PRINCIPLES OF PHYSICS

PRINCIPLES OF PHYSICS

HANS C. OHANIAN

Rensselaer Polytechnic Institute

W.W. NORTON & COMPANY

New York London

Printed in the United States of America

First Edition

The text of this book is composed in ITC New Baskerville
with the display set in Huxley Vertical
Composition by New England Typographic Service, Inc.
Maufacturing by Von Hoffmann Press, Inc.
Book design by M space

Library of Congress Cataloging-in-Publication Data

Ohanian, Hans C.
Principles of physics / Hans C. Ohanian.
p. cm.
1. Physics. I. Title.
QC23.04 1994
530—dc20 93-25636

ISBN 0-393-95773-x

W. W. Norton & Company, Inc., 500 Fifth Avenue, New York, N.Y. 10110
W. W. Norton & Company Ltd., 10 Coptic Street, London WC1A 1PU

1 2 3 4 5 6 7 8 9 0

For Mara

BRIEF CONTENTS

CONTENTS

CHAPTER 1

CHAPTER 2

* This section is optional.

CHAPTER 5

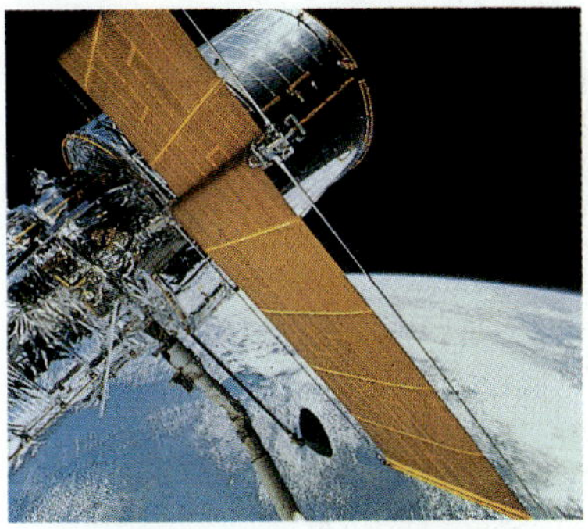

GRAVITATION 159

CHAPTER 6

MOMENTUM AND COLLISIONS 184

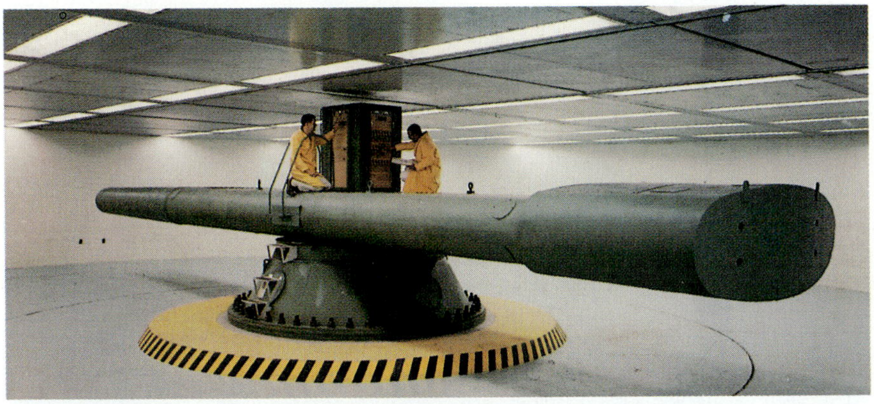

CHAPTER 7

ROTATION OF A RIGID BODY 219

* This section is optional.

* This section is optional.

C H A P T E R 1 1

SOUND 320

C H A P T E R 1 2

MOTION OF FLUIDS 346

* This section is optional.

* This section is optional.

* This section is optional.

* This section is optional.

C H A P T E R 2 2

ELECTROMAGNETIC INDUCTION 619

C H A P T E R 2 3*

AC CIRCUITS 645

C H A P T E R 2 4

ELECTROMAGNETIC WAVES 666

C H A P T E R 2 5

REFLECTION, REFRACTION, AND OPTICS 692

* This section is optional.

C H A P T E R 2 6

INTERFERENCE AND DIFFRACTION 729

C H A P T E R 2 7

THE THEORY OF SPECIAL RELATIVITY 757

C H A P T E R 2 8

QUANTA OF LIGHT 786

C H A P T E R 2 9

SPECTRAL LINES AND BOHR'S THEORY 810

* This section is optional.

CHAPTER 30

QUANTUM STRUCTURE OF ATOMS, MOLECULES, AND SOLIDS 835

CHAPTER 31

NUCLEI 861

* This section is optional.

CHAPTER 32

ELEMENTARY PARTICLES 894

APPENDICES

PREFACE

Principles of Physics is a textbook for a one year algebra-based introductory physics course. The book is intended for students in the life sciences, the premedical curriculum, the earth and environmental sciences, and the liberal arts.

Although the program of fundamental laws and concepts covered in *Principles of Physics* is somewhat similar to that of a typical calculus-based textbook, the execution of the program is quite different. Apart from the difference in mathematical level, there is much less emphasis on derivations and more emphasis on simple, qualitative explanations. Since for many students this is their first encounter with physics, I introduce new concepts by qualitative and intuitive discussions whenever possible. And to motivate the students and to maintain their enthusiasm, I draw on examples from sports, biology, medicine, and everyday life, chosen to match the background and special interests of the students. In the same spirit, the book includes discussions of the biomolecular basis of muscular contraction, automobile collisions, biomechanics of the limbs of animals, the ear, the vocal cords, ultrasound imaging, the heart and blood flow, mechanical efficiency of the human body, electrocardiography, hazards of electric currents, lightning, the eye, microscopy, X-Rays, CAT, MRI, and other forms of medical imaging. Some of these topics are included in the chapters; others are included in separate supplements called Physics in Action (see below). By means of such special examples and topics, I seek to strike a balance between presenting physics for its own sake and presenting physics as a tool for practical applications.

I believe that the same assortment of fundamental laws and concepts of physics needs to be taught to all students, whether they are pursuing careers in engineering or in the life sciences. To gain a clear view of physics, the students must learn to see the material world through the eye of a physicist and appreciate the priorities that govern the thinking of a physicist. Some algebra-based texts give students a distorted view of physics by skipping some fundamental laws, such as Maxwell's equations, that are close to the heart of every physicist. The notion that Maxwell's laws for electromagnetism should be taught to engineers but not to biologists strikes me as just as absurd as the notion that, say, the Renaissance should be included in a European history course for one group of students but not for another. Although Maxwell's laws, written with integrals, are beyond the algebra level, these laws can be reexpressed as statements about the *average values* of the magnitudes of electric and magnetic fields along paths or over surfaces. In this simplified form, these laws remain within the algebra level, and they are accessible to all students.

Furthermore, I believe that students should always be encouraged to keep in mind the atomic structure of matter and to regard the material world as a multitude of restless particles. This point of view is introduced in the very first pages, in the Prelude which presents a brief survey of the world of matter, from galaxies and the entire universe to atoms, nuclei, and quarks. In the mechanics chapters, I emphasize that all macroscopic bodies are systems of particles, and in the electricity chapters, I introduce the concepts of positive and negative electric charge by referring to protons and electrons, not by referring to the antiquated procedure of rubbing glass rods with silk rags or with cats.

The following paragraphs briefly summarize some aspects of the text.

Mathematical Level. Students are expected to have some knowledge of algebra (Appendix 2 provides a quick refresher). Algebraic steps are spelled out in great detail, especially in the early chapters. Acquaintance with trigonometry is not a prerequisite, but would be helpful. The necessary trigonometric concepts and formulas are gradually introduced in a self-contained manner, beginning with a box in Chapter 2 giving the definitions of the trigonometric functions (Appendix 4 presents a systematic survey of trigonometry). Students are expected to know how to use an electronic calculator, equipped with trigonometric functions and their inverses.

Units. SI units are used throughout the book. British engineering units for length, mass and force are mentioned, but used only in a handful of problems in the first chapters. In accord with the practice adopted by the American Institute of Standards (ANSI), the Institute of Electrical and Electronic Engineers (IEEE), and the United States Department of Defense, as well as *A Physicist's Desk Reference* published by the American Institute of Physics, the pound (lb) is regarded as the unit of mass of the engineering system, and the pound-force (lbf) as the unit of force.

Key Statements, Boxed Equations, and Side Heads. Boldface italics, colored boxes, and colored side heads help the students identify the key ideas in each chapter and quickly find items needed for solving problems.

Color Diagrams and Photographs. The colors of lines and of vectors, the labels, and the style of the diagrams have been carefully adjusted for consistency and clarity. Color photographs have been selected to highlight ideas presented in the text.

Solved Examples. A generous number of solved examples are scattered through each chapter. These serve to reinforce the student's grasp of the concepts, and they also provide models for the solution of problems. Many of the examples contain a paragraph or two of COMMENTS AND SUGGESTIONS, which draw attention to pitfalls in the solution and indicate what general lessons can be extracted from the solution.

Problems. Over 2,000 questions and problems are included at the ends of the chapters. The problems are grouped by sections and by topic within each section, and, where possible, they are arranged in order of increasing difficulty. The levels of difficulty of the problems are roughly indicated by no star (easy and straightforward, mostly of the plug-in type), one star * (of medium difficulty, requiring the combination of several concepts or the manipulation of several formulas), and two stars ** (difficult and challenging, demanding considerable thought and insight). Many of these problems are based on realistic data extracted from *The Guinness Book of World Records,* repair or maintenance manuals, newspapers, scientific journals, radio reports, etc. I hope that these problems will convince students that physics is fun or, at least, that physics is relevant in the real, everyday world. Answers to the even-numbered problems are given in Appendix 9.

Problem-solving Help. A section in Chapter 1 gives some general rules for the solution of problems. Many of the solved Examples include COMMENTS AND SUGGESTIONS that provide problem-solving help. In the appropriate sections, general guidance is provided for dealing with problems in kinematics, problems involving "free-body" diagrams, and problems involving energy conservation and momentum conservation.

Qualitative Questions. These questions, at the ends of the chapters, serve to stimulate thinking and to provoke discussions.

Summaries. Brief summaries of the crucial definitions and equations are attached to the chapters. These summaries are intended to serve for quick reference; they are not intended to serve as recapitulations of the contents of the chapters. In class trials, students have commented favorably on the brevity of these summaries.

Prelude. The Prelude introduces powers-of-ten notation and presents a panoramic survey of the world of matter, with emphasis on the atomic structure of matter.

Optional Sections. Some sections are marked as optional in the text. Later sections of the book do not rely on such optional sections, which can therefore be omitted without subsequent complications.

Physics in Action Supplements. These twelve supplements, inserted after relevant chapters, illustrate some of the important practical applications of fundamental principles of physics. The twelve supplements deal with a broad range of topics, from sports to molecular biology to MRI. The twelve Supplements are *Long Jumps, Muscular Contraction, Automobile Collisions, Animal Mechanics, Ultrasound Images, Blood Flow, The Electrocardiograph, Thunderstorms and Lightning, Microscopy, X-Ray Images, Laser Light,* and *Medical Imaging.* All of these supplements are, of course, optional.

ANCILLARIES

The following items are available from the publisher:

Solutions Manual, by Professor Carl Bromberg, contains the complete step-by-step solutions to all the problems. (Available to teachers only.)

Answers Pamphlet lists the answers for both the even-numbered and odd-numbered problems. (Available for purchase by students, but only with permission of the instructor.)

Computerized test bank, by Professor Miles Dresser, contains over 1,000 multiple-choice questions in printed form and on diskettes. Many of these are "generator-questions," which can be transformed into new questions by a few keystrokes.

Principles of Physics Videodisc contains animated demonstrations of key physical concepts provided by Physics Curriculum and Instruction and hundreds of still images from the text. Available free to departments ordering a minimum number of copies of *Principles of Physics.*

Interactive Physics II by Knowledge Revolution is an advanced motion simulation program that allows instructors to exhibit the physics of moving objects by drawing on the computer screen—a powerful lecture demonstration tool. Available free to departments ordering a minimum number of copies of *Principles of Physics.*

VHS tapes from the "Physics Demonstration and Concepts" series by Physics Curiculum and Instruction are designed to present physical phenomena not easily observed in the classroom. Two tapes are available free to departments ordering a minimum number of copies of *Principles of Physics.*

Full-color transparencies of more than 180 diagrams in the book.

Instructor's Resource Manual, by Professor Chris Roddy, contains useful advice on using the *Principles of Physics* ancillary package, teaching strategies, and a guide to available films, software, and other pedagogical aids.

Study Guide, by Professor Van E. Neie, includes for each chapter, a brief survey, a list of key terms for review, commentaries on the main ideas, and a large variety of sample problems; both fully solved, and guided (which provide step-by-step schemes that lead students to the solution).

Spreadsheet Exercises and Solutions for Principles of Physics, by Professor Peter Drago, contains approximately 150 spreadsheet exercises tied to problems in the book. These exercises use numerical methods and explore "what-if" scenarios through thoughtful questions, tabulations of results, and graphs.

Sensei Physics by Broderbund Software is a computer program for the Macintosh that presents basic concepts, animated demonstrations, and interactively-solved problems. The program serves as the student's own personal review guide to the full year course.

The Norton Student Version of Pro Solv by Problem Solving Concepts is a computer program for Windows that allows students to design their own problems from a large set of mathematical and physical principles. Under the "computer solve" option, abundant tutorial help is available for solving problems, and the program specifically covers more than 120 problems from *Principles of Physics.*

ACKNOWLEDGEMENTS

I am deeply indebted to the many reviewers of the book for their thoughtful comments: Professors Carl Bromberg, Michigan State University; Alex F. Burr, New Mexico State University; Lawrence B. Coleman, University of California, Davis; Miles J. Dresser, Washington State University; Nate O. Folland, Kansas State University; Robert J. Friauf, University of Kansas; Lothar Frommhold, University of Texas, Austin; John A. Gilreath, Clemson University; Laurent Hodges, Iowa State University; Richard W. Huggett, Louisiana State University; Sanford Kern, Colorado State University; Ian M. Littlewood, University of Central Florida; Elihu Lubkin, University of Wisconsin-Milwaukee; Y. Daines Lund, Utah State University; Peter W. Mathews, University of British Columbia; Marvin Morris, San Jose State University; Richard A. Morrow, University of Maine; Van E. Neie, Purdue University; Paul M. Parker, Michigan State University; Harvey Picker, Trinity College; Raymond S. Robinson, Colorado State University; Dennis K. Ross, Iowa State University; Larry Rowan, University of North Carolina at Chapel Hill; Jay D. Strieb, Villanova University; George Strobel, University of Georgia; Gerald E. Tripard, Washington State University; Gilbert Ward, Pennsylvania State University; and David M. Wolfe, University of New Mexico. And I am equally indebted to the reviewers of the Physics in Action Supplements: Professors Thomas J. Beck, Johns Hopkins Hospital (*X-Ray Images* and *Medical Imaging*); Barbara C. Boyer, Union College (*Microscopy*); Richard Graetzer, Pennsylvania State University (all Supplements); Allen Scher, University of Washington (*Blood Flow* and *The Electrocardiograph*); Steven Vogel, Duke University (all Supplements); Ronald Wakai, University of Wisconsin, Madison (*The Electrocardiograph, Microscopy, Laser Light, X-Rays Images,* and *Medical Imaging*). I was glad to adopt most of the corrections and improvements suggested by all these reviewers, but, of course, there were some instances where I decided to persist in my own way of doing things.

Several versions of the book were given class trials by Professors Ian Littlewood at the University of Central Florida, Nate O. Folland at Kansas State University, Lothar Frommhold at the University of Texas at Austin, and Harvey Picker at Trinity College. I thank them for their efforts and their comments, and I also thank their students for their forbearance.

And I am grateful to the staff of W. W. Norton & Company, Inc. for their meticulous work on this book. Drake McFeely, Editor and Vice President, guided the book through its first few drafts. Stephen R. Mosberg read and reread the manuscript and suggested many additional explanations and other improvements. He picked up enough physics along the way to make many valuable contributions to the clarity of the text. Mary Walsh kept the production process on track and on schedule. Debra McKay excelled in copyediting the manuscript, making this as painless for me as possible; she checked the stylistic consistency of the text and even spotted some mistakes in some calculations. JoAnn Schambier displayed exceptional ingenuity and persistence in searching for the color photos that adorn the pages, and Gyodi Reid enthusiastically completed the search.

H. C. O.
August 1993

The World of Physics

Physics is the study of matter. In a quite literal sense, physics is the greatest of all natural sciences: it encompasses the smallest particles, such as electrons and quarks; and it also encompasses the largest bodies, such as galaxies and the entire Universe. The smallest particles and the largest bodies differ in size by a factor of more than ten thousand billion billion billion billion! In the pictures on the following pages we will survey the world of physics and attempt to develop some rough feeling for the sizes of things in this world. This preliminary survey sets the stage for our explanations of the mechanisms that make things behave in the way they do. Such explanations are at the heart of physics, and they are the concern of the later chapters of this book.

Since the numbers we will be dealing with in this prelude and in the later chapters are often very large or very small, we will find it convenient to employ the ***scientific notation*** for these numbers. In this notation, numbers are written with powers of 10; thus, hundred is written as 10^2, thousand is written as 10^3, ten thousand is written as 10^4, and so on. A tenth is written as 10^{-1}, a hundredth is written as 10^{-2}, a thousandth is written as 10^{-3}, and so on. The following table lists some powers of ten:

$10 = 10^1$	$0.1 = 1/10 = 10^{-1}$
$100 = 10^2$	$0.01 = 1/100 = 10^{-2}$
$1000 = 10^3$	$0.001 = 1/1000 = 10^{-3}$
$10{,}000 = 10^4$	$0.0001 = 1/10{,}000 = 10^{-4}$
$100{,}000 = 10^5$	$0.00001 = 1/100{,}000 = 10^{-5}$
$1{,}000{,}000 = 10^6$	$0.000001 = 1/1{,}000{,}000 = 10^{-6}$,

etc.

Note that the power of 10, or the exponent on the 10, simply tells us how many zeros follow the 1 in the number (if the power of 10 is positive) or how many zeros follow the 1 in the denominator of the fraction (if the power of 10 is negative).

In scientific notation, a number that does not coincide with one of the powers of 10 is written as a product of a decimal number and a power of 10. For example, in this notation, 1,500,000,000 is written as 1.5×10^9. Alternatively, this number could be written as 15×10^8 or as 0.15×10^{10}; but in scientific notation it is customary to place the decimal point immediately after the first nonzero digit. The same rule applies to numbers smaller than 1; thus, 0.000015 is written as 1.5×10^{-5}.

The pictures on the following pages fall into two sequences. In the first sequence we zoom out: we begin with a picture of a woman's face and proceed step by step to pictures of the entire Earth, the Solar System, the Galaxy, and the Universe. This ascending sequence contains 27 pictures, with the scale decreasing in steps of factors of 10.

Most of our pictures are photographs. Many of these have become available only in recent years; they were taken by high-flying U-2 aircraft, Landsat satellites, astronauts on the Moon, or sophisticated electron microscopes. For some of our pictures no photographs are available and we have to rely, instead, on drawings.

Part I: The Large-Scale World

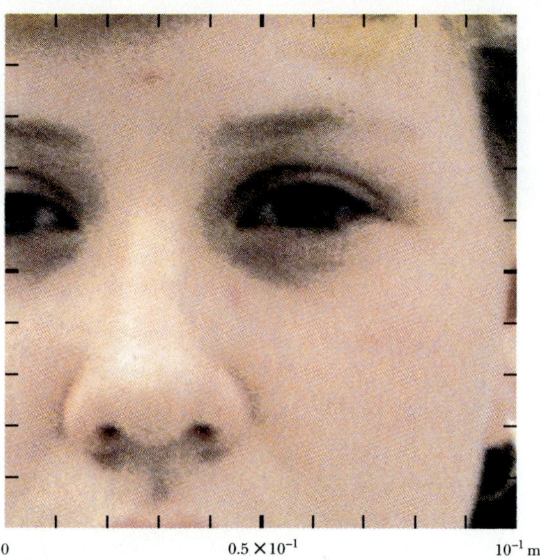

SCALE 1:1.5 This is Clare, an intelligent biped of the planet Earth, Solar System, Orion Spiral Arm, Milky Way Galaxy, Local Group, Local Supercluster. Clare belongs to the Phylum Chordata, Class Mammalia, Order Primates, Family Hominidae, Genus *Homo,* Species *sapiens.* She is made of 4.8×10^{27} atoms, with 1.7×10^{28} electrons, the same number of protons, and 1.3×10^{28} neutrons.

SCALE 1:1.5 × 10 Clare has a height of 1.6 meters and a mass of 52 kilograms. Her chemical composition (by mass) is 65% oxygen, 18.5% carbon, 9.5% hydrogen, 3.3% nitrogen, 1.5% calcium, 1% phosphorus, and 0.35% of other elements.

The matter in Clare's body and the matter in her immediate environment occurs in three states of aggregation: ***solid, liquid,*** and ***gas.*** All these forms of matter are made of atoms and molecules, but solid, liquid, and gas are qualitatively different because the arrangements of the atomic and molecular building blocks are different.

In a solid, each building block occupies a definite place. When a solid is assembled out of molecular or atomic building blocks, these blocks are locked in place once and for all, and they cannot move or drift about except with great difficulty. This rigidity of the arrangement is what makes the aggregate hard — it makes the solid "solid." In a liquid, the molecular or atomic building blocks are not rigidly connected. They are thrown together at random and they move about fairly freely, but there is enough adhesion between neighboring blocks to prevent the liquid from dispersing. Finally, in a gas, the molecules or atoms are almost completely independent of one another. They are distributed at random over the volume of the gas and are separated by appreciable distances, coming in touch only occasionally during collisions. A gas will disperse spontaneously if it is not held in confinement by a container or by some restraining force.

The molecules of a gas are forever moving around at high speed. For example, at a temperature of 0°C the average speed of a molecule of nitrogen in air is about 450 meters/second, faster than the speed of sound. The speed of the molecules is directly related to the temperature: the speed increases if the air is heated and decreases if the air is cooled. The molecules of a liquid also move; their speeds are not very different from those of the molecules of a gas. However, since there is little space between the molecules in a liquid, the motion is continually interrupted by collisions. Because of these frequent collisions, the path of a molecule consists of a series of random zigzags, and the molecule takes a long time to wander from one part of the liquid to another. Even in a solid there is some motion of the building blocks. But the motion of each atom or molecule is merely a vibration around its assigned position — the atom behaves as if kept on a short leash and although it moves back and forth at a high speed, it never strays beyond some tight limits.

If we regard the motion of the atoms of a gas in a container as analogous to the bouncing of a few dice in a shaker, then the motion of atoms in a liquid is analogous to the random wandering of the dice in a shaker that has been loosely but completely filled with dice; and the motion of atoms in a solid is analogous to the impotent rattling of the dice in a shaker that has been tightly and regularly packed full of dice.

The eternal, dancing motion of the atoms is the key to

the transformations of state from solid to liquid to gas and vice versa. If we heat a solid, the vibrational motion becomes more violent and the atoms or molecules finally shake themselves out of place — the solid softens and melts, turning into a liquid. If we heat this resulting liquid further, the motion ultimately becomes so violent that the adhesion between the atoms or molecules cannot prevent some of them from escaping from the surface of the liquid. Gradually more and more escape — the liquid evaporates, turning into a gas.

SCALE 1:1.5 × 10² The building behind Clare is the New York Public Library, one of the largest libraries on Earth. This library holds 1.4×10^{10} volumes, containing roughly 10% of the total accumulated knowledge of our terrestrial civilization.

0 0.5×10^1 10^1 m

SCALE 1:1.5 × 10³ The New York Public Library is located at the corner of Fifth Avenue and 42nd Street, in the middle of New York City.

0 0.5×10^2 10^2 m

SCALE 1:1.5 × 10⁴ This aerial photograph shows an area of 1 kilometer × 1 kilometer in the vicinity of the New York Public Library. The streets in this part of the city are laid out in a regular rectangular pattern. The library is the building in the park in the upper middle of the picture. The Empire State Building shows up in the lower left of the picture. This skyscraper is 381 meters high and it casts a long shadow.

This photograph was taken from an airplane flying at an altitude of a few thousand meters. North is at the top of the photograph.

SCALE 1:1.5 × 10^5 This photograph shows a large portion of New York City. We can recognize the library and its park as a small rectangular patch slightly above the center of the picture. The central mass of land is the island of Manhattan, with the Hudson River on the left and the East River on the right.

This photograph was taken from a U-2 aircraft flying at an altitude of about 20,000 meters.

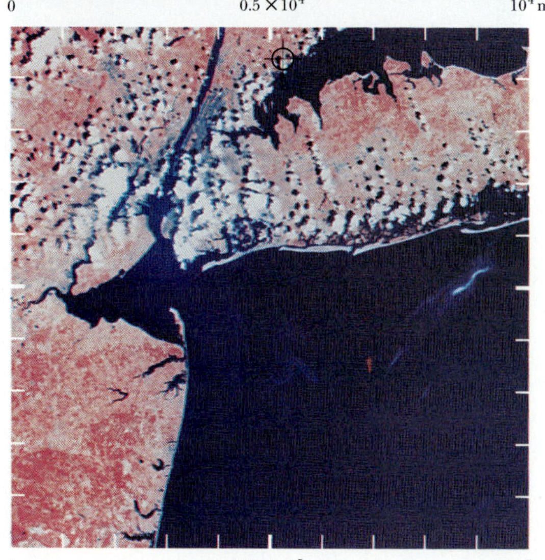

SCALE 1:1.5 × 10^6 In this photograph, Manhattan is in the upper left quadrant. On this scale, we can no longer distinguish the pattern of streets in the city. The vast expanse of water in the lower right of the picture is part of the Atlantic Ocean. The mass of land in the upper right is Long Island. Parallel to the south shore of Long Island we can see a string of very narrow islands; they almost look man-made. These are barrier islands; they are heaps of sand piled up by ocean waves in the course of thousands of years.

This photograph was taken by a Landsat satellite orbiting the Earth at an altitude of 920 kilometers.

SCALE 1:1.5 × 10⁷ Here we see the eastern coast of the United States, from Cape Cod to Cape Fear. Cape Cod is the hook near the northern end of the coastline, and Cape Fear is the promontory near the southern end of the coastline. If we move along the coast starting at the north, we first come to Long Island; then to Delaware Bay and Chesapeake Bay, two deep indentations in the coastline; and then to Cape Hatteras, at the extreme end of the large bulge of land thrusting eastward into the Atlantic. Note that on this scale no signs of human habitation are visible. However, at night the lights of large cities would stand out clearly.

This picture is a mosaic, assembled by joining together many Landsat photographs such as the one above.

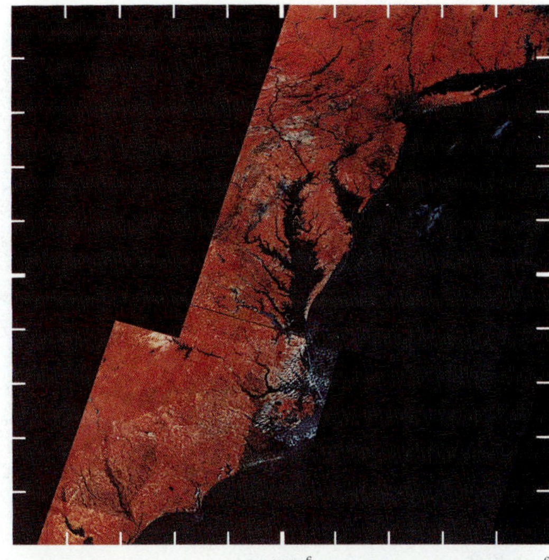

SCALE 1:1.5 × 10^8 In this photograph, taken by the Apollo 16 astronauts during their trip to the Moon, we see a large part of the Earth. Through the gap in the clouds in the lower middle of the picture, we can see the coast of California and Mexico. We can recognize the peninsula of Baja California and the Gulf of California. In the middle right of the photograph we can recognize the Gulf of Mexico. Clare's location, the East Coast of the United States, is covered by a big system of swirling clouds in the upper right of the photograph.

Note that a large part of the area visible in this photograph is ocean. About 71% of the surface of the Earth is ocean; only 29% is land. The atmosphere covering this surface is about 100 kilometers thick; on the scale of this photograph, its thickness is about 0.7 millimeter. Seen from a large distance, the predominant colors of the planet Earth are blue (oceans) and white (clouds).

SCALE 1:1.5 × 10^9 This photograph of the Earth was taken by the Apollo 16 astronauts standing on the surface of the Moon. Sunlight is striking the Earth from the top of the picture.

As is obvious from this and from the preceding photograph, the Earth is a sphere. Its radius is 6.38×10^6 meters and its mass is 5.98×10^{24} kilograms.

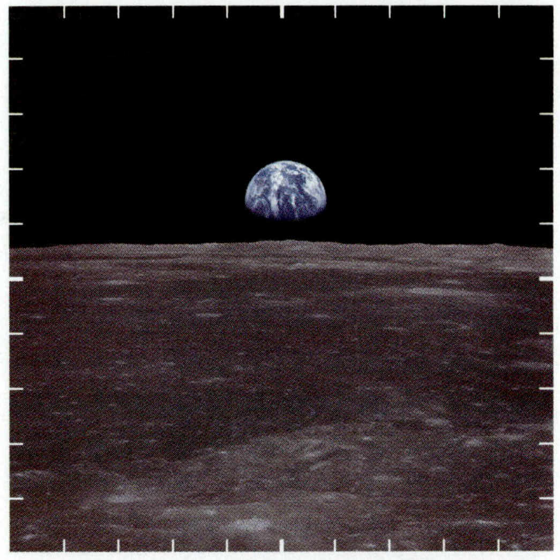

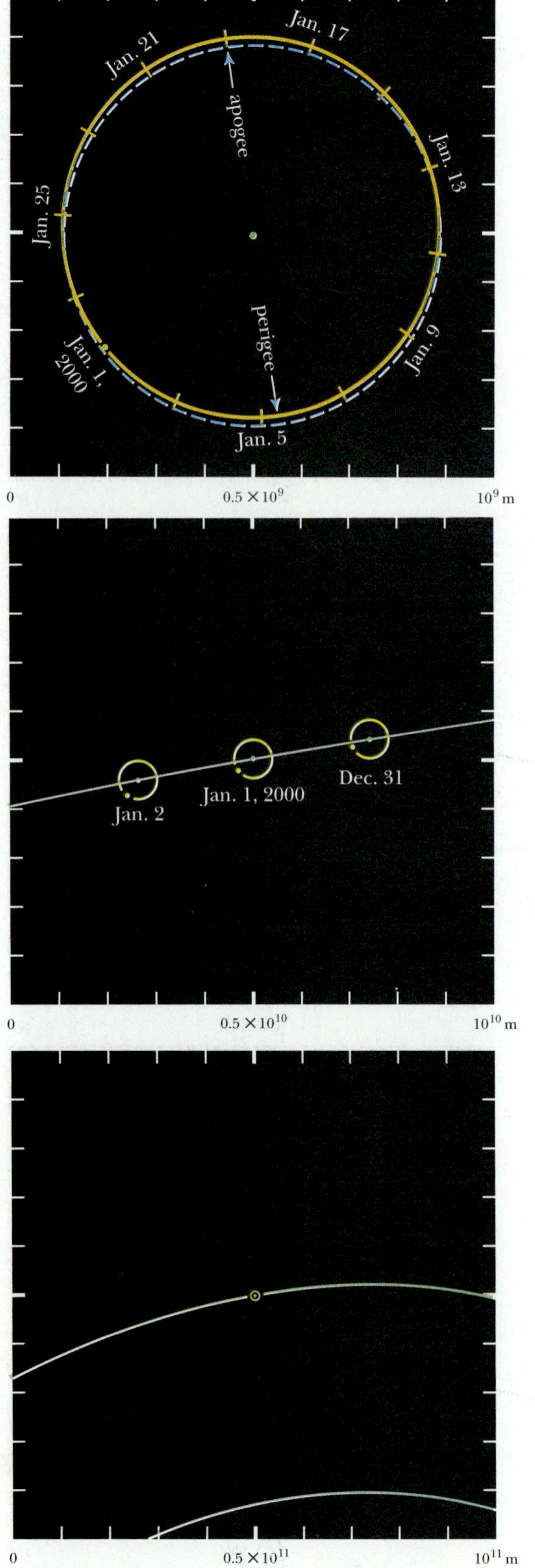

SCALE 1:1.5 × 10^{10} In this picture, we see the Earth and the Moon. The solid line indicates the orbit of the Moon around the Earth (many of the pictures on the following pages are also drawings). As in the preceding picture, the Sun is far below the bottom of the picture. The position of the Moon is that of January 1, 2000.

The orbit of the Moon around the Earth is an ***ellipse,*** but an ellipse that is very close to a circle. The solid curve in the picture is the orbit of the Moon and the dashed curve is a circle; by comparing these two curves we can see how little the ellipse deviates from a circle. The point on the ellipse closest to the Earth is called the ***perigee*** and the point farthest from the Earth is called the ***apogee.*** The distance between the Moon and the Earth is roughly 30 times the diameter of the Earth. The Moon takes 27.3 days to travel once around the Earth.

SCALE 1:1.5 × 10^{11} This picture shows the Earth, the Moon, and portions of their orbits around the Sun. On this scale, both the Earth and the Moon look like small dots. Again, the Sun is far below the bottom of the picture. In the middle, we see the Earth and the Moon in their positions for January 1, 2000. On the right and on the left we see, respectively, their positions for 1 day before and 1 day after this date.

Note that the net motion of the Moon consists of the combination of two simultaneous motions: the Moon orbits around the Earth, which in turn orbits around the Sun.

SCALE 1:1.5 × 10^{12} Here we see the orbits of the Earth and of Venus. However, Venus itself is beyond the edge of the picture. The small circle is the orbit of the Moon. The dot representing the Earth is much larger than what it should be, although the draftsman has drawn it as minuscule as possible. On this scale, even the Sun is quite small; if it were included in this picture, it would be only 1 millimeter across.

SCALE 1:1.5 × 10^{13} This picture shows the positions of the Sun and the inner planets: Mercury, Venus, Earth, and Mars. The positions of the planets are those of January 1, 2000. The orbits of all these planets are ellipses, but they are close to circles. The point of the orbit nearest to the Sun is called the ***perihelion*** and the point farthest from the Sun is called the ***aphelion.*** The Earth reaches perihelion about January 3 and aphelion about July 6 of each year.

All the planets travel around their orbits in the same direction: counterclockwise in our picture. The marks along the orbit of the Earth indicate the successive positions at intervals of 10 days.

Beyond the orbit of Mars, a large number of asteroids orbit around the Sun; these have been omitted to prevent excessive clutter. Furthermore, a large number of comets orbit around the Sun. Most of these have pronounced elliptical orbits. The comet Halley has been included in our picture.

The Sun is a sphere of radius 6.96 × 10^8 meters. On the scale of the picture, the Sun looks like a very small dot, even smaller than the dot drawn here. The mass of the Sun is 1.99 × 10^{30} kilograms.

The matter in the Sun is in the ***plasma*** state, sometimes called the fourth state of matter. Plasma is a very hot gas in which violent collisions between the atoms in their random thermal motion have fragmented the atoms, ripping electrons off them. An atom that has lost one or more electrons is called an ***ion.*** Thus, plasma consists of a mixture of electrons and ions, all milling about at high speed and engaging in frequent collisions. These collisions are accompanied by the emission of light, making the plasma luminous.

SCALE 1:1.5 × 10^{14} This picture shows the positions of the outer planets of the Solar System: Jupiter, Saturn, Uranus, Neptune, and Pluto. On this scale, the orbits of the inner planets are barely visible. As in our other pictures, the positions of the planets are those of January 1, 2000.

The outer planets move slowly and their orbits are very large; thus they take a long time to go once around their orbit. The extreme case is that of Pluto, which takes 248 years to complete one orbit.

Uranus, Neptune, and Pluto are so far away and so faint that their discovery became possible only through the use of telescopes. Uranus was discovered in 1781, Neptune in 1846, and the tiny Pluto in 1930.

SCALE 1:1.5 × 10^{15} We now see that the Solar System is surrounded by a vast expanse of space. Although this space is shown empty in the picture, the Solar System is encircled by a large cloud of millions of comets whose orbits crisscross the sky in all directions. Furthermore, the interstellar space in this picture and in the succeeding pictures contains traces of gas and of dust. The interstellar gas is mainly hydrogen; its density is typically 1 atom per cubic centimeter.

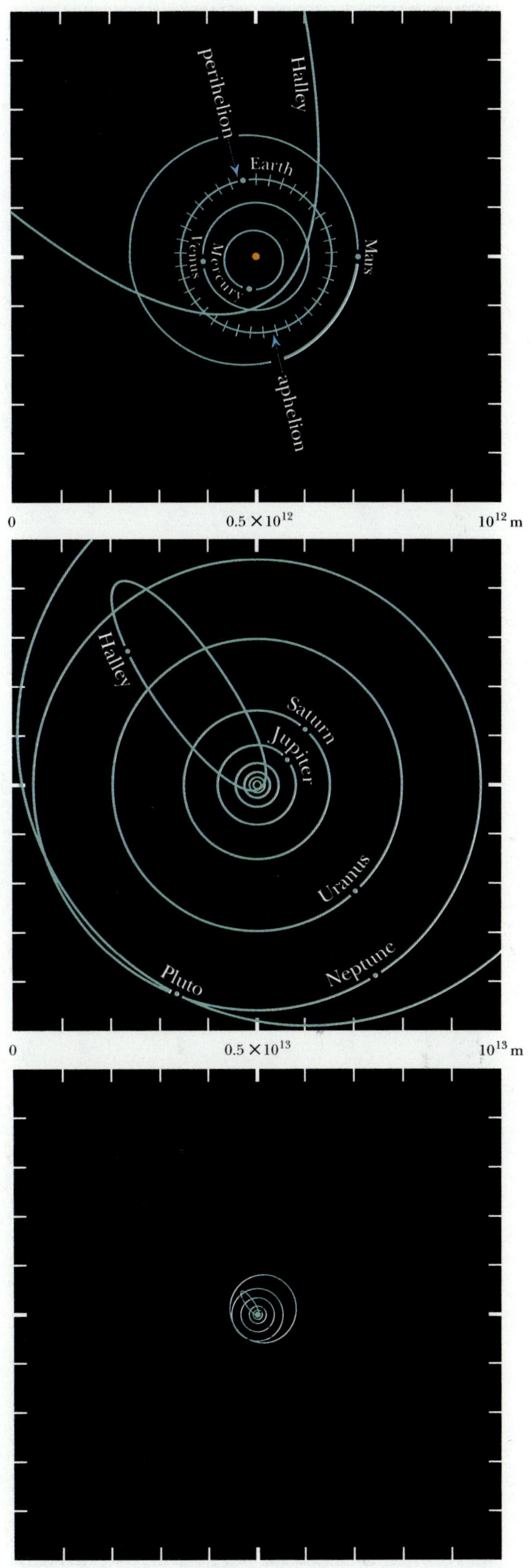

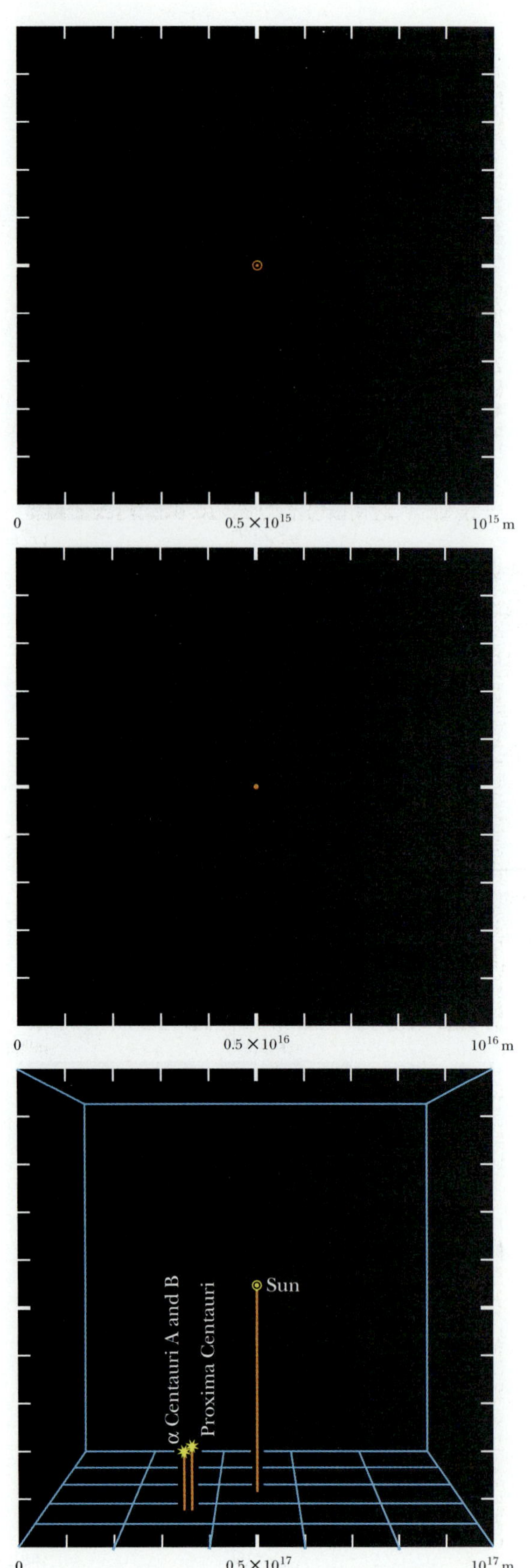

SCALE 1:1.5 × 10¹⁶ More interstellar space. The small circle is the orbit of Pluto.

SCALE 1:1.5 × 10¹⁷ And more interstellar space. On this scale, the Solar System looks like a minuscule dot, 0.1 millimeter across.

SCALE 1:1.5 × 10¹⁸ Here, at last, we see the stars nearest to the Sun. The picture shows all the stars within a cubical box 10^{17} meters × 10^{17} meters × 10^{17} meters centered on the Sun: Alpha Centauri A, Alpha Centauri B, and Proxima Centauri. All three are in the constellation Centaurus, in the southern sky.

The star closest to the Sun is Proxima Centauri. This is a very faint, reddish star (a "red dwarf"), at a distance of 4.0×10^{16} meters from the Sun. Astronomers like to express stellar distances in light-years: Proxima Centauri is 4.2 light-years from the Sun, which means light takes 4.2 years to travel from this star to the Sun.

Proxima Centauri is too faint to be seen by the naked eye. The nearest stars that can be seen by the naked eye are Alpha Centauri A and Alpha Centauri B. These two stars are so close together that we need to use a telescope to distinguish between them. They form a double star, continually orbiting around each other.

SCALE 1:1.5 × 10^{19} This picture displays the brightest stars within a cubical box 10^{18} meters × 10^{18} meters × 10^{18} meters centered on the Sun. There are many more stars in this box besides those shown — the total number of stars in this box is close to 2000.

Sirius is the brightest of all the stars in the night sky. If it were at the same distance from the Earth as the Sun, it would be 28 times brighter than the Sun.

SCALE 1:1.5 × 10^{20} Here we expand our box to 10^{19} meters × 10^{19} meters × 10^{19} meters, again showing only the brightest stars and omitting many others. The total number of stars within this box is about 2 million. We recognize several clusters of stars in this picture: the Pleiades Cluster, the Hyades Cluster, the Coma Berenices Cluster, and the Perseus Cluster. Each of these has hundreds of stars crowded into a fairly small patch of sky.

SCALE 1:1.5 × 10^{21} Now there are so many stars in our field of view that they appear to form clouds of stars. There are about a million stars in this photograph, and there are many more stars too faint to show up distinctly. Although this photograph is not centered on the Sun, it simulates what we would see if we could look toward the Solar System from very far away. The photograph shows a view of the Milky Way in the direction of the constellation Sagittarius. When we look from our Solar System in this direction, we see the clouds of stars in the neighboring spiral arm of our Galaxy (the next picture shows a galaxy and its spiral arms). This neighboring arm is the Sagittarius Spiral Arm; the cloud of stars in which our Sun is located belongs to the ***Orion Spiral Arm.***

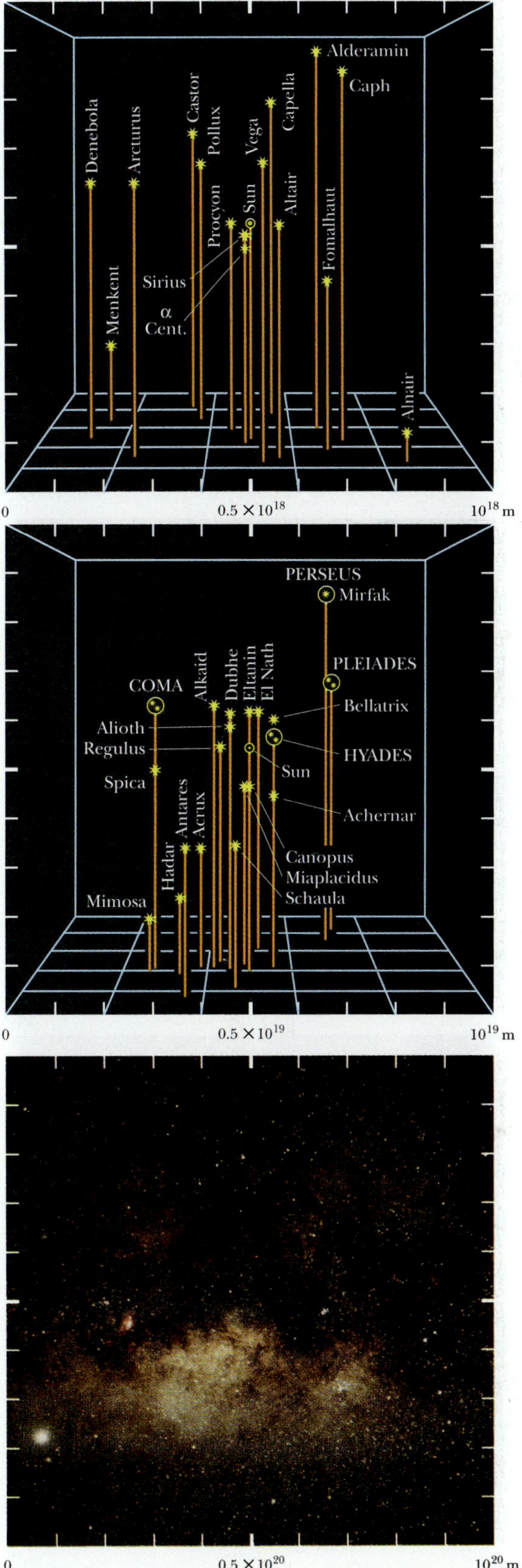

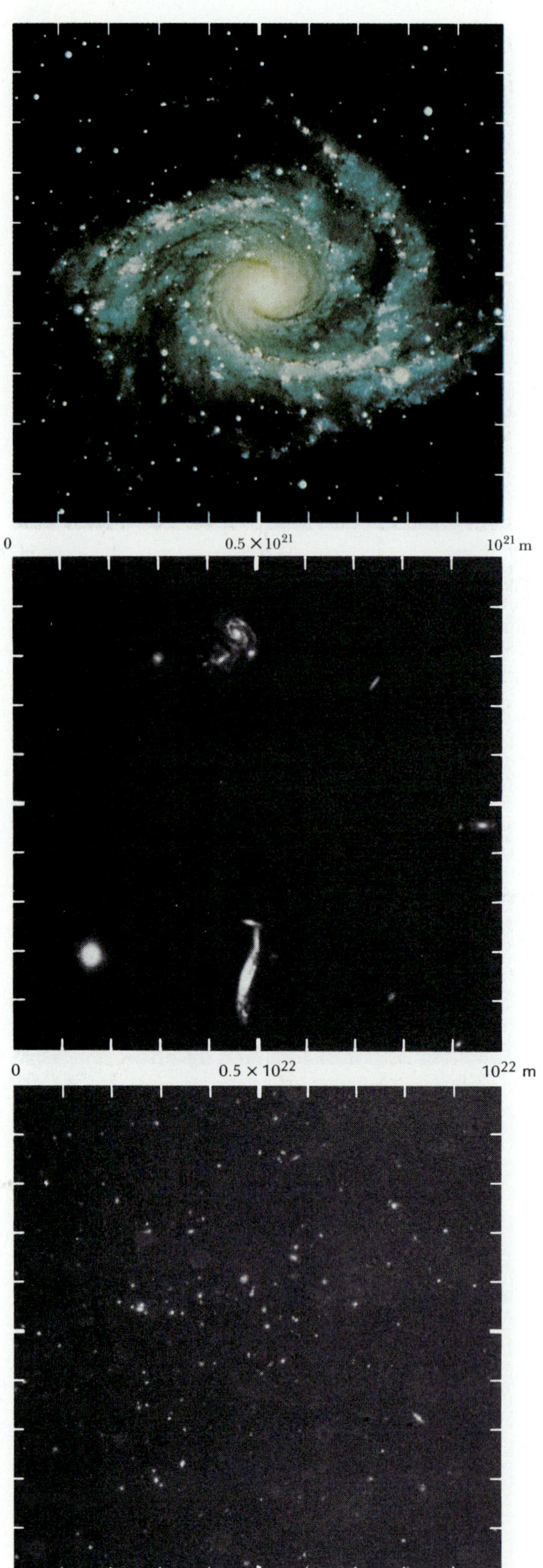

SCALE 1:1.5 × 10²² This is the spiral galaxy NGC 2997. Its clouds of stars are arranged in spiral arms wound around a central bulge. The bright central bulge is the nucleus of the galaxy; it has a more or less spherical shape. The surrounding region, with the spiral arms, is the disk of the galaxy. This disk is quite thin; it has a thickness of only about 3% of its diameter. The stars making up the disk circle around the galactic center in a clockwise direction.

Our Sun is in a spiral galaxy of roughly similar shape and size: the ***Galaxy of the Milky Way.*** The total number of stars in this galaxy is about 10^{11}. The Sun is in one of the spiral arms, roughly one-third inward from the edge of the disk toward the center.

SCALE 1:1.5 × 10²³ Galaxies are often found in clusters of several galaxies. Some of these clusters consist of just a few galaxies, others of hundreds or even thousands. The photograph shows a modest cluster, or group, of four galaxies beyond the constellation Hercules. The group contains an elliptical galaxy like a luminous egg (lower left), two spiral galaxies close together (top), and a spiral with a bar (bottom).

Our Galaxy is part of another modest cluster, the ***Local Group,*** consisting of our own Galaxy, the great Andromeda Galaxy, the Triangulum Galaxy, the Large Magellanic Cloud, plus 16 other small galaxies.

According to recent investigations, the dark, apparently empty, space near galaxies contains some form of distributed matter, with a total mass several times as large as the mass in the luminous, visible galaxies. But the composition of this dark, invisible, extragalactic matter is not known.

In this and the following photograph, the stars in the foreground have been erased to prevent their confusion with galaxies.

SCALE 1:1.5 × 10²⁴ The Local Group lies on the fringes of a very large cluster of galaxies, called the ***Local Supercluster.*** This is a cluster of clusters of galaxies. At the center of the Local Supercluster is the Virgo Cluster with several thousand galaxies. Seen from a large distance, our supercluster would present a view comparable to this photograph, which shows a rich cluster beyond the constellation Hydra, a cluster that is at a large distance from us.

All the distant galaxies are moving away from us and away from each other. For instance, the Hydra Cluster is moving away from our Galaxy at the rate of 6030 kilometers per second. This motion of recession of the galaxies is analogous to the outward motion of, say, the fragments of a grenade after its explosion. The motion of the galaxies suggests that the Universe began with a big explosion, the ***Big Bang,*** that launched the galaxies away from each other.

SCALE 1:1.5 × 10^{25} On this scale a galaxy equal in size to our own Galaxy would look like a fuzzy dot, 0.1 millimeter across. Thus, the galaxies are too small to show up clearly on a photograph. Instead we must rely on a plot of the positions of the galaxies. The plot shows the positions of about 200 of the brightest galaxies in an angular patch of the sky, 40° by 40°. The dense cluster of galaxies in the lower half of the picture is the Virgo Cluster.

Since we are looking into a volume of space, some of the galaxies are in the foreground, some are in the background; but our plot takes no account of perspective. The distance scale has been computed for those galaxies that are at a middle distance, at about 1.4 × 10^8 light-years from the Earth; the scale is not valid for galaxies in the foreground or in the background.

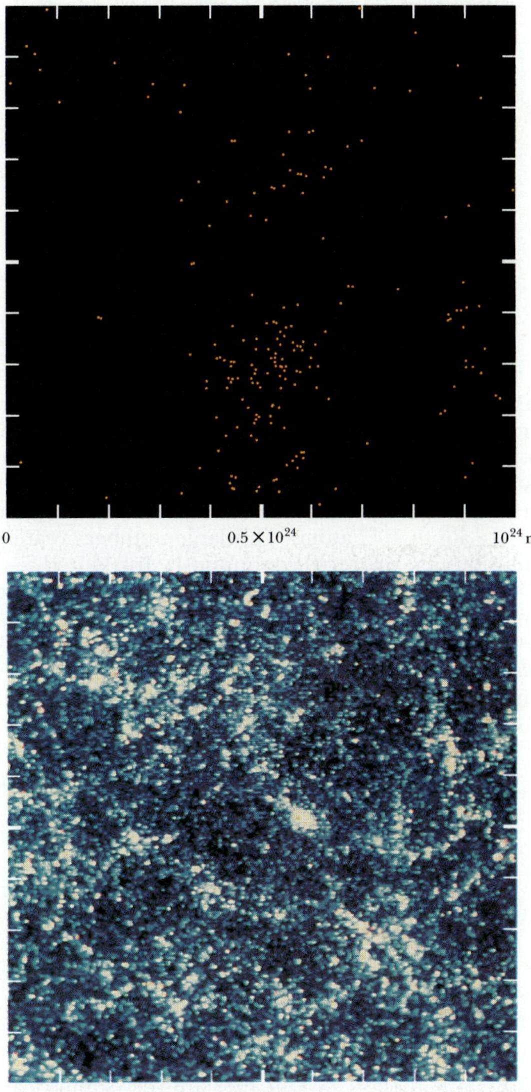

SCALE 1:1.5 × 10^{26} Now there are so many galaxies in our field of view that it is impractical to plot the positions of individual galaxies. In this picture, the sky has been divided into small squares, and the brightness of each square has been adjusted so it is proportional to the number of galaxies in that square. The picture includes about 2.3 × 10^5 galaxies in an angular patch of the sky, 40° by 40°. The distance scale has been computed for those galaxies that are at a middle distance, at about 1.4 × 10^9 light-years from the Earth.

This is the last of our pictures in the ascending series. We have reached the limits of zoom-out. If we wanted to draw another diagram, 10 times larger than this, we would need to know the shape and the size of the entire Universe. We do not yet know that.

Part II: The Small-Scale World

SCALE 1:1.5 We now return to Clare and zoom in on her eye. The surface of her skin appears smooth and firm. But this is an illusion. Matter appears continuous because the number of atoms in each cubic centimeter is extremely large. In a cubic centimeter of human tissue there are about 10^{23} atoms. This large number creates the illusion that matter is continuously distributed — we see only the forest and not the individual trees. The solidity of matter is also an illusion. The atoms in our bodies are mostly vacuum. As we will discover in the following pictures, within each atom the volume actually occupied by subatomic particles is only a very small fraction of the total volume.

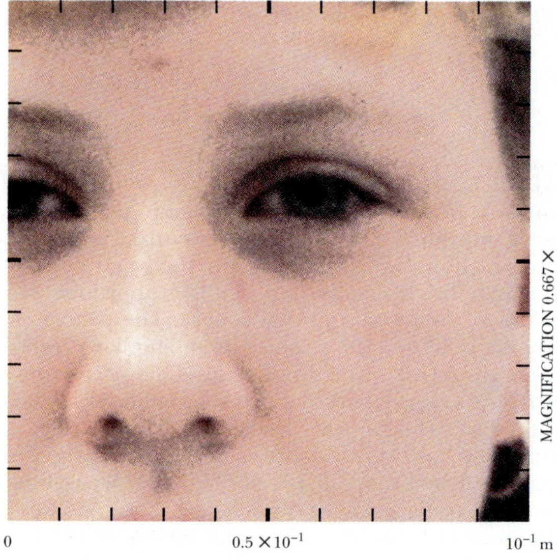

SCALE 1:1.5 × 10^{-1} Our eyes are very sophisticated sense organs; they collect more information than all our other sense organs taken together. The photograph shows the pupil and the iris of Clare's eye. Annular muscles in the iris change the size of the pupil and thereby control the amount of light that enters the eye. In strong light the pupil automatically shrinks to about 2 millimeters; in very weak light it expands to as much as 7 millimeters.

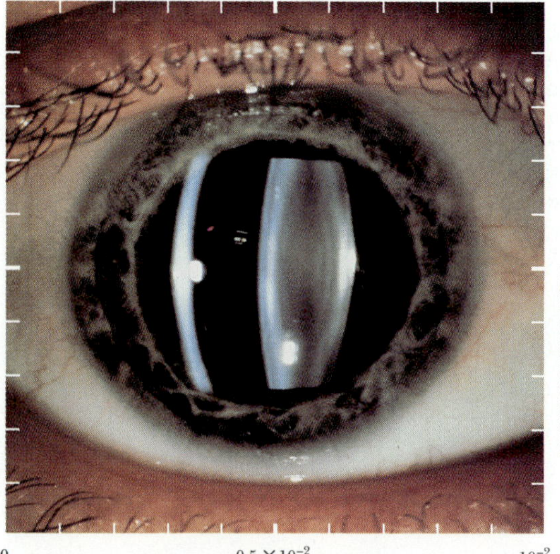

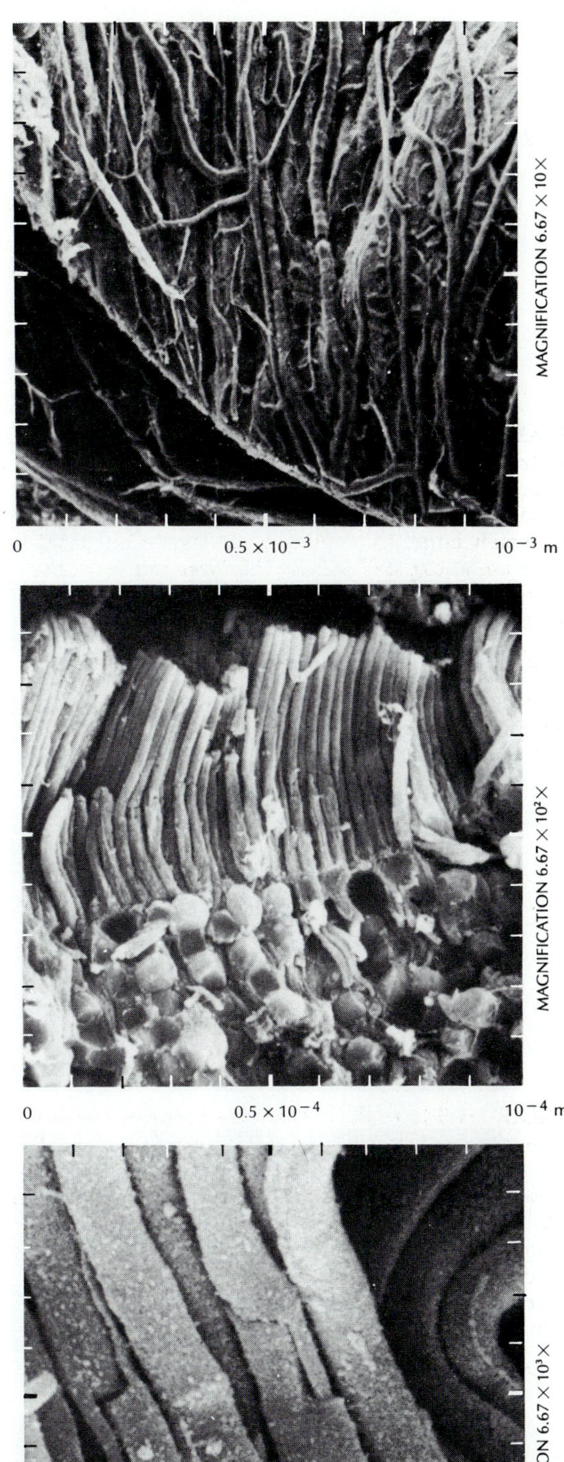

SCALE 1:1.5 × 10^{-2} This photograph shows the delicate network of blood vessels and nerve fiber bundles on the front surface of the retina, the light-sensitive membrane lining the interior of the eyeball. The rear surface of the retina is densely packed with two kinds of cells that sense light: cone cells and rod cells. In a human retina there are about 6 million cone cells and 120 million rod cells. The cone cells distinguish colors; the rod cells distinguish only brightness and darkness, but they are more sensitive than the cone cells and therefore give us vision in faint light ("night vision").

This and the following photographs were made with various kinds of ***electron microscopes.*** An ordinary microscope uses a beam of light to illuminate the object; an electron microscope uses a beam of electrons. Electron microscopes can achieve much sharper contrast and much higher magnification than ordinary microscopes.

SCALE 1:1.5 × 10^{-3} Here we have a clear photograph of rod cells prepared with a scanning electron microscope (SEM). For this photograph, the retina was cut apart and the microscope was aimed at the edge of the cut. In the top half of the picture we see tightly packed rods. Each rod is connected to the main body of a cell. In the bottom part of the picture we can distinguish tightly packed cell bodies. The round balls lying about in the middle are the nuclei of some cells. The cutting of the retina has broken some cells apart and has exposed their nuclei.

SCALE 1:1.5 × 10^{-4} This is a close-up view of a few rods, showing the scarfed joints between the upper and the lower portions. The upper portions of the rods contain a special pigment — visual purple — which is very sensitive to light. The absorption of light by this pigment initiates a chain of chemical reactions that finally trigger nerve pulses from the eye to the brain.

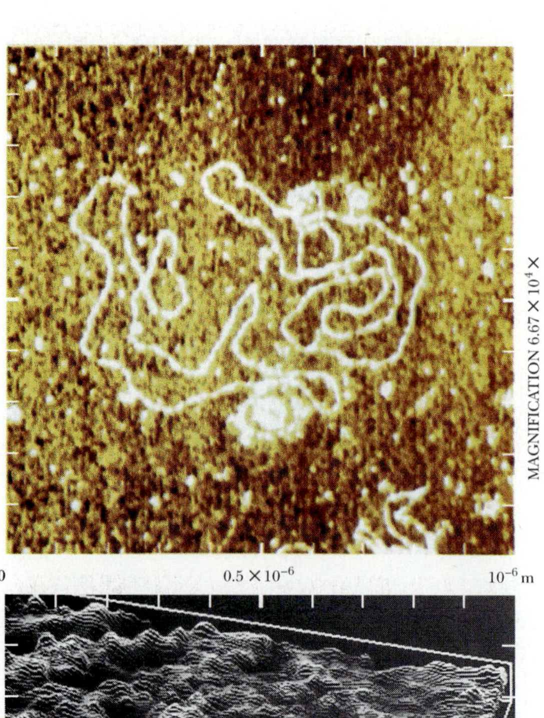

SCALE 1:1.5 × 10^−5^ These are strands of DNA, or deoxyribonucleic acid, as seen with a transmission electron microscope (TEM) at very high magnification. DNA is found in the nuclei of cells. It is a long molecule made by stringing together a large number of nitrogenous base molecules on a backbone of sugar and phosphate molecules. The base molecules are of four kinds, the same in all living organisms. But the sequence in which they are strung together varies from one organism to another. This sequence spells out a message — the base molecules are the "letters" in the "words" of this message. The message contains all the genetic instructions governing the metabolism, growth, and reproduction of the cell.

The strands of DNA in the photograph are encrusted with a variety of small protein molecules. At intervals, the strands of DNA are wrapped around larger protein molecules that form lumps looking like the beads of a necklace.

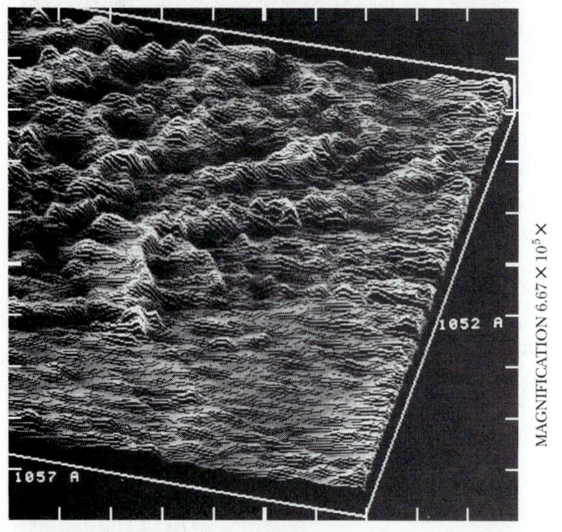

SCALE $1{:}1.5 \times 10^{-6}$ The highest magnifications are attained by a new kind of electron microscope, the scanning tunneling microscope (STM). This picture and the next two pictures were prepared with such a microscope. The picture shows strands of DNA deposited on a substrate of graphite. In contrast to the strands of the preceding picture, these strands are uncoated, that is, they are without protein encrustations.

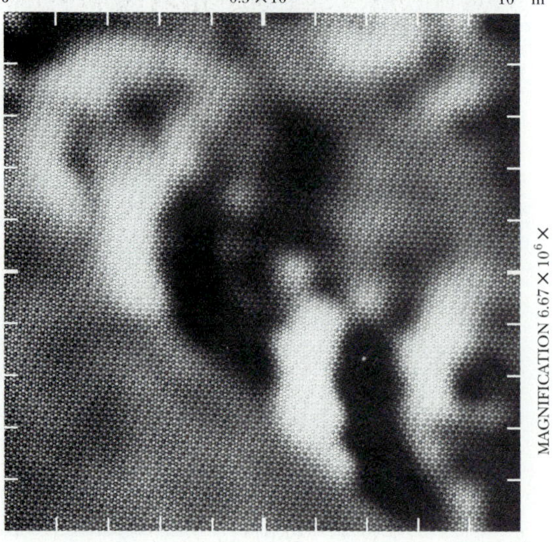

SCALE $1{:}1.5 \times 10^{-7}$ This close-up picture of a strand of DNA reveals the helical structure of this molecule. The strand consists of a pair of helical coils (white) wrapped around each other. The narrow groove (black) indicates the gap between the two coils in the pair.

SCALE 1:1.5 × 10^{-8} This picture shows the atoms on the surface of a small grain of gold. Here we have visual evidence of the atomic structure of matter. The gold atoms are arranged in a symmetric, repetitive pattern. Materials with such regular arrangements of atoms are called ***crystals.***

Each of the gold atoms is approximately a sphere, about 3×10^{-10} meter across. However, the atom does not have a sharply defined boundary; its surface is somewhat fuzzy. Atoms of other elements are also approximately spheres, with sizes that range from 2×10^{-10} to 4×10^{-10} meter across.

At present we know of more than 100 kinds of atoms or chemical elements. The lightest atom is hydrogen, with a mass of 1.67×10^{-27} kilogram; the heaviest is element 109, with a mass about 266 times as large.

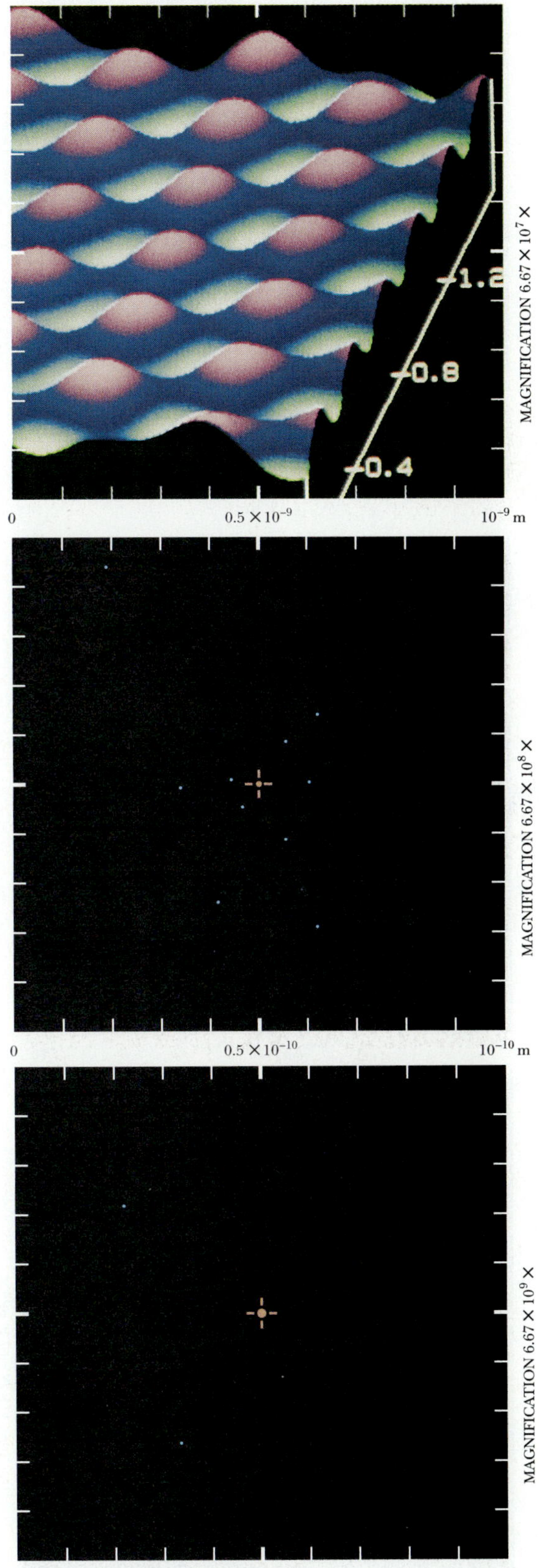

SCALE 1:1.5 × 10^{-9} The drawing shows the interior of an atom of neon. This atom consists of 10 electrons orbiting around a nucleus. In the drawing, the electrons have been indicated by small dots, and the nucleus by a slightly larger dot at the center of the picture. These dots have been drawn as small as possible, but even so the size of these dots does not give a correct impression of the actual sizes of the electrons and of the nucleus. The electron is smaller than any other particle we know; maybe the electron is truly pointlike and has no size at all. The nucleus has a finite size, but this size is much too small to show up on the drawing. Note that the electrons tend to cluster near the center of the atom. However, the overall size of the atom depends on the distance to the outermost electron; this electron defines the outer edge of the atom.

The electrons move around the nucleus in a very complicated motion. The drawing shows the electrons as they would be seen at one instant of time with a ***Heisenberg microscope.*** This is a hypothetical microscope that employs gamma rays instead of light rays to illuminate an object; no such microscope has yet been built.

The mass of each electron is 9.1×10^{-31} kilogram, but most of the mass of the atom is in the nucleus; the 10 electrons of the neon atom have only 0.03% of the total mass of the atom.

SCALE 1:1.5 × 10^{-10} Here we are closing in on the nucleus. We are seeing the central part of the atom. Only two electrons are in our field of view; the others are beyond the margin of the drawing. The size of the nucleus is still much smaller than the size of the dot at the center of the drawing.

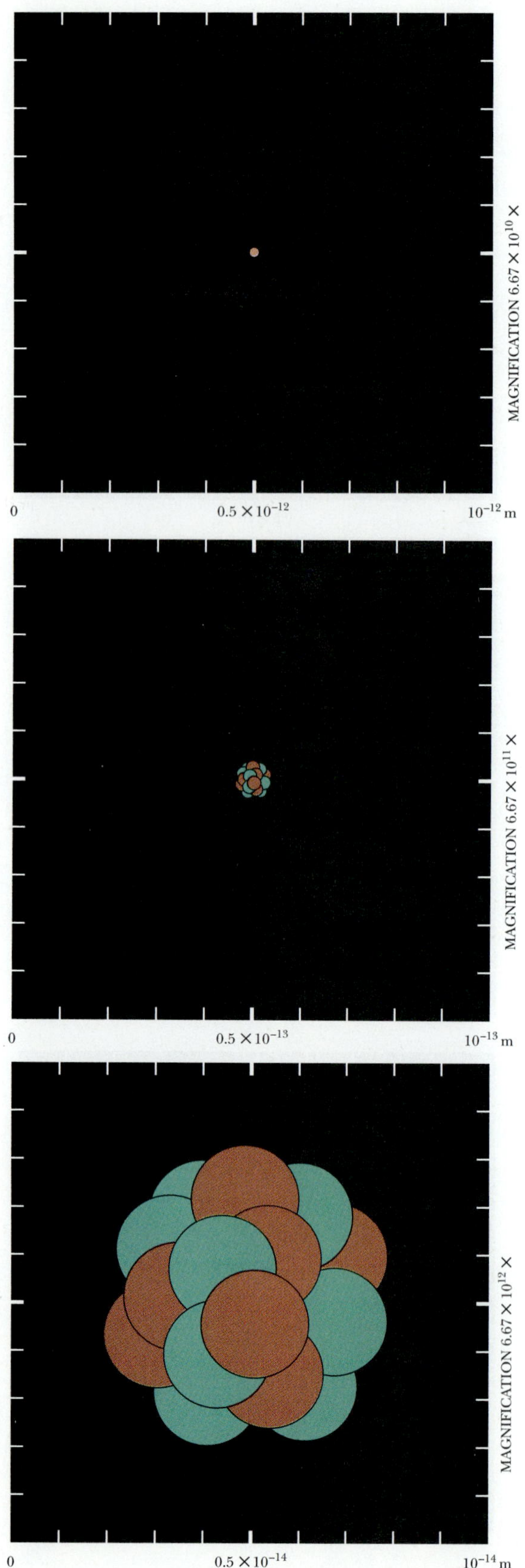

SCALE 1:1.5 × 10^{-11} In this drawing we finally see the nucleus in its true size. At this magnification, the nucleus of the neon atom looks like a small dot, 0.5 millimeter in diameter. Since the nucleus is extremely small and yet contains most of the mass of the atom, the density of the nuclear material is enormous. If we could assemble a drop of pure nuclear material of a volume of 1 cubic centimeter, it would have a mass of 2.3×10^{11} kilograms, or 230 million metric tons!

Our drawings show clearly that most of the volume within the atom is empty space. The nucleus occupies only a very small fraction of this volume.

SCALE 1:1.5 × 10^{-12} We can now begin to distinguish the nuclear structure. The nucleus has a nearly spherical shape, but its surface is slightly fuzzy.

SCALE 1:1.5 × 10^{-13} At this extreme magnification we can see the details of the nuclear structure. The nucleus of the neon atom is made up of 10 protons (white balls) and 10 neutrons (gray balls). Each proton and each neutron is a sphere with a diameter of about 2×10^{-15} meter, and a mass of 1.67×10^{-27} kilogram. In the nucleus, these protons and neutrons are tightly packed together, so tightly that they almost touch. The protons and neutrons move around the volume of the nucleus at high speed in a complicated motion.

SCALE 1:1.5 × 10^{-14} This final picture shows three pointlike bodies within a proton. These pointlike bodies are ***quarks*** — each proton and each neutron is made of three quarks. Recent experiments have told us that the quarks are much smaller than protons, but we do not yet know their precise size. Hence the dots in the drawing probably do not give a fair description of the size of the quarks. The quarks within protons and neutrons are of two kinds, called ***up*** and ***down.*** The proton consists of two *up* quarks and one *down* quark joined together; the neutron consists of one *up* quark and two *down* quarks joined together.

This final picture takes us to the limits of our knowledge of the subatomic world. As a next step we would like to zoom in on the quarks and show what they are made of. But we do not yet know whether they are made of anything else.

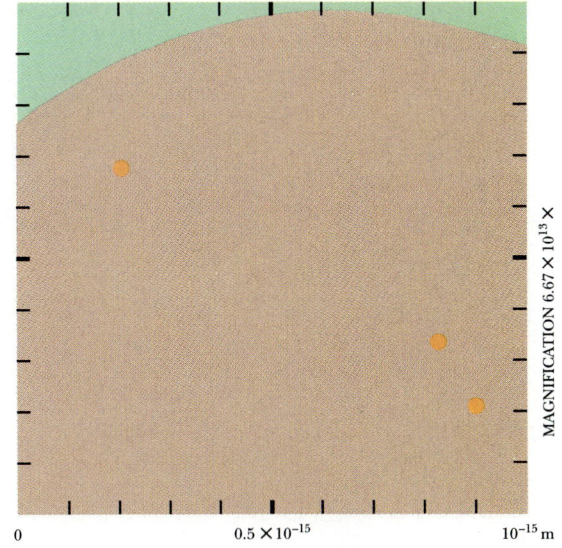

CHAPTER 1

On this straight railroad track, the handcar moves along a straight line. The motion is one dimensional.

MOTION ALONG A STRAIGHT LINE

The branch of physics that studies the motion of bodies is called ***mechanics.*** In antiquity, the science of mechanics (from the Greek *mechane,* machine) was the study of machines, and this is still what we have in mind when we call an automobile repairperson a mechanic. But physicists soon recognized that the essential aspect of the study of machines is motion, and mechanics thus became the study of motion. Broadly speaking, mechanics is divided into kinematics and dynamics. ***Kinematics*** deals with the mathematical description of motion in terms of position, velocity, and acceleration; ***dynamics*** deals with force as the cause of changes in motion.

◄ **Kinematics and dynamics**

In this chapter, and in the next six chapters, we will be concerned only with the translational motion of a particle, that is, a change of position of the particle with time. In the case of an ideal particle — a pointlike mass of infinitesimal size — ***the dependence of position on time provides a complete description of the motion of the particle.*** In the case of a more complicated body — an automobile, a ship, or a planet — the dependence of position on time does not provide a complete description. Such a complicated body can rotate, or change its orientation in space; furthermore, the body has many internal parts which can move in relation to one another. But insofar as we are not interested in the size, shape, orientation, and internal structure of a complicated body, we many find it useful to concentrate on its translational motion and ignore all rotational or internal motions. Under these circumstances we may pretend that the motion of the complicated body is particle motion.

We begin this chapter with a brief discussion of how physicists describe and measure position and motion.

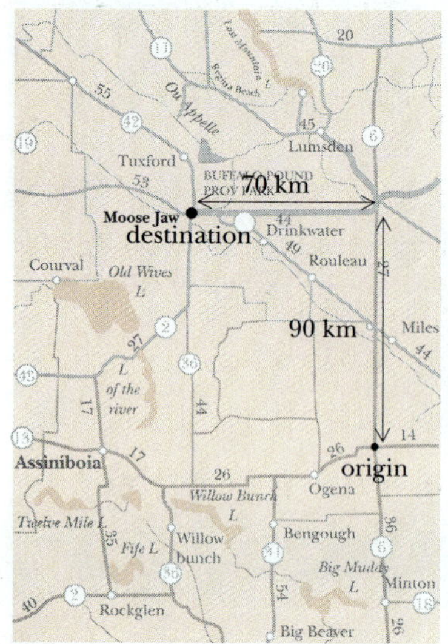

FIGURE 1.1

To reach Moose Jaw, Canada, the automobile has to travel 90 km north and then 70 km west.

1.1 MEASUREMENT OF SPACE AND TIME; UNITS

If you are lost somewhere on the highways in Canada and you stop at a service station to ask for directions to Moose Jaw, the attendant might instruct you to go 90 kilometers north on Route 6 and then 70 kilometers west on Route 1 (see Fig. 1.1). In giving these instructions, the attendant is taking the service station as ***origin,*** and he is specifying the position of Moose Jaw relative to this origin. To achieve a precise, quantitative description of the position of a particle, physicists use much the same procedure. They first take some convenient point of space as origin and then specify the position of the particle relative to this origin. For this purpose, they imagine a grid of lines around the origin and give the location of the particle within this grid; that is, they imagine that the ground is covered with graph paper, and they specify the position of the particle by means of coordinates read off this graph paper.

Rectangular coordinates ▶

The most common kind of coordinates are rectangular coordinates x ***and*** y***, which rely on a rectangular grid.*** Figure 1.2 shows such a rectangular grid. The mutually perpendicular lines through the origin O are called the x and y axes. The coordinates of the grid point P, where the particle is located, simply tell us how far we must move parallel to the corresponding axis in order to go from the origin O to the point P. For example, the point P shown in Fig. 1.2a has coordinates $x = 3$ units and $y = 5$ units. If we move in a direction opposite to that indicated by the arrow on the axis, then the coordinate is negative; thus, the point P shown in Fig. 1.2b has a negative x coordinate, $x = -3$ units.

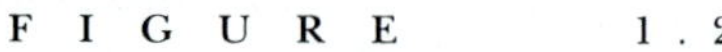

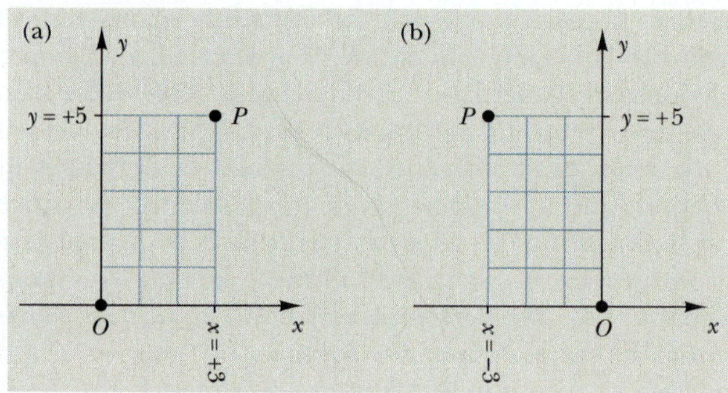

FIGURE 1.2

Rectangular coordinates x and y of a point P. In (a) both coordinates are positive; in (b), the x coordinate is negative.

The two-dimensional grid shown in Fig. 1.2 is adequate when we want to describe the two-dimensional (east–west and north–south) motion of an automobile traveling on flat ground or the motion of a ship on the (nearly) flat surface of the water of a harbor. However, if we want to describe the three-dimensional (east–west, north–south, and up–down) motion of an aircraft flying through the air, then we need a three-dimensional grid, with x, y, and z axes. And if we want to describe the motion of an automobile along a straight road, then we need only a one-dimensional grid, that is, we need only the x axis, which we imagine placed along the road.

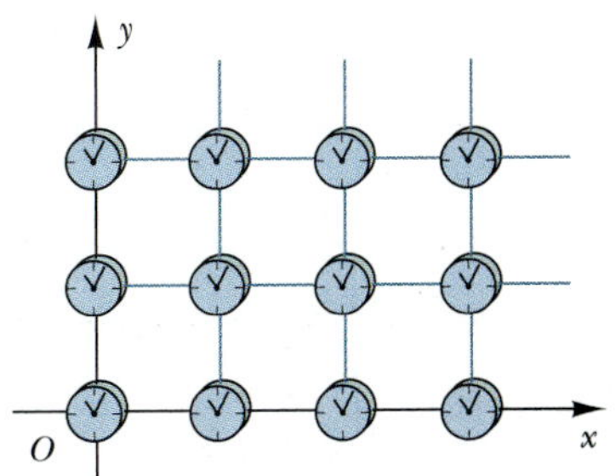

FIGURE 1.3

A reference frame consisting of a coordinate grid and synchronized clocks.

To achieve a complete description of the motion, we must specify both the position of the particle and the time at which it has this position. For the determination of the time, we use a set of synchronized clocks which we imagine arranged at regular intervals along the coordinate grid (see Fig. 1.3). When a particle passes through a grid point P, the coordinates give us the position of the particle in space, and the time registered by the nearby clock gives us the time t. ***Such a coordinate grid with an array of synchronized clocks is called a reference frame.***

◂ **Reference frame**

In order to make numerical records of our measurements of position and time, we need to select a unit of length and a unit of time, so we can express our measurements as numerical multiples or fractions of these units. The system of units most widely used by scientists is the ***metric system,*** which is based on ***the meter as the unit of length, the second as the unit of time, and the kilogram as the unit of mass*** (we will deal with mass in Chapter 3).

Originally, the standard of length that defined the size of one meter (1 m) was the standard meter bar kept at the Bureau International des Poids et Mesures (International Bureau of Weights and Measures) at Sèvres, France. This is a bar made of platinum–iridium alloy with a fine scratch mark near each end (see Fig. 1.4). By definition, the distance between these scratch marks was taken to be exactly one meter.

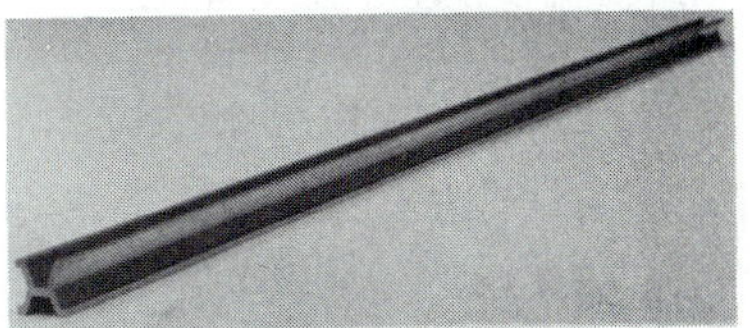

FIGURE 1.4

International standard meter bar.

The precision of the standard meter bar is limited by the coarseness of the scratch marks at the ends. For the sake of higher precision, physicists developed other, improved, definitions of the standard of length. The most recent improvement emerged from the development of stabilized lasers (see Fig. 1.5). This kind of laser emits light waves of extreme unifor-

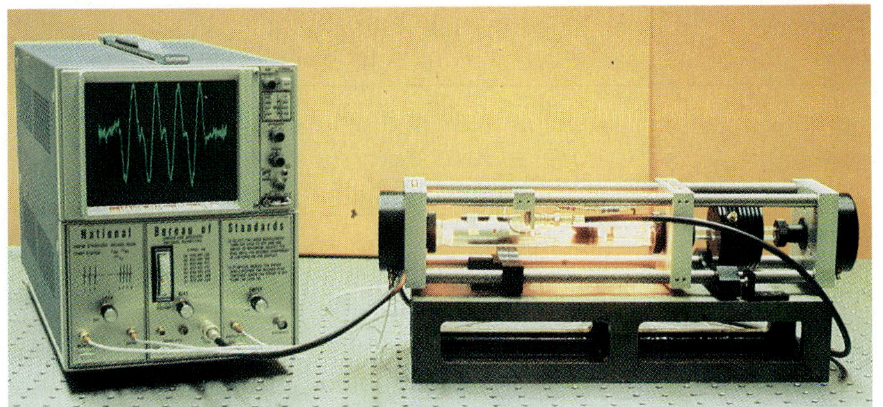

FIGURE 1.5

Stabilized laser at the National Institute of Standards and Technology.

mity which made it possible to determine the speed of light with extreme precision. This led to a new definition, officially adopted in 1983, for the length of the meter in terms of the speed of light: ***the meter is the length traveled by a light wave in vacuum in a time interval of 1/299,792,458 second.*** Note that, since the meter is adjusted so that light travels exactly one meter in 1/299,792,458 second, the speed of light is exactly

Meter (m) ▶

$$[\text{speed of light}] = 299{,}792{,}458 \text{ meters per second} \tag{1}$$

TABLE 1.1	SOME DISTANCES AND SIZES
Distance to boundary of observable Universe	$\simeq 1 \times 10^{26}$ m
Distance to Andromeda galaxy	2.1×10^{22} m
Diameter of our Galaxy	7.6×10^{20} m
Distance to nearest star (Proxima Centauri)	4.0×10^{16} m
Earth–Sun distance	1.5×10^{11} m
Radius of Earth	6.4×10^{6} m
Wavelength of radio wave (AM band)	$\simeq 3 \times 10^{2}$ m
Length of ship *Queen Elizabeth*	3.1×10^{2} m
Height of man (average male)	1.8 m
Diameter of 5¢ coin	2.1×10^{-2} m
Diameter of red blood cell (human)	7.5×10^{-6} m
Wavelength of visible light	$\simeq 5 \times 10^{-7}$ m
Diameter of smallest virus (potato spindle)	2×10^{-8} m
Diameter of atom	$\simeq 1 \times 10^{-10}$ m
Diameter of atomic nucleus (iron)	8×10^{-15} m
Diameter of proton	2×10^{-15} m

TABLE 1.2	MULTIPLES AND SUBMULTIPLES OF THE METER
kilometer	1 km = 10^{3} m
meter	1 m
centimeter	1 cm = 10^{-2} m
millimeter	1 mm = 10^{-3} m
micron	1 μm = 10^{-6} m
nanometer	1 nm = 10^{-9} m
Ångstrom	1 Å = 10^{-10} m
fermi	1 fm = 10^{-15} m

TABLE 1.3	MULTIPLES AND SUBMULTIPLES OF THE FOOT
mile	1 mi = 5280 ft = 1609.38 m
yard	1 yd = 3 ft = 0.9144 m
foot	1 ft = 0.3048 m
inch	1 in. = $\frac{1}{12}$ ft = 2.540 cm

Thus, the new definition of the meter amounts to the adoption of the speed of light as a standard of speed.

Table 1.1 lists a few distances and sizes, from the largest to the smallest. Many of the quantities listed in the table have already been mentioned in the Prelude. Quantities indicated with an $\simeq$ sign are not precisely defined; these quantities are rough approximations.

Table 1.2 lists some multiples and submultiples of the meter and their abbreviations.

In the British system of units — abandoned by Britain and almost all other countries, but regrettably still in use in the United States — the unit of length is the ***foot*** (ft), which is exactly 0.3048 m. Table 1.3 gives the multiples and submultiples of the foot, but we will have little need for this table because we will hardly ever use British units in this book.

The unit of time is the ***second.*** Originally one second (1 s) was defined as $1/(60 \times 60 \times 24)$, or 1/86,400 of a mean solar day. The solar day is the time interval required for the Earth to complete one rotation relative to the Sun. The length of the solar day depends on the rate of rotation of the Earth, which is subject to a host of minor variations, both seasonal and long-term, which make the rotation of the Earth an imperfect timekeeper.

FIGURE 1.6

Cesium atomic clock at the National Institute of Standards and Technology.

To avoid any variation in the unit of time, we now use an atomic standard of time. This standard is the period of one vibration of an atom of cesium. ***The second is defined as the time needed for 9,192,631,770 vibrations of a cesium atom.*** Figure 1.6 shows one of the atomic clocks at the National Institute of Standards and Technology in Boulder, Colorado; in this clock, the feeble vibrations of the cesium atoms are amplified to a level that permits them to control the dial of the clock. Good cesium clocks are very, very good — they lose or gain no more than one second in 300,000 years.

◀ **Second (s)**

Precise time signals keyed to the cesium atomic clocks of the National Institute of Standards are continuously transmitted by radio station WWV, Fort Collins, Colorado. These time signals can be picked up worldwide on shortwave receivers tuned to 2.5, 5, 10, 15, or 20 megahertz. Precise time signals are also announced continuously by telephone — the telephone number is (303)499-7111. The time announced on the radio and on the telephone is Coordinated Universal Time (Greenwich time), which is exactly 5 hours ahead of Eastern Standard Time.

Table 1.4 lists some typical time intervals, and Table 1.5 gives multiples and submultiples of the second.

▼ **Derived units**

The meter, the second, and the kilogram are the fundamental units, or ***base units, for length, time, and mass. Any other physical quantity can be measured by introducing a derived unit constructed from some combination of the base units.*** For example, ***area*** can be measured with a derived unit that is the square of the unit of length; thus, in the metric system, the unit of area is the square meter, which is the area of a square one meter on a side ($1\ \text{m}^2 = 1\ \text{m} \times 1\ \text{m}$; see Fig. 1.7a). And ***volume*** can be measured with a derived unit that is the cube of the unit of length; in the metric system, the unit of volume is the cubic meter, which is the volume of a cube one meter on a side ($1\ \text{m}^3 = 1\ \text{m} \times 1\ \text{m} \times 1\ \text{m}$; see Fig. 1.7b). Tables 1.6 and 1.7 give multiples and submultiples of these units.

(a)

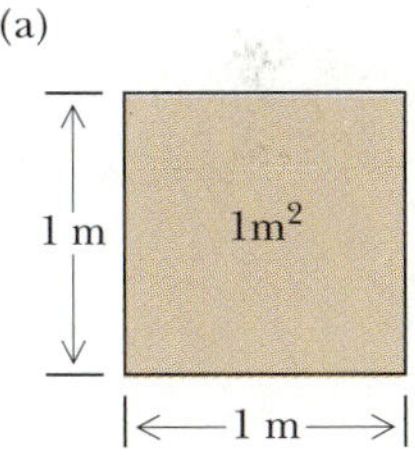

(b)

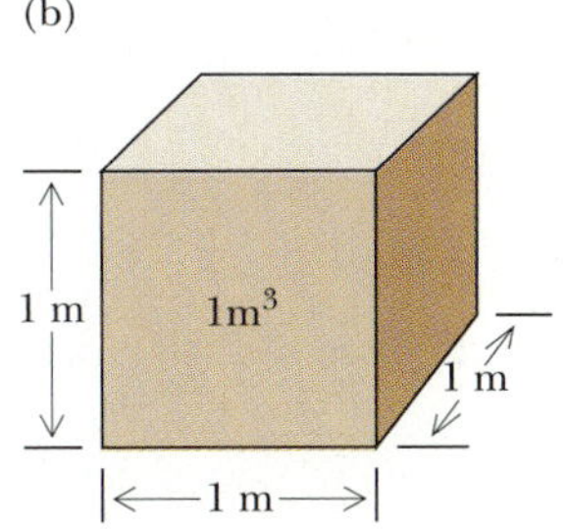

FIGURE 1.7

(a) One square meter. (b) One cubic meter.

TABLE 1.4	SOME TIME INTERVALS
Age of the Universe	$\simeq 4 \times 10^{17}$ s
Age of the Solar System	1.4×10^{17} s
Age of the oldest written records (Sumerian)	1.6×10^{11} s
Life-span of man (average)	2.2×10^{9} s
Travel time for light from nearest star	1.4×10^{8} s
Revolution of Earth (1 year)	3.2×10^{7} s
Rotation of Earth (1 day)	8.6×10^{4} s
Life-span of free neutron (average)	9.2×10^{2} s
Travel time for light from Sun	5×10^{2} s
Travel time for light from Moon	1.3 s
Period of heartbeat (human)	$\simeq 0.9$ s
Period of sound wave (middle C)	3.8×10^{-3} s
Period of radio wave (AM band)	$\simeq 1 \times 10^{-6}$ s
Period of light wave	$\simeq 2 \times 10^{-15}$ s
Life-span of shortest lived, unstable particle	$\simeq 10^{-24}$ s

TABLE 1.5	MULTIPLES AND SUBMULTIPLES OF THE SECOND
century	1 century = 100 yr = 3.156×10^{9} s
year	1 year = 3.156×10^{7} s = 365.25 days
day	1 day = 86,400 s
hour	1 h = 3600 s
minute	1 min = 60 s
millisecond	1 ms = 10^{-3} s
microsecond	1 μs = 10^{-6} s
nanosecond	1 ns = 10^{-9} s
picosecond	1 ps = 10^{-12} s

TABLE 1.6	MULTIPLES AND SUBMULTIPLES OF THE SQUARE METER
square meter	1 m^2
square kilometer	1 $km^2 = 10^6\ m^2$
square centimeter	1 $cm^2 = 10^{-4}\ m^2$
square millimeter	1 $mm^2 = 10^{-6}\ m^2$

TABLE 1.7	MULTIPLES OF SUBMULTIPLES OF THE CUBIC METER
cubic meter	1 m^3
cubic kilometer	1 $km^3 = 10^9\ m^3$
liter	1 liter = $10^{-3}\ m^3 = 10^3\ cm^3$
cubic centimeter	1 $cm^3 = 10^{-6}\ m^3$
cubic millimeter	1 $mm^3 = 10^{-9}\ m^3$

EXAMPLE 1. Your heart pumps 92 cm^3 of blood per second (when you are resting). How much blood does it pump per day? Express the answer in m^3.

SOLUTION: A day has 24 hours of 60 minutes of 60 seconds, that is, $24 \times 60 \times 60\ s = 8.64 \times 10^4$ s. Given that in each second the heart pumps 92 cm^3, in a day it pumps

$$92\ cm^3 \times 8.64 \times 10^4 = 7.9 \times 10^6\ cm^3 \tag{2}$$

Since $1\ cm = 10^{-2}$ m, we find that

$$1\ cm^3 = 1\ cm \times 1\ cm \times 1\ cm = 10^{-2}\ m \times 10^{-2}\ m \times 10^{-2}\ m = 10^{-6}\ m^3$$

(This result agrees with Table 1.7, and we could have copied it from there.) The volume of blood pumped in one day then becomes

$$7.9 \times 10^6\ cm^3 = 7.9 \times 10^6 \times (10^{-6}\ m^3) = 7.9\ m^3$$

COMMENTS AND SUGGESTIONS: In Eq. (2), the result of the multiplication has been rounded off to two significant figures, since the data for the problem have two significant figures. As mentioned in the Prelude, in scientific notation, the digits, or ***significant figures, in a number indicate the precision to within which the number is known.*** As a general rule, the final result of a calculation cannot be more precise than the data used in the calculation; hence you should always round off the final result in a manner consistent with the data. In a multiplication or a division, consistency means that the result should be rounded to the least number of significant figures that appear in the multiplicands; for example, $2.00 \times 1.41 = 2.82$, but $2.0 \times 1.41 = 2.8$. In an addition or subtraction, consistency means that the result should be rounded to the last significant figure that is present in *both* of the addends, when they are written in decimal notation without powers of ten; for example, $2.00 + 1.41 = 3.41$, but $2.0 + 1.41 = 3.4$, and $(2.0 \times 10^{-3}) + 1.41 = 0.0020 + 1.41 = 1.41$. Note that in this last example the 0.0020 is too small to make a "significant" contribution when added to 1.41.

◂ **Significant figures**

1.2 AVERAGE SPEED

If you travel the 160 km of highway from the service station to Moose Jaw in 2 hours, you would say that your average speed for the trip is 80 kilometers per hour. But this statement relies on the implicit assumption that the motion of the automobile can be regarded as particle motion. To make the definition and the measurement of the average speed precise and unambiguous, we must pick some marker point on the automobile, which we imagine painted on the automobile. For instance, we might take

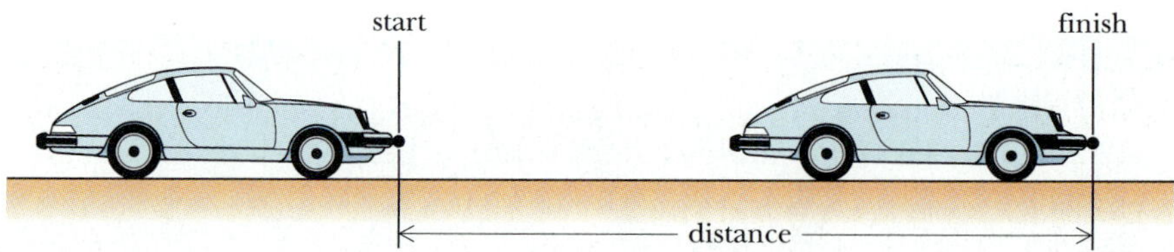

FIGURE 1.8 For a precise measurement, the departure and the arrival of the automobile must be reckoned according to when the front end of the bumper crosses the starting line and the finishing line.

the center of the front bumper as our marker point. We reckon departure and arrival according to when the marker point crosses the starting line and the finishing line, and we reckon distance according to the path traced out by the marker point (see Fig. 1.8). We can then treat the motion of the automobile as particle motion.

Average speed ▶

For a particle that travels along some path, straight or curved, the distance traveled and the time taken to travel this distance have a precise, unambiguous meaning. ***The average speed of the particle is defined as the ratio of this distance to the magnitude of the time interval.*** Expressing this as a formula, we can write

$$[\text{average speed}] = \frac{[\text{distance traveled}]}{[\text{time taken}]} \tag{3}$$

Thus, the average speed is the distance traveled per unit time. We see from Eq. (3) that the unit of speed is the unit of length divided by the unit of time. ***In the metric system, the unit of speed is the meter per second*** (m/s). In practice, the speed of automobiles, aircraft, and other everyday bodies is often measured in kilometers per hour (km/h),

$$1 \text{ km/h} = \frac{1000 \text{ m}}{3600 \text{ s}} = 0.278 \text{ m/s}$$

Table 1.8 gives some examples of typical speeds.

FIGURE 1.9 Carl Lewis winning the 100-m race in 1991.

EXAMPLE 2. The world record set by Carl Lewis for the 100-m run is 9.86 s (Fig. 1.9). What average speed, in m/s, did he attain while setting this record?

SOLUTION: According to Eq. (3),

$$[\text{average speed}] = \frac{[\text{distance traveled}]}{[\text{time taken}]} = \frac{100 \text{ m}}{9.86 \text{ s}} = 10.1 \text{ m/s}$$

Here the final result has been rounded off to three significant figures, since the data for the problem have three significant figures.

EXAMPLE 3. Express the average speed of the runner in km/h.

SOLUTION: With the substitutions 1 m = (1/1000) km and 1 s = (1/3600) h, the average speed becomes

$$10.1\ \frac{\text{m}}{\text{s}} = 10.1\ \frac{(1/1000)\ \text{km}}{(1/3600)\ \text{h}} = 36.4\ \text{km/h}$$

COMMENTS AND SUGGESTIONS: An alternative method for the conversion of units from one set of units is the following: since 1 km = 1000 m, we have the identity

$$1 = \frac{1\ \text{km}}{1000\ \text{m}}$$

and similarly,

$$1 = \frac{3600\ \text{s}}{1\ \text{h}}$$

This means that any quantity can be multiplied by (1 km)/(1000 m) or (3600 s)/(1 h) without changing its value. Thus, starting with 10.1 m/s, we obtain

$$10.1\ \frac{\text{m}}{\text{s}} = 10.1\ \frac{\cancel{\text{m}}}{\cancel{\text{s}}} \times \frac{1\ \text{km}}{1000\ \cancel{\text{m}}} \times \frac{3600\ \cancel{\text{s}}}{1\ \text{h}}$$

Multiplying this out, and canceling the m and s (as indicated by the colored slashes), we find

$$10.1\ \frac{\text{m}}{\text{s}} = 10.1 \times \frac{3600\ \text{km}}{1000\ \text{h}} = 36.4\ \text{km/h}$$

Factors such as (1 km)/(1000 m) or (3600 s)/(1 h), consisting of the ratio of two equal quantities, are called ***conversion factors.*** To change the units of a quantity, simply multiply the quantity by whatever conversion factors will bring about the cancellation of the old units and leave the result in the new, desired units.

◄ Conversion factors

Motion and speed are relative; the value of the speed depends on the frame of reference with respect to which it is calculated. Example 2 gave the speed of the runner relative to a reference frame attached to the surface of the Earth. However, relative to the reference frame of a bicyclist riding the same direction as the runner, the speed will be different (Fig. 1.10). For instance, if the groundspeed of the bicyclist is also 10.1 m/s, then relative to her reference frame, the runner will have zero speed.

FIGURE 1.10
A bicyclist and her reference frame. If both the bicyclist and the runner are moving toward the right at the same speed, then the runner is at rest relative to the reference frame of the bicyclist.

Thus, questions regarding speed are meaningless unless the reference frame is first specified. In everyday language, "speed" often means speed relative to a reference frame attached to the Earth's surface. If the speed is reckoned relative to some other reference frame, this will usually be clear from the context. For example, in Table 1.8 the speed of the rifle

TABLE 1.8 SOME SPEEDS

Light	3.0×10^8 m/s
Recession of fastest known quasar	2.8×10^8 m/s
Electron around nucleus (hydrogen)	2.2×10^6 m/s
Earth around Sun	3.0×10^4 m/s
Rifle bullet (muzzle velocity) **(a)**	$\simeq 7 \times 10^2$ m/s
Random motion of molecules in air (average)	4.5×10^2 m/s
Sound	3.3×10^2 m/s
Jet airliner (Boeing 747, maximum airspeed)	2.7×10^2 m/s
Cheetah (maximum) **(b)**	28 m/s
Federal highway speed limit (55 mi/h)	25 m/s
Man (maximum)	12 m/s
Man (walking briskly)	1.3 m/s
Snail	$\simeq 10^{-3}$ m/s
Glacier **(c)**	$\simeq 10^{-6}$ m/s
Rate of growth of hair (human)	3×10^{-9} m/s
Continental drift	$\simeq 10^{-9}$ m/s

(a)

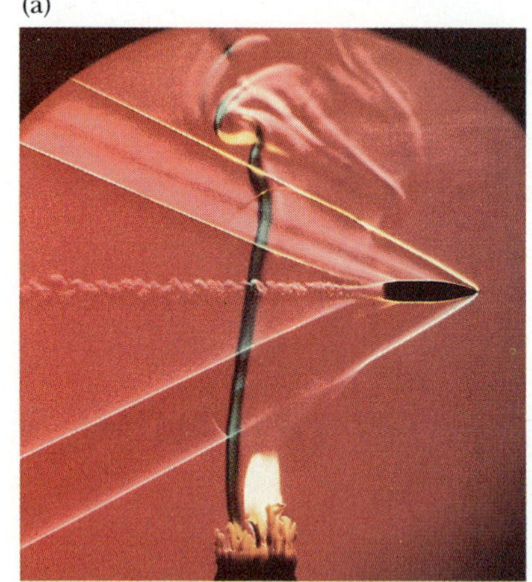

(b)

(c)

bullet is reckoned relative to the reference frame of the rifle (which may be in motion relative to the Earth). We will have to be careful to specify the reference frame whenever it is not clear from the context.

1.3 AVERAGE VELOCITY FOR MOTION ALONG A STRAIGHT LINE

For the rest of this chapter, we will consider the special case of motion along a straight line, that is, motion in one dimension. For convenience, we will assume that the straight line coincides with the x axis (see Fig. 1.11). We can then give a complete description of the motion by specifying the x coordinate at each instant of time. In mathematical language,

FIGURE 1.11
An automobile moving along a straight line. The x axis coincides with this straight line, and the position of the automobile can be completely specified by its x coordinate.

this means that the x coordinate is a function of time. Graphically, we can represent the motion by means of a plot of the x coordinate vs. the time coordinate. For example, Fig. 1.12 is such a plot of the position coordinate x vs. the time coordinate t for an automobile that starts from rest, accelerates along a straight road for 10 seconds, and then brakes and comes to a full stop 4.3 seconds later. The plot is based on data from a road test of a Maserati sports car; in this test, the driver floored the accelerator for 10 seconds, and then he slammed on the brakes. The position is measured from the starting point on the road to a marker point painted on the automobile. The plot shown in this figure gives a complete description of the (translational) motion of the automobile.

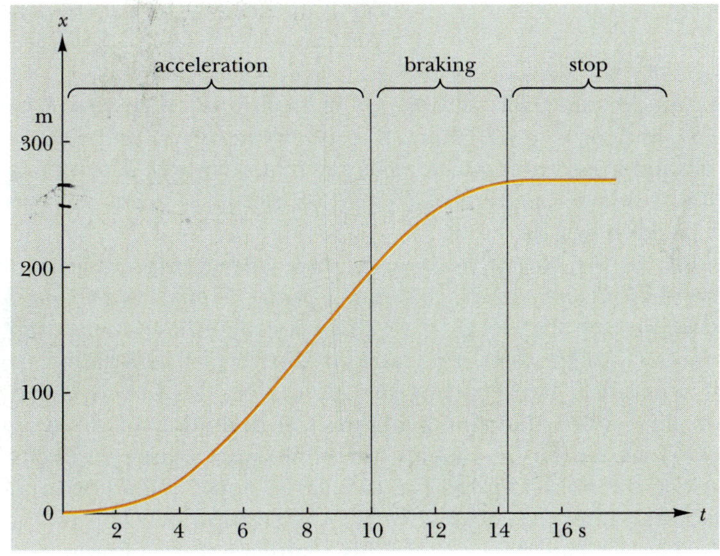

FIGURE 1.12
Plot of position vs. time for an automobile that accelerates ($0\text{ s} < t < 10\text{ s}$), then brakes ($10\text{ s} < t < 14.3\text{ s}$), and then stops ($t > 14.3\text{ s}$) (based on data from a road test of a Maserati Bora by *Road & Track* magazine).

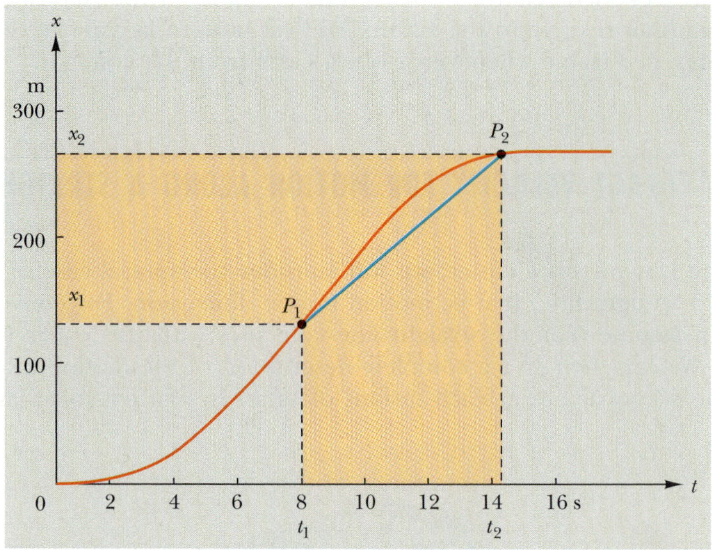

FIGURE 1.13
The average velocity for the interval from $t_1 = 8.0$ s to $t_2 = 14.3$ s is the slope of the straight line P_1P_2.

Suppose that at time t_1 the automobile is at position x_1 and at a subsequent time t_2 the automobile is at position x_2 (see Fig. 1.13). Then $x_2 - x_1$ is the change of position that occurred in the time interval $t_2 - t_1$. ***The ratio of this change of position to the time interval is called the average velocity,***

Average velocity ▶

$$\bar{v} = \frac{x_2 - x_1}{t_2 - t_1} \tag{4}$$

Here, the overbar on the symbol for velocity is a standard notation used in physics to indicate an average quantity. The expression (4) can also be written as

$$\bar{v} = \frac{\Delta x}{\Delta t} \tag{5}$$

where Δx and Δt stand for the *changes* in position and in time, that is, $\Delta x = x_2 - x_1$ and $\Delta t = t_2 - t_1$ (here, the Greek letter Δ, or "delta," is a standard notation used to indicate a change in a quantity). Thus, the average velocity is the average rate of change of the position, or the average change of position per unit time.

Graphically, ***in the plot of position vs. time, the average velocity is the ratio of the vertical separation between the points P_1 and P_2 and the horizontal separation*** (see Fig. 1.13). If we draw a straight line connecting the points P_1 and P_2 in the plot, the ratio of the vertical separation to the horizontal separation between any two points on this line is called the ***slope*** of the line. Note that this mathematical definition of slope agrees with the everyday notion of slope: a steep line, with a large vertical separation between the points P_1 and P_2, has a large slope; and a horizontal or nearly horizontal line, with a small vertical separation between the points P_1 and P_2, has a small slope. With this mathematical definition of the slope, we can say that the average velocity equals the slope of the straight line connecting the points P_1 and P_2. For instance, if $t_1 = 8.0$ s and $t_2 = 14.3$ s, then the average velocity, or the slope of the straight line connecting the points P_1 and P_2 in Fig. 1.13, is

$$\bar{v} = \frac{[\text{vertical separation}]}{[\text{horizontal separation}]} = \frac{x_2 - x_1}{t_2 - t_1}$$

$$= \frac{272\text{ m} - 130\text{ m}}{14.3\text{ s} - 8.0\text{ s}} = \frac{142\text{ m}}{6.3\text{ s}} = 23\text{ m/s} \qquad (6)$$

According to the general formula (4), the velocity is positive or negative depending on whether x_2 is larger or smaller than x_1, that is, depending on whether the x coordinate increases or decreases in the time interval $t_2 - t_1$. This means that the sign of the velocity depends on the direction of motion. If the motion is in the positive x direction — as in the example plotted in Fig. 1.13 — the velocity is positive; if the motion is in the negative x direction, the velocity is negative. Thus, according to the precise terminology used in physics, speed [defined by Eq. (3)] and velocity [defined by Eq. (4)] are not the same thing, because ***speed is always positive, whereas velocity is positive or negative depending on the direction of motion.*** Furthermore, if the motion has one portion in the positive x direction and another portion in the negative x direction, then it is possible for the average velocity to be zero even though the average speed is not zero, as the following example shows.

EXAMPLE 4. A runner runs 100 m on a straight track in 11.0 s and then walks back in 80.0 s. What are the average velocity and the average speed for each part of the motion and for the entire motion?

SOLUTION: The plot of position vs. time for the runner is shown in Fig. 1.14. The motion has two parts: the run (from $t = 0$ to $t = 11.0$ s) and the walk (from $t = 11.0$ s to $t = 91.0$ s). The average velocity for the run is

$$\bar{v} = \frac{\Delta x}{\Delta t} = \frac{+100\text{ m}}{11.0\text{ s}} = +9.09\text{ m/s}$$

The average velocity for the walk is

$$\bar{v} = \frac{-100\text{ m}}{80.0\text{ s}} = -1.25\text{ m/s}$$

(Here, the minus sign in -100 m indicates that the change of position is in the negative x direction.) The average velocity for the entire motion is

$$\bar{v} = \frac{0\text{ m}}{91.0\text{ s}} = 0\text{ m/s}$$

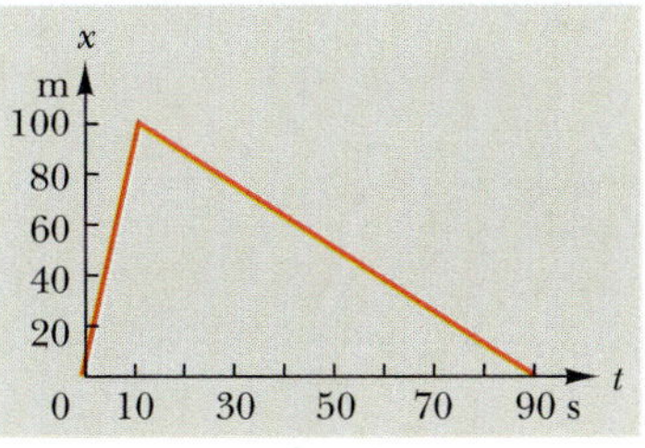

FIGURE 1.14
Plot of position vs. time for a runner.

This average velocity is zero because the net change of position is zero.

The average speeds for the run and the walk are, respectively, 9.09 m/s and 1.25 m/s (here, there is no minus sign, since the distance used in the calculation of speed is always reckoned as positive). The average speed for the entire motion is the total distance traveled divided by the time taken,

$$\frac{200\ \text{m}}{91.0\ \text{s}} = 2.20\ \text{m/s}$$

The average *speed* differs from the average *velocity* because the distance traveled (200 m) differs from the net change of position (zero).

1.4 INSTANTANEOUS VELOCITY

If, during the road test of the Maserati, the driver was keeping an eye on the speedometer, he would have seen the needle gradually climb from zero to a value well above the legal limit while he was stepping on the accelerator, and then quickly fall to zero when he slammed on the brakes. The speedometer of an automobile displays the instantaneous velocity, that is, the velocity at one instant of time (except that the speedometer does not make a distinction between positive velocity and negative velocity — it does not care whether you are driving eastward or westward along a street, and it displays only the *magnitude* of the instantaneous velocity, that is, it displays the instantaneous *speed*). But although we are all familiar with speedometers in automobiles, this familiarity does not help us with the mathematical definition of instantaneous velocity, because the design and the calibration of speedometers hinge on knowing the definition of what we are trying to measure.

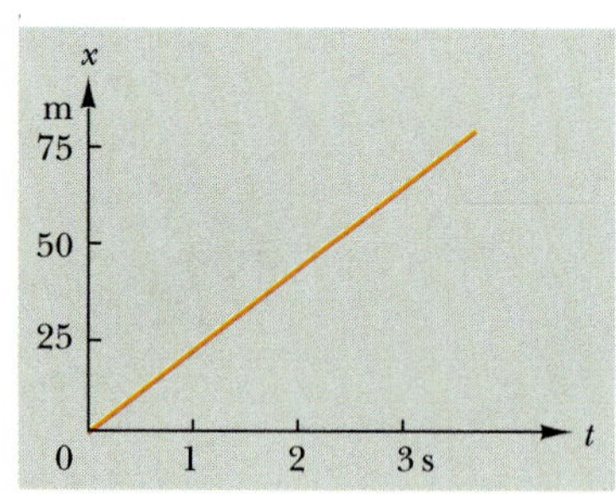

FIGURE 1.15

Plot of position vs. time for an automobile moving at constant velocity.

Since a moving automobile or a moving particle does not cover any distance in one instant, it is not immediately obvious how we should define the instantaneous velocity. Only in the exceptional case of a particle moving uniformly, with constant velocity, is the instantaneous velocity obvious — it then coincides with the average velocity. We can see this from an examination of the plot of position vs. time. If a particle moves at constant velocity, the plot of position vs. time is a straight line, with a slope equal to the velocity. For example, Fig. 1.15 shows a plot of position vs. time for an automobile moving along a straight road at a constant velocity of 25 m/s. This plot is a straight line of constant slope — the slope in any time interval is equal to the slope in any other time interval. Thus, the average velocity is the same for all time intervals — it is always 25 m/s. Since the velocity is the same for all time intervals, we may regard the instantaneous velocity for this motion as identical to the average velocity.

If a particle moves with a varying velocity, the plot of position vs. time is a curve. The plot of position vs. time for the accelerating automobile shown in Fig. 1.12 gives us an example of this: the automobile first accelerates and then brakes, and the plot of position vs. time is a curve of vary-

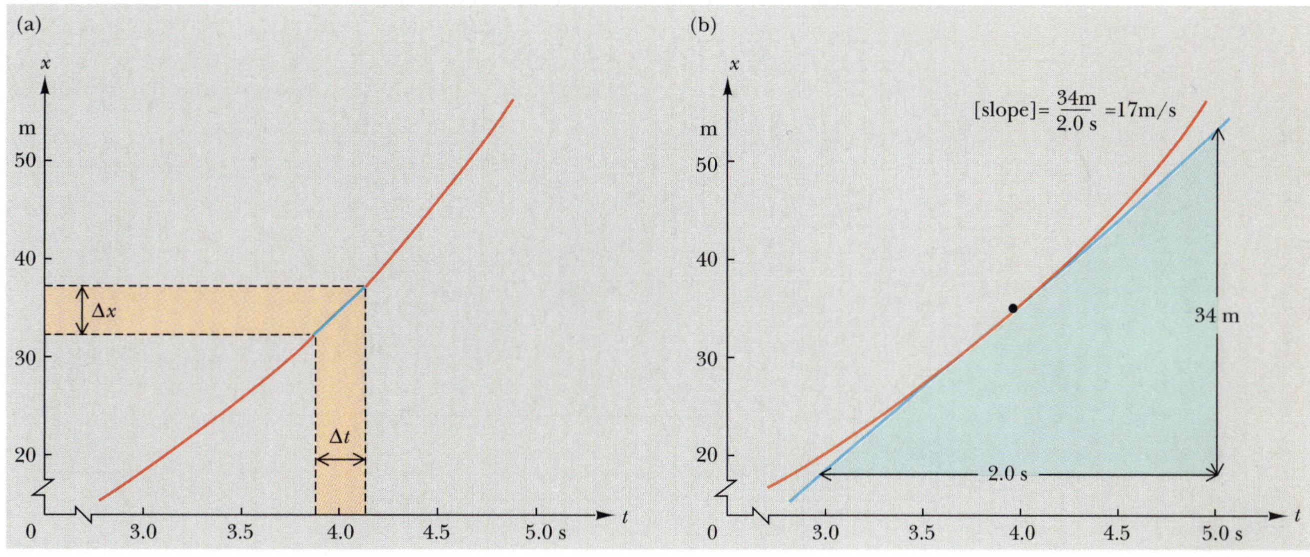

FIGURE 1.16

(a) Plot of position vs. time for an automobile moving with varying velocity. Over a small time interval, we can approximate the plot by a short straight line (blue). (b) The straight line tangent to the plot at 4.0 s rises by 34 m in 2.0 s; thus, the line has a slope of 34 m/2.0 s = 17 m/s.

ing slope. How can we construct a definition of the instantaneous velocity of the automobile on the basis of this plot?

To formulate a definition of the instantaneous velocity, consider the instant $t = 4$ s. We can find an *approximate* value for the velocity at this instant by taking a small time interval of, say, 0.001 s centered on 4 s, that is, a time interval from 3.9995 s to 4.0005 s. In this time interval the automobile covers some small distance Δx, and we can approximate the actual (curved) plot of position vs. time by a straight line segment connecting the endpoints of the interval (see Fig. 1.16a). According to the discussion at the beginning of this section, the instantaneous velocity associated with a straight plot of position vs. time is simply the slope of the plot; hence the instantaneous velocity at $t = 4$ s can be evaluated approximately as the slope of the short line segment shown in Fig. 1.16a. Whether this is a good approximation depends on how closely the straight line segment coincides with the actual curved plot. The approximation can be improved by taking a shorter time interval, 0.0001 s or even less. In the limiting case of extremely small time intervals (infinitesimal time intervals), the line segment has the direction of the tangent line that touches the plot at the point $t = 4$ s. Hence ***the instantaneous velocity at a given time equals the slope of the tangent that touches the plot at that time.*** For example, drawing the tangent that touches the plot at $t = 4$ s (Fig. 1.16b) and measuring its slope on the graph, we readily find that this slope is 17 m/s; hence the instantaneous velocity at $t = 4$ s is 17 m/s.

◀ **Instantaneous velocity**

By drawing tangents at other points of the plot and measuring their slopes (see Fig. 1.17), we can obtain a complete table of values of the instantaneous velocity of the automobile at different times. After we assemble such a table of values of the instantaneous velocity, we can prepare a plot of the instantaneous velocity vs. time. Figure 1.18 is such a plot of the instantaneous velocity. It is instructive to compare this plot of instantaneous velocity with the plot of position shown in Fig. 1.17. The velocity is initially zero (zero slope in Fig. 1.17), then increases to a maximum of

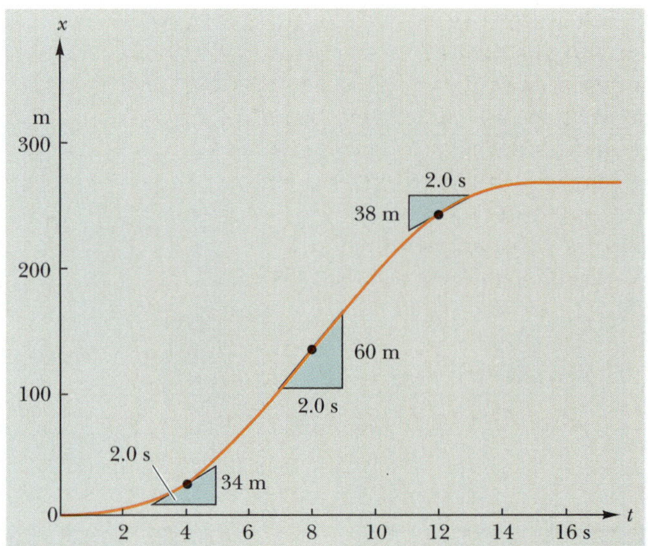

$t = 4$ s [slope] $= \dfrac{34\text{ m}}{2.0\text{ s}} = 17$ m/s

$t = 8$ s [slope] $= \dfrac{60\text{ m}}{2.0\text{ s}} = 30$ m/s

$t = 12$ s [slope] $= \dfrac{38\text{ m}}{2.0\text{ s}} = 19$ m/s

FIGURE 1.17 To find the instantaneous velocities at different times, we draw tangents to the plot at these times and measure their slopes.

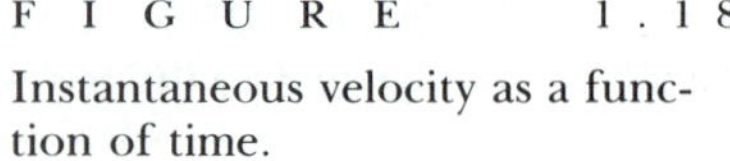

FIGURE 1.18 Instantaneous velocity as a function of time.

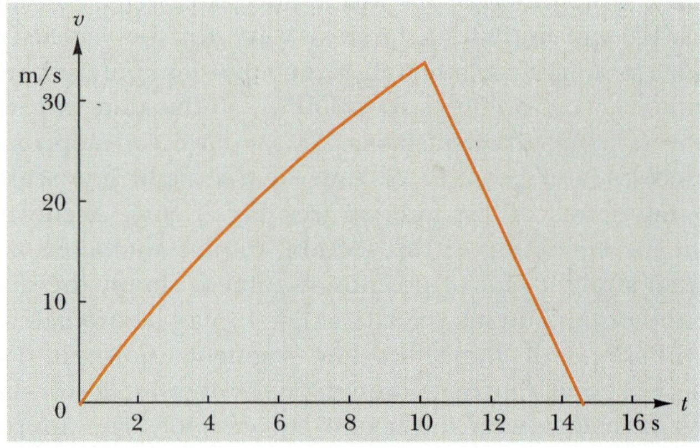

34.9 m/s at $t = 10$ s (maximum slope in Fig. 1.17), and finally decreases to zero at $t = 14.3$ s (zero slope in Fig. 1.17).

The graphical method for the determination of the instantaneous velocity has the disadvantage that we must first prepare a careful plot of position vs. time. If we have available a complete table of values of the position at different times or an exact mathematical formula for the position as a function of time, then we can calculate the instantaneous velocity by a numerical method, without bothering with a plot. We recall that in the graphical method we began with some small time interval centered on the relevant time, and then took an even smaller time interval, and ultimately considered what happens to the slope in the limiting case of extremely small time intervals (infinitesimal time intervals). Likewise, in the numerical method, we begin with some small time interval Δt, find the change of position Δx, and evaluate the ratio $\Delta x/\Delta t$. Then we take a smaller time interval and repeat the calculation. And then we take an even smaller time interval and repeat the calculation again, and so on. Ultimately, we discover that in the limiting case of extremely small time intervals, the result of the calculation does not depend on the size of the time interval, that is, we obtain the same result for an extremely small time interval and for an even smaller time interval. ***This limiting result for the ratio $\Delta x/\Delta t$ is the correct value of the instantaneous velocity.*** Thus, we can write the following formula for the instantaneous velocity:

$$v = \frac{\Delta x}{\Delta t} \quad \text{with } \Delta t \text{ extremely small} \tag{7}$$

1.5 Acceleration

Any motion with a change of velocity is accelerated motion. Thus, the motion of an automobile that speeds up is accelerated motion; but so is the motion of an automobile that slows down while braking — in both cases there is a *change* of velocity. If an automobile or a particle has velocity v_1 at time t_1 and velocity v_2 at time t_2, then ***the average acceleration for this time interval is defined as the change of velocity divided by the change of time,***

◀ **Average acceleration**

$$\bar{a} = \frac{v_2 - v_1}{t_2 - t_1} \tag{8}$$

or

$$\bar{a} = \frac{\Delta v}{\Delta t} \tag{9}$$

where $\Delta v = v_2 - v_1$ and $\Delta t = t_2 - t_1$. Accordingly, the average acceleration is the average rate of change of the velocity, or the average change of velocity per unit time. The unit of acceleration is the unit of velocity divided by the unit of time. Consequently, ***in the metric system, the unit of acceleration is the meter per second per second, or meter per second squared*** [(m/s)/s, or m/s^2]. Table 1.9 gives some typical values of accelerations.

The acceleration can be positive or negative, depending on the sign of the velocity change $v_2 - v_1$ in the formula (8). If the velocity is positive and increasing in magnitude, the acceleration is positive; if the velocity is positive and decreasing in magnitude, the acceleration is negative. However, note that if the velocity is negative (motion in the negative x direction) and increasing in magnitude, that is, becoming more negative, the acceleration is *negative.* Thus, an automobile speeding up while moving in the negative x direction has a negative acceleration; conversely, an auto-

TABLE 1.9	SOME ACCELERATIONS
Baseball struck by bat	3×10^4 m/s^2
Soccer ball struck by foot	3×10^3 m/s^2
Rat flea, starting a jump	2×10^3 m/s^2
Automobile crash (60 mi/h into fixed barrier)	1×10^3 m/s^2
Parachute opening (extreme)	3.2×10^2 m/s^2
Free fall on surface of Sun	2.7×10^2 m/s^2
Explosive seat ejection from aircraft (extreme)	1.5×10^2 m/s^2
Loss of consciousness of man ("blackout")	70 m/s^2
Free fall on surface of Earth	9.8 m/s^2
Braking of automobile	$\simeq$8 m/s^2
Free fall on surface of Moon	1.7 m/s^2

mobile braking while moving in the negative x direction has positive acceleration! These quirks of the formula for the acceleration must always be kept in mind.

On a plot of velocity vs. time, the average acceleration is the slope of the straight line connecting the points Q_1 and Q_2 corresponding to $t = t_1$ and $t = t_2$. For instance, in the case of the accelerating Maserati discussed above, the plot of velocity as a function of time is shown in Fig. 1.19, and the average acceleration for the time interval from $t = 0$ s to $t = 10.0$ s is the ratio of the vertical separation to the horizontal separation between the two points Q_1 and Q_2 on the plot (see Fig. 1.19):

$$\bar{a} = \frac{v_2 - v_1}{t_2 - t_1} = \frac{34.9 \text{ m/s} - 0 \text{ m/s}}{10.0 \text{ s} - 0 \text{ s}} = 3.49 \text{ m/s}^2$$

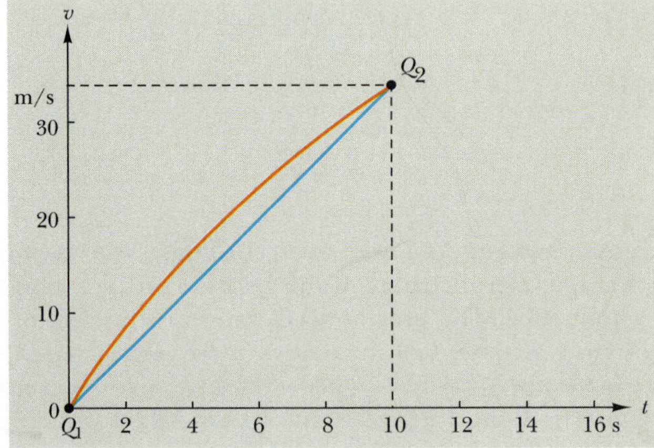

FIGURE 1.19
The average acceleration for the interval from $t_1 = 0$ to $t_2 = 10.0$ s is the slope of the straight line Q_1Q_2.

EXAMPLE 5. While starting a jump, a locust pushes itself off the ground with its legs and thereby accelerates its body upward. Slow-motion photography shows that the locust pushes for 0.013 s and attains a speed of 3.0 m/s when its legs lose contact with ground. What is its average acceleration during this push-off?

SOLUTION: The change of velocity is $\Delta v = v_2 - v_1 = 3.0 \text{ m/s} - 0 \text{ m/s} = 3.0 \text{ m/s}$ and the change of time is $\Delta t = t_2 - t_1 = 0.013 \text{ s} - 0 \text{ s} = 0.013 \text{ s}$. The average acceleration is

$$\bar{a} = \frac{\Delta v}{\Delta t} = \frac{3.0 \text{ m/s}}{0.013 \text{ s}} = 2.3 \times 10^2 \text{ m/s}^2$$

◀ **Instantaneous acceleration**

The instantaneous acceleration at some specified time is the slope of the tangent drawn on the plot of velocity vs. time. As in the case of velocity, the instantaneous acceleration can also be calculated from the ratio of small increments,

$$a = \frac{\Delta v}{\Delta t} \quad \text{with } \Delta t \text{ extremely small} \tag{10}$$

By drawing tangents at different points on the plot of velocity vs. time and measuring their slopes, we can prepare a complete table of values of the instantaneous acceleration at different times. Figure 1.20 shows the values of the instantaneous acceleration obtained from the plot of the velocity given in Fig. 1.18, from our road test of the Maserati. At the initial instant $t = 0$, the acceleration is large (large slope in Fig. 1.18); as the automobile gains velocity, the acceleration gradually drops (decreasing slope in Fig. 1.18); at $t = 10$ s, the driver applies the brakes, which produces a large negative acceleration of -8.0 m/s^2 (negative slope in Fig. 1.18); this negative acceleration, or ***deceleration,*** is maintained until the automobile comes to a halt. Note that when the driver steps on the brakes at time $t = 10$ s, the acceleration of the automobile suddenly switches from positive to negative; thus, the plot in Fig. 1.20 has a jump at this point.

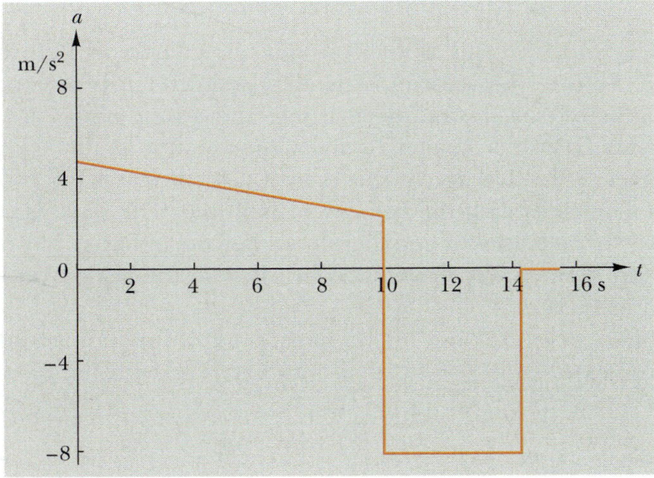

FIGURE 1.20
Instantaneous acceleration as a function of time.

1.6 Motion with Constant Acceleration

Constant acceleration implies a constant slope in the plot of velocity vs. time; thus, the plot is a straight line. In this case, the velocity increases (or decreases) by equal amounts in each 1-second time interval. A familiar example is a braking automobile, for which the velocity usually decreases by equal amounts in each 1-second interval (for instance, we found that for the braking automobile mentioned in the preceding section, the velocity decreases by 8.0 m/s in each second).

In the case of constant acceleration, we can obtain some simple relations among acceleration, velocity, position, and time. Suppose that the initial velocity at time zero is v_0 and that the velocity increases at a constant rate given by the constant acceleration a. After a time t has elapsed, the velocity will have increased by an amount at, and it will have attained the value

Constant acceleration: *v*, *a*, and *t*

$$v = v_0 + at \tag{11}$$

[Note that this equation is essentially the same as Eq. (8), slightly rearranged, and with the substitutions $t_1 = 0$, $t_2 = t$, $v_1 = v_0$, and $v_2 = v$.]

Suppose that the initial position is x_0 at time zero. After a time t has elapsed, the position will have changed by an amount equal to the product of the average velocity multiplied by the time, that is, the position will have changed from the initial value x_0 to

$$x = x_0 + \overline{v}t \tag{12}$$

The average velocity coincides with the velocity at halftime, that is, the velocity at one-half the time t,

$$\overline{v} = v_0 + a(\tfrac{1}{2}t) \tag{13}$$

To understand why this is so, consider what happens before and after halftime. Before, the velocity is smaller, and after it is larger. If we consider a time of, say, 1 s before halftime and a time of 1 s after halftime, the velocity before is smaller by the same amount as the velocity after is larger. Hence the average of this pair of velocities at 1 s before and after halftime coincides with the velocity of halftime. The same is true for any other pair of velocities at equal intervals before and after halftime. Hence the overall average velocity must coincide with the velocity at halftime, as asserted in Eq. (13).

To obtain an expression for the position as a function of time, we substitute the expression $\overline{v} = v_0 + \frac{1}{2}at$ into Eq. (12), with the result

$$x = x_0 + \overline{v}t = x_0 + (v_0 + \tfrac{1}{2}at)t \tag{14}$$

that is

$$x = x_0 + v_0 t + \tfrac{1}{2}at^2 \tag{15}$$

Thus, with constant acceleration, the change in position is

$$x - x_0 = v_0 t + \tfrac{1}{2}at^2 \tag{16}$$

◀ **Constant acceleration: *x*, *a*, and *t***

The right side of this equation consists of two terms: the term $v_0 t$ represents the change in position that the particle would suffer if moving at constant velocity v_0, and the term $\frac{1}{2}at^2$ represents the effect of the acceleration.

Equations (11) and (16) express velocity and position in terms of time. By eliminating the time t between these two equations, we obtain a direct relation between position and velocity, which is sometimes useful. According to Eq. (11),

$$t = \frac{v - v_0}{a}$$

and if we substitute this into Eq. (16), we obtain

$$\begin{aligned} x - x_0 &= v_0\left(\frac{v - v_0}{a}\right) + \tfrac{1}{2}a\left(\frac{v - v_0}{a}\right)^2 \\ &= \frac{v_0 v - v_0^2}{a} + \frac{\frac{1}{2}(v^2 - 2vv_0 + v_0^2)}{a} \\ &= \frac{\frac{1}{2}(v^2 - v_0^2)}{a} \end{aligned}$$

which we can rearrange as follows:

$$a(x - x_0) = \tfrac{1}{2}(v^2 - v_0^2) \tag{17}$$

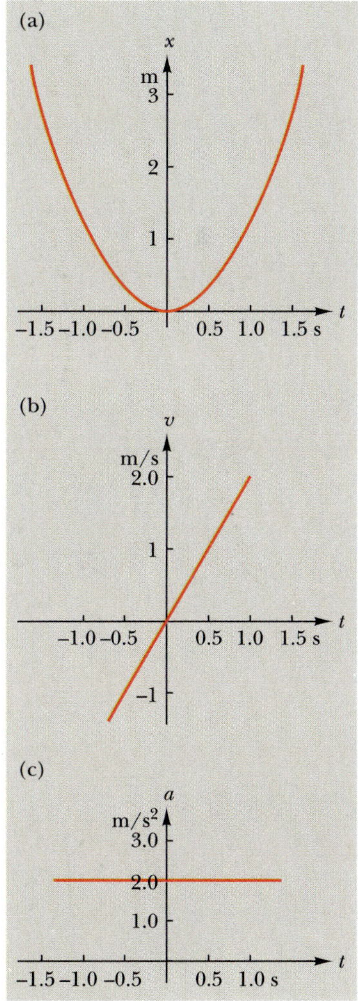

FIGURE 1.21

(a) Plot of position vs. time for a particle moving with a constant acceleration of 2.0 m/s^2; the plot is a parabola. (b) Velocity vs. time; this plot is a straight line of slope 2.0 m/s^2. (c) Acceleration vs. time; this plot shows a constant value of 2.0 m/s^2.

Figure 1.21 shows graphs of position, velocity, and acceleration for motion with constant acceleration. In these graphs the motion for negative values of t (instants before zero) has also been included, with the assumption that the acceleration a always has the same constant value. Note that the plot in Fig. 1.21a has the shape of a parabola; this shape of the plot is a distinctive characteristic of motion with constant acceleration.

Equations (11), (16), and (17) contain the acceleration a, the time t, and the instantaneous and initial positions and velocities x, x_0, v, v_0. In a typical problem of motion with constant acceleration, some of these quantities will be known and the others will be unknown, to be calculated from the equations. Which of these equations are the most useful will depend on the problem.

EXAMPLE 6. A subway train leaving a station accelerates along a straight track at 0.25 m/s^2 (see Fig. 1.22). How long does the train take to reach its cruising speed of 15 m/s? How far from the station will the train have traveled by this time?

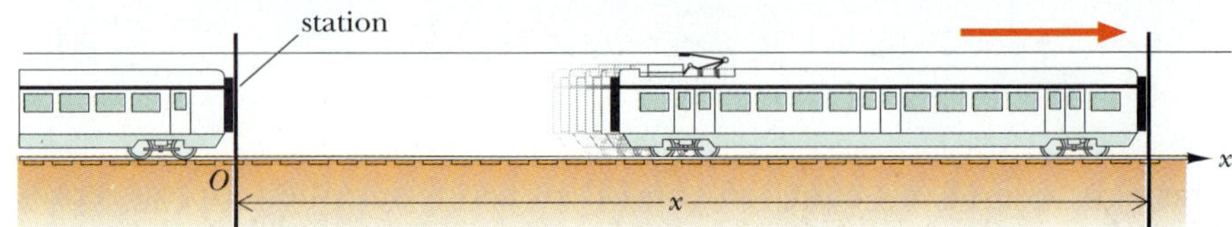

FIGURE 1.22 An accelerating subway train. The origin of coordinates is at the station ($x_0 = 0$).

SOLUTION: For motion with constant acceleration, the relevant equations are Eqs. (11), (16), and (17). To solve this problem, we must decide which of these equations we need. For the first part of the problem (calculation of the time), the unknown quantity is the time, and the known quantities are the acceleration and the final and initial speeds. To keep this information in sight, it is helpful to make a list:

Unknown	Known
t	$a = 0.25\ \text{m/s}^2$
	$v = 15\ \text{m/s}$
	$v_0 = 0$

Both Eqs. (11) and (16) contain the unknown quantity t. Equation (11) contains no unknown quantities except t, whereas Eq. (16) also contains the unknown quantity x. Hence Eq. (11) is our best choice. To solve this equation for t, we set $v_0 = 0$ and we divide both sides by a:

$$t = \frac{v}{a} = \frac{15\ \text{m/s}}{0.25\ \text{m/s}^2} = 60\ \text{s}$$

For the second part of the problem (calculation of distance), the unknown quantity is the distance, and the known quantities are the acceleration, the final and initial speeds, and the time. For convenience, we assume that the origin of our coordinates is at the station, so $x_0 = 0$ (see Fig. 1.22). Then the known and the unknown quantities are

Unknown	Known
x	$a = 0.25\ \text{m/s}^2$
	$v = 15\ \text{m/s}$
	$v_0 = 0$
	$x_0 = 0$
	$t = 60\ \text{s}$

Equations (16) and (17) contain the unknown quantity x, and they contain other quantities, all of which are known. We can use either of these equations to find x. The simplest choice is Eq. (16), since it is already solved for x:

$$x = x_0 + v_0 t + \tfrac{1}{2}at^2 = 0 + 0 + \tfrac{1}{2} \times \left(0.25 \,\frac{\text{m}}{\cancel{\text{s}^2}}\right) \times (60\,\cancel{\text{s}})^2 \tag{18}$$

$$= 450 \text{ m} \tag{19}$$

COMMENTS AND SUGGESTIONS: Note that in the denominator and numerator of Eq. (18), one factor of s^2 cancels against another factor of s^2, leaving the final unit m in Eq. (19). Thus the final unit in this equation is that of distance, as it should be. It is a general rule that in any calculation with the equations of physics, ***the units can be multiplied and divided as though they were algebraic quantities, and this automatically yields the correct units for the final result*** — if it does not, there is some mistake in the calculation. Thus, we will always find it worthwhile to carry the units along in all our calculations, because this provides us with extra protection against costly mistakes. Any failure of the expected cancellations is a sure sign of trouble!

EXAMPLE 7. An automobile is traveling at 88 km/h on a straight road when the driver spots a wreck ahead and slams on the brakes. The automobile decelerates at 8.0 m/s^2. What is the stopping distance?

SOLUTION: If we take the origin of coordinates at the point at which braking begins, then $x_0 = 0$ (see Fig. 1.23). The stopping distance is unknown, and the acceleration and the initial and final velocites are known:

Unknown	Known
x	$a = -8.0 \text{ m/s}^2$ $v = 0$ $v_0 = 88 \text{ km/h}$ $x_0 = 0$

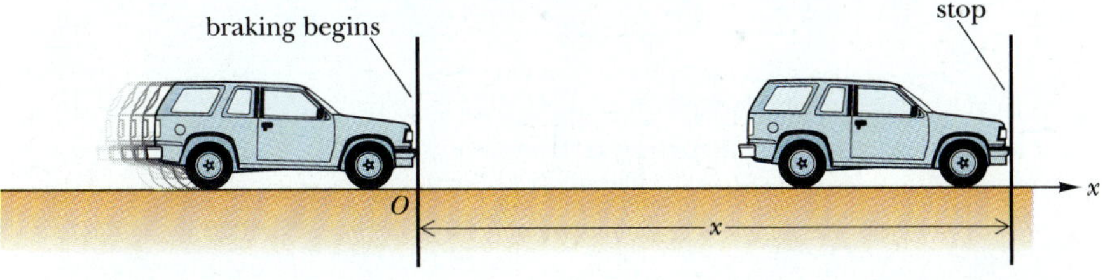

FIGURE 1.23 A braking automobile. The origin of coordinates is at the point at which braking begins.

Here, the acceleration is negative, since the automobile is decelerating. The most suitable equation for the solution of this problem is Eq. (17), since it contains the unknown quantity, and all the other quantities in it are known. Solving this equation for x, we find

$$x = x_0 + \frac{v^2 - v_0^2}{2a}$$

But before we substitute the known quantities into this equation, we had better convert 88 km/h into m/s; for this we use the conversion factors $1 = 10^3$ m/(1 km) and $1 = 1$ h/(3600 s):

$$88\,\frac{\text{km}}{\text{h}} = 88\,\frac{\text{km}}{\not{\text{h}}} \times \frac{10^3\text{ m}}{1\text{ km}} \times \frac{1\,\not{\text{h}}}{3600\text{ s}} = 24\text{ m/s}$$

Substituting this value of v_0 into our equation for x, we find that the stopping distance is

$$x = 0 + \frac{0 - (24\text{ m/s})^2}{2(-8.0\text{ m/s}^2)} = 36\text{ m}$$

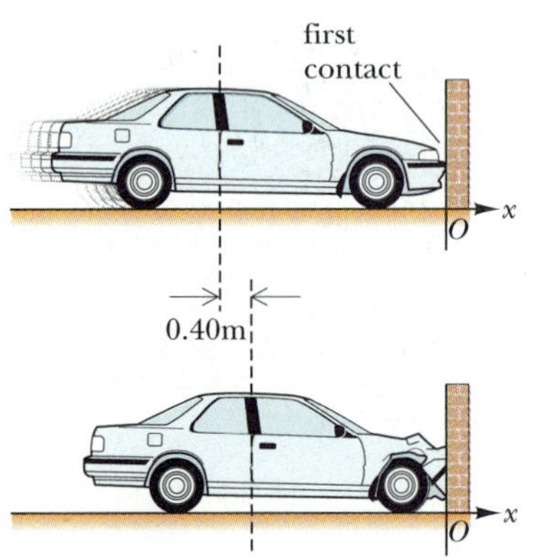

FIGURE 1.24
Deformation of the front end of an automobile crashing into a barrier. The origin of coordinates is at the barrier. While the front end crumples, the passenger compartment advances 0.40 m.

EXAMPLE 8. In a crash test, an automobile initially traveling at 50 km/h crashes into a stationary, rigid barrier. The passenger compartment comes to rest after the front end of the automobile crumples by 0.40 m (see Fig. 1.24). Assuming constant deceleration during the crash, what is the value of the deceleration of the passenger compartment? How long does it take to stop?

SOLUTION: The known quantities are the initial velocity ($v_0 = 50$ km/h just before the automobile comes into contact with the barrier), the final velocity ($v = 0$ when the automobile comes to rest), and the change in position of the passenger compartment during the crash ($x - x_0 = 0.40$ m; see Fig. 1.24):

Unknown	Known
a	$x - x_0 = 0.40$ m
t	$v_0 = 50$ km/h
	$v = 0$

To find the unknown acceleration, it will be best to use Eq. (17), since there the acceleration appears as the *only* unknown quantity.

If we divide both sides of Eq. (17) by $(x - x_0)$, we find

$$a = \frac{v^2 - v_0^2}{2(x - x_0)}$$

Upon substituting the known quantities into this equation, with $v_0 = 50\ \text{km/h} = 50 \times 1000\ \text{m}/3600\ \text{s} = 13.9\ \text{m/s}$, we find

$$a = \frac{v^2 - v_0^2}{2(x - x_0)} = \frac{(-13.9\ \text{m/s})^2}{2(0.40\ \text{m})} = -240\ \text{m/s}^2$$

This is a large acceleration; a passenger involved in such a crash would suffer severe injuries, unless well restrained by a snug seat belt or an air bag.

We can next use Eq. (11) to calculate the time the automobile takes to stop:

$$t = \frac{v - v_0}{a} = \frac{-13.9\ \text{m/s}}{-240\ \text{m/s}^2} = 0.058\ \text{s}$$

Galileo Galilei, *1564–1642, Italian mathematician, astronomer, and physicist. Galileo demonstrated experimentally that all bodies fall with the same acceleration, and he deduced that the trajectory of a projectile is a parabola. With a telescope of his own design, he discovered the satellites of Jupiter and sun spots. He vociferously defended the heliocentric system of Copernicus, for which he was sentenced by the Inquisition.*

1.7 THE ACCELERATION OF FREE FALL

A body released near the surface of the Earth will accelerate downward under the influence of the pull of gravity. If the frictional resistance of the air has been eliminated (by placing the body in an evacuated container), then the body is in ***free fall,*** and the downward motion proceeds with constant acceleration. It is a remarkable fact that ***the value of this downward acceleration is exactly the same for all bodies released at the same location*** — the value of the acceleration is completely independent of the speeds, masses, sizes, shapes, chemical compositions, etc., of the bodies. Figure 1.25 shows a simple experiment that verifies this equality of the rates of free fall for two bodies of different masses. The universality of the rates of free fall is one of the most precisely and rigorously tested laws of nature; a long series of careful experiments have tested the equality of the rates of free fall of different bodies to within 1 part in 10^{10}, and in some special cases even to within 1 part in 10^{12}.

The acceleration of free fall near the surface of the Earth is usually denoted by g. The numerical value of g is approximately

$$g \simeq 9.81\ \text{m/s}^2 \tag{20}$$

◄ **Acceleration of free fall**

The exact value of the acceleration of free fall varies somethat from location to location on the Earth, and it varies with altitude. However, for any location on the surface of the Earth, this variation amounts to a mere fraction of a percent, and we will neglect it.

For the description of free-fall motion, we can use the formulas for motion with constant acceleration developed in the preceding section. To apply these formulas, we will assume that the x axis is in the upward direction.* With this choice for the direction of the x axis, the acceleration of a freely-falling particle is negative, $a = -g$, and Eqs. (11), (16), and (17) become

Free-fall motion ▶

$$v = v_0 - gt \tag{21}$$

$$x = x_0 + v_0 t - \tfrac{1}{2} g t^2 \tag{22}$$

$$-g(x - x_0) = \tfrac{1}{2}(v^2 - v_0^2) \tag{23}$$

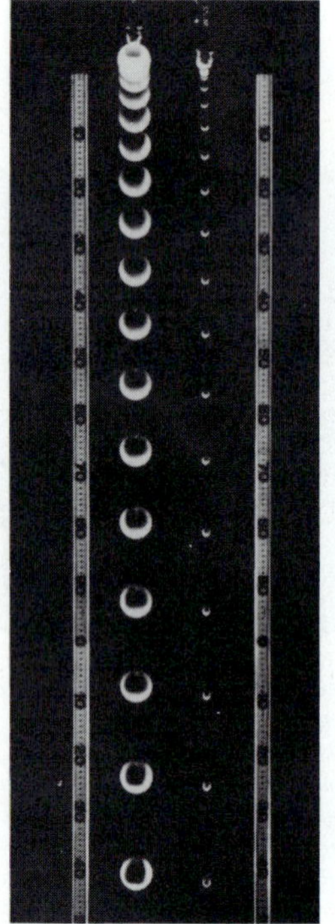

FIGURE 1.25

Stroboscopic photograph of two freely-falling balls of unequal masses released simultaneously. This photograph was made by leaving the shutter of the camera open, and triggering a flash at regular intervals.

Strictly, these equations are valid only for bodies falling in a vacuum, where there is no frictional resistance. But they are also good approximations for dense bodies, such as chunks of metal or stone, released in air. For such bodies the frictional reistance offered by the air is unimportant as long as the speed is low (the exact restriction to be imposed on the speed depends on the mass, size, and shape of the body, and on the desired accuracy). In all the following calculations we will ignore the resistance of air, even when the speeds are not all that low.

EXAMPLE 9. At Acapulco, professional divers amuse tourists by jumping from a 36-m-high cliff into the sea (Fig. 1.26). How long do they fall? What is their impact velocity on the water?

SOLUTION: For this problem the relevant equations are Eqs. (21), (22), and (23). The known quantities are the change of position and the initial velocity; the unknown quantities are the time of fall and the final velocity. The initial velocity is $v_0 = 0$ (we assume that the diver merely drops from his initial position, without pushing himself up or down when he jumps off the cliff). The change of position is $x - x_0 = -36$ m (this is negative because the final position is below the initial position, that is, the motion is in the negative x direction; see Fig. 1.26):

Unknown	Known
t	$x - x_0 = -36$ m
v	$v_0 = 0$
	$g = 9.81$ m/s^2

To calculate the time from the known quantities, we will use Eq. (22), in which the time is the only unknown. With $v_0 = 0$, Eq. (22) yields

$$(x - x_0) = -\tfrac{1}{2} g t^2$$

* This is a matter of choice. We could equally well assume that the x axis is in the downward direction, so the acceleration of a freely-falling particle is positive, $a = +g$. The choice is unimportant, but once we make a choice of direction of the x axis, we must adhere to it throughout the entire calculation.

which we can solve for t by dividing both sides by $-\frac{1}{2}g$, and taking the square root of both sides:

$$t = \sqrt{\frac{-2(x - x_0)}{g}} \tag{24}$$

$$= \sqrt{\frac{-2(-36 \text{ m})}{9.81 \text{ m/s}^2}} = 2.7 \text{ s}$$

From Eq. (21), the impact velocity is then

$$v = 0 - gt = -9.81 \text{m/s}^2 \times 2.7 \text{ s}$$
$$= -26 \text{ m/s}$$

This is about 94 km/h.

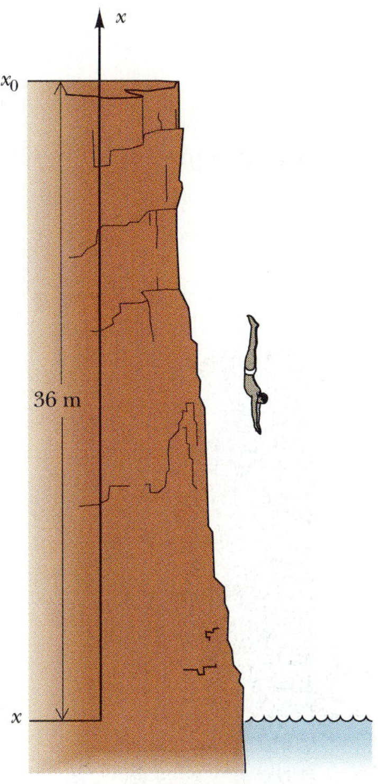

FIGURE 1.26

Jump of a diver. The change of position is negative ($x - x_0 < 0$).

EXAMPLE 10. A powerful bow, such as those used to establish world records in archery, can launch an arrow at a velocity of 90 m/s. How high will such an arrow rise if aimed vertically upward? How long will it take to return to the ground? What will be its velocity when it hits the ground?

SOLUTION: At the ground, the initial velocity is positive, $v_0 = 90$ m/s (see Fig. 1.27). At the highest point of the motion, the arrow ceases to move upward and is momentarily at rest; hence at this point the instantaneous velocity is zero, $v = 0$. For the upward motion, we can therefore regard the initial and final velocities as known. The height reached and the time are unknown:

Unknown	Known
$x - x_0$	$v_0 = 90$ m/s
t	$v = 0$
	$g = 9.81$ m/s^2

Equation (23) relates the height to the known quantities. Dividing this equation by $-g$, we find

$$x - x_0 = \frac{-(v^2 - v_0^2)}{2g} = \frac{-0 + (90 \text{ m/s})^2}{2 \times 9.81 \text{ m/s}^2} = 4.1 \times 10^2 \text{ m}$$

Equation (21) relates the time for the upward motion to the known quantities. Solving this equation for t, we find the time for the upward motion:

$$t = \frac{v_0 - v}{g} = \frac{90 \text{ m/s} - 0}{9.81 \text{ m/s}^2} = 9.2 \text{ s} \tag{25}$$

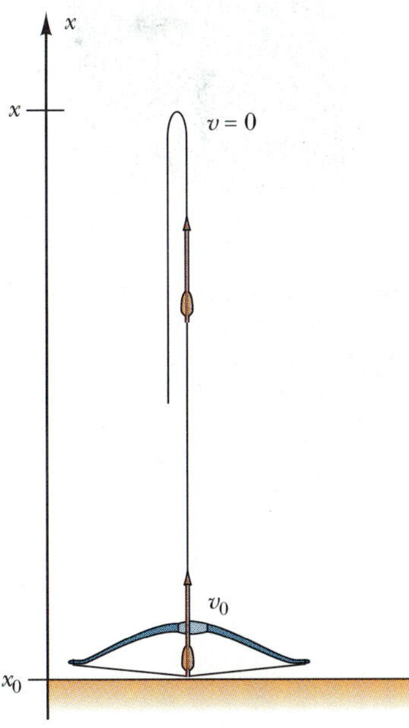

FIGURE 1.27

An ascending arrow. The change of position is positive ($x - x_0 > 0$).

The downward motion is simply the reverse of the upward motion. The downward motion therefore takes exactly as long as the upward motion, and the time for the arrow to return to the ground is 2×9.2 s $= 18.4$ s.

The velocity of the arrow when it hits the ground must be the reverse of the initial velocity. Thus, it must be -90 m/s.

COMMENTS AND SUGGESTIONS: Note that although the instantaneous velocity of the arrow is zero at the highest point of the motion, the acceleration is still the same as that at any other point, $a = -g$. The arrow is momentarily at rest, but it is still accelerating!

Finally, some brief comments on the effects of the frictional resistance of air on bodies falling at high speeds. By holding your hand out of the window of a speeding automobile, you can readily feel that air offers a substantial frictional resistance to motion at speeds in excess of a few tens of kilometers per hour. This frictional resistance increases with speed (roughly in proportion to the square of the speed). Hence a falling, accelerating body will experience a larger and larger frictional resistance as its speed increases. Ultimately. this resistance becomes so large that it counterbalances the pull of gravity — the body ceases to accelerate and attains a constant speed. This ultimate speed is called the ***terminal speed.*** The precise value of the terminal speed depends on the mass of the body, its size, and its shape; for instance, a sky diver with a closed parachute (see Fig. 1.28) attains a terminal speed of about 200 km/h, whereas a sky diver with an open parachute attains a terminal speed of only about 18 km/h.

Terminal speed ▶

FIGURE 1.28
Sky divers falling with closed parachutes.

SUMMARY

- **Average speed:** $\dfrac{\text{[distance traveled]}}{\text{[time taken]}}$

- **Average velocity:** $\bar{v} = \dfrac{\Delta x}{\Delta t}$

- **Instantaneous velocity:** $v = \dfrac{\Delta x}{\Delta t}$ with Δt extremely small

- **Average acceleration:** $\bar{a} = \dfrac{\Delta v}{\Delta t}$

- **Instantaneous acceleration:** $a = \dfrac{\Delta v}{\Delta t}$ with Δt extremely small

- **Motion with constant acceleration:** $v = v_0 + at$

$$x = x_0 + v_0 t + \tfrac{1}{2} at^2$$

$$a(x - x_0) = \tfrac{1}{2}(v^2 - v_0^2)$$

- **Acceleration of free fall:** $g \simeq 9.81 \text{ m/s}^2$

- **Free-fall motion** (with positive direction upward): $v = v_0 - gt$

$$x = x_0 + v_0 t - \tfrac{1}{2}gt^2$$

$$-g(x - x_0) = \tfrac{1}{2}(v^2 - v_0^2)$$

ADVICE ON SOLVING PROBLEMS. The solving of problems is an art; there is no simple recipe for obtaining the solutions. Most of the problems in this and the following chapters are applications of the concepts and principles developed in the text. The examples scattered throughout each chapter illustrate typical cases of problem solving. Sometimes you will be able to solve a problem by imitating one of these examples. But if you can't see how to begin the solution, try the following steps:

1. Carefully read and reread the problem and prepare a complete list of the given (known) and sought (unknown) quantities.

2. Draw a sketch of the situation described in the problem and label the known and unknown quantities.

3. If the problem deals with some kind of motion, try to visualize the progress of the motion in time, as though you were watching a movie.

4. Ask yourself what physical conditions characterize the situation sketched and visualized in the preceding steps, and ask yourself what physical principles are applicable. For instance, does the

motion proceed with constant velocity? With constant acceleration? Does the principle of universality of the rate of free fall apply?

5. Having decided what physical conditions and principles are applicable, examine the formulas that are valid under the given conditions. Then try to spot a formula that permits you to express the unknowns in terms of the known quantities (see Examples 6, 7, and 8). Be discriminating in your selection of formulas — sometimes a formula will tempt you because it displays all the desired quantities, but it will be an invalid formula if the assumptions that went into its derivation are not satisfied in your problem.

6. You will sometimes find that you seem to have too many unknowns and too few equations. Then ask yourself: Are there any quantities that you can calculate from the known quantities? Do these calculated quantities bring you nearer to the answer?

7. It is good practice to solve all the equations by algebraic manipulations, and substitute numbers only at the very end; this makes it easier to spot and correct mistakes.

8. When you substitute numbers, also include the units of these numbers. The units in your equations should then combine or cancel in such a way as to give the correct units for the final result. If the units do not combine or cancel in the expected way, something has gone wrong with your algebra.

9. After you have finished your calculations, always check whether the answer is plausible. For instance, if your calculation yields the result that a diver jumping off a cliff hits the water at 3000 km/h, then somebody has made a mistake somewhere!

10. Last, remember to round off your final answer to the same number of significant figures as given in the data for the problem.

QUESTIONS

1.–Try to estimate by eye the lengths, in centimeters or meters, of a few objects in your immediate environment. Then measure them with a ruler or meter stick. How good were your estimates?

2.–How close is your watch to standard time right now? Roughly how many minutes does your watch gain or lose per month?

3.–Mechanical clocks (with pendulums) were not invented until the tenth century A.D. What clocks were used by the ancient Greeks and Romans?

4.–By counting aloud "One, two, three, . . ten, one, . . ." at a reasonable rate you can measure seconds fairly accurately. Try to measure 30 seconds in this way. How good a timekeeper are you?

5.–Pendulum clocks are affected by the temperature and pressure of air. Why?

6.–In 1761 an accurate chronometer built by John Harrison was tested aboard H.M.S. *Deptford* during a voyage at sea for 5 months. During this voyage, the chronometer accumulated an error of less than 2 minutes. For this achievement, Harrison was ultimately awarded a prize of £20,000 that the British government had offered for the discovery of an accurate method of the determination of geographical longitude at sea. Explain how the navigator of a ship uses a chronometer and observation of the position of the Sun in the sky to find longitude.

Harrison's chronometer H.4.

7.–Lecky's *Wrinkles in Practical Navigation,* a famous nineteenth-century textbook of celestial navigation, recommends that each ship carry three chronometers for accurate timekeeping. What can the navigator do with three chronometers that he cannot do with two?

8.–Suppose that by an "act of God" (or by the act of a thief) the standard kilogram (kept at Sèvres, near Paris) were destroyed. Would this destroy the metric system?

9.–Consider the piece of paper on which this sentence is printed. If you had available suitable instruments, what physical quantities could you measure about this piece of paper? Make the longest list you can and give the units. Are all these units derived from the meter, second, and kilogram?

10.–The motion of a runner can be regarded as particle motion, but the motion of a gymnast cannot. Explain.

11.–Do our sense organs permit us to feel velocity? Acceleration?

12.–What is your velocity at this instant? Is this a well-defined question? What is your acceleration at this instant?

13.–Suppose that at one instant of time the velocity of a body is zero. Can this body have a nonzero acceleration at this instant? Give an example.

14.–Give an example of a body in motion with instantaneous velocity and acceleration of the same sign. Give an example of a body in motion with instantaneous velocity and acceleration of opposite signs.

15.–Experienced drivers recommend that when driving in traffic you should stay at least 2 s behind the car in front of you. This is equivalent to a distance of about two car lengths for every 10 mi/h. Why is it necessary to leave a larger distance between the cars when the speed is larger?

16.–In the seventeenth century Galileo Galilei measured the acceleration of gravity by rolling balls down an inclined plane. Why did he not measure the acceleration directly by dropping a stone from a tower?

17.–Why did astronauts find it easy to jump on the Moon? If an astronaut can jump to a height of 15 cm on the Earth (with his spacesuit), how high can he jump on the Moon?

18.–An elevator is moving upward with a constant velocity of 5 m/s. If a passenger standing in this elevator drops an apple, what will be the acceleration of the apple relative to the elevator?

19.–Some people are fond of firing guns into the air when under the influence of drink or patriotic fervor. What happens to the bullets? Is this practice dangerous?

20.–A woman riding upward in an elevator drops a penny in the elevator shaft when she is passing by the third floor. At the same instant, a man standing at the elevator door at the third floor also drops a penny in the elevator shaft. Which coin hits the bottom first? Which coin hits with the higher speed? Neglect friction.

21.–Suppose that you drop a $\frac{1}{2}$-kg packet of sugar and a $\frac{1}{2}$-kg ball of lead from the top of a building. Taking air friction into account, which will take the shorter time to reach the ground? Suppose that you place the sugar and the lead in identical sealed glass jars before dropping them. Which will now take the shorter time?

22.–Is air friction important in the falling motion of a raindrop? If a raindrop were to fall without friction from a height of 1000 m and hit you, what would it do to you?

23.–An archer shoots an arrow straight up. If you consider the effects of the air friction, would you expect the arrow to take a longer time to rise or to fall?

PROBLEMS

Section 1.1: Measurement of Space and Time; Units

1.——What is your height in feet? In meters?

2.——With a ruler, measure the thickness of this book, excluding the cover. Deduce the thickness of each of the sheets of paper making up the book.

3.——A football field measures 100 yd $\times$ $53\frac{1}{3}$ yd. Express this in meters.

4.——Express the last three entries in Table 1.1 in inches.

5.——Express the following fractions of an inch in millimeters: $\frac{1}{2}$, $\frac{1}{4}$, $\frac{1}{8}$, $\frac{1}{16}$, $\frac{1}{32}$, and $\frac{1}{64}$ in. Round your answers off to three significant figures.

6.——The tallest man was Robert Wadlow, who continued to grow throughout his life and attained 8 feet 11.1 inches a few days before his death in 1940. Express his height in meters. How many significant figures are there in your result?

7.——If each step you take in 0.60 m, how many steps do you need to cover 1.0 km?

8.——The Earth is approximately a sphere of radius 6.37×10^6 m. Calculate the distance from the pole to the equator, measured along the surface of the Earth.

9.*——Analogies can often help us to imagine the very large or very small distances that occur in astronomy or in atomic physics. Use the information given in the Prelude and Table 1.1 to answer the following questions:

(a) If the Sun were the size of a grapefruit, how large would the Earth be? How far away would the nearest star be?

(b) If your head were the size of Earth, how large would an atom be? How large would a red blood cell be?

10.*——As seen from the Earth, the Sun has an angular diameter of 0.53°. The distance between the Earth and the Sun is 1.5×10^{11} m. From this, calculate the radius of the Sun.

11.*——A nautical mile (nmi) equals 1.151 mi, or 1852 m. Show that a distance of 1 nmi along a meridian of the Earth corresponds to a change of latitude of 1 minute of arc.

12.——What is your age in days? In seconds?

13.——The age of the Earth is 4.5×10^9 years. Express this in seconds.

14.——How many days is a million seconds?

15.——How many hours are there in a week? How many seconds?

16.——Your heart beats 70 times per minute. How often does it beat in a year?

17.——The longest officially verified human life-span was attained by the Japanese man Shigechiyo Izumi, who died in 1986 at an age of 120 years and 237 days. Express this age in seconds. How many significant figures are there in your result?

18.*——In 1985 Carlos Lopez set a world record of 2 hours, 7 minutes, and 11.0 seconds for running the marathon. In scientific notation, this record becomes 7.6310×10^3 s, which is a number with five significant figures. If we round the number to four significant figures, it becomes 7.631×10^3 s; to three significant figures, it becomes 7.63×10^3 s; to two significant figures, it becomes 7.6×10^3 s; and to one significant figure, it becomes 8×10^3 s.

(a) The time was measured to five significant figures with a stopwatch. If instead of a stopwatch, the referees had used an ordinary wristwatch with a seconds hand, how many significant figures could they have reported?

(b) If they had used an ordinary wristwatch without a seconds hand (but with a minute hand), how many significant figures could they have reported?

(c) If they had used a "designer" wristwatch without numbers on its face, how many significant figures could they have reported?

19.——How many square feet are there in a square meter?

20.——How many cubic feet are there in a cubic meter?

21.——A football field measures 100 yd. $\times$ $53\frac{1}{3}$ yd. Calculate the area of this field; express your result in square meters.

22.——A tennis court measures 78 ft $\times$ 27 ft. Calculate the area of this court; express your result in square meters.

23.——Normal human blood contains 5.1×10^6 red blood cells per cubic millimeter. The total volume of blood in a man of 73 kg is 5.5 liters. How many red blood cells does this man have?

24.——As stated in Example 1, your heart pumps 92 cubic centimeters of blood per second. If your total volume of blood is 5.5 liters, what is the average travel time for your blood to complete one trip around your circulatory system?

25.*——Smokestacks in the United States spew out 8×10^6 metric tons of fly ash per year. If this stuff settles uniformly over all of the area of the United States (9.4×10^6 km^2), how many kilograms of fly ash will be deposited per square meter per year?

Section 1.2: Average Speed

26.——In setting a world record for speed, the horse Big Racket ran 402 m ($\frac{1}{4}$ mile) in 23.8 s. What was its average speed?

27.——The speed of nerve impulses in mammals is typically 10^2 m/s. If a shark bites the tail of a 30-m-long whale, roughly how long will it take before the whale knows of this?

28.——The world record for the 100-yard run is 9.0 seconds. What is the corresponding average speed in km per hour?

29.——In 1958 the nuclear-powered submarine *Nautilus* took 6 days and 12 hours to travel submerged 5068 km across the Atlantic from Portland, England, to New York City. What was the average speed (in kilometers per hour) for this trip?

30.——The speed-skating record for women is held by Bonnie Blair, who completed a 500-m race in 39.10 s. What is the corresponding average speed? Give the answer in m/s and in km/h.

Florence Joyner wins the race.

31.——The fastest speed ever measured for a tennis ball served by a player was 263 km/h. At this speed, how long does the tennis ball take to go from one end of the court to the other, a distance of 23.8 m?

32.——The distance from New York to Belém, Brazil, is 5280 km. How long does it take you to travel this distance by airliner, at 900 km/h? How long does it take you by ship, at 35 km/h?

33.——At the 1988 Olympics, Florence Griffith Joyner ran 100 m in 10.54 s, and she ran 200 m in 21.34 s. What was her average speed in each case?

34.——A galaxy beyond the constellation Corona Borealis is moving directly away from our Galaxy at the rate of 21,600 km/s. This galaxy is now at a distance of 1.4×10^9 light-years from our Galaxy. Assuming that the galaxy has always been moving at a constant speed, how many years ago was it right on top of our Galaxy?

35.——In the case of the closest verdict for a championship ski race, the winner of a cross-country race reached the finishing line one-hundredth of a second ahead of his closest competitor. If both were moving at a speed of 6 m/s, what was the distance between them at the finish?

36.*——You travel in your car for 30 minutes at 35 km/h, and then for another 30 minutes at 85 km/h.

(a) What distance do you cover in the first 30 minutes? In the second 30 minutes?

(b) What is your average speed for the entire trip?

37.*——You make a 100-km trip in an automobile, traveling the first 50 km at 60 km/h, and the second 50 km at 80 km/h. What is your average speed for this trip? Explain why the average speed is not 70 km/h.

38.*——You are traveling on a highway at 80 km/h and you are overtaking a car traveling in the same direction at 50 km/h. How long does it take you to go from 10 m behind this car to 10 m ahead of this car? The length of the car is 4 m.

39.*——On one occasion a tidal wave (tsunami) originating near Java was detected in the English Channel 32 h later. Roughly measure the distance from Java to England by sea (round the Cape of Good Hope) on a map of the world and calculate the average speed of the tidal wave.

40.*——A hunter shoots an arrow at a deer running directly away from him. When the arrow leaves the bow, the deer is at a distance of 40 m. When the arrow strikes, the deer is at a distance of 50 m. The speed of the arrow is 65 m/s. What must have been the speed of the deer? How long did the arrow take to travel to the deer?

41.*——In 1971, Francis Chichester, in the yacht *Gypsy Moth V*, attempted to sail the 4000 nautical miles (nmi) from Portuguese Guinea to Nicaragua in no more than 20 days.

(a) What minimum average speed (in nautical miles per hour) does this require?

(b) After sailing 13 days, he still had 1720 nmi to go. What minimum average speed did he require to reach his goal in the remaining 7 days? Knowing that his yacht could at best achieve a maximum speed of 10 nmi/h, what could he conclude at this point?

42.**——The fastest land animal is the cheetah, which runs at a speed of up to 28 m/s. The second fastest is the antelope, which runs at a speed of up to 24 m/s.

(a) Suppose that a cheetah begins to chase an antelope. If the antelope has a head start of 50 m, how long does it take the cheetah to catch the antelope? How far will the cheetah have traveled by this time?

(b) The cheetah can only maintain its top speed for about 20 s (and then has to rest), whereas the antelope can do it for a considerably longer time. What is the maximum head start the cheetah can allow the antelope?

Section 1.3: Average Velocity for Motion along a Straight Line
Section 1.4: Instantaneous Velocity

43.—— A runner runs 100 m in 10 s, rests 60 s, and returns at a walk in 80 s. What is is average speed for the complete motion? What is the average velocity?

44.—— Suppose you throw a baseball straight up so it reaches a maximum height of 8.0 m and returns to you 2.5 s after you throw it. What is the average speed for this motion of the ball? What is the average velocity?

45.*—— Consider the automobile with the plot of position vs. time given in Fig. 1.12. What is the average velocity for the interval from $t = 0$ to $t = 10$ s? From $t = 10$ s to $t = 14.3$ s? From $t = 0$ to $t = 14.3$ s?

Section 1.5: Acceleration

46.—— A football player kicks a stationary ball and sends it flying. Slow-motion photography shows that the ball is in contact with the foot for 8.2×10^{-3} s and leaves with a speed of 25 m/s. What is the average acceleration of the ball while in contact with the foot?

Kicking a football.

47.—— The highest speed attained by a cyclist on level ground in 105 km/h. To attain this speed, the cyclist used a streamlined recumbent bike. Starting from rest, he gradually built up his speed by pedaling furiously over a distance of 3.2 km. If his acceleration was uniform over this distance, what was the acceleration and how long did the cyclist take to build up his final speed?

48.—— A Porsche racing car takes 2.2 s to accelerate from 0 to 96 km/h (60 mi/h). What is the average acceleration?

49.—— In an experiment with a water-braked rocket sled, an Air Force volunteer (?) was subjected to an acceleration of 810 m/s^2 for 0.04 s. What was his change of speed in this time interval?

50.—— The driver of an automobile traveling at 80 km/h suddenly slams on the brakes and stops in 2.8 s. What is the average deceleration during braking?

51.*—— (a) The solid line in Fig. 1.29 is a plot of velocity vs. time for a Triumph sports car undergoing an acceleration test. By drawing tangents to the velocity curve, find the accelerations at time $t = 10$ and 40 s.

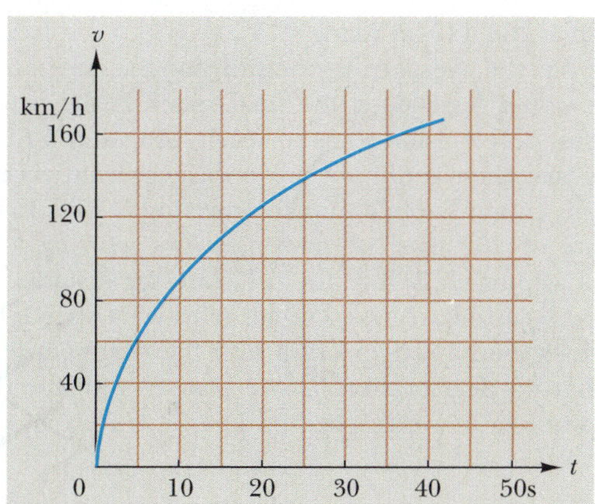

FIGURE 1.29
Instantaneous velocity for a Triumph sports car undergoing an acceleration test.

(b) The dashed line in Fig. 1.29 is a plot of velocity vs. time for the same car when coasting with its gears in neutral. Find the acceleration at time $t = 10$ and 40 s.

Section 1.6: Motion with Constant Acceleration

52.——The takeoff speed of a jetliner is 360 km/h. If the jetliner is to take off from a runway of length 2100 m, what must be its acceleration along the runway (assumed constant)?

53.——A British 6-inch naval gun has a barrel 6.63 m long. The muzzle speed of a projectile fired from this gun is 657 m/s. Assuming that upon detonation of the explosive charge the projectile moves along the barrel with constant acceleration, what is the magnitude of this acceleration? How long does it take the projectile to travel the full length of the barrel?

54.——In an accident on motorway M.1 in England, a Jaguar sports car made skid marks 290 m long while braking. Assuming that the deceleration was 10 m/s^2 during this skid (this is approximately the maximum deceleration that a car with rubber wheels can attain on ordinary pavements), calculate the initial speed of the car before braking.

55.——A jet-powered car racing on the Salt Flats of Utah went out of control and made skid marks 9.5 km long. If the inital speed of the car was 800 km/h, what must have been the (uniform) deceleration during the skid?

56.——In a "drag" race, a car with a rocket engine attained a final speed of 377.8 mi/h at the end of a 440-yd run, starting from rest. Assuming that the acceleration was constant during this run, find the value of the acceleration and find the time required to travel the given distance.

57.——The operation manual of a passenger automobile states that the stopping distance is 50 m when the brakes are fully applied at 96 km/h. What is the deceleration?

58.*—The nearest star is Proxima Centauri, at a distance of 4.2 light-years from the Sun. Suppose we wanted to send a spaceship to explore this star. To keep the astronauts comfortable, we want the spaceship to travel with a constant acceleration of 9.8 m/s^2 at all times (this will simulate ordinary gravity within the spaceship). If the spaceship accelerates at 9.8 m/s^2 until it reaches the midpoint of its trip and then decelerates at 9.8 m/s^2 until it reaches Proxima Centauri, how long will the one-way trip take? What will be the speed of the spaceship at the midpoint? Do your calculations according to Newtonian physics (actually, the speed is so large that the calculation should be done according to relativistic physics; see Chapter 27).

59.*—In a "drag" race a car starts at rest and attempts to cover 440 yd in the shortest possible time. The world record for a piston-engined car is 5.637 s; while setting this record, the car reached a final speed of 250.69 mi/h at the 440-yd mark.

(a) What was the average acceleration for the run?

(b) What would have been the final speed is the car had moved with constant acceleration so as to reach 440 yd in 5.637 s?

(c) Prove that the car did not move with constant acceleration.

60.*—In a collision, an automobile initially traveling at 50 km/h decelerates at a constant rate of 200 m/s^2. A passenger not wearing a seat belt crashes against the dashboard. Before the collision, the distance between the passenger and the dashboard was 0.60 m. With what speed, relative to the automobile, does the passenger crash into the dashboard? Assume that the passenger has no deceleration before contact with the dashboard. (Hint: Solve this problem in a reference frame that travels at a constant velocity of 50 km/h.)

61.*—In a large hotel, a fast elevator takes you from the ground floor to the 21st floor. The elevator takes 17 s for this trip: 5 s at constant acceleration, 7 s at constant velocity, and 5 s at constant deceleration. Each floor in the hotel has a height of 2.5 m. Calculate the values of the acceleration and deceleration (assume they are equal). Calculate the maximum speed of the elevator.

62.*—The driver of an automobile traveling at 95 km/h perceives an obstacle on the road and slams on the brakes.

(a) Calculate the total stopping distance (in meters). Assume that the reaction time of the driver is 0.75 s (so there is a time interval of 0.75 s during which the automobile continues at constant speed while the driver gets ready to apply the brakes) and that the deceleration of the automobile is 7.8 m/s^2 when the brakes are applied.

(b) Repeat the calculation of Part (a) for initial speeds of 15, 30, 45, 60, and 75 km/h. Make a plot of stopping distance vs. initial speed.

63.*—An automobile is traveling at 90 km/h on a country road when the driver suddenly notices a cow in the road 30 m ahead. The driver attempts to brake the automobile, but the distance is too short. With what velocity does the automobile hit the cow? Assume that, as in Problem 62, the reaction time of the driver is 0.75 s and that the deceleration of the automobile is 7.8 m/s^2 when the brakes are applied.

Section 1.7: The Acceleration of Free Fall

64.—An engineer standing on a bridge drops a penny toward the water and sees the penny splashing into the water 3.0 s later. How high is the bridge?

65.—An apple drops from the top of the Empire State Building, 380 m above street level. How long does the apple take to fall? What is its impact velocity on the street? Ignore air resistance.

66.—Peregrine falcons dive on their prey with speeds of up to 130 km/h. From what height must a falcon fall freely to achieve this speed? Ignore air resistance.

67.—The muzzle speed of a 22-caliber bullet fired from a rifle is 360 m/s. If there were no air resistance, how high would this bullet rise when fired straight up?

68.—Cats are known for their ability to survive falls from buildings several floors high. If a cat falls three floors down, a distance of 8.7 m, what is its speed when it hits the ground?

69.—The world record for a high jump from a standing position is 1.90 m. With what speed must the jumper leave the ground to attain this height? (Treat this as particle motion, although it really is not.)

70.—The volcano Loki on Io, one of the moons of Jupiter, ejects debris to a height of 200 km. What must be the initial ejection velocity of the debris? The acceleration of gravity on Io is 1.80 m/s^2. There is no atmosphere on Io, hence no air resistance.

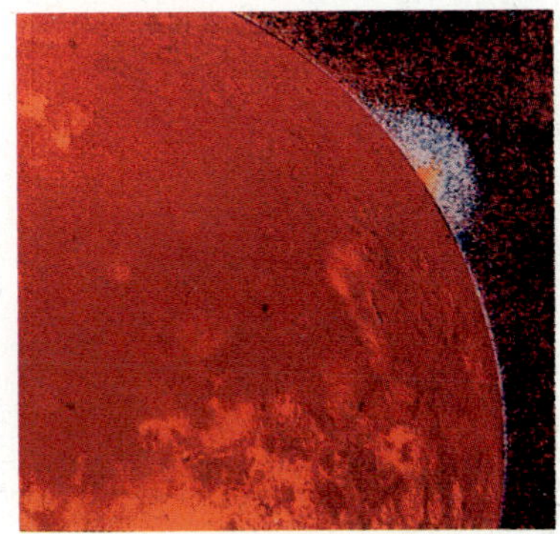

The volcano Loki.

71.*—In diving competitions, a diver jumping from a platform 10 m high performs $1\frac{1}{2}$ somersaults (head up, to down, to up, to down) while falling toward the water. How much time does the diver have for each $\frac{1}{2}$ somersault (head up to down, or down to up)?

72.*—The nozzle of a fire hose discharges water at the rate of 280 liters/min at a speed of 26 m/s. How high will the stream of water rise if the nozzle is aimed straight up? How many liters of water will be in the air at any given instant?

73.*—According to an estimate, a man who survived a fall from a 56-m cliff took

0.015 s to stop upon impact on the ground. What was his speed just before impact? What was his average deceleration during impact?

74.*—A particle is initially at rest at some height. If the particle is allowed to fall freely, what distance does it cover in the time from $t = 0$ s to $t = 1$ s? From $t = 1$ s to $t = 2$ s? From $t = 2$ s to $t = 3$ s? Show that these successive distances are in the ratios 1:3:5:7. . . .

75.*—A golf ball released from a height of 1.5 m above a concrete floor bounces back to a height of 1.1 m. If the ball is in contact with the floor for 6.2×10^{-4} s, what is the average acceleration of the ball while in contact with the floor?

76.*—In 1978 the stuntman A. J. Bakunas died when he jumped from the 23rd floor of a skyscraper and hit the pavement. The air bag that was supposed to cushion his impact ripped.

(a) The height of his jump was 96 m. What was his impact speed?

(b) The air bag was 3.6 m thick. What would have been the man's deceleration had the air bag not ripped? Assume that his deceleration would have been uniform over the 3.6-m interval.

77.*—The crash of an automobile into a solid barrier can be simulated by dropping the automobile, nose down, from some height on a hard floor (some of the early crash tests were performed in this manner). From what height must you drop the automobile to simulate a crash at 45 km/h? At 75 km/h? At 105 km/h? To gain a better appreciation of what these heights mean, express each height as a number of floors up a typical apartment building, with 2.9 m per floor (2.9 m is one floor up, etc.).

78.**—From a window on the fifth floor of a building, 13 m up, you drop two tennis balls with an interval of 1 s between the first and the second. Where is the first ball when you release the second? Where is the second ball when the first hits the ground?

79.**—A sky diver jumps out of an airplane at a height of 1000 m. Calculate how long she takes to fall to the ground. Assume that she falls freely with a downward acceleration g until she reaches the terminal speed of 200 km/h, and that she then continues to fall without acceleration until very near the ground (where she opens her parachute at the last moment).

CHAPTER 2

These wakes mark the paths of several boats maneuvering on the surface of a lake. The surface of the lake is (approximately) a flat plane, and the motion of the boats is two dimensional.

MOTION IN A PLANE

In this chapter we will deal with translational motion in a plane, such as the motion of an automobile traveling on the crisscrossing and curving streets of a flat city or the motion of a boat maneuvering on the flat surface of a lake. This is two-dimensional motion, and it is a simple generalization of the case of one-dimensional motion we studied in the preceding chapter. In essence, ***two-dimensional motion consists of two one-dimensional motions occurring simultaneously.*** Thus, we will have to apply the formulas of the preceding chapter separately to each of these one-dimensional motions. We could further generalize to the case of three-dimensional motion, such as the motion of an aircraft or the motion of an automobile on a mountain road, consisting of three one-dimensional motions. But we will not bother with this, since the case of two dimensions is adequate for most of the physical problems discussed in later chapters.

The mathematical concept of ***vector*** turns out to be very useful for the description of the position, velocity, and acceleration in two- or three-dimensional motion. The second half of this chapter is an introduction to vectors and their addition and subtraction. We will see that a vector description of the motion permits us to give precise meaning to the *direction* of the velocity and acceleration in two or three dimensions. And we will see in later chapters that the vector concept is also useful for the description of many other physical quantities, such as force and momentum, which have both a magnitude and a direction.

2.1 COMPONENTS OF VELOCITY AND ACCELERATION

To describe the translational motion of a particle in a plane, we need two coordinates, say, an x coordinate and a y coordinate. For instance, if the particle is an automobile traveling on the streets of a (flat) city, we can describe the position by choosing an origin at, say, the library, and laying out an x axis in the eastward direction and a y axis in the northward direction (Fig. 2.1). The x and y coordinates measured with respect to these axes then provide a complete description of the two-dimensional translational motion of the automobile.

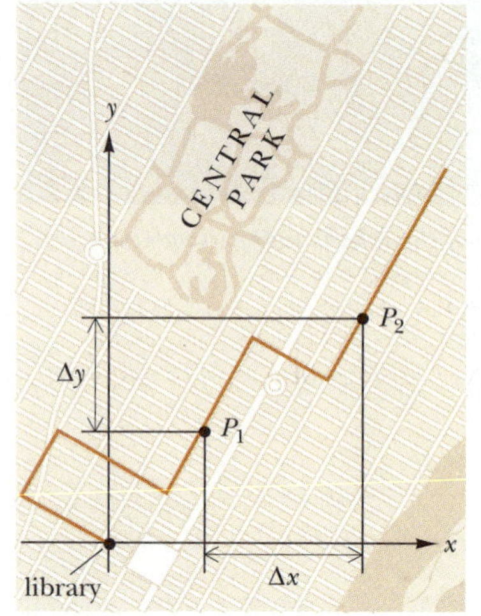

FIGURE 2.1
Path of an automobile along the streets of New York City. In a time interval Δt, the x and the y displacements of the automobile are Δx and Δy.

The x coordinate and the y coordinate of a particle are often called the x component of the position and the y component of the position. For a moving particle, both of these components of the position change with time. Correspondingly, there is an x component of the velocity and a y component of the velocity. In the case of the average velocity, we can define these x and y components by analogy with Eq. (1.5):

$$\bar{v}_x = \frac{\Delta x}{\Delta t} \tag{1}$$

and

$$\bar{v}_y = \frac{\Delta y}{\Delta t} \tag{2}$$

where $\Delta x = x_2 - x_1$ and $\Delta y = y_2 - y_1$ are the changes in the x and y components of the position in the time interval $\Delta t = t_2 - t_1$ (see Fig. 2.1). Stated in words: the x component of the velocity is the rate of change of the x coordinate, and the y component of the velocity is the rate of change of the y coordinate.

Likewise, in the case of the of the instantaneous velocity, we can define the x and y components by analogy with Eq. (1.7):

Components of the instantaneous velocity ▶

$$v_x = \frac{\Delta x}{\Delta t} \tag{3}$$

$$v_y = \frac{\Delta y}{\Delta t} \quad \text{with } \Delta t \text{ extremely small} \tag{4}$$

As we saw in the preceding chapter, for one-dimensional motion the instantaneous speed (such as the speed indicated by the speedometer of an automobile) equals the magnitude of the instantaneous velocity. For two-dimensional motion, the velocity has two components, and the relationship between speed and velocity is not that obvious. To discover what the relationship is, consider the motion of a particle in a small time interval Δt. In this time interval the particle travels a distance P_1P_2 from the point P_1 to the point P_2 (see Fig. 2.2a), and the changes in its x and y coordinates are Δx and Δy. The distances Δx and Δy form the sides of a small right triangle, and the distance P_1P_2 is the hypotenuse of this triangle (see Fig. 2.2b). According to the Pythagorean theorem, the hypotenuse is the square root of the sum of the squares of the sides,

$$[\text{distance traveled}] = P_1P_2 = \sqrt{(\Delta x)^2 + (\Delta y)^2} \tag{5}$$

Hence the instantaneous speed is

$$\begin{aligned}[\text{instantaneous speed}] &= \frac{[\text{distance traveled}]}{[\text{time taken}]} \\ &= \frac{P_1P_2}{\Delta t} \quad \text{with } \Delta t \text{ extremely small} \\ &= \frac{\sqrt{(\Delta x)^2 + (\Delta y)^2}}{\Delta t} = \sqrt{\frac{(\Delta x)^2 + (\Delta y)^2}{(\Delta t)^2}} \\ &= \sqrt{\left(\frac{\Delta x}{\Delta t}\right)^2 + \left(\frac{\Delta y}{\Delta t}\right)^2} \qquad (6) \\ &= \sqrt{v_x^2 + v_y^2} \qquad (7)\end{aligned}$$

Thus, ***the instantaneous speed is the square root of the sum of the squares of the components of the instantaneous velocity.*** If we use the letter v (without any subscript) to represent the instantaneous speed in two-dimensional motion, we can write Eq. (7) as

$$v = \sqrt{v_x^2 + v_y^2} \tag{8}$$

With the help of some trigonometry, we can derive two other useful expressions relating the speed to the x and y components of the velocity. Comparing the speed $P_1P_2/\Delta t$, the x component of the velocity $\Delta x/\Delta t$, and the y component of the velocity $\Delta y/\Delta t$, we see that they all share a common factor $1/\Delta t$; discarding this common factor, we find that the speed, the x component of the velocity, and the y component of the velocity are in the same proportions as the distances P_1P_2, Δx, and Δy. Hence, if we draw a triangle with sides representing the speed, the x component of velocity, and the y component of velocity (see Fig. 2.3), then this triangle of velocities will be similar to the triangle of distances (see Fig. 2.2b); that is, these two triangles have the same angles and their sides are in the same proportions. In view of this similarity of the triangles, if the direc-

(a)

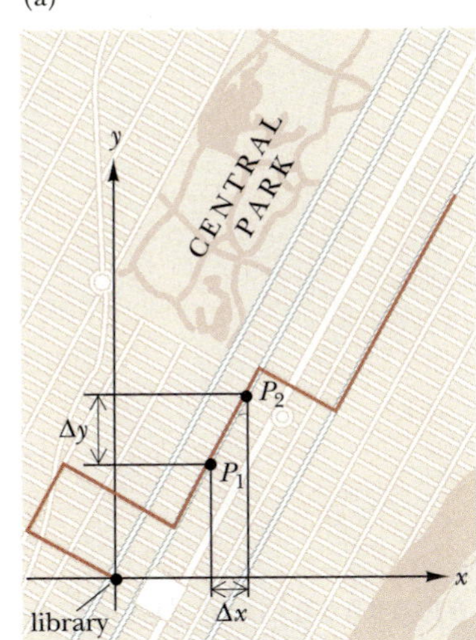

(b)

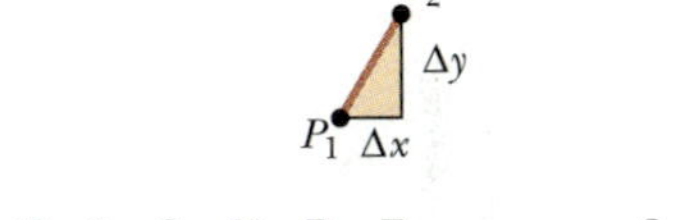

FIGURE 2.2

(a) In a small time interval Δt, the x and the y displacements of the automobile are Δx and Δy. (b) The displacements Δx and Δy are sides of a right triangle.

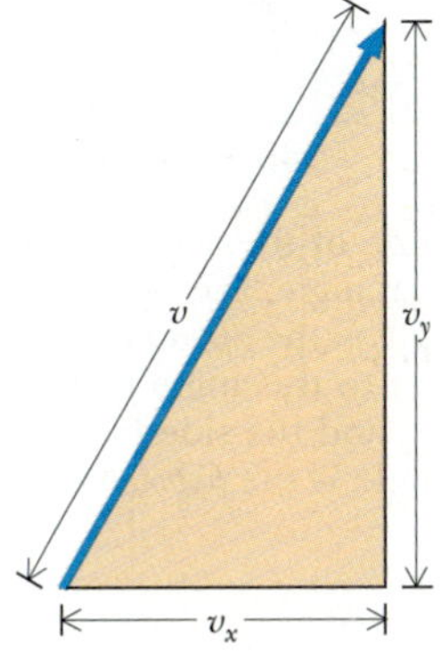

FIGURE 2.3

The sides of this triangle represent the speed, the x component of velocity, and the y component of velocity. The blue arrow indicates the direction of motion.

tion of motion of the particle makes some angle with the x direction, then the hypotenuse of the velocity triangle must make the same angle with the side representing v_x. In the velocity triangle, the hypotenuse is v, the adjacent side is v_x, and the opposite side is v_y. We will denote the angle of the triangle by θ, the Greek letter theta. According to the definitions of the cosine and the sine of an angle in a right triangle (see box), the cosine is the adjacent side divided by the hypotenuse, and the sine is the opposite side divided by the hypotenuse; that is,

$$\cos\theta = \frac{v_x}{v} \quad \text{and} \quad \sin\theta = \frac{v_y}{v}$$

Solving these equations for v_x and v_y, we obtain

Components in terms of angle ▶

$$v_x = v\cos\theta \tag{9}$$

and

$$v_y = v\sin\theta \tag{10}$$

These equations permit us to calculate the x and y components of velocity if we know the speed and the direction of motion.

Trigonometry ▶

TRIGONOMETRY OF THE RIGHT TRIANGLE*

$$\sin\theta = \frac{[\text{opposite side}]}{[\text{hypotenuse}]}$$

$$\cos\theta = \frac{[\text{adjacent side}]}{[\text{hypotenuse}]}$$

$$\tan\theta = \frac{[\text{opposite side}]}{[\text{adjacent side}]}$$

Pythagorean theorem:

$$[\text{hypotenuse}]^2 = [\text{opposite side}]^2 + [\text{adjacent side}]^2$$

This theorem implies that $1 = \sin^2\theta + \cos^2\theta$.

(In principle, the numerical value of the sine or cosine of any angle can be found by laying out a right triangle with this angle, measuring its sides, and evaluating the ratios given in the definitions. In practice, numerical values of cosines and sines are obtained from trigonometric tables or from electronic pocket calculators.)

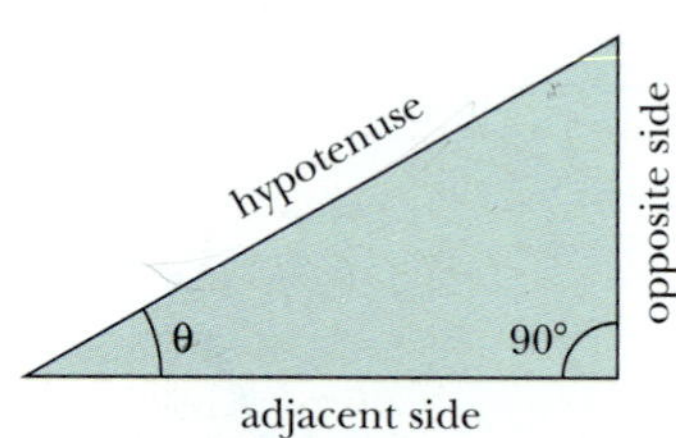

If θ is one of the acute angles of a right triangle, the side facing this angle is the *opposite side,* the side next to the angle is the *adjacent side,* and the side facing the right angle is the *hypotenuse*.

* Appendix 4 gives a review of trigonometry.

EXAMPLE 1. A golfer launches a ball in the eastward direction with an initial speed of 30.0 m/s at an upward angle of 34° with the horizontal (see Fig. 2.4). What are the components of the initial velocity of the ball in the horizontal and vertical directions? Assume the x axis is horizontal and eastward and the y axis is vertical and upward.

SOLUTION: According to Eqs. (9) and (10), with $v = 30.0$ m/s and $\theta = 34°$, the components of the initial velocity are

$$v_x = v \cos\theta = 30.0 \text{ m/s} \times \cos 34°$$

$$v_y = v \sin\theta = 30.0 \text{ m/s} \times \sin 34°$$

Our electronic pocket calculator tells us that $\cos 34° = 0.829$ and $\sin 34° = 0.559$, so

$$v_x = 30.0 \text{ m/s} \times 0.829 = 24.9 \text{ m/s}$$

$$v_y = 30.0 \text{ m/s} \times 0.559 = 16.8 \text{ m/s}$$

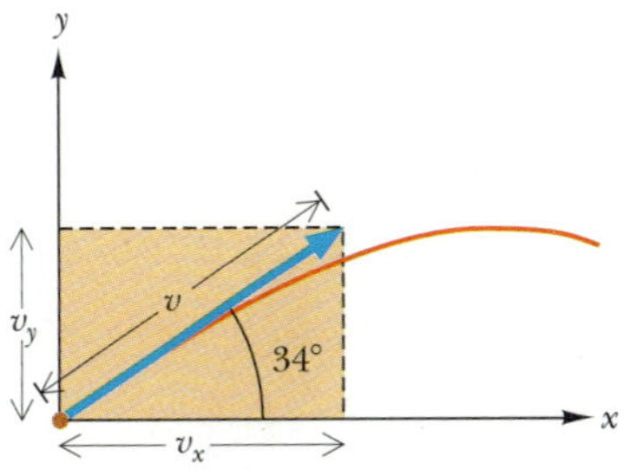

FIGURE 2.4 The blue arrow indicates the direction of motion of the golf ball at launch.

The definitions of the components of the acceleration in two dimensions can be formulated in much the same way as the definitions of the velocity components. For the average acceleration we have

$$\bar{a}_x = \frac{\Delta v_x}{\Delta t} \tag{11}$$

and

$$\bar{a}_y = \frac{\Delta v_y}{\Delta t} \tag{12}$$

where, of course, Δv_x and Δv_y are the changes in the x and y components of the velocity. For the instantaneous acceleration we have

◀ **Components of the instantaneous acceleration**

$$a_x = \frac{\Delta v_x}{\Delta t} \tag{13}$$

and

$$a_y = \frac{\Delta v_y}{\Delta t} \quad \text{with } \Delta t \text{ extremely small} \tag{14}$$

It is an important consequence of these definitions that ***there is an acceleration whenever any of the components of the velocity changes.*** This means that not only is there an acceleration when an automobile increases or decreases speed, but there is also an acceleration when the automobile travels around a curve at constant speed, as we will see in the following example.

EXAMPLE 2. An automobile enters a 90° curve at a constant speed of 25 m/s and emerges from this curve 6.0 s later. What are the components of the average acceleration for this time interval?

SOLUTION: Figure 2.5 shows the path of the automobile and the orientation of the axes. The initial velocity has an x component ($v_x = 25$ m/s) and no y component. The final velocity has a y component ($v_y = -25$ m/s) and no x component. Hence the changes in the velocity components are

$$\Delta v_x = [\text{final } x \text{ velocity}] - [\text{initial } x \text{ velocity}] = 0 - 25 \text{ m/s} = -25 \text{ m/s}$$

and

$$\Delta v_y = [\text{final } y \text{ velocity}] - [\text{initial } y \text{ velocity}] = -25 \text{ m/s} - 0 = -25 \text{ m/s}$$

The components of the average acceleration are then

$$\bar{a}_x = \frac{\Delta v_x}{\Delta t} = \frac{-25 \text{ m/s}}{6.0 \text{ s}} = -4.2 \text{ m/s}^2$$

and

$$\bar{a}_y = \frac{\Delta v_y}{\Delta t} = \frac{-25 \text{ m/s}}{6.0 \text{ s}} = -4.2 \text{ m/s}^2$$

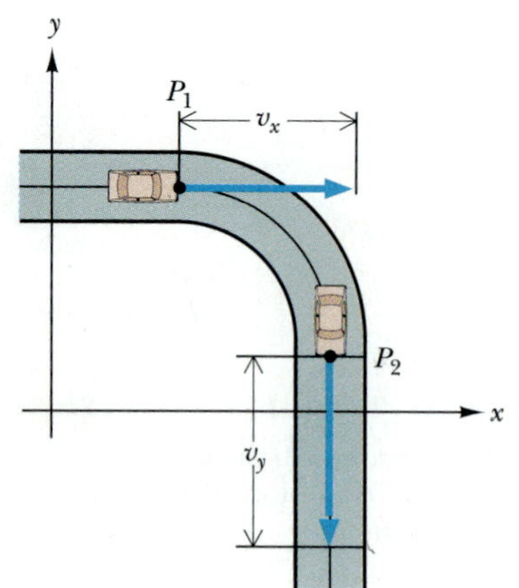

FIGURE 2.5

Path of an automobile through a 90° curve. The arrows indicate the directions of motion of the automobile at the points P_1 and P_2 before and after the curve.

2.2 MOTION WITH CONSTANT ACCELERATION

For a particle moving in two dimensions with constant acceleration, we can derive equations relating acceleration, velocity, position, and time analogous to the equations that hold in one dimension (see Section 1.6). If the x component of the acceleration is a_x and the y component a_y, then, by analogy with Eq. (1.11), we see that the components of the velocity at time t will be

$$v_x = v_{0x} + a_x t \tag{15}$$

$$v_y = v_{0y} + a_y t \tag{16}$$

where, as always, the subscript "0" indicates the values at the initial time zero. Furthermore, a mathematical argument analogous to that used in the one-dimensional case [see Eqs. (1.13)–(1.16)] leads us to the following expressions for the change in the position:

$$x - x_0 = v_{0x}t + \tfrac{1}{2}a_x t^2 \tag{17}$$

$$y - y_0 = v_{0y}t + \tfrac{1}{2}a_y t^2 \tag{18}$$

Equations (15)–(18) state that the x and y components of the motion proceed completely independently of one another. Thus, the x acceleration affects only the x velocity, and the change in the x position is completely determined by the x acceleration and the initial x velocity. Figure 2.6a shows an experimental demonstration of this independence between the x and y components of the motion. Two balls were released simultaneously from a platform above the surface of the Earth; one was merely dropped from rest, the other was launched with an initial horizontal velocity. From our discussion of the acceleration of free fall in Chapter 1, we know that the vertical free-fall accelerations of the two balls are the same. According to Eq. (18), the vertical motions (y motions) of the two balls should then be the same, even though their horizontal motions (x motions) differ. Furthermore, according to Eq. (17), the horizontal motion of the second ball should simply proceed with uniform horizontal velocity, even though this ball has a vertical acceleration.

The stroboscopic images of the balls recorded on the photograph at equal intervals of time confirm these predictions. The balls indeed fall downward in unison, reaching equal heights at the same instants of time. The red grid lines drawn in Fig. 2.6b permit us to verify that the vertical components of the positions of the two balls are always exactly the same, even though the horizontal components differ. Furthermore, the blue

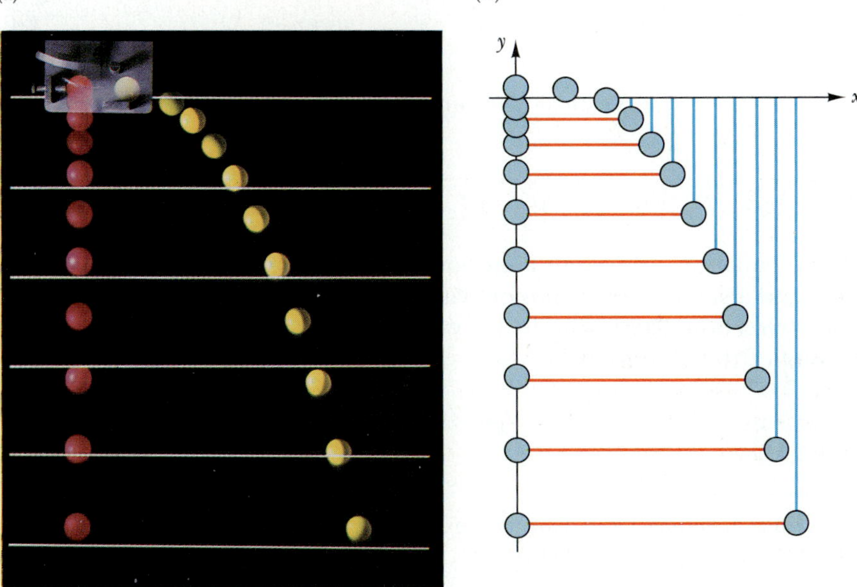

FIGURE 2.6
(a) Stroboscopic photograph showing multiple exposures of two balls that have been released simultaneously from a platform; one ball has a horizontal velocity, the other does not. The time interval between the exposures is 1/40 s. (b) For the analysis of the photograph, we draw horizontal (red) and vertical (blue) grid lines through the positions of the balls. The red grid lines verify that the vertical components of the positions of the two balls always coincide. The blue grid lines, by their uniform spacing, verify that the horizontal component of the velocity of the second ball remains constant.

grid lines permit us to verify that the horizontal component of the velocity of the second ball is constant.

2.3 THE MOTION OF PROJECTILES

We know that near the surface of the Earth, the pull of gravity gives a freely-falling body a downward acceleration of about 9.81 m/s^2. If we ignore friction with the air, this is the only acceleration that the body experiences when launched from some initial position with some initial velocity. Thus, the motion of a baseball thrown by hand or hit by a bat is ***motion with constant vertical acceleration and zero horizontal acceleration.*** This kind of motion is called ***projectile motion,*** or ballistic motion. In most cases of projectile motion, we deliberately launch the body with an initial upward component of velocity, so that it flies farther before striking the ground. But we can also launch the projectile horizontally, without any initial upward component of velocity (as in the case of the ball whose motion we examined in Fig. 2.6), and we can even launch the projectile with an initial downward component of velocity.

In any case, the initial velocity of the projectile can be characterized by its vertical and horizontal components. If we take the y axis in the upward direction and the x axis in the direction of the initial horizontal velocity, we have $a_x = 0$, $a_y = -g = -9.81$ m/s^2.* Furthermore, let us assume that the origin of coordinates coincides with the initial position of the projectile, so $x_0 = 0$ and $y_0 = 0$. The components of the velocity and position at the time t will then be, according to Eqs. (15)–(18),

Motion of projectile ▸

Horizontal Motion: $$v_x = v_{0x} \tag{19}$$

$$x = v_{0x}t \tag{20}$$

Vertical Motion: $$v_y = v_{0y} - gt \tag{21}$$

$$y = v_{0y}t - \tfrac{1}{2}gt^2 \tag{22}$$

These equations represent a motion with constant velocity in the x direction and a simultaneous motion with constant (downward) acceleration in the y direction. According to Eq. (1.17), the vertical position and velocity also obey the identity

$$-gy = \tfrac{1}{2}(v_y^2 - v_{0y}^2) \tag{23}$$

---

* Note that in Section 1.7 we took the x axis in the upward direction, whereas now we are taking the y axis in the upward direction.

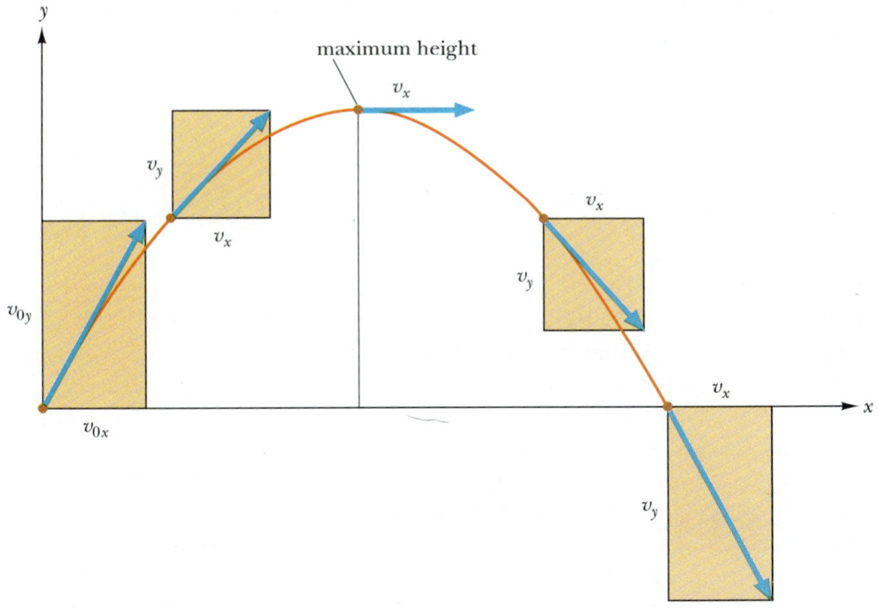

FIGURE 2.7

Motion of a projectile. The horizontal component of the velocity remains constant, and the vertical component of the velocity decreases, reaches zero at the instant of maximum height, and then becomes negative.

If the motion is initially upward, the projectile will ascend to some maximum height and then begin its descent. At the instant of maximum height, the vertical component of the velocity is zero (see Fig. 2.7), since at this one instant the projectile has ceased to move upward, and has not yet begun to move downward. We can find the instant of maximum height and the value of the maximum height by inserting the condition $v_y = 0$ into Eqs. (21) and (23), respectively.

EXAMPLE 3. Consider the golf ball launched as described in Example 1. At what time does this projectile reach its maximum height? What is this maximum height?

SOLUTION: The relevant equations are (19)–(22). The known quantities are the components of the initial velocity; according to Example 1, these components are $v_{0x} = 24.9$ m/s and $v_{0y} = 16.8$ m/s. Furthermore, we know that when the projectile reaches maximum height, $v_y = 0$ (see Fig. 2.7). Since the problem asks no questions about the horizontal motion, we can ignore this motion altogether, and we can concentrate on the vertical motion. The known and unknown quantities for the vertical motion are

Unknown	Known
t	$v_{0y} = 16.8$ m/s
y	$v_y = 0$
	$g = 9.81$ m/s^2

With these values of v_{0y} and v_y, Eq. (21) determines the instant of maximum height,

$$0 = v_{0y} - gt$$

or, solving this for t,

$$t = \frac{v_{0y}}{g} = \frac{16.8 \text{ m/s}}{9.81 \text{ m/s}^2} = 1.71 \text{ s} \tag{24}$$

With this value of the time, we can then calculate the maximum height from Eq. (22):

$$\begin{aligned} y &= v_{0y}t - \tfrac{1}{2}gt^2 \\ &= 16.8 \text{ m/s} \times 1.71 \text{ s} - \tfrac{1}{2} \times 9.81 \text{ m/s}^2 \times (1.71 \text{ s})^2 \\ &= 14.4 \text{ m} \end{aligned}$$

In our calculation we have ignored the effects of air resistance. For a projectile with a fairly low velocity, this is a good approximation.

EXAMPLE 4. In low-level bombing (at "smokestack level") a bomber releases a bomb at a height of 50 m above the surface of the sea while in horizontal flight at a constant speed of 300 km/h. How long does the bomb take to fall to the surface of the sea? How far ahead (horizontally) of the point of release is the point of impact?

SOLUTION: It is convenient to place the origin of coordinates at the point of release, 50 m above the level of the sea, with the x axis along the horizontal path of the bomber (see Fig. 2.8). Since the bomb merely drops out of the bomber, without being pushed, the initial velocity of the bomb is the same as that of the bomber. Thus, the initial horizontal component of the velocity is

$$v_{0x} = 300 \text{ km/h} = \frac{300 \times 1000 \text{ m}}{3600 \text{ s}} = 83.3 \text{ m/s}$$

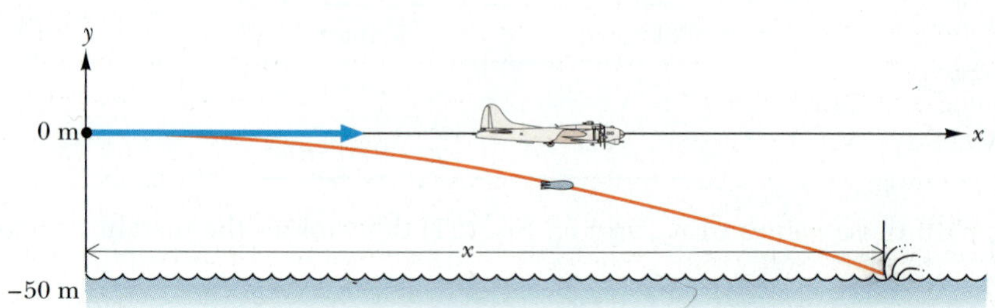

FIGURE 2.8

Trajectory of a bomb. The initial direction of motion is the same as that of the bomber, and the initial direction of motion is horizontal (blue arrow).

and the initial vertical component of the velocity is

$$v_{0y} = 0$$

When the bomb reaches the level of the sea, its vertical position is $y = -50$ m (that is, the bomb is 50 m below our origin of coordinates).

For the vertical motion, the unknown quantity is the time of impact:

Unknown	Known
t	$v_{0y} = 0$ $y = -50$ m $g = 9.81$ m/s^2

With these known quantities, we can use Eq. (22) to calculate the time of impact:

$$y = v_{0y}t - \tfrac{1}{2}gt^2 = -\tfrac{1}{2}gt^2$$

Upon dividing both sides of this equation by $-\frac{1}{2}g$ and taking the square root of both sides, we find

$$t = \sqrt{-\frac{2y}{g}} \tag{25}$$

$$= \sqrt{\frac{2 \times 50 \text{ m}}{g}} = \sqrt{\frac{2 \times 50 \text{ m}}{9.81 \text{ m/s}^2}} = 3.19 \text{ s}$$

At this time, the horizontal position of the bomb is

$$x = v_{0x}t = 83.3 \text{ m/s} \times 3.19 \text{ s} = 266 \text{ m} \tag{26}$$

Note that the bomber moves exactly the same *horizontal* distance in this time; that is, the bomb always remains directly below the bomber because both have exactly the same horizontal velocity $v_{0x} = 83.3$ m/s. Figure 2.9 shows bombs released by a bomber at successive instants of time, and demonstrates that they remain directly below the bomber.

FIGURE 2.9

A "string" of bombs released from a bomber. The bombs continue to move forward with the same horizontal velocity as that of the bomber.

If air resistance is negligible, ***the path of any projectile launched with some horizontal and vertical velocity is a parabola or a portion of a parabola*** (see Fig. 2.10). We can recognize this parabolic shape in the path of the ball photographed in Fig. 2.6, which shows a portion of half a parabola, from the apex downward. In the case of a bomb dropped from a speeding airplane, the path is a similar portion of a parabola, which begins at the apex and descends to the level of the target. In the case of a ball or stone launched by hand or a shot or bullet fired from a gun, the relevant portion begins at the level of the launch point or muzzle, rises to

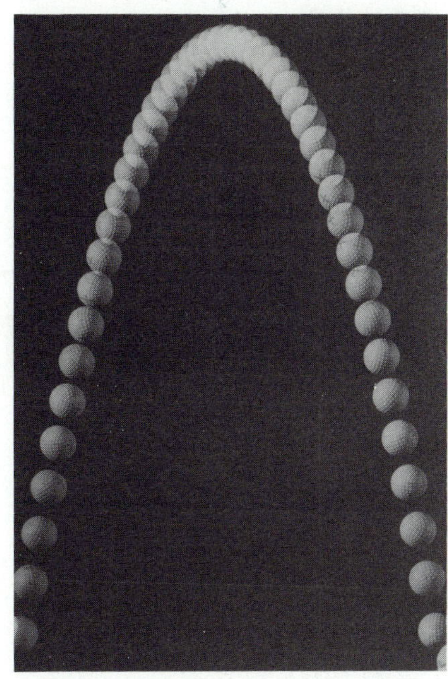

FIGURE 2.10

Stroboscopic photograph showing the path of a projectile (a golf ball) launched upward at an angle. The path is a parabola. At the apex of the parabola (the point of maximum height), the vertical component of the velocity is instantaneously zero.

the apex, and then descends to the target. In the latter case, it is often important to calculate the ***maximum height*** reached, and also the ***time of flight*** (time between the instants of launch and impact) and the ***range*** (distance between the points of launch and impact). This kind of calculation is illustrated in the following example.

FIGURE 2.11
A champion discus thrower.

EXAMPLE 5. A champion discus thrower (Fig. 2.11) throws a discus with an initial speed of 26 m/s. If the discus is thrown upward at an angle of 45°, what maximum height and what range will the discus attain? What is its time of flight?

SOLUTION: For the sake of simplicity we ignore the height of the hand over the ground, so the launch point and the impact point are at the same height. We take the origin of coordinates at the launch point. Equations (19)–(22) are then applicable to this problem. However, to extract the answers we want, we must manipulate these equations somewhat.

Since Eqs. (19)–(22) depend on the components v_{0x} and v_{0y} of the initial velocity, we begin by calculating these components. From Fig. 2.12 we see that:

$$v_{0x} = v_0 \cos\theta \tag{27}$$

$$= v_0 \cos 45^\circ = 26 \text{ m/s} \times \cos 45^\circ = 26 \text{ m/s} \times 0.707 = 18.4 \text{ m/s}$$

and

$$v_{0y} = v_0 \sin\theta \tag{28}$$

$$= v_0 \sin 45^\circ = 26 \text{ m/s} \times \sin 45^\circ = 26 \text{ m/s} \times 0.707 = 18.4 \text{ m/s}$$

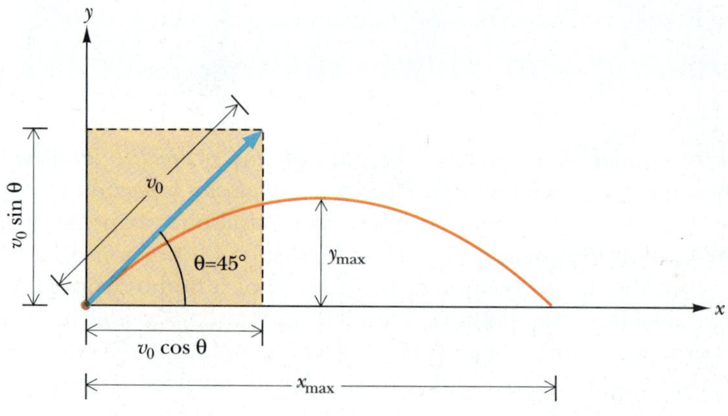

FIGURE 2.12
The blue arrow indicates the initial direction of motion of the discus. The angle θ between this direction and the horizontal is the elevation angle.

the time it spends in flight; an angle larger than 45° increases the time it spends in flight, but gives it a smaller x velocity. By inspection of Fig. 2.13 we see that any two angles that are equal amounts above or below 45° yield equal ranges; for instance, 60° (15° above 45°) and 30° (15° below 45°) yield equal ranges.

In Fig. 2.13, as in all calculations of this section, air resistance has been neglected. For a high-speed projectile, such as a rifle bullet, air resistance is quite important, and the parabolic trajectory is distorted into a trajectory of more complicated shape, called a ***ballistic curve.*** If we attempt to use the simple formulas of this section in the calculation of the motion of a high-speed projectile, our results will bear only a very vague resemblance to reality.

2.4 VECTORS

The concept of vector provides us with a precise and elegant mathematical description of the position, velocity, and acceleration in two- or three-dimensional motion. By means of vectors we can give mathematical meaning to the intuitive notion that velocity and acceleration have *direction* as well as magnitude: the velocity and the acceleration of a particle can point north, or east, or up, or down, or in any direction in between. As mentioned in the introduction, the vector concept is also useful for the description of many other physical quantities, such as force and momentum, which have both magnitude and direction.

◂ **Displacement vector**

We begin our discussion of vectors with the concept of displacement. The ***displacement*** of a particle is simply a change of its position. If a particle moves from a point P_1 to a point P_2, we can represent the change of position graphically by an arrow, or directed line segment, from P_1 to P_2. ***The directed line segment is the displacement vector*** of the particle. For example, if a ship moves from Liberty Island to the Battery in New York Harbor, then the displacement vector is as shown in Fig. 2.14. Note that

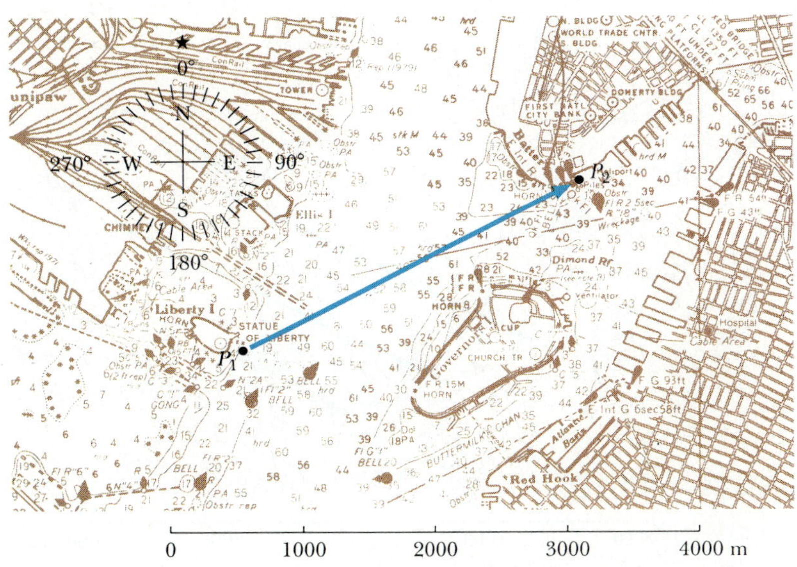

FIGURE 2.14

Displacement vector for a ship moving from Liberty Island to the Battery in New York Harbor. The length of this vector is 2890 m; its direction is 65° east of north. (Excerpt from National Ocean Survey Chart 12328.)

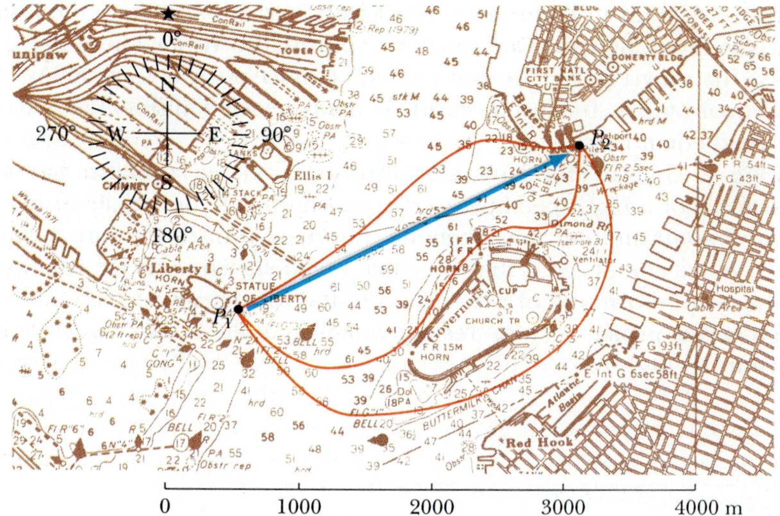

FIGURE 2.15
Several alternative paths (red) from Liberty Island to the Battery. All of these result in the same final displacement.

the displacement vector tells us only where the final position (P_2) is in relation to the initial position(P_1); it does not tell us what path the ship followed between the two positions. Thus, any of the paths shown by the red lines in Fig. 2.15 results in the same final displacement vector.

The displacement vector has a length and a direction. Instead of describing the vector graphically by drawing a picture, we can describe it numerically by giving the numerical value of its length (in, say, meters) and the numerical value of the angle (in, say, degrees) it makes with some reference direction. For example, we can specify the displacement vector in Fig. 2.14 by stating that it is 2890 m long and points at an angle of 65° east of north.

Since the displacement vector describes a *change* in position, any two line segments of identical length and direction represent equal vectors, regardless of whether the endpoints of the line segments are the same. Thus, the parallel directed line segments show in Fig. 2.16 do not represent different vectors; both involve the same *change* of position (same distance and same direction), and both represent equal displacement vectors.

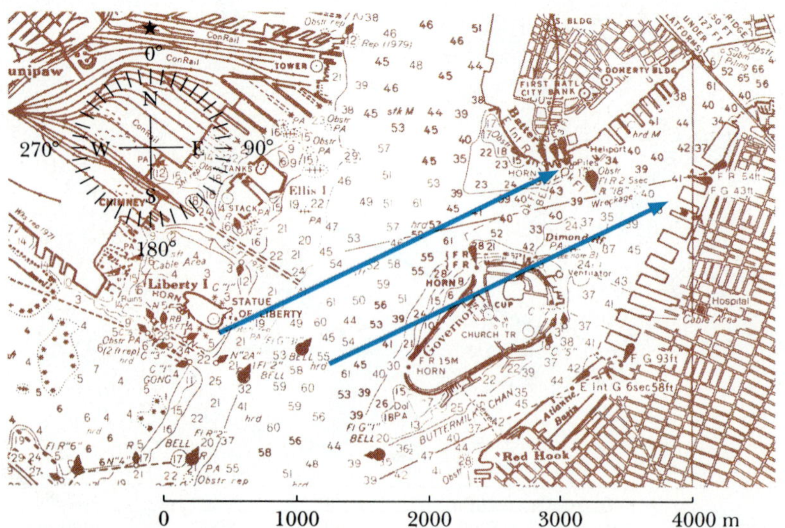

FIGURE 2.16
These two displacement vectors are equal.

In printed books, vectors are usually indicated by boldface letters, such as **A,** and we will follow this convention. In handwritten calculations, an alternative notation consisting of either a small arrow, such as $\vec{A}$, or a wavy underline, such as $\underset{\sim}{A}$, is usually more convenient. We will use an ordinary italic letter, such as *A*, to denote the length, or the magnitude, of a displacement vector. For instance, if **A** stands for the displacement vector shown in Fig. 2.14, then *A* stands for its length, that is, $A = 2890$ m.

The displacement vector serves as prototype for all other vectors. To decide whether some quantity endowed with both magnitude and direction is a vector, we compare its mathematical properties with those of the displacement vector. ***Any quantity that has magnitude and direction and that has the mathematical behavior of the displacement vector is a vector.*** For example, velocity, acceleration, and force are vectors; they can be represented graphically by directed line segments of a length equal to the magnitude of the velocity, acceleration, or force (in some suitable units) and a corresponding direction.

◀ General vector

In contrast, ***any quantity that has a magnitude but no direction is called a scalar.*** For example, length, time, mass, area, volume, density, and energy are scalars; they can be completely specified by their numerical magnitude. Note that the length of a displacement vector, such as the length $A = 2890$ m of the displacement vector in Fig. 2.14, is a quantity which has a magnitude but no direction; that is, this length is a scalar.

◀ Scalar

2.5 Vector Addition and Subtraction

Since by definition all vectors have the mathematical properties of displacement vectors, we can investigate all the mathematical operations with vectors by looking at displacement vectors. The most important of these mathematical operations is ***vector addition.***

◀ Vector addition

Two displacements carried out in succession result in a net displacement, which can be regarded as the sum of the individual displacements. For example, Fig. 2.17 shows a displacement vector **A** (from P_1 to P_2) and a dis-

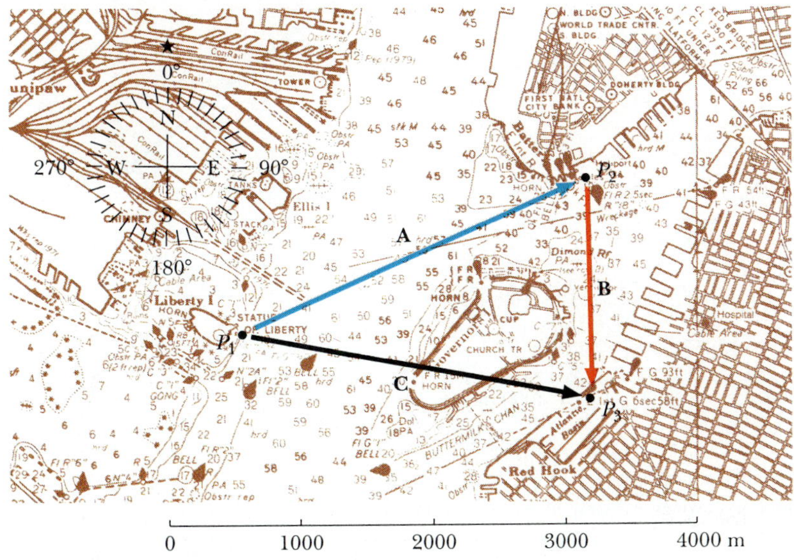

FIGURE 2.17
The displacement **A** is followed by the displacement **B.** The net displacement is **C.**

placement vector **B** (from P_2 to P_3). The net displacement vector is the directed line segment from P_1 to P_3; this net displacement vector is denoted by **C** in Fig. 2.17. This vector **C** can be regarded as the sum of the individual displacements,

$$\mathbf{C} = \mathbf{A} + \mathbf{B} \tag{38}$$

Resultant ▸ The sum of two vectors is usually called the ***resultant*** of these vectors. Thus **C** is called the resultant of **A** and **B**.

EXAMPLE 6. A ship moves from Ellis Island in New York Harbor to the Battery and from there to the Atlantic Basin (see Fig. 2.18a). The first displacement vector is 2180 m due east, and the second is 1790 m due south. What is the resultant?

SOLUTION: The resultant of the two displacement vectors **A** and **B** is the vector **C**, from the tail of **A** to the tip of **B**. For a graphical determination of **C**, we can measure the length of **C** directly on the chart using the scale of length marked on the chart, and we can measure the direction of **C** with a protractor (the way the navigator of the ship would solve the problem). This yields a length of about 2800 m and an angle of about 39° for the resultant vector **C**.

For a more precise, numerical, determination of **C**, we note that **A**, **B**, and **C** form a right triangle. We can therefore find **C** by using the standard trigonometric methods for the solution of triangles.* The lengths of the two known sides of this triangle are $A = 2180$ m and $B = 1790$ m (see Fig. 2.18b). The unknown side is the hypotenuse of

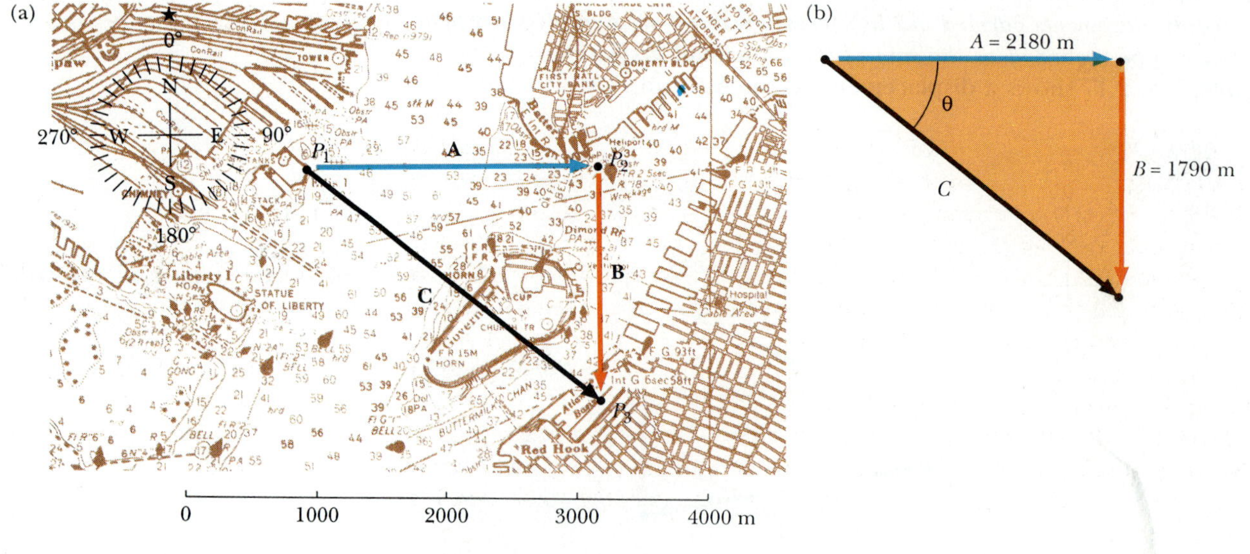

FIGURE 2.18

(a) The vectors **A** and **B** represent the displacements of the ship moving from Ellis Island to the Battery and from the Battery to the Atlantic Basin. The vector **C** is the sum of these two vectors. (b) The vectors **A**, **B**, and **C** form a right triangle. Two of the sides are known, and the hypotenuse is unknown.

* Appendix 4 gives a review of trigonometry.

the triangle. By the Pythagorean theorem, the length C of the hypotenuse is

$$C = \sqrt{A^2 + B^2}$$

from which

$$C = \sqrt{(2180\ \text{m})^2 + (1790\ \text{m})^2} = 2820\ \text{m}$$

The tangent of the angle θ is the ratio of the opposite side B to the adjacent side A,

$$\tan\theta = \frac{B}{A} = \frac{1790\ \text{m}}{2180\ \text{m}} = 0.821 \tag{39}$$

Using an electronic pocket calculator, we find that the angle whose tangent is 0.821 is

$$\theta = 39.4°$$

Thus, the resultant vector **C** has a length of 2820 m at an angle 39.4° south of east, which is consistent with values obtained by the graphical method.

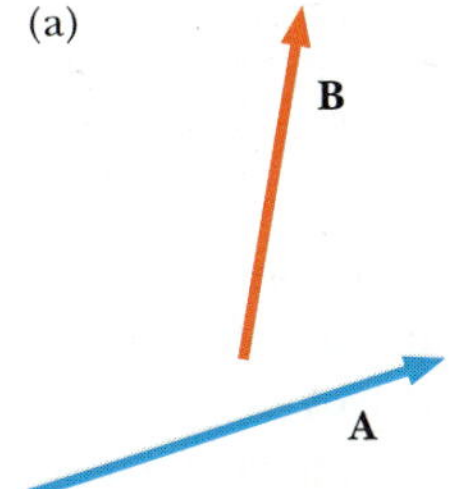

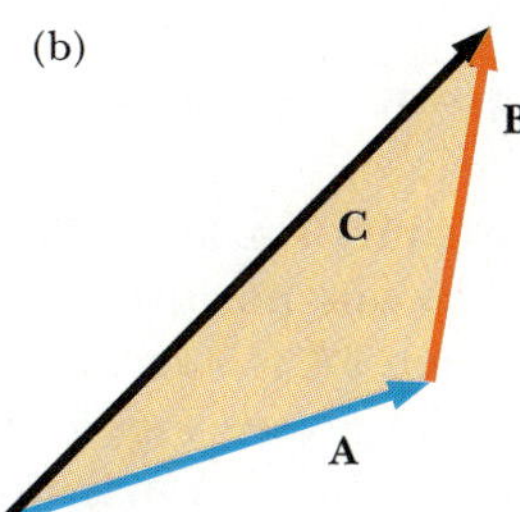

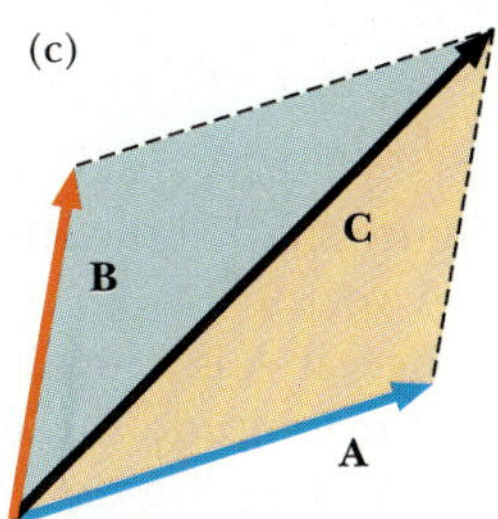

FIGURE 2.19 The vector sum **A** + **B**; the resultant is **C.** (a) The two vectors **A** and **B.** (b) Addition of **A** and **B** by the tail-to-head method. (c) Addition of **A** and **B** by the parallelogram method.

The procedure for the addition of all other vectors — such as velocity, acceleration, and force vectors — mimics that for displacement vectors. All such vectors can be represented by arrows. If **A** and **B** are two arbitrary vectors (see Fig. 2.19a), then their resultant can be obtained by placing the tail of **B** on the head of **A;** ***the directed line segment connecting the tail of A to the head of B is the resultant*** (see Fig. 2.19b). Alternatively, the resultant can be obtained by placing the tail of **B** on the tail of **A** and drawing a parallelogram with **A** and **B** as two of the sides; ***the diagonal of the parallelogram is then the resultant*** (see Fig. 2.19c).

◀ **Tail-to-head method**

◀ **Parallelogram method**

Note that the order in which the two vectors are added makes no difference to the final result. Whether we place the tail of **B** on the head of **A** or the tail of **A** on the head of **B,** the resultant is the same (see Fig. 2.20). Hence

$$\mathbf{A} + \mathbf{B} = \mathbf{B} + \mathbf{A} \tag{40}$$

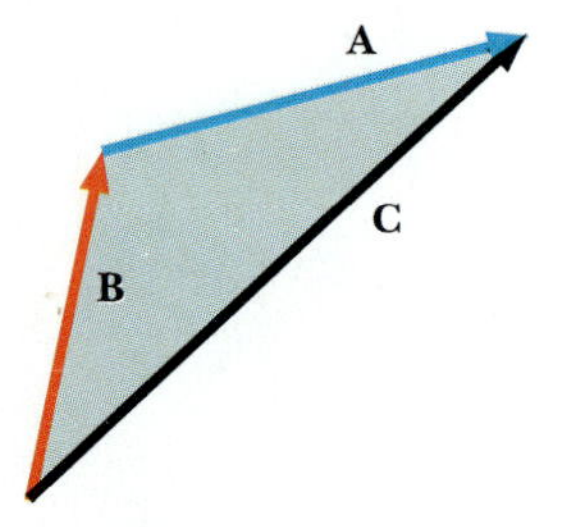

FIGURE 2.20
The resultant for the vector sum **B** + **A** is the same as for the vector sum **A** + **B** (compare Fig. 2.19b).

This identity indicates that, just as in ordinary addition of numbers, the order of the terms is irrelevant.

The magnitude of the resultant of two vectors is usually less than the sum of the magnitudes of the vectors. Thus, if

$$\mathbf{C} = \mathbf{A} + \mathbf{B} \tag{41}$$

then

$$C \leq A + B \tag{42}$$

This inequality simply expresses the fact that in a triangle (see Fig. 2.20) the length of any side is less than the sum of the lengths of the other two sides. Only in the special case where **A** and **B** are parallel (see Fig. 2.21) will the magnitude of **C** equal the sum of the magnitudes of **A** and **B;** it can never exceed this sum.

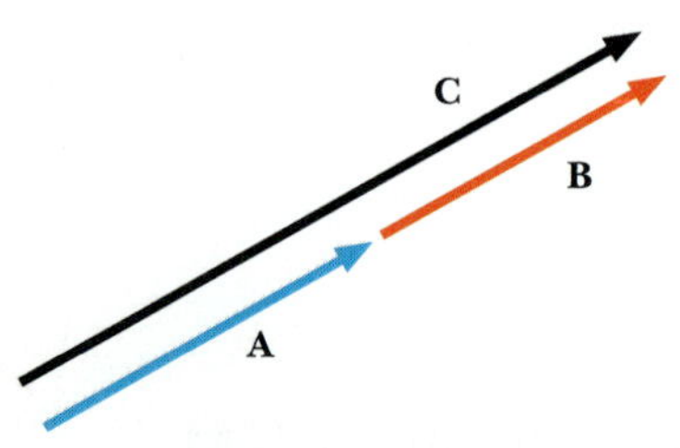

FIGURE 2.21
Parallel vectors **A** and **B,** and their resultant **C.**

The negative of a given vector* A *is a vector of the same magnitude, but opposite direction; this new vector is denoted by −**A** (see Fig. 2.22). Obviously, the sum of a vector and its negative gives a vector of zero magnitude,

$$\mathbf{A} + (-\mathbf{A}) = 0 \tag{43}$$

Vector subtraction ▸

***The subtraction of two vectors* A *and* B *is defined as the sum of* A *and* −B,**

$$\mathbf{A} - \mathbf{B} = \mathbf{A} + (-\mathbf{B}) \tag{44}$$

Figure 2.23 shows this vector sum of **A** and −**B.**

A vector can be multiplied by any positive or negative number. For instance, if **A** is a given vector, then 3**A** is a vector of the same direction and of a magnitude three times as large (see Fig. 2.24a); and −3**A** is a vector of the opposite direction and, again, of a magnitude three times as large (see Fig. 2.24b). In particular, if we multiply a vector by −1, we obtain the negative of that vector:

$$(-1)\mathbf{A} = -\mathbf{A} \tag{45}$$

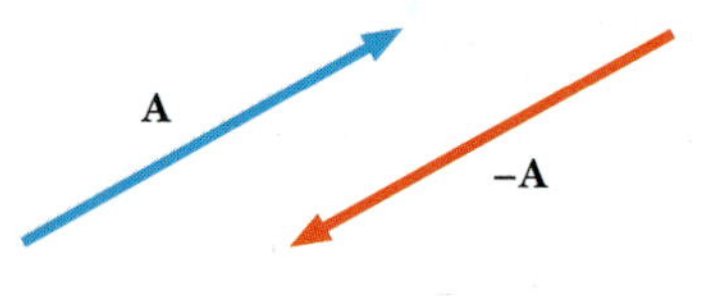

FIGURE 2.22
The negative of the vector **A** has the same length as the vector **A,** but the opposite direction.

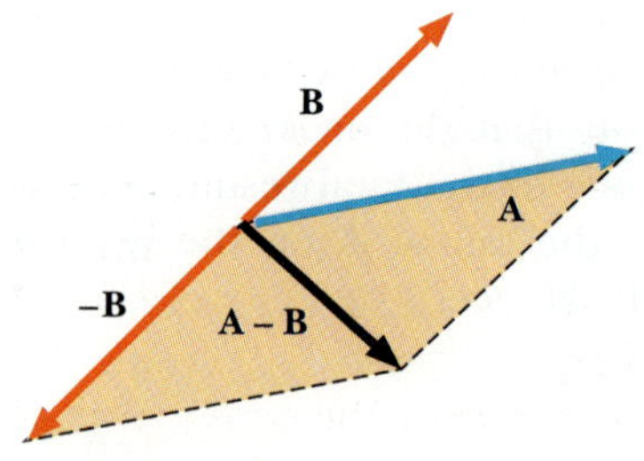

FIGURE 2.23
To subtract the vector **B** from **A,** we first draw the vector −**B,** and then construct the vector sum **A** + (−**B**) by the parallelogram method.

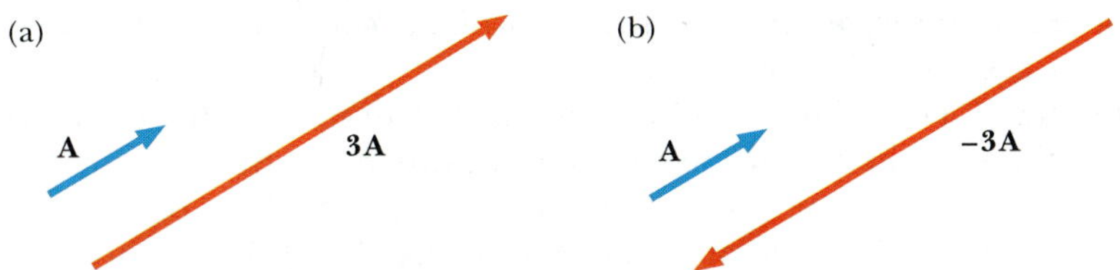

FIGURE 2.24 (a) The vector 3**A** has the same direction as the vector **A**, and is three times as long. (b) The vector −3**A** has a direction opposite to that of **A**, and is three times as long.

2.6 THE POSITION, VELOCITY, AND ACCELERATION VECTORS

In Section 1.1, we saw that to describe the position of a point, we must choose an origin and construct a coordinate grid. If the grid is rectangular, then the position of a point (in two dimensions) will be specified by the rectangular coordinates x and y. Alternatively, we can describe the position by means of the displacement vector from the origin to the point. Figure 2.25 shows the point P, its coordinates x and y, and the displacement vector from the origin to the point. This displacement vector is called the ***position vector,*** usually denoted by **r.**

◀ Position vector

As indicated in Fig. 2.25, the position vector of a point has a length r and it makes an angle θ with the x axis. Consider the right triangle shown in Fig. 2.25. The length r of the position vector is the hypotenuse of this triangle, and the x and y coordinates form the sides. From the Pythagorean theorem, the length of the position vector is

$$r = \sqrt{x^2 + y^2} \tag{46}$$

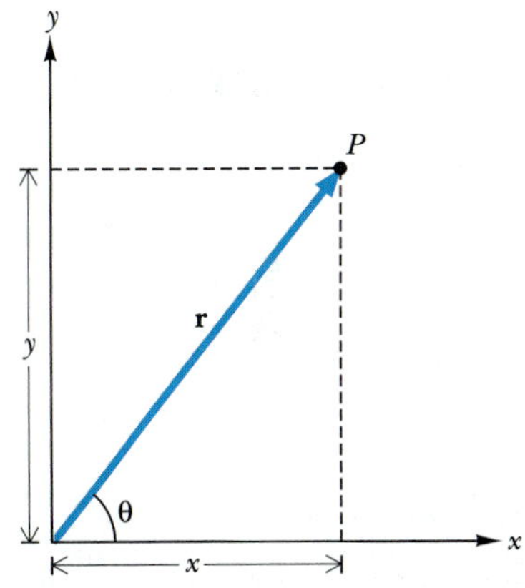

FIGURE 2.25 The position vector **r** in two dimensions and its components.

And from the definitions of the cosine and the sine of the angle θ,

$$\cos\theta = \frac{x}{r}$$

$$\sin\theta = \frac{y}{r}$$

Hence

$$x = r\cos\theta \tag{47}$$

$$y = r\sin\theta \tag{48}$$

These equations show how the x and y coordinates of a point are related to the length and the direction of the position vector. According to the terminology introduced in the preceding chapter, x and y are the components of the position. We will now adopt a more precise terminology according to which x ***and*** y ***are the components of the position vector.*** Note that graphically the components of the position vector are determined by

◀ Components of the position vector

dropping perpendiculars from the head of the vector to the x and y axes (see Fig. 2.25).

In general, for an arbitrary vector **A**, the definition of the components is analogous to the definition of the components of the position vector. We place the tail of the vector at the origin, and we drop perpendiculars from the head of the vector to the x and y axes. The intercepts of these perpendiculars with the axes (which may be positive or negative) give us the x and y components of the vector **A** (see Fig. 2.26). Designating these components by A_x and A_y, we see from Fig. 2.26 that the magnitude of the vector is

Magnitude in terms of components ▶

$$A = \sqrt{A_x^2 + A_y^2} \tag{49}$$

and that the components are

Components in terms of angle ▶

$$A_x = A \cos\theta \tag{50}$$

$$A_y = A \sin\theta \tag{51}$$

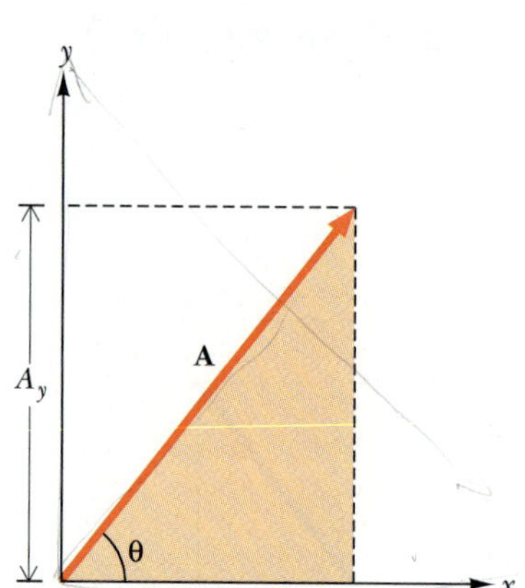

FIGURE 2.26
An arbitrary vector **A** in two dimensions and its components.

These equations are analogous to Eqs. (47) and (48).

Note that the components A_x and A_y uniquely specify the vector — if we know the components, we can find the magnitude and the direction of the vector from Eqs. (49)–(51).

The addition or subtraction of two vectors can be performed by adding or subtracting their components. Thus, if A_x, A_y and B_x, B_y are the components of the vectors **A** and **B**, then the components of the resultant of these two vectors are

$$C_x = A_x + B_x$$

$$C_y = A_y + B_y$$

This component method for the calculation of a vector sum or difference is often the most convenient.

EXAMPLE 7. The eye of a hurricane is 200 km from Miami on a bearing of 30° south of east. A reconnaissance airplane is 100 km due north of Miami. What displacement vector will bring the plane to the eye of the hurricane?

SOLUTION: In Example 6 we saw how to use a graphical method and a trigonometric method for finding an unknown vector. Here we will see how to use the component method.

For this calculation, we need to make a choice of coordinate system, that is, a choice of origin and of axes. In Fig. 2.27, we have placed the origin on Miami, with the x axis eastward and the y axis northward. In this coordinate system, the airplane has a position vector **A** with components

$$A_x = 0 \text{ km} \quad A_y = 100 \text{ km}$$

The hurricane has a position vector **B** with components [see Eqs. (50) and (51); Fig. 2.27 b]

$$B_x = 200 \text{ km} \times \cos 30^\circ = 173 \text{ km}$$

$$B_y = -200 \text{ km} \times \sin 30^\circ = -100 \text{ km}$$

The displacement vector **C** from the airplane to the hurricane is the *difference* between these position vectors, that is, **C** = **B** − **A** (see Fig. 2.27a). Subtraction of two vectors is equivalent to subtraction of their components; thus

$$C_x = B_x - A_x = 173 \text{ km} - 0 \text{ km} = 173 \text{ km}$$

$$C_y = B_y - A_y = -100 \text{ km} - 100 \text{ km} = -200 \text{ km}$$

The magnitude of the vector **C** is then, according to Eq. (49),

$$C = \sqrt{C_x^2 + C_y^2} = \sqrt{(173 \text{ km})^2 + (-200 \text{ km})^2} = 264 \text{ km}$$

and the angle between **C** and the x axis is given by [see Eq. (50); Fig. 2.27c]

$$\cos\theta = \frac{173}{264} = 0.655$$

Our electronic calculator then tells us that

$$\theta = 49^\circ$$

Hence a displacement of 264 km at 49° south of east will bring the airplane to the hurricane.

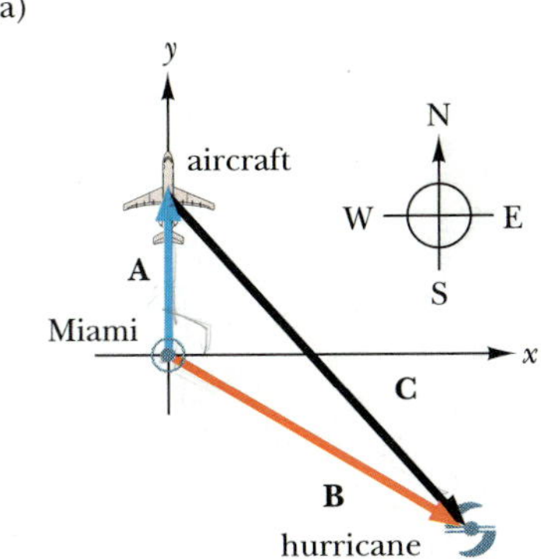

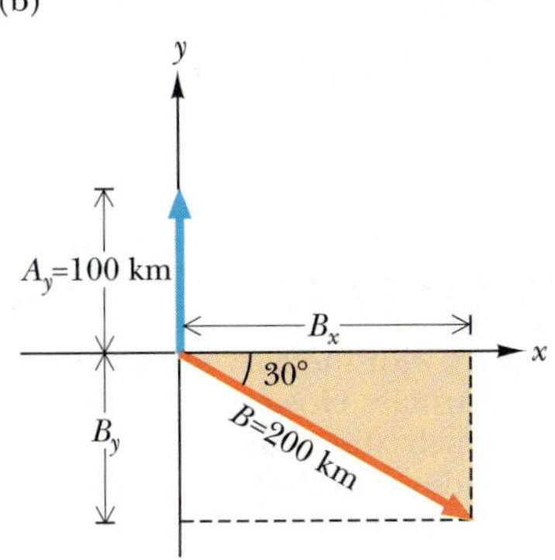

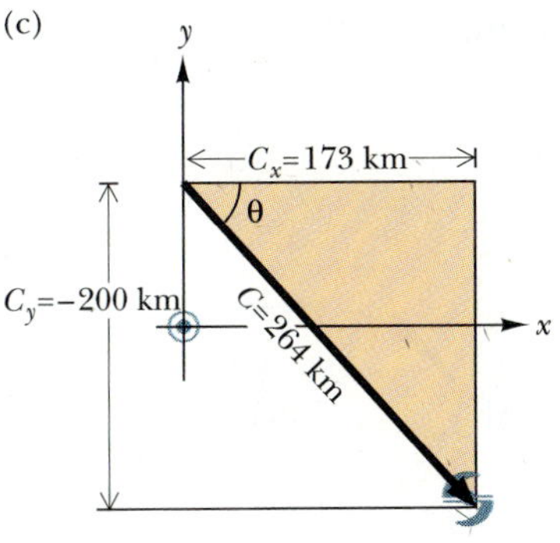

FIGURE 2.27

(a) Position vectors of the aircraft **(A)** and of the hurricane **(B).** The displacement vector from the aircraft to the hurricane is the difference between these vectors, **C** = **B** − **A.** (b) The x and y components of **A** and **B.** (c) The x and y components of **C.**

According to the discussion in Section 2.1, the x and y components of the (instantaneous) velocity of a particle moving in two dimensions are

$$v_x = \frac{\Delta x}{\Delta t} \tag{52}$$

$$v_y = \frac{\Delta y}{\Delta t} \quad \text{with } \Delta t \text{ extremely small} \tag{53}$$

Since Δx and Δy are the components of the vector $\Delta\mathbf{r}$ representing the change in the position vector in the small time interval Δt, it is clear that

v_x and v_y in Eqs. (52) and (53) are the x and y components of the vector $\Delta\mathbf{r}/\Delta t$, which is obtained from the vector $\Delta\mathbf{r}$ by dividing by Δt. The vector $\Delta\mathbf{r}/\Delta t$ is the ***velocity vector.*** We will use the notation $\mathbf{v}$ for this vector,

Velocity vector ▶

$$\mathbf{v} = \frac{\Delta\mathbf{r}}{\Delta t} \quad \text{with } \Delta t \text{ extremely small} \tag{54}$$

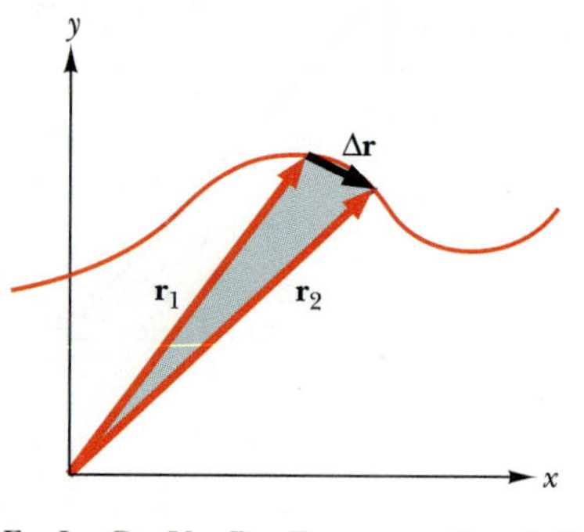

FIGURE 2.28

In a small time interval Δt, the particle has a displacement $\Delta\mathbf{r}$. Since $\Delta\mathbf{r}$ is a straight line segment, it cannot coincide exactly with the curved path of the particle. However, if $\Delta\mathbf{r}$ is extremely short, then it almost coincides with the path, as shown in this figure. In this case, the direction of $\Delta\mathbf{r}$ is tangent to the path, and so is the direction of $\Delta\mathbf{r}/\Delta t$.

This equation simply says that the velocity vector is the rate of change of the position vector.

The direction of the velocity vector is the direction of the small displacement $\Delta\mathbf{r}$. As shown in Fig. 2.28, in the limiting case of an extremely small time interval Δt, the direction of $\Delta\mathbf{r}$, and hence also ***the direction of the velocity $\Delta\mathbf{r}/\Delta t$, is tangent to the path of the particle.*** If the particle moves along a curved path, then, in general, both the magnitude and the direction of the velocity change with time. For example, Fig. 2.29 shows the instantaneous velocity vectors at different times for a projectile launched upward at some angle; at each instant, the velocity is tangent to the path.

The equations relating the components of the velocity vector to the magnitude and the direction of this vector have already been stated in Section 2.1 [see Eqs. (8)–(10)]:

$$v = \sqrt{v_x^2 + v_y^2}$$

$$v_x = v\cos\theta$$

$$v_y = v\sin\theta$$

These equations are special instances of Eqs. (49)–(51), which are valid for any vector.

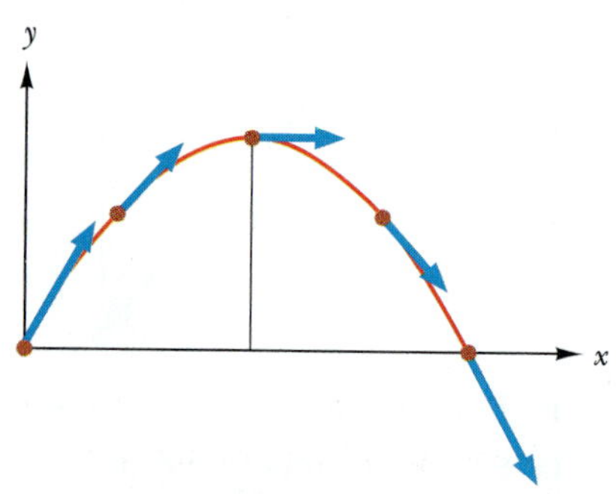

FIGURE 2.29

Instantaneous velocity vectors of a projectile at different times.

Similar arguments lead us to introduce the acceleration vector. We begin, again, with the definitions of the components of the (instantaneous) acceleration from Section 2.1,

$$a_x = \frac{\Delta v_x}{\Delta t} \tag{55}$$

$$a_y = \frac{\Delta v_y}{\Delta t} \quad \text{with } \Delta t \text{ extremely small} \tag{56}$$

Here Δv_x and Δv_y are the components of the vector $\Delta\mathbf{v}$ representing the change in the velocity vector. We therefore conclude that a_x and a_y are the components of an ***acceleration vector,*** which we will designate by **a,**

Acceleration vector ▶

$$\mathbf{a} = \frac{\Delta\mathbf{v}}{\Delta t} \quad \text{with } \Delta t \text{ extremely small} \tag{57}$$

The acceleration vector is the rate of change of the velocity vector. The direction of the acceleration vector is the direction of the change of the velocity vector. In the next section, we will see an example of how to find the direction of the instantaneous change of the velocity vector.

2.7 UNIFORM CIRCULAR MOTION

◀ Uniform circular motion

Uniform circular motion is motion with constant speed along a circular path, such as the motion of an automobile around a traffic circle or the motion of a planet around the Sun. Figure 2.30 shows the positions at different times for a particle in uniform circular motion. The velocity vector at any instant is tangent to the path, that is, it is tangent to the circle. All the velocity vectors shown have the same magnitude (same speed), but they differ in direction. Because of this change of direction, ***uniform circular motion is accelerated motion.***

Suppose that the constant speed of the particle is v and the radius of the circle is r. To find the value of the instantaneous acceleration, we must look at the velocity change in a very short time interval Δt. Figure 2.31a shows the particle at two positions $\mathbf{r}_1$ and $\mathbf{r}_2$ a short time apart; the difference between these vectors is $\Delta\mathbf{r} = \mathbf{r}_2 - \mathbf{r}_1$. The figure also shows the two velocity vectors $\mathbf{v}_1$ and $\mathbf{v}_2$; both velocity vectors have the same magnitude v; Fig. 2.31b shows the two velocity vectors tail to tail and their difference $\Delta\mathbf{v} = \mathbf{v}_2 - \mathbf{v}_1$. The position vectors make a small angle $\Delta\theta$ with each other; the velocity vectors are always perpendicular to the position vectors, and the angle between the velocity vectors is therefore the same as the angle between the position vectors. Since the angles $\Delta\theta$ in Figs. 2.31a and b are equal, the triangles formed by the position vectors and by the velocity vectors are similar — hence the ratio of the short sides of these triangles equals the ratio of their long sides, that is, the ratio of the magnitudes of the vectors $\Delta\mathbf{r}$ and $\Delta\mathbf{v}$ equals the ratio of v and r:

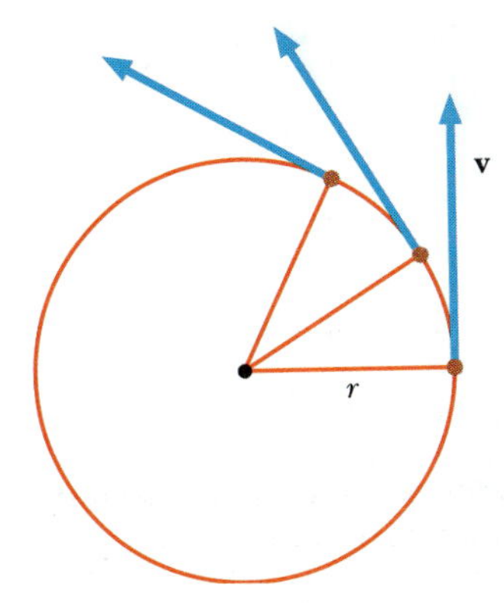

FIGURE 2.30

Instantaneous velocity vectors for a particle in uniform circular motion.

$$\frac{[\text{magnitude of } \Delta\mathbf{v}]}{[\text{magnitude of } \Delta\mathbf{r}]} = \frac{v}{r}$$

We can rewrite this as

$$[\text{magnitude of } \Delta\mathbf{v}] = \frac{v}{r} \times [\text{magnitude of } \Delta\mathbf{r}] \tag{58}$$

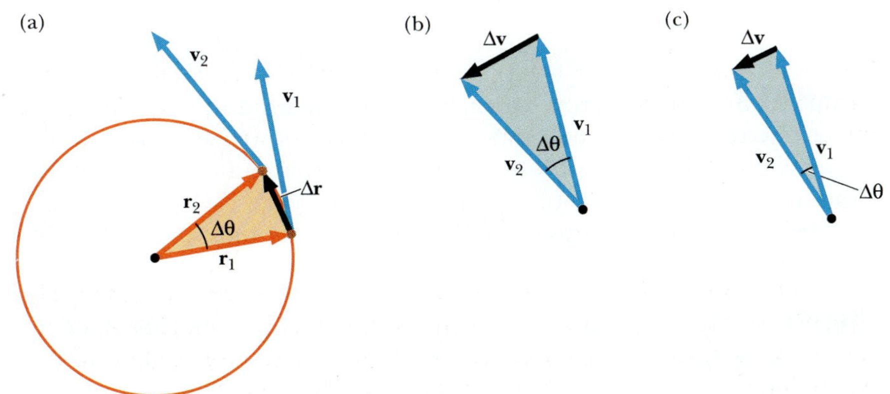

FIGURE 2.31

(a) Position vectors $\mathbf{r}_1$ and $\mathbf{r}_2$ at two times with an interval Δt. The difference between these two position vectors is $\Delta\mathbf{r}$. (b) Velocity vectors $\mathbf{v}_1$ and $\mathbf{v}_2$ at the same two times. The difference between these velocity vectors is $\Delta\mathbf{v}$. The colored triangle is similar to the colored triangle in Part (a). (c) If the angle $\Delta\theta$ is very small, then the two velocity vectors will be nearly parallel, and their difference $\Delta\mathbf{v}$ will be nearly perpendicular to both vectors.

If Δt is very small, then $\Delta\theta$ will also be very small, and the straight line segment $\Delta\mathbf{r}$ will approximately coincide with the circular arc from the tip of $\mathbf{r}_1$ to the tip of $\mathbf{r}_2$. The latter length is simply the distance traveled by the particle in the time Δt,

$$[\text{magnitude of } \Delta\mathbf{v}] \simeq \frac{v}{r} \times [\text{distance traveled in time } \Delta t] \tag{59}$$

The magnitude of the acceleration is the magnitude of $\Delta\mathbf{v}$ divided by the time Δt:

$$a = \frac{[\text{magnitude of } \Delta\mathbf{v}]}{\Delta t} \simeq \frac{v}{r} \times \frac{[\text{distance traveled in time } \Delta t]}{\Delta t} \tag{60}$$

But the distance traveled divided by the time Δt is the speed v, so

$$a \simeq \frac{v}{r} \times v \tag{61}$$

This equation becomes exact if the time Δt is extremely small, giving us the result

Centripetal acceleration ▸

$$a = \frac{v^2}{r} \tag{62}$$

The *direction* of this acceleration remains to be determined. From Fig. 2.31c it is clear that, for the case of very small Δt, the direction of $\Delta\mathbf{v}$ will be perpendicular to the velocity vectors $\mathbf{v}_1$ and $\mathbf{v}_2$ (which will be nearly parallel in this limiting case). Hence the instantaneous acceleration is perpendicular to the instantaneous velocity. Since the velocity vector corresponding to circular motion is tangential to the circle, ***the acceleration vector points along the radius, toward the center of the circle.*** Figure 2.32 shows the velocity and acceleration vectors at two positions along the circular path. The acceleration of uniform circular motion, with a magnitude given by Eq. (62), is called ***centripetal acceleration,*** because it is directed toward the center of the circle.

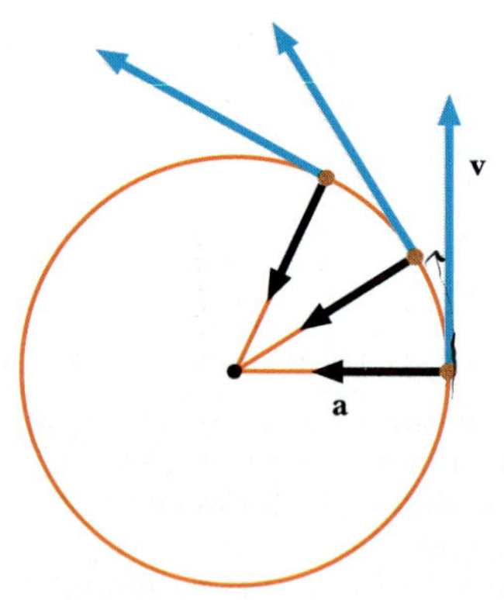

FIGURE 2.32
Instantaneous acceleration vectors for a particle in uniform circular motion.

EXAMPLE 8. In tests of the effects of high acceleration on the human body, astronauts at the NASA Manned Spacecraft Center in Houston are placed in a capsule that is whirled around a circular path of radius 15 m at the end of a revolving girder (see Fig. 2.33). If the girder makes 24 revolutions per minute, what is the acceleration of the capsule?

SOLUTION: The circumference of the circular path is $2\pi \times [\text{radius}] = 2\pi \times 15$ m. Since the capsule makes 24 revolutions per minute, or 24 revolutions per 60 seconds, the capsule takes 60/24 s to go around this circumference. The speed is therefore

$$v = \frac{2\pi \times 15 \text{ m}}{60/24 \text{ s}} = 38 \text{ m/s}$$

From Eq. (62), the centripetal acceleration is then

$$a = \frac{v^2}{r} = \frac{(38 \text{ m/s})^2}{15 \text{ m}} = 96 \text{ m/s}^2 \tag{63}$$

FIGURE 2.33

Centrifuge at the Manned Spacecraft Center in Houston.

EXAMPLE 9. An automobile drives around a traffic circle of radius 30 m. If the wheels of the automobile can withstand a maximum transverse acceleration of 8.0 m/s² without skidding, what is the maximum permissible speed?

SOLUTION: With a radius of 30 m and a centripetal acceleration of 8.0 m/s², Eq. (62) gives

$$v^2 = ar \tag{64}$$

$$= 8.0 \text{ m/s}^2 \times 30 \text{ m} = 240 \text{ (m/s)}^2$$

Hence

$$v = \sqrt{240 \text{ (m/s)}^2} = 15 \text{ m/s}$$

This is 56 km/h. If the driver tries to go around the circle faster than this, she will skid out of the circle.

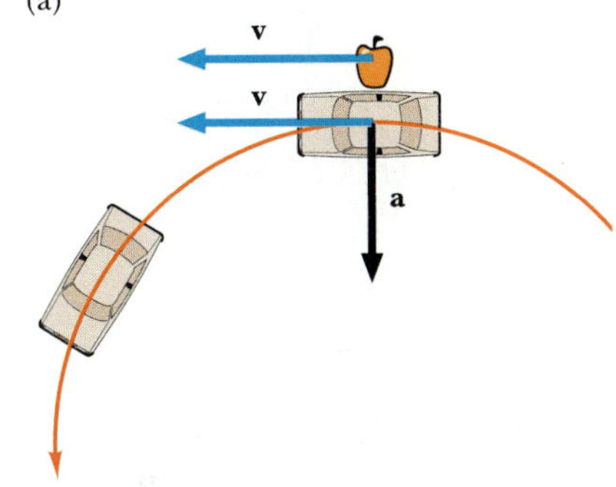

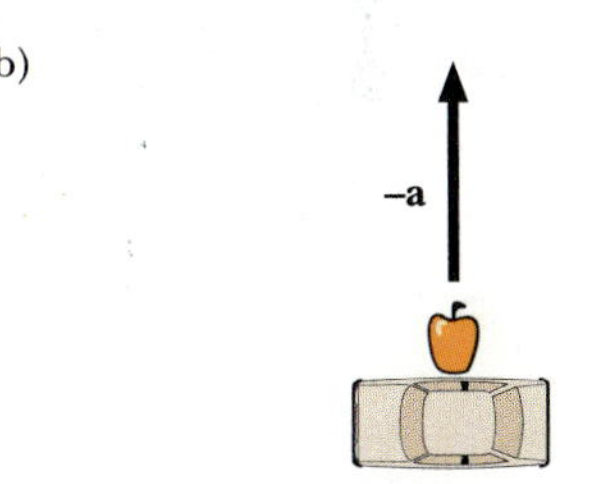

FIGURE 2.34

(a) Automobile in uniform circular motion. The driver has just released an apple, which now moves with constant velocity **v** relative to the ground. The automobile accelerates away from the apple with the centripetal acceleration **a.** (b) Relative to the automobile, the apple has a centrifugal acceleration −**a.**

Note that the acceleration of a particle, or an automobile, in uniform circular motion is always centripetal (toward the center), never centrifugal (away from the center). However, with respect to the reference frame of the accelerating automobile, any body free to move in a horizontal direction — such as an apple that the driver of the automobile has placed on the (frictionless) dashboard — slides away and accelerates toward the outer side of the automobile with a centrifugal acceleration of magnitude equal to the centripetal acceleration of the automobile with respect to the ground (see Fig. 2.34). This centrifugal acceleration of the apple exists only in the reference frame of the automobile; it does not exist in the reference frame of the ground, where the apple merely continues its horizontal motion with constant velocity while the dashboard of the automobile accelerates sideways away from it.

2.8* THE RELATIVITY OF MOTION; ADDITION OF VELOCITIES

Motion is relative — the values of the position, velocity, and acceleration of a particle depend on the frame of reference in which these quantities are measured. For example, consider one reference frame attached to the shore and a second reference frame attached to a ship moving away from the shore due east at a constant velocity of 5 m/s. Suppose that observers in both reference frames measure and plot the position vector of a fast motorboat passing by. The observers will then find different results for the position and the velocity of the motorboat. If the velocity of the motorboat is 12 m/s due east in the reference frame of the shore, it will be 7 m/s due east in the reference frame of the ship. The velocities of the motorboat in the two reference frames are related by a simple addition rule: the velocity relative to the shore is the velocity relative to the ship plus the velocity of the ship relative to the shore, that is, 12 m/s = 7 m/s + 5 m/s. This simple addition rule seems intuitively obvious, but to see where it comes from, we need to examine the position vectors of the motorboat in the two reference frames.

The position vector measured in the first reference frame (shore) will be denoted by $\mathbf{r}$; that measured in the second reference frame (ship) will be denoted by $\mathbf{r}'$. These two position vectors are different (see Fig. 2.35, where the position vector $\mathbf{r}$ is red and the position vector $\mathbf{r}'$ is blue). The velocity vector of the ship relative to the shore will be denoted by $\mathbf{V}_O$. For the sake of simplicity, let us assume that the velocity of the ship relative to the shore is constant, and that the ship started at the shore at time $t = 0$. The position vector of the ship relative to the shore is then $\mathbf{R} = \mathbf{V}_O t$. The vectors $\mathbf{r}$, $\mathbf{r}'$, and $\mathbf{R}$ form a vector triangle (see Fig. 2.35), and by inspec-

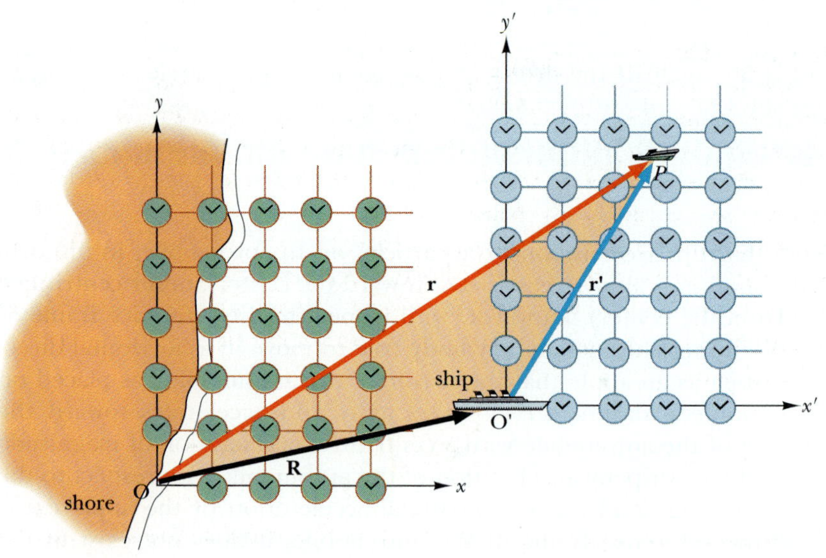

FIGURE 2.35

The reference frame x'–y' of the ship moves relative to the reference frame x–y of the shore.

* This section is optional.

tion of this triangle we recognize that

$$\mathbf{r} = \mathbf{r}' + \mathbf{R}$$

or

$$\mathbf{r} = \mathbf{r}' + \mathbf{V}_O t \qquad (65)$$

This equation merely says that the position vector of the motorboat relative to the shore is the position vector relative to the ship plus the position vector of the ship relative to the shore.

To extract the addition rule for velocities from Eq. (65), we contemplate what happens in a small time interval Δt. In such a time interval, the position vectors $\mathbf{r}$ and $\mathbf{r}'$ change by $\Delta\mathbf{r}$ and $\Delta\mathbf{r}'$. According to Eq. (65), these small changes are related by

$$\Delta\mathbf{r} = \Delta\mathbf{r}' + \mathbf{V}_O \Delta t \qquad (66)$$

Dividing this by Δt, we find

$$\frac{\Delta\mathbf{r}}{\Delta t} = \frac{\Delta\mathbf{r}'}{\Delta t} + \mathbf{V}_O \qquad (67)$$

But $\Delta\mathbf{r}/\Delta t$ is the instantaneous velocity of the motorboat relative to the shore, and $\Delta\mathbf{r}'/\Delta t$ is the instantaneous velocity of the motorboat relative to the ship. Designating these velocities by $\mathbf{v}$ and $\mathbf{v}'$, respectively, we obtain

$$\mathbf{v} = \mathbf{v}' + \mathbf{V}_O \qquad (68)$$

◀ **Addition rule for velocities**

This is the ***addition rule for velocities,*** sometimes called the Galilean addition rule for velocities, because Galileo first considered how changes of reference frame affect velocities and accelerations. Equation (68) is a vector equation, which tells us both the magnitude and the direction of the velocity: ***the velocity vector* $\mathbf{v}$ *measured in the first reference frame (shore) equals the velocity vector* $\mathbf{v}'$ *measured in the second reference frame (ship) plus the velocity vector* $\mathbf{V}_O$ *of the second reference frame relative to the first.***

Note that since the velocities in the two reference frames differ by only a constant quantity $\mathbf{V}_O$, the *changes* in these velocities are the same in the two reference frames. Thus, the accelerations in the two reference frames are the same,

$$\mathbf{a}' = \mathbf{a} \qquad (69)$$

If the motorboat is accelerating at, say, 3 m/s^2 relative to the shore, it will be accelerating at the same rate relative to the ship. This means that for reference frames in uniform motion relative to one another, ***acceleration is an absolute quantity.***

EXAMPLE 10. Off the coast of Miami, the Gulf Stream current has a velocity of 4.8 km/h in a direction due north. The captain of a motorboat wants to travel on a straight course from Miami to Bimini Island, due east of Miami. His boat has a speed of 18 km/h relative to the water. In what direction must he head his boat? What is his speed relative to the shore?

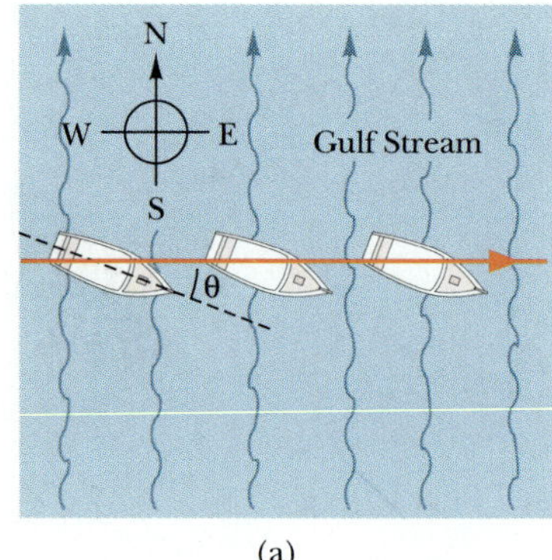

(a)

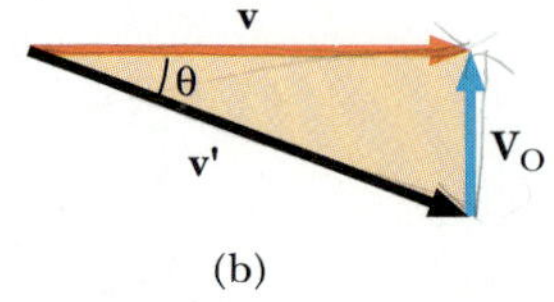

(b)

FIGURE 2.36

(a) The path of the boat is toward the east, but the heading of the boat is south of east. (b) The velocity vector $\mathbf{V}_O$ of the water relative to the shore, and the velocity vector $\mathbf{v}'$ of the boat relative to the water. The sum of these vectors equals the velocity $\mathbf{v}$ of the boat relative to the shore.

SOLUTION: Obviously, the captain must head his boat somewhat to the south to compensate for the Gulf Stream current. Figure 2.36a shows the path of the boat relative to the shore and the heading of the boat. Figure 2.36b shows the velocity vector $\mathbf{V}_O$ of the water relative to the shore and the velocity vector $\mathbf{v}'$ of the boat relative to the water. According to Eq. (68), the velocity vector $\mathbf{v}$ of the boat relative to the shore is the sum of $\mathbf{v}'$ and $\mathbf{V}_O$,

$$\mathbf{v} = \mathbf{v}' + \mathbf{V}_O \tag{70}$$

By hypothesis, this vector $\mathbf{v}$ points due east (see Fig. 2.36b). Since the vector triangle is a right triangle, the sine of the angle between $\mathbf{v}'$ and the eastward direction is

$$\sin\theta = \frac{V_O}{v'} \tag{71}$$

$$= \frac{4.8 \text{ km/h}}{18 \text{ km/h}} = 0.27$$

With an electronic calculator, we find that if the sine is 0.27, the angle is $\theta = 15^\circ$. Thus, the boat must head 15° south of east.

The speed of the boat relative to the shore is

$$v = v' \cos\theta = (18 \text{ km/h}) \times \cos 15^\circ = 17 \text{ km/h}$$

SUMMARY

- **Average velocity:** $\bar{v}_x = \dfrac{\Delta x}{\Delta t} \quad \bar{v}_y = \dfrac{\Delta y}{\Delta t}$
- **Instantaneous velocity:** $v_x = \dfrac{\Delta x}{\Delta t} \quad v_y = \dfrac{\Delta y}{\Delta t}$ with Δt extremely small
- **Average acceleration:** $\bar{a}_x = \dfrac{\Delta v_x}{\Delta t} \quad \bar{a}_y = \dfrac{\Delta v_y}{\Delta t}$
- **Instantaneous acceleration:** $a_x = \dfrac{\Delta v_x}{\Delta t} \quad a_y = \dfrac{\Delta v_y}{\Delta t}$ with Δt extremely small

- **Motion of a projectile:** $v_x = v_{0x} = v_0 \cos\theta$
$$v_y = v_{0y} - gt = v_0 \sin\theta - gt$$
$$x = v_{0x}t$$
$$y = v_{0y}t - \tfrac{1}{2}gt^2$$
- **Vector:** Quantity with magnitude and direction; it behaves like a displacement.
- **Addition of vectors:** Use the parallelogram method or the tail-to-head method; alternatively, add components.
- **Components of a vector:** $A_x = A \cos\theta$
$$A_y = A \sin\theta$$
$$A = \sqrt{A_x^2 + A_y^2}$$
- **Centripetal acceleration of uniform circular motion:** $a = \dfrac{v^2}{r}$
- **Addition rule for velocities:** $\mathbf{v} = \mathbf{v}' + \mathbf{V}_\mathrm{O}$

QUESTIONS

1.–Can an automobile have eastward instantaneous velocity and northward instantaneous acceleration? Give an example.

2.–A projectile is launched over level ground. Its initial velocity has a horizontal component v_{0x} and a vertical component v_{0y}. What is the average velocity of the projectile between the instants of launch and of impact?

3.–What is the acceleration of a projectile when it reaches the top if its trajectory?

4.–Are the velocity and the acceleration of a projectile ever perpendicular? Parallel?

5.–If you throw a crumpled piece of paper, its trajectory is not a parabola. How does it differ from a parabola and why?

6.–If a projectile is subject to air resistance, then the elevation angle for maximum range is not 45°. Do you expect the angle to be larger or smaller than 45°?

7.–Figure 2.37 shows the trajectory of a high-speed projectile subject to strong air friction. The trajectory is not symmetric about its highest point. Why not? Would you expect the ascending or the descending portion of the trajectory to take more time?

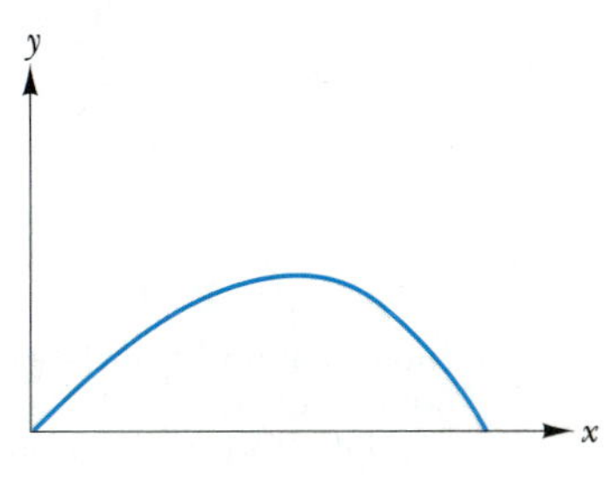

FIGURE 2.37
Trajectory of a projectile with air friction.

8.–Pitchers are fond of throwing curveballs. How does the trajectory of such a ball differ from the simple parabolic trajectory we studied in this chapter?

9.–A large oil tanker proceeds from Kharg Island (Persian Gulf) to Rotterdam via the Cape of Good Hope. A small oil tanker proceeds from Kharg Island to Rotterdam via the Suez Canal. Are the displacement vectors of the two tankers equal? Are the distances covered equal?

10.–An airplane flies from Boston to Houston and back to Boston. Is the displacement zero in the reference frame of the Earth? In the reference frame of the Sun?

11.–Does a vector of zero magnitude have a direction? Does it matter?

12.–Can the magnitude of a vector be negative? Zero?

13.–Two vectors have nonzero magnitude. Under what conditions will their sum be zero? Their difference?

14.–Is it possible for the sum of two vectors to have the same magnitude as the difference of the two vectors?

15.–Three vectors have the same magnitude. Under what conditions will their sum be zero?

16.–The magnitude of a vector is never smaller than the magnitude of any one component of the vector. Explain.

17.–A particle travels once around a circle with uniform circular motion. What is the average velocity and the average acceleration?

18.–A pendulum is swinging back and forth. Is this uniform circular motion? Draw a diagram showing the direction of the velocity and the acceleration at the top of the swing. Draw a similar diagram at the bottom of the swing.

19.–Why do raindrops fall down at a pronounced angle with the vertical when seen from the window of a speeding train? Is this angle necessarily the same as that of the path of a water drop sliding down along the outside surface of the window?

Sailboat on a beat.

20.–When a sailboat is sailing to windward ("beating"), the wind feels much stronger than when the sailboat is sailing downwind ("running"). Why?

21.–Rain is falling vertically. If you run through the rain, at what angle should you hold your umbrella? If you don't have an umbrella, should you bend forward while running?

22.–In the reference frame of the ground, the path of a sailboat beating to windward makes an angle of 45° with the direction of the wind. In the reference frame of the sailboat, the angle is somewhat smaller. Explain.

23.–Suggest a method for measuring the speed of falling raindrops. (Hint: When seen from a moving automobile, the direction of fall of a raindrop makes an angle with the vertical.)

24.–According to a theory proposed by Galileo, the tides on the oceans are caused by the Earth's rotational motion about its axis combined with its translational motion around the Sun. At midnight these motions are in the same direction; at noon they are in opposite directions (Fig. 2.38). Thus, any

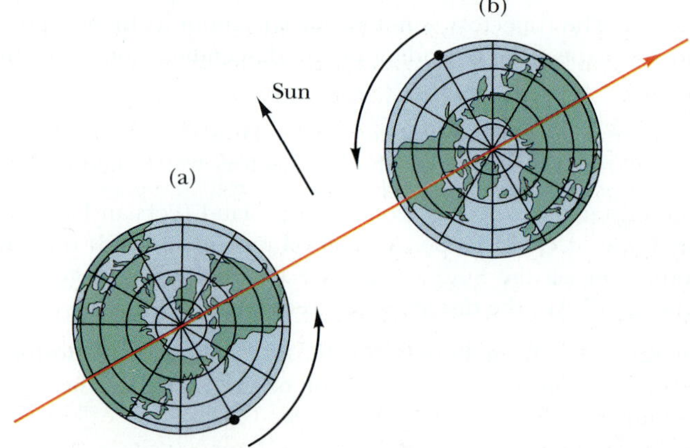

FIGURE 2.38

Rotational and translational motions of the Earth: (a) at midnight and (b) at noon.

point on the Earth alternately speeds up and slows down. Galileo was of the opinion that the speeding up and slowing down of the ocean basins would make the water slosh back and forth, thus giving rise to tides. What is wrong with this theory? (Hint: What is the acceleration of a point on the surface of the Earth? Does this acceleration depend on the translational motion?)

PROBLEMS

Section 2.1: Components of Velocity and Acceleration

1.——At the entrance of Ambrose Channel at New York Harbor, the maximum tidal current has velocity of 4.2 km/h in a direction 20° south of east. What is the component of this velocity in the east direction? In the north direction?

2.——A blimp motoring at a constant altitude has a velocity component of 15 km/h in the north direction and a velocity component of 15 km/h in the east direction. What is the speed of the blimp? What is the direction of motion of the blimp?

3.——With its engine cut off, a small airplane glides downward at an angle of 15° below the horizontal at a speed of 240 km/h.

(a) What are the horizontal and the vertical components of its velocity?

(b) If the airplane is initially at a height of 2000 m above the ground, how long does it take to crash into the ground?

4.*——Suppose an automobile enters a 180° curve at a constant speed of 25 m/s and emerges from this curve 12 s later. What are the components of the average acceleration for this time interval?

5.*——An automobile with a drunken driver at the wheel travels round and round a traffic circle at 30 km/h. The automobile takes 80 s to go once around the circle. At $t = 0$, the automobile is at the east of the traffic circle; at $t = 20$ s it is at the north; at $t = 40$ s it is at the west, etc. What are the components of the velocity of the automobile at $t = 0$, $t = 10$ s, $t = 20$ s, $t = 30$ s, and $t = 40$ s? The x axis points eastward and the y axis points northward.

Section 2.2: Motion with Constant Acceleration
Section 2.3: The Motion of Projectiles

6.——An archer shoots and arrow over level ground. The arrow leaves the bow at a height of 1.5 m with an initial velocity of 60 m/s in a horizontal direction.

(a) How long does this arrow take to fall to the ground?

(b) At what horizontal distance does this arrow strike the ground?

7.——Consider the bomb dropped from the bomber described in Example 4.

(a) What are the final horizontal and the vertical components of the velocity of the bomb when it strikes the surface of the sea?

(b) What is the final speed of the bomb? Compare this with the initial speed of the bomb.

8.——The fastest recorded speed of a baseball thrown by a pitcher is 162.4 km/h, achieved by Nolan Ryan in 1974 at Anaheim Stadium. If the baseball leaves the pitcher's hand with a horizontal velocity of this magnitude, how far will the ball have fallen vertically by the time it has traveled 20 m horizontally?

9.——At Acapulco, professional divers jump from a 36-m-high cliff into the sea (compare Example 1.9). At the base of the cliff, a rocky ledge sticks out

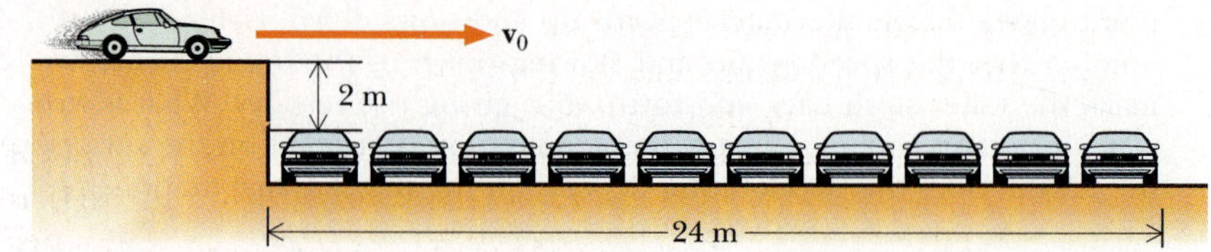

FIGURE 2.39
A stunt.

for a horizontal distance of 6.4 m. With what minimum horizontal velocity must the divers jump off if they are to clear this ledge?

10.—A stunt driver wants to make his car jump over 10 cars parked side by side below a horizontal ramp (Fig. 2.39). With what minimum speed must he drive off the ramp? The vertical height of the ramp is 2 m and the horizontal distance he must clear is 24 m.

11.—A large stone-throwing engine designed by Archimedes could throw a 77-kg stone over a range of 180 m. What must have been the initial speed of the stone if thrown at an initial angle of 45° with the horizontal?

12.—In a circus act at the Ringling Bros. and Barnum & Bailey Circus, a "human cannonball" was fired from a large cannon with a muzzle velocity of 86 km/h. Assume that the firing angle was 45° from the horizontal. How many seconds did the human cannonball take to reach maximum height? How high did he rise? How far from the cannon did he land?

13.—According to a reliable report, in 1795 a member of the Turkish embassy in England shot an arrow to a distance of 441 m. According to a less reliable report, a few years later the Turkish Sultan Selim shot an arrow to 889 m. In each of these cases calculate what must have been the minimum initial speed of the arrow.

14.*—A golfer claims that a golf ball launched with an elevation angle of 12° can reach a horizontal range of 250 m. Ignoring air friction, what would the initial speed of such a golf ball have to be? What maximum height would it reach?

15.*—A gunner wants to fire a gun at a target at a horizontal distance of 12,500 m from his position.

(a) If his gun fires with a muzzle velocity of 700 m/s and if $g = 9.81\ \text{m/s}^2$, what is the correct elevation angle? Pretend that there is no air resistance.

(b) If the gunner mistakenly assumes $g = 9.80\ \text{m/s}^2$, by how many meters will he miss the target?

16.*—According to the *Guinness Book of World Records,* during a catastrophic explosion at Halifax on December 6, 1917, William Becker was thrown through the air for some 1500 m and was found, still alive, in a tree. Assume that Becker left the ground and returned to the ground (ignore the height of the tree) at an angle of 45°. With what speed did he leave the ground? How high did he rise? How long did he stay in flight?

17.*—The motion of a ballistic missile can be regarded as the motion of a projectile, because along the greatest part of its trajectory the missile is in free fall. Suppose that a missile is to strike a target 1000 km away. What minimum speed must the missile have at the beginning of its trajectory? What maximum height does it reach when launched with this minimum speed? How long does it take to reach its target? For these calculations assume that $g = 9.8\ \text{m/s}^2$ everywhere along the trajectory.

18.*—Volcanoes on the Earth eject rocks at speeds of up to 700 m/s. Assume that the rocks are ejected in all directions; ignore the height of the volcano and ignore air friction.

(a) What is the maximum height reached by rocks?
(b) What is the maximum horizontal distance reached by rocks?
(c) Is it reasonable to ignore air friction in these calculations?

19.*—The nozzle of a fire hose ejects 570 liters of water per minute at a speed of 26 m/s. How far away will the stream of water land if the nozzle is aimed at an angle of 35° with the horizontal? How many liters of water are in the air at any given instant?

Firehose ejecting water.

20.*—According to an ancient Greek source, a stone-throwing machine on one occasion achieved a range of 730 m. If this is true, what must have been the minimum initial speed of the stone as it was ejected from the engine? When ejected with this speed, how long would the stone have taken to reach its target?

21.*—The world record for the javelin throw by a woman, established in 1976 by Ruth Fuchs in Berlin, was 69.11 m (226 ft 9 in.). If Fuchs had thrown her javelin with the same initial velocity in Buenos Aires rather than in Berlin, how much farther would it have gone? The acceleration of gravity is 9.8128 m/s^2 in Berlin and 9.7967 m/s^2 in Buenos Aires. Pretend that air resistance plays no role in this problem.

22.*— (a) A golfer wants to drive a ball to a distance of 240 m. If he launches the ball with an elevation angle of 14°, what is the appropriate initial speed? Ignore air resistance.
(b) If the speed is too great by 1 m/s, how much farther will the ball travel when launched at the same angle?
(c) If the elevation angle is 0.5° larger than 14°, how much farther will the ball travel if launched with the speed calculated in Part (a)?

23.*—Show that for a projectile launched with an elevation angle of 45°, the maximum height reached is one-quarter of the range.

24.**—During a memorable jump in Richmond, Virginia, in 1903, the horse Heatherbloom with its rider jumped over an obstacle 2.64 m high while covering a horizontal distance of 11 m. At what angle and with what speed did the horse leave the ground? Make the (somewhat doubtful) assumption that the motion of the horse is particle motion. (Hint: Begin by calculating the time of flight for the jump.)

25.**—A hay-baling machine throws each finished bundle of hay 2.5 m up in the air so it can land on a trailer waiting 5 m behind the machine. What must be the speed with which the bundles are launched? What must be the angle of launch? Assume that the launch point and the landing point are at the same height.

Section 2.4: Vectors
Section 2.5: Vector Addition and Subtraction

26.—A ship moves from the Golden Gate Bridge in San Francisco Bay to Alcatraz Island and from there to Point Blunt. The first displacement vector is 10.2 km due east, and the second is 5.9 km due north. What are the magnitude and the direction of the resultant displacement vector?

27.—To reach Moose Jaw, Canada, you drive your automobile due north 90 km, and then due west 70 km. What are the magnitude and the direction of your displacement vector? What is the distance you traveled?

28.*—An aircraft flies 250 km in a direction 30° east of south, and then 250 km in a direction 30° west of south. What are the magnitude and the direction of the resultant displacement vector?

29.*—The displacement vector **A** has a length of 350 m in the direction 70° west of north; the displacement vector **B** has a length of 120 m in the di-

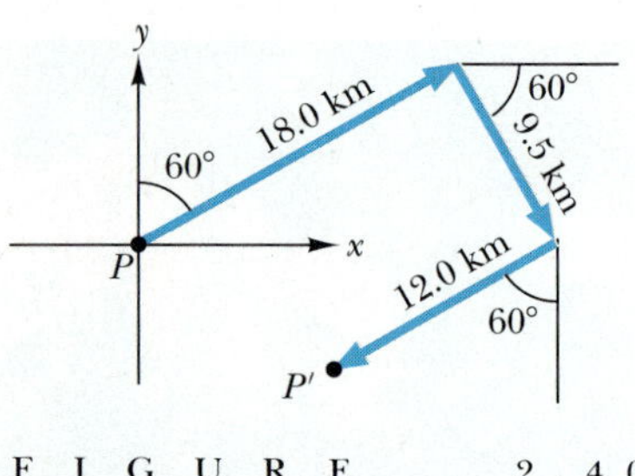

FIGURE 2.40
Successive displacement vectors of an aircraft.

rection due north. Find the magnitude and direction of the resultant of these vectors.

30.*—Figure 2.40 shows the successive displacements of an aircraft flying a search pattern. The initial position of the aircraft is P and the final position is P'. What is the net displacement (magnitude and direction) between P and P'? Find the answer graphically by carefully drawing a page-size diagram with protractor and ruler and measuring the resultant.

31.*—A sailboat tacking against the wind moves as follows: 3.2 km at 45° east of north, 4.5 km at 45° west of north, 2.6 km at 45° east of north. What is the net displacement for the entire motion?

32.*—Three displacement vectors **A, B,** and **C** are, respectively, 4 cm at 30° west of north, 8 cm at 30° east of north, and 3 cm due north. Carefully draw these vectors on a sheet of paper. Find **A** + **B** + **C** graphically. Find **A** + **B** − **C** graphically.

33.**—The Earth moves around the Sun in a circle of radius 1.50×10^{11} m at (approximately) constant speed.

(a) Taking today's position of the Earth as origin, draw a diagram showing the position vector 3 months, 6 months, 9 months, and 12 months later.

(b) Draw the displacement vector between the 0-month and the 3-month position; the 3-month and the 6-month position, etc. Calculate the magnitude of the displacement vector for one of these 3-month intervals.

34.**—Both Singapore and Quito are (nearly) on the Earth's equator; the longitude of Singapore is 104° East and that of Quito is 78° West. What is the magnitude of the displacement vector between these cities? What is the distance between them measured along the equator?

35.**—By a method known as "doubling the angle on the bow," the navigator of a ship can determine his position relative to a fixed point, such as a lighthouse. Figure 2.41 shows the (straight) track of a ship passing by a lighthouse. At the point P, the navigator measures the angle α between the line of sight to the lighthouse and the direction of motion of the ship. He then measures how far the ship advances through the water until the angle between the line of sight and the direction of motion is twice as large as it was initially. Prove that the magnitude of the displacement vector PP' equals the magnitude of the position vector AP' of the ship relative to the lighthouse.

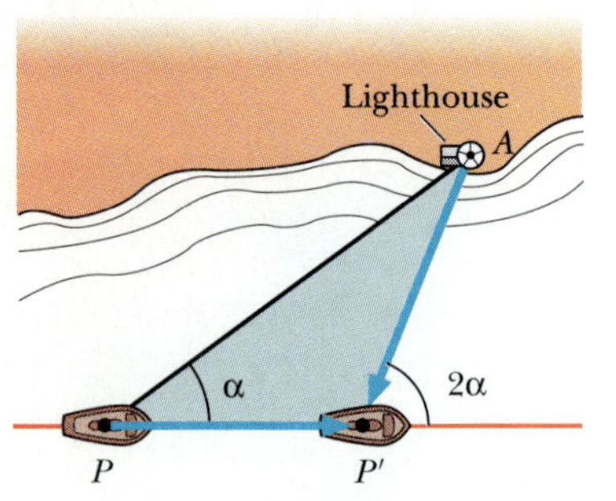

FIGURE 2.41
"Doubling the angle on the bow."

Section 2.6: The Position, Velocity, and Acceleration Vectors

36.—A vector of length 5.0 m is in the x–y plane at an angle of 30° with the x axis. What is the x component of this vector? The y component?

37.—A displacement vector has a magnitude of 12.0 km in the direction 40° west of north. What is the north component of this vector? The west component?

38.—A vector in the x–y plane has a magnitude of 8.0 units. The angle between the vector and the x axis is 52°. What are the x and y components of this vector?

39.—A vector has a magnitude of 2.0 and makes an angle of 30° with the vertical direction. What are the vertical and horizontal components of this vector?

40.—A vector of magnitude 4.0 points vertically downward. What is the component of this vector along a line sloping upward at 25°? (Hint: Place your x axis along the sloping line.)

41.—A vector **A** has components $A_x = 3$ and $A_y = 2$. What is the magnitude of this vector?

42.—A vector has components $A_x = 5.0$ and $A_y = -3.0$. What is the magnitude of this vector? What is the angle between this vector and the x axis? The y axis?

43.*—Air traffic controllers usually describe the position of an aircraft relative to the airport by altitude, horizontal distance, and bearing. Suppose an aircraft is at altitude 500 m, distance 15 km, and bearing 35° east of north. What are the x, y, and z components (in meters) of the position vector? The x axis is east, the y axis is north, and the z axis is vertical.

44.*—The displacement vectors **A** and **B** are in the x–y plane. Their components are $A_x = 3$ cm, $A_y = 2$ cm, $B_x = -1$ cm, $B_y = 3$ cm.

(a) Draw a diagram showing these vectors.

(b) Calculate the resultant of **A** and **B**. Draw the resultant in your diagram.

45.*—Suppose that the vectors **A** and **B** have components $A_x = -5$, $A_y = -3$, $B_x = 2$, and $B_y = 1$. Calculate the following:

(a) $\mathbf{A} + \mathbf{B}$

(b) $\mathbf{A} - \mathbf{B}$

(c) $2\mathbf{A} - 3\mathbf{B}$

46.**—A vector **A** has components $A_x = 2$ and $A_y = -1$. Find a vector (give its components) that has the same direction as **A** but a magnitude of 1 unit.

47.—In one-half year, the Earth moves halfway around its orbit, a circle of radius 1.50×10^{11} m centered on the Sun. What is the average speed and what is the magnitude of the average velocity for this time interval?

48.*—A sailboat tacking against the wind moves as follows: 5.8 km at 45° east of north, and then 4.5 km at 50° west of north. The entire motion takes 1 h 15 min.

(a) What is the total displacement for this motion?

(b) What is the average velocity for this motion?

(c) What is the speed if it is assumed to be constant?

49.**—The fastest bird is the spine-tailed swift, which reaches speeds of 170 km/h. Suppose that you wish to shoot such a bird with a 22-caliber rifle that fires a bullet with a speed of 270 m/s. If you fire at the instant when the bird is 30 m directly overhead, how many meters ahead of the bird must you aim the rifle? Ignore gravity in this problem.

Section 2.7: Uniform Circular Motion

50.—A phonograph record rotates at 33.3 revolutions per minute. The radius of the record is 15 cm. What is the speed of a point at its rim?

51.—The Earth rotates once around its axis in (approximately) 24 hours. What is the rotational speed of a point on the equator of the Earth? What is the centripetal acceleration of this point?

52.—An ultracentrifuge spins a small test tube in a circle of radius 10 cm at 1000 revolutions per second. What is the centripetal acceleration of the test tube?

53.—The blade of a circular saw has a diameter of 20 cm. If this blade rotates at 7000 revolutions per minute (its maximum safe speed), what is the speed and what is the centripetal acceleration of a point on the rim?

54.—At the Fermilab accelerator (one of the world's largest atom smashers), protons are forced to travel in an evacuated tube in a circular orbit of diameter 2.0 km. The protons have a speed nearly equal to the speed of light (99.99995% of the speed of light). What is the centripetal acceleration of these protons?

The main accelerator ring at Fermilab.

55.—An automobile travels at a steady 90 km/h along a road leading over a small hill. The top of the hill is rounded so that, in the vertical plane, the road approximately follows an arc of circle of radius 70 m. What is the centripetal acceleration of the automobile at the top of the hill?

56.——The Earth moves around the Sun in a circular path of radius 1.50×10^{11} m at uniform speed. What is the magnitude of the centripetal acceleration of the Earth toward the Sun?

57.*—An automobile has wheels of diameter 0.65 m. What is the centripetal acceleration of a point on the rim of this wheel when the automobile is traveling at a constant velocity of 95 km/h?

58.*—The table printed on the endpapers lists the radii of the orbits of the planets ("mean distance from the Sun") and the times required for moving around the orbit ("period of revolution"). Assume that the planets move along circular orbits at constant speed. Calculate the centripetal acceleration for each of the first three planets (Mercury, Venus, and Earth). Verify that the centripetal accelerations are in proportion to the inverses of the squares of the orbital radii.

Section 2.8: The Relativity of Motion; Addition of Velocities

59.——In an airport, a moving sidewalk has a speed of 1.5 m/s relative to the ground. What is the speed, relative to the ground, of a passenger running forward on this sidewalk at 4 m/s? What is the speed, relative to the ground, of a passenger running backward on this sidewalk at 4 m/s?

60.——You are driving an automobile at a steady speed of 90 km/h along a straight highway. Ahead of you is a 10-m-long truck traveling at a steady speed of 60 km/h. You decide to pass the truck, and you switch into the left-hand lane when 40 m behind the truck.

(a) What is your speed in the reference frame of the truck?

(b) How long do you take to pass the truck, starting 40 m behind the truck and ending 40 m ahead of the truck? (Hint: Calculate this time in the reference frame of the truck.)

61.——A white automobile is traveling at a constant speed of 90 km/h on a highway. The driver notices a red automobile 1.0 km behind, traveling in the same direction. Two minutes later, the red automobile passes the white automobile.

(a) What is the average speed of the red automobile relative to the white?

(b) What is the speed of the red automobile relative to the ground?

62.——On a rainy day, a steady wind is blowing at 30 km/h. In the reference frame of the *air,* the raindrops are falling vertically with a speed of 10 m/s. What are the magnitude and direction of the velocity of the raindrops in the reference frame of the ground?

63.——Raindrops are falling vertically with a speed of 7.0 m/s. What is the angle with the vertical that the trajectories of these raindrops make in the reference frame of an airplane flying horizontally at 300 km/h?

64.——On a rainy day, raindrops are falling with a vertical velocity of 10 m/s. If an automobile drives through the rain at 25 m/s, what is the velocity (magnitude and direction) of the raindrops relative to the automobile?

65.——A wind of 30 m/s is blowing from the west. What will be the speed, relative to the ground, of a sound signal traveling due west? Due east? Due north? The speed of sound, relative to air, is 330 m/s.

66.——On a windy day, a hot-air balloon is ascending at a rate of 0.5 m/s relative to the air. Simultaneously, the air is moving with a horizontal velocity of 12 m/s. What is the velocity (magnitude and direction) of the balloon relative to the ground?

67.*—A blimp is motoring at constant altitude. The airspeed indicator on the blimp shows that its speed relative to the air is 20 km/h and the compass shows that the heading of the blimp is due north. If the air is moving over the ground with a velocity of 15 km/h due east, what is the velocity (magnitude and direction) of the blimp relative to the ground? For an observer on the ground, what is the angle between the longitudinal axis of the blimp and the direction of motion?

68.*—A sailboat is moving in a direction due east at a speed of 8.5 km/h. The wind measured by an instrument aboard the sailboat has an apparent (relative to the sailboat) speed of 20 km/h coming from an apparent direction of 10° east of north. Find the true (relative to the ground) speed and direction of the wind.

69.*—A lump of concrete falls off a crumbling overpass and strikes an automobile traveling on a highway below. The lump of concrete falls 5 m before impact and the automobile has a speed of 90 km/h.

(a) What is the speed of impact of the lump in the reference frame of the automobile?

(b) What is the angle of impact?

70.*—A wind is blowing at 50 km/h from a direction 45° west of north. The pilot of an airplane wishes to fly on a route due north from an airport. The airspeed of the airplane is 250 km/h.

(a) In what direction must the pilot point the nose of the airplane?

(b) What will be the airplane's speed relative to the ground?

PHYSICS IN ACTION

LONG JUMPS

Many of our most popular sports involve projectile motion. In games of baseball, basketball, football, soccer, tennis, golf, etc., players launch balls by hand or by foot, or with bats, racquets, or clubs. The player has to launch the ball with a speed and an elevation angle that will send the ball to some desired target at the end of its trajectory. This demands delicate judgment and fine muscular control. The intense concentration displayed by basketball players, tennis players, golf players, and other ball players during a launch suggests that a considerable, though subconscious, mental effort is needed to select the correct launch speed and elevation angle to strike the target.

In athletic events of long jumping, high jumping, and high diving, the athlete launches his or her own body as a projectile. If we focus on the motion of the center of the athlete's body (more precisely, the center of mass; see Chapter 6), the translational motion of the human projectile can be described as particle motion, and this motion obeys the equations we developed in Section 2.3. However, besides this fairly simple translational motion, the athlete's body also has a deformational motion — the athlete usually changes the configuration of his body while in flight. Such changes of configuration play an essential role in high jumps, where the athlete attempts to "flow" over the bar, keeping all parts of his body as low as possible, except for the part that is at imminent risk of coming in contact with the bar. And they also play an essential role in high dives, where the athlete is required to perform a variety of exacting pirouettes while in flight.

But in long jumps, changes of configuration while in flight are less important. The athlete needs to perform no drastic changes in his configuration, except when pushing off with his legs at the beginning of the jump and when landing at the end of the jump, where it is important to land with the feet as far forward as possible, and then fall forward. (The distance in a long jump is measured from a takeoff board at the end of the "runway" to the nearest point of contact with the ground, and if the athlete falls backward upon landing, the officially measured distance will suffer.) Thus, except for minor corrections at the beginning and the end of the jump, we can treat the long jump as pure translational motion. We can therefore directly apply the equations of Section 2.3 to the jump, and we will see that this permits us to draw an interesting conclusion about the maximum range that can be attained in a long jump.

We know from Section 2.3 that, for maximum range, the elevation angle for the launch of the projectile should be 45°, which gives the projectile equal horizontal and vertical velocities at launch (this assumes that air resistance is absent; a more realistic calculation taking into account air resistance indicates that the best choice of elevation angle for long jumpers is *slightly* smaller than 45°). However, long jumpers always launch themselves with elevation angles much smaller than 45°, typically 20° or so (see Fig. 1). The reason is that the long jumper finds it much easier to attain a large horizontal velocity than a large vertical velocity. The jumper attains his horizontal velocity by running toward the

F I G U R E 1 Multiple-exposure photograph of long jump.

takeoff board as fast as he can. When he reaches the takeoff board, the jumper propels himself upward. Thus, he attains his vertical velocity only at the last moment, as the result of only a single downward push of his legs. Not surprisingly, his resulting vertical velocity is considerably smaller than his horizontal velocity, and the launch angle is therefore smaller than 45°. The only way a jumper could attain equal vertical and horizontal velocities, and a launch angle of 45°, would be by deliberately slowing his run. But this would be counterproductive — if the vertical velocity is fixed, a slower horizontal run will always give a smaller range than a faster horizontal run. Clearly, the best strategy is to run as fast as possible and to push off vertically as hard as possible.

Since the jumper acquires his horizontal and vertical velocities more or less independently, we can deduce a theoretical maximum range for the long jump by combining independent data for running and for vertical jumping (standing jump). All long jumpers are good runners, and many were champion runners before they became long jumpers. The maximum speed that has been attained by champion runners is 10 m/s. The maximum height that has been attained in standing jumps is 1.76 m. However, this is the height measured from the floor to the bar, whereas we really need the height by which the center of the jumper rises during the jump, since this determines the time of flight. In the "standing" jump, the jumper actually begins the jump in a crouch, with the legs contracted. In this configuration, the center of the jumper is about 0.50 or 0.60 m from the ground; subtracting this from 1.76 m, we find that the height by which the center of the jumper rises is about 1.2 m.

We can now set a limit for the maximum range for the long jump by assuming that a long jumper runs as fast as a champion runner and also succeeds in jumping upward as fast as a champion "standing" jumper. For the calculation of the range, we proceed as in the calculation of the range of the discus in Example 2.5. First we find the time of flight, and from this, the range. The time required for the jumper to rise by 1.2 m is given by the familiar equation $y - y_0 = \frac{1}{2}gt^2$. This yields

$$t = \sqrt{\frac{2 \times (y - y_0)}{g}}$$

$$= \sqrt{\frac{2 \times 1.2 \text{ m}}{9.81 \text{ m/s}^2}} = 0.495 \text{ s} \qquad (1)$$

The time of flight, for the jumper to rise and to descend, is twice as large:

$$t_{\text{flight}} = 0.99 \text{ s} \qquad (2)$$

If the horizontal velocity of the jumper has the maximum value of 10 m/s, then in a time of 0.99 s, he covers a horizontal distance

$$x - x_0 = v_{0x} t_{\text{flight}} = 9.9 \text{ m} \qquad (3)$$

This is the distance traveled by the center of his body. But since the jumper lands with his legs extended forward, the point at which his feet make contact with the ground is a bit farther, typically 0.6 m farther. The theoretical prediction for the maximum attainable distance for the long jump is therefore about 9.9 m + 0.6 m = 10.5 m.

How does this theoretical prediction compare with experimental data? The three best long jumpers have been the Americans Bob Beamon, Carl Lewis, and Mike Powell. At the Olympics in Mexico City, in 1968, Beamon set a world record with a spectacular jump of 8.90 m (see Fig. 2). His jump was regarded as an astounding achievement, since it exceeded the previous record by more than half a meter. Beamon's record remained unbroken for 23 years, becoming the longest-enduring record in track-and-field. Finally, at the 1991 World Championships in Tokyo, this record was broken in a dramatic competition between Lewis and Powell. First

FIGURE 2 Bob Beamon in midair in his spectacular jump of 8.90 m (29 ft, $2\frac{1}{2}$ in.) during the 1968 Olympics in Mexico City.

F I G U R E 3 Carl Lewis in his best jump of 8.91 m (29 ft, $2\frac{3}{4}$in.) during the 1991 World Championships in Tokyo.

F I G U R E 4 Mike Powell setting a new record of 8.95 m (29 ft, $4\frac{1}{2}$ in.) during the 1991 World Championships.

Lewis jumped 8.91 m (see Fig. 3); but this jump was disallowed because the wind was slightly stronger than permitted by the rules. Then Powell jumped 8.95 m, and he thereby established a new world record (see Fig. 4).

The close agreement between the maximum distance attained by Beamon, Lewis, and Powell suggests that these jumpers are approaching the physical limits for the long jump. If we compare Powell's "experimental result" of 8.95 m with our calculated value of 10.5 m, we see that Powell is about 1.5 m below the predicted theoretical maximum. Thus, our prediction is in satisfactory accord with the data. Presumably, Powell did not reach the theoretical maximum because he did not run quite as fast as a championship runner, and he did not jump quite as high as a championship standing jumper — but he came very close.

Although the rules for long jumps impose limits on the strongest permitted wind speed, so as to prevent an unfair advantage for a jumper, the rules do not restrict other environmental conditions. When Beamon jumped in Mexico City, he may have been significantly helped by special conditions at that site. The acceleration of free fall at Mexico City is about 0.28% smaller than the standard value $g = 9.81\ \text{m/s}^2$. Since the range of a projectile is inversely proportional to g [see Eq. (2.37)], a 0.28% decrease in g implies a 0.28% *increase* in the range, for a fixed launch speed and elevation angle. Furthermore, Mexico City is at high altitude, and the pressure and density of air are about 23% smaller than the pressure and density at sea level. This means that the air resistance is smaller, which permits the jumper to run faster while approaching the takeoff board. Like all champion jumpers, Beamon was an excellent runner. He was capable of running 100 yards in 9.5 s (an average speed of 9.6 m/s); but it is quite likely that in the thin air of Mexico City he achieved an even higher speed, perhaps higher by 2%, which would have produced a 2% increase in the range of his jump. Furthermore, the smaller air resistance also helps during the actual jump. Calculations indicate that, at sea level, air resistance reduces the range of a long jump by about 1.9% relative to an ideal jump without air resistance. But at the altitude of Mexico City, the reduction is only 1.4%; hence Beamon gained a relative advantage of about 0.5% by jumping at Mexico City. The net gain resulting from all these favorable special conditions is about 2.8%, or 0.25 m.

A gain of 0.25 m resulting from special conditions is not much, but it is enough to bring into question the fairness of comparing world records set at different sites. In Mexico City in 1968, not only did Beamon set a record that remained unbroken for many years, but so did competitors in the 100-m race and the 400-m race. Some sport experts have proposed that world records for jumping, for running, and for some other athletic events should be corrected (handicapped) according to the special conditions at the site. The effect of a change in g on projectile motion is easily calculated and easily corrected. But the effect of a change of air pressure and air resistance on projectile motion and on running motion is much harder to calculate and to correct. The correction would be different for runners of different build and for runners who adopt somewhat different postures while running. The only way to ensure fair comparisons among world records would be to hold the Olympics at the same site each time, perhaps at Olympia in Greece.

REFERENCES

Alexander, R. McN., *Animal Mechanics* (Blackwell Scientific Publications, Oxford, 1983).

Alexander, R. McN., *The Human Machine* (Columbia University Press, New York, 1992).

Armenti, A., ed., *The Physics of Sports* (American Institute of Physics, New York, 1992.)

Brancazio, P. J., *Sport Science* (Simon and Schuster, New York, 1984).

Gardner, R., *Science and Sports* (Franklin Watts, New York, 1988).

Vogel, S., *Life's Devices* (Princeton University Press, Princeton, N.J., 1988).

C H A P T E R 3

Tugboats pushing a supertanker. In the absence of any push, the tanker would remain at rest. The pushes, or the forces, exerted by the tugboats cause it to accelerate.

NEWTON'S LAWS

So far we have dealt only with the mathematical description of motion — the definitions of position, velocity, and acceleration and the relationships among these quantities. We did not inquire what causes a body to accelerate. In this chapter we will see that ***the cause of acceleration is a force exerted on the body by some external agent.*** The fundamental properties of force and the relationship between force and acceleration are given by Newton's three laws of motion. The first of these laws describes the natural state of motion of a body on which no external forces are acting, whereas the other two laws deal with the behavior of bodies under the influence of external forces.

The first law was actually discovered by Galileo Galilei early in the seventeenth century, but it remained for Isaac Newton, in the second half of the seventeenth century, to formulate a coherent theory of forces and to lay down a complete set of equations from which the motion of bodies under the influence of arbitrary forces can be calculated. The study of forces and their effects on the motion of bodies is called ***dynamics,*** and Newton's laws of motion are sometimes called the laws of dynamics.

◂ Dynamics

Sir Isaac Newton, *1642–1727, English mathematician and physicist, widely regarded as the greatest scientist of all time. His brilliant discoveries in mechanics were published in 1687 in his book* Principia Mathematica, *one of the glories of the Age of Reason. In this book, Newton laid down the laws of motion and the law of universal gravitation, and he demonstrated that planets in the sky as well as bodies on the Earth obey the same mathematical equations. For over 2000 years, Newton's laws stood as the unchallenged basis of all our attempts at a scientific explanation of the physical world.*

3.1 NEWTON'S FIRST LAW

Everyday experience seems to suggest that a force — a push or a pull — is needed to keep a body moving at constant velocity. For example, if the wind pushing a sailboat suddenly ceases, the boat will coast along for some distance, but it will gradually slow down, stop, and remain stopped until a new gust of wind comes along. However, what actually slows down the sailboat is not the *absence* of a propulsive force but, rather, the *presence* of a friction force. The water rubs against the hull of the sailboat, and this friction force resists the motion of the boat and gradually slows it down. ***Under ideal frictionless conditions, a body left to itself would continue to move forever with uniform velocity.*** Experiments with pucks or gliders riding on a cushion of air on a low-friction air table or air track give some hint of the persistence of motion; but in order to eliminate friction entirely, it is best to use bodies moving in a vacuum, without even air against which to rub. Observations on particles moving in evacuated tubes show that a body left to itself, on which no external forces are acting, persists indefinitely in its state of uniform motion.

In this context, an external force is any force exerted on the body by some *other* body. By contrast, internal forces are those exerted by one part of the body on another part of the *same* body. For instance, the forces that the screws or bolts in the sailboat exert on its planks are internal forces; such internal forces do not affect the motion of the boat.

Newton's First Law summarizes experiments and observations on the motion of bodies on which no external forces are acting:

> ***A body at rest remains at rest, and a body in motion continues to move at constant velocity unless acted upon by an external force.***

◂ Newton's First Law

The tendency of a body to continue in its initial state of motion (a state of rest or a state of uniform velocity) is called its ***inertia.*** Accordingly, the First Law is often called the law of inertia.

◂ Inertia

A crucial restriction on Newton's First Law concerns the choice of reference frame: the law is not valid in all reference frames but only in certain special frames. Whenever this law is valid in some given reference frame, then it cannot be valid in a second reference frame that has an accelerated motion relative to the first. For example, in the reference frame of the ground, a ball at rest on the floor of a train station remains at rest, but in the reference frame of an accelerating train leaving the station, a ball initially at rest on the (frictionless) floor of one of the cars will begin to slip backward, and this "spontaneous" acceleration toward the rear of the car is in contradiction to Newton's First Law (see Fig. 3.1). ***Those special reference frames in which the law is valid are called inertial reference***

◂ Inertial reference frames

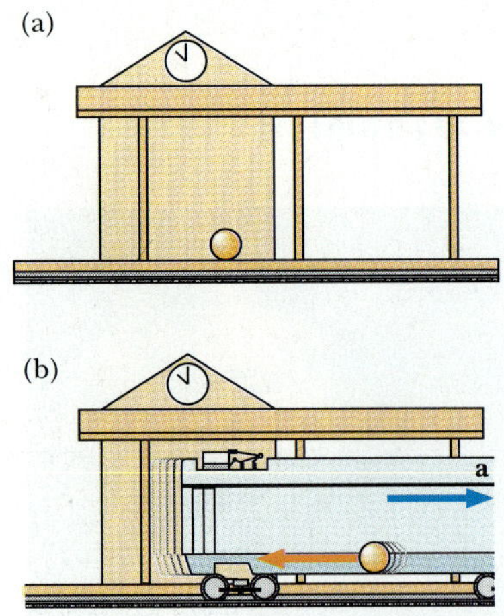

FIGURE 3.1
(a) In the reference frame of the ground, a ball at rest on the floor of a train station remains at rest when no force acts on it. (b) In the reference frame of an accelerating train, a ball initially at rest on the (frictionless) floor of the car begins to slip backward; that is, it accelerates backward, even though no force acts on it.

frames. Thus, the reference frame of the ground is an inertial reference frame, but that of the accelerating train is not.

If some given reference frame is inertial, any other reference frame in uniform translational motion relative to the first will also be inertial, but any other reference frame in accelerated motion relative to the first will not be inertial. Thus, any two inertial reference frames can differ only by some constant relative velocity; they cannot differ by an acceleration. This implies that, ***as measured with respect to inertial reference frames, acceleration is absolute:*** when a particle has some acceleration in one inertial reference frame, then it will have exactly the same acceleration in any other. By contrast, the velocity of the particle is relative; the velocities are related by vector addition [see Eq. (2.68)].

3.2 NEWTON'S SECOND LAW

Newton's Second Law of motion establishes the relationship between the force acting on a body and the acceleration caused by this force. This law summarizes the results of experiments and observations on bodies moving under the action of external forces. ***Qualitatively, a force is any push or pull exerted on a body.*** It is intuitively obvious that such a push or pull has a direction as well as a magnitude — in fact, force is a vector quantity, and it can be represented graphically by an arrow (see Fig. 3.2). For the sake of simplicity, we assume that only one force is acting on the body, but we will relax this assumption in the next section. ***Newton's Second Law*** states:

Newton's Second Law ▶

> ***An external force acting on a body gives it an acceleration that is in the direction of the force and has a magnitude directly proportional to the magnitude of the force and inversely proportional to the mass of the body,***

$$\mathbf{a} = \mathbf{F}/m \tag{1}$$

or

$$m\mathbf{a} = \mathbf{F} \tag{2}$$

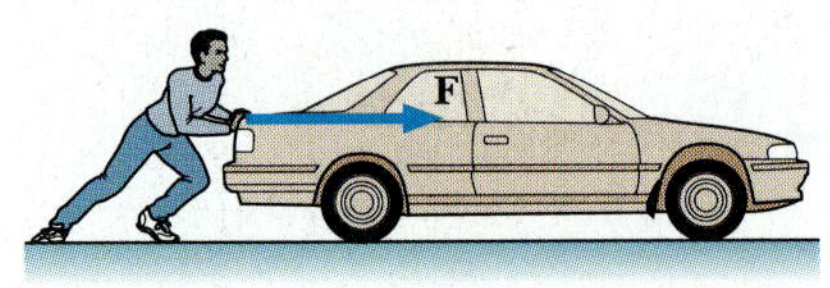

FIGURE 3.2
The force, or push, you exert on the car has a magnitude and a direction.

According to Eq. (1), the acceleration vector equals the force vector divided by the mass; thus, this equation specifies both the magnitude and the direction of the acceleration, as demanded by the verbal statement of the law.

The Second Law is subject to the same restrictions as the First Law: ***it is valid only in inertial reference frames.***

Before we deal with applications of the Second Law, we must give the precise definitions of mass and of force. These definitions are contained in the Second Law itself; that is, the Second Law plays a dual role as law of nature and as definition of mass and force.

The definition of mass hinges on comparing the unknown mass with a standard mass, which is assumed known. To compare the two masses, we exert forces of identical magnitudes on each, and we measure the accelerations that these forces produce on each. For instance, we might attach identical rubber bands or springs to the unknown mass and to the standard mass, and stretch these rubber bands or springs by identical amounts, thereby producing forces of identical magnitudes. According to the Second Law, ***if two bodies of different masses are subjected to forces of identical magnitudes, the accelerations will be in the inverse ratio of the masses.*** If we designate the acceleration of the unknown mass by a and that of the standard mass by a_s and their masses by m and m_s, we can express this inverse ratio of masses and accelerations as

$$\frac{m}{m_s} = \frac{a_s}{a} \tag{3}$$

◀ **Definition of mass**

This relation serves to define the unknown mass m in terms of the standard mass m_s. The relation says that the unknown mass is large if its acceleration is small. This is of course quite reasonable. A large mass is hard to accelerate — it has a large inertia. If we pull a baseball bat with our rubber band, it will accelerate readily; but if we pull a supertanker, it will hardly accelerate at all. The precise definition given by Eq. (3) expresses the intuitive notion that mass is a measure of the resistance that the body offers to changes in its velocity.

The standard of mass used in the metric system of units is the standard kilogram, a cylinder of platinum–iridium alloy kept at the Bureau International des Poids et Mesures (see Fig. 3.3). By definition, the mass of this cylinder is exactly one ***kilogram*** (1 kg). In the British system of units, the unit of mass is the ***pound*** (lb), which equals 0.4536 kg. Table 3.1 gives

◀ **Kilogram (kg)**

TABLE 3.1	SOME MASSES
Observable Universe	$\simeq 10^{53}$ kg
Galaxy	4×10^{41} kg
Sun	2.0×10^{30} kg
Earth	6.0×10^{24} kg
Ship *Queen Elizabeth*	7.6×10^{7} kg
Jet airliner (Boeing 747, empty)	1.6×10^{5} kg
Automobile	1.5×10^{3} kg
Man (average male)	73 kg
Apple	0.2 kg
5¢ coin	5.2×10^{-3} kg
Raindrop	2×10^{-6} kg
Red blood cell	9×10^{-14} kg
Smallest virus (potato spindle)	4×10^{-21} kg
Atom (iron)	9.5×10^{-26} kg
Proton	1.7×10^{-27} kg
Electron	9.1×10^{-31} kg

FIGURE 3.3
International standard kilogram.

TABLE 3.2	MULTIPLES AND SUBMULTIPLES OF THE KILOGRAM
metric ton (tonne)	$1\ \text{t} = 10^3$ kg
kilogram	1 kg
gram	$1\ \text{g} = 10^{-3}$ kg
milligram	$1\ \text{mg} = 10^{-6}$ kg
atomic mass unit	$1\ \text{u} = 1.66 \times 10^{-27}$ kg

some examples of masses expressed in kilograms; Table 3.2 lists multiples and submultiples of the kilogram.

The definition of force also relies on the Second Law. To measure a given force — say, the force generated by a spring that has been stretched a certain amount — we apply this force to the standard kilogram. If the resulting acceleration of the standard kilogram is a_s, then the force has a magnitude

$$F = m_s a_s = 1\ \text{kg} \times a_s \tag{4}$$

After the standard of mass has been used to measure the force, any other masses to which this same force is applied will be found to obey the Second Law. In regard to these other masses, the Second Law is an assertion about the physical world that can be verified by experiments — it is a law of nature.

Newton (N) ▸

In the metric system of units, the unit of force is the newton (N); this is the force that will give a mass of 1 kg an acceleration of 1 m/s^2,

$$1\ \text{newton} = 1\ \text{N} = 1\ \text{kg} \cdot \text{m/s}^2 \tag{5}$$

In the British system of units, the unit of force is the ***pound-force*** (lbf), which equals 4.448 N. In everyday usage, the pound-force is often simply called pound, but we must be careful not to confuse the pound-force (a unit of force) with the pound (a unit of mass). The widespread confusion between the two kinds of pounds stems from their close relationship — the pound-force is the weight (see below) of a pound-mass.*

Table 3.3 lists the magnitudes of some typical forces.

FIGURE 3.4
Spirit of America on the Salt Flats of Utah.

EXAMPLE 1. The jet-engined car *Spirit of America* (see Fig. 3.4), which set a world record for speed on the Salt Flats of Utah, had a mass of 4100 kg, and its engine could develop up to 68,000 N of thrust. What acceleration could this car achieve?

---

* Compounding this confusion, some physics textbooks define the pound as a unit of force, and they introduce a separate unit of mass, called the slug. But in engineering usage (see, for example, the references mentioned in the Preface), the pound is a unit of mass, and since the British system of units is an engineering system, physicists ought to accept engineering usage.

SOLUTION: According to Newton's Second Law, the magnitude of the acceleration is

$$a = \frac{F}{m} = \frac{68{,}000\ \text{N}}{4100\ \text{kg}} = 17\ \text{m/s}^2 \qquad (6)$$

EXAMPLE 2: Some small animals — locusts, beetles, and fleas — attain very large accelerations while starting a jump. The rat flea attains an acceleration of about 2×10^3 m/s^2. Calculate what force the hind legs of the flea must exert on the body while pushing it off with this acceleration. The mass of the flea is about 6×10^{-11} kg; neglect the mass of the legs.

SOLUTION: The magnitude of the force is

$$F = ma = (6 \times 10^{-11}\ \text{kg}) \times (2 \times 10^3\ \text{m/s}^2) = 1.2 \times 10^{-7}\ \text{N}$$

This brings us to the question of the practical measurement of mass and force. In laboratories on the Earth, the most common and most precise mass measurements are carried out with beam balances that compare the weights of the masses (see Fig. 3.5). Weight is the downward pull that the gravity of the Earth exerts on a body. The precise meaning of "weight" will be further discussed in Section 3.5; as we will see, measurements of mass via weight give results consistent with those obtained by the primary procedure based on Eq. (3). Likewise, measurements of force are often carried out by matching the unknown force with a known weight.

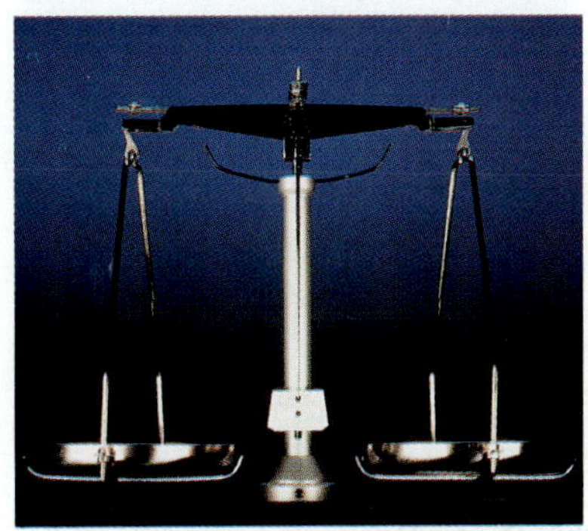

FIGURE 3.5
Beam balance.

TABLE 3.3 SOME FORCES	
Gravitational pull of Sun on Earth	3.5×10^{22} N
Thrust of Saturn 5 rocket engines	3.3×10^{7} N
Pull of large tugboat	1×10^{6} N
Thrust of jet engines (Boeing 747)	7.7×10^{5} N
Pull of large locomotive	5×10^{5} N
Decelerating force on automobile during braking	1×10^{4} N
Force between two protons in a nucleus	$\simeq 10^{4}$ N
Accelerating force on automobile	7×10^{3} N
Gravitational pull of Earth on man	7.3×10^{2} N
Maximum upward force exerted by forearm (isometric)	2.7×10^{2} N
Gravitational pull of Earth on apple	2 N
Gravitational pull of Earth on 5¢ coin	5.1×10^{-2} N
Force between electron and nucleus of atom (hydrogen)	8×10^{-8} N

TABLE 3.4 THE MASSES OF THE ELECTRON, PROTON, AND NEUTRON

Particle	Mass
Electron	9.110×10^{-31} kg
Proton	1.673×10^{-27}
Neutron	1.675×10^{-27}

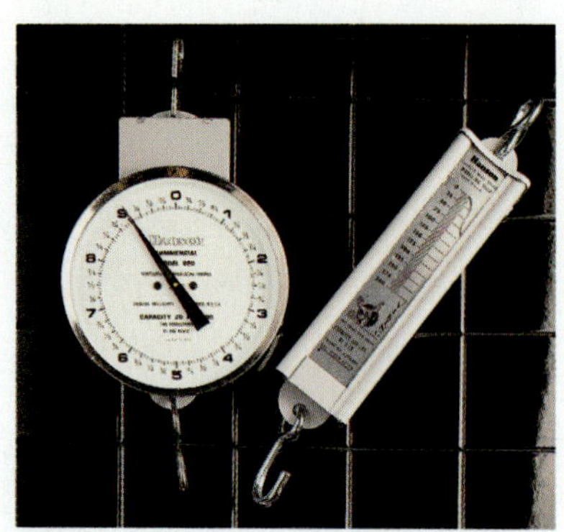

FIGURE 3.6
Spring balances.

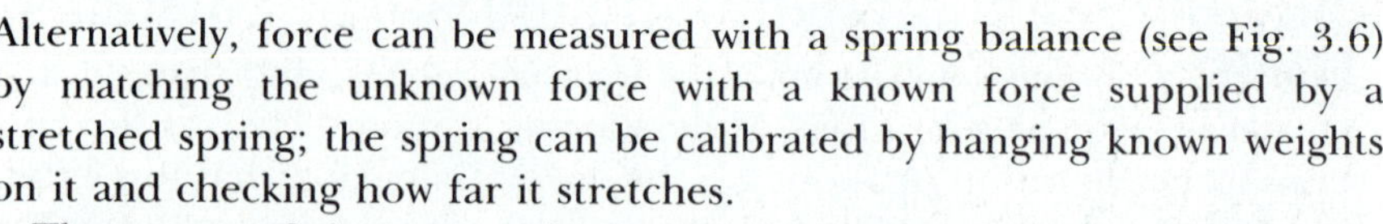

Alternatively, force can be measured with a spring balance (see Fig. 3.6) by matching the unknown force with a known force supplied by a stretched spring; the spring can be calibrated by hanging known weights on it and checking how far it stretches.

The masses of electrons and protons and the masses of ions (atoms with missing electrons or extra electrons) are too small to be measured by their weight. Instead, they are measured with a procedure based on Eq. (2). The procedure involves applying a known force to the particle, measuring the resulting acceleration, and then calculating the mass from Eq. (2). The device used for such mass determinations is called a mass spectrometer. It consists of an evacuated vessel into which is shot a steady succession of particles, so they form a beam, like the beam of electrons in a TV tube. In the absence of forces, this beam would proceed along a straight path. But the spectrometer applies known electric and magnetic forces to the particles, causing a deflection of the beam. The measured value of the deflection indicates the acceleration, and Eq. (2) then permits the calculation of the masses of the particles. Table 3.4 lists the masses of the electron, the proton, and the neutron.

The most severe empirical test of the Second Law is supplied by the study of the motion of celestial bodies. The observed motions of the planets, satellites, and comets have been found to agree with this law with extreme precision. Another severe empirical test of the law is supplied by experiments with electrons, protons, and ions when under the influence of electric and magnetic forces. The motion of such particles is in accord with the Second Law as long as the speeds of the particles are low compared with the speed of light, but the motion displays deviations from the Second Law when the speed is comparable with the speed of light. These deviations from the Second Law at high speeds indicate a failure of Newtonian physics. As we will see in Chapter 27, high-speed motion belongs to the realm of the theory of Special Relativity, where the physics of Einstein supersedes the physics of Newton. However, the motion of ordinary bodies with ordinary speeds — such as the motion of baseballs, automobiles, ships, aircraft, or space shuttles — is in complete agreement with Newton's Second Law, and we can always trust this law when dealing with such bodies.

EXAMPLE 3: During the Skylab mission, three astronauts were kept for about 2 months in weightless conditions. Since an ordinary balance will not work under these conditions, scientists who wanted a daily record of the astronaut's masses had to invent a new mass measurement device. Figure 3.7 shows the device used aboard Skylab. It consisted of a small chair that could be accelerated back and forth by a spring attached to it. Instruments connected to the chair measured the acceleration.* Suppose that with a known standard mass of 66.9 kg placed in the chair, the bent spring produced an acceleration of 0.0262 m/s². With the standard mass removed and with astronaut J. R. Lousma sitting in the chair, the bent spring (with the same amount of bending) produced an acceleration of 0.0204 m/s². Deduce the mass of Lousma. Ignore the mass of the chair.

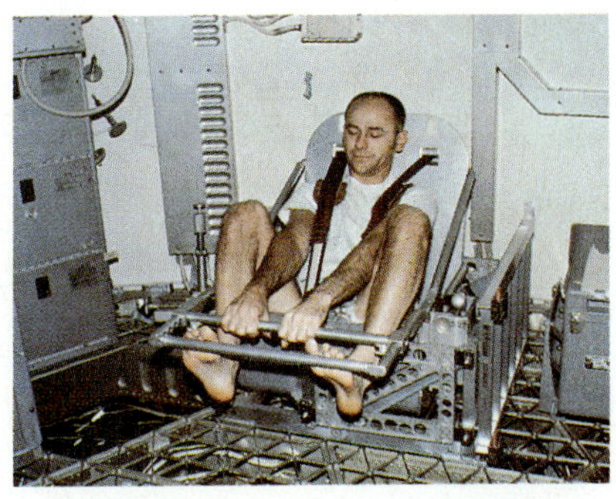

FIGURE 3.7

An astronaut on the body-mass measurement device (BMMD) on Skylab.

SOLUTION: The bent spring provides the same force F when the standard mass is placed in the chair and when the astronaut is placed in the chair. Hence, the accelerations will be in the inverse ratio of the masses, as in Eq. (3),

$$\frac{m}{m_s} = \frac{a_s}{a}$$

We then find

$$m = \frac{a_s}{a} m_s = \frac{0.0262 \text{ m/s}^2}{0.0204 \text{ m/s}^2} \times 66.9 \text{ kg} = 85.9 \text{ kg} \tag{7}$$

for the mass of Lousma.

3.3 THE COMBINATION OF FORCES

More often than not, a body will be subjected to the simultaneous action of several forces. For example, Fig. 3.8 shows a barge under tow by two tugboats. The forces acting on the barge are the pull of the first tow rope,

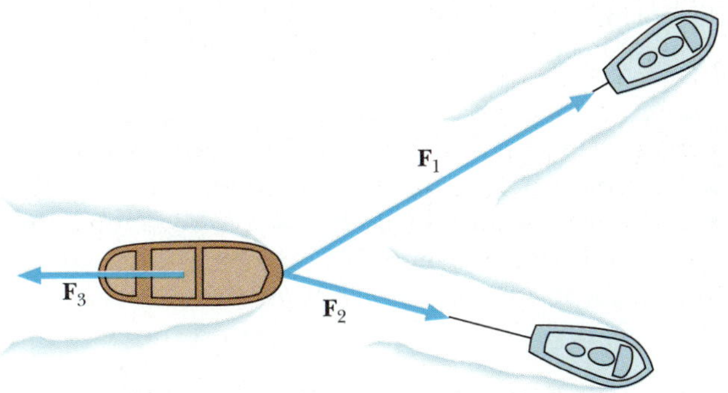

FIGURE 3.8

A barge under tow by two tugboats.

* The instruments actually measured the period of the back-and-forth oscillations of the chair; this amounts to an indirect measurement of the acceleration.

the pull of the second tow rope, and the frictional resistance of the water*; these forces are indicated by the arrows in Fig. 3.8. Newton's Second Law tells us what each of these forces would do if it acted by itself. The question now is: How can we calculate the simultaneous effect of two or more forces? The answer is supplied by the principle of Combination of Forces, or principle of ***Superposition of Forces:***

Superposition of Forces ▶

If several forces $\mathbf{F}_1$, $\mathbf{F}_2$, $\mathbf{F}_3$, . . . act simultaneously on a body, then the acceleration is the same as that produced by a single force $\mathbf{F}_{\text{net}}$ given by the vector sum of the individual forces,

$$\mathbf{F}_{\text{net}} = \mathbf{F}_1 + \mathbf{F}_2 + \mathbf{F}_3 + \ldots \tag{8}$$

The single force $\mathbf{F}_{\text{net}}$ that has the same effect as the combination of the individual forces is called the ***net force,*** or the ***resultant force.*** In terms of the net force, Newton's Second Law becomes

$$m\mathbf{a} = \mathbf{F}_{\text{net}} \tag{9}$$

Equation (8) can be interpreted as follows: Each force ($\mathbf{F}_1$, $\mathbf{F}_2$, $\mathbf{F}_3$, . . .) acting by itself produces its own acceleration ($\mathbf{a}_1 = \mathbf{F}_1/m$, $\mathbf{a}_2 = \mathbf{F}_2/m$, $\mathbf{a}_3 = \mathbf{F}_3/m$, . . .); according to Eq. (8) all the forces acting together produce a net acceleration $\mathbf{a}$ equal to

$$\mathbf{a} = \frac{1}{m}\mathbf{F}_{\text{net}} = \frac{1}{m}(\mathbf{F}_1 + \mathbf{F}_2 + \mathbf{F}_3 + \ldots) = \mathbf{a}_1 + \mathbf{a}_2 + \mathbf{a}_3 + \ldots \tag{10}$$

This is simply the sum of the individual accelerations. Thus, the principle of Superposition of Forces is equivalent to the assertion that each force produces an acceleration independently of the presence or absence of other forces.

We must emphasize that this principle is a law of nature, which has the same status as Newton's laws. The most precise empirical test of this principle emerges from the study of planetary motion. There it is found that the net force on a planet is indeed the vector sum of all the gravitational pulls exerted by the Sun and by the other planets. Somewhat less precise tests of this principle can be performed in laboratory experiments by pulling on a body with known forces in known directions.

EXAMPLE 4. Suppose that the two tow ropes in Fig. 3.8 pull with forces of 2.5×10^5 N and 1.0×10^5 N, respectively, and that these forces make angles of 30° and 15° with the long axis of the barge (see Fig. 3.9). Suppose that the friction force is zero. What are the magnitude and direction of the net force on the barge?

---

* These are the horizontal forces. There are also vertical forces: the downward pull of gravity (the weight) and the upward pressure of the water (the buoyancy). These vertical forces will be ignored for now.

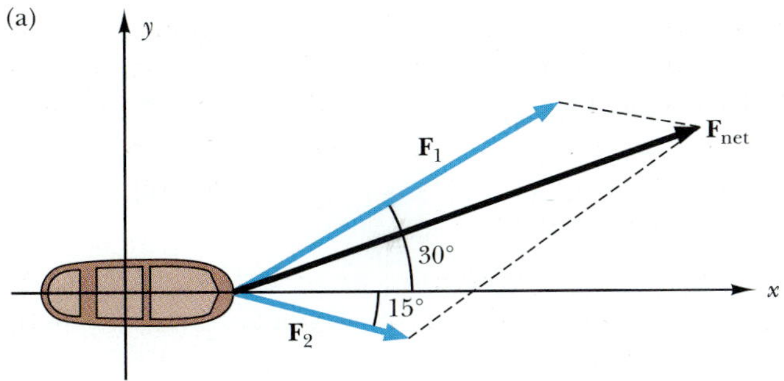

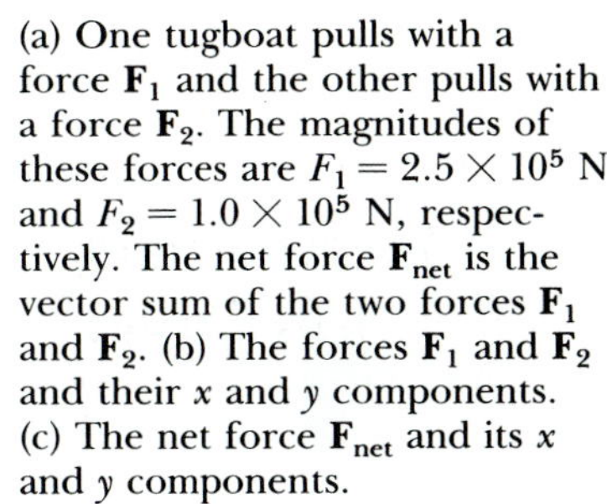

FIGURE 3.9

(a) One tugboat pulls with a force $\mathbf{F}_1$ and the other pulls with a force $\mathbf{F}_2$. The magnitudes of these forces are $F_1 = 2.5 \times 10^5$ N and $F_2 = 1.0 \times 10^5$ N, respectively. The net force $\mathbf{F}_{\text{net}}$ is the vector sum of the two forces $\mathbf{F}_1$ and $\mathbf{F}_2$. (b) The forces $\mathbf{F}_1$ and $\mathbf{F}_2$ and their x and y components. (c) The net force $\mathbf{F}_{\text{net}}$ and its x and y components.

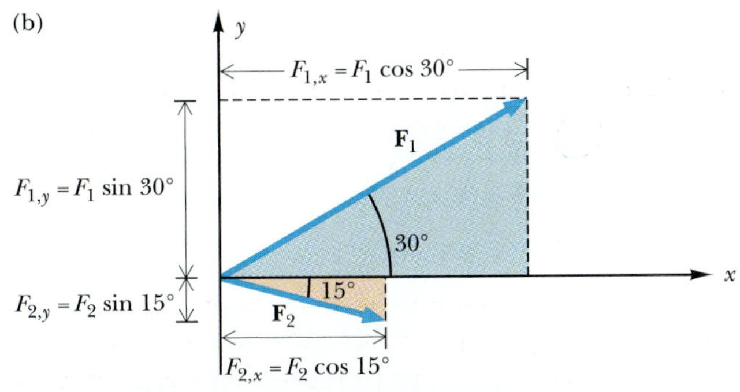

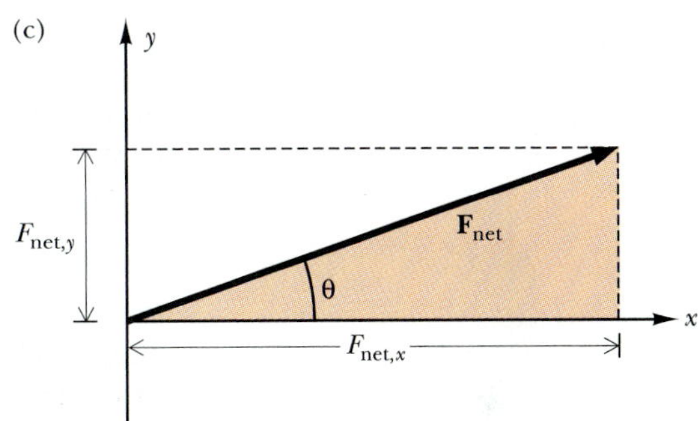

SOLUTION: The net force is the vector sum

$$\mathbf{F}_{\text{net}} = \mathbf{F}_1 + \mathbf{F}_2 \tag{11}$$

where $\mathbf{F}_1$ is the force of the first tow rope and $\mathbf{F}_2$ that of the second. The net force is shown in Fig. 3.9a. With the x and y axes arranged as in Fig. 3.9a, the x component of the net force is the sum of the x components of the individual forces (see Fig. 3.9b),

$$\begin{aligned} F_{\text{net},x} &= F_{1,x} + F_{2,x} \qquad (12) \\ &= 2.5 \times 10^5 \text{ N} \times \cos 30° + 1.0 \times 10^5 \text{ N} \times \cos 15° \\ &= 2.5 \times 10^5 \text{ N} \times 0.866 + 1.0 \times 10^5 \text{ N} \times 0.966 = 3.13 \times 10^5 \text{ N} \end{aligned}$$

Likewise, the y component of the net force is the sum of the y components of the individual forces (see Fig. 3.9b),

$$\begin{aligned} F_{\text{net},y} &= F_{1,y} + F_{2,y} \qquad (13) \\ &= 2.5 \times 10^5 \text{ N} \times \sin 30° - 1.0 \times 10^5 \text{ N} \times \sin 15° \\ &= 2.5 \times 10^5 \text{ N} \times 0.500 - 1.0 \times 10^5 \text{ N} \times 0.259 = 0.99 \times 10^5 \text{ N} \end{aligned}$$

(The y components of the individual forces are of opposite signs because one tugboat pulls to the left and the other to the right.)

The components $F_{\mathrm{net},x}$ and $F_{\mathrm{net},y}$ uniquely specify the net force, and we could end our calculation of the net force with these components. However, the problem asks for the magnitude and the direction of the net force, and we therefore have to take our calculation a step further. According to Eq. (2.49), the magnitude of the net force is the square root of the sum of squares of the components,

$$F_{\mathrm{net}} = \sqrt{(F_{\mathrm{net},x})^2 + (F_{\mathrm{net},y})^2} \tag{14}$$

$$= \sqrt{(3.13 \times 10^5\ \mathrm{N})^2 + (0.99 \times 10^5\ \mathrm{N})^2} = 3.3 \times 10^5\ \mathrm{N}$$

The direction of the net force makes an angle θ with the x axis (see Fig. 3.9c). According to Eq. (2.50), this angle is given by

$$\cos\theta = \frac{F_{\mathrm{net},x}}{F_{\mathrm{net}}} \tag{15}$$

$$= \frac{3.13 \times 10^5\ \mathrm{N}}{3.3 \times 10^5\ \mathrm{N}} = 0.95$$

With our calculator, we find that the angle with this cosine is 18°.

(a)

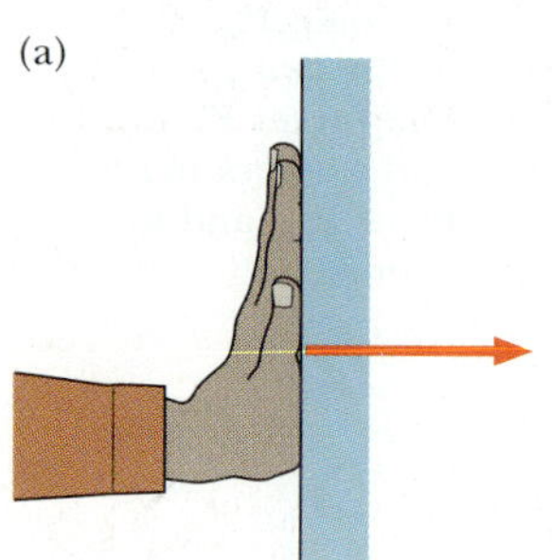

(b)

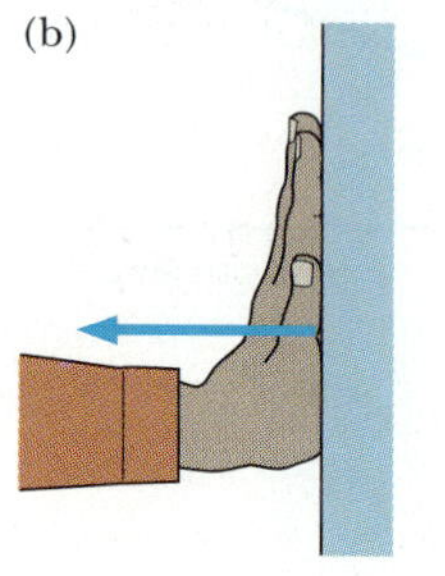

FIGURE 3.10
(a) Hand pushes on wall (action). (b) Wall pushes on hand (reaction).

3.4 NEWTON'S THIRD LAW

Action and reaction ▸

When you push with your hand against a table or wall, you can feel the wall pushing back at you. Thus, the mutual interaction of your hand and the wall involves two forces: the "action" force of the hand on the wall and the "reaction" force of the wall on the hand (see Fig. 3.10). These forces are said to form an ***action–reaction pair.*** Which of the forces is regarded as "action" and which as "reaction" is irrelevant. It may seem reasonable to regard the push of the hand as an action; then the push of the wall is a reaction. However, it is equally valid to regard the push the wall exerts on the hand as an action, and then the push of the hand on the wall is a reaction. The important point is that forces always occur in pairs; each of them cannot exist without the other.

Newton's Third Law gives the quantitative relationship between the action force and the reaction force:

Newton's Third Law ▸

> ***Whenever a body exerts a force on another body, the latter exerts a force of equal magnitude and opposite direction on the former.***

For instance, if the push of your hand on the wall has a magnitude of 60 N, then the push of the wall on your hand also has a magnitude of 60 N.

This equality of the magnitudes of action and reaction holds even if the body you push against is not held in a fixed position (like a wall), but is free to move. Thus, if you push on a cart with a force of 60 N, the cart will push back on you with a force of 60 N (see Fig. 3.11). Note that although these action and reaction forces are of equal magnitudes, they act on different bodies and their effects are quite different: the first force gives an acceleration to the cart (if there is no other force acting on the cart), whereas the second force merely slows your hand and prevents it from accelerating as much as it would if the cart were not there. ***Thus, although action and reaction are forces of equal magnitudes and of opposite directions, their effects do not cancel because they are acting on different bodies.***

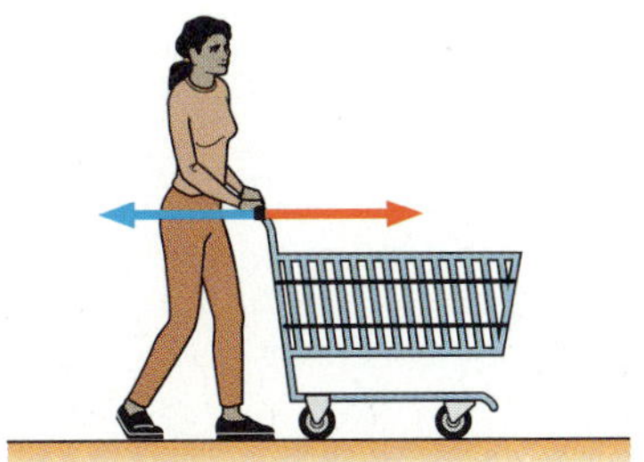

FIGURE 3.11
Student pushing a cart.

Reaction forces play a crucial role in all machines that produce locomotion by pushing against the ground, water, or air. For example, an automobile moves by pushing backward on the ground with its wheels; the reaction of the ground then pushes the automobile forward (see Fig. 3.12). A man walks by pushing backward on the ground; the reaction of the ground then pushes the man forward (see Fig. 3.13). A tugboat moves by pushing backward on the water with its propeller; the reaction of the water on the propeller then pushes the tugboat forward (Fig. 3.14). Even the propulsion of a rocket relies on reaction forces. The rocket engine expels exhaust gases, pushing them out in the rearward direction; the reaction of the exhaust gases then pushes the engine and the rocket in the forward direction (Fig. 3.15). Note that this rocket-propulsion mechanism works just as well in empty space — the rocket does not need to push against the atmosphere, but only against its exhaust gases. The atmosphere is of no help in rocket propulsion; rather it is a hindrance, since it exerts a frictional resistance on the rocket.

FIGURE 3.12
Automobile pushes on ground; ground pushes on automobile.

Reaction forces can exist even if the two interacting bodies are not in direct contact, so the forces between them must bridge the intervening empty space. For instance, consider an apple in free fall at some height above the ground. The Earth pulls on the apple by means of gravity. If this pull has a magnitude of, say, 2 N, then the Third Law requires that the apple pull on the Earth with an opposite force of 2 N (see Fig. 3.16). This reaction force is also a form of gravity — it is the gravity that the apple exerts on the Earth. However, the effect of the apple on the motion of the Earth is insignificant because the mass of the Earth is so large that a force of only 2 N causes only an insignificant acceleration.

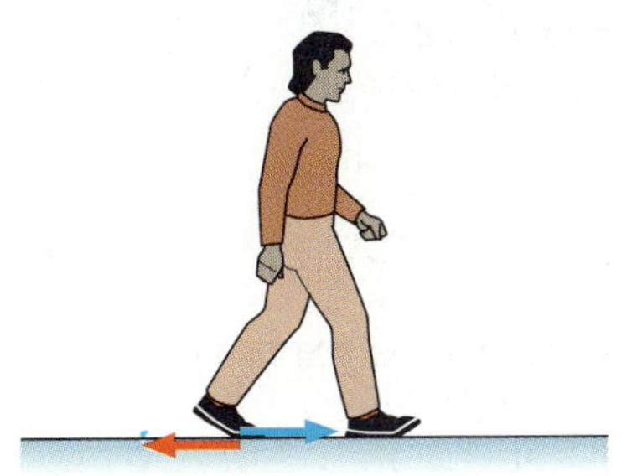

FIGURE 3.13
Man pushes on ground; ground pushes on man.

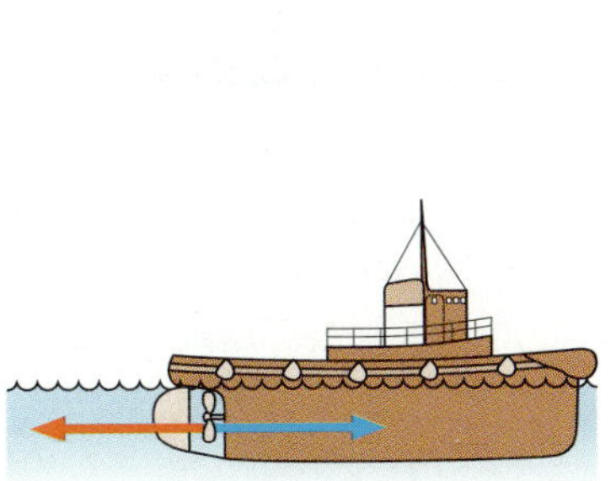

FIGURE 3.14
Propeller pushes on water; water pushes on propeller.

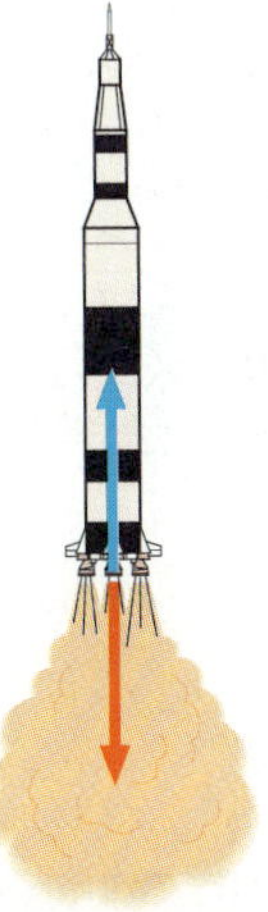

FIGURE 3.15
Rocket pushes on exhaust gases; exhaust gases push on rocket.

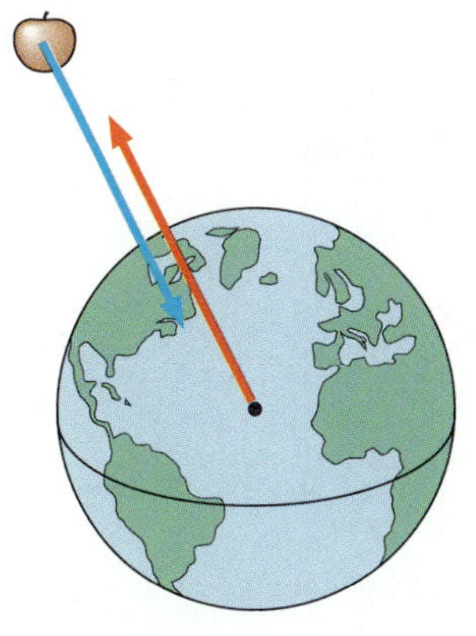

FIGURE 3.16
Earth pulls on apple; apple pulls on Earth.

EXAMPLE 5. A tugboat tows a barge of mass 50,000 kg by means of a cable (see Fig. 3.17). If the tugboat exerts a pull of 6000 N on the cable, what is the acceleration of the barge? What is the tension in the cable at its forward end? At its rearward end? Assume that the mass of the cable can be neglected (that is, assume the cable is practically massless), and ignore the friction of the water on the barge.

SOLUTION: The force of 6000 N exerted by the tugboat must accelerate both the barge and the cable. The accelerated mass is 50,000 kg. Hence, the resulting acceleration is

$$a = \frac{F}{m} = \frac{6000\ \text{N}}{50{,}000\ \text{kg}} = 0.12\ \text{m/s}^2 \tag{16}$$

Tension ▶

By tension is meant the force with which the cable pulls on what is attached to it. Since the tugboat pulls on the cable with a force of 6000 N, Newton's Third Law requires that the cable pull on the tugboat with an equally large force, and consequently the tension at the forward end of the cable is 6000N.

The tension at the rearward end of the cable must also be 6000 N, since the cable must exert a pull of this magnitude on the barge to produce the acceleration of 0.12 m/s^2.

COMMENTS AND SUGGESTIONS: Note that although the tensions at the forward and rearward ends of the cable are two forces of equal magnitudes and of opposite direction, these two forces are *not* an action–reaction pair. The tension at the forward end forms an action–reaction pair with the pull exerted by the tugboat on the cable, and the tension at the backward end forms an action–reaction pair with the pull exerted by the barge on the cable (see Fig. 3.18).

FIGURE 3.17
A tugboat tows a barge by means of a steel cable.

FIGURE 3.18
Forces exerted on the ends of the cable (blue) and forces exerted by the ends of the cable (red).

In the preceding example, the equality of the magnitudes of the tensions at the forward and the rearward ends of the cable is a consequence of neglecting the mass of the cable. If we were to take into account the mass of the cable, then we would need a net force to accelerate the cable. This means that the force pulling on the forward end of the cable would have to be *larger* than the force pulling on the rearward end. Correspondingly, by the Third Law, the force with which the forward end of the cable pulls on the tugboat would have to be larger than the force with which the rearward end of the cable pulls on the barge — that is, the tensions would be of unequal magnitudes.

For practical purposes, the mass of a cable, rope, string, or chain can often be neglected compared with the mass of the body to which it is attached. Under these conditions, ***the cable transmits the tension without any significant change of magnitude, and the force that one body exerts on another body via the cable is as though the bodies were in direct contact.*** The rule that a cable transmits the magnitude of the tension without change also holds for a cable that runs over a pulley, provided that the mass of the pulley and the friction on its axle can be neglected; only the *direction* of the tension changes when the cable passes over such a pulley. In our calculations we will hereafter always neglect the masses of cables and pulleys unless it is explicitly stated otherwise.

EXAMPLE 6. Figure 3.19a shows a traction apparatus used in hospitals to exert a steady pull on a broken leg, in order to keep the bones aligned. The middle pulley is attached to the cast, and the other two pulleys are attached to the bed or the wall. A flexible wire passes over these pulleys, and a heavy brick hanging from this wire provides a tension. The upper and the lower portions of the wire lead, respectively, upward and downward from the middle pulley at angles of 35°. If the horizontal pull on the leg is to be 50 N, what vertical pull must the brick provide at the end of the wire?

FIGURE 3.19

(a) Traction apparatus for fractured leg. (b) Forces exerted by the wire on the middle pulley. The upper portion of the wire pulls with a force $\mathbf{F}_1$ and the lower portion pulls with a force $\mathbf{F}_2$. The magnitudes of these two forces are equal, $F_1 = F_2 = w$. The resultant is in the x direction. (c) Components of the forces $\mathbf{F}_1$ and $\mathbf{F}_2$.

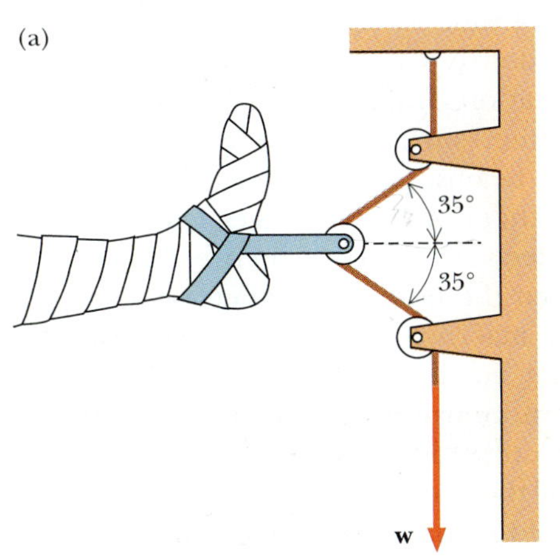

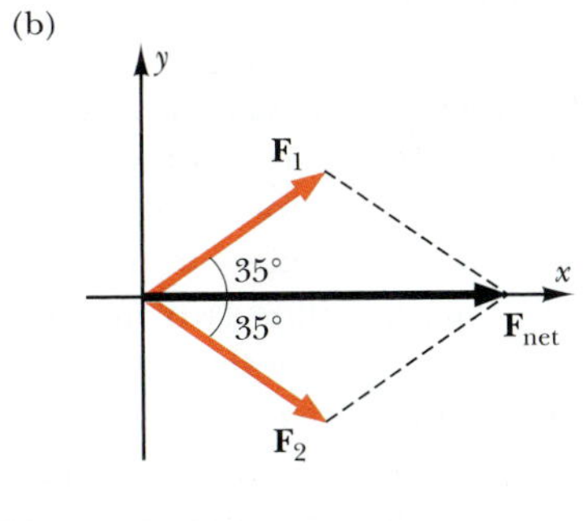

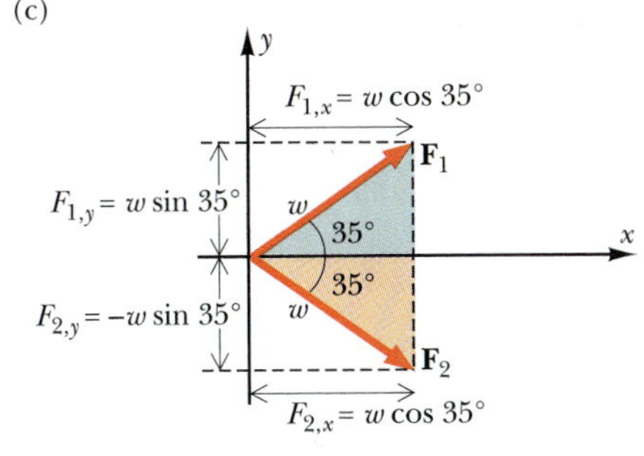

SOLUTION: As discussed above, the tension is constant along the entire length of the wire. The magnitude of the tension at the lower end of the wire equals the magnitude w of the pull of the brick; hence, the magnitude of the tension at all other points of the wire must also be w. Under static conditions, the upper and the lower portions of the wire may be regarded as attached to the middle pulley at the points of first contact. Thus the upper portion of the wire pulls upward at an angle of 35° with a force of magnitude w, and the lower portion pulls downward at an angle of 35° with a force of equal magnitude (see Fig. 3.19b). The vertical components of these pulls therefore cancel, since they have opposite signs and equal magnitudes. The horizontal components add, since they both have positive signs and equal magnitudes. The net force is therefore in the x direction.

The components of $\mathbf{F}_1$ and $\mathbf{F}_2$ in the x direction are $w\cos 35°$ (see Fig. 3.19c). Hence the x component of the net force is

$$F_{\text{net},x} = w\cos 35° + w\cos 35°$$

$$= 2w\cos 35°$$

Since the magnitude of $F_{\text{net},x}$ is supposed to be 50 N, we obtain

$$w = \frac{F_{\text{net},x}}{2\cos 35°}$$

$$= \frac{50\ \text{N}}{2 \times 0.819} = 31\ \text{N}$$

Therefore the brick must provide a vertical pull of 31 N.

3.5 Kinds of Forces; Weight and Normal Force

In everyday experience we encounter a wide variety of forces: the gravity of the Earth which pulls all bodies downward, contact forces between rigid bodies that resist their interpretation, friction forces that resist the motion of a surface sliding over another surface, elastic forces that oppose the deformation of springs and beams, pressure forces exerted by air or water on bodies immersed in them, adhesive forces exerted by a layer of glue bonding two surfaces, electrostatic forces between two electrified bodies, magnetic forces between the poles of magnets, and so on.

Besides these forces that act in the macroscopic world of everyday experience, there are many others that act in the microscopic world of atomic and nuclear physics. There are intermolecular forces that attract or repel molecules to or from each other, interatomic forces that bind atoms into molecules or repel them if they come too close to each other, atomic forces within the atom that hold its parts together, nuclear forces

(a)

(b)

(c)

(a) The gravitational force is responsible for the free-fall motion of this diver. (b) Electric forces between the atoms give rise to the contact forces and elastic forces that act in this impact of a club on a golf ball. (c) The "strong" force brings about the thermonuclear fusion reactions in this explosion of a hydrogen bomb. (d) The "weak" force causes the decays of several elementary particles and the creation of new particles, which made these tracks in a bubble chamber at an accelerator laboratory

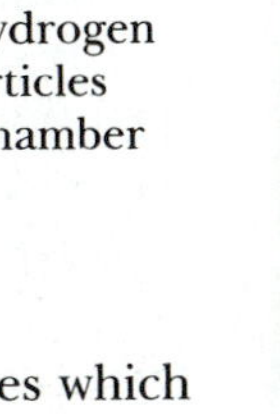

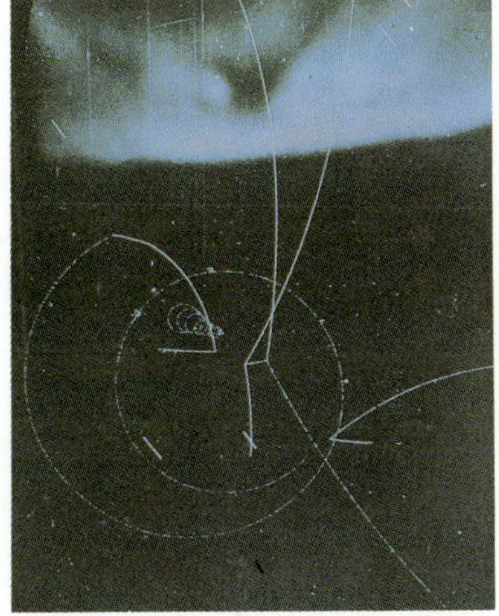

(d)

that act on the parts of the nucleus, and even more esoteric forces which act only for a brief instant when subnuclear particles are made to suffer violent collisions in high-energy experiments performed in accelerator laboratories.

Yet this bewildering variety of macroscopic and microscopic forces involves only four different fundamental kinds of force. ***The four fundamental forces are the gravitational force, the electromagnetic force, the "strong" force, and the "weak" force.***

◀ **Gravitational force**

The gravitational force is a mutual attraction between all masses. The gravitational force is the weakest of the fundamental forces. Between two masses of a few kilograms, such as your mass and that of your roommate, the gravitational force is so weak as to be imperceptible. On the surface of the Earth, we feel the downward pull of gravity only because the mass of the Earth is very large.

◀ **Electromagnetic force**

The electromagnetic force is an attraction or repulsion between electric charges. The electric and the magnetic forces, once considered to be separate, are now grouped together because they are closely related: the magnetic force is an extra electric force that acts whenever charges are in motion. Of all the forces, the electric force plays the largest role in our lives. With the exception of the Earth's gravity, every force in our everyday, macroscopic environment is electric. Contact forces between rigid bodies, elastic forces, pressure forces, adhesive forces, friction forces, etc., are nothing but electric forces between the charged particles in the atoms of one body and those in the atoms of another.

◀ **"Strong" force**

The "strong" force acts mainly within the nuclei of atoms. It plays the role of a nuclear glue that prevents the pieces of the nucleus from flying apart. This nuclear force is called "strong" because it is the strongest of the four fundamental forces. But because this force is hidden away within the nucleus, we do not notice it directly.

"Weak" force ▶

Finally, ***the "weak" force manifests itself only in special reactions among elementary particles, such as the reactions that occur in radioactive decay,*** involving the spontaneous breakup of a particle into several other particles (we will discuss radioactive decay in Chapter 31). This force is called "weak" because it is very weak, although not as weak as the gravitational force.

In the next sections and chapters, we will not concern ourselves with the fundamental aspects of forces. We will adopt a macroscopic point of view, and deal only with those forces of practical importance in our everyday environment. The most prominent among these forces are the gravity of the Earth, the contact forces that act between the adjacent surfaces of two bodies when they are pressed together, and elastic forces, such as the elastic force of a spring.

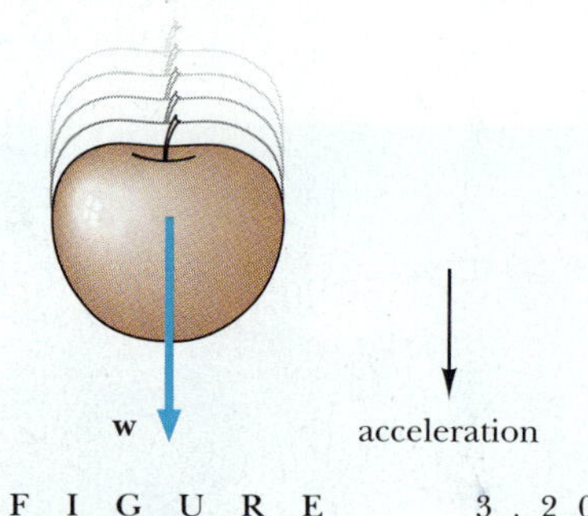

FIGURE 3.20
A body in free fall. The weight force **w** produces a downward acceleration g.

The gravity of the Earth is the most familiar of all forces. When we hold a body in our hands, we can directly feel the downward pull of gravity on the body; and if we release the body, we can see it accelerating under the influence of this pull. In the terminology of physics, ***the pull of gravity on a body is called the weight of the body.*** Thus, weight is a force, it is a vector quantity — it has a direction (downward) as well as a magnitude. The unit of weight is the unit of force; that is, the unit of weights is the newton (N).

The magnitude of the weight force is directly proportional to the mass of the body. To understand this, consider a body of mass m in free fall near the surface of the Earth — say, an apple you have released from your hand (see Fig. 3.20). The body has a downward acceleration g. Since we attribute this acceleration to the weight force, Newton's Second Law tells us that the magnitude of the weight force acting on the body must be

$$F = ma$$

$$= mg \tag{17}$$

We will denote the weight of the vector symbol **w.** According to Eq. (17), the magnitude of the weight is

Weight ▶

$$w = mg \tag{18}$$

If the body is not in free fall but is held in a stationary position by some support, then the weight is of course still the same as given by Eq. (18); however, the support balances the downward weight force and prevents it from producing a downward motion.

EXAMPLE 7. What is the weight of a 54-kg woman? Assume that $g = 9.81\ \text{m/s}^2$.

SOLUTION: By Eq. (18), the weight is

$$w = mg = 54 \text{ kg} \times 9.81 \text{ m/s}^2 = 530 \text{ N}$$

COMMENTS AND SUGGESTIONS: Since the value of g depends on location, the weight of a body also depends on its location. For example, if the 54-kg woman travels from London ($g = 9.81$ m/s^2) to Hong Kong ($g = 9.79$ m/s^2), her weight will decrease from 530 N to 529 N, a difference of 1 N. And if this woman were to travel to the Moon ($g = 1.62$ m/s^2), her weight would decrease to 87 N!

The preceding example illustrates an essential distinction between mass and weight: ***mass is an intrinsic property of a body,*** measuring the inertial resistance with which the body opposes changes in its motion; the definition of mass is designed in such a way that a given body has the same mass regardless of its position in the universe. ***Weight is an extrinsic property of a body,*** measuring the pull of gravity on the body; it depends on the (gravitational) environment in which the body is located, and therefore depends on position.

FIGURE 3.21

This soccer player has jumped into the air. Both he and the ball are in free fall.

A body deep in intergalactic space, far from the gravitational pull of any star or planet, will experience hardly any gravitational pull — the weight of the body will be nearly zero. ***This condition of weightlessness can be simulated in the vicinity of the Earth by means of a freely-falling reference frame.*** Consider an observer in free fall, such as the soccer player in Fig. 3.21, who has jumped into the air to kick a ball. The player and the ball both accelerate downward at the same rate; thus, the ball does not accelerate relative to the player. In the reference frame of the player (a freely-falling reference frame accelerating downward with the acceleration g), the freely-falling ball or any other freely-falling body continues to move with constant velocity, as though there were no force acting on it. This means that in such a reference frame the gravitational pull is *apparently* zero; the weight is *apparently* zero. Of course, this simulated weightlessness arises from the accelerated motion of the reference frame — in the unaccelerated, inertial reference frame of the ground, the weight of the ball is certainly not zero. Nevertheless, if the player insists on looking at things from his own reference frame, he will judge the weight of the ball, and the weight of his own body, as zero. This sensation of weightlessness is also simulated within an airplane flying along a parabola, imitating the motion of a (frictionless) projectile (see Fig. 3.22); and it is also simulated in a spacecraft orbiting the Earth (see Fig. 3.23). Both of these motions are free-fall motions.

FIGURE 3.22

Astronauts training in an airplane.

The gravity of the Earth reaches from the Earth to any other body, even a body placed high above the surface. Gravity bridges empty space, and requires no perceptible medium for its transmission. In contrast, most of the other forces familiar from everyday experience require direct contact between the bodies. You cannot exert a push on a box unless your hand is in contact with the box; and the box cannot exert a push on the floor, unless it is in contact with the floor. ***The push that the surface of a***

FIGURE 3.23
Astronaut floating in the cargo bay of the space shuttle.

Contact force ▶

body exerts on the adjacent surface of another body is called a contact force. If the two bodies are solid, the contact forces between their adjacent surfaces are of two kinds: the compressional force, or normal force, that arises when the surfaces are pressed together perpendicularly, and the friction force that arises when the surfaces are sliding, or trying to slide, past each other. We will deal with friction forces in Section 3.7.

Normal force ▶

The normal force represents the resistance that the solid bodies offer to interpenetration. When we try to push two bodies together, their surfaces begin to repel as soon as they come into contact. For instance, consider a solid box sitting on a floor (see Fig. 3.24). There are two forces acting on this box: the downward pull of gravity, and the upward push of the floor against the bottom surface of the box. The upward push of the floor is distributed uniformly over the bottom surface of the box, but in Fig. 3.24 this push is shown as though acting only at the center of the surface. The upward push of the floor is called the normal force, because it is "normal," or perpendicular, to the floor (of course, there is also a normal force of equal magnitude and opposite direction that the box exerts on the floor; but Fig. 3.24 shows only the forces that act on the box). This normal force is a repulsive force with which the floor supports the box and prevents it from passing through the floor.

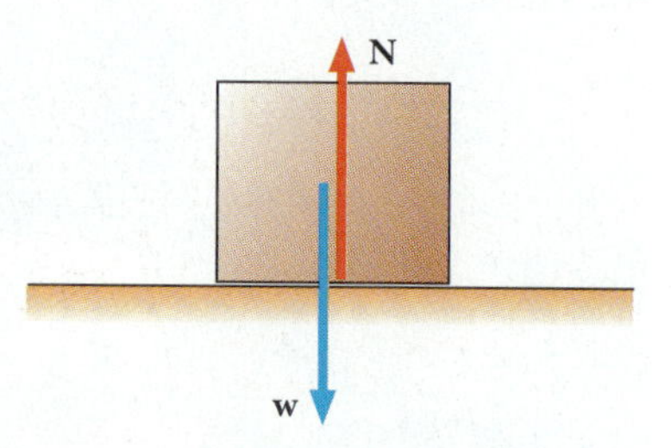

FIGURE 3.24
A box sitting on a floor. The weight **w** pulls downward on the box, and the normal force **N** exerted by the floor pushes upward on the box.

This force arises from the contact between the atoms of the box and the atoms in the floor; the atoms of the floor and the atoms on the surface of the box exert repulsive forces on each other, which prevent their interpenetration. You can feel this kind of repulsive contact force when you push against any solid surface. For instance, if you push your foot down on the floor, you can feel the floor pushing up against the sole of your foot, stopping your foot from penetrating the floor. How does the floor succeed in preventing your foot from penetrating the floor, regardless of how hard you push? The resistance offered by the floor results from a slight compression of the material of the floor. The atoms in the material in the floor behave like an array of miniature springs; these atomic springs compress slightly when you push your foot against them, but the force that these springs exert on your foot increases with the amount of compression. Hence your foot sinks into the floor until the increasing force of the atomic springs matches your downward push. If the material of the floor is hard — for example, concrete — the amount of compression is so slight as to be unnoticeable.

When the normal force plays the role of a supporting force, as in the case of our box sitting on the floor, then the magnitude of the normal force is always such as to achieve a balance with whatever other forces are pressing the bodies together. In the case of the box, the only force pressing the box down is its weight $w = mg$. Hence the magnitude of the normal force must also be $N = mg$, so that the box has no net force acting on it and remains at rest. If there were an extra force pressing the box down, say, if you were pushing down on the box, then the normal N force would be larger than mg (we will see an example of this sort in the next section).

3.6 Motion with a Constant Force

◀ Equation of motion

Newton's Second Law is often called the ***equation of motion.*** If the force on a particle is known, then ***the Second Law determines the acceleration, and from this the complete motion of the particle can be calculated.*** Thus, in principle, the motion of the particle is completely predictable.

If the force acting on a particle is constant, then the acceleration is also constant. The motion is then given by the formulas we developed for motion with constant acceleration in Chapter 2 [Eqs. (2.15)–(2.18)].

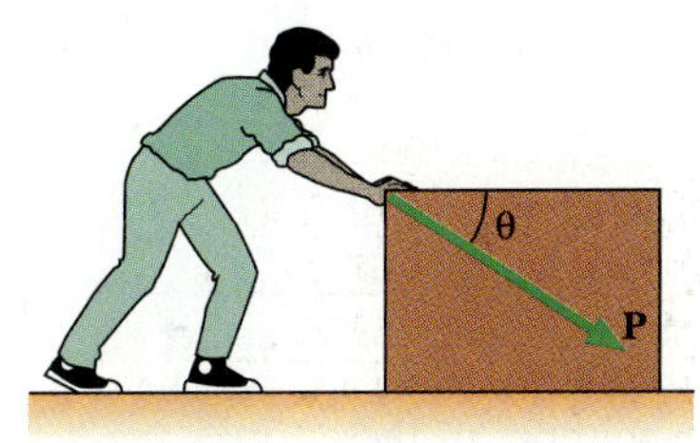

FIGURE 3.25
Man pushing a box.

As a simple example of motion with a constant force, consider a rectangular cardboard box of mass m being pushed along a smooth, frictionless floor. Figure 3.25 shows the box and the man pushing it. The push can be represented by a vector **P** of magnitude P pointing at an angle θ with the horizontal direction. Besides this push, there are two other forces acting on the box: the weight **w** of the box pointing downward in the vertical direction, and the normal force of the floor **N** pointing upward in the vertical direction. Figure 3.26a shows only the box and the three forces **P, w,** and **N** acting on it. The man and the floor on which the box slides are not shown in this diagram — the effects of the man and of the floor are entirely contained in the push **P** and the normal force **N**. ***Such a diagram, showing the body and all the forces acting on it, but not showing the surrounding bodies that exert these forces, is called a "free-body" diagram.*** (In this context *free* does not mean free of forces; it means that the surroundings are represented by the exerted forces.) Such a diagram eliminates clutter and helps us to focus on the body and on the forces needed to formulate the equation of motion of the body.

◀ "Free-body" diagram

The net force on our box is the vector sum of **P, w,** and **N;** and the acceleration of the box is related to this net force by Newton's Second Law. Although the motion of the box is one dimensional (horizontal, along the

(a)

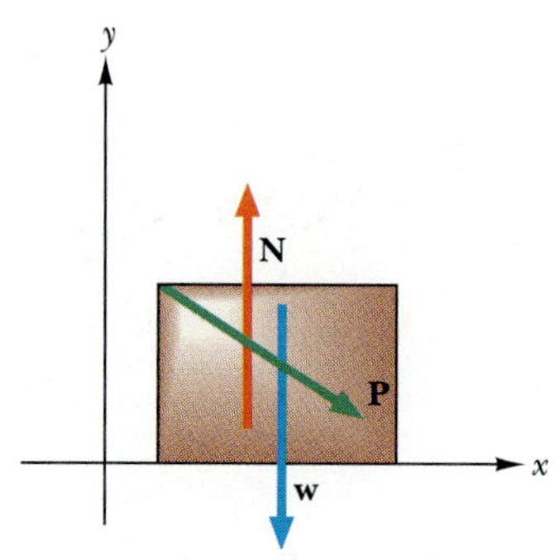

(b)

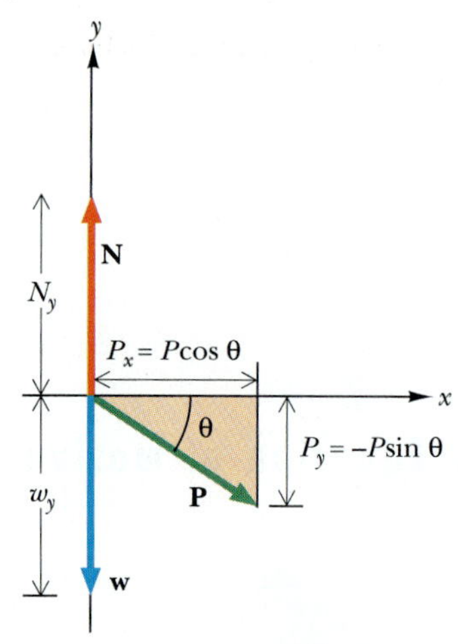

FIGURE 3.26
(a) "Free-body" diagram for the box. The three forces acting on the box are **P, w,** and **N.** The magnitudes of these forces are *P, mg,* and *N.* (b) The *x* and *y* components of the forces **P, w,** and **N.**

floor), the forces acting on it are not one dimensional, and we must consider both the horizontal and the vertical components of these forces. For this purpose, it is convenient to take the x axis in the horizontal direction and the y axis in the vertical direction. Looking at Fig. 3.26b, we then find that the x and the y components of the individual forces are

$$P_x = P\cos\theta \qquad P_y = -P\sin\theta \tag{19}$$

$$w_x = 0 \qquad w_y = -mg \tag{20}$$

$$N_x = 0 \qquad N_y = N \tag{21}$$

Here N is the (unknown) magnitude of the normal force.

The components of the net force are

$$F_x = P_x + w_x + N_x = P\cos\theta + 0 + 0$$

$$= P\cos\theta \tag{22}$$

$$F_y = P_y + w_y + N_y$$

$$= -P\sin\theta - mg + N \tag{23}$$

For the calculation of the translational motion of the box, we can regard the box as a particle. The x and y components of Newton's equation of motion are $ma_x = F_x$ and $ma_y = F_y$, from which we obtain the components of the acceleration:

$$a_x = \frac{F_x}{m} = \frac{P\cos\theta}{m} \tag{24}$$

$$a_y = \frac{F_y}{m} = \frac{-P\sin\theta - mg + N}{m} \tag{25}$$

Equation (24) says that the acceleration of the box along the floor is $P\cos\theta/m$. This determines the motion of the box along the floor, since the magnitude and direction (P and θ) of the push exerted by the man are assumed known. To find the velocity and the position at any time we need only to substitute the acceleration a_x into our old equations (1.11) and (1.16) for uniformly accelerated motion.

Equation (25) can be used to evaluate the magnitude of the normal force. Since the motion is necessarily along the floor, the acceleration a_y in the direction perpendicular to the floor is zero; hence

$$0 = (-P\sin\theta - mg + N)/m \tag{26}$$

We can solve this for N, with the result

$$N = mg + P\sin\theta \tag{27}$$

Note that if the push is horizontal ($\theta = 0$ and $\sin\theta = 0$) or if the push is absent ($P = 0$), then the normal force is $N = mg$. Thus, under these conditions, the normal force simply balances the weight; this is, of course, exactly what we would expect for a box sitting on a floor when there is no push. Furthermore, note that if the push is vertical ($\theta = 90°$ and $\sin\theta = 1$, $\cos\theta = 0$), the normal force is $N = mg + P$, and the horizontal acceleration is zero.

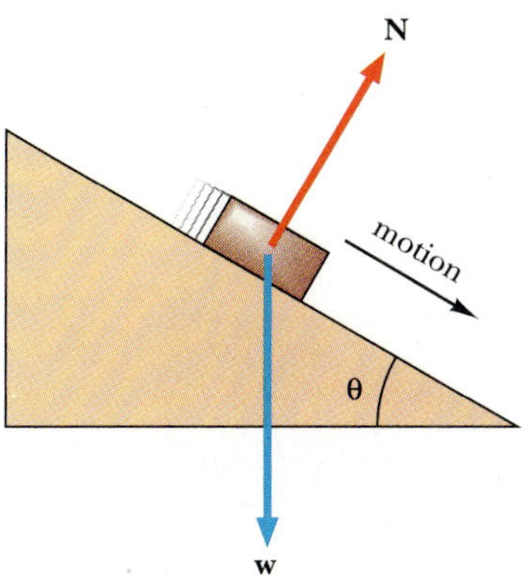

FIGURE 3.27
Block sliding down an inclined plane.

Another example of motion with constant force is the motion of a body sliding on a ramp, or an inclined plane. Figure 3.27 shows a block of mass m sliding down a smooth, frictionless plane inclined at an angle θ with respect to the horizontal direction. There are two forces acting on the block: the weight **w** pointing vertically downward and the normal force **N** pointing in a direction perpendicular to the inclined plane. Figure 3.28 shows these two forces on a "free-body" diagram. The net force acting on the block is the vector sum of **w** and **N.** For the calculation of the components of these forces, it is convenient to take the x axis parallel to the inclined plane and the y axis perpendicular; this simplifies the calculation of the motion, since the velocity and the acceleration are then entirely along the x axis. With this choice of axes, we find that the components of the two forces are

$$N_x = 0 \qquad N_y = N \tag{28}$$

$$w_x = mg\sin\theta \qquad w_y = -mg\cos\theta \tag{29}$$

and the components of the net force are

$$F_x = N_x + w_x = mg\sin\theta \tag{30}$$

$$F_y = N_y + w_y = N - mg\cos\theta \tag{31}$$

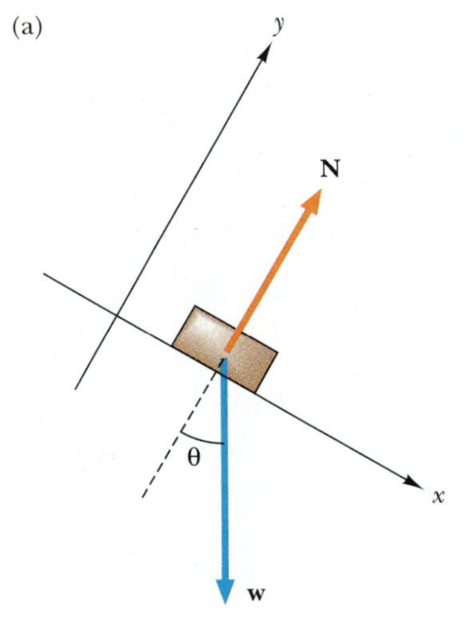

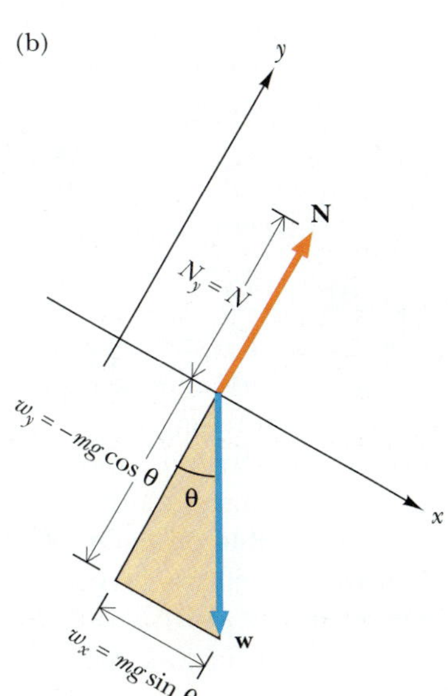

FIGURE 3.28
(a) "Free-body" diagram showing the forces **N** and **w.** (b) The x and y components of the forces **N** and **w.**

Newton's equation of motion then gives us the corresponding components of the acceleration:

$$a_x = \frac{F_x}{m} = \frac{mg \sin\theta}{m} = g \sin\theta \tag{32}$$

$$a_y = \frac{F_y}{m} = \frac{N - mg \cos\theta}{m} \tag{33}$$

Equation (32) tells us that the acceleration of the block along the inclined plane is $g \sin \theta$. In the case of a horizontal plane ($\theta = 0$ and $\sin\theta = 0$), there is no acceleration; in the case of a vertical plane ($\theta = 90°$ and $\sin\theta = 1$), the acceleration is that of free fall. Both of these extreme cases are as we would expect them to be.

Equation (33) can be used to evaluate the normal force. Since the motion is along the plane, the acceleration perpendicular to the plane must be identically zero, and thus

$$0 = \frac{N - mg \cos\theta}{m} \tag{34}$$

From this we find

$$N = mg \cos\theta \tag{35}$$

In the case of a horizontal plane ($\theta = 0$), the normal force has a magnitude mg, that is, the magnitude of the normal force matches the weight. In the case of a vertical plane ($\theta = 90°$), the normal force vanishes.

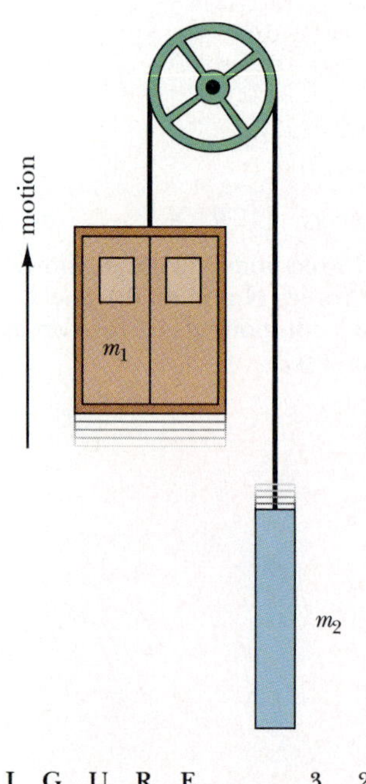

FIGURE 3.29
Elevator with counterweight.

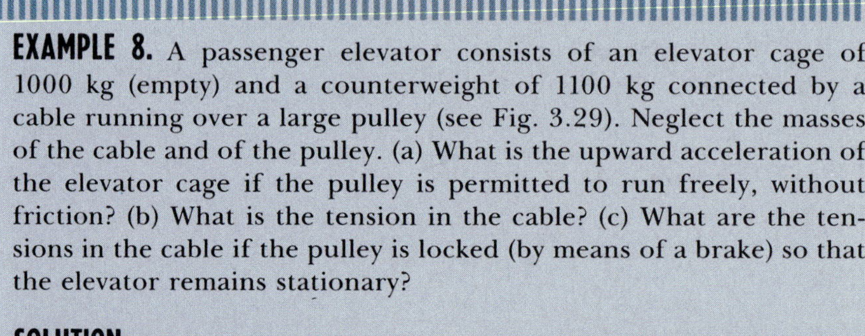

EXAMPLE 8. A passenger elevator consists of an elevator cage of 1000 kg (empty) and a counterweight of 1100 kg connected by a cable running over a large pulley (see Fig. 3.29). Neglect the masses of the cable and of the pulley. (a) What is the upward acceleration of the elevator cage if the pulley is permitted to run freely, without friction? (b) What is the tension in the cable? (c) What are the tensions in the cable if the pulley is locked (by means of a brake) so that the elevator remains stationary?

SOLUTION: (a) Since the elevator cage and the counterweight are linked by the cable, it is necessary to solve the equations of motion for these two bodies simultaneously. Designate the masses of the elevator cage and the counterweight by m_1 and m_2, respectively. The cable merely transmits the tension from one mass to the other, without change of its magnitude; consequently, the upward tension forces exerted by the ends of the cable on each mass are exactly equal.

Figure 3.30 shows "free-body" diagrams for the masses m_1 and m_2. For a system consisting of several bodies, such as the system of two bodies we are dealing with here, the "free-body" diagrams are especially helpful, since they permit us to view each body in isolation and they give us a clear picture of what happens to each individual body. The vector **T** in Fig. 3.30 represents the tension force, and $\mathbf{w}_1$ and $\mathbf{w}_2$ represent the weights. The vertical force acting on m_1 is $F_1 = T - w_1$; the vertical force on m_2 is $F_2 = T - w_2$. Hence, the equation for the vertical motion of each mass is

$$m_1 a_1 = F_1 = T - w_1 \tag{36}$$

$$m_2 a_2 = F_2 = T - w_2 \tag{37}$$

where force and acceleration are regarded as positive when directed upward. Since the two masses are connected by a fixed length of cable, their accelerations a_1 and a_2 are always of the same magnitudes and in opposite directions, that is,

$$a_1 = -a_2 \tag{38}$$

With this equation and with $w_1 = m_1 g$ and $w_2 = m_2 g$, we obtain from Eqs. (36) and (37)

$$m_1 a_1 = T - m_1 g \tag{39}$$

$$-m_2 a_1 = T - m_2 g \tag{40}$$

These are two simultaneous equations for the two unknowns a_1 and T. To solve these equations, we subtract each side of the second equation from each side of the first equation:

$$m_1 a_1 - (-m_2 a_1) = T - m_1 g - (T - m_2 g) \tag{41}$$

Here the unknown T cancels out, leaving us with an equation for a_1:

$$m_1 a_1 + m_2 a_1 = -m_1 g + m_2 g \tag{42}$$

We can then factor a_1 on the left side and g on the right side:

$$a_1(m_1 + m_2) = (m_2 - m_1)g \tag{43}$$

If we divide this by $(m_1 + m_2)$, we obtain an expression for a_1:

$$a_1 = \frac{m_2 - m_1}{m_1 + m_2} g \tag{44}$$

This equation tells us the acceleration. With $m_1 = 1000$ kg and $m_2 = 1100$ kg, we obtain

$$a_1 = \frac{1100 \text{ kg} - 1000 \text{ kg}}{1000 \text{ kg} + 1100 \text{ kg}} g = \frac{100}{2100} g$$

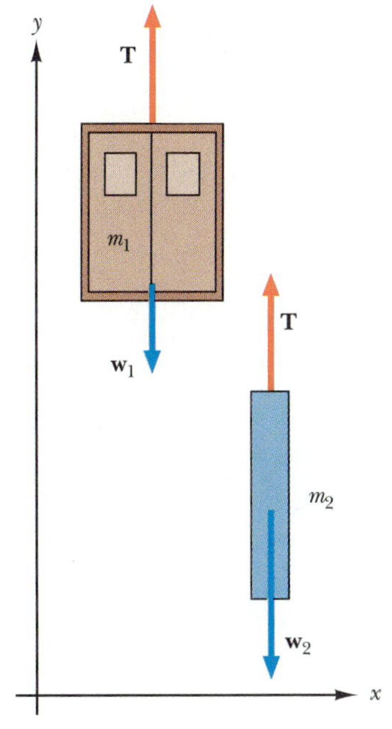

FIGURE 3.30
"Free-body" diagrams for the elevator (m_1) and the counterweight (m_2).

$$= 0.0476g = 0.0476 \times 9.81 \text{ m/s}^2 = 0.467 \text{ m/s}^2 \qquad (45)$$

(b) Next, we must find the tension in the cable. Substituting the result for a_1 into Eq. (39), we obtain an equation for T:

$$m_1 \frac{m_2 - m_1}{m_1 + m_2} g = T - m_1 g \qquad (46)$$

which leads to

$$T = m_1 \frac{m_2 - m_1}{m_1 + m_2} g + m_1 g \qquad (47)$$

If we combine the two terms on the right side by means of a common denominator, we find

$$T = \frac{m_1(m_2 - m_1)g + (m_1 + m_2)m_1 g}{m_1 + m_2} = \frac{2gm_1m_2}{m_1 + m_2} \qquad (48)$$

which tells us the tension in the cable. Numerically,

$$T = \frac{2gm_1m_2}{m_1 + m_2} = \frac{2 \times 9.81 \text{ m/s}^2 \times 1000 \text{ kg} \times 1100 \text{ kg}}{1000 \text{ kg} + 1100 \text{ kg}} \qquad (49)$$

$$= 1.03 \times 10^4 \text{ N}$$

(c) If the pulley is locked and the elevator is stationary, the tension in the cable on either side of the pulley must match the weight hanging on that side. Thus

$$T_1 = m_1 g = 1000 \text{ kg} \times 9.81 \text{ m/s}^2 = 9.81 \times 10^3 \text{ N} \qquad (50)$$

and

$$T_2 = m_2 g = 1100 \text{ kg} \times 9.81 \text{ m/s}^2 = 1.08 \times 10^4 \text{ N} \qquad (51)$$

COMMENTS AND SUGGESTIONS: As we might have expected, Eq. (44) shows that if the two masses m_1 and m_2 are equal, the acceleration is zero — the two masses are then in equilibrium, and they either remain at rest or move with uniform velocity (until the cable runs out).

Note that when the masses are unequal and the pulley is locked, the tensions in the two parts of the cable are *not* equal. This is because the locked pulley exerts extra forces on the cable, and the two portions of the cable now behave as though they were suspended independently.

From the examples analyzed in this section, we see that the solution of a problem of motion with several forces acting on a body proceeds in a sequence of steps:

1. [illegible] ep is always a careful enumeration of all the forces. Make a [illegible] st of these forces, and label each with a vector symbol.

2. I[illegible] ee-body" diagram in which each force is represented by an arrow. Remember that the only forces to be included in the "free-body" diagram are the forces that act *on* the body, not the forces exerted *by* the body.

3. Then draw coordinate axes on the diagram, preferably placing one of the axes along the direction of motion. If the motion proceeds along a sloping ramp, it is convenient to use tilted coordinate axes, with one axis along the ramp, as in Fig. 3.27b.

4. Next, examine the components of the individual forces and the components of the net force.

5. And, as a final step, apply Newton's Second Law to calculate the components of the acceleration (if the acceleration is unknown), or the components of some force (if some force is unknown).

Leonardo da Vinci, *1452–1519, Italian artist, engineer, and scientist. Famous for his brilliant achievements in painting, sculpture, and architecture, Leonardo also made pioneering contributions to science. But Leonardo's investigations of friction were forgotten, and the laws of friction were rediscovered two hundred years later by Guillaume Amontons, a French physicist.*

3.7 FRICTION

Friction forces, which we have ignored up to now, play an important role in our environment and provide us with many interesting examples of motion with constant force. For instance, if the driver of a moving automobile suddenly slams on the brakes, the wheels will lock and begin to skid on the pavement. The skidding wheels experience an (approximately) constant friction force that opposes the motion and decelerates the automobile at an (approximately) constant rate of 8 to 10 m/s^2. The magnitude of the deceleration depends on the characteristics of the tires and the pavement; besides, the heavy friction of rubber wheels on a typical pavement is accompanied by abrasion of the wheels, which introduces additional complications.

For the sake of simplicity, let us concentrate on an idealized case of friction, involving a solid block of metal sliding on a flat surface of metal. Figure 3.31 shows a block of steel, in the shape of a brick, sliding on a horizontal steel plate. If we give the block some initial velocity and then let it coast, friction will decelerate it. The forces acting on the block are the weight **w**, the normal force **N**, and the friction force **f**. The weight **w** acts downward with a magnitude *mg*. The normal force **N** exerted by the plate on the block acts upward. The magnitude of this normal force must be *mg*, so that it balances the weight. The friction force **f** acts horizontally, parallel to the plate, in a direction opposite to the motion. This force, just like the normal force, is a contact force which acts over the entire bottom surface of the block; however, in Fig. 3.31 it is shown as though acting at the center of the surface.

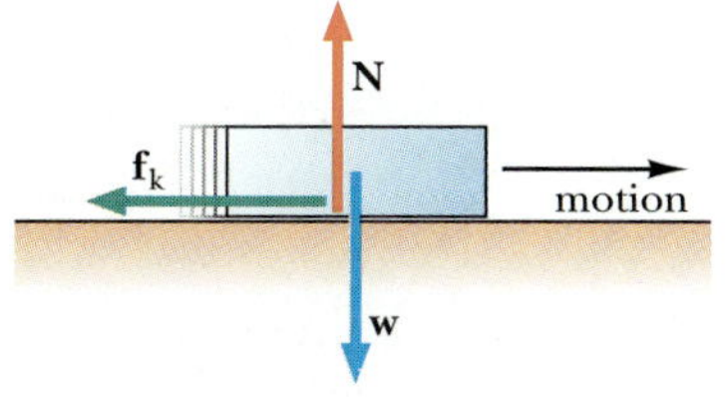

FIGURE 3.31
Forces on a block sliding on a plate. The friction force is in the direction opposite to the motion.

Friction involving surfaces in relative motion is called sliding friction, or kinetic friction. The friction force arises from adhesion between the two

◀ **Kinetic friction**

pieces of metal: the atoms in the block form bonds with the atoms in the plate, and when the block slides, these bonds are continually ruptured and formed again. The macroscopic friction force represents the effort required to rupture the microscopic bonds. Although at the microscopic level the phenomenon of friction is very complicated, at the macroscopic level the resulting friction force can often be described adequately by a simple empirical law, first enunciated by Leonardo da Vinci:

The magnitude of the friction force between unlubricated, dry surfaces sliding one over the other is proportional to the normal force acting on the surfaces and is independent of the area of contact and of the relative speed.

According to this law, the magnitude of the force of kinetic friction can be written mathematically as

$$f_k = \mu_k N \tag{52}$$

Coefficient of kinetic friction ▶

where μ_k is the ***coefficient of kinetic friction,*** a constant characteristic of the material involved. Table 3.5 lists typical friction coefficients for various materials.

Note that Eq. (52) states that the magnitudes of the friction force and the normal force are proportional. ***The directions of these forces are, however, quite different:*** the normal force $\mathbf{N}$ is perpendicular to the surface of contact, whereas the friction force $\mathbf{f}_k$ is parallel to this surface, in a direction opposite to that of the motion.

The above simple "law" of friction lacks the general validity of, say, Newton's laws. It is only approximately valid and it is phenomenological, which means that it is merely a descriptive summary of empirical observations which does not rest on any detailed theoretical understanding of the mechanism that causes friction. Deviations from this simple law occur at high speeds and at low speeds. However, we can ignore these deviations

TABLE 3.5 KINETIC AND STATIC FRICTION COEFFICIENTS[a]

Materials	μ_k	μ_s
Steel on steel	0.6	0.7
Steel on lead	0.9	0.9
Steel on copper	0.4	0.5
Copper on cast iron	0.3	1.1
Copper on glass	0.5	0.7
Waxed ski on snow		
at −10°C	0.2	—
at 0°C	0.05	—
Rubber on concrete	≃1	≃1

[a] The friction coefficient depends on the condition of the surfaces. The values in this table are typical for dry surfaces but not entirely reliable.

in many everyday engineering problems in which the speeds are not extreme. The simple friction law is then a reasonably good approximation for a wide range of materials, and it is at its best for metals.

The fact that the friction force is independent of the area of contact means that the friction force of the block sliding on the plate is the same whether the block slides on a large face or on one of the small faces (see Fig. 3.32). This may seem surprising at first — we might expect the friction force to be larger when the block slides on the larger face, with more area in contact with the plate. However, the normal force is then distributed over a larger area, and is therefore less effective in pressing the atoms together; the net result is that the friction force is independent of the area of contact.

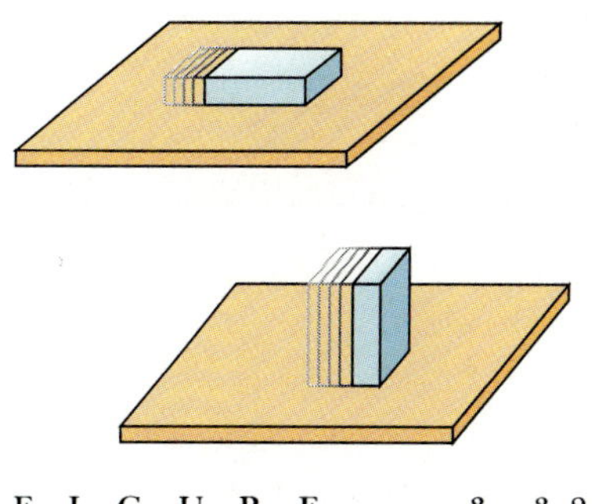

FIGURE 3.32
Steel block on a steel plate, sliding on a large face or on a small face.

EXAMPLE 9. Suppose that the coefficient of kinetic friction of the hard rubber of an automobile tire sliding on the pavement of a street is $\mu_k = 0.8$. What is the deceleration of an automobile on a flat street if the driver brakes sharply, so all the wheels are locked and skidding?

SOLUTION: Figure 3.33 shows a "free-body" diagram with all the forces on the automobile. These forces are the weight **w**, the normal force **N** exerted by the street, and the friction force $\mathbf{f}_k$. The normal force must balance the weight; hence the magnitude of the normal force is the same as the magnitude of the weight, $N = mg$. According to Eq. (52), the magnitude of the friction force is then

$$f_k = \mu_k N = 0.8 \times mg$$

Since this friction force is the only horizontal force on the automobile, the deceleration of the automobile along the street is

$$a_x = -\frac{f_k}{m} = -\frac{0.8 \times mg}{m} = -0.8 \times g$$

$$= -0.8 \times 9.81 \text{ m/s}^2 = -8 \text{ m/s}^2$$

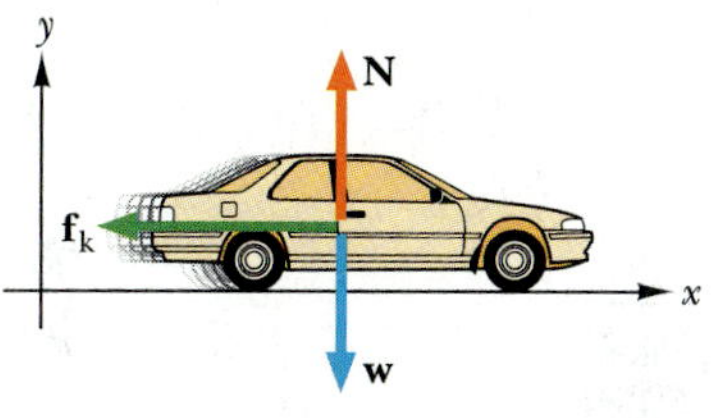

FIGURE 3.33
"Free-body" diagram for an automobile skidding with locked wheels. A sliding friction force acts on each wheel; but in the diagram all these forces are shown combined in a single force $\mathbf{f}_k$.

EXAMPLE 10. A ship is launched toward the water on a slipway making an angle of 5° with the horizontal direction (see Fig. 3.34a). The coefficient of kinetic friction between the bottom of the ship and the slipway is $\mu_k = 0.08$. What is the acceleration of the ship along the slipway?

SOLUTION: Figure 3.34b is a "free-body" diagram for the ship. The forces shown are the weight **w**, the normal force **N**, and the friction

(a)

(b)

(c)

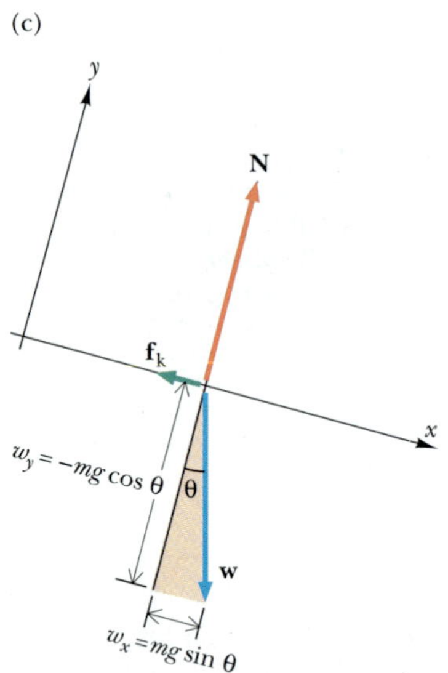

FIGURE 3.34

(a) Ship on slipway. (b) "Free-body" diagram for the ship. (c) The x and y components of the force **w**.

force $\mathbf{f}_k$. The magnitude of the weight is

$$w = mg$$

Since there is no motion in the direction perpendicular to the slipway, we find, as in Eq. (35), that the normal force is

$$N = mg\cos\theta \tag{53}$$

And the magnitude of the friction force is

$$f_k = \mu_k N = \mu_k mg\cos\theta \tag{54}$$

With the x axis parallel to the slipway, the x component of the net force (Fig. 3.34c) is

$$F_x = w_x - f_k = mg\sin\theta - \mu_k mg\cos\theta$$

Hence the acceleration of the ship along the slipway is

$$a_x = \frac{F_x}{m} = \frac{mg\sin\theta - \mu_k mg\cos\theta}{m}$$

$$= (\sin\theta - \mu_k\cos\theta)g \tag{55}$$

Note that in this equation the mass has canceled — the acceleration is the same for a large ship and a small ship. With $\theta = 5^0$ and $\mu_k = 0.08$, Eq. (55) gives

$$a_x = (\sin 5^0 - 0.08 \times \cos 5^0) \times 9.81 \text{ m/s}^2$$

$$= 0.073 \text{ m/s}^2$$

Friction forces also act between two surfaces at rest. If we exert a force against the side of, say, a steel block initially at rest on a steel plate, the block will not move unless the force is sufficiently large to overcome the friction that holds it in place. ***Friction between surfaces at rest is called static friction.*** The maximum magnitude of the static friction force, that is, the magnitude that this force attains when the lateral push is just about to start the motion, can be described by an empirical law quite similar to that for the kinetic friction force:

> ***The magnitude of the maximum friction force between unlubricated, dry surfaces at rest with respect to each other is proportional to the normal force and independent of the area of contact.***

Mathematically,

Static friction ▶

$$f_{s,\max} = \mu_s N \tag{56}$$

Here μ_s is a constant of proportionality, called the ***coefficient of static friction,*** which depends on the material. The direction of the static friction force is parallel to the surface, so as to oppose the lateral push that tries to move the body (see Fig. 3.35).

◂ **Coefficient of static friction**

The force in Eq. (56) is labeled with the subscript "max" because it represents the largest friction force that the surfaces can support without beginning to slide; in other words, $f_{s,max}$ is the force at the "breakaway" point, when the lateral push is just about to start the motion. Of course, ***if the lateral push is less than this critical value, then the static friction force*** f_s ***is less than*** $f_{s,max}$ ***and exactly matches the magnitude of the lateral push.***

Table 3.5 includes some typical values of the coefficient of static friction. ***For most materials*** μ_s ***is larger than*** μ_k, and therefore the maximum static friction force is larger than the kinetic friction force. This implies that if the lateral push applied to the block is large enough to overcome the static friction and start the block moving, it will more than compensate for the subsequent kinetic friction, and it will therefore accelerate the block continuously.

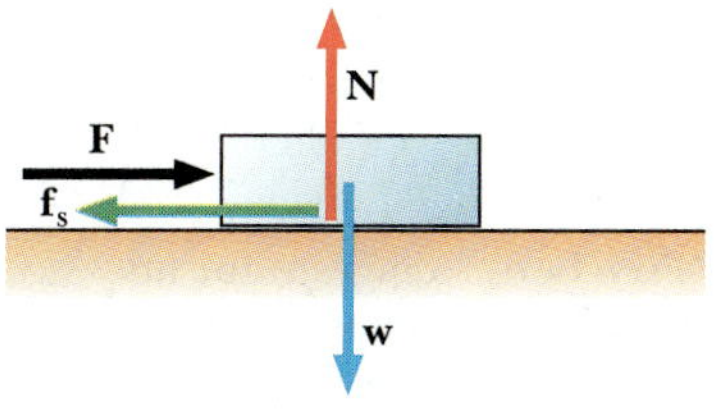

FIGURE 3.35

Forces on a steel block at rest on a steel plate. The friction force is in a direction opposite to the push.

EXAMPLE 11. The coefficient of static friction of the hard rubber of automobile tires on a street surface is $\mu_s = 0.9$. What is the steepest slope of a street on which an automobile with such tires (and locked wheels) can rest without slipping?

SOLUTION: The "free-body" diagram is shown in Fig. 3.36. The angle θ is assumed to be at its maximum value, so that the friction force has its maximum value $f_{s,max} = \mu_s N$. As in Example 10, $N = mg\cos\theta$, and hence $f_{s,max} = \mu_s mg\cos\theta$. With the x axis parallel to the street surface, we then find that the x component of the net force is

$$F_x = w_x - f_{s,max} = mg\sin\theta - \mu_s mg\cos\theta \tag{57}$$

This component of the force determines the motion along the street. If the automobile is to remain stationary, F_x must be zero:

$$0 = mg\sin\theta - \mu_s mg\cos\theta$$

Hence

$$\sin\theta = \mu_s\cos\theta$$

or, dividing this equation by $\cos\theta$,

$$\tan\theta = \mu_s \tag{58}$$

With $\mu_s = 0.9$, this gives $\tan\theta = 0.9$. Thus, the slope of the street is 0.9, or 9:10. (With a calculator, we find that the angle with this tangent of 0.9 is 42°.)

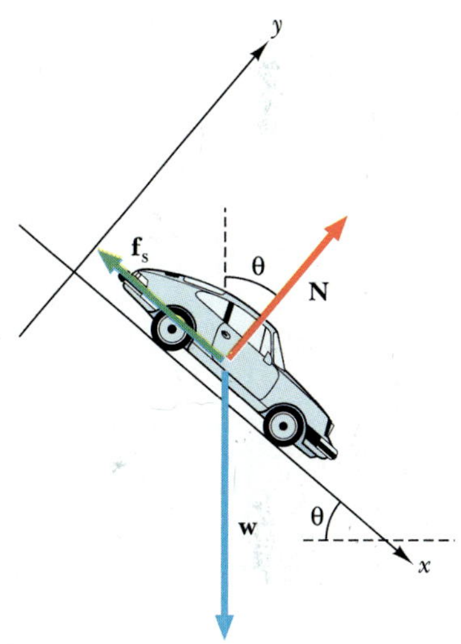

FIGURE 3.36

"Free-body" diagram for an automobile parked on a very steep street. A static friction force acts at the point of contact of each wheel with the ground; these forces are shown combined in a single force $\mathbf{f}_s$.

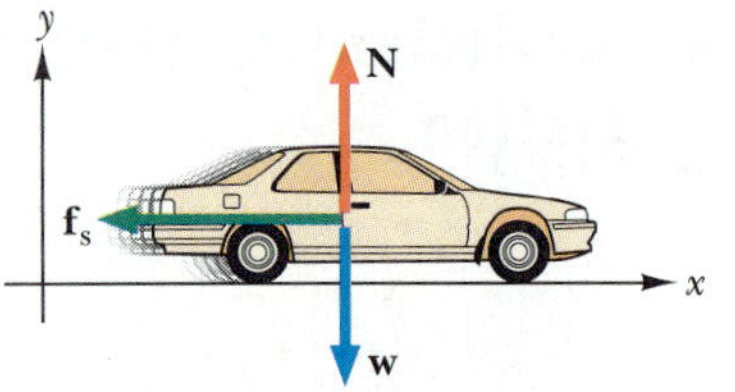

FIGURE 3.37
"Free-body" diagram for an automobile braking without skidding. A static friction force acts at the instantaneous point of contact of each wheel with the ground; again, these forces are shown combined in a single force $\mathbf{f}_s$.

EXAMPLE 12. An automobile is braking on a level road. What is the maximum deceleration that the automobile can achieve without skidding? As in the preceding example, assume that the tires of the automobile have a coefficient of static friction $\mu_s = 0.9$.

SOLUTION: Figure 3.37 shows the "free-body" diagram. If the wheels are rolling without skidding, their rubber surface *does not slide* on the street surface (the point of contact between the rolling wheel and the street is instantaneously at rest on the street; you can easily convince yourself of this by rolling any round object on a tabletop). Since there is no sliding, the relevant friction force is the *static* friction force. The maximum value of this force is

$$f_{s,\mathrm{max}} = \mu_s N \tag{59}$$

Here the magnitude of the normal force N is simply mg, since the normal force must balance the weight. The deceleration is then given by

$$a_x = -\frac{f_{s,\mathrm{max}}}{m} = -\mu_s \frac{mg}{m} = -\mu_s g \tag{60}$$

which yields

$$a_x = -0.9 \times 9.81 \text{ m/s}^2 = -9 \text{ m/s}^2$$

3.8 RESTORING FORCE OF A SPRING; HOOKE'S LAW

Elastic body ▶

A body is said to be ***elastic if it suffers a deformation when subjected to a stretching or compressing force, and if it returns to its original shape when the force is removed.*** For example, suitable forces can stretch a coil spring or a rubber band, and they can bend a flexible rod or a beam of metal or wood. Even bodies normally regarded as rigid, such as the balls of a ball bearing made of hardened steel, are somewhat elastic — they will deform if subjected to a sufficiently large force.

The force with which a body resists deformation is called its restoring force. If we stretch a spring by pulling with a hand at one end (see Fig. 3.38), we can feel the restoring force opposing our pull.

FIGURE 3.38
The restoring force of the spring opposes the pull exerted by the hand.

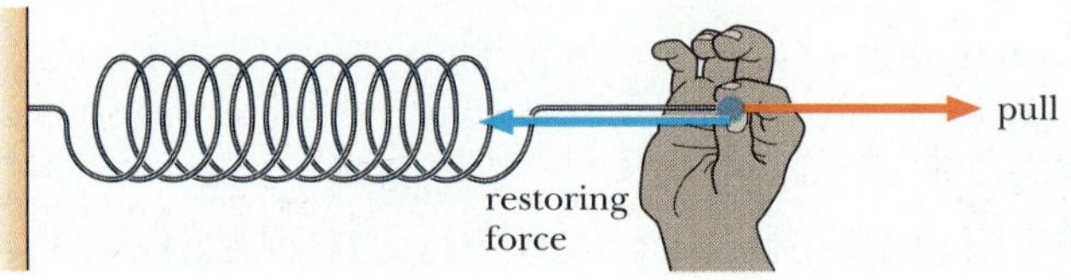

Under static conditions, the restoring force with which an elastic body opposes whatever pulls on it often obeys a simple empirical law known as ***Hooke's Law:***

◀ **Hooke's Law**

> ***The magnitude of the restoring force is directly proportional to the deformation.***

This is not a general law of physics — the exact restoring force produced by the deformation of an elastic body depends in a complicated way on the shape of the body and on the detailed properties of the material of the body. Hooke's Law is only an approximate description of the restoring force. However, it is often a quite good approximation, provided the restoring force and the deformation are small.

As an example, consider a coil spring. Figure 3.39a shows such a spring in its relaxed state; it is loosely coiled and can therefore be compressed as well as stretched. Suppose we attach the left end of the spring to a wall and we apply a stretching or compressing force to the right end. Under the influence of this force, the spring will settle into a new equilibrium configuration such that the restoring force exactly balances the externally applied force. We can measure the deformation of the spring by the displacement that the right end undergoes relative to its initial position. In Figs. 3.39b and 3.39c, this displacement is denoted by x. Clearly, x is simply the change in the length of the spring. A positive value of x corresponds to an elongation of the spring, and a negative value of x corresponds to a compression.

(a)

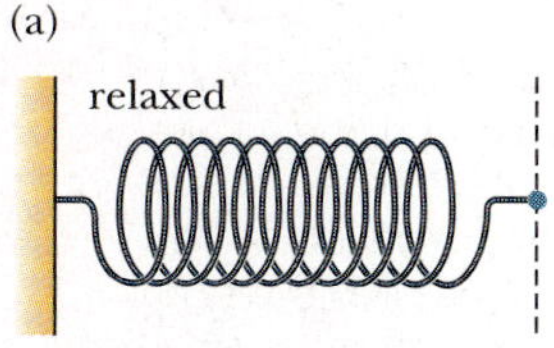

(b)

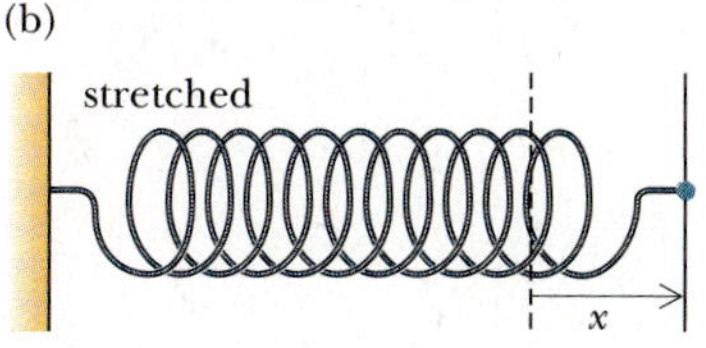

(c)

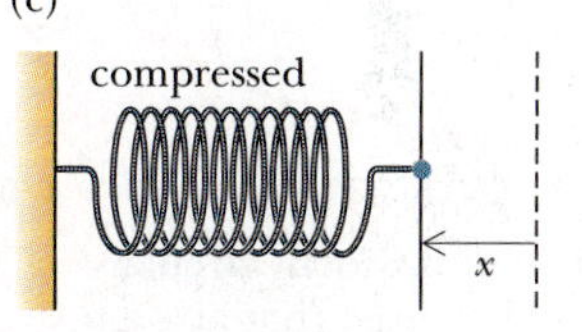

FIGURE 3.39

(a) Spring, relaxed. (b) Spring, stretched by a length x. (c) Spring, compressed by a length x.

Expressed mathematically, Hooke's Law then says that ***the restoring force is directly proportional to the displacement*** x,

$$F = -kx \tag{61}$$

The constant of proportionality k is the ***spring constant;*** it is a positive number characteristic of the spring. The spring constant is a measure of the stiffness of the spring — a stiff spring has a high value of k, and a soft spring has a low value of k. The units for the spring constant are newtons per meter (N/m). The negative sign in Eq. (61) indicates that the restoring force opposes the deformation: if the spring is elongated (positive x), then the restoring force is negative and opposes the external stretching force; if the spring is compressed (negative x), then the restoring force is positive and opposes the external compressing force.

EXAMPLE 13. The manufacturer's specifications for the coil spring for the front suspension of a Triumph sports car call for a spring with a relaxed length of 0.316 m, and a length of 0.205 m when under a load of 399 kg. What is the spring constant?

SOLUTION: The weight of 399 kg is $w = mg = 399\ \text{kg} \times 9.81\ \text{m/s}^2 = 3.91 \times 10^3$ N. The magnitude of the restoring force that will balance this weight must then also be 3.91×10^3 N. The corresponding change of length is 0.205 m − 0.316 m = −0.111 m. Hence, Eq. (61) gives us

$$k = -\frac{F}{x} = -\frac{3.91 \times 10^3 \text{ N}}{-0.111 \text{ m}} = 3.52 \times 10^4 \text{ N/m} \tag{62}$$

3.9 FORCE FOR UNIFORM CIRCULAR MOTION

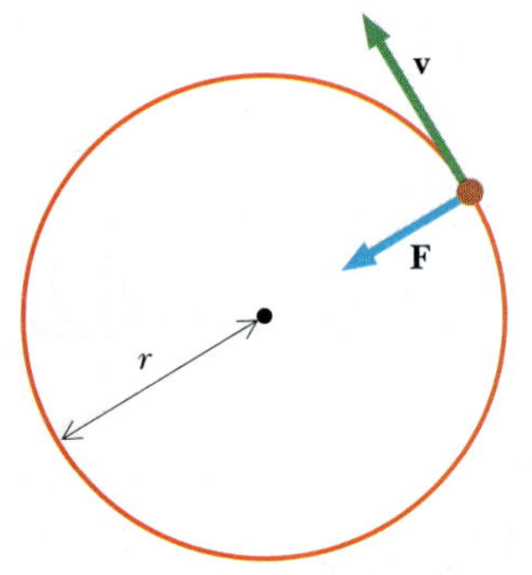

FIGURE 3.40

A particle in uniform circular motion. The force that acts on the particle is directed toward the center of the circle.

As we saw in Section 2.7, uniform circular motion of a particle is accelerated motion with a centripetal acceleration. If the motion proceeds with speed v around a circle of radius r, Eq. (2.62) tells us that the magnitude of the centripetal acceleration is

$$a = v^2/r \tag{63}$$

According to Newton's Second Law, this acceleration must be caused by a force having the same direction as that of the acceleration; that is, ***the direction of the force must be toward the center of the circle*** (Fig. 3.40). For instance, the centripetal acceleration of a stone being whirled around a circle at the end of a string is caused by the pull of the string toward the center of the circle (see Fig. 3.41).

The magnitude of the force required to maintain uniform circular motion is

Force required for uniform circular motion ▶

$$F = ma = \frac{mv^2}{r} \tag{64}$$

This equation can be used to calculate the magnitude of the force if the speed of the motion is known, or it can be used to calculate the speed if the force is known. The following examples illustrate such calculations with different kinds of forces.

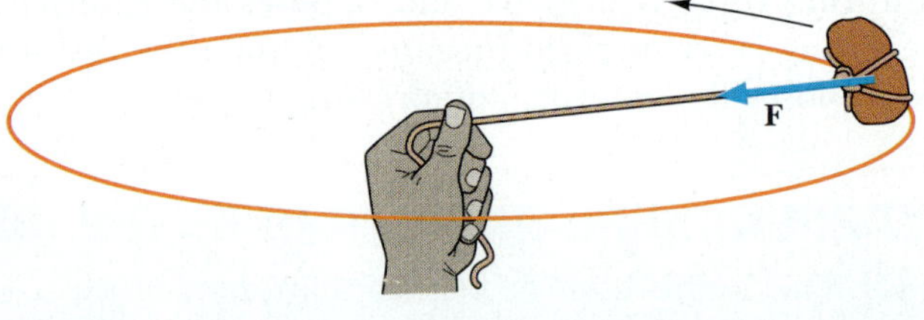

FIGURE 3.41

The string must exert a pull toward the center of the circle to produce a centripetal acceleration and to keep the stone in uniform circular motion.

EXAMPLE 14. In the hammer throw, an athlete launches a "hammer" consisting of a heavy metal ball attached to a handle by a steel cable. Just before launching the hammer, the athlete swings it around several times in a circle. The mass of the ball is 7.3 kg, the distance from the hammer to the center of its circular motion is 1.9 m (including the length of the athlete's arms; see Fig. 3.42). The speed of the hammer is 27 m/s. What is the centripetal force that the athlete must exert to keep the hammer moving in its circle?

SOLUTION: According to Eq. (64), the magnitude of the force must be

$$F = ma = mv^2/r = [7.3 \text{ kg} \times (27 \text{ m/s})^2]/(1.9 \text{ m}) = 2.8 \times 10^3 \text{ N} \qquad (65)$$

This is a rather large force! Hammer throwing requires great physical strength, and hammer throwers must be of hefty build.

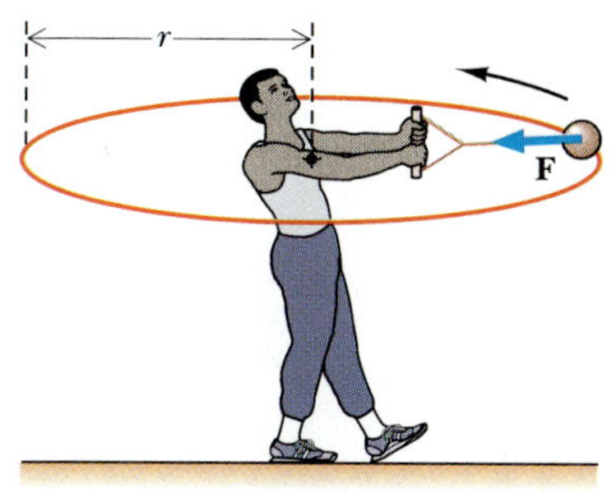

FIGURE 3.42
Circular motion of the hammer.

EXAMPLE 15. What is the maximum speed with which an automobile can round a curve of radius 100 m without skidding? Assume that the road is flat and that the coefficient of static friction between the tires and road surface is $\mu_s = 0.8$.

SOLUTION: The "free-body" diagram for the automobile is given in Fig. 3.43. The forces on the automobile are the weight **w**, the normal force **N**, and the friction force $\mathbf{f}_s$. The weight balances the normal force, that is, $N = mg$. The horizontal friction force must provide the centripetal force; hence the magnitude of the friction force must be

$$f_s = ma = mv^2/r \qquad (66)$$

The friction is *static* because, by assumption, there is no lateral slippage. At the maximum possible speed, the friction force has its maximum value

$$f_s = f_{s,\max} = \mu_s mg$$

and consequently

$$\mu_s mg = mv^2/r$$

We can cancel the masses on both sides of this equation and take the square root of both sides of the equation; this yields

$$\sqrt{\mu_s g} = \frac{v}{\sqrt{r}}$$

from which

$$v = \sqrt{\mu_s g r} \qquad (67)$$

$$= \sqrt{0.8 \times 9.81 \text{ m/s}^2 \times 100 \text{ m}} = 28 \text{ m/s}$$

This is about 100 km/h.

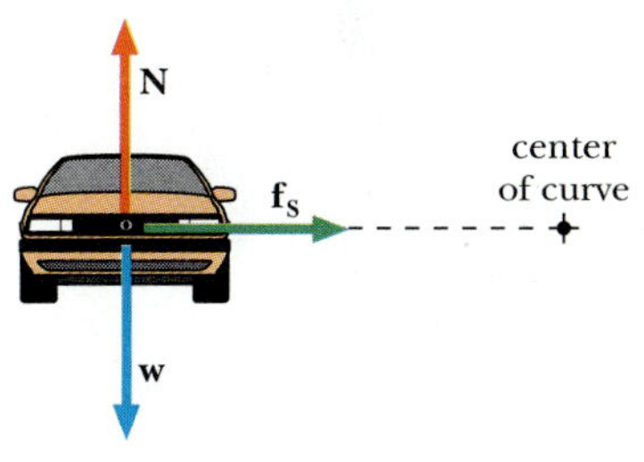

FIGURE 3.43
"Free-body" diagram for an automobile rounding a curve.

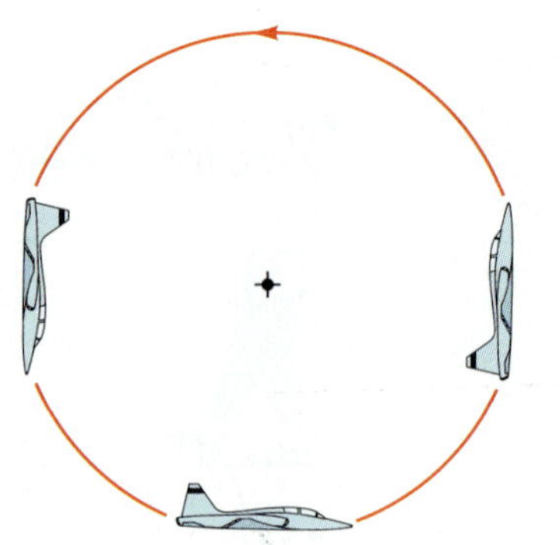

FIGURE 3.44
Jet looping the loop.

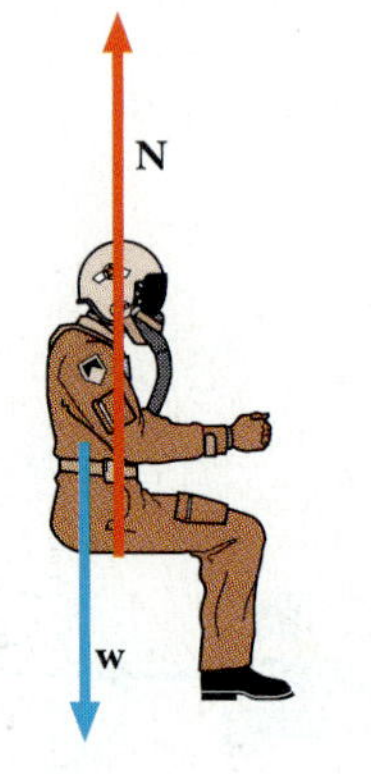

FIGURE 3.45
"Free-body" diagram for jet pilot at the bottom of the loop.

EXAMPLE 16. A pilot in a fast jet aircraft loops the loop (see Fig. 3.44). The radius of the loop is 1.5 km, and the aircraft has a speed of 200 m/s when it passes through the bottom of the loop. What is the apparent weight that the pilot feels; in other words, what is the force with which the seat of his pants presses against his chair? Express the answer as a multiple of his normal weight.

SOLUTION: Figure 3.45 shows a "free-body" diagram for the pilot at the bottom of the loop. The forces acting on him are the (true) weight **w** and the normal force **N** exerted by the chair. The net vertical upward force is $N - mg$, and this must provide the centripetal acceleration:

$$N - mg = ma = \frac{mv^2}{r} \tag{68}$$

Note that here we used the familiar formula v^2/r for the centripetal acceleration, even thought the speed is not constant (the speed of the aircraft increases somewhat as it goes down the loop, and the speed decreases as it goes up the loop). Such a change of speed along the loop implies that there is an extra acceleration *along* the loop. But this extra acceleration does not affect the centripetal acceleration — the two accelerations are at right angles, and they are independent. For the purposes of our problem, we do not need to pay any attention to the extra acceleration along the loop.

Solving Eq. (68) for N, we obtain

$$N = mg + \frac{mv^2}{r} = mg\left(1 + \frac{v^2}{gr}\right) \tag{69}$$

With $v = 200$ m/s and $r = 1500$ m, we then find

$$N = mg\left(1 + \frac{(200\text{ m/s})^2}{9.81\text{ m/s}^2 \times 1500\text{ m}}\right)$$

$$= mg \times 3.7$$

This is the force with which the chair presses against the pilot, and it is therefore the apparent weight of the pilot. The apparent weight equals his normal weight multiplied by a factor of 3.7. In these circumstances, the pilot would say he is experiencing 3.7 "gee," because he feels as through gravity had been magnified by a factor of 3.7.

This example shows that in the reference frame of a body in circular motion, ***the centripetal acceleration generates an effective increase of weight*** — the apparent weight of a body in such a rotating reference frame is larger than its normal weight. The operation of centrifuges and ultracentrifuges hinges on this effective increase of weight. For instance, consider a test tube with some liquid, which is being spun in a horizontal circle in a centrifuge (see Fig. 3.46). Suppose the liquid contains some particles in suspension; for instance, the liquid might be blood, consisting of a suspension of red blood corpuscles (and other corpuscles) in liquid blood plasma. The horizontal force required to keep a corpuscle of mass m suspended in a fixed position relative to the test tube is mv^2/r in the centripetal direction. Therefore, the apparent weight with which the corpuscle pushes against the plasma is mv^2/r in the outward, or centrifugal, direction (besides this apparent weight mv^2/r, there is also a true weight mg, in the vertical direction; but in a high-speed centrifuge, the true weight is negligible compared with the apparent weight). ***The apparent weight mv^2/r is called the centrifugal force.*** Since the corpuscle is more dense than the plasma, its large apparent weight will enable it to shoulder the plasma aside, and to settle quickly against the outermost wall of the test tube. Thus, the centrifuge accomplishes a quick segregation of the contents of the test tube into layers of different density, just as though the test tube had been subjected to a manifold enhancement of gravity. Ultracentrifuges (see Fig. 3.47) spinning at up to 100,000 revolutions per minute generate apparent centrifugal forces of up to 500,000 times the true weight; they are used in chemical and biochemical research. Special ultracentrifuges can separate even different isotopes of chemical elements, and they are used to separate the isotopes of uranium.

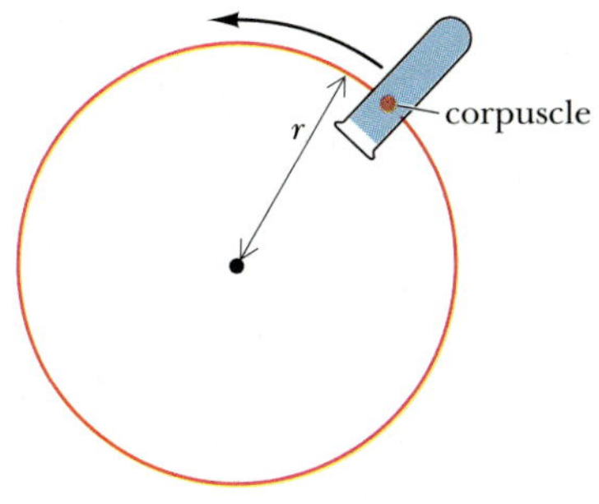

FIGURE 3.46

The test tube rotates around a circle of radius r. A corpuscle of mass m is suspended in the liquid in the test tube. A centripetal force mv^2/r is required to keep this particle moving on the circular path.

FIGURE 3.47

An ultracentrifuge.

SUMMARY

- **Newton's First Law:** In an inertial reference frame, a body at rest remains at rest and a body in motion continues to move at constant velocity unless acted upon by an external force.
- **Definition of mass:** $m/m_s = a_s/a$
- **Newton's Second Law:** $m\mathbf{a} = \mathbf{F}_{net}$
- **Superposition of Forces:** $\mathbf{F}_{net} = \mathbf{F}_1 + \mathbf{F}_2 + \mathbf{F}_3 + \ldots$
- **Newton's Third Law:** Whenever a body exerts a force on another body, the latter exerts a force of equal magnitude and opposite direction on the former.
- **The four fundamental forces:** gravitational, "weak," electromagnetic, "strong"
- **Weight:** magnitude: $w = mg$
  direction: downward
- **Kinetic friction:** $f_k = \mu_k N$
- **Static friction:** $f_{s,max} = \mu_s N$
- **Restoring force of a spring (Hooke's Law):** $F = -kx$
- **Force required for uniform circular motion:** magnitude: mv^2/r
  direction: centripetal

QUESTIONS

1.–Estimate the masses, in grams or kilograms, of a few bodies in your environment. Check the masses with a balance if you have one available.

2.–Why does a pound of feathers give the impression of being less massive than a pound of lead?

3.–If a glass stands on a smooth table on top of a sheet of paper, you can remove the paper without touching the glass by jerking the paper away very sharply. Explain why the glass more or less stays put.

4.–Suppose that a (strange) body has negative mass. Suppose you connect this body to a body of positive mass of the same magnitude by means of a rubber band stretched between the bodies. Describe the motion of the two bodies.

5.–Does the magnitude or the direction of a force depend on the frame of reference?

6.–A fisherman wants to reel in a large dead shark hooked on a thin fishing line. If he jerks the line, it will break; but if he reels it in very gradually and smoothly, it will hold. Explain.

7.–If a body crashes into a water surface at high speed, the impact is almost as hard as on a solid surface. Explain.

8.–In a tug-of-war, two teams of children pull on a rope (Fig. 3.48). Is the tension constant along the entire length of the rope? Along what portion of the rope is it constant?

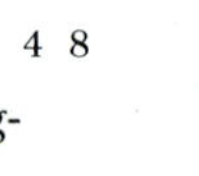

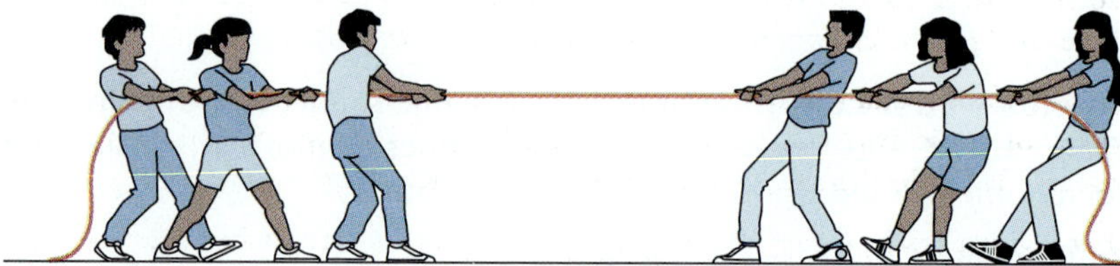

FIGURE 3.48
Two teams of children in a tug-of-war.

9.–When you are standing on the Earth, your feet exert a force (push) against the surface. Why does the Earth not accelerate away from you?

10.–On a windless day, a sailor puts an electric fan, powered by a battery, on the stern of the boat and blows a stream of air into his sail. Will the boat move forward?

11.–When an automobile accelerates on a level road, the force that produces this acceleration is the push of the road on the wheels (see Fig. 3.12). If so, why does the automobile need an engine?

12.–You are in a small boat in the middle of a calm lake. You have no oars, and you cannot put your hands in the water because the lake is full of piranhas. The boat carries a large load of coconuts. How can you get to the shore?

13.–According to the adherents of parapsychology, some people are endowed with the supernormal power of psychokinesis—that is, spoon-bending-at-a-distance via mysterious psychic forces emanating from the brain. Physicists are confident that the only forces acting between pieces of matter are those listed in Section 3.5, none of which are implicated in psychokinesis. Given that the brain is nothing but a (very complicated) piece of matter, what conclusions can a physicist draw about psychokinesis?

14.–If you carry a spring balance from London to Hong Kong, do you have to recalibrate it? If you carry a beam balance?

15.–An automobile is parked on a street. (a) Draw a "free-body" diagram showing the forces acting on the automobile. What is the net force? (b) Draw a "free-body" diagram showing the forces that the automobile exerts on the Earth. Which of the forces in diagrams (a) and (b) are action–reaction pairs?

16.–Some old-time roofers claim that when walking on a rotten roof, it is important to "walk with a light step so that your full weight doesn't rest on the roof." Can you walk on a roof with less than your full weight? What is the advantage of a light step?

17.–The label on a package of sugar claims that the contents are "1 lb or 454 g by weight." What is wrong with this statement?

18.–A physicist stands on a bathroom scale in an elevator. When the elevator is stationary, the scale reads 60 kg. Describe qualitatively how the reading of the scale will fluctuate while the elevator makes a trip to a higher floor.

19.–How could you use a pendulum suspended from the roof of your automobile to measure its acceleration?

20.–Seen from a reference frame moving with the wave, the motion of a surfer is analogous to the motion of a skier down a mountain.* If the wave were to last forever, could the surfer ride it forever? In order to stay on the wave as long as possible, in what direction should the surfer ski the wave?

21.–Some drivers like to spin the wheels of their automobiles for a quick start. Does this give them greater acceleration? (Hint: $\mu_s > \mu_k$.)

22.–Cross-country skiers like to use a ski wax that gives their skis a large coefficient of static friction, but a low coefficient of kinetic friction. Why is this useful? How do "waxless" skis achieve the same effect?

23.–For a given initial speed, the stopping distance of a train is much longer than that of a truck. Why?

24.–Why does the traction on snow or ice of an automobile with rear-wheel drive improve when you place extra weight over the rear wheels?

25.–Why are wet streets slippery?

26.–In order to stop an automobile on a slippery street in the shortest distance, it is best to brake as hard as possible without initiating a skid. Why does skidding lengthen the stopping distance? (Hint: $\mu_s > \mu_k$.)

27.–Suppose that in a panic stop, a driver locks the wheels of his automobile and leaves skid marks on the pavement. How can you deduce his initial speed from the length of the skid marks?

28.–When rounding a curve in your automobile, you get the impression that a force tries to pull you toward the outside of the curve. Is there such a force?

---

* There is, however, one complication: surf waves grow higher as they approach the beach. Ignore this complication.

(a)

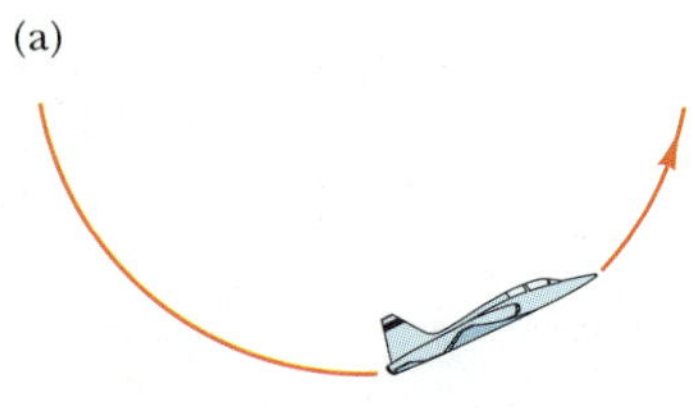

(b)

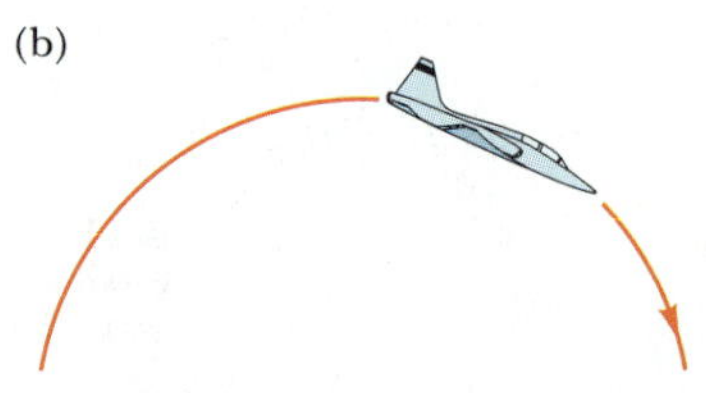

FIGURE 3.49
Aircraft (a) pulling out of a dive (b) beginning a dive.

Pilot in a G suit.

29.–(a) If a pilot in a fast aircraft very suddenly pulls out of a dive (Fig. 3.49a), he will suffer "blackout" caused by loss of blood pressure in the brain. If he suddenly begins a dive while climbing (Fig. 3.49b), he will suffer "redout" caused by excessive blood pressure in the brain. Explain.
(b) A pilot wearing a G suit—a tightly fitting garment that squeezes the tissues of the legs and abdomen—can tolerate 50 m/s² while pulling out of a dive. How does the G suit prevent blackout? A pilot can tolerate no more than −20 m/s² while beginning a dive. Why does the G suit not help against redout?

30.–If the Earth were to stop spinning (other things remaining equal), the value of g at all points of the surface except the poles would become slightly larger. Why?

PROBLEMS

Section 3.2: Newton's Second Law

1.—What is your mass in pounds? In kilograms? What is your weight in pounds-force? In newtons?

2.—What percentage of the mass of the Solar System is in the planets? What percentage is in the Sun? Use the data given in the table printed on the endpapers of the book.

3.—According to the *Guinness Book of Records,* the heaviest man ever had a confirmed mass of 975 pounds. Express this in kilograms.

4.*—The hydrogen atom consists of one proton and one electron. What is the mass of one hydrogen atom? How many hydrogen atoms are there in 1 kg of hydrogen gas?

5.*—The oxygen atom consists of 8 protons, 8 neutrons, and 8 electrons. What is the mass of one oxygen atom? How many oxygen atoms are there in 1 kg of oxygen gas?

6.*—On a flat road, a Maserati sports car can accelerate from 0 to 80 km/h (0 to 50 mi/h) in 5.8 s. The mass of the car is 1620 kg. What are the average acceleration and the average force on the car?

7.*—Pushing with both hands, a sailor standing on a pier exerts a horizontal force of 300 N on a destroyer of 1700 metric tons. Assuming that the mooring ropes do not interfere and that the water offers no resistance, what is the acceleration of the ship? How far does the ship move in 60 s?

8.*—A woman of 57 kg is held firmly in the seat of her automobile by a lap-and-shoulder seat belt. During a collision, the automobile decelerates from 50 to 0 km/h in 0.12 s. What is the average force that the seat belt exerts on the woman?

9.*—A heavy freight train has a total mass of 16,000 metric tons. The locomotive exerts a pull of 670,000 N on this train. What is the acceleration? How long does it take to increase the speed from 0 to 50 km/h?

10.*—With brakes fully applied, a 1500-kg automobile decelerates at the rate of 8.0 m/s² on a flat road. What is the braking force acting on the automobile? Draw a diagram showing the direction of motion of the automobile and the direction of the braking force.

11.—Consider the impact of the automobile on a barrier described in Example 1.8. If the mass of the automobile is 1400 kg, what is the force acting on the automobile during the deceleration?

12.*—The projectile fired by the gun described in Problem 1.53 has a mass of 45 kg. What is the average force on this projectile as it moves along the barrel?

13.**—Figure 1.28 shows the plot of velocity vs. time for a Triumph sports car coasting along with its gears in neutral. From the value of the deceleration at the time $t = 40$ s (see Problem 1.51b), calculate the friction force that the car experiences at this time. The mass of the car is 1160 kg.

Section 3.3: The Combination of Forces

14.—While braking, an automobile of mass 1200 kg decelerates along a level road at 8.0 m/s^2. What is the horizontal force that the road exerts on the automobile? The vertical force? What are the magnitude and direction of the net force that the road exerts on the automobile? Ignore the friction of the air.

15.—In 1978, in an accident at a school in Harrisburg, Pennsylvania, several children lost parts of their fingers when a nylon rope suddenly snapped during a giant tug-of-war among 2300 children. The rope was known to have a breaking tension of 5.8×10^4 N. Each child can exert a pull of approximately 130 N. Was it safe to employ this rope in this tug-of-war?

16.*—The Earth exerts a gravitational pull of 2.0×10^{20} N on the Moon; the Sun exerts a gravitational pull of 4.3×10^{20} N on the Moon. What is the net force on the Moon when the angular separation between the Earth and the Sun is 90° as seen from the Moon?

17.*—A sailboat is propelled through the water by the combined action of two forces: the push ("lift") of the wind on the sail and the push of the water on the keel. Figure 3.50 shows the magnitudes and the directions of these forces acting on a medium-sized sailboat (this oversimplified diagram does not include the drag of wind and water). What is the resultant of the forces in Fig. 3.50?

18.**—A boat is tied to a dock by four (horizontal) ropes. Two ropes, with a tension of 270 N each, are at right angles to the dock. Two other ropes, with a tension of 360 N each, are at an angle of 20° with the dock (Fig. 3.51). What is the resultant of these forces?

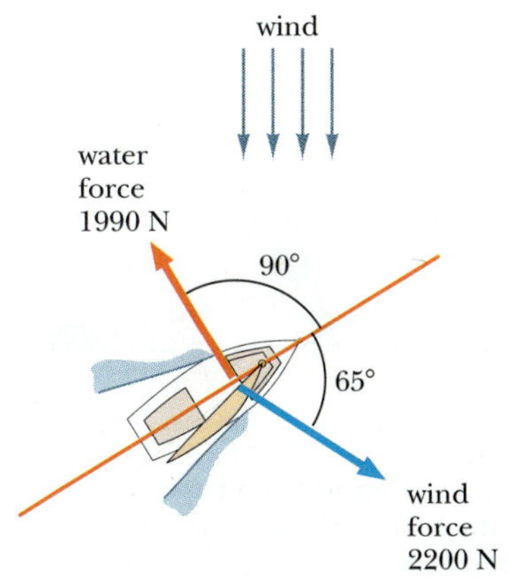

FIGURE 3.50

Forces on a sailboat. The angles are measured relative to the line of motion.

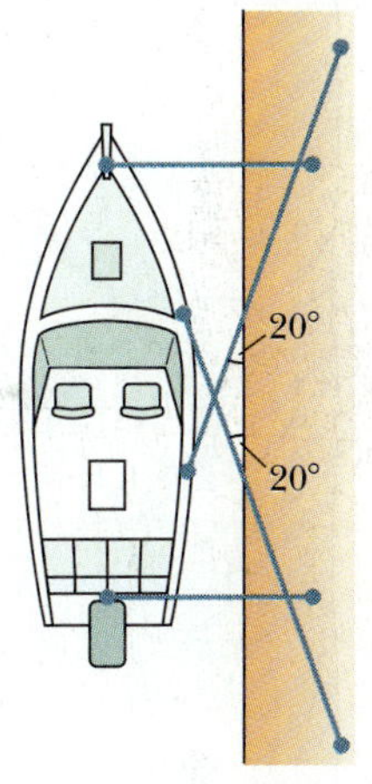

FIGURE 3.51

Ropes holding a boat to a dock.

Section 3.4: Newton's Third Law

19.*—A long freight train consists of 250 cars each of mass 64 metric tons. The pull of the locomotive accelerates this train at the rate of 0.043 m/s^2 along a level track. What is the tension in the coupling that holds the first car to the locomotive? What is the tension in the coupling that holds the last car to the next-to-last car? Ignore friction.

20.*—Two heavy boxes of masses 20 kg and 30 kg sit on a smooth, frictionless surface. The boxes are in contact and a force of 60 N pushes horizontally against the smaller box (Fig. 3.52). What is the acceleration of the two boxes? What is the force that the smaller box exerts on the larger box? What is the force that the larger box exerts on the smaller box?

21.*—An archer pulls the string of her bow back with her hand with a force of 180 N. If the two halves of the string above and below her hand make an angle of 120° with each other, what is the tension in each half of the string?

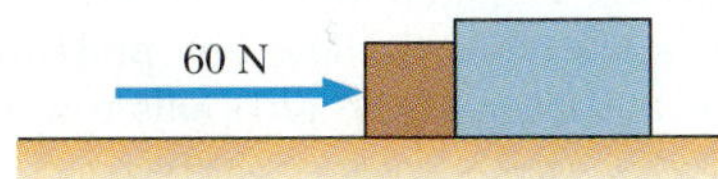

FIGURE 3.52

Two boxes in contact.

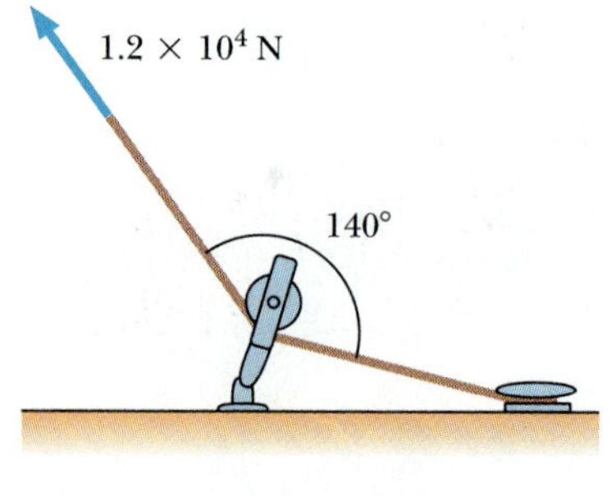

FIGURE 3.53
The left end of the rope is attached to the sail; the right end is attached to the cleat.

22.*—On a sailboat, a rope holding the foresail passes through a block (a pulley) and is made fast on the other side to a cleat (Fig. 3.53). The two parts of the rope make an angle of 140° with each other. The sail pulls on the rope with a force of 12,000 N. What is the force that the rope exerts on the block?

23.**—A crate of mass 2000 kg is hanging from a crane at the end of a cable 12 m long. If we attach a horizontal rope to this crate and gradually apply a pull of 1800 N, what angle will the cable finally make with the vertical?

24.**—A horse, walking along the bank of a canal, pulls a barge. The horse exerts a pull of 300 N on the barge at an angle of 30° (Fig. 3.54). The bargeman relies on the rudder to steer the barge on a straight course parallel to the bank. What transverse force (perpendicular to the bank) must the rudder exert on the barge?

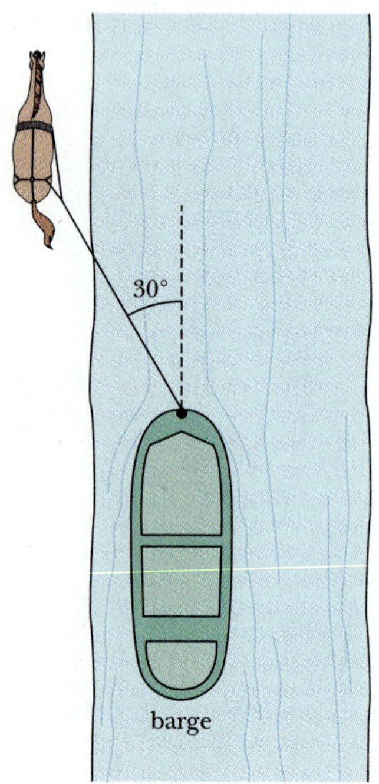

FIGURE 3.54
Horse pulling a barge.

Section 3.5: Kinds of Forces; Weight and Normal Force

25.—A bar of gold of mass 500 g is transported from Paris ($g = 9.8094$ m/s²) to San Francisco ($g = 9.7996$ m/s²). What is the decrease of the weight of the gold? Does this mean that the bar of gold is worth less in San Francisco?

26.—A man carrying a 20-kg sack on his shoulder rides in an elevator. What is the force the sack exerts on his shoulder when the elevator is accelerating upward at 2.0 m/s²?

27.—A woman stands on a chair. Her mass is 60 kg and the mass of the chair is 20 kg. What is the force that the chair exerts on the woman? What is the force that the floor exerts on the chair?

28.—A diver of mass 75 kg is in free fall after jumping off a high platform.
(a) What is the force that the Earth exerts on the diver? What is the force that the diver exerts on the Earth?
(b) What is the acceleration of the diver? What is the acceleration of the Earth?

29.*—A 60-kg woman stands on a bathroom scale placed on the floor of an elevator.
(a) What does the scale read when the elevator is at rest?
(b) What does the scale read when the elevator is accelerating upward at 1.8 m/s²?
(c) What does the scale read when the elevator is moving upward with constant velocity?
(d) What does the scale read if the cable of the elevator is cut (and the brakes have not yet engaged), so the elevator is in free fall?

30.*—At lift off, the Saturn 5 rocket used for the Appolo mission has a mass of 2.45×10^6 kg.
(a) What is the minimum thrust that the rocket engines must develop to achieve lift-off?
(b) The actual thrust that the engines develop is 3.3×10^7 N. What is the vertical acceleration of the rocket at lift-off?
(c) At burnout, the rocket has spent its fuel and its remaining mass is 0.75×10^6 kg. What is the acceleration just before burnout? Assume that the motion is still vertical and that the strength of gravity is the same as when the rocket is on the ground.

31.*—A boy and a girl are engaged in a tug-of-war. Suppose that they are evenly matched, and neither moves. The boy pulls toward the left on the rope with a force of 250 N, and the girl pulls toward the right with a force of 250 N.

(a) Draw separate "free-body" diagrams for the boy, the girl, and the rope. In each of these diagrams include all the appropriate *horizontal* forces (ignore the vertical forces).
(b) What is the horizontal force that the ground exerts on the boy? On the girl?
(c) What is the tension in the rope?
(d) The girl ties her end of the rope to a stout tree and walks away, while the boy continues pulling as before. What is the tension in the rope now?

32.*—The world's steepest railroad track, found in Guatemala, has a slope of 1:11 (see Fig. 3.55). A boxcar of 20 metric tons is being pulled up this track.
(a) Draw a "free-body" diagram for the boxcar.
(b) What force (along the track) is required to move the boxcar up the track at constant speed? Ignore friction and treat the motion as particle motion.

FIGURE 3.55
Boxcar on a steep railroad track.

Section 3.6: Motion with a Constant Force

33.—A boy on a skateboard rolls down a hill of slope 1:5. What is his acceleration? What speed will he reach after rolling for 50 m? Ignore friction.

34.—A skier of mass 75 kg is sliding down a frictionless hillside inclined at 35° to the horizontal.
(a) Draw a "free-body" diagram showing all the forces on the skier (regarded as a particle); draw a separate diagram showing the resultant of these forces.
(b) What is the magnitude of each force? What is the magnitude of the resultant?
(c) What is the acceleration of the skier?

35.*—Suppose that the last car of a train becomes uncoupled while the train is moving upward on a slope of 1:12 at a speed of 45 km/h. How far will the car coast up the slope before it stops? Ignore friction.

36.*—A bobsled slides down an icy track making an angle of 30° with the horizontal. How far must the bobsled slide in order to attain a speed of 90 km/h if initially at rest? When will it attain this speed? Assume that the motion is frictionless.

37.*—Figure 3.56a shows two masses hanging from a string running over a pulley. Such a device can be used to measure the acceleration of gravity; it is then called ***Atwood's machine*** (Fig. 3.56b). If the masses are nearly equal, then the acceleration a of the masses will be much smaller than g; this makes it convenient to measure a and then calculate g by means of Eq. (44). Suppose that an experimenter using masses $m_1 = 400.0$ g and $m_2 = 402.0$ g finds that the masses move a distance of 0.50 m in 6.4 s starting from rest. What value of g does this imply? Assume the pulley is massless and frictionless.

(a)

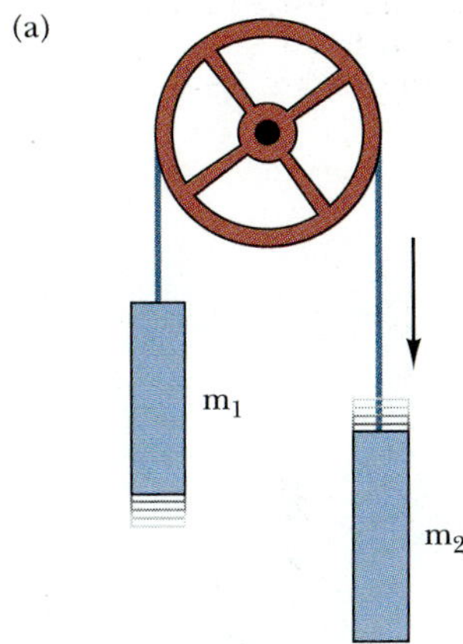

(b)

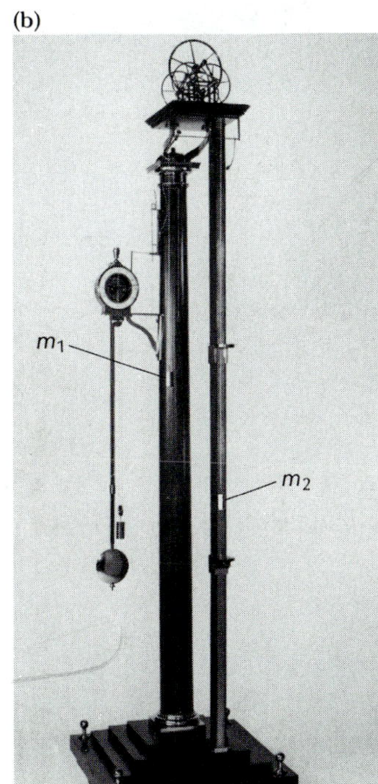

FIGURE 3.56
(a) Two masses and pulley.
(b) Atwood's machine.

Section 3.7: Friction

38.—The ancient Egyptians moved large stones by dragging them across the sand in sleds. How many Egyptians were needed to drag an obelisk of 700 metric tons? Assume that $\mu_k = 0.3$ for the sled on sand and that each Egyptian exerted a horizontal force of 350 N.

39.—The base of a winch is bolted to a mounting plate with four bolts. The base and the mounting plate are flat surfaces made of steel; the friction coefficient of these surfaces in contact is $\mu_s = 0.4$. The bolts provide a normal force of 2600 N each. What maximum static friction force will act between the steel surfaces and help to oppose lateral slippage of the winch on its base?

40.* According to tests performed by the manufacturer, an automobile with an initial speed of 65 km/h has a stopping distance of 20 m on a level road. Assuming that no skidding occurs during braking, what is the value of μ_s between the wheels and the road required to achieve this stopping distance?

41.* A crate sits on the load platform of a truck. The coefficient of friction between the crate and the platform is $\mu_s = 0.4$. If the truck stops suddenly, the crate will slide forward and crash into the cab of the truck. What is the maximum braking deceleration that the truck may have if the crate is to stay put?

42.* When braking (without skidding) on a dry road, the stopping distance of a sports car with a high initial speed is 38 m. What would have been the stopping distance of the same car with the same initial speed on an icy road? Assume that $\mu_s = 0.95$ for the dry road and $\mu_s = 0.20$ for the icy road.

43.* If the coefficient of static friction between the tires of an automobile and the road is $\mu_s = 0.8$, what is the minimum distance the automobile needs in order to stop without skidding from an initial speed of 88 km/h?

44.* In a remarkable accident on motorway M.1 (in England), a Jaguar car initially speeding at 160 km/h skidded 290 m before coming to rest. Assuming that the wheels were completely locked during the skid, find the value of the coefficient of kinetic friction between the wheels and the road.

45.* Because of a failure of its landing gear, an airplane has to make a belly landing on the runway of an airport. The landing speed of the airplane is 90 km/h and the coefficient of kinetic friction between the belly of the airplane and the runway is $\mu_k = 0.6$. How far will the airplane slide along the runway?

46.* The "Texas" locomotives of the old T&P railway had a mass of 200,000 kg, of which 140,000 kg rested on the driving wheels. What maximum acceleration could such a locomotive attain (without slipping) when pulling a train of 100 boxcars of 18,000 kg each on a level track? Assume that the coefficient of static friction between the driving wheels and the track is 0.25.

47.* A 40-kg crate falls off a truck traveling at 80 km/h on a level road. The crate slides along the road and gradually comes to a halt. The coefficient of kinetic friction between the crate and the road is 0.80.

(a) Draw a "free-body" diagram for the crate sliding on the road.
(b) What is the normal force the road exerts on the crate?
(c) What is the friction force the road exerts on the crate?
(d) What is the weight force on the crate? What is the net force on the crate?
(e) What is the deceleration of the crate? How far does the crate slide before coming to a halt?

48.* A 20-kg box rests on an inclined plane which makes an angle of 30° with the horizontal. The coefficient of static friction between the box and the plane is 0.90.

(a) Draw a "free-body" diagram for the box.
(b) What is the normal force that the inclined plane exerts on the box?
(c) What is the friction force the inclined plane exerts on the box?
(d) What is the net force the inclined plane exerts on the box? What is the direction of this force?

49.*——The friction force (including air friction and rolling friction) acting on an automobile traveling at 65 km/h amounts to 500 N. What slope must a road have if the automobile is to roll down this road at a constant speed of 65 km/h (with its gears in neutral)? The mass of the automobile is 1.5×10^5 kg.

50.*——In a downhill race, a skier slides down a 40° slope. Starting from rest, how far must she slide down the slope in order to reach the speed of 130 km/h? How many seconds does it take to reach this speed? The friction coefficient between her skis and the snow is $\mu_k = 0.1$. Ignore the resistance offered by the air.

51.*——To measure the coefficient of static friction of a block of plastic on a plate of steel, an experimenter places the block on the plate and then gradually tilts the plate. The block suddenly begins to slide when the plate makes an angle of 38° with the horizontal. What is the value of μ_s?

52.*——A block of wood rests on a sheet of paper lying on a table. The coefficient of static friction between the block and the paper is $\mu_s = 0.7$ and that between the paper and the table is $\mu_s = 0.5$. If you tilt the table, at what angle will the block begin to move?

Section 3.8: Restoring Force of a Spring; Hooke's Law

53.——Attempting to measure the force constant of a spring, an experimenter clamps the upper end of the spring in a vise and suspends a mass of 1.5 kg from the lower end. This stretches the spring by 0.20 m. What is the force constant of the spring?

54.——A spring with a force constant $k = 150$ N/m has a relaxed length of 0.15 m. What force must you exert to stretch this spring to twice its length? What force must you exert to compress this spring to one-half its length?

55.——The body of an automobile is held above the axles of the wheels by means of four springs, one near each wheel. Assume that the springs are vertical and that the forces on all the springs are the same. The mass of the automobile is 1200 kg and the spring constant of each spring is 2.0×10^4 N/m. When the automobile is stationary on a level road, how far are the springs compressed from their relaxed length?

56.——A rubber band of relaxed length 6.3 cm stretches to 10.2 cm under a force of 1.0 N, and to 16.5 cm under a force of 2.0 N. Does this rubber band obey Hooke's Law?

Section 3.9: Force for Uniform Circular Motion

57.——A swing consists of a seat supported by a pair of ropes 5 m long. A 60-kg woman sits in the swing. Suppose that the speed of the woman is 5.0 m/s at the instant the swing goes through its lowest point. What is the tension in each of the two ropes? Ignore the masses of the seat and the ropes.

58.——The Moon moves around the Earth in a circular orbit of radius 3.8×10^8 m in 27 days. The mass of the Moon is 7.3×10^{22} kg. From these data, calculate the magnitude of the force required to keep the Moon in its orbit.

59.——The Earth moves around the Sun in a circular orbit of radius 1.50×10^{11} m. The speed of the Earth is 3.0×10^4 m/s. Calculate the magnitude of the force required to maintain this motion.

60.*——A few copper coins are lying on the (flat) dashboard of an automobile. The coefficient of static friction between the copper and the dashboard is 0.5. Suppose the automobile rounds a curve of radius 100 m. At what speed of the automobile will the coins begin to slide? The curve is *not* banked.

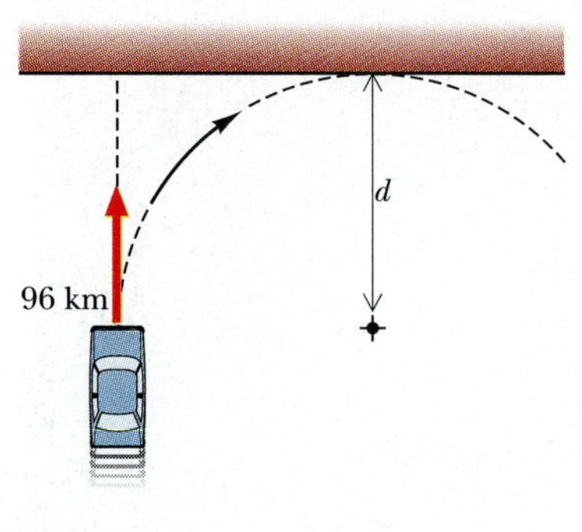

FIGURE 3.57
Automobile approaching a brick wall.

61.* A man of 80 kg is standing in the cabin of a Ferris wheel of radius 30 m rotating at 1 rev/min. What is the force that the feet of the man exert on the floor of the cabin when he reaches the highest point? The lowest point?

62.* The highest part of a road over the top of a hill follows an arc of a vertical circle of radius 50 m. With what minimum speed must you drive an automobile along this road if its wheels are to lose contact with the road at the top of the hill?

63.* A man holds a pail full of water by the handle and whirls it around in a vertical circle at constant speed. The radius of this circle is 0.9 m. What is the minimum speed that the pail must have at the top of its circular motion if the water is not to spill out of the upside-down pail?

64.** An automobile traveling at speed v on a level surface approaches a brick wall (Fig. 3.57). When the automobile is at a distance d from the wall, the driver suddenly realizes that he must either brake or turn. If the coefficient of static friction between the tires and the surface is μ_s, what is the minimum distance that the driver needs to stop (without turning)? What is the minimum distance that the driver needs to complete a 90° turn (without braking)? What is the safest tactic for the driver?

CHAPTER 4

This mountain biker performs hard work while pedaling his bike up a steep slope.

WORK AND ENERGY

◀ Conservation laws

Conservation laws play an important role in the world of matter. Such laws assert that some quantity is conserved, which means that some quantity remains constant even when matter suffers drastic changes involving motions, collisions, and reactions. One familiar example of a conservation law is the conservation of mass. Expressed in its simplest form, this law asserts that the mass of a given particle remains constant. In the preceding two chapters we took this conservation law for granted, and we treated the particle mass appearing in Newton's Second Law ($m\mathbf{a} = \mathbf{F}$) as a constant, time-independent quantity. In everyday life and in commercial and industrial operations, we always rely implicitly on the conservation of mass. For instance, in the chemical plants that reprocess the uranium fuel for nuclear reactors, the batches of uranium compounds are carefully weighed at several checkpoints during the reprocessing operation to make sure that none of the uranium is diverted for nefarious purposes. This procedure would make no sense if the particle masses were changeable, so the net mass of a batch could increase or decrease spontaneously.

Other conservation laws involving mechanical quantities are the conservation of energy, momentum, and angular momentum. Physicists take advantage of these conservation laws to ***make predictions about some aspects of the motion of a particle or of a system of particles*** when it is undesirable or impossible to calculate the full details of the motion from Newton's Second Law. This is especially helpful in those cases where the forces are not known exactly (we will see some examples of this kind in Chapter 6).

This chapter deals with the conservation of energy. This conservation law is one of the most fundamental laws of nature. Although we will derive this law from Newton's laws, it is actually much more general than Newton's laws, and it remains valid even when we step outside of the realm of Newtonian physics and enter the realm of relativistic physics or atomic physics, where Newton's laws fail. No violation of the law of conservation of energy has ever been discovered.

But before we can deal with energy and its conservation, we must introduce the concept of work. Energy and work are closely related. We will see that the work done by a force on a body is related to the change of the kinetic energy (the energy of motion) of the body.

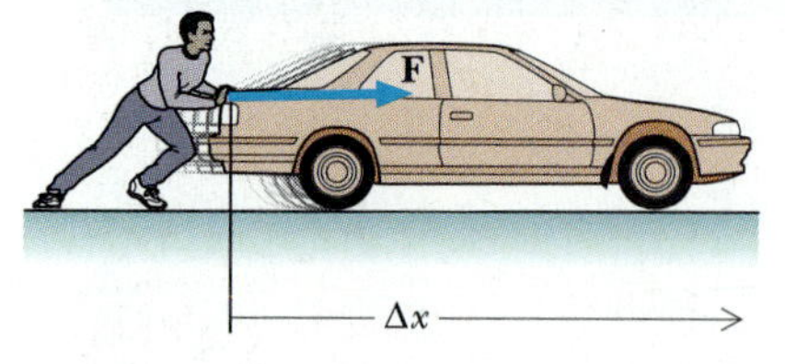

FIGURE 4.1

You do work while pushing an automobile along a road with a horizontal force F_x.

4.1 WORK

To introduce the definition of work done by a force, we begin with the simple case of motion along a straight line, with the force along the line of motion, and then we will proceed to the general case of motion along some arbitrary curved path, with the force in some arbitrary direction. Consider a particle moving along a straight line, say, the x axis, and suppose that a constant force F_x, directed along the same straight line, acts on the particle. For example, the particle might be a stalled automobile that you are pushing along a road (see Fig. 4.1). Then ***the work done by the force F_x on the particle as it moves some given distance Δx is defined as the product of the force and the displacement,***

Work done by a force ▶

$$W = F_x \Delta x \tag{1}$$

This rigorous definition of work is consistent with our intuitive notion of what constitutes "work" — if you push the stalled automobile, the work that you perform is proportional to the magnitude of the force you have to exert, and it is also proportional to the distance you move the automobile.

Note that in Eq. (1), F_x is reckoned as positive if the force is in the positive x direction and negative if in the negative x direction; that is, F_x is the component of the force along the x axis. Thus, according to Eq. (1), ***the work is positive if the force and the displacement are in the same direction*** (both positive, or both negative), ***and the work is negative if they are in opposite directions*** (one positive, the other negative). When pushing the automobile, you do positive work on the automobile if you push in the direction of the motion, so your push tends to accelerate the automobile (Fig. 4.2a); but you do negative work on the automobile if you push in the direction opposite to the motion, so your push tends to decelerate the automobile (Fig. 4.2b).

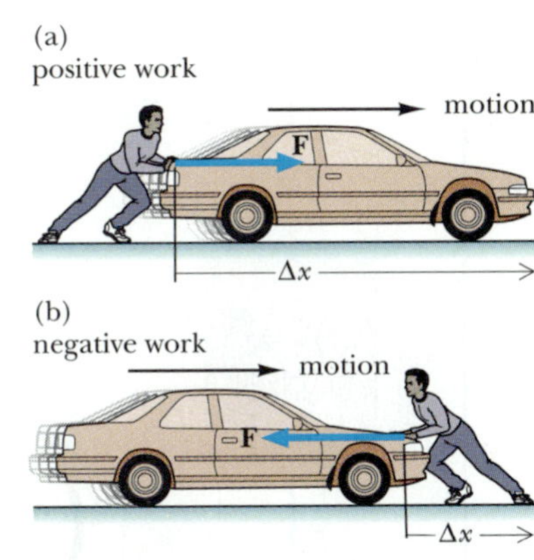

FIGURE 4.2

(a) The work you do on the automobile is positive if you push in the direction of motion. (b) The work you do on the automobile is negative if you push in the direction opposite to the motion.

Equation (1) gives the work done by one of the forces acting on the particle. If several forces act, then Eq. (1) can be used to calculate the work done by each force. If we add the amounts of work done by all the forces acting on the particle, we obtain the net amount of work done by all these forces together. This net amount of work can also be directly calculated from the net force,

$$W = F_{\text{net},x}\Delta x$$

In the metric system, the unit of work is the joule (J), which is the work done by a force of 1 N during a displacement of 1 m. Thus,

◂ **Joule (J)**

$$1 \text{ joule} = 1 \text{ J} = 1 \text{ N}\cdot\text{m}$$

EXAMPLE 1. Suppose you push your stalled automobile along a straight road (see Fig. 4.1). If the force required to overcome friction and to keep the automobile moving is 500 N, how much work must you do to push the automobile 30 m?

SOLUTION: With $F_x = 500$ N and $\Delta x = 30$ m, Eq. (1) gives

$$W = F_x\Delta x = 500 \text{ N} \times 30 \text{ m} = 15{,}000 \text{ J} \qquad (2)$$

James Prescott Joule (jool), *1818–1889, English physicist. He established experimentally that heat is a form of mechanical energy, and he made the first direct measurement of the mechanical equivalent of heat. By a series of meticulous mechanical, thermal, and electrical experiments, Joule provided empirical proof of the general law of conservation of energy.*

EXAMPLE 2. A 1000-kg elevator cage descends 400 m within a skyscraper. (a) What is the work done by gravity on the elevator cage during this displacement? (b) Assuming that the elevator cage descends at constant velocity, what is the work done by the tension of the suspension cable?

SOLUTION: (a) With the x axis arranged vertically upward (see Fig. 4.3), the displacement is negative, $\Delta x = -400$ m, and the x component of the weight is also negative, $w_x = -mg = -1000 \text{ kg} \times 9.81 \text{ m/s}^2 = -9810$ N. Hence the work is

$$W = F_x\Delta x = w_x\Delta x = (-9810 \text{ N}) \times (-400 \text{ m}) = 3.92 \times 10^6 \text{ J} \quad (3)$$

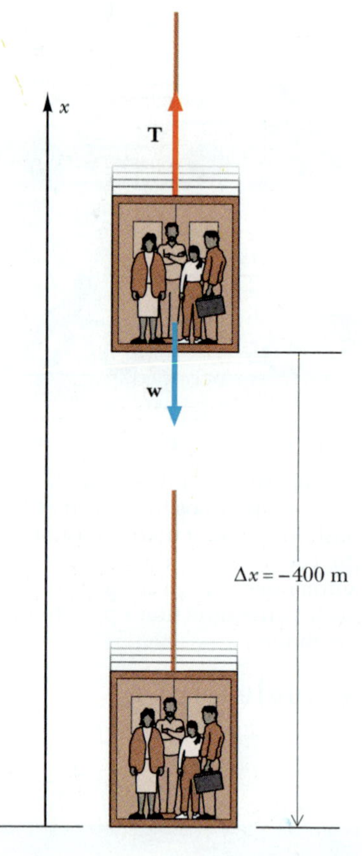

FIGURE 4.3

Gravity does work on a descending elevator. Since the positive x axis is directed upward, the displacement of the elevator is negative, $\Delta x = -400$ m.

Note that the work done by gravity is completely independent of the details of the motion; the work depends on the total vertical displacement and on the weight, but not on the velocity or the acceleration of the motion.

(b) For motion at constant velocity, the tension force must exactly balance the weight. Therefore, the tension force has the same magnitude as the weight, but the opposite direction,

$$T_x = +mg = 9810 \text{ N}$$

The work done by this force is then

$$W = F_x \Delta x = T_x \Delta x = (9810 \text{ N}) \times (-400 \text{ m}) = -3.92 \times 10^6 \text{ J} \quad (4)$$

This work is negative because the tension force and the displacement are in opposite directions.

COMMENTS AND SUGGESTIONS: In this example, the work done by the tension is exactly the opposite of the work done by gravity, and thus the net work done by both forces together is zero (we can also see this by examining the work done by the net force; the net force $F_{\text{net},x} = w_x + T_x$ is zero, and hence the net work $F_{\text{net},x}\Delta x$ is zero). However, the result (4) for the work done by the tension depends implicitly on the assumptions made about the motion. Only for unaccelerated motion does the tension force remain constant at 9810 N. For instance, if the elevator cage were allowed to fall freely with the acceleration of gravity, then the tension would be zero; the work done by the tension would then also be zero, whereas the work done by gravity would still be 3.92×10^6 J.

Although the rigorous definition of work given in Eq. (1) agrees to some extent with our intuitive notion of what constitutes "work," the rigorous definition clashes with our intuition in some instances. For example, consider a man holding a bowling ball in a fixed position in his outstretched hand (see Fig. 4.4). Our intuition suggests that the man does work — yet Eq. (1) indicates that no work is done on the ball, since the ball does not move, and the displacement Δx is zero. The resolution of this conflict hinges on the observation that, although the man does no work *on the ball,* he does work *within his own muscles* and, consequently, grows tired of holding the ball. A contracted muscle is never in a state of complete rest; within it, atoms, cells, and muscle fibers engage in complicated chemical and mechanical processes which involve motion and work. This means that work is done, and wasted, internally within the muscle, while no work is done externally on the bone to which the muscle is attached or on the bowling ball supported by the bone.

FIGURE 4.4

Man holding a bowling ball. The displacement of the ball is zero; hence the work is zero.

Another conflict between our intuition and the rigorous definition of work arises when we consider a body in motion. Suppose that the man with the bowling ball in his hand rides in an elevator moving upward at constant velocity (Fig. 4.5). In this case, the displacement is not zero, and the force (push) exerted by the hand on the ball does work — the displacement and the force are in the same direction, and consequently the man continuously does positive work on the ball. Nevertheless, to the man the ball feels no different when riding in the elevator than when standing on the ground. The man plays a passive role: the elevator floor does work on his feet, and the man transmits a suitable fraction of this work to the bowling ball. This example illustrates that ***the amount of work done on a body depends on the reference frame.*** In the reference frame of the ground, the ball is moving upward and work is done on it; in the reference frame of the elevator, the ball is at rest, and no work is done on it. The lesson we learn from this is that before proceeding with a calculation of work, we must be careful to specify the reference frame.

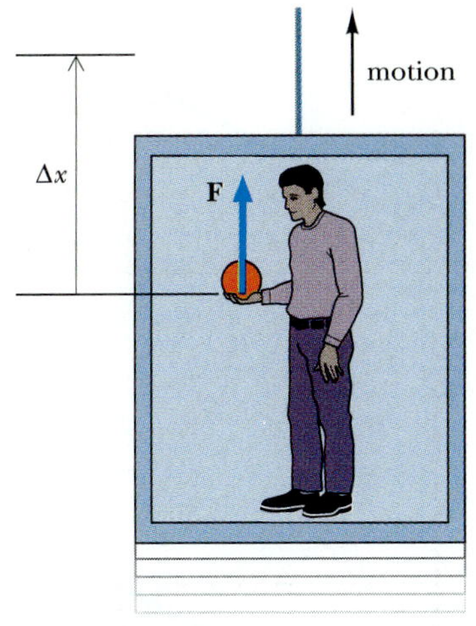

FIGURE 4.5

The man holding the ball rides in an elevator. The displacement of the ball is not zero; hence the work is not zero.

If the motion of the particle and the force are not along the same line, then the simple definition of work given in Eq. (1) must be generalized. Consider a particle moving along some arbitrary curved path, and suppose that the force that acts on the particle is constant. The force can then be represented by a vector **F** (see Fig. 4.6a) which is constant in magnitude and direction. The work done by this constant force during a (vector) displacement **s** is defined as

$$W = Fs\cos\theta \tag{5}$$

where F is the magnitude of the force, s is the length of the displacement, and θ is the angle between the force and the direction of the displacement. As can be seen from Fig. 4.6b, this expression can be regarded as the magnitude of the displacement (s) multiplied by the component of the force along the direction of the displacement ($F\cos\theta$). If the force is parallel to the direction of the displacement ($\theta = 0$ and $\cos\theta = 1$), then the work is simply Fs, this coincides with the case of motion along a straight line. If the force is perpendicular to the direction of motion ($\theta = 90°$ and $\cos\theta = 0$), then the work is zero. For instance, if a woman holding a bowling ball walks along a level road, she does not do any work on the ball, since the force she exerts on the ball is perpendicular to the direction of motion (Fig. 4.7a). However, if the woman climbs up some stairs while holding the ball, then she does work on the ball, since now the force she exerts has a component along the direction of motion (Fig. 4.7b).

(a)

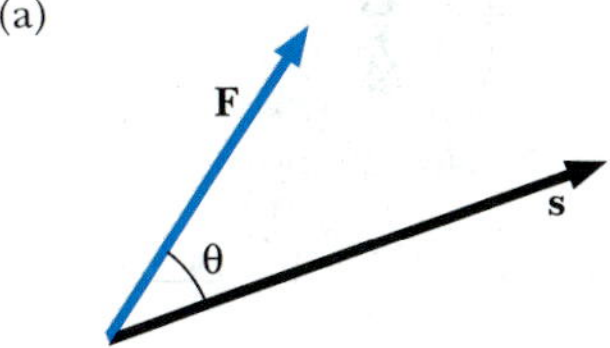

(b)

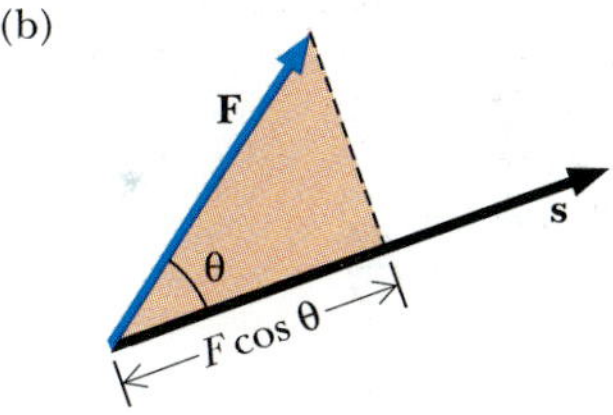

FIGURE 4.6

(a) A constant force **F** acts during a displacement **s**. The force makes an angle θ with the displacement. (b) The component of the force along the direction of the displacement is $F\cos\theta$.

(a)

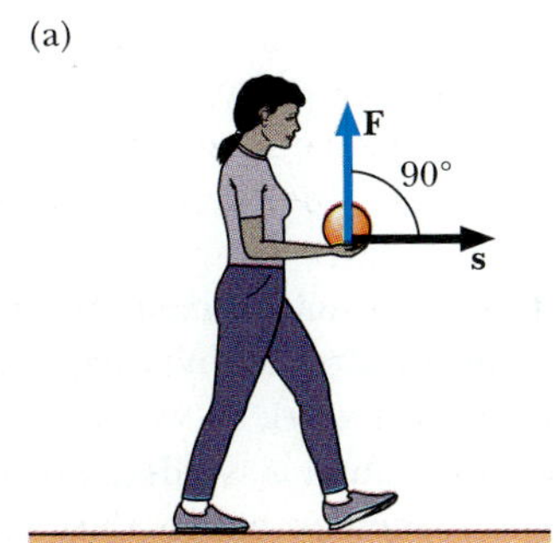

(b)

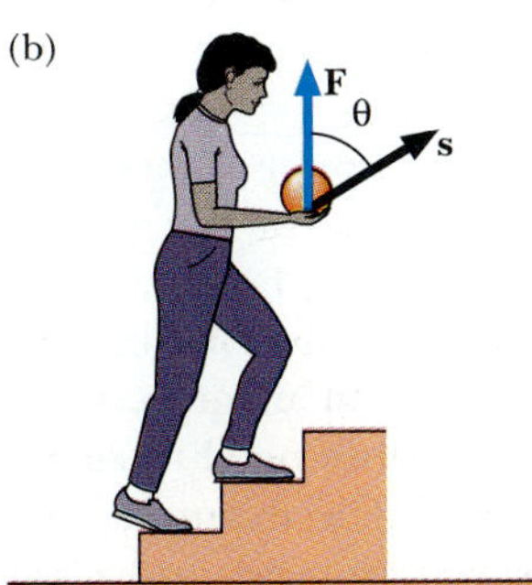

FIGURE 4.7

(a) The force exerted by the woman is perpendicular to the displacement. (b) The force exerted by the woman is not perpendicular to the displacement.

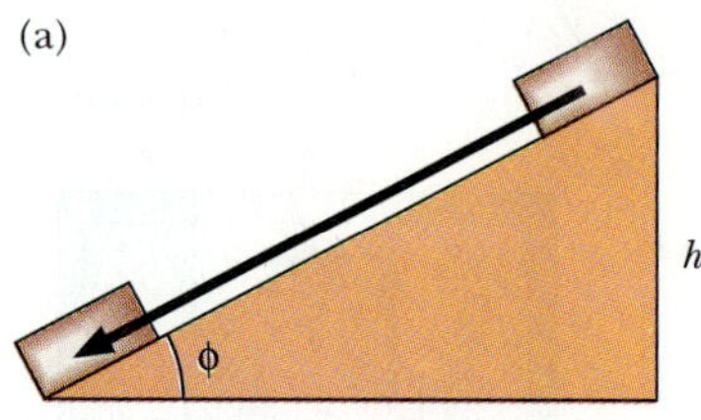

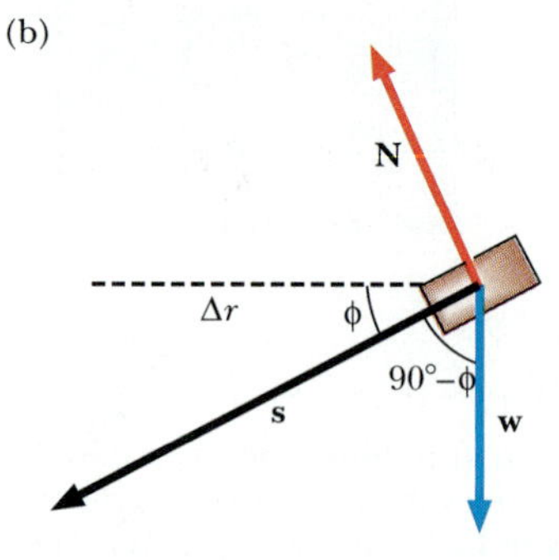

FIGURE 4.8
(a) A block undergoing a displacement along an inclined plane.
(b) "Free-body" diagram showing the weight, the normal force, and the displacement of the block.

EXAMPLE 3. What is the work done by gravity on a block of mass m that slides down an inclined plane from a height h? What is the work done by the normal force?

SOLUTION: Figure 4.8a shows the inclined plane; the block moves down the full length of this plane. By inspection of the right triangle formed by the plane and the ground, we see that the displacement of the block has a magnitude

$$s = h/\sin\phi.$$

[Here we use the label ϕ (Greek phi) for the angle of the inclined plane to distinguish it from the angle θ appearing in Eq. (5).] Figure 4.8b shows a "free-body" diagram for the block; the forces acting on the block are the normal force **N** and the weight **w.** The weight makes an angle of $90° - \phi$ with the displacement. Therefore the work done by the weight is

$$W = Fs\cos\theta = mg \times \frac{h}{\sin\phi}\cos(90° - \phi) \tag{6}$$

Since $\cos(90° - \phi) = \sin\phi$, the work is

$$W = mg \times \frac{h}{\sin\phi} \times \sin\phi = mgh \tag{7}$$

Note that this result is independent of the angle of the inclined plane — it is the same as though the block had moved down vertically through the height h.

The work done by the normal force is zero, since this force makes an angle of 90° with the displacement.

COMMENTS AND SUGGESTIONS: Note that the result of zero work for the normal force is quite general. The normal force **N** acting on any body sliding on any kind of surface never does work on the body, since this force is always perpendicular to the displacement.

4.2 WORK FOR A VARIABLE FORCE

The definition of work in the preceding section assumed that the force was constant (in magnitude and in direction). But many forces are not constant, and we need to revise our definition of work so we can deal with such forces. For example, suppose that you push a stalled automobile along a straight road, and suppose that the force you exert is not constant

— as you move along the road, you sometimes push hard and sometimes less hard. Figure 4.9 shows how the force might vary with position. (The reason why you sometimes push harder is irrelevant — maybe the automobile passes through the muddy portion of the road and requires more of a push, or maybe you get impatient and want to hurry the automobile along; all that is relevant for the calculation of the work is the value of the force at different positions, as shown in the plot.) ***The work done on a particle by such varying force is defined as the product of the average value of the force and the displacement,***

$$W = \overline{F}_x \Delta x \tag{8}$$

◂ **Work done by a variable force**

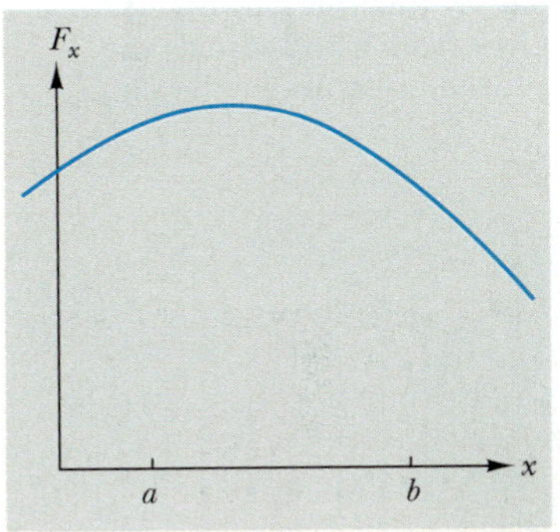

FIGURE 4.9

Plot of F_x vs. x for a force that varies with position.

In this equation, the overbar on the symbol for force is the standard notation used in physics to indicate an average value. Graphically, the average force $\overline{F}_x$ is the average height of the points plotted in Fig. 4.9. A precise evaluation of this average height is quite difficult if the plot has a complicated shape.

A simple case of a force that depends on position is the force exerted by a spring on a particle attached to it. The force is directly proportional to the distance measured from the equilibrium position [see Eq. (3.61)],

$$F_x = -kx \tag{9}$$

The plot of this force vs. x is simply a straight line of slope $-k$ (see Fig. 4.10). Suppose the particle moves from $x = a$ to $x = b$, and we want to evaluate the work done by the spring during this displacement. According to Eq. (5), we need the average force, or the average height of the points plotted in Fig. 4.10 (this average "height" is negative, since all the points are below the horizontal axis). At the starting point $x = a$, the "height" is $F_x = -kx = -ka$; and at the endpoint $x = b$, the "height" is $F_x = -kx = -kb$. Since the force varies uniformly between a and b, the average value of the force is simply the average of the values at the start and at the end; thus, the average force is

$$\overline{F}_x = \tfrac{1}{2}(-ka - kb) = -\tfrac{1}{2}k(a + b) \tag{10}$$

The displacement is $\Delta x = b - a$. According to Eq. (8), the work is then

$$W = \overline{F}_x\, \Delta x = -\tfrac{1}{2}k(a + b) \times (b - a) = -\tfrac{1}{2}k(ab - a^2 + b^2 - ba)$$

$$= -\tfrac{1}{2}k(b^2 - a^2) \tag{11}$$

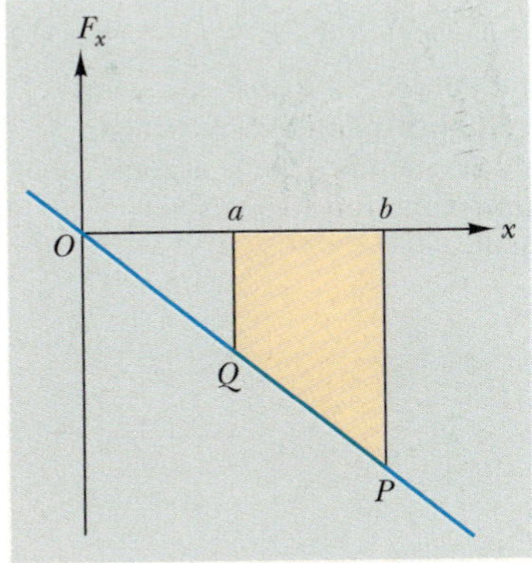

FIGURE 4.10

The plot of the force $F_x = -kx$ is a straight line. The work done by the force as the particle moves from a to b equals the (colored) quadrilateral area under this plot.

This result has an interesting graphical interpretation: ***the work is the area between the plot of the force and the*** x ***axis, between*** a ***and*** b. This area is shown colored in Fig. 4.10. To verify this, note that the colored quadrilateral area $abPQ$ in Fig. 4.10 is the difference between the triangular areas ObP and OaQ. The former area is $\frac{1}{2}$[base] × [height] $= \frac{1}{2}b \times (-kb)$ and the latter area is $\frac{1}{2}$[base] × [height] $= \frac{1}{2}a \times (-ka)$; both of these areas are neg-

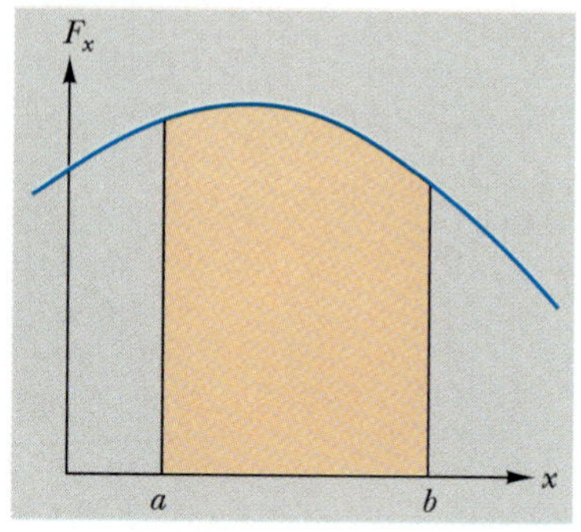

FIGURE 4.11

Plot of F_x vs. x for a force that varies with position. The work done by this force as the particle moves from a to b equals the (colored) area under the plot.

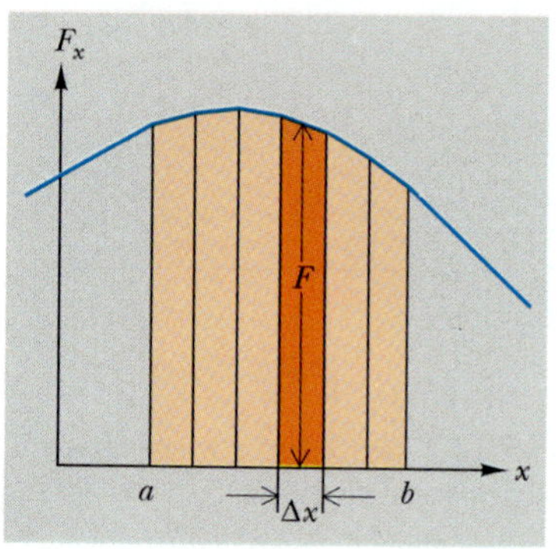

FIGURE 4.12

The curved plot of F_x vs. x has been approximated by straight line segments. This is a good approximation if Δx is very small.

ative, because their "heights" are negative. The difference between these two areas is

$$(-\tfrac{1}{2}kb^2) - (-\tfrac{1}{2}ka^2) = -\tfrac{1}{2}k(b^2 - a^2) \tag{12}$$

which shows that the quadrilateral area *abPQ* agrees with the work calculated in Eq. (11).

Although we have here obtained this graphical interpretation in the special case of a force directly proportional to the distance, the graphical interpretation is of general validity. For any force varying with position in some arbitrary way (see Fig. 4.11), the work is equal to the area between the plot of the force and the x axis, between the starting point and the endpoint of the motion (this area is shown colored in Fig. 4.11). For a proof, we approximate the exact curved plot of force vs. position by a plot consisting of straight line segments, each spanning a small interval of width Δx along the x axis (see Fig. 4.12). Within each of these intervals, the work and the area can be calculated as we did above for the case of the spring force, because in each of these intervals, the plot of the force is a straight line, and the area between the plot and the x axis has the same quadrilateral shape as for the spring force (compare the colored areas in Figs. 4.10 and 4.12). Thus, as in the case of the spring force, in each of these intervals the area and the work are equal. The net work for the entire motion is the sum of the small amounts of work done in all the small intervals, and therefore the work done by the force plotted in Fig. 4.12 equals the area under the plot.

We might have anticipated this equality of work and area under the plot from Eq. (8), since it seems self-evident that the product of the average height of the plot and the width of the plot should equal the area. This suggests we could have saved ourselves the trouble of going through the above lengthy geometrical arguments with areas. However, for a complicated plot of force, there is no direct way of reckoning the average height of the plot, and the geometrical calculation of areas by the approximation method described in Fig. 4.12 is then the best way of calculating the work.

4.3 Kinetic Energy

In everyday language, *energy* means a capacity for vigorous activities and hard work. Likewise, in the language of physics, energy is a capacity for performing work. Energy is "stored" work, or latent work, which can be converted into actual work under suitable conditions. For instance, a body in motion, such as a speeding arrow, has energy of motion, or kinetic energy. This kinetic energy will be converted into work when the arrow strikes a target, such as the trunk of a tree. The tip of the arrow then performs work on the wood, prying apart and cutting the wood fibers. The arrow continues to perform work and to penetrate the wood for a few centimeters, until all of its kinetic energy has been exhausted. A high-speed arrow has a deeper penetration and delivers a larger amount of work to the target. Thus, we see that the kinetic energy of the arrow, or the kinetic energy of any kind of particle, must be large if the speed is large. To discover the dependence of kinetic energy on the speed, we must examine how work performed by or on a particle is related to changes of the speed of the particle.

For convenience, we will consider the work done *on* a particle by the net external force acting on it, rather than the work done *by* the particle (the work done *by* the particle on the external system is difficult to calculate directly if the external system suffers deformations and disruptions, as in the case of the arrow penetrating into a target of wood; for such a calculation, we would need to know the motions of all the wood fibers and wood particles). When the net force acts, it accelerates the particle, does work on the particle, and "stores" this work in the particle (or removes "stored" work, as in the case of the penetrating arrow). If the acceleration has a component along the direction of motion of the particle, it will result in a change of the speed of the particle. We can establish an important identity between the work done by the force and the change of speed it produces. Let us do this for the simple case of a constant net force F_x acting on a particle moving along a straight line (see Fig. 4.13). In this case, the work done on the particle during a displacement Δx from a starting point x_1 to an endpoint x_2 is

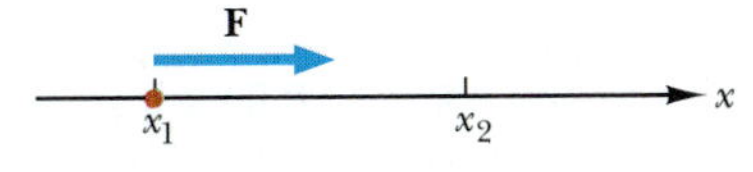

FIGURE 4.13

A particle moves on a straight line from x_1 to x_2 while a force F_x acts on it.

$$W = F_x \Delta x = F_x(x_2 - x_1) \tag{13}$$

Since the force is constant, it produces a constant acceleration, $a = F_x/m$. For such a motion with constant acceleration, Eq. (1.17) tells us that the acceleration is related to the change in the square of the speed,

$$v_2^2 - v_1^2 = 2a(x_2 - x_1) \tag{14}$$

or, since $a = F_x/m$,

$$v_2^2 - v_1^2 = 2\frac{F_x}{m}(x_2 - x_1) \tag{15}$$

The product $F_x(x_2 - x_1)$ is the work. Hence Eq. (15) becomes

$$v_2^2 - v_1^2 = \frac{2}{m} W \tag{16}$$

which we can rearrange to read

$$\tfrac{1}{2}mv_2^2 - \tfrac{1}{2}mv_1^2 = W \tag{17}$$

This shows that ***the change in the square of the speed is proportional to the work done by the force.***

Although we have here obtained the result (17) for the simple case of a constant force, it can be shown that the same result is valid for a variable force. Furthermore, the result is valid not only for motion along a straight line, but for motion along any curve and for any force of variable magnitude and direction.

According to Eq. (17), whenever we perform positive work on the particle, we increase the "amount of $\frac{1}{2}mv^2$" in the particle; and whenever we perform negative work on the particle (that is, when we let the particle perform work on us), we decrease the "amount of $\frac{1}{2}mv^2$" in the particle. Thus, ***the quantity $\frac{1}{2}mv^2$ is the amount of work stored in the particle, or the kinetic energy of the particle.*** We represent the kinetic energy by the symbol K:

Kinetic energy ▶

$$K = \tfrac{1}{2}mv^2 \tag{18}$$

With this notation, Eq. (18) states that ***the change of kinetic energy equals the net work done on the particle,*** that is,

$$K_2 - K_1 = W \tag{19}$$

or

Work-energy theorem ▶

$$\Delta K = W \tag{20}$$

This result is called the ***work-energy theorem.*** Keep in mind that the work in Eqs. (16), (17), and (20) must be evaluated with the *net* force, that is, all the forces that do work on the particle must be included in the calculation.

When a force does positive work on a particle initially at rest, the kinetic energy of the particle increases. The particle then has a capacity to do work: if the moving particle subsequently is allowed to push against some obstacle, then this obstacle does negative work on the particle and simultaneously the particle does positive work on the obstacle. When the particle does work, its kinetic energy decreases. The total amount of work the particle can deliver to the obstacle is equal to its kinetic energy. Thus, ***the kinetic energy represents the capacity of a particle to do work by virtue of its speed.***

The acquisition of kinetic energy through work and the subsequent production of work by this kinetic energy are neatly illustrated in the operation of a waterwheel driven by falling water. In a flour mill of an old Spanish Colonial design, the water runs down from a reservoir in a steep, open channel (see Fig. 4.14). The motion of the water particles is essentially that of particles sliding down an inclined plane. If we ignore friction, then the only force that does work on the water particles is gravity. This work is positive; consequently, the kinetic energy of the water increases and it attains a maximum value at the lower end of the channel. The stream of water emerges from this channel with high kinetic energy

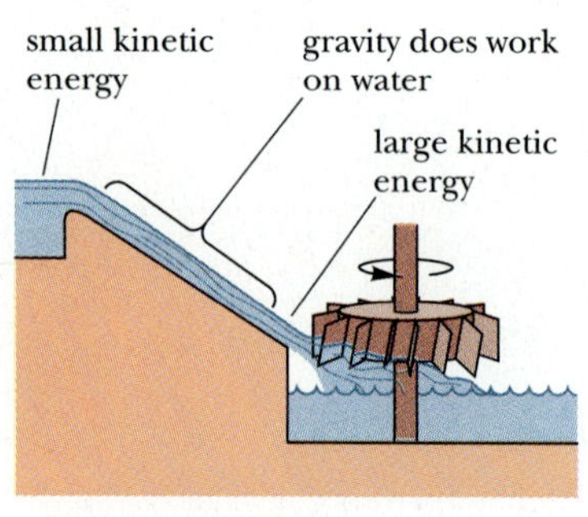

FIGURE 4.14
Water pushing on a horizontal waterwheel.

TABLE 4.1 SOME KINETIC ENERGIES	
Orbital motion of Earth	2.6×10^{33} J
Ship *Queen Elizabeth* (at cruising speed)	9×10^{9} J
Jet airliner (Boeing 747 at maximum speed)	7×10^{9} J
Automobile (at 90 km/h)	5×10^{5} J
Rifle bullet	4×10^{3} J
Man walking	60 J
Falling raindrop	4×10^{-5} J
Proton from large accelerator (Fermilab)	1.6×10^{-7} J
Electron in atom (hydrogen)	2.2×10^{-18} J
Air molecule (at room temperature)	6.2×10^{-21} J

and hits the blades of the waterwheel. The water pushes on the wheel, turns it, and gives up its kinetic energy while doing work — and the wheel runs the millstones and does useful work on them. Thus, the work that gravity does on the descending water is ultimately converted into useful work, with the kinetic energy playing an intermediate role in this process.

The unit of kinetic energy is the joule, the same as the unit of work. Table 4.1 lists some typical kinetic energies.

EXAMPLE 4. During a baseball game, the pitcher throws the ball with a speed of 30 m/s (Fig. 4.15). The mass of the ball is 0.15 kg. What is the kinetic energy of the ball when it leaves his hand? How much work did his hand do on the ball during the throw?

SOLUTION: The final speed of the ball, when it leaves the hand at the end of the throwing motion, is $v_2 = 30$ m/s. The final kinetic energy of the ball is

$$K_2 = \tfrac{1}{2}mv_2^2 = \tfrac{1}{2} \times 0.15\ \text{kg} \times (30\ \text{m/s})^2 = 68\ \text{J} \tag{21}$$

According to the work–energy theorem, the work done by the hand on the ball equals the change of kinetic energy. Since the initial kinetic energy at the beginning of the throwing motion is zero ($v_1 = 0$), the change of kinetic energy equals the final kinetic energy, and the work is

$$W = K_2 = 68\ \text{J}$$

Note that for this calculation of the work we did not need to know the (complicated) details of how the force varies during the throwing motion. The work–energy theorem gives us the answer directly.

FIGURE 4.15

Pitcher throwing a ball. The ball leaves his hand with a speed of 30 m/s.

4.4 POTENTIAL ENERGY

As we saw in the preceding section, the kinetic energy represents the capacity of a particle to do work by virtue of its speed. We will now become acquainted with another form of energy that represents the capacity of the particle to do work by virtue of its position in space. This is the ***potential energy.*** In this section, we will examine the special case of gravitational potential energy for a particle moving under the influence of the constant gravitational force, and we will formulate a law of conservation of energy for such a particle. In the next section we will examine other cases of potential energy and formulate the general law of conservation of energy.

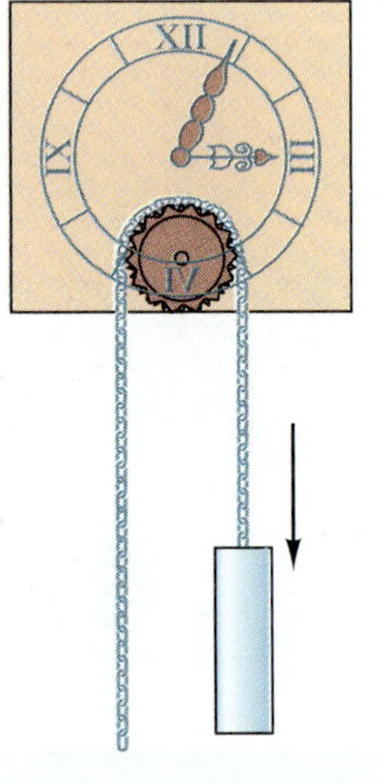

FIGURE 4.16
The descending weight of the grandfather clock pulls on the chain and does work on the wheel of the clock.

The gravitational potential energy represents the capacity of the particle to do work by virtue of its height above the surface of the Earth. When we lift a particle to some height above the surface, we have to do work against gravity, and we thereby store work in the particle. Thus, a particle high above the surface is endowed with a large amount of latent work, which can be exploited and converted into actual work by allowing the particle to push against some obstacle as it descends. A good example of such an exploitation of gravitational potential energy is found in a grandfather clock, where a weight hanging on a chain drives the wheel of the clock (Fig. 4.16). The weight does work on the wheel, and gradually converts all of its potential energy into work (in a typical grandfather clock, the weight takes about a week to sink down from the top to the bottom, and you must then rewind the clock, by lifting the weight).

To obtain a general expression for the gravitational potential energy of a particle moving on a straight or a curving path, we first consider a particle moving on an inclined plane. According to Eq. (7), when a particle of mass m descends a distance h along an inclined plane, the work done by gravity is

$$W = mgh \tag{22}$$

As already remarked in Example 3, this result is independent of the angle of inclination of the plane — it depends only on the change of height. More generally, for a curved path, the result is independent of the shape of the path that the particle follows from its starting point to its endpoint. For instance, the curved path and the straight sloping path in Fig. 4.17a lead to exactly the same result for the work done by gravity. To recognize this, we simply approximate the curved path by small straight segments (see Fig. 4.17b). Each such small segment can be regarded as a small inclined plane, and therefore the work is mg times the small change of height. The net amount of work for all the small segments taken together is then mg times the net change of height, in agreement with Eq. (22).

If the vertical coordinate of the starting point is y_1 and the vertical coordinate of the endpoint is y_2 (see Fig. 4.17), then $h = y_1 - y_2$ and Eq. (22) reads

$$W = mg(y_1 - y_2)$$

or

$$W = -(mgy_2 - mgy_1) \tag{23}$$

According to Eq. (23), whenever gravity performs positive work on the particle (a descending particle), the "amount of mgy" of the particle decreases; and whenever gravity performs negative work on the particle (an ascending particle), the "amount of mgy" increases. Thus, ***the quantity*** mgy ***is the amount of stored, or latent, gravitational work, that is, it is the gravitational potential energy.*** We will adopt the notation U for the gravitational potential energy:

◂ **Gravitational Potential Energy**

$$U = mgy \tag{24}$$

In terms of this potential energy, Eq. (23) for the work done by gravity becomes

$$W = -U_2 + U_1 \tag{25}$$

which says that the work equals the negative of the change in potential energy.

EXAMPLE 5. What is the kinetic energy and what is the gravitational potential energy (relative to the ground) of a jet airliner of mass 73,000 kg cruising at 240 m/s at an altitude of 9000 m?

SOLUTION: The kinetic energy is

$$K = \tfrac{1}{2}mv^2 = \tfrac{1}{2} \times 7.3 \times 10^4 \text{ kg} \times (240 \text{ m/s})^2 = 2.1 \times 10^9 \text{ J}$$

The gravitational potential energy is $U = mgy$. If we measure the y coordinate from the ground level, then $y = 9000$ m for our airliner, and

$$U = mgy = 7.3 \times 10^4 \text{ kg} \times 9.81 \text{ m/s}^2 \times 9 \times 10^3 \text{ m} = 6.4 \times 10^9 \text{ J}$$

We see that the airliner has more potential energy than kinetic energy.

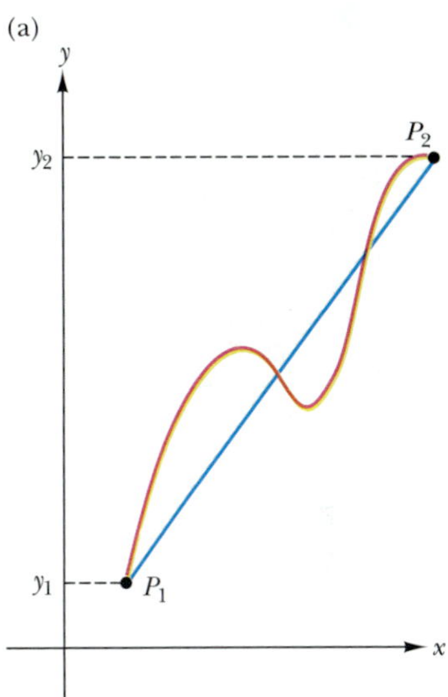

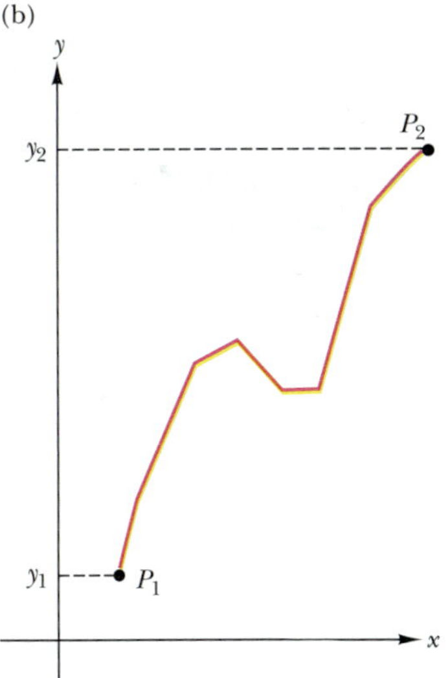

FIGURE 4.17
(a) A curved path and a straight path from point P_1 to point P_2. (b) The curved path can be approximated by short straight segments.

If we let particle push or pull on some obstacle (such as the wheel of the grandfather clock) during its descent from y_1 to y_2, then the total amount of work that we can extract during this descent is equal to the work done by gravity, that is, it is equal to $-U_2 + U_1$, or the negative of the change of potential energy. Of course, the work extracted in this way really arises from the Earth's gravity — the particle can do work on the obstacle because gravity is doing work on the particle. Hence, ***the gravita-***

Christiaan Huygens (hoigens), *1629–1695, Dutch mathematician and physicist. He invented the pendulum clock, made improvements in the manufacture of telescope lenses, and discovered the rings of Saturn. Huygens investigated the theory of collisions of elastic bodies and the theory of oscillations of the pendulum, and he stated the law of conservation of mechanical energy for motion under the influence of gravity.*

tional potential energy is really a joint property of both the particle and the Earth; it is a property of the configuration of the particle–Earth system.

If the only force acting in the particle is gravity, then by combining Eqs. (20) and (25) we can obtain a relation between potential energy and kinetic energy. Since $K_2 - K_1 = W$ by Eq. (20), and $W = -U_2 + U_1$ by Eq. (25), we see that

$$K_2 - K_1 = -U_2 + U_1 \tag{26}$$

which we can rewrite as follows:

$$K_2 + U_2 = K_1 + U_1 \tag{27}$$

This equality indicates that the quantity $K + U$ is a constant of the motion, that is, it has the same value at the endpoint as it had at the starting point. We can express this as

$$K + U = [\text{constant}] \tag{28}$$

The sum of the kinetic and potential energies is called the mechanical energy of the particle. It is usually designated by the symbol E,

Mechanical energy ▶

$$E = K + U \tag{29}$$

This energy represents the total capacity of the particle to do work by virtue of both its speed and its position.

Equation (28) show that if the only force acting on the particle is gravity, then the mechanical energy remains constant,

Law of conservation of mechanical energy ▶

$$E = K + U = [\text{constant}] \tag{30}$$

This is the ***law of conservation of mechanical energy.***

Apart from its practical significance in terms of work, the mechanical energy is very helpful in the study of the motion of a particle. Since the sum of the potential and kinetic energies must remain constant during the motion, an increase in one must be compensated by a decrease in the other; this means that ***during the motion, kinetic energy is converted into potential energy and vice versa.*** For instance, if we launch an arrow straight upward from ground level ($y_1 = 0$), the initial kinetic energy is large and the initial potential energy is zero. As the arrow rises, its potential energy increases and, correspondingly, its kinetic energy decreases, so as to keep the sum of the kinetic and potential energies constant. When the arrow reaches its maximum height, its potential energy has the largest value, and the kinetic energy is (instantaneously) zero. As the arrow falls, its potential energy decreases, and its kinetic energy increases.

If we make use of the formulas for K and U, Eq. (30) becomes

$$E = \tfrac{1}{2}mv^2 + mgy = [\text{constant}] \tag{31}$$

This shows explicitly how the arrow, or any other particle moving under the influence of gravity, trades speed for height during the motion: whenever y increases, v must decrease (and conversely), so as to keep the sum of the two terms in Eq. (31) constant.

An important facet of Eq. (31) is that it is valid not only for a particle in free fall (a projectile), but also for a particle sliding on a surface or a track of arbitrary shape, provided that there is no friction. Of course, under these conditions, besides the gravitational force there also acts the normal force; but this force does no work, and hence does not affect Eq. (23), or any of the equations following after it. The next examples illustrate how these results can be applied to simplify the study of fairly complicated motions, which would be extremely difficult to investigate by direct calculation with Newton's Second Law. These examples give us a glimpse of the elegance and power of the law of conservation of mechanical energy.

EXAMPLE 6. A bobsled run leading down a hill at Lake Placid descends 148 m from its highest point to its lowest. Suppose that a bobsled, initially at rest at the highest point, slides down this run without friction. What speed will the bobsled attain at the lowest point?

SOLUTION: The coordinates of the highest and the lowest points are $y_1 = 148$ m and $y_2 = 0$ m, respectively (see Fig. 4.18). According to Eq. (31), the energy at the start of the motion is

$$E = \tfrac{1}{2}mv_1^2 + mgy_1 = 0 + mgy_1 \tag{32}$$

and the energy at the end of the motion is

$$E = \tfrac{1}{2}mv_2^2 + mgy_2 = \tfrac{1}{2}mv_2^2 + 0 \tag{33}$$

The conservation of energy implies that the right sides of Eqs. (32) and (33) are equal,

$$\tfrac{1}{2}mv_2^2 = mgy_1 \tag{34}$$

We divide both sides of this equation by $\frac{1}{2}m$ and then take the square root of both sides. We then find

$$v_2 = \sqrt{2gy_1} \tag{35}$$

which gives

$$v_2 = \sqrt{2 \times 9.81\ \text{m/s}^2 \times 148\ \text{m}} = 54\ \text{m/s}$$

COMMENTS AND SUGGESTIONS: This example illustrates how energy conservation can be exploited to answer a question about motion. To

A bobsled starting to go downhill.

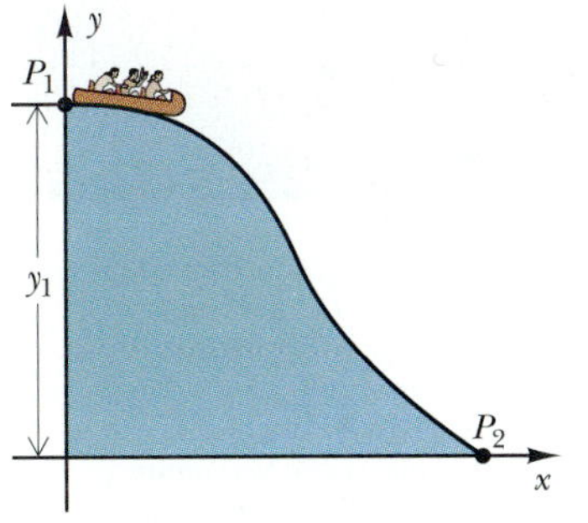

FIGURE 4.18

A bobsled run. The bobsled descends from P_1 to P_2.

obtain the final speed by direct computation of forces and accelerations would have been extremely difficult — it would have required detailed knowledge of the shape of the path down the hill. With the law of conservation of energy we can bypass these complications.

The use of energy conservation in a problem of motion typically involves three steps: (1) First write an expression for the energy at one point of the motion [Eq. (32)]; (2) then write an expression for the energy at another point [Eq (33)], and then rely on energy conservation to equate the two expressions [Eq. (34)]. This yields one equation, which can be solved for the unknown final speed or the unknown final position (if the final speed is known).

Note that the value of the gravitational potential energy mgy depends on the level from which you measure the y coordinate. However, the *change* in the potential energy does not depend on the choice of this level, and therefore any choice will lead to the same result for the change in the kinetic energy. Thus, you can make any choice of zero level, but you must stick to this choice throughout the entire calculation. You will usually find it most convenient to place the zero level for the y coordinate either at the final position of the particle (as in the example) or at the initial position. And always remember that the formula mgy for the gravitational potential energy assumes that the y axis is directed vertically upward.

Joseph Louis, Comte Lagrange (lagranj), *1736–1813, French mathematician and theoretical astronomer. In his elegant mathematical treatise* Analytical Mechanics, *Lagrange formulated Newtonian mechanics in the language of advanced mathematics and introduced the general definition of the potential-energy. Lagrange is also known for his calculations of the motion of planets and for his influential role in securing the adoption of the metric system of units.*

4.5 Potential Energy of a Conservative Force

In the preceding section we saw how to formulate a law of conservation of mechanical energy for a particle moving under the influence of the Earth's gravity. Now we will seek to formulate such a conservation law when other forces — such as the force exerted by a spring — act on the particle.

To deal with some general force, we proceed as in the case of gravitational potential energy. We begin with the work–energy theorem [see Eq. (19)],

$$K_2 - K_1 = W \tag{36}$$

and we express the work W as a difference between two potential energies [see Eq. (25)],

$$W = U_1 - U_2 \tag{37}$$

This gives us

$$K_2 - K_1 = U_1 - U_2$$

from which we immediately find the conservation law for the sum of the kinetic and potential energies:

$$K_2 + U_2 = K_1 + U_1 \tag{38}$$

As an illustration of this general procedure for the construction of the conservation law for mechanical energy, let us deal with the case of a particle moving under the influence of the force exerted by a spring attached to the particle. This force is

$$F_x = -kx \tag{39}$$

where x is, as always, measured from the equilibrium position of the particle. The crucial step in the construction of the conservation law is to express the work W as a difference of two potential energies. For this purpose, we take advantage of the result established in Section 4.2 [see Eq. (11)], according to which the work done by the spring force during a displacement from $x_1 = a$ to $x_2 = b$ is

$$W = \tfrac{1}{2}kx_1^2 - \tfrac{1}{2}kx_2^2 \tag{40}$$

This shows that if we identify the potential energy as

$$U = \tfrac{1}{2}kx^2 \tag{41}$$

◂ **Potential energy of a spring**

then the work is, indeed, the difference between two potential energies, $U_1 = \frac{1}{2}kx_1^2$ and $U_2 = \frac{1}{2}kx_2^2$.

The potential energy $U = \frac{1}{2}kx^2$ ***of the spring represents the capacity of the spring to do work by virtue of its deformation.*** When we compress a spring, we store latent work in it, which we can recover at a later time by letting the spring push against something. An old-fashioned watch, operated by a wound spring, illustrates this storage of energy in a spring (however, the springs in watches are not coil springs, but spiral springs, which are compressed by turning the knob of the watch).

As in the case of the particle moving under the influence of gravity, from the expression (40) for work done by the spring we conclude that the conserved energy for a particle moving under the influence of the spring force is

$$E = K + U = \tfrac{1}{2}mv^2 + \tfrac{1}{2}kx^2 = [\text{constant}] \tag{42}$$

This equation gives us some information about the general features of the motion of the particle; it shows that, say, an increase of x requires a decrease of v so as to keep the sum $\frac{1}{2}mv^2 + \frac{1}{2}kx^2$ constant.

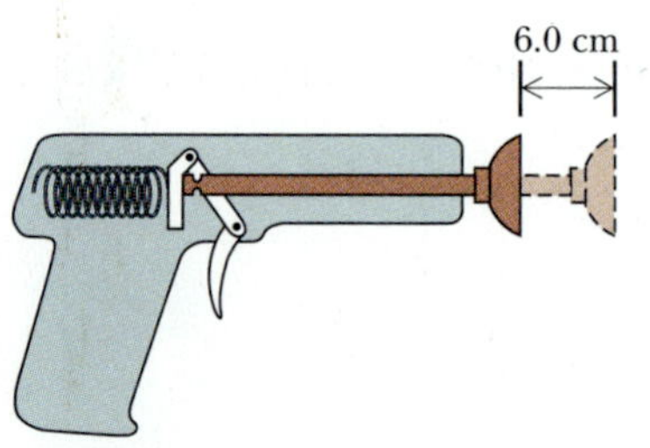

FIGURE 4.19

A toy gun. The spring is initially compressed 6.0 cm.

EXAMPLE 7. A child's toy gun shoots a dart by means of a compressed spring. The constant of the spring is $k = 320$ N/m and the mass of the dart is 8.0 g. Before shooting, the spring is compressed by 6.0 cm and the dart is placed in contact with the spring (see Fig. 4.19). The spring is then released. What will be the speed of the dart when the spring reaches its equilibrium length?

SOLUTION: The dart can be regarded as a particle moving under the influences of a force $F_x = -kx$. The initial value of x is negative ($x_1 = -6.0$ cm). According to Eq. (42), the initial energy is

$$E = \tfrac{1}{2}mv_1^2 + \tfrac{1}{2}kx_1^2 = 0 + \tfrac{1}{2}kx_1^2 \tag{43}$$

When the spring reaches its equilibrium length ($x_2 = 0$), the energy will be

$$E = \tfrac{1}{2}mv_2^2 + \tfrac{1}{2}kx_2^2 = \tfrac{1}{2}mv_2^2 + 0 \tag{44}$$

Conservation of energy demands that the right sides of Eqs. (43) and (44) be equal:

$$\tfrac{1}{2}mv_2^2 = \tfrac{1}{2}kx_1^2 \tag{45}$$

If we cancel the factors of $\frac{1}{2}$ in this equation, divide both sides by m, and take the square root of both sides, we find that the speed of the dart as it leaves the spring at $x_2 = 0$ is

$$v_2 = \sqrt{\frac{k}{m}x_1^2} \tag{46}$$

$$= \sqrt{\frac{320\ \text{N/m}}{0.008\ \text{kg}} \times (-0.060\ \text{m})^2} = 12\ \text{m/s}$$

COMMENTS AND SUGGESTIONS: Note that the use of energy conservation in the solution of this example involved the three steps outlined in Example 6. We began with an expression for the energy at one point [Eq. (43)] and an expression for the energy at another point [Eq. (44)], and then we used energy conservation to equate these expressions [Eq. (45)].

To formulate the law of conservation of mechanical energy for a particle moving under the influence of any other force, with some more complicated dependence on position, we want to imitate the above construction. We will be able to do this, if, and only if, the work performed by the force can be expressed as a difference between two potential energies, as in Eq. (37). ***If the force meets this requirement (and therefore permits the construction of a conservation law), the force is called conservative.*** Thus, the force of gravity and the force of a spring are conservative forces. Note that for any such force, the work done when the particle starts at the point x_1 and *returns* to the same point is necessarily zero, since, with $x_2 = x_1$, Eq. (37) implies

◀ Conservative force

$$W = U_1 - U_1 = 0 \tag{47}$$

This simply means that for a roundtrip that starts and ends at x_1, the work the force does during the outward portion of the trip is exactly the negative of the work the force does during the return portion of the trip, and therefore the net work for the roundtrip is zero (see Fig. 4.20). Thus, the energy supplied by the force is recoverable: the energy supplied by the force during motion in one direction is restored during the return motion in the opposite direction. For instance, when a particle moves downward from some starting point, gravity performs positive work; and when the particle moves upward, returning to its starting point, gravity performs negative works of a magnitude exactly equal to that of the positive work.

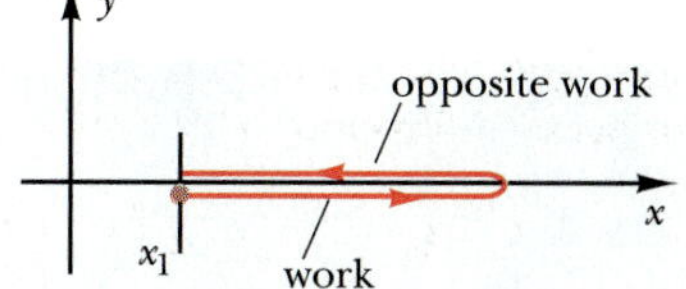

FIGURE 4.20
A particle starts at a point x_1 and returns to the point x_1 after completing some roundtrip. If the force is conservative, the work done is zero, because the work for the outward portion of the trip is opposite to that for the inward portion.

The requirement of zero work for a roundtrip can be used to discriminate between conservative and nonconservative forces. Friction is an example of a nonconservative force. If we slide a metal block through some distance along a table, and then slide the block back to its starting point, the net work is not zero (see Fig. 4.21). The work performed by the friction force during the outward portion of the motion is negative, and the work performed by the friction force during the return portion of the trip is also negative — the friction always opposes the motion, and the work done by the friction force is always negative. Thus, the work done by the friction force cannot be expressed as a difference between two potential energies, and we cannot formulate a law of conservation of mechanical energy if friction forces are acting.

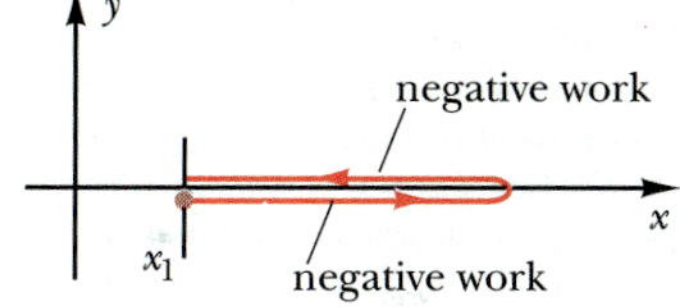

FIGURE 4.21
If a frictional force acts on the particle, the work is negative for all portions of the roundtrip.

If several conservative forces simultaneously act on a particle, then the net potential energy is the sum of all potential energies of all these forces. The total mechanical energy is the sum of the kinetic energy and the net potential energy; this total mechanical energy is, of course, conserved.

4.6 OTHER FORMS OF ENERGY

If the forces acting on a particle are conservative, then the mechanical energy of the particle is conserved. But if some of the forces acting on the particle are not conservative, then the mechanical energy of the particle — consisting of the sum of the kinetic energy and the net potential energy of all the conservative forces acting on the particle — will not remain constant. For instance, if friction forces are acting, they do negative work and thereby decrease the mechanical energy of the particle.

However, it is a remarkable fact about our physical universe that ***whenever mechanical energy is lost by a body, this energy never disappears — it is merely transmuted into other forms of energy.*** Thus, in the case of friction, the mechanical energy lost by the body is transformed into kinetic and potential energy of the atoms in the body and in the surface against which it is rubbing. The energy that the atoms acquire in the rubbing process is disorderly kinetic and potential energy — it is spread out among the atoms in an irregular, random fashion. This disorderly kinetic and potential energy of the atoms of a body is what is meant by ***heat,*** or ***thermal energy.*** Hence, friction produces heat.

Thermal energy ▸

Hermann von Helmholtz, *1821–1894, Prussian surgeon, biologist, mathematician, and physicist. His scientific contributions ranged from the invention of the ophthalmoscope and studies of the physiology and physics of vision and hearing to the measurement of the speed of light and studies in theoretical mechanics. Helmholtz formulated the general law of conservation of energy, treating it as a consequence of the basic laws of mechanics and electricity.*

Heat is a form of energy, but whether it is to be regarded as a new form of energy or not depends on the point of view we wish to adopt. If we take a macroscopic point of view, we ignore the atomic motions; then heat is to be regarded as distinct from mechanical energy. If we take a microscopic point of view, we recognize heat as kinetic and potential energy of the atoms; then heat is to be regarded as mechanical energy. (We will further discuss heat and thermal energy in Chapter 14.)

Chemical energy and nuclear energy are two other forms of energy. The former is kinetic and potential energy of the electrons within the atoms; the latter is kinetic and potential energy of the protons and neutrons within the nuclei of atoms. As in the case of heat, whether these are to be regarded as new forms of energy depends on the point of view.

Electric and magnetic energy are forms of energy associated with electric charges and with light and radio waves. (We will examine these forms of energy in Chapters 18 and 24.)

Table 4.2 lists some examples of different forms of energy. All the energies in Table 4.2 have been expressed in joules, the unit of energy in the metric system. However, for reasons of tradition and convenience, some other energy units are often used in specialized areas of physics and engineering.

TABLE 4.2 SOME ENERGIES	
Nuclear fuel in the Sun	1×10^{45} J
Explosion of a supernova	1×10^{44} J
Fossil fuel available on Earth	2.0×10^{23} J
Yearly energy expenditure of the United States	8×10^{19} J
Volcanic explosion (Krakatoa)	6×10^{18} J
Annihilation of 1 kg of matter–antimatter	9.0×10^{16} J
Explosion of thermonuclear bomb (1 megaton)	4.2×10^{15} J
Gravitational potential energy of airliner (Boeing 747 at 10,000 m)	2×10^{10} J
Combustion of 1 gal. of gasoline	1.3×10^{8} J
Daily food intake of man (3000 kcal)	1.3×10^{7} J
Explosion of 1 kg of TNT	4.6×10^{6} J
Metabolization of one apple (110 kcal)	4.6×10^{5} J
One push-up	3×10^{2} J
Fission of one uranium nucleus	3.2×10^{-11} J
Energy of ionization of hydrogen atom	2.2×10^{-18} J

The energy of atomic and subatomic particles is usually measured in ***electron-volts*** (eV), where

$$1 \text{ electron-volt} = 1 \text{ eV} = 1.60 \times 10^{-19} \text{ J} \tag{48}$$

The energy supplied by electric power plants is usually expressed in ***kilowatt-hours*** (kW · h),

$$1 \text{ kilowatt-hour} = 1 \text{ kW} \cdot \text{h} = 3.60 \times 10^{6} \text{ J} \tag{49}$$

◀ **Kilowatt-hour (kW · h)**

And the thermal energy supplied by the combustion of fuels or by other chemical reactions is expressed in ***kilocalories*** (kcal),

$$1 \text{ kilocalorie} = 1 \text{ kcal} = 4.178 \times 10^{3} \text{ J} \tag{50}$$

We will learn more about these units in later chapters.

All these forms of energy can be transformed into one another. For example, in an internal-combustion engine, chemical energy of the fuel is transformed into heat and kinetic energy; in a hydroelectric power station, gravitational potential energy of the water is transformed into electric energy; in a nuclear reactor, nuclear energy is transformed into heat, light, kinetic energy, etc. However, in any such transformation process, the sum of all the energies of all the pieces of matter involved in the process remains constant: ***the form of the energy changes, but the total amount of energy does not change.*** This is the ***general law of conservation of energy.***

◀ **General law of conservation of energy**

EXAMPLE 8. The "calorie" used by dieticians to express the energy equivalents of different foods is actually a kilocalorie, or a "large" calorie. To measure the energy equivalent of some kind of food — for instance, sugar — a sample is placed in a bomb calorimeter, a closed vessel filled with oxygen at high pressure (see Fig. 4.22). The sample is ignited and burnt completely (complete oxidation). The number of calories released in this chemical reaction — for instance, 4.1 kcal for 1 g of sugar — tells us the maximum amount of energy that can be extracted from this food. The human body does not necessarily "burn" food quite as completely, and the muscles do not convert all of the available chemical energy into mechanical energy. However, energy conservation tells us that from one gram of sugar the body cannot produce more than 4.1 kcal of mechanical work. If you eat one spoonful (4 g) of sugar, what is the maximum height to which this permits you to climb stairs? Assume your mass is 70 kg.

SOLUTION: The energy equivalent of 4 g of sugar is 4×4.1 kcal $= 16.4$ kcal $= 16.4 \times 4.178 \times 10^{3}$ J $= 6.85 \times 10^{4}$ J. When you climb the stairs to a height y, this energy becomes gravitational potential energy:

$$mgy = 6.85 \times 10^{4} \text{ J}$$

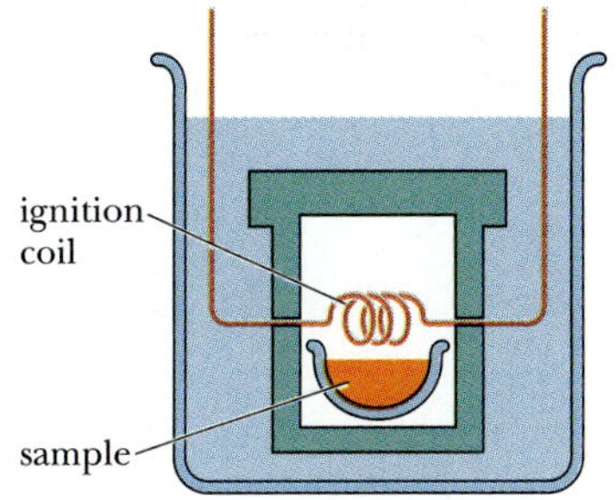

FIGURE 4.22

A bomb calorimeter. The sample is ignited electrically, by a glowing wire.

From which

$$y = \frac{6.85 \times 10^4 \text{ J}}{70 \text{ kg} \times 9.81 \text{ m/s}^2} = 100 \text{ m} \tag{51}$$

In practice, because of the limited efficiency of your body, only about 20% of the chemical energy of food is converted into mechanical energy; thus, the actual height you can climb is only about 20 m. (Because of the strong musculature of the human leg, stair climbing is one of your most efficient activities; most of your other physical activities are rather less efficient in converting chemical energy into mechanical energy.)

James Watt, *1736–1819, Scottish inventor and engineer. He modified and improved an earlier steam engine. Watt introduced the* horsepower *as a unit of mechanical power.*

4.7 Power

When we use an automobile engine to move a car up a hill or when we use an electric motor to lift an elevator cage, the important characteristic of the engine is not how much force it can exert, but rather how much work it can perform in a given amount of time. The force is only of secondary importance, because by shifting to a low gear we can make sure that even a "weak" engine exerts enough force on the wheels to propel the automobile uphill. But the work performed in a given amount of time, or the rate of work, is crucial since it determines how fast the engine can propel the car up the hill. While the car moves uphill, the gravitational force takes energy from the car, that is, it performs negative work on the car. To keep the car moving, the engine must perform an equal amount of positive work. If the engine is able to perform this work at a fast rate, it can propel the car uphill at a fast speed.

The rate at which a force does work on a body is called the power delivered by the force. If the force does an amount of work W in an interval of time Δt, then the ***average power*** is the ratio of W and Δt,

Power ▸

$$\overline{P} = \frac{W}{\Delta t} \tag{52}$$

The ***instantaneous power*** is defined by a procedure analogous to that involved in the definition of the instantaneous velocity. We consider the small amount of work W done in the small interval of time Δt and take the ratio of these small quantities:

$$\overline{P} = \frac{W}{\Delta t} \quad \text{with } \Delta t \text{ extremely small} \tag{53}$$

According to these definitions, the engine of your automobile delivers high power if it performs a large amount of work on the wheels (or, rather, the driveshaft) in a short time. The maximum power delivered by the engine determines the maximum speed of which this automobile is capable when moving up a hill, and also when moving on a flat road. Although a speeding automobile on a flat road loses no energy to the gravitational force, it loses energy to air resistance at a prodigious rate, and this loss has to be made good by the engine. You might also expect that the power of the engine determines the maximum acceleration of which the automobile is capable. But the acceleration is determined by the maximum force exerted by the engine on the wheels, and this is not directly related to the power as defined above.

In the metric system, the unit of power is the watt (W), which is the rate of work of one joule per second,

$$1 \text{ watt} = 1 \text{ W} = 1 \text{ J/s} \tag{54}$$

◀ **Watt (W)**

In engineering practice, power is often measured in ***horsepower*** (hp) units, where

$$1 \text{ horsepower} = 1 \text{ hp} = 746 \text{ W} \tag{55}$$

◀ **Horsepower (hp)**

This is roughly the rate at which a (very strong) horse can do work.

Note that multiplication of a unit of power by a unit of time gives a unit of energy. An example of this is the kilowatt-hour (kW · h), already mentioned in Section 4.6:

$$1 \text{ kilowatt-hour} = 1 \text{kW} \cdot \text{h} = 1 \text{ kW} \times 1 \text{ h} = 1000 \text{ W} \times 3600 \text{ s}$$

$$= 3.6 \times 10^6 \text{ J} \tag{56}$$

This unit is commonly used to measure the electric energy delivered to homes and factories.

EXAMPLE 9. An elevator cage has a mass of 1000 kg. How many horsepower must the motor deliver to the elevator if it is to raise the elevator cage at the rate of 2.0 m/s? The elevator has no counterweight (see Fig. 4.23).

SOLUTION: The weight of the elevator is $w = mg = 1000 \text{ kg} \times 9.81 \text{ m/s}^2 \simeq 9800$ N. By means of the elevator cable, the motor must exert an upward force equal to the weight to raise the elevator at a steady rate. If the elevator moves up a distance Δy, the work done by the force is

$$W = F\Delta y \tag{57}$$

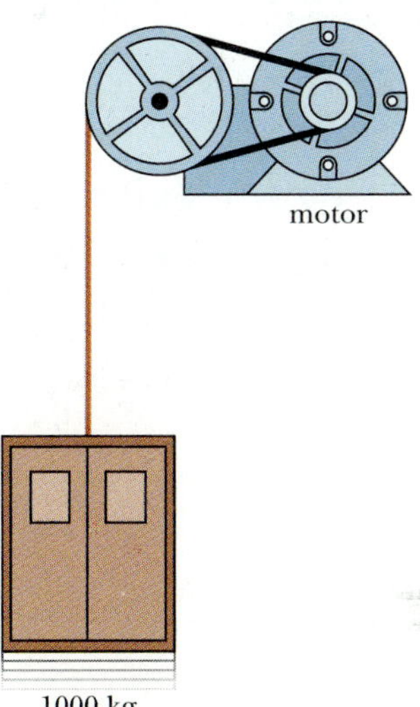

FIGURE 4.23
Elevator cage and motor.

To obtain the power, or the rate of work, we divide this by the time interval Δt:

$$P = \frac{W}{\Delta t} = \frac{F\Delta y}{\Delta t} = Fv \tag{58}$$

where $v = \Delta y/\Delta t$ is the speed of the elevator. With $F = 9800$ N and $v = 2.0$ m/s, we then find

$$P = Fv = 9800 \text{ N} \times 2.0 \text{ m/s} = 2.0 \times 10^4 \text{ W}$$

Since 1 hp = 746 W, this equals

$$P = \frac{2.0 \times 10^4}{746} \text{ hp} = 27 \text{ hp} \tag{59}$$

Equation (58) is a special instance of the following general formula, which expresses the instantaneous power in terms of force and speed. When the body moves a small distance Δs, the force **F** will perform an amount of work

$$W = F\Delta s \cos\theta \tag{60}$$

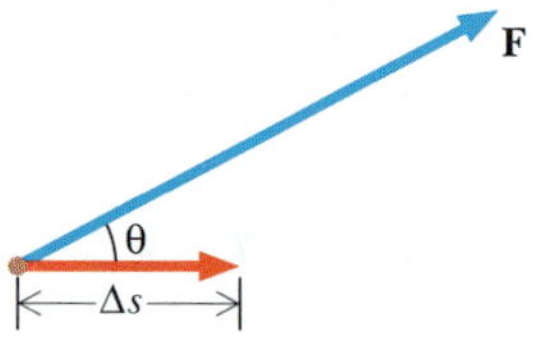

FIGURE 4.24
The force **F** makes an angle θ with the direction of motion.

where θ is the angle between the direction of the force and the direction of motion (see Fig. 4.24). The instantaneous power delivered by this force is then

$$P = \frac{W}{\Delta t} = F\frac{\Delta s}{\Delta t}\cos\theta \tag{61}$$

Since $\Delta s/\Delta t$ is the speed, the general expression for the power is

Power delivered by a force ▸

$$P = Fv\cos\theta \tag{62}$$

EXAMPLE 10. A horse pulls a sled up a steep snow-covered street of slope 1:7 (see Fig. 4.25a). The sled has a mass of 300 kg and the force of friction between the sled and the snow is 350 N. If the horse pulls parallel to the surface of the street and delivers a power of 1.0 hp, what is the maximum (constant) speed with which the horse can pull the sled uphill? What fraction of the horse's power is expended against gravity? What fraction against friction?

SOLUTION: Figure 4.25b is a "free-body" diagram for the sled, showing the weight ($w = mg$), the normal force ($N = mg\cos\phi$), the friction force ($f_k = \mu_k N$), and the pull of the horse (T). With the x axis along the street and the y axis perpendicular, the components of these forces are (see Fig. 4.25c)

$$w_x = -mg\sin\phi \qquad w_y = -mg\cos\phi$$

$$N_x = 0 \qquad N_y = mg\cos\phi$$

$$f_{k,x} = 350\text{ N} \qquad f_{k,y} = 0$$

$$T_x = T \qquad T_y = 0$$

Since the acceleration along the street is zero, the sum of the x components of these forces must be zero,

$$-mg\sin\phi + 0 - 350\text{ N} + T = 0 \tag{63}$$

This tells us what the pull of the horse is:

$$T = mg\sin\phi + 350\text{ N} \tag{64}$$

This simply says that the pull of the horse must balance the component of the weight along the street plus the friction force. The direction of this pull is parallel to the direction of motion of the sled. Hence, in Eq. (62) we have $\theta = 0$, and the power delivered by the horse is

$$P = Tv = (mg\sin\phi + 350\text{ N})v \tag{65}$$

Solving this equation for v, we find

$$v = \frac{P}{mg\sin\phi + 350\text{ N}} \tag{66}$$

For a slope of 1:7, the tangent of the angle of inclination is $\tan\phi = 1/7$, and, with our calculator, we find $\phi = 8°$. Hence

$$v = \frac{746\text{ W}}{300\text{ kg} \times 9.81\text{ m/s}^2 \times \sin 8° + 350\text{ N}} = \frac{746\text{ W}}{409\text{ N} + 350\text{ N}}$$

$$= 0.98\text{ m/s}$$

The weight of the sled makes an angle of 90° + 8° with the direction of motion of the sled (see Fig. 4.25b). The power delivered by the weight of the sled is given by Eq. (62), with $F = mg$ and $\cos\theta = \cos 98°$,

$$P_{\text{gravity}} = mgv\cos 98° = 300\text{ kg} \times 9.81\text{ m/s}^2 \times 0.98\text{ m/s} \times \cos 98°$$

$$= -401\text{ W} = -0.54\text{ hp}$$

Thus, 54% of the horse's power is expended against gravity, and the remaining 46% against friction.

(a)

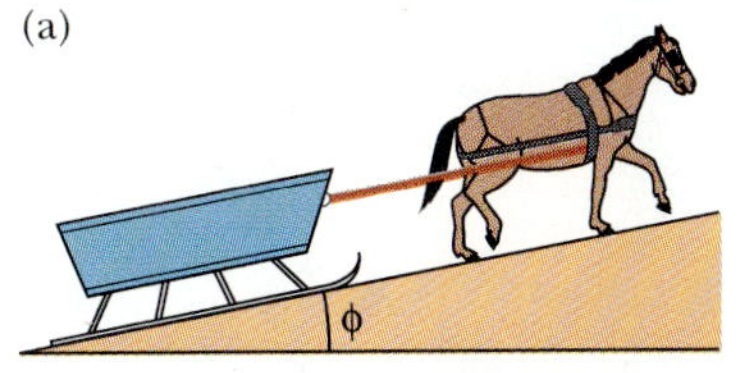

(b)

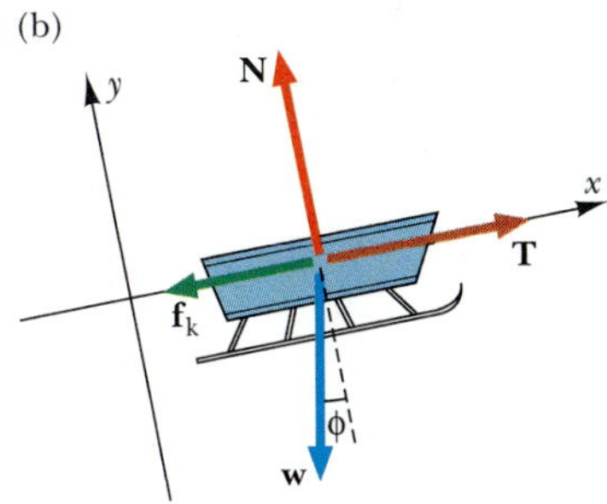

(c)

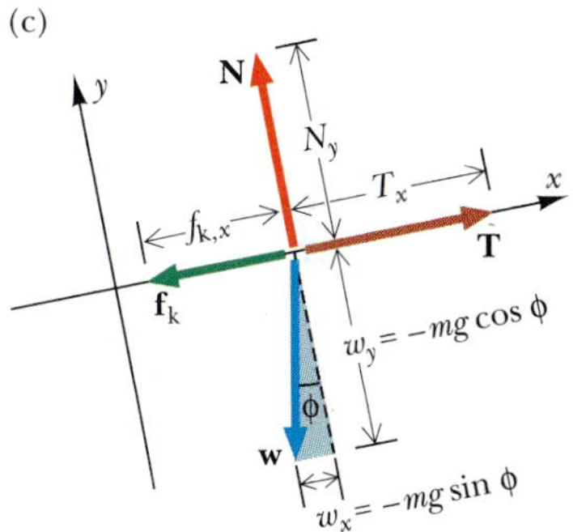

FIGURE 4.25

(a) Horse dragging a sled up a street, (b) "free-body" diagram for the sled, and (c) the components of these forces.

(a)

(b)

(c)

TABLE 4.3	SOME POWERS
Light and heat emitted by the Sun	3.9×10^{26} W
Mechanical power generated by hurricane **(a)**	2×10^{13} W
Total power used in United States (average)	2×10^{12} W
Large electric power plant	$\simeq 10^{9}$ W
Jet airliner engines (Boeing 747) **(b)**	2.1×10^{8} W
Automobile engine	1.5×10^{5} W
Solar light and heat per square meter at Earth	1.4×10^{3} W
Electricity used by toaster	1×10^{3} W
Work output of man (athlete at maximum)	2×10^{2} W
Electricity used by light bulb	1×10^{2} W
Basal metabolic rate for man (average)	88 W
Heat and work output of bumblebee (in flight) **(c)**	2×10^{-2} W
Atom radiating light	$\simeq 10^{-10}$ W

The above equations all refer to *mechanical* power. ***In general, power is the rate at which energy is converted from one form of energy to another or the rate at which energy is transported from one place to another.*** For instance, an automobile engine converts chemical energy of fuel into mechanical energy. A nuclear power plant converts nuclear energy into electric energy. And a high-voltage power line transports electric energy from one place to another. Table 4.3 gives some examples of different kinds of power.

The basal metabolic rate listed in Table 4.3 is for a man of average mass (73 kg) resting in a prone position.* The metabolic rate represents the rate of consumption of internal chemical energy (extracted from food by physiological processes). Under conditions of rest, the body of the man produces no external mechanical work, and the net result of the chemical and physiological processes within the body is merely the conversion of 88 W of internal chemical energy into an equal amount of heat. Thus, the basal metabolic rate of 88 W equals the rate of production of heat in the body of the man.

The metabolic rate is much higher when the man is performing heavy

* The basal metabolic rate is often expressed per kilogram of body mass, 1.2 W/kg; but since the metabolic rate is not proportional to the body mass (rather, it is proportional to the body mass raised to the power 0.67), the metabolic rate per kilogram is *not* the same for bodies of large and of small mass, and it is not a very meaningful quantity.

work. For instance, during strenuous bicycling, an athlete delivers 220 W of mechanical power to the pedals of the bicycle and, simultaneously, generates about 760 W of waste heat. The metabolic rate is then 220 W + 760 W, or 980 W.

SUMMARY

- **Work done by a constant force aligned with the displacement:** $W = F_x \Delta x$
- **Work done by a constant force not aligned with the displacement:** $W = Fs \cos\theta$
- **Work done by a variable force:** $W = \overline{F}_x \Delta x$
- **Kinetic energy:** $K = \frac{1}{2}mv^2$
- **Work–energy theorem:** $\Delta K = W$
- **Gravitational potential energy:** $U = mgy$
- **Mechanical energy:** $E = K + U$
- **Conservation of mechanical energy:** $E = K + U = [\text{constant}]$
- **Conservative force:** The work done by the force is zero for any roundtrip.
- **Work done by conservative force:** $W = U_1 - U_2$
- **Potential energy of a spring:** $U = \frac{1}{2}kx^2$
- **Average power:** $\overline{P} = \dfrac{W}{\Delta t}$
- **Instantaneous power:** $P = \dfrac{W}{\Delta t}$ with Δt extremely small
- **Mechanical power delivered by a force:** $P = Fv \cos\theta$

QUESTIONS

1.–Does the work of a force on a body depend on the frame of reference in which it is calculated? Give some examples.

2.–Does your body do work (external or internal) when standing at rest? When walking steadily along a level road?

3.–Consider a pendulum swinging back and forth. During what part of the motion does the weight do positive work? Negative work?

4.–Consider a woman steadily climbing a flight of stairs. The external forces on the woman are her weight and the normal force of the stairs against her feet. During the climb, the weight does negative work, while the normal force does no work. Under these conditions how can the kinetic energy of the woman remain constant? (Hint: the entire woman cannot be regarded as particle, since her legs are not rigid; but the upper part of her body can be regarded as a particle, since it is rigid. What is the force of her legs against the upper part of her body? Does this force do work?)

5.–An automobile increases its speed from 80 to 100 km/h. What is the percent increase of kinetic energy? What is the percent reduction of travel time for a given distance?

6.–Why do elevators have counterweights? (See Fig. 3.29.)

7.–A parachutist jumps out of an airplane, opens a parachute, and lands safely on the ground. Is the mechanical energy for this motion conserved?

8.–If you release a tennis ball at some height above a hard floor, it will bounce up and down several times, with a gradually decreasing amplitude. Where does the ball suffer a loss of mechanical energy?

9.–Two ramps, one steeper than the other, lead from the floor to a loading platform (Fig. 4.26). It takes more force to push a (frictionless) box up the steeper ramp. Does this mean it takes more work to raise the box from the floor to the platform?

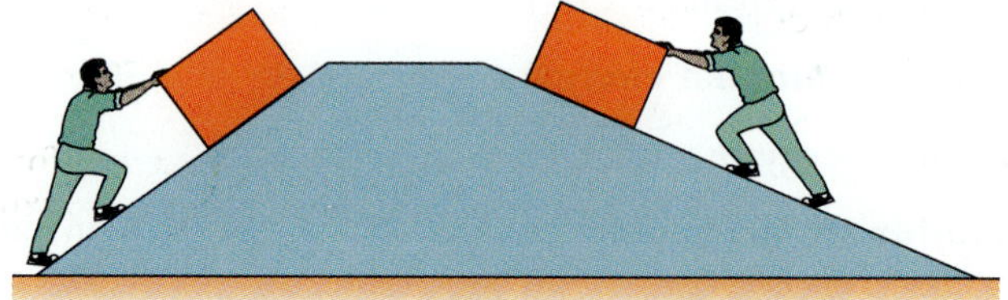

FIGURE 4.26
Two ramps of different steepness.

10.–Consider the two ramps described in the preceding question. Taking friction into account, which ramp requires less work for raising a box from the floor to the platform?

11.–If you stretch a spring so far that it suffers a permanent deformation, is the force exerted by the spring during this operation conservative?

12.–What happens to the kinetic energy of an automobile during braking without skidding? With skidding?

13.–Consider a stone thrown vertically upward. If we take friction against the air into account, we see that $\frac{1}{2}mv^2 + mgy$ must *decrease* as a function of time. From this, prove that the stone will take longer for the downward motion than for the upward motion.

14.–An automobile travels down a road leading from a mountain peak to a valley. What happens to the gravitational potential energy of the automobile? How is it dissipated?

15.–Suppose you wind up a watch and then place it into a beaker full of nitric acid and let it dissolve. What happens to the potential energy stored in the spring of the watch?

16.–News reporters commonly speak of "energy consumption." Is it accurate to say that energy is *consumed?* Would it be more accurate to say that energy is *dissipated?*

17.–The explosive yield of thermonuclear bombs (Fig. 4.27) is usually reported in kilotons or megatons of TNT. Would the explosion of a 1-megaton hydrogen bomb really produce the same effects as the explosion of 1-megaton on TNT (a mountain of TNT several hundred feet high)?

FIGURE 4.27
Thermonuclear explosion.

18.–It takes about 5000 hp to keep a 26-m yacht moving at its top speed of 89 km/h (50 knots). What happens to this power?

19.–In order to travel at 128 km/h, an automobile of average size needs an engine delivering about 40 hp to overcome the effects of air friction, road friction, and internal friction (in the transmission and drive train). Why do most drivers think they need an engine of 150 or 200 hp?

PROBLEMS

Section 4.1: Work

1.——If it takes a horizontal force of 300 N to push a stalled automobile along a level road, how much work must you do to push this automobile a distance of 5.0 m?

2.——In an overhead lift, a champion weight lifter raises 254 kg from the floor to a height of 2.0 m. How much work does he do?

3.——An apple falls down 35 m from the fifth floor of an apartment building to the street. The mass of the apple is 0.20 kg. How much work does gravity do on the apple during this fall?

4.——Suppose that the force required to push a saw back and forth through a piece of wood is 40 N. If you push this saw back and forth 30 times, moving it forward 12 cm and back 12 cm each time, how much work do you do?

5.*——A man pushes a crate along a flat concrete floor. The mass of the crate is 120 kg and the coefficient of friction between the crate and the floor is $\mu_k = 0.5$. How much work does the man do if, pushing horizontally, he moves the crate 15 m?

6.*——A woman slowly lifts a 20-kg package of books from the floor to a height of 1.8 m, and then slowly returns it to the floor. How much work does she do on the package while lifting? How much work does she do on the package while lowering? What is the total work she does on the package? From the information given, can you tell how much work she expends internally in her muscles — that is, how many calories she expends?

7.*——An automobile of 1200 kg is traveling at 25 m/s when the driver suddenly applies the brakes so as to lock the wheels and cause the automobile to skid to a stop. The coefficient of sliding friction between the tires and the road is 0.9.

(a) What is the deceleration of the automobile, and what is the stopping distance?

(b) What is the friction force of the road on the wheels, and what is the amount of work that this friction force does during the stopping process?

8.*——An elevator consists of an elevator cage and a counterweight attached to the ends of a cable which runs over a pulley (Fig. 4.28). The mass of the cage (with its load) is 1200 kg and the mass of the counterweight is 1000 kg. The elevator is driven by an electric motor attached to the pulley. Suppose that the elevator is initially at rest at the first floor of the building, and the motor makes the elevator accelerate upward at the rate of 1.5 m/s^2.

(a) What is the tension in the part of the cable attached to the elevator cage? What is the tension in the part of the cable attached to the counterweight?

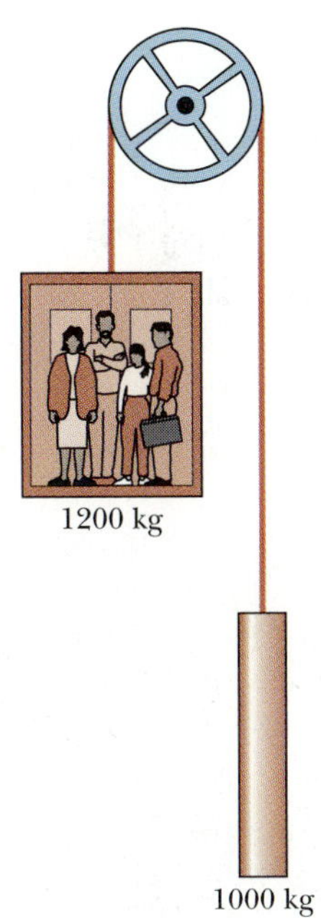

FIGURE 4.28
Elevator cage and counterweight.

(b) The acceleration lasts exactly 1.0 s. How much work has the electric motor done in this interval? Ignore friction forces and ignore the mass of the pulley.

(c) After the acceleration interval of 1.0 s, the motor pulls the elevator upward at constant speed until it reaches the third floor, exactly 10.0 m above the first floor. What is the total amount of work that the motor has done up to this point?

9.**—A man pushes a heavy box up an inclined ramp making an angle of 30° with the horizontal. The mass of the box is 60 kg and the coefficient of kinetic friction between the box and the ramp is 0.45. How much work must the man do to push the box to a height of 2.5 m? Assume that the man pushes on the box in a direction parallel to the surface of the ramp.

10.*—A woman pulls a sled by a rope. The rope makes an upward angle of 45° with the ground, and the woman exerts a pull of 150 N on the rope. How much work does the woman do if she pulls this sled 20 m?

11.*—Suppose you push on a block sliding on a table. Your push has a magnitude of 50 N and makes a downward angle of 60° with the direction of motion. What is the work you do on the block while the block moves a distance of 1.6 m?

12.*—Consider the barge being pulled by two tugboats, as described in Example 3.4. The pull of the first tugboat is 2.5×10^5 N at 30° to the left, and the pull of the second tugboat is 1.0×10^5 N at 15° to the right (see Fig. 3.9). What is the work done by each tugboat on the barge while the barge moves 100 m forward (in the direction of the x axis in Fig. 3.9)? What is the total work done by both tugboats on the barge?

13.—The spring used in the front suspension of a Triumph sports car has a spring constant $k = 3.5 \times 10^4$ N/m. How much work must you do to compress this spring by 0.10 m from its relaxed condition?

14.*—A spring has a force constant of 2×10^3 N/m. The spring is initially relaxed. How much work must you do to stretch the spring by 3 cm? How much work must you do to stretch the spring a further 3 cm?

Section 4.3: Kinetic Energy

15.—Calculate the kinetic energy that the Earth has owing to its motion around the Sun.

16.—The electron in a hydrogen atom has a speed of 2.2×10^6 m/s. What is the kinetic energy of this electron?

17.—A 1500-kg automobile is traveling at 20 m/s on a level road. How much work must be done on the body of the automobile to accelerate it from 20 m/s to 25 m/s? From 25 m/s to 30 m/s?

18.—According to statistical data, the probability that an occupant of an automobile suffers a lethal injury when involved in a crash is proportional to the square of the speed of the automobile. At a speed of 50 mi/h, the probability is approximately 3%. What are the probabilities at 60 mi/h, 70 mi/h, and 80 mi/h?

19.—The fastest skier is Michael Prufer, who attained 223.75 km/h on a steep slope. The fastest runner is Robert Hayes, who briefly attained 44.88 km/h on a level track. Assume that the skier and the runner each have a mass of 73 kg. What is the kinetic energy of each? By what factor is the kinetic energy of the skier larger than that of the runner?

20.—The Skylab satellite disintegrated when it reentered the atmosphere. Among the pieces that crashed down on the surface of the Earth, one of the heaviest was a lead-lined film vault of 1770 kg which had an estimated impact speed of 120 m/s on the surface. What was its kinetic energy? How many kilograms of TNT would we have to explode to release the same amount of energy? (One kilogram of TNT releases 4.6×10^6 J.)

21.* An automobile of mass 1600 kg is traveling along a straight road at 80 km/h.

(a) What is the kinetic energy of this automobile in the reference frame of the ground?

(b) What is the kinetic energy in the reference frame of a motorcycle traveling in the same direction at 60 km/h?

(c) What is the kinetic energy in the reference frame of a truck traveling in the opposite direction at 60 km/h?

22.* The velocity of small bullets can be roughly measured with ballistic putty. When the bullet strikes a slab of putty, it penetrates a distance that is roughly proportional to the kinetic energy. Suppose that a bullet of velocity 160 m/s penetrates 0.8 cm into the putty and a second, identical bullet fired from a more powerful gun penetrates 1.2 cm. What is the velocity of the second bullet?

23.* With the brakes fully applied, a 1500-kg automobile decelerates at the rate of 8.0 m/s^2.

(a) What is the braking force acting on the automobile?

(b) If the initial speed is 90 km/h, what is the stopping distance?

(c) What is the work done by the braking force in bringing the automobile to a stop from 90 km/h?

24.** A large stone-throwing engine designed by Archimedes could throw a 77-kg stone over a range of 180 m. Assume the stone is thrown at an initial angle of 45° with the vertical.

(a) Calculate the initial kinetic energy of this stone.

(b) Calculate the kinetic energy of the stone at the highest point of the trajectory.

Section 4.4: Potential Energy

25. It has been reported that at Cherbourg, France, waves smashing on the coast lifted a boulder of 3200 kg over a 6-m wall. What minimum energy must the waves have given to the boulder?

26. A small aircraft, of mass 1200 kg, is cruising at 250 km/h at an altitude of 3000 m. What is the kinetic energy and what is the gravitational potential energy of this aircraft?

27. A 75-kg man walks up the stairs from the first to the third floor of a building, a height of 10 m. How much work does he do against gravity? Compare your answer with the food energy he acquires by eating an apple (see Table 4.2).

28.* A golf ball of mass 50 g released from a height of 1.5 m above a concrete floor bounces back to a height of 1.0 m.

(a) What is the kinetic energy of the ball just before contact with the floor begins? Ignore air friction.

(b) What is the kinetic energy of the ball just after contact with the floor ends?

(c) What is the loss of energy during contact with the floor?

29.* Surplus energy from an electric power plant can be temporarily stored as gravitational energy by using this surplus energy to pump water from a river into a reservoir at some altitude above the level of the river. If the reservoir is 250 m above the level of the river, how much water (in kilograms) must we pump in order to store 2×10^{13} J?

A pole vaulter.

30.* The track of a cable car on Nob Hill in San Francisco rises more than 60 m from its lowest point. Suppose that a car is ascending at 13 km/h along the track when it breaks away from its cable at a height of exactly 60 m. It will then coast up the hill some extra distance, stop, and begin to race down the hill. What speed does the car attain at the lowest point of the track? Ignore friction.

31.* In pole vaulting, the jumper achieves great height by converting his kinetic energy of running into gravitational potential energy. The jumper uses the pole for this conversion. Suppose that a jumper runs at a speed of 10 m/s. If the jumper converts all of the corresponding kinetic energy into gravitational potential energy, how high will his center of mass rise? The actual height reached by pole vaulters is 5.7 m (measured from the ground). Is this consistent with your calculation?

32.* Because of a brake failure, a bicycle with its rider careens down a steep hill 45 m high. If the bicycle starts from rest and if there is no friction, what is the final speed attained at the bottom of the hill?

33.* Under favorable conditions, an avalanche can reach extremely great speeds because the snow rides down the mountain on a cushion of trapped air that makes the sliding motion nearly frictionless. Suppose that a mass of 2×10^7 kg of snow breaks loose from a mountain and slides down into a valley 500 m below the starting point. The explosion of 1 short ton (2000 lb) of TNT releases 4.2×10^9 J. How many tons of TNT release the same energy as the avalanche?

34.* A roller coaster near St. Louis is 34 m high at its highest point.
  (a) What is the maximum speed that a car can attain by rolling down from the highest point if initially at rest? Ignore friction.
  (b) Some people claim that cars reach a maximum speed of 100 km/h. If this is true, what must be the initial speed of a car at the highest point?

35.* A pendulum consists of a mass m tied to one end of a string of length l. The other end of the string is attached to a fixed point on the wall (see Fig. 4.29). Suppose that the pendulum is initially held at an angle of 90° with the vertical. If the pendulum is released from this position, what will be the speed of the pendulum at the instant it passes through its lowest point?

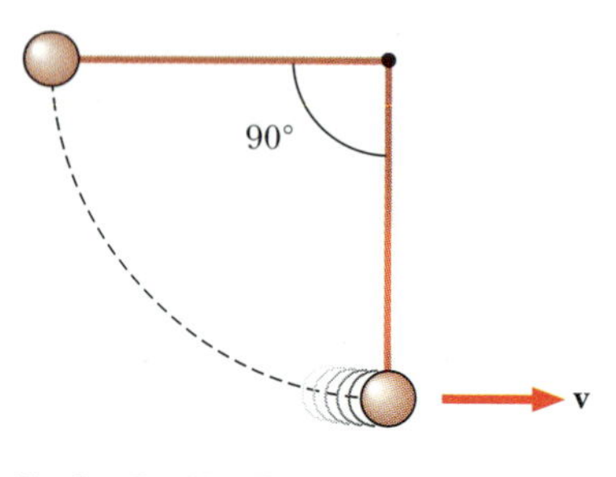

FIGURE 4.29
A pendulum.

36.* In a roller coaster, a car starts on the top of a 30-m-high mountain. It rolls down into a valley, and then up a 20-m-high mountain. What is the speed of the car at the bottom of the valley, at ground level? What is the speed of the car at the top of the second mountain?

37.* A parachutist of mass 60 kg jumps out of a helicopter at an altitude of 800 m. Her parachute opens and she lands on the ground with a speed of 5.0 m/s. How much energy has been lost to air friction in this jump?

38.** In some barge canals built in the nineteenth century, barges were lifted from a low level of the canal to a higher level by means of wheeled carriages. In a French canal, barges of 70 metric tons were placed on a carriage of 35 metric tons which was pulled, by a wire rope, to a height of 12 m along an inclined track 500 m long.
  (a) What was the tension in the wire rope?
  (b) How much work was done to lift the barge and carriage?
  (c) If the cable had broken just as the carriage reached the top, what would have been the final speed of the carriage when it crashed at the bottom? Ignore friction.

39.** A center fielder throws a baseball of mass 0.17 kg with an initial speed of 28 m/s and elevation angle of 30°. What is the kinetic energy and what is the potential energy of the baseball when it reaches the highest point of its trajectory? Ignore friction.

Section 4.5: Potential Energy of a Conservative Force

40.——The spring from an automobile suspension has a spring constant 3.53×10^4 N/m (see Example 3.13). How much potential energy do you store in this spring if you compress it from its relaxed length of 0.316 m to 0.205 m?

41.*——A mass m hangs at rest on a vertical spring of spring constant k.
(a) How far will this hanging mass have stretched the spring from its relaxed length?
(b) If you now push up on the mass and lift it until the spring reaches its relaxed length, how much work will you have done against gravity? Against the spring?

42.*——A 3.0-kg block is released from a compressed spring whose force constant is 120 N/m. After leaving the spring, it travels over a horizontal surface, with coefficient of friction 0.20 for a distance of 8.0 m before coming to rest (Fig. 4.30).
(a) What was its maximum kinetic energy?
(b) How far was the spring compressed before being released?

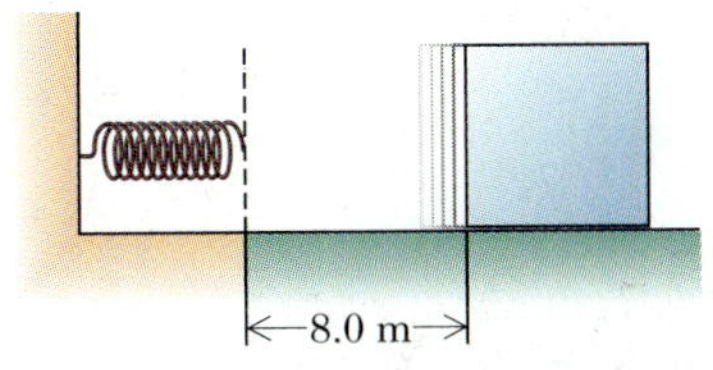

FIGURE 4.30
Block released from a spring.

43.*——A bow may be regarded mathematically as a spring. The archer stretches this "spring" and then suddenly releases it so that the bowstring pushes against the arrow. Suppose that when the archer stretches the "spring" 0.50 m, he must exert a force of 160 N to hold the arrow in this position. If he now releases the arrow, what will be the speed of the arrow when the "spring" reaches its equilibrium position? The mass of the arrow is 0.020 kg. Pretend that the "spring" is massless.

44.**——The four wheels of an automobile of mass 1200 kg are suspended below the body by vertical springs of constant $k = 7.0 \times 10^4$ N/m each. If the forces on all wheels are the same, what will be the maximum instantaneous deformation of the springs if the automobile is lifted by a crane and dropped on the street from a height of 0.8 m? [The height of 0.8 m is measured from the initial position of the automobile to the final position, where the automobile is (instantaneously) at rest.]

45.**——A rope can be regarded as a long spring; when under tension, it stretches and stores elastic potential energy. Consider a nylon rope similar to that which snapped during a giant tug-of-war at a school in Harrisburg (see Problem 3.15). Under a tension of 5.8×10^4 N (applied at its ends), the rope of initial length 300 m stretches to 390 m. What is the elastic energy stored in the rope at this tension? What happens to this energy when the rope breaks?

Section 4.6: Other Forms of Energy

46.——Express the last three entries in Table 4.1 in electron-volts.

47.——The mass of a TNT molecule is 227.0 u. The explosion of 1 kg of TNT releases 4.6×10^6 J. Calculate the energy released per molecule of TNT. Express your answer in electron-volts.

48.——Using the data of Table 4.2, calculate the amount of gasoline that would be required if all of the energy requirements of the United States were to be met by the consumption of gasoline. How many gallons per day would have to be consumed?

49.*——The following table lists the fuel consumption and the passenger capacity of several vehicles. Assume that the energy content of the fuel is that of gasoline (see Table 4.2). Calculate the amount of energy used by each vehicle per passenger per mile. Which is the most energy-efficient vehicle? The least energy efficient?

Vehicle	Passenger capacity	Fuel consumption
Motorcycle	1	60 mi/gal.
Snowmobile	1	12
Automobile	4	12
Intercity bus	45	5
Jetliner	360	0.1
Concorde SST	110	0.12

Section 4.7: Power

50.——The heart of a resting person delivers a mechanical power of about 1.5 W for pumping blood. Express this power in horsepower.

51.——If a 60-W light bulb is left on for 24 hours each day, how many kilowatt-hours of electric power does it use in one year? If the electric energy costs you 8 cents per kilowatt-hour, what is your cost for 1 year?

52.——While an automobile is cruising at a steady speed of 65 km/h, its engine delivers a mechanical power of 20 hp. How much energy does the engine deliver per hour?

53.——A large windmill delivers 10 kW of mechanical power. How much energy does the windmill deliver in a working day of 8 hours?

54.——For an automobile traveling at a steady speed of 65 km/h, the friction of the air and the rolling friction of the ground on the wheels provide a total external friction force of 500 N. At what rate does this force remove energy from the automobile?

55.——In 1979, B. Allen flew a very lightweight propeller airplane across the English Channel. His legs, pushing bicycle pedals, supplied the power to turn the propeller. To keep the airplane flying, he had to supply about 0.30 hp. How much energy did he supply for the full flight lasting 2 h 49 min? Express your answer in kilocalories.

56.——The ancient Egyptians and Romans relied on slaves as a source of mechanical power. One slave, working desperately by turning a crank, can deliver about 200 W of mechanical power (at this power the slave would not last long). How many slaves would be needed to match the output of a modern automobile engine (150 hp)? How many slaves would an ancient Egyptian have to own in order to command the same amount of power as the average per capita power used by residents of the United States (about 10 kW)?

57.——An electric clock uses 2 W of electric power. How much electric energy (in kilowatt-hours) does this clock use in 1 year? What happens to this electric energy?

58.——Nineteenth-century engineers reckoned that a laborer turning a crank can do steady work at the rate of 110 W. Suppose that four laborers working a manual crane attempt to lift a load of 8 metric tons. If there is no friction, what is the rate at which they can lift this load? How long will it take them to lift the load 5 m?

59.——The driver of an automobile traveling on a straight road at 80 km/h pushes forward with his hands on the steering wheel with a force of 50 N. What is the rate at which his hands do work on the steering wheel in the reference frame of the ground? In the reference frame of the automobile?

60.——A crane is powered by an electric motor delivering 60 hp. What is the maximum speed with which this crane can raise a load of 10 metric tons? Assume that 28% of the power is lost to friction within the crane.

61.——A horse walks along the bank of a canal and pulls a barge by means of a long horizontal towrope making an angle of 30° with the bank (see Fig. 3.54). The horse walks at the rate of 5 km/h and the tension in the rope is 400 N. What horsepower does the horse deliver?

62.*—The movement of a grandfather clock is driven by a 5-kg weight which drops a distance of 1.5 m in the course of a week. What is the power delivered by the weight to the movement?

63.——A six-cylinder internal-combustion engine, such as used in an automobile, delivers an average power of 150 hp while running at 3000 rev/min. Each of the cylinders fires once every two revolutions. How much energy does each cylinder deliver each time it fires?

64.*—The ancient Egyptians moved large stones by dragging them across the sand in sleds. Suppose that 6000 Egyptians are dragging a sled with a coefficient of sliding friction $\mu_k = 0.3$ along a level surface of sand.

(a) If each Egyptian exerts a force of 350 N, what is the maximum mass they can move at constant speed?

(b) If each Egyptian delivers a mechanical power of 0.20 hp, what is the maximum speed with which they can move this mass?

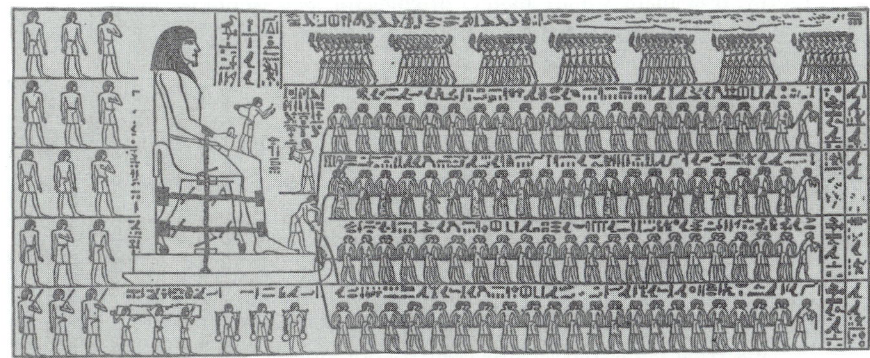

Ancient Egyptian wall mural from 1900 B.C.

65.*—An automobile engine typically has an efficiency of about 25%; that is, it converts about 25% of the chemical energy available in gasoline into mechanical energy. Suppose that an automobile engine has a mechanical output of 110 hp. At what rate (in gallons per hour) will this engine consume gasoline? See Table 4.2 for the energy content of gasoline.

66.*—In a braking test, a 990-kg automobile takes 2.1 s to come to a full stop from an inital speed of 60 km/h. What is the amount of energy dissipated in the brakes? What is the average power dissipated in the brakes? Ignore external friction in your calculation and express the power in horsepower.

67.*—The takeoff speed of a DC-3 airplane is 100 km/h. Starting from rest, the airplane takes 10 s to reach this speed. The mass of the (loaded) airplane is 11,000 kg. What is the average power delivered by the engines to the airplane during takeoff?

68.——A 900-kg automobile accelerates from 0 to 80 km/h in 7.5 s. What are the initial and the final kinetic energies of the automobile? What is the average power delivered by the engine to the automobile in this time interval? Express your answer in horsepower.

69.*—In a waterfall on the Alto Paraná river (between Brazil and Paraguay), the height of fall is 34 m and the average rate of flow is 18,000 m^3 of water per second. What is the power wasted by this waterfall?

70.*—Suppose that in the waterwheel shown in Fig. 4.14 the stream of water against the blades of the wheel has a speed of 15 m/s, and the amount of water is 30 liters per second (or 30 kg per second). If the water gives all of its kinetic energy to the blades (and then drips down with zero horizontal speed), how much mechanical power does the water deliver to the wheel?

71.*—At Niagara Falls, 6200 m^3 of water per second fall down a height of 49 m.

(a) What is the rate (in watts) at which gravitational potential energy is wasted by the falling water?

(b) What is the amount of energy (in kilowatt-hours) wasted in 1 year?

(c) Power companies get paid about 8 cents per kilowatt-hour of electric energy. If all of the gravitational potential energy wasted at Niagara Falls could be converted into electric energy, how much money would this be worth?

72.*—A pump sitting on the shore of a pond takes in 0.80 kg of water per second and squirts it out of a nozzle at 50 m/s. What mechanical power does this pump supply to the water?

73.*—The energy of sunlight arriving at the surface of the Earth amounts to about 1 kW per square meter of surface (facing the Sun). If all of the energy incident on a collector of sunlight could be converted into useful energy, how many square meters of collector area would we need to satisfy all of the energy demands in the United States? See Table 4.2 for the energy expenditure of the United States.

74.**—A record of 203.1 km/h (126.2 mi/h) for speed skiing set by Franz Weber at Velocity Peak, Colorado, was achieved on a mountain slope inclined downward at 51°. At this speed, the force of friction (air and sliding friction) balances the pull of gravity along the slope, so that the motion proceeds at constant velocity.

(a) What is the rate at which gravity does work on the skier? Assume that the mass of the skier is 75 kg.

(b) What is the rate at which sliding friction does work? Assume that the coefficient of friction is $\mu_k = 0.03$.

(c) What is the rate at which air friction does work?

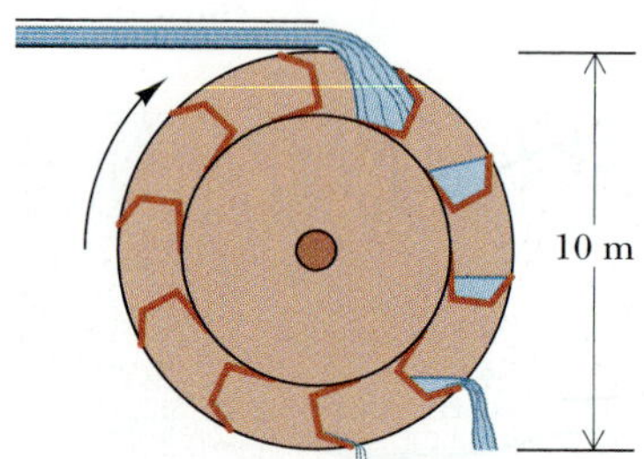

FIGURE 4.31
Overshot waterwheel.

75.**—Figure 4.31 shows an overshot waterwheel, in which water flowing onto the top of the wheel fills buckets whose weight causes the wheel to turn. The water descends in the buckets to the bottom, and there it is spilled out, so the ascending buckets are always empty. If in a waterwheel of diameter 10 m the amount of water carried down by the wheel is 20 liters per second (or 20 kg per second), what is the mechanical power that the descending water delivers to the wheel? Assume that the water flowing onto the top of the wheel has roughly the same speed as the wheel, and exerts no horizontal push on the wheel. [Hint: The kinetic energy of the water is the same when the water enters the bucket and when it spills out (since the speed of the bucket is constant); hence the kinetic energy of the water does not affect the answer.]

76.**—A 27,000-kg truck has a 550-hp engine. What is the maximum speed with which this truck can move up a 10° slope?

77.**—An electric water pump is rated at 15 hp. If this water pump is to lift water to a height of 30 m, how many kilograms of water can it lift per second? How many liters? Neglect the kinetic energy of the water.

PHYSICS IN ACTION

MUSCULAR CONTRACTION

When your muscles contract, they convert chemical energy from food into mechanical energy of motion. In some regards a muscle is similar to a steam engine or an automobile engine, which converts the chemical energy of fuel into mechanical energy. But steam engines and automobile engines are very inefficient; they first convert the chemical energy of fuel into heat, and they then convert this heat into mechanical energy — they are heat engines. Muscle is much more efficient; it converts chemical energy *directly* into mechanical energy by a molecular process that we cannot duplicate technologically.

The chemical fuel that muscle cells use to operate their contractile machinery is adenosine triphosphate (ATP), a high-energy molecule which releases its chemical energy when it splits into low-energy molecules of adenosine diphosphate (ADP) and inorganic phosphate (P_i). Metabolic reactions in the muscle cells immediately recycle the ADP and P_i, recombining them into high-energy ATP. The energy for recycling the ADP comes from the "raw" chemical fuel that the muscle cells absorb from the bloodstream. The preferred raw fuel is glucose, supplied by the liver and the digestive system; but if your body exhausts its supply of glucose, the muscle cells can metabolize fats and proteins. The metabolic reactions in the cell oxidize, or "burn," the glucose, leaving carbon dioxide and water as waste products, which the cells dump into the bloodstream.

To understand how the contractile machinery in the muscle cell converts the chemical energy of ATP into mechanical energy, we need to examine the microscopic and molecular structure of muscle. In the following description we will concentrate on skeletal, or striated, muscle, such as the muscles that produce the voluntary motions of your limbs or your trunk. This kind of muscle achieves the fastest and most efficient contractions.

Striated muscle is a bundle of long ***muscle fibers,*** each fiber about 20–80 μm in diameter. Each fiber consists of a single cell, but in contrast to ordinary cells, the fiber cell contains many nuclei because it is originally formed by the fusion of many cells. The fiber, in turn, is a bundle of a large number of long cylindrical units, called ***myofibrils,*** each about 1 μm in diameter (see Fig. 1). When examined with an electron microscope, the myofibrils have a banded, or striated, appearance (see Fig. 2). The dark bands

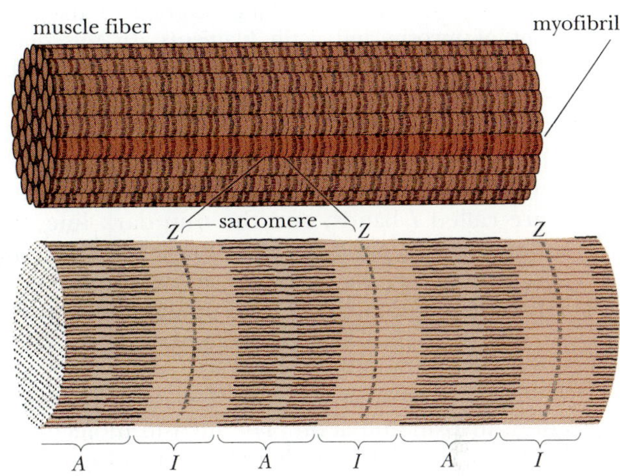

FIGURE 1 The structure of muscle.

FIGURE 2 Myofibrils of rabbit leg muscle. The myofibrils run diagonally across the picture. Note the dark and light striations.

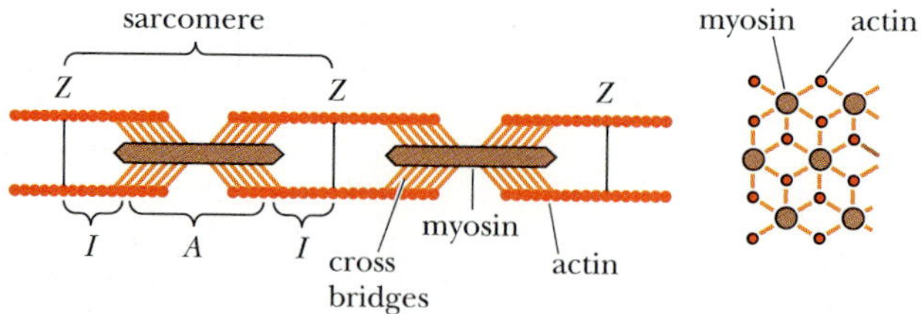

FIGURE 3 Schematic diagram of the arrangement of myosin and actin filaments. In this diagram, the transverse dimensions have been exaggerated by a factor of about 10, for clarity (compare Fig. 1). The sarcomere is shown in its relaxed state. Cross bridges extend outward from the myosin filament and link it to the actin filaments.

across the myofibril are called *A* bands, the light bands are called *I* bands, and the thin dark line in the center of the *I* band is called the *Z* line. These bands repeat at regular intervals along the myofibril. The basic repeating unit, extending from one *Z* line to the next, is called a ***sarcomere*** (from the Greek *sarco,* flesh, and *mere,* part).

The myofibril consists of two kinds of protein molecules, myosin and actin. The myosin molecules are arranged in thick parallel filaments which extend across the *A* band, from one edge of the *A* band to the other. The actin is arranged in thin parallel filaments which extend from the *Z* line in both directions across the *I* band, and slightly beyond the edge of the *I* band into the *A* band (see Fig. 3). Thus, the ends of the myosin filaments fit into slots formed by actin filaments. If we view the myofibril in cross section, end on, we find that the parallel filaments of myosin and actin are packed in a regular, hexagonal pattern (see Fig. 4).

The myofibrils are the contractile machinery of the muscle cell. The myofibrils contract when immersed in a supply of ATP. During their contraction, the myofibrils convert the chemical energy of ATP into mechanical energy. The myosin and the actin filaments in the myofibril do not change their lengths during contraction, but they slide past each other, so the ends of the myosin filaments fit

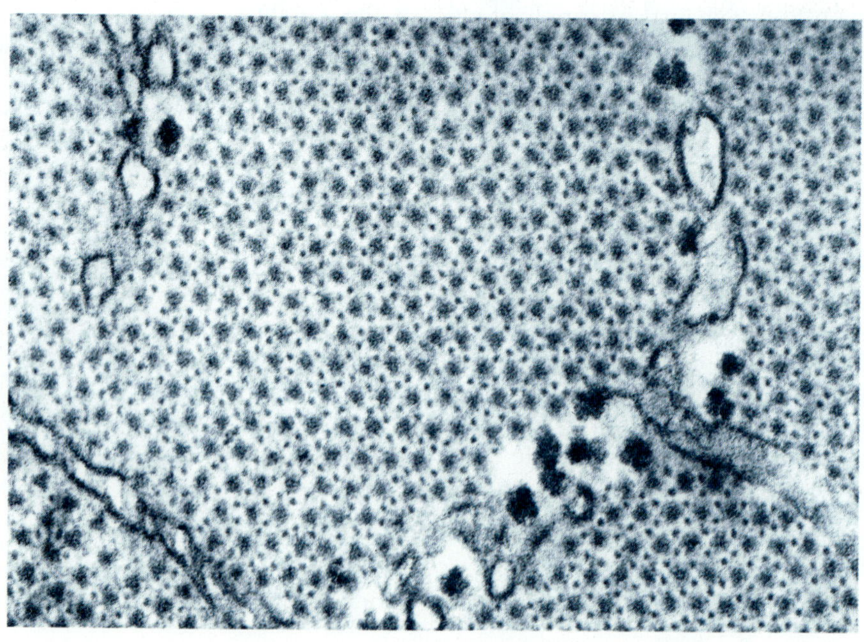

FIGURE 4 Regular hexagonal packing of myosin and actin filaments in a myofibril. Note that each myosin filament (thick) is surrounded by six actin filaments (thin) to which it is linked by cross bridges. The schematic diagram in Fig. 3 shows only two of these actin filaments.

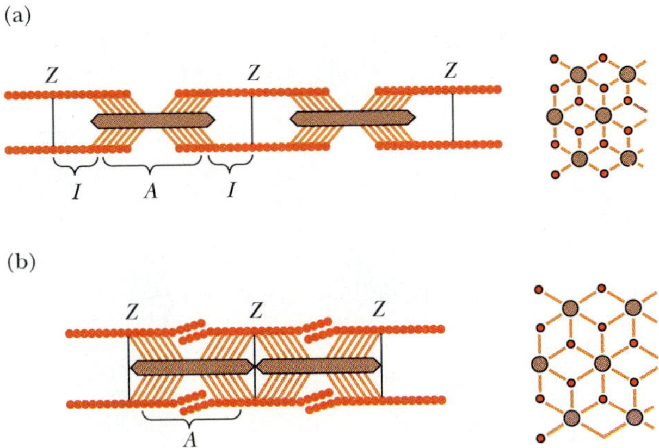

FIGURE 5 Contraction of myofibril. (a) Sarcomere in its relaxed state. (b) Sarcomere in its extreme contracted state. Note that the *I* band has disappeared.

deeper into the slots of actin filaments (see Fig. 5). This results in a shortening of each sarcomere. Viewed in a microscope, this shortening shows up as a shortening of the *I* band; the *A* band stays at a constant length (compare Figs. 5a and b). At the same time, the transverse dimension of the myofibril expands, so the volume of the myofibril remains constant (the bulk of muscle tissue is water, and the overall density of the tissue has to remain constant during contraction). This expansion of the transverse dimension is what makes your biceps bulge when you tense it.

The mechanism that brings about the sliding motion relies on cross bridges that link adjacent myosin and actin filaments. These cross bridges are the ends of individual myosin molecules that stick out of the sides of the myosin filament like little oars. These oars swing forward, then grip the actin filaments, and then swing back, thereby pulling the myosin filament forward, and sliding it deeper into the slot formed by the actin filaments. The oars then release the actin, again swing forward, again grip the actin, etc.; and they go on repeating this pulling motion several times per second.

If the muscle is contracting while under a load, the cross bridges must do mechanical work while pulling the myosin filaments forward. The cross bridges acquire the energy for this work by interacting with ATP. At the end of a pull, the cross bridge reacts with an ATP molecule, which binds to it. The interaction between the cross bridge and the ATP bends the cross bridge forward and also splits the ATP into ADP and P_i, depositing energy in the bent cross bridge. The cross bridge then grips the actin filament, releases the ADP and P_i, and uses the energy deposited within it to swing back and to pull the actin filament forward.

The efficiency of conversion of the chemical energy of ATP into mechanical energy is amazingly high. For instance, experiments with muscle fibers from mammalian soleus muscle (in the calf of the leg) indicate that, under favorable conditions, up to 75% of the chemical energy of ATP is converted into mechanical work; the remaining 25% is wasted as heat. We can gain a better appreciation of this efficiency by noting that an automobile engine typically converts only 20% of the chemical energy of fuel into work and wastes 80% as heat.

However, muscle fibers attain the high efficiency of 75% only at an intermediate speed of contraction. If the muscle contracts too fast or too slowly, the efficiency suffers. As an extreme illustration of this dependence of the efficiency on the speed of contraction, we note that the efficiency becomes zero if the muscle fiber has zero speed of contraction, which happens if the muscle fiber merely holds a load at a fixed position. In such an isometric condition, the muscle fiber still consumes ATP; but the muscle fiber does no external work, and all the energy of the ATP is wasted as heat, resulting in zero efficiency. The isometric muscle consumes ATP because the cross bridges still go through their repeated pulling motions, just as in a contraction at nonzero speed. But fewer cross bridges participate at any given instant, and the pulls of these few participating cross bridges are insufficient to make the myosin filament slide forward, into the channel of actin filaments. The pulls merely stretch portions of the filaments momentarily, but the stretched portions of the filaments snap back when the pulls end, and, on the average, the length of the muscle fiber remains constant. Thus, the work done by the cross bridges is dissipated by the internal friction within the stretching filaments, and the muscle fiber performs no external work.

Furthermore, the high value of 75% for the conversion of chemical energy into mechanical energy is somewhat misleading. The overall efficiency of muscle fibers depends not only on the contraction process, but also on the conversion of the energy of the "raw" fuel (glucose) into the energy of ATP. Unfortunately, this process is rather inefficient — about 50% of the energy in the glucose is wasted as heat. Hence the overall efficiency of mammalian muscle fibers is no more than 37%. Besides, if we want to compare the muscle power of an animal with the power of an engine, we also need to take into account the inefficiencies in the digestion of food, the inefficiencies in the transport of oxygen, glucose, and carbon dioxide in the circulatory system, and the inefficiencies resulting from mechanical friction in the muscles, tendons, and joints. We may also want to take into account the inefficiency associated with the basal metabolic rate of the animal. In contrast to an engine, which does not require any fuel when it is switched off, the animal needs some minimum amount of food to support its life functions, even when resting, without doing any external work. Depending on context, it may be desirable to take some or all of these inefficiencies into account. The overall efficiency for the production of mechanical power by the animal therefore depends on the kind of food consumed, the kind of motion performed, the speed of the motion, and other conditions; and it is meaningless to ascribe an efficiency to the animal without specifying the conditions under which the efficiency is measured.

The maximum force that muscles can exert is about 40 newtons per square centimeter of cross section. For instance, a human biceps muscle, with a cross-sectional area of 40 cm^2, can exert a pull of about 1600 newtons (see Fig. 6). The maximum mechanical power that muscles can deliver is about 200 watts per kilogram. For comparison, note that typical automobile engines deliver about 300 watts per kilogram (at lower efficiency). But most muscles cannot continue to deliver their maximum power for an extended period of time. The flight muscles of birds and of insects are exceptional in this regard — they are capable of delivering high power for extended periods.

FIGURE 6 Weightlifter during lift.

REFERENCES

Alberts, B., Bray, D., Lewis, J., Raff, M., Roberts, K., and Watson, J. D., *Molecular Biology of the Cell* (Garland Publishers, New York, 1983).

Alexander, R. McN., and Goldspink, G., eds., *Mechanics and Energetics of Animal Locomotion* (Chapman and Hill, London, 1977).

Bendall, J. R., *Muscles, Molecules and Movement* (Elsevier Publishing Company, New York, 1969).

Kleiber, M., *The Fire of Life* (Wiley, New York, 1961).

Saturn and its rings, in an image acquired by the Voyager 2 spacecraft while approaching the planet. The rings consist of fragments of ice, held in circular orbits by the pull of Saturn's gravity.

GRAVITATION

Within the Solar System, planets orbit around the Sun, and satellites orbit around the planets. These circular, or nearly circular, motions require a centripetal force pulling the planets toward the Sun and the satellites toward the planets. It was Newton's great discovery that ***this interplanetary force holding the celestial bodies in their orbits is of the same kind as the force of gravity that causes apples, and other things, to fall downward near the surface of the Earth.*** Newton found that a single formula, his law of universal gravitation, encompasses both the gravitational force acting between celestial bodies and the gravitational force acting on bodies near the surface of the Earth.

Calculations based on Newton's laws of motion and on Newton's law of universal gravitation permit us to predict the motions of the planets, satellites, and comets; these predicted motions agree very precisely with astronomical observations. For example, the predicted planetary angular positions agree with the observed positions to within better than a thousandth of a degree, even after a lapse of tens of years.

By the nineteenth century, Newton's theory of gravitation had proved itself so trustworthy that when astronomers noticed an irregularity in the motion of Uranus, they could not bring themselves to believe that the theory was at fault. Instead, they suspected that a new, unknown planet caused these irregularities by its gravitational pull on Uranus. J. C. Adams and U. J. J. Leverrier proceeded to calculate the expected position of this hypothetical planet — and the new planet, later named Neptune, was promptly found at just about the expected position. This discovery of Neptune was a spectacular success of Newton's theory of gravitation.

Newton's theory has had other great successes, but it has also had a minor, though significant, failure. Leverrier discovered an enigmatic, very small discrepancy in the motion of Mercury which could not be explained by Newton's theory. It was finally explained by Einstein's theory of gravitation (General Relativity), a new theory based on a radical revision of our fundamental concepts of space and time.

Apart from the barely noticeable defect in the motion of Mercury, Newton's theory of gravitation has stood up remarkably well to the test of centuries; and it remains one of the most accurate and successful theories in all of physics, and in all of science.

5.1 NEWTON'S LAW OF UNIVERSAL GRAVITATION

Newton proposed that just as the Earth gravitationally attracts bodies placed near its surface and causes them to fall downward, the Earth also attracts more distant bodies, such as the Moon, or the Sun, or other planets. In turn, the Earth is gravitationally attracted by all these bodies. More generally, *every* body in the Universe attracts *every* other body with a gravitational force that depends on their masses and on their distances. The gravitational force that two bodies exert on each other is large if their masses are large, and small if their masses are small. The gravitational force decreases if we increase the distance between the bodies. The ***law of universal gravitation*** formulated by Newton can be stated most easily for the case of particles:

> ***Every particle attracts every other particle with a force directly proportional to the product of their masses and inversely proportional to the square of the distance between them.***

Expressed mathematically, the magnitude of the gravitational force that two particles of masses M and m separated by a distance r exert on each other is

Law of universal gravitation ▶

$$F = \frac{GMm}{r^2} \tag{1}$$

where G is a universal constant of proportionality, the same for all pairs of particles. This constant G is known as the ***gravitational constant.*** In the metric system of units its value is

$$G = 6.67 \times 10^{-11}\ \mathrm{N \cdot m^2/kg^2} \tag{2}$$

◀ **Gravitational constant**

Figure 5.1 shows the directions of the forces on each particle. Note that the gravitational forces on the two particles are of equal magnitudes and opposite directions; they form an action–reaction pair, in accordance with Newton's Third Law.

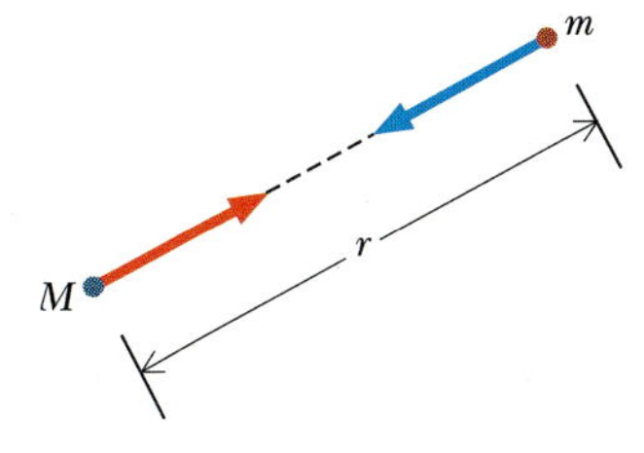

FIGURE 5.1

Two particles attract each other gravitationally. The forces are of equal magnitudes and of opposite directions.

The gravitational force of Eq. (1) is an inverse-square force: it decreases by a factor of 4 when the distance increases by a factor of 2; it decreases by a factor of 9 when the distance increases by a factor of 3; it decreases by a factor of 16 when the distance increases by a factor of 4, and so on. Figure 5.2 is a plot of the magnitude of the gravitational force as a function of the distance. Although the force decreases with distance, it never quite reaches zero. Thus, every particle in the Universe continually attracts every other particle at least a little bit, even if the distance between them is very, very large.

The gravitational force does not require any contact between the interacting particles. While reaching from one remote particle to another, the gravitational force somehow bridges the empty space between the particles. In Chapter 17 we will discuss how forces manage to reach through empty space.

It is also quite remarkable that ***the gravitational force between two particles is unaffected by the presence of intervening masses.*** For example, a particle in Washington, D.C., attracts a particle in Beijing with exactly the force given by Eq. (1), even though all of the bulk of the Earth lies between Washington and Beijing. This means that it is impossible to shield a particle from the gravitational attraction of another particle.

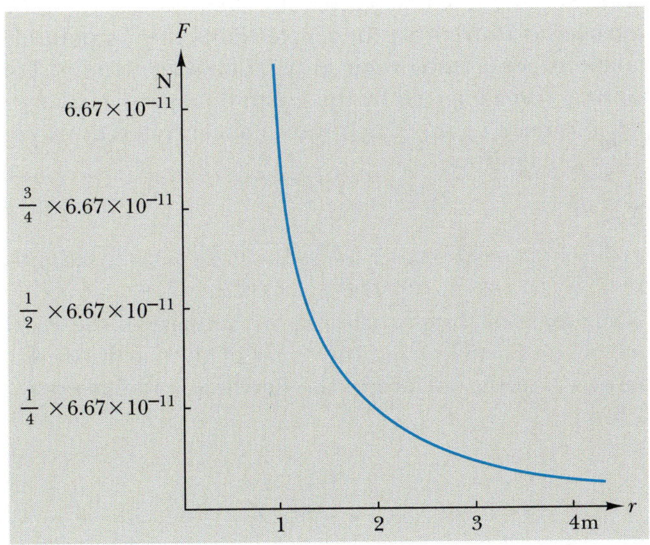

FIGURE 5.2

Magnitude of the gravitational force exerted by a particle of mass 1 kg on another particle of mass 1 kg.

EXAMPLE 1. What is the gravitational force between two bowling balls of 7.2 kg each, separated by a distance of 1.0 m? Treat both balls as particles.

SOLUTION: From Eq. (1)

$$F = \frac{GMm}{r^2}$$

$$= \frac{6.67 \times 10^{-11}\ \text{N} \cdot \text{m}^2/\text{kg}^2 \times 7.2\ \text{kg} \times 7.2\ \text{kg}}{(1.0\ \text{m})^2}$$

$$= 3.6 \times 10^{-9}\ \text{N}$$

This is a very small force, but as we will see in the next section, the measurement of such small forces is not beyond the reach of sensitive instruments.

Since the gravitational attraction between two particles is completely independent of the presence of other particles, it follows that the net gravitational force between two bodies (for example, the Earth and the Moon or the Earth and an apple) is merely the vector sum of the individual forces between all the particles making up the bodies — that is, the gravitational force obeys the principle of linear superposition of forces (see Section 3.3). As a consequence of this simple vector summation of the gravitational forces of the individual particles in a body, it can be shown that ***the net gravitational force between two spherical bodies acts just as though each body were concentrated at the center of its respective sphere.*** This important result is known as ***Newton's theorem.*** Since the Sun, the planets, and most of their satellites are almost exactly spherical, this theorem permits us to treat all these celestial bodies as pointlike particles in all calculations concerning their gravitational attractions. For instance, the gravitational force exerted by the Earth on a particle above its surface is as though the mass of the Earth were concentrated at its center; thus, this force has a magnitude

Newton's theorem ▶

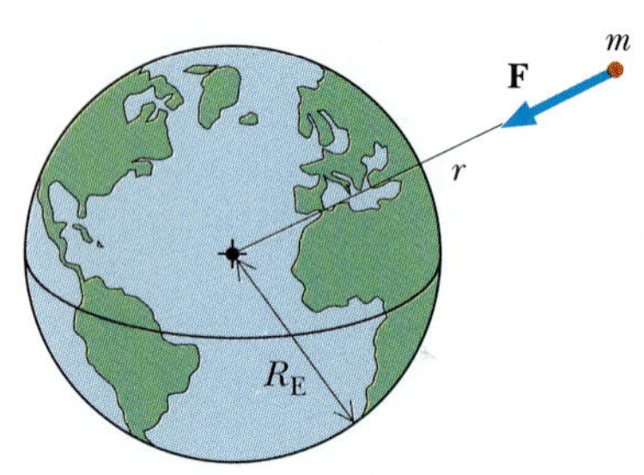

FIGURE 5.3
The gravitational force exerted by the Earth on a particle is directed toward the center of the Earth.

$$F = \frac{GM_E m}{r^2} \tag{3}$$

where m is the mass of the particle, M_E is the mass of the Earth, and r is the distance of the particle from the *center* of the Earth (see Fig. 5.3).

If the particle is at the surface of the Earth, at a radius $r = R_E$, then Eq. (3) gives a force

$$F = \frac{GM_E m}{R_E^2} \tag{4}$$

The corresponding acceleration of the mass m is

$$a = \frac{F}{m} = \frac{GM_E}{R_E^2} \tag{5}$$

But this acceleration is what we usually call the acceleration of free fall; thus

$$g = \frac{GM_E}{R_E^2} \tag{6}$$

◀ **Acceleration of free fall**

This equation establishes the connection between the ordinary force of gravity we experience at the surface of the Earth and Newton's law of universal gravitation. [We could substitute the values of the mass and the radius of the Earth into this formula, and we would then find the expected result $g = 9.81\ \text{m/s}^2$, but this is not a useful calculation, because we need Eq. (6) for the calculation of the mass of the Earth; see the next section.]

Note that although we obtained Eq. (6) for a particle at the surface of the Earth (at sea level), the same equation is approximately valid for a particle a few meters, or a few hundred meters, or even a few thousand meters above the surface. For such a particle the actual distance to the center of the Earth is only a small fraction of a percent larger than R_E, and hence the value of g is almost the same as on the surface of the Earth. The decrease of gravity that occurs when we climb a tower or a mountain is too small to be detectable, except with very sensitive instruments.

An equation analogous to Eq. (6) relates the acceleration of gravity at the surface of any (spherical) celestial body to the mass and the radius of that body. For example, we can calculate the acceleration of free fall on the surface of the Moon from its mass and radius.

EXAMPLE 2. The mass of the Moon is 7.35×10^{22} kg and its radius is 1.74×10^6 m. Calculate the acceleration of free fall on the Moon and compare it with the acceleration of free fall on the Earth.

SOLUTION: For the Moon, the formula analogous to Eq. (6) is

$$g_{\text{Moon}} = \frac{GM_{\text{Moon}}}{R_{\text{Moon}}^2} = \frac{6.67 \times 10^{-11}\ \text{N} \cdot \text{m}^2/\text{kg}^2 \times 7.35 \times 10^{22}\ \text{kg}}{(1.74 \times 10^6\ \text{m})^2}$$

$$= 1.62\ \text{m/s}^2$$

This is about 1/6 of the acceleration of free fall on the surface of the Earth. If you can jump upward to a height of 0.5 m on the Earth, then this same jump will take you to a height of 3 m on the Moon.

5.2 THE MEASUREMENT OF G

Cavendish torsion balance ▸

The gravitational constant G is rather difficult to measure with precision. The trouble is that gravitational forces between masses of laboratory size are extremely small, and thus a very delicate apparatus is needed to detect these forces. Measurements of G are usually done with a ***Cavendish torsion balance*** (see Fig. 5.4). Two equal, small spherical masses m, m' are attached to a lightweight horizontal beam which is suspended at its middle by a thin vertical fiber. When the beam is left undisturbed, it will settle into an equilibrium position such that the fiber is completely untwisted. If two equal, large masses M, M' are brought near the small masses, the gravitational attraction between each small mass and this neighboring large mass tends to rotate the beam clockwise (as seen from above). The twist of the fiber opposes this rotation, and the net result is that the beam settles into a new equilibrium position in which the force on the beam generated by the gravitational attraction between the masses is exactly balanced by the force exerted by the twisted fiber. The gravitational constant can then be calculated from the measured values of the angular displacement between the two equilibrium positions, the values of the masses, their distances, and the characteristics of the fiber.

Henry Cavendish, *1731–1810, English experimental physicist and chemist. His torsion balance for the absolute measurement of the gravitational force was based on an earlier design used by Coulomb for the measurement of the electric force.*

Several modern methods for the measurement of G rely on clever modifications of the basic Cavendish balance.

Note that the mass of the Earth can be calculated from Eq. (6) using the known values of G, R_E, and g:

$$M_E = \frac{R_E^2 g}{G} = \frac{(6.38 \times 10^6\,\text{m})^2 \times 9.81\,\text{m/s}^2}{6.67 \times 10^{-11}\,\text{N}\cdot\text{m}^2/\text{kg}^2}$$

$$= 5.98 \times 10^{24}\,\text{kg}$$

This calculation would seem to be a rather roundabout way to arrive at the mass of the Earth, but there is no direct route, since we cannot place the Earth on a balance. Because the calculation requires a prior measure-

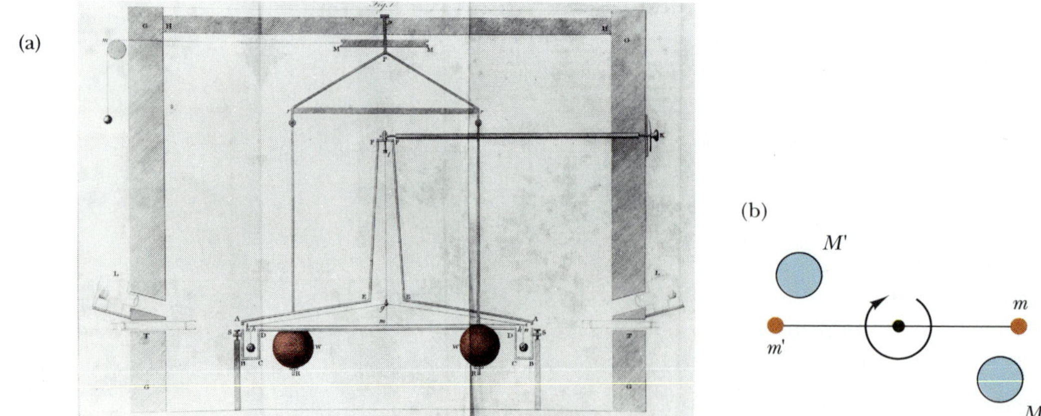

FIGURE 5.4

(a) Torsion balance used by Cavendish. The small masses m, m' are attached to a beam suspended by a thin fiber at the middle. (b) Top view of the balance, showing the small masses m, m' and the large masses M, M'. The gravitational attraction between m and M and between m' and M' tends to rotate the beam clockwise.

ment of the value of G, the Cavendish experiment has often been described figuratively as "weighing the Earth."

Nicholas Copernicus, *1473–1543, Polish astronomer. In his book* De Revolutionibus Orbium Coelestium *he formulated the heliocentric system for the description of the motion of the planets, according to which the Sun is immovable and the planets orbit around it. This new system — ardently defended by Galileo — gradually supplanted the old Ptolemaic system, according to which the Earth is immovable at the center of the Universe.*

5.3 CIRCULAR ORBITS

The gravitational force is responsible for holding the Solar System together; it makes the planets orbit around the Sun, and it makes the satellites orbit around the planets. Although the mutual gravitational forces of the Sun on a planet and of the planet on the Sun are of equal magnitudes, the mass of the Sun is much larger than the mass of a planet, and hence its acceleration is much smaller. It is therefore an excellent approximation to regard the Sun as fixed and immovable, and it then only remains to investigate the motion of the planet. If we designate the masses of the Sun and the planet by M_S and m, respectively, and their separation by r, then the magnitude of the gravitational force on the planet is

$$F = \frac{GM_S m}{r^2} \tag{7}$$

This force points toward the center of the Sun, that is, the center of the Sun is the center of force (see Fig. 5.5). For a particle moving under the influence of such a central force, the simplest conceivable motion is uniform circular motion, with the gravitational force acting as centripetal force. The motion of the planets in our Solar System is somewhat more complicated than that — as we will see in the next section, the planets move along ellipses, instead of circles. However, none of these planetary ellipses deviates very much from a circle, and as a first approximation we can pretend that the planetary orbits are circles.

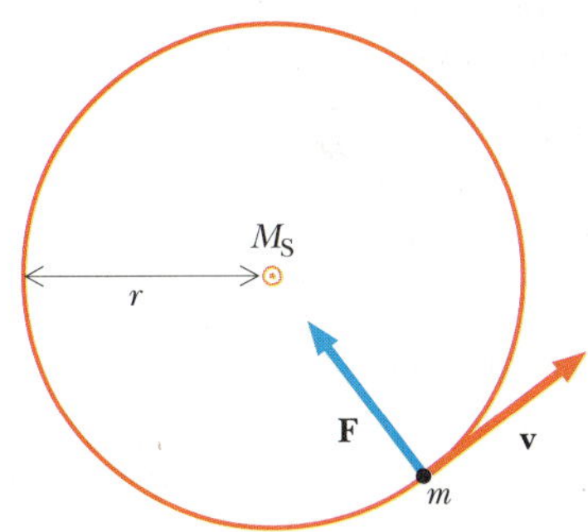

FIGURE 5.5 Circular orbit of a planet around the Sun.

By combining Newton's equation of motion with Newton's law of universal gravitation we can find a relation between the radius of the circular orbit and the speed. If the speed of the planet is v, then the centripetal acceleration is v^2/r [see Eq. (2.62)], and the equation of motion, $ma = F$, becomes

$$\frac{mv^2}{r} = F \tag{8}$$

or

$$\frac{mv^2}{r} = \frac{GM_S m}{r^2} \tag{9}$$

We can cancel a factor of m and a factor of $1/r$ on each side of this equation, and we obtain

$$v^2 = \frac{GM_S}{r} \tag{10}$$

Speed for circular orbit

or

$$v = \sqrt{\frac{GM_S}{r}} \tag{11}$$

EXAMPLE 3. The mass of the Sun is 1.99×10^{30} kg and the radius of the Earth's orbit around the Sun is 1.50×10^{11} m. What is the orbital speed of the Earth?

SOLUTION: According to Eq. (11), the orbital speed is

$$v = \sqrt{\frac{GM_S}{r}} = \sqrt{\frac{6.67 \times 10^{-11}\ \text{N}\cdot\text{m}^2/\text{kg}^2 \times 1.99 \times 10^{30}\ \text{kg}}{1.50 \times 10^{11}\ \text{m}}}$$

$$= 3.0 \times 10^4\ \text{m/s} = 30\ \text{km/s}$$

Period ▸ ***The time the planet takes to travel once around the Sun, or the time for one revolution, is called the period of the planet.*** We will designate the period by T. The speed of the planet is equal to the circumference $2\pi r$ of the orbit divided by the time T,

$$v = \frac{2\pi r}{T} \tag{12}$$

With this expression for the speed, Eq. (10) becomes

$$\frac{4\pi^2 r^2}{T^2} = \frac{GM_S}{r}$$

which can be rearranged to read

$$T^2 = \frac{4\pi^2}{GM_S} r^3 \tag{13}$$

This says that ***the square of the period is proportional to the cube of the radius of the orbit,*** with a constant proportionality depending on the mass of the central body. As we will see in the next section, Eq. (13) is one of Kepler's laws. We can use this equation to find the relative sizes of the planetary orbits from their periods.

EXAMPLE 4. Both Venus and the Earth have approximately circular orbits around the Sun. The period of the orbital motion of Venus is

0.615 year and the period of the Earth is 1 year. According to Eq. (13), by what factor do the sizes of the two orbits differ?

SOLUTION: If we take the cube root of both sides of Eq. (13), we see that the orbital radius is proportional to the 2/3 power of the period. Hence we can set up the following proportion for the orbital radii of the Earth and Venus:

$$\frac{r_E}{r_V} = \frac{T_E^{2/3}}{T_V^{2/3}} \tag{14}$$

$$= \frac{(1\text{ year})^{2/3}}{(0.615\text{ year})^{2/3}} = 1.38$$

Of course, an equation analogous to Eq. (13) also applies to the circular motion of a moon or artificial satellite around a planet. In this case, the planet plays the role of the central body and, in Eq. (13), its mass replaces the mass of the Sun.

EXAMPLE 5. A communications satellite is placed in a circular orbit around the Earth. The orbit of the satellite is in the equatorial plane, and the direction of its motion is eastward. The period of the orbit of the satellite is exactly 1 day, so that the satellite always hovers in a fixed position relative to the rotating Earth. What must be the radius of such a "synchronous," or "geostationary," orbit?

A communications satellite about to be released from the Space Shuttle.

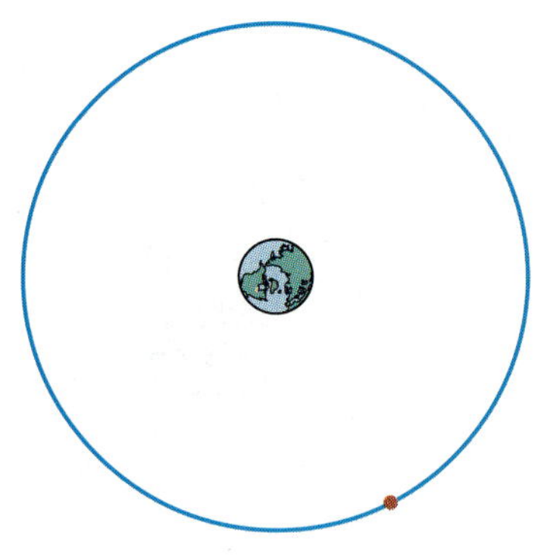

FIGURE 5.6
Orbit of a "geostationary" satellite around the Earth.

SOLUTION: Since the central body is the Earth, the equation analogous to Eq. (13) is

$$T^2 = \frac{4\pi^2}{GM_E} r^3 \tag{15}$$

or

$$r^3 = \frac{GM_E T^2}{4\pi^2} \tag{16}$$

Taking the cube root of both sides of this equation we find

$$r = \left(\frac{GM_E T^2}{4\pi^2}\right)^{1/3} \tag{17}$$

$$= \left[\frac{6.67 \times 10^{-11}\ \text{N} \cdot \text{m}^2/\text{kg}^2 \times 5.98 \times 10^{24}\ \text{kg} \times (24 \times 60 \times 60\ \text{s})^2}{4\pi^2}\right]^{1/3}$$

$$= 4.23 \times 10^7\ \text{m}$$

The orbit is shown in Fig. 5.6, which is drawn to scale. Several communications satellites, such as the Telstar, Galaxy and Syncom satellites, have been placed in this geostationary orbit. These satellites routinely relay radio and TV signals over large distances.

Johannes Kepler, *1571–1630, German astronomer and mathematician. Kepler relied on the theoretical framework of the Copernican system, and he extracted his three laws by a meticulous analysis of the observational data on planetary motions collected by the great Danish astronomer Tycho Brahe.*

5.4 ELLIPTICAL ORBITS; KEPLER'S LAWS

Although the orbits of the planets around the Sun are approximately circular, none of these orbits is *exactly* circular. We will not attempt the general solution of the equation of motion for such noncircular orbits. A complete calculation shows that with the inverse-square force of Eq. (1), the planetary orbits are ellipses. This is ***Kepler's First Law:***

Kepler's First Law ▶

The orbits of the planets are ellipses with the Sun at one focus.

Figure 5.7 shows an elliptical planetary orbit (for the sake of clarity, the elongation of this ellipse has been exaggerated; actual planetary orbits have only very small elongations). The point closest to the Sun is called the ***perihelion;*** the point farthest from the Sun is called the ***aphelion.*** The sum of the perihelion and the aphelion distances is the ***major axis*** of the ellipse. The distance from the center of the ellipse to the perihelion (or aphelion) is the ***semimajor axis;*** this distance equals the average of the perihelion and aphelion distances.

Kepler originally discovered his First Law and his other two laws (see below) early in the seventeenth century, by direct analysis of the available

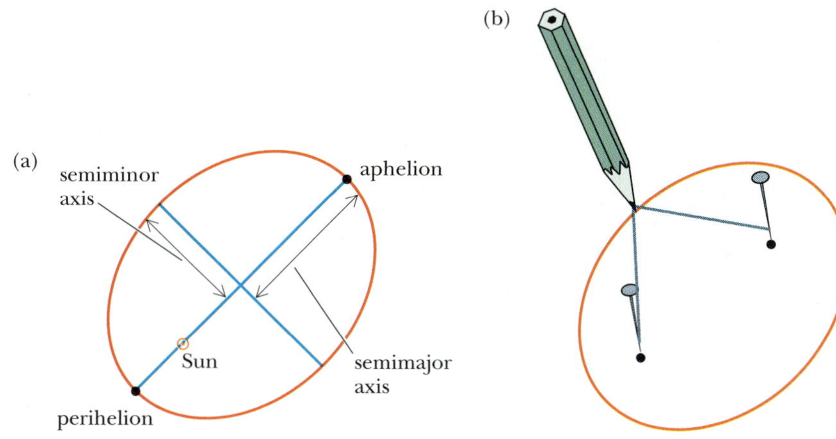

F I G U R E 5.7

(a) Orbit of a planet around the Sun. The orbit is an ellipse, with the Sun at one focus.
(b) An ellipse is defined geometrically by the condition that the sum of the distance from one focus of the ellipse and the distance from the other focus is the same for all points on the ellipse. This geometrical condition leads to a simple method for the construction of an ellipse: Stick pins into the two foci and tie a length of string to these pins. Stretch the string tautly to the point of a pencil, and move this pencil around the foci while keeping the string taut.

observational data on planetary motions. Hence, Kepler's laws were originally purely phenomenological statements; that is, they described the phenomenon of planetary motion but did not explain its causes. The explanation only came later, when Newton laid down his laws of motion and his law of universal gravitation and deduced the features of planetary motion from these fundamental laws.

Kepler's Second Law

Kepler's Second Law describes the variation in the speed of the motion:

> ***The radial line segment from the Sun to the planet sweeps out equal areas in equal times.***

Figure 5.8 illustrates this law. The two colored areas are equal and the planet takes equal times to move from P to P' and from Q to Q'. From Fig. 5.8 we see that the speed of the planet is larger when it is near the Sun (at Q) than when it is far from the Sun (at P).

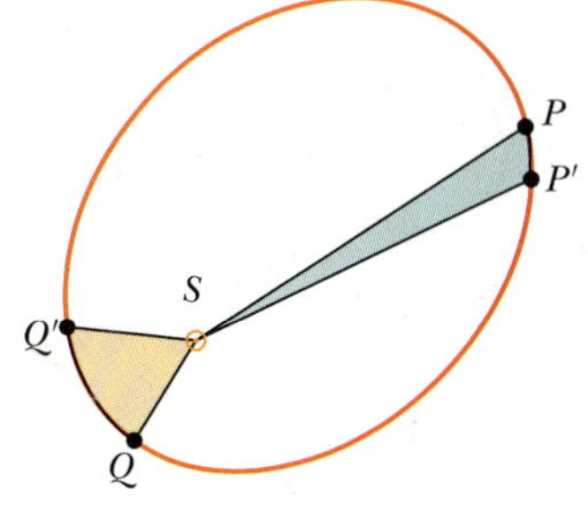

F I G U R E 5.8

For equal time intervals, the areas $QQ'S$ and $PP'S$ are equal. The distance QQ' is larger than the distance PP'.

This Second Law, also called the law of areas, is a direct consequence of the central direction of the gravitational force. This means that Kepler's Second Law is valid not only for planetary motion, but also for motion with any kind of central force.

Let us examine what Kepler's Second Law has to say about the speeds of a planet at aphelion and at perihelion. Figure 5.9 shows the triangular area SPQ swept out by the radial line in one second at, or near, aphelion. The height PQ of this triangle equals the speed v_1 at aphelion; hence the area of the triangle is $\frac{1}{2}r_1v_1$. Likewise, the triangular area $SP'Q'$ swept out by the radial line in one second at, or near, perihelion is $\frac{1}{2}r_2v_2$. By Kepler's law these two areas must be equal,

$$\tfrac{1}{2}r_1v_1 = \tfrac{1}{2}r_2v_2 \qquad (18)$$

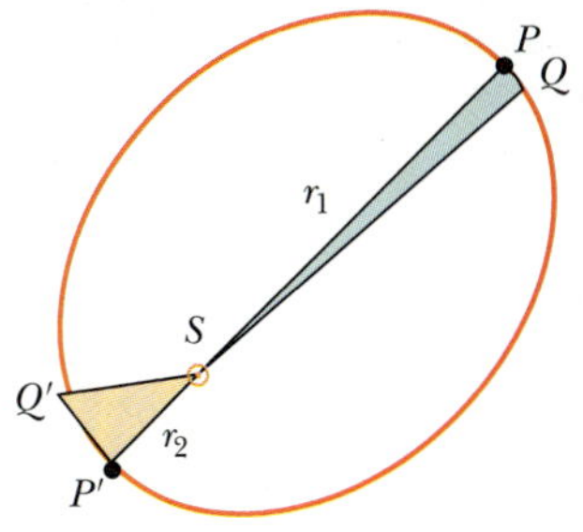

F I G U R E 5.9

Triangular area SPQ swept out in one second after aphelion and triangular area $SP'Q'$ swept out in one second after perihelion.

According to this equation, the ratio of the aphelion and perihelion speeds is the inverse of the ratio of the distances.

In Chapter 7 we will become acquainted with the ***angular momentum***, which, for a planet at aphelion or at perihelion, is defined as the product mvr. Apart from an irrelevant factor of $\frac{1}{2}$ and an irrelevant factor of m, Eq. (18) therefore asserts that the angular momentum at aphelion equals the angular momentum at perihelion. Thus, Kepler's Second Law is equivalent to a conservation law for angular momentum.

EXAMPLE 6. The perihelion and aphelion distances for Mercury are 45.9×10^9 m and 69.8×10^9 m, respectively. The speed of Mercury at aphelion is 3.88×10^4 m/s. What is the speed at perihelion?

SOLUTION: From Eq. (18),

$$v_2 = \frac{r_1}{r_2} v_1 = \frac{69.8 \times 10^9 \text{ m}}{45.9 \times 10^9 \text{ m}} \times 3.88 \times 10^4 \text{ m/s} \tag{19}$$

$$= 5.91 \times 10^4 \text{ m/s}$$

Kepler's Third Law ▶

Kepler's Third Law relates the period of the orbit to the size of the orbit:

> ***The square of the period is proportional to the cube of the semimajor axis of the planetary orbit.***

This Third Law, or law of periods, is nothing but a generalization of Eq. (13) to elliptical orbits.

Table 5.1 lists the orbital data for the planets of the Solar System. The mean distance listed in this table is the average of the perihelion and aphelion distances, that is, it is the semimajor axis of the ellipse. The difference between the perihelion and aphelion distances gives an indication of the elongation of the ellipse. Figure 5.10 shows the orbits of the planets Mercury, Venus, Earth, Mars, Jupiter, and Saturn on scale diagrams. The orbits of Uranus, Neptune, and Pluto are considerably larger and would have to be plotted on a separate diagram (see the Prelude). Inspection of these diagrams makes it obvious that the orbits of Mercury and Mars are noticeably different from circles.

TABLE 5.1 THE PLANETS

Planet	Mass	Mean distance from Sun (semimajor axis)	Perihelion distance	Aphelion distance	Period
Mercury	3.30×10^{23} kg	57.9×10^6 km	45.9×10^6 km	69.8×10^6 km	0.241 year
Venus	4.87×10^{24}	108	107	109	0.615
Earth	5.98×10^{24}	150	147	152	1.00
Mars	6.42×10^{23}	228	207	249	1.88
Jupiter	1.90×10^{27}	778	740	816	11.9
Saturn	5.67×10^{26}	1430	1350	1510	29.5
Uranus	8.70×10^{25}	2870	2730	3010	84.0
Neptune	1.03×10^{26}	4500	4460	4540	165
Pluto	1.50×10^{22}	5890	4410	7360	248

A montage of images of planets obtained by Pioneer, Mariner, Viking, and Voyager spacecraft. The relative sizes of these images are not to scale. From above center: Mercury, Venus, Earth and Moon, Mars, Jupiter, Saturn, Uranus, and Neptune.

FIGURE 5.10

(Below) (a) Orbits of Mercury, Venus, Earth, and Mars. The colored dots indicate the positions of these planets on January 1, 2000. The tick marks indicate the positions at intervals of 10 days. (b) Orbits of Jupiter and Saturn. The tick marks indicate the positions at intervals of 1 year.

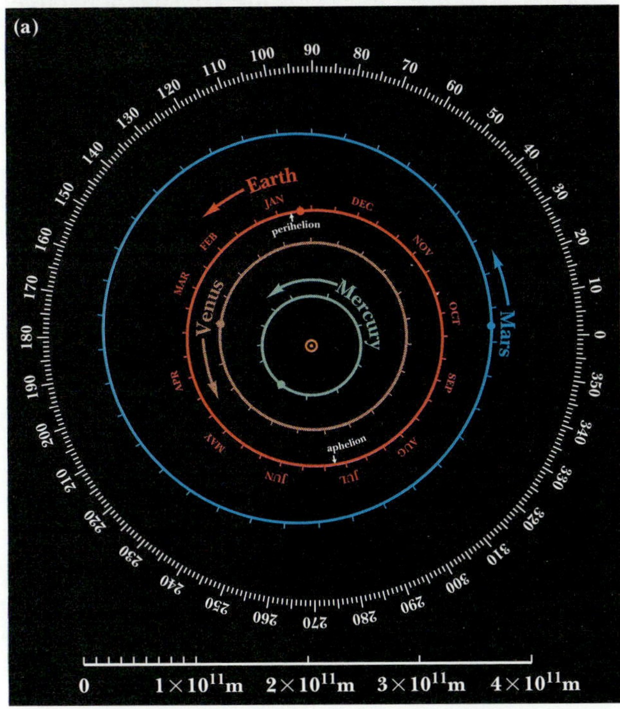

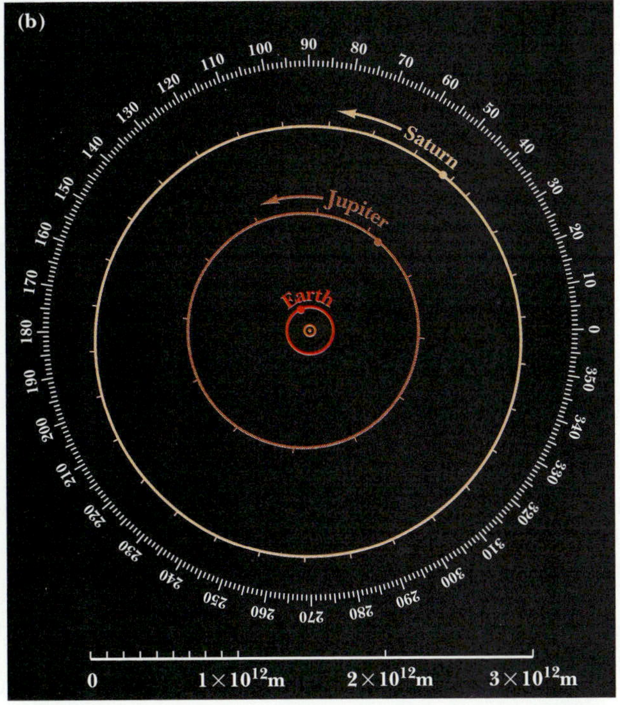

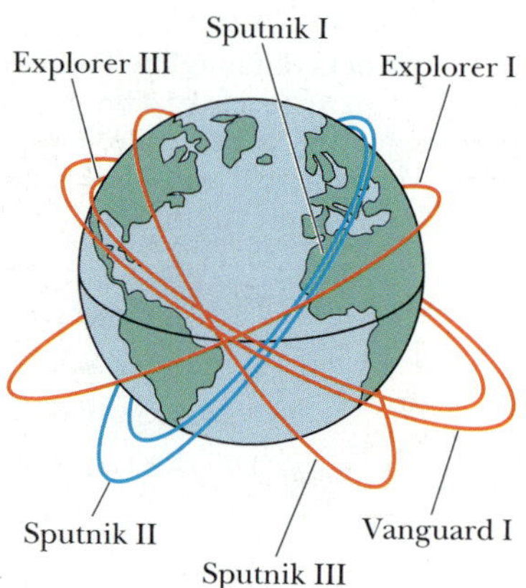

FIGURE 5.11
Orbits of the first artificial satellites of the Earth.

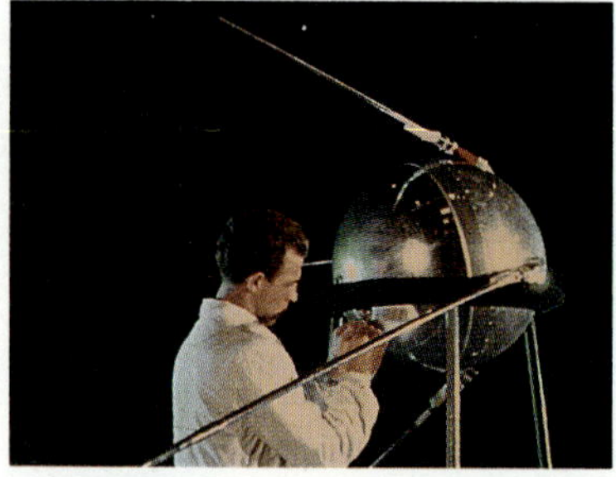

FIGURE 5.12
Sputnik, the first artificial satellite of the Earth. This satellite had a mass of 83 kg.

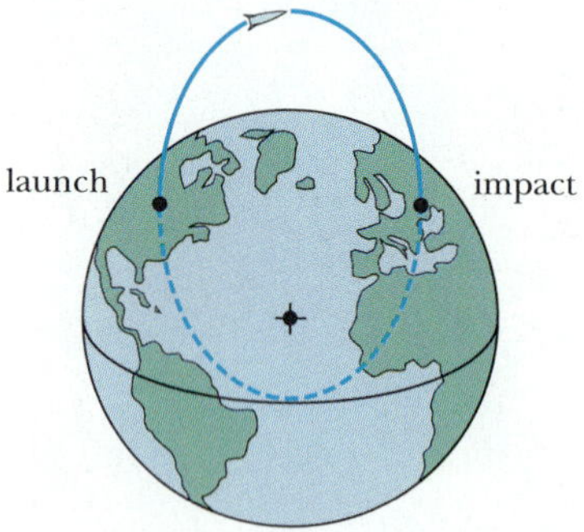

FIGURE 5.13
Trajectory of an intercontinental ballistic missile (ICBM).

Kepler's three laws apply not only to planets, but also to satellites and to comets. For example, Fig. 5.11 shows the orbits of a few of the many artificial satellites of the Earth. All these orbits are ellipses. The early artificial satellites were quite small, with masses below 100 kg (see Fig. 5.12). Nowadays, satellites with masses of several tons are not unusual. All of the early artificial satellites burned up in the atmosphere after a few months or a few years because they were not sufficiently far from the Earth to avoid the effects of residual atmospheric friction.

Kepler's laws also apply to the motion of a projectile near the Earth. For instance, Fig. 5.13 shows the trajectory of an intercontinental ballistic missile (ICBM). During most of its trajectory, the only force acting on the missile is the gravity of the Earth, because the thrust of the engines and the friction of the atmosphere act only during the relatively short initial and final segments of the trajectory (on the scale of Fig. 5.13, these initial and final segments of the trajectory are too small to be noticed). The trajectory is a portion of an elliptical orbit cut short by the impact on the Earth. Likewise, the motion of an ordinary low-altitude projectile, such as a cannonball, is also a portion of an elliptical orbit (if we ignore atmospheric friction). In Chapter 2 we assumed that gravity was constant in magnitude and direction; with these approximations we found that the orbit of a projectile was a parabola. Although the exact orbit of a projectile is an ellipse, the parabola approximates this ellipse quite well over the relatively short distances and low altitudes involved in ordinary projectile motion (see Fig. 5.14).

The connection between projectile motion and orbital motion was neatly illustrated by Newton by means of an imaginary experiment, or as it is called these days, a *Gedankenexperiment.** Imagine that we fire a can-

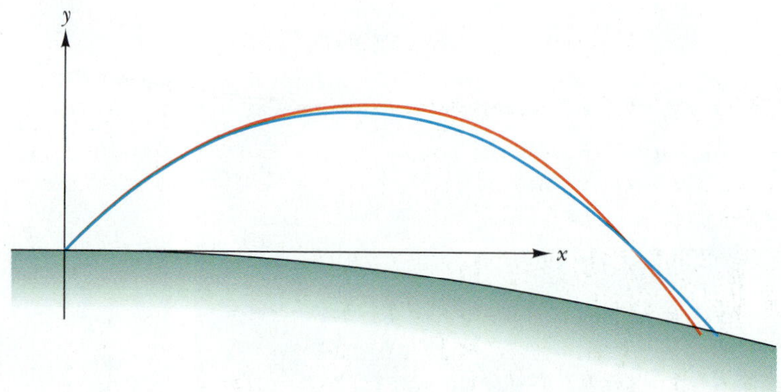

FIGURE 5.14 Exact elliptical orbit (red) and approximate parabolic orbit (blue) for a projectile without air friction. In this example, the parabola nearly coincides with the ellipse. For an orbit of greater height or longer range, the deviation between the parabola and the ellipse would be larger.

* *Gedankenexperiment* is German for "thought experiment." This word is used by physicists for an experiment that can be done in principle but that has never been done in practice, and whose outcome can be discovered by thought.

nonball horizontally from a gun emplaced on a high mountain (see Fig. 5.15). If the muzzle speed is fairly low, the cannonball will arc toward the Earth and strike near the base of the mountain. The trajectory is a segment of a parabola, or, more precisely, a segment of an ellipse. If we increase the muzzle speed, the cannonball will describe larger and larger arcs. Finally, if the muzzle speed is just large enough, the rate at which the trajectory curves downward is precisely matched by the curvature of the surface of the Earth — the cannonball never hits the Earth and keeps on falling forever while moving in a circular orbit. This example makes it very clear that orbital motion is free-fall motion (if we ignore atmospheric friction).

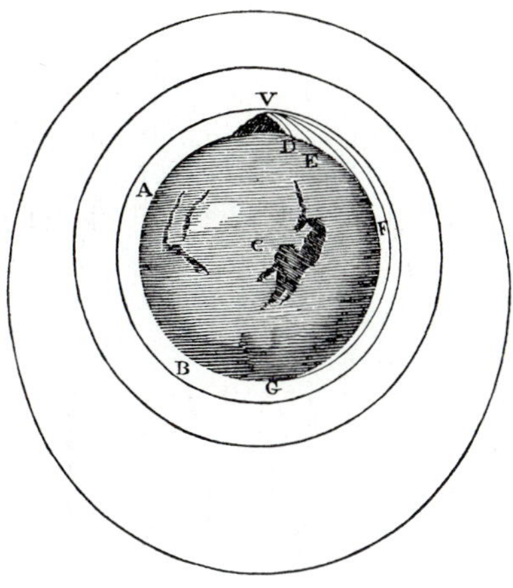

FIGURE 5.15
An imaginary experiment by Newton. The trajectory of a fast cannonball is a circular orbit.

Finally, we note that in our mathematical description of planetary motion we have neglected the gravitational forces that the planets exert on one another. These forces are much smaller than the force exerted by the Sun, but in a precise calculation they must be taken into account. The net force on any planet then depends on the positions of all the other planets. This means that the motions of all the planets are coupled together, and the calculation of the motion of one planet requires the simultaneous calculation of the motions of all the other planets. This makes the precise mathematical treatment of planetary motion extremely complicated. Kepler's simple laws neglect the complications introduced by the interplanetary forces; these laws therefore do not provide an exact description of planetary motions, but only a reasonably good first approximation.

5.5* GRAVITATIONAL POTENTIAL ENERGY

The gravitational force $F = GMm/r^2$ is a conservative force; that is, the work done by this force on a particle moving from some point P_1 to some other point P_2 can be expressed as a difference between two potential energies, and the work done in any roundtrip starting and ending at some given point is zero. By a mathematical argument, it can be shown that the potential energy associated with the gravitational force is

$$U = -\frac{GMm}{r} \tag{20}$$

◀ **Gravitational potential energy**

This potential energy is always negative, and its magnitude is inversely proportional to r. Figure 5.16 gives a plot of this potential energy. If the distance r is small, the potential energy is low (the potential energy is much below zero); if the distance r is large, the potential energy is higher (the potential energy is still negative, but not so much below zero). Thus, the potential energy *increases* with distance; it increases from a large negative value to a smaller negative value, nearer to zero. Such an increase of po-

* This section is optional.

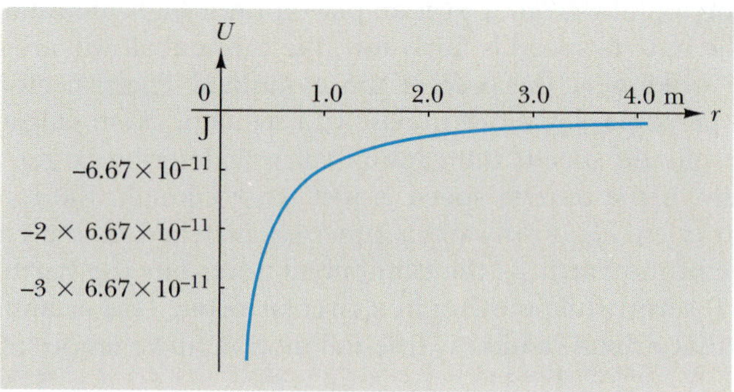

FIGURE 5.16
Gravitational potential energy for a particle of mass 1 kg gravitationally attracted by another particle of mass 1 kg.

tential energy with distance is characteristic of an attractive force. For instance, if we want to lift a communications satellite from a low initial orbit (just above the Earth's atmosphere) into a high final orbit (such as the geostationary orbit described in Example 5), we must do work on this satellite (by means of a rocket engine). The work we do while lifting the satellite increases the gravitational potential energy from a large negative value (much below zero), to a smaller negative value (not so much below zero).

The total mechanical energy is the sum of the potential energy and the kinetic energy. Since we are assuming that the mass M is stationary, the kinetic energy is entirely due to the motion of the mass m, and the total mechanical energy takes the form

Mechanical energy ▶

$$E = K + U = \tfrac{1}{2}mv^2 - \frac{GMm}{r} \tag{21}$$

If the only force acting on the mass m is the gravitational force (no rocket engine or other external force!), then this total energy remains constant during the motion. For instance, the energy (21) is constant for a planet orbiting the Sun, and for a satellite or a spacecraft (with rocket engines shut off) orbiting the Earth. As we saw in Chapter 4, the examination of the energy reveals some general features of the motion. Equation (21) implies that whenever r increases during the motion of the mass m, then v must decrease, so that the sum of the two terms $\frac{1}{2}mv^2$ and $-GMm/r$ remains constant. Conversely, if r decreases, v must increase.

Let us now investigate the possible orbits around, say, the Sun from the point of view of their energy. For a circular orbit, the square of the orbital speed is [see Eq. (10)]

$$v^2 = \frac{GM_S}{r} \tag{22}$$

and the kinetic energy is

$$K = \tfrac{1}{2}mv^2 = \frac{GM_S m}{2r} \tag{23}$$

Hence the total mechanical energy is

$$E = \tfrac{1}{2}mv^2 - \frac{GM_S m}{r} = \frac{GM_S m}{2r} - \frac{GM_S m}{r}$$

or

$$E = -\frac{GM_S m}{2r} \tag{24}$$

Consequently, ***the total energy for a circular orbit is negative and is exactly one-half of the potential energy.*** Note that small orbits, with a small radius r, have low energy (their energy is much below zero); large orbits, with a large radius r, have a higher energy (their energy is still negative, but not so much below zero). This increase of energy with orbital size agrees with the intuitive notion that we must supply energy to lift the orbiting mass from a small (or low) orbit to a large (or high) orbit. The maximum possible energy of a circular orbit is zero; this corresponds to an orbit of infinite radius ($r = \infty$; here, the mathematical symbol ∞ stands for infinity).

For an elliptical orbit, the total energy is also negative. It can be proven that the energy can still be written in the form of Eq. (24), but the quantity r must be taken equal to the semimajor axis of the ellipse. ***The total energy does not depend on the shape of the ellipse, but only on its semimajor axis.*** Figure 5.17 shows several orbits of different shapes but with exactly the same total energy. All these orbits have the same semimajor axis.

If the energy is nearly zero, then the size of the orbit is very large. Such orbits are characteristic of comets, many of which have elliptical orbits that extend far beyond the edge of the Solar System (see Fig. 5.18). If the energy is exactly zero, then the "ellipse" extends all the way to infinity and never closes; such an "open ellipse" is actually a parabola (see Fig. 5.19). Equation (21) indicates that if the energy is zero, the comet will reach infinite distance with zero velocity (at $r = \infty$, the potential energy is zero, hence the kinetic energy is zero, and hence v is zero). By considering the reverse of this motion, we recognize that a comet initially at rest at a very large distance from the Sun will fall along this type of parabolic orbit.

If the energy is positive, then the obit again extends all the way to infinity and again fails to close; such an open orbit is a hyperbola (see Fig. 5.20). Equation (21) indicates that if the energy is positive, the comet will

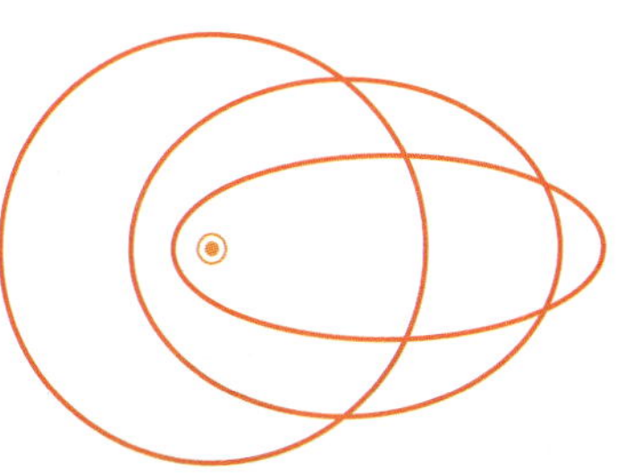

FIGURE 5.17

Orbits of the same total energy. All these orbits have the same semimajor axis.

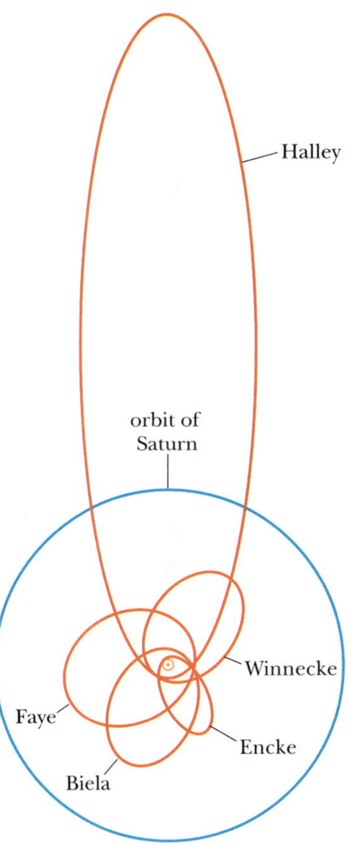

FIGURE 5.18

Orbits of some periodic comets.

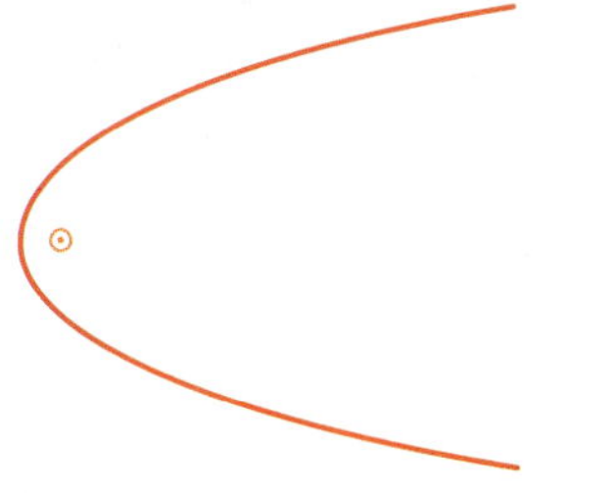

FIGURE 5.19

Orbit of zero energy — a parabola.

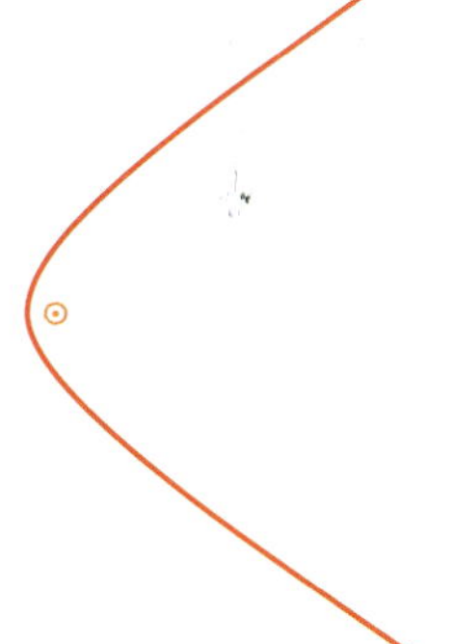

FIGURE 5.20

Orbit of positive energy — a hyperbola.

reach infinite distance with some nonzero velocity (at $r = \infty$, the potential energy is zero, hence the total energy equals the kinetic energy, which is not zero, and hence v is nonzero).

This meteorite is what remains of a meteoroid that entered the atmosphere of the Earth and partially melted before striking the ground.

EXAMPLE 7. A meteoroid (a chunk of rock) is initially at rest in interplanetary space at a large distance from the Sun. Under the influence of gravity, the meteoroid begins to fall toward the Sun along a straight radial line. With what speed does it strike the Sun?

SOLUTION: This radial motion is an extreme case of parabolic motion; if we were to push the meteoroid ever so slightly away from its radial line, its orbit would become a very narrow, elongated parabola. The energy of the meteoroid is

$$E = \frac{1}{2}mv^2 - \frac{GM_S m}{r} = [\text{constant}] \tag{25}$$

Initially, both the kinetic and potential energies are zero ($v = 0$ and $r \simeq \infty$). Hence at any later time

$$\frac{1}{2}mv^2 - \frac{GM_S m}{r} = 0$$

or

$$\frac{1}{2}mv^2 = \frac{GM_S m}{r} \tag{26}$$

If we cancel a factor of m on both sides of this equation, take the square root of both sides, and substitute $r = R_S$, we find the speed at the moment of impact:

$$v = \sqrt{\frac{2GM_S}{R_S}} \tag{27}$$

This is called the escape velocity (see below). With $R_S = 6.96 \times 10^8$ m and $M_S = 1.99 \times 10^{30}$ kg, we obtain

$$v = \sqrt{\frac{2 \times 6.67 \times 10^{-11}\ \text{N} \cdot \text{m}^2/\text{kg} \times 1.99 \times 10^{30}\ \text{kg}}{6.96 \times 10^8\ \text{m}}}$$

$$= 6.18 \times 10^5\ \text{m/s} = 618\ \text{km/s}$$

Escape velocity ▶

The quantity given by Eq. (27) is called the ***escape velocity*** because it is the minimum initial velocity with which a body must be launched upward from the surface of the Sun if it is to escape and never fall back. We can recognize this by looking at the motion of the meteoroid in Example 7 in reverse; it starts with a velocity of 618 km/s at the surface of the Sun and gradually slows as it rises, but never quite stops until it reaches a very large distance ($r \simeq \infty$).

Note that the direction in which the escaping body is launched is immaterial — the body will succeed in its escape whenever the direction of launch is above the horizon. Of course, the path that the body takes will depend on the direction of launch (see Fig. 5.21).

The escape velocity for a body launched from the surface of the Earth can be calculated from a formula analogous to Eq. (27), provided that we ignore atmospheric friction and the pull of the Sun on the body. Atmospheric friction will be absent if we launch the body from just above the atmosphere, and the pull of the Sun has only a small effect on the velocity of escape from the Earth if we contemplate a body that "escapes" to a distance of, say, $r = 100\ R_E$ or $200\ R_E$ rather than $r = \infty$. For such a body, the escape velocity is approximately $\sqrt{2GM_E/R_E} \simeq 11.2$ km/s.

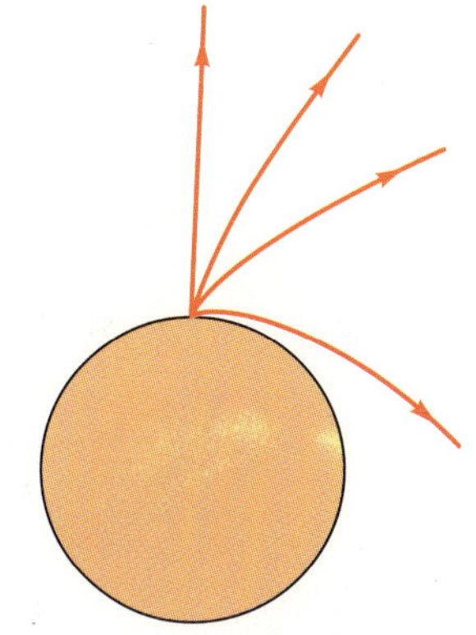

FIGURE 5.21
Different parabolic orbits with the same starting point and initial speed.

SUMMARY

- **Law of universal gravitation:** $F = \dfrac{GMm}{r^2}$
- **Gravitational constant:** $G = 6.67 \times 10^{-11}\ \text{N} \cdot \text{m}^2/\text{kg}^2$
- **Acceleration of free fall on Earth:** $g = \dfrac{GM_E}{R_E^2}$
- **Speed for circular orbit around Sun:** $v^2 = \dfrac{GM_S}{r}$
- **Kepler's First Law:** The orbits of the planets are ellipses with the Sun at one focus.
- **Kepler's Second Law:** The radial line segment from the Sun to the planet sweeps out equal areas in equal times.
- **Kepler's Third Law:** The square of the period is proportional to the cube of the semimajor axis of the planetary orbit.
- **Gravitational potential energy:** $U = -\dfrac{GMm}{r}$
- **Total mechanical energy:** $E = \frac{1}{2}mv^2 - \dfrac{GMm}{r}$

QUESTIONS

1.–Can you directly feel the gravitational pull of the Earth with your sense organs? (Hint: Would you feel anything if you were in free fall?)

2.–According to a tale told by Professor R. Lichtenstein, some apple trees growing in the mountains of Tibet produce apples of negative mass. In what direction would such an apple fall if it fell off its tree? How would such an apple hang on the tree?

3.–Eclipses of the Moon can occur only at full Moon. Eclipses of the Sun can occur only at new Moon. Why?

4.–Explain why the sidereal day (the time of rotation of the Earth relative to the stars, or 23 h 56 min 4 s) is shorter than the mean solar day (the time between successive passages of the Sun over a given meridian, or 24 h). (Hint: The rotation of the Earth around its axis and the revolution of the Earth around the Sun are in the same direction.)

5.–Communications satellites are usually placed in an orbit or radius of 4.2×10^7 m so that they remain stationary above a point on the Earth's equator (see Example 5). Can we place a satellite in an orbit so it remains stationary above the North Pole?

6.–When an artificial satellite — such as the ill-fated Skylab — experiences friction against the residual atmosphere of the Earth, the radius of the orbit decreases while at the same time the speed of the satellite *increases.* Explain.

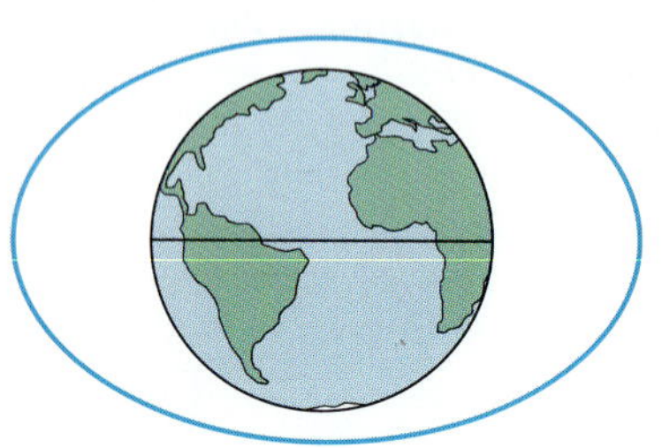

FIGURE 5.22
Proposed orbit for a communications satellite.

7.–Suppose that an airplane flies around the Earth along the equator. If this airplane flies *very* fast, it would not need wings to support itself. Why not?

8.–The mass of Pluto was not known until 1978 when a moon of Pluto was finally discovered. How did the discovery of this moon help?

9.–It is easier to launch an Earth satellite into an eastward orbit than into a westward orbit. Why?

10.–Would it be advantageous to launch rockets into space from a pad at very high altitude on a mountain? Why has this not been done?

11.–Describe how you could play tennis by yourself on a small, round asteroid. What rules of the game would you want to lay down?

12.–According to an NBC news report of April 5, 1983, a communications satellite launched from the space shuttle went into an orbit as shown in Fig. 5.22. Is this believable?

FIGURE 5.23
The jump of an Apollo astronaut.

13.–If we take into account both the gravitational pull exerted by the Sun and that exerted by other planets, is Kepler's Second Law valid for the Earth?

14.–Why were the Apollo astronauts able to jump much higher on the Moon than on Earth (Fig. 5.23)? If they had landed on a small asteroid, could they have launched themselves into a parabolic or hyperbolic orbit by a jump?

15.–The Earth reaches perihelion on January 3, and aphelion on July 6. Why is it not warmer in January than in July?

16.–The gravitational force that a hollow spherical shell of mass exerts on a particle in its interior in zero. Does this mean that such a shell acts as a gravity shield?

17.–Consider an astronaut launched in a rocket from the surface of the Earth and then placed in a circular orbit around the Earth. Describe the astronaut's weight (measured in an inertial reference frame) at different times during this trip. Describe the astronaut's *apparent* weight (measured in his own reference frame) at different times.

18.–Several of our astronauts suffered severe motion sickness while under conditions of apparent weightlessness aboard Skylab. Since the astronauts were not being tossed about (as in an airplane or a ship in a storm), what caused this motion sickness? What other difficulties does an astronaut face in daily life under conditions of weightlessness?

19.–An astronaut on Skylab lights a candle. Will the candle burn like a candle on Earth?

20.–Astrology is an ancient superstition according to which the planets influence phenomena on the Earth. The only force that can reach over the large distances between the planets and act on pieces of matter on the Earth is gravitation (planets do not have electric charge and they therefore do not exert electric forces; some planets do have magnetism, but their magnetic forces are too weak to reach the Earth). Given that the Earth is in free fall under the action of the net gravitational force of the planets and the Sun, is there any way that the gravitational forces of the planets can affect what happens on the Earth?

PROBLEMS

Section 5.1: Newton's Law of Universal Gravitation
Section 5.2: The Measurement of *G*

1.——Two supertankers, each with a mass of 7×10^8 kg, are separated by a distance of 2.0 km. What is the gravitational force that each exerts on the other? Treat them as particles.

2.——What is the gravitational force between two protons separated by a distance equal to their diameter, 2.0×10^{-15} m?

3.——What is the magnitude of the gravitational force that the Sun exerts on you? What is the magnitude of the gravitational force that the Moon exerts on you? The masses of the Sun and the Moon and their distances are given on the endpapers; assume that your mass is 70 kg. Compare these forces with your weight. Why don't you feel these forces? (Hint: You and the Earth are in free fall toward the Sun and the Moon.)

4.——Calculate the gravitational force that the Earth exerts on an astronaut of mass 75 kg in a space capsule at a height of 1000 km above the surface of the Earth. Compare with the gravitational force that this astronaut would experience if on the surface of the Earth.

5.——The masses used in the Cavendish experiment typically are a few kilograms for the large masses and a few tens of grams for the small masses. Suppose that a "large" spherical mass of 8.0 kg is at a center-to-center distance of 10 cm from a "small" spherical mass of 30 g. What is the magnitude of the gravitational force?

6.——A 70-kg man and a 60-kg woman are separated by a distance of 10 m. Treating them as particles, find the magnitude of the gravitational force between them.

7.*——If the Earth were a Styrofoam ball (of the same radius), what would be the value of the acceleration of free fall at the surface of this ball? The density of Styrofoam is 30 kg/m^3.

8.*——Somewhere between the Earth and the Moon there is a point where the gravitational pull of the Earth on a particle exactly balances that of the Moon. At what distance from the Earth is this point?

9.——Calculate the values of the acceleration of gravity at the surfaces of Venus, Mercury, and Mars. Use the data on planetary masses and radii given in the table printed on the endpapers.

10.——The asteroid Ceres has a diameter of 1100 km and a mass of (approximately) 7×10^{20} kg. What is the value of the acceleration of gravity at its surface? What would be the weight of a 75-kg astronaut standing on this asteroid?

11.——Jupiter is the largest planet; its mass is 1.90×10^{27} kg and its radius is 7.14×10^{7} m. Calculate the acceleration of free fall on the surface of Jupiter. By what factor would your weight on Jupiter be larger than your weight on Earth? Would you find it difficult to remain standing on Jupiter?

Section 5.3: Circular Orbits

12.——The orbital speed of the Earth is 30 km/s. What would be the orbital speed of a (hypothetical) planet in a circular orbit twice as large as the orbit of the Earth? Half as large as the orbit of the Earth?

13.——The radius of the orbit of Neptune around the Sun is 4.5×10^{12} m. What is the orbital speed of Neptune?

14.——Suppose we place a spacecraft in a tight circular orbit around the Sun, with an orbital radius of just twice the radius of the Sun. What would be the orbital speed of such a spacecraft? What would be the orbital period of such a spacecraft?

15.——The radius of the Moon's orbit around the Earth is 3.84×10^{8} m. Suppose that the Earth had a second moon, with an orbit twice as large. Calculate the orbital speed of this second moon around the Earth. Calculate the orbital period.

16.——With what speed would an aircraft have to fly at a constant altitude a few kilometers above the surface of the Earth to make the passengers feel weightless?

17.——The Midas 2 spy satellite was launched into a circular orbit at a height of 500 km above the surface of the Earth. Calculate the orbital period and the orbital speed of this satellite.

18.——The Sun is moving in a circular orbit around the center of our Galaxy. The radius of this orbit is 3×10^{4} light-years. Calculate the period of the orbital motion and calculate the orbital speed of the Sun. The mass of our Galaxy is 4×10^{41} kg and all of this mass can be regarded as concentrated at the center of the Galaxy.

19.——The asteroid belt of the Solar System consists of chunks of rock orbiting around the Sun in approximately circular orbits. The mean distance of the asteroid belt from the Sun is about 2.9 times the distance of the Earth. What is the mean period of the orbital motion of the asteroids?

20.——Consider the communications satellite described in Example 5. What is the speed of this satellite?

21.——Imagine that somewhere in interstellar space a small pebble is in a circular orbit around a spherical asteroid of mass 1000 kg. If the radius of the circular orbit is 1 km, what is the period of the motion?

22.——Instead of calculating the mass of the Earth from the measured values of g and G, as in Section 5.2, you can calculate this mass from the measured value of G and the known radius and period of the orbit of the Moon around the Earth, 3.84×10^{8} m and 27.3 days. What value for the mass of the Earth do you deduce from these data?

23.*—Calculate the orbital speed of Venus from the data given in Examples 3 and 4.

24.*—Europa is a moon of Jupiter. Astronomical observations show that this moon is in a circular orbit of radius 6.71×10^{8} m with a period of 3.55 days. From these data deduce the mass of Jupiter.

Europa, one of Jupiter's moons.

25.*—Suppose that a baseball player standing on the surface of the Moon throws a baseball horizontally. If the baseball has a high enough speed and does not strike any mountain, it can orbit around the Moon in a circular orbit (and, after completing one orbit, strike the baseball player from behind). What is the speed with which the baseball player must launch the ball for such a circular orbit? How long does the ball take to complete one orbit? See Example 2 for data on the Moon.

26.*—Suppose that an artificial satellite is in a circular orbit in the equatorial plane of the Earth. The satellite has an eastward direction of motion, and it has a period of one-half day. What is the radius of the orbit of this satellite? How often per day does the satellite pass overhead as seen from Quito, Ecuador, on the equator?

27.*—The following table lists some of the moons of Saturn. Their orbits are circular.

Moon	Distance from Saturn	Period	Orbital speed
Tethys	2.95×10^5 km	1.89 days	—
Dione	3.77	—	—
Rhea	5.27	—	—
Titan	12.22	—	—
Iapetus	35.60	—	—

(a) From the information given, calculate the periods and orbital speeds of all these moons.
(b) Calculate the mass of Saturn.

28.**—The Discoverer 2 satellite had an approximately circular orbit passing over both poles of the Earth. The radius of the orbit was about 6.67×10^5 km. Taking the rotation of the Earth into account, if the satellite passed over New York City at one instant, over what point of the United States would it pass after completing one more orbit?

Section 5.4: Elliptical Orbits; Kepler's Laws

29.—Which of the planetary elliptical orbits is the most elongated (most eccentric)? Which is the least elongated?

30.—Halley's comet orbits the Sun in an elliptical orbit (the comet reached perihelion in 1986). When the comet is at perihelion, its distance from the Sun is 8.75×10^{10} m, and its speed is 5.46×10^4 m/s. When the comet is at aphelion, its distance is 5.26×10^{12} m. What is the speed at aphelion?

31.—Explorer 1, the first American artificial satellite, had an elliptical orbit around the Earth with a perigee distance of 6.74×10^6 m and an apogee distance of 8.91×10^6 m. The speed of this satellite was 6.21×10^5 m/s at apogee. Calculate the speed at perigee.

32.—The orbit of the Earth around the Sun is not quite a circle but, rather, an ellipse of very small elongation (small eccentricity). At perihelion, the Earth–Sun distance is 1.47×10^{11} m; at aphelion, the Earth–Sun distance is 1.52×10^{11} m. By what factor is the speed of the Earth at perihelion greater than the speed at aphelion?

33.—Vanguard 1, the second American artificial satellite, moved in an elliptical orbit around the Earth with a perigee distance of 7.02×10^6 m and an apogee distance of 10.3×10^6 m. At perigee, the speed of this satellite was 8.22×10^3 m/s. What was the speed at apogee?

34.—What would be the mean distance from the Sun of a hypothetical planet of period 50 years? 100 years?

35.*—The Explorer 10 satellite had an orbit with perigee 175 km and apogee 181,200 km above the surface of the Earth. What was the period of this satellite?

Tethys, one of Saturn's moons.

Comet Halley, photographed in 1986.

The Vanguard 1 satellite.

36.*—Calculate the orbital period of Sputnik 1 from its apogee distance of 7.33×10^3 km and perigee distance of 6.60×10^3 km.

37.*—What would be the period of a hypothetical asteroid of perihelion distance 3.7×10^{11} m and aphelion distance 5.2×10^{11} m?

38.*—Suppose we place a spacecraft into an elliptical orbit of period equal to that of the Earth (1 year), but of perihelion distance half as large as the Earth–Sun distance. What would the aphelion distance of this spacecraft be?

39.*—Use the data in Table 5.1 to calculate the ratio T^2/r^3 (where r is the mean distance from the Sun) for each planet. Are all the values the same, as required by Kepler's Third Law? What is a likely reason for discrepancies in your values?

40*—For a trip to Mars, we want to place a spacecraft into an elliptical orbit such that its perihelion coincides with the mean Earth–Sun distance, but its aphelion coincides with the mean Mars–Sun distance. What is the period of such an elliptical orbit? How long does the spacecraft take to coast from the perihelion (at the Earth) to the aphelion (at Mars)?

Section 5.5: Gravitational Potential Energy

41.*—What is the kinetic energy and what is the gravitational potential energy for the orbital motion of the Earth around the Sun? What is the total energy?

42.*—The Voskhod 1 satellite, which carried Yuri Gagarin into space in 1961, had a mass of 4.7×10^3 kg. The radius of the orbit was (aproximately) 6.6×10^3 km. What were the orbital speed and the orbital energy of this satellite?

43.**—The motor of a Scout rocket uses up all the fuel and stops when the rocket is at an altitude of 200 km above the surface of the Earth and is moving vertically at 8.50 km/s. How high will this rocket rise? Ignore any residual atmospheric friction.

44.**—Neglect the gravity of the Moon, neglect the atmospheric friction, and neglect the rotational speed of the Earth in the following problem: A long time ago, Jules Verne, in his book *From Earth to the Moon* (1865), suggested sending an expedition to the Moon by means of a projectile fired from a gigantic gun.

(a) With what muzzle speed must a projectile be fired vertically from a gun on the surface of the Earth if it is to (barely) reach the distance of the Moon?

(b) Suppose that the projectile has a mass of 2000 kg. What energy must the gun deliver to the projectile? The explosion of 1 ton (2000 lb) of TNT releases 4.2×10^9 J. How many tons of TNT are required for firing this gun?

(c) If the gun barrel is 500 m long, what must be the average acceleration of the projectile during firing?

45.**—An artificial satellite of 3500 kg made of aluminum is in a circular orbit at a height of 100 km above the surface of the Earth. Atmospheric friction removes energy from the satellite and causes it to spiral downward so it ultimately crashes into the ground.

(a) What is the initial orbital energy (gravitational plus kinetic) of the satellite? What is the final energy when the satellite comes to rest on the ground? What is the energy change? (Neglect the rotational speed of the Earth and the ground.)

(b) Suppose that all of this energy is absorbed in the form of heat by the material of the satellite. Is this enough heat to melt the material of the satellite? To vaporize it? The heats of fusion and of vaporization of aluminum are given in Table 14.4.

46.*—The Pons–Brooks comet had a speed of 47.30 km/s when it reached its perihelion, 1.160×10^8 km from the Sun. Is the orbit of this comet elliptic, parabolic, or hyperbolic?

47.**—At a radial distance of 2.00×10^7 m from the center of the Earth, three artificial satellites (1, 2, 3) are ejected from a rocket. The three satellites 1, 2, 3 are given initial speeds of 5.47 km/s, 4.47 km/s, and 3.47 km/s, respectively; the initial velocities are all in the tangential direction.

(a) Which of the satellites 1, 2, 3 will have a circular orbit? Which will have elliptical orbits? Explain your answer.

(b) Draw the circular orbit. Also, on the top of the same diagram draw the elliptical orbits of the other satellites; label the orbits with the numbers of the satellites. (Note: You need not calculate the exact size of the ellipses, but your diagram should show whether the ellipses are larger or smaller than the circle.)

48.**—(a) Since the Moon has no atmosphere, it is possible to place an artificial satellite in a circular orbit that skims along the surface of the Moon (provided that the satellite does not hit any mountains!). Suppose that such a satellite is to be launched from the *surface* of the Moon by means of a gun that shoots the satellite in a horizontal direction. With what velocity must the satellite be shot out from the gun? How long does the satellite take to go once around the Moon?

(b) Suppose that a satellite is shot from the gun with a horizontal velocity of 2.00 km/s. Make a rough sketch showing the Moon and the shape of the satellite's orbit; indicate the position of the gun on your sketch.

(c) Suppose that a satellite is shot from the gun with a horizontal velocity of 3.00 km/s. Make a rough sketch showing the Moon and the shape of the satellite's orbit. Is this a closed orbit?

49.**—Sputnik 1, the first Russian satellite (1957), had a mass of 83.5 kg; its orbit reached perigee at a height of 225 km and apogee at 959 km. Explorer 1, the first American satellite (1958), had a mass of 14.1 kg; its orbit reached perigee at a height of 368 km and apogee at 2540 km. What was the orbital energy of each of these satellites?

50.**—The orbits of most meteoroids around the Sun are nearly parabolic.

(a) With what speed will a meteoroid reach a distance from the Sun equal to the distance of the Earth from the Sun? (Hint: In a parabolic orbit the speed at any radius equals the escape velocity at that radius. Why?)

(b) Taking into account the Earth's orbital speed of 30 km/s, what will be the speed of the meteroid *relative to the Earth* in a head-on collision with the Earth? In an overtaking collision? Ignore the effect of the gravitational pull of the Earth on the meteroid.

51.—Calculate the escape velocity for a projectile launched from the surface of Jupiter. The mass and radius of Jupiter are 1.9×10^{27} kg and 7.1×10^7 m, respectively.

52.—What is the escape velocity for a projectile launched from the surface of our Moon?

53.*—Compare the escape velocity given by Eq. (27) with the velocity required for a circular orbit of radius R_S, according to Eq. (10). By what factor is the escape velocity larger than the velocity for the circular orbit?

54.**—A communications satellite of mass 700 kg has been placed in a circular orbit of radius 4.23×10^7 m around the Earth.

(a) What is the total orbital energy of this satellite?

(b) How much extra energy would we have had to give this satellite to put it into a parabolic orbit that permits it to escape to infinite distance from the Earth?

CHAPTER 6

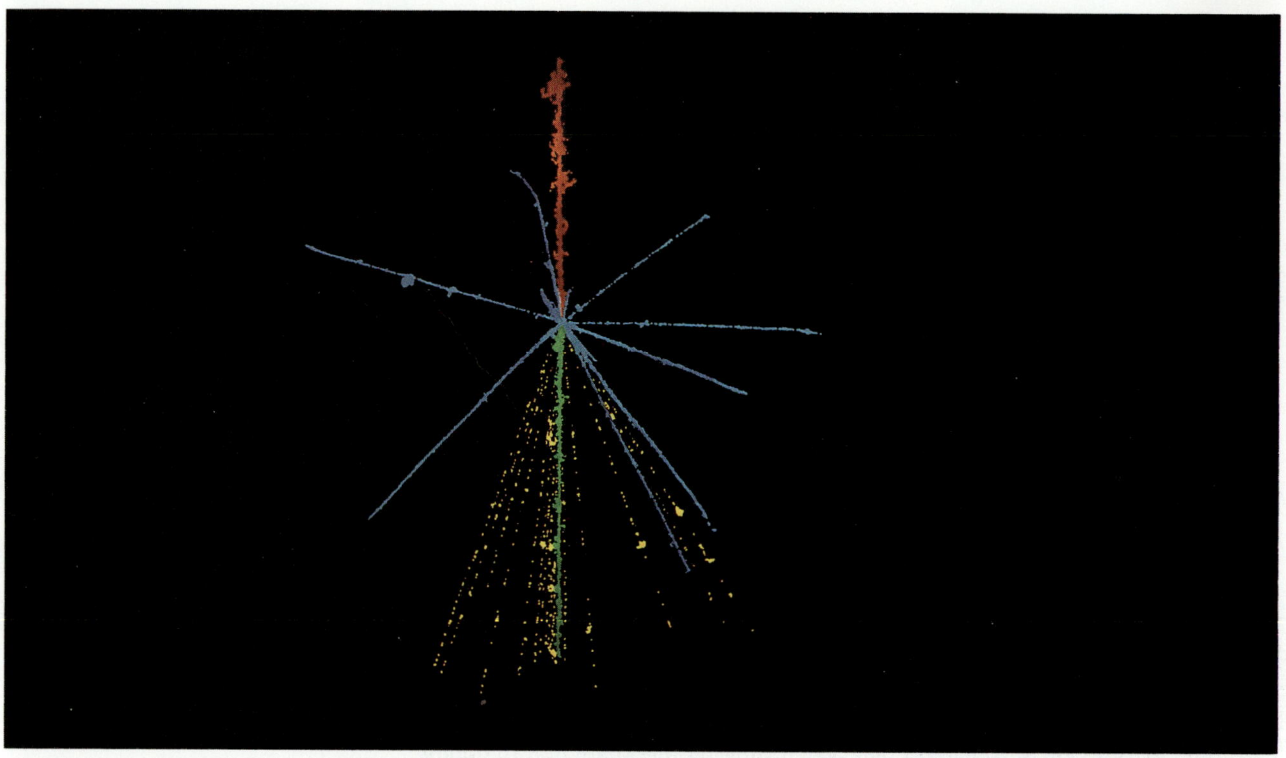

A collision between subatomic particles is here made visible by the fine tracks that these particles produce when passing through the emulsion of a photographic plate. In this enlarged and color-coded view of a small piece of the emulsion, we see the track (red) of a high-speed nucleus that entered the emulsion from above and collided with one of the nuclei in the emulsion. This violent collision fragmented both nuclei, giving rise to a multitude of new particles (blue, green, yellow).

MOMENTUM AND COLLISIONS

So far we have dealt almost exclusively with the motion of a single particle. Now we will begin to study systems of several particles interacting with each other via some forces. This means we must examine, and solve, the equations of motion of all these particles simultaneously.

Since chunks of ordinary matter are made of particles — electrons, protons, and neutrons — all the macroscopic bodies that we encounter in our everyday environment are in fact many-particle systems containing a very large number of particles. However, for most practical purposes, it is not desirable to adopt such an extreme microscopic point of view, and in the preceding chapters we treated the motion of a macroscopic body, such as an automobile, as motion of a single particle. Likewise, when dealing with a system consisting of several macroscopic bodies, we will often find it convenient to treat these bodies as particles and ignore the internal structure of the bodies. For example, when investigating a collision between two automobiles, we may find it convenient to pretend that each of the automobiles is a particle — we then regard the colliding automobiles as a system of two particles which exert forces on each other when in contact. And when investigating the Solar System, we may find it convenient to pretend that each planet and each satellite is a particle — we then regard the Solar System as a system of such planet and satellite particles loosely held together by gravitation and orbiting around the Sun and around each other.

The equations of motion of a system of several particles are often hard, and sometimes impossible, to solve. It is therefore necessary to make the most of any information that can be extracted from the general conservation laws. One of these conservation laws is the conservation of energy. Another is the conservation of momentum, which we will discuss in this chapter.

6.1 MOMENTUM

Newton's laws can be expressed very neatly in terms of momentum, a vector quantity of great importance in physics. ***The momentum of a particle is defined as the product of its mass and velocity:***

$$\mathbf{p} = m\mathbf{v} \tag{1}$$

◀ Momentum

Thus, the momentum $\mathbf{p}$ is a vector that has the same direction as the velocity vector, but a magnitude that is m times as large. ***The unit of momentum is the kilogram-meter per second*** (kg·m/s); this is the momentum of a mass of 1 kg when moving at 1 m/s.

The mathematical definition of momentum is consistent with the intuitive, everyday notion of "momentum." If two cars have equal masses, but one has twice the velocity of the other, it has twice the momentum. And if a truck has three times the mass of a car and the same velocity, it has three times the momentum. During the nineteenth century physicists argued as to whether momentum or kinetic energy was the best measure of the "amount of motion" in a body. They finally decided that the answer depends on the context — as we will see in the examples in this chapter, sometimes momentum is the most relevant quantity, sometimes the energy, and sometimes both are relevant.

Expressed in terms of momentum, Newton's First Law states that, ***in the absence of external forces, the momentum of a particle remains constant,***

$$\mathbf{p} = [\text{constant}] \tag{2}$$

◀ First Law in terms of momentum

Thus, we can say that the momentum of the particle is conserved. Of course, we could equally well say that the velocity of this particle is conserved; but the deeper significance of momentum will emerge when we study the motion of a system of several particles exerting forces on one another. We will find that the total momentum of such a system is conserved — any momentum lost by one particle is compensated by a momentum gain of some other particle or particles.

To express the Second Law in terms of momentum, we note that any change in the momentum of the particle must be attributed to a change in the velocity, since the mass is a constant and cannot contribute to a change of momentum. Thus, the rate of change of the momentum $m\mathbf{v}$ is simply m times the rate of velocity $\mathbf{v}$,

$$\frac{\Delta \mathbf{p}}{\Delta t} = m\frac{\Delta \mathbf{v}}{\Delta t}$$

or, since $\Delta \mathbf{v}/\Delta t$ is the acceleration $\mathbf{a}$,

$$\frac{\Delta \mathbf{p}}{\Delta t} = m\mathbf{a}$$

Second Law in terms of momentum ▶

But, according to the Second Law [see Eq. (3.2)], $m\mathbf{a}$ equals the force. Hence, ***the rate of change of the momentum equals the force,***

$$\frac{\Delta \mathbf{p}}{\Delta t} = \mathbf{F} \qquad (3)$$

This equation gives the Second Law a concise and elegant form.

(a)

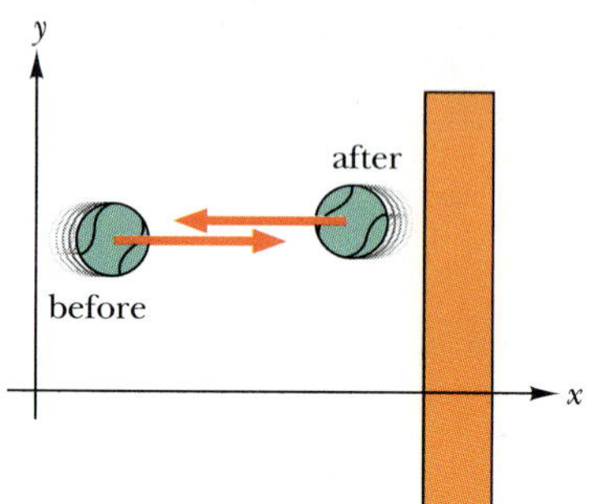

(b)

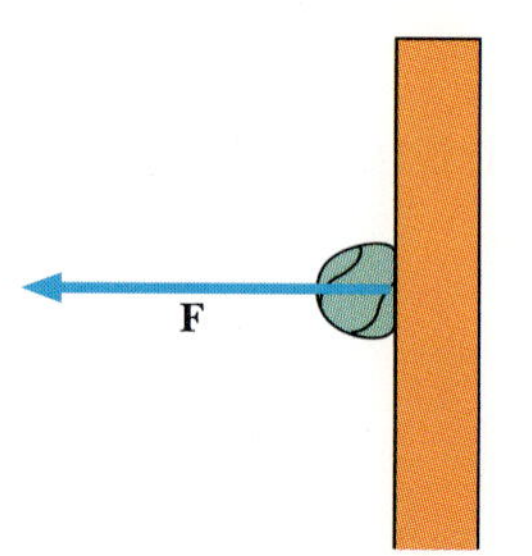

FIGURE 6.1

(a) A tennis ball bounces off a wall. (b) At the instant of impact, the wall exerts a large force on the ball.

EXAMPLE 1. A tennis player smashes a ball of mass 0.060 kg at a vertical wall. The ball hits the wall perpendicularly with a speed of 40 m/s and bounces straight back with the same speed. What is the change of momentum during the impact?

SOLUTION: Take the positive x axis along the direction of the initial motion of the ball (see Fig. 6.1a). The momentum of the ball before impact is then in the positive direction and the x component of the momentum is

$$p_x = mv_x = 0.060 \text{ kg} \times 40 \text{ m/s} = 2.4 \text{ kg}\cdot\text{m/s}$$

The momentum of the ball after impact has the same magnitude but the opposite direction:

$$p'_x = -2.4 \text{ kg}\cdot\text{m/s}$$

(Throughout this chapter, the primes on mathematical quantities indicate that these quantities are evaluated *after* the collision.) The change of momentum is

$$\Delta p_x = p'_x - p_x = -2.4\ \text{kg·m/s} - 2.4\ \text{kg·m/s} = -4.8\ \text{kg·m/s}$$

This change of momentum is produced by the (large) force that acts on the ball during impact on the wall (see Fig. 6.1b). The change of momentum is negative because the force is negative.

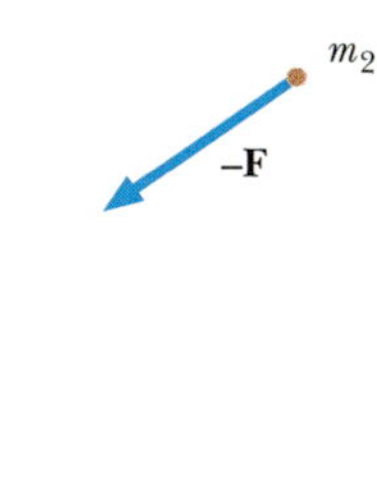

FIGURE 6.2
Two particles exerting mutual forces on each other.

We can also express Newton's Third Law in terms of momentum. Since the action force is exactly opposite to the reaction force, the rate of change of momentum generated by the action force on one body is exactly opposite to the rate of change of momentum generated by the reaction force on the other body. Hence, we can state the Third Law as follows: ***Whenever two bodies exert forces on each other, the resulting changes of momentum are of equal magnitudes and of opposite directions.*** This balance in the changes of momentum leads us to a general law of conservation of the total momentum for a system of particles.

◀ **Third Law in terms of momentum**

The total momentum of a system of n particles — such as the Solar System, consisting of the Sun, the planets, and their satellites — is simply the (vector) sum of all the individual momenta of all the particles. Thus, if $\mathbf{p}_1 = m_1\mathbf{v}_1$, $\mathbf{p}_2 = m_2\mathbf{v}_2$, . . . , $\mathbf{p}_n = m_n\mathbf{v}_n$ are the individual momenta of the particles, then the ***total momentum of the system*** is

$$\mathbf{P} = \mathbf{p}_1 + \mathbf{p}_2 + \ldots + \mathbf{p}_n \tag{4}$$

◀ **Momentum of a system**

The simplest of all many-particle systems consists of just two particles exerting some mutual forces on each other (see Fig. 6.2). Let us assume that the two particles are isolated from the rest of the Universe so that, apart from their mutual forces, they experience no extra forces of any kind. According to the above formulation of the Third Law, the rates of change of $\mathbf{p}_1$ and $\mathbf{p}_2$ are then exactly opposite. The rate of change of the sum $\mathbf{p}_1 + \mathbf{p}_2$ is therefore zero, since the rate of change of the first term in this sum is canceled by the rate of change of the second term. This means that the sum $\mathbf{p}_1 + \mathbf{p}_2$ is a constant of the motion,

$$\mathbf{p}_1 + \mathbf{p}_2 = [\text{constant}] \tag{5}$$

◀ **Conservation of momentum**

This is the ***law of conservation of momentum.*** Note that Newton's Third Law is an essential ingredient for establishing the conservation of momentum: ***the total momentum is constant because the equality of action and reaction keeps the momentum changes of the two particles exactly equal in magnitude but opposite in direction — the particles merely exchange some momentum by means of their mutual forces.***

Conservation of momentum is a powerful tool which permits us to calculate some general features of the motion even when we are ignorant of the detailed properties of the interparticle forces. The following examples illustrate how we can use conservation of momentum to solve some problems of motion.

EXAMPLE 2. An automobile of mass 1500 kg traveling at 25 m/s crashes into a similar parked automobile. The two automobiles remain joined together after the collision. What is the velocity of the wreck immediately after the collision? Neglect friction against the road, since this force is insignificant compared with the large mutual forces that the automobiles exert on each other.

SOLUTION: Under the assumptions of the problem, the only horizontal forces are the mutual forces of one automobile on the other. Thus, momentum conservation applies to the horizontal component of the momentum — the value of this component must be the same before and after the collision. Before the collision, the (horizontal) velocity of the moving automobile is $v_1 = 25$ m/s and that of the other is $v_2 = 0$. With the x axis along the direction of motion (see Fig. 6.3a), the total momentum is therefore

$$P_x = m_1 v_1 + m_2 v_2 = m_1 v_1$$

After the collision, both automobiles have the same velocity (see Fig. 6.3b). We will designate the velocities of the automobiles after the collision by v_1' and v_2', respectively (as in Example 1, the primes on mathematical quantities indicate that these quantities are evaluated *after* the collision). With $v_1' = v_2'$, the total momentum is

$$P_x' = m_1 v_1' + m_2 v_2' = (m_1 + m_2) v_1'$$

By momentum conservation, the momenta P_x and P_x' before and after the collision must be equal,

$$m_1 v_1 = (m_1 + m_2) v_1' \tag{6}$$

(a)

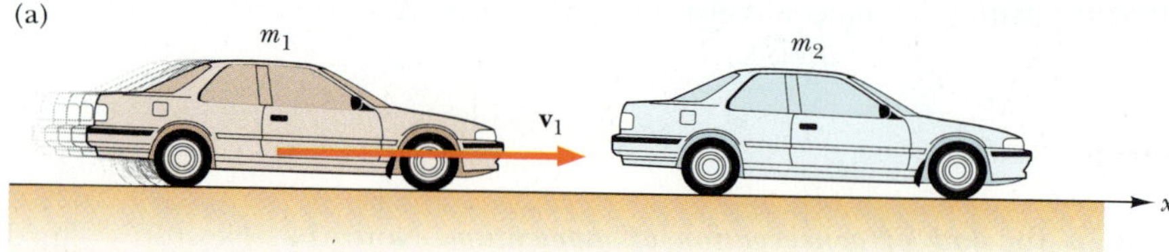

(b)

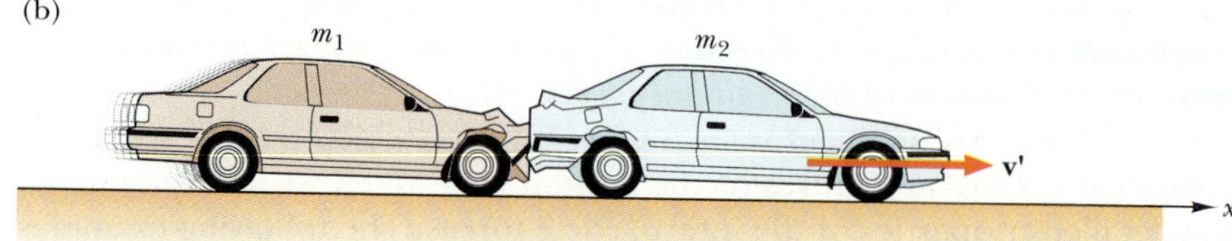

FIGURE 6.3

(a) Initially, the red automobile has a speed of 25 m/s, and the blue automobile is at rest.
(b) After the collision, both automobiles are in motion with velocity v_1'.

When we solve this for v'_1, we find

$$v'_1 = \frac{m_1 v_1}{m_1 + m_2} \tag{7}$$

$$= \frac{1500\ \text{kg} \times 25\ \text{m/s}}{1500\ \text{kg} + 1500\ \text{kg}} = 12.5\ \text{m/s}$$

COMMENTS AND SUGGESTIONS: Note that the solution of this example involves three steps similar to those we used in examples of energy conservation: First write an expression for the momentum before the collision, then write an expression for the momentum after the collision, and then equate these two expressions.

The forces acting during the collision are extremely complicated, but momentum conservation permits us to bypass these complications and directly obtain the answer for the final velocity. Incidentally: It is easy to check that kinetic energy is *not* conserved in this collision; some of this energy is used up to produce changes in the shapes of the automobiles.

EXAMPLE 3. A gun used on board an eighteenth-century warship is mounted on a carriage which allows the gun to roll back each time it is fired (Fig. 6.4). The mass of the gun, including the carriage, is 2200 kg. The gun fires a 6-kg shot horizontally with a velocity of 500 m/s. What is the recoil velocity of the gun?

SOLUTION: The total momentum of the shot plus the gun must be the same before the firing and just after the firing. Before, the total momentum is zero. After, the (horizontal) velocity of the shot is $v'_1 = 500$ m/s and that of the gun is v'_2; hence the total momentum is $P'_x = m_1 v'_1 + m_2 v'_2$, where $m_1 = 6$ kg is the mass of the shot and $m_2 = 2200$ kg is the mass of the gun (including the carriage). Thus, momentum conservation tells us

$$0 = m_1 v'_1 + m_2 v'_2$$

from which

$$v'_2 = -\frac{m_1}{m_2} v'_1$$

$$= -\frac{6\ \text{kg}}{2200\ \text{kg}} \times 500\ \text{m/s} = -1.4\ \text{m/s}$$

The negative sign indicates that the recoil velocity of the gun is opposite to the velocity of the shot.

(a)

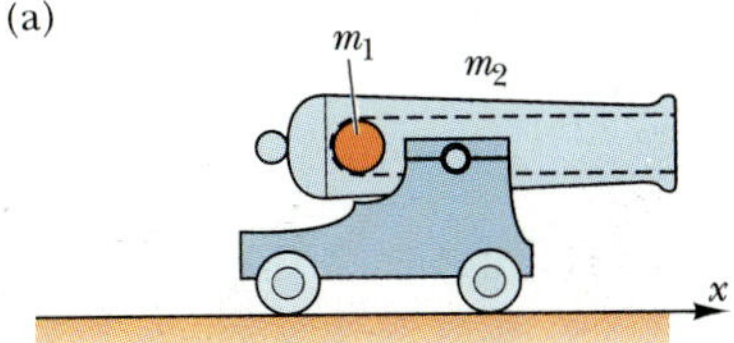

(b)

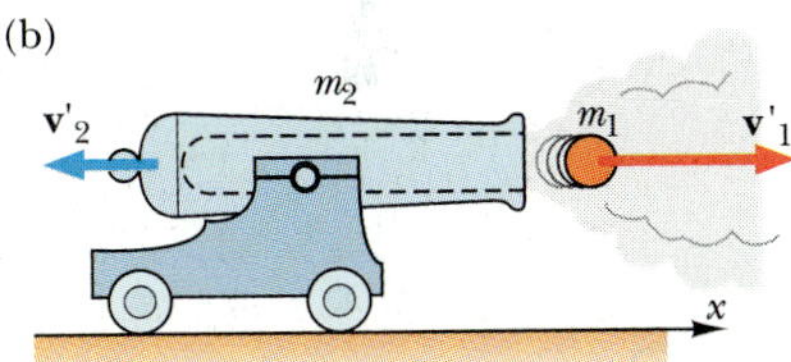

FIGURE 6.4

(a) Initially, the gun and the shot are at rest. (b) After the firing, the gun recoils toward the left (the velocity v'_2 of the gun is negative).

The conservation law for momentum depends on the absence of "extra" forces. If the particles are not isolated from the rest of the Universe, then besides the mutual forces exerted by one particle on the other, there are also forces exerted by other bodies not belonging to the particle system. The former forces are called ***internal forces*** of the system and the latter ***external forces.*** For example, for the colliding automobile of Example 2, the gravity of the Earth, the normal force of the road, and the friction of the road are external forces. In Example 2, we ignored these external forces, because gravity and the normal force cancel each other, and the friction force was assumed to be insignificant by comparison with the much larger impact force that the automobiles exert on each other. But if the external forces are significant, we must take them into account, and we must modify Eqs. (3) and (4). If the external force on particle 1 is $\mathbf{F}_{1,\mathrm{ext}}$, then the total force on this particle is $\mathbf{F} + \mathbf{F}_{1,\mathrm{ext}}$ and its equation of motion will be

$$\frac{\Delta \mathbf{p}_1}{\Delta t} = \mathbf{F} + \mathbf{F}_{1,\mathrm{ext}} \tag{8}$$

Likewise

$$\frac{\Delta \mathbf{p}_2}{\Delta t} = -\mathbf{F} + \mathbf{F}_{2,\mathrm{ext}} \tag{9}$$

If we add the left sides of these equations and the right sides, the contributions from the internal forces cancel, and what remains is

$$\frac{\Delta \mathbf{p}_1}{\Delta t} + \frac{\Delta \mathbf{p}_2}{\Delta t} = \mathbf{F}_{1,\mathrm{ext}} + \mathbf{F}_{2,\mathrm{ext}} \tag{10}$$

The sum of the rates of change of the momenta is the same as the rate of change of the sum of the momenta; hence,

$$\frac{\Delta (\mathbf{p}_1 + \mathbf{p}_2)}{\Delta t} = \mathbf{F}_{1,\mathrm{ext}} + \mathbf{F}_{2,\mathrm{ext}} \tag{11}$$

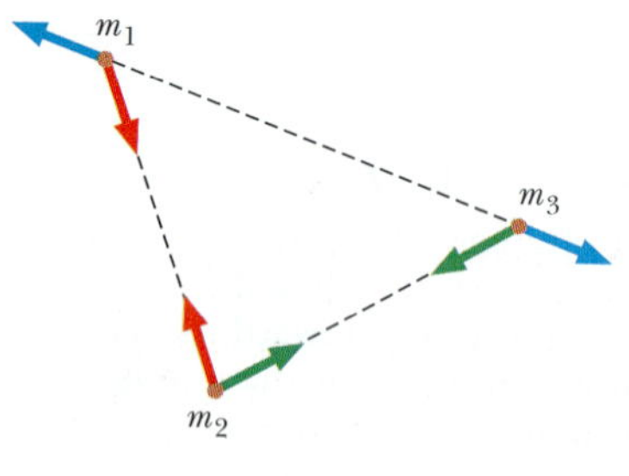

FIGURE 6.5
Three particles exerting forces on each other. As in the case of two particles, the mutual forces between pairs of particles merely exchange momentum between them.

The sum $\mathbf{p}_1 + \mathbf{p}_2$ is the total momentum and the sum $\mathbf{F}_{1,\mathrm{ext}} + \mathbf{F}_{2,\mathrm{ext}}$ is the total external force on the particle system. Thus, Eq. (11) states that the rate of change of the total momentum of the two-particle system equals the total *external* force.

For a system containing more than two particles we can obtain similar results. If the system is isolated so there are no external forces, then the mutual interparticle forces acting between pairs of particles merely transfer momentum from one particle of the pair to the other, just as in the case of two particles. Since all the internal forces necessarily arise from such forces between pairs of particles, these internal forces cannot change the total momentum (see Fig. 6.5). Consequently, the total momentum of an isolated system obeys the conservation law

$$\mathbf{P} = [\text{constant}] \tag{12}$$

If, besides the internal forces, there are external forces, then the latter will change the momentum. The rate of change can be calculated in essentially the same way as for the two-particle system and, again, the rate of change of the total momentum is equal to the total external force. We can write this as

$$\frac{\Delta \mathbf{P}}{\Delta t} = \mathbf{F}_{\text{ext}} \qquad (13)$$

◀ Second Law for a system

where $\mathbf{F}_{\text{ext}}$ is the total external force, that is, the sum of all the external forces acting on all the particles in the system.

Equations (12) and (13) have exactly the same mathematical form as Eqs. (2) and (3), and they may be regarded as the generalizations for a system of particles of Newton's First and Second Laws. As we will see in Section 6.3, Eq. (13) is an equation of motion for the system of particles — it determines the overall translational motion of the system.

6.2 Center of Mass

In our study of kinematics and dynamics in the preceding chapters we always ignored the size of the bodies; even when analyzing the motion of a large body — an automobile or a ship — we pretended that the motion could be treated as particle motion, with position being described by reference to some marker point which we imagine painted on the body. In reality, large bodies are systems of particles, and their motion obeys Eq. (13) for a system of particles. This equation can be converted into an equation of motion, containing just one acceleration rather than the rate of change of momentum of the entire system, by making reference to one special point on the body: the ***center of mass.*** The equation that describes the motion of this special point has the same mathematical form as the equation of motion of a single particle, that is, the motion of the center of mass mimics single-particle motion (see, for example, Fig. 6.6).

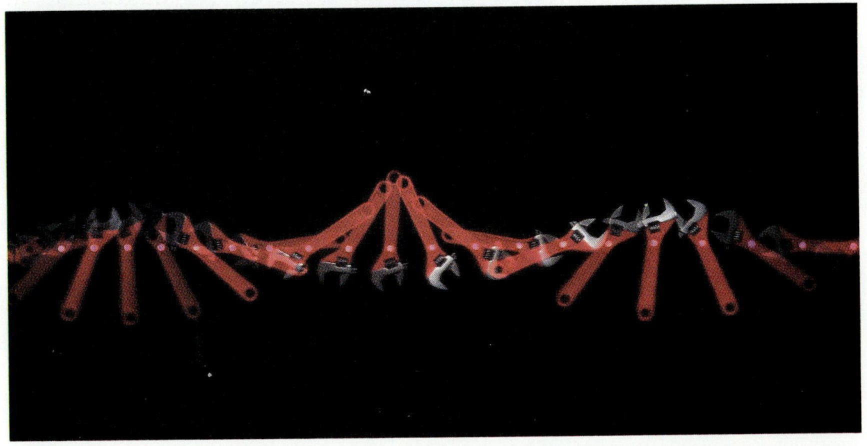

FIGURE 6.6

A wrench moving freely in the absence of external forces. The center of mass, marked with a white dot, moves with uniform velocity along a straight line (you can check this by laying a ruler along the white dots).

The position of the center of mass is merely the average position of the mass of the system. For instance, if the system consists of two particles, each of mass 1 kg, then the center of mass is halfway between them (see Fig. 6.7). In any system consisting of n particles of equal mass — such as a piece of pure metal with atoms of only one kind — the coordinate of the center of mass is simply the average of the coordinates of all the particles,

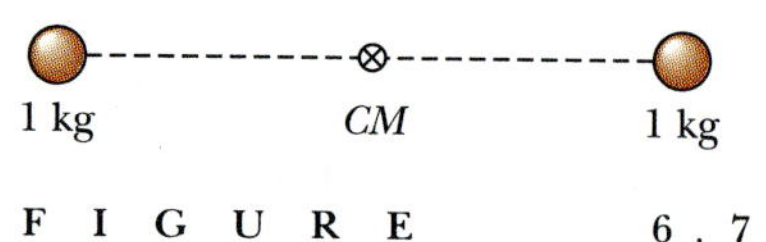

FIGURE 6.7

Two particles of equal masses, and their center of mass.

$$x_{CM} = \frac{x_1 + x_2 + \ldots x_n}{n} \tag{14}$$

with similar equations for the y and the z coordinates.

If the system consists of particles of unequal masses, then the position of the center of mass can be calculated by first subdividing the particles into fragments of equal mass. For instance, if the system consists of two particles of masses 2 kg and 1 kg, respectively, then we can pretend that we have *three* particles of equal masses 1 kg, two of which are located at the same position. The coordinate of the center of mass is then

$$x_{CM} = \frac{x_1 + x_1 + x_2}{3} \tag{15}$$

We can also write this in the equivalent form

$$x_{CM} = \frac{m_1 x_1 + m_2 x_2}{m_1 + m_2} \tag{16}$$

where $m_1 = 2$ kg and $m_2 = 1$ kg. The formula (16) is actually valid for any values of the masses m_1 and m_2. The formula simply asserts that in the average position, the position of particle 1 is included m_1 times and the position of particle 2 is included m_2 times — that is, the number of times the position of each particle is included in the average is in direct proportion to its mass.

EXAMPLE 4. A 50-kg woman and an 80-kg man sit on the two ends of a seesaw of length 3.0 m (see Fig. 6.8). Treating them as particles,

FIGURE 6.8

A woman and a man on a seesaw.

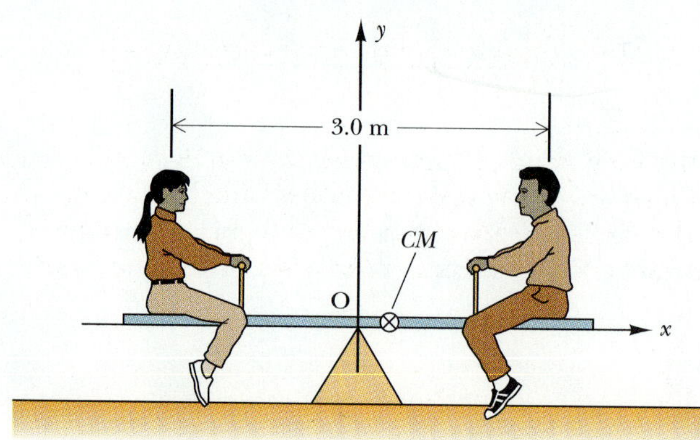

and ignoring the mass of the seesaw, find the center of mass of this system.

SOLUTION: In Fig. 6.8, the origin of coordinates is at the center of the seesaw; hence the woman has a negative x coordinate ($x = -1.5$ m) and the man a positive x coordinate ($x = +1.5$ m). According to Eq. (16), the coordinate of the center of mass is

$$x_{CM} = \frac{m_1x_1 + m_2x_2}{m_1 + m_2} = \frac{50 \text{ kg} \times (-1.5 \text{ m}) + 80 \text{ kg} \times 1.5 \text{ m}}{50 \text{ kg} + 80 \text{ kg}}$$

$$= 0.35 \text{ m}$$

COMMENTS AND SUGGESTIONS: Note that the distance of the woman from the center of mass is 1.5 m + 0.35 m = 1.85 m and the distance of the man from the center of mass is 1.5 m − 0.35 m = 1.15 m. The ratio of these distances is 1.85/1.15 = 1.6, which coincides with the inverse of the ratio of the masses, 50/80 = 1/1.6. This is a general rule: ***the distances of two particles from their center of mass are in the inverse ratio of the masses.***

As we will see in a later chapter, if the point of support (the fulcrum) of the seesaw coincides with the center of mass, then the seesaw will be in equilibrium; that is, it will remain at rest if initially at rest.

If the system consists of n particles of masses $m_1, m_2, \ldots, m_n$, then we apply the same prescription: the number of times the position of each particle is included in the average is in direct proportion to its mass. This leads to the following general expression for the x coordinate of the center of mass:

$$x_{CM} = \frac{m_1x_1 + m_2x_2 + \ldots + m_nx_n}{m_1 + m_2 + \ldots + m_n} \tag{17}$$

or

$$x_{CM} = \frac{m_1x_1 + m_2x_2 + \ldots + m_nx_n}{M} \tag{18}$$

◀ Center of mass

where M is the total mass of the system, $M = m_1 + m_2 + \ldots + m_n$.

If the particles of our system are spread out in two or three dimensions, then the center of mass has not only an x coordinate, but also a y and z coordinate. The formulas for these coordinates are similar to Eq. (18):

$$y_{CM} = \frac{m_1y_1 + m_2y_2 + \ldots + m_ny_n}{M} \tag{19}$$

$$z_{CM} = \frac{m_1 z_1 + m_2 z_2 + \ldots + m_n z_n}{M} \tag{20}$$

The position of the center of mass of a solid body can, in principle, be calculated from Eqs. (18)–(20), since a solid body is a collection of atoms, each of which can be regarded as a particle. However, it would be awkward to deal with 10^{23} or so atoms that make up a chunk of matter the size of, say, a coin. It is more convenient to pretend that matter in bulk has a smooth and continuous distribution of mass over its entire volume.

Density ▸

The mass distribution can then be described by the ***density,*** that is, the amount of mass per unit volume (the number of kilograms per cubic meter). For a body of uniform density (with the same number of kilograms in every cubic meter), the amount of mass in any given volume is directly proportional to the amount of volume. To find the position of the center of mass of such a body, we therefore can subdivide the body into small, equal volume elements and compute the average position of all these small volume elements. ***Thus, the position of the center of mass is the average position of all the small volume elements making up the volume of the body.*** If the body has a symmetric shape, this average position will often be obvious by inspection. For instance, a sphere of uniform density, or a ring, or a circular plate, or a cylinder, or a parallelepiped will have its center of mass at the geometrical center (see Fig. 6.9).

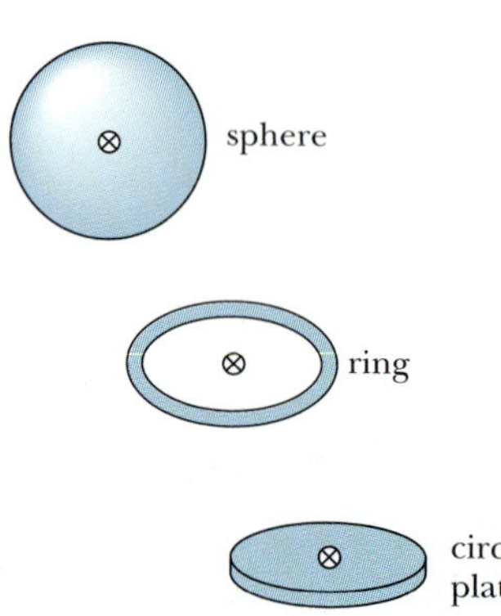

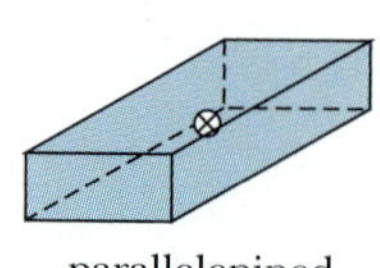

FIGURE 6.9

Several bodies for which the center of mass coincides with the geometric center.

EXAMPLE 5. A meter stick of aluminum is bent at its midpoint, so the two halves are at right angles (see Fig. 6.10). Where is the center of mass of this bent stick?

SOLUTION: When dealing with a body of complicated shape, it is advantageous to regard the body as composed of simpler pieces. We can regard the bent stick as consisting of two straight pieces, each of 0.50 m. The centers of mass of these straight pieces are at their midpoints, 0.25 m from their ends (see Fig. 6.10). The center of mass of the entire stick is the average position of the centers of mass of the two halves. With the coordinate axes arranged as in Fig. 6.10, the x and y coordinates of the center of mass are then, according to Eq. (14),

$$x_{CM} = \frac{0.25\ \text{m} + 0}{2} = 0.125\ \text{m}$$

$$y_{CM} = \frac{0.25\ \text{m} + 0}{2} = 0.125\ \text{m}$$

Note that the center of mass of this bent stick is *outside* of the stick, that is, it is not in the volume of the stick. If you find this surprising, remember that the center of mass of two equal particles separated by some distance is at a point in (empty) space, halfway between the particles; and the center of mass of a ring (see Fig. 6.9) is in (empty) space, at the center of the ring. There are many other examples of systems in

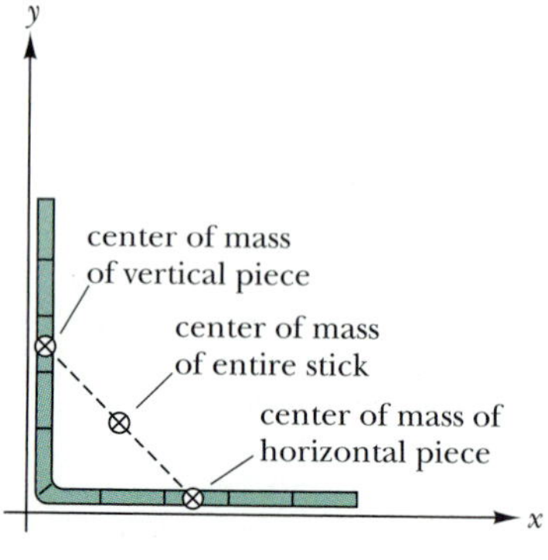

FIGURE 6.10

A meter stick, bent through 90° at its midpoint.

which the center of mass is in empty space, outside of the region occupied by the mass of the system.

The position of the center of mass enters into the calculation of the gravitational potential energy of an extended body located near the surface of the Earth. According to Eq. (4.24), the potential energy of a single particle of mass m at a height y above the ground is mgy. For a system of particles, the total gravitational potential energy is the sum of the potential energies of all the particles:

$$U = m_1gy_1 + m_2gy_2 + \ldots + m_ngy_n$$

$$= (m_1y_1 + m_2y_2 + \ldots + m_ny_n)g \tag{21}$$

Comparison with Eq. (19) shows that the quantity in parentheses is My_{CM}. Hence, Eq. (21) becomes

$$U = Mgy_{CM} \tag{22}$$

This expression for the gravitational potential energy of the system has the same mathematical form as for a single particle — it is as though the entire mass of the system were located at the center of mass.

For a human body standing upright, the position of the center of mass is in the middle of the trunk, at about the height of the navel. This is therefore the height to be used in the calculation of the gravitational potential energy of the body. However, if the body is bent in a downward arc, the position of the center of mass is below and outside of the trunk. Olympic jumpers take advantage of this shift of the center of mass to reduce the gravitational potential energy they need to supply for a high jump. By adopting a bent position as they pass over the bar, they permit their center of mass to pass *below* the bar (see Fig. 6.11), and they thereby reduce the required gravitational potential energy by a significant amount.

FIGURE 6.11
High jumper passing over the bar.

Figure 6.12a gives the centers of mass of the body segments of a man of average proportions standing upright. Figure 6.12b shows the hinge

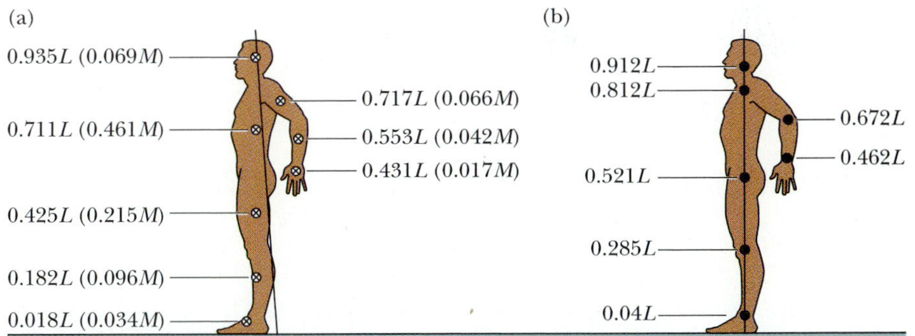

FIGURE 6.12
(a) Centers of mass of the body segments of an average male of mass M and height L standing upright. The numbers give the heights of the centers of mass of the body segments from the floor and (in parentheses) the masses of the body segments; right and left limbs are shown combined. (b) Hinge points of the body. The numbers give the heights of the joints from the floor.

points at which these body segments are joined. From the data in this figure, we can calculate the location of the center of mass when the body adopts any other position, and we can calculate the work done against gravity to change the position of any segment.

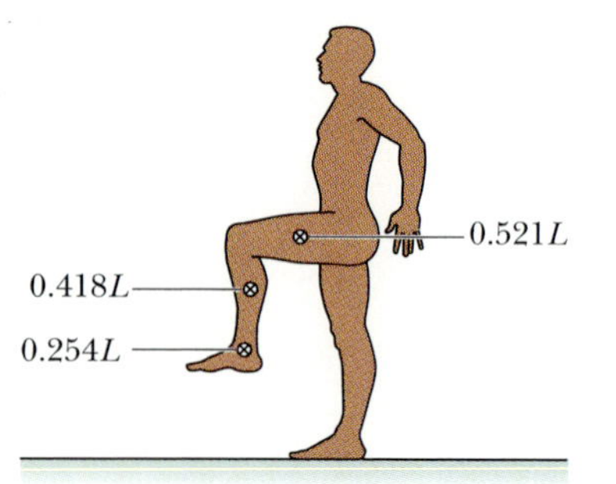

FIGURE 6.13
Centers of mass of the segments of a raised leg.

EXAMPLE 6. Suppose the man lifts one leg, so the thigh is horizontal (see Fig. 6.13). How much work must he do against gravity? Assume the mass of the man is $M = 73$ kg and his height $L = 1.75$ m.

SOLUTION: According to Fig. 6.12, the masses of one thigh, one lower leg, and one foot are $\frac{1}{2} \times 0.215M = 0.108M$, $\frac{1}{2} \times 0.096M = 0.048M$, and $\frac{1}{2} \times 0.034M = 0.017M$, respectively. The heights of the centers of mass of these segments above ground are $0.425L$, $0.182L$, and $0.018L$, respectively, in the upright position shown in Fig. 6.12a. Thus, the center of mass of the leg is at

$$y_{CM} = \frac{(0.108 \times 0.425 + 0.048 \times 0.182 + 0.017 \times 0.018)ML}{(0.108 + 0.048 + 0.017)M}$$

$$= 0.318L$$

With the leg in the lifted position shown in Fig. 6.13, the height of the center of mass of the thigh is the height of the hip joint, $0.521L$. The centers of mass of the lower leg and the foot are lifted through a distance equal to the length of the thigh, which according to Fig. 6.12b is $0.521L - 0.285L = 0.236L$. Thus, the heights of the lower leg and foot become, respectively, $0.182L + 0.236L = 0.418L$ and $0.018L + 0.236L = 0.254L$. The center of mass of the lifted leg is then at

$$y_{CM} = \frac{(0.108 \times 0.521 + 0.048 \times 0.418 + 0.017 \times 0.254)ML}{(0.108 + 0.048 + 0.017)M}$$

$$= 0.466L$$

The change in the height of the center of mass of the leg is therefore $\Delta y_{CM} = 0.466L - 0.318L = 0.148L$ and, according to Eq. (22), the work done against gravity is

$$W = \Delta U = M_{\text{leg}} g \Delta y_{CM} \qquad (23)$$

$$= (0.108 + 0.048 + 0.017)M \times g \times 0.148L = 0.0256MgL$$

$$= 0.0256 \times 73\ \text{kg} \times 9.81\ \text{m/s}^2 \times 1.75\ \text{m} = 32.1\ \text{J}$$

6.3 THE MOTION OF THE CENTER OF MASS

When a system of particles moves, so does the center of mass. We will now obtain an equation for the motion of the center of mass, an equation which relates the acceleration of the center of mass to the external force. This equation will permit us to calculate the overall translational motion of a system of particles.

According to Eq. (18), if the positions of the particles change by small amounts Δx_1, Δx_2, . . . ,Δx_n, then the position of the center of mass changes by

$$\Delta x_{CM} = \frac{1}{M}(m_1 \Delta x_1 + m_2 \Delta x_2 + \ldots + m_n \Delta x_n)$$

with Δx extremely small (24)

Dividing this by the time Δt taken for these changes of position, we obtain

$$\frac{\Delta x_{CM}}{\Delta t} = \frac{1}{M}\left(m_1 \frac{\Delta x_1}{\Delta t} + m_2 \frac{\Delta x_2}{\Delta t} + \ldots + m_n \frac{\Delta x_n}{\Delta t}\right) \tag{25}$$

The left side of this equation is the x component of the instantaneous velocity of the center of mass, and the rates of change on the right side are the x components of the instantaneous velocities of the individual particles; thus

$$v_{x,CM} = \frac{m_1 v_{x,1} + m_2 v_{x,2} + \ldots + m_n v_{x,n}}{M}$$

Note that this equation has the same mathematical form as Eq. (18); that is, the velocity of the center of mass is an average over the particle velocities, and the number of times the velocity of each particle is included is directly proportional to its mass.

Since similar equations apply to the y and z components of the velocity, we are entitled to write a vector equation for the velocity of the center of mass:

$$\mathbf{v}_{CM} = \frac{m_1 \mathbf{v}_1 + m_2 \mathbf{v}_2 + \ldots + m_n \mathbf{v}_n}{M} \tag{26}$$

The quantity in the numerator is simply the total momentum [compare Eq. (4)]; hence Eq. (26) says

$$\mathbf{v}_{CM} = \frac{\mathbf{P}}{M} \tag{27}$$

or

$$\mathbf{P} = M\mathbf{v}_{CM} \tag{28}$$

◂ **Momentum and velocity of center of mass**

This equation expresses the total momentum of a system of particles as the product of the total mass and the velocity of the center of mass. Obviously, this equation is analogous to the familiar equation $\mathbf{p} = m\mathbf{v}$ for the momentum of a single particle.

Since we know, from Eq. (13), that the rate of change of the total momentum equals the net external force on the system, we see from Eq. (28) that the rate of change of $M\mathbf{v}_{CM}$ must equal the net external force. The mass M is constant; hence, the rate of change of $M\mathbf{v}_{CM}$ is M times the rate of change of $\mathbf{v}_{CM}$, which is M times the acceleration of the position of the center of mass. Consequently:

Equation of motion ▶ of center of mass

$$\mathbf{F}_{\text{ext}} = M\mathbf{a}_{CM} \tag{29}$$

This equation of motion for a system of particles is the analog of Newton's equation of motion for a single particle. ***The equation asserts that the center of mass moves as though it were a particle of mass M under the influence of a force*** $\mathbf{F}_{\text{ext}}$.

This result justifies some of the approximations we made in previous chapters. For instance, in Example 3.10 we treated a ship sliding down a slipway as a particle. Equation (29) shows that this treatment is legitimate: the center of mass of the ship, under the influence of the external forces (gravity and friction), moves parallel to the slipway, just as though it were a particle on an inclined plane with the external forces acting directly on it.

If the net external force vanishes, the acceleration of the center of mass vanishes; the center of mass then remains at rest or it moves with uniform velocity.

EXAMPLE 7. During a "space walk," an astronaut floats in space 8.0 m from his Gemini spacecraft orbiting the Earth. He is tethered to the spacecraft by a long umbilical cord; to return he pulls himself in by this cord. How far does the spacecraft move toward him? The mass of the spacecraft is 3500 kg and the mass of the astronaut, including his space suit, is 140 kg.

(a)

SOLUTION: In the absence of external forces, the center of mass of the astronaut–spacecraft system remains at rest. Astronaut and spacecraft exert equal and opposite forces on each other (via the cord); the astronaut is pulled toward the spacecraft and the spacecraft is pulled toward the astronaut. Thus, the spacecraft and the astronaut both move toward the center of mass and there they meet.

With the x axis as in Fig. 6.14, the x coordinate of the center of mass is

$$x_{CM} = \frac{m_1 x_1 + m_2 x_2}{m_1 + m_2} \tag{30}$$

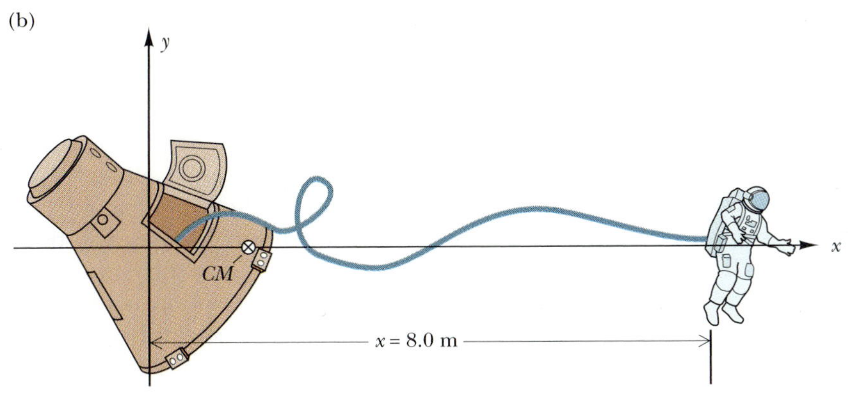

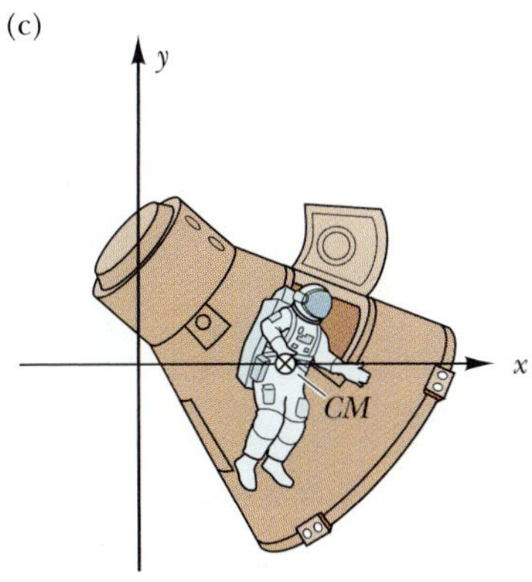

FIGURE 6.14

(a) Astronaut on a "space walk" during the Gemini 4 mission. (b) Initial position of the astronaut and the spacecraft. The center of mass is between them. (c) Final position of the astronaut and the spacecraft. They are both at the center of mass.

where $m_1 = 3500$ kg is the mass of the spacecraft and $m_2 = 140$ kg is the mass of the astronaut. Strictly, the coordinates x_1 and x_2 of the spacecraft and of the astronaut should correspond to the centers of mass of these bodies, but, for the sake of simplicity, we neglect their sizes and treat both as particles. The initial values of the coordinates are $x_1 = 0$ and $x_2 = 8.0$ m; hence

$$x_{CM} = \frac{0 + 140\text{ kg} \times 8.0\text{ m}}{3500\text{ kg} + 140\text{ kg}} = 0.31\text{ m}$$

During the pulling in, the spacecraft will move from $x_1 = 0$ to $x_1 = 0.31$ m; simultaneously, the astronaut will move from $x_2 = 8.0$ m to $x_2 = 0.31$ m. (The astronaut moves a large distance, and the spacecraft a small distance, in inverse proportion of their masses.)

Finally, a comment about the kinetic energy of a system of particles. The total kinetic energy of a system of particles is simply the sum of the individual kinetic energies of all the particles,

Kinetic energy of a system

$$K = \tfrac{1}{2}m_1v_1^2 + \tfrac{1}{2}m_2v_2^2 + \ldots + \tfrac{1}{2}m_nv_n^2 \qquad (31)$$

Since Eq. (28) for the momentum of a system of particles resembles the expression for the momentum of a single particle, we might be tempted to guess that the equation for the kinetic energy of a system of particles also can be expressed in the form $K = \frac{1}{2}Mv_{CM}^2$ resembling the kinetic energy of a single particle. But this is wrong! The total kinetic energy of a system of particles is usually larger than $\frac{1}{2}Mv_{CM}^2$. We can see this in the following simple example: Consider two automobiles of equal masses moving toward each other at equal speeds. The velocity of the center of mass is then zero, $v_{CM} = 0$, and consequently $\frac{1}{2}Mv_{CM}^2 = 0$. However, since each automobile has a positive kinetic energy, the total kinetic energy is *not* zero.

FIGURE 6.15 Crash test of a Mercedes-Benz automobile. The photographs show an impact at 49 km/h into a rigid barrier. The first photograph was taken 5×10^{-3} s after the initial contact; the others were taken at intervals of 20×10^{-3} s. The automobile remains in contact with the barrier for 0.120 s; it then recoils from the barrier with a speed of 4.7 km/h. The checkered bar on the ground has a length of 2 m.

6.4 COLLISIONS

The collision between two bodies — an automobile and solid wall, a ship and an iceberg, a molecule of oxygen and a molecule of nitrogen — involves a violent change of the motion, a change brought about by very strong forces that begin to act suddenly when the bodies come into contact, last a short time, and then cease just as suddenly when the bodies separate. The forces that act during a collision usually vary in a rather complicated way, so their complete theoretical description is impossible (for example, in an automobile collision) or at least very difficult (for example, in an atomic collision). However, even without exact knowledge of the force law, we can make some predictions about the collision by taking advantage of the general laws of conservation of momentum and energy. In the following sections we will see what constraints these laws impose on the motion of the colliding bodies.

The force that two colliding bodies exert on each other acts only for a short time, giving a brief but strong push. Such a force is called an impulsive force. During the collision, the impulsive force is much stronger than any other forces that may be present; consequently the impulsive force produces a large change in the motion while the other forces produce only small and insignificant changes. For instance, during the automobile collision shown in Fig. 6.15, the only important force is the push of the wall on the front end of the automobile; the effects produced by gravity and by the friction force of the road during the collision are insignificant.

Suppose the collision lasts some short time Δt, say, from $t = 0$ to $t = \Delta t$, and that during this time an impulsive force $\mathbf{F}$ acts on one of the colliding bodies. This force is zero before $t = 0$ and it is zero after $t = \Delta t$, but it is large between these times. For example, Fig. 6.16 shows a plot of the force experienced by an automobile in a collision with a solid wall lasting 0.120 s. The force is zero before $t = 0$ and after $t = 0.120$ s, and varies in a complicated way between these times.

We can establish a simple equation relating the average value of the impulsive force to the momentum change produced by the collision. If the velocity before the collision is $\mathbf{v}$ and the velocity after the collision is $\mathbf{v}'$, then the average acceleration during the collision is

$$\bar{\mathbf{a}} = \frac{\mathbf{v}' - \mathbf{v}}{\Delta t} \tag{32}$$

Multiplying this by the mass of the colliding body, we obtain

$$m\bar{\mathbf{a}} = \frac{m\mathbf{v}' - m\mathbf{v}}{\Delta t} \tag{33}$$

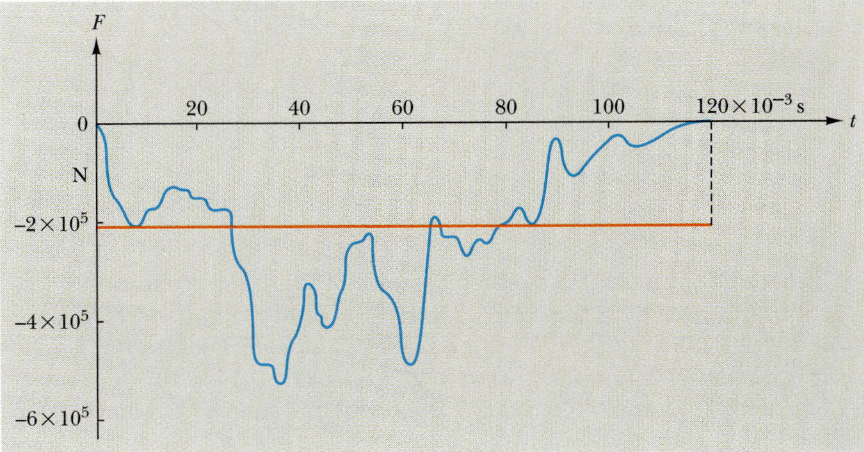

FIGURE 6.16

Force on the automobile as a function of time during the impact shown in Fig. 6.15. The red horizontal line indicates the time-average force. (From data supplied by Mercedes-Benz of North America, Inc.)

The left side of this equation is the product of the mass and the average acceleration; by Newton's Second Law this is the average force. The right side is the difference between the final momentum ($\mathbf{p}' = m\mathbf{v}'$) and the initial momentum ($\mathbf{p} = m\mathbf{v}$) divided by the time interval Δt of the collision. Hence, Eq. (33) is equivalent to

$$\overline{\mathbf{F}} = \frac{\mathbf{p}' - \mathbf{p}}{\Delta t} \tag{34}$$

◀ Average force

Note that the average force in this equation is an average over time (in contrast to the average force we used in the definition of work, which was an average over position).

Equation (34) gives a convenient estimate of the magnitude of the impulsive force if the collision interval and the momentum change are known.

EXAMPLE 8. The collision between the automobile and the wall shown in Fig. 6.15 lasts 0.120 s. The mass of the automobile is 1700 kg and the initial and final velocities are $v = 13.6$ m/s and $v' = -1.3$ m/s, respectively (the final velocity is negative because the automobile recoils, or bounces back). From these data, evaluate the average force that acts on the automobile during the collision. Evaluate the average force that acts on the wall.

SOLUTION: With the x axis along the direction of the initial motion, the change of momentum is

$$p'_x - p_x = mv' - mv \tag{35}$$

$$= 1700 \text{ kg} \times (-1.3 \text{ m/s}) - 1700 \text{ kg} \times 13.6 \text{ m/s}$$

$$= -2.53 \times 10^4 \text{ kg} \cdot \text{m/s}$$

According to Eq. (34), the average force is then

$$\overline{F_x} = \frac{p'_x - p_x}{\Delta t} \tag{36}$$

$$= \frac{-2.53 \times 10^4 \text{ kg} \cdot \text{m/s}}{0.120 \text{ s}} = -2.11 \times 10^5 \text{ N}$$

Since the mutual forces on two bodies engaged in a collision are an action–reaction pair, Newton's Third Law tells us that the forces on the automobile and on the wall are of equal magnitudes and of opposite directions. Thus, the force on the wall is $+2.11 \times 10^5$ N. This is quite a large force — it equals the weight of about 2×10^4 kg, or 20 tons.

EXAMPLE 9. When an automobile collides with an obstacle and suddenly stops, a passenger not restrained by a seat belt will not stop simultaneously with the automobile, but instead will continue traveling at nearly constant speed until he hits the dashboard and the windshield. The collision of the passenger's head with the windshield often results in severe or fatal injuries. Consider a human head striking a windshield at 15 m/s and stopping in a time of 0.015 s (this time is considerably shorter than the time of about 0.12 s for stopping the automobile because the front end of the automobile crumples gradually and cushions the collision to some extent; there is no such cushioning for the head striking the windshield). What is the average force on the head during impact on the windshield? What is the average deceleration? Treat the head as a body of mass 5.0 kg, moving independently of the neck and trunk.

SOLUTION: The initial momentum of the head is

$$p_x = mv_x = 5.0 \text{ kg} \times 15 \text{ m/s} = 75 \text{ kg} \cdot \text{m/s}$$

The final momentum is zero. Hence the average force is

$$\overline{F_x} = \frac{p'_x - p_x}{\Delta t} = -\frac{p_x}{\Delta t}$$

$$= -\frac{75 \text{ kg} \cdot \text{m/s}}{0.015 \text{ s}} = -5.0 \times 10^3 \text{ N}$$

The average acceleration is

$$\overline{a_x} = \frac{\overline{F_x}}{m} = -\frac{5.0 \times 10^3 \text{ N}}{5.0 \text{ kg}} = -1.0 \times 10^3 \text{ m/s}^2$$

Such a crash of the head against the windshield will cause a severe concussion, and it is also likely to cause a fracture of the cranial bones.

Often it is not possible to calculate the motion of the colliding bodies by direct solution of Newton's equation of motion because the impulsive forces that act during the collision are not known in sufficient detail. We must then glean whatever information we can from the general laws of conservation of momentum and energy, which do not depend on the details of these forces. In some simple instances, these general laws permit the deduction of the motion after the collision from what is known about the motion before the collision.

◂ Elastic collision

A collision in which the kinetic energy before and after the collision is the same is called elastic. (The usage of the word *elastic* in the context of collisions must not be confused with the usage we encountered previously when discussing the restoring force of a deformable body. However, there is a connection between these usages of *elastic:* if the colliding bodies exert purely elastic restoring forces on each other during the collision, then the kinetic energy before and after the collision will indeed be the same; that is, the collision will be elastic.) Collisions between macroscopic bodies are usually not elastic — during the collision some of the kinetic energy is transformed into heat by the internal friction forces and some is expended in doing work to change the internal configuration of the bodies. For example, the automobile collision shown in Fig. 6.15 is highly inelastic; almost the entire initial kinetic energy is expended in doing work on the automobile parts, changing their shape. On the other hand, the collision of a Superball and a hard wall or the collision of two billiard balls comes pretty close to being elastic (there is some internal friction associated with the deformations that the balls suffer during the collision, and therefore the collision is not *exactly* elastic).

Collisions between "elementary" particles — such as electrons, protons, and neutrons — are often elastic. These particles have no internal friction forces which could dissipate kinetic energy. A collision between such particles can be inelastic only if it involves the creation of new particles; such new particles may arise either by conversion of some of the available kinetic energy into mass or by transmutation of the old particles by means of a change in their internal structure.

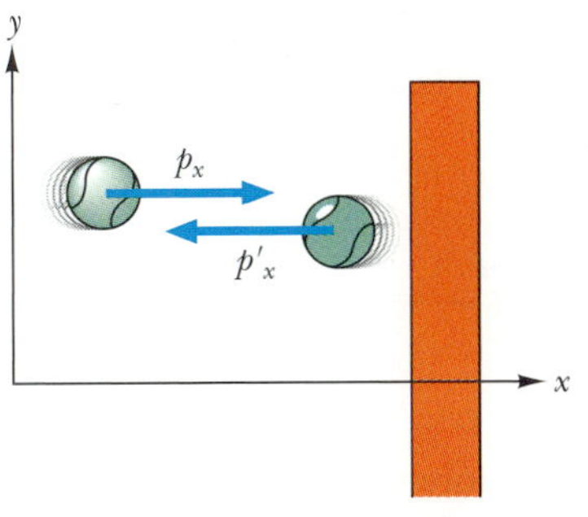

FIGURE 6.17
The initial momentum p_x of the ball is positive; the final momentum p'_x is negative.

EXAMPLE 10. A Superball made of a rubberlike plastic is thrown against a hard, smooth wall. The ball strikes the wall perpendicularly with speed v. Assuming that the collision is elastic, find the speed of the ball after the collision.

SOLUTION: The only force on the ball is the normal force exerted by the wall; this force reverses the motion of the ball (see Fig. 6.17). Since the wall is very massive, the reaction force of the ball will not give it any appreciable velocity. Hence the kinetic energy of the system, both before and after the collision, is merely the kinetic energy of the ball. Conservation of this kinetic energy then requires that the ball rebound with a speed v equal to the incident speed.

COMMENTS AND SUGGESTIONS: Note that although the kinetic energy of the ball is the same before and after the collision, the momentum is not the same before and after the collision (see also Example 1). If the x axis is in the direction of the initial motion, then the momentum of the ball before the collision is $p_x = mv$, and after the collision it is $p'_x = -mv$. Hence the change of momentum is $\Delta p_x = p'_x - p_x = -2mv$. The wall suffers an equal and opposite momentum change of $+2mv$, so that the total momentum of the system is conserved. (The wall can acquire the momentum $2mv$ without acquiring any appreciable velocity because its mass is large and it is attached to the ground.)

The study of collisions is an important tool in the experimental investigations of atoms, nuclei, and elementary particles. All subatomic bodies are too small to be made visible with any kind of microscope. Just as a blind man uses a cane to feel his way around a room, a physicist who cannot see the interior of an atom uses probes to "feel" for subatomic structures. The probe used by physicists in the exploration of subatomic structures is simply a stream of fast-moving particles — electrons, protons, alpha particles, or others. These projectiles are aimed at a target containing a sample of the atoms, nuclei, or elementary particles under investigation. From the manner in which the projectiles collide and react with the target, physicists can deduce some of the properties of the subatomic structures in the target.

6.5 Elastic Collisions on a Straight Line

The collision of two boxcars on a railroad track is an example of a collision on a straight line. More generally, the collision of any two bodies that approach head-on and recoil along their original line of motion is a collision along a straight line. Such collisions will occur only under exceptional circumstances; nevertheless, we find it instructive to study such collisions because they display in a simple way some of the important features of more complicated collisions.

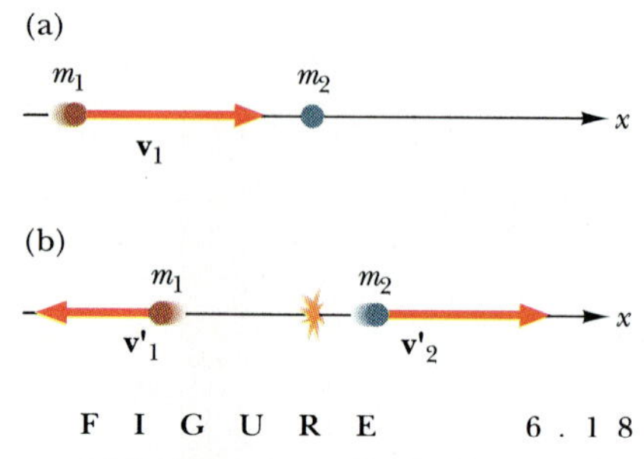

FIGURE 6.18
(a) Before the collision, particle 2 is at rest, and particle 1 has velocity $\mathbf{v}_1$. (b) After the collision, particle 1 has velocity $\mathbf{v}'_1$, and particle 2 has velocity $\mathbf{v}'_2$.

In an elastic collision of two particles moving along a straight line, the laws of conservation of momentum and energy completely determine the final velocities in terms of the initial velocities. In the following calculations, we will assume that one particle (the "target") is initially at rest, and the other (the "projectile") is initially in motion.

Figure 6.18a shows the particles before the collision, and Fig. 6.18b shows them after; the x axis is along the direction of motion. We will designate the x components of the velocity of particle 1 ("projectile") and particle 2 ("target") before the collision by v_1 and v_2, respectively. We will designate these velocities after the collision by v_1' and v_2'.

Since particle 2 is initially at rest, $v_2 = 0$. The initial momentum is therefore simply the momentum $m_1 v_1$ of particle 1. The final momentum, after the collision, is $m_1 v_1' + m_2 v_2'$. Conservation of momentum then states that

$$m_1 v_1 = m_1 v_1' + m_2 v_2' \tag{37}$$

The initial kinetic energy is $\frac{1}{2} m_1 v_1^2$ and the final kinetic energy is $\frac{1}{2} m_1 v_1'^2 + \frac{1}{2} m_2 v_2'^2$. Conservation of energy states that

$$\tfrac{1}{2} m_1 v_1^2 = \tfrac{1}{2} m_1 v_1'^2 + \tfrac{1}{2} m_2 v_2'^2 \tag{38}$$

In these equations, we can regard the initial velocities v_1 and v_2 as known, and the final velocities v_1' and v_2' as unknown. We therefore want to solve these equations for the unknown quantities. For this purpose, it is convenient to rearrange the two equations somewhat. If we subtract $m_1 v_1'$ from both sides of Eq. (37), we obtain

$$m_1 (v_1 - v_1') = m_2 v_2' \tag{39}$$

If we multiply both sides of Eq. (38) by 2 and subtract $m_1 v_1'^2$ from both sides, we obtain

$$m_1 (v_1^2 - v_1'^2) = m_2 v_2'^2 \tag{40}$$

With the identity $(v_1^2 - v_1'^2) = (v_1 - v_1')(v_1 + v_1')$ this becomes

$$m_1 (v_1 - v_1')(v_1 + v_1') = m_2 v_2'^2 \tag{41}$$

Now divide Eq. (41) by Eq. (39); that is, divide the left side of Eq. (41) by the left side of Eq. (39), and the right side of Eq. (41) by the right side of Eq. (39). The result is

$$v_1 + v_1' = v_2' \tag{42}$$

This trick gets rid of the bothersome squares in Eq. (40) and leaves us with two equations — Eqs. (39) and (42) — without squares. To complete the solution for our unknowns, we take the value $v_2' = v_1 + v_1'$ given by Eq. (42) and substitute it into the right side of Eq. (39):

$$m_1 (v_1 - v_1') = m_2 (v_1 + v_1') \tag{43}$$

This can be solved immediately for the unknown v_1', with the result

$$v_1' = \frac{m_1 - m_2}{m_1 + m_2} v_1 \tag{44}$$

Finally, we substitute this value of v_1' into the expression $v_2' = v_1 + v_1'$ and we find

$$v_2' = \frac{2m_1}{m_1 + m_2} v_1 \tag{45}$$

These equations give us the final velocities v_1' and v_2' in terms of the initial velocity v_1.

EXAMPLE 11. An empty boxcar of mass $m_1 = 20{,}000$ kg rolling on a straight track at 5 m/s collides with a loaded stationary boxcar of mass $m_2 = 65{,}000$ kg (see Fig. 6.19). Assuming that the cars bounce off each other elastically, find the velocities after the collision.

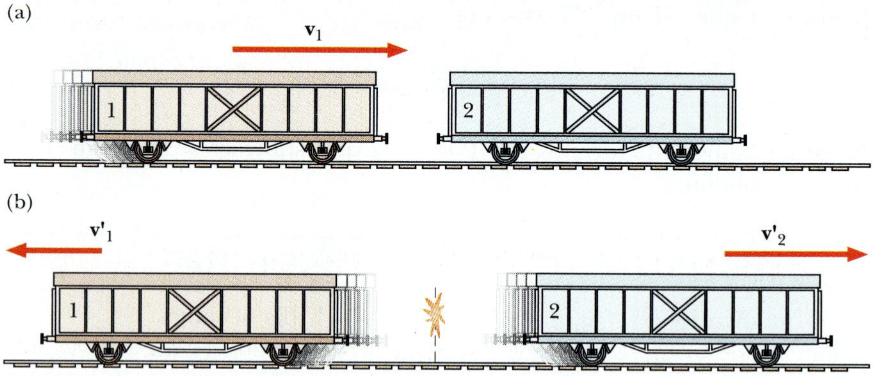

FIGURE 6.19 (a) Initially, boxcar 1 is moving toward the right, and boxcar 2 is stationary. (b) After the collision, boxcar 1 is moving toward the left, and boxcar 2 is moving toward the right.

SOLUTION: With $m_1 = 20{,}000$ kg and $m_2 = 65{,}000$ kg, Eqs. (44) and (45) yield

$$v_1' = \frac{20{,}000\ \text{kg} - 65{,}000\ \text{kg}}{20{,}000\ \text{kg} + 65{,}000\ \text{kg}} \times 5\ \text{m/s} = -2.6\ \text{m/s}$$

$$v_2' = \frac{2 \times 20{,}000\ \text{kg}}{20{,}000\ \text{kg} + 65{,}000\ \text{kg}} \times 5\ \text{m/s} = 2.4\ \text{m/s}$$

If the mass of the target is much larger than the mass of the projectile, then m_1 can be neglected compared with m_2. Equation (44) then becomes

$$v_1' \simeq -\frac{m_2}{m_2} v_1 = -v_1 \tag{46}$$

and Eq. (45) becomes

$$v_2' \simeq \frac{2m_1}{m_2} v_1 \simeq 0 \tag{47}$$

This means the projectile bounces off the target with a reversed velocity, and the target remains nearly stationary. Conversely, if the mass of the projectile is much larger than the mass of the target, then m_2 can be neglected compared with m_1, and Eqs. (44) and (45) become

$$v_1' \simeq \frac{m_1}{m_1} v_1 = v_1 \tag{48}$$

and

$$v_2' \simeq \frac{2m_1}{m_1} v_1 = 2v_1 \tag{49}$$

This means that the projectile plows along with unchanged velocity, and the target bounces off with *twice* the speed of the incident projectile. For example, when a (heavy) golf club strikes a golf ball, the ball bounces away with approximately twice the speed of the club (see Fig. 6.20).

FIGURE 6.20
Impact of club on golf ball. By inspection of this multiple-exposure photograph, we see that the speed of the ball is larger than the initial speed of the club.

6.6 INELASTIC COLLISIONS ON A STRAIGHT LINE

◀ Totally inelastic collision

If the collision is inelastic, with some loss of kinetic energy during the collision, then the only conservation law that is applicable is the conservation of momentum. This, by itself, is insufficient to calculate the velocities of both particles after the collision. However, ***if the collision is totally inelastic, so a maximum amount of kinetic energy is lost, then the conservation of momentum suffices to determine the velocities of both particles after the collision.***

In a totally inelastic collision, the particles do not bounce off each other at all; instead, they stick together, like two automobiles that form a single mass of interlocking wreckage after a collision, or two railroad boxcars that couple together. Under these conditions, it is clear that the velocities of both particles must coincide with the velocity of the center of mass. But the velocity of the center of mass *after* the collision is the same as the velocity of the center of mass *before* the collision, since, in the absence of external forces, the acceleration of the center of mass is zero [see Eq. (29)]. Before the collision, the velocity of the target particle is zero ($v_2 = 0$), and the general equation [Eq. (26)] for the velocity of the center of mass yields

$$v_{CM} = \frac{m_1 v_1}{m_1 + m_2} \tag{50}$$

This must then be the final velocity of both particles after a totally inelastic collision:

$$v_1' = v_2' = v_{CM} = \frac{m_1 v_1}{m_1 + m_2} \tag{51}$$

We have already come across an instance of this formula in Example 2.

EXAMPLE 12. Suppose that the two boxcars of Example 11 couple during the collision and remain locked together (see Fig. 6.21). What is the velocity of the combination after the collision? How much kinetic energy is dissipated during the collision?

SOLUTION: With $m_1 = 20{,}000$ kg, $m_2 = 65{,}000$ kg, and $v_1 = 5$ m/s, Eq. (50) gives us the velocity of the center of mass:

$$v_{CM} = \frac{20{,}000 \text{ kg} \times 5 \text{ m/s}}{20{,}000 \text{ kg} + 65{,}000 \text{ kg}} = 1.2 \text{ m/s}$$

and this must be the velocity of the coupled cars after the collision.

The kinetic energy before the collision is

$$\tfrac{1}{2} m_1 v_1^2 = \tfrac{1}{2} \times 20{,}000 \text{ kg} \times (5 \text{ m/s})^2 = 2.5 \times 10^5 \text{ J}$$

and the kinetic energy after the collision is

$$\tfrac{1}{2} m_1 v_{CM}{}^2 + \tfrac{1}{2} m_2 v_{CM}{}^2 = \tfrac{1}{2}(m_1 + m_2) v_{CM}{}^2 \tag{52}$$

$$= \tfrac{1}{2}(20{,}000 \text{ kg} + 65{,}000 \text{ kg}) \times (1.2 \text{ m/s})^2 = 0.6 \times 10^5 \text{ J}$$

Thus, the loss of kinetic energy is

$$2.5 \times 10^5 \text{ J} - 0.6 \times 10^5 \text{ J} = 1.9 \times 10^5 \text{ J}$$

This energy is absorbed by friction (in the bumpers) during the coupling of the boxcars.

(a)

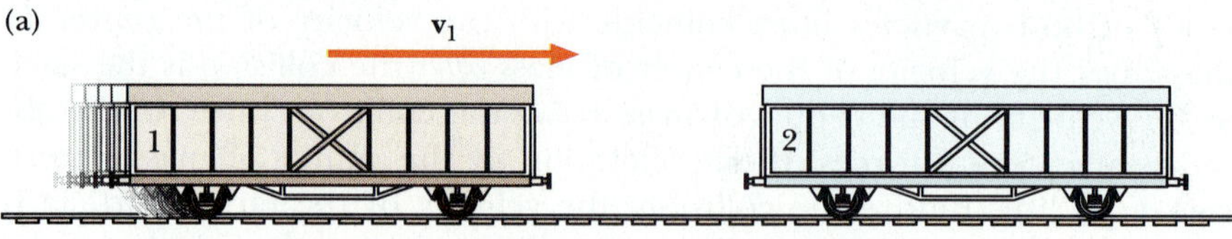

(b)

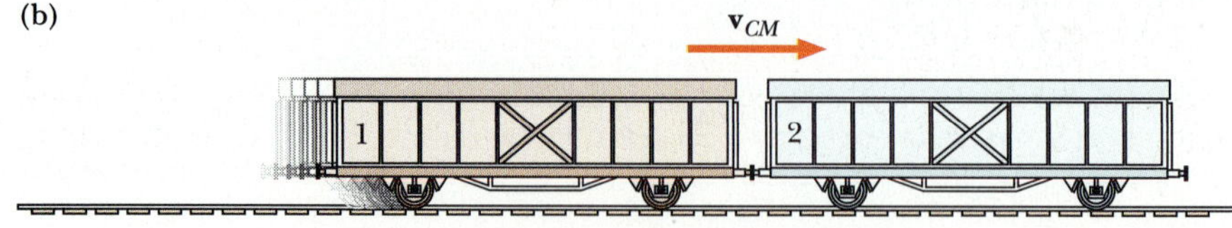

FIGURE 6.21

(a) Initially, boxcar 1 is moving toward the right, and boxcar 2 is stationary, as in Fig. 6.19. (b) After the collision, the boxcars remain locked together. Their common velocity must then be the velocity of the center of mass.

EXAMPLE 13. Figure 6.22a shows a ***ballistic pendulum,*** a device used in the past century to measure the speeds of bullets. The pendulum consists of a large block of wood of mass m_2 suspended from thin wires. Initially, the pendulum is at rest. The bullet, of mass m_1, strikes the block horizontally and remains stuck in it. The impact of the bullet puts the block in motion, causing it to swing upward to a height h (see Fig. 6.22b). In a test of a Springfield rifle firing a bullet of 9.7 g, a ballistic pendulum of 4.0 kg swung up to a height of 19 cm. What was the speed of the bullet before impact?

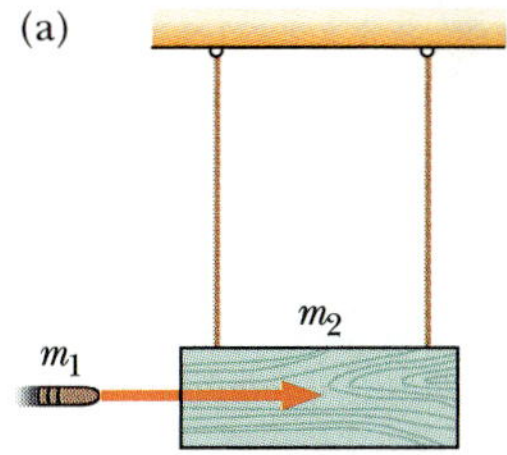

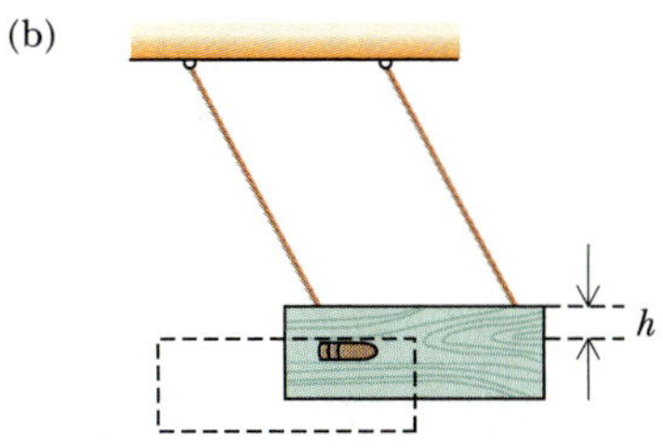

FIGURE 6.22

(a) Before the bullet strikes, the block of wood is at rest. (b) After the bullet strikes, the block, with the embedded bullet, moves toward the right and swings upward to a height h.

SOLUTION: For the analysis of this motion, we consider two stages: the collision and the subsequent swinging motion. The collision of the bullet with the wood is totally inelastic. Hence, immediately after the collision, bullet and block move horizontally with the velocity of the center of mass,

$$v_{CM} = \frac{m_1 v_1}{m_1 + m_2} \tag{53}$$

During the subsequent swinging motion of the pendulum, the total mechanical energy (kinetic plus potential) is conserved. At the bottom of the swing, the energy is kinetic, $\frac{1}{2}(m_1 + m_2)v_{CM}^2$; and at the top of the swing, it is potential, $(m_1 + m_2)gh$. Hence, conservation of energy tells us that

$$\tfrac{1}{2}(m_1 + m_2)v_{CM}^2 = (m_1 + m_2)gh \tag{54}$$

If we divide this by $(m_1 + m_2)$ and take the square root of both sides, we find

$$v_{CM} = \sqrt{2gh} \tag{55}$$

Substitution of this into Eq. (53) yields

$$\sqrt{2gh} = \frac{m_1 v_1}{m_1 + m_2} \tag{56}$$

which we can solve for v_1, with the result

$$v_1 = \frac{m_1 + m_2}{m_1}\sqrt{2gh} \tag{57}$$

$$= \frac{0.0097\text{ kg} + 4.0\text{ kg}}{0.0097\text{ kg}}\sqrt{2 \times 9.81\text{ m/s}^2 \times 0.19\text{ m}}$$

$$= 800\text{ m/s}$$

COMMENTS AND SUGGESTIONS: Note that during the collision, momentum is conserved but not energy (the collision is inelastic); and that during the swinging motion, energy is conserved, but not momentum (the swinging motion proceeds under the influence of the "external" forces of gravity and the tensions in the wires).

SUMMARY

- **Momentum of a system of particles:** $\mathbf{P} = \mathbf{p}_1 + \mathbf{p}_2 + \ldots + \mathbf{p}_n$
- **Rate of change of momentum:** $\dfrac{\Delta \mathbf{P}}{\Delta t} = \mathbf{F}_{\text{ext}}$
- **Conservation of momentum (in the absence of external forces):** $\mathbf{P} = [\text{constant}]$
- **Center of mass:** $x_{CM} = \dfrac{m_1 x_1 + m_2 x_2 + \ldots + m_n x_n}{m_1 + m_2 + \ldots + m_n}$

$$y_{CM} = \frac{m_1 y_1 + m_2 y_2 + \ldots + m_n y_n}{m_1 + m_2 + \ldots + m_n}$$

$$z_{CM} = \frac{m_1 z_1 + m_2 z_2 + \ldots + m_n z_n}{m_1 + m_2 + \ldots + m_n}$$

- **Momentum and velocity of center of mass:** $\mathbf{P} = M\mathbf{v}_{CM}$
- **Equation of motion of the center of mass:** $M\mathbf{a}_{CM} = \mathbf{F}_{\text{ext}}$
- **Kinetic energy of a system of particles:** $K = \frac{1}{2}m_1 v_1^2 + \frac{1}{2}m_2 v_2^2 + \ldots + \frac{1}{2}m_n v_n^2$
- **Average force in collision:** $\overline{\mathbf{F}} = \dfrac{\mathbf{p}' - \mathbf{p}}{\Delta t}$
- **Elastic collision:** Kinetic energy is conserved.
- **Speeds in one-dimensional elastic collision:**
  before: $v_1 \neq 0$, $v_2 = 0$

  after: $v_1' = \dfrac{m_1 - m_2}{m_1 + m_2} v_1, \qquad v_2' = \dfrac{2m_1}{m_1 + m_2} v_1$
- **Speeds in totally inelastic collision:**
  before: $v_1 \neq 0$, $v_2 = 0$

  after: $v_1' = v_2' = v_{CM} = \dfrac{m_1 v_1}{m_1 + m_2}$

QUESTIONS

1.–When the nozzle of a fire hose is discharging a large amount of water at high speed, several strong firemen are needed to hold the nozzle steady. Explain.

2.–When firing a shotgun, a hunter always presses it tightly against his shoulder. Why?

3.–Where is the center of mass of this book when it is closed? Mark the center of mass with a cross.

4.–Roughly, where is the center of mass of this book when it is open, as it is at this moment?

5.–A fountain shoots a stream of water into the air (see Fig. 6.23). Roughly, where is the center of mass of the water that is in the air at one instant? Is the center of mass higher or lower than the middle height?

FIGURE 6.23
Stream of water from a fountain.

6.–Consider the moving wrench shown in Fig. 6.6. If the center of mass on this wrench has not been marked, how could you have found it by inspection of this photograph?

7.–Within the Mexican jumping bean, a small insect larva jumps up and down. How does this lift the bean off the table?

8.–An elephant jumps off a cliff. Does the Earth move upward while the elephant falls?

9.–A juggler stands on a balance, juggling five balls (Fig. 6.24). On the average, will the balance register the weight of the juggler plus the weight of the five balls? More than that? Less?

FIGURE 6.24
Juggler on a balance.

10.–Suppose you fill a rubber balloon with air and then release it so that the air spurts out of the nozzle. The balloon will fly across the room. Explain.

11.–The combustion chamber of a rocket engine is closed at the front and at the sides, but it is open at the rear (Fig. 6.25). Explain how the pressure of the gas on the walls of this combustion chamber gives a net forward force that propels the rocket.

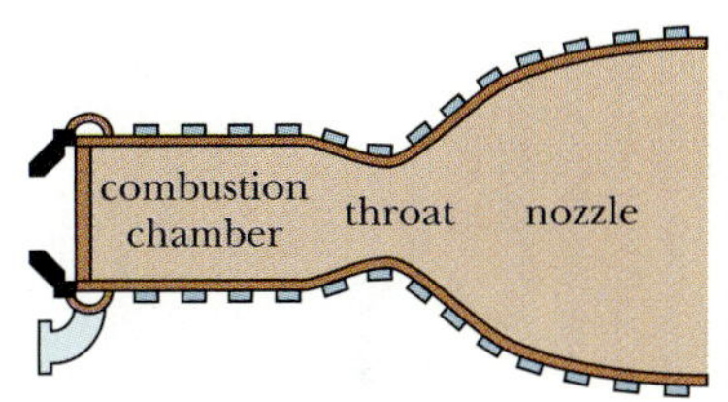

FIGURE 6.25
Combustion chamber of a rocket engine.

12.–In experimental simulations of automobile collisions, baboons were used as passengers. When wearing only lap belts, 50% of the animals suffered fatal injuries in collisions of 300 m/s^2. When wearing lap belts and shoulder harnesses, 50% of the animals suffered fatal injuries in collisions of 1000 m/s^2. When protected by air bags, the animals suffered only minor injuries in collisions of up to 1200 m/s^2. (Accelerations of 300 and 1000 m/s^2 are typical of automobile impacts into a solid wall at 35 and 60 mi/h, respectively.) Why are air bags better than seats belts (in head-on collisions)? What are some of the advantages and disadvantages of air bags and seat belts?

13.–A (foolish) stuntman wants to jump out of an airplane at high altitude without a parachute. He plans to jump while tightly encased in a strong safe which can withstand the impact on the ground. How would you convince the stuntman to abandon this project?

14.–Each year 50,000 people die in automobile accidents in the United States; even more suffer severe injuries sometimes resulting in amputation, blindness, or paralysis. Why have automobile manufacturers failed to design an automobile that would permit its occupants to walk away from a 60-mi/h crash? Is this technologically impossible?

15.–Is the collision shown in Fig. 6.15 totally inelastic? For the sake of safety, would it be desirable to design automobiles so that their collisions are elastic or inelastic?

16.–Statistics show that, on the average, the occupants of a heavy ("full-size") automobile are more likely to survive a crash than those of a light ("compact") automobile. Why would you expect this to be true?

FIGURE 6.26
A pendulum toy, sometimes called "Newton's cradle."

17.–Give an example of a collision between two bodies in which *all* of the kinetic energy is lost to inelastic processes.

18.–Explain the operation of the pendulum toy shown in Fig. 6.26.

19.–In order to split a log with a small ax, you need a greater impact speed than you would need with a large ax. Why? If the energy required to split the log is the same in both cases, why is it more tiring to use a small ax? (Hint: Think about the kinetic energy of your arms.)

20.–If you throw an (elastic) baseball at an approaching train, the ball will bounce back at you with an increased speed. Explain.

PROBLEMS

Section 6.1: Momentum

1.——What is the momentum of a rifle bullet of mass 15 g and speed 600 m/s? An arrow of mass 40 g and speed 80 m/s?

2.——Using the entries listed in Tables 1.8 and 3.1, find the magnitude of the momentum for each of the following: Earth moving around Sun, jet airliner at maximum airspeed, automobile at federal speed limit, man walking.

3.——What is the momentum of an automobile of 1000 kg moving at 40 km/h? If a truck of 8000 kg is to have the same momentum as the automobile, what must be its speed?

4.*——An electron of mass 9.1×10^{-31} kg is moving in the x–y plane; its speed is 2.0×10^5 m/s and its direction of motion makes an angle of 25° with the x axis. What are the components of the momentum of the electron?

5.*——A sky diver of mass 75 kg is in free fall (without air friction). What is the rate of change of his momentum?

6.*——Calculate the change of kinetic energy in the collision between the two automobiles described in Example 2.

7.*——A woman of 55 kg jumps off a small rowing boat of 40 kg. The woman jumps off with a horizontal velocity of 5 m/s. What is the recoil speed of the boat?

8.*——Two automobiles, moving at 25 km/h in opposite directions, collide head-on. One automobile has a mass of 680 kg, the other a mass of 1500 kg. After the collision both remain joined together. What is the velocity of the wreck? What is the change of the velocity of each automobile during the collision?

9.*——The nucleus of an atom of radium (mass 3.77×10^{-25} kg) suddenly ejects an alpha particle (mass 6.68×10^{-27} kg) of an energy of 7.26×10^{-16} J. What is the velocity of recoil of the nucleus? What is the kinetic energy of the recoil?

10.**——A lion of mass 120 kg leaps at a hunter with a horizontal velocity of 12 m/s. The hunter has an automatic rifle firing bullets of mass 15 g with a muzzle velocity of 630 m/s and he attempts to stop the lion in midair. How many bullets would the hunter have to fire into the lion to stop its horizontal motion? Assume the bullets stick inside the lion.

11.**——A vase falls off a table and hits a smooth floor, shattering into three fragments of equal mass. Two of the fragments leave the point of impact with horizontal velocities of equal magnitude v at right angles. What are the magnitude and direction of the horizontal velocity of the third fragment?

12.**—Find the recoil velocity for the gun described in Example 3 if the gun is fired with an elevation angle of 20°.

13.**—An automobile of mass 1500 kg and a truck of 3500 kg collide at an intersection. Just before the collision the automobile was traveling north at 80 km/h and the truck was traveling east at 50 km/h. After the collision both vehicles remain joined together.

(a) What is the velocity (magnitude and direction) of the vehicles immediately after collision?

(b) How much kinetic energy is lost during the collision?

14.**—The solar wind sweeping past the Earth consists of a stream of particles, mainly hydrogen ions of mass 1.7×10^{-27} kg. There are about 10^7 ions per cubic meter and their speed is 4×10^5 m/s. What force does the impact of the solar wind exert on an artificial Earth satellite that has an area of 1.0 m^2 facing the wind? Assume that upon impact the ions at first stick to the surface of the satellite.

15.**—The nozzle of a fire hose ejects 750 liters/min of water at a speed of 26 m/s. Estimate the recoil force on the nozzle. By yourself, can you hold this nozzle steady in your hands?

16.**—Grain is being loaded into an almost full railroad car from an overhead chute (see Fig. 6.27). If 500 kg per second fall freely from a height of 4 m to the top of the car, what downward push does the impact of the grain exert on the car?

17.**—A boy holds the end of a garden hose and lets water stream out of the hose, in an almost vertical direction (see Fig. 6.28). The hose has a diameter of 2.0 cm and the stream of water rises to a height of 1.5 m.

(a) What is the speed of the water as it leaves the hose?

(b) What is the amount of water streaming out of the hose per second?

(c) What is the recoil force that the water exerts on the hose?

(d) What is the mass of the rising column of water (from the end of the hose to the top; ignore the descending stream)? Compare the weight of the ascending column with the recoil force.

18.*—A spaceship of frontal area 25 m^2 passes through a cloud of interstellar dust at a speed of 1.0×10^6 m/s. The density of dust is 2.0×10^{-18} kg/m^3. If all the particles of dust that impact on the spaceship stick to it, find the average decelerating force that the impact of the dust exerts on the spaceship.

Section 6.2: Center of Mass

19.—A penny coin lies on a table at a distance of 30 cm from a stack of four penny coins. Where is the center of mass of the five coins?

20.—Three firemen of equal masses are climbing a long ladder. When the first fireman is 20 m up the ladder, the second is 15 m up, and the third is 5 m up. Where is the center of mass of the three firemen?

21.—Consider the system Earth–Moon; use the data in the table printed on the endpapers. How far from the center of the Earth is the center of mass of this system?

22.*—Four identical books are arranged on the vertices of an equilateral triangle of side 1.0 m. Two of the books are together at one vertex of the triangle, and the other two are at the other two vertices. Where is the center of mass of this arrangement?

23.—The distance between the centers of the atoms of potassium and bromine in a potassium bromide (KBr) molecule is 0.282 nm (Fig. 6.29). Treating the atoms as particles, find the center of mass.

24.*—Two uniform squares of sheet metal of dimension $L \times L$ are joined at a

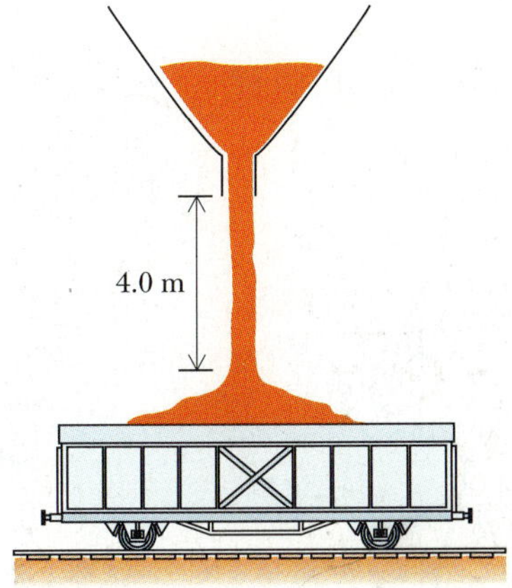

FIGURE 6.27

Grain falls into a railroad car from a chute.

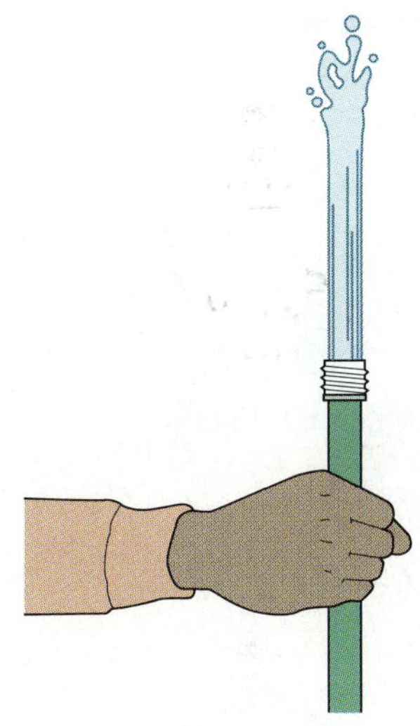

FIGURE 6.28

Water streaming from garden hose.

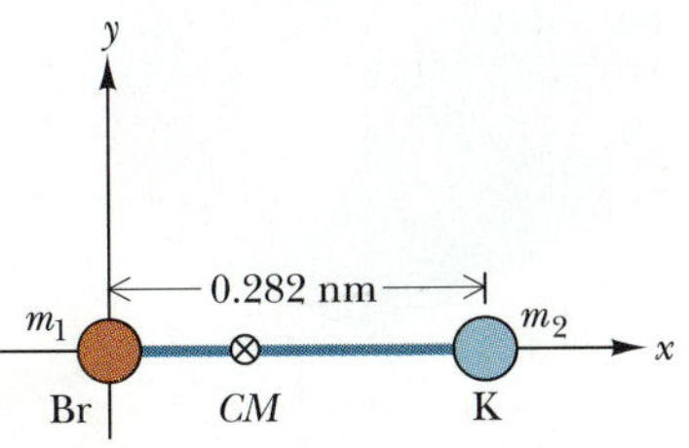

FIGURE 6.29

A potassium bromide molecule. The atoms of bromine (Br) and potassium (K) are regarded as particles.

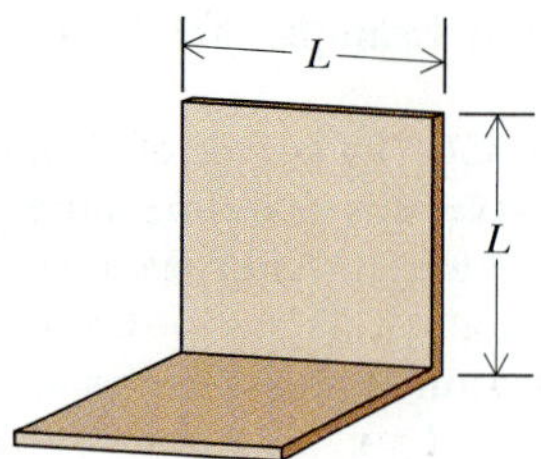

FIGURE 6.30

Two uniform squares of sheet metal joined along one edge.

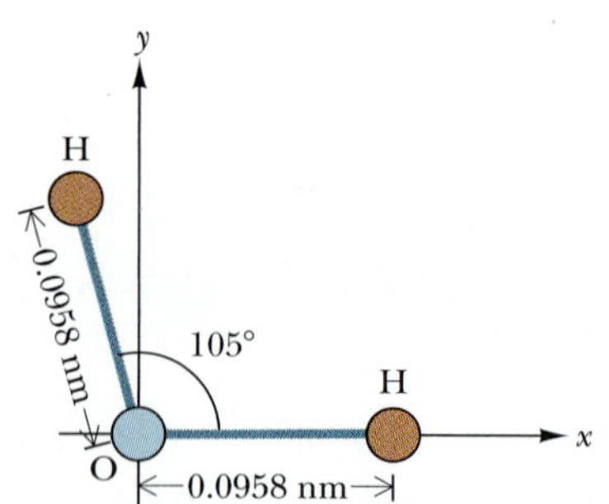

FIGURE 6.31

A water molecule.

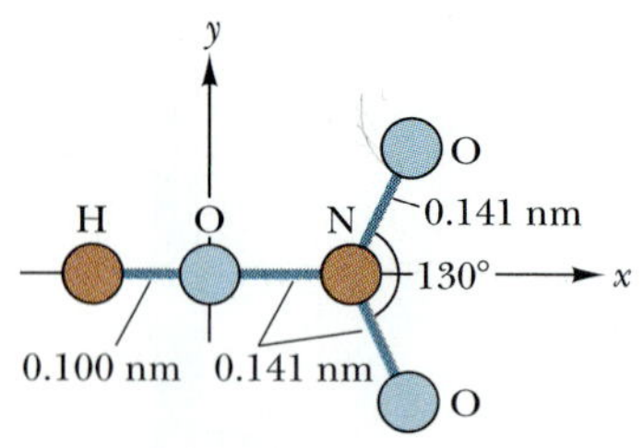

FIGURE 6.32

A molecule of nitric acid.

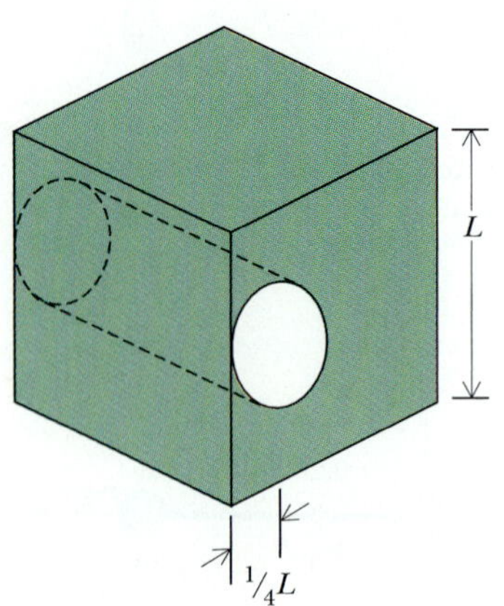

FIGURE 6.33

Iron cube with a hole.

right angle along one edge (Fig. 6.30). Find the center of mass of the combined squares.

25.**—The distance between the oxygen and each of the hydrogen atoms in a water (H_2O) molecule is 0.0958 nm; the angle between the two oxygen–hydrogen bonds is 105° (Fig. 6.31). Treating the atoms as particles, find the center of mass.

26.**—Figure 6.32 shows the shape of a nitric acid (HNO_3) molecule and its dimensions. Treating the atoms as particles, find the center of mass of this molecule.

27.**—A cube of iron has dimensions $L \times L \times L$. A hole of radius $\frac{1}{4}L$ has been drilled all the way through the cube, so one side of the hole is tangent to one face along its entire length (Fig. 6.33). Where is the center of mass of the drilled cube?

28.*—A lock on the Champlain canal is 72 m long and 9 m wide; the lock has a lift of 3.6 m, that is, the difference between the water levels of the canal on one side of the lock and on the other side is 3.6 m. How much gravitational potential energy is wasted each time the lock goes through one cycle (involving the filling of the lock with water from the high level and then the spilling of this water to the low level)?

29.*—Suppose that a man of mass 75 kg and height 1.75 m runs in place, raising his legs high, as in Fig. 6.13. If he runs at the rate of 80 steps per minute (each leg), what power does he expend in raising his legs?

Section 6.3: The Motion of the Center of Mass

30.—A boy and a girl are engaged in a tug-of-war on smooth, frictionless ice. The mass of the boy is 40 kg, and that of the girl is 30 kg; their separation is initially 4.0 m. Each pulls with a force of 200 N on the rope. What is the acceleration of each? If they keep pulling, where will they meet?

31.—A tugboat of mass 4×10^6 kg and a ship of mass 28×10^6 kg are joined by a long towrope of 400 m. Both vessels are initially at rest in the water. If the tugboat reels in 200 m of towrope, how far does the ship move relative to the water? The tugboat? Ignore the resistance that the water offers to the motion.

32.*—A fisherman in a boat catches a great white shark with a harpoon. The shark struggles for a while and then becomes limp when at a distance of 300 m from the boat. The fisherman pulls in the shark by the rope attached to the harpoon. During this operation, the boat (initially at rest) moves 50 m in the direction of the shark. The mass of the boat is 5400 kg. What is the mass of the shark? Pretend that the water exerts no friction.

33.*—A 75-kg man climbs the stairs from the ground to the fourth floor of a building, a height of 15 m. How far does the Earth recoil in the opposite direction as the man climbs?

34.*—A 6000-kg truck stands on the deck of an 80,000-kg ferryboat. Initially the ferry is at rest and the truck is located at its front end. If the truck now drives 15 m along the deck toward the rear of the ferry, how far will the ferry move forward relative to the water? Pretend that the water has no effect on the motion.

35.*——In a molecule, such as the potassium bromide (KBr) molecule of Problem 23, the atoms usually execute a rapid vibrational motion about their equilibrium positions. Suppose that in an isolated KBr molecule the speed of the potassium atom is 5.0×10^3 m/s at one instant (relative to the center of mass). What is the speed of the bromine atom at the same instant?

36.——A proton of energy 1.6×10^{-13} J is moving toward a proton at rest. What is the velocity of the center of mass of the system?

37.——An automobile of 1200 kg and an automobile of 1500 kg are traveling in the same direction on a straight road. The speeds of the two automobiles are 60 km/h and 80 km/h, respectively. What is the velocity of the center of mass of the two-automobile system?

Section 6.4: Collisions

38.——A stunt man of mass 77 kg "belly-flops" on a shallow pool of water from a height of 11 m. When he hits the pool, he comes to rest in about 0.05 s. What is the average braking force that the water and the bottom of the pool exert on his body during this time interval?

39.——A large ship of 7×10^8 kg steaming at 20 km/h runs aground on a reef, which brings it to a halt in 5.0 s. What is the average force on the ship? What is the average deceleration?

40.*——Suppose that a seat-belted mother riding in an automobile holds a 10-kg baby in her arms. If the automobile crashes at 50 km/h into a rigid barrier and comes to a stop in 0.10 s, what impulsive force must the mother exert on the baby to keep it in her arms? Will she be able to hold on?

41.*——According to federal safety standards, the steering wheel of an automobile must exert a force of no more than 1.1×10^4 N when the chest of the driver strikes it at 24 km/h (15 mi/h). Estimate how much the steering wheel must yield to meet this standard. Pretend that the chest is rigid.

42.*——Slow-motion photography shows that when a golf club hits a golf ball, the club and the ball typically remain in contact for 1.0×10^{-3} s and the ball acquires a speed of 70 m/s. The mass of the ball is 45 g. Estimate the magnitude of the force that the club exerts on the ball.

43.*——The "land divers" of Pentecost Island (New Hebrides) jump from platforms 21 m high. Long liana vines tied to their ankles jerk them to a halt just short of the ground. If the pull of the liana takes 0.02 s to halt the diver, what is the average acceleration of the diver during this time interval? If the mass of the diver is 64 kg, what is the corresponding average force on his ankles?

44.*——In a remarkable incident, a 52-kg woman jumped from the 10th floor of a building, fell 28 m, and landed on her side on soft earth in a freshly dug garden. She fractured her wrist and a rib, but remained conscious and fully alert, and recovered fully after some time in a hospital. The earth was depressed 15 cm by her impact.

(a) What was her impact speed?

(b) Assuming constant deceleration upon contact with the ground, what was her deceleration?

(c) What was the force of the ground on her body during deceleration?

Section 6.5: Elastic Collisions on a Straight Line

45.——An automobile approaching an intersection at 10 km/h bumps into the rear of another automobile standing at the intersection with its brakes off and its gears in neutral. The mass of the moving automobile is 1200 kg and that of the stationary automobile is 700 kg. If the collision is elastic, find the velocities of both automobiles after the collision.

46.—The impact of the head of a golf club on a golf ball can be approximately regarded as an elastic collision. The mass of the head of the golf club is 0.15 kg and that of the ball is 0.045 kg. If the ball is to acquire a speed of 60 m/s in the collision, what must be the speed of the club before impact?

47.*—The impact of a hammer on a nail can be regarded as an elastic collision between the head of the hammer and the nail. Suppose that the mass of the head of the hammer is 0.50 kg and it strikes a nail of mass 12 g with an impact speed of 5 m/s. How much energy does the nail acquire in this collision?

48.**—When a baseball bat strikes a ball, the impact can be approximately regarded as an elastic collision (the hands of the hitter have little effect in the short time the bat and the ball are in contact). Suppose that a bat of 0.85 kg moving horizontally at 30 m/s encounters a ball of 0.15 kg moving at 40 m/s in the opposite direction. We cannot directly apply the results of Section 6.5 to this collision, since *both* particles are in motion before collision ($v_1 = 40$ m/s and $v_2 = -30$ m/s). However, we can apply these results if we use a reference frame that moves at a velocity of $V_O = -30$ m/s in the direction of the initial motion of the bat; in this reference frame, the initial velocity of the bat is zero ($v_2 = 0$).

(a) What is the initial velocity of the ball in this reference frame?

(b) What are the final velocity of the ball and the final velocity of the bat, just after the collision?

(c) What are these final velocities in the reference frame of the ground?

49.**—A boy throws a baseball at another baseball sitting on a 1.5-m-high fence. The collision of the balls is elastic. The thrown ball moves horizontally at 20 m/s just before the collision.

(a) What are the velocities of the two balls just after the collision?

(b) Where do the two balls land on the ground?

Section 6.6: Inelastic Collisions on a Straight Line

50.—A small automobile of 500 kg traveling at 45 km/h strikes a deer of 120 kg standing on the road. Assume that the collision is totally inelastic (the deer remains draped over the front end of the automobile). What is the speed of the automobile immediately after this collision?

51.—When Wilhelm Tell shot the apple off his son's head, the arrow remained stuck in the apple, which means the collision between arrow and apple was totally inelastic. Suppose that the speed of the arrow was 80 m/s before it hit, the mass of the arrow was 40 g, and the mass of the apple was 200 g. Calculate the speed of the apple and arrow immediately after the collision.

52.—In karate, the fighter makes his hand collide at high speed with the target; this collision is inelastic and a large portion of the kinetic energy of the hand becomes available to do damage in the target. According to a crude estimate, the energy required to break a concrete block (28 cm $\times$ 15 cm $\times$ 1.9 cm supported only at its short edges) is of the order of 10 J. Suppose the fighter delivers a downward hammer-fist strike with a speed of 12 m/s to such a concrete block. In principle, is there enough energy to break the block? Assume that the fist has a mass of 0.4 kg.

FIGURE 6.34
Meteor crater in Arizona.

53.*—Meteor Crater in Arizona (Fig. 6.34), a hole 180 m deep and 1300 m across, was gouged in the surface of the Earth by the impact of a large meteorite. The mass and speed of this meteorite have been estimated at 2×10^9 kg and 10 km/s, respectively, before impact.

(a) What recoil velocity did the Earth acquire during this (totally inelastic) collision?

(b) How much kinetic energy was released for inelastic processes during the collision? Express this energy in the equivalent of tons of TNT; 1 ton of TNT releases 4.2×10^9 J upon explosion.

54.* It has been reported (fallaciously) that the deer botfly can attain a maximum airspeed of 818 mi/h, that is, 366 m/s. Suppose that such a fly, buzzing along at this speed, strikes a stationary hummingbird and remains stuck in it. What will be the recoil velocity of the hummingbird? The mass of the fly is 2 g; the mass of the hummingbird is 50 g.

55.** A ship of 3.0×10^7 kg steaming at 40 km/h strikes an iceberg of 8.0×10^8 kg. If the collision is totally inelastic, what fraction of the initial kinetic energy of the ship is converted into inelastic energy? What fraction remains as kinetic energy of the ship–iceberg system?

56.** Two automobiles of 540 kg and 1400 kg collide head-on while moving at 80 km/h in opposite directions. After the collision the automobiles remain locked together.

(a) Find the velocity of the wreck immediately after the collision.

(b) Find the kinetic energy of the two-automobile system before and after the collision.

57.** You can make a fairly accurate measurement of the speed of a bullet by shooting it horizontally into a block of wood sitting on a fence. The collision of the bullet and the block is inelastic, and the block will fall off the fence and land on the ground at some distance from its bottom. The speed of the bullet is proportional to this distance. Suppose that a bullet of mass 15 g fired into a block of 4.0 kg sitting on a 1.8-m-high fence causes the block to land 1.4 m from the bottom of the fence. Calculate the speed of the bullet.

58.** According to a tall tale told by Baron Münchhausen, on one occasion, while cannon shots were being exchanged between a besieged city and the enemy camp, he jumped on a cannonball as it was being fired from the city, rode the cannonball toward the enemy camp, and then, in midair, jumped onto an enemy cannonball and rode back to the city. The collision of Münchhausen and the enemy cannonball must have been inelastic, since he held on to it. Suppose that his speed just before hitting the enemy cannonball was 150 m/s southward, and the speed of the enemy cannonball was 300 m/s northward. The mass of Münchhausen was 90 kg, and the mass of the enemy cannonball was 20 kg. What must have been the speed just after the collision?

Collision of two hockey players.

59.** Two hockey players of mass 80 kg collide while skating at 7.0 m/s. The angle between their initial directions of motion is 90°.

(a) Suppose that the collision is totally inelastic. What is their velocity immediately after collision?

(b) Suppose that the collision lasts 0.030 s. What is the magnitude of the average acceleration of each player during the collision?

60.** A speeding automobile hits the rear of a parked automobile. After the impact, the two automobiles remain joined together, and they skid along the pavement with all their wheels locked. An investigation of this accident establishes that the length of the skid marks made by the automobiles after the impact was 18 m; the mass of the moving automobile was 2200 kg and that of the parked automobile was 1400 kg, and the coefficient of friction between the wheels and the pavement was 0.95.

(a) What was the speed of the two automobiles immediately after impact?

(b) What was the speed of the moving automobile before impact?

61.** Two automobiles of equal mass collide head-on at an intersection. After the collision, they remain joined together and skid, with locked wheels, be-

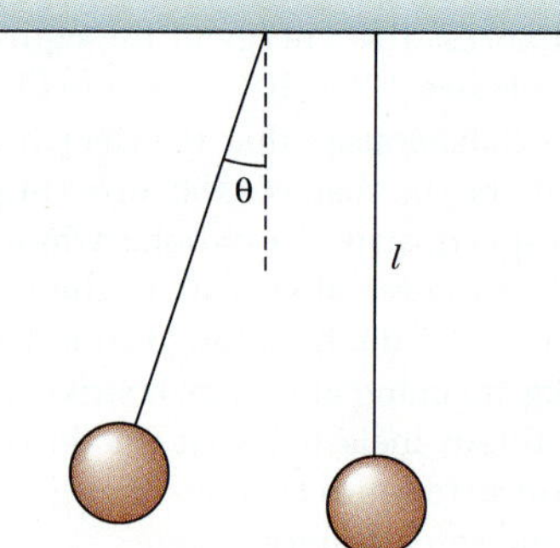

FIGURE 6.35
Two steel balls.

fore coming to rest. The length of the skid marks is 18 m and the coefficient of friction between the locked wheels and the pavement is 0.80. Each driver claims his speed was less than 14 m/s (30 mi/h) before the collision. Prove that at least one driver is lying.

62.** (a) Two identical steel balls are suspended from strings of length l so they touch when in their equilibrium position (Fig. 6.35). If we pull one of the balls back until its string makes an angle θ with the vertical and then let go, it will collide elastically with the other ball. How high will the other ball rise?

(b) Suppose that instead of steel balls we use putty balls. They will then collide inelastically and remain stuck together. How high will the balls rise?

PHYSICS IN ACTION

AUTOMOBILE COLLISIONS

Motor vehicle accidents are the leading cause of death for young and middle-aged people in the industrialized nations. In the United States, 40,000 to 50,000 people are killed each year in such accidents (Fig. 1). Besides this high toll of life, the carnage on the roads exacts a high toll of injuries. In the United States, some 4,000,000 people are injured each year in motor vehicle accidents. Among these, about 70,000 suffer permanent brain damage or spinal-cord damage, often leading to epilepsy, paraplegia, or quadruplegia. We can bring this stupefying toll of deaths and injuries into sharper focus by asking what these numbers mean for a single individual. An infant born in the United States today faces a chance of 2 in 3 of suffering an injury in a motor vehicle accident at some time during his life, and he faces a chance of 1 in 60 of ending his life in such an accident.

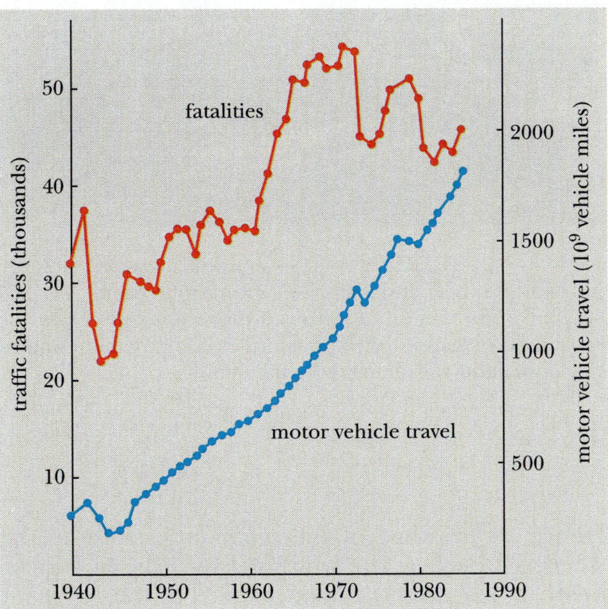

FIGURE 1 Deaths and motor vehicle travel in the United States. Note the sharp dip in the number of deaths in 1974, when the federal speed limit of 55 mi/h was imposed. Although some portion of this dip reflects a temporary decrease of vehicle use caused by the oil shortage, most of it reflects live saved by the lower driving speed.

An automobile of 1500 kg traveling at 90 km/h (56 mi/h) has a kinetic energy of 4.7×10^5 J. We can better appreciate the magnitude of this energy by noting that it is equivalent to the potential energy released in dropping the automobile from a height of 32 m, or eleven floors down from the top of an apartment building. Most automobile collisions are totally inelastic, or nearly so. During a collision, all, or almost all, of this large kinetic energy is dissipated in the body of the automobile (and in the bodies of the occupants), with a consequent large amount of deformation and breakage.

Throughout the following discussion we will concentrate on head-on collisions, since these are the most frequent and usually the most severe, and they are also the easiest to analyze. Head-on collisions between automobiles and stationary, rigid barriers are routinely used in experimental tests of automobile safety. In the United States, such barrier crash tests are required for all new models of automobiles. The sequence of photographs in Fig. 6.15 was obtained in one of these crash tests. The automobile was towed into the barrier, a massive concrete wall, at a speed of 49 km/h (30 mi/h). The impulsive force acting during the collision is quite large. As we saw in Example 6.8, the average impulsive force for this crash is 2.1×10^5 N, but the maximum instantaneous force is more than twice as large.

During the collision, the front end of the automobile is crushed by the impulsive force. This crushing dissipates the kinetic energy. In a collision at 49 km/h, the typical crush distance for the front part of the automobile is 0.6 to 0.7 m. The crush distance increases with impact speed.

Measurements of the durations of automobile-barrier collisions show that the duration is always about 0.1 s, regardless of the impact speed. For instance, in Fig. 6.16 we see that the collision lasts 0.12 s, from the moment of zero acceleration (just before contact with the barrier) to the moment of zero acceleration (just after contact). The duration of the collision is not greatly affected by the impact speed, because the crush distance for a fast-moving automobile is larger than for a slow-moving automobile, and hence both automobiles take about the same time to travel through their crush distances and come to a halt. Heavy, large automobiles tend to have somewhat larger crush durations and crush distances than small automobiles. In such automobiles, there are larger gaps between the separate parts, and there is more space available for crushing. Thus, in the large automobile, the crushing of the front provides better cushioning of the collision than in the small automobile (Fig. 2).

A head-on collision between two automobiles of equal masses and of equal speeds is roughly equivalent to a head-on collision with a barrier. The only difference is that in a collision between two automobiles the front ends interpenetrate to some extent, which cannot happen in a collision with a rigid, flat barrier. A head-on collision between two automobiles of unequal masses results in unequal accelerations and unequal velocity changes. Since the mutual forces on the two automobiles are of equal magnitudes, the accelerations are inversely proportional to the masses, and the velocity changes are also inversely proportional to the masses. Hence, the collision is more severe for the smaller automobile.

Injury to the occupant of an automobile usually results from the collision of the occupant with some part of the interior of the automobile, such as the steering wheel, the instrument panel, or the windshield. This collision of the occupant with the interior of the automobile is called the ***secondary collision,*** or the human collision. It occurs some time after the beginning of the primary collision of the front of the automobile. Consider, for instance, an automobile crash with an impact speed of 56 km/h (35 mi/h). If the occupant is not restrained by a seat belt, he will continue to move forward at a constant speed of 56 km/h while the body of the automobile decelerates. The occupant continues to move forward at this speed until he collides with the steering wheel, instrument panel, or windshield. In the typical case, the time the occupant spends flying through the air is comparable with the time the body of the automobile takes to come to a full stop. Hence, at the time of the secondary collision, the steering wheel or the windshield will be at rest or nearly at rest, and the impact speed of the occupant on these obstacles will therefore be comparable with the primary impact speed of 56 km/h. The crush distance, or yield distance, of a steering wheel struck by a human chest or a windshield struck by a head is considerably shorter than the crush distance of the front of the automobile. Thus, the secondary collision has less cushioning than the primary colli-

(a)

(b)

FIGURE 2 (a) Collision between a large automobile (red) and a smaller automobile (blue). (b) The smaller automobile has suffered more damage. Note the strong deformation of the passenger compartment and of the door, and the shattered windshield.

sion, and it generates larger decelerations. The large forces involved in this secondary collision result in serious or fatal injuries to the occupant.

We can better appreciate the effects of the secondary collision on the human body if we compare the impact speed with the speed attained by a body in free fall from some height. The impact of a head on the windshield at 56 km/h is equivalent to falling four floors down from an apartment building and landing head first on a hard surface. Our intuition tells us that this is likely to be fatal. Since our intuition about the dangers of heights is much better than our intuition about the dangers of speeds, it is often instructive to compare impact speeds with equivalent heights of fall. Figure 3 displays impact speeds and equivalent heights, expressed as the number of floors the automobile has to fall down to acquire the same speed (keep in mind that according to the American way of labeling floors of buildings, you have to climb to the "fifth" floor if you want to fall down four floors).

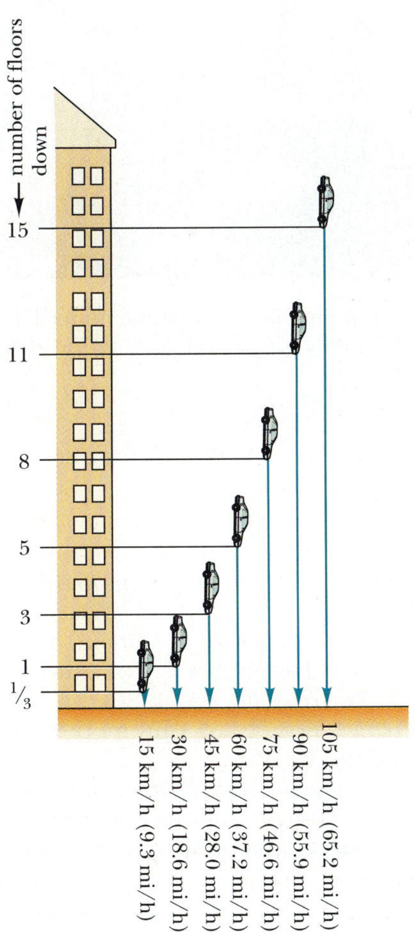

FIGURE 3 Comparison of impact speeds and heights of fall. Each floor of the apartment building shown here is 2.9 m.

If the occupants of a colliding automobile are held snugly by seat belts, they will not suffer a secondary collision. Instead they will decelerate at the same rate as the body of the automobile and come to a stop at the same time as the automobile. Of course, the seat belt must supply the force to decelerate the occupant, and this seat-belt force can itself cause some injuries; but such injuries are usually minor compared with the severe injuries produced by the secondary collision of an unrestrained occupant.

At present, safety standards for American automobiles require lap-and-shoulder belts for the two outboard front seats, but require no more than lap belts for the middle front seat (if any) or the rear seats. Lap belts alone are much less effective in preventing a secondary collision than lap-and-shoulder belts. During the collision, the lap belt restrains the hips, but the torso jackknifes forward and down, and the head of the occupant suffers a violent secondary collision with the instrument panel. Besides, the sudden snubbing of a slack lap belt is concentrated entirely across the hips and the abdomen, and can lead to pelvic and lumbar fractures and to ruptures of the abdominal organs.

Many automobile manufacturers now offer air-bag restraint systems. This restraint system consists of a plastic bag packed into a compartment in the hub of the steering wheel or in the glove compartment, a gas generator for inflating the bag, and an electronic crash sensor installed in the body of the automobile. When the sensor experiences a deceleration in excess of a preset threshold level, it triggers the explosive combustion of a solid propellant in the gas generator. This combustion produces nitrogen gas, which inflates the air bag in about 0.04 s. The bag balloons out of its compartment and slams into the face and chest of the driver, braking his forward motion and preventing a secondary collision with the steering wheel or the instrument panel (see Fig. 4). Crash tests with volunteers have shown that such an air-bag system prevents injuries in a frontal 48-km/h crash into a barrier. However, the air bag offers no protection in a lateral crash or in a rollover. For full protection, the driver must still wear a seat belt, even if the automobile has an air-bag system.

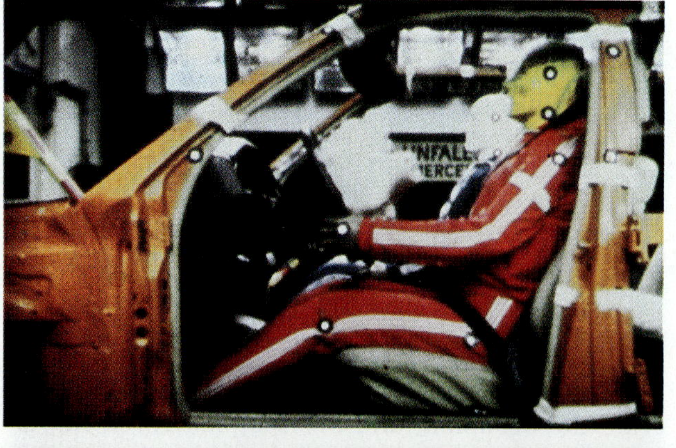

FIGURE 4 Test of the air-bag system on a Mercedes-Benz automobile. A computer linked to a small electronic accelerometer senses the beginning of a crash and triggers the inflation of the bag if the deceleration exceeds a preset level.

REFERENCES

Damask, A. C., "Forensic Physics of Vehicle Accidents," *Physics Today,* March 1987.

Evans, L., *Traffic Safety and the Driver* (Van Nostrand Reinhold, New York, 1991).

Evans, L., and Schwing, R. C., eds., *Human Behavior and Traffic Safety* (Plenum, New York, 1985).

Selzer, M. L., Gikas, P. W., and Huelke, D. F., *The Prevention of Highway Injury* (Highway Safety Institute, University of Michigan, Ann Arbor, 1967).

Wolf, R. A., "Four Facets of Automotive Crash Injury Research," *New York State Journal of Medicine,* July 1966.

Human Tolerance to Impact Conditions as Related to Motor Vehicle Design (Society of Automotive Engineers, Warrendale, PA, 1986)

Crash Protection (Society of Automotive Engineers, Warrendale, PA 1982).

Automobile Occupant Crash Protection (U.S. Department of Transportation, National Highway Traffic Safety Administration, Washington, D.C.)

This time-exposure photograph of a spinning ferris wheel illustrates rotational motion.

ROTATION OF A RIGID BODY

◀ **Rigid body**

A body is rigid if the particles in the body do not move relative to one another. Thus, the body has a fixed shape, and all its parts have a fixed position relative to one another. A hammer is a rigid body, and so is a baseball bat. A baseball is not rigid — when struck a blow by the bat, the ball suffers a substantial deformation. However, the baseball can be regarded as a rigid body while it flies through the air — the air resistance is not sufficiently large to produce an appreciable deformation of the ball. This example indicates that whether a body can be regarded as rigid depends on the circumstances. None of the bodies in our everyday environment is absolutely rigid; when subjected to a sufficiently large force, any body will suffer some deformation or perhaps even break into several pieces. In this chapter, we will ignore such deformations produced by the forces acting on bodies. We will examine the motion of bodies under the assumption that rigidity is a good approximation.

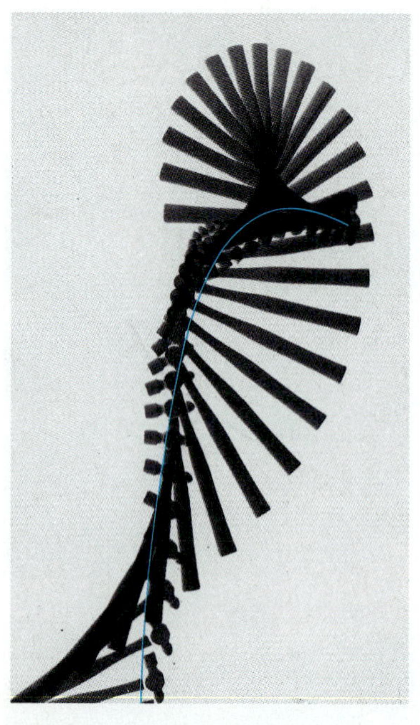

FIGURE 7.1
A hammer in free fall under the influence of gravity. The center of mass of the hammer, marked by the blue line, moves with constant vertical acceleration g, just like a particle in free fall.

A rigid body can simultaneously have two kinds of motion: ***it can change its position in space and it can change its orientation in space.*** Change of position is translational motion; as we saw in Chapter 6, this motion can be conveniently described as motion of the center of mass. Change in orientation is rotational motion, that is, it is rotation about some axis.

As an example, consider the motion of a hammer thrown upward (see Fig. 7.1). The orientation of the hammer changes relative to fixed coordinates attached to the ground. Instantaneously, the hammer rotates about a horizontal axis, say, a horizontal axis that passes through the center of mass. In Fig. 7.1, this horizontal axis sticks out of the plane of the page and it moves upward with the center of mass. The complete motion can then be described as a simultaneous rotation of the hammer about this axis and a translation of the axis along a parabolic path.

We will take advantage of this separability of the motion into rotation and translation, and we begin this chapter by concentrating on the rotation.

7.1 Rotation about a Fixed Axis

The simplest case of rotational motion is rotation about a fixed axis, such as the rotational motion of a phonograph turntable, a roulette wheel, or a revolving door (see Fig. 7.2). For the sake of notational uniformity, we

FIGURE 7.2 Some examples of rotational motion with a fixed axis (phonograph turntable, roulette wheel, revolving door).

will assume that the fixed axis coincides with the z axis. Figure 7.3 shows a rigid body of some arbitrary shape rotating about the z axis. During this rotational motion, each point of the body remains at a given distance from this axis and moves along a circle centered on the axis. To describe the orientation of the body at any instant, we select one particle in the body and use it as a marker; any particle will do as marker, provided that it is not on the axis of rotation. ***The circular motion of this marker particle*** (labeled P in Fig. 7.3) ***is then representative of the rotational motion of the entire body, and the angular position of this particle along its circular path is representative of the angular orientation of the entire body.***

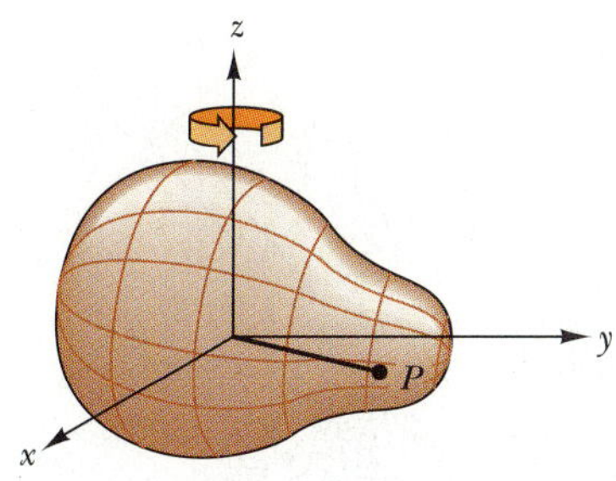

FIGURE 7.3
Rotation of a rigid body about a fixed axis.

Figure 7.4 shows the rotating rigid body as seen from along the axis of rotation. The coordinates in Fig. 7.4 have been chosen so the z axis coincides with the axis of rotation whereas the x and y axes are in the plane of the circle traced out by the motion of the marker particle. The angular position of the marker particle P — and hence the angular position of the entire rigid body — can be described by the position angle ϕ between the radial line OP and the x axis (ϕ is the Greek letter phi, often used for angles). Conventionally, the angle ϕ is taken as positive when reckoned in a counterclockwise direction, as in Fig. 7.4. We will always measure this position angle in radians, rather than degrees. By definition, ***the angle ϕ in radians is the length s of the circular arc divided by the radius R,***

◀ **Radian**

$$\phi = \frac{s}{R} \tag{1}$$

In Fig. 7.4, the length s of the circular arc is the distance traveled by the marker particle from the x axis to the point P. Note that if the length s is the circumference of a full circle, then $s = 2\pi R$ and $\phi = s/R = 2\pi R/R = 2\pi$. Thus, there are 2π radians in a full circle; that is, there are 2π radians in 360°,

$$2\pi \text{ radians} = 360°$$

Accordingly, 1 radian equals $360°/2\pi$, or

$$1 \text{ radian} = 57.3°$$

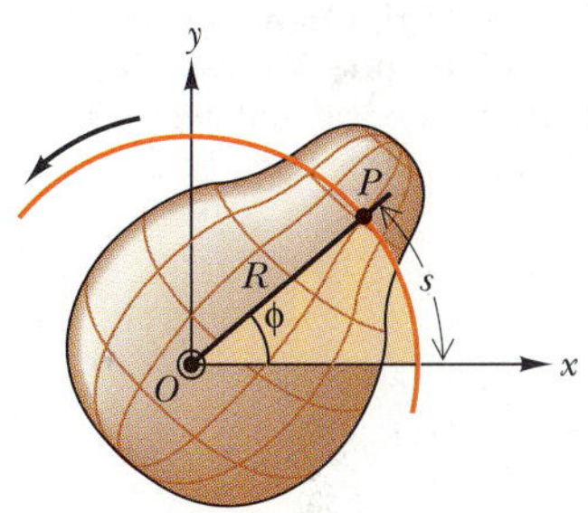

FIGURE 7.4
Motion of the "marker" particle P in a rigid body. The radius of the circle traced out by the motion of the "marker" particle is R.

EXAMPLE 1. The accuracy of the guidance system of the Hubble Space Telescope is such that if the telescope were sitting in New York, the guidance system could aim it at a dime placed on top of the Washington Monument, at a distance of 320 km. The width of a dime is 1.8 cm. What angle does the dime subtend when seen from New York?

SOLUTION: In Fig. 7.5, the radius of the arc of circle subtended by the dime is 320 km. For a small angle, such as in this figure, the length s of the arc between one side of the dime and the other is approximately

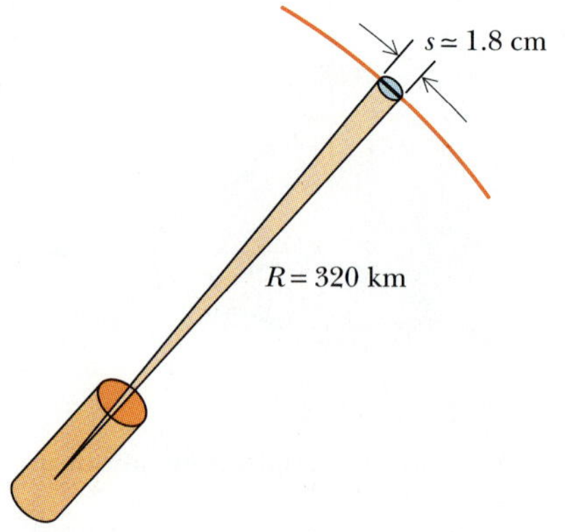

FIGURE 7.5

A dime placed at a distance of 320 km from the telescope. The length of the small arc s is approximately the width of the dime.

the same as the length of the straight line from one side to the other, which is the width of the dime. Hence, the angle in radians is

$$\phi = \frac{s}{R} = \frac{1.8\text{ cm}}{320\text{ km}} = \frac{0.018\text{ m}}{3.2 \times 10^5\text{ m}} = 5.6 \times 10^{-8}\text{ radian}$$

Expressed in degrees, this becomes

$$\phi = 5.6 \times 10^{-8}\text{ radian} \times \frac{360°}{2\pi\text{ radian}} = 3.2 \times 10^{-6}\text{ degree}$$

When a rigid body rotates, the position angle ϕ changes with time. The body then has an angular velocity. The definition of the angular velocity for rotational motion is mathematically analogous to the definition of velocity for translational motion (see Sections 1.3 and 1.4). Suppose that in a time Δt, the body rotates through an angle $\Delta\phi$. The ***average angular velocity*** is then defined as

Angular velocity ▸

$$\overline{\omega} = \frac{\Delta\phi}{\Delta t} \qquad (2)$$

Here ω is the Greek letter omega, commonly used to denote the angular velocity. Likewise, the ***instantaneous angular velocity*** is defined as

$$\omega = \frac{\Delta\phi}{\Delta t} \quad \text{with } \Delta t \text{ extremely small} \qquad (3)$$

According to these definitions, the angular velocity is the rate of change of the angle with time. ***The unit of angular velocity is the radian per second*** (1 radian/s). The radian is the ratio of two lengths [compare Eq. (1)], and hence it is a pure number, without units of meters, seconds, or kilograms; thus, 1 radian/s is the same thing as 1/s. However, to prevent confusion, it is often useful to retain the vacuous label *radian.* Table 7.1 gives some examples of angular velocities, in radians per second.

If the body rotates with constant angular velocity, then we can also measure the rate of rotation in terms of the ***frequency, or the number of revolutions per second.*** Since each complete revolution comprises a change of ϕ by 2π radians, the number of revolutions equals the number of radians divided by 2π; hence, the frequency equals the angular velocity divided by 2π. Denoting the frequency by f, we can express the frequency in terms of the angular velocity as follows:

Frequency ▸

$$f = \frac{\omega}{2\pi} \qquad (4)$$

TABLE 7.1	SOME ANGULAR VELOCITIES
Circular saw	7×10^2 radians/s
Electric blender blades	5×10^2 radians/s
Jet engine	4×10^2 radians/s
Magnetic storage disk of computer	3×10^2 radians/s
Airplane propeller	3×10^2 radians/s
Automobile engine	2×10^2 radians/s
Small fan	60 radians/s
Helicopter rotor	40 radians/s
Phonograph turntable	3.5 radians/s
Neutron star (pulsar) rotation	0.1 radian/s
Earth rotation	7.3×10^{-5} radian/s
Earth revolution about Sun	2.0×10^{-7} radian/s

The unit of frequency is the revolution per second (1 rev/s). Like the radian, the revolution is a pure number, and hence 1 rev/s is the same thing as 1/s. But we will keep the label *rev* to prevent confusion between rev/s and radian/s.

As in the case of planetary motion, ***the time per revolution is called the period of the motion.*** If the number of revolutions per second is f, then the time per revolution is $1/f$; that is,

$$T = \frac{1}{f} \tag{5}$$

◄ **Period**

EXAMPLE 2. A phonograph turntable rotates at 33.3 revolutions per minute. What is the frequency of revolution? What is the angular velocity? What is the period of the motion?

SOLUTION: Each minute is 60 s; hence 33.3 revolutions per minute amount to 33.3 revolutions in 60 s, that is

$$f = \frac{33.3\ \text{rev}}{60\ \text{s}} = 0.555\ \text{rev/s}$$

Since each revolution comprises 2π radians, the angular velocity is

$$\omega = 2\pi f = 2\pi \times 0.555\ \text{rev/s} = 3.49\ \text{radians/s}$$

Note that here we have dropped a label *rev* in the third step and inserted a label *radians.* As remarked above, these labels merely serve to prevent confusion, and they can be inserted and dropped at will once they have served their purpose.

The period of the motion is

$$T = \frac{1}{f} = \frac{1}{0.555 \text{ rev/s}} = 1.80 \text{ s} \tag{6}$$

If the angular velocity of the rigid body is changing, the body has an angular acceleration. The rotational motion of a phonograph turntable that is gradually building up speed immediately after being switched on is an example of accelerated rotational motion. The mathematical definition of the ***average angular acceleration*** is, again, analogous to the definition of acceleration for translational motion. If the angular velocity changes by $\Delta\omega$ in a time Δt, then the angular acceleration is

$$\overline{\alpha} = \frac{\Delta\omega}{\Delta t} \tag{7}$$

The Greek letter α is commonly used to denote the angular acceleration. Likewise, the ***instantaneous angular acceleration*** is defined as

$$\alpha = \frac{\Delta\omega}{\Delta t} \quad \text{with } \Delta t \text{ extremely small} \tag{8}$$

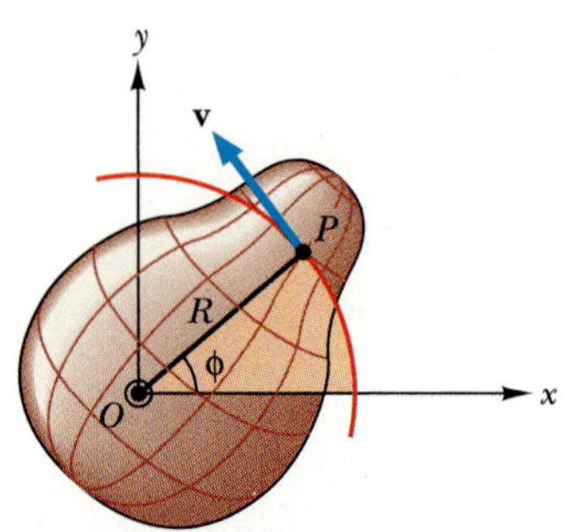

FIGURE 7.6
The instantaneous translational velocity of a particle in a rotating rigid body is tangent to the circular path.

Thus, the angular acceleration is the rate of change of the angular velocity. The unit of angular acceleration is the radian per second per second, or radian per second squared (1 radian/s^2).

Equations (3) and (8) give the angular velocity and acceleration of the rigid body, and they therefore give the angular velocity and acceleration of every particle in the body. It is interesting to focus on one of these particles and evaluate its *translational* speed and acceleration as it moves along its circular path around the axis of rotation of the rigid body. If the particle moves along a circle of radius R around the axis of rotation (see Fig. 7.6), then the length along the circular path of the particle is, according to Eq. (1),

$$s = \phi R \tag{9}$$

Since the radius R of the circle is a constant, the rate of change of s is entirely due to the rate of change of ϕ, that is,

$$\frac{\Delta s}{\Delta t} = R\frac{\Delta\phi}{\Delta t} \tag{10}$$

Here $\Delta s/\Delta t$ is the speed v with which the particle moves along its circular path and $\Delta\phi/\Delta t$ is the angular velocity ω; hence Eq. (10) is equivalent to

Translational speed ▸

$$v = R\omega \tag{11}$$

This says that ***the translational speed of the particle along its circular path around the axis is directly proportional to the radius:*** the farther a particle of the rigid body is from the axis, the faster it moves. We can understand this by comparing the motions of two particles, one on a circle of small radius R, and the other on a circle of large radius R. For each revolution of the rigid body, both of these particles complete one trip around their circles. But the particle on the larger circle has to travel a larger distance, and hence must move with a larger speed.

If the rotation of the body is accelerating, then the speed v with which the particle moves along its circle will be increasing. It follows from Eq. (11) that the rate of change of v is entirely due to the rate of change of ω; the rate of change of v is proportional to the rate of change of ω,

$$\frac{\Delta v}{\Delta t} = R\frac{\Delta \omega}{\Delta t} \tag{12}$$

A rate of change of the speed along the circle implies that the particle has an acceleration along the circle, called a tangential acceleration. According to Eq. (12), this tangential acceleration is

$$a_{\text{tan}} = R\alpha \tag{13}$$

◂ **Tangential acceleration**

Note that, besides this tangential acceleration directed along the circle, the particle also has a centripetal acceleration directed toward the center of the circle. From Section 2.7, we know that this centripetal acceleration is

$$a_{\text{cent}} = \frac{v^2}{R} \tag{14}$$

◂ **Centripetal acceleration**

With $v = \omega R$, this becomes

$$a_{\text{cent}} = R\omega^2 \tag{15}$$

The net acceleration of the particle is the vector sum of the tangential and the centripetal accelerations (see Fig. 7.7).

Although we have here introduced the concept of tangential acceleration in the context of the rotational motion of a rigid body, this concept is also applicable to the purely translational motion of a particle. For instance, consider an automobile (regarded as a particle) traveling around a curve. If the driver steps on the accelerator (or on the brake), the automobile will suffer a change of speed as it travels around the curve. It will then have both a tangential and a centripetal acceleration.

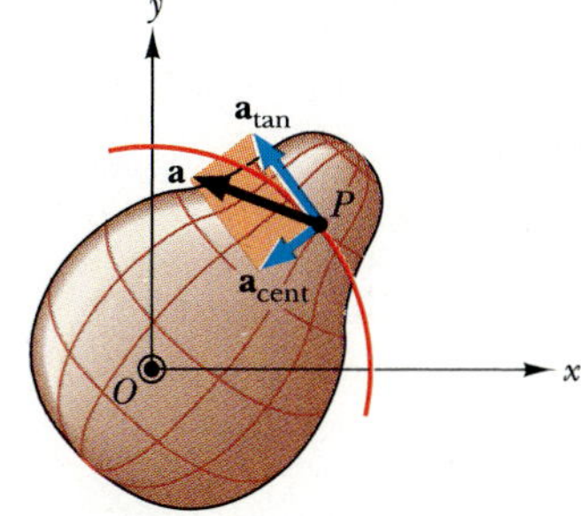

FIGURE 7.7
A particle in a rotating rigid body with an angular acceleration has both a centripetal acceleration $\mathbf{a}_{\text{cent}}$ and a tangential acceleration $\mathbf{a}_{\text{tan}}$. The net instantaneous translational acceleration $\mathbf{a}$ is then the vector sum of $\mathbf{a}_{\text{cent}}$ and $\mathbf{a}_{\text{tan}}$.

EXAMPLE 3. The blade of a circular saw is initially rotating at 7000 revolutions per minute. Then the motor is switched off, and the

The blade of a circular saw.

blade coasts to a stop in 8.0 s. What is the average angular acceleration?

SOLUTION: In radians per second, 7000 rev/min corresponds to an initial angular velocity $\omega_1 = 7000 \times 2\pi$ radians/min, or

$$\omega_1 = 7000 \times \frac{2\pi \text{ radians}}{60 \text{ s}} = 7.3 \times 10^2 \text{ radians/s}$$

The final angular velocity is $\omega_2 = 0$. Hence the average angular acceleration is

$$\bar{\alpha} = \frac{\Delta\omega}{\Delta t} = \frac{\omega_2 - \omega_1}{t_2 - t_1} = \frac{0 - 7.3 \times 10^2 \text{ radians/s}}{8.0 \text{ s} - 0}$$

$$= -91 \text{ radians/s}^2$$

7.2 Motion with Constant Angular Acceleration

The kinematical equations describing rotational motion with constant angular acceleration are mathematically analogous to the equations describing translational motion with constant acceleration (see Section 2.5), and they can be derived by the same methods. If the rigid body rotates with a constant angular acceleration α, then the angular velocity increases at a constant rate and, after a time t has elapsed, the angular velocity will attain the value

Motion with constant angular acceleration ▶

$$\omega = \omega_0 + \alpha t \tag{16}$$

where ω_0 is the initial value at $t = 0$.

The angular position ϕ can be calculated from this angular velocity by the arguments used in Section 1.6 to calculate x from v_0 and a [see Eqs. (1.12)–(1.16)]. The result is

$$\phi = \phi_0 + \omega_0 t + \tfrac{1}{2}\alpha t^2 \tag{17}$$

Furthermore, the arguments of Section 1.6 lead to an identity between acceleration, position, and velocity [see Eq. (1.17)]:

$$\alpha(\phi - \phi_0) = \tfrac{1}{2}(\omega^2 - \omega_0^2) \tag{18}$$

Note that all these equations have exactly the same mathematical form as the equations of Section 1.5, with ϕ taking the place of x, ω the place of v, and α the place of a.

The analogy between rotational and translational quantities can serve as a useful mnemonic for remembering the equations for rotational motion. Table 7.2 displays analogous equations.

◀ **Analogies between rotational and translational quantities**

TABLE 7.2 ANALOGIES BETWEEN ROTATIONAL AND TRANSLATIONAL QUANTITIES

$v = \Delta x/\Delta t$	$\rightarrow$	$\omega = \Delta\phi/\Delta t$
$a = \Delta v/\Delta t$	$\rightarrow$	$\alpha = \Delta\omega/\Delta t$
$v = v_0 + at$	$\rightarrow$	$\omega = \omega_0 + \alpha t$
$x = x_0 + v_0 t + \frac{1}{2}at^2$	$\rightarrow$	$\phi = \phi_0 + \omega_0 t + \frac{1}{2}\alpha t^2$
$a(x - x_0) = \frac{1}{2}(v^2 - v_0^2)$	$\rightarrow$	$\alpha(\phi - \phi_0) = \frac{1}{2}(\omega^2 - \omega_0^2)$

EXAMPLE 4. The cable supporting an elevator runs over a wheel of radius 0.36 m (see Fig. 7.8). If the elevator ascends with an upward acceleration of 0.60 m/s^2, what is the angular acceleration of the wheel? How many turns does the wheel make if this accelerated motion starts from rest and lasts 5.0 s? Assume that the cable runs over the wheel without slipping.

SOLUTION: If there is no slipping, the speed of the cable must always coincide with the speed of a point on the rim of the wheel. The acceleration $a = 0.60$ m/s^2 of the cable must then coincide with the tangential acceleration of a point on the rim of the wheel,

$$a = a_{\tan} = R\alpha \tag{19}$$

where $R = 0.36$ m is the radius of the wheel. Hence

$$\alpha = \frac{a}{R} = \frac{0.60\ \text{m/s}^2}{0.36\ \text{m}} = 1.67\ \text{radians/s}^2$$

To calculate the angular displacement, we use Eq. (17) with $\alpha =$ 1.76 radians/s^2 and $t = 5.0$ s:

$$\phi - \phi_0 = \omega_0 t + \tfrac{1}{2}\alpha t^2 \tag{20}$$

$$= 0 + \tfrac{1}{2} \times 1.67\ \text{radians/s}^2 \times (5.0\ \text{s})^2$$

$$= 21\ \text{radians}$$

Each revolution comprises 2π radians; thus, 21 radians is the same as $21/2\pi$ revolutions, or 3.3 revolutions.

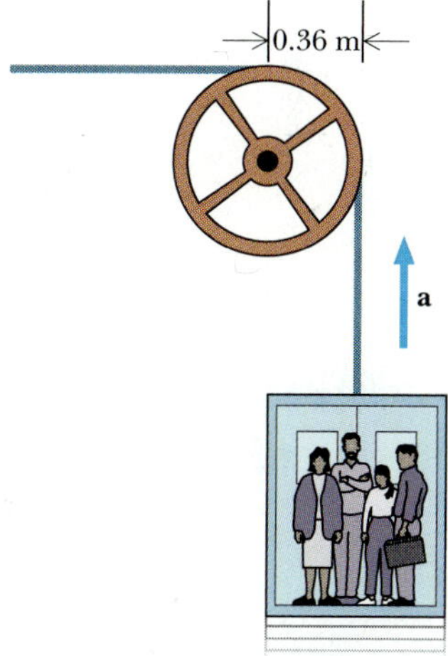

FIGURE 7.8
The cable supporting an elevator runs over a rotating wheel.

7.3 KINETIC ENERGY OF ROTATION; MOMENT OF INERTIA

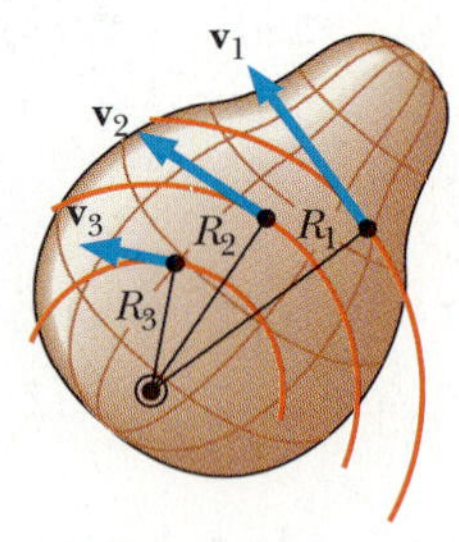

FIGURE 7.9
Several particles in a rigid body rotating about a fixed axis. The speeds are proportional to the radial distances.

A rigid body is a system of particles, and, as for any system of particles, the total kinetic energy of a rotating rigid body is simply the sum of the individual kinetic energies of all the particles. If the particles in the rigid body have masses $m_1, m_2, m_3, \ldots$ and speeds $v_1, v_2, v_3, \ldots$ (see Fig. 7.9), then the kinetic energy is

$$K = \tfrac{1}{2}m_1v_1^2 + \tfrac{1}{2}m_2v_2^2 + \tfrac{1}{2}m_3v_3^2 + \ldots \qquad (21)$$

For a rigid body rotating about a fixed axis, all the particles move with the same angular velocity ω along circular paths. From Eq. (11) we know that the speeds of the particles along their circular paths are proportional to their radial distances:

$$v_1 = R_1\omega \quad v_2 = R_2\omega \quad v_3 = R_3\omega, \ldots \qquad (22)$$

and hence the total kinetic energy is

$$K = \tfrac{1}{2}m_1R_1^2\omega^2 + \tfrac{1}{2}m_2R_2^2\omega^2 + \tfrac{1}{2}m_3R_3^2\omega^2 + \ldots$$

We can write this conveniently as

Kinetic energy of rotation ▶

$$K = \tfrac{1}{2}I\omega^2 \qquad (23)$$

where the quantity

Moment of inertia ▶

$$I = m_1R_1^2 + m_2R_2^2 + m_3R_3^2 + \ldots \qquad (24)$$

is called the ***moment of inertia*** of the rotating body about the given axis. The unit of moment of inertia is $\text{kg}\cdot\text{m}^2$.

Note that Eq. (23) has a mathematical form reminiscent of the familiar expression $\frac{1}{2}mv^2$ for the kinetic energy of a single particle — the moment of inertia replaces the mass, and the angular velocity replaces the translational velocity. As we will see later in this chapter, this analogy between moment of inertia and mass is of general validity. ***The moment of inertia is a measure of the resistance that a body offers to changes in its rotational motion,*** just as mass is a measure of the resistance that a body offers to changes in its translational motion.

Equation (24) shows that the moment of inertia — and consequently the kinetic energy for a given value of ω — is large if most of the mass of the body is at a large distance from the axis of rotation. This is quite reasonable: for a given value of ω, particles at a large distance from the axis move with high speeds, and therefore have large kinetic energies.

EXAMPLE 5. A 50-kg woman and an 80-kg man sit on a massless seesaw of length 3.0 m. The seesaw rotates about a pivot placed at the center of mass of the system (see Fig. 7.10). If the (instantaneous) angular velocity of the seesaw is 0.40 radian/s, calculate the kinetic energy. Treat both the woman and the man as particles.

SOLUTION: According to Example 6.4, the center of mass of this system is at a distance of $R_1 = 1.85$ m from the woman and at a distance of $R_2 = 1.15$ m from the man. Hence the moment of inertia for rotation about an axis through the center of mass is

$$I = m_1R_1^2 + m_2R_2^2 \tag{25}$$

$$= 50 \text{ kg} \times (1.85 \text{ m})^2 + 80 \text{ kg} \times (1.15 \text{ m})^2 = 277 \text{ kg}\cdot\text{m}^2$$

The kinetic energy for the rotational motion is

$$K = \tfrac{1}{2}I\omega^2$$

$$= \tfrac{1}{2} \times 277 \text{ kg}\cdot\text{m}^2 \times (0.40 \text{ radian/s})^2 = 22 \text{ J} \tag{26}$$

This kinetic energy could equally well have been obtained by first calculating the individual speeds of the woman and the man ($v_1 = R_1\omega$ and $v_2 = R_2\omega$) and then adding the corresponding individual kinetic energies ($\frac{1}{2}m_1v_1^2$ and $\frac{1}{2}m_2v_2^2$).

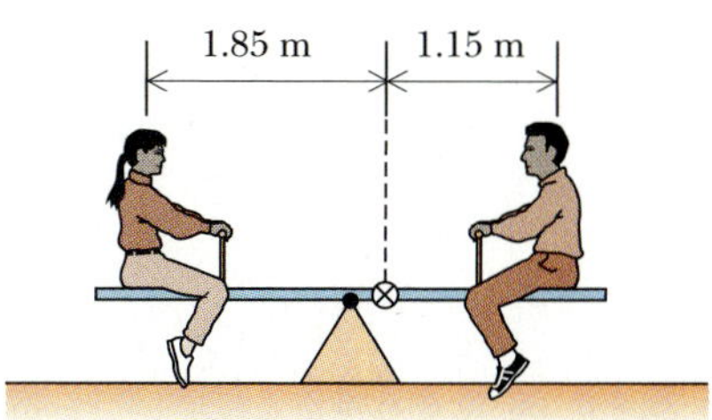

FIGURE 7.10
Woman and man on a seesaw.

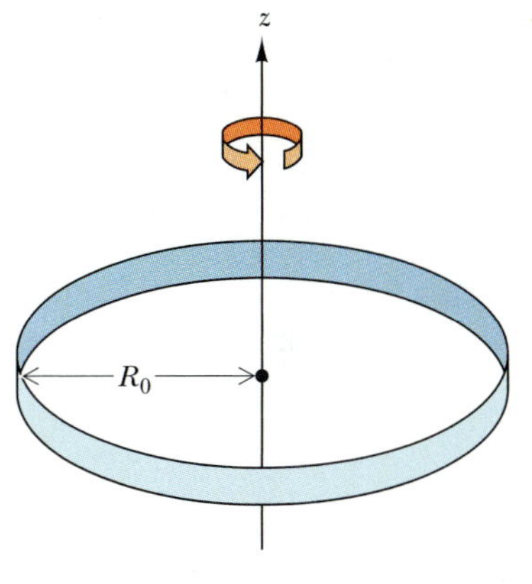

FIGURE 7.11
A thin hoop rotating about its axis of symmetry.

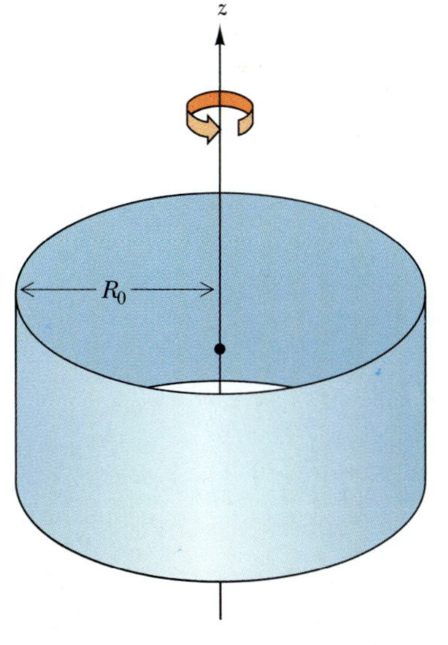

FIGURE 7.12
A thin cylindrical shell rotating about its axis of symmetry.

If we regard the mass of a solid body as continuously distributed throughout its volume, then we can calculate the moment of inertia by the same method we used for the calculation of the center of mass: we subdivide the body into small volume elements and add the moments of inertia contributed by all these small amounts of mass. Since the contributions of the small volume elements are proportional to their values of R^2, the moment of inertia of the body is the total mass multiplied by the average value of R^2; thus, the calculation of the moment of inertia hinges, in essence, on a calculation of the average value of R^2.

The calculation of the moment of inertia usually requires some fancy mathematics. However, in a few simple cases, it is possible to find the moment of inertia without much trouble. For example, if the rigid body is a thin hoop (see Fig. 7.11) or a thin cylindrical shell (see Fig. 7.12) of radius R_0 rotating about its axis of symmetry, then *all* of the mass of body is at the same distance from the axis of rotation — the moment of inertia is then simply the total mass M of the hoop or shell multiplied by its radius R_0 squared,

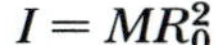

$$I = MR_0^2$$

TABLE 7.3 SOME MOMENTS OF INERTIA

Body		Moment of inertia
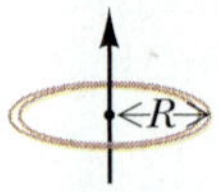	Thin hoop about symmetry axis	MR^2
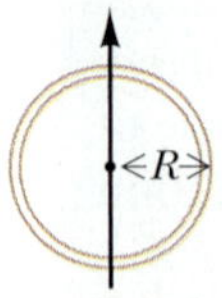	Thin hoop about diameter	$\frac{1}{2}MR^2$
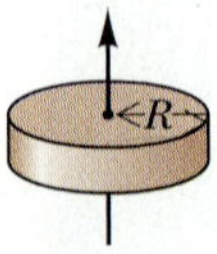	Disk or cylinder about symmetry axis	$\frac{1}{2}MR^2$
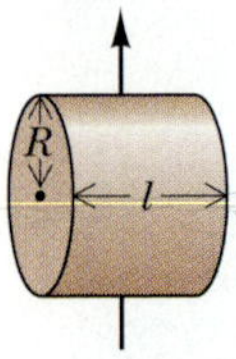	Cylinder about diameter through center	$\frac{1}{4}MR^2 + \frac{1}{12}Ml^2$
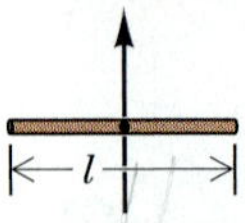	Thin rod about perpendicular axis through center	$\frac{1}{12}Ml^2$
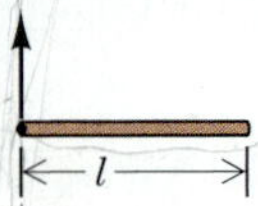	Thin rod about perpendicular axis through end	$\frac{1}{3}Ml^2$
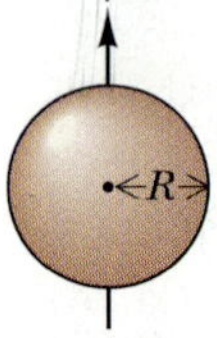	Sphere about diameter	$\frac{2}{5}MR^2$
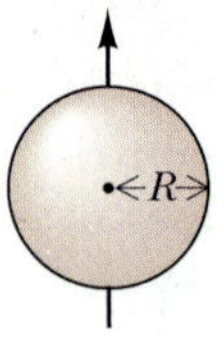	Thin spherical shell about diameter	$\frac{2}{3}MR^2$

Table 7.3 lists the moments of inertia of a variety of rigid bodies about the axes indicated in the diagrams; all the bodies are assumed to have uniform density. The moments of inertia depend on the total mass of the bodies and on their dimensions.

EXAMPLE 6. Calculate the rotational kinetic energy of the Earth. Treat the Earth as a rigid sphere of (approximately) uniform density, with a radius of 6.37×10^6 m and with a mass of 5.98×10^{24} kg.

SOLUTION: For a uniform sphere, Table 7.3 gives the following formula for the moment of inertia:

$$I = \tfrac{2}{5}MR^2$$

If we substitute the mass and the radius of the Earth into this formula, we find

$$I = \tfrac{2}{5} \times 5.98 \times 10^{24}\ \text{kg} \times (6.37 \times 10^6\ \text{m})^2$$

$$= 9.7 \times 10^{37}\ \text{kg} \cdot \text{m}^2$$

The rate of rotation of the Earth is one revolution in 24 hr, that is 2π radians in 24 hr, or

$$\omega = \frac{2\pi}{24 \times 60 \times 60\ \text{s}} = 7.3 \times 10^{-5}\ \text{radian/s}$$

From Eq. (23), we then find a kinetic energy

$$K = \tfrac{1}{2}I\omega^2 = \tfrac{1}{2} \times 9.7 \times 10^{37}\ \text{kg} \cdot \text{m}^2 \times (7.3 \times 10^{-5}\ \text{radian/s})^2$$

$$= 2.6 \times 10^{29}\ \text{J}$$

In this example we have ignored the orbital motion of the Earth around the Sun; if we take this motion into account, then the Earth has both rotational kinetic energy (due to its rotation about its axis) and translational kinetic energy (due to its motion around the Sun). The total kinetic energy of the Earth, or of any other rigid body with simultaneous rotational and translational motions, can be shown to be the sum of the rotational kinetic energy $\frac{1}{2}I\omega^2$ and the translational kinetic energy $\frac{1}{2}Mv_{CM}^2$, provided the rotational kinetic energy is reckoned for an axis passing through the center of mass.

7.4 WORK, ENERGY, AND POWER IN ROTATIONAL MOTION; TORQUE

In the case of translational motion, Newton's Second Law tells us that any acceleration of the body is caused by an external force. In the case of rotational motion, any angular acceleration is also caused by an external force. However, the relationship between the angular acceleration and the external force is somewhat complicated, and the best way to discover this relationship is to begin with a calculation of the work done by an external force on a rigid body rotating about a fixed axis.

Figure 7.13 shows the body as seen from along the axis; the force **F** is applied at some point of the body at a distance R from the axis of rota-

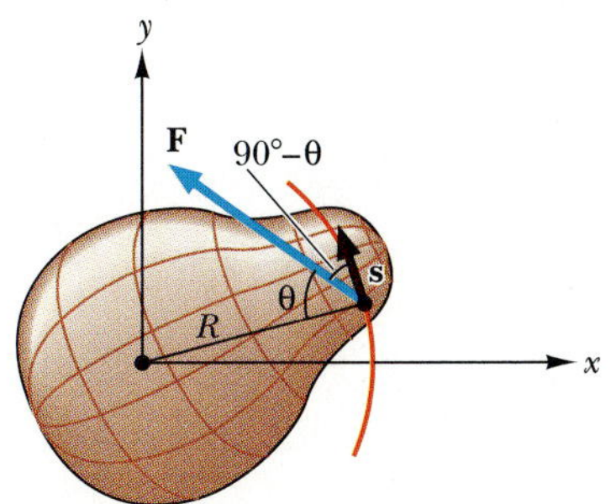

FIGURE 7.13

Force applied to a rigid body rotating about a fixed axis. The force **F** makes an angle θ with the radial line and an angle $90° - \theta$ with the instantaneous displacement **s**.

tion. For a start, we will assume that the force has no component parallel to the axis; any such component would be of no interest in the present context, since the body does not move in the direction parallel to the axis, and a force parallel to the axis can then do no work. In Fig. 7.13, the force is shown entirely in the plane of the page.

The work done by this force during a small displacement of the point at which the force acts is the product of the magnitude F of the force, the magnitude s of the displacement and the cosine of the angle between the force and the displacement [see Eq. (5.5)]. If the angle between the force and the radial line is θ, as in Fig. 7.13, then the angle between the force and the displacement is $90° - \theta$. The cosine of the angle $90° - \theta$ is equal to the sine of the angle θ, according to the general trigonometric identity $\cos(90° - \theta) = \sin\theta$. Hence, we can write the work as

$$W = Fs\sin\theta \tag{27}$$

If the body rotates through a small angle $\Delta\phi$, the magnitude of the displacement is $s = R\Delta\phi$, and hence

$$W = FR\Delta\phi\sin\theta \tag{28}$$

The product $FR\sin\theta$ ***is called the torque of the force* F, *usually designated by the symbol*** τ,

Torque ▶

$$\tau = FR\sin\theta \tag{29}$$

With this notation, the work done by the force is simply

Work done by torque ▶

$$W = \tau\Delta\phi \tag{30}$$

This is the rotational analog of the familiar equation $W = F\Delta x$ for work done in translational motion. The torque is analogous to the force, and the angular displacement $\Delta\phi$ is analogous to the translational displacement Δx. The analogy between torque in rotational motion and the force in translational motion extends beyond the equation for the work. As we will see in the next section, a torque applied to a rigid body causes angular acceleration, just as a force applied to a particle causes translational acceleration.

The unit of torque is the unit of force multiplied by the unit of distance; ***the unit of torque is therefore the newton-meter*** (N · m).

Note that according to Eq. (29), for a force of given magnitude, the torque is largest if the force acts at right angles to the radial line ($\theta = 90°$) and if the force acts at a large distance from the axis of rotation (large R). This dependence of the torque and of the work on the distance from the axis of rotation agrees with your everyday experiences in pushing doors open or shut. A door is a rigid body, which rotates about a vertical axis through the hinges. If you push against the edge of the door farthest from the hinge (largest R; see Fig. 7.14a), you produce a large torque, which does work on the door, increases its kinetic energy, and swings the door quickly. If, with the same force, you push at a point near the hinge (small R; see Fig. 7.14b), the door responds more sluggishly.

You produce a smaller torque, and you have to push through a larger angle to do the same amount of work and attain the same amount of kinetic energy and the same final angular velocity.

(a)

(b)

FIGURE 7.14

(a) A push against the door far from the hinge produces a large angular acceleration. (b) The same push near the hinge produces a small angular acceleration.

EXAMPLE 7. Suppose that while opening a 1-m-wide door, you push against the edge farthest from the hinge with a steady force of 0.9 N at right angles to the surface of the door. How much work do you do on the door during an angular displacement of 30°?

SOLUTION: Equation (30) tells us that the work done by the torque is

$$W = \tau \Delta\phi$$

With $R = 1$ m and $\theta = 90°$, Eq. (29) tells us that the torque is

$$\tau = FR \sin 90° = 0.9 \text{ N} \times 1 \text{ m} \times 1 = 0.9 \text{ N} \cdot \text{m}$$

The angular displacement must be expressed in radians, $\Delta\phi = 30° \times (2\pi \text{ radians}/360°) = 0.52$ radian. Then

$$W = \tau\Delta\phi = 0.9 \text{ N} \cdot \text{m} \times 0.52 \text{ radian}$$

$$= 0.5 \text{ J}$$

The equation for the power in rotational motion and the equations that express the work–energy theorem and the conservation law for energy in rotational motion are analogous to the equations we formulated for translational motion in Chapter 4. To find the power delivered by the torque, we divide both sides of Eq. (30) by Δt, with the result

$$P = \frac{W}{\Delta t} = \tau \frac{\Delta\phi}{\Delta t} \tag{31}$$

or

$$P = \tau\omega \tag{32}$$

◀ **Power delivered by torque**

This equation is analogous to the equation $P = Fv$ for the power in translational motion.

The work done by the torque changes the rotational kinetic energy of the body. The work–energy theorem for rotational motion says that the work done on the body equals the change in rotational kinetic energy:

$$W = K_2 - K_1 = \tfrac{1}{2} I\omega_2^2 - \tfrac{1}{2} I\omega_1^2 \tag{33}$$

If the force acting on the body is conservative — such as the force of gravity or the force of a spring — then the work equals the negative of the change in potential energy, and Eq. (33) becomes

$$-U_2 + U_1 = \tfrac{1}{2}I\omega_2^2 - \tfrac{1}{2}I\omega_1^2 \qquad (34)$$

or

$$\tfrac{1}{2}I\omega_1^2 + U_1 = \tfrac{1}{2}I\omega_2^2 + U_2 \qquad (35)$$

This expresses the law of ***conservation of energy in rotational motion: the sum of the kinetic and potential energies is constant,***

Conservation of energy in rotational motion

$$E = \tfrac{1}{2}I\omega^2 + U = [\text{constant}] \qquad (36)$$

EXAMPLE 8. A meter stick is initially standing vertically on the floor. If the meter stick falls over, with what angular velocity will it hit the floor? Assume that the end in contact with the floor does not slip.

SOLUTION: We can solve this problem by exploiting energy conservation. The motion of the meter stick is rotation about a fixed axis passing through the point of contact with the floor (see Fig. 7.15). The stick is a uniform rod of mass M and length $l = 1$ m. Its moment of inertia about the end is $I = Ml^2/3$ (see Table 7.3), and its rotational kinetic energy is therefore $K = \frac{1}{2}I\omega^2 = \frac{1}{2} \times Ml^2/3 \times \omega^2 = Ml^2\omega^2/6$. The gravitational potential energy is Mgy_{CM}, where y_{CM} is the height of the center of mass above the floor. When the meter stick is standing vertically, $y_{CM} = l/2$, and the gravitational potential energy therefore is $Mgl/2$. The initial energy is purely potential:

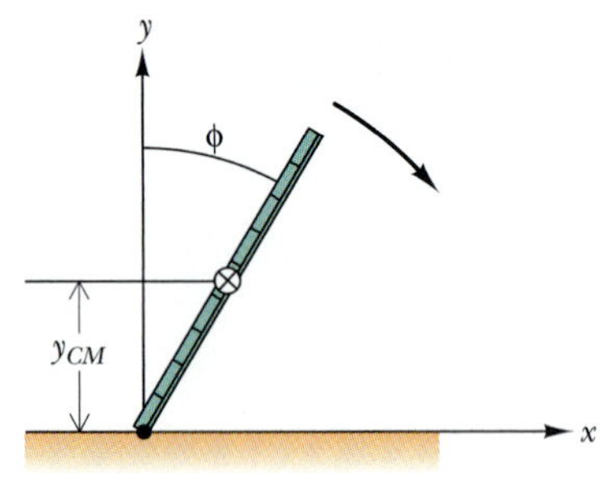

FIGURE 7.15
Meter stick rotating about its lower end.

$$E = \tfrac{1}{6}Ml^2\omega_1^2 + Mgy_{CM} = 0 + Mgl/2 \qquad (37)$$

When the meter stick hits the floor, $y_{CM} = 0$, and thus the final energy is purely kinetic:

$$E = \tfrac{1}{6}Ml^2\omega_2^2 + Mgy_{CM} = \tfrac{1}{6}Ml^2\omega_2^2 + 0 \qquad (38)$$

Conservation of energy tells us that the initial energy equals the final energy:

$$\tfrac{1}{6}Ml^2\omega_2^2 = Mgl/2 \qquad (39)$$

From this we obtain

$$\omega_2^2 = 3g/l$$

And taking the square root of both sides, we find

$$\omega_2 = \sqrt{\frac{3g}{l}} = \sqrt{\frac{3 \times 9.81\ \text{m/s}^2}{1.0\ \text{m}}} = 5.4\ \text{radians/s}$$

COMMENTS AND SUGGESTIONS: Note that the procedure for the solution of a problem involving energy conservation in rotational motion is the same as the procedure for translational motion: First write an expression for the energy at one point in the motion [Eq. (37)], then write an expression for the energy at another point [Eq. (38)], and then use energy conservation to equate the two expressions [Eq. (39)].

7.5 THE EQUATION OF ROTATIONAL MOTION

Our intuition tells us that a torque acting on a wheel or some other body free to rotate about an axis will produce an angular acceleration. For instance, the push of your hand against a crank on a wheel exerts a torque, or "twist," that starts the wheel turning. The angular acceleration depends on the magnitude of your push on the crank and also on its direction. Your push will be most effective if exerted tangentially, at right angles to the radius (at $\theta = 90°$ and $\sin\theta = 1$; see Fig. 7.16a), because this is the direction in which the crank is free to move. It will be less effective if exerted at a smaller or larger angle (see Fig. 7.16b). And it will be completely ineffective if exerted parallel to the radius (at $\theta = 0°$ or $180°$ and $\sin\theta = 0$; see Fig. 7.16c) — such a push in the radial direction produces no rotation at all. These qualitative considerations are in agreement with the definition of torque,

$$\tau = FR\sin\theta \tag{40}$$

According to this equation, the torque or "twist" provided by a force of a given magnitude F is maximum if the force is at right angles to the radius ($\theta = 90°$), and it is zero if the force is parallel to the radius ($\theta = 0°$ or $180°$).

The quantity $R\sin\theta$ appearing in Eq. (40) has a simple geometric interpretation: it is ***the perpendicular distance between the line of action of the***

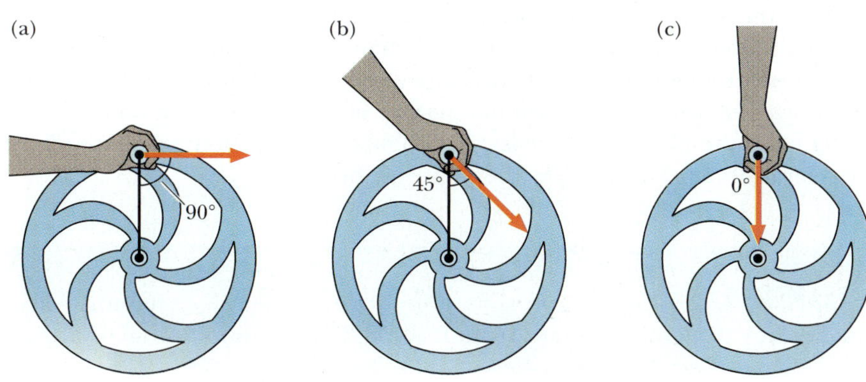

FIGURE 7.16

(a) A push at right angles to the radius is most effective in producing rotation. (b) A push at 45° is less effective. (c) A push parallel to the radius produces no rotation.

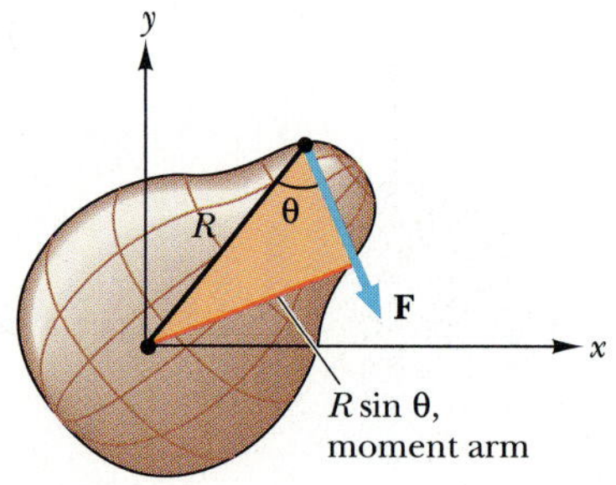

FIGURE 7.17
Force applied to a rigid body rotating about a fixed axis. The distance between the center of rotation and the point of application of the force is R. The perpendicular distance between the center of rotation and the line of action of the force is $R \sin\theta$; this distance is the moment arm.

force and the axis of rotation (see Fig. 7.17); ***this perpendicular distance is called the moment arm of the force.*** Hence, Eq. (40) states that the torque equals the magnitude of the force multiplied by the moment arm.

To find a quantitative relationship between torque and angular acceleration, we recall that the power delivered by a torque acting on a body is

$$\frac{W}{\Delta t} = \tau\omega \tag{41}$$

The work–energy theorem tells us that the work W must equal the change of kinetic energy in the small time interval Δt. If the kinetic energy is $\frac{1}{2}I\omega^2$ at the beginning of this time interval, it will be $\frac{1}{2}I(\omega + \Delta\omega)^2$ at the end of the time interval. The difference between these two kinetic energies must equal the work W:

$$\begin{aligned} W &= \tfrac{1}{2}I(\omega + \Delta\omega)^2 - \tfrac{1}{2}I\omega^2 \\ &= \tfrac{1}{2}I[\omega^2 + 2\omega\Delta\omega + (\Delta\omega)^2] - \tfrac{1}{2}I\omega^2 \\ &= I\omega\Delta\omega + \tfrac{1}{2}I(\Delta\omega)^2 \end{aligned} \tag{42}$$

Since the increment $\Delta\omega$ in the angular velocity in a small time interval Δt is small, the square of $\Delta\omega$ is even smaller, and it can be neglected. Thus, $W \simeq I\omega\Delta\omega$. Inserting this into the left side of Eq. (41), we find

$$\frac{I\omega\Delta\omega}{\Delta t} = \tau\omega \tag{43}$$

Here, we can cancel a factor of ω on both sides of the equation, which leaves us with

$$I\frac{\Delta\omega}{\Delta t} = \tau \tag{44}$$

But $\Delta\omega/\Delta t$ is the angular acceleration; thus

Equation for rotational motion ▸

$$I\alpha = \tau \tag{45}$$

This is the equation for rotational motion. As we might have expected, this equation says that ***the angular acceleration is directly proportional to the torque.*** Equation (45) is mathematically analogous to Newton's Second Law, $ma = F$, for the translational motion of a particle; the moment of inertia takes the place of the mass, the angular acceleration the place of the acceleration, and the torque the place of the force.

EXAMPLE 9. A phonograph turntable driven by an electric motor accelerates at a constant rate from 0 to 33.3 rev/min in a time of 2.0 s.

The turntable is a uniform disk of metal, of a mass 1.5 kg and a radius 13 cm. What torque is required to drive this turntable? If the driving wheel makes contact with the turntable at its outer rim (see Fig. 7.18), what is the force that it must exert?

SOLUTION: The torque can be calculated from the equation for the rotational motion, Eq. (45). The final angular velocity is $2\pi \times 33.3$ radians/60 s = 3.49 radians/s and therefore the angular acceleration is

$$\alpha = \frac{\Delta\omega}{\Delta t} = \frac{\omega_2 - \omega_1}{t_2 - t_1} \tag{46}$$

$$= \frac{3.49 \text{ radians/s} - 0}{2.0 \text{ s} - 0} = 1.75 \text{ radians/s}^2$$

The moment of inertia of the turntable is that of a disk (see Table 7.3):

$$I = \tfrac{1}{2}MR^2 \tag{47}$$

$$= \tfrac{1}{2} \times 1.5 \text{ kg} \times (0.13 \text{ m})^2 = 1.27 \times 10^{-2} \text{ kg}\cdot\text{m}^2$$

Hence the required torque is, according to Eq. (45),

$$\tau = I\alpha = 1.27 \times 10^{-2} \text{ kg}\cdot\text{m}^2 \times 1.75 \text{ radians/s}^2$$

$$= 2.22 \times 10^{-2} \text{ N}\cdot\text{m}$$

The driving force is perpendicular to the radius (see Fig. 7.18), so $\sin\theta = 1$ and Eq. (29) reduces to $\tau = FR$, which yields

$$F = \frac{\tau}{R} = \frac{2.22 \times 10^{-2} \text{ N}\cdot\text{m}}{0.13 \text{ m}} = 0.171 \text{ N}$$

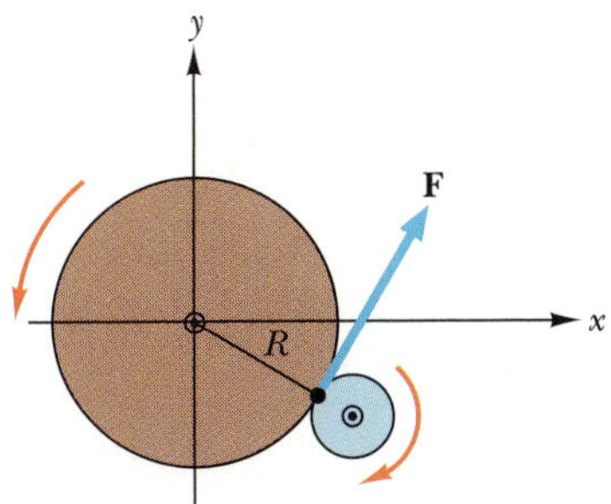

FIGURE 7.18

The driving wheel exerts a tangential force on the turntable.

EXAMPLE 10. An automobile with rear-wheel drive is accelerating at 5.0 m/s² along a straight road. Consider one of the front wheels of this automobile (see Fig. 7.19a). The axle pushes the wheel forward, providing an acceleration of 5.0 m/s². Simultaneously, the friction force of the road pushes the bottom of the wheel rearward, providing a torque that gives the wheel an angular acceleration. The wheel has a radius of 0.38 m and a mass of 25 kg. Assume that the wheel is (approximately) a uniform disk, and assume it rolls without slipping. Find the magnitude of the rearward force that friction exerts on the bottom of the wheel.

SOLUTION: Figure 7.19b shows a "free-body" diagram of the wheel, with the horizontal forces acting on it. Besides these horizontal forces, there is also a vertical downward push exerted by the axle and a verti-

(a)

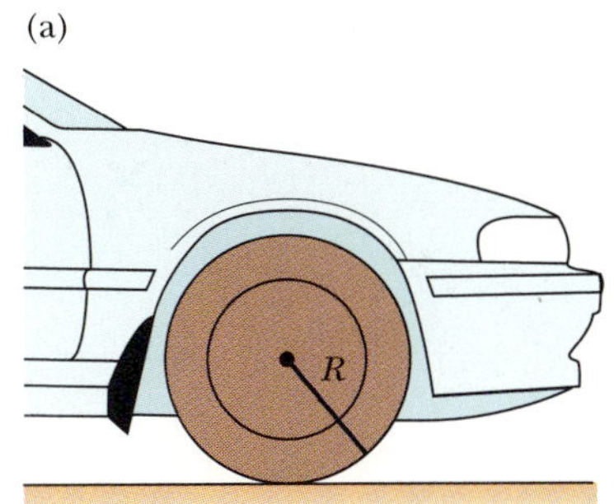

(b)

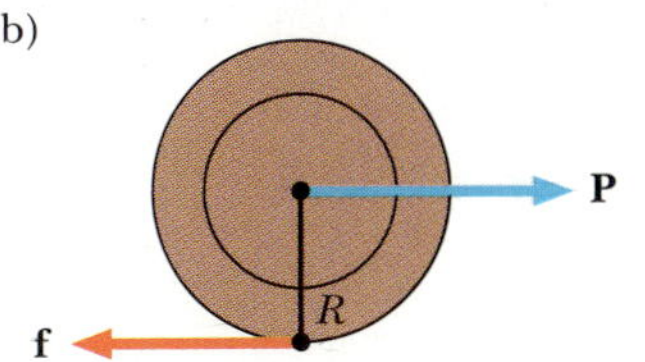

FIGURE 7.19

(a) Front wheel of an automobile. (b) "Free-body" diagram for the wheel. The friction force of the road pushes the wheel backward. The axle pushes the wheel forward.

cal upward normal force exerted by the road; these forces cancel, and they need not concern us here.* The forward push of the axle is **P** and the rearward push of the friction exerted by the road is **f.** The force **P,** acting at the center of the wheel, exerts no torque, the force **f,** acting at the rim, exerts a torque $\tau = Rf$. Thus, the equation for the rotational motion of the wheel is

$$I\alpha = Rf \tag{48}$$

or, since $I = \frac{1}{2}MR^2$,

$$\tfrac{1}{2}MR\alpha = f \tag{49}$$

To evaluate the force f from this equation, we first need to find the angular acceleration α. This angular acceleration is related to the translational acceleration by $a = \alpha R$. We can establish this relationship most conveniently by viewing the motion of the wheel in the reference frame of the automobile. In this reference frame, the ground is moving backward at speed v, and the bottom point of the rotating wheel is moving rearward at the rotational speed ωR. Since the wheel is supposed to move without slipping, the speed of the ground must match the speed of the bottom point of the wheel; that is, $v = \omega R$. This proportionality for v and ω implies the same proportionality for the accelerations, and therefore establishes the relationship $a = \alpha R$. With this, the equation (48) for rotational motion reduces to

$$\tfrac{1}{2}Ma = f \tag{50}$$

from which

$$f = \tfrac{1}{2}Ma = \tfrac{1}{2} \times 25 \text{ kg} \times 5.0 \text{ m/s}^2$$

$$= 63 \text{ N}$$

COMMENTS AND SUGGESTIONS: In this problem, the axis of rotation of the wheel is not a fixed axis, and we had to consider both the rotational motion of the wheel and the translational motion. Problems with moving axes of rotation are usually quite difficult, but in our example the rotational and the translational motions can be handled separately, which makes possible the separate calculation of the friction force f. (If we want, we can next calculate the magnitude of the push P by writing the equation for the translational motion of the wheel, $Ma = P - f$; see Problem 45.)

Note that in our calculation we implicitly assumed that the equation of rotational motion (45) remains valid even though the axis of rota-

* Note that this "free-body" diagram applies to the front wheels only. For the rear wheels (the "driving" wheels), the forces are quite different; furthermore, the rear wheels are bolted to their axles, and the engine provides a torque on them.

tion of the wheel is in accelerated motion and is *not a fixed axis.* This assumption can be justified by some further arguments that demonstrate that ***Eq. (45) remains valid for rotation about an axis in accelerated translational motion, provided the axis passes through the center of mass of the rotating body.***

7.6 Conservation of Angular Momentum

In Chapter 6 we saw how to express the equation for translational motion in terms of the momentum: the rate of change of the momentum equals the force ($\Delta \mathbf{p}/\Delta t = \mathbf{F}$). Likewise, we can express the equation for rotational motion in terms of angular momentum. ***The angular momentum of a body rotating about a fixed axis is defined as the product of moment of inertia and angular velocity,***

$$L = I\omega \tag{51}$$

◀ **Angular momentum**

This equation for angular momentum is analogous to the equation $p = mv$ for translational momentum. The unit of angular momentum is kg · m^2/s, which can also be written in the alternative form J · s. Table 7.4 gives some examples of typical values of angular momenta.

To express the equation for rotational motion in terms of angular momentum, we proceed as we did in the translational case. We note that if the change of angular velocity is $\Delta\omega = \omega_2 - \omega_1$, then $I\omega_2 - I\omega_1 = L_2 - L_1 = \Delta L$. Thus, the left side of Eq. (45) can be expressed as a change of angular momentum, and the equation becomes

$$\frac{\Delta L}{\Delta t} = \tau \tag{52}$$

TABLE 7.4	SOME ANGULAR MOMENTA
Orbital motion of Earth	2.7×10^{40} J · s
Rotation of Earth	5.8×10^{33} J · s
Helicopter rotor (320 rev/min)	5×10^{4} J · s
Automobile wheel (90 km/h)	1×10^{2} J · s
Electric fan	1 J · s
Frisbee	1×10^{-1} J · s
Toy gyroscope	1×10^{-1} J · s
Phonograph record (33.3 rev/min)	6×10^{-3} J · s
Bullet fired from rifle	2×10^{-3} J · s
Orbital motion of electron in atom	1.05×10^{-34} J · s
Spin of electron	0.53×10^{-34} J · s

(a)

(b)

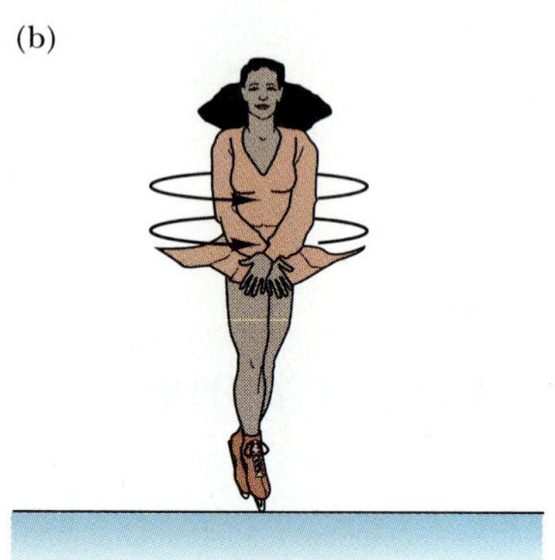

FIGURE 7.20
Figure skater performing a pirouette. (a) Arms extended. (b) Arms folded against body.

TABLE 7.5	FURTHER ANALOGIES BETWEEN ROTATIONAL AND TRANSLATIONAL QUANTITIES	
$W = F\Delta x$	$\rightarrow$	$W = \tau\Delta\phi$
$P = Fv$	$\rightarrow$	$P = \tau\omega$
$K = \frac{1}{2}mv^2$	$\rightarrow$	$K = \frac{1}{2}I\omega^2$
$ma = F$	$\rightarrow$	$I\alpha = \tau$
$p = mv$	$\rightarrow$	$L = I\omega$
$\Delta p/\Delta t = F$	$\rightarrow$	$\Delta L/\Delta t = \tau$

This says that the rate of change of angular momentum equals the torque. Obviously, this equation is analogous to the equation $\Delta p/\Delta t = F$ for translational motion.

We now see that the analogy between rotational and translational quantities mentioned in Section 7.2 can be extended to angular momentum and translational momentum. Table 7.5 lists further analogous quantities, including the quantities for work, power, and kinetic energy.

If there is no torque acting on the rotating body, $\tau = 0$ and therefore $\Delta L/\Delta t = 0$, which means that the angular momentum does not change,

$$L = [\text{constant}] \tag{53}$$

This is the law of ***conservation of angular momentum.*** Since $L = I\omega$, we can also write this law as

$$I\omega = [\text{constant}] \tag{54}$$

A pirouette performed by a figure skater on ice provides a nice illustration of the conservation of angular momentum. The skater begins the pirouette by spinning about her vertical axis with her arms extended horizontally (see Fig. 7.20a); in this configuration, the arms are far from the axis of rotation, and therefore the arms have a large moment of inertia. She then brings her arms close to her body (see Fig. 7.20b), suddenly decreasing her moment of inertia. Since the ice is nearly frictionless, the external torque on the skater is nearly zero, and therefore the angular momentum is conserved. According to Eq. (54), a decrease of I requires an increase of ω to keep the angular momentum constant. Thus, the change of configuration of her arms causes the skater to whirl around her vertical axis with a dramatic increase of angular velocity (see Fig. 7.21).

Like the law of conservation of translational momentum, the law of conservation of angular momentum is often useful in the solutions of problems in which the forces are not known in detail.

FIGURE 7.21
Photo of figure skater.

EXAMPLE 11. Suppose that the phonograph turntable described in Example 9 is coasting (with the motor disengaged) at 33.3 rev/min when a stack of 10 phonograph records suddenly drops down on it. What is the angular velocity after the drop? Each record has a mass of 0.170 kg and a radius of 15.2 cm. Ignore the (small) friction force in the axle of the turntable.

SOLUTION: Since there is no external torque on the system of turntable and records, the angular momentum of this system is conserved. The angular momentum before the drop is $I\omega$ (where ω is the initial angular velocity and I the moment of inertia of the turntable), and the angular momentum after the drop is $I'\omega'$ (where ω' is the final angular velocity and I' the moment of inertia of turntable and records combined). Hence

$$I\omega = I'\omega' \tag{55}$$

from which we find

$$\omega' = \frac{I}{I'}\,\omega \tag{56}$$

The moment of inertia of the turntable is (see Example 9)

$$I = 1.27 \times 10^{-2}\ \text{kg}\cdot\text{m}^2$$

and the moment of inertia of the 10 records is that of a disk with 10 times the mass of one record,

$$I_r = \tfrac{1}{2}M_r R_r^2 = \tfrac{1}{2} \times (10 \times 0.170\ \text{kg}) \times (0.152\ \text{m})^2 \tag{57}$$

$$= 1.96 \times 10^{-2}\ \text{kg}\cdot\text{m}^2$$

Accordingly,

$$\omega' = \frac{I}{I'}\,\omega = \frac{I}{I + I_r}\,\omega \tag{58}$$

$$= \frac{1.27 \times 10^{-2}\ \text{kg}\cdot\text{m}^2}{1.27 \times 10^{-2}\ \text{kg}\cdot\text{m}^2 + 1.96 \times 10^{-2}\ \text{kg}\cdot\text{m}^2} \times 3.49\ \text{radians/s}$$

$$= 1.37\ \text{radians/s}$$

As already mentioned in Chapter 5, the law of conservation of angular momentum also applies to a single particle moving in an orbit under the influence of a central force. Such a force is always directed along the radial line, and it therefore exerts no torque. If the particle is (instantaneously) moving with speed v along a circle of a radius r (see Fig. 7.22), its moment of inertia is mr^2 and its angular velocity is $\omega = v/r$ [see Eq. (11)].

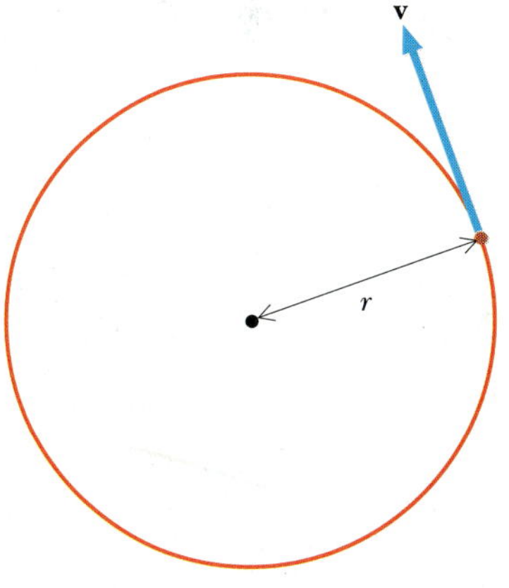

FIGURE 7.22
A particle moving with speed v along a circle of radius r. The moment of inertia of this particle with respect to the center of the circle is $I = mr^2$.

Hence $I\omega = mr^2 \times v/r = mvr$, and the angular momentum of the particle is

Orbital angular momentum ▶

$$L = mvr \tag{59}$$

This formula is valid not only for a circular orbit, but also for the perihelion and aphelion points of an elliptical orbit, where the instantaneous velocity is perpendicular to the radius. In Chapter 5 we took advantage of the conservation of the angular momentum $L = mvr$ to compare the speeds of a planet at perihelion and at aphelion.

The angular momentum (59) is called the ***orbital angular momentum*** to distinguish it from the ***spin angular momentum*** of a body rotating about its own axis. For instance, the Earth has both an orbital angular momentum (due to its motion around the Sun) and a spin angular momentum (due to its rotation about its axis). Table 7.4 includes examples of both kinds of angular momentum.

SUMMARY

- **Average angular velocity:** $\overline{\omega} = \dfrac{\Delta\phi}{\Delta t}$
- **Instantaneous angular velocity:** $\omega = \dfrac{\Delta\phi}{\Delta t}$ with Δt extremely small
- **Frequency:** $f = \dfrac{\omega}{2\pi}$
- **Average angular acceleration:** $\overline{\alpha} = \dfrac{\Delta\omega}{\Delta t}$
- **Instantaneous angular acceleration:** $\alpha = \dfrac{\Delta\omega}{\Delta t}$ with Δt extremely small
- **Speed of a particle on rotating body:** $v = R\omega$
- **Acceleration of a particle on rotating body:** $a_{\text{tan}} = R\alpha$
  $a_{\text{cent}} = R\omega^2$
- **Motion with constant angular acceleration:**
  $\omega = \omega_0 + \alpha t$
  $\phi = \phi_0 + \omega_0 t + \frac{1}{2}\alpha t^2$
  $\alpha(\phi - \phi_0) = \frac{1}{2}(\omega^2 - \omega_0^2)$
- **Moment of inertia:** $I = m_1R_1^2 + m_2R_2^2 + m_3R_3^2 + \ldots$
- **Kinetic energy of rotation:** $K = \frac{1}{2}I\omega^2$
- **Torque:** $\tau = FR\sin\theta$
- **Work done by torque:** $W = \tau\Delta\phi$
- **Power delivered by torque:** $P = \tau\omega$

- **Conservation of energy in rotational motion:**
  $E = \frac{1}{2}I\omega^2 + U = [\text{constant}]$
- **Equation for rotational motion (fixed axis):** $I\alpha = \tau$
- **Angular momentum of rotation:** $L = I\omega$
- **Conservation of angular momentum:** $I\omega = [\text{constant}]$
- **Orbital angular momentum of particle:** $L = mvr$

QUESTIONS

1.–A spinning flywheel in the shape of a disk suddenly shatters into many small fragments. Draw the trajectories of a few of these small fragments; assume that the fragments do not interfere with each other.

2.–You may have noticed that in some old movies the wheels of carriages or stagecoaches seem to rotate backward. How does this come out?

3.–Relative to an inertial reference frame, what is your angular velocity right now about an axis passing through your center of mass?

4.–When engineers design the teeth for gears, they usually make the sides of the teeth rounded (Fig. 7.23). Can you guess why?

FIGURE 7.23
Gears.

5.–The hands of a watch are small rectangles with a common axis passing through one end. The minute hand is long and thin; the hour hand is short and thicker. Both hands have the same mass. Which has the larger moment of inertia? Which has the greater kinetic energy and angular momentum?

6.–What configuration and what axis would you choose to give your body the smallest possible moment of inertia? The greatest?

7.–About what axis through the center of mass is the amount of inertia of this book largest? Smallest? Assume the book is closed.

8.–Automobile engines and other internal-combustion engines have flywheels attached to their crankshafts. What is the purpose of these flywheels? (Hint: Each explosive combustion in one of the cylinders of such an engine gives a sudden push to the crankshaft. How would the crankshaft respond to this push if it had no flywheel?)

9.–Many farmers have been injured when their tractors suddenly flipped over backward while pulling a heavy piece of farm equipment. Can you explain how this happens?

10.–According to popular belief, a falling piece of bread always lands on its buttered side. Why or why not?

11.–Rifle bullets are given a spin about their axis by spiral grooves ("rifling") in the barrel of the gun. What is the advantage of this?

12.–You are standing on a frictionless turntable (like a phonograph turntable, but sturdier). How can you turn 180° without leaving the turntable or pushing against any exterior body?

FIGURE 7.24
A tightrope walker.

(a)

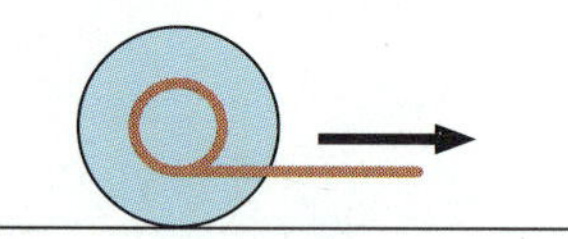

(b)

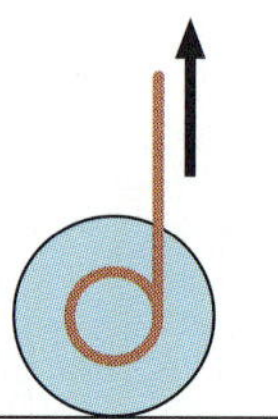

FIGURE 7.25
Yo-yo resting on a table.
(a) String pulls horizontally.
(b) String pulls vertically.

13.–If you give a hard-boiled egg resting on a table a twist with your fingers, it will continue to spin. If you try doing the same with a raw egg, it will not. Why?

14.–A tightrope walker uses a balancing pole to keep steady (Fig. 7.24). How does this help?

15.–Why do helicopters need a small vertical propeller on their tail?

16.–The rotation of the Earth is subject to small seasonal variations (see Section 1.1). Does this mean that angular momentum is not conserved?

17.–Why does the front end of an automobile dip down when the automobile is braking sharply?

18.–The friction of the tides against the ocean coasts and the ocean shallows is gradually slowing down the rotation of the Earth. What happens to the lost angular momentum? (Hint: What causes the tides?)

19.–An automobile is traveling on a straight road at 90 km/h. What is the speed, relative to the ground, of the lowermost point on one of its wheels? The topmost point? The midpoint?

20.–A sphere and a hoop of equal masses roll down an inclined plane without slipping. Which will get to the bottom first? Will they have equal kinetic energies when they reach the bottom?

21.–A yo-yo rests on a table (Fig. 7.25). If you pull the string horizontally, which way will it move? If you pull vertically?

22.–Stand a pencil vertically on its point on a table and let go. The pencil will topple over.

(a) If the table is very smooth, the point of the pencil will slip in the direction opposite to that of the toppling. Why?

(b) If the table is somewhat rough, or covered with a piece of paper, the point of the pencil will jump in the direction of the toppling. Why? (Hint: During the early stages of toppling, friction holds the point of the pencil fixed; thus the pencil requires a horizontal momentum.)

PROBLEMS

Section 7.1: Rotation about a Fixed Axis

1.——An electric drill rotates at 5000 rev/min. What is the frequency of rotation (in rev/s)? What is the time for one revolution? What is the angular velocity (in radians/s)?

2.——The rim of a phonograph record is at a distance of 15 cm from the center, and the rim of the paper label on the record is at a distance of 5 cm from the center.

(a) When this record is rotating at 33.3 rev/min, what is the translational speed of a point on the rim of the record? The translational speed of a point on the rim of the paper label?

(b) What are the centripetal accelerations of these points?

3.——A jet engine is spinning at 3000 revolutions per minute when the tip of one of its fan blades breaks off. The radius of the fan is 0.7 m. What is the speed with which the tip of the blade moves away?

4.——A merry-go-round has a diameter of 10 m. During a ride lasting 3 minutes, the merry-go-round turns through 28 revolutions. What is its average angular velocity? What is the average speed of a point on its rim?

5.——A bicycle has wheels of radius 0.36 m. How often does each wheel turn if the bicycle travels 1 km? How fast (in rev/min) do the wheels turn if the bicycle travels at a speed of 20 km/h?

6.*——The minute hand of a wall clock has a length of 20 cm. What is the speed of the tip of this hand?

7.*——Quito is on the Earth's equator; New York is at latitude 41° North. What is the angular velocity of each city about the Earth's axis of rotation? What is the linear speed of each?

8.*——An automobile has wheels with a radius of 0.30 m. What is the angular velocity (in radians per second) and the frequency (in revolutions per second) of the wheels when the automobile is traveling at 88 km/h?

9.*——An aircraft passes directly over you with a speed of 900 km/h at an altitude of 10,000 m. What is the angular velocity of the aircraft (relative to you) when directly overhead?

10.*——An automobile has wheels of diameter 0.63 m. If the automobile is traveling at 80 km/h, what is the instantaneous velocity vector (relative to the ground) of a point at the top of a wheel? At the bottom? At the front?

A merry-go-round.

Section 7.2: Motion with Constant Angular Acceleration

11.——The blade of a circular saw of diameter 20 cm accelerates uniformly from rest to 7000 rev/min in 1.2 s.

(a) What is the angular acceleration?

(b) How many revolutions will the blade have made by the time it reaches full speed?

12.——When you turn off the motor, a phonograph turntable initially rotating at 33.3 rev/min makes 25 revolutions before it stops. Calculate the angular deceleration of this turntable; assume it is constant.

13.——The shaft of an automobile engine accelerates at a constant rate from 200 rev/min to 3000 rev/min in 7.0 s and then runs at constant speed.

(a) Find the angular velocity and the angular acceleration at $t = 0$ (just after acceleration begins) and at $t = 7.0$ s (just before acceleration ends).

(b) A flywheel with a radius of 18 cm is attached to the shaft of the engine. Calculate the tangential and the centripetal accelerations of a point on the rim of the flywheel at the times given above.

14.*——An automobile accelerates uniformly from 0 to 80 km/h in 6.0 s. The automobile has wheels of radius 30 cm. What is the angular acceleration of the wheels about their axles?

Section 7.3: Kinetic Energy of Rotation; Moment of Inertia

15.——Estimate the moment of inertia of a human body spinning rigidly about its longitudinal axis. Treat the body as a uniform cylinder of mass 70 kg, length 1.7 m, and average diameter 23 cm.

16.——Find the moment of inertia of an orange of mass 300 g and diameter 9.0 cm. Treat the orange as a uniform sphere.

17.——What is the moment of inertia of a broomstick of mass 0.5 kg, length 1.5 m, and diameter 2.5 cm about its longitudinal axis? About an axis at right angles to the broomstick, passing through its center?

18.——According to spectroscopic measurements, the moment of inertia of an oxygen molecule about an axis through the center of mass and perpendicular to the line joining the atoms is $1.95 \times 10^{-46}\ \mathrm{kg \cdot m^2}$. The mass of an

An airplane propeller.

oxygen atom is 2.66×10^{-26} kg. What is the distance between the atoms? Treat the atoms as pointlike particles.

19.—What is the moment of inertia (about its axle) of a bicycle wheel of mass 4.0 kg and radius 0.33 m? Neglect the mass of the spokes.

20.—The drilling pipe of an oil rig is 2 km long, 15 cm in diameter, and has a mass of 20 kg per meter of length. Assume that the wall of the pipe is very thin.

(a) What is the moment of inertia of this pipe rotating about its longitudinal axis?

(b) What is the kinetic energy when rotating at 1 rev/s?

21.—Engineers have proposed that large flywheels be used for the temporary storage of surplus energy generated by electric power plants. A suitable flywheel would be 3.6 m in diameter, have a mass of 300 metric tons, and spin at 3000 revolutions per minute. What is the kinetic energy of rotation of this flywheel? Give the answer in both joules and kilowatt-hours. Assume that the moment of inertia of the flywheel is that of a uniform disk.

22.*—An airplane propeller consists of three radial blades, each of length 1.8 m and mass 20 kg. What is the kinetic energy of this propeller when rotating at 2500 rev/min? Assume that each blade is (aproximately) a uniform rod.

23.*—Problem 6.26 gives the dimension of a molecule of nitric acid (HNO_3). What is the moment of inertia of this molecule when rotating about the axis passing through the H, O, and N atoms? Treat the atoms as pointlike particles.

24.*—An empty beer can has a mass of 50 g, a length of 12 cm, and a radius of 3.3 cm. Find the moment of inertia of the can about its longitudinal axis. Assume that the can is a perfect cylinder of sheet metal with no ridges, indentations, or holes.

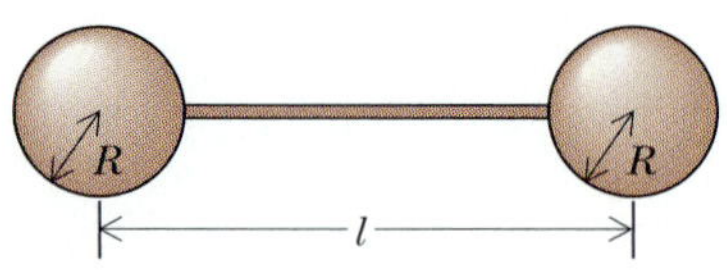

FIGURE 7.26
A dumbbell.

25.*—A dumbbell consists of two uniform spheres of mass M and radius R joined by a thin rod of mass m and length l (Fig. 7.26). What is the moment of inertia of this device about an axis through the center of the rod perpendicular to the rod? About an axis along the rod?

26.*—In order to increase her moment of inertia about a vertical axis, a spinning figure skater stretches out her arms horizontally; in order to reduce her moment of inertia, she brings her arms down vertically along her sides. Calculate the change of moment of inertia between these two configurations of the arms. Assume that each arm is a thin, uniform rod of length 0.60 m and mass 2.8 kg hinged at the neck, on the axis of rotation (that is, ignore the distance of the shoulder joint from the axis of rotation).

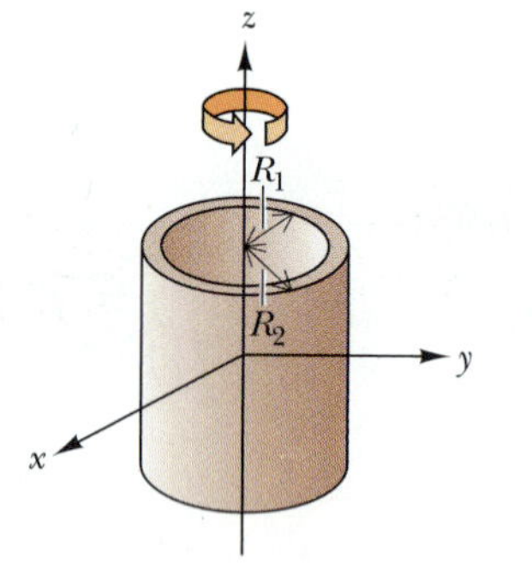

FIGURE 7.27
A piece of pipe.

27.**—A piece of steel pipe (Fig. 7.27) has an inner radius R_1, an outer radius R_2, and a mass M. Find the moment of inertia about the axis of the pipe.

28.**—An automobile of mass 1400 kg has wheels 0.75 cm in diameter weighing 27 kg each. Taking into account the rotational kinetic energy of the wheels about their axles, what is the total kinetic energy of the automobile when traveling at 80 km/h? What percentage of the kinetic energy belongs to the rotational motion of the wheels about their axles? Pretend that each wheel has a mass distribution equivalent to that of a uniform disk.

Section 7.4: Work, Energy, and Power in Rotational Motion; Torque

29.—You hold a 10-kg book in your hand with your arm extended horizontally in front of you. What is the torque that the weight of this book exerts about your shoulder joint, at a distance of 0.60 m from the book?

30.——A door is 0.8 m wide. What is the torque you exert about the axis passing through the hinges if you push against this door with a perpendicular force of 200 N at its middle? What if you push at the edge? A wind is blowing against the other side of the door and trying to push it open. Where should you push to keep the door closed?

31.——A 2-kg trout hangs from one end of a 2-m-long stiff fishing pole which the fisherman holds with one hand by the other end. If the pole is horizontal, what is the torque that the weight of the trout exerts about the end the fisherman holds? If the pole is tilted upward at an angle of 60°?

32.——The repair manual for an automobile specifies that the cylinder-head bolts are to be tightened to a torque of 200 N · m. If a mechanic uses a wrench of length 20 cm on such a bolt, what perpendicular force must he exert on the end of this wrench to achieve the correct torque?

33.——If you bend over, so your trunk is horizontal, the weight of your trunk exerts a rather strong torque about the sacrum, where your backbone is pivoted on your pelvis. Assume that the mass of your trunk (including arms and head) is 48 kg, and that the weight effectively acts at a distance of 0.40 m from the sacrum. What is the torque that this weight exerts?

34.——An elevator of mass 900 kg is being lifted at constant speed by a cable wrapped around a wheel (see Fig. 7.28). The radius of the wheel is 0.35 m. What torque does the cable exert on the wheel?

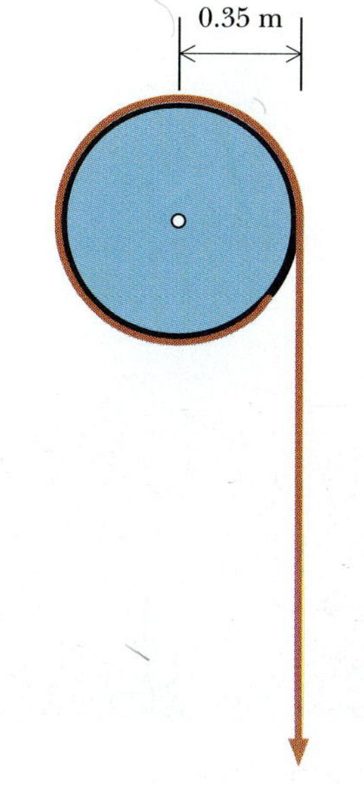

FIGURE 7.28

Elevator cable attached to a wheel.

35.——While tightening a screw, a carpenter exerts a torque of 0.5 N · m on the screwdriver. How much work does the carpenter do on the screwdriver while turning it through 360°?

36.——A manual winch has a crank of length (radius) 0.25 m. If a laborer pushes against this handle tangentially with a force of 200 N, how much work does he do while turning the crank through 10 revolutions?

37.——A cyclist is turning the pedals of her bicycle at the rate of 1 revolution per second. With each foot, she exerts an average force of 80 N in the tangential direction on each pedal of radius 0.16 m. What is the mechanical power she delivers to the pedals?

38.——The engine of an automobile delivers a maximum torque of 670 N · m when running at 4600 rev/min and it delivers a maximum power of 142 hp when running at 5750 rev/min. What power does the engine deliver when running at maximum torque? What torque does it deliver when running at maximum power?

39.——The Wright Cyclone engine on a DC-3 airplane delivers a power of 850 hp with the propeller revolving steadily at 2100 rev/min. What is the torque exerted by air resistance on the propeller?

40.*——The flywheel of a motor is connected to the flywheel of a pump by a drive belt (Fig. 7.29). The first flywheel has a radius R_1, and the second a radius R_2. The pump offers resistance to the rotational motion, and the motor has to exert a constant torque to drive the pump. While the motor wheel is rotating at a constant angular velocity ω_1, the tensions in the upper and the lower portions of the drive belt are T and T', respectively. Assume that the drive belt is massless.

(a) What is the angular velocity of the pump wheel?

(b) What is the torque of the drive belt on each wheel?

(c) By taking the product of torque and angular velocity, calculate the power delivered by the motor to the drive belt, and the power removed by the pump from the drive belt. Are these powers equal?

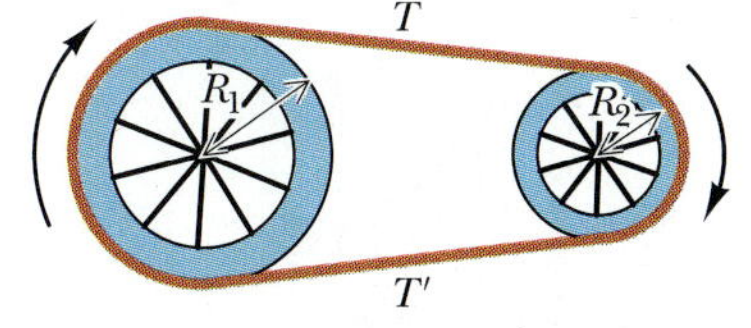

FIGURE 7.29

Motor and pump wheels connected by a drive belt.

Section 7.5: The Equation of Rotational Motion

41.——While starting up a roulette wheel, the croupier exerts a torque of 100 N · m with his hand on the spokes of the wheel. What angular accelera-

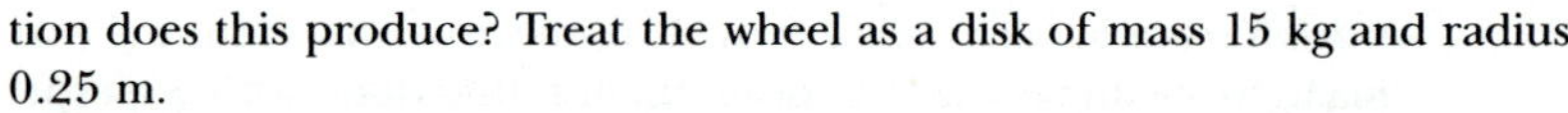

tion does this produce? Treat the wheel as a disk of mass 15 kg and radius 0.25 m.

42.—With your bicycle upside down on the ground, and the wheel free to rotate, you grasp the front wheel at the top and give it a horizontal push of 20 N. What is the instantaneous angular acceleration of the wheel? The wheel is a hoop of mass 4.0 kg and radius 0.33 m; ignore the mass of the spokes.

43.—Because of friction, the turntable of a phonograph gradually slows down when the motor is disengaged and finally stops. The initial angular speed of the turntable is 33.3 rev/min and the time it takes to stop is 1.2 min. The turntable is a disk of mass 2.0 kg and radius 15 cm. What is the average frictional torque on the turntable?

The original Ferris wheel.

44.—The original Ferris wheel, built by George Ferris, had a radius of 38 m and a mass of 9.7×10^5 kg. Assume that all of its mass was uniformly distributed along the rim of the wheel. If the wheel was initially rotating at 0.5 rev/min, what constant torque had to be applied to bring it to a full stop in 30 s? What force exerted on the rim of the wheel would have given such a torque?

45.*—Consider the automobile wheel discussed in Example 10. From the data given in this example, calculate the magnitude of the horizontal push P that the axle exerts on the wheel. Compare your result with the push that would be required if the bottom of the wheel were to slip on a frictionless road, and the wheel were to accelerate without rotating.

46.*—The center span of a revolving drawbridge consists of a uniform steel girder of mass 300 metric tons and length 25 m. This girder can be regarded as a uniform thin rod. The bridge opens by rotating about a vertical axis through its center. What torque is required to open this bridge in 60 s? Assume that the bridge first accelerates uniformly through an angular interval of 45° and then decelerates uniformly through an angular interval of 45° so as to come to rest after rotating by 90°.

47.**—A uniform disk of mass M is free to rotate about a fixed horizontal axis. A string is wrapped around the rim of this disk and a mass m is attached to this string (Fig. 7.30). What is the downward acceleration of the mass?

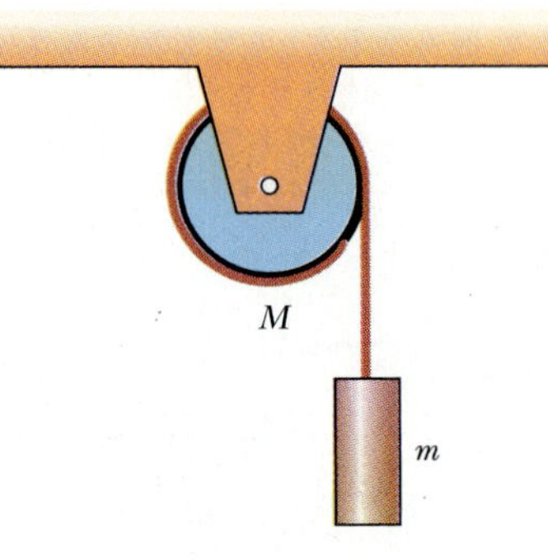

FIGURE 7.30
A mass m attached to a disk.

Section 7.6: Conservation of Angular Momentum

48.—What is the angular momentum of a Frisbee spinning at 20 rev/s about its axis of symmetry? Treat the Frisbee as a uniform disk of mass 200 g and radius 15 cm.

49.—You spin a hard-boiled egg on a table, at 5 rev/s. What is the angular momentum of the egg? Treat the egg as a sphere of mass 70 g and mean diameter 5 cm.

50.—The wheel of an automobile has a mass of 25 kg and a diameter of 70 cm. Assume that the wheel can be regarded as a uniform disk. What is the angular momentum of the wheel about its axle when the automobile is traveling at 25 m/s (55 mi/h) on a straight road?

51.—A phonograph turntable is a uniform disk of radius 15 cm and mass 1.4 kg. If this turntable accelerates from 0 rev/min to 78 rev/min in 2.5 s, what is the average rate of change of the angular momentum in this time interval?

52.—The propeller shaft of a cargo ship has a diameter of 8.8 cm, a length of 26 m, and a mass of 1200 kg. What is the rotational kinetic energy of this propeller shaft when it is rotating at 200 rev/min? What is the angular momentum?

53.—At the Fermilab accelerator, protons of momentum 5.3×10^{-16} kg·m/s travel around a circular path of diameter 2.0 km. What is the orbital angular momentum of one of these protons? Assume that the origin is at the center of the circle.

54.——Prior to launching a stone from a sling, a Bolivian native whirls the stone at 3.0 revolutions per second around a circle of radius 0.75 m. The mass of the stone is 0.15 kg. What is the angular momentum of the stone relative to the center of the circle?

55.——A communications satellite of mass 100 kg is in a circular orbit of radius 4.22×10^7 m around the Earth. The orbit is in the equatorial plane of the Earth and the satellite moves along it from west to east with a speed of 4.90×10^2 m/s. What is the magnitude of the angular momentum of this satellite?

56.——According to Bohr's (oversimplified) theory, the electron in the hydrogen atom moves in one or another of several possible circular orbits around the nucleus. The radii and the orbital velocities of the three smallest orbits are, respectively, 0.529×10^{-10} m, 2.18×10^6 m/s; 2.12×10^{-10} m, 1.09×10^6 m/s; and 4.76×10^{-10} m, 7.27×10^5 m/s. For each of these orbits calculate the orbital angular momentum of the electron, with the origin at the center. How do these angular momenta compare?

57.*—According to a simple (but erroneous) model, the proton is a uniform rigid sphere of mass 1.67×10^{-27} kg and radius 1.0×10^{-15} m. The spin angular momentum of the proton is 5.3×10^{-35} J·s. According to this model, what is the angular velocity of rotation of the proton? What is the linear velocity of a point on its equator? What is the rotational kinetic energy?

58.**—There are 1.1×10^8 automobiles in the United States, each of an average mass of 2000 kg. Suppose that one morning all these automobiles simultaneously start to move in an eastward direction and accelerate to a speed of 80 km/h.

(a) What change in the total angular momentum about the axis of the Earth do all these automobiles contribute together? Assume that the automobiles travel at an average latitude of 40°.

(b) How much will the rate of rotation of the Earth change because of the action of these automobiles? Assume that the axis of rotation of the Earth remains fixed. The moment of inertia of the Earth is 8.1×10^{37} kg·m^2.

59.**—A flywheel rotating freely on a shaft is suddenly coupled by means of a drive belt to a second flywheel sitting on a parallel shaft (Fig. 7.31). The initial angular velocity of the first flywheel is ω; that of the second is zero. The flywheels are uniform disks of masses M_1, M_2, and of radii R_1, R_2, respectively. The drive belt is massless and the shafts are frictionless and exert no torques.

(a) Calculate the final angular velocity of each flywheel.

(b) Calculate the kinetic energy lost during the coupling process. What happens to this energy?

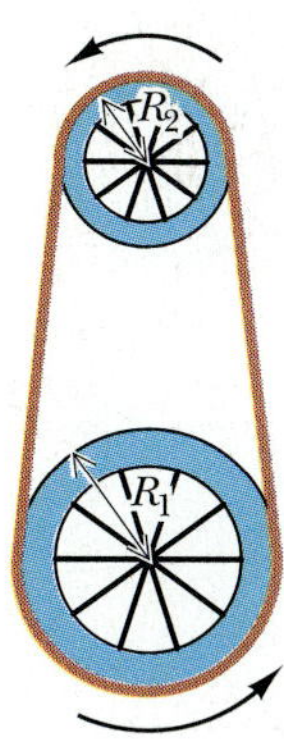

FIGURE 7.31
Two flywheels coupled by a drive belt.

60.**—A woman stands in the middle of a small rowboat. The rowboat is floating freely and experiences no friction against the water. The woman is initially facing east. If she turns around 180° so that she faces west, through what angle will the rowboat turn? Assume that the woman performs her turning movement at constant angular velocity and that her moment of inertia remains constant during this movement. The moment of inertia of the rowboat about the vertical axis is 20 kg·m^2 and that of the woman is 0.80 kg·m^2.

Chapter 8

The Pont du Gard at Nîmes, France was built by Roman engineers in the 1st century A.D. as a combined bridge and aqueduct. The roadway (lower level) still carries traffic.

STATICS AND ELASTICITY

Engineers and architects concerned with the design of bridges, buildings, and other structures need to know under what conditions a body will remain at rest, even when forces act on it. For instance, the designer of a railroad bridge must make sure that the bridge will not tip over or break when a heavy train passes over it. ***A body that remains at rest, even though several forces act on it, is said to be in equilibrium.*** The branch of physics that studies the conditions for the equilibrium of a body is called ***statics.*** Statics is the oldest branch of physics. The ancient Egyptians, Greeks, and Romans had a good grasp of the basic principles of statics, as is evident from their construction of elegant arches for doorways and bridges. The oldest surviving physics textbook is a treatise on the statics of ships by Archimedes.

In the first three sections of this chapter, we will rely on the assumption that the "rigid" structural members — such as beams and columns — indeed remain rigid; that is, they do not deform. In essence, this means that we assume that the forces are not so large as to produce a significant bending or compression of the beams or columns. However, in the last section, we will take a brief look at the phenomenon of the deformation of solid bodies when subjected to the action of large forces.

8.1 STATICS OF RIGID BODIES

If a rigid body is to remain at rest, its translational and rotational accelerations must be zero. Hence, the condition for the static equilibrium of a rigid body is that ***the sum of external forces and the sum of external torques on the body must be zero.*** This means that the forces and the torques are in balance; each force is compensated by some other force or forces, and each torque is compensated by some other torque or torques. For example, when a bowling ball rests in the palm of your hand (Fig. 8.1), the external forces on the ball are its (downward) weight and your (upward) push. If the ball is to remain at rest, these two forces must add to zero, which tells us that the forces are of equal magnitudes. Likewise, the torques of these forces must add to zero. If we choose an axis passing through the center of the ball, each of the two external forces has zero torque about this axis, and the sum of torques is zero plus zero, so the equilibrium condition for the torques is trivially satisfied. But if we choose as axis passing through some point in space, at some horizontal distance from the center of the ball (see Fig. 8.1), the individual torques of the two external forces are not zero. Nevertheless, the sum of these torques is still zero, because, with the choice of axis in Fig. 8.1, the weight of the ball exerts a counterclockwise (positive) torque and the push of your hand exerts a clockwise (negative) torque of the same magnitude.

◀ **Condition for static equilibrium**

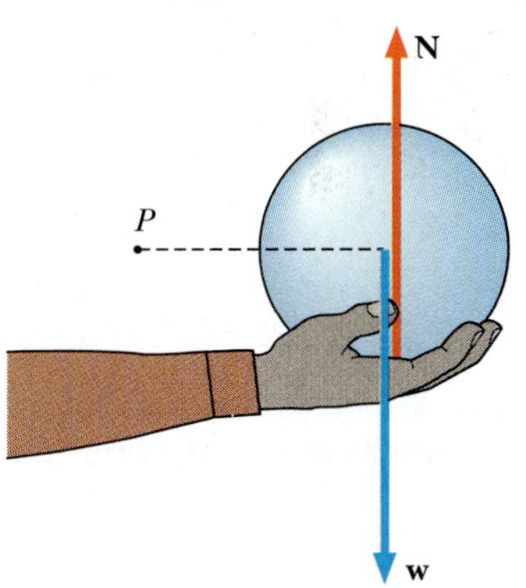

FIGURE 8.1
A bowling ball at rest in your hand. The external forces add to zero. The external torques about any axis, such as an axis passing through the point P and perpendicular to the page, also add to zero.

Note that for the purposes of statics, ***any line through the body or any line passing at some distance from the body can be thought of as a conceivable axis of rotation; and the torque about every such axis must be zero.*** This means we have complete freedom in the choice of the axis of rotation, and we can make whatever choice seems convenient. With some practice, one learns to recognize which choice of axis will be most useful for the solution of a problem in statics.

The force of gravity plays an important role in many problems of statics. The force of gravity on a body is distributed over all parts of the body, each part being subjected to a force proportional to its mass. However, for the calculation of the torque exerted by gravity on a rigid body, ***the entire gravitational force may be regarded as acting on the center of mass.*** The proof of this rule is easy: suppose that we release the rigid body and permit it to fall freely from an initial condition of rest. Since all the particles in the body fall at the same rate, the body will not change its orientation as it falls. The absence of angular acceleration implies that gravity does not generate any torque about the center of mass. Hence, if we want to simulate gravity by a single force acting at one point of the rigid body, that point will have to be the center of mass, so this single force

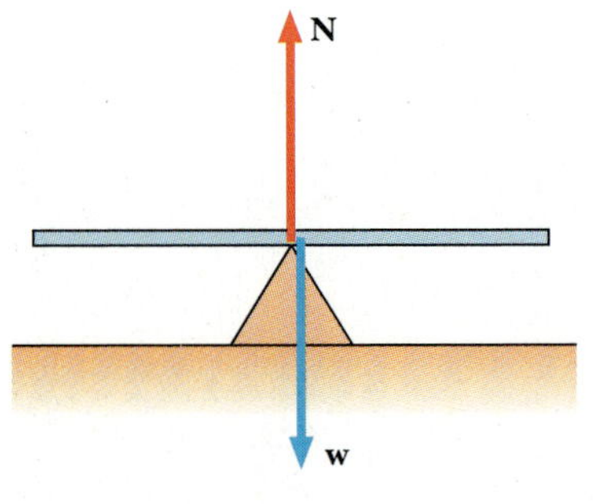

FIGURE 8.2

A seesaw supported at its center of mass is in equilibrium.

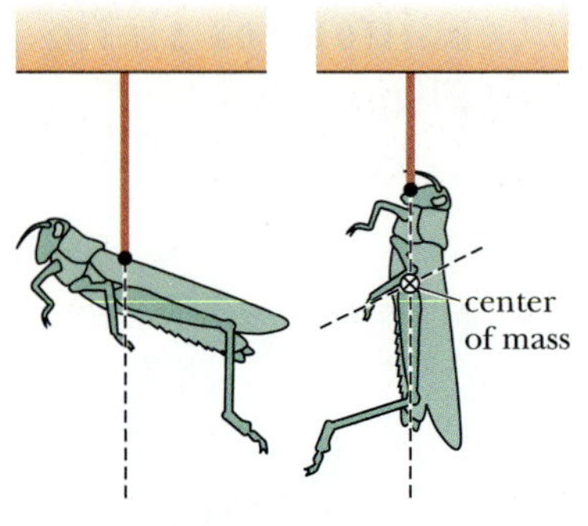

FIGURE 8.3

A locust suspended from a string glued to its exoskeleton. The two diagrams illustrate the equilibrium position for two different points of attachment of the string.

does not generate any torque either. Both gravity and the single force that replaces it then produce exactly the same rotational motion about the center of mass (namely, no motion), and they are therefore equivalent in regard to the equations of rotational motion of the body.

Given that in a rigid body the force of gravity effectively acts on the center of mass, we see that a rigid body supported at the center of mass is in equilibrium. For instance, a seesaw supported at its center of mass is in equilibrium (Fig. 8.2), because both the downward forces of gravity and the upward supporting force provided by the pivot act at the same point, and they exert no torque.

More generally, a rigid body supported by an upward force acting at a point on the vertical line through its center of mass is in equilibrium, since the support force is then collinear with the effective force of gravity, and such collinear forces of equal magnitudes and opposite directions exert no torque. This gives us a simple method for the experimental determination of the center of mass of a body of complicated shape: Suspend the body from a string attached to a point of its surface (Fig. 8.3); the body will then settle into an equilibrium position such that the center of mass is on the vertical downward prolongation of the string (this vertical prolongation is marked dashed in Fig. 8.3). Next, suspend the body from a string attached at another point of its surface, and mark the new vertical prolongation of the string. The center of mass is then at the intersection of the new and the old prolongations of the string.

Stable, unstable, and neutral equilibrium ▶

A body suspended from a point above its center of mass, as in Fig. 8.4a, is in stable equilibrium. If we turn this body through some angle, so the center of mass is not any more vertically below the point of support, the force of gravity and the supporting force will produce a torque that tends to return the body to the equilibrium position. In contrast, ***if this body is supported by a force applied at a point below the center of mass, as in Fig. 8.4b, the body is in unstable equilibrium.*** If we turn the body ever so slightly, the force of gravity and the supporting force will produce a torque that tends to turn the body farther away from the equilibrium position — the body tends to topple over. Finally, ***a body supported at its center of mass is in neutral equilibrium.*** If we turn such a body, it remains in equilibrium in its new position, and exhibits no tendency to return to its original position or to turn farther away.

Similar stability criteria apply to a body resting on a surface. A body is in stable equilibrium if it resists small disturbances and tends to return to its original position when the disturbances cease. A car sitting on flat ground is an example of this kind of equilibrium; if we lift one corner of the car and then let go, the car falls back to its original position. A body is in unstable equilibrium if it offers no resistance to a disturbance and tends to move away from its original position when disturbed. An upright

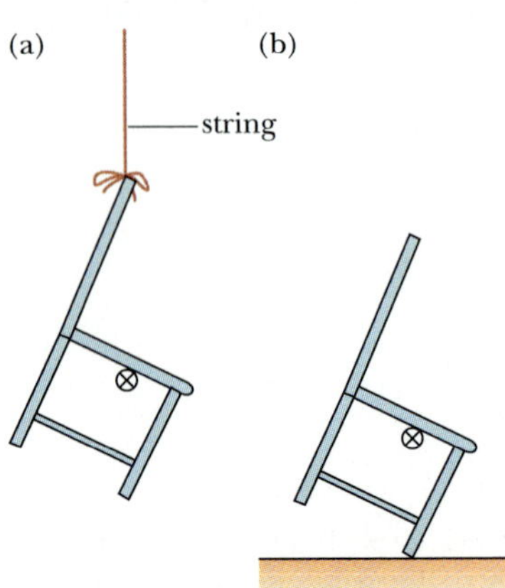

FIGURE 8.4 (a) A chair suspended from a point directly above its center of mass is in stable equilibrium. (b) A chair supported at a point directly below its center of mass is in unstable equilibrium.

bicycle is an example of this second kind of equilibrium (see Fig. 8.5). If we push the cyclist from a side, he topples over (unless he steadies himself by some nonrigid motion, by turning the front wheel or shifting his body to restore the balance).

FIGURE 8.5

An upright bicycle is in unstable equilibrium.

8.2 EXAMPLES OF STATIC EQUILIBRIUM

The following are some examples of solutions of problems in statics. In these examples, the condition of a zero sum of external forces and of external torques is used either to find the magnitudes of the forces that hold the body in equilibrium or to find whether the body can achieve equilibrium at all.

EXAMPLE 1. A locomotive of 90,000 kg is one-third of the way across a bridge 90 m long. The bridge consists of a uniform iron girder of 900,000 kg, which rests on two piers (see Fig. 8.6a). What is the load on each pier?

SOLUTION: The body whose equilibrium we must investigate is the bridge. Figure 8.6b is a "free-body" diagram for the bridge, showing all the forces acting on it. The weight of the bridge is shown acting at the center of mass. The bridge is static, and hence the net torque on the bridge reckoned about any point must be zero. Let us first consider the torques about the point P_2, at the right pier. These torques are generated by the weight of the bridge acting at a distance of 45 m, the weight of the locomotive acting at a distance of 30 m, and the upward thrust $\mathbf{F}_1$ of the pier at P_1 acting at a distance of 90 m (the upward thrust $\mathbf{F}_2$ generates no torque about P_2). The weight of the bridge is $m_{\text{brid}}g = 9 \times 10^5 \text{ kg} \times g$, and the weight of the locomotive is $m_{\text{loc}}g = 9 \times 10^4 \text{ kg} \times g$. Since each of the forces acts at right angles to the (horizontal) line from P_2 to the point of application of the force, the torque for each force is simply the product of the distance and the force. According to the equilibrium condition, we must set the sum of the three torques equal to zero:

$$-45\text{m} \times 9 \times 10^5\text{kg} \times g - 30\text{m} \times 9 \times 10^4\text{kg} \times g + 90\text{m} \times F_1 = 0 \qquad (1)$$

Here, the first two torques have been reckoned as negative since they tend to produce clockwise rotation, and the last torque has been reckoned as positive since it tends to produce counterclockwise rotation. Solving this equation for F_1, we find

$$F_1 = \frac{(45 \text{ m} \times 9 \times 10^5 \text{ kg} + 30 \text{ m} \times 9 \times 10^4 \text{ kg}) \times g}{90 \text{ m}} = 4.8 \times 10^5 \text{ kg} \times g$$

$$= 4.8 \times 10^5 \text{ kg} \times 9.81 \text{ m/s}^2 = 4.7 \times 10^6 \text{ N}$$

(a)

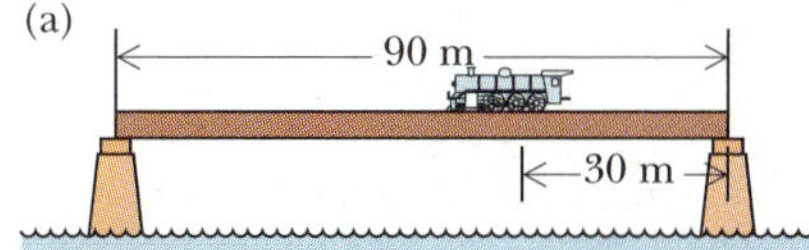

(b)

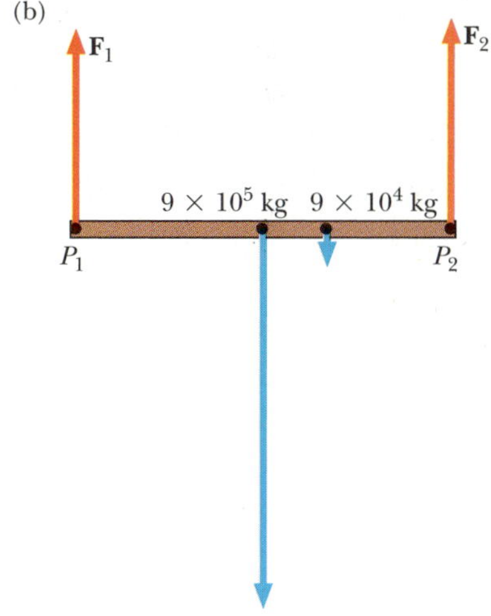

FIGURE 8.6

(a) Bridge with a locomotive on it. (b) "Free-body" diagram for the bridge.

Next, consider the torques about the point P_1. These torques are generated by the weight of the bridge, the weight of the locomotive, and the upward thrust $\mathbf{F}_2$ at point P_2 (the upward thrust of $\mathbf{F}_1$ generates no torque about P_1). Setting the sum of these three torques about the point P_1 equal to zero, we have

$$45\text{ m}\times 9\times 10^5\text{ kg}\times g + 60\text{ m}\times 9\times 10^4\text{ kg}\times g - 90\text{ m}\times F_2 = 0 \quad (2)$$

from which

$$F_2 = 5.0\times 10^6\text{ N}$$

COMMENTS AND SUGGESTIONS: Note that the net vertical upward force exerted by the piers is $F_1 + F_2 = 9.7\times 10^6$ N. It is easy to check that this matches the sum of the weights of the bridge and the locomotive; thus, the condition for zero net vertical force, as required for translational static equilibrium, is automatically satisfied. This automatic result for the equilibrium of vertical forces came about because we used the condition for rotational equilibrium twice. Instead, we could have used the condition for rotational equilibrium once [Eq. (1)], and then evaluated F_2 by means of the condition for translational equilibrium. The result for zero net torque about the point P_1 would then have emerged automatically.

From this example we see that the steps in the solution of a problem of statics resemble the steps we employed in Chapter 3:

1. The first step is the selection of the body which is to obey the equilibrium conditions.
2. Next, list all the external forces that act on this body, and display these forces on a "free-body" diagram.
3. If the forces have different directions, draw coordinate axes on the diagram, and resolve the forces into x and y components.
4. For each component, apply the static equilibrium condition: the sum of forces is zero.
5. Make a choice of axis rotation, calculate the torque of each force about this axis, and apply the static equilibrium condition: the sum of torques is zero. As mentioned in Section 8.1, any line through the body or any line passing near the body can be thought of as an axis of rotation; and the torque about every such axis must be zero. From the above example we see that sometimes it is convenient to consider two different axes of rotation, and to examine the separate equilibrium conditions of the torques for each of these axes.

EXAMPLE 2. The mast of a sailboat is held by two steel cables attached as shown in Fig. 8.7a. The front cable has a tension of 5.0×10^3 N. What is the tension in the rear cable? What force does the foot of the mast exert on the sailboat? Assume that the weight of the mast can be neglected and that the foot of the mast is hinged (and therefore exerts no torque).

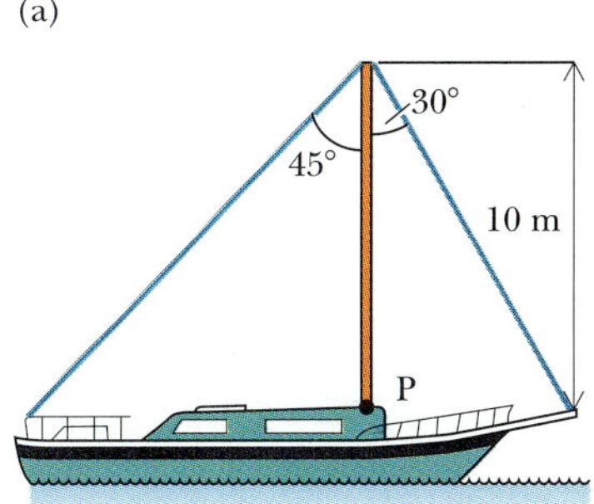

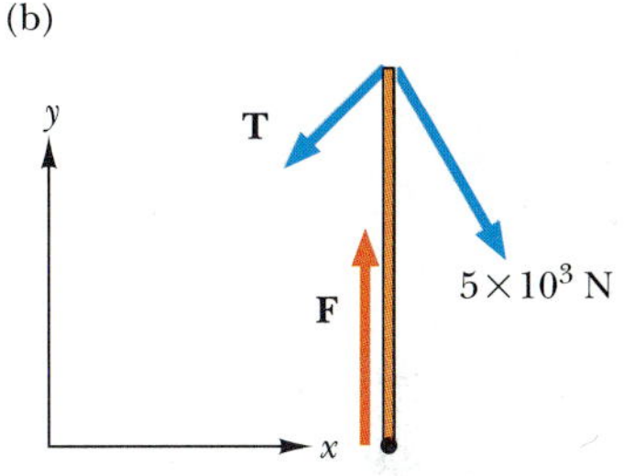

FIGURE 8.7

(a) Steel cables staying a mast.
(b) "Free-body" diagram for the mast.

SOLUTION: Figure 8.7b is a "free-body" diagram of the mast displaying the forces acting on it. To find the tension **T** in the rear cable, it is convenient to reckon the net torque on the mast about the point P, the foot of the mast. The force **F** exerts no torque about this point. The front cable pulls at a distance of 10 m at an angle of 30° (see Fig. 8.7a); hence it exerts a torque of $10 \text{ m} \times 5.0 \times 10^3 \text{ N} \times \sin 30°$. The rear cable pulls at the same distance, at an angle of 45°, and exerts a torque of $10 \text{ m} \times T \times \sin 45°$. According to the equilibrium condition, the sum of these torques must be zero:

$$-10 \text{ m} \times T \times \sin 45° + 10 \text{ m} \times 5.0 \times 10^3 \text{ N} \times \sin 30° = 0$$

From this we find

$$T = \frac{10 \text{ m} \times 5.0 \times 10^3 \text{ N} \times \sin 30°}{10 \text{ m} \times \sin 45°} \tag{3}$$

$$= 3.5 \times 10^3 \text{ N}$$

The net force on the mast must also be zero. The force **F** exerted by the boat against the mast has horizontal and vertical components F_x and F_y (however, as our calculations will show, the horizontal component is zero, and Fig. 8.7 has been drawn in anticipation of this result). Taking into account the horizontal and vertical components of the pulls exerted by the two cables, we find that the net horizontal force is

$$F_x + 5.0 \times 10^3 \text{ N} \cos 60° - 3.5 \times 10^3 \text{ N} \cos 45° = 0 \tag{4}$$

and the net vertical force is

$$F_y - 5.0 \times 10^3 \text{ N} \cos 30° - 3.5 \times 10^3 \text{ N} \cos 45° = 0 \tag{5}$$

From the first of these equations we find

$$F_x = 0$$

and from the second

$$F_y = 6.8 \times 10^3 \text{ N}$$

Thus, the force exerted by the boat on the mast is vertically upward, of magnitude 6.8×10^3 N. The force exerted by the mast on the boat is opposite to this (it is the reaction force); it is vertically downward, of the same magnitude.

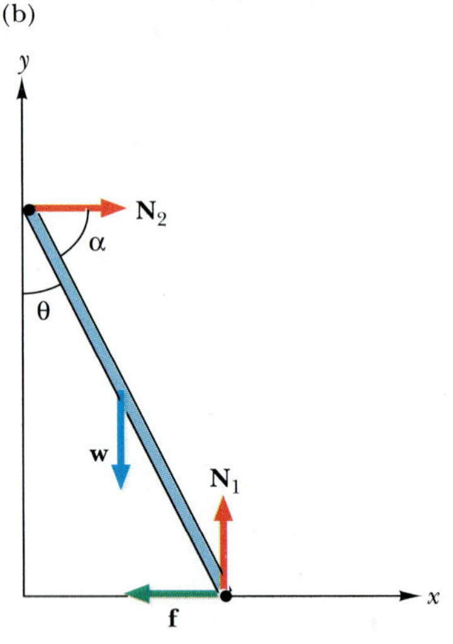

FIGURE 8.8
(a) Ladder leaning against a wall. (b) "Free-body" diagram for the ladder.

EXAMPLE 3. The bottom of a short ladder rests on the floor and the top rests against a wall (see Fig. 8.8a). If the coefficient of static friction between the ladder and the floor is $\mu_s = 0.4$ and the wall is frictionless, what is the maximum angle that the ladder can make with the wall without slipping?

SOLUTION: Figure 8.8b shows a "free-body" diagram for the ladder, with all the forces. If the ladder is about to slip, the friction force at the floor has its maximum magnitude

$$f = \mu_s N_1 \tag{6}$$

The weight of the ladder acts downward at the center of mass.

The sum of the torques must be zero. If we reckon the torques about the point of contact with the floor, the normal force $\mathbf{N}_1$ and the friction force $\mathbf{f}$ exert no torques about this point. The weight $w = Mg$ acting at the center of mass exerts a torque $(l/2) \times Mg \times \sin\theta$, and the normal force $\mathbf{N}_2$ of the wall exerts a torque $l \times N_2 \times \sin\alpha$, where α is the angle between the ladder and the normal force (see Fig. 8.8b); since $\alpha = 90° - \theta$, the sine of α equals the cosine of θ, and the torque equals $l \times N_2 \times \cos\theta$. Hence,

$$-\tfrac{l}{2}Mg \sin\theta + lN_2 \cos\theta = 0 \tag{7}$$

or, equivalently,

$$\tfrac{1}{2}Mg \sin\theta = N_2 \cos\theta \tag{8}$$

Dividing both sides of this equation by $\frac{1}{2}Mg \cos\theta$, we find

$$\frac{\sin\theta}{\cos\theta} = \frac{2N_2}{Mg}$$

or, since $\sin\theta/\cos\theta = \tan\theta$,

$$\tan\theta = \frac{2N_2}{Mg} \tag{9}$$

To evaluate the angle θ we still need to determine N_2. For this, we use the condition for translational equilibrium: the net vertical and the net horizontal forces must be zero:

$$N_1 - Mg = 0 \tag{10}$$

$$N_2 - \mu_s N_1 = 0 \tag{11}$$

From the first of these equations, $N_1 = Mg$; therefore, from the second equation, $N_2 = \mu_s Mg$. Inserting this into our expression (9) for the tangent of the angle θ, we obtain the final result

$$\tan\theta = \frac{2\mu_s Mg}{Mg} = 2\mu_s \tag{12}$$

With $\mu_s = 0.4$, this yields $\tan\theta = 0.8$. With our calculator, we find that the angle with this tangent is $\theta = 39°$. For any angle larger than this, the ladder will slip.

EXAMPLE 4. A uniform rectangular box 2 m high, 1 m wide, and 1 m deep stands on a flat floor. You push the upper end of the box to one side and then release it (see Fig. 8.9a). At what angle of release will the box begin to topple over on its side?

SOLUTION: The forces on the box when it has been released are as shown in Fig. 8.9b. Both the normal force **N** and the friction force **f** act at the bottom corner, which is the only point of contact of the box with the floor. The weight **w** acts at the center of mass, which is at the center of the box.

Since the box rotates about the bottom corner, let us consider the torque about this point. The only force that produces a torque about the bottom corner is the weight. This torque can be expressed as $d \times Mg$, where d is the perpendicular distance from the bottom corner to the vertical line through the center of mass (see Fig. 8.9b). This torque produces counterclockwise rotation if the center of mass is to the left of the bottom corner, and it produces clockwise rotation if the

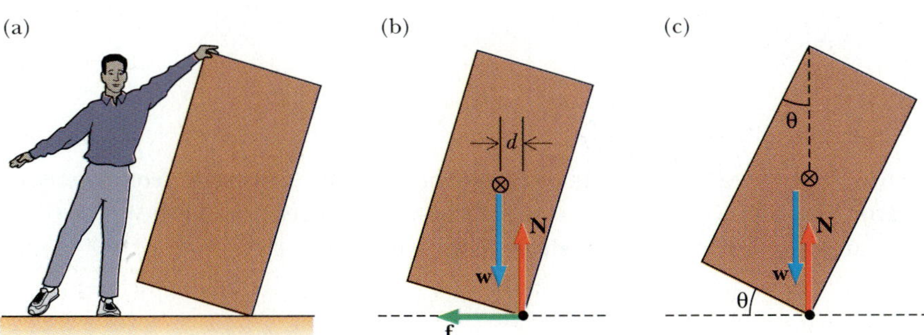

FIGURE 8.9
(a) Box standing on edge.
(b) "Free-body" diagram for the box. (c) "Free-body" diagram if the box is tilted at the critical angle. The center of mass is directly above the corner.

center of mass is to the right of the bottom corner. This means that in the former case, the box returns to its initial position, and in the latter case it topples over on its side. Thus, the critical angle corresponds to vertical alignment of the bottom corner and the center of the box (Fig. 8.9c). This critical angle equals the angle between the side of the box and the diagonal. The tangent of this angle is the ratio of the width and the height of the box,

$$\tan\theta = \frac{1.0 \text{ m}}{2.0 \text{ m}} = 0.5 \tag{13}$$

Our calculator tells us that the critical angle is then $\theta = 27°$.

COMMENTS AND SUGGESTIONS: In this example we found that the box begins to topple over if its inclination is such that the center of mass is vertically aligned with the bottom corner. This is a special instance of the general rule that a rigid body resting on a surface (flat or otherwise) becomes unstable when its center of mass is vertically aligned with the outermost point of support.

8.3* LEVERS AND PULLEYS

Lever ▸

A ***lever*** consists of a rigid bar swinging on a pivot (see Fig. 8.10). If we apply a force **F** at the long end, the short end of the bar pushes against a load with a larger force **F′**. Thus, the lever permits us to lift a larger load than we could with our bare hands. The relationship between the magnitudes of the forces **F** and **F′** follows from the condition for static equilibrium of the lever: the net torque about the pivot point P must be zero. Since, for the arrangement shown in Fig. 8.10, both forces **F** and **F′** are at right angles to the distances l and l', the condition on the net torque is

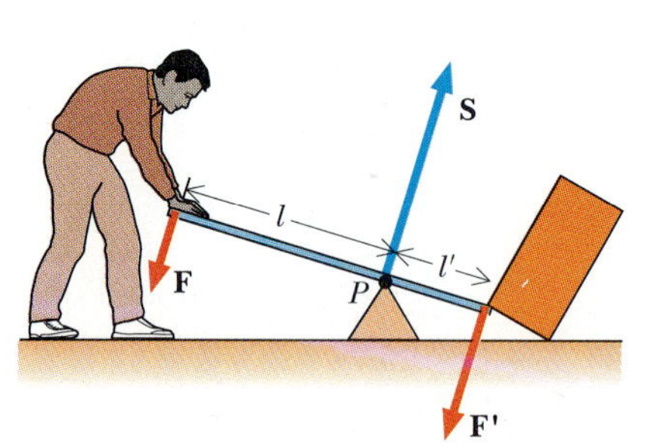

FIGURE 8.10

A lever. The vectors show the forces acting on the lever; **F′** is our push, **F** is the push of the load, and **S** is the supporting force of the pivot. The force exerted by the lever on the load is of the same magnitude as **F′**, but of opposite direction.

$$Fl - F'l' = 0 \tag{14}$$

from which we find

$$\frac{F'}{F} = \frac{l}{l'} \tag{15}$$

Thus, ***the forces are in the inverse ratio of the distances from the pivot point.*** For a powerful lever, we must make the lever arm l as long as possible and the lever arm l' as short as possible. ***The ratio F'/F of the magni-***

* This section is optional.

tudes of the force delivered by the lever and the force we must supply is called the mechanical advantage.

◂ **Mechanical advantage**

Apart from its application in the lifting of heavy loads, the principle of the lever finds application in many hand tools, such as pliers and bolt cutters. The handles of these tools are long, and the working ends are short, yielding an enhancement of the force exerted by the hand (see Fig. 8.11). A simple manual winch also relies on the principle of the lever. The handle of the winch is long and the drum of the winch, which acts as the short lever arm, is small (see Fig. 8.12). The force the winch delivers to the rope attached to the drum is then larger than the force exerted by the hand pushing on the handle. Compound winches, used for trimming sails on sailboats, have internal sets of gears which provide a larger mechanical advantage; in essence, such compound winches stagger one winch within another, so the force ratio generated by one winch is further multiplied by the force ratio of the other.

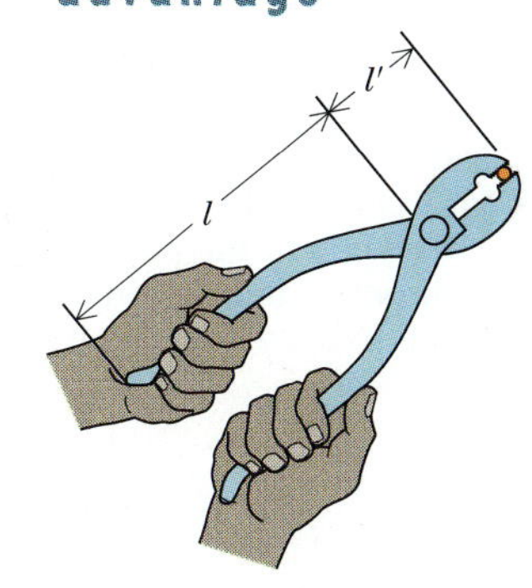

FIGURE 8.11
A pair of pliers serves as lever.

In the human body, many bones play the roles of levers which permit muscles or groups of muscles to support or to move the body. For example, Fig. 8.13 shows the bones of the foot; these act as a lever, hinged at the ankle. The rear end of this lever, at the heel, is tied to the muscles of the calf by the Achilles tendon, and the front end of the lever is in contact with the ground, at the ball of the foot. When the muscle contracts, it rotates the heel about the ankle, and presses the ball of the foot against the ground, thereby lifting the entire body on tiptoe. Note that the muscle is attached to the short end of this lever — the muscle must provide a larger force than the force generated at the ball of the foot. At first sight, it would seem advantageous to install a longer projecting spur at the heel of the foot and attach the Achilles tendon to the end of this spur; but this would require that the contracting muscle move through a longer distance. Muscle is good at producing large forces, but not so good at contracting over long distances, and the attachment of the Achilles tendon at the short end of the lever represents the best compromise. In most of the levers found in the human skeleton, the muscle is attached to the short end of the lever.

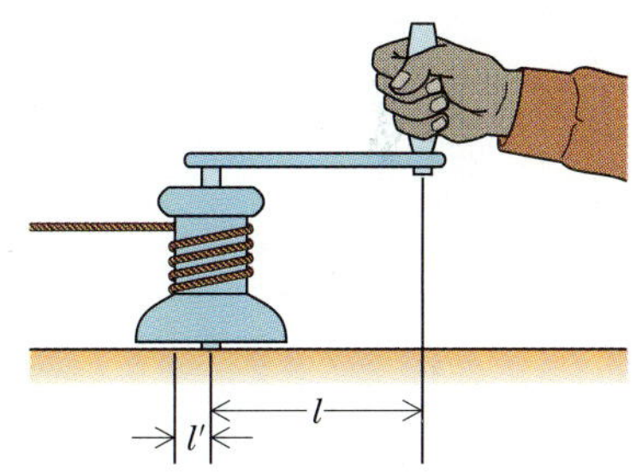

FIGURE 8.12
A manual winch.

Equation (15) is valid only if the forces are applied at right angles to the lever. A similar equation is valid if the forces are applied at some other angle, but instead of the lengths l and l' of the lever, we must substitute the lengths of the moment arms of the forces, that is, the perpendicular distances betwween the pivot point and the lines of action of the forces. These moment arms play the role of effective lengths of the lever.

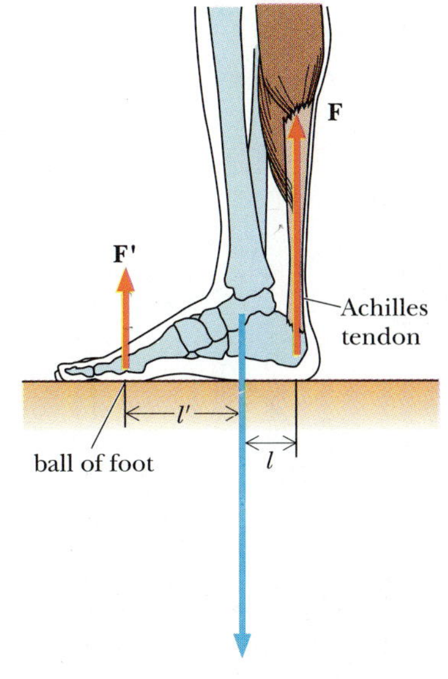

FIGURE 8.13
Bones of the foot acting as a lever.

EXAMPLE 5. When you bend over to pick up something from the floor, your backbone acts as a lever pivoted at the sacrum (see Fig. 8.14). The weight of the trunk pulls downward on this lever, and the muscles attached along the upper part of the backbone pull upward. The actual arrangement of the muscles is rather complicated, but for a simple mechanical model we can pretend that the muscles are equivalent to a string attached to the backbone at an angle of about 12° at a point beyond the center of mass (the other end of the

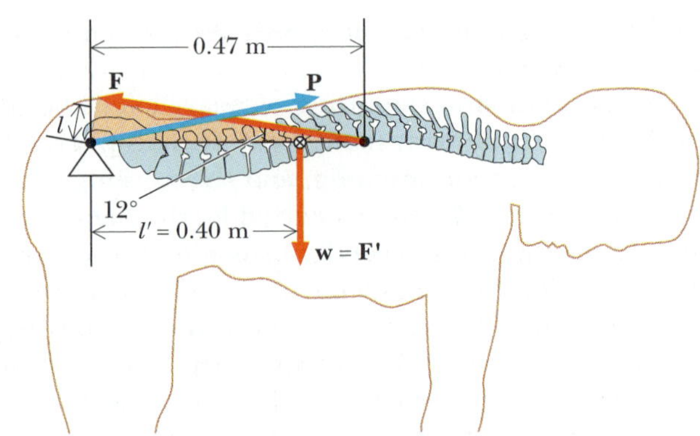

FIGURE 8.14
"Free-body" diagram for the backbone acting as lever. The forces on the backbone are the weight **w** of the trunk (including the weight of the backbone), the pull **F** of the muscles, and the thrust **P** of the pelvis acting as pivot.

"string" is attached to the pelvis). Assume that the mass of the trunk, including head and arms, is 48 kg, and that the dimensions are as shown in the diagram. What force must the muscles exert to balance the weight of the trunk when bent over horizontally?

SOLUTION: Figure 8.14 shows a "free-body" diagram for the backbone, with all the forces acting on it. Since the weight **w** of the trunk acts at right angles to the backbone, the lever arm for this weight is equal to the distance $l' = 0.40$ m between the pivot and the center of mass of the trunk. The lever arm for the muscle is the (small) distance l, which equals $l = 0.47\ \text{m} \times \sin 12° = 0.10$ m. According to Eq. (15), the magnitude of the force **F** exerted by the muscles is then

$$F = \frac{l'}{l}F' = \frac{l'}{l}Mg = \frac{0.40\ \text{m}}{0.10\ \text{m}} \times Mg = 4.0 \times Mg$$

$$= 4.0 \times 48\ \text{kg} \times 9.81\ \text{m/s}^2 = 1.9 \times 10^3\ \text{N}$$

This is a quite large force, 4.0 times larger than the weight of the trunk. Thus, bending over horizontally puts a severe stress on the muscles of the back. Furthermore, it puts an almost equally large compressional stress on the backbone, pulling it hard against the sacrum. The stresses are even larger if you try to lift a load from the floor while your body is bent over in this position. To avoid damage to the muscles and to the lumbosacral disk, it is best to lift by bending the knees, keeping the backbone vertical.

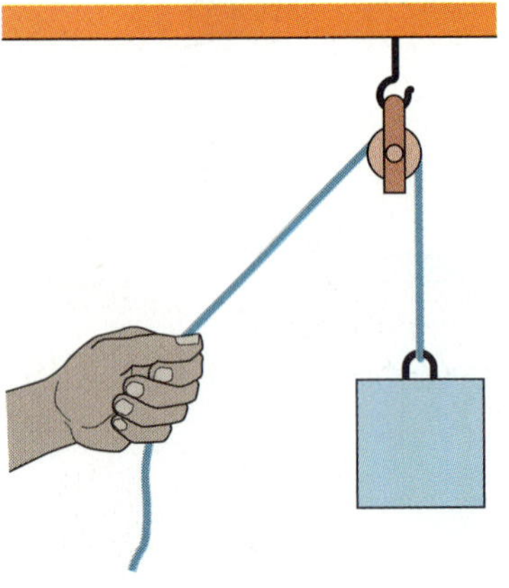

FIGURE 8.15
A single pulley.

Often, a force is applied to a load by means of a flexible rope, or a string. A pulley is then sometimes used to change the direction of the string or rope, and the direction of the force exerted on the body. If the pulley is frictionless, the tension at each point of a flexible rope passing over the pulley is the same. For instance, if we want to lift a load with a rope passing over a single pulley attached to the ceiling (see Fig. 8.15), the force we must exert on the rope has the same magnitude as the weight of the load. Thus, there is no gain of mechanical advantage in such an arrangement of a single pulley; the only benefit is that it permits us to pull more comfortably than if we attempted to lift the load directly.

However, an arrangement of several pulleys linked together, called block and tackle, can provide a large gain of mechanical advantage. For example, consider the arrangement of three pulleys shown in Fig. 8.16a; the axles of the two upper pulleys are bolted together, and they are linked to each other and to the third pulley by a single rope. If the rope segments linking the pulleys are parallel and there is no friction, then the mechanical advantage of this arrangement is 3; that is, the magnitudes of the forces **F** and **F′** are in the ratio of 1 to 3. This can be most easily understood by drawing the "free-body" diagram for the lower portion of the pulley system (Fig. 8.16b). In this diagram, the three ropes leading upward have been cut off and replaced by the forces exerted on them by the external (upper) portions of ropes. Since the tension is the same everywhere along the rope, the forces pulling upward on each of the three rope ends shown in the "free-body" diagram all have the same magnitude F, and thus the net upward force is $3F$.

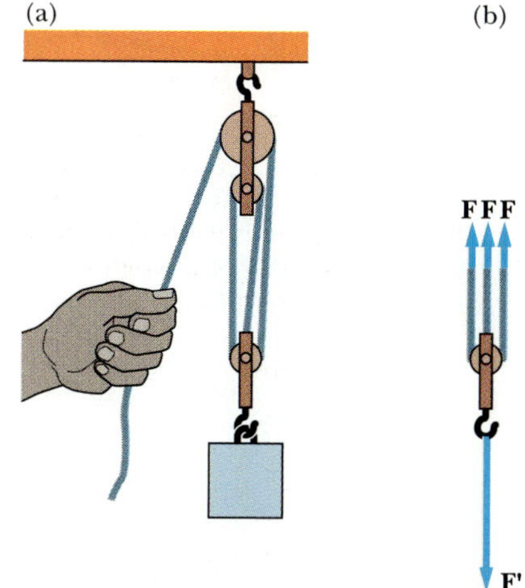

FIGURE 8.16
(a) Block and tackle. (b) "Free-body" diagram for the lower portion of the pulley system.

Block-and-tackle arrangements have many practical applications. They use used to provide the proper tension in overhead power cables for electric trains and trams (see Fig. 8.17); without such an arrangement, the cables would sag on warm days when thermal expansion increases their length, and they would be stretched excessively tight and perhaps snap on cold days, when they contract. One common cause of power failures on cold winter nights is the snapping of power lines lacking such compensating pulleys.

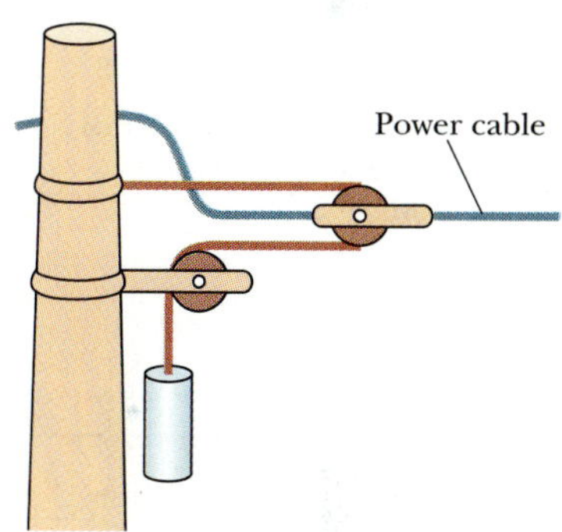

FIGURE 8.17
Block and tackle used for tensioning power line.

Another practical application of block and tackle is found in the traction devices used in hospitals to immobilize and align fractured bones, especially leg bones. A typical arrangement is shown in Fig. 8.18; here the pull applied to the leg is twice as large as the magnitude of the weight attached on the lower end to the rope.

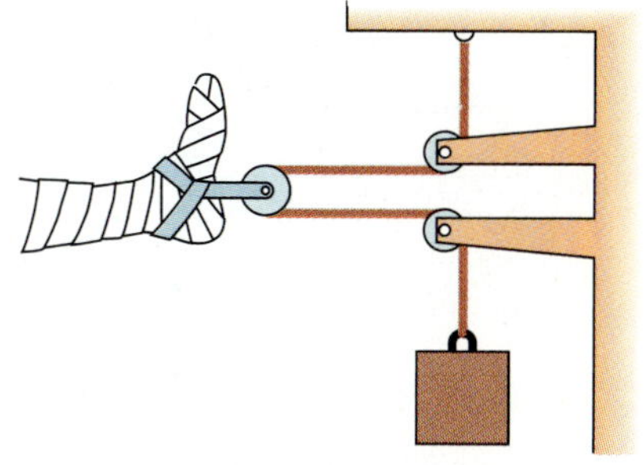

FIGURE 8.18
Block and tackle in traction apparatus for fractured leg.

The mechanical advantage provided by levers, arrangements of pulleys, or other devices can be calculated in a general and elegant way by appealing to the law of conservation of energy. A lever merely transmits the work we supply at one end to the load at the other end. We can express this equality of work input and work output by

$$F's' = Fs \tag{16}$$

where s is the displacement of our hand and s' the displacement of the load. According to this equation, the forces F' and F are in the inverse ratio of the displacements,

$$\frac{F'}{F} = \frac{s}{s'} \tag{17}$$

Consider, now, the rotation of the lever by a small angle (see Fig. 8.19). Since the two triangles included between the initial and final positions of the lever are similar, the distances s and s' are in the same ratio as the lever arms l and l'; thus, we immediately recognize from Eq. (17) that the mechanical advantage of the lever is l/l'.

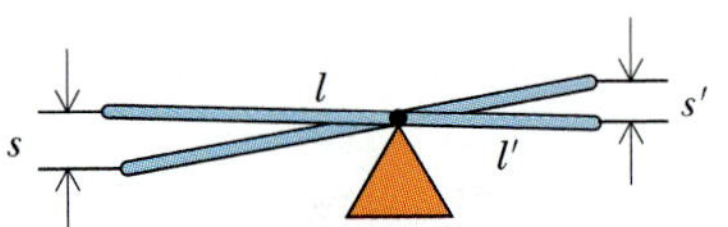

FIGURE 8.19

Rotation of the lever by a small angle produces displacements s and s' of the ends.

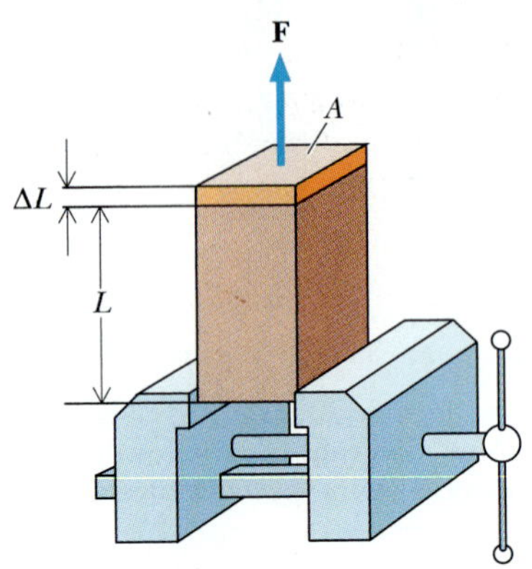

FIGURE 8.20

Tension applied to the end of a block of material causes elongation.

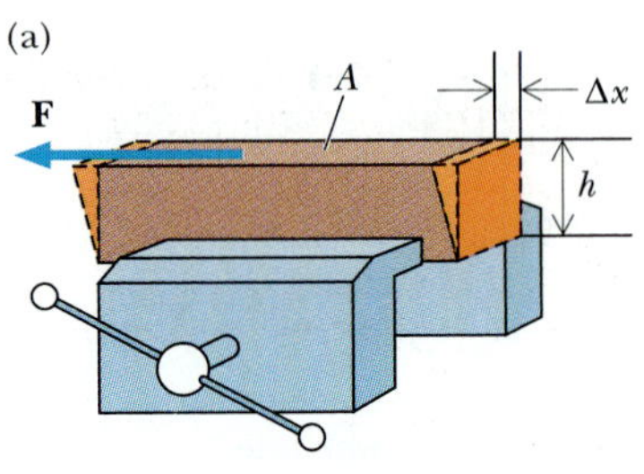

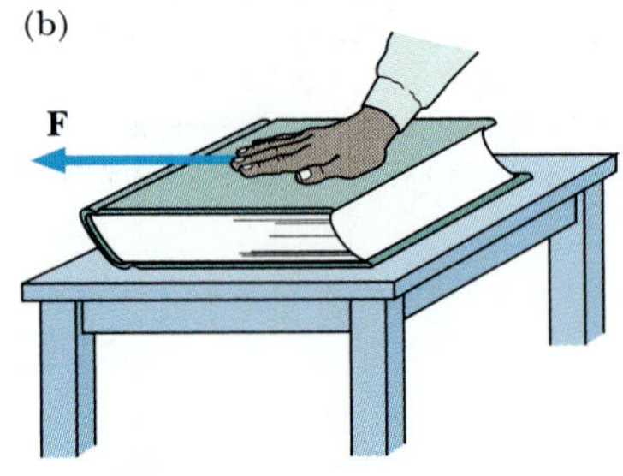

FIGURE 8.21

(a) Tangential force applied to the side of a block of material causes shear. (b) When such a tangential force is applied to the cover of a book, the pages slide past one another.

Likewise, we immediately recognize from Eq. (17) that the mechanical advantage of the arrangement of pulleys shown in Fig. 8.16 is 3, since whenever our hand pulls a length s of rope out of the upper pulley, the load moves upward by a distance of only $s/3$.

8.4* Elasticity of Materials

In our examples of bridges, masts of sailboats, etc., we assumed that the bodies on which the forces act are rigid; that is, they do not deform. Although solid bodies, such as bars or blocks of steel, are nearly rigid, they are not exactly rigid, and they will deform if a large enough force is applied to them. A solid bar may be thought of as a very stiff spring. If the force is fairly small, this "spring" will suffer only an insignificant deformation, but if the force is large, it will suffer a noticeable deformation. Provided that the force and the deformation remain within some limits, ***the deformation of a solid body is elastic, which means that the body returns to its original shape once the force ceases to act.*** Such elastic deformations of a solid body usually obey Hooke's Law: the deformation is proportional to the force. But the constant of proportionality is small, giving a small deformation unless the force is large.

A solid block of material can suffer several kinds of deformation, depending on how the force is applied. If one end of the body is held fixed, and the force pulls on the other end, the deformation is a simple ***elongation*** of the body (see Fig. 8.20). If one side of the body is held fixed, and the force pushes tangentially along the other side, then the deformation is a ***shear,*** which changes the shape of the body from a rectangular parallelepiped to a rhomboidal parallelepiped (see Fig. 8.21a). During this deformation, the parallel layers of the body slide past one another just as the pages of a book slide past one another when we push along its cover (see Fig. 8.21b). If the force is applied from all sides simultaneously, by subjecting the body to the pressure of a fluid in which the body is immersed, then the deformation is a ***compression*** of the volume of the body, without any change of the geometrical shape (see Fig. 8.22).

In all of these cases, ***the fractional deformation is directly proportional to the applied force and inversely proportional to the area over which the force is distributed.*** For instance, if a given force produces an elongation of 1% when pulling on the end of a block, then the same force pulling on the end of another block of, say, twice the cross-sectional area will produce an elongation of $\frac{1}{2}$%. This can be readily understood if we think of the block as consisting of parallel rows of atoms linked by springs, which represent the interatomic forces that hold the atoms in their places (see Fig. 8.23). When we pull on the end of the block with a given force, we stretch the interatomic springs by some amount; and when we pull on another block of twice the cross-sectional area, we have to stretch twice as many springs, and therefore the force acting on each spring is only half as large

* This section is optional.

and produces only half the elongation in each spring. Furthermore, since the force applied to the end of a row of atoms is communicated to all the interatomic springs in that row, a given force produces a given elongation in each spring in a row. The net elongation of the block is therefore proportional to the number of atoms in the row, which means it is proportional to the (initial) length of the block. Thus, if a block elongates by 0.1 mm when subjected to a given force, then another block, say, twice as long will elongate by 0.2 mm when subjected to the same force, which means the fractional elongation, or the percent elongation, remains the same.

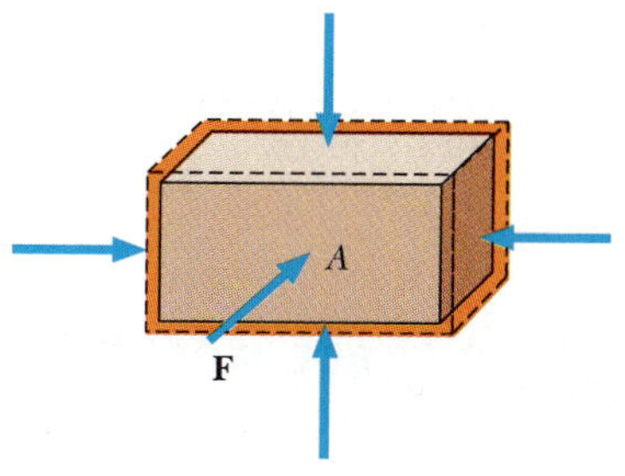

FIGURE 8.22
Pressure applied to all sides of a block of material causes compression.

To express the relationships among elongation, force, and area mathematically, consider a block of initial length L and cross-sectional area A (see Fig. 8.20). If a force F pulls on the end of this block, the elongation is ΔL, and the fractional elongation is $\Delta L/L$. This fractional elongation is directly proportional to the force and inversely proportional to the area A:

$$\frac{\Delta L}{L} = \frac{1}{Y}\frac{F}{A} \qquad (18)$$

◀ **Elongation**

Here the quantity Y is the constant of proportionality. In Eq. (18), this constant is traditionally written as $1/Y$, so it divides the right side, instead of multiplying it; thus, a stiff material, such as steel, that elongates by only a small amount, has a large value of Y. The constant Y is called ***Young's modulus.*** Table 8.1 lists values of Young's moduli for a few solid materials. Note that if, instead of exerting a pull on the end of the block, we exert a push, then F in Eq. (18) must be reckoned as negative, and the change ΔL of length will then likewise be negative — the block becomes shorter.

◀ **Young's modulus**

In engineering language, ***the fractional deformation is usually called the strain, and the force per unit area is called the stress.*** In this terminology, Eq. (18) simply states that the strain is proportional to the stress.

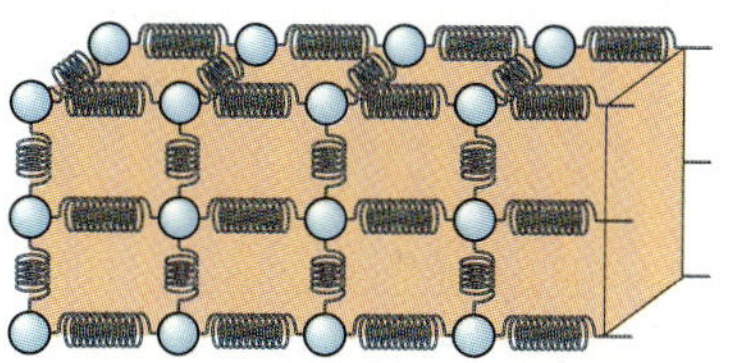

FIGURE 8.23
Microscopically, a block of solid material may be thought of as rows of atoms linked by springs. The springs stretch when a tension is applied to the block.

This proportionality of strain and stress is also valid for shearing deformations and compressional deformations, provided we adopt a suitable definition of strain, or fractional deformation, for these cases. For shear, the fractional deformation is defined as the ratio of the sideways displacement Δx of the edge of the block to the height h of the block (see Fig. 8.21a). This fractional deformation is directly proportional to the force F and inversely proportional to the area A (note that the relevant area A is

TABLE 8.1 ELASTIC MODULI OF SOME MATERIALS

Material	Young's modulus	Shear modulus	Bulk modulus
Steel	22×10^{10} N/m^2	8.3×10^{10} N/m^2	16×10^{10} N/m^2
Iron, cast	15	6.0	11
Aluminum	7.0	2.5	7.8
Nylon	0.36	0.12	0.59
Bone (long)	3.2	1.2	3.1
Water	—	—	0.22
Glycol	—	—	0.27

now the top area of the block, where the force is applied):

Shear ▶

$$\frac{\Delta x}{h} = \frac{1}{S}\frac{F}{A} \tag{19}$$

Shear modulus ▶

Here, the constant of proportionality S is called the ***shear modulus.*** Table 8.1 includes values of shear moduli of solids.

For compression, the fractional deformation is defined as the ratio of the change ΔV of the volume to the initial volume, and this fractional deformation is, again, proportional to the force F pressing on each face of the block and inversely proportional to the area A of that face:

Compression ▶

$$\frac{\Delta V}{V} = -\frac{1}{B}\frac{F}{A} \tag{20}$$

Bulk modulus ▶

In this equation, the minus sign indicates that ΔV is negative; that is, the volume decreases. The constant of proportionality B in the equation is called the ***bulk modulus.*** Table 8.1 includes values of bulk moduli for solids. This table also includes values of bulk moduli for two liquids. The formula (20) is equally valid for solids and for liquids — when we squeeze a liquid from all sides, it will suffer a compression. Note that Table 8.1 does *not* include values of Young's moduli and of shear moduli for liquids; ***elongation and shear stress are not supported by a liquid*** — we can elongate or shear a block of liquid as much as we please without having to exert any significant force.

EXAMPLE 6. A steel piano wire of length 1.8 m and radius 0.3 mm is subjected to a tension of 70 N by means of a weight attached to its lower end (see Fig. 8.24). By how much does this wire stretch in excess of its initial length?

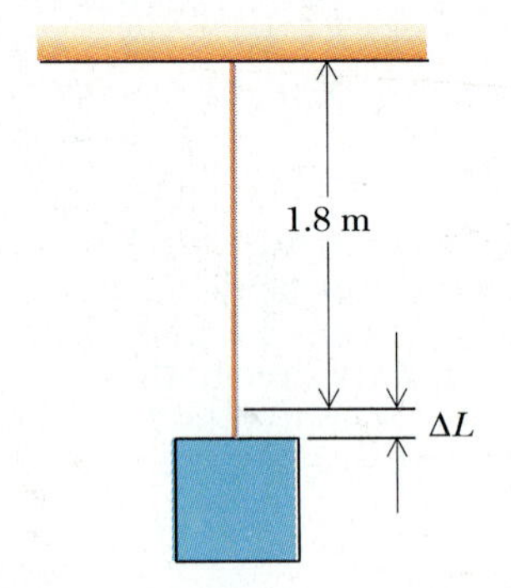

FIGURE 8.24
Elongation of a piano wire.

SOLUTION: The cross-sectional area of the wire is

$$A = \pi r^2 = \pi \times (0.0003 \text{ m})^2 = 2.8 \times 10^{-7} \text{ m}^2$$

and the force per unit area is

$$\frac{F}{A} = \frac{70 \text{ N}}{2.8 \times 10^{-7} \text{ m}^2} = 2.5 \times 10^8 \text{ N/m}^2$$

According to Table 8.1, the Young's modulus of steel is 22×10^{10} N/m². Hence, Eq. (18) yields

$$\frac{\Delta L}{L} = \frac{1}{Y}\frac{F}{A} = \frac{1}{22 \times 10^{10} \text{ N/m}^2} \times 2.5 \times 10^8 \text{ N/m}^2$$

$$= 1.1 \times 10^{-3}$$

The change of length is therefore

$$\Delta L = 1.1 \times 10^{-3} \times L = 1.1 \times 10^{-3} \times 1.8 \text{ m}$$

$$= 2.0 \times 10^{-3} \text{ m} = 2.0 \text{ mm}$$

EXAMPLE 7. What pressure must you exert on a sample of water if you want to compress its volume by 0.1%?

SOLUTION: By Eq. (20), the pressure, or the force per unit area, is

$$\frac{F}{A} = -B\frac{\Delta V}{V}$$

We want to achieve a fractional change of volume of $\Delta V/V = -0.001$. Since the bulk modulus of water is 0.22×10^{10} N/m^2, the required pressure is

$$\frac{F}{A} = 0.22 \times 10^{10} \text{ N/m}^2 \times 0.001 = 2.2 \times 10^6 \text{ N/m}^2$$

The simple uniform deformations of elongation, shear, and compression described above require a rather special arrangement of forces. In general, the forces applied to a solid body will produce non-uniform elongation, shear, and compression. For instance, a beam supported at its ends and sagging in the middle because of its own weight, or the weight of a load placed on it, will elongate along its lower edge and compress along its upper edge.

Finally, note that the formulas (18)–(20) are valid only as long as the deformation is reasonably small — a fraction of a percent or so. If the deformation is excessive, the material will be deformed beyond its elastic limit; that is, the material will suffer a permanent deformation and will *not* return to its original size and shape when the force ceases. If the deformation is even larger, the material will break apart or crumble. For instance, steel will break apart if the tensile stress exceeds 5×10^8 N/m^2, or if the shearing stress exceeds 2.5×10^8 N/m^2, and it will crumble if the compressive stress exceeds 5×10^8 N/m^2.

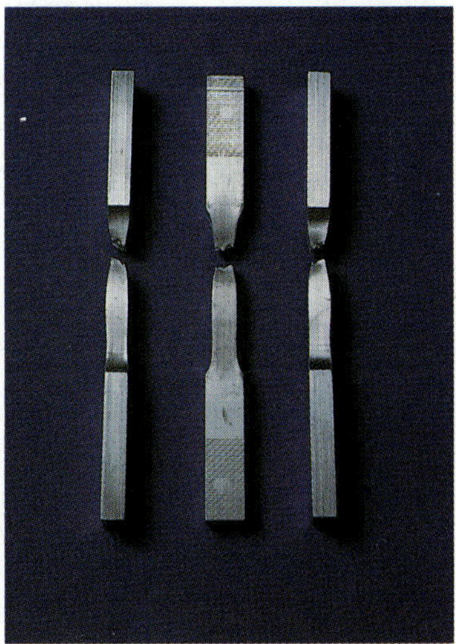

These steel rods broke apart when a large tension was applied.

SUMMARY

- **Static equilibrium:** The sums of the external forces and of the external torques on a rigid body are zero. Gravity effectively acts at the center of mass.

▶ **Mechanical advantage of lever:** $\dfrac{F'}{F} = \dfrac{l}{l'}$

▶ **Elongation of elastic material:** $\dfrac{\Delta L}{L} = \dfrac{1}{Y}\dfrac{F}{A}$

▶ **Shear:** $\dfrac{\Delta x}{h} = \dfrac{1}{S}\dfrac{F}{A}$

▶ **Compression of volume:** $\dfrac{\Delta V}{V} = -\dfrac{1}{B}\dfrac{F}{A}$

QUESTIONS

1.–In order to pull an automobile out of the mud in which it is stuck, a man stretches a rope tautly from the front end of the automobile to a stout tree. He then pushes sideways against the rope at the midpoint (see Fig. 8.25). Explain why the resulting force on the automobile is much larger than if the man were pushing directly on the automobile.

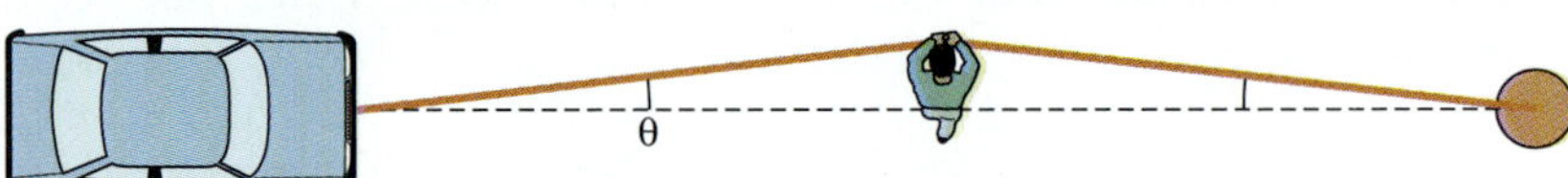

FIGURE 8.25
The rope is stretched between the automobile and the tree. The man pushes at the midpoint.

2.–The collapse of several "skywalks" at the Hyatt Regency hotel in Kansas City on July 17, 1981, with the loss of 114 lives, was due to a defective design of the suspension system. Instead of suspending the beams of the skywalks directly from single, long steel rods anchored at the top of the building, some incompetent engineers decided to use several short steel rods joining the beams of each skywalk to those of the skywalk above (Fig. 8.26). Criticize this design, keeping in mind that the beams are made of a much weaker material than the rods.

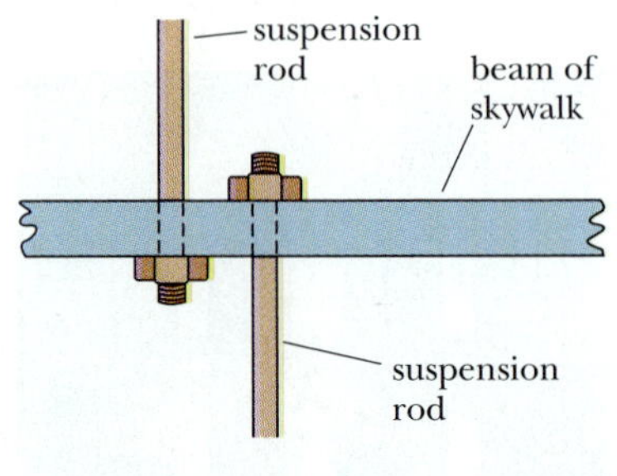

FIGURE 8.26
Beam of skywalk and suspension rods.

3.–If the legs of a table are of exactly the same length and if the floor is exactly flat, then the weight of the table will be equally distributed over all four legs. But if there are small deviations from exactness, then the weight will not be equally distributed. Is it possible for all of the weight to rest on three legs? On two?

4.–Consider a cone (a) lying on its side, (b) standing on its base, and (c) standing on its apex on a flat table (Fig. 8.27). For which of these positions is the equilibrium stable, unstable, or neutral?

5.–List as many examples as you can of joints in the human skeleton that act as pivots for levers. Do any of these levers in the human skeleton have a mechanical advantage larger than 1?

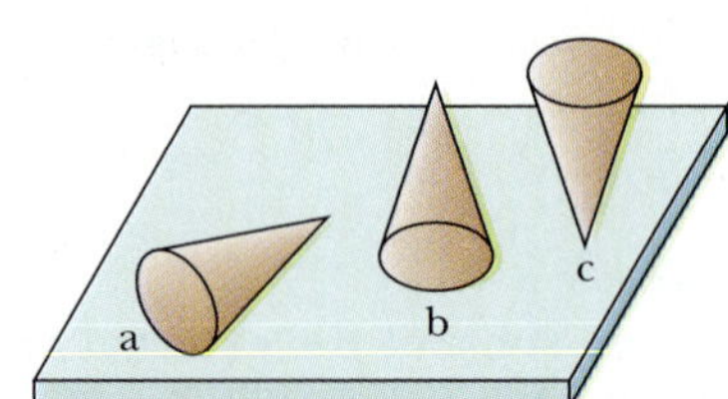

FIGURE 8.27
A cone lying on its side (a), standing on its base (b), and standing on its apex (c).

6.–Design a block and tackle with a mechanical advantage of 4.

7.–Figure 8.28 shows a differential windlass consisting of two rigidly joined drums around which the two ends of a rope are wound in opposite directions. By turning the crank, the operator raises the pulley and the load hanging from it. Explain why this device gives a very large mechanical advantage if the radii R_1 and R_2 of the two drums are nearly equal.

8.–A steel rod is much less flexible than a woven steel rope of the same diameter. Explain this.

9.–A carpenter wants to support the (flat) roof of a building with horizontal beams of wood of rectangular cross section. To achieve maximum strength of the roof (least sag), should he install the beams with their narrow side up or with their wide side up?

10.–The long bones in the limbs of vertebrates have the shape of hollow pipes. If the same amount of bone tissue had been assembled in a solid rod (of correspondingly smaller cross section), would the bone have been more rigid or less rigid?

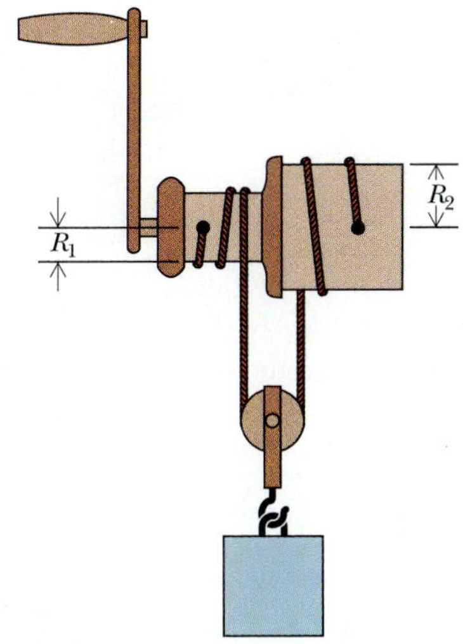

FIGURE 8.28
Differential windlass.

PROBLEMS

Section 8.2: Examples of Static Equilibrium

1.——The operating instructions for a small crane specify that when the boom is at an angle of 20° above the horizontal (Fig. 8.29), the maximum safe load for the crane is 500 kg. Assuming that this maximum load is determined by the maximum torque that the pivot can withstand, what is the maximum safe load for 40°? For 60°?

2.——A traffic lamp of 25 kg hangs from a wire stretched between two posts. The traffic lamp hangs at the middle of the wire, and the two halves of the wire sag downward at an angle of 20° (Fig. 8.30). What is the tension in the wire? Assume the wire is massless.

3.——A tractor pulls a trailer along a street (Fig. 8.31). The rear wheels, which are connected to the engine by means of the axle, have a radius of 0.60 m. Draw a "free-body" diagram for one of the rear wheels; be sure to include the forces and the torque exerted by the axle on the wheel, but neglect the weight of the wheel. If the tractor is to provide a pull of 8,000 N (a pull of 4,000 N from each wheel), what torque must the axle exert on each rear wheel?

4.——One end of a string is tied to a meter stick at the 80-cm mark and the other end is tied to a hook in the ceiling. You push against the bottom edge of the meter stick at the 30-cm mark, so the stick is held horizontally (Fig. 8.32). The mass of the meter stick is 0.24 kg. What is the magnitude of the force you must exert? What is the tension in the string?

5.——A beam of steel hangs from a crane by means of cables attached to the upper corners of the beam making an angle of 60° with each other. The

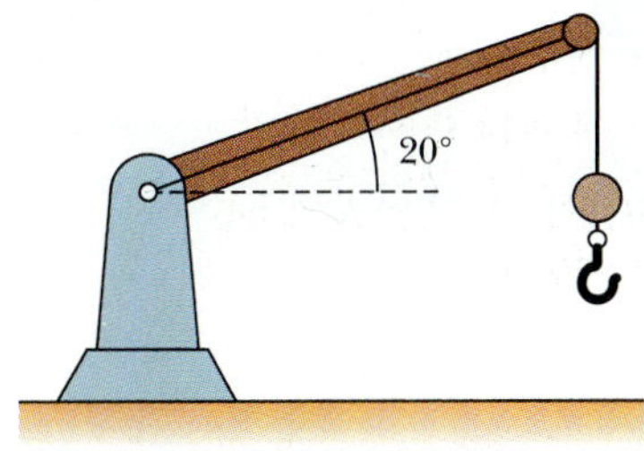

FIGURE 8.29
Small crane.

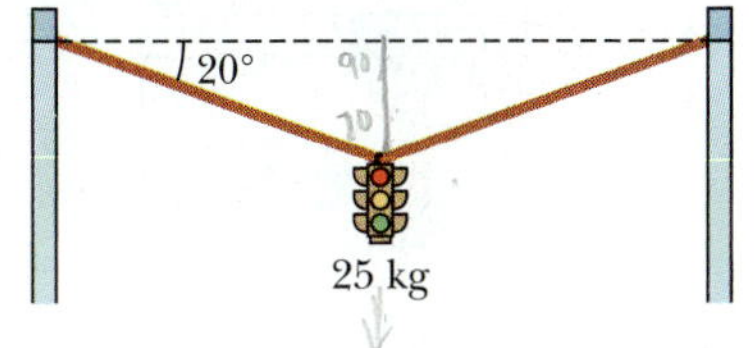

FIGURE 8.30
Traffic lamp.

FIGURE 8.31
Tractor.

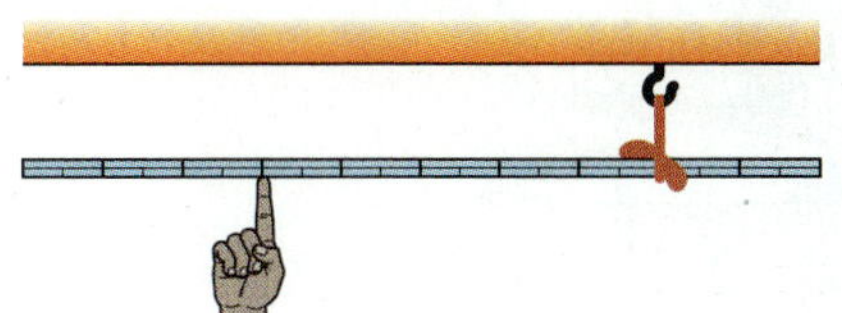

FIGURE 8.32
Meter stick tied to a hook.

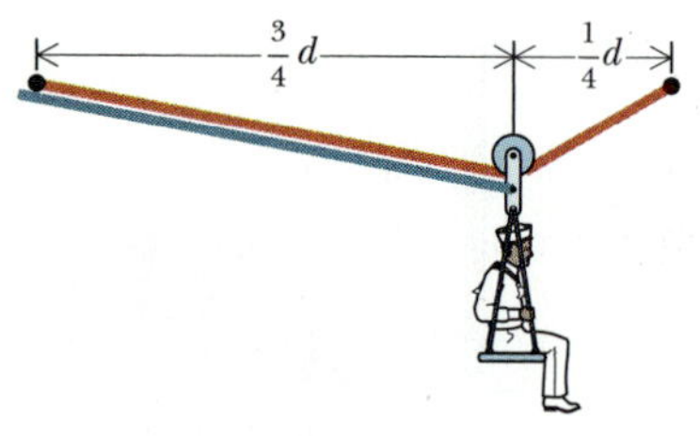

FIGURE 8.33
Sailor in bosun's chair.

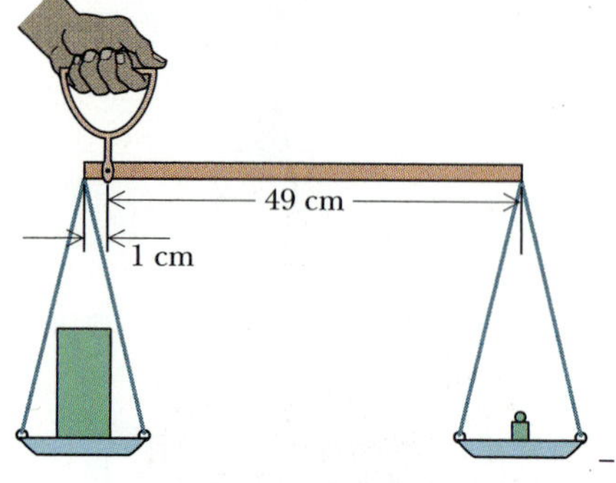

FIGURE 8.34
Unequal-arm balance.

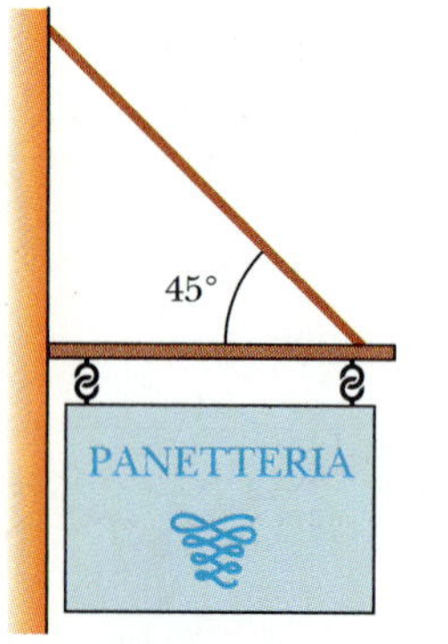

FIGURE 8.35
Sign hanging from a boom.

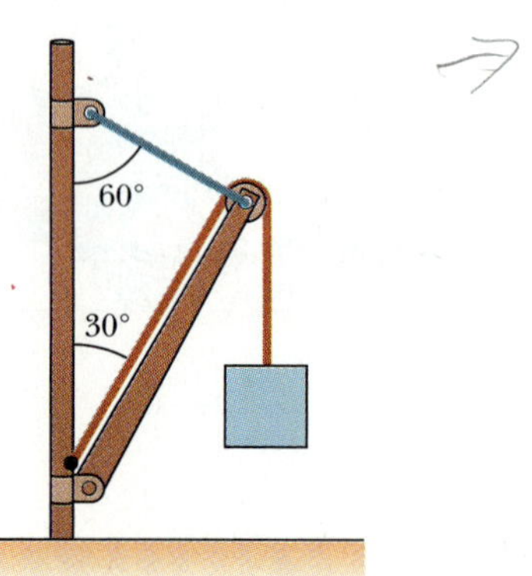

FIGURE 8.36
Cargo hanging from a boom.

mass of the beam is M. Find the tensions in the cables and the compressional force in the beam.

6.—A sailor is being transferred from one ship to another by means of a bosun's chair (Fig. 8.33). The chair hangs from a rope strung between the two ships. The mass of the sailor and chair is 90 kg. If the two parts of the rope make angles of 40° and 20° with the horizontal, what are the tensions in the two parts?

7.—In an unequal-arm balance, the beam is pivoted at a point near one end. With such a balance, large loads can be balanced with small standard weights. Figure 8.34 shows such a balance with an arm of 50 cm swinging on a pivot 1.0 cm from one end. When a package of sugar is deposited in the balance pan, equilibrium is attained with a standard mass of 0.12 kg in the other pan. What is the mass of the sugar? Neglect the masses of the pans.

8.—A 50-kg log of uniform thickness lies horizontally on the ground.
(a) What vertical force must you exert on one end of the log to barely lift this end off the ground?
(b) If you continue to exert a purely vertical force on the end of the log, what is the magnitude of the force required to hold the log at an angle of 30° to the ground? At an angle of 60°? At an angle of 85°?
(c) If instead you exert a force at right angles to the length of the log, what is the magnitude of the force required to hold the log at an angle of 30° to the ground? At an angle of 60°? At an angle of 85°?

9.—A heavy shop sign hangs from a boom sticking out horizontally from a building (Fig. 8.35). The boom is hinged at the building and is supported by a diagonal wire, making an angle of 45° with the boom. The mass of the sign is 50 kg, and the boom and wire are massless. What is the tension in the wire? What is the force with which the end of the boom pushes against the building?

10.—Consider a long, thick cable of diameter d and density ρ hanging vertically down the side of a building. The length of the cable is l. What is the tension in the cable at its upper end? At its midpoint?

11.—Figure 8.36 shows cargo hanging from the loading boom of a ship. If the boom is inclined at an angle of 30° and the cargo has a mass of 2500 kg, what is the tension in the upper cable? What is the compressional force in the boom? Neglect the mass of the boom.

12.—Consider the bridge with the locomotive described in Example 1 and suppose that, besides the first locomotive at 30 m from the right end, there is a second locomotive, also of 90,000 kg, at 80 m from the right end. What is the load on each pier in this case?

13.—One end of a uniform beam of mass 50 kg and length 3.0 m rests on the ground, the other end is held above the ground by a pivot placed 1.0 m

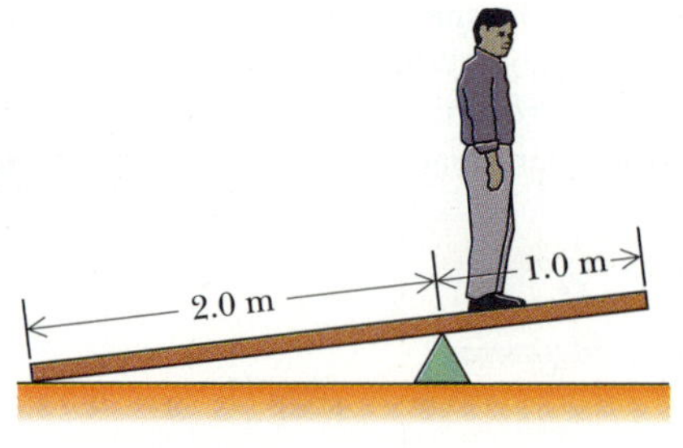

FIGURE 8.37
Man on a beam.

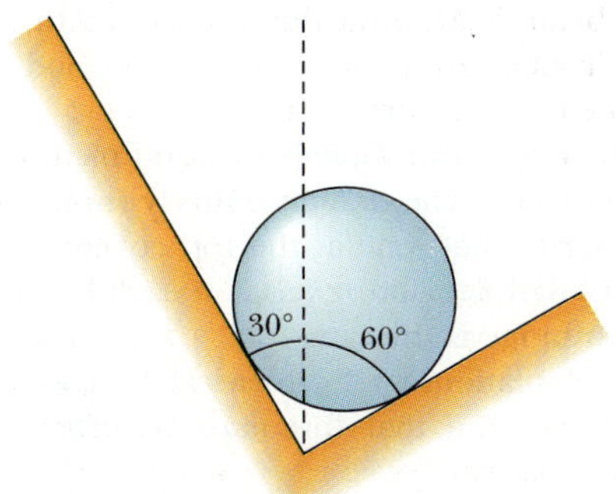

FIGURE 8.38
A bowling ball in a groove.

from that end (see Fig. 8.37). An 80-kg man walks along the beam, from the low end toward the high end. How far beyond the pivot can the man walk before the high end of the beam swings down?

14.* A bowling ball of mass 10 kg rests in a groove with smooth, perpendicular walls, inclined at angles of 30° and 60° with the vertical, as shown in Fig. 8.38. Calculate the magnitudes of the normal forces at the points of contact.

15. Sheerlegs are widely used to suspend loads. They consist of two rigid beams leaning against each other, like the legs of the letter "**A**" (Fig. 8.39). The load is suspended by a cable from the apex of the "**A.**" Suppose that a pair of sheerlegs, each at an angle of 30° with the vertical, is used to suspend an automobile engine of mass 400 kg. What is the compressional force in each leg? What are the horizontal and vertical forces that each leg exerts on the ground?

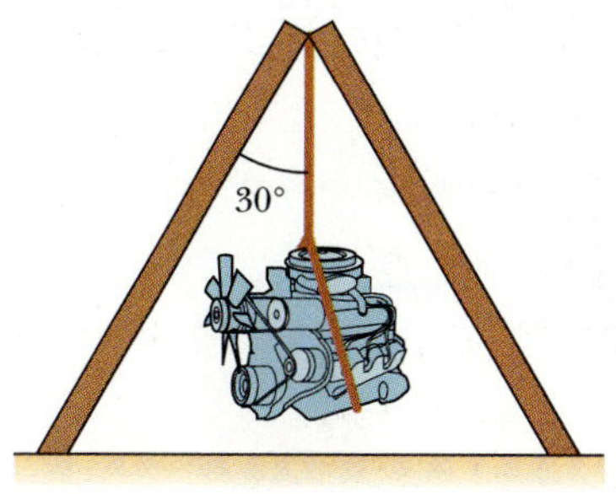

FIGURE 8.39
Sheerlegs supporting a load.

16.* Figure 8.40 shows the arrangement of wheels on a locomotive of the Caledonian Railway. The numbers give the distances between the wheels in meters and the downward forces that each wheel exerts on the track in thousands of newtons (the numbers for the forces include both the right and the left wheels). From the information given, find how far the center of mass of the locomotive is behind the front wheel.

17.* A meter stick of wood of 0.40 kg is nailed to a wall at the 75-cm mark. If the stick is free to rotate about the nail, what horizontal force must you exert at the upper (short) end to deflect the stick 30° to one side?

18.* The plant of the foot of an average male is 26 cm and the height of the center of mass above the floor is 1.03 m. When standing upright, the center of mass is vertically aligned with the ankle, 18 cm from the tip of

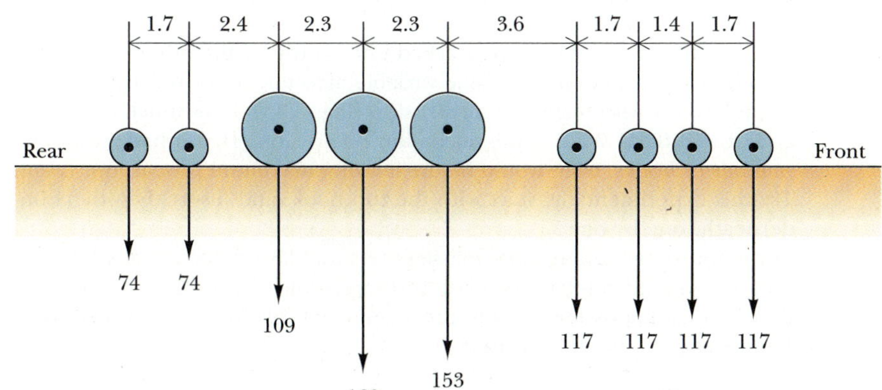

FIGURE 8.40
Wheels of a locomotive.

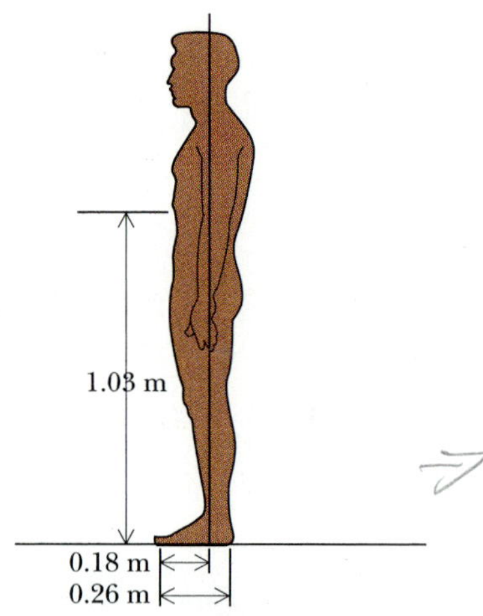

FIGURE 8.41
Man standing on stiff feet.

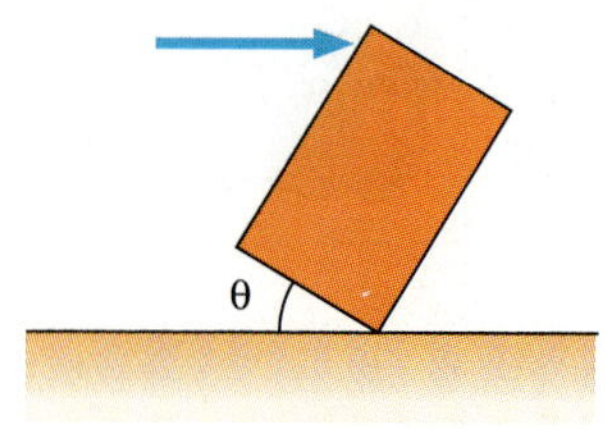

FIGURE 8.42
Tilted box.

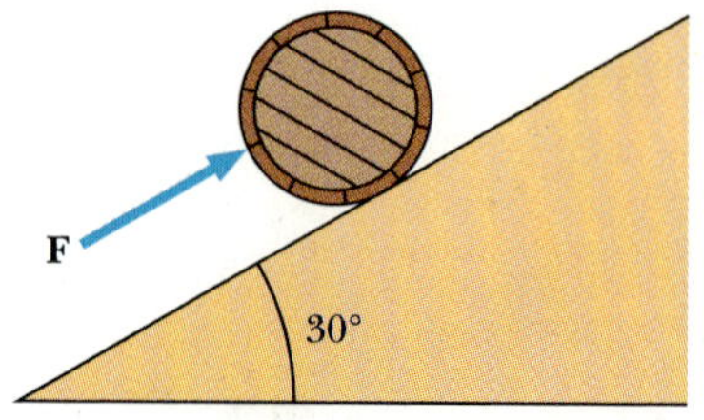

FIGURE 8.43
Barrel on a ramp.

FIGURE 8.44
A stack of books.

the foot (Fig. 8.41). Without losing his equilibrium, how far can the man lean forward or backward while keeping his body straight and his feet stiff and immobile?

19.*—A door made of a uniform piece of wood measures 0.90 m by 2.10 m and has a mass of 18 kg. The door is entirely supported by two hinges, one at the bottom corner and one at the top corner. Find the force (magnitude and direction) that the door exerts on each hinge. Assume that the *vertical* force on each hinge is the same.

20.*—Repeat the calculations of Problem 11, but assume that the mass of the boom is 800 kg, and that this mass is uniformly distributed along the length of the boom.

21.*—A wooden box, filled with a material of uniform density, stands on a concrete floor. The box has a mass of 75 kg and is 0.5 m wide, 0.5 m long, and 1.5 m high. The coefficient of friction between the box and the floor is $\mu_s = 0.80$. If you exert a (sufficiently strong) horizontal push against the side of the box, it will either topple over or start sliding without toppling over, depending on how high above the level of the floor you push. What is the maximum height at which you can push if you want the box to slide? What is the magnitude of the force you must exert to start the sliding?

22.*—An automobile with a wheelbase (distance from the front wheels to the rear wheels) of 3.0 m has its center of mass at a point midway between the wheels at a height of 0.65 m above the road. When the automobile is on a level road, the force with which each wheel presses on the road is 3100 N. What is the normal force with which each wheel presses on the road when the automobile is standing on a steep road of slope 3:10 with all the wheels locked?

23.**—A wooden box is filled with material of uniform density. The box (with its contents) has a mass of 80 kg; it is 0.6 m wide, 0.6 m deep, and 1.2 m high. The box stands on a level floor. By pushing against the box, you can tilt it over (Fig. 8.42). Assume that when you do this, one edge of the box remains in contact with the floor without sliding.

(a) Plot the gravitational potential energy of the box as a function of the angle θ between the bottom of the box and the floor.

(b) What is the critical angle beyond which the box will topple over when released?

(c) How much work must you do to push the box to this critical angle?

24.**—Consider the ladder leaning against a wall described in Example 3. If the ladder makes an angle of 30° with the wall, how hard can you push down vertically on the top of the ladder with your hand before slipping begins?

25.**—A 100-kg barrel is placed on a 30° ramp (Fig. 8.43). What push, parallel to the ramp, must be exerted against the middle of the barrel to keep it from rolling down? Assume that the friction between the barrel and the ramp prevents slipping of the barrel; that is, the barrel would roll without slipping if released.

26.**—Five identical books are to be stacked one on top of the other. Each book is to be shifted sideways by some variable amount, so as to form a curved leaning tower with maximum protrusion (Fig. 8.44). How much must each book be shifted? What is the maximum protrusion? If you had an infinite number of books, what would be the limiting maximum protrusion? (Hint: Try this experimentally; start with the top book, and insert the others underneath, one by one.)

27.**—You want to pick up a nearly massless rectangular cardboard box by grabbing its top and side between you forefinger and thumb (Fig. 8.45). Show that this is impossible unless the coefficient of friction between your fingers and the box is at least 1.

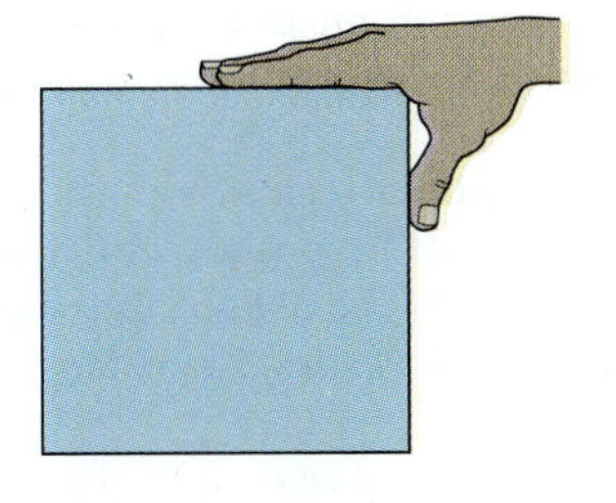

FIGURE 8.45
Holding a box.

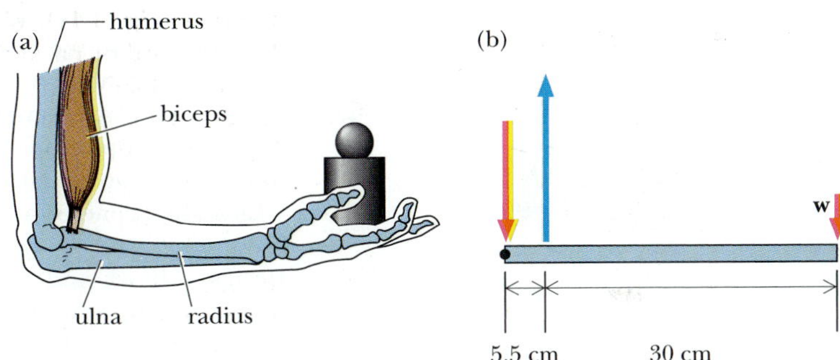

FIGURE 8.46 Forearm as lever.

Section 8.3: Levers and Pulleys

28.——The human forearm (including the hand) can be regarded as a lever pivoted at the joint of the elbow and pulled upward by the tendon of the biceps (Fig. 8.46a). The dimensions of this lever are given in Fig. 8.46b. Suppose that a load of 25 kg rests in the hand. What upward force must the biceps exert to keep the forearm horizontal? What is the downward force at the elbow joint? Neglect the weight of the forearm.

29.——Repeat the preceding problem if, instead of being horizontal, the forearm is angled upward at 45°.

30.——A simple manual winch consists of a drum of radius 4.0 cm to which is attached a handle of radius 25 cm (Fig. 8.47). When you turn the handle, the rope winds up on the drum and pulls the load. Suppose that the load carried by the rope is 2500 N. What force must you exert on the handle to hold this load?

31.——A rope hoist consists of four pulleys assembled in two pairs with rigid straps, with a rope wrapped around as shown in Fig. 8.48. A load of 300 kg hangs from the lower pair of pulleys. What tension must you apply to the rope to hold the load steady? Treat the pulleys and the rope as massless, and ignore any friction in the pulleys.

32.——Figure 8.49 shows a pair of pliers and its dimensions. If you push against the handles of the pliers with a force of 200 N from each side, what is the force that the jaws of the pliers exert against each other?

33.——A man of 73 kg stands on one foot, resting all of his weight on the ball of the foot. As described in Section 8.3, the bones of the foot play the role

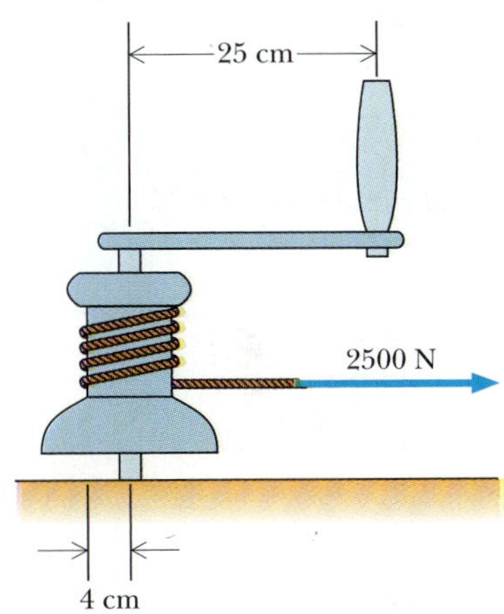

FIGURE 8.47
Manual winch.

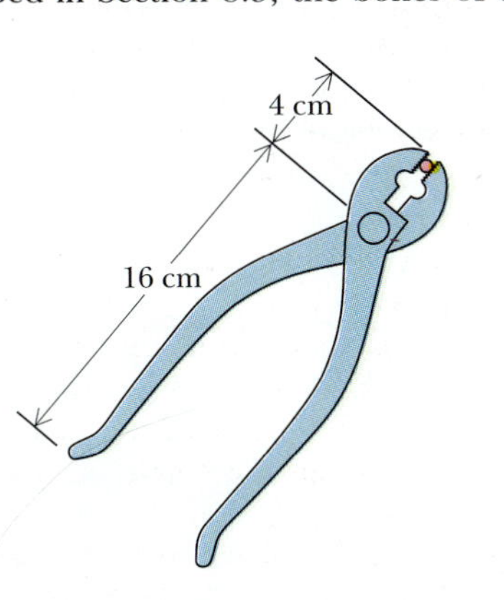

FIGURE 8.49
Pliers.

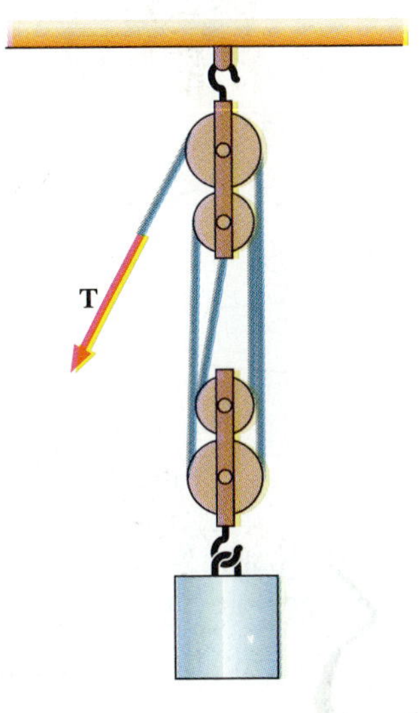

FIGURE 8.48
Rope hoist.

of a lever. The short end of the lever (to the heel) measures 5.0 cm and the long end (to the ball of the foot) 14 cm. Calculate the force exerted by the Achilles tendon and the force at the ankle.

34.—To help his horses drag a heavy wagon up a hill, a teamster pushes forward at the top of one of the wheels (Fig. 8.50). If he pushes with a force of 600 N, what forward force does he generate on the axle of the wagon? (Hint: The diameter of the wheel can be regarded as a lever pivoted at the ground.)

35.—Design a block and tackle with a mechanical advantage of 2. Design a block and tackle with a mechanical advantage of 5. If you connect these two block and tackles in tandem, so the second pulls on the free end of rope of the first, what mechanical advantage do you get?

36.*—A parbuckle is a simple device used by laborers for raising or lowering a barrel or some other cylindrical object along a ramp. It consists of a loop of rope wrapped around the barrel (Fig. 8.51). One end of the rope is tied to the top of the ramp, and the laborer pulls on the other end. Suppose that the laborer exerts a pull of 500 N on the rope, parallel to the ramp. What is the force that the rope exerts on the barrel? What is the mechanical advantage of the parbuckle?

37.**—Consider the differential windlass illustrated in Fig. 8.28. Calculate what clockwise torque must be applied to the handle to lift a load m. What tangential force must be exerted on the handle? What is the mechanical advantage of this windlass?

38.**—Figure 8.52 shows a compound bolt cutter. If the dimensions are as indicated in this figure, what is the mechanical advantage?

39.**—The drum of a winch is rigidly attached to a concentric large gear, which is driven by a small gear attached to a crank. The dimensions of the drum, the gears, and the crank are given in Fig. 8.53. What is the mechanical advantage of this geared winch?

FIGURE 8.50
Teamster pushing on a wheel.

FIGURE 8.51
Parbuckle used to move a barrel up a ramp.

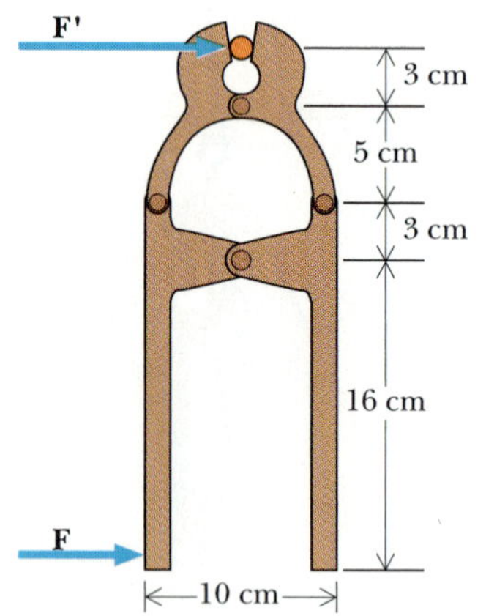

FIGURE 8.52
Compound bolt cutter.

Section 8.4: Elasticity of Materials

40.—The anchor rode of a sailboat is a nylon rope of length 60 m and diameter 1.3 cm. While anchored during a storm, the sailboat momentarily pulls on this rope with a force of 1.8×10^4 N. How much does the rope stretch?

41.—If a steel rope and a nylon rope of equal lengths are to stretch by equal amounts when subjected to equal tensions, what must be the ratio of their diameters?

42.—The length of the femur of a woman is 38 cm and the average cross section is 10 cm^2. How much will the femur be compressed in length if the woman lifts another woman of 68 kg and carries her piggyback? Assume that, momentarily, all of the weight rests on one leg.

43.—In a skyscraper, an elevator is suspended from three equal, parallel,

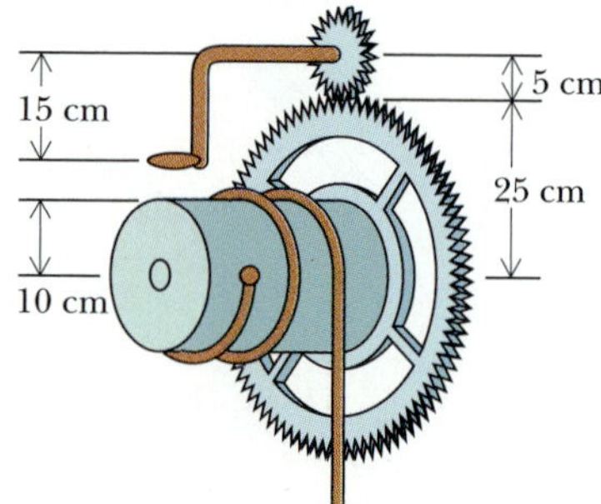

FIGURE 8.53
A geared winch.

300-m-long steel cables, each of diameter 1.0 cm. How much do these cables stretch if the mass of the elevator is 1000 kg?

44.——A concrete column supporting one corner of a building is 6.0 m high and 0.60 m in diameter. The mass resting on the top of this column is 50,000 kg. How much does the column compress under the weight of this mass? Young's modulus for concrete is 2×10^{10} N/m^2.

45.——According to (somewhat oversimplified) theoretical considerations, Young's modulus, the shear modulus, and the bulk modulus are related by

$$Y = \frac{9BS}{3B + S}$$

Check this for the first four materials listed in Table 8.1.

46.——Suppose you drop an aluminum sphere of radius 10 cm into the ocean and it sinks to a depth of 5000 m, where the pressure is 5.7×10^7 N/m^2. Calculate by how much the diameter of this sphere will shrink.

47.——The piano wire described in Example 6 can be regarded as a spring. What is the effective spring constant of this spring?

48.——At the bottom of the Marianas trench in the Pacific Ocean, at a depth of 10,900 m, the pressure is 1.24×10^8 N/m^2. What is the percent increase of the density of water at this depth as compared to the density at the surface?

49.*——A slab of stone of mass 1200 kg is attached to the wall of a building by two bolts of iron of diameter 1.5 cm (Fig. 8.54). The distance between the wall and the slab of stone is 1.0 cm. Calculate by how much the bolts will sag downward because of the shear stress they are subjected to.

50.**——A rope of length 12 m consists of an upper half of nylon of diameter 1.9 cm spliced to a lower half of steel wire of diameter 0.95 cm. How much will this rope stretch if a mass of 4000 kg is suspended from it? Young's modulus for the steel wire is 19×10^{10} N/m^2.

51.**——When a bar of steel is heated, it expands in length by 0.0012% for each degree centigrade of temperature increase. If the length of the heated bar is to be reduced to its original value, a compressive stress must be applied to it. The compressive stress required to cancel the thermal expansion is called ***thermal stress.*** What is its value for a cylindrical bar of cross section 4 cm^2 heated by 150°C?

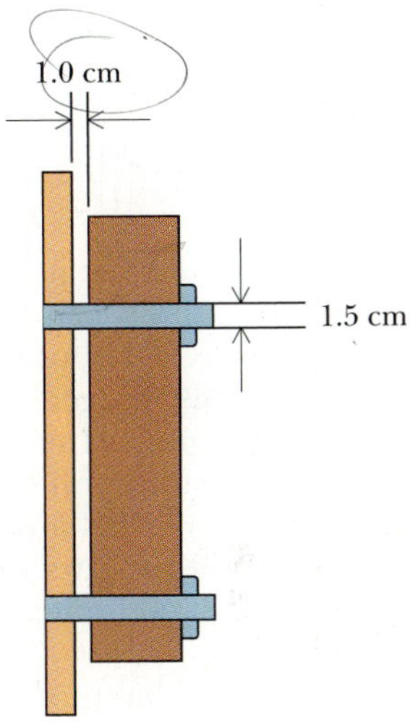

FIGURE 8.54
A slab of stone held by bolts.

PHYSICS IN ACTION

ANIMAL MECHANICS

When an animal runs, or jumps, or grabs at a prey, the motion of its limbs can be regarded as the motion of a machine. The bones of the limb play the role of rigid beams, hinged at the joints. The muscles and the tendons of the animal play the role of motors and ropes that pull on the beams and exert torques on them. We can analyze the motion of the animal by applying the laws we learned in our study of the motion of rigid bodies, such as the law of the lever.

Among the many levers in the human body is the forearm (Fig. 1). The two bones of the forearm, the radius and the ulna, are tied together by ligaments, and they move together, as a single rigid beam. This beam is hinged at the elbow joint, which is the pivot for the beam. The tendon of the biceps muscle is attached to the forearm at a short distance in front of the elbow. If your forearm is extended horizontally and you hold a weight in your hand, there are three forces acting on the forearm (including the hand): the upward pull of the biceps (**F**), the downward push of the weight (**F′**), and the downward push of the joint (besides, there is the weight of the forearm, acting at the center of mass; but, for the sake of simplicity, we will neglect this weight). Figure 1b displays these forces and their points of application. The upward pull of the biceps acts at a distance of 5.5 cm from the joint, and the downward push of the weight acts at a distance of 35.5 cm. According to the law of the lever, for equilibrium, the magnitudes of the forces F and F' must be in the inverse ratio of these distances,

$$\frac{F'}{F} = \frac{l}{l'} = \frac{5.5 \text{ cm}}{35.5 \text{ cm}} = 0.15$$

Thus, the mechanical advantage of your forearm is considerably less than 1 — your forearm machine operates with a mechanical *dis*advantage. Your biceps must exert a much larger force than the weight you hold in your hand. For instance, if the weight in your hand is $F' = 250$ N, then the force your biceps must exert is $F = 1600$ N. A force of 250 N corresponds to the weight of 25 kg, which is pretty near the maximum you can expect to support in your hand (if you are strongly built); but a force of 1600 N corresponds to the weight of two fully grown persons hanging on the tendon of your biceps! Besides this large force that your biceps

(a)

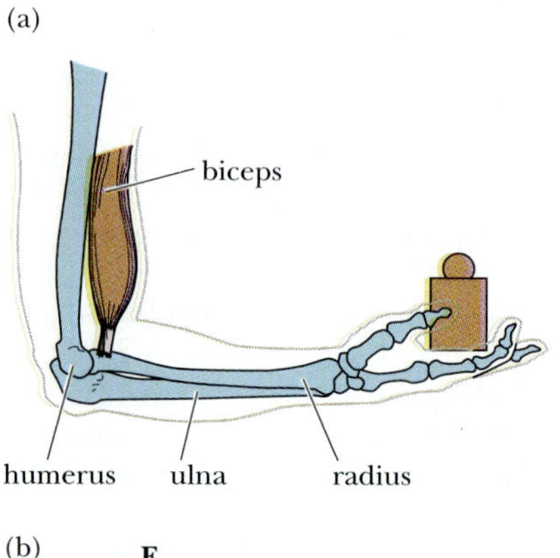

(b)

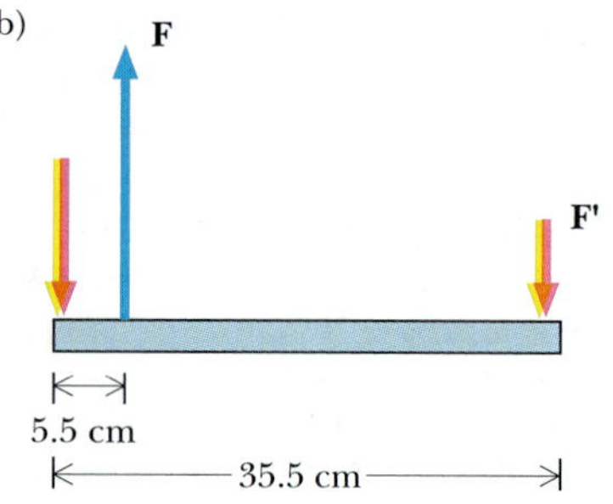

FIGURE 1 (a) Forearm and hand holding a weight in equilibrium. The downward push of the weight is balanced by the upward pull of the tendon of the biceps. (b) "Free-body" diagram for the forearm, idealized as a straight beam, or lever, hinged at the elbow.

must exert, there is also a large reaction force in the elbow joint, about 1400 N.

Mechanical advantages less than 1 are the rule throughout the bone and muscle machinery in the bodies of animals. Limbs often have a low mechanical advantage for one direction of motion, and an even lower mechanical advantage for the opposite direction. In human limbs, the lowest mechanical advantages are found in the muscles that straighten (extend) a limb after some other muscle has bent (flexed) the limb. Muscle can only pull on the tendon; it cannot push the tendon back to its original position. After your biceps has flexed your forearm, your triceps has to straighten it out. The biceps and the triceps are said to be antagonists; the former is a ***flexor,*** and the latter is an ***extensor.*** Of course, if you are holding a weight, you can just let your lower arm drop, so gravity straightens the arm, without need of the triceps. But if instead of holding a weight, you are trying to pull downward on, say, a rope, you do need your triceps. Figure 2 shows the forces on your forearm and their points of application when your hand is pulling downward on a rope. The upward pull of the triceps acts at a distance of only 1.4 cm from the elbow joint. The mechanical advantage for the triceps is only 1.4 cm/35.5 cm = 0.039, even smaller than for the biceps.

From an engineering point of view, the low mechanical advantage of the levers in the machinery of animals poses a puzzle. It would seem that the human machine could be made much more efficient by shifting the attachment point of the tendon of the biceps forward by a few centimeters; if the attachment point were twice as far from the joint, the same load of 25 kg could be supported by a biceps muscle of half the cross-sectional size. But such a change in the design of the human machine would impose a penalty on the range of the motion of the forearm and on the speed of motion. Muscle has a limited range of contraction; usually it contracts by no more than about 30% of its original length. If the attachment point of the biceps were twice as far from the joint, the biceps could move the forearm over only half its arc of swing. Furthermore, muscle has a limited speed of contraction; the biceps needs about 0.1 s to contract. If in this time the forearm moves over only half its arc, the speed of the hand would be only half as large. Obviously, a halving of the speed of motion of the limbs would be a severe handicap to an animal engage in physical combat, in hunting, or in swinging from trees. Thus, the arrangement of the tendons in the machinery of animals is designed to attain a suitable compromise between force and speed.

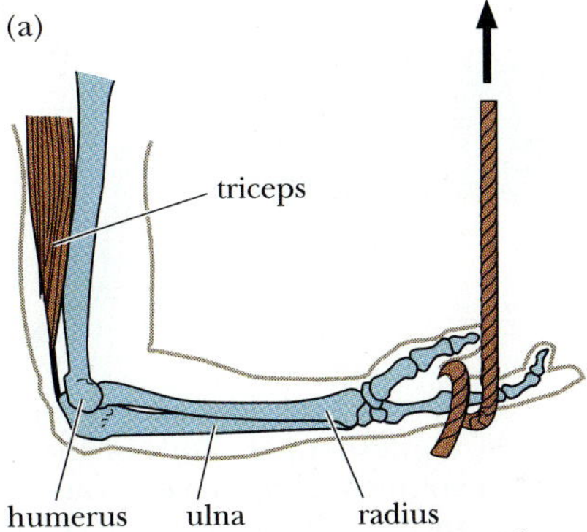

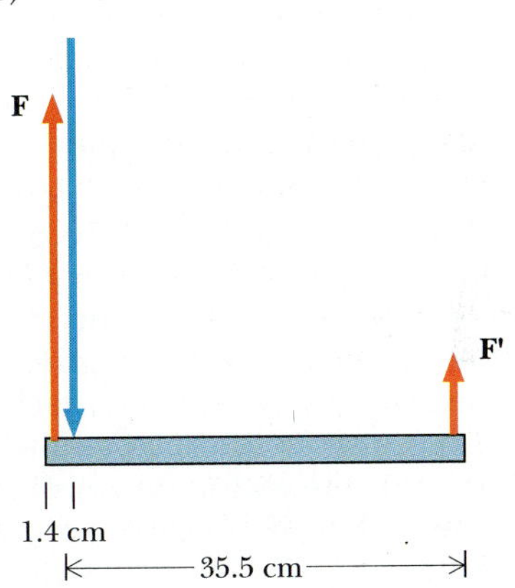

FIGURE 2 (a) Forearm and hand holding a rope. The upward pull of the rope is balanced by the upward pull of the tendon of the triceps. (b) "Free-body" diagram for the forearm.

Muscle does not really need a large lever arm to enhance its contractile force; but it does need a short lever arm so it can generate high speeds of motion.

Animals with exoskeletons, such as insects and crustaceans, often cannot accommodate adequate lever arms within the confines of their skeleton. For instance, Fig. 3 shows the hind leg of a grasshopper. The leg is long, and the space available for the hinge and the attachment of the extensor and the flexor muscles is very narrow, giving an extremely low mechanical advantage. If the muscles were of the same kind as in your biceps, with long parallel muscle fibers, the insect could not pack enough fibers into the available space. Instead, the muscles are of a different kind, with short fibers arranged at a slant; each of these short fibers is connected to the exoskeleton at its outer end and to a long central strap at its inner end (see Fig. 3). This kind of muscle is called pinnate, or featherlike. The slanted

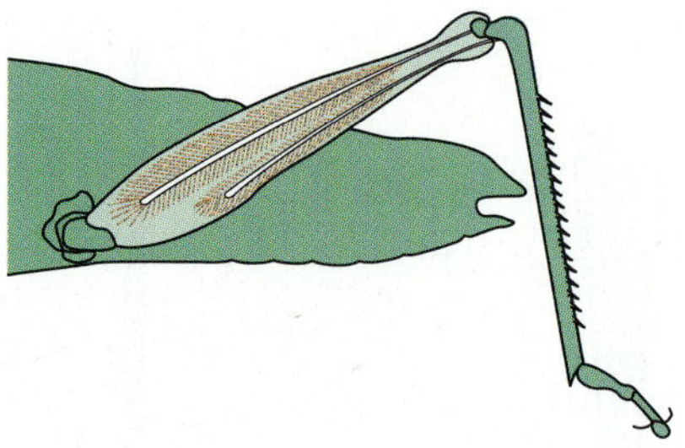

F I G U R E 3 Extensor and flexor muscles in the hind leg of a grasshopper. Both of these muscles are attached to extremely short lever arms. Note the featherlike appearance of the muscle fibers. (Based on Vogel, *Life's Devices.*)

arrangement of short fibers in pinnate muscle permits a much larger number of fibers to be packed into the available space, and therefore permits this kind of muscle to exert a much larger force than a parallel-fiber muscle of equal volume (the force a fiber exerts is independent of its length). Note that the net force exerted by two fibers attached at opposite sides of the strap is the vector sum of the individual forces, and this vector sum is of somewhat smaller magnitude than if the fibers were parallel — the components of the forces at right angles to the strap are in opposition, and they cancel. Just as you waste energy in an isometric exercise, these opposing components of the forces waste energy within the fibers. Another disadvantage is that the pinnate muscle can contract only over a short distance; the more it contracts, the steeper the slant of the fibers, and the less the net forward pull. But this is not a serious defect, since the very short lever arm can in any case not move far.

As mentioned above, levers of low mechanical advantage require not only large muscle forces, but they also generate large reaction forces in the joints. The elbow joints of your arms and the knee joints of your legs have to be massive to withstand these forces. As an alternative to massive, heavy joints, some animals have evolved special arrangements of muscles to stabilize joints and relieve the stress upon them. The arrangement of the jaw muscles of carnivores is a good example of this. Figure 4 shows the jaw of a marten, a small carnivore that feeds on squirrels and similar prey. The jaw is an **L**-shaped lever, hinged to the skull at the corner of the **L.** Figure 4 shows the forces on the jaw when the marten is chewing a piece of bone or meat with its hind teeth (maybe you have seen your cat or your dog doing this with a bone). The force (**F**) exerted by the piece of bone acts vertically down, on the large carnassial tooth. The jaw is pulled up by the combined pulls of two muscles: the masseter pulling diagonally up and forward (**F**′), and the temporalis pulling diagonally up and backward (**F**″). As can be seen from Fig. 4b, the sum of the forces exerted by these muscles matches the force on the tooth — the hinge does not have to exert any force. For comparison, Fig. 5 shows what the forces would be if the masseter muscle were disabled. Now the equilibrium would require a very large reaction force at the hinge of the jaw. If the hinge fails to provide this large forward force, the jaw moves backward and becomes dislocated. In essence, the masseter muscle relieves this large hinge force and prevents dislocation of the jaw.

The human jaw also has temporalis and masseter

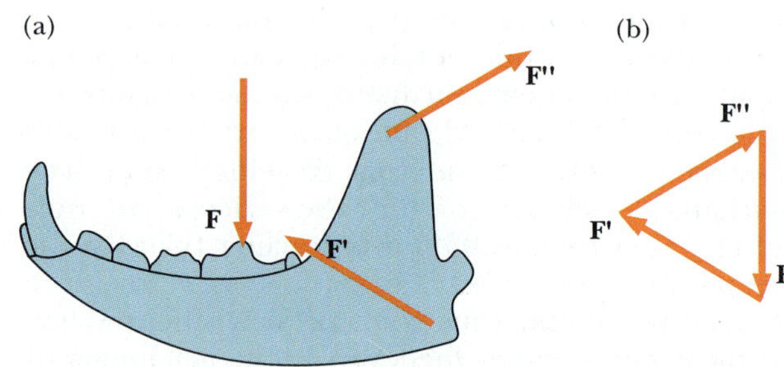

F I G U R E 4 (a) Forces acting on the lower jaw of a marten when chewing on a piece of bone. The force exerted by the bone is **F,** and the forces exerted by the masseter and temporalis muscles are **F**′ and **F**″. (b) The vector sum of these forces is zero, and the jaw is in equilibrium even though the hinge of the jaw contributes no force. (Based on Alexander, *Animal Mechanics.*)

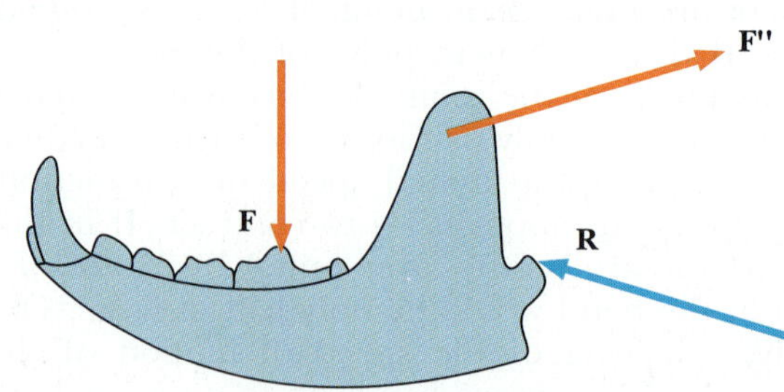

F I G U R E 5 Forces acting on the jaw if the masseter muscle were disabled. Now the hinge of the jaw must exert a large reaction force **R** for equilibrium. (Based on Alexander, *Animal Mechanics.*)

muscles; but the muscles are not as strongly developed as in carnivores, and they are not angled backward and forward as much. We do not chew on bones with the violence of a carnivore, and we do not need as much protection against damage to our jaw hinge.

In our discussion we have so far treated tendons as passive ropes, which merely transmit the pull from the muscle to the bone. But this does not do justice to the tendons — in many circumstances the elasticity of the tendon makes an important contribution to the operation of the machinery. The hopping motion of the kangaroo is a spectacular example of the active role of the tendon. When they want to move quickly, kangaroos don't walk or run — they hop, and they go on hopping (Fig. 6). Typically, a large kangaroo hops 6 m forward and about 0.4 m up (in a pinch, they can hop farther and higher). What is impressive about this is not so much the length or the height of the hop — human athletes can jump as far and as high — but the fact that the kangaroo can go on making such hops for a long time, for hours, according to some reports.

FIGURE 6 Hopping kangaroo.

Let us examine the energy requirements for such a long-term hopping motion. To lift a kangaroo of 70 kg to a height of 0.4 m requires an energy $mgh =$ 270 J. A simple calculation shows that for each up-and-down motion (free-fall motion!) to a height of 0.4 m, the kangaroo takes a time $\Delta t = 0.57$ s. Thus, it would seem that as long as the kangaroo goes on hopping, its leg muscles must deliver an average mechanical power of $mgh/\Delta t = 270\text{ J}/0.57\text{ s} = 470$ watts. But this is far in excess of what the animal's metabolism can provide. For comparison, a human athlete of about the same mass riding a bicycle at maximum effort can deliver a long-term mechanical power of only 200 watts (in short spurts of a few minutes, the athlete can deliver a higher power, and so can the kangaroo). So, where does the kangaroo get the energy to operate its hopping machinery? The secret of the kangaroo is in its Achilles tendon. The kangaroo does not make an excessive muscular effort while it hops — it merely bounces off the ground like a rubber ball. Its Achilles tendon is very thick and very elastic. The kangaroo's leg muscles have to provide the energy for the first jump, but when the kangaroo lands on the ground, it converts most of the kinetic energy of its downward motion into elastic potential energy, stored in the stretched Achilles tendon. For the next jump, the kangaroo simply uses this stored potential energy. Thus, the motion of a hopping kangaroo is analogous to the motion of a bouncing rubber ball, or the motion of a child on a pogo stick (but the Achilles tendon stores elastic energy when stretched, whereas the rubber ball or the spring of the pogo stick stores elastic energy when compressed). The kangaroo's muscles do not have to provide a fresh supply of energy for each jump; they only have to provide the fraction of the energy that is lost to friction.

REFERENCES

Alexander, R. McN., *Animal Mechanics* (Blackwell Scientific Publications, Oxford, 1983).

Alexander, R. McN., *The Human Machine* (Columbia University Press, New York, 1992).

Schmidt-Nielsen, K., *How Animals Work* (Cambridge University Press, Cambridge, 1972).

Vogel, S., *Life's Devices* (Princeton University Press, Princeton, N.J., 1988).

White, D. C. S., "Muscle Mechanics," *in* Alexander, R. McN., and Goldspink, G., eds., *Mechanics and Energetics of Animal Locomotion* (Chapman and Hill, London, 1977).

CHAPTER 9

Foucault pendulum at the Smithsonian Institution in Washington, D.C. The period of oscillation of this pendulum is 8 s.

OSCILLATIONS

Periodic motion ▶ ***The motion of a particle or of a system of particles is periodic, or cyclic, if it repeats again and again at regular intervals of time.*** The orbital motion of a planet around the Sun, the uniform rotational motion of a phonograph turntable, the back-and-forth motion of a piston in an automobile engine or in a water pump, the swinging motion of a pendulum bob, and the vibration of a guitar string are examples of periodic motions. **Oscillation ▶** ***If the periodic motion is a back-and-forth or swinging motion, it is called an oscillation.*** Thus, the motion of the piston is an oscillation, and so are the motion of the pendulum and the motion of the individual particles of the guitar string.

In this chapter we will examine in some detail the motion of a mass oscillating back and forth under the push and pull exerted by an ideal, massless spring. The equations that we will develop for the description of this mass–spring system are of great importance because analogous equations also occur in the description of many other oscillating systems.

9.1 SIMPLE HARMONIC MOTION

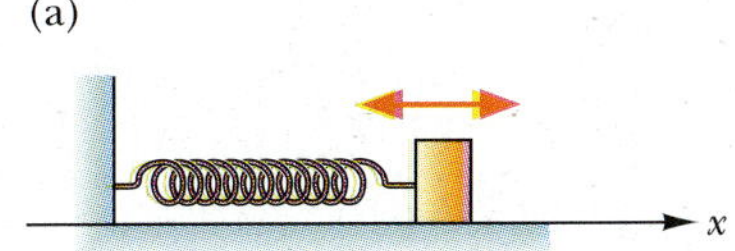

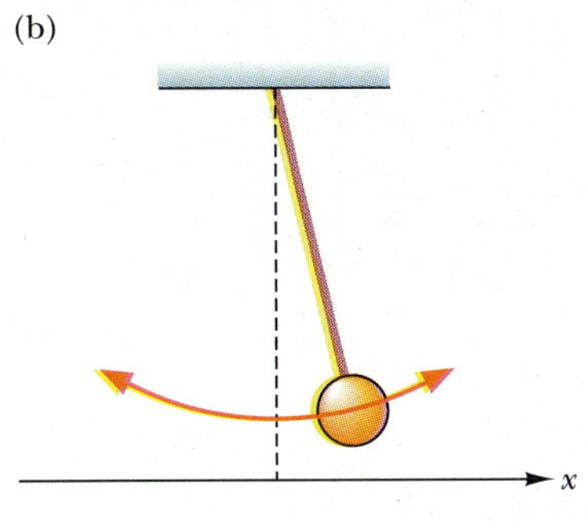

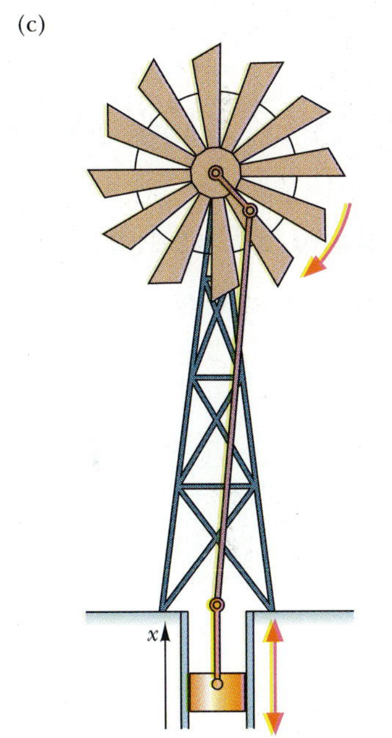

FIGURE 9.1

(a) The motion of a mass oscillating back and forth in response to the push and pull of a spring is simple harmonic. (b) The motion of a pendulum bob is approximately simple harmonic. (c) The motion of a piston of a water pump being pushed up and down by a long rod connected to a crank on a windmill is approximately simple harmonic.

Simple harmonic motion is a special kind of one-dimensional periodic motion. In any kind of one-dimensional periodic motion, the particle moves back and forth along a straight line, repeating the same motion again and again. In the special case of ***simple harmonic motion, the particles position can be expressed as a cosine or a sine function of time.*** As we will see later, the motion of a mass oscillating back and forth under the push and pull of a spring is simple harmonic (Fig 9.1a), and so is the motion of a pendulum bob swinging back and forth (Fig. 9.1b), and so is the up-and-down motion of the piston in a water pump (provided the piston rod is very long, as shown in Fig. 9.1c). However, for a start we want to explore the mathematical description of simple harmonic motion, and we will postpone the question of what causes the motion.

As a numerical example of simple harmonic motion, suppose that the piston of the water pump in Fig. 9.1c moves up and down between $x = -0.3$ m and $x = +0.3$ m (where the x axis is assumed to be vertical). The motion of the piston may be regarded as particle motion; Fig. 9.2 gives a plot of the position of the piston as a function of time. The plot in Fig. 9.2 has the mathematical form

$$x = 0.3 \cos\left(\frac{\pi t}{4}\right) \tag{1}$$

where it is assumed that distance is measured in meters and time in seconds, and it is assumed that the "angle" $\pi t/4$ in the cosine function is reckoned in radians. For instance, if the time is 3.0 seconds, then $\pi t/4 = \pi \times 3.0/4 = 2.36$, and with our calculator we find $x = 0.3 \cos(2.36 \text{ radians}) = 0.3 \times (-0.707) = -0.212$, that is, $x = -0.212$ m.

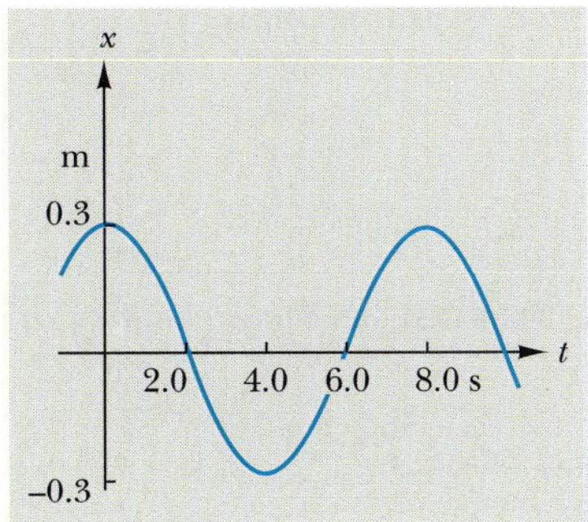

FIGURE 9.2

Plot of position vs. time for a particle oscillating along the x axis with simple harmonic motion.

Cosines and sines are called ***harmonic functions,*** which is why we call the motion of the particle *harmonic.* At $t = 0$, the particle is at its maximum distance from the origin ($x = 0.3$ m) and is just starting to move; at $t = 2.0$ s, it passes through the origin ($x = 0$); at $t = 4.0$ s, it reaches maximum distance, but on the negative side of the origin ($x = -0.3$ m); at $t = 6.0$ s, it again passes through the origin. Finally, at $t = 8.0$ s, the particle returns to its maximum distance on the positive side of the origin, exactly as at $t = 0$ — it has completed one cycle of the motion and is ready to begin the next cycle. Thus, the period, or the repeat time, of the motion is

$$T = 8.0 \text{ s} \tag{2}$$

and the frequency of the motion, or the rate of repetition of the motion, is

$$f = \frac{1}{T} = \frac{1}{8.0 \text{ s}} = 0.125/\text{s} \tag{3}$$

Turning points ▶

The points $x = 0.3$ m and $x = -0.3$ m, at which the x coordinate attains its maximum and minimum values, are the ***turning points*** of the motion; and the point $x = 0$ is the midpoint, or the ***equilibrium point.***

Equation (1) is a specific example of simple harmonic motion. More generally, the motion of a particle is simple harmonic if the dependence of position on time has the form of a cosine or a sine function, such as

$$x = A\cos(\omega t) \tag{4}$$

Amplitude ▶

Angular frequency ▶

The quantities A and ω are constants. The quantity A is called the ***amplitude*** of the motion; it is simply the distance between the midpoint ($x = 0$) and either of the turning points ($x = +A$ or $x = -A$). The quantity ω is called the ***angular frequency;*** it is directly related to the period and to the frequency of the motion. To establish this relationship, note that if we increase the time by $2\pi/\omega$, we increase t to $t + 2\pi/\omega$, and we increase ωt to $\omega t + 2\pi$. This changes the right side of Eq. (4) from $A\cos(\omega t)$ to $A\cos(\omega t + 2\pi)$. But this is the same as $A\cos(\omega t)$, since increasing an angle by 2π radians means increasing the angle by a full circle, which is really no change at all. We therefore conclude that after the lapse of a time $2\pi/\omega$, the particle returns to the same point; thus, $2\pi/\omega$ must be the ***period*** of the motion,

Period ▶

$$T = \frac{2\pi}{\omega} \tag{5}$$

Frequency ▶

and $\omega/2\pi$ must be the ***frequency*** of the motion,

$$f = \frac{1}{T} = \frac{\omega}{2\pi} \tag{6}$$

The units of angular frequency are radians per second. The units of frequency are cycles per second (cycle/s). Like the label *revolution* that we used in revolutions/s in rotational motion, the label *cycle* in cycle/s can be omitted in the course of a calculation, and so can the label *radian* in radian/s. But it is useful to retain these labels whenever there is a chance of confusion. The unit of frequency is called a ***hertz*** (Hz):

Hertz (Hz) ▶

$$1 \text{ hertz} = 1 \text{ Hz} = 1 \text{ cycle/s} \tag{7}$$

For instance, if the frequency is $f = 0.125$ Hz $= 0.125/\text{s}$, as in the example of the piston of our water pump, then the angular frequency is

$$\omega = 2\pi f = 2\pi \times 0.125/\text{s} = 0.785 \text{ radian/s}$$

Here, in the last step of the calculation, the label *radian* has been inserted, so as to distinguish the angular frequency ω from the ordinary frequency f.

Note that the above equations connecting *angular frequency,* period, and frequency are formally the same as the equations connecting *angular velocity,* period, and frequency of uniform rotational motion [see Eqs. (7.4) and (7.5)]. This coincidence arises from a special geometrical relationship between simple harmonic motion and uniform circular motion. Suppose that a particle moves with simple harmonic motion according to Eq. (4), with amplitude A and angular frequency ω; and consider a "satellite" particle that is constrained to move in uniform circular motion with angular velocity ω along a circle of radius A, centered on the midpoint of the harmonic motion. Figure 9.3 shows this circle, called the ***reference circle.*** At time $t = 0$, both the particle and its satellite are on the x axis. After this time, the particle moves along the x axis, so its position is

$$x = A\cos(\omega t) \tag{8}$$

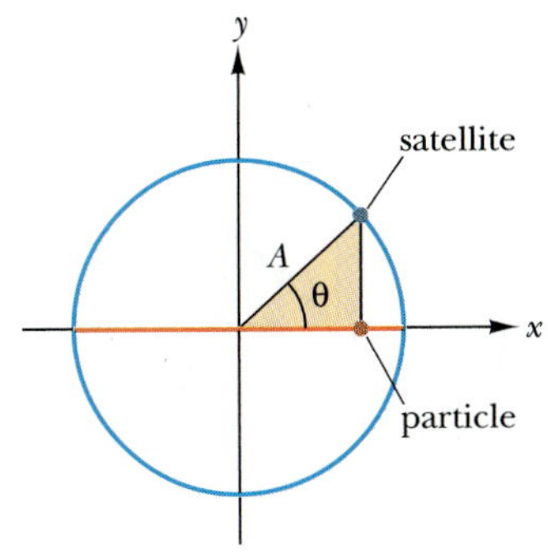

FIGURE 9.3

Particle oscillating along the x axis and satellite particle moving around reference circle. The particle and the satellite are always aligned vertically.

Meanwhile, the satellite moves around the circle, and its angular position is

$$\theta = \omega t \tag{9}$$

Now note that the x coordinate of the satellite is the adjacent side of the triangle shown in Fig. 9.3:

$$x_{\text{sat}} = A\cos\theta = A\cos(\omega t) \tag{10}$$

Comparing this with Eq. (8), we see that the x coordinate of the satellite always coincides with the x coordinate of the particle — that is, the particle and the satellite always have exactly the same x motion. This means that in Fig. 9.3 the satellite always is on that point of the reference circle directly above or directly below the particle.

This geometric relationship between simple harmonic motion and uniform circular motion can be used to generate simple harmonic motion from uniform circular motion. Figure 9.4 shows a simple mechanism for accomplishing this by means of a slotted arm placed over a peg which is attached to a wheel in uniform circular motion. The slot is vertical and the arm is constrained to move horizontally. The peg plays the role of "satellite," and the midpoint of the slot in the arm plays the role of "particle." The peg drags the arm back and forth and makes it move with simple harmonic motion.

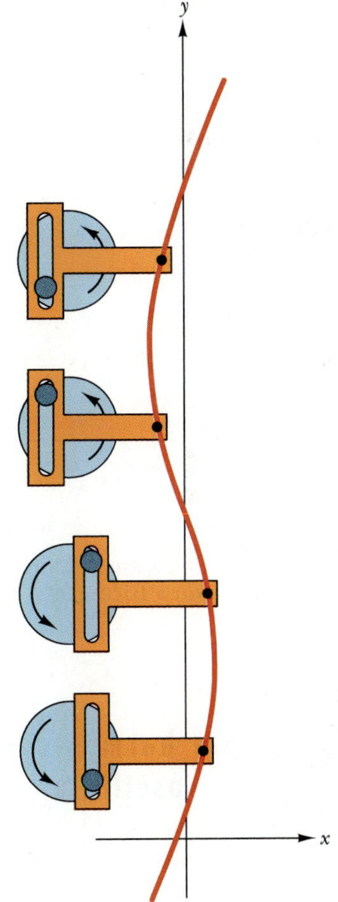

FIGURE 9.4

Rotating wheel with a peg driving a slotted arm back and forth. The drawing shows several "snapshots" of the wheel at different times.

Finally, let us exploit the relationship between simple harmonic motion and uniform circular motion to calculate the instantaneous acceleration in simple harmonic motion. Since the x coordinate of the satellite always

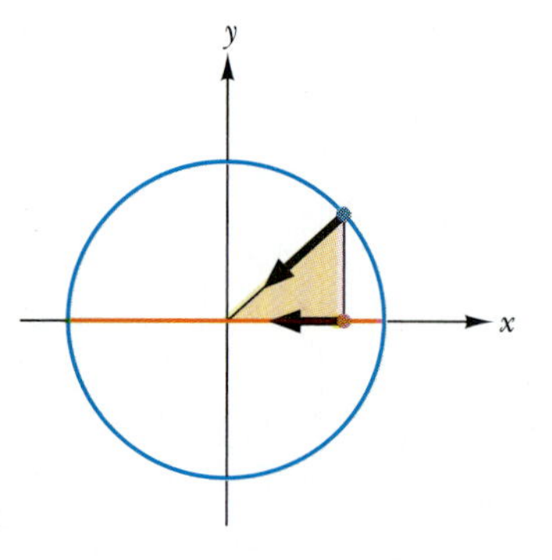

FIGURE 9.5
Acceleration of particle and acceleration of satellite. The x components of these acceleration are equal, and they are both negative.

coincides with the x coordinate of the particle, the x component of the acceleration of the satellite must coincide with the acceleration of the particle. From Chapter 7 [see Eq. (7.15)] we know that the acceleration of the satellite in its uniform circular motion is centripetal, of magnitude

$$a_{\text{cent}} = \omega^2 A \tag{11}$$

From Fig. 9.5 we see that the component of this acceleration in the x direction is

$$-a_{\text{cent}} \cos\theta = -\omega^2 A \cos\theta \tag{12}$$

Here, we have included a negative sign because, as shown in Fig. 9.5, the x acceleration is in the negative x direction. This x acceleration of the satellite must coincide with the acceleration of the particle; thus the acceleration of the particle must be

$$a = -\omega^2 A \cos\theta \tag{13}$$

Sine $\theta = \omega t$, we can express this as

$$a = -\omega^2 A \cos(\omega t) \tag{14}$$

Alternatively, by means of Eq. (4), we can express this as

$$a = -\omega^2 x \tag{15}$$

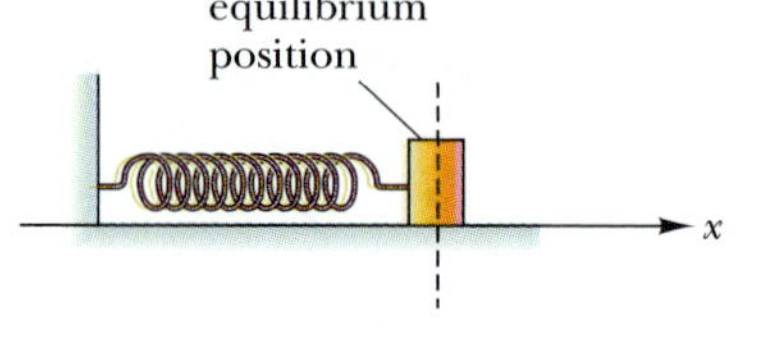

FIGURE 9.6
Mass attached to a spring sliding back and forth on a frictionless surface.

This equation shows that ***the instantaneous acceleration of a particle in simple harmonic motion is proportional to the instantaneous distance x, but is in the opposite direction.*** This proportionality is a characteristic feature of simple harmonic motion, a fact that will be useful in the next section.

9.2 THE SIMPLE HARMONIC OSCILLATOR

Simple harmonic oscillator ▸

The simple harmonic oscillator consists of a mass coupled to an ideal, massless spring which obeys Hooke's Law, that is, a spring that provides a force proportional to the elongation or compression of the spring. One end of the spring is attached to the mass, and the other end is held fixed (see Fig. 9.6). We will ignore gravity and friction, so the spring force is the only force acting on the mass. The system has an equilibrium position corresponding to the relaxed length of the spring. If the mass is initially at some distance from this equilibrium position, then the stretched spring supplies a restoring force which pulls the mass toward the equilibrium position. The mass speeds up as it moves toward the equilibrium position, and it overshoots the equilibrium position. Then, the mass begins to compress the spring and slows down, coming to rest at the other side of the equilibrium position, at a distance equal to its initial distance. The compressed spring then pushes the mass back toward the equilibrium position. The mass again speeds up, overshoots the equilibrium position, and so on. The result is that the mass oscillates back and forth — forever if there is no friction.

A practical example of such a mass–spring system is the device used aboard Skylab for the daily measurement of the masses of the astronauts. This consists of a spring coupled to a chair into which the astronaut is strapped (see Fig. 9.7). Pushed by the spring, the chair with the astronaut oscillates back and forth. As we will see below, the frequency of oscillation of the mass–spring system depends on the mass, and therefore the frequency can serve as an indicator of the mass of the astronaut.

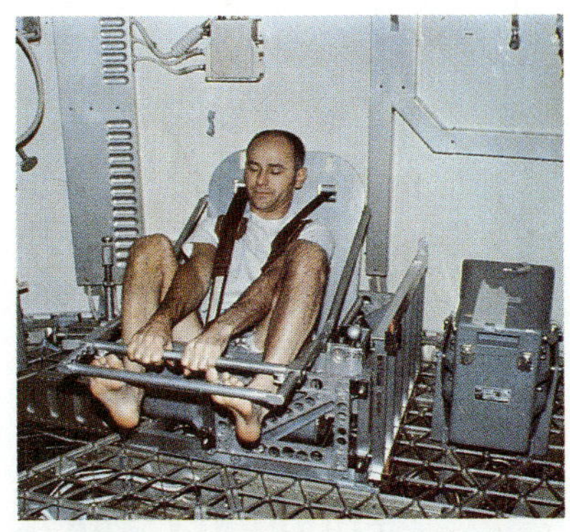

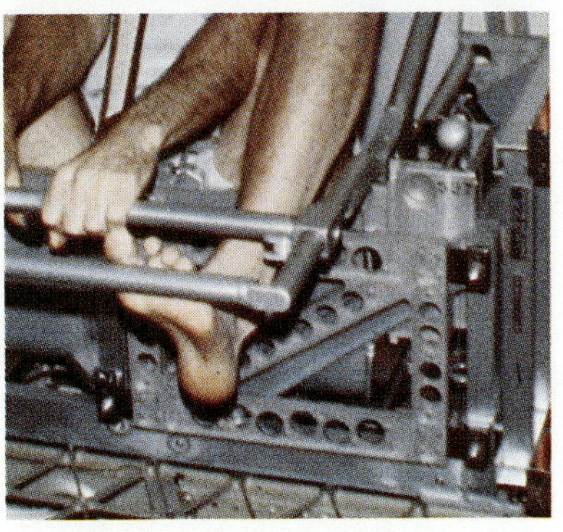

FIGURE 9.7

Body-mass measurement device on Skylab (top). The leaf spring is attached to the corners of the triangular frame on the right (bottom).

Aside from such practical applications, the great importance of the simple harmonic oscillator is that many physical systems are mathematically equivalent to simple harmonic oscillators; that is, these systems have an equation of motion of the same mathematical form as the simple harmonic oscillator. A pendulum, the balance wheel of a watch, a tuning fork, the air in an organ pipe, and the atoms in a diatomic molecule are systems of this kind; the restoring force and the inertia are of the same mathematical form in these systems as in the simple harmonic oscillator, as we can transcribe the general mathematical results from the latter to the former.

To obtain the equation of motion of the simple harmonic oscillator, we begin with Hooke's Law for the force exerted by the spring on the mass [compare Eq. (3.61)]:

$$F = -kx \tag{16}$$

Here the displacement is measured from the equilibrium position, which corresponds to $x = 0$. The constant k is the spring constant. Note that the force is negative if x is positive (stretched spring; see Fig. 9.8a); and the force is positive if the displacement is negative (compressed spring; see Fig. 9.8b).

With the force as given by Eq. (16), the equation of the motion of the mass is

$$ma = -kx \tag{17}$$

This equation says that the acceleration of the mass is always proportional to the distance x, but is in the opposite direction. We now recall, from Eq. (15), that such a proportionality of acceleration and distance is characteristic of simple harmonic motion, and we therefore conclude that the motion of a mass coupled to a spring must be simple harmonic motion. By comparing Eqs. (17) and (15), we see that these equations become identical if

$$\omega^2 = \frac{k}{m} \tag{18}$$

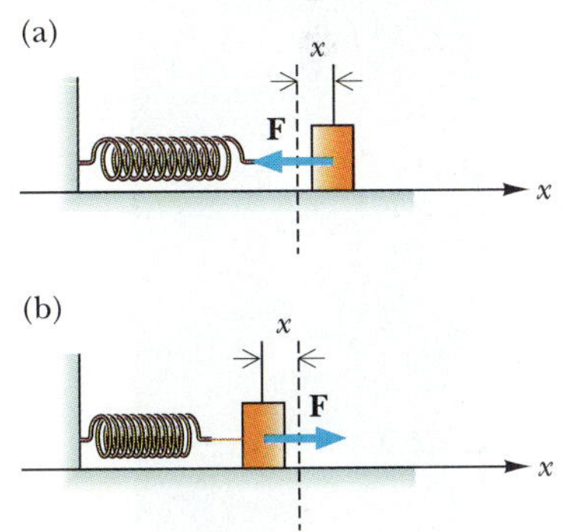

FIGURE 9.8

(a) Positive displacement of the mass; the force exerted by the spring is negative. (b) Negative displacement of the mass; the force exerted by the spring is positive.

and we therefore conclude that the angular frequency of the oscillation of the mass on a spring is

$$\omega = \sqrt{\frac{k}{m}} \tag{19}$$

Consequently, the frequency and the period are

Frequency and period for simple harmonic oscillator ▶

$$f = \frac{\omega}{2\pi} = \frac{1}{2\pi}\sqrt{\frac{k}{m}} \tag{20}$$

and

$$T = \frac{1}{f} = 2\pi\sqrt{\frac{m}{k}} \tag{21}$$

With the value (19) for the angular frequency, the expression (4) for the position as a function of time becomes

$$x = A\cos\left(\sqrt{\frac{k}{m}}t\right) \tag{22}$$

This formula, like the formulas in the preceding section, assumes that the mass is at the turning point $x = A$ at time $t = 0$. If, instead, the mass is passing (in the negative direction) through the equilibrium point $x = 0$ at time $t = 0$, then we must replace the time t in Eq. (22) by $t + \frac{1}{4}T$, which is equivalent to starting our clock earlier by one quarter cycle.

According to Eq. (21) the frequency of the motion of the simple harmonic oscillator depends *only* on the spring constant and on the mass. ***The frequency of the oscillator will always be the same, regardless of the amplitude with which it has been set swinging.*** This property of the oscillator is called ***isochronism.***

Note that the period is long if the mass is large and the spring constant is small. This is as expected, since in each period the spring must accelerate and decelerate the mass, and a weak spring will give a large mass only little acceleration.

Figure 9.9 shows a multiple-exposure picture of the oscillations of a mass on a spring. The picture illustrates the variations of speed in simple harmonic motion: the mass moves at high speed near the midpoint of the motion, and at low speed near the endpoints.

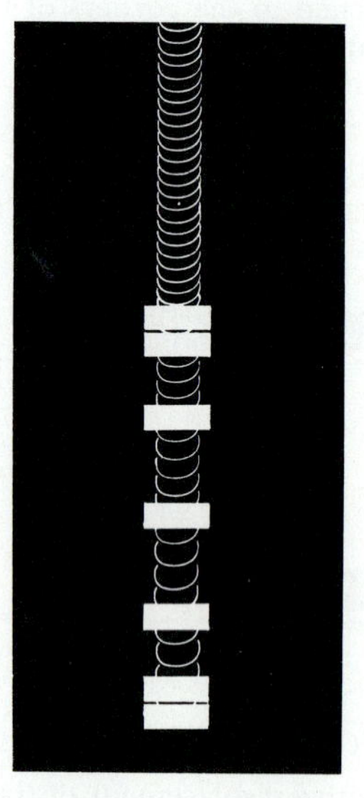

FIGURE 9.9
Stroboscopic photograph of an oscillating mass on a spring. Note that the mass moves slowly at the extremes of its motion.

EXAMPLE 1. You place a heavy encyclopedia, of mass 8 kg, on a kitchen scale (a spring scale; see Fig. 9.10). You notice that before coming to equilibrium, the pointer of the scale oscillates back and forth around the equilibrium position a few times with a period of 0.4 s. What is the spring constant of the internal spring of the kitchen scale?

FIGURE 9.10
A heavy book on a spring scale oscillates up and down.

SOLUTION: The mass of 8 kg in conjunction with the internal spring of the scale form a mass-and-spring system, to which we can apply Eq. (21). If we square both sides of this equation, we obtain

$$T^2 = 4\pi^2 \frac{m}{k}$$

which gives us

$$k = \frac{4\pi^2 m}{T^2}$$

With $m = 8$ kg and $T = 0.4$ s, this becomes

$$k = 4\pi^2 \times \frac{8 \text{ kg}}{(0.4 \text{ s})^2} = 2 \times 10^3 \text{ N/m}$$

COMMENTS AND SUGGESTIONS: In this example, there is not only the force of the spring acting on the mass, but also the weight of the mass. The weight determines where the spring will reach equilibrium, but the weight has no direct effect on the frequency of oscillation around equilibrium; this frequency depends exclusively on the spring constant.

EXAMPLE 2. A particle of 4.0 kg is moving along the x axis under the influence of the force of a spring with $k = 3.5 \times 10^2$ N/m. There are no other forces acting on the particle. The equilibrium point is at $x = 0$. Suppose that at $t = 0$, the particle is (instantaneously) at rest at $x = 0.20$ m. Where will the particle be at $t = 0.05$ s? At $t = 0.10$ s? What will be the speed of the particle when it passes through the equilibrium point?

SOLUTION: Since the particle is initially at rest at $x = 0.20$ m, this must be one of the turning points of the motion; thus, the amplitude of the motion must be $A = 0.20$ m. At time $t = 0.05$ s, the position of the particle will be

$$x = A\cos(\omega t) = 0.20 \text{ m} \times \cos(\omega \times 0.05 \text{ s})$$

To evaluate this, we need the angular frequency of the oscillation,

$$\omega = \sqrt{\frac{k}{m}} = \sqrt{\frac{3.5 \times 10^2 \text{ N/m}}{4.0 \text{ kg}}}$$

$$= 9.4 \text{ radians/s}$$

With this value of ω, we obtain

$$x = 0.20 \text{ m} \times \cos(9.4 \text{ radians/s} \times 0.05 \text{ s})$$

$$= 0.20 \text{ m} \times \cos(0.47 \text{ radian}) = 0.20 \text{ m} \times 0.89 = 0.18 \text{ m}$$

Likewise, at time $t = 0.10$ s, the position is

$$x = A\cos(\omega t) = 0.20 \text{ m} \times \cos(9.4 \text{ radians/s} \times 0.10 \text{ s})$$

$$= 0.20 \text{ m} \times \cos(0.94 \text{ radian}) = 0.20 \text{ m} \times 0.59 = 0.12 \text{ m}$$

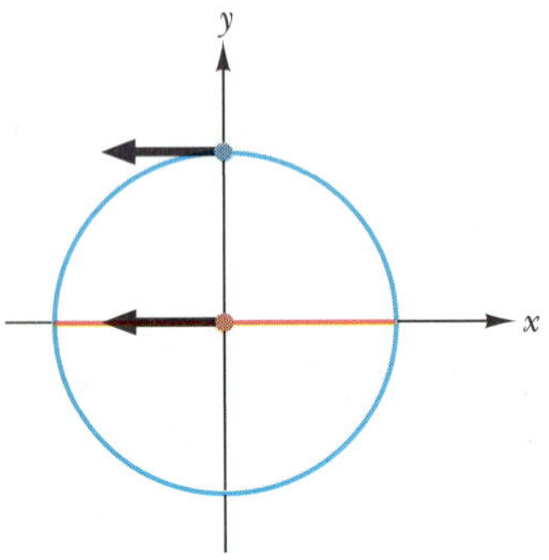

To find the speed when the particle passes through the equilibrium point, we again use the relationship between simple harmonic motion and uniform circular motion discussed in Section 9.1. When the particle passes through the equilibrium point, its motion is instantaneously parallel to the motion of the satellite particle (see Fig. 9.11). Thus, the speeds of the particle and the satellite are equal at this time. The speed of the satellite is the circumference of the reference circle divided by the period of the motion:

$$v = \frac{2\pi A}{T}$$

In view of Eqs. (20) and (21), the speed can be expressed as

$$v = 2\pi f A = \omega A \tag{23}$$

This must then also be the speed of the particle. With the numbers specified in the problem, the speed becomes

$$v = 9.4 \text{ radians/s} \times 0.20 \text{ m} = 1.9 \text{ m/s}$$

FIGURE 9.11
When the particle passes through the equilibrium point, the satellite passes through the y axis, and its direction of motion is parallel to that of the particle.

The relationship between simple harmonic motion and uniform circular motion we used in Example 2 to obtain the velocity at the equilibrium point can also be used to obtain the velocity at any point. It can be shown that the general result for the instantaneous velocity in simple harmonic motion is

$$v = -\omega A \sin(\omega t) \tag{24}$$

As expected, the instantaneous velocity calculated from this formula is zero at $t = 0$, when the mass is at the turning point; and the instantaneous velocity attains a maximum magnitude of ωA when the mass passes through the equilibrium point [note that the maximum magnitude of $\sin(\omega t)$ is 1].

9.3 KINETIC ENERGY AND POTENTIAL ENERGY

We know from Section 4.5 that the force exerted by a spring is a conservative force, for which we can construct a potential energy. With this potential energy, we obtain a law of conservation of the mechanical energy: the sum of the kinetic energy and the potential energy is a constant,

$$E = K + U = [\text{constant}]$$

In this section we will see how to calculate the kinetic energy and the potential energy of the simple harmonic oscillator at each instant of time, and we will verify explicitly that the sum of these energies is constant.

The kinetic energy of a moving mass is

$$K = \tfrac{1}{2}mv^2 \tag{25}$$

For simple harmonic motion, the speed is given by Eq. (24), and the kinetic energy becomes

$$K = \tfrac{1}{2}mv^2 = \tfrac{1}{2}m[-\omega A \sin(\omega t)]^2 \tag{26}$$

$$= \tfrac{1}{2}m\omega^2 A^2 \sin^2(\omega t)$$

Since $m\omega^2 = k$, we can also write this as

$$K = \tfrac{1}{2}kA^2 \sin^2(\omega t) \tag{27}$$

The potential energy associated with the force $F = -kx$ is [see Eq. (4.41)]

$$U = \tfrac{1}{2}kx^2 \tag{28}$$

For simple harmonic motion, this becomes

$$U = \tfrac{1}{2}k[A \cos(\omega t)]^2$$

$$= \tfrac{1}{2}kA^2 \cos^2(\omega t) \tag{29}$$

The kinetic energy and the potential energy both depend on time. According to Eqs. (27) and (29), each oscillates between a minimum value of zero and a maximum value of $\tfrac{1}{2}kA^2$. At the initial time $t = 0$, the mass is at maximum distance from the equilibrium point and its instantaneous speed is zero; thus, the potential energy is at its maximum value, and the kinetic energy is zero. A quarter of a cycle later, the mass passes through the equilibrium point and attains its maximum speed; thus, the kinetic energy is at its maximum value and the potential energy is zero. Figure 9.12 plots the oscillations of the kinetic energy and the potential energy as a function of time.

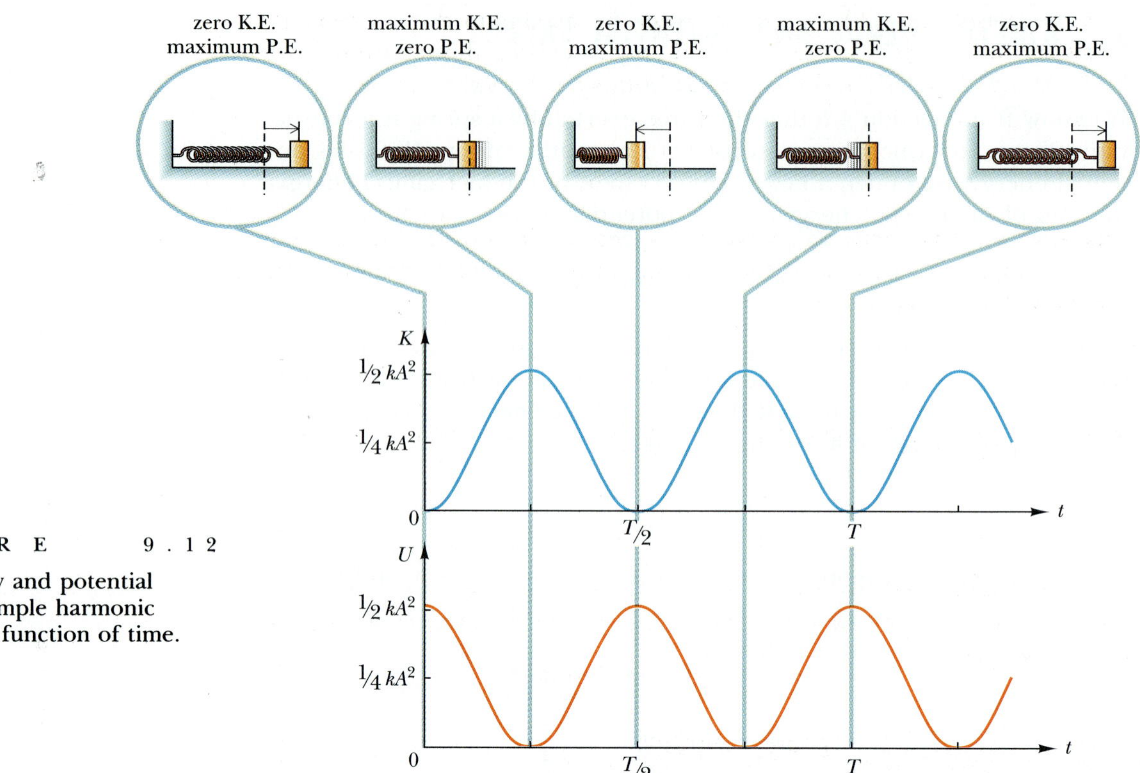

FIGURE 9.12
Kinetic energy and potential energy of a simple harmonic oscillator as a function of time.

Since the force $F = -kx$ is conservative, the total mechanical energy $E = K + U$ is a constant of the motion. To verify this conservation law for the energy explicitly, we take the sum of Eqs. (27) and (29),

$$E = K + U \tag{30}$$

$$= \tfrac{1}{2}kA^2 \sin^2(\omega t) + \tfrac{1}{2}kA^2 \cos^2(\omega t)$$

$$= \tfrac{1}{2}kA^2[\sin^2(\omega t) + \cos^2(\omega t)] \tag{31}$$

We can simplify this expression if we use the trigonometric identity $\sin^2\beta + \cos^2\beta = 1$, which is valid for any angle β.* With this identity, we find that the right side of Eq. (31) is simply $\frac{1}{2}kA^2$:

Energy of simple harmonic oscillator ▶

$$E = \tfrac{1}{2}kA^2 \tag{32}$$

This shows that the energy of the motion is constant and is proportional to the square of the amplitude of oscillation.

---

* For a demonstration of this identity, consider a right triangle with an angle β and a hypotenuse of length 1. The opposite side in this triangle is $\sin\beta$, and the adjacent side is $\cos\beta$. The identity $\sin^2\beta + \cos^2\beta = 1$ is then simply the Pythagorean theorem — the sum of the squares of the two sides equals the square of the hypotenuse.

By means of Eq. (32), we can express the maximum displacement in terms of the energy. For this, we need only to extract the square root of both sides of Eq. (32), and then divide both sides by $\sqrt{k/2}$:

$$x_{\text{max}} = A = \sqrt{2E/k} \tag{33}$$

Likewise, we can express the maximum speed in terms of the energy. For this, we note that when the particle passes through the equilibrium point, the energy is purely kinetic,

$$E = \tfrac{1}{2}mv_{\text{max}}^2 \tag{34}$$

Again, we extract the square root of both sides, and then divide both sides by $\sqrt{m/2}$:

$$v_{\text{max}} = \sqrt{2E/m} \tag{35}$$

These equations tell us that both the maximum displacement and the maximum speed increase with the energy — they both increase in proportion to the square root of the energy.

EXAMPLE 3. For the particle of 4.0 kg moving under the influence of a spring with $k = 3.5 \times 10^2$ N/m, as described in Example 2, what is the total mechanical energy? What is the kinetic energy and what is the potential energy at $t = 0$? What is the kinetic energy and what is the potential energy at $t = 0.10$ s?

SOLUTION: The total mechanical energy is

$$E = \tfrac{1}{2}kA^2 = \tfrac{1}{2} \times 3.5 \times 10^2 \text{ N/m} \times (0.20 \text{ m})^2$$

$$= 7.0 \text{ J}$$

At $t = 0$, the particle is at rest at $x = 0.20$ m. The kinetic energy is zero, and the potential energy is at its maximum,

$$U = \tfrac{1}{2}kA^2 = 7.0 \text{ J}$$

At $t = 0.10$ s, the kinetic energy is, by Eq. (27),

$$K = \tfrac{1}{2}mv^2 = \tfrac{1}{2}kA^2 \sin^2(\omega t)$$

$$= \tfrac{1}{2} \times 3.5 \times 10^2 \text{ N/m} \times (0.20 \text{ m})^2 \times \sin^2(9.4 \text{ radians/s} \times 0.10 \text{ s})$$

$$= \tfrac{1}{2} \times 3.5 \times 10^2 \text{ N/m} \times (0.20 \text{ m})^2 \times (0.81)^2$$

$$= 4.6 \text{ J}$$

and the potential energy is, by Eq. (29),

$$U = \tfrac{1}{2}kA^2 \cos^2(\omega t)$$

$$= \tfrac{1}{2} \times 3.5 \times 10^2 \text{ N/m} \times (0.20 \text{ m})^2 \times \cos^2(9.4 \text{ radians/s} \times 0.10 \text{ s})$$

$$= \tfrac{1}{2} \times 3.5 \times 10^2 \text{ N/m} \times (0.20 \text{ m})^2 \times (0.59)^2$$

$$= 2.4 \text{ J}$$

Note that the sum of these kinetic and potential energies is $K + U = 4.6 \text{ J} + 2.4 \text{ J} = 7.0 \text{ J}$, which agrees with our result for the total mechanical energy.

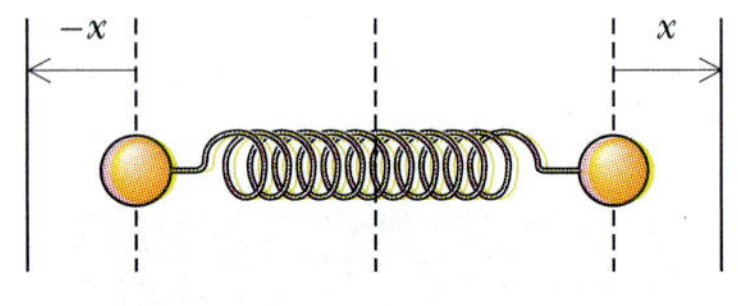

FIGURE 9.13

A hydrogen molecule, represented as two masses joined by a spring. The masses move symmetrically relative to the center of mass.

EXAMPLE 4. The hydrogen molecule (H_2) may be regarded as two masses joined by a spring (see Fig. 9.13). The center of the spring is the center of mass of the molecule. This point can be assumed to remain fixed, so the molecule consists of two identical simple harmonic oscillators vibrating in opposite directions. The spring constant for each of these oscillators is 1.13×10^3 N/m, and the mass of each hydrogen atom is 1.67×10^{-27} kg. Suppose that the vibrational energy of the molecule is 1.3×10^{-19} J. Find the corresponding amplitude of oscillation and the maximum speed.

SOLUTION: Each atom has half the energy of the molecule; thus, the energy per atom is

$$E = \tfrac{1}{2} \times 1.3 \times 10^{-19} \text{ J} = 6.5 \times 10^{-20} \text{ J}$$

According to Eqs. (33) and (35), the amplitude of oscillation and the maximum speed of each atom are then

$$x_{\text{max}} = \sqrt{\frac{2E}{k}} \tag{36}$$

$$= \sqrt{\frac{2 \times 6.5 \times 10^{-20} \text{ J}}{1.13 \times 10^3 \text{ N/m}}} = 1.1 \times 10^{-11} \text{ m}$$

and

$$v_{\text{max}} = \sqrt{\frac{2E}{m}} \tag{37}$$

$$= \sqrt{\frac{2 \times 6.5 \times 10^{-20} \text{ J}}{1.67 \times 10^{-27} \text{ kg}}} = 8.8 \times 10^3 \text{ m/s}$$

9.4 THE SIMPLE PENDULUM

A ***simple pendulum*** consists of a bob suspended by a string or a rod (see Fig 9.14). The bob is assumed to behave like a particle of mass m, and the string is assumed to be massless. Gravity acting on the bob provides a restoring force. When in equilibrium, the pendulum hangs vertically, just like a plumb line. When released at some angle with the vertical, the pendulum will swing back and forth along an arc of circle (see Fig. 9.15). The motion is two-dimensional; however, the position of the pendulum can be completely described by a single parameter: the angle between the string and the vertical (see Fig. 9.14). We will reckon this angle as positive on the right side of the vertical, and as negative on the left side.

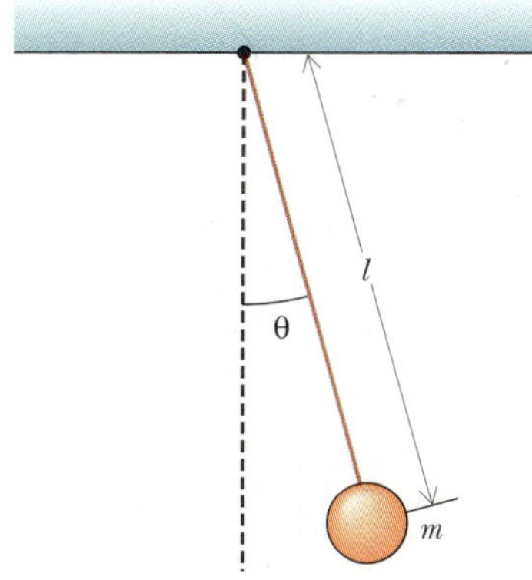

FIGURE 9.14
A pendulum. The angle θ is reckoned as positive if the deflection of the pendulum is toward the right, as in this figure.

Since the bob and the string swing as a rigid unit, the motion can be regarded as rotation about a horizontal axis through the point of suspension, and the equation of motion is that of a rigid body [see Eq. (7.45)]:

$$I\alpha = \tau \tag{38}$$

Here the moment of inertia I and the torque τ are reckoned about the horizontal axis through the point of suspension.

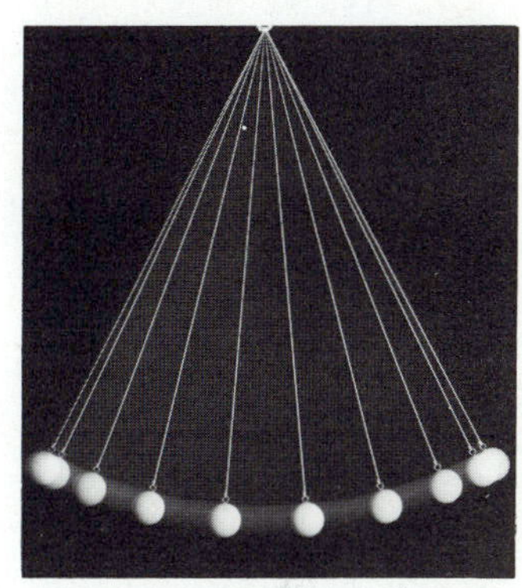

FIGURE 9.15
Stroboscopic photograph of a swinging pendulum.

Figure 9.16 shows a "free-body" diagram for the string–bob system with all the external forces. These external forces are the weight **w** acting on the mass m and the suspension force **S** acting on the string at the point of support. The suspension force exerts no torque, since its point of application is on the axis of rotation (its moment arm is zero). The weight exerts a torque

$$\tau = -mgl \sin\theta \tag{39}$$

where l is the length of the pendulum, measured from the point of suspension to the center of the bob. The minus sign in Eq. (39) indicates that this is a restoring torque, which tends to pull the pendulum to its equilibrium position.

The moment of inertia I is simply that of a particle of mass m at a distance l from the axis of rotation:

$$I = ml^2 \tag{40}$$

Hence the equation of motion becomes

$$ml^2\alpha = mgl \sin\theta \tag{41}$$

or

$$l\alpha = -g \sin\theta \tag{42}$$

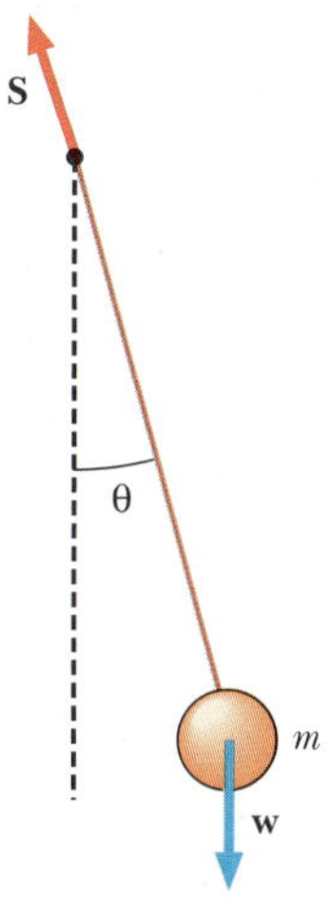

FIGURE 9.16
"Free-body" diagram for the string–pendulum system.

We will solve this equation of motion only in the special case of small oscillations about the equilibrium position. If θ is small, we can make the

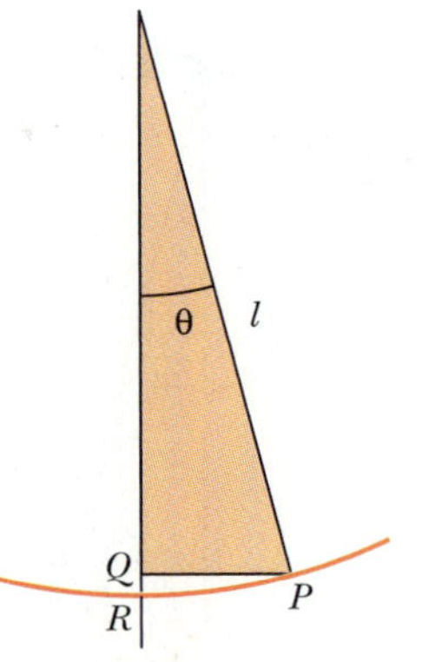

FIGURE 9.17
If the angle θ is small, the length of the straight line PQ is approximately the same as the length of the circular arc PR.

approximation

$$\sin\theta \simeq \theta \tag{43}$$

where the angle is measured in radians. To understand how this approximation comes about, consider the small angle θ shown in Fig. 9.17. The sine of this angle is $\sin\theta = PQ/l$. If θ is small, the length of the straight line PQ is approximately the same as the length of the curved circular are PR. Thus, $\sin\theta \simeq PR/l$. But the ratio PR/l is the angle θ expressed in radians. Thus, $\sin\theta \simeq \theta$. This approximation is usually satisfactory if θ is less than 10°, or less than 0.17 radian.

With this approximation, the equation of motion becomes

$$l\alpha = -g\theta \tag{44}$$

This equation has the same mathematical form as Eq. (17). Comparing these two equations, we see that the angle θ replaces the distance x, the angular acceleration α replaces the linear acceleration a, l replaces m, and g replaces k. Hence the angular motion is simple harmonic. Making the appropriate replacements in Eqs. (4) and (19), we find that the motion is described by the equation

$$\theta = A\cos(\omega t) \tag{45}$$

with an angular frequency

$$\omega = \sqrt{\frac{g}{l}} \tag{46}$$

The frequency and the period of the pendulum are then

Frequency and period of pendulum ▸

$$f = \frac{\omega}{2\pi} = \frac{1}{2\pi}\sqrt{\frac{g}{l}}$$

and

$$T = \frac{2\pi}{\omega} = 2\pi\sqrt{\frac{l}{g}} \tag{47}$$

Note that these expressions for the frequency and the period depend only on the length of the pendulum and on the acceleration of gravity; they do not depend on the mass of the pendulum bob or on the amplitude of oscillation (but, of course, our calculation depends on the assumption that the angle θ is small).

Like the simple harmonic oscillator, the pendulum has the property of ***isochronism — its frequency is (approximately) independent of the ampli-***

tude with which it is swinging. This property can be easily verified by swinging two pendulums of equal lengths side by side, with different amplitudes. The pendulums will continue to swing in step for a long while.

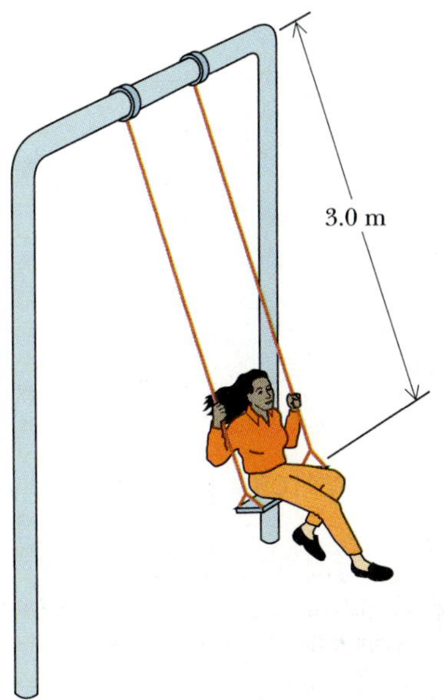

FIGURE 9.18
Woman on a swing.

EXAMPLE 5. A woman sits in a swing of length 3.0 m (see Fig. 9.18). What is the period of oscillation of this swing?

SOLUTION: We can regard the swing as a pendulum of an approximate length 3.0 m. From Eq. (47) we then find

$$T = 2\pi\sqrt{\frac{l}{g}}$$

$$= 2\pi\sqrt{\frac{3.0\text{ m}}{9.81\text{ m/s}^2}} = 3.5\text{ s}$$

FIGURE 9.19
Pendulum clock of astronomical observatory.

EXAMPLE 6. The "seconds" pendulum in a pendulum clock built for an astronomical observatory has a period of exactly 2.0 s, so each one-way motion of the pendulum takes exactly 1.0 s (see Fig. 9.19). What is the length of such a "seconds" pendulum at a place where the acceleration of gravity is $g = 9.81\text{ m/s}^2$? At a place where the acceleration of gravity is 8.79 m/s^2?

SOLUTION: If we square both sides of Eq. (47) and then solve for l, we find

$$l = \left(\frac{T}{2\pi}\right)^2 g \qquad (48)$$

With $g = 9.81\text{ m/s}^2$, this gives

$$l = \left(\frac{2.0\text{ s}}{2\pi}\right)^2 \times 9.81\text{ m/s}^2 = 0.994\text{ m}$$

With $g = 8.79\text{ m/s}^2$, it gives

$$l = \left(\frac{2.0\text{ s}}{2\pi}\right)^2 \times 8.79\text{ m/s}^2 = 0.992\text{ m}$$

The most familiar application of pendulums is the construction of pendulum clocks. Up to about 1950, the most accurate clocks were pendulum clocks of a special design, which were kept inside airtight flasks placed in deep cellars to protect them from disturbances caused by the variations of the atmospheric pressure and temperature (see Fig. 9.20). The best of these high-precision pendulum clocks were accurate to within a few thousandths of a second per day. Later, such pendulums were superseded by quartz clocks and then by atomic clocks.

Another important application of pendulums is the measurement of the acceleration of gravity g. For this purpose it is only necessary to time the swings of a pendulum of known length; the value of g can then be calculated from Eq. (47). The pendulums used in precise determinations of g usually consist of a solid bar swinging about one end, instead of a bob on a string. Such a pendulum consisting of a swinging rigid body is called a ***physical pendulum;*** its period is related to its size and shape. The pendulum method for the determination of the acceleration of gravity is very precise and very convenient, but modern electronic instrumentation permits the attainment of even higher precision by the direct timing of the free-fall motion of small projectiles in an evacuated chamber.

Physical pendulum ▶

Finally, we must emphasize that the approximation contained in Eq. (43) is valid only for small angles. If the amplitude of oscillation of the pendulum is more than a few degrees — say, 10 degrees — the approximation of Eq. (43) begins to fail, and the motion of the pendulum begins to deviate from simple harmonic motion. At large amplitudes, the period of the pendulum depends on the amplitude — the larger the amplitude, the larger the period. For instance, a pendulum oscillating with an amplitude of 30° has a period 1.7% longer than the value given by Eq. (47).

FIGURE 9.20

This electromechanical clock, regulated by a pendulum, served as the U.S. frequency standard in the 1920s. Its master pendulum is enclosed in the canister at right.

9.5* DAMPED OSCILLATIONS AND FORCED OSCILLATIONS

So far we have proceeded on the assumption that the only force acting on a simple harmonic oscillator or a pendulum is the restoring force $F = -kx$ or the restoring torque $\tau = -mgl \sin\theta$. However, in a real oscillator or a real pendulum, there is always some extra force caused by friction. For instance, if the pendulum starts its swinging motion with some initial amplitude, the friction against the air and against the point of support will gradually brake the pendulum, reducing its amplitude of oscillation.

Figure 9.21 is a plot of the position as a function of time for a harmonic oscillator with fairly strong friction. The amplitude of oscillation suffers a noticeable decrease from one cycle to the next. ***Such a gradually decreasing oscillation is called damped harmonic motion.***

Damped harmonic motion ▶

Since the oscillator must do work against the friction, the mechanical energy gradually decreases. The energy loss per cycle is usually a constant fraction of the energy E that the oscillator has at the beginning of the cycle. If we represent the energy change per cycle by ΔE, then ΔE is pro-

* This section is optional.

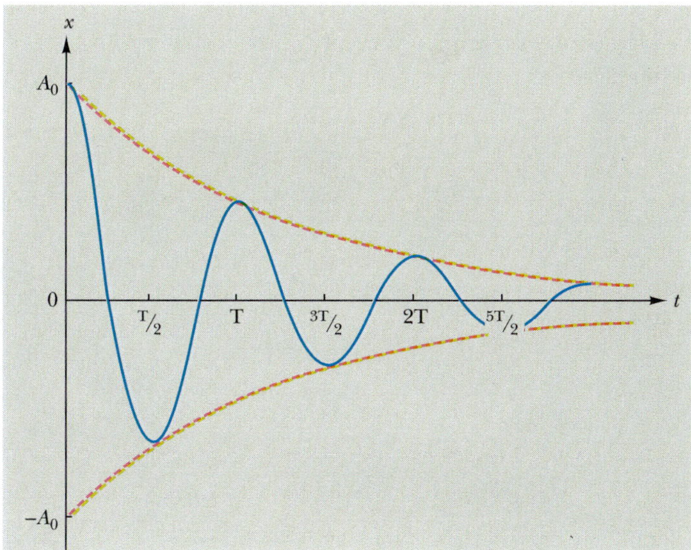

FIGURE 9.21
Plot of position vs. time for a particle with damped harmonic motion.

portional to E,

$$\Delta E = -\left(\frac{2\pi}{Q}\right)E \tag{49}$$

Here, the constant of proportionality has been written in the somewhat complicated form $2\pi/Q$, which is the form usually adopted in engineering. The quantity Q is called the "quality factor" of the oscillator. An oscillator with low friction as a high value of Q, and a small energy loss per cycle; an oscillator with high friction has a low value of Q, and a large energy loss per cycle. The value of Q roughly coincides with the number of cycles the oscillator completes before the oscillations damp away significantly. Mechanical oscillators of low friction, such as tuning forks or piano strings, have Q values of a few thousand; that is, they "ring" for a few thousand cycles before their oscillations fade noticeably.

To maintain the oscillations of a damped harmonic oscillator at a constant level, it is necessary to exert some extra force on the oscillator, so the energy fed into the oscillator by this new force compensates for the energy lost to friction. An extra force is also needed to start the oscillations of any oscillator, damped or not, by supplying the initial energy for the motion. Any such extra force exerted on an oscillator is called ***driving force.*** A familiar example of a driving force is the "pumping" force that you must exert on a swing (a pendulum) to start it moving and to keep it moving at a constant amplitude. This is an example of a periodic driving force.

◀ **Driving force**

If the period of the driving force coincides with the period of the natural oscillations of the oscillator, then even a quite small driving force can gradually build up large amplitudues. Essentially, what happens is that under these conditions the driving force steadily feeds energy into the oscillations, and the amplitude of these grows until the friction becomes so large that it inhibits further growth. Thus, the ultimate amplitude

FIGURE 9.22
Collapse of the bridge at Angers, as illustrated in a contemporary newspaper.

reached depends on friction; in an oscillator of low friction, or high Q, this ultimate amplitude can be extremely large. ***The buildup of a large amplitude by the action of a driving force in tune with the natural frequency of an oscillator is called resonance, or sympathetic oscillation.***

Resonance ▶

The phenomenon of resonance plays a crucial role in many pieces of industrial machinery — if one vibrating part of a machine is driven at resonance by a perturbing force originating from some other part, then the amplitude of oscillation can build up to violent levels and shake the machine apart. Such dangerous resonance effects can occur not only in moving pieces of machinery, but also in structures that are normally regarded as static. In a famous accident that took place in 1850 in Angers (France), the stomping of 487 soldiers marching over a suspension bridge excited a resonant swinging motion of the bridge (Fig. 9.22); the motion quickly rose to a disastrous level and broke the bridge apart, causing the death of 226 of the soldiers.

SUMMARY

- **Simple harmonic motion:** $x = A\cos(\omega t)$
- **Period:** $T = 2\pi/\omega$
- **Frequency:** $f = 1/T = \omega/2\pi$
- **Equation of motion of simple harmonic oscillator:** $ma = -kx$
- **Angular frequency and period of simple harmonic oscillator:**

$$\omega = \sqrt{\frac{k}{m}} \qquad T = 2\pi\sqrt{\frac{m}{k}}$$

- **Energy of simple harmonic oscillator:** $E = \frac{1}{2}kA^2$
- **Angular frequency and period of simple pendulum:**

$$\omega = \sqrt{\frac{g}{l}} \qquad T = 2\pi\sqrt{\frac{l}{g}}$$

FIGURE 9.23
Pendulum of grandfather clock, with adjusting screw.

QUESTIONS

1.–If the motion of the piston of the water pump shown in Fig. 9.1 is to be approximately simple harmonic, the rod connecting the piston to the crank must be long. Why does the length of the rod matter?

2.–In our calculation of the frequency of the simple harmonic oscillator, we ignored the mass of the spring. Qualitatively, how does the mass of the spring affect the frequency?

3.–A grandfather clock is regulated by a pendulum. If the clock is running late, how must we readjust the length of the pendulum (Fig. 9.23)?

4.–Figure 9.24 shows the escapement of a pendulum clock, that is, the linkage that permits the pendulum to control the rotation of the wheels of the clock. Explain how the wheel turns as the pendulum swings back and forth.

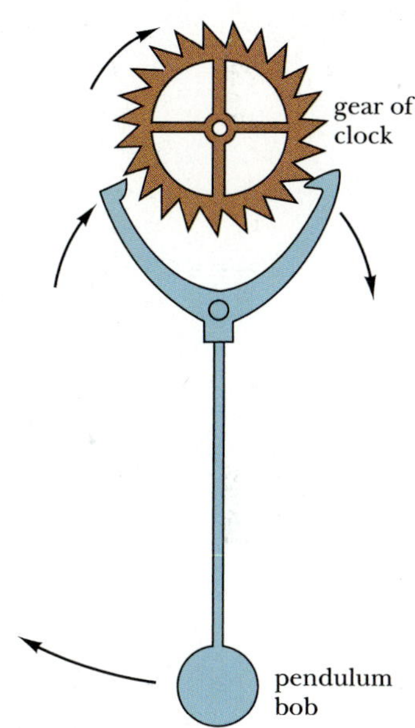

FIGURE 9.24
Escapement mechanism of a pendulum clock. At the instant shown, the tooth at the left is just escaping from the left arm, and the tooth on the right is starting to push against the right arm.

5.–Would a pendulum clock keep good time on a ship?

6.–Galileo claimed that the oscillations of a pendulum are isochronous, even for an amplitude of oscillation as large as 30°. What is your opinion of this claim?

7.–Why would you expect a pendulum oscillating with an amplitude of nearly (but not quite) 180° to have a very long period?

8.–Can a pendulum oscillate with an amplitude of more than 180°?

9.–Figure 9.25 shows a "tilted pendulum" designed by Christian Huygens in the seventeenth century. When the pendulum is tilted, its period is longer than when the pendulum is vertical. Explain.

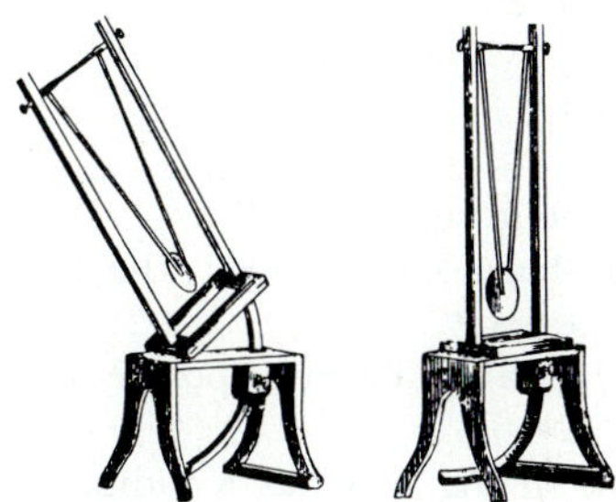

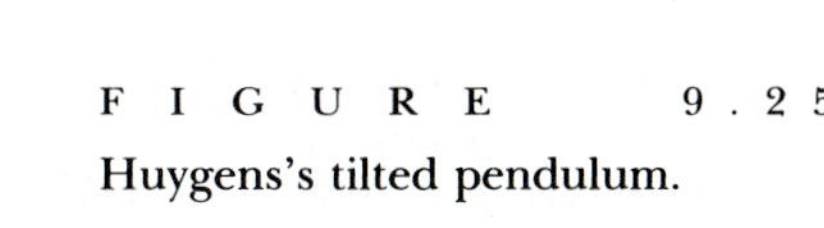

FIGURE 9.25
Huygens's tilted pendulum.

10.–Galileo described an experiment to compare the acceleration of gravity of lead and of cork:

> I took two balls, one of lead and one of cork, the former being more than a hundred times as heavy as the latter, and suspended them from two equal thin strings, each four or five bracchia long. Pulling each ball aside from the vertical, I released them at the same instant, and they, falling along the circumferences of the circles having the strings as radii, passed thru the vertical and returned along the same path. This free oscillation, repeated more than a hundred times, showed clearly that the heavy body kept time with the light body so well that neither in a

> hundred oscillations, nor in a thousand, will the former anticipate the latter by even an instant, so perfectly do they keep step.

Since air friction affects the cork ball much more than the lead ball, do you think Galileo's results are credible?

Newton reported a more careful experiment that avoided the inequality of friction:

> I tried the thing in gold, silver, lead, glass, sand, common salt, wood, water, and wheat. I provided two equal wooden boxes. I filled the one with wood, and suspended an equal weight of gold (as exactly as I could) in the centre of oscillation of the other. The boxes, hung by equal threads of 11 feet, made a couple of pendulums perfectly equal in weight and figure . . . and, placing the one by the other, I observed them to play together forwards and backwards for a long while, with equal vibrations. . . . And by these experiments, in bodies of the same weight, one could have discovered a difference of matter less than the thousandth part of the whole.

Explain how Newton's experiment was better than Galileo's.

11.–A simple pendulum hangs below a table, with its string passing through a small hole in the tabletop. Suppose you gradually pull the string while the pendulum is swinging. What happens to the frequency of oscillation? To the (angular) amplitude?

12.–Shorter people have a shorter length of stride, but a higher rate of step while walking "naturally." Explain.

13.–Suppose that the spring in the front-wheel suspension of an automobile has a natural frequency of oscillation equal to the frequency of rotation of the wheel at, say, 80 km/h. Why is this bad?

14.–When marching soldiers are about to cross a bridge, they break step. Why?

PROBLEMS

Section 9.1: Simple Harmonic Motion

1.——A particle moves as follows as a function of time:

$$x = 3.0\cos(2.0t)$$

where distance is measured in meters and time in seconds.

(a) What is the amplitude of this simple harmonic motion? The frequency? The angular frequency? The period?

(b) At what time does the particle reach the equilibrium point? The turning point?

2.——Suppose that the rotating wheel illustrated in Fig. 9.4 has a radius of 4.0 cm. The wheel turns at a rate of 600 rev/min. What is the amplitude of the simple harmonic motion of the slotted arm? What are the period, the frequency, and the angular frequency?

3.——A particle performs simple harmonic motion along the x axis with an amplitude of 0.20 m and a period of 0.8 s. At $t = 0$, the particle is at maximum distance from the origin, that is, $x = 0.20$ m.

(a) What is the equation that describes the position of the particle as a function of time?

(b) Calculate the position of the particle at $t = 0.1$ s, 0.2 s, 0.3 s, and 0.4 s.

4.——In an electric saber saw, the rotational motion of the electric motor is converted into a back-and-forth motion of the saw blade by a mechanism similar to that shown in Fig. 9.4. Suppose the peg of the rotating wheel moves around a circle of diameter 3.0 cm at 4000 revolutions per minute and thereby moves the slotted arm to which the saw blade is bolted. What is the amplitude and the frequency of the back-and-forth simple harmonic motion of the blade?

5.——Consider the particle executing simple harmonic motion according to Eq. (1).

(a) What is the speed of the satellite particle for this case?

(b) At $t = 2.0$ s, the particle is at the equilibrium point and its instantaneous velocity is parallel to that of the satellite particle. What speed can you deduce from this for the particle?

6.*——Suppose that two particles are performing simple harmonic motion with a period of 8.0 s. The first particle moves according to the equation

$$x = 0.3 \cos\left(\frac{\pi t}{4}\right)$$

and the second particle according to the equation

$$x = 0.3 \sin\left(\frac{\pi t}{4}\right)$$

where the distance is measured in meters and the time in seconds.

(a) When does the first particle reach the equilibrium point? The turning point? Draw a diagram showing the particle and its satellite particle at these times.

(b) When does the second particle reach the equilibrium point? The turning point? Draw a diagram showing the particle and its satellite particle at these times.

(c) By some argument establish that whenever the first particle passes through a point on the x axis, the second particle passes through this same point 2.0 s later.

7.*——A particle moves back and forth along the x axis between the points $x = 0.20$ m and $x = -0.20$ m. The period of the motion is 1.2 s, and it is simple harmonic. At the time $t = 0$, the particle is at $x = 0.20$ m and its velocity is zero.

(a) What is the frequency of the motion? The angular frequency?

(b) What is the amplitude of the motion?

(c) At what time will the particle reach the point $x = 0$? At what time will it reach the point $x = -0.20$ m?

(d) What is the acceleration of the particle when it is at $x = 0$? What is the acceleration of the particle when it reaches the point $x = -0.20$ m?

8.*——The motion of the piston in an automobile engine is approximately simple harmonic (Fig. 9.26). Suppose that the piston travels back and forth over a distance of 8.50 cm and that the piston has a mass of 1.2 kg. What is its maximum acceleration and what is its maximum speed if the engine is turning over at its highest safe rate of 6000 rev/min? What is the maximum force on the piston?

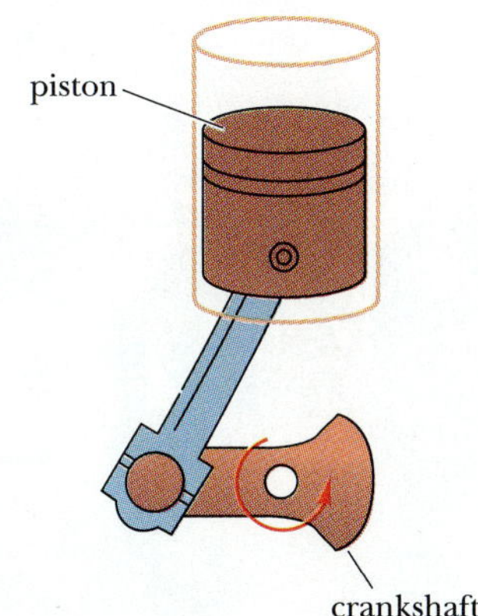

FIGURE 9.26
Piston of automobile engine.

9.*——A given point on a guitar string (say, the midpoint of the string) executes simple harmonic motion with a frequency of 440 Hz and an amplitude of 1.2 mm. What is the maximum acceleration of this motion?

10.* Experience shows that from one-third to one-half of the passengers in an airliner can be expected to suffer motion sickness if the airliner bounces up and down with a peak acceleration of 4 m/s^2 and a frequency of about 0.3 Hz. Assume that this up-and-down motion is simple harmonic. What is the amplitude of the motion?

11.* The frequency of oscillation of a mass attached to a spring is 3.0 Hz. At time $t = 0$, the mass has an initial displacement of 0.20 m and an initial velocity of zero.

(a) What is the position of the mass as a function of time?

(b) When will the mass first reach the equilibrium point? What will be its acceleration at that time?

12.* Consider a particle moving with simple harmonic motion as in Problem 1, with

$$x = 3.0 \cos(2.0\ t)$$

where distance is measured in meters and time in seconds.

(a) To generate this simple harmonic motion with a satellite particle, what must be the radius of the reference circle? What must be the frequency (in rev/s) of the satellite particle?

(b) What is the centripetal acceleration of the satellite particle? What is the corresponding acceleration of the particle (not the satellite!) when it reaches $x = 3.0$ m? When it passes through $x = 1.5$ m?

13.** A particle is performing simple harmonic motion along the x axis according to the equation

$$x = 0.6 \cos\left(\frac{\pi t}{2}\right)$$

where the distance is measured in meters and the time in seconds.

(a) Calculate the position x of the particle at $t = 0$, $t = 0.50$ s, and $t = 1.00$ s.

(b) Calculate the instantaneous velocity of the particle at these times.

(c) Calculate the instantaneous acceleration of the particle at these times.

Section 9.2: The Simple Harmonic Oscillator

14. A simple harmonic oscillator, consisting of a mass of 0.25 kg sliding on a frictionless track while pushed and pulled by a spring, has a period of oscillation of 0.5 s. What will be the new period of oscillation if we decrease the mass to 0.15 kg?

15. A simple harmonic oscillator consists of a mass sliding on a frictionless surface under the influence of a force exerted by a spring connected to the mass. The frequency of this harmonic oscillator is 8.0 Hz. If we connect a second, identical spring to the mass, what will be the new frequency of oscillation?

16. The body-mass measurement device used aboard Skylab consists of a chair supported by a spring. The device was calibrated before the space flight by placing a standard mass of 66.91 kg in the chair; with this mass the period of oscillation of the chair was 2.088 s. During the space flight, astronaut Lousma sat in the chair; the period of oscillation was then 2.299 s. What was the mass of the astronaut? Ignore the mass of the chair.

FIGURE 9.27
Men on pogo sticks.

17. A man of mass 70 kg is bouncing up and down on a pogo stick (see Fig. 9.27). He finds that if he just holds on rigidly and lets the stick do the bouncing (after getting it started), the period of the up-and-down motion

is 0.70 s. What is the spring constant of the spring in the pogo stick? Assume that the bottom of the stick remains in touch with the floor and ignore the mass of the stick.

18.——The body of an automobile of mass 1100 kg is supported by four vertical springs attached to the axles of the wheels. In order to test the suspension, a man pushes down on the body of the automobile and then suddenly releases it. The body rocks up and down with a period of 0.75 s. What is the spring constant of each of the springs? Assume that all the springs are identical and that the compressional force on each spring is the same; also assume that the shock absorbers of the automobile are completely worn out so they do not affect the oscillation frequency.

19.——A Small Mass Measurement Instrument (SMMI) was used on Skylab to measure the masses of biological samples, small animals, chemicals, and other such items used in life-sciences experiments while in orbit (see Fig. 9.28). The sample to be measured is strapped to a tray supported by leaf springs, and the mass is determined from the observed period of oscillation of the tray-and-mass. To calibrate this instrument, a test mass of 1.00 kg is first placed on the tray; the period of oscillation is them 1.08 s. Suppose that when the test mass is removed, and an unknown sample is placed on the tray, the period becomes 1.75 s. What is the mass of the sample? Assume that the mass of the tray (and the straps) is 0.40 kg.

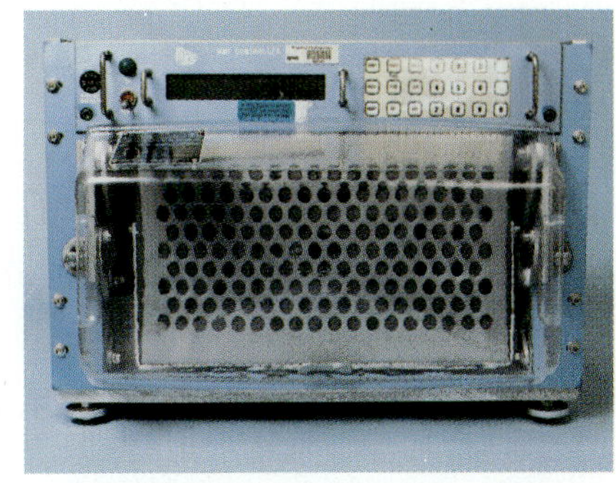

FIGURE 9.28
Small Mass Measurement Instrument.

20.——Ropes used by mountain climbers are quite elastic, and they behave like springs. A rope of 10 m has a spring constant $k = 4.9 \times 10^3$ N/m. Suppose that a mountain climber of 80 kg hangs on this rope, which is stretched vertically down. What is the frequency of up-and-down oscillations of the mountain climber?

21.——In response to a sound wave, the middle of your eardrum oscillates back and forth with a frequency of 4000 Hz and an amplitude of 1.0×10^{-5} m. What is the maximum speed of the eardrum?

22.——Deuterium is an isotope of hydrogen. The mass of the deuterium atom is 1.998 times larger than the mass of the hydrogen atom. Given that the frequency of vibration of the H_2 molecule is 1.31×10^{14} Hz, calculate the frequency of vibration of the D_2 molecule. Assume the "spring" connecting the atoms is the same in H_2 and D_2.

23.——A physicist of 55 kg stands on a bathroom scale (a spring scale, with an internal spring). She observes that when she mounts the scale suddenly, the pointer of the scale at first oscillates back and forth a few times with a frequency of 2.4 Hz.

(a) What value of the spring constant can she deduce from these data?
(b) If she then takes a child of 20 kg in her arms and again stands on the scale, what will be the new frequency of oscillation of the pointer?

24.*——A mass of 6.0 kg is executing simple harmonic motion along the x axis under the influence of a spring. The mass moves according to the equation

$$x = 0.2 \cos(3.0t)$$

where x is measured in meters and t in seconds.

(a) What is the frequency of the motion? What is the spring constant of the spring? What is the maximum speed of the motion?
(b) Suppose we replace the mass by a new mass of 2.0 kg (but we keep the same spring), and suppose we start the motion with the same amplitude of 0.2 m. What will be the new frequency of the motion? What will be the new maximum speed?

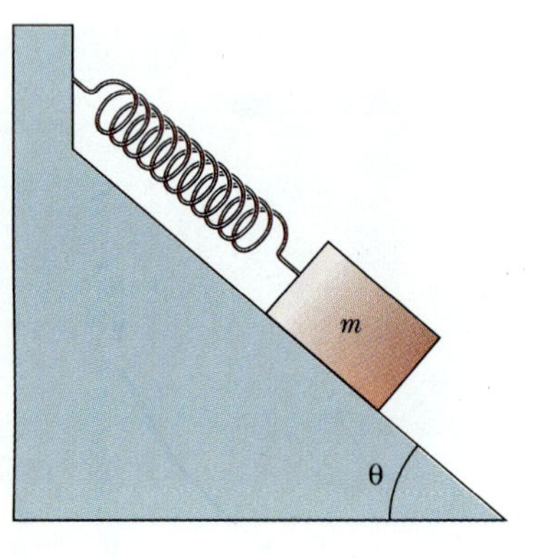

FIGURE 9.29
Mass sliding on an inclined frictionless plane.

25.*—The wheel of a sports car suspended below the body of the car by a vertical spring with a spring constant 1.1×10^4 N/m. The mass of the wheel is 14 kg and the diameter of the wheel is 61 cm.

(a) What is the frequency of up-and-down oscillations of the wheel? Regard the wheel as a mass on one end of a spring and regard the body of the car as a fixed support for the other end of the spring.

(b) Suppose that the wheel is slightly out of round, having a bump on one side. As the wheel rolls on the street it receives a periodic push each time the bump comes in contact with the street. At what speed of the translational motion of the car will the frequency of this push coincide with the natural frequency of the up-and-down oscillations of the wheel? What will happen to the car at this speed? (Note: This problem is not quite realistic because the elasticity of the tire also contributes a restoring force to the up-and-down motion of the wheel.)

26.**—A mass $m = 2.5$ kg hangs from the ceiling by a spring with $k = 90$ N/m. Initially, the spring is in its unstretched configuration and the mass is held at rest by your hand. If, at time $t = 0$, you release the mass, what will be its position as a function of time? (Hint: How far is the equilibrium position below the initial position of the mass?)

27.**—A mass m slides on a frictionless plane inclined at an angle θ with the horizontal. The mass is attached to a spring, parallel to the plane (Fig. 9.29); the spring constant is k. How much is the spring stretched at equilibrium? What is the frequency of the oscillations of this mass up and down the plane?

Section 9.3: Kinetic Energy and Potential Energy

28.—A mass of 8.0 kg is attached to a spring and oscillates with an amplitude of 0.25 m and a frequency of 0.60 Hz. What is the energy of the motion?

29.—A simple harmonic oscillator consists of a mass of 3.0 kg sliding back and forth along a horizontal frictionless track while pushed and pulled by a spring with $k = 6.0 \times 10^2$ N/m. Suppose that initially the mass is released from rest at a distance of 0.25 m from the equilibrium point. What is the energy of this harmonic oscillator? What is the maximum speed it attains when passing through the equilibrium point?

30.—A simple harmonic oscillator consists of a mass of 2.0 kg sliding back and forth along a horizontal frictionless track while pushed and pulled by a spring with $k = 8.0 \times 10^2$ N/m. Suppose that when the mass is at the equilibrium point, it has an instantaneous speed of 3.0 m/s. What is the energy of this harmonic oscillator? What is the amplitude of oscillation?

31.—A simple harmonic oscillator of mass 0.60 kg oscillates with a frequency of 3.0 Hz and an amplitude of 0.15 m. Suppose that, while the mass is instantaneously at rest at its turning point, we quickly attach another mass of 0.60 kg to it. How does this change the amplitude of the motion? The frequency? The energy? The maximum speed? The maximum acceleration?

32.—A simple harmonic oscillator of mass 0.80 kg oscillates with a frequency of 2.0 Hz and an amplitude of 0.12 m. Suppose that, while the mass is instantaneously at rest at its turning point, we quickly shift the fixed end of the spring to a new fixed point, 0.12 m farther away from the mass. How does this change the amplitude of the motion? The frequency? The energy? The maximum speed? The maximum acceleration?

33.—Suppose that a particle of mass 0.24 kg acted upon by a spring undergoes simple harmonic motion with the parameters given in Problem 1.

(a) What is the total energy of this motion?

(b) At what time is the kinetic energy zero? At what time is the potential energy zero?

(c) At what time is the kinetic energy equal to the potential energy?

34.—The separation between the equilibrium positions of the two atoms of a hydrogen molecule is 1.0×10^{-10} m. Using the data given in Example 4, calculate the value of the vibrational energy that corresponds to an amplitude of vibration of 5.0×10^{-11} m for each atom. Is it valid to treat the motion as a small oscillation if the energy has this value?

35.*—Consider a mass m moving along the x axis under the influence of a spring of spring constant k. The equilibrium point is at $x = 0$, and the amplitude of the motion is A.

(a) At what point x is the kinetic energy of the mass equal to its potential energy?

(b) When the mass reaches the point $x = \frac{1}{2}A$, what fraction of its energy is potential, and what fraction is kinetic?

Section 9.4: The Simple Pendulum

36.—The longest pendulum in existence is a 27-m Foucault pendulum in Portland, Oregon. What is the period of this pendulum?

37.—Use your calculator to make a table of sin θ vs. θ for $\theta = 0$ to $\theta = 0.2$ radian at intervals of 0.02 radian. Verify that $\sin\theta \simeq \theta$.

38.—At a construction site, a bucket full of concrete hangs from a crane. You observe that the bucket slowly swings back and forth, 8 times per minute. What is the length of the cable from which the bucket hangs?

39.—A pendulum has a length of 2.0 m. What is the period of this pendulum? If you want to construct a pendulum with exactly half this period, how long would it have to be?

40.—A "seconds" pendulum is a pendulum that has a period of exactly 2.0 s: each one-way swing of the pendulum therefore takes exactly 1.0 s. What is the length of the "seconds" pendulum in Paris ($g = 9.809$ m/s^2), Buenos Aires ($g = 9.797$ m/s^2), and Washington, D.C. ($g = 9.801$ m/s^2)?

41.—A mass suspended from a parachute (which is descending at constant velocity) can be regarded as a pendulum. What is the frequency of pendulum oscillations of a human body suspended 7 m below a parachute?

42.—The elevator cage of a skyscraper hangs from a 300-m-long steel cable. The elevator cage is guided within the elevator shaft by railings. If we remove these railings, and we let the elevator cage swing from side to side (with small amplitude), what is its period of oscillation?

43.—On the Earth, a pendulum of length 0.994 m has a period of 2.00 s (compare Example 6). If we take this pendulum to the surface of Jupiter, where $g = 24.8$ m/s^2, what will be its period?

44.*—The pendulum of a grandfather clock has a length of 0.994 m. If the clock runs late by 1 minute per day, how much must you shorten the pendulum to make it run on time?

45.*—A grandfather clock controlled by a pendulum of length 0.9932 m keeps good time in New York ($g = 9.803$ m/s^2).

(a) If we take this clock to Austin ($g = 9.793$ m/s^2), how many minutes per day will it fall behind?

(b) In order to adjust the clock, by how many millimeters must we shorten the pendulum?

46.*—To test that the acceleration of gravity is the same for a piece of iron and a piece of brass, an experimenter takes a pendulum of length 1.800 m with an iron bob and another pendulum of the same length with a brass bob and starts them swinging in unison. After swinging for 12 min, the two pendulums are no more than one-quarter of a (one-way) swing out of step. What is the largest difference between the values of g for iron and for brass consistent with these data? Express your answer as a fractional difference.

47.*—At the National Institute of Standards and Technology in Washington, D.C., the value of the acceleration of gravity is 9.80095 m/s^2. Suppose that at this location a very precise pendulum, designed for measurements of the acceleration of gravity, has a period of 2.10356 s. If we take this pendulum to a new location at the U.S Coast and Geodetic Survey, also in Washington, D.C., it has a period of 2.10354 s. What is the value of the acceleration of gravity at this new location? What is the percentage change of the acceleration between the two locations?

48.**—The pendulum of a pendulum clock consists of a rod of length 0.99 m with a bob of mass 0.40 kg. The pendulum bob swings back and forth along an arc of length 20 cm.

(a) What is the maximum velocity of the pendulum bob along the arc?

(b) What is the force that the pendulum exerts on its support when it is at the midpoint of its swing? Neglect the mass of the rod in your calculations.

49.**—The pendulum of a regulator clock consists of a mass of 120 g at the end of a (massless) wooden stick of length 44 cm.

(a) What is the total energy (kinetic plus potential) of this pendulum when oscillating with an amplitude of 4°?

(b) What is the speed of the mass when at its lowest point?

Section 9.5: Damped Oscillations and Forced Oscillations

50.——Roughly, what is the frequency of stomping of soldiers on the march? What must have been the resonant frequency of the bridge at Angers that broke when soldiers marched across it?

51.——A swing of length 2 m hangs from a horizontal branch of a tree. With what frequency should you rock the branch to build up the oscillations of the pendulum by resonance?

52.*—Consider the motion of the damped harmonic oscillator plotted in Fig. 9.21.

(a) According to this plot, what fraction of its amplitude does the oscillator lose in its first oscillation?

(b) What fraction of its energy does the oscillator lose in its first oscillation?

(c) According to Eq. (49), what is the value of Q for this oscillator?

Large waves produced by a storm in the Pacific Ocean. The wave crests and wave troughs form a regular, repetitive pattern.

WAVES

A wave is a vibrational, trembling motion in an elastic, deformable body. The wave is initiated by some external force that acts on some part of the body and deforms it. The elastic restoring forces within the body communicate this initial disturbance from one part of the body to the next, adjacent part. The disturbance therefore gradually spreads outward through the entire elastic body. For instance, Fig. 10.1 shows a familiar example of wave motion: a stone has been dropped into a pond, and it has disturbed the calm surface of the water. The disturbance propagates outward in the form of concentric water waves.

FIGURE 10.1

Water waves spreading out on the surface of a pond. The wave fronts are concentric circles.

The elastic body in which the wave propagates is called the ***medium.*** Thus, water is the medium for water waves, air is the medium for sound waves, the crust of the Earth is the medium for seismic waves, the stretched string of a guitar is the medium for vibrational "string" waves, and so on. When a wave propagates through a medium, ***the particles in***

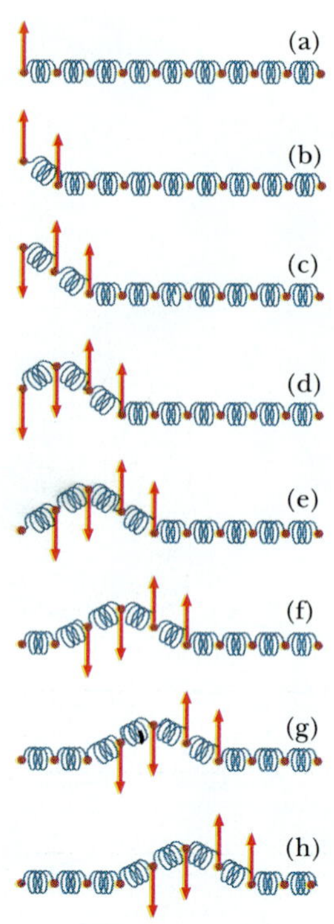

FIGURE 10.2

Particles joined by springs. A transverse disturbance propagates from left to right. The diagrams show snapshots at successive times. The particles move up and down.

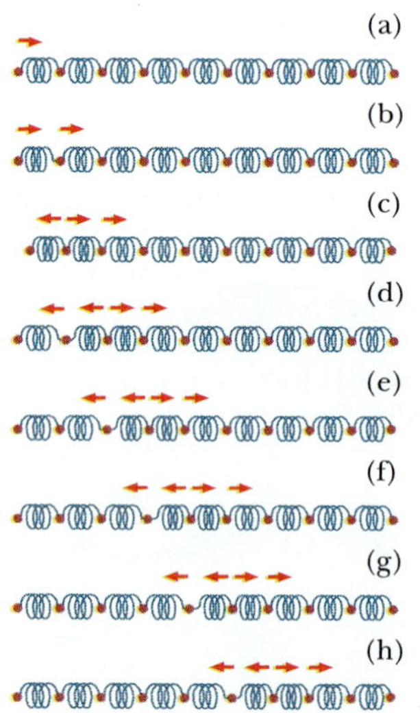

FIGURE 10.3

A longitudinal disturbance propagates from left to right. The particles move back and forth.

the medium vibrate back and forth, but the medium as a whole does not perform translational motion. This is obvious in the case of a wave on a guitar string, where we can see the string vibrate, and we know that the string cannot travel anywhere, since it is tied down at its ends. But the lack of motion of the medium as a whole is not so obvious for water waves—when watching ocean waves, we often gain the impression that the water travels with the wave, especially when the waves are large and when we see them crashing against a seawall or some other obstruction. But we can check that the water does not flow with the wave if we watch a chip of wood or some other flotsam on the water. Such a chip of wood only bobs up and down and rocks back and forth; it does not travel forward with the wave.

For the sake of simplicity, in this chapter we will concentrate on the motion of waves on a string. However, most of our mathematical results also apply to wave motion in other elastic bodies. In the next chapter we will examine some features of wave motion in air, that is, sound waves.

10.1 TRANSVERSE AND LONGITUDINAL WAVE MOTION

Consider a tightly stretched elastic string, such as a long rubber cord. If we shake one end of the string up and down with the flick of the wrist, a disturbance travels along the string. Figure 10.2 shows in detail how such a traveling disturbance comes about. The string may be regarded as a row of particles connected by small, massless springs. Initially, the particles are at their equilibrium positions, evenly spaced along the string. When we jerk the first particle to one side, it will pull the second particle to the same side, and this will pull the third, and so on. If we then jerk the first particle back to its original position, it will pull the second particle back, and this will likewise pull the third, and so on. As the motion is transmitted from one particle to the next particle, the disturbance propagates along the row of particles. Such a disturbance is called a ***transverse*** wave pulse.

Alternatively, we can generate a disturbance by suddenly pushing the first particle toward the second. Figure 10.3 shows such a compressional disturbance propagating along the row of particles. This kind of disturbance is called a ***longitudinal*** wave pulse.

Note that although the wave pulse travels along the full length of the string, the particles do not—they merely move back and forth around their equilibrium positions. Also note that in the region of the wave pulse, the string has kinetic energy (due to the back-and-forth motion of the particles) and potential energy (due to the stretching of the springs between the particles). Hence, a wave pulse traveling along the string carries energy with it—the wave transports energy from one end of the string to the other.

Wave motion in air, water, or any other medium displays the same general features. The waves are propagating disturbances in the medium communicated by pushes and pulls from one particle to the next. The waves transport energy without transporting particles. The waves are longitudinal, transverse, or both. A sound wave in air is longitudinal; the air molecules move forward and backward, parallel to the direction of propagation of the wave. But a wave on the surface of the ocean is both longitudinal and transverse; the water molecules move up and down and, simultaneously, forward and backward—the net result of these simultaneous motions is that each water molecule traces out an elliptical path. Seismic waves in the crust of the Earth can be either longitudinal (P waves) or transverse (S waves). These two kinds of seismic waves have different speeds, and their relative intensities depend on the characteristics of the earthquake that generated them.

10.2 PERIODIC WAVES

If we shake the end of a long string up and down, and we continue shaking it steadily, we will generate a ***periodic wave*** on the string. Such a wave can be regarded as consisting of a steady succession of positive and negative wave pulses, which repeat at regular intervals. Figure 10.4a shows a periodic wave at one instant of time. The high points of the wave are called the ***wave crests,*** and the low points are called the ***wave troughs. The distance from one crest to the next or from one trough to the next is called the wavelength, designated by the symbol*** λ (Greek lambda). The wavelength is the repeat distance of the wave pattern—a shift of the wave pattern by one wavelength to the right (or the left) reproduces the original wave pattern.

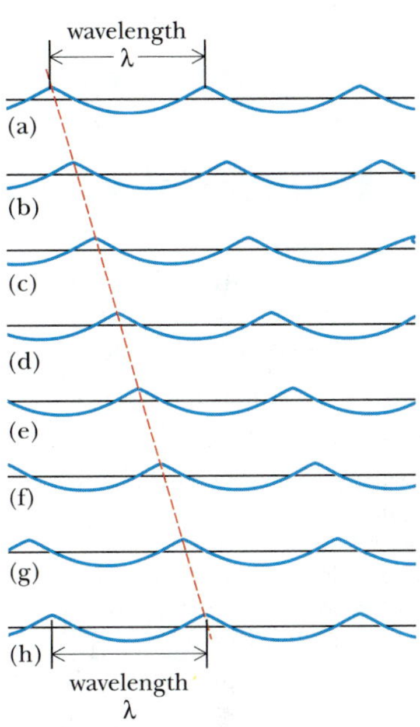

FIGURE 10.4

A periodic wave traveling to the right. The diagrams show snapshots of the wave at successive instants of time. The wave pattern (h) coincides with the wave pattern (a) because the wave has moved exactly one wavelength to the right.

With the passing of time, the wave crests and wave troughs travel toward the right at a speed v. As the wave travels, the entire wave pattern shifts toward the right; that is, the wave pattern (but not the string) performs a rigid translation motion. Figures 10.4b–10.4h show the wave at successive instants of time. These pictures span one ***period*** of the wave; that is, they span the interval of time required for the wave pattern to travel exactly one wavelength to the right. The period is the repeat time of the wave pattern—after one period, each wave crest or wave trough will have traveled to the position previously occupied by the adjacent wave crest or wave trough, and the wave will have attained exactly the same configuration as it had at the initial time.

Since in one period, the wave travels a distance equal to one wavelength, the ratio of wavelength to period must equal the wave speed,

$$\frac{\lambda}{T} = v \qquad (1)$$

◀ **Wavelength**

As in the case of simple harmonic motion, we define the frequency of the wave as the inverse of the period,

$$f = \frac{1}{T} \qquad (2)$$

The frequency of the wave is simply the number of wave crests arriving at some point on the string per second. The unit for the frequency is cycles per second, or Hz. For example, if the period of the wave is 0.1 s, then in one second there will be 10 wave crests arriving at some point on the string, and consequently the frequency of the wave is 10 cycles per second, or 10 Hz.

In terms of the frequency, Eq. (1) becomes

Wavelength, frequency, and speed ▸

$$\lambda f = v \tag{3}$$

This equation permits us to calculate the frequency from the wavelength, or the wavelength from the frequency, provided we know the speed of the wave.

Harmonic wave ▸

Amplitude of wave ▸

An important special case of a periodic wave is a ***harmonic wave. This kind of a wave has the shape of a sine curve*** (see Fig. 10.5). The height of the wave crests (or the depth of the wave troughs) of this sine curve is called the ***amplitude*** of the wave, designated by A. As the wave passes a point of the string, the particle at this point moves up a distance A; then,

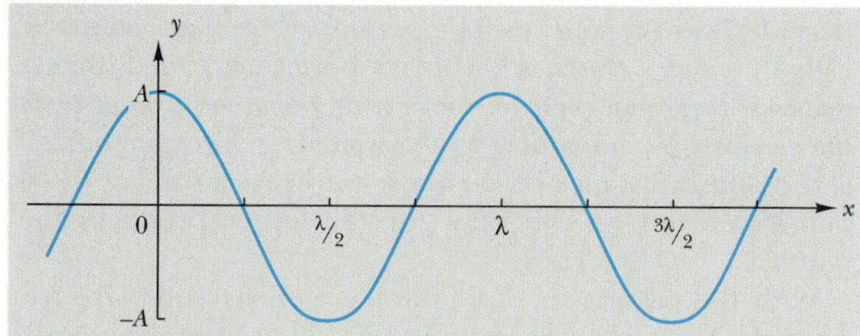

FIGURE 10.5

A harmonic wave.

half a cycle later, it moves down a distance A; and then, another half cycle later, it again moves up a distance A. Thus, ***the particle executes simple harmonic motion of a frequency and an amplitude equal to the frequency and the amplitude of the wave*** (see Fig. 10.6). Harmonic waves play a central role in the study of wave motion because, as we will see in the next section, any periodic wave of arbitrary shape can be regarded as a superposition, or sum, of several harmonic waves of suitably chosen amplitudes and wavelengths. Thus, if we understand the motion of harmonic waves,

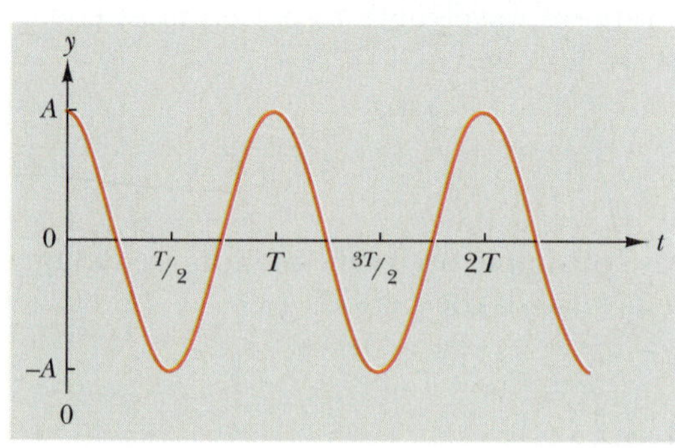

FIGURE 10.6

Each particle on the string executes simple harmonic motion with a period equal to the period of the wave, and an amplitude equal to the amplitude of the wave.

we understand the motion of any kind of periodic wave. Hereafter, we will concentrate on harmonic waves.

The speed of the wave depends on the characteristics of the string. The relevant characteristics are the tension F in the string, its mass M, and its length L. By examining the implications of Newton's Second Law for a small deformed segment of string being accelerated by the tension forces acting on the ends of the segment, we can demonstrate that the speed of the wave on a string is

$$v = \sqrt{\frac{F}{M/L}} \tag{4}$$

◂ **Speed of wave on a string**

The characteristic of the string that enters into Eq. (4) is the ratio of mass to length (M/L), or the mass per unit length. If we use a given kind of string, of some given thickness, the mass per unit length will be the same, regardless of whether we use a short length of string or a long length in our experiments. Note that the speed of the wave is large if the tension is large and the mass per unit length is small (a thin string). This is intuitively reasonable, since a large tension can accelerate a small mass very quickly, and therefore produces a quick back-and-forth motion of the mass elements in the string; that is, it produces a high frequency of the motion. According to Eq. (3), for a given wavelength, a high frequency implies a large speed of the wave.

Of course, Eq. (4) is valid not only for strings, but also for tightly stretched wires and cables.

EXAMPLE 1. A long piece of piano wire of a mass of 3.9×10^{-3} kg per meter is under a tension of 1.0×10^{3} N. What is the speed of transverse waves on this wire? What is the wavelength of a harmonic wave on this wire if its frequency is 262 Hz?

SOLUTION: If the mass is 3.9×10^{-3} kg per meter, the ratio M/L in Eq. (4) is 3.9×10^{-3} kg/m, and hence the speed of the wave is

$$v = \sqrt{\frac{F}{M/L}} \tag{5}$$

$$= \sqrt{\frac{1.0 \times 10^{3}\ \text{N}}{3.9 \times 10^{-3}\ \text{kg/m}}} = 510\ \text{m/s}$$

Consequently, with $f = 262$ Hz $= 262/\text{s}$, the wavelength is

$$\lambda = \frac{v}{f} = \frac{510\ \text{m/s}}{262/\text{s}} = 1.9\ \text{m}$$

Although a wave on a string is a rather special case of wave motion, the mathematical description of other kinds of waves is similar to that of waves on a string. The instantaneous configuration of the wave can always be described by a plot of the wave disturbance vs. position, such as the plot in Fig. 10.5, but the vertical axis of the plot must be adapted to the physical properties of the wave. For instance, to describe a sound wave in air, we can plot the pressure disturbance produced by the wave vs. the position.

Our Eq. (4) for the wave speed applies only to waves on a string. But this equation exhibits a general feature of wave propagation: in broad terms, this equation states that the speed of the wave depends on the restoring force and on the inertia of the medium in which the wave is propagating. This is true for all kinds of waves. In all cases, some force within the medium opposes its deformation—tension tends to keep the string straight, the pressure within a gas tends to keep the density of the gas uniform, gravity tends to keep the surface of the sea smooth, and so on. But if something provides an initial disturbance, then the restoring force will cause it to propagate, as in Fig. 10.2, with a speed depending on the magnitude of the restoring force and on the amount of inertia or, equivalently, the amount of mass in the medium. In general, the speed will be large if the restoring force is large and the amount of mass in the medium is small.

10.3 THE SUPERPOSITION OF WAVES

Superposition principle ▶

Waves on a string and waves in other elastic bodies usually obey a ***superposition principle:*** When two or more waves arrive at any given point simultaneously, ***the resultant instantaneous deformation is the sum of the individual instantaneous deformations.*** Such a superposition means that the waves do not interact; they have no effect on one another. Each wave propagates as though the other were not present, and the contribution that each makes to the displacement of a particle in the elastic body is as though the other were not present. For instance, if the sound waves from a violin and a flute reach us simultaneously, then each of these waves produces a displacement of the air molecules just as though it were acting alone, and the net displacement of the air molecules is simply the (vector) sum of these individual displacements.

For waves of low amplitude on a string and for sound waves of ordinary intensity in air, the superposition principle is very well satisfied. However, for waves of very large amplitude or intensity, the superposition principle fails. When a wave of very large amplitude is propagating on a string, it alters the tension of the string, and therefore affects the behavior of a second wave propagating on the same string. Likewise, a very intense sound wave (a shock wave, such as the loud bang from a nearby explosion) produces significant alterations of the temperature and the pressure of the air, and therefore affects the behavior of a second wave propagat-

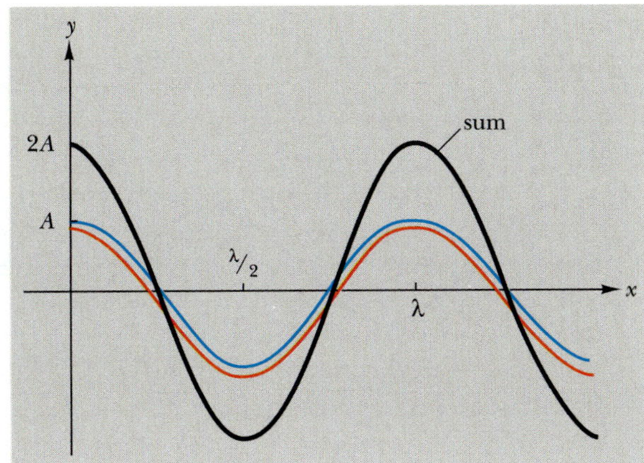

FIGURE 10.7. Constructive interference of two waves.

ing through this same region. In this section, we will not worry about such extreme conditions, and we will assume that the superposition principle is a good approximation.

As a first example of superposition, let us consider two waves propagating in the same direction with the same frequency and amplitude. If the wave crests and the wave troughs of the two waves coincide, the waves are said to be ***in phase.*** The superposition of these two waves yields a wave of twice the amplitude of the individual waves (see Fig. 10.7). ***Such a reinforcement of one wave by another is called constructive interference.*** If the wave crests of one wave match the wave troughs of the other, the waves are said to be ***out of phase,*** or to differ in phase by half a cycle. The superposition of these two waves yields a wave of zero amplitude (see Fig. 10.8). ***Such a cancellation of one wave by another is called destructive interference.****

◂ **Constructive interference**

◂ **Destructive interference**

If the two waves are out of phase, but their amplitudes are not equal,

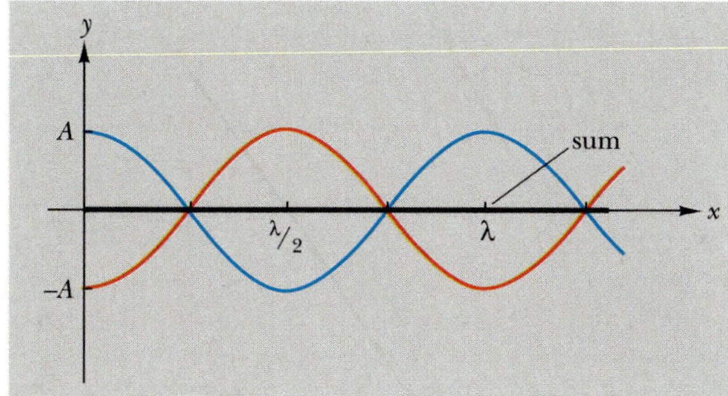

FIGURE 10.8 Destructive interference of two waves; the waves cancel everywhere.

* This cancellation raises a question: If the waves cancel, what happens to the energy they carry? To answer this question we must examine in detail how the two waves were brought together. For instance, if the two waves were initially propragating on two separate strings which merge into a single string at a junction or knot, then a cancellation of the waves beyond the junction is impossible unless there are strong backward reflections of the two incident waves at the junction. The waves reflected backward from the junction account for the missing energy.

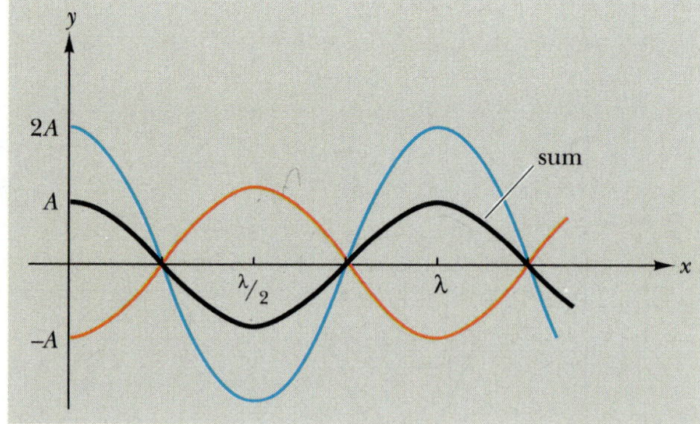

FIGURE 10.9

Destructive interference of two waves of different amplitudes. The sum is small, but not zero.

then their cancellation will not be total; some portion of the wave that has the larger amplitude will be left over (see Fig. 10.9).

As another example of superposition, let us consider two waves of the same amplitude, but slightly different frequencies and, therefore, slightly different wavelengths. Figure 10.10 shows the two waves at one instant of time and their superposition. At $x = 0$, the waves are in phase, and they interfere constructively, giving a large net amplitude. But farther along the x axis, the difference in wavelengths gradually causes the waves to acquire a phase difference. At the point P, the waves are out of phase by half a cycle, and they interfere destructively, giving a net amplitude of zero. Beyond this point, the phase difference exceeds one half cycle. At the point Q the phase difference has grown to one cycle; but since a phase difference of one cycle means that the crests of the two waves coincide, they interfere constructively, again giving a large amplitude, and so on. Thus, the superposition of the two waves displays regularly alternating re-

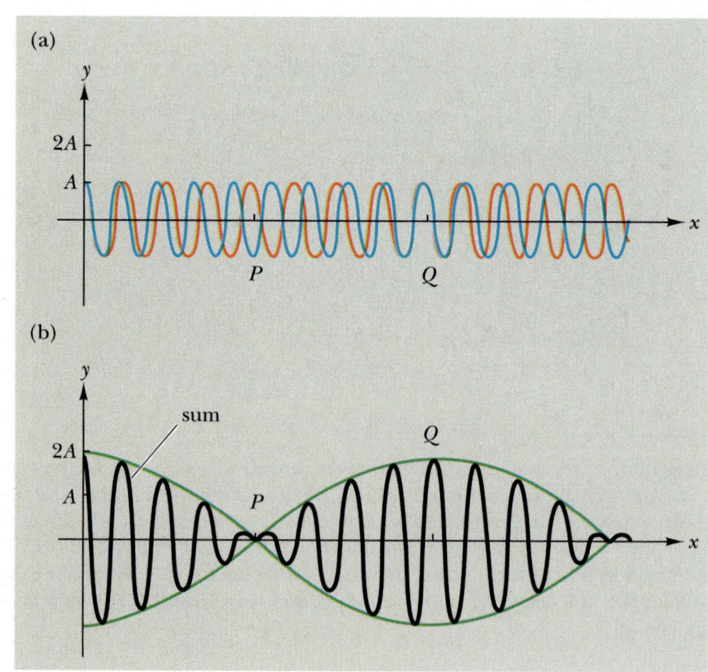

FIGURE 10.10

Superposition of two waves of slightly different wavelengths and frequencies. (a) The two waves before addition. (b) The sum of the two waves. The green line shows the wave envelope, or the average amplitude.

gions of constructive and destructive interference, that is, alternating regions of large amplitude and small amplitude (see Fig. 10.10b).

With the passing of time, the entire pattern in Fig. 10.10b moves rigidly toward the right with the wave speed. The periodic variations in the amplitude of the wave pattern give rise to the phenomenon of ***beats.*** At any given position, the amplitude of the wave pulsates—first the amplitude is large, then it becomes small, then again large, and so on. ***The frequency with which the amplitude pulsates is called the beat frequency.*** It is easy to check that the beat frequency is simply the difference between the frequencies of the two waves,

◂ **Beat frequency**

$$f_{\text{beat}} = f_2 - f_1 \tag{6}$$

To establish this result, we note that one of the waves oscillates through f_1 cycles per second, and the other wave through f_2 cycles per second. Therefore, in a time interval $1/(f_2 - f_1)$, the former wave oscillates through $f_1/(f_2 - f_1)$ cycles, and the latter wave through $f_2/(f_2 - f_1)$ cycles. The difference between these numbers of cycles is

$$\frac{f_2}{f_2 - f_1} - \frac{f_1}{f_2 - f_1} = \frac{f_2 - f_1}{f_2 - f_1} = 1 \tag{7}$$

This means that in this time interval $1/(f_2 - f_1)$, the waves accumulate a difference of exactly one cycle, and the amplitude of the superposition goes through one pulsation: from maximum intensity, to zero intensity, back to maximum intensity. Thus, $1/(f_2 - f_1)$ must be the period of the beats, and $f_2 - f_1$ must be the frequency of the beats.

EXAMPLE 2. Suppose that two flutes generate sound waves of frequency 264 Hz and 262 Hz, respectively. What is the beat frequency?

SOLUTION: According to Eq. (6),

$$f_{\text{beat}} = 264 \text{ Hz} - 262 \text{ Hz} = 2 \text{ Hz} \tag{8}$$

Hence, a listener will hear a tone of average frequency 263 Hz, but with a loudness pulsating 2 times per second.

Beats are a sensitive indication of small frequency differences, and they are very useful in the tuning of musical instruments. For example, to bring the two flutes of the preceding example in tune, the musicians listen to the beats and, by trial and error, adjust one of the flutes so as to reduce the beat frequency; when the beats disappear entirely (zero beat frequency), the two flutes will be generating waves of exactly the same frequencies.

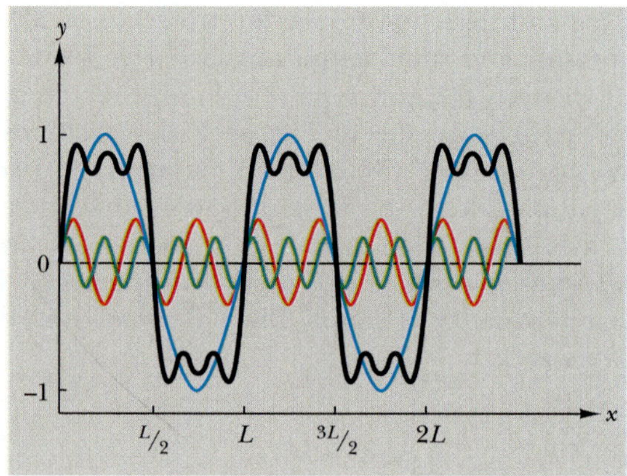

FIGURE 10.11
A wave (black) constructed by superposition of three harmonic waves (red, blue, green) of wavelengths L, $L/3$, and $L/5$.

By the superposition of harmonic waves of different amplitudes and wavelengths, we can construct some rather complicated wave shapes. For example, Fig. 10.11 shows a periodic wave constructed by the superposition of three harmonic waves of wavelengths L, $L/3$, and $L/5$ whose amplitudes are in the ratio $1:\frac{1}{3}:\frac{1}{5}$. It can be shown that ***any arbitrary periodic wave can be constructed by the superposition of a sufficiently large number of harmonic waves.*** This is called **Fourier's theorem.** As already mentioned in the preceding section, such a superposition implies that we can regard any arbitrary periodic wave as a sum of harmonic waves; and it implies that the study of periodic waves can, in essence, be reduced to the study of harmonic waves.

10.4 Standing Waves

Next, we will consider the superposition of two waves of the same amplitude and frequency, but of opposite directions of propagation. Figure 10.12a shows the two waves and their sum at an initial instant of time. At this instant, the waves are in phase, and their sum is a wave of twice the amplitude of each. At a slightly later time, one wave has moved to the right and the other to the left (see Fig. 10.12b). At the points P, Q, R, . . . the waves were initially zero; now, one wave has a positive value at these points, and the other wave has an equally large negative value. Thus, the sum of the two waves still yields a result of zero at these points. Furthermore, the two waves will continue to cancel at these points at all times (see Figs. 10.12c and d). These points at which the sum of the waves is zero are called ***nodes.*** They are one half wavelength apart.

Nodes ▶

Midway between the nodes we find points at which the sum of the two waves is maximum (positive or negative). These points are called ***antinodes.*** Figure 10.12 shows that with the passing of time, the height of the wave crests at the antinodes oscillates, but the positions of these wave crests remain fixed. Thus, ***the superposition of two waves traveling in opposite directions is a standing wave.*** This wave travels neither to the right nor to the left; its wave crests remain at fixed positions while the entire wave increases and decreases in unison. The frequency of this pulsation of the

Antinodes ▶

Standing wave ▶

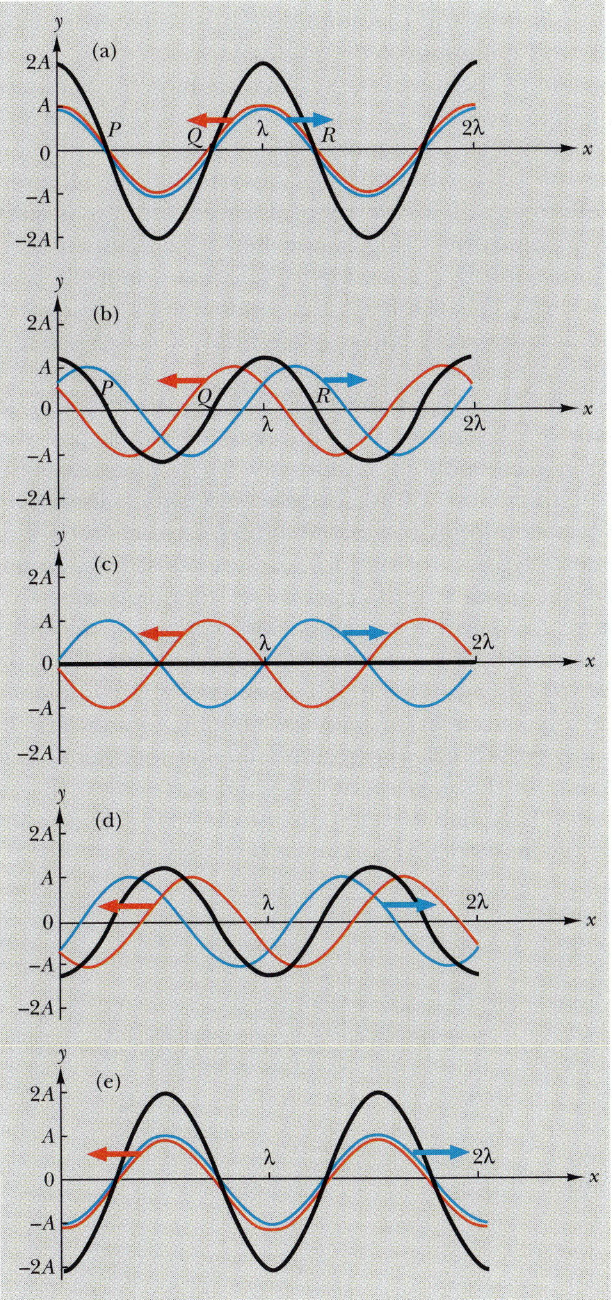

FIGURE 10.12

Superposition (black) of two waves (red and blue) of the same amplitude but of opposite directions of propagation. (a) At $t = 0$ the waves are in phase. (b) At $t = (\frac{1}{8}$ of a period), one wave has moved $\frac{1}{8}$ of a wavelength to the right, and the other the same distance to the left. (c) At $t = (\frac{1}{4}$ of a period), one wave has moved $\frac{1}{4}$ of a wavelength to the right and the other the same distance to the left; the waves have therefore moved apart $\frac{1}{2}$ wavelength, and they are out of phase and they cancel every where. (d) At $t = (\frac{3}{8}$ of a period), the waves have moved $\frac{3}{8}$ of a wavelength to the right and the left, respectively. (e) At $t = (\frac{1}{2}$ period), the waves are again in phase, and so on.

standing wave is the same as the frequency of the two underlying traveling waves.

If this standing wave is a wave on a string, then each particle of this string executes an up-and-down simple harmonic motion. However, in contrast to the case of a traveling wave, where the amplitudes of the harmonic oscillations of all the particles are the same, the amplitudes of

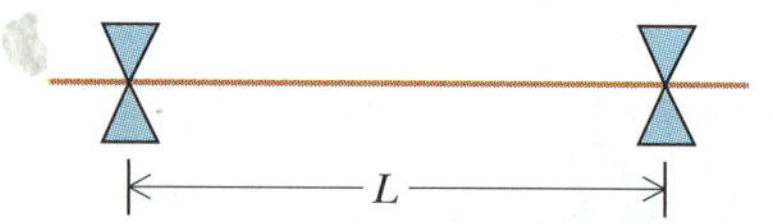

FIGURE 10.13

A tightly stretched string with fixed ends. The next figure shows possible standing waves on this string.

oscillation now depend on position: the amplitude of oscillation is maximum at the antinodes, and minimum at the nodes.

So far, in our discussion of the waves on a string, we have assumed that the string is very long, and we have ignored the endpoints of the string. When a traveling wave arrives at an endpoint, something drastic will have to happen to it: either the wave will have to be absorbed at the endpoint or it will have to be reflected, with a reversal of its direction of motion. If the endpoint is a fixed point (the string is attached to a rigid support), then the endpoint cannot absorb the energy of the wave, and the wave will be completely reflected. This results in the simultaneous presence of two waves of equal amplitudes and opposite directions of travel—that is, it results in a standing wave.

If the other endpoint of the string is also fixed (see Fig. 10.13), then the possible standing waves on this string are subject to the restriction that the wave must be zero at each endpoint at all times. Such a restriction on what happens at the endpoints of a wave is called a ***boundary condition.*** Obviously, the boundary condition for our standing wave requires that the endpoints are nodes. Figures 10.14 and 10.15 show possible standing waves on a string of some given length L; all these standing waves have nodes at the endpoints. ***The possible standing-wave motions of the string shown in Fig. 10.14 are called the normal modes.*** Figure 10.14a shows the ***fundamental mode;*** Fig. 10.14b and Fig. 10.14c show the next two modes.

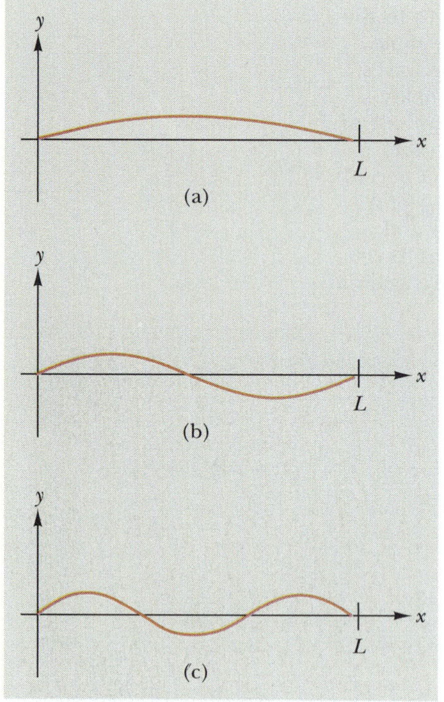

FIGURE 10.14

(a) The fundamental mode.
(b) The first overtone.
(c) The second overtone.

In all these modes, some number of half wavelengths exactly fits the length of the string (see Fig. 10.14). In the fundamental mode, one half wavelength fits the string; in the next mode, two half wavelengths fit the string; in the next mode, three half wavelengths fit the string, and so on. Thus, the wavelengths of the modes are

$$\lambda_1 = 2L, \quad \lambda_2 = \frac{2L}{2} = L, \quad \lambda_3 = \frac{2L}{3}, \text{ etc.} \tag{9}$$

The frequencies of oscillation of the modes are related to the wavelengths in the usual way ($f = v/\lambda$):

$$f_1 = \frac{v}{2L}, \quad f_2 = \frac{2v}{2L} = \frac{v}{L}, \quad f_3 = \frac{3v}{2L}, \text{ etc.} \tag{10}$$

FIGURE 10.15

Time-exposure photograph of standing waves on a string.

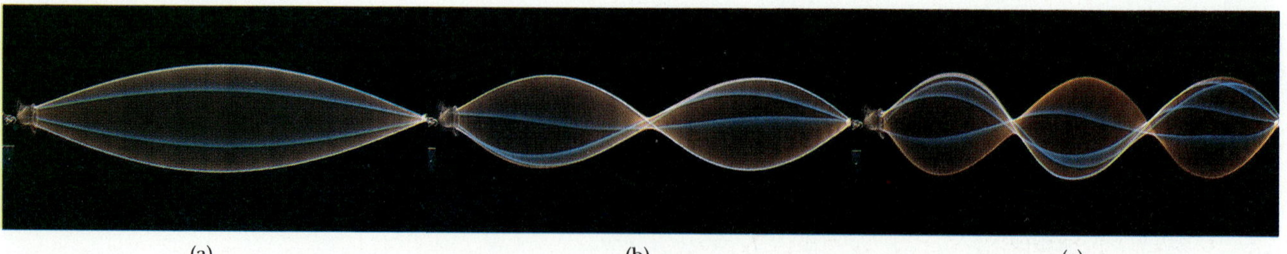

The frequencies of these modes are called the normal frequencies, proper frequencies, or ***eigenfrequencies*** of the string. In musical terminology, these frequencies are called the ***fundamental frequency*** and the ***overtones.*** Note that in the case of our vibrating string, all the frequencies are multiples of the fundamental frequency: the first overtone is twice the frequency of the fundamental, the second overtone is three times the frequency of the fundamental, and so on.

◀ Eigenfrequencies of normal modes on a string

In general, any arbitrary motion of a freely vibrating string (with fixed endpoints) will be some superposition of several of the above normal modes. Which modes will be present in the superposition depends on how the motion is started. For instance, when a guitar player plucks a string on her guitar near the middle, she will excite the fundamental frequency and also the second overtone and, to a lesser extent, some of the higher even-numbered overtones.

EXAMPLE 3. The low E string on the guitar vibrates with a frequency of 82.4 Hz when excited in its fundamental mode. What are the frequencies of the second and the fourth overtones of this string?

SOLUTION: In our notation, the frequency of the second overtone is f_3, and the frequency of the fourth overtone is f_5. According to Eq. (10), $f_3 = 3f_1$, and $f_5 = 5f_1$. Hence the frequencies of the second and fourth overtones are, respectively, 3×82.4 Hz and 5×82.4 Hz; that is, 247 Hz and 412 Hz.

The normal modes of vibration of a long, thin elastic rod or a beam fixed at both ends are mathematically similar to the normal modes of a string. However, such an elastic body can experience both transverse deformations (like those of a string) and rotational, or "torsional," deformations. Figure 10.16 shows a spectacular example of a torsional standing wave in the span of a bridge at Tacoma, Washington. This standing wave was excited by a wind blowing across the bridge, which generated vortices in the air in resonance with one of the normal modes of vibration of the span. The bridge vibrated for several hours, with increasing amplitude, and then broke apart.

FIGURE 10.16

Standing wave on the deck of the Tacoma Narrows Bridge, July 1, 1940. The bridge broke apart a short time after this picture was taken.

SUMMARY

- **Wavelength, period, frequency and wave speed:** $\dfrac{\lambda}{T} = \lambda f = v$
- **Speed of wave on a string:** $v = \sqrt{\dfrac{F}{M/L}}$
- **Beat frequency:** $f_{\text{beat}} = f_2 - f_1$
- **Wavelengths of normal modes of string:**

$$\lambda_1 = 2L, \quad \lambda_2 = \frac{2L}{2}, \quad \lambda_3 = \frac{2L}{3}, \quad \text{etc.}$$

- **Eigenfrequencies:** $f_1 = \dfrac{v}{2L}, \quad f_2 = \dfrac{v}{L}, \quad f_3 = \dfrac{3v}{2L}, \quad \text{etc.}$

Questions

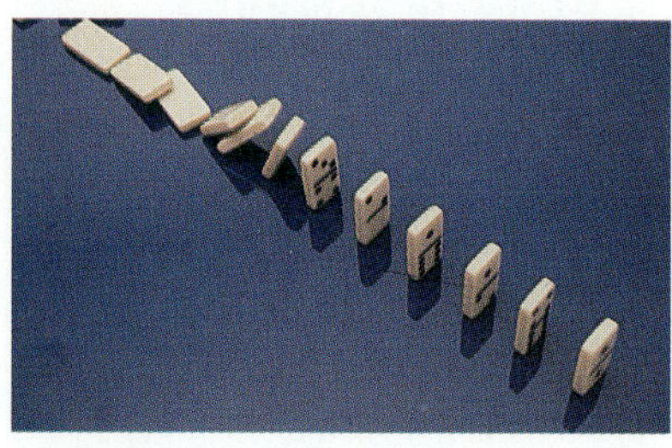

FIGURE 10.17
Toppling dominoes.

1.–You have a long, thin steel rod and a hammer. How must you hit the end of the rod to generate a longitudinal wave? A transverse wave?

2.–Some people enjoy arranging long rows of dominoes on the floor, so the toppling of one domino triggers the toppling of all the others, by a chain reaction (see Fig. 10.17). The propagation of the disturbance along such a chain of dominoes has some of the properties of a wave pulse. In what way is it similar to a wave pulse? In what way is it different?

3.–A wave pulse on a string transports energy. Does it also transport momentum? To answer this question, imagine a washer loosely encircling the string at some place; what happens to the washer when the wave pulse strikes it?

FIGURE 10.18
Wire wheel for auto.

4.–The strings of a guitar are made of wires of different thicknesses (the thickest wires are manufactured by wrapping copper or brass wire around a strand of steel). Why is it impractical to use wire of the same thickness for all the strings?

5.–According to Eq. (4), the speed of a wave on a string increases by a factor of 2 if we increase the tension by a factor of 4. However, in the case of a rubber string, the speed increases by more than a factor of 2 if we increase the tension by a factor of 4. Why are rubber strings different?

6.–A harmonic wave is traveling along a string. Where in this wave is the kinetic energy at maximum? The potential energy? The total energy?

7.–Figure 10.11 shows a standing wave on a string. At time $t = \frac{1}{4}T$, the amplitude of the wave is everywhere zero. Does this mean the wave has zero energy at this instant?

8.–After an arrow has been shot from a bow, the bowstring will oscillate back and forth, forming a standing wave. Which of the overtones shown in Fig. 10.14 do you expect to be present?

9.–In tuning a guitar or violin, by what means do you change the frequency of the string?

10.–A mechanic can make a rough test of the tension in the spokes of a wire wheel (Fig. 10.18) by striking the spokes with a wrench or a small hammer. A spoke under tension will ring, but a loose spoke will not. Explain.

11.–What is the purpose of the frets on the neck of a guitar or a mandolin (see Fig. 10.19)?

FIGURE 10.19
Close-up of frets on stem of guitar.

Problems

Section 10.2: Periodic Waves

1.——An ocean wave has a wavelength of 120 m and a period of 8.77 s. Calculate the frequency, angular frequency, and speed of this wave.

2.——Light waves have a speed of 3.00×10^8 m/s. Red light has a wavelength of 7.0×10^{-7} m, and violet light has a wavelength of 4.0×10^{-7} m. What is the frequency of red light? Of violet light?

3.——Radio waves have the same speed as light waves, 3.00×10^8 m/s. What is the wavelength of an FM radio wave, of frequency 105 MHz?

4.——The speed of tidal waves in the Pacific is about 740 km/h.

(a) How long does a tidal wave take to travel from Japan to San Francisco, a distance of 8000 km?

(b) If the wavelength of the wave is 300 km, what is its frequency?

5.——A harmonic wave on a string has an amplitude of 2.0 cm, a wavelength of 1.2 m, and a speed of 6.0 m/s. What are the period, frequency, and angular frequency of this wave?

6.——Ocean waves smash into a breakwater at the rate of 12 crests per minute. The wavelength of these waves is 39 m. What is their speed?

7.——The velocity of sound in fresh water at 15°C is 1440 m/s and at 30°C it is 1530 m/s. Suppose that a sound wave of frequency 440 Hz penetrates from a layer of water at 30°C into a layer of water at 15°C. What will be the change in the wavelength? Assume that the frequency remains unchanged.

8.——A light wave of frequency 5.5×10^{14} Hz penetrates from air into water. What is its wavelength in air? In water? The speed of light is 3.0×10^8 m/s in air and 2.3×10^8 m/s in water; assume that the frequency remains the same.

9.——In deep water (when the depth is much larger than the wavelength), the speed of waves is given by the formula $v = \sqrt{g\lambda/2\pi}$. Calculate the speed of short water waves, with $\lambda = 1.0$ m. Calculate the speed of long waves with $\lambda = 300$ m.

10.——In shallow water (when the depth is shorter than a wavelength), the speed of waves is given by the formula $v = \sqrt{gh}$, where h is the depth of the water.

(a) Calculate the speed of water waves in a shallow pond with a depth of 2.0 m.

(b) For ocean waves of extremely long wavelength, such as tidal waves, the oceanic basins can be treated as shallow ponds, since their depth is small compared with the wavelength. Calculate the speed of a tidal wave in the Pacific Ocean, where the mean depth is 4.3 km.

11.——To determine the speed and the frequency of periodic waves on a lake, the owner of a motorboat first runs the boat in the direction of the waves, and he finds that when his boat keeps up with a wave crest the speed indicator shows 16 m/s. He then anchors the boat and finds that the waves make it bounce up and down 6 times per minute. What is the speed, frequency, and wavelength of the waves?

12.——A string is stretched along the x axis (horizontally) and it oscillates in the y direction (vertically). At one instant of time, a transverse wave on this string is described by the mathematical formula

$$y = 0.03 \cos(1.2x)$$

where y and x are measured in meters.

(a) What is the amplitude of the wave?

(b) What is the wavelength of this wave?

(c) Where are the first three wave crests and the first three wave troughs on the x axis nearest the origin?

13.——A transverse wave travels along a stretched string with a wave speed of 14 m/s. A particle at a fixed location on this string oscillates up and down as follows as a function of time:

$$y = 0.02 \cos(9.0t)$$

where the displacement y is measured in meters and the time t is measured in seconds.

(a) What is the amplitude of the wave?
(b) What is the frequency of the wave?
(c) What is the wavelength of the wave?

14.*—In the crust of the Earth, seismic waves of the P type have a speed of almost 5 km/s; waves of the S type have a speed of about 3 km/s. Suppose that after an earthquake, a seismometer placed at some distance first registers the arrival of P waves and 9 min later the arrival of S waves. What is the distance between the seismometer and the source of the waves?

15.*—Ocean waves of wavelength 100 m have a speed of 6.2 m/s; ocean waves of wavelength 20 m have a speed of 2.8 m/s. Suppose that a sudden storm at sea generates waves of all wavelengths. The long-wavelength waves travel fastest and reach the coast first. A fisherman standing on the coast first notices the arrival of 100-m waves; 10 hours later he notices the arrival of 20-m waves. How far is the storm from the coast?

16.**—A motorboat is speeding at 12 m/s through a group of periodic ocean waves. When the motorboat travels in the same direction as the waves, it smashes into 6.5 waves per minute. When the motorboat is traveling in the direction opposite to the waves, it smashes into 30 waves per minute. Calculate the speed, frequency, and wavelength of the ocean waves.

17.—A transverse harmonic wave on a stretched string has an amplitude of 1.2 cm, a speed of 8.0 m/s, and a wavelength of 2.2 m.

(a) What is the maximum transverse speed attained by a particle on the string? Does the particle attain this maximum speed when a wave crest passes the particle or at some other time?
(b) What is the maximum transverse acceleration attained by a particle on the string? Does the particle attain this maximum acceleration when a wave crest passes the particle or at some other time?

18.—When a periodic transverse wave travels along a clothesline, a ladybug sitting on the line experiences a maximum transverse velocity of 0.20 m/s and a maximum transverse acceleration of 4.0 m/s^2. Deduce the amplitude and the frequency of the wave.

19.*—A giant, freak wave encountered by a weather ship in the North Atlantic was 23.5 m high from trough to crest; its wavelength was 350 m and its period 15.0 s. Calculate the maximum vertical acceleration of the ship as the wave passed underneath; calculate the maximum vertical velocity. Assume that the motion of the ship was purely vertical and simple harmonic.

20.*—Many inhabitants of Tangshan, China, reported that during the catastrophic earthquake of July 28, 1976, they were thrown 2 m into the air as if by a "huge jolt from below."

(a) With what speed must a body be thrown upward to reach a height of 2 m?
(b) Assume that the vertical wave motion of the ground was simple harmonic with a frequency of 1 Hz. What amplitude of the vertical motion is required to generate a speed equal to that calculated in Part (a)?

21.*—A passenger in an airplane flying over an anchored ship notices that ocean waves are smashing into the ship regularly at the rate of 10 per minute. He knows an ocean wave of this frequency has a speed of 9.4 m/s. He also notices that the length of the ship is the same as about three wavelengths. Deduce the length of the ship from this information.

22.**—In the open sea, a tsunami usually has an amplitude less than 30 cm and a wavelength longer than 80 km. Assume that the speed of the tsunami is 740 km/h. What is the maximum vertical velocity and acceleration that such a tsunami will give to a ship floating on the water? Will the crew of the ship notice the passing of the tsunami?

23.** —The National Ocean Survey has deployed buoys off the Atlantic Coast to measure ocean waves. Such a buoy detects waves by the vertical acceleration that it experiences as it is lifted and lowered by the waves. In order to calibrate the device that measures the acceleration, scientists placed the buoy on a Ferris wheel at an amusement park. If the *vertical* acceleration (as a function of time) of a buoy riding on a Ferris wheel of a radius 6.1 m rotating at 6 rev/min is to simulate the vertical acceleration of a buoy riding a wave, what would be the amplitude and frequency of this wave? Assume that the waves are in deep water and that the buoy always rides on the surface of the water.

24.—An elevator cage of 2000 kg hangs from a steel cable. The mass per unit length of this cable is 0.60 kg/m. What is the speed of transverse waves on this cable?

25.—While an anchored sailboat pulls on its anchor rope, the tension in the rope is 5×10^3 N. The anchor rope is nylon, of diameter 0.92 cm. The density of nylon is 1.1×10^3 kg/m^3. What is the mass per unit length for this rope? What is the speed of transverse waves on this rope?

26.—A clothesline of length 10 m is stretched between a house and a tree. The clothesline is under a tension of 50 N and it has a density of 6.0×10^{-2} kg/m. How long does a wave pulse take to travel from the house to the tree and back?

27.—A wire rope used to support a radio mast has a length of 20 m and a density of 0.8 kg/m (see Fig. 10.20). When you give the wire rope a sharp blow at the lower end and generate a wave pulse, it takes 1 s for this wave pulse to travel to the upper end and to return. What is the tension in the wire rope?

28.—A nylon rope of length 24 m is under a tension of 1300 N. The total mass of this rope is 2.7 kg. If a wave pulse starts at one end of this rope, how long does it take to reach the other end?

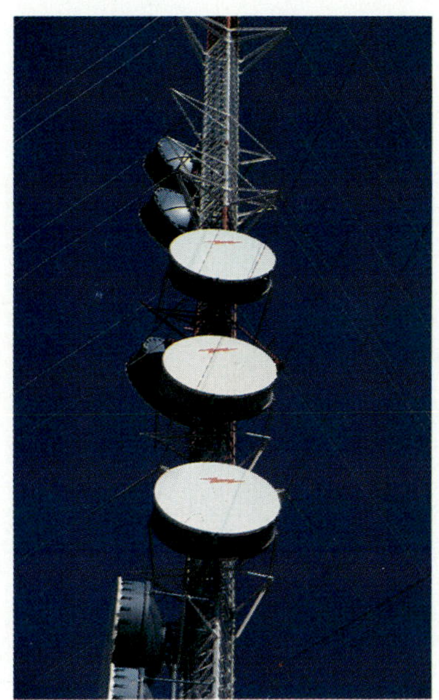

FIGURE 10.20
Radio mast showing guy ropes.

Section 10.3: The Superposition of Waves

29.—At one instant of time two transverse waves are traveling in the same direction along a stretched string. The instantaneous shapes of the waves are represented by

$$y = 0.02 \cos(4.0x) \quad \text{and} \quad y = 0.03 \cos(4.0x)$$

where the transverse displacements y and the position x are measured in meters.

(a) Are these waves in phase or out of phase?
(b) What is the amplitude and the wavelength of the net wave?

30.—At one instant of time two transverse waves are traveling in the same direction along a stretched string. The instantaneous shapes of the waves are represented by

$$y = 0.03 \cos(4.0x) \quad \text{and} \quad y = 0.03 \sin(4.0x)$$

where the transverse displacements y and the position x are measured in meters.

(a) What is the phase difference between these waves?
(b) Find the position x nearest the origin where the net wave has a wave crest. What is the amplitude of the net wave? (Hint: The wave crest of the net wave is somewhere between the wave crest of the cosine wave and the wave crest of the sine wave. You can find the wave crest of the net wave by plotting the net wave, or by trial and error, or by using a trigonometric identity.)

31.*—Consider two transverse harmonic waves of different wavelengths traveling in the same direction along a stretched string. At one instant of time, the shapes of the waves are given by

$$y = 0.012 \cos(3.0x) \quad \text{and} \quad y = 0.030 \cos(5.0x)$$

where the transverse displacements y and the position x are measured in meters.

(a) Is the superposition of these waves a harmonic wave? Is it a periodic wave?

(b) What is the wavelength of the net wave formed by the superposition of the two waves? What is the amplitude (the maximum transverse displacement) of the net wave?

32.*—Since there are two independent transverse directions of motion for a string, there are two possible directions of oscillation for transverse waves. For instance, if the string is horizontal, the two possible directions of oscillation (sometimes called directions of polarization) are horizontal and vertical. If both these kinds of transverse waves are present simultaneously on a string, the net displacement of a particle on the string is the vector sum of the horizontal and the vertical displacements.

(a) Suppose that horizontal and vertical transverse waves of equal amplitudes 2.0 cm are in phase (the crests of the two waves are in step). Describe the motion of a particle at a fixed location on the string. What is the amplitude of the net wave?

(b) Suppose that horizontal and vertical transverse waves of equal amplitudes are 180° out of phase (the crests of one wave are in step with the troughs of the other wave). Describe the motion of a particle at a fixed location on the string. What is the amplitude of the net wave?

(c) Suppose that horizontal and vertical transverse waves of equal amplitudes are 90° out of phase (the crests and the troughs of one wave are in step with the zeros of the other wave). Describe the motion of a particle at a fixed location on the string. What is the amplitude of the net wave?

33.—Two cars, of identical make, have horns that generate sound waves of slightly different frequencies, 600 Hz and 612 Hz. What beat frequency do you hear if both of these cars are blowing their horns?

34.—A pianist simultaneously plays the notes C and C# on his piano. What is the beat frequency he hears? See Table 11.1 for the frequencies of these notes.

35.—Two ocean waves with $\lambda = 100$ m, $f = 0.125$ Hz and $\lambda = 90$ m, $f = 0.132$ Hz arrive at a seawall simultaneously. What is the beat frequency of these waves?

36.*—A guitar player attempts to tune her instrument with the help of a tuning fork. If the guitar player sounds the tuning fork and a string on her guitar simultaneously, she perceives beats at a frequency of 4 Hz. The tuning fork is known to have a frequency of 294 Hz. By what factor must she increase (or decrease) the tension of the guitar string to bring the guitar in tune with the tuning fork? From the available information, can you tell whether an increase or decrease of tension is required?

Section 10.4: Standing Waves

37.—An elevator cage of 3000 kg hangs at rest from a steel cable. The length of the cable is 300 m and its mass is 180 kg. What are the wavelengths and the frequencies of the first three possible standing waves on this cable?

38.——The fundamental mode of the G string of a violin has a frequency of 196 Hz. What are the frequencies of the first, second, third, and fourth overtones?

39.——Suppose that a vibrating mandolin string of length 0.34 m vibrates in a mode with five nodes (including the nodes at the ends) and four antinodes. What overtone is this? What is its wavelength?

40.——A telegraph wire made of copper is stretched tightly between two telephone poles 50 m apart. The tension in the wire is 500 N and the mass of the segment of wire is 1.0 kg. What is the frequency of the fundamental mode? The first overtone?

41.——Some automobiles are equipped with wire wheels. The spokes of these wheels are made of short segments of thick wire installed under large tension. Suppose that one of these wires is 9.0 cm long, 0.40 cm in diameter, and under a tension of 2200 N. The wire is made of steel; its mass is 8.8×10^{-3} kg. To check the tension, a mechanic gives the spoke a light blow with a wrench near its middle. With what frequency will the spoke ring? Assume that the frequency is that of the fundamental mode.

42.——The fundamental mode of the G string in a mandolin has a frequency of 196 Hz. The length of this string is 0.34 m and its mass is 1.4×10^{-3} kg. What is the tension on this string?

43.*——The middle C string of a piano is supposed to vibrate at 261.6 Hz when exited in its fundamental mode. A piano tuner finds that in a piano that has tension of 900 N on this string, the frequency of vibration is too low (flat) by 15 Hz. By how much must he increase the tension of the string to achieve the correct frequency?

44.*——A car is being towed by means of a rope which has a mass per unit length of 0.080 kg/m. The length of the rope is 3.0 m and the tension in the rope is 2.2×10^3 N. What is the frequency for a standing wave on this rope, in the fundamental mode?

45.*——The A string on a violin has a length of 0.32 m and a mass (in this length) of 2.4×10^{-4} kg. The fundamental mode of the string has a frequency of 440 Hz. What is the tension in this string?

46.*——The D string of a violin is vibrating in its fundamental mode with a frequency of 294 Hz and an amplitude of 2.0 mm. What are the maximum velocity and the maximum acceleration of the midpoint of the string?

47.*——Many men enjoy singing in shower stalls because their voice resonates in the cavity of the show stall. Consider a shower stall measuring 1 m $\times$ 1 m $\times$ 2.5 m. What are the four lowest resonant frequencies of standing sound waves in such a shower stall?

48.*——A violin has four strings; all the strings have (approximately) equal tensions and lengths but they have different masses so that when excited in their fundamental modes they vibrate at different frequencies. The fundamental frequencies of the four strings are 196, 294, 440, and 659 Hz. What must be the ratios of the masses of the strings?

49.*——A light wave of wavelength 5.0×10^{-7} m strikes a mirror perpendicularly. The reflection of the wave by the mirror makes a standing wave with a node at the mirror. At what distance from the mirror is the nearest antinode? The nearest node?

50.*——A wave on the surface of the sea with a wavelength of 3.0 m and a period of 4.4 s strikes a seawall perpendicularly to its path. The reflection of the wave by the seawall sets up a standing wave. For such a wave, there is an antinode at the seawall. How far from the seawall will there be nodes?

CHAPTER 11

Anechoic chamber at the Naval Research Laboratory, Washington. The cones of foam rubber seen in the foreground absorb the sound waves incident on them, and thereby eliminate reflections.

SOUND

In the preceding chapter we dealt with waves propagating in one dimension, such as waves propagating on a string. These waves can be represented graphically by a plot of the displacement of the string vs. the distance x. Now we want to deal with sound waves propagating through air, in three dimensions. When sound waves emerge from a source — a human mouth, a loudspeaker, or the horn of a car — they spread outward in all directions, filling the three-dimensional volume of air surrounding the source. Since we cannot prepare a multidimensional plot of the wave vs. all the three distances x, y, and z of three-dimensional space, we need some other graphical method to represent the wave. The simplest such graphical representation of three-dimensional waves is by their *wave fronts,* that is, the locations of the wave crests at a given instant of time. For example, Fig. 11.1 shows the wave fronts of a sound wave emerging from the earpiece of a telephone handset. The sound wave

Wave fronts ▸

emerging from a pointlike source, such as the earpiece shown in Fig. 11.1, is a spherical wave; the wave fronts of this wave are concentric spherical surfaces.

As time passes, the wave fronts spread outward, expanding as they move away from the source. This spreading of the waves is a characteristic feature of wave propagation in two or three dimensions. It implies that ***the amplitude at a given wave front decreases as the wave front increases in size.*** Figure 10.1 illustrates this decrease of the wave amplitude with distance for the case of circular water waves spreading out from a pointlike disturbance caused by the impact of a pebble on the water's surface. As a given circular wave front spreads outward, its circumference increases, and its energy is distributed along this larger circumference; thus, the amplitude of the wave front must decrease in accord with the decreased concentration of its energy. Likewise, for a spherical sound wave spreading outward from a pointlike source, the wave amplitude decreases with distance.

At a very large distance from the source, the spherical wave fronts of a sound wave can be regarded as nearly flat, provided we concentrate our attention on a small region (see Fig. 11.2). ***Such waves with flat, parallel wave fronts are called plane waves.***

FIGURE 11.1

A sound wave spreading out in air. The wave fronts are concentric spheres, seen here in cross section. The sound wave has been made visible by means of a small electric light bulb attached to a microphone that controlled the brightness of the bulb. The bulb and microphone were swept through the space in front of the telephone earpiece along arcs, as indicated by the fine pattern of ridges. Wherever the sound had a large amplitude, the microphone made the light bulb bright, and this brightness was recorded on a time-exposure photograph.

11.1 SOUND WAVES IN AIR

A sound wave in air consists of alternating zones of low density and high density (or, equivalently, zones of low pressure and of high pressure). Such zones of alternating density are generated by the vibrating diaphragm of a loudspeaker or the vibrating prong of a tuning fork which exerts successive pushes on the air that is in contact with it. Figure 11.3 shows a sound wave in air at one instant of time. The alternating zones of low density and high density travel to the right, away from the source. However, although these density disturbances travel, the air as a whole does not travel — the air molecules merely oscillate back and forth.

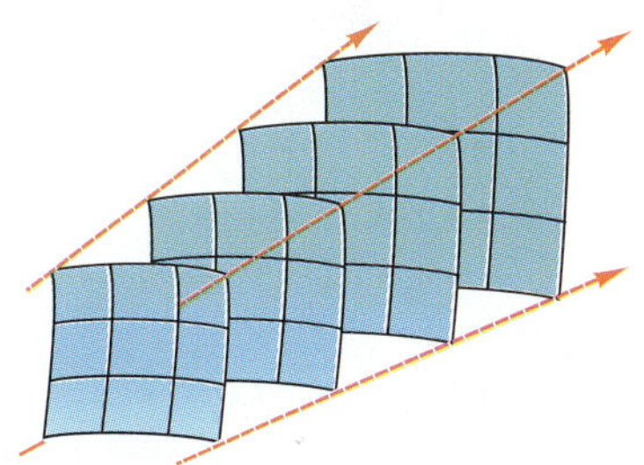

FIGURE 11.2

Parallel, nearly flat wave fronts.

The pushes of the loudspeaker or of the tuning fork on the air are longitudinal, and the sound wave itself is also longitudinal. The air molecules oscillate back and forth along the direction of propagation of the sound wave. The restoring force that drives these oscillations is the pressure of air. Wherever the density of molecules is higher than normal, the pressure also is higher than normal and pushes the molecules apart; wherever the density of molecules is lower than normal, the pressure also is lower than normal, and therefore the higher pressure of the adjacent regions pushes these molecules together. Thus, the pressure in air plays the same role as the tension in a string.

Note that at the centers of the zones of high density in the wave, the molecules are instantaneously at rest while molecules from the right and the left have converged on them. At the centers of the zones of low density, the molecules also are instantaneously at rest, but molecules on the right and the left have moved away. Thus, ***the zones of maximum and minimum density in the wave coincide with zones of zero displacement of the mol-***

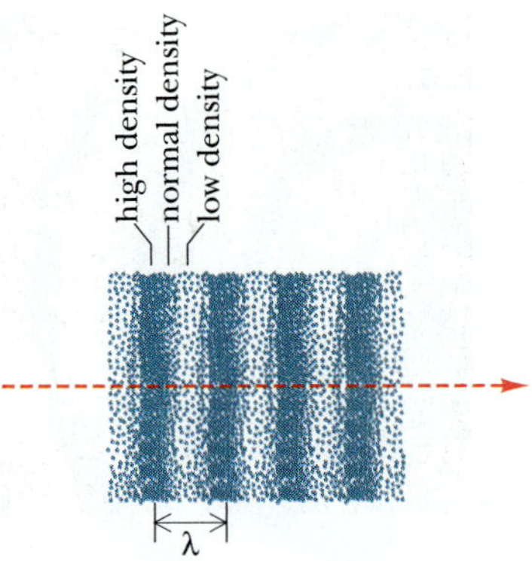

FIGURE 11.3

The sound wave consists of alternating zones of high and low density of air.

ecules. Conversely, the zones of zero density disturbance in the wave coincide with zones of maximum (positive or negative) displacement of the molecules.

In Fig. 11.3, the density disturbances are shown much exaggerated. Even in an extremely intense sound wave, such as that produced by the engines of a jet airliner at takeoff, the displacements of the molecules are only about a tenth of a millimeter and the density enhancements only about 1%.

The frequency of the sound wave determines the pitch we hear; that is, it determines whether the tone is perceived as high or low by our ear (pitch is to sound what color is to light).

According to Fourier's theorem, mentioned in Section 10.3, a periodic sound wave of arbitrary shape can be regarded as a superposition of harmonic waves. The relative strengths of the harmonic waves in this superposition determine the perceived timbre, or quality, of the sound. Pure noise, or ***white noise,*** consists of a mixture of harmonic waves of all frequencies with equal strengths. White noise sounds like air rushing through a hole; to produce something like white noise, blow air out of your mouth, making a strong shushing sound (or turn up the volume on your TV set after selecting an inactive channel). In contrast, the musical tones emitted by a musical instrument consist of a mixture of just a few harmonic waves: the fundamental and its first few overtones. Figure 11.4 shows the wave forms emitted by a violin, a trumpet, and a clarinet when the musical note C is played on these instruments. In all cases the wave is periodic, repeating at the rate of 261.7 cycles per second; but the shape of the wave and the strengths of the harmonics are quite different in each case. It is because of this difference in the shapes of the waves that our ear can distinguish between diverse musical instruments.

Table 11.1 lists the frequencies of the 12 notes of the chromatic musical scale. These frequencies are based on the system of equal temperament: successive frequencies in the scale differ by a factor of $2^{1/12} = 1.059$. The

TABLE 11.1 THE CHROMATIC MUSICAL SCALE

Note	Frequency[a]
C	261.7 Hz
C#	277.2
D	293.7
D#	311.2
E	329.7
F	349.2
F#	370.0
G	392.0
G#	415.3
A	440.0
A#	466.2
B	493.9

[a] Based on a frequency of 440 Hz for A.

(a)

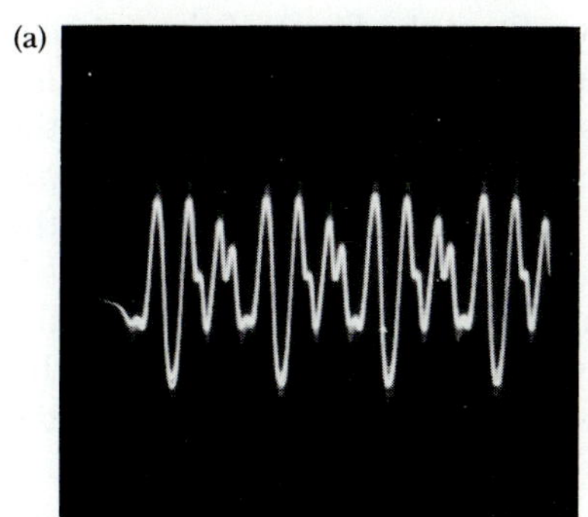

(b)

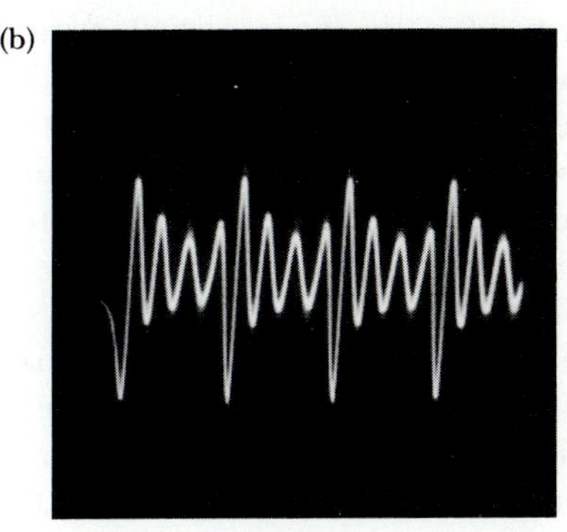

(c)

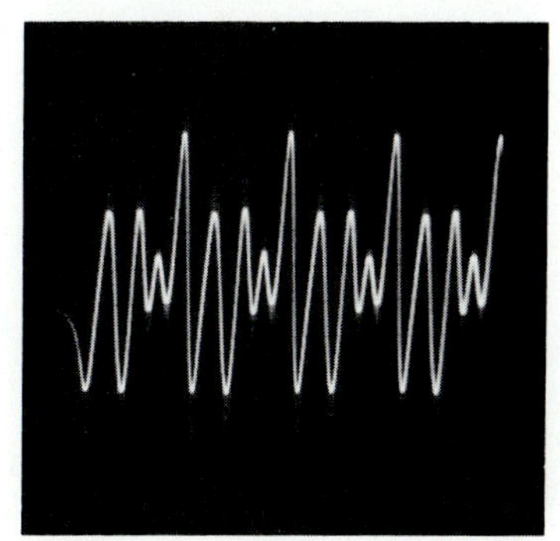

F I G U R E 11.4

Wave forms emitted by (a) a violin, (b) a trumpet, and (c) a clarinet playing the note C. The wave pattern repeats 261.7 times per second.

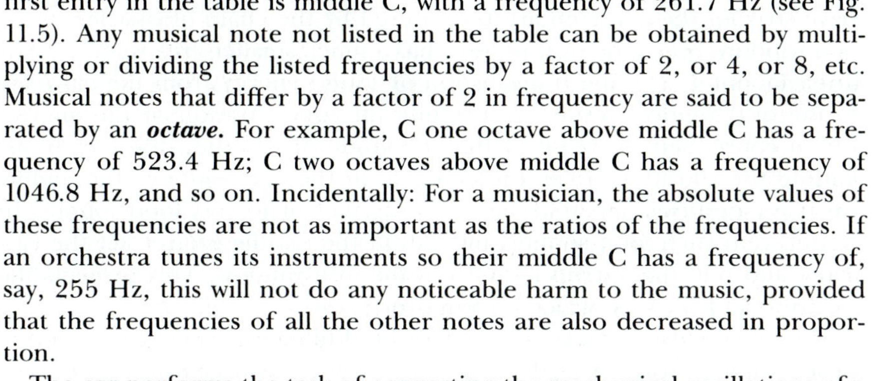

first entry in the table is middle C, with a frequency of 261.7 Hz (see Fig. 11.5). Any musical note not listed in the table can be obtained by multiplying or dividing the listed frequencies by a factor of 2, or 4, or 8, etc. Musical notes that differ by a factor of 2 in frequency are said to be separated by an ***octave.*** For example, C one octave above middle C has a frequency of 523.4 Hz; C two octaves above middle C has a frequency of 1046.8 Hz, and so on. Incidentally: For a musician, the absolute values of these frequencies are not as important as the ratios of the frequencies. If an orchestra tunes its instruments so their middle C has a frequency of, say, 255 Hz, this will not do any noticeable harm to the music, provided that the frequencies of all the other notes are also decreased in proportion.

F I G U R E 11.5

Middle C.

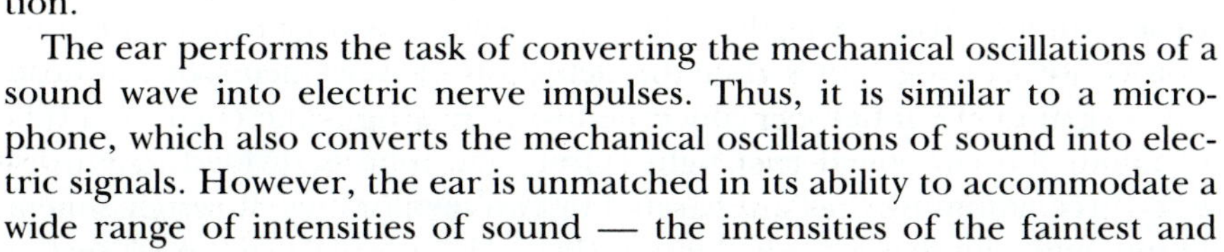

The ear performs the task of converting the mechanical oscillations of a sound wave into electric nerve impulses. Thus, it is similar to a microphone, which also converts the mechanical oscillations of sound into electric signals. However, the ear is unmatched in its ability to accommodate a wide range of intensities of sound — the intensities of the faintest and the loudest sounds acceptable to the ear differ by a factor of 10^{12}!

Figure 11.6 displays the structure of the human ear with its three main parts: the outer ear, the middle ear, and the inner ear. The ***outer ear*** consists of the auricle and the ear canal. The auricle serves to funnel sound

◀ **Outer ear**

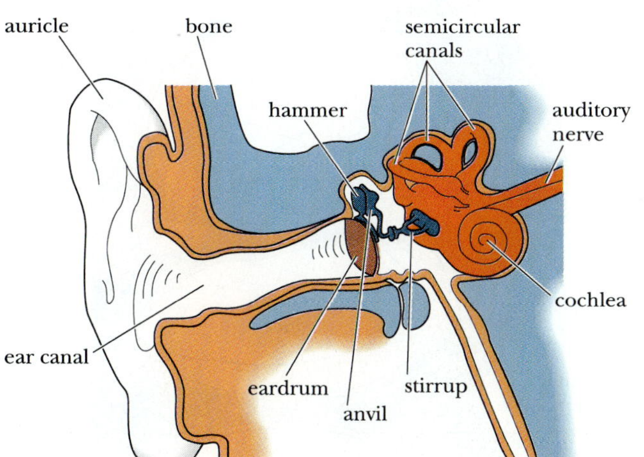

F I G U R E 11.6

Anatomy of the human ear.

waves into the ear, especially waves arriving from the front of the listener. The ear canal is a tube, about 2.7 cm long, closed off at the inner end by the eardrum, or tympanum. The ear canal guides sound waves toward the eardrum, and also enhances sound waves of a frequency of a few thousand hertz, which are in resonance with the standing-wave modes of the air column in the canal.

Middle ear ▶

The ***middle ear*** is an air-filled cavity in the temporal bone of the skull. The cavity is connected to the nasopharynx by the Eustachian tube; this tube permits equalization of the air pressure in the middle ear with the external atmospheric pressure. The middle ear contains three small bones, or ossicles: the hammer, the anvil, and the stirrup (malleus, incus, and stapes). These ossicles are arranged in a chain from the eardrum to the oval window of the inner ear. The vibrations generated by a sound wave striking the eardrum are transmitted by the chain of ossicles to the oval window. Since the oval window has a much smaller area than the eardrum (about 1/25), the transmission of sound energy from the eardrum to the oval window results in a significant concentration of the energy, with a consequent increase of the amplitude of vibration. Besides transmitting the vibrations from the eardrum to the oval window, the middle ear plays a crucial role in accommodating the ear to very loud sounds. In response to such loud sounds, muscles in the middle ear retract the eardrum and pull the stirrup away from the oval window. This protects the eardrum and the oval window from overloads.

Inner ear ▶

The ***inner ear*** is a complex system of fluid-filled cavities in the temporal bone. Among these cavities are the three semicircular canals, whose function is not hearing, but rather the detection of movements of the head. The organ of the inner ear concerned with hearing is the cochlea, a tube of about 3.5 cm coiled in a tight spiral. The tube is divided lengthwise into three adjoining ducts, separated by two membranes, Reissner's membrane and the basilar membrane (Fig. 11.7 displays a cross section through the tube of the cochlea, showing these three ducts). The sensory receptor of the inner ear is the organ of Corti, consisting of thousands of

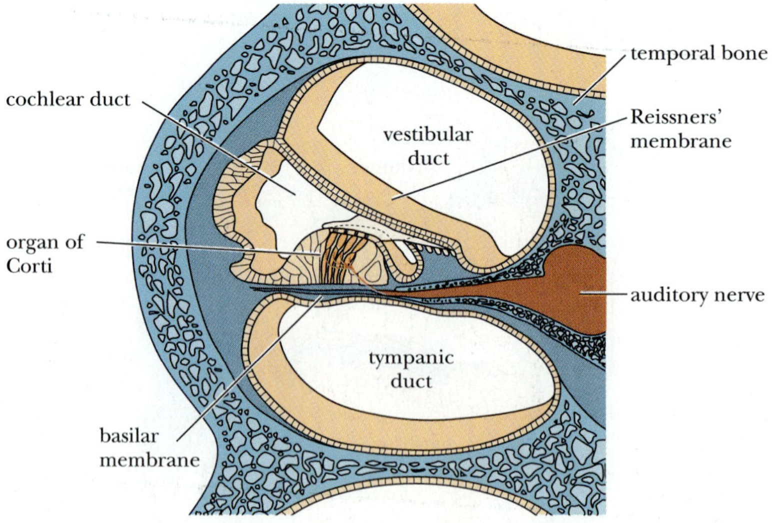

FIGURE 11.7
Cross section through the tube of the cochlea.

hair cells which sit on the basilar membrane. The vibrations of the oval window excite a wave motion in the fluid of the cochlea, which shakes the basilar membrane. The hair cells detect this motion of the basilar membrane and convert the mechanical energy into electric nerve impulses. The basilar membrane is stiff at the end near the oval window, and soft at the distant end. Because of this, the near part of the membrane responds most readily to high-frequency disturbances, and the distant part responds most readily to low-frequency disturbances. Thus, there is a correlation between the frequency of the incident sound and the regional pattern of motion of the basilar membrane. The hair cells, and the brain, recognize these regional patterns and thereby discriminate among frequencies.

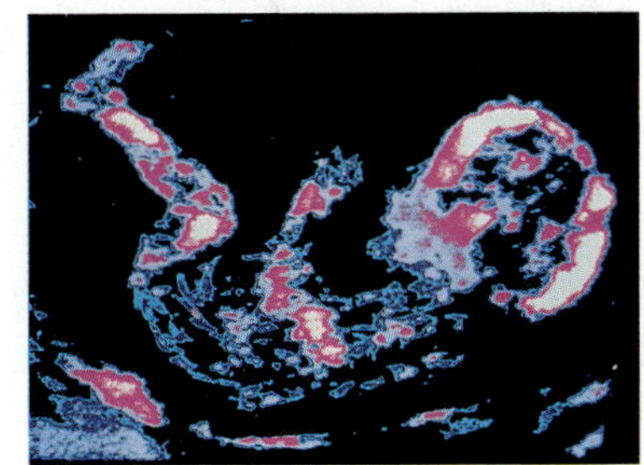

FIGURE 11.8
Sonogram of fetus.

The range of frequencies audible to the human ear extends from 20 Hz to 20,000 Hz. These limits are somewhat variable; for instance, the ears of older people are less sensitive to high frequencies. Sound waves above 20,000 Hz are called ***ultrasound;*** some animals — dogs, cats, bats, and dolphins — can hear these frequencies.

◀ Ultrasound

Ultrasonic waves of very high frequency do not propagate well in air — they are rapidly dissipated and absorbed by air molecules. However, these waves propagate readily through liquids and solids; and, in recent years, this property has been exploited in the development of some interesting practical applications of ultrasound. For instance, such waves are now used in place of X rays to take pictures of the interior of the human body. This technique, called ***sonography,*** permits examination of the fetus in the body of a pregnant woman, and avoids the damage that X rays might do to the very sensitive tissues of the fetus (see Fig. 11.8). The ultrasonic "cameras" that take such pictures employ sound waves of a frequency of about 10^6 Hz. Further development of this technique has led to the construction of acoustic microscopes. The most powerful of these devices employ ultrasound waves of a frequency in excess of 10^9 Hz to make highly magnified pictures of small samples of materials. The wavelength of sound waves of such extremely high frequency is about 10^{-6} m, roughly the same as the wavelengths of ordinary light waves. The micrographs made by experimental acoustic microscopes compare favorably with micrographs made by ordinary optical microscopes (see Fig. 11.9).

FIGURE 11.9
Acoustic micrograph of a portion of a transistor, at a magnification of 100×. This picture was made with sound waves of 2.7×10^9 Hz.

11.2 Intensity of Sound

A sound wave is intense and loud if it has a large amplitude. However, the amplitude of a sound wave is hard to measure directly, and it is more convenient to reckon the intensity of a sound wave by the energy it carries. ***The intensity of a sound wave is defined as the energy per second transported by this wave per square meter of wave front,*** that is, the power transported by this wave per square meter. Thus, to measure the intensity, we have to erect an area facing the wave, and we have to check how much energy the wave delivers in this area per second (Fig. 11.10). It can be shown that the intensity of a sound wave is proportional to the square of

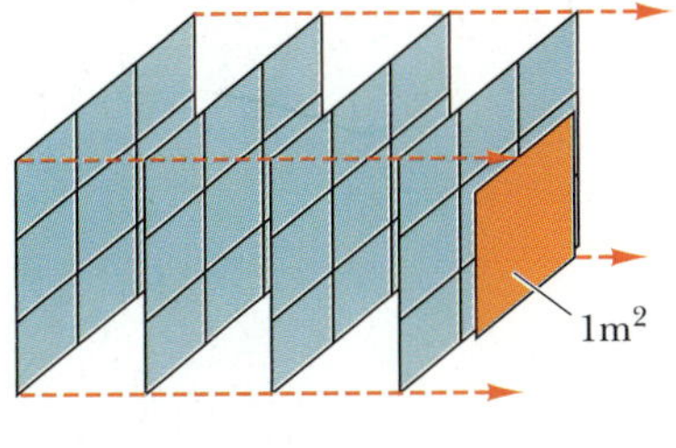

FIGURE 11.10
Area of 1 m² facing a wave. The intensity of the wave equals the energy incident on this area in 1 s.

the pressure disturbance it produces in the air; equivalently, the intensity is proportional to the square of the density disturbance.

The unit of intensity is the watt per meter squared (W/m^2). At a frequency of 1000 Hz, the minimum intensity audible to the human ear is about 2.5×10^{-12} W/m^2. This intensity is called the threshold of hearing. There is no upper limit for the audible intensity of sound; however, an intensity above 1 W/m^2 produces a painful sensation in the ear.

Note that since the eardrum has an area of about 4×10^{-5} m^2, the energy delivered per second by a sound wave of minimum intensity is only about $2.5 \times 10^{-12}\ J/m^2 \times 4 \times 10^{-5}\ m^2 \simeq 10^{-16}$ J; this is a very small amount of energy, and it testifies to the extreme sensitivity of the ear.

Decibel (dB) ▶

The intensity of sound is often expressed on a logarithmic scale called the intensity level. The unit of intensity level is the ***decibel*** (dB); like the radian, this unit is a pure number, without any dimensions of m, s, or kg. The definition of intensity level is as follows: We take an intensity of 1×10^{-12} W/m^2 as our standard of intensity, which corresponds to 0 dB.* An intensity 10 times as large corresponds to 10 dB; an intensity 100 times as large corresponds to 20 dB; an intensity 1000 times as large corresponds to 30 dB, and so on. This scale of intensity level is intended to agree with our subjective perception of the loudness of sounds. We tend to underestimate increments in the intensity of sound — our ears perceive a sound of 100×10^{-12} W/m^2 (20 dB) as only twice as loud as a sound of 10×10^{-12} W/m^2 (10 dB).

Mathematically, the relationship between the intensity in W/m^2 and the intensity level in dB is given by a formula involving a logarithm:

Intensity level ▶

$$[\text{intensity level in dB}] = (10\text{ dB}) \times \log\left(\frac{\text{intensity in W/m}^2}{1 \times 10^{-12}\text{ W/m}^2}\right) \qquad (1)$$

The logarithm in this formula is the common logarithm, with base 10. According to this formula, a sound of intensity 1×10^{-12} W/m^2 has an intensity level of 10 dB × log(1), which is 0 dB, since log(1) = 0. And a sound of intensity 10×10^{-12} W/m^2 has an intensity level of 10 dB × log(10), which is 10 dB, since log(10) = 1, and so on. Thus, the mathematical formula agrees with the assertions made in the preceding paragraph.

Table 11.2 gives some examples of sounds of different intensities.

EXAMPLE 1. Express the threshold of hearing (2.5×10^{-12} W/m^2) and the threshold of pain (1 W/m^2) in decibel.

---

* More precisely, the standard of intensity, which corresponds to 0 dB, is 0.937×10^{-12} W/m^2. But it will be adequate for our purposes to round this off to 1×10^{-12} W/m^2.

SOLUTION: From Eq. (1), we find that the intensity level corresponding to 2.5×10^{-12} W/m² is

$$(10 \text{ dB}) \times \log\left(\frac{2.5 \times 10^{-12} \text{ W/m}^2}{1 \times 10^{-12} \text{ W/m}^2}\right) = (10 \text{ dB}) \times \log(2.5) \qquad (2)$$

With our calculator, we obtain log(2.5) = 0.40, and hence the intensity level is 4.0 dB.

Likewise, we find that the intensity level corresponding to 1 W/m² is

$$(10 \text{ dB}) \times \log\left(\frac{1 \text{ W/m}^2}{1 \times 10^{-12} \text{ W/m}^2}\right) = (10 \text{ dB}) \times \log(10^{12}) \qquad (3)$$

Since $\log(10^{12}) = 12$, the intensity level is 120 dB.

TABLE 11.2 SOME SOUND INTENSITIES

Sound	Intensity level	Intensity
Rupture of eardrum	160 dB	1.0×10^4 W/m²
Jet engine (at 30 m)	130	10
Threshold of pain	120	1.0
Rock music	115	0.30
Thunder (loud)	110	0.10
Subway train (New York City)	100	1.0×10^{-2}
Heavy street traffic	70	1.0×10^{-5}
Normal conversation	60	1.0×10^{-6}
Whisper	20	1.0×10^{-10}
Normal breathing	10	1.0×10^{-11}
Threshold of hearing	4	2.5×10^{-12}

As a sound wave spreads out from its source, its intensity falls off because the area of the wave front grows larger, and therefore the wave energy per unit area grows smaller. Figure 11.11 helps to make this clear; it shows a spherical wave front at successive instants of time. The wave front grows from an old radius r_1 to a new radius r_2; correspondingly, the area of the wave front grows from $4\pi r_1^2$ to $4\pi r_2^2$. The total power carried by the wave front remains the same; hence the power per unit area, or the intensity, is inversely proportional to the area. If the intensity at r_1 is I_1, and at r_2 is I_2, then

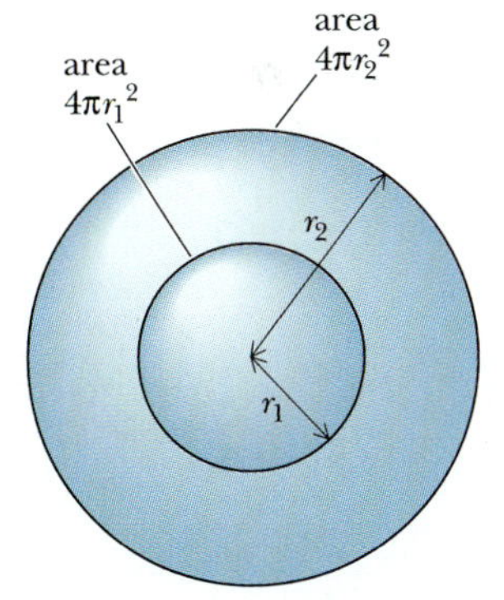

FIGURE 11.11
Concentric spherical wave fronts of a sound wave in air.

$$I_1 \propto 1/4\pi r_1^2 \quad \text{and} \quad I_2 \propto 1/4\pi r_2^2 \qquad (4)$$

which says that ***the intensity is inversely proportional to the square of the***

distance. Taking the ratio of the two proportionalities in Eq. (4), we obtain

Decrease of intensity with distance

$$\frac{I_2}{I_1} = \frac{r_1^2}{r_2^2} \tag{5}$$

Although Fig 11.11 shows the wave front spreading out uniformly in all directions, Eq. (5) remains valid if the wave is a beam, such as the beam emitted by a loudspeaker with a horn, aimed in some preferential direction. In this case, the intensity falls off as the inverse square of the distance along the direction of the beam.

EXAMPLE 2. At a distance of 60 m from a jet airliner engaged in takeoff, the intensity of sound is 1 W/m^2. What is the intensity at a distance of 180 m in the same direction?

SOLUTION: With $r_1 = 60$ m and $r_2 = 180$ m, Eq. (5) gives

$$I_2 = \frac{r_1^2}{r_2^2} I_1 \tag{6}$$

$$= \frac{(60 \text{ m})^2}{(180 \text{ m})^2} \times 1 \text{ W/m}^2 = 0.11 \text{ W/m}^2$$

11.3 THE SPEED OF SOUND

As in the case of a wave on a string, the speed of sound in air depends on the restoring force and on the amount of mass. Since the restoring force is the pressure of air, and since the amount of mass is proportional to the density of air, we expect the speed of sound to depend on the pressure and on the density of air. A somewhat involved calculation shows that the theoretical formula for the speed of sound in air is

$$v = \sqrt{1.40 \frac{p_0}{\rho_0}} \tag{7}$$

where p_0 designates the atmospheric pressure and ρ_0 the atmospheric density. Note that this formula does display the expected increase of speed with restoring force and decrease with mass — for air of higher pressure or air of lower density, the speed is higher.

Under so-called standard conditions, the pressure of air is $p_0 = 1.01 \times 10^5$ N/m^2 and the density is $\rho_0 = 1.29$ kg/m^3 (at a temperature of 0°C);

under these conditions, the speed of sound is

$$v = \sqrt{1.40 \frac{p_0}{\rho_0}} = \sqrt{1.40 \times \frac{1.01 \times 10^5 \text{ N/m}^2}{1.29 \text{ kg/m}^3}} = 331 \text{ m/s} \qquad (8)$$

◀ Speed of sound

The speed of sound in liquids and in solids is considerably higher than in air because the restoring force is much larger — liquids and solids offer much more opposition to compression than gases. For instance, the speed of sound in water is about 4 times as large as in air, and the speed of sound in iron or steel is about 15 times as large as in air. Table 11.3 gives the values of the speed of sound in some materials.

TABLE 11.3 THE SPEED OF SOUND IN SOME MATERIALS

Material	v
Air	
0°C, 1 atm	331 m/s
20°C, 1 atm	344
100°C, 1 atm	386
Helium, 0°C, 1 atm	965
Water (distilled)	1497
Water (sea)	1531
Aluminum	5104
Iron	5130
Glass	5000–6000
Granite	6000

A simple method for the measurement of the speed of sound in air takes advantage of standing waves in a tube open at one end and closed at the other (see Fig. 11.12). As in the case of a standing wave on a string, the standing sound wave in the column of air in the tube consists of two traveling waves, of opposite directions of travel. ***The standing sound wave in the tube must have a displacement node at the closed end,*** since the wall at this end forbids the motion of the molecules. ***The wave must have a displacement antinode at the open end.*** This is not so obvious, but can be understood by first considering the pressure variation in the wave. The pressure excess, relative to normal atmospheric pressure, is large where the displacement of the molecules is small, and small where the displacement of the molecules is large (compare the discussion of Fig. 11.3 in Section 11.1). Thus, the pressure nodes are displacement antinodes, and conversely. At the open end of our tube, the pressure must remain constant, because the open end is accessible to the atmosphere, and hence any incipient decrease or increase of the pressure would immediately lead to an inflow or outflow of air from the surrounding atmosphere, canceling the pressure change. Thus, the atmosphere tends to behave as a "reservoir" of constant pressure, and the pressure excess of the standing

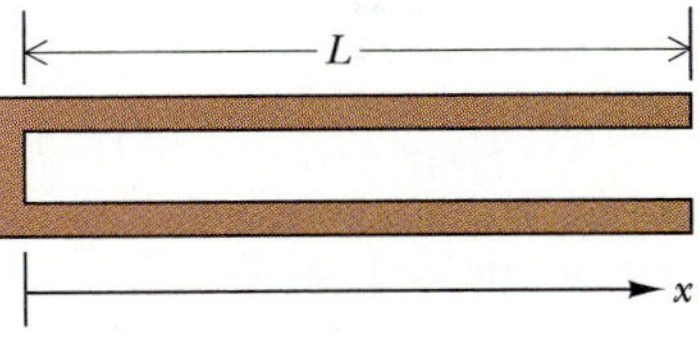

FIGURE 11.12
A tube open at one end and closed at the other.

wave must have a node at the open end of the tube. This implies that the open end is a displacement antinode.

With these boundary conditions for the closed end and the open end of the tube, the possible standing waves, or normal modes, are as shown in Fig. 11.13. There is either one quarter of a wavelength in the tube, or three quarters of a wavelength, or five quarters of a wavelength, etc. If the length of the tube is L, then the wavelengths for these normal modes are

Wavelengths and eigenfrequencies of normal modes in a tube ▶

$$\lambda_1 = 4L, \quad \lambda_2 = \frac{4L}{3}, \quad \lambda_3 = \frac{4L}{5}, \ldots \tag{9}$$

The eigenfrequencies of these normal modes are

$$f_1 = \frac{v}{4L}, \quad f_2 = \frac{3v}{4L}, \quad f_3 = \frac{5v}{4L}, \ldots \tag{10}$$

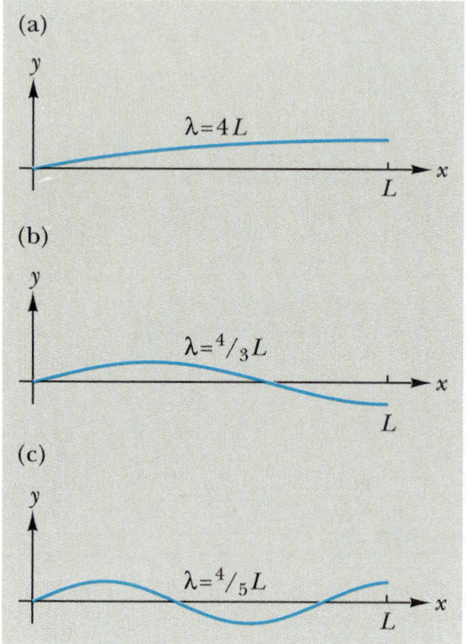

FIGURE 11.13
Possible standing waves in tube open at one end. (a) The fundamental mode. (b) The first overtone. (c) The second overtone.

Note that the expressions (9) for the wavelengths of the normal modes of sound in a tube differ from the expressions (10.9) for the wavelengths of the normal modes of a string. This is due to the difference in boundary conditions; the tube has a node at one end and an antinode at the other end, whereas the string has nodes at both ends.

The speed of sound can be determined by means of Eq. (10) by measuring the resonant frequency of a tube of known length.

Standing waves play a crucial role in most musical instruments. Organs, flutes, trumpets, trombones, and other wind instruments are essentially tubes within which standing waves are excited by a stream of air blown across or into a blowhole or mouthpiece. In organs and flutes, the blowhole acts as an open end of the tube, and therefore the normal modes are those of a tube with *two* open ends. In trumpets and trombones, the lips of the player act approximately as a closed end, and the normal modes are those of a tube with one closed and one open end. The eigenfrequencies of the tube depend on its length. In many wind instruments — flutes, trumpets, French horns — the effective length of the tube can be varied by opening or closing valves, thereby changing the eigenfrequencies.

The excitation of vibrations in the tube of an organ or a flute by a steady stream of air blown across the blowhole depends on a rotational motion that develops behind the edges of the blowhole when the velocity of flow is high. As the air streams past the edge, it forms a vortex (see Fig. 11.14). This vortex soon breaks away from the edge and is replaced by another vortex, and another, and so on. The regular succession of vortices constitutes a vibration of the stream of air, and this excites standing waves in the tube of the organ by resonance.

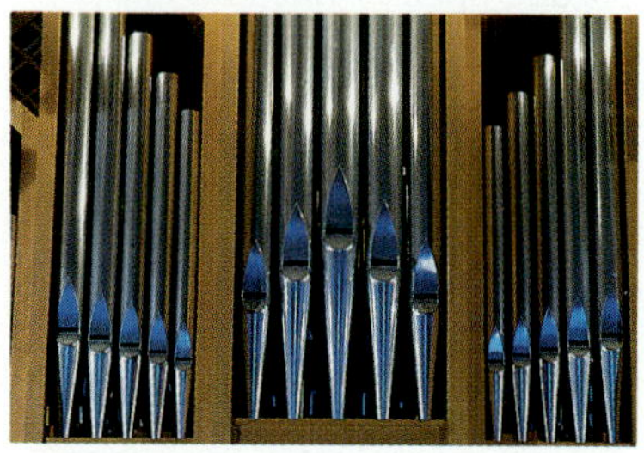

FIGURE 11.14
Organ pipes.

The excitation of vibrations in the tube of a trumpet or trombone involves a different mechanism. These instruments have a cup-shaped mouthpiece, across which the player stretches his lips, which then behave somewhat like a pair of strings under tension, with a natural period of vibration. The vibration of the lips is triggered by the stream of air that the player blows out of his mouth. If the lips are initially close together and the gap between them is small, the pressure in the mouth builds up. This high pressure pushes the lips apart. But when the gap between them becomes wide, air rushes out and the pressure decreases. This permits the lips to snap back to their initial configuration. The gap between the lips therefore periodically widens and narrows, with a natural frequency that is determined by their tension. The periodic puffs of air produced by this vibration excite standing waves in the tube of the trumpet.

Stringed instruments — violins, guitars, mandolins — use a resonant cavity to amplify and modify the sound produced by the string. The cavity is mechanically coupled to the string, and the vibrations of the latter excite resonant vibrations in the former. The resonant vibrations not only involve standing waves in the air in the cavity, but also standing waves in the solid material (wood) of the walls. Because the area of the body of, say, a violin is much larger than the area of its strings, the body pushes against much more air and radiates sound more efficiently than the strings. Hence, most of the sound from a violin emerges from its body.

The human voice organ is roughly analogous to a trumpet. The throat and the oral and nasal cavities form the "tube" of this trumpet, and the larynx forms its "mouthpiece." Standing waves are excited in these cavities by the vibration of the ***vocal cords,*** or vocal folds, stretched across the larynx (see Fig. 11.15). The space between each vocal cord and the side of the larynx is closed off by a membrane. Hence, the vocal cords form a pair of "lips" that can open and close, leaving between them a variable gap called the glottis. Laryngeal muscles control the tension in the vocal cords and therefore their natural frequency of vibration. As in the case of the trumpet, the vibration of the vocal cords, or "lips," is triggered by the stream of air blown through the glottis; this vibration periodically opens and closes the glottis and generates periodic puffs of air which excite standing waves in the oral and nasal cavities.

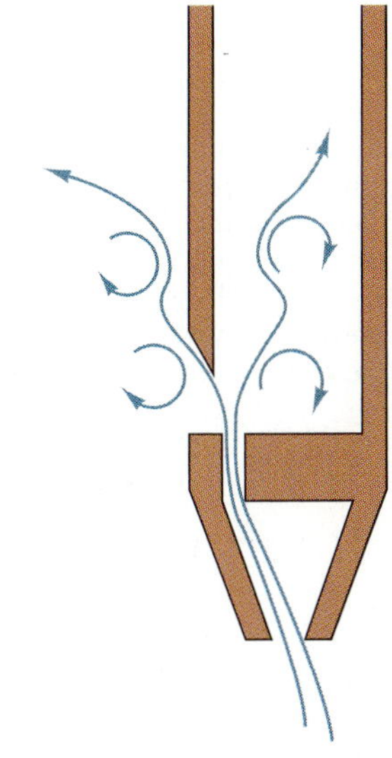

FIGURE 11.15
Vortices at the blowhole of an organ pipe.

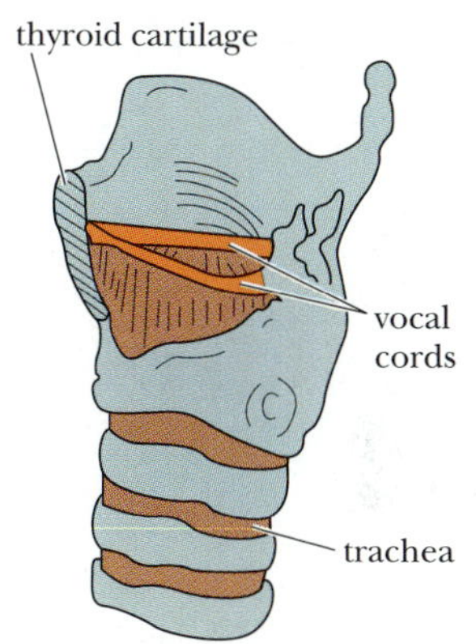

FIGURE 11.16
Cutaway view of the human larynx, showing the two vocal cords.

11.4 THE DOPPLER EFFECT

The speed of a sound wave in air is 331 m/s when measured in a reference frame at rest in the air. But when measured in a reference frame moving through the air, the speed of the sound wave will be larger or smaller, depending on the direction of motion of the reference frame. For example, if a train moving at 30 m/s approaches a stationary siren which is emitting sound waves (see Fig. 11.17), the speed of the sound waves relative to the train will be 361 m/s. And if the train moves away from the siren, the speed of the sound waves relative to the train will be 301 m/s.

The motion of the train affects not only the speed of the sound waves relative to the train, but also their frequency. For instance, if the train ap-

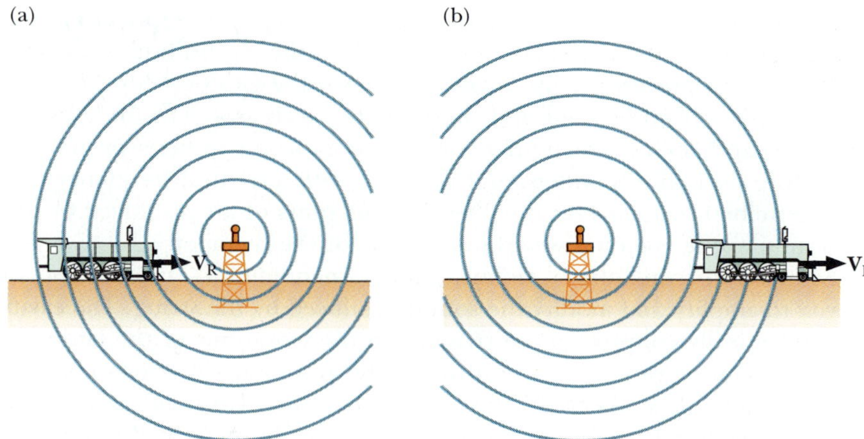

FIGURE 11.17
(a) Train approaching a siren. The train encounters *more* wave fronts per unit time than when stationary. (b) Train receding from the siren. The train encounters *fewer* wave fronts per unit time than when stationary.

proaches the siren, it runs head on into the sound waves (see Fig. 11.17a) and hence encounters more wave fronts per second than if it were stationary; and if the train recedes from the siren, it runs with the sound waves (see Fig. 11.17b) and hence encounters fewer wave fronts per second. Consequently, ***a receiver on the train will detect a higher frequency when approaching the siren, and a lower frequency when receding.*** This frequency change caused by the motion of the receiver (or by motion of the source; see below) is called the ***Doppler shift.***

Doppler shift ▸

To calculate the frequency shift, we note that in the reference frame of the ground, we have the usual relation between the frequency, speed, and wavelength of the sound wave:

$$f = v/\lambda \tag{11}$$

and in the reference frame of the train, we have a corresponding relation:

$$f' = v'/\lambda \tag{12}$$

The wavelengths in Eqs. (11) and (12) are exactly the same because the distance between the wave crests does not depend on the reference frame. Dividing Eq. (12) by Eq. (11), we obtain

$$\frac{f'}{f} = \frac{v'}{v} \tag{13}$$

We will designate by V_R the speed of the train acting as receiver of sound waves. In the reference frame of the train, the speed of sound is then $v' = v \pm V_R$, where the positive sign corresponds to motion of the train toward the source of sound and the negative sign to motion of the train away from the source of sound. With this expression for v', Eq. (13) yields

Frequency at moving receiver, for stationary emitter ▸

$$f' = f\left(1 \pm \frac{V_R}{v}\right) \qquad \begin{array}{l} + \text{ for approaching receiver} \\ - \text{ for receding receiver} \end{array} \tag{14}$$

EXAMPLE 3. Suppose that a stationary siren emits a tone of frequency 440 Hz as the train moves away from it at 30 m/s. What is the frequency received on the train?

SOLUTION: Since the train is moving away, we must use Eq. (14) with the negative sign:

$$f' = f\left(1 - \frac{V_R}{v}\right) \tag{15}$$

$$= 440 \text{ Hz} \times \left(1 - \frac{30 \text{ m/s}}{331 \text{ m/s}}\right) = 400 \text{ Hz}$$

Note that if the receiver is moving away from the source at a speed equal to the speed of sound ($V_R = v$), then the frequency f' is zero; this simply means that the receiver is moving exactly with the waves, and therefore no wave fronts catch up with it. If the receiver is moving away at a speed greater than the speed of sound ($V_R > v$), then Eq. (14) gives a negative frequency; this means that the receiver overruns the wave fronts from behind. A receiver speed equal to or larger than the speed of sound can be achieved only by mounting the receiver on a supersonic aircraft or some other fast-flying object, an arrangement of little practical interest. However, Eq. (14) applies not only to sound waves, but also to water waves and other kinds of waves. For water waves, it is not at all hard to arrange for a "receiver" with a speed V_R in excess of the speed v of the water waves.

Christian Doppler, *1803–1853, Austrian physicist. After Doppler discovered his formula for the frequency shift of sound, he recognized that light from a moving source should also be subject to a frequency shift, resulting in a change of color of the received light. However, unless the speed of the source is extremely large, such changes of the color of light are too slight to be perceived by the eye.*

EXAMPLE 4. A motorboat speeding at 15 m/s is moving in the same direction as a group of water waves of frequency 0.17 Hz and speed 9.3 m/s (relative to the water; see Fig. 11.18). What is the frequency with which the wave crests pound on the motorboat?

SOLUTION: Since the motorboat is moving away from the source of waves, we again use Eq. (14) with the negative sign:

$$f' = f\left(1 - \frac{V_R}{v}\right) \tag{16}$$

$$= 0.17 \text{ Hz} \times \left(1 - \frac{15 \text{ m/s}}{9.3 \text{ m/s}}\right) = -0.10 \text{ Hz}$$

The negative sign indicates that the motorboat overtakes the waves at the rate of 0.10/s, that is, one wave every 10 seconds.

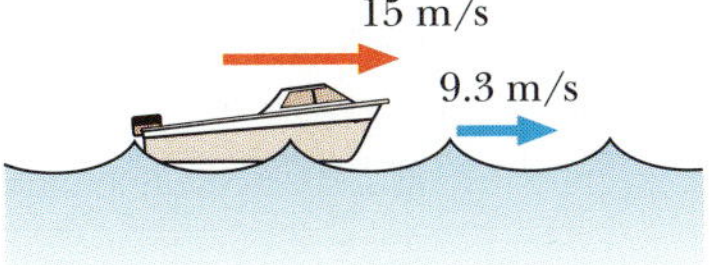

FIGURE 11.18

Motorboat overtaking waves on the surface of the water.

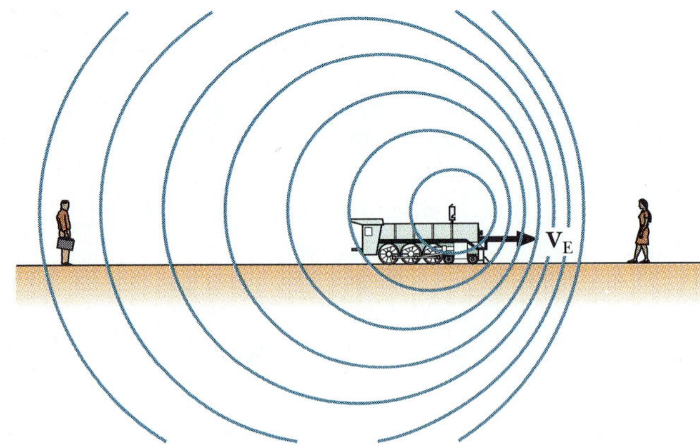

FIGURE 11.19

Train emitting sound waves while in motion. The wavelength ahead of the train is shorter and that behind the train is longer than when the train is stationary.

A shift between the frequency emitted by the source of sound waves (or other waves) and the frequency detected by the receiver will also occur if the source is in motion and the receiver is at rest. For example, if a train approaching a station blows a whistle, the successive wave fronts emitted by the whistle are centered at intervals along the path of the whistle, and they will be crowded together in the forward direction and spaced apart in the rearward direction (see Fig. 11.19). Consequently, ***a stationary receiver will detect a higher frequency when in front of the train, and a lower frequency when behind the train.*** As the train rushes by the stationary receiver, the detected frequency suddenly changes from high to low. This explains the sudden drop in pitch that you hear when standing next to a railroad track as a whistling train passes by, or when standing next to a street as an ambulance with its siren passes by.

We will designate by V_E the speed of the train when it acts as an emitter of sound. To calculate the frequency change produced by the motion of the emitter, we begin by noting that in the time $1/f$ corresponding to one period, the train travels a distance $(1/f)V_E$ between emitting one wave front and the next, and hence the wavelength is shortened or lengthened from its normal value λ to a new value $\lambda' = \lambda \mp V_E/f$, where the negative sign corresponds to motion of the train toward the receiver and the positive sign to motion away from the receiver. The new frequency is therefore

$$f' = \frac{v}{\lambda'} = \frac{v}{\lambda \mp V_E/f} = \frac{v}{v/f \mp V_E/f} \tag{17}$$

If we multiply both the numerator and the denominator of the right side of this equation by f/v, we obtain

Frequency at stationary receiver, for moving emitter

$$f' = f\left(\frac{1}{1 \mp V_E/v}\right) \quad \begin{array}{l} - \text{ for approaching emitter} \\ + \text{ for receding emitter} \end{array} \tag{18}$$

EXAMPLE 5. Suppose that the whistle of a train emits a tone of frequency 440 Hz as the train recedes from a stationary observer at 30 m/s. What frequency does the observer hear?

SOLUTION: For a receding whistle, we must use the plus sign in Eq. (18):

$$f' = f\left(\frac{1}{1 + V_E/v}\right) \tag{19}$$

$$= \frac{440\ \text{Hz}}{1 + (30\ \text{m/s})/(331\ \text{m/s})} = 403\ \text{Hz}$$

If we compare the results of Examples 3 and 5, we see that motion of the source and motion of the receiver have nearly the same effect on the frequency — in both examples the frequency is decreased by about 10%, from 440 Hz to 400 or 403 Hz. This symmetry of the Doppler shift has to do with the low speed of the motion. When the speed is low compared with the speed of sound, the effects of motion of the source and motion of the receiver are approximately the same; but when the speed is high, the effects of motion of the source and motion of the receiver are quite different.

Ernst Mach (makh), *1838–1916, Austrian philosopher and physicist. Mach obtained visual evidence for the Mach cone by photographing projectiles in flight. His book* The Science of Mechanics *is a profound critical examination of the historical and logical foundations of mechanics.*

Finally, let us consider the case of an emitter, such as a fast aircraft, moving at a speed nearly equal to the speed of sound. If the aircraft emits sound of some frequency f, then Eq. (18) indicates that the frequency received at points just ahead of the aircraft is very large — in the limiting case of a speed V_E almost equal to the speed of sound, the frequency becomes extremely large. This is because all the wave fronts are infinitely bunched together, and they all arrive at almost the same instant as the aircraft (see Fig. 11.20). If the speed of the aircraft exceeds the speed of sound, then the aircraft will overtake the wave fronts (see Fig. 11.21). In this case ***the sound is confined to a conical region that has the aircraft at its apex and moves with the aircraft at the speed*** V_E; ahead of this region, the air has not yet been disturbed, although it will be disturbed at a later time, when the aircraft moves sufficiently far to the right. The cone is called the ***Mach cone.***

◀ **Mach cone**

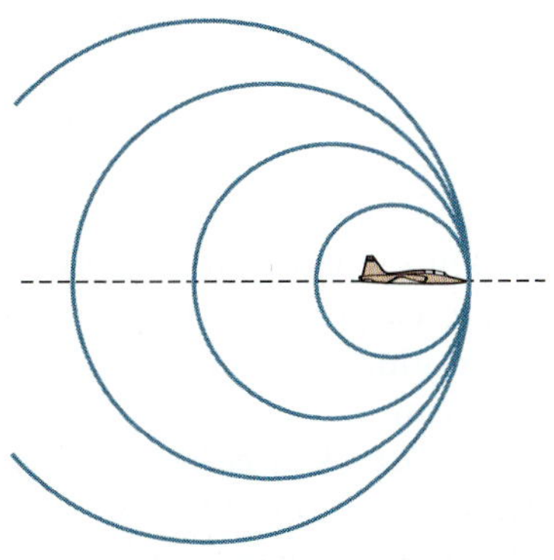

FIGURE 11.20
Sound waves emitted by a subsonic aircraft at a speed almost equal to the speed of sound.

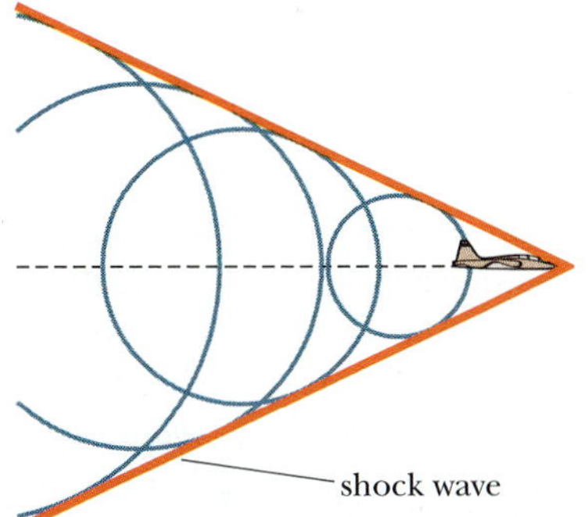

FIGURE 11.21
Sound waves emitted by a supersonic aircraft.

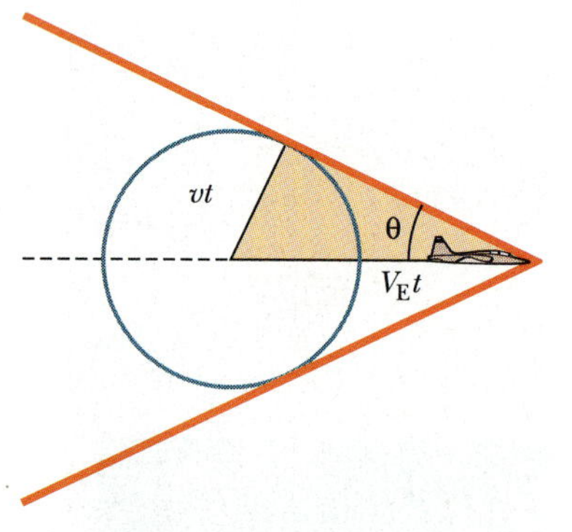

FIGURE 11.22
Mach cone. In a time *t*, the aircraft moves a distance $V_E t$ and the initial sound wave moves a distance *vt*.

FIGURE 11.23
A bullet and its Mach cone. The bullet is passing through a soap bubble filled with argon gas.

FIGURE 11.24
Concorde SST.

The half angle of the apex of the Mach cone is given by the formula

$$\sin\theta = v/V_E \tag{20}$$

This can readily be seen from Fig. 11.22, which shows the aircraft at a time t and the wave front that was emitted by the aircraft at time zero. In the time t the sound travels a distance vt, and, simultaneously, the aircraft travels a distance $V_E t$. Thus, the radius of the wave front is vt. This radius is the opposite side of a right triangle of hypotenuse $V_E t$. Consequently,

$$\sin\theta = vt/V_E t \tag{21}$$

which leads to Eq. (20) if we cancel the factor t.

Any supersonic aircraft (or other body) moving through air will generate a Mach cone, regardless of whether or not it carries an artificial source of sound aboard (see Fig. 11.23). The motion of the body through the air creates a pressure disturbance which spreads outward with the speed of sound and forms the cone. The cone trails behind the body much as a wake trails behind a ship. The sharp pressure disturbance at the surface of the cone is heard as a loud bang whenever the cone sweeps over the ear. This is the ***sonic boom.*** For a large aircraft, such as the Concorde SST (see Fig. 11.24), the noise level of the sonic boom reaches the pain threshold even if the aircraft is 20 km away.

11.5* Diffraction

It is a characteristic feature of waves that they will deflect around the edges of an obstacle placed in their path and penetrate into the "shadow" zone behind the obstacle. For example, Fig. 11.25 shows water waves striking a breakwater at the entrance of a harbor. The region directly behind the breakwater is out of the direct path of the waves, but nevertheless waves reach this region because each wave front spreads sideways once it has passed the entrance to the harbor. This lateral spreading of the wave fronts can be easily understood: The breakwater cuts a segment out of each wave front, and such a segment of wave front cannot just keep moving straight on as though nothing had happened. The end of the segment is a vertical wall of water where the breakwater has chopped off the wave front, and the water at the ends will immediately begin to spill out sideways, producing a disturbance at the edge of the segment. This disturbance continues to spread out and gradually forms curved wave fronts to the left and the right of the main beam of the wave.

Diffraction ▸

Such a deflection of waves by the edge of an obstacle is called diffraction. It is a general rule that the amount of diffraction suffered by a wave passing through an aperture depends on the ratio of wavelength to the size of

* This section is optional.

the aperture. An increase of the wavelength (or a decrease of the size of the aperture) makes the diffraction effects more pronounced; a decrease of the wavelength (or an increase of the size of the aperture) makes the diffraction effects less pronounced. For instance, Fig. 11.26a shows diffraction of waves of relatively long wavelength by a small aperture; the waves spread out very strongly, forming divergent, fanlike beams of concentric wave fronts. Figure 11.26b shows diffraction of waves of shorter wavelength — here the wave spreads out only slightly; most of the wave remains within a straight beam of nearly parallel wave fronts. Note that in Fig. 11.26b, the beam has a fairly well-defined edge — the region in the "shadow" of the breakwater remains nearly undisturbed while the region facing the aperture receives the full impact of the waves.

FIGURE 11.25

Ocean waves incident on a breakwater with an aperture.

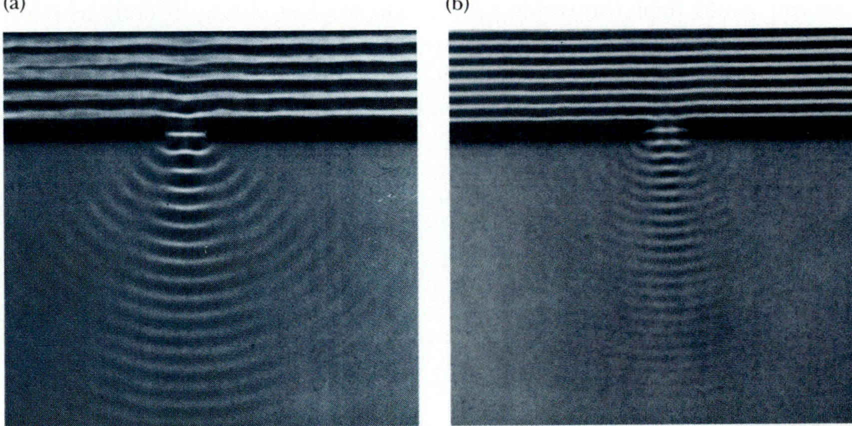

FIGURE 11.26

(a) Water waves of fairly long wavelength in a ripple tank exhibit strong diffraction when passing through an aperture. (b) Water waves of shorter wavelength exhibit less diffraction.

The fanlike beams of waves spreading out from the aperture constitute a ***diffraction pattern.*** In Fig. 11.26a, the diffraction pattern consists of a central beam and two clearly recognizable secondary beams on each side. The beams are separated by nodal lines along which the wave amplitude is zero. Figures 11.27a and 11.27b shows the diffraction patterns generated by a small island. Note that if the wavelength is large compared with the size of the island, then there exists no shadow zone; instead, the island merely produces some distortion of the waves.

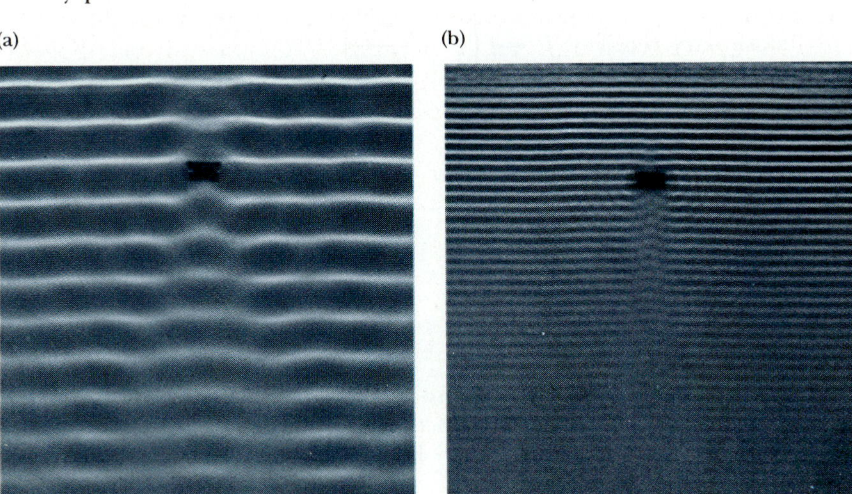

FIGURE 11.27

(a) Water waves of long wavelength diffracted by a small island. (b) Water waves of shorter wavelength.

FIGURE 11.28
Hi-fi sound system with separate loudspeakers for low, intermediate, and high frequencies.

Diffraction plays an important role in the propagation of sound. We can hear a person whose mouth is out of our direct line of sight — say, a person facing away from us or a person talking in an adjacent room — because the sound waves diffract through the open mouth or the open door and spread to fill the entire room.

Diffraction must be taken into account in the design of loudspeakers for hi-fi systems. For strong reproduction of the low-frequency sounds, hi-fi systems use large speakers ("woofers"), as much as 40 cm in diameter. The low-frequency sound waves have wavelengths in excess of 40 cm, and when these waves emerge from the speaker, they diffract and spread to fill the entire room. However, the higher-frequency waves have shorter wavelengths, and if they were emitted by the woofers, they would not diffract as much. These waves would then form a more or less narrow beam, and you could hear the treble only if you were standing directly in front of the woofer. To achieve a better spreading of the sound waves (and for other reasons), the middle frequencies in the music are routed to mid-range "drivers," and the high frequencies are routed to "tweeters." The drivers are about 20 cm in diameter, whereas the tweeters are just a few centimeters across (see Fig. 11.28). These choices of the sizes of the speakers ensure that all the waves are strongly diffracted.

SUMMARY

- **Decrease of intensity of sound wave with distance:** $\dfrac{I_2}{I_1} = \dfrac{r_1^2}{r_2^2}$
- **Speed of sound in air:** 331 m/s
- **Standing wave in tube open at one end:**
  $\lambda_1 = 4L$, $\lambda_2 = \frac{4}{3}L$, $\lambda_3 = \frac{4}{5}L$, etc.
- **Doppler shift:**

  stationary emitter: $f' = f\left(1 \pm \dfrac{V_R}{v}\right)$

  stationary receiver: $f' = f\left(\dfrac{1}{1 \mp V_E/v}\right)$
- **Mach cone:** $\sin\theta = v/V_E$

QUESTIONS

1.–Could an astronaut be heard playing the violin while standing on the surface of the Moon?

2.–A hobo can hear a very distant train by placing an ear against the rail. How does this help? (Hint: Ignoring frictional losses, how does the intensity of sound decrease with distance in air? In the rail?)

3.–If you speak while standing in a corner with your face toward the wall, you will sometimes notice that your voice sounds unusually loud. Explain.

4.–Why does the wind whistle in the rigging of a ship or in the branches of a tree?

5.–How does a flutist play different musical notes on a flute?

6.–When inside a boat, you can often hear the engine noises of another boat much more clearly than when on deck. Can you guess why?

7.–Many men like singing in shower stalls because the stall somehow enhances their voice. How does this happen? Would the effect be different for men and women?

FIGURE 11.29
Large pipes on an organ.

8.–The pipes that produce the lowest frequencies in a great organ are very long, usually 4.88 m (see Fig. 11.29). Why must they be long?

9.–Some of the old European opera houses and concert halls renowned for their acoustic excellence have very irregular walls, heavily encrusted with an abundance of stucco ornamentation which reflects sound waves in almost all directions. (Fig. 11.30). How does the sound reaching a listener in such a hall differ from the sound reaching a listener in a modern concert hall with four flat, plain walls?

FIGURE 11.30
European opera house, interior.

10.–Electric guitars simplify the sound of the strings electronically. Do such guitars need a body?

11.–Does the temperature of the air affect the pitch of a flute? A guitar?

12.–The human auditory system is very sensitive to small differences between the arrival times of a sound signal at the right and left ears. Explain how this permits us to perceive the direction from which a sound signal arrives.

13.–The depth finder (or "fish finder") on a boat sends a pulse of sound toward the bottom and measures the time an echo takes to return (Fig. 11.31). The dial of the depth finder displays this echo time on a scale directly calibrated in distance units. Experienced operators can tell whether the bottom is clean rock, or rock covered by a layer of mud, or whether a school of fish is swimming somewhere above the bottom. What echo times would you expect to see displayed on the dial of the depth finder in each of these instances?

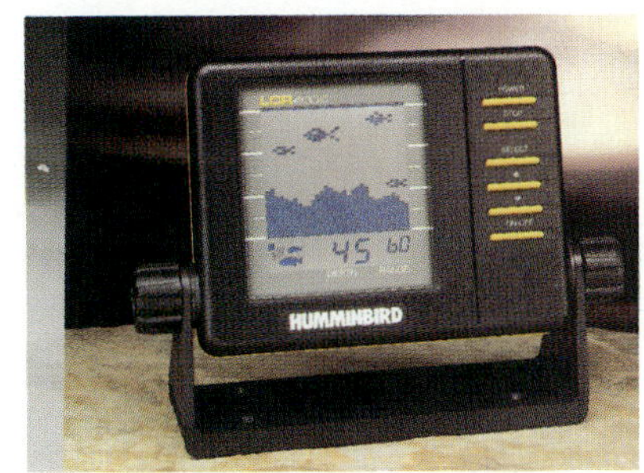

FIGURE 11.31
Fish finder.

14.–The helmsman of a fast motorboat heading toward a cliff sounds his horn. A woman stands on the top of the cliff and listens. Compare the frequency of the horn, the frequency heard by the woman, and the frequency heard by the helmsman in the echo from the cliff. Which of these three is the highest frequency? Which is the lowest?

15.–Two automobiles are speeding in opposite directions while sounding their horns. Describe the changes of pitch that each driver hears as they pass by one another.

16.–A man is standing north of a woman while a strong wind is blowing from the south. If the man and the woman yell at each other, how does the wind affect the pitch of the voice of each as heard by the other?

17.–A Concorde SST passing overhead at an altitude of 20 km produces a sonic boom with an intensity level of 120 dB lasting about half a second. How does this compare with some other loud noises? Would it be acceptable to let this aircraft make regular flights over populated areas?

18.–Many people have reported seeing UFOs traveling through air noiselessly at speeds much greater than the speed of sound. If the UFO consisted of a solid impenetrable body, would you expect its motion to produce a sonic boom? What can you conclude from the absence of sonic booms?

19.–Occasionally ocean waves passing by a harbor entrance will excite very high standing waves ("seiche") within the harbor. Under what conditions will this happen?

20.–The amplitude of an ocean wave initially decreases as the wave travels outward from its point of origin; but when the wave has traveled a quarter of the distance around the Earth, its amplitude *increases.* Explain how this comes about. (Hint: If the wave were to travel half the distance around the Earth, it would converge on a point, if no continents block its progress.)

21.–If you are standing on the south side of a house, you can speak to a friend standing on the east side, out of sight around the corner. How do your sound waves reach into the shadow zone?

PROBLEMS†

Section 11.1: Sound Waves in Air

1.—The range of frequencies audible to the human ear extends from 20 to 20,000 Hz. What is the corresponding range of wavelengths?

2.—The lowest musical note available on a piano is A, three octaves below that listed in Table 11.1; and the highest note available is C, four octaves above that listed in Table 11.1. What are the frequencies of these notes?

3.—When vibrating in its lowest mode, the C string of the cello produces the musical note C that is two octaves below middle C. What is the frequency of this note? What is the wavelength of the sound wave?

4.*—Estimate the wavelength and the frequency of the sound waves made visible in Fig. 11.1. Assume that the telephone handset is of standard size, like those found on pay telephones.

5.*—The lowest note that can be played on a guitar is E one octave below the middle E listed in Table 11.1. This note is produced by the thickest string when vibrating in its fundamental mode. The length of the string is 0.62 m and its mass per unit length is 5.4×10^{-3} kg/m. What must be the tension in this string?

6.*—A violin has four strings, each of them 0.326 m long. When vibrating in their fundamental modes, the four strings have frequencies of 196, 294, 440, and 659 Hz, respectively.

(a) What is the wavelength of the standing wave on each string? What is the wavelength of the sound wave generated by the string?

(b) What are the frequency and the wavelength of the first overtone on each string? What is the corresponding wavelength of the sound wave generated by each string?

(c) According to Table 11.1, to what musical tones do the frequencies calculated above correspond?

7.**—A mandolin has strings 34.0 cm long fixed at their ends. When the mandolin player plucks one of these strings, exciting its fundamental mode,

† In all the problems assume that the speed of sound in air is 331 m/s, unless otherwise stated.

this string produces the musical note D (293.7 Hz; see Table 11.1). In order to produce other notes of the musical scale, the player shortens the string by holding a portion of the string against one or another of several frets (small transverse metal bars) placed underneath the string. The player shortens the string by one fret to produce the note D#, by two frets to produce the note E, by three frets to produce the note F, etc. Calculate the correct spacing between the successive frets of the mandolin for one complete octave. Assume that the string always vibrates in its fundamental mode and assume that the tension in the string is always the same.

FIGURE 11.32
Pneumatic hammer (jackhammer).

Section 11.2: Intensity of Sound

8.—Suppose we turn the volume control of a loudspeaker up, and we increase the intensity of the sound reaching our ears by a factor of 2. What is the corresponding increase in intensity level, in dB?

9.—At a distance of 15 m from a pneumatic hammer (Fig. 11.32), the sound intensity is about 1×10^{-3} W/m^2. What is the intensity level in decibel?

10.—The noise level in a quiet automobile is 50 dB. Find the sound intensity in W/m^2.

11.—What is the energy incident per second on your eardrum if exposed to a sound wave of 160 dB?

12.*—The intensity level of sound near a loud rock band is 120 dB. What is the intensity level of sound near two such rock bands playing together? (Hint: Be careful; the answer is not 240 dB.)

13.*—In a screaming contest, a Japanese woman achieved 115 dB. How many such women would have to scream at you to bring you to the threshold of pain, 120 dB?

14.*—Suppose that a whisper has an intensity level of 20 dB at a distance of 0.5 m from the speaker's mouth. At what distance will this whisper be below your threshold of hearing?

15.*—A loudspeaker receives 8 W of electric power from an audio amplifier and converts 3% of this power into sound waves. Assuming that the loudspeaker radiates the sound uniformly over a hemisphere (a vertical and horizontal angular spread of 180°), what will be the intensity and the intensity level at a distance of 10 m in front of the loudspeaker?

16.*—An old-fashioned hearing trumpet has the shape of a flared funnel, with a diameter of 8 cm at its wide end and a diameter of 0.7 cm at its narrow end. Suppose that all of the sound energy that reaches the wide end is funneled into the narrow end. By what factor does this hearing trumpet increase the intensity of sound, measured in W/m^2? By how many dB does it increase the intensity level?

Section 11.3: The Speed of Sound

17.—Spectators at soccer matches often notice that they hear the sound of the impact of the ball on the player's foot (or head) sometime after seeing this impact. If a spectator notices that the delay time is about 0.5 s, how far is he from the player?

18.—In the past century a signal gun was fired at noon at most harbors so that the navigators of the ships at anchor could set their chronometers. This method is somewhat inaccurate, because the sound signal takes some time to travel the distance from gun to ship. If this distance is 3.0 km, how long does the signal take to reach the ship? Can you suggest a better method for signaling noon?

19.—Sound waves used for medical ultrasound scans of the soft tissues in the human body have frequencies in the range 0.8 to 15 MHz. What wave-

FIGURE 11.33
Lightning.

lengths correspond to these frequencies? The speed of sound in the soft tissues of the human body is 1500 m/s.

20.—Designers of hi-fi systems usually call sound waves of frequency below 800 Hz low frequency; from 800 Hz to 3500 Hz, middle frequency; and above 3500 Hz, high frequency. What are the wavelengths that correspond to these frequencies?

21.—The peak speed v_{max} acquired by particles of air when exposed to a sound wave is related to the intensity I by

$$I = \frac{v}{2}\rho_0 v_{\text{max}}^2$$

where ρ_0 is the density of air, 1.29 kg/m³. Calculate the maximum speed acquired by the particles in a sound wave of intensity 1.0×10^4 W/m².

22.—While standing at some distance from a large stone cliff, you notice that if you clap your hands, an echo of the clap returns to you about 1.5 s later. What is the distance to the cliff?

23.—You can estimate your distance from a bolt of lightning (Fig. 11.33) by counting the seconds between seeing the flash and hearing the thunder, and then dividing by 3 to obtain the distance in kilometers (or by 5 to obtain the distance in miles). Verify this rule.

24.—The pitch of the vowels produced by the human voice is determined by the frequency of standing waves in several resonant cavities (larynx, pharynx, mouth, and nose). In an amusing demonstration experiment, a volunteer inhales helium gas and then speaks a few words. As long as his resonant cavities are filled with helium gas, the pitch of his voice will be much higher than normal. Given that the speed of sound in helium is about three times as large as in air, calculate the factor by which the eigenfrequencies of his resonant cavities will be higher than normal.

25.*—In fresh water, sound travels at a speed of 1460 m/s. In air, sound travels at a speed of 331 m/s. Suppose that an explosive charge explodes on the surface of a lake. A woman with her head in the water hears the bang of the explosion and, lifting her head out of the water, she hears the bang again 5 s later. How far is she from the site of the explosion?

FIGURE 11.34
Camera with ultrasonic rangefinder.

26.*—The ultrasonic rangefinder on a new automatic camera sends a pulse of sound to the target and determines the distance by the time an echo takes to return (Fig. 11.34).

(a) If the rangefinder is to determine a distance of 50 cm with an error no larger than ± 2 cm, how accurately (in seconds) must it measure the travel time?

(b) If you aim this camera at an object placed beyond a sheet of glass (a window or a glass door), on what will the camera focus?

27.*—A bat can sense its distance from the wall of a cave (or whatever) by emitting a sharp ultrasonic pulse that is reflected by the wall. The bat can tell the distance from the time the echo takes to return.

(a) If a bat is to determine the distance of a wall 10 m away with an error of less than ± 0.5 m, how accurately must it sense the time interval between emission and return of the pulse?

(b) Suppose that a bat flies into a cave filled with methane (swamp gas). By what factor will this gas distort the bat's perception of distances? The speed of sound in methane is 432 m/s.

28.*—The sound waves used for medical ultrasound scans of the human body typically have a frequency of 10 MHz and an amplitude of 8×10^{-8} m (this is the amplitude of the simple harmonic motion of the particles in the tissues through which the wave passes). When subjected to such a sound wave, what is the maximum speed of the motion of a particle? What is the maximum acceleration of the particle? Compare this acceleration with g, the acceleration of free fall.

29.*—As described in Problem 26, the ultrasonic rangefinder on a camera sends a pulse of sound to the target and determines the distance by the time an echo takes to return. Suppose that, after waterproofing this camera somehow, you try to use it underwater, in a swimming pool. If you aim the camera at a target 5 m away, what distance will the rangefinder indicate? The speed of sound in water is 1500 m/s.

30.*—Consider a tube of length L open at both ends. Are the ends displacement nodes or antinodes? Show that the eigenfrequencies of standing sound waves in this tube are

$$\nu_n = n\frac{v}{2L} \qquad n = 1, 2, 3, \ldots$$

Draw diagrams similar to those of Fig. 11.13 showing the displacement amplitude for each of the first four standing waves.

31.*—The human ear canal is approximately 2.7 cm long. The canal can be regarded as a tube open at one end and closed at the other. What are the eigenfrequencies of standing waves in this tube? The ear is most sensitive at a frequency of about 3000 Hz. Would you expect that resonance plays a role in this?

32.*—Consider a tube of length L closed at both ends. Show that the eigenfrequencies of standing sound waves in this tube are

$$\nu_n = n\frac{v}{2L} \qquad n = 1, 2, 3, \ldots$$

Draw diagrams similar to those of Fig. 11.13 showing the displacement amplitude for each of the first four standing waves.

33.*—The largest pipes in a great organ usually have a length of about 4.88 m. These pipes are open at both ends so that a standing sound wave will have a displacement antinode at each end. What is the frequency of the fundamental mode of such a pipe?

34.*—The mass per unit length of a steel wire of diameter 1.3 mm is 0.010 kg/m and the yield strength, of maximum tension that the wire can withstand without breaking, is 3.6×10^3 N. Is it possible to apply enough tension to this wire so the speed of transverse waves on the wire exceeds the speed of sound in the steel, 5000 m/s?

35.*—The Bay of Fundy (Nova Scotia) is about 250 km long. The speed of water waves of long wavelength in the bay is about 30 m/s.

(a) What are the frequency and the period of the fundamental mode of oscillation of the bay? Treat the bay as a long, narrow tube open at one end and closed at the other.

(b) The period of the tidal pull exerted by the Moon is about 12 h. Would you expect that the very large tidal oscillations (with a height of up to 15 m) observed in the Bay of Fundy are due to resonance?

36.**—A flute can be regarded as a tube open at both ends. It will emit a musical note if the flutist excites a standing wave in the air column in the tube.

(a) The lowest musical note that can be played on a flute is C (261.7 Hz; see Table 11.1). What must be the length of the tube? Assume that the air column is vibrating in its fundamental mode (see Problem 30).

(b) In order to produce higher musical notes, the flutist opens valves arranged along the side of the tube. Since the holes in these valves are large, an open valve has the same effect as shortening the tube. The flutist opens one valve to play C#, two valves to play D, etc. Calculate the successive spacings between the valves of a flute for one complete octave. (The actual spacings used on flutes differ slightly from the results of this simple theoretical evaluation because the mouth cavity of the flutist also resonates and affects the frequency.)

37.**—The tube of a flute has a sliding joint that can be used to change the length, to tune the flute. Suppose that a flute has been tuned to perfect pitch while outdoors, where the temperature is 0°C and the speed of sound is 331 m/s. For this perfectly tuned flute, the frequency of middle C, which corresponds to the fundamental mode of the tube of the flute, is 261.7 Hz. Suppose that this flute is then taken indoors, where the temperature is 20°C and the speed of sound is 344 m/s. What will the frequency of the flute's middle C be now? To restore the flute to perfect pitch, how much must we increase the length of the tube of the flute? Express your answer as a percentage of the length.

38.**—In order to measure the depth of a ravine, a physicist standing on a bridge drops a stone and counts the seconds between the instant she releases the stone and the instant she hears it strike some rocks at the bottom. If this time interval is 6.0 s, how deep is the ravine? Take into account the travel time of the sound signal, but ignore air friction.

39.**—Because the human auditory system is very sensitive to small differences between the arrival times of a sound signal at the right and left ears, we can perceive the direction from which a sound signal arrives to within about 5°. Suppose that a source of sound (a ringing bell) is 10 m in front of and 5° to the left of a listener. What is the difference in the arrival times of sound signals at the left and right ears? The separation between the ears is about 15 cm.

Section 11.4: The Doppler Effect

40.——The horn of a stationary automobile emits a sound wave of 580 Hz. What frequency will you hear if you are driving toward this automobile at 80 km/h?

41.——In an experiment performed shortly after the discovery of the Doppler effect by the Austrian physicist Christian Doppler in 1842, several trumpeters were placed on a train and told to play a steady musical tone. As the train sped by a listener standing on the side of the track, the pitch perceived by the listener suddenly dropped by a noticeable amount. Suppose that the train had a speed of 60 km/h and that the trumpets on the train sounded the note of E (329.7 Hz; see Table 11.1). What was the frequency of the note perceived by the listener on the ground when the train was approaching? When the train was receding? Approximately to what musical notes do these Doppler-shifted frequencies correspond?

42.——Ocean waves with a wavelength of 100 m have a period of 8.0 s. A motorboat, with a speed of 9.0 m/s, heads directly into such waves. What is the speed of the waves relative to the motorboat? With what frequency do wave crests hit the front of the motorboat?

43.——The horn of an automobile emits a tone of frequency 520 Hz. What frequency will a pedestrian hear when the automobile is approaching at a speed of 85 km/h? Receding at the same speed?

44.*—Two automobiles are driving on the same road in opposite directions. The speed of the first automobile is 90.0 km/h and that of the second is 60.0 km/h. The horns of both automobiles emit tones of frequency 524 Hz. Calculate the frequency that the driver of each automobile hears coming from the other automobile. Assume that there is no wind blowing along the road.

45.*—A train approaches a mountain at a speed of 75 km/h. The train's engineer sounds a whistle that emits a frequency of 420 Hz. What will be the frequency of the echo that the engineer hears reflected off the mountain?

46.*—Suppose that a moving train carries a source of sound and also a receiver of sound so that both have the same velocity relative to the air. Show that in this case the Doppler shift due to motion of the source cancels the Doppler shift due to motion of the receiver — the frequency detected by the receiver is the same as the frequency generated by the source.

47.**—The helmsman of a fast motorboat heading toward a cliff sounds his horn. The speed of the motorboat is 15 m/s, and the frequency of the sound emitted by the horn, when at rest, is 660 Hz. A woman stands on the top of the cliff and listens.

(a) What is the frequency heard by the helmsman?
(b) What is the frequency heard by the woman?
(c) What is the frequency heard by the helmsman in the echo from the cliff?

48.**—The whistle on a train generates a tone of 440 Hz as the train approaches a station at 30 m/s. A wind blows at 20 m/s in the same direction as the motion of the train. What is the frequency that an observer standing at the station will hear?

49.—A rifle bullet has a speed of 674 m/s. What is the half angle of the Mach cone generated by this bullet?

50.—Figure 11.23 shows the shock wave of a bullet speeding through air. Measure the angle of the Mach cone and calculate the speed of the bullet.

51.—According to tradition, Superman flies faster than a speeding bullet. The speed of a typical bullet is 700 m/s, so let us guess that the speed of Superman is 800 m/s. At this speed, what will be the half angle of the Mach cone that Superman produces?

52.*—The Concorde SST has a cruising speed of 2160 km/h.

(a) What is the half angle of the Mach cone generated by this aircraft?
(b) If the aircraft passes directly over your head at an altitude of 12,000 m, how long after this instant will the shock wave strike you?

PHYSICS IN ACTION

ULTRASOUND IMAGES

Modern medicine relies heavily on a variety of imaging techniques that generate pictures of the interior of the human body for diagnostic purposes. The oldest of these imaging techniques, and still the most widely used, is radiography, which generates pictures by irradiating the body with X rays and recording the shadows of the internal anatomical structures on photographic film (see *X-Ray Images*, pp. 809A–809D). X rays give us sharp and clear shadows of the bones, but they are not very well suited for imaging the soft tissues of the body. X rays cannot discriminate between tissues of approximately equal densities. For instance, when they pass through the heart, X rays do not discriminate between heart muscle and the blood filling the heart cavities — an X-ray picture of the heart is merely a blob, which does not reveal the details of the heart's anatomy.

Several of the newer imaging techniques generate better pictures of soft tissues. Among these newer techniques is ultrasound imaging, which is second only to X rays in popularity among medical practitioners. Ultrasound imaging produces pictures of the anatomical structures by "illuminating" the body with sound waves of high frequency, far above the threshold of hearing. The sound waves used by medical ultrasound equipment have frequencies between 10^6 Hz and 10^7 Hz and wavelengths (in soft tissue) between 1.5 mm and 0.15 mm. In contrast to X rays, which produce a picture by penetrating through the body and shadowing the organs, the ultrasonic waves produce a picture by reflecting at the walls of organs.

Ultrasound imaging exploits an echo-detection scheme similar to that of sonar and radar. The sonar equipment in a ship detects an underwater target by sending a short pulse of sound waves into the water and listening for the echo returned by the target. The time delay for the return of the echo indicates the distance to the target. Likewise, medical ultrasound equipment detects the anatomical structures within the body by sending a short pulse of sound waves into the body and listening for the echo. The speed of sound in soft body tissues is about the same as that in water, $v = 1500$ m/s (this is not surprising, since soft tissues are mostly water). If the time delay for the return of the echo is Δt, then the roundtrip distance traveled by the pulse is $v\Delta t$, and the actual distance to the point of reflection at the wall of the heart or some other organ is $v\Delta t/2$. For instance, if $\Delta t = 10^{-4}$ s, then the distance is $v\Delta t/2 = 1500 \text{ m/s} \times 10^{-4} \text{ s}/2 = 0.07$ m. Since the travel times for ultrasound echoes are quite short, the imaging equipment must be able to measure time with high precision.

The pulses of ultrasound are generated by a transducer, which converts a high-frequency electric signal into a sound signal. In principle, such a transducer is similar to the loudspeaker in your radio, which also converts an electric signal into a sound wave. For this purpose, the loudspeaker uses a small electromagnet to push a diaphragm back and forth, thereby communicating periodic pushes to the air in front of the diaphragm. However, the diaphragm of a loudspeaker cannot move fast enough to produce ultrasound. Instead, the ultrasound transducer uses a piezoelectric (pronounced py-ezo) material, such as PZT (lead zirconate titanate), which contracts when an electric signal is applied to it (Fig. 1). A layer of such material can easily vibrate at frequencies of 10^7 Hz. The ultrasound transducer is mounted at the tip of a probe which is placed in contact with the surface of the body (see Fig. 2). For good contact between probe and skin, the end of the probe is smeared with a

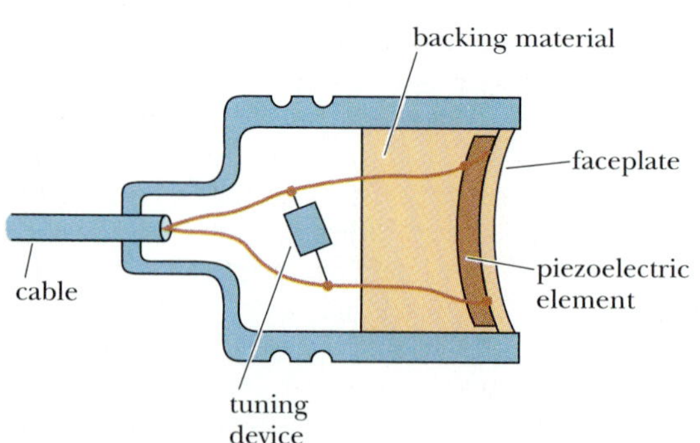

FIGURE 1 Schematic diagram of a piezoelectric transducer. The layer of piezoelectric material in the transducer has a curved shape to concentrate and focus the ultrasound in a narrow beam.

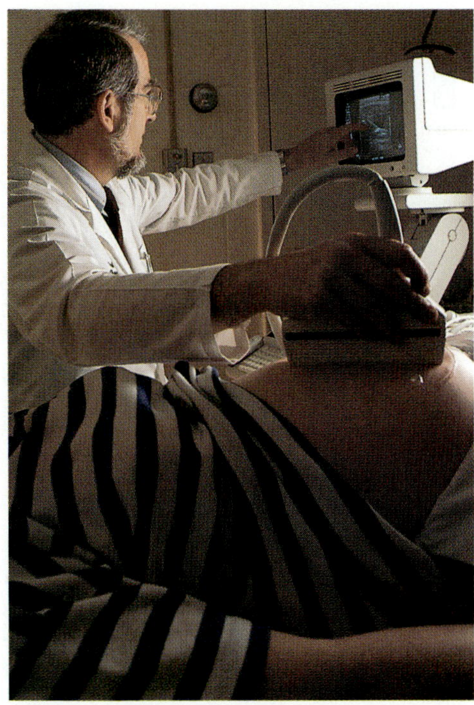

FIGURE 2 Technician using an ultrasound probe on a patient.

jelly; this reduces reflection of the ultrasound by the skin.

The transducer serves not only as the emitter of sound waves, but also as the detector of the echo. The piezoelectric material contracts when an electric signal is applied to it; and, conversely, it generates an electric signal when it is compressed by an arriving sound wave. Thus, the material can serve both as emitter and as receptor of sound waves. In this regard this material is similar to the antenna of your cordless telephone, which acts both as emitter of radio waves (when you speak) and as receptor (when you listen). In a typical ultrasound imaging system, the transducer automatically switches back and forth between its emission and reception modes. It first acts as emitter, sending out a short pulse of sound; then it acts as receptor, listening to echoes; then it again acts as emitter, etc. Typically, the transducer switches back and forth between its two modes about a thousand times per second, sending out about a thousand pulses per second.

The individual pulses must be very short and sharp, since otherwise the length of the pulse would introduce a significant uncertainty in the timing of the echo. Typically, each individual pulse lasts only 10^{-6} s, and it comprises only a few complete wave periods. Note that if we were to attempt to construct a pulse lasting 10^{-6} s with ordinary sound waves, we could not even fit a single wave period into the pulse, that is, we could not construct such a pulse (the shortest wave period for ordinary sound is about 5×10^{-5} s). This is one of the reasons why ultrasound is needed for imaging. Another reason is diffraction. The aperture of the probe acts like the aperture of a loudspeaker — the waves spread out laterally and broaden the narrow focused beam of ultrasound. In the case of a loudspeaker, this lateral spreading of the beam of sound is desirable (see Section11.5); but in the case of an ultrasound probe, the spreading makes the image unsharp, since a broad beam cannot sense fine detail. In order to minimize the loss of detail resulting from diffraction, we need to use waves of extremely short wavelength, which are less susceptible to diffraction.

To build up a picture of the interior of the body, the beam of ultrasound must be swept through the body, either by moving the probe back and forth over the body or by swinging it from side to side, through some angle. For each position or direction of the probe, the time delays of the echoes must be recorded. Such a sweeping search for echoes is called a ***scan.*** The time-delay information for a complete scan is accumulated in the memory of a computer, which later processes this information and displays a picture on a video monitor.

Figure 3 shows an ultrasound image of the face

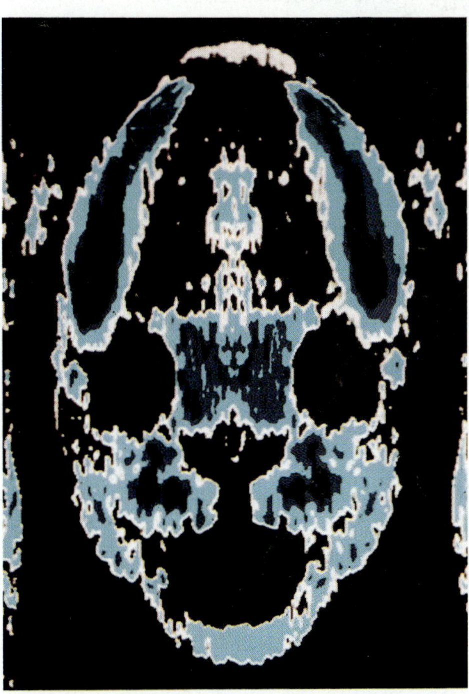

FIGURE 3 Ultrasound image of fetus. The mouth of the fetus is wide open.

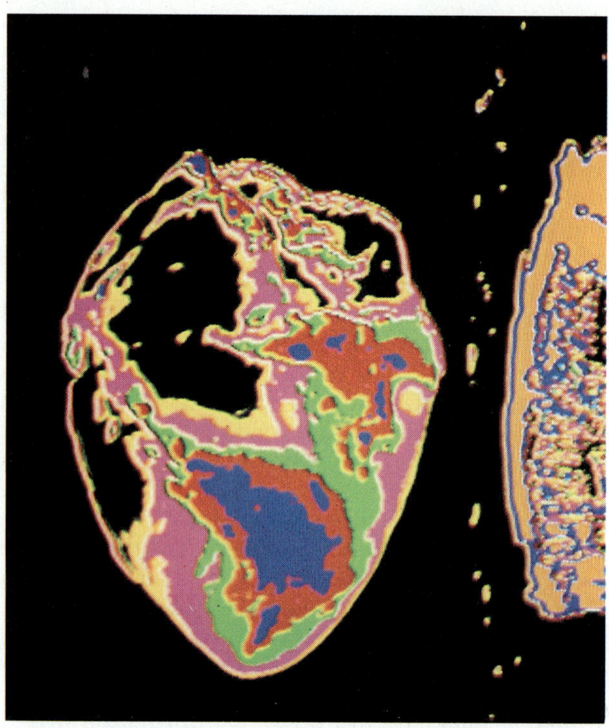

FIGURE 4 Ultrasound image of the heart.

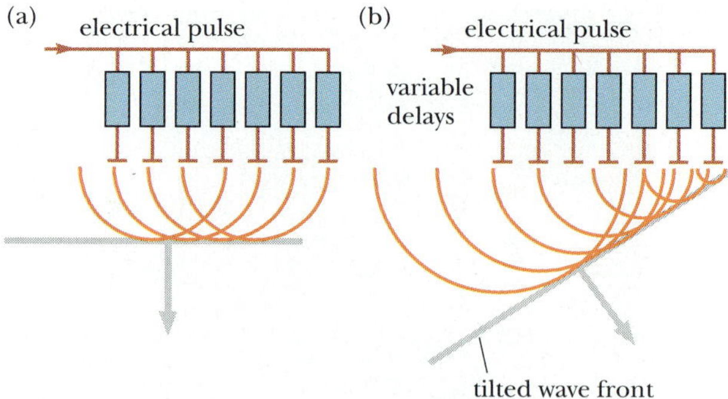

FIGURE 5 (a) Several transducers emitting sound waves in phase. (b) Several transducers with a phase delay from one side to the other. The wave front is tilted to one side.

of a fetus. Ultrasound is now routinely used to examine pregnant women, to detect fetal abnormalities at an early stage. X rays are unsuitable for this purpose, because they tend to produce genetic damage, to which the fetus is especially susceptible.

Figure 4 shows an ultrasound image of the heart (echocardiogram); this image clearly displays the four chambers of the heart and the valves between them. In the ultrasound probe used for producing such a picture of the heart, several transducers are placed side by side, so the individual waves form a single wave front by superposition (see Fig. 5a). The advantage of this arrangement, called a phased array, is that the wave front can be steered to the left or the right by introducing a delay in the electric signal applied to the transducers on the left or the right; such a delay tilts the wave front, and alters its direction of advance (Fig. 5b). A probe with a phased array can automatically sweep the direction of the ultrasound beam across the heart, and we do not need to move the probe itself. To perform a scan, the operator merely holds the probe steady against the body, and the electronic circuitry quickly sweeps the beam of ultrasound from one side to the other. Fast scanning is especially valuable in echocardiography, because it is desirable to sweep the beam across the entire heart in a time so short that the heart hardly moves at all. Such a fast scan produces a snapshot of the heart, and by repeating these snapshots in quick succession, we can make a movie that shows the motion of the heart in real time.

The images shown in Figs. 3 and 4 are called ***B-scans*** (*B*rightness scans), because they represent the strength of the echo by brightness on the screen. Several other scanning methods are also in use. In the ***A-scan,*** the probe and the beam of sound are held fixed, and the time delays of echoes returning from the interior of the body are used for precise measurements of the distances to interior structures. This kind of scan gives no picture, but is useful for precise measurements of the diameter of the eye, in preparation for the surgical implantation of a new lens. In the ***M-scan*** (*M*otion scan) the probe and the beam of sound are also held fixed, but they are aimed at a moving target, such as the leaflets of one of the valves of the heart. This provides a measurement of the positions of the valve leaflets as a function of time (Fig. 6), which permits diagnosis of abnormalities in the operation of the valves, such as mitral valve prolapse (failure of the valve to close tightly, or "floppy valve syndrome") or mitral valve stenosis (narrowing of the valve).

Yet another scan is the ***Doppler scan,*** which measures not the position of a moving target, but its speed. This scan exploits the Doppler effect. When a sound wave is reflected by a moving target of speed V_T, the reflected wave acquires a Doppler shift $(f' - f) = 2fV_T/v$ [see Eq. (11.15)]. Here, an

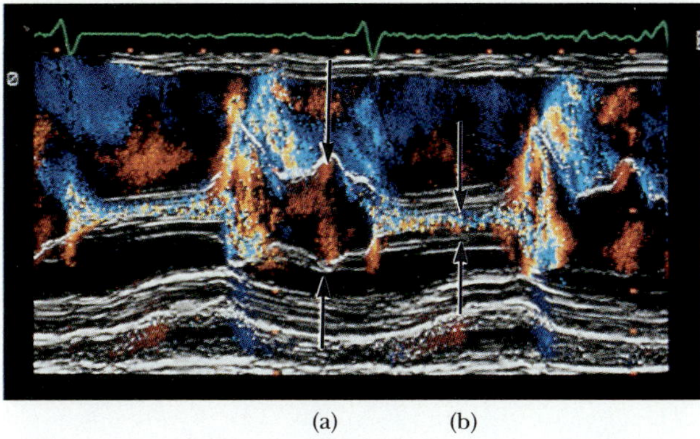

(a) (b)

F I G U R E 6 M-scan of the mitral valve of the heart. The two middle traces plot the positions of the two valve leaflets as a function of time. At the instant (a), the valve is open (leaflets separated); at the instant (b), it is closed (leaflets joined).

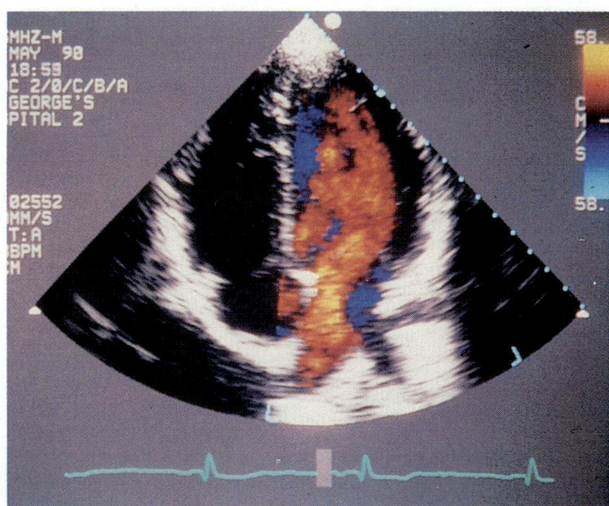

F I G U R E 7 False-color image indicating blood flow in the chambers of the heart, as revealed by an ultrasound Doppler scan. In this image, the transducer is at the top, and the heart is shown upside down; the ventricles are at the top, and the atria at the bottom. The red color indicates a flow velocity toward the transducer. The blood is streaming from the left atrium (bottom) into the left ventricle (top).

extra factor of 2 has been inserted into the usual Doppler formula, because the sound wave acquires a Doppler shift $(f' - f) = fV_T/v$ when it strikes the moving target (the target acts as moving "receiver"), and another, approximately equal, Doppler shift when the sound wave is reflected (the target acts as moving "emitter"). In Doppler scans of the heart, the moving target is the blood squirting through the valves, which has speeds of the order of several meters per second. Figure 7 shows the speed of flow of the blood in the heart, as measured by a Doppler scan. This scan clearly reveals the high speed of flow of the blood leaking through the mitral valve.

R E F E R E N C E S

Gramiak, R., and Waag, R. C., eds., *Cardiac Ultrasound* (Mosby, St. Louis, 1975).

Putman, C. E., and Ravin, C. E., *Textbook of Diagnostic Imaging* (W. B. Saunders, Philadelphia, 1988).

Webb, S., ed., *The Physics of Medical Imaging* (Hilger, Bristol and Philadelphia, 1988).

Chapter 12

The ALVIN deep submergence vessel can dive to a depth of 4000 m where the water pressure is 400 atmospheres.

MOTION OF FLUIDS

Fluid ▸

A fluid is a system of particles loosely held together by their own cohesive forces or by the restraining forces exerted by the walls of a container. Both liquids and gases are fluids — liquids are held together by their cohesive forces, and gases are held together by the restraining forces of a container (or, in the case of atmospheric air, by the weight of the atmosphere). In contrast to the particles in a rigid body, which are permanently locked into fixed positions, the particles in a fluid body are more or less free to wander about within the volume of the fluid body (see Fig. 12.1). A fluid will flow, that is, it will change its shape in response to external forces — a body of water or a body of air will change its shape in response to the forces exerted by gravity and by the container. The difference between these two kinds of fluids is that liquids are incompressible, or nearly so, whereas gases are compressible. This means that a body of water has a constant volume independent of the container, whereas a

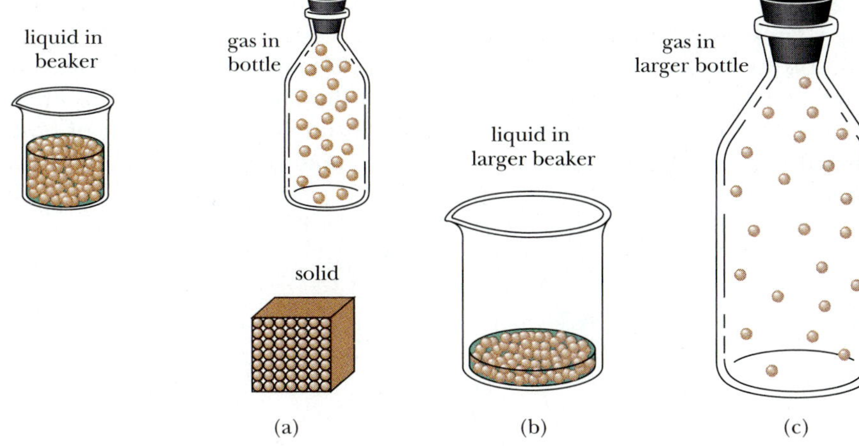

FIGURE 12.1

(a) Molecules in a solid are tightly packed together and locked into a rigid array. (b) Molecules of a liquid are loosely packed, and they have some freedom to wander about, although they suffer frequent collisions. (c) Molecules in a gas are widely dispersed. The molecules have complete freedom to move, and the gas is spread out over all the available volume.

body of air has a variable volume — the air always spreads out so as to fill the container entirely, and it can be made to expand or contract by increasing or decreasing the size of the container.

Throughout most of this chapter we will neglect the viscosity of fluids. Viscosity is an internal friction or stickiness within the fluid that offers resistance to its flow. For instance, honey is a fluid of high viscosity, whereas water is a fluid of fairly low viscosity. In general, the magnitude of the viscous frictional force within a fluid increases with the speed of flow. Thus, the viscosity of water can be neglected at low speeds of flow, but it cannot be neglected at high speeds. A fluid that is flowing sufficiently slowly so its viscosity can be neglected is said to be a ***perfect fluid. Such a fluid does not offer any resistance to flow except through its inertia.***

◄ **Perfect fluid**

12.1 DENSITY AND FLOW VELOCITY

Although a fluid is a system of particles, the number of particles in, say, a cubic centimeter of water is so large that it is not feasible to describe the state of the fluid in terms of the masses, positions, and velocities of all the individual particles in the system. Instead, ***we will describe the fluid in terms of its velocity of flow, density, and pressure.*** These quantities give us a macroscopic description of the fluid — they tell us the average behavior of the particles in regions within the fluid. For example, if the velocity of flow of water in a firehose is 4 m/s, this does not mean that all the water molecules have this velocity. The water molecules have a high-speed thermal motion of about 600 m/s; they move in short zigzags because they frequently collide with one another. This thermal motion of a water molecule is random; the motion is as likely to be in a direction opposite to the flow as along the flow (see Fig. 12.2). The flow of water molecules along the firehose at 4 m/s represents a slow drift superimposed on the much faster random zigzag motion. However, on a macroscopic scale, we notice only the drift and not the random small-scale motion — we notice only the *average* motion of the water molecules.

FIGURE 12.2

Motion of a molecule in water. The straight segments of the motion are typically 10^{-10} m long.

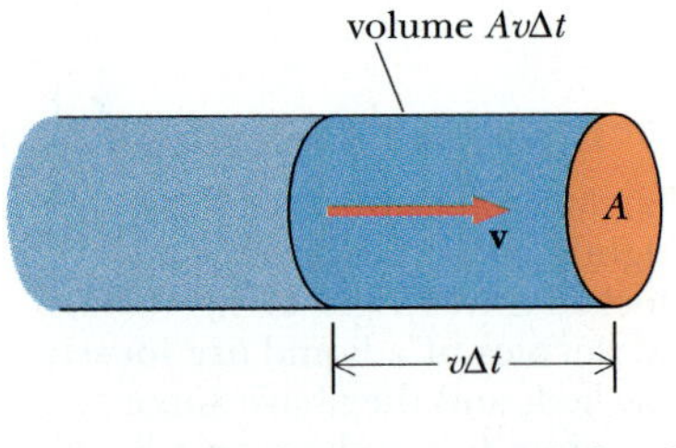

FIGURE 12.3

Flow of a fluid across an area A. In a time Δt, the fluid within a distance $v\Delta t$ reaches the area A, but fluid farther to the left does not.

From the magnitude v of the flow velocity of the fluid we can calculate the volume of fluid that flows across an area A perpendicular to the direction of the velocity per unit time. Figure 12.3 helps to make this clear; it shows a (stationary) area A and a volume of fluid about to cross this area. The fluid that crosses the area in a time Δt is initially in a cylinder of base A and length $v\Delta t$. The amount of fluid volume that crosses the area is therefore

$$\Delta V = Av\Delta t \tag{1}$$

and the amount that crosses per unit time is

$$\frac{\Delta V}{\Delta t} = Av \tag{2}$$

Density ▸

The density is the amount of mass per unit volume. The unit of density is the kilogram per cubic meter (kg/m^3). Table 12.1 lists the densities of a few liquids and gases. In the table, the density is designated by the customary symbol ρ, the Greek letter rho. The densities of gases depend on the temperature and the pressure (this dependence will be discussed in Chapter 13); unless otherwise noted, the values of the densities listed in Table 12.1 are for standard temperature and standard pressure (0°C and 1 atm). The densities of liquids depend only slightly on pressure, but they do depend appreciably on temperature. For instance, water has a maximum density at about 4°C.

TABLE 12.1 DENSITIES OF SOME FLUIDS[a]

Fluid	ρ
Water	
0°C	999.8 kg/m^3
4°C	1000.0
20°C	998.2
100°C	958.4
Sea water, 15°C	1025
Mercury	13,600
Sodium, liquid at 98°C	929
Texas crude oil, 15°C	875
Gasoline, 15°C	739
Olive oil, 15°C	920
Human blood, 37°C	1060
Air	
0°C	1.29
20°C	1.20
Water vapor, 100°C	0.598
Hydrogen	0.0899
Helium	0.178
Nitrogen	1.25
Oxygen	1.43
Carbon dioxide	1.98
Propane	2.01

[a] At 0°C and 1 atm, unless otherwise noted.

EXAMPLE 1. The water in a firehose of diameter 6.4 cm, or 0.064 m, has a flow velocity of 4.0 m/s. At what rate does this hose deliver water? Give the answer in both cubic meters per second and kilograms per second.

SOLUTION: The radius of the hose is 0.032 m. The cross-sectional area of the hose is $A = \pi \times (0.032\ \text{m})^2 = 0.0032\ \text{m}^2$. Hence, by Eq. (2), the rate of delivery in terms of volume is

$$\frac{\Delta V}{\Delta t} = Av \tag{3}$$

$$= 0.0032\ \text{m}^2 \times 4.0\ \text{m/s} = 0.013\ \text{m}^3/\text{s}$$

To find the rate of delivery in terms of the mass of water, we must multiply $\Delta V/\Delta t$ by the density of water, $\rho = 1000\ \text{kg/m}^3$:

$$\frac{\Delta m}{\Delta t} = \rho\frac{\Delta V}{\Delta t} \tag{4}$$

$$= 1000\ \text{kg/m}^3 \times 0.013\ \text{m}^3/\text{s} = 13\ \text{kg/s}$$

12.2 INCOMPRESSIBLE STEADY FLOW; STREAMLINES

In most of the examples in this chapter we will deal with ***steady flow, for which the velocity at any given point of space remains constant in time.*** Thus, in steady flow, each small parcel of fluid that starts at any given point follows exactly the same path as a small parcel that passes through the same point at an earlier (or later) time. For example, Fig. 12.4 shows an example of velocity vectors for the steady flow of water around a cylindrical obstacle, say, the flow of the water of a broad river around a cylindrical piling placed in the middle. The water enters the picture in a broad stream from the left, and disappears in a similar broad stream toward the right.

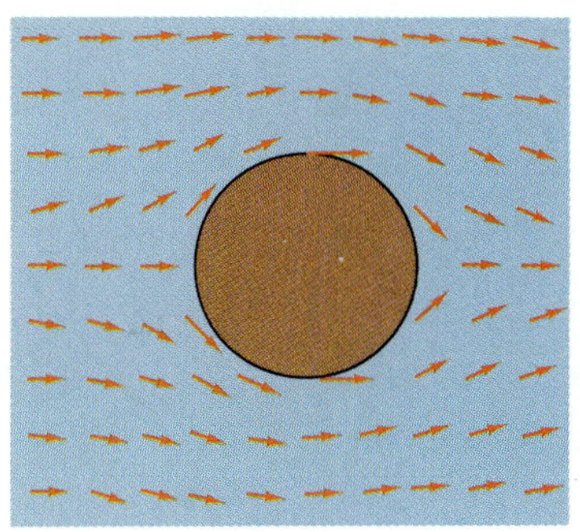

FIGURE 12.4
Velocity vectors for water flowing around a cylinder.

For the steady flow of an incompressible fluid, such as water, the picture of velocity vectors can be replaced by an alternative graphical representation. Suppose we focus our attention on a small parcel of water, say, 1 mm^3 of water, and we observe the path of this 1 mm^3 from the "source" of water to the "sink," that is, from the place where the motion of the water begins to where it ends. ***The path traced out by the small parcel of fluid is called a streamline.*** Neighboring small parcels will trace out neighboring streamlines. Figure 12.5 displays the pattern of streamlines for the same steady flow of water that already was represented in Fig. 12.4 by means of velocity vectors. The streamlines on the far left (and far

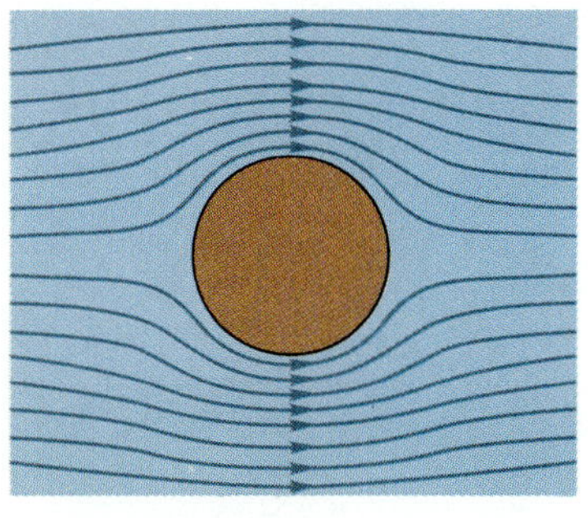

FIGURE 12.5
Streamlines for water flowing around a cylinder.

right) of Fig. 12.5 are evenly spaced to indicate the uniform and parallel flow in this region.

Streamline flow ▶

The steady flow of an incompressible fluid is often called ***streamline flow.*** Note that streamlines never cross. A crossing of two streamlines would imply that a small parcel of water moving along one of these streamlines has to penetrate through the small parcel of water moving along the other streamline. This is impossible — it would lead to disruption of both the small parcels and to destruction of the steadiness of flow. Because the streamlines for steady incompressible flow never cross, such flow is also called ***laminar.***

If we know the velocity of flow throughout the fluid, we can trace out the motions of small parcels of fluid and therefore construct the streamlines. But the converse is also true — if we know the streamlines, we can reconstruct the velocity of flow. We can do this by means of the following rule: ***The direction of the velocity at any one point is tangent to the streamline, and the magnitude of the velocity is proportional to the density of streamlines.***

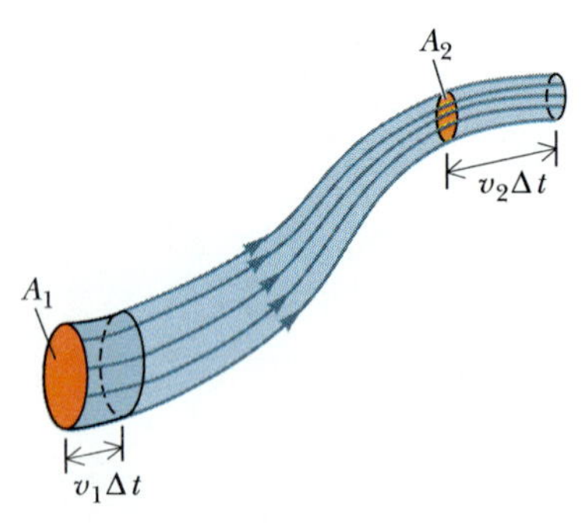

FIGURE 12.6
A stream tube.

The first part of this rule is self-evident, since the direction of motion of a small parcel of fluid is tangent to the streamline. To establish the second part, consider a bundle of streamlines forming a pipelike region, called a ***stream tube.*** Any fluid inside the stream tube will have to move along the tube; it cannot cross the surface of the tube because streamlines never cross. The tube therefore plays the same role as a pipe made of some impermeable material — it serves as a conduit for the fluid. If we assume that the tube is very narrow, so its cross-sectional area is very small, the velocity of flow will vary along the length of the tube, but it will be nearly the same at all points on a given cross-sectional area. For instance, on the area A_1 (see Fig. 12.6) the velocity is v_1, and on the area A_2 the velocity is v_2. In a time Δt, the fluid volume that enters across the area A_1 is $v_1 A_1 \Delta t$, and the fluid volume that leaves across the area A_2 is $v_2 A_2 \Delta t$. The amount of fluid that enters must match the amount that leaves, since, under steady conditions, fluid cannot accumulate in the segment of tube between A_1 and A_2. Hence,

$$v_1 A_1 \Delta t = v_2 A_2 \Delta t \tag{5}$$

or, canceling the factor Δt on both sides of the equation,

Continuity equation ▶

$$v_1 A_1 = v_2 A_2 \tag{6}$$

This relation is called the ***continuity equation.*** It shows that along any thin stream tube, ***the speed of flow is inversely proportional to the cross-sectional area of the stream tube.***

The density of streamlines inside the stream tube is the number of such lines divided by the cross-sectional area; since the number of streamlines entering A_1 is necessarily the same as that leaving A_2, the density of streamlines is inversely proportional to the cross-sectional area. This implies that the speed at any point in the fluid is directly proportional to the density of streamlines at that point. For example, in Fig. 12.5, the speed of the water is large at the top and bottom of the obstacle (large density of streamlines) and smaller to the left and right (smaller density of streamlines).

In experiments on fluid flow, the streamlines of a fluid can be made directly visible by several clever techniques. If the fluid is water, the streamlines can be made visible by placing grains of dye at diverse points within the volume of water; the dye will then be carried along by the flow, and it will mark the streamlines. The photograph of Fig. 12.7 shows a pattern of streamlines made visible by this technique. The water emerges from a pointlike source on the left and disappears into a pointlike sink on the right. The streamers of dye were created by small grains of potassium permanganate while dissolving in the water.

FIGURE 12.7

Streamers of dye indicate the streamlines in water flowing from a source (left) to a sink (right).

If the fluid is air, the streamlines can be made visible by releasing smoke from small jets at diverse points within the flow of air. The photograph of Fig. 12.8 shows fine trails of smoke marking the streamlines in air flowing past a scale model of the wing of an airplane in a wind tunnel. The experimental investigation of such streamline patterns plays an important role in airplane design. Incidentally: The flow of air can be regarded as nearly incompressible provided that the speed of flow is well below the speed of sound (331 m/s). Although the air will suffer some changes of density as it flows around obstacles, the changes are usually small enough to be neglected.

Finally, Fig 12.9 shows an example of turbulent flow. In the region behind the wing, the streamers of smoke become twisted and chaotic. This is due to the generation of eddies, or vortices, in this region. As the eddies form, grow, break away, and disappear in quick succession, the velocity of flow fluctuates violently. The flow of the fluid becomes unsteady and irregular. The formation of eddies and the onset of turbulence have to do

FIGURE 12.8

(Left). Fine trails of smoke indicate the streamlines in air flowing around the wing of an aircraft.

FIGURE 12.9

(Right). Here, the wing is in a partial stall, and the flow behind the wing has become turbulent.

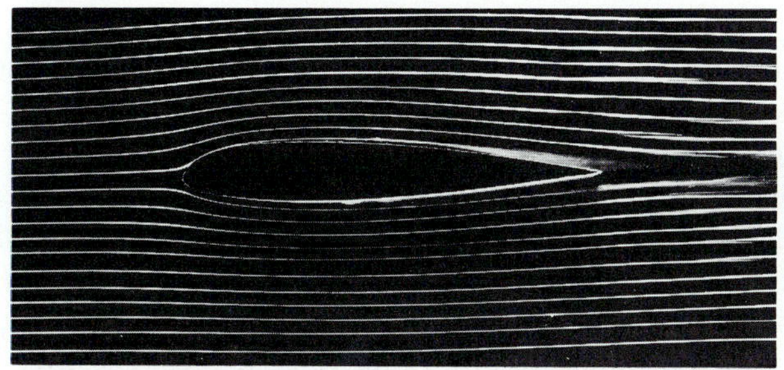

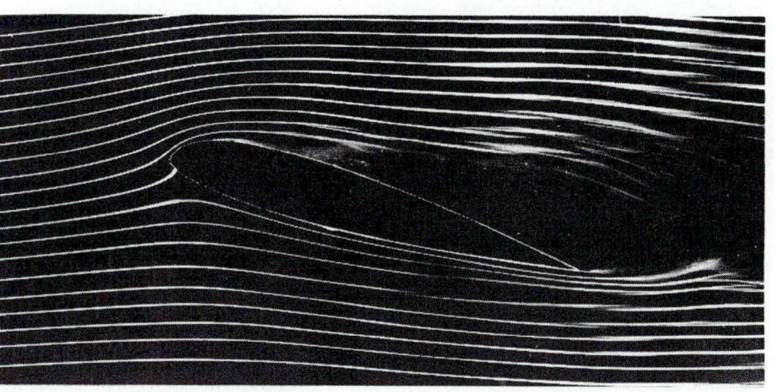

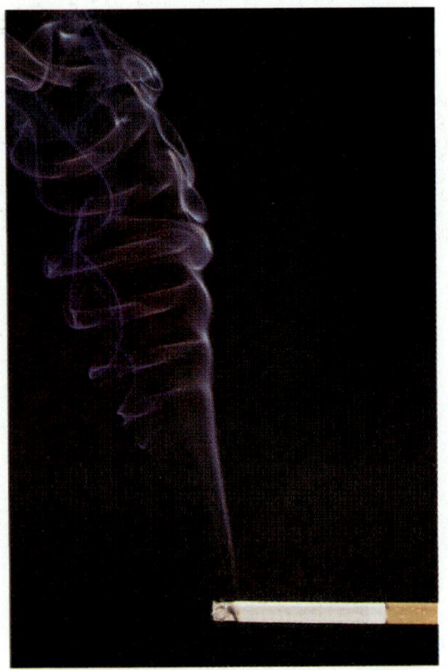

FIGURE 12.10

Ascending smoke from a cigarette. At first the flow of the smoke is steady and smooth. Then it becomes turbulent.

with viscosity in the fluid; it is a general rule that vortices and turbulence will develop in a fluid of given viscosity whenever the velocity of flow, the size of the obstacle, or both exceed a certain limit. We can see the transition from steady flow to turbulent flow in the ascending smoketrail from a cigarette (Fig. 12.10). The flow starts out steady, with smoke particles moving along well-defined streamlines; but at some height above the cigarette, where the flow velocity exceeds the critical limit, the flow becomes turbulent.

EXAMPLE 2. In the human circulatory system, the blood flows out of the heart via the aorta which is connected to other arteries which branch out into a multitude of small capillaries (see Fig. 12.11). In the average adult, the aorta has a radius of 1.2 cm, or 0.012 m, and the speed of flow of the blood is 0.20 m/s. The radius of each capillary is about 0.0003 cm, or 3×10^{-6} m, and the number of open capillaries, under conditions of rest, is 1×10^{10}. Calculate the speed of flow of the blood in the capillaries.

SOLUTION: The cross-sectional area of the aorta is

$$A_1 = \pi r_1^2 = \pi \times (0.012 \text{ m})^2 = 4.5 \times 10^{-4} \text{ m}^2$$

and the net cross-sectional area of all the capillaries is

$$A_2 = [\text{number of capillaries}] \times [\text{area of each}]$$

$$= 1 \times 10^{10} \times \pi r_2^2 = 1 \times 10^{10} \times \pi \times (3 \times 10^{-6} \text{ m})^2$$

$$= 2.8 \times 10^{-1} \text{ m}^2$$

From the continuity equation, with $v_1 = 0.20$ m/s, we then find that the speed of flow in the capillaries is

$$v_2 = \frac{A_1}{A_2} v_1 = \frac{4.5 \times 10^{-4} \text{ m}^2}{2.8 \times 10^{-1} \text{ m}^2} \times 0.20 \text{ m/s}$$

$$= 3 \times 10^{-4} \text{ m/s} = 0.3 \text{ mm/s}$$

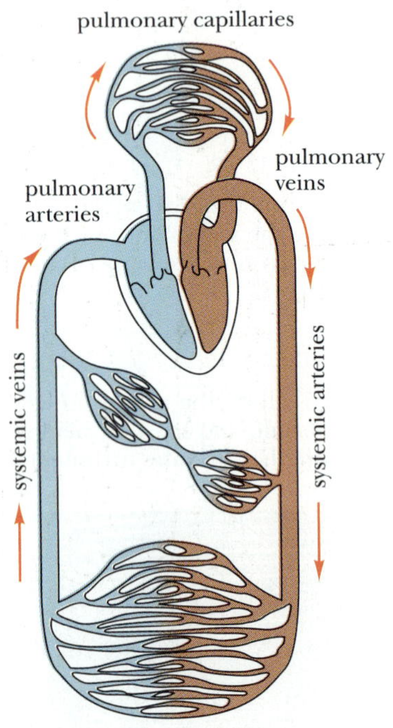

FIGURE 12.11

The human circulatory system.

EXAMPLE 3. The world's tallest fountain (at Fountain Hills, Arizona; see Fig. 12.12) shoots water to a height of 170 m at the rate of 26,000 liters/min. If we ignore friction, the motion of small parcels of water is projectile motion, and the deceleration of the ascending small parcels of water is 9.81 m/s^2. The speed therefore decreases with height in the same way as for a projectile. From this, it is easy to check that the speed of flow must be 58 m/s at the base, and 37 m/s at a height of 100 m. Given these speeds, calculate the cross-sectional area of the water column at the base and at a height of 100 m.

SOLUTION: To find the cross-sectional area, we use Eq. (2),

$$A = \frac{1}{v} \frac{\Delta V}{\Delta t}$$

The rate of delivery of the fountain is

$$\frac{\Delta V}{\Delta t} = \frac{2.6 \times 10^4 \text{ liters}}{1 \text{ min}} = \frac{26 \text{ m}^3}{60 \text{ s}} = 0.43 \text{ m}^3/\text{s}$$

Therefore, at the base,

$$A = \frac{1}{58 \text{ m/s}} \times 0.43 \text{ m}^3/\text{s} = 0.0074 \text{ m}^2 = 74 \text{ cm}^2$$

and at a height of 100 m,

$$A = \frac{1}{37 \text{ m/s}} \times 0.43 \text{ m}^3/\text{s} = 0.0116 \text{ m}^2 = 116 \text{ cm}^2$$

Note that the water column is narrow at the base and widens at the top. The photograph in Fig. 12.12 does not show this widening because the column of rising water is hidden in a curtain of falling water.

FIGURE 12.12
The world's tallest fountain.

12.3 PRESSURE

The ***pressure*** within a fluid is defined in terms of the force that a small volume of fluid exerts on an adjacent volume or on the adjacent wall of a container. Figure 12.13 shows two small adjacent volumes of fluid of cubical shape. The cube of fluid on the left presses against the cube on the right, and vice versa. Suppose that the magnitude of the perpendicular force between the two cubes is F and that the area of one face of one of the cubes is A; ***then the pressure p is defined as the magnitude F of the force divided by the area A:***

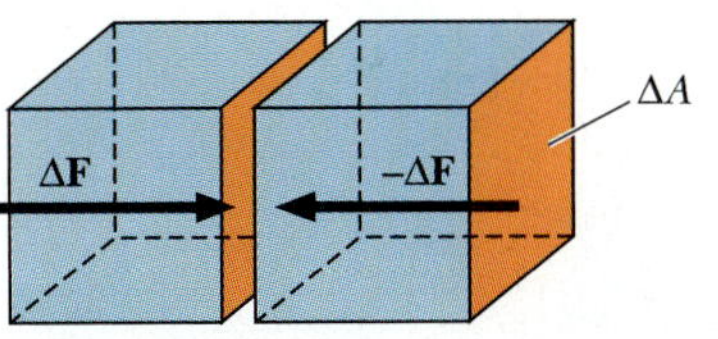

FIGURE 12.13
Adjacent small cubes of fluid exerting forces on each other.

$$p = \frac{F}{A} \qquad (7)$$

◄ **Pressure**

According to this definition, pressure is simply the magnitude of the force per unit area. Note that, in contrast to the force, ***the pressure is a quantity without direction.*** We cannot associate a direction with the pressure, because a small volume of fluid, or any small body immersed in the fluid, experiences pressures from all directions, and the pressures from all directions are equal. For instance, your hand is subjected to the pressure

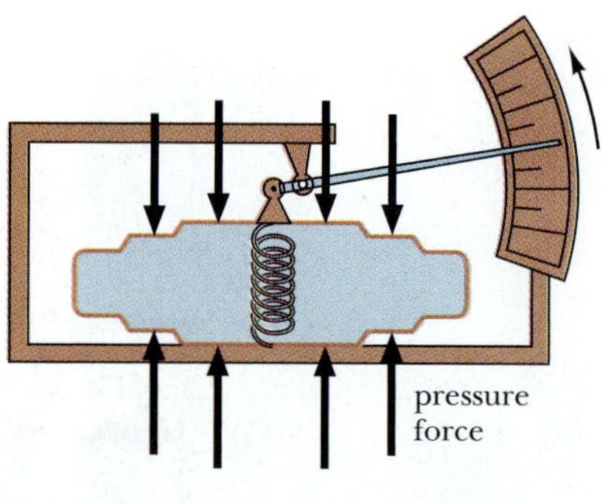

FIGURE 12.14

Device for the measurement of atmospheric pressure consisting of a hermetically sealed cylindrical can made of thin sheet metal. An internal spring holds the flexible bases apart, and prevents the atmospheric pressure from collapsing the can. The upper flexible base of the can is linked to a pointer that indicates the atmospheric pressure.

of the air in which you are immersed; but this pressure pushes from all sides, and it has no preferential direction.

A simple mechanical device for the measurement of pressure consists of a hermetically sealed, evacuated cylindrical can with corrugated flexible bases (see Fig. 12.14). If this can is immersed in a fluid at high pressure, the bases will be compressed inward; if it is immersed in a fluid at low pressure, the bases will bulge outward. Thus, the deformation of the bases of the can serves as an indicator of the pressure, and a pointer linked to one base can be calibrated to read the pressure. This device is widely used in aneroid barometers for the measurement of atmospheric pressure.

Pascal (Pa) ▸

In the metric system, the unit of pressure is the newton per square meter (N/m^2), ***which has been given the name pascal*** (Pa):

$$1 \text{ Pa} = 1 \text{ N/m}^2 \tag{8}$$

Atmosphere (atm) ▸

Another unit in common use is the ***atmosphere*** (atm),

$$1 \text{ atm} = 1.01 \times 10^5 \text{ N/m}^2 \tag{9}$$

This is the average value of pressure of air at sea level. Note that this is quite a large pressure. For instance, the force that the atmospheric pressure of 1 atm exerts on the palm of your hand, of approximate area 0.006 m^2, is

$$F = Ap = 0.006 \text{ m}^2 \times 1.01 \times 10^5 \text{ N/m}^2 = 600 \text{ N}$$

This is roughly the weight of 60 kg, but you do not notice this pressure force because an equal pressure force of opposite direction acts on the back of your hand, leaving the hand in equilibrium (you don't even notice that these opposed forces squeeze your hand, because the external pressure forces directed against your skin are compensated by the internal pressure forces exerted by your body fluids).

Table 12.2 gives various examples of values of pressures.

Blaise Pascal, *1623–1662, French scientist. He made important contributions to mathematics and is regarded as the founder of modern probability theory. In physics, he performed experiments on atmospheric pressure and on the equilibrium of fluids.*

TABLE 12.2 SOME PRESSURES	
Core of neutron star	1×10^{38} N/m^2
Center of Sun	2×10^{16} N/m^2
Highest sustained pressure achieved in laboratory	5×10^{11} N/m^2
Center of Earth	4×10^{11} N/m^2
Bottom of Pacific Ocean (5.5-km depth)	6×10^{7} N/m^2
Water in core of nuclear reactor	1.6×10^{7} N/m^2
Overpressure[a] in automobile tire	2×10^{5} N/m^2
Air at sea level	1.0×10^{5} N/m^2
Overpressure at 7 km from 1-megaton explosion	3×10^{4} N/m^2
Air in funnel of tornado	2×10^{4} N/m^2
Overpressure in human heart	
systolic	1.6×10^{4} N/m^2
diastolic	1.1×10^{4} N/m^2
Lowest vacuum achieved in laboratory	10^{-14} N/m^2

[a] The *overpressure* is the amount of pressure in excess of normal atmospheric pressure.

EXAMPLE 4. In 1934, C. W. Beebe and O. Barton descended to a depth of 923 m below the surface of the ocean in a steel bathysphere (see Fig. 12.15). This bathysphere had an observation porthole of diameter 0.50 m. The pressure at this depth is 9.3×10^6 N/m^2. Under these conditions, what is the force pressing against the outside of the porthole?

SOLUTION: The area of the porthole is $A = \pi \times (0.25\ m)^2 = 0.20\ m^2$. Hence the force is

$$F = A \times p = 0.20\ m^2 \times 9.3 \times 10^6\ N/m^2 = 1.9 \times 10^6\ N$$

This is the weight of about 190 tons!

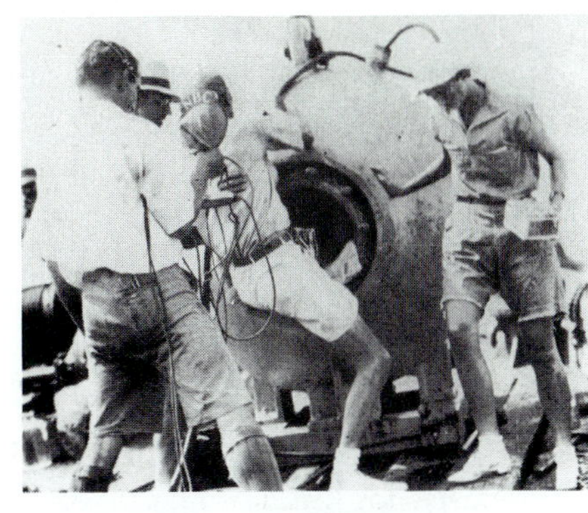

FIGURE 12.15

Explorer C.W. Beebe entering the bathysphere.

12.4 Pressure in a Static Fluid

A fluid is said to be in static equilibrium when the flow velocity is everywhere zero. An example of such a static fluid is the air in a closed room with no air currents. At first, we will neglect gravity and pretend that the only forces acting on the fluid are those exerted by the walls of the container. Under these conditions, ***the pressure at all points within the fluid must be the same.*** To see that this is so, consider two points P_1 and P_2, and imagine a long, thin parallelepiped of fluid with bases at these two points (see Fig. 12.16). The fluid outside the parallelepiped exerts pressure forces on the fluid inside the parallelepiped. The components of these forces along the long direction of the parallelepiped are entirely due to the forces on the bases at P_1 and P_2. If the parallelepiped of fluid is to remain static, these forces on the opposite bases must be equal in magnitude. Hence the pressure at P_1 and P_2 must be equal. For example, the pressure of the air is the same at all points of a room — if the pressure is 1 atm in one corner of the room, it will be the same at any other point of the room.

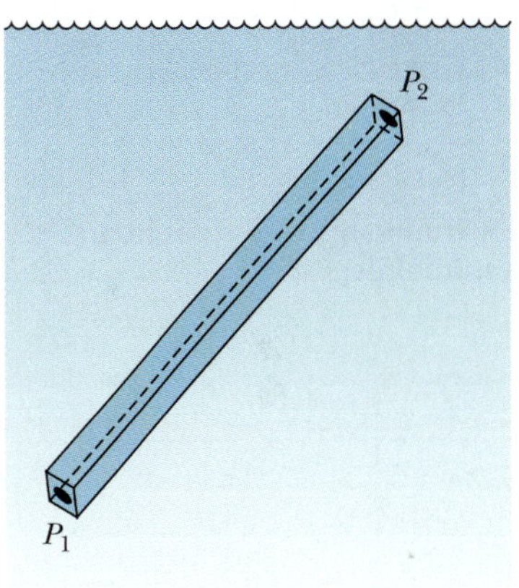

FIGURE 12.16

A long, thin parallelepiped within a static fluid.

The uniformity of pressure throughout a static fluid implies that if we apply a pressure to some part of the surface of a confined fluid by means of a piston or a weight pushing against the surface, ***then this pressure will be transmitted without change to all parts of the fluid.*** This rule for the transmission of pressure in a static fluid is called ***Pascal's Principle,*** and it finds widespread application in the design of hydraulic presses, jacks, and remote controls. Figure 12.17 is a schematic diagram of a hydraulic press consisting of two cylinders with pistons, one small and one large. The cylinders are filled with an incompressible fluid, and they are connected by a pipe. By pushing down on the small piston, we increase the pressure in the fluid; this increases the force on the large piston. Since the pressures on both pistons are the same, the forces on the pistons are in the ratio of the areas of their faces; thus, a small force on the small piston will gener-

◄ **Pascal's Principle**

FIGURE 12.17
Hydraulic press.

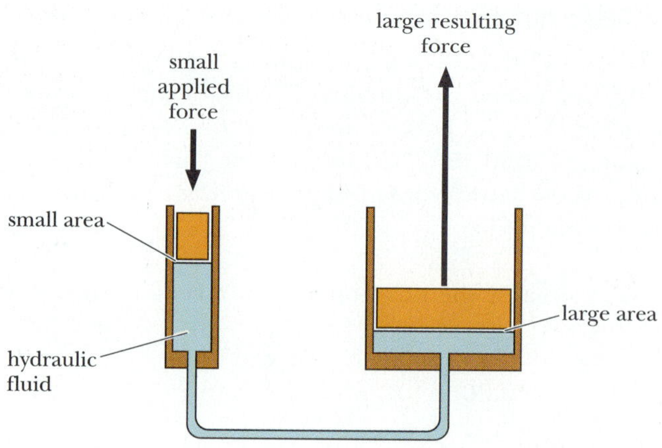

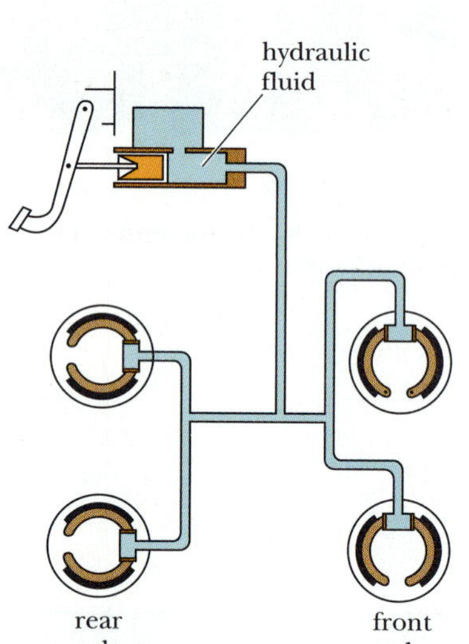

FIGURE 12.18
Hydraulic brake system in an automobile.

ate a large force on the large piston. The brake systems and other control systems on automobiles, trucks, and aircraft employ such arrangements of cylinders. Figure 12.18 shows the hydraulic brake system of an automobile. The brake pedal pushes on the small master piston, and the brake fluid communicates the resultant pressure to the large slave piston and activates the brakes.

EXAMPLE 5. The diameter of the small piston in Fig. 12.17 is 1.5 cm and that of the large piston is 7.0 cm. If you exert a force of 300 N on the master piston, what force will this generate on the slave piston?

SOLUTION: The forces are in the ratio of the areas of the pistons, and these areas are in the ratio of the squares of the diameters:

$$F_2 = F_1 \frac{A_2}{A_1} \tag{10}$$

$$= 300 \text{ N} \times \frac{(7.0 \text{ cm})^2}{(1.5 \text{ cm})^2} = 6.5 \times 10^3 \text{ N}$$

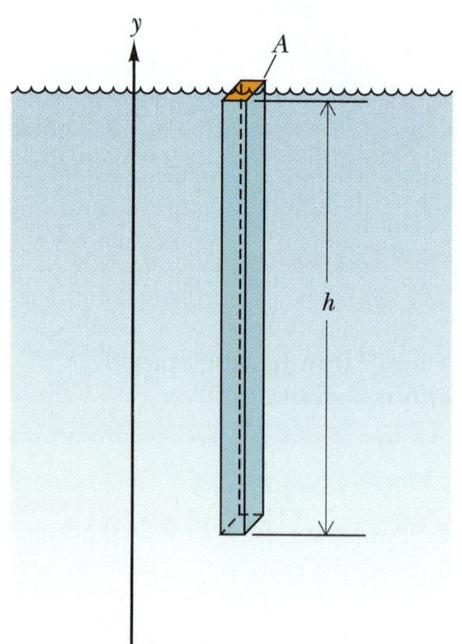

FIGURE 12.19
A thin parallelepiped of fluid from the surface of a lake to a depth h.

Next, we want to take into account the effect of gravity on the pressure in a fluid. For a static fluid subjected to gravity, such as the water of a calm lake, the pressure force at any given depth must support the weight of the overlaying mass of fluid; consequently, the pressure must increase with depth. To derive a formula for the dependence of this ***hydrostatic pressure*** on depth, consider a long, thin, vertical parallelepiped, extending from the surface of the lake to a depth h (see Fig. 12.19). The area of each of the two small horizontal faces of the parallelepiped is A. The pressure of the fluid on the bottom face is p, and the pressure on the top face is the atmospheric pressure p_0. Hence, the upward force acting on the bottom face of the parallelepiped is Ap, and the downward force acting on the top face is Ap_0. The mass of fluid in the parallelepiped is $\rho \times$ [volume] $= \rho Ah$, and the weight of this mass is ρAhg. If the parallelepiped is to remain in equilibrium, the difference between the vertical pressure forces on its bottom and top faces must match the weight:

$$Ap - Ap_0 = \rho Ahg \tag{11}$$

or, canceling the factor of A,

$$p - p_0 = \rho gh \tag{12}$$

◀ **Hydrostatic pressure**

This shows that ***the increment in pressure is directly proportional to the depth.***

EXAMPLE 6. What is the pressure of a depth of 10 m below the surface of a lake? Assume that the pressure of air at the surface of the lake is 1.0 atm.

SOLUTION: With $p_0 = 1$ atm $= 1.01 \times 10^5$ N/m², $\rho = 1000$ kg/m³, and $h = 10$ m, Eq. (12) gives

$$p = p_0 + \rho gh \tag{13}$$

$$= 1.01 \times 10^5 \text{ N/m}^2 + 1000 \text{ kg/m}^3 \times 9.81 \text{ m/s}^2 \times 10 \text{ m}$$

$$= 1.99 \times 10^5 \text{ N/m}^2$$

Thus, the pressure is 1 atm at the surface of the lake and about 2 atm at a depth of 10 m, that is, the pressure increases by about 1 atm per 10 m of water.

Several simple instruments for the measurement of pressure make use of a column of liquid. Figure 12.20 shows a mercury barometer consisting of a tube of glass, about 1 m long, closed at the upper end and open at the lower end. The tube is filled with mercury, except for a small empty space at the top. The bottom of the tube is immersed in an open bowl filled with mercury. The atmospheric pressure acting on the exposed surface of mercury in the bowl prevents the mercury from flowing out of the tube. At the level of the exposed surface, the pressure exerted by the column of mercury is ρgh [see Eq. (12)], where $\rho = 1.36 \times 10^4$ kg/m³ is the density of mercury and h the height of the mercury column. For equilibrium, this pressure must match the atmospheric pressure:

$$p_0 = \rho gh \tag{14}$$

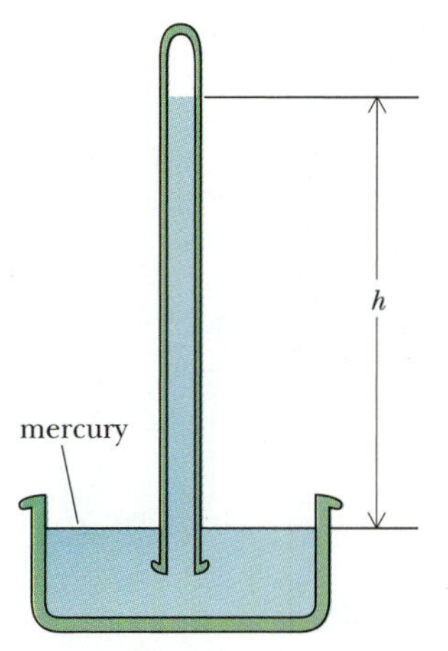

FIGURE 12.20
A mercury barometer.

This equation permits a simple determination of the atmospheric pressure from a measurement of the height of the mercury column.

In view of the direct correspondence of the atmospheric pressure and the height of the mercury column, the pressure is often quoted in terms of this height, usually expressed in millimeters of mercury (mm-Hg). ***The average value of the atmospheric pressure at sea level is 760 mm-Hg, which by definition is one atmosphere*** (atm). Hence,

Atmospheric pressure ▶

$$1 \text{ atm} = 760 \text{ mm-Hg} = \rho \times g \times 0.760 \text{ m} \tag{15}$$

$$= 1.36 \times 10^4 \text{ kg/m}^3 \times 9.81 \text{ m/s}^2 \times 0.760 \text{ m}$$

$$= 1.01 \times 10^5 \text{ N/m}^2 \tag{16}$$

This value of the atmosphere has already been mentioned in Eq. (9).

EXAMPLE 7. What is the change in atmospheric pressure between the basement of a house and the attic, at a height of 10 m above the basement? Express the result in mm-Hg. Assume that the density of air has its standard value 1.29 kg/m^3.

SOLUTION: Although air is a compressible fluid, the change in its density is small if the change of altitude (and pressure) is small, as it is in the present example. Therefore, Eq. (12) is a good approximation; however, we need to insert an extra minus sign in this equation because the "depth" h is now measured upward:

$$p - p_0 = -\rho g h \tag{17}$$

$$= 1.29 \text{ kg/m}^3 \times 9.81 \text{ m/s}^2 \times 10 \text{ m} = -1.3 \times 10^2 \text{ N/m}^2$$

Since 1.01×10^5 N/m^2 equals 760 mm-Hg, 1.3×10^2 N/m^2 must equal

$$760 \text{ mm-Hg} \times \frac{1.3 \times 10^2 \text{ N/m}^2}{1.01 \times 10^5 \text{ N/m}^2} = 0.98 \text{ mm-Hg}$$

Hence the pressure decreases by about 1 mm-Hg in 10 m. This decrease of pressure can be readily detected by carrying an ordinary barometer from the basement to the attic of the house. Note that the pressure change associated with 10 m of air is only about one-thousandth of the pressure change associated with 10 m of water, calculated in Example 6. This reflects the lower density of air.

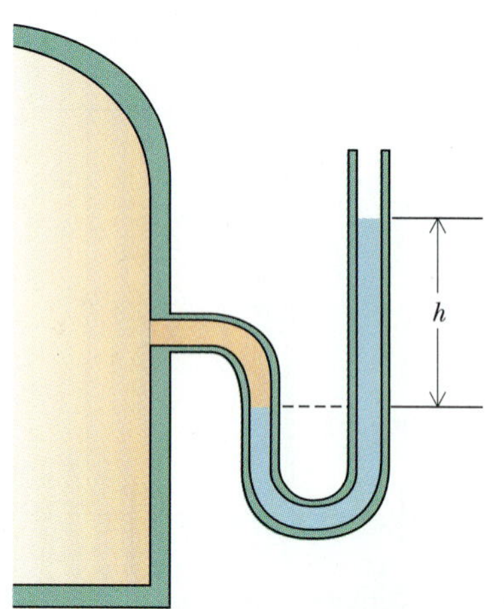

FIGURE 12.21
Open-tube manometer.

Figure 12.21 shows an open-tube ***manometer,*** a device for the measurement of the pressure of a fluid, such as that contained in the tank shown on the left. The tube contains mercury, or water, or oil. One side of the tube is in contact with the fluid in the tank; the other is in contact with the air. The fluid in the tank therefore presses down on one end of the mercury column and the air presses down on the other end. The difference h in the heights of the levels of mercury at the two ends gives the difference in the pressure at the two ends,

$$p - p_0 = \rho g h \tag{18}$$

Hence, this kind of manometer indicates the amount of pressure in the tank in excess of the atmospheric pressure. This excess is called the ***overpressure,*** or ***gauge pressure.*** It is well to keep in mind that many of the pressure gauges used in engineering practice are calibrated in terms of the overpressure rather than absolute pressure. For instance, the pressure gauges used for automobile tires read overpressure.

◂ **Overpressure**

Among the many practical applications of manometers is the ***sphygmomanometer*** used to measure cardiac blood pressure. This consists of a manometer connected to an air sac in the form of a cuff (see Fig. 12.22). The air sac is wrapped around the upper arm of the patient, and is then inflated by means of a hand pump until the pressure of the sac against the arm collapses the brachial artery and cuts off the blood flow. The air is then slowly allowed to leak out of the sac. When the pressure drops to a value equal to the systolic (maximum) cardiac pressure, blood will intermittently squirt through the artery with each heart beat. This initiation of intermittent blood flow can be readily detected by listening to the noise of the rushing blood with a stethoscope placed just below the cuff. As the pressure drops further, the intervals of intermittent blood flow become longer; and when the pressure has dropped to a value equal to the minimum (diastolic) cardiac pressure, the blood flow becomes continuous, with a steady rushing noise. Thus, the onset of intermittent noise and onset of steady rushing noise indicate, respectively, the systolic and the diastolic cardiac pressures. Typical values of these pressures in healthy adults are 120 mm-Hg and 80 mm-Hg, usually reported in the abbreviated notation 120/80 (these pressures are overpressures, that is, the atmospheric pressure has been discounted). Note that in order to obtain an accurate reading of the cardiac pressure, the cuff must be placed at the level of the heart; if it were placed lower (say, on a leg) or higher (say, on a raised arm), then there would be a pressure difference $\rho g h$ between the height of the heart and the height of the cuff. This pressure difference would amount to about 10 mm-Hg for every 10 cm of height difference — a significant discrepancy.

◂ **Sphygmomanometer**

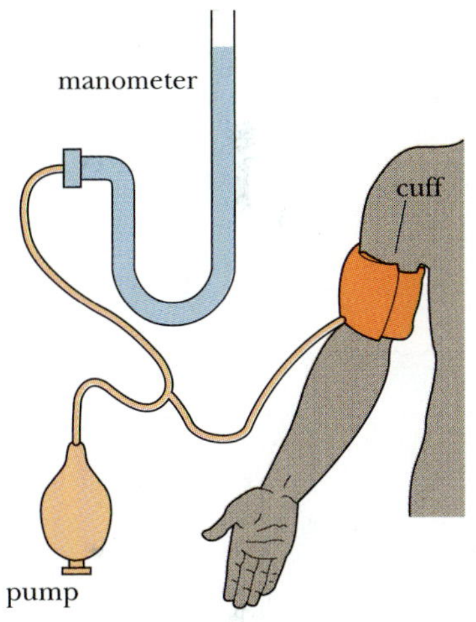

FIGURE 12.22
A sphygmomanometer.

12.5 ARCHIMEDES' PRINCIPLE

If you try to push a beach ball below the surface of the water, you notice that the water exerts a strong upward push on the ball. You can barely force the ball under, and if you release it, the ball pops out of the water with violence. ***The upward force that water or some other fluid exerts on a***

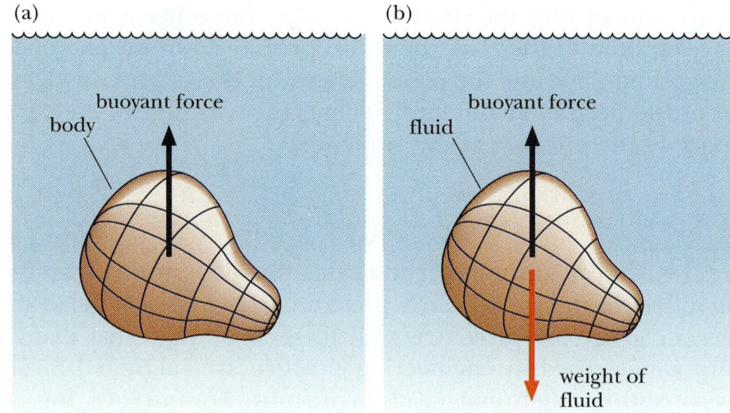

FIGURE 12.23
(a) Submerged body.
(b) Volume of fluid of same shape as the body.

partially or totally immersed body is called the buoyant force. This force results from the pressure difference between the bottom and the top of the body. In a fluid in equilibrium under the influence of gravity, the pressure increases with depth; hence the pressure of the fluid at the bottom of the body is larger than that at the top of the body, and there is more force pressing the body up than down. The magnitude of the buoyant force is given by ***Archimedes' Principle:***

Archimedes' Principle

> ***The buoyant force on an immersed body has the same magnitude as the weight of the fluid displaced by the body.***

The proof of this famous principle is very simple. Imagine that we replace the immersed volume of the body by an equal volume of fluid of the same shape (see Fig. 12.23). The volume of fluid will then be in static equilibrium. Obviously, this requires a balance between the weight of the fluid and the resultant of all the pressure forces acting on the surface enclosing this volume of fluid. But the pressure forces acting on the surface of the original immersed body are exactly the same as the pressure forces acting on the surface of the volume of fluid by which we have replaced it. Hence, the magnitude of the resultant of the pressure forces acting on the original body must equal the weight of the displaced fluid.

If a body of density less than that of water floats in water, equilibrium of the weight and of the buoyant force is achieved when the body is partially submerged — the weight of the body must match the weight of the water displaced by the submerged part of the body. For instance, the human body has an average density slightly below that of water; if inert, it floats with only the top of the head sticking out. But the equilibrium is quite delicate, and some people can make themselves sink by merely exhaling air, thereby reducing their chest volume and increasing their average density, Most species of bony fishes use a similar method to adjust their buoyancy; they have an internal swim bladder filled with gas, and they preserve neutral buoyancy by making adjustments to the volume of this swim bladder. Submarines use much the same method to move down or up in the water. They have ballast tanks filled partially with water and partially with pressurized air (Fig. 12.24). To dive, they pump air out of the tanks, into storage cylinders, and allow water to flood the tanks, thereby effectively increasing the mass and the density of the submarine.

(a)

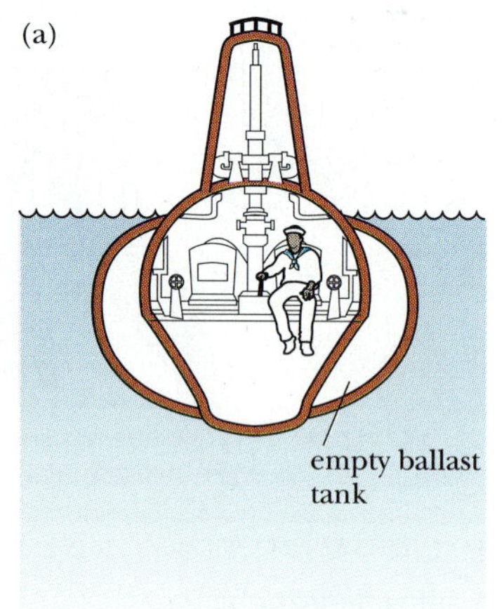

(b)

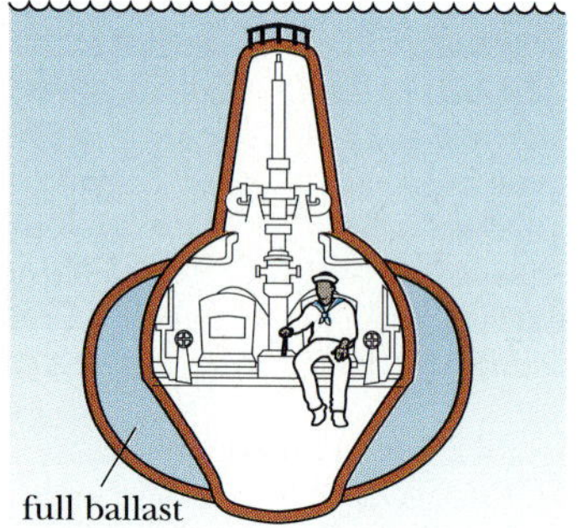

FIGURE 12.24
(a) Submarine afloat.
(b) Submarine diving.

To surface, they blow high-pressure air into the tanks, and drive out the water.

EXAMPLE 8. A hot-air balloon (Fig. 12.25) has a volume of 2.20×10^3 m^3. What is the buoyant force that the surrounding cold air exerts on the balloon? Assume that the density of the surrounding air is $1.29\ kg/m^3$.

SOLUTION: The mass of the cold air displaced by the balloon is $m =$ [density] $\times$ [volume] $= 1.29\ kg/m^3 \times 2.20 \times 10^3\ m^3 = 2.84 \times 10^3$ kg. The weight of this air is $mg = 2.84 \times 10^3\ kg \times 9.81\ m/s^2 = 2.79 \times 10^4$ N. By Archimedes' Principle, this weight gives us the buoyant force,

$$F = 2.79 \times 10^4\ N$$

If the balloon is to stay aloft, its weight (including the weight of the hot air inside it) must be less than the buoyant force of 2.79×10^4 N.

FIGURE 12.25
Hot-air balloon.

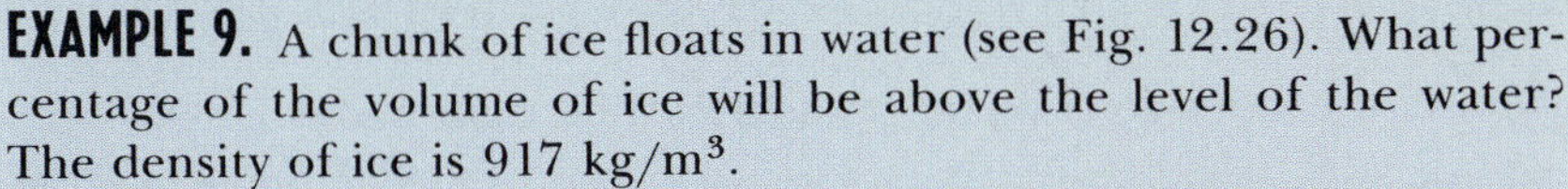

EXAMPLE 9. A chunk of ice floats in water (see Fig. 12.26). What percentage of the volume of ice will be above the level of the water? The density of ice is $917\ kg/m^3$.

SOLUTION: If the mass of the chunk of ice is, say, 1000 kg, it must displace an amount of water of the same weight, that is, 1 m^3 of water. The volume of ice below the water level is then 1 m^3, whereas the total volume of ice is [mass]/[density] $= (1000\ kg)/(917\ kg/m^3) = 1.091\ m^3$. The fraction of ice above the water level is therefore

$$\frac{0.091\ m^3}{1.091\ m^3} = 0.083, \text{ or } 8.3\%.$$

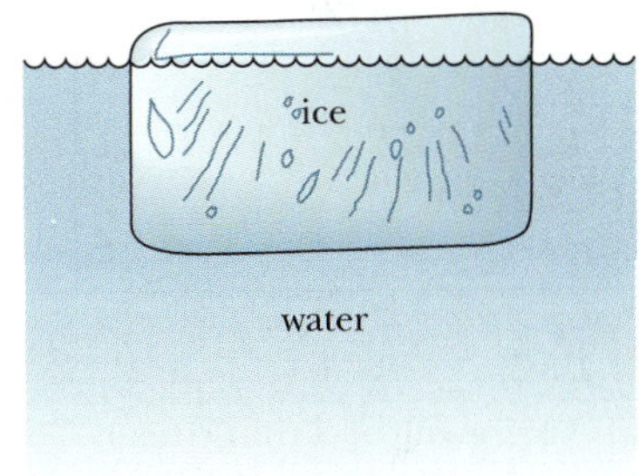

FIGURE 12.26
Ice floating in water.

12.6 BERNOULLI'S EQUATION

When moving air encounters an obstacle that slows down its motion, the air exerts an extra pressure on the obstacle. You can feel the push of this extra pressure if you stand in a strong wind, or if you put a hand out of the window of a speeding car. The pressure changes that occur when air flows around obstacles or when water or some other fluid flows through pipes of varying cross sections can be calculated in a simple way by exploiting the conservation law for the mechanical energy. In this section, we will formulate the conservation law for energy for the special case of

Daniel Bernoulli (bernooyee), *1700–1782, Swiss physician, physicist, and mathematician. His great treatise* Hydrodynamica *included the equation named after him. Several other members of the Bernoulli family made memorable contributions to mathematics and physics.*

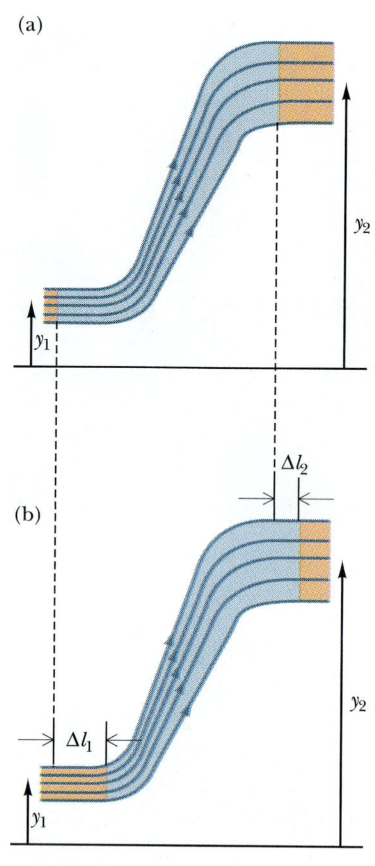

FIGURE 12.27
(a) A segment of a thin stream tube. The beginning and the end of the segment are marked by the dashed lines. (b) Motion of the fluid along the stream tube. The fluid enters the segment from the left and emerges on the right.

steady flow of an incompressible fluid without viscosity. This conservation law is called Bernoulli's equation.

We know from Section 12.2 that steady incompressible flow can be described by streamlines. As in the derivation of the equation of continuity, we consider a bundle of streamlines forming a thin stream tube. The fluid flows inside this tube as though the surface of the tube were an impermeable pipe. Figure 12.27a shows a segment of this "pipe" extending from a point 1 to a point 2; this segment contains some mass of fluid. Figure 12.27b shows the same mass of fluid at a slightly later time—the fluid has moved toward the right. During this movement, the pressures at the left and at the right ends of the segment exert forces and do some work on the mass of fluid. By energy conservation, this work must equal the change of kinetic and potential energy. To express this mathematically, we begin by calculating the work done by the pressure. As the left end of the mass of fluid moves through a distance Δl_1, the work done by the pressure is the force $A_1 p_1$ multiplied by the distance Δl_1,

$$W_1 = A_1 p_1 \Delta l_1$$

Since the product $A_1 \Delta l_1$ is the volume V vacated by the retreat of the fluid on the left end, we can also write this as

$$W_1 = p_1 V \tag{19}$$

Likewise, the work done by the pressure at the right end is

$$W_2 = -p_2 V \tag{20}$$

This is negative because the force is opposite to the displacement. Note that the *same* volume V appears in Eqs. (19) and (20) — since the fluid is incompressible, the volume vacated by the retreat of the fluid at one end must equal the volume occupied by the advance of the fluid at the other end. The net work done by the pressure is then

$$W = W_1 + W_2 = p_1 V - p_2 V \tag{21}$$

The change in kinetic and potential energy is entirely due to the changes at the ends of the mass of fluid; everywhere else, the shift of the fluid merely replaces fluid of some kinetic energy and potential energy with fluid of exactly the same kinetic and potential energy. The change at the ends involves replacing a mass m of fluid, of speed v_1 at height y_1, by an equal mass m, of speed v_2 at the height y_2. The corresponding change of kinetic and potential energy is

$$\Delta K + \Delta U = \tfrac{1}{2}mv_2^2 - \tfrac{1}{2}mv_1^2 + mgy_2 - mgy_1 \tag{22}$$

This change of mechanical energy must match the work done by the pressure:

$$\tfrac{1}{2}mv_2^2 - \tfrac{1}{2}mv_1^2 + mgy_2 - mgy_1 = p_1 V - p_2 V \tag{23}$$

If we divide both sides of this equation by V, and we move all terms with subscript "2" to the left side, and all terms with subscript "1" to the right side, we obtain

$$\frac{1}{2}\frac{m}{V}v_2^2 + \frac{m}{V}gy_2 + p_2 = \frac{1}{2}\frac{m}{V}v_1^2 + \frac{m}{V}gy_1 + p_1 \tag{24}$$

or, since m/V is the density of the fluid,

$$\frac{1}{2}\rho v_2^2 + \rho gy_2 + p_2 = \frac{1}{2}\rho v_1^2 + \rho gy_1 + p_1 \tag{25}$$

◀ **Bernoulli's equation**

This is ***Bernoulli's equation.*** Note that $\frac{1}{2}\rho v^2$ is the density of kinetic energy (amount of kinetic energy per unit volume) and ρgy is the density of potential energy (potential energy per unit volume). Hence, Bernoulli's equation states that for any two points on a streamline, the sums of the density of kinetic energy, the density of potential energy, and the pressure are the same. Since the points are arbitrary, the quantity $\frac{1}{2}\rho v^2 + \rho gy + p$ must be a constant of the motion:

$$\frac{1}{2}\rho v^2 + \rho gy + p = [\text{constant}] \tag{26}$$

Note that if the fluid is static, so $v = 0$, Eq. (26) reduces to

$$p + \rho gy = [\text{constant}] \tag{27}$$

This is equivalent to Eq. (12) for the pressure in a static fluid.

According to Bernoulli's equation, in any region in which the hydrostatic pressure ρgy is constant or approximately constant, ***the pressure along any given streamline must decrease wherever the velocity increases.*** Intuitively, we might expect that where the velocity is large, the pressure is large; but energy conservation demands exactly the opposite. This inverse relation between pressure and velocity plays an important role in the design of wings for airplanes. Figure 12.28 shows an airfoil and the streamlines of air flowing around it, as seen in the reference frame of the airfoil. The shape of the airfoil has been designed so that along its upper part the velocity is large (high density of streamlines), and along its lower part the velocity is small (low density of streamlines). Consider now one streamline passing just over the airfoil and one just under it. At a large distance to the left, the fluid on all streamlines has the same pressure and the same velocity. Bernoulli's equation applied to each of the two streamlines then tells us that in the region just above the airfoil, the pressure is low, and in the region just below the airfoil, the pressure is high. This leads to a net upward force, or ***lift,*** on the airfoil — this lift force supports the airplane in flight.

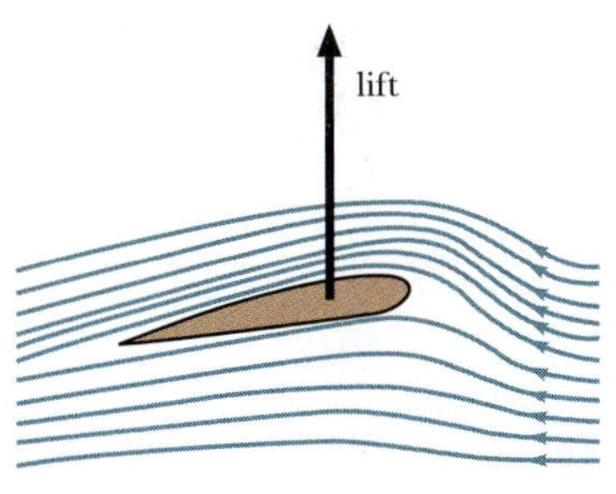

FIGURE 12.28
Flow of air around an airfoil.

A similar lift force acts on the sail and on the keel of a sailboat. The combination of these forces permits the sailboat to move upwind, at an angle as close as 40° or 45° relative to the wind direction (see Fig. 12.29). The sail acts like the wing of the airplane, and the wind streaming over the sail generates a lift force at right angles to the sail (see Fig. 12.30). At

FIGURE 12.29
Sailboat sailing upwind.

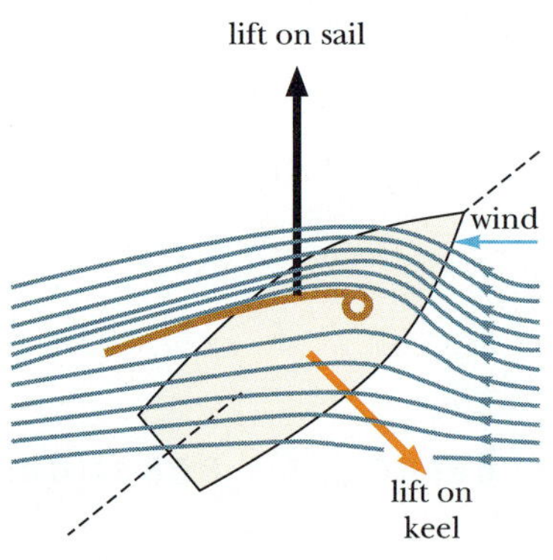

FIGURE 12.30

Lift forces on the sail and on the keel of a sailboat sailing upwind. The resultant force is forward, roughly parallel to the centerline of the sailboat.

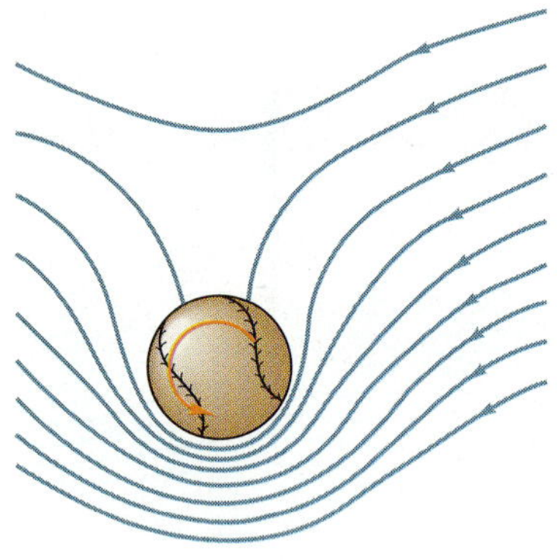

FIGURE 12.31

Flow of air around a spinning baseball, as seen in the reference frame of the baseball. For the direction of spin shown in the diagram, the flow velocity is large at the bottom and smaller at the top.

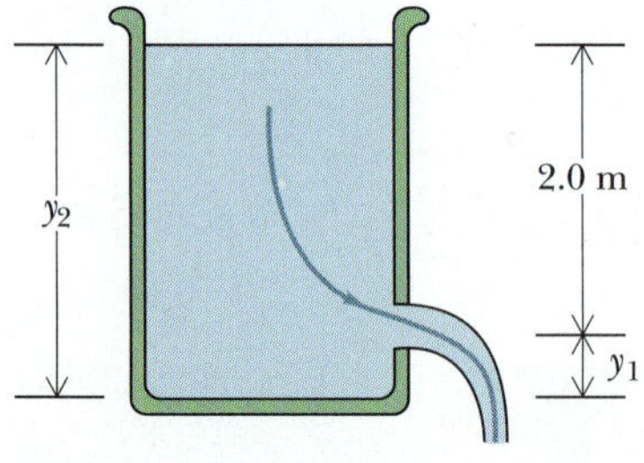

FIGURE 12.32

Streamline for a parcel of water flowing out of a tank.

the same time, the water streaming over the keel generates a lift force at right angles to the keel. The resultant is a propulsive force, roughly parallel to the centerline of the sailboat.

The lift force also explains the puzzling deviations from simple projectile motion observed in baseballs, golf balls, and tennis balls. For instance, baseball pitchers are adept at throwing "curveballs" whose trajectory curves downward at a rate significantly in excess of the rate expected for parabolic projectile motion. The keys to this deviation of the ball are its spin and the viscosity of the air. To curve down, the ball must be thrown with a quick rotational motion so its top is rotating in the forward direction and its bottom in the backward direction. Because of the viscous drag of the air, the top surface of the ball then slows the air flow over the top, and the bottom surface speeds up the air flow over the bottom. In the reference frame of the ball, the streamlines would look as in Fig. 12.5, if the air flow over the top and the bottom were the same. But when the air flow is disturbed by the spinning ball, the streamlines become asymmetric, as in Fig. 12.31. Since the flow velocity of air is large at the bottom and smaller at the top, Bernoulli's equation tells us that the pressure is large at the top and smaller at the bottom — the ball experiences a negative lift, which curves it down.

Note that ***Bernoulli's equation is not valid for the flow of a fluid through a pump or a turbine wheel with a moving piston or moving blades, which do work on the fluid and add energy or remove energy.*** Since Bernoulli's equation expresses energy conservation for the fluid, it cannot be valid when such an external device adds energy to or removes energy from the fluid. Mathematically, the failure of Bernoulli's equation under these conditions arises from the failure of the assumption of steady flow, which entered in the above derivation. The motion of the piston of the pump or the motion of the blades of the turbine requires a time-dependence of the pattern of flow of the fluid; this means the flow is *not* steady.

The following examples illustrate some applications of Bernoulli's equation.

EXAMPLE 10. A water tank has a (small) hole near its bottom at a depth of 2.0 m from the top surface (see Fig. 12.32). What is the speed of the stream of water emerging from the hole?

SOLUTION: Qualitatively, one of the streamlines for the water flowing out of the tank will look as shown in Fig. 12.32. Since the hole is small, the water level at the top of the tank drops only very slowly; at the top, we can therefore take $v_1 = 0$ in Eq. (25). Furthermore, the pressures at the top and in the emerging stream of water are the same; both are equal to the atmospheric pressure p_0. Thus, $p_1 = p_2 = p_0$. With this, Eq. (25) becomes

$$\tfrac{1}{2}\rho v_2^2 + \rho g y_2 + p_0 = \rho g y_1 + p_0 \qquad (28)$$

We can cancel the terms p_0, and we can move the term $\rho g y_2$ to the right side of the equation:

$$\tfrac{1}{2}\rho v_2^2 = \rho g y_1 - \rho g y_2 \tag{29}$$

If we here divide both sides by $\frac{1}{2}\rho$ and extract the square root of both sides, we find

$$v_2 = \sqrt{2g(y_1 - y_2)} \tag{30}$$

With $y_1 - y_2 = 2.0$ m, the speed is

$$v_2 = \sqrt{2 \times 9.81\ \text{m/s}^2 \times 2.0\ \text{m}} = 6.3\ \text{m/s}$$

COMMENTS AND SUGGESTIONS: The use of Bernoulli's equation in a problem of motion of a fluid involves the same steps as the use of energy conservation in earlier chapters: First write an expression for $\frac{1}{2}\rho v_1^2 + \rho g y_1 + p_1$ at one point on a streamline, then write an expression for $\frac{1}{2}\rho v_2^2 + \rho g y_2 + p_2$ at another point of the streamline, and then equate these two expressions [as in Eq. (28)]. This yields one equation, which can be solved for the unknown speed or the unknown final pressure (if the final speed is specified).

Note that in our example, the speed of the water emerging from the tank is exactly what it would be if the water were to fall freely from a height of 2.0 m [compare Eq. (2.23)]. This, of course, expresses conservation of energy: when a drop of water flows out at the bottom, the loss of potential energy of the water in the tank is equivalent to the removal of a drop of water from the top; the conversion of this potential energy into kinetic energy will give the drop the speed of free fall.

EXAMPLE 11. The pressure in a firehose of diameter 6.4 cm is 3.5×10^5 N/m^2 and the speed of flow in 4.0 m/s. The firehose ends in a metal tip of diameter 2.5 cm (see Fig. 12.33). What are the velocity and the pressure of water in the tip?

SOLUTION: The magnitudes of the velocities in the hose and in the tip are related to the cross-sectional areas by the continuity equation [Eq. (6)],

$$v_2 = v_1 \frac{A_1}{A_2} \tag{31}$$

The ratio of the cross-sectional areas is equal to the ratio of the

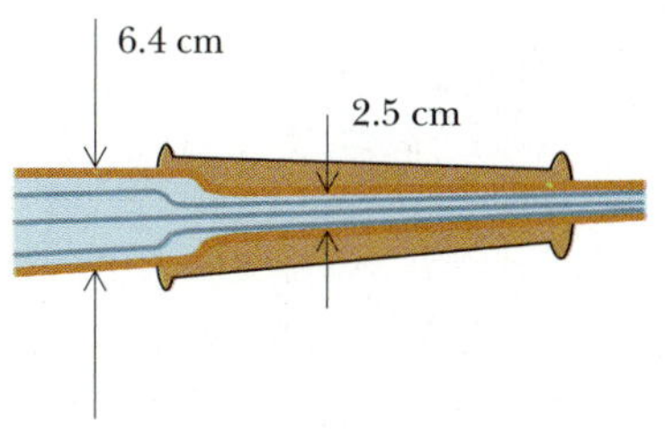

FIGURE 12.33
A firehose tip with streamlines of the water flowing out of the tip.

squares of the diameters; hence the velocity in the tip is

$$v_2 = 4.0 \text{ m/s} \times \frac{(6.4 \text{ cm})^2}{(2.5 \text{ cm})^2} = 26.2 \text{ m/s}$$

With $y_1 = y_2$, Bernoulli's equation becomes

$$\tfrac{1}{2}\rho v_2^2 + p_2 = \tfrac{1}{2}\rho v_1^2 + p_1$$

From this, we find the pressure of the water in the tip:

$$p_2 = \tfrac{1}{2}\rho v_1^2 - \tfrac{1}{2}\rho v_2^2 + p_1 \tag{32}$$

$$= \tfrac{1}{2} \times 1000 \text{ kg/m}^3 \times (4.0 \text{ m/s})^2 - \tfrac{1}{2} \times 1000 \text{ kg/m}^3 \times (26.2 \text{ m/s})^2 + 3.5 \times 10^5 \text{ N/m}^2$$

$$= 1.4 \times 10^4 \text{ N/m}^2$$

Venturi flowmeter ▶

EXAMPLE 12. The ***Venturi flowmeter*** is a device that measures the velocity of a fluid flowing in a pipe. It consists of a constriction in the pipe with a cross-sectional area A_2 that is smaller than the cross-sectional area A_1 of the pipe itself (see Fig. 12.34). Small holes in the constriction and in the pipe permit the measurement of the pressures at these points by means of a manometer. Express the velocity of flow in terms of the pressure difference registered by the manometer.

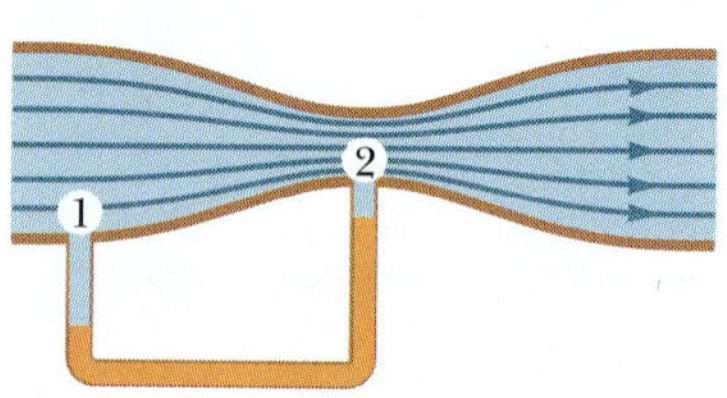

FIGURE 12.34
Venturi flowmeter. The manometer measures the pressure difference between the points 1 and 2.

SOLUTION: As in the preceding example, the magnitudes of the velocities at points 1 and 2 are related to the cross-sectional areas by the continuity equation,

$$v_2 = v_1 \frac{A_1}{A_2} \tag{33}$$

With $y_1 = y_2$, Bernoulli's equation then gives us:

$$\tfrac{1}{2}\rho v_2^2 + p_2 = \tfrac{1}{2}\rho v_1^2 + p_1$$

Which leads to the following equation for the pressure difference:

$$p_1 - p_2 = \tfrac{1}{2}\rho v_2^2 - \tfrac{1}{2}\rho v_1^2$$

$$= \tfrac{1}{2}\rho (v_1 A_1/A_2)^2 - \tfrac{1}{2}\rho v_1^2$$

$$= \tfrac{1}{2}\rho v_1^2 [(A_1/A_2)^2 - 1] \tag{34}$$

Taking the square root of both sides of this equation and solving for v_1, we find

$$v_1 = \sqrt{\frac{2(p_1 - p_2)}{\rho[(A_1/A_2)^2 - 1]}} \tag{35}$$

This says that the flow velocity is proportional to the square root of the pressure difference.

12.7* Viscous Flow

◀ Viscosity

Viscosity is an internal friction or stickiness in the fluid. The layers of a viscous fluid stick to the adjacent layers of the fluid and to the adjacent walls of the pipe or conduit in which the fluid is flowing, and this tends to stop the flow. For instance, molasses and pitch are two fluids of high viscosity — they will flow, but only slowly and gradually, and they will stop flowing almost immediately if the external forces pushing them are removed. Water has a much lower viscosity, and in the examples of the preceding sections we neglected the viscous forces in water, because these forces could be regarded as small compared with other forces acting on the water. This is an adequate approximation for water flowing in large pipes at fairly low speeds. But if the water is flowing in a narrow pipe or a thin tube, then the viscous forces become important.

In viscous flow, ***the thin layer of fluid immediately adjacent to the wall of the pipe adheres to the wall, that is, the layer has zero velocity.*** The next layer, a bit farther away from the wall, has a small velocity. The next layer has a somewhat larger velocity, and so on. The fluid at the center of the pipe has the largest velocity (Fig. 12.35). Each layer of fluid exerts frictional forces on the adjacent layers. For instance, in the lower half of the pipe in Fig. 12.35, each layer exerts a frictional force on the next layer immediately above it, which tends to slow down this next layer. These frictional forces remove mechanical energy from the layers of fluid, converting it into heat. If the flow is to continue, the loss of mechanical energy must be made good by a pump or a plunger that does work on the fluid while pushing it into the pipe from the left side.

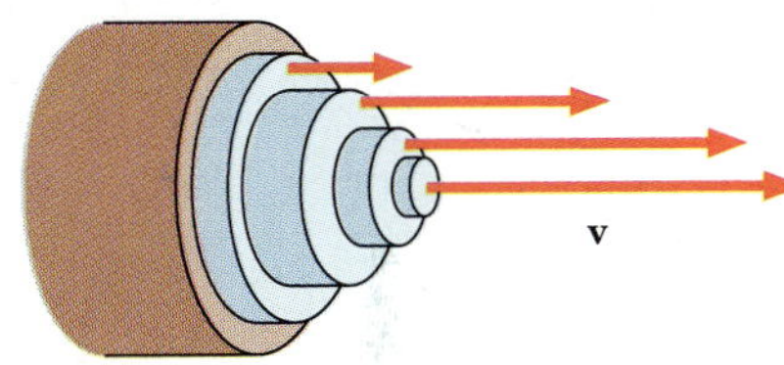

FIGURE 12.35

Flow of a viscous fluid through a cylindrical pipe. The concentric cylindrical layers of fluid shown in this diagram have different velocities. The velocity is lowest (zero) at the wall and it is highest at the center of the pipe.

For a quantitative measurement of the viscous resistance, consider a flat layer of fluid between two parallel flat plates (Fig. 12.36). The bottom plate is stationary, and the top plate is made to move with velocity v toward the right, by means of some machinery (not shown in Fig. 12.36). The fluid in contact with the bottom plate will then be at rest, and the fluid in contact with the top plate will have a velocity v toward the right. The viscous friction within the fluid generates a force that opposes the

* This section is optional.

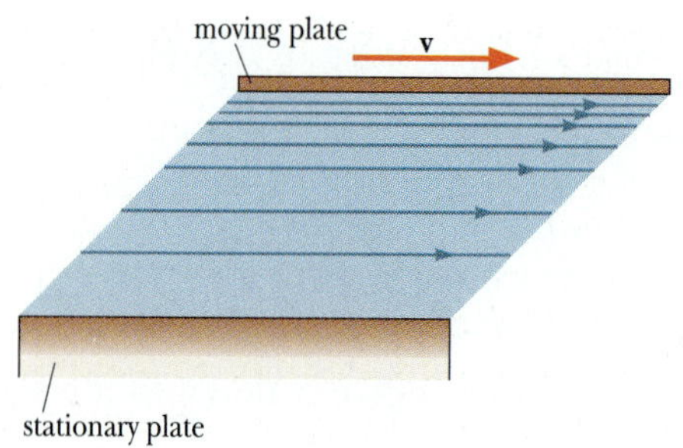

FIGURE 12.36
Flow of a flat layer of viscous fluid between two parallel plates. The bottom plate is stationary, and the top plate is in motion with velocity v toward the right.

motion of the top plate. Experimentally, this force is directly proportional to the velocity v and to the area A of the plate in contact with the fluid, and it is inversely proportional to the distance between the plates,

Viscous friction ▶

$$F = \eta \frac{vA}{h} \tag{36}$$

Coefficient of viscosity ▶

The constant of proportionality η (Greek eta) in this equation is called the ***coefficient of viscosity.***

Table 12.3 lists values of this coefficient for a few fluids. Sticky fluids, such as glucose, have very high values of η. Note that the value of η depends on the temperature; the values in Table 12.3 are for a temperature of 20°C, except as noted. In most fluids, the viscosity increases when the temperature drops. For instance, during a cold winter night, the motor oil in an automobile engine becomes much more viscous, and the large friction then makes it difficult or impossible to start the engine.

If a perfect fluid, with negligible viscous resistance, is initially flowing in a long horizontal pipe, the fluid will continue to flow forever, at constant velocity, in accordance with Newton's First Law. Furthermore, from Bernoulli's equation, we see that the pressure at all points within such a fluid is the same. However, if a real fluid, with viscous resistance, is flowing in

TABLE 12.3 VISCOSITIES OF SOME FLUIDS

Fluid	Coefficient of viscosity
Air	1.8×10^{-5} s · N/m^2
Water	1.0×10^{-3}
Alcohol	1.2×10^{-3}
Blood (37°C)	2.3×10^{-3}
Motor oil, No. 10	2.0×10^{-1}
Glycerin	1.5
Glucose	6.6×10^{10}

such a pipe, it will gradually slow down and stop, because of the frictional resistance generated by the viscous forces. To keep the fluid moving, we have to provide a pressure difference between one end of the pipe and the other, so the excess pressure at the upstream end pushes the fluid along and overcomes the frictional resistance.

The pressure difference required to maintain the flow is directly proportional to the average speed of the flow. Since the average speed of flow is directly proportional to the rate of delivery of the fluid (in cubic meters per second), the mathematical relation for the pressure difference required to maintain the flow is usually stated in terms of the rate of delivery:

$$[\text{rate of delivery}] = \frac{\Delta V}{\Delta t} = \frac{\pi r^4 (p_1 - p_2)}{8\eta l} \tag{37}$$

◂ **Poiseuille's equation**

where r is the radius of the pipe, l is the length of the pipe, and $p_1 - p_2$ is the pressure difference between its ends (Fig. 12.37). Equation (37) is called ***Poiseuille's equation*** (pronounced *pwasöl*).

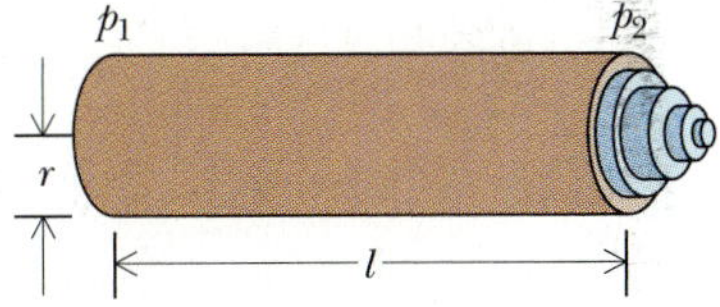

FIGURE 12.37

Flow in a cylindrical pipe with a pressure difference $p_1 - p_2$ along its length.

According to this equation, the rate of delivery is inversely proportional to the viscosity of the fluid and to the length of the pipe; this is not surprising, since high viscosity and long length result in large friction. What is surprising is the dependence on the fourth power of the radius of the pipe. Naively, we might have expected that the rate of delivery is proportional to the cross-sectional area of the pipe ($A = \pi r^2$), and therefore proportional to the square of the radius.

The dependence on the fourth power of the radius implies that a much larger pressure difference is required to push a viscous fluid through a small pipe than through a large pipe. Even a slight reduction in the radius of the pipe requires a substantial increase in the pressure difference, if a given rate of delivery is to be maintained. For instance, a decrease of the radius by a mere 10% (or by a factor of 0.9) requires an increase of the pressure difference by a factor of $1/(0.9)^4 = 1.5$. This explains why the blood flow in the human circulatory system is so sensitive to slight reductions of the diameter of the arteries. Even a slight buildup of plaque in the interior of the arteries requires a substantial increase of blood pressure, if a fixed rate of delivery of blood is to be maintained. The heart then has to work harder, to provide this higher pressure.

Note that if the rate of delivery of fluid is small, then the pressure difference required to maintain this rate of delivery is also small. This means that ***for low speeds of flow, the viscous resistance is low.*** Hence Poiseuille's equation confirms that at sufficiently low speeds of flow, any fluid will behave like a perfect fluid.

EXAMPLE 13. A hypodermic syringe has a needle of length 4 cm and internal radius 0.2 mm (Fig. 12.38). The syringe contains 4 cm^3 of distilled water. What pressure must you apply to the water to squirt it out of the syringe in 10 s?

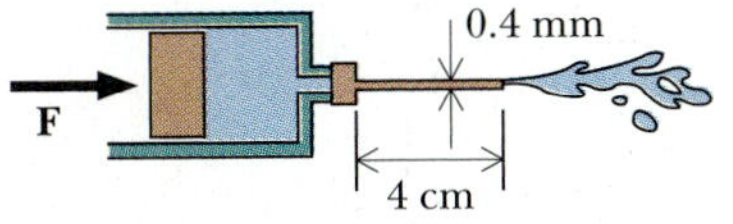

FIGURE 12.38

The needle of a syringe is a cylindrical pipe.

SOLUTION: Since the radius of the needle is much smaller than the radius of the body of the syringe, most of the viscous resistance is in the needle, and we can regard the pressure in the body of the syringe as constant. The rate of delivery is $\Delta V/\Delta t = 4 \times 10^{-6}\ \text{m}^3/10\ \text{s} = 4 \times 10^{-7}\ \text{m}^3/\text{s}$. According to Poiseuille's equation, the required pressure difference from one end of the needle to the other is then

$$p_1 - p_2 = \frac{8\eta l}{\pi r^4} \times \frac{\Delta V}{\Delta t}$$

$$= \frac{8 \times 1.0 \times 10^{-3}\ \text{s} \cdot \text{N/m}^2 \times 0.04\ \text{m}}{\pi \times (0.0002\ \text{m})^4} \times 4 \times 10^{-7}\ \text{m}^3/\text{s}$$

$$= 2.5 \times 10^4\ \text{N/m}^2$$

The pressure p_2 at the outer end of the needle is atmospheric pressure. Hence $p_1 - p_2 = 2.5 \times 10^4\ \text{N/m}^2$ is the overpressure; that is, it is the pressure in excess of atmospheric pressure, and it is the pressure you must supply by pushing on the plunger of the syringe.

SUMMARY

- **Continuity equation:** $v_1 A_1 = v_2 A_2$
- **Definition of pressure:** $p = \dfrac{F}{A}$
- **Atmospheric pressure:** 1 atm = 760 mm-Hg = $1.01 \times 10^5\ \text{N/m}^2$ or Pa
- **Hydrostatic pressure (for incompressible fluid):** $p - p_0 = \rho g h$
- **Archimedes' Principle:** The buoyant force has the same magnitude as the weight of the displaced fluid.
- **Bernoulli's equation:** $\frac{1}{2}\rho v^2 + \rho g y + p = [\text{constant}]$
- **Viscous friction:** $F = \eta(vA/h)$
- **Poiseuille's equation:** $\dfrac{\Delta V}{\Delta t} = \dfrac{\pi r^4 (p_1 - p_2)}{8\eta l}$

QUESTIONS

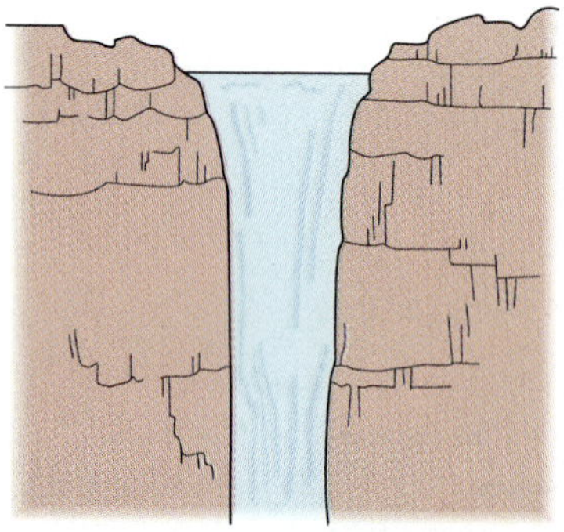

FIGURE 12.39
Sheet of water on a smooth waterfall, narrowing at bottom.

1.–According to popular belief, blood is denser than water. Is this true?

2.–The sheet of water of a waterfall is thick at the top and thin at the bottom (Fig. 12.39). Explain.

3.–If you place a block of wood on the bottom of a swimming pool, why does the pressure of the water not keep it there?

4.–Explain what holds a suction cup on a smooth surface.

5.–Newton gave the following description of his classic experiment with a rotating bucket:

> If a vessel, hung by a long cord, is so often turned about that the cord is strongly twisted, then filled with water, and held at rest together with the water; thereupon, by the sudden action of another force, it is whirled about the contrary way, and while the cord is untwisting itself, the vessel continues for some time in this motion; the surface of the water will at first be plain, as before the vessel began to move; but after that, the vessel, by gradually communicating its motion to the water, will make it begin sensibly to revolve, and recede by little and little from the middle, and ascend to the sides of the vessel, forming itself into a concave figure (as I have experienced), and the swifter the motion becomes, the higher will the water rise, till at last, performing its revolutions in the same times with the vessel, it becomes relatively at rest in it.

Explain why hydrostatic equilibrium requires that the rotating water be higher at the rim of the bucket than at the center.

6.–Scuba divers have survived short intervals of free swimming at depths of 430 m. Why does the pressure of the water at this depth not crush them?

7.–Figure 12.40 shows a glass vessel with vertical tubes of different shapes. Explain why the water level in all the tubes is the same.

8.–In a celebrated experiment, Blaise Pascal attached a long metal funnel to a tight cask (Fig. 12.41). When he filled the cask by pouring water into this funnel, the cask burst. Explain.

9.–Face masks for scuba divers have two indentations at the bottom into which the diver can stick thumb and forefinger to pinch his nose shut. What is the purpose of this arrangement?

10.–Figure 12.42 shows grain silos held together by circumferential steel bands. Why has the farmer placed more bands near the bottom than near the top?

11.–The level of a large oil slick on the sea is slightly higher than the level of the surrounding water. Explain.

12.–Flooding on the low coasts of England and the Netherlands is most severe when the following three conditions are in coincidence: onshore wind, full or new moon, and low barometric pressure. Can you explain this?

13.–Why do some men or women float better than others? Why do they all float better in salt water than in plain water?

14.–Would you float if your lungs were full of water?

15.–Will a stone float in a tub full of mercury?

16.–An ice cube floats in a glass full of water. Will the water level rise or fall when the ice melts?

17.–The density of a solid body can be determined by first weighing the body in air and then weighing it again when it is immersed in water (Fig. 12.43). How can you deduce the density from these two measurements?

18.–A buoy floats on the water. Will the flotation level of the buoy change when the atmospheric pressure changes?

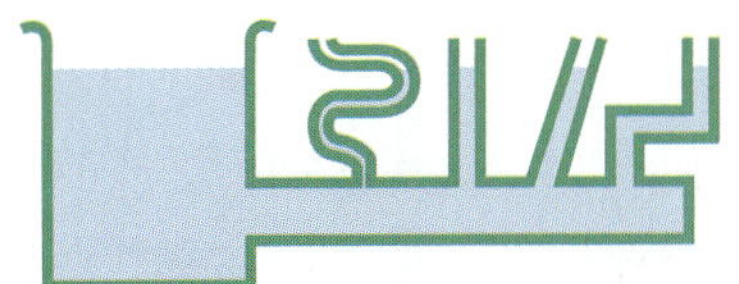

FIGURE 12.40
Glass vessel with vertical tubes.

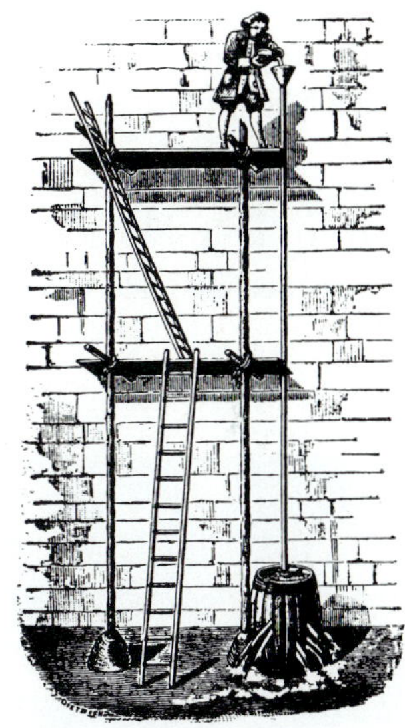

FIGURE 12.41
Blaise Pascal's experiment.

FIGURE 12.42
Grain silos.

FIGURE 12.43
Weighing a body underwater. (Above.)

FIGURE 12.44
Astronauts floating underwater. (Right.)

FIGURE 12.45
The bathyscaphe *Trieste*.

FIGURE 12.46
A Cartesian diver.

19.–Will the water level in a canal lock rise or fall if a ship made of steel sinks in the lock? What if the ship is made of wood?

20.–While training for the conditions of weightlessness they would encounter in an orbiting spacecraft, NASA astronauts were made to float submerged in a large water tank (Fig. 12.44). Small weights attached to their spacesuits gave the astronauts neutral buoyancy. To what extent does such simulated weightlessness imitate true weightlessness?

21.–A girl standing in a subway car holds a helium balloon on a string. Which way will the balloon move when the car accelerates?

22.–A slurry of wood chips is used in some tanning operations. What would happen to a man were he to fall into a vat filled with such slurry?

23.–The bathyscaphe *Trieste*, which set a record of 10,917 m in a deep dive in the Marianas Trench in 1960, consists of a large tank filled with gasoline below which hangs a steel sphere for carrying the crew (Figure 12.45). Can you guess the purpose of the tank of gasoline?

24.–Gasoline vapor from, say, a small leak in the fuel system poses a very serious hazard in a motorboat, but only a minor hazard in an automobile. Explain. (Hint: Gasoline vapor is denser than air.)

25.–How does a submarine dive? Ascend?

26.–How does a balloonist control the ascent and descent of a hot-air balloon?

27.–When you release a bubble of air while underwater, the bubble grows in size as it ascends. Explain.

28.–A "Cartesian diver" consists of a small inverted bottle floating inside a larger bottle whose mouth is covered by a rubber membrane (Fig. 12.46). By depressing the membrane, you can increase the water pressure in the large bottle. How does this affect the buoyancy of the small bottle?

29.–Why are the continuity equation and Bernoulli's equation only *approximately* valid for the flow of air?

30.–To throw a curveball, the baseball pitcher gives the ball a spinning motion about a vertical axis. The air on the left and right sides of the ball will then have slightly different speeds. Using Bernoulli's equation, explain how this creates a lateral deflecting force on the ball.

31.–Hurricanes with the lowest central barometric pressures have the highest wind speeds. Observations show that the square of the wind speed in a hurricane is roughly proportional to the difference between the barometric pressure outside the hurricane and the barometric pressure in the hurricane. Show that this proportionality is expected from Bernoulli's equation. (Hint: The streamlines of air start outside the hurricane and gradually spiral in toward the "eye.")

32.–If you place a Ping-Pong ball in the jet of air from a vacuum cleaner hose aimed vertically upward, the Ping-Pong ball will be held in stable equilibrium within this jet. Explain this by means of Bernoulli's equation. (Hint: The speed of air is maximum at the center of the jet.)

PROBLEMS

Section 12.1: Density and Flow Velocity
Section 12.2: Incompressible Steady Flow; Streamlines

1.——The average rate of flow of blood in the human aorta is 92 cm^3/s. The radius of the aorta is 1.2 cm. From this, calculate the average speed of the blood in the aorta.

2.——The following table, taken from a firefighter's manual, lists the rate of flow of water (in liters per minute) through a firehose of diameter 3.81 cm connected to a nozzle of given diameter; the listed rate of flow will maintain a pressure of 3.4 atm in the nozzle:

RATE OF FLOW FOR A 3.81-CM HOSE

Rate of flow	Nozzle diameter	Nozzle pressure
95 liters/min	0.95 cm	3.4 atm
190	1.27	3.4
284	1.59	3.4

For each case calculate the velocity of flow (in meters per second) in the hose and in the nozzle.

3.*——A fountain shoots a stream of water vertically upward. Assume that the stream is inclined very slightly to one side so that the descending water does not interfere with the ascending water. The upward velocity at the base of the column of water is 15 m/s.

(a) How high will the water rise?

(b) The diameter of the column of water is 7.0 cm at the base. What is the diameter at the height of 5 m? At the height of 10 m?

Section 12.3: Pressure

4.——A metallic can is filled with a carbonated soft drink. The can is cylindrical, of radius 3.2 cm. The overpressure of the carbon dioxide in the liquid is 0.8 atm. What is the force acting on one of the circular bases of the can?

5.——Within the funnel of a tornado, the air pressure is much lower than normal — about 0.20×10^5 N/m^2 as compared to the normal value of 1.01×10^5 N/m^2. Suppose that such a tornado suddenly envelops a house; the air pressure inside the house is 1.01×10^5 N/m^2 and the pressure outside suddenly drops to 0.20×10^5 N/m^2. This will cause the house to burst explosively. What is the net outward force the pressure exerts on a 12 m $\times$ 3 m wall of this house? Is the house likely to suffer less damage if all the windows and doors are open?

6.——What is the downward force that air pressure (1 atm) exerts on the upper surface of a sheet of paper ($8\frac{1}{2}$ in. $\times$ 11 in.) lying on a table? Why does this force not squash the paper against the table?

7.——At a distance of 7 km from a 1-megaton nuclear explosion, the blast wave has an overpressure of 3×10^4 N/m^2. Calculate the force that this blast wave exerts on the front of a standing man; the frontal area of the man is 0.7 m^2. (The actual force on a man exposed to the blast wave is larger than the result of this simple calculation because the blast wave will be reflected by the man and this leads to a substantial increase of pressure.)

8.——The overpressure in the tires of a 1300-kg automobile is 2.4×10^5 N/m^2. If each tire supports one-fourth the weight of the automobile, what must be the area of each tire in contact with the ground? Pretend the walls of the tires are completely flexible.

9.——The shape of the wing of an airplane is carefully designed so that, when the wing moves through the air, a pressure difference develops between the bottom surface of the wing and the top surface; this supports the weight of the airplane. A fully loaded DC-3 airplane has a mass of 10,900 kg. The (bottom) surface area of its wings is 92 m^2. What is the average pressure difference between the top and bottom surfaces when the airplane is in flight?

10.——Pressure gauges used on automobile tires read the overpressure, that is, the amount of pressure in excess of atmospheric pressure. If a tire has an overpressure of 2.4×10^5 N/m^2 on a day when the barometric pressure is 724 mm-Hg, what will be the overpressure when the barometric pressure increases to 770 mm-Hg? Assume that the pressure in the tire remains constant.

11.——Commercial jetliners have pressurized cabins enabling them to carry passengers at a cruising altitude of 10,000 m. The air pressure at this altitude is 210 mm-Hg. If the air pressure inside the jetliner is 760 mm-Hg, what is the net outward force on a 1 m $\times$ 2 m door in the wall of the cabin?

12.——The baggage compartment of the DC-10 airliner is under the floor of the passenger compartment. Both compartments are pressurized at a normal pressure of 1 atm. In a disastrous accident near Orly, France, in 1974, a faulty lock permitted the baggage compartment door to pop open in flight, depressurizing this compartment. The normal pressure in the passenger compartment then caused the floor to collapse, jamming the control cables. At the time the airliner was flying at an altitude of 3800 m, where the air pressure is 0.64 atm. What was the net pressure force on a 1 m $\times$ 1 m square of the floor?

13.*—In 1654 Otto von Guericke, the inventor of the air pump, gave a public demonstration of air pressure. He took two hollow hemispheres of copper, whose rims fitted tightly together, and evacuated them with his air pump. Two teams of 15 horses each, pulling in opposite directions, were unable to separate these hemispheres. If the evacuated sphere had a radius of 40 cm and the pressure inside it was nearly zero, what force would each team of horses have to exert to pull the hemispheres apart?

Section 12.4: Pressure in a Static Fluid

14.——Porpoises dive to a depth of 500 m. What is the water pressure at this depth?

15.——Figure 12.47 shows a device for the measurement of venous pressure. A hypodermic needle is inserted into the vein, and saline solution from the syringe is pushed into the vertical manometer, until the hydrostatic pressure of the column of saline solution matches the pressure of the blood (there is then no blood flow or saline flow through the needle). If the height of the equilibrium column of saline is 103 mm, what is the blood pressure, in mm-Hg? The density of saline is the same as that of blood.

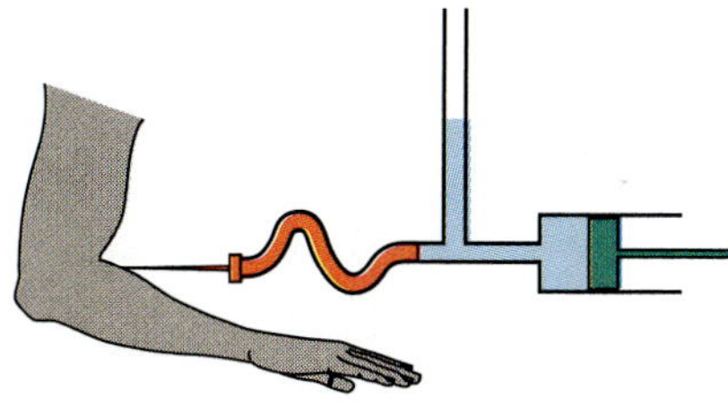

FIGURE 12.47 Measurement of venous pressure.

16.——You can use a barometer as an altimeter. Suppose that when you carry the barometer up a hill, its reading decreases by 8 mm-Hg. What is the height of the hill?

17.*——(a) What is the arterial systolic pressure at your wrist if your arm is hanging straight down? Assume that the pressure is 120 mm/Hg at heart level, and that the wrist is 38 cm below this level.
(b) What is the pressure if you raise your arm straight up? Assume that the wrist is then 78 cm above heart level.

18.*——A diver attempts to breathe by a long snorkel, that is, a tube connecting his mouth to a float on the surface of the water. If the diver stays near the surface, he can breathe through his snorkel. But suppose the diver descends to a depth of 2 m. What is the water pressure on the outside of his chest? What is the air pressure on the inside of his chest? What is the net force on the front of his chest, of area 0.1 m^2? Can he breathe? (Hint: Could you breathe if two hefty football players were sitting on your chest?)

19.*——A tanker is full of oil of density 880 kg/m^3. The flat bottom of the hull is at a depth of 26 m below the surface of the surrounding water. Inside the hull, oil is stored with a depth of 30 m (Fig. 12.48). What is the pressure of the water on the bottom of the hull? The pressure of the oil? What is the net vertical force the pressure exerts on 1 m^2 of bottom?

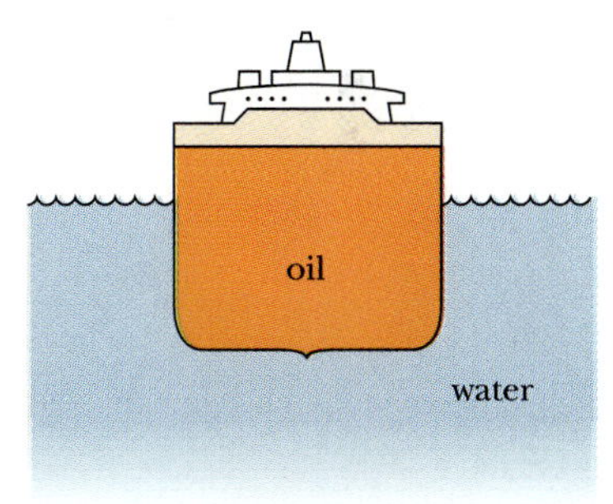

FIGURE 12.48 Cross section of a tanker.

20.*——A pencil sharpener is held to the surface of a desk by means of a rubber "suction" cup measuring 6 cm $\times$ 6 cm. The air pressure under the suction cup is zero and the air pressure above the suction cup is 1 atm.
(a) What is the magnitude of the pressure force pushing the cup against the table?
(b) If the coefficient of static friction between the rubber and the table is 0.9, what is the maximum transverse force the suction cup can withstand?

21.*——(a) Calculate the mass of air in a column of base 1 m^2 extending from sea level to the top of the atmosphere. Assume that the pressure at sea level is 760 mm-Hg and that the value of the acceleration of gravity is 9.81 m/s^2, independent of height.
(b) Multiply your result by the surface area of the Earth to find the total mass of the entire atmosphere.

22.*——Suppose that a zone of low atmospheric pressure (a "low") is at some place on the surface of the sea. The pressure at the center of the "low" is 64 mm-Hg less than the pressure at a large distance from the center. By how much will this cause the water level to rise at the center?

23.*—In a deep bowl a layer of olive oil is floating on top of a layer of water. The thickness of the layer of olive oil is 6.0 cm and the thickness of the layer of water is 5.0 cm. The density of the olive oil is 918 kg/m^3. What is the pressure at the bottom of the bowl?

24.*—A large slick of very viscous oil dumped by a tanker floats on the surface of the sea. The density of the oil is 950 kg/m^3. If the slick is 10 cm thick at its center, how high is the center above the normal level of the sea?

25.*—(a) Under normal conditions the human heart exerts a pressure of 120 mm-Hg on the arterial blood. What is the arterial blood pressure in the feet of a man standing upright? What is the blood pressure in the brain? The feet are 140 cm below the level of the heart; the brain is 40 cm above the level of the heart; the density of human blood is 1060 kg/m^3.

(b) Under conditions of stress the human heart can exert a pressure of up to 190 mm-Hg. Suppose that an astronaut were to land on the surface of a large planet where the acceleration of gravity is 61 m/s^2. Could the astronaut's heart maintain a positive blood pressure in his brain while he is standing upright? Could the astronaut survive?

26.*—A ***"suction" pump*** consists of a piston in a cylinder with a long pipe leading down into a well (Fig. 12.49). What is the maximum height to which such a pump can "suck" water?

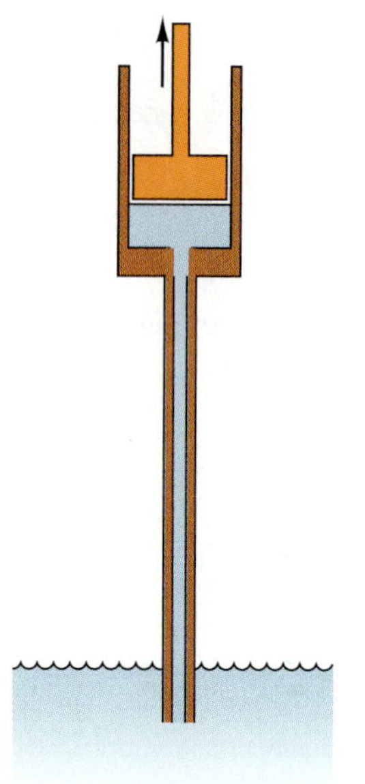

FIGURE 12.49
Suction pump.

Section 12.5: Archimedes' Principle

27.—What is the buoyant force on a human body of volume 7.4×10^{-2} m^3 when totally immersed in air? In water?

28.—Icebergs commonly found floating in the North Atlantic are 30 m high (above the water) and 400 m $\times$ 400 m across. The density of ice is 920 kg/m^3.

(a) What is the total volume of such an iceberg (including the volume below the water)?

(b) What is the total mass?

29.—A rectangular block of wood has a density of 600 kg/m^3. What fraction of this block will be submerged if this block floats in water?

30.—A water trough is 1 m deep, 1 m wide, and 2 m long.

(a) You slip a log of mass 200 kg and density 500 kg/m^3 into this trough. How much will the water level in the trough rise?

(b) You slip a rock of mass 200 kg and density 2000 kg/m^3 into the trough. How much will the water level rise?

31.*—You can walk on water if you wear very large shoes shaped like boats. Calculate the length of the shoes that will support you; assume that each shoe is 0.3 m $\times$ 0.3 m in cross section.

32.*—A gasoline barrel, made of steel, has a mass of 20 kg when empty. The barrel is filled with 0.12 m^3 of gasoline with a density of 730 kg/m^3. Will the full barrel float in water? Neglect the volume of the steel.

FIGURE 12.50
Fully loaded and empty supertankers.

33.*—The supertanker *Globtik London* has a mass of 2.2×10^8 kg when empty and it can carry up to 4.4×10^8 kg of oil when fully loaded. Assume that the shape of its hull is approximately that of a rectangular parallelepiped 380 m long, 60 m wide, and 40 m high (Fig. 12.50).

(a) What is the draft of the empty tanker; that is, how deep is the hull submerged in the water? Assume that the density of (sea) water is 1.02×10^3 kg/m^3.

(b) What is the draft of the fully loaded tanker?

34.*—A supertanker has a draft (submerged depth) of 30 m when in sea water (density $\rho = 1.02 \times 10^3$ kg/m^3). What will be the draft of this tanker when it enters a river estuary with fresh water (density $\rho = 1.00 \times 10^3$ kg/m^3)? Assume that the sides of the ship are vertical.

35.*——In an experiment to measure the average density of the tissues of the human body, a man was first weighed in air, and then he was weighed while immersed in water. In air, his weight was 816.2 N; in water (after he expelled as much air from his lungs as he could, which caused him to sink), his weight was 42.2 N. The estimated residual volume of air in his lungs was 1.20 liters. What density can you deduce from these data?

36.*——A rectangular barge measures 5 m wide, 20 m long, and 3 m deep. Its mass is 5×10^4 kg when empty. What is the maximum load that this barge can carry, before the water swamps its deck?

37.*——The mean density of the tissues in an average human body is 1.071×10^3 kg/m^3 (this value of the density excludes the volume of the lungs; that is, the air in the lungs is not counted as part of the body). What volume of air must a man of mass 80 kg take into his lungs if he wants to remain (barely) afloat in water?

38.**——A "gold" bracelet is made of a gold and copper alloy. The bracelet has a mass of 0.0900 kg. To determine the amount of gold in the alloy, you weigh the bracelet when immersed in water. Its weight is then 0.820 N. What is the percentage of gold in the alloy? The density of gold is 19,300 kg/m^3 and the density of copper is 8900 kg/m^3.

Section 12.6: Bernoulli's Equation

39.——Suppose that a brisk wind of 18 m/s is blowing over your house. The outside pressure is 1 atm. What is the pressure inside your house? (Hint: Consider a streamline that passes through your house, entering and leaving through some crack, and assume that the speed of flow inside your house is nearly zero.)

40.——If you blow a stream of air with a speed of 7 m/s out of your mouth, what must be the overpressure in your mouth?

41.——A thin stream of water emerges vertically from a small hole on the side of a water pipe and ascends to a height of 1.2 m. What is the pressure inside the pipe? Assume the water inside the pipe is nearly static.

42.——A tank full of water, 2.0 m deep, has a circular opening of radius 1.0 cm at its bottom. What is the rate (in m^3 per second) at which water flows out of this opening?

43.*——A pump has a horizontal intake pipe at a depth of 8.0 m below the surface of a lake. What is the maximum speed for steady flow of water into this pipe? (Hint: Inside the pipe the maximum speed of flow corresponds to zero pressure.)

44.*——To fight a fire on the fourth floor of a building, firemen want to use a hose of diameter 6.4 cm to shoot 950 liters/min of water to a height of 12 m.

(a) With what minimum speed must the water leave the nozzle of the firehose if it is to ascend 12 m?

(b) What pressure must the water have inside the firehose? Ignore friction.

45.*——Streams of water from firehoses are sometimes used to disperse crowds. Suppose that the stream of water emerging from the firehose described in Example 1 impinges horizontally on a man. The collision of the water with the man is totally inelastic.

(a) What is the force that the stream of water exerts on the man, that is, what is the rate at which the water delivers momentum to the man?

(b) What is the rate at which the water delivers energy?

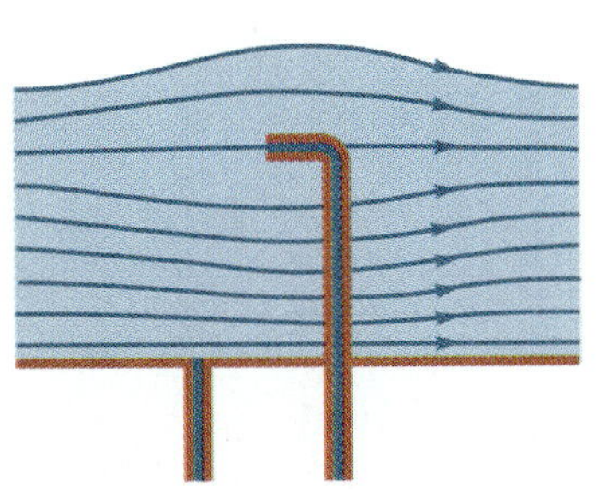

FIGURE 12.51
Pitot tube.

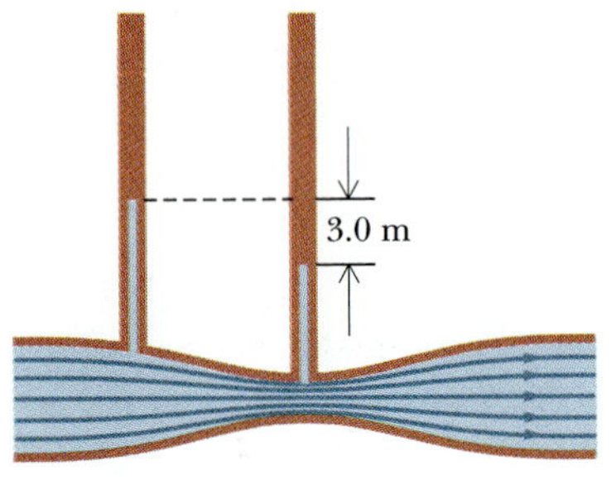

FIGURE 12.52
Venturi flowmeter.

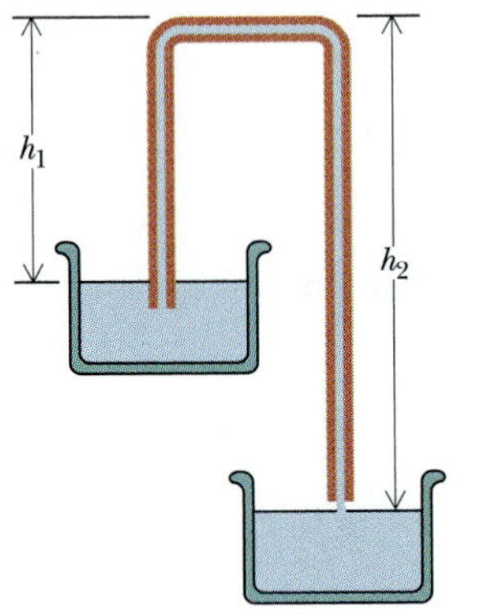

FIGURE 12.53
Siphon.

*—An airplane has a wing of area (bottom area) 92 m^2. When in flight, the speed of the air along the bottom of the wing is 85 m/s, and along the top of the wing it is 95 m/s. Approximately what lift force does this wing provide? The density of air is 1.29 kg/m^3.

*—The ***Pitot tube*** is used for the measurement of the flow speeds of fluids, such as the flow speed of air past the fuselage of an airplane. It consists of a bent tube protruding into the airstream (Fig. 12.51) and another tube opening flush with the fuselage. The pressure difference between the air in the two tubes can be measured with a manometer. Show that in terms of the pressure p_1 and p_2 in the two tubes, the speed of airflow is

$$v = \sqrt{\frac{2(p_2 - p_1)}{\rho}}$$

where ρ is the density of air. (Hint: p_1 is simply the static air pressure. To find p_2, consider a streamline reaching the opening of the bent tube; at this point, the velocity of the air is zero.)

48.*—A Venturi flowmeter in a water main of diameter 30 cm has a constriction of diameter 10 cm. Vertical pipes are connected to the water main and to the constriction (Fig. 12.52); these pipes are open at their upper ends and the water level within them indicates the pressure at their lower ends. Suppose that the difference in the water levels in these two pipes is 3.0 m. What is the velocity of flow in the water main? What is the rate (in liters per second) at which water is delivered?

49.**—The pump of a fire engine draws 1100 liters of water per minute from a pond at the same level as the pump and discharges this water into a firehose of diameter 6.4 cm at an overpressure of 5.5×10^5 N/m^2. In the absence of friction, what power (in hp) does this pump require? (Hint: Consider the work the pump must do to push the water into the hose, against the pressure.)

50.**—A ***siphon*** is an inverted U-shaped tube that is used to transfer liquid from a container at a high level to a container at a low level (Fig. 12.53).

(a) Using the lengths shown in Fig. 12.53 and the density ρ of the liquid, find a formula for the speed with which the liquid emerges from the lower end of the siphon.
(b) Find the pressure at the highest point of the siphon.
(c) By setting this pressure equal to zero, find the maximum height h_2 with which the siphon can operate.

Section 12.7: Viscous Flow

51.—The human aorta has a radius of 1.2 cm; this radius is approximately constant over most of its length. The average rate of flow of blood in the aorta is 9.2×10^{-5} m^3/s. What is the pressure loss attributable to the viscosity of blood in a length of 10 cm?

52.—A horizontal firehose of diameter 6.4 cm delivers water at the rate of 1.3×10^{-2} m^3 per second. What is the pressure drop per meter of length of the firehose caused by the viscous friction in the water?

53.—A typical capillary in the human circulatory system has a radius of 3.0×10^{-6} m and a length of 7.5×10^{-4} m. The rate of flow of the blood is 8.5×10^{-15} m^3/s. What is the pressure difference between one end of the capillary and the other?

54.—For a blood transfusion, a 4-cm-long needle of inner radius 0.2 mm is inserted into a patient's vein. The needle is connected by tubing to a bag of whole blood suspended above the patient. How far above the patient should the bag be suspended if the rate of delivery of the blood is to be 10 cm^3 per minute? The pressure of the blood in the vein is 8 mm-Hg (overpressure), and the temperature of the blood in the needle is 37°C (body temperature). Assume that the viscous friction in the tubing can be neglected.

55.*—Figure 12.54 shows an apparatus used for the practical measurement of the viscosity of a fluid. A solid metallic cylinder is placed inside a slightly larger cylindrical metallic can, and the gap between them is filled with the fluid. The cylinder is then rotated about its axis at a constant angular speed, which makes the fluid flow in circles around the axis. The torque on the cylinder required to maintain this rotation is measured, and from this torque the viscosity can be calculated. Suppose that the radius of the solid cylinder is 9.0 cm and that of the can is 9.5 cm, so the thickness of the layer of fluid is 0.5 cm. The depth of the fluid is 25 cm. If a torque of 7.2×10^{-2} N·m is required to maintain rotation at one revolution per second, what is the value of η? (Hint: In this problem, the outer can plays the role of the stationary plate, and the inner rotating cylinder plays the role of the moving plate.)

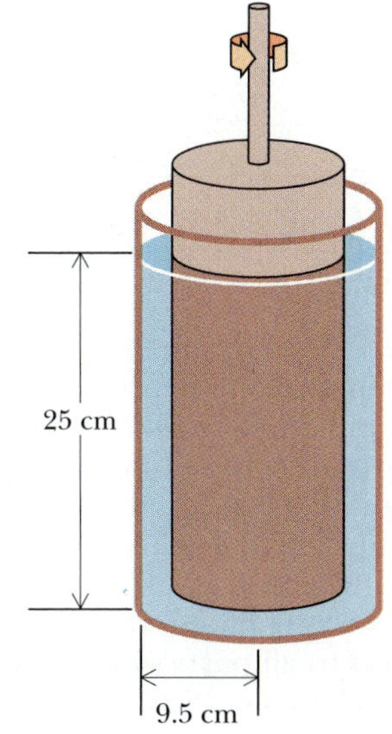

FIGURE 12.54

Apparatus for the measurement of viscosity.

56.*—A tube of radius r branches into two equal tubes of smaller radius r' (see Fig. 12.55). A viscous liquid flows within the tube. If the pressure drop per unit length is to be the same along the first tube and the two branching r tubes, what must be the ratio of the radii r and r'?

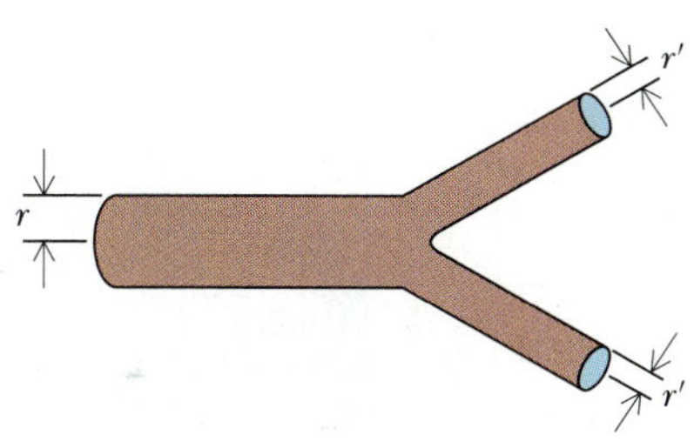

FIGURE 12.55

A tube branches into two tubes.

57.*—At the site of a fire, a firehose of diameter 3.8 cm and length 30 m has been laid from a pump on the ground to the third floor of a house, 10 m high. The pump delivers 0.4 m^3 of water per minute at an overpressure of 6×10^5 N/m^2 into the lower end of the firehose. What is the overpressure at the upper end of the firehose? Take into account both the height difference and the viscosity of water.

58.**—A long straight pipe of radius r leads downward at a slant from a large tank full of viscous liquid (see Fig. 12.56). If the viscosity and the flow rate are not excessively large, then gravity can provide the force necessary to overcome the viscous resistance, and the fluid can flow down the pipe even when there is *no pressure difference* between the ends of the pipe. Show that for such flow with zero pressure difference, the downward angle of the pipe must be given by

$$\sin\theta = \frac{8\eta}{\pi r^4 \rho g}\frac{\Delta V}{\Delta t}$$

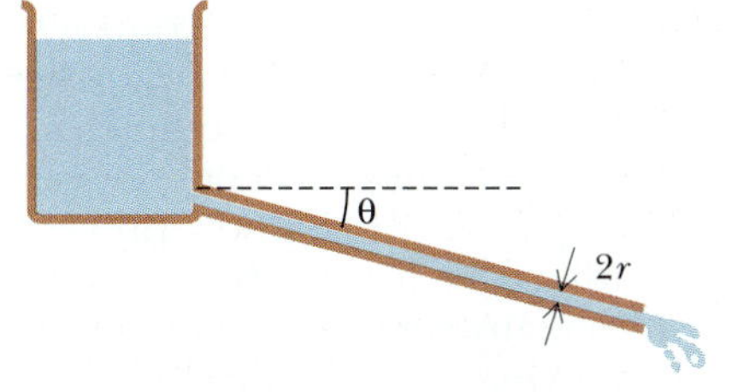

FIGURE 12.56

A long straight pipe connected to a tank. The pipe is directed downward at an angle θ.

PHYSICS IN ACTION

BLOOD FLOW

We are accustomed to thinking of the heart as a pump, but the heart is actually two pumps working in tandem. The first pump, on the right side of the heart, receives oxygen-depleted, dark-red blood from all parts of the body via the vena cava, and pumps this blood to the lungs. The second pump, on the left side of the heart, receives oxygen-rich, bright red blood from the lungs, and pumps this blood to all parts of the body via the aorta. Figure 1 is a schematic diagram of the cardiovascular circuit.

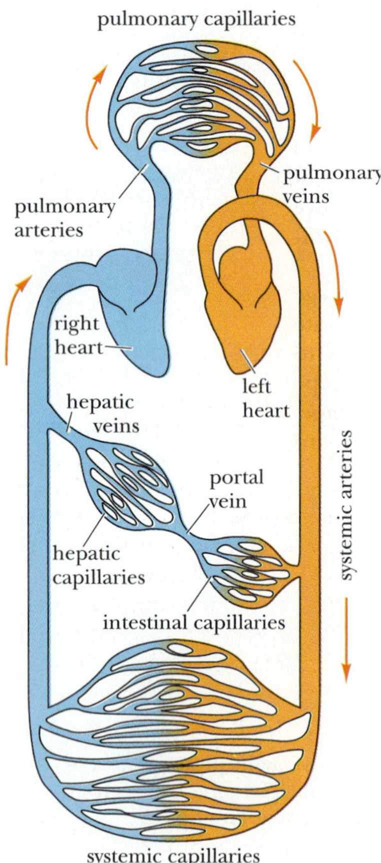

FIGURE 1 Schematic diagram of the cardiovascular circuit. The right and the left halves of the heart are two pumps, and they are shown separated in the diagram. The diagram shows the heart and blood vessels as you would see them in somebody else's body, looking into the body from the front.

The circuit is divided into two segments: the pulmonary segment, from the right heart, to the lungs, to the left heart; and the systemic segment, from the left heart, to all parts of the body, to the right heart.

Each of the pumps in the heart has two chambers. The upper chamber is called the ***atrium,*** and the lower chamber the ***ventricle*** ("belly"). The atria accumulate the blood arriving at the heart and then inject it into the ventricles. The ventricles eject the blood from the heart into the arteries; they perform most of the work required for the pumping. Each pump has two flap valves (see Fig. 2). The right pump has the tricuspid valve (between the atrium and the ventricle) and the pulmonary valve (at the beginning of the pulmonary artery). The left pump has the mitral valve (between the atrium and the ventricle) and the aortic valve (at the beginning of the aorta). These flap valves operate passively; their leaflets bend open to permit the flow of blood in the forward direction; but they flip shut when there is an incipient flow in the backward direction.

We can best understand the operation of the chambers and the valves of the heart by tracing the flow of a parcel of blood through the heart and around the entire circulatory system. The blood arrives at the heart via the inferior and the superior vena cava and enters the right atrium, where it accumulates until the atrium begins to contract and the tricuspid valve opens. The blood then flows into the relaxed right ventricle, filling and expanding it. When the ventricle begins to contract a moment later, the leaflets of the tricuspid valve flip into their shut position, while the leaflets of the pulmonary valve open. The contraction of the ventricle then propels the blood into the pulmonary artery. When the ventricle completes its contraction and begins to relax, the pulmonary valve shuts behind the parcel of blood, preventing any backflow. Within the lungs, the pulmonary artery branches into many small arterioles and even smaller capillaries. While passing through these pulmonary capillaries, the blood sheds its load of carbon dioxide and absorbs oxygen. The capillaries connect to veins, which merge into the main pulmonary veins. These veins carry the blood to the left atrium of the heart, where it accumulates until the mitral valve

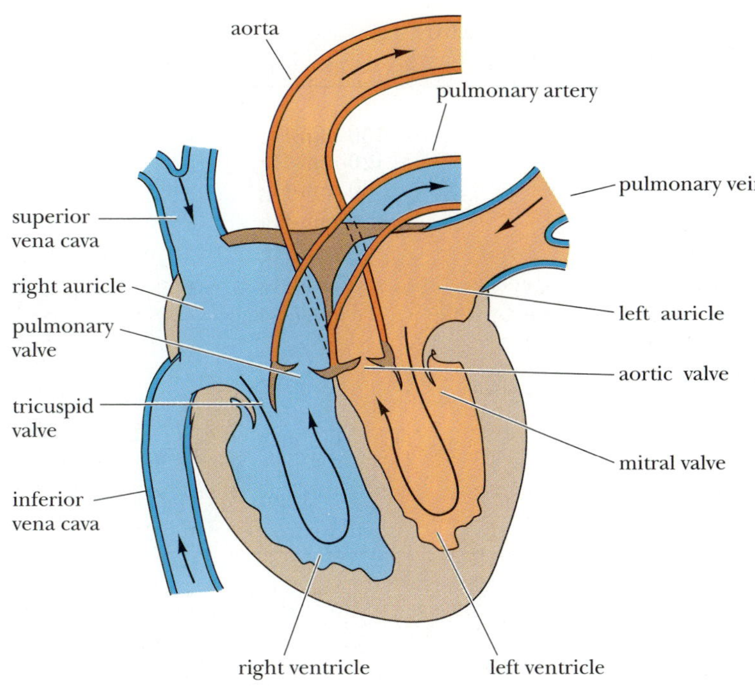

FIGURE 2 The chambers and the valves of the heart.

opens. The blood then flows into the relaxed left ventricle, filling and expanding it. The ventricle begins to contract, and the mitral valve shuts, while the aortic valve opens. The ventricle completes its contraction, pushing the blood into the aorta. When the ventricle begins to relax, the aortic valve shuts, preventing backflow of the parcel of blood. The aorta branches out into arteries, arterioles, and ultimately into capillaries, which distribute the blood throughout the body. While passing through these capillaries, the blood delivers oxygen to the cells and absorbs carbon dioxide dumped by the cells during their metabolic activity. The systemic capillaries merge into veins, and finally the blood returns to the heart via the inferior and the superior vena cava, completing the circuit.

The walls of the heart consist of layers of muscle, wound several times around the atria and the ventricles in a complicated arrangement. The layers of muscle around the ventricles are thicker than those around the atria. The heart muscle pumps blood by its cyclic contractions and relaxations. The contracted phase of the heart is called ***systole,*** and the relaxed phase is called ***diastole.*** Each contraction begins in the walls of the atria, and this squeezes the blood from the atria into the ventricles; a moment later, the walls of the ventricles contract, squeezing the blood out of the heart into the arteries. Since the walls of both atria contract jointly, and the walls of both ventricles also contract jointly, the right and the left pumps in the heart always operate in unison. The layers of muscle around the left ventricle are much thicker than those around the right ventricle (see Fig. 3). The left ventricle generates the highest pressures and does most of the mechanical work.

Under conditions of rest, the heart typically goes through 70 cycles of contraction and relaxation per minute, that is, 70 heartbeats per minute. Each contraction lasts about 0.3 s, and each relaxation about 0.5 s. However, under conditions of heavy exercise or stress, the heart rate may be as high as 180 heartbeats per minute (see Table 1).

The rate of flow of blood through each side of the heart is 5.5 liters per minute, at rest. Trained athletes attain a rate of flow of up to 35 liters per minute, during heavy exercise. The total volume of

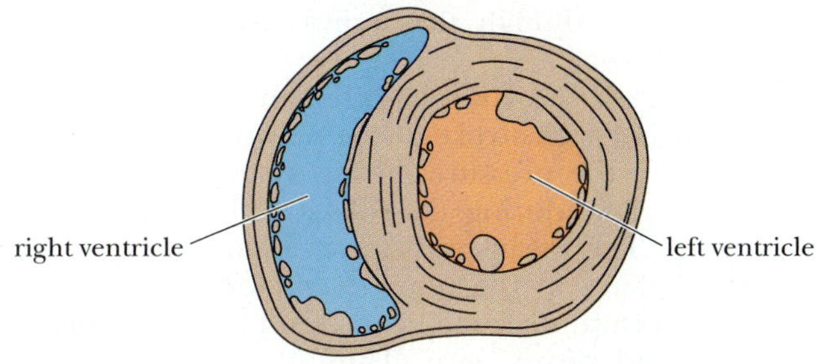

FIGURE 3 Cross section through the heart. Note the heavy musculature of the left ventricle.

TABLE 1: SOME DATA FOR THE HUMAN CIRCULATORY SYSTEM	
Pressure in aorta while resting,	
mean	100 mm-Hg
systole	120 mm-Hg
diastole	80 mm-Hg
Pressure in vena cava	$\simeq$0 mm-Hg
Pressure in pulmonary artery, mean	15 mm-Hg
Pressure in pulmonary vein	8 mm-Hg
Volume pumped with each heartbeat	78 cm^3
Heartbeat rate	70/min
Radius of aorta	1.2 cm
Speed of blood in aorta,	
mean	0.2 m/s
maximum	0.7 m/s
Radius of capillary	3 μm
Speed of blood in capillary, mean	0.3 mm/s

blood in the human body is 5 to 6 liters, and therefore a rate of flow of 5.5 liters per minute implies that, on the average, a parcel of blood takes just about a minute to travel around the complete cardiovascular circuit. Since the blood flows in a closed circuit, without any loss of blood, the rate of flow must be the same everywhere along the circuit. Thus, if 5.5 liters per minute flow through the caval veins, then the same amount must flow through the lungs, and through the aorta, and through the systemic capillaries. However, the speed of flow is different at different points of the circuit, as required by the equation of continuity (see Fig. 4). The mean speed is about 0.2 m/s in the aorta, but it is much lower in the capillaries, only about 0.3 mm/s. As we saw in Example 12.2, this low speed in the capillaries comes about because there are very many capillaries in the human body, about 10^{10}, and the total cross-sectional area of all these capillaries together is much larger than the cross-sectional area of the aorta, even though the individual capillaries are very narrow.

The pressure of the blood in the arteries fluctuates during each stroke of the pump. At the base of the aorta, the pressure during systole reaches 120 mm-Hg. and during diastole it falls to about 80 mm-Hg. The mean pressure is about 100 mm-Hg. The pressure in the veins fluctuates much less; in the vena cava, near the heart, the pressure is nearly steady and nearly zero. Thus, the mean pressure difference across the systemic segment of the circulatory system is about 100 mm-Hg. A pressure difference is required to balance the viscous resistance of blood and to maintain a more or less steady rate of flow. Blood has a viscosity of $2.1 \times 10^{-3}\ \text{s}\cdot\text{N/m}^2$, about three times that of water. According to Poiseuille's equation, the pressure drop along a tube, such as an artery or a vein, is directly proportional to the viscosity, the length, and the rate of flow, and inversely proportional to the fourth power of the radius of the tube,

$$p_1 - p_2 = \frac{8l\eta}{\pi r^4}\frac{\Delta V}{\Delta t} \tag{1}$$

Most of the pressure drop actually occurs in the small arteries, or arterioles, that connect to the capillaries. In the capillaries themselves, the pressure drop is not so large, even though these are the smallest vessels in the circulatory system, with a radius of only 3×10^{-6} m. The arterioles have a larger radius than the capillaries, but they also have a larger length and a higher rate of flow, and because of this they contribute a larger pressure drop than the capillaries (see Fig. 5).

The net power lost to viscous friction in the systemic segment of the circulatory system can be calculated by taking the difference between the mechanical power supplied by the heart to the arterial blood and the mechanical power returned by the venous blood to the heart. When the heart pumps a parcel of blood of volume ΔV into the

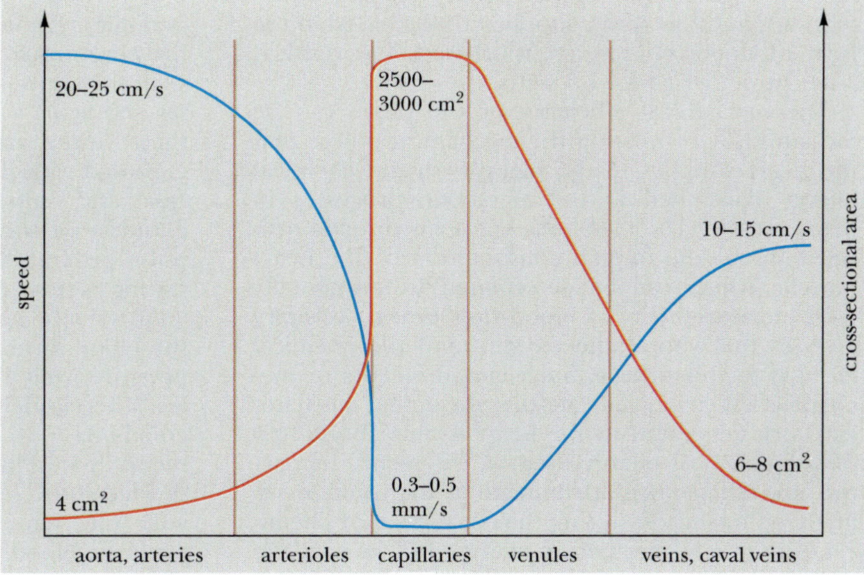

FIGURE 4 Mean speed of flow at different positions along the systemic segment of the circulatory system.

aorta at a mean pressure P_a, the work it does is $P_a \Delta V$. And when the blood in the vena cava pushes an equal parcel of blood ΔV into the heart at a pressure P_v, the work it does is $P_v \Delta V$; since the venous pressure is nearly zero, this work is nearly zero. The net rate of work, or the power lost to viscous friction in the systemic segment, is therefore

$$\frac{\Delta W}{\Delta t} = P_a \frac{\Delta V}{\Delta t} \tag{1}$$

$$= 100 \text{ mm-Hg} \times \frac{5.5 \text{ liters}}{\text{minute}}$$

$$= 1.3 \times 10^4 \text{ N/m}^2 \times \frac{5.5 \times 10^{-3} \text{ m}^3}{60 \text{ s}} = 1.2 \text{ watts}$$

The heart must supply at least this much power to keep the blood flowing in the systemic segment of the circulatory system. To this power we must add the power supplied to the blood flowing in the pul-

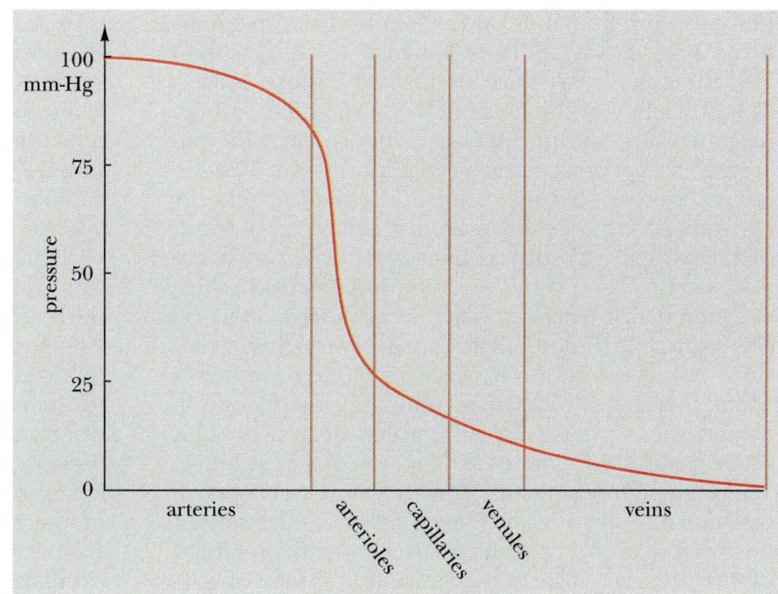

FIGURE 5 Mean pressure at different positions along the systemic segment of the circulatory system. Note the steep pressure drop in the arterioles.

monary segment of the circulatory system, about 0.2 watt, and the power supplied to the blood in the form of kinetic energy, about 0.1 watt. The result is a net power of about 1.5 watts.

The mechanical efficiency of the heart can be measured by comparing the mechanical power that the heart supplies to the blood with the chemical energy consumed in the metabolic processes in heart muscle. The metabolic energy is directly proportional to the oxygen consumption of the heart muscle, which can be determined experimentally from measurements of blood flow in the coronary arteries that supply the oxygen (and the chemical energy) to this muscle. Such measurements of mechanical efficiency indicate that the actual mechanical work generated by the heart is only about 10% of the chemical energy used by the heart. In part this low efficiency is attributable to the usual losses incurred by muscle during the conversion of chemical energy into mechanical energy. But in part the low efficiency results from energy wasted within the heart muscle to maintain a tension in the muscle, a tension that is required to balance the pressure of blood in the chambers of the heart. The tension must be maintained regardless of how fast or how slow the heart pushes the blood out of the chamber. This means that in addition to the external work on the blood, the heart muscle must expend some internal work, analogous to the isometric work that your biceps expends internally when you tense the muscle while holding the arm steady.

Note that from Eq. (2) we can immediately understand why an abnormally high blood pressure places severe demands on the heart. If the mean arterial pressure is 180 mm-Hg instead of 100 mm-Hg, then the heart must deliver 80% more power. The efficiency of the heart is about the same at high pressure and at low pressure; hence an 80% increase in the power delivered requires an 80% increase in the oxygen and the chemical energy consumed. High blood pressure occurs as a result of physical stress, such as heavy exercise. But it can also occur as a result of emotional stress; thus, emotional disturbances place a *mechanical* load on the heart. Chronic high blood pressure (hypertension) is a common occurrence in old age, and it is often correlated with an accumulation of plaque in the arteries (atherosclerosis), which decreases their inner radius and increases the viscous resistance. The high blood pressure and the increased viscous resistance tend to balance each other — the rate of flow of the blood remains unaffected. Chronic high blood pressure also occurs as a side effect of various diseases, and it sometimes occurs for no discernible reason. Whatever its origin, high blood pressure means that the heart must work harder.

In the above discussion we have treated the arteries and veins as rigid, inert tubes. But the walls of these vessels are actually elastic, and they are also endowed with thin layers of muscle, which can contract and reduce the diameter of the vessel. The elasticity of the arteries smooths out the pressure pulse generated by the heart. The arteries distend during systole (you can feel this in the arteries at your wrist or neck), and their elastic force squeezes the blood during diastole, maintaining an adequate pressure while the blood flows out into the capillaries. The elasticity of the veins permits them to accommodate to the somewhat irregular, sluggish blood flow within them. In your limbs, the veins are divided into compartments by one-way valves, and your muscular movements squeeze the veins and push the blood from one compartment to the next. Thus, to a large extent, the blood is pushed along these veins by incidental movements of the limbs. If you sit or stand for an extended interval of time without moving your legs, an excess of blood accumulates in the veins, and the pressure of this pooled blood causes fluid to leak out of the capillaries, which swells your feet and ankles, and often produces some discomfort. The elasticity of the veins not only permits them to adapt to an increase in the volume of blood, but also a decrease. For instance, if your body suffers a loss of blood through an injury, the veins simply shrink and adapt to the smaller blood volume. Thus, by their elasticity, the arteries act as a "pressure buffer," and the veins as a "volume buffer" for the circulatory system.

Control over the distribution of blood flow within the body is provided by the contraction of layers of muscle in the walls of arterioles and venules and also by the contraction of small rings of muscle (pre-capillary sphincters) at the entrances of capillaries. These muscles contract or relax in response to signals from the central nervous system, or in response to hormones or other chemicals in the blood, thereby directing the blood flow to the parts of the body where the need is greatest. The capillaries play a large role in this control of the blood distribution. At rest, only about 25% to 35% of the capillaries in the circulatory system are open; the remainder are shut off, but they open when the tissues develop a need for more oxygen. The capillaries in the skin also serve to regulate the heat loss of the body. For instance, when your body is in a cold environment, the blood flow through the skin capillaries is shut off to prevent heat loss. And when your body is in a warm environment, the skin capillaries are widely dilated to bring heat to the surface where you can dissipate it by sweating. Vasodilator drugs, such as alcohol, will open the skin capillaries even when your body is in a cold environment. This produces a pleasant rush of heat to the skin, but sabotages the mechanism by which your body is attempting to protect itself against heat loss.

REFERENCES

Folkow, B., and Neil, E., *Circulation* (Oxford University Press, New York, 1971).

Rushmer, R. F., ed., *Cardiovascular Dynamics* (W. B. Saunders, Philadelphia, 1970).

Shepherd, J. T., and Vanhoutte, P. M., *The Human Cardiovascular System* (Raven Press, New York, 1979).

Vogel, S., *Vital Circuits: On Pipes, Pumps, and the Workings of Circulatory Systems* (Oxford University Press, New York, 1991).

CHAPTER 13

This photograph, taken from the Space Shuttle, shows that the atmosphere of the Earth is concentrated in a relatively thin layer (about 100 km deep) parallel to the Earth's surface. When a gas is not confined by a vessel, it usually disperses; but the gas of the atmosphere is held in place by the Earth's gravity.

THE IDEAL GAS

The particles of a gas — that is, the individual molecules of the gas — are well separated, and they fly about quite independently. The gas would disperse if it were not restrained by the forces exerted by the wall of the container. The gas molecules collide only occasionally with each other, and if we ignore these intermolecular collisions, we can regard the gas as a system of free particles. This means that motion of the gas particles is very simple: they move with uniform velocity on straight lines, except when they collide with the walls of the container (Fig. 13.1). However, in spite of this simplicity of the motion of the gas particles, we cannot keep track of their motion in detail, because there are too many of them. For instance, one cubic centimeter of air contains about 2.7×10^{19} molecules. We have no way of ascertaining the initial positions and velocities of each individual molecule; and even if we had, the calculation of the simultaneous motions of such an enormous number of molecules is far beyond the capabilities of even the fastest conceivable computer.

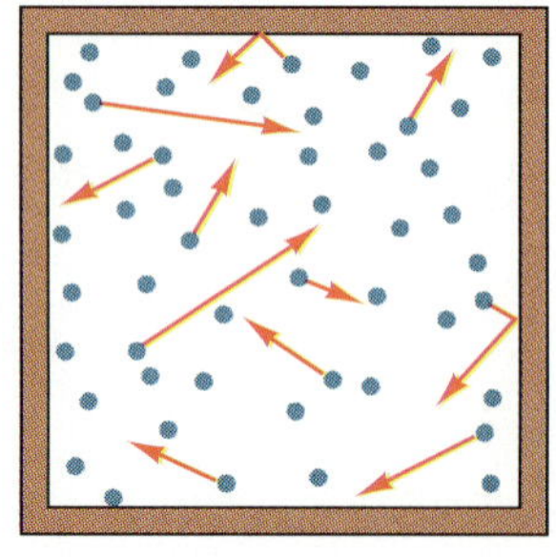

FIGURE 13.1

Random motion of the molecules of gas in a container. Between collisions, the motion of each molecule is along a straight line.

In lack of a microscopic description involving the individual positions and velocities of the molecules of the gas, we must be satisfied with a macroscopic description involving just a few variables that characterize the *average* conditions in the volume of gas. For example, we will regard the mass as such a macroscopic variable. The assertion that the mass of a cubic centimeter of air is 1.29×10^{-6} kg does not mean that every cubic centimeter of air has exactly the same mass as every other cubic centimeter. The air molecules wander about and sometimes a few more, sometimes a few less, molecules will be within any given cubic centimeter of space. We can say only what the mass will be on the average. In practice, the fluctuations of the mass in a cubic centimeter of air are quite small — typically about $10^{-8}\%$.

Other macroscopic parameters characterizing the average conditions of a gas that can be measured with large-scale laboratory instruments are the number of moles, the volume, the density, the pressure, and the temperature.

In this chapter we will study the macroscopic properties of gases, and we will see how these macroscopic properties are related to the average microscopic properties of the molecules of the gas.

13.1 THE IDEAL-GAS LAW

The pressure, volume, and temperature of a gas obey some simple laws. Before we state these laws, let us recall the definition of pressure from the preceding chapter. Imagine that the gas is divided into small adjacent cubical volumes. The pressure is the magnitude of the force that one of these cubes exerts on an adjacent cube, or on an adjacent wall, divided by the area of one face of the cube; that is, the pressure is the magnitude of the force per unit area. As we know from Section 12.4, the pressure is the same throughout the entire volume of a container of gas. (Within a container of gas, gravity causes a small decrease of pressure from the bottom to the top, but this decrease of pressure can usually be ignored.)

Consider now a given amount of gas, that is, a given mass of gas. Instead of reckoning the amount of gas by mass, we will find it convenient to reckon the amount by the number of moles, a concept borrowed from chemistry. ***One mole of a chemical element (or a chemical compound) is that amount that contains exactly as many atoms (or molecules) as there are atoms in 12* g *of the isotope carbon-12.*** This isotope is the most common of the forms of carbon found in nature. For the definition of the mole, carbon has been adopted as the standard chemical element with which others are to be compared because carbon forms compounds with a large variety of other elements, and this helps in making comparisons among elements. ***The number of atoms (or molecules) in one mole is called Avogadro's number.*** Experimentally, this number can be determined by counting the number of molecules in one mole of some crystal, such as diamond (C) or salt (NaCl): since the molecules in the crystal are arranged in evenly spaced rows and columns, they are fairly easy to count. The experimental

◀ **Mole**

data yield the following value for Avogadro's number.

Avogadro's number ▶

$$N_A = 6.02 \times 10^{23} \text{ molecules/mole} \tag{1}$$

Atomic mass ▶

The mass of one mole of a chemical element is called the atomic mass. A table of atomic masses is included in Appendix 8. According to this table, one mole of carbon atoms (C) has a mass of 12.0 grams, one mole of hydrogen atoms (H) has a mass of 1.0 gram, one mole of oxygen atoms (O) has a mass of 8.0 grams, one mole of hydrogen molecules (H_2) has a mass of 2.0 grams, one mole of oxygen molecules (O_2) has a mass of 16.0 grams, and so on. The advantage of measuring the amount of a chemical substance by moles rather than by kilograms is that one mole of any substance has exactly the same number of atoms (or molecules) as one mole of any other substance — when we reckon the amount of a gas by moles, we are reckoning the amount by the number of molecules.

Ideal-gas law ▶

Suppose we take n moles of a gas and place them in a container of volume V at a temperature T. The gas will then exert a pressure p. Experiments show that — to a good approximation — ***the pressure p, the volume V, and the temperature T of the n moles of gas are related by the ideal-gas law:***

$$pV = nRT \tag{2}$$

Universal gas constant ▶

Kelvin ▶

Here R is the ***universal gas constant*** with the value

$$R = 8.31 \text{ J/K} \tag{3}$$

From the ideal-gas law we can calculate one of the three quantities that characterize the state of the gas (pressure, volume, temperature) if the other two are known.

The temperature in Eq. (2) is measured on the ***absolute temperature scale,*** and the unit of temperature is the ***kelvin,*** or ***degree absolute,*** abbreviated K. We have not previously given the definition of temperature because Eq. (2) plays a dual role: it is a law of physics and also serves for the definition of temperature. This is by now a familiar story — in Chapter 3 we already came across laws that play such a dual role.

We will give the details concerning the definition of temperature in the next section. For now it will suffice to note that the freezing point of water corresponds to a temperature of 273.15 K, and the boiling point of water corresponds to 373.15 K; hence, there is an interval of exactly 100 K between the freezing and the boiling points. The zero of temperature on the absolute scale is the absolute zero, $T = 0$ K . According to Eq. (2), the pressure of the gas vanishes at this point. Actually, the gas will liquefy or even solidify before the absolute zero of temperature can be reached; when this happens, Eq. (2) becomes inapplicable.

William Thomson, Lord Kelvin, *1824–1907, British physicist and engineer. Besides inventing the absolute temperature scale, he made many other contributions to the theory of heat. He was first to state the principle of dissipation of energy incorporated in the Second Law of Thermodynamics.*

The ideal-gas law is a simple relation between the macroscopic parameters that characterize a gas. At normal densities and pressures, real gases obey this law quite well; but if a real gas is compressed to an excessively high density, then its behavior will deviate from this law. We will not deal with any examples of gases under such extreme conditions, and we will therefore neglect any deviations from Eq. (2). A gas that obeys Eq. (2) exactly is called an ***ideal gas.*** The ideal gas is a limiting case of a real gas when the density and the pressure of the latter tend to zero. The ideal gas may be thought of as consisting of atoms of infinitesimal size, exerting no forces on each other or on the walls of the container, except for instantaneous impact forces exerted during collisions.

◀ **Ideal gas**

EXAMPLE 1. What is the volume of 1 mole of air at a temperature of 273 K and a pressure of 1 atm, called ***standard temperature and pressure*** (STP)?

◀ **Standard temperature and pressure (STP)**

SOLUTION: In metric units, 1 atm is 1.01×10^5 N/m². For 1 mole ($n = 1$), the ideal-gas law then gives us

$$V = \frac{nRT}{p} = \frac{1 \times 8.31\ \text{J/K} \times 273\ \text{K}}{1.01 \times 10^5\ \text{N/m}^2}$$

$$= 2.24 \times 10^{-2}\ \text{m}^3 = 22.4\ \text{liters}$$

COMMENTS AND SUGGESTIONS: Note that it makes no difference whether the gas in this calculation is air or something else — one mole of *any* gas at STP has a volume of 22.4 liters (see Fig. 13.2)

EXAMPLE 2. Suppose you heat 1 kg of water and convert it into steam at the boiling temperature of water, 373 K, and at normal atmospheric pressure, 1 atm, or 1.01×10^5 N/m². What is the volume of the steam?

SOLUTION: The molecular mass of water (H_2O) is the sum of the atomic masses of two hydrogen atoms and one oxygen atom, that is, 1.0 g + 1.0 g + 16.0 g = 18.0 g. Thus 1 mole of steam is 18.0 g, and the number of moles in 1 kg of steam is

$$n = \frac{1\ \text{kg}}{18.0\ \text{g}} = \frac{1000\ \text{g}}{18.0\ \text{g}} = 55.6\ \text{moles}$$

According to Eq. (2), the volume of the steam is then

$$V = \frac{nRT}{p} = \frac{55.6 \times 8.31\ \text{J/K} \times 373\ \text{K}}{1.01 \times 10^5\ \text{N/m}^2}$$

$$= 1.71\ \text{m}^3$$

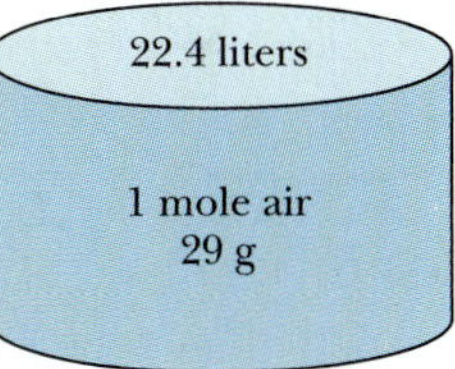

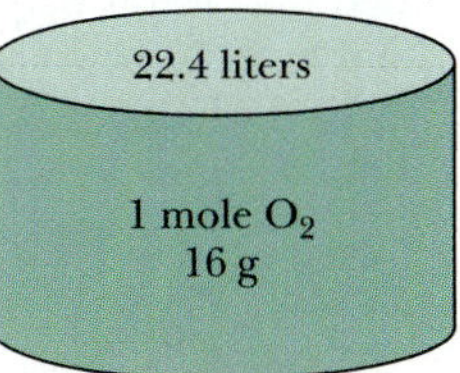

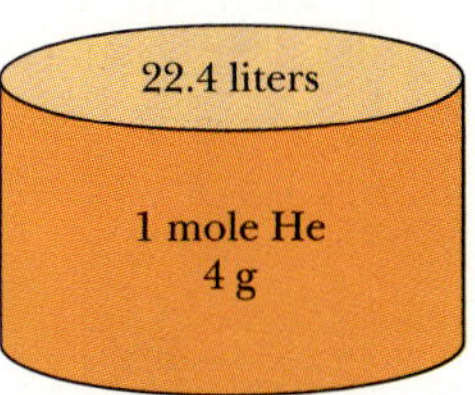

FIGURE 13.2

At a standard temperature of 273 K and pressure of 1 atm (STP conditions) a mole of any kind of gas occupies a volume of 22.4 liters.

Boyle's Law ▶

The ideal-gas law incorporates two laws: ***Boyle's Law*** and ***Gay-Lussac's Law.*** Boyle's Law asserts that if the temperature is held constant, then the product of pressure and volume must remain constant as a given amount of gas is compressed or expanded,

$$pV = [\text{constant}] \text{ for } T = [\text{constant}] \tag{4}$$

Gay-Lussac's Law ▶

Gay-Lussac's Law asserts that if the pressure is held constant, the ratio of volume to temperature remains constant as a given amount of gas is heated or cooled,

$$V/T = [\text{constant}] \text{ for } p = [\text{constant}] \tag{5}$$

The relation given by Eq. (4) can be tested experimentally by placing a sample of gas in a cylinder with a movable piston, surrounded by some substance at a fixed temperature, say, a water bath (see Fig. 13.3). The relation given by Eq. (5) can be tested with a similar cylinder–piston arrangement with a fixed weight mounted on the piston and a heat source below the cylinder (Fig. 13.4). Together, the experimental tests of Eqs. (4) and (5) amount to a test of the ideal-gas law.

Note that the ideal-gas law also incorporates another law: if the volume is held constant, then the ratio of pressure to temperature remains constant,

$$P/T = [\text{constant}] \tag{6}$$

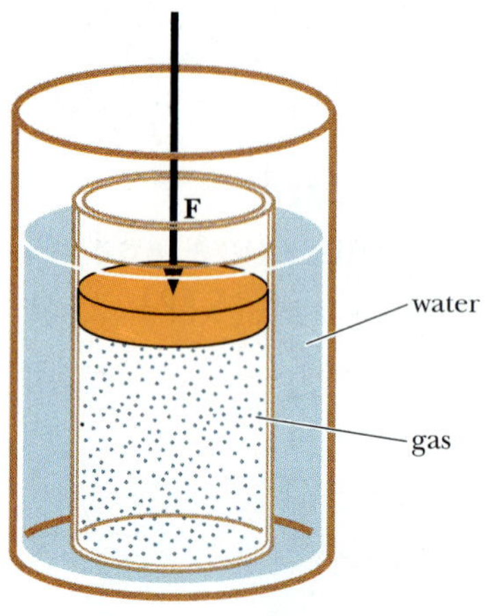

FIGURE 13.3

Compression of a gas by a force applied to the piston. The container with the gas is surrounded by a water bath that keeps the temperature constant.

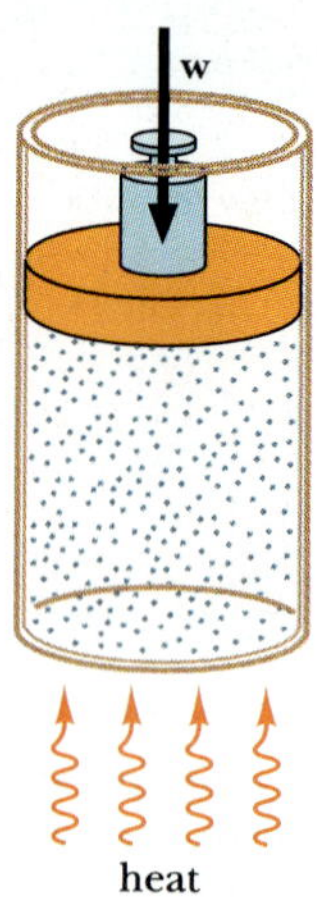

FIGURE 13.4

Expansion of a gas kept at constant pressure by a weight on the piston. The gas is being heated.

EXAMPLE 3. Early in the morning, at the beginning of a trip, the tires of an automobile are cold (280 K) and their air is at a pressure of 3.0 atm. Later in the day, after a long trip on hot pavements, the tires are hot (330 K). What is the pressure? Assume that the volume of the tires remains constant.

SOLUTION: At constant volume, the pressure is proportional to the temperature [see Eq. (2) or Eq. (6)]. Hence

$$p_2 = p_1 \times \frac{T_2}{T_1}$$

$$= 3.0 \text{ atm} \times \frac{330 \text{ K}}{280 \text{ K}} = 3.5 \text{ atm}$$

Note that the pressure gauges for automobile tires are commonly calibrated to read ***overpressure,*** that is, the excess above atmospheric pressure. Thus, the pressure gauge would read 2.0 atm in the morning, and 2.5 atm later in the day (if we assume that the atmospheric pressure remains constant at 1 atm).

The ideal-gas law can also be written in terms of the number of molecules, instead of the number of moles. The number of molecules equals the number of moles times the number of molecules per mole, that is, times Avogadro's number:

$$N = N_A n = 6.02 \times 10^{23}\ n \tag{7}$$

With $n = N/N_A$, the ideal-gas law becomes

$$pV = \frac{N}{N_A} RT \tag{8}$$

or

$$pV = NkT \tag{9}$$

where

$$k = \frac{R}{N_A} = \frac{8.31\ \text{J/K}}{6.02 \times 10^{23}} = 1.38 \times 10^{-23}\ \text{J/K} \tag{10}$$

◀ Boltzmann's constant

The constant k is called ***Boltzmann's constant.*** As we will see, this constant tends to make an appearance in equations relating macroscopic quantities (such as p or V) to microscopic quantities (such as the number N of molecules).

Robert Boyle, *1627–1691, English experimental physicist. He invented a new air pump, with which he performed the experiments on gases that led to discovery of the law named after him.*

Ludwig Boltzmann, *1844–1906, Austrian theoretical physicist. He made crucial contributions in the kinetic theory of gases and in statistical mechanics.*

EXAMPLE 4. (a) What is the number of molecules in 1 cm^3 of air at a temperature of 273 K and a pressure of 1 atm (STP)? (b) What is the mass of air in 1 cm^3 under these conditions? The mean molecular mass of air is 29.0 g.*

SOLUTION: (a) In metric units, 1 atm is 1.01×10^5 N/m^2 and 1 cm^3 is 10^{-6} m^3. Hence, by Eq. (9), the number of molecules in 1 cm^3 is

$$N = \frac{pV}{kT} = \frac{1.01 \times 10^5\ \text{N/m}^2 \times 10^{-6}\ \text{m}^3}{1.38 \times 10^{-23}\ \text{J/K} \times 273\ \text{K}}$$

$$= 2.68 \times 10^{19}\ \text{molecules}$$

Note that this result is valid for any kind of gas — the number of molecules in 1 cm^3 of any kind of gas under STP conditions if 2.68×10^{19}.

---

* Air consists, on the average, of 76% nitrogen, 23% oxygen, and 1% argon by mass.

(b) The number of moles in 1 cm^3 equals the number of molecules divided by Avogadro's number:

$$\frac{2.68 \times 10^{19}}{N_A} = \frac{2.68 \times 10^{19}}{6.02 \times 10^{23}} = 4.45 \times 10^{-5}$$

Since the mass per mole is 29.0 g, the amount of mass is

$$4.45 \times 10^{-5} \times 29.0 \text{ g} = 1.29 \times 10^{-3} \text{ g} = 1.29 \times 10^{-6} \text{ kg}$$

This implies that the mass in 1 m^3 of air is 1.29 kg, in agreement with the density of air listed in Table 12.1.

The mass of one atom or molecule is the "atomic mass" (or the "molecular mass") divided by N_A. For instance, the mass of one carbon atom is

$$\frac{12 \text{ g}}{N_A} = \frac{12 \text{ g}}{6.02 \times 10^{23}} = 1.99 \times 10^{-23} \text{ g} = 1.99 \times 10^{-26} \text{ kg}$$

Masses of atom are often measured in terms of the ***atomic mass unit*** (u), which is exactly 1/12 of the mass of a carbon atom,

$$1 \text{ atomic mass unit} = 1 \text{ u} = \frac{1.99 \times 10^{-26} \text{ kg}}{12}$$

that is,

$$1 \text{ u} = 1.66 \times 10^{-27} \text{ kg}$$

Note that with this definition of the atomic mass unit, the "atomic mass," or the number of grams in one mole, necessarily has the same numerical value as the mass of one atom expressed in u.

13.2 THE TEMPERATURE SCALE

To use the ideal-gas law for the definition of temperature, we take a fixed amount of some gas, such as helium, and place it an airtight, nonexpanding container, such as a Pyrex glass bulb. According to Eq. (2), the pressure for a gas kept in such a constant volume is directly proportional to the temperature. Thus, a simple measurement of pressure gives us the temperature.

Triple point of water ▶

To calibrate the scale of this thermometer, we must choose a standard reference temperature. The standard adopted in the metric system of units is the temperature of the ***triple point of water, that is, the temperature at which water, ice, and water vapor coexist when placed in a closed***

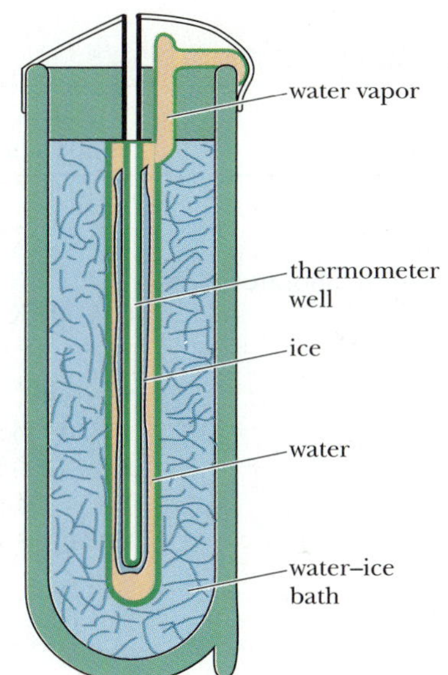

FIGURE 13.5

Triple-point cell of the National Institute of Standards and Technology. The inner tube contains water, water vapor, and ice in equilibrium.

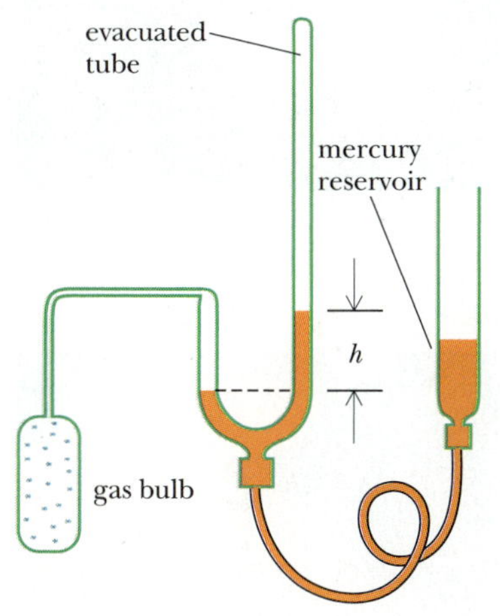

FIGURE 13.6

Constant-volume gas thermometer.

vessel. Figure 13.5 shows a triple-point cell used to achieve the standard temperature. This standard temperature has been assigned the value of 273.16 kelvin, or 273.16 K. If the bulb of the gas thermometer is placed in thermal contact with this cell so that it attains a temperature of 273.16 K, it will read some pressure p_{tri}. If the bulb is then placed in thermal contact with some body at an unknown temperature T, it will read a pressure p which is greater or smaller than p_{tri} by some factor. The unknown temperature T is then greater or smaller than 273.16 K by this same factor; for instance, if the pressure p is half as large as p_{tri}, then $T = \frac{1}{2} \times$ 273.16 K. ***The temperature scale defined by this procedure is called the ideal-gas temperature scale, or the absolute temperature scale.***

◀ **Absolute temperature scale**

When connecting a pressure gauge to the bulb of gas, we must take special precautions to ensure that the operation of the pressure gauge does not alter the volume available to the gas. Figure 13.6 shows a device that will serve our purposes; this device is called a ***constant-volume gas thermometer.*** The pressure gauge used in this thermometer consists of a closed-tube manometer; one branch of the manometer is connected to the bulb of gas, and the other branch consists of a closed, evacuated tube. The difference h in the heights of the levels of mercury in these two branches is proportional to the pressure of the gas. The manometer is also connected to a mercury reservoir. During the operation of the thermometer, this reservoir must be raised or lowered so that the level of mercury in the left branch of the manometer tube always remains at a constant height; this keeps the gas in the bulb at a constant volume. The bulb of the thermometer may be put in contact with any body whose temperature we wish to measure, and the pressure registered by the manometer then gives us the absolute temperature.

◀ **Constant-volume gas thermometer**

Table 13.1 lists some examples of temperatures of various bodies.

For everyday and industrial use, the ideal-gas thermometer is somewhat inconvenient, and is often replaced by mercury-bulb thermometers, bimetallic strips, electrical-resistance thermometers, or thermocouples (see

(a)

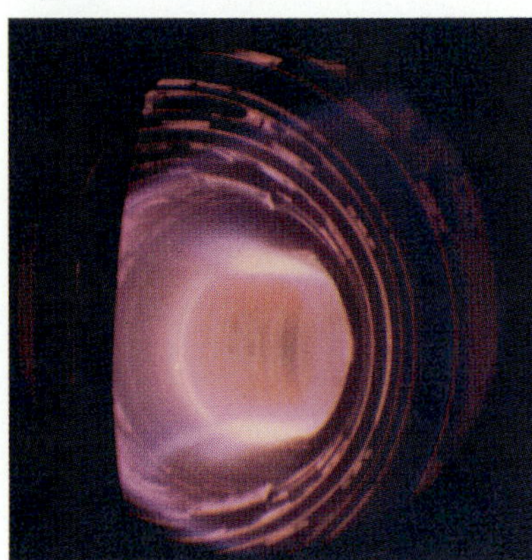

(b)

TABLE 13.1 SOME TEMPERATURES		
	Kelvin temperature	Celsius temperature
Interior of hottest stars	10^9 K	10^9°C
Center of H-bomb explosion	10^8 K	10^8°C
Highest temperature attained in laboratory (plasma) **(a)**	6×10^7 K	6×10^7°C
Center of Sun	1.5×10^7 K	1.5×10^7°C
Surface of Sun	4.5×10^3 K	4.2×10^3°C
Center of Earth	4×10^3 K	3.7×10^3° C
Acetylene flame	2.9×10^3 K	2.6×10^3°C
Melting of iron **(b)**	1.8×10^3 K	1.5×10^3°C
Melting of lead	6.0×10^2 K	3.3×10^2°C
Boiling of water	373 K	100°C
Human body	310 K	37°C
Surface of Earth (average)	287 K	14°C
Freezing of water	273 K	0°C
Liquefaction of nitrogen **(c)**	77 K	−196°C
Liquefaction of hydrogen	20 K	−253°C
Liquefaction of helium **(d)**	4.2 K	−269°C
Interstellar space	3 K	−270°C
Lowest temperature attained in laboratory	3×10^{-8} K	$\simeq -273.15$°C

(c)

(d)

Figs. 13.7–13.10). These must be calibrated in terms of the ideal-gas thermometer if they are to read absolute temperature.

Although the absolute temperature scale is the only scale of fundamental significance, several other temperature scales are in practical use. The ***Celsius scale*** (formerly known as the centigrade scale) is shifted 273.15 K relative to the absolute scale,

Celsius scale ▶

$$T_C = T - 273.15°C \tag{11}$$

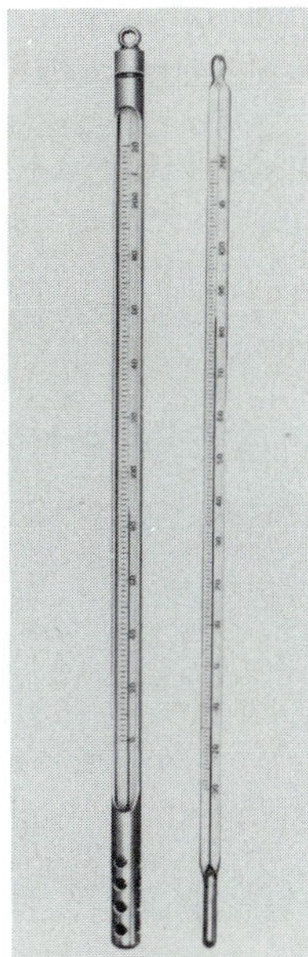

FIGURE 13.7

Mercury-bulb thermometers. The thermal expansion of the mercury in the bulb causes it to rise in the capillary tube, indicating the temperature.

FIGURE 13.8

Bimetallic-strip thermometer. The stem of the thermometer contains a helix consisting of joined bands of different metals. With an increase of temperature, the bands expand by different amounts, which tends to coil the helix, and turns the pointer.

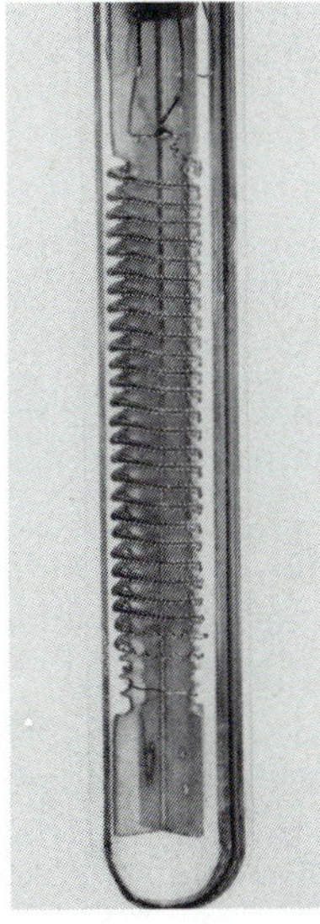

FIGURE 13.9

Platinum-resistance thermometer. The resistance that the fine coiled wire of platinum offers to an electric current serves as an indicator of the temperature.

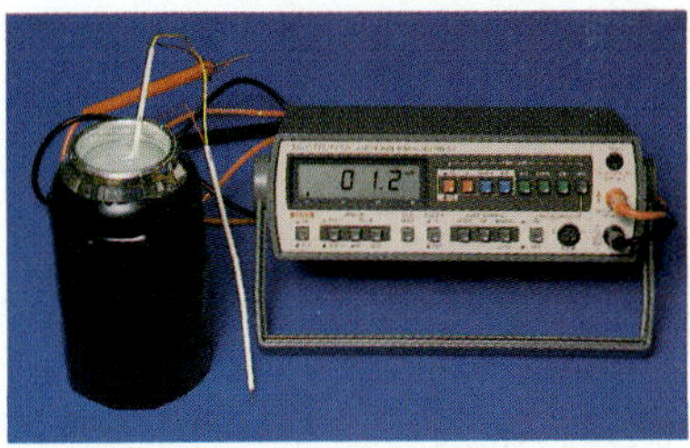

FIGURE 13.10

Thermocouple thermometer. Each of the three thermocouples shown here consists of one wire of platinum and one wire of platinum–rhodium alloy joined at their ends. If the other ends (not shown) of the wires are connected to an external electric circuit, the thermocouple will generate a weak electric current, of a magnitude that depends on the temperature.

Note that on the Celsius scale, absolute zero is at $-273.15°C$. The triple point of water is then at $0.01°C$, the freezing point at $0°C$, and the boiling point at $100°C$.

◀ **Fahrenheit scale**

The ***Fahrenheit scale,*** in common use in the United States, is shifted relative to the Celsius scale and, furthermore, uses degrees of smaller size, each degree Fahrenheit corresponding to $\frac{5}{9}$ degree Celsius:

$$T_F = \tfrac{9}{5}\,T_C + 32°F \qquad (12)$$

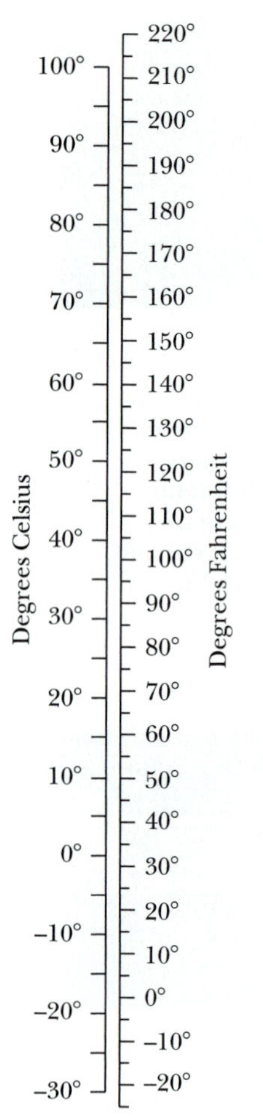

FIGURE 13.11
The correspondence between the Fahrenheit scale and the Celsius scale.

On this scale, the freezing point of water is at 32°F, and the boiling point at 212°F. Figure 13.11 can be used for a rough conversion between the Fahrenheit and Celsius scales.

Keep in mind that neither the Celsius temperature nor the Fahrenheit temperature is compatible with the ideal-gas law. The temperature must always be converted to absolute temperature before it is substituted into the ideal-gas law.

EXAMPLE 5. The normal temperature of the human body is 98°F. Express this in °C.

SOLUTION: According to Eq. (12),

$$T_C = \frac{5}{9}(T_F - 32°\text{F}) = \frac{5}{9}(98°\text{F} - 32°\text{F}) = 37°\text{C}$$

13.3 A MICROSCOPIC VIEW OF PRESSURE

The pressure of a gas against the walls of its container is due to the impacts of the molecules on the walls. We can calculate this pressure by considering the average motion of the molecules of gas. We will assume that the container is a cube of side L, that all the gas molecules have the same mass m, and that the gas molecules collide only with the walls but not with each other, and that the collisions are elastic. These assumptions are not necessary, but they simplify the calculations.

Figure 13.12 shows the container filled with gas molecules. The motion of each molecule can be resolved into x, y, and z components. Consider one molecule, and consider the component of its motion in the x direction. The component of the velocity in this direction is v_x, and the magnitude of this velocity remains constant, since the collisions with the wall are elastic. The time that the molecule takes to move from the face of the cube at $x = 0$ to $x = L$ and back to $x = 0$ is

$$t = 2L/v_x \tag{13}$$

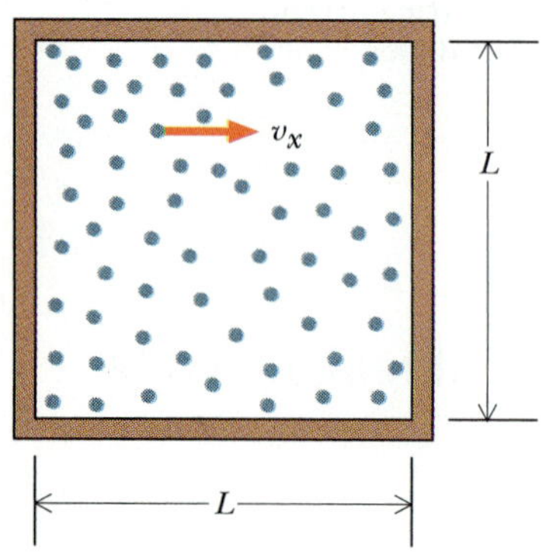

FIGURE 13.12
Gas molecules in a container.

This is therefore the time between one collision with the face at $x = 0$ and the next collision with the same face. When the molecule strikes the face, its x velocity is reversed from $-v_x$ to $+v_x$. Hence, during each collision at $x = 0$, the x momentum of the molecule changes from $-mv_x$ to $+mv_x$, a net change of $+2mv_x$. The average rate at which the molecule transfers momentum to the face at $x = 0$ is then the momentum change per collision divided by the time between collisions,

$$\frac{2mv_x}{t} = \frac{2mv_x}{2L/v_x} = \frac{mv_x^2}{L} \tag{14}$$

This is therefore the average force that the impacts of this one molecule exert on the wall. To find the total force exerted by the impacts of all the N molecules, we must multiply the force given in Eq. (14) by N; and to find the pressure, we must divide by the area L^2 of the face of the cube. This leads to a pressure

$$p = \frac{N}{L^2} \frac{mv_x^2}{L} \tag{15}$$

or, in terms of the volume $V = L^3$ of the gas,

$$p = \frac{Nmv_x^2}{V} \tag{16}$$

In this calculation we made the implicit assumption that all the molecules have the same velocity. This is, of course, not true; the molecules of the gas have a distribution of velocities — some have high velocities, and some have low velocities. To account for this spread of velocities, we must replace the force in Eq. (14) due to one given molecule by the average over all the molecules. Consequently, we must replace v_x^2 by an average over all the molecules in the container. We will designate the average by an overbar, $\overline{v_x^2}$. Equation (16) then becomes

$$p = \frac{Nm\overline{v_x^2}}{V} \tag{17}$$

To proceed further, we note that on the average, molecules are just as likely to move in the x, y, or z direction. Hence, the average values of v_x^2, v_y^2, and v_z^2 are equal:

$$\overline{v_x^2} = \overline{v_y^2} = \overline{v_z^2} \tag{18}$$

The sum of the squares of the components of the velocity is the square of the magnitude of the velocity,

$$\overline{v_x^2} + \overline{v_y^2} + \overline{v_z^2} = \overline{v^2} \tag{19}$$

and therefore each of the terms on the left side of Eq. (19) must equal $\frac{1}{3}\overline{v^2}$. We can then write Eq. (17) as

$$p = \frac{Nm\overline{v^2}}{3V} \tag{20}$$

Let us now compare this result for the pressure with the ideal-gas law [Eq. (9)], according to which the pressure is

$$p = \frac{NkT}{V} \tag{21}$$

The agreement between these two expressions (20) and (21) for the pressure demands

$$\frac{Nm\overline{v^2}}{3V} = \frac{NkT}{V}$$

or, if we cancel the factor N/V on both sides,

$$\frac{m\overline{v^2}}{3} = kT \qquad (22)$$

This equation shows that the average square of the molecular speed is proportional to the temperature. ***The square root of $\overline{v^2}$ is called the root-mean-square speed, or the rms speed, and it is usually designated by v_{rms}.*** If we divide both sides of Eq. (22) by $m/3$ and extract the square root of both sides, we find

Root-mean-square ▶ speed

$$v_{rms} = \sqrt{\overline{v^2}} = \sqrt{\frac{3kT}{m}} \qquad (23)$$

The root-mean-square speed may be regarded as the typical speed of the molecules of the gas. According to Eq. (23), this typical speed is directly proportional to the square root of the temperature and inversely proportional to the square root of the mass — at a given temperature, the molecules of the lowest mass have the highest speeds. Incidentally: There are other ways of calculating a typical speed; for example, we may want to know the average of all the molecular speeds, or the most probable of all the molecular speeds. These other typical speeds turn out to be approximately the same as v_{rms}, but their calculation requires some knowledge of the distribution of molecular speeds.

EXAMPLE 6. What is the root-mean-square speed of nitrogen molecules in air at 0°C? Of oxygen molecules?

SOLUTION: The molecular mass of N_2 molecules is 28.0 g, twice the atomic mass of nitrogen. Hence, the mass of one molecule is

$$m = \frac{28.0\text{ g}}{N_A} = \frac{28.0 \times 10^{-3}\text{ kg}}{6.02 \times 10^{23}} = 4.65 \times 10^{-26}\text{ kg}$$

and Eq. (23) yields

$$v_{rms} = \sqrt{\frac{3kT}{m}} \qquad (24)$$

$$= \sqrt{\frac{3 \times 1.38 \times 10^{-23}\ \text{J/K} \times 273\ \text{K}}{4.65 \times 10^{-26}\ \text{kg}}} = 493\ \text{m/s}$$

Likewise, for O_2 molecules, the molecular mass is 32.0 g, twice the atomic mass of oxygen. Proceeding as in the case of nitrogen, we find that the mass of one oxygen molecule is 5.32×10^{-26} kg, and v_{rms} is 461 m/s.

COMMENTS AND SUGGESTIONS: Note that the rms speed of nitrogen molecules is slightly larger than that of oxygen molecules. In general, Eq. (23) shows that the rms speed is inversely proportional to the square root of the mass of the molecule — at a given temperature, the molecules of lowest mass have the highest speeds.

13.4* THE INTERNAL ENERGY OF AN IDEAL GAS

Since a gas is a system of particles, its energy is the sum of the energies of all these particles. To calculate the energy of the gas, we begin with the kinetic energy of one molecule. According to Eq. (23), the average value of the square of the speed of a molecule in an ideal gas is

$$\overline{v^2} = 3\,\frac{kT}{m}$$

Hence the average kinetic energy of a molecule is

$$\tfrac{1}{2}\,m\overline{v^2} = \tfrac{1}{2} \times 3kT = \tfrac{3}{2}kT \tag{25}$$

If we multiply this by the total number of molecules, we obtain the total translational kinetic energy of all the molecules jointly. In an ideal gas, the molecules exert no forces on one another, and hence there is no intermolecular potential energy. The kinetic energy is then the total energy,

$$E = K = \tfrac{3}{2}NkT$$

◂ **Thermal energy of ideal gas**

This formula tells us how much energy is stored in the microscopic thermal motion of the gas. This energy is called the ***thermal energy*** of the gas. It is sometimes also called the ***internal energy,*** because it is stored (and hidden) in the microscopic motions in the interior of the gas, instead of being manifest in an overall macroscopic translational motion of the entire body

* This section is optional.

of gas. Since $Nk = nR$ [see Eqs. (7) and (10)], we can also write the formula for the energy as

$$E = \tfrac{3}{2}nRT \tag{27}$$

Note that in our calculation we have assumed that the molecules behave like pointlike particles — each molecule has translational kinetic energy, but no rotational kinetic energy or other energy associated with motions within the molecule. Monoatomic gases, such as helium, argon, and krypton, behave in this way, and Eq. (27) is applicable to these gases.

EXAMPLE 7. What is the thermal kinetic energy in 1 kg of helium gas at 0°C? How much extra energy must be supplied to this gas to increase its temperature to 60°C (at constant volume)?

SOLUTION: The atomic mass of helium is 4.0 g; hence the number of moles in 1 kg of helium is

$$n = \frac{1 \text{ kg}}{4.0 \text{ g}} = 250 \text{ moles}$$

According to Eq. (27), the energy at a temperature of 273 K is

$$E = \tfrac{3}{2}nRT = \tfrac{3}{2} \times 250 \times 8.31 \text{ J/K} \times 273 \text{ K}$$

$$= 8.5 \times 10^5 \text{ J}$$

The extra energy needed to increase the temperature to 60°C is

$$\Delta E = \tfrac{3}{2}nR\Delta T = \tfrac{3}{2} \times 250 \times 8.31 \text{ J/K} \times 60 \text{ K}$$

$$= 1.9 \times 10^5 \text{ J}$$

FIGURE 13.13
A diatomic molecule represented as two pointlike particles joined by a rod.

Diatomic gases, such as N_2 and O_2, store additional amounts of energy in the internal motions of the atoms within each molecule. The molecules of these gases may be regarded as two pointlike particles rigidly connected together (a dumbbell; see Fig. 13.13). If such a molecule collides with another molecule or with the wall of the container, it will usually start rotating about its center of mass. We therefore expect that, on the average, an appreciable fraction of the energy of the gas will be in the form of this kind of rotational kinetic energy.

The molecule may rotate about either of the two axes through the center of mass perpendicular to the line joining the atoms (see Fig. 13.13). If the moments of inertia about these axes are I_1 and I_2, and if the corresponding angular velocities are ω_1 and ω_2, then the kinetic energy for these two rotations is

$$\tfrac{1}{2}I_1\omega_1^2 + \tfrac{1}{2}I_2\omega_2^2 \tag{28}$$

The average of this kinetic energy is

$$\tfrac{1}{2}I_1\overline{\omega_1^2} + \tfrac{1}{2}I_2\overline{\omega_2^2} \tag{29}$$

where, as in the preceding section, the overbars denote the average over all the molecules of the gas.

To discover the values of these average rotational energies, let us return to Eq. (25) and write it in terms of the x, y, and z components of velocity:

$$\tfrac{1}{2}m\overline{v_x^2} + \tfrac{1}{2}m\overline{v_y^2} + \tfrac{1}{2}m\overline{v_z^2} = \tfrac{3}{2}kT \tag{30}$$

We know that the average x, y, and z speeds are equal. Hence, in Eq. (30) the three terms on the left side are equal, and the value of each must be $\tfrac{1}{2}kT$, so the sum of these three terms is $\tfrac{3}{2}kT$:

$$\tfrac{1}{2}m\overline{v_x^2} = \tfrac{1}{2}kT, \quad \tfrac{1}{2}m\overline{v_y^2} = \tfrac{1}{2}kT, \quad \tfrac{1}{2}m\overline{v_z^2} = \tfrac{1}{2}kT \tag{31}$$

It turns out that this is true not only for the components of the translational motion, but also for the components of the rotational motion. The general result, which we will not prove, is known as the ***equipartition theorem:***

◀ **Equipartition theorem**

> ***Each translational or rotational component of the random thermal motion of a molecule has an average kinetic energy of*** $\tfrac{1}{2}kT$.

According to this theorem, each of the terms in Eq. (29) has a value $\tfrac{1}{2}kT$. The total average rotational energy is then kT, and when we add this to the translational kinetic energy $\tfrac{3}{2}kT$, we obtain a total kinetic energy

$$kT + \tfrac{3}{2}kT = \tfrac{5}{2}kT \tag{32}$$

for one molecule. The energy of all the molecules of the diatomic gas taken together is then

$$E = \tfrac{5}{2}NkT \tag{33}$$

Note that in this calculation we have ignored the possibility of rotation about the longitudinal axis of the molecule. This means we have ignored the rotation of the atoms about an axis through them, just as we have ignored this kind of rotation of the atoms in a monoatomic gas. Furthermore, we have ignored the vibrational motion of the atoms of the diatomic molecule. The interatomic forces do not really hold these atoms in a rigid embrace; rather, the forces act somewhat like springs (see Example 9.3), and they permit a restricted back-and-forth vibration of the atoms about their equilibrium positions. The reason why we have ignored these motions in our calculation lies beyond the realm of classical physics; it lies in the realm of quantum physics. There it is established that the rotation of atoms about their own axis and the vibration of atoms in a molecule do not occur unless the temperature is rather high, 400°C or more. As we will see in Section 14.5, the energies calculated from Eqs. (27) and (33) actually agree quite well with experiments, provided we do not exceed this temperature limit.

SUMMARY

- **Ideal-gas law:** $pV = nRT$, $R = 8.31$ J/K
  $pV = NkT$, $k = 1.38 \times 10^{-23}$ J/K

- **Temperature scales:**
  Absolute: T
  Celsius: $T_C = T - 273.15°C$
  Fahrenheit: $T_F = \frac{9}{5}T_C \times 32°F$

- **Root-mean-square speed:** $v_{rms} = \sqrt{\overline{v^2}} = \sqrt{\dfrac{3kT}{m}}$

- **Thermal energy of ideal monoatomic gas:** $K = \frac{3}{2}NkT$

- **Equipartition theorem:** Each translational or rotational component of the random thermal motion of a molecule has an average kinetic energy of $\frac{1}{2}kT$.

QUESTIONS

1.–Why do meteorologists usually measure the temperature in the shade rather than in the sun?

2.–Why are there no negative temperatures on the absolute temperature scale?

3.–The temperature of the ionized gas in the ionosphere of the Earth is about 2000 K, but the density of this gas is extremely low, only about 10^5 gas particles per cubic centimeter. If you were to place an ordinary mercury thermometer in the ionosphere, would it register 2000 K? Would it melt?

4.–The temperature of the intergalactic space is 3 K. How can empty space have a temperature?

5.–At the airport of La Paz, Bolivia, one of the highest in the world, pilots find it preferable to take off early in the morning or late at night, when the air is very cold. Why?

6.–If you release a rubber balloon filled with helium, it will rise to a height of a few thousand meters and then remain stationary. What determines the height reached? Is there an optimum pressure to which you should inflate the balloon to reach greatest height?

7.–How can you use a barometer as an altimeter?

8.–Explain why a real gas behaves like an ideal gas at low densities but not at high densities.

9.–Helium and neon approach the behavior of an ideal gas more closely than do any other gases. Why would you expect this?

10.–If you open a bottle of perfume in one corner of a room, it takes a rather long time for the smell to reach the opposite corner (assuming that there are no air currents in the room). Explain why the smell spreads slowly, even though the typical speeds of perfume molecules are 300–400 m/s.

11.–Ultrasound waves of extremely short wavelength cannot propagate in air. Why not?

12.–Prove that it is impossible for all of the molecules in a gas to have the same speeds and to keep these speeds forever. (Hint: Consider an elastic collision between two molecules with equal speeds. Will the speeds remain equal if the initial lines of motion are not parallel?)

13.–If you increase the absolute temperature by a factor of 2, by what factor will you increase the average speed of the molecules of gas?

14.–Air consists of a mixture of nitrogen (N_2), oxygen (O_2), and argon (A). Which of these molecules has the highest average speed? The lowest?

15.–Equipartition of energy applies not only to atoms and molecules, but also to macroscopic "particles" such as, for example, golf balls. If so, why do golf balls remain at rest on the ground instead of flying through the air like molecules? (Hint: Compare the rms speeds of a golf ball and a molecule.)

PROBLEMS

Section 13.1: The Ideal-Gas Law
Section 13.2: The Temperature Scale

1.——Express the last six temperatures listed in Table 13.1 in terms of degrees Fahrenheit.

2.——The hottest place on Earth is Al' Aziziyah, Libya, where the temperature has soared to 136.4°F. The coldest place is Vostok, Antarctica, where the temperature has plunged to −126.9°F. Express these temperatures in degrees Celsius and in kelvin.

3.——In summer when the temperature is 30°C, the pressure within an automobile tire is 3.0 atm. What will be the pressure within this tire in winter when the temperature is 0°C? Assume that no air is added to the tire and that no air leaks from the tire; assume that the volume of the tire remains constant.

4.——In the Middle Ages, physicians applied suction cups to the skin, to draw out "bad humors" (Fig. 13.14). The cups produced "suction" by means of hot air. Suppose that a hot suction cup, at a temperature of 85°C, is applied to the skin, and its rim makes an airtight seal against the skin. The cup initially contains air at 85°C and at atmospheric pressure, 1.0 atm. What will be the underpressure generated in the cup (that is, the difference between the pressure in the cup and atmospheric pressure) when the cup and the air trapped inside it cool from 85°C to 30°C?

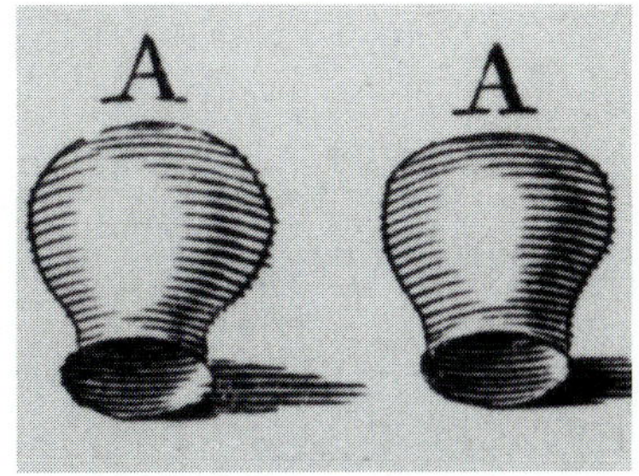

FIGURE 13.14
Suction cups as illustrated in an old textbook of surgery.

5.——A bicycle pump is a cylinder of diameter 2.5 cm and length 30 cm. Initially, when the piston is fully pulled out, the cylinder is filled with air at 25°C and 1.0 atm. If you slowly compress the air to half its initial volume, what force must you exert on the piston to hold it in the compressed position? Assume that the valve is blocked, and no air escapes.

6.——A scuba diver releases an air bubble of diameter 1.0 cm at a depth of 15 m below the surface of a lake. What will be the diameter of this bubble when it reaches the surface? Assume that the temperature of the bubble remains constant.

7.——The storage tank of a small air compressor holds 0.3 m^3 of air at a pressure of 5.0 atm and a temperature of 20°C. How many moles of air is this?

FIGURE 13.15
Cloud of interstellar gas.

8.——The lowest pressure attained in a "vacuum" in a laboratory on the Earth is 1×10^{-17} atm, at a temperature of 4.1 K. What is the number of molecules per cubic centimeter in this vacuum?

9.——Clouds of interstellar hydrogen gas have densities of up to 10^{10} atoms/m^3 and temperatures of up to 10^4 K (Fig. 13.15). What is the pressure in such a cloud?

10.——On a warm day, the outdoor temperature is 32°C and the indoor temperature in an air-conditioned house is 17°C. What is the difference between the densities of the air outdoors and indoors? Assume the pressure is 1 atm.

11.——A paper clip has a mass of 0.5 g. The paper clip is made of iron. How many atoms of iron are in this paper clip?

12.——What is the number of sodium and of chlorine atoms in one spoonful (10 g) of salt, NaCl?

13.——The atomic mass of fissionable uranium is 235.0 g. What is the mass of a single uranium atom? Express your answer in kilograms and in atomic mass units.

14.——How many water molecules are there in 1.0 kg of water? How many oxygen atoms? Hydrogen atoms?

15.——What is the molecular mass of methanol, CH_3OH? What is the number of molecules in 1.0 kg of methanol?

16.——The following table gives the pressure and density of the Earth's upper atmosphere as a function of altitude:

Altitude	Pressure	Density
20,000 m	56×10^2 N/m^2	9.2×10^{-2} kg/m^3
40,000	3.2×10^2	4.3×10^{-3}
60,000	0.28×10^2	3.8×10^{-4}
80,000	0.013×10^2	2.5×10^{-5}

Calculate the temperature at each altitude. The mean molecular mass for air is 29.0 g.

17.——What is the density (in kilograms per cubic meter) of helium gas at 1 atm at the temperature of boiling helium liquid (see Table 12.1)?

18.——Suppose you pour 10 g of water into a 1-liter jar and seal it tightly. You then place the jar into an oven and heat it to 500°C (a dangerous thing to do!). What will be the pressure of the vaporized water?

19.——Consider the automobile tire described in Example 3. If the volume of this tire is 2.5×10^{-2} m^3, what is the mass of air inside it? The mean molecular mass of air is 29.0 g.

20.*—During the volcanic eruption of Mt. Pelée on the island of Martinique in 1902, a *nuée ardente* (burning cloud) of very hot gas rolled down the side of the volcano and killed the 30,000 inhabitants of Saint-Pierre. The temperature in the cloud has been estimated at 700°C. Assume that this cloud consisted of a gas of high molecular mass. What must have been this molecular mass to make the cloud as dense as, or denser than, the surrounding air (at 20°C)?

21.*—An airplane flies through air at a temperature of 5°C. The lift force generated by the flow of air over the wings is 1.2×10^3 N. What would be the lift force if the airplane were flying through air at a temperature of 35°C, other conditions remaining equal?

22.* The helium atom has a volume of about 3×10^{-30} m^3. What fraction of a volume of helium gas at STP is actually occupied by atoms?

23.* How much does the frequency of middle C (see Table 11.1) played on a flute change when the air temperature drops from 20°C to −10°C? [Hint: The speed of sound in air is given by Eq. (11.7).]

24.* At high altitudes, pilots and mountain climbers must breathe an enriched mixture containing more oxygen than the standard concentration of 23% found in ordinary air at sea level. At an altitude of 11,000 m, the atmospheric pressure is 0.22 atm. What oxygen concentration is required at this altitude if with each breath the same number of oxygen molecules is to enter the lungs as for ordinary air at sea level?

25. A carbon dioxide (CO_2) fire extinguisher has an interior volume of 2.8×10^{-3} m^3. The extinguisher has a mass of 5.9 kg when empty and a mass of 8.2 kg when fully loaded with CO_2. At a temperature of 20°C, what is the pressure of CO_2 in the extinguisher?

26.* (a) When you heat the air in a house, some air escapes because the pressure inside the house must remain the same as the pressure outside. Suppose you heat the air from 10°C to 30°C. What fraction of the mass of air originally inside will escape?

(b) If the house were completely airtight, the pressure would have to increase as you heat the house. Suppose that the initial pressure inside the house is 1.0 atm. What is the final pressure? What force does the excess inside pressure exert on a window 1.0 m high and 1.0 m wide? Do you think the window can withstand this force?

27.* In one method for the determination of the average density of the tissues of the human body, the subject is locked in a hermetic chamber of known volume V_c (see Fig. 13.16) containing an unknown volume V of air at an initial pressure p. Then, by means of a small piston, the volume of this chamber is reduced by an amount ΔV. This causes an increase of pressure Δp (at constant temperature). Show that the volume of *air* in the chamber is approximately given by

$$V = \frac{p}{\Delta p}\Delta V$$

and that therefore the volume of the body of the subject is given by

$$V_s = V_c - \frac{p}{\Delta p}\Delta V$$

{Hint: $pV = [\text{constant}]$, and hence $pV = (p + \Delta p)(V + \Delta V)$.}

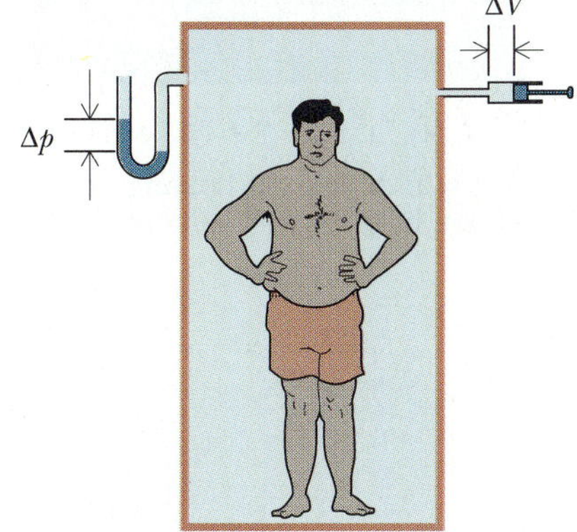

FIGURE 13.16
A volunteer locked in a hermetic chamber.

28.* An oxygen cylinder for medical use contains oxygen at a pressure of 140 atm, at room temperature (20°C). The cylinder measures 20 cm in diameter and 110 cm in length. How many kilograms of oxygen does this cylinder contain? What will be the volume of the oxygen if it is allowed to expand to normal atmospheric pressure, at room temperature?

29.* The volume of air in the fully expanded human lungs is 5 liters. How many molecules are in the lungs? How many molecules of oxygen and how many of nitrogen? Assume that air is 76% nitrogen and 24% oxygen by mass.

30.* Assume that air is 76% nitrogen and 24% oxygen by mass. What is the percentage composition of air by number of molecules?

FIGURE 13.17
A research balloon floating at 40 km in the upper atmosphere.

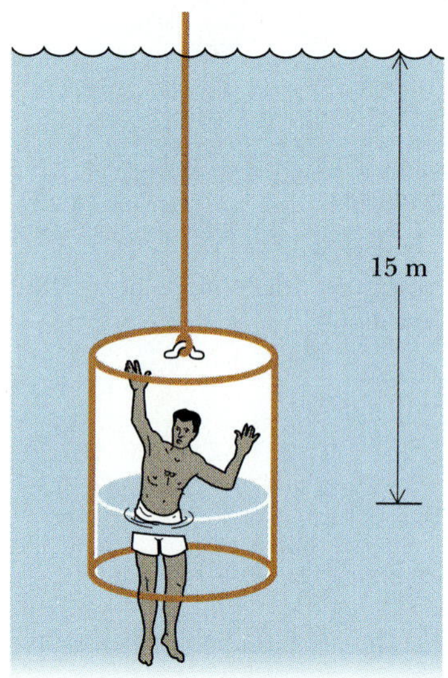

FIGURE 13.18
Submerged diving bell.

31.**—A typical hot-air balloon has a volume of 2200 m^3 and a mass of 730 kg (including balloon, gondola, four passengers, and a propane tank). Since the balloon is open at the bottom, the pressures of the internal and the external air are (approximately) equal. If the temperature of the external air is 20°C, what must be the minimum temperature of the internal air in the balloon to achieve lift-off? The density of the external air is 1.20 kg/m^3.

32.**—A research balloon ascends to an altitude of 40 km and floats in equilibrium (Fig. 13.17). The pressure (outside and also inside the balloon) is 3.2×10^2 N/m^2 and the temperature is −13°C. The volume of the balloon is 8.5×10^5 m^3 and it is filled with helium. What payload (including the mass of the fabric but excluding the helium) can this balloon carry? What was the volume of the balloon on the ground (at STP), before it was released?

33.**—A sunken ship of steel is to be raised by making the upper part of the hull airtight and then pumping compressed air into it while letting the water escape through holes in the bottom. The mass of the ship is 5×10^7 kg and it is at a depth of 60 m. How much compressed air (in kilograms) must be pumped into the ship? The temperature of the air and the water is 15°C.

34.**—A ***diving bell*** is a cylinder closed at the top and open at the bottom; when it is immersed in the water, any air initially in the cylinder remains trapped in the cylinder. Suppose that such a diving bell, 2 m high and 1.5 m across, is immersed to a depth of 15 m measured from water level to water level (see Fig. 13.18).

(a) How high will the water have risen within the diving bell?

(b) If compressed air is pumped into the bell, water will be expelled from the bell. How much air (in kilograms) must be pumped into the bell, and at what pressure, to get rid of all the water? Assume that the temperature of the air is 15°C.

Section 13.3: A Microscopic View of Pressure

35.—What is the rms speed of molecules of water vapor in air at 0°C.?

36.—What is the rms speed of hydrogen atoms on the surface of the Sun, where the temperature is 4.5×10^3 K? At the center of the Sun, where the temperature is 1.5×10^7 K?

37.—In Example 6 we calculated the rms speeds of nitrogen and of oxygen molecules in air at 0°C. If we want to increase these rms speeds by a factor of 2, what temperature do we need?

38.—A sample of gas has some initial pressure p, volume V, and temperature T. By what factor does the rms speed of its molecules increase or decrease if we do one of the following:

(a) Increase the temperature to $2T$?

(b) Decrease the volume to $V/3$?

(c) Increase the volume to $4V$ and simultaneously decrease the pressure to $p/2$?

(d) Decrease the temperature to $T/4$ and simultaneously decrease the pressure to $p/3$?

39.—Consider separate samples of nitrogen gas and of oxygen gas. The temperature of the nitrogen gas is 20°C. What must be the temperature of the oxygen gas if the rms speed of the oxygen molecules is to equal the rms speed of the nitrogen molecules?

40.—What is the average kinetic energy of an oxygen molecule in air at STP? A nitrogen molecule?

41.—What is the rms speed of a helium atom at 0°C? At −269°C?

42.——At the top of the stratosphere, at an altitude of 30 km, the temperature is $-38°C$. What is the rms speed of an oxygen molecule at $-38°C$? Of an ozone (O_3) molecule? What are the average kinetic energies for these molecules?

43.——According to Eq. (11.7) the speed of sound in air is $\sqrt{1.4p/\rho}$.

(a) Show by means of the ideal-gas law that this expression equals $\sqrt{1.4kT/m}$, where m is the average mass per molecule of air.

(b) Show that, in terms of the rms speed, the latter expression equals $\sqrt{1.4/3}v_{rms}$ or $0.68v_{rms}$.

(c) Calculate the speed of sound in air at temperatures of 0°C, 10°C, 20°C, and 30°C.

44.——The rms speed of nitrogen molecules in air at some temperature is 493 m/s. What is the rms speed of hydrogen molecules in air at the same temperature? The atomic mass of nitrogen is 14.0 g and that of hydrogen is 1.0 g.

45.——One method for the separation of the rare isotope ^{235}U (used in nuclear bombs and reactors) from the abundant isotope ^{238}U relies on diffusion through porous membranes. Both isotopes are first made into a gas of uranium hexafluoride (UF_6). The molecules of $^{235}UF_6$ have a higher rms speed and they will diffuse faster through a porous membrane than the molecules of $^{238}UF_6$. The molecular masses of these two molecules are 349 g and 352 g, respectively. What is the percentage difference between their rms speeds at a given temperature?

46.*—Estimate the number of impacts of air molecules on the palm of your hand per second. Assume that the air is at 20°C and 1 atm, and assume that it consists entirely of nitrogen molecules.

Section 13.4: The Internal Energy of an Ideal Gas

47.——What is the thermal kinetic energy of 1 mole of helium gas at 300 K? How much does this kinetic energy increase if we increase the temperature by 20 K? If we increase the pressure by 3.0 atm (at fixed temperature)?

48.——What is the thermal kinetic energy in 1 kg of oxygen gas at a temperature of 20°C? What fraction of this energy is translational? What fraction is rotational?

49.——Assume that air consists of the diatomic gases O_2 and N_2. How much must we increase the thermal energy of 1 kg of air in order to increase its temperature by 1°C?

50.*—A container is divided into two equal compartments by a partition. One compartment is initially filled with helium at a temperature of 250 K; the other is filled with oxygen at a temperature of 310 K. Both gases are at the same pressure. If we remove the partition and allow the gases to mix, what will be their final temperature?

CHAPTER 14

Molten steel pours out of a bucket into forms.

HEAT

In everyday language, heat is what makes things hot. When we place a kettle full of water on a stove, the water absorbs heat from the stove and it becomes hot. But in the precise language of physics, what the water absorbs from the stove is not called heat but ***thermal energy, that is, the kinetic and potential energy of the random microscopic motions of molecules, atoms, ions, electrons, and other particles.*** When the water is in contact with the hot stove, the atoms of the stove communicate some of their violent random microscopic motions to the water molecules. Thus, the thermal energy of the water molecules increases — they bounce around more violently than before. At the macroscopic level, such an increase of the

Thermal energy ▶

energy of the random microscopic motions manifests itself as an increase of the temperature of the water.

We examined the connection between the random microscopic motions and the temperature in the preceding chapter, where we saw that the increase of the kinetic energy of the random microscopic motions of the molecules in a gas is directly proportional to the increase of temperature [see Eq. (13.25)]. In a liquid or a solid, the kinetic energy of the random microscopic motions also increases with the temperature. Furthermore, the atoms and molecules in a liquid or a solid have potential energies associated with the forces they exert on one another; these potential energies also increase with temperature. Thus, the microscopic view of thermal energy as kinetic and potential energy of the random motions of atoms and molecules agrees with the intuitive notion that the absorption of thermal energy should lead to an increase of temperature.

◄ Heat

In the language of physics, ***heat is the transfer of thermal energy from a hot body to a colder body.*** The relationship of heat to thermal energy is analogous to the relationship of work to mechanical energy we studied in Chapter 4. Work done on a particle increases the mechanical energy of the particle. Work has the same units as energy, but work is not energy — rather it is a transfer of energy. Likewise, heat has the same units as thermal energy, but it is not thermal energy — rather it is a transfer of thermal energy. This analogy between heat and work is not merely formal. In fact, heat can be regarded as microscopic work done by the particles in the hotter body on the particles in the colder body, and this microscopic work accomplishes the transfer of thermal energy. Although in a strict sense heat is a transfer of thermal energy, physicists sometimes use the word *heat* in a loose sense as a synonym for *thermal energy*. Thus, we speak of heat flow, heat storage, heat loss, etc., when it would be more accurate to say flow of thermal energy, storage of thermal energy, loss of thermal energy, etc.

Benjamin Thompson, Count Rumford, *1753–1814, American-British scientist. On the basis of experimental observations that he collected while supervising the boring of cannon, Rumford argued against the prevailing view that heat is a substance, and he proposed that heat is nothing but the random microscopic motion of the particles within a body.* **Robert von Mayer,** *1814–1878, German physician and physicist, calculated the mechanical equivalent of heat by comparing the work done on a gas during compression with the consequent increase of temperature. Finally,* **J. P. Joule** *measured this quantity directly by means of his famous experiment.*

The ambiguity in the usage of the word *heat* arises from historical roots. Until well into the nineteenth century, scientists did not have a clear understanding of the concept of energy, and they thought that heat was an invisible, weightless fluid, which they called "caloric." The first experiments to give conclusive evidence of the nature of heat were performed by Count Rumford, who showed that the mechanical energy lost in friction is converted into heat. You can verify such a frictional conversion of mechanical energy into heat by rubbing your hands against each other — a few seconds of rubbing produce a noticeable warming.

The practical development of steam engines for the industrial generation of (macroscopic) mechanical energy from heat motivated a careful examination of the theoretical principles underlying the operation of such engines and led to the discovery of the law of conservation of energy and to the recognition that heat is a form of energy transfer. Steam engines and other heat engines do not create energy; they merely convert thermal energy into mechanical energy, which can be used to perform useful work. For example, the steam engine of an old-fashioned locomotive converts thermal energy from the combustion of coal into mechanical energy, and the engine of an automobile converts thermal energy from the combustion of gasoline into mechanical energy.

14.1 Heat as a Form of Energy Transfer

Calorie (cal) ▶

Long before physicists recognized that heat is a form of energy transfer resulting from the random microscopic motion of atoms, they defined heat in terms of the temperature changes it produces in a body. ***The traditional unit of heat is the calorie*** (cal), which is the amount of heat needed to raise the temperature of 1 g of water by 1°C. The kilocalorie is 1000 cal:

$$1 \text{ kcal} = 1000 \text{ cal} \tag{1}$$

Incidentally: The "calories" marked on some packages of food in grocery stores are actually kilocalories, sometimes also called large calories.

Specific heat ▶

The heat necessary to raise the temperature of 1 kg ***of a material by*** 1°C ***is called the specific heat capacity, or the specific heat, usually designated by the symbol*** c. Thus, by definition, water has a specific heat of

$$c = 1 \text{ kcal/kg}\cdot{}^\circ\text{C} \tag{2}$$

The specific heat varies from substance to substance; Table 14.1 lists the specific heats of some common substances. The specific heat also varies with temperature. For example, the specific heat of water varies by about 1% between 0°C and 100°C, reaching a minimum of 35°C.* Finally, the

TABLE 14.1 SOME SPECIFIC HEATS[a]

Substance	C
Aluminum	0.214 kcal/kg · °C
Brass	0.092
Copper	0.092
Iron, steel	0.11
Lead	0.031
Tin	0.054
Silver	0.056
Mercury	0.033
Water	1.00
Ice (−10°C)	0.530
Ethyl alcohol	0.581
Glycol	0.571
Mineral oil	0.5
Glass, thermometer	0.20
Marble	0.21
Granite	0.19
Sea water	0.93

[a] At room temperature (20°C) and 1 atm, unless otherwise noted.

* This variation must be taken into account for a precise definition of the calorie: a calorie is the heat needed to raise the temperature of 1 g of water from 14.5°C to 15.5°C.

specific heat depends on the pressure to which the material is subjected during the heating. All the values listed in Table 14.1 are applicable at room temperature (20°C) and at a constant pressure of 1.0 atm.

By inspection of Table 14.1, we see that the specific heat of water is exceptionally large. This exceptional capacity of water to absorb heat with only a relatively small change of temperature accounts for the thermal stability of lakes and of oceans. Whereas the temperature of air might vary by 10°C in the course of a day, the temperature of the water of a lake varies only little — the only significant temperature change is a gradual warming up and cooling down of the lake in the course of the seasons.

The values in Table 14.1 tell us the amount of heat required to increase the temperature of 1 kg of the given substance by 1°C. The amount of heat Q required to increase the temperature of a mass m by ΔT is proportional to m and to ΔT:

$$Q = mc\Delta T \tag{3}$$

This merely says that a large mass or a large temperature change requires more heat, in proportion to the mass and the temperature change.

EXAMPLE 1. You pour 0.1 kg of water at 20°C into an aluminum pot of 0.2 kg at the same temperature. How much heat must you supply to bring the water and the pot to a temperature of 100°C?

SOLUTION: The temperature change is $\Delta T = 80°\text{C}$. Hence the heat absorbed by the water is

$$Q = 0.1\text{ kg} \times 1\text{ kcal/kg}\cdot°\text{C} \times 80°\text{C} = 8.0\text{ kcal}$$

The specific heat of aluminum is 0.214 kcal/kg·°C. Hence the heat absorbed by the aluminum is

$$Q = 0.2\text{ kg} \times 0.214\text{ kcal/kg}\cdot°\text{C} \times 80°\text{C} = 3.4\text{ kcal}$$

The net heat is

$$8.0\text{ kcal} + 3.4\text{ kcal} = 11.4\text{ kcal}$$

Since heat is a form of work, it can be transformed into macroscopic mechanical work, and vice versa. The transformation of heat into macroscopic work is accomplished by a steam engine, a steam turbine, or a similar machine; we will examine the theory of such heat engines in the next chapter. The transformation of macroscopic work into heat requires no special machinery — any kind of friction will convert work into heat.

Since heat is a form of work, the calorie is a unit of energy and it must be possible to express it in joules. The conversion factor between these units is called the ***mechanical equivalent of heat.*** The traditional method for the measurement of the mechanical equivalent of heat is ***Joule's experiment.*** A set of falling weights drives a paddle wheel which churns the water in a thermally insulated bucket (see Fig. 14.1). The churning raises the temperature of the water by a measurable amount, converting a known amount of gravitational potential energy into a known amount of heat. The best available experimental results give

Joule's experiment ▶

Mechanical equivalent of heat ▶

$$1 \text{ cal} = 4.186 \text{ J} \tag{4}$$

for the mechanical equivalent of heat.

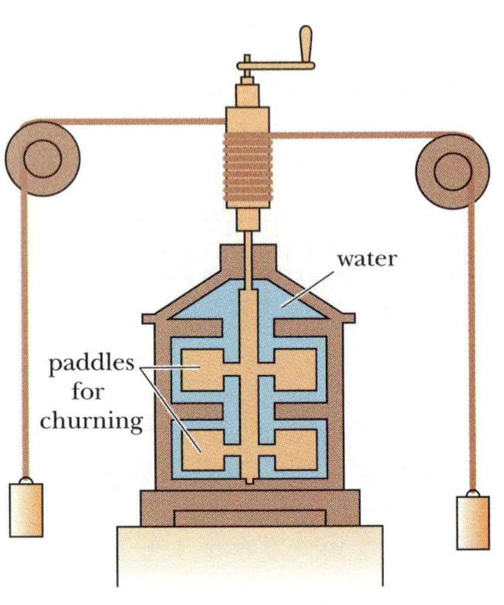

FIGURE 14.1
Joule's apparatus.

EXAMPLE 2. When an automobile is braking, the friction between the brake drums and the brake shoes converts translational kinetic energy into heat. If a 2000-kg automobile brakes from 25 m/s to 0 m/s, how much heat is generated in the brakes? If each of the four brake drums has a mass of 9.0 kg of iron of specific heat 0.11 kcal/kg · °C, how much does the temperature of the brake drums rise? Assume that all the heat accumulates in the brake drums (there is not enough time for the heat to leak away into the air).

SOLUTION: The initial kinetic energy of the automobile is

$$K = \tfrac{1}{2}mv^2 = \tfrac{1}{2} \times 2000 \text{ kg} \times (25 \text{ m/s})^2 = 6.2 \times 10^5 \text{ J}$$

Expressed in kilocalories, this gives an amount of heat

$$Q = 6.2 \times 10^5 \text{ J} \times \frac{1 \text{ cal}}{4.19 \text{ J}} = 1.5 \times 10^5 \text{ cal} = 1.5 \times 10^2 \text{ kcal}$$

The total mass of iron in the four brake drums is 4×9.0 kg. Hence the temperature increase of this mass of iron is, according to Eq. (3),

$$\Delta T = \frac{Q}{mc} \tag{5}$$

$$= \frac{1.5 \times 10^2 \text{ kcal}}{4 \times 9.0 \text{ kg} \times 0.11 \text{ kcal/kg} \cdot {}^\circ\text{C}} = 38^\circ\text{C}$$

14.2 Thermal Expansion of Solids and Liquids

As we saw in the preceding chapter, if the pressure is held constant, the volume of a given amount of gas will increase with the temperature [see

Eq. (13.5)]. Such an increase of volume with temperature also occurs for solids and liquids; this phenomenon is called ***thermal expansion.*** However, the thermal expansion of solids and of liquids is much less than that of gases. For example, if we raise the temperature of a piece of iron by 100°C, we will increase its volume by only 0.36%. During the expansion, the solid retains its shape, but all its dimensions increase in proportion. Figure 14.2a illustrates the expansion of a piece of metal; for the sake of clarity, the expansion has been exaggerated. An expanding liquid does not, of course, retain its shape; the liquid will merely fill more of the container that holds it. Figure 14.2b illustrates the thermal expansion of a liquid.

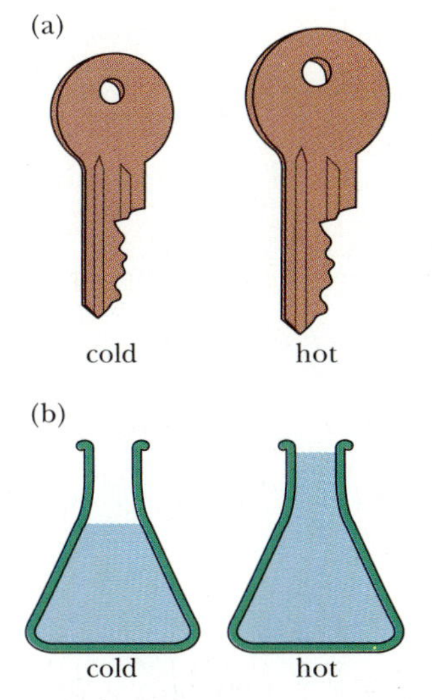

FIGURE 14.2

(a) Thermal expansion of a solid. (b) Thermal expansion of a liquid. The expansion of the flask has been neglected.

From a microscopic point of view, the thermal expansion of solids and liquids is due to the increase of thermal motion caused by the increase of temperature — in a solid the amplitude of the back-and-forth motions of the atoms about their equilibrium positions increases with temperature, and in a liquid the bouncing zigzag motions of the molecules increase with temperature. This increase of the random motions tends to push the atoms or molecules farther apart, and therefore leads to an increase of the volume of the solid or liquid.

The thermal expansion of a solid can be best described mathematically by the increase in the linear dimensions of the solid (see Fig. 14.3). ***The increment in the length is directly proportional to the increment of temperature and to the original length,***

$$\Delta L = \alpha L \Delta T \tag{6}$$

◀ **Thermal expansion in length**

The constant of proportionality α in this equation is called the ***coefficient of linear expansion.*** Table 14.2 lists the value of this coefficient for a few materials.

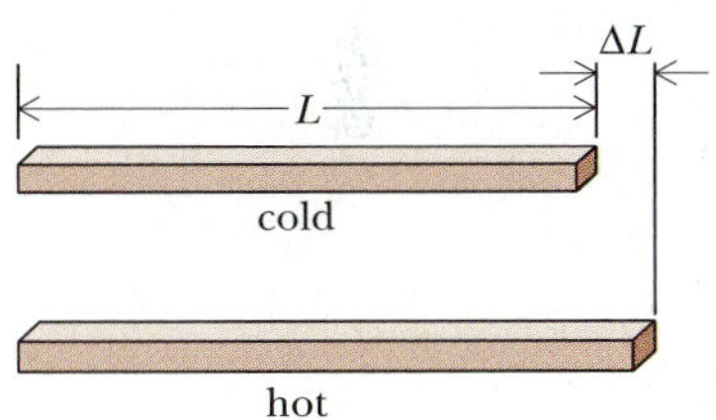

FIGURE 14.3

Increase of the length of a solid rod by thermal expansion.

TABLE 14.2 COEFFICIENTS OF EXPANSION[a]

Solids	α	Liquids	β
Lead	$29 \times 10^{-6}/°C$	Alcohol, ethyl (99%)	$1.01 \times 10^{-3}/°C$
Aluminum	24	Carbon tetrachloride	1.18
Brass	19	Ether	1.51
Copper	17	Gasoline	0.95
Iron, steel	≃ 12	Glycerine	0.49
Concrete	≃ 12	Olive oil	0.68
Glass		Mercury	0.18
crown	9.0		
Pyrex	3.6		
Quartz, fused	0.50		

[a] At room temperature (20°C).

EXAMPLE 3. The highest tower in the world is the steel radio mast of Warsaw Radio in Poland, which has a height of 646 m. How much does its height increase between a cold winter day when the temperature is $-35°C$ and a hot summer day when the temperature is $+35°C$?

SOLUTION: The increment of temperature is $\Delta T = 70°C$. With a value of $\alpha = 12 \times 10^{-6}°C$ for steel, we then find

$$\Delta L = \alpha L \Delta T = \frac{12 \times 10^{-6}}{°C} \times 646 \text{ m} \times 70°C = 0.54 \text{ m}$$

Thermal expansion in volume

The increment in the volume of a solid is directly proportional to the increment of temperature and to the original volume,

$$\Delta V = \beta V \Delta T \tag{7}$$

Here the constant of proportionality β is called the ***coefficient of volume expansion.*** This coefficient is three times the coefficient of linear expansion,

$$\beta = 3\alpha \tag{8}$$

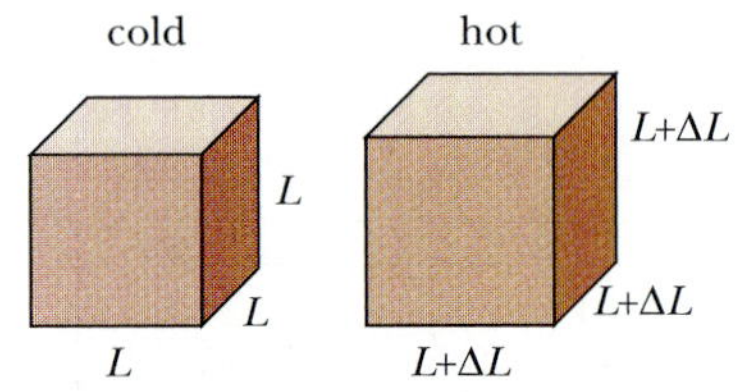

FIGURE 14.4
Thermal expansion of a solid cube. The initial volume is $L \times L \times L$; the final volume is $(L + \Delta L) \times (L + \Delta L) \times (L + \Delta L)$.

To see how this relationship comes about, suppose that the solid has the shape of a cube of side L (see Fig. 14.4). Since each side expands to $L + \Delta L$, the volume expands to $(L + \Delta L)^3$, which equals

$$(L+\Delta L)^3 = L^3 + 3L^2\Delta L + 3L(\Delta L)^2 + (\Delta L)^3 \tag{9}$$

Since ΔL is a small quantity, $(\Delta L)^2$ is a doubly small quantity, and $(\Delta L)^3$ is a triply small quantity. Thus, the last two terms in Eq. (9) can be neglected compared with the first two terms, and

$$(L + \Delta L)^3 \simeq L^3 + 3L^2\Delta L \tag{10}$$

Here, L^3 is the original volume, and $3L^2\Delta L$ therefore represents the increment of volume. Hence, from Eq. (6), we obtain

$$\Delta V = 3L^2\Delta L = 3L^2(\alpha L \Delta T) = 3\alpha L^3 \Delta T = 3\alpha V \Delta T \tag{11}$$

Comparing this result for ΔV with Eq. (7), we see that, indeed, $\beta = 3\alpha$.

The increment in the volume of a liquid can be described by the same equation [Eq. (7)] as the increment in the volume of a solid. Table 14.2 lists values of coefficients of volume expansion for some liquids.

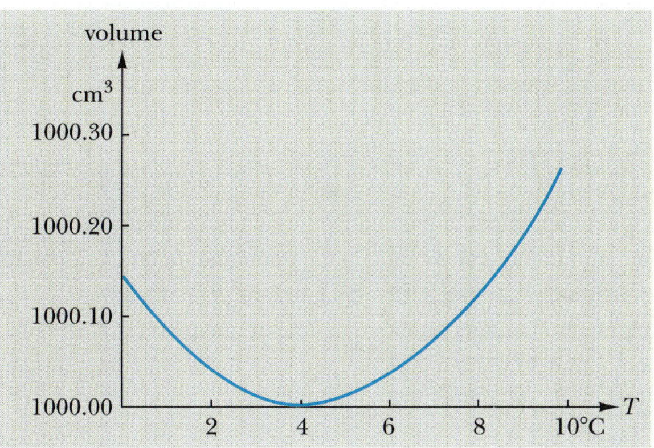

FIGURE 14.5
Volume of 1 kg of water as a function of temperature.

Water has not been included in this table because its behavior is quite peculiar: from 0°C to 4°C, the volume *decreases* with temperature, but not uniformly; above 4°C, the volume increases with temperature. Figure 14.5 gives the volume of 1 kg of water for temperatures ranging from 0°C to 10°C. The strange behavior of the density of water at low temperatures can be traced to the crystal structure of ice. Water molecules have a rather angular shape that prevents a tight fit of these molecules; when they assemble in a solid, they adopt a very complicated crystal structure with large gaps. As a result, ice has a lower density than water — the density of ice is 917 kg/m^3, and the volume of 1 kg of ice is 1091 cm^3, in contrast to 1000 cm^3 for an equal mass of water. At a temperature slightly above the freezing point, water is liquid, but some of the water molecules have already assembled themselves into microscopic (and ephemeral) ice crystals; these microscopic ice crystals give the cold water an excess volume.

The maximum in the density of water at about 4°C has an important consequence for the ecology of lakes. In winter, the layer of water on the surface of the lake cools, becomes denser than the lower layer, and sinks to the bottom. This process continues until the temperature of the entire body of the lake reaches 4°C. Beyond this point, the cooling of the surface layer will make it *less dense* than the lower layers; thus, the cold surface layer stays in place, floating on the top of the lake. Ultimately, this surface layer freezes, becoming a solid sheet of ice, while the body of the lake remains at 4°C. The sheet of ice inhibits the heat loss from the lake, especially if covered with an insulating blanket of snow. Besides, any further heat loss merely causes some thickening of the sheet of ice, without disturbing the deeper layers of water, which remain at a stable temperature of 4°C — fish and other aquatic life can survive the winter in reasonable comfort.

EXAMPLE 4. A glass vessel of volume 200 cm^3 is filled to the rim with mercury. Taking into account the expansions of the mercury and the glass, how much of the mercury will overflow the vessel if we raise the temperature by 30°C?

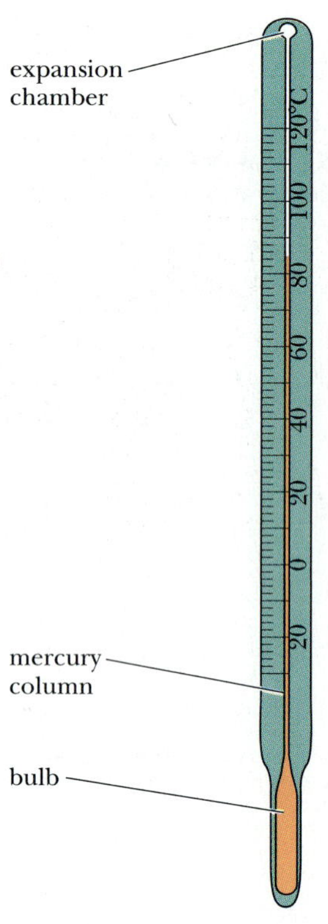

FIGURE 14.6

Mercury-bulb thermometer.

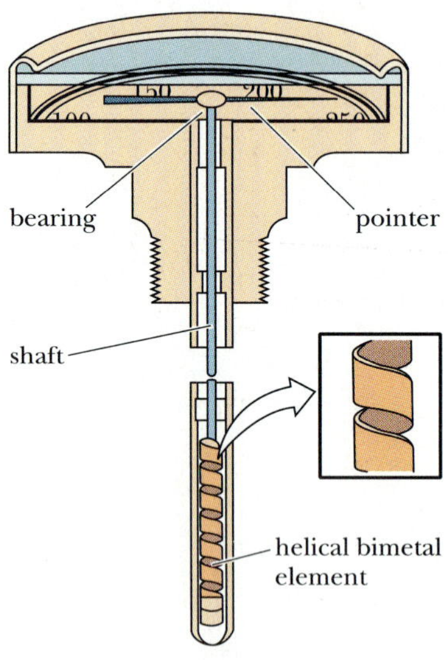

FIGURE 14.7

Bimetallic-strip thermometer.

SOLUTION: The volume of mercury will increase by

$$\Delta V = \beta_{Hg} V \Delta T \tag{12}$$

$$= 0.18 \times 10^{-3}/^{\circ}\mathrm{C} \times 200\ \mathrm{cm}^3 \times 30^{\circ}\mathrm{C} = 1.08\ \mathrm{cm}^3$$

The volume of the glass vessel will increase just as though all of the vessel were filled with glass (as in Fig. 14.2, the hole, or cavity, in the vessel expands as though it were filled with glass); hence

$$\Delta V_{glass} = \beta_{glass} V \Delta T = 3\alpha_{glass} V \Delta T \tag{13}$$

$$= 3 \times 9.0 \times 10^{-6}/^{\circ}\mathrm{C} \times 200\ \mathrm{cm}^3 \times 30^{\circ}\mathrm{C} = 0.16\ \mathrm{cm}^3$$

The difference $1.08\ \mathrm{cm}^3 - 0.16\ \mathrm{cm}^3 = 0.92\ \mathrm{cm}^3$ is the volume of mercury that will overflow.

Ordinary thermometers and thermostats make use of thermal expansion to sense changes in temperature. The mercury-bulb thermometer already mentioned in the preceding chapter (see Fig. 14.6) consists of a glass bulb filled with mercury connected to a thin capillary tube. Thermal expansion makes the mercury overflow into the capillary tube and increases the length of the mercury column; this length indicates the temperature. The bimetallic-strip thermometer (see Fig. 14.7) consists of two parallel strips of different metals — such as aluminum and iron — welded together and curled into a spiral or a helix. The differential thermal expansion increases the length of one side of the welded strip more than that of the other side; this causes the strip to curl up more tightly and rotates the upper end of the helix relative to the lower end; a pointer attached to the upper end indicates the temperature.

Thermal expansion must be taken into account in the design of long structures, such as bridges or railroad tracks. The decks of bridges usually have several expansion joints with gaps (see Fig. 14.8) that permit changes

FIGURE 14.8

Expansion joints in deck of a bridge.

of length and prevent the bridge from buckling. Likewise, gaps are left between the segments of rail in a railroad track; but if the temperature changes exceed the expectations of the designers, the results can be disastrous (see Fig. 14.9).

Incidentally: Our ability to erect large buildings and other structures out of reinforced concrete hinges on the fortuitous coincidence of the coefficients of expansion of iron and concrete (see Table 14.2). Reinforced concrete is made by pouring concrete around an array of iron rods. The iron rods strengthen the concrete, and prevent it from cracking. If the coefficients of expansion for these two materials were appreciably different, the daily and seasonal temperature changes would cause the iron rods to move relative to the concrete — ultimately, the iron rods would loosen, and the reinforcement would come to an end.

FIGURE 14.9

Buckling of railroad rails due to heat.

14.3 CONDUCTION OF HEAT

If you stick one end of an iron poker or rod into the fire and hold the other end in your hand, you will feel the end in your hand gradually become warmer. This is an example of heat transfer by ***conduction.*** The atoms and electrons in the hot end of the rod have greater kinetic and potential energies than those in other parts of the rod. In random collisions, these energetic atoms and electrons share some of their energy with their less energetic neighbors; these, in turn, share their energy with their neighbors, and so on. The result is a gradual diffusion of the thermal energy from the hot end to the cold end (Fig. 14.10).

FIGURE 14.10

Heat flows from the hot end of a rod to the cold end.

Metals are excellent conductors of heat, and also excellent conductors of electricity. The high thermal and electric conductivities of a metal are due to an abundance of "free" electrons within the volume of the metal; these are electrons that have become detached from their atoms — they move at high speeds and they wander all over the volume of the metal with little hindrance, but they are held back by the surface of the metal. The free electrons behave like particles of a gas, and the metal acts like a bottle holding this gas. Typically, a free electron will move past a few hundred atoms before it suffers a collision. Because the electrons move such fairly large distances between collisions, they can travel very quickly from one end of a metallic rod to the other. Thus, the motion of the free electrons transports the thermal energy much more efficiently than does the back-and-forth vibrational motion of the atoms.

◀ **Heat flow**

Quantitatively, we can describe the transport of heat by the ***heat flow,*** or the heat current; ***this is the amount of heat that passes by some given place on the rod per unit time.*** We will use the symbol $Q/\Delta t$ for heat flow. In the metric system, the unit of heat flow is the joule per second (J/s); however, in practice, the preferred unit usually is the calorie per second (cal/s).

Consider a rod of cross-sectional area A and length Δx (see Fig. 14.11). Assume that the cold end of the rod is kept at a constant temperature T_1 and the hot end at a constant temperature T_2, so the difference of temperature between the ends is $\Delta T = T_2 - T_1$. For instance, one end of the

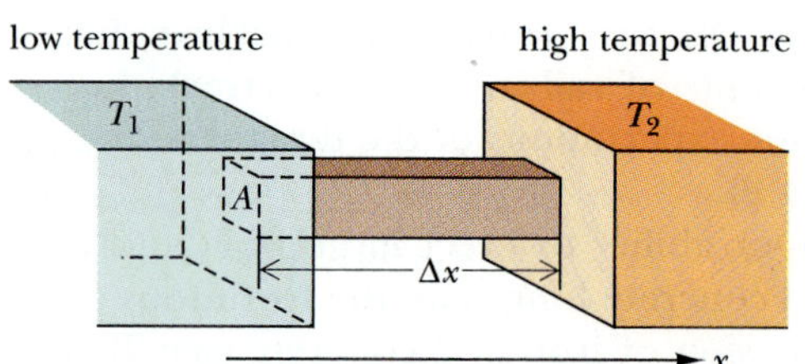

FIGURE 14.11
A rod of cross-sectional area A conducting heat from a high-temperature reservoir (T_2) to a low-temperature reservoir (T_1).

rod might be placed in contact with a block of ice, and the other end might be immersed in boiling water. If the ends are kept at constant temperatures for a while, then the temperatures at all other points of the rod will settle to final steady values. Under such steady-state conditions, the heat flow of the rod is found to be directly proportional to the temperature difference ΔT and to the cross-sectional area A, and inversely proportional to the length:

Conduction of heat ▶

$$\frac{Q}{\Delta t} = kA\frac{\Delta T}{\Delta x} \tag{14}$$

Thermal conductivity ▶

The direction of the heat flow is, of course, from the hot end of the rod toward the cold end. The constant of proportionality k in our equation is called the ***thermal conductivity.*** Table 14.3 lists values of k for some materials.*

Equation (14) can be regarded as an empirical law that has been veri-

TABLE 14.3 SOME THERMAL CONDUCTIVITIES[a]

Substance	k
Aluminum	49 cal/(s · m · °C)
Copper	92
Iron, cast	11
Steel	11
Lead	8.3
Silver	97
Ice, 0°C	0.3
Snow, 0°C, compact	0.05
Glass, crown	0.25
Porcelain	0.25
Concrete	0.2
Styrofoam	0.002
Wood, pine	0.03
Fiberglas (batten)	0.01
Down	0.005

[a] At room temperature (20°C), unless otherwise noted.

* The thermal conductivity must not be confused with the Boltzmann constant; both these quantities are designated with the same letter k, but they are not related.

fied by many experiments. Alternatively, this equation can be derived from a detailed study of the process of diffusion of the thermal energy along the rod. Note that Eq. (14) also applies to heat conduction through a slab or plate; such a piece of material can be regarded as a rod of very short length and very large cross section.

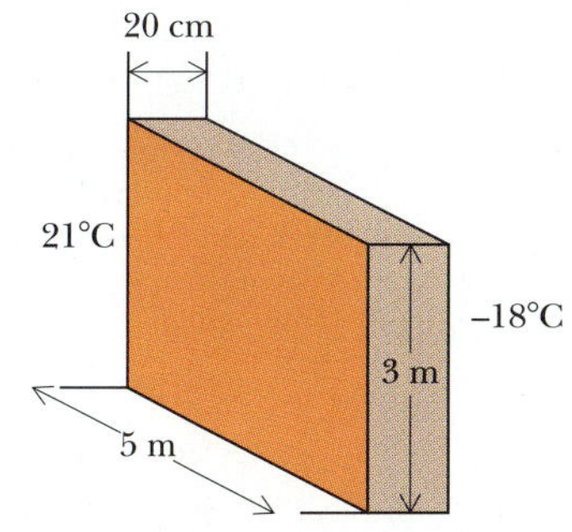

FIGURE 14.12
Heat flow through the wall of a house.

EXAMPLE 5. A house is built of bricks, with walls 20 cm thick. A wall in one of the rooms of this house measures 5 m × 3 m (see Fig. 14.12). What is the heat flow through this wall if the inside temperature is 21°C and the outside temperature −18°C?

SOLUTION: The temperature change across the wall is $\Delta T = 39°\text{C}$, the thickness of the wall is $\Delta x = 20\text{ cm} = 0.20\text{ m}$, and the area of the wall is 15 m^2. Hence, with $k = 0.15\text{ cal/(s·m·°C)}$, Eq. (14) gives

$$\frac{Q}{\Delta t} = kA\frac{\Delta T}{\Delta x}$$

$$= 0.15\text{ cal/(s·m·°C)} \times 15\text{ m}^2 \times \frac{39°\text{C}}{0.20\text{ m}}$$

$$= 440\text{ cal/s} \qquad (15)$$

This is a rather large, intolerable heat loss. Obviously, the house needs insulation. If a 12-cm layer of Fiberglas insulation is placed over the brick, the heat loss will be reduced by a factor of about 10, and the heat loss becomes tolerable.

Besides conduction, there are two other mechanisms of heat transfer: convection and radiation. ***In convection, the heat is stored in a moving fluid, and it is carried from one place to another by the motion of this fluid. In radiation, the heat is carried from one place to another by electromagnetic waves*** — light waves, infrared waves, or radio waves. All three mechanisms of heat transfer are neatly illustrated by the operation of a hot-water heating system in a house. In this system, the heat is carried from the boiler to the radiators in the rooms by means of water flowing in pipes (convection); the heat then diffuses through the metallic walls of the radiators (conduction); and finally spreads from the surface of the radiators into the volume of the room (radiation, supplemented by some convection of air heated by direct contact with the radiators).

Radiation is the only mechanism of heat transfer that can carry heat through a vacuum; for instance, the heat of the Sun reaches the Earth by radiation. We will study thermal radiation in Chapter 28.

(a)

(b)

(c)

(a) The hot water rises from the hot bottom of the kettle and transfers heat by convection. (b) The rod held by the blacksmith transfers heat from its hot end to its cold end by conduction. (c) The coil of this heater emits infrared radiation, which transfers heat from the coils to the surroundings.

14.4 Changes of State

Heat absorbed by a body will not only increase the temperature, but it may also bring about a change of state from solid to liquid or from liquid to gas when the body reaches its melting point or its boiling point. At the melting temperature or the boiling temperature, the thermal motion of the atoms and molecules becomes so violent that the bonds holding them in the solid or liquid loosen or break. The loosening of the bonds in a solid transforms it into a liquid, and the breaking of bonds in a liquid transforms it into a gas.

While the body is melting or boiling, it absorbs some amount of heat without any increase of temperature. This heat represents the energy required to loosen or break the bonds that hold the atoms inside the solid or liquid. ***The heat absorbed during the change of state is called the heat of transformation, or, more specifically, the heat of fusion or the heat of vaporization, for the change of state from a solid to a liquid or a liquid to a gas, respectively.*** Table 14.4 lists the heats of fusion and vaporization for a few substances (at a pressure of 1 atm).

Heat of transformation ▶

The values listed in Table 14.4 depend on the pressure. The decrease of the boiling point of water with a decrease of atmospheric pressure is a phenomenon familiar to people living at high altitude; for instance, in Denver, Colorado, at an altitude of 1600 m, the mean pressure is 0.96 atm, and the boiling point of water is 98.9°C.

TABLE 14.4 HEATS OF FUSION AND VAPORIZATION[a]

Substance	Melting point	Heat of fusion	Boiling point	Heat of vaporization
Water	0°C	79.7 kcal/kg	100°C	539 kcal/kg
Nitrogen	−210	6.2	−196	47.8
Oxygen	−218	3.3	−183	51
Helium	—	—	−269	5.97
Hydrogen	−259	15.0	−253	107
Aluminum	660	95.3	2467	2520
Copper	1083	48.9	2567	1240
Iron	1535	65	2750	1620
Lead	328	6.8	1740	203
Tin	232	14.2	2270	463
Silver	962	23.7	2212	563
Tungsten	3410	44	5660	1180
Mercury	−39	2.7	357	69.7
Carbon dioxide[b]	−79	—	—	138

[a] At a pressure of 1 atm.

[b] Undergoes direct vaporization (sublimation) from solid to gas.

EXAMPLE 6. How many ice cubes (at 0°C) must be added to a bowl containing a liter of boiling water at 100°C so that the resulting mixture reaches a temperature of 40°C? Assume that each ice cube has a mass of 20 g and that the bowl and the environment do not exchange heat with the water.

SOLUTION: Since all of the heat released by the water is absorbed by the ice, the amount of heat released by the hot water during cooling must equal the amount of heat absorbed by the ice during melting and during the subsequent heating of the molten ice from 0°C to 40°C.

The heat released by the hot water during cooling from 100°C to 40°C is

$$Q = 1\ \text{kg} \times 1\ \text{kcal/(kg}\cdot{}^\circ\text{C)} \times 60^\circ\text{C} = 60\ \text{kcal} \tag{16}$$

If the total mass of ice is m, then the heat absorbed by this mass during melting and subsequent heating from 0°C to 40°C is

$$\begin{aligned} Q &= m \times 79.7\ \text{kcal/kg} + m \times 1\ \text{kcal/(kg}\cdot{}^\circ\text{C)} \times 40^\circ\text{C} \\ &= m \times (79.7\ \text{kcal/kg} + 40\ \text{kcal/kg}) = m \times (119.7\ \text{kcal/kg}) \end{aligned} \tag{17}$$

Since these amounts of heat must be equal, we obtain the equation

$$m \times (119.7\ \text{kcal/kg}) = 60\ \text{kcal}$$

Solving this for m, we find

$$m = 0.50\ \text{kg}$$

Since each ice cube has a mass of 0.020 kg, this is 0.50/0.020 = 25 ice cubes.

14.5* THE SPECIFIC HEAT OF A GAS

If you heat a gas, the increase of temperature causes an increase of the pressure, and this tends to bring about an expansion of the gas. You can observe this expansion if you leave a loosely inflated beach ball or plastic bag in the sun; the beach ball or bag soon becomes taut because the air warms and expands. The value of the specific heat of a gas depends on whether the container permits expansion during heating. If the container is perfectly rigid, the heating proceeds at constant volume (see Fig. 14.13). For gases it is customary to reckon the specific heat per mole,

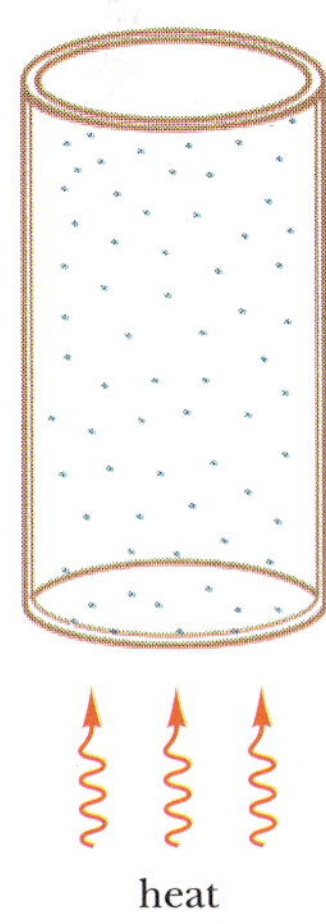

FIGURE 14.13
A gas kept at constant volume while being heated.

* This section is optional.

Specific heat at constant volume, C_V ▶

rather than per kilogram. The ***specific heat at constant volume*** is designated by C_V; it is the amount of heat needed to raise the temperature of 1 mole of gas by 1 °C. The amount of heat Q required to increase the temperature of n moles by ΔT is proportional to n and to ΔT:

$$Q = nC_V\Delta T \tag{18}$$

This equation resembles Eq. (3), but the number of moles (n) appears instead of the mass (m), because now we are reckoning the specific heat per mole.

Specific heat at constant pressure, C_p ▶

If the container is fitted with a vertical piston carrying a load whose weight presses down on the gas, the heating proceeds at a constant pressure p determined by the combined weight of the piston and the load and the area A of the piston (see Fig. 14.14). The ***specific heat at constant pressure*** is designated by C_p. For n moles of gas, the heat absorbed and the temperature increase are related by

$$Q = nC_p\Delta T \tag{19}$$

We expect C_p to be larger than C_V because, if we supply some amount of heat to the container in Fig. 14.14, only part of this heat will go into a temperature increase of the gas; the rest will be converted into work as the expanding gas lifts the piston. Let us calculate the difference between the two heat capacities.

At constant volume, the gas does no work, since there is no moving piston. Hence all the heat supplied to the gas will go to increase the internal energy of the gas and increase its temperature. If we supply a small amount of heat Q, the energy increment ΔE must match Q:

$$Q = \Delta E \tag{20}$$

or, according to Eq. (18),

$$nC_V\Delta T = \Delta E \tag{21}$$

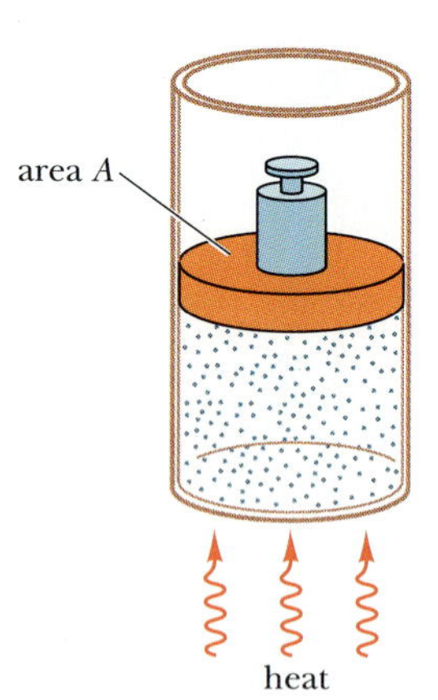

FIGURE 14.14

A gas kept at constant pressure while being heated. The pressure p equals the combined weight of the piston and the load divided by the area A.

At constant pressure, the gas does work against the moving piston. Suppose that the piston is displaced a small distance Δx (see Fig. 14.15). The force of the gas on the piston is Ap, and the work done by the gas is

$$W = Ap\Delta x \tag{22}$$

The product $A\Delta x$ is simply the small change ΔV of the volume of the gas. Hence

$$W = p\Delta V \tag{23}$$

The heat supplied to the gas must provide both the energy increase of the gas and the work done by the gas:

$$Q = \Delta E + W = \Delta E + p\Delta V \tag{24}$$

or, according to Eq. (19),

$$nC_p\Delta T = \Delta E + p\Delta V \tag{25}$$

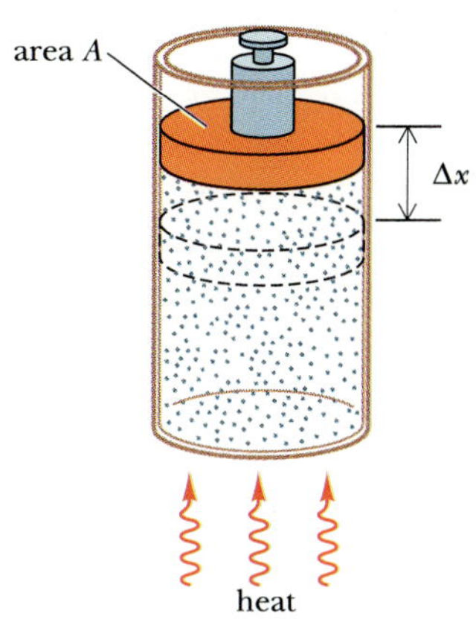

FIGURE 14.15
Displacement of the piston by the expanding gas.

In an ideal gas, the internal energy E depends on the temperature only [see Eqs. (13.27) and (13.33)]; consequently, if the temperature increment at constant pressure has the same value as the temperature increment at constant volume, the increase ΔE of the energy must be the same. We can therefore insert the expression for ΔE given by Eq. (25), and we obtain

$$nC_p\Delta T = nC_V\Delta T + p\Delta V \tag{26}$$

From the ideal-gas law [Eq. (13.2)], we find that, at constant pressure, the changes of volume and of temperature are related by

$$p\Delta V = nR\Delta T \tag{27}$$

With this, Eq. (26) becomes

$$nC_p\Delta T = nC_V\Delta T + nR\Delta T \tag{28}$$

and, canceling the factors $n\Delta T$ in this equation, we obtain the final result

$$C_p = C_V + R \tag{29}$$

◂ **Relation between specific heats of a mole of ideal gas**

The numerical value of R is 8.31 J/K, or 1.99 cal/K. Hence, Eq. (29) shows that C_p is larger than C_V by about 2 cal/K·mole — ***it takes about* 2 *more calories to heat a mole of gas by* 1 K *at constant pressure than at constant volume.***

Note that although the above general argument did permit us to evaluate the difference between C_p and C_V, it does not permit us to find the individual values of C_p and C_V. For this, we must know something about the energy stored in the internal rotational or vibrational motions of the molecules. For a monoatomic gas there is no such extra energy, and according to Eq. (13.27),

$$\Delta E = \tfrac{3}{2}nR\Delta T \tag{30}$$

By the definition of C_V [Eq. (21)], this leads to

$$C_V = \tfrac{3}{2}R \tag{31}$$

and by Eq. (29)

$$C_p = C_V + R = \tfrac{3}{2}R + R = \tfrac{5}{2}R \tag{32}$$

For a diatomic gas, the rotational energy of the molecules results in a larger value of ΔE [see Eq. (13.33)], and consequently a larger value of C_V

and C_p:

$$C_V = \tfrac{5}{2}R \tag{33}$$

and

$$C_p = \tfrac{7}{2}R \tag{34}$$

Table 14.5 lists the experimental values of the specific heats of some gases. In the cases of monoatomic and diatomic gases, these values are in reasonable agreement with the theoretical predictions given by Eqs. (31)–(34), although there are some minor deviations because the gases are not quite ideal gases. Note that in all cases the difference $C_p - C_V$ agrees quite precisely with Eq. (29). Thus, our simple theory of the microscopic behavior of molecules is successful in explaining some of the macroscopic properties of gases.

TABLE 14.5 SPECIFIC HEATS OF SOME GASES[a]

Gas	C_V	C_p	$C_p - C_V$
Helium (He)	3.00 cal/K·mole	4.98 cal/K·mole	1.98 cal/K·mole
Argon (A)	3.00	5.00	2.00
Nitrogen (N_2)	4.96	6.95	1.99
Oxygen (O_2)	4.96	6.95	1.99
Carbon monoxide (CO)	4.93	6.95	2.02
Carbon dioxide (CO_2)	6.74	8.75	2.01
Methane (CH_4)	6.48	8.49	2.01

[a] At STP.

EXAMPLE 7. During a sunny day, the sunlight warms the ground, which in turn warms the air in contact with it. How many calories must the ground supply to heat an initial volume of 1 m^3 of air from 0°C to 10°C? The atmospheric pressure is steady at 1 atm.

SOLUTION: Since the atmosphere surrounding the given amount of air provides a constant pressure, the relevant specific heat is C_p, the specific heat at constant pressure. Air is mostly N_2 and O_2, and Table 14.5 tells us that for N_2 and for O_2, $C_p = 6.95$ cal/K·mole; and this must then also be the right value of C_p for any mixture of these two gases.

During the heating, the volume of air expands, but we can calculate the number of moles from the initial volume and temperature. For this calculation, we can use either the ideal-gas law or, more simply, the known volume of 22.4 liters for 1 mole at STP. By proportions, a volume of 1 m^3, or 1000 liters, contains $n = 1000/22.4$ moles $= 44.6$

moles. Hence the amount of heat absorbed by the air is

$$Q = nC_p\Delta T = 44.6 \text{ moles} \times 6.95 \text{ cal/K} \cdot \text{mole} \times 10 \text{ K}$$

$$= 3.10 \times 10^3 \text{ cal} = 3.10 \text{ kcal}$$

This is a fairly small amount of heat (for comparison, if we wanted to heat a volume of 1 m^3 of water by 10°C, we would need to supply 10^4 kcal).

SUMMARY

- **Specific heat of water:** $c = 1 \text{ kcal/kg} \cdot {}^\circ\text{C}$
- **Mechanical equivalent of heat:** $1 \text{ cal} = 4.186 \text{ J}$
- **Thermal expansion:** $\Delta L = \alpha L \Delta T$
  $\Delta V = \beta V \Delta T$
- **Conduction of heat:** $\dfrac{Q}{\Delta t} = kA\dfrac{\Delta T}{\Delta x}$
- **Relation between specific heats of a mole of ideal gas:** $C_p = C_V + R$

QUESTIONS

1.–Can the body heat from a crowd of people produce a significant temperature increase in a room?

2.–The expression "cold enough to freeze the balls off a brass monkey" originated aboard ships of the British Navy where cannonballs of lead were kept in brass racks ("monkeys"). Can you guess how the balls might fall off a "monkey" on a very cold day?

3.–If the metal lid of a glass jar is stuck, it can usually be loosened by running hot water over the lid. Explain.

4.–On hot days, bridges expand. How do bridge designers prevent this expansion from buckling the road?

5.–At regular intervals, oil pipelines have lateral loops (shaped like a U; see Fig. 14.16). What is the purpose of these loops?

6.–When you heat soup in a metal pot, sometimes the soup rises at the rim of the pot and falls at the center. Explain.

7.–A sheet of glass will crack if heated in one spot. Why?

8.–When aluminum wiring is used in electric circuits, special terminal connectors are required to hold the ends of the wires securely. If an ordinary brass screw were used to hold the end of an aluminum wire against a brass plate, what is likely to happen during repeated heating and cooling of the circuit?

FIGURE 14.16
Loop in an oil pipeline.

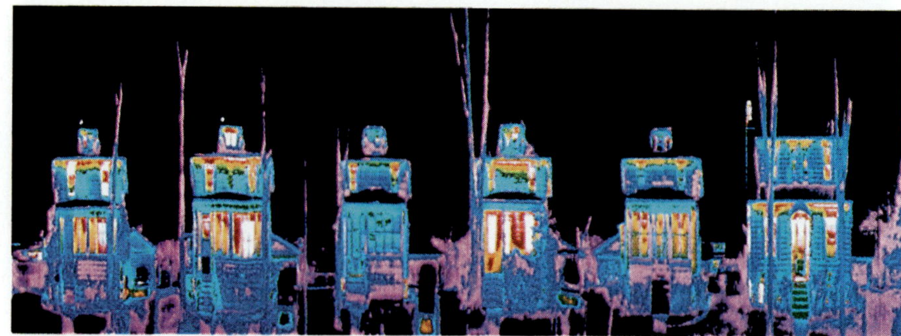

FIGURE 14.17
Thermal photograph of houses. Bright regions indicate high heat loss.

9.–Suppose that a piece of metal and a piece of wood are at the same temperature. Why does the metal feel colder to the touch than the wood?

10.–In lack of better, some nineteenth-century explorers in Africa measured altitude by sticking a thermometer into a pot of boiling water. Explain.

11.–In the cooling system of an automobile, how is the heat transferred from the combustion cylinder to the cooling water (conduction, convection, or radiation)? How is the heat transferred from the water in the engine to the water in the radiator? How is the heat transferred from the radiator to the air?

12.–A fan installed near the ceiling of a room blows air down toward the floor. How does such a fan help to keep you cool in summer and warm in winter?

13.–It is often said that an open fireplace sends more heat up the chimney than it delivers to the room. What is the mechanism for heat transfer to the room? For heat transfer up the chimney?

14.–A large fraction of the heat lost from a house escapes through the windows (Fig. 14.17). This heat is carried to the windowpane by convection — hot air at the top of the room descends along the windowpane as it cools, giving up its heat to the windowpane. Suppose that the windows are equipped with venetian blinds. In order to minimize the heat loss, should you close the blinds so the slats are oriented down and away from the window or up and away?

15.–Fiberglas insulation used in the walls of houses has a shiny layer of aluminum foil on one side (Fig. 14.18). What is the purpose of this layer?

16.–Is it possible to add heat to a system without changing its temperature? Give an example.

17.–Why is boiling oil much more likely to cause severe burns on skin than boiling water?

18.–A very cold ice cube, fresh out of the freezer, tends to stick to the skin of your fingers. Why?

19.–What is likely to happen to the engine of an automobile if there is no antifreeze in the cooling system and the water freezes?

20.–If an evacuated glass vessel, such as a TV tube, fractures and implodes, the fragments fly about with great violence. From where does the kinetic energy of these fragments come?

21.–When you boil water and convert it into water vapor, is the heat you supply equal to the change of the internal energy of the water?

22.–If you let some air out of the valve of an automobile tire, it feels cold. Why?

FIGURE 14.18
Fiberglas insulation.

23.–A sample of ideal gas is initially confined in a bottle at some given temperature. If we break the bottle and let the gas expand freely into an evacuated chamber of larger volume, will the temperature of the gas change?

PROBLEMS

Section 14.1: Heat as a Form of Energy Transfer

1.——The immersible electric heating element in a coffee maker converts 620 W of electric power into heat. How long does this coffee maker take to heat 1.0 liter of water from 20°C to 100°C? Assume that no heat is lost to the environment.

2.——The body heat released by children in a school makes a contribution toward heating the building. How many kilowatts of heat do 1000 children release? Assume that the daily food intake of each child has a chemical energy of 2000 kcal and that this food is burned at a steady rate throughout the day.

3.——In 1847 Joule attempted to measure the frictional heating of water in a waterfall near Chamonix in the French Alps. If the water falls 120 m and all of its gravitational energy is converted into thermal energy, how much does the temperature of the water increase? (Actually, Joule found no increase of temperature because the falling water cools by evaporation.)

4.——A nuclear power plant takes in 5×10^6 m^3 of cooling water per day from a river and exhausts 1200 megawatts of waste heat into this water. If the temperature of the inflowing water is 20°C, what is the temperature of the outflowing water?

5.——Your metabolism extracts about 100 kcal of chemical energy from one apple. If you want to get rid of all this energy by jogging, how far must you jog? At a speed of 12 km/h, jogging requires about 750 kcal/h.

6.——For basic subsistence a human body requires a diet with about 2000 kcal/day. Express this power in watts.

7.——By turning a crank, you can do mechanical work at the steady rate of 0.15 hp. If the crank is connected to paddles churning 4 liters of water, how long must you churn the water to raise its temperature by 5°C?

8.——You can warm the surfaces of your hands by rubbing one against the other. If the coefficient of friction between your hands is 0.6 and if you press your hands together with a force of 60 N while rubbing them back and forth at an average relative speed of 0.5 m/s, at what rate (in calories per second) do you generate heat on the surfaces of your hands?

9.——A simple gadget for heating water for showers consists of a black plastic bag holding 10 liters of water. When hung in the sun, the bag absorbs heat. On a clear, sunny day, the power delivered by sunlight per unit area facing the Sun is 1.0×10^3 W/m^2. The bag has an area of 0.10 m^2 facing the sun. How long does it take for the water to warm from 20°C to 50°C? Assume that the bag loses no heat.

10.*—A solar collector consists of a flat plate that absorbs the heat of sunlight. A water pipe attached to the back of the plate carries away the absorbed heat (Fig. 14.19). Assume that the solar collector has an area of 4.0 m^2 facing the Sun and that the power per unit area delivered by sunlight is 1.0×10^3 W/m^2. What is the rate at which water must circulate through the pipe if the temperature of the water is to increase by 40°C as it passes through the collector?

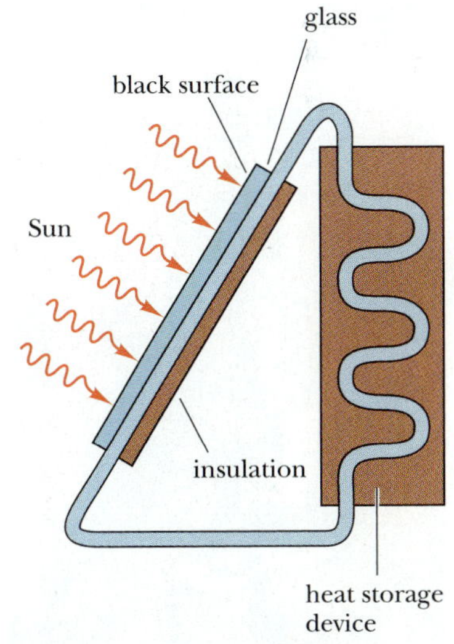

FIGURE 14.19
Collector of solar heat.

11.*—In the cooling system of the engine of a boat, water is pumped from the outside through the engine and then returned to the outside. The engine produces 3.0 kW of waste heat, and the pump circulates 12 liters of water per minute through the engine. If all of the waste heat is carried away by the water, what is the increase in the temperature of the water?

12.*—The first quantitative determination of the mechanical equivalent of heat was made by Robert von Mayer, who compared available data on the amount of mechanical work needed to compress a gas and the amount of heat generated during the compression. From this comparison, Mayer deduced that the energy required for warming 1 kg of water by 1°C is equivalent to the potential energy released when a mass of 1 kg falls from a height of 365 m. By how many percent does Mayer's result differ from the modern result given in Section 14.1?

FIGURE 14.20
Beam dump at SLAC.

13.*—On a hot summer day, the use of air conditioners raises the consumption of electric power in New York City to 22,400 megawatts. All of this electric power ultimately produces heat. Compare the heat produced in this way with the solar heat incident on the city. Assume that the incident flux of solar energy is 1 kilowatt per m^2 and the area of the city is 850 km^2. Would you expect that the consumption of electric power significantly increases the ambient temperature?

14.*—The beam dump at the Stanford Linear Accelerator consists of a large tank with 12 m^3 of water into which the accelerated electrons can be aimed when they are not wanted elsewhere (Fig. 14.20). The beam carries 3.0×10^{14} electrons/s; the kinetic energy per electron is 3.2×10^{-9} J. In the beam dump this energy is converted into heat.

(a) What is the rate of production of heat?

(b) If the water in the tank is stagnant and does not lose any heat to the environment, what is the rate of increase of temperature of the water?

(c) To prevent overheating, cooling water is pumped through the tank at the rate of 2.0 m^3/min; this carries away the heat. If the temperature of the inflowing water is 20°C, what is the temperature of the outflowing water?

FIGURE 14.21
Sears Tower, Chicago.

Section 14.2: Thermal Expansion of Solids and Liquids

15.—The tallest building in the world is the Sears Tower in Chicago, which is 443 m high (Fig. 14.21). It is made of concrete and steel. How much does its height change between a day when the temperature is 35°C and a day when the temperature is −29°C?

16.—The height of the Eiffel Tower is 321 m. How much does its height change when the temperature changes from −20°C to 35°C?

17.—Machinists use gauge blocks of steel as standards of length. A one-inch gauge block is supposed to have a length of 1 in., to within $\pm 10^{-6}$ in. In order to keep the length of the block within this tolerance, how precisely must the machinist control the temperature of the block?

18.—A mechanic wants to place a sleeve (pipe) of copper around a rod of steel. At a temperature of 18°C, the sleeve of copper has an inner diameter of 0.998 cm and the rod of steel has a diameter of 1.000 cm. To what temperature must the mechanic heat the copper to make it fit around the steel?

19.—(a) Segments of steel railroad rails are laid end to end. In an old railroad, each segment is 18 m long. If they are originally laid at a temperature of −7°C, how much of a gap must be left between adjacent segments if they are to just barely touch at a temperature of 43°C?

(b) In a modern railroad, each segment is 790 m long, with a special expansion joint at each end. How much of a gap must be left between adjacent segments in this case?

20.——The largest ship is the supertanker *Seawise Giant,* with a length of 458 m and a beam of 69 m. By how many meters does this tanker expand in length and in width when it travels from the wintry North Atlantic ($-20°C$) to the hot Persian Gulf ($+40°C$)? By how many square meters does its deck area increase? Assume that the deck is approximately a rectangle, $458 \text{ m} \times 69 \text{ m}$.

21.——The supertanker *Seawise Giant* (see also Problem 20) has an enclosed volume of $1.8 \times 10^6 \text{ m}^3$. By how many cubic meters does its volume expand when it travels from the wintry North Atlantic ($-20°C$) to the hot Persian Gulf ($+40°C$)?

22.*——Suppose you heat a 1-kg cube of iron from 20°C to 80°C while it is surrounded by air at a pressure of 1 atm. How much work does the iron do against the atmospheric pressure while expanding? Compare this work with the heat absorbed by the iron. (The density of iron is $7.9 \times 10^3 \text{ kg/m}^3$.)

23.*——When a solid expands, the increment of the area of one of its faces is directly proportional to the increment of temperature and to the original area. Show that the coefficient of proportionality for this expansion of area is two times the coefficient of linear expansion.

24.*——A spring made of steel has a relaxed length of 0.316 m at a temperature of 20°C. By how much will the length of this spring increase if we heat it to 150°C? What compressional force must we apply to the hot spring to bring it back to its original length? The spring constant is $3.5 \times 10^4 \text{ N/m}$.

25.*——The pendulum (rod and bob) of a pendulum clock is made of brass.

(a) What will be the fractional increment in the length of this pendulum if the temperature increases by 20°C? What will be the fractional increase in the period of the pendulum?

(b) The pendulum clock keeps good time when its temperature is 15°C. How much time (in seconds per day) will the clock lose when its temperature is 35°C?

26.*—— (a) The density of gasoline is 730 kg/m^3 when the temperature is 0°C. What will be the density of gasoline when the temperature is 30°C?

(b) The price of gasoline is 120 cents per gallon. What is the price per kilogram at 0°C? What is the price per kilogram at 30°C? (Note that $1 \text{ gal.} = 3.80 \times 10^{-3} \text{ m}^3$.) Is it better to buy cold gasoline or warm gasoline?

27.*——An ordinary mercury thermometer consists of a glass bulb to which is attached a fine capillary tube. As the mercury expands, it rises up the capillary tube. Given that the bulb has a volume of 0.20 cm^3 and that the capillary tube has a diameter of $7.0 \times 10^{-3} \text{ cm}$, how far will the mercury column rise up the capillary tube for a temperature increase of 10°C? Ignore the expansion of the glass and ignore the expansion of the mercury in the capillary tube.

Section 14.3: Conduction of Heat

28.——The walls of an igloo are made of compacted snow, 30 cm thick. What thickness of Styrofoam would provide the same insulation as the snow?

29.——The bottom of a tea kettle consists of a layer of steel 0.080 cm thick. The area of the bottom of the kettle is 300 cm^2. Below the steel is a hot plate at a temperature of 101.2°C, and above the steel is boiling water at 100.00°C. What is the rate of heat transfer through the bottom of the kettle?

30.——A rod of steel 0.70 cm in diameter is surrounded by a tight copper sleeve of inner diameter 0.70 cm and outer diameter 1.00 cm. What will be the heat flow along this compound rod if the temperature gradient along the rod is 50°C/cm? What fraction of the heat flows in the copper? What fraction in the steel?

31.——A window in a room measures 1 m $\times$ 1.5 m. It consists of a single sheet of glass of thickness 2.5 mm. What is the heat flow through this window if the temperature difference between the inside surface of the glass and the outside is 39°C? Compare the heat loss through the window with the heat loss through the wall calculated in Example 5.

32.——A pan of aluminum, filled with boiling water, sits on a hot plate. The bottom area of the pan is 300 cm^2 and the thickness of the aluminum is 0.10 cm. If the hot plate supplies 2000 W of heat to the bottom of the pan, what must be the temperature of the upper surface of the hot plate?

33.*—A Styrofoam box, used for the transportation of medical supplies, is filled with dry ice (carbon dioxide) at a temperature of -79°C. The box measures 30 cm $\times$ 30 cm $\times$ 40 cm and its walls are 4 cm thick. If the outside surface of the box is at a temperature of 20°C, what is the rate of loss of dry ice by vaporization?

34.*—The icebox on a sailboat measures 60 cm $\times$ 60 cm $\times$ 60 cm. The contents of this icebox are to be kept at a temperature of 0°C for 4 days by the gradual melting of a block of ice of 20 kg, while the temperature of the outside of the box is 30°C. What minimum thickness of Styrofoam insulation is required for the walls of the icebox?

35.*—A man has a skin area of 1.8 m^2; his skin temperature is 34°C. On a cold winter day, the man wears a whole-body suit insulated with down. The temperature of the outside surface of his suit is -25°C. If the man can stand a heat loss of no more than 100 kcal/h, what is the minimum thickness of down required for his suit?

Section 14.4: Changes of State

36.——While jogging on a level road, your body generates heat at the rate of 750 kcal/h. Assume that evaporation of sweat removes 50% of this heat and convection and radiation the remainder. The evaporation of 1 kg (or 1 liter) of sweat requires 580 kcal. How many kilograms of sweat do you evaporate per hour?

37.——You place 1 kg of ice (at 0°C) in a pot and heat it until the ice melts and the water boils off, making steam. How much heat must you supply to accomplish this?

38.——Thunderstorms obtain their energy by condensing the water vapor contained in humid air. Suppose that a thunderstorm succeeds in condensing *all* the water vapor in 10 km^3 of air.

(a) How much heat does this release? Assume the air is initially at 100% humidity and that each cubic meter of air at 100% humidity (at 20°C and 1 atm) contains 1.74×10^{-2} kg of water vapor. The heat of vaporization of water is 585 kcal/kg at 20°C.

(b) The explosion of an A bomb releases an energy of 2×10^{10} kcal. How many A bombs does it take to make up the energy of one thunderstorm?

39.——The Mediterranean loses a large volume of water by evaporation. This loss is made good, in part, by currents flowing into the Mediterranean through the straits joining it to the Atlantic and the Black Sea. Calculate the rate of evaporation (in km^3/h) of the Mediterranean on a clear summer day from the following data: the area of the Mediterranean is 2.9×10^6 km^2, the power per unit area supplied by sunlight is 1×10^3

W/m², and the heat of vaporization of water is 580 kcal/kg (at a temperature of 21°C). Assume that all the heat of sunlight is used for evaporation.

40.*——If you pour 0.50 kg of molten lead at 328°C into 2.5 liters of water at 20°C, what will be the final temperature of the water and the lead? The specific heat of (solid) lead has an average value of 3.4×10^{-2} kcal/kg · °C over the relevant temperature range.

41.*——During a rainstorm lasting 2 days, 7.6 cm of rain fell over an area of 2.6×10^3 km².

(a) What is the total mass of the rain (in kilograms)?

(b) Suppose that the heat of vaporization of water in the rain clouds is 580 kcal/kg. How many calories of heat are released during the formation of the total mass of rain by condensation of the water vapor in the rain clouds?

(c) Suppose that the rain clouds are at a height of 1500 m above the ground. What is the gravitational potential energy of the total mass of rain before it falls? Express your answer in calories.

(d) Suppose that the raindrops hit the ground with a speed of 10 m/s. What is the total kinetic energy of all the raindrops taken together? Express your answer in calories. Why does your answer to Part (c) not agree with this?

42.*——The heat of vaporization of water at 100°C and 1 atm is 539 kcal/kg. How much of this energy is due to the work the water vapor does against atmospheric pressure? What would be this work at a pressure of 0.1 atm? At (nearly) zero pressure?

43.*——Suppose you drop a cube of titanium of mass 0.25 kg into a Dewar (a thermos bottle) full of liquid nitrogen at −196°C. The initial temperature of the titanium is 20°C. How many kilograms of nitrogen will boil off as the titanium cools from 20°C to −196°C? The specific heat of titanium is 8.2×10^{-2} kcal/kg · °C.

Section 14.5: The Specific Heat of a Gas

44.——A TV tube of glass with zero pressure inside and atmospheric pressure outside suddenly cracks and implodes. The volume of the tube is 2.5×10^{-2} m³. During the implosion, the atmosphere does work on the fragments of the tube and on the layer of air immediately adjacent to the tube. This amount of work represents the energy released in the implosion. Calculate this energy. If all of this energy is acquired by the fragments of glass, what will be the mean speed of the fragments? The total mass of the glass is 2.0 kg.

45.——How much heat must we supply to heat 1 mole of argon gas from 30°C to 100°C at constant volume? At constant pressure?

46.——If we heat 1.00 kg of hydrogen gas from 0°C to 50.0°C in a cylinder with a piston keeping the gas at a constant pressure of 1.0 atm, we must supply 1.69×10^2 kcal of heat. How many kilocalories of heat must we supply to heat the same amount of gas from 0°C to 50.0°C in a container of constant volume?

47.——The rear end of an air conditioner dumps 3000 kcal/h of waste heat into the air outside a building. A fan assists in the removal of this heat. The fan draws in 15 m³/min of air at a temperature of 30°C and ejects this air after it has absorbed the waste heat. With what temperature does the air emerge?

48.——Table 14.5 gives the specific heat C_V per mole for several gases. Calculate the specific heat per kilogram for each gas. Which gas has the highest value of the specific heat per kilogram? The lowest?

49.——To discover whether an unknown gas is monoatomic or diatomic, an experimenter takes a 2.0-liter sample of this gas at STP and heats this sample to 100°C at constant volume.

(a) The heat that the gas absorbs during this process is 44 cal. Is the gas monoatomic or diatomic? Assume it is an ideal gas.

(b) The experimenter weighs the sample, and finds that its mass is 2.5 g. Can you tell what gas it is?

50.——Suppose we heat 1 mole of oxygen gas at a constant pressure of 1 atm from 20°C to 80°C, and then cool it at constant volume from 80°C back to 20°C.

(a) How much heat is absorbed by the gas during the first step?

(b) How much heat is released by the gas during the second step?

(c) What is the volume of the gas at the end of the first step? What is the pressure at the end of the second step?

(d) How much work does the gas perform during the first step? During the second step?

51.——As in Example 7, consider an initial volume of 1 m^3 of air that is warmed by sunlight from 0°C to 10°C at a constant pressure of 1 atm. What is the final volume of this air? During its expansion, how much work does the air do against the pressure of the surrounding atmosphere?

52.——What are the specific heats C_V and C_p for air consisting of 76% nitrogen, 23% oxygen, and 1% argon at STP? Use the values for C_V and C_p of nitrogen, oxygen, and argon listed in Table 14.5.

53.——A helium balloon consists of a large bag loosely filled with 600 kg of helium at an initial temperature of 10°C. While exposed to the heat of the Sun, the helium gradually warms to a temperature of 30°C. The heating proceeds at a constant pressure of 1.0 atm. How much heat does the helium absorb during this temperature change?

54.*—On a winter day you inhale cold air at a temperature of −30°C and at 0% humidity. The amount of air you inhale is 0.45 kg per hour. Inside your body you warm and humidify the air; you then exhale the air at a temperature of 37°C and 100% relative humidity. At a temperature of 37°C, each kilogram of air at 100% relative humidity contains 0.041 kg of water vapor. How many calories are carried out of your body by the air that passes through your lungs in one hour? Take into account both the heat needed to warm the air at constant pressure and the heat needed to vaporize the moisture that the exhaled air carries out of your body. The specific heat of air at constant pressure is 0.25 kcal/kg·°C; the heat of vaporization of water at 37°C is 576 kcal/kg.

CHAPTER 15

This steam-driven locomotive is a heat engine—it converts the thermal energy supplied by burning coal into mechanical energy.

THERMODYNAMICS

◂ Thermodynamics

Thermodynamics deals with the conversion of one form of energy into another, especially the conversion of thermal energy into other forms of energy. These conversions are governed by the two fundamental laws of thermodynamics. The first of these is essentially a general statement of the law of conservation of energy, and the second is a statement about the maximum efficiency attainable in the conversion of heat into work.

Thermodynamics describes physical processes in terms of purely macroscopic parameters. Such a macroscopic, large-scale description is necessarily somewhat crude, since it overlooks all of the small-scale, microscopic details. However, in practical applications these details are often irrelevant. For instance, an engineer investigating the behavior of the combustion gases in an automobile engine can get by reasonably well with such macroscopic quantities such as temperature, pressure, density, and heat capacity, without knowing the microscopic behavior of the gases.

The study of thermodynamics was inaugurated by nineteenth-century engineers who wanted to know what ultimate limitations the laws of physics impose on the operation of steam engines and other machines that generate mechanical energy. They soon established that perpetual-motion machines, which earlier inventors had sought to build, are impossible. A ***perpetual-motion machine of the first kind*** is a (hypothetical) device that supplies an endless output of work without any input of fuel or any other input of energy. Figure 15.1 shows a proposed design for such a machine. Weights are attached to the rim of a wheel by short pivoted rods resting against pegs. With the rods in the position shown, there is an imbalance in the weight distribution causing a clockwise torque on the wheel; as the wheel turns, the rod coming to the top presumably flips over, maintaining the imbalance. This perpetual torque would not only keep the wheel turning, but would also continually deliver energy to the axle of the wheel. However, a detailed analysis demonstrates that the machine will not perform as intended — the wheel actually settles in a static equilibrium configuration such that the top rod just barely fails to flip over. The First Law of Thermodynamics, or the law of conservation of energy, directly tells us of the failure of this machine: after one revolution of the wheel, the masses all return to their initial positions, their potential energy returns to its initial value, and they will not have delivered net energy to the motion of the wheel.

Perpetual-motion machine of the first kind ▶

Perpetual-motion machine of the second kind ▶

A ***perpetual-motion machine of the second kind*** is a device that extracts thermal energy from some heat source, such as air or the water of the ocean, and converts it into mechanical energy. Such a device is not forbidden by conservation laws. The oceans are enormous reservoirs of thermal energy; if we could extract this thermal energy, a temperature drop of just 1°C of the oceans would supply the energy needs of the United States for the next 50 years. But, as we will see, the Second Law of Thermodynamics tells us that conversion of heat into work requires not only a heat source, but also a heat sink. Heat flows out of a warm body only if there is a cooler body that can absorb it. If we want heat to flow from the ocean into our machine, we must provide a low-temperature heat sink toward which the heat will tend to flow spontaneously. If no low-temperature sink is available, the extraction of heat from the ocean is impossible, and we cannot build a perpetual-motion machine of the second kind.

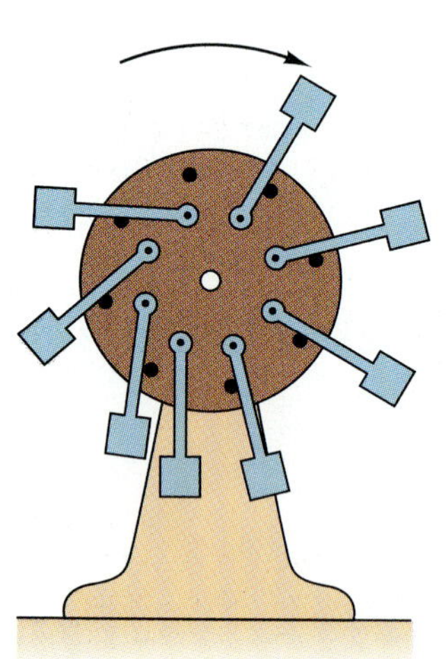

FIGURE 15.1
A hypothetical perpetual-motion machine.

15.1 THE FIRST LAW OF THERMODYNAMICS

Consider some amount of gas with a given initial volume V_1, pressure p_1, and temperature T_1. The gas is in a container fitted with a piston (see Fig. 15.2). Suppose we push on the piston and compress the gas to some smaller volume V_2, and we also cool the gas and lower its temperature to some smaller value T_2; the pressure will then reach some new value p_2. (Such a compression and cooling process is of practical importance in the liquefaction of gases, such as oxygen or nitrogen — before the gas can be liquefied it must be compressed and cooled.) We can reach the new state V_2, p_2, and T_2 from the old state V_1, p_1, and T_1 in a variety of ways. For instance, we may first compress the gas and then cool it. Or else, we may first cool it and then compress it. Or we may go through small alternating steps of compressing and cooling. In order to compress the gas, we must do work on it [see Eq. (14.23)]; and in order to cool the gas, we must remove heat from it. The work done on or by the gas and the heat transferred from or to the gas result in a change of the internal energy of the gas. We can express this change of internal energy as

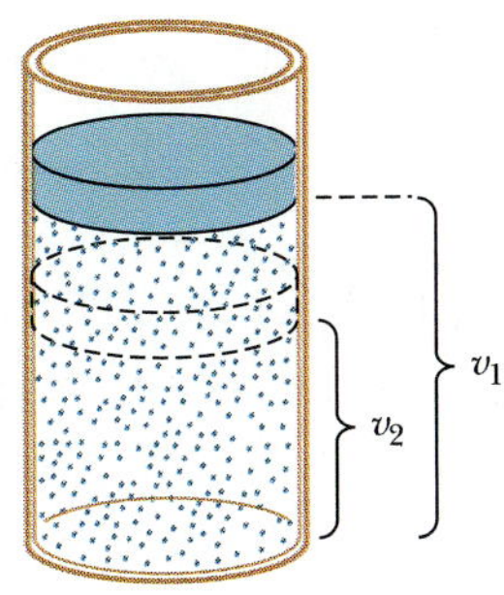

FIGURE 15.2

Compression of a gas by a piston.

$$\Delta E = Q - W \tag{1}$$

where Q is the amount of heat supplied to the gas, and W is the amount of work performed by the gas. Note the sign conventions in this equation: Q is positive if we add heat to the gas and negative if we remove heat; W is positive if the gas does work on us and negative if we do work on the gas.

The values of Q and W depend on the process. If we first compress the gas and subsequently cool it, then during the first step W is negative and Q zero; and during the second step W is zero and Q is negative. If we first cool the gas and then compress it, the values of W and Q will be quite different. Yet, it turns out that regardless of what sequence of operations we use to transform the gas from its initial state V_1, p_1, T_1 to its final state V_2, p_2, T_2, the net change ΔE in the internal energy is always the same: Q and W vary, but the sum of Q and $-W$ remain fixed. This is the ***First Law of Thermodynamics:***

◀ **First Law of Thermodynamics**

> ***Whenever we employ some process involving heat and work to change a system from an initial state characterized by certain values of the macroscopic parameters to a final state characterized by new values of the macroscopic parameters, the change in the internal energy of the system has a fixed value***
>
> $$\Delta E = Q - W \tag{2}$$
>
> ***which does not depend on the details of the process.***

The sign conventions in this equation are as in Eq. (1) (see Fig. 15.3).

Note that the First Law tells us that the energy is conserved — the change of internal energy of the system equals the input of heat and work; that is, it equals the input of microscopic work and macroscopic work. But the First Law tells us more than that. If we describe a system in terms of the detailed microscopic positions and velocities of all its constituent particles, then energy conservation is uncontestable — it is a theorem of mechanics. But if we describe a system in terms of nothing but macroscopic parameters, then it is not at all obvious that we have available enough information to determine the energy and to formulate a conservation law. The First Law of Thermodynamics tells us that a knowledge of the macroscopic parameters is indeed sufficient to determine the energy of the system.

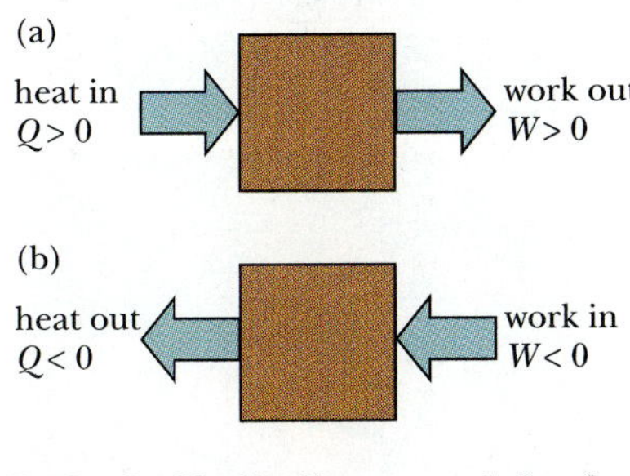

FIGURE 15.3

(a) If the system receives heat from its surroundings, and performs work on its surroundings, Q is positive and W is positive. (b) If the system delivers heat to its surroundings, and the surroundings perform work on the system, Q is negative and W is negative.

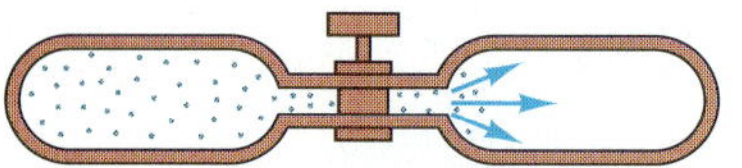

FIGURE 15.4
Free expansion of a gas.

EXAMPLE 1. Some amount of gas at temperature T_1 is stored in a thermally insulated bottle. By means of a pipe with a stopcock, we connect this bottle to another insulated bottle which is evacuated (see Fig. 15.4). If we suddenly open the stopcock, the gas will rush from the first bottle into the second until the pressures are equalized. This is called a free expansion of the gas, because the gas expands without pushing against anything. What does the First Law say about the change of the internal energy of the gas in this process?

SOLUTION: The expansion process neither adds nor removes heat from the gas, that is, $Q = 0$. Furthermore, the expansion process involves no work (the gas does not push against any piston), that is, $W = 0$. Consequently, Eq. (2) tells that the internal energy of the gas does not change,

$$\Delta E = Q - W = 0 \tag{3}$$

We can take this conclusion a step further if we assume that the gas behaves like an ideal gas. If so, the internal energy E depends on the temperature only [E is directly proportional to T; see Eqs. (13.27) and (13.33)]. Since the energy does not change, we can then conclude that the temperature does not change. Thus, in the free expansion of the gas, the temperature remains constant.

FIGURE 15.5
Steam engine and auto engine.

15.2 HEAT ENGINES; THE CARNOT ENGINE

Steam engines and automobile engines convert thermal energy into mechanical energy (Fig. 15.5). The steam engine obtains heat from the combustion of coal or oil in a boiler; the automobile engine obtains heat from the (explosive) combustion of gasoline in its cylinders. If an engine is continually to convert thermal energy into mechanical energy, it must operate cyclically. At the end of each cycle, it must return to its initial configuration, so it can repeat the process of conversion of heat into work over and over again. Steam engines and automobile engines are obviously cyclic — after one revolution (or sometimes two), they return to their initial configuration. These engines are not 100% efficient. The condenser of the steam engine and the radiator and exhaust of the automobile engine eject a substantial amount of heat into the environment; this waste heat represents lost energy.

Heat engine ▶

Any device that converts heat into work by means of a cyclic process is called a heat engine. During each cycle of operation, the engine absorbs heat from a heat reservoir at high temperature, converts this heat partially into work, and ejects the remainder as waste heat into a reservoir at low temperature. In this context, a ***heat reservoir*** is simply a body that remains at constant temperature, even when heat is removed from or added to it. In practice, the high-temperature heat reservoir is often a boiler whose temperature is kept constant by the controlled combustion of some fuel, and the low-temperature reservoir is usually a condenser in contact with a body of water or in contact with the atmosphere of the Earth, whose large volume permits it to absorb the waste heat without appreciable change of temperature.

◀ **Heat reservoir**

Figure 15.6 is a flow chart for the energy, showing the heat Q_1 flowing into the engine from the high-temperature reservoir, the heat Q_2 (waste heat) flowing out of the engine into the low-temperature reservoir, and the work generated. The work generated is the difference between Q_1 and Q_2,

$$W = Q_1 - Q_2 \tag{4}$$

◀ **Efficiency of a heat engine**

The efficiency of the engine is defined as the ratio of this work to the heat absorbed from the high-temperature reservoir,

$$e = \frac{W}{Q_1} = \frac{Q_1 - Q_2}{Q_1} = 1 - \frac{Q_2}{Q_1} \tag{5}$$

This says that if there were no waste heat ($Q_2 = 0$), then the efficiency would be $e = 1$, or 100%. If so, the engine would convert the high-temperature heat *totally* into work. As we will see later, this extreme efficiency is unattainable. Even under ideal conditions, the engine will produce some waste heat.

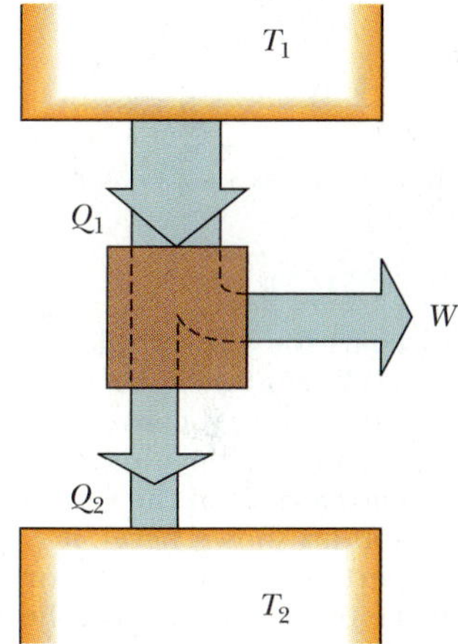

FIGURE 15.6

Flow chart for a heat engine. The square box represents the heat engine.

EXAMPLE 2. The steam engine of a locomotive delivers 5.4×10^8 J of work per minute and receives 3.6×10^9 J of heat per minute from its boiler. What is the efficiency of this engine? How much heat is wasted per minute?

SOLUTION: From Eq. (5),

$$e = \frac{W}{Q_1} = \frac{5.4 \times 10^8 \text{ J}}{3.6 \times 10^9 \text{ J}} = 0.15 \tag{6}$$

Expressed in percent, this is 15%.

The wasted heat is the difference between the heat received and the work:

$$Q_2 = Q_1 - W = 3.6 \times 10^9 \text{ J} - 5.4 \times 10^8 \text{ J} = 3.1 \times 10^9 \text{ J} \tag{7}$$

FIGURE 15.7
Athlete on stationary bike.

EXAMPLE 3. During strenuous bicycling, an athlete delivers 220 W of mechanical power to the pedals of the bicycle and, simultaneously, generates 760 W of waste heat (Fig. 15.7). What overall efficiency is implied by these data?

SOLUTION: A power of 1 W is equal to 1 J per second. Thus, in one second, the mechanical work delivered by the athlete is 220 J and the waste heat generated is 760 J. Accordingly, the thermal (chemical) energy flowing into the athlete's muscles must be $Q_1 = Q_2 + W = 760\text{ J} + 220\text{ J} = 980\text{ J}$, which implies an efficiency of

$$e = \frac{W}{Q_1} = \frac{220\text{ J}}{980\text{ J}} = 0.22$$

COMMENTS AND SUGGESTIONS: Physiologists sometimes subtract the basal metabolic rate of 88 W from the rate of production of waste heat before calculating the efficiency, on the excuse that the basal metabolic rate serves to support the life of the organism, and has nothing to do with the external work. If we include this modification in our calculation of the efficiency, then

$$e = \frac{W}{Q_1} = \frac{220\text{ J}}{980\text{ J} - 88\text{ J}} = 0.25$$

However, engineers and physicists are not in the habit of making such allowances for the energy wasted by an engine while "idling."

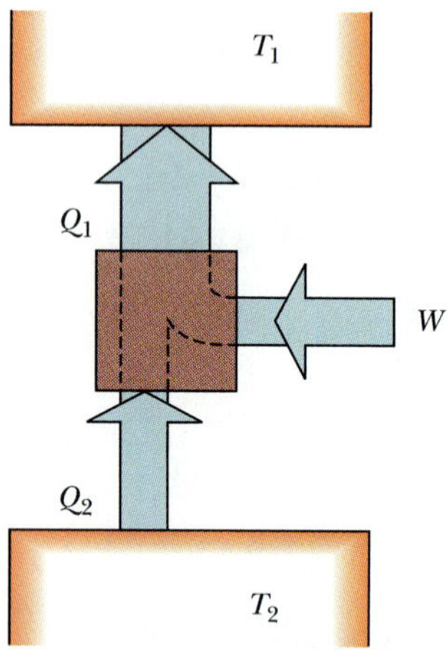

FIGURE 15.8
Flow chart for a reversible engine. When the engine moves in one direction, it converts some amount of heat into work. When it operates in the reverse direction (as shown here), it reconverts this amount of work into the original amount of heat.

We will now calculate the efficiency of an ideal heat engine that converts heat into work with maximum efficiency. As we will see in the next section, for maximum efficiency, the thermodynamic process within the engine should be *reversible*. ***This means that the engine can, in principle, be operated in reverse, and it then converts work into heat at the same rate as it converts heat into work when operating in the forward direction*** (see Fig. 15.8).

Reversible engine ▶

Carnot engine ▶

The simplest kind of reversible engine is the ***Carnot engine,*** consisting of some amount of ideal gas enclosed in a cylinder with a piston (see Fig. 15.9). We can alternately place the cylinder in thermal contact with a high-temperature reservoir (where it absorbs heat) or a low-temperature reservoir (where it dumps waste heat). The gas delivers work when it pushes the piston outward, and the gas absorbs work when we push the piston inward. To achieve reversibility with this engine, the motion of the piston must be sufficiently slow, so that the gas is always in an equilibrium configuration. If the piston were to have a sudden motion, a pressure disturbance would travel through the gas, and the motion of this pressure disturbance could not be reversed by giving the piston a sudden motion in the opposite direction — this would merely create a second pressure disturbance. Furthermore, the temperature of the gas must coincide with the temperature of the heat reservoir during contact. If the gas were to have, say, a lower temperature than that of the heat reservoir with which it is in contact, heat would rush from the reservoir into the gas, and this flow of heat could not be reversed by any manipulation of the piston. In practice, we cannot attain exact reversibility; but it is nevertheless worthwhile to consider the ideal Carnot engine with exact reversibility, because this tells us what is the best we can hope for when attempting to convert heat into work.

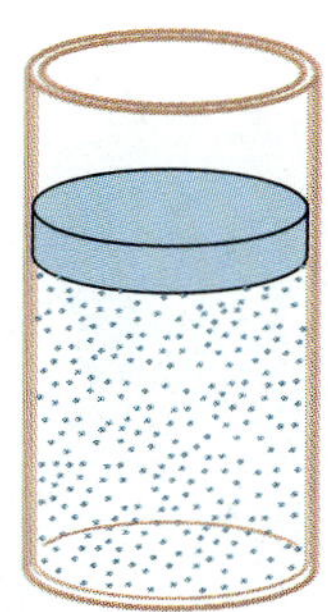

FIGURE 15.9
Carnot engine: a gas-filled cylinder with a piston.

The operation of the Carnot engine takes the gas through a sequence of four steps with varying volume and pressure, but at the end of the last step the gas returns to its initial volume and pressure. The four steps are illustrated in Fig. 15.10. This sequence of four steps is called the ***Carnot cycle:***

◂ Carnot cycle

a. We begin the cycle by placing the cylinder in contact with the high-temperature heat reservoir, which maintains the temperature of the gas at the constant value T_1. The gas is now allowed to expand from the initial volume V_1 to a new volume V_2. This is called an ***isothermal expansion.*** During the expansion, the gas does work on the piston; that is, the engine absorbs heat from the reservoir and converts it into work.

◂ Isothermal expansion

b. When the gas has reached volume V_2 and pressure p_2, we remove it from the heat reservoir and allow it to continue the expansion on its own, in thermal isolation. The expansion of a thermally isolated gas, which neither receives heat from its surroundings nor loses any, is called an ***adiabatic expansion.*** During this expansion, the temperature of the gas decreases. (You can readily notice such a decrease of temperature of an expanding gas when you let air rush out of the valve of an automobile tire — the expanding air feels quite cool. This expansion is approximately adiabatic, because the rushing air, although not isolated from its surroundings, expands so quickly that it does not have time to exchange heat with the surrounding atmospheric air.

◂ Adiabatic expansion

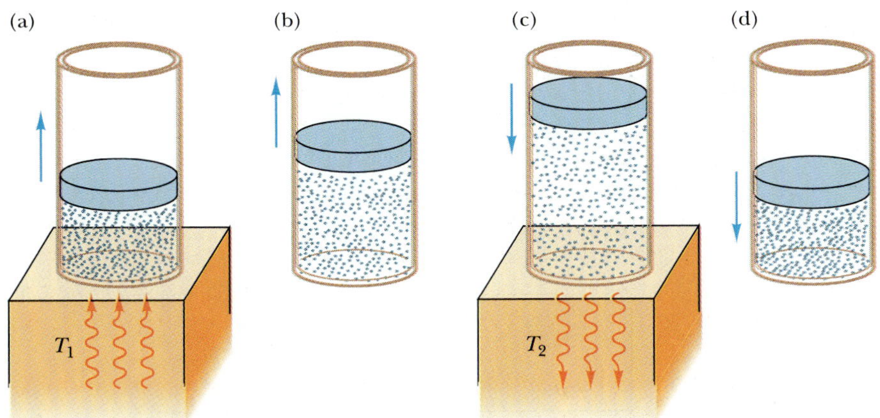

FIGURE 15.10
The Carnot cycle. The arrows indicate the displacements of the piston. (a) Expansion at constant temperature T_1 while in contact with a high-temperature heat reservoir. (b) Adiabatic expansion. (c) Compression at constant temperature T_2 while in contact with a low-temperature heat reservoir. (d) Adiabatic compression to the initial volume and pressure.

Sadi Carnot (karno), *1796–1832, French engineer and physicist. In his book* On the Motive Power of Heat *he formulated the theory of the conversion of heat into work.*

Conversely, you can notice the increase of temperature during an adiabatic compression of a gas when you operate a manual bicycle pump. The compression of the air in the barrel of the pump produces a detectable warming of the pump.)

c. When the temperature of the gas has decreased to the temperature T_2 of the low-temperature reservoir, we stop the piston and place the gas in contact with this low-temperature reservoir. The volume at this instant is V_3 and the pressure p_3. We now begin to push the piston back toward its starting position, that is, we compress the gas isothermally. This means that the engine converts work to heat and ejects this heat into the low-temperature reservoir.

d. When the gas has reached volume V_4 and pressure p_4, we remove it from contact with the low-temperature reservoir and continue to compress it adiabatically until the volume and the pressure return to their initial values. During this adiabatic compression, the temperature increases from T_2 to its initial value T_1.

To describe the operation of this engine mathematically, it is best to use a p–V diagram (see Fig. 15.11). Each point in this diagram represents an equilibrium configuration of the gas; p and V can be read directly from the diagram, and T can then be calculated from the ideal-gas law.

To find the efficiency, we need to calculate the heat Q_1 that the engine absorbs from the high-temperature reservoir in step (a) and the heat Q_2 that it ejects into the low-temperature reservoir in step (c). These calculations are a bit difficult because the volume and the pressure vary in a somewhat complicated way along the curved lines in Fig. 15.11. But Carnot showed that the final result of the calculation is very simple: the amounts of heat Q_1 and Q_2 are in direct proportion to the respective temperatures T_1 and T_2 of the reservoirs. Thus, the ratio of Q_2 and Q_1

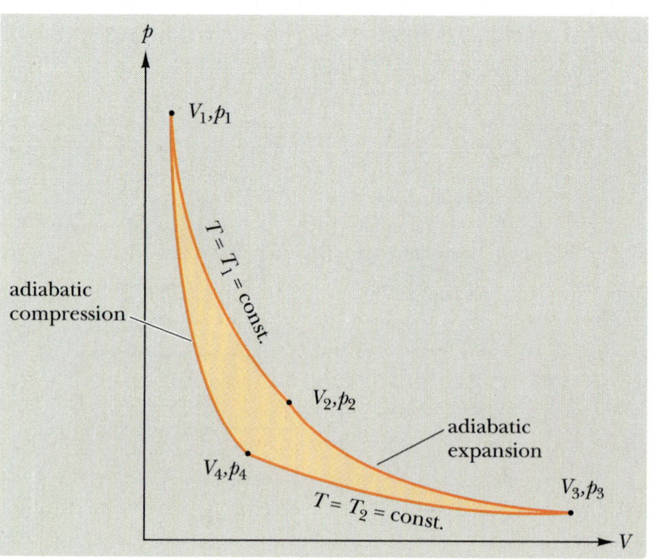

FIGURE 15.11
The Carnot cycle displayed on a p–V diagram.

equals the ratio of T_2 and T_1,

$$\frac{Q_2}{Q_1} = \frac{T_2}{T_1} \tag{8}$$

With this expression for Q_2/Q_1, the formula (5) for the efficiency tells us that the efficiency of the Carnot engine is

$$e = 1 - \frac{Q_2}{Q_1} = 1 - \frac{T_2}{T_1} \tag{9}$$

◂ **Efficiency of a Carnot engine**

This shows that ***the efficiency depends only on the temperatures of the heat reservoirs.***

Note that an efficiency of $e = 1$ (or 100%) can be achieved only if $T_2 = 0$, that is, if the low-temperature reservoir is at the absolute zero of temperature. Unfortunately, we have no such absolutely cold reservoir available on Earth.

EXAMPLE 4. The boiler of a steam engine produces steam at a temperature of 500°C. The engine exhausts its waste heat into the atmosphere where the temperature is 20°C. The actual efficiency of this steam engine is 0.15. Compare this with the efficiency of a Carnot engine operating between the same temperatures.

SOLUTION: According to Eq. (9), the efficiency of a Carnot engine with $T_1 = (500 + 273)$ K and $T_2 = (20 + 273)$ K is

$$e = 1 - \frac{293\text{ K}}{773\text{ K}} = 0.62 \tag{10}$$

Thus, the ideal Carnot engine is more efficient than the actual steam engine by 0.47, or 47%.

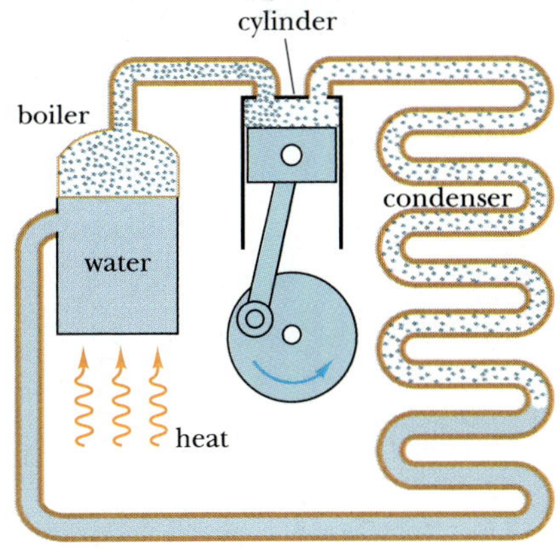

FIGURE 15.12
A steam engine (schematic).

◂ **Steam engine**

The basic steps in the operation of a practical steam engine resemble the steps of the Carnot cycle. The ***steam engine*** has a cylinder and a piston that performs a cyclic motion of expansion and compression. However, in contrast to the Carnot engine, which uses gas as working fluid, practical steam engines use gas and liquid (steam and water) as working fluid, and the cycle of a steam engine is more complicated than the Carnot cycle. Figure 15.12 is a schematic diagram of the main parts of a simple steam engine with boiler, cylinder, and condenser. The boiler produces hot, high-pressure steam which enters the cylinder and pushes against the piston doing work. The low-pressure, spent steam is then exhausted from the cylinder and sent to a condenser where an external coolant (air or flowing water) condenses the steam into liquid water. This liquid water is pumped back to the boiler. Each completed circulation of the fluid through the circuit can be regarded as one cycle of operation of the steam engine. Such simple steam engines have efficiencies of about 5–18%. Most modern steam engines employ a turbine wheel instead of the cylinder and piston; large engines of this kind achieve efficiencies of up to 40%.

Refrigerator ▸

If the Carnot engine is operated in reverse, it uses up work to transfer heat from the low-temperature reservoir to the high-temperature reservoir. This is the principle involved in the operation of ***refrigerators,*** air conditioners, and "heat pumps" (see Fig. 15.13). The amount of work required to operate a Carnot engine in reverse can be calculated from Eq. (8).

FIGURE 15.13
Heat pump.

EXAMPLE 5. Suppose a homeowner uses a Carnot engine operating in reverse as a heat pump to extract heat from the outside air and inject it into her home. If the outside temperature is −10 °C, and the inside temperature is 20 °C, what is the amount of work that must be supplied to pump 1 kcal of heat from the outside to the inside?

SOLUTION: For a Carnot engine, the ratio of the heats exchanged at the low-temperature reservoir and at the high-temperature reservoir is given by Eq. (8), with $T_1 = (20 + 273)$ K and $T_2 = (-10 + 273)$ K:

$$\frac{Q_2}{Q_1} = \frac{T_2}{T_1} = \frac{263\ \text{K}}{293\ \text{K}} = 0.90 \qquad (11)$$

According to this equation, if $Q_2 = 1$ kcal, then $Q_1 = Q_2/0.90 = 1.11$ kcal. The difference $Q_1 - Q_2$ represents the work that must be supplied [see Eq. (4)]; hence the work is 0.11 kcal.

Note that by the expenditure of 0.11 kcal of work, the heat pump delivers a total of 1.11 kcal of heat into the house. This is obviously a more economical heating method than the expenditure of 1.11 kcal of fuel or electric energy in a conventional furnace or electric heater.

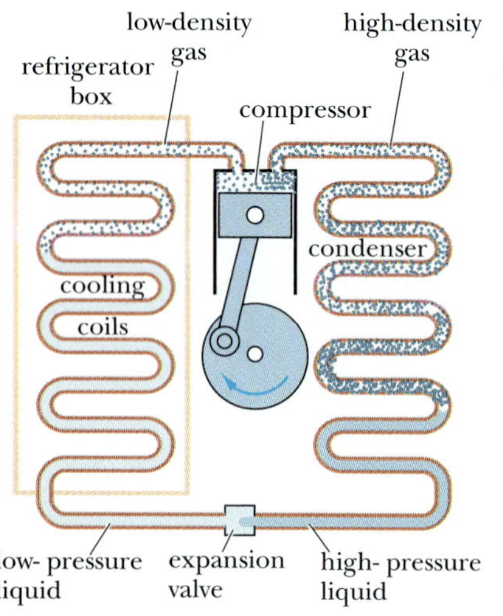

FIGURE 15.14
A refrigerator (schematic).

Practical refrigerators use gas and liquid freon (dichlorodifluoromethane) as working fluid, and their cycle differs from the Carnot cycle. Freon and similar substances employed as refrigerants have a boiling point near room temperature when at high pressure, but a boiling point below 0 °C when at low pressure. Figure 15.14 is a schematic diagram of the parts of a practical refrigerator. Liquid freon at low pressure enters the cooling coils in the refrigerator box and absorbs heat while evaporating into freon gas (Fig. 15.15). This gas flows to the compressor, where its pressure and density are increased by the push of a piston. The high-pressure gas then circulates through the condenser coils, which are exposed to the atmospheric air. The freon gas loses its heat and condenses into a liquid. This high-pressure liquid then passes through an expansion valve (a small orifice) where its pressure is reduced to match the low pressure in the

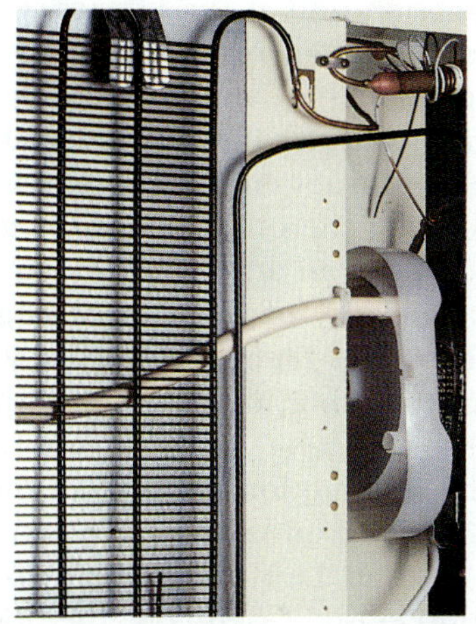

FIGURE 15.15
Cooling coils, motor, and compressor of a refrigerator.

cooling coils. This return of the fluid to the cooling coils completes the cycle.

An air conditioner employs a similar refrigeration cycle; and so does a "heat pump," which is merely an air conditioner turned around, so its cold end is outdoors and its warm end indoors.

15.3 THE SECOND LAW OF THERMODYNAMICS

As we saw in the preceding section, a heat engine operating with an ideal gas as working fluid has a limited efficiency — it fails to convert all of the heat into work and instead produces some waste heat. The ***Second Law of Thermodynamics*** asserts that this is a limitation from which all heat engines suffer. As formulated by Lord Kelvin and Max Planck, this law simply states:

> ***An engine operating in a cycle cannot transform heat into work without some other effect on its surroundings.***

◂ Second Law of Thermodynamics

It is an immediate corollary of this law that ***the efficiency of any heat engine operating between two heat reservoirs of high and low temperature is never greater than the efficiency of a Carnot engine; furthermore, the efficiency of any reversible engine equals the efficiency of a Carnot engine.*** The proof of these statements, known as ***Carnot's theorem,*** is by contradiction. Imagine that a heat engine more efficient than a Carnot engine converts heat from a reservoir at a high temperature into work and ejects only a small amount of waste heat into a reservoir at low temperature. We can then use the work output of this engine to drive a Carnot engine in reverse, pumping the waste heat from the low-temperature reservoir to the high-temperature reservoir. By hypothesis, the given engine is more efficient than the Carnot engine; hence only part of its work output will be needed to drive the reversed Carnot engine and return all of the waste heat to the high-temperature reservoir. The remainder of the output constitutes available work (see Fig. 15.16 for a flow chart). The net effect of the joint operation of both engines is the complete conversion of heat into work, without waste heat, in contradiction to the Second Law. We

◂ Carnot's theorem

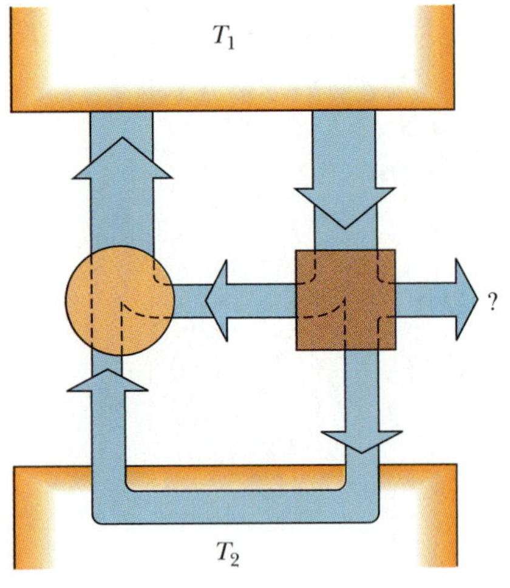

FIGURE 15.16

Flow chart for an arbitrary reversible engine (box) and a Carnot engine (circle) connected together. The arbitrary engine drives the Carnot engine in reverse.

can avoid this contradiction only if the efficiency of any engine is never greater than that of a Carnot engine.

To prove that the efficiency of any reversible engine equals that of a Carnot engine, we again consider the net effect of the joint operation of the two engines, with the Carnot engine running in the forward direction and its work output driving the other reversible engine in the backward direction (see Fig. 15.17). By an argument similar to that given above, it now follows that the efficiency of the Carnot engine cannot be greater than that of the other engine. Thus, the efficiency of each engine can be no greater than that of the other, that is, they both must have exactly the same efficiency.

We recall that a perpetual-motion machine of the second kind is an engine that takes heat energy from a reservoir and completely converts it into work. Thus, the Second Law asserts that no perpetual-motion machine of the second kind exists. Essentially, the operation of any heat engine hinges on the temperature difference between two heat reservoirs. The heat in the high-temperature reservoir has a high "potential" for work, and that in the low-temperature reservoir has a low "potential." If we were to place the two reservoirs in thermal contact, heat would rush from the high-temperature reservoir to the low-temperature reservoir. By interposing a heat engine in the path of this rush of heat, we can force the heat to do useful work. Thus, heat engines depend on the tendency of heat to flow from a hot reservoir to a cooler reservoir. The Second Law can be formulated in an alternative form that is based on this characteristic of the flow of heat. This formulation, due to Rudolph Clausius, states:

An engine operating in a cycle cannot transfer heat from a cold reservoir to a hot reservoir without some other effect on its surroundings.

The Clausius and Kelvin–Planck formulations of the Second Law are equivalent — each implies the other. The proof is again by contradiction, and relies on examining the result of a joint operation of two engines: a Carnot engine and a second engine that violates one of the two formulation of the Second Law. It is then easy to show that the joint operation violates the other formulation of the Second Law.

We recall that, according to Eq. (9), the efficiency of a Carnot engine is

$$e = 1 - \frac{T_2}{T_1} \tag{12}$$

This must then also be the efficiency of any other reversible engine, since the efficiency of any such engine equals that of a Carnot engine.

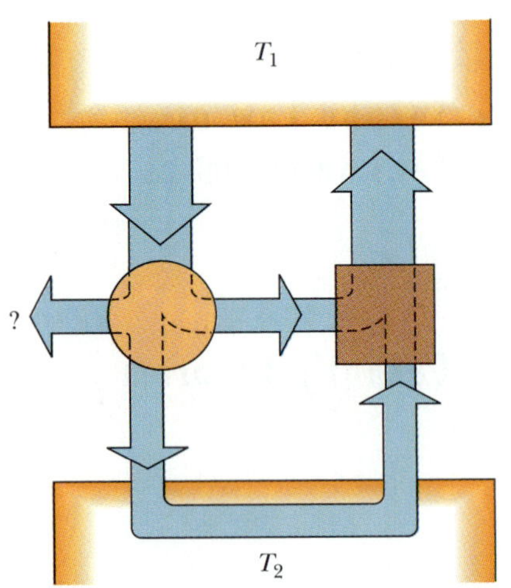

FIGURE 15.17

Flow chart for an arbitrary reversible engine (box) and a Carnot engine (circle) connected together. The Carnot engine drives the arbitrary engine in reverse.

15.4 The Entropy

In the operation of a Carnot engine, or any other reversible engine, the ratio of the heats absorbed and ejected at the high- and low-temperature reservoirs is, according to Eq. (8),

$$\frac{Q_2}{Q_1} = \frac{T_2}{T_1}$$

We can rewrite this equation in a suggestive way by multiplying both sides by Q_1/T_2 and then moving all terms to the same side of the equation:

$$\frac{Q_1}{T_1} - \frac{Q_2}{T_2} = 0 \tag{13}$$

If, as in Section 15.1, we regard the heat as positive when it enters the system and negative when it leaves the system, then we must replace Q_2 by $-Q_2$, since this heat leaves the system. With this sign convention, Eq. (13) says that, for a Carnot cycle, the sum of all the heat exchanges divided by their respective temperatures is zero:

$$\frac{Q_1}{T_1} + \frac{Q_2}{T_2} = 0 \tag{14}$$

Rudolph Clausius, *1822–1888, German mathematical physicist. He was one of the creators of the science of thermodynamics. He contributed the concept of entropy, as well as the restatement of the Second Law of Thermodynamics.*

This result can also be shown to be valid for any other reversible cycle, because any other reversible cycle can be approximated by a large number of small Carnot cycles (in the same way as any given area in a plane can be approximated by a large number of small rectangles). A general reversible cycle may have a large number of successive heat exchanges, each at some different temperature. Some of the amounts of heat in these exchanges are positive (heat enters the system), and some are negative (heat leaves the system). For such a cycle, the generalized version of Eq. (14) is

$$\frac{Q_1}{T_1} + \frac{Q_2}{T_2} + \frac{Q_3}{T_3} + \frac{Q_4}{T_4} + \ldots = 0 \tag{15}$$

Thus, the negative heat exchanges divided by their temperatures exactly compensate the positive heat exchanges divided by their temperatures. This result is called the ***theorem of Clausius.***

◀ **Theorem of Clausius**

The importance of the theorem of Clausius lies in that it permits us to define a new physical quantity called the ***entropy,*** or more precisely, the entropy difference. Given two states A and B of a system, ***the entropy difference*** ΔS ***between them is defined as the sum of the heat exchanges divided by the temperature differences for some reversible process that takes the system from the initial state*** A ***to the final state*** B:

$$\Delta S = \frac{Q_1}{T_1} + \frac{Q_2}{T_2} + \frac{Q_3}{T_3} + \ldots \tag{16}$$

◀ **Entropy difference**

Since there is usually a wide choice of different reversible processes that take the system from the given initial state to the given final state, the definition (16) would make no sense if these different processes gave different values to the sum (16). The theorem of Clausius guarantees that all these reversible processes lead to the same value for (16). To understand the connection between the definition of entropy and the Clausius theorem, consider the example of the ideal gas in the Carnot engine, for which different states are plotted in the p–V diagram in Fig. 15.11. If the initial state A corresponds to the point 1 in Fig. 15.11, and the final state B to the point 3, then one possible reversible path from 1 to 3 consists of the steps (a) and (b) of the Carnot cycle. The heat exchanged in step (a) is Q_1, and the heat exchanged in step (b) is zero; thus Eq. (16) tells us that the entropy difference between points 1 and 3 is

$$\Delta S = \frac{Q_1}{T_1} \tag{17}$$

But another reversible path from 1 to 3 consists of step (d) in *reverse* followed by step (c) in *reverse* — this also takes the system from point 1 to 3. The heat exchanged in step (d) is zero, and the heat exchanged in step (c) is $-Q_2$. Hence Eq. (16) now tells us

$$\Delta S = -\frac{Q_2}{T_2} \tag{18}$$

But Eq. (14) — which is the theorem of Clausius for the Carnot cycle — immediately guarantees that the two expressions (17) and (18) for the entropy difference are equal.

The unit of entropy is the unit of energy divided by the unit of temperature: joule per kelvin (J/K) or, alternatively, calories per kelvin (cal/K).

The entropy in thermodynamics plays a role somewhat analogous to that of the potential energy in mechanics. Just as the potential energy allows us to make some predictions about possible motions of a mechanical system, the entropy allows us to make some predictions about the possible behavior of a thermodynamic system. For instance, it can be shown from the Second Law of Thermodynamics that the entropy of a closed system — which is thermally and mechanically isolated from its surroundings — can never decrease. Either the entropy of such a system remains constant (if only reversible processes are occurring in the system) or it increases (if irreversible processes are occurring). ***The assertion that the entropy of a closed system can never decrease may be regarded as an alternative formulation of the Second Law of Thermodynamics.***

The increase of entropy in closed systems is illustrated in the following examples.

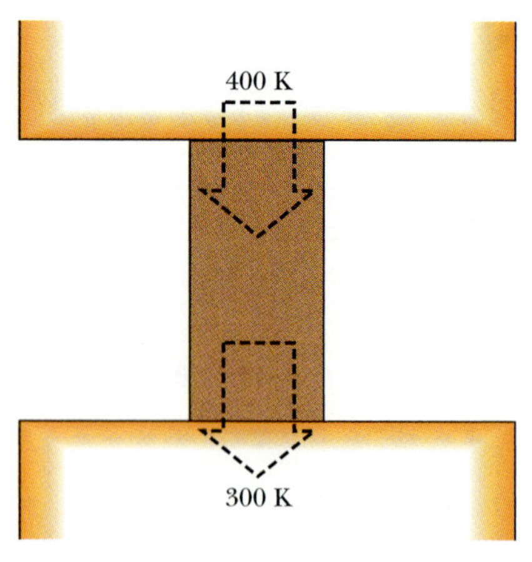

FIGURE 15.18

A hot reservoir (400 K) is in thermal contact with a colder reservoir (300 K). Heat flows through a conducting rod connecting the two reservoirs, from the hot reservoir into the colder reservoir.

EXAMPLE 6. A heat reservoir at a temperature of $T_1 = 400$ K is briefly put in thermal contact with a reservoir at $T_2 = 300$ K (see Fig. 15.18). If 1 cal of heat flows from the hot reservoir to the cold reservoir, what is the change of the entropy of the system consisting of both reservoirs?

SOLUTION: The heat flow from the hot to the cold reservoir is irreversible — once the heat has flowed to the cold reservoir, it cannot flow back to the hot reservoir. Since Eq. (16) is restricted to reversible processes, it is not directly applicable to our heat flow. To evaluate the entropy change by means of Eq. (16), we must first imagine some process that reversibly takes the system from the initial state to the final state.

We can imagine that 1 cal flows from the hot reservoir into an auxiliary reservoir of a temperature just barely below 400 K, and that simultaneously 1 cal flows from another auxiliary reservoir of a temperature just barely above 300 K into the cold reservoir. Then all processes are reversible, and Eq. (16) gives us the change of entropy:

$$\Delta S = \frac{Q_1}{T_1} + \frac{Q_2}{T_2} = -\frac{1\ \text{cal}}{400\ \text{K}} + \frac{1\ \text{cal}}{300\ \text{K}} \tag{19}$$

$$= 8.3 \times 10^{-4}\ \text{cal/K}$$

EXAMPLE 7. A large stone, of mass 80 kg, slides down a hill of a vertical height of 100 m and is stopped by friction at the bottom. What is the increase of the entropy of the stone plus the environment? Assume that the temperature of the environment (hill and air) is 270 K.

SOLUTION: In this process, all of the initial mechanical energy of the stone is converted into heat,

$$Q = mgh = 80\ \text{kg} \times 9.81\ \text{m/s}^2 \times 100\ \text{m}$$

$$= 7.8 \times 10^4\ \text{J} = 1.87 \times 10^4\ \text{cal}$$

This heat is delivered to the environment at a temperature of 270 K.

Clearly, the process described is irreversible. To calculate the entropy with Eq. (16), we must imagine a reversible process that brings the stone down the hill and delivers the heat into the environment. In principle, we can imagine that we use an elevator to let the stone down slowly without friction and extract work while removing the potential energy; and afterward we use a heat reservoir of a temperature barely above 270 K to supply reversibly the correct amount of heat (1.87×10^4 cal) to the stone's environment. The first of these steps makes no contribution to the entropy, and the second makes a contribution

$$\Delta S = \frac{Q}{T} = \frac{1.87 \times 10^4\ \text{cal}}{270\ \text{K}} = 69\ \text{cal/K} \tag{20}$$

From a microscopic point of view, the increase of entropy of a system is an increase of disorder. This can be seen in the preceding example of the conversion of mechanical energy into heat by friction. The translational motion of a macroscopic body, such as a sliding stone, is ordered energy — all the particles in the body move in the same direction with the same

Walther Hermann Nernst, *1864–1941, German physicist and chemist. He was a pioneer in physical chemistry and received the Nobel Prize in chemistry in 1920 for his discovery of the Third Law of Thermodynamics.*

speed. Thermal energy is disordered energy — the particles in the heated body move in random directions with a mixture of speeds.

The increase of disorder also holds true in our other example of spontaneous flow of heat from a hot to a cold reservoir. When a given amount of heat is added to a reservoir of low temperature, it causes more disorder than when the same amount of heat is added to a reservoir of high temperature; in both reservoirs, the added heat generates extra random motion and extra disorder, but in the cold reservoir the percent increment of the random motions is larger than in the hot reservoir, and consequently the extra disorder is larger.

The connection between entropy and disorder can be given a precise mathematical meaning, but the details would require a discussion of statistical mechanics and information theory — and here we cannot deal with this. The Second Law can be reformulated to say that the processes in a closed system always tend to increase the disorder.

The human activities on the Earth, like any other processes in nature, cause an increase of disorder. Of course, some of our activities result in an increase of order in some portion of the system. For instance, when we extract dispersed bits of metal from ores, manufacture gears and springs from them, and meticulously assemble these into a watch, we are obviously increasing the order of the bits of metal. However, we can do this only by simultaneously generating disorder somewhere else — the smelting of ores and the machining of metals demand an input of energy which is converted into waste heat, and this increases the disorder of the environment. The net result is always an increase of disorder. All of our activities depend on a supply of highly ordered energy in the form of chemical or nuclear fuels or light from the Sun that can "soak up" disorder while they burn and release degraded waste heat into the environment. We continually convert useful, ordered energy into unusable, disordered energy.

We end this chapter with a brief statement of the ***Third Law of Thermodynamics.*** This law, as formulated by Walther Nernst, asserts:

Third Law of Thermodynamics ▶

The entropy of a system at absolute zero is zero.

We can understand this law in terms of the connection between entropy and disorder, mentioned above. If we lower the temperature of a system, we decrease the random thermal motions, and we decrease the disorder. The random thermal motions cease completely, or almost completely, when the temperature of the system approaches absolute zero. The system then tends to settle into a state of minimum disorder, that is, a state of minimum entropy.

SUMMARY

▶ **First Law of Thermodynamics:** $\Delta E = Q - W$

▶ **Efficiency of heat engine:** $e = 1 - \dfrac{Q_2}{Q_1}$

▶ **Efficiency of Carnot engine:** $e = 1 - \dfrac{T_2}{T_1}$

- **Carnot's theorem:** The efficiency of any engine cannot exceed that of a Carnot engine; the efficiency of any reversible engine equals that of a Carnot engine.
- **Second Law of Thermodynamics:** An engine operating in a cycle cannot transform heat into work without some other effect on its surroundings.
- **Entropy difference (reversible process):** $\Delta S = \frac{Q_1}{T_1} + \frac{Q_2}{T_2} + \frac{Q_3}{T_3} + \ldots$
- **Third Law of Thermodynamics:** At absolute zero the entropy is zero.

QUESTIONS

1.–Figure 15.19 shows a perpetual-motion machine designed by M. C. Escher. Exactly at what point is there a defect in this design?

FIGURE 15.19
Waterfall. Lithograph by M. C. Escher.

2.–An inventor proposes the following scheme for the propulsion of ships on the ocean without an input of energy: cover the hull of the ship with copper sheets and suspend an electrode of zinc in the water at some distance from the hull; since sea water is an electrolyte, the hull of the ship and the electrode will then act as the terminals of a battery which can deliver energy to an electric motor propelling the ship. Will this scheme work? Does it violate the First Law of Thermodynamics?

3.–The lowest-temperature heat sink available in nature is interstellar and intergalactic space with a temperature of 3 K. Why don't we use this heat sink in the operation of a Carnot engine?

4.–In some showrooms, salesmen demonstrate air conditioners by simply plugging them into an outlet without bothering to install them in a window or a wall. Does such an air conditioner cool the showroom or heat it?

5.–Can mechanical energy be converted completely into heat? Give some examples.

6.–If you leave the door of a refrigerator open, will it cool the kitchen?

7.–Which of the following processes are irreversible, which are reversible? (a) Burning of a piece of paper. (b) Slow descent of a frictionless elevator attached to a perfectly balanced counterweight. (c) Breaking of a windowpane. (d) Explosion of a stick of dynamite. (e) Electric pump lifting water from a low-level reservoir to a high-level reservoir. (f) Block being slowly pushed up a frictionless inclined plane.

8.–A gas is in a cylinder fitted with a piston. Does it take more work to compress the gas isothermally or adiabatically?

9.–According to the result of Section 6.5, when a particle of small mass collides elastically with a body of very large mass, the particle gains kinetic energy if the body of large mass was approaching the particle before the collision. Using this result, explain how the collisions between the particles of gas and the moving piston lead to an increase of temperature during an adiabatic compression.

10.–The air near the top of a mountain is usually cooler than that near the bottom. Explain this by considering the adiabatic expansion of a parcel of air carried from the bottom to the top by an air current.

11.–If we measure the efficiency of a Carnot engine directly, we can use Eq. (9) to calculate the temperature of one of the heat reservoirs. What are the advantages and the disadvantages of such a thermodynamic determination of temperature?

12.–Without a heat sink, energy cannot be extracted from an ocean at uniform temperature. However, within the oceans there are small temperature differences between the warm water on the surface and the cooler water in the depths. Discuss how we could use a Carnot engine to exploit this temperature difference.

13.–Why are heat pumps used for heating houses in mild climates but not in very cold climates?

14.–Does the Second Law of Thermodynamics forbid a spontaneous flow of heat between two bodies of equal temperature?

15.–The inside of an automobile parked in the sun becomes much hotter than the surrounding air. Does this contradict the Second Law of Thermodynamics?

16.–According to a story by George Gamow,* on one occasion Mr. Tompkins was drinking a highball when all of a sudden one small part at the surface of the liquid became hot and boiled with violence, releasing a cloud of steam, while the remainder of the liquid became cooler. Is this consistent with the First Law of Thermodynamics? With the Second Law?

17.–A vessel is divided into two equal volumes by a partition. One of these volumes contains helium gas, and the other contains argon at the same temperature and pressure. If we remove the partition and allow the gases to mix, does the entropy of the system increase?

18.–Does the motion of the planets around the Sun generate entropy?

19.–Does static friction generate entropy?

20.–Consider the process of emission of light by the surface of the Sun followed by absorption of this light by the surface of the Earth. Does this entail an increase of entropy?

21.–Suppose that a box contains gas of extremely low density, say, only 50 molecules of gas altogether. The molecules move at random and it is possible that once in a while all of the 50 molecules are simultaneously in the left half of the box, leaving the right half empty. At this instant the entropy of the gas is less than when the gas is more or less uniformly distributed throughout the box. Is this a violation of the Second Law of Thermodynamics? What are the implications for the range of validity of this law?

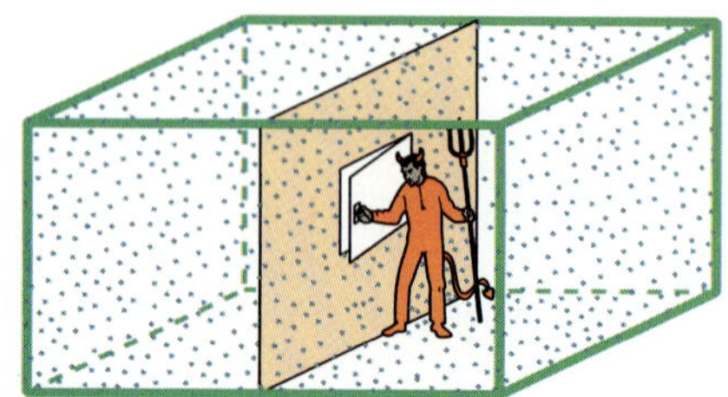

FIGURE 15.20
Maxwell's demon.

22.–***Maxwell's demon*** is a tiny hypothetical creature that can see individual molecules. The demon can make heat flow from a cold body to a hot body as follows: Suppose that a box initially filled with gas at uniform temperature and pressure is divided into two equal volumes by a partition, equipped with a small door which is closed but can be opened by the demon (Fig. 15.20). Whenever a molecule of above-average speed approaches the door from the left, the demon quickly opens the door and lets it through. Whenever a mole-

* G. Gamow, *Mr. Tompkins in Paperback.*

cule of below-average speed approaches from the right, the demon also lets it through. This selective action of the demon accumulates hot gas in the right volume, and cool gas in the left volume. Does this violate the Second Law of Thermodynamics? Do any of the activities of the demon involve an *increase* of entropy?

23.–According to one cosmological model based on Einstein's theory of General Relativity, the Universe oscillates — it expands, then contracts, then expands, and then contracts, and so on. If the Second Law of Thermodynamics is valid, can each cycle of oscillation be the same as the preceding cycle?

24.–***Negentropy*** is defined as the negative of the entropy ([negentropy] = $-S$). Explain the following statement: "In our everyday activities on the Earth, we do not consume energy, but we consume negentropy."

25.–If you tidy up a messy room you are producing a decrease of disorder. Does this violate the Second Law?

PROBLEMS

Section 15.1: The First Law of Thermodynamics

1.——A cylinder with a piston contains 1 mole of helium gas. Suppose we heat the cylinder and allow the gas to expand, so it performs work on the piston, at constant pressure. If, during this expansion, the gas absorbs 3000 J of heat and performs 2000 J of work on the piston, what is the change ΔE in the internal energy of the gas? What is the change of its temperature?

2.——A sample of gas in a cylinder with a piston is in thermal contact with a heat reservoir at a temperature of 353 K. While keeping this gas at this constant temperature and at a constant pressure of 1.01×10^5 N/m^2, we permit the gas to expand by 1.5×10^{-3} m^3. How much heat does the gas absorb during this process?

3.——We place a 1.0-mole sample of an ideal monoatomic gas in a cylinder with a piston and we heat the gas so it expands and performs work against the piston. Suppose the temperature of the gas increases by 90°C while at the same time it performs 800 J of work. What is the change of the internal energy of the gas in this process? How much heat does the gas absorb during this process?

4.——A ball of lead of mass 0.25 kg drops from a height of 0.80 m, hits the floor, and remains there at rest. Assume that all the heat generated during the impact remains within the lead. What are the values of ΔQ, ΔW, and ΔE, for the lead during this process? What is the increase of temperature of the lead?

5.——A cylinder with a piston contains 1 mole of air (regarded as an ideal diatomic gas). The cylinder is insulated, so no heat can escape from it. If we push the piston inward and compress the gas, the temperature of the gas will increase. How much does the temperature increase if we do 2500 J of work on the gas during the compression?

6.——The freezing of 1 kg of water converts 1 liter of liquid water at 100°C into a volume of ice at the same temperature. The ice occupies a larger volume (1.091 liter) than the water; hence, during freezing, the water must do work on the surrounding atmosphere while it expands. How much work does 1 kg of water do while it expands into ice, at 1 atm? Given that the heat of fusion of ice is 79.7 kcal/kg, what is the change of internal energy of the water during freezing?

7.——The vaporization of 1 kg of water converts 1 liter of liquid water at 100°C into a volume of water vapor at the same temperature. The water vapor occupies a larger volume than the water; hence, during vaporization, the water must do work on the surrounding atmosphere while it expands.
   (a) How much work does 1 kg of water do while it expands into vapor, at a pressure of 1 atm? Given that the heat of vaporization of water is 539 kcal/kg, what is the change of internal energy of the water during vaporization?
   (b) How much work does 1 kg of water do while it expands into vapor, at a pressure of 0.1 atm? Deduce the heat of vaporization of water at 0.1 atm (and 100°C). Assume that the change of the density of liquid water can be neglected in this calculation.

8.——A 4.0-liter sample of a monoatomic ideal gas is initially at STP. We first heat this gas at constant volume, until its pressure is doubled. We then continue heating at constant pressure, and allow the gas to expand until its volume is also doubled.
   (a) What are the values of ΔE, Q, and W for the first step of this process?
   (b) What are the values of ΔE, Q, and W for the second step of this process?

9.*——When pressurized air rushes out of the nozzle of a tire, the air must do work against the surrounding atmosphere while it expands. This means that the air suffers a loss of internal energy, and its temperature therefore drops (you can easily feel that the emerging air is quite cold). Suppose that each mole of the emerging air does 1800 J of work, and suppose that the air rushes out so fast that it has no time to absorb heat from the surrounding atmosphere (adiabatic expansion). If the initial temperature of the air in the tire is 20°C, what is the temperature of the expanded air? Treat the air as an ideal diatomic gas.

10.*——In the cylinder of a diesel engine, the piston compresses the air–fuel mixture and does work on it. This work increases the internal energy of the mixture, and therefore heats it. The temperature attained by the compression is sufficient to ignite the mixture, without any need of a spark plug. How much work must the piston do on a parcel of gas, 0.030 mole, to heat it from 40°C to 790°C? Assume that the compression is so fast that the gas loses no heat to the surroundings (adiabatic compression), and assume the gas behaves like a diatomic ideal gas.

11.*——A large, tightly sealed plastic bag contains 0.10 m^3 of an unknown gas at an initial temperature of 10°C and at the same pressure as the surrounding atmosphere, 1.0 atm. You place this bag in the sun and let the gas warm up to 38°C and expand to 0.11 m^3. During this process, the gas absorbs 840 cal of heat. Assume the bag is large enough so it remains only loosely filled with gas, and the pressure of the gas remains at 1.0 atm.
   (a) How many moles of gas are in the bag?
   (b) What is the work done by the gas in the bag against the atmosphere during the expansion?
   (c) What is the change in the internal energy of the gas in the bag?
   (d) Is the gas a monoatomic gas? A diatomic gas?

12.*——A tightly sealed plastic bag contains 4.0 liters of air at 0°C. The plastic bag is only loosely filled, and it exerts no compression on the air; the pressure of the air is therefore the pressure of the surrounding atmosphere, 1 atm. We place this plastic bag in the sun, and let it warm up to 60°C.
   (a) What is the new volume of the air? Assume that the plastic bag remains only loosely filled with air, so the pressure remains at 1 atm.
   (b) How much work has the air in the bag done on the surrounding atmosphere while it expanded? What is the change in the internal en-

ergy of the air? What is the amount of the heat the air has absorbed? Assume that the air behaves like an ideal diatomic gas.

13.*—A submerged scuba diver, 10 m below the surface of the water, blows a bubble of air from the valve of her air tank. The final volume of the bubble is 4.0 cm^3, and the temperature of the air is that of the water, 15°C. Calculate how much heat the bubble absorbs as it forms, and how much work it does on the surrounding water. Assume that the temperature of the air in the bubble remains equal to the water temperature.

14.**—At a pressure of 1 atm, the heat of vaporization of water is 539 kcal/kg; this is the heat required to convert 1 kg of water at 100°C into water vapor at the same temperature. Given that the specific heat of water is $c = 1$ kcal/kg·°C and that of water vapor is (approximately†) $c_p = 0.48$ kcal/kg·°C, use the First Law of Thermodynamics to calculate the heat of vaporization of water at 20°C. [Hint: Instead of directly converting water at 20°C into water vapor, we can first heat the water to 100°C, then vaporize it, and then cool the vapor to 20°C (without condensation). The net work done against atmospheric pressure is the same during this indirect process as during direct vaporization (why?); hence, according to the First Law, the heat absorbed during each process is also the same.]

Section 15.2: Heat Engines; the Carnot Engine

15.—Suppose that a heat engine takes 3.0×10^4 J of heat from the high-temperature reservoir to produce 2.0×10^4 J of mechanical work. What is the efficiency of this engine? How much waste heat does it produce?

16.—A heat engine takes 5.0×10^3 kcal of heat from the high-temperature reservoir and ejects 2.0×10^3 kcal of heat into the low-temperature reservoir. How much work does it produce? What is its efficiency?

17.—Electric motors convert electric energy into mechanical energy with an efficiency of 95%. If the electric current supplies a power of 3 kW to such an electric motor, what mechanical power will the motor produce? How much heat (in kW) will the motor produce?

18.—An electric power plant consists of a coal-fired boiler that makes steam, a turbine, and an electric generator. The boiler delivers 90% of the heat of combustion of the coal to the steam; the turbine converts 50% of the heat of the steam into mechanical energy; and the electric generator converts 99% of this mechanical energy into electric energy. What is the overall efficiency of generation of electric power?

19.—A coal-burning power plant uses thermal energy at a rate of 850 megawatts and produces 300 megawatts of mechanical power for the generation of electricity. What is the efficiency of this power plant?

20.—Although a sprinter running at a steady speed on level ground performs no external work (except for a small amount of work against air resistance), the sprinter performs a considerable amount of work to accelerate and decelerate his own limbs during each stride and to lift his limbs against gravity. The following list gives the mechanical power used by a sprinter for motion of limbs and body during a sprint (the power was calculated from data obtained from photographic analysis of the motion).

Acceleration of limbs, 1.5 hp
Deceleration of limbs, 0.67 hp
Work against gravity, 0.1 hp
Speed changes of body, 0.5 hp

---

† The specific heat of water is somewhat dependent on temperature.

What is the total mechanical power expended in these motions of the sprinter? According to a measurement of the oxygen consumption of the sprinter, his expenditure of chemical energy was 13 hp during the sprint. What is his efficiency of conversion of chemical energy into mechanical energy?

21.——In an automobile proceeding at medium speed, the engine delivers 20 hp of mechanical power. The engine burns gasoline which provides thermal energy at the rate of 15 kilocalories per second. What is the efficiency of the engine under these conditions? What is the rate at which the engine ejects waste heat?

22.——Each of the two engines of a DC-3 airplane produces 1100 hp. The engines consume gasoline; the combustion of 1 kg of gasoline yields 44×10^6 J. If the efficiency of the engines is 20%, at what rate do the two engines consume gasoline?

23.——A nuclear power plant generates 1000 megawatts of electric (or mechanical) power. If the efficiency of this plant is 33%, at what rate does the plant generate waste heat? If this waste heat is to be removed by passing water from a river through the plant, and if the water is to suffer a temperature increase of at most 8°C, how many cubic meters of water per second are required?

24.——To produce 120 hp of mechanical power, an automobile engine requires a supply of heat of 105 kcal per second from combustion of the fuel. What is the efficiency of this engine?

25.——In an experiment on the work efficiency of horses, a tracheotomized horse connected to an oxygen supply was made to do work on a treadmill (see Fig. 15.21). When the horse was delivering 869 watts of work to the treadmill, its rate of oxygen consumption was 10.8 liters per minute. The horse's metabolism yields 5 kcal of chemical energy per liter of oxygen consumed. Calculate the efficiency of conversion of chemical energy into external mechanical work.

26.——A Carnot engine operates between a high-temperature reservoir at 100°C and a low-temperature reservoir at 0°C. How much energy must the engine take from the high-temperature reservoir to produce 5.0×10^4 J of work? How much waste heat does it produce?

27.——Consider a Carnot engine operating between heat reservoirs at 400 K and at 300 K. What is the efficiency of this Carnot engine? If we want to increase the efficiency by 10%, by how much must we increase the temperature of the high-temperature reservoir? Alternatively, by how much must we decrease the temperature of the low-temperature reservoir?

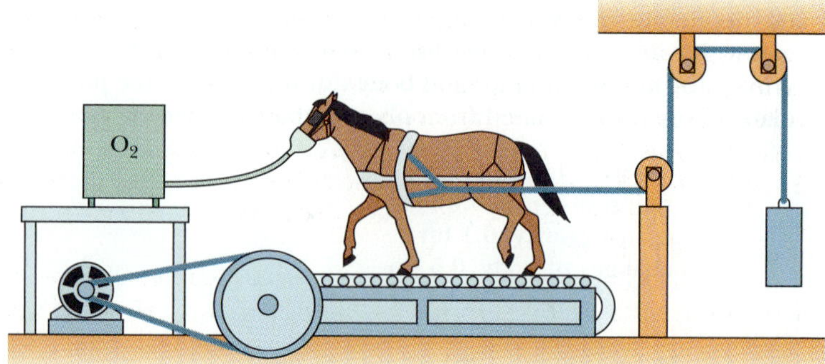

FIGURE 15.21
Horse doing work on a treadmill.

28.——The efficiency of striated muscle is typically 37%, that is, the muscle converts 37% of the chemical energy reaching it (in the form of glucose) into mechanical work, and converts the remainder into waste heat. If your biceps muscle performs mechanical work at the rate of 50 W, what is the rate at which it consumes chemical energy, and what is the rate at which it produces waste heat? Given that the oxidation of glucose yields 3.7×10^3 kcal/kg, what is the rate at which this muscle consumes glucose?

29.——A geothermal power plant at Wairakei, New Zealand, uses hot underground water at a temperature of 300°C as heat source and it uses the atmosphere at a temperature of, say, 25°C as heat sink (Fig. 15.22). What would be the efficiency of a Carnot engine operating between these temperatures? If this Carnot engine produces 10,000 kW of mechanical power, what power does it eject as waste heat?

FIGURE 15.22
Geothermal power plant at Wairakei, New Zealand.

30.——A Carnot engine operates between two heat reservoirs of temperature 500°C and 30°C, respectively.

(a) What is the efficiency of this engine?
(b) If the engine generates 1.5×10^3 J of work (in some time interval), how many calories of heat does it absorb from the hot reservoir? Eject into the cold reservoir?

31.——In principle, nuclear reactions can achieve temperatures of the order of 10^{11} K. What is the efficiency of a Carnot engine taking in heat from such a nuclear reaction and exhausting waste heat at 300 K?

32.——An automobile engine takes heat from the combustion of gasoline, converts part of this heat into mechanical work, and ejects the remainder into the atmosphere. The temperature attainable by the combustion of gasoline is about 2100°C and the temperature of the atmosphere is 20°C. What is the maximum theoretical efficiency of an engine operating between these temperatures? (The actual efficiency attained by an automobile engine is typically 0.2, much lower than the theoretical maximum.)

33.——According to some naive speculations of the last century, the human body was supposed to be a heat engine in which the combustion of food produces body heat which is then somehow converted into mechanical work by the muscles. If this were true, what would be the maximum (ideal) efficiency of the human body? The temperature of the "hot" human body (the heat source) is 37°C and the temperature of the heat sink is that of the surrounding environment, about 20°C. How many kilocalories of food energy would you need to consume to climb up one flight of stairs, to a height of 3 m? (Actually, muscles convert chemical energy directly into work; they are not heat engines.)

34.——Suppose we operate a Carnot engine in reverse to "pump" heat from a heat reservoir at 0°C into a heat reservoir at 100°C. How much mechanical work must we supply per joule of heat removed from the 0°C reservoir?

35.——The electric motor of your refrigerator uses 2.5×10^3 J of electric energy to remove 6.0×10^3 J of heat from the refrigerator compartment. How much heat does this refrigerator dump into your kitchen?

36.——A refrigerator is operated by a Carnot engine. The temperature inside the refrigerator is 4°C and the temperature of the surrounding room is 20°C. How much mechanical work must the refrigerator's electric motor perform to remove 1 kcal of heat from the inside of the refrigerator and dump it into the room?

37.——On a hot day a house is kept cool by an air conditioner. The outside temperature is 32°C and the inside temperature is 21°C. Heat leaks into the house at the rate of 9000 kcal/h. What is the minimum mechanical power that the air conditioner requires to hold the inside temperature constant?

38.——A scheme for the extraction of energy from the oceans attempts to take advantage of the temperature difference between the upper and lower layers of ocean water. The temperature at the surface in tropical regions is about 25°C; the temperature at a depth of 300 m is about 5°C.

(a) What is the efficiency of a Carnot engine operating between these temperatures?

(b) If a power plant operating at the maximum theoretical efficiency generates 1 megawatt of mechanical power, at what rate does this power plant release waste heat?

(c) The power plant obtains the mechanical power and the waste heat from the surface water by cooling this water from 25°C to 5°C. At what rate must the power plant take in surface water?

39.——In a nuclear power plant, the reactor produces steam at 520°C and the cooling tower eliminates waste heat into the atmosphere at 30°C. The power plant generates 500 megawatts of electric (or mechanical) power.

(a) If the efficiency is that of a Carnot engine, what is the rate of release of waste heat (in megawatts)?

(b) Actual efficiencies of nuclear power plants are about 33%. For this efficiency, what is the rate of release of heat?

40.——An air conditioner removes 2000 kcal/h of heat from a room at a temperature of 21°C and ejects this heat into the ambient air at a temperature of 27°C. This air conditioner requires 950 W of electric power.

(a) How much mechanical power would a Carnot engine, operating in reverse, require to remove this heat at the same rate?

(b) By what factor is the power required by the air conditioner larger than that required by the Carnot engine?

41.**—The boiler of a power plant supplies steam at 540°C to a turbine which generates mechanical power. The steam emerges from the turbine at 260°C and enters a steam engine that generates extra mechanical power. The steam is finally released into the atmosphere at a temperature of 38°C. Assume that the conversion of heat into work proceeds with the maximum efficiency permitted by Carnot's theorem.

(a) What is the efficiency of the turbine? Of the steam engine?

(b) What is the net efficiency of both engines acting together? How does it compare with the efficiency of both engines acting together? How does it compare with the efficiency of a single engine operating between 540°C and 38°C?

42.**—An ice-making plant consists of a reversed Carnot engine extracting heat from a well-insulated icebox. The temperature in the icebox is -5°C and the temperature of the ambient air is 30°C. Water, of an initial temperature of 30°C, is placed in the icebox and allowed to freeze and to cool to -5°C. If the ice-making plant is to produce 10,000 kg of ice per day, what mechanical power is required by the Carnot engine?

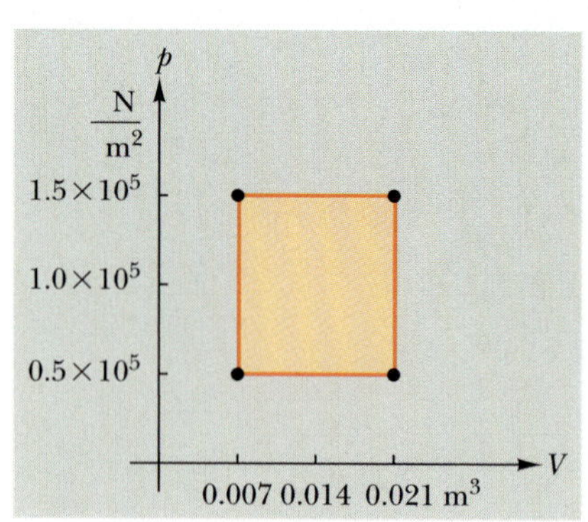

FIGURE 15.23

A cyclic process for a gas displayed in a p–V diagram.

43.**—One mole of ideal gas is taken through a cyclic process consisting of four steps: an expansion at constant pressure, followed by a reduction of pressure at constant volume, followed by a compression at constant pressure, followed by an increase of pressure at constant volume. Figure 15.23 displays these steps on a p–V diagram and gives relevant numbers. During each of these steps heat is added or removed from the gas as required. For each of the four steps calculate the work W done by the gas, the change of internal energy ΔE, and the heat Q that must be added or removed.

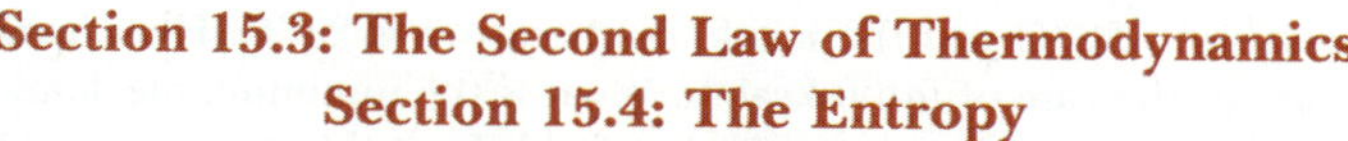

Section 15.3: The Second Law of Thermodynamics
Section 15.4: The Entropy

44.——On a winter day heat leaks out of a house at the rate of 2.5×10^4 kcal/h. The temperature inside the house is 21°C and the temperature outside is $-5°C$. At what rate does this process produce entropy?

45.——Your body generates about 2000 kcal of heat per day. Estimate how much entropy you generate per day. Neglect the (small) amount of entropy that enters your body in the food you consume.

46.——A steam engine operating between reservoirs at temperatures of 480°C and 27°C has an efficiency of 40%. The engine delivers 2000 hp of mechanical power. At what rate does this engine generate entropy?

47.——Suppose that 1.0 kg of water freezes while at 0°C. What is the change of entropy of the water during this freezing process?

48.——For an automobile moving at an constant speed of 65 km/h on a level road, rolling friction, air friction, and friction in the drive train absorb a mechanical power of 12 kW. At what rate do these processes generate entropy? The temperature of the environment is 20°C.

49.——Consider the refrigerator described in Problem 36. Calculate the rate of increase of entropy contributed by the operation of this refrigerator.

50.——What is the increase of entropy of 1.0 kg of water when it vaporizes at 100°C and 1 atm?

51.——At a temperature of −79°C, solid carbon dioxide ("dry ice") transforms into gas by sublimation (that is, direct vaporization from solid to gas). From the heat of transformation given in Table 14.4, calculate the increase of entropy per kilogram of carbon dioxide during sublimation.

52.——Table 14.4 gives data for the melting and the vaporization of lead. What is the increase of entropy per kilogram of lead during melting? During vaporization?

53.——An automobile of 2100 kg moving at 80 km/h brakes to a stop. In this process the kinetic energy of the automobile is first converted into thermal energy of the brake drums; this thermal energy later leaks away into the ambient air. Suppose that the temperature of the brake drums is 60°C when the automobile stops, and that the temperature of the air, and the final temperature of the brake drums, is 20°C.

(a) How much entropy is generated by the conversion of mechanical energy into thermal energy of the brake drums?

(b) How much extra entropy is generated as the heat leaks away into the air?

54.——A parachutist of 80 kg descends at a constant speed of 5 m/s. What is the rate of increase of entropy of the parachute and the environment? The air temperature is 20°C.

55.——At Niagara Falls, 5700 m^3/s of water fall through a vertical distance of 50 m, dissipating all of their gravitational energy (Fig. 15.24). Calculate the rate of increase of entropy contributed by this falling water. The temperature of the environment is 20°C.

FIGURE 15.24
Niagara Falls.

56.——The amount of energy dissipated in the United States per year is 8×10^{19} J. Roughly, what is the entropy increase that results from this dissipation?

57.**—It can be demonstrated that when the temperature of a mass m of solid or liquid is changed from some initial value T_1 to a new value T_2, the change of entropy is given by the formula

$$\Delta S = mc \ln \left(\frac{T_2}{T_1}\right)$$

where c is the specific heat and ln stands for the logarithm to base e (the "natural" logarithm). Suppose we mix 1 liter of water at 20°C with 1 liter of water at 100°C. What is the increase of entropy?

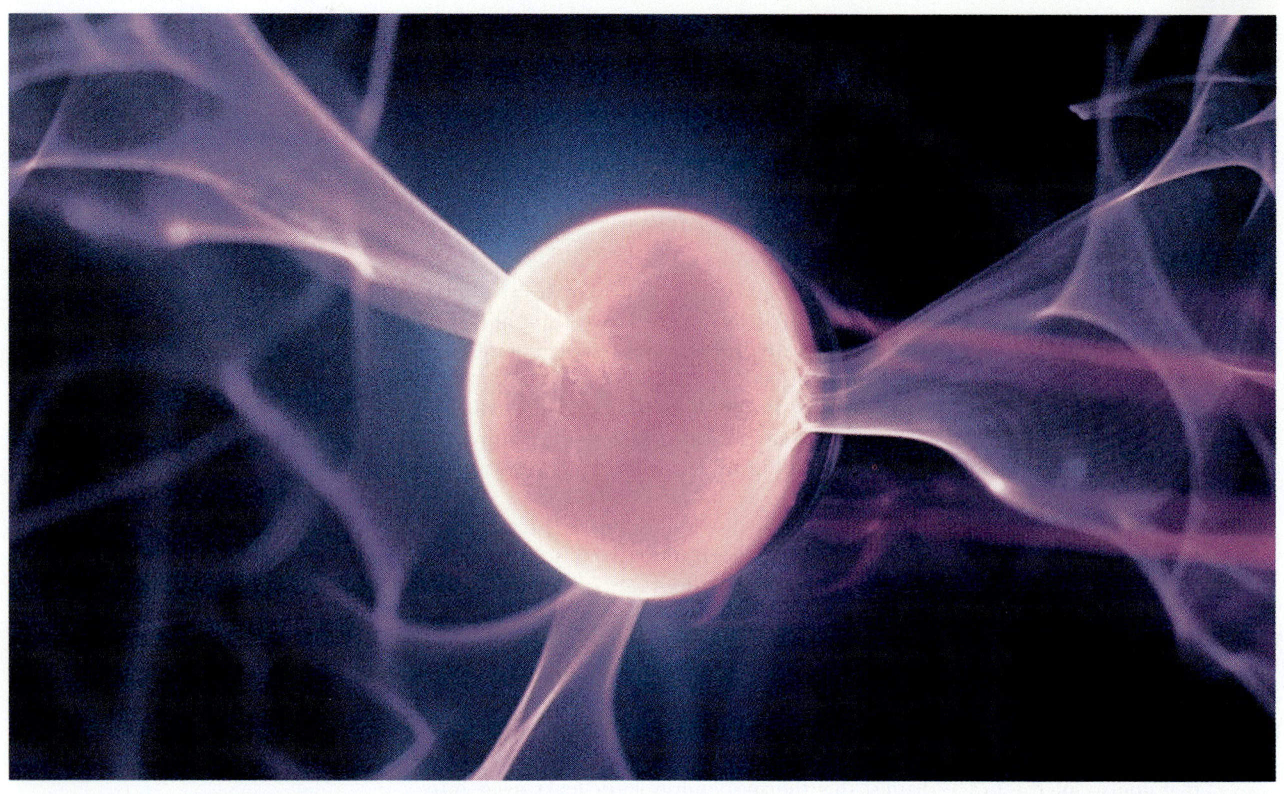

Electric discharge in a glass bulb filled with argon gas. A generator delivers electric charges to the terminal at the center, and these charges then leak away, through the gas, producing the bright pink glow.

ELECTRIC FORCE AND ELECTRIC CHARGE

Our society is dependent on electricity. An electric power failure demonstrates our dependence — subways and elevators stop, traffic lights, streetlights, and the lights in our homes go out, refrigerators fail, food can't be cooked, homes can't be heated, radios, TVs, and computers can't be operated. But our dependence on electricity runs even deeper than our reliance on electrical machinery and gadgetry would suggest. Electricity is an essential ingredient in all the atoms in our bodies and in our environment. The forces that hold the parts of an atom together are electric forces. Furthermore, so are the forces that bind atoms in a molecule and hold these building blocks together in large-scale macroscopic structures, such as a rock, a tree, a human body, a skyscraper, or a supertanker. All the mechanical "contact" forces of everyday experience — the push of a hand against a door, the pull of an elevator cable, the pressure of water against the hull of a ship — are nothing but the combined electric forces of many atoms. Thus, ***our immediate environment is dominated by electric forces.***

In the following chapters, we will study electric forces and their effects. For a start (Chapters 16–20), we will assume that the particles exerting these forces are at rest or moving only very slowly. The electric forces exerted under these conditions are called ***electrostatic forces.*** Later on (Chapters 21–23), we will consider the electric forces when the particles are moving with uniform velocity or nearly uniform velocity. Then the electric forces are modified — besides the electrostatic force there arises a ***magnetic force,*** which depends on the velocities of the particles. The combined electrostatic and magnetic forces are called ***electromagnetic forces.*** Finally we will consider the forces exerted when the particles are moving with accelerated motion (Chapter 24). The electromagnetic forces are then further modified with a drastic consequence, that is, the emission of electromagnetic waves, such as light and radio waves.

Electricity was first discovered through friction. The ancient Greeks noticed that rods of amber ("elektron" in Greek), when rubbed with a cloth or fur, gave off sparks and attracted small bits of straw or feathers. You can easily duplicate this ancient discovery by rubbing a plastic comb on a shirt or a sweater (preferably use a material with synthetic fibers and wait for a day of low humidity); in the dark, you can then see a multitude of small sparks produced by this rubbing process, and the electrified comb will attract small bits of paper or lint. In the nineteenth century, technical applications of electricity were gradually developed, but it was only in the twentieth century that the pervasive presence of interatomic electric forces holding together all the matter of our environment was recognized.

Benjamin Franklin, *1706–1790, American scientist, statesman, and inventor. Although he is most often remembered for his hazardous experiments with a kite in a thunderstorm, which demonstrated that lightning is an electric phenomenon, and for his invention of lightning rods, Franklin also made other significant contributions to the experimental and theoretical studies of electricity, and he was admired and honored by the leading scientific associations in Europe. Among these contributions was his formulation of the law of conservation of electric charge and his introduction of the modern notation for plus and minus charges, which he regarded as an excess or deficiency of "electric fluid."*

16.1 Electric Charge

Ordinary matter — solids, liquids, and gases — consists of atoms, each with a nucleus surrounded by a swarm of electrons. For example, Fig. 16.1 shows the structure of an atom of neon. At the center of this atom there is a nucleus made of ten protons and ten neutrons packed very tightly together — the diameter of the nucleus is only about 6×10^{-15} m. Moving around this nucleus are ten electrons; these electrons are confined to a roughly spherical region about 1×10^{-10} m across.

The atom somewhat resembles the Solar System, with the nucleus as Sun and the electrons as planets. In the Solar System, the force that holds a planet near the Sun is the gravitational force. In the atom, the force that holds an electron near the nucleus is the ***electric force*** of attraction between the electron and the protons in the nucleus. This electric force resembles gravitation in that it decreases in proportion to the inverse

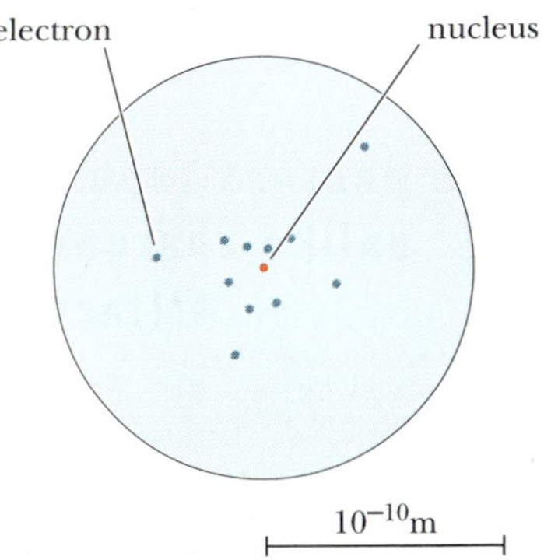

FIGURE 16.1

Neon atom. This drawing shows the electrons at one instant of time, as they would appear in a hypothetical, extremely powerful microscope.

TABLE 16.1	ELECTRIC FORCES (QUALITATIVE)
Particles	**Force**
Electron and proton	Attractive
Electron and electron	Repulsive
Proton and proton	Repulsive
Neutron and anything	Zero

square of the distance. But the electric force is much stronger than the gravitational force. The electric attraction between an electron and a proton (at any given distance) is about 2×10^{39} times as strong as the gravitational attraction. Thus, the electric force is by far the strongest force felt by an electron in an atom.

The other great difference between the gravitational force and the electric force is that the gravitational force between two particles is always attractive, whereas electric forces can be attractive, repulsive, or zero, depending on what two particles we consider. The electron–proton electric force is attractive; but the electron–electron and the proton–proton electric forces are repulsive; and the electron–neutron and the proton–neutron electric forces are zero. Table 16.1 gives a qualitative summary of the electric forces between the fundamental particles.

Electric charge ▶

Particles that exert electric forces are said to have an ***electric charge;*** particles that do not exert electric forces are said to have no electric charge. Thus, ***electric charge is thought of as the source of electric force, just as mass is the source of gravitational force.*** Electrons and protons have electric charge, but neutrons have no electric charge. Since the electron–proton force, the electron–electron force, and the proton–proton force all have the same magnitudes (for a given distance), the strengths of the sources on electrons and protons are of equal magnitudes; that is, their electric charges are of equal magnitudes. For the mathematical formulation of the law of electric force, we assign a positive charge to the proton and a negative charge (of equal magnitude) to the electron. We designate these charges of the proton and the electron by $+e$ and $-e$, respectively. Table 16.2 summarizes these values of the charges.

Like charges repel; ▶ unlike charges attract

In terms of these electric charges, we can then state that ***the electric force between charges of like sign is repulsive and the electric force between charges of unlike sign is attractive.***

TABLE 16.2	ELECTRIC CHARGES OF PROTONS, ELECTRONS, AND NEUTRONS
Particle	**Charge**
Proton, p	$+e$
Electron, e	$-e$
Neutron, n	0

The numerical value of the charge e of the proton depends on the system of units. In the metric, or SI, system of units, the electric charge is measured in ***coulombs*** (C), and the corresponding numerical values of the charges of the proton and of the electron are

◀ **Coulomb (C)**

$$e = 1.60 \times 10^{-19}\ \text{C for proton;} \qquad (1)$$

$$-e = -1.60 \times 10^{-19}\ \text{C for electron}$$

◀ **Charges of proton and of electron**

The net electric charge of a body containing some number of electrons and protons is the (algebraic) sum of the electron and proton charges. For instance, the net electric charge of an atom containing equal numbers of electrons and protons is zero; that is, the atom is electrically neutral. Sometimes atoms lose an electron, and sometimes they gain an extra electron. Such ***atoms with missing electrons or with extra electrons are called ions.*** They have a net positive charge if they have lost electrons, and a net negative charge if they have gained electrons. The positive or negative charge on a macroscopic body — such as on a plastic comb electrified by rubbing — arises in the same way, from a deficiency or an excess of electrons.

◀ **Ions**

The electric forces between two neutral atoms tend to cancel; each electron in one atom is attracted by the protons in the nucleus of the other atom and simultaneously it is repelled by the equal number of electrons of that atom. However, the cancellation of these electric attractive and repulsive forces among the electrons and the protons in the two atoms is sometimes not complete. For instance, the "contact" force between two atoms close together arises from an incomplete cancellation of the attractive and repulsive forces. The force between the atoms depends on the relative locations of the electrons and the nuclei. If the distributions of the electrons are somewhat distorted so, on the average, the electrons in one atom are closer to the nucleus of the neighboring atom than to its electrons, then the electrons in the first atom will be more strongly attracted by the nucleus of the neighboring atom than they are repelled by its electrons — the net force between these atoms will be attractive. Figure 16.2a shows such a distortion that leads to an attractive net force; the distortion may either be intrinsic to the structure of the atom or induced by the presence of the neighboring atom. Figure 16.2b shows a distortion that leads to a repulsive force.

Likewise, the electric forces between two neutral macroscopic bodies separated by some appreciable distance tend to cancel. For example, if the macroscopic bodies are a baseball and a tennis ball separated by a distance of 2 m, then each electron of the baseball is attracted by the protons of the tennis ball, but simultaneously it is repelled by the electrons of the tennis ball; and these forces cancel each other. Only when the surfaces of the two balls are very near each other ("touching") will the atoms in the surface exert a net electric force on those in the other surface.

This cancellation of the electric forces between neutral macroscopic bodies explains why we do not see large electric attractions or repulsions between the macroscopic bodies in our environment, even though the

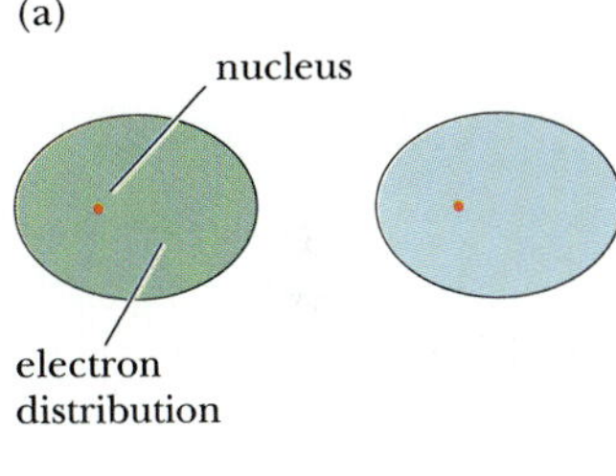

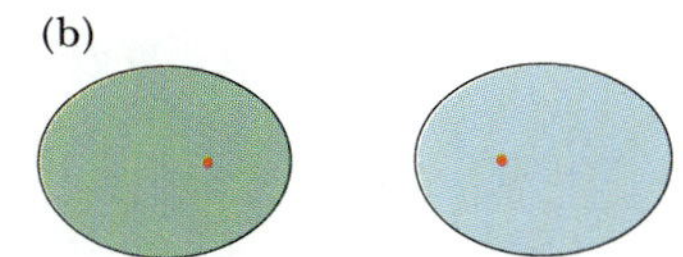

FIGURE 16.2

(a) Two neighboring distorted atoms. The colored regions represent the average distribution of the electrons. The electrons of the left atom are closer to the nucleus of the right atom than to its electrons.
(b) The nucleus of the left atom is closer to the nucleus of the right atom than to its electrons.

electric forces between individual electrons and protons are much stronger than the gravitational forces. Most such macroscopic bodies are electrically neutral, and they therefore exert no net electric forces on each other, except for contact forces.

16.2 COULOMB'S LAW

As already mentioned above, the electric force between two particles decreases with the inverse square of the distance, just as does the gravitational force. The dependence of the electric force on distance was discovered through experiments by Coulomb, who investigated the repulsion between small balls charged by rubbing. His experimental results are summarized in ***Coulomb's Law:***

Coulomb's Law ▶

> ***The magnitude of the electric force that a particle exerts on another particle is directly proportional to the product of their charges and inversely proportional to the square of the distance between them. The direction of the force is along the line from one particle to the other.***

To express this law mathematically, we denote the charges on the particles by q and q', and the distance by r. Coulomb's Law is then represented by the formula

$$F = [\text{constant}] \times \frac{q'q}{r^2} \tag{2}$$

This formula gives not only the magnitude of the force, but also the direction, if we interpret a positive value of the force F as repulsive and a negative value as attractive. For instance, for the case of the force exerted by a proton on an electron, the charges are $q' = e$ and $q = -e$, and the formula (2) yields

$$F = [\text{constant}] \times \frac{e \times (-e)}{r^2} \tag{3}$$

which is negative, indicating attraction.

The electric force that the particle of charge q exerts on the particle of charge q' has the same magnitude as the force exerted by q' on q, but the opposite direction. These mutual forces are an action–reaction pair (see Fig. 16.3).

In SI units, the constant of proportionality in Coulomb's Law has the value

$$[\text{constant}] = 8.99 \times 10^9 \,\frac{\text{N} \cdot \text{m}^2}{\text{C}^2}$$

This constant is traditionally written in the form

$$[\text{constant}] = \frac{1}{4\pi\epsilon_0} \tag{4}$$

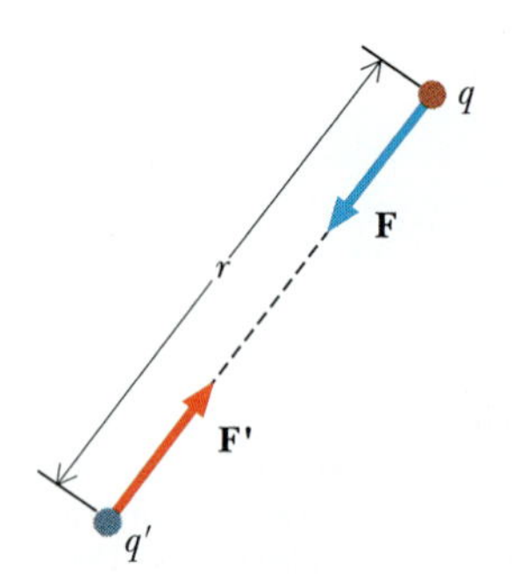

FIGURE 16.3

Two charged particles q and q'. The mutual electric forces that these particles exert on each other lie along the line joining the particles.

with

$$\epsilon_0 = 8.85 \times 10^{-12}\ \mathrm{C^2/(N \cdot m^2)} \quad (5)$$

◂ **Permittivity constant**

The quantity ϵ_0 (epsilon nought) is called the ***permittivity constant.*** In terms of the permittivity constant, Coulomb's Law for the force that a particle of charge q' exerts on a particle of charge q becomes

$$F = \frac{1}{4\pi\epsilon_0}\frac{q'q}{r^2} = 8.99 \times 10^9\ \frac{\mathrm{N \cdot m^2}}{\mathrm{C^2}}\frac{q'q}{r^2} \quad (6)$$

Although the second expression on the right side of Eq. (6) is most convenient for numerical calculations of the Coulomb force, the first expression with $1/4\pi\epsilon_0$ is generally used in manipulations involving formulas. Of course, the two expressions are mathematically equivalent, and they give the same results.

Charles Augustin de Coulomb (koolom), *1736–1806, French physicist. with the torsion balance, which he invented, he established that the electric force between small charged balls obeys an inverse-square law.*

Coulomb's Law applies to particles — electrons and protons — and also to any small charged bodies, provided that the sizes of these bodies are much smaller than the distance between them; such bodies are called ***point charges.*** Equation (6) clearly resembles Newton's law for the gravitational force, $F = GMm/r^2$. The constant $1/4\pi\epsilon_0$ is analogous to the gravitational constant G, and the electric charges are analogous to the gravitating masses.

In the SI system, the coulomb is defined in terms of a standard electric current: ***one coulomb is the amount of electric charge that a current of one ampere delivers in one second.*** Unfortunately, the definition of the ampere involves the use of magnetic fields, and we will therefore have to postpone the details of the precise definition of ampere and coulomb to Chapter 21.

◂ **Definition of coulomb**

EXAMPLE 1. Compare the magnitudes of the electric force of attraction and the gravitational force of attraction between the electron and the proton in a hydrogen atom. According to Newtonian mechanics, what is the acceleration of the electron? The mass of an electron is 9.11×10^{-31} kg; the mass of a proton is 1.67×10^{-27} kg. Assume that the distance between these particles is 0.53×10^{-10} m.

SOLUTION: The magnitude of the charge of an electron or a proton is $e = 1.60 \times 10^{-19}$ C. The magnitude of the electric force is

$$\frac{1}{4\pi\epsilon_0}\frac{e \times e}{r^2} = 8.99 \times 10^9\ \mathrm{N \cdot m^2/C^2} \times \frac{(1.60 \times 10^{-19}\ \mathrm{C}) \times (1.60 \times 10^{-19}\ \mathrm{C})}{(0.53 \times 10^{-10}\ \mathrm{m})^2}$$

$$= 8.2 \times 10^{-8}\ \mathrm{N}$$

The magnitude of the gravitational force is

$$G\frac{m \times M}{r^2} = 6.67 \times 10^{-11}\ \mathrm{N \cdot m^2/kg^2}$$

$$\times \frac{(9.11 \times 10^{-31}\ \mathrm{kg}) \times (1.67 \times 10^{-27}\ \mathrm{kg})}{(0.53 \times 10^{-10}\ \mathrm{m})^2}$$

$$= 3.6 \times 10^{-47}\ \mathrm{N}$$

The ratio of these forces is $8.2 \times 10^{-8}\ \mathrm{N}/3.6 \times 10^{-47}\ \mathrm{N} = 2.3 \times 10^{39}$.

Since the gravitational force is insignificant compared with the electric force, it can be neglected. The acceleration of the electron is then

$$a = \frac{F}{m} = \frac{8.2 \times 10^{-8}\ \mathrm{N}}{9.11 \times 10^{-31}\ \mathrm{kg}} = 9.0 \times 10^{22}\ \mathrm{m/s^2}$$

EXAMPLE 2. How much negative charge and how much positive charge is there on the electrons and the protons of a cup of water (0.25 kg)?

SOLUTION: The molecular mass of water is 18 g (16 g from the oxygen atom in the molecule, and 2 g from the two hydrogen atoms). Hence, 1 mole of water is 18 g, and the number of moles in 250 g is 250/18 moles. Each mole has 6.0×10^{23} molecules, giving $6.0 \times 10^{23} \times (250/18)$ molecules in the cup. Each molecule consists of two hydrogen atoms (1 electron apiece) and one oxygen atom (8 electrons). Thus, there are 10 electrons in each molecule, and the total negative charge on all the electrons together is $-6.0 \times 10^{23} \times (250/18) \times 10 \times 1.6 \times 10^{-19}\ \mathrm{C} = -1.3 \times 10^{7}\ \mathrm{C}$. The positive charge on the protons is the opposite of this.

EXAMPLE 3. What is the magnitude of the attractive force exerted by the electrons in a cup of water on the protons in a second cup of water at a distance of 10 m?

SOLUTION: According to the preceding example, the charge on the electrons is $-1.3 \times 10^{7}\ \mathrm{C}$ and the charge on the protons is $+1.3 \times 10^{7}\ \mathrm{C}$. If we treat both of these charges as point charges, the electric force is

$$F = \frac{1}{4\pi\epsilon_0}\frac{qq'}{r^2} \tag{7}$$

$$= 8.99 \times 10^9 \text{ N} \cdot \text{m}^2/\text{C}^2$$

$$\times \frac{(-1.3 \times 10^7 \text{ C}) \times (1.3 \times 10^7 \text{ C})}{(10 \text{ m})^2}$$

$$= -1.5 \times 10^{22} \text{ N}$$

This is approximately the weight of 10^{18} tons! This enormous attractive force is precisely canceled by an equally large repulsive electric force exerted by the protons in one cup on the protons in the other cup. Thus, the cups exert no net electric force on each other.

16.3 THE SUPERPOSITION OF ELECTRIC FORCES

The electric force, like any other force, has a magnitude and a direction; that is, the electric force is a vector. According to Coulomb's Law, the *magnitude* of the electric force exerted by a point charge q' on a point charge q is

$$F = \frac{1}{4\pi\epsilon_0} \frac{qq'}{r^2} \tag{8}$$

The *direction* of this force is along the line from one charge to the other. As illustrated in Fig. 16.4, this force can be represented by a vector **F** pointing along the line from one charge to the other.

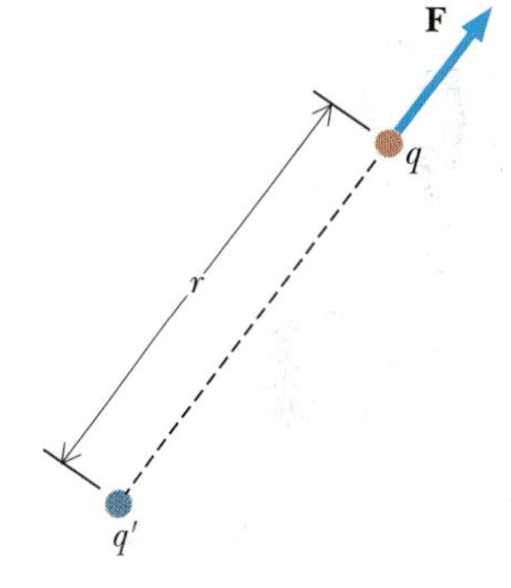

FIGURE 16.4

A charge q' exerts an electric force **F** on the charge q.

If several point charges q_1, q_2, q_3, . . . simultaneously exert electric forces on the charge q, then ***the net force on q is obtained by taking the vector sum of the individual forces*** (see Fig. 16.5). Thus, if the vectors representing the individual electric forces produced by q_1, q_2, q_3, . . . are $\mathbf{F}_1$, $\mathbf{F}_2$, $\mathbf{F}_3$, . . . , respectively, then the net force is

$$\mathbf{F} = \mathbf{F}_1 + \mathbf{F}_2 + \mathbf{F}_3 + \ldots \tag{9}$$

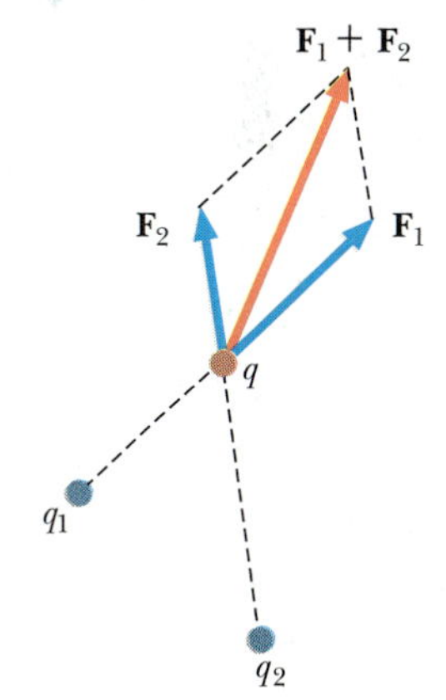

FIGURE 16.5

Two point charges q_1 and q_2 exert electric forces $\mathbf{F}_1$ and $\mathbf{F}_2$ on the point charge q. The net force on q is the vector sum of these forces.

Equation (9) expresses the ***principle of superposition*** of electric forces. According to Eq. (9), the force contributed by each charge is independent of the presence of the other charges. For instance, the charge q_2 does not affect the interaction of q_1 with q, it merely adds its own interaction with q. This simple combination law is an important *empirical* fact about electric forces. Since the contact forces (pushes and pulls) of everyday experience are electric forces, they will likewise obey the superposition principle, and they can be combined by simple vector addition. Incidentally: The gravitational forces on the Earth and within the Solar System also obey the superposition principle. Thus, all the forces in our immediate environment obey this principle. We have already made much use of the principle of superposition of forces in our study of mechanics; now we recognize that ***the superposition of mechanical forces, such as "contact" forces, hinges on the superposition of electric forces.***

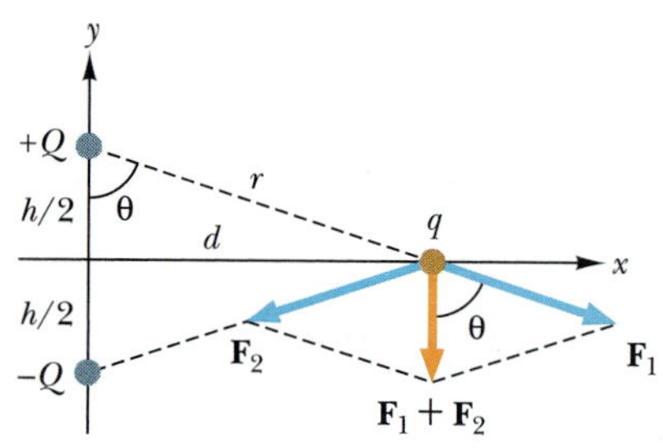

FIGURE 16.6
The charges $+Q$ and $-Q$ exert forces $\mathbf{F}_1$ and $\mathbf{F}_2$ on the charge q.

EXAMPLE 4. Point charges $+Q$ and $-Q$ are separated by a distance h. A point charge q is equidistant from these charges, at a distance d from their midpoint (see Fig. 16.6). What is the electric force on q?

SOLUTION: As illustrated in Fig. 16.6, the charge $+Q$ produces a repulsive force on the charge q, and the charge $-Q$ produces an attractive force. Thus, the vector $\mathbf{F}_1$ points away from $+Q$ and the vector $\mathbf{F}_2$ points toward $-Q$. The net force on q is the vector sum $\mathbf{F}_1+\mathbf{F}_2$. With the choice of axes shown in Fig. 16.6, the distance from each of the charges $+Q$ and $-Q$ to the charge q is $r=\sqrt{d^2+h^2/4}$. Hence the magnitudes of the individual Coulomb forces exerted by $+Q$ and $-Q$ are equal,

$$F_1=F_2=\frac{1}{4\pi\epsilon_0}\frac{qQ}{r^2}=\frac{1}{4\pi\epsilon_0}\frac{qQ}{d^2+h^2/4} \tag{10}$$

From Fig. 16.6 we see that in the vector sum $\mathbf{F}_1+\mathbf{F}_2$ for the net force, the horizontal components (x components) of $\mathbf{F}_1$ and $\mathbf{F}_2$ cancel, and the vertical components (y components) add, giving a net vertical component twice as large as each individual vertical component. Thus, the net force $\mathbf{F}$ has a y component but no x component. In terms of the angle θ shown in Fig. 16.6, the y component of $\mathbf{F}_1$ is $F_1\cos\theta$ and the y component of $\mathbf{F}_2$ is $F_2\cos\theta$. Thus, the y component of the net force is

$$F_y=-F_1\cos\theta-F_2\cos\theta=-2F_1\cos\theta$$

$$=-\frac{2}{4\pi\epsilon_0}\frac{qQ}{d^2+h^2/4}\cos\theta$$

From Fig. 16.6, we see that $\cos\theta=\frac{1}{2}h/\sqrt{d^2+h^2/4}$, and therefore

$$F_y=-\frac{1}{4\pi\epsilon_0}\frac{qQh}{(d^2+h^2/4)^{\frac{3}{2}}} \tag{11}$$

COMMENTS AND SUGGESTIONS: Note that if the charge q is at a large distance from the two charges $\pm Q$, then h^2 can be neglected compared with d^2, so $(d^2+h^2/4)^{\frac{3}{2}}\simeq(d^2)^{\frac{3}{2}}=d^3$. The force F_y is then proportional to $1/d^3$, that is, the force decreases in inverse proportion to the cube of the distance. Thus, although the force contributed by each point charge $\pm Q$ is an inverse-square force, the net force has a quite different behavior, because at large distance the force contributed by one charge tends to cancel the force contributed by the other.

16.4 CHARGE QUANTIZATION AND CHARGE CONSERVATION

Not only electrons and protons exert electric forces on each other, but so do many other particles. The magnitudes of these electric forces are given by Eq. (7) with the appropriate values of the electric charges. Table 16.3 lists the electric charges of some particles; a more complete list will be found in Chapter 32. The charges of antiparticles are always opposite to those of the corresponding particles; for example, the antielectron (or positron) has charge $+e$, the antiproton has charge $-e$, the antineutron has charge 0, and so on.

TABLE 16.3 ELECTRIC CHARGES OF SOME PARTICLES

Particle	Charge
Electron, e	$-e$
Muon, μ	$-e$
Pion, π^0	0
Pion, π^+	$+e$
Pion, π^-	$-e$
Proton, p	$+e$
Neutron, n	0
Delta, Δ^+	$+e$
Delta, Δ^{++}	$+2e$

◀ Charge quantization

All the known particles have charges that are some integer multiple of the fundamental charge, that is, the charges are always 0, $\pm e$, $\pm 2e$, $\pm 3e$, etc. Why no other charges exist is a mystery for which classical physics offers no explanation. Much effort has been expended on experimental searches for charges of $\frac{1}{3}e$ and $\frac{2}{3}e$ (these are the charges of the quarks, the elementary constituents supposedly contained inside protons and neutrons; see Chapter 32). But neither such fractional charges nor any other fractional charges have ever been found in isolation.

Since charges exist in discrete packets, we say that ***charge is quantized*** — the fundamental charge e is called the quantum of charge. However, in a description of the charge distribution on macroscopic bodies, the discrete nature of charge can often be ignored and it is usually sufficient to treat the charge as a continuous "fluid" with a charge density (C/m^3) that varies more or less smoothly over the volume of the charged body. This is analogous to describing the mass distribution of a solid, liquid, or a gas by a smooth mass density (kg/m^3) which ignores the fact that, on a microscopic scale, the mass is concentrated in atoms. A solid, a liquid, or a gas seems smooth because there are very many atoms in each cubic millimeter, and the distances between atom are very small. Likewise, charge distributions placed on wires or other conductors seem smooth because they consist of very many electrons (or protons) in each cubic millimeter.

◀ Charge conservation

The electric charge is a ***conserved quantity: in any reaction involving charged particles, the total charges before and after the reaction are always the same.*** Even in a reaction in which particles are destroyed — such as the annihilation of matter and antimatter — the net electric charge remains constant. An example of such a reaction is the annihilation of an electron and an antielectron:

$$\begin{array}{lccccc} & [\text{electron}] & + & [\text{antielectron}] & \rightarrow & 2\,[\text{photons}] \\ \textit{charges:} & -e & + & e & \rightarrow & 0 \end{array} \tag{12}$$

No reaction that creates or destroys net electric charge has ever been discovered.

Charge is of course also conserved in chemical reactions. For instance, in a lead–acid battery (automobile battery), plates of lead and of lead dioxide are immersed in an electrolytic solution of sulfuric acid (Fig. 16.7). The reactions that take place on these plates involve sulfate ions (SO_4^{--}) and hydrogen ions (H^+); the reactions release electrons at the lead plate and they absorb electrons at the lead-dioxide plate:

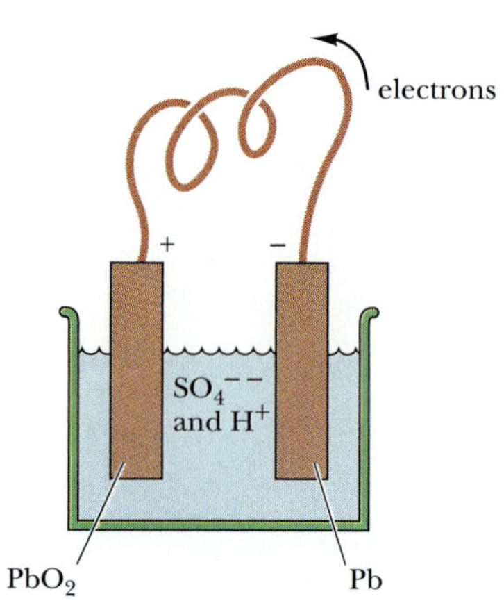

FIGURE 16.7
Schematic diagram of lead–acid battery.

at lead plate:

$$\begin{array}{lccccc} & Pb + SO_4^{--} & \rightarrow & PbSO_4 & + & 2\,[\text{electrons}] \\ \textit{charges:} & 0 + (-2e) & \rightarrow & 0 & + & (-2e) \end{array} \tag{13}$$

at lead-dioxide plate:

$$\begin{array}{lccccc} & PbO_2 + 4H^+ + SO_4^{--} + 2\,[\text{electrons}] & \rightarrow & PbSO_4 + 2H_2O \\ \textit{charges:} & 0 \;+\; 4e \;+ (-2e) + \;(-2e) & \rightarrow & 0 \;+\; 0 \end{array} \tag{14}$$

The plates of such a battery are connected by an external circuit (a wire), and the electrons released by the reaction (13) travel from one plate to the other, forming an electric current (see Fig. 16.7).

EXAMPLE 5. A fully "charged" battery contains a large amount of sulfuric acid in an electrolytic solution (H_2SO_4 in the form of SO_4^{--} and H^+ ions). As the battery delivers electric charge to the external circuit connecting its terminals, the amount of sulfuric acid in solution gradually decreases. Suppose that while discharging completely, the positive terminal of an automobile battery delivers an electric charge of 1.8×10^5 C through the external circuit. How many grams of sulfuric acid will be used up in this process?

SOLUTION: Since each electron has a charge of -1.6×10^{-19} C, the number of electrons in -1.8×10^5 C is $1.8 \times 10^5 / 1.6 \times 10^{-19}$, or 1.1×10^{24}. According to the reactions (13) and (14), whenever two electrons are transferred from the lead to the lead-dioxide plate, two sulfate ions are absorbed (one at each plate). Thus, 1.1×10^{24} sulfate ions will be used up, which means 1.1×10^{24} molecules of sulfuric acid will be used up. The required number of moles of sulfuric acid is therefore $1.1 \times 10^{24}/6.0 \times 10^{23} = 1.9$ and, since the molecular mass of sulfuric acid is 98 g per mole, the required mass of sulfuric acid is $1.9 \times 98\text{ g} = 183$ g.

The conservation of electric charge in chemical reactions, such as the reactions (13) and (14), is an immediate consequence of the conservation of electrons and protons. All such reactions involve nothing but a rearrangement of the electrons and protons in the molecules; during this rearrangement, the numbers of electrons and of protons remain constant. Obviously, the net electric charge must then also remain constant.

The same argument applies to all macroscopic electric processes, such as the operation of electrostatic machines and generators, the flow of currents on wires, the storage of charge in capacitors, the electric discharge of thunderclouds, etc. All such processes involve nothing but a rearrangement of electrons and protons. Consequently, the net electric charge must remain constant.

16.5 CONDUCTORS AND INSULATORS; FRICTIONAL ELECTRICITY

◀ Conductor
◀ Insulator

A conductor — such as copper, aluminum, or iron — is a material that permits the motion of electric charges through its volume. An insulator — such as glass, porcelain, rubber, or nylon — is a material that does not permit the motion of electric charges. Thus, when we deposit some electric charge on one place on a conductor, it immediately spreads out over the entire conductor until it finds an equilibrium distribution*; or, if the conductor is connected to some other conductor, the charge leaks away. In contrast, when we deposit some charge on one place of an insulator, it stays in that place. This is why you can electrify a nylon comb by rubbing, but you cannot electrify a metal comb — any charge that the rubbing produces on the metal comb immediately leaks away through your hand.

◀ Free electrons

All metals are good conductors. The motion of charge in metal is due to the motion of electrons. ***In a metal, some of the electrons of each atom are free,*** that is, they are not bound to any particular atom although they are bound to the metal as a whole. The free electrons come from the outer parts of the atoms. The outer electrons of the atom are not very strongly attached and readily come loose; the inner electrons are firmly bound to the nucleus of the atom and are likely to stay put. The free electrons wander through the entire volume of the metal, suffering occasional collisions, but they experience a restraining force only when they encounter the surface of the metal. The electrons are held inside the metal in much the same way as particles of a gas are held inside a container — the particles of gas can wander through the volume of the container, but they are restrained by the walls. In view of this analogy, electrons in a metallic conductor are often said to form a ***free-electron gas.*** If one end of a metallic conductor has an excess or deficit of electrons, the motion of the

◀ Free-electron gas

* We will study the conditions for the equilibrium of electric charge on a conductor in the next chapter (Section 17.4). It turns out that when the extra charges deposited on a conductor finally reach equilibrium, they will all sit on the surface of the conductor.

free-electron gas will quickly distribute this excess or deficit to other parts of the metallic conductor.

As already mentioned in Section 16.1, the charging of a macroscopic body is usually accomplished by the removal or the addition of electrons. A body will acquire a net positive charge if electrons are removed and a net negative charge if electrons are added. Thus, positive charge on a body of metal is simply a deficit of electrons, and negative charge an excess of electrons.

Liquids containing ions (atoms or molecules with missing electrons or with excess electrons) also are good conductors. For instance, a solution of common salt in water contain ions of Na^+ and Cl^-. The motion of charges through the liquid is due to the motion of these ions. ***Liquid conductors with an abundance of ions are called electrolytes.***

Electrolytes ▶

Incidentally: Very pure distilled water is a poor conductor because it lacks ions. But ordinary water is a good conductor because it contains some ions contributed by dissolved impurities. The ubiquitous water in our environment makes many substances into conductors. For example, earth (soil) is a reasonably good conductor, mainly because of the presence of water. Furthermore, on a humid day many insulators acquire a microscopic surface film of water, and this permits electric charge to leak away along the surface of the insulator; thus, on humid days, it is difficult to store electric charge on bodies supported by insulators. The soft tissues of the human body are good conductors, since they contain a large amount of water, with an abundance of ions. However, dry human skin is a fairly good insulator, and this provides some protection against the hazards of accidental contact with live electric wires.

FIGURE 16.8
Lightning.

Ordinary gases are insulators, but ionized gases are good conductors. For example, ordinary air is an insulator, but the ionized air found in a lightning bolt is a good conductor (see Fig. 16.8). This ionized air contains a mixture of positive ions and free electrons; the motion of charge in such a mixture is due mainly to the motion of the electrons. ***Such an ionized gas is called a plasma.***

Plasma ▶

Ionization of a gas occurs whenever the gas molecules are subjected to large electric forces arising from a strong concentration of electric charge, such as the concentration of electric charge in a thundercloud. The electric forces on the air molecules produce a sudden, catastrophic ionization of the air within a fairly narrow channel between the cloud and the ground. The electric charge of the thundercloud then quickly spills through this channel to the ground, forming the lightning bolt.

Although lightning gives us the most spectacular evidence of electrical activity in a thunderstorm, the largest part of the electrical activity in a thunderstorm proceeds silently and almost imperceptibly in the form of ***corona discharge.*** The electric charge in the thundercloud attracts opposite charge from the ground. This charge on the ground concentrates at any sharp points sticking out of the ground — such as the tips of the leaves of trees. The concentration of charge causes a steady, nearly imperceptible ionization of the air near the sharp points. Charge can then leak into this ionized air and flow through the air between the thundercloud and the ground. This kind of ionization of air is called corona discharge because it can sometimes be seen as a glowing halo surrounding the pointed object (see Fig. 16.9).

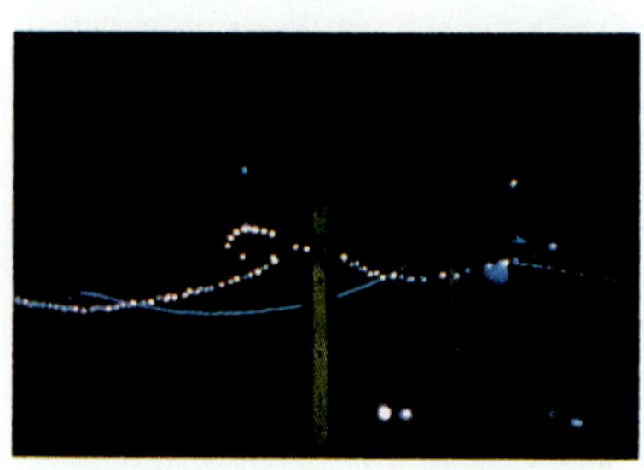

FIGURE 16.9
Corona discharge from power lines.

Lightning rods, invented by Benjamin Franklin, were originally intended to dissipate the electric charge of thunderclouds by corona discharge, and they were therefore supposed to inhibit lightning. However, the traditional lightning rod, with a single sharp point, cannot produce a sufficient amount of corona discharge. The lightning rod is connected to ground by a hefty conducting cable, and its main benefit is that this cable conducts harmlessly to the ground the electric charge that arrives during a lightning stroke. Lightning rods of modern design are much better at producing corona discharge. These dissipative lightning rods end in a tassel of fine wires (see Fig. 16.10). Thus, the lightning rod has a multitude of sharp points, and it can dissipate a substantial fraction of the total charge of a thundercloud passing overhead.

◀ **Frictional electricity**

We will end this chapter with a few brief comments on ***frictional electricity.*** It is easy to accumulate electric charge on a glass rod merely by rubbing it with a piece of silk. The glass becomes positively charged, and the silk negatively charged — the rubbing motion between the surfaces of the glass and the silk rips charges off one of these surfaces and makes them stick to the other. It is believed that what is usually involved is a transfer of ions from one surface to the other. Contaminants residing on the rubbed surfaces play a crucial role in frictional electricity. If glass is rubbed with an absolutely clean piece of silk or other textile material, the glass becomes negatively charged rather than positively charged. Ordinary pieces of silk apparently have such a large amount of dirt on their surfaces that the charging process is dominated by the dirt rather than by the silk.

Once we have accumulated some charge on, say, a rod of glass, we can produce charges on other bodies by a process of ***induction,*** as follows: First we bring the glass rod near a metallic body supported on an insulating stand. The positive charge on the rod will then attract free electrons to the near side of the metallic body and leave a deficit of free electrons on the far side (see Fig. 16.11a); thus, the near side will acquire negative charge and the far side positive charge. If we next momentarily connect the far side to the ground, the positive charge will be neutralized by an influx of electrons from the ground (in effect, the positive charge leaks away; see Fig. 16.11b). This leaves the metallic body with a net negative charge. When we finally withdraw the glass rod, this charge will remain on the metallic body, distributing itself over its entire surface (see Fig. 16.11c).

FIGURE 16.10

Modern dissipative lightning rod. The rod ends in a tassel of fine wires, which provide many sharp points for corona discharge.

(a)

(b)

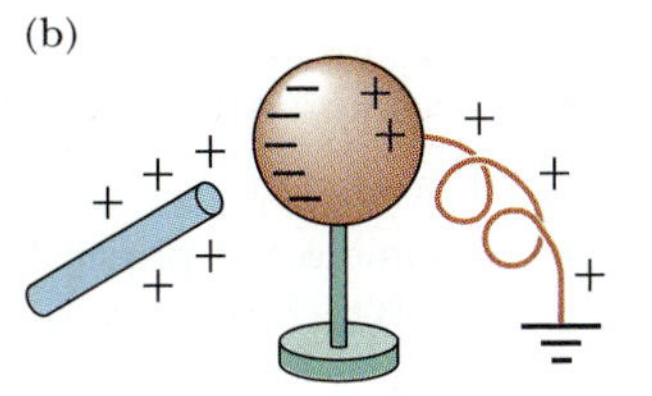

(c)

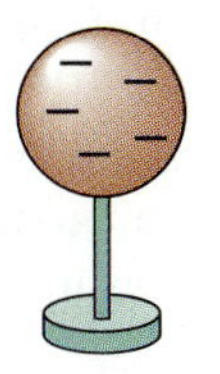

FIGURE 16.11

(a) The positively charged glass rod induces a charge distribution on the metallic sphere. (b) When the far side of the sphere is connected to the ground by means of a wire, the positive charge leaks away. (c) Finally, only negative charge remains on the sphere.

SUMMARY

- **Electric charges may be positive, negative, or zero; like charges repel, unlike charges attract.**
- **Coulomb's Law:** $F = \frac{1}{4\pi\epsilon_0}\frac{qq'}{r^2}$
- **Permittivity constant:** $\epsilon_0 = 8.85 \times 10^{-12}\ \mathrm{C^2/(N \cdot m^2)}$

$$\frac{1}{4\pi\epsilon_0} = 8.99 \times 10^9\ \mathrm{N \cdot m^2/C^2}$$

- **Superposition principle:** Electric forces produced by different charges combine by vector addition.
- **Charge of proton:** $e = 1.60 \times 10^{-19}$ C
- **Charge of electron:** $-e = -1.60 \times 10^{-19}$ C
- **Charge conservation:** In any reaction or process, the net electric charge remains constant.
- **Conductor:** Permits the motion of charge.
- **Insulator:** Does not permit the motion of charge.

QUESTIONS

1.–Suppose that the Sun has a positive electric charge and each of the planets a negative electric charge, and suppose that there is no gravitational force. In what way would the motions of the planets predicted by this "electric" model of the Solar System differ from the observed motions?

2.–The protons in the nucleus of an atom repel each other electrically. What holds the protons (and the neutrons) together, and prevents the nucleus from bursting apart?

3.–Describe how you would set up an experiment to determine whether the electric charges on an electron and a proton are exactly the same.

4.–If we were to assign a positive charge to the electron and a negative charge to the proton, would this affect the mathematical statement [Eq. (6)] of Coulomb's Law?

5.–In the cgs, or Gaussian, system of units, Coulomb's Law is written as $F = qq'/r^2$, without the factor $1/4\pi\epsilon_0$. In terms of grams, centimeters, and seconds, what are the units of electric charge in this system?

6.–Could we use the electric charge of an electron as an atomic standard of electric charge to define the coulomb? What would be the advantages and disadvantages of such a standard?

7.–Besides electric charge, list some other physical quantities that are conserved in reactions among particles. Which of these quantities are quantized?

8.–Since the free electrons in a piece of metal are free to move any which way, why don't they all fall to the bottom of the piece of metal under the influence of the pull of gravity?

9.–If the surface of a piece of metal acts like a container in confining the free electrons, why can't we cause these electrons to spill out by drilling a hole in the surface?

10.–If you rub a plastic comb, it will attract hairs or bits of paper, even though they have no net electric charge. Explain.

11.–Some old-fashioned physics textbooks define positive electric charge as the kind of charge that accumulates on a glass rod when rubbed with silk. What is wrong with this definition?

12.–When you rub your shoes on a carpet, you sometimes pick up enough electric charge to feel an electric shock if you subsequently touch a radiator or some other metallic body connected to the ground. Why is this more likely to happen in winter than in summer?

13.–Some automobile operators hang a conducting strap on the underside of their automobile, so this strap drags on the street. What is the purpose of this arrangement?

14.–Some electric charge has been deposited on a ping pong ball. How could you find out whether the charge is positive or negative?

15.–Two aluminum spheres of equal radii hang from the ceiling on insulating threads. You have a glass rod and a piece of silk. How can you give these two spheres exactly equal amounts of electric charge?

An electric charge on this plastic comb attracts small bits of paper.

PROBLEMS

Section 16.1 Electric Charge

1.——A lightning stroke typically deposits −25 C on the ground. How many electrons is this?

2.——The mass of the electron cannot be measured directly, since macroscopic amounts of mass always contain a combination of electrons, protons, and neutrons, never pure electrons. Instead, the mass is calculated from a measurement of the electric charge $-e$ of the electron and a measurement of the charge-to-mass ratio $-e/m_e$. The best values for these quantities are $-e = -1.602177 \times 10^{-19}$ C and $-e/m_e = -1.758820 \times 10^{11}$ C/kg. What best value of the mass of the electron can you deduce from this?

3.——Suppose that during a thunderstorm, the corona discharge from a dissipative lightning rod into the surrounding air amounts to 1×10^{-4} C of positive charge per second. If this discharge goes on more or less steadily for an hour, how much electric charge flows out of the lightning rod? How many electrons flow into the lightning rod from the surrounding air?

4.——How many electrons do you need to remove from an initially neutral bowling ball to give it a positive electric charge of 10^{-6} C?

5.——The electric charge in 1 mole of protons is called *Faraday's constant.* What is its numerical value?

6.——How many electrons are in a paper clip of iron of mass 0.3 g?

7.——What is the number of electrons and of protons in a human body of mass 73 kg? The chemical composition of the body is roughly 70% oxygen, 20% carbon, and 10% hydrogen (by mass).

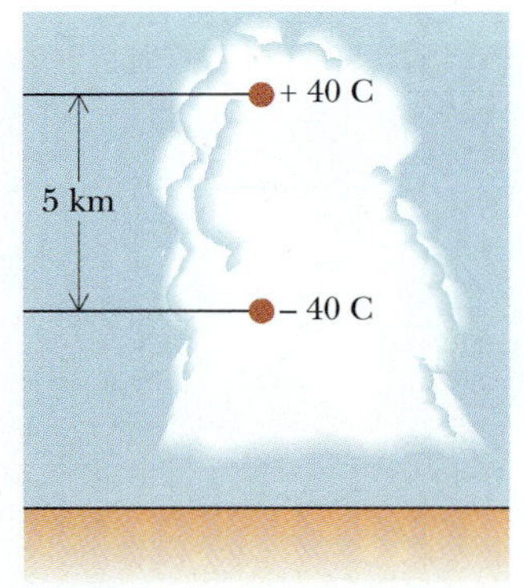

FIGURE 16.12
Charges in a thundercloud.

Section 16.2: Coulomb's Law

8.——By rubbing a small glass ball against a small nylon ball, you deposit a charge of 6×10^{-11} C on the glass ball and a charge of -6×10^{-11} C on the nylon ball. If you then separate the balls to a distance of 20 cm, what is the attractive electric force between them?

9.——Within a typical thundercloud there are electric charges of -40 C and $+40$ C separated by a vertical distance of 5 km (see Fig. 16.12). Treating these charges as pointlike, find the magnitude of the electric force of attraction between them.

10.——According to Example 1, the electric force of attraction between an electron and a proton separated by a distance of 0.53×10^{-10} m is 8.2×10^{-8} N. What is the force if the separation is twice as large? Three times as large? Four times as large?

11.——In an HCl molecule, the nuclei of the H and the Cl atoms, with charges $+e$ and $+17e$, respectively, are separated by a distance of 1.28×10^{-10} m. What is the electric force of repulsion between these nuclei?

12.——Consider two protons separated by a distance of 1.0×10^{-12} m.

(a) What is the gravitational force of attraction between the protons?

(b) What is the electric force of repulsion between these protons? What is the ratio of the electric force and the gravitational force?

(c) Consider a second pair of protons separated by a larger distance, so their electric repulsion matches the gravitational attraction calculated in Part (a). How far apart would these protons have to be?

13.——Suppose that two grains of dust of equal masses each have a single electron charge. What must be the masses of the grains if their gravitational attraction is to balance their electric repulsion?

14.——A crystal of NaCl (common salt) consists of a regular arrangement of ions of Na^+ and Cl^-. The distance from one ion to its neighbor is 2.82×10^{-10} m. What is the magnitude of the electric force of attraction between the two ions? Treat the ions as point charges.

15.——Suppose that the two protons in the nucleus of a helium atom are at a distance of 2×10^{-15} m from each other. What is the magnitude of the electric force of repulsion that they exert on each other? What would be the acceleration of each if this were the only force acting on them? Treat the protons as point charges.

16.——An alpha particle (charge $+2e$) is launched at high speed toward a nucleus of uranium (charge $+92e$). What is the magnitude of the electric force on the alpha particle when it is at a distance of 5×10^{-14} m from the nucleus? What is the corresponding instantaneous acceleration of the alpha particle?

17.——According to recent theoretical and experimental investigations, the subnuclear particles are made of quarks and of antiquarks (see Chapter 32). For example, a positive pion is made of a u quark and a d antiquark. The electric charge on the u quark is $\frac{2}{3}e$ and the electric charge on the d antiquark is $\frac{1}{3}e$. Treating the quarks as classical particles, calculate the electric force between the quarks in the pion if the distance between them is 1.0×10^{-15} m.

18.——The electric charge flowing through an ordinary 115-volt, 150-watt light bulb is 1.3 C/s. How many electrons per second does this amount to?

19.——Two tiny chips of plastic of mass 5×10^{-5} g are separated by a distance of 1 mm. Suppose that they carry opposite electrostatic charges of equal magnitudes. What must the magnitude of the charge be if the electric attraction between them is to equal their weight?

20.——In the lead atom, the nucleus has an electric charge $82e$. The innermost electron in this atom is typically at a distance of 6.5×10^{-13} m from the nucleus. What is the electric force that the nucleus exerts on such an electron? What is the acceleration that this force produces on the electron? Treat the electron as a classical particle.

21.—A maximum electric charge of 7.5×10^{-6} C can be placed on a metallic sphere of radius 15 cm before the surrounding air suffers electric breakdown (sparks). How many excess electrons (or missing electrons) does the sphere have when breakdown is about to occur?

22.*—A small charge of 2×10^{-8} C is at the point $x = 2$ m, $y = 0$ on the x axis. A second small charge of -3×10^{-6} C is at the point $x = 0$, $y = -3$ m on the y axis (see Fig. 16.13). What is the electric force that the first charge exerts on the second? What is the force that the second charge exerts on the first? Express your answers as vectors, with x and y components.

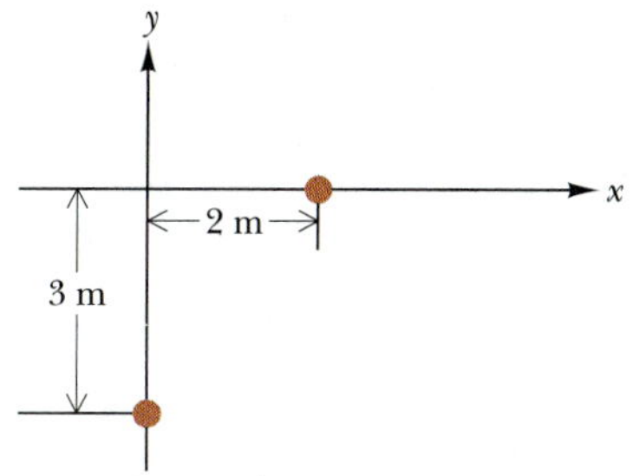

FIGURE 16.13
Two pointlike charges.

23.*—Two small balls of plastic carry equal charges of opposite signs and of unknown magnitudes. When the balls are separated by a distance of 18 cm, the attractive electric force between them is 0.3 N. What is the excess of electrons on one ball and the deficit of electrons on the other?

24.*—A proton is at the origin of coordinates. An electron is at the point $x = 4.0 \times 10^{-11}$ m, $y = 2.0 \times 10^{-11}$ m in the x–y plane (see Fig. 16.14). What are the x and y components of the electric force that the proton exerts on the electron? That the electron exerts on the proton?

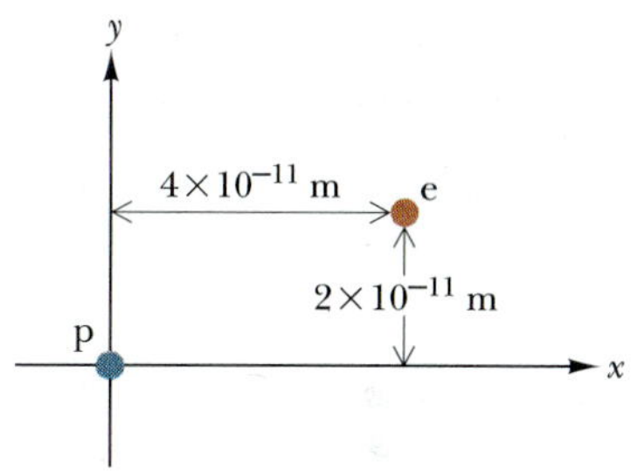

FIGURE 16.14
A proton and an electron.

25.*—Deimos is a small moon of Mars, with a mass of 2×10^{15} kg. Suppose that an electron is at a distance of 100 km from Deimos. What is the gravitational attraction acting on the electron? What negative electric charge would have to be placed on Deimos to balance this gravitational attraction? How many electron charges does this amount to? Treat the mass and the charge distributions as pointlike in your calculations.

26.*—Under the influence of the electric force of attraction, the electron in a hydrogen atom orbits around the proton on a circle of radius 0.53×10^{-10} m. What is the orbital speed? What is the orbital period?

27.*—Suppose that you remove one electron from each of the atoms in a copper coin of mass 2.7 g and place these electrons at a distance of 2.0 m from the remaining copper ions. What would be the electric force of attraction on the electrons?

28.*—Suppose that a proton and an electron are initially separated by a distance of 5.0×10^{-9} m, and they are initially at rest.

(a) What is the instantaneous acceleration of the electron toward the proton? Of the proton toward the electron? Assume that the only force acting on the particles is their mutual electric attraction.

(b) Some time later, the particles have come closer together, and the electron has attained an instantaneous speed of 3×10^{6} m/s. What is the instantaneous speed of the proton at this time?

29.*—At a place directly below a thundercloud, the induced electric charge on the surface of the Earth is $+10^{-7}$ coulomb per square meter of surface. How many singly charged positive ions per square meter does this represent? The number of atoms on the surface of a solid is typically 3×10^{20} per square meter. What fraction of these atoms must be ions to account for the above electric charge?

30.*—How many extra electrons would we have to place on the Earth and on the Moon so that the electric repulsion between these bodies cancels their gravitational attraction? Assume that the numbers of extra electrons on the Earth and on the Moon are the same.

31.**—Precise experiments have established that the magnitudes of the electric charges of an electron and a proton are equal to within an experimental error of $\pm 10^{-21}e$ and that the electric charge of a neutron is zero to within $\pm 10^{-21}e$. Making the worst possible assumption about the combi-

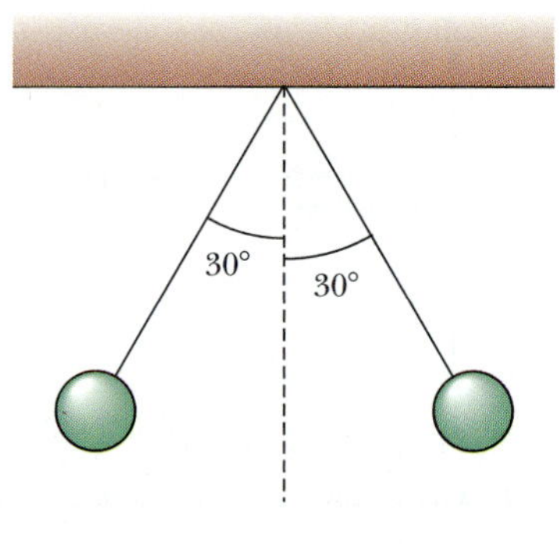

FIGURE 16.15
Balls suspended from strings.

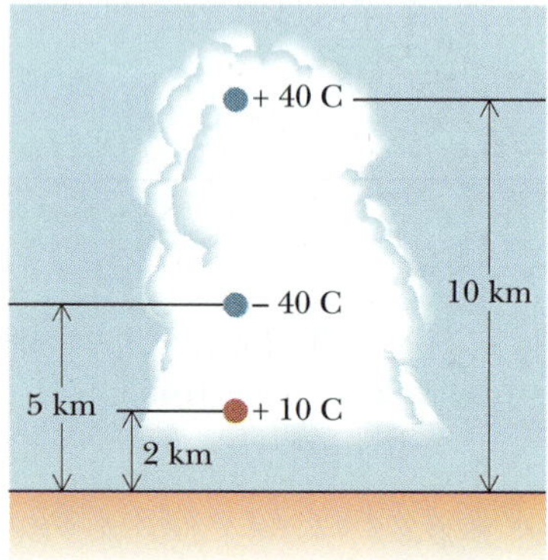

FIGURE 16.16
Charges in a thundercloud.

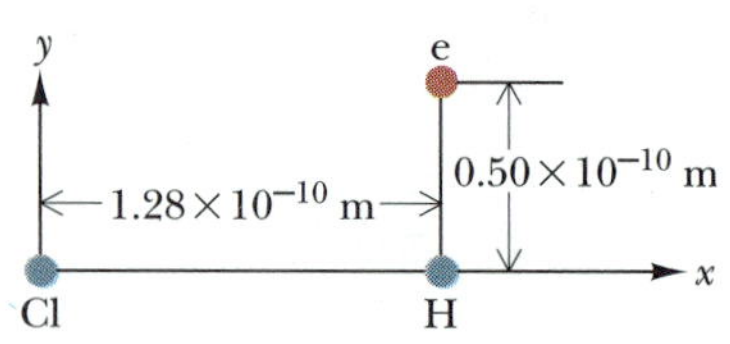

FIGURE 16.17
The positive nuclear charges in chlorine and hydrogen exert electric forces on an electron.

nation of errors, what is the largest conceivable electric charge of an oxygen atom consisting of 8 electrons, 8 protons, and 8 neutrons? Treating the atoms as point particles, compare the electric force between two such oxygen atoms with the gravitational force between these atoms. Is the net force attractive or repulsive?

32.** Two small Styrofoam balls, of mass 1.5×10^{-3} kg each, are suspended from the same point by strings 10 cm long (see Fig. 16.15). Equal positive charges are placed on the balls. What must be the magnitudes of these positive charges if the balls are to remain in equilibrium at angles of 30° from the vertical?

Section 16.3: The Superposition of Electric Forces

33. The distribution of electric charges in a thundercloud can be approximated by several pointlike charges placed at different heights. Suppose that a thundercloud has electric charges of +10 C, −40 C, and +40 C at altitudes of 2 km, 5 km, and 10 km, respectively (see Fig. 16.16). Treating these charges as pointlike, find the net electric force that the two charges of ±40 C exert on the charge of +10 C.

34. Figure 16.17 shows the arrangement of nuclear charges (positive charges) in an HCl molecule. The magnitudes of these charges are e and $17e$, respectively, and the distance between them is 1.28×10^{-10} m. What is the net electric force that these charges exert on an electron placed 0.50×10^{-10} m above the H nucleus?

35.* Point charges $+Q$ and $-2Q$ are separated by a distance d. A point charge q is equidistant from these charges, at a distance x from their midpoint (see Fig. 16.18). What is the electric force on q?

36.* Three positive point charges $+Q$ are placed at three corners of a square, and a negative point charge $-Q$ is placed at the fourth corner (see Fig. 16.19). The side of the square is L. Calculate the net electric force that the positive charges exert on the negative charge.

37.* Four point charges of $\pm Q$ are arranged on the corners of a square on side L, as illustrated in Fig. 16.20. What is the net electric force that these charges exert on a point charge q placed at the center of the square?

38.* Figure 16.21 shows the approximate charge distribution in a thundercloud consisting of a pointlike charge of +40 C at a height of 10 km and a pointlike charge of −30 C at a height of 4 km. What is the force that these two charges exert on an electron located at a height of 10 km and at a horizontal distance of 4 km to the right?

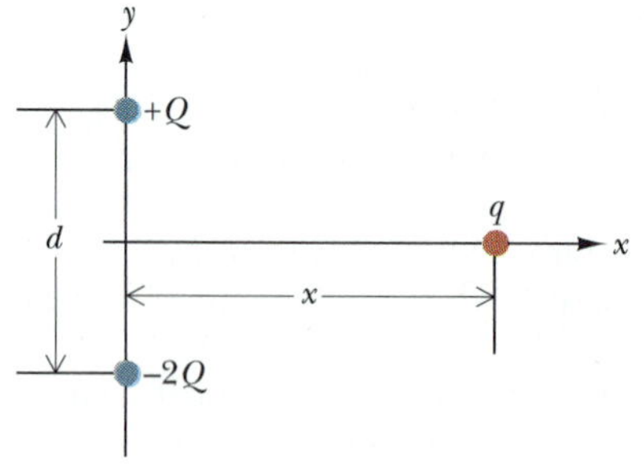

FIGURE 16.18
Charges $+Q$ and $-2Q$ exert forces on a charge q.

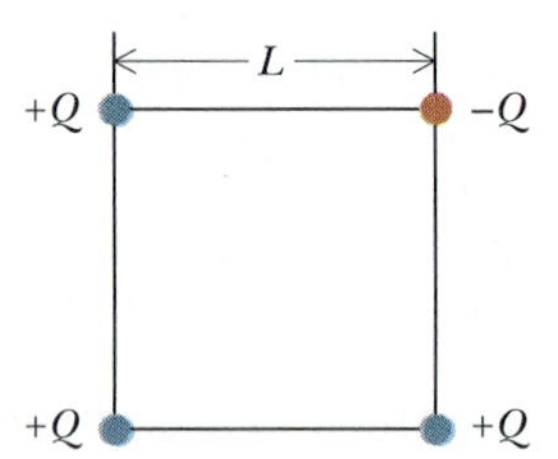

FIGURE 16.19
Three positive point charges and one negative point charge.

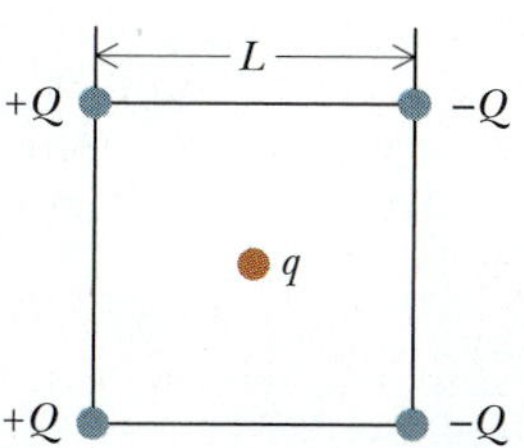

FIGURE 16.20
Four point charges.

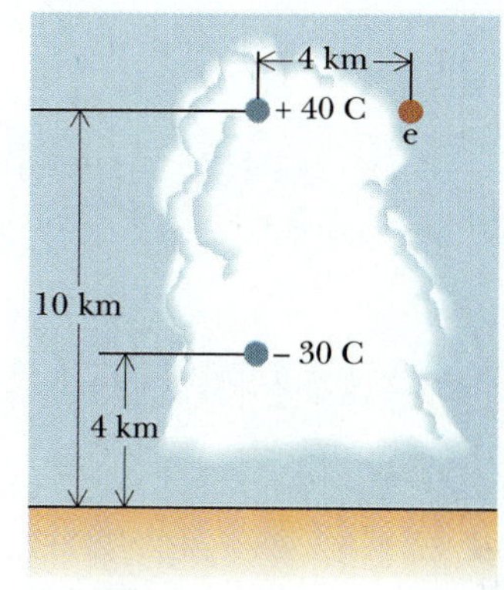

FIGURE 16.21
Charges in a thundercloud.

39.** Repeat the preceding problem but with the electron at a height of 7 km and at a horizontal distance of 4 km to the right.

Section 16.4: Charge Quantization and Charge Conservation

40.— Consider the reaction

$$Ni^{++} + 4H_2O \rightarrow NiO_4^{--} + 8H^+ + [\text{electrons}]$$

How many electrons does this reaction release?

41.— Consider the following hypothetical reactions involving the collision between a high-energy proton (from an accelerator) and a stationary proton (in the nucleus of a hydrogen atom serving as a target):

$$\mathrm{p} + \mathrm{p} \rightarrow \mathrm{n} + \mathrm{n} + \pi^+$$
$$\mathrm{p} + \mathrm{p} \rightarrow \mathrm{n} + \mathrm{p} + \pi^0$$
$$\mathrm{p} + \mathrm{p} \rightarrow \mathrm{n} + \mathrm{p} + \pi^+$$
$$\mathrm{p} + \mathrm{p} \rightarrow \mathrm{p} + \mathrm{p} + \pi^0 + \pi^0$$
$$\mathrm{p} + \mathrm{p} \rightarrow \mathrm{n} + \mathrm{p} + \pi^0 + \pi^-$$

where the symbols p, n, π^+, π^-, and π^0 stand for proton, neutron, positively charged pion, negatively charged pion, and neutral pion, of charges $+e$, 0, $+e$, $-e$, and 0, respectively. Which of these reactions are impossible, because they violate the conservation of charge?

42.* We can silver-plate a metallic object, such as a spoon, by immersing the spoon and a bar of silver in a solution of silver nitrate ($AgNO_3$). If we then connect the spoon and the silver bar to an electric generator and make a current flow from one to the other, the following reactions will occur at the immersed surfaces (Fig. 16.22):

$$Ag^+ + [\text{electron}] \rightarrow Ag_{(\text{metal})}$$
$$Ag_{(\text{metal})} \rightarrow Ag^+ + [\text{electron}]$$

The first reaction deposits silver on the spoon and the second removes silver from the silver bar. How many electrons must we make flow from the silver bar to the spoon in order to deposit 1 g of silver on the spoon?

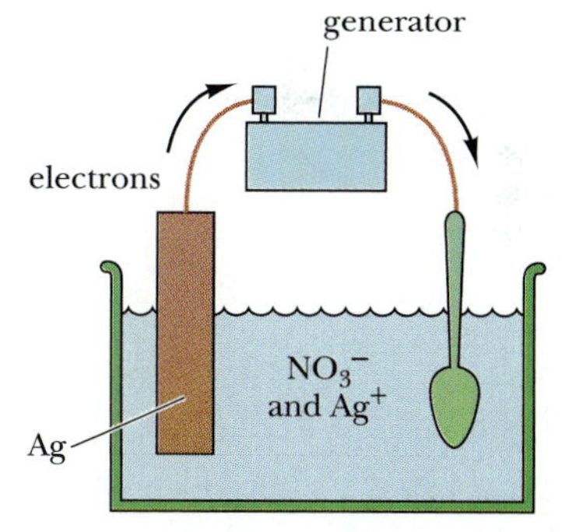

FIGURE 16.22
Silver-plating a spoon.

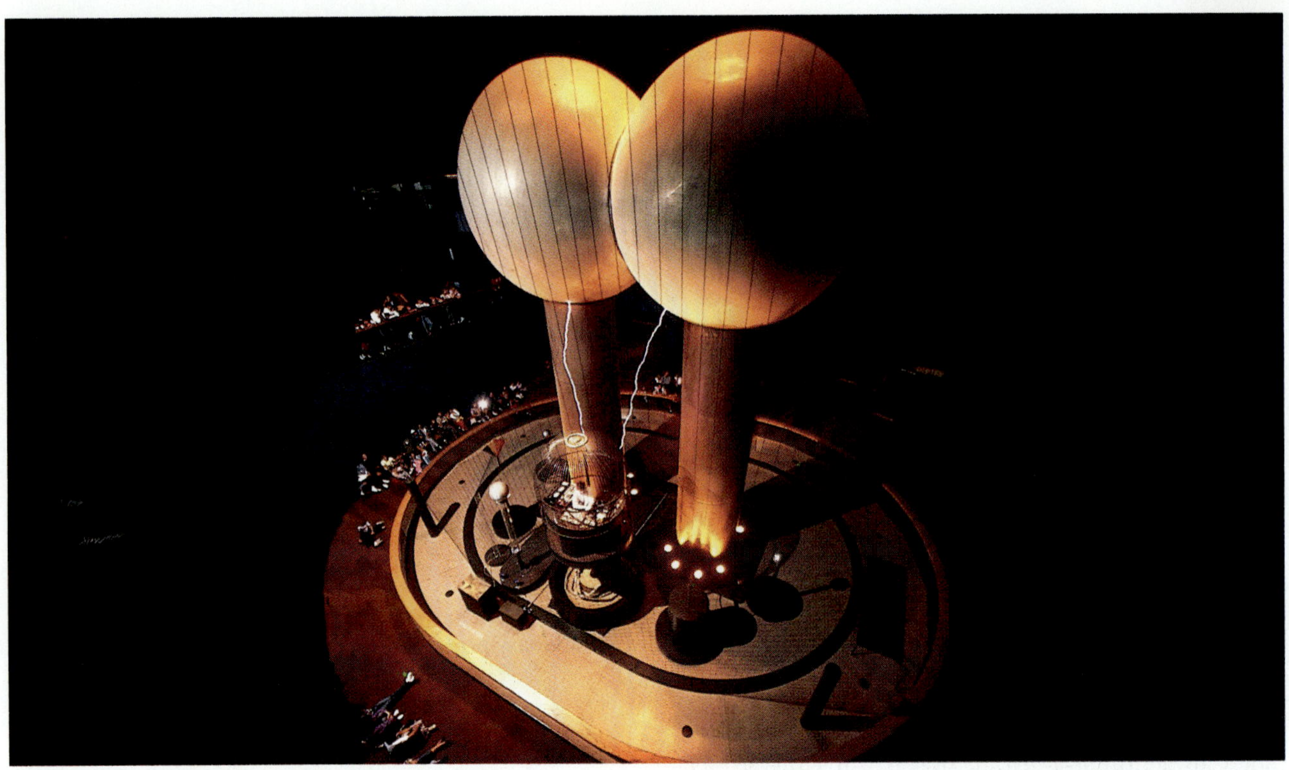

This Van de Graaff generator at the Boston Science Museum accumulates electric charge on two large metallic spheres. The concentration of charge produces a strong electric field in the vicinity of the spheres. The resulting strong electric forces on the air molecules cause electric breakdown of the air—sparks jump from the spheres.

THE ELECTRIC FIELD

Action-at-a-distance ▶

Up to here we have taken the view that the gravitational forces and the electric forces between particles involve ***action-at-a-distance,*** that is, a particle exerts a direct gravitational or electric force on another particle even though these particles are not touching. Although such an interpretation of gravitational and electric forces as a ghostly tug-of-war between distant particles is suggested by Newton's law of gravitation and by Coulomb's Law, Newton himself had considerable misgivings about this interpretation. In his own words:

> It is inconceivable that inanimate brute matter, should, without the mediation of something else, which is not material, operate upon and affect other matter without mutual contact. That Gravity should be innate, inherent and essential to Matter so that one Body may act upon another at a Distance thro' a *Vacuum* without the Mediation of anything else, by and through which their Action and Force may be conveyed from one to an-

> other, is to me so great an Absurdity that I believe no Man who has in philosophical Matters a competent Faculty of thinking can ever fall into it.

According to the modern view, there is indeed an entity that acts as mediator of force, conveying the force over the distance from one body to another. This entity is the ***field.*** A gravitating or electrically charged body generates a gravitational or electric field which permeates the (apparently) empty space around the body, and forms an invisible disturbance in the space around the body. This field exerts pushes or pulls whenever it comes in contact with another body. Thus, fields convey forces from one body to another through ***action-by-contact.***

◀ **Action-by-contact**

In the present chapter, we will become acquainted with the electric field that conveys the electric force from one body to another body and we will see how to represent the electric field graphically by field lines.

17.1 THE ELECTRIC FIELD

According to the naive action-at-a-distance interpretation of Coulomb's Law, a charge q' exerts a direct force on a charge q even though these charges are separated by a large distance and are not touching. However, such an action-at-a-distance interpretation of electric force leads to serious difficulties in the case of moving charges. Suppose that we suddenly move the charge q' somewhat nearer to the charge q; then the electric force has to increase. But the required increase cannot occur instantaneously — the increase can be regarded as a signal from q' and q, and it is a fundamental principle of physics, based on the theory of relativity, that no signal can propagate faster than the speed of light. This suggests that, when we suddenly move the charge q' (see Fig. 17.1), some kind of disturbance propagates through space from q' to q and adjusts the electric force to the new increased value. Thus, ***charges exert forces on one another by means of disturbances that they generate in the space surrounding them. These disturbances are called electric fields.***

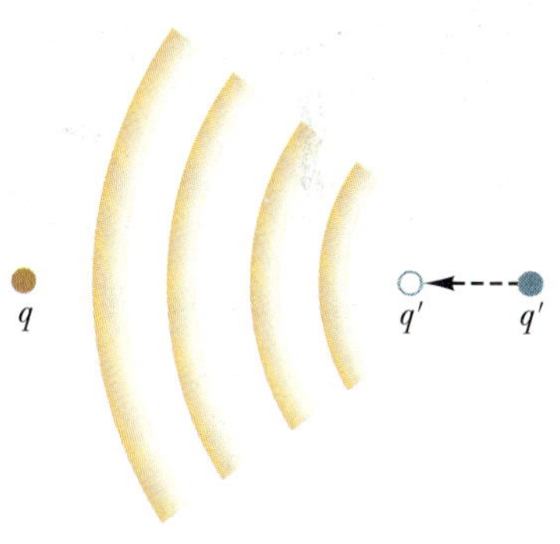

FIGURE 17.1
A disturbance emanates from the charge q' and reaches the charge q.

◀ **Electric fields**

Fields are a form of matter — they are endowed with energy and momentum, and they therefore exist in a material sense. If we think of solids, liquids, gases, and plasmas as the first four states of matter, then fields are the fifth state of matter. In the context of the above example, it is easy to see why the disturbance, or field, generated by the sudden displacement of q' must possess momentum: when we suddenly move q' toward q, the force on q' immediately increases according to Coulomb's Law, but the increase in the force on q will be delayed until a signal has had time to propagate from q' carrying the information regarding the changed position of q'. Thus, action and reaction will be temporarily out of balance. We know from Chapter 6 that the conservation of momentum in a system of particles hinges on the balance of action and reaction — any imbalance implies that the momentum of our system of two charged particles is not conserved. In order to maintain an overall momentum conservation, the momentum missing from the particles q and q' must be transferred to the electric field, that is, the field must acquire momentum.

Although the above arguments for the existence of fields arose from the problem of charges in motion, we will now adopt the very natural view that the forces on charges at rest involve the same mechanism. We suppose that each charge at rest generates a permanent, static disturbance in the space surrounding it, and that this disturbance exerts forces on other charges when they come in contact with the disturbance. Thus, we take the view that the electric interaction between charges is ***action-by-contact: a charge q' generates an electric field which permeates the surrounding space and exerts a force on any other charge that it touches.*** The electric field serves as the mediator of forces according to the scheme

Action-by-contact provided by electric field ▶

$$\text{charge } q' \rightarrow \text{electric field of charge } q' \rightarrow \text{force on charge } q \qquad (1)$$

To translate this conceptual scheme into mathematical language, we start with Coulomb's Law for the force exerted by the charge q' on the charge q,

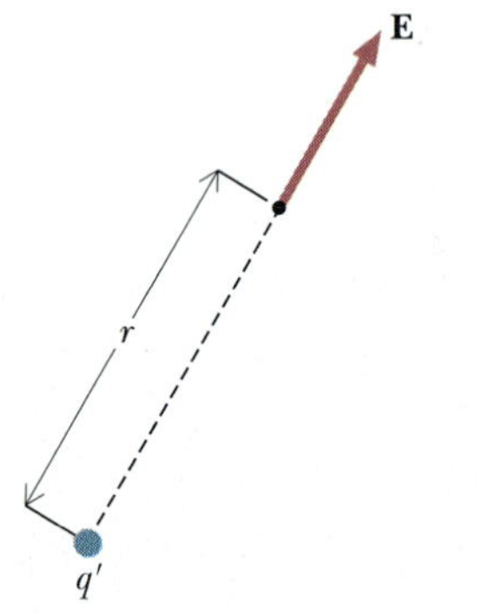

FIGURE 17.2
A charge q' generates an electric field **E** at some distance r. The direction of this electric field is along the radial line.

$$F = \frac{1}{4\pi\epsilon_0} \frac{qq'}{r^2} \qquad (2)$$

and we separate this expression into a product of two factors: a factor q characteristic of the point charge on which the force is being exerted, and a factor q'/r^2 characteristic of the point charge that exerts the force and its distance r. We also include the constant of proportionality $1/4\pi\epsilon_0$ in the second factor, so the expression for the force becomes

$$F = q \times \left(\frac{1}{4\pi\epsilon_0} \frac{q'}{r^2} \right) \qquad (3)$$

The second factor is defined to be the electric field of the point charge q'; we designate this electric field by E:

Electric field of point charge ▶

$$E = \frac{1}{4\pi\epsilon_0} \frac{q'}{r^2} \qquad (4)$$

The force on the charge q is then simply

$$F = qE$$

Like the force, the electric field is a vector, and it can be represented by an arrow (see Fig. 17.2). The direction of the electric field depends on the sign of the charge q'. ***The electric field is directed radially outward if q' is positive, and radially inward if q' is negative.*** In vector language, the force on the charge q becomes

$$\mathbf{F} = q\mathbf{E} \qquad (5)$$

With this special case of the electric field of a point charge as guidance, we can proceed to the definition of the electric field for the general case of an arbitrary charge distribution. To find the electric field at a given position, we take a point charge q (a "test charge") and place it at that position. The charge q will then experience an electric force **F**. The relation between the force **F** and the electric field **E** in this general case ought to be the same as the relation (5) we found in the special case of the electric field of the point charge q'. Accordingly, ***we define the electric field* E *as the electric force* F *divided by the magnitude* q *of the charge,***

$$\mathbf{E} = \mathbf{F}/q \tag{6}$$

◀ General definition of electric field

This means that the electric field is the force per unit charge. Note that with this definition the electric field is independent of the magnitude of the test charge [the force **F** is proportional to q, and hence in the ratio $\mathbf{F}/q$ the factor of q included in **F** cancels against the q in the denominator of Eq. (6)]. Thus, the electric field **E** depends on the magnitudes and positions of the charges that produce the electric field, but it does not depend on the magnitude of the test charge q used to detect it.

The unit of electric field is the newton/coulomb (N/C).* Table 17.1 gives the magnitudes of some typical electric fields.

◀ Newton/Coulomb (N/C)

TABLE 17.1	SOME ELECTRIC FIELDS
At surface of pulsar	$\simeq 10^{14}$ N/C
At orbit of electron in hydrogen atom	6×10^{11} N/C
In X-ray tube	5×10^{6} N/C
Electrical breakdown of air	3×10^{6} N/C
In Van de Graaff accelerator	2×10^{6} N/C
Within lightning bolt	$\simeq 10^{4}$ N/C
Under thundercloud	1×10^{4} N/C
Near radar transmitter (FPS-6)	7×10^{3} N/C
In sunlight (rms)	1×10^{3} N/C
In atmosphere (fair weather)	1×10^{2} N/C
In beam of small laser (rms)	1×10^{2} N/C
In fluorescent lighting tube	10 N/C
In radio wave	$\simeq 10^{-1}$ N/C
Within household wiring	$\simeq 3 \times 10^{-2}$ N/C

EXAMPLE 1. In a TV tube, an electron is placed in an electric field of 3.0×10^4 N/C generated by a nearby distribution of charges on the deflecting plates of the tube. The direction of the electric field is vertically upward (see Fig. 17.3). What is the acceleration of the electron?

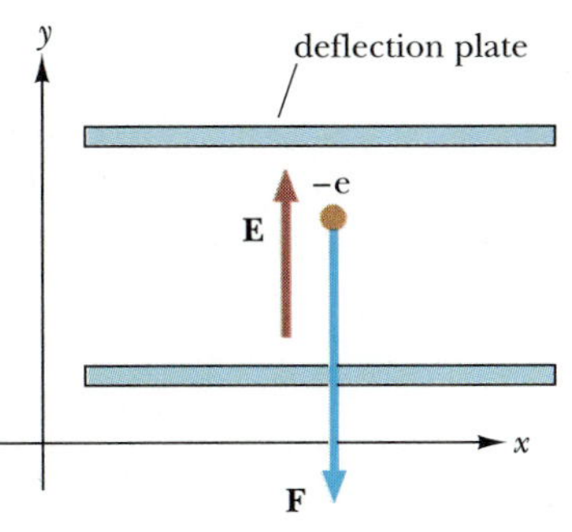

FIGURE 17.3

The upward electric field **E** exerts a downward force **F** on the electron.

* As we will see in Section 18.1, newton/coulomb is the same thing as volt/meter.

SOLUTION: The charge of the electron is $q = -e$. According to Eq. (6), the electric force that the electric field exerts on the electron is then

$$\mathbf{F} = q\mathbf{E} = (-e) \times \mathbf{E}$$

If the y axis is vertically upward, $E_y = 3.0 \times 10^4$ N/C, and the force has only a y component,

$$F_y = (-e) \times E_y = (-1.6 \times 10^{-19}\ \text{C}) \times (3.0 \times 10^4\ \text{N/C})$$

$$= -4.8 \times 10^{-15}\ \text{N}$$

Newton's Second Law then implies that this force gives the electron an acceleration

$$a_y = \frac{F_y}{m_e} = \frac{-4.8 \times 10^{-15}\ \text{N}}{9.1 \times 10^{-31}\ \text{kg}} = -5.3 \times 10^{15}\ \text{m/s}^2$$

The negative sign tells us that the direction of this acceleration is downward, *opposite* to the direction of the electric field (see Fig. 17.3). Note that in this calculation we have neglected the acceleration due to gravity; since this acceleration is much smaller than that due to the electric field, its neglect is justified.

Superposition principle for electric fields ▶

The net electric field generated by any distribution of point charges with specified positions can be calculated by forming the vector sum of the individual electric fields of the point charges, where the magnitude of each individual electric field is given by an equation like Eq. (4). This procedure is justified by the principle of superposition for electric forces stated in the preceding chapter [see Eq. (16.9)].

EXAMPLE 2. The distribution of electric charge in a thundercloud can usually be approximated by several pointlike charges placed at different heights. Figure 17.4a shows such an approximate charge distribution consisting of a charge of +40 C at a height of 10 km and a charge of −30 C at a height of 4 km in the thundercloud. What are the horizontal and the vertical components of the electric field that these two charges produce at a point P at a height of 10 km and at a horizontal distance of 6 km to the right?

SOLUTION: At the point P, the charge $Q_1 = +40$ C produces an electric field that is directed away from this charge, and the charge $Q_2 = -30$ C produces an electric field that is directed toward that charge. The net electric field is the vector sum of these two individual electric fields.

If h represents the height difference between the charges and d the horizontal position of the point P, then the distances from the charges to P are, respectively, d and $\sqrt{d^2+h^2}$. According to Eq. (4), the magnitudes of the two individual electric fields are

$$E_1 = \frac{1}{4\pi\epsilon_0}\frac{Q_1}{d^2}$$

and

$$E_2 = \frac{1}{4\pi\epsilon_0}\frac{|Q_2|}{d^2+h^2}$$

Here $|Q_2|$ stands for the absolute value of Q_2, that is, the magnitude of Q_2, without regard to its negative sign. From Fig. 17.4b we see that the horizontal components (x components) of the two individual electric fields are E_1 and $-E_2\cos\theta$, and the vertical components (y components) are 0 and $-E_2\sin\theta$, respectively. We therefore obtain the following components of the net electric field:

$$E_x = \frac{1}{4\pi\epsilon_0}\left(\frac{Q_1}{d^2} - \frac{|Q_2|}{d^2+h^2}\cos\theta\right) \tag{7}$$

and

$$E_y = -\frac{1}{4\pi\epsilon_0}\frac{|Q_2|}{d^2+h^2}\sin\theta \tag{8}$$

With $d = h = 6$ km, and with $\theta = 45°$, we find

$$E_x = 8.99\times10^9\,\frac{\mathrm{N\cdot m^2}}{\mathrm{C^2}}\left(\frac{40\ \mathrm{C}}{(6\times10^3\ \mathrm{m})^2} - \frac{30\ \mathrm{C}}{2\times(6\times10^3\ \mathrm{m})^2}\cos45°\right)$$

$$= 7.3\times10^3\ \mathrm{N/C}$$

and

$$E_y = -8.99\times10^9\,\frac{\mathrm{N\cdot m^2}}{\mathrm{C^2}}\,\frac{30\ \mathrm{C}}{2\times(6\times10^3\ \mathrm{m})^2}\sin45°$$

$$= -2.7\times10^3\ \mathrm{N/C}$$

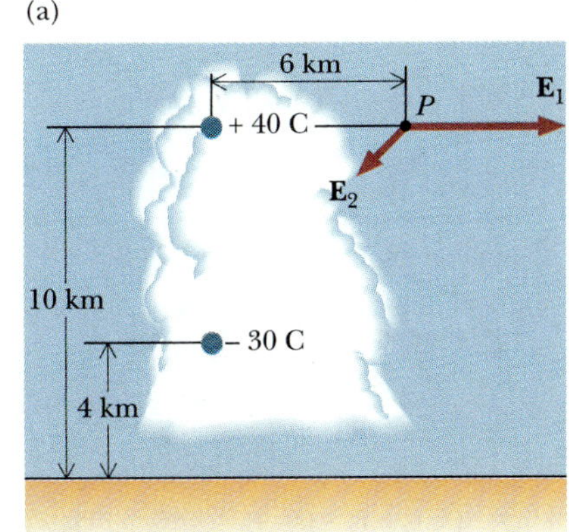

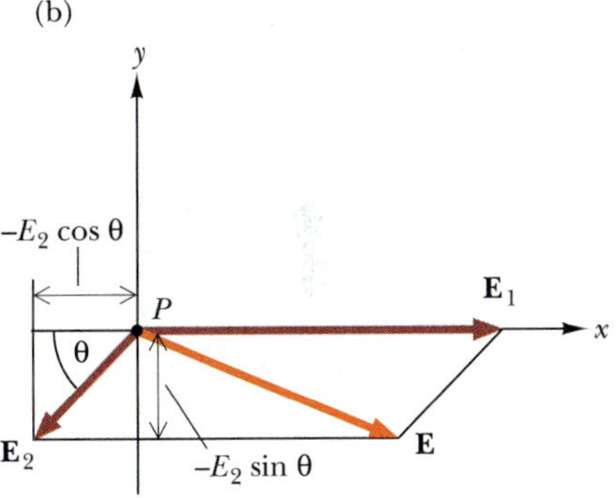

FIGURE 17.4
(a) Distribution of positive and negative electric charges in a thundercloud. (b) The net electric field is the vector sum of the individual electric fields $\mathbf{E}_1$ and $\mathbf{E}_2$.

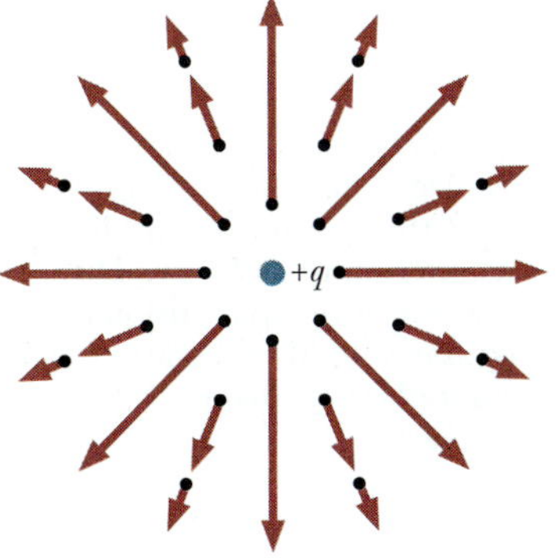

FIGURE 17.5
Electric field vectors surrounding a positive point charge. The field vectors are directed radially outward. The positions to which the field vectors belong are marked in black.

17.2 LINES OF ELECTRIC FIELD

The electric field can be represented graphically by drawing, at any given point of space, a vector whose magnitude and direction are those of the electric field at that point. Figure 17.5 shows the electric field vectors in

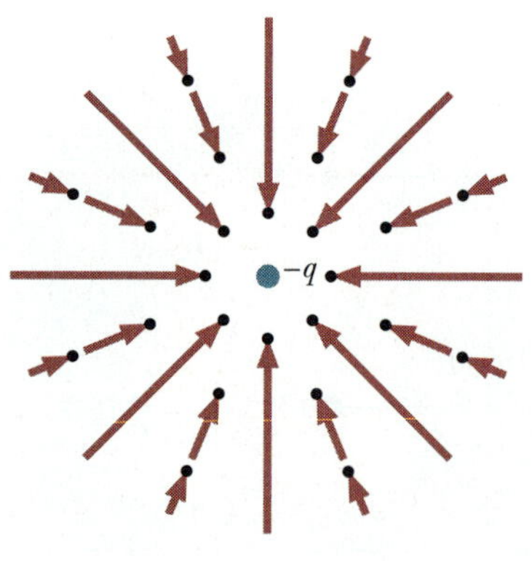

FIGURE 17.6
Electric field vectors surrounding a negative point charge. The field vectors are directed radially inward.

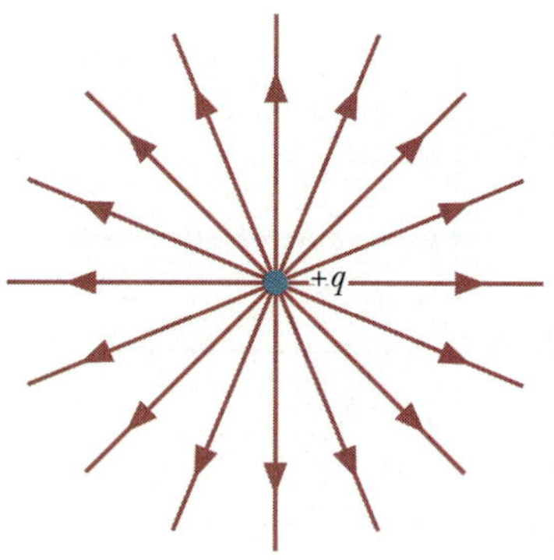

FIGURE 17.7
Electric field lines of a positive point charge. Note that in three dimensions the lines spread out in all three directions of space, whereas the diagram shows the lines spreading out in only the two directions within the page. This gives a misleading impression of the density of field lines as a function of distance. This limitation of two-dimensional diagrams should be kept in mind when looking at diagrams of field lines.

the space surrounding a positive point charge. Figure 17.6 shows the electric field vectors in the space surrounding a negative point charge. Note that, as demanded by Eq. (4), the magnitudes of these vectors decrease in inverse proportion to the squares of their distances from the point charge.

Alternatively, the electric field can be represented graphically by ***field lines.*** These lines are drawn in such a way that, at any given point, ***the tangent to the field line has the direction of the electric field. Furthermore, the density of lines is directly proportional to the magnitude of the electric field,*** that is, where the lines are closely spaced the electric field is strong, and where the lines are far apart the electric field is weak. Figure 17.7 shows the electric field lines of a positive point charge, and Fig. 17.8 shows those of a negative point charge. The arrows on these lines indicate the direction of the electric field along each line.

When we draw a pattern of field lines, we have to continue each line indefinitely, except where it begins on a positive charge or where it ends on a negative charge. Since the magnitude of the electric field is directly proportional to the amount of electric charge, the number of field lines that we draw emerging from a positive point charge must be proportional to the amount of charge. Figure 17.9 shows the electric field lines of a positive point charge twice as large as that of Fig. 17.7. We will adopt the convention that the number of electric field lines emerging from a charge q is q/ϵ_0; hence the number of lines emerging from one coulomb of charge is $1/\epsilon_0 = 1/(8.85 \times 10^{-12}) = 1.13 \times 10^{11}$. This convention, or "normalization," for the number of field lines is very convenient for performing computations with field lines — it makes the density of field lines not only proportional to but exactly equal to the magnitude of the electric field (we will establish this equality later in this section). However, this normalization is not always practical for making drawings — depending on the magnitude of the charge, it sometimes yields an enormous number of field lines so drawing them becomes an unbearable chore, and sometimes a fractional number, so drawing them becomes altogether meaningless. In case of need, we can alter the normalization, but we must be careful to

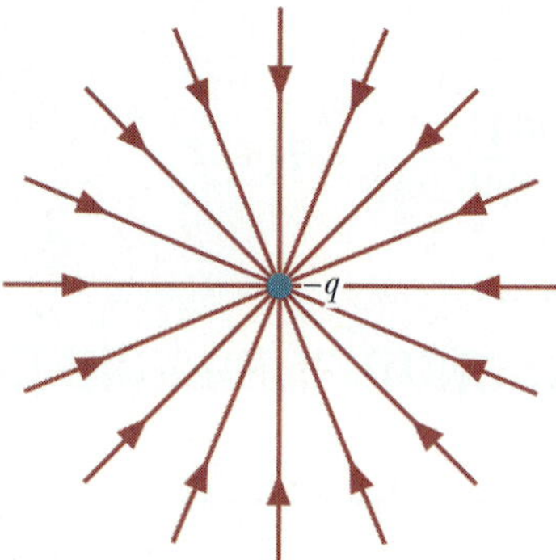

FIGURE 17.8
Electric field lines of a negative point charge.

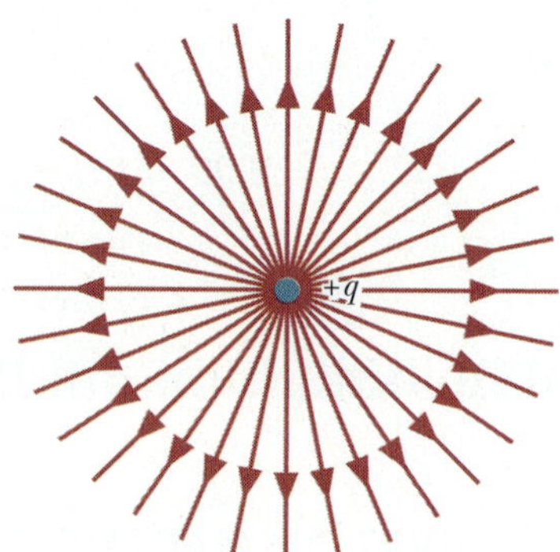

FIGURE 17.9
Electric field lines of a positive point charge twice as large as in Fig. 17.7.

maintain a fixed normalization throughout any given computation or series of drawings.

Figure 17.10 shows the field lines generated jointly by a pair of one positive charge and one negative charge of equal magnitudes. Figure 17.11 shows the field lines of a pair of equal positive charges, and Fig. 17.12 those of a pair of unequal positive and negative charges.

Note that in all cases the field lines start on positive charges and end (if they end) on negative charges — ***the positive charges are sources of field lines, and the negative charges are sinks.*** Also, note that the field lines never intersect (except where they start or end on point charges). If the lines ever were to intersect, the electric field would have *two* directions at the point of intersection; this is impossible.

The above pictures of field lines help us to develop some intuitive feeling for the spatial dependence of the electric fields surrounding diverse arrangements of electric charges. But we must not fall into the trap of thinking of the field lines as physical objects. The electric field is a form of matter, that is, it is a physical object, whereas the field lines are merely mathematical crutches to aid our imagination.

Besides providing us with a pictorial representation of the electric field, the field lines also are useful in some computations of the electric fields of given charge distributions. The following example illustrates how the concept of field lines can be exploited in the computation of the electric field of a large, flat, charged sheet.

Suppose that positive charge is uniformly distributed over a very large flat horizontal sheet of paper. The amount of charge on the sheet is σ coulombs per square meter (the Greek letter σ, or sigma, is commonly used to represent an amount of charge per unit area). To find the electric field in the space above and below the sheet, let us examine the pattern of field lines generated by this charge distribution.

The field lines must start on the charges on the sheet, and some of the field lines must go in the upward direction, some in the downward direction (see Fig. 17.13). We can take advantage of symmetry arguments to deduce the arrangement of the field lines. Such arguments are based on the rule that if the charge distribution makes no distinction between up

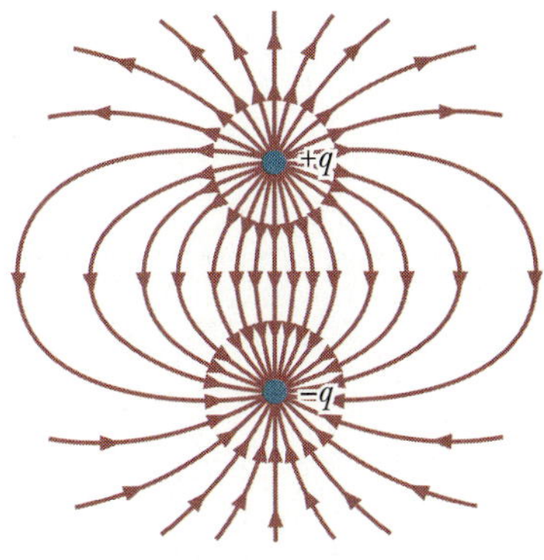

FIGURE 17.10

Field lines generated by positive and negative charges of equal magnitudes.

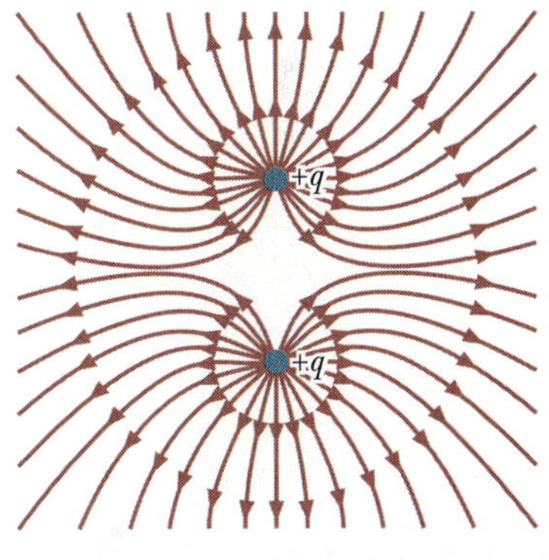

FIGURE 17.11

Field lines generated by two positive charges of equal magnitudes.

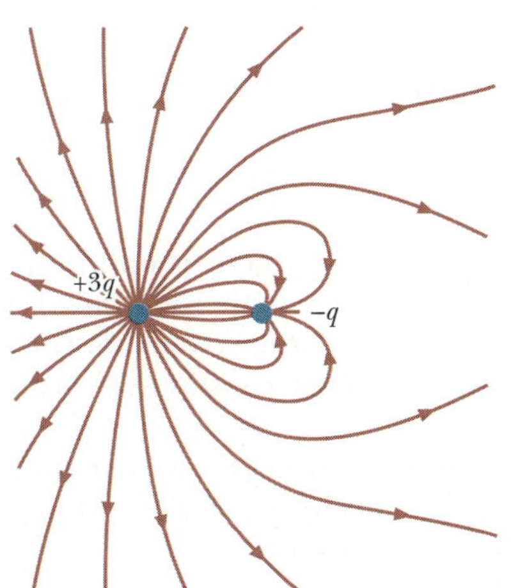

FIGURE 17.12

Field lines generated by positive and negative charges of unequal magnitudes. The positive charge has three times the magnitude of the negative charge.

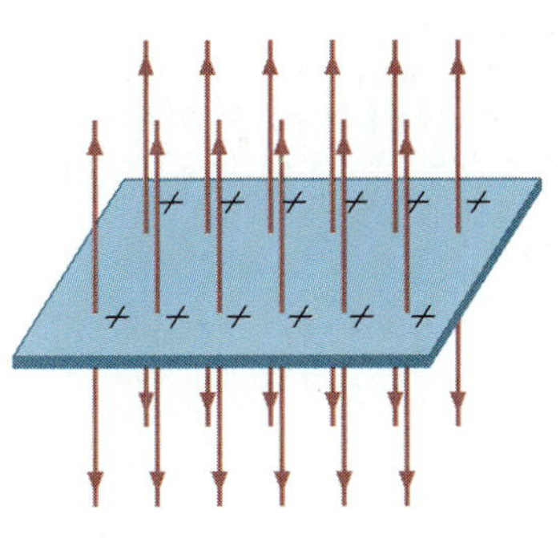

FIGURE 17.13

Field lines of a very large sheet of charge.

and down or left and right, then the pattern of field lines must, likewise, not make any such distinction; that is, the pattern of field lines must respect the symmetry of the charge distribution. Since the upper and the lower surfaces of the sheet are physically equivalent, symmetry requires that the pattern of field lines in the space above the sheet be the mirror image of the pattern below the sheet, with the same number of field lines above and below the sheet. Therefore one-half of the field lines must leave the sheet in the upward direction, and one-half in the downward direction. Next, consider one of the field lines starting on the sheet and going in the, say, upward direction. Since, for an infinite sheet, the portions of the sheet to the right of the field line and the left of the field line are physically equivalent, symmetry requires that the field line bend neither to the right nor to the left. Thus, we conclude that every field line must be a straight vertical line. Finally, since the charge is uniformly distributed, the field lines must also be uniformly distributed. We therefore recognize that the pattern of field lines must consist of uniformly spaced vertical lines, as shown in Fig. 17.13. Corresponding to this uniform arrangement of the field lines, the electric field is of uniform magnitude throughout all of space, and its direction is vertically upward in the space above the sheet and vertically downward in the space below.

To discover the magnitude of the electric field, we make use of the equality between this magnitude and the density of field lines. Consider an area A of the sheet. The amount of charge within this area is σA, the product of the amount of charge per unit area and the area. Therefore the number of field lines starting on this amount of charge is $\sigma A/\epsilon_0$. The number of field lines going in the upward direction is half of this, or $\sigma A/2\epsilon_0$. The density of these field lines is the number of lines per unit area, or the number of lines intercepted by the area A divided by this area A,

$$[\text{density of lines}] = \frac{[\text{number of intercepted lines}]}{A} \tag{9}$$

$$= \frac{\sigma A/2\epsilon_0}{A} = \frac{\sigma}{2\epsilon_0} \tag{10}$$

This density equals the magnitude of the electric field:

Electric field of flat sheet ▶

$$E = \frac{\sigma}{2\epsilon_0} \tag{11}$$

Thus, the electric field generated by the flat sheet is directly proportional to the charge per unit area. Although this result is strictly valid only for the case of an infinitely large sheet, it is also a good approximation for a sheet of finite size, provided we stay near the sheet and we stay away from the vicinity of the edges.

Large charged flat sheets or plates are often used in physics laboratories to generate uniform electric fields. In practice, two parallel charged sheets with charges of opposite signs are preferred over a single sheet.

The individual electric fields of the two sheets are then in the same direction in the space between the sheets (see Fig. 17.14a); they therefore reinforce each other giving a net field twice as large as the field of one sheet, that is, an electric field

$$E = \frac{\sigma}{\epsilon_0} \tag{12}$$

The individual electric fields of the two sheets are in opposite directions in the space above or below the two sheets; thus, these electric fields cancel, and the space above and below has zero electric field. The net result of this superposition of the individual electric fields of the sheets is that all the field lines starting on the positively charged sheet end on the negatively charged sheet. Figure 17.14b shows the net electric field of the two sheets.

The inverse-square law for the electric field of a point charge can be "derived" from the picture of field lines — it is easy to show that the density of lines necessarily obeys an inverse-square law. Consider a point charge q; there will be q/ϵ_0 lines emerging from this point charge. Since the point charge is spherically symmetric and makes no distinction between one radial direction and another, symmetry arguments tell us that the arrangement of field lines must also be spherically symmetric, with a uniform distribution over all radial directions. At a distance r from the point charge, these lines are uniformly distributed over the area $4\pi r^2$ of a concentric sphere; consequently, there are $(q/\epsilon_0)/4\pi r^2$ lines per unit area. Hence the density of lines decreases in proportion to the inverse square of the distance. Furthermore, if we compare our expression $(q/\epsilon_0)/4\pi r^2$ for the density of lines with the expression for the magnitude of the electric field [Eq. (4)], we see that the density of lines is not only proportional to but exactly *equal* to the magnitude of the electric field. This justifies our use of the equality between the density of lines and the magnitude of the electric field in the example of the large flat sheet.

Note that this "derivation" of the inverse-square law from the picture of field lines is really no more than a consistency check — the electric field can be represented by field lines only because it is an inverse-square field; any other dependence on distance would make it impossible to draw continuous field lines that start and end only on charges.

One new result we can extract from this derivation is that ***the electric field outside of any spherical charge distribution of net charge q is given by the same formula as for a point charge,***

$$E = \frac{1}{4\pi\epsilon_0}\frac{q}{r^2} \tag{13}$$

The reason is that such a charge distribution has the same spherical symmetry as a point charge, and therefore outside of the charge distribution the arguments given in the above paragraph apply equally to the case of the point charge and to the case of such a spherical charge distribution.

Thus, the electric field outside of a hollow spherical shell (see Fig. 17.15) is the same as that of a point charge, and the electric field outside

(a)

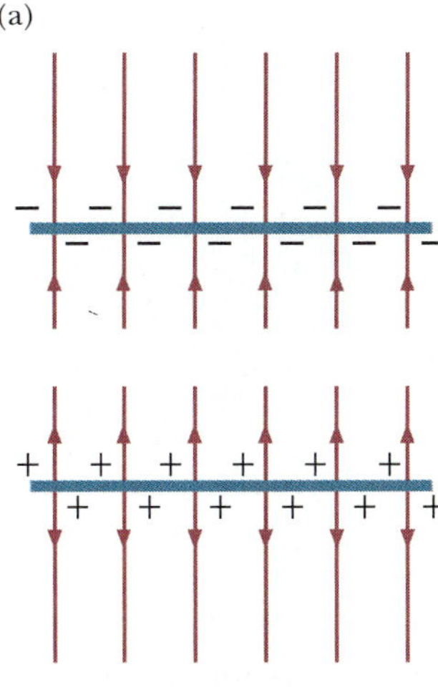

(b)

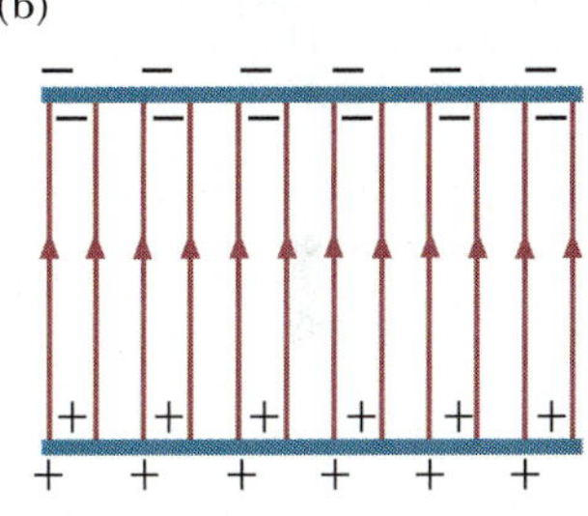

FIGURE 17.14

(a) Individual electric fields of two sheets of charge of opposite signs. (b) The net electric field of the two sheets.

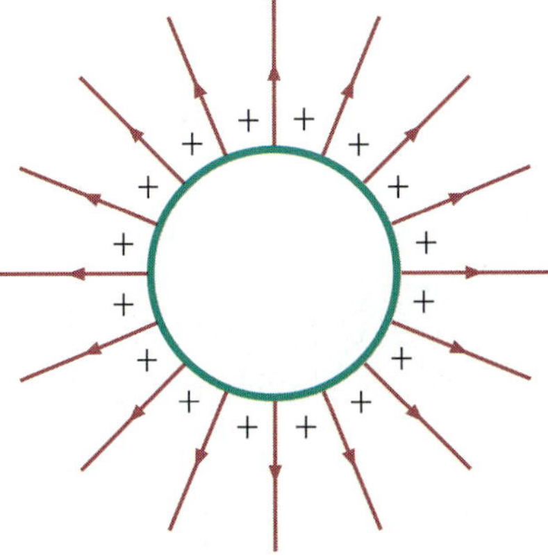

FIGURE 17.15

A hollow spherical shell with a uniform distribution of charge over its surface. The field lines are radial.

of a solid charged sphere (see Fig. 17.16) is also the same as that of a point charge. Outside of any such spherical charge distribution, the electric field is as though all of the charge of the sphere were concentrated at its center. This result is similar to Newton's famous theorem, mentioned in Section 5.1, concerning the gravitational force exerted by a planet: the gravitational force exerted by a spherically symmetric planet can be calculated as though all the mass were concentrated in a point at the planetary center. This similarity between electricity and gravitation reflects the similarity of the laws of force. Incidentally: It can be shown that a spherical charge distribution mimics a point charge not only in regard to the electric field that it produces, but also in regard to the force that it experiences when placed in some arbitrary external electric field. Thus, the electric force that an arbitrary electric field exerts on a spherical charge distribution can be calculated as though all of the charge were concentrated in a point at the center.

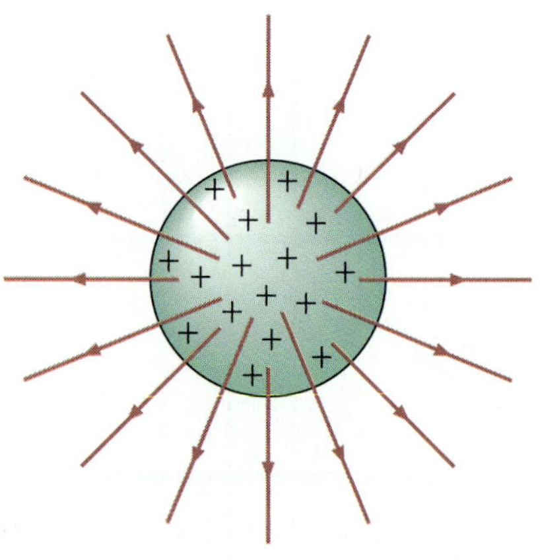

FIGURE 17.16
A solid sphere with a uniform distribution of charge throughout its volume. The field lines are, again, radial.

EXAMPLE 3. A proton is (approximately) a spherically symmetric ball of charge of radius 1.0×10^{-15} m. (a) What is the electric field at the surface of the proton? (b) If a second proton is brought within touching distance of the first, what is the repulsive electric force?

SOLUTION: (a) With $q = e = 1.6 \times 10^{-19}$ C and $r = 1.0 \times 10^{-15}$ m, we find that the electric field produced by one proton is

$$E = \frac{1}{4\pi\epsilon_0}\frac{q}{r^2} = 8.99 \times 10^9\ \mathrm{N \cdot m^2/C^2} \times \frac{1.6 \times 10^{-19}\ \mathrm{C}}{(1.0 \times 10^{-15}\ \mathrm{m})^2}$$

$$= 1.4 \times 10^{21}\ \mathrm{N/C}$$

(b) If the protons are touching (see Fig. 17.17), the center-to-center distance is $2 \times 1.0 \times 10^{-15}$ m. The electric field generated by one proton at this distance is one-quarter of the above value, that is, $E = 3.5 \times 10^{20}$ N/C. The force exerted by one proton on the other is then

$$F = qE = eE = 1.6 \times 10^{-19}\ \mathrm{C} \times 3.5 \times 10^{20}\ \mathrm{N/C} \tag{14}$$

$$= 56\ \mathrm{N}$$

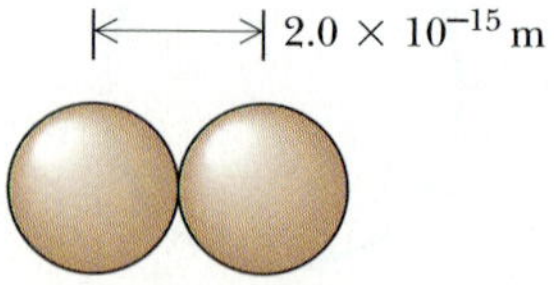

In the nucleus of an atom, this large repulsive electric force between neighboring, tightly packed protons is more than compensated by an even larger binding force — the "strong" force — that holds the nucleus together.

2.0×10^{-15} m

FIGURE 17.17
Two protons in contact. Each proton may be regarded as a spherical ball of positive charge.

The picture of field lines can also be exploited for the computation of the electric field of a long straight charged line, say, a tightly stretched line of silk with charge distributed uniformly along its length (see Fig. 17.18). This charge distribution has cylindrical symmetry around the line of charge — the charge distribution makes no distinction between the front part of the line of charge and the rear, or between one radial direction away from the line and another. The only directions for the field lines consistent with this cylindrical symmetry of the charge distribution are the radial directions (within the planes in Fig. 17.19) — if the field lines had any other direction (in front of or behind the planes in Fig. 17.19), they would make an unacceptable distinction between the front and the rear parts of the line of charge. Furthermore, the field lines must be uniformly distributed over all radial directions.

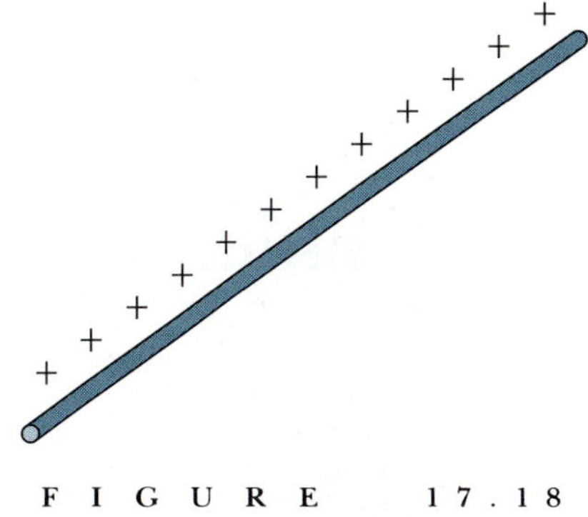

FIGURE 17.18

A long straight line with a uniform distribution of charge.

As a next step, by examining the density of field lines, we can find the magnitude of the electric field produced by the straight charged line. The result is

$$E = \frac{1}{2\pi\epsilon_0}\frac{\lambda}{r} \tag{15}$$

◀ **Electric field of a charged line**

where λ is the amount of charge per unit length along the line, in coulombs per meter, and r is the radial distance from the line. This result shows that the electric field of a long line of charge decreases as the inverse of the distance, not the inverse square of the distance.

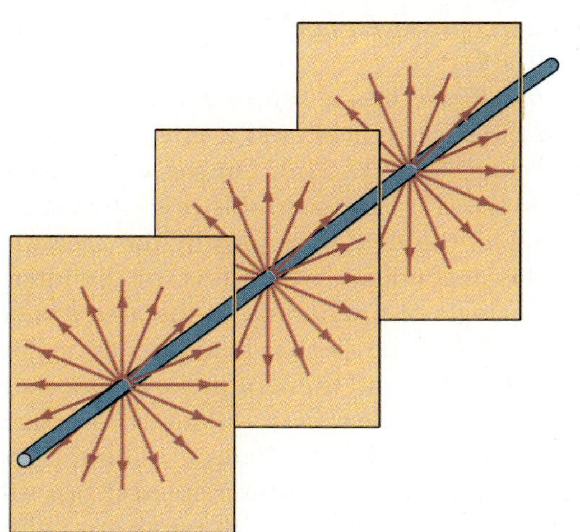

FIGURE 17.19

Field lines of a long straight charged line. The field lines are uniformly distributed over all radial directions.

17.3 ELECTRIC FLUX AND GAUSS' LAW

The mathematical relationship between the electric charge and the electric field generated by this charge can be given an elegant and concise form by means of the concept of electric flux. Consider a mathematical (that is, imagined) surface in the shape of a rectangle of area A. Suppose that this surface is immersed in a constant electric field **E** (see Fig. 17.20). The electric field makes an angle with the surface; the electric field vector has a component tangential to the surface and a component normal (that

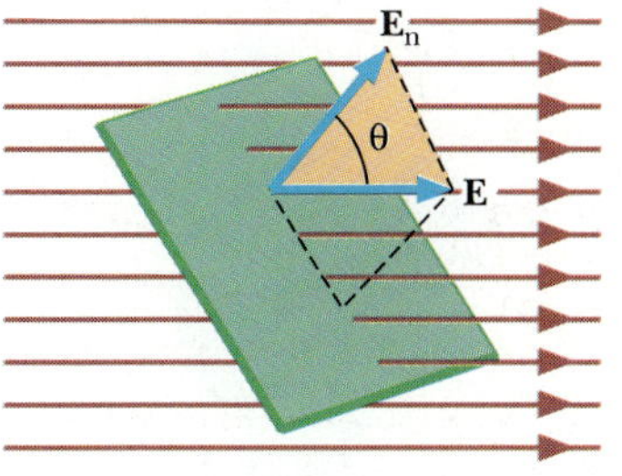

FIGURE 17.20

Flat rectangular surface immersed in a uniform electric field. The perpendicular to the surface makes an angle θ with the field lines.

is, perpendicular) to the surface. ***The electric flux Φ through the surface is defined as the product of the area A by the magnitude of the normal component of the electric field.*** If we represent the normal component of the electric field by E_n, then

Electric flux ▶

$$\Phi = E_n A \tag{16}$$

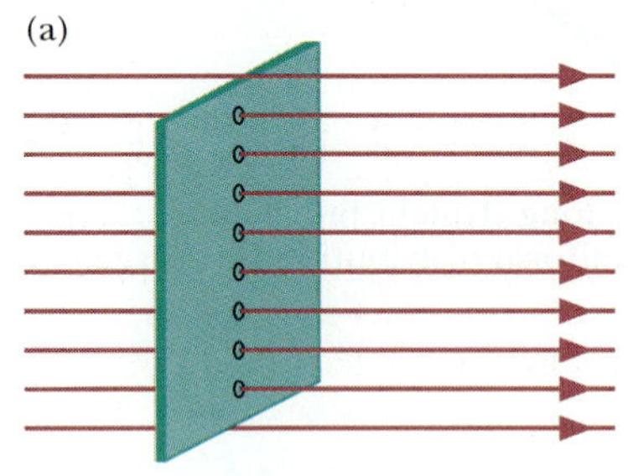

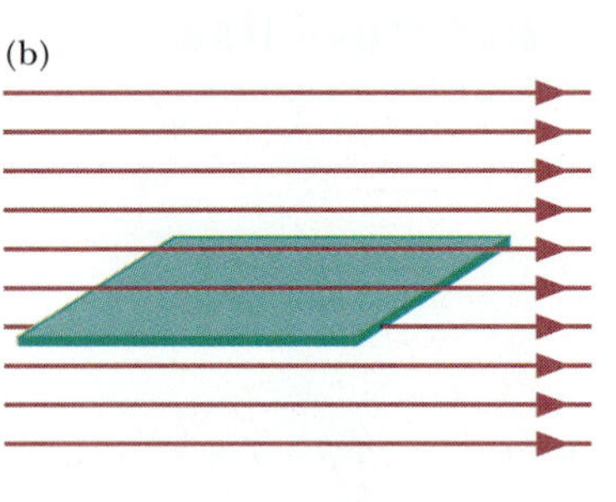

F I G U R E 17.21
(a) The area A is perpendicular to the electric field lines. It intercepts an electric flux EA. (b) The area A is parallel to the field lines. It intercepts no electric flux.

According to the usual rule for calculating a component, the normal component E_n can also be written as $E \cos\theta$, where θ is the angle between **E** and the perpendicular erected on the surface (see Fig. 17.20). Hence, Eq. (16) can be written as

$$\Phi = EA \cos\theta \tag{17}$$

The product $A \cos\theta$ can be interpreted as the projection of the area A onto a plane perpendicular to the electric field; that is, $A \cos\theta$ can be regarded as that portion of the area that effectively faces the electric field. Thus, Eq. (17) states that the flux is the magnitude E of the electric field multiplied by the area facing the electric field. According to Section 17.2, the magnitude E of the electric field is numerically equal to the number of field lines intercepted by a unit area facing the electric field. Hence, the electric flux through the area A is numerically equal to the number of field lines intercepted by the area.* For instance, in Fig. 17.20, the rectangular surface intercepts seven field lines, that is, the electric flux through the surface is $\Phi = 7$.

Note that if $\theta = 0$, the flux is simply EA; in this case, the area A is exactly face-on to the electric field, and it intercepts the maximum possible number of field lines (Fig. 17.21a). On the other hand, if $\theta = 90°$, then the flux given by Eq. (17) is zero; in this case, the area A is parallel to the electric field, and all the field lines pass by the area without interception (Fig. 17.21b). This dependence of the flux, or the intercepted number of field lines, on the angle of orientation of the area can be understood by a simple analogy: Think of the electric field lines as analogous to the trajectories of falling raindrops and think of the area as a sheet of paper. If you hold the paper face-on to the rain, it will intercept a maximum number of drops; if you hold it parallel to the rain, it will intercept no drops. At intermediate angles, the number of intercepted drops will be proportional to $\cos\theta$, as in Eq. (17). This analogy with raindrops falling along field lines also explains the curious terminology: the electric "flux" is analogous to the flow of raindrops against the paper.

For a mathematical surface of arbitrary, curved shape immersed in an arbitrary, nonuniform electric field (see Fig. 17.22), the electric flux

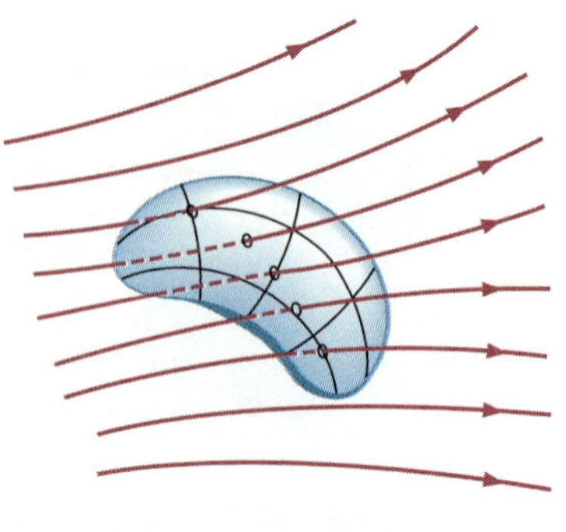

F I G U R E 17.22
An arbitrary surface immersed in an arbitrary, nonuniform electric field. The flux through the surface is $\Phi = 5$.

* This equality hinges on the normalization adopted in Section 17.2 — flux and number of intercepted field lines are equal if and only if electric field and number of field lines per unit area are equal, and the latter is true if and only if we adopt the normalization that q/ϵ_0 lines emerge from each charge q.

through the surface is defined by

$$\Phi = \overline{E}_n A \tag{18}$$

Here A is the area of the surface and $\overline{E}_n$ is the *average* value of the normal component of the electric field (as in earlier chapters, we use an overbar to indicate an average value). This average is to be evaluated by averaging over the normal components at all points of the surface (as in the calculation of work in Section 4.2, the average can be obtained by subdividing the surface into small pieces and summing the contributions associated with all these small pieces).

According to the arguments given above, the electric flux is again equal to the number of field lines intercepted by the surface. Note that lines going through the surface from one side make a positive contribution to the flux, whereas lines going through from the opposite side make a negative contribution (see Fig. 17.23 for an example). This means that the normal component E_n must be reckoned as positive if the electric field points into the surface from one side, and negative when it points into the surface from the opposite side.

The definition (18) is also valid for a *closed* surface, such as the surface of a sphere or the surface of a cube. The electric flux through this surface is given by Eq. (18), where the average is to be performed over all points of the closed surface. ***The normal component E_n is reckoned as positive if the direction of the electric field is outward from the surface; and negative if it is inward, into the surface.*** Thus, the electric flux through the closed surface is simply the *net* number of field lines emerging from the surface. The electric flux will be positive or negative depending on whether more field lines emerge from the surface or enter the surface.

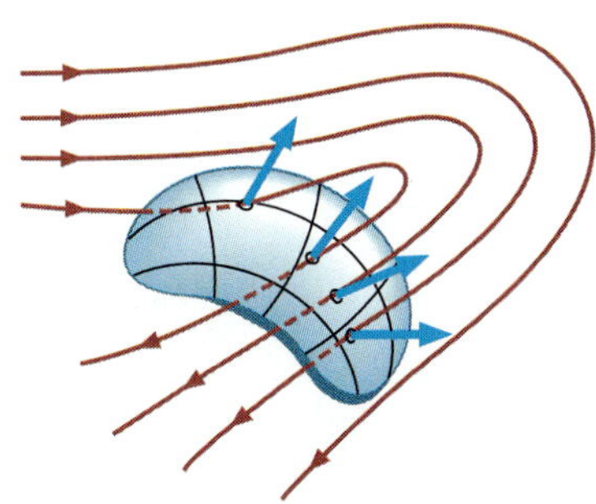

FIGURE 17.23

Arbitrary surface immersed in arbitrary electric field. The small black arrows are the perpendiculars to the surface. The flux through the surface is $\Phi = -2$.

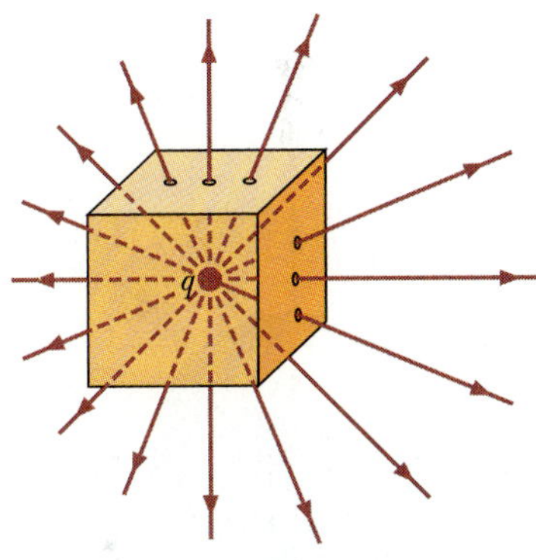

FIGURE 17.24

A cubical Gaussian surface encloses a charge q.

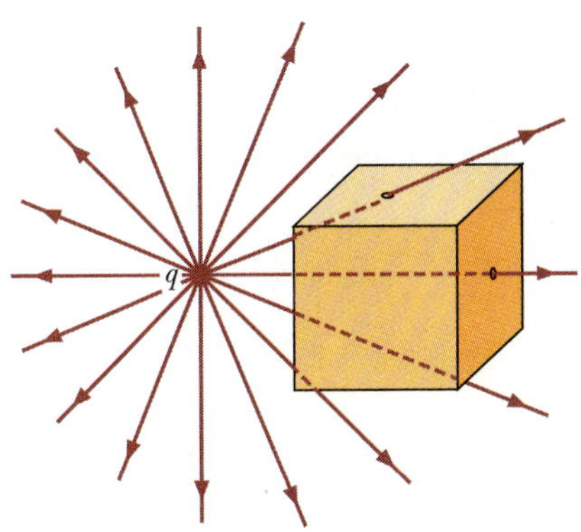

FIGURE 17.25

The cubical Gaussian surface is adjacent to the charge q.

EXAMPLE 4. (a) Consider a cubical surface enclosing a positive point charge q (see Fig. 17.24). What electric flux does the electric field of this point charge produce through the surface? (b) Consider a cubical surface adjacent to (but not enclosing) a positive point charge q (see Fig. 17.25). What electric flux does the electric field of this point charge produce through this surface?

SOLUTION: (a) A positive charge q is the starting point of q/ϵ_0 outward field lines. As shown in Fig. 17.24, the surface surrounds the point charge, and therefore all of the field lines must emerge through the surface. Since the electric flux equals the net number of field lines emerging from the surface, the electric flux must be q/ϵ_0.

(b) As shown in Fig. 17.25, any field line that enters the face of the cube nearest the point charge will leave again through some other face. Thus, each such field line makes a negative contribution to the flux at the nearest face, and an equal positive contribution at the other face — the net contribution to the flux is zero.

Note that by appealing to the concept of field lines, we were able to get the answer for the flux without having to perform the explicit evaluation of the average electric field in Eq. (18). For our cubical surface, such an explicit evaluation would have been quite difficult.

We saw in Example 4 that a point charge enclosed within a surface produces an electric flux q/ϵ_0 through this surface. This result can be generalized for an arbitrary charge distribution enclosed within an arbitrary surface. The general result is called ***Gauss' Law:***

Gauss' Law ▶

If the volume within an arbitrary closed surface holds a net electric charge Q, then the electric flux through the surface is Q/ϵ_0, that is,

$$\Phi = Q/\epsilon_0 \tag{19}$$

Gaussian surface ▶

The closed surface involved in this law is usually called the ***Gaussian surface*** (see Fig. 17.26).

The proof of Gauss' Law is easy, if we exploit the concept of field lines. The electric field appearing in Eq. (18) is a sum of the individual electric fields of some number of point charges. Some of the point charges are outside the closed surface and some are inside the closed surface. Let us consider the contribution to the flux from each individual electric field of each individual point charge. The individual electric field of a charge outside the closed surface produces no net flux through the closed surface — any field line of this field either does not touch the surface or else enters it at one point and leaves at another; neither case makes any net contribution to the flux. However, the individual electric field of a positive or negative point charge enclosed within the surface does contribute to the flux. For example, a positive point charge q has q/ϵ_0 outward field lines, all of which will pierce the closed surface — such a charge therefore contributes a flux q/ϵ_0 (as in Fig. 17.24). Taking into account that positive charges produce positive flux and negative charges negative flux, we see that the net number of field lines piercing the surface, or the net flux, is equal to the net charge divided by ϵ_0. This completes the proof of Gauss' Law.

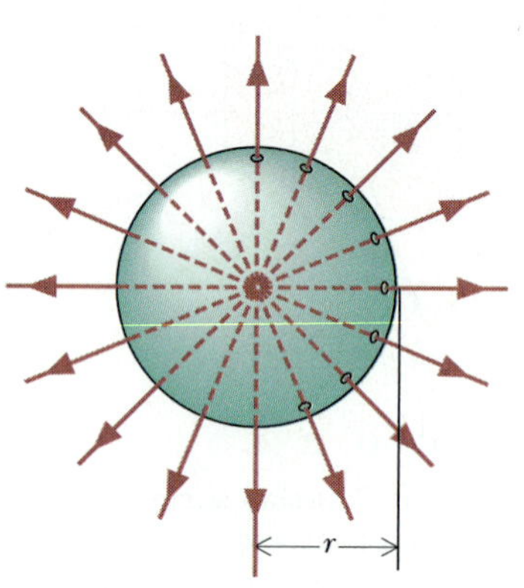

FIGURE 17.26
A spherical Gaussian surface surrounds a positive point charge concentrically.

Gauss' Law can be used to calculate the electric field generated by a charge distribution provided the charge distribution has a high degree of symmetry. The first step in such a calculation is a determination of the *direction* of the electric field from symmetry arguments, in much the same way as in our calculations with field lines in Section 17.2. The second step involves the exploitation of Eq. (19) to extract the *magnitude* of the electric field.

EXAMPLE 5. Use Gauss' Law to deduce the electric field of a point charge.

SOLUTION: For the purposes of this example we will pretend that we do not know the electric field of a point charge. We begin with a symmetry argument: Since the point charge is spherically symmetric, the electric field must also be spherically symmetric. Hence, at all points on the surface of a mathematical sphere of radius r centered on the charge q, the electric field must be in the radial direction and it must have the same magnitude.

If the Gaussian surface is taken to coincide with the sphere of radius r (see Fig. 17.26), then the normal component of the electric field coincides with the magnitude of the electric field, and the average value of the normal component over this sphere coincides with the magnitude of the electric field at any point. Thus, Gauss' Law reduces to

$$\Phi = EA = q/\epsilon_0 \tag{20}$$

But, for a sphere, the area is $A = 4\pi r^2$, so

$$E \times 4\pi r^2 = q/\epsilon_0 \tag{21}$$

from which we immediately obtain

$$E = \frac{1}{4\pi\epsilon_0}\frac{q}{r^2} \tag{22}$$

Karl Friedrich Gauss, *1777–1855, German mathematician, physicist, and astronomer. Gauss was an indefatigable calculator, and he loved to perform enormously complicated computations, which today would be regarded as impossible without an electronic computer. He developed many new methods for calculations in celestial mechanics; later he became interested in electric and magnetic phenomena, which he researched in collaboration with* **W. Weber,** *1804–1891.*

The proof of Eq. (19) establishes that Gauss' Law is a consequence of the field-line concept, that is, Gauss' Law is a consequence of Coulomb's Law. Conversely, Example 5 establishes that Coulomb's Law is a consequence of Gauss' Law. These two laws are therefore equivalent, at least in regard to the electric fields of charges at rest.

17.4 CONDUCTORS AND ELECTRIC FIELDS

As we pointed out in Section 16.5, in metallic conductors — such as copper, silver, aluminum — some of the electrons are free, that is, they can move without restraint within the volume of the metal. If such a conductor is immersed in an electric field, the free electrons move in response to the electric force. As an excess of electrons accumulates on one part of the conductor, a deficit of electrons will appear on another part of the conductor; thus, negative and positive charges are induced on the conductor. Within the volume of the conductor, the electric field of the induced charges tends to cancel the original electric field in which the conductor was immersed (see Fig. 17.27). The accumulation of negative and positive charge on the surface of the conductor continues until the electric field generated by these charges exactly cancels the original elec-

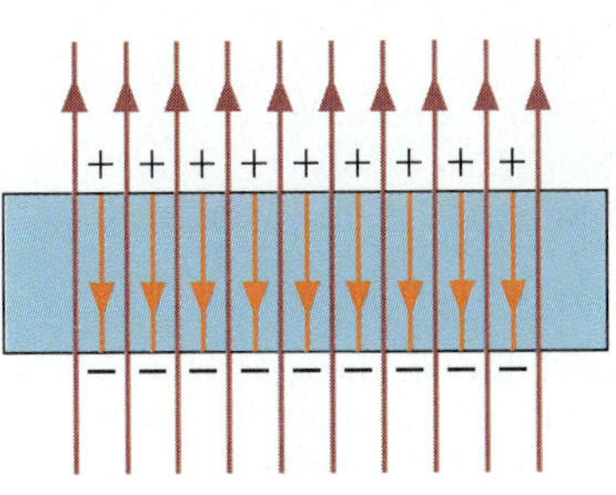

FIGURE 17.27

A very large slab of conductor, immersed in a uniform electric field. The uniform distribution of charge that has accumulated on its surface generates an electric field (black) opposite to the original electric field (red).

tric field. Consequently, ***when the charge distribution on a conductor reaches electrostatic equilibrium, the net electric field within the material of the conductor is exactly zero.*** The proof of this statement is by contradiction: if the electric field were different from zero, the free electrons would continue to move and the charge distribution would *not* (yet) be in equilibrium. For a good conductor (copper, aluminum, etc.), the equilibrium is reached in a fairly short time, a small fraction of a second.

Electrostatic equilibrium ▶

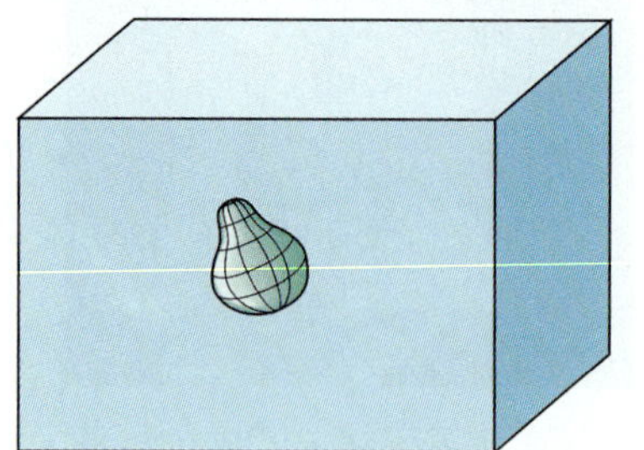

FIGURE 17.28
A closed Gaussian surface (green) inside a volume of conducting material.

Furthermore, ***for a conductor in electrostatic equilibrium, any (extra) electric charge deposited on the conductor resides on the surface of the conductor.*** We can prove this by means of Gauss' Law: Consider a small closed Gaussian surface inside the conducting material (Fig. 17.28). Since $\mathbf{E} = 0$ everywhere in this material, the left side of Eq. (19) vanishes for this surface, and therefore the right side must also vanish — which means that the charge enclosed by *any* arbitrary small surface is zero, that is, the charge in *any* small volume of the conductor is zero. If the charges are not in the volume of the conductor, they must be on the surface.

Finally, we can say something about the electric field just outside a conductor: ***the electric field at the surface of a conductor in electrostatic equilibrium is normal to the surface.*** The proof is again by contradiction: if the electric field had a component tangential to the surface of the conductor, the free electrons would move along the surface and the charge distribution would *not* be in equilibrium.

Note that this argument does not exclude an electric field perpendicular to the surface of the conductor; such an electric field merely pushes the free electrons against the surface, where they are held in equilibrium by the combination of the force exerted by the electric field and the restraining force exerted by the surface of the conductor. Figure 17.29 displays an experimental demonstration of electric fields perpendicular to the surfaces of conductors. The flat plate and the cylinder shown in this figure are conductors, and we see that the field lines meet the surfaces of these conductors at right angles.

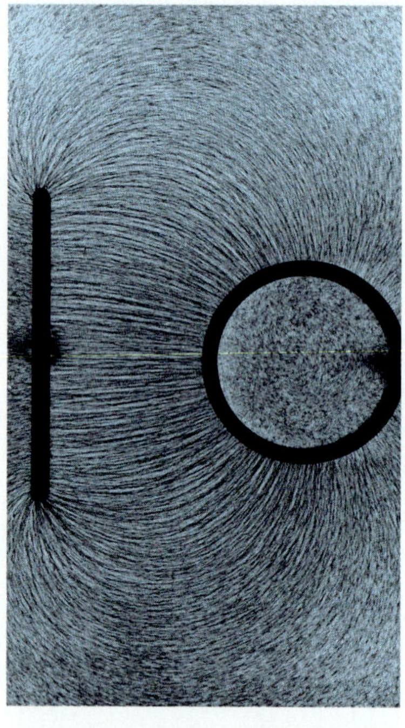

FIGURE 17.29
Field lines in the space surrounding a charged flat plate and cylinder, made visible by small bits of thread suspended in oil.

EXAMPLE 6. Find the electric field outside a very large flat conducting slab (Fig. 17.30) on which there is a surface charge density of σ coulomb per square meter.

SOLUTION: The calculation of this electric field can be performed by relying on field lines and by exploiting the symmetry of the charge distribution, in the same way as in the calculation of the electric field of the large flat charged sheet of paper (see Section 17.2). The only difference is that the electric field inside the conducting slab is zero, and hence no field lines can penetrate into the slab, that is, no field lines can penetrate in the downward direction in Fig. 17.30. Thus, all the field lines that start on the charges on the surface of the slab must go in the *upward* direction, whereas in the case of the sheet of paper, one-half of the field lines must go upward, one-half downward. This means that the density of the field lines above the slab is twice as large as in the case of the sheet of paper, that is, the electric field is twice as strong as that given by Eq. (11):

$$E = \sigma/\epsilon_0 \tag{23}$$

COMMENTS AND SUGGESTIONS: Over a small region, any smooth curved conducting surface can be approximated by a flat surface. Hence the expression (23) can be used to find the electric field in a region very near any smooth curved conducting surface. It is not even necessary that σ be constant — it can be some smooth function of position.

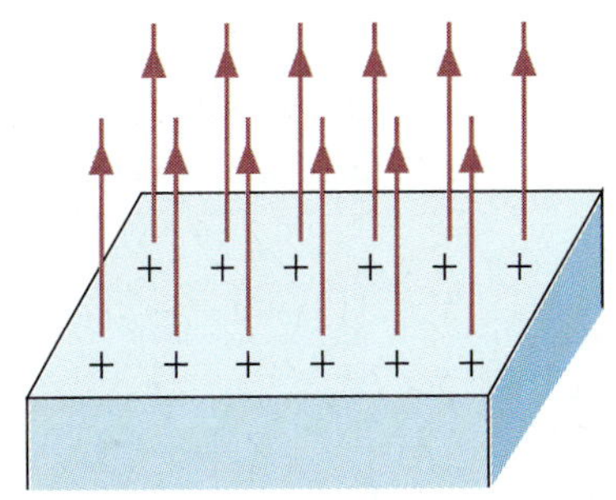

FIGURE 17.30
A large flat conducting slab with a uniform distribution of charge on its surface.

EXAMPLE 7. At the ground directly below a thundercloud, the electric field is 2×10^4 N/C and points upward. What is the surface-charge density on the ground?

SOLUTION: For the purposes of this problem, we treat the ground as a good conductor. Equation (23) gives the relation between electric field and surface-charge density for points at or near the charged surface,

$$\sigma = \epsilon_0 E = 8.85 \times 10^{-12} \frac{C^2}{N \cdot m^2} \times 2 \times 10^4 \text{ N/C}$$

$$= +1.8 \times 10^{-7} \text{ C/m}^2$$

17.5* THE ELECTRIC DIPOLE

A body that has a positive electric charge on one end and a negative electric charge of equal magnitude on the other end is called an electric dipole. Many asymmetric molecules have an excess of electrons on one end of the molecule and a corresponding deficit on the other. This means that the molecule has an accumulation of negative charge on one end and of positive charge on the other, and any such molecule is an electric dipole. For example, Figs. 17.31 and 17.32 show the structures of the carbon monoxide molecule (CO) and the water molecule (H_2O). In both of these molecules, the electrons tend to concentrate on the oxygen atom; hence, in Figs. 17.31 and 17.32 the left sides of the molecules are negatively charged, and the right sides positively charged. Because of this separation of the average positions of the negative and positive charges, the molecules are electric dipoles.

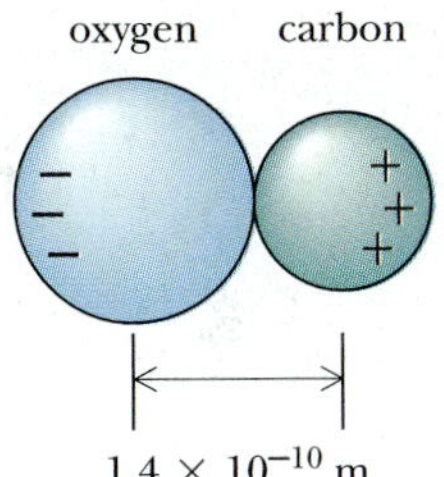

FIGURE 17.31
Structure of the CO molecule.

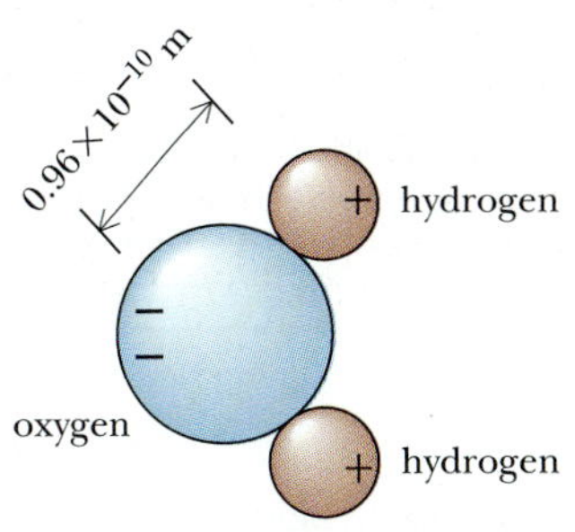

FIGURE 17.32
Structure of the H_2O molecule.

* This section is optional.

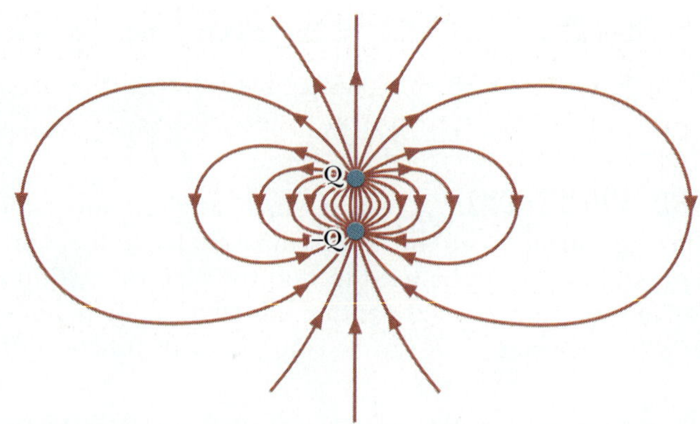

FIGURE 17.33
Field lines of an electric dipole.

Figure 17.33 shows the electric field lines of an electric dipole. For instance, the water molecule has an electric field with such field lines. By means of this electric field, the water molecule can act on other molecules in its vicinity, and exert forces on the electric charges in these molecules. The force exerted by the electric dipole field of the water molecule is what makes water such a good solvent.

Apart from the force that an electric dipole exerts on other charges, we may also be interested in the force that other charges exert on the dipole. Suppose that the other charges produce some electric field **E** at the position of the dipole. We can then calculate the force that this electric field exerts on the dipole. At first sight, it would seem that the force exerted on the dipole is zero, since the positive and negative charges in the dipole are of equal magnitudes, and the net electric charge of the dipole is zero. We expect that when a body of zero electric charge is placed in an electric field, the body experiences no net electric force. However, this expectation fails if the electric force on the positive charges is larger or smaller than that on the negative charges; the body then experiences a net force. Such an imbalance of the forces on the positive and negative charges will happen if the electric field is stronger or weaker at the location of the positive charges than at the location of the negative charges. For example, the dipole shown in Fig. 17.34, with positive charges on the left end and negative charges on the right end, will be pushed toward the left because the electric field that acts on the body is stronger at the location of the negative charges. Note that this electric field — indicated by the field lines in Fig. 17.34 — is not the field generated by the body; rather it is an electric field generated by some other charges (not shown in fig. 17.34). The electric field that acts on a body is often called the ***external field.*** In contrast, the field generated by the body itself is called the ***self-field.*** The self-field exerts only internal forces within the body and does not contribute to the net force acting on the body from the outside.

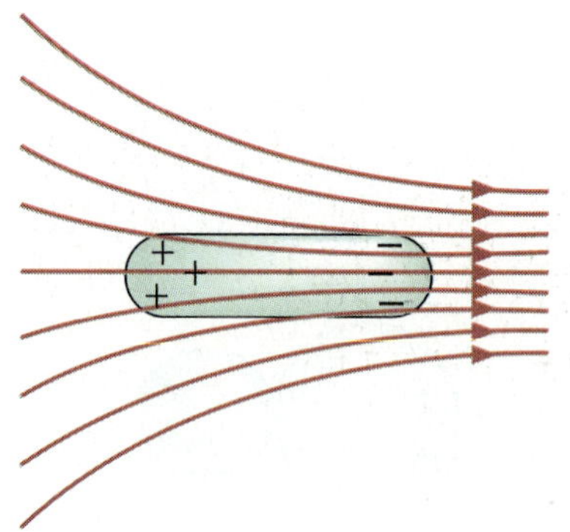

FIGURE 17.34
An electric dipole placed in a nonuniform electric field.

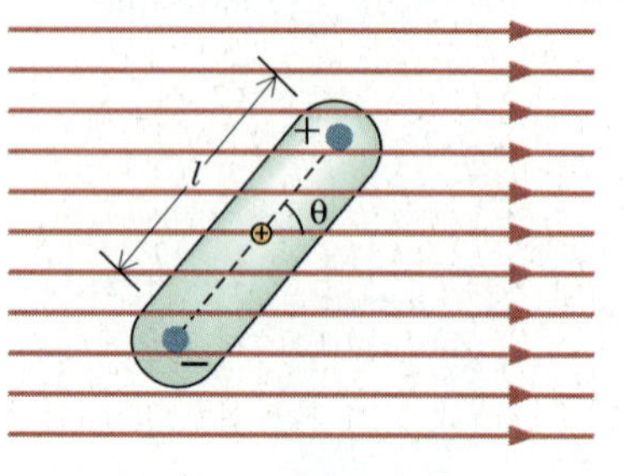

FIGURE 17.35
Electric dipole placed in a uniform electric field. The cross indicates the center of mass of the dipole.

If the external electric field is uniform, such as the electric field generated by the large flat charged slab discussed in the preceding section, then the forces on the positive and negative charges in a neutral body cancel, and there is no net force. Thus, the dipole shown in Fig. 17.35, experiences no net force in a uniform electric field. However, this dipole experiences a torque. The charges $\pm Q$ on the body are separated by a distance l. Since the force on the positive charge is toward the right and the force on the negative charge is toward the left, these combined forces tend to rotate the dipole. The torque of each force about the center of the dipole is the product of the magnitude of the force (QE), the distance from the center ($\frac{1}{2}l$), and the sine of the angle between the line from the center and the force ($\sin\theta$). Thus, the torque of each force is $-\frac{1}{2}lQE\sin\theta$, and the net torque of both forces together is

$$\tau = -lQE\sin\theta \tag{24}$$

The minus sign in this equation indicates that the torque is clockwise, in the sense of negative angles. ***The torque tends to align the dipole with the electric field.***

We can write Eq. (24) as

$$\tau = -pE\sin\theta \tag{25}$$

◀ **Torque on dipole**

where

$$p = lQ \tag{26}$$

◀ **Dipole moment**

The quantity p is called the ***dipole moment*** of the dipole; it is simply the charge at each end multiplied by the separation between the charges. The units of dipole moment are coulomb-meter (C · m). For example, the dipole moment of a CO molecule is $p = 8.0 \times 10^{-30}$ C · m, and the dipole moment of a water molecule is $p = 6.1 \times 10^{-30}$ C · m.

EXAMPLE 8. A CO molecule is placed in an electric field of 2.0×10^5 N/C. What is the torque that this electric field exerts on the molecule if its orientation is perpendicular to the electric field?

SOLUTION: When the orientation of the dipole is perpendicular to the electric field, the angle θ in Eq. (25) is 90° and the torque is

$$\tau = -pE\sin 90^\circ = -pE = -8.0 \times 10^{-30}\ \text{C}\cdot\text{m} \times 2.0 \times 10^5\ \text{N/C}$$

$$= -1.6 \times 10^{-24}\ \text{N}\cdot\text{m}$$

Molecules or atoms that do not have a permanent dipole moment may acquire a temporary dipole moment when placed in an electric field. The opposite electric forces on the positive and negative charges can distort the molecule and produce a charge separation. Such a dipole moment, which lasts only as long as the molecule is immersed in the electric field, is called an ***induced dipole moment.*** The magnitude of the induced dipole moment is approximately directly proportional to the magnitude of the electric field causing it.

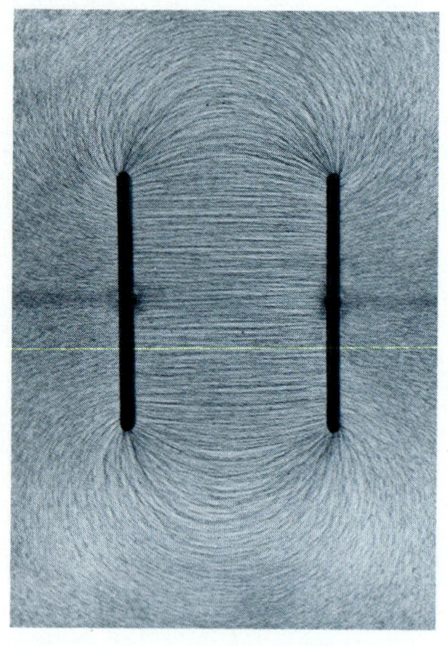

FIGURE 17.36

Field lines between a pair of charged parallel plates, made visible by small bits of thread aligned with the field lines. Note the fringing, or spreading, of the field lines near the edges of the plates.

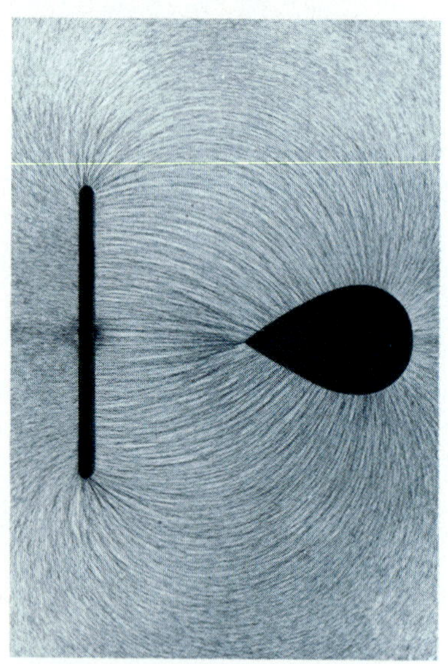

FIGURE 17.37

Field lines between a charged pointed body and a charged flat plate, made visible by small bits of thread aligned with the field lines. Note the strong concentration of field lines at the sharp point, indicating a strong electric field.

The tendency for alignment of a dipole with an electric field can be exploited to make the field lines visible. For this purpose, small bits of thread are suspended in oil in a container placed in the electric field; alternatively, small grass seeds are scattered on a sheet of paper placed in the electric field. The electric field induces a dipole moment along the long axis of the bit of thread or the grass seed, and the torque on this dipole then aligns the bit of thread or the grass seed with the electric field. Figures 17.36 and 17.37 show two photographs of field lines made visible by bits of thread. (Figures 19.1 and 19.2 show photographs of field lines made visible with grass seeds.)

SUMMARY

- **Definition of electric field:** $\mathbf{E} = \mathbf{F}/q$
- **Electric field of point charge:** $E = \dfrac{1}{4\pi\epsilon_0}\dfrac{q'}{r^2}$
- **Electric field of large, uniformly charged flat sheet:** $E = \sigma/2\epsilon_0$
- **Electric field of charged line:** $E = \dfrac{1}{2\pi\epsilon_0}\dfrac{\lambda}{r}$
- **Electric flux through a surface:** $\Phi = \overline{E}_{\mathrm{n}}\, A$
- **Gauss' Law (for closed surface):** $\Phi = Q/\epsilon_0$
- **Conductor in electrostatic equilibrium:**
  The electric field within the conductor is zero.
  The charge resides on the surface.
  The electric field at the surface is perpendicular and of magnitude $E = \sigma/\epsilon_0$.
- **Torque on dipole:** $\tau = -pE\sin\theta$
- **Electric dipole moment:** $p = lQ$

QUESTIONS

1.–Does it make any difference whether the value of the charge q in the equation defining the electric field [Eq. (2)] is positive or negative?

2.–During days of fair weather, the Earth has an atmospheric electric field that points vertically down. This electric field is due to charges on the surface of the Earth. What must be the sign of these charges?

3.–Figure 17.38 shows diagrams of hypothetical field lines corresponding to some static charge distributions, which are beyond the edge of the diagram. What is wrong with these field lines?

4.–If a positive point charge is released from rest in an electric field, will its orbit coincide with a field line? What if the point charge has zero mass?

5.–A ***tube of force*** is the volume enclosed between a bundle of adjacent field lines (Fig. 17.39; such a tube of force is analogous to a flow tube in hydrodynamics, and the field lines are analogous to streamlines). Along such a tube of force, the magnitude of the electric field varies in inverse proportion to the cross-sectional area of the tube. Explain.

6.–A negative point charge $-q$ sits in front of a very large flat sheet with a uniform distribution of positive charge. Make a rough sketch of the pattern of field lines of the net electric field produced jointly by the charge $-q$ and the sheet. Is there any point where the electric field is zero?

7.–A large, flat, thick slab of insulator has positive charge uniformly distributed over its volume. Sketch the field lines on both sides and inside the slab; pay careful attention to the starting points of the field lines.

8.–How could you build a "compass" that indicates the direction of the electric field?

9.–When a neutral metallic body, insulated from the ground, is placed in an electric field, it develops a charge separation, acquiring positive charge on one end and negative charge on the other. This means the body acquires an induced dipole moment. How is the direction of this dipole moment related to the direction of the electric field?

10.–By inspection of Fig. 17.37, make a rough, qualitative plot of the electric field strength as a function of position along a horizontal line midway between the charges.

11.–One electric dipole is at the origin, oriented parallel to the y axis. Another electric dipole is at some distance on the x axis. The electric field of the first dipole then exerts a torque on the second dipole. For what orientation of the second dipole is the potential energy at minimum?

12.–A point charge Q is inside a spherical Gaussian surface, which is enclosed in a larger cubical Gaussian surface. Compare the fluxes through these two surfaces.

13.–Figure 17.40 shows a Gaussian surface and lines of electric field entering and leaving this surface. Assuming that the number of lines has been normalized according to the convention of Section 17.2, what can you say about the magnitude and the sign of the electric charge within the surface?

14.–A Gaussian surface contains an electric dipole, and no other charge. What is the electric flux through this surface?

15.–Suppose we drop a charged Ping Pong ball into a cookie tin and quickly close the lid. What happens to the portions of the electric field lines that are outside of the cookie tin when we close the lid?

16.–The free electrons belonging to a metal are uniformly distributed over the entire volume of the metal. Does this contradict the result we derived in Section 17.4, according to which the charges are supposed to reside on the surface of a conductor?

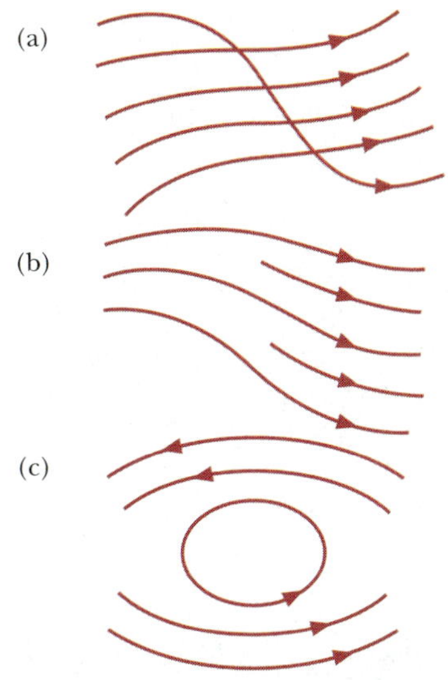

FIGURE 17.38
Hypothetical field lines.

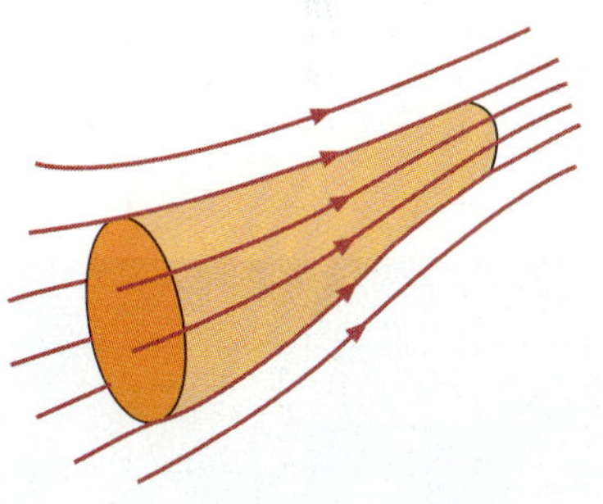

FIGURE 17.39
A tube of force.

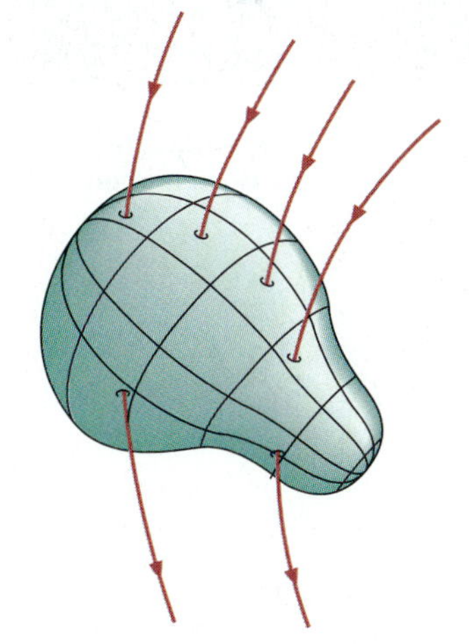

FIGURE 17.40
Gaussian surface and some field lines.

PROBLEMS

Section 17.1: The Electric Field

1.——Electric fields as large as 3.4×10^5 N/C have been measured by airplanes flying through thunderclouds. What is the force on an electron exposed to such a field? What is its acceleration?

2.——An electron moving through an electric field is observed to have an acceleration of 10^{16} m/s^2 in the x direction. What must be the magnitude and the direction of the electric field that produces this acceleration?

3.——Example 2 gives the x and y components of the electric field near a thundercloud. What are the magnitude and direction of this electric field?

4.——The Earth has not only a magnetic field, but also an atmospheric electric field. During days of fair weather (no thunderclouds), this atmospheric electric field has a strength of about 100 N/C and it points vertically downward. Taking into account this electric field and also gravity, what will be the acceleration (magnitude and direction) of a grain of dust of mass 1.0×10^{-18} kg carrying a single electron charge?

5.——***Millikan's experiment*** measures the elementary charge e by the observation of the motion of small oil droplets in an electric field. The oil droplets are charged with one or several elementary charges and, if the (vertical) electric field has the right magnitude, the electric force on the droplet will balance its weight, holding the drop suspended in midair. Suppose that an oil droplet of radius 1.0×10^{-4} cm carries a single elementary charge. What electric field is required to balance the weight? The density of oil is 0.80 g/cm^3.

6.——In an X-ray tube, electrons are exposed to an electric field of 8×10^5 N/C. What is the force on an electron? What is its acceleration?

7.——According to a theoretical estimate, at the surface of a neutron star of mass 1.4×10^{30} kg and radius 1.0×10^4 m there is an electric field of magnitude 6×10^3 N/C pointing vertically up. Show that the corresponding electric force on a proton more than balances the gravitational force on the proton.

8.——A long hair, taken from a girl's braid, has a mass of 1.2×10^{-3} g. The hair carries a charge of 1.3×10^{-9} C distributed along its length. If we want to suspend this hair in midair, what (uniform) electric field do we need?

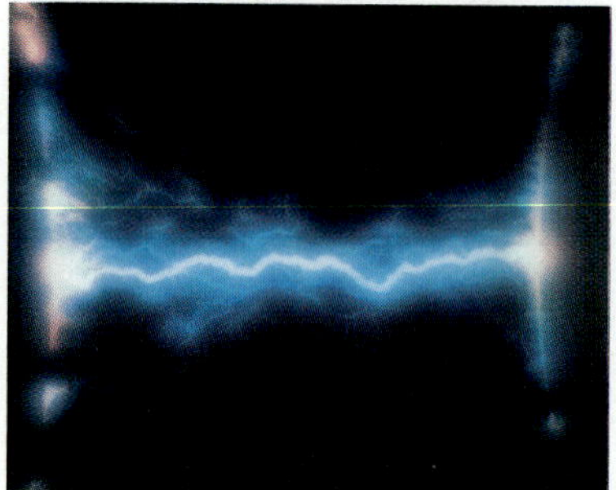

An electric spark jumps through the air between two charged bodies.

9.*——The electric field in the electron gun of a TV tube is supposed to accelerate electrons uniformly from 0 to 3.3×10^7 m/s within a distance of 1.0 cm. What electric field is required?

10.*——Electric breakdown (sparks) occurs in air if the electric field reaches 3×10^6 N/C. At this field strength, free electrons present in the atmosphere are quickly accelerated to such large velocities that upon impact on atoms they knock electrons off the atom and thereby generate an avalanche of electrons. How far must a free electron move under the influence of the above electric field if it is to attain a kinetic energy of 3×10^{-19} J (which is sufficient to produce ionization)?

11.——The nuclei of the atoms in a chunk of metal lying on the surface of the Earth would fall to the bottom of the metal if their weight were the only force acting on them. Actually, within the interior of any metal exposed to gravity there exists a very small electric field that points vertically up. The corresponding electric force on a nucleus just balances the weight of the nucleus. Show that for a nucleus of atomic number Z, mass m, the required field has a magnitude $mg/(Ze)$. What is the numerical value of this electric field in a chunk of iron?

12.——The hydrogen atom has a radius of 0.53×10^{-10} m. What is the magnitude of the electric field that the nucleus of the atom (a proton) produces at this radius?

13.——What is the strength of the electric field at the surface of a uranium nucleus? The radius of the nucleus is 7.4×10^{-15} m and the electric charge is $92e$. For the purposes of this problem the electric charge may be regarded as concentrated at the center.

14.——Suppose that in a hydrogen atom the electron is (instantaneously) at a distance of 2.1×10^{-10} m from the proton. What is the net electric field that

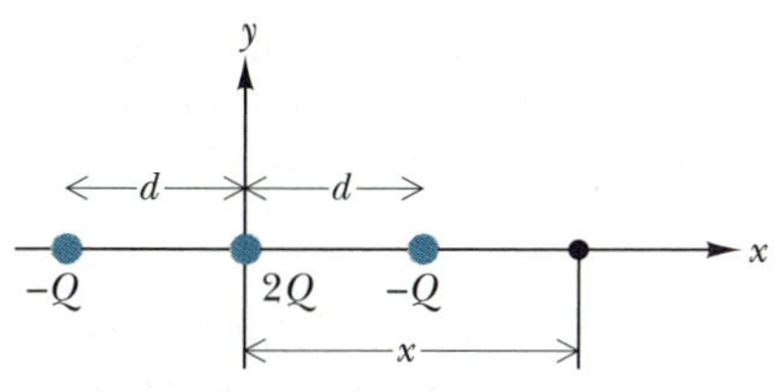

FIGURE 17.41
Three point charges.

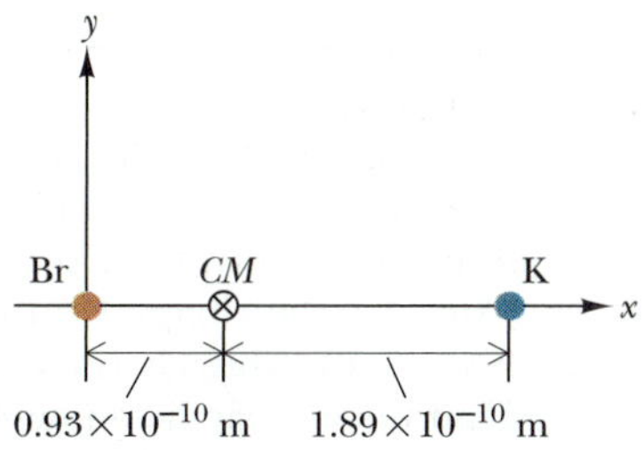

FIGURE 17.42
The positive (nuclear) charges in a KBr molecule.

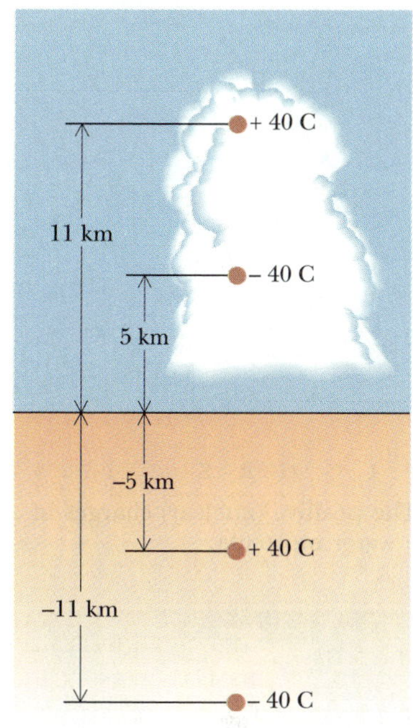

FIGURE 17.43
Charges in a thundercloud and image charges below the ground.

the electron and the proton produce jointly at a point midway between them?

15.—Three point charges $-Q$, $2Q$, and $-Q$ are arranged on a straight line, as illustrated in Fig. 17.41. What is the electric field that these charges produce at a distance x to the right of the central charge?

16.—Figure 17.42 shows the arrangement of nuclear charges (positive charges) of a KBr molecule. Find the electric field that these charges produce at the center of mass at a distance of 0.93×10^{-10} m from the Br atom.

17.*—Suppose that the charge distribution of a thundercloud can be approximated by two point charges, a negative charge of -40 C at a height of 5 km (above ground) and a positive charge of 40 C at a height of 11 km. To find the electric field strength at the ground, we must take into account that the ground is a conductor and that the charge of the thundercloud induces charges on the ground. It can be shown that the effect of the induced charges can be simulated by a point charge of 40 C at 5 km below ground and a point charge of -40 C at 11 km below ground (Fig. 17.43); these fictitious charges are called ***image charges.*** By adding the electric field of the image charges to that of the two real charges in the thundercloud, calculate the magnitude of the electric field at a point on the ground directly below the thundercloud charges.

18.*—Three point charges $-Q$, $2Q$, and $-Q$ are arranged on a straight line, as illustrated in Fig. 17.44. What is the electric field that these charges produce at a distance y above the central charge?

19.*—Three charges $\pm Q$ are arranged at the vertices of an equilateral triangle of side L, as shown in Fig. 17.45. Calculate the electric field that these charges produce at the center of the triangle.

20.*—If a 1.0×10^{-10} C charge is placed on the x axis 0.15 m from the origin of a coordinate system (see Fig. 17.46), what is the magnitude of the electric field at a point 0.10 m up the y axis?

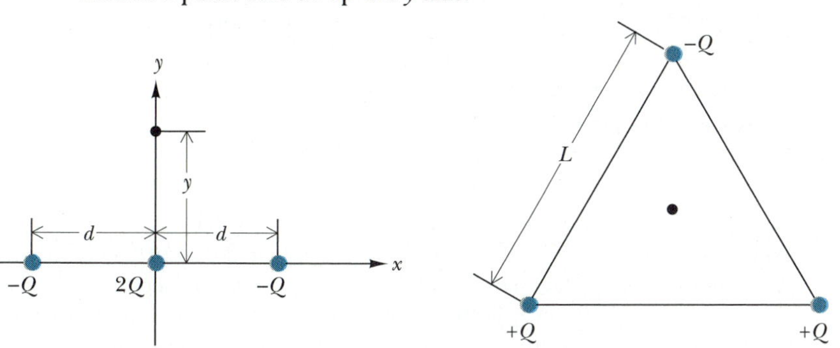

FIGURE 17.44

FIGURE 17.45

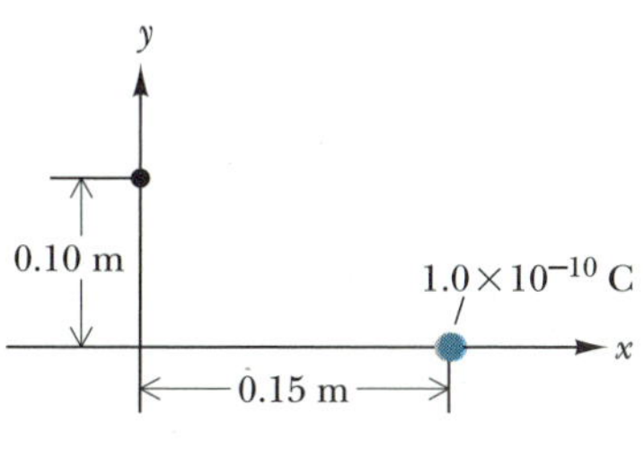

FIGURE 17.46

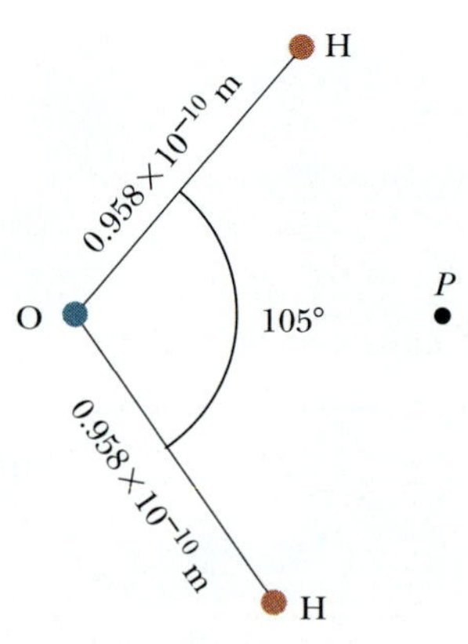

FIGURE 17.47

The positive (nuclear) charges in a water molecule.

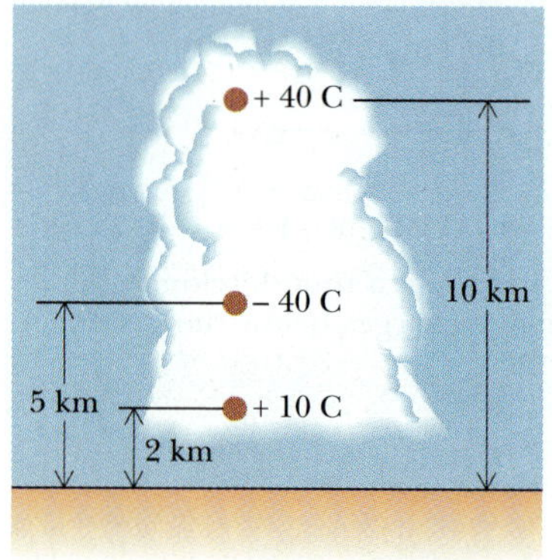

FIGURE 17.48

Charges in a thundercloud.

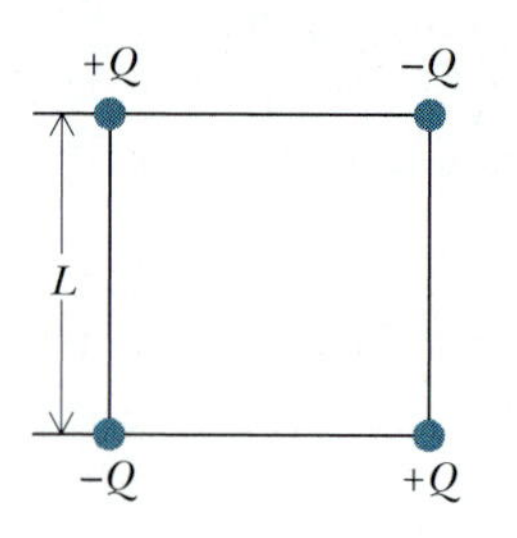

FIGURE 17.49

21.*—The distance between the oxygen nucleus and each of the hydrogen nuclei in an H_2O molecule is 0.958×10^{-10} m; the angle between the two hydrogen atoms is 105° (Fig. 17.47). Find the electric field produced by the nuclear charges (positive charges) at the point P at a distance of 1.2×10^{-10} m to the right of the oxygen nucleus.

22.*—Consider the charge distribution within a thundercloud as illustrated in Fig. 17.48. This charge distribution consists of three pointlike charges. Find the electric field (magnitude and direction) that this charge distribution produces at a height of 10 km and a horizontal distance of 5 km to the right.

23.*—Four point charges $\pm Q$ are arranged at the corners of a square of side L, as shown in Fig. 17.49. Consider the midpoint of each of the four sides of the square. At each midpoint, find the magnitude and the direction of the electric field.

24.**—To measure the magnitude of a horizontal electric field, an experimenter attaches a small charged cork ball to a string and suspends this device in the electric field. The electric force pushes the cork ball to one side, and the ball attains equilibrium when the string makes an angle of 35° with the vertical (see Fig. 17.50). The mass of the ball is 3×10^{-5} kg, and the charge on the ball is 4×10^{-7} C. What is the magnitude of the electric field?

25.**—Consider eight of the ions of Cl^- and Na^+ in a crystal lattice of common salt. The ions are located at the vertices of a cube measuring 2.82×10^{-10} m on an edge (see Fig. 17.51). What is the electric field that these eight ions produce at the center of the cube? What is the electric field that they produce at the center of each face of the cube?

26.**—Figure 17.48 shows the charge distribution within a thundercloud. There is a charge of 40 C at a height of 10 km, −40 C at 5 km, and 10 C at 2 km. Treating these charges as pointlike, find the electric field (magnitude and direction) that they produce at a height of 5 km and a horizontal distance of 3 km to the right.

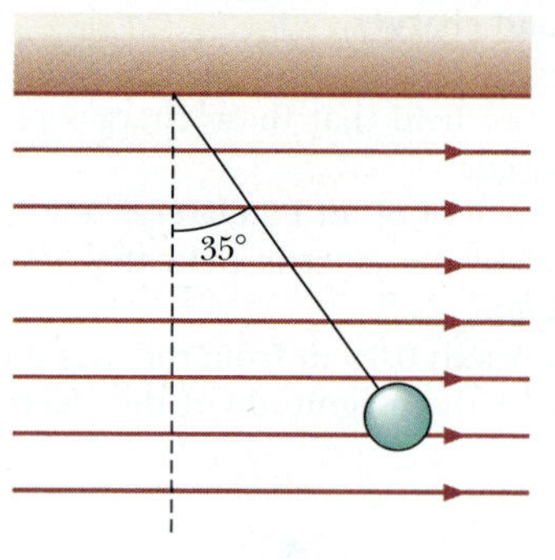

FIGURE 17.50

Cork ball suspended in an electric field.

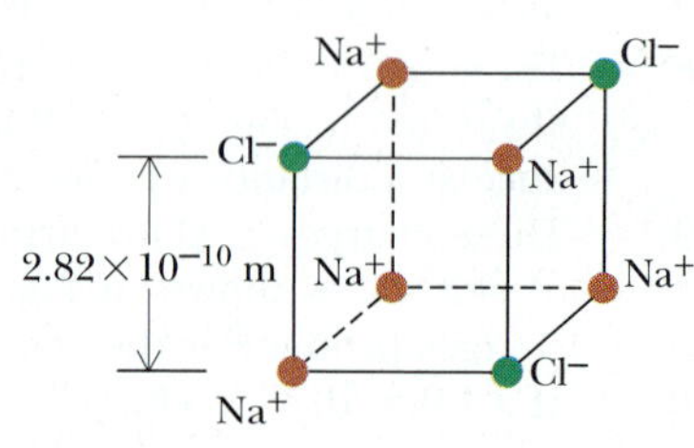

FIGURE 17.51

Ions of Cl^- and Na^+ in a salt crystal.

Section 17.2: Lines of Electric Field

27.—You wish to generate a uniform electric field of 2.0×10^5 N/C in the space between two flat, parallel plates of metal placed face-to-face. The plates measure $0.30 \text{ cm} \times 0.30 \text{ cm}$. How much electric charge must you put on each plate? Assume that the gap between the plates is small so that the charge distribution and the electric field are approximately uniform, as for infinite plates.

28.* Consider two large parallel plates with uniform, opposite charge distributions, as in Fig. 17.52a. Suppose that the magnitude of the charge density on each plate is 2×10^{-5} C/m^2. The upper plate is positive and the lower negative.
   (a) What is the magnitude of the electric field in the region between the plates?
   (b) You now insert a large parallel metallic plate in the space between the two charged plates (see Fig. 17.52b). Suppose that this plate is thick, say, 1 cm thick. What is the magnitude of the electric field inside the thick plate? What is the magnitude of the electric field in the remaining space above the thick plate? Below the thick plate? What is the charge density on the upper surface of the thick plate? On the lower surface?

29.* Electric charge is uniformly distributed over each of three large, parallel sheets of paper (Fig. 17.53). The charges per unit area on the sheets are 2×10^{-6} C/m^2, 2×10^{-6} C/m^2, and -2×10^{-6} C/m^2, respectively. The distance between one sheet and the next is 1.0 cm. Find the strength of the electric field **E** above the sheets, below the sheets, and in the space between the sheets. Find the direction of **E** at each place.

30.* Each of two very large, flat sheets of paper carries a uniform positive charge distribution of 3.0×10^{-4} C/m^2. The two sheets of paper intersect at an angle of 45° (Fig. 17.54). What are the magnitude and the direction of the electric field at a point between the two sheets?

31.* Two large sheets of paper intersect at right angles. Each sheet carries a uniform distribution of positive charge (Fig. 17.55). The charge per unit area on the sheets is 3×10^{-6} C/m^2. Find the magnitude of the electric field in each of the four quadrants. Draw the field lines in each quadrant.

32. How many electrons must be added to a copper sphere with an initial charge of -5.00×10^{-6} C to triple the magnitude of the electric field at a fixed location outside of the sphere?

33. The diameter of a proton is about 1.0×10^{-15} m. How many diameters apart are two protons if the Coulomb force between them is 1.0 N?

34. Charge is placed on a small metallic sphere which is surrounded by air. If the radius of the sphere is 0.5 cm, how much charge can be placed on the sphere before the air near the sphere suffers electric breakdown? The critical electric field strength that leads to breakdown in air is 3×10^6 N/C.

35. In symmetric fission, a uranium nucleus splits into two equal pieces each of which is a palladium nucleus. The palladium nucleus is spherical with a radius of 5.9×10^{-15} m and a charge of $46e$ uniformly distributed over its volume. Suppose that immediately after fission, the two palladium nuclei

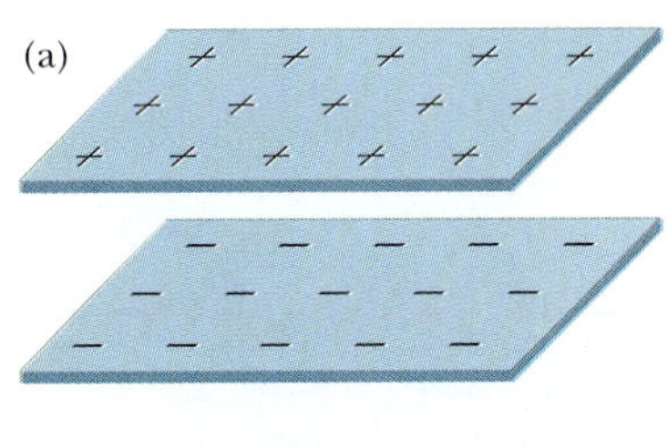

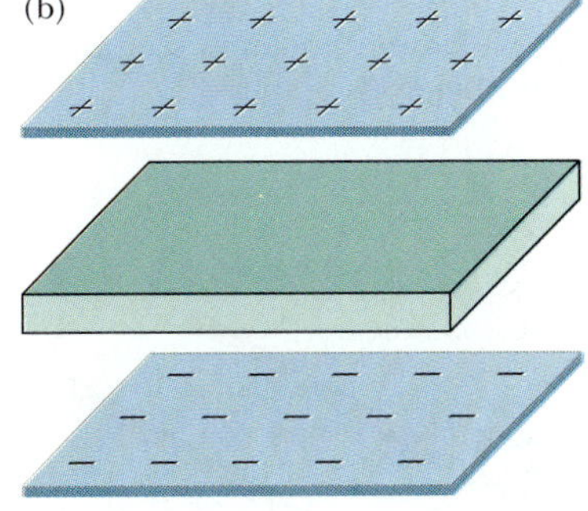

FIGURE 17.52

(a) Parallel plates with opposite charges. (b) Another plate inserted between the other two.

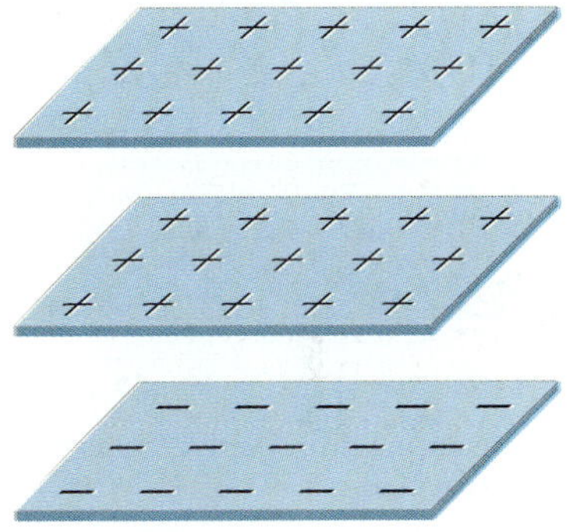

FIGURE 17.53

Three parallel charged sheets of paper.

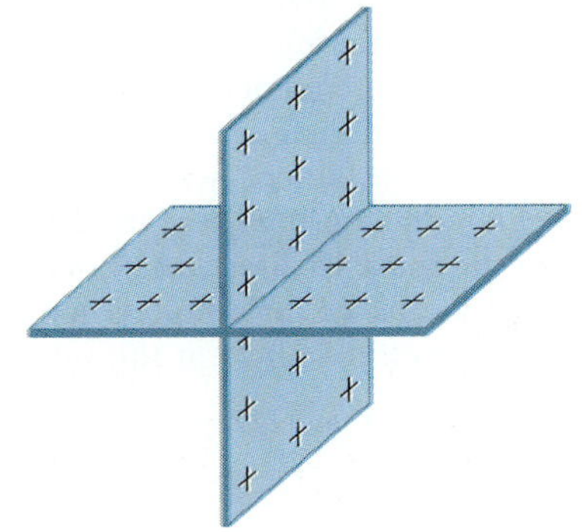

FIGURE 17.55

Another two intersecting charged sheets of paper.

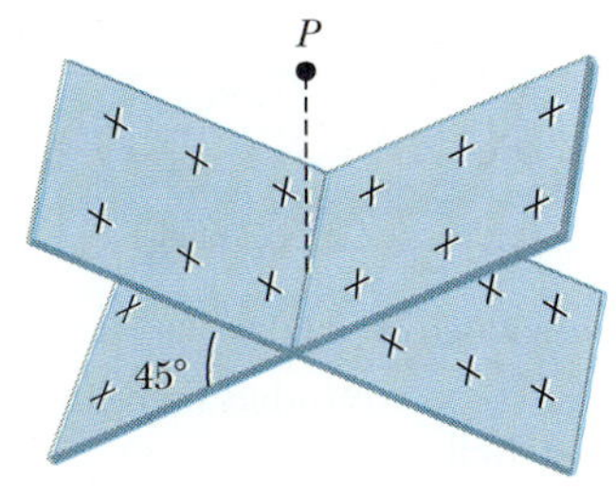

FIGURE 17.54

Two intersecting charged sheets of paper.

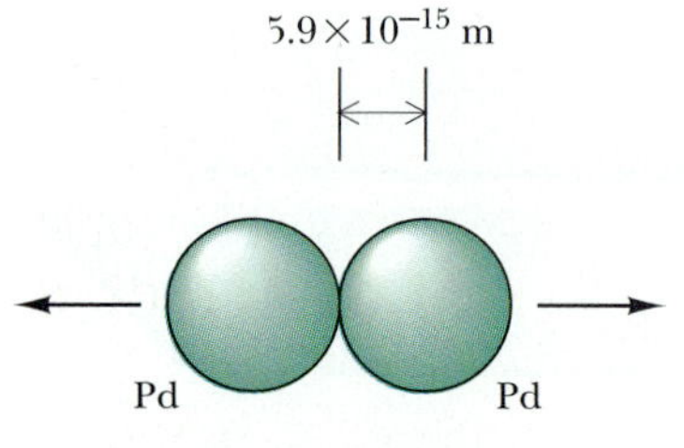

FIGURE 17.56

Two palladium nuclei in contact. Each nucleus is a sphere.

are barely touching (Fig. 17.56). What is the value of the total electric field at the center of each? What is the repulsive force between them? What is the acceleration of each? The mass of a palladium nucleus is 1.99×10^{-25} kg.

36.—A long straight string carries a uniform distribution of electric charge of 2×10^{-10} C/m. What is the electric field at a distance of 0.5 m from this string? At 1.0 m? At 1.5 m?

37.—What is the maximum amount of electric charge per unit length that we can place on a long, straight, human hair of diameter 8×10^{-3} cm if the surrounding air is not to suffer electrical breakdown? The air will suffer breakdown if the electric field exceeds 3×10^6 N/C.

38.—In a xerographic copier, during the copying process, electric charge is deposited on the sheet of paper by means of a corona discharge from a thin wire carrying a concentration of charge (in some copiers you can see this corona wire if you open the top of the copier). The wire has a radius of 2×10^{-5} m. To produce the corona discharge, an electric field of 3×10^6 N/C is required near the surface of the wire. What amount λ of charge per unit length must be placed on the wire to attain this electric field?

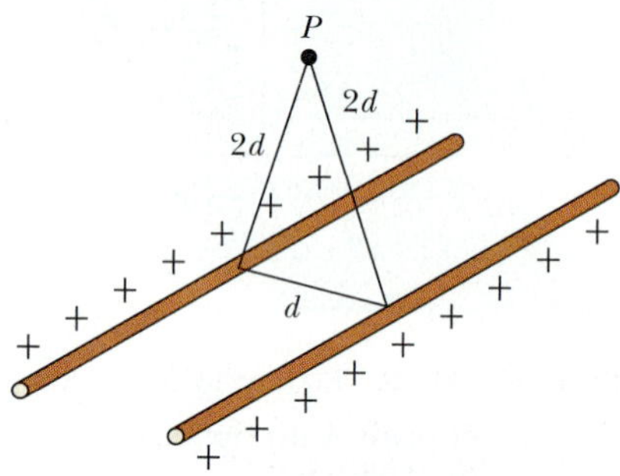

FIGURE 17.57

Two parallel charged lines.

39.*—Each of two very long, straight, parallel lines carries a positive charge of λ coulomb per meter. The distance between the lines is d. Find the electric field at a point equidistant from the lines, with a distance $2d$ from each line (see Fig. 17.57). Draw a diagram showing the direction of the electric field.

40.*—The surface of a long, cylindrical copper pipe has a charge of λ coulomb per meter (Fig. 17.58). What is the electric field outside the pipe?

41.*—Each of two very long, straight, parallel lines carries a positive charge of 1.0×10^{-9} C/m. One line lies along the x axis, the other along the y axis. Find the electric field (magnitude and direction) at the point $x = 0.50$ m, $y = 0.20$ m.

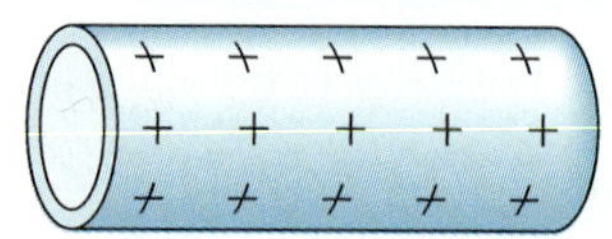

FIGURE 17.58

A long pipe with positive charge on its surface.

42.*—According to Eq. (15), the magnitude of the electric field produced by a long thin straight line of charge is

$$E = \frac{1}{2\pi\epsilon_0}\frac{\lambda}{r}$$

To obtain this result, consider a segment of the straight charged line, of length h.

(a) What is the amount of electric charge on this segment?
(b) How many field lines start on this segment?
(c) Consider a cylindrical mathematical surface of radius r, concentric with the straight charged line. What is the area of this surface? What is the density of field lines intercepted by this surface?
(d) What is the magnitude of the electric field at this surface?

43.*—A spherical rubber balloon has a uniform distribution of charge over its surface. Show that the electric field that this charge produces in the (empty) interior of the balloon is exactly zero. (Hint: If there were an electric field in the interior, what would the field lines look like?)

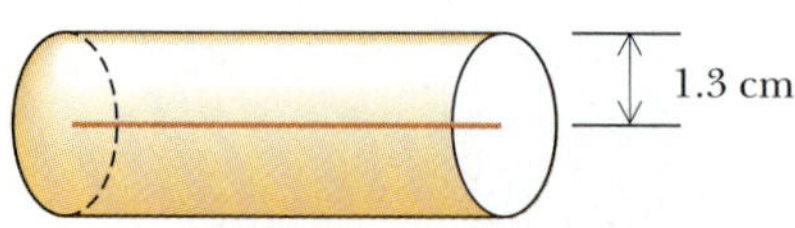

FIGURE 17.59

Thin wire surrounded by cylindrical shell.

44.*—The tube of a Geiger counter consists of a thin conducting wire of radius 1.3×10^{-3} cm stretched along the axis of a conducting cylindrical shell of radius 1.3 cm (Fig. 17.59). The wire and the cylinder have equal and opposite charges of 7.2×10^{-10} C distributed along their length of 9.0 cm. Find a formula for the electric field in the space between the wire and the cylinder; pretend that the electric field is that of an infinitely long wire and cylinder. What is the magnitude of the electric field at the surface of the wire?

Section 17.3: Electric Flux and Gauss' Law

45.——A point charge of 2×10^{-12} C is located at the center of a cubical Gaussian surface. What is the electric flux through *each* of the faces of the cube?

46.——Consider a point charge q located just above a very large, plane surface. What is the electric flux through this surface?

47.——Consider the thundercloud described in Problem 22 and Fig. 17.48. What is the total electric flux coming out of the surface of the cloud?

48.——Two very large plane surfaces intersect at 90°, forming a floor and a wall (see Fig. 17.60).

(a) If a point charge q is placed just outside the edge formed by the intersecting surfaces, what is the electric flux through the combined surfaces?

(b) If the point charge q is placed just inside the edge formed by the intersecting surfaces, what is the electric flux through the combined surfaces?

49.——Consider a cubical Gaussian surface of edge 5 cm. You do not know the electric charge or the electric field inside the cube, but you do know the electric field at the surface: at the top of the cube the electric field has a magnitude of 5×10^5 N/C and points perpendicularly out of the cube; at the bottom of the cube, the electric field has a magnitude of 2×10^5 N/C and points perpendicularly into the cube; on all other faces, the electric field is tangential to the surface of the cube.

(a) How much electric charge is inside the cube?

(b) Can you guess what charge distribution inside (and outside) the cube would generate this kind of electric field?

50.——A small square surface, 1 cm × 1 cm, is placed at a distance of 1 m from a point charge of 3×10^{-9} C. What is the electric flux through this square if it is face-on to the electric field (see Fig. 17.61)? If it is tilted by 30°? If it is tilted by 60°?

51.*——Consider the (infinite) midplane between two charges of equal magnitudes and opposite signs (see Fig. 17.10). If the charges are $\pm q$, what is the electric flux through this midplane?

52.*——A point charge of 6×10^{-8} C sits on the wood floor of a room. You hold a small cardboard square measuring 1 cm × 1 cm face down, and carry it across the room at a height of 2 m, passing over the position of the point charge. Roughly plot the flux through the square as a function of position.

53.*——A spherical Gaussian surface of radius 1 m has a small hole of radius 0.5 cm. A point charge of 2×10^{-9} C is placed at the center of this spherical surface (see Fig. 17.62). What is the flux through the surface?

54.*——Defining the gravitational field as the gravitational force per unit mass, formulate a Gauss' Law for gravity. Check that your law implies Newton's law of universal gravitation.

55.**——Consider a point charge q located at one vertex of a cubical Gaussian surface, just inside the cube (see Fig. 17.63). What is the electric flux through each face of the cube?

56.**——Prove that if the electric field is uniform in some region, then the charge density must be zero in that region. (Hint: Use Gauss' Law.)

Section 17.4: Conductors and Electric Fields

57.——On a clear day, the Earth's atmospheric electric field near the ground has a magnitude of 100 N/C and points vertically down. Inside the ground, the electric field is zero, since the ground is a conductor. Consider a

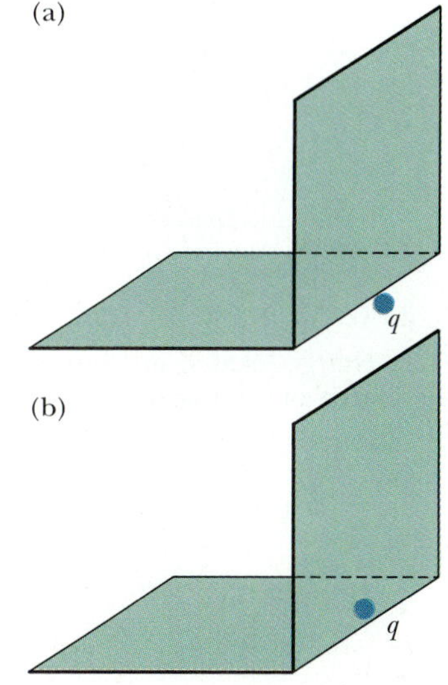

FIGURE 17.60

(a) Point charge located just outside of the edge formed by the intersecting surfaces.
(b) Point charge located just inside of the edge formed by the intersecting surfaces.

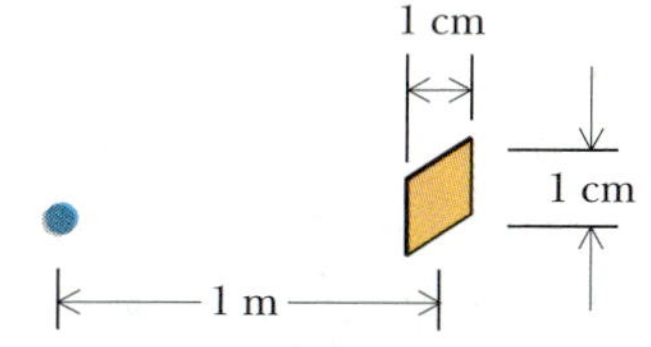

FIGURE 17.61

Point charge and small square surface.

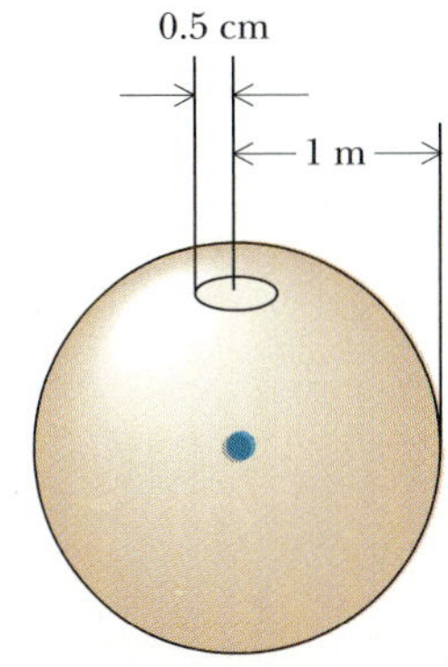

FIGURE 17.62

Spherical Gaussian surface with a small hole.

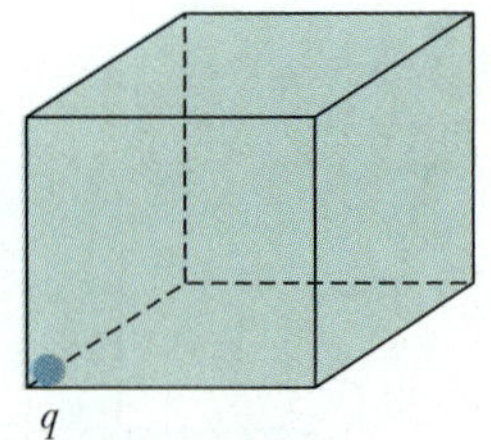

FIGURE 17.63
Point charge located just inside the vertex of a cubical Gaussian surface.

Gaussian surface in the shape of a box of 1 m × 1 m × 1 m, half below the ground and half above. What is the electric flux through each of the square sides of this box? What is the charge enclosed by the box?

58.——A point charge of 1.0×10^{-8} C is placed inside an uncharged metallic can (say, a closed beer can) insulated from the ground. How many flux lines will emerge from the surface of the can when the point charge is inside?

59.——Suppose we suspend a small ball carrying a charge of 1.0×10^{-6} C in the middle of a safe and lock the door. The safe is made of solid steel; it has inside dimensions 0.3 m × 0.3 m × 0.3 m and outside dimensions 0.4 m × 0.4 m × 0.4 m. What is the electric flux through a cubical surface measuring 0.2 m × 0.2 m × 0.2 m centered on the ball? A cubical surface measuring 0.35 m × 0.35 m × 0.35 m? A cubical surface measuring 0.5 m × 0.5 m × 0.5 m?

Section 17.5: The Electric Dipole

60.——The two charges of ±40 C in the thundercloud of Fig. 17.48 form a dipole. What is the dipole moment?

61.——In a hydrogen atom, the electron is at a distance of 0.53×10^{-10} m from a proton.

(a) What is the instantaneous dipole moment of this system?

(b) Taking into account that the electron moves around the proton on a circular orbit, what is the time-average dipole moment of this system?

62.——(a) Pretend that the HCl molecule consists of (pointlike) ions of H^+ and Cl^- separated by a distance of 1.3×10^{-10} m. If so, what would be the dipole moment of this system?

(b) The observed dipole moment is 3.4×10^{-30} C·m. Can you suggest a reason for this discrepancy?

63.——The dipole moment of an HCl molecule is 3.4×10^{-30} C·m. Calculate the magnitude of the torque that an electric field of 2.0×10^{6} N/C exerts on this molecule when the angle between the electric field and the longitudinal axis of the molecule is 45°.

64.——A small, straight bit of thread, such as used to make the pictures of field lines in Fig. 17.36, has a charge of $+1 \times 10^{-14}$ C at one end and a charge of -1×10^{-14} C at the other. The length of the thread is 2 mm.

(a) What is the dipole moment?

(b) What is the torque on this thread if it is placed in an electric field of 6×10^{5} N/C at right angles to the field?

Banks of capacitors at Lawrence Livermore Laboratory store electric energy for the operation of the powerful NOVA laser used for fusion research.

ELECTROSTATIC POTENTIAL AND ENERGY

From our study of mechanics we know that if we want to formulate a law of conservation of energy for a particle moving under the influence of some force, we have to construct a potential energy corresponding to this force. In this chapter we will construct the electrostatic potential energy for a charged particle moving in the electric field generated by a static charge distribution. Like the potential energies we examined in Chapter 4, the electrostatic potential energy is helpful in the calculation of the motion of the particle. Furthermore, as we will see in the next chapter, the electrostatic potential energy is needed for the analysis of the electric currents that flow in circuits connected to batteries or similar devices.

Besides formulating the electric potential energy for a single particle in an electric field, we will formulate the mutual electric potential energy for a configuration of several charged particles and for an arrangement of conductors with charges placed on them. Since the charges exert electric forces on each other, we have to perform a certain amount of work to bring the charges together into their final configuration, starting with an initial configuration of very large (infinite) separation. This work is the potential energy of the configuration. It represents energy stored in the configuration during its assembly. We will see that this energy is actually stored in the electric field. The energy is concentrated in those regions of space where the electric field is strong.

Arrangements of isolated conductors, such as a pair of parallel metallic plates mounted on nonconducting supports, can be used to store electric charge. Any such arrangement of conductors used for the storage of charge is called a ***capacitor.*** When the capacitor stores electric charge, it also stores electric potential energy. In most capacitors, the gaps between the plates or other conductors are filled with an insulating material, usually called ***dielectric*** material, which affects the electric field. In the final sections of this chapter, we will examine the mutual effects between the electric field and dielectric materials. These effects are of practical importance in the use of capacitors, but they are also interesting in their own right. For instance, air is a dielectric material and, since most of the electric fields we deal with are in air, we ought to inquire how the electric field in air differs from that in vacuum.

18.1 THE ELECTROSTATIC POTENTIAL

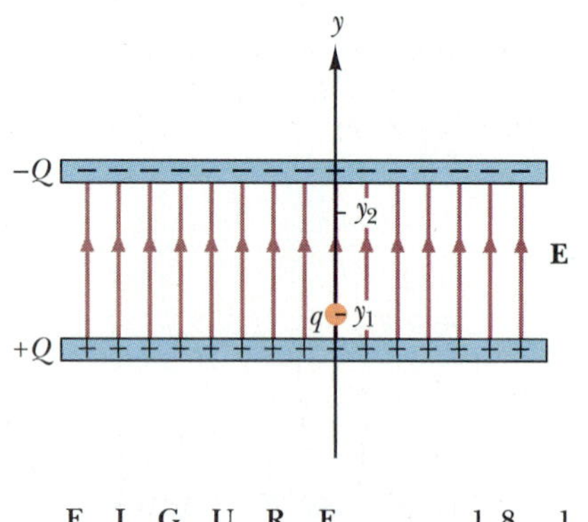

FIGURE 18.1
Uniform electric field in the space between two charged parallel plates.

In Chapter 4 we learned that if a force is conservative, then the work done by this force on a particle during a displacement can be expressed as a difference between two potential energies, one for the starting point and one for the end point of the displacement. If we know the potential energy that corresponds to the force, we can immediately construct the conserved mechanical energy, which is simply the sum of the kinetic energy and the potential energy.

The electric force that a static distribution of charges exerts on a point charge is a conservative force. For instance, consider the electric force exerted by two uniform distributions of positive and of negative charges placed on two large parallel sheets or plates (see Fig. 18.1). We know from Section 17.2 that such parallel charged plates generate a uniform electric field E in the space between them; the magnitude of this electric field is directly proportional to the amount of charge per unit area on each plate [see Eq. (17.11)]. The force that this electric field exerts on a point charge q is qE, and the work done by this force during a displacement from the point y_1 to the point y_2 illustrated in Fig. 18.1 is

$$W = F\Delta y = qE(y_2 - y_1) \qquad (1)$$

or

$$W = -qEy_1 + qEy_2 \tag{2}$$

This shows that if we identify the potential energy as

$$U = -qEy \tag{3}$$

then the work is the difference between the potential energies corresponding to the points y_1 and y_2:

$$W = -qEy_1 + qEy_2 = U_1 - U_2 \tag{4}$$

The conserved mechanical energy is the sum of the kinetic energy $K = \frac{1}{2}mv^2$ and the potential energy $U = -qEy$,

$$E = \tfrac{1}{2}mv^2 - qEy = [\text{constant}] \tag{5}$$

Note that the potential energy (3) is directly proportional to the distance y from the lower plate in Fig. 18.1. This direct proportionality of potential energy and height is reminiscent of the gravitational potential energy mgy. Mathematically, the electric potential energy and the gravitational potential energy are similar because both involve a constant force. [But note that the signs of these two potential energies are opposite, because the electric force that the plates in Fig. 18.1 exert on a (positive) charge q is upward, whereas the gravitational force is downward; the signs of these two potential energies would be the same if the electric field in Fig. 18.1 were downward or if the charge q were negative.]

We recall from Section 4.5 that a general criterion for a conservative force is that the work the force performs on any roundtrip must be zero. When we attempt to apply this criterion to the electric force exerted by the parallel charged plates in Fig. 18.1, we run into a paradox. Consider the roundtrip shown in Fig. 18.2, which consists of a straight segment from the lower plate to the upper; followed by a long horizontal segment to the left, beyond the end of the plate; followed by a downward segment on the far left; and, finally, a long horizontal segment toward the right, to the starting point. If the electric field lines are as shown in Fig. 18.2, then the work done by the electric field along the horizontal segments is zero, since the electric field is perpendicular to these segments; and the work done along the downward segment on the far left is also zero, since there is no electric field there. Thus, the only segment that contributes to the work is the first upward segment, between the plates. Since the contribution to the work from this segment is positive, it would seem we cannot achieve zero work for the roundtrip!

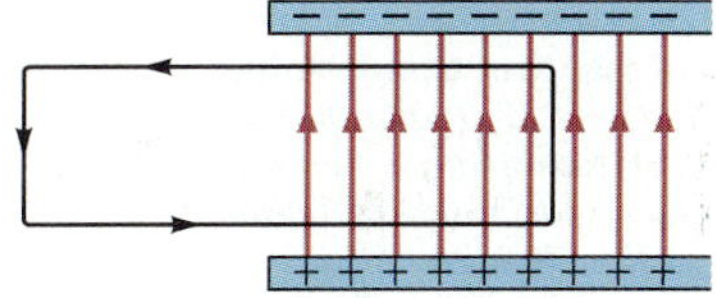

FIGURE 18.2

A closed path with an upward segment between the plates and a downward segment outside of the plates.

The resolution of this paradox hinges on the existence of electric fields beyond the edges of the plates. An exact calculation of the electric field shows that near the edges of the plates the electric field is not uniform, and the electric field spills out into the space beyond the plates. The actual distribution of field lines is as shown in Fig. 18.3. The field that spills out beyond the edges of the plates is called the ***fringing field.*** The portions of the path that pass through this fringing field contribute to the

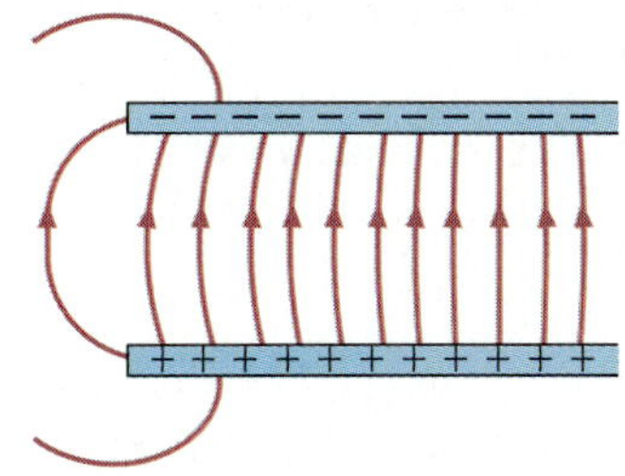

FIGURE 18.3

The fringing field that extends beyond the space between the plates.

work, and the net work for the complete roundtrip turns out to be zero, as required by our criterion for a conservative force. More generally, the electric force generated by any distribution of static charges can be shown to be a conservative force.

We found it useful to define the electric field as the electric force divided by the charge q on which this force acts. Likewise, we find it useful to ***define the electrostatic potential V as the electric potential energy U divided by the charge q,***

Electrostatic potential ▸

$$V = U/q \tag{6}$$

In brief, the electrostatic potential is the potential energy per unit charge. For example, in the case of a uniform electric field, with a potential energy $U = -qEy$, the electrostatic potential is

$$V = -Ey \tag{7}$$

The unit of electrostatic potential is the volt (V),*

$$1 \text{ volt} = 1 \text{ V} = 1 \text{ joule/coulomb} = 1 \text{ J/C} \tag{8}$$

The unit of electric field we have employed in the preceding chapters is the N/C. This unit can be expressed in terms of volts as follows:

$$1\,\frac{\text{N}}{\text{C}} = 1\,\frac{\text{N}\cdot\text{m}}{\text{C}\cdot\text{m}} = 1\,\frac{\text{J}}{\text{C}}\frac{1}{\text{m}} = 1\,\frac{\text{V}}{\text{m}} \tag{9}$$

Thus, N/C and V/m are equal units; in practice, volt per meter is the preferred unit for the electric field. Table 18.1 gives some values of electrostatic potential.

Alessandro, Conte Volta, *1745–1827, Italian physicist. Volta established that the "animal electricity" observed by* **Luigi Galvani,** *1737–1798, in experiments with frog muscle tissue placed in contact with dissimilar metals, was not due to any exceptional property of animal tissues, but was also generated whenever any wet body was sandwiched between dissimilar metals. This led him to develop the first "voltaic pile," or battery, consisting of a large stack of moist disks of cardboard (electrolyte) sandwiched between disks of metal (electrodes).*

EXAMPLE 1. Suppose that near the ground directly below a thundercloud, the electric field is constant of magnitude 2×10^4 V/m and points upward. What is the potential difference between the ground and a point in the air, 50 m above ground?

SOLUTION: From Eq. (7),

$$V = -Ey = -2 \times 10^4 \text{ V/m} \times 50 \text{ m} = -1 \times 10^6 \text{ volt}$$

COMMENTS AND SUGGESTIONS: This calculation assumes that the ground is flat, without protuberances, such as trees or buildings. It would be

---

* Note that the same letter "V" is used in physics both as a symbol for potential and as the abbreviation for *volt*. This leads to confusing equations such as $V = 3.0$ V (which is intended to mean $V = 3.0$ volt). If there is a possibility of confusion, it is best not to abbreviate *volt*.

wrong to conclude that the potential at the top of a 50-m-high building is -1×10^6 volt. Buildings are usually made of conducting materials, and, as we will see, the potential at all points of a conductor is the same. Thus, the building effectively acts as part of the ground, and the potential difference between the top of the building and the ground is zero. The presence of the building or of some other conducting protrusion modifies the electric field, so the above calculation does not apply.

For a nonuniform electric field, we can calculate the electric potential energy of a point charge q by the same method as for the uniform electric field: we express the work done by the electric field during a displacement from one point to another as the difference of two terms, one of which depends on the position of the starting point, and the other on the position of the end point. These two terms are the potential energies U_1 and

TABLE 18.1 SOME POTENTIALS AND POTENTIAL DIFFERENCES

Thundercloud to ground		5×10^7 V
Van de Graaff generator	**(a)**	10^7 V
High-voltage power line	**(b)**	5×10^5 V
At nucleus of uranium atom		2×10^5 V
Power supply for X-ray tube		10^5 V
Power supply for TV tube		2×10^4 V
Automobile ignition		10^4 V
Power supply for neon tube		2×10^3 V
Household outlet (Europe)		220 V
Household outlet (USA)		115 V
At electron orbit in hydrogen atom		26 V
Automobile battery	**(c)**	12 V
Dry cell		1.5 V
Single solar cell	**(d)**	0.6 V
Resting potential across nerve membrane		0.09 V
Potential changes on skin (EKG and EEG)		5×10^{-5} V

(a)

(b)

(d)

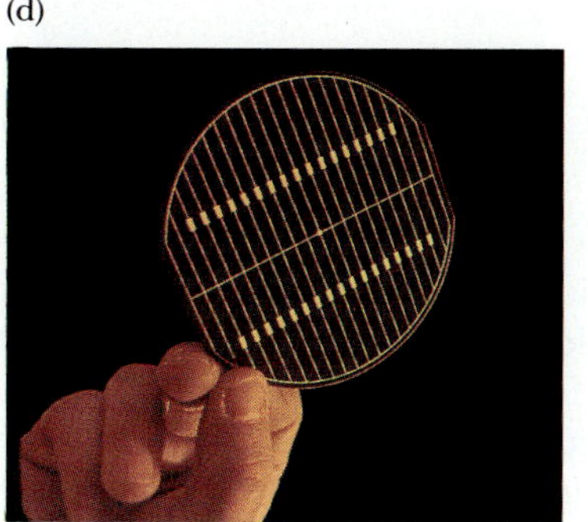

(c)

U_2 corresponding to these two points. The work is then

$$W = U_1 - U_2 \tag{10}$$

and the conserved mechanical energy is

$$E = K + U = \tfrac{1}{2}mv^2 + U = [\text{constant}] \tag{11}$$

or, with $U = qV$,

$$E = \tfrac{1}{2}mv^2 + qV = [\text{constant}] \tag{12}$$

The electric field of a fixed point charge q' is an important case of a nonuniform electric field. The calculation of the potential energy of a charge q placed in this electric field is slightly difficult, but the final result for this electric energy is simple:

Potential energy of two interacting point charges ▶

$$U = \frac{1}{4\pi\epsilon_0}\frac{qq'}{r} \tag{13}$$

This electric potential energy is really a joint property of the two point charges q and q'. We can equally well regard it as the potential energy of the charge q placed in the electric field of the charge q', or as the potential energy of the charge q' placed in the electric field of the charge q. ***For two charges of equal signs, the electric potential energy (13) is positive*** and it decreases in inverse proportion to the distance. A decrease of potential energy with distance is characteristic of an attractive force. ***For two charges of opposite signs, the electric potential energy is negative,*** and this negative potential energy *increases* with distance (it increases from a large negative value to zero).

Note that the dependence of the electric potential energy of two point charges on distance is the same as the dependence of the gravitational potential energy of two point masses on distance [see Eq. (5.20)] — both are inversely proportional to the distance. This was to be expected, since the electric force is mathematically similar to the gravitational force — both are inversely proportional to the square of the distance.

The electrostatic potential produced by the point charge q' is $V = U/q$, that is,

Electrostatic potential of point charge ▶

$$V = \frac{1}{4\pi\epsilon_0}\frac{q'}{r} \tag{14}$$

Figure 18.4a is a plot of the electrostatic potential of a positive point charge q' vs. distance. Figure 18.4b is a plot of the electrostatic potential of a negative point charge q' vs. distance.

Equation (14) also gives the electrostatic potential outside of a hollow charged spherical shell and the electrostatic potential outside of a solid

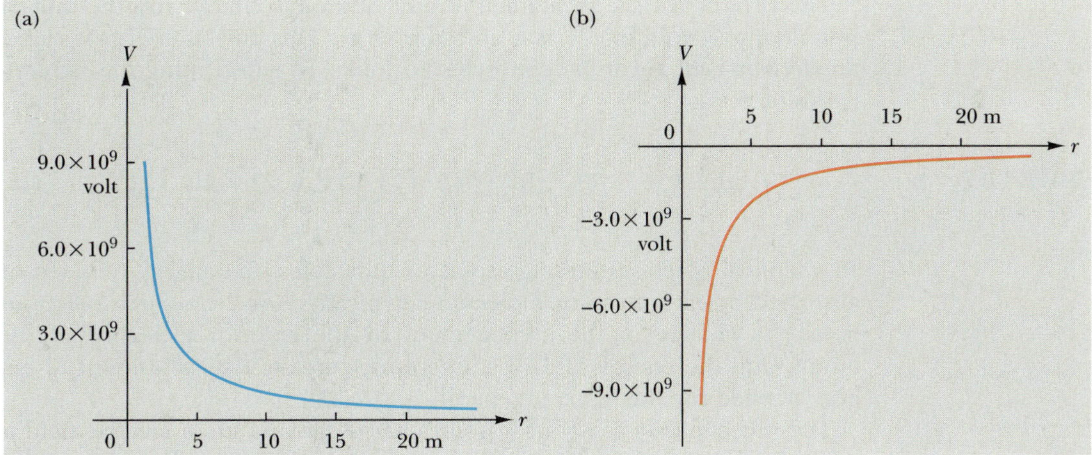

charged sphere. The electric field outside of such spherical charge distributions is the same as that of a point charge, and therefore the electrostatic potential is also the same as for a point charge.

FIGURE 18.4

(a) Electrostatic potential of a charge q' vs. distance. For this plot, the magnitude of the charge is take as $q' = 1$ C. (b) Electrostatic potential for a negative charge, $q' = -1$ C.

EXAMPLE 2. The electron in a hydrogen atom is at a distance of 0.53×10^{-10} m from the proton (see Fig. 18.5). The proton is a small ball of charge with $q' = e = 1.6 \times 10^{-19}$ C. What is the electrostatic potential generated by the proton at this distance? What is the potential energy of the electron?

SOLUTION: According to Eq. (14), the electrostatic potential produced by the proton is

$$V = \frac{1}{4\pi\epsilon_0}\frac{q'}{r} \tag{15}$$

$$= \frac{1}{4\pi \times 8.85 \times 10^{-12}\ \mathrm{C^2/N \cdot m^2}} \times \frac{1.6 \times 10^{-19}\ \mathrm{C}}{0.53 \times 10^{-10}\ \mathrm{m}}$$

$$= 27 \text{ volts}$$

The charge of the electron is $q = -e = -1.6 \times 10^{-19}$ C. From Eq. (6), the potential energy of the electron is then

$$U = qV = -e \times 27 \text{ volts} \tag{16}$$

$$= -1.6 \times 10^{-19}\ \mathrm{C} \times 27 \text{ volts} = -4.3 \times 10^{-18}\ \mathrm{J} \tag{17}$$

For the purposes of atomic physics, the joule is a rather large unit of energy and it is more convenient to leave the answer as in Eq. (16),

$$U = -27e \times \text{volt}$$

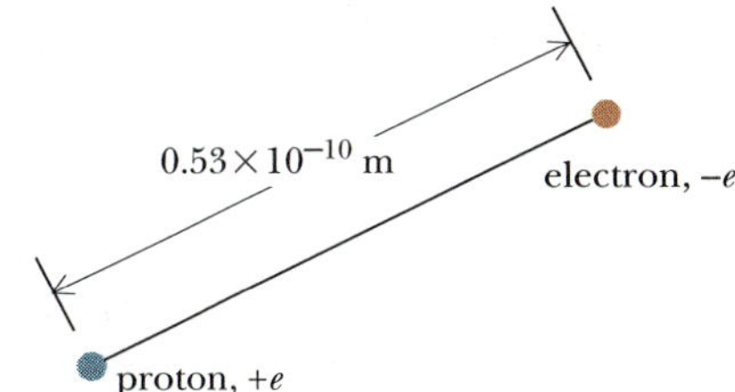

FIGURE 18.5

Proton ($+e$) and electron ($-e$) separated by a distance of 0.53×10^{-10} m.

The product of the fundamental unit of atomic charge and the unit of potential, $e \times$ volt, or eV, is a unit of energy. This unit of energy is called an ***electron-volt.*** It can be converted to joules by substituting the numerical value for e,

Electron-volt (eV) ▶

$$1 \text{ eV} = (1.60 \times 10^{-19} \text{ C}) \times (1 \text{ V}) = 1.60 \times 10^{-19} \text{ J} \tag{18}$$

In chemical reactions among atoms or molecules, the energy released or absorbed by each atom or molecule is typically 1 or 2 eV. Such reactions involve a change in the arrangement of the exterior electrons of the atoms, and the energy of 1 or 2 eV represents the typical amount of energy needed for this rearrangement.

The mechanical energy of a point charge moving in an electric field is the sum of the electric potential energy and the kinetic energy. The ***law of conservation of energy*** for the motion of a point charge q in the electric field of a fixed point charge q' takes the form

Conservation of energy ▶

$$E = K + U = \tfrac{1}{2}mv^2 + \frac{1}{4\pi\epsilon_0}\frac{qq'}{r} = [\text{constant}] \tag{19}$$

This total energy remains constant during the motion. As we saw in Chapter 4, the examination of the energy reveals some general features of the motion. Obviously, if qq' is negative (opposite charges, attractive Coulomb force), then Eq. (19) implies that whenever r increases, v must decrease, and conversely.

EXAMPLE 3. An electron is initially at rest at a very large distance from a proton. Under the influence of the electric attraction, the electron falls toward the proton, which remains (approximately) at rest. What is the speed of the electron when it has fallen to within 0.53×10^{-10} m of the proton? (See Fig. 18.6.)

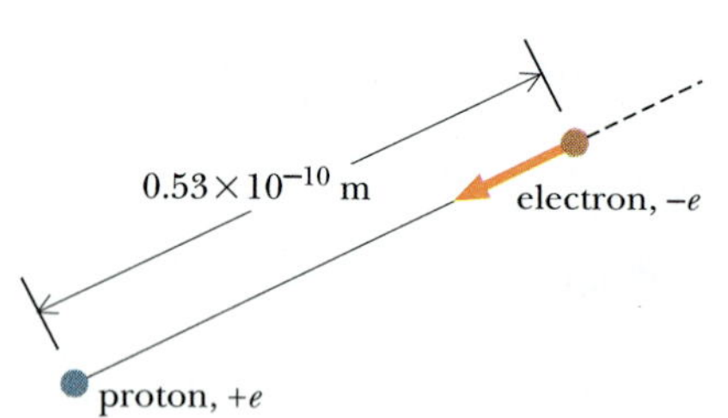

FIGURE 18.6
The approaching electron is instantaneously at a distance of 0.53×10^{-10} m from the proton.

SOLUTION: The potential energy of the electron is $qV = -eV$. The total energy is the sum of the potential and kinetic energies,*

$$E = \tfrac{1}{2}m_e v^2 - eV \tag{20}$$

This total energy is conserved. The initial value of the energy is zero; hence the final value of the energy must also be zero:

$$\tfrac{1}{2}m_e v^2 - eV = 0 \tag{21}$$

* Do not confuse the expression eV (for $e \times$ [potential]) with eV (for $e \times$ volt, or electron-volt).

Solving this for v we find

$$v = \sqrt{\frac{2eV}{m_e}} \tag{22}$$

According to Example 2, $V = 27$ volts for $r = 0.53 \times 10^{-10}$ m; with $m_e = 9.1 \times 10^{-31}$ kg, we then obtain

$$v = \sqrt{\frac{2 \times 1.6 \times 10^{-19}\ \mathrm{C} \times 27\ \mathrm{V}}{9.1 \times 10^{-31}\ \mathrm{kg}}}$$

$$= 3.1 \times 10^{6}\ \mathrm{m/s}$$

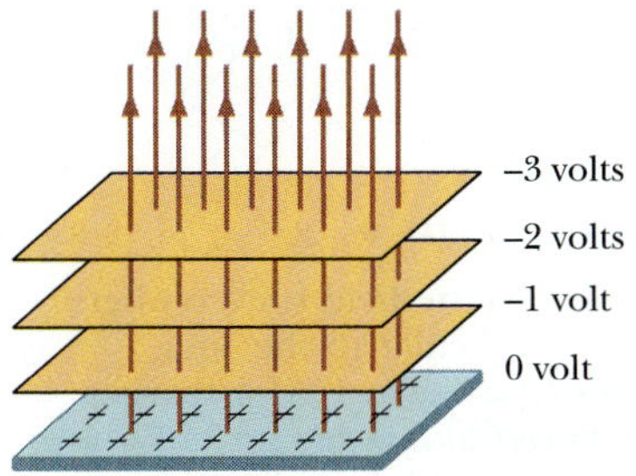

FIGURE 18.7

Equipotential surfaces for a very large sheet with a uniform distribution of charge. The equipotentials are flat parallel planes. This indicates that all points at the same height above the sheet are at the same potential.

18.2 EQUIPOTENTIALS

A mathematical surface on which the electrostatic potential has a fixed, constant value is called an ***equipotential surface.*** Figure 18.7 shows the equipotential surfaces belonging to the potential of a uniformly charged flat sheet — the equipotential surfaces are parallel planes. Figure 18.8 shows the equipotential surfaces belonging to the potential of a positive point charge — the equipotential surfaces are concentric spheres. The equipotential surfaces associated with a given potential provide a graphical representation of the potential.

Note that ***the electric field is everywhere perpendicular to the equipotentials*** (see Fig. 18.9). This is an immediate consequence of Eq. (10): if there were an electric field parallel to the equipotential surface, then this electric field would do work on a charge q transported along the surface, and, according to Eq. (10), the potential energy U would change along the surface. Since $\Delta V = \Delta U/q$, this would conflict with the requirement that the electrostatic potential V must not change along the equipotential surface. Consequently, the electric field in the direction parallel to the surface must be zero, and the electric field must be entirely perpendicular to the surface.

Conversely, if the electric field is everywhere perpendicular to a given surface, then this surface must be an equipotential surface. This, also, is a consequence of Eq. (10): if the electric field is zero in the direction parallel to the surface, then for a displacement parallel to the surface, the work is zero, and hence $\Delta U = 0$. This implies $\Delta V = 0$, and thus the potential is constant. Since we already know (from Section 17.4) that along the surface of any conductor in electrostatic equilibrium the electric field is perpendicular to the surface, we conclude that any conducting surface is an equipotential surface.

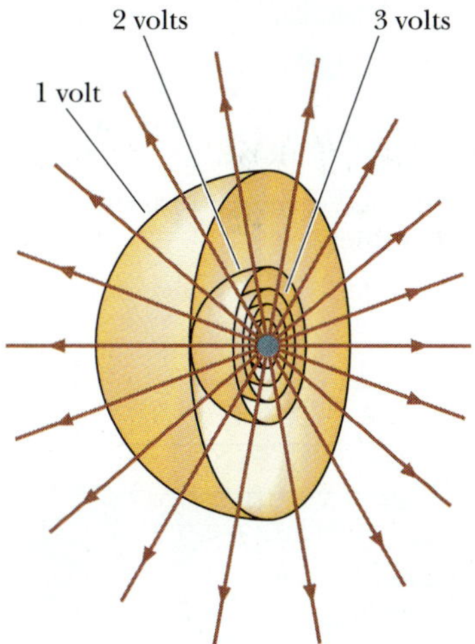

FIGURE 18.8

Equipotential surfaces for a positive point charge. The equipotentials are concentric spheres. This indicates that points at the same distance from the central charge are at the same potential.

Since the electric field in a conducting body in electrostatic equilibrium is zero, no work is performed by the electric field as a charge moves from

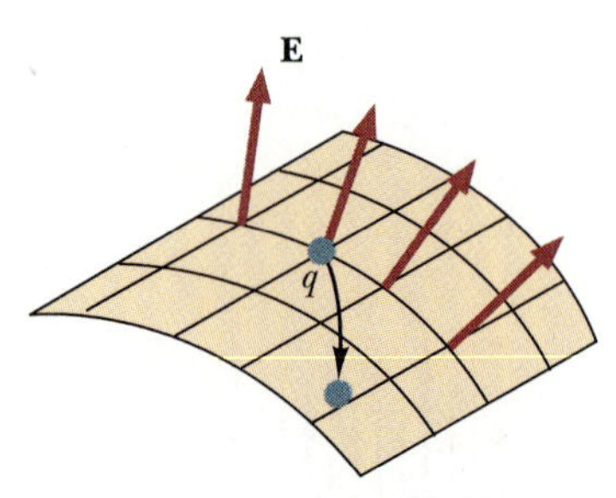

FIGURE 18.9
The electric field is everywhere perpendicular to the equipotential surface. If a charge q moves along this surface, the electric field does no work on it.

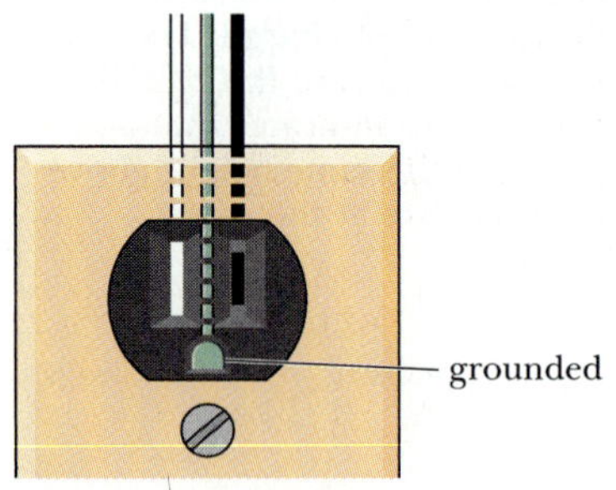

FIGURE 18.10
Grounded terminal of a three-hole outlet.

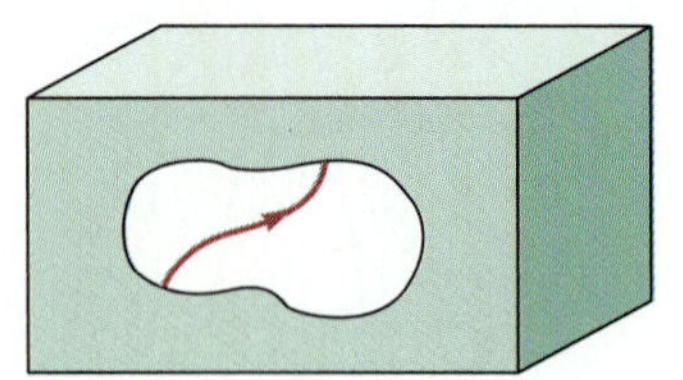

FIGURE 18.11
Empty cavity in a volume of conductor. The red line is a hypothetical field line in the cavity.

one point to another within the conducting body. Accordingly, Eq. (10) implies that the potential difference between any two points within a conducting body is zero. Thus, ***all points within a conducting body are at the same electrostatic potential.*** For instance, since the ground is a conductor, all points on the surface of the Earth or within the Earth are at the same electrostatic potential; that is, the surface of the Earth is an equipotential. In experiments with electric circuits, it is usually convenient to adopt the convention that the potential of the surface of the Earth is zero, $V = 0$. The surface of the Earth is said to be the ***electric ground,*** and any conductors connected to it are said to be grounded. For instance, the third terminal (round hole) in an ordinary household electric outlet (see Fig. 18.10) is grounded — it is connected to a plate or a rod buried in the ground outside the house. This grounded terminal is intended as a safety feature — if the insulation in your electric drill or some other appliance fails, the electric current from the "live" (flat) terminals can leak away harmlessly into the ground instead of leaking into your hands.

Given that a conducting body is at constant potential, we can prove an interesting theorem about the electric field in an empty, completely enclosed cavity within a conducting body (see Fig. 18.11): ***In a closed, empty cavity within a homogeneous conductor, the electric field is exactly zero.*** The proof of the theorem is by contradiction. If there were an electric field inside this cavity, then there would have to be field lines in the cavity. Consider one of these field lines. Since the cavity is empty and contains no charge, the field line cannot end or begin within the space of the cavity — it must therefore begin and end on the surface of the cavity, as shown in Fig. 18.11 (note that the field line cannot penetrate the conducting material, since the electric field is zero in this material). Now, suppose that we transport a positive charge q on a path that follows the field line from the point where it begins to the point where it ends. Since the electric field is everywhere tangent to the field line, the work that the electric field does on the charge q is positive; according to Eq. (10), then $U_1 > U_2$ and $V_1 > V_2$. But the electrostatic potentials at the beginning and the end of the field line cannot be different, since these potentials are evaluated at the surface of the conductor, and all points on a conductor are necessarily at the same potential. This contradiction establishes the theorem.

The absence of electric fields in closed cavities surrounded by conductors has important practical applications. Delicate electric instruments can be shielded from atmospheric electric fields, and other stray electric fields, by placing them in a box made of sheet metal. Such a box is called a ***Faraday cage.*** Often, the box is made of a mesh of fine copper wire rather than sheet metal; although such wire mesh does not have the perfect shielding properties of solid sheet metal, it provides good enough shielding for most purposes. In the same way, the sheet metal of an automobile provides fairly good shielding against atmospheric electric fields, such as the electric fields of lightning. The windows of the automobile leave some gaps in this shield and permit the penetration of some field lines, but the strength of the external atmospheric fields is attenuated to such an extent that the occupants are quite safe from atmospheric electrical discharges.

18.3 ENERGY OF A SYSTEM OF POINT CHARGES

The electrical potential energy of two point charges q and q' separated by a distance r is

$$U = \frac{1}{4\pi\epsilon_0} \frac{qq'}{r} \tag{23}$$

This potential energy can be regarded as the work required to move the charge q from infinite distance to within a distance r of the charge q'. It is a *mutual* potential energy which belongs to both q and q', that is, it is an energy associated with the relative configuration of the pair q, q'.

For configurations consisting of more than two charges, the net potential energy can be calculated by writing down a term similar to that in Eq. (23) for *each pair* of charges. For instance, if we are dealing with three charges q_1, q_2, and q_3 (see Fig. 18.12), we have three possible pairs (q_1, q_2), (q_2, q_3), and (q_1, q_3), and so the net potential energy is

$$U = \frac{1}{4\pi\epsilon_0} \frac{q_1 q_2}{r_{12}} + \frac{1}{4\pi\epsilon_0} \frac{q_2 q_3}{r_{23}} + \frac{1}{4\pi\epsilon_0} \frac{q_1 q_3}{r_{13}} \tag{24}$$

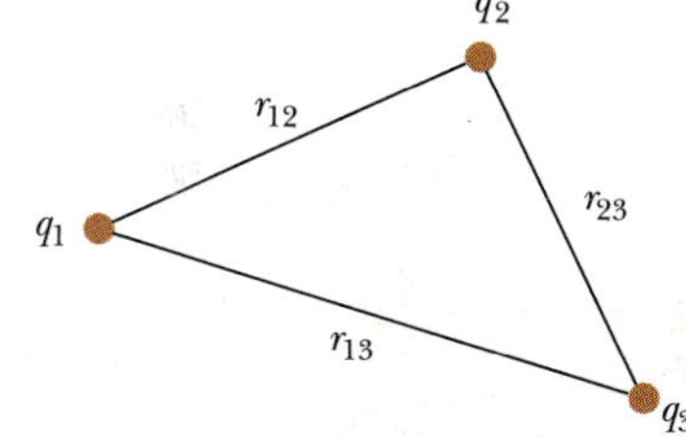

FIGURE 18.12
Three point charges.

where r_{12}, r_{23}, and r_{13} are the distances indicated in Fig. 18.12. This potential energy is the work required to assemble the charges into the final configuration shown in Fig. 18.12, starting from an initial condition of infinite separation of all charges.

Note that Eq. (24) is identically equal to

$$U = \frac{1}{2}\left(\frac{1}{4\pi\epsilon_0}\frac{q_2}{r_{12}} + \frac{1}{4\pi\epsilon_0}\frac{q_3}{r_{13}}\right)q_1 + \frac{1}{2}\left(\frac{1}{4\pi\epsilon_0}\frac{q_1}{r_{12}} + \frac{1}{4\pi\epsilon_0}\frac{q_3}{r_{23}}\right)q_2 \tag{25}$$

$$+ \frac{1}{2}\left(\frac{1}{4\pi\epsilon_0}\frac{q_1}{r_{13}} + \frac{1}{4\pi\epsilon_0}\frac{q_2}{r_{23}}\right)q_3$$

Here, the potential energy of each pair of charges appears twice, each time with a factor of $\frac{1}{2}$. The terms in parentheses are the electric potentials produced by the individual charges; for instance, the term in the first parentheses on the right side of Eq. (25) is the sum of the potentials produced by charge 2 and charge 3 at the position of charge 1. Equation (25) then leads to the following simplified expression for the energy in terms of potentials:

$$U = \tfrac{1}{2} q_1 V_{\text{others at 1}} + \tfrac{1}{2} q_2 V_{\text{others at 2}} + \tfrac{1}{2} q_3 V_{\text{others at 3}} \tag{26}$$

Where $V_{\text{others at 1}}$ is the electric potential produced at the position of charge 1 by the *other* charges (charges 2 and 3), that is,

$$V_{\text{others at 1}} = \frac{1}{4\pi\epsilon_0}\frac{q_2}{r_{12}} + \frac{1}{4\pi\epsilon_0}\frac{q_3}{r_{13}} \tag{27}$$

and similarly for $V_{\text{others at 2}}$ and $V_{\text{others at 3}}$.

By means of a generalization of this argument, we can easily demonstrate that for a configuration consisting of any number of point charges, the electric potential energy is

Potential energy of a system of point charges ▶

$$U = \tfrac{1}{2}q_1 V_{\text{others at 1}} + \tfrac{1}{2}q_2 V_{\text{others at 2}} + \tfrac{1}{2}q_3 V_{\text{others at 3}} + \tfrac{1}{2}q_4 V_{\text{others at 4}} + \ldots \tag{28}$$

This expression gives the work that must be done to bring the point charges to their final positions, starting from initial positions at very large distances from one another. This work is energy that has been stored in the configuration of the charges, energy which can be recovered by allowing the charges to move back to their initial positions at very large distances.

18.4 Energy of a System of Charged Conductors

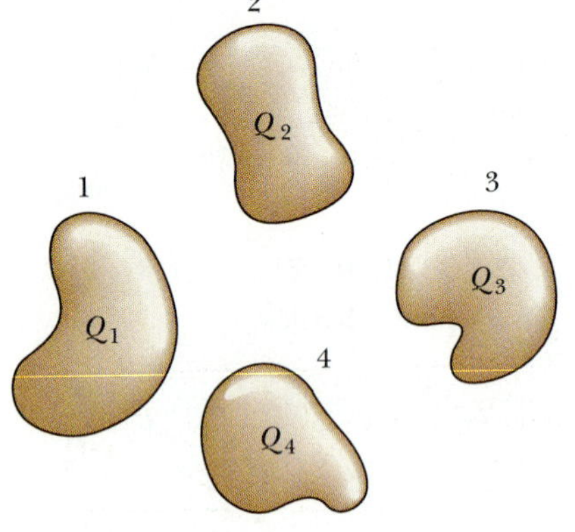

FIGURE 18.13 Four conducting bodies carrying electric charges.

In practical applications, the point charges are usually placed on conductors, each conductor carrying many point charges. By a slight modification of Eq. (28), we can calculate the potential energy of a system of charges placed on conductors. For instance, Fig. 18.13 shows several conducting bodies — such as bodies of metal — with charges Q_1, Q_2, Q_3, . . . distributed over their surfaces. We recall that on each conducting body, the potential has a constant value over the entire volume of the body. Suppose that these electrostatic potentials of the bodies are V_1, V_2, V_3, . . . The charge Q_1 on conductor 1 consists of many small pieces of charge distributed over the surface of conductor 1. Each of these small pieces of charge is then at potential V_1. This potential V_1 acting on a given small piece of charge can be regarded as due to the *other* small pieces of charge (on conductor 1 and on the other conductors), because the given piece of charge makes only an insignificant contribution to the potential V_1. Hence, according to Eq. (28), the electric potential energy associated with each small piece of charge ΔQ_1 on conductor 1 is $\frac{1}{2}(\Delta Q_1)V_1$, and therefore the potential energy associated with all the small pieces of charge on conductor 1 together is $\frac{1}{2}Q_1V_1$. Since similar arguments apply to the other conductors, we conclude that the net electric potential energy is

Potential energy of a system of charged conductors ▶

$$U = \tfrac{1}{2}Q_1V_1 + \tfrac{1}{2}Q_2V_2 + \tfrac{1}{2}Q_3V_3 + \ldots \tag{29}$$

EXAMPLE 4. The metallic sphere on the top of a small Van de Graaff generator has a radius of 0.15 m. Suppose that the sphere carries a charge of 2.5×10^{-6} C uniformly distributed over its surface. How much electric energy is stored in this charge distribution?

SOLUTION: In this problem, there is only one conductor, the metallic sphere. Accordingly, we need only the first term in Eq. (29):

$$U = \tfrac{1}{2}Q_1V_1$$

or, with $Q_1 = Q$ and $V_1 = V$,

$$U = \tfrac{1}{2}QV \tag{30}$$

The potential outside a spherically symmetric charge distribution is given by Eq. (14). At $r = R$, this potential is

$$V = \frac{1}{4\pi\epsilon_0}\frac{Q}{R}$$

According to Eq. (30), the electric potential energy is then

$$U = \tfrac{1}{2}QV = \tfrac{1}{2}Q\frac{1}{4\pi\epsilon_0}\frac{Q}{R} = \frac{1}{8\pi\epsilon_0}\frac{Q^2}{R} \tag{31}$$

With $Q = 2.5 \times 10^{-6}$ C and $R = 0.15$ m, this gives

$$U = \frac{1}{8\pi \times 8.85 \times 10^{-12}\,\text{C}^2/\text{N}\cdot\text{m}^2}\frac{(2.5 \times 10^{-6}\,\text{C})^2}{0.15\ \text{m}} = 0.19\ \text{J}$$

EXAMPLE 5. Two very large, parallel metallic plates of area A are separated by a distance d. Charges $+Q$ and $-Q$ are placed on the plates, respectively (see Fig. 18.14). What is the electric potential energy?

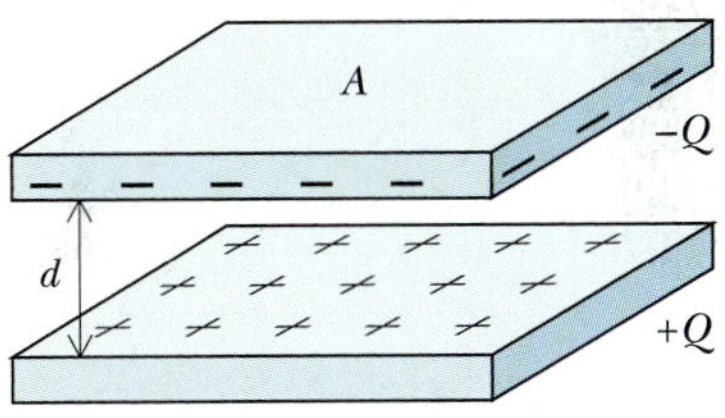

FIGURE 18.14
Two very large, parallel metallic plates with opposite electric charges.

SOLUTION: Here, we are dealing with two conductors, with $Q_1 = Q$ and $Q_2 = -Q$. Thus, we need two terms in Eq. (29):

$$U = \tfrac{1}{2}Q_1V_1 + \tfrac{1}{2}Q_2V_2 = \tfrac{1}{2}QV_1 - \tfrac{1}{2}QV_2 \tag{32}$$

$$= \tfrac{1}{2}Q(V_1 - V_2)$$

To proceed, we must evaluate the potential difference $V_1 - V_2$ between the plates. The electric field in the region between the plates is approximately the field of a pair of infinite charged sheets. According to Eq. (17.12), the electric field is

$$E = \frac{\sigma}{\epsilon_0}$$

where σ is the charge per unit area on the plate. Since the total charge on a plate is Q and the area is A, the charge per unit area is $\sigma = Q/A$

and

$$E = \frac{Q}{\epsilon_0 A} \tag{33}$$

This expression fails near the edges of the plates, where there is an electric fringing field that is not constant. But if the plates are very large, then the edge region is only a very small fraction of the total region between the plates and we can ignore this region without introducing excessive errors in our calculation.

With the constant electric field (33) we can calculate the potential difference between two plates, using Eq. (7) for the potential,

$$V_1 - V_2 = -Ey_1 + Ey_2 = E(y_2 - y_1) = Ed$$

from which

$$V_1 - V_2 = \frac{Q}{\epsilon_0 A} d \tag{34}$$

Substituting this into Eq. (32), we obtain the result

$$U = \tfrac{1}{2} Q \times \frac{Q}{\epsilon_0 A} d = \frac{1}{2} \frac{Q^2 d}{\epsilon_0 A} \tag{35}$$

Equation (35) can be rewritten in the following interesting way, in terms of the electric field:

$$U = \tfrac{1}{2} \epsilon_0 \left(\frac{Q}{\epsilon_0 A} \right)^2 Ad \tag{36}$$

$$= \tfrac{1}{2} \epsilon_0 E^2 \times Ad \tag{37}$$

Here the product Ad is the volume of the space between the plates, that is, the volume filled with electric field. Thus, the electric potential energy has the form

$$U = \tfrac{1}{2} \epsilon_0 E^2 \times [\text{volume}] \tag{38}$$

This formula suggests that the energy is distributed over the regions of space in which there is electric field, with an energy density, or amount of energy per unit volume, of $\frac{1}{2}\epsilon_0 E^2$. According to this formula, ***the energy is concentrated in the regions of space where the electric field is strong.***

Note that although Eq. (38) expresses the energy in terms of the electric field and suggests that the energy is located in the electric field, Eq. (29) expresses the energy in terms of the electric charges and suggests that the energy is located on these charges. Thus, these two equations, which are mathematicaly equivalent, suggest conflicting physical interpretations. To decide which alternative is correct, we need some extra information. The clue is the existence of electric fields that are independent of electric charges. As we will see in Chapter 24, radio waves and light waves consist of electric and magnetic fields traveling through space. Such fields are originally created by electric charges, but they persist even when the charges are eliminated. For instance, a radio wave or a light beam continues to travel through space long after the radio transmitter has been shut down or the candle has been snuffed out — obviously, the energy of a radio wave resides in the radio wave itself, in its electric and magnetic fields, and not in the electric charges in the antenna of the radio transmitter. We can then argue that if energy is associated with the traveling electric fields of a radio wave, energy should also be associated with the electric fields of a static charge distribution. A more detailed calculation confirms that the expression we found above for the energy density in the special case of a uniform electric field is valid in general, for uniform or nonuniform electric fields. In any electric field (in otherwise empty space) the energy density is

$$[\text{energy density}] = \tfrac{1}{2}\epsilon_0 E^2 \qquad (39)$$

◀ **Energy density in the electric field**

EXAMPLE 6. What is the energy density in the electric field of a thundercloud, where $E = 2 \times 10^6$ V/m?

SOLUTION: According to Eq. (39),

$$[\text{energy density}] = \tfrac{1}{2}\epsilon_0 E^2$$

$$= \tfrac{1}{2} \times 8.85 \times 10^{-12}\ \text{C}^2/\text{N}\cdot\text{m}^2 \times (2 \times 10^6\ \text{V/m})^2$$

$$= 18\ \text{J/m}^3$$

18.5 CAPACITANCE

Any arrangement of conductors that is used to store electric charge is called a capacitor, or condenser. Since work must be done during the charging process, the capacitor will also store electric potential energy whenever it stores electric charge. In our electronic technology, capacitors find widespread application — they are part of the circuitry of radios, computers, automobile ignition systems, and so on.

As a first example of a capacitor, consider an isolated metallic sphere of radius R (see Fig. 18.15). Charge can be stored on this sphere. If the amount of charge placed on the sphere is Q, then the potential of the sphere will be

$$V = \frac{1}{4\pi\epsilon_0}\frac{Q}{R} \qquad (40)$$

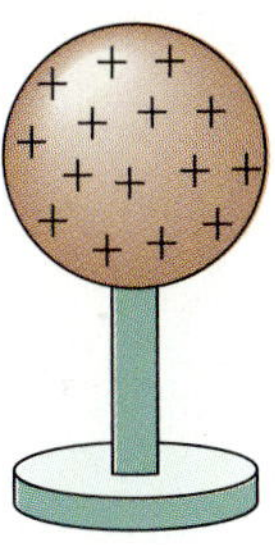

FIGURE 18.15
An isolated metallic sphere used to store electric charge.

Thus, the amount of charge stored on the sphere is directly proportional to the potential.

This proportionality holds in general for any conductor of arbitrary shape. The charge on the conductor produces an electric field whose strength is directly proportional to the amount of charge (twice the charge produces twice the field strength), and the electric field yields a potential which is directly proportional to the field strength (twice the field strength yields twice the potential); hence, charge and potential are proportional. We write this relationship as

Capacitance ▶

$$Q = CV \tag{41}$$

where C is the constant of proportionality. This constant is called the ***capacitance*** of the conductor. ***The capacitance is large if the conductor is capable of storing a large amount of charge at a low potential.*** For instance, according to Eq. (40), the capacitance of a spherical conductor is

$$C = \frac{Q}{V} = \frac{Q}{(1/4\pi\epsilon_0)\, Q/R} = 4\pi\epsilon_0 R \tag{42}$$

Thus, the capacitance of the sphere increases with its radius.

The unit of capacitance is the farad (F):

Farad (F) ▶

$$1 \text{ farad} = 1 \text{ F} = 1 \text{ coulomb/volt} \tag{43}$$

This unit of capacitance is rather large; in practice, electrical engineers prefer the ***microfarad*** and the ***picofarad.*** A microfarad equals 10^{-6} farad ($1\ \mu\text{F} = 10^{-6}$ F) and a picofarad equals 10^{-12} farad ($1 \text{ pF} = 10^{-12}$ F).

Note that $1 \text{ F} = 1 \text{ C/V} = 1 \text{ C}^2/\text{N}\cdot\text{m}$, so the constant ϵ_0 can be written

$$\epsilon_0 = 8.85 \times 10^{-12} \frac{\text{C}^2}{\text{N}\cdot\text{m}^2} = 8.85 \times 10^{-12} \text{ F/m} \tag{44}$$

The latter expression is the one usually listed in tables of physical constants.

EXAMPLE 7. What is the capacitance of an isolated metallic sphere of radius 20 cm?

SOLUTION: According to Eq. (42),

$$C = 4\pi\epsilon_0 R = 4\pi \times 8.85 \times 10^{-12} \text{ F/m} \times 0.20 \text{ m}$$

$$= 2.2 \times 10^{-11} \frac{\text{coulomb}}{\text{volt}} = 2.2 \times 10^{-11} \text{ F} = 22 \text{ pF}$$

EXAMPLE 8. What is the capacitance of the Earth, regarded as a conducting sphere?

SOLUTION: The radius of the Earth is 6.4×10^6 m; therefore

$$C = 4\pi\epsilon_0 R = 4\pi \times 8.85 \times 10^{-12} \text{ F/m} \times 6.4 \times 10^6 \text{ m}$$

$$= 7.1 \times 10^{-4} \text{ F}$$

As capacitances go, this is a rather large capacitance. However, note that it takes a charge of only about 10^{-3} coulomb to alter the potential of the Earth by 1 volt.

The most common variety of capacitor consists of *two* metallic conductors, insulated from each other and carrying opposite amounts of charge $\pm Q$. The capacitance of such a pair of conductors is defined in terms of the *difference* of potential ΔV between the two conductors:

$$Q = C\Delta V \tag{45}$$

◀ Capacitance of a pair of conductors

In this expression, both Q and ΔV are taken as positive quantities. Note that the quantity Q is *not* the net charge in the capacitor, but the magnitude of the charge on each plate. The net charge in the two-conductor capacitor is zero, since the two conductors carry opposite amounts of charge.

Figure 18.16 shows such a two-conductor capacitor consisting of two very large, parallel metallic plates, each of area A, separated by a distance d. The plates carry charges $+Q$ and $-Q$, respectively, on their facing surfaces. The potential difference between the plates is [see Eq. (34)]

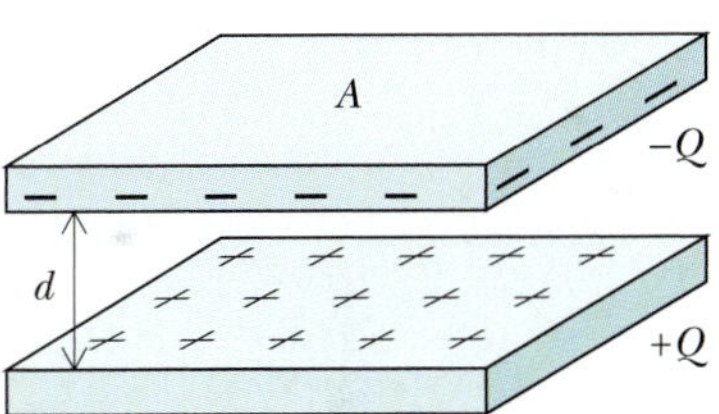

FIGURE 18.16

Two parallel plates, with charges $+Q$ and $-Q$.

$$\Delta V = Ed = \frac{Qd}{\epsilon_0 A} \tag{46}$$

Hence the capacitance of this configuration is

$$C = \frac{Q}{\Delta V} = \frac{Q}{Qd/\epsilon_0 A}$$

or

$$C = \frac{\epsilon_0 A}{d} \tag{47}$$

◀ Capacitance of a parallel-plate capacitor

From this formula we recognize that to store a large amount of charge at a low potential, we want a large plate area A, but a small plate separation d.

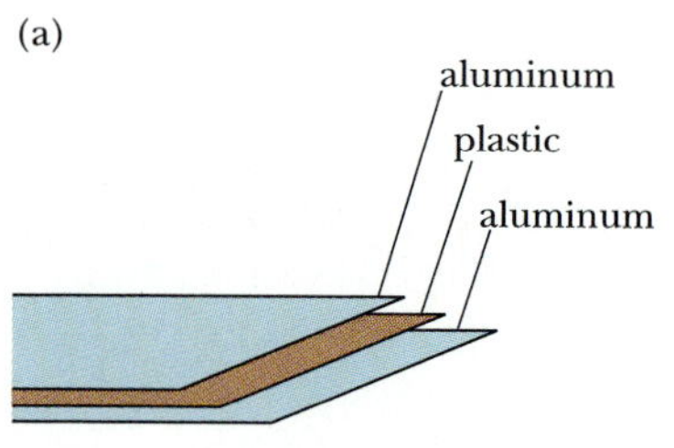

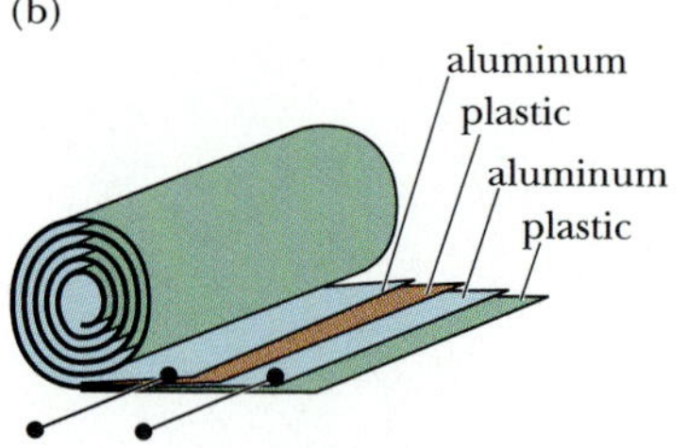

FIGURE 18.17

(a) Sheets of aluminum foil separated by a sheet of plastic. (b) Rolled capacitor.

Parallel-plate capacitors are usually manufactured out of two parallel sheets of aluminum foil, a few centimeters wide, but several meters long. The sheets are placed very close together, but kept from contact by a thin sheet of plastic sandwiched between them (see Fig. 18.17a). For convenience, the entire sandwich is covered with another sheet of plastic and rolled up like a roll of toilet paper (see Fig. 18.17b).

EXAMPLE 9. A parallel-plate capacitor consists of two strips of aluminum foil, each with an area of 0.20 m², separated by a distance of 0.10 mm. The space between the foils is empty. The two strips are connected to the poles of a battery which produces a potential difference of 200 volts between them. What is the capacitance of this capacitor? What is the electric charge on each plate? What is the strength of the electric field between the plates?

SOLUTION: According to Eq. (47), the capacitance is

$$C = \frac{\epsilon_0 A}{d} = \frac{8.85 \times 10^{-12}\ \text{F/m} \times 0.20\ \text{m}^2}{1.0 \times 10^{-4}\ \text{m}} = 1.8 \times 10^{-8}\ \text{F}$$

$$= 0.018\ \mu\text{F}$$

The charge on each plate is

$$Q = C\Delta V = 1.8 \times 10^{-8}\ \text{F} \times 200\ \text{volts} = 3.6 \times 10^{-6}\ \text{coulomb}$$

and the electric field between the plates is

$$E = \frac{\Delta V}{d} = \frac{200\ \text{volts}}{1.0 \times 10^{-4}\ \text{m}} = 2.0 \times 10^{6}\ \text{volts/m}$$

As already mentioned above, capacitors store not only electric charge, but also electric potential energy. Consider a two-conductor capacitor, with charges $\pm Q$ on its plates. The electric potential energy can be calculated directly from Eq. (32):

$$U = \tfrac{1}{2}Q(V_1 - V_2) \tag{48}$$

where V_1 and V_2 are the potentials of the plates. Thus, the potential energy can be expressed in terms of the charge and the potential difference,

Potential energy in capacitor ▸

$$U = \tfrac{1}{2}Q\Delta V \tag{49}$$

By means of the definition of capacitance, $Q = C\Delta V$, this can be put in the alternative forms

$$U = \tfrac{1}{2}C(\Delta V)^2 \tag{50}$$

or

$$U = \tfrac{1}{2}Q^2/C \tag{51}$$

EXAMPLE 10. What is the amount of electric potential energy in the capacitor described in Example 9?

SOLUTION: With $Q = 3.6 \times 10^{-6}$ coulomb and $\Delta V = 200$ volts,

$$U = \tfrac{1}{2}Q\Delta V = \tfrac{1}{2} \times 3.6 \times 10^{-6} \text{ coulomb} \times 200 \text{ volts} = 3.5 \times 10^{-4} \text{ J}$$

Variable capacitors are used in the tuning circuits of radios. These capacitors consist of a fixed plate (or plates) and a movable plate (or plates). When you turn the tuning knob, you shift the moving plate parallel to the fixed plate, and you decrease or increase the amount of overlap of the plates, and thereby change the capacitance. Another special kind of capacitor is used in the ***capacitor microphone,*** illustrated in Fig. 18.18. The flexible diaphragm of this microphone forms one plate of the capacitor and a rigid disk forms the other plate. When a sound wave impinges on the diaphragm, the periodic fluctuations of the air pressure alternately push and pull the diaphragm toward and away from the rigid plate. The change of distance between the plates produces a change in capacitance. Since the plates are connected to a battery which maintains a steady potential difference between the plates, the change in capacitance results in a change in the amount of electric charge on the plates. The charge that leaves the capacitor plates flows along the wires forming an electric current. Thus, the capacitor microphone transforms a sound signal into an electric signal, which can be fed into an amplifier and from there into a loudspeaker or a tape recorder. This kind of microphone has a good sensitivity to a wide range of frequencies, and it is often used in recording studios.

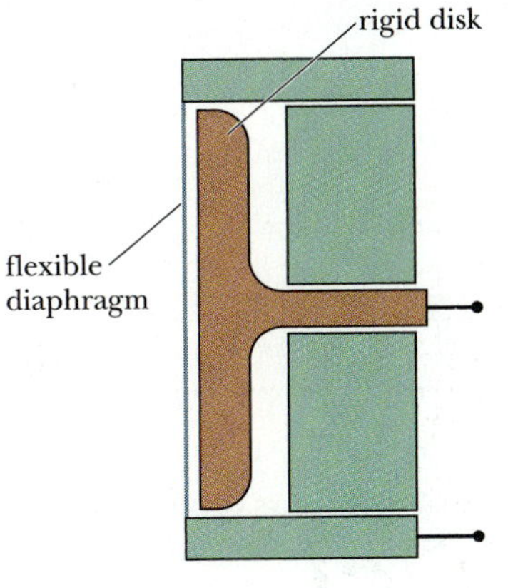

FIGURE 18.18
A capacitor microphone.

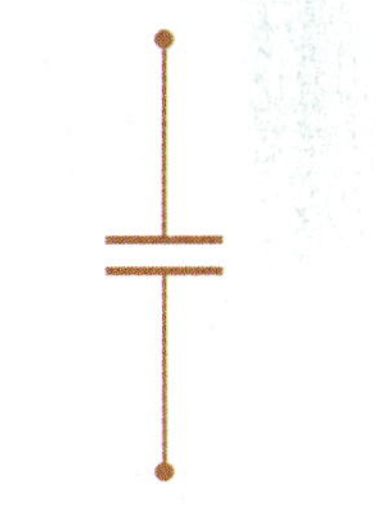

FIGURE 18.19
Symbol for a capacitor in a circuit diagram.

18.6 CAPACITORS IN COMBINATION

Capacitors used in practical applications in electric circuitry commonly are of the two-conductor variety. Schematically, such capacitors are represented in a circuit diagram as two parallel lines with terminals attached to their middles (see Fig. 18.19). In a circuit, several such capacitors are often wired together, and it is then necessary to calculate the net capacitance of the combination. The two simplest ways of wiring capacitors together are in *parallel* and in *series.*

Figure 18.20 shows two capacitors connected in ***parallel.*** If charge is fed into this combination via the two terminals, some of the charge will be stored on the first capacitor and some on the second. The net capacitance of the combination can be found as follows: Since the corresponding

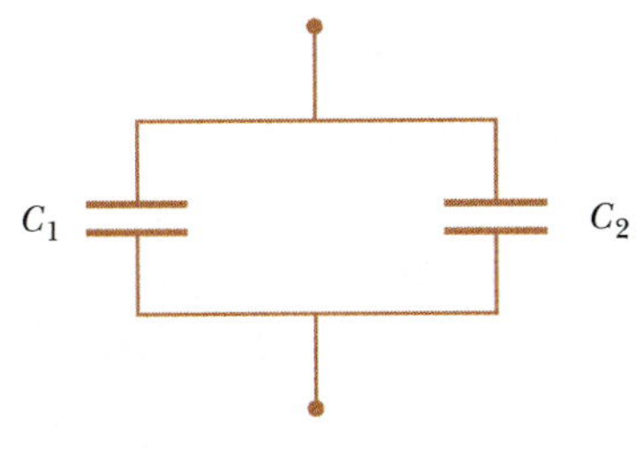

FIGURE 18.20
Two capacitors connected in parallel.

plates of the capacitors are joined by conductors, the potentials of corresponding plates must be equal and the potential differences across both capacitors must also be equal,

$$\Delta V = \frac{Q_1}{C_1} \quad \text{and} \quad \Delta V = \frac{Q_2}{C_2} \tag{52}$$

Therefore the net charge for the capacitor combination can be expressed as

$$Q = Q_1 + Q_2 = C_1 \Delta V + C_2 \Delta V \tag{53}$$

that is,

$$Q = (C_1 + C_2)\Delta V \tag{54}$$

Comparing this with the definition for capacitance given in Eq. (45), we recognize that the combination is equivalent to a single capacitor of capacitance

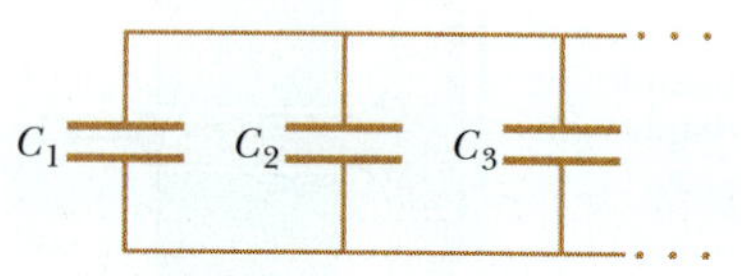

FIGURE 18.21
Several capacitors connected in parallel.

$$C = C_1 + C_2 \tag{55}$$

Thus, the net capacitance of the parallel combination is simply the sum of the individual capacitances.

It is easy to obtain a similar result for any number of capacitors connected in parallel (see Fig. 18.21). The net capacitance for such a parallel combination is

Parallel combination of capacitors ▶

$$C = C_1 + C_2 + C_3 + \ldots \tag{56}$$

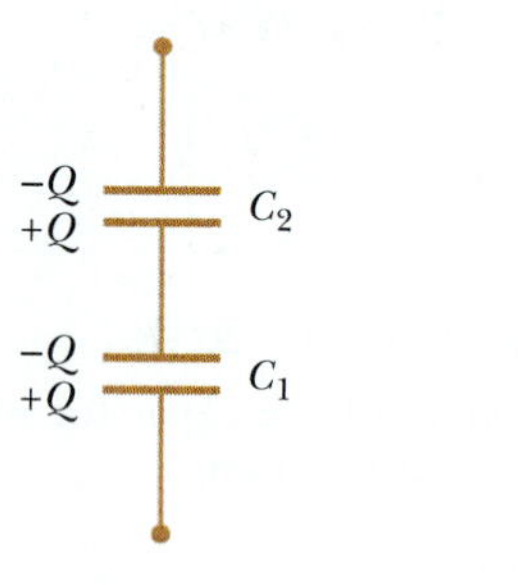

FIGURE 18.22
Two capacitors connected in series.

Next, we consider the alternative way of connecting capacitors. Figure 18.22 shows two capacitors connected in ***series.*** Any charge fed into this combination via the two outside terminals will have to remain on the *outside* plates [the lower plate of the first capacitor (C_1) and the upper plate of the second (C_2); see Fig. 18.22]. Thus, the bottom plate will have a charge $+Q$ and the top plate a charge $-Q$. But these charges on the outside plates will induce charges on the inside plates (the upper plate of the first capacitor and the lower plate of the second). The charge $+Q$ on the bottom plate will attract electrons to the facing plate and a charge $-Q$ will accumulate on this plate. Corresponding to the excess electrons on the upper plate of the first capacitor, there will be a deficit of electrons on the lower plate of the second capacitor, and a charge $+Q$ will accumulate there. The capacitance of the combination can then be found as follows. The individual potential differences across the two capacitors are

$$\Delta V_1 = \frac{Q}{C_1} \quad \text{and} \quad \Delta V_2 = \frac{Q}{C_2} \tag{57}$$

The net potential difference between the external terminals is the sum of the individual potential differences across the two capacitors, since the net work that the electric force does on a charge that we transport across both capacitors is the sum of the individual amounts of work. Thus,

$$\Delta V = \Delta V_1 + \Delta V_2 = \frac{Q}{C_1} + \frac{Q}{C_2}$$

from which

$$\Delta V = Q\left(\frac{1}{C_1} + \frac{1}{C_2}\right) \qquad (58)$$

Comparing this, again, with the definition of capacitance given by Eq. (45), we see that the combination is equivalent to a single capacitor with

$$\frac{1}{C} = \frac{1}{C_1} + \frac{1}{C_2} \qquad (59)$$

or

$$C = \frac{C_1 C_2}{C_1 + C_2} \qquad (60)$$

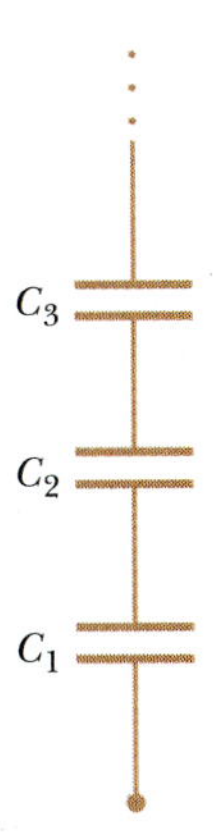

FIGURE 18.23
Several capacitors connected in series.

According to Eq. (59), the inverse of the net capacitance of the series combination is obtained by taking a sum of inverses. Note that the net capacitance is always *less* than the individual capacitances; for example, if $C_1 = C_2$, then $C = \frac{1}{2}C_1 = \frac{1}{2}C_2$.

A similar result applies to any number of capacitors connected in series (see Fig. 18.23). The net capacitance of such a series combination is given by

$$\frac{1}{C} = \frac{1}{C_1} + \frac{1}{C_2} + \frac{1}{C_3} + \ldots \qquad (61)$$

◄ **Series combination of capacitors**

EXAMPLE 11. Two capacitors of 4 μF each are connected in parallel and then a third capacitor of 6 μF is connected in series with the first two (see Fig. 18.24a). What is the net capacitance of this combination?

SOLUTION: First we calculate the combined capacitance of the two 4-μF capacitors. Since they are connected in parallel, Eq. (55) tells us that their combined capacitance is the sum 4 μF + 4 μF = 8μF.

Next, we consider the combined capacitance of this effective 8-μF capacitor connected in series with the 6-μF capacitor (see Fig. 18.24b). For this series combination, Eq. (59) tells us that the net capacitance is given by

$$\frac{1}{C} = \frac{1}{8\,\mu\text{F}} + \frac{1}{6\,\mu\text{F}} = \frac{6\,\mu\text{F} + 8\,\mu\text{F}}{8\,\mu\text{F} \times 6\,\mu\text{F}} = \frac{14}{48\,\mu\text{F}}$$

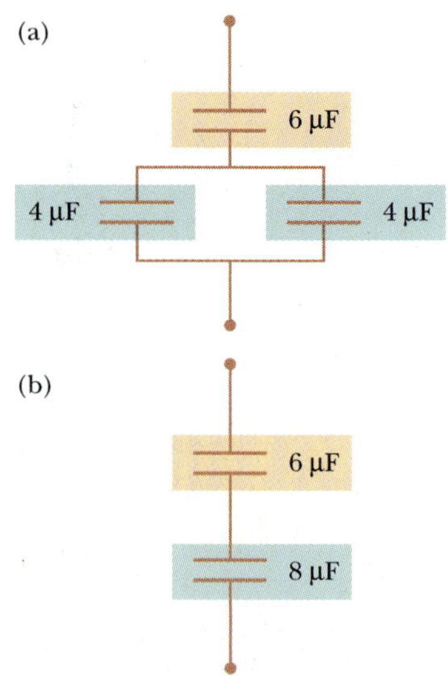

FIGURE 18.24

(a) The two 4-μF capacitors (blue) are connected in parallel, and the 6-μF capacitor (orange) is connected in series with these two. (b) The two 4-μF capacitors are equivalent to a single 8-μF capacitor, which is connected in series to the 6-μF capacitor.

so

$$C = \frac{48}{14}\ \mu\text{F} = 3.4\ \mu\text{F}$$

This is the net capacitance of the entire combination.

COMMENTS AND SUGGESTIONS: When dealing with a circuit containing several capacitors connected in some complicated manner, proceed in two steps. In the first step, look for groups of capacitors that form simple parallel or series combinations (such as the group of the two 4-μF capacitors in Fig. 18.24a). Evaluate the capacitance of each such group. In the second step, see how these effective group capacitances are connected to each other (as in Fig. 18.24b), and evaluate the net capacitance of the combination of the groups.

18.7 Dielectrics

So far, in dealing with problems in electrostatics, we have assumed that the space surrounding the electric charges consisted of a vacuum, which has no effect on the electric field, or of air, which has only an insignificant effect on the electric field. However, in dealing with capacitors, we must take the effects of the medium into account. The space between the plates of the capacitor is usually filled with an electrical insulator, or ***dielectric,*** which drastically changes the electric field from what it would be in a vacuum: ***the dielectric reduces the strength of the electric field.***

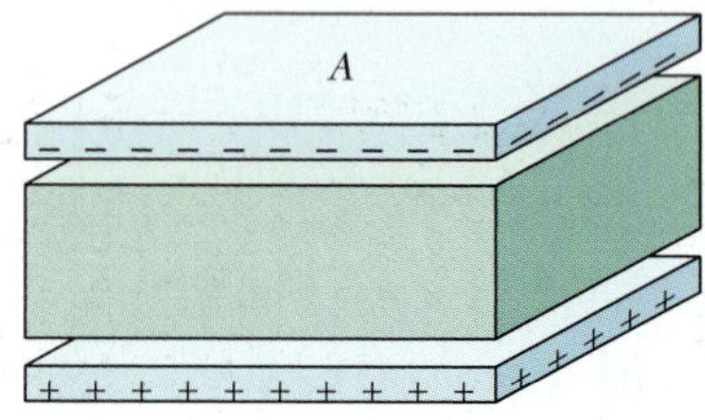

FIGURE 18.25

A slab of dielectric between the plates of a capacitor.

To understand this, consider a parallel-plate capacitor whose plates carry some amount of charge per unit area. Suppose that a slab of dielectric, such as glass or nylon, fills most of the space between the plates (see Fig. 18.25). This dielectric contains a large number of atomic nuclei and electrons but, of course, these positive and negative charges balance each other, so the material is electrically neutral. ***In an insulator, all the charges are bound*** — the electrons are confined within their atoms or molecules and they cannot wander about as in a conductor. Nevertheless, in response to the force exerted by the electric field, the charges will move very slightly without leaving their atoms. The positive charges in the atom move in the direction of the electric field; the negative charges move in a direction opposite to that of the electric field. These opposite displacements slightly separate the positive and negative charges and thereby create electric dipoles within the dielectric. In most dielectrics, the magnitudes of the charge separations and the magnitudes of the dipole moments are directly proportional to the strength of the electric field; such dielectrics are said to be ***linear.***

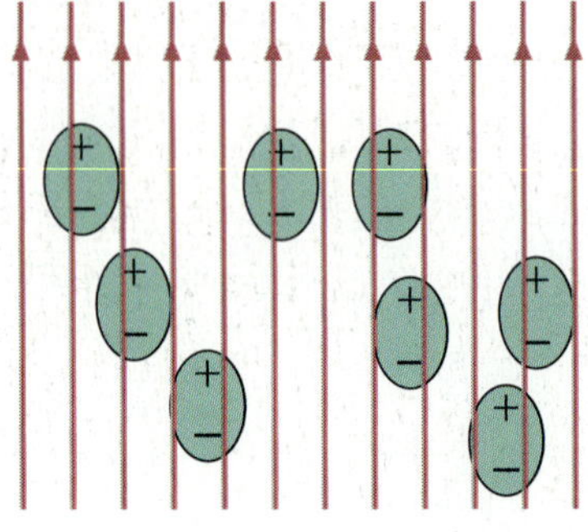

FIGURE 18.26

The electric field produces a distortion of the molecules.

The details of the mechanism of displacement and separation of charge depend on the dielectric. In some dielectrics — such as glass, nylon, and other solids — the creation of dipole moments involves a distortion of the molecules or atoms. By tugging on the electrons and protons in opposite directions, the electric field stretches the molecule while producing a charge separation within it (see Fig. 18.26). In other dielectrics — such as distilled water* or carbon dioxide — the charge separation results mainly from a realignment of existing dipoles. In such dielectrics, the molecules have permanent dipole moments which are randomly oriented when the dielectric is left to itself. The randomness of the orientation of the dipoles means that, on the average, there is no charge separation in the dielectric. But when the dielectric is placed in an electric field, the permanent dipoles experience a torque that tends to align them with the electric field (see Fig. 18.27). Random thermal motions oppose this alignment, and the molecules achieve an average equilibrium state in which the average amount of alignment is (approximately) proportional to the strength of the electric field. This average alignment is equivalent to an average charge separation.

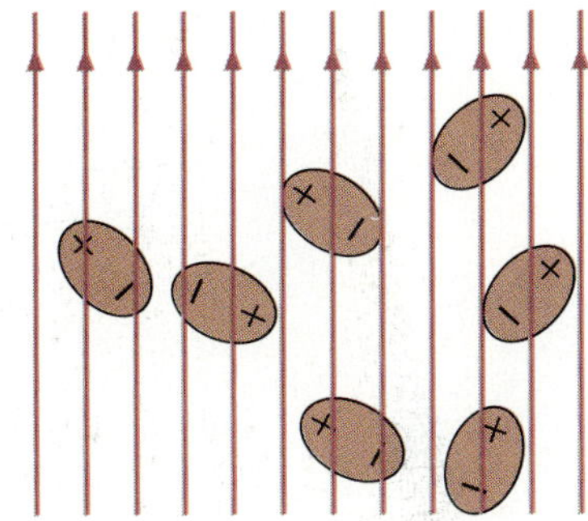

FIGURE 18.27

The electric field produces a (partial) alignment of already distorted molecules.

The displacement of the negative and the positive charges in opposite directions implies that the positive and negative charge distributions of the dielectric cease to overlap precisely (see Fig. 18.28). Consequently, there will be an excess of positive charge on one surface of the slab of dielectric and an excess of negative charge on the opposite surface. The slab of dielectric is then said to be ***polarized.*** These surface charges act just like a pair of parallel sheets of positive and negative charge; between the sheets, the charges generate an electric field that is *opposite* to the original applied electric field. The total electric field, consisting of the sum of the field of the free charges on the conducting plates plus the field of the bound charges on the dielectric surfaces, is therefore *smaller* than the field of the free charges alone (see Fig. 18.29).

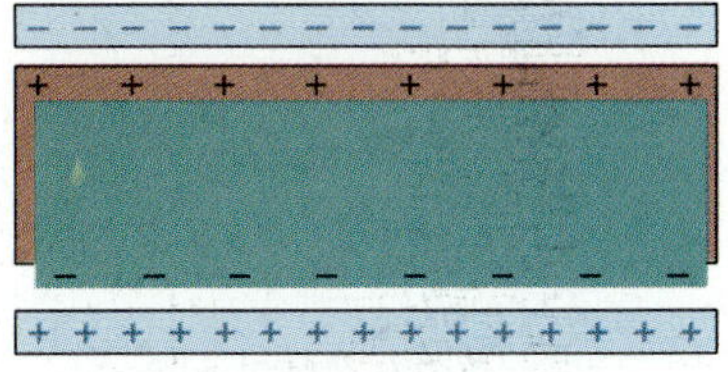

FIGURE 18.28

The distributions of positive charge (red) and of negative charge (green hatched) of the slab of dielectric fail to overlap precisely.

◀ **Dielectric constant κ**

In a linear dielectric, the amount by which the dielectric reduces the strength of the electric field can be characterized by the ***dielectric constant*** κ. This constant is merely the factor by which the electric field in the dielectric between the parallel plates is reduced; that is, if E_{free} is the electric field that the free charges produce by themselves and E the electric field that the free charges and the bound charges produce together, then

$$E = \frac{1}{\kappa} E_{\text{free}} \tag{62}$$

where κ is larger than 1.

Table 18.2 lists the values of the dielectric constants of some materials. Note that air has a value very near 1, that is, the dielectric properties of air are not very different from those of a vacuum, and the electric fields produced by (free) charges placed in air are almost the same as those produced in vacuum. This justifies our policy of neglecting the presence of air in many of the problems in preceding chapters, and we will continue this policy.

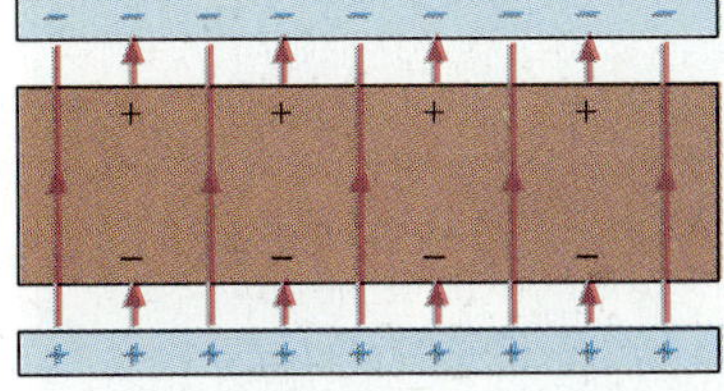

FIGURE 18.29

Some electric field lines stop on the negative charges at the bottom of the slab of dielectric. The density of field lines is smaller in the dielectric than in the empty gaps adjacent to the plates.

If the slab of dielectric fills the space between the plates entirely, then the formula (62) for the reduction of the strength of the electric field ap-

* Remember that distilled water is an insulator.

TABLE 18.2 DIELECTRIC CONSTANTS OF SOME MATERIALS[a]

Material	κ
Vacuum	1
Air	1.00054
Carbon dioxide	1.00098
Polyethylene	2.3
Rubber, hard	2.8
Transformer oil	$\simeq 3$
Plexiglas	3.4
Nylon	3.5
Epoxy resin	3.6
Paper	$\simeq 4$
Glass	$\simeq 6$
Porcelain	$\simeq 7$
Water, distilled	80

[a] At room temperature (20°C) and 1 atm.

plies throughout all of this space. Since the potential difference between the capacitor plates is directly proportional to the strength of the electric field, it follows that, for a given amount of free charge on the plates, the presence of the dielectric also reduces the potential difference by the factor κ. If ΔV_0 is the potential difference in the absence of the dielectric, the potential difference in the presence of the dielectric is

$$\Delta V = \frac{1}{\kappa} \Delta V_0 \tag{63}$$

From Eq. (45) it then follows that the dielectric increases the capacitance by a factor κ. If C_0 is the capacitance in the absence of the dielectric, the capacitance in the presence of the dielectric is

$$C = \frac{Q}{\Delta V} = \kappa \frac{Q}{\Delta V_0} = \kappa C_0 \tag{64}$$

For example, the capacitance of a parallel-plate capacitor filled with Plexiglas is

$$C = \kappa C_0 = 3.4\ C_0 = 3.4 \times \frac{\epsilon_0 A}{d} \tag{65}$$

FIGURE 18.30 Electric breakdown in a Plexiglas block exposed to a very strong electric field caused minute perforations in the block and created this beautiful arboreal pattern.

Thus, by filling the space between the capacitor plates with Plexiglas, we obtain a gain of capacitance by a factor of 3.4. Furthermore, the dielectric can prevent electric breakdown in the space between the plates. If this space contains air, sparking will occur between the plates when the electric field reaches a value of about 3×10^6 V/m, and the capacitor will discharge spontaneously. Some dielectrics are better insulators than air and they will tolerate an electric field that is appreciably larger than 3×10^6 V/m. For instance, Plexiglas will tolerate an electric field of up to 40×10^6 V/m before it suffers electric breakdown (see Fig. 18.30).

EXAMPLE 12. A parallel-plate capacitor, such as found in a radio, is made of two strips of aluminum foil with a plate area of 0.75 m^2. The plates are separated by a layer of polyethylene 2.0×10^{-5} m thick. Suppose that a potential difference of 30 V is applied to this capacitor. What is the magnitude of the free charge on each plate? What is the electric field in the dielectric?

SOLUTION: For polyethylene, $\kappa = 2.3$ (see Table 18.2). As in Eq. (65), we find that the capacitance is

$$C = \kappa \frac{\epsilon_0 A}{d} = 2.3 \times \frac{8.85 \times 10^{-12}\ \mathrm{F/m} \times 0.75\ \mathrm{m}^2}{2.0 \times 10^{-5}\ \mathrm{m}}$$

$$= 7.6 \times 10^{-7}\ \mathrm{F}$$

The magnitude of the free charge on each plate is then

$$Q_{\text{free}} = C\Delta V = 7.6 \times 10^{-7}\ \mathrm{F} \times 30\ \mathrm{V} = 2.3 \times 10^{-5}\ \text{coulomb}$$

The electric field produced by the free charges on the plates (in the absence of dielectric) would be

$$E_{\text{free}} = \frac{Q}{\epsilon_0 A}$$

and the field in the dielectric is therefore, according to Eq. (62),

$$E = \frac{1}{\kappa} E_{\text{free}} = \frac{1}{\kappa} \frac{Q}{\epsilon_0 A} = \frac{1}{2.3} \times \frac{2.3 \times 10^{-5}\ \text{coulomb}}{8.85 \times 10^{-12}\ \mathrm{F/m} \times 0.75\ \mathrm{m}^2}$$

$$= 1.5 \times 10^6\ \mathrm{V/m}$$

The simple formula (62) for the reduction of the electric field by a dielectric applies to any configuration in which the dielectric and the distribution of free charges have the same symmetry — for instance, a pair of flat charged plates with a flat slab of dielectric (as in the parallel-plate capacitor) or a spherical distribution of charge surrounded by a concentric spherical shell of dielectric. However, if the symmetries of the charge distribution and the dielectric differ — for instance, a parallel-plate capacitor with a spherical ball of dielectric placed between its plates — then the simple formula (62) is not applicable, and the reduction of the electric field in the dielectric becomes rather more difficult to calculate.

EXAMPLE 13. What is the electric field of a point charge q surrounded by a large volume of dielectric — for instance, a point charge placed in a large volume of gas?

SOLUTION: If the volume of the surrounding dielectric is large, then the electric field in the neighborhood of the point charge is not significantly influenced by the shape of the (remote) surfaces of the dielectric. Thus, the dielectric can be regarded as providing a spherically symmetric environment for the spherically symmetric point charge. Consequently, Eq. (62) is applicable to this problem and

$$E = \frac{1}{\kappa} E_{\text{free}} = \frac{1}{\kappa} \frac{q}{4\pi\epsilon_0 r^2} \tag{66}$$

where, as usual, r is the distance from the point charge.

Note that this inverse-square electric field produced by the point charge in the large volume of dielectric differs from the electric field of a point charge in vacuum only by the factor $1/\kappa$.

EXAMPLE 14. Consider the parallel-plate capacitor filled with polyethylene, described in Example 12. What is the stored potential energy in this capacitor? What is the energy density?

SOLUTION: By Eq. (49),

$$U = \tfrac{1}{2} Q \Delta V = \tfrac{1}{2} \times 2.3 \times 10^{-5} \text{ coulomb} \times 30 \text{ volts} = 3.4 \times 10^{-4} \text{ J}$$

The energy density is the energy divided by the volume between the plates,

$$[\text{energy density}] = \frac{U}{Ad} = \frac{3.4 \times 10^{-4} \text{ J}}{0.75 \text{ m}^2 \times 2.0 \times 10^{-5} \text{ m}} = 23 \text{ J/m}^3$$

Note that here we calculated the energy density directly from the energy and the volume, instead of using the formula $\frac{1}{2}\epsilon_0 E^2$ from Section 18.4. If we had used this formula, we would have obtained a wrong answer, because the formula from Section 18.4 is not applicable in a dielectric. It is easy to check that in a dielectric material, the correct formula for the energy density is $\frac{1}{2}\kappa\epsilon_0 E^2$.

SUMMARY

- **Definition of electrostatic potential:** $V = U/q$
- **Potential energy of two point charges:** $U = \dfrac{1}{4\pi\epsilon_0} \dfrac{qq'}{r}$

- **Electrostatic potential of point charge:** $V = \frac{1}{4\pi\epsilon_0} \frac{q'}{r}$
- **Potential energy of a system of point charges:**

$$U = \tfrac{1}{2}q_1 V_{\text{others at 1}} + \tfrac{1}{2}q_2 V_{\text{others at 2}} + \tfrac{1}{2}q_3 V_{\text{others at 3}} + \ldots$$

- **Potential energy of a system of conductors:**

$$U = \tfrac{1}{2}Q_1 V_1 + \tfrac{1}{2}Q_2 V_2 + \tfrac{1}{2}Q_3 V_3 + \ldots$$

- **Energy density in electric field:** $\frac{1}{2}\epsilon_0 E^2$
- **Capacitance of a pair of conductors:** $C = Q/\Delta V$
- **Capacitance of parallel plates:** $C = \epsilon_0 A/d$
- **Potential energy in capacitor:** $U = \frac{1}{2}Q\Delta V$
- **Parallel combination of capacitors:** $C = C_1 + C_2 + C_3 + \ldots$
- **Series combination of capacitors:** $1/C = 1/C_1 + 1/C_2 + 1/C_3 + \ldots$
- **Electric field in dielectric between parallel plates:** $E = \frac{1}{\kappa} E_{\text{free}}$

QUESTIONS

1.–The potential difference between the poles of an automobile battery is 12 volts. Explain what this means in terms of the definition of potential as work per unit charge.

2.–An old-fashioned word for electrostatic potential is electrostatic *tension.* Is it reasonable to think of the potential as analogous to mechanical tension?

3.–If the electric field is zero in some region, must the potential also be zero? Give an example.

4.–A bird sits on a high-voltage power line which is at a potential of 345,000 volts. Does this harm the bird?

Bird sitting on a high-voltage line.

5.–Consider an electron placed in the vicinity of a proton. Where is the electrostatic potential produced by the proton highest? Where is the potential energy of the electron highest?

6.–Consider a sphere of radius R with a charge Q uniformly distributed over its volume. Where does the potential produced by the sphere have a maximum?

7.–If the potential in a three-dimensional region of space is known to be constant, what can you conclude about the electric field in this region? If the potential on a two-dimensional surface is known to be constant, what can you conclude about the electric field on this surface?

8.–In many calculations it is convenient to assign a potential of 0 volt to the ground. If so, what is the potential at the top of the Eiffel Tower? What is the potential at the top of your head? (Hint: Your body is a conductor.)

9.–If a high-voltage power cable falls on top of your automobile, you will probably be safest if you remain inside the automobile. Why?

10.–Suppose that several separate solid metallic bodies have been placed near a charge distribution. Is it necessarily true that all of these bodies will have the same potential?

11.–If we surround some region with a conducting surface, we shield it from external electric fields. Why can we not shield a region from gravitational fields by a similar method?

12.–A cavity is completely surrounded by conducting material. Can you create an electric field inside this cavity?

13.–Show that different equipotential surfaces cannot intersect.

14.–Consider the patterns of field lines shown in Figs. 17.10 and 17.11. Roughly, sketch some of the equipotential surfaces for each case.

15.–Suppose we have a system of electric point charges with the electric potential energy given by Eq. (28). By what factor will this energy change if we increase the values of all the electric charges by a factor of 2?

16.–Consider a metallic sphere carrying a given amount of charge. Explain why the electric energy is large if the radius of the sphere is small. Would you expect a similar inverse proportion between the electric energy and the size of a conductor of arbitrary shape?

17.–Suppose that we increase the separation between the metallic plates described in Example 5. Does this change the electric energy density? The net electric energy?

18.–Consider the electric fields shown in Figs. 17.10 and 17.11. In what regions of the latter figure is the electric energy density larger than in the former? Can you guess which of these electric fields has a larger energy density on the average?

19.–Figure 18.31 shows a sequence of deformations of a nucleus about to undergo fission. The volume of the nucleus and the electric charge remain constant during these deformations. Which configuration has the highest electric energy? The lowest?

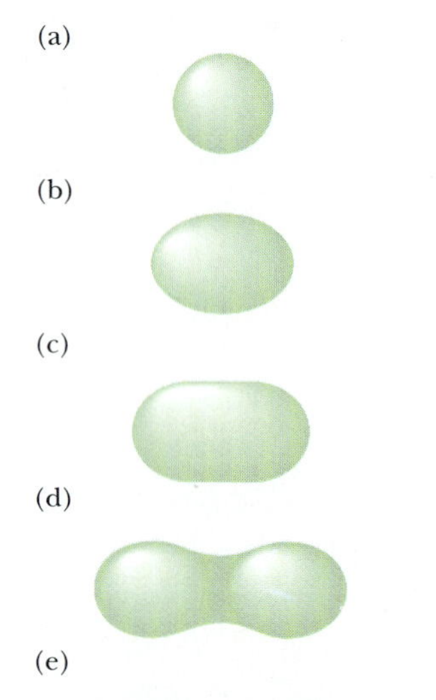

FIGURE 18.31
Fission of a nucleus.

20.–Commercially available large capacitors have a capacitance of several farads. How is it possible that the capacitance of such a device is larger than the capacitance of the Earth?

21.–Suppose we enclose the entire Earth in a conducting shell of a radius slightly larger than the Earth's radius. Explain why this would make the capacitance of the Earth much larger than the value calculated in Example 8.

22.–Equation (47) shows that $C \to \infty$ as $d \to 0$. In practice, why can we not construct a capacitor of arbitrarily large C by making d sufficiently small? (Hint: What happens to E as $d \to 0$ while ΔV is held constant?)

23.–If you put more charge on one plate of a parallel-plate capacitor than on the other, what happens to the extra charge?

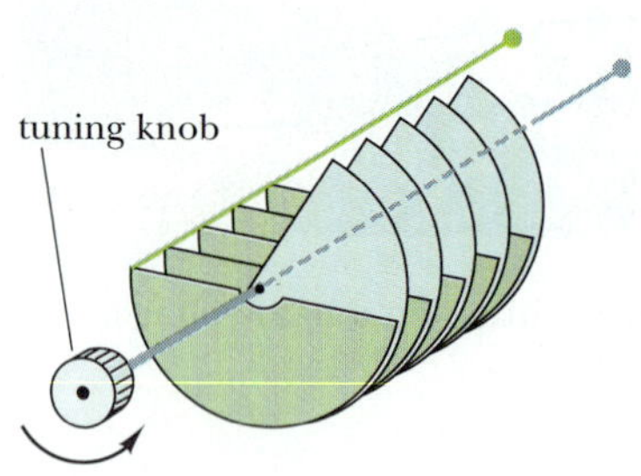

FIGURE 18.32
The blue plates are connected together, and the green plates are connected together. By turning the knob and the attached blue plates, we change the overlap between the blue and the green plates.

24.–Figure 18.32 shows the design of an adjustable capacitor used in the tuning circuit of a radio. This capacitor can be regarded as several connected capacitors. Are these several capacitors connected in series or in parallel? If we turn the tuning knob (and the attached plates) counterclockwise, does the capacitance increase or decrease?

25.–Consider a parallel-plate capacitor. Does the capacitance change if we insert a thin conducting sheet between the two plates, parallel to them?

26.–Suppose we insert a thick slab of metal between the plates of a parallel-plate capacitor, parallel to the plates and not touching them. Does the capacitance increase or decrease?

27.–Consider a fluid dielectric that consists of molecules with permanent dipole moments. Will the dielectric constant increase or decrease as a function of temperature?

PROBLEMS

Section 18.1: The Electrostatic Potential

1.——The electric potential difference between the positive and negative poles of an automobile battery is 12 volts. In order to charge the battery fully, the charging device must force $+2.0 \times 10^5$ coulombs from the negative terminal of the battery to the positive terminal. How much work must the charging device do during this process?

2.——A proton sits at the origin of coordinates. How much work must you do against the electric force of the proton to push an electron from the point $x = 1.0 \times 10^{-10}$ m, $y = 0$ in the x–y plane to the point $x = 2.5 \times 10^{-10}$ m, $y = 2.5 \times 10^{-10}$ m?

3.——On days of fair weather, the atmospheric electric field of the Earth is about 100 V/m; this field points vertically downward (compare Problem 17.4). What is the electric potential difference between the ground and an airplane flying at 600 m? What is the potential difference between the ground and the tip of the Eiffel Tower? Treat the ground as a flat conductor.

4.——Consider the arrangement of parallel sheets of charge described in Problem 17.29. Find the potential difference between the upper sheet and the lower sheet.

5.——A proton is decelerated from an initial speed of 6.9×10^6 m/s to rest by a uniform electric field of 2.5×10^6 V/m. How far did the proton travel during the motion? What is the difference in electrostatic potential between the initial and final positions of the proton?

6.——A proton is accelerated from rest through a potential of 2.50×10^5 V. What is its final speed?

7.——The gap between the electrodes of a spark plug in an automobile is 0.64 mm. In order to produce an electric field of 3×10^6 V/m (required to initiate an electric spark), what minimum potential difference must you apply to the spark plug?

8.——Prove that the plane midway between a positive and a negative point charge of equal magnitudes is an equipotential surface. Is this also true if both charges are positive?

9.——The nucleus of lead has a charge of $82e$ uniformly distributed over a spherical region of radius 7.1×10^{-15} m. What is the electrostatic potential at the nuclear surface?

10.*——The potential difference between the two poles of an automobile battery is 12.0 V. Suppose that you place such a battery in empty space and you release an electron at a point next to the negative pole of the battery. The electron will then be pushed away by the electric force and move off in some direction.

(a) If the electron strikes the positive pole of the battery, what will be its impact speed?

The Stanford Linear Accelerator (SLAC).

(b) If instead the electron moves away toward infinity, what will be its ultimate speed?

11.*—At the Stanford Linear Accelerator (SLAC), electrons are accelerated from an energy of 0 eV to 20×10^9 eV as they travel in a straight evacuated tube 1600 m in length. The acceleration is due to a strong electric field pushing the electrons along. Assume that the electric field is uniform. What must be its strength?

12.*—An alpha particle of kinetic energy 1.7×10^{-12} J is shot directly toward a platinum nucleus. What will be the distance of closest approach? The electric charge of the alpha particle is $2e$ and that of the platinum nucleus is $78e$. Treat the alpha particle and the nucleus as spherical charge distributions and disregard the motion of the nucleus.

13.*—What is the minimum kinetic energy with which an alpha particle must be launched toward a plutonium nucleus if it is to make contact with the nuclear surface? The plutonium nucleus is a sphere of radius 7.5×10^{-15} m with a charge of $94e$ uniformly distributed over the volume. For the purpose of this problem, the alpha particle may be regarded as a particle (of negligible radius) with a charge of $2e$.

14.*—A thorium nucleus emits an alpha particle according to the reaction

$$\text{thorium} \rightarrow \text{radium} + \text{alpha}$$

Assume that the alpha particle is pointlike and that the residual radium nucleus is spherical with a radius of 7.4×10^{-15} m. The charge on the alpha particle is $2e$ and that on the radium nucleus is $88e$.

(a) At the instant the alpha particle emerges from the nuclear surface, what is its electrostatic potential energy?

(b) If the alpha particle has no initial kinetic energy, what will be its final kinetic energy and speed when far away from the nucleus? Assume that the radium nucleus does not move. The mass of the alpha particle is 6.7×10^{-27} kg.

15.*—Consider again the arrangement of charges within the thundercloud of Fig. 17.48. Find the electric potential due to these charges at a point which is at a height of 7.5 km and on the vertical line passing through the charges. Find the electric potential at a second point that is at the same height and has a horizontal distance of 5 km from the first point.

16.*—A positive point charge q, with mass m, is released at a distance d from a fixed positive point charge Q. How fast is the charge q moving when the distance has grown to three times the initial value?

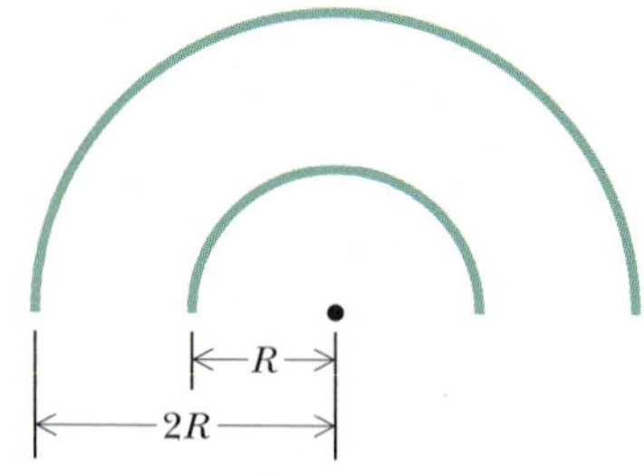

FIGURE 18.33
Two concentric semicircular rods.

17.*—Two semicircular rods are arranged as shown in Fig. 18.33. The rods carry a charge of λ coulombs per unit length. Calculate the potential at the center of this configuration.

18.**—A nucleus of carbon (charge $6e$) and one of helium (charge $2e$) are separated by a distance of 1.2×10^{-13} m and are instantaneously at rest. The center of mass of this system is at a distance of 0.4×10^{-13} m from the carbon nucleus. Take this point as origin and take the x axis along the line joining the nuclei, with the carbon nucleus on the negative x axis.

(a) Find the potential produced jointly by these charges at the point $x = 0$. Repeat for the point $x = 0.02 \times 10^{-13}$ m.

(b) From this, evaluate the (approximate) rate of change $\Delta V/\Delta x$ of the potential at the point $x = 0.01 \times 10^{-13}$ m, halfway between the two points in Part (a).

(c) Calculate the electric field produced by the two nuclei at this point, and compare with the value you obtained for $\Delta V/\Delta x$ in Part (b).

Section 18.3: Energy of a System of Point Charges

19.——Consider once more the distribution of charges within the thundercloud shown in Fig. 17.48. What is the electric potential energy of this charge distribution? Treat the charges as pointlike.

20.——Figure 17.32 describes the arrangement of nuclear charges (positive charges) in a water molecule. Treating the nuclei as point charges, calculate the electric potential energy of this arrangement of three charges.

21.——Suppose that at one instant the electrons and the nucleus of a helium atom occupy the positions shown in Fig. 18.34; at this instant, the electrons are at a distance of 0.20×10^{-10} m from the nucleus. What is the electric potential energy of this arrangement? Treat the electrons and the nucleus as point charges.

22.*—According to the alpha-particle model of the nucleus, some nuclei consist of a regular geometric arrangement of alpha particles. For instance, the nucleus of ^{12}C consists of three alpha particles arranged on an equilateral triangle (Fig. 18.35). Assuming that the distance between pairs of alpha particles is 3.0×10^{-15} m, what is the potential energy (in eV) of this arrangement of alpha particles? Treat the alpha particles as pointlike.

23.**—According to the alpha-particle model (see also the preceding problem), the nucleus of ^{16}O consists of four alpha particles arranged on the vertices of a tetrahedron (Fig. 18.36). If the distance between pairs of alpha particles is 3.0×10^{-15} m, what is the potential energy (in eV) of this configuration of alpha particles? Treat the alpha particles as pointlike.

24.**—Four equal particles of positive charges q and masses m are initially held at the four corners of a square of side L. If these particles are released simultaneously, what will be their speeds when they have separated by a very large distance?

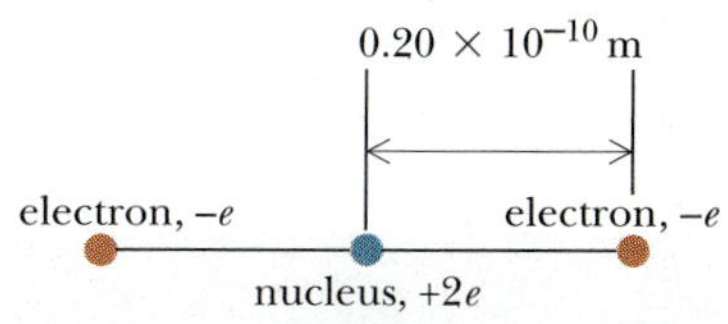

FIGURE 18.34
The nucleus ($+2e$) and the electrons ($-e$) of an atom of helium at one instant of time.

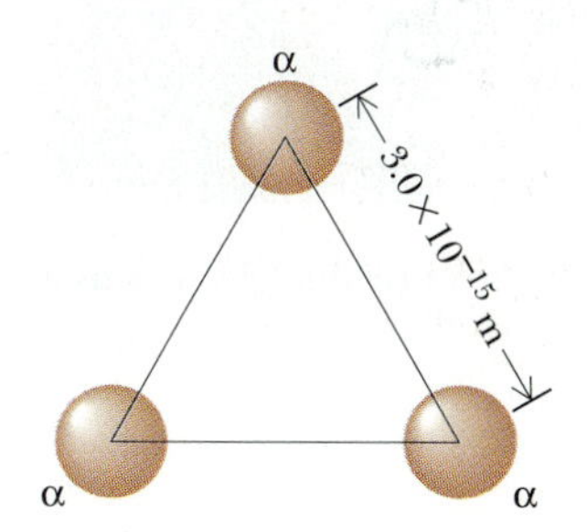

FIGURE 18.35
Three alpha particles.

Section 18.4: Energy of a System of Charged Conductors

25.——A pair of parallel plates, each measuring 30 cm $\times$ 30 cm, are separated by a gap of 1.0 mm. How much work must you do against the electric forces to charge these plates with $+1.0 \times 10^{-6}$ C and -1.0×10^{6} C, respectively?

26.——A charge of 7.5×10^{-6} C can be placed on a metallic sphere of radius 15 cm before the surrounding air suffers electrical breakdown. What is the electric potential energy of the sphere with this charge?

27.*—Pretend that an electron is a conducting sphere of radius R with a charge e distributed uniformly over its surface. In terms of e and the mass m_e of the electron, what must be the radius R if the electric potential energy is to equal the energy $m_e c^2$, called the rest-mass energy of the electron? Numerically, what is the value of R?

28.——Near the surface of the nucleus of a lead atom, the electric field has a strength of 3.4×10^{21} V/m. What is the energy density in this field?

29.——The atmospheric electric field near the surface of the Earth has a strength of 100 V/m.

(a) What is its energy density?

(b) Assuming that the field has the same magnitude everywhere in the atmosphere up to a height of 10 km, what is the corresponding total energy?

30.——Calculate the energy density in each of the electric fields listed in the first four entries of Table 17.1.

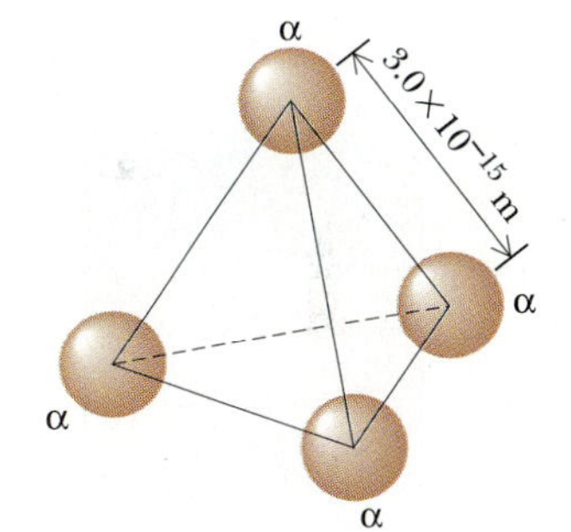

FIGURE 18.36
Four alpha particles.

Section 18.5: Capacitance

31.——Consider an isolated metallic sphere of radius R and another isolated metallic sphere of radius $3R$. If both spheres are at the same potential, what

The head of this girl acts as a spherical capacitor.

is the ratio of their charges? If both spheres carry the same charge, what is the ratio of their potentials?

32.——The collector of an electrostatic machine is a metal sphere of radius 18 cm.

(a) What is the capacitance of this sphere?

(b) How many coulombs of charge must you place on this sphere to raise its potential to 2.0×10^5 V?

33.——Your head is (approximately) a conducting sphere of radius 10 cm. What is the capacitance of your head? What will be the charge on your head if, by means of an electrostatic machine, you raise your head (and your body) to a potential of 100,000 V?

34.——A parallel-plate capacitor has a plate area of 900 cm^2 and a plate separation of 0.50 cm. The space between the plates is empty.

(a) What is the capacitance?

(b) What is the potential difference if the charges on the plates are $\pm 6.0 \times 10^{-8}$ C?

(c) What is the electric field between the plates?

(d) The energy density?

(e) The total potential energy?

35.——Two parallel conducting plates of area 0.5 m^2 placed in a vacuum have a potential difference of 2.0×10^5 V when charges of $\pm 4.0 \times 10^{-3}$ C are placed on them, respectively:

(a) What is the capacitance of the pair of plates?

(b) What is the distance between them?

(c) What is the electric field between them?

(d) What is the potential energy?

36.——A TV receiver contains a capacitor of 10 μF charged to a potential difference of 2×10^4 V. What is the amount of charge stored in this capacitor? The amount of energy?

37.——How much energy is stored in a 3×10^3-μF capacitor charged to 100 volts?

38.*—What is the electric field in a 3.0-μF capacitor with parallel plates of area 15 m^2 charged to 4.4 volts?

39.*—A 4.00-μF capacitor has been charged by a 9.00-volt battery. How many electrons must be moved from the negative plate to the positive plate of the capacitor to reverse the electric field inside of the capacitor?

40.*—A primitive capacitor, originally called a ***Leyden jar,*** consists of a glass bottle filled with water and wrapped on the outside with metal foil (see Fig. 18.37). The foil plays the role of one plate of the capacitor, and the surface of the water facing the foil plays the role of the other plate; these "plates" are separated by the layer of glass (dielectric). Suppose that, as shown in Fig. 18.37, the wrapped portion of the bottle is cylindrical, of height 15 cm and diameter 15 cm (the bottom is not wrapped). The thickness of the glass is 0.2 cm. What is the capacitance of this contraption? (Hint: Treat this as a parallel-plate capacitor.)

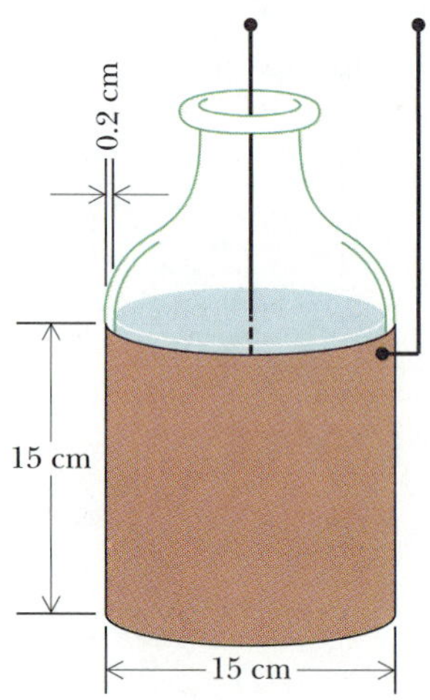

FIGURE 18.37
A Leyden jar.

41.*—In many computer keyboards, the switches under the keys consist of small parallel-plate capacitors (see Fig. 18.38). The key is attached to the upper plate, which is movable. When you push the key down, you push the upper plate toward the lower plate, and you alter the plate separation d and the capacitance. The capacitor is connected to an external circuit that maintains a constant potential difference ΔV across the plates. The change of capacitance therefore triggers a transfer of charge between the capacitor and the computer circuit. Suppose that the initial plate separation is 5.0 mm and the initial capacitance is 6×10^{-13} F. The final plate separation (with the key full depressed) is 0.2 mm. The constant potential difference is 8 V. What is the change in capacitance when you depress the key?

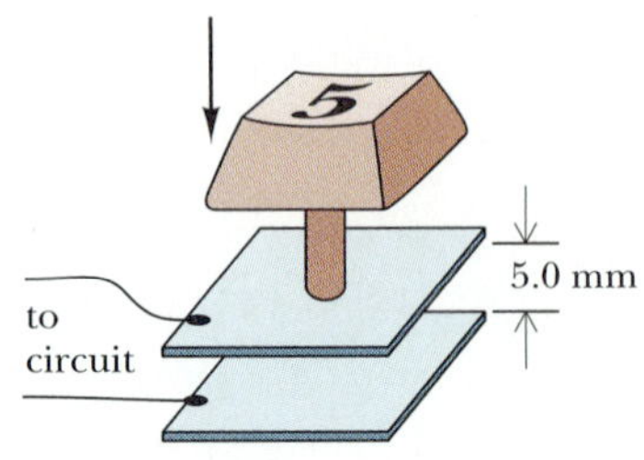

FIGURE 18.38
Capacitor attached to a key in a computer keyboard.

What is the amount of electric charge transferred between the capacitor and the computer circuit?

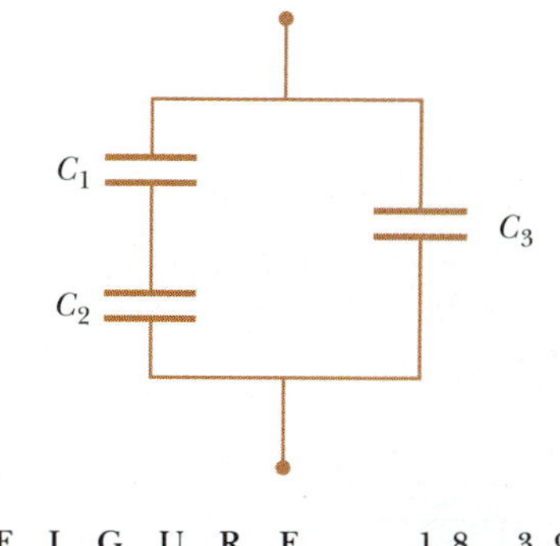

FIGURE 18.39

Section 18.6: Capacitors in Combination

42.——What is the combined capacitance if three capacitors of 3.0, 5.0, and 7.5 μF are connected in parallel? What is the combined capacitance if they are connected in series?

43.——Three capacitors with capacitances $C_1 = 5.0\ \mu$F, $C_2 = 3.0\ \mu$F, and $C_3 = 8.0\ \mu$F are connected as shown in Fig. 18.39. Find the combined capacitance.

44.——Two capacitors of 5.0 μF and 8.0 μF are connected in series to a 24-V battery. What is the potential energy stored in the capacitors?

45.——What is the total charge stored on the three capacitors connected to a 30-V battery as shown in Fig. 18.40?

46.*——Six identical capacitors of capacitance C are connected as shown in Fig. 18.41. What is the total capacitance of the combination?

47.*——Six identical capacitors of capacitance C are connected as shown in Fig. 18.42. What is the total capacitance of the combination?

48.*——Three capacitors are connected as shown in Fig. 18.43. Their capacitances are $C_1 = 2.0\ \mu$F, $C_2 = 6.0\ \mu$F, and $C_3 = 8.0\ \mu$F. If a voltage of 200 V is applied to the two free terminals, what will be the charge on each capacitor? What will be the potential energy in each?

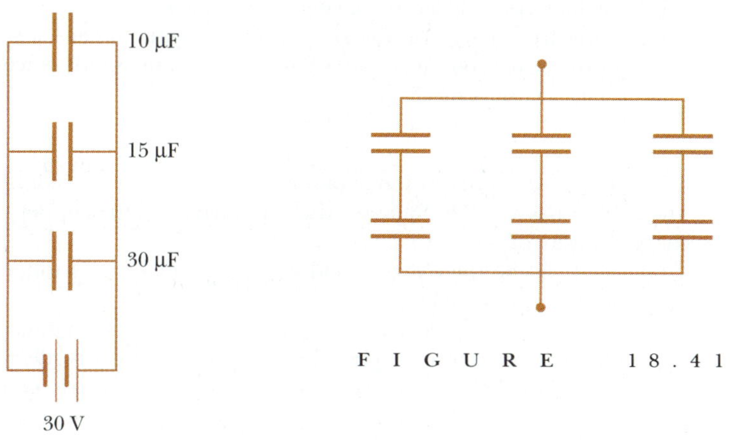

FIGURE 18.40

FIGURE 18.41

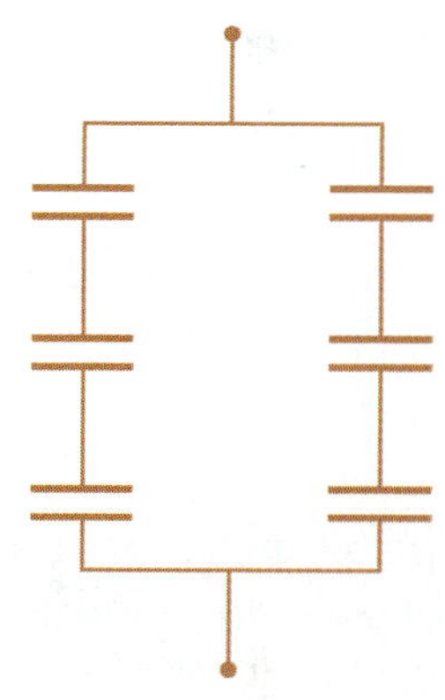
FIGURE 18.42

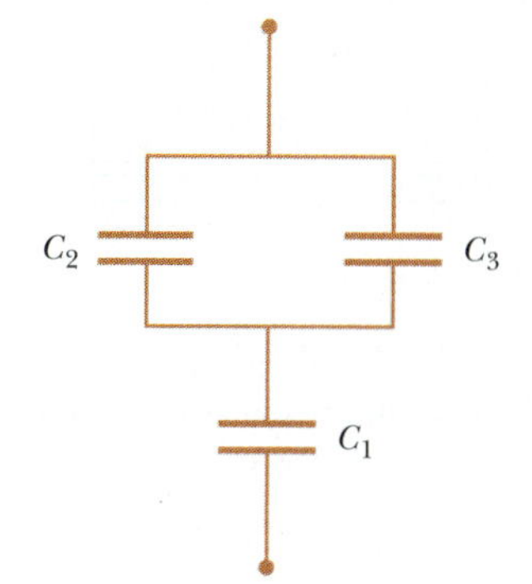

FIGURE 18.43

Section 18.7: Dielectrics

49.——You wish to construct a capacitor out of a sheet of polyethylene of thickness 5×10^{-2} mm and $\kappa = 2.3$ sandwiched between two aluminum sheets. If the capacitance is to be 3.0 μF, what must be the area of the sheets?

50.——What is the capacitance of a sphere of radius R immersed in a large volume of gas of dielectric constant κ?

51.*——In order to measure the dielectric constant of a dielectric material, a slab of this material 2.0 cm thick is slowly inserted between a pair of parallel conducting plates separated by a distance of 2.0 cm. Before insertion of the dielectric, the potential difference across these capacitor plates is 3.0×10^5 V. During insertion, the charge on the plates remains constant. After insertion, the potential difference is 1.8×10^5 V. What is the value of the dielectric constant?

52.*——A parallel-plate capacitor is filled with carbon dioxide at 1 atm pressure. Under these conditions the capacitance is 0.5 μF. We charge the capacitor

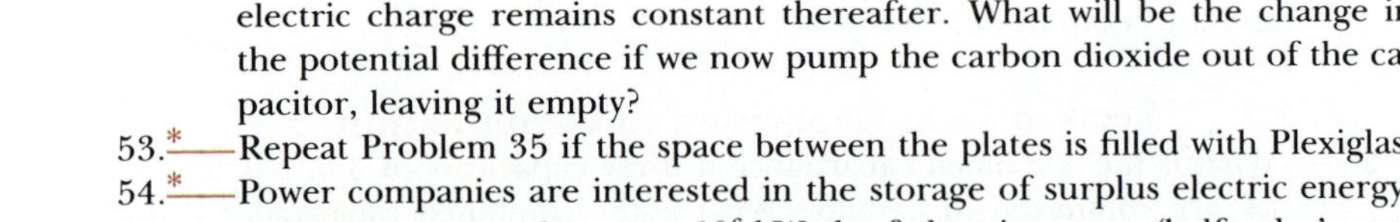
by means of a 48-V battery and then disconnect the battery so that the electric charge remains constant thereafter. What will be the change in the potential difference if we now pump the carbon dioxide out of the capacitor, leaving it empty?

53.* Repeat Problem 35 if the space between the plates is filled with Plexiglas.

54.* Power companies are interested in the storage of surplus electric energy. Suppose we wanted to store 10^6 kW · h of electric energy (half a day's output for a large power plant) in a large parallel-plate capacitor filled with a plastic dielectric with $\kappa = 3.0$. If the dielectric can tolerate a maximum electric field of 5×10^7 V/m, what is the minimum total volume of dielectric needed to store this energy?

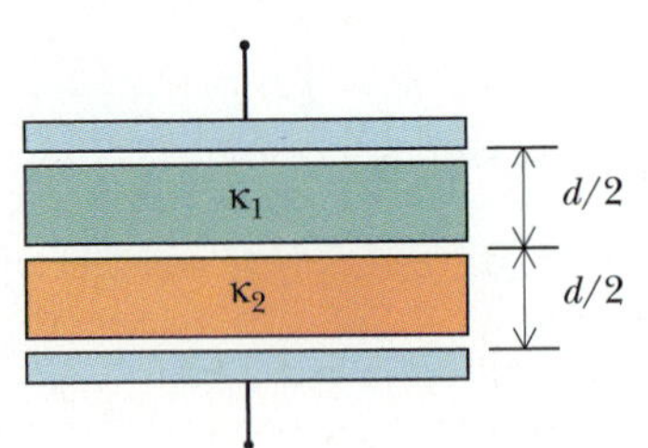

FIGURE 18.44

Parallel-plate capacitor with two slabs of dielectric.

55.* The plates of a parallel-plate capacitor are movable and are initially separated by an air gap of width d. A piece of dielectric of dielectric constant κ and a thickness of $3d$ is inserted between the plates. If the ratio of the capacitance before inserting the dielectric to the capacitance after inserting the dielectric is 1.5, what is the value of κ?

56.* Starting from the general expression $\frac{1}{2} Q\Delta V$ for the electric energy in a parallel-plate capacitor (equally valid for a capacitor with or without a dielectric), show that the energy density in the electric field between the plates of a capacitor with a dielectric is $\frac{1}{2}\kappa\epsilon_0 E^2$.

57.* A spherical capacitor consists of a metallic sphere of radius R_1 surrounded by a concentric metallic shell of radius R_2. The space between R_1 and R_2 is filled with dielectric having a constant κ.

(a) If the free charge on the surface of the inner sphere is Q and that on the outer sphere is $-Q$, what is the potential difference between the two conductors?

(b) What is the capacitance of this spherical capacitor?

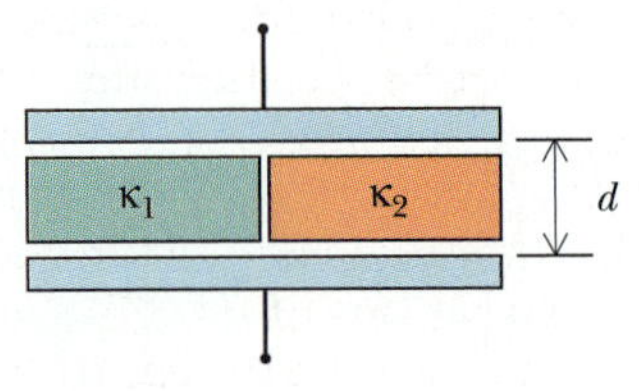

FIGURE 18.45

Parallel-plate capacitor with two slabs of dielectric side by side.

58.* A long cylindrical copper wire of radius 0.20 cm is surrounded by a cylindrical sheath of rubber of inner radius 0.20 cm and outer radius 0.30 cm. The rubber has $\kappa = 2.8$. Suppose that the surface of the copper has a free charge density of 4.0×10^{-6} C/m^2.

(a) What will be the electric field in the rubber near its inner surface? Near its outer surface?

(b) What will be the electric field just outside the rubber sheath?

59.* A parallel-plate capacitor of place area A and spacing d is filled with two parallel slabs of dielectric of equal thickness with dielectric constants κ_1 and κ_2, respectively (Fig. 18.44). What is the capacitance? (Hint: Check that the configuration of Fig. 18.44 is equivalent to two capacitors in series.)

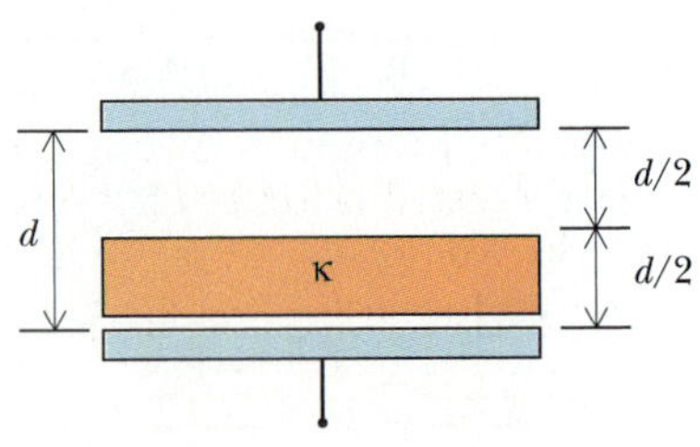

FIGURE 18.46

Parallel-plate capacitor, partially filled with dielectric.

60.* A capacitor with two large parallel plates of area A separated by a distance d is filled with two equal slabs of dielectric side by side (Fig. 18.45). The dielectric constants are κ_1 and κ_2. What is the capacitance?

61.* A parallel-plate capacitor of plate area A and separation d contains a slab of dielectric of thickness $d/2$ and constant κ (see Fig. 18.46). What is the capacitance of this capacitor? (Hint: Regard this capacitor as two capacitors in series, one with dielectric, one without.)

PHYSICS IN ACTION

THE ELECTROCARDIOGRAPH

The human heart beats about 70 times per minute. Each beat begins with a contraction of the atria which is followed a moment later by the contraction of the ventricles (see *Blood Flow,* pp. 379A–379F). The contractions of the heart muscle, like the contractions of other muscles, are triggered by electric signals. But in contrast to other muscles, where the electric signals travel along nerve fibers, the electric signals in heart muscle travel along the muscle fibers. These electric signals involve changes of the electric potential in and around the muscle fibers. The changes of the electric potential associated with the heartbeats are strongest in the immediate vicinity of the muscle fibers; but, with a sensitive detector, the changes in the potential can be measured at some distance from the heart, on the surface of the skin. The detector used to measure such changes of potential is called ***electrocardiograph;*** it is widely used by physicians to monitor the operation of the heart and to discover defects in the heart muscle and in its nerve conduction.

The mechanism for the propagation of electric signals along muscle fibers is almost exactly the same as for the propagation along nerve fibers. Each muscle fiber is a single, long cell, sheathed in a membrane (see Fig. 1). When the muscle fiber is at rest (inactive), the membrane carries a layer of positive charge on its outside surface and a layer of negative charge on its inside surface. A membrane with such layers of opposite charges is said to be ***polarized.*** The positive charges are mainly K^+ ions, and the negative charges are mainly Cl^- ions. Between these layers of charge there is a uniform electric field (like the electric field between oppositely charged parallel plates); consequently, there is a potential difference between the outside and the inside of the cell. The potential inside of the cell is negative; this ***resting potential*** is about -90 millivolts.

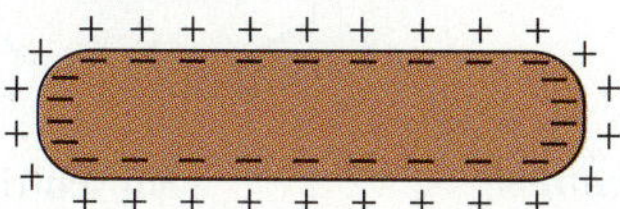

FIGURE 1 A muscle fiber and the positive and negative charges on the outside and inside surfaces of its membrane.

When a muscle fiber is stimulated by an electric signal from an adjacent muscle fiber, positive charges (mostly Na^+ ions) from the interstitial fluid surrounding the fiber flow through the membrane, into the fiber. The accumulation of these positive charges reverse the potential difference, from -90 millivolts to about $+30$ millivolts. The membrane is then said to be ***depolarized*** (actually, polarized in the reverse direction). The depolarization propagates along the length of the muscle fiber and constitutes the electric signal. This depolarization triggers the contraction process of the fiber (see *Muscular Contraction,* pp. 158A–158D, for a discussion of the contractile machinery in muscle fibers). Within a short time, chemical processes within the fiber pump the positive charges out of the fiber and restore the charge distribution to the initial, resting state. The repolarized fiber is then ready for the next depolarization and the next contraction.

The depolarization of muscle fibers can be detected at some distance from the fibers by the changes in the electric potential. Consider a muscle fiber that is initially polarized, and progressively becomes depolarized, from left to right (see Fig. 2). Initially the positive charge distribution is centered on the negative charge distribution. The potential at a point M beyond one end of the fiber is nearly zero, since the average positions of the positive and the negative charge distributions in Fig. 2a coincide, and electric fields they produce at any exterior point M cancel almost exactly. But when the fiber is partially depolarized, the average positions of the positive and the negative charge distributions are slightly different — the average position of the positive charge distribution in Fig. 2b is slightly to the right of the average position of the negative charge distribution. The separation between the average positions of the charge distributions implies that the muscle fiber has a ***dipole moment*** (see Section 17.5). The electric fields then do not cancel, and the potential at the exterior point M is not zero. Since, on the average, the positive charges are slightly nearer to the point M than the negative charges,

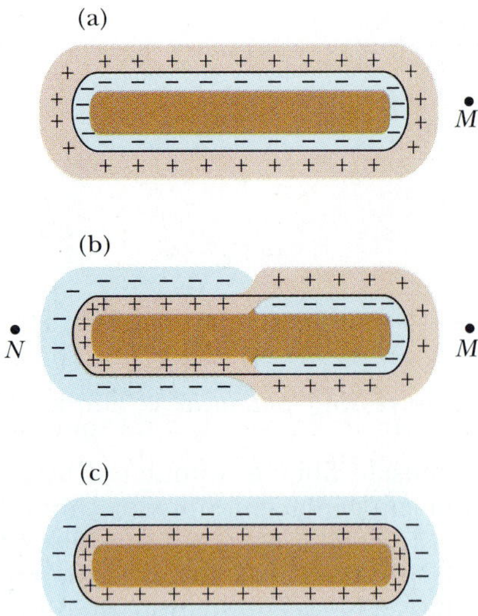

FIGURE 2 Progressive depolarization of a muscle fiber. (a) Initially the fiber is polarized. The average position, or the center, of the positive charge distribution (red) coincides with the average position of the negative charge distribution (blue). (b) At a later time, the fiber is partially depolarized. The average position of the positive charge distribution (red) is slightly to the right of the average position of the negative charge distribution (blue). (c) Finally, the fiber is completely depolarized. Now the average positions of the positive and the negative charge distributions again coincide.

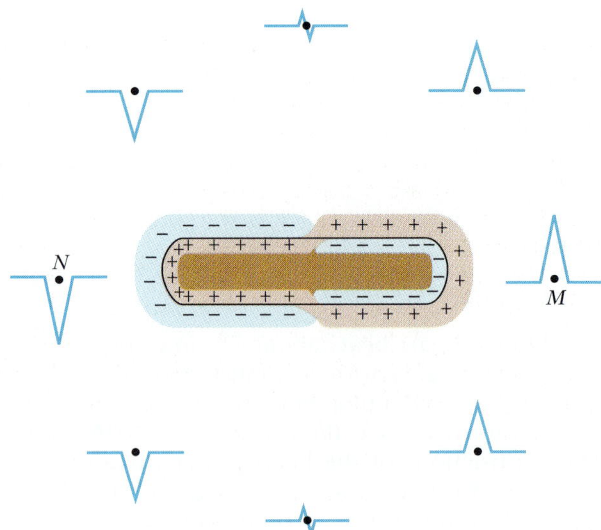

FIGURE 3 Plots of the potential vs. time, at different points, all at about the same distance from the fiber.

the potential at the point M is positive. On the other hand, the potential at the point N, beyond the other end of the fiber, is negative. Finally, when the fiber is completely depolarized, the potential will again be zero, since the average positions of the positive and the negative charge distributions will then again coincide. Thus, during the progressive depolarization of the muscle fiber, the potential at the point M first increases, then decreases, and then returns to zero. Figure 3 gives a qualitative plot of the potential vs. time for the point M, and also similar plots for the point N and for several other points at different directions around the fiber.

In the heart, the depolarization and contraction of the muscle fibers of the atria and the ventricles are initiated, respectively, by the ***sinus node*** and the ***atrioventricular node*** (A-V node). These nodes are small clumps of muscle fibers located, respectively, at the base of the superior vena cava and in the septum, or the interior dividing wall, of the heart (see Fig. 4). The sinus node is the primary pacemaker of the heart. Its muscle cells depolarize and contract spontaneously, at regular intervals. The electric signals produced during this activity are transmitted to the muscles of the atria and initiate their contraction. When the electric signals reach the A-V node, it relays these signals to the ventricles and thereby initiates their contraction.

To observe the potential changes generated during the depolarization of the muscle fibers in the heart, electrodes (contacts) are placed on the skin and connected by wire leads to a sensitive voltmeter. The potential changes on the skin are of the order of a few millivolts, and to detect such small potentials, the voltmeters used in the electrocardiograph are equipped with amplifier circuits. The standard procedure for electrocardiography is to attach electrodes to the left wrist, the right wrist, and the left ankle (other electrodes are attached

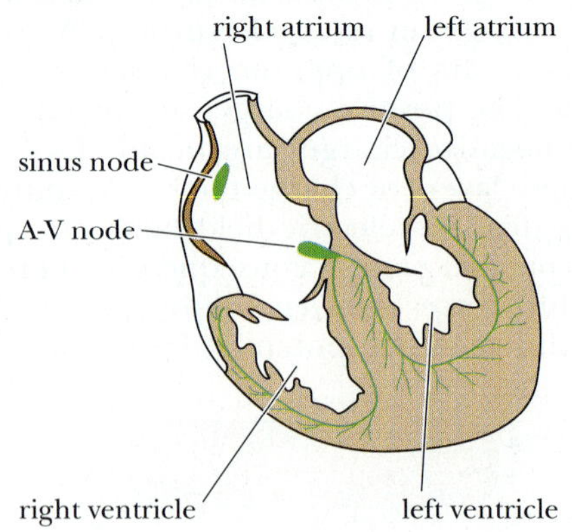

FIGURE 4 The sinus node and the A-V node. The sinus node is the primary pacemaker of the heart.

to the chest wall, but we will ignore these). The electrodes are covered with a salt paste, for good electric contact. The choice of location of the electrodes on the arms and legs is a matter of convenience. The arms and the legs act as conducting segments, and attaching electrodes to the ends of the arms or legs is therefore equivalent to attaching electrodes to the shoulders or the groin, where the arms or legs merge with the trunk. Thus, measurements with the wrist and ankle electrodes are roughly equivalent to measurements at the vertices of a triangle centered on the heart (see Fig. 5), called the ***Einthoven triangle.***

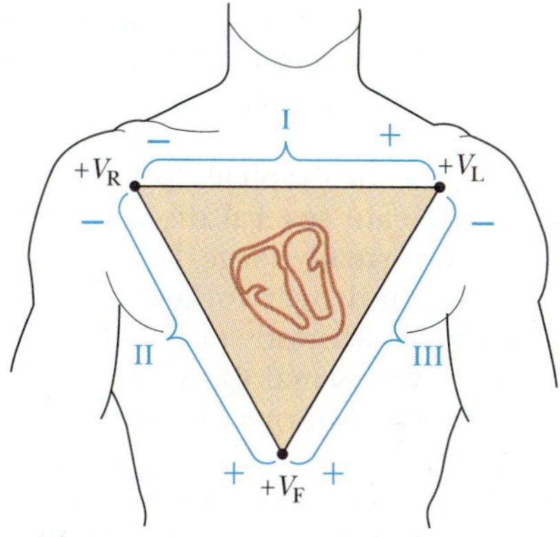

FIGURE 5 Measurements of the potential at the wrists and the left leg are roughly equivalent to measurements at the vertices of a triangle centered on the heart. The potentials at these points are labeled V_R, V_L, and V_F.

With these three electrodes at the vertices of the triangle, six measurements are routinely made and plotted by the electrocardiograph: three "unipolar" measurements, and three "bipolar" measurements. The ***unipolar*** measurements simply give the values of the potential at the three vertices of the triangle. These potentials are designated V_R (for right wrist), V_L (for left wrist), and V_F (for left ankle or foot). The plus signs included next to these symbols in Fig. 5 indicate that the + terminal of the voltmeter is connected to this electrode during the measurement. The − terminal of the voltmeter is connected to a reference terminal, or "indifferent terminal," whose potential is zero.

The ***bipolar*** measurements give the potential differences between adjacent vertices. These potential differences are designated I, II, and III, as indicated in Fig. 5. The plus and minus signs included with symbols in Fig. 5 indicate where the + and − terminals of the voltmeter are connected during the measurement. Note that I, II, and III can be expressed as differences between V_R, V_L, and V_F:

$$\mathrm{I} = V_L - V_R$$

$$\mathrm{II} = V_F - V_R$$

$$\mathrm{III} = V_F - V_L$$

Hence the information contained in I, II, and III is actually redundant, but it helps in a quick interpretation of the data. From the dipole character of the potential, it can be shown that the potential differences I, II, and III coincide with $\sqrt{3}$ times the potentials at three points lying at equal distances from the heart, on radial lines parallel to the sides of the triangle (see Fig. 6). Thus, the three unipolar and the three bipolar measurements give us the potentials at six points around the heart. The potentials at the points at equal distances in the opposite directions are the opposites of these potentials.*

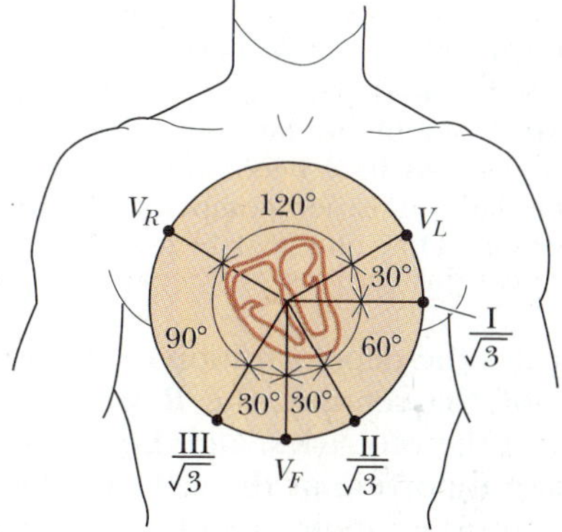

FIGURE 6 The three unipolar measurements V_R, V_L, and V_F and the three bipolar measurements I, II, and III determine the potentials at six points lying at equal distances from the heart.

Figure 7 is an example of an ***electrocardiogram*** (EKG) measures on a healthy individual. This EKG displays the potentials I, II, III, V_R, V_L, and V_F as a function of time. Figure 8 is an enlarged sketch of the potential II. The main features of this plot are the maxima and minima, called the ***P, Q, R, S, and T "waves".*** These waves indicate the potentials generated by depolarizations and repolarizations of muscle fibers in different parts of the heart. The P

* Medical textbooks often cavalierly omit the factor of $\sqrt{3}$. But they usually include a factor of 1.5 (see below) with each of the unipolar values. The resulting relative variation from one point to another in Fig. 6 is roughly the same as when the factor of $\sqrt{3}$ is included.

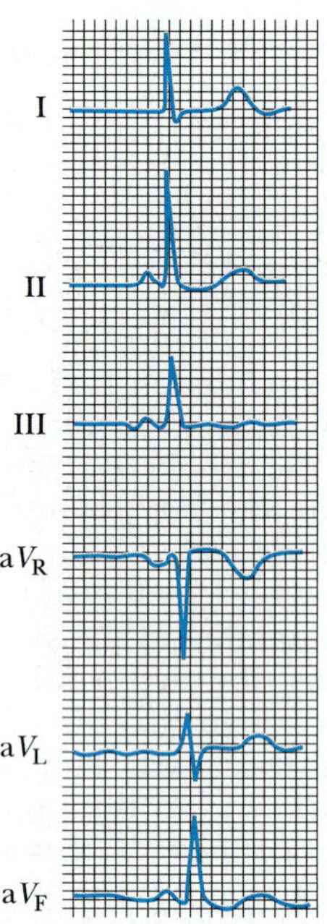

F I G U R E 7 An electrocardiogram displaying the potentials I, II, III, V_R, V_L, and V_F vs. time (the labels "a" in front of the last three potentials indicate that these potentials have been amplified by a factor of 1.5, for the sake of easier comparison with the other three potentials). Only one heartbeat cycle is shown.

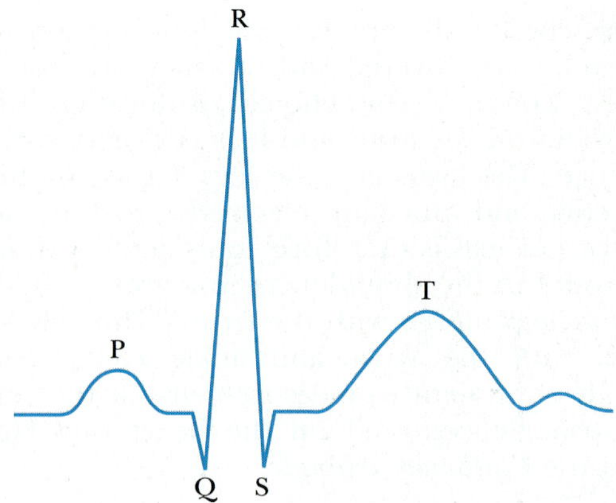

F I G U R E 8 Schematic diagram illustrating the main features in a "typical" plot of the potential II.

wave indicates the depolarization of the atria; the Q wave that of the septum, the R wave that of the main mass of the ventricles, and the S wave that of a smaller, lateral portion of the right ventricle. The T wave indicates the repolarization of the ventricles.

These waves are of different magnitudes because the muscle mass is different (largest for ventricle, smallest for septum), and because the directions of the progressive depolarization of the muscle fibers are different [the direction of depolarization in the ventricle is, on the average, aligned with the direction to which the potential II belongs (see Fig. 6), and therefore this depolarization gives a large contribution to this potential].

Deviations from the "typical" potential illustrated in Fig. 8 can indicate several diseases of the heart. For instance, abnormally large R and S waves indicate an enlarged and overworked ventricle (hypertrophy), commonly found in patients with high blood pressure. Abnormal Q, S, and T waves indicate dead patches (infarcts) in the heart muscle, caused by heart attacks. And an excessively long time interval between the P wave and the Q, R, S waves indicates an A-V block, that is, a failure of the A-V node to transmit the electric signal.

Comparisons among the different potential plots in the EKG in Fig. 7 are useful for determining the orientation of the main "electrical axis" of the heart, that is, the direction of depolarization of the main mass of muscle fibers in the ventricle. The direction of this axis is simply the direction in which we detect the largest R wave. For example, in the EKG in Fig. 7, the largest R wave shows up in the potential II, and according to Fig. 6, this potential belongs to a direction of 60° below the horizontal, in the lower "left" quadrant. This is the normal quadrant for the main electrical axis. An electrical axis outside of this quadrant may indicate some form of heart disease.

R E F E R E N C E S

Chung, E. K., *Electrocardiography* (Harper & Row, Hagerstown, 1980).

Owen, S. G., *Electrocardiography* (Little, Brown, and Company, Boston, 1973).

Glasser O., et al., *Medical Devices* (Year Book Medical Publisher, Inc., Chicago, updated yearly).

Scher, A. M., "The Electrocardiogram," *Scientific American*, November 1961.

C H A P T E R **19**

Electric currents flow in these wires connecting the circuits of a large computer at Fermilab.

CURRENTS AND OHM'S LAW

Under static conditions there can exist no electric field inside a conductor. But suppose that we suddenly deposit opposite amounts of electric charge on the opposite ends of a long metallic conductor, such as a copper wire connected to the opposite plates of a capacitor (see Fig. 19.1). The conductor will then not be in electrostatic equilibrium, and the

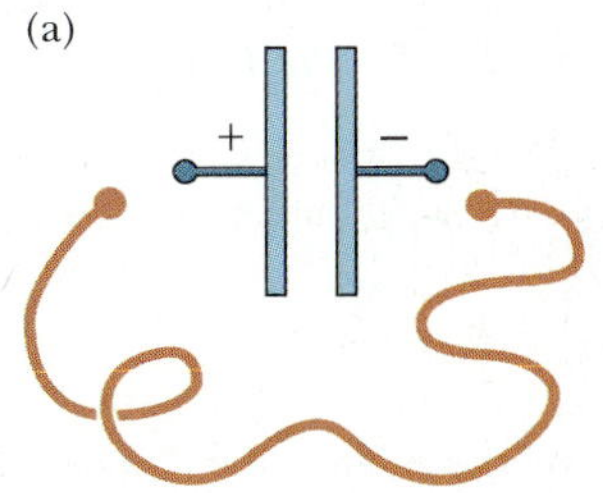

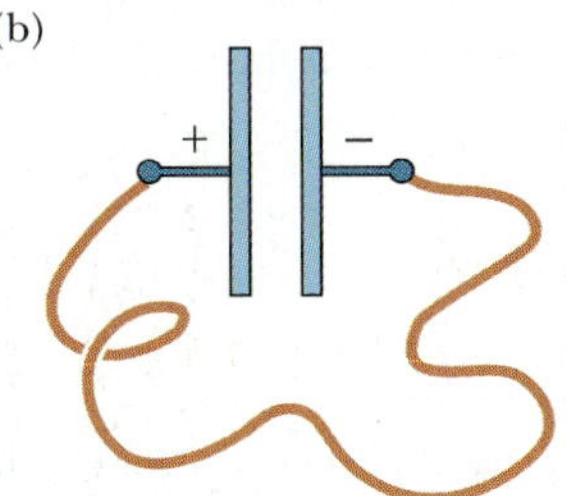

FIGURE 19.1

(a) A charged capacitor and a separate, uncharged piece of copper wire. Both the capacitor and the wire are in electrostatic equilibrium. (b) We suddenly connect the ends of the wire to the capacitor plates. The wire and the plates now form a single conductor, with opposite amounts of charge on its ends. This conductor is not in electrostatic equilibrium.

FIGURE 19.2
Electric field lines in and near a straight conductor not in equilibrium. The conductor consists of a strip of metallic paint on a paper surface. The field lines have been made visible by sprinkling grass seeds on the paper. (From O. Jefimenko, *Am. J. Phys.* **30,** 19, 1962.)

FIGURE 19.3
Electric field lines in and near a rectangular conductor carrying an electric current. (From O. Jefimenko, *Am. J. Phys.* **30,** 19, 1962.)

charges will generate an electric field along and inside the conductor (see Fig. 19.2). This electric field propels the charges toward each other. When the charges meet, they cancel, and the electric field then disappears — the conductor reaches equilibrium.

For a good conductor, such as copper, the approach to equilibrium is fairly rapid; typically, the time required to achieve equilibrium is a small fraction of a second. However, we can keep a conductor in a permanent state of disequilibrium if we continually supply more electric charge to its ends. For example, we can connect the two ends of a copper wire to the terminals of a battery or an electric generator. The terminals of such a device act as a source and sink of electric charge, just like the outlet and the intake of a pump act as source and sink of water. Under these conditions electric charge will continually flow from one terminal to the other, forming an electric current.*

19.1 ELECTRIC CURRENT

When a wire is connected between the two terminals of a battery or generator, the electric charges are propelled from one end of the wire to the other by the electric field that exists along and within the wire. Most of the field lines originate at the terminals of the battery or generator, but some field lines originate at charges that have accumulated on the wire itself. As Figs. 19.2 and 19.3 show, the field lines tend to concentrate within the conductor, and they tend to follow the conductor. If the conductor has no sharp kinks, the field lines are uniformly distributed over the cross-sectional area of the conductor. For instance, if the conductor is a more or less straight wire of constant thickness, then the electric field inside the wire will be of constant magnitude and of a direction parallel to the wire. If the length of the wire is l and if the battery or generator maintains a difference of potential ΔV across its ends, then this constant electric field in the wire has a magnitude

$$E = \Delta V / l \tag{1}$$

This electric field causes the flow of charge, or ***electric current,*** from one end of the wire to the other. Before we can explore the dependence of the current on the electric field, we need a precise definition of the current. Suppose that an amount of charge Δq flows past some given place on the wire (for example, the end of the wire) in a time Δt; then ***the electric current is defined as the charge divided by the time,***

$$I = \frac{\Delta q}{\Delta t} \tag{2}$$

* We will discuss the inner workings of batteries, generators, and other "pumps" of electricity in the next chapter.

Note that if the sides of the wire do not leak (good insulation), then the conservation of electric charge requires that the current be the same everywhere along the wire, that is, the current is simply the rate at which charge enters the wire at one end or the rate at which charge leaves at the other end.

The SI unit of current is the ampere (A); this is a flow of charge of one coulomb per second,

$$1 \text{ ampere} = 1 \text{ A} = 1 \text{ C/s} \tag{3}$$

◀ **Ampere (A)**

Table 19.1 gives the values of selected currents.

(a)

(b)

(c)

(d)

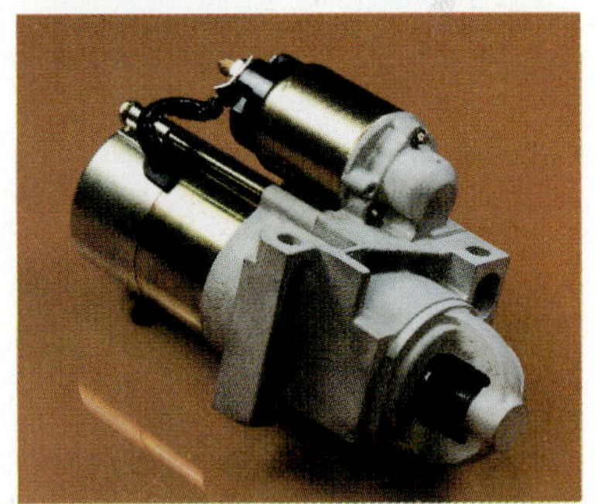

(e)

TABLE 19.1 SOME CURRENTS	
Lightning stroke **(a)**	10^4 A
High-tension power line **(b)**	10^3 A
Large transformer **(c)**	10^3 A
Large electromagnet	200 A
Starter motor of automobile **(d)**	100 A
Alternator of automobile	30 A
Fuse blows	30 A
Air conditioner	12 A
Hair dryer	10 A
Defibrillation treatment for heart	1 A
Ordinary light bulb	1 A
Flashlight bulb	0.5 A
Lethal fibrillation of heart	0.1 A
Barely perceptible by skin	1×10^{-3} A
Electronic calculator **(e)**	1×10^{-4} A

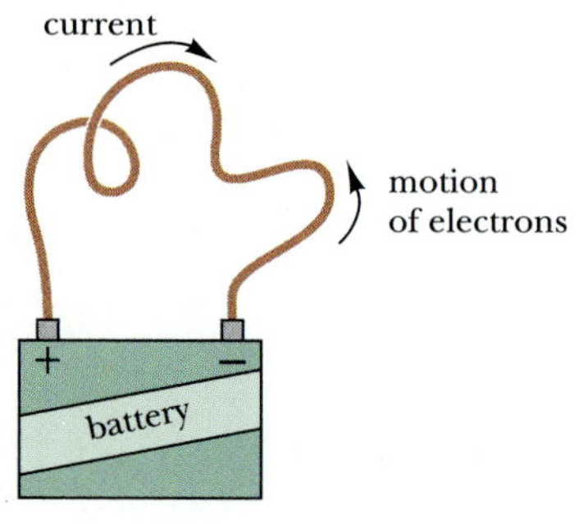

FIGURE 19.4

Flow of charge in a wire connected to a battery. According to our convention, the direction of the current is from the positive terminal to the negative, even though the actual motion of electrons in the wire is from the negative terminal to the positive.

In metallic conductors, the charge carriers are electrons — a current in a metal is nothing but a flow of electrons. In electrolytes, such as salt water, the charge carriers are positive ions, negative ions, or both — a current in such a conductor is a flow of ions. For the sake of mathematical uniformity, whenever we need to indicate the direction of the current along a conductor, ***we will adopt the convention that the current has the direction of an assumed positive flow of charge.*** This means that we pretend that the moving charges are always positive charges. Of course, in metals the moving charges are actually negative charges (electrons), and hence the above convention assigns to the current a direction opposite to that of the true motion of the charges (see Fig. 19.4). However, as regards the transfer of charge, the transport of negative charge in one direction is equivalent to the transfer of positive charge in the opposite direction. Our convention of labeling the direction of the current takes advantage of this equivalence.

EXAMPLE 1. In the headlamp of an automobile, a current of 8 A flows through the filament of the light bulb. How much electric charge flows through the filament in a minute? How many electrons?

SOLUTION: The amount of charge that flows through the filament in 1 min is

$$\Delta q = I\Delta t = 8\text{ A} \times 1\text{ min} = 8\text{ A} \times 60\text{ s} = 480\text{ C}$$

A charge of 480 C moving through the light bulb in one direction is equivalent to an electron charge of -480 C moving in the opposite direction. The number of electrons equals this charge divided by the charge per electron,

$$[\text{number of electrons}] = \frac{-480\text{ C}}{-e} = \frac{-480}{-1.6 \times 10^{-19}\text{ C}}$$

$$= 3.0 \times 10^{21}\text{ electrons}$$

19.2 RESISTANCE AND OHM'S LAW

We will now examine in detail the behavior of a current in a metallic conductor. Such conductors contain a vast number of free electrons; for example, copper has about 8×10^{22} free electrons per cubic centimeter. These electrons form a gas which fills the entire volume of the metal. Of course, in an electrically neutral conductor, the negative charge of the free electrons is exactly balanced by the positive charge of the ions that make up the crystal lattice of the metal. A current in this metallic conductor is simply a flow of the gas of electrons, while the ions remain at rest.

The flow of the gas of electrons along a metallic wire is analogous to the flow of water along a canal leading down a gentle slope. In such a canal, the force of gravity acting on the water has a component along the canal; this component pushes the water along. But the water does not accelerate — the friction between the water and the walls of the canal opposes the motion, and the water moves at a constant speed because the friction exactly matches the push of gravity.

Likewise, the electric field in the wire pushes the gas of electrons along. But the gas of electrons does not accelerate — friction between the gas and the crystal lattice of the wire opposes the motion, and the gas moves at a constant speed because the friction exactly matches the push of the electric field.

The analogy between the motion of water and the motion of the electron gas extends to the motion of the individual water molecules and individual electrons. Although the water in a canal usually has a fairly low speed, perhaps a few meters per second, the individual molecules within the water have a rather high speed — the typical speed of the random thermal motion of water molecules is about 600 m/s at ordinary temperature. But since this thermal motion consists of rapid random zigzags which are just as likely to move the molecule backward as forward, this high speed does not contribute to the net downhill motion of water. Figure 19.5 shows the motion of a water molecule in the canal; on a microscopic scale, this motion consists of rapid zigzags on which is superimposed a much slower "drift" along the canal.

FIGURE 19.5

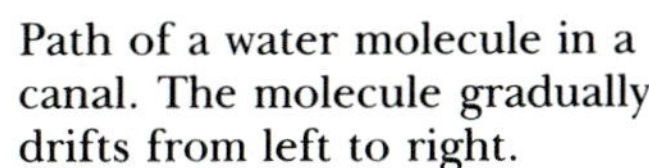

Path of a water molecule in a canal. The molecule gradually drifts from left to right.

Likewise, the electron gas moves along the wire at a rather low speed, perhaps 10^{-2} m/s, but the individual electrons have a much higher speed — the typical speed of the random motion of electrons in a metal is about 10^6 m/s (this very high speed is due to quantum-mechanical effects, which we cannot discuss here). Thus, the net motion of an electron also consists of rapid zigzags on which is superimposed a much slower drift motion along the wire. Qualitatively, the motion resembles the path of a water molecule shown in Fig. 19.5, but the amount of drift per zigzag is even less than shown in this figure.

The friction between the electron gas and the wire is caused by collisions between the electrons and the ions of the crystal lattice of the wire. An electron moving through a wire of copper will suffer about 10^{14} collisions with ions per second. Each collision slightly slows the electron down and deflects the electron from its original direction of motion. Because of the disturbing effects of these collisions, the electron never gains much velocity from the electric field that is attempting to accelerate it. The collisions dissipate the kinetic energy that the electron acquires from the electric field. This dissipated kinetic energy of the electrons remains in the crystal lattice in the form of random kinetic and potential energy of the ions, that is, it remains in the form of thermal energy or heat. In some instances, the amount of heat produced in a wire is so large as to make the wire glow. The bright glow of a light bulb is produced in this way; and so is the dull red glow of the coils of an electric range.

Georg Simon Ohm, *1787–1854, German physicist. Ohm was led to his law by an analogy between the conduction of electricity and the conduction of heat.*

The average velocity, or drift velocity, that an electron attains in the electric field is proportional to the strength of the electric field,

$$v_{\mathrm{d}} \propto E \tag{4}$$

This proportionality merely reflects the fact that if the electric field is strong, the electron gains more velocity between one collision and the next, and therefore attains a large average velocity. The electric current carried by the wire is proportional to the average velocity of the electrons, and it is therefore proportional to the electric field,

$$I \propto v_{\text{d}} \propto E \tag{5}$$

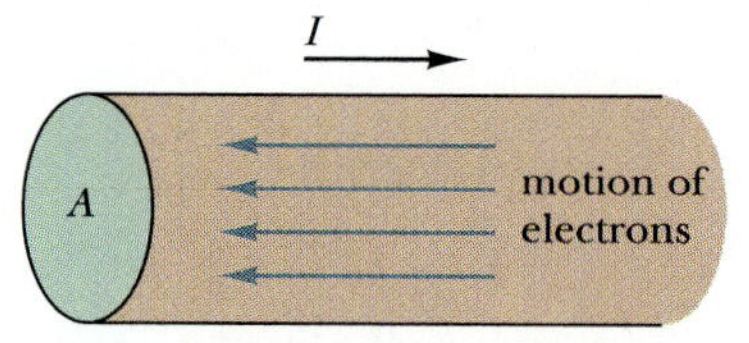

FIGURE 19.6
The current carried by the wire is uniformly distributed over its entire cross section. A wire of twice the cross-sectional area would carry twice the current (with the same electric field).

The current is also proportional to the cross-sectional area A of the wire (see Fig. 19.6), because a large cross-sectional area means that more electrons participate in the transport of charge. Since the current is proportional to both the area A and the electric field E, we can write

$$I \propto AE \tag{6}$$

With $E = \Delta V/l$, this proportionality becomes

$$I \propto A\frac{\Delta V}{l} \tag{7}$$

To transform this into an equality, we write it as

$$I = \frac{1}{\rho}\frac{A}{l}\Delta V \tag{8}$$

Resistivity ▶

where ρ is a constant of a proportionality that depends on the characteristics of the material of the wire. This constant is called the ***resistivity*** of the wire.

It is customary to define the ***resistance*** of the wire as

Resistance in terms of resistivity ▶

$$R = \rho\frac{l}{A} \tag{9}$$

Equation (8) can then be expressed in the convenient form

Ohm's Law ▶

$$I = \frac{\Delta V}{R} \tag{10}$$

This equation is called ***Ohm's Law.*** It asserts that ***the current is proportional to the potential difference between the ends of the conductor.*** Note that in Eq. (10), the resistance R plays the role of a constant of proportionality. For a wire of uniform cross section, the resistance can be calculated from the simple formula (9). But Ohm's Law is also valid for conductors of arbitrary shape — such as wires of nonuniform cross section — for which the resistance must be calculated from a more complicated formula tailored to the shape and the size of the conductor.

Ohm's Law is valid for metallic conductors and also for many nonmetallic conductors (for example, carbon) in which the current is carried by a flow of electrons. It is also valid for plasma and for electrolytes, in which the current is carried by a flow of both electrons and ions. However, we ought to keep in mind that in spite of its wide range of applicability, Ohm's Law is not a general law of nature, such as Gauss' Law. In many materials, Ohm's Law fails when the current is large; and in inhomogeneous materials, such as the layered materials used in semiconductor devices, Ohm's Law fails even when the current is small (in fact, the operation of semiconductor devices usually hinges on the exploitation of this non-ohmic behavior). Thus, Ohm's Law is not of universal applicability; it is merely an assertion about the electrical properties of some conducting materials.

EXAMPLE 2. To measure the resistance of a long piece of wire, a physicist connects this wire between the poles of a 6-volt battery. He finds that this produces a current of 30 A in the wire. What is the resistance of the wire?

SOLUTION: According to Ohm's Law,

$$R = \frac{\Delta V}{I}$$

Since the potential difference across the wire is $\Delta V = 6$ volts and the current is 30 A, we obtain

$$R = \frac{6\ \text{V}}{30\ \text{A}} = 0.2\ \text{V/A}$$

The unit V/A is called the ohm (see the next section); thus, the resistance is 0.2 ohm.

19.3 The Resistivity of Materials

As we saw in the preceding section, the resistance of a wire of uniform cross section is related to the resistivity by the formula

$$R = \rho\frac{l}{A} \tag{11}$$

We can use this formula to calculate the resistance if the resistivity of the material is known, and we can also use it to calculate the resistivity if the resistance has been measured experimentally. The latter calculation is important in the experimental determination of the resistivity of a material, which is done by measuring the potential difference and current in a wire of given length and cross section made of a sample of the material.

TABLE 19.2 RESISTIVITIES OF METALS[a]

Material	ρ	α
Silver	$1.6 \times 10^{-8}\ \Omega \cdot \text{m}$	$3.8 \times 10^{-3}/°\text{C}$
Copper	1.7×10^{-8}	3.9×10^{-3}
Aluminum	2.8×10^{-8}	3.9×10^{-3}
Brass	$\simeq 7 \times 10^{-8}$	2×10^{-3}
Nickel	7.8×10^{-8}	6×10^{-3}
Iron	10×10^{-8}	5×10^{-3}
Steel	$\simeq 11 \times 10^{-8}$	4×10^{-3}
Constantan	49×10^{-8}	1×10^{-5}
Nichrome	100×10^{-8}	4×10^{-4}

[a] At a temperature of 20°C.

As is obvious from Ohm's Law, ***the unit of resistance is 1 volt/ampere; this unit is called ohm*** (Ω),

Ohm (Ω) ▶

$$1 \text{ ohm} = 1\ \Omega = 1\ \text{V/A} \tag{12}$$

The unit of resistivity is the ohm-meter. Table 19.2 lists the resistivities of some metallic conductors.

EXAMPLE 3. A wire commonly used for electrical installations in homes is No. 10 copper wire, which has a radius of 0.129 cm. What is the resistance of a piece of this wire 30 m long? What is the potential difference along this wire if it carries a current of 10 A?

SOLUTION: The cross-sectional area of the wire is

$$A = \pi r^2 = \pi \times (0.129 \times 10^{-2}\ \text{m})^2 = 5.2 \times 10^{-6}\ \text{m}^2$$

The resistivity of copper is $1.7 \times 10^{-8}\ \Omega \cdot \text{m}$. By Eq. (11), the resistance is

$$R = \rho \frac{l}{A} = 1.7 \times 10^{-8} \Omega \cdot \text{m} \times \frac{30\ \text{m}}{5.2 \times 10^{-6}\ \text{m}^2} = 0.098\ \Omega$$

For a current of 10 A, Ohm's Law then gives a potential difference

$$\Delta V = IR = 10\ \text{A} \times 0.098\ \Omega = 0.98\ \text{volt} \tag{13}$$

The resistivity of materials depends somewhat on temperature. In ordinary metals, the ***resistivity increases slightly with temperature.*** This is due to an increase of the rate of collision between the moving electrons and the atoms of the lattice — at high temperature the atoms jump violently about their positions in the lattice, and they are then more likely to disturb the motion of the electrons. The numbers in Table 19.2 give the resistivity at room temperature (20 °C).

For a small increment in temperature, the increase in the resistivity and in the resistance is directly proportional to the increment in temperature. Mathematically, we can express this increase of resistance as

$$\Delta R = \alpha R_0 \Delta T \tag{14}$$

◀ **Increase of resistance with temperature**

where R_0 is the resistance at the initial temperature. [Note that this equation is reminiscent of the equation for the thermal expansion of the length of a solid; see Eq. (14.6).] The constant of proportionality α is called the ***temperature coefficient of resistance.*** Table 19.2 lists values of this coefficient.

◀ **Temperature coefficient of resistance**

EXAMPLE 4. Suppose that because of a current overload, the temperature of the copper wire in Example 3 increases from 20 °C to 50 °C. How much does the resistance increase?

SOLUTION: The temperature increment is $\Delta T = 30\,°\text{C}$. According to Table 19.2, the temperature coefficient of resistance for copper is $\alpha = 3.9 \times 10^{-3}/°\text{C}$. The change of resistance is therefore

$$\Delta R = 3.9 \times 10^{-3}/°\text{C} \times 0.098\ \Omega \times 30\,°\text{C}$$

$$= 0.011\ \Omega$$

The new resistance of the wire will then be

$$0.098\ \Omega + 0.011\ \Omega = 0.109\ \Omega$$

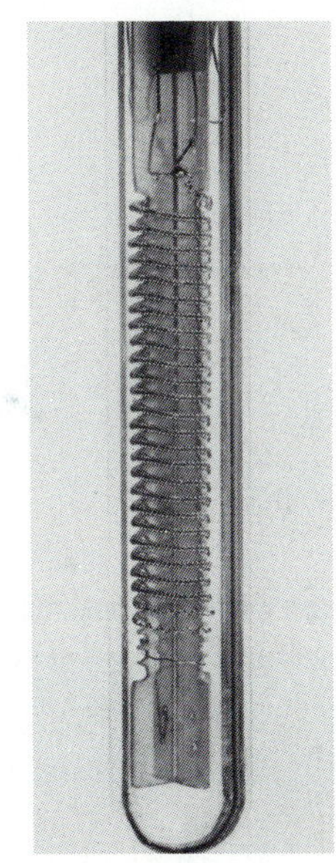

FIGURE 19.7
A coil of fine platinum wire serves as sensor in a resistance thermometer.

The change of electrical resistance with temperature is exploited in the operation of the resistance thermometer. Figure 19.7 shows such a thermometer, made of a coil of fine platinum wire. We can calibrate this thermometer in the same way as, say, a mercury thermometer by first immersing it into an ice–water mixture (0 °C) and then into boiling water (100 °C). Measurements of the resistance at these two temperatures tell us how many ohms correspond to 100 °C, and, by extrapolation, how many ohms correspond to any other temperature.

At very low temperatures, the resistivity of a metal will be substantially less than at room temperature. Some metals such as lead, tin, zinc, and niobium, exhibit the phenomenon of ***superconductivity: their resistance vanishes completely at some critical temperature above absolute zero.*** For example, Fig. 19.8 displays a plot of resistivity vs. temperature for tin; at an absolute temperature of 3.72 K, the resistivity abruptly vanishes. In one

◀ **Superconductivity**

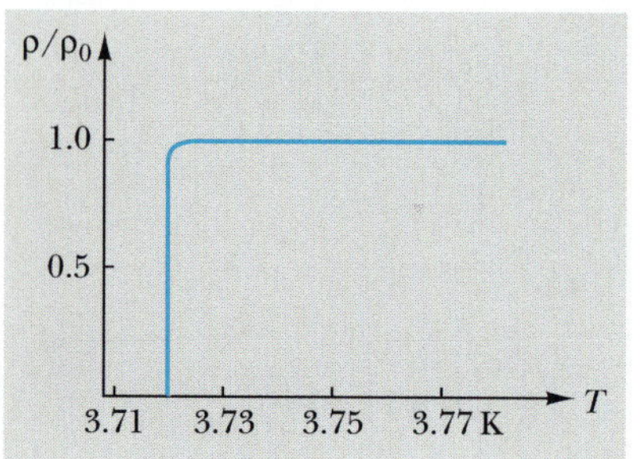

FIGURE 19.8

Resistivity of tin as a function of temperature. Below 3.72 K, the resistivity is zero. For the purpose of this plot, the resistivity has been expressed as a fraction of the resistivity at 4.2 K, the temperature of liquefaction of helium.

experiment, a current of several hundred amperes was started in a superconducting ring; the current continued on its own with undiminished strength for over a year, without any battery or generator to maintain it.

According to the definition we gave in Section 16.5, an ideal insulator is a material that does not permit any motion of electric charge. Real insulators, such as porcelain or glass, do permit some very slight motion of charge. What distinguishes them from conductors is their enormously large resistivity. Typically, the resistivities of insulators are more than 10^{20} times as large as those of conductors (see Table 19.3). This means that even when we apply a high voltage to a piece of glass, the flow of current will be insignificant (provided, of course, that the material does not suffer electrical breakdown). In fact, on a humid day it is likely that more current will flow along the microscopic film of water that tends to form on the surface of the insulator than will flow through the insulator itself. The resistivities of semiconductors, such as carbon and silicon, are between those of conductors and insulators (see Table 19.4).

TABLE 19.3 RESISTIVITIES OF INSULATORS

Material	ρ
Polyethylene	$2 \times 10^{11}\ \Omega \cdot \mathrm{m}$
Glass	$\simeq 10^{12}$
Porcelain, unglazed	$\simeq 10^{12}$
Rubber, hard	$\simeq 10^{13}$
Epoxy	$\simeq 10^{15}$

TABLE 19.4 RESISTIVITIES OF SEMICONDUCTORS

Material	ρ
Silicon	$2.6 \times 10^{3}\ \Omega \cdot \mathrm{m}$
Germanium	4.2×10^{-1}
Carbon (graphite)	3.5×10^{-5}

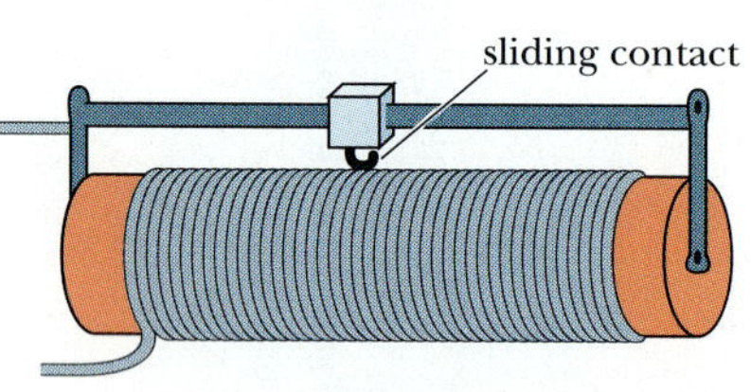

FIGURE 19.9

A rheostat consisting of a long coiled wire with a sliding contact.

19.4 RESISTANCES IN COMBINATION

Resistors ▶

The metallic wires of any electric circuit have some resistance. But in electronic instruments — radios, televisions, amplifiers — the main contribution to the resistance is usually due to devices that are specifically designed to have a high resistance. These devices are called ***resistors,*** and they are used to control and modify the currents. For instance, the volume controls on radio and TV receivers are adjustable resistors. Such a volume control is made of a long, coiled piece of high-resistance wire on which rests a sliding contact (see Fig. 19.9); by moving the sliding contact, you increase or decrease the length of wire that lies between the two external terminals, and you increase or decrease the resistance, and thereby modify the current. Adjustable resistors of this kind are called rheostats, potentiometers, or simply "pots."*

Resistors used in the circuits of electronic devices are often made of a piece of pure carbon (graphite) connected between two terminals (Fig. 19.10a). Carbon has a high resistivity and hence a small piece of carbon can have a higher resistance than a long piece of metallic wire. Such resistors obey Ohm's Law (current proportional to potential difference) over a wide range of values of the current; but if the resistor is overloaded with current, it will heat up, possibly even burn, and Ohm's Law will fail.

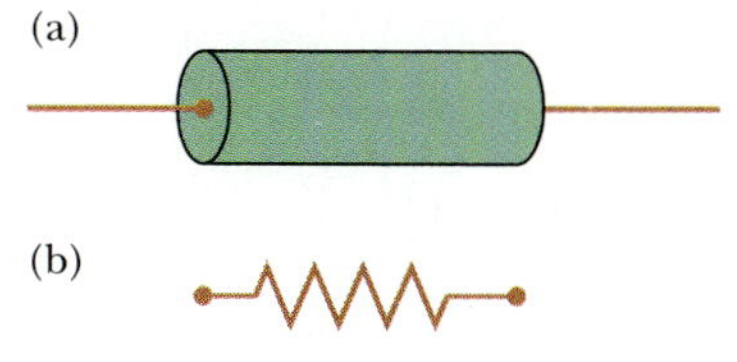

FIGURE 19.10

(a) A resistor consisting of a cylinder of carbon with two terminals attached. (b) Symbol for a resistor in a circuit diagram.

In circuit diagrams the symbol for resistor is a zigzag line, reminiscent of the path of an electron inside a conducting material (see Fig. 19.10b).

Even circuits without resistors, such as the wiring of a house, often have the main contribution to their resistance concentrated, or lumped, in one place. For instance, when an electric toaster or a light bulb is plugged into an electric outlet, the current flowing through the circuit encounters the most resistance in the relatively short piece of wire within the toaster or the light bulb. As concerns the flow of current, the short, high-resistance piece of wire in the appliance has the same effect as a carbon resistor, and in a circuit diagram it can be treated as though it were a resistor.

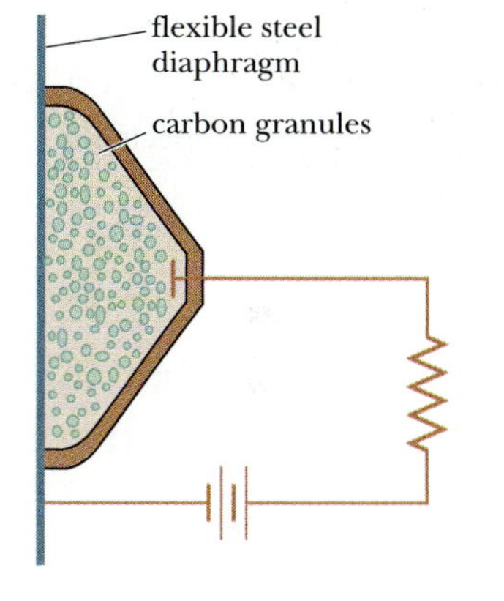

FIGURE 19.11

A resistance microphone.

A special kind of carbon resistor is used as the sensitive element in a ***resistance microphone,*** or carbon-button microphone, illustrated in Fig. 19.11. The carbon button consists of small carbon granules, packed together. The resistance of the packed granules varies with the pressure exerted on them — if the button is compressed by an external force, the granules come into better contact, and the resistance across the button decreases. In the microphone, the external force is provided by the diaphragm, which moves in response to an incident sound wave. The button rests against the diaphragm of the microphone. When the diaphragm moves, it compresses the granules and reduces the resistance of the button. If the microphone is connected to a battery, as in Fig. 19.11, the change of resistance results in a change in the current in the circuit — the microphone transforms the sound signal into an electric signal. Resistance microphones are cheap and rugged, and they are widely used in telephones. But their sensitivity to high frequencies is poor, and because of this, voices on the telephone have a flat, nonresonant quality.

FIGURE 19.12

Two resistors connected in series.

For resistors, as for capacitors, the two simplest ways of connecting several resistors are in *series* and in *parallel.* Figure 19.12 shows two resistors connected in ***series.*** Since each of these resistors offers a resistance to the current entering the wire, our intuition suggests that the net resistance of this combination is the sum of the individual resistances,

$$R = R_1 + R_2 \tag{15}$$

The formal derivation of this result begins with the observation that if the potential differences across the individual resistors are ΔV_1 and ΔV_2, then the net potential difference across the combination is

$$\Delta V = \Delta V_1 + \Delta V_2 \tag{16}$$

* The name *potentiometer* is also used in a more technical sense for a device used to measure potential, which incorporates a sliding contact similar to that shown in Fig. 19.9.

This additivity of the potentials is a direct consequence of the definition of the potential as work per unit charge — the net work done by the electric field on a unit charge that moves through the first resistor and then through the second resistor is simply the sum of the work done in the first and the work done in the second. Furthermore, the currents in both resistors are exactly the same, since any charge that flows through the first resistor continues to flow through the second. Hence, using Ohm's Law, we find

$$\Delta V = \Delta V_1 + \Delta V_2 = IR_1 + IR_2 = I(R_1 + R_2)$$

From this it is clear that the resistance of the combination is equivalent to a single resistance given by Eq. (15).

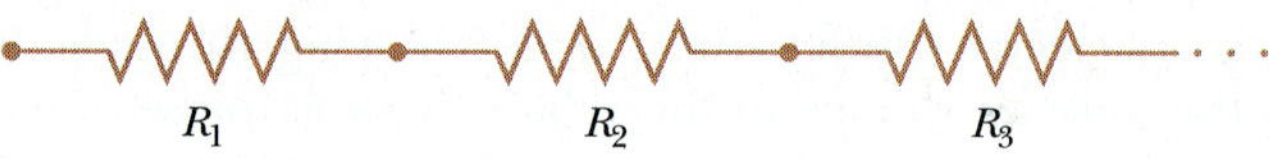

FIGURE 19.13
Several resistors connected in series.

We can easily generalize this result to any number of resistors in series (see Fig. 19.13). The net resistance of the series combination is

Series combination of resistors ▶

$$R = R_1 + R_2 + R_3 + \dots \tag{17}$$

Figure 19.14 shows two resistors in ***parallel.*** The potential difference across each resistor is the same as the potential difference across the combination. Hence, from Ohm's Law, the currents are

$$I_1 = \frac{\Delta V}{R_1} \quad \text{and} \quad I_2 = \frac{\Delta V}{R_2} \tag{18}$$

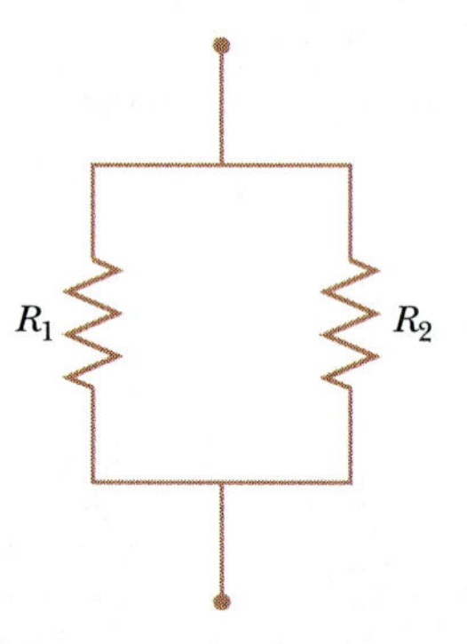

FIGURE 19.14
Two resistors connected in parallel.

The total current through the combination is the sum of the individual parallel currents,

$$I = I_1 + I_2 = \frac{\Delta V}{R_1} + \frac{\Delta V}{R_2} = \left(\frac{1}{R_1} + \frac{1}{R_2}\right)\Delta V \tag{19}$$

The resistance of the combination is therefore equivalent to a single resistance given by

$$\frac{1}{R} = \frac{1}{R_1} + \frac{1}{R_2} \tag{20}$$

Note that the resistance of the parallel combination is less than each of the individual resistances. For example, if $R_1 = R_2 = 1.0\ \Omega$, then $R = 0.5\ \Omega$.

We can also generalize this to any number of resistors in parallel (see Fig. 19.15). The net resistance of the parallel combination is given by

Parallel combination of resistors ▶

$$\frac{1}{R} = \frac{1}{R_1} + \frac{1}{R_2} + \frac{1}{R_3} + \dots \tag{21}$$

EXAMPLE 5. Two resistors, with $R_1 = 10\ \Omega$ and $R_2 = 20\ \Omega$, are connected in parallel (see Fig. 19.14). A total current of 1.8 A flows through this combination. What is the potential difference across the combination? What is the current in each resistor?

SOLUTION: According to Eq. (21), the net resistance is given by

$$\frac{1}{R} = \frac{1}{R_1} + \frac{1}{R_2} = \frac{1}{10\ \Omega} + \frac{1}{20\ \Omega} = \frac{30}{200\ \Omega}$$

which yields $R = 200\ \Omega/30 = 6.67\ \Omega$. Hence, Ohm's Law applied to the combination as a whole tells us that the potential difference is

$$\Delta V = IR = 1.8\ \text{A} \times 6.67\ \Omega = 12.0\ \text{volts}$$

and Ohm's Law applied to the two resistors tells us that the individual currents are

$$I_1 = \frac{\Delta V}{R_1} = \frac{12.0\ \text{volts}}{10\ \Omega} = 1.20\ \text{A}$$

$$I_2 = \frac{\Delta V}{R_2} = \frac{12.0\ \text{volts}}{20\ \Omega} = 0.60\ \text{A}$$

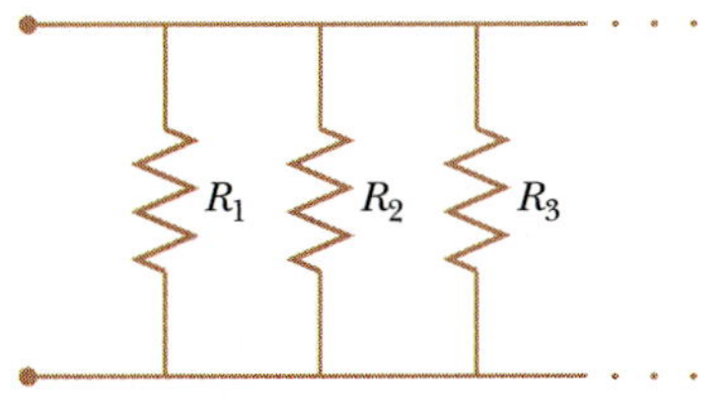

FIGURE 19.15

Several resistors connected in parallel.

In the calculations of the preceding example, we neglected the resistance of the wires connecting the resistors. This is a good approximation if the resistance of these wires is small compared with the resistance of the resistors. If this is not so, then we must take the resistance of the wires into account in our calculation. In circuit diagrams, it is customary to represent the resistance of a wire schematically by an equivalent resistor of suitable magnitude; this means we pretend that all of the resistance of a wire is concentrated, or lumped, in one place. The line segments connecting these equivalent resistors to the rest of the circuit can then be assumed to have zero resistance exactly. For instance, if the resistance of each of the two wires connecting the resistors in Fig. 19.14 is 2 Ω, then the schematic circuit diagram that includes the resistance of the wires is as shown in Fig. 19.16. Note that in this diagram, we placed the resistors of 2 Ω that represent the resistances of the wires immediately above the resistors R_1 and R_2. We could equally well place these equivalent resistors immediately below R_1 and R_2, or one-half of each above and one-half below — all these arrangements are effectively equivalent, and lead to the same distribution of current among the two branches of the circuit. Also note that we neglected the resistance of the short leads sticking out of the circuit above and below; if these have a significant resistance, then we would also need to take these into account.

(a)

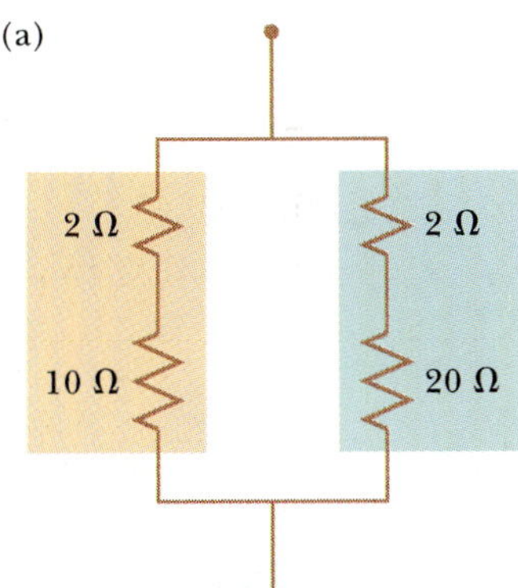

(b)

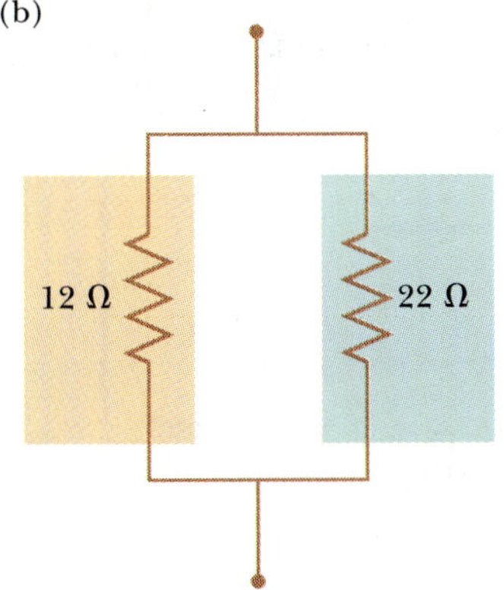

FIGURE 19.16

(a) In this circuit diagram, the resistances of the wires are represented by the two small resistors. (b) The right (blue) and the left (orange) branches of the circuit are equivalent to two resistors of 12 Ω and 22 Ω connected in parallel.

EXAMPLE 6. If, as in the preceding example, a current of 1.8 A flows through the combination of resistors and resistive wires shown schematically in Fig. 19.16, what is the potential difference across the combination? What is the current in each resistor?

SOLUTION: To solve this problem we proceed in two steps. First, we find the resistances of the left and the right branches of the circuit. The left branch consists of a resistor R_1 of 10 Ω in series with 2 Ω; the right branch consists of a resistor R_2 of 20 Ω in series with 2 Ω. Hence, the resistance of the left branch of the circuit is $R_1' = R_1 + 2\,\Omega = 12\ \Omega$ and that of the right branch is $R_2' = R_2 + 2\ \Omega = 22\ \Omega$. Next, we find the net resistance of the circuit by combining R_1' and R_2' in parallel (Fig. 19.16b). This parallel combination of the resistances of the left and the right branches gives:

$$\frac{1}{R} = \frac{1}{R_1'} + \frac{1}{R_2'} = \frac{1}{12\ \Omega} + \frac{1}{22\ \Omega} = \frac{34}{264\ \Omega}$$

From this, we find $R = 7.76\ \Omega$ for the net resistance of the circuit. Hence, by Ohm's Law, the potential difference across the combination is

$$\Delta V = IR = 1.8\ \text{A} \times 7.76\ \Omega = 14.0\ \text{V}$$

and the individual currents in the branches are

$$I_1 = \frac{\Delta V}{R_1'} = \frac{14.0\ \text{volts}}{12\ \Omega} = 1.16\ \text{A}$$

$$I_2 = \frac{\Delta V}{R_2'} = \frac{14.0\ \text{volts}}{22\ \Omega} = 0.64\ \text{A}$$

COMMENTS AND SUGGESTIONS: When seeking the net resistance of a circuit containing several resistors connected in some complicated manner, proceed in two steps, as in the case of a circuit with capacitors. In the first step, look for groups of resistors that form simple parallel or series combinations (such as the group of two resistors in the left branch and the group of two resistors in the right branch in Fig. 19.16a). Evaluate the resistance of each such group. In the second step, see how these effective group resistances are connected to each other (as in Fig. 19.16b), and evaluate the net resistance of the combination of the groups.

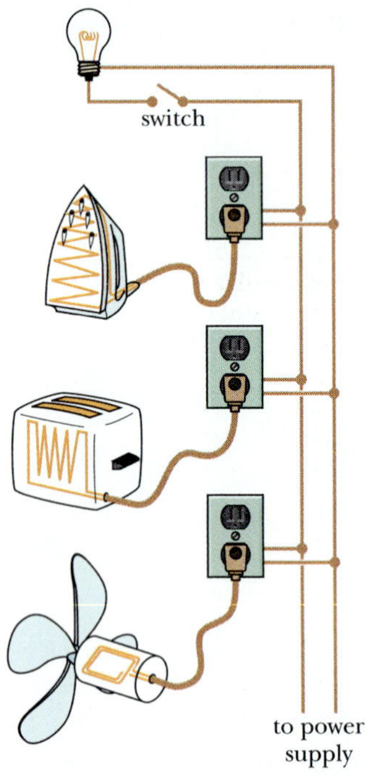

FIGURE 19.17
Several appliances plugged into the power outlets in a house.

Ordinary appliances plugged into the power outlets in your home are connected in parallel (see Fig. 19.17). The potential difference across each of these appliances is 115 V, which is the potential difference supplied across the wires that feed electric power into the house.* Some of these appliances — such as light bulbs, electric toasters, irons, hot plates, heaters, and blankets — simply consist of a piece of resistive wire in which the dissipation of the electric energy produces heat. More complicated appliances — such as refrigerators, fans, washers, radios, TV receivers, videocassette recorders, and tape decks — contain electric motors and a variety of electronic devices. Such complicated devices usually do not obey Ohm's Law; for instance, if you plug your TV into a 120-V outlet instead of a 115-V outlet, the current will not increase in proportion to the potential. However, for purposes of comparison, it is often useful to assign to a complicated device an effective resistance $R = \Delta V/I$, as though Ohm's Law were valid. Since this effective resistance of a complicated device is *not* constant, it can be used only in calculations at the given, fixed potential difference of, say, 115 V.

(a)

115 V

(b)

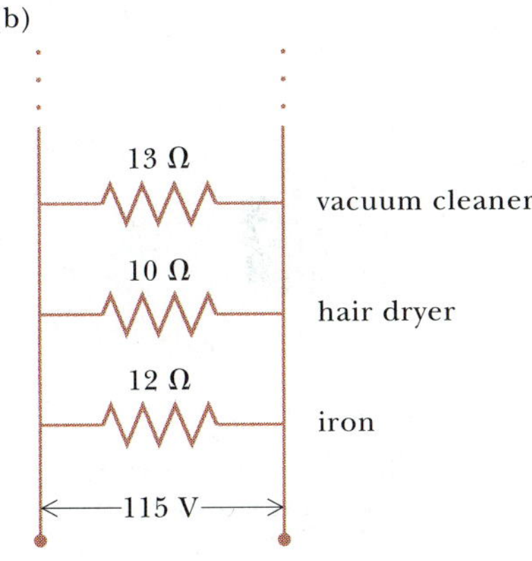

FIGURE 19.18

(a) Three appliances plugged into the same outlet.
(b) Schematic diagram showing how the resistances are connected.

EXAMPLE 7. A vacuum cleaner, a hair dryer, and an electric iron are simultaneously plugged into a single outlet (see Fig. 19.18a). Their resistances are 9.0 Ω, 10 Ω, and 12 Ω, respectively. The outlet supplies a potential difference of 115 V. What is the net current flowing through the outlet? If the maximum safe current that the outlet can carry is 30 A, is it advisable to plug all of these appliances into the outlet?

SOLUTION: All the appliances are connected in parallel (see Fig. 19.18b); thus the potential difference across each is 115 V. By Ohm's Law, the individual currents are then, respectively,

$$I_1 = 115\ \text{V}/9.0\ \Omega = 12.8\ \text{A}$$

$$I_2 = 115\ \text{V}/10\ \Omega = 11.5\ \text{A}$$

$$I_3 = 115\ \text{V}/12\ \Omega = 9.6\ \text{A}$$

The net current is the sum 12.8 A + 11.5 A + 9.6 A = 33.9 A. This is more than the outlet can carry safely. These devices should not be plugged into the same outlet, or into several outlets supplied by the same wire.

* The potential difference supplied across the power wires is actually an oscillating potential difference, which periodically reverses sign. We will deal with the details of such oscillating, or alternating, potentials and currents in Chapter 23.

SUMMARY

- **Electric field in uniform wire:** $E = \Delta V/l$
- **Electric current:** $I = \Delta q/\Delta t$
- **Resistance in terms of resistivity:** $R = \rho\frac{l}{A}$
- **Increase of resistance with temperature:** $\Delta R = \alpha R_0 \Delta T$
- **Ohm's Law:** $I = \frac{\Delta V}{R}$
- **Series combination of resistors:** $R = R_1 + R_2 + R_2 + \ldots$
- **Parallel combination of resistors:** $1/R = 1/R_1 + 1/R_2 + 1/R_3 + \ldots$

QUESTIONS

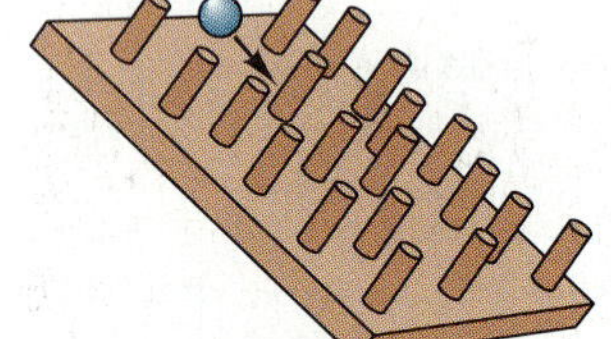

FIGURE 19.19
Mechanical analog of electric conductor with resistance.

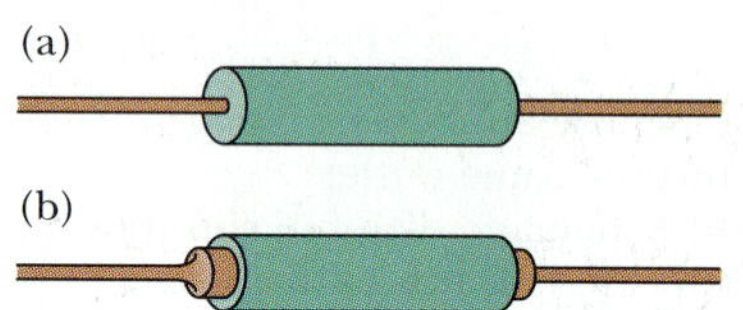

FIGURE 19.20
(a) The wires touch the carbon cylinder directly. (b) The wires end in contact plates which touch the carbon cylinder.

FIGURE 19.21
End of wire wrapped around terminal post.

1.–A wire is carrying a current of 15 A. Is this wire in electrostatic equilibrium?

2.–Can a current flow in a conductor when there is no electric field? Can an electric field exist in a conductor when there is no current?

3.–Figure 19.19 shows a putative mechanical analog of an electric conductor with resistance. A marble rolling down the inclined plane is stopped every few centimeters by a collision with a pin and therefore maintains a constant average velocity v_d. Is this a good analog — that is, is the average velocity v_d proportional to g?

4.–By what factor must we increase the diameter of a wire to decrease its resistance by a factor of 2?

5.–What derivations from Ohm's Law do you expect if the current is very large?

6.–An automobile battery has a potential difference of 12 V between its terminals even when there is no current flowing through the battery. Does this violate Ohm's Law?

7.–Figure 19.20 shows two alternative experimental arrangements for determining the resistance of a carbon cylinder. A known potential difference is applied via the copper terminals, the current is measured, and the resistance is calculated from $R = \Delta V/I$. The arrangement in Fig. 19.20a yields a higher resistance than that in Fig. 19.20b. Why?

8.–Why is it bad practice to operate a high-current appliance off an extension cord?

9.–The installation instructions for connecting an outlet to the wiring of a house recommended that the wire be wrapped at least three-quarters of the way around the terminal post (Fig. 19.21). Explain.

10.–Aluminum wire should never be connected to terminals designed for copper wire. Why not? (Hint: Aluminum has a considerably higher coefficient of thermal expansion than copper.)

11.–A wire of copper and a wire of silver are connected in parallel. Both wires have the same length and the same diameter. Which carries more current?

PROBLEMS

Section 19.1: Electric Current

1.——The electric current in the light bulb of a flashlight is 0.5 A. How much electric charge flows through the light bulb in one hour? How many electrons pass through the light bulb?

2.——The filament of a light bulb consists of a piece of tungsten wire, 5.0 cm long. If the potential difference across this filament is 115 V, what is the electric field in the wire?

3.——In a typical lightning stroke, the electric current is about 20,000 A and it lasts about 10^{-4} s. The direction of the current is from the ground to the cloud. What is the charge (magnitude and sign) that this stroke deposits on the ground?

4.——A fully charged automobile battery delivers 40 A for 1 hour before it runs down. How much electric charge flows through the battery during this time? How many electrons pass through the battery?

5.*——A 40-μF capacitor is initially charged with a 9-V battery. To reverse the voltage on the capacitor, how long must a constant 3-A current flow from the positive to the negative plate of the capacitor?

6.*——What is the capacitance of a capacitor which charges to 1.4 V in 0.50 μs by a constant current of 25 mA?

Section 19.2: Resistance and Ohm's Law

7.——A conducting wire of length 2.0 m is connected between the terminals of 12-V battery. The resistance of the wire is 3.0 Ω. What is the electric current in the wire? What is the electric field in the wire?

8.——When a thin copper wire is connected between the poles of a 1.5-V battery, the current in the wire is 0.50 A. What is the resistance of this wire? What will be the current in the wire if it is connected between the terminals of a 7.5-V battery?

9.——The resistance of a 150-W, 115-V light bulb is 0.73 Ω when the light bulb is at its operating temperature. What current passes through this light bulb when in operation? How many electrons per second does this amount to?

10.——The resistance of the wire in the windings of an electric starter motor for an automobile is 3.0×10^{-2} Ω. The motor is connected to a 12-V battery. What current will flow through the motor when it is stalled (does not turn)?

11.——The resistance of the filament of the light bulb of a flashlight is 8 Ω (at its operating temperature). The two batteries of the flashlight, of 1.5 V each, are connected in series, and they therefore supply 3 V. What is the current through the filament? What will be the current if you rewire this flashlight so it operates on a single battery? Assume that the resistance of the filament remains the same (actually, with less current, the filament will be cooler, and its resistance will be lower).

12.*——Two copper wires are connected in parallel. Both wires have the same length, but one has twice the diameter of the other. What fraction of the total current flows in each wire?

13.* An aluminum wire has a resistance of 0.10 Ω. If you draw this wire through a die, making it thinner and twice as long, what will be its new resistance?

14.* Show that for a wire of given length made of a given material, the resistance is inversely proportional to the mass of the wire.

Section 19.3: The Resistivity of Materials

15. According to Example 3, the resistance of a copper wire of radius 0.129 cm and length 30 m is 0.098 Ω. What is the resistance of a similar aluminum wire?

16. When the starter motor of an automobile is in operation, the cable connecting it to the battery carries a current of 80 A. This cable is made of copper and is 0.50 cm in diameter. What is the electric field in the cable?

17. The electromagnet of a bell is constructed by winding copper wire around a cylindrical core, like thread on a spool. The diameter of the copper wire is 0.45 mm, the number of turns in the winding is 260, and the average radius of a turn is 5.0 mm. What is the resistance of the wire?

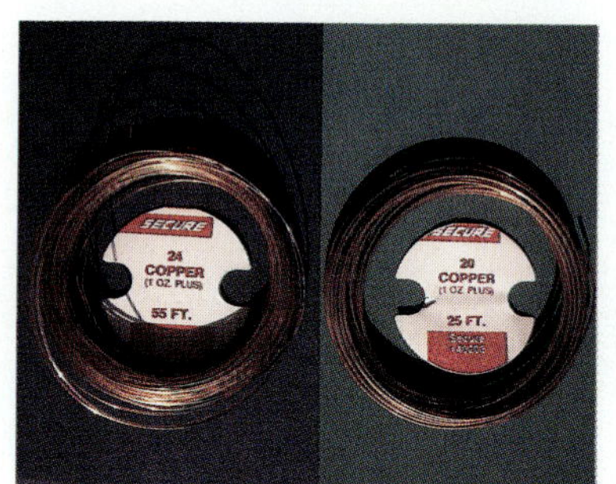

Samples of copper wires of different diameters.

18. The following is a list of some types of copper wire manufactured in the United States:

Gauge No.	Diameter
10	0.2588 cm
11	0.2305
12	0.2053

For each type of wire, calculate the resistance for a 100-m segment.

19. To measure the resistivity of a metal, an experimenter takes a wire of this metal of diameter 0.500 mm and length 1.10 m and applies a potential difference of 12.0 V to the ends. She finds that the resulting current is 3.75 A. What is the resistivity?

20. A high-voltage transmission line has an aluminum cable of diameter 3.0 cm, 200 km long. What is the resistance of this cable?

21. A lightning rod of iron has a diameter of 0.80 cm and a length of 0.50 m. During a lightning stroke, it carries a current of 1.0×10^4 A. What is the potential drop along the rod?

22. You want to make a resistor of 1.0 Ω out of a carbon rod of diameter 1.0 mm. How long a piece of carbon do you need?

23. The copper cable connecting the positive pole of a 12-V automobile battery to the starter motor is 0.60 m long and 0.50 cm in diameter.
   (a) What is the resistance of this cable?
   (b) When the starter motor is stalled, the current in the cable may be as much as 600 A. What is the potential drop along the cable under these conditions?

24. The electromagnet of a bell is wound with 8.2 m of copper wire of diameter 0.45 mm. What is the resistance of the wire? What is the current through the wire if the electromagnet is connected to a 12-V source?

25.* The air conditioner in a home draws a current of 12 A.
   (a) Suppose that the pair of wires connecting the air conditioner to the fuse box are No. 10 copper wire with a diameter of 0.259 cm and a length of 25 m each. What is the potential drop along each wire? Suppose that the voltage delivered to the home is exactly 115 V at the fuse box. What is the voltage delivered to the air conditioner?
   (b) Some older homes are wired with No. 12 copper wire with a diameter of 0.205 cm. Repeat the calculation of Part (a) for this wire.

26.*—A copper cable in a high-voltage transmission line has a diameter of 3.0 cm and carries a current of 750 A. What is the electric field in the wire?

27.—Consider the aluminum cable described in Problem 20. If the temperature of this cable increases from 20°C to 50°C, how much will its resistance increase?

28.—What increase of temperature will increase the resistance of a nickel wire from 0.5 Ω to 0.6 Ω?

29.—The resistivity of a certain material is found to increase by 4.45% when heated from 20.0°C to 70.0°C. What will be the resistivity of the material at 120°C?

30.*—An aluminum wire of length 15 m is to carry a current of 25 A with a potential drop of no more than 5 V along its length. What is the minimum acceptable diameter of this cable?

31.*—According to the National Electrical Code, the maximum permissible current in a No. 12 copper wire (diameter 0.205 cm) with rubber insulation is 25 A.

(a) What is the potential drop along a 1-meter segment of the wire carrying this current?

(b) What is the electric field in the wire?

32.*—A parallel-plate capacitor with a plate area of 8.0×10^{-2} m^2 and a plate separation of 1.0×10^{-4} m is filled with polyethylene. If the potential difference between the plates is 2.0×10^4 V, what will be the current flowing through the polyethylene from one plate to the other?

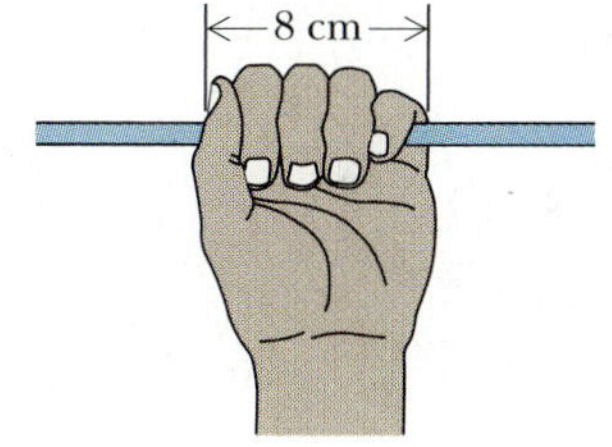

FIGURE 19.22
Contact between hand and wire.

33.*—The resistance of a square centimeter of dry human epidermis is about 10^5 Ω. Suppose that a (foolish) man firmly grasps two wires in his fists (see Fig. 19.22). The wires have a radius of 0.13 cm, and the skin of each hand is in full contact with the surface of the wire over a length of 8 cm.

(a) Calculate the resistance the man offers to a current flowing through his body from one wire to the other. In this calculation you can neglect the resistance of the internal tissues of the human body, because the body fluids are reasonably good conductors, and their resistance is small compared with the skin resistance.

(b) What current will flow through the body of the man if the potential difference between the wires is 12 V? If it is 110 V? If it is 240 kV? What is your prognosis in each case? See Section 20.8 for the effects of electric currents on the human body.

34.*—Although aluminum has a somewhat higher resistivity than copper, it has the advantage of having a considerably lower density. Find the weight of a 100-m segment of aluminum cable 3 cm in diameter. Compare this weight with that of a copper cable of the same length and the same resistance. The densities of aluminum and of copper are 2700 kg/m^3 and 8950 kg/m^3, respectively.

Section 19.4: Resistances in Combination

35.—Three resistors of 4 Ω, 6 Ω, and 8 Ω, respectively, are connected in series. What is the resistance of this combination? If this combination is connected to a 12-V battery, what is the current?

36.—Three resistors of 5 Ω, 7 Ω, and 9 Ω, respectively, are connected in parallel. What is the resistance of this combination? If this combination is connected to a 12-V battery, what is the net current? What is the current in each resistor?

37.—A brass wire and iron wire of equal diameters and of equal lengths are connected in parallel. Together they carry a current of 6.0 A. What is the current in each?

38.—Ten equal strands of thin wire are held together side by side to form a thick wire of resistance R. Six of the strands are then separated and tied together end-to-end to form a new wire six times as long as the original wire. What is the resistance of the long thin wire?

39.—A long thin wire of resistance R is cut into 8 pieces. Four of these pieces are then placed side by side to form a new wire $\frac{1}{8}$ of the original length. What is the resistance of the new wire?

40.—Consider the brass and iron wires described in Problem 37. What is the current in each if they are at a temperature of 90°C instead of a temperature of 20°C?

41.*—An electric cable of length 12.0 m consists of a copper wire of diameter 0.30 cm surrounded by a cylindrical layer of rubber insulation of thickness 0.10 cm. A potential difference of 6.0 V is applied to the ends of the cable.

(a) What will be the current in the copper?

(b) Taking into account the conductivity of the rubber (see Table 19.3), what will be the current in the rubber?

42.*—A water pipe is made of iron with an outside diameter of 2.5 cm and an inside diameter of 2.0 cm. The pipe is used to ground an electric appliance. If a current of 20 A flows from the appliance into the water pipe, what fraction of this current will flow in the iron? What fraction in the water? Assume that water has a resistivity of 0.01 $\Omega \cdot$m.

43.*—A copper wire, of length 0.50 m and diameter 0.259 cm, has been accidentally cut by a saw. The region of the cut is 0.4 cm long, and in this region the remaining wire has a cross-sectional area of only one-quarter of the original area. What is the percentage increase of the resistance of the wire caused by this cut?

44.*—The windings of high-current electromagnets are often made of copper pipe. The current flows in the walls of the pipe and cooling water flows in the interior of the pipe. Suppose the copper pipe has an outside diameter of 1.20 cm and an inside diameter of 0.80 cm. What is the resistance of 30 m of this copper pipe? What voltage must be applied to it if the current is to be 600 A?

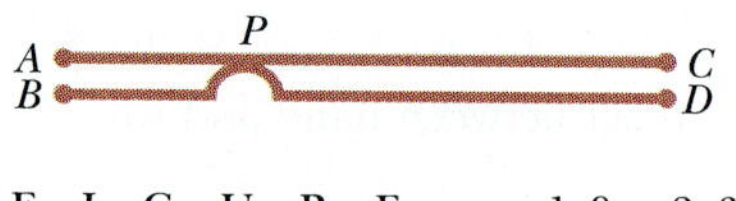

FIGURE 19.23

A pair of wires with a short at the point P (the wires are joined at P).

45.*—An underground telephone cable, consisting of a pair of wires, has suffered a short somewhere along its length (Fig. 19.23). The telephone cable is 5 km long and in order to discover where the short is, a technician first measures the resistance across the terminals AB; then he measures the resistance across the terminals CD. The first measurement yields 30 Ω; the second 70 Ω. Where is the short?

46.*—Three resistors with resistances of 3.0 Ω, 5.0 Ω, and 8.0 Ω are connected in parallel. If this combination is connected to a 12.0-V battery, what is the current through each resistor? What is the current through the combination?

47.*—A flexible wire for an extension cord for electric appliances is made of 24 strands of fine copper wire, each of diameter 0.053 cm, tightly twisted together. What is the resistance of a length of 1.0 m of this kind of wire?

48.*—Two copper wires of diameters 0.26 cm and 0.21 cm, respectively, are connected in parallel. What is the current in each if the combined current is 18 A?

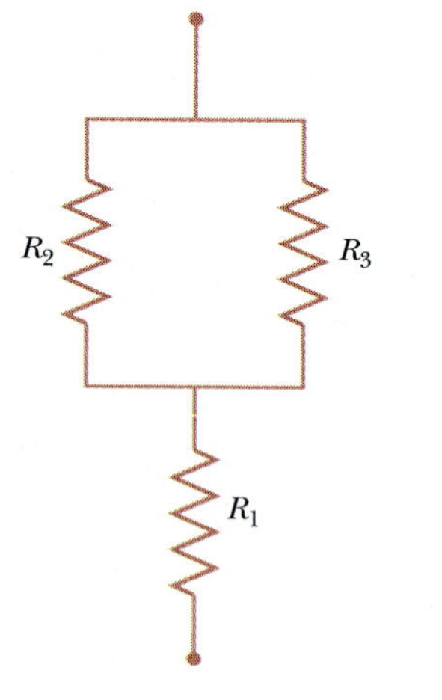

FIGURE 19.24

49.*—Three resistors with $R_1 = 2.0\ \Omega$, $R_2 = 4.0\ \Omega$, and $R_3 = 6.0\ \Omega$ are connected as shown in Fig. 19.24.

(a) Find the net resistance of the combination.

(b) Find the current that passes through the combination if a potential difference of 8.0 V is applied to the terminals.

(c) Find the potential difference and the current for each individual resistor.

50.* —Three resistors with $R_1 = 4.0\ \Omega$, $R_2 = 6.0\ \Omega$, and $R_3 = 8.0\ \Omega$ are connected as shown in Fig. 19.25.

(a) Find the net resistance of the combination.

(b) Find the current that passes through the combination if a potential difference of 12.0 V is applied to the terminals.

(c) Find the potential difference and the current for each individual resistor.

51.** —Commercially manufactured superconducting cables consist of filaments of superconducting wire embedded in a matrix of copper. As long as the filaments are superconducting, all the current flows in them, and no current flows in the copper. But if superconductivity suddenly fails because of a temperature increase, the current can spill into the copper; this prevents damage to the filaments of the superconductor. Calculate the resistance per meter of length of the copper matrix. The diameter of the copper matrix is 0.7 mm, and each of the 2100 filaments has a diameter of 0.01 mm.

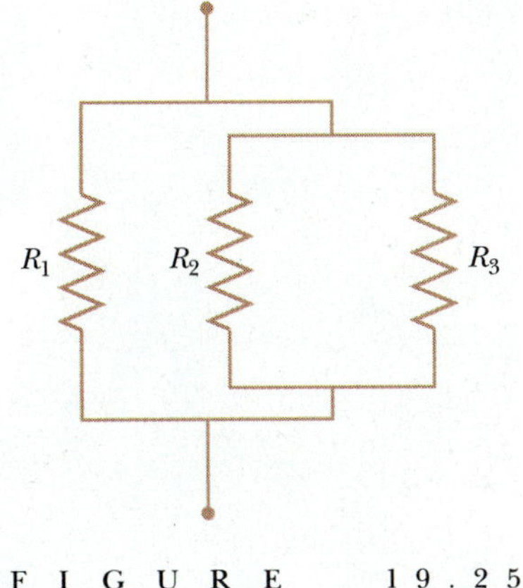

FIGURE 19.25

In a light bulb, electric energy is converted into thermal energy, and this heats the thin tungsten filament to incandescence. The filament is normally enclosed in an evacuated glass bulb, so the filament will not burn. Here the bulb has been broken, but the current is low, and the filament barely glows and does not burn.

DC CIRCUITS

Direct currents (DC) and alternating currents (AC) ▸

The electric circuits installed in your automobile and the electric circuits installed in your home carry two different kinds of currents: ***direct currents (DC)*** and ***alternating currents (AC).*** The direct current flows steadily along the wires of the circuit; it remains constant, except when it is switched on or off. The alternating current periodically reverses its direction of flow along the wire; like a pendulum swinging back and forth, the alternating current oscillates sinusoidally from one direction (positive) to the opposite direction (negative).

To keep any current flowing in the wires of the circuit, we must connect the ends of the wire to a "pump of electricity," a device that continuously supplies electric charges to one end of the wire and removes them from the other. What kind of current flows in the wire depends on what kind of pump we use. A steady pump — such as an automobile battery — produces a steady current. An alternating pump — such as the generator of the power station to which the circuits in your home are connected — produces an alternating current. In this chapter we will deal with direct currents produced by batteries or by other pumps of electricity that behave like batteries.

20.1 ELECTROMOTIVE FORCE

Figure 20.1 shows a simple circuit consisting of a single (resistive) wire connected to the terminals of a battery. A steady, time-independent current will then flow around this circuit; in Fig. 20.1, we have indicated the direction of the current according to our convention that it is the direction of flow of (hypothetical) positive charges. The current flowing in this circuit is an example of a direct current — as long as the "strength" of the battery and the resistance of the wire remain constant, the current will also remain constant.

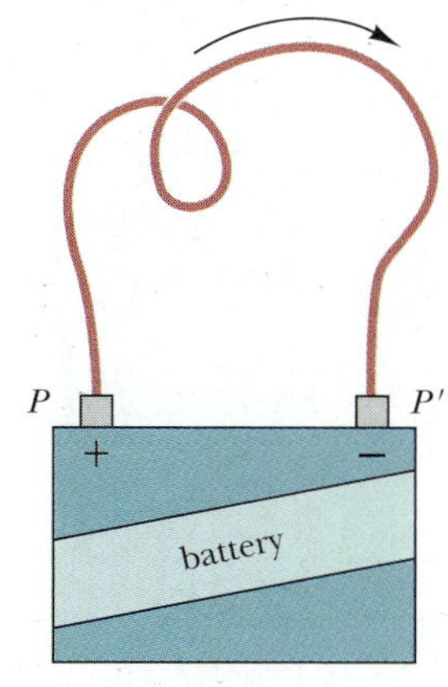

FIGURE 20.1
A wire connected between the terminals of a battery.

The battery must do work on the charges in order to keep them flowing around the circuit. Suppose that a (hypothetical) positive charge is originally at the point P, at one terminal of the battery. Pushed along by the electric field, the charge moves along the wire. On the average, the kinetic energy that the charge gains from the electric field is dissipated by friction within the wire, and the charge reaches the point P', at the other terminal of the battery, with its original kinetic energy.

Thus, the kinetic energy does not change. But the potential energy of the charge does change. The electric field is directed along the wire, and it does work on the charge; hence the electric potential steadily decreases with distance along the wire, and the charge reaches the point P' with a potential energy lower than its original potential energy. In order to keep the current flowing, the battery must "pump" the charge from the low-potential terminal to the high-potential terminal; that is, the battery must supply electric potential energy to the charge.

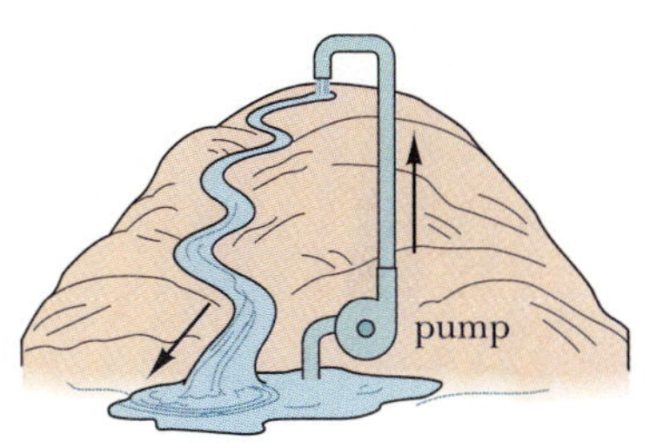

FIGURE 20.2
Mechanical analog of the battery–wire circuit of Fig. 20.1.

The role of the battery is analogous to that of a hydraulic pump that lifts water from the bottom to the top of a hill, and thereby increases the gravitational potential energy of the water (see Fig. 20.2). The wire is analogous to a channel by means of which the water runs down the hill returning to the pump. The water then flows in a closed hydraulic circuit, just as charge flows in a closed electric circuit. The hydraulic pump of Fig. 20.2 can be regarded as a source of gravitational potential energy — it produces this energy from an external supply of chemical or mechanical energy. Likewise, the "pump of electricity" of Fig. 20.1 can be regarded as a source of electric potential energy — it produces this energy from a supply of chemical energy.

◀ **Electromotive force (emf)**

To characterize the "strength" of a source of electric potential energy, we introduce the concept of ***electromotive force,*** or ***emf.*** The emf of a source of electric potential energy is defined as ***the amount of electric energy delivered by the source per coulomb of positive charge as this charge passes through the source from the low-potential terminal to the high-potential terminal.*** Since the emf is energy per unit charge, its units are volts. Note that the electromotive "force" is not a force, but an energy per coulomb — the confusing name became attached to it a long time ago, when physicists were not yet making a sharp distinction between force and energy. Because the units of emf are volts, the emf is often simply called the ***voltage*** of the source.

If a steady, time-independent current carries one coulomb of charge around the circuit of Fig. 20.1 from P to P' along the wire and from P' to P through the source of emf, then the energy that this charge receives from the source of emf must exactly match the energy it loses within the wire. If so, the charge returns to its starting point with exactly the same energy it had originally, and it can repeat this roundtrip again and again, in exactly the same manner. We can write this energy balance as

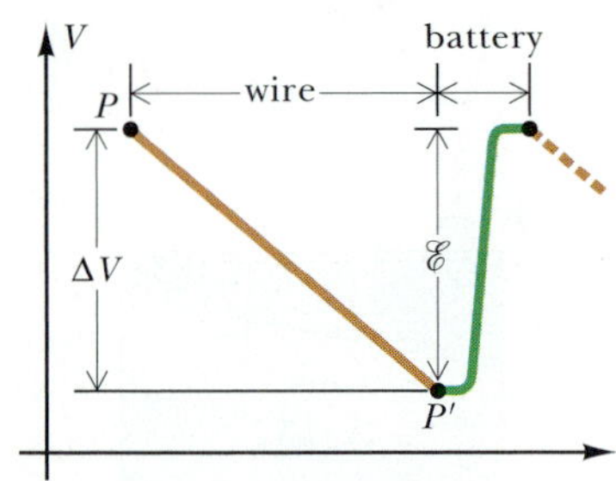

FIGURE 20.3
Plot of the electric potential, or the potential energy per coulomb of charge, as a function of position along the wire for the circuit illustrated in Fig. 20.1. The potential is highest at the point P (the positive pole of the battery), and lowest at the point P' (the negative pole).

$$\mathscr{E} + \Delta V = 0 \tag{1}$$

where $\mathscr{E}$ represents the emf, or the increase of potential energy, due to the source and ΔV represents the decrease of potential energy along the wire ($\mathscr{E}$ is positive and ΔV is negative; see Fig. 20.3).

According to Eq. (1), the emf $\mathscr{E}$ has the same magnitude as the potential drop in the external circuit connected between the terminals of the source of emf. For example, a battery with an emf of 1.5 V (or 1.5 J/C) connected to an external circuit will do 1.5 J of work on a coulomb of positive charge that passes through the battery in the forward direction (from the − terminal to the + terminal), and the resistors and other devices in the external circuit will do −1.5 J of work on the charge as it flows around this circuit (from the + terminal to the − terminal).

EXAMPLE 1. A fresh flashlight battery with a voltage of 1.5 V will deliver a current of 1 A for about 1 h before running down. How much work does the battery do in this time interval?

SOLUTION: The battery does 1.5 J of work on each coulomb that passes through. If the current is 1 A, the charge that passes through in 1 h is 1 A × 3600 s = 3600 C. The total work is then the work per coulomb multiplied by the number of coulombs, that is, 1.5 J/C × 3600 C = 5400 J.

Several kinds of small batteries.

Note that if one coulomb of positive charge is forced through the battery in the *reverse* direction (from the + terminal to the − terminal), then the charge will yield up electric potential energy to the battery. The charge will then emerge from the battery at a potential that is 1.5 volts lower than the potential with which it entered. Either the energy yielded up by the charge will be stored within the battery (if it is a reversible, or rechargeable, battery) or else it will merely be wasted as heat within the battery (if it is an irreversible battery).

20.2 SOURCES OF ELECTROMOTIVE FORCE

The most important kinds of sources of emf are batteries, electric generators, fuel cells, and solar cells. We will now briefly discuss some of these.

Batteries

◂ Lead-acid battery

These sources of emf convert chemical energy into electric energy. A very common type of battery is the ***lead–acid battery*** which finds widespread use in automobiles. In its simplest form, this battery consists of two plates of lead — the positive electrode and the negative electrode — immersed in a solution of sulfuric acid (see Fig. 20.4). The positive electrode is covered with a layer of lead dioxide, PbO_2. When the external circuit is closed, the sulfuric acid reacts with the immersed surfaces of the electrodes. As already mentioned in Section 16.4, the reactions that occur at the negative and the positive electrodes are, respectively:

$$Pb + SO_4^{--} \rightarrow PbSO_4 + 2e^- \tag{2}$$

$$PbO_2 + SO_4^{--} + 4H^+ + 2e^- \rightarrow PbSO_4 + 2H_2O \tag{3}$$

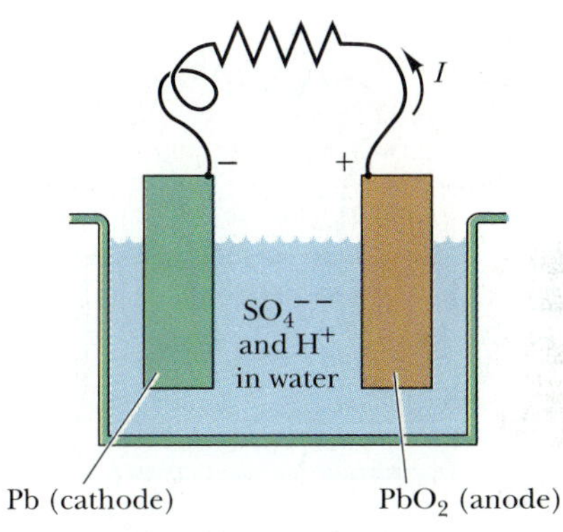

FIGURE 20.4
A lead–acid battery.

These reactions deposit electrons on the negative electrode and absorb electrons from the positive electrode. Thus, the battery acts as a pump for electrons — the negative electrode is the outlet, the positive electrode the intake, and the electrons flow from one to the other via the external circuit.

The reactions (2) and (3) deplete the sulfuric acid in the solution and they deposit lead sulfate on the electrodes. The depletion of sulfuric acid finally halts the reaction — the battery is then "discharged."

The lead–acid battery can be "charged" by simply passing a current through it in the backward direction. This reverses the reactions (2) and (3) and restores the sulfuric-acid solution. Note that what is stored in the battery during the "charging" process is not electric charge, but chemical energy. The number of positive and negative electric charges (protons and electrons) in the battery remains constant; what changes is the concentration of chemical compounds. A "charged" battery contains chemical compounds (lead, lead dioxide, sulfuric acid) of relatively high internal energy; a "discharged" battery contains chemical compounds (lead sulfate, water) of lower internal energy. Charging a battery is analogous to winding up the spring of a clock or pumping water into a reservoir on a high hill — in all these cases we are storing energy which we can subsequently release upon demand.

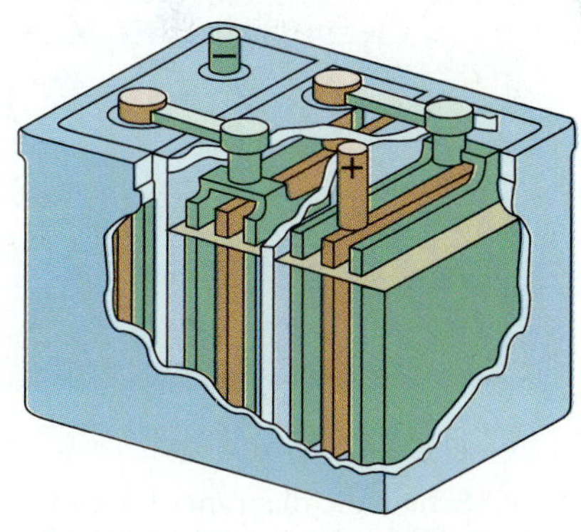

FIGURE 20.5
An automobile battery, with several cells connected in series.

The single-cell battery shown in Fig. 20.4 has an emf of 2.0 V. In an automobile battery (see Fig. 20.5) six such cells are stacked together and connected in series to give an emf of 12.0 V. The energy stored in such a battery is typically about 0.5 kW · h.

◂ Dry cell

Another familiar type of battery is the ***dry cell,*** or flashlight battery. The positive electrode consists of a carbon rod and the negative electrode of a zinc cylinder. The electrolyte in which these electrodes are "im-

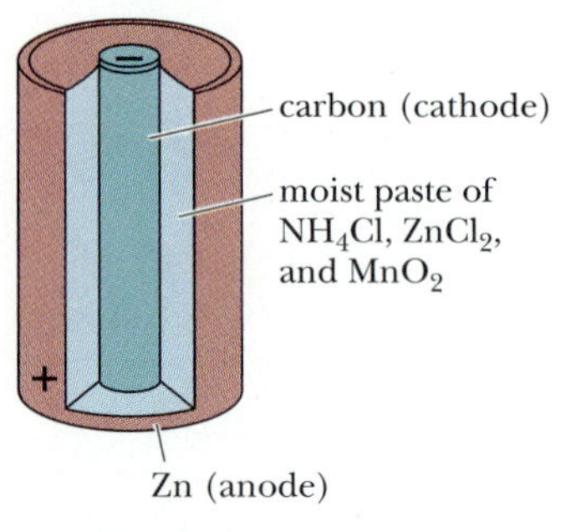

FIGURE 20.6
A dry cell.

mersed" is a moist paste of ammonium chloride, zinc chloride, and manganese dioxide (see Fig. 20.6). The chemical reactions at the electrodes convert chemical energy into electric energy and pump electrons from one electrode to the other via the external circuit. The emf of such a dry cell is 1.5 V. Since there is no liquid to slosh around, these batteries are particularly suitable for portable devices. The energy stored in a flashlight battery is typically of the order of 2×10^{-3} kW·h.

Electric Generators

Generators convert mechanical energy (kinetic energy) into electric energy. Their operation involves magnetic fields and the phenomenon of induction. We will leave the description of electric generators for Section 22.3.

Fuel Cells

These resemble batteries in that they convert chemical energy into electric energy. However, in contrast to a battery, neither the high-energy chemicals nor the low-energy reaction products are stored inside the fuel cell. The former are supplied to the fuel cell from external tanks and the latter are ejected. Essentially, the fuel cell acts as a combustion chamber in which a controlled chemical combustion takes place. The fuel cell "burns" a high-energy fuel, but produces electric energy rather than thermal energy.

Figure 20.7a shows a fuel cell that burns a hydrogen–oxygen fuel. The electrodes of the fuel cell are hollow cylinders of porous carbon; oxygen at high pressure is pumped into the positive electrode and hydrogen into the negative electrode. The electrodes are immersed in a potassium-hydroxide electrolyte. The reactions at the negative and positive electrode are, respectively,

$$2H_2 + 4OH^- \rightarrow 4H_2O + 4e^- \tag{4}$$

$$O_2 + 2H_2O + 4e^- \rightarrow 4OH^- \tag{5}$$

These reactions deposit electrons on the negative electrode and remove electrons from the positive electrode. This pumps electrons from one electrode to the other via the external circuit.

Note that the net result of the sequence of reactions (4) and (5) is the conversion of oxygen and hydrogen into water. This reaction is the reverse of the electrolysis of water (decomposition of water by an electric current). The excess water is removed from the cell in the form of water vapor.

All fuel cells produce a certain amount of waste heat. The best available fuel cells convert about 45% of the chemical energy of the fuel into electric energy and they waste the remainder. Fuel cells are still at an experimental stage, but they have already been put to use as practical power sources aboard the Apollo spacecraft and on Skylab (see Fig. 20.7b). They are compact and clean; on Skylab, the waste water eliminated from the fuel cell was used both for drinking and for washing.

(a)

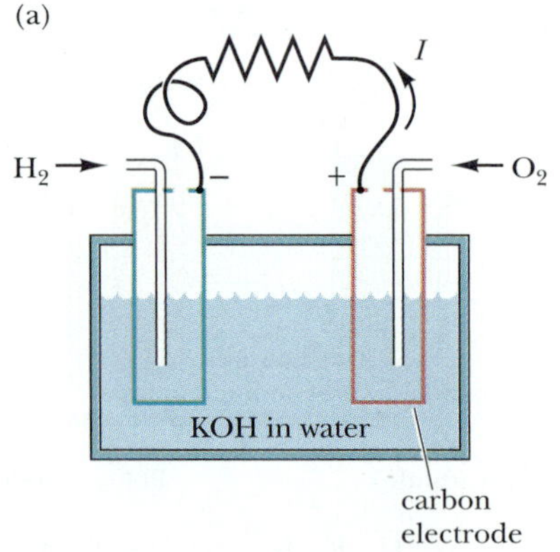

(b)

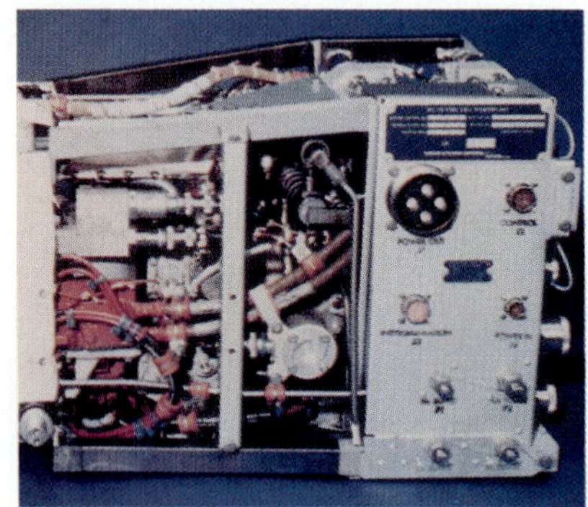

FIGURE 20.7
(a) Schematic diagram of a fuel cell. (b) Fuel cell used on Skylab.

Solar Cells

Solar cells convert the energy of sunlight directly into electric energy. They are made of thin wafers of a semiconductor, such as silicon. We will discuss semiconductors and their applications in Chapter 30.

20.3 SINGLE-LOOP CIRCUITS

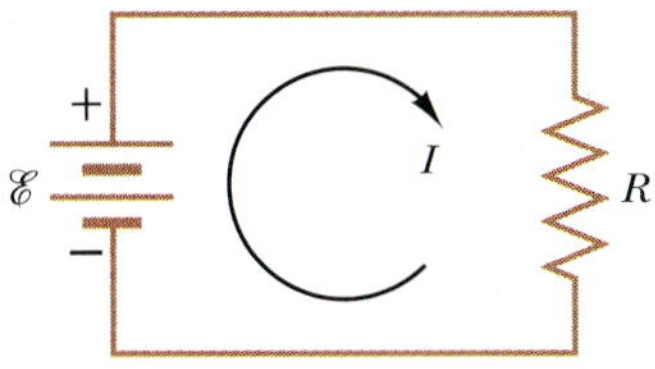

FIGURE 20.8
A simple circuit with a source of emf and a resistor.

The electric circuits in automobiles and in battery-operated appliances or tools, such as flashlights, vacuum cleaners, and electric drills, contain one or several batteries or other sources of emf connected by wires to lightbulbs, electric motors, buzzers, etc. In schematic diagrams of electric circuits, the latter devices can be represented by their resistances. Hence the schematic circuit diagram consists of one or several sources of emf connected to one or several resistors. In such a diagram, any source with a time-independent emf is represented by a stack of parallel short and long lines suggesting the plates of a lead–acid battery (see Fig. 20.8). The high-potential terminal is represented by a long line (marked with a plus sign) and the low-potential terminal by a short line (marked with a minus sign). If the terminals of such a source are connected to a network with resistances, a direct current, or DC, will flow through the network. In this section and the next we will learn how to calculate the currents that flow through the different branches of a circuit. We begin with simple circuits, consisting of a single closed loop.

Figure 20.8 shows a schematic circuit diagram for a single-loop circuit consisting of a source of emf, such as a battery, connected to a resistor. The emf of the battery is $\mathscr{E}$ and the resistance of the resistor is R. The wires from the resistor to the battery are assumed to have negligible resistance (if higher accuracy is required, the resistance of the wires must be included in the circuit diagram). The resistance R in Fig. 20.8 could equally well represent a carbon resistor or some other device, such as a light bulb, endowed with an electrical resistance; the resistance R could even represent the resistance of a wire by itself connected between the poles of the battery. To find the current that flows through the circuit, we note that, according to Ohm's Law, the potential change across the resistor must be

$$\Delta V = -IR \tag{6}$$

Here the potential change has been reckoned in the direction of the arrow (see Fig. 20.8), from the upper end to the lower end of the resistor. The negative sign on the right side of Eq. (6) means that for a positive charge that moves around the circuit in the direction of the arrow, the potential decreases across the resistor. According to Eq. (1), the emf plus the potential change must equal zero; hence

$$\mathscr{E} - IR = 0 \tag{7}$$

Gustav Robert Kirchhoff, (keerkh-hoff), *1824–1887, German physicist. Mainly known for his development of spectroscopy, he also made many important contributions to mathematical physics, among them his first and second rules for circuits.*

from which

$$I = \frac{\mathscr{E}}{R} \tag{8}$$

Kirchhoff's Rule ▶

Equation (7) is an instance of ***Kirchhoff's Rule,*** which states that ***around any closed loop in a circuit the sum of all the emfs and all the potential changes across resistors and other circuit elements must equal zero.**** In this sum, the emf of a source is reckoned as positive if the current flows through the source in the forward direction and negative if in the backward direction.

The proof of this general rule is similar to the proof of Eq. (1). If one coulomb of positive charge flows once around a closed loop in a circuit with one or several sources of emf and resistors, it will gain potential energy while passing through each source of emf and lose potential energy while passing through each resistor. Under steady conditions, the sum of gains and losses must equal zero, since the charge must return to its starting point with no change of energy. In this sum, the emf must be reckoned as positive (gain of potential energy) whenever the charge flows through a source of emf in the forward direction, from the − terminal to the + terminal; and as negative (loss of potential energy) whenever the charge flows through a source of emf in the backward direction, from the + terminal to the − terminal.

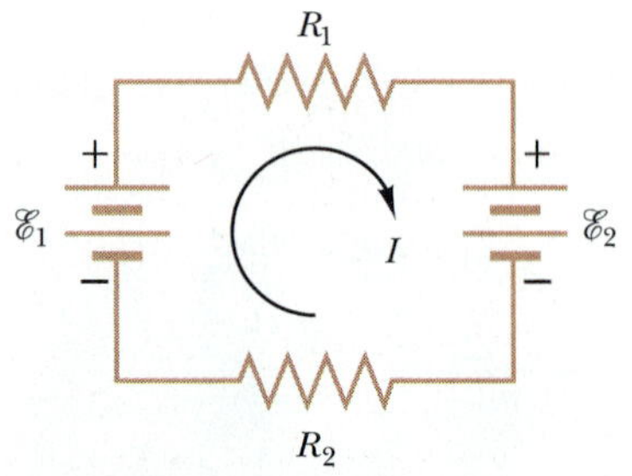

FIGURE 20.9
Two sources of emf and two resistors.

EXAMPLE 2. Figure 20.9 shows a circuit with two batteries and two resistors. The emfs of the batteries are $\mathscr{E}_1 = 12$ V and $\mathscr{E}_2 = 15$ V; the resistances are $R_1 = 4\ \Omega$ and $R_2 = 2\ \Omega$. What is the current in the circuit?

SOLUTION: To apply Kirchhoff's Rule, we must decide in which direction the current flows around the loop. We will arbitrarily assume that the current flows in the clockwise direction. If this hypothesis is wrong, our calculation of the current will yield a negative value, and this will indicate that the direction of the actual current is opposite to the hypothetical current.

According to Kirchhoff's Rule, the sum of all emfs and all potential changes across resistors is zero,

$$\mathscr{E}_1 - IR_1 - \mathscr{E}_2 - IR_2 = 0 \tag{9}$$

Note that $\mathscr{E}_2$ enters with a negative sign into this equation since the hy-

* This is often called Kirchhoff's second rule. Kirchhoff's first rule states that if a current branches out from one wire into several wires at a junction, the sum of the currents leaving the junction must equal the current entering. This rule is sometimes used in calculations with multiloop circuits; but the method of calculation described in the next section allows us to bypass this rule, and we will have no need for it.

pothetical current passes through this source of emf in the backward direction. The solution of Eq. (9) leads to

$$I = \frac{\mathscr{E}_1 - \mathscr{E}_2}{R_1 + R_2} \tag{10}$$

or

$$I = \frac{12\ \text{V} - 15\ \text{V}}{4\ \Omega + 2\ \Omega} = -0.5\ \text{A} \tag{11}$$

Here, the negative sign indicates that the current is *not* clockwise, but counterclockwise.

COMMENTS AND SUGGESTIONS: We could have guessed the direction of the current, since it is obvious that the stronger battery on the right will force the current backward through the weaker battery on the left. But in more complicated circuits the direction of the current will not be so obvious, and we will have to discover the direction by carefully keeping track of the signs in our calculations.

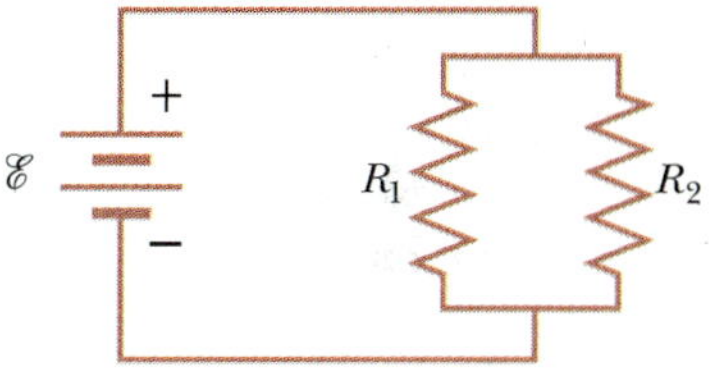

FIGURE 20.10

A circuit containing two resistances in parallel.

A circuit that contains two or more resistors in parallel, such as the circuit shown in Fig. 20.10, is not a genuine single-loop circuit, since the parallel resistors and their connecting wires form additional loops. The general method for dealing with multiloop circuits will be discussed in the next section. However, circuits with parallel resistors, such as the circuit in Fig. 20.10, can be handled by the same method as single-loop circuits, because, as we know from the preceding chapter, the parallel resistances are effectively equivalent to a single resistance [see Eq. (19.23)].

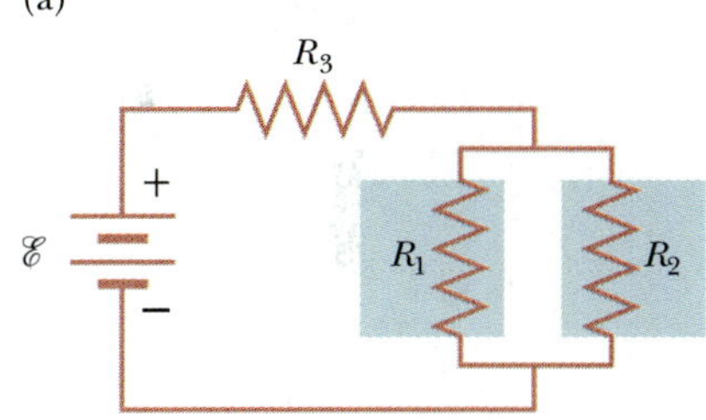

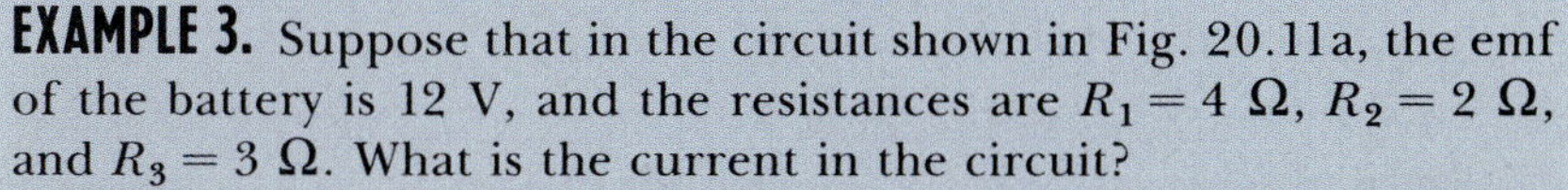

EXAMPLE 3. Suppose that in the circuit shown in Fig. 20.11a, the emf of the battery is 12 V, and the resistances are $R_1 = 4\ \Omega$, $R_2 = 2\ \Omega$, and $R_3 = 3\ \Omega$. What is the current in the circuit?

SOLUTION: The net resistance of the parallel combination is given by Eq. (19.21),

$$\frac{1}{R} = \frac{1}{R_1} + \frac{1}{R_2} = \frac{1}{4\ \Omega} + \frac{1}{2\ \Omega} = \frac{3}{4\ \Omega}$$

or

$$R = 4/3\ \Omega$$

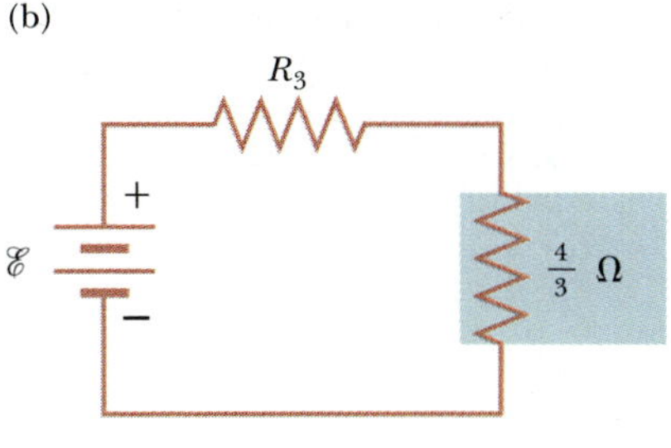

FIGURE 20.11

(a) A circuit containing two resistances in parallel (blue). (b) The two parallel resistances are equivalent to a single resistance of $4/3\ \Omega$.

The circuit shown in Fig. 20.11a is therefore effectively equivalent to the single-loop circuit shown in Fig. 20.11b. Kirchhoff's Rule for this single-loop circuit with resistances of $R = 4/3\ \Omega$ and $R_3 = 3\ \Omega$ in series with an emf of 12 V tells us

$$\mathscr{E} - IR - IR_3 = 0$$

This equation has the solution

$$I = \frac{\mathscr{E}}{R + R_3} = \frac{12\ \text{V}}{4/3\ \Omega + 3\ \Omega} = 2.8\ \text{A}$$

COMMENTS AND SUGGESTIONS: The currents in the individual parallel resistors can be found from this total current in the same way as in Example 19.5.

20.4* MULTILOOP CIRCUITS

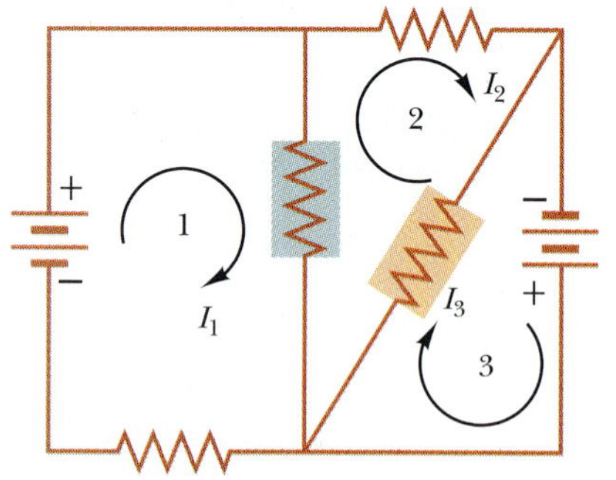

FIGURE 20.12

A multiloop circuit. The currents I_1, I_2, and I_3 are regarded as flowing in closed loops. Note that loops 1 and 2 share the middle resistor (blue), and loops 2 and 3 share the diagonal resistor (orange). The net current in the middle resistor is $I_1 - I_2$, and the net current in the diagonal resistor is $I_2 - I_3$.

If several sources of emf and several resistors are connected in some complicated circuit with several branches, then the currents will flow along several alternative paths. For instance, Fig. 20.12 shows a complicated circuit with several branches. Such ***a circuit can always be regarded as consisting of several single-loop circuits which overlap in some way.*** Thus, the circuit in Fig. 20.12 can be regarded as consisting of three single-loop circuits (one rectangle and two triangles). Loops 1 and 2 share the middle resistor and its connecting wires; loops 2 and 3 share the diagonal resistor and its connecting wires. To find the currents in such a complicated circuit, we must solve a simultaneous set of several equations with the currents as unknowns.

The procedure for obtaining the necessary equations is as follows:

1. Regard the given circuit as a collection of closed current loops. The loops may overlap, but each loop must have at least some portion that does not overlap with other loops.

2. Label the currents in the loops I_1, I_2, I_3, . . . and arbitrarily assign a direction to each of these currents.

3. Apply Kirchhoff's Rule to each loop: the sum of all the emfs and all the potential changes across resistors must add to zero around each loop. Note that when calculating the potential change across a resistor, we must take the product of the resistance and the *net* current through the resistor; if the resistor belongs to two adjacent current loops, then the *net* current is the algebraic sum of the two loop currents.

---

* This section is optional.

This procedure, called the ***loop method,*** will result in the right number of equations for the unknown currents I_1, I_2, I_3, . . . We can then solve the equations for these unknowns by the standard mathematical methods for the solution of simultaneous equations for several unknowns (see Appendix A2.5). If a current turns out to be negative, its direction is opposite to the direction assigned in step 2.

The following example illustrates this method for the two-loop circuit shown in Fig. 20.13.

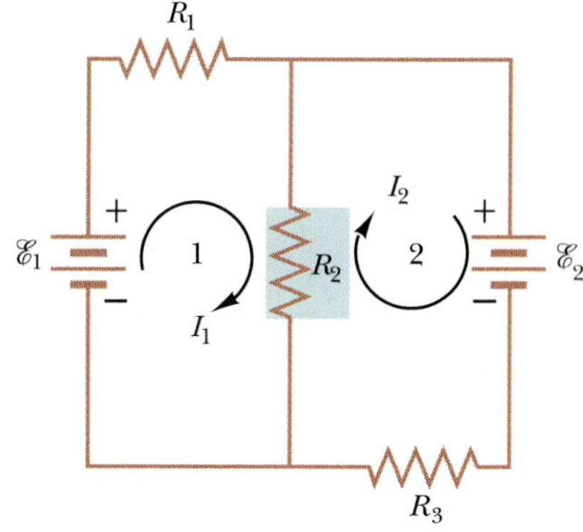

FIGURE 20.13

The loops 1 and 2 share the resistor R_2 (blue). The currents I_1 and I_2 flow through R_2 in opposite directions; the net current through R_2 is therefore $I_1 - I_2$.

EXAMPLE 4. Suppose that the emfs and the resistors in the circuit shown in Fig. 20.13 are $\mathscr{E}_1 = 12.0$ V, $\mathscr{E}_2 = 8.0$ V, $R_1 = 4.0\ \Omega$, $R_2 = 4.0\ \Omega$, and $R_3 = 2.0\ \Omega$. Find the current in each of the resistors.

SOLUTION: Obviously, this circuit can be regarded as consisting of the two loops indicated by the arrows in Fig. 20.13. We label the currents in the loops I_1 and I_2; these symbols have been written next to the arrows that indicate the directions. When calculating the potential change in the resistor R_2 (which is included in both loops), we must take into account that both loop currents flow through R_2 simultaneously, in opposite directions. The net current through R_2 in the direction of the arrow of loop 1 is therefore $I_1 - I_2$. The only emf in loop 1 is $\mathscr{E}_1$, and the potential changes in the resistors are $-I_1R_1$ and $-(I_1 - I_2)R_2$. With this, Kirchhoff's Rule for loop 1 yields

$$\mathscr{E}_1 - I_1R_1 - (I_1 - I_2)R_2 = 0. \tag{12}$$

Likewise, the net current through R_2 in the direction of the arrow of loop 2 is $I_2 - I_1$. The emf in this loop is $-\mathscr{E}_2$ (this emf enters with a negative sign since the hypothetical current passes through it in the backward direction), and the potential changes in the resistors are $-I_2R_3$ and $-(I_2 - I_1)R_2$. Kirchhoff's Rule for loop 2 yields

$$\mathscr{E}_2 - I_2R_3 - (I_2 - I_1)R_2 = 0 \tag{13}$$

This gives us two equations for the two unknown currents I_1 and I_2.

Before proceeding to the solution of these equations, it is convenient to substitute the numerical values for the known quantities $\mathscr{E}$ and R. With these numerical values, Eqs. (12) and (13) become

$$12 - 8I_1 + 4I_2 = 0 \tag{14}$$

$$-8 - 6I_2 + 4I_1 = 0 \tag{15}$$

If we solve the first of these equations for I_2, we find

$$I_2 = \frac{8I_1 - 12}{4} = 2I_1 - 3 \tag{16}$$

Substituting this into Eq. (15), we obtain

$$-8 - 6(2I_1 - 3) + 4I_1 = 0$$

or

$$-8I_1 + 10 = 0$$

This immediately gives

$$I_1 = \frac{10}{8} = 1.25$$

And then Eq. (16) gives

$$I_2 = 2I_1 - 3 = 2 \times 1.25 - 3 = -0.50$$

These currents are, of course, measured in amperes, that is $I_1 = 1.25$ A and $I_2 = -0.50$ A. The negative sign on I_2 indicates that the current in the second loop is opposite to the direction shown in Fig. 20.13.

The current in the resistor R_1 is then $I_1 = 1.25$ A; the current in the resistor R_2 is $I_1 - I_2 = 1.25\text{ A} - (-0.50\text{ A}) = 1.75$ A; the current in the resistor R_3 is $I_2 = -0.50$ A.

COMMENTS AND SUGGESTIONS: In setting up the equations for the several loops in the circuit, you must handle the directions of the currents and the sign conventions for the currents and potentials consistently. To maintain this consistency, first assign a direction for each of the loops by drawing an arrow, then go around each loop in the direction of its arrow and apply Kirchhoff's Rule. Each emf contributes a positive term if the direction of the arrow is the forward direction for this emf, and it contributes a negative term if the direction of the arrow is the backward direction. Each resistor contributes a negative term of the form −[resistance] × [net current], where the net current is the algebraic sum of all the loop currents that flow through that resistor. In this algebraic sum, the current of the loop under consideration is to be regarded as positive, and the current of any other loop sharing the resistor is to be regarded as positive or as negative depending on the direction of its arrow.

Keep in mind that the currents you obtain by setting up and solving the loop equations are the *loop currents.* If you need to know the net current through a given emf or a given resistor, you must calculate these afterward by forming the algebraic sum of the loop currents that simultaneously flow through this emf or resistor.

And if you need to know the potential difference between two given points in the circuit, simply sum all the emfs and voltage changes in resistors along a path from one point to another, complying with the same sign conventions as for a closed loop.

20.5 ENERGY IN CIRCUITS; JOULE HEAT

As we saw in Section 20.1, to keep a current flowing in a circuit, the batteries or other sources of emf must do work. If an amount of charge Δq passes through a source of emf, the amount of work it does will be

$$\Delta W = \mathscr{E} \Delta q \tag{17}$$

Hence the rate at which the source does work is

$$\frac{\Delta W}{\Delta t} = \mathscr{E} \frac{\Delta q}{\Delta t} \tag{18}$$

The rate of work is the power; the rate of flow of charge is the current. Equation (18) therefore asserts that the electric power delivered by the source of emf to the current is

$$P = I\mathscr{E} \tag{19}$$

◂ **Power delivered by source of emf**

Thus, the power delivered by the "pump of electricity" is large if it pumps a large current through a large potential difference. In terms of our analogy with a hydraulic pump, this merely means that the power delivered by a hydraulic pump is large if it pumps water at a fast rate (large water current) and lifts the water to a large height (large gravitational potential difference).

Note that in Eq. (19) we have not yet taken into account the algebraic sign of the power. We will have to attach a positive sign to the power if the current passes through the source in the forward direction and a negative sign if the current passes through in the backward direction. In the former case, the source delivers energy to the current and in the latter case the source receives energy from the current.

EXAMPLE 5. What power do the two batteries described in Example 4 deliver?

SOLUTION: The current through the first battery is 1.25 A and its emf is 12.0 V. This current passes through the battery in the forward direction; hence the power delivered by the battery is

$$P = 1.25 \text{ A} \times 12.0 \text{ V} = 15.0 \text{ W}$$

Likewise, the power delivered by the other battery is

$$P = 0.50 \text{ A} \times 8.0 \text{ A} = 4.0 \text{ W}$$

The net power delivered by both batteries is 19.0 W.

The electric potential energy acquired by the charges is carried along the circuit to the resistors, and it is continually dissipated in the resistors. If within a given resistor the charge Δq suffers a potential change ΔV (regarded as a positive quantity), then the loss of potential energy is $\Delta U = \Delta V \Delta q$, and the rate at which energy is dissipated is

$$\frac{\Delta U}{\Delta t} = \Delta V \frac{\Delta q}{\Delta t}$$

Hence the power dissipated in the resistor is

Power dissipated by resistor ▶

$$P = (\Delta V) I \tag{20}$$

By means of Ohm's Law, $\Delta V = IR$, we can also write the power as

$$P = I^2 R \tag{21}$$

or as

$$P = \frac{(\Delta V)^2}{R} \tag{22}$$

Joule heating ▶

The energy lost by the charges during their passage through a resistor generates thermal energy, that is, it generates disordered, microscopic kinetic and potential energy of the atoms in the resistor. This ***conversion of electric energy into thermal energy in a resistor is called Joule heating.***

Many simple electric appliances — such as electric toasters, heaters, hot plates, irons, blankets — rely on Joule heating. All these devices merely convert electric energy into thermal energy by means of a heating element consisting of a coiled wire of fairly high resistance. For instance, in a typical electric toaster, the ribbon of wire you can see wound around the plates of insulating material inside the bread slots has a resistance of 10 or 15 Ω, which is much larger than the resistance of the wires connecting the toaster to the outlet, or the wires connecting the outlet to the fuse box of your house. The resistance of the internal wire in the toaster must be kept larger than the resistance of the outside wire because otherwise the outside wire would produce more heat than the internal wire, which would be wasteful and hazardous, since overheating of the outside wire could start a fire.

Incandescent light bulbs also rely on Joule heating. Within the light bulb, the current passes through a very fine tungsten filament which acts like a miniature electric heater. The temperature of the tungsten filament reaches 3000°C, and the filament glows, white hot. Since the light bulb is evacuated, the heat cannot escape by conduction or convection (except for a small amount conducted by the wires connected to the filament); thus, the heat escapes by radiation, in the form of infrared light and visible light.

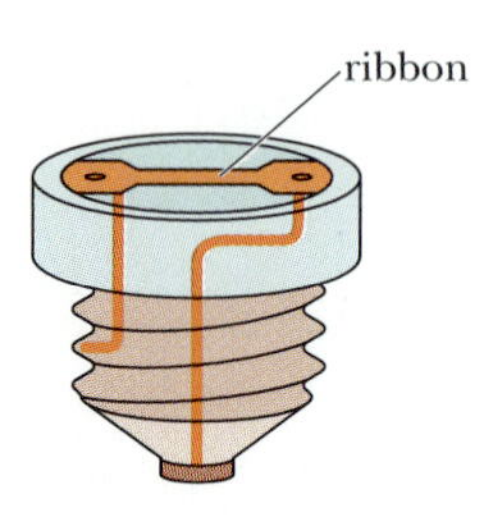

FIGURE 20.14
An ordinary fuse.

Joule heating can also be used for a crude control of currents by means of fuses and circuit breakers. These safety devices are designed to cut off the current if it exceeds a preset level. In an ordinary fuse (see Fig. 20.14), the current passes through a thin ribbon of metal of low melting point. The current heats this ribbon, and if the current is excessive, the ribbon melts and cuts off the current. In a circuit breaker, the current sensor is a bimetallic strip (see Fig. 20.15), which bends when heated by the current. If the bimetallic strip bends far enough up, it permits a contact arm to flip to one side. The contact arm acts as a switch, and when it flips to one side, it opens the contact points and cuts the current. After the bimetallic strip cools, the contact arm can be pushed back into place, so as to close the contact points, and again permit the current to flow.

Maximum currents permitted for wires of different thicknesses are specified in the National Electrical Code. For instance, the maximum current permitted for the typical wire (No. 10) installed in modern houses is 30 A, and the fuses protecting such wire are therefore designed to cut off the current if it exceeds this level.

In any circuit consisting of several sources of emf and several resistors with steady currents, the total power delivered by the sources of emf must equal the total power dissipated in the resistors. This equality can be shown to be mathematically equivalent to Kirchhoff's Rule. The net result of the flow of current in such a circuit is therefore a conversion of energy of the sources of emf into an equal amount of energy of heat.

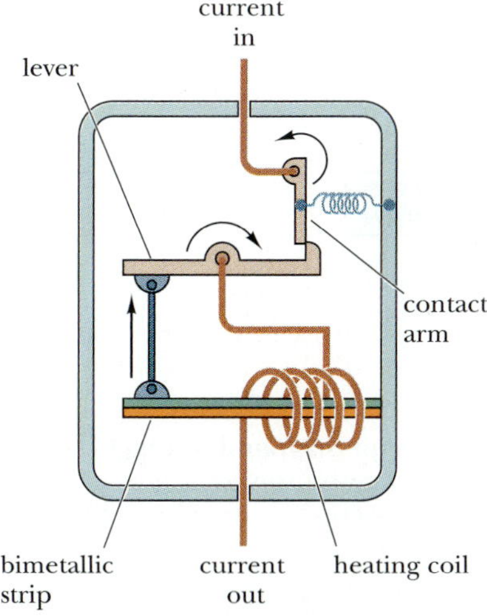

FIGURE 20.15

A circuit breaker. When heated, the bimetallic strip bends up and pushes the lever, which releases the spring-loaded contact arm, and breaks the circuit.

EXAMPLE 6. What is the rate at which Joule heat is produced in the resistors of Example 4?

SOLUTION: The resistances are $R_1 = 4.0\ \Omega$, $R_2 = 4.0\ \Omega$, and $R_3 = 2.0\ \Omega$; the corresponding currents are 1.25 A, 1.75 A, and 0.50 A. Equation (21) then gives the power in each resistor:

$$P_1 = (1.25\ \text{A})^2 \times 4.0\ \Omega = 6.25\ \text{W}$$

$$P_2 = (1.75\ \text{A})^2 \times 4.0\ \Omega = 12.25\ \text{W}$$

$$P_3 = (0.50\ \text{A})^2 \times 2.0\ \Omega = 0.50\ \text{W}$$

COMMENTS AND SUGGESTIONS: Note that the net power dissipated is 19.0 W, which agrees with the net power delivered by the batteries (compare with Example 5).

EXAMPLE 7. A high-voltage transmission line that connects a city to a power plant consists of a pair of copper cables, each with a resistance of 4 Ω. The current flows to the city along one wire and back along the other. (a) The transmission line delivers to the city 1.7×10^5 kW of power at 2.3×10^5 V. What is the current in the transmission line? How much power is lost as Joule heat in the transmission line? (b) If the transmission line were to deliver the

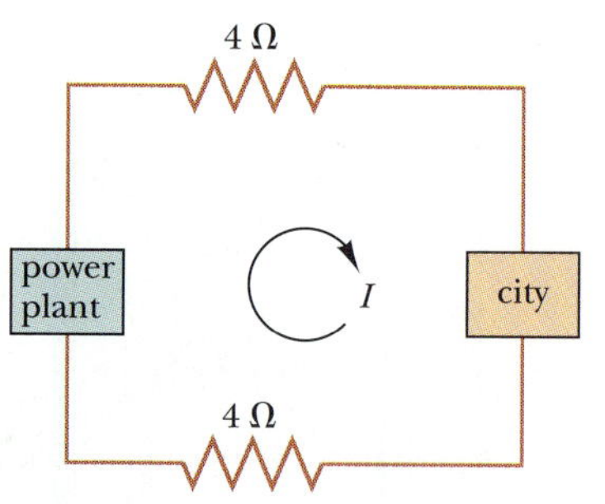

FIGURE 20.16
Circuit diagram for a high-voltage transmission line connecting a city to a power plant.

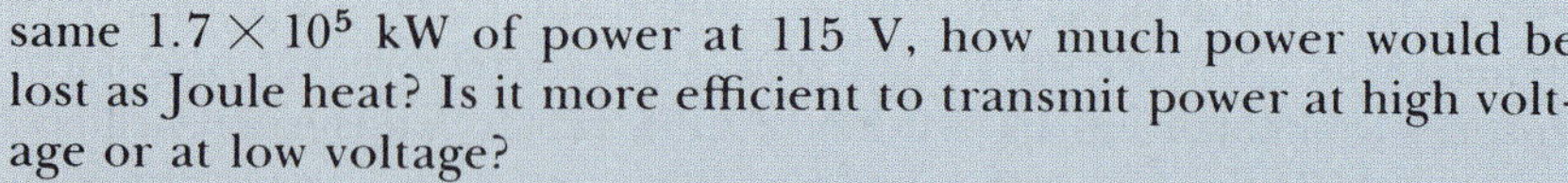

same 1.7×10^5 kW of power at 115 V, how much power would be lost as Joule heat? Is it more efficient to transmit power at high voltage or at low voltage?

SOLUTION: (a) Figure 20.16 shows the circuit consisting of power plant, transmission line, and city. In terms of the power and the voltage delivered to the city, the current through the city is, according to Eq. (20),

$$I = \frac{P_{\text{delivered}}}{\Delta V_{\text{delivered}}} = \frac{1.7 \times 10^8 \text{ W}}{2.3 \times 10^5 \text{ V}} = 7.4 \times 10^2 \text{ A}$$

The current in both portions of the transmission line must be the same, since any current that flows toward the city must return to the power plant. The combined resistance of both wires is $4\ \Omega + 4\ \Omega = 8\ \Omega$, and hence the power lost in the transmission line is, according to Eq. (21),

$$P_{\text{lost}} = I^2R = (7.4 \times 10^2 \text{ A})^2 \times 8\ \Omega = 4.4 \times 10^6 \text{ W}$$

Thus, the power lost is about 3% of the power delivered.
(b) For $\Delta V_{\text{delivered}} = 115$ V, the current is

$$I = \frac{P_{\text{delivered}}}{\Delta V_{\text{delivered}}} = \frac{1.7 \times 10^8 \text{ W}}{115 \text{ V}} = 1.5 \times 10^6 \text{ A}$$

and the power lost is

$$P_{\text{lost}} = I^2R = (1.5 \times 10^6 \text{ A})^2 \times 8\ \Omega = 1.8 \times 10^{13} \text{ W}$$

Thus, the power lost is much larger than the power delivered! Comparing the results of (a) and (b), we see that transmission at high voltage is much more efficient than transmission at low voltage.

Analog and digital multimeters; these can be used both as voltmeters or as ammeters.

20.6* ELECTRICAL MEASUREMENTS

Measurements of currents, potentials, and resistances in electrical circuits require various specialized instruments. Here we will briefly discuss some of the instruments used in DC circuits.

Ammeter and Voltmeter

Most electrical measurements are performed with ammeters and voltmeters. The ammeter measures the electric current flowing into its termi-

* This section is optional.

nals, and the voltmeter measures the potential difference applied to its terminals.

The internal mechanisms of the ammeter and the voltmeter are similar. In the traditional moving-coil instruments, the sensitive element is a small coil of wire, delicately suspended between the poles of a magnet, which suffers a deflection when a current passes through it (see Figs. 21.43 and 21.44 for a view of the internal mechanism of these instruments). Thus, ammeters and voltmeters both respond to an electric current passing through the instrument. The difference is that ***the ammeter has a low internal resistance and permits the passage of whatever current enters its terminals with little hindrance, whereas the voltmeter has a very large internal resistance and draws only an extremely small current,*** even when the potential difference applied to its terminals is large.

◂ **Internal resistances of ammeter and voltmeter**

Figure 20.17a shows an example of an electric circuit. To measure the electric current at, say, the point P in this circuit, the experimenter must cut the wire apart and insert the ammeter. Figure 20.17b shows the correct way of connecting the ammeter (for comparison, Fig. 20.17c shows a wrong way). Since the ammeter has a very low internal resistance, its insertion in the circuit usually has an insignificant effect on the current. However, if the resistances in the circuit are very small, then the insertion of an ammeter can have a significant inhibiting effect on the current. For the accurate measurement of the current, the experimenter must select an ammeter whose internal resistance is much lower than the resistances in the circuit.

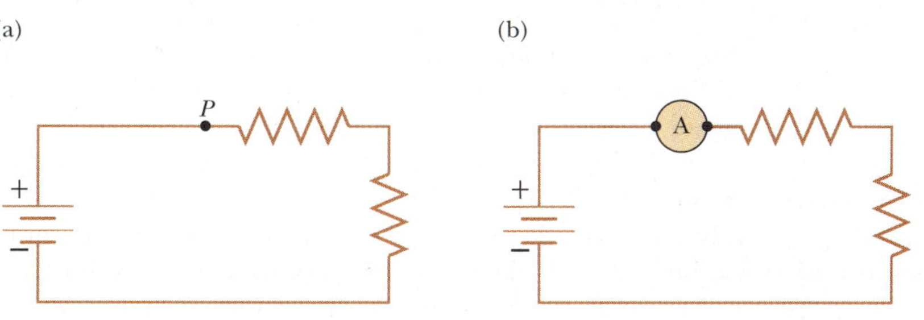

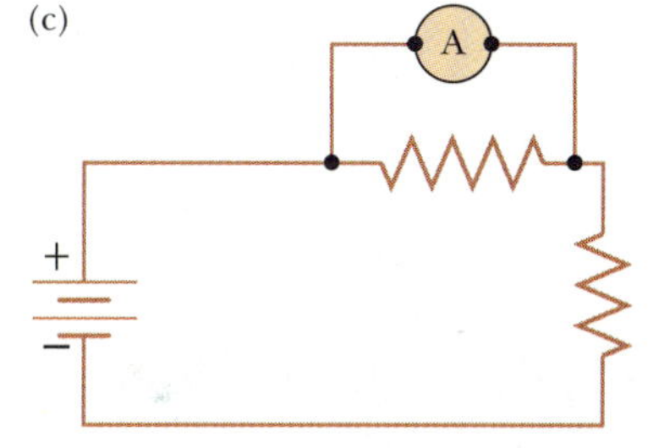

FIGURE 20.17

(a) An electric circuit.
(b) Correct connection of the ammeter. The symbol for the ammeter is a circle labeled *A*, with two terminals. The current to be measured enters through one terminal of the ammeter and leaves through the other.
(c) Incorrect connection of the ammeter.

To measure the potential difference between, say, the points P and P' in a circuit, the experimenter must connect the terminals of the voltmeter to these points. Figure 20.18a shows the correct way of connecting the voltmeter (for comparison, Fig. 20.18b shows a wrong way). Since the voltmeter has a very large internal resistance, it draws only a very small current, and the alteration in the flow of current through the resistance

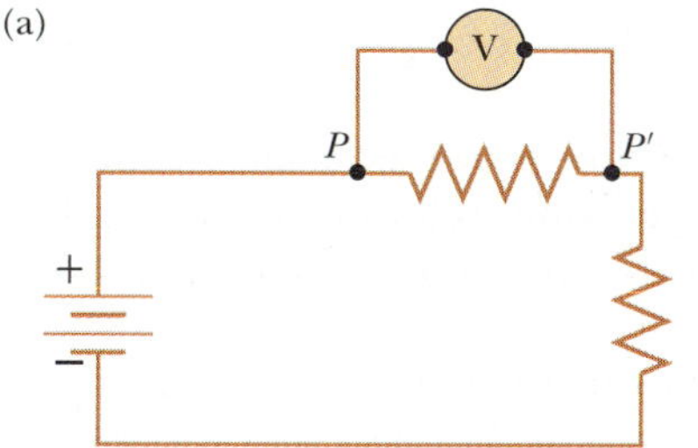

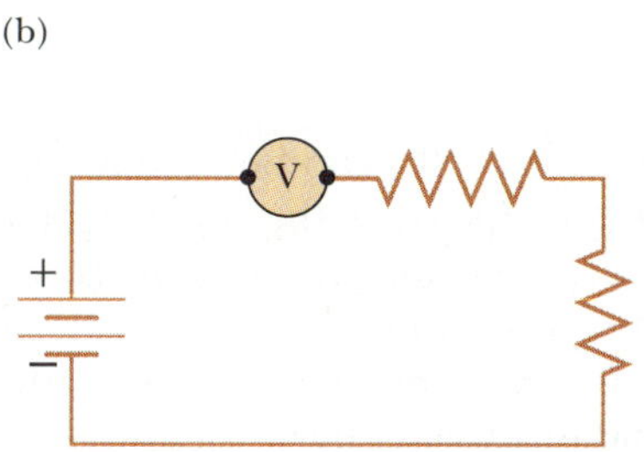

FIGURE 20.18

(a) Correct connection of the voltmeter. The symbol for the voltmeter is a circle labeled *V*, with two terminals. One terminal is connected to the point *P;* the other is connected to the point P'. (b) Incorrect connection of the voltmeter.

PP' and the consequent alteration of the potential difference between P and P' are usually insignificant. However, if the resistance PP' across which the voltmeter is connected is large, then the voltmeter can draw a significant fraction of the current, with a consequent decrease of potential across the resistance PP'. For an accurate measurement of the potential, the experimenter must select a voltmeter with an internal resistance much larger than the resistance PP'.

Potentiometer ▶

Potentiometer

A voltmeter can be used for a direct measurement of the emf of a battery, or other source of emf. However, batteries usually have some internal resistance, and when the voltmeter is connected across the terminals of a battery and draws a current, the potential across the terminals is smaller than the emf by the potential drop produced within this internal resistance. Thus, the potential measured by the voltmeter depends on how much current it draws. To avoid this difficulty, it is desirable to measure the emf when no current is flowing. The potentiometer is a precision instrument used for such measurements of the emf at zero current, or the "open-circuit voltage."

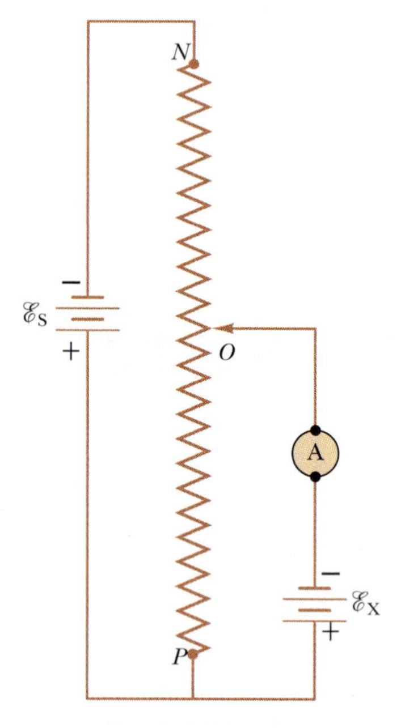

FIGURE 20.19
Schematic diagram for a potentiometer.

The potentiometer is an electrical balance — it compares an unknown emf $\mathscr{E}_X$ with a known reference emf $\mathscr{E}_S$. In order to attain the condition of zero current for both the unknown emf and the known emf, the comparison is made indirectly, in two steps, by means of a third, auxiliary emf. A schematic diagram of the potentiometer is shown in Fig. 20.19. The auxiliary emf $\mathscr{E}_0$ is connected to a long uniform wire of a metal of fairly high resistivity; in Fig. 20.19, this wire is represented by the long resistance PN. A steady current will then flow in the wire, and the potential along the wire will be a linearly decreasing function of the length of wire measured from the point P. In the first step of the measurement, the experimenter connects the unknown source of emf $\mathscr{E}_X$ to the end P of the wire and to a sensitive ammeter A. The experimenter then moves the sliding contact O until the ammeter A reads zero current. Under these conditions, the emf of the unknown source exactly matches the potential drop along the length PO of wire. The potentiometer is then said to be balanced. In the second step of the measurement, the experimenter replaces the unknown source of emf $\mathscr{E}_X$ by the known reference source $\mathscr{E}_S$ and again balances the potentiometer. The emf of the reference source will match the potential drop along some length PO' of the wire. The ratio of the lengths PO and PO' is equal to the ratio of the unknown emf $\mathscr{E}_X$ and the known emf $\mathscr{E}_S$,

$$\frac{\mathscr{E}_X}{\mathscr{E}_S} = \frac{PO}{PO'} \tag{23}$$

From this, $\mathscr{E}_X$ can be readily calculated. Note that the result is completely independent of the auxiliary emf $\mathscr{E}_0$; the only relevant property of this emf is that it must remain constant during the steps of the operation of the potentiometer.

The advantage of the potentiometer over the voltmeter is not only that the unknown emf is measured at zero current, but also that the standard cell, used as reference emf, maintains a precisely fixed value of its emf, which provides a more reliable calibration of the potential than is to be had with a voltmeter.

Wheatstone Bridge

◀ Wheatstone bridge

The Wheatstone bridge is used for precise comparisons of an unknown resistance with a known reference resistance. Its operation is based on much the same principles as the potentiometer. Figure 20.20 shows a schematic diagram of a Wheatstone bridge. Again, a long uniform wire is connected across an auxiliary source of emf $\mathscr{E}_0$. One end of the unknown resistance R_X is connected to the point P, and one end of a known reference resistance R_S is connected to the point N. The other ends of these two resistances are joined at the ammeter A. To balance the bridge, the experimenter moves the sliding contact O until the ammeter reads zero current. Thus, steady currents I_1 and I_2 flow along the parallel paths of the circuit, and no current crosses the bridge AO. Under these conditions, the potential drop I_2R_X must equal the potential drop across the segment PO of the wire, and the potential drop I_2R_S must equal the potential drop across the segment ON. Since the potential drops in the wire are in proportion to the lengths,

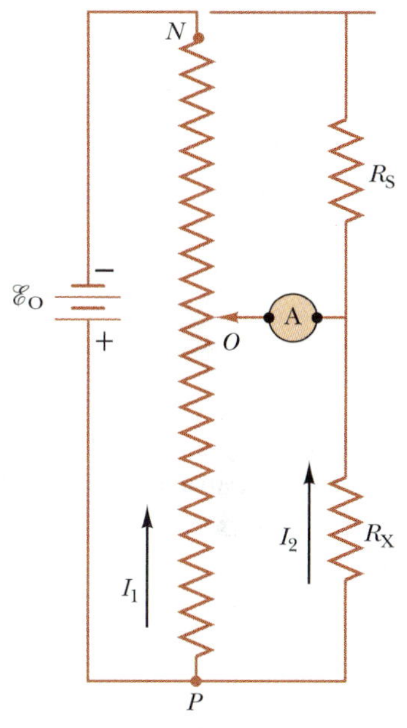

FIGURE 20.20
Schematic diagram for a Wheatstone bridge with a slide wire.

$$I_2R_X \propto PO \quad \text{and} \quad I_2R_S \propto ON \tag{24}$$

Taking the ratio of these proportions, we obtain

$$\frac{R_X}{R_S} = \frac{PO}{ON} \tag{25}$$

This equation permits the calculation of the unknown resistance. Again, the result is independent of the auxiliary emf.

In essence, the operation of the Wheatstone bridge hinges on an adjustment of the ratio of resistances in the circuit. In some variants of the Wheatstone bridge, the lengths (and resistances) PO and ON are held fixed, and the ratio R_X/R_S is varied by using an adjustable, calibrated resistance R_S. If $PO = ON$, then the bridge will balance when $R_X = R_S$; in this case, the dial on the adjustable, calibrated resistance directly tells us the value of the unknown resistance.

20.7* THE RC CIRCUIT

◀ Kirchhoff's Rule for time-dependent currents

Throughout this chapter we have dealt only with steady, time-independent currents. However, ***Kirchhoff's Rule and the methods for solving circuits we have developed in this chapter apply also to time-dependent currents.*** The only restriction is that the emfs and the currents in the circuit must not vary too quickly. (A rough criterion for the applicability of

* This section is optional.

Kirchhoff's Rule is that the currents and emfs must not change significantly in an interval equal to the travel time for a light signal around the circuit.)

In Chapter 23 we will deal with a variety of circuits with time-dependent currents. Here we will deal with the simple case of a time-dependent current in a circuit consisting of a resistor and a capacitor connected in series and being charged by a battery. Figure 20.21 shows a schematic diagram for such an RC circuit. We assume that the capacitor is initially uncharged, and that the battery is suddenly connected at time $t = 0$. Initially, the potential difference across the capacitor is then zero. When the battery is connected, charge flows from the terminals of the battery to the plates of the capacitor. As the charge accumulates on the plates, the potential difference across the plates gradually increases. The flow of charge will come to a halt when the potential difference across the plates matches the emf of the battery. This qualitative discussion of the charging process indicates that the current is initially large, but gradually tapers off, and ultimately approaches zero.

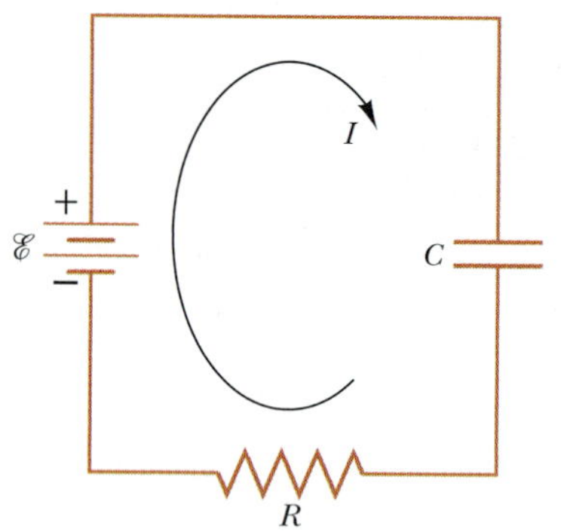

FIGURE 20.21
An RC circuit, consisting of a resistor and a capacitor connected in series to a battery.

For a mathematical treatment of the current in the circuit we turn to Kirchhoff's Rule: The sum of all the emfs and voltage drops around the circuit must be zero. The emf of the battery is $\mathscr{E}$. If the current at some instant is I, the potential drop across the resistor is IR. And if the charge on the capacitor plates at some instant has a magnitude Q, the potential difference across the plates is Q/C. Hence

$$\mathscr{E} - IR - \frac{Q}{C} = 0 \tag{26}$$

At the initial time, $Q = 0$ and Eq. (26) then tells us that the initial current is $I = \mathscr{E}/R$. At the end of the charging process, $I = 0$ and Eq. (26) then tells us that $Q = C\mathscr{E}$. These conclusions are in accord with our qualitative discussion.

The calculation of the current and the charge at intermediate times is somewhat complicated. but we can make a rough estimate of the time required for the charging process if we ignore the decrease of the current, and pretend that the current is approximately $\mathscr{E}/R$ at all times. Upon dividing the final charge $C\mathscr{E}$ by the current $\mathscr{E}/R$, we obtain an estimate of the time required to complete the charging process (this is analogous to dividing distance by speed to obtain the travel time). We will designate this estimated charging time by τ (the Greek letter tau):

Characteristic time for RC circuit ▶

$$\tau = \frac{C\mathscr{E}}{\mathscr{E}/R} = RC \tag{27}$$

This time is called the ***characteristic time*** for the RC circuit.

An exact calculation, which takes into account the gradual decrease of the current, shows that the time to complete the charging is a very long time (infinite time), but most of the charging occurs within a time of about RC. Figure 20.22 gives plots of the current in the RC circuit and of the charge in the capacitor. ***At the characteristic time $\tau = RC$, the charge on the capacitor reaches 63% of its final value.***

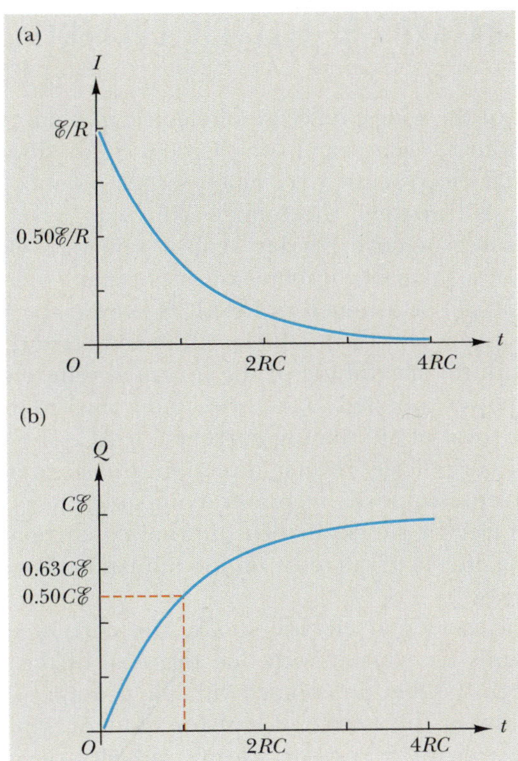

FIGURE 20.22

(a) Current in the RC circuit as a function of time. The initial current is $\mathscr{E}/R$, and the final current (at $t=\infty$) is zero.
(b) Charge on the capacitor as a function of time. The initial charge is zero, and the final charge (at $t=\infty$) is $C\mathscr{E}$.

EXAMPLE 8. Suppose that in the circuit illustrated in Fig. 20.21, the resistance is $R = 8.0 \times 10^3\ \Omega$, the capacitance is $C = 2.0\ \mu\text{F}$, and the emf of the battery is $\mathscr{E} = 1.5$ V. What is the initial value of the current, at the instant after the battery is connected? What is the final value of the charge on the capacitor? At what time is the charge 63% of its final value?

SOLUTION: The initial value of the current is

$$I = \frac{\mathscr{E}}{R} = \frac{1.5\ \text{V}}{8.0 \times 10^3\ \Omega} = 1.9 \times 10^{-4}\ \text{A}$$

The final value of the charge is

$$Q = C\mathscr{E} = 2.0 \times 10^{-6}\ \text{F} \times 1.5\ \text{V} = 3.0 \times 10^{-6}\ \text{C}$$

The time at which the charge is 63% of its final value is the characteristic time τ,

$$\tau = RC = 8.0 \times 10^3\ \Omega \times 2.0 \times 10^{-6}\ \text{F} = 1.6 \times 10^{-2}\ \text{s}$$

20.8* THE HAZARDS OF ELECTRIC CURRENTS

As a side effect of the widespread use of electric machinery and devices in factories and homes, each year in the United States about 1000 people die by accidental electrocution. A much larger number suffer nonfatal electric shocks. Fortunately, the human skin is a fairly good insulator, which provides a protective barrier against injurious electric currents. The resistance of a square centimeter of dry human epidermis in contact with a conductor can be as much as $10^5\ \Omega$. However, the resistance varies in a rather sensitive way with the thickness, moisture, and temperature of the skin, and with the magnitude of the potential difference.†

The electric power supplied to factories and homes in the United States is usually in the form of alternating currents, or AC. These are oscillating currents, which periodically reverse direction (the standard period of the alternating current supplied by power companies is $\frac{1}{60}$ second). Since most accidental electric shocks involve alternating currents, the following discussion of the effects of currents on the human body will emphasize alternating currents.

In the typical accidental electric shock, the current enters the body through the hands (in contact with one terminal of the source of emf) and exits through the feet (in contact with the ground, which constitutes the other terminal of the source of emf for most AC circuits). Thus, the body plays the role of a resistor, closing an electric circuit (see Fig. 20.23).

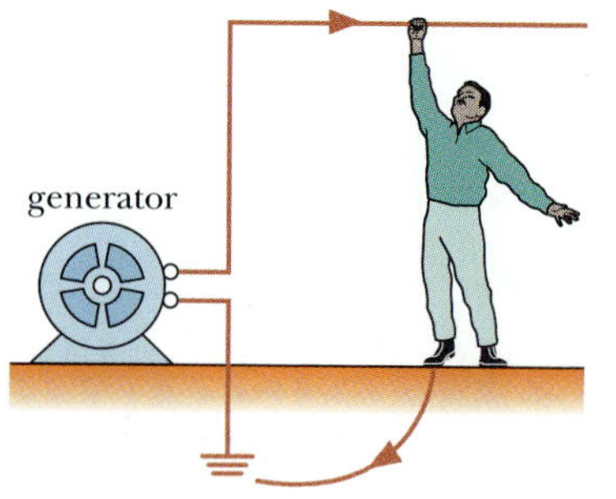

FIGURE 20.23
A human body closing an electric circuit. The current enters through the hand and exits through the feet.

The damage to the body depends on the magnitude of the current passing through it. An alternating current of about 0.001 A produces only a barely detectable tingling sensation. Higher currents produce pain and strong muscular contractions. If the victim has grasped an electric conductor — such as an exposed power cable — with the hand, the muscular contraction may prevent the victim from releasing the hold on the conductor. The magnitude of the "let-go" current, at which the victim can just barely release the hold on the conductor, is about 0.01 A. Higher currents lock the victim's hand to the conductor. Unless the circuit is broken within a few seconds, the skin in contact with the conductor will then suffer burns and blisters. Such damage to the skin drastically reduces its resistance, which can lead to a fatal increase of the current.

An alternating current of about 0.02 A flowing through the body from the hands to the feet produces a contraction of the chest muscles that halts breathing; this leads to death by asphyxiation if it lasts for a few minutes. A current of about 0.1 A lasting just a few seconds induces fibrillation of the heart muscles, with cessation of the natural rhythm of the heartbeat and cessation of the pumping of blood. Fibrillation usually continues even when the victim is removed from the electric circuit; the consequences are fatal unless immediate medical assistance is available. The

* This section is optional.

† The variation of resistance with potential difference implies that skin does not obey Ohm's Law.

treatment for fibrillation involves the deliberate application of a severe electric shock to the heart by means of electrodes placed against the chest; this arrests the motion of the heart completely. When the shock ends, the heart usually resumes beating with its natural rhythm.

A current of a few amperes produces a seizure of the nervous system and paralysis of the respiratory muscles. Victims of such currents can sometimes be saved by prompt recourse to artificial respiration. At these high values of the current, the effects of AC and DC are not very different. But at lower values, a DC current poses less of a hazard than the comparable AC current, because the DC current does not trigger the strong muscular contractions triggered by the AC current.

In the above we assumed that the path of the current through the body is from the hands to the feet. If the current enters through one leg and exits through the other, no vital organs lie in its path, and the threat to life is lessened. However, an intense current through a limb tends to kill the tissue through which it passes, and may ultimately require the surgical excision of large amounts of dead tissue, and even the amputation of the limb.

Other things being equal, a higher voltage will result in a higher current. The hazard posed by contact with high-voltage sources is therefore obvious. But under exceptional circumstances, even sources of low voltage can be hazardous. Several cases of electrocution by contact with sources of a voltage as low as 12 V have been reported. It seems that in these cases death resulted from an unusually sensitive response of the nervous system; it is also conceivable that an unusually small skin resistance was a contributing factor. Thus, it is advisable to treat even sources of low voltage with respect!

◀ First aid for victims of electric shock

Aid to victims of electric shock should begin with switching off the current. When no switch, plug, or fuse for cutting off the current is accessible, the victim must be pushed or pulled away from the electric conductor by means of a piece of insulating material, such as a piece of* dry *wood or a rope. The rescuer must be careful to avoid electric contact. If the victim is not breathing, artificial respiration must be started at once. If there is no heartbeat, cardiac massage must be applied by trained personnel until the victim can be treated with a defibrillation apparatus.

SUMMARY

- **Kirchhoff's (second) Rule:** The sum of emfs and potential changes across resistors around any closed loop in a circuit must equal zero.
- **Power delivered by a source of emf:** $P = I\mathscr{E}$
- **Power dissipated by a resistor:** $P = I\Delta V$
- **Characteristic time for RC circuit:** $\tau = RC$

QUESTIONS

1.–Can we use a capacitor as a pump of electricity in a circuit? In what way would such a pump differ from a battery?

2.–Does a fully charged battery have the same emf as a partially charged battery?

3.–The emf of a battery is often called the "open-circuit voltage." Explain.

4.–Kirchhoff's second rule is equivalent to energy conservation in the electric circuit. Explain.

5.–What would happen if we were to connect an ammeter incorrectly, as shown in Fig. 20.17c?

6.–What would happen if we were to connect a voltmeter incorrectly, as shown in Fig. 20.18b?

7.–A mechanic determines the internal resistance of an automobile battery by connecting a rugged, high-current ammeter (of nearly zero resistance) directly across the poles of the battery. The internal resistance is inversely proportional to the ammeter reading, $R_i \propto 1/I$. Explain.

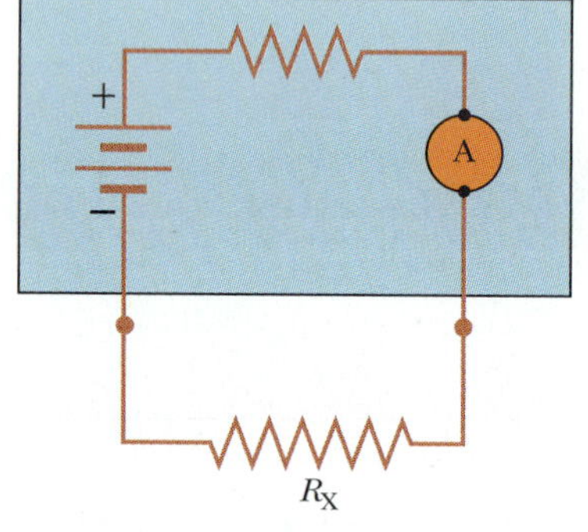

FIGURE 20.24
An ohmmeter, consisting of a battery, a reference resistance, and an ammeter enclosed in a box (blue).

8.–An ***ohmmeter*** consists of a reference source of emf (a battery), connected in series with a reference resistance, and an ammeter. When the terminals of the ohmmeter are connected to an unknown resistor (Fig. 20.24), the current registered by the ammeter permits the evaluation of the unknown resistance. Explain.

9.–At $t = 0$ you connect a battery and a resistor to a capacitor, as shown in Fig. 20.21. The capacitor is initially charged. Qualitatively, describe the current as a function of time.

10.–A homeowner argues that he should not pay his electric bill since he is not keeping any of the electrons that the power company delivers to his home — any electron that enters the wiring of his home sooner or later leaves and returns to the power station. How would you answer?

11.–The spiral heating elements commonly used in electric ranges *appear* to be made of solid metal. Why do they not short-circuit when you place an iron pot on them?

Spiral heating element of an electric range.

12.–What are the advantages and what are the disadvantages of high-voltage power lines compared with low-voltage power lines?

13.–In many European countries, electric power is delivered to homes at 220 V, instead of the 115 V customary in the United States. What are the advantages and what are the disadvantages of 220 V?

PROBLEMS

Section 20.1: Electromotive Force

1.——The smallest batteries have a mass of 1.5 g and store an electric energy of about 5×10^{-6} kW · h. The largest batteries (used aboard submarines) have a mass of 2.7×10^{5} kg and store an electric energy of 5×10^{3} kW · h. What is the amount of energy stored per kg of battery in each case?

2.——A heavy-duty 12-V battery for a truck is rated at 160 A · h, that is, this battery will deliver 1 A for 160 h (or a larger current for a correspondingly shorter time). What is the amount of electric energy that this battery will deliver?

3.——The electric starter motor in an automobile equipped with a 12-V battery draws a current of 80 A when in operation.

(a) Suppose it takes the starter motor 3.0 s to start the engine. What amount of electric energy has been withdrawn from the battery?

(b) The automobile is equipped with a generator that delivers 5.0 A to the battery when the engine is running. How long must the engine run so that the generator can restore the energy in the battery to its original level?

4.——A size D flashlight battery will deliver 1.2 A·h at 1.5 V (that is, it will deliver 1.2 A for 1 h or a larger or smaller current for a correspondingly shorter or longer time). An automobile battery will deliver 55 A·h at 12 V. The flashlight battery is cylindrical with a diameter of 3.3 cm, a length of 5.6 cm, and a mass of 0.086 kg. The automobile battery is rectangular with dimensions 30 cm × 17 cm × 23 cm and a mass of 23 kg.

(a) What is the available electric energy stored in each battery?

(b) What is the amount of energy stored per cubic centimeter of each battery?

(c) What is the energy stored per kilogram of each battery?

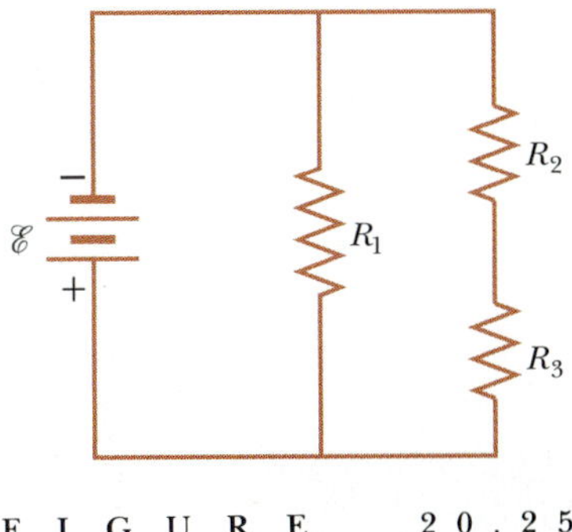

FIGURE 20.25

Section 20.3: Single-Loop Circuits

5.——Two small light bulbs are connected in parallel to a 6-V lantern battery. If the resistance of each light bulb, at its operating temperature, is 2 Ω, what is the current in the circuit? What is the current in each light bulb?

6.——An array of solar cells is to be used to charge a 12-V lead–acid battery. If each solar cell generates 0.6 V, how many cells must we connect in series to charge the battery?

7.——In a flashlight, two 1.5-V batteries are connected in series. An 8-Ω light bulb closes the circuit. What is the current in the circuit? What is the electric energy delivered to the light bulb in 1 hour?

8.——Consider the circuit shown in Fig. 20.25. When the resistance of R_2 is decreased, do the currents in R_1 and R_3 increase or decrease?

9.*——In order to reduce the effective internal resistance of an ammeter, a physicist connects a resistor in parallel across its terminals (see Fig. 20.26). If the resistance of this parallel resistor, or ***shunt*** resistor, is 1/10 of the resistance of the ammeter, by what factor does the shunt reduce the effective resistance of the modified ammeter? How must the physicist recalibrate the dial of the ammeter?

10.*——Four resistors, with $R_1 = 25\ \Omega$, $R_2 = 15\ \Omega$, $R_3 = 40\ \Omega$, and $R_4 = 20\ \Omega$, are connected to a 12-V battery as shown in Fig. 20.27.

(a) Find the combined resistance of the four resistors.

(b) Find the current in each resistor.

11.*——Four resistors are connected in parallel to a voltage source as shown in Fig. 20.28. The 25-Ω resistor carries a current of 5 A. What is the total current supplied by the battery to the resistors?

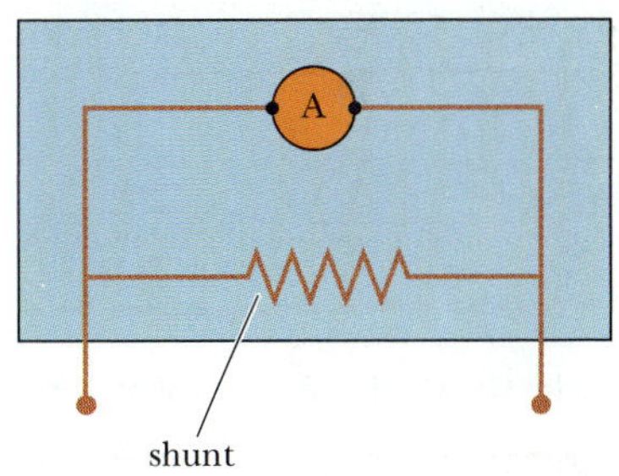

FIGURE 20.26

An ammeter with a parallel resistor, or shunt resistor.

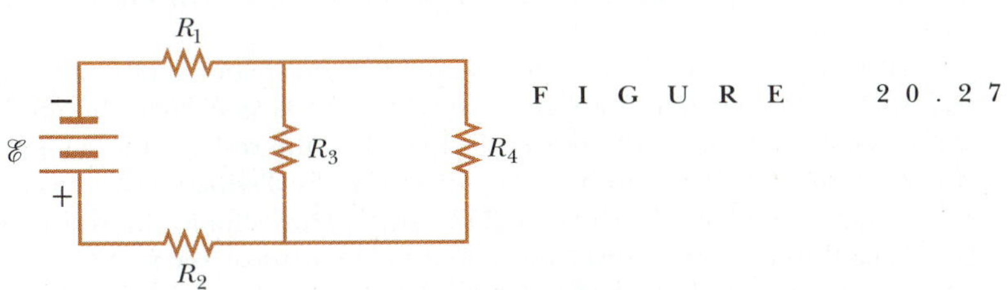

FIGURE 20.27

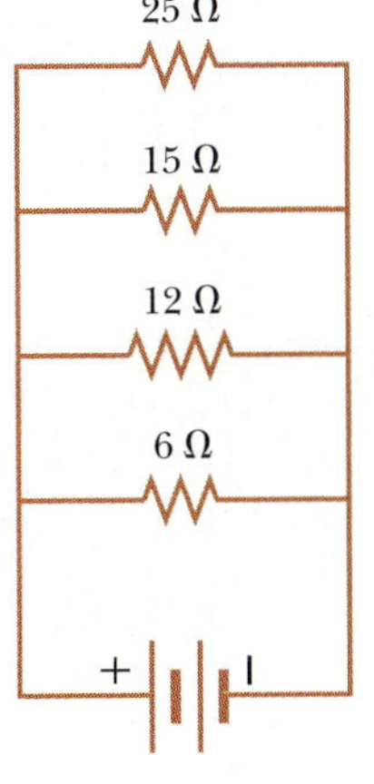

FIGURE 20.28

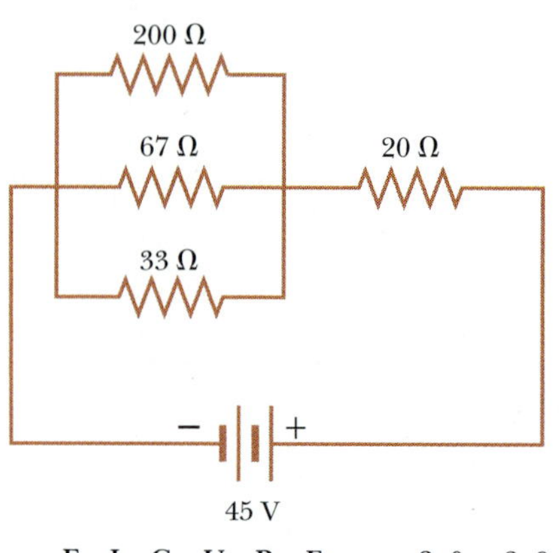

FIGURE 20.29

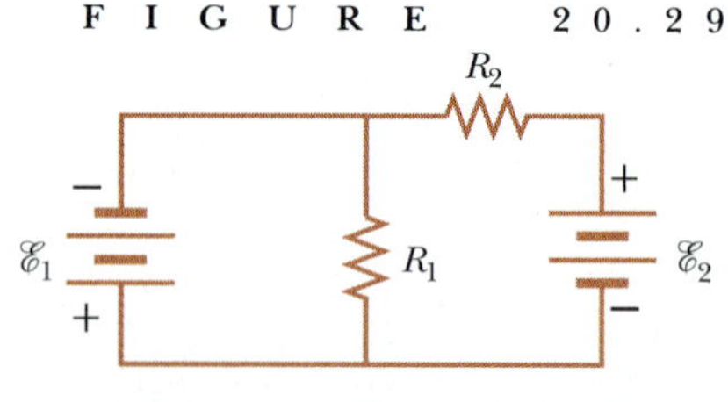

FIGURE 20.30

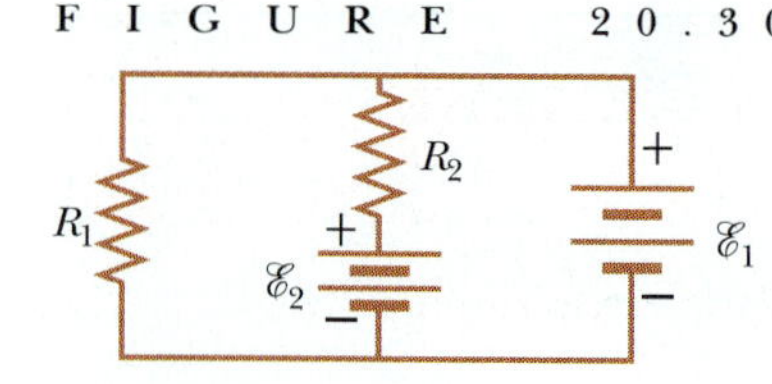

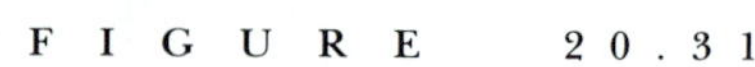

FIGURE 20.31

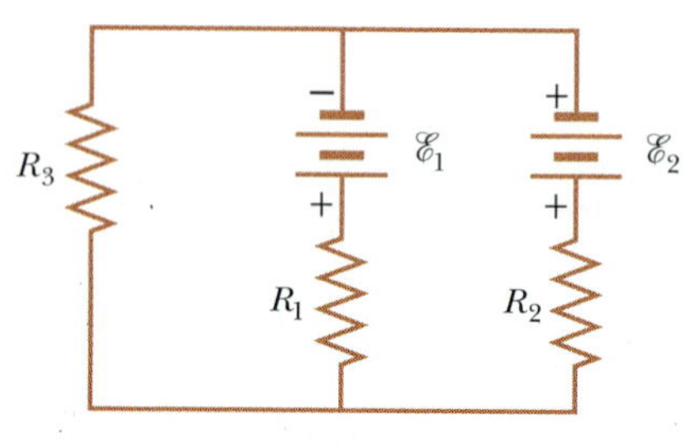

FIGURE 20.32

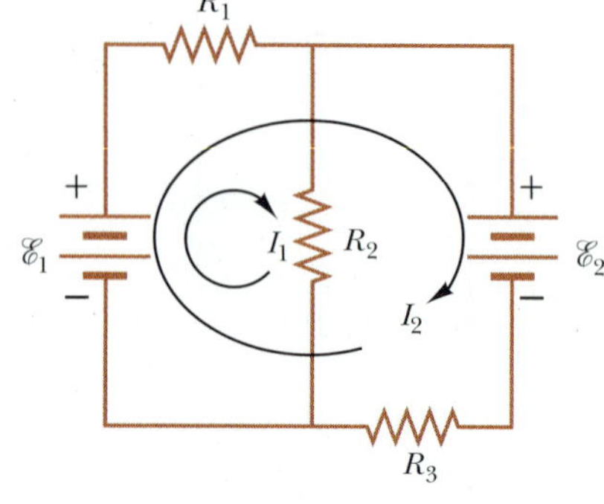

FIGURE 20.33
Alternative choice for current loops.

12.*—What is the current through the 200-Ω resistor in the circuit shown in Fig. 20.29?

Section 20.4: Multiloop Circuits

13.*—Consider the circuit shown in Fig. 20.30. Given that $\mathscr{E}_1 = 6.0$ V, $\mathscr{E}_2 = 10$ V, and $R_1 = 2.0\ \Omega$, what must be the value of the resistance R_2 if the current through this resistance is to be 2.0 A?

14.*—Find the current in the two resistors shown in Fig. 20.31. Find the power delivered by the battery $\mathscr{E}_2$. The resistances are $R_1 = 10\ \Omega$, $R_2 = 8\ \Omega$; the emfs are $\mathscr{E}_1 = 10$ V and $\mathscr{E}_2 = 12$ V.

15.*—Consider the circuit shown in Fig. 20.13 with the given resistances and emfs (see Example 4). Suppose we replace the emf $\mathscr{E}_1 = 12.0$ V by a larger emf. How large must we make $\mathscr{E}_1$ if the current I_2 is to charge the battery $\mathscr{E}_2$?

16.*—Find the currents in the circuit shown in Fig. 20.13 if the battery on the right is reversed.

17.*—Two batteries and three resistors are connected as shown in Fig. 20.32. Given that $R_1 = 0.25\ \Omega$, $R_2 = 0.20\ \Omega$, $R_3 = 0.50\ \Omega$, $\mathscr{E}_1 = 6.0$ V, and $\mathscr{E}_2 = 12.0$ V, find the currents in the resistors R_1 and R_3.

18.*—When using the loop method to obtain the equations for a circuit, we can make several choices for the loops. In Fig. 20.13, we made one possible choice and obtained Eqs. (12) and (13). Suppose that instead we make the choice shown in Fig. 20.33. What are the two loop equations in this case? Show that the new loop equations lead to the same result as the old ones.

Section 20.5: Energy in Circuits; Joule Heat

19.—An electronic calculator operates from a 3.0-V battery. The calculator uses a current of 2.0×10^{-4} A. What power does it use?

20.—An electric water heater of resistance 8 Ω draws 15 A of current when connected to the voltage supply. What is the cost of operating the water heater for 4 hours if the electric company charges 8¢ per kilowatt-hour?

21.—How much does it cost you to operate a 100-watt light bulb for 24 hours? The price of electric energy is 8¢ per kilowatt-hour.

22.—An electric toaster uses 1200 W at 115 V. What is the current through the toaster? What is the resistance of its heating coils?

23.—While cranking the engine, the starter motor of an automobile draws 80 A at 12.0 V for a time interval of 2.5 s. What is the electric power used by the starter motor? How many horsepower does this amount to? What is the electric energy used up in the given time interval?

24.—An air conditioner operating on 115 V uses 1500 W of electric energy. What is the electric current through the air conditioner? What is the resistance of the air conditioner?

25.—A small electric motor operating on 115 V delivers 0.75 hp of mechanical power. Ignoring friction losses within the motor, what current does this motor require?

26.*—A resistor R, connected to a source of emf $\mathscr{E}$, is dissipating a power P. The resistor is then changed to a resistance value 20% larger. How should the emf be changed to keep the power dissipated by the resistor the same?

27.*—A resistor R, connected to a source of emf $\mathscr{E}$, is dissipating a power P. The voltage is increased to a value 25% larger. How should the resistance be changed to keep the power dissipated by the resistor the same?

28.*—To heat one gram of water (density 1 g/cm³) 1°C takes 4.18 joules. How long does it take to heat one liter of water by 50°C with an electric heating element of resistance 2.0 Ω carrying a current of 32 A?

29.*—The rate of flow of water over Niagara Falls is 2800 m^3/s; this water falls a vertical distance of 51 m. At night, one-half of the water is diverted to a power plant. If the plant converts all of the potential energy of this diverted water into electric power, what is the electric power in kilowatts? If this power is fed into a power line at 240 kV, what is the current?

30.*—A cyclotron accelerator produces a beam of protons of an energy of 700 million eV. The average current of this beam is 1.0×10^{-6} A. What is the number of protons per second delivered by the accelerator? What is the corresponding power delivered by the accelerator?

31.*—The aluminum cable of a high-voltage transmission line carries a current of 600 A. The cable is 60 km long, and it has a diameter of 2.5 cm. What is the power lost to Joule heat in this cable?

32.*—Two wires of equal lengths and equal diameters are connected in parallel to a source of emf. One wire is made of copper, the other of aluminum. What is the ratio of the Joule heats produced in these two wires?

An electric automobile.

33.*—An engineer wants to design a high-voltage transmission line 100 km long.
(a) The aluminum cable used in this transmission line is to have a resistance of no more than 6 Ω. What diameter of cable is required?
(b) For a cable of this diameter carrying a current of 750 A, what is the power lost to Joule heat?

34.*—An electric automobile is equipped with an electric motor supplied by a bank of sixteen 12-V batteries. When fully charged, each battery stores an energy of 2.2×10^6 J.
(a) What current is required by the motor when it is delivering 12 hp? Ignore friction losses.
(b) With the motor delivering 12 hp, the car has a speed of 65 km/h (on a level road). How far can the car travel before its batteries run down?

35.*—The banks of batteries in a submarine store an electric energy of 5×10^3 kW·h. If the submarine has an electric motor developing 1000 hp, how long can it run on these batteries?

36.*—An electric toothbrush draws 7 watts. If you use it 4 minutes per day and if electric energy costs you 8¢/kW·h, what do you have to pay to use your toothbrush for one year?

37.*—In a small electrostatic generator, a rubber belt transports charge from the ground to a spherical collector at 2.0×10^5 V. The rate at which the belt transports charge is 2.5×10^{-6} C/s. What is the rate at which the belt does work against the electrostatic forces?

38.*—A solar panel (an assemblage of solar cells) measures 58 cm × 53 cm. When facing the sun, this panel generates 2.7 A at 14 V. Sunlight delivers an energy of 1.0×10^3 W/m^2 to an area facing it. What is the efficiency of this panel, that is, what fraction of the energy in sunlight is converted into electric energy?

39.*—A 40-m cable connecting a lightning rod on a tower to the ground is made of copper with a diameter of 7 mm. Suppose that during a stroke of lightning the cable carries a current of 1×10^4 A.
(a) What is the potential drop along the cable?
(b) What is the rate at which Joule heat is produced?

40.*—The maximum recommended current for a No. 10 copper wire of diameter 0.259 cm is 30 A. For such a wire with this current, what is the rate of production of Joule heat per meter of wire? What is the potential drop per meter of wire? What is the electric field in the wire?

41.*—The cable connecting the electric starter motor of an automobile with the 12.0-V battery is made of copper with a diameter of 0.50 cm and a length of 0.60 m. If the starter motor draws 500 A (while stalled), what is the rate at which Joule heat is produced in the cable? What fraction of the power delivered by the battery does this Joule heat represent?

42.*—Two heating coils have resistances of 12.0 Ω and 6.0 Ω, respectively.
(a) What is the Joule heat generated in each if they are connected in parallel to a source of emf of 115 V?
(b) What if they are connected in series?

43.*—An electric clothes dryer operates on a voltage of 220 V and draws a current of 20 A. How long does the dryer take to dry a full load of clothes? The clothes weigh 6.0 kg when wet and 3.7 kg when dry. Assume that all the electric energy going into the dryer is used to evaporate water (the heat of evaporation is 539 kcal/kg).

44.**—A large electromagnet draws a current of 200 A at 400 V. The coils of the electromagnet are cooled by a flow of water passing over them. The water enters the electromagnet at a temperature of 20°C, absorbs the Joule heat, and leaves at a higher temperature. If the water is to leave with a temperature no higher than 80°C, what must be the minimum rate of flow of water (in liters per minute) through the electromagnet?

45.**—You want to design a hot-water heater for use in a house. The heater is to operate at 115 V, and it is to heat 150 liters of water from an initial temperature of 10°C to a final temperature of 60°C. If the heater is to take no more than 1 h to reach this final temperature, what electric power is required? What must be the resistance of the electric heating coil in this heater? Assume that no water is taken out of the heater before it reaches its final temperature, and assume that the heater is well insulated and loses no heat to the environment.

46.**—A hair dryer intended for travelers operates at 115 V and also at 220 V. A switch on the dryer adjusts the dryer for the voltage. At each voltage, the dryer delivers 1000 W of heat. What must be the resistance of the heating coils for each voltage? For such a dryer, design a circuit consisting of two identical heating coils connected to a switch and to the power outlet.

Section 20.6: Electrical Measurements

47.—A battery of unknown emf is being measured with a potentiometer. When this battery is inserted in the potentiometer, the balance is achieved with $PO = 30.2$ cm. When a standard cell of emf 1.50 V is inserted in the potentiometer, the balance is achieved at $PO = 44.5$ cm. What is the emf of the battery?

48.—An unknown resistance is being measured with a Wheatstone bridge equipped with a slide wire of length 1.00 m and a reference resistance of 200 Ω. The bridge balances at $PO = 68.4$ cm. What is the value of the unknown resistance?

49.*—A circuit consists of a resistor of 3.0 Ω connected to a (resistanceless) battery. To measure the current in this circuit, you insert an ammeter of internal resistance 2×10^{-3} Ω. This ammeter then reads 3.955 A. What was the current in the circuit before you inserted the ammeter?

50.**—A voltmeter of internal resistance 5000 Ω is connected across the poles of a battery of internal resistance 0.2 Ω. The voltmeter reads 1.4993 volts. What is the actual zero-current emf of the battery?

51.**—To measure the internal resistance of a battery, a physicist places the battery in a potentiometer and finds that it balances at a length $PO = 55.2$ cm. Without removing the battery from the potentiometer, he then connects a 2.0-Ω resistor across the terminals of the battery and, while a current is flowing through the resistor–battery circuit, he again balances the potentiometer. The new balance is at 53.4 cm. What is the internal resistance of the battery?

Section 20.7: The RC Circuit

52.——Consider the circuit shown in Fig. 20.34. How long does it take after the switch is closed to charge the capacitor to within 63% of its final voltage?

53.——A capacitor charged through a 75-Ω resistor takes 2.7 ms to come within 63% of its final voltage. What is its capacitance?

54.——How long does it take for a fully charged 4-μF capacitor to decay to 37% of its initial charge when discharged through a 100-Ω resistor?

55.——For a laboratory demonstration, you want to construct an RC circuit with a time constant of 15 s. You have available a capacitor of 20 μF. What resistance do you need?

56.——Two capacitors of 2.0 μF and 4.0 μF and a resistor of 8×10^3 Ω are connected to a battery as shown in Fig. 20.35. What is the time constant of this circuit?

57.——Ideal capacitors have an infinite internal resistance between their plates (that is, the material between the plates is a perfect insulator). However, real capacitors have a finite internal resistance, and consequently the charge will gradually leak from one plate to the opposite, and the capacitor will gradually discharge when left to itself. If a capacitor of 8.0 μF has an internal resistance of 5.0×10^8 Ω, how long does it take for 63% of its initial charge to leak away?

58.*——A capacitor with $C = 20$ μF and a resistor with $R = 100$ Ω are suddenly connected in series to a battery with $\mathscr{E} = 6.0$ V.

(a) What is the charge on the capacitor at $t = 0$? At $t = 0.002$ s?

(b) What is the final value of the charge?

(c) What is the rate of increase of the charge at $t = 0$?

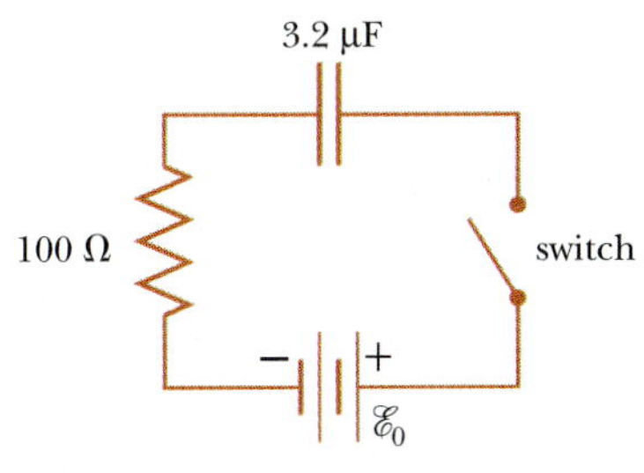

FIGURE 20.34

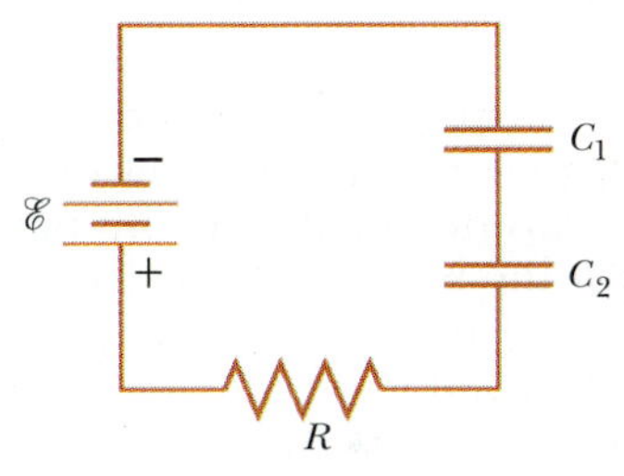

FIGURE 20.35

PHYSICS IN ACTION

THUNDERSTORMS AND LIGHTNING

The atmosphere acts as a great electric machine. Thunderstorms and lightning are the most spectacular manifestations of electric activity in the atmosphere (see Fig. 1), but even in fair weather, the atmosphere is endowed with electric fields and electric currents. The thunderstorms act as giant electrostatic generators; they deliver negative charge to the ground, and thereby pump a positive current into the upper level of the atmosphere. This upper level of the atmosphere, called the ionosphere, is a good conductor, and the current reaching it quickly spreads laterally, over the entire globe. In fair-weather regions, this current gradually leaks down to the ground, completing the atmospheric electric circuit.

FIGURE 1 Thunderstorm with lightning.

There are roughly 2000 thunderstorms in action all over the Earth at any given time. The time-average current generated by each thunderstorm is about 1 A (the instantaneous current can be much larger — up to 20,000 A in a stroke of lightning). Thus, at any given time, all the thunderstorms together contribute an average of 2000 A to the current in the atmospheric circuit.

Figure 2 is a schematic diagram of the ***atmospheric electric circuit.*** Thunderstorms act as the source of emf in this circuit, and the fair-weather atmosphere acts as a resistor. Both the ionosphere and the ground are good conductors, and each is an equipotential surface. The potential difference between them is about 300,000 V. The resistance of the entire fair-weather atmosphere between ionosphere and ground is about 200 Ω. Most of this resistance is concentrated in the dense, low regions of the atmosphere; correspondingly, most of the 300,000-V potential drop occurs in the low regions of the atmosphere, within a few kilometers from the ground. On the average, the total electric power delivered to the atmospheric circuit by thunderstorm activity is about $2000\ \text{A} \times 300{,}000\ \text{V} = 6 \times 10^8\ \text{W}$, that is, nearly a million kW.

The electric current that flows through the fair-weather atmosphere from the ionosphere to the ground is carried by ions, that is, atoms and molecules with positive or negative electric charges. Such ions are always present in air, in small concentrations. The current consists of a downward motion of positive ions and an upward motion of negative ions. The driving force that maintains this motion is provided by an atmospheric electric field, which points vertically downward; this electric field is analogous to the electric field that exists in a current-carrying wire and drives the electrons along the wire. Near the surface of the Earth, over open ground, in fair weather, the strength of the atmospheric electric field is 100 to 200 V/m.

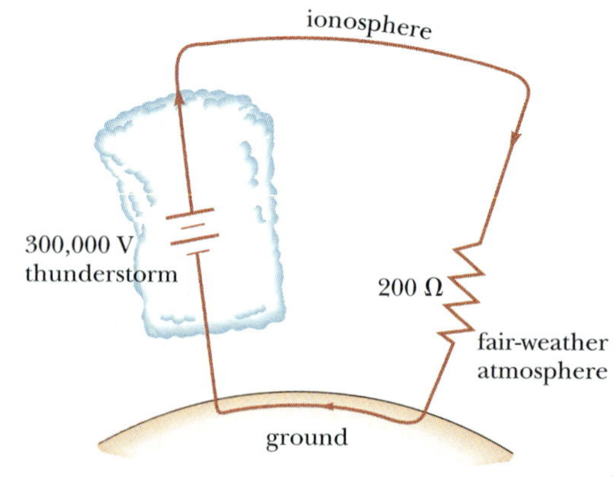

FIGURE 2 Schematic diagram of the atmospheric electric circuit.

Like any source of emf, thunderstorms convert nonelectric energy into electric energy. Thunderstorms obtain the energy for their violent mechanical and electric activity from humid air. Thunderstorms are heat engines, and the humid air is their heat reservoir. They condense the water vapor in humid air and convert the heat from this into mechanical and electric energy. One cubic kilometer of air at 100% relative humidity contains about 10^7 kg of water vapor; condensation of all of this water vapor would release about 10^{10} kcal of heat. Although the efficiency for the energy conversion is low, the release of just a small fraction of the energy stored in humid air suffices to account for the violent activity of a thunderstorm. Tornadoes and hurricanes also obtain their energy from humid air, and their destructive power reflects the large amount of stored energy.

A thunderstorm is made of several ***cells,*** or thunderclouds, within each of which air moves upward or, in later stages of the cell's development, downward. The mechanism that drives these updrafts and downdrafts hinges on the instability of humid air. If, because of some disturbance, a parcel of humid air is pushed upward to slightly higher altitude, its expansion and consequent reduction of temperature leads to condensation of a fraction of the water vapor. The heat released by this condensation tends to rewarm the air. The parcel of air will then be warmer and less dense than the surrounding dry air that resides at the higher altitude. Thus, the parcel of humid air becomes buoyant and it continues to rise. The humid air is unstable — once an upward motion starts, it will continue faster and faster until all the water vapor has condensed, and even then the upward motion can continue, since the water can supply further amounts of heat by freezing into ice.

In a young thunderstorm cell, the air rises as in a chimney, drawing in more and more humid air at the base while the top of the cell grows upward at speeds of 10 m/s or more. A mature thunderstorm cell may extend to a height of 12,000 to 18,000 m (see Fig. 3). The updrafts within the cell are typically 15 m/s (updrafts and downdrafts of such a magnitude are dangerous to aircraft). Inside the thunderstorm cell, raindrops and hail form and begin to fall. This downward motion drags some air along and starts a downdraft. Once the air begins to descend, it will continue descending — humid air is unstable both for upward motion and for downward motion. An old, dissipating thunderstorm is dominated by downdrafts.

Exactly how electric charges are generated in a thunderstorm cell remains somewhat of a mystery.

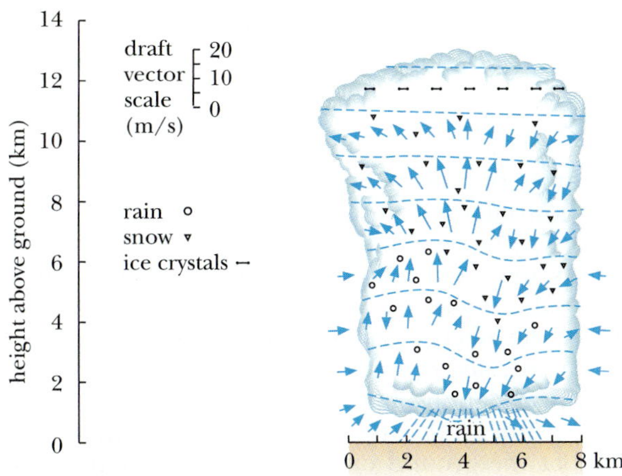

FIGURE 3 A mature thunderstorm cell. The vectors indicate the updrafts and the downdrafts.

The charge distribution that accumulates in a mature thunderstorm cell is typically as shown in Fig. 4. There is a positive charge in the upper part of the cell and a negative charge in the lower part; furthermore, at the very bottom of the cell, there is often an extra, small positive charge. Several mechanisms have been proposed to explain this accumulation of positive charge in the upper part of the cell and negative charge in the lower part. According to most meteorologists, updrafts play a crucial role in the charging mechanism. Within the cell, some kind of charge-separation process deposits positive and negative charges on raindrops or ice crystals of different sizes. In this process, the positive charge is attached to raindrops or ice crystals of

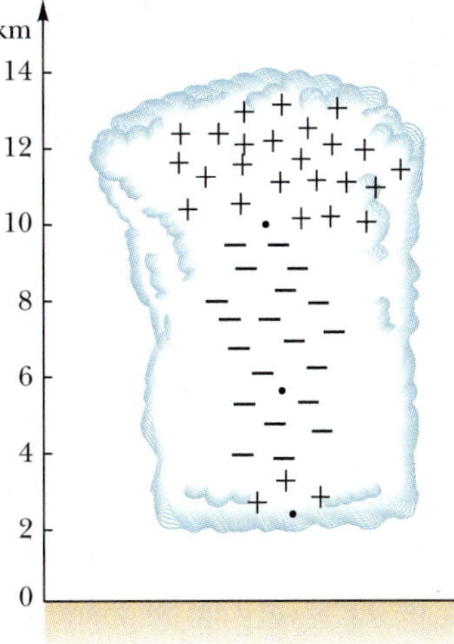

FIGURE 4 Charge distribution in a mature thunderstorm cell.

small size and the negative charge to those of larger size. The updrafts then carry those of small size (and positive charge) toward the top of the cell, while those of large size (and negative charge) fall toward the bottom of the cell. Typically, the amounts of charge that accumulate at the top and the bottom are 40 C and −40 C, respectively.

The negative electric charge at the bottom of the cloud induces a positive electric charge on the ground, immediately below the thundercloud. Thus, the electric field under the thundercloud is in the vertically upward direction (opposite to the downward atmospheric electric field that prevails in fair-weather regions). The strength of the electric field at the ground under the thundercloud is typically 10^4 V/m. However, at any sharp protuberance, the strength of the field will be much larger. For instance, Fig. 5 shows the electric field lines near a

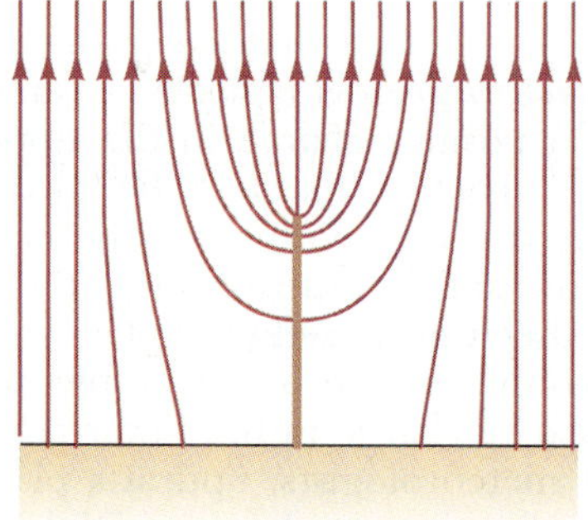

FIGURE 5 Electric field lines near a thin vertical spike sticking out of the ground.

thin conducting spike sticking out of the ground; the electric field is very strong at the tip of the spike, where the field lines converge. The strong electric fields at sharp spikes or edges sticking out of the ground produce ***point discharge*** — the air becomes a conductor, and a current flows from the spike into the atmosphere. The conductivity of air results from electric breakdown of air molecules. In the intense electric field, the normally neutral molecules lose one or two electrons and they become ions. Point discharges in such ionized air are not visible, and yet, on the average, the total point discharges from all the protuberances under the thundercloud carry most of the electric current that the thundercloud pumps into the atmospheric electric circuit. Lightning makes an impressive visual display, and it carries a very large instantaneous current; but the time-average current associated with lightning is several times smaller than the time-average current associated with the total of point discharges.

Near the concentration of charge in the thundercloud, the electric field is very intense. If this electric field near the charge concentration attains a strength of about 10^6 V/m, it causes a catastrophic electric breakdown of air, resulting in lightning. The sequence of events resulting in a lightning stroke has been investigated by high-speed photography. The electric breakdown begins in a faintly luminous channel that grows downward from the thundercloud. The channel follows a tortuous path, often with lateral branches. This channel is called the ***step leader,*** because the tip of this channel advances downward in steps of about 50 m, with short pauses between one step and the next. Once the tip of the step leader reaches the ground, it forms a continuous conducting channel from cloud to ground, and charge can flow from cloud to ground with little resistance. The electrons at the bottom of the channel are the first to move — they drain out into the ground, giving a large current. Then electrons from progressively higher positions drain down. Thus, a "drainage front" speeds upward along the channel toward the thundercloud. This drainage front starts an intense flow of current in the channel, as much as 10,000 to 20,000 A, called the ***return stroke*** (see Fig. 6). The return stroke lasts only a fraction of a millisecond. The first return stroke is often succeeded by several more return strokes. A typical flash of lightning contains three to five return strokes. Each of the succeeding strokes is preceded by a leader of its own, called a ***dart leader,*** which speeds from the cloud to the ground in a single step.

The total charge delivered to the ground by a complete flash of lightning with its several return strokes amounts to about −25 C. The intense cur-

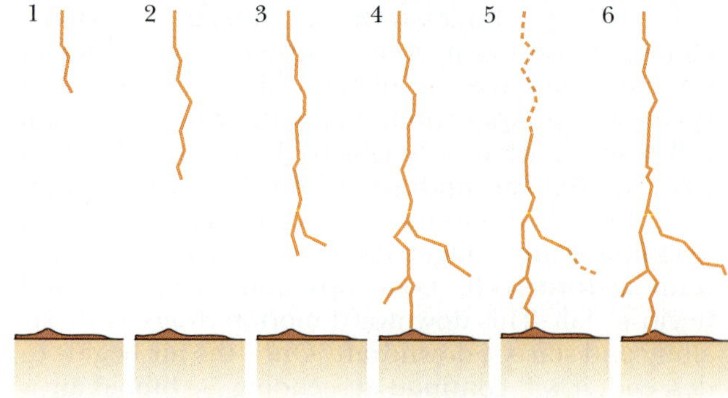

FIGURE 6 This sequence of diagrams, based on high-speed photographs, shows the downward advance of the step leader (first four diagrams), and the upward advance of the return stroke (last two diagrams). The return stroke advances much faster than the leader.

rent in the return stroke heats air to an extremely high temperature. The air becomes totally ionized — it becomes a plasma. We see this incandescent channel of plasma as a lightning bolt. The sudden heating of the air also results in a high pressure in the channel. The high-pressure air expands explosively, generating a shock wave in the surrounding atmospheric air. This shock wave dissipates quickly, and it changes into a sound wave. When the sound pulse reaches our ears, we hear it as thunder. The rumbling noise in thunder is due to successive arrivals of sound pulses from different portions of the lightning bolt, which are at different distances from our ears.

At some distance from the lightning bolt, we will hear the sound some seconds after we see the flash of lightning. Because of the high speed of light, the flash of lightning arrives almost instantaneously, while the sound takes an appreciable time to cover the distance. Since the speed to sound is about ⅓ km/s, we can use the following rule of thumb to reckon the distance (in kilometers) to the lightning bolt: Count the seconds between flash and thunder and divide by 3.

REFERENCES

Battan, L. J., *The Nature of Violent Storms* (Doubleday, Garden City, 1961).

Schonland, B. F. J., *The Flight of Thunderbolts* (Oxford Univ. Press, NY, 1964).

Viemeister, P. E., *The Lightning Book* (Doubleday, Garden City, 1961).

Vonnegut, B., "Thundercloud Electricity," *Discovery*, March 1965.

Handel, P. H., "Bolts from the Blue," *The Sciences*, July/August 1986.

CHAPTER 21

Magnetic forces repel a small magnet from the surface of a superconductor (below) and hold it suspended in midair.

MAGNETIC FORCE AND FIELD

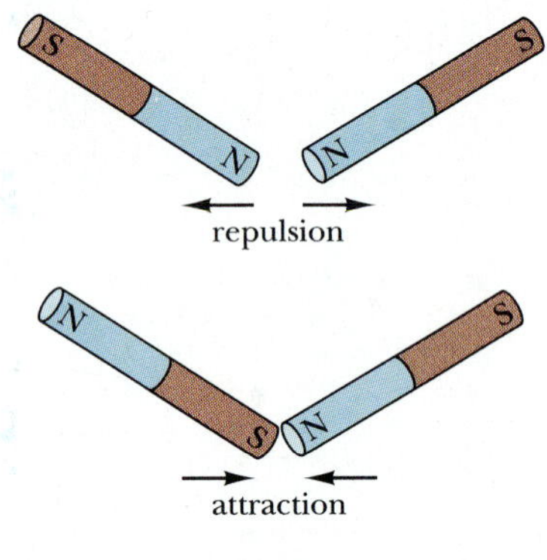

FIGURE 21.1
Equal ends of two bar magnets repel; opposite ends attract.

The magnetic forces most familiar from everyday experience are the forces that permanent magnets exert on each other or on pieces of iron or other "magnetic" materials. If you bring two bar magnets together, they will attract or repel. Each magnet has two distinct ends, called its north pole and its south pole. If you bring together equal poles, they repel; and if you bring together opposite poles, they attract (see Fig. 21.1). This attraction or repulsion between permanent magnets also accounts for the behavior of a compass needle, which is simply a small magnet free to rotate on a pivot. When we place a compass needle near the, say, south pole of a bar magnet, this south pole will attract the north pole of the compass needle, and the needle will settle into an equilibrium configuration, pointing roughly toward the south pole of the bar magnet (see Fig. 21.2). When we remove the compass needle from the disturbing influence of nearby bar magnets, it will respond to the magnetic force exerted by the Earth, and it will point northward. The core of the Earth acts like a large permanent magnet, whose poles roughly coincide with the geographic poles. (Note, however, that since the north pole of the compass needle is attracted by the geographic north pole of the Earth, this geographic north pole is actually the magnetic south pole of the Earth; see Fig. 21.3.)

The magnetic forces between permanent magnets were known for many centuries, but only during the nineteenth century did experimenters discover that electric currents also exert magnetic forces on permanent magnets and that electric currents exert magnetic forces on one another. Finally, physicists came to understand that the magnetic force is simply an extra electric force acting between charges in motion. Whether the charges are moving along a wire, forming currents, or the charges are moving through empty space, on their own, makes no essential difference; in both cases ***moving charges exert magnetic forces on one another.***

The magnetic forces between permanent magnets involve the same fundamental mechanism as the magnetic forces between moving charges or between currents. The magnetic forces between permanent magnets arise from microscopic currents in the atoms of the magnet. Within these atoms, currents flow around in closed loops. Although the individual microscopic rings of current are too weak to produce a noticeable force, the magnet contains many atoms and many rings of current, and the combined effect of all these microscopic rings gives a noticeable macroscopic force on another nearby magnet or on a nearby moving charge or current. You can observe the force exerted by a magnet on a moving charge if you place the magnet near your TV tube.* The magnetic force on the moving electrons in the tube will then deflect the electrons and produce grotesque distortions of the image on your TV screen (see Fig. 21.4).

FIGURE 21.2

A compass needle placed near a bar magnet.

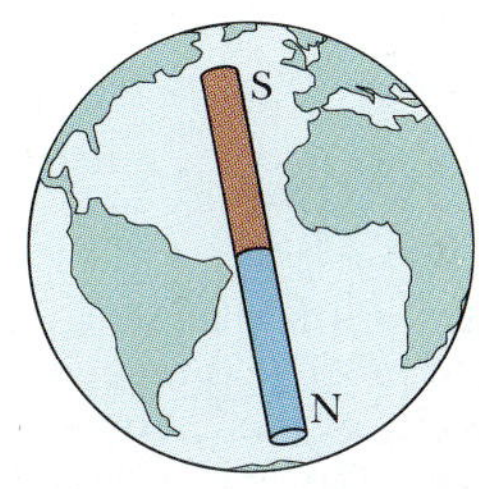

FIGURE 21.3

The magnetic effects of the Earth are like those of a large permanent magnet placed within the Earth.

FIGURE 21.4

A magnet placed near a TV tube distorts the image.

Hans Christian Oersted (örstad), *1777–1851, Danish physicist and chemist. He observed that a compass needle suffers a deflection when placed near a wire carrying an electric current. This discovery gave the first empirical evidence of a connection between electric and magnetic phenomena.*

21.1 THE MAGNETIC FORCE

In principle, it would be desirable to begin the study of magnetic forces with the law for the magnetic force between two moving point charges, just as we began our study of electric forces with the law for the electric force between two point charges at rest (Coulomb's Law). However, the

* You can do this experiment with a black-and-white TV tube, but don't do it with a color TV tube; you are likely to cause a permanent discoloration of the screen.

magnitude and direction of the magnetic force that two moving charges exert on each other depend in a complicated manner on the magnitudes and directions of their velocities, and the mathematical formula for the force is rather messy.

Instead of dealing with this complicated case, we find it easier to begin with the formula for the magnetic force exerted on one moving point charge q by a steady current I flowing in a long, straight wire. The formula for this magnetic force is somewhat similar to the formula for the electric force exerted on a point charge by a long, straight line of charge. According to Eq. (17.15), the electric force that such a line of charge exerts on the point charge is inversely proportional to the distance r between the point charge and the line of charge. This inverse proportionality to the distance r also holds for the magnetic force that a current in a straight wire exerts on a moving point charge. However, in contrast to the electric force, the magnetic force also depends on the velocity **v** of the charge — the magnitude and the direction of the magnetic force **F** depend on the magnitude and direction of the velocity **v**. To spell out this dependence in detail, it is useful to consider three separate cases:

(a) v parallel to the current: As illustrated in Fig. 21.5a, the current I is in the x direction and the velocity **v** of the point charge q is in the same direction. In this case, the magnetic force is in the radial direction, toward or away from the current. ***The magnitude of the magnetic force is directly proportional to the product of the current*** I***, the charge*** q***, and the velocity*** v***, and it is inversely proportional to the distance*** r***:***

$$F = -[\text{constant}] \times \frac{qvI}{r} \tag{1}$$

The minus sign in this formula indicates that the force is attractive if the charge q, the velocity v, and the current I are all positive, or, more generally, if the product qvI is positive. Thus, in contrast to the electric force, which is repulsive between positive charges, the magnetic force exerted on a moving positive charge q by the moving positive charges that constitute the current on the wire is attractive if the velocities of the charges are

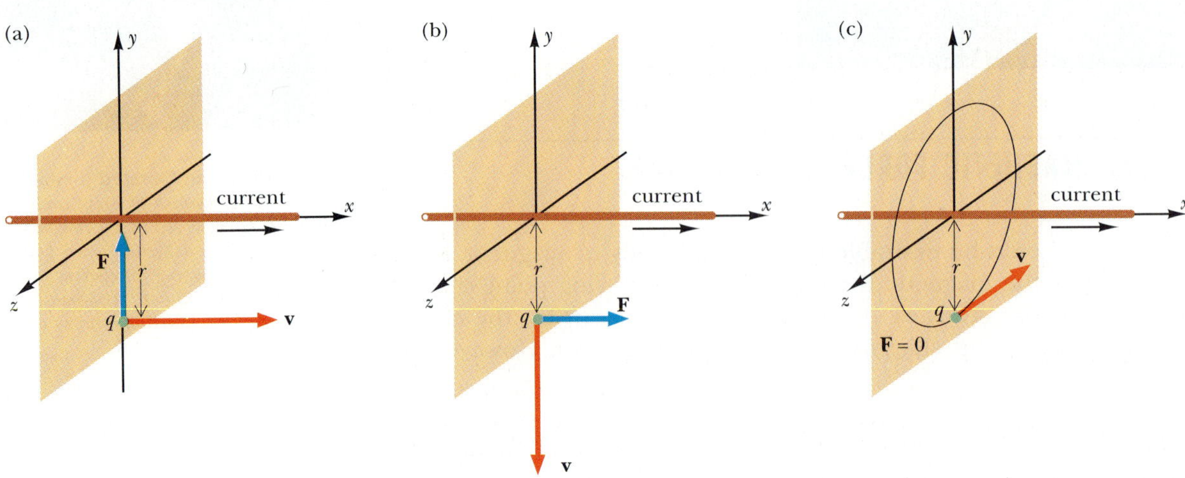

FIGURE 21.5 Direction of the magnetic force for three possible directions of the velocity of the charge q (in these diagrams, q is assumed to be positive). (a) The current on the wire is in the x direction, and the velocity of the charge q is in the same, parallel, direction. The magnetic force is in the radial direction, toward the current. (b) The velocity of the charge q is in the radial direction, perpendicularly away from the current. The magnetic force is in the x direction, parallel to the current. (c) The velocity of the charge q is in the tangential direction. The magnetic force is then zero.

parallel. However, the force is repulsive if the velocities are antiparallel — that is, if these charges are moving in opposite directions. Note that the magnetic force is zero unless both the velocity v and the current I are different from zero — the charge q and the charges on the wire must *both* be in motion if there is to be a magnetic force between them.

In SI units, the numerical value of the constant of proportionality in Eq. (1) is

$$[\text{constant}] = 2.00 \times 10^{-7}\ \text{N} \cdot \text{s}^2/\text{C}^2 \tag{2}$$

This constant is conventionally written in the form

$$[\text{constant}] = \frac{\mu_0}{2\pi} \tag{3}$$

with

$$\mu_0 = 4\pi \times 10^{-7}\ \text{N} \cdot \text{s}^2/\text{C}^2 = 1.26 \times 10^{-6}\ \text{N} \cdot \text{s}^2/\text{C}^2 \tag{4}$$

◀ Permeability constant

The quantity μ_0 (pronounced "mu-nought") is called the ***permeability constant.*** With this constant, our equation for the magnetic force on the point charge q becomes

$$F = -\frac{\mu_0}{2\pi}\frac{qvI}{r} \tag{5}$$

◀ Magnetic force; charge moving parallel to current

(b) v in the radial direction: This case is illustrated in Fig. 21.5b. The velocity **v** is directed radially away from the direction of the current. The magnetic force is now in the x direction, parallel to the direction of the current. The magnitude of the magnetic force is the same as in the first case:

$$F = +\frac{\mu_0}{2\pi}\frac{qvI}{r} \tag{6}$$

◀ Magnetic force; charge moving radially away from current

The + sign indicates that the force is in the positive x direction if the product qvI is positive (here, v is reckoned as positive if the motion of the charge is radially away from the current).

(c) v in the tangential direction: This final case is illustrated in Fig. 21.5c. The velocity **v** is now directed tangentially to the circumference of a circle concentric with the current. If the position of the charge q is as shown in Fig. 21.5c, this direction of the velocity is perpendicular to the plane of the page. In this case, the magnetic force is *zero* (and has no direction).

If the direction of the velocity **v** is not as described in one of the three basic cases (a), (b), and (c) above, then the velocity can be split into three components along the directions described in these three basic cases, and the magnetic force can be obtained by calculating the contributions for each direction separately and then taking the vector sum of the separate magnetic forces. Thus, in principle, we can calculate the magnetic force on a charge q moving in some arbitrary direction by taking a suitable combination of the three basic cases discussed above (in the next section we will learn about a more convenient way of calculating the force on a charge moving in an arbitrary direction, by means of the magnetic field). Note that according to the three basic cases, ***the magnetic force is perpendicular to the velocity of the point charge,*** whenever the magnetic force is not zero; this perpendicularity of force and velocity is a general feature of the magnetic force.

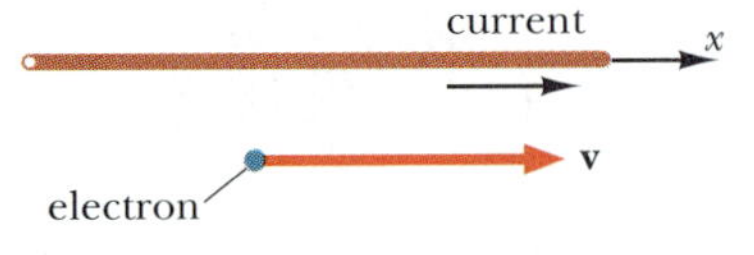

FIGURE 21.6
An electron moves parallel to a wire carrying a current. The magnetic force is directed perpendicularly away from the current.

EXAMPLE 1. A long, straight wire carries a current of 50 A. An electron of speed 2.0×10^6 m/s is (instantaneously) moving parallel to this wire at a distance of 0.030 m. What is the magnetic force on the electron?

SOLUTION: This is essentially case (a) above, but the charge q is negative, $q = -e$. If the current I is in the x direction, as in Fig. 21.6, and the velocity of the electron is also in the x direction, then according to Eq. (5) the magnetic force is repulsive (positive), of magnitude

$$F = -\frac{\mu_0}{2\pi}\frac{qvI}{r} = +\frac{\mu_0}{2\pi}\frac{evI}{r} \tag{7}$$

$$= 2.00 \times 10^{-7}\,\frac{\text{N}\cdot\text{s}^2}{\text{C}^2} \times \frac{1.6 \times 10^{-19}\,\text{C} \times 2.0 \times 10^6\,\text{m/s} \times 50\,\text{A}}{0.030\,\text{m}}$$

$$= 1.1 \times 10^{-16}\,\text{N}$$

The formulas (5) and (6) for the magnetic force exerted by a current on a moving charge are to be regarded as basic laws of physics, rooted in experiments with currents and with moving charges. In our study of magnetism, these laws for the magnetic force play a role analogous to that played in electricity by Coulomb's Law.

Another instance of a magnetic force is illustrated in Fig. 21.7, which shows two long, straight, parallel wires carrying currents. The conditions here are similar to those described in case (a). The moving charges of one wire exert magnetic forces on the moving charges of the other wire. If the currents are in the same direction, as in Fig. 21.7, these magnetic forces are attractive. If the currents are in opposite directions (antiparallel), then the magnetic forces are repulsive. In Section 21.6 we will see how to calculate the magnitude of the magnetic force between the currents on such wires.

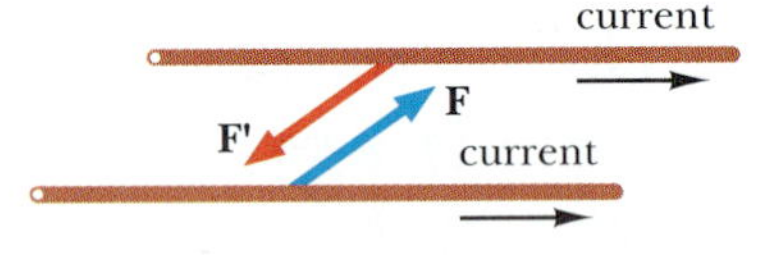

FIGURE 21.7
Two long, straight, parallel wires carrying currents in the same direction. The magnetic force that each wire exerts on the other is attractive.

Historically, the first evidence for the magnetic force between moving charges came from experiments with such currents on two or more wires. Our introduction to magnetic forces on the basis of a single moving charge and a current therefore goes counter to the historical development. But the formulas for the magnetic force between two wires of arbitrary orientations are prohibitively complicated, whereas the formulas (5) and (6) are fairly simple.

21.2 THE MAGNETIC FIELD

In Section 17.1 we saw that the electric force is communicated from one charge to another by action-by-contact, through an electric field. Likewise, ***the magnetic force is communicated from one moving charge to another through a magnetic field.*** Starting with our formulas for the magnetic force exerted by a current, we define the magnetic field by separating the expression for the force into two factors: one factor comprising quantities associated with the point charge (its charge q, its velocity $\mathbf{v}$), and another factor comprising quantities associated with the current (its magnitude I and its distance r from the point charge). Thus, in the case of a point charge moving parallel to the current, we write

$$F = (qv) \times \left(\frac{\mu_0 I}{2\pi r}\right) \tag{8}$$

Nikola Tesla, *1856–1943, American electrical engineer and inventor. He made many brilliant contributions to high-voltage technology, ranging from new motors and generators to transformers and a system for radio transmission. Tesla designed the power-generating station at Niagara Falls.*

We identify the second factor (including the constant of proportionality μ_0) as the magnetic field of the current, and we designate this magnetic field by B:

$$B = \frac{\mu_0 I}{2\pi r} \tag{9}$$

◂ **Magnetic field of current on straight wire**

The expression for the magnitude of the magnetic force then becomes

$$F = qvB \tag{10}$$

The direction of the magnetic field of the current is defined as tangent to a circle concentric with the current. Figure 21.8 shows the direction of the magnetic field at several positions around the current. Note that the direction of the magnetic field coincides with the direction of the velocity $\mathbf{v}$ for which the magnetic force on the moving charge is zero [compare case (c) in Section 21.1].

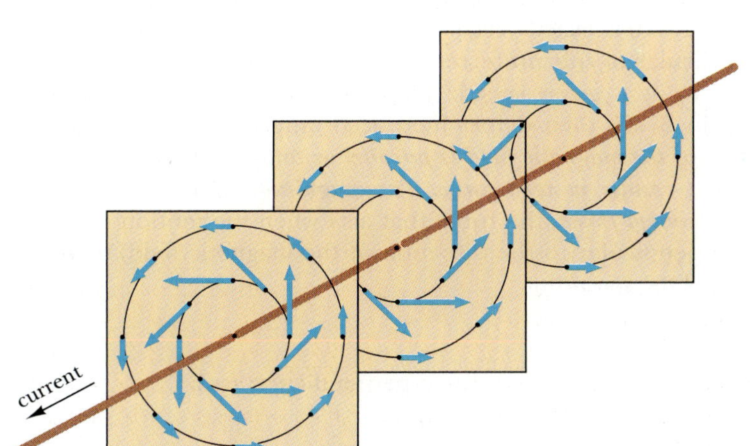

FIGURE 21.8
Direction of the magnetic field surrounding a current on a wire.

Right-hand rule for magnetic field ▶

The direction of the magnetic field of a current on a wire can be remembered by a simple ***right-hand rule for the magnetic field: If you place the thumb of your right hand along the direction of the current, then the fingers will curl around the wire in the direction of the magnetic field*** (see Fig. 21.9). This rule is consistent with the direction of the magnetic field shown in Fig. 21.8.

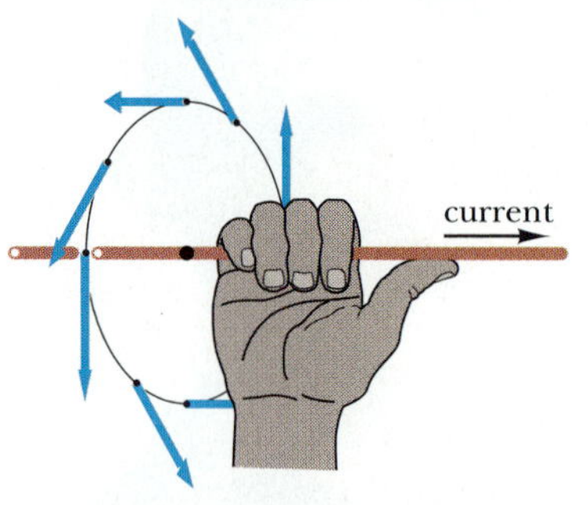

FIGURE 21.9
A right-hand rule indicates the direction of the magnetic field of a current on a wire. If the thumb of your right hand is placed along the direction of the current, then the fingers will curl around the wire in the direction of the magnetic field.

According to Eq. (9), the magnetic field of the current on a long, straight wire is inversely proportional to the distance r. In this regard the magnetic field of the long, straight wire is similar to the electric field of a long, straight line of charge [see Eq. (17.15)]. But the directions of these fields differ — the magnetic field is in the tangential direction, whereas the electric field is in the radial direction.

Equation (10) is valid for the case of a point charge moving parallel to the current. But we can easily generalize this equation for all the other cases of motion of the point charge by recognizing that in cases (a) and (b) the point charge is moving perpendicularly to the magnetic field, and in case (c) the point charge is moving parallel (or, rather, antiparallel) to the magnetic field (see Fig. 21.8). The absence of a magnetic force in the case of motion parallel (or antiparallel) to the magnetic field means that if the point charge moves at some general angle α with respect to the magnetic field (see Fig. 21.10), only the component of its velocity perpendicular to the magnetic field generates a magnetic force; the component parallel to the magnetic field does not. But the component of the velocity perpendicular to the magnetic field is $v\ \sin\alpha$ (see Fig. 21.10). Hence, a general expression for the magnitude of the magnetic force is

Force exerted by magnetic field ▶

$$F = qvB\ \sin\alpha \tag{11}$$

This expression automatically gives zero if $\alpha = 0$ (parallel motion) and it gives $F = qvB$ if $\alpha = 90°$ [perpendicular motion; see Eq. (10)].

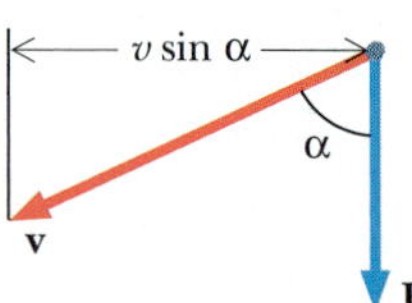

FIGURE 21.10
A point charge q moving at some angle α with respect to the direction of the magnetic field. The component of the velocity perpendicular to the magnetic field is $v\ \sin\alpha$.

Using the definition of the magnetic field produced by a current on a long, straight wire as guidance, we can now proceed to define the magnetic field produced by any general distribution of moving charges and/or currents. To discover the magnetic field at a given position, we use a test charge q and let it move repeatedly through that position, with different velocities. If the force acting on the test charge is a purely magnetic force, there will be one direction of motion (and the opposite direction) for which the force is exactly zero. Furthermore, for any direction of motion perpendicular to this preferred direction, the force will have a maximum strength, at a given speed. The direction and magnitude of the magnetic field **B** are then determined according to the following rules: ***The direction of the magnetic field is parallel or antiparallel to the direction of motion that results in zero force. The magnitude of the magnetic field equals the maximum force*** (the force that acts when the motion of the test charge is perpendicular to the direction of the magnetic field) ***divided by the charge and the speed,***

General definition of magnetic field ▶

$$B = \frac{F}{qv} \quad \text{for } \mathbf{v} \text{ perpendicular to } \mathbf{B} \tag{12}$$

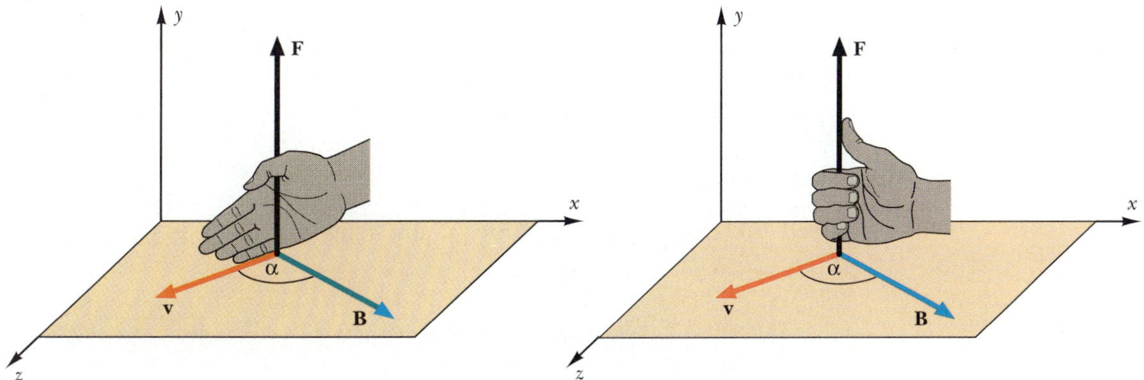

FIGURE 21.11

This right-hand rule for the magnetic force on a moving point charge relates the directions of the magnetic field **B**, the magnetic force **F**, and the velocity **v**. The vector **F** is always perpendicular to the plane defined by the vectors **B** and **v** (horizontal plane in this figure), and points in the direction indicated by the thumb of your right hand if the fingers are placed along **v** [Left] and curled toward **B** [Right]. This rule determines the direction of **B** (if **F** is known), or it determines the direction of **F** (if **B** is known).

◀ **Right-hand rule for magnetic force**

The remaining ambiguity in the direction of the magnetic field is resolved by the ***right-hand rule for the magnetic force***: ***If you place the fingers of your right hand along the direction of the velocity* v *and curl the fingers toward the direction of the magnetic field* B *through the smallest angle between* v *and* B, *the thumb will lie along the direction of the force* F** for a positive test charge (see Fig. 21.11). This is a general rule relating the directions of the magnetic field, the velocity, and the magnetic force for a positive test charge (for a negative test charge, the magnetic force is opposite to that for a positive test charge). It is easy to check that the direction of the magnetic force exerted by the current on a wire [see cases (a) and (b) in Section 21.1] agrees with the right-hand rule.

According to Eq. (12), the magnetic field is the force per unit charge and unit velocity. The SI unit of magnetic field is N/(C · m/s), the unit of force divided by the unit of charge and of velocity; this unit is called the ***tesla*** (T),

$$1 \text{ tesla} = 1 \text{ T} = 1 \text{ N/(C} \cdot \text{m/s)} \tag{13}$$

Table 21.1 lists the values of some typical magnetic fields.

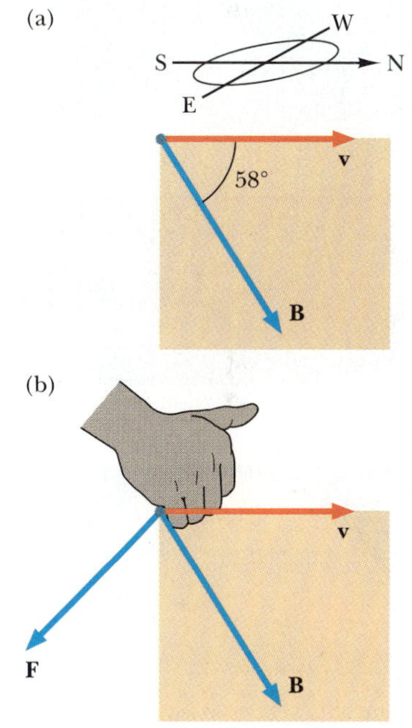

FIGURE 21.12

(a) Direction of the magnetic field of the Earth in Florida. An electron is moving south to north in this magnetic field. (b) Right-hand rule for the magnetic force on a *positive* charge. For the *negative* electron the magnetic force (blue) is opposite to the direction given by the right-hand rule.

EXAMPLE 2. In Florida, the Earth's magnetic field is in the north–south vertical plane, but directed downward from the horizontal at an angle of 58° (see Fig. 21.12a). The magnitude of this magnetic field is 5.3×10^{-5} T. Suppose that an electron in a TV tube is moving with an (instantaneous) horizontal velocity of 2×10^{6} m/s in the south to north direction. What are the magnitude and the direction of the force that the magnetic field of the Earth exerts on this electron?

SOLUTION: According to the right-hand rule, the magnetic force would be horizontal toward the west for a *positive* charge (see Fig. 21.12b). Hence for the *negative* electron, the magnetic force is horizontal toward the east. The magnitude of this magnetic force is given by Eq. (11),

$$F = evB \sin\alpha = 1.6 \times 10^{-19}\ \text{C} \times 2 \times 10^{6}\ \text{m/s} \times 5.3 \times 10^{-5}\ \text{T} \times \sin 58^\circ$$

$$= 1.4 \times 10^{-17}\ \text{N}$$

(a)

TABLE 21.1 SOME MAGNETIC FIELDS	
At surface of pulsar	$\simeq 10^{8}$ T
Maximum achieved in laboratory:	
Explosive compression of field lines	1×10^{3} T
Steady	30 T
A large bubble-chamber magnet **(a)**	2 T
In MRI magnet	1.5 T
In sunspot **(b)**	$\simeq 0.3$ T
Near small ceramic magnet	$\simeq 2 \times 10^{-2}$ T
At surface of Sun	$\simeq 10^{-2}$ T
Near household wiring **(c)**	$\simeq 10^{-4}$ T
At surface of Earth	$\simeq 5 \times 10^{-5}$ T
In sunlight (rms)	3×10^{-6} T
In Crab Nebula **(d)**	$\simeq 10^{-8}$ T
In radio wave (rms)	$\simeq 10^{-9}$ T
In interstellar galactic space	$\simeq 10^{-10}$ T
Produced by human body	3×10^{-10} T
In shielded antimagnetic chamber	2×10^{-14} T

(b)

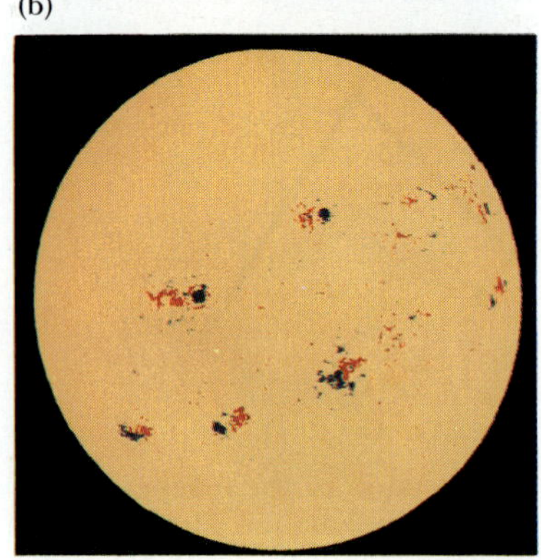

(c)

(d)

The magnetic field can be represented graphically by field lines. As in the case of the electric field, the tangent to the field lines indicates the direction of the field, and the density of field lines indicates the strength of the field. Figure 21.13 shows the pattern of magnetic field lines for the magnetic field produced by a current on a long, straight wire. The de-

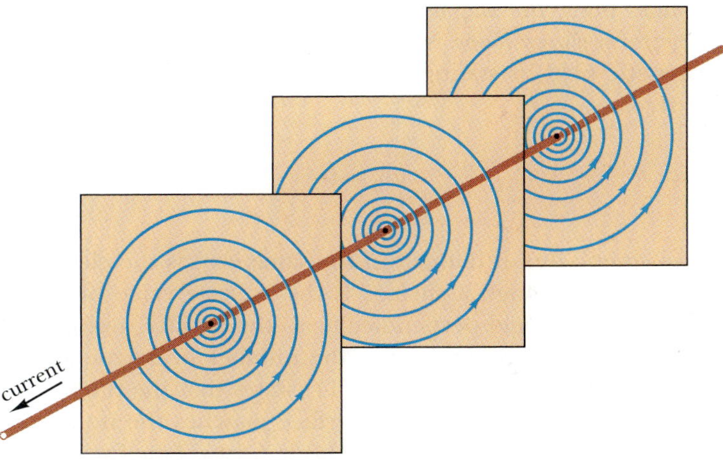

FIGURE 21.13
Magnetic field lines around a current on a long, straight wire.

FIGURE 21.14
Magnetic field lines around a current on a long, straight wire, made visible by iron filings sprinkled on a sheet of paper.

crease of the strength of the magnetic field with distance is indicated by the decrease of density of the field lines. The magnetic field of the straight wire can be made visible by sprinkling small iron filings on a sheet of paper placed around the wire (see Fig. 21.14). The iron filings behave like small compass needles, and they align in the direction of the magnetic field. Figures 21.15 and 21.16 show the pattern of magnetic field lines for the magnetic field produced by a bar magnet. ***The end of the magnet from which the field lines emerge is the north pole of the magnet, and the end into which the field lines enter is the south pole.***

For computations with field lines we will adopt the convention that the number of lines per unit area equals the magnitude of the magnetic field. However, for the purpose of making drawings, this normalization is sometimes unwieldy, and we will alter it in case of need.

Note that the magnetic field lines in Figs. 21.13 and 21.15 form closed loops, that is, the magnetic field lines do not begin or end anywhere in the way that electric field lines begin and end on positive and negative charges. Since there is no "magnetic charge" that acts as source or sink of

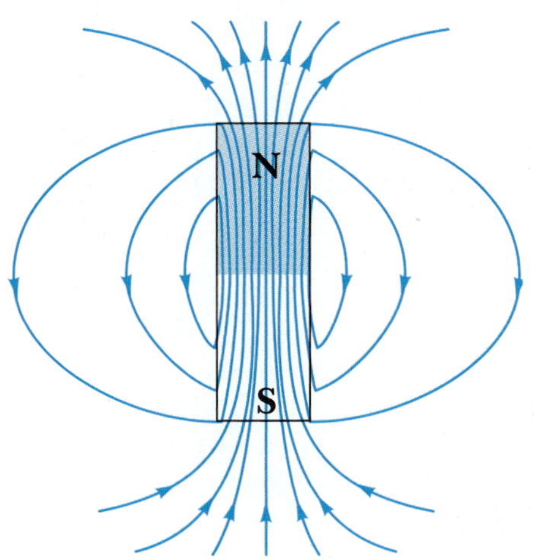

FIGURE 21.15
Magnetic field lines of a bar magnet.

FIGURE 21.16
Magnetic field lines of a bar magnet made visible by iron filings sprinkled on a sheet of paper placed next to the magnet.

magnetic field lines, ***the magnetic field lines of any kind of magnetic field must always form closed loops.*** Mathematically, we can express this feature of the magnetic field by stating that ***the magnetic flux through any closed surface is zero:***

Gauss' Law for magnetic field ▶

$$\Phi_B = 0 \tag{14}$$

Here the magnetic flux is defined in the same way as the electric flux — the magnetic flux through a surface is the average value of the component of the magnetic field perpendicular to the surface multiplied by the amount of area or, equivalently, it is the number of magnetic field lines intercepted by the surface. Thus, Eq. (14) states that the number of magnetic field lines entering any closed surface matches the number leaving the surface (see Fig. 21.17). Equation (14) is ***Gauss' Law*** for the magnetic field.

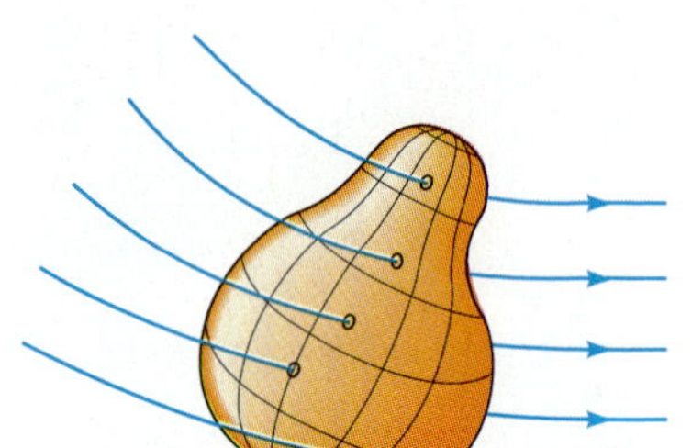

FIGURE 21.17

For any closed surface, the number of magnetic field lines entering the surface equals the number leaving the surface.

21.3 AMPÈRE'S LAW

The magnetic fields produced by currents flowing on wires are of great practical interest, because many applications of magnetism, such as electromagnets and electric motors, rely on currents on wires to produce magnetic fields. However, the arrangements of wires used in practical applications are usually much more complicated than the single long, straight wire we dealt with in the preceding sections. We therefore need to develop a more general method for the calculation of magnetic fields.

Superposition principle ▶

The net magnetic field produced by several wires or other current distributions, each of which produces an individual magnetic field, obeys the ***principle of superposition.*** This principle states that ***the net magnetic field produced by several currents is the (vector) sum of the individual magnetic fields of the individual currents.*** The following example illustrates this principle of superposition.

(a)

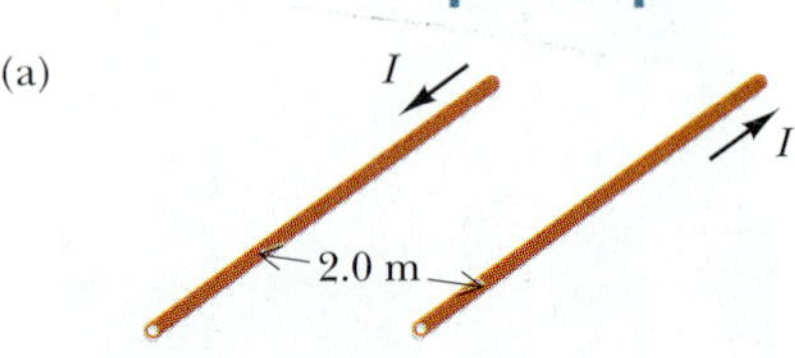

(b)

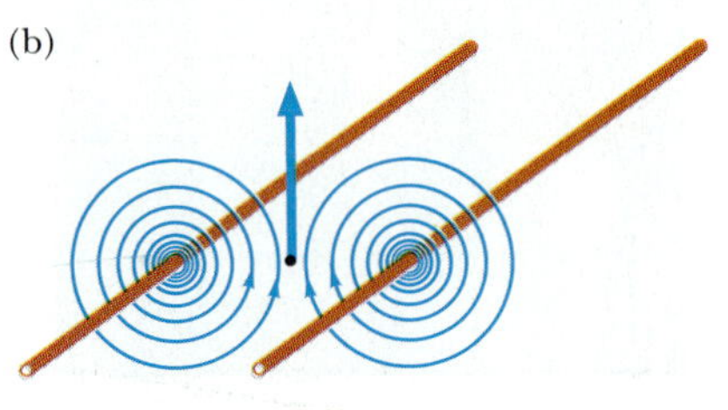

FIGURE 21.18

(a) Two long, parallel wires carrying opposite currents.
(b) Magnetic field lines of the wires.

EXAMPLE 3. A high-voltage transmission line consists of two long parallel wires separated by a distance of 2.0 m. The wires carry currents of 800 A in opposite directions (see Fig. 21.18a). What is the net magnetic field that these wires produce jointly at a point midway between these wires?

SOLUTION: Figure 21.18b shows the magnetic field lines of each wire. Since the currents are opposite, the field lines curl around the two wires in opposite directions. At the midpoint between the wires, the individual magnetic fields of the two wires are parallel. These individual magnetic fields therefore add, and the net magnetic field is twice as large as the individual magnetic field of each wire. By Eq. (9), the individual magnetic field of each wire is $\mu_0 I/2\pi r$, and hence the net magnetic field is

$$B = 2 \times \frac{\mu_0}{2\pi}\frac{I}{r} = 2 \times \frac{1.26 \times 10^{-6}\,\mathrm{N \cdot s^2/C^2}}{2\pi} \times \frac{800\ \mathrm{A}}{1.0\ \mathrm{m}}$$

$$= 3.2 \times 10^{-4}\ \mathrm{T}$$

This magnetic field is about ten times as strong as that of the Earth.

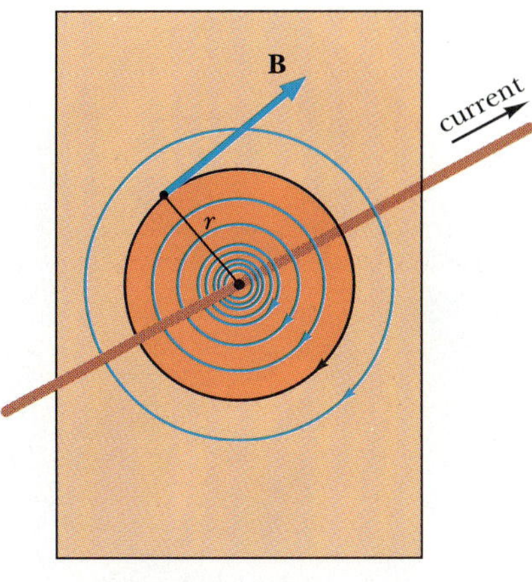

F I G U R E 21.19

Circular path of radius r around a long straight wire carrying a current I. This circular path coincides with a magnetic field line, and the magnetic field is everywhere tangent to this path.

The expression (9) for the magnetic field produced by a current on a long, straight wire leads to an interesting relationship between field and current. We can rewrite Eq. (9) as

$$2\pi r B = \mu_0 I \tag{15}$$

or as

$$Bl = \mu_0 I \tag{16}$$

where $l = 2\pi r$ is the circumference of the circle of radius r (see Fig. 21.19). In words, Eq. (16) states that the magnetic field along a circular path multiplied by the length of the path equals μ_0 times the current intercepted by the area within the circular path.

In a slightly modified form, this statement is valid for a closed path of arbitrary shape in the magnetic field of an arbitrary distribution of currents. Consider some closed mathematical path (see Fig. 21.20), and designate by B_t the component of the magnetic field tangent to the path. Then ***Ampère's Law*** states:

The average component of the magnetic field tangent to the direction of a closed path multiplied by the length of the path equals μ_0 times the current intercepted by the area within the path:

$$\overline{B_t}\, l = \mu_0 I \tag{17}$$

◀ **Ampère's Law**

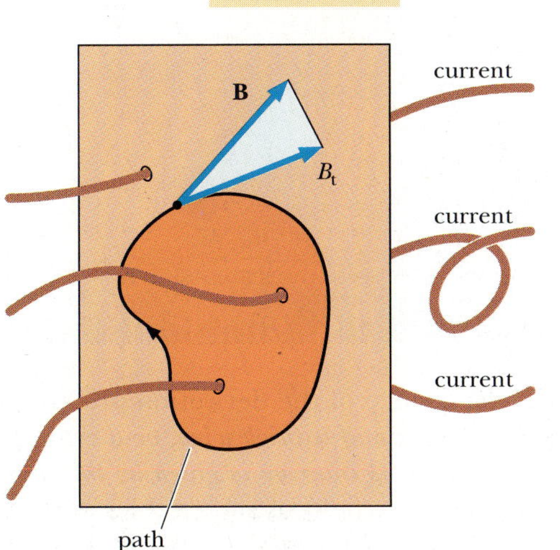

F I G U R E 21.20

A closed path in a magnetic field. The area within the path (orange) intercepts some currents flowing on wires or on other conductors.

André Marie Ampère, *1775–1836, French physicist and mathematician. After Oersted's discovery of the generation of magnetic fields by electric currents, Ampère demonstrated experimentally that currents exert magnetic forces on each other. He carefully investigated the relationship between currents and magnetic fields, and he established that a magnet is equivalent to a distribution of currents.*

Ampère's Law is one of the fundamental and general laws of magnetism. The only restriction on this law is that the currents must be steady, that is, they must be constant in time.

Ampère's Law can be used to calculate the magnetic field of a given distribution of currents, provided that the distribution has a high degree of symmetry. The technique for calculating magnetic fields by means of Ampère's Law is similar to the technique for calculating electric fields by means of Gauss' Law. As in Section 17.4, the calculation involves two steps: first determine the *direction* of the magnetic field by appealing to symmetry arguments, and then determine the *magnitude* of the magnetic field by evaluating Ampère's Law along some suitable, cleverly chosen path.

The following example illustrates this procedure in the simple case of a current on a long, straight wire. For the purposes of this example, we will pretend that we do not know the magnetic field [Eq. (9)] of the straight wire, and we will see how this magnetic field can be deduced from Ampère's Law.

EXAMPLE 4. Deduce the magnetic field of a current on a very long, straight wire from Ampère's Law.

SOLUTION: The arrangement of magnetic field lines has to match the symmetry of the current. Since the current has cylindrical symmetry, the arrangement of magnetic field lines must also have cylindrical symmetry. Thus, the field lines must be either concentric circles around the wire, radial lines, or parallel lines in the same direction as the wire. Radial lines would require that the field lines start on the axis, which is impossible, since the field lines must form closed loops. Parallel lines in the direction of the wire are likewise inconsistent with closed loops. Thus, the field lines must necessarily be concentric circles. Furthermore, by symmetry, the magnetic field must have a constant magnitude along each circle.

Now take a path that follows one of these circles, of radius r as in Fig. 21.19. For this path, $B_t = B$, and the left side of Ampère's Law is $2\pi r \times B$. The right side is $\mu_0 I$; hence

$$2\pi r B = \mu_0 I$$

As expected, this yields

$$B = \frac{\mu_0}{2\pi}\frac{I}{r} \tag{18}$$

An interesting consequence of the derivation presented in Example 4 is that ***the magnetic field outside of a thick wire, or a cylinder, with a uniform distribution of current over its interior is given by the same formula (18) as for a thin wire.*** The magnetic field of such a thick wire has the same symmetry as for a thin wire, and therefore the calculation presented in Example 4 is equally valid for a thick wire and for a thin wire.

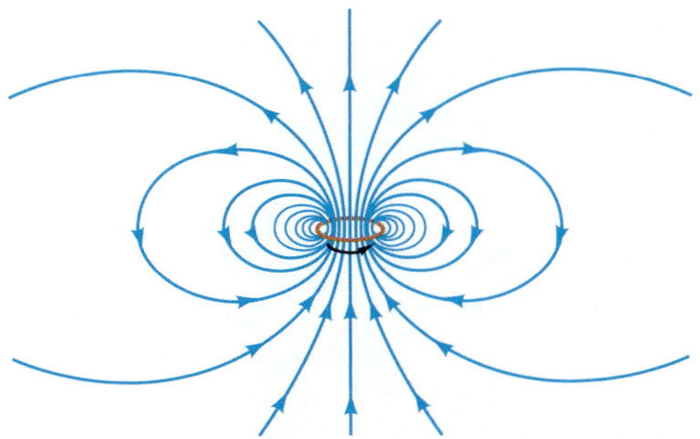

FIGURE 21.21

Magnetic field lines of a ring of current.

FIGURE 21.22

Magnetic field lines of a ring of current made visible with iron filings.

Although circular concentric field lines and a magnetic field with the strength given by the formula (18) are characteristic of a current on a long, straight wire (thin or thick), this formula also provides an approximation for the magnetic field in the immediate vicinity of any segment of wire, straight or curved, except where the wire has a sharp kink. From points very near the wire, any wire looks almost straight, and the magnetic field at such points is then approximately that of a long, straight wire. The magnetic field of a current flowing on a ring illustrates this behavior. The field lines for such a ring are shown in Figs. 21.21 and 21.22; very near the wire, the field lines are almost circles, concentric with the wire. However, far from the wire, the pattern of field lines is much more complex.

The calculation of the complete magnetic field of a ring of current at points near and far from the ring is quite difficult, and we will not attempt to do this calculation. It can be shown that the magnitude of the magnetic field at the center of a ring of radius R carrying a current I is

$$B = \frac{\mu_0 I}{2R} \tag{19}$$

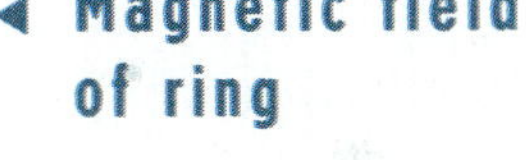

The direction of the magnetic field at this point is perpendicular to the plane of the ring (see Fig. 21.23).

Note that the pattern of magnetic field lines at large distance from the ring of current resembles the pattern of electric field lines of an electric dipole (compare Fig. 17.33). This resemblance is also found in the pattern of magnetic field lines at large distance from a closed loop of current of any other shape. At large distance, square loops, rectangular loops, and oval loops all produce the same pattern of field lines as a circular ring — the exact shape of the loop has little effect on the distant magnetic field. For example, in the central core of the Earth, currents flow in loops of some kind, and the magnetic field that these currents produce at the sur-

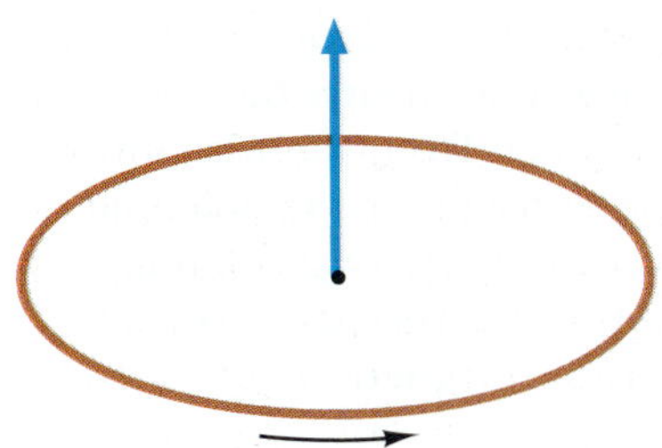

FIGURE 21.23

Direction of the magnetic field at the center of a ring of current.

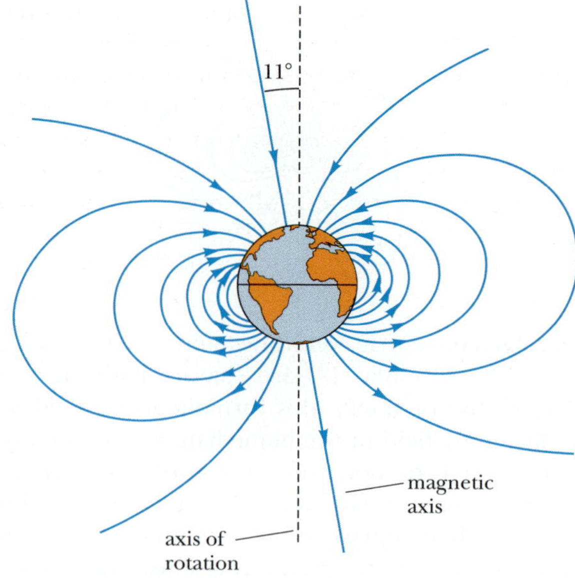

FIGURE 21.24
Magnetic field of the Earth. The axis of the magnetic field makes an angle of about 11° with the axis of rotation of the Earth.

face of the Earth and in the space above is pretty much that of a ring of current (see Fig. 21.24).

21.4 SOLENOIDS AND MAGNETS

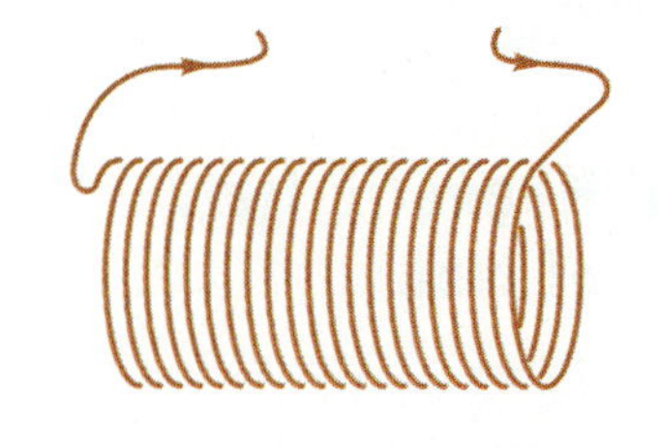

FIGURE 21.25
A solenoid.

A solenoid is a conducting wire wound in a tight helical coil of many turns (see Fig. 21.25). A current in this wire will produce a strong magnetic field within the coil. Because of the similarity of the current distributions, such a tight coil produces essentially the same magnetic field as a large number of rings stacked next to one another. The calculation of the magnetic field of such a solenoid of finite length is fairly difficult, and we will not attempt it here. Instead, we will calculate the magnetic field of an ***ideal solenoid,*** that is, a very long (infinitely long) solenoid with very tightly wound coils, so the current distribution on the surface of the solenoid is nearly uniform.

To find this magnetic field, we begin with an appeal to symmetry, as in Example 4. The ideal solenoid has translational symmetry (along the axis of the solenoid) and rotational symmetry (around the axis). For consistency with these symmetries, the magnetic field lines inside the solenoid will then have to be either concentric circles, or radial lines, or lines parallel to the axis. Concentric circles and radial lines are unacceptable; the former would require the presence of a current along the axis (compare Example 4), and the latter would require that the field lines begin on the axis, which is impossible, since magnetic field lines cannot begin or end anywhere. Thus, the field lines inside the solenoid must all be parallel to the axis. These lines emerge from the (distant) end of the solenoid, and they curve around the exterior of the solenoid and return to the other (distant) end. For an ideal, very long solenoid, these external field lines will then be spread out over a very large region of space; hence, the density of field lines, and the magnetic field, outside the solenoid is nearly zero.

We can now use Ampère's Law to determine the magnitude of the magnetic field. To evaluate the left and the right sides of Ampère's Law, we must choose a closed path. A convenient choice is the square path shown in Fig. 21.26, which is partially inside the solenoid and partially outside. The magnetic field has a component tangent to the path only along the lower side of the square, within the solenoid; along this side $B_t = B$. Along all other sides, either the magnetic field is zero or the component of the magnetic field tangent to the path is zero. Hence the value of B_t averaged over all four sides is $\frac{1}{4}B$. The total length of the path is $l = 4h$. Hence the left side of Ampère's Law is $\frac{1}{4}B \times 4h = Bh$. The net current intercepted by the area of the square is $I_0 \times N$, where I_0 is the current in one wire and N is the number of wires intercepted by the area of the square. Thus, Ampère's Law becomes

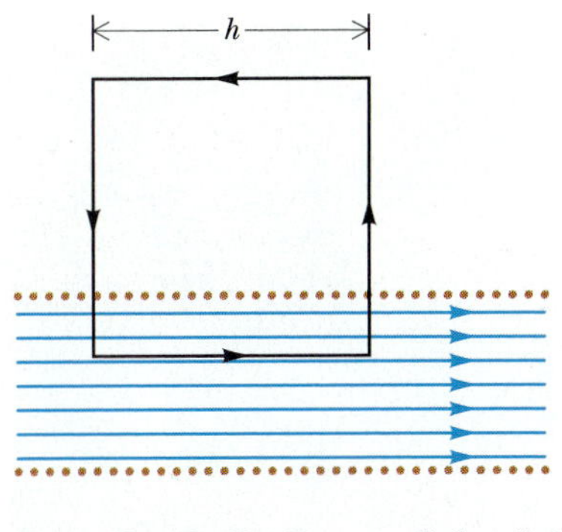

FIGURE 21.26
Magnetic field lines of a very long solenoid.

$$Bh = \mu_0 I_0 N$$

from which

$$B = \mu_0 I_0 (N/h) \tag{20}$$

The ratio N/h is the number of turns of wire per unit length of the solenoid, commonly designated by n. Thus,

$$B = \mu_0 n I_0 \tag{21}$$

◂ **Magnetic field of solenoid**

This shows that to obtain a large magnetic field, we want a solenoid with a large current and a large number of turns of wire per unit length (a densely wound solenoid).

Note that this result is independent of the "depth" to which the square is immersed in the solenoid. Hence ***the magnetic field has the same magnitude everywhere within the solenoid.*** This means that the magnetic field within an ideal solenoid is perfectly uniform.

EXAMPLE 5. A solenoid used for research consists of 180 turns of wire wound on a narrow cardboard tube 19 cm long. The current in the wire is 5.0 A. What is the strength of the magnetic field within the tube?

SOLUTION: The number of turns per unit length is

$$n = 180/0.19 \text{ m} = 9.5 \times 10^2/\text{m}$$

Although this solenoid is not infinitely long, its length is large compared with its width, and therefore the magnetic field at all points except those near the ends will be approximately given by Eq. (21):

$$B = \mu_0 n I_0 = 1.25 \times 10^{-6} \text{ N}\cdot\text{s}^2/\text{C}^2 \times 9.5 \times 10^2/\text{m} \times 5.0 \text{ A} \tag{22}$$
$$= 6.0 \times 10^{-3} \text{ T}$$

FIGURE 21.27
Magnetic field lines of a solenoid made visible with irons filings. Note that inside the solenoid, the distribution of field lines is nearly uniform.

For a more precise calculation of the magnetic field of a short solenoid, we must take the sum of the magnetic fields of the individual rings in this solenoid. But even a solenoid of just a few rings produces a fairly uniform magnetic field in its interior. For instance, Fig. 21.27 shows a photograph of the magnetic field lines, made visible with iron filings, of a short solenoid of just five turns. The magnetic field in the interior is not far from uniform.

An ***electromagnet*** is essentially a solenoid with a gap, or what amounts to the same thing, a pair of solenoids with their ends placed close together (see Fig. 21.28a). Magnetic field lines come out of one solenoid and go into the other solenoid (of course, field lines will also have to come out of the solenoids at their other ends, curve around, and close on themselves). The first solenoid is called the north pole of the electromagnet, and the second the south pole. If the gap is small, then the magnetic field in this gap is almost the same as inside the solenoids.

In most electromagnets, the space inside the solenoids is filled with an iron core (Fig. 21.28b). Iron and other ***ferromagnetic materials*** enhance the magnetic field, making it much stronger than the value given by Eq. (21). It is not unusual for the magnetic field to be enhanced by a factor of several thousand. This enhancement arises from an alignment of electron spins.

Electromagnet ▶

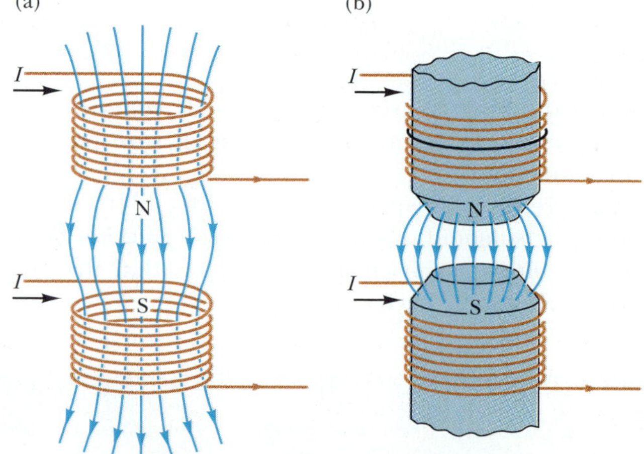

FIGURE 21.28
(a) An electromagnet with two coils. The coil from which the field lines emerge into the gap is called the north pole; the coil into which the field lines enter is called the south pole. (b) An electromagnet with iron pole pieces.

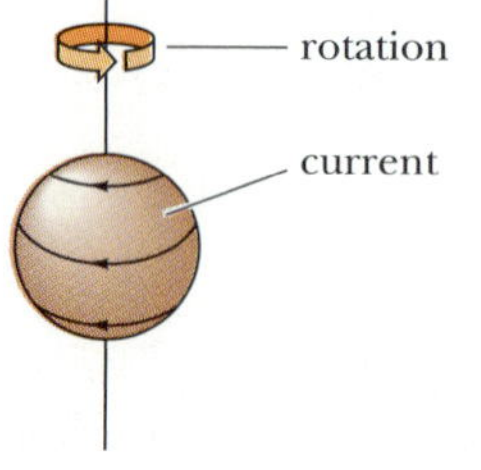

FIGURE 21.29
A small ball of negative charge spinning about an axis.

Crudely, we can picture the electron as a small ball of negative electric charge spinning about an axis (see Fig. 21.29). The rotational motion of the charge is equivalent to currents flowing in rings around the axis of spin, and the magnetic field surrounding the electron is therefore that of a small ring of current. When a piece of iron is left to itself, the electron spins are oriented at random, and the magnetic fields of the electrons average to zero. But when the iron is placed in an external magnetic field, such as the field of a solenoid, the spins of the outer electrons of the atoms align with the magnetic field; although this alignment is triggered by the external magnetic field, which exerts a torque (see Section 21.7) on the small current loops within the electrons, the amount of alignment is strongly magnified by a natural tendency for spin alignment in iron and other ferromagnetic materials. ***The combined magnetic fields of***

all these aligned electrons add to the original magnetic field and enhance it drastically.

◀ **Ferromagnetic materials**

◀ **Permanent magnet**

A piece of iron with aligned electron spins is said to be magnetized. If the iron is removed from the solenoid, it will retain some of its magnetization and act as a ***permanent magnet.*** Some ferromagnetic materials, such as Alnico (an alloy of iron, aluminum, and nickel), retain more magnetization than pure iron, and they make better permanent magnets. Figure 21.30 shows two kinds of permanent magnets: bar magnets and horseshoe magnets.

Electromagnets are used in many kinds of electric machinery, such as electric motors (see Section 21.7), electric generators, magnetic-tape recorders, videocassette recorders (VCRs), computer drives, etc. For example, in the recording head of a magnetic-tape recorder, a small electromagnet produces a magnetic field whose strength varies in proportion to the amplitude of the sound signal arriving at the microphone (see Fig. 21.31). The tape passing over the recording head contains a magnetizable material. Initially, the tape is blank and unmagnetized. When the tape is subjected to the magnetic field of the recording head, it becomes magnetized. The strength of the induced magnetization in the tape is proportional to the strength of the magnetic field of the recording head, which is proportional to the intensity of the sound signal. Thus, as the tape passes over the recording head, it stores the information about the varying sound signal in the form of varying zones of high and low magnetization. The gap between the poles of the electromagnet in the recording head must be extremely narrow — typically no more than 10^{-5} m — so an instant of the sound signal is recorded on a very narrow, almost pointlike spot or band on the tape. The quality of the recording improves if the speed of the tape over the recording head is high, since this leads to a sharper separation between the magnetizations recorded at successive instants. The hi-fi tape recorders in broadcasting studios typically use a tape speed of 15 in./s. One of the important advantages of magnetic tape over the old-fashioned phonograph record is that the tape can be erased easily. For this purpose the tape is run over a recording head that applies a high-frequency oscillating magnetic field which disrupts and randomizes the magnetization of the tape, leaving it unmagnetized and blank.

FIGURE 21.30

Different kinds of permanent magnets.

(a)

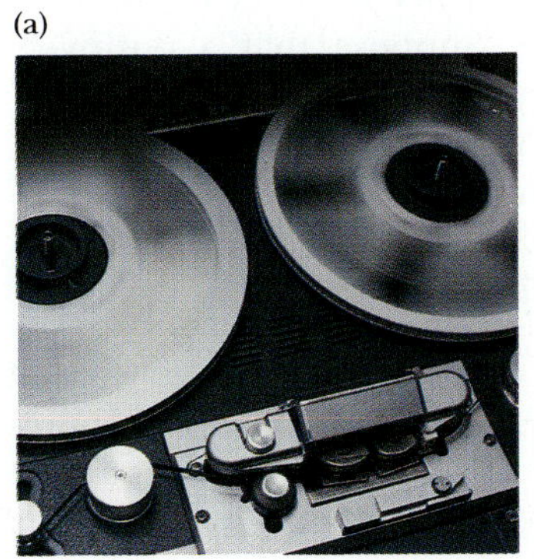

(b)

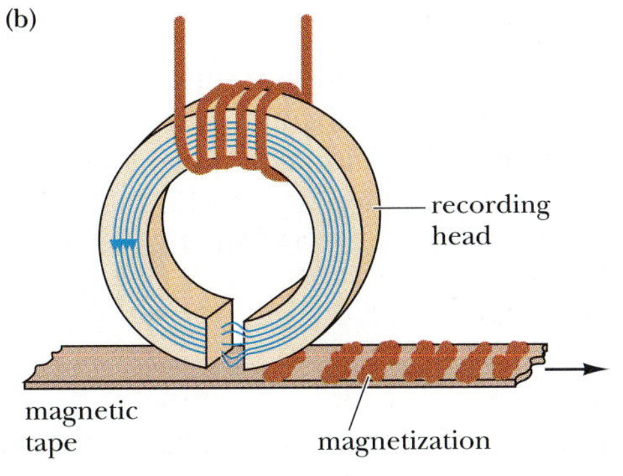

FIGURE 21.31

(a + b) Recording heads of magnetic-tape recorder.

21.5 MOTION OF CHARGES IN MAGNETIC FIELDS

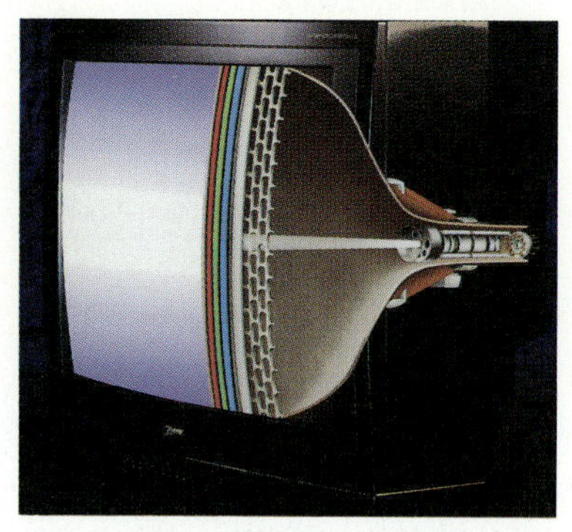

FIGURE 21.32
Cutaway view of a TV tube. The beam of electrons from the cathode at the rear end of the tube is deflected by the coils in the stem of the tube.

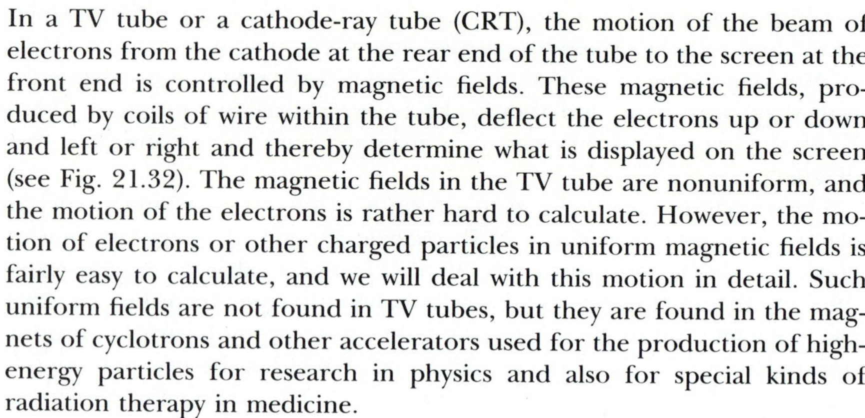

In a TV tube or a cathode-ray tube (CRT), the motion of the beam of electrons from the cathode at the rear end of the tube to the screen at the front end is controlled by magnetic fields. These magnetic fields, produced by coils of wire within the tube, deflect the electrons up or down and left or right and thereby determine what is displayed on the screen (see Fig. 21.32). The magnetic fields in the TV tube are nonuniform, and the motion of the electrons is rather hard to calculate. However, the motion of electrons or other charged particles in uniform magnetic fields is fairly easy to calculate, and we will deal with this motion in detail. Such uniform fields are not found in TV tubes, but they are found in the magnets of cyclotrons and other accelerators used for the production of high-energy particles for research in physics and also for special kinds of radiation therapy in medicine.

The direction of the force exerted by a magnetic field on a moving charged particle is always perpendicular to both the magnetic field and the velocity, as specified by the right-hand rule (see Fig. 21.9). Since the magnetic force is always perpendicular to the velocity, the acceleration — and the small change of velocity in a small time interval — is perpendicular to the velocity. If the change of velocity is always perpendicular to the velocity, then the velocity can never change in magnitude, but only in direction (see Fig. 21.33). We therefore recognize that ***the magnetic force acting on a particle will deflect the particle, keeping the speed constant.*** We can also recognize this in another way: the force is always perpendicular to the displacement, therefore does no work, and therefore the kinetic energy and the speed remain constant.

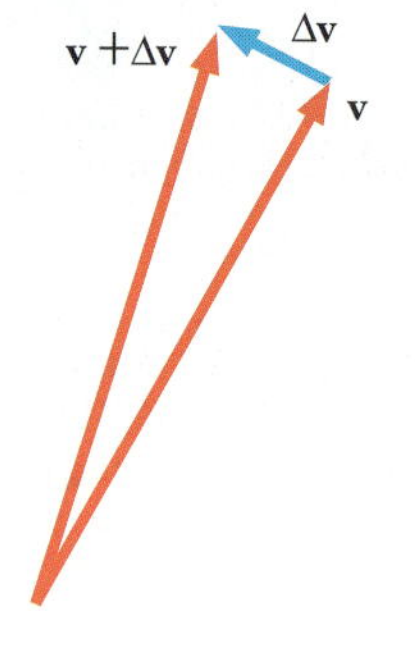

FIGURE 21.33
If the (small) change of the instantaneous velocity is perpendicular to the velocity, then the velocity vector turns through some (small) angle while maintaining a constant magnitude.

According to Eq. (11), the magnitude of the magnetic force is

$$F = qvB \sin\alpha \tag{23}$$

where α is the angle between the magnetic field **B** and the velocity **v.** This force is zero if the direction of the velocity is parallel to the magnetic field. The force is not zero if the velocity is perpendicular to the magnetic field. In the latter case, the force is

$$F = qvB \tag{24}$$

Figure 21.34 shows a region with a uniform magnetic field, directed perpendicularly into the plane of the page. Suppose that a positively charged particle has an initial velocity in the plane of the page; this initial velocity is perpendicular to the magnetic field. The magnetic force is then in the plane of the page, perpendicular to both the velocity and the magnetic field; its direction is shown in Fig. 21.33. According to Eq. (24), the acceleration caused by this force has a magnitude

$$a = \frac{F}{m} = \frac{qvB}{m} \tag{25}$$

and its direction is perpendicular to the velocity. Such an acceleration is characteristic of uniform circular motion. Thus, the particle will move in a circle of some radius r, and the acceleration given by Eq. (25) will play the role of the centripetal acceleration v^2/r, that is,

$$\frac{qvB}{m} = \frac{v^2}{r} \tag{26}$$

This leads to the following formula for the radius of the circular orbit of a particle in a uniform magnetic field:

$$r = \frac{mv}{qB} \tag{27}$$

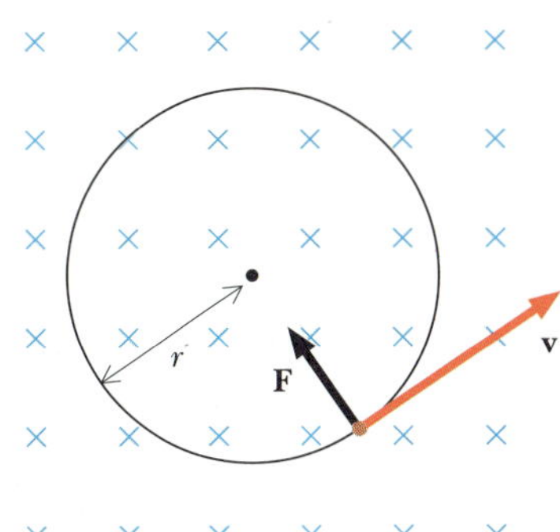

FIGURE 21.34

Positively charged particle in a uniform magnetic field. The magnetic field points perpendicularly into the plane of the page; the crosses show the tails of the magnetic field vectors. The velocity is in the plane of the page, and the force is also in the plane of the page and perpendicular to the velocity.

Figure 21.35 is a photograph of a beam of electrons executing such uniform circular motion in a cathode-ray tube placed in a magnetic field.

The frequency of the circular motion is

$$f = \frac{[\text{speed}]}{[\text{circumference}]} = \frac{v}{2\pi r}$$

or

$$f = \frac{v}{2\pi mv/qB} = \frac{qB}{2\pi m} \tag{28}$$

◀ **Circular orbit in magnetic field**

◀ **Cyclotron frequency**

This is called the ***cyclotron frequency***, because the operation of cyclotrons (described below) involves particles moving with this frequency in a magnetic field. Note that the cyclotron frequency is independent of the speed of the circular motion — in a uniform magnetic field, slow particles and fast particles (of a given charge and mass) move around circles at the same frequency, but the slow particles move along smaller circles than the fast particles.

The ***cyclotron*** is a device for the acceleration of protons or other ions. It consists of an evacuated cavity placed between the poles of a large electromagnet; within the cavity there is a flat metallic can cut into two D-shaped pieces, or ***Dees*** (Fig. 21.36). An oscillating high-voltage generator

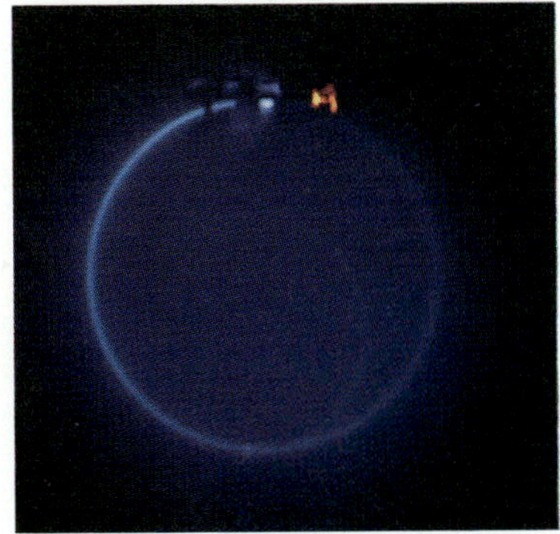

FIGURE 21.35

Electrons moving in a circle in a cathode-ray tube in a magnetic field. The tube contains a gas at a very low pressure, and the atoms of the gas glow under the impact of the electrons; this makes the electron beam visible.

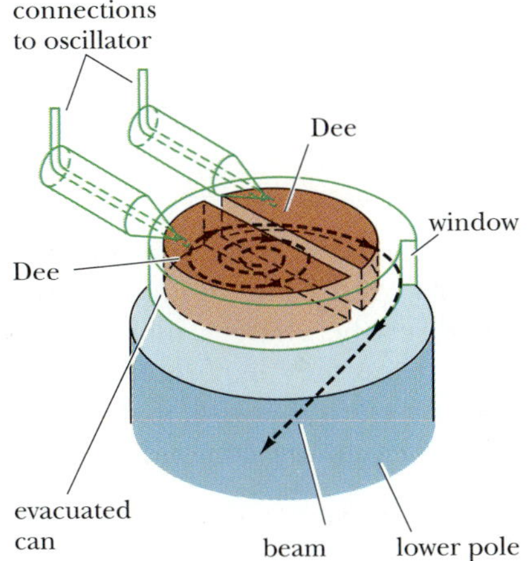

FIGURE 21.36

Trajectory of a particle within the Dees of a cyclotron. In this diagram, the upper pole of the electromagnet has been omitted for the sake of clarity.

is connected to the Dees; this creates an oscillating electric field in the gap between the Dees. The frequency of the voltage generator is adjusted so that it coincides with the cyclotron frequency of Eq. (28). An ion source at the center of the cyclotron releases protons or other ions. The electric field in the gap between the Dees gives each of these protons a push, and the uniform magnetic field in the cyclotron then makes the proton travel on a semicircle inside the first Dee. When the proton returns to the gap after one-half period, the high-voltage generator will have reversed the electric field in the gap; the proton therefore receives an additional push which sends it into the second Dee. There, it travels on a semicircle of slightly larger radius corresponding to its slightly larger energy, and so on. Each time the proton crosses the gap between the Dees, it receives an extra push and extra energy. The proton therefore travels along arcs of circles of stepwise increasing radius. When the protons reach the outer edge of the Dees they leave the cyclotron as a high-energy beam.

EXAMPLE 6. One of the first cyclotrons, built by E. O. Lawrence at Berkeley in 1932, had Dees with a diameter of 28 cm and its magnet was capable of producing a magnetic field of 1.4 T. What was the maximum energy of the protons accelerated by this cyclotron?

SOLUTION: When the proton reaches its maximum energy, its orbit has a radius of 14 cm. Since the magnetic field is 1.4 T, the speed of such a proton is, according to Eq. (27), with $q = e$ and $m = m_{\rm p} = 1.67 \times 10^{-27}$ kg,

$$v = \frac{qBr}{m} = \frac{eBr}{m_{\rm p}} = \frac{1.6 \times 10^{-19}\ \text{C} \times 1.4\ \text{T} \times 0.14\ \text{m}}{1.67 \times 10^{-27}\ \text{kg}}$$

$$= 1.9 \times 10^{7}\ \text{m/s}$$

and the energy is

$$K = \tfrac{1}{2} m_{\rm p} v^2 = \tfrac{1}{2} \times 1.67 \times 10^{-27}\ \text{kg} \times (1.9 \times 10^{7}\ \text{m/s})^2 = 3.0 \times 10^{-13}\ \text{J}$$

21.6 FORCE ON A WIRE

If a wire carrying a current is placed in a magnetic field, the moving charges within the wire will experience a force. Since the motion of the charges in the wire is constrained by the wire (the charges must move along the wire, regardless of the magnetic force they experience), any force acting on these charges is merely transferred to the wire, and therefore the wire as a whole will experience a force equal to that acting on the charges.

Consider a segment Δl of a wire on which flows a current I. To find the amount of moving charge in this segment, suppose that the speed of these moving charges is v; then they take a time $\Delta l/v$ to move out of the segment (see Fig. 21.37). Since the amount of charge is the product of the time and the current, we find that

$$\Delta q = (\Delta l/v)I \tag{29}$$

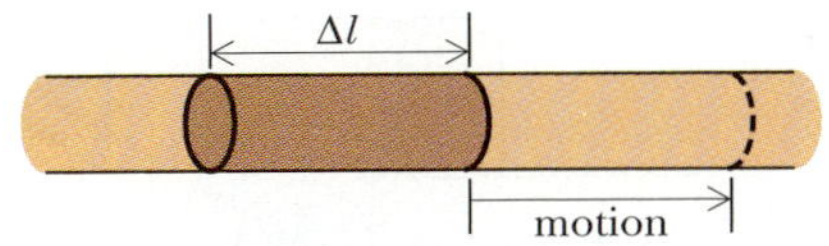

FIGURE 21.37

In a time $\Delta l/v$, the moving charge moves a distance Δl toward the right. Hence, in this time, all the moving charge in the segment Δl of the wire moves out of this segment toward the right (this charge is replaced by charge entering the segment from the left).

where, as always, we pretend that the moving charge is positive. If the wire segment is oriented at right angles to the magnetic field, then the magnetic force is

$$\Delta F = \Delta q v B = (\Delta l/v) I v B = I B \Delta l \tag{30}$$

The direction of this force is related to the direction of the current and the direction of the magnetic field by the right-hand rule. For instance, the force on the wire segment oriented perpendicularly to the magnetic field, as illustrated in Fig. 21.38a, is perpendicular to the wire and to the magnetic field, out of the plane of the page.

If the wire segment is oriented parallel to the magnetic field, then the force is zero. Other orientations of the wire segment gives forces of a magnitude somewhere between zero and the maximum magnitude $IB\Delta l$. We can find the dependence of the force on the orientation of the wire segment by noting that for a charged particle moving at an angle α relative to the magnetic field the force is reduced by a factor $\sin\alpha$. Hence this same factor must appear in the general expression for the force on a wire segment oriented at an angle α to the magnetic field (Fig. 21.38b):

$$\Delta F = I B \Delta l \sin\alpha \tag{31}$$

◀ **Force on wire segment**

This equation contains Eq. (30) as a special case (with $\alpha = 90°$).

To find the force on an entire wire, we must (vectorially) add the contributions (31) from all the small segments of this wire.

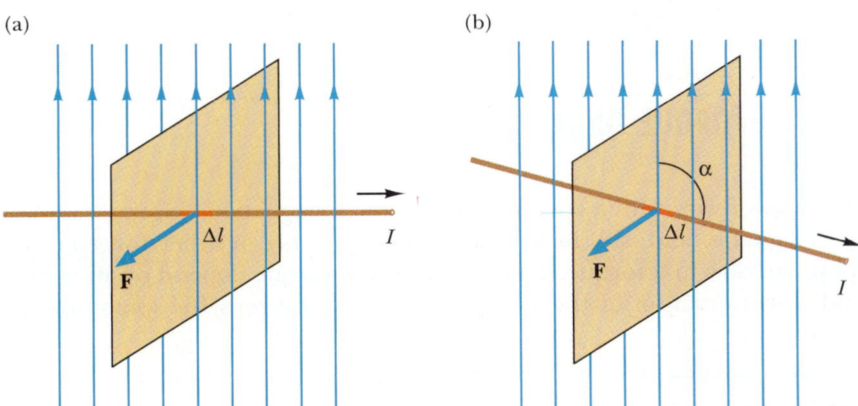

FIGURE 21.38

(a) A wire with a current is oriented at right angles to a uniform magnetic field. The force on the small segment of wire is perpendicularly out of the plane of the page. (b) A wire with a current is oriented at an angle α to a uniform magnetic field. The magnetic force is, again, perpendicularly out of the plane of the page.

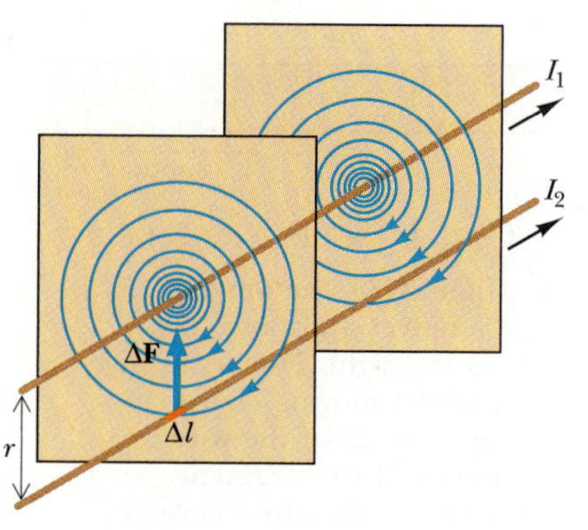

FIGURE 21.39

A long straight wire carrying a current I_2 in the magnetic field of a long straight wire carrying a current I_1.

EXAMPLE 7. Two very long, parallel wires separated by a distance r carry currents I_1 and I_2, respectively. Find the magnetic force that each wire exerts on a segment Δl of the other wire.

SOLUTION: Each wire generates a magnetic field, which exerts a force on the other wire. Figure 21.39 shows the magnetic field $\mathbf{B}_1$ that the current I_1 produces in the vicinity of the current I_2. By Eq. (9), this magnetic field has a magnitude

$$B_1 = \frac{\mu_0}{2\pi}\frac{I_1}{r}$$

and it is perpendicular to the wire 2. By Eq. (30), the force on a segment Δl of the wire 2 has a magnitude

$$\Delta F = I_2 B_1 \Delta l = \frac{\mu_0}{2\pi}\frac{I_2 I_1}{r}\Delta l \tag{32}$$

The force on a segment of the wire 1 is of the same magnitude, but opposite direction. The force between the wires is attractive if the currents in the wires are parallel and repulsive if they are antiparallel.

SI unit of current ▶

The official definition of the ***SI unit of current*** is based on the force per meter of length between two long parallel wires. This force can be measured very precisely by holding one wire stationary and suspending the other from a balance; the wires are connected in series so that the currents are exactly equal ($I_1 = I_2$). The force per unit length and the distance r can be measured experimentally, and the value of I_1 (or I_2) calculated from Eq. (32) is then the current in amperes. The constant μ_0 appearing in Eq. (32) is assigned the value $\mu_0 = 4\pi \times 10^{-7}\ \mathrm{N \cdot s^2/C^2}$ by definition.

SI unit of charge ▶

The official definition of the ***SI unit of charge*** is based on the unit of current. ***The coulomb is defined as the amount of charge that a current of one ampere delivers in one second.***

21.7* TORQUE ON A LOOP

If a loop of wire with a current is placed in a magnetic field, the action of the magnetic field will, in general, result not only in a net force on the loop, but also in a torque. The force and the torque depend on the shape and the orientation of the loop, and they can be zero for some orienta-

* This section is optional.

tions of the loop. For instance, let us consider the simple case of a loop oriented perpendicularly to the magnetic field, illustrated in Fig. 21.40. The forces on the four sides of the loop are given by Eq. (30). The forces on opposite sides are opposite (because the currents are opposite); hence the forces cancel in pairs, and the loop will be in equilibrium.

Figure 21.41 shows a loop oriented parallel to the magnetic field. Again the forces on opposite sides are opposite, but the lines of action of the forces pulling on the left and on the right do not coincide — these forces exert a torque which tends to rotate the loop. Since the magnetic field is constant, Eq. (30) tells us that the magnitude of the forces on the top and bottom sides is

$$F = IaB \tag{33}$$

where a is the length of each of these sides. If, as shown in Fig. 21.41, the length of each of the other sides is b, then the moment arm about the center of the loop is $b/2$, and the torque due to the pair of forces IaB is

$$\begin{aligned}\tau &= F\,b/2 + F\,b/2 \\ &= IabB \end{aligned}\tag{34}$$

It is easy to see that for a loop oriented at an angle θ to the magnetic field, as shown in Fig. 21.42, the moment arm is $(b/2)\sin\theta$ instead of $b/2$, and the torque is

$$\tau = IabB\sin\theta \tag{35}$$

The product of the area of the loop and the current flowing around it is called the ***magnetic dipole moment*** of the loop,*

$$\mu = Iab \tag{36}$$

Note that if the loop consists of several turns of wire, then Eq. (36) must be evaluated with the net current flowing around the loop. This net current equals the number N of turns of wire times the current I_0 in one turn, so

$$\mu = NI_0ab \tag{37}$$

In any case, the torque is

$$\tau = \mu B\sin\theta \tag{38}$$

The direction of the torque is such that it tends to twist the loop into an orientation perpendicular to the magnetic field. The formula (38) can be

* Do not confuse the magnetic dipole moment with the permeability constant. Both are represented by the same letter μ. But they have nothing to do with each other.

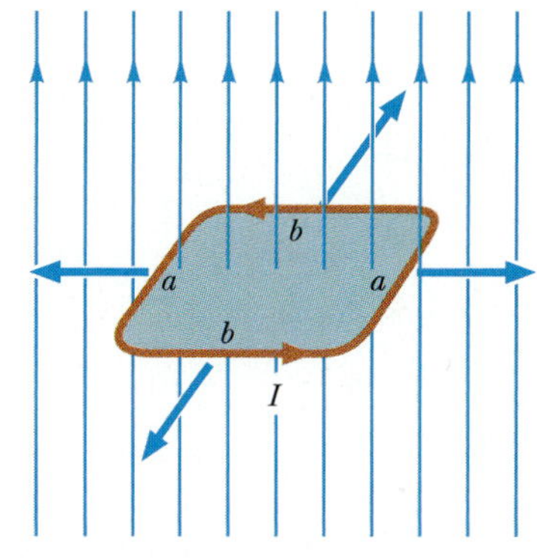

FIGURE 21.40
Forces on a rectangular loop of current oriented perpendicularly to a uniform magnetic field.

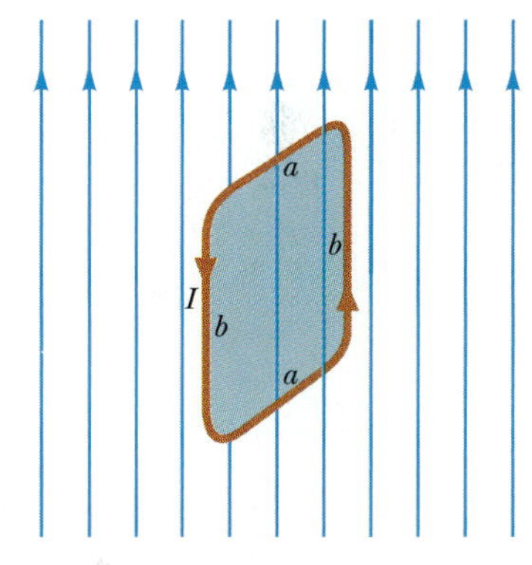

FIGURE 21.41
Forces on a rectangular loop of current oriented parallel to the magnetic field.

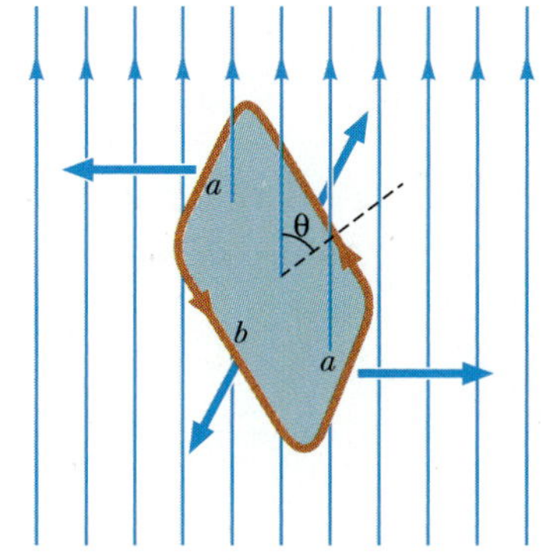

FIGURE 21.42
Forces on a rectangular loop of current oriented at an angle θ with the magnetic field.

shown to be valid not only for the rectangular loop, but also for loops of any other shape. The magnetic moment of a (flat) loop of arbitrary shape is the product of the area and the net current around the loop.

A loop of current suitably pivoted on an axis acts as a compass needle; the normal to the loop seeks to align itself with the magnetic field. This similarity is no accident. A compass needle is a small permanent magnet and, as we saw in Section 21.4, such a permanent magnet contains a large number of electrons acting as small current loops. The mechanism underlying the alignment of a compass needle with a magnetic field is therefore the same as for a current loop.

Electric motor ▸

The torque on a current loop pivoted on an axis and placed in a strong magnetic field is exploited in ***electric motors***. This torque brings about the rotational motion of the current loop and provides the means of converting electric energy into mechanical energy of rotational motion. In small electric motors, the magnetic field is produced by permanent magnets; but in large motors, the magnetic field is produced by electromagnets. The following example illustrates the operation of a simple electric motor.

EXAMPLE 8. A simple electric motor consists of a rectangular coil of wire that rotates on a longitudinal axle in a magnetic field of 0.50 T (see Fig. 21.43). The coil measures 10 cm × 20 cm; it has 40 turns of wire and the current in the wire is 8.0 A. (a) In terms of the angle θ between the magnetic field and the normal to the coil, what is the torque that the magnetic field exerts on the coil? (b) In order to keep the sign of the torque constant, a switch (commutator) mounted on the axle reverses the current in the coil whenever θ passes through 0° and 180°. Plot this torque, vs. the angle θ.

SOLUTION: (a) According to Eq. (38), the torque on the coil is

$$\tau = NI_0abB\sin\theta$$

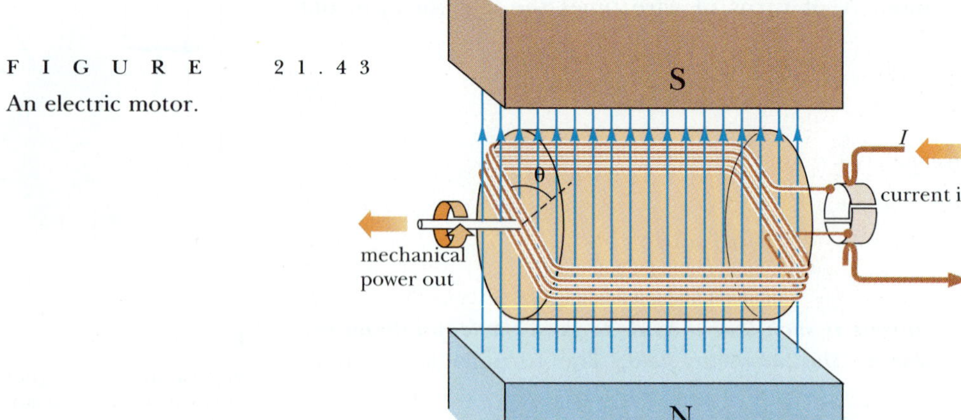

FIGURE 21.43
An electric motor.

$$= 40 \times 8.0 \text{ A} \times 0.10 \text{ m} \times 0.20 \text{ m} \times 0.50 \text{ T} \times \sin\theta$$

$$= 3.2 \text{ N}\cdot\text{m} \sin\theta$$

(b) If the torque always has the same sign (say, positive), then the plot of the torque for the interval 180° to 360° merely repeats the plot for the interval 0° to 180°. Figure 21.44 shows this plot.

COMMENTS AND SUGGESTIONS: Practical electric motors consist of many such coils, each with its commutator, arranged at regular intervals around the axle. The plot of the net torque of this arrangement is a sum of plots such as shown in Fig. 21.44, but with different starting angles. This averages out the ups and downs of Fig. 21.44 and yields a torque that is nearly constant at all angles.

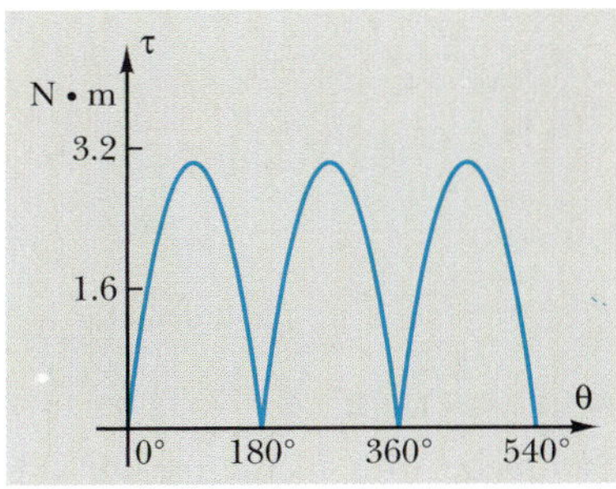

FIGURE 21.44

Torque as a function of angle, according to Example 8.

The torque on a current loop is also exploited in instruments for the measurement of current, generally called ***galvanometers.*** In such an instrument, the current to be measured is sent into a coil of several turns pivoted on an axis and placed in the magnetic field of a permanent magnet (see Fig. 21.45). A fine spiral spring attached to the moving coil provides a restoring torque that opposes the magnetic torque. Under the influence of these two opposing torques, the moving coil attains an angular deflection that increases with the magnitude of the current. A pointer attached to the moving coil indicates the magnitude of the current on a calibrated scale.

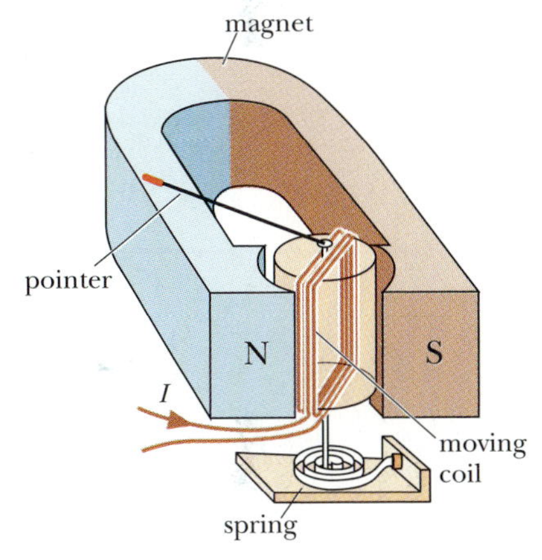

FIGURE 21.45

Mechanism of a galvanometer.

Galvanometers are the sensitive elements in ammeters and voltmeters. Since the moving coil in the galvanometer is made of very fine wire, it cannot tolerate a large current, and the galvanometer inside an ammeter or a voltmeter must be protected from excessively large currents or voltages.

In the ammeter, this protection is accomplished by a ***shunt*** resistor connected in parallel to the galvanometer coil (see Fig. 21.46). This shunt resistor has a much smaller resistance than the resistance of the coil. When a current enters the external terminals of the ammeter, most of the current passes through the shunt, and only a small fraction passes through the coil. Furthermore, since the resistance of the shunt is low, it ensures that the net internal resistance between the terminals of the ammeter is low, and therefore the ammeter does not significantly disturb the flow of current in the external circuit into which it is inserted.

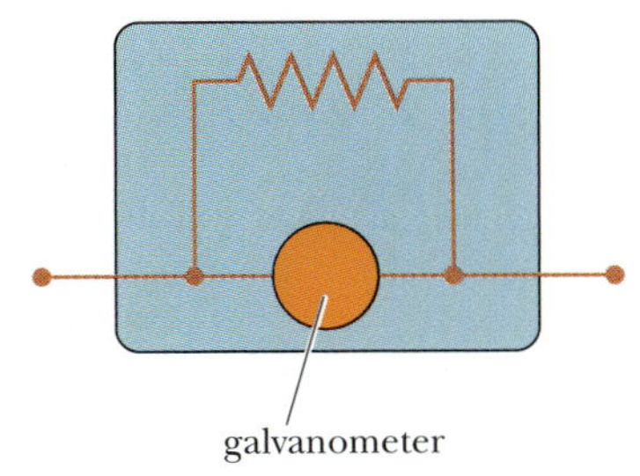

FIGURE 21.46

Ammeter with a shunt resistor, which permits most of the current to bypass the galvanometer.

In the voltmeter, the protection of the galvanometer coil is accomplished by placing a large resistance in series with the coil (see Fig. 21.47). When a voltage is applied to the external terminals of the voltmeter, this large resistance prevents any large flow of current through the coil. Furthermore, with this large internal resistance between its terminals, the voltmeter does not significantly disturb the currents and the potentials in the external circuit to which it is connected.

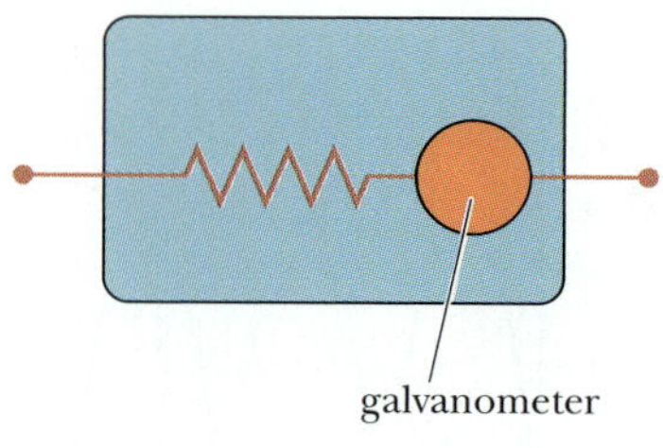

FIGURE 21.47
Voltmeter with a large resistor placed in series with the galvanometer.

SUMMARY

- **Magnetic force exerted on a point charge by a current:**

$$F = \pm \frac{\mu_0}{2\pi} \frac{qvI}{r}$$

- **Magnetic field of a current:** $B = \dfrac{\mu_0 I}{2\pi r}$
- **Permeability constant:** $\mu_0 = 1.26 \times 10^{-6}\ \text{N} \cdot \text{s}^2/\text{C}^2$
- **Force exerted by magnetic field:** $F = qvB \sin\alpha$
- **Gauss' Law for magnetism:** $\Phi_B = 0$
- **Ampère's Law:** $\overline{B}_t l = \mu_0 I$
- **Magnetic field at center of ring of current:** $B = \dfrac{\mu_0 I}{2R}$
- **Magnetic field of ideal solenoid:** $B = \mu_0 n I_0$
- **Circular orbit in magnetic field:** $r = \dfrac{mv}{qB}$
- **Cyclotron frequency:** $f = \dfrac{qB}{2\pi m}$
- **Force on wire segment:** $\Delta F = IB\Delta l \sin\alpha$
- **Magnetic dipole moment of current loop:** μ = [current] × [area]
- **Torque on current loop:** $\tau = \mu B \sin\theta$

QUESTIONS

1.–How would the magnetic field lines shown in Fig. 21.13 differ if the current on the wire were in the opposite direction?

2.–Theoretical physicists have proposed the existence of ***magnetic monopoles,*** which are sources and sinks of magnetic field lines, just as electric charges are sources and sinks of electric field lines. What would the pattern of magnetic field lines of a positive magnetic monopole look like?

3.–The Earth's magnetic field at the equator is horizontal, in the northward direction. What is the direction of the magnetic force on an electron moving vertically up?

4.–An electron with a vertical velocity passes through a magnetic field without suffering any deflection. What can you conclude about the magnetic field?

5.–At an initial time, a charged particle is at some point P in a magnetic field and it has an initial velocity. Under the influence of the magnetic field, the particle moves to a point P'. If you now reverse the velocity of the particle, will it retrace its orbit and return to the point P?

6.–An electron moving northward in a region of space is deflected toward the east by a magnetic field. What is the direction of the magnetic field?

7.–A Faraday cage shields electric fields. Does it also shield magnetic fields?

8.–Strong electric fields are hazardous — if you place some part of your body in a strong electric field you are likely to receive an electric shock. Are strong magnetic fields hazardous? Do they produce any effect on your body?

9.–Figure 21.24 shows the magnetic field of the Earth. What must be the direction of the currents flowing in loops inside the Earth to give this magnetic field?

10.–The needle of an ordinary magnetic compass indicates the direction of the horizontal component of the Earth's magnetic field. Explain why the magnetic compass in unreliable when used near the poles of the Earth.

11.–A ***dip needle*** is a compass needle that swings about a horizontal axis. If the axis is oriented east–west, then the equilibrium direction of the dip needle is the direction of the Earth's magnetic field. The ***dip angle*** of the dip needle is the angle that it makes with the horizontal. How does the dip angle vary as you transport a dip needle along the surface of the Earth from the South Pole to the North Pole?

12.–In order to eliminate or reduce the magnetic field generated by the pair of wires that connect a piece of electric equipment to an outlet, a physicist twists these wires tightly about each other. How does this help?

13.–Consider a circular loop of wire carrying a current. Describe the direction of the magnetic field at different points in the plane of the loop, both inside and outside of the loop.

14.–Consider a long solenoid and a long straight wire along its axis, both carrying some current. Describe the field lines of the net magnetic field within the solenoid.

15.–Cosmic rays are high-speed charged particles — mostly protons — that crisscross interstellar space and strike the Earth from all directions. Why is it easier for the cosmic rays to penetrate through the magnetic field of the Earth near the poles than anywhere else?

16.–If we want a proton to orbit all the way around the equator in the Earth's magnetic field, must we send it eastward or westward?

17.–If a strong current flows through a thick wire, it tends to cause a compression of the wire. Explain.

18.–A horizontal wire carries a current in the eastward direction. The wire is in a uniform magnetic field. What must be the direction of this field if the magnetic force is to balance the weight of the wire?

19.–The ***tangent galvanometer*** is an old form of ammeter, consisting of an ordinary magnetic compass mounted at the center of a coil whose axis is horizontal and oriented along the east–west line (Fig. 21.48). If there is no current in the coil, the compass needle points north. Explain how the compass needle will deviate from north when there is a current in the coil.

20.–A simple indicator of electric current, first used by H. C. Oersted in his early experiments, consists of a compass needle placed below a wire stretched in the northward direction. Explain how the angle of the compass needle indicates the electric current in the wire.

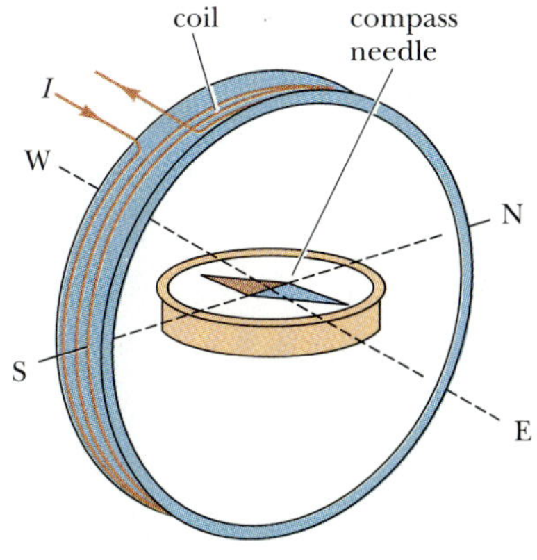

FIGURE 21.48
Tangent galvanometer.

PROBLEMS

Section 21.1: The Magnetic Force

1.——Suppose that, instead of moving parallel to the wire, the electron in Example 1 is moving radially away from the wire. In this case, what is the magnitude of the magnetic force on this electron? What is its direction?

2.——A proton is instantaneously at a distance of 0.10 m from a long straight wire carrying a current of 30 A. The speed of the proton is 5.0×10^6 m/s, and its direction of motion is radially toward the wire. What is the magnitude of the magnetic force on the proton? What is the magnitude of the instantaneous acceleration of the proton? Draw a diagram displaying the direction of the current, and the proton's velocity and acceleration.

3.——In a TV tube, electrons are moving at 2.0×10^7 m/s on a straight path from the back of the tube to the front. This TV tube is placed near a (single) straight cable carrying a current of 12 A parallel to the path of the electrons, at a radial distance of 0.30 m from the path of the electrons. What is the magnetic force on each electron? What is the transverse acceleration caused by this force?

4.——Since the magnetic force is proportional to the speed of the charged particle, it is largest if the particle has the largest possible speed — that is, the speed of light. Suppose that an electron of a speed almost equal to the speed of light is moving radially toward a long straight wire carrying a current of 15 A. What is the magnitude of the magnetic force on this electron when it is at a distance of 0.050 m from the wire?

5.——In terms of m, s, and kg, what are the units of $1/\sqrt{\epsilon_0 \mu_0}$?

Section 21.2: The Magnetic Field

6.——A wire of superconducting niobium, 0.20 cm in diameter, can carry a current of up to 1900 A. What is the strength of the magnetic field just outside of the wire when it carries this current?

7.——The current in a lightning bolt may be as much as 2×10^4 A. What is the magnetic field at a distance of 1.0 m from a lightning bolt? The bolt can be regarded as a straight line of current.

8.——The cable of a high-voltage power line is 25 m above the ground and carries a current of 1.8×10^3 A.

(a) What magnetic field does this current produce at the ground?

(b) The strength of the magnetic field of the Earth is 0.60×10^{-4} T at the location of the power line. By what factor do the fields of the power line and of the Earth differ?

9.——An electromagnet produces a magnetic field of 0.60 T in the vertical downward direction. An electron of speed 6.0×10^6 m/s is moving through this magnetic field. What are the magnitude and the direction of the magnetic force on the electron if the electron is moving:

(a) Vertically downward

(b) Horizontally toward the east

(c) Horizontally toward the south

(d) Toward the south and downward at an angle of 30°?

10.——A proton is moving through a vertical magnetic field. The (instantaneous) velocity of the proton is 8.0×10^5 m/s horizontally in the north direction. The (instantaneous) acceleration produced by the magnetic force is 3.2×10^{14} m/s in the west direction. What is the magnitude of the magnetic field? Is the direction of this field up or down?

11.——The magnetic field surrounding the Earth typically has a strength of 5×10^{-5} T. Suppose that a cosmic-ray electron of energy 3×10^4 eV is instantaneously moving in a direction perpendicular to the lines of this magnetic field. What is the force on this electron?

12.——At a location where the strength of the Earth's magnetic field is 0.60×10^{-4} T, what must be the minimum speed of an electron if the magnetic force on it is to exceed its weight?

13.——A charge of q is traveling with a velocity of v at an angle θ with respect to the direction of a magnetic field **B** which points along the x axis (see Fig. 21.49). For what angle θ is the magnitude of the magnetic force one-third of the maximum magnetic force?

14.——Suppose you want to balance the downward gravitational force on a proton by a magnetic force. If the proton is moving horizontally in the east direction with a speed of 6.0×10^4 m/s, what magnetic field do you need (in magnitude and direction)?

15.——An electron is traveling with a velocity of 2×10^5 m/s at an angle of 120° to the direction of a 0.33-T magnetic field which points along the x axis (see Fig. 21.50). What are the magnitude and direction of the force on the electron?

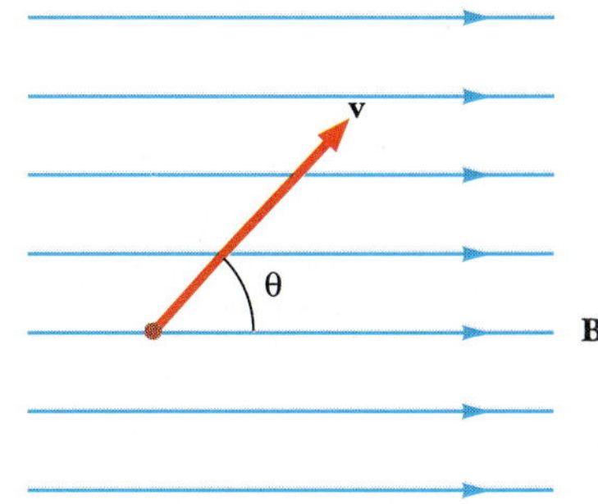

FIGURE 21.49
Magnetic field and velocity vector of a charge q.

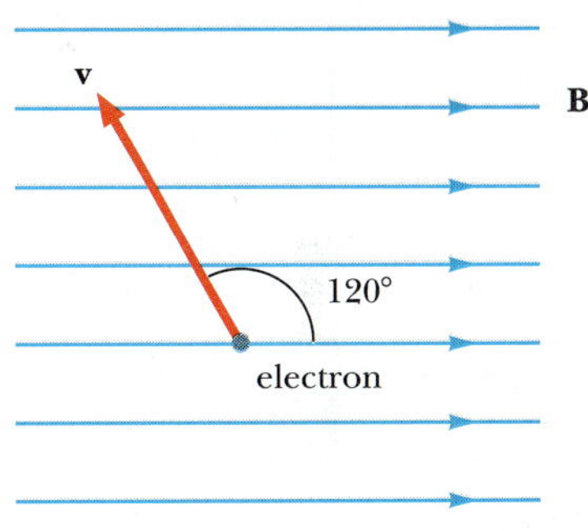

FIGURE 21.50
Magnetic field and velocity vector of an electron.

16.*——In New York, the magnetic field of the Earth has a vertical (down) component of 0.60×10^{-4} T and a horizontal (north) component of 0.17×10^{-4} T. What are the magnitude and direction of the magnetic force on an electron of velocity 1.0×10^6 m/s moving (instantaneously) in an east to west direction in a television tube?

17.*——At the surface of a pulsar, or neutron star, the magnetic field may be as strong as 10^8 T. Consider the electron in a hydrogen atom on the surface of such a neutron star. The electron is at a distance of 0.53×10^{-10} m from the proton and has a speed of 2.2×10^6 m/s. Compare the electric force that the proton exerts on the electron with the magnetic force that the magnetic field of the neutron star exerts on the electron. Is it reasonable to expect that the hydrogen atom will be strongly deformed by the magnetic field?

18.*——The electric field of a long straight line of charge with λ coulombs per meter is [see Eq. (17.15)]

$$E = \frac{1}{2\pi\epsilon_0}\frac{\lambda}{r}$$

Suppose that we move this line of charge parallel to itself at speed v'.

(a) The moving line of charge constitutes an electric current. What is the magnitude of the current?

(b) What is the magnitude of the magnetic field produced by this current? Show that the magnitude of the magnetic field is proportional to the magnitude of the electric field,

$$B = \mu_0\epsilon_0 v' E$$

(However, the direction of the electric and magnetic fields differ. The electric field is radial, whereas the magnetic field is tangential.)

19.*——According to the preceding problem, when a line of charge is made to move at speed v' parallel to itself, it produces a magnetic field of magnitude proportional to its electric field, $B = \mu_0\epsilon_0 v'E$. Use this result to find the magnetic field of a large charged flat sheet of paper with a charge of σ coulombs per square meter moving at speed v' in a direction parallel to itself.

20.*——Two very long parallel wires separated by a distance of 1.0 cm carry opposite currents of 8.0 A.

(a) Find the magnetic field at the midline between the wires.

(b) Find the magnetic field in the plane of the wires, at a distance of 2.0 cm from the midline.

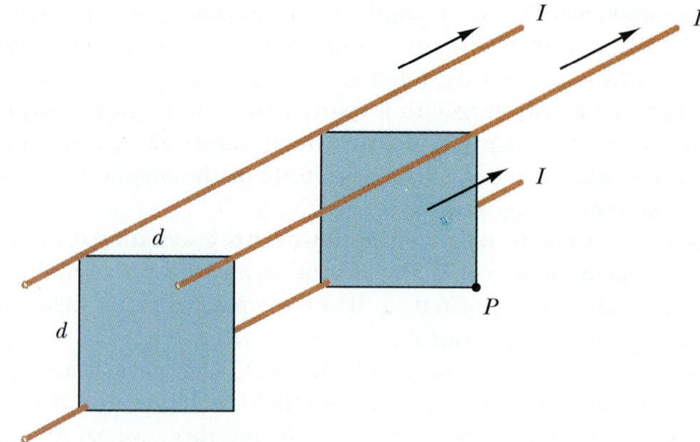

FIGURE 21.51
Three long, parallel wires. The wires pass through three corners of a square.

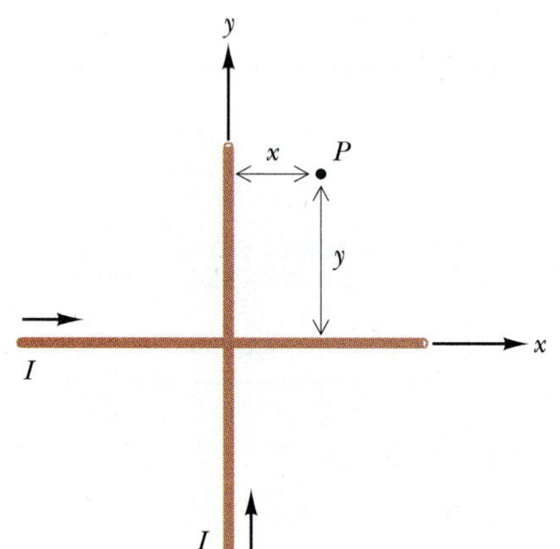

FIGURE 21.52
Two long, straight wires.

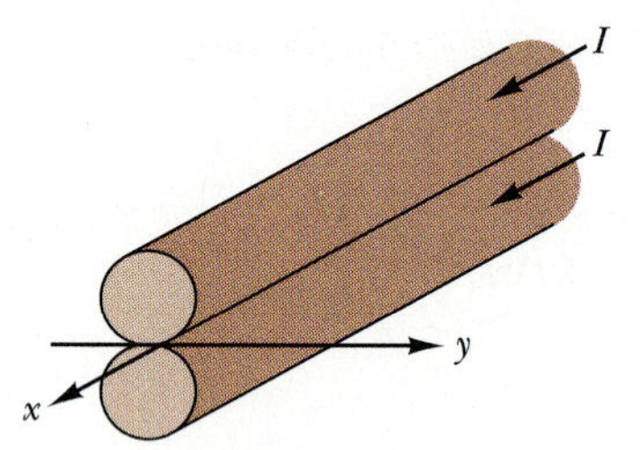

FIGURE 21.53
Two long, straight, thick wires in contact.

21.*—Two very long, straight, parallel wires separated by distance d carry currents of magnitude I in opposite directions. Find the magnetic field at a point equidistant from the lines, with a distance $2d$ from each line. Draw a diagram showing the direction of the magnetic field.

22.*—Three parallel wires are spaced as shown in Fig. 21.51. The wires carry equal currents in the same direction. What is the magnetic field at the point P? Draw a diagram giving the direction of the magnetic field.

23.*—In a proton accelerator, protons of velocity 3×10^8 m/s form a beam of current of 2×10^{-3} A. Assume that the beam has a circular cross section of radius 1 cm and that the current is uniformly distributed over the cross section. What is the magnetic field that the beam produces at its edge? What is the magnetic force on a proton at the edge of the beam?

24.*—A long straight wire of copper with a radius of 1 mm carries a current of 20 A. What are the instantaneous magnetic force and the corresponding acceleration on one of the conduction electrons moving at 10^6 m/s along the surface of the wire in a direction opposite to that of the current? What is the direction of the acceleration?

25.*—Two very long straight wires carry currents I are right angles. One of the wires lies along the x axis; the other lies along the y axis (see Fig. 21.52). Find the magnetic field at a point P in the first quadrant.

26.**—Two long parallel wires of copper of radius R are in contact along their full length (see Fig. 21.53). The wires carry equal currents I, in the same direction. The currents are uniformly distributed over the volumes of the wires. Find the magnetic field in the midplane (x–y plane) of the wires, as a function of the distance y from the point of contact. Where is this magnetic field maximum and what is the value of the maximum magnetic field?

27.**—Repeat the preceding problem if the currents carried by the wires are in opposite directions.

Section 21.3: Ampère's Law

28.—A ring of superconducting wire carries a current of 2.0 A. The radius of the ring is 1.5 cm. What is the magnitude of the magnetic field at the center of the ring?

29.—A circular coil consists of 60 turns of wire wound around the rim of a plywood disk of radius 0.10 m. If a current of 2.0 A is sent through this coil, what is the magnetic field produced at the center of the disk?

30.—A circular ring of wire of diameter 0.60 m carries a current of 35 A. What acceleration will the magnetic field generated by this ring give to an electron that is passing through the center of the ring with a velocity of 1.2×10^6 m/s in the plane of the ring?

31.—A ring of radius 8.0 cm is placed concentrically around a ring of radius 6.0 cm (see Fig. 21.54). Each ring carries a current of 4.0 A. What is the magnitude of the magnetic field at the center of the rings if the two currents flow in the same direction? If they flow in opposite directions?

32.*—A charge of 2.0×10^{-6} C is distributed uniformly around the rim of a cardboard disk of radius 8.0 cm, and this disk is made to spin about its axis at the rate of 15 revolutions per second. What is the current generated by the motion of the charge? What is the magnetic field that this current produces at the center of the disk?

33.*—Two rings of the same radius 0.20 m are placed at right angles (see Fig. 21.55). The rings carry equal currents of 10 A. What is the magnitude of the magnetic field at the center of this arrangement of rings? Draw a diagram showing the direction of this magnetic field.

34.*—A long, straight wire is bent into a circular loop of radius R near its midpoint (see Fig. 21.56). The wire carries a current I. What are the magnitude and direction of the magnetic field at the center of the loop?

35.*—A long, straight wire is bent into a circular loop of radius r near its midpoint, and this loop is twisted so its plane is perpendicular to the wire (see Fig. 21.57). The wire carries a current I. What are the magnitude and direction of the magnetic field at the center of the loop?

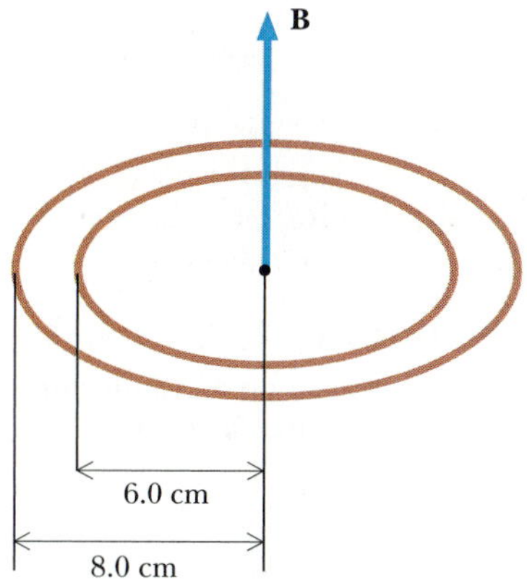

FIGURE 21.54
Two concentric rings.

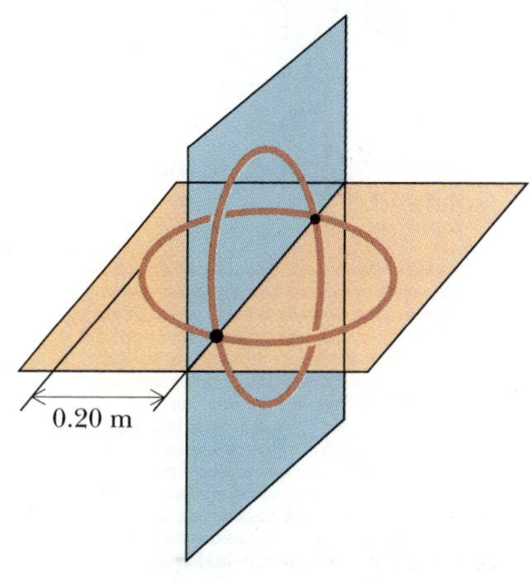

FIGURE 21.55
Two rings at right angles.

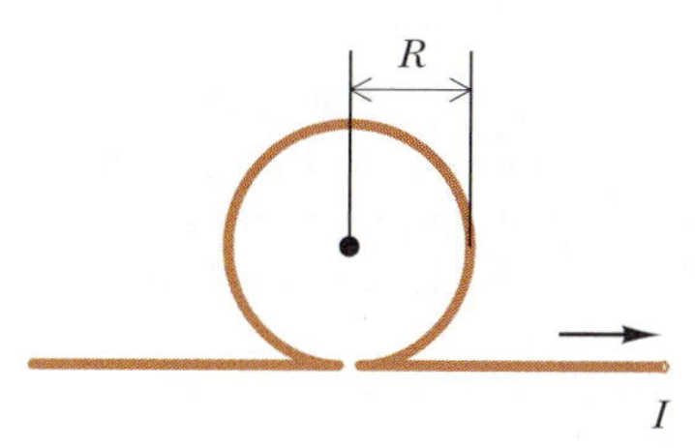

FIGURE 21.56
Long, straight wire deformed into a loop in the same plane.

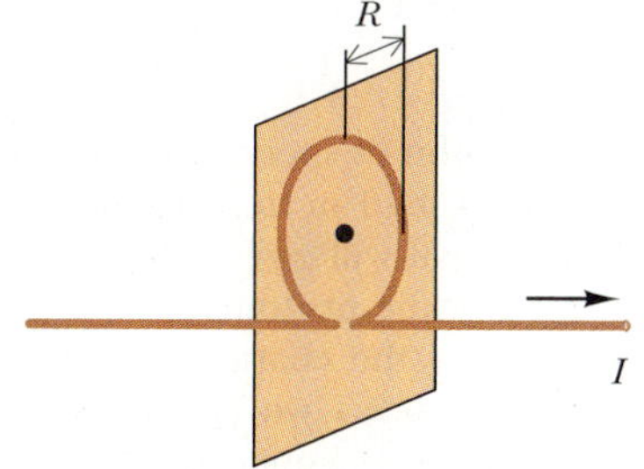

FIGURE 21.57
Long, straight wire deformed into a perpendicular loop.

Section 21.4: Solenoids and Magnets

36.—A long stolenoid has 15 turns per centimeter. What current must we put through its winding if we wish to achieve a magnetic field of 5.0×10^{-2} T in its interior?

37.—The electromagnet of a small electric bell is a solenoid with 260 turns in a length of 2.0 cm. What magnetic field will this solenoid produce if the current is 8.0 A?

38.—For two cases (a) and (b) in Fig. 21.58 decide if the two magnets will attract or repel each other.

39.*—A long solenoid has been placed inside another long solenoid of larger radius (see Fig. 21.59). The solenoids are coaxial and both have the same number n of turns per unit length and the same current I. What is the formula for the magnetic field in the region within the smaller solenoid? Between the smaller and the larger solenoid?

(a)

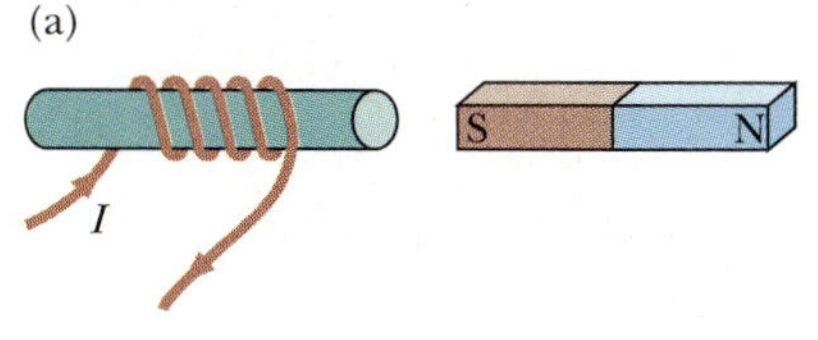

(b)

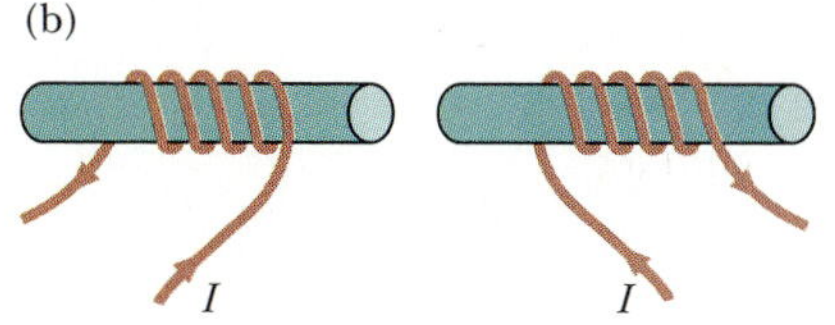

FIGURE 21.58
(a) Electromagnet and permanent magnet. (b) Two electromagnets.

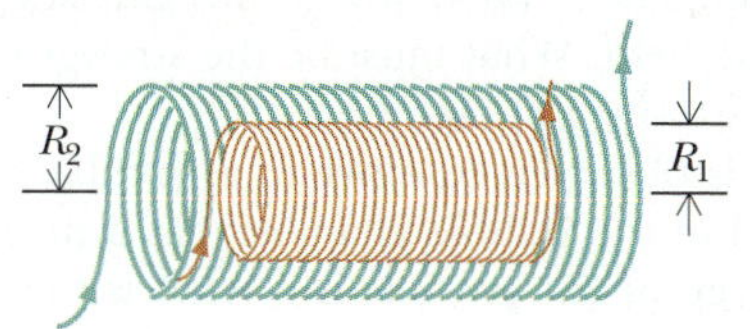

FIGURE 21.59
Two coaxial solenoids.

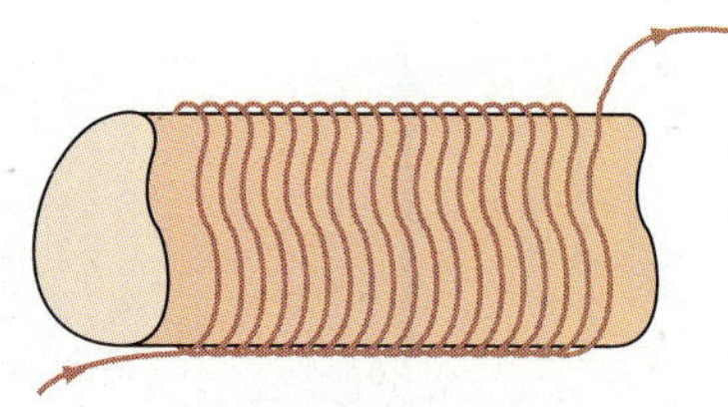

FIGURE 21.60
Solenoid wound on a cylinder of arbitrary, noncircular cross section.

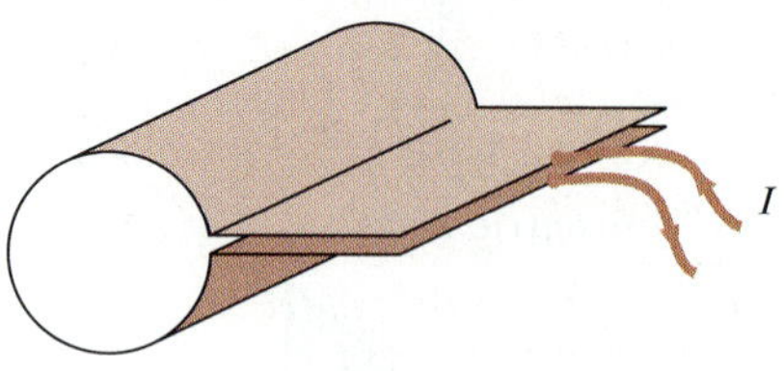

FIGURE 21.61
A sheet of copper shaped into a solenoid.

FIGURE 21.62
Long, straight, parallel wires.

40.*—Figure 21.60 shows a solenoid of arbitrary cross section; this solenoid is a (noncircular) cylinder. Suppose that there are n turns of wire per unit length and that the current in the wire is I. Use Ampère's Law to show that the magnetic field in the solenoid is $\mu_0 nI$, the same as for a circular cylinder.

41.*—Figure 21.61 shows a "solenoid" made of one turn of a sheet of copper. The solenoid has a length of 20 cm, and the current flowing through it is 2×10^3 A. What is the magnetic field in this solenoid? Assume that the current is uniformly distributed over the sheet of copper, and treat the solenoid as very long.

42.**—A long, straight wire carrying a current I is placed along the axis of a circular solenoid with n turns per unit length, also carrying a current I. What is the magnetic field at some point in the solenoid, at a distance r from the axis? Describe the shape of the magnetic field lines.

43.**—A large number of thin, long, straight wires are laid parallel to each other on a flat plane (see Fig. 21.62). The number of wires per unit length measured along the plane perpendicularly to the wires is n, and each carries a current I. Use Ampère's Law to find the magnetic field produced by this current distribution. Describe the direction of the magnetic field. Explain why the magnetic field differs in strength from that inside of a solenoid [Eq. (21)].

Section 21.5: Motion of Charges in Magnetic Fields

44.—A proton of speed 4.0×10^5 m/s is observed to move in a circular orbit of radius 0.40 m in a magnetic field. What is the strength of the magnetic field that will give such circular motion? What is the frequency of the motion of the proton?

45.—What is the charge-to-mass ratio of a particle that orbits in a magnetic field of 0.2 T with an angular frequency of 3.51×10^{10} rad/s?

46.—A bubble chamber, used to make the tracks of protons and other charged particles visible, is placed between the poles of a large electromagnet that produces a uniform magnetic field of 2.0 T. A high-energy proton passing through the bubble chamber makes a track that is an arc of a circle of radius 3.5 m. According to Eq. (27), what is the momentum of the proton?

47.—A proton of energy 1.0×10^7 eV moves in a circular orbit in the magnetic field near the Earth. The strength of the field is 0.50×10^{-4} T. What is the radius of the orbit?

48.—In principle, a proton of the right energy can orbit the Earth in an equatorial orbit under the influence of the Earth's magnetic field. If the orbital radius is to be 6.5×10^3 km and the magnetic field at this radius is 0.29×10^{-4} T, what must be the momentum of the proton?

49.—In the Crab Nebula (the remnant of a supernova explosion), electrons of a momentum of up to about 10^{-16} kg·m/s orbit in a magnetic field of about 10^{-8} T. What is the orbital radius of such electrons?

50.—At the Fermilab accelerator, protons of momentum 5.3×10^{-16} kg·m/s are held in a circular orbit of diameter 2.0 km by a vertical magnetic field. What is the strength of the magnetic field required for this?

51.—Some astrophysicists believe that the radio waves of 10^9 Hz reaching us from Jupiter are emitted by electrons of fairly low (nonrelativistic) energies orbiting in Jupiter's magnetic field. What must be the strength of this field if the cyclotron frequency is to be 10^9 Hz?

52.—In a bubble chamber, a particle is observed to move in a circular orbit of radius 1.2 m. The magnetic field in the bubble chamber has a magnitude of 2.0 T. Assuming that the charge of the particle is e, what is the magnitude of its momentum?

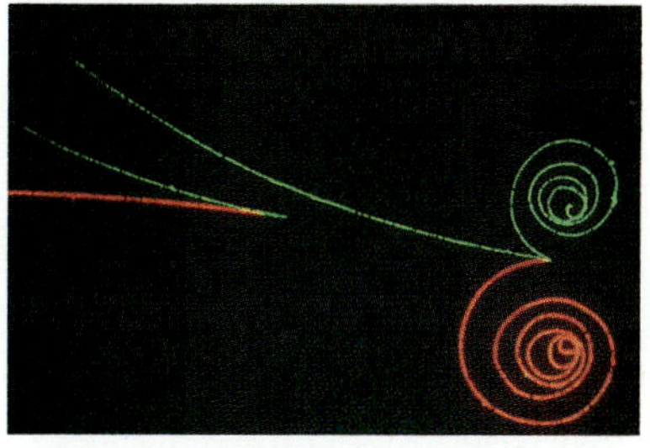

FIGURE 21.63

Tracks of an electron and an antielectron in a bubble chamber. The tracks spiral because the particles suffer a loss of energy as they pass through the liquid. For the purposes of Problem 55, concentrate on the initial portions of the tracks.

53.*——A beam containing the nuclei of two isotopes of carbon, ^{12}C (6 protons and 6 neutrons) and ^{14}C (6 protons and 8 neutrons), at the same momentum enters a magnetic field which bends the trajectories of these two kinds of particles into circles of radii r_{12} and r_{14}, respectively. What is the ratio of the two radii?

54.*——You want to confine an electron of energy 3.0×10^4 eV by making it circle inside a solenoid of radius 10 cm under the influence of the force exerted by the magnetic field. The solenoid has 120 turns of wire per centimeter. What minimum current must you put through the wire if the electron is not to hit the wall of the solenoid?

55.*——Figure 21.63 shows the tracks of an electron (charge $-e$) and an antielectron (charge $+e$) created in a bubble chamber. When the particles made these tracks, they were under the influence of a magnetic field of 1.0 T and a direction perpendicular to and into the plane of the figure. What is the momentum of each particle? Assume that they are moving in the plane of the figure and that this figure is $\frac{1}{10}$ natural size. Which is the track of the electron and which is the track of the antielectron?

56.*——Two positively charged particles separated by a distance d with a charge q and mass m are initially moving with a velocity v in opposite directions perpendicular to the line joining them (see Fig. 21.64). A magnetic field applied perpendicularly to the plane of the page will bend the paths of the particles into circles. What strength of magnetic field is necessary to make them collide head-on midway between the two starting points? (Ignore the electric forces between the charges.)

57.*——In a ***mass spectrometer,*** ions of a precisely selected speed are sent into a uniform magnetic field where they move along a semicircle and then strike a photographic plate, making a mark. A careful measurement of the distance of the mark from the entrance slit gives us the diameter (and the radius) of the circular motion of the ion in the magnetic field, and from this radius the mass can be calculated by means of Eq. (27). Suppose that the magnetic field in the mass spectrometer is 0.050 T. Suppose that an ion of charge e and speed 4.0×10^4 m/s is found to strike the photographic plate at a distance of 0.332 m from the entrance slit. What is the mass of this ion?

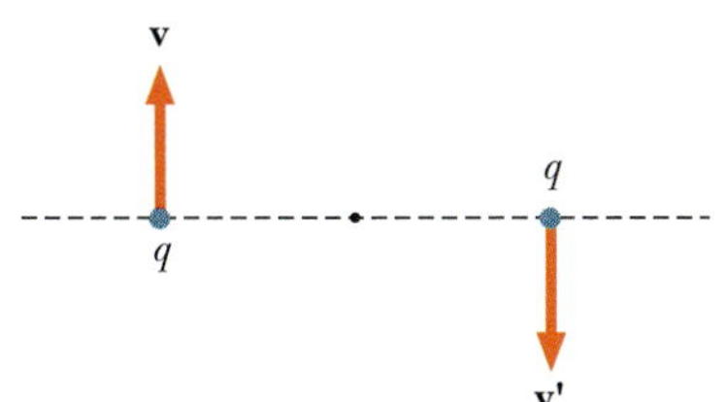

FIGURE 21.64

Two positively charged particles in a magnetic field.

Section 21.6: Force on a Wire

58.——An electromagnet produces a magnetic field of 0.60 T in the vertical downward direction. A straight wire placed in this magnetic field carries a current of 3.0 A. Calculate the magnitude and direction of the magnetic force on a 0.10-m segment of this wire if the orientation of the wire is:

(a) Vertically downward
(b) Horizontally toward the east
(c) Horizontally toward the south
(d) Toward the south and downward at an angle of 30°

59.——Figure 21.65 shows a balance used for the measurement of a magnetic field. A loop of wire carrying a precisely known current is partially im-

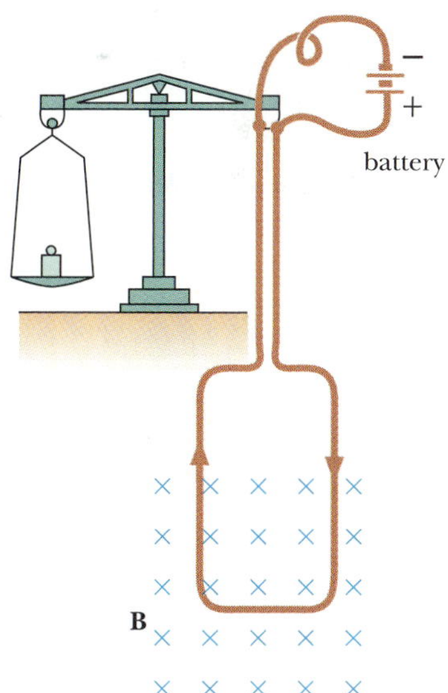

FIGURE 21.65

A current balance. The magnetic field **B** is perpendicular to the plane of the page. The crosses show the tails of the magnetic field vectors.

mersed in the magnetic field. The force that the magnetic field exerts on the loop can be measured with the balance and this permits the calculation of the strength of the magnetic field. Suppose that the short side of the loop measures 10.0 cm, the current in the wire is 0.225 A, and the magnetic force is 5.35×10^{-2} N. What is the strength of the magnetic field?

60.—The electric cable supplying an electric clothes dryer consists of two long straight wires separated by a distance of 1.2 cm. Opposite currents of 20 A flow on these wires. What is the magnetic force experienced by a 1.0-cm segment of wire due to the entire length of the other wire?

61.—Two parallel cables of a high-voltage power line carry opposite currents of 1.8×10^3 A. The distance between the cables is 4.0 m. What is the magnetic force pushing on a 50-m segment of one of these cables? Treat both cables as very long, straight wires.

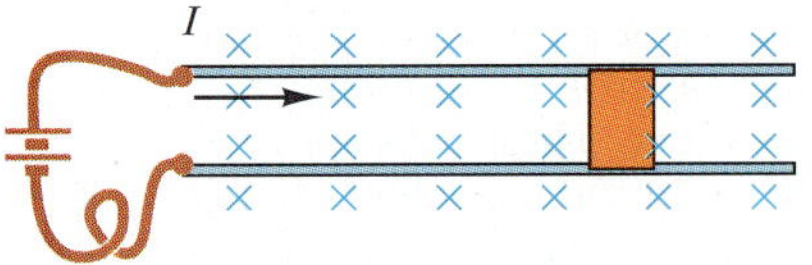

FIGURE 21.66
Electromagnetic launcher.

62.*—A straight wire is placed in a uniform magnetic field; the wire makes an angle of 30° with the magnetic field. The wire carries a current of 6.0 A and the magnetic field has a strength of 0.40 T. Calculate the force on a 10-cm segment of this wire. Show the direction of the force in a diagram.

63.*—An electromagnetic launcher, or rail gun, consists of two parallel conducting rails across which is laid a conducting bar, which serves as projectile. To launch this projectile, the rails are immersed in a magnetic field, and a current is sent through the rails and the bar (see Fig. 21.66). The magnetic force on the current in the bar then accelerates the bar (in actual rail guns of this kind, the magnetic field is itself produced by the current in the rails; but let us ignore this complication). Suppose that the magnetic field has a strength of 0.2 T and the bar has a length of 0.1 m and a mass of 0.2 kg. Ignore friction. What current must you send through the bar to give it an acceleration of 1×10^5 m/s^2?

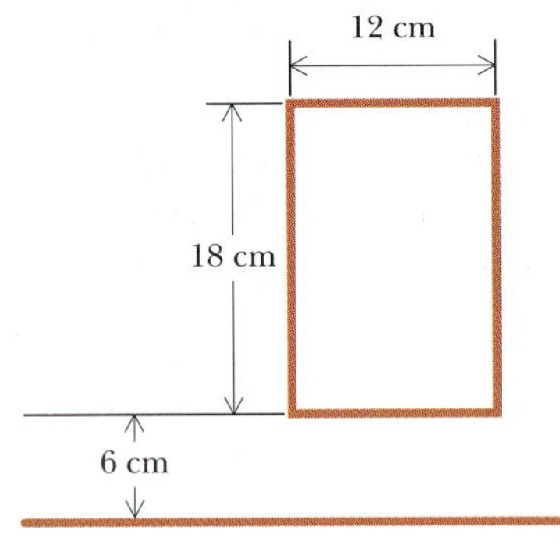

FIGURE 21.67
Long, straight wire and rectangular loop.

64.**—A rectangular loop of wire of dimensions 12 cm × 18 cm is near a long, straight wire. One of the short sides of the rectangle is parallel to the straight wire and at a distance of 6.0 cm; the long sides are perpendicular to the straight wire (Fig. 21.67). A current of 40 A flows on the straight wire and a current of 60 A flows around the loop. What are the magnitude and direction of the next magnetic force that the straight wire exerts on the loop?

Section 21.7: Torque on a Loop

65.—A circular coil consists of 60 turns of wire wound around the rim of a plastic disk of radius 2.0 cm. What is the magnetic moment of this coil if the current in the wire is 0.30 A?

66.—The proton has a magnetic moment of 1.41×10^{-26} A·m^2. If this magnetic moment makes an angle of 45° with a uniform magnetic field of 0.80 T, what is the torque on the proton?

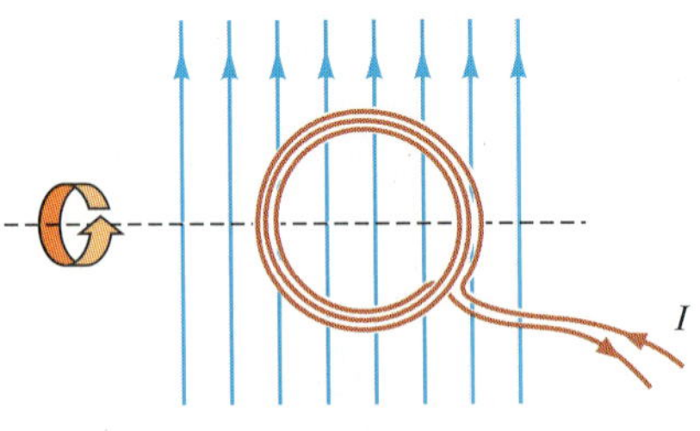

FIGURE 21.68
Circular coil in magnetic field.

67.—The coil in the mechanism of an ammeter is a rectangular loop, measuring 1 cm × 2 cm, with 120 turns of wire. The coil is immersed in a magnetic field of 0.01 T. What is the torque on this coil when it is parallel to the magnetic field and carries a current of 0.001 A?

68.—A horizontal circular loop of wire of radius 20 cm carries a current of 25 A. At the location of the loop, the magnetic field of the Earth has a magnitude of 0.39×10^{-4} T and points down at an angle of 16° with the vertical. What is the magnitude of the torque that this magnetic field exerts on the loop?

69.*—A 25-turn circular coil has a mass of 0.050 kg. The coil is immersed in a uniform 0.20-T magnetic field, with the plane of the coil oriented parallel to the magnetic field (see Fig. 21.68). The coil carries 5.0 A of current. When released, what will be the angular acceleration of the coil?

CHAPTER 22

Hydraulic turbines drive these large electric generators at the Hoover Dam on the Colorado River. The generators deliver 1,300 megawatts of electric power.

ELECTROMAGNETIC INDUCTION

In this chapter we will discover that electric fields can be generated not only by charges, but also by changing magnetic fields. For instance, if we increase the currents in the windings of an electromagnet and thereby increase the strength of the magnetic field, this changing magnetic field generates an electric field, called an ***induced electric field.*** This kind of electric field exerts the usual electric forces on charges — in this regard the induced electric field does not differ from an ordinary electrostatic electric field.

However, the induced and the electrostatic electric fields differ in that the forces exerted by the electrostatic field are conservative, whereas the forces exerted by the induced field are nonconservative. We recall that, according to the requirement laid down in Chapter 4, what discriminates between conservative and nonconservative forces is the work done during a roundtrip. For conservative forces, such as the forces exerted by electrostatic electric fields, the work done during a roundtrip is zero; for

Michael Faraday, *1791–1867, English physicist and chemist. Faraday's earliest research lay in chemistry, but he soon turned to research in electricity and magnetism, making contributions of the greatest significance. His discovery of electromagnetic induction was no accident, but arose from a systematic experimental investigation of whether magnetic fields can generate electric currents. Although Faraday was essentially an experimenter, with no formal training in mathematics, he made an important theoretical contribution by introducing the concept of field lines and by recognizing that electric and magnetic fields are physical entities.*

nonconservative forces, such as the forces exerted by induced electric fields, the work done during a roundtrip is nonzero. This means that when a charge moves around a closed circuit, the induced electric fields deliver work to the charge, and they constitute a source of emf. Like the emf of a battery, this ***induced emf*** is capable of driving a current around the circuit. As we will see, a celebrated law formulated by Michael Faraday asserts that the magnitude of the induced emf is directly proportional to the rate at which magnetic flux sweeps across the circuit. Such induced emf's find practical application in electromagnetic generators, widely used for the generation of electric power.

We begin with a discussion of the induced emf produced by the motion of a conductor in a constant magnetic field. Such a moving conductor sweeps across magnetic flux and thereby generates an induced emf, called a motional emf.

22.1 Motional EMF

Suppose that we push a rod of metal with some velocity **v** through a uniform magnetic field, such as the magnetic field of a large electromagnet. The free electrons in the metal will then also have a velocity **v,** and they will experience a magnetic force. For motion perpendicular to the magnetic field, the magnitude of the magnetic force on an electron is evB [see Eq. (21.11)]. If the rod and the velocity **v** of the rod are perpendicular to each other and to the magnetic field, then the direction of this magnetic force on the free electrons is parallel to the rod, as indicated by the right-hand rule (see Fig. 22.1). The electrons will therefore flow along the rod, accumulating negative charge on the upper end and leaving positive charge on the lower end. The flow of charge will stop when the electric repulsion generated by the accumulated charges balances the magnetic force evB. However, if the ends of the rod are in sliding contact with a pair of long wires that provide a (stationary) return path, then the electrons will flow continually around the circuit (see Fig. 22.2). ***Thus, the moving rod acts as a "pump of electricity," or a source of emf.*** In Fig. 22.2, the upper end of the rod is the negative terminal of this source, and the lower end is the positive terminal.

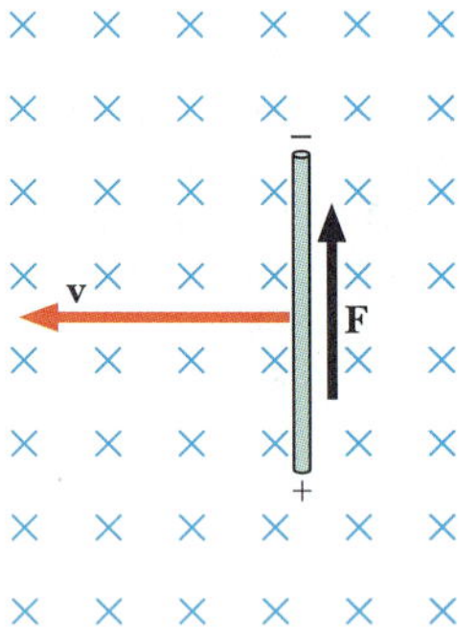

FIGURE 22.1
Conducting rod moving with velocity **v** through a uniform magnetic field (the magnetic field is directed into the plane of the page; the crosses indicate the tails of the magnetic field vectors). A free electron within this rod experiences a force in the upward direction.

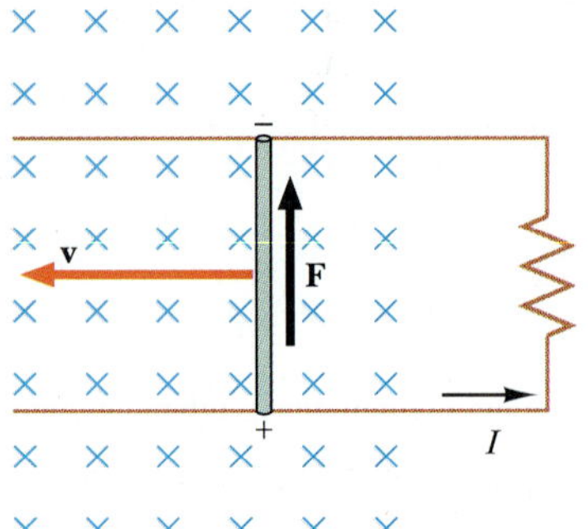

FIGURE 22.2
If the ends of the rod are in sliding contact with a pair of long wires, a current will flow around the circuit.

The emf associated with the rod is defined as the work done by the driving force per unit positive charge that passes from the negative end of the rod to the positive. Since the magnitude of the force on an electron is evB, the magnitude of the force per unit charge is $evB/e = vB$. If the length of the rod is l, the work per unit charge equals l times the force per unit charge, and therefore

$$\mathscr{E} = lvB \tag{1}$$

This is called a ***motional emf,*** because it is generated by the motion of the rod through the magnetic field.

◀ Motional emf

EXAMPLE 1. Suppose you drop an aluminum rod of length 1.0 m out of a window at a place where the horizontal magnetic field of the Earth is 2.0×10^{-5} T. The rod is oriented horizontally, at right angles to the magnetic field (see Fig. 22.3). What is the induced emf between the ends of the rod when its instantaneous downward velocity reaches 12 m/s?

SOLUTION: The magnetic field of the Earth has both a horizontal and a vertical component. In our derivation of Eq. (1) for the motional emf we assumed that the magnetic field B is perpendicular to the rod and to its velocity. For a vertically falling rod, as in Fig. 22.3, the horizontal component of the Earth's magnetic field is perpendicular to the rod and to its velocity, and hence generates a motional emf. By contrast, the vertical component of the Earth's magnetic field exerts no force on the free electrons in the rod (remember that the magnetic force is zero when the velocity is parallel to the magnetic field), and therefore generates no motional emf.

With $B = B_x = 2.0 \times 10^{-5}$ T, Eq. (1) gives

$$\mathscr{E} = lvB_x = 1.0 \text{ m} \times 12 \text{ m/s} \times 2.0 \times 10^{-5} \text{ T} = 2.4 \times 10^{-4} \text{ V}$$

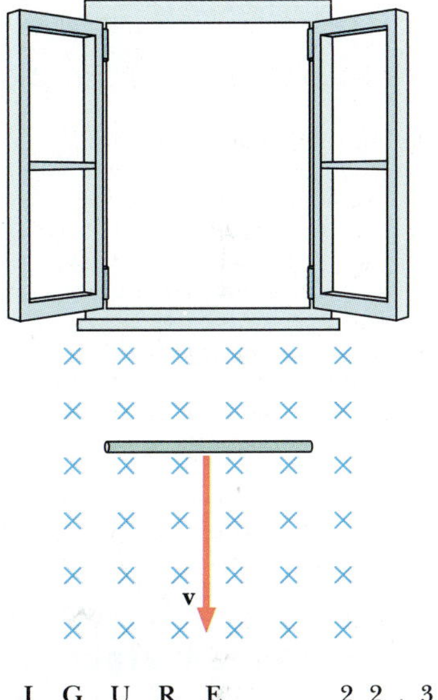

FIGURE 22.3

A rod has been dropped out of a window. The horizontal component of the magnetic field of the Earth is directed into the plane of the page.

A practical application of motional emf's is the ***electromagnetic flowmeter*** used to measure the speed of flow of a conducting liquid, such as blood, detergent, liquid sodium, tomato pulp, beer, and so on. In this device, the liquid is made to flow in a nonconducting pipe placed at right angles to a magnetic field, and a motional emf of the magnitude given by Eq. (1) will then appear across the diameter of the column of liquid, because the diameter acts like a conducting rod. Electrodes inserted into the pipe on opposite sides pick up the emf, which can be measured on a sensitive voltmeter. The velocity of flow can then be calculated from the measured emf, the strength of the magnetic field, and the dimension of the pipe.

◀ Electromagnetic flowmeter

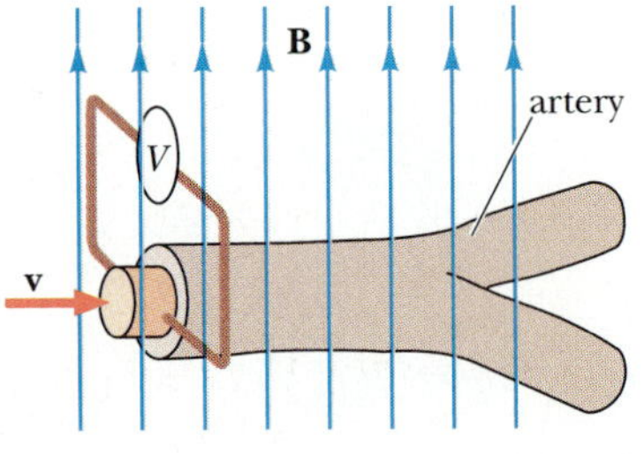

FIGURE 22.4
The moving column of blood in an artery is in contact with two electrodes at diametrically opposite points.

EXAMPLE 2. To measure the velocity of the flow of blood in the mesenteric artery in the abdomen of a dog, a researcher places the animal in a magnetic field of 3×10^{-2} T, inserts electrodes through the wall of the artery from each side, and measures the emf with a voltmeter (see Fig. 22.4). The inner diameter of the artery is 0.30 cm and the measured emf is 1.8×10^{-6} V. What is the velocity of flow?

SOLUTION: According to Eq. (1), with $l = 0.30$ cm,

$$v = \frac{\mathscr{E}}{lB} = \frac{1.8 \times 10^{-6}\,\text{V}}{0.30 \times 10^{-2}\,\text{m} \times 3 \times 10^{-2}\,\text{T}} = 2 \times 10^{-2}\ \text{m/s}$$

It is instructive to reexamine the generation of the induced emf from the point of view of a reference frame in which the rod is at rest. In such a reference frame moving with the rod, the free charges have no velocity, and there is *no magnetic force.* However, the free charges must experience some other kind of force that pushes them along the rod. The question is then: What is this new kind of force in the moving reference frame that produces the same effect as the magnetic force in the stationary reference frame? Since the only kinds of forces that act on electric charges are the magnetic force and the electric force, the "new" kind of force must be due to a "new" kind of electric field, an electric field that exists in the moving reference frame, but not in the stationary reference frame (see Fig. 22.5).

Induced electric field ▶

The "new" electric field that exists in the reference frame of the rod moving through a magnetic field is called an induced electric field. We can determine the magnitude of the induced electric field from a consistency requirement: the "new" electric force qE' in the moving reference frame must coincide with the magnetic force qvB in the stationary reference frame. This tells us that the electric field in the moving reference frame must have a magnitude

$$E' = vB \tag{2}$$

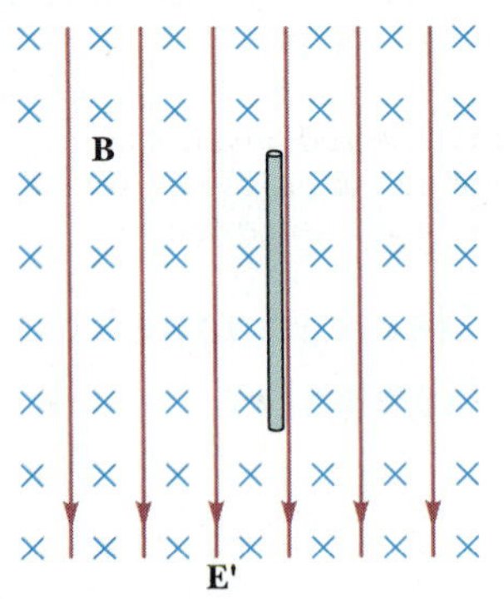

FIGURE 22.5
In the moving reference frame of the rod, there is a magnetic field **B** (blue crosses) and also an electric field **E**′ (red lines).

This electric field does work on the free charges and therefore represents a source of emf. The work done on a unit positive charge that passes through the rod is

$$\mathscr{E} = E'l \tag{3}$$

In view of Eq. (2), this value of the emf coincides with the value that we obtained in Eq. (1). Thus, the motional emf can be calculated either in a stationary reference frame or in a moving reference frame; in the former case it arises from a magnetic field and in the latter from an electric field.

22.2 FARADAY'S LAW

The quantity lvB appearing on the right side of Eq. (1) can be given an interesting interpretation in terms of the magnetic flux. As already mentioned in Section 21.2, the magnetic flux is defined in the same way as the electric flux. For any open or closed surface of area A, ***the magnetic flux Φ_B is defined as the average value of the normal component of the magnetic field multiplied by the area A,***

◀ **Magnetic flux**

$$\Phi_B = \overline{B}_n A \tag{4}$$

The magnetic flux is simply the net number of magnetic field lines that are intercepted by the surface, lines crossing the surface in one direction being reckoned as positive, and lines in the opposite direction as negative. Note that for a flat surface in a constant magnetic field, the magnetic flux is simply

$$\Phi_B = BA\cos\theta \tag{5}$$

where θ is the angle between the magnetic field and the perpendicular erected on the surface (see Fig. 22.6).

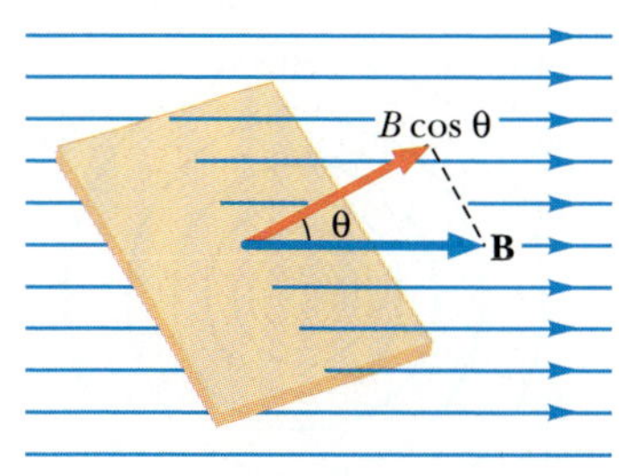

FIGURE 22.6

Flat rectangular surface in a magnetic field. The component of the magnetic field perpendicular to the surface is $B\cos\theta$.

Now, consider the rod moving through the magnetic field at velocity **v**. In a time interval Δt, the rod advances a distance $v\Delta t$ and therefore sweeps through an area $l \times v\Delta t$, perpendicular to the magnetic field (see Fig. 22.7). The product of this area and the magnetic field is the magnetic flux that the rod sweeps through, that is,

$$(lv\Delta t)B = \Delta\Phi_B$$

Thus,

$$lvB = \frac{\Delta\Phi_B}{\Delta t}$$

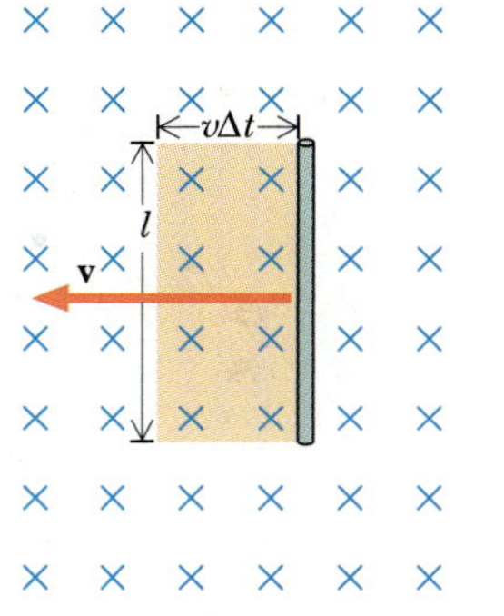

FIGURE 22.7

When the rod advances a distance $v\Delta t$, it sweeps through an area $l \times v\Delta t$ (orange).

which tells us that the quantity lvB is the rate at which the rod sweeps through magnetic flux. We can then rewrite Eq. (1) as

$$\mathscr{E} = \frac{\Delta\Phi_B}{\Delta t} \tag{6}$$

The advantage of this interpretation of the induced emf in terms of magnetic flux is that Eq. (6) is of more general validity than Eq. (1) — as we will see, Eq. (6) is a general law relating the induced emf to the rate at which the rod sweeps across flux, and it is valid regardless of how the sweeping of flux comes about. For instance, instead of moving the rod relative to the magnet that produces the magnetic field, we can hold the rod fixed and move the magnet. Flux will then sweep across the rod, and Eq. (6) suggests that an emf should be induced along the rod. Experi-

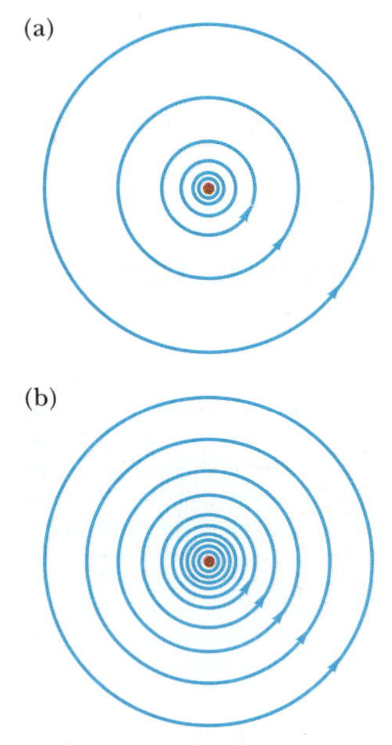

FIGURE 22.8
(a) Magnetic field lines of a current on a very long wire. The wire is perpendicular to the plane of the page. (b) Magnetic field lines of a stronger current.

ments do indeed confirm that in both cases the induced emf is exactly the same.

But there is another way in which the rod will sweep across magnetic flux: we can hold the rod fixed and increase or decrease the strength of the magnetic field. To understand why flux will be swept across the rod under these conditions, we must first take a look at what happens to the field lines of a current when the current increases or decreases. Figure 22.8a shows the field lines produced by a current on a very long, straight wire. If we increase the current, the magnetic field increases, that is, the number of field lines increases. Figure 22.8b shows the field lines of a stronger current. Where do the extra field lines come from? Obviously, the current has to make them. When the current slowly increases, it makes new, small circles of field lines in its immediate vicinity; meanwhile, the circles that already exist in Fig. 22.8a expand and move outward, just as ripples on the surface of a pond that are created when we drop a stone into the pond. Thus, the pattern shown in Fig. 22.8a gradually grows into the new pattern shown in Fig. 22.8b. Note that the pattern grows from the inside out.

If we decrease the current, field lines must disappear. But this is not quite the reverse of the creation of field lines — the circles of Fig. 22.8b do not contract and disappear at the center. The pattern cannot change from the outside in; it must first change *near* the current. When the current slowly decreases, it makes new small circles of field lines of *opposite* direction (negative field lines) in its immediate vicinity. These opposite circles expand and move outward, gradually canceling the original field lines.

In any case, a change in the strength of the magnetic field involves moving field lines. If a rod is located in the vicinity of the current, the moving field lines will sweep across the rod, that is, magnetic flux will sweep across the rod. Hence an increase or decrease of the strength of the magnetic field can cause flux to sweep across a *stationary* rod.

Experiments show that such a sweeping of flux across a stationary rod induces an emf which is given by Eq. (6), the same formula as for the case of the sweeping of flux across a moving rod. More generally, from the emf induced in a rod, we can infer the emf induced in a wire of arbitrary shape, since we can regard any such wire as consisting of short straight segments, or short rods, joined one to another. The net emf induced between the two ends of the wire is then the sum of the emf's induced in all these short rods, and this net emf equals the net rate at which magnetic flux sweeps across the wire.

More generally, we can make an assertion about the emf induced along an arbitrary moving or fixed mathematical path immersed in a constant or changing magnetic field. An emf will exist between the ends of this path regardless of whether we place a rod or a wire along the path — whenever a unit positive charge moves along this path, it will gain an amount $\mathscr{E}$ of energy from the induced electric field, regardless of whether the charge moves on a rod or through empty space. The rod or wire merely serves as a convenient conduit for the flow of charge. Of course, for the practical exploitation of the emf, we usually find it convenient to provide a rod, wire, or some other conductor along which the charge can flow, and we will then also have to provide a return path for the charge.

The general statement about the induced emf is known as ***Faraday's Law of Induction:***

The induced emf along any moving or fixed path in a constant or changing magnetic field equals the rate at which magnetic flux sweeps across the path,

$$\mathscr{E} = -\frac{\Delta \Phi_B}{\Delta t} \tag{7}$$

The minus sign that we have inserted into Eq. (7) indicates how the polarity of the induced emf is related to the change of flux; we will discuss the determination of this polarity in the next section.

For a closed path, or a closed circuit, Faraday's Law can be expressed in terms of the flux intercepted by the surface within the path. For this purpose we imagine that the path is spanned (or closed off) by some smooth mathematical surface, such as shown in Fig. 22.9. Since the magnetic field lines are continuous (without sources or sinks), the exact shape of the surface is of no consequence; we can use any flat or curved surface that has the path as boundary — any such surface intercepts the same flux. If we move the path or if the magnetic field changes, flux will sweep across the path. But any flux that sweeps across the path represents a gain or a loss of flux intercepted by the surface, since any field line that moves across the path moves either into the surface or out of the surface. Therefore, the rate $\Delta\Phi_B/\Delta t$ at which flux sweeps across the closed path equals the rate of change of the flux intercepted by the surface, and we can interpret Eq. (7) as stating that the induced emf equals the rate of change of this intercepted flux. This means that for a closed path Faraday's Law can be stated as follows:

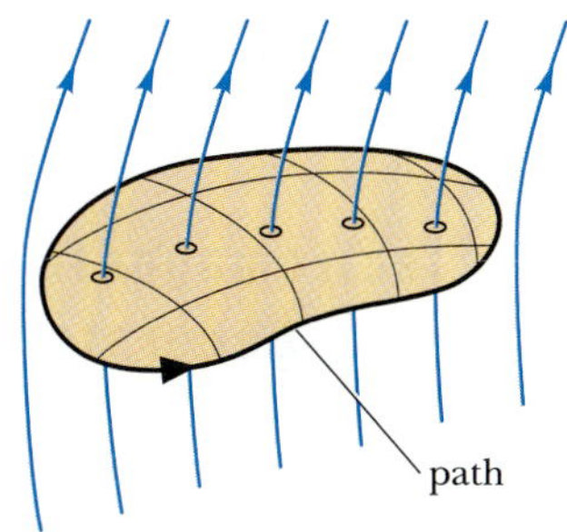

FIGURE 22.9
A closed path spanned by a surface. The magnetic field lines are intercepted by this surface.

◂ **Faraday's Law for closed circuit**

The induced emf around a closed path in a magnetic field is equal to the rate of change of the magnetic flux intercepted by the surface within the path.

Note that the magnetic flux intercepted by the surface within a closed path, say, a wire loop, can be made to change in a variety of ways (see Fig. 22.10). We can move the wire loop in or out of the magnetic field, or deform it so as to increase or decrease its area, or change its orientation, or change the strength of the magnetic field by changing the current in the magnet that produces the field — but in all these cases the emf induced around the loop is related to the change of flux by Eq. (7).

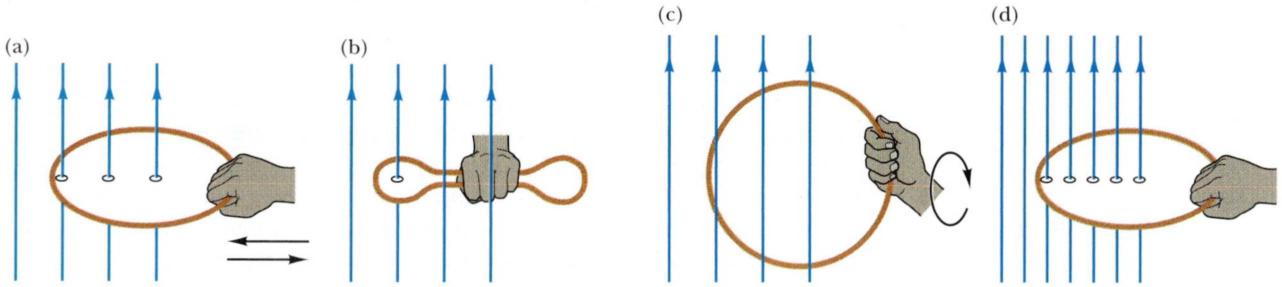

FIGURE 22.10
The flux through the loop changes when we (a) move the loop in or out of the magnetic field, (b) squeeze the loop, altering its shape, (c) rotate the loop in the magnetic field, or (d) increase the strength of the magnetic field.

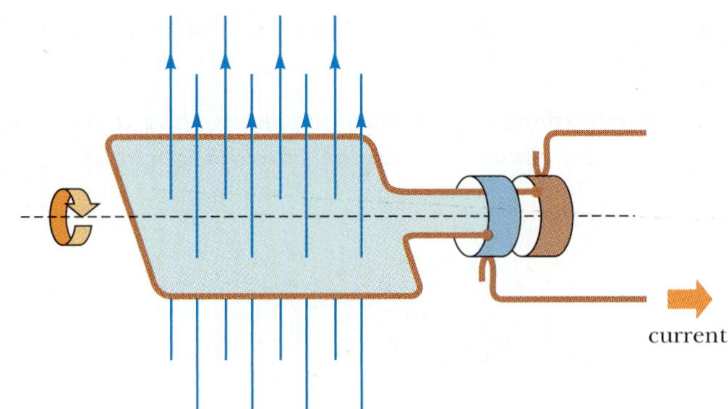

FIGURE 22.11
A simple electromagnetic generator consisting of a loop of wire rotating in a uniform magnetic field. Sliding contacts connect the loop to the external circuit.

Electromagnetic generator

The most important practical application of induction is the ***electromagnetic generator,*** widely used to generate the electric power that we need in our homes and factories. In its simplest form, the electromagnetic generator consists of a loop of wire that is made to rotate in a magnetic field (see Fig. 22.11). When the loop is face-on to the magnetic field, the flux intercepted by the surface within the loop is large; when the loop is edge-on to the magnetic field, the flux is zero. Thus, the rotation of the loop about its axis leads to a rate of change of flux, which induces an emf and drives a current through the loop and through the external circuit connected to the ends of the loop.

In essence, the generator is an electric motor operating in "reverse." In a motor, we convert electric power into mechanical power — we send a current through a loop in a magnetic field, and we produce rotational motion. In a generator, we convert mechanical power into electric power — we rotate a loop in a magnetic field, and we produce an emf and a current in the loop and in the external circuit. Small- or medium-size generators are driven by gasoline or diesel engines; large-size generators at electric power plants are driven by steam turbines or by hydraulic turbines.

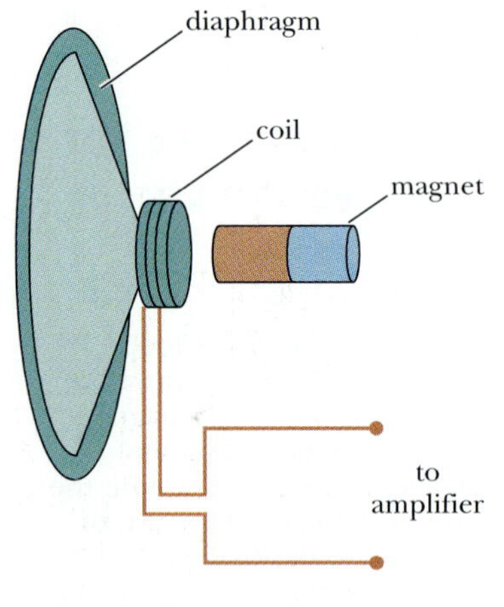

FIGURE 22.12
An induction microphone.

Another interesting practical application is the ***induction microphone,*** or moving-coil microphone. This consists of a flexible diaphragm with an attached small coil of wire, which is placed near a bar magnet (see Fig. 22.12). When a sound wave strikes the diaphragm, the pressure fluctuations move the diaphragm and the coil back and forth, and this motion of the coil produces a changing flux and induces an emf. Thus, the microphone transforms pressure fluctuations into an electric signal, which can be fed into the electric circuit of an audio amplifier, and from there into a tape recorder or a loudspeaker. A loudspeaker is simply an induction microphone operating in reverse — a current fed into the coil produces a magnetic field, which pushes or pulls the coil away or toward the bar magnet and moves the diaphragm back and forth, producing sound waves. Since the basic mechanism of the loudspeaker is the same as that of the induction microphone, it is actually possible to use a loudspeaker as a microphone—simply hook the loudspeaker into an amplifier and speak into it, and the loudspeaker will transform your sound signal into an electric signal and act as a (somewhat poor) microphone.

For later use in Chapter 24, we need to reformulate Faraday's Law in terms of the induced electric field. As we saw in the example of the rod moving in a magnetic field, the induced emf for a moving straight path sweeping through a constant magnetic field is related to the induced electric field by $\mathscr{E} = E'l$. More generally, the induced emf for an arbitrary moving or fixed path in a constant or changing magnetic field is related to the induced electric field (measured in the reference frame of the path) by

$$\mathscr{E} = \overline{E}_t l \tag{8}$$

where $\overline{E}_t$ is the average component of the induced electric field tangent to the path, and l is the length of the path. This permits us to express Faraday's Law as

$$\overline{E}_t l = -\frac{\Delta \Phi_B}{\Delta t} \tag{9}$$

This version of Faraday's Law can be used to calculate the induced electric field. However, in most of the problems of this chapter we will be concerned only with the induced emf, and the somewhat simpler version (7) of Faraday's Law will then suffice.

22.3 Some Examples; Lenz' Law

In this section we will look at some examples of the calculation of induced emf's. We begin by laying down a simple rule for finding the polarity of the induced emf, a rule known as ***Lenz' Law:***

> ***The induced current is always such as to oppose the change of flux that generated it.***

◂ Lenz' Law

For instance, consider the moving rod illustrated in Fig. 22.2, which generates a counterclockwise current. By the right-hand rule, within the area enclosed by the circuit, the magnetic field of this current is opposite to the original magnetic field (the magnetic field of the current in the wire is out of the plane of the page in Fig. 22.2, whereas the original magnetic field is into this plane). Thus, the flux contributed by the induced current opposes the increasing flux contributed by the original magnetic field, in agreement with Lenz' Law.

In practical applications, induced emf's are usually generated in coils consisting of many turns of wire, that is, many loops. In such a coil, each individual loop has a rate of change of flux and generates its own emf. The emf's of the individual loops of wire add together, and for a coil of N loops, the net emf is N times the emf of one loop.

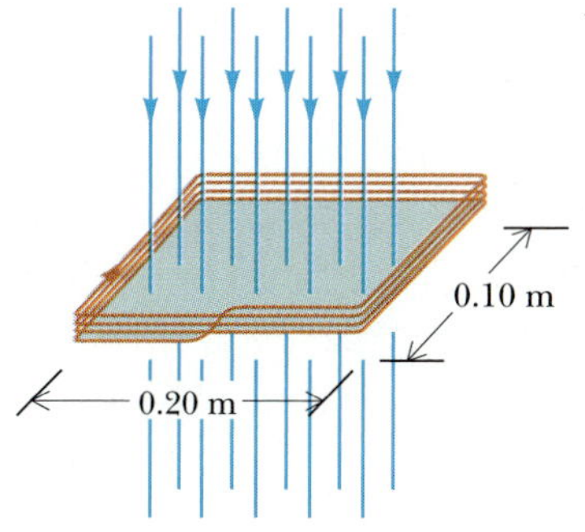

FIGURE 22.13
A rectangular coil in a uniform, decreasing magnetic field.

EXAMPLE 3. A rectangular coil of 150 loops forming a closed circuit measures 0.20 m × 0.10 m. The resistance of the coil is 5.0 Ω. The coil is placed between the poles of an electromagnet, face-on to the magnetic field (see Fig. 22.13). Suppose that when we switch the electromagnet off, the strength of the magnetic field decreases at the rate of 20 T per second. What is the induced emf in the coil? What is the direction of the induced current?

SOLUTION: The flux intercepted by each loop in the coil is AB, where A is the area of the loop and B the magnetic field. The net flux intercepted by all the 150 loops is therefore $\Phi_B = 150AB$. If the magnetic field changes by ΔB, the flux changes by $\Delta\Phi_B = 150A\Delta B$, and Faraday's Law tells us that the induced emf for the entire coil is

$$\mathscr{E} = -\frac{\Delta\Phi_B}{\Delta t}$$

$$= -150A\frac{\Delta B}{\Delta t} = -150 \times (0.20 \text{ m} \times 0.10 \text{ m}) \times (-20 \text{ T/s}) = 60 \text{ V}$$

According to Ohm's Law, the magnitude of the induced current is then

$$I = \frac{\mathscr{E}}{R} = \frac{60 \text{ V}}{5.0\ \Omega} = 12 \text{ amperes}$$

To determine the direction of the induced current, we appeal to Lenz' Law. The magnetic field shown in Fig. 22.13 and its flux are decreasing. Hence, the current in the coil ought to provide a magnetic field that compensates (in part) for this decreasing magnetic flux. This requires a clockwise current around the loop, so the magnetic field of this current is downward (according to the right-hand rule).

EXAMPLE 4. An electromagnetic generator consists of a rectangular coil of N loops of wire that rotates about an axis perpendicular to a constant magnetic field. Sliding contacts connect the coil to an external circuit (see Fig. 22.14). What emf does the coil deliver to the

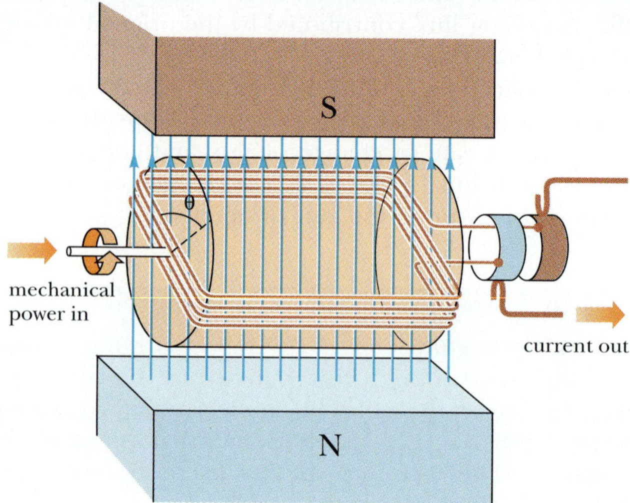

FIGURE 22.14
An electromagnetic generator.

external circuit? The coil has an area A and rotates with an angular frequency ω.

SOLUTION: The perpendicular to the coil makes an angle $\theta = \omega t$ with the magnetic field. The component of the magnetic field perpendicular to the coil is therefore $B_n = B \cos\theta = B \cos\omega t$. The magnetic flux through the coil is then $\Phi_B = NAB \cos\omega t$. To find the induced emf, we need to evaluate the rate of change of this flux, which means we need to evaluate the rate of change of $\cos\omega t$. We can do this by recalling a mathematical result from Chapter 9: for a particle moving in simple harmonic motion with a position $\cos\omega t$, the rate of change of the position, or the velocity, is $-\omega \sin\omega t$ [see Eq. (9.24)]. This is a general mathematical result for the rate of change of $\cos\omega t$, and it remains valid regardless of whether $\cos\omega t$ represents position or some other physical quantity. Thus, if the magnetic flux is $\Phi_B = NAB \cos\omega t$, the rate of change of this magnetic flux is $\Delta\Phi_B/\Delta t = -NAB\omega \sin\omega t$, and the induced emf is

$$\mathscr{E} = -\frac{\Delta\Phi_B}{\Delta t} = NAB\omega \sin\omega t \tag{10}$$

At the instant shown in Fig. 22.14, the magnetic flux through the coil is decreasing; according to Lenz' Law, the induced current must therefore flow around the coil in a clockwise direction so as to contribute an extra magnetic flux that opposes the decrease of flux.

◀ **Alternating emf, or AC voltage**

◀ **Commutator**

The emf given by Eq. (10) is an ***alternating emf, or AC voltage,*** that oscillates sinusoidally between positive and negative values. A plot of this emf is shown in Fig. 22.15. To obtain a purely positive emf from a generator, we can use a slotted sliding contact, or ***commutator,*** arranged so that the emf on the external terminals always has the same sign (see Fig. 22.16). A plot of the emf produced by such a generator is shown in Fig. 22.17. This emf oscillates between zero and a maximum value, but always remains positive. To eliminate or reduce the oscillation of the emf, we can connect several generators in series, each with its coil at a slightly differ-

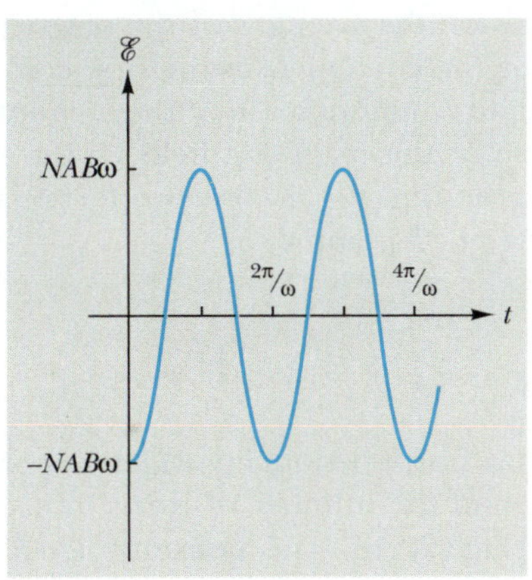

FIGURE 22.15
Emf of the generator; this is an alternating voltage (AC).

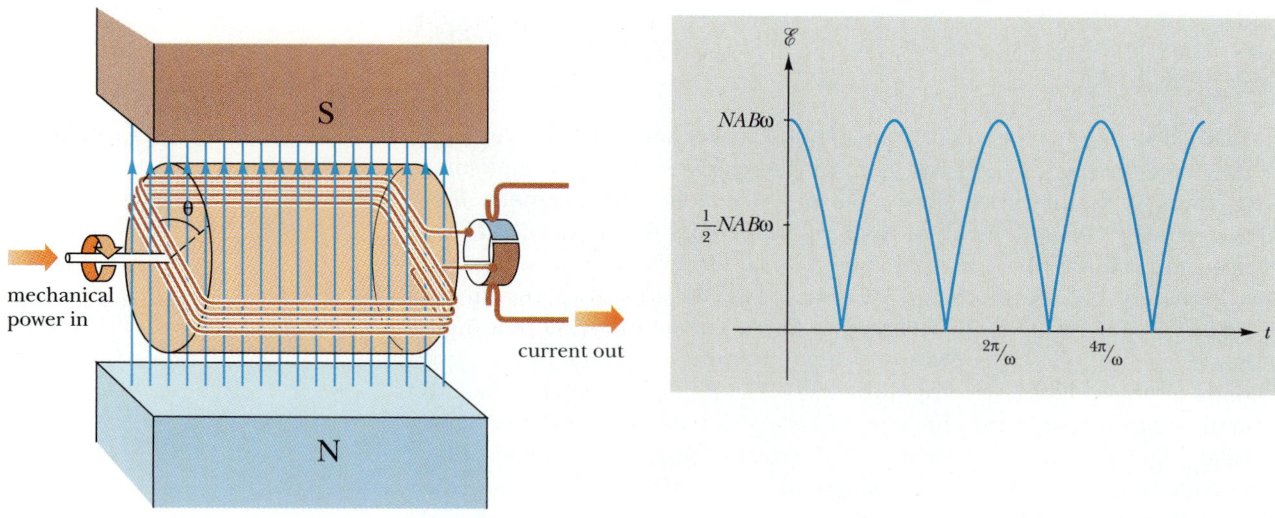

FIGURE 22.16
Electromagnetic generator with commutator. (Left).

FIGURE 22.17
Emf of the generator with commutator. (Right).

ent initial angle to the magnetic field. In the sum of the emfs, the oscillations then tend to average out.

22.4 Inductance

If a conductor carrying a time-dependent current is near some other conductor, then the changing magnetic field of the former can induce an emf in the latter. Thus, a time-dependent current in one conductor can induce a current in another, nearby conductor. For instance, consider the two coils in Fig. 22.18. The first of these coils is connected to some alternating source of emf, such as an AC generator, and it carries a time-dependent current. This coil therefore produces a time-dependent magnetic field $\mathbf{B}_1$. The changing magnetic flux Φ_{B_1} through the second coil induces an emf in this coil. The emf in the second coil is

$$\mathscr{E}_2 = -\frac{\Delta\Phi_{B_1}}{\Delta t} \tag{11}$$

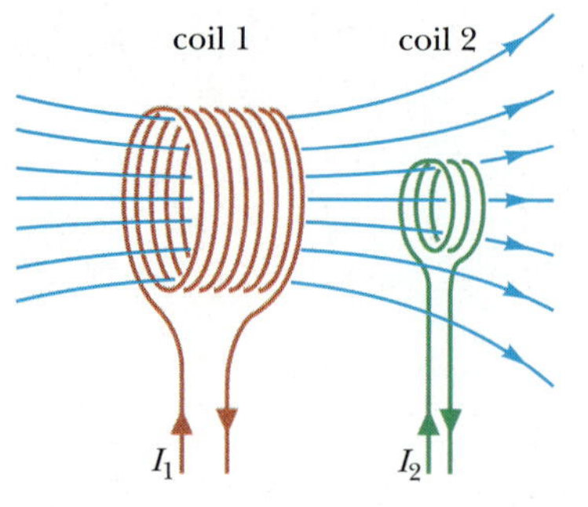

FIGURE 22.18
Coil 1 creates a magnetic field. Some of the field lines pass through coil 2.

The flux Φ_{B_1} depends on the strength of the magnetic field $\mathbf{B}_1$ in the second coil produced by the current I_1 in the first coil. This field strength is directly proportional to I_1 (an increase of the current by, say, a factor of 2 results in an increase of the magnetic field by the same factor; see Ampère's Law). Hence, the flux Φ_{B_1} is also proportional to I_1. We can write the relationship between Φ_{B_1} and I_1 as

$$\Phi_{B_1} = LI_1 \tag{12}$$

where L is a constant of proportionality which depends on the size of the coils, their distance, and the number of turns in each, that is, L depends on the geometry of Fig. 22.18. This constant is called the ***mutual induc-***

tance of the coils. If we change the current by some small amount ΔI_1, the flux will change by $L\Delta I_1$, and hence Eq. (11) becomes

$$\mathscr{E}_2 = -L\frac{\Delta I_1}{\Delta t} \tag{13}$$

◀ **Mutually induced emf**

This equation states that the emf induced in coil 2 is proportional to the rate of change of the current in coil 1.

The converse is also true; if coil 2 carries a current, then the emf induced in coil 1 is proportional to the rate of change of the current in coil 2,

$$\mathscr{E}_1 = -L\frac{\Delta I_2}{\Delta t} \tag{14}$$

The constants of proportionality appearing in Eqs. (13) and (14) are the same. Although we will accept this statement without proof, we note that the result is quite reasonable: the mutual inductance reflects the geometry of the *relative* arrangement of the coils, and that is of course the same in both cases.

The SI unit of inductance is called the henry (H),

$$1 \text{ henry} = 1 \text{ H} = 1 \text{ V}\cdot\text{s/A} \tag{15}$$

◀ **Henry (H)**

Incidentally: The permeability constant is commonly expressed in terms of this unit of inductance,

$$\mu_0 = 1.26 \times 10^{-6} \text{ H/m} \tag{16}$$

Mutual inductance finds an important application in the operation of transformers, used to step up or step down the emf supplied by an AC generator. The transformer consists of two coils — the primary and the secondary — arranged close together. The emf supplied to the primary produces a current which induces an emf in the secondary. This induced emf will be larger or smaller than the original emf, depending on the number of turns in the coils (we will examine transformers in detail in Section 23.5).

Joseph Henry, *1797–1878, American experimental physicist. He made important improvements in electromagnets by winding coils of insulated wire around iron pole pieces, and invented an electromagnetic motor and a new, efficient telegraph. He discovered self-induction and investigated how currents in one circuit induce currents in another.*

EXAMPLE 5. Suppose that the mutual inductance of the two coils in Fig. 22.18 is 2.0×10^{-7} H. If the first coil is connected to an alternating source of emf and the instantaneous rate of change of the current in the first coil is 40 A/s, what emf will be induced in the second coil?

SOLUTION: With $L = 2.0 \times 10^{-7}$ H, Eq. (13) gives

$$\mathscr{E}_2 = -L\frac{\Delta I_1}{\Delta t} = -2.0 \times 10^{-7} \text{ H} \times 40 \text{ A/s} = -8.0 \times 10^{-6} \text{ V}$$

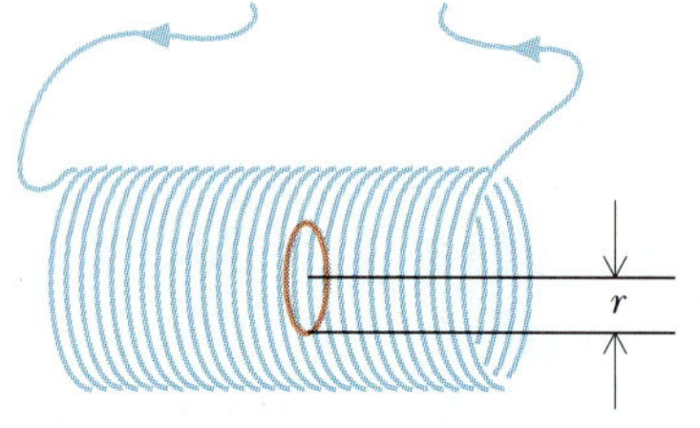

FIGURE 22.19
A ring of wire inside a solenoid.

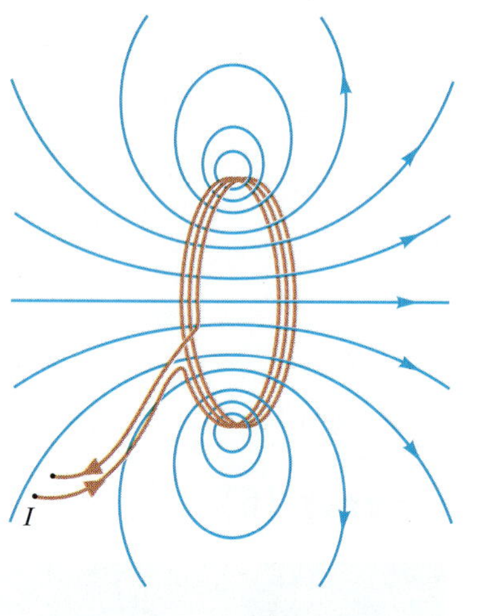

FIGURE 22.20
A coil and its magnetic field.

EXAMPLE 6. A long solenoid has n turns per unit length. A ring of wire of radius r is placed within the solenoid, perpendicular to the axis (see Fig. 22.19). What is the mutual inductance?

SOLUTION: If the current in the solenoid windings is I_1, the magnetic field is $B_1 = \mu_0 n I_1$ [see Eq. (21.21)], and the flux through the ring is

$$\Phi_{B_1} = [\text{area}] \times B_1 = \pi r^2 \times \mu_0 n I_1$$

According to Eq. (12), the inductance is

$$L = \Phi_{B_1}/I_1 = \pi r^2 \mu_0 n \tag{17}$$

A conductor by itself has a ***self-inductance.*** Consider a coil with a time-dependent current. The magnetic field of this coil (see Fig. 22.20) will then produce a time-dependent magnetic flux and, by Faraday's Law, an induced emf. The net emf acting on the coil is then the sum of the external emf (supplied by some generator connected to the terminals in Fig. 22.20) and the self-induced emf. This means that whenever the current is time-dependent, the coil will act back on the current and modify it (we will see how to calculate the net resultant current in Section 23.3). For this reason the self-induced emf is called a ***back emf.*** From Lenz' Law, we immediately recognize that ***the self-induced emf always acts in such a direction to oppose the change in the current, that is, it attempts to maintain the current constant.***

In terms of the flux through the circuit, the definition of the self-inductance is of the same form as Eq. (12),

Self-inductance ▶

$$\Phi_B = LI \tag{18}$$

and therefore the self-induced emf is

Self-induced emf ▶

$$\mathscr{E} = -L\frac{\Delta I}{\Delta t} \tag{19}$$

EXAMPLE 7. A long solenoid has n turns per unit length and a radius R. What is its self-inductance per unit length?

SOLUTION: The magnetic field inside the solenoid is $B = \mu_0 n I$. The number of loops in a length l is nl; each of these loops has a flux $\pi R^2 B$. Hence the flux through all the loops in a length l is

$$\Phi_B = \pi R^2 B \times [\text{number of turns}] = \pi R^2 B \times nl = \mu_0 n^2 \pi R^2 I l$$

and the self-inductance for the length l is

$$L = \Phi_B/I = \mu_0 n^2 \pi R^2 l \quad (20)$$

The self-inductance per unit length is therefore

$$L/l = \mu_0 n^2 \pi R^2 \quad (21)$$

22.5 MAGNETIC ENERGY

Inductors store magnetic energy, just as capacitors store electric energy. When we connect an external source of emf to an inductor and start a current through the inductor, the back emf will oppose the increase of the current, and the external emf must do work in order to overcome this opposition and establish the flow of current. This work is stored in the inductor, and it can be recovered by removing the external source of emf from the circuit. As the current begins to decrease, the inductor will supply a back emf which tends to keep the current flowing for a while (at a decreasing rate). Thus, the inductor delivers energy to the current while the current decreases.

To calculate the amount of stored energy, we note that when the current increases at the rate of $\Delta I/\Delta t$, the back emf is

$$\mathscr{E} = -L\frac{\Delta I}{\Delta t}$$

The inductor does work on the current at a rate given by the usual formula for the electric power [see Eq. (20.20)]:

$$I\mathscr{E} = -IL\frac{\Delta I}{\Delta t} \quad (22)$$

Here the negative sign implies that the energy is delivered by the current to the inductor rather than vice versa. In a time Δt, the energy stored in the inductor is therefore

$$\Delta U = -I\mathscr{E}\Delta t = LI\Delta I \quad (23)$$

To find the net energy stored in the inductor, we have to sum all the small increments ΔU in the energy. We can do this by noting that Eq. (23) has the same mathematical form as the equation for a small increment in the position of a particle with uniform acceleration, $\Delta x = at\Delta t$. By comparing these equations we see that L is mathematically analogous to a, and I is mathematically analogous to t. For the accelerated particle, the sum of all the small increments Δx is $x = \frac{1}{2}at^2$. According to our analogy, the sum of all the small increments ΔU must then be

Magnetic energy in inductor ▶

$$U = \tfrac{1}{2}LI^2 \tag{24}$$

Thus, ***the total energy stored in the inductor is proportional to the square of the current.***

EXAMPLE 8. A solenoid has a radius of 2.0 cm; its winding has one turn of wire per millimeter. A current of 10 A flows through the winding. What is the amount of energy stored per unit length of the solenoid?

SOLUTION: According to Eq. (21), the inductance per unit length is

$$L/l = \mu_0 n^2 \pi R^2$$

and the energy per unit length is

$$U/l = \tfrac{1}{2}(L/l)I^2 = \tfrac{1}{2}\mu_0 n^2 \pi R^2 I^2 \tag{25}$$

$$= \tfrac{1}{2} \times 1.26 \times 10^{-6}\ \text{H/m} \times (10^3/\text{m})^2 \times \pi \times (0.02\ \text{m})^2 \times (10\ \text{A})^2$$

$$= 7.9 \times 10^{-2}\ \text{J/m}$$

The energy stored in a solenoid can be expressed in terms of the magnetic field. Consider a length l of the solenoid. From Eq. (25), the energy associated with this portion is

$$U = \tfrac{1}{2}\mu_0 n^2 \pi R^2 I^2 l \tag{26}$$

Since for the solenoid $B = \mu_0 nI$, we can write this as

$$U = \frac{1}{2\mu_0} B^2 \pi R^2 l \tag{27}$$

And since $\pi R^2 l$ is the volume filled with magnetic field, the energy is

$$U = \frac{1}{2\mu_0} B^2 \times [\text{volume}] \tag{28}$$

According to this equation, the quantity $(1/2\mu_0)B^2$ can be regarded as the magnetic energy per unit volume. Thus, the energy density in the magnetic field is

Energy density in magnetic field ▶

$$[\text{energy density}] = \frac{1}{2\mu_0} B^2 \tag{29}$$

Although we have derived this equation only for the special case of a long solenoid, it turns out to be generally valid (in vacuum and nonmagnetic materials). The magnetic field, just as the electric field, stores energy. The magnetic energy density is proportional to B^2, just as the electric energy density is proportional to E^2 [compare Eq. (18.39)]. Accordingly, ***the magnetic energy is concentrated in the regions of space where the magnetic field is strong.***

EXAMPLE 9. Near the surface, the Earth's magnetic field typically has a strength of 0.3×10^{-4} T and the Earth's atmospheric electric field typically has a strength of 100 V/m. What is the energy density in each field?

SOLUTION: The magnetic energy density is

$$\frac{1}{2\mu_0} B^2 = \frac{(0.3 \times 10^{-4}\ \text{T})^2}{2 \times 1.26 \times 10^{-6}\ \text{H/m}} = 3.6 \times 10^{-4}\ \text{J/m}^3$$

and the electric energy density is

$$\frac{\epsilon_0}{2} E^2 = \frac{8.85 \times 10^{-12}\ \text{F/m}}{2} \times (100\ \text{V/m})^2$$

$$= 4.4 \times 10^{-8}\ \text{J/m}^3$$

Thus, the magnetic energy density in our immediate environment is much larger than the electric energy density.

22.6* THE RL CIRCUIT

The RL circuit consists of a resistor and an inductor connected in series to a battery or some other source of emf. This circuit provides a good illustration of the effects of self-inductance on a current. Figure 22.21 is the schematic diagram for this circuit. In the diagram, the inductor is represented by a coiled line, reminiscent of a coil of wire. The inductance is L, the resistance is R, and the emf of the battery is $\mathscr{E}$. The inductor is supposed to be resistanceless; if the wire in the coils of the inductor has some resistance, then this resistance must be included in R.

We assume that the current in the circuit is initially zero, and that the battery is suddenly connected at the initial time $t = 0$. The current then starts to increase. But the self-inductance will generate an emf across the inductor, which, by Lenz' Law, opposes the increase of the current. Because of this self-induced emf, the current in the circuit cannot increase suddenly; it can increase only gradually. The self-induced emf continues to oppose the current and to restrain its growth until this current attains

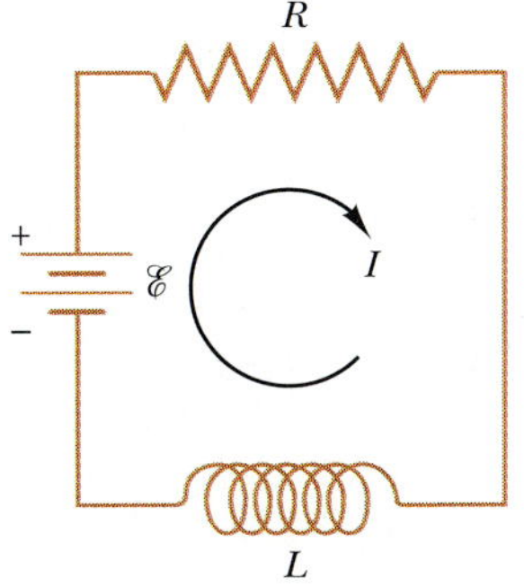

FIGURE 22.21
An RL circuit, consisting of a resistor and an inductor connected in series to a battery.

* This section is optional.

its final, steady value. This steady, final value of the current is simply $I = \mathscr{E}/R$, as though the inductor were absent, because when the current becomes steady, the inductor ceases to contribute to the emf in the circuit; that is, it ceases to affect the circuit. Qualitatively, the gradual growth of current in an RL circuit is analogous to the increase of charge in an RC circuit. As we will see, the plot of the current vs. time in the RL circuit is mathematically similar to the plot of the charge vs. time in the RC circuit.

For an estimate of the time required for the current to attain its final, steady value, let us apply Kirchhoff's Rule to the RL circuit at the first instant, when the battery has just been connected. The current is then still zero (it has not yet had time to build up). For zero current, the potential drop across the resistor is zero; consequently, only the battery and the inductor enter into Kirchhoff's Rule. The emf of the battery is $\mathscr{E}$, and the emf across the inductor is $-L\Delta I/\Delta t$. The sum of these two emf's must therefore be zero:

$$\mathscr{E} - L\frac{\Delta I}{\Delta t} = 0 \tag{30}$$

This tell us the rate of change of the current at the initial instant:

$$\frac{\Delta I}{\Delta t} = \frac{\mathscr{E}}{L} \tag{31}$$

We obtain a rough estimate of the time required for the current to increase from zero to its final value $\mathscr{E}/R$ if we divide the increase of the current by the rate of change of the current (this is analogous to dividing distance by speed to obtain the travel time). We will designate this estimate of the time by τ:

Characteristic time for RL circuit ▶

$$\tau = \frac{\mathscr{E}/R}{\mathscr{E}/L} = \frac{L}{R} \tag{32}$$

This time τ is called the ***characteristic time*** for the RL circuit.

As in the case of the RC circuit, our calculation of the time required for the current to increase to its final value $\mathscr{E}/R$ is not exact, because the rate of change of the current is not constant. The rate of change of the current is large at first, but gradually decreases toward zero; and the current takes a very long time (infinite time) to reach its final value $\mathscr{E}/R$. Figure 22.22 is a plot of the current vs. time, based on an exact calculation. ***At the characteristic time $\tau = L/R$, the current reaches only 63% of its final value.***

Although in the schematic diagram in Fig. 22.21 the inductance and the resistance are shown separated, this diagram can equally well represent a coil of resistive wire, such as the coil making up the windings of an electromagnet, which has both inductance and resistance. For purposes of calculation, the inductance and the resistance of such a coil may be regarded as placed in series, since each produces its own change of potential in the circuit. From our results for the RL circuit we therefore see

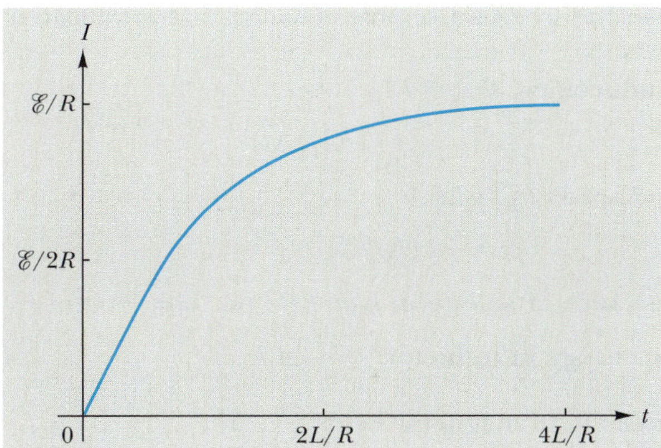

FIGURE 22.22

Growing current in the RL circuit as a function of time. The current is initially zero, and it increases toward a final value of $\mathscr{E}/R$ (at $t = \infty$).

that whenever an electromagnet is suddenly switched on, the current and the magnetic field in the magnet take a while to build up to a steady value.

EXAMPLE 10. The windings of a large electromagnet have an inductance of 10 H and a resistance of 8 Ω. This electromagnet is connected to an external emf of 230 V. After the electromagnet is switched on, how long does the current take to build up to 63% of its final value? What is the final, steady value of the current that is attained after a fairly long time?

SOLUTION: The characteristic time for this electromagnet is

$$\tau = \frac{L}{R} = \frac{10\text{ H}}{8\ \Omega} = 1.3\text{ s}$$

This is the time required to reach 63% of the final value of the current.

The final, steady value of the current is

$$I = \frac{\mathscr{E}}{R} = \frac{230\text{ V}}{8\ \Omega} = 29\text{ A}$$

This value of the current is attained after a long time, that is, a time long compared with 1.3 s.

SUMMARY

- **Motional emf in rod:** $\mathscr{E} = lvB$
- **Magnetic flux:** $\Phi_B = \overline{B}_n A$
- **Faraday's Law:** $\mathscr{E} = -\dfrac{\Delta \Phi_B}{\Delta t}$

▶ **Lenz' Law:** Induced emf opposes change that produced it.

▶ **Mutual inductance:** $\Phi_{B_1} = LI_1$

$$\mathscr{E}_2 = -L(\Delta I_1/\Delta t)$$

▶ **Self-inductance:** $\Phi_B = LI$

$$\mathscr{E} = -L(\Delta I/\Delta t)$$

▶ **Self-inductance of solenoid:** $\mu_0 n^2 \pi R^2$ per unit length

▶ **Magnetic energy in inductor:** $U = \frac{1}{2}LI^2$

▶ **Energy density in magnetic field:** $\frac{1}{2\mu_0} B^2$

▶ **Time constant of RL circuit:** $\tau = L/R$

QUESTIONS

1.–At the latitude of the United States, the magnetic field of the Earth has a downward component, larger than the northward component. Suppose that an airplane is flying due west in this magnetic field. Will there be an emf between its wingtips? Which wingtip will be positive? Will there be a flow of current?

2.–What is the magnetic flux that the magnetic field of the Earth produces through the surface of the Earth?

3.–A long straight wire carries a steady current. A square conducting loop is in the same plane as the wire. If we push the loop toward the wire, how is the direction of the current induced in the loop related to the direction of the current in the wire?

4.–A ***flip coil*** serves to measure the strength of a magnetic field. It consists of a small coil of many turns connected to a sensitive ammeter. The coil is placed face-on in the magnetic field and then suddenly flipped over. How does this indicate the presence of the magnetic field?

5.–The ***magneto*** used in the ignition system of old automobile engines consists of a permanent magnet mounted on the flywheel of the engine. As the flywheel turns, the magnet passes by a stationary coil, which is connected to the spark plug. Explain how this device produces a spark.

6.–A long straight wire carries a current that is increasing as a function of time. A rectangular loop is near the wire, in the same plane as the wire. How is the direction of the current induced in the loop related to the direction of the current in the wire?

7.–Consider two adjacent rectangular circuits, in the same plane. If the current in one circuit is suddenly switched off, describe the direction of the current induced in the other circuit.

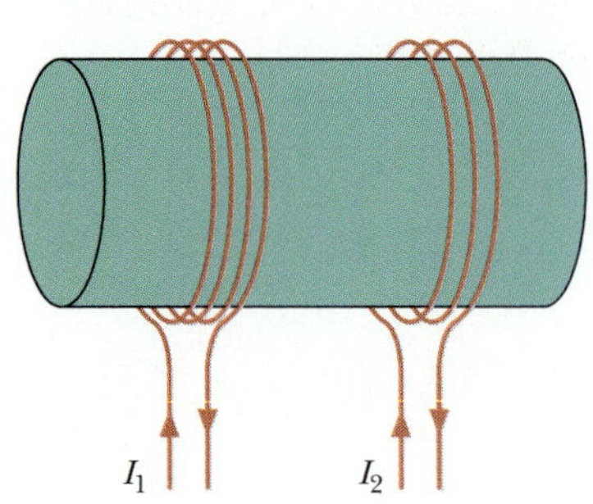

FIGURE 22.23
Two adjacent coils of wire.

8.–Figure 22.23 shows two coils of wire wound around a plastic cylinder. If the current in the left coil is made to increase, what is the direction of the current induced in the coil on the right?

9.–A circular conducting ring is being pushed toward the north pole of a bar magnet. Describe the direction of the current induced in the ring.

10.–A sheet of aluminum is being pushed between the poles of a horseshoe magnet (Fig. 22.24). Describe the direction of flow of the induced currents, or ***eddy currents,*** in the sheet. Explain why there is a strong friction force that opposes the motion of the sheet.

11.–A bar magnet, oriented vertically, is dropped toward a flat horizontal copper plate. Describe the direction of flow of the induced currents (eddy currents) in the copper plate. Explain why there will be a repulsive force between the bar magnet and the copper plate. Is this an elastic force — that is, will the bar magnet bounce if it is very strong?

12.–A conducting ring is falling toward a bar magnet (Fig. 22.25). Explain why there will be a repulsive force between the ring and the magnet. Explain why there will be no such force if the ring has a slot cut through it (see dashed lines in Fig. 22.25.

13.–You have two coils, one of slightly smaller radius than the other. To achieve maximum mutual inductance, should you place these coils face-to-face or one inside the other?

14.–Two circular coils are separated by some distance. Qualitatively, describe how the mutual inductance varies as a function of the orientation of the coils.

15.–If a strong current flows in a circuit that includes some coils of wire, and you suddenly break the circuit (by opening a switch), a large spark is likely to jump across the switch. Explain.

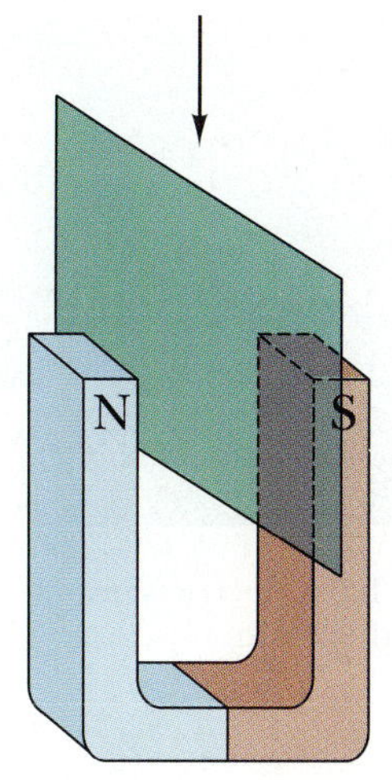

FIGURE 22.24
Sheet of aluminum and horseshoe magnet.

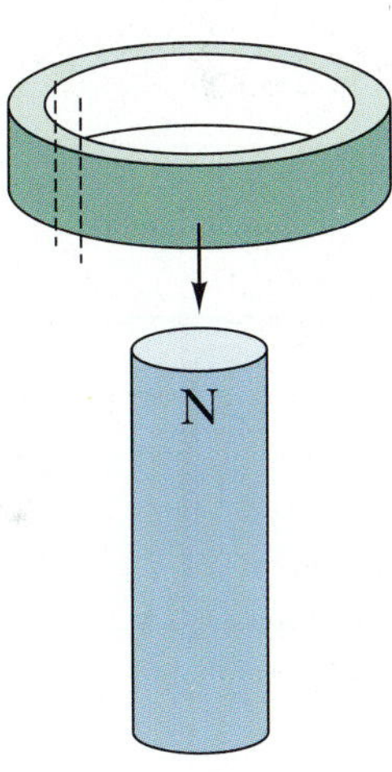

FIGURE 22.25
A conducting ring falling toward a bar magnet. The dashed lines indicate where a slot will be cut through the ring.

PROBLEMS

Section 22.1: Motional Emf

1.——An automobile travels at 88 km/h along a level road. The vertical downward component of the Earth's magnetic field is 0.58×10^{-4} T. What is the induced emf between the right and the left door handles separated by a distance of 2.1 m? Which side is positive and which negative?

2.——The DC-10 jet aircraft has a wingspan of 47 m. If such an aircraft is flying horizontally at 960 km/h at a place where the vertical component of the Earth's magnetic field is 0.60×10^{-4} T, what is the induced emf between its wingtips?

3.——In order to detect the movement of water in the ocean, oceanographers sometimes rely on the motional emf generated by this movement of the water through the magnetic field of the Earth. Suppose that, at a place where the vertical magnetic field is 0.70×10^{-4} T, two electrodes are immersed in the water separated by a distance of 200 m measured perpendicularly to the movement of the water. If a sensitive voltmeter connected to the electrodes indicates a potential difference of 7.0×10^{-3} V, what is the speed of the water?

4.——The rate of flow of a conducting liquid, such as detergent, tomato pulp, beer, liquid sodium, sewage, etc., can be measured with an electromagnetic flowmeter that detects the emf induced by the motion of the liquid in a magnetic field. Suppose that a plastic pipe of diameter 10 cm carries beer with a speed of 1.5 m/s. The pipe is in a transverse magnetic field of 1.5×10^{-2} T. What emf will be induced between the opposite sides of the column of liquid?

5.——An experiment attempted with the Space Shuttle, but not completed because of mechanical difficulties, was designed to obtain electric power from the motional emf induced by the motion of the Space Shuttle through the Earth's magnetic field. While in orbit around the Earth at an

Sharks have electric sensors below their skin.

altitude of 296 km at a speed of 7.7 km/s, a 20-km long wire was to be stretched radially outward between the Space Shuttle and a small "tethered satellite." The magnetic field of the Earth at the altitude of the Space Shuttle is 2.7×10^{-5} T. Calculate the magnitude of the motional emf induced between the ends of such a 20-km wire in this magnetic field. Assume that the motion of the wire is at right angles to the magnetic field.

6.——A boxcar of a train is 2.5 m wide. The boxcar travels at 60 km/h on a level track at a place where the vertical component of the Earth's magnetic field is 0.62×10^{-4} T. What is the induced emf between the sides of the boxcar?

7.*——Sharks have delicate sensors on their bodies that permit them to sense small differences of potential. They can sense electrical disturbances created by other fish and they can also sense the Earth's magnetic field and use this for navigation. Suppose that a shark is swimming horizontally at 25 km/h at a place where the magnetic field has a strength of 4.7×10^{-5} T and points down at an angle of 40° with the vertical. Treat the shark as a cylinder of diameter 30 cm. What is the largest induced emf between diametrically opposite points on the sides of the shark when heading north?

8.*——A ***homopolar generator*** consists of a disk of metal spinning in a magnetic field perpendicular to the face of the disk. The disk has a radius R and spins with angular velocity ω; and the magnetic field has a strength B. What is the motional emf induced between the axis of the disk and its rim? (Hint: Any radius of the disk can be regarded as a conducting rod.)

9.*——A helicopter has blades of length 4.0 m rotating at 3 rev/s in a horizontal plane. If the vertical component of the Earth's magnetic field is 0.65×10^{-4} T, what is the induced emf between the tip of the blade and the hub?

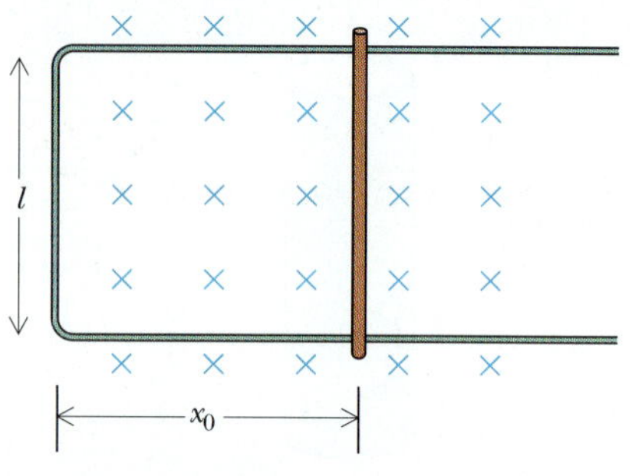

FIGURE 22.26

Rod sliding on parallel tracks. The crosses show the tails of the magnetic field vectors.

10.**——A metal rod of length l and mass m is free to slide, without friction, on two parallel metal tracks. The tracks are connected at one end so that they and the rod form a closed circuit (Fig. 22.26). The rod has a resistance R, and the tracks have negligible resistance. A uniform magnetic field is perpendicular to the plane of this circuit. The magnetic field is increasing at a constant rate $\Delta B/\Delta t$. Initially the magnetic field has strength B_0 and the rod is at rest at a distance x_0 from the connected end of the rails. Express the acceleration of the rod at this instant in terms of the given quantities.

Section 22.2: Faraday's Law
Section 22.3: Some Examples; Lenz' Law

11.——The plane of a coil is initially at right angles to the direction of an applied magnetic field. By what angle must the coil be turned to *reverse* the direction of the flux and to reduce the magnitude of the flux through the coil to 30% of its initial value?

12.——In Idaho, the magnetic field of the Earth points downward at an angle of 69° below the horizontal. The strength of the magnetic field is 0.59×10^{-4} T. What is the magnetic flux through 1 m² of ground in Idaho?

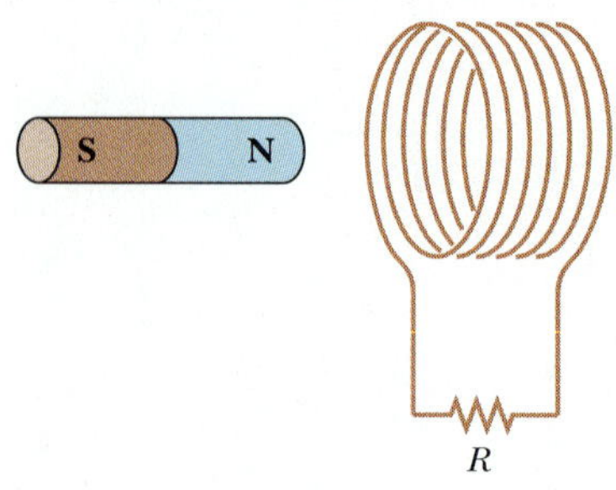

FIGURE 22.27

Bar magnet and coil.

13.——The bar magnet shown in Fig. 22.27 is thrust through the coil at a constant speed. Consider the following stages: (i) when the north pole approaches the coil, (ii) when the magnet is centered in the coil, and (iii) when the south pole leaves the right side of the coil. What is the direction of the current in the resistor at each of these stages?

14.——An electromagnetic generator of the kind described in Example 4 has a coil of area 2×10^{-4} m² with 300 turns of wire. What is the amplitude of the alternating emf delivered by this coil when rotating at a rate of 2000 revolutions per minute in a magnetic field of 0.02 T?

15.——An electric generator consists of a rectangular loop of wire rotating about its longitudinal axis which is perpendicular to a magnetic field of 2.0×10^{-2} T. The loop measures 10.0 cm $\times$ 20.0 cm and it has 120 turns of wire. The ends of the wire are connected to an external circuit. At what speed (in rev/s) must you rotate this loop in order to induce an alternating emf of amplitude 12.0 V between the ends of the wire?

16.*——A 25-turn coil of wire has an area of 4.0×10^{-3} m^2 and it is oriented perpendicularly to a magnetic field **B**. The coil has a resistance of 15 Ω. At what rate must the magnitude of **B** change for an induced current of 5.0 mA to appear in the coil?

17.*——(a) A long solenoid has 300 turns of wire per meter and it has a radius of 3.0 cm. If the current in the wire is increasing at the rate of 50 amperes per second, at what rate does the strength of the magnetic field in the solenoid increase?

(b) The solenoid is surrounded by a coil of wire with 120 turns (Fig. 22.28). The radius of this coil is 6.0 cm. What induced emf will be generated in this coil while the current in the solenoid is increasing?

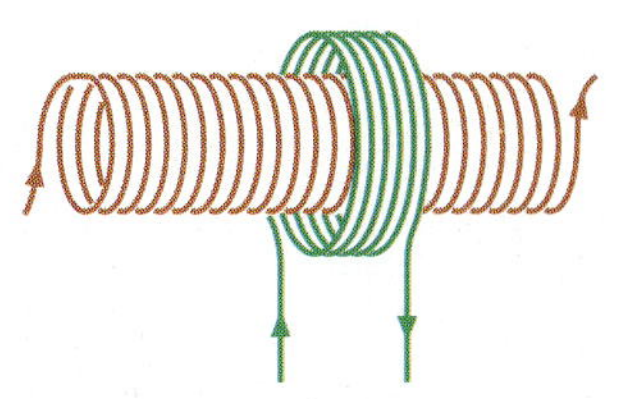

FIGURE 22.28
A long solenoid surrounded by a circular coil.

18.*——A girl uses a jump rope made of flexible wire. She holds the ends of the rope in her extended hands, so the rope has approximately the shape of a semicircle of radius 0.7 m, and she whirls the rope at the rate of 1 revolution per second at a place where the magnetic field of the Earth is nearly vertical, of magnitude 5×10^{-5} T. The motion of the wire rope generates an alternating emf between the ends of the rope. What is the amplitude of oscillation of this alternating emf?

19.*——A circular loop of wire is placed in a magnetic field of 0.30 T while the free ends of the wire are attached to a 15-Ω resistor as shown in Fig. 22.29. When you squeeze the loop, the area of the loop is reduced from 200 to 100 cm^2 in 0.020 s. What are the magnitude and direction of the current in the resistor?

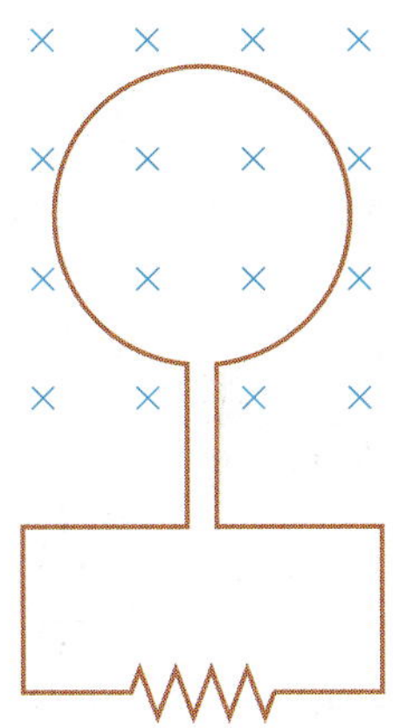

FIGURE 22.29
Circular loop in magnetic field.

20.*——Large superconducting magnets are used in some hospitals to produce pictures of the interior of the body by magnetic resonance imaging (MRI). For this purpose, the patient is shoved within the coils of the magnet, where the magnetic field is 1.5 T (see Fig. 22.30). Suppose that the patient is shoved into the magnetic field in a time of 10 s. Estimate the emf induced around the patient's trunk, 0.90 m in circumference. Should the patient be shoved into the magnetic field more slowly?

21.*——A washer (annulus) of aluminum is lying on top of a vertical solenoid (see Fig. 22.31). When the current in the solenoid is suddenly switched on, the washer flies upward. Carefully explain why the end of the solenoid exerts a repulsive force on the washer under these conditions. (Hint: Take into account that, at the end of the solenoid, the magnetic field lines spread out.)

22.*——A square loop of dimension 8.0 cm $\times$ 8.0 cm is made of copper wire of

FIGURE 22.30
Large magnet used for MRI.

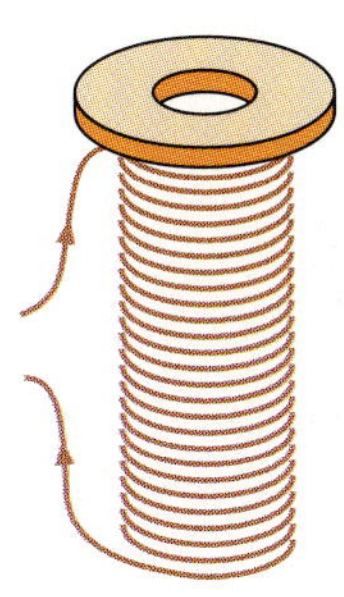

FIGURE 22.31
Washer lying on top of a solenoid.

radius 0.10 cm. The loop is placed face-on in a magnetic field which is increasing at the constant rate of 80 T/s. What induced current will flow around the loop? Draw a diagram showing the direction of the field and the induced current.

23.*——A very long solenoid with 20 turns per centimeter of radius 5.0 cm is surrounded by a rectangular loop of copper wire. The rectangular loop measures 10 cm $\times$ 30 cm and its wire has a radius of 0.050 cm. The resistivity of copper is $1.7 \times 10^{-8}\ \Omega \cdot \text{m}$. Find the induced current in the rectangular loop if the current in the solenoid is increasing at the rate of 5.0×10^4 A/s.

24.**——A very long straight wire carries a current that increases at a steady rate $\Delta I/\Delta t$.

(a) What is the rate of increase of the magnetic field at a radial distance r?

(b) What is the induced emf around the rectangular loop shown in Fig. 22.32? Assume that the width of the loop is small so the magnetic field, and its rate of change, is approximately the same at all points within the loop.

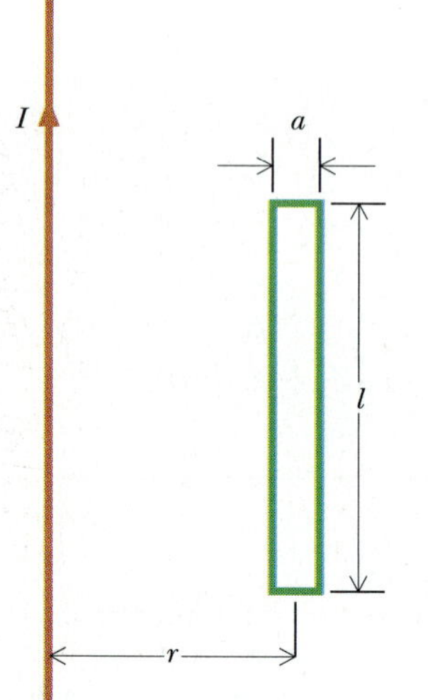

FIGURE 22.32

Long, straight wire and rectangular loop.

25.**——A circular coil of insulated wire has a radius of 9.0 cm and contains 60 turns of wire. The ends of the wire are connected in series with a 15-Ω resistor closing the circuit. The normal to the loop is initially parallel to a constant magnetic field of 5.0×10^{-2} T. If the loop is flipped over, so that the direction of the normal is reversed, a pulse of current will flow through the resistor. What amount of charge will flow through the resistor? Assume that the resistance of the wire is negligible compared with that of the resistor. (Hint: Suppose the flipping takes a time Δt. What is the average rate of change of the magnetic flux? The average emf? The average current?)

26.**——A flux meter used to measure magnetic fields consists of a small coil of radius 0.80 cm wound with 200 turns of fine wire. The coil is connected by means of a pair of tightly twisted trailing wires to a galvanometer that measures the electric charge flowing through the coil. The resistance of the coil–wire–galvanometer circuit is 6.0 Ω. Suppose that when the coil is quickly moved from a place outside the magnetic field to a place inside the magnetic field the galvanometer registers a flow of charge of 0.30 C. What is the magnetic flux through the coil? What is the component of **B** perpendicular to the face of the coil? Assume that **B** is uniform over the area of the coil. (Hint: Suppose that the coil is moved in a time Δt. What is the average rate of change of the magnetic flux? The average induced emf? The average current?)

Section 22.4: Inductance

27.——A loop of wire carrying a current of 100 A generates a magnetic flux of $50\ \text{T} \cdot \text{m}^2$ through the area bounded by the loop.

(a) What is the self-inductance of the loop?

(b) If the current is decreased at the rate of 20 A/s, what is the induced emf?

28.——Two coils are arranged face-to-face, as in Fig. 22.18. Their mutual inductance is 2.0×10^{-2} H. The current in coil 1 is increasing at the rate of 4.5×10^3 A/s.

(a) What is the magnetic flux that this current generates in coil 2?

(b) What is the induced emf that this current induces in coil 2?

(c) What is the direction of the induced current in coil 2, according to Lenz' Law? Assume that the positive direction for the current I_1 is as shown by the arrows in the figure.

29.——A current of 15 A in a coil produces a magnetic flux of $0.10\ \text{T} \cdot \text{m}^2$

through each of the turns of an adjacent coil of 60 turns. What is the mutual inductance?

30.*——A long solenoid of radius R has n turns per unit length. A circular coil of wire of radius R' with 200 turns surrounds the solenoid (Fig. 22.33). What is the mutual inductance? Does the shape of the coil of wire matter?

31.*——A long solenoid has 400 turns per meter. A coil of wire of radius 1.0 cm with 30 turns of insulated wire is placed inside the solenoid, its axis parallel to the axis of the solenoid. What is the mutual inductance? What emf will be induced around the coil if the current in the solenoid windings changes at the rate of 200 A/s?

32.*——Two long concentric solenoids of n_1 and n_2 turns per unit length have radii R_1 and R_2, respectively (Fig. 22.34). What is the mutual inductance per unit length of the solenoids? Assume $R_1 < R_2$.

33.——What is the self-inductance of a solenoid 0.1 m long, 1 cm in radius, with 2×10^4 turns per meter of length?

34.——A long solenoid has 2000 turns per meter and a radius of 2.0 cm.
   (a) What is the self-inductance for a 1.0-m segment of this solenoid?
   (b) What back emf will this segment generate if the current in the solenoid is changing at the rate of 3.0×10^2 A/s?

35.*——The resistance of the coils of the electric motor of a record player is 2.0 Ω when the motor is not in operation (not rotating). When the motor is connected to 115 V and is rotating at full speed, it draws a current of 0.1 A. Deduce the back emf that the coils of the motor produce when rotating.

36.*——The current in a 115-volt electric motor jumps from 4 A to 36 A when the motor coils are suddenly halted by a brake. What is the back emf when the motor is rotating at full speed?

37.*——A solenoid of self-inductance 2.2×10^{-3} H in which there is initially no current is suddenly connected in series with the poles of a 24-V battery. What is the instantaneous rate of increase of the current in the solenoid?

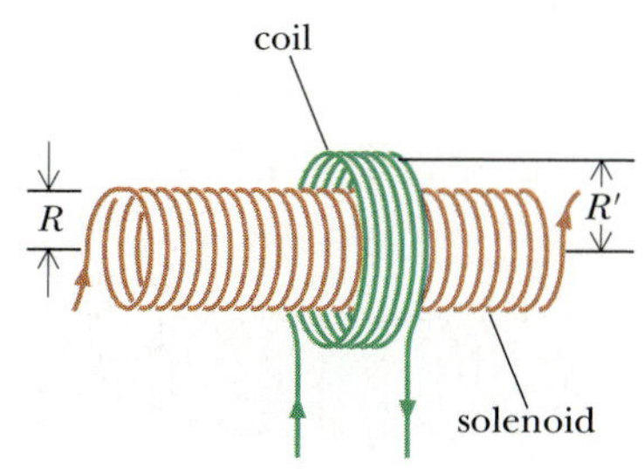

FIGURE 22.33
A long solenoid and a circular coil.

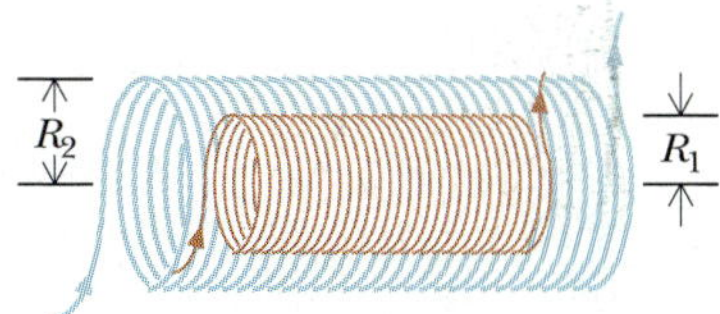

FIGURE 22.34
Two long concentric solenoids.

Section 22.5: Magnetic Energy

38.——A ring of thick wire has a self-inductance of 4.0×10^{-6} H. How much work must you do to establish a current of 25 A in this ring?

39.——Suppose that the magnetic energy stored in an inductor is 2.0×10^{-3} J when the current in this inductor is 30 A. What will be the magnetic energy of this same inductor if the current is 60 A? 90 A?

40.——The self-inductance of the aluminum cable of a high-voltage transmission line is 6.6×10^{-4} H per kilometer of length. If the cable carries a current of 800 A, what is the energy in the magnetic field of the cable, per kilometer?

41.——The strongest magnetic field achieved in a laboratory is 10^3 T. This field can be produced only for a short instant by compressing the magnetic field lines with an explosive device. What is the energy density in this field?

42.——For each of the first four entries in Table 21.1, calculate the energy density in the magnetic field.

43.——In a region of vacuum containing a magnetic field of 1 T and an electric field of 1 V/m, which field has the larger energy density?

44.——A large toroid built in the early days of plasma research in the former Soviet Union had a major radius of 1.50 m and a minor radius of 0.40 m, and a volume of 5.7 m^3 (Fig. 22.35). If the average magnetic field within such a toroid is 4.0 T, what is the magnetic energy?

45.*——For a crude estimate of the energy in the Earth's magnetic field, pretend that this field has a strength of 0.5×10^{-4} T from the ground up to an altitude of 6×10^6 m above ground. What is the total magnetic energy in this region?

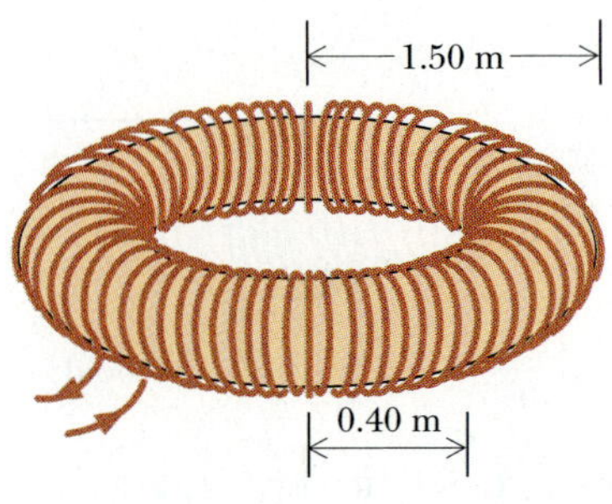

FIGURE 22.35
A toroid.

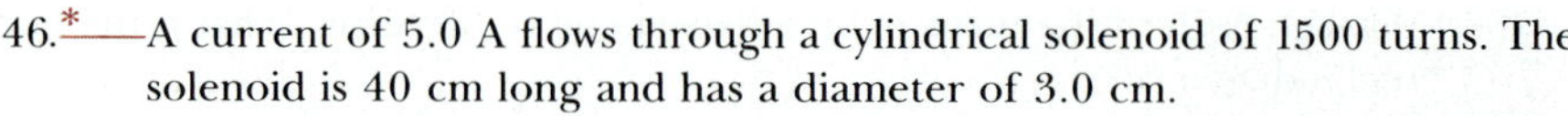

46.* A current of 5.0 A flows through a cylindrical solenoid of 1500 turns. The solenoid is 40 cm long and has a diameter of 3.0 cm.

(a) Find the magnetic field in the solenoid. Treat the solenoid as very long.

(b) Find the energy density in the magnetic field and find the magnetic energy stored in the space within the solenoid.

47.* According to one proposal, the surplus energy from a power plant could be temporarily stored in the magnetic field within a very large toroid. If the strength of the magnetic field is 10 T, what volume of the magnetic field would we need to store 1.0×10^5 kW·h of energy? If the toroid has roughly the proportions of a doughnut, roughly what size would it have to be? (Hint: The volume of a toroid equals the cross-sectional area multiplied by the average radius.)

Section 22.6: The RL Circuit

48. An inductor with $L = 2.0$ H and a resistor with $R = 100\ \Omega$ are suddenly connected in series to a battery with $\mathscr{E} = 6.0$ V.

(a) What is the current at $t = 0$?

(b) What is the final, steady value of the current?

(c) What is the rate of increase of the current at $t = 0$?

49. Design an RL circuit with an arrangement of switches so that the battery can be suddenly switched out of the circuit and the current in the inductor can be suddenly fed into the resistor. If the self-inductance of the inductor is 0.2 H, what resistance do you need to obtain a time constant of 10 s in your circuit?

50.* An inductor with an initial current of 10 A is suddenly connected in series to a resistor with $R = 10\ \Omega$. After a time of 6.0×10^{-2} s, the current has dropped to 3.7 A. What is the self-inductance?

51.* An RL circuit with $L = 0.50$ H and $R = 0.025\ \Omega$ is initially connected to a battery of 1.2 V. When the current reaches its maximum, steady value, the battery is suddenly switched out of the circuit and the current is switched into the resistor.

(a) What is the maximum value of the current? At what time will the current in the inductor drop to 37% of its maximum value?

(a) What is the maximum value of the energy stored in the inductor? What percentage of the energy remains when the current has dropped to 37% of its maximum value?

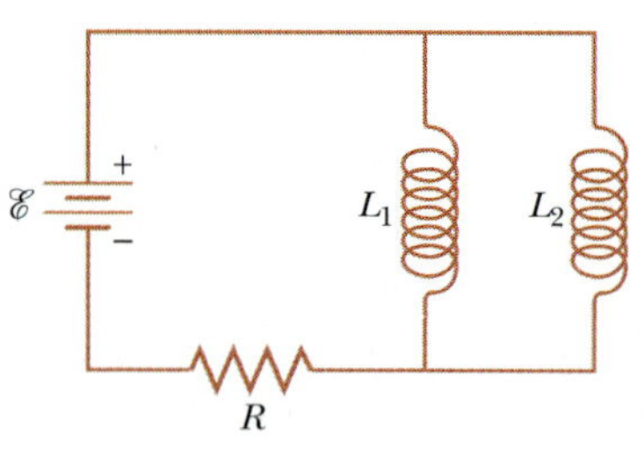

FIGURE 22.36

52.* An RL circuit consists of two inductors of self-inductances $L_1 = 4$ H and $L_2 = 2$ H connected in parallel to each other, and connected in series to a resistor of 6 Ω and a battery of 3 V (Fig. 22.36). Assume that the inductors have no mutual inductance.

(a) When the battery is suddenly connected, what is the initial rate of change of the current in each inductor?

(b) What is the final, steady current in the resistor? What are the final, steady currents in each inductor?

53.* To measure the self-inductance and the internal resistance of an inductor, a physicist first connects the inductor across a 3.0-V battery. Under these conditions, the final, steady current in the inductor is 24 A. The physicist then suddenly short-circuits the inductor with a thick (resistanceless) wire placed across its terminals. The current then decreases from 24 A to 8.9 A in 0.22 s. What are the self-inductance and the internal resistance of the inductor?

CHAPTER 23*

These cables, suspended from ceramic insulators, deliver high-voltage AC power from a generating station to a factory.

AC CIRCUITS

The current delivered by power companies to homes and factories is an oscillating, time-dependent current. It periodically flows forward and backward, 60 times per second. This is called ***alternating current, or AC.*** Power companies prefer alternating currents to direct currents because of the ease with which alternating voltages can be stepped up or down by means of transformers. This makes it possible to step up the output of a power plant to several hundred thousand volts, transmit the power along a high-voltage line that minimizes the Joule heating losses, and finally step down the power to 230-volt AC or 115-volt AC just before delivery to the consumer (see Fig. 23.1).

◂ **Alternating current (AC)**

All the appliances connected to ordinary outlets in homes therefore involve circuits with oscillating currents. Furthermore, electronic devices — radio receivers, TV receivers, computers — involve a variety of circuits

* This chapter is optional.

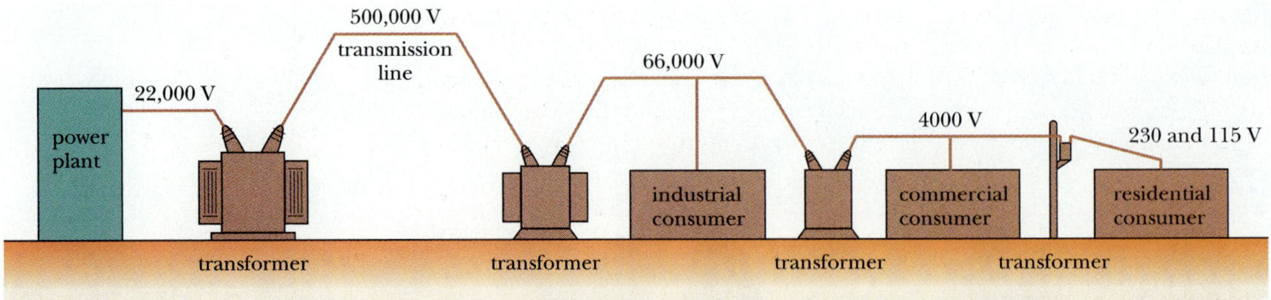

FIGURE 23.1

Typical voltage transformations during the transmission of electric power. The small transformers used to step the power down to 115 V for residential use are often attached to telephone poles.

with oscillating currents of high frequency. Many of these circuits have a natural frequency of oscillation. Such circuits exhibit the phenomenon of resonance when the natural frequency matches the frequency of a signal applied to the circuit. For instance, the tuning of a radio relies on an oscillating circuit whose frequency of oscillation is adjusted by means of a variable capacitor (attached to the tuning knob), so that it matches the frequency of the radio signal.

The calculations of currents in AC circuits are similar to the calculations in DC circuits. As we will see, the starting point for these calculations is again Kirchhoff's Rule.

23.1 Resistor Circuit

In the United States, the alternating emf supplied by the power companies to electric outlets in private homes is an oscillating emf with an amplitude $\mathscr{E}_{\text{max}} = 163$ V and a frequency of 60 Hz. Thus the angular frequency is $\omega = 2\pi \times 60$ radians/s and the time dependence of the emf is given by the trigonometric function

$$\mathscr{E} = \mathscr{E}_{\text{max}} \cos \omega t = 163 \text{ V} \times \cos \omega t \tag{1}$$

Figure 23.2 shows a plot of this emf as a function of time. This kind of voltage is usually called "115-V AC" for reasons that will become clear

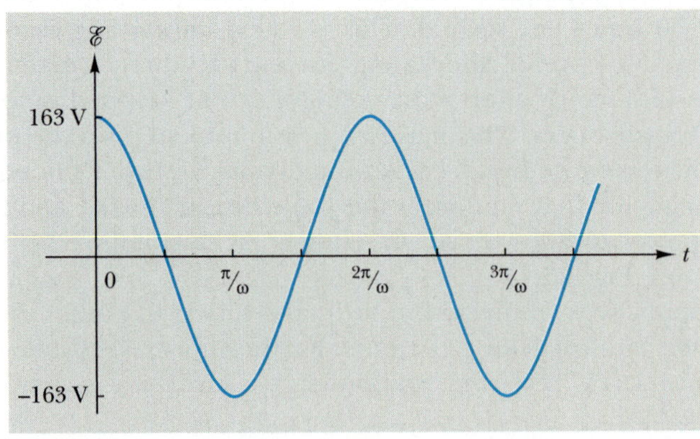

FIGURE 23.2

Emf supplied by an outlet as a function of time.

shortly. In many European countries, the amplitude of the AC voltage supplied to outlets in homes is twice as large; it is "230-V AC." An appliance designed for 115-V AC is likely to overheat and burn out if plugged into 230-V AC. To prevent this, the plugs for 115-V and 230-V appliances are built in different shapes, so they cannot fit into the wrong kind of outlet.

FIGURE 23.3
Resistor connected to a source of alternating emf.

The simplest possible AC circuit consists of a pure resistor connected to an oscillating source of emf. In the circuit diagram (see Fig. 23.3), the source of emf is symbolized by a wavy line enclosed in a circle. This circuit might represent an electric heater or an incandescent lamp plugged into an ordinary wall outlet.

According to Kirchhoff's Rule, the sum of emf's and potential changes across resistors in any circuit must be zero. Although we first developed this law for DC circuits, it is equally valid for AC circuits. For the circuit shown in Fig. 23.3, this tells us that, at any instant of time,

$$\mathscr{E} - IR = 0 \tag{2}$$

from which we find

◂ Current in resistor circuit

$$I = \frac{\mathscr{E}}{R} = \frac{\mathscr{E}_{\max} \cos \omega t}{R} \tag{3}$$

Thus, ***the current oscillates with the same frequency*** ω ***as the emf, but the amplitude of the current is*** $\mathscr{E}_{\max}/R$. The maxima and the minima of the current and the emf occur simultaneously, and the current is said to be ***in phase*** with the emf. Figure 23.4 compares plots of the emf and the current.

The instantaneous electric power dissipated in the resistor is the product of the instantaneous current and the emf, as in the case of a DC circuit,

$$P = I\mathscr{E} = \frac{\mathscr{E}_{\max} \cos \omega t}{R} \times \mathscr{E}_{\max} \cos \omega t$$

$$= \frac{\mathscr{E}_{\max}^2 \cos^2 \omega t}{R} \tag{4}$$

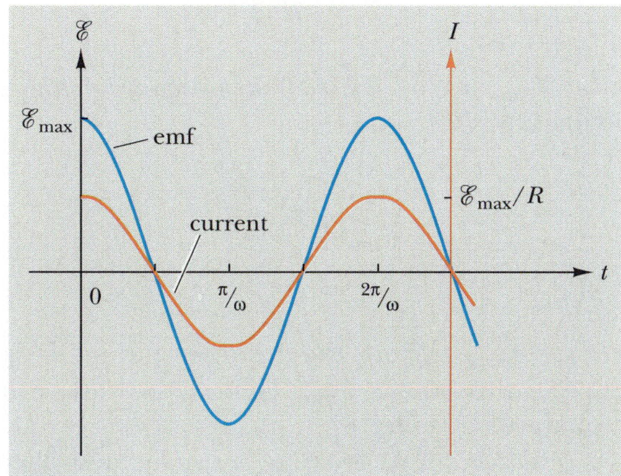

FIGURE 23.4
The emf (blue) and the current (red) in the resistor circuit as a function of time. The maxima and the minima of the emf and the current occur simultaneously.

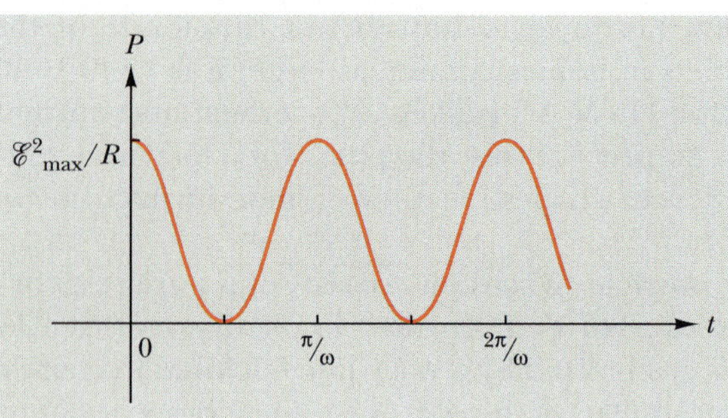

FIGURE 23.5
Instantaneous power dissipated in the resistor as a function of time. Note that the frequency of oscillation of the power is twice that of the emf (in one cycle of the emf, the power goes through two cycles).

Although the emf and the current both are negative for one-half of each cycle, the power is always positive, because the emf and the current become negative simultaneously (the emf and the current are in phase), and therefore their product is always positive. Figure 23.5 shows how this power oscillates between zero and a maximum value $\mathscr{E}^2_{max}/R$.

To determine the long-term power dissipated by the resistor, we do not care about the oscillations of the power; we care only about the ***time-average power.*** This time-average power can be obtained by averaging the quantity $\cos^2\omega t$ in Eq. (4) over one cycle of oscillation. This average can be evaluated by noting that the average of $\cos^2\omega t$ is the same as the average of $\sin^2\omega t$, because both of these quantities have the same number of ups and downs in one cycle. Using overbars to indicate averages, we therefore find

$$\overline{\cos^2\omega t} = \tfrac{1}{2}\,(\overline{\cos^2\omega t} + \overline{\cos^2\omega t}) = \tfrac{1}{2}\,(\overline{\cos^2\omega t} + \overline{\sin^2\omega t}) \tag{5}$$

But $\cos^2\omega t + \sin^2\omega t = 1$, and hence Eq. (5) tells us that $\overline{\cos^2\omega t} = \frac{1}{2}$. The time-average power is then

$$\overline{P} = \frac{\mathscr{E}^2_{max}}{2R} \tag{6}$$

This is usually written in the form

Time-average power absorbed by resistor ▶

$$\overline{P} = \frac{\mathscr{E}^2_{rms}}{R} \tag{7}$$

where the quantity $\mathscr{E}_{rms}$, called the ***root-mean-square voltage,*** is the maximum voltage divided by $\sqrt{2}$,

Root-mean-square voltage ▶

$$\mathscr{E}_{rms} = \frac{\mathscr{E}_{max}}{\sqrt{2}} \tag{8}$$

Note that since $\mathscr{E}^2 = \mathscr{E}^2_{max}\cos^2\omega t$, the average value of $\mathscr{E}^2$ is $\mathscr{E}^2_{max}/2$ and, according to Eq. (8), this equals $\mathscr{E}^2_{rms}$. Hence, the square of $\mathscr{E}_{rms}$ is the average of the square of $\mathscr{E}$.

In engineering practice, an AC voltage is usually described in terms of $\mathscr{E}_{rms}$. For example, if $\mathscr{E}_{max} = 163$ V, then $\mathscr{E}_{rms} = 163/\sqrt{2}$ V $= 115$ V. An oscillating voltage with this value of $\mathscr{E}_{max}$ is described as "115-volt AC." Ordinary voltmeters and ammeters are calibrated so as to display the rms voltage and the rms current when connected to an AC circuit. Thus, a voltmeter plugged into a wall outlet will read 115 V.

Comparison of Eq. (7) and Eq. (20.22) shows that the average AC power dissipated in the resistor is equal to the DC power dissipated in the same resistor when connected to a steady DC voltage of magnitude $\mathscr{E}_{rms}$. Thus, ***a voltage of 115-volt AC*** (with $\mathscr{E}_{max} = 163$ V) ***delivers the same average power to the resistor as 115-volt DC.*** This means that in any calculations involving the electric power dissipated by resistors in such an AC circuit we can pretend that we are dealing with a 115-volt DC circuit.

EXAMPLE 1. A 115-V AC incandescent light bulb is rated at 150 W. What is the resistance of this light bulb (when at its operating temperature)?

SOLUTION: We have $\mathscr{E}_{rms} = 115$ V and $\overline{P} = 150$ W. Hence Eq. (7) gives us

$$R = \frac{\mathscr{E}_{rms}^2}{\overline{P}} = \frac{(115\text{ V})^2}{150\text{ W}} = 88\ \Omega \tag{9}$$

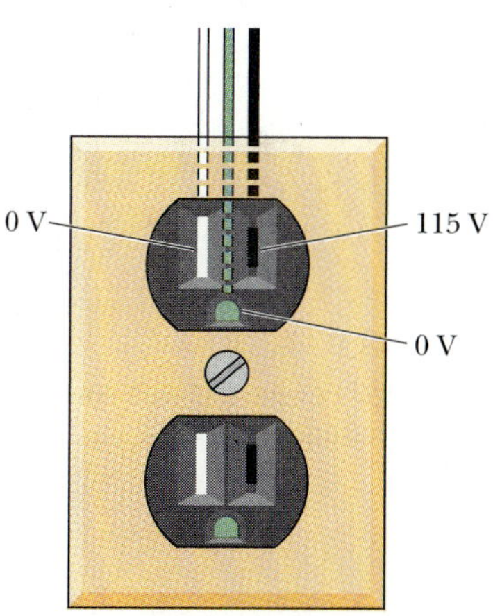

FIGURE 23.6
Standard power outlet for three-prong plug.

When several appliances are plugged into the same outlet in a house, or into different outlets that are part of the same circuit in the house, all the appliances will be connected in parallel across the supplied emf of 115-volt AC. In ordinary wall outlets for three-prong plugs, such as illustrated in Fig. 23.6, the small slot is at a potential of 115-volt AC, the large slot is at a potential of zero, and the round slot is also at a potential of zero. The two flat slots are connected to the two wires coming from the power station, and, when an appliance is plugged into the outlet, the current flows into the appliance through one slot and out through the other. The round slot is connected to the ground, usually via a grounding plate buried just outside the house. Under normal circumstances, this slot carries no current. It comes into play only when there is a failure in the electric insulation of the appliance; then the ground slot permits the leaking currents to spill away into the ground (instead of spilling into your hands).

◀ **Open-circuit potentials**

Note that the potentials indicated in Fig. 23.6 are ***open-circuit potentials,*** that is, they are the potentials before an appliance is plugged into the outlet. When an appliance drawing a heavy current — such as an electric heater or an air conditioner — is plugged into the outlet, the potential at the small flat slot may drop by several volts, and the potential at the large flat slot will rise by an equal amount. These changes in potential are due to the resistance of the wires connecting the outlets to the power station. The wires and the appliance form a series circuit, and the total potential drop of 115 V provided by the power station is distributed among these wires and the appliance in direct proportion to their resistances, as required by Ohm's Law.

23.2 Capacitor Circuit

A capacitor connected to an alternating source of emf behaves very differently from a capacitor connected to a constant source of emf, such as a battery. When we connect the capacitor to the battery, there is an initial, brief surge of current that charges the capacitor. The current deposits positive charge on one plate of the capacitor, and negative charge on the other, and thereby increases the voltage across the capacitor. But once the voltage across the charged capacitor matches the emf, no further current flows, and the capacitor thereafter remains in equilibrium.

In contrast, when we connect a capacitor to an alternating source of emf, the current begins to deposit positive charges on one plate and negative charges on the other. But half a cycle later, the emf reverses, and so does the current. Thus, the current now removes the charges it deposited earlier, and begins to deposit reversed charges on the plates. This reversal of the current and of the charges repeats each half cycle. The current flows back and forth along the wires from the emf to the plates, and the signs of the charges on the plates alternate with the same frequency as the emf. Note that the capacitor permits this periodic back-and-forth flow of the current on the wires connected to its plates, even though no current can cross the gap between the plates.

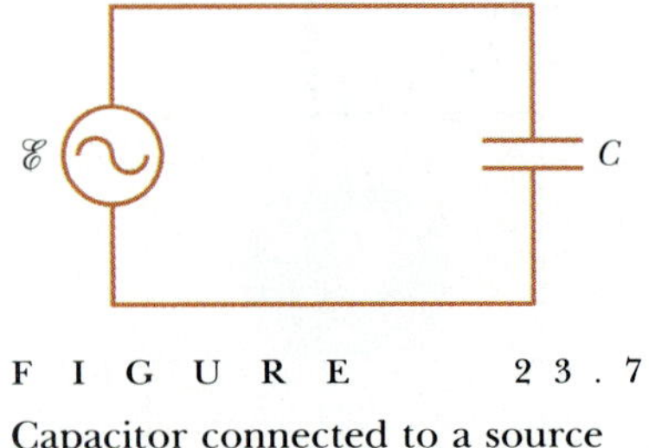

FIGURE 23.7
Capacitor connected to a source of alternating emf.

Figure 23.7 shows the circuit diagram for a capacitor connected to our oscillating source of emf. The instantaneous voltage across the capacitor is Q/C and therefore Kirchhoff's Rule gives

$$\mathscr{E} - Q/C = 0 \tag{10}$$

With the expression (1) for $\mathscr{E}$, this yields

$$Q = C\mathscr{E} = C\mathscr{E}_{\text{max}} \cos\omega t \tag{11}$$

The current in the circuit is the rate of change of the charge, $I = \Delta Q/\Delta t$. This means we must evaluate the rate of change of $\cos\omega t$. We have already done so in Example 22.4, where we found that the rate of change of $\cos \omega t$ is $-\omega \sin \omega t$. Hence

$$I = -\omega C\mathscr{E}_{\text{max}} \sin\omega t \tag{12}$$

According to this equation ***the current oscillates sinusoidally with a frequency ω and with an amplitude $\omega C\mathscr{E}_{\text{max}}$***. Figure 23.8 shows plots of the emf and the current. Note that the current has a maximum magnitude ($\omega C\mathscr{E}_{\text{max}}$) when the emf has a minimum magnitude (0); and the current has minimum magnitude (0) when the emf has maximum magnitude ($\mathscr{E}_{\text{max}}$). Because of this, the current and the emf are said to be ***90° out of phase.***

It is customary to write Eq. (12) as

Current in capacitor circuit ▶

$$I = -\frac{\mathscr{E}_{\text{max}} \sin\omega t}{X_C} \tag{13}$$

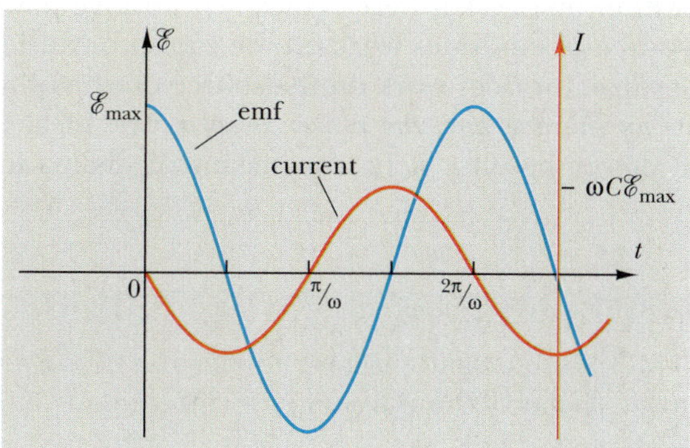

F I G U R E 23.8

The emf (blue) and the current (red) in the capacitor as a function of time. The maxima of the emf coincide with zeroes of the current.

where

$$X_C = \frac{1}{\omega C} \tag{14}$$

◀ **Capacitive reactance**

is called the ***capacitive reactance. This quantity X_C plays roughly the same role for a capacitor in an AC circuit as does the resistance for a resistor*** [compare Eqs. (3) and (13)]. Note, however, that the reactance depends not only on the characteristics of the capacitor, but also on the frequency at which we are operating the circuit. The unit of reactance is the ohm, as it is for resistance.

From Eq. (12) or (13) we see that the current is small if the frequency ω is low. This was to be expected, since it means that the response of the capacitor to a low-frequency emf is nearly the same as its response to a constant emf (DC). As we saw in the discussion at the beginning of this section, for a constant emf, the capacitor blocks the flow of current.

The instantaneous power delivered to the capacitor is, again, the product of the instantaneous current and the emf,

$$P = I\mathscr{E} = -\omega C\mathscr{E}_{max}^2 \sin\omega t \cos\omega t \tag{15}$$

The time dependence of this expression is contained in the factor $\sin\omega t$ $\cos\omega t$. According to a standard trigonometric identity, this equals $\frac{1}{2}\sin 2\omega t$, which shows that the power oscillates at a frequency of 2ω. But the important point is that the average power delivered is zero — as Fig.

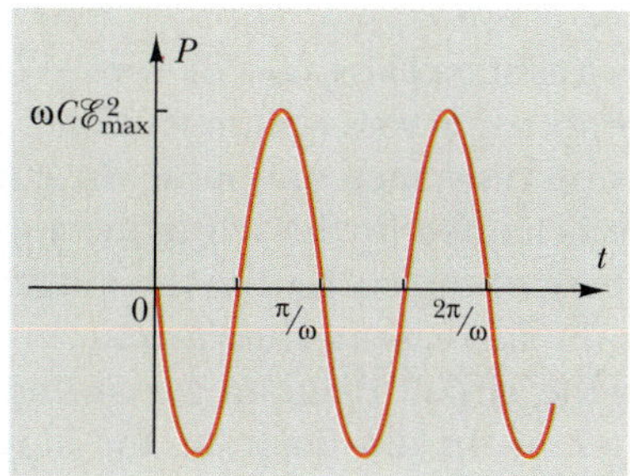

F I G U R E 23.9

Instantaneous power delivered to the capacitor as a function of time.

23.9 shows, within one cycle, there is as much positive power as negative power. The source of emf does work on the capacitor during parts of the cycle, but the capacitor does work on the source during other parts of the cycle, so that, ***on the average, the power is zero.*** The ideal capacitor does not consume power because it has no means of dissipating electric energy.

EXAMPLE 2. Suppose you plug the terminals of a 20-pF capacitor into a 115-V outlet. What is the maximum instantaneous current?

SOLUTION: The maximum instantaneous emf is $\mathscr{E}_{\text{max}} = 115\text{ V} \times \sqrt{2} = 163\text{ V}$. From Eq. (13) we see that the maximum instantaneous current occurs when $\sin \omega t = 1$ or -1, and it has the value

$$\frac{\mathscr{E}_{\text{max}}}{X_{\text{C}}} = \omega C \mathscr{E}_{\text{max}}$$

$$= (2\pi \times 60/\text{s}) \times (20 \times 10^{-12}\text{ F}) \times 163\text{ V} = 1.2 \times 10^{-6}\text{ A}$$

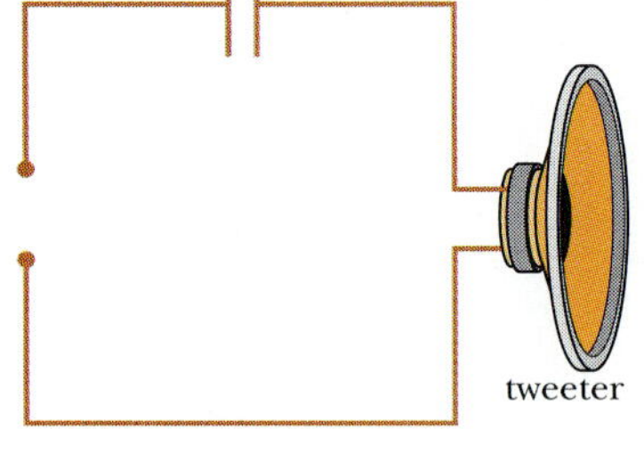

FIGURE 23.10
Capacitor connected in series with tweeter.

The selective blocking of low-frequency currents by a capacitor is exploited in hi-fi systems. As discussed in Section 11.5, hi-fi systems route high-frequency signals into small loudspeakers ("tweeters") and low-frequency signals into large loudspeakers ("woofers"). For this purpose, a capacitor is connected in series with the tweeter (see Fig. 23.10); this capacitor acts as a filter which blocks the low-frequency currents, but permits the passage of high-frequency currents. As we will see in the next section, an inductor has the opposite effect of a capacitor: the current through the inductor is small if the frequency is high. In the hi-fi system, an inductor is connected in series with the woofer (see Fig. 23.11); this inductor acts as a filter which blocks high-frequency currents, but permits the passage of low-frequency currents. Combinations of capacitors and inductors are used in many other kinds of electronic circuits that filter frequencies.

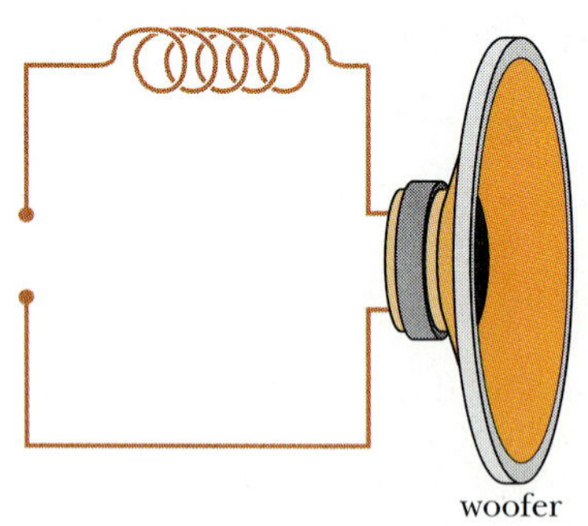

FIGURE 23.11
Inductor connected in series with woofer.

23.3 INDUCTOR CIRCUIT

An ideal inductor is a resistanceless coil of wire. We might expect that when such a coil is connected across a source of emf, a very large current will flow, since there is no resistance to oppose the current. However, the current in the inductor is limited by the self-inductance. As we know from Section 22.4, any increase of the current in the inductor generates a back emf, and this opposes the increase in the current.

Figure 23.12 shows the circuit diagram for an inductor connected to a source of alternating emf. In this diagram, the inductor is represented by a coiled line. The induced emf in the inductor, or the back emf, is $-L\Delta I/\Delta t$. By Kirchhoff's Rule, the sum of this back emf and the emf of the source must be zero,

FIGURE 23.12
Inductor connected to a source of alternating emf.

$$\mathscr{E} - L\frac{\Delta I}{\Delta t} = 0 \tag{16}$$

This equation gives us

$$\frac{\Delta I}{\Delta t} = \frac{\mathscr{E}}{L} = \frac{\mathscr{E}_{\text{max}} \cos \omega t}{L} \tag{17}$$

This tells us the rate of change of the current. To discover the current, we note that $\omega \cos\omega t$ is the rate of change of $\sin\omega t$ (this can be established by an argument similar to that which gave us $-\omega \sin\omega t$ for the rate of change of $\cos\omega t$; see Example 22.4); thus, the quantity $\cos\omega t$ appearing in Eq. (17) is the rate of change of $(1/\omega) \sin\omega t$. Equation (17) therefore asserts that the rate of change of the current equals a constant times the rate of change of $\sin\omega t$. But if the rates of change of two quantities are equal, and if the quantities are equal at some initial time, then the two quantities must themselves be equal. Hence the current must be a constant times $\sin\omega t$:

$$I = \frac{\mathscr{E}_{\text{max}} \sin \omega t}{\omega L} \tag{18}$$

This current oscillates sinusoidally with a frequency ω and an amplitude $\mathscr{E}_{\text{max}}/\omega L$.

Figure 23.13 gives plots of the current and the emf. Again, comparison of these plots shows that the current is 90° out of phase with the emf — when the current reaches its maximum magnitude ($\mathscr{E}_{\text{max}}/\omega L$), the emf reaches its minimum magnitude (0).

We can write Eq. (18) as

$$I = \frac{\mathscr{E}_{\text{max}} \sin \omega t}{X_L} \tag{19}$$

◀ **Current in inductor circuit**

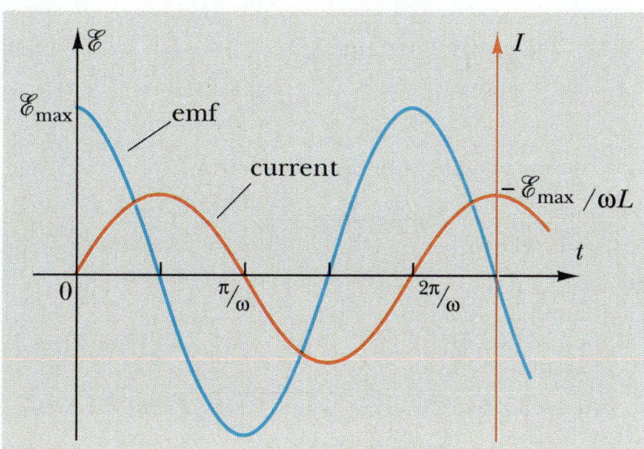

FIGURE 23.13
The emf (blue) and the current (red) in the inductor as a function of time. The maxima of the emf coincide with zeroes of the current.

where

Inductive reactance ▸

$$X_L = \omega L \tag{20}$$

is the ***inductive reactance.*** The unit of this reactance is, again, the ohm.

Note that according to Eq. (18) or (19), the amplitude of the current is small if the frequency ω is high, and the amplitude of the current is large if the frequency ω is low. This merely means that the response of the inductor to a low-frequency emf is close to its response to a steady emf — for a steady emf (DC), the inductor permits a very large current, since it has no resistance.

The instantaneous power delivered to the inductor is

$$P = I\mathscr{E} = \frac{1}{\omega L}\mathscr{E}_{max}^2 \sin\omega t \cos\omega t \tag{21}$$

As in the case of the capacitor, ***the average power is zero.***

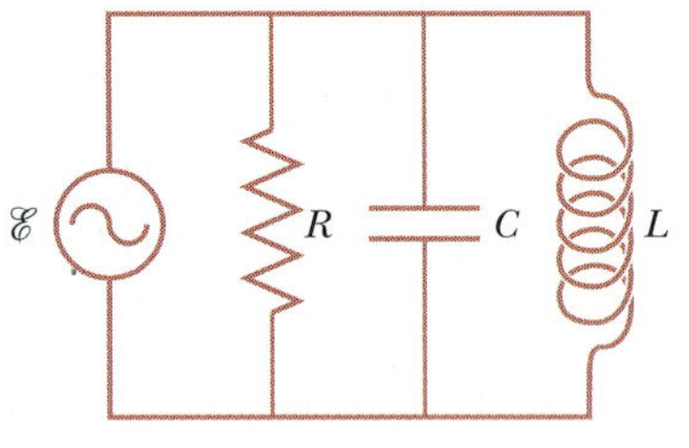

FIGURE 23.14
Resistor, capacitor, and inductor connected in parallel to a source of oscillating emf.

EXAMPLE 3. A resistor, a capacitor, and an inductor are connected in parallel (see Fig. 23.14) to a source of oscillating emf of frequency $\omega = 6 \times 10^3$ radians/s and amplitude 1×10^{-3} V. The resistance of the resistor is 200 Ω. (a) What is the maximum instantaneous current in the resistor? (b) If we want to make the maximum instantaneous currents in the capacitor and in the inductor equal to that in the resistor, what values of the capacitance and of the inductance must be select?

SOLUTION: (a) In a parallel connection, the source of emf supplies the same voltage across each of the circuit elements. Thus, the maximum instantaneous voltage across the resistor is 1×10^{-3} V, and the maximum instantaneous current in the resistor is, by Eq. (3),

$$\frac{\mathscr{E}_{max}}{R} = \frac{1 \times 10^{-3}\ \text{V}}{200\ \Omega} = 5 \times 10^{-6}\ \text{A}$$

(b) According to Eqs. (13) and (19), if the maximum instantaneous currents in the capacitor and the inductor are to match that in the resistor, their reactances must match the resistance:

$$X_C = X_L = 200\ \Omega$$

Thus,

$$\frac{1}{\omega C} = 200\ \Omega \quad \text{and} \quad \omega L = 200\ \Omega$$

which gives

$$C = \frac{1}{6 \times 10^3/\text{s} \times 200\ \Omega} = 8.3 \times 10^{-7}\ \text{F}$$

and

$$L = \frac{200\ \Omega}{6 \times 10^3/\text{s}} = 3.3 \times 10^{-2}\ \text{H}$$

COMMENTS AND SUGGESTIONS: Note that with this choice of C and L, the maximum currents are equal at the frequency of 6×10^3 radians/s only. At any other frequency, this choice of C and L will not result in equal maximum currents.

23.4 FREELY OSCILLATING LC CIRCUIT

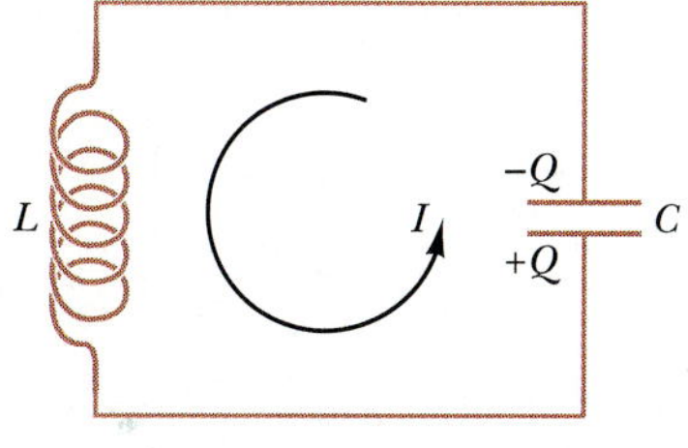

FIGURE 23.15
Inductor and capacitor connected in series.

An LC circuit consists of an inductor and a capacitor connected in series (see Fig. 23.15). The circuit has no source of emf; nevertheless, a current will flow in this circuit, provided that the capacitor is *initially charged*. The potential on one plate is then initially high and that on the other plate low. A current will begin to flow around the circuit from the positive plate to the negative. If the circuit had no inductance, the current would merely neutralize the charge on the plates, that is, the capacitor would discharge, and this would be the end of the current. But the inductor makes a difference: the inductor initially opposes the buildup of the current, but once the current has become established, the inductor will keep it going for some extra time. Hence *more* charge flows from one capacitor plate to the other than required for neutrality, and reversed charges accumulate on the capacitor plates. When the current finally stops, the capacitor will again be fully charged, with reversed charges. And then a reversed current will begin to flow, and so on. Thus, the charge sloshes back and forth around the circuit.

The LC system is analogous to a mass–spring system. ***The inductor is analogous to the mass — it tends to keep the current constant and provides "inertia." The charged capacitor is analogous to the stretched spring — it tends to accelerate the current and provides a "restoring force."***

The equation of motion of the LC system follows from Kirchhoff's Rule: the sum of emf's and other voltage changes around the circuit must add to zero. Going around the circuit in the direction of the arrow shown in Fig. 23.15, we find that the induced emf in the inductor (back emf) is

$$-L\frac{\Delta I}{\Delta t}$$

and the voltage across the capacitor is

$$-\frac{Q}{C}$$

Hence

$$-L\frac{\Delta I}{\Delta t}-\frac{Q}{C}=0 \tag{22}$$

or, eliminating the minus signs,

$$L\frac{\Delta I}{\Delta t}+\frac{Q}{C}=0 \tag{23}$$

Note that here Q is reckoned as positive when the charge on the lower plate is positive, and I is reckoned as positive when the charge on the lower plate is increasing.

Equation (23) has exactly the same mathematical form as the equation of motion for the simple harmonic oscillator [see Eq. (9.17)],

$$m\frac{\Delta v}{\Delta t}+kx=0$$

Comparing this with Eq. (23), we see that Q plays the role of x, whereas L replaces m, and $1/C$ replaces k. The current ($I=\Delta Q/\Delta t$) plays the role of the velocity ($v=\Delta x/\Delta t$). Hence the solution of Eq. (23) can be found by recalling the solution for the simple harmonic oscillator [see Eq. (9.22)],

$$x=A\cos\left(\sqrt{\frac{k}{m}}t\right)$$

With the above replacements for x, m, and k, we immediately obtain

$$Q=Q_0\cos\left(\frac{1}{\sqrt{LC}}t\right) \tag{24}$$

where Q_0 is the amount of charge on the positive plate at time $t=0$. Furthermore, according to the equation for the velocity of the simple harmonic oscillator [see Eq. (9.24)],

$$v=\frac{\Delta x}{\Delta t}=-\sqrt{\frac{k}{m}}A\sin\left(\sqrt{\frac{k}{m}}t\right)$$

With the above replacements, we then find that the current in the LC circuit must be

$$I=\frac{\Delta Q}{\Delta t}=-\frac{Q_0}{\sqrt{LC}}\sin\left(\frac{1}{\sqrt{LC}}t\right) \tag{25}$$

According to Eqs. (24) and (25), the charge and the current oscillate with

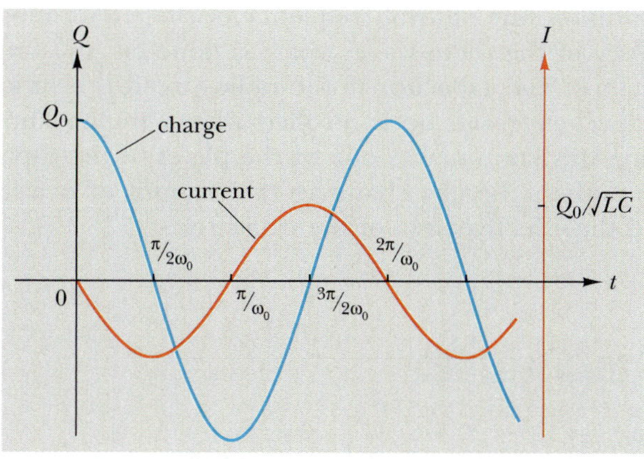

FIGURE 23.16
Charge (blue) on the capacitor and current (red) in the inductor as a function of time.

a natural frequency

$$\omega_0 = \frac{1}{\sqrt{LC}} \tag{26}$$

Figure 23.16 is a plot of the charge and the current in an LC circuit oscillating according to Eqs. (24) and (25).

EXAMPLE 4. A primitive radio transmitter, such as those used in the early days of "wireless telegraphy," consists of an LC circuit oscillating at high frequency (Fig. 23.17a). The circuit is inductively coupled to an antenna (Fig. 23.17b), so the oscillating current in the circuit induces an oscillating current on the antenna; the latter current then radiates radio waves. Suppose that the inductance in the circuit in Fig. 23.17a is 20 μH. What capacitance do we need if we want to produce oscillations of a frequency of 1.5×10^6 Hz?

SOLUTION: The angular frequency is $2\pi \times 1.5 \times 10^6$/s. Hence, from Eq. (26),

$$C = \frac{1}{\omega_0^2 L} = \frac{1}{(2\pi \times 1.5 \times 10^6/\text{s})^2 \times 20 \times 10^{-6}\,\text{H}}$$

$$= 5.6 \times 10^{-10}\,\text{F} = 560\,\text{pF}$$

(a)

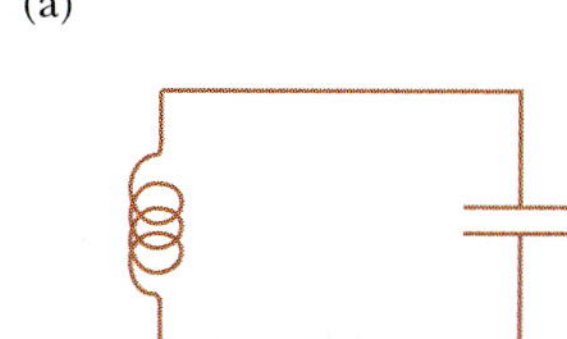

(b)

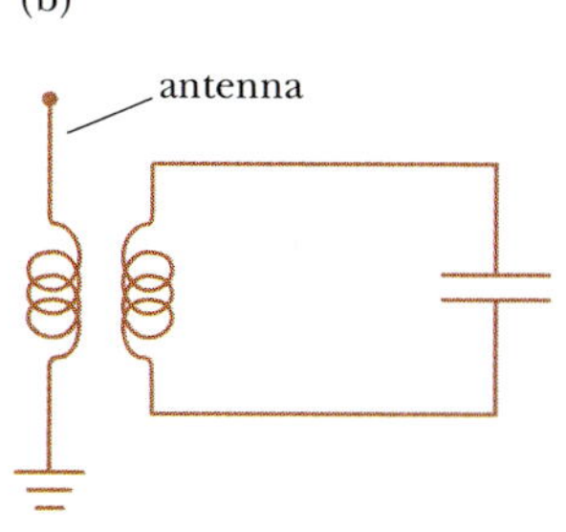

FIGURE 23.17
(a) An LC circuit in a radio.
(b) The LC circuit is coupled to the antenna by the mutual inductance of the two inductors.

A radio receiver employs much the same circuit as shown in Fig. 23.17b to pick up radio signals reaching its antenna. When a radio wave — consisting of oscillating electric and magnetic fields — reaches the antenna, it causes an oscillating current to flow along the antenna, which induces a current in the LC circuit. The current in the circuit will build up to a relatively large value if the frequency of the driving force supplied by the antenna matches the natural frequency of the circuit. To attain this

FIGURE 23.18
Adjustable capacitor for a radio. The capacitor consists of two sets of parallel, semicircular plates. A tuning knob controls the amount of overlap of the plates.

resonance condition, the natural frequency of the circuit must be tuned to the frequency of the radio wave, which is done by adjusting the value of the capacitance. The capacitor in the radio circuit is a variable capacitor, whose capacitance can be controlled with a tuning knob (see Fig. 23.18). Turning this knob makes one of the plates of the capacitor move parallel to the other, thereby changing the amount of overlap between the plates and the effective area of the capacitor.

23.5 THE TRANSFORMER

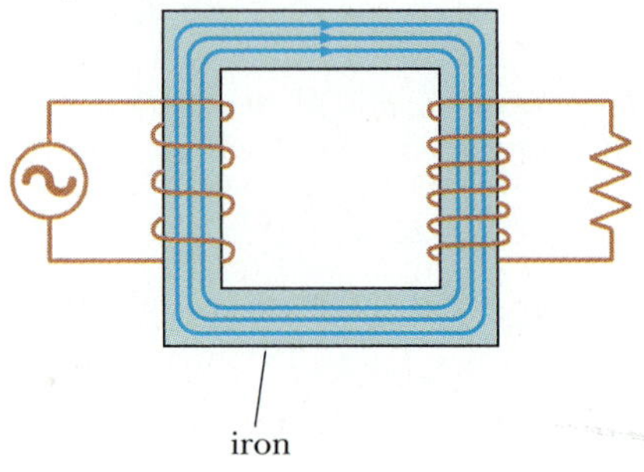

FIGURE 23.19
A transformer.

A transformer consists of two coils arranged in such a way that (almost) all the magnetic flux lines generated by one of them pass through the other. This can be achieved by winding both coils on a common iron core (see Fig. 23.19). As we saw in Section 21.4, the iron increases the strength of the magnetic field in its interior by a large factor. Since the field is much stronger in the iron then outside, most of the field lines have to stay inside the iron; this means that ***the iron tends to keep the field lines together and acts as a conduit for the field lines from one coil to the other.***

Each coil is part of a separate electric circuit (see Fig. 23.20). The ***primary circuit*** has a source of alternating emf, and the ***secondary circuit*** has a resistance or some other "load" that consumes electric power. The alternating current in the primary circuit induces an alternating emf in the secondary circuit. We will show that the emf $\mathscr{E}_2$ produced in the secondary circuit is related as follows to the emf $\mathscr{E}_1$ in the primary circuit:

Emf's of transformer ▶

$$\mathscr{E}_2 = \mathscr{E}_1 \frac{N_2}{N_1} \tag{27}$$

where N_1 and N_2 are, respectively, the numbers of turns in the primary and the secondary coils.

To prove Eq. (27), we begin with Kirchhoff's Rule as it applies to the primary circuit: the emf $\mathscr{E}_1$ of the source must equal the induced emf $\mathscr{E}_{1,\text{ind}}$ across the primary coil. But by Faraday's Law, the induced emf equals the rate of change of the flux,

$$\mathscr{E}_1 = \mathscr{E}_{1,\text{ind}} = -\frac{\Delta \Phi_1}{\Delta t} \tag{28}$$

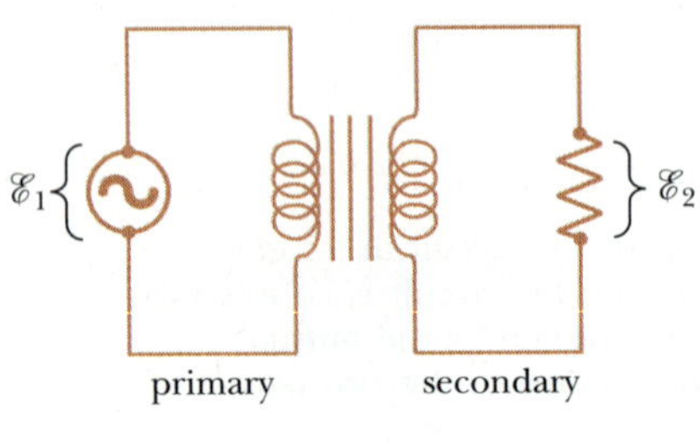

FIGURE 23.20
Circuit diagram for the transformer. The parallel lines represent the mutual inductance.

Likewise, the emf $\mathscr{E}_2$ delivered to the load must equal the induced emf $\mathscr{E}_{2,\text{ind}}$, which, in turn, equals the rate of change of the flux in that coil,

$$\mathscr{E}_2 = \mathscr{E}_{2,\text{ind}} = -\frac{\Delta \Phi_2}{\Delta t} \tag{29}$$

The ratio of these equations is

$$\frac{\mathscr{E}_2}{\mathscr{E}_1} = \frac{\Delta \Phi_2/\Delta t}{\Delta \Phi_1/\Delta t} \tag{30}$$

Since the same numbers of magnetic field lines pass through both coils, the fluxes and their rates of change necessarily are in the ratio N_2/N_1,

$$\frac{\Delta\Phi_2/\Delta t}{\Delta\Phi_1/\Delta t} = \frac{N_2}{N_1} \tag{31}$$

Inserting this into Eq. (30), we find

$$\frac{\mathscr{E}_2}{\mathscr{E}_1} = \frac{N_2}{N_1} \tag{32}$$

which is equivalent to Eq. (27).

If $N_2 > N_1$, we have a ***step-up transformer,*** and if $N_2 < N_1$, we have a ***step-down transformer.***

◂ **Step-up and step-down transformers**

EXAMPLE 5. Door bells and buzzers usually are designed for 12-volt AC and they are powered by small transformers which step down 115-volt AC to 12-volt AC. Suppose that such a transformer has a primary winding with 1500 turns. How many turns are there in the secondary winding?

SOLUTION: Equation (27) applies to the instantaneous voltages. It is therefore also valid for the rms voltages. With the appropriate numerical values, Eq. (27) gives

$$N_2 = N_1 \frac{\mathscr{E}_2}{\mathscr{E}_1} = 1500 \times \frac{12 \text{ V}}{115 \text{ V}} = 157 \text{ turns}$$

As long as the secondary circuit is open and carries no current ($I_2 = 0$), ***an ideal transformer does not consume electric power.*** Under these conditions, the primary circuit consists of nothing but the source of emf and an inductor — it is a pure L circuit. In such a circuit, the power delivered by the source of emf to the inductor averages to zero (see Section 23.3).

If the secondary circuit is closed, a current will flow ($I_2 \neq 0$). This current contributes to the magnetic flux in the transformer and induces a current in the primary circuit. The current in the latter is then different from that in a pure L circuit, and the power will *not* average to zero over a cycle. ***In an ideal transformer, the electric power that the primary circuit takes from the source of emf exactly matches the power that the secondary circuit delivers to the load.*** Since the power is the product of the current and the emf, we can express the quality of these average powers as

$$I_1\mathscr{E}_1 = I_2\mathscr{E}_2 \tag{33}$$

where the currents and emf's can be taken to be either maximum values or rms values. Good transformers approach this ideal condition of conservation of electric energy fairly closely: about 99% of the power supplied to the input terminals emerges at the output terminals; the difference is lost as heat in the iron core and the windings.

Transformers play a large role in our electric technology. As we saw in Example 20.7, transmission lines for electric power operate much more efficiently at high voltage, since this reduces the Joule heating losses caused by the resistance of the wires. To take advantage of this high efficiency, power lines are made to operate at several hundred kilovolts. The voltage must be stepped up to this value at the power plant and, for safety's sake, it must be stepped down just before it reaches the consumer (see Fig. 23.1). For these operations, large banks of transformers are needed at both ends. Transformers are also used in many electric and electronic devices, such as TV tubes and X-ray machines, which require high voltage to accelerate beams of electrons.

SUMMARY

- **Rms voltage:** $\mathscr{E}_{\text{rms}} = \dfrac{\mathscr{E}_{\text{max}}}{\sqrt{2}}$
- **Average AC power absorbed by resistor:** $\overline{P} = \dfrac{\mathscr{E}^2_{\text{max}}}{2R} = \dfrac{\mathscr{E}^2_{\text{rms}}}{R}$
- **Reactance of capacitor circuit:** $X_{\text{C}} = \dfrac{1}{\omega C}$
- **Reactance of inductor circuit:** $X_{\text{L}} = \omega L$
- **Natural frequency of LC circuit:** $\omega_0 = \dfrac{1}{\sqrt{LC}}$
- **Emf's of transformer:** $\mathscr{E}_2 = \mathscr{E}_1 \dfrac{N_2}{N_1}$

QUESTIONS

1.–You can perceive the 120-Hz flicker (two peaks of intensity per AC cycle) in a fluorescent light tube (by sweeping your eye quickly across the tube), but you cannot perceive any such flicker in an incandescent light bulb. Explain.

2.–Do the electrons from the power station ever reach the wiring of your house?

3.–Some electric motors operate only on DC, others only on AC. What is the difference between these motors? (Hint: Consider the commutators.)

4.–If you connect a capacitor across a 115-V outlet, does any current flow through the connecting wires? Through the space between the capacitor plates? Does the outlet deliver instantaneous electric power? Average electric power?

5.–It is sometimes said that a capacitor becomes a short circuit at high frequencies, and an inductor becomes an open circuit at high frequencies. Explain.

6.–Can you blow a fuse by connecting a very large capacitor across an ordinary 115-V outlet?

7.–How could you use an LC circuit to measure the capacitance of a capacitor?

8.–Why can we not use a transformer to step up the voltage of a battery?

9.–Does an electric motor absorb more electric power when pulling a mechanical load than when running freely?

PROBLEMS

An electric heater.

Section 23.1: Resistor Circuit

1.——An electric heater plugged into a 115-V AC outlet uses an average electric power of 1200 W.

(a) What are the rms current and the maximum instantaneous current through the heater?

(b) What are the maximum instantaneous power and the minimum instantaneous power?

2.——A resistor of 100 Ω is connected to an AC source with an amplitude of 163 V. What is the instantaneous current in the resistor when the instantaneous voltage is 163 V? What is the instantaneous current one-quarter cycle later? One-half cycle later? Three-quarters cycle later?

3.——A high-voltage power line operates on an rms voltage of 230,000 volts AC and delivers an rms current of 740 A.

(a) What are the maximum instantaneous voltage and current?

(b) What are the maximum instantaneous power and the average power delivered?

4.——An electric heater operating with a 115-V AC power supply delivers 1200 W of heat.

(a) What is the rms current through this heater?

(b) What is the maximum instantaneous current?

(c) What is the resistance of this heater?

5.——The GG-1 electric locomotive develops 4600 hp; it runs on an AC voltage of 1100 V.

(a) What rms current does this locomotive draw?

(b) Why is it advantageous to supply the electric power for locomotives at high voltage (and fairly low current)?

6.——An immersible heating element used to boil water consumes an (average) electric power of 400 W when connected to a source of 115 volts AC. Suppose that you connect this heating element to a source of 115 volts DC. What power will it consume?

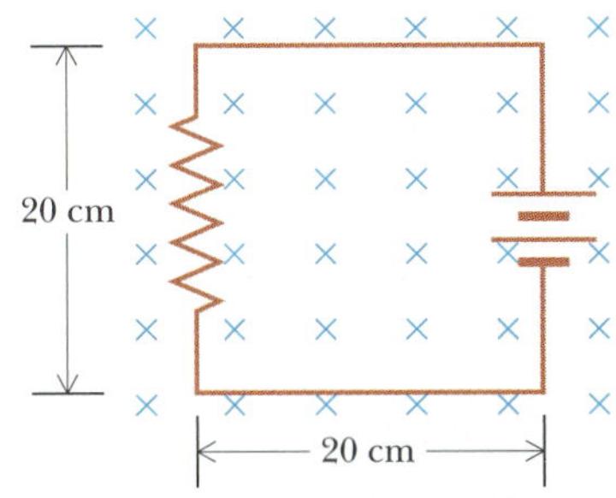

FIGURE 23.21 Circuit placed in magnetic field.

7.** A circuit consists of a resistor connected in series to a battery; the resistance is 5 Ω and the emf of the battery is 12 V. The wires (of negligible resistance) connecting these circuit elements are laid out along a square of 20 cm × 20 cm (Fig. 23.21). The entire circuit is placed face-on in an oscillating magnetic field. The instantaneous value of the magnetic field is

$$B = B_0 \sin \omega t$$

with $B_0 = 0.15$ T and $\omega = 360$ radians/s.

(a) Find the instantaneous current in the resistor.

(b) Find the average power dissipated in the resistor.

Section 23.2: Capacitor Circuit

8.——A capacitor of 2.2 μF is connected to a generator supplying 12-V AC. What is the rms current if the angular frequency of the generator is 1.2×10^3 radians/s? What is the rms current if the angular frequency of the generator is 2.4×10^3 radians/s?

9.——A capacitor has a capacitance of 0.15 μF. What is the reactance when connected to a circuit operating at $\omega = 6 \times 10^3$ radians/s? At $\omega = 1.8 \times 10^4$ radians/s?

10.——A capacitor has a reactance of 3×10^6 Ω at a frequency of 2 kHz. What is its reactance at 3 kHz? At 4 kHz?

11.——A capacitor of 600 pF is connected to an audio generator of adjustable frequency. At what frequency is the reactance of this capacitor 2×10^5 Ω?

12.——A capacitor with $C = 8.0 \times 10^{-7}$ F is connected to an oscillating source of emf. This source provides an emf $\mathscr{E} = \mathscr{E}_{\max} \cos \omega t$, with $\mathscr{E}_{\max} = 0.20$ V and $\omega = 6.0 \times 10^3$ radians/s.

(a) What is the reactance of the capacitor?
(b) What is the maximum current in the circuit?
(c) What is the current at time $t = 0$? At time $t = \pi/4\omega$?

13.——A capacitor of 0.4 μF is connected to an AC source of amplitude 12 V and angular frequency 3.0×10^3 radians/s. What is the instantaneous current in the circuit when the instantaneous voltage is 12 V? What is the instantaneous current one-quarter cycle later? One-half cycle later? Three-quarters cycle later?

14.——To measure the capacitance of a capacitor, a physicist connects it to an oscillating source of emf of angular frequency 2×10^3 radians/s and of amplitude 4×10^{-3} V. She finds that the maximum instantaneous current in the capacitor is 0.8×10^{-6} A. What is the capacitance?

15.——We want to generate a maximum current of 5×10^{-5} A in a circuit consisting of a capacitor of 2 pF connected to a source of oscillating emf of angular frequency of 4×10^4 radians/s. What amplitude of oscillation of the emf do we need?

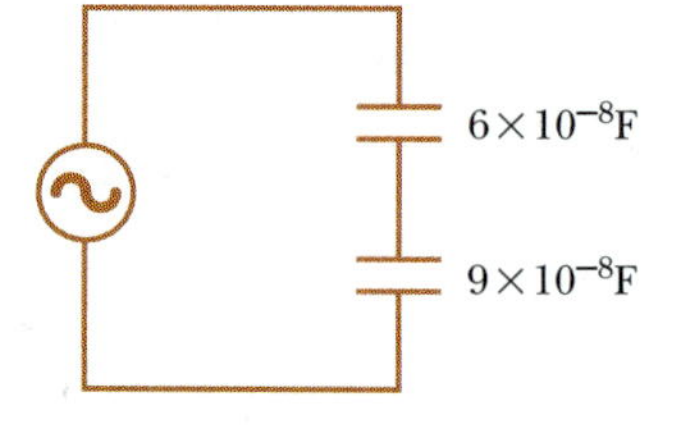

FIGURE 23.22
Two capacitors in series connected to a source of emf.

16.*—A parallel-plate capacitor has plates of area 0.3 m^2 separated by an air gap of 0.002 cm. This capacitor is connected to a generator supplying 12-V AC at an angular frequency of 9.0×10^3 radians/s.

(a) What is the capacitance?
(b) What is the rms current flowing into the plates?
(c) What is the rms charge in the capacitor?
(d) What is the rms energy stored in the capacitor?

17.*—A circuit consists of two capacitors of 6.0×10^{-8} F and 9.0×10^{-8} F connected in series to an oscillating source of emf (Fig. 23.22). This source delivers a cosinusoidal emf $\mathscr{E} = 1.8 \cos(120\pi t)$, where $\mathscr{E}$ is in volts and t in seconds.

(a) Find the charge on each capacitor as a function of time.
(b) At what time is the charge on the capacitors maximum? At what time minimum?
(c) What is the maximum energy in the capacitors? What is the time-average energy?

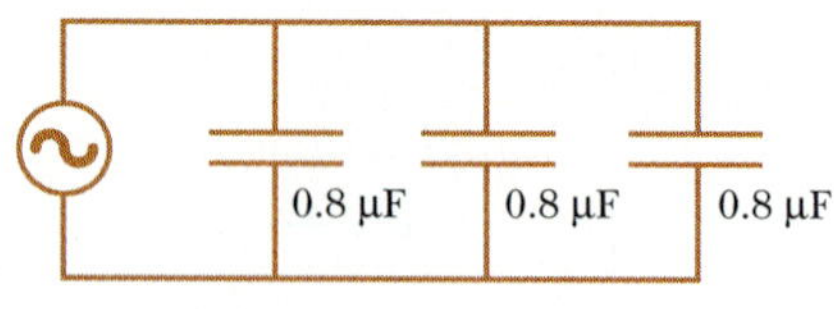

FIGURE 23.23
Three capacitors in parallel connected to a source of emf.

18.*—A circuit consists of three capacitors of 0.80 μF each, connected in parallel to the terminals of a generator that delivers 24-V AC at an angular frequency of 1.8×10^3 radians/s (see Fig. 23.23). What is the net rms current in this circuit?

Section 23.3: Inductor Circuit

19.——What is the rate of change of the current in a coil with an inductance of 250 μH which has an instantaneous voltage of 25 V across it?

20.——An inductor has an inductance of 0.25 H. What is the reactance when connected to a circuit operating at $\omega = 6 \times 10^3$ radians/s? At $\omega = 1.8 \times 10^4$ radians/s?

21.——An inductor of 0.20 H is connected to an audio generator of adjustable frequency. At what frequency is the reactance of this inductor $2 \times 10^3\ \Omega$?

22.——A 1-km segment of the aluminum cable of a high-voltage transmission line has a resistance of $7.3 \times 10^{-2}\ \Omega$, a capacitance of 4.7×10^{-8} F, and a self-inductance of 6.6×10^{-4} H. At the standard frequency of 60 Hz, what are the capacitive and the inductive reactances of this segment? Compare the reactances X_C, X_L, and the resistance R. Which of these quantities is largest? Smallest?

23.——An inductor of 0.30 H is connected to a generator supplying 12-V AC. What is the rms current if the angular frequency of the generator is 1.2×10^3 radians/s? What is the rms current if the angular frequency of the generator is 2.4×10^3 radians/s?

24.——An inductor of 0.0030 H is connected to an AC source of amplitude 8.0 V and angular frequency of 3.0×10^3 radians/s. What is the instantaneous current in the inductor when the instantaneous voltage is 8.0 V? What is the instantaneous current one-quarter cycle later? One-half cycle later? Three-quarters cycle later?

25.——We want to generate a maximum current of 5×10^{-5} A in an inductor of 1×10^{-2} H connected to a source of oscillating emf of angular frequency 4×10^4 radians/s. What amplitude of oscillation of the emf do we need?

26.——The primary winding of a transformer has an inductance of 6.2 H. What rms current will flow in this winding if it is connected to an outlet supplying 115-V AC?

27.——To measure the inductance of an inductor, an experimenter connects it to an oscillating source of emf of angular frequency 3×10^3 radians/s and of amplitude 8×10^{-3} V. He finds that the maximum instantaneous current in the inductor is 0.2×10^{-6} A. What is the inductance?

28.*——An inductor of 1.6×10^{-3} H is connected to a source of alternating emf. The current in the inductor is $I = I_0 \sin \omega t$, with $I_0 = 180$ A and $\omega = 120\pi$ radians/s.

(a) What is the potential difference across the inductor at time $t = 0$? At time $t = 1/240$ s?

(b) What is the energy in the inductor at time $t = 0$? At time $t = 1/240$ s?

(c) What is the instantaneous power delivered by the source of emf to the inductor at time $t = 0$? At time $t = 1/240$ s?

29.*——An inductor with $L = 4.0 \times 10^{-2}$ H is connected to a source of alternating emf. This source provides an emf $\mathscr{E} = \mathscr{E}_{max} \cos \omega t$, with $\mathscr{E}_{max} = 0.20$ V and $\omega = 6.0 \times 10^3$ radians/s.

(a) What is the reactance of the inductor?

(b) What is the maximum current in the circuit?

(c) What is the current at time $t = 0$? At time $t = \pi/4\omega$?

30.*——A 2.0-μF capacitor and a 3.0-mH inductor are connected in parallel with an AC power supply. At what frequency will the inductive reactance be three times the capacitive reactance?

31.*——Consider the circuit shown in Fig. 23.24. The emf of the source is of the form $\mathscr{E}_0 \cos \omega t$. In terms of this emf and the capacitance C and inductance L, find the instantaneous currents through the capacitor and the inductor. Find the instantaneous current and the instantaneous power delivered by the source of emf.

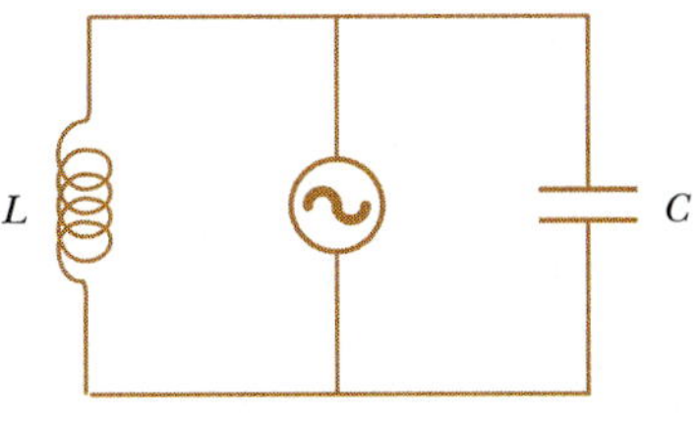

FIGURE 23.24
Inductor and capacitor connected to a source of emf.

32.*——In Example 3 we examined the current in a circuit containing a resistor, a capacitor, and an inductor connected in parallel to a source of oscillating emf of an angular frequency 6×10^3 radians/s and amplitude 1×10^{-3} V.

How would the answers to this example change if we reduce the angular frequency to 3×10^3 radians/s?

33.*—Suppose that in Example 3 the emf is of the form $\mathscr{E} = \mathscr{E}_{max} \cos \omega t$. What is the net instantaneous current in the circuit at $t = 0$? At $t = \pi/2\omega$? At $t = \pi/\omega$?

34.*—An inductor of 4.5×10^{-2} H and a capacitor of 0.25 μF are connected in parallel to a source of alternating emf of frequency ω. For what value of ω will the rms currents in the inductor and the capacitor be of equal magnitudes?

35.*—An inductor is built in the shape of a solenoid of radius 0.20 cm, length 4.0 cm, with 1000 turns. This inductor is connected to a generator supplying 1.2×10^{-4}-V AC at an angular frequency of 9.0×10^3 radians/s.

(a) What is the inductance? Assume that the solenoid can be treated as very long.
(b) What is the rms current flowing through the inductor?
(c) What is the rms energy stored in the inductor?

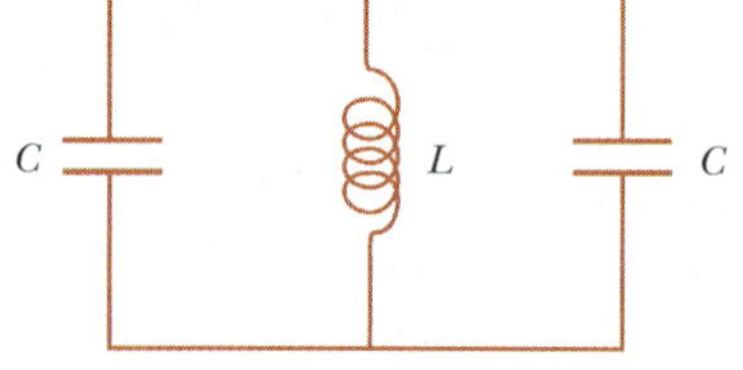

FIGURE 23.25
Two equal capacitors connected to an inductor.

Section 23.4: Freely Oscillating LC Circuit

36.—What is the natural frequency for an LC circuit consisting of a 2.2×10^{-6} F capacitor and an 8.0×10^{-2} H inductor?

37.—You want to construct an LC circuit of natural frequency 8×10^3 Hz. You have available a capacitor of 0.2 μF. What inductor do you need?

38.—What is the natural frequency of oscillation of the circuit shown in Fig. 23.25? The capacitances are 2.4×10^{-5} F each and the inductance is 1.2×10^{-3} H.

39.—A radio receiver contains an LC circuit whose natural frequency of oscillation can be adjusted, or tuned, to match the frequency of incoming radio waves. The adjustment is made by means of a variable capacitor. Suppose that the inductance of the circuit is 15 μH. Over what range of capacitances must the capacitor be adjustable if the frequencies of oscillation of the circuit are to span the range from 530 kHz to 1600 kHz?

40.*—The circuit of Fig. 23.26a is oscillating with the switch S closed. The graph of current vs. time is shown in Fig. 23.26b.

(a) At time t_1 the switch S is suddenly opened. Is the frequency of oscillation increased, decreased, or unchanged? In the space on the right in Fig. 23.26b sketch a rough, qualitative graph of current for times after t_1.
(b) At time t_2 the switch S is closed. Sketch the graph of current after this time.

(a)

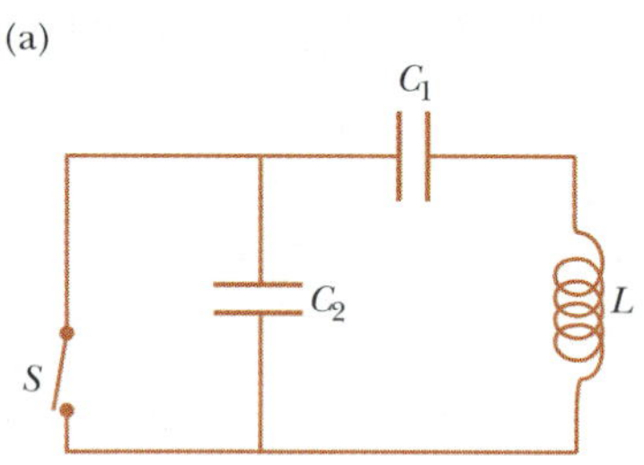

(b)

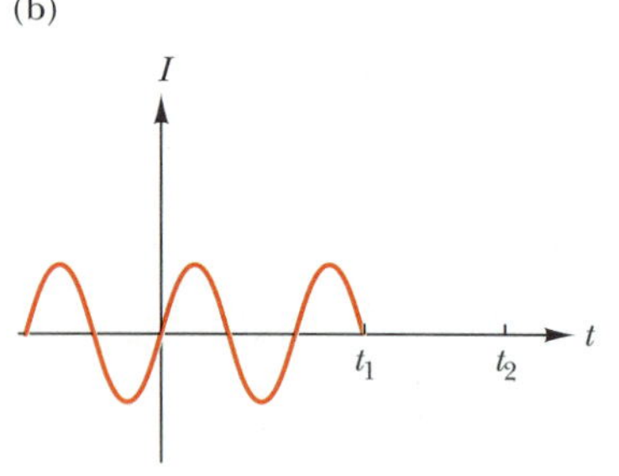

FIGURE 23.26
(a) Inductor and two capacitors in a circuit. The switch S is initially closed. (b) Current in the circuit as a function of time. At $t = t_1$, the switch S is opened.

41.*—An LC circuit has an inductance of 5.0×10^{-2} H and a capacitance of 5.0×10^{-6} F. At $t = 0$, the capacitor is fully charged so $Q_0 = 1.2 \times 10^{-4}$ C. What is the energy in this circuit? At what time, after $t = 0$, will the energy be purely magnetic? At what *later* time will it be purely electric?

Section 23.5: The Transformer

42.—The primary winding of a transformer has 1200 turns and the secondary winding has 80 turns. If the emf supplied to the primary is 115-V AC, what is the emf delivered by the secondary?

43.—A transformer used to step up 115 V to 5000 V has a primary coil of 100 turns. What must be the number of turns in the secondary coil?

44.—The primary winding of a transformer has 140 turns and the secondary has 2200 turns. This transformer is connected to a second, identical transformer, so the output of the first transformer serves as input for the second transformer. If the primary of the first transformer is connected to a source of 115-volt AC, what is the voltage delivered by the secondary of the second transformer?

45.——Consider each of the four transformers illustrated in Fig. 23.1; assume that the last of these delivers 115-volt AC. What is the ratio N_1/N_2 of the numbers of turns in the primary and secondary windings of each of these transformers?

46.——How much current is flowing in the 1200-V secondary of a transformer if the primary has a voltage of 12 V and a current of 3.0 A?

47.——The generators of a large power plant deliver an electric power of 2000 MW at 22 kilovolts AC. For transmission, this voltage is stepped up to 400 kV by a transformer. What is the rms current delivered by the generators? What is the rms current in the transmission line? Assume that the transformer does not waste any power.

48.——A transformer operating on a primary voltage of 115 volts AC delivers a secondary voltage of 6.0 volts AC to a small electric buzzer. If the current in the secondary circuit is 3.0 A, what is the rms current in the primary circuit? Assume that no electric power is lost in the transformer.

49.——In a transformer, the ratio of turns of primary and secondary is 10:1. If the primary voltage is 120 V and the secondary current is 3.0 A, how much power is being absorbed in the primary circuit of the transformer?

50.——The largest transformer ever built handles a power of 1.50×10^9 W. This transformer is used to step down 765 kV to 345 kV. What is the rms current in the primary? What is the current in the secondary? Assume that no electric power is lost by the transformer.

51.*——A power station feeds 1.0×10^8 W of electric power at 760 kV into a transmission line. Suppose that 10% of this power is lost in Joule heat in the transmission line. What percentage of the power would be lost if the power station were to feed 340 kV into the transmission line instead of 760 kV, other things being equal?

A transformer.

CHAPTER 24

Antennas for the reception (large dishes) and the emission (horns, at lower left) of radio waves on an Intelsat communications satellite.

ELECTROMAGNETIC WAVES

We already know that a changing magnetic field induces an electric field. In this chapter, we will discover that the converse is also true: a changing electric field induces a magnetic field. The law describing this induction effect of electric fields was formulated by James Clerk Maxwell, who thereby achieved a wide-ranging unification of the laws of electricity and magnetism. These laws became known as Maxwell's equations. The next two chapters in this book are, in essence, nothing but applications of Maxwell's equations.

The mutual induction of electric and magnetic fields gives rise to the phenomenon of self-supporting electromagnetic oscillations in empty space. If, initially, there exists an oscillating electric field, it will induce a magnetic field, and this will induce a new electric field, and so on. Thus, these fields can perpetuate each other. Of course, an oscillating charge or current is needed to get the fields started, but after this initiation the fields continue on their own. These self-supporting oscillations are ***elec-***

tromagnetic waves, either traveling waves or standing waves. Among such electromagnetic waves are radio waves, microwaves, light, and X rays. All of these kinds of waves are qualitatively the same — they consist of oscillating electric and magnetic fields. The only difference between, say, a radio wave and light is in their wavelengths — the radio wave has a much longer wavelength than light. In this chapter we will examine the properties of electromagnetic waves, and see how they are generated by accelerated electric charges.

24.1 INDUCTION OF MAGNETIC FIELDS; MAXWELL'S EQUATIONS

As we saw in Chapter 22 [Eq. (22.9)], Faraday's Law for the electric field induced by a changing magnetic field can be expressed in the form

$$\overline{E}_t l = -\frac{\Delta \Phi_B}{\Delta t} \tag{1}$$

where $\overline{E}_t$ is the average component of the induced electric field tangent to a closed path (Fig. 24.1), l is the length of the path, and $\Delta\Phi_B/\Delta t$ is the rate of change of the magnetic flux intercepted by the surface within the path.

The magnetic field induced by a changing electric field is described by an equation analogous to Faraday's Law. On the left side of the equation that describes this new kind of induction there appears the induced magnetic field tangent to a closed path (Fig. 24.2), and on the right side there appears the rate of change of the electric flux:

$$\overline{B}_t l = \mu_0 \epsilon_0 \frac{\Delta \Phi}{\Delta t} \tag{2}$$

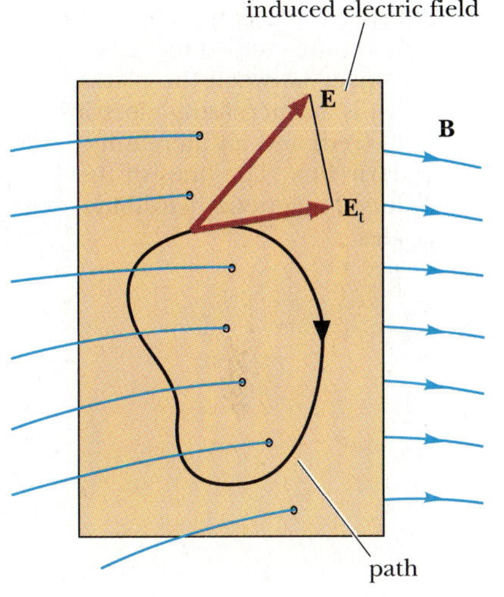

FIGURE 24.1

A closed path placed in a changing magnetic field. E_t is the component of the induced electric field tangent to the path.

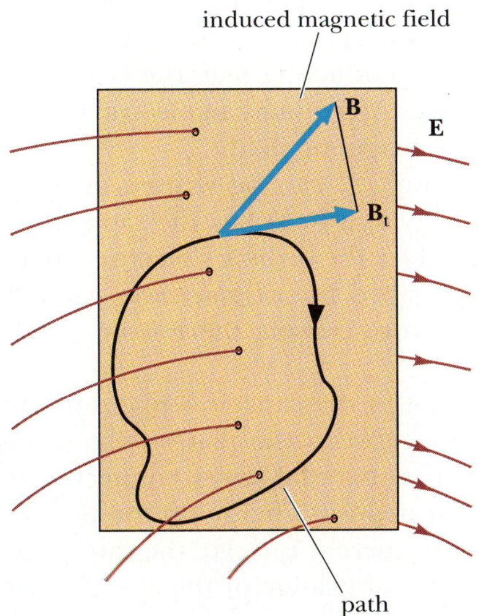

FIGURE 24.2

A closed path placed in a changing electric field. B_t is the component of the induced magnetic field tangent to the path.

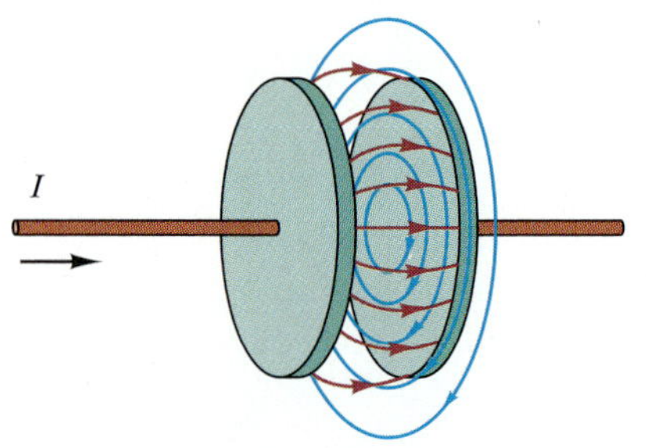

FIGURE 24.3

A capacitor is being charged by wires connected to its plates. In the space between the plates, there is an increasing electric field (red) and an increasing electric flux. The lines of the induced magnetic field (blue) are circles.

Apart from the minus sign in Eq. (1) and the constant $\mu_0\epsilon_0$ in Eq. (2), these two equations are "mirror images" of each other; they can be obtained from each other by exchanging electric and magnetic fields. Hence these equations indicate a certain symmetry in the effects of electric and magnetic fields on one another. Equation (1) states that a rate of change of magnetic flux induces an electric field. Equation (2) states that ***a rate of change of electric flux induces a magnetic field.*** This induction of magnetic fields was first proposed by Maxwell on purely theoretical considerations; the experimental demonstration of this kind of induction came only later.

For a simple example of a system with a rate of change of electric flux and an induced magnetic field, consider a capacitor that is being charged by currents flowing into its plates (see Fig. 24.3). In the space between the plates there is then an increasing electric field, and there is increasing electric flux. If the plates of the capacitor are circular, as shown in Fig. 24.3, the induced magnetic field lines will have the shape of concentric circles, just like the field lines surrounding a long wire carrying a current. The strength of the magnetic field can be calculated in the manner familiar from the calculations with Ampère's Law in Chapter 21: take a path that follows one of the circular field lines and evaluate each side of Eq. (2) for this choice of path. According to Eq. (2), we then find that the strength of the induced magnetic field is directly proportional to the rate of change of the electric flux, or to the rate of change of the electric field between the plates.

Although the magnetic fields of currents (described by Ampère's Law) and the magnetic fields induced by electric fields [described by Eq. (2)] have different sources, the equations describing these fields both involve the tangential component of the field. Hence Ampère's Law and Eq. (2) can be combined into one single equation:

Maxwell-Ampère's Law ▶

$$\overline{B}_t l = \mu_0 I_0 + \mu_0 \epsilon_0 \frac{\Delta \Phi}{\Delta t} \tag{3}$$

This equation covers all conceivable cases of production of magnetic fields — by currents, by changing electric fields, or by any combination of both. For example, if the space between the capacitor plates in Fig. 24.3 is filled with a slightly conducting material (a leaky capacitor), then there will be both an electric current and an electric flux in this space, and both will contribute to the magnetic field.

The right side of Eq. (3) can be written as $\mu_0(I + \epsilon_0 \Delta\Phi/\Delta t)$, which shows that the quantity $\epsilon_0 \Delta\Phi/\Delta t$ has the same effect for the production of magnetic fields as does the ordinary electric current I. Accordingly, the quantity $\epsilon_0 \Delta\Phi/\Delta t$ is called the ***displacement current*** (although there is a good reason for the word *current,* there is no good reason for the word *displacement*).

Displacement current ▶

Note that in the case of the capacitor plates illustrated in Fig. 24.3 the displacement current between the plates is large whenever the ordinary current flowing on the external wires connected to the plates is large. This ordinary current deposits charge on the plates of the capacitor, and whenever the ordinary current is large, the rate of change of the charge is large, and so is the rate of change of the electric field between the plates, and so is the rate of change of the electric flux. It is not hard to show that the magnitude of the displacement current between the plates is equal to the magnitude of the ordinary current flowing on the external wires. Thus, ***the displacement current between the capacitor plates can be viewed as a continuation of the ordinary current "by other means."***

Equation (3) is Maxwell's modification of Ampère's Law. The great importance of this equation lies in its general validity — Maxwell boldly proposed that this equation is valid not only for the electric and magnetic fields associated with capacitors, wires, and other such devices, but also for the fields associated with electromagnetic waves.

Equation (3) is the last of the fundamental laws that we need for a complete mathematical description of the behavior of electric and magnetic fields. There are four fundamental laws, as follows:

Gauss' Law of electricity [Eq. (17.19)],

$$\Phi = Q/\epsilon_0 \tag{4}$$

◂ Maxwell's equations

Gauss' Law for magnetism [Eq. (21.14)],

$$\Phi_B = 0 \tag{5}$$

Faraday's Law [Eq. (22.9)],

$$\overline{E_t} l = -\frac{\Delta \Phi_B}{\Delta t} \tag{6}$$

Maxwell–Ampère's Law [Eq. (3)],

$$\overline{B_t} l = \mu_0 I + \mu_0 \epsilon_0 \frac{\Delta \Phi}{\Delta t} \tag{7}$$

James Clerk Maxwell, *1831–1879, Scottish physicist. Maxwell at first attempted to explain the behavior of electric and magnetic fields in terms of a complicated mechanical model, according to which all of space was filled with an elastic medium, or ether. But ultimately Maxwell discarded the mechanical model and treated the electric and magnetic fields as physical entities that exist in their own right, without any need for an underlying medium. He published his electromagnetic theory in 1873 in his celebrated* Treatise on Electricity and Magnetism. *In this book he laid down a complete and consistent set of laws for all electromagnetic phenomena and thereby accomplished for electricity and magnetism what Newton had accomplished for mechanics.*

The physical basis for each of these four laws may be briefly summarized as follows. ***Gauss' Law for electricity is based on Coulomb's Law describing the forces of attraction and repulsion between stationary charges. Gauss' Law for magnetism asserts that there are no sources or sinks of magnetic field lines. Faraday's Law describes the induction of an electric field by motion or by a changing magnetic field. And finally, Maxwell–Ampère's Law is based on the law of magnetic force between moving charges and it also contains the induction of a magnetic field by a changing electric field.***

Taken as a whole, the four laws (4)–(7) are known as ***Maxwell's equations,*** because Maxwell supplied the missing link between the magnetic and the electric fields [Eq. (3)], and thereby placed the capstone on electromagnetic theory. Maxwell recognized that these equations imply a dynamic interplay between electric and magnetic fields, an interplay that couples and unifies electric and magnetic phenomena.

Maxwell's equations provide a complete description of the interactions among charges, currents, electric fields, and magnetic fields. All the properties of the fields can be deduced by mathematical manipulation of these equations. If the distribution of charges and currents is given, then these equations uniquely determine the corresponding fields. Even more important, Maxwell's equations uniquely determine the time evolution of the fields, starting from a given initial condition for these fields. Thus, ***these equations accomplish for the dynamics of electromagnetic fields what Newton's equations of motion accomplish for the dynamics of particles.***

Although the experimental foundations on which we based the development of Maxwell's equations were restricted to charges at rest or charges in uniform motion, these equations also govern the fields of accelerated charges and the fields of light and radio waves. In the remaining sections of this chapter we will calculate the fields of electromagnetic waves from our equations, and we will see that the results are in agreement with the observed properties of light and radio waves.

24.2* THE ELECTROMAGNETIC WAVE PULSE

We started our study of electricity and magnetism with the electric field of a point charge at rest — this is the Coulomb field given by Eq. (17.4). Later, we dealt with the fields of a current, consisting of charges in motion with uniform velocity — in addition to the Coulomb field, such uniformly moving charges have a magnetic field. Now we will investigate the fields of a charge with *accelerated motion.* We will find that in this case there are extra electric and magnetic fields that spread outward from the position of the charge, like ripples on a pond in which a stone has been dropped, and carry away energy and momentum. ***These spreading fields are called radiation fields, or wave fields. These fields constitute an electromagnetic wave.***

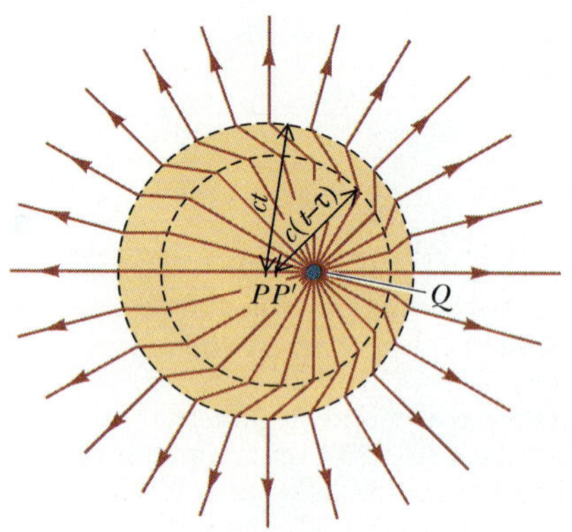

FIGURE 24.4
Electric field lines of a charge that has suffered an acceleration. Here Q is the present position of the charge; P is the initial position of the charge. Between P and P' the charge suffered a constant acceleration; between P' and Q the charge moved at constant velocity. The outer dashed sphere (outer edge of the kink) has radius ct and is centered on P; the inner dashed sphere (inner edge of kink) has radius $c(t-\tau)$ and is centered on P'.

We can gain some insight into the fields spreading out from an accelerated charge by the following argument. Consider a charge that is initially at rest, then is quickly accelerated for some short time interval, and then continues to move with a constant final velocity. The initial electric field lines originate at the initial position of the charge. But the field lines at some later time, after the acceleration ceases, must originate on the new position of the charge. The field lines cannot change from their initial configuration to their final configuration instantaneously; rather a disturbance must travel outward from the position of the charge and gradually change the field lines. The disturbance takes the form of a kink connecting the old and the new field lines.

The disturbance travels at some speed c (we will see in the next section that the speed of the disturbance is actually the speed of light, and our notation anticipates this result). Figure 24.4 shows the situation at some time after the acceleration has ended. Suppose that the acceleration lasts from time $t = 0$ to time $t = \tau$. Then the disturbance of the electric field lines begin at time $t = 0$ and ends at time $t = \tau$. The leading edge of the disturbance (outer edge of kink) travels outward from the initial position

* This section is optional.

of the charge and in a time t it reaches out to a distance ct. Beyond the sphere of radius ct, the electric field is still the old field with field lines centered on the initial position of the charge.

The disturbance ceases as soon as the acceleration ceases. The field in the vicinity of the uniformly moving charge then settles into the new radial configuration centered on the new position of the charge. The trailing edge of the disturbance (inner edge of kink), marking the cessation of the acceleration, travels outward from the position that the charge has at the time $t = \tau$, when the acceleration ceases, and by some later time t it reaches out to a distance $c(t - \tau)$. Within the sphere of radius $c(t - \tau)$, the electric field is the new radial field centered on the new position of the uniformly moving charge.

The disturbance produced by the accelerated charge is confined to the space between the larger and the smaller spheres in Fig. 24.4. The field lines in this zone must connect the lines of the new field of the uniformly moving charge with the lines of the old field of the stationary charge. As shown in Fig. 24.4, each connecting segment of field line makes an angle with the radial line to the position of the charge. Thus, ***the electric field in the zone of the kink has both a radial component and a tangential, or transverse, component.*** This transverse component is the radiation field, or the wave field, of the accelerated charge.

◀ **Transverse electric field**

The traveling kink is a single wave pulse, analogous to the single wave pulse we might produce on a stretched string by flicking its end just once. If we want to produce a periodic electromagnetic wave, we have to accelerate the charge back and forth periodically, and thereby generate a succession of kinks in the field lines. This picture of an electromagnetic wave as a succession of kinks in the field lines explains how such waves are able to travel in a vacuum, a puzzle that greatly worried the physicists of the nineteenth century. The electromagnetic wave does not really travel in a vacuum — it travels on the field lines of the charge, that is, it travels in the electric field of the charge. The space surrounding the charge is not really a vacuum; the space is filled with the electric field of the charge, and this electric field is the medium in which the electromagnetic wave propagates.

As the kink travels outward, the angle between the radial direction and the direction of the electric field in the zone of the kink increases; thus, the kink becomes more pronounced, and the electric field becomes more transverse; Fig. 24.5 illustrates this for one field line. Thus, when the kink

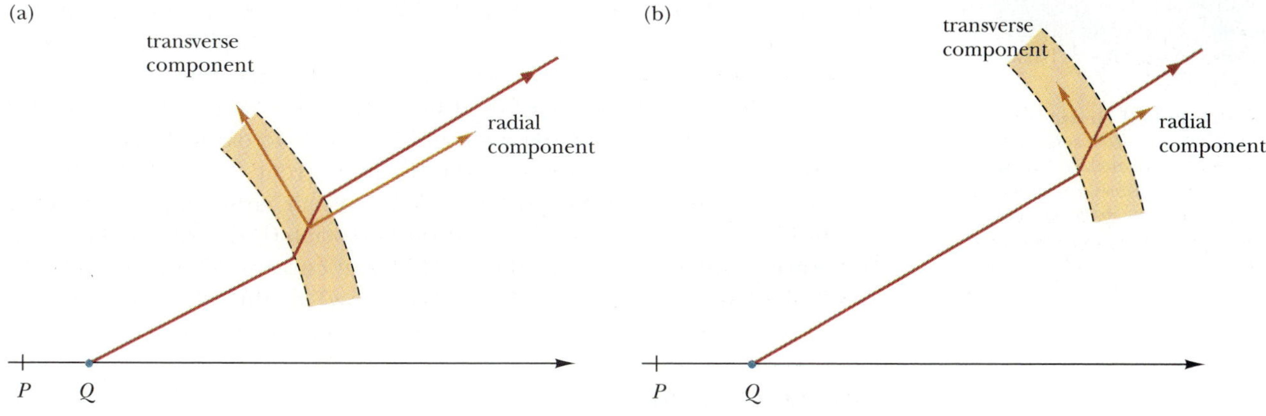

FIGURE 24.5

(a) One of the field lines of Fig. 24.4. In the zone of the kink the electric field has both a radial and a transverse component. (b) The same field line at a later time, when the kink has traveled farther outward. The kink is more pronounced because the radial component of the electric field has decreased more than the transverse component has decreased.

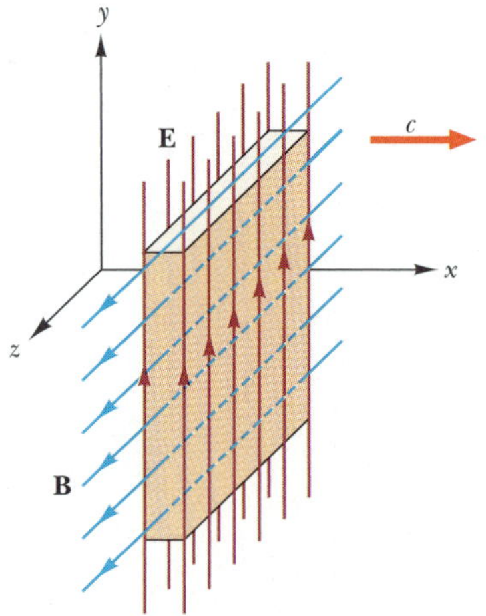

FIGURE 24.6
Electric field lines (red) of a plane wave pulse consisting of a region of uniform electric field. The wave front, or front surface of the pulse, is a flat plane. The magnetic field lines (blue) are at right angles to the electric field lines. The entire pattern of field lines travels to the right with the wave speed *c*.

Heinrich Rudolf Hertz, *1857–1894, German physicist. He supplied the first experimental evidence for the electromagnetic waves predicted by Maxwell's theory. Hertz generated these waves by means of an electric spark, measured their speed and wavelength, and established their similarity to light waves in the phenomena of reflection, refraction, and polarization.*

reaches a large distance from the charge, the electric field in the zone of the kink will be almost entirely transverse. Such ***an electric field at right angles to the direction of propagation is a characteristic feature of electromagnetic waves.***

The transverse kink propagating in the outward direction is a spherical wave pulse, with spherical wave fronts. Often we will be interested in the behavior of the wave over only a relatively small interval of distances, an interval small compared with the radius of the spherical wave. For example, we might be interested in the wave that the moving charges on the antenna of a distant radio transmitter produce in the room in which we are sitting. If the length of the room is, say, 5 m, and the distance to the transmitter is 20 km, then the portion of the spherical wave pulse that passes through the room will look nearly flat, and we can approximate the spherical wave pulse by a plane wave pulse, with plane wave fronts and with straight electric field lines at right angles to the direction of propagation.

Figure 24.6 shows such a flat wave pulse propagating in the x direction; the transverse electric field, at right angles to the direction of propagation, is represented by the straight field lines drawn parallel to the y direction. For the sake of simplicity, we will assume that the electric field in the wave pulse is of constant magnitude (the magnitude of the electric field in the wave pulse depends on the acceleration of the charge that generated the pulse — constant magnitude of the electric field requires constant magnitude of the acceleration).

Since the wave pulse sweeps through the room at the speed c, which, as we will see, is the speed of light, it lasts only some short time at any one point of the room. This means the electric field of the wave pulse is time-dependent: initially it is zero, then it suddenly increases to some constant value E when the leading edge of the pulse arrives, then it remains constant for a while, and finally it drops to zero when the rear edge of the pulse passes.

Besides the transverse electric field, ***the wave pulse is also endowed with a transverse magnetic field, at right angles to the electric field and to the direction of propagation.*** In Fig. 24.6, the magnetic field is represented by magnetic field lines parallel to the z direction (out of the plane of the page). This magnetic field arises by induction, from the time-dependent electric field and the time-dependent electric flux. The induced magnetic field can be calculated from the time-dependent electric field by means of the Maxwell–Ampère Law. We will not attempt to calculate the induced magnetic field, but let us try to understand why its direction is perpendicular to the electric field. For this purpose, consider a small stationary rectangle perpendicular to the electric field. In Fig. 24.7, this rectangle is in the x–z plane. This rectangle registers a rate of change of electric flux when the leading edge of the wave pulse sweeps across it; thus there is a rate of change of flux, or a displacement current, associated with the leading edge of the wave pulse. Likewise, there is an opposite rate of change of flux and displacement current associated with the rear edge of the wave pulse. In Fig. 24.7, the displacement current is shown by parallel vertical lines indicating the direction of "flow" of this current. Since the effect of such a displacement current is the same as that of a real current, the magnetic field produced by this displacement current is the same as that of an array of many adjacent parallel straight wires lying on the front surface of the wave pulse, and a similar array with opposite current lying on the rear surface of the wave pulse. According to the usual right-hand rule for current and magnetic field, such wires produce a magnetic field perpendicular to the current, that is, a magnetic field in the z direction, as shown in Fig. 24.7. Note that the directions of the magnetic field and the electric field are related by a simple right-hand rule: ***If you place the fingers of the right hand along the direction of the electric field and curl them toward the magnetic field, the thumb will lie along the direction of propagation of the wave.***

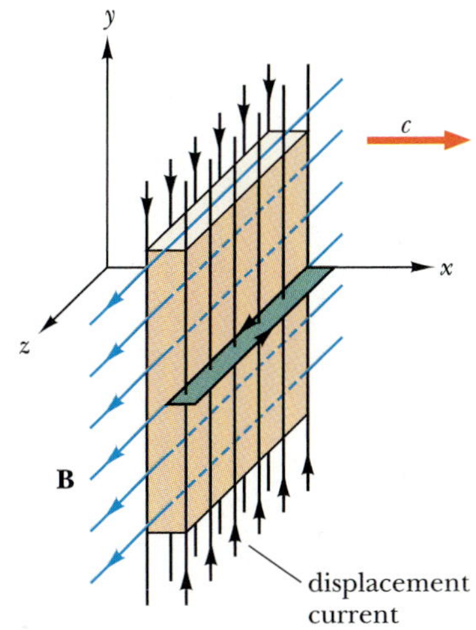

FIGURE 24.7
A small stationary rectangle (green) perpendicular to the electric field lines registers a rate of change of electric flux at the instant the front edge of the wave pulse sweeps across the rectangle, and again at the instant the rear edge of the wave pulse sweeps across the rectangle. Hence there is a sheet of displacement current at the front edge of the wave pulse, and another sheet of displacement current at the rear edge. The black lines indicate the direction of the displacement current. The blue lines indicate the magnetic field produced by these currents.

Like the electric field of the wave pulse, the magnetic field is time-dependent: initially it is zero, then it suddenly increases to some constant value B when the leading edge of the pulse arrives, then it remains constant for a while, and then it drops to zero when the rear edge of the pulse passes. According to Faraday's Law, this time-dependent magnetic field and its time-dependent magnetic flux give rise to an induced electric field. By means of Faraday's Law, it can be verified that this induced electric field coincides with the original electric field E of the wave pulse. Thus, the electric and magnetic fields of the wave pulse mutually induce each other — the electric field induces the magnetic field, and the magnetic field induces the original electric field. These fields thereby become self-supporting, and they continue to exist and to propagate independently of what happens to the charge that initially generated these fields.

24.3 PLANE WAVES; POLARIZATION

In this section, we will concentrate on periodic electromagnetic waves, characterized by a frequency and a wavelength. ***A periodic electromagnetic wave consists of alternating zones of positive and negative electric field and magnetic field.*** The positive and negative electric fields define the wave crests and wave troughs of the wave, respectively.

As described in the preceding section, ***the electric and magnetic fields of the electromagnetic wave mutually induce each other,*** and the wave thereby becomes self-supporting. The wave must initially be generated by some accelerated charges; but once it has been generated, it propagates on its own, independent of the accelerated charges that generated it.

By a detailed theoretical analysis, based on this mutual induction of the electric and magnetic fields according to Maxwell–Ampère's Law and Faraday's Law, we can establish that the speed c of the electromagnetic wave is given by the expression

$$c = \frac{1}{\sqrt{\mu_0 \epsilon_0}} \tag{8}$$

◂ **Speed of electromagnetic wave**

This result was first obtained by Maxwell. Numerically, Maxwell's theoretical expression for the speed of electromagnetic waves yields

$$c = \frac{1}{\sqrt{\mu_0 \epsilon_0}} = \frac{1}{\sqrt{1.26 \times 10^{-6}\ \mathrm{H/m} \times 8.85 \times 10^{-12}\ \mathrm{F/m}}}$$

$$= 3.00 \times 10^8\ \mathrm{m/s} \tag{9}$$

This value coincides with the measured value of the speed of light in vacuum, a coincidence that led Maxwell to propose that ***light waves are electromagnetic waves,*** consisting of self-supporting, mutually induced electric and magnetic fields.

Maxwell's theory of the propagation of electromagnetic waves and of their generation by accelerated charges received direct experimental confirmation at the hands of Heinrich Hertz, who generated the first artificial radio waves by means of sparks triggered in a gap in a high-frequency LC circuit.

We recall from Chapter 10 that the wavelength of the wave is the distance between one wave crest and the next, and the frequency is the number of wave crests arriving at some fixed point per second. As for any wave, the product of the wavelength and the frequency of an electromagnetic wave equals the speed of the wave, that is, the speed of light,

Wavelength and frequency ▶

$$\lambda f = c \tag{10}$$

EXAMPLE 1. The wavelength of green light is 5.5×10^{-7} m. What is the frequency of this kind of light?

SOLUTION: From Eq. (10),

$$f = \frac{c}{\lambda} = \frac{3.00 \times 10^8\ \mathrm{m/s}}{5.5 \times 10^{-7}\ \mathrm{m}} = 5.5 \times 10^{14}\ \mathrm{Hz}$$

The most precise modern method for the determination of the speed of light relies on separate measurements of the wavelength and the frequency of the light emitted by a stabilized laser. The speed can then be evaluated as the product of these measured quantities, as in Eq. (10). The value of the speed obtained by these means is

$$c = 299{,}792{,}458\ \mathrm{m/s} \tag{11}$$

As stated in Chapter 1, this value of the speed of light was adopted as a standard of speed in 1983, and it is now used as the basis for the definition of the meter.*

---

* According to modern practice, Eq. (9) is then used to calculate the value of ϵ_0 from the value of the speed of light and the value of μ_0.

A periodic wave may be regarded as a succession of positive and negative wave pulses. Therefore the arrangement of the electric and magnetic fields in the periodic electromagnetic wave is the same as in the electromagnetic wave pulse, discussed in Section 24.2. ***The directions of the electric and magnetic fields are perpendicular to the direction of propagation and they are perpendicular to each other.*** The direction of the magnetic field is related to the direction of the electric field by the right-hand rule already stated in Section 24.2. ***If you place the fingers of the right hand along the direction of* E *and curl them toward the direction of* B, *the thumb will point in the direction of propagation.***

◀ **Right-hand rule for electric and magnetic fields in wave**

The theoretical analysis of the mutual induction of the electric and magnetic fields in the wave shows that the magnitude of the magnetic field differs from that of the electric field by a factor of c,

$$B = \frac{1}{c} E \tag{12}$$

◀ **Magnetic field of wave**

Thus, the magnetic field of the wave is large wherever the electric field is large; that is, the wave crests (and wave troughs) of the magnetic field coincide with the wave crests (and wave troughs) of the electric field.

Figure 24.8a shows the electric and magnetic field lines for such a periodic wave, with the direction of propagation parallel to the x axis, the direction of the electric field parallel to the y axis, and the direction of the magnetic field parallel to the z axis. Figure 24.8b gives plots of the strengths of the electric and magnetic fields in a harmonic wave, that is, a wave with a sinusoidal dependence on position. The maximum strength of the electric field of such a wave, or the strength of the electric field at the wave crests, is called the ***amplitude*** of the wave, usually designated by E_0. Note that ***the electric and magnetic fields of the wave are always in phase*** — the wave crests of the magnetic field are always at the same position as the wave crests of the electric field.

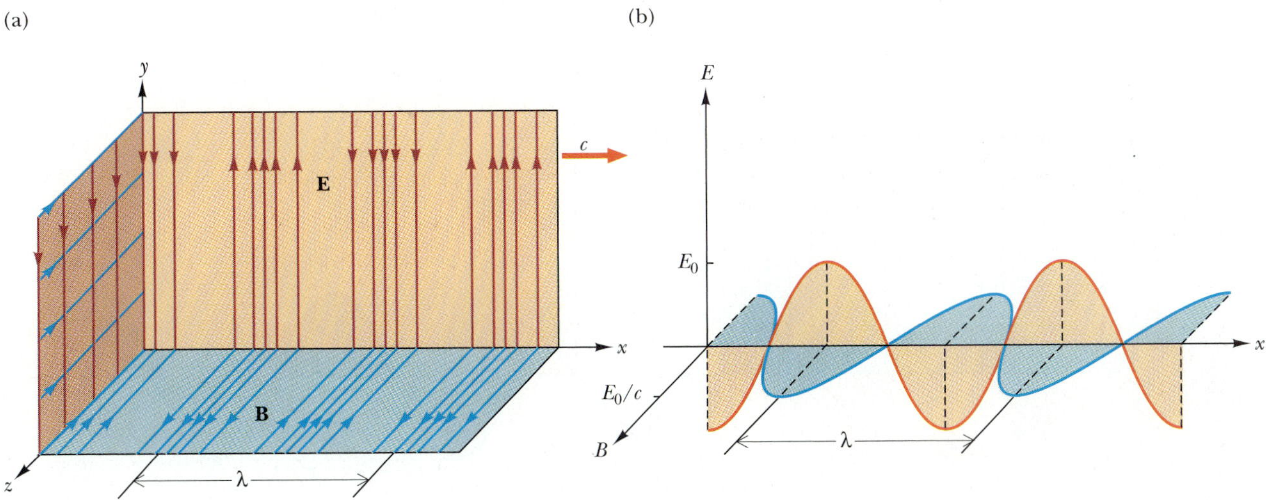

FIGURE 24.8

(a) Electric (red) and magnetic (blue) field lines for a plane wave traveling toward the right, shown at one instant of time. The electric field is vertical, and the magnetic field is horizontal. Only the electric field lines in the x–y plane are shown. There are many more electric and magnetic field lines parallel to those shown. These lines fill slabs perpendicular to the x axis. (b) Plots of the strengths of the electric and magnetic fields as a function of x.

EXAMPLE 2. The wave reaching a point at some distance from a radio transmitter has an electric field with an amplitude of 2.0×10^{-3} V/m. What is the amplitude of the magnetic field?

SOLUTION: By Eq. (12), the amplitude of the magnetic field is

$$B_0 = \frac{E_0}{c} = \frac{2.0 \times 10^{-3}\ \text{V/m}}{3.0 \times 10^{8}\ \text{m/s}} = 6.7 \times 10^{-12}\ \text{T}$$

The wave shown in Fig. 24.8 is a ***plane wave,*** with uniform electric and magnetic fields over any plane perpendicular to the direction of propagation, such as the plane shown at the far left in Fig. 24.8a. Such perpendicular planes, drawn at the location of the wave crests, where the electric field is maximum, are the wave fronts of the wave.

Polarization ▶

The direction of the electric field is called the direction of polarization of the wave. Thus the wave shown in Fig. 24.8 is polarized in the y direction. By contrast, the wave shown in Fig. 24.9 is polarized in the z direction. It is of course also possible to construct waves polarized in some intermediate direction, say, at 45° to the y and z axes; but such waves are superpositions of those shown in Figs. 24.8 and 24.9, and they therefore are nothing essentially new. Hence, electromagnetic waves (of a given direction of propagation) have ***only two independent directions of polarization.*** In this regard, electromagnetic waves are analogous to transverse waves on a string; if the string is stretched horizontally, then there are only two independent transverse waves, since we can shake the string either up and down (vertical polarization) or right and left (horizontal polarization). Schematically, the direction of polarization of a light wave is often represented by a double-headed arrow which indicates the direction of the positive and negative electric fields of the wave (see Fig. 24.10).

Although an individual light wave, like any other kind of electromagnetic wave, is always polarized in some direction or another, the light beams produced by ordinary light sources — the Sun, a light bulb, a can-

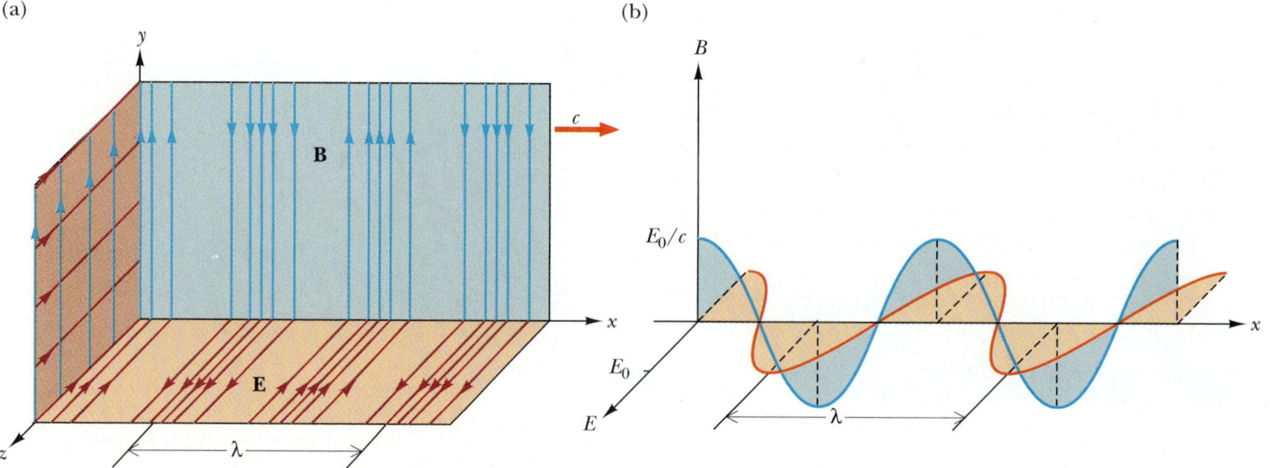

FIGURE 24.9
(a) Electric (red) and magnetic (blue) field lines of another plane wave traveling toward the right, shown at one instant of time. The electric field is horizontal, and the magnetic field is vertical. (b) Plots of the strengths of the electric and magnetic fields as a function of x.

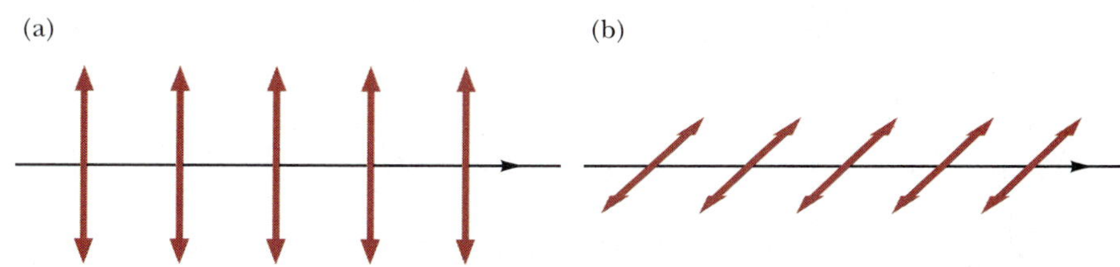

FIGURE 24.10

Schematic representation of the direction of polarization of a light wave. (a) Vertical polarization. (b) Horizontal polarization.

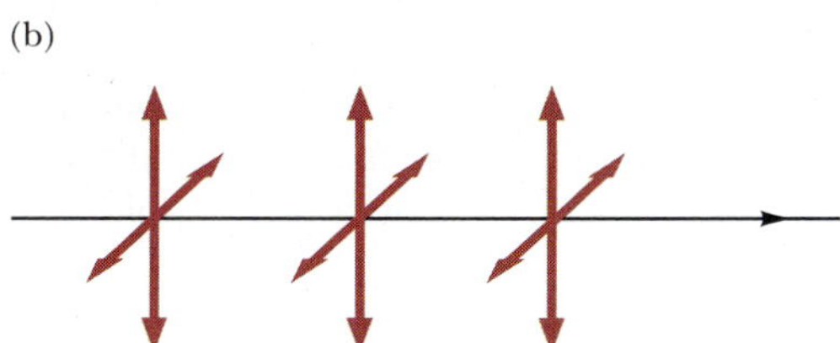

FIGURE 24.11

(a) Unpolarized sunlight consists of a superposition of many plane waves with random directions of polarization. (b) Such light can be represented as a random mixture of horizontally and vertically polarized waves.

dle — do not exhibit any noticeable polarization. Such ***"unpolarized" light consists of a superposition of a very large number of plane waves with random directions of polarization*** (see Fig. 24.11a). Hence, on the average, there is no polarization in the beam. Note that, in Fig. 24.11a, any of the plane waves of an intermediate direction of polarization can be regarded as a superposition of waves of horizontal and of vertical polarizations; hence we can equally well represent an unpolarized light beam as a random mixture of horizontally and vertically polarized waves (see Fig. 24.11b).

Unpolarized light can be given a polarization by passing it through a ***polarizer,*** such as a sheet of Polaroid, that permits only the passage of the electric field component lying parallel to the preferential direction in the sheet and absorbs the electric field component perpendicular to the preferential direction. The Polaroid sheet contains long chains of organic molecules arranged parallel to each other; and the preferential direction that permits passage of the electric field of a wave is *perpendicular* to the direction of alignment of these molecules. An analogous polarizer for microwaves, or radio waves of short wavelength, can be constructed out of a number of thin conducting rods or wires arranged parallel to each other (see Fig. 24.12). The preferential direction of polarization that permits the passage of the electric field of a wave is then *perpendicular* to the direction of the wires, because the wires have very little effect on a perpendicular electric field; on the other hand, an electric field parallel to the wires causes strong currents to flow along the wires, which both reflect the wave and dissipate its energy.

Polaroid is widely used in sunglasses. On the average, unpolarized sunlight consists of an equal mixture of both directions of polarization, paral-

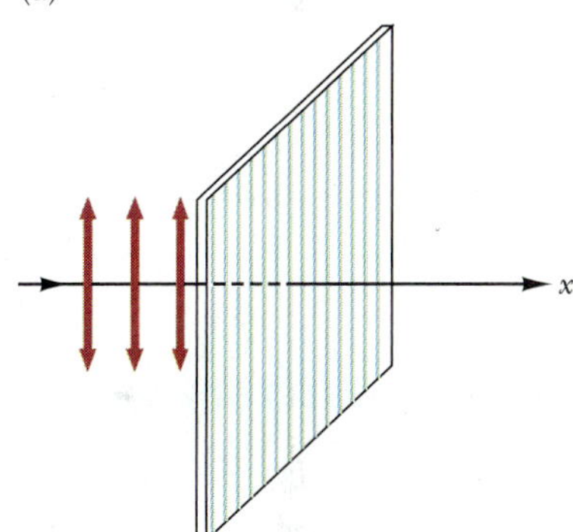

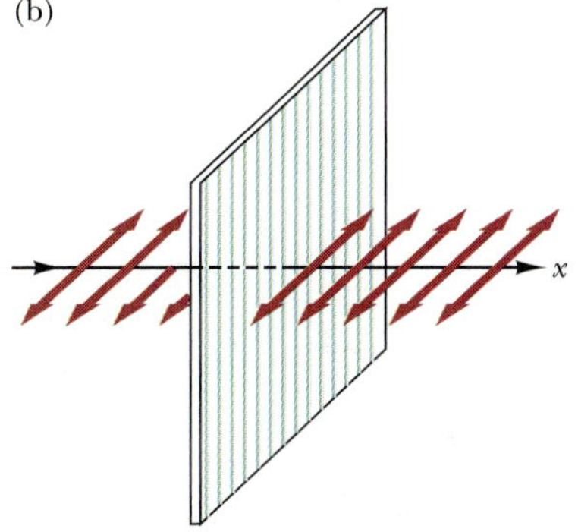

FIGURE 24.12

An array of vertical wires (a) blocks the passage of a microwave of vertical polarization but (b) permits the passage of a microwave of horizontal polarization.

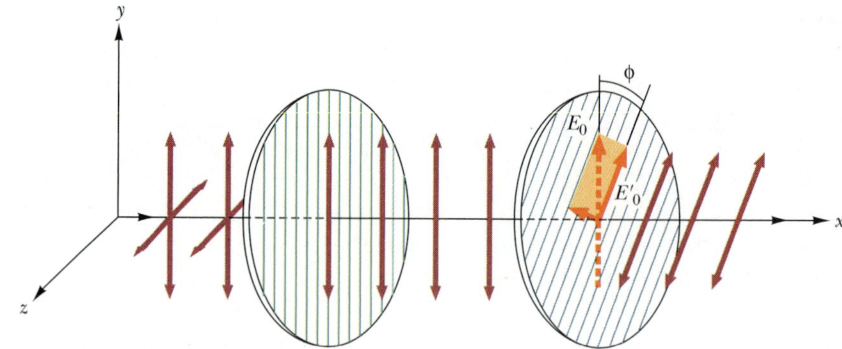

FIGURE 24.13
Two polarizers arranged in tandem. The first polarizer has its preferential direction oriented vertically; the second (analyzer) has its preferential direction inclined at an angle ϕ.

lel to the preferential direction of the Polaroid and perpendicular to it. Hence the Polaroid will absorb half the sunlight (actually somewhat more, since the Polaroid in sunglasses is slightly tinted). But the important advantage of Polaroid sunglasses over ordinary sunglasses is that they strongly attenuate the glare of reflected sunlight, such as the sunlight reflected by water or by a road. In distinction to ordinary sunlight, the reflected sunlight is somewhat polarized in the horizontal direction, and Polaroid sunglasses have their preferential direction in the vertical plane (unless you tilt your head!), and they block the horizontally polarized light.

A variety of interesting experiments can be performed with two or more polarizers arranged in tandem. For instance, Fig. 24.13 shows a simple arrangement of two polarizers. The first polarizer has its preferential direction oriented vertically; the second has its preferential direction inclined at an angle ϕ. Unpolarized light is incident on the first polarizer, which selects waves of vertical polarization and allows them to pass; the light emerging from this first polarizer is therefore vertically polarized. When this light is incident on the second polarizer, or ***analyzer,*** its electric field vector makes an angle ϕ with the preferential direction. We can regard a light wave with such an electric field as a superposition of two light waves, whose electric fields are, respectively, parallel and perpendicular to the preferential direction of the analyzer. The analyzer permits the passage of the former wave but blocks the latter. If the amplitude of the wave incident on the analyzer is E_0, then the amplitude of the parallel wave is

Polarizers and analyzers ▶

$$E'_0 = E_0 \cos\phi \tag{13}$$

Since the intensity of a light wave, like that of any other kind of wave, is proportional to the square of the amplitude,* the intensity of the wave transmitted by the analyzer is smaller than that of the wave incident on the analyzer by a factor of $\cos^2\phi$.

Law of Malus ▶

$$[\text{transmitted intensity}] = [\text{incident intensity}] \times \cos^2\phi \tag{14}$$

* For a precise definition of the intensity of a light wave we just examine the energy flux in the wave (that is, the energy carried by the wave per square meter per second); we will deal with this precise definition of the intensity in Section 24.5.

This relation between the intensity incident on the analyzer and the intensity transmitted by the analyzer is called the ***Law of Malus.*** Note that if $\phi = 90°$, the transmitted intensity is zero; that is, such "crossed" polarizers block the light completely.

This blocking of light by crossed polarizers can be readily demonstrated by means of two Polaroid sunglasses (see Fig. 24.14). However, these sunglasses cannot be used for a quantitative test of Eq. (14), because the reduction of intensity of light by the sunglasses is caused not entirely by polarization but also by the color in the glass.

FIGURE 24.14

The light that passes through the distant sunglasses becomes polarized in the vertical direction, because this is the preferential direction of the Polaroid in the sunglasses. If the near sunglasses are oriented parallel to the distant sunglasses ($\phi = 0$), the polarized light can pass and reach us. But if the near sunglasses are oriented perpendicular to the distant sunglasses ($\phi = 90°$), the polarized light is blocked.

EXAMPLE 3. Suppose that the preferential direction of the second polarizer makes an angle of 30° with the preferential direction of the first. If unpolarized light is incident on the first polarizer from the left, what fraction of this incident light will pass through both polarizers and emerge on the right?

SOLUTION: In unpolarized light, on the average, one-half of the light waves are polarized in the vertical direction and one-half in the horizontal direction. Hence, one-half of the light will be able to pass through the first polarizer. According to Eq. (14), a fraction $\cos^2\phi = \cos^2 30° = 0.75$ of this light will then pass through the second polarizer. Hence the fraction of the light that passes through both polarizers is $0.5 \times 0.75 = 0.375$.

24.4 THE GENERATION OF ELECTROMAGNETIC WAVES

As discussed in some detail in Section 24.2, an accelerated point charge creates an electromagnetic wave pulse which spreads outward from the charge. In essence, this wave pulse is a disturbance of the familiar electric and magnetic fields with which we began our study of electricity and magnetism. As long as the charge moves with uniform velocity, these fields accompany the charge — they move as though they were rigidly attached to the charge. But if the charge is forced to accelerate, then parts of the fields break away — they become independent of the charge and they travel outward as an electromagnetic wave pulse. If the charge moves back and forth with periodic motion and periodic acceleration, then it will radiate a periodic wave, with a frequency equal to that of the motion of the charge. The wavelength of the wave is related to its frequency by Eq. (10),

$$\lambda = \frac{c}{f} \tag{15}$$

Antennas of a radio station.

For instance, the electric charges on the antenna of an FM radio station typically oscillate back and forth with a frequency of 10^8 Hz (or 100

MHz); correspondingly, the wavelength of the radiation emitted by these accelerated charges has a wavelength of

$$\lambda = \frac{c}{f} = \frac{3.0 \times 10^8 \text{ m/s}}{10^8/\text{s}} = 3.0 \text{ m} \tag{16}$$

The oscillations of the charges on the antenna are produced by means of a resonating LC circuit coupled to the antenna by a mutual inductance (see Fig. 23.17). In essence, this is the method used to generate ***long waves, medium waves*** (AM), and ***short waves*** (including FM), as well as ***TV waves.*** Such radio waves span a wavelength range from 10^5 m to a few centimeters.

The frequencies of AM or FM radio waves indicated on the dials of radio receivers refer to the frequency of the carrier wave, which is the steady, periodic wave emitted by the radio station during moments of silence, when there is no audio signal (no voice signal and no musical signal). AM and FM stations use different methods for imprinting the audio signal on the carrier wave. In the AM (*a*mplitude *m*odulation) method, the amplitude of the carrier wave is altered in accordance with the amplitude of the audio signal to be transmitted; thus, the carrier amplitude is increased or decreased when the amplitude of the audio signal increases or decreases (see Fig. 24.15). In the FM (*f*requency *m*odulation) method, the frequency of the carrier wave is altered in accordance with the amplitude of the audio signal; thus, the carrier frequency is increased or decreased when the amplitude of the audio signal increases or decreases (see Fig. 24.16), but the amplitude of the carrier wave is kept constant. FM attains a higher fidelity than AM because it is quite insensitive to disturbances in the strength of the carrier wave. Any such disturbances in the strength of the FM carrier wave have no effect on its frequency, and hence do not alter the audio signal imprinted on the wave. By contrast, any disturbance in the strength of an AM carrier wave leads to a distortion of the audio signal.

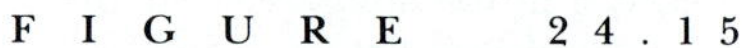
FIGURE 24.15

(a) Carrier radio wave. (b) Audio, or sound, signal. (c) AM radio wave. The amplitude of the wave is modulated according to the audio signal.

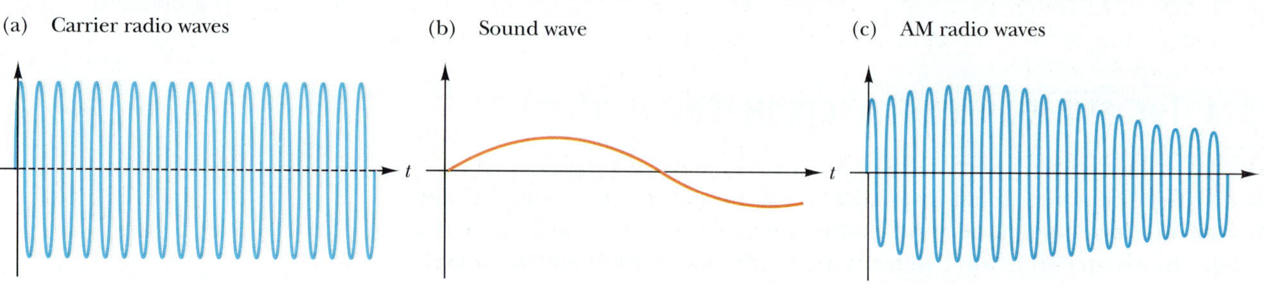

Waves of a wavelength shorter than that of ordinary radio waves, called ***microwaves,*** are best generated by a resonating electromagnetic cavity, consisting of an empty metallic can in which a standing electromagnetic wave is set up by an electron beam passing through, much as a standing sound wave is set up in an organ pipe by a stream of air passing over the blowhole. The antenna that radiates the microwaves is simply a horn con-

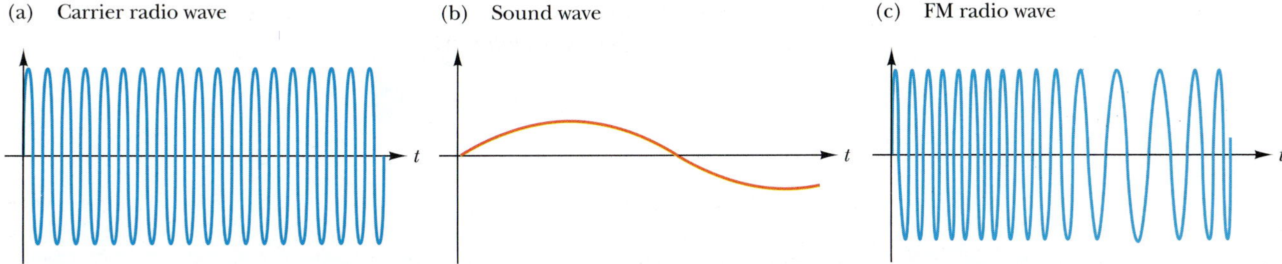

FIGURE 24.16

(a) Carrier radio wave. (b) Audio signal. (c) FM radio wave. The frequency of the wave is modulated according to the audio signal.

nected to the electromagnetic cavity by a metallic pipe, or waveguide, which permits the waves to spill out into space (see Fig. 24.17). This method can be used to generate waves as short as about a millimeter. Shorter wavelengths cannot be generated with currents oscillating in macroscopic laboratory equipment; however, short wavelengths can be easily generated by electrons oscillating within molecules and atoms subjected to stimulation by heat or by an electric current. Depending on the details of the motion, the electrons in molecules and in atoms will emit ***infrared radiation, visible light, ultraviolet radiation,*** or ***X rays;*** the corresponding wavelengths range from 10^{-3} m to 10^{-11} m. X rays can also be generated by the acceleration that high-speed electrons suffer during impact on a target; this is ***Bremsstrahlung*** (German for *braking radiation*). Radiations of even shorter wavelengths are emitted by protons and neutrons moving within a nucleus; these are ***gamma rays,*** with wavelengths as short as 10^{-13} m. However, the motion of subatomic particles and their emission of radiation cannot be calculated by classical mechanics or classical electricity and magnetism; such calculations require quantum mechanics.

FIGURE 24.17

Horns of microwave antennas.

◂ **Incoherent and coherent waves**

In an ordinary light source, such as an incandescent light bulb, the individual atoms or molecules radiate independently. The emerging light consists of a superposition of many individual light waves with random phase differences, random directions of polarization, and diverging directions of propagation; such ***light waves with random, unpredictable phase differences are said to be incoherent.*** In a laser, the atoms or molecules radiate in unison, by a quantum-mechanical phenomenon called ***stimulated emission.*** The emerging light is a superposition of light waves with exactly the same phases, the same directions of polarization, and the same directions of propagation; such ***light waves with no phase differences, or with predictable phase differences, are said to be coherent.*** Since the individual light waves in this kind of light combine constructively, the light emerging from the laser is very intense, and it also forms a very narrow, sharp beam.

Another important mechanism for the generation of electromagnetic waves is ***cyclotron emission.*** This involves high-speed electrons undergoing centripetal acceleration while spiraling in a magnetic field (see Section 21.5). Depending on the speed of the electrons and the strength of the magnetic field, the radiation may consist of radio waves, X rays, or anything in between (including visible light). Most of the radio waves reaching us from stars, pulsars, and radio galaxies are generated by this process.

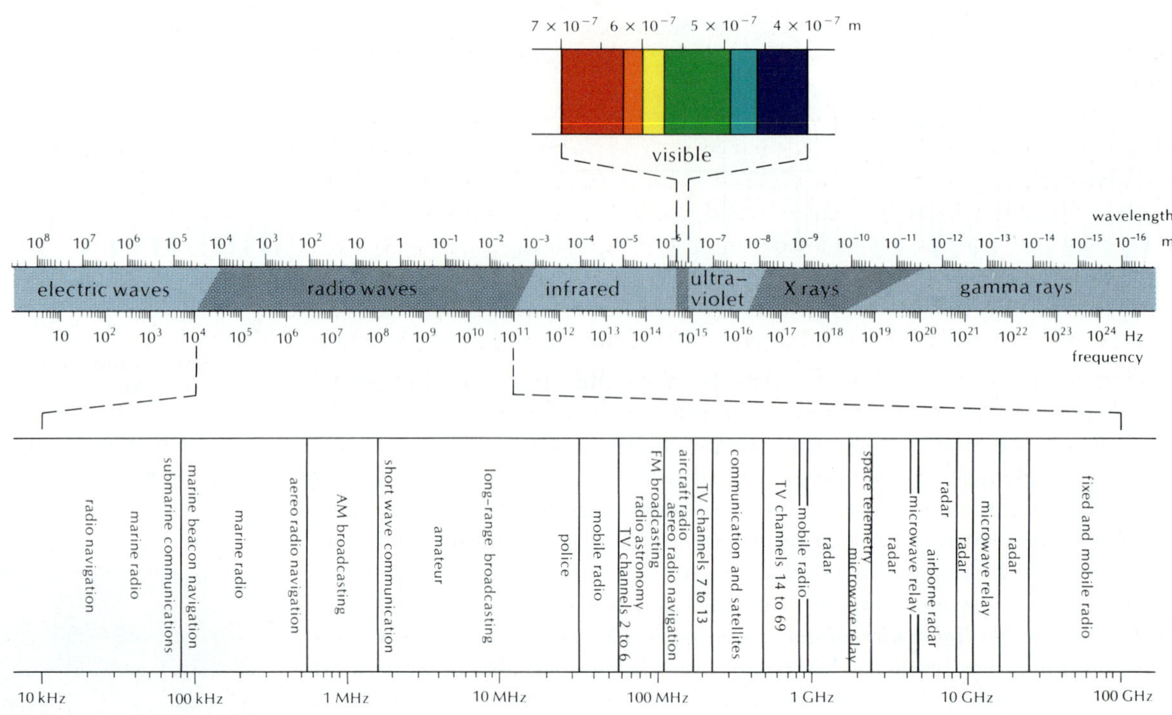

F I G U R E 24.18

Wavelength and frequency bands of electromagnetic radiation.

Figure 24.18 displays the wavelength and the frequency bands of electromagnetic radiation. The bands overlap to some extent, because the names assigned to the different ranges of wavelengths depend not only on the value of the wavelength, but also on the method used to generate and/or detect the radiation. For example, radiation of a wavelength of a tenth of a millimeter will be called a radio wave (microwave) if detected by a radio receiver, but it will be called infrared radiation if detected by a heat sensor.

The wavelengths of visible light range from about 7×10^{-7} m to 4×10^{-7} m. The wavelength of light is usually expressed in nanometers (1 nm $= 10^{-9}$ m), in terms of which ***the range of wavelengths of visible light extends from 400*** nm ***to 700*** nm. This is the range of wavelengths over which our eyes are sensitive to light. We perceive different wavelengths within the visible region as having different colors. Figure 24.19 is a plot

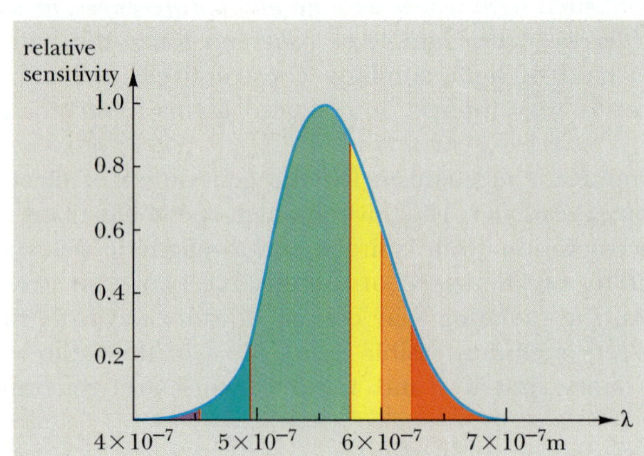

F I G U R E 24.19

Sensitivity of the human eye at different wavelengths and colors in bright light. The sensitivity is maximum at about 550 nm (yellow-green light) and decreases to about 1% of this maximum at 690 nm (deep red) and at 430 nm (violet). In dim light, the sensitivity changes; it becomes maximum at about 500 nm (green).

of the sensitivity of the human eye and shows how colors are correlated with wavelength.

Incidentally: Our eyes are almost completely insensitive to the polarization of light waves. We can detect the polarization only with special equipment, such as Polaroid sunglasses. Radio and TV antennas are of course very sensitive to the direction of polarization of radio waves, and they must have the proper orientation to pick up a strong signal. For instance, a simple antenna in the form of a straight wire or rod is sensitive to a radio wave polarized in the same direction as the antenna, but insensitive to a radio wave polarized at right angles to the antenna. Some TV antennas are equipped with an electric motor so that they can be easily rotated to the optimal orientation for each TV station you want to receive.

24.5 ENERGY OF A WAVE

The electric and magnetic fields of an electromagnetic wave contain energy. As the wave moves along, so does this energy — the wave transports energy.

Let us calculate the flow of energy in a plane wave, or an approximately plane portion of a spherical wave. Suppose the wave propagates in the positive x direction, with its electric fields and magnetic fields in the y and z directions, respectively. Figure 24.20 shows a slab of the wave within which the electric and magnetic fields are nearly uniform. The densities of electric and magnetic energy are $(\epsilon_0/2)E^2$ and $(1/2\mu_0)B^2$ [see Eqs. (18.39) and (22.29)], so the energy density in the wave is

$$\frac{\epsilon_0}{2}E^2 + \frac{1}{2\mu_0}B^2 \tag{17}$$

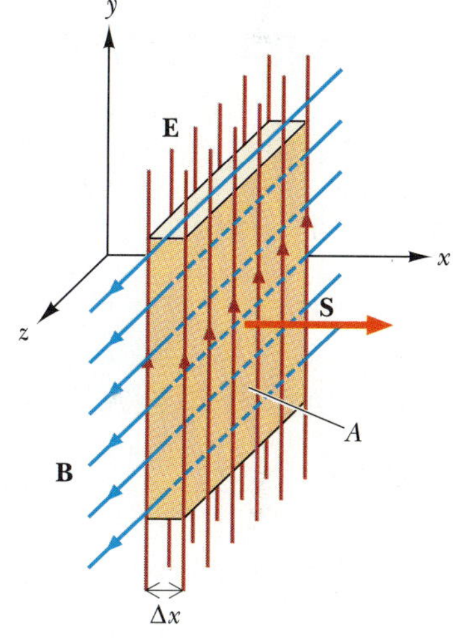

FIGURE 24.20
A slab of electric (red) and magnetic (blue) fields in a plane wave propagating toward the right. The slab of thickness Δx and frontal area A moves with the wave, at speed c.

If the thickness of the wave slab is Δx and the frontal area is A (see Fig. 24.20), then the volume of the slab is $A\Delta x$, and the total amount of energy in the slab is

$$\Delta U = \left(\frac{\epsilon_0}{2}E^2 + \frac{1}{2\mu_0}B^2\right) \times A\Delta x \tag{18}$$

Since $B = E/c$ and $\epsilon_0 = 1/\mu_0 c^2$, we can write this energy as

$$\Delta U = \left[\frac{1}{2\mu_0 c^2}E^2 + \frac{1}{2\mu_0}\left(\frac{E}{c}\right)^2\right] A\Delta x \tag{19}$$

$$= \frac{1}{\mu_0 c^2}E^2 A\Delta x \tag{20}$$

Note that the two terms on the right side of Eq. (19) are equal; that is, the electric and magnetic energy densities in an electromagnetic wave are equal.

The forward speed of the wave slab is c, and hence the amount of energy ΔU moves out of the (stationary) volume $A\Delta x$ in a time $\Delta t = \Delta x/c$. The rate of flow of energy, or the power, is therefore

$$\frac{\Delta U}{\Delta t} = \frac{\Delta U}{\Delta x/c} = \frac{1}{\mu_0 c} AE^2 \tag{21}$$

Energy flux in a wave ▶

In Fig. 24.20, the energy flows toward the right, along the direction of propagation of the wave. ***The energy flux in the wave is defined as the rate of energy flow per unit frontal area,*** that is, $(1/A)\Delta U/\Delta t$. The rate of energy flow is the power; hence the energy flux is the power carried per unit area of wave front. The units for this energy flux are watts per square meter (W/m^2). According to Eq. (21), the energy flux for our plane wave is

$$\frac{1}{A}\frac{\Delta U}{\Delta t} = \frac{1}{\mu_0 c} E^2 \tag{22}$$

This is called the ***Poynting flux,*** usually designated by S,

Poynting flux ▶

$$S = \frac{1}{\mu_0 c} E^2 \tag{23}$$

The energy flux, or the Poynting flux, provides us with a precise measure of the intensity of the wave. A wave of high intensity — such as the wave produced by a powerful laser — is a wave of large energy flux.

Since the electric field oscillates in time, so does the energy flux. For a harmonic wave, such as shown in Fig. 24.8b, with a sinusoidal dependence on time, the average of the square of the electric field is one-half of the square of the amplitude of the electric field, $\overline{E^2} = E_0^2/2$ (compare the calculation of the average of the square of the emf in Section 23.1). Hence the time-average energy flux in the wave is

$$\overline{S} = \frac{1}{2\mu_0 c} E_0^2 \tag{24}$$

EXAMPLE 4. At a distance of 8.5 km from a radio transmitter, the amplitude of the oscillating electric field in the radio wave is $E_0 = 0.13$ V/m. What is the time-average energy flux? What is the total power radiated by the radio transmitter? Assume that the transmitter radiates uniformly in all directions.

SOLUTION: From Eq. (24) we find

$$\overline{S} = \frac{1}{2\mu_0 c} E_0^2 = \frac{(0.13 \text{ V/m})^2}{2 \times 1.26 \times 10^{-6} \text{ H/m} \times 3 \times 10^8 \text{ m/s}}$$

$$= 2.2 \times 10^{-5} \text{ W/m}^2$$

To obtain the total power, we must multiply the power per unit area by the area over which the radio wave spreads. This area is a sphere of radius r and of area $4\pi r^2$, with $r = 8.5$ km (see Fig. 24.21). Therefore the total power is

$$\overline{P} = [\text{area}] \times [\text{energy flux}] = 4\pi r^2 \overline{S} \tag{25}$$

$$= 4\pi \times (8.5 \times 10^3 \text{ m})^2 \times 2.2 \times 10^{-5} \text{ W/m}^2$$

$$= 2.0 \times 10^4 \text{ W} = 20 \text{ kW}$$

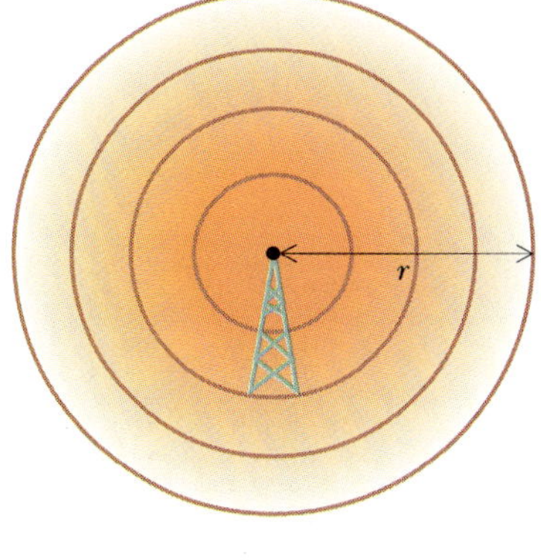

FIGURE 24.21

The energy of the radio wave is spread out over a sphere of radius $r = 8.5$ km.

Note that, according to Eq. (25), ***the energy flux $\overline{S}$ for a spherical wave spreading out from a source in all directions is inversely proportional to the square of the distance,***

$$\overline{S} = \frac{\overline{P}}{4\pi r^2} \tag{26}$$

We must take this dependence of the flux on distance into account whenever we want to investigate the spreading of the wave over a large range of distances (but we can ignore this dependence over a small range of distances, where the wave can be approximated as a plane wave of constant energy flux).

SUMMARY

- **Maxwell's equations:** $\Phi = Q/\epsilon_0$ (Gauss' Law for electricity)

  $\Phi_B = 0$ (Gauss' Law for magnetism)

  $\overline{E}_t l = -\dfrac{\Delta \Phi_B}{\Delta t}$ (Faraday's Law)

  $\overline{B}_t l = \mu_0 I + \mu_0 \epsilon_0 \dfrac{\Delta \Phi}{\Delta t}$ (Maxwell–Ampère's Law)

- **Speed of electromagnetic wave:** $c = 1/\sqrt{\mu_0 \epsilon_0}$
- **Magnetic field of wave:** $B = E/c$
- **Time-average energy flux in plane wave:** $\overline{S} = \dfrac{1}{2\mu_0 c} E_0^2$

QUESTIONS

1.–Consider the electric field of a single positive electric charge moving at constant velocity. What is the direction of the displacement current intercepted by a circular area perpendicular to the velocity in front of the charge? Behind the charge?

2.–Which of Maxwell's equations permits us to deduce the electric Coulomb field of a static charge? Which of Maxwell's equations permits us to deduce the magnetic field of a charge moving with uniform velocity?

3.–Suppose that there exist magnetic monopoles, that is, positive and negative magnetic charges that act as sources and sinks of magnetic field lines, analogous to positive and negative electric charges. Which of Maxwell's equations would have to be modified to take into account such monopoles? Qualitatively, what are the required modifications?

4.–Some microwave ovens have rotating turntables that continually turn the food while cooking. What is the purpose of this arrangement? (Hint: Microwave ovens are cavities with standing electromagnetic waves.)

5.–A charged particle moving around a circular orbit at uniform speed has a centripetal acceleration and therefore produces a radiation field. However, a uniform current flowing around a circular loop does *not* produce a radiation field. Is this a contradiction? Explain.

6.–Consider transverse waves on a string. How many independent directions of polarization do such waves have? How could you construct a mechanical polarization filter that permits only the passage of a wave polarized in a preferential direction?

7.–If the preferential directions of two adjacent sheets of Polaroid are at right angles, no light will pass through. However, if you now slip a third sheet of Polaroid between the other two and orient its preferential direction so that it lies between the directions of the other two, then some light will pass through the three sheets. Explain.

8.–It has been proposed that we could eliminate the glare of the headlights of approaching automobiles by covering the windshields and the headlights with sheets of Polaroid. What orientation should we pick for the sheets of Polaroid installed on windshields and on headlights so that the light of every approaching automobile is blocked out, but our own light is not?

9.–Some small-boat sailors like to wear Polaroid sunglasses because these make disturbances of the water surface stand out with exceptional contrast, and thereby make it easier to spot approaching gusts of wind. Why are Polaroid sunglasses better for this purpose than ordinary sunglasses? (Hint: The amount of polarization of reflected sunlight depends on the angle of reflection.)

10.–Figure 24.19 shows the sensitivity of the human eye as a function of the wavelength of light. The sensitivity is maximum at about 5.5×10^{-7} m and drops to about 1% at 6.9×10^{-7} m and at 4.3×10^{-7} m. Suppose that the sensitivity of your eye were constant over the entire interval of wavelengths shown in Fig. 24.19. How would this alter your visual perception of some of the things you see in your everyday life?

11.–Why does the radio reception fade in the receiver of your automobile when you enter a tunnel?

12.–Short-wave radio waves are reflected by the ionosphere of the Earth; this makes them very useful for long-range communication. Explain.

13.–In the seventeenth century, the Danish astronomer Ole Roemer noticed that the orbital periods of the moons of Jupiter, as observed from the Earth, exhibit some systematic irregularities: the periods are slightly longer when the Earth is moving away from Jupiter, and slightly shorter when the Earth is moving toward Jupiter. Roemer attributed this apparent irregularity to the finite speed of propagation of light, and used it to make the first determination of the speed of light. Explain how the lengthening or the shortening of the observed period arises from the travel time of light from the moon to the Earth. (This shift of period may be regarded as the earliest discovery of a Doppler shift.)

PROBLEMS

Section 24.1: Induction of Magnetic Fields; Maxwell's Equations

1.——A parallel-plate capacitor of plate area A is being charged by a current I flowing into its plates via external wires. At one instant, the charge on the capacitor plates is Q.
   (a) Assume that the electric field between the plates is uniform. Show that the electric field between the plates at this instant is $E = Q/\epsilon_0 A$.
   (b) Show that the electric flux crossing the mathematical midplane between the plate surfaces is Q/ϵ_0.
   (c) What is the displacement current in this capacitor?
   (d) Show that this displacement current matches the ordinary current I flowing into the plates.

2.——A capacitor has circular plates, such as illustrated in Fig. 24.3, of radius 0.15 m. Between these plates there is a uniform electric field. Suppose that this electric field is increasing at the rate of 3.8×10^{13} V/m per second. What is the displacement current between the plates? What is the magnitude of the magnetic field at the edge of the capacitor, halfway between the plates?

3.*——A parallel-plate capacitor is being charged by a current of 4.0 A.
   (a) What is the displacement current between its plates?
   (b) What is the rate of change of the electric flux intercepted by each plate?

4.**——The space between the plates of a leaky capacitor is filled with a material of resistance $5.0 \times 10^5\ \Omega$. The capacitor has a capacitance of 2.0×10^{-6} F, its plates are circular, with a radius of 0.30 m, and its electric field is uniform. At time $t = 0$, the initial voltage across the capacitor is zero.
   (a) What is the displacement current if we increase the voltage at the steady rate of 1.0×10^3 V/s?
   (b) At what time will the real current leaking through the capacitor equal the displacement current?

5.**——A parallel-plate capacitor has circular plates of area A separated by a distance d. A thin straight wire of length d lies along the axis of the capacitor and connects the two plates (Fig. 24.22); this wire has a resistance R. The exterior terminals of the plates are connected to a source of alternating emf with a voltage $V = V_0 \sin \omega t$.
   (a) What is the current in the thin wire?
   (b) What is the displacement current through the capacitor?
   (c) What is the current arriving at the outside terminals of the capacitor?

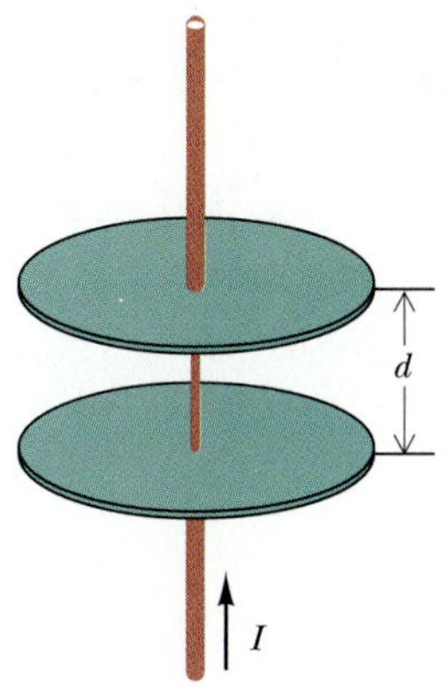

FIGURE 24.22
Parallel-plate capacitor with a thin wire connecting the inside faces of the plates.

6.**——Suppose that the parallel-plate capacitor discussed in Section 24.1 (see Fig. 24.3) is filled with a slab of dielectric with a dielectric constant κ. Assume that, as in the case of the empty capacitor, the displacement current can be viewed as a continuation of the ordinary current. Show that this implies that the displacement current in the filled capacitor must be

$$\kappa\epsilon_0 \frac{\Delta\Phi}{\Delta t}$$

and Maxwell's modification of Ampère's Law must be

$$\overline{B}_t l = \mu_0 I + \kappa\mu_0\epsilon_0 \frac{\Delta\Phi}{\Delta t}$$

Section 24.2: The Electromagnetic Wave Pulse

7.——Roughly sketch the electric field lines for a positive charge that is initially moving at uniform velocity and suddenly stops.

8.——In a collision with an atom, an electron suddenly stops. Describe the directions of the electric and magnetic radiation fields at some distance from the electron at right angles to the acceleration.

9.——Draw a diagram analogous to Fig. 24.4, showing the electric field lines of a charge that has suffered an acceleration toward the left (that is, an acceleration opposite to that involved in Fig. 24.4).

10.*—Suppose that a charge is initially at rest, then is accelerated for a short time interval from $t = 0$ to $t = \tau$, then is decelerated from $t = \tau$ to $t = 2\tau$, and then remains at rest after $t = 2\tau$. Carefully draw a diagram analogous to Fig. 24.4, showing the electric field lines of such a charge, at some time after the deceleration has ended.

11.——Figure 24.6 shows the electric and magnetic field lines for a flat wave pulse propagating in the positive x direction. Draw an analogous diagram for a flat wave pulse propagating in the negative x direction.

FIGURE 24.23
A laser rangefinder.

Section 24.3: Plane Waves; Polarization

12.——A severe limitation on the speed of computation of large electronic computers is imposed by the speed of light because the electric signals on the connecting wires within the computer are electromagnetic waves ("guided waves"), which travel at a speed roughly equal to the speed of light. If the computer measures about 1 m across, what is the minimum travel time required for a typical signal sent from one end of the computer to another? What is the maximum number of signals that can be sent back and forth (sequentially) per second? Is there any way to avoid the limitation imposed by the travel time of signals?

13.——A light-year is defined as the distance that light travels in one year. Calculate how much this is in meters.

14.——Laser rangefinders used by surveyors (see Fig. 24.23) determine the distance to a reflecting target by means of a pulse at laser light, which travels from the rangefinder to the target and back. The distance is automatically calculated from the travel time of this pulse. If such a rangefinder is to determine a distance of 100 m to within 1 cm, what is the maximum permitted error in the measurement of the travel time?

15.——When the American astronauts on the Moon were in conversation with Mission Control on Earth, there was a noticeable delay between questions and answers. What is the roundtrip travel time for a radio signal from Earth to Moon and back? The distance to the Moon is 3.8×10^8 m.

16.——An electromagnetic wave is traveling vertically upward. The instantaneous electric field at some point in this wave is eastward, of magnitude 150 V/m. What are the magnitude and the direction of the instantaneous magnetic field?

17.—(a) One type of antenna for a radio receiver consists of a short piece of straight wire; when the electric field of a radio wave strikes this wire it makes currents flow along it, which are detected and amplified by the receiver. Suppose that the electric field of a radio wave is vertical. What must be the orientation of the wire for maximum sensitivity?

(b) Another type of antenna consists of a circular loop; when the magnetic field of a radio wave strikes this loop it induces currents around it. Suppose that the magnetic field of a radio wave is horizontal. What must be the orientation of the loop for maximum sensitivity?

18.—A plane electromagnetic wave travels in the eastward direction. At one instant the electric field at a given point has a magnitude of 0.60 V/m and points down. What are the magnitude and direction of the magnetic field at this instant? Draw a diagram showing the electric field, the magnetic field, and the direction of propagation.

19.—Linearly polarized light is incident on a polarizer whose preferential direction is inclined at an angle of 20° relative to the plane of polarization of the incident light. What fraction of the incident light is transmitted? Repeat for an angle of 40°, and repeat for an angle of 60°.

20.*—Suppose that an unpolarized light beam is incident from the left on the arrangement of two polarizers illustrated in Fig. 24.13. If the intensity of the light emerging on the right is 30% of the incident intensity, what must be the angle between the preferential directions of the polarizers?

21.*—A light wave is polarized in the vertical plane. This wave is incident on a polarizer whose preferential direction is inclined at 60° to the vertical, and then on a second polarizer whose preferential direction is vertical. What fraction of the incident light emerges from the second polarizer?

22.*—The preferential directions of two adjacent sheets of Polaroid make an angle of 45°. A beam of polarized light, whose direction of polarization coincides with the preferential direction of the *second* sheet, is incident on the *first* sheet. By what factor is the intensity of the transmitted beam emerging from the second sheet reduced compared with the intensity of the incident beam? Assume that the sheets act as ideal polarizing filters.

23.*—Two sheets of Polaroid are placed on top of one another. Unpolarized light is perpendicularly incident on the sheets. By what factor is the intensity of the emerging light reduced (relative to the incident light) if the preferential directions of the sheets differ by an angle of 30°? 45°? 60°?

Section 24.4: The Generation of Electromagnetic Waves

24.—At many coastal locations, radio stations of the National Weather Service transmit continuous weather reports at a frequency of 162.5 MHz. What is the wavelength of these transmissions?

25.—The shortest microwaves have a wavelength of about 1 mm. What is the frequency of such waves?

26.—Radio station WWV of the National Institute of Standards, Fort Collins, Colorado, transmits precise time signals at radio frequencies of 2.5, 5, 10, 15, and 20 MHz. What are the wavelengths of these transmissions?

27.—An ordinary radio receiver, such as found in homes across the country, has an AM dial and an FM dial. The AM dial covers a range from 530 to 1600 kHz and the FM dial a range from 88 to 108 MHz. What is the range of wavelengths for AM? For FM?

28.—Hydrogen atoms in interstellar clouds of gas emit radio waves of wavelength 21 cm. What is the frequency of these waves?

29.—For radio communication with submerged submarines, the U.S. Navy uses ELF (extremely low frequency) radio waves of wavelength 4000 km; such waves can penetrate for some distance below water. What is the frequency of such waves?

AM and FM dials on an ordinary radio.

30.——A radar antenna emits radio waves of a frequency of 1.1×10^{10} Hz. What is the wavelength of these waves?

31.——Figure 24.19 gives the sensitivity of the human eye as a function of the wavelength of light. For what color is the sensitivity maximum? For what color is the sensitivity one-half of the maximum? One-quarter of the maximum?

Laser used as torch to cut plates of metal.

Section 24.5: Energy of a Wave

32.——At one point in an electromagnetic wave, the instantaneous electric field has a magnitude of 80 V/m. What is the energy density?

33.——At a distance of several kilometers from a radio transmitter, the electric field of the emitted radio wave has a magnitude of 0.12 V/m at one instant of time. What is the energy density in this electric field? What is the energy density in the magnetic field of the radio wave?

34.——A radio wave has an instantaneous magnetic field of 2×10^{-10} T. What is the magnitude of the instantaneous Poynting vector?

35.——The average energy flux of sunlight incident on the top of the Earth's atmosphere is 1.4×10^3 W/m^2. What are the corresponding amplitudes of oscillation of the electric and magnetic fields?

36.——A plane electromagnetic wave travels in the northward direction. At one instant, the electric field at a given point has a magnitude of 0.50 V/m and is in the eastward direction. What are the magnitude and direction of the magnetic field at the given point? What are the magnitude and direction of the Poynting vector?

37.——A laser used as a torch, to cut plates of metal, produces a light beam with an energy flux of 10^9 W/m^2. What are the magnitudes of the rms electric and magnetic fields in such a light beam?

38.——Starlight arriving at the Earth from the star Capella has an rms energy flux of 1.2×10^{-8} W/m^2. The distance of this star is 4.3×10^{17} m. Calculate the power radiated by this star.

39.——The beam of light produced by a small laser is cylindrical, of diameter 2.5 mm. The rms power that the laser feeds into this beam is 1.2 W. Calculate the rms values of the electric and magnetic fields.

40.——The Sun emits radiation uniformly in all directions. At the Earth, at a distance of 1.5×10^{11} m, the energy flux of sunlight is 1.4×10^3 W/m^2. Calculate the power radiated by the Sun.

Solar cells.

41.——The beam of a powerful laser has a diameter of 0.2 cm and carries a power of 6 kW. What is the time-average Poynting vector in this beam? What are the rms values of the electric and magnetic fields?

42.——A TV transmitter emits a spherical wave, that is, a wave spreading out uniformly in all directions. At a distance of 5 km from the transmitter, the amplitude of the wave is 0.22 V/m. What is the magnitude of the time-average Poynting vector at this distance? What is the time-average power emitted by the transmitter?

43.*—A silicon solar cell of frontal area 13 cm^2 delivers 0.20 A at 0.45 V when exposed to full sunlight of energy flux 1.0×10^3 W/m^2. What is the efficiency for conversion of light energy into electric energy?

44.*—A magnifying glass of diameter 10 cm focuses sunlight into a spot of diameter 0.50 cm. The energy flux in the sunlight incident on the lens is 0.10 W/cm^2.

(a) What is the energy flux in the focal spot? Assume that all points in the spot receive the same flux.

(b) Will newspaper ignite when placed at the focal spot? Assume that the flux required for ignition is 2 W/cm^2.

45.*—Binoculars are usually marked with their magnification and lens size. For instance, 7×50 binoculars magnify angles by a factor of 7 and their collecting lenses have an aperture of diameter 50 mm. Your pupil, when dark adapted, has an aperture of diameter 7.0 mm. When observing a distant pointlike light source at night, by what factor do these binoculars increase the energy flux penetrating your eye? Neglect reflection of light by the lenses.

Binoculars.

46.*—In the United States, the accepted standard for the safe maximum level of continuous whole-body exposure to microwave radiation is 10 milliwatts/cm^2.

(a) For this energy flux, what are the corresponding amplitudes of oscillation of the electric and magnetic fields?

(b) Suppose that a man of frontal area 1.0 m^2 completely absorbs microwaves with an intensity of 10 milliwatts/cm^2 incident on this area and that the microwave energy is converted to heat within his body. What is the rate (in calories per second) at which his body develops heat?

47.*—A radio transmitter emits a time-average power of 5 kW in the form of a radio wave with uniform intensity in all directions. What are the amplitudes of the electric and magnetic fields of this radio wave at a distance of 10 km from the transmitter?

48.*—A radio receiver has a sensitivity of 2×10^{-4} V/m. At what maximum distance from a radio transmitter emitting a time-average power of 10 kW will this radio receiver still be able to detect a signal? Assume that the transmitter radiates uniformly in all directions.

49.*—At a distance of 6.0 km from a radio transmitter, the amplitude of the electric radiation field of the emitted radio wave is $E_0 = 0.13$ V/m. Taking into account the decrease of the amplitude of the wave with distance, what will be the amplitude of the radio wave when it reaches a distance of 12.0 km? A distance of 18.0 km?

50.**—Two plane-wave pulses of the kind described in Section 24.3 are traveling in opposite directions. Their polarizations are parallel and the magnitudes of their electric fields are 2×10^{-3} V/m.

(a) What are the electric energy density and the magnetic energy density in each pulse?

(b) Suppose that at one instant the two pulses overlap. What are the magnitudes of the electric field and the magnetic field in this superposition?

(c) What are the electric energy density and the magnetic energy density in this superposition?

51.**—At night, the naked, dark-adapted eye can see a star provided the energy flux reaching the eye is at least 8.8×10^{-11} W/m^2.

(a) Under these conditions, how many watts of power enter the eye? The diameter of the dark-adapted pupil is 7.0 mm.

(b) Assume that in our neighborhood there are, on the average, 3.5×10^{-3} stars per cubic light-year and that each of these emits the same amount of light as the Sun (3.9×10^{26} W). If so, what is the maximum distance at which we could see a star with the naked eye? How many stars could we see in the sky?

CHAPTER 25

A large concave glass mirror, 1.8 m across, for an astronomical telescope. Large mirrors are easier to manufacture than large lenses, and most astronomical telescopes use mirrors.

REFLECTION, REFRACTION, AND OPTICS

So far we have examined the propagation of light waves and other electromagnetic waves only in a vacuum. There, a plane wave will simply propagate in a fixed direction at the constant speed c. But if the wave encounters the surface of a region filled with matter — a sheet of metal, a pane of glass, or a layer of water — then the wave will interact with the matter and suffer changes in speed, direction, intensity, and polarization. One part of the wave will be ***reflected*** by the surface; that is, it will bounce off. The other part will be ***refracted;*** that is, it will penetrate into the matter-filled region and continue to propagate, with some change of speed

Reflected wave and refracted wave ▶

and of direction. You can notice this partial reflection of light by a water surface when you look at the light from streetlamps or other bright sources reflected by the surface of a calm dark pond or puddle; the reflected light is not as strong as the direct light. You can notice the refraction of light by looking at your fingers through a drinking glass filled with water; your fingers appear distorted, because the rays of light traveling from your fingers to your eye suffer a change of direction when they enter the drinking glass, and they suffer another change of direction when they leave the drinking glass.

Since light is an electromagnetic wave, the changes in speed, direction, intensity, and polarization of the light wave can be calculated from Maxwell's equations, taking into account the motion of the electric charges and the flow of current caused by the action of the wave on the matter. But the complete calculation of all the changes in the wave is rather complicated, and, furthermore, Maxwell's equations often tell us more than we want to know. For instance, if a light wave encounters a water surface, we may wish to compute the angle at which it penetrates the water, but we do not always need to know the changes in intensity or polarization.

FIGURE 25.1
Rays of sunlight.

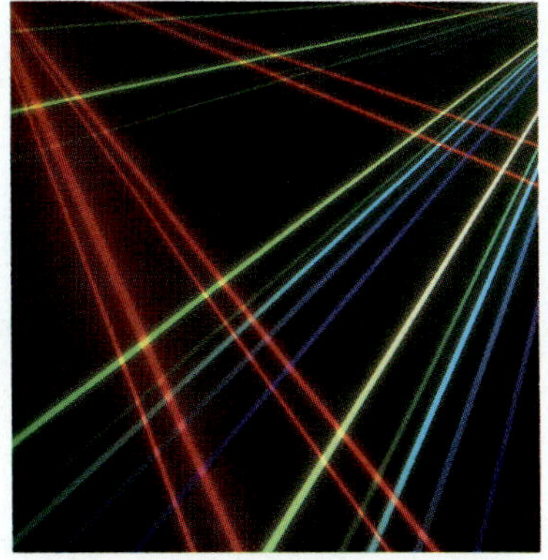

FIGURE 25.2
Beams of laser light.

In this chapter we will see that much can be learned by considering merely one aspect of Maxwell's equations, namely, their implications for the speed of light in matter-filled regions. We will see that in such regions the speed of light is reduced, and that this reduction of the speed of light leads to the change of direction of propagation during refraction, when the wave penetrates from vacuum into a matter-filled region or when the wave penetrates from one matter-filled region into another.

The law governing the change of direction during refraction can be deduced from the change of speed by means of Huygens' Construction, a simple geometric construction that reveals how the wave fronts advance when they encounter the surface of a region filled with matter. Huygens' Construction also reveals how the wave fronts are altered during reflection, and it allows us to deduce the law governing the change of direction during reflection.

◀ Geometric optics

These laws of reflection and refraction are the basis for ***geometric optics,*** which studies the propagation of light under the assumption that ***light propagates in a fixed direction, along a straight line (rectilinearly), while in a uniform medium, and suffers changes of direction only when it encounters the surface (or interface) separating two different media.*** You can observe this straight-line propagation in the beam of a strong searchlight aimed at the dark sky, or in beams of sunlight piercing through holes in a cloud (Fig. 25.1). But the most dramatic demonstration of straight-line propagation is provided by laser beams, which look like fine straight lines in space (Fig. 25.2).

25.1 HUYGENS' CONSTRUCTION

◀ Wave fronts

The propagation of a light wave or some other electromagnetic wave can be conveniently described by means of the ***wave fronts,*** or wave crests, that is, the points at which the electric field of the wave has maximum

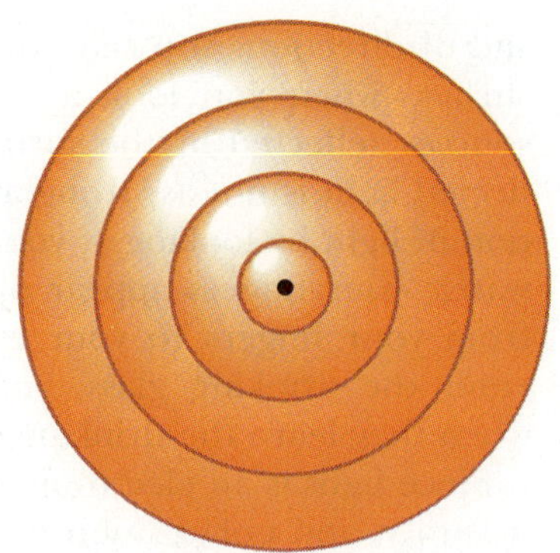

FIGURE 25.3

Spherical wave fronts at one instant of time. At a later time, each of these wave fronts will have moved outward by some distance.

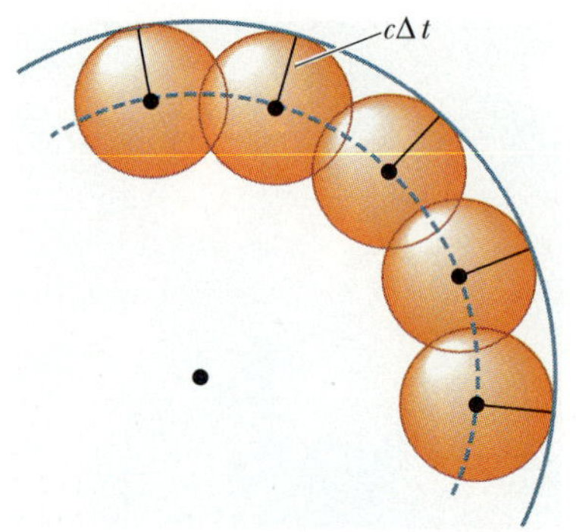

FIGURE 25.4

Huygens' Construction for the propagation of a spherical wave front. The inner (blue dashed) arc shows the wave front at time t; the outer (blue solid) arc shows the propagated wave front at time $t + \Delta t$.

strength at some instant of time. For example, Fig. 25.3 shows the instantaneous wave fronts of the spherical radio wave emitted by a radio station. The interval between one wave front and the next is one wavelength. With the passing of time, each of these wave fronts spreads in the outward direction.

The rule governing the propagation of wave fronts is ***Huygens' Construction:***

> ***To find the change of position of a wave front in a small time interval*** Δt, ***draw many small spheres of radius [wave speed]*** $\times$ Δt ***with centers on the old wave fronts. The new wave front is the surface that touches the leading edges of these small spheres.***

Wavelets ▸

The small spheres employed in this construction are called ***wavelets.*** Figure 25.4 shows how Huygens' Construction applies to the propagation of the spherical wave fronts in Fig. 25.3. The wave speed in this example is simply c, and hence the radius of the wavelets is $c\Delta t$. Erecting wavelets of this radius on the old wave front, we find the new wave front that touches the outer edges of these wavelets; since, in this example, all the wavelets have the same radius, the new wave front is concentric with the old wave front.

Figure 25.5 shows a similar construction for the case of the propagation of a plane radio wave or a plane light wave. When we erect the wavelets on the plane wave front, the result is another plane wave front, parallel to the first. This means that the plane wave continues to propagate in the same direction, without deviating to one side or another. A light beam, such as the light beam from a laser, consists of plane wave fronts, and Huygens' Construction therefore accounts for the straight-line propagation of such a light beam.

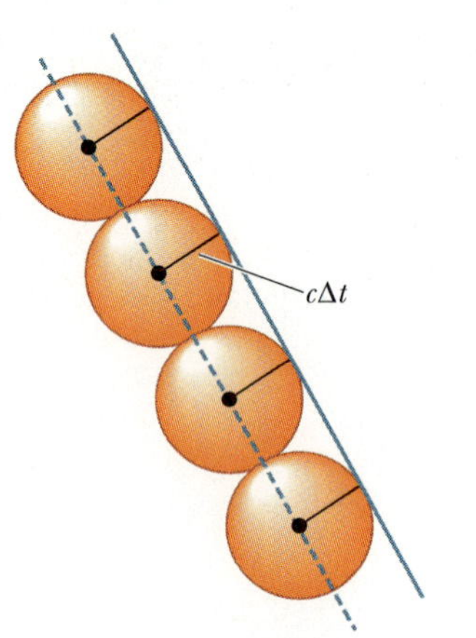

FIGURE 25.5

Huygens' Construction for the propagation of a plane wave front.

Huygens' Construction applies not only to the propagation of light waves in a vacuum, but also to their propagation in any transparent material, such as air, glass, or water. As we will see in the following two sections, this construction permits us to derive the laws of reflection and refraction. Although our emphasis will be on the propagation of light, Huygens' Construction is a general feature of wave propagation — it applies just as well to sound waves and water waves. The laws of reflection and refraction for all such waves are essentially the same, and many of our results for light waves can be readily generalized to such other waves.

25.2 REFLECTION

When a light wave encounters the surface of a transparent material — such as the surface of a pane of glass, or the surface of a pond — part of the wave penetrates the surface and part is reflected. When a light wave encounters the surface of a very smoothly polished metal — such as the silvered surface of a mirror — almost all of the wave is reflected. In this section we will deal with the reflected part of the wave; in the next section we will deal with the part of the wave that penetrates from one transparent material into the other.

◀ Law of reflection

The ***law of reflection*** for a wave incident on a flat surface at an angle has been known since ancient times: ***the angle of incidence equals the angle of reflection.*** Figure 25.6 gives an experimental demonstration of this law of reflection with a strong light beam shining down on a mirror. To derive this law of reflection from Huygens' Construction, we begin with Fig. 25.7a, which shows wave fronts approaching a reflecting surface; at the instant shown, one edge of the leading wave front is barely touching the surface at the point P. Figure 25.7b shows some Huygens' wavelets a short time later, when the second wave front has moved down to take the place previously occupied by the leading wave front. The portions of the wavelets below the reflecting surface have been omitted as irrelevant. The new leading wave front constructed on these wavelets touches the surface at the point P'. Obviously, to the right of the point P', the new wave front is simply parallel to the old wave front, that is, this part of the wave has not yet been reflected. To find the new wave front to the left of the point P', we draw a straight line that starts at P' and is tangent to the wavelet centered on P. This straight line represents the part of the wave that has already been reflected. To see that the incident wave front and the re-

FIGURE 25.6

Reflection of a beam of light by a mirror. The angle of incidence equals the angle of reflection.

(a)

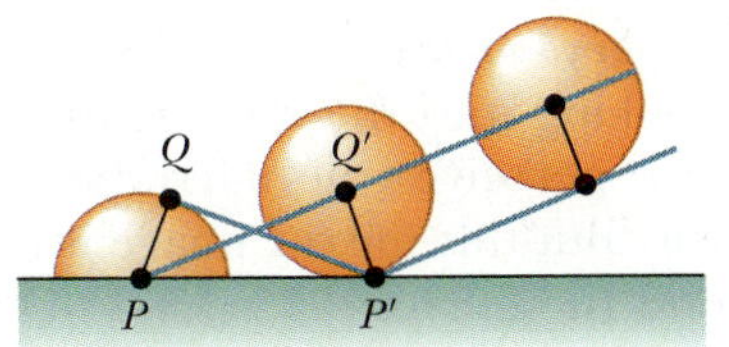

(c)

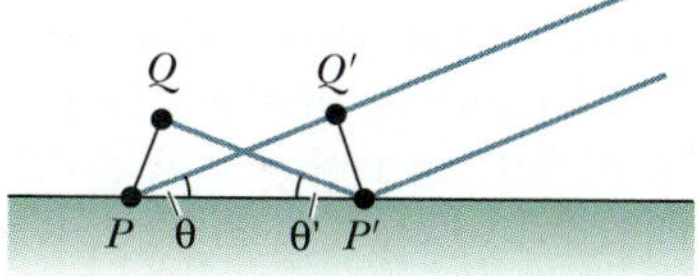

FIGURE 25.7

(a) Wave front approaching a reflecting surface. The leading wave front barely touches the reflecting surface. (b) Huygens' wavelets erected on the leading wave front of Part (a).
(c) The incident wave front PQ' makes an angle θ with the reflecting surface; the reflected wave front QP' makes an angle θ' with this surface.

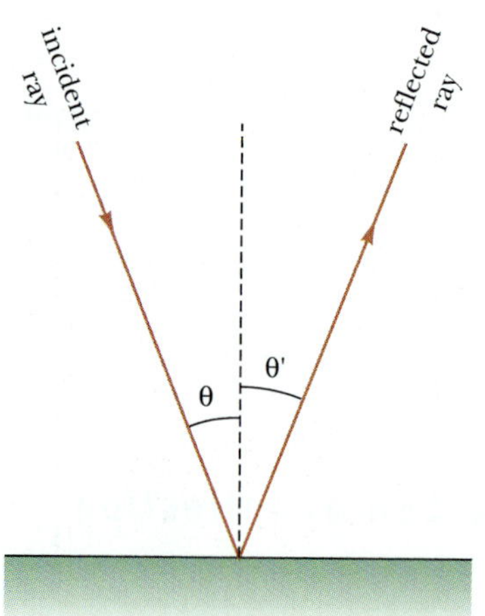

FIGURE 25.9
The angle of incidence θ and the angle of reflection θ'. These angles are the same as in Fig. 25.8.

FIGURE 25.8
The rays of the wave are perpendicular to the wave fronts.

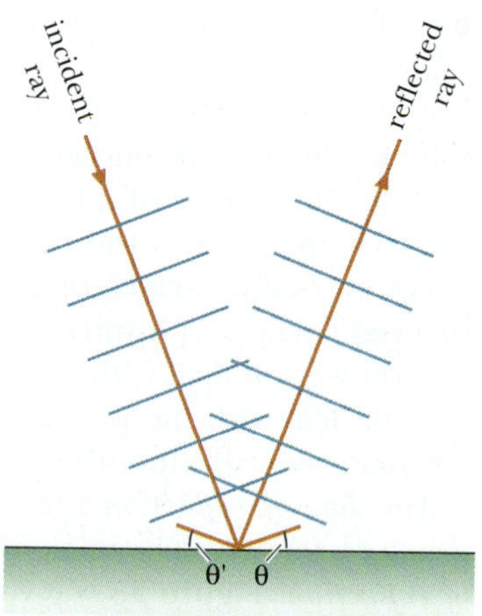

flected wave front make the same angle with the reflecting surface, we appeal to Fig. 25.7c. The right triangles $PQ'P'$ and $P'QP$ are identical since they have a common long side (PP') and their short sides (PQ and $P'Q'$) are equal. Hence the angles θ and θ' are equal.

The direction of propagation of a wave is commonly described by the rays of the wave. These ***rays are lines perpendicular to the wave fronts.*** For example, Fig. 25.8 shows the rays associated with the incident and the reflected wave fronts.

Angle of incidence ▶
Angle of reflection ▶

The angle θ (or θ') between the wave front and the reflecting surface is equal to the angle between the ray and the normal to the surface. The angles θ and θ' are called the ***angles of incidence and of reflection*** (see Fig. 25.9). Thus, from the Huygens' Construction, we have deduced that the angle of incidence equals the angle of reflection. This is the law of reflection.

Image ▶

When light from some source strikes a flat mirror, the reflection of the light leads to formation of an image of the source. Figure 25.10 shows a point source of light and the rays emerging from it; the figure also shows the reflected rays. If we extrapolate the reflected rays to the far side of the mirror, we find that they ***all appear to come from a point source of light placed beyond the mirror. This apparent point source is the image.*** To an eye looking into the mirror, the image looks like the original source — the eye perceives the mirror image as existing in the space beyond the mirror. But this mirror image is an illusion; the light does not come from beyond the mirror. This kind of illusory image that gives the impression that light rays emerge from where they do not is called a ***virtual image.***

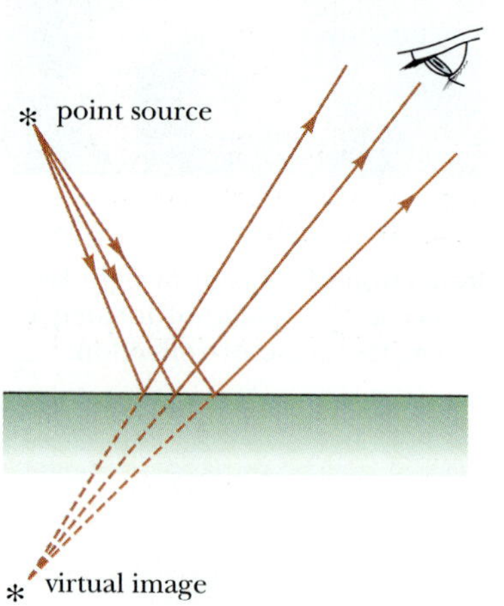

FIGURE 25.10
Rays emerging from a point source are reflected by a mirror. The extrapolated rays (dashed) appear to come from a point source beyond the mirror.

If, instead of a single luminous point, our light source consists of an extended object — such as a book, a hand, or a face — then the mirror image will also be an extended object. If the object is illuminated by sunlight or by a lamp, each point on the surface of the object scatters light and acts as a luminous point. Each such point forms an image in the mirror, and the net result is a mirror image of the entire object. Note that the mirror image of an object is a mirror-reversed object. For instance, Fig. 25.11 shows some written letters and their mirror images. This reversal is commonly referred to as a reversal of left to right. However, it is more accurately described as a reversal of front to back — mirror writing is ordinary writing seen from behind. And the mirror image of, say, a hand facing north is a hand facing south (the reversal is not an ordinary "about face," but involves passing the front of the hand through its back, thereby converting a right hand into a left hand, and vice versa; see Fig. 25.12).

FIGURE 25.11
Some letters and their images in a mirror.

25.3 Refraction

The speed of light in a transparent material — such as air, water, or glass — differs from the speed of light in vacuum. We can recognize this immediately by recalling the theoretical formula for the speed of light derived from Maxwell's equations,

$$c = \frac{1}{\sqrt{\epsilon_0 \mu_0}} \tag{1}$$

We know from Chapter 18 that in a material with given dielectric characteristics, the quantity ϵ_0 in the equation for the electric field gets replaced by $\kappa\epsilon_0$ [see Eq. (18.66)].* Likewise, the formula (1) for the speed of light gets replaced by

$$v = \frac{1}{\sqrt{\kappa \epsilon_0 \mu_0}} \tag{2}$$

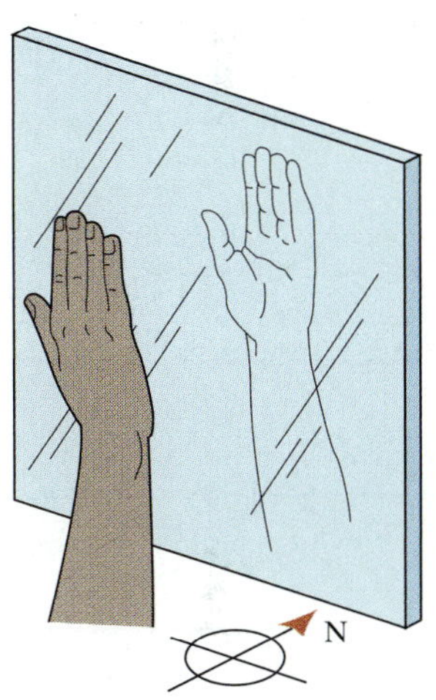

FIGURE 25.12
A hand facing north and its image in a mirror.

This is usually written as

$$v = \frac{c}{n} \tag{3}$$

where $c = 1/\sqrt{\epsilon_0 \mu_0}$ is the standard speed of light in vacuum and

$$n = \sqrt{\kappa} \tag{4}$$

◀ **Index of refraction**

The quantity n is called the ***index of refraction*** of the material. The index of refraction is larger than 1; according to Eq. (3), ***the speed of light in the material is less than the speed of light in vacuum.*** In connection with Eq. (4), it is important to keep in mind that the value of the dielectric constant depends on the frequency of the electric field. Hence, the values of the dielectric constants from Table 18.1 cannot be inserted into Eq. (4), because the former values apply only to static fields, whereas we are now concerned with the high-frequency fields of a light wave.

* In a material with magnetic properties, μ_0 must also be multiplied by a factor; but this factor is very near 1 except in ferromagnetic materials (where light does not propagate because these materials are not transparent). We will therefore ignore any correction factor for μ_0.

TABLE 25.1 INDICES OF REFRACTION OF SOME MATERIALS[a]

Material	n
Air, 1 atm, 0°C	1.00029
1 atm, 15°C	1.00028
1 atm, 30°C	1.00026
Water	1.33
Ethyl alcohol	1.36
Castor oil	1.48
Quartz, fused	1.46
Glass, crown	1.52
light flint	1.58
heavy flint	1.66

[a] For light of wavelength $\simeq$ 550 nm.

Table 25.1 gives the values of the indices of refraction of a few materials. For instance, water has $n = 1.33$, and the speed of light in water is

$$v = \frac{c}{n} = \frac{c}{1.33} = \frac{3.00 \times 10^8 \text{ m/s}}{1.33} = 2.26 \times 10^8 \text{ m/s}$$

The values in the table apply to light waves of medium frequency (yellow-green light). The index of refraction is slightly larger for blue light and slightly smaller for red light; we will deal with this complication later in this section. Note that the index of refraction of air is very close to 1; consequently, in most calculations we can ignore the distinction between air and vacuum.

With the wave speed $v = c/n$, the relation between frequency and wavelength becomes

$$\lambda f = v = \frac{c}{n} \tag{5}$$

or

$$\lambda = \frac{1}{n}\frac{c}{f} \tag{6}$$

Since c/f is the wavelength that the wave would have in vacuum [see Eq. (24.10)], we can also write Eq. (6) as

$$\lambda = \frac{\lambda_{\text{vac}}}{n} \tag{7}$$

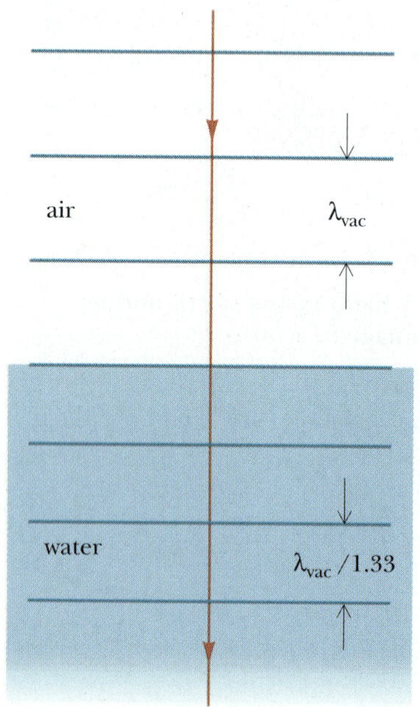

FIGURE 25.13
Change of wavelength of a light wave as it penetrates from air into water.

For example, if a wave penetrates from vacuum or from air into water, where $n = 1.33$, its speed is reduced by a factor of 1.33, but its ***frequency remains constant.*** Consequently, Eq. (7) shows that its wavelength will be reduced by a factor of 1.33 (see Fig. 25.13). The fact that the frequency

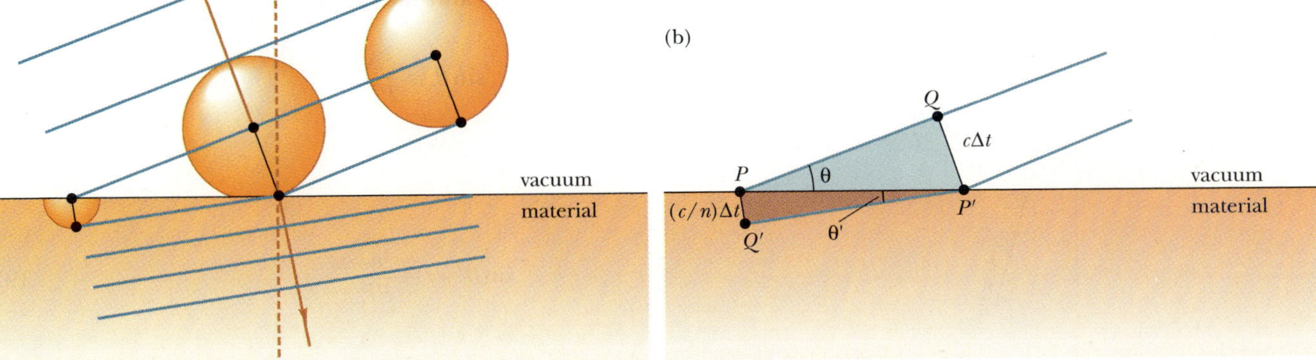

FIGURE 25.14

(a) Huygens' wavelets erected on a wave front whose edge barely touches the dielectric surface. (b) The incident wave front makes an angle θ with the dielectric surface; the refracted wave front makes an angle θ'.

of the wave remains constant can be understood in terms of the atomic mechanism underlying the interaction of the wave with the material. When the wave strikes the water surface, it shakes the electrons of the water molecules; this acceleration of electric charges produces extra waves, which combine with the original wave and result in a slowed-down wave. The frequency of the combined, slowed wave is the same as that of the original wave, because the shaking of the electrons proceeds at the original frequency, and the extra waves produced by the electrons are therefore also of the same frequency as the original wave.

Incidentally: Eq. (6) does not imply that a light source changes color when immersed in water. ***The color we perceive depends on the frequency of the light reaching our eyes;*** and this frequency is independent of whether the light source, our eyes, or both are immersed in air or water.

When a wave strikes the surface of a transparent material, part of it is reflected and part of it penetrates into the material. In the preceding section we have investigated the direction of propagation of the reflected wave; now let us investigate the penetrating wave. Again, we will use Huygens' Construction to find out what the wave does when its strikes the surface of the material. In vacuum the speed of light is c; in the material it is c/n. Figure 25.14a shows the approaching wave fronts at one instant of time. The left edge of one wave front barely touches the surface. Figure 25.14a also shows the Huygens' wavelets that determine the position of this wave front at a later time; only the forward portions of these wavelets are relevant. Above the surface, the wavelet has a radius $c\Delta t$; below the surface, in the material, the wavelet has a smaller radius $(c/n)\Delta t$. The reduction of the speed of propagation of one side of the wave front causes

Christiaan Huygens, *1629–1695, Dutch mathematician and physicist. Besides making important contributions to mechanics, he refined and extended the wave theory of light first proposed by Hooke. Huygens placed this theory on a firm foundation by showing that reflection and refraction could be explained in terms of wave propagation.*

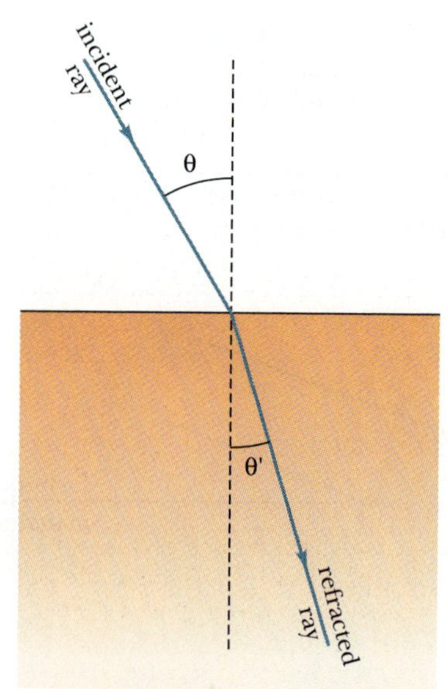

FIGURE 25.15
The angle of incidence θ and the angle of refraction θ'.

the wave front to swing around, changing its direction of advance. This is analogous to the change of the direction of advance of a row of marching soldiers when the soldiers on one edge slow down by taking short steps, while the soldiers on the other edge take longer steps.

To obtain the law of refraction, we need to examine the triangles in Fig. 25.14b. The right triangles $PP'Q$ and $PP'Q'$ have the long side PP' in common. In terms of the length PP' of this common side, the sines of the angles between the wave fronts and the surface are

$$\sin\theta = \frac{c\Delta t}{PP'} \tag{8}$$

and

$$\sin\theta' = \frac{(c/n)\Delta t}{PP'} \tag{9}$$

The ratio of the sines is therefore

$$\frac{\sin\theta}{\sin\theta'} = \frac{c}{c/n} \tag{10}$$

from which

Law of refraction ▸

$$\sin\theta = n\sin\theta' \tag{11}$$

This equation describes the change of direction of the wave upon penetrating into a material. ***This change of direction is called refraction, and Eq. (11) is called the law of refraction, or Snell's Law.*** The angle θ is the ***angle of incidence,*** and θ' is the ***angle of refraction.*** It is usually convenient to measure these angles between the rays and the normal to the surface. Figure 25.15 shows the incident and refracted rays, and the angles of incidence and refraction. As can be seen from this figure, the ray in the material is bent toward the normal (θ' is smaller than θ).

Note that we can also use the formula (11) for a ray of light that emerges from the material into vacuum. The formula is valid for the ray shown in Fig. 25.15 and also for the ray propagating in the reverse direction, provided we always assume that θ is the angle in vacuum and θ' is the angle in the material, regardless of the direction of propagation.

Our formula (11) describing refraction at the interface between vacuum and a material is a special case of a general formula describing refraction at the interface of two different materials. If the indices of refraction are n_1, n_2 and the angles between the rays and the normal are θ_1, θ_2, respectively, then

$$n_1\sin\theta_1 = n_2\sin\theta_2 \tag{12}$$

EXAMPLE 1. A ray of light enters a thick plate of glass of index of refraction $n = 1.52$ at an angle of incidence of 45° (see Fig. 25.16). (a) What is the angle of refraction of the ray at the upper surface of the glass? (b) When the ray reaches the lower surface of the glass it is refracted again and it emerges into air. What is the angle at which it emerges?

SOLUTION: (a) With $n = 1.52$ and $\theta = 45^\circ$ the law of refraction gives us

$$\sin 45^\circ = 1.52 \sin\theta' \tag{13}$$

from which

$$\sin\theta' = \frac{\sin 45^\circ}{1.52} = \frac{0.707}{1.52} = 0.465 \tag{14}$$

With our electronic calculator we then find that the angle of refraction is $\theta' = 28^\circ$.

(b) Within the glass, the ray travels at 28° with respect to the normal, and this will be the angle of incidence at the lower surface. Hence refraction at the lower surface (where the ray proceeds from glass into air) is simply the reverse of the refraction at the upper surface (where the ray proceeds from air into glass). The angle at which the ray emerges is therefore 45°.

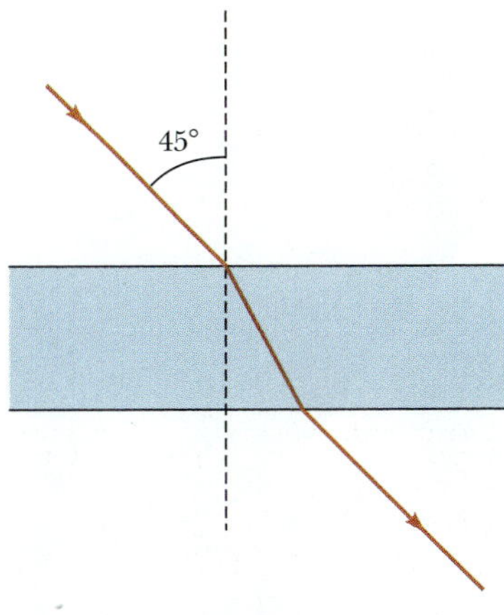

FIGURE 25.16

A ray of light passes through a thick plate of glass. The angle of incidence on the upper surface of glass is 45°.

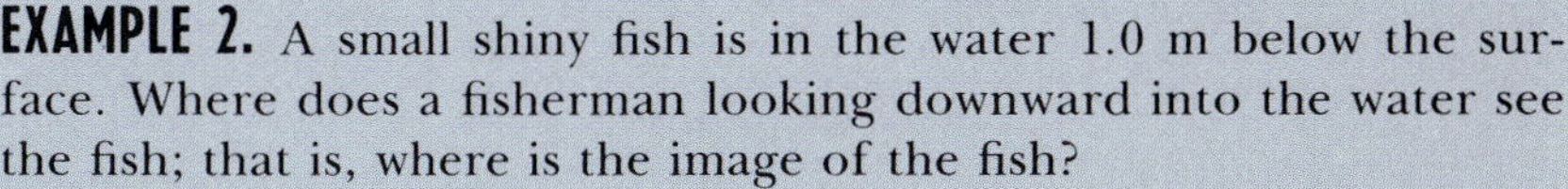

EXAMPLE 2. A small shiny fish is in the water 1.0 m below the surface. Where does a fisherman looking downward into the water see the fish; that is, where is the image of the fish?

SOLUTION: Figure 25.17 shows two light rays from the fish to the eye of the fisherman. The first light ray is perpendicular to the surface and is not bent. The second light ray is bent away from the normal when it emerges from the water. With the assumption that θ' is the angle in water and θ the angle in air, Eq. (11) can be applied to this ray emerging from the water. Since the angles are small, we can use the familiar approximation that the sine of the angle is approximately equal to the angle expressed in radians (see Section 9.4), $\sin\theta \simeq \theta$ and $\sin\theta' \simeq \theta'$. Hence Eq. (11) becomes approximately

$$\theta \simeq n\theta' = 1.33\theta'$$

When the fisherman's eye extrapolates the refracted ray back into the water, it seems to intersect the vertical ray at the point P', above the point P. Hence the image is above the object. The image distance OP' and the object distance OP are related as follows (see Fig. 25.17):

$$OP \tan\theta' = OP' \tan\theta$$

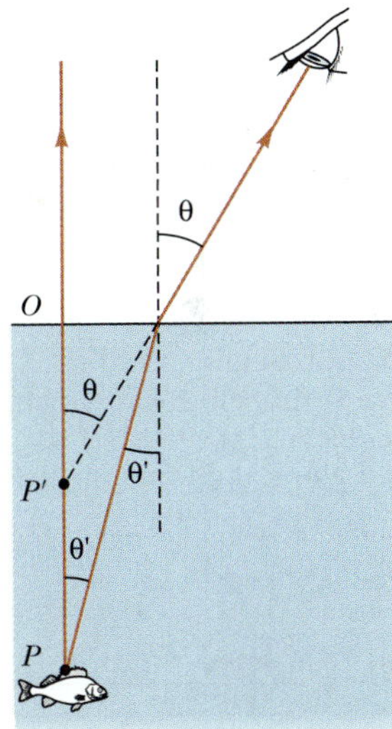

FIGURE 25.17

A small shiny fish as a source of light. Note that the direction of propagation of the light is opposite to that shown in Fig. 25.15. This does not affect the validity of Eq. (11). The extrapolated ray (dashed) appears to come from the point P'.

FIGURE 25.18
A rod partially immersed in water appears bent.

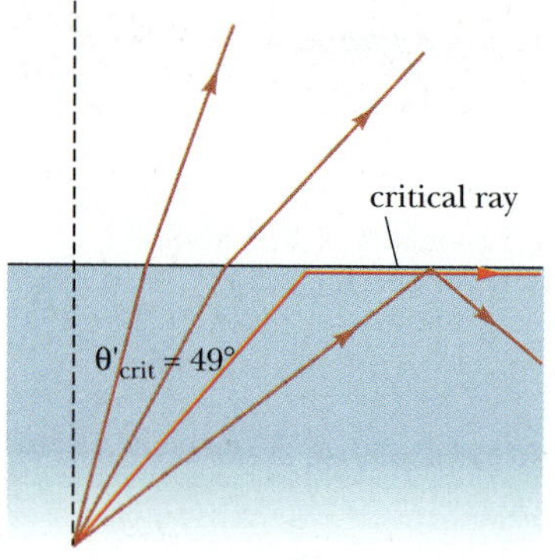

FIGURE 25.19
A ray approaching a water surface from below with an angle of incidence $\theta'_{\rm crit} = 49°$ is refracted along the water surface.

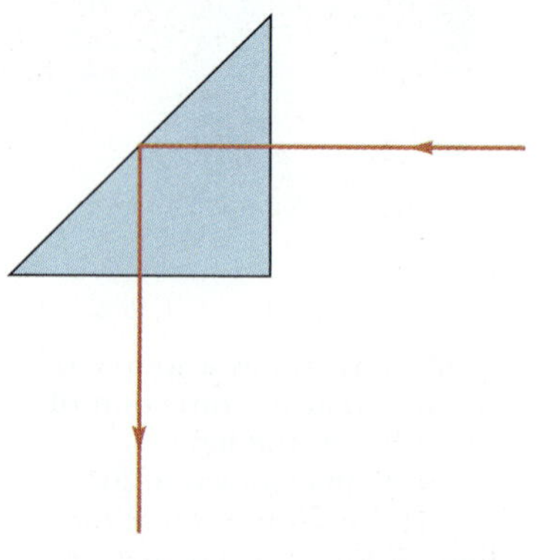

FIGURE 25.20
Internal reflection in a prism.

For small angles, we can make the approximations $\tan\theta \simeq \theta$ and $\tan\theta' \simeq \theta'$. Hence

$$\frac{OP'}{OP} \simeq \frac{\theta'}{\theta} \simeq \frac{1}{1.33}$$

This shows that the image distance is smaller than the object distance by a factor of 1.33. For instance, if $OP = 1.0$ m, then $OP' = 0.75$ m. The fish seems to be nearer to the surface than it is (if the fisherman wants to spear the fish, he must shove his spear down deeper than where the fish *seems* to be).

COMMENTS AND SUGGESTIONS: The apparent bending that you perceive when you look at a straight rod partially immersed in water results from this apparent shrinking of the vertical distance. All the immersed portions of the rod seem to be nearer the water surface than they are, and the rod seems to have a kink at the place where it enters the water (see Fig. 25.18).

For a ray attempting to leave water, there is a critical angle beyond which refraction is impossible. As the light ray emerges into air, it is bent *away* from the vertical; in the extreme case, it is bent so much that it lies almost along the water surface (see Fig. 25.19); this extreme case corresponds to $\theta = 90°$ in Eq. (11),

$$n \sin\theta' = \sin 90° = 1 \tag{15}$$

The ***critical angle*** for this extreme form of refraction is therefore given by

$$\sin\theta'_{\rm crit} = \frac{1}{n} \tag{16}$$

With $n = 1.33$, we then find

$$\theta'_{\rm crit} = 49° \tag{17}$$

If a light ray strikes a water surface from below at an angle larger than this, refraction is impossible. Instead, the ray suffers a complete reflection — the water surface behaves like a perfect mirror. This phenomenon is called ***total internal reflection.*** It can occur whenever the index of refraction of the medium containing the light ray is larger than the index of refraction of the adjacent medium.

Total internal reflection has many important practical applications in optics. For instance, in periscopes the light is reflected down the tube by internal reflection in a prism (see Fig. 25.20); because the reflection is total, without any loss of light, this gives a much better image than reflection in a mirror. Such reflecting prisms are also used in binocular telescopes to reflect the path of the rays of light back and forth, and increase the effective length of the telescope.

Optical fibers are another application of total internal reflection. In an optical fiber, light moves along a thin rod made of transparent material; the light zigzags back and forth between the walls of the rod, undergoing a sequence of total internal reflections — the fiber acts as a pipe for light (see Fig. 25.21). Such optical fibers are being used to replace telephone cables. The electrical pulses normally carried on a wire cable are converted into pulses of infrared laser light which can be transmitted in an optical fiber. The efficiency of optical telephone lines is very high because an optical fiber can carry many telephone conversations simultaneously; in modern telephone systems, single optical fibers are being used to carry several hundred telephone conversations simultaneously.

Optical fibers also find application in flexible endoscopes, used by physicians to examine the interior of the intestine and the stomach. This device is a bundle of optical fibers, sometimes over a meter in length, one end of which is inserted into the patient's stomach or intestine while the physician looks into the other end. A small light bulb at the leading end provides illumination (see Fig. 25.22).

In most materials, the index of refraction depends somewhat on the wavelength of the light. Usually, the index of refraction increases as the wavelength decreases. For instance, Fig. 25.23 displays a plot of the index of refraction of light in water over a range of wavelengths (the wavelengths plotted along the horizontal axis of Fig. 25.23 are measured in air, before the light penetrates the water). When a ray of light containing several wavelengths, or colors, is refracted by a material with an index of refraction that depends on wavelength, the refracted rays of different colors will emerge at somewhat different angles. ***The separation of a ray by refraction into distinct rays of different colors is called dispersion.***

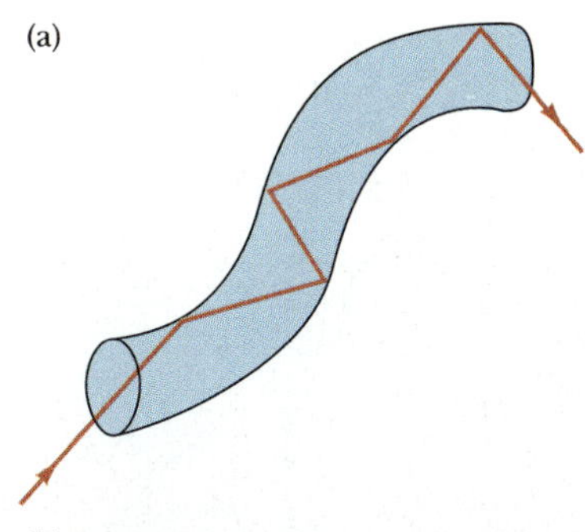

FIGURE 25.21

(a) Internal reflection in an optical fiber. (b) Light enters the optical fiber at the top left and emerges at the center.

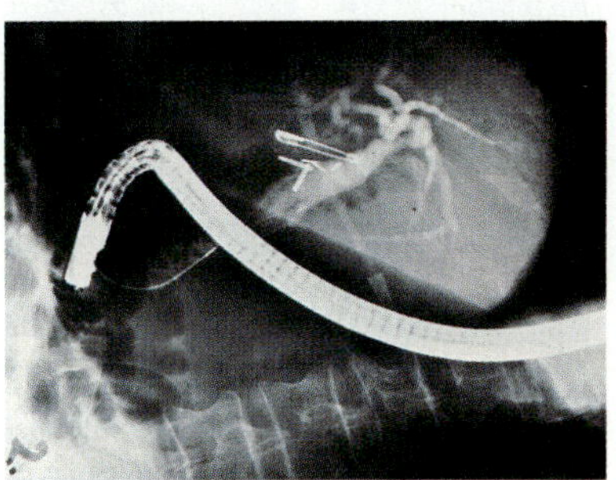

FIGURE 25.22

Endoscope inside gall bladder patient.

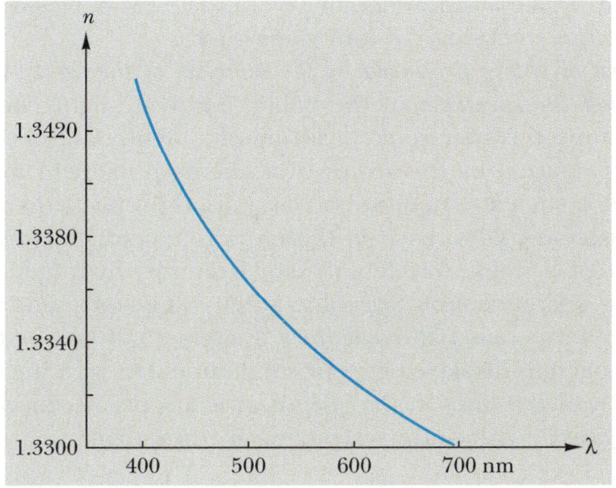

FIGURE 25.23

Index of refraction of light in water as a function of wavelength. The index of refraction varies by about 1% over the range of visible wavelengths.

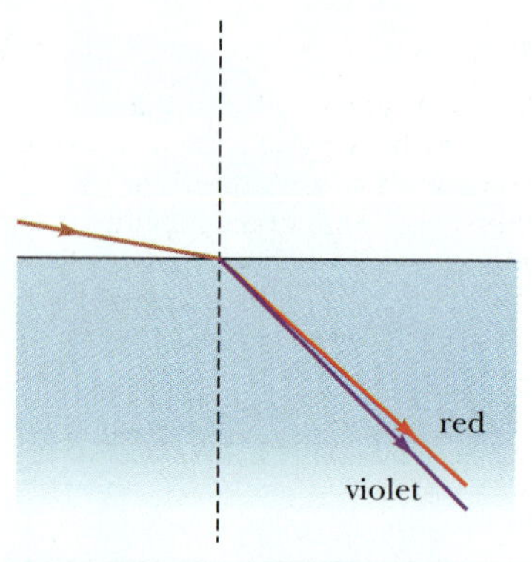

FIGURE 25.24
Refraction of red and of violet light in water. The difference between the angles of the refracted rays has been exaggerated for the sake of clarity.

EXAMPLE 3. The index of refraction for red light in water is 1.330 and for violet light it is 1.342. Suppose that a ray of light approaches a water surface with an angle of incidence of 80°. What are the angles of refraction for red light and for violet light?

SOLUTION: For $n = 1.330$ Eq. (11) yields

$$\sin\theta' = \frac{\sin\theta}{n} = \frac{\sin 80^\circ}{1.330} = 0.740$$

for which our calculator gives us $\theta' \simeq 48^\circ$.

For $n = 1.342$ Eq. (11) yields

$$\sin\theta' = \frac{\sin 80^\circ}{1.342} = 0.734$$

and $\theta' \simeq 47^\circ$. The violet light is bent more toward the vertical than the red light (see Fig. 25.24). For $\theta = 80^\circ$, the difference in the angles of refraction is roughly 1° (a more precise calculation shows that the difference in angles is 0.6°). Thus, refraction in water separates light rays according to colors. A beautiful demonstration of this effect is found in rainbows, which are produced by the refraction of sunlight in water droplets.

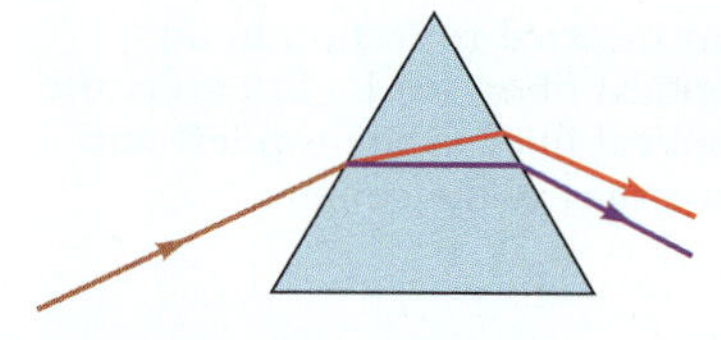

FIGURE 25.25
Refraction of red and of violet light by a prism.

A ***prism*** is the traditional device employed for the separation of light rays into their constituent colors. The basic mechanism is the same as discussed in Example 3; the glass in the prism has slightly different indices of refraction for light of different wavelengths, and hence it bends rays of different colors by different amounts (see Fig. 25.25). In passing through a prism, light is refracted twice, first when it enters the glass, and then when it leaves the glass. Under normal operating conditions, a good prism will introduce a difference of several degrees between the angular directions of the emerging red and violet rays.

The pattern of colors produced by the analysis of the light by means of a prism is called the spectrum of the light. The white light emitted by the Sun has a continuous spectrum consisting of a blend of all the colors. The colored light emitted by the atoms of a chemical element in an electric discharge tube, such as a neon tube, has a discrete spectrum consisting of just a few discrete colors. Each of these discrete colors is absolutely pure — it is light of a single wavelength. For example, hydrogen atoms emit the following discrete colors: red, blue-green, blue-violet, and violet. Such discrete colors are called ***spectral lines.*** Figure 25.26 shows the spectral lines of hydrogen as displayed by a prism illuminated with the light from a fine slit. Each of the lines in the spectrum is a separate image of the slit made by a separate color after refraction by the prism.

Each of these two prisms separates a ray of white light into its constituent colors.

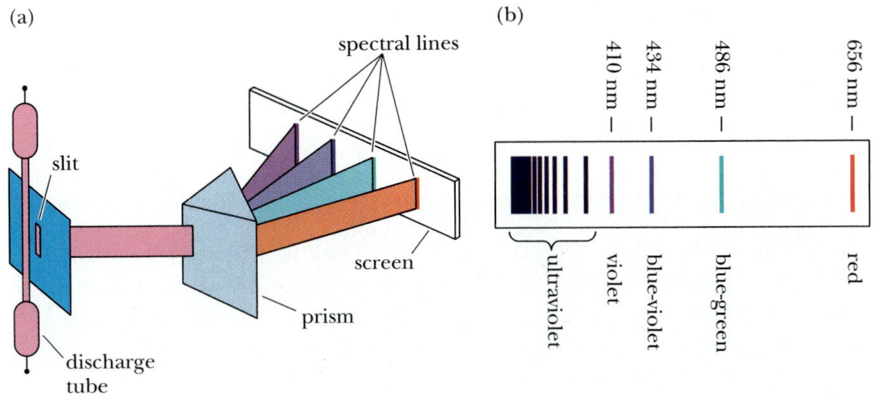

FIGURE 25.26

(a) Arrangement for the analysis of light by a prism. Each discrete color forms a spectral line on the screen or photographic plate. (b) The spectral lines of hydrogen. (A color print of these spectral lines is included on page 834E.)

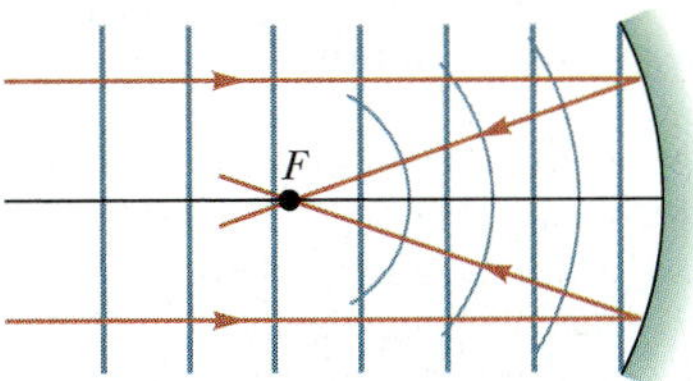

FIGURE 25.27

A concave spherical mirror focuses an incident plane wave to a point.

25.4 SPHERICAL MIRRORS

◀ **Concave mirror**

The law of reflection is also valid for a wave incident on a small portion of a curved reflecting surface, since such a portion can be approximated by a flat, tangent surface. Figure 25.27 shows ***concave*** mirror with a surface curved like the inner surface of a sphere. When a wave is incident on this mirror, each portion of the wave is reflected according to the law of reflection, with equal angles of incidence and reflection. However, the angles of incidence for different portions of the wave are different; for instance, the angle of incidence is 90° for the portion of the wave that strikes the exact center of the mirror, but the angle of incidence is less than 90° for the portions of the wave that strike above or below the center. Since the angles of reflection for different portions of the wave are different, the plane wave incident on the spherical mirror becomes a wave with curved wave fronts.

As can be seen in Fig. 25.27, the reflected wave fronts converge toward a point, the ***focal point*** of the mirror. We can describe the direction of propagation of the wave fronts by rays; Fig. 25.28 shows the incident and reflected rays. The reflected rays intersect at the focal point.

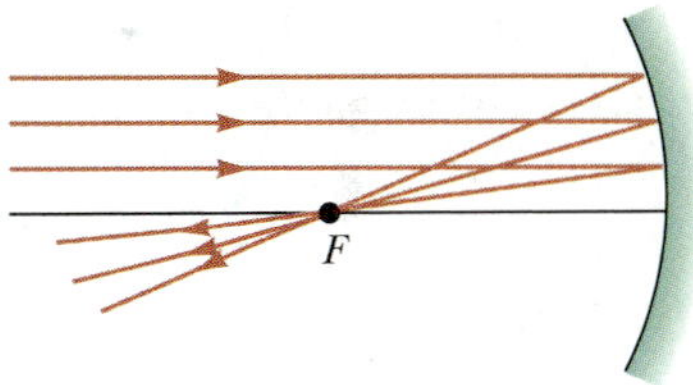

FIGURE 25.28

Reflection of parallel rays by a concave spherical mirror.

The focal point of the spherical mirror is halfway between the mirror and the center of the spherical surface. For a proof, we use Fig. 25.29, which shows the path of a single ray of light. The focal point is the intersection of the ray with the axial line CA. To find the distance FA, called the ***focal length,*** we begin with the observation that in the isosceles triangle CFQ the length CF equals FQ. Under the assumption that angle θ is small (equivalently, that the incident ray is near the axial line), the length FQ is approximately equal to FA. Hence.

$$CF = FA \tag{18}$$

and thus the point F is halfway between the mirror (A) and the center (C). This means the focal length is one-half of the radius of the spherical sur-

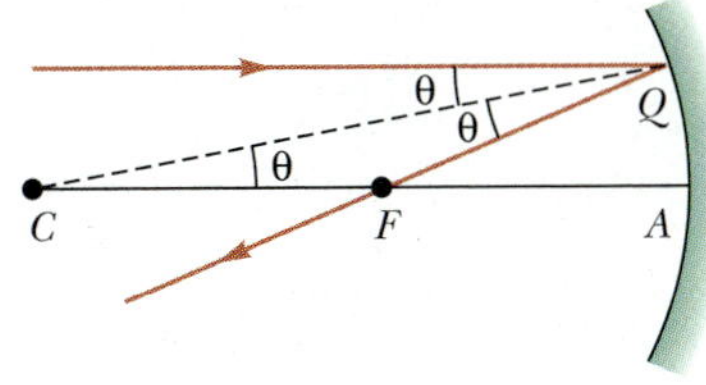

FIGURE 25.29

Reflection of a single ray.

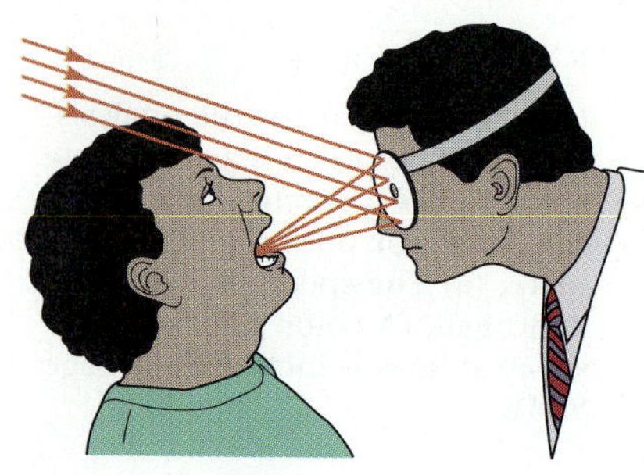

FIGURE 25.30
Mirror used to throw light into the patient's mouth.

face. Designating the former by f and the latter by R, we can write

$$f = \tfrac{1}{2}R \tag{19}$$

EXAMPLE 4. A dentist uses a concave mirror to throw light into the patient's mouth (see Fig. 25.30). If a parallel beam of light supplied by a lamp is to be concentrated at a distance of 25 cm from the mirror, what should be the radius of curvature of the mirror?

SOLUTION: For strong concentration of the light, we want the focal point of the mirror to be at the distance of 25 cm from the mirror. According to Eq. (19) the radius in the mirror must be

$$R = 2f = 2 \times 25\text{ cm} = 50\text{ cm}$$

Convex mirror ▶

Figure 25.31 shows a ***convex*** spherical mirror. This kind of mirror is also curved like the surface of a sphere; but, in contrast to the concave mirror, the reflecting surface is the outer surface of the sphere. Parallel rays incident on this mirror diverge upon reflection. If we extrapolate the divergent rays to the far side of the mirror, they all seem to come from a single point, the focal point of the convex mirror. An argument similar to that given above demonstrates that the focal length is again one-half of the radius of the spherical surface,

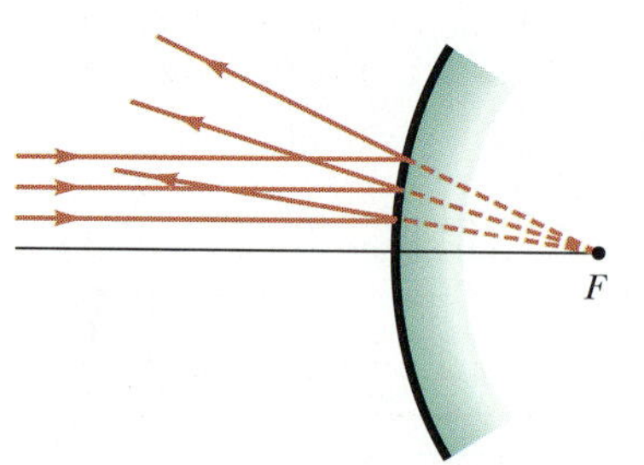

FIGURE 25.31
Reflection of parallel rays by a convex mirror.

$$f = -\tfrac{1}{2}R \tag{20}$$

A negative sign has been inserted in Eq. (20) to indicate that the focal point is on the far side of the spherical surface. This negative sign will prove useful in algebraic calculations with mirrors (see below).

Both concave and convex mirrors will form images of objects placed in front of them. Figure 25.32 shows a point source of light P in front of a concave mirror. To find the position of the image, we must trace some rays of light and find where they intersect. The three rays that are easiest to trace are the three rays shown in Fig. 25.32, called the ***principal rays:***

Principal rays for mirror ▶

1. The first of these rays (PC in Fig. 25.32a) passes through the center of the spherical surface of the mirror. It then strikes the mirror perpendicularly and is therefore reflected on itself.
2. The second ray (PR in Fig. 25.32b) begins parallel to the axial line and passes through the focus after being reflected.
3. The third ray (PF in Fig. 25.32c) passes through the focal point. Upon reflection, this ray emerges parallel to the axial line (this behavior of the third ray becomes obvious if we recognize that the third ray

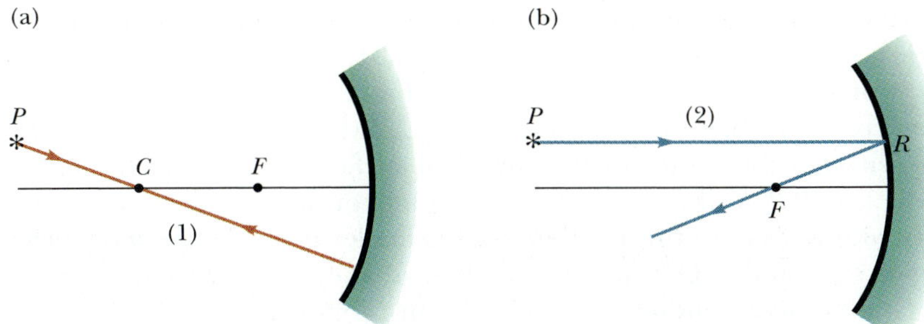

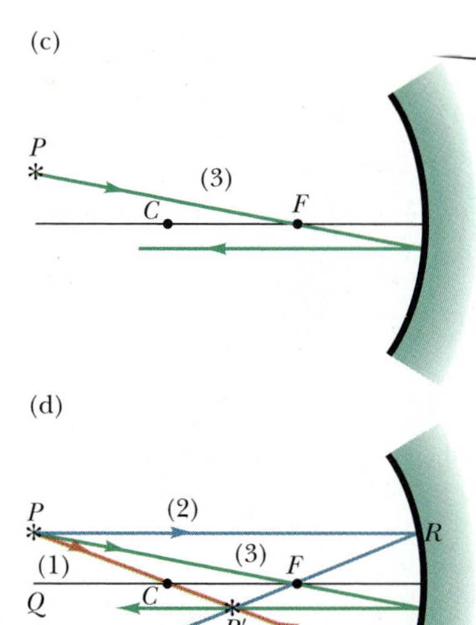

FIGURE 25.32

(a) Ray from P through center of mirror. (b) Ray from P parallel to axial line passes through focus. (c) Ray from P through focus emerges parallel to axial line. (d) The rays (1), (2), and (3) intersect at the image P'.

is merely the reverse of a ray that arrives at the mirror from a direction parallel to the axial line).

All these rays, and any other rays originating from the point source P, come together at P' (see Fig. 25.32d). This point P' is the image of the point source P. Note that to locate the image, two out of the three rays mentioned above are already sufficient — the third is redundant but serves as a useful check.

If the source of light is an extended object, then we must find the image of each of its points. For instance, a luminous object in the shape of an arrow has an image as shown in Fig. 25.33. We can easily verify this by drawing the rays that emerge from, say, the tip of the arrow, the midpoint of the arrow, the tail of the arrow, and so on. All the rays emerging from the tip of the arrow intersect at the tip of the image; all the rays emerging from the midpoint of the arrow intersect at the midpoint of the image, and so on. However, for finding the position of the image, it often suffices to draw the rays emerging from one or two points of the object. For instance, in Fig. 25.33, we need only the rays emerging from the tip, since we know that the tail of the arrow and the image of the tail both sit on the axial line.

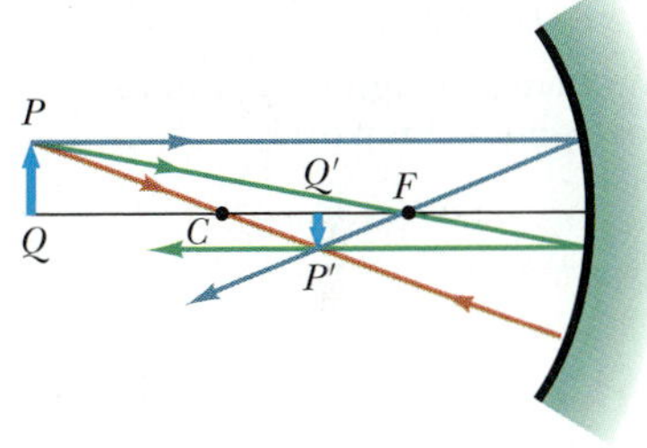

FIGURE 25.33

An object PQ in the shape of an arrow and its image $P'Q'$ formed by a concave mirror.

The ray-tracing technique summarized in Fig. 25.33 is a graphical method for finding the image of a source. This method also applies to convex mirrors; an example of this is shown in Fig. 25.34.

Although ray-tracing is useful to locate the image roughly and to gain a qualitative understanding of how the positions of object, mirror, and image are related, its precision is severely limited. For higher precision, we must calculate the position of the image algebraically by means of the ***mirror equation***,

$$\frac{1}{s}+\frac{1}{s'}=\frac{1}{f} \tag{21}$$

◄ **Mirror equation**

Here s is the distance from the object to the mirror and s' is the distance from the image to the mirror. The distance s or s' is positive if the object or image is in front of the mirror; the distance s or s' is negative if the object or image is behind the mirror (the object can be behind the mirror if what serves as "object" for the mirror is actually an image produced by another mirror or lens). As already mentioned above, f is positive for a

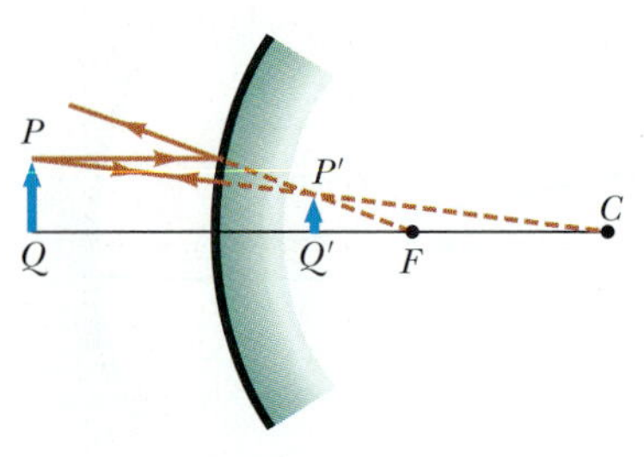

FIGURE 25.34

An object PQ and its image $P'Q'$ formed by a convex mirror.

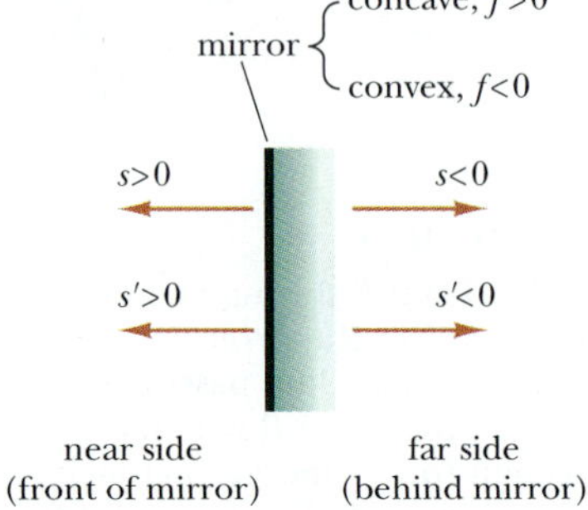

FIGURE 25.35

Summary of sign conventions for spherical mirrors.

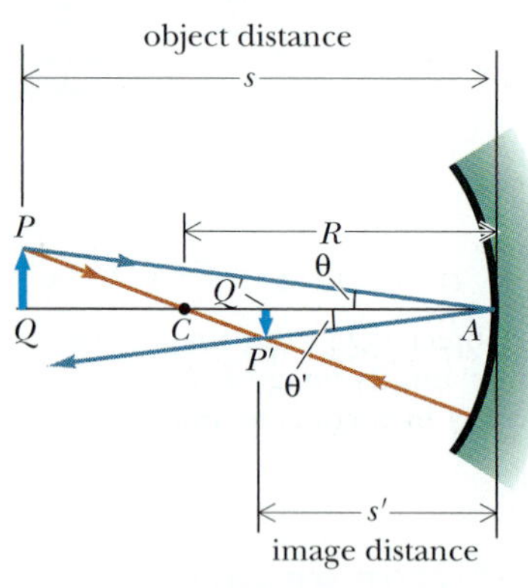

FIGURE 25.36

The angles θ and θ' are equal, hence the right triangles PQA and $P'Q'A$ are similar.

concave mirror, negative for a convex mirror. These sign conventions for spherical mirrors are summarized in Fig. 25.35.

For a proof of Eq. (21), we make use of Fig. 25.36, which shows a concave mirror, an object PQ, its image $P'Q'$, and two rays. The ray PCP' passes through the center of the spherical surface and is reflected on itself; the ray PAP' strikes the center of the mirror and is reflected symmetrically with respect to the axial line so the angles θ and θ' are equal. The triangles PQA and $P'Q'A$ are similar, that is, they have the same angles. Hence their corresponding sides must be in proportion,

$$\frac{PQ}{P'Q'} = \frac{s}{s'} \tag{22}$$

The triangles PQC and $P'Q'C$ are also similar; hence

$$\frac{PQ}{P'Q'} = \frac{QC}{Q'C} \tag{23}$$

or, since $QC = s - R$ and $Q'C = R - s'$,

$$\frac{PQ}{P'Q'} = \frac{s-R}{R-s'} \tag{24}$$

Combining Eqs. (22) and (24), we find

$$\frac{s}{s'} = \frac{s-R}{R-s'} \tag{25}$$

We can rearrange this equation as

$$\frac{R-s'}{s'} = \frac{s-R}{s}$$

or

$$\frac{R}{s'} - 1 = 1 - \frac{R}{s}$$

If we divide both sides of this by R and shift terms from one side of the equation to the other, we obtain

$$\frac{1}{s'} + \frac{1}{s} = \frac{2}{R} \tag{26}$$

Since $f = R/2$, this is the same as Eq. (21).

EXAMPLE 5. A candle is placed 41 cm in front of a concave spherical mirror of radius 60 cm. Where is the image?

SOLUTION: With $s = 41$ cm and $f = 30$ cm, Eq. (21) gives

$$\frac{1}{41\text{ cm}} + \frac{1}{s'} = \frac{1}{30\text{ cm}}$$

or

$$\frac{1}{s'} = \frac{1}{30\text{ cm}} - \frac{1}{41\text{ cm}} = \frac{41-30}{41 \times 30}\frac{1}{\text{cm}}$$

from which $s' = 112$ cm. The positive value of s' indicates that the image is on the same side of the mirror as the object (see Fig. 25.37).

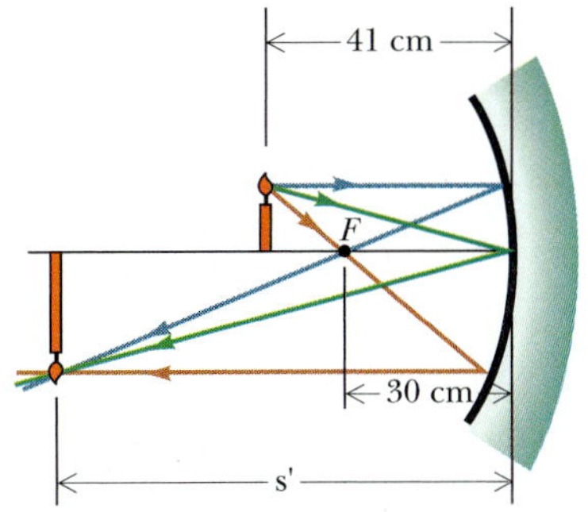

FIGURE 25.37

A candle and its image. This is a *real* image; it lies in front of the mirror.

The image in the preceding example is a ***real image. The light rays that we see when we look toward the concave mirror not only seem to come from this image, but they actually do.*** As Fig. 25.37 shows, the light rays pass through the real image and diverge from it, just as they diverge from the object. The real image is in front of the mirror. Visually, it gives the impression of a ghostly replica of the object floating in midair. In contrast, the virtual images that we found in earlier examples were behind the surfaces of the mirrors. Light rays do not actually pass through a virtual image; they merely *seem* to come from such an image.

25.5 THIN LENSES

A lens made of a refracting material with two spherical surfaces will focus the parallel rays in a beam of light to a point (see Fig. 25.38). For a thin lens, the focal length is given by the ***lens-maker's formula***

$$\frac{1}{f} = (n-1)\left(\frac{1}{R_1} + \frac{1}{R_2}\right) \qquad (27)$$

◀ **Lens-maker's formula**

where n is the index of refraction of the material of the lens, and R_1 and R_2 are the radii of the two spherical surfaces making up the lens. This equation is based on the assumption that the lens is thin (its thickness is small compared with R_1 and R_2) and that the incident rays are near the axial line.

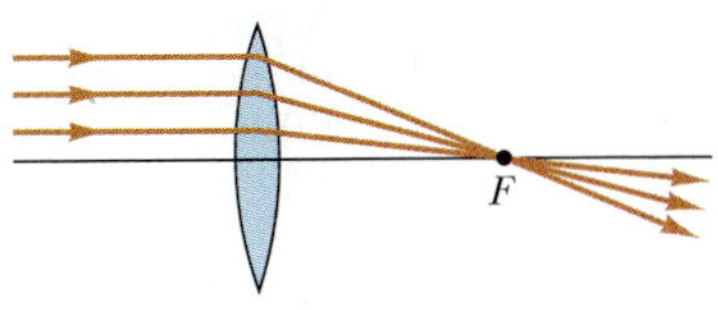

FIGURE 25.38

Refraction of rays by a convex lens. In Eq. (27), the radii R_1 and R_2 are reckoned as positive. The focal length is positive.

Equation (27) can be verified by tracing rays through the lens, taking into account their refraction at the two curved surfaces. We will not perform this tedious calculation and note only that the focusing depends on the fact that parallel rays at some distance from the axial line strike the surface of the lens with larger angles of incidence than rays near the axial line, and consequently they are bent more sharply toward the axial line.

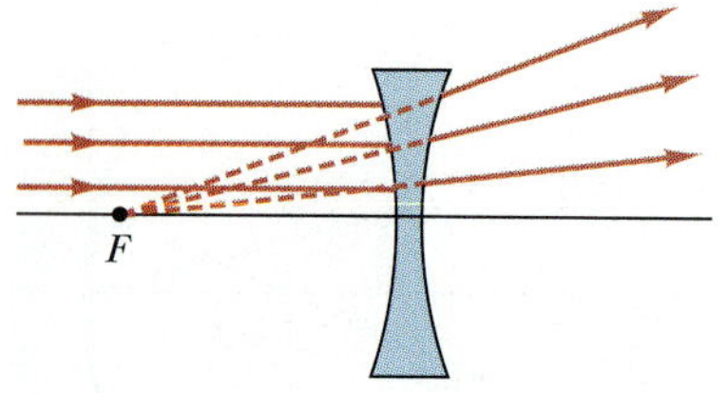

FIGURE 25.39

Refraction of rays by a concave lens. In Eq. (27), the radii R_1 and R_2 are reckoned as negative ($R_1 < 0$ and $R_2 < 0$). The focal length is negative.

This is of course exactly what is required to make these distant rays cross the axial line at the same point (the focus) as the near rays.

Equation (27) may also be applied to a concave lens (see Fig. 25.39). In this case, the radii R_1 and R_2 must be reckoned as *negative* and the focal distance f is then also negative. The meaning of a negative focal length is the same as in the case of mirrors: parallel rays incident on the lens diverge when they emerge from the lens, and the focal point is the point at which the extrapolated rays appear to intersect (see Fig. 25.39). Furthermore, Eq. (27) can be applied to a concave–convex lens, such as shown in Fig. 25.40. Whether such a lens produces net convergence or divergence depends on whether the positive radius (convex) or the negative radius (concave) is smaller; for instance, the lens of Fig. 25.40 produces convergence.

Principal rays for lens ▶

Note that ***a lens has two focal points at equal distances to the right and left of the lens.*** The point on the right of a converging lens is the focus of a parallel beam coming from the left and, conversely, the point on the left is the focus of a parallel beam coming from the right.

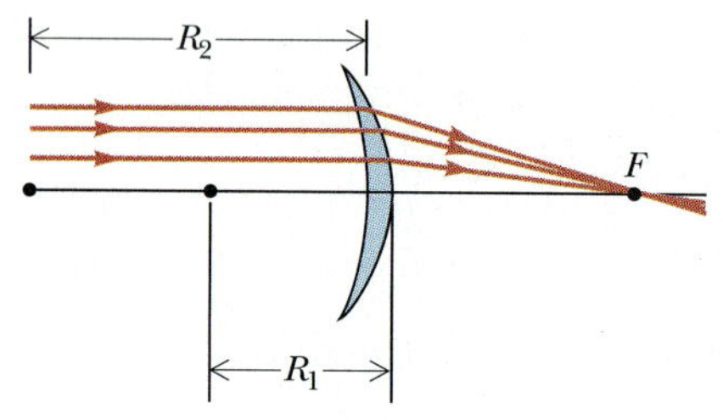

FIGURE 25.40

Refraction of rays by a concave–convex lens. The convex surface (right surface) has a smaller radius than the concave surface (left surface). In Eq. (27), the radius of the former surface is reckoned as positive ($R_1 > 0$), and the radius of the latter surface is reckoned as negative ($R_2 < 0$). For this lens, the sum $1/R_1 + 1/R_2$ is positive.

To find the image of an object placed near the lens, we can use a ray-tracing technique similar to that used for mirrors. Figure 25.41 shows the three rays that are easy to trace, or the ***principal rays:***

1. The first ray (PQP' in Fig. 25.41) that starts parallel to the axial line. This ray is deflected by the lens so it passes through the focal point F.

2. The second ray (PCP' in Fig. 25.41) that passes through the center of the lens. This ray continues straight through the lens, without deflection.

3. The third ray ($PQ'P'$ in Fig. 25.41) that passes through the focal point F'. That ray is deflected by the lens so it emerges parallel to the axial line (this becomes obvious if we recognize that the third ray is merely the reverse of a ray that arrives at the lens from the right, parallel to the axial line). All these rays intersect at the point P', the image point.

As in the case of mirrors, two of the above three rays are already sufficient to locate the image. And, of course, the same ray-tracing technique can be applied to a concave lens. Figure 25.42 gives an example with such a concave lens.

The ***lens equation*** to be used for the algebraic calculation of the image

FIGURE 25.41

A point source of light P in front of a convex lens. The rays (1), (2), and (3) intersect at the image P'.

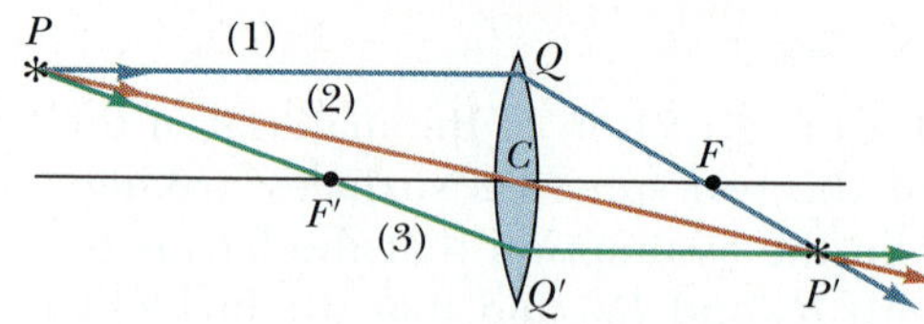

FIGURE 25.42

A point source of light P in front of a concave lens, and the image P'.

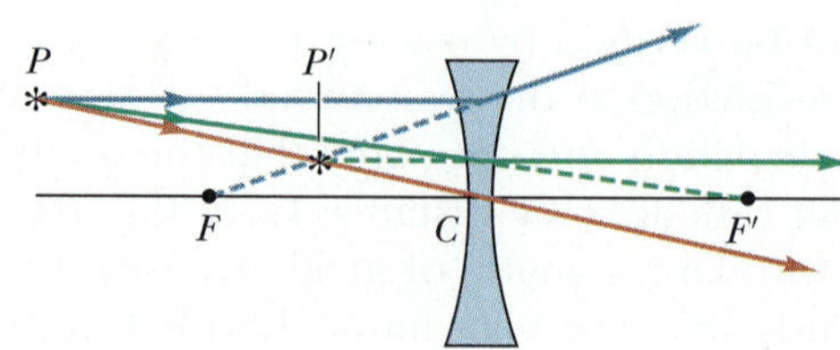

distances is the same as Eq. (21),

$$\frac{1}{s}+\frac{1}{s'}=\frac{1}{f} \tag{28}$$

◂ **Lens equation**

but the sign conventions are slightly different. The object distance s is positive if the object is on the near side of the lens and negative if it is on the far side; the image distance s' is positive if the image is on the far side of the lens and negative if it is on the near side. In this context, the "near" side is the side from which the light rays are incident on the lens (left side in the above figures), and the "far" side is the other side (right side in the above figures). The sign of f is positive for a convex lens, and negative for a concave lens. These sign conventions for lenses are summarized in Fig. 25.43.

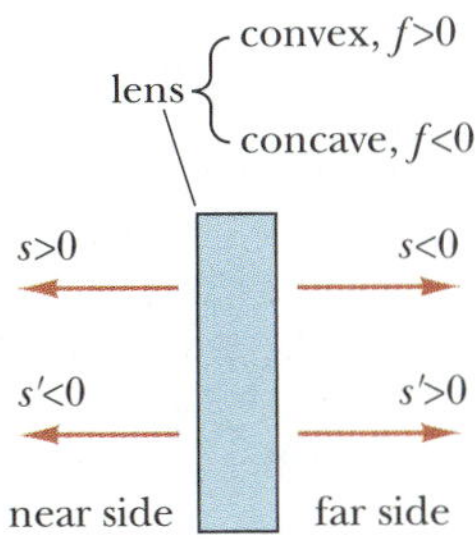

FIGURE 25.43 Summary of sign conventions for lenses.

Although the derivation of the lens equation (28) can be based on a geometric argument similar to that used for the mirror formula (21), we can bypass this labor by a trick. We begin by noting that a concave mirror is equivalent to one-half of a convex lens placed directly in front of a flat mirror. If the concave mirror and the (entire) convex lens have the same focal length, then the two arrangements shown in Figs. 25.44a and b have exactly the same optical properties — in both cases the focal lengths are the same, and the image distances are therefore the same. If we now remove the flat mirror in Fig. 25.44b and replace the one-half lens by the entire lens, the image distances will remain the same, but the image will form on the opposite side of the lens, that is, the sign of the image distance will be reversed. Consequently, the same equation (21) must apply to the concave mirror and to the convex lens; the only difference is that the sign of the image distance is reversed. This reversal of sign has already been taken into account in our description of the sign convention associated with Eqs. (21) and (28) — for the mirror, s' is taken as *positive* if the image is on the near side of the mirror, whereas for the lens, s' is taken as *negative* if it is on the near side of the lens. There is, of course, a similar correspondence between a convex mirror and a concave lens.

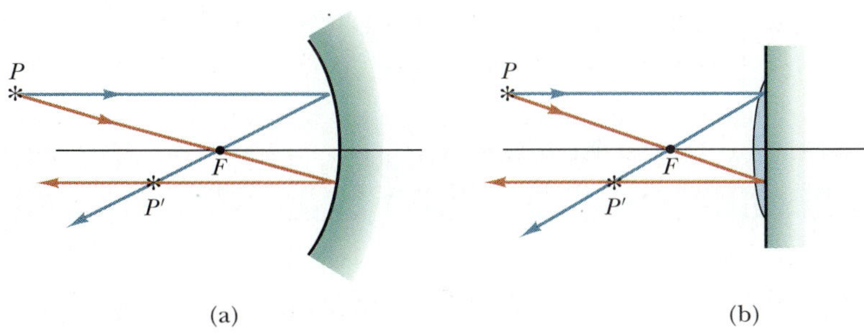

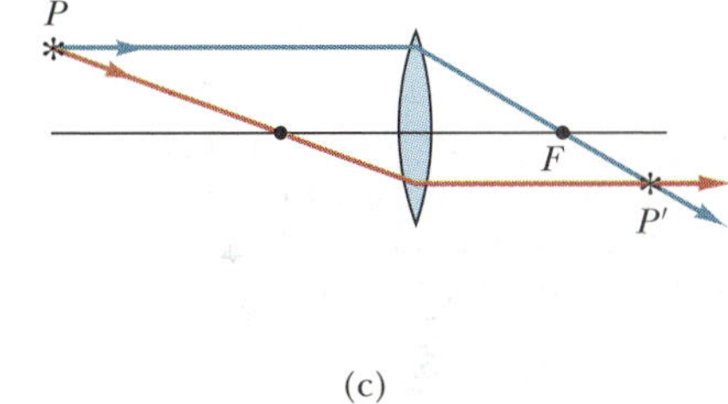

FIGURE 25.44 (a) A point source of light and its image formed by a concave mirror. (b) A similar image is formed by one-half of a convex lens placed in front of a flat mirror. Note that each ray has to pass through the one-half lens twice: once before reflection by the mirror, once after. The deflection suffered by a ray in two passages through one-half of a lens is the same as that in a single passage through the entire lens. (c) A similar image is also formed by the entire lens, but the image is now on the other side of the lens.

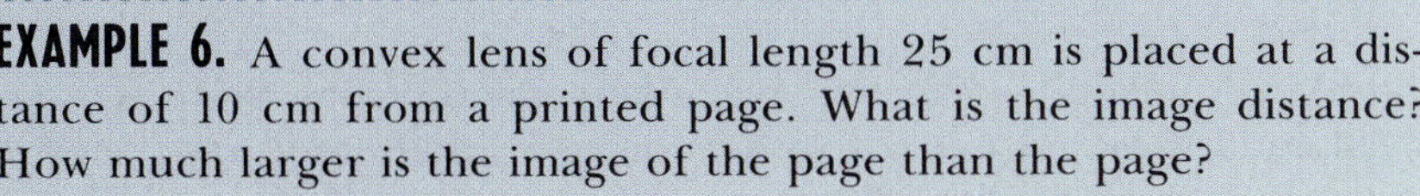

EXAMPLE 6. A convex lens of focal length 25 cm is placed at a distance of 10 cm from a printed page. What is the image distance? How much larger is the image of the page than the page?

SOLUTION: With $s = 10$ cm and $f = 25$ cm, Eq. (28) gives

$$\frac{1}{10\text{ cm}} + \frac{1}{s'} = \frac{1}{25\text{ cm}}$$

from which

$$\frac{1}{s'} = \frac{1}{25\text{ cm}} - \frac{1}{10\text{ cm}} = \frac{10-25}{25\times 10}\,\frac{1}{\text{cm}}$$

and

$$s' = -16.7\text{ cm}$$

The negative sign indicates that the image is on the near side of the lens (see Fig. 25.45). The image is virtual.

Since the triangles $P'Q'C$ and PQC are similar, the sizes of image and object are in the ratio

$$\frac{P'Q'}{PQ} = \frac{Q'C}{QC} = -\frac{s'}{s} \qquad (29)$$

$$= \frac{16.7\text{ cm}}{10\text{ cm}} = 1.67$$

That is, the image is larger than the object by a factor of 1.67. This is the principle involved in the magnifying glass.

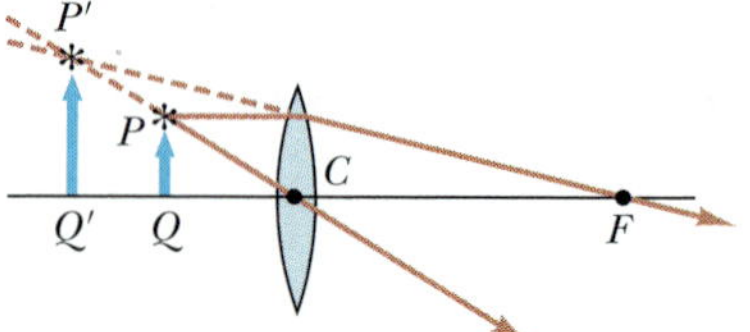

FIGURE 25.45
An object PQ and its image $P'Q'$.

Equation (29) is a general result for the magnification produced by a lens or by a mirror. The size of the image always equals the size of the object multiplied by a factor $-s'/s$. This factor is called the ***magnification:***

Magnification ▶

$$M = -\frac{s'}{s} \qquad (30)$$

The sign of M tells us something about the character of the image. If M is positive, then the image is upright and virtual, as in Example 6. If M is negative, then the image is inverted and real.

25.6 Optical Instruments

Optical instruments — cameras, magnifiers, microscopes, and telescopes — employ lenses and mirrors to form images in different ways. Most of the lenses in high-quality optical systems are compound lenses, that is, combinations of several simple lenses chosen in such a way that the optical deficiencies of one lens are canceled by those of another. However, in the following discussion of optical instruments, we will ignore this complication, and we will schematically represent the compound lenses in the instrument by single thin lenses of suitable focal lengths.

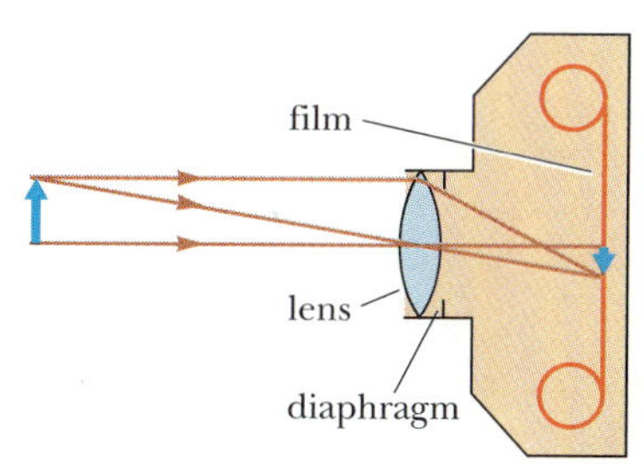

FIGURE 25.46
A photographic camera.

The Photographic Camera

The lens of the camera focuses a real image of the object on the photographic film and thereby imprints the image on the film (see Fig. 25.46). The distance between the lens and the film is adjustable, so the image can always be made to fall on the film, regardless of the object distance. A shutter controls the exposure time during which light is admitted to the camera. For a given exposure time, the amount of light entering the camera is proportional to the area of the lens. The size of the camera lens is commonly labeled by the **f *number,*** which is defined as the ratio of the focal length of the lens to its diameter. For instance, a lens of focal length 55 mm and a diameter of 32 mm has an f number of 55/32, or 1.7. A lens of small f number is said to be "fast," because it collects sufficient light for a photograph in a short exposure time.

Good cameras have an adjustable iris diaphragm that can be used to block part of the area of the lens and thereby alter its effective f number. If the diaphragm is closed down so only a small central portion of the lens remains unblocked, the camera will require a long exposure time, but it will have a large depth of field — it simultaneously forms sharp images for objects spanning a large range of object distances. This is so because the rays emitted by any object then enter the camera at small angles, and they therefore intersect the image at small angles; thus the rays will stay close together on the photographic film, even if the position of the image does not fall exactly on the film (see Fig. 25.47).

(a)

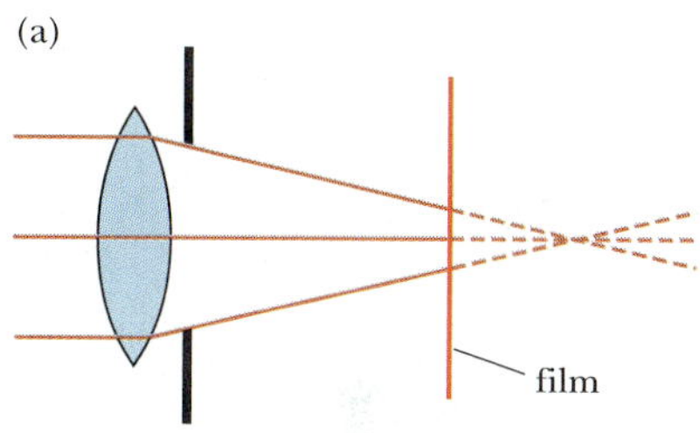

(b)

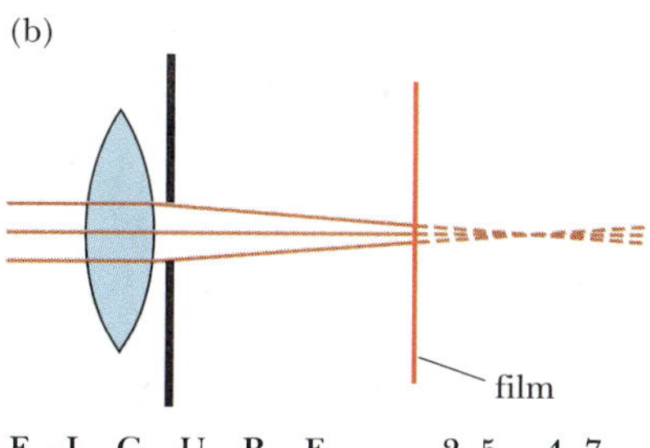

FIGURE 25.47
(a) When the aperture is large, the rays are spread over a large angle. If the focal point is beyond the film, the rays make a large spot on the film. (b) When the aperture is small, the rays are confined to a narrow range of angles. Even if the focal point is beyond the film, the rays make only a small spot on the film.

The Eye

In principle, the eye is similar to a camera. The lens of the eye forms a real image on the retina, a delicate membrane packed with light-receptor cells which send nerve impulses to the brain. Figure 25.48 shows a horizontal section through the human eye. The diameter of the eyeball is typically 2.3 cm.

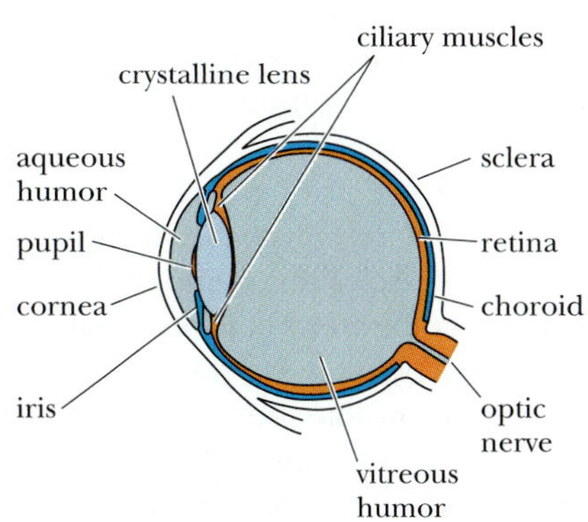

FIGURE 25.48
The human eye (natural size). The space between the cornea and the crystalline lens is filled with a transparent jelly, the aqueous humor. The main body of the eye is also filled with a transparent jelly, the vitreous humor. The indices of refraction of the humors are 1.34, nearly the same as for water. The index of refraction of the crystalline lens is 1.44. The sclera is the thick white outer casing of the eye. The choroid is a pigmented black membrane, like the black paint inside cameras.

The cornea and the aqueous humor act as a lens; they provide most of the refraction for rays entering the eye. The crystalline lens merely provides the fine adjustments of focal length required to make the image of an object fall on the retina, regardless of the object distance. This adjustment is called accommodation. The crystalline lens is flexible, and its focal length is controlled by the ciliary muscles. If the eye is viewing a distant object, the muscles are relaxed and the lens is fairly flat, with a long focal length. If the eye is viewing a nearby object, the muscles are contracted and the lens is more rounded, with a short focal length.

The shortest attainable focal length determines the shortest distance at which an object can be placed from the eye and still be seen sharply. This shortest distance is called the ***near point*** of the eye. For a normal young adult, the near point is typically 25 cm. With advancing age, the lens loses its flexibility and the near point recedes; for instance, at an age of 60 years, the near point is typically around 200 cm.

The iris serves the same function as the diaphragm of the camera — it controls the size of the pupil and the amount of light admitted to the eye. The pupil contracts to a diameter of 2 mm in bright light, and expands to 7 mm in faint light, a change of area by a factor of about 12.

The receptor cells lining the retina are of two kinds: about 120 million rods and 6 million cones. The rods are more sensitive to light than the cones, but they do not discriminate among colors. The cones are blind at low light levels, but they give us color perception.

The cones also give us high-acuity vision, because some 2000 of them are densely packed into a small spot, the fovea, which collects the most detailed visual information. The fovea is a fraction of a millimeter across, and it is located near the center of the retina. When the eye is aimed at an object, the image falls on the fovea, where the densely packed cones provide sharp resolution of fine details. For maximum exploitation of the information collected at the fovea, the cone cells in this region are connected to an equal number of nerve fibers, whereas elsewhere in the retina one nerve fiber typically services more than a hundred receptor cells.

×   ●

FIGURE 25.49

Hold the book at a distance of half a meter. Close your left eye and steadily stare at the cross. If you gradually bring the book closer to your eye, the dot will disappear when its image falls on your blind spot.

The density of rods cells is maximum at a distance of about 20° from the fovea. For this reason, very faint light, such as the light of a faint star, is best perceived peripherally, by looking to one side of the light source, instead of staring directly at it.

The bundle of nerve fibers connecting the receptor cells to the brain leaves the eye at a place somewhat to one side of the fovea. This is called the blind spot, because it lacks receptor cells. You are normally not aware of this blind spot, because the brain fills in the visual picture from information acquired during eye movements; but you can demonstrate the existence of the blind spot by means of Fig. 25.49.

Nearsighted eye ▸

The three most common optical defects of the eye are nearsightedness, farsightedness, and astigmatism. ***In a nearsighted (myopic) eye, the focal length is excessively short,*** even when the ciliary muscles are completely relaxed. Thus, parallel rays from a distant object come to a focus in front of the retina (see Fig. 25.50) and fail to form a sharp image on the retina — vision of distant objects is blurred. This condition can be corrected by eyeglasses with divergent lenses.

Farsighted eye ▸

In a farsighted (hyperopic) eye, the focal length is excessively long, even

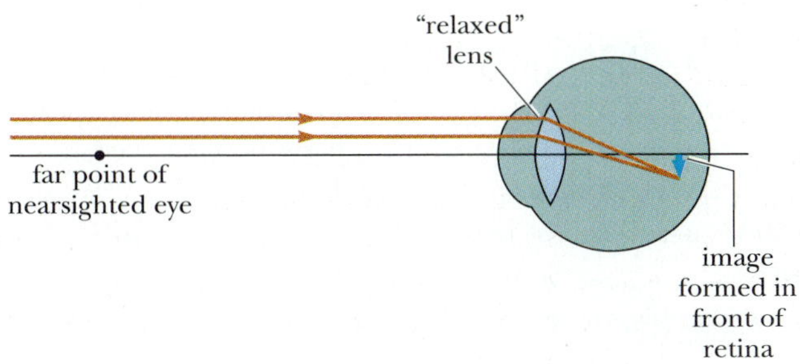

FIGURE 25.50

Nearsighted eye forms an image in front of the retina.

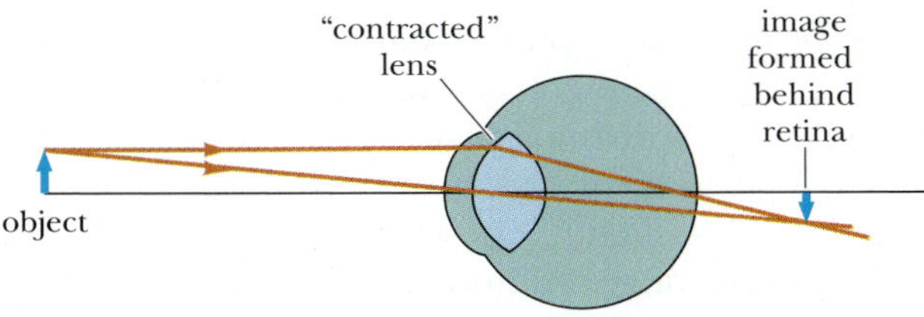

FIGURE 25.51

Farsighted eye forms an image beyond the retina.

when the ciliary muscles are fully contracted (in other words, the near point of the eye is farther away than normal). Hence rays from nearby objects converge toward an image beyond the retina (see Fig. 25.51) and fail to form a sharp image on the retina — vision of nearby objects is blurred. The condition can be corrected by eyeglasses with convergent lenses. In old age, both nearsightedness and farsightedness often occur simultaneously, through loss of flexibility of the crystalline lens and the weakening of the ciliary muscles, and the consequent failure of the eye to accommodate over the full range of distances (presbyopea). The correction then requires bifocal lenses, with a lower convergent portion for near vision and an upper divergent portion for far vision.

EXAMPLE 7. A farsighted person has a near point of 200 cm. What kind of glasses does this person need to be able to read a newspaper held at 25 cm from the eyes?

SOLUTION: The object distance is 25 cm (we ignore the small difference between distances measured from the eye and from the glasses). Since the nearest point at which the person can see distinctly is 200 cm away, the lens must form an image of the newspaper at this distance, that is, $s' = -200$ cm. The lens equation then tells us

$$\frac{1}{f} = \frac{1}{s} + \frac{1}{s'} = \frac{1}{25\text{ cm}} + \frac{1}{-200\text{ cm}} = \frac{8-1}{200}\,\frac{1}{\text{cm}}$$

and

$$f = 29\text{ cm}$$

Thus, the lenses in these eyeglasses must be convex, of focal length 29 cm.

Refractive power ▶
Diopters ▶

COMMENTS AND SUGGESTIONS: Optometrists customarily describe lenses by the value of $1/f$ rather than the value of f. The quantity $1/f$ is called the ***refractive power*** of the lens, and it is measured in units of inverse meters, called ***diopters.*** Thus, a lens with $f = 29$ cm $= 0.29$ m is said to have a refractive power of $1/f = 1/(0.29\text{ m}) = 3.4\text{ m}^{-1}$, or 3.4 diopters.

Astigmatism ▶

Astigmatism is an inability to focus simultaneously light rays arriving in different planes, for instance, light rays arriving in the vertical plane and light rays arriving in the horizontal plane. This is caused by a slight horizontal or vertical flattening of the cornea — instead of being curved spherically, the cornea is slightly out of round, with more curvature in one direction than in the other. Although astigmatism is quite common, it is often so slight as to be unnoticeable. Figure 25.52 provides a simple test for astigmatism. Correction of this condition requires a lens with a cylindrical surface, which focuses rays of light in the, say, vertical plane, but does not deflect rays of light in the horizontal plane.

FIGURE 25.52
Close one eye and focus the other on the central ends of the radial lines. If some of the lines appear less sharp or less black than others, your eye is astigmatic. (To detect mild astigmatism, you may find it helpful to perform this test on each eye and notice the difference, if any.)

The Magnifier

In order to see fine detail with the naked eye, we must bring the object very close to the eye, so the angular size of the object is large and, correspondingly, the image on the retina is large (see Fig. 25.53). This means we want to bring the object to the near point, at a typical distance of 25 cm for the eye of a young adult. To see finer detail, we need a magnifier. This consists of a strong convergent lens placed adjacent to the eye, as shown in Fig. 25.54.* Such a lens permits us to bring the object closer to the eye than 25 cm, and thereby increase the size of the image on the retina.

The angular magnification of the magnifier is defined as the ratio of the angular size of the image produced by the magnifier at infinity (as in Fig. 25.54) to the angular size of the object seen by the naked eye at the standard distance of 25 cm (as in Fig. 25.53). Since the angle θ in Fig. 25.53 is small, we can use the familiar formula that this angle (in radians) approximately equals the sine of the angle, or the size h divided by the distance of 25 cm,

$$\theta = \frac{h}{25\text{ cm}}$$

Likewise, the angle θ' in Fig. 25.54 is small and approximately equals the

Jeweler using a magnifier.

* Note that such a magnifier is not the same thing as a magnifying glass. In common use, the magnifying glass is placed at an appreciable distance from the eye, near the object to be magnified, because this maximizes the magnification. A magnifying glass can be regarded as a magnifier (in the technical sense of the word) only if it is placed near the eye.

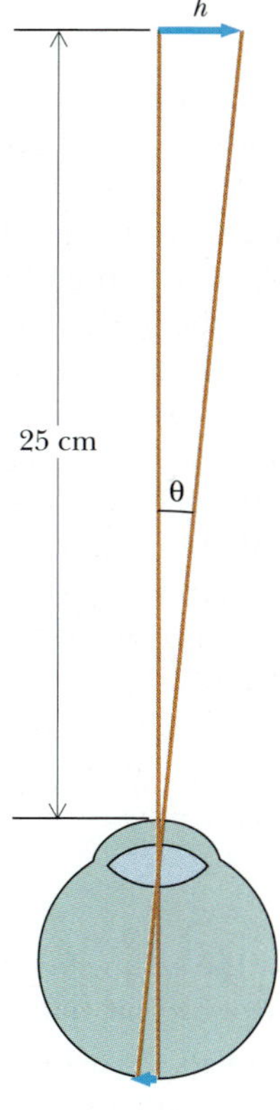

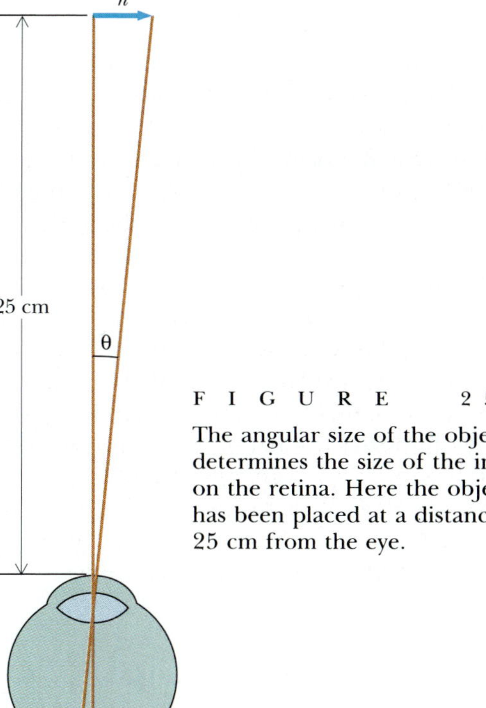

FIGURE 25.53
The angular size of the object determines the size of the image on the retina. Here the object has been placed at a distance of 25 cm from the eye.

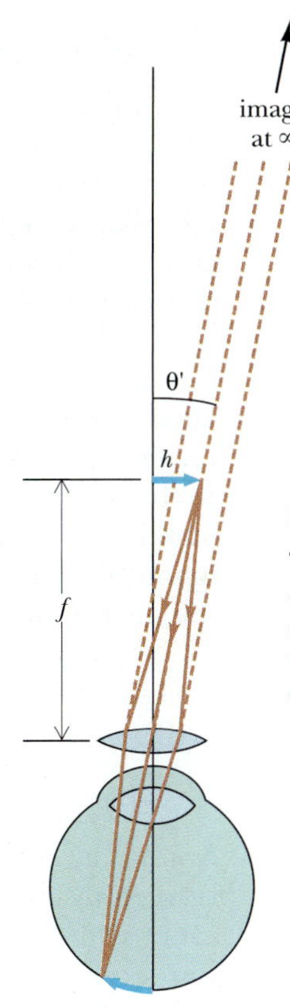

FIGURE 25.54
The magnifier is adjacent to the eye. The object has been placed at a distance slightly shorter than the focal distance, so that the eye sees the image at infinity.

size h divided by the distance f,

$$\theta' = \frac{h}{f}$$

Taking the ratio of these two angles, we find the angular magnification of the magnifier,

$$[\text{angular magnification}] = \frac{\theta'}{\theta} = \frac{25\text{ cm}}{f} \qquad (31)$$

◂ Angular magnification of magnifier

This tells us the magnification relative to that of the typical naked eye, that is, it tells us how much better the magnifier is than the naked eye. For example, a magnifier with $f = 5$ cm has an angular magnification of 25 cm/5 cm = 5. Note that this result is valid only under the assumption that the magnifier is placed adjacent to the eye, and that the object is placed near the focus of the magnifier, so the image is at infinity. The second of these assumptions is not crucial — if the object is placed somewhat closer than the focus, the magnification will be changed only slightly. But the first assumption is crucial — if the magnifier is placed at some appreciable distance from the eye, then the magnification will be quite different.

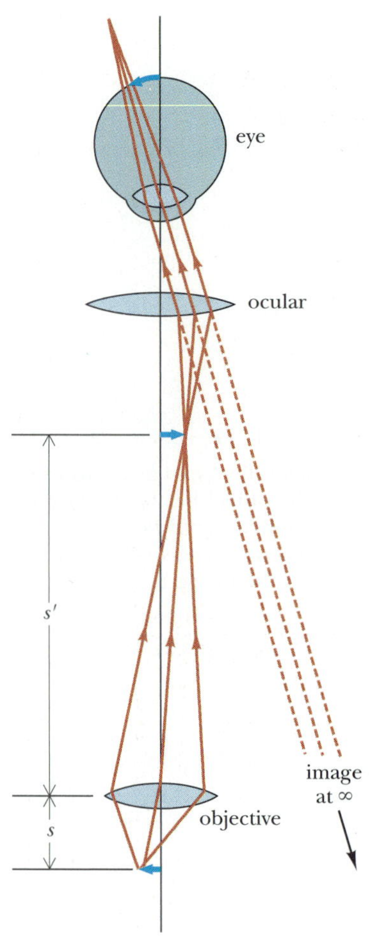

FIGURE 25.55

Arrangement of lenses in a microscope. The object to be magnified is below the objective, and the eye is just above the ocular. The ocular forms a virtual image at infinity, and the lens of the eye focuses on the retina the parallel rays emerging from the ocular.

The Microscope

The microscope consists of two lenses: the objective and the ocular, or eyepiece. Both of these lenses have very short focal lengths. The objective is placed near the object, and it forms a real, magnified image of the object. This image serves as object for the ocular, which acts as a magnifier and forms a virtual image at infinity (see Fig. 25.55). Thus both the objective and the ocular contribute to the magnification of the microscope. The net angular magnification of the microscope is the angular magnification of the ocular multiplied by the magnification of the objective. The angular magnification of the ocular is given by Eq. (31); and the magnification of the objective is given by Eq. (30), where s and s' are, respectively, the object and image distances for the objective (these distances are marked in Fig. 25.55). If we combine these two magnifications, and if we ignore the sign in Eq. (30), we find that the net angular magnification of the microscope is

$$[\text{angular magnification}] = \frac{25\text{ cm}}{f_{\text{oc}}} \times \frac{s'}{s} \tag{32}$$

Note that, as in the case of the magnifier, this tells us the magnification relative to the (typical) naked eye. Good microscopes operate at magnifications of up to 1400. Although higher magnifications can be achieved, this serves little purpose because diffraction of the light (see next chapter) at the objective limits the details that can be resolved. To overcome this limitation, we need to use waves of shorter wavelength than light waves, such as the electron waves used in electron microscopes.

The Ophthalmoscope

The ophthalmoscope is widely used by physicians to examine the interior of the eyeball. It consists of a light source that throws a strong beam of light into the patient's eye with a mirror and a magnifier that permits the physician to view the illuminated portion of the retina (see Fig. 25.56). The light from the illuminated portion of the retina travels through the lens of the patient's eye and through the magnifier lens. These two lenses act like the objective and the ocular lens of a (low-power) microscope, giving the physician an enlarged view of the retinal surface. Figure 25.57 is a photograph of the retina as seen through an ophthalmoscope; the blood vessels on the surface of the retina stand out clearly.

The Telescope

A simple astronomical telescope consists of an objective of very long focal length and an ocular of short focal length. These two lenses are separated by a distance (nearly) equal to the sum of their individual focal lengths, so their focal points coincide. The objective forms a real image FP' of a distant object. This image serves as object for the ocular, which forms a magnified virtual image at infinity (see Fig. 25.58).

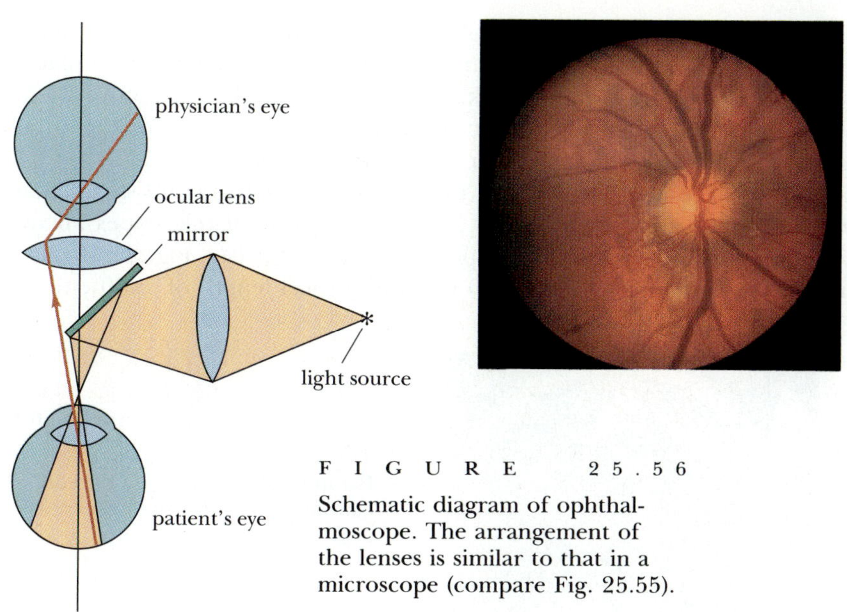

FIGURE 25.56

Schematic diagram of ophthalmoscope. The arrangement of the lenses is similar to that in a microscope (compare Fig. 25.55).

FIGURE 25.57

View of retina through an ophthalmoscope.

To find the angular magnification produced by this telescope, we begin by noting that the lens equation with $s = \infty$ applied to the objective gives

$$\frac{1}{\infty} + \frac{1}{s'} = \frac{1}{f_{\text{ob}}} \tag{33}$$

But $1/\infty = 0$, and hence

$$s' = f_{\text{ob}} \tag{34}$$

This means that the image is at the focal distance. Accordingly, in Fig. 25.58, the image FP' is shown located at the focal point F. We can now use the geometric relationships contained in this figure. The angular magnification is the ratio of the angles θ and θ' that represent, respectively, the angular sizes of the object and the final image viewed by the eye. Since both these angles are small, their values (in radians) are approximately the transverse distance FP' divided by the distance to the appropriate lens,

$$\frac{\theta'}{\theta} = \frac{FP'/FB}{FP'/FA} = \frac{FA}{FB} = \frac{f_{\text{ob}}}{f_{\text{oc}}}$$

and the angular magnification is

$$[\text{angular magnification}] = \frac{\theta'}{\theta} = \frac{f_{\text{ob}}}{f_{\text{oc}}} \tag{35}$$

For example, an astronomical telescope with $f_{\text{ob}} = 120$ cm and $f_{\text{oc}} = 2.5$ cm has an angular magnification of 120 cm/2.5 cm = 48.

Many astronomical telescopes are reflecting telescopes in which a con-

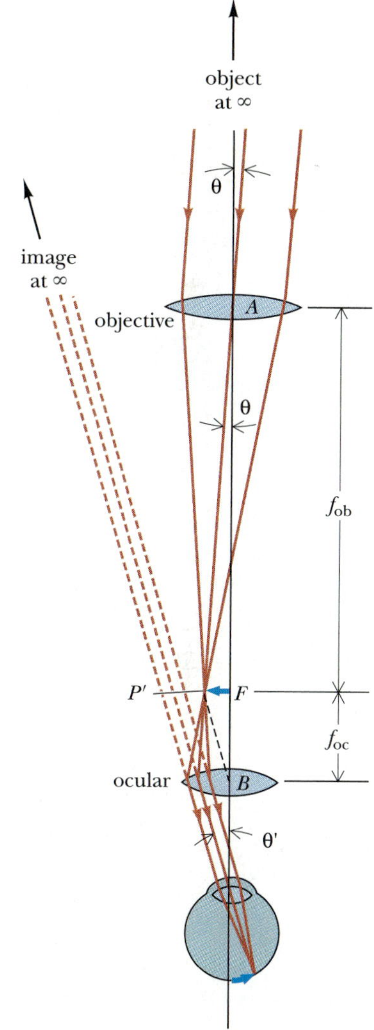

FIGURE 25.58

An astronomical telescope. The object (a star) is at a large distance above. The observer's eye is below the ocular. The ocular forms a virtual image at infinity, and the lens of the eye focuses on the retina the parallel rays emerging from the ocular.

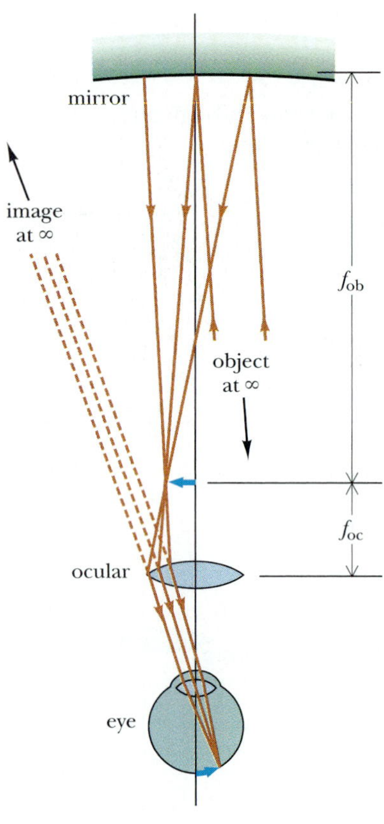

FIGURE 25.59

A reflecting telescope. The object has been placed below, to facilitate comparison with Fig. 25.58.

cave mirror plays the role of the objective. The mirror forms a real image that serves as object for the ocular. Of course, the ocular must be placed in front of the mirror (and blocks out some of the light). Figure 25.59 shows this arrangement of mirror and ocular. The formula for the angular magnification of such a mirror telescope is, again, Eq. (35).

Because large mirrors of good quality are easier to manufacture than large lenses of good quality, the largest astronomical telescopes all use mirrors. For instance, the telescope on Mt. Palomar (see Fig. 25.60) uses a mirror of diameter 510 cm (200 in.) and a focal length of 1680 cm.

FIGURE 25.60

The 200-in. telescope on Palomar Mountain. The mirror can be seen under the observer cage.

SUMMARY

- **Huygens' Construction:** The new wave front is the surface that touches the wavelets erected on the old wave front.
- **Law of reflection:** The angle of incidence equals the angle of reflection.
- **Index of refraction:** $v = c/n$
- **Law of refraction:** $\sin\theta = n \sin\theta'$
- **Critical angle for total internal reflection:** $\sin\theta'_{\text{crit}} = 1/n$
- **Focal length of spherical mirror:** $f = \pm\frac{1}{2}R$ (f is positive for concave mirror, negative for convex)
- **Mirror equation:** $\frac{1}{s} + \frac{1}{s'} = \frac{1}{f}$ (s or s' is positive if object or image is in front of mirror, negative if behind)
- **Lens-maker's formula:** $\frac{1}{f} = (n-1)\left(\frac{1}{R_1} + \frac{1}{R_2}\right)$ (f is positive for convex lens, negative for concave)
- **Lens equation:** $\frac{1}{s} + \frac{1}{s'} = \frac{1}{f}$ (s is positive if object is on near side of lens, negative if on far side; s' is positive if image is on far side of lens, negative if on near side)

- **Magnification:** $M = -s'/s$
- **Angular magnification of magnifier:** $25\ \text{cm}/f$
- **Angular magnification of microscope:** $(25\ \text{cm}/f_{oc}) \times (s'/s)$
- **Angular magnification of telescope:** f_{ob}/f_{oc}

QUESTIONS

FIGURE 25.61

1.–When light is incident on a smooth surface—a glass surface, a painted surface, a water surface—the reflection is strongest if the angle of incidence is near 90° (grazing incidence). Can Huygens' Construction explain this?

2.–In celestial navigation, the navigator measures the angle between the Sun, or some other celestial body, and the horizon with a sextant. If the navigator is on dry land, where the horizon is not visible, he can measure instead the angle between the Sun and its reflection in a pan full of water, and divide this angle by two. Explain.

3.–Suppose we release a short flash of light in the space between two parallel mirrors placed face-to-face. Why does this light flash not travel back and forth between the two mirrors forever?

4.–Artists are notorious for making mistakes when drawing or painting mirror images. What is wrong with the position and orientation of the mirror images shown in the cartoon in Fig. 25.61?

5.–Two parallel mirrors are face-to-face. Describe what you see if you stand between these two mirrors.

6.–If you immerse one-half of a stick in the water, it will appear bent. Explain.

7.–At sunset, the image of the Sun remains visible for some time after the actual position of the Sun has sunk below the horizon. Explain.

8.–After a navigator measures the angle between the Sun and the horizon with a sextant, he must make a correction for the refraction of sunlight by the atmosphere of the Earth. Does this refraction increase or decrease the apparent angle between the Sun and the horizon?

9.–At amusement parks you find mirrors that make you look very short and fat or very tall and thin. What kinds of mirrors achieve these effects?

10.–Storeowners often install convex mirrors at strategic locations in their stores to supervise the customers. What is the advantage of a convex mirror over a flat mirror?

Convex mirrors used to supervise customers in a store.

11.–You look toward a lens or a mirror and you see the image of an object. How can you tell whether this image is real or virtual?

12.–Hand mirrors are sometimes concave, but never convex. Why?

13.–If you place a book in front of a concave mirror, will it show you mirror writing? Does the answer depend on the distance of the book from the mirror?

14.–How could you make a lens that focuses sound waves?

15.–If you place a small luminous light bulb at the focus of a convex lens and look toward the lens from the other side, what will you see?

(a)

(b)

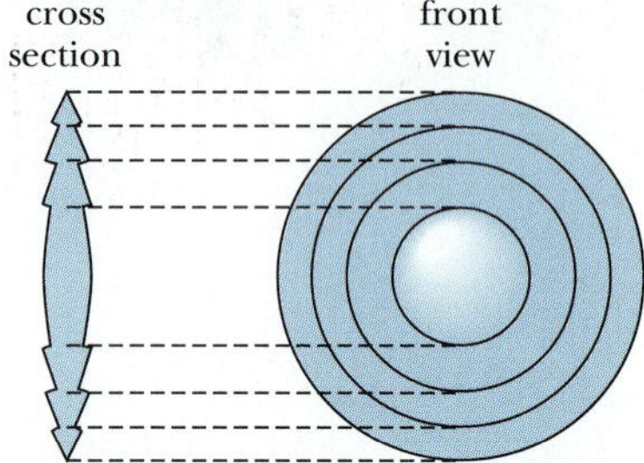

FIGURE 25.62

(a) Fresnel lenses of the lighthouse at Heceta Head, Oregon. (b) Detail of one of these lenses.

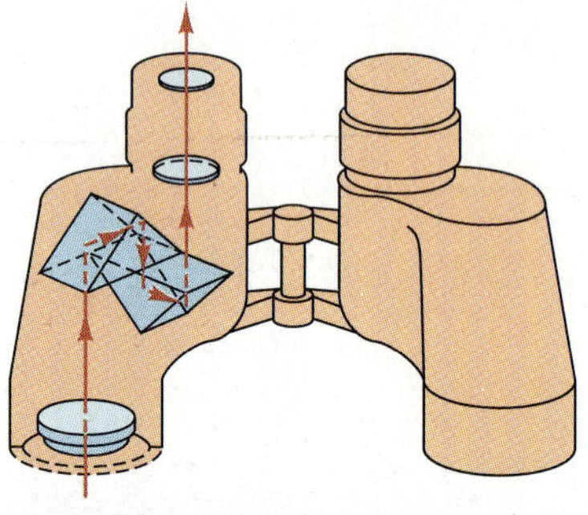

FIGURE 25.63

Arrangement of prisms in a binocular.

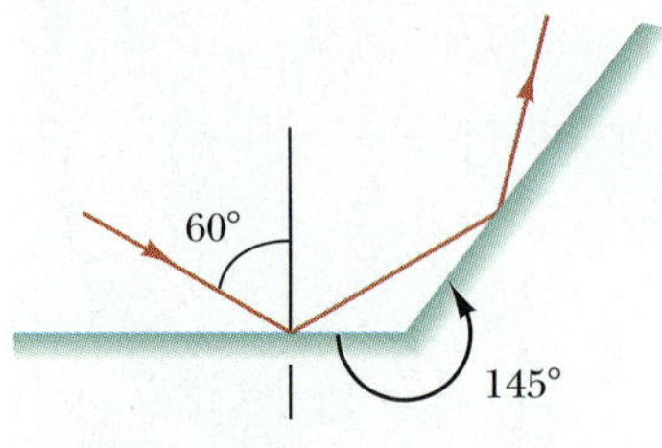

FIGURE 25.64

Two adjacent flat mirrors at an angle.

16.–Consider Fig. 25.41. How do we know that the ray $PQ'P'$ emerges parallel to the axis of the lens?

17.–Suppose you place a magnifying glass against a flat mirror and look into the glass. What do you see if your face is very near the glass? If it is not very near?

18.–Figure 25.62 shows large **Fresnel lenses** used in the lantern of a lighthouse. The lens consists of annular segments, each with a curved surface similar to the curved surface of an ordinary lens. Why is this arrangement better than a single curved surface?

19.–Binoculars use prisms to reflect the light back and forth (Fig. 25.63). What is the purpose of these prisms, and what is their advantage over mirrors?

PROBLEMS

Section 25.2: Reflection

1.——According to an ancient report, Archimedes set fire to the Roman ships besieging Syracuse by focusing the light of the Sun on them with mirrors. Suppose that Archimedes used flat mirrors. How many flat mirrors must simultaneously reflect sunlight at a piece of canvas if it is to catch fire? The energy flux of the sunlight at the surface of the Earth is 0.1 W/cm^2 and the energy flux required for ignition of canvas if 4 W/cm^2. Assume that the mirrors reflect the sunlight without loss.

2.——Two flat mirrors are arranged at an angle of 145° with respect to each other (see Fig. 25.64). A ray of light strikes one of the mirrors at an angle of incidence of 60°, it is reflected, and it then strikes the other mirror. What is the angle of reflection at the second mirror?

3.——A ray of light is incident on a flat mirror at an angle of θ. If we then rotate the mirror by 10° (see Fig. 25.65) while keeping the incident ray fixed, by what angle will we rotate the reflected ray?

4.*—Two flat mirrors are joined along their edges so they make an angle of 90° with each other (see Fig. 25.66). A cat sits in the angle formed by the mirrors. Draw a diagram showing the location and orientation of all the images of the cat in this system of mirrors.

5.*—Two mirrors meeting at a right angle make a corner reflector (see Fig. 25.67). Prove that a ray of light reflected successively by both mirrors will emerge on a path antiparallel to its original path.

6.**—A vertical mirror, oriented toward the Sun, throws a rectangular patch of sunlight on the floor in front of the mirror. The size of the mirror is 0.5 m × 0.5 m and its bottom rests on the floor. If the Sun is 50° above the horizon, what is the size of the patch of sunlight on the floor?

7.**—What is the minimum size of a mirror hanging on a wall such that you can see your entire body when standing in front of it? Assume that your height is 1.8 m.

Section 25.3†: Refraction

8.——A ray of sunlight approaches the surface of a smooth pond at an angle of incidence of 40°. What is the angle of refraction?

9.——What is the speed of light in crown glass? Light flint? Heavy flint?

† In all problems, assume that the index of refraction of air is 1, unless otherwise stated.

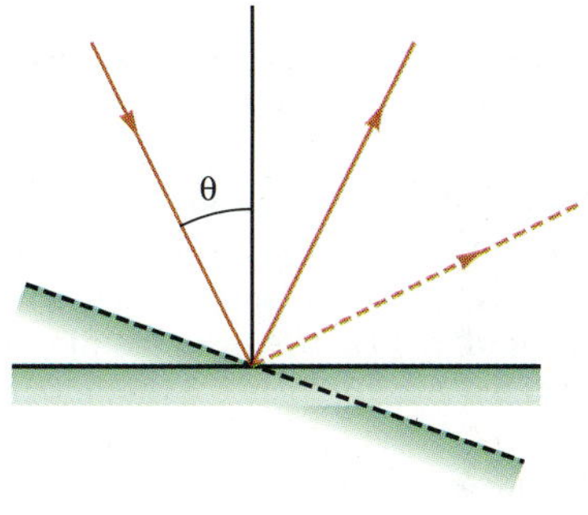

FIGURE 25.65

A flat mirror, before and after rotation.

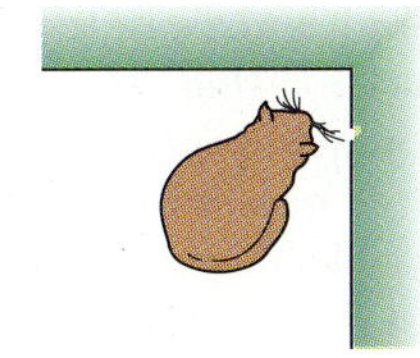

FIGURE 25.66

Cat looking at two mirrors at an angle of 90°.

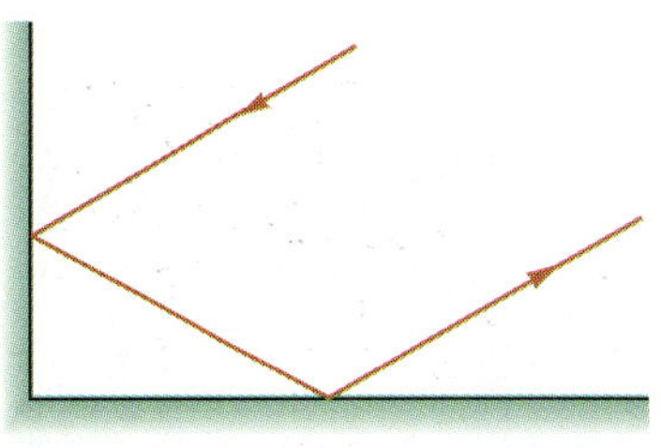

FIGURE 25.67

Two flat mirrors at an angle of 90°.

10.——The speed of sound in air is 340 m/s and in water it is 1500 m/s. If a sound wave in air approaches a water surface with an angle of incidence of 10°, what will be the angle of refraction?

11.——Make a plot of the angle of incidence vs. the angle of refraction for light rays incident on a water surface. What is the maximum angle of refraction?

12.——A ship's navigator observes the position of the Sun with his sextant and measures that the Sun is exactly 39° away from the vertical. Taking into account the refraction of the Sun's light by air, what is the true angular position of the Sun? For the purpose of this problem assume that the Earth is flat and that the atmosphere can be regarded as a flat, transparent plate of uniform density and an index of refraction of 1.0003.

13.——A ray of light strikes a plate of window glass of index of refraction 1.5 with an angle of incidence of 50°.

(a) What is the angle of refraction?

(b) At what angle does the ray of light emerge on the other side of the glass?

14.*——A prism of glass has two angles of 45° and one of 90°. To determine the index of refraction of the glass in this prism, an experimenter aims a beam of light perpendicularly into a face of this prism and measures the angle at which the ray emerges on the far side (see Fig. 25.68). If this angle is 72°, what is the index of refraction?

15.*——The bottom corner of a swimming pool is 2.0 m below the surface of the water. If a ray of light travels upward from the corner and reaches the water surface at an angle of 40° with the vertical, what will be the angle the ray makes with the vertical when it emerges from the water? If your eye intercepts this ray, how deep will the swimming pool seem to you?

16.*——A prism has three equal angles of 60°. The prism is made of flint glass of index of refraction 1.66. At what angle should a ray of light enter this prism so it is refracted symmetrically (see Fig. 25.69)?

17.*——The walls of a (filled) aquarium are made of glass with an index of refraction of 1.5. If a ray strikes the glass from the inside at an angle of incidence of 45°, what is the angle at which it emerges into the air?

18.*——One of the angles of a prism is 90°. The prism is made of glass of index of refraction 1.58. If a ray of light enters this prism at an angle of incidence of 70°, as shown in Fig 25.70, what is the angle at which this ray emerges?

19.*——The bottom half of a beaker of depth 20 cm is filled with water ($n = 1.33$), and the top half is filled with oil ($n = 1.48$). If you look into this beaker from above, how far below the upper surface of the oil does the bottom of the beaker seem to be?

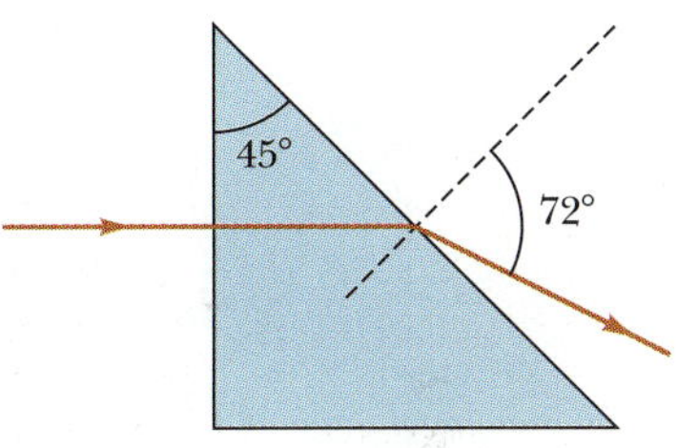

FIGURE 25.68

A prism.

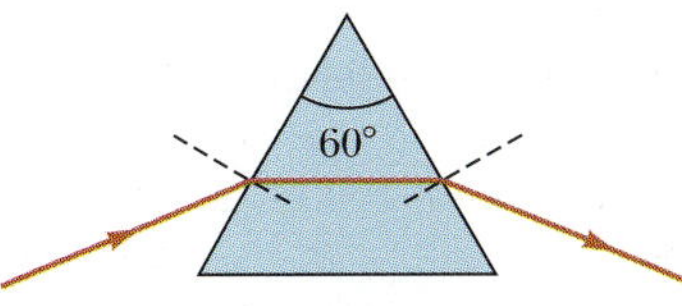

FIGURE 25.69

A symmetric prism.

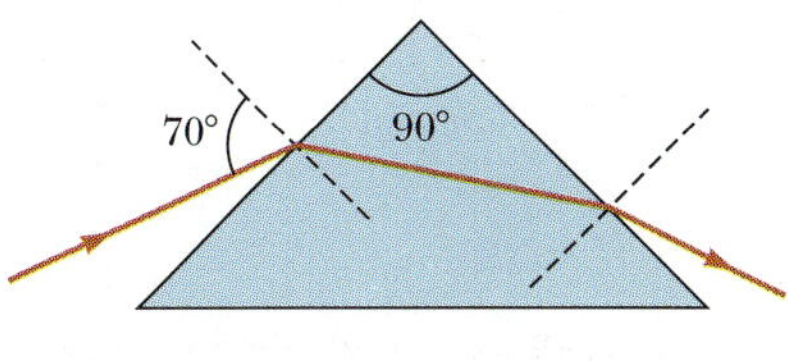

FIGURE 25.70

Another prism.

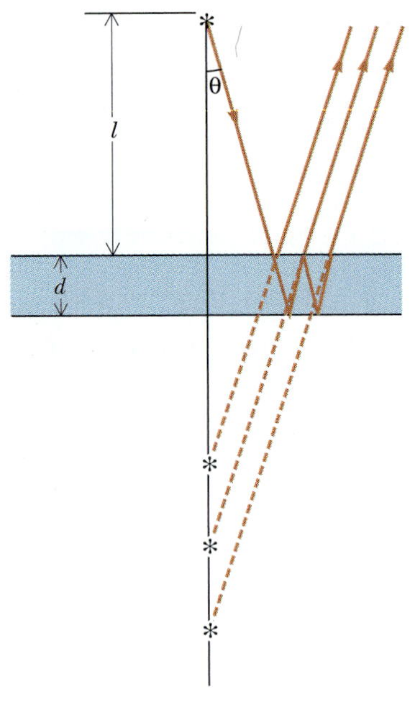

FIGURE 25.71

Ray of light reflected and refracted by a thick plate of glass.

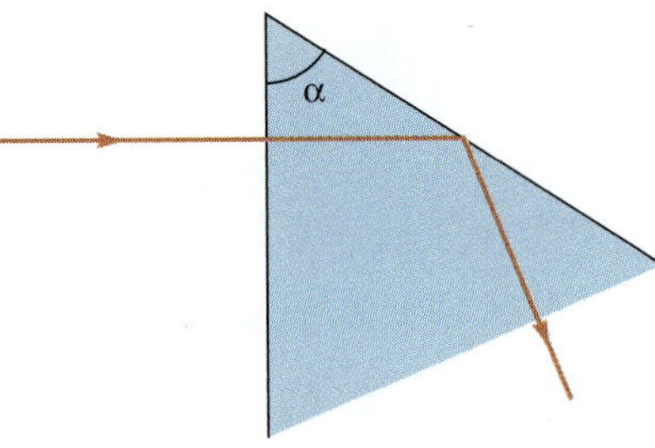

FIGURE 25.72

Total internal reflection in a prism.

20.** —A point source of light is placed above a thick plate of glass of index of refraction n (Fig. 25.71). The distance from the source to the upper surface of the plate is l, and the thickness of the plate is d. A ray of light from the source may suffer either a single reflection at the upper surface, or a single reflection at the lower surface, or multiple alternating reflections at the lower and upper surfaces. Thus, each ray splits into several rays, giving rise to multiple images. In terms of l and d, find the distance of the first and second images below the upper surface of the plate. Assume that the angle of incidence of the ray is small.

21.—To determine the index of refraction of a liquid, an experimenter sends a ray of light through the liquid from below and measures the critical angle for total internal reflection. She finds that this angle is 56°. What is the index of refraction?

22.—An optical fiber is made of a thin strand of glass of index of refraction 1.5. If a ray of light is to remain trapped within this fiber, what is the largest angle it may make with the surface of the fiber?

23.* —A signal rocket explodes at a height of 200 m above a ship on the surface of a smooth lake. The explosion sends out sound waves in all directions. Since the speed of sound in water (1500 m/s) is larger than the speed of sound in air (340 m/s), a sound wave can suffer total reflection at a water surface if it strikes at a sufficiently large angle of incidence. At what minimum distance from the ship will the sound wave from the explosion suffer total reflection?

24.* —A prism is to be used for total internal reflection of a ray of light perpendicularly incident on one of its faces (see Fig. 25.72). The index of refraction of the glass is 1.60. What must be the angle of the vertex of this prism to accomplish this total internal reflection?

25.* —(a) A transparent medium of index of refraction n_1 adjoins a transparent medium of index of refraction n_2. Assuming $n_2 > n_1$, show that the critical angle for total internal reflection of a ray attempting to leave the second medium is

$$\sin \theta'_{\text{crit}} = \frac{n_1}{n_2}$$

(b) A layer of kerosene ($n_1 = 1.2$) floats on a water surface. In this case, what is the critical angle for total internal reflection for a light ray attempting to leave the water?

26.* —What is the critical angle for a ray of light to be internally reflected inside a plate of crown glass? What if this plate is immersed in water?

27.* —An indoor swimming pool has a depth of 2.0 m. The ceiling of the room is 3.0 m above the surface of the water. A waterproof electric light is installed in the bottom of the pool. What is the radius of the illuminated circle that this electric light throws on the ceiling?

28.* —A seagull sits on the (smooth) surface of the sea. A shark swims toward the seagull at a constant depth of 5 m. How close (measured horizontally) can the shark approach before the seagull can see it clearly?

29.** —To discover the percentage of sucrose (cane sugar) in an aqueous solution, a chemist determines the index of refraction of the solution very precisely and then finds the percentage in a table giving the dependence of index of refraction on sucrose concentration. He determines the index of refraction by immersing a glass prism in the sucrose solution and measuring the critical angle for total internal reflection of a light ray inside the glass prism.

(a) Suppose that with a prism of index of refraction 1.6640 the critical angle is 57.295°. Use the result of Problem 25a to find the index of refraction of the sucrose solution.

(b) Use the following table, interpolating if necessary, to find the concentraton of sucrose to four significant figures.

Concentration	n
40.00%	1.3997
40.10	1.3999
40.20	1.4001
40.30	1.4003

30.**—Consider a light wave incident on a dielectric medium of index of refraction n. It can be shown that if the light is polarized in the plane of incidence (plane of the page in Fig. 25.73) and the refracted ray is perpendicular to the reflected ray, then the intensity of the reflected wave is zero. Show that the angle of incidence that makes the refracted ray perpendicular to the reflected ray is given by $\tan\theta = n$. This relation is called **Brewster's Law.**

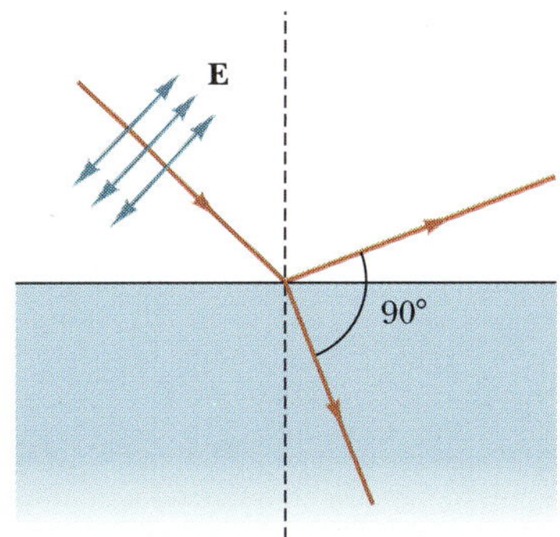

FIGURE 25.73

The light wave is polarized in the plane of the page. The refracted ray is perpendicular to the reflected ray.

31.*—A prism of flint glass has an acute angle of 30°. A ray of white light is perpendicularly incident on one face of this prism (see Fig. 25.74). What are the angles at which red, green, and violet light emerge? The indices of refraction for red (660 nm), green (550 nm), and violet (400 nm) light are 1.650, 1.660, and 1.690, respectively.

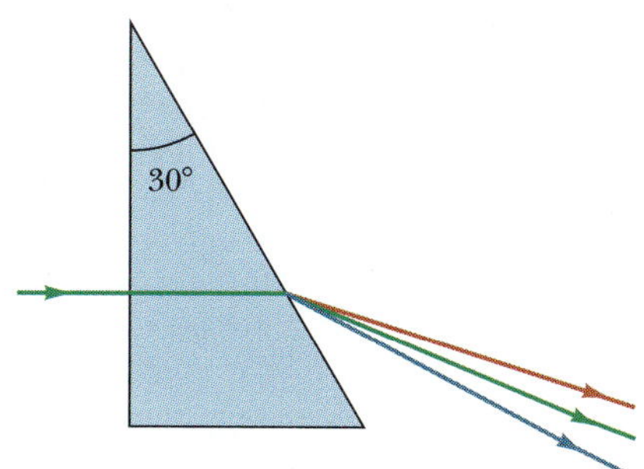

FIGURE 25.74

Spectrum formed by a prism.

Outside rear view mirror of an automobile.

Section 25.4: Spherical Mirrors

32.—A candle is placed at a distance of 15 cm in front of a concave mirror of radius of curvature 40 cm. Where is the image of the candle? Draw a diagram showing the candle, the mirror, and the image. Show the three principal rays.

33.—A light bulb is placed at a distance of 15 cm in front of a convex mirror of radius of curvature 40 cm. Where is the image of the light bulb? Draw a diagram showing the light bulb, the mirror, and the image. Show the three principal rays.

34.—The outside rearview mirror on an automobile is convex. When you look into this mirror, you notice that the image of a distant truck or some other distant object is about 1.0 m beyond the mirror. What is the radius of curvature of the mirror?

35.—In the head lamp of an automobile, a light bulb placed in front of a concave spherical mirror generates a parallel beam of light. If the distance between the light bulb and the mirror is 4 cm, what must the radius of the mirror be?

36.—The mirror of the Hubble Space Telescope is approximately spherical, of radius of curvature of 11.04 m. As shown in Fig. 25.75, a technician stands at a distance of 4.0 m in front of this mirror. What is the location of the image?

FIGURE 25.75

Mirror of the Hubble Space Telescope.

37.——You place an object in front of a concave mirror, at a distance smaller than the focal length. Is the image real or virtual? Erect or inverted? Magnified or reduced? What if the distance is greater than the focal length? What if the mirror is convex?

38.*——In a grocery store, a convex mirror with a radius of curvature of 4.0 m is used for surveillance. Suppose a person of height 1.7 m is 12 m from such a mirror. Where is the image? What is the size of the image?

39.*——At what distance from a concave mirror of radius R must you place an object if the image is to be at the same position as the object?

40.*——A concave mirror has a radius of curvature R. If you want to form a real image, within what range of distances from the mirror must you place the object? If you want to form a virtual image, within what range of distances must you place the object?

41.*——A woman's hand mirror is to show a (virtual) image of her face magnified 1.5 times when held at a distance of 20 cm from the face. What must be the radius of curvature of a spherical mirror that will serve the purpose? Must it be concave or convex?

42.*——The surface of a highly polished convex doorknob of brass has a radius of curvature of 4.5 cm. If you hold this doorknob 15 cm away from your face, where is the image that you see? By what factor does the size of the image differ from the size of your face?

43.*——A concave mirror of radius 30 cm faces a second concave mirror of radius 24 cm. The distance between the mirrors is 80 cm and their axes coincide. A light bulb is suspended between the mirrors, at a distance of 20 cm from the first mirror.

(a) Where does the first mirror form an image of the light bulb?

(b) Where does the second mirror form an image of this image?

44.**——A concave mirror of radius 60 cm faces a convex mirror of the same radius. The distance between the mirrors is 50 cm, and their axes coincide. A candle is held between the mirrors, at a distance of 10 cm from the convex mirror. Consider rays of light that first reflect off the concave mirror and then off the convex mirror. Where do these rays form an image?

Section 25.5: Thin Lenses

45.——The crystalline lens of a human eye has two convex surfaces with radii of curvature of 10 mm and 6.0 mm. The index of refraction of its material is 1.44. Treating it as a thin lens, what is its focal length when it is removed from the eye and placed in air?

46.——A thin lens of flint glass with $n = 1.58$ has one concave surface of radius 15 cm and one flat surface.

(a) What is the focal length of this lens?

(b) If you place this lens at a distance of 40 cm from a candle, where will you find the image of the candle?

47.——A thin, symmetric, convex lens of crown glass with index of refraction $n = 1.52$ is to have a focal length of 20 cm. What are the correct radii of the spherical surfaces of the lens?

48.——A convex lens is made of plastic of index of refraction 1.2. If you immerse this lens in water, will it produce convergence or divergence of incident parallel rays?

49.——A light bulb is placed at a distance of 15 cm in front of a convex lens of focal length 30 cm. Where is the image of the light bulb? Draw a diagram showing the light bulb, the lens, and the image. Show the three principal rays.

50.——A candle is placed at a distance of 50 cm in front of a convex lens of focal length 30 cm. Where is the image of the candle? Draw a diagram showing the candle, the lens, and the image. Show the three principal rays.

51.——A light bulb is placed at a distance of 15 cm in front of a concave lens of focal length 30 cm. Where is the image of the light bulb? Draw a diagram showing the light bulb, the lens, and the image. Show the three principal rays.

52.——A candle is placed at a distance of 50 cm in front of a concave lens of focal length 30 cm. Where is the image of the candle? Draw a diagram showing the candle, the lens, and the image. Show the three principal rays.

53.——You place an object in front of a convex lens at a distance smaller than the focal length. Is the image real or virtual? Erect or inverted? Magnified or reduced? What if the distance is greater than the focal length? What if the lens is concave?

54.——A slide projector has a lens of focal length 13 cm. The slide is at a distance of 2.0 m from the screen. What must be the distance from the slide to the lens if a sharp image of the slide is to be seen on the screen?

55.——If you place a convex lens of focal length 18 cm at a distance of 30 cm from a small light bulb, where will you find the image of the light bulb? Is this a real or virtual image? Is it upright or inverted?

56.——The convex lens of a magnifying glass has a focal length of 20 cm. At what distance from a postage stamp must you hold this lens if the image of the stamp is to be twice as large as the stamp?

57.*—A convex lens of focal length 25 cm is at a distance of 60 cm from a concave mirror of focal length 20 cm. A light bulb is 80 cm from the lens (Fig. 25.76).

(a) Where does the lens form an image of the light bulb?

(b) Where does the mirror form an image of this image?

58.*—Two lenses, one concave and one convex, have equal focal lengths of 30 cm. The lenses are separated by a distance of 10 cm. A candle is 20 cm from the convex lens (Fig. 25.77).

(a) Where does the convex lens form an image?

(b) Where does the concave lens form an image of this image?

59.**—A light bulb is 15 cm in front of a convex mirror of radius 10 cm. A convex lens of focal length 25 cm is 5 cm beyond the light bulb (Fig. 25.78). Where do you see the light bulb if you look through the convex lens at the mirror?

Section 25.6: Optical Instruments

60.——The lens of a 35-mm camera has a focal length of 50 mm. The distance of the lens from the film is adjustable over a range from 50 mm to 56 mm. Over what range of object distances (measured from the lens) is this camera capable of producing sharp pictures?

61.——A miniature Minox camera has a lens of focal length 15 mm. This camera can be focused on an object as close as 20 cm, or as far away as infinity. What must be the distance from the lens to the film if the camera is set for 20 cm? What if the camera is set for infinity?

62.——The light meter of a 35-mm camera with a lens of f number 1.7 indicates that the correct exposure time for a photograph is $\frac{1}{250}$ s. If the iris diaphragm is closed down so that the f number becomes 4, what will be the correct exposure time?

63.——Pretend that the cornea and the crystalline lens of the human eye act together as a single thin lens placed at a distance of 2.2 cm from the retina (Fig. 25.79). This lens is deformable; it can change its focal length by changing its shape.

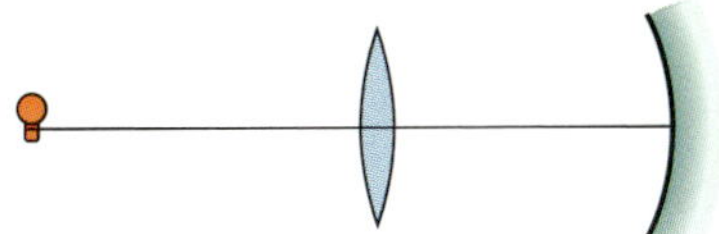

FIGURE 25.76

Light bulb, convex lens, and concave mirror.

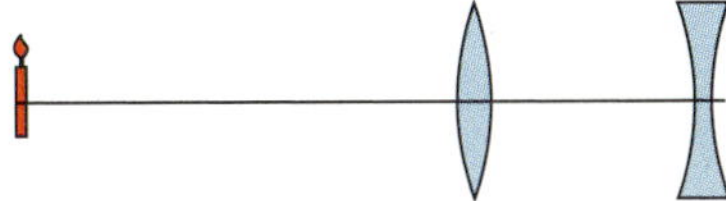

FIGURE 25.77

Candle, convex lens, and concave lens.

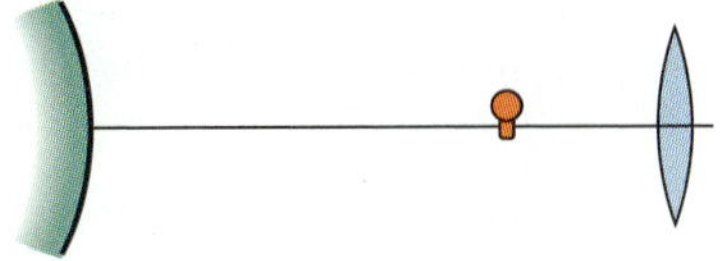

FIGURE 25.78

Light bulb, convex mirror, and convex lens.

Lens of a 35-mm camera.

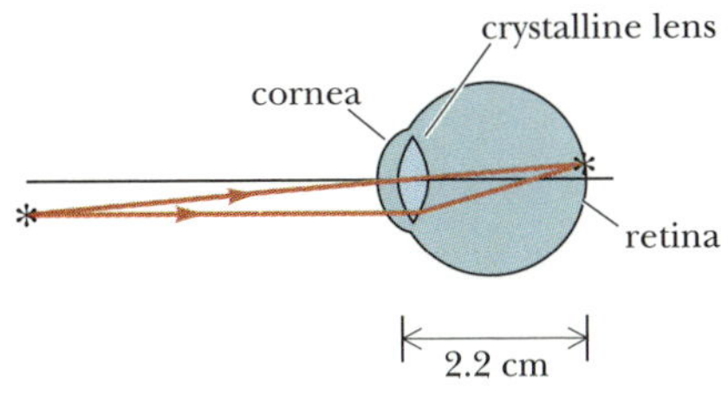

FIGURE 25.79

Lens of human eye forms image on retina.

(a) What must be the focal length if the eye is viewing an object at a very large distance?

(b) What must be the focal length if the eye is viewing an object at a distance of 25 cm?

64.—A microscope has an objective of focal length 4.0 mm. This lens forms an image at a distance of 224 mm from the lens. If we want to attain a net angular magnification of 550, what choice must we make for the angular magnification of the ocular?

65.—A microscope has an objective of focal length 1.9 mm and an ocular of focal length 25 mm. The distance between these lenses is 180 mm.

(a) At what distance must the object be placed from the objective so that the ocular forms an image at infinity, as shown in Fig. 25.55?

(b) What is the net angular magnification of this microscope?

66.—A telescope has an objective of focal length 160 cm and an ocular of focal length 2.5 cm. If you look into the *objective* (that is, into the wrong end) of this telescope, you will see distant objects *reduced* in size. By what factor will the angular size of objects be reduced?

The Multiple-Mirror Telescope on Mt. Hopkins, Arizona.

67.—An amateur astronomer uses a telescope with an objective of focal length 90.0 cm and an ocular of focal length 1.25 cm. What is the angular magnification of this telescope?

68.—The large reflecting telescope on Mt. Palomar has a mirror of focal length 1680 cm. If this telescope is operated with an ocular of focal length 1.25 cm, what is the angular magnification?

69.—It has been proposed that the six 1.8-m mirrors in the Multiple-Mirror Telescope be replaced by a single 6.5-m mirror, which would fit into the existing mounting. By what factor would this replacement enhance the amount of light collected by the telescope?

70.**—Equation (31) gives the angular magnification for a magnifier if the object is so placed that the image is at infinity. Show that if the object is so placed that the image is at a distance of 25 cm, then the angular magnification is $1 + 25\ \text{cm}/f$.

CHAPTER 26

The Very Large Array (VLA) radiotelescope in New Mexico consists of 27 separate dish antennas connected together, so the radio signals arriving at them can interfere constructively or destructively. Two of these dish antennas are seen here.

INTERFERENCE AND DIFFRACTION

Geometric optics relies on the assumption that the propagation of light is rectilinear, except when it is reflected or refracted at the interface between two different materials. This means that in the calculations with mirrors and lenses in the preceding chapter we could adequately describe the behavior of light by its rays, and we did not have to worry about the wave properties of light, such as wavelength, frequency, wave amplitude, etc. The wave properties of light do not show up explicitly in systems of mirrors or lenses, although these wave properties entered into our derivations of the laws of reflection and refraction.

The absence of any explicit wave effects during the propagation of light through a system of mirrors and lenses is related to the small wavelength of light. In comparison with the sizes of typical mirrors and lenses, the wavelength of light is extremely small (about 5×10^{-7} m). We cannot detect this small wavelength and the small-scale wave behavior of light by means of ordinary mirrors and lenses for the same reason that we cannot detect ripples on the surface of the ocean by means of a supertanker — our instruments are too crude. If we want to detect ripples on the surface of the ocean, we need to use small pieces of flotsam, preferably of a size comparable with the wavelength of the ripples. Likewise, to detect the wave behavior of light, we need to let the light interact with a body or an obstacle of a size comparable with a wavelength. In that case, light will display its wave properties by the phenomena of interference and diffraction. As mentioned in Chapters 10 and 11, ***interference is the constructive or destructive combination of two or more waves meeting at one place; diffraction is the bending and spreading of waves around obstacles. Wave optics, or physical optics, deals with interference and diffraction of light,*** and other phenomena that directly involve the wave properties of light.

Interference and diffraction ▶

Wave optics ▶

In this chapter we will begin by examining the interference of light waves and other electromagnetic waves. Like all electric and magnetic fields, the fields of electromagnetic waves obey the principle of superposition: if two waves meet at some point, the resultant electric or magnetic field is simply the vector sum of the individual fields. If two waves of equal amplitude meet crest to crest, they combine and produce a wave of double amplitude; if they meet crest to trough, they cancel and give a wave of zero amplitude. The former case is called ***constructive interference*** and the latter ***destructive interference.*** We will encounter both cases of interference in the following sections.

Constructive and destructive interference ▶

26.1 THIN FILMS

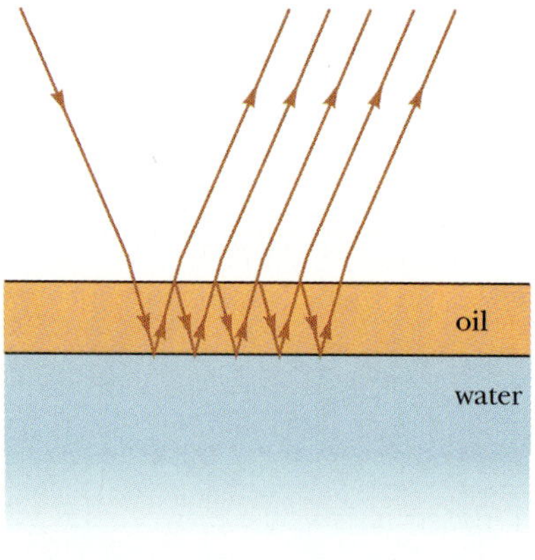

The interference between light waves is difficult to observe. For instance, the simplest case of interference is that between two waves traveling in opposite directions, giving a standing wave, with nodes and antinodes. Such a standing wave can be set up by shining light perpendicularly at a mirror, so the incident wave interferes with the reflected wave. However, since the wavelength of light is quite small, our eyes cannot perceive the individual nodes and antinodes — we see only the average intensity, without noticeable interference effects.

Beautiful and spectacular interference effects may become visible when a light wave is reflected by a thin film, such as a thin film of oil floating on water. When the wave strikes the upper surface of the film (see Fig. 26.1), it will set up a multitude of reflected waves due to reflection at the upper surface, reflection at the lower surface, and multiple zigzags produced by reflections between the surfaces. These reflected waves all travel in the same direction, and they can interfere destructively or constructively over a large region of space. Note that we are now interested only in the interference between the several reflected waves — the incident wave plays no direct role in this.

FIGURE 26.1
Incident ray and multiple reflected and refracted rays produced by a thin film.

To find the conditions for constructive and destructive interference between the waves reflected by a thin film, let us make the simplifying assumption that the direction of propagation of the wave is nearly perpendicular to the surface of the film. The two most intense waves are those that suffer only one reflection: the wave that reflects only at the upper surface and the wave that reflects only at the lower surface. Figure 26.2 shows the rays corresponding to these two waves. Under what conditions will these waves interfere constructively in the space above the film? The wave that is reflected from the lower surface has to travel an extra distance to emerge from the film. If the thickness of the film is d, and if the direction of propagation is nearly perpendicular to the film, then the extra distance the wave has to travel is $2d$, that is, down through the film and then back up. Provided that this extra distance is equal to one, two, three, etc., wavelengths, the wave reflected from the lower surface will meet crest to crest with the wave reflected from the upper surface. ***The condition for constructive interference is therefore***

$$2d = \lambda,\ 2\lambda,\ 3\lambda,\ \ldots \tag{1}$$

Likewise, if the extra distance is equal to one-half wavelength, three-halves wavelengths, five-halves wavelengths, etc., the wave reflected from the lower surface will meet crest to trough with the wave reflected from the upper surface. ***The condition for destructive interference is therefore***

$$2d = \tfrac{1}{2}\lambda,\ \tfrac{3}{2}\lambda,\ \tfrac{5}{2}\lambda,\ \ldots \tag{2}$$

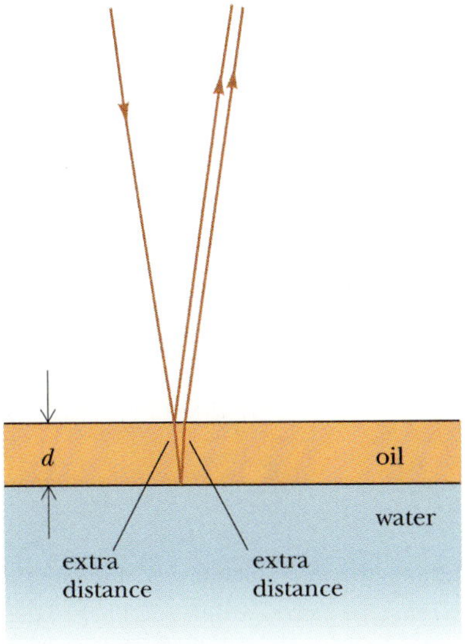

FIGURE 26.2

Incident ray and reflected rays for nearly perpendicular incidence and reflection. The wave reflected at the lower surface has to travel an extra distance of approximately $d + d$, or $2d$.

Condition for constructive and destructive interference

Thus, depending on the thickness of the film and on the wavelength of the light, the reflected wave can be either very strong or very weak. Note that the wavelength λ in Eqs. (1) and (2) is the wavelength of the light within the film; the wavelength outside the film will differ from this by a factor depending on the index of refraction (see Example 1).

The shimmering color displays seen in oil slicks and on soap bubbles arise from such interference effects. Different portions of an oil or soap film usually have different thicknesses, and they therefore give constructive interference for different wavelengths. When white light, consisting of a mixture of all visible colors, is incident on such a film, some portions of the film will display strong colors in the reflected light, by constructive interference. The result is a pattern of bright colored bands, or colored ***fringes*** (Fig. 26.3).

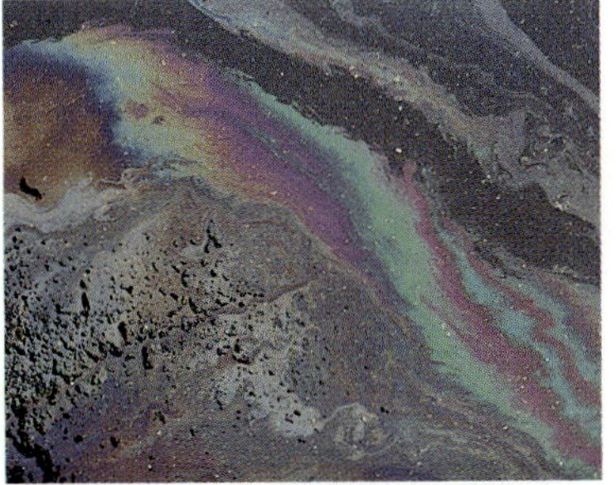

FIGURE 26.3

Colored fringes in the light reflected by a thin film.

EXAMPLE 1. A film of kerosene 450 nm thick floats on water. White light, a mixture of all visible colors, is vertically incident on this film. Which of the wavelengths contained in the white light will give maximum intensity upon reflection? Which will give minimum intensity? The index of refraction of kerosene is 1.2.

SOLUTION: For maximum intensity we need constructive interference, or, according to Eq. (1),

$$\lambda = 2d,\ \tfrac{2}{2}d,\ \tfrac{2}{3}d,\ \ldots \tag{3}$$

$$= 900\ \text{nm},\ 450\ \text{nm},\ 300\ \text{nm},\ \ldots$$

These are the wavelengths in kerosene; to obtain the wavelengths in air, we must multiply by the index of refraction of kerosene, $n = 1.2$. Of the resulting wavelengths, the only one in the visible region is 1.2×450 nm $= 540$ nm.

For minimum intensity we need destructive interference, or, according to Eq. (2),

$$\lambda = 4d, \tfrac{4}{3}d, \tfrac{4}{5}d, \ldots \tag{4}$$

$$= 1800 \text{ nm}, 600 \text{ nm}, 360 \text{ nm}, \ldots$$

Upon multiplication by 1.2, we find that the only wavelength in air in the visible region is 1.2×360 nm $= 432$ nm.

Note that the wavelength of 540 nm for constructive interference corresponds to a yellow-green color. The film will therefore be seen to have this color in the reflected light.

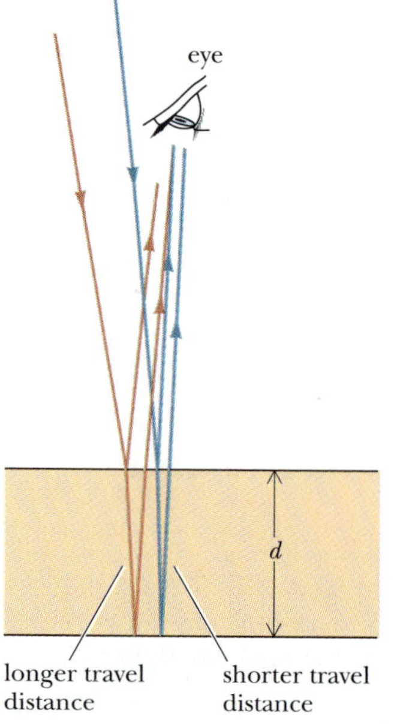

FIGURE 26.4

Two rays arriving at slightly different angles at a thick film. For the ray with the smaller angle of incidence (blue), the travel distance through the film and back is, say, 5000 wavelengths; for the ray with the larger angle of incidence (red), the travel distance is, say, 5000.5 wavelengths. The former ray gives constructive interference; the latter gives destructive interference. The interference effects tend to average out.

If the oil film is thick, no interference fringes will be visible. In a thick film, the distance from the top of the film to the bottom is many wavelengths. Thus, even a very small change in the angle at which the ray of light enters the film results in a change of travel distance by many wavelengths. The interference maxima and minima for rays of slightly different angles therefore overlap, and they tend to average out. The rays emitted by an ordinary source, such as the Sun or a lamp, and the rays entering your eye, always include a range of different angles; thus, you see only an average intensity when observing such rays reflected by a thick film, and you do not see interference fringes (see Fig. 26.4).

We can also use Eqs. (1) and (2) for a soap film suspended in air, instead of an oil film floating on water; but we must take into account a minor complication. ***Whenever an electromagnetic wave propagating in one medium is reflected at the surface of another medium of larger index of refraction, the wave suffers a reversal of its electric field, that is, a change of phase by 180°*** (this change of phase can be deduced from a detailed examination of the process of reflection). In the case of the oil film floating on water, the wave reflected at the upper surface of the film suffers such a change of phase, and so does the wave reflected at the lower surface; the net result is that the relative phase between the two waves is not altered, and the extra phase changes introduced by the process of reflection can be ignored. However, in the case of a soap film, or some other film, suspended in air, only the wave reflected at the upper surface suffers the extra change of phase. Thus, besides the phase difference introduced by the thickness of the film, there is an extra phase difference of 180° between the two waves. This means that the waves reflected from the upper and the lower surfaces will be out of phase if the distance $2d$ is equal to one, two, three, etc., wavelengths; and Eq. (1) is now the condition for destructive interference. Conversely, Eq. (2) is now the condition for constructive interference.

Similar interference effects can also arise in a narrow gap between two adjacent glass surfaces. Figure 26.5 shows a photograph of the interference fringes produced by the thin film of air between a flat glass plate and a lens of large radius of curvature. The convex surface of the lens is in contract with the plate at the center, but leaves a gradually widening air gap, or air film, for increasing distances from the center (Fig. 26.6). The photograph was taken with monochromatic light. At each of the bright fringes, the width of the air gap is such as to give constructive interference of the light reflected at the upper and the lower surfaces of the air gap. At each of the dark fringes, the width is such as to give destructive interference. The rings in Fig. 26.5 are called ***Newton's rings.***

Thin films are of great importance in the manufacture of optical instruments. The lenses of high-quality instruments are often coated with thin films of dielectric material, so that undesirable reflections of light are eliminated. Of course, with one thin film, we can achieve destructive interference at only one wavelength; but with several layers of thin films, we can achieve destructive interference at several wavelengths.

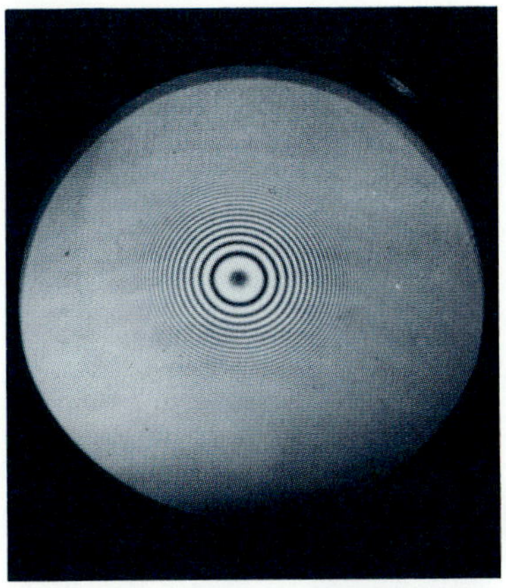

FIGURE 26.5

Fringes of constructive and destructive interference seen in the light reflected in the gap between a flat glass plate and a spherical lens in contact.

26.2* THE MICHELSON INTERFEROMETER

The Michelson interferometer takes advantage of the interference between two light waves to accomplish an extremely precise comparison between two lengths. Figure 26.7 is a schematic diagram of the essential parts of such an interferometer. The apparatus consists of two arms at the ends of which are mounted mirrors M_1 and M_2. Light waves from a monochromatic source S fall on a semitransparent mirror M (a half-silvered mirror). This mirror splits the light wave into two parts: one part continues straight ahead and reaches mirror M_1, the other part is reflected and reaches mirror M_2. These mirrors reflect the waves back toward the central mirror M and, upon reflection or transmission by this mirror, the waves emerge from the interferometer. When they emerge, they interfere constructively or destructively. Suppose that the lengths MM_1 and MM_2

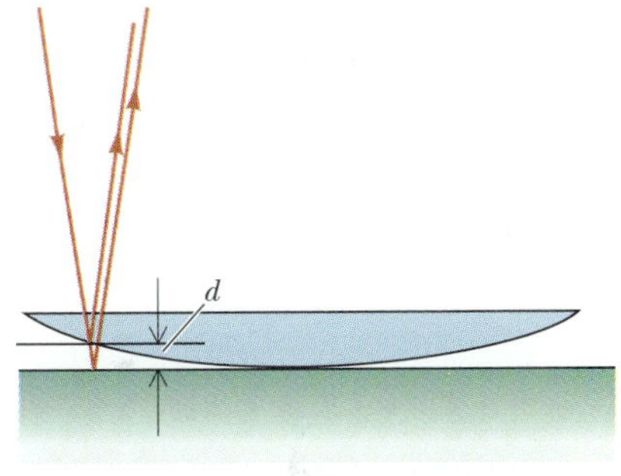

FIGURE 26.6

Rays reflected at the top and the bottom of the air gap between the flat plate and the lens.

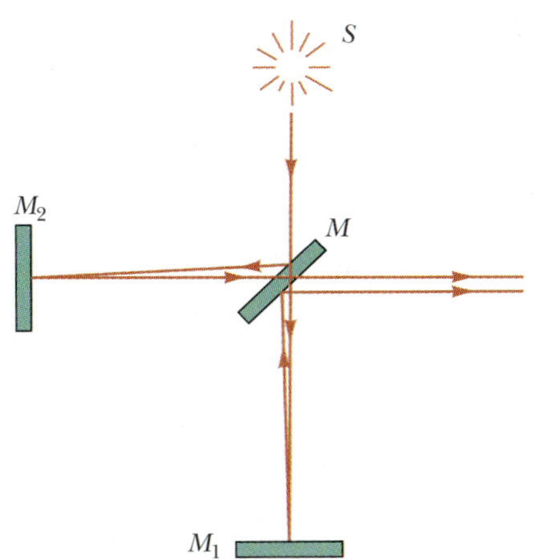

FIGURE 26.7

Paths of rays in a Michelson interferometer. For the sake of clarity, rays are shown reaching the mirrors M_1 and M_2 with a small angle between them; they are actually parallel, and they overlap.

* This section is optional.

differ by d. Then one of the waves must travel an extra distance $2d$ and the condition for constructive interference is

Conditions for constructive and destructive interference in interferometer

$$2d = 0,\ \lambda,\ 2\lambda,\ \ldots \tag{5}$$

and for destructive interference it is

$$2d = \tfrac{1}{2}\lambda,\ \tfrac{3}{2}\lambda,\ \tfrac{5}{2}\lambda,\ \ldots \tag{6}$$

FIGURE 26.8

Special interferometer used at the BIPM for the comparison of lengths. The interferometer automatically counts the wavelengths when the carriage C moves through a given distance.

To achieve this interference, the mirrors M_1 and M_2 must be very precisely aligned, so that the image of M_1 seen in M is exactly parallel to M_2. The alignment can be achieved by means of adjusting screws on the backs of the mirrors.

The mirror M_1 is usually mounted on a carriage that can be driven along a track by means of an accurately machined screw. If the distances MM_1 and MM_2 are exactly equal, the emerging light waves interfere constructively. If we displace the carriage outward or inward by $\frac{1}{4}\lambda$, then $2d = \frac{1}{2}\lambda$, and the interference becomes destructive. Upon another displacement by $\frac{1}{4}\lambda$, the interference becomes constructive. If we continue to displace the mirror slowly outward or inward, the interference of the emerging waves will continue to alternate back and forth between constructive and destructive whenever the mirror is displaced by $\frac{1}{4}$ wavelength — and the intensity of the emerging light will change back and forth between maxima and minima, which are seen as bright and dark fringes. Thus, the displacement of the mirror can be measured very precisely by counting fringes and fractions of a fringe. This measurement expresses the displacement in terms of the wavelength of the light. Modern interferometers used for length measurements, such as that illustrated in Fig. 26.8, are designed to count fringes automatically with a photoelectric sensor.

Albert Abraham Michelson, *1852–1931, American experimental physicist. He first performed the "Michelson–Morley" experiment on his own in 1881 and then repeated it several times in collaboration with* **E. W. Morley,** *1838–1923, with increasing accuracy. Michelson also made precise measurements of the speed of light. He received the Nobel Prize in 1907.*

Interferometers have played an important role in the test for the dependence of the speed of light on the motion of the Earth. This test is the famous ***Michelson–Morley experiment,*** first performed in 1881. The idea behind this experiment is as follows: If light were to propagate in a manner analogous to sound, then we would expect that the motion of the Earth toward or away from a light wave propagating through the "stationary" interplanetary space would affect the speed of the light relative to the Earth, just as the motion of a train toward or away from a sound wave propagating through the stationary air affects the speed of the sound relative to the train. Such an alteration of the speed of light could be detected with an interferometer by orienting one of the arms parallel to the direction of motion of the Earth and the other arm perpendicular. A difference in the speeds of light c along the arms would entail a difference in the corresponding wavelengths ($\lambda = c/f$), and alter the conditions (5) and (6) for bright and dark fringes. The easiest way to detect the speed difference is by rotating the interferometer, so the arm that has been parallel to the motion becomes perpendicular and vice versa. If the speeds were different in the two directions, this rotation could shift the fringes from bright to dark or vice versa.

Michelson and Morley found that there was no observable fringe shift to within the accuracy of their experiments. Taking into account possible experimental errors, they established that the effect of the motion of the Earth on the speed of light was at most ± 5 km/s. This value is substantially less than the speed of the Earth around the Sun ($\simeq 30$ km/s) and proved beyond all reasonable doubt that the propagation of light through space is *not* analogous to the propagation of sound through air. Figure 26.9 shows Michelson and Morley's interferometer.

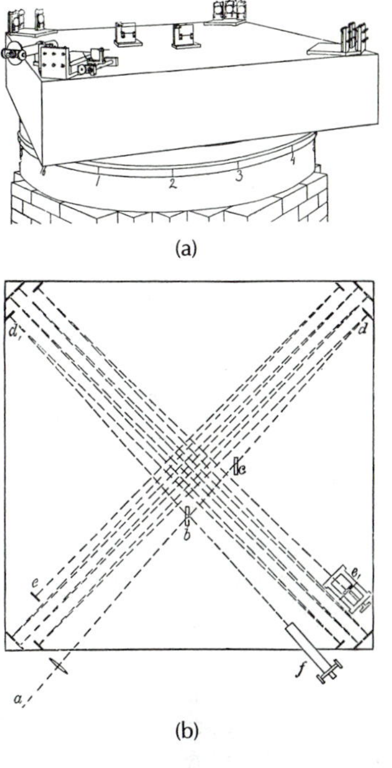

FIGURE 26.9

(a) The interferometer used in the experiment of Michelson and Morley. (b) The many mirrors reflect the light beams back and forth several times, increasing the path length of the light.

26.3 INTERFERENCE FROM TWO SLITS

A very clear experimental demonstration of interference effects in light can be performed with two small light sources placed near each other. ***The light waves spread out from the sources, run into one another, and interfere constructively or destructively, giving rise to a pattern of bright and dark zones*** (see Fig. 26.10). These interference effects were first discovered by Thomas Young around 1800, and, in conjunction with diffraction effects (see Section 26.5), they settled a long-standing controversy regarding the nature of light. Before Young performed his experiments, some physicists — among them Newton — had argued in favor of a particle nature of light; they had conjectured that light is a stream of particles. But Young's demonstration of interference gave unmistakable evidence in favor of waves. However, the question of what these waves were made of was not settled until much later, when Maxwell formulated his equations and established that light is a wave consisting of electric and magnetic fields.

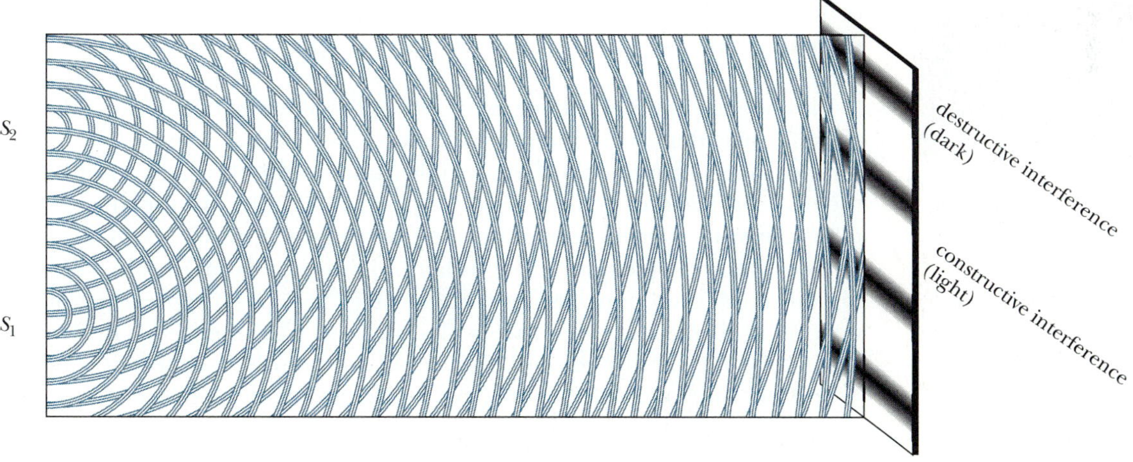

FIGURE 26.10

This diagram by Thomas Young illustrates how the interference of light waves emerging from two small light sources produces a pattern of light and dark zones. Along the directions indicated by the colored lines, the wave crests of the two waves coincide, giving constructive interference. At intermediate directions, the wave crests of one wave fall on the wave troughs of the other, giving destructive interference. If the light is intercepted by a screen, these directions of constructive and destructive interference give rise to bright and dark bands, or interference fringes.

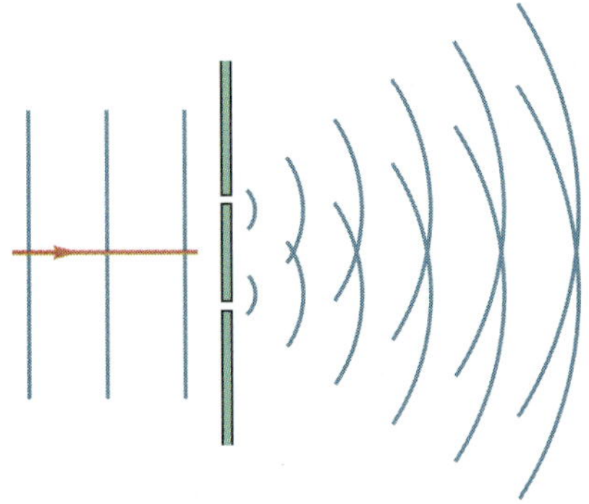

FIGURE 26.11

A plane light wave strikes a plate with two very narrow slits. The slits act as two coherent light sources. Light waves diverge from the two slits.

If the interference pattern produced by two light sources is to remain stationary in space, ***the two sources must be coherent, that is, they must emit waves of the same frequency and the same phase*** (a *constant* phase difference is also acceptable; what is crucial is that the phase difference must not fluctuate in time). Such coherent sources are easily manufactured by aiming a monochromatic wave with plane or spherical wave fronts at an opaque plate with two small slits or holes; the waves diverging from the two slits are then coherent because they arise from a single original wave (see Fig. 26.11). An extended object, such as an ordinary light bulb, can be used to illuminate the slits provided it is placed at a *very large* distance. The light bulb will then effectively act as a point source, illuminating the slits with light waves that consist of a succession of nearly plane wave fronts.*

Note that if we use *two* separate light bulbs as sources, then we will not see any interference pattern. The light from two separate light bulbs is ***incoherent;*** it consists of a mixture of a large number of light waves with variable phase differences. We can understand the distinction between coherent and incoherent light sources by means of an analogy with water waves. If we simultaneously drop two large stones into a pond, they act as coherent sources of water waves. The waves from these two sources will interfere constructively at some places, and destructively at others (we can observe the interference effects of the waves by means of bits of cork scattered on the water; on the midline between the sources, the waves interfere constructively, and the bits of cork bob up and down with twice the amplitude they would have had if only one stone had been dropped; at some other places, where the waves interfere destructively, the bits of cork remain at rest). But if instead of dropping two large stones, we dribble two handfuls of small pebbles into the pond, then the two sources are incoherent. There will then be no distinguishable interference pattern on the surface of the pond (all the bits of cork will bob up and down, more or less at random, although on the average the energy of their bobbing motion will be twice as large as if we had dropped a single handful of pebbles).

Thomas Young, *1773–1829, English physicist, physician, and Egyptologist. Young worked on a wide variety of scientific problems, ranging from the structure of the eye and the mechanism of vision to the decipherment of the Rosetta stone. He revived the wave theory of light and recognized that interference phenomena provide proof of the wave properties of light.*

In modern practice, a laser is often used to illuminate the two slits, because it provides a very intense plane wave, which makes even faint interference and diffraction effects visible (lasers and the light emitted by them are discussed in *Laser Light,* pp. 834A–834D). If the slits in the plate are very narrow, less than one wavelength in width, then the portions of the wave passing through them will diverge from the slits, spreading out in all directions, as illustrated in Fig. 26.11. (The spreading of the wave is a diffraction effect; we will study this in some detail in Section 26.5.) This means that the two slits act as two pointlike, coherent sources.

To find the zones of constructive and destructive interference, or the zones of maximum and minimum light intensity, in the space beyond the

* If the light bulb is placed too close to the slits, then light waves arrive at the slits from several directions at once, and this tends to wash out the interference pattern because waves of different directions of incidence will produce overlapping zones of brightness and darkness. For convenience, the light bulb is sometimes placed fairly near the slits; but it must then be covered with a shield perforated by a single pinhole, so the emerging light comes from just one portion on the surface of the light bulb; this again makes the light source into a point source.

slits, we need to examine the path difference between the rays from each of the slits to some point in the space beyond. Figure 26.12 shows the light wave incident on the plate from the left and the rays QP and $Q'P$ leading from the slits to the point P on the right. We will assume that the light source that generates the incident wave is either a laser or else some other light source placed very far from the plate, so the incident wave is a plane wave. The waves emerging from the slits and reaching P interfere constructively if the difference between the lengths QP and $Q'P$ is zero, or one wavelength, or two wavelengths, etc.; and they interfere destructively if this difference is one-half wavelength, or three-halves wavelengths, or five-halves wavelengths, etc.

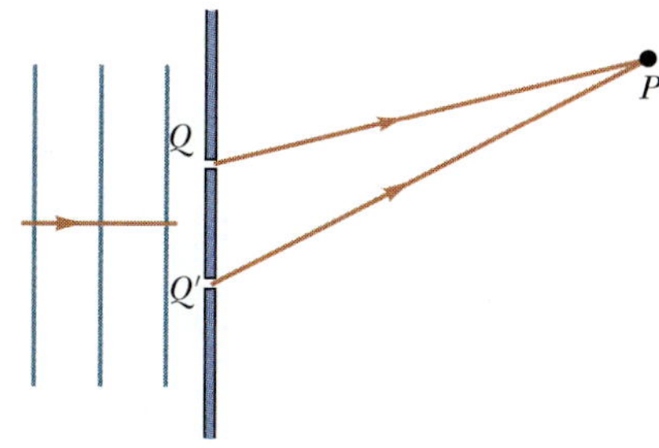

FIGURE 26.12

The waves reaching the point P have different path lengths QP and $Q'P$.

We can obtain a simple formula for the angular positions of the maxima and minima if we make the additional assumption that the point P is at a very large distance from the plate (to be precise, QP is assumed to be very large compared with QQ'). If so, then the rays QP and $Q'P$ are nearly parallel, as illustrated in Fig. 26.13. The difference between the lengths QP and $Q'P$ is then the distance $Q'R$, which is the short side of the right triangle $QQ'R$. Since the distance d between the slits is the hypotenuse of this triangle, we can express $Q'R$ in terms of d and the angle θ:

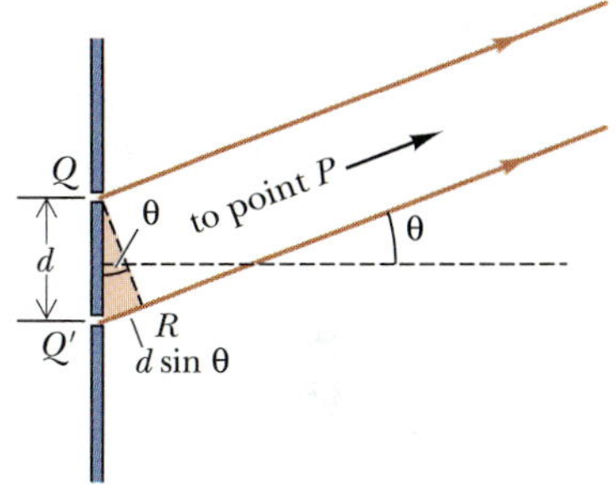

FIGURE 26.13

If P is far away, QP and $Q'P$ are nearly parallel. The lengths QP and $Q'P$ then differ by $d\sin\theta$.

$$Q'R = d\sin\theta \tag{7}$$

The angle θ in the triangle equals the angle between the line toward P and the perpendicular midline of the plate (see Fig. 26.13); thus, θ represents the angular position of the point P.

Our condition for maximum intensity is then that the distance $Q'R$ equals zero, one, two, etc., wavelengths:

◀ **Conditions for maxima and minima in interference pattern**

$$d\sin\theta = 0,\ \lambda,\ 2\lambda,\ \ldots \tag{8}$$

Likewise, the condition for minimum intensity is

$$d\sin\theta = \tfrac{1}{2}\lambda,\ \tfrac{3}{2}\lambda,\ \tfrac{5}{2}\lambda,\ \ldots \tag{9}$$

These equations give the angular positions of the interference maxima and minima. Note that the same equations apply whether θ is measured up or down from the centerline in Fig. 26.13 — whenever there is a maximum (or a minimum) at some upward angle, there is also a maximum (or a minimum) at an equal downward angle.

The zones of high intensity have the shape of beams fanning out in the space beyond the slits. The central beam (corresponding to $\sin\theta = 0$, or $\theta = 0$) is called the ***central maximum,*** or the ***zero-order maximum.*** The beams on each side of this (corresponding to $d\sin\theta = \lambda$, where θ is measured either up or down) are called the ***first-order maxima,*** etc. The beams of high intensity are separated by lines of zero intensity; these are nodal lines, analogous to the nodal points produced by destructive interference in the standing wave on a string (see Section 10.4). The lines of zero intensity between the zero-order maximum and the first-order maxima (corresponding to $d\sin\theta = \frac{1}{2}$) are called the ***first-order minima,*** etc.

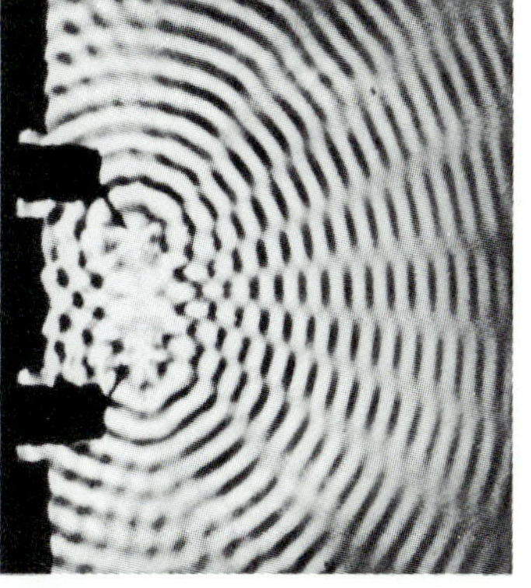

FIGURE 26.14

Interference between water waves spreading out from two coherent pointlike sources in a ripple tank.

Figure 26.14 is a photograph of such a pattern of fanlike beams pro-

(a)

(b)

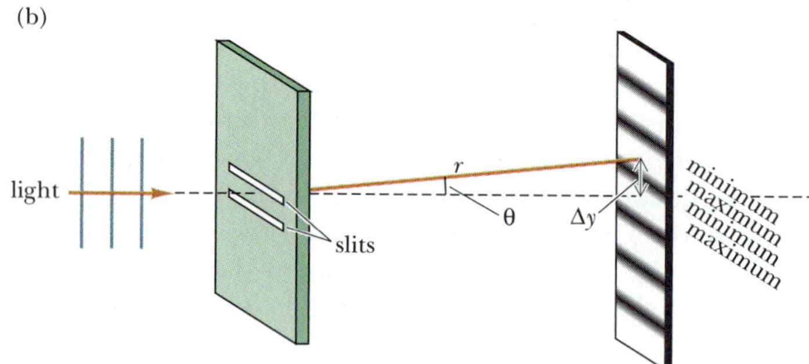

FIGURE 26.15

(a) A photographic film placed beyond two illuminated narrow slits records a regular pattern of bright and dark fringes. (Courtesy C. C. Jones, Union College.) (b) Placement of the photographic film beyond the slits.

duced by the interference of water waves spreading out from two point-like sources [the formulas (8) and (9) apply to light waves, water waves, and to any other kinds of waves]. With light waves, it is difficult to photograph the entire pattern of beams at once; instead we must be content with the photograph in Fig. 26.15, which shows the pattern of bright and dark fringes recorded on a photographic film that intercepts the beam at some fixed distance beyond the slits.

EXAMPLE 2. Two narrow slits separated by a distance of 0.12 mm are illuminated with light of wavelength 589 nm from a sodium lamp. What is the angular position of the first lateral maximum in the interference pattern? If the light is intercepted by a photographic film placed 2.00 m beyond the slits, what is the distance on the film between this maximum and the central maximum?

SOLUTION: With $d = 1.2 \times 10^{-4}$ m, and $\lambda = 5.89 \times 10^{-7}$ m, Eq. (8) gives

$$\sin\theta = \frac{\lambda}{d} = \frac{5.89 \times 10^{-7}\text{ m}}{1.2 \times 10^{-4}\text{ m}} = 4.9 \times 10^{-3}$$

For small angles, we can make the approximation that the sine of the angle is approximately equal to the angle in radians. Thus,

$$\theta \simeq \sin\theta = 4.9 \times 10^{-3}\text{ radian} \tag{10}$$

The distance between the points with $\theta = 0$ and $\theta = 4.9 \times 10^{-3}$ radian on the photographic film is approximately equal to the radial distance r from the slit to the film (see Fig. 26.15b) multiplied by the angle 4.9×10^{-3} radian,

$$\Delta y \simeq r\theta \tag{11}$$

$$\simeq 2.00\text{ m} \times 4.9 \times 10^{-3} = 9.8 \times 10^{-3}\text{ m} = 0.98\text{ cm}$$

Thus, the maxima are separated by nearly 1 cm.

26.4 INTERFERENCE FROM MULTIPLE SLITS

It is easy to see that the condition (8) for maximum intensity also applies to the case of a plane wave incident on multiple slits. For instance, Fig. 26.16 shows an opaque plate with three evenly spaced slits. If waves emerging from *adjacent* slits interfere constructively, then the waves emerging from *all* three slits interfere constructively. Hence the condition for maximum intensity is the same as Eq. (8),

$$d\sin\theta = 0, \lambda, 2\lambda, \ldots \tag{12}$$

◀ **Condition for principal maxima, multiple slits**

and this same equation also applies to more than three slits. However, the condition for minimum intensity is *not* Eq. (9). For instance, in the case of three slits, destructive interference of waves from adjacent slits will lead to cancellation of the waves originating from one pair of slits, but the wave from the remaining slit will not suffer cancellation. Thus, the condition for minima is somewhat more complicated. It turns out that between the maxima given by Eq. (12), called ***principal maxima,*** there are several minima and several ***secondary maxima.*** The secondary maxima result from a partial constructive interference: the waves from a majority of the slits contribute constructively, but the waves from the remaining slits contribute destructively. Figure 26.17 shows a photograph illustrating the strong principal maxima and the weaker secondary maxima of the diffraction pattern of three slits. Hereafter, we will concentrate our attention on the principal maxima, because the secondary maxima are quite weak, especially if the number of slits is large.

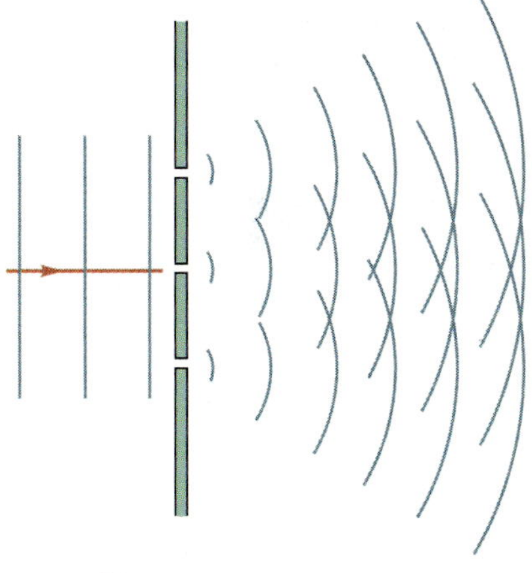

FIGURE 26.16
A plane wave strikes a plate with three very narrow slits.

Arrangements of multiple slits, with a large number of slits, are called ***gratings.*** They are commonly used in spectroscopy laboratories to analyze light into its colors. If a light beam containing several wavelengths passes through such a grating, the maxima for these different wavelengths will form beams at different angles — the beams of long-wavelength light will be found at larger angles than the beams of short-wavelength light [compare Eq. (12)]. Thus, the system of slits separates the light according to color and produces a spectrum in much the same way that a prism does. There is, however, one important difference between the spectra formed by a prism and by a grating: in the prism the long-wavelength light (red)

◀ **Gratings**

FIGURE 26.17
A photographic film placed beyond the slits shows the strong principal maxima and the weaker secondary maxima. (Courtesy C. C. Jones, Union College.)

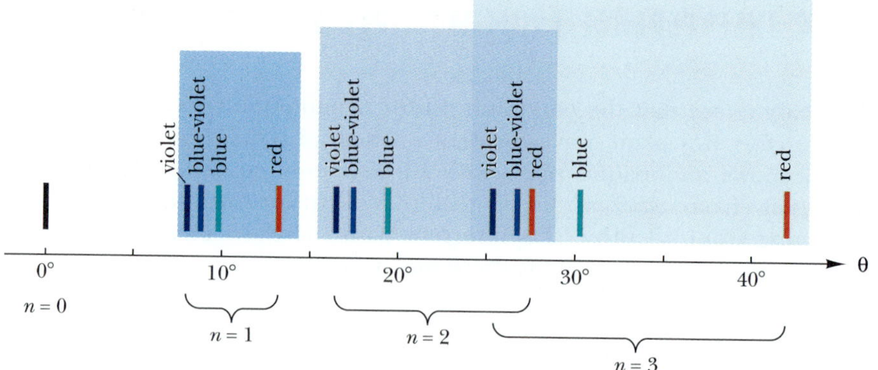

FIGURE 26.18

First-, second-, and third-order spectra of hydrogen light produced by a system of N slits. The lines correspond to the principal maxima; the secondary maxima are weak and can be ignored. Each spectrum consists of a violet, blue-violet, blue, and red spectral line (compare Fig. 25.26). The pattern of spectral lines for negative values of θ (negative n) is similar.

suffers the least deflection; in the grating the long-wavelength light suffers the most deflection.

The grating will produce one complete spectrum for each of the possible alternatives listed for the right side in Eq. (12), that is, each principal maximum, except the central maximum, gets spread out by color. These spectra are called the first-order spectrum [$d \sin\theta = \lambda$ in Eq. (12)], the second-order spectrum ($d \sin\theta = 2\lambda$), and so on. Sometimes these spectra overlap, for example, the red end of the second-order spectrum may show up at the same angle as the blue end of the third-order spectrum (see Fig. 26.18).

EXAMPLE 3. Among the spectral lines emitted by atoms of iron are spectral lines of wavelength $\lambda = 500.57$ nm and $\lambda = 500.61$ nm. Suppose we analyze the spectrum of iron with a grating with $d = 2 \times 10^{-6}$ m. What is the angular separation between the maxima for these wavelengths in the first-order spectrum produced by this grating?

SOLUTION: For small angles, the sine of the angle is approximately equal to the angle in radians. Hence the equation $d \sin\theta = \lambda$ for the angular position of the maximum in the first-order spectrum becomes

$$d\,\theta = \lambda \tag{13}$$

Thus, the angle is proportional to the wavelength, and the difference in angle is therefore proportional to the difference in wavelength,

$$d\,\Delta\theta = \Delta\lambda$$

With $\Delta\lambda = 500.61 \text{ nm} - 500.57 \text{ nm} = 0.04$ nm, we find

$$\Delta\theta = \frac{\Delta\lambda}{d} = \frac{0.04 \times 10^{-9}\text{ m}}{2 \times 10^{-6}\text{ m}} = 2 \times 10^{-5} \text{ radian}$$

The best gratings have a very large number of slits, because this produces sharp, narrow maxima and enhances the ability of the grating to discriminate between spectral lines of nearly identical wavelengths. Since it is difficult to cut a large number of slits in an opaque plate, gratings are usually manufactured by cutting fine parallel grooves in a glass or metal surface with a diamond stylus guided by a special ruling machine. When illuminated by a plane light wave, the edges of the grooves act as coherent light sources, in much the same way as slits. High-quality gratings used in spectroscopy have 100,000 or more grooves with distances of about 10^{-6} m between them.

You can verify the effects of a grating on light by a simple experiment with an LP or a CD record. The grooves on these records make them behave like crude gratings. You can see spectral colors with these crude gratings if you let them reflect the light from some narrow light source, such as sunlight coming through a crack between the blinds of a window or light from the edge of a bright lamp. CD records have tighter grooves than LP records, and they give more pronounced spectral colors.

◂ Radiotelescopes

Radiotelescopes, consisting of regularly spaced arrays of antennas, act as "gratings" for radio waves. The same formulas for the directions of the maxima apply to these as to ordinary gratings used with light. Of course, there is a difference between the operation of a radiotelescope and that of an ordinary grating: in the former the radio waves from a distant source *enter* the antennas and interfere constructively or destructively within the radio receiver, whereas in the latter the light waves *emerge* from the slits and travel to a distant point where they interfere. Nevertheless, the condition for, say, maximum intensity can still be expressed in terms of the direction of the entering wave by Eq. (12), because this condition only hinges on the phase relationships among the waves, and these relationships are independent of the direction of propagation (for instance, in Fig. 26.13, if the waves traveling away from the slits in some direction reach a distant point in phase, then waves traveling in the reverse direction toward the slits from this point will arrive at the slits in phase).

EXAMPLE 4. One branch of the VLA (Very Large Array) radiotelescope at Socorro, New Mexico, has nine antennas arranged on a straight line with distances of 2.0 km between one antenna and the next (see Fig. 26.19). These antennas are all connected to a single radio receiver by waveguides of equal lengths.* The receiver then registers a maximum intensity if the radio waves incident on all antennas are in phase. Radio waves of wavelength 21 cm from a point-like source in the sky strike this radiotelescope. When a source is at the zenith, the intensity in the radio receiver is maximum. What angular displacement of the source from the zenith will also result in maximum intensity?

FIGURE 26.19

The VLA radiotelescope. The branch in the center consists of nine antennas. The distance between the antennas can be adjusted by rolling them along the track. Here the antennas are shown close together.

* The waveguides used with these antennas are hollow tubes, which conduct radio waves, just as the coaxial cable used with your TV antenna conducts TV waves.

SOLUTION: The zenith direction corresponds to the central maximum, $\theta = 0$. To find the angular direction of the first maximum next to this central maximum, we again use the approximation that, for small angles, the sine of the angle equals the angle in radians. Thus, Eq. (12) gives, for the maximum next to the central maximum,

$$d\,\theta = \lambda$$

Hence

$$\theta = \frac{\lambda}{d} = \frac{0.21\ \text{m}}{2.0 \times 10^{3}\ \text{m}} = 1.1 \times 10^{-4}\ \text{radian}$$

FIGURE 26.20

Diffraction of water waves in a ripple tank. Beyond the narrow aperture, the waves spread out in a fanlike pattern.

26.5 DIFFRACTION BY A SINGLE SLIT

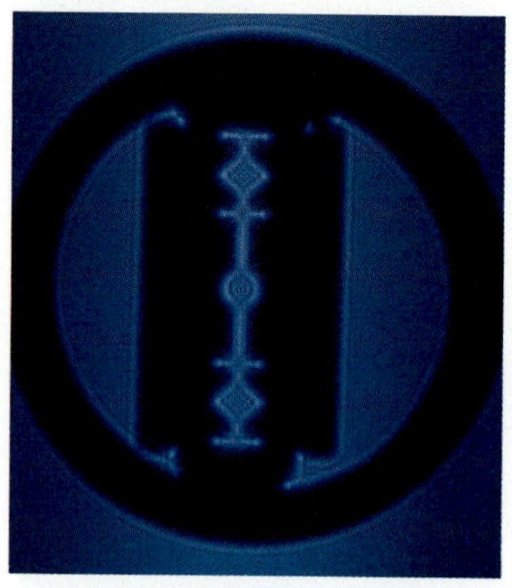

FIGURE 26.21

Diffraction of light around a razor blade. The diffracted light generates a complex pattern of fine fringes at the edges of the shadow. This photograph was prepared by illuminating the razor blade with a distant light source, and throwing its shadow on a distant screen.

Like any wave, ***light displays diffraction effects — it bends and spreads around obstacles.*** We can easily see diffraction effects in water waves. For instance, Fig. 26.20 shows the diffraction of water waves by a narrow aperture; beyond the aperture, the waves spread out in a wide fanlike pattern. The diffraction of these water waves is quite pronounced because their wavelength is fairly large; it is about as large as the aperture. Diffraction effects with light waves are much harder to observe, because the wavelength of the light waves is so small. The diffraction becomes observable only when light passes through extremely narrow slits, or when it strikes extremely small opaque obstacles, or when we examine the fine detail at the boundaries of shadows produced by sharp edges (Fig. 26.21).

Diffraction effects with light were first noticed in the seventeenth century. Nevertheless, Newton and some of his contemporaries held to the belief that light is of a corpuscular nature, and consists of a stream of particles. The importance of diffraction effects in establishing the wave nature of light was not fully appreciated until 1818, when Augustin Fresnel mathematically formulated the theory of diffraction and demonstrated that diffraction is a general and distinctive characteristic of waves. The brilliant success of this theory finally convinced physicists that light is a wave. In this section we will discuss the sample case of diffraction by a single narrow slit.

Figure 26.22 shows a plane light wave approaching a slit in an opaque plate. To find the light distribution in the space beyond the slit, we use the following prescription, called the ***Huygens–Fresnel Principle:***

Huygens-Fresnel Principle ▶

Pretend that each point of the wave front reaching the slit can be regarded as a point source of light emitting a spherical wave; the net wave in the region beyond the slit is simply the superposition of all these waves.

This prescription has some obvious similarities to Huygens' Construction. However, the latter is merely a geometric construction for finding the

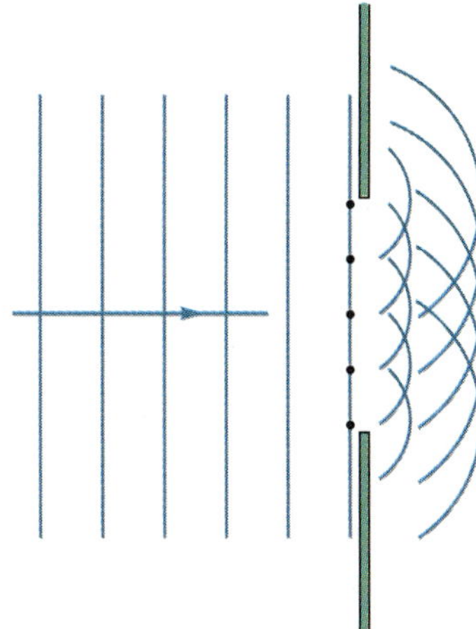

FIGURE 26.22

A plane wave approaches a slit in an opaque plate. Each point on the wave front at the slit gives rise to a spherical wave that spreads out beyond the slit.

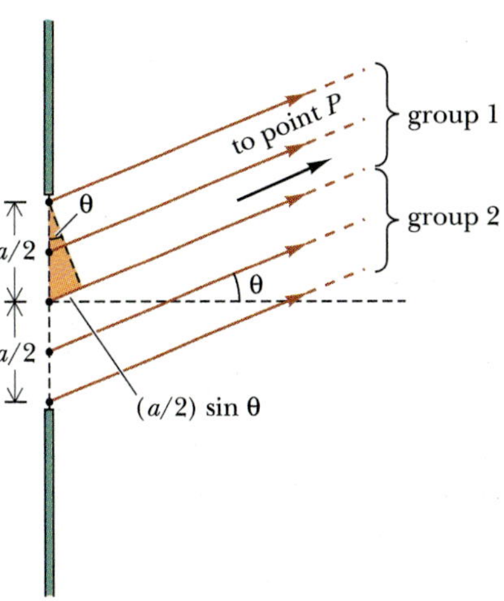

FIGURE 26.23

Two groups of rays from points in the slit to the point P. The path difference between the uppermost ray of the first group and the uppermost ray of the second group is $(a/2)\sin\theta$.

successive positions of the wave fronts and does not yield any information about the distribution of intensity in different directions, whereas the aim of our new prescription is precisely the calculation of this distribution of intensity. (Although the new prescription has a strong intuitive appeal, it turns out that it is not all that easy to justify rigorously. The trouble is that light waves are not really sources of light waves — only accelerated charges are sources of light waves. The intensity pattern on the far side of the slit does not arise from waves radiated by the original wave, but from the superposition of the original wave with all the waves radiated by the electric charges sitting in the opaque plate when accelerated by the original wave. Thus, our prescription lacks a simple physical basis. Nevertheless, it gives the right answer, or almost the right answer; this can be shown by a somewhat more sophisticated mathematical argument based on Maxwell's equations.)

We can use the Huygens–Fresnel Principle to find the positions of the minima in the diffraction pattern of the slit. Figure 26.23 shows some rays associated with spherical waves that spread out from the points of a wave front at the slit. All the rays lead to a point P in the space beyond the slit. As in our calculation for the case of two slits, we will assume that P is very far away, so all the rays are nearly parallel. We can divide these rays into two equal groups: those that come from the upper part of the slit and those that come from the lower (the middle ray belongs equally to both these groups; see Fig. 26.23). Rays from the first group have a shorter distance to travel to the point P than rays from the second group. Consider the uppermost ray from the first group and the uppermost ray from the second group. The path difference between these is $(a/2)\sin\theta$, where a is the width of the slit. If this path difference is $\frac{1}{2}\lambda$, the two rays will interfere destructively. Furthermore, pairs of rays that originate at an equal distance below each of the two uppermost rays in the two groups will also interfere destructively. This establishes that all the waves cancel in pairs when

$$\frac{a}{2}\sin\theta = \tfrac{1}{2}\lambda \tag{14}$$

Augustin Fresnel (frenel), *1788–1827, French physicist and engineer. His brilliant experimental and mathematical investigations firmly established the wave theory of light. Fresnel was commissioner of lighthouses, for which he designed large compound lenses (Fresnel lenses) as a replacement for systems of mirrors.*

or

$$a \sin\theta = \lambda \tag{15}$$

This is the condition for the first minimum. [Note that this equation is mathematically similar to the condition for the first *maximum* in the interference pattern of a double slit. This mathematical similarity between Eqs. (15) and (8) merely reflects the similarities between the geometries in Figs. 26.23 and 26.13.]

To find the next minimum, we divide the rays into four equal groups and consider the uppermost rays from the first and second groups (see Fig. 26.24). These rays, and other pairs of rays, will interfere destructively if their path difference is $\frac{1}{2}\lambda$:

$$\frac{a}{4}\sin\theta = \tfrac{1}{2}\lambda \tag{16}$$

or

$$a \sin\theta = 2\lambda \tag{17}$$

By continuing this argument, we find a general condition for minima:

Condition for minima in diffraction pattern ▶

$$a \sin\theta = \lambda,\ 2\lambda,\ 3\lambda,\ \ldots \tag{18}$$

As regards the maxima, there is of course a strong central maximum, at $\theta = 0$. The secondary maxima cannot be found by any simple argument; roughly, their position is halfway between the successive minima. On a photographic film placed at some distance, the maxima and minima show up as a pattern of bright and dark fringes (see Fig. 26.25).

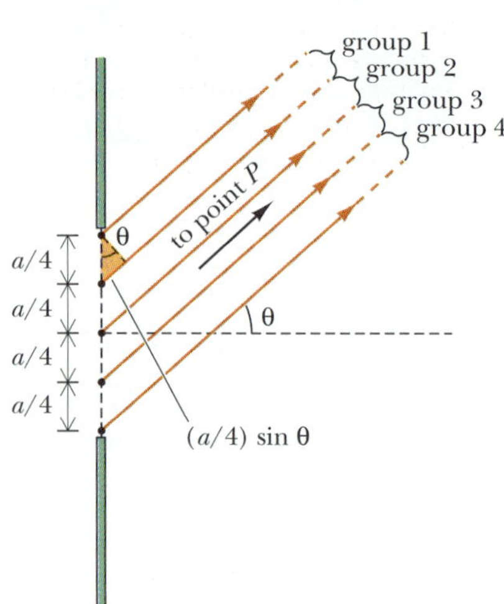

FIGURE 26.24

Four groups of rays. The path difference between the uppermost ray of the first group and the uppermost ray of the second group is $(a/4)\sin\theta$.

FIGURE 26.25

A photographic film placed beyond an illuminated slit records a strong central maximum and successively weaker secondary maxima. (Courtesy C. C. Jones, Union College.)

EXAMPLE 5. Equation (18) applies not only to light waves, but also to radio waves and other waves. Suppose that radio waves from a TV transmitter with a wavelength of 0.80 m strike the wall of a large building. In this wall there is a very wide window with a height of 1.4 m. The wall is opaque to radio waves and the window is transparent and acts as a horizontal single slit. What is the angular width of the central maximum of the diffraction pattern formed by the waves inside the building?

SOLUTION: According to Eq. (18), the first minimum is at an angle such that

$$a \sin\theta = \lambda$$

which yields

$$\sin\theta = \frac{\lambda}{a} = \frac{0.80 \text{ m}}{1.4 \text{ m}} = 0.57$$

Our calculator tells us that the angle θ is then 35°. The central maximum extends from $-35°$ to $+35°$, that is, the width is 70°.

26.6 DIFFRACTION BY A CIRCULAR APERTURE; RAYLEIGH'S CRITERION

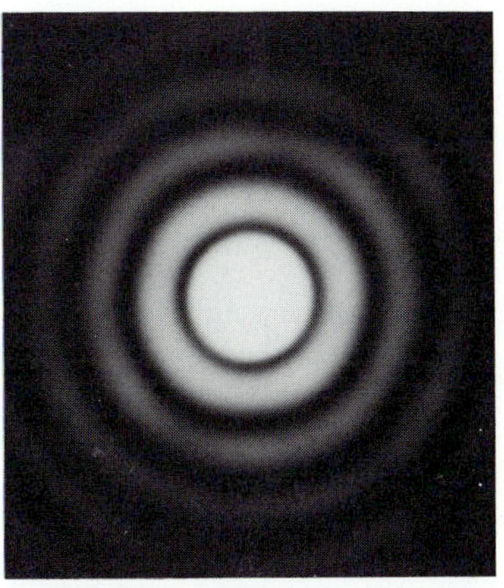

FIGURE 26.26
A photographic film placed beyond a circular aperture records a strong central maximum and annular secondary maxima. (Courtesy C. C. Jones, Union College.)

The diffraction of light by a circular aperture is in principle no different from the diffraction by a slit (a very long rectangular aperture). To find the maxima and minima in the diffraction pattern of the light emerging from such a circular aperture, we must apply the Huygens–Fresnel Principle, which tells us to sum the spherical waves originating from all points of a wave front at the circular aperture. This calculation is fairly complicated and we will not attempt it. Figure 26.26 is a photograph that displays the central maximum and the bright and dark fringes generated by diffraction at the circular aperture. The angular position of the first minimum is given by the simple formula

◂ **First minimum in diffraction pattern of circular aperture**

$$\sin\theta = \frac{1.22\,\lambda}{a} \tag{19}$$

where a is the diameter of the circular aperture.

Many optical instruments — telescopes, microscopes, cameras, etc. — have circular apertures, and these will diffract light. For instance, the objective lens of an astronomical telescope will act like a circular opening in a plate; the parallel wave fronts arriving from some distant star will suffer diffraction effects at this aperture and produce an intensity distribution

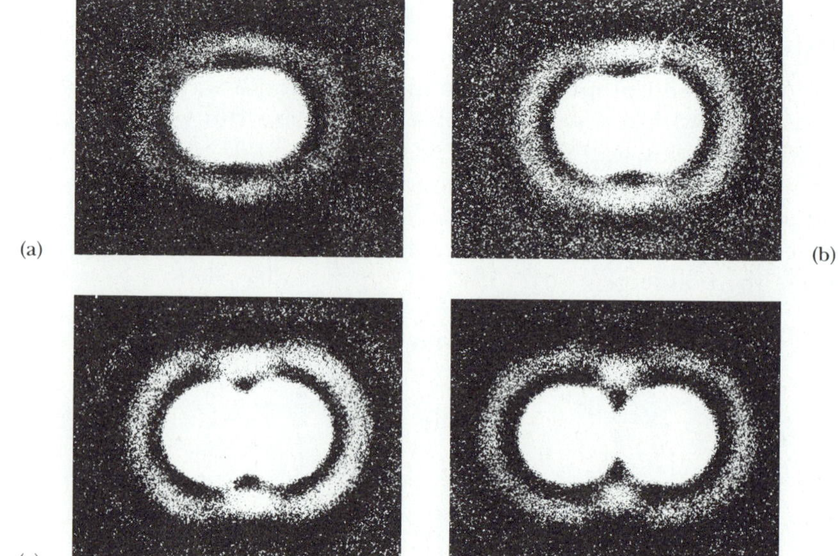

FIGURE 26.27
When the light waves from two pointlike sources arrive at a circular aperture simultaneously, each set of light waves will produce its own diffraction pattern. If the angular separation between the two sources is small, the diffraction patterns overlap. In photograph (a), the angular separation is small. In photographs (b), (c), and (d), the angular separation is progressively larger. In the first photograph, the two light sources are not resolved (not distinct). In the second photograph, they are just barely resolved, according to Rayleigh's criterion. (Courtesy C. C. Jones, Union College.)

John William Strutt, 3rd Baron Rayleigh, *1842–1919, English physicist. He is best known for his extensive mathematical investigations of sound and of light. He also investigated the behavior of gases at high densities, and he discovered argon; for this he was awarded the Nobel Prize in 1904.*

such as that shown in Fig. 26.26. The image of the star as seen through this telescope will then not be a bright point, but a disk surrounded by concentric rings, as in Fig. 26.26. For example, as seen through a telescope with an objective lens 6 cm in diameter, stars look like small disks, about 2×10^{-5} radian (or 4 seconds of arc) across.

This spreading out of the image puts a limit on the detail that can be perceived through the telescope. If two stars are very close together, their images tend to merge, and it may be impossible to tell them apart. Figures 26.27a–d show the images produced by a pair of pointlike light sources upon diffraction by a circular aperture. In the first of these figures, the angular separation of the sources is so small that the two images look like one image. In the second figure, the angular separation is large enough to give a clear indication of the existence of two separate images.

The angular separation of the sources in Fig. 26.27b is such that the central maximum of the diffraction pattern of one source coincides with the minimum of the diffraction pattern of the other source. Since we are now dealing with small angles, we can approximate $\sin\theta$ by θ, and Eq. (19) then tells us that the angular position of the first minimum, or the angular separation between the sources, is

$$\theta = 1.22\,\frac{\lambda}{a} \tag{20}$$

Rayleigh's criterion ▸

We will regard this as the critical angle that decides whether the two sources are clearly distinguishable: the telescope (or other optical instrument) can resolve the sources if their angular separation is larger than that in Eq. (20), and it cannot resolve them if the separation is smaller. This is ***Rayleigh's criterion.***

The ability of a telescope to resolve stars or other objects of small angular separation improves with the size of the telescope — a telescope of

EXAMPLE 6. The star ζ Orionis is a binary star, that is, it consists of two stars very close together. The angular separation between the stars is 2.8 seconds of arc. Can the stars be resolved with a telescope having an objective lens 6 cm in diameter? Assume that the wavelength of the starlight is 550 nm.

SOLUTION: According to Rayleigh's criterion, a telescope of this aperture can resolve stars as close as

$$\theta = 1.22\,\frac{\lambda}{a} = 1.22 \times \frac{550 \times 10^{-9}\text{ m}}{0.06\text{ m}}$$

$$= 1.1 \times 10^{-5}\text{ radian}$$

Since one second of arc is 4.8×10^{-6} radian (see the conversion factors in Appendix 6), we find that 1.1×10^{-5} radian is 2.3 seconds of arc. Hence the telescope can resolve these double stars.

FIGURE 26.28
The Hubble Space Telescope.

30-cm diameter can resolve angular separations as small as 0.5 second of arc. However, beyond 30 cm, the resolution of an Earth-bound telescope does not improve further with size, because fluctuations in the density of the atmosphere introduce irregularities in the path of the light coming down from the sky; this smears out the light and prevents the telescope from reaching its full potential.

The Hubble Space Telescope, launched in 1990, is placed above the atmosphere of the Earth and it is therefore not affected by atmospheric fluctuations. This telescope (see Fig. 26.28) has an aperture of 2.4 m and it was expected to attain an angular resolution of about 0.1 second of arc, close to the limit set by Rayleigh's criterion. Unfortunately, through a stupid error committed by the manufacturer, the mirror of the telescope is slightly out of shape, and the telescope has failed to achieve its expected resolution. It does achieve a somewhat better resolution than Earth-bound optical telescopes, but not much better. However, it is able to observe at ultraviolet and infrared wavelengths, which are blocked by the atmosphere.

For Earth-bound radio telescopes, the atmosphere poses no problem — radio waves do not suffer from the effects of atmospheric density fluctuations. Hence, with increasing size, the resolution of a radio telescope improves indefinitely. Figure 26.29 shows the large radiotelescope at Arecibo, Puerto Rico. The concave "mirror" of this telescope has an aperture of 300 m and a radius of curvature which is also 300 m. The shortest wavelength at which this telescope has been operated is 4 cm. For this wavelength, Rayleigh's criterion gives a limiting angular resolution

FIGURE 26.29
The Arecibo radiotelescope.

$$\theta = 1.22\,\frac{\lambda}{a} = 1.22 \times \frac{0.04\text{ m}}{300\text{ m}} = 1.6 \times 10^{-4}\text{ radian}$$

which is about 30 seconds of arc.

SUMMARY

- **Constructive interference:** Waves meet crest to crest.
- **Destructive interference:** Waves meet crest to trough.
- **Two-slit interference pattern:**
  maxima: $d \sin\theta = 0, \lambda, 2\lambda, \ldots$
  minima: $d \sin\theta = \frac{1}{2}\lambda, \frac{3}{2}\lambda, \frac{5}{2}\lambda, \ldots$
- **Multiple-slit interference pattern:**
  principal maxima: $d \sin\theta = 0, \lambda, 2\lambda, \ldots$
- **Single-slit diffraction pattern:**
  minima: $a \sin\theta = \lambda, 2\lambda, 3\lambda, \ldots$
- **First minimum for circular aperture:** $a \sin\theta = 1.22\lambda$
- **Rayleigh's criterion:** $\theta = 1.22\lambda/a$

(a)

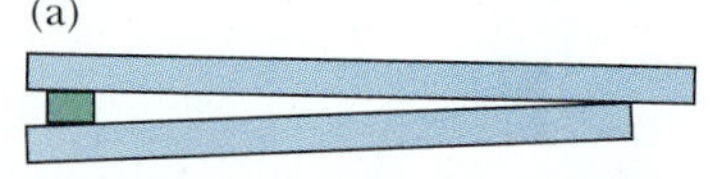

(b)

FIGURE 26.30
(a) Two flat plates of glass, separated by a thin wedge of air. (b) Interference fringes.

QUESTIONS

1.–When two waves interfere destructively at one place, what happens to their energy?

2.–When two coherent waves of equal intensity interfere constructively at one place, the energy density at that place becomes four times as large as the energy density of each individual wave. Is this a violation of the law of conservation of energy?

3.–Suppose that N waves of equal intensity meet at one place. If the waves are coherent, the net intensity is N^2 times that of each individual wave. If the waves are incoherent, the (average) intensity is N times that of each individual wave. Explain, and give an example of each case.

4.–When light strikes a windowpane, some rays will be reflected back and forth between the two glass surfaces. Why do these reflected rays not produce visible colored interference fringes?

5.–Suppose that a lens is covered with an antireflective coating that eliminates the reflection of perpendicularly incident light of some given color. Will this coating also eliminate the reflection of light incident at an angle?

6.–Explain how Newton's rings may be used as a sensitive test of the rotational symmetry of a lens.

7.–Two flat plates of glass are in contact at one edge and separated by a thin spacer at the other edge (Fig. 26.30a). Explain why we see parallel interference fringes in the reflected light if we illuminate these plates from above (Fig. 26.30b).

8.–In the experiment that sought to test the dependence of the speed of light on the motion of the Earth, Michelson and Morley used an interferometer with very long arms (about 11 m, obtained by multiple reflections back and forth between sets of mirrors). Why does this make the instrument more sensitive?

9.–If you stand next to your TV receiver, your body will sometimes affect the reception. Why?

10.–Consider (1) sunlight, (2) sunlight passed through a monochromatic filter selecting one wavelength, (3) light from a neon tube, (4) light from a laser, (5) starlight passed through a monochromatic filter, and (6) radio waves emitted by a radio station. Which of these kinds of light or electromagnetic radiation is sufficiently coherent so that when it is used to illuminate two slits such as shown in Fig. 26.11, it will give rise to an interference pattern?

11.–When installing a pair of stereo loudspeakers, the terminals of the loudspeakers should be connected to the amplifier in the same way, so the loudspeakers are in phase. What would happen to sound waves of long wavelength if the loudspeakers were out of phase?

12.–Suppose that a plane wave is incident at an angle on a plate with two narrow slits (Fig. 26.31). In what direction will we then find the central maximum?

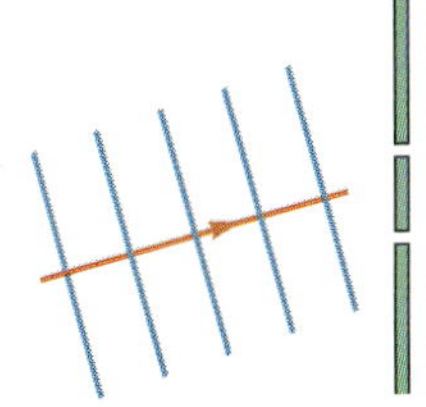

FIGURE 26.31
Plane wave obliquely incident on the plate.

13.–Several radio antennas are arranged at regular intervals along a straight line; the antennas are connected to the same radio transmitter, so they radiate coherently. How can this array be used to concentrate the radio emission in a selected direction?

14.–Gratings used for analyzing light are often called diffraction gratings, but it would be more accurate to call them interference gratings. Why?

15.–Figure 11.25 shows the diffraction of water waves at a breakwater at the entrance of a harbor. Could this diffraction be eliminated by making the entrance smaller?

16.–What would you expect the diffraction pattern of a rectangular aperture to look like?

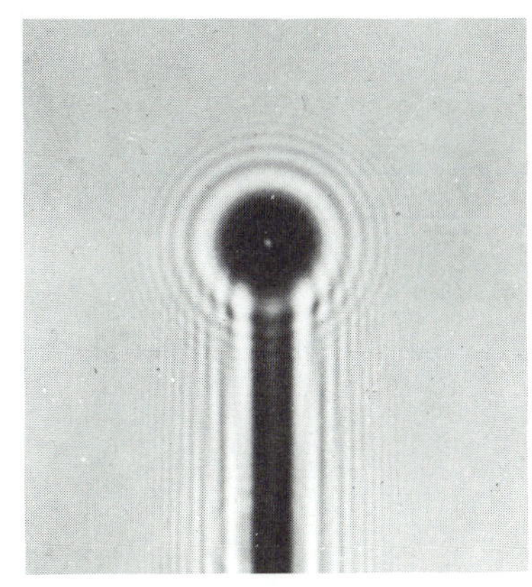

FIGURE 26.32
The Poisson spot.

17.–In the center of the shadow of a disk or a sphere there is a small bright spot, called the ***Poisson spot*** (Fig. 26.32). This spot is very faint near the disk, but becomes more noticeable at large distances. Qualitatively, explain why the diffraction of the light waves around the edges of the disk gives rise to this spot.

18.–Besides good angular resolution, what other advantage does a telescope of large aperture have over a telescope of small aperture?

19.–Spy satellites use cameras with lenses of very large diameter, 30 cm or more. Why are such large diameters necessary?

20.–In order to beam a sound wave sharply in one direction with a loud-hailer, should the horn of the loud-hailer have a small aperture or a large aperture?

21.–The maximum useful magnification of an optical microscope is determined by diffraction effects at the objective lens. Explain.

22.–Other things being equal, how much resolution can you gain by operating an optical microscope with blue light instead of red light? Why can you not operate the microscope with ultraviolet light?

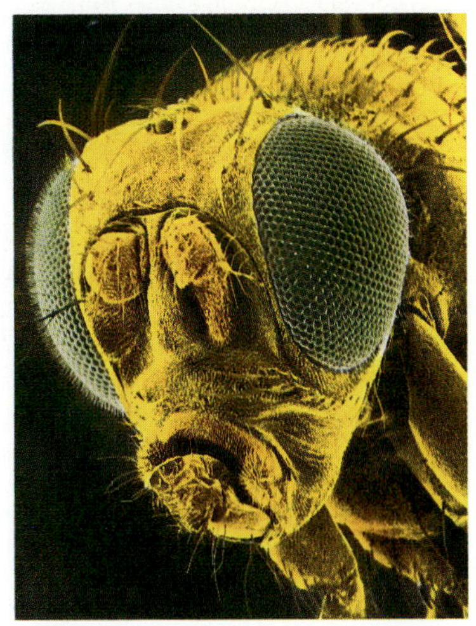

FIGURE 26.33
The compound eyes of a fruit fly. These eyes consist of about 28,000 ommatidia each.

23.–The compound eye of insects consists of a large number of small eyes, or ommatidia (Fig. 26.33). Each ommatidium is typically 0.03 mm across; it does not form an image, but merely acts as a sensor of the intensity of light arriving from a narrow cone of directions. What are some of the advantages and disadvantages of such a compound eye as compared with the camera eye of vertebrates?

24.–The diffraction of sunlight by small droplets of water in clouds produces a bright ring surrounding the Sun, called the ***corona.*** What is the color of the outer edge of the corona? The inner edge?

25.–Antennas used for the transmission of microwaves in communication links consist of metal dishes. What factors determine the size of these dishes?

PROBLEMS

Section 26.1: Thin Films

1.——The wall of a soap bubble floating in air has a thickness of 400 nm. If sunlight strikes the wall perpendicularly, what colors in the reflected light will be strongly enhanced as seen in air? The index of refraction of the soap film is 1.35.

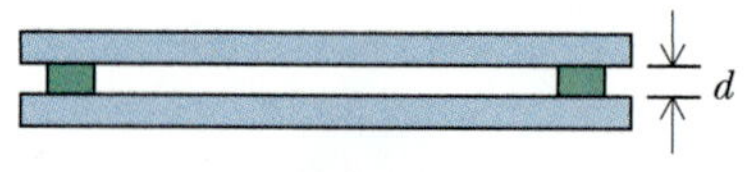

FIGURE 26.34
Two parallel, flat plates of glass.

2.——Two flat, parallel plates of glass are separated by thin spacers so as to leave a gap of width d (Fig. 26.34). If light of wavelength λ is normally incident on these plates, what is the condition for constructive interference between the rays reflected by the lower surface of the top plate and the upper surface of the bottom plate?

3.——When you look at the point on a spherical soap bubble nearest your eye (where the rays of light reaching your eye are perpendicular to the bubble's surface), you see strongly reflected red light, of a wavelength of 650 nm. What can you conclude about the thickness of the wall of the soap bubble? The index of refraction of the soapy water is 1.35.

4.*—When you look at a spherical soap bubble, under the conditions described in the preceding problem, you see strongly reflected light of a wavelength of 650 nm. Suppose you now inflate this bubble (without adding liquid) until the strongly reflected light becomes blue, of a wavelength of 430 nm. By what factor must you inflate the diameter of the bubble to accomplish this? The index of refraction of the soapy water is 1.35.

5.*—A thin oil slick, of index of refraction 1.3, floats on water. When a beam of white light strikes this film vertically, the only colors enhanced in the reflected beam seen in air are orange-red (about 650 nm) and violet (about 430 nm). From this information, deduce the thickness of the oil slick.

6.*—A lens made of flint glass with an index of refraction of 1.61 is to be coated with a thin layer of magnesium fluoride with an index of refraction of 1.38.

(a) How thick should the layer be so as to give destructive interference for the perpendicular reflection of light of wavelength 550 nm seen in air?

(b) Does your choice of thickness permit *constructive* interference for the reflection of light of some other wavelength in the visible spectrum? (If it does, you ought to make a better choice.)

7.*—A layer of oil of thickness 200 nm floats on top of a layer of water of thickness 400 nm resting on a flat, metallic mirror. The index of refraction of the oil is 1.24 and that of the water 1.33. A beam of light is normally incident on these layers. What must be the wavelength of the beam if the light reflected by the top surface of the oil is to interfere destructively with the light reflected by the mirror?

8.*—Consider the light *transmitted* by a thin film. For a light wave with a direction of propagation perpendicular to the surface of the film, what is the condition for constructive interference between the direct wave and the wave that is reflected once by the lower surface and once by the upper surface? Would you expect that this condition for constructive interference in transmission coincides with the condition of destructive interference in reflection?

Section 26.2: The Michelson Interferometer

9.——The interferometer of the Bureau International des Poids et Mesures can count 19,000 bright fringes (maxima) per second. To achieve this count rate, what must be the speed of motion of the moving mirror (mirror M_1 in Fig. 26.7)? Assume that the interferometer operates with krypton light of wavelength $\lambda = 605.8$ nm.

10.*——The ***Fabry–Perot interferometer*** consists of two parallel half-silvered mirrors. A ray of light entering the space between the mirrors may pass straight through, or be reflected once or several times by each mirror (Fig. 26.35). Show that the condition for constructive interference of the emerging light (at large distance from the mirrors) is

$$2d \cos\theta = 0, \lambda, 2\lambda, \ldots$$

where d is the distance between the mirrors, λ the wavelength of light, and θ the angle of incidence of the light.

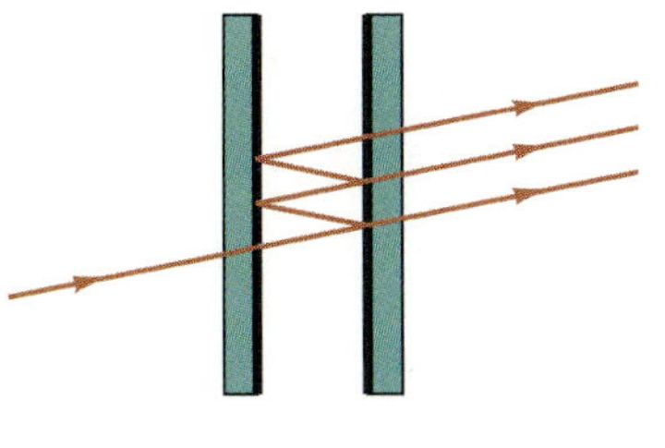

FIGURE 26.35
Fabry–Perot interferometer.

11.**——The index of refraction of air for monochromatic light of some given wavelength can be determined very precisely with a Michelson interferometer one of whose arms (MM_1 in Fig. 26.7) is in air while the other (MM_2) is in a vacuum tank. With the vacuum tank filled with air, the arms are first adjusted so they are exactly equal. Then the vacuum tank is slowly evacuated, while at the same time the length of the arm is gradually increased so as to maintain constructive interference (this increase of length maintains a fixed number of wavelengths in the arm). Show that the index of refraction of air can be expressed in terms of the fractional change of length:

$$n - 1 = \frac{\Delta L}{L}$$

Suppose that for light of wavelength 580 nm, an interferometer with arms of length 40.000 cm requires a readjustment of 0.01106 cm to maintain constructive interference while the vacuum tank is being evacuated. What is the index of refraction of air at this wavelength?

Section 26.3: Interference from Two Slits

12.——Suppose that the two slits, the light source and the film, described in Example 2 are immersed in water. What will now be the distance between the central maximum and the first lateral maximum?

13.——A piece of aluminum foil with two narrow slits is being illuminated with red light of wavelength 694.3 nm from a laser. This yields a row of evenly spaced bright bands on a screen placed 3.00 m beyond the slits. The interval between the bright bands is 1.4 cm. What is the distance between the two slits?

14.——Consider the water waves shown in Fig. 26.14. With a ruler, measure the wavelength of the waves and the distance between the sources; with a protractor, measure the angular positions of the nodal lines (minima). Do these measured quantities satisfy Eq. (9)?

15.——Microwaves of wavelength 2.0 cm from a radio transmitter are aimed at two narrow parallel slits in an aluminum plate. The slits are separated by a distance of 5.0 cm. At what angles at some distance beyond the plate will we find maxima in the interference pattern?

16.——When a beam of monochromatic light is incident on two narrow slits separated by a distance of 0.15 mm, the angle between the central beam and the third lateral maximum in the interference pattern is 0.52°. What is the wavelength of the light?

17.——An interferometric radiotelescope consists of two antennas separated by a distance of 1.0 km. The two antennas feed their signals into a common receiver tuned to a frequency of 2300 MHz. The receiver will detect a maximum (constructive interference) if the wave from a radio source in the sky arrives at the two antennas with the same phase. What possible angular propositions of the radio source will result in such a maximum? Measure the angular position from the vertical line erected at the midpoint of the antennas. Treat the antennas as pointlike.

18.——Interference effects can be detected with sound waves passing through a picket fence. Suppose that the fence consists of vertical boards or rods separated by a distance of 20 cm. Suppose you stand on one side of the fence, and a friend stands on the other side of the fence, at some reasonably large distance at an angle of 30° away from the direction perpendicular to the fence. If you clap your hands, producing white noise, what wavelengths and frequencies will your friend hear strongly?

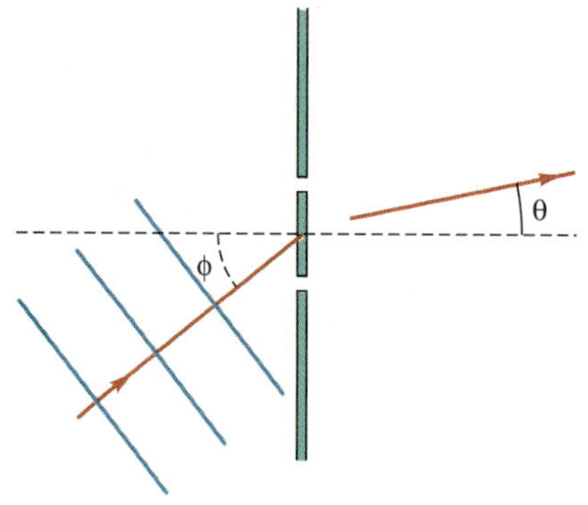

FIGURE 26.36
Plane wave obliquely incident on the plate.

19.*—Two radio beacons are located on an east–west line and separated by a distance of 6.0 km. The radio beacons emit synchronous (in phase) sinusoidal waves of a frequency of 1.0×10^5 Hz. The navigator of a ship wants to determine his position relative to the radio beacons. His radio receiver indicates zero signal strength at the position of the ship. What is the angular bearing of the ship relative to the radio beacons? Assume that the distance between the ship and the radio beacons is much larger than 6 km. Note: This problem has *several* answers.

20.**—Two radio beacons emit waves of frequency 2.0×10^5 Hz. The beacons are on a north–south line, separated by a distance of 3.0 km. The southern beacon emits waves $\frac{1}{4}$ of a cycle later than the northern beacon. Find the angular directions for constructive interference. Reckon the angles relative to the east–west line, and assume that the distance between the beacons and the point of observation is large.

21.**—Light of wavelength λ is obliquely incident on a plate with two narrow slits separated by a distance d. The angle of incidence of the light on the plate is ϕ (Fig. 26.36). Show that the light emerging at an angle θ interferes constructively if

$$d \sin\theta - d \sin\phi = 0,\ \lambda,\ 2\lambda,\ \ldots$$

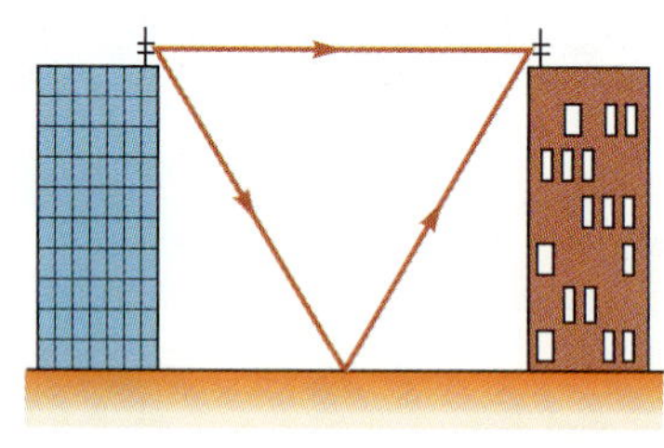

FIGURE 26.37
Antennas on two buildings.

22.**—The radio wave from a transmitter to a receiver may follow either a direct path or else an indirect path involving a reflection on the ground (Fig. 26.37). This can lead to destructive interference of the two waves and a consequent fading of the radio signal at certain locations. Suppose that a transmitter and a receiver operating at a wavelength λ are at a height h on tall buildings with bare ground between. The distance from the emitter to the receiver is d.

(a) Show that the condition for destructive interference is

$$\sqrt{d^2 + 4h^2} - d = 0,\ \lambda,\ 2\lambda,\ \ldots$$

Note that in this calculation you must take into account the reversal, or change of phase, of the wave during reflection on the ground.

(b) If $h = 60$ m and $d = 2300$ m, what wavelengths will lead to destructive interference?

Section 26.4: Interference from Multiple Slits

23.——A grating has 5000 lines per centimeter. What are the angular positions of the principal maxima produced by this grating when illuminated with light of wavelength 650 nm?

24.——Sodium light with wavelengths 588.99 nm and 589.59 nm is incident on a grating with 5500 lines per centimeter. A screen is placed 3.0 m beyond the grating. What is the distance between the two spectral lines in the first-order spectrum on the screen? In the second-order spectrum?

25.——The red line in the spectrum of hydrogen has a wavelength of 656.3 nm; the blue line in this spectrum has a wavelength of 434.0 nm. If hydrogen light falls on a grating with 6000 slits per centimeter, what will be the angular separation (in degrees) of these two spectral lines as seen in the first-order spectrum?

26.——A thin curtain of fine batiste consists of vertical and horizontal threads of cotton forming a net that has a regular array of square holes. While looking through this curtain at the red (670 nm) taillight of an automobile, a physicist notices that the taillight appears as a multiple array of images (an array of principal maxima). The angular separation between adjacent images is 2×10^{-3} radian. From this information deduce the spacing between the threads in the batiste curtain.

27.——For the most detailed mapping of the sky, radioastronomers use radiotelescopes placed on different continents, with (straight-line) separations of several thousand kilometers. Suppose that two such radiotelescopes are separated by an east–west distance of 5000 km, and they are connected to a single radio receiver tuned to a wavelength of 21 cm. If a source is symmetrically located above these radiotelescopes, the radio waves reaching the receiver will be in phase, and the receiver will register maximum intensity. What westward angular displacement of the source from this location will, again, result in maximum intensity?

Section 26.5: Diffraction by a Single Slit

28.——Consider the diffraction pattern shown in Fig. 26.25. How would this pattern change if the slit, the light source, and the film were immersed in water?

29.——Light of wavelength 632.8 nm from a He–Ne laser illuminates a single slit of width 0.10 mm. What is the width of the central maximum formed on a screen placed 2.0 m beyond the slit?

30.——Consider the water waves shown in Fig. 26.38. With a ruler, measure the wavelength of the waves and the length of the gap; with a protractor, measure the angular positions of the two nodal lines (minima). Check whether these quantities satisfy Eq. (17).

FIGURE 26.38
Diffraction of water waves by an aperture.

31.——A sound wave of frequency 820 Hz passes through a doorway of width 1.0 m. What are the angular directions of the minima of the diffraction pattern?

32.——Water waves of wavelength 20 m approach a harbor entrance 50 m across at right angles to their path. What is the angular width of the central beam of diffracted waves beyond the entrance?

33.——The beam of a ruby laser, with light of wavelength 693 nm, is aimed at a slit of width 0.050 mm cut in an aluminum sheet. The diffracted light is intercepted by a screen placed 2.5 m beyond the slit. How far from the centerline on this screen will the first-order minimum be found? The second-order minimum?

34.*——A slit of width 0.11 mm cut in a sheet of metal is illuminated with light of wavelength 577 nm from a mercury lamp. A screen is placed 4.0 m beyond the slit.

(a) Find the width of the central maximum in the diffraction pattern on the screen, that is, find the distance between the first minimum on the left and on the right.

(b) Find the width of the second-order maximum, that is, find the distance between the first minimum and the second minimum on the same side.

35.*—You want to prepare a photograph, such as Fig. 26.25, showing diffraction by a single slit. Suppose you use the light of wavelength 577 nm from a mercury lamp, and you place your photographic film 2.0 m behind the slit. If the width of the central maximum is to be at least 0.5 cm on your film, what width of slit do you require?

36.*—Suppose you want to detect diffraction fringes in the intensity pattern produced by *sunlight* passing through a fine slit cut in the blinds covering a window. You have available a filter that blocks all wavelengths except 550 nm, and you place this filter over the slit.

(a) If the width of the slit is 0.1 mm, will you be able to detect diffraction fringes? Take into account that the Sun's angular diameter is about $\frac{1}{2}°$.

(b) If the width of the slit is 0.01 mm?

(c) If you remove the filter?

Section 26.6: Diffraction by a Circular Aperture; Rayleigh's Criterion

37.—When the eye looks at a star (a point of light), diffraction at the pupil spreads the image of the star on the retina into a small disk.

(a) When opened to maximum size, the diameter of the pupil of a human eye is 7.0 mm. Assuming the starlight has a wavelength of 550 nm, what is the angular size of the image on the retina?

(b) The distance from pupil to retina is 23 mm. What is the linear size of the image of the star?

(c) At the midpoint on the retina (fovea), there are 150,000 light-sensitive cells (rods) per mm^2. How many of these cells are illuminated when the eye looks at a star?

38.—A sailor uses a speaking trumpet to concentrate his voice into a beam. The opening at the front end of the speaking trumpet has a diameter of 25 cm. If the sailor emits a sound of wavelength 15 cm (this is very roughly the wavelength a man emits when yelling "eeeee . . ."), what is the angular width of the central maximum of the beam of sound?

39.—The antenna of a small radar transmitter operating at 1.5×10^{10} Hz consists of a circular dish of diameter 1.0 m. What is the angular width of the central maximum of the radar beam? What is the linear width at a distance of 5.0 km from the transmitter?

40.—The radiotelescope at Jodrell Bank (England) is a dish with a circular aperture of diameter 76 m. What angular resolution can this radiotelescope achieve when operating at a wavelength of 21 cm?

41.—The Hubble Space Telescope placed into an orbit above the atmosphere of the Earth has an aperture of 2.4 m. According to Rayleigh's criterion, what angular resolution can this telescope achieve with visible light of wavelength 550 nm? With ultraviolet light of wavelength 120 nm? How much better is this than the angular resolution of 0.5 second of arc achieved by telescopes on the surface of the Earth?

42.—For an optically perfect lens, the size of the focal spot is limited only by diffraction effects. Suppose that a lens of diameter 10 cm and focal length 18 cm is illuminated with parallel light of wavelength 550 nm. What is the angular width of the central maximum in the diffraction pattern? What is the corresponding linear width at the focal distance?

43.—According to a recent proposal, solar energy is to be collected by a large power station on an artificial satellite orbiting the Earth at an altitude of 35,000 km. The power is to be beamed down to the surface of the Earth in the form of microwaves. If the microwaves have a wavelength of 10 cm and if the antenna emitting the microwaves is 1.5 km in diameter, what is the angular width of the central beam emerging from this antenna? What will be the transverse dimension of the beam when it reaches the surface of the Earth?

44.—Rumor has it that a photographic camera on a spy satellite can read the license plate of an automobile on the ground.

(a) If the altitude of the satellite is 160 km, roughly what angular resolution does the camera need to read a license plate? Assume that the reading requires a linear resolution of about 5 cm.

(b) To attain this angular resolution, what must be the diameter of the aperture of the camera?

45.*—At night, on a long stretch of straight road in Nevada, a truckdriver sees the distant headlights of another truck. How close must he be to the other truck in order for his eyes to resolve the two headlights? Assume that the pupils of the truckdriver have a diameter of 5.0 mm, that the headlights are separated by 1.8 m, and that the light has a wavelength of 550 nm.

46.*—(a) According to Rayleigh's criterion, what is the angular resolution that the human eye can achieve for light of wavelength 550 nm? The fully distended pupil of the human eye has a diameter of 7.0 mm.

(b) Even during steady fixation, the eye has a spontaneous tremor that swings it through angles of 20 or 30 seconds of arc. Compare this angular tremor with the angular resolution that you found in Part (a). Would the elimination of the tremor greatly improve the acuity of the eye?

47.*—According to newspaper reports, a photographic camera on a "Blackbird" reconnaissance jet flying at an altitude of 27 km can distinguish detail on the ground as small as the size of a man.

(a) Roughly, what angular resolution does this require?

(b) According to Rayleigh's criterion, what minimum diameter must the lens of the camera have?

48.*—Some spy satellites carry cameras with lenses 30 cm in diameter and with a focal length of 2.4 m.

(a) What is the angular resolution of the camera according to Rayleigh's criterion? Assume that the wavelength of light is 550 nm.

(b) If such a satellite looks down on the Earth from a height of 150 km, what is the distance between two points on the ground that the camera can barely resolve?

(c) The lens projects images of the two points on a film at the focal plane of the lens. What is the distance between the two images projected on the film?

49.*—Kalahari bushmen are said to be able to see the four moons of Jupiter with the naked eye. According to Rayleigh's criterion, what must be the minimum separation between two small light sources placed at the distance of Jupiter if they are to be resolved by the human eye? Compare this with the separations between the moons of Jupiter. Does the limit on the resolving power of the human eye prevent *you* from seeing the moons? The (average) distance to Jupiter is 630×10^6 km; the separations between the moons are typically 400×10^3 km. Assume that the diameter of the pupil of the eye is 7 mm and that the wavelength of the light is 550 nm.

50.* A microwave antenna used to relay communication signals has the shape of a circular dish of diameter 1.5 m. The antenna emits waves with $\lambda =$ 4.0 cm.

(a) What is the width of the central maximum of the beam of this antenna at a distance of 30 km?

(b) The power emitted by the antenna is 1.5×10^3 W. What is the energy flux directly in front of the antenna? What is the energy flux at a distance of 30 km? Assume that the power is evenly distributed over the width of the central beam.

51.* When exposed to strong light, the pupil of the eye of a cat narrows to a fine slit, about 0.3 mm across. Suppose that the cat is looking at two white mice 20 m away and separated by a distance of 5 cm. Can the cat distinguish one mouse from the other?

52.* Mars has a radius of 3400 km; when Mars is at its closest to the Earth, its distance is 78×10^6 km. Calculate the angular size of Mars as seen from Earth. Estimate the diameter of the objective lens of the telescope of smallest size that will permit you to tell that Mars has a disk, that is, that the image of Mars is wider than the image of a star.

PHYSICS IN ACTION

MICROSCOPY

Microscopes are widely used in biology and in medicine to study microbes, tissues, cells, and structures within cells. Light microscopes used for these purposes operate at magnifications of up to 1300×. Although electron microscopes can attain much higher magnifications — more than 10^5× — such high magnifications are not needed for most of the routine work in medicine and biology. Besides, some special techniques available for use with light microscopes can often reveal structures that an electron microscope cannot detect.

Many of the objects that biologists and medical practitioners want to examine are transparent or nearly transparent to light, and such objects are nearly invisible when examined by transmitted light in an ordinary, or "bright-field," microscope. Among the techniques that have been developed by microscopists to enhance the contrast and to render such transparent objects visible, the simplest and the oldest is staining with some chemical dye that selectively darkens different structures within the object. More sophisticated techniques exploit the diffraction, interference, and polarization of light to enhance the contrast in the image, without staining the object. We will briefly discuss the main techniques of microscopy: ***bright-field, dark-field, phase-contrast, differential interference-contrast, polarization,*** and ***fluorescence.***

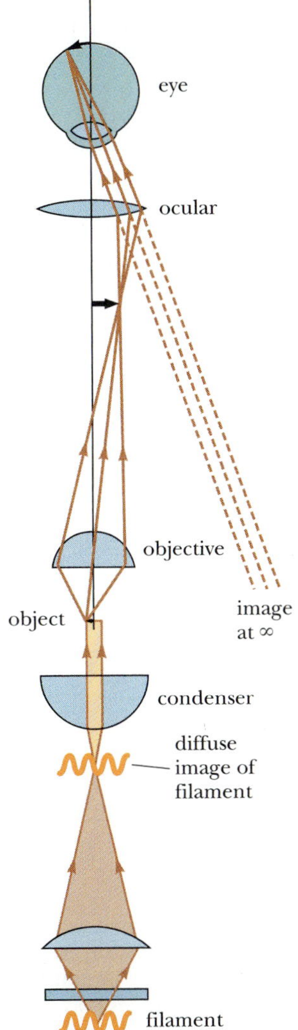

FIGURE 1 Schematic diagram of a bright-field microscope. Illumination of the object is provided by light rays that originate from the incandescent filament of a light bulb and are made parallel by a condenser lens (Köhler illumination).

Bright-field

In the bright-field microscope, the object is illuminated with a beam of light from below, and the image is formed by transmitted light rays that pass through the object and enter the objective lens placed above the object (see Fig. 1). The image depends on the opacity of the object — opaque portions of the object show up darkly, against a bright background (see Fig. 2). The bright-field microscope is routinely used for examination of fixed and

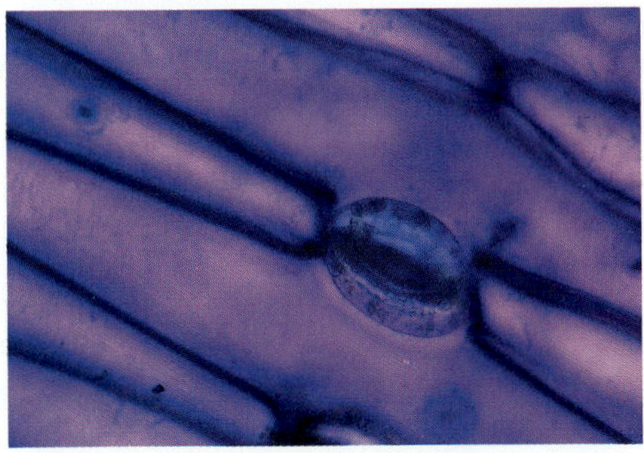

FIGURE 2 Image produced by a bright-field microscope. The object is the epidermis of an amaryllis leaf, stained with methylene blue to enhance the veins and stomates (pores). The magnification is about 300×. In Figs. 5, 6, 7, and 9 the object is the same, but unstained.

stained slices of tissues, stained cells, and stained bacteria.

The background in the field of view of the microscope should be of uniform brightness, since any nonuniformity could be misinterpreted as a structure within the object. The requisite uniformity of the background is attained by a condenser lens, with the light source (or the image of the light source) placed at its focus (see Fig. 1). The light rays emerging from the condenser are then parallel rays, uniformly distributed over the object plane, even if the light source (filament of a light bulb) is not of uniform brightness. This method of illumination is called ***Kōhler illumination.***

As we discussed in Section 25.6, the magnification attained by the microscope is the product of the magnifications of the objective lens and the ocular lens [see Eq. (25.32)].* Large magnifications can be attained by using oculars of short focal length. However, excessively large magnifications lead to blurry images and loss of detail. The smallest detail that can be distinguished by the microscope is limited by diffraction effects. The aperture of a microscope, like the aperture of a telescope, produces a diffraction pattern, and the image of a pointlike object is a small diffraction disk. This spreading of the image limits the detail that can be seen through the microscope — if two pointlike objects are very close together, their images tend to overlap and merge, like the images of two close stars seen through a telescope. However, unlike the telescope, the microscope is afflicted with extra diffraction at the object, because the object is not self-luminous, but is illuminated by coherent or partially coherent light from a single light source. This light is diffracted when it passes through the object, and the emerging light consists not only of a zero-order, central beam, but also a multitude of higher-order, lateral beams. To discover the net resolution in the image of the microscope, we need to take into account both this diffraction at the object and the diffraction at the aperture of the microscope. As always, we can calculate the diffraction by using Huygens' wavelets, but instead of starting these wavelets at the aperture of the microscope, we must start them at the object, where the first diffraction occurs. If we then track the wavelets through the microscope, the additional diffraction by the microscope will be automatically included.

* In all modern microscopes these lenses are actually compound lenses, consisting of several elements; but they are effectively equivalent to single lenses of suitable focal lengths.

We can obtain a simple formula for the size of the finest detail that can be resolved by a microscope if we consider a test object made out of many thin, parallel bright and dark lines (a test object of this kind is also often used to test the resolution of a TV screen). Such a test object is a grating, and when we illuminate it with plane light waves (parallel rays), the diffraction and interference effects generated by the slits in this grating produce a zero-order, central beam and higher-order, lateral beams. If the spacing between adjacent bright lines in the grating is d, then the angles θ at which the lateral beams emerge is given by Eq. (26.12),

$$d \sin\theta = \lambda,\ 2\lambda,\ 3\lambda,\ \ldots \tag{1}$$

The microscope will form a perfect image of the test object if it captures all the lateral beams and focuses them in the image plane, so they can recombine and reproduce the original intensity distribution. But the objective lens of the microscope actually cannot capture all the lateral beams, since some emerge at large angles θ and miss the lens. Such lost beams imply a loss of sharpness of the image. If the microscope is to reveal the presence of the bright and dark lines in our test object, at the very least the first-order lateral beam must enter the objective lens (see Fig. 3). This requires

$$d = \frac{\lambda}{\sin\theta_0} \tag{2}$$

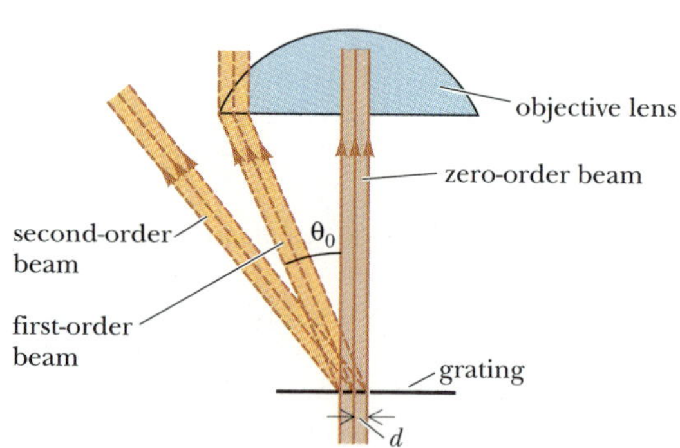

F I G U R E 3 The microscope is aimed at a grating of bright and dark lines, which serves as test object. The first-order lateral beam (barely) enters the objective lens; the higher-order lateral beams are lost. The first-order lateral beam suffices to reveal the presence of bright and dark lines in the test object and their spacing; the loss of the higher-order lateral beams implies that the microscope will not see the lines very sharply (their edges will be blurred somewhat).

where θ_0 is the angle between the axial line of the microscope and the line from the object to the rim of the objective lens. This formula tells us the closest distance d for which the microscope can still reveal the presence of the bright and dark lines. Note that d is proportional to the wavelength λ and inversely proportional to the sine of θ_0. For high resolution, we want a large angle θ_0. But a large angle θ_0 leads to aberrations in the lenses, and this limits the best microscope lenses to about $\theta_0 \simeq 60^0$.* With a typical wavelength of light, $\lambda = 550$ nm, the closest resolvable distance is then

$$d = \frac{550 \text{ nm}}{\sin 60^0} = 640 \text{ nm} = 0.64\ \mu\text{m} \qquad (3)$$

Any details smaller than that will be totally blurred.

The resolution of the microscope can be improved somewhat by filling the space between the object and the objective lens with oil or some other transparent liquid. The index of refraction of the oil used for this ***immersion*** technique is about 1.5, and the wavelength in Eq. (2) is therefore decreased by a factor of 1.5, which leads to a corresponding decrease in d. Furthermore, the oil prevents the total internal reflection of rays within the cover glass that is usually placed on top of the object, between the object and the objective lens. Such total internal reflection is undesirable, since it limits the maximum angle of acceptance to a smaller value than the 60^0 tolerated by the lens (see footnote above). Microscopes intended for the immersion technique have special objective lenses whose mountings will not be damaged by oil.

Phase-contrast

A perfectly transparent object does not affect the intensity of the light passing through it; such an object would therefore not be visible in a bright-field microscope. However, if the index of refraction of the object is different from 1, then the object will produce diffracted beams, even though the net intensity of the light passing through the object is not

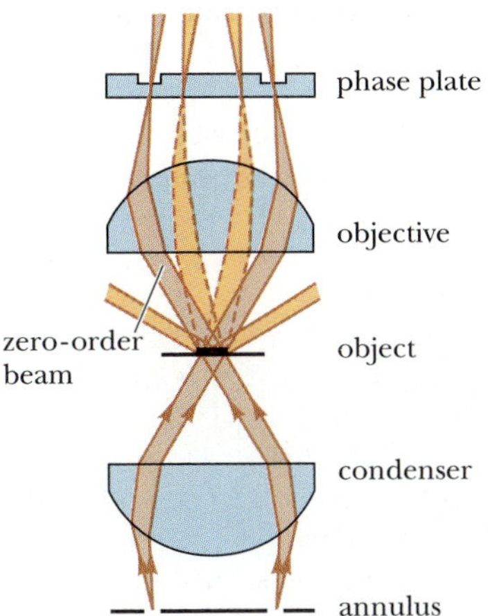

FIGURE 4 Schematic diagram of a phase-contrast microscope. The phase plate has an annular notch, so it is thinner at the rim than at its center.

altered. The phase-contrast microscope takes advantage of these diffracted beams to produce an image. For this purpose, the microscope artificially introduces a phase difference between the zero-order beam and the higher-order diffracted beams that emerge from the object. Figure 4 shows how this phase difference is introduced. The illuminating light reaches the object obliquely; consequently the zero-order beam leaves the object obliquely, at the same angle as the incident beam, and it enters the objective near its rim. But the first-order beam enters the objective near its center. The phase difference between the zero-order beam and the first-order beam is introduced by a ***phase plate*** of glass placed behind the objective lens. This phase plate is thinner at the rim (where the zero-order beam passes) than at its center (where the first-order beam passes). Since the wavelength in glass is shorter than in air, a beam that passes through the thick part accumulates a larger phase than the beam that passes through the thin part. When these beams are brought together in the image plane of the microscope, they will interfere and produce an image whose bright and dark regions correspond to constructive and destructive interference. Note that if we were to remove the object, there would be no higher-order beams, and no interference between the zero-order beam and the higher-order beams, and no image. If we were to remove the phase plate, then the interference between the zero-order beam and the higher-order beams would merely reconstruct the (transparent) object, which would remain invisible. In essence, the phase plate destroys

---

* Specimens used in biology and medicine are usually mounted under a cover glass, and total internal reflection of light within this cover glass actually limits the maximum angle of acceptance θ_0 to considerably less than 60°. This limitation can be overcome by immersion of the cover glass and the objective in oil; see the next paragraph.

this constructive interference between the beams, and permits some parts of the object to show up darkly (see Fig. 5).

The phase-contrast microscope is often used to view living cells, which would be harmed by staining.

FIGURE 5 Image obtained with phase-contrast microscope.

FIGURE 6 Image obtained with dark-field microscope.

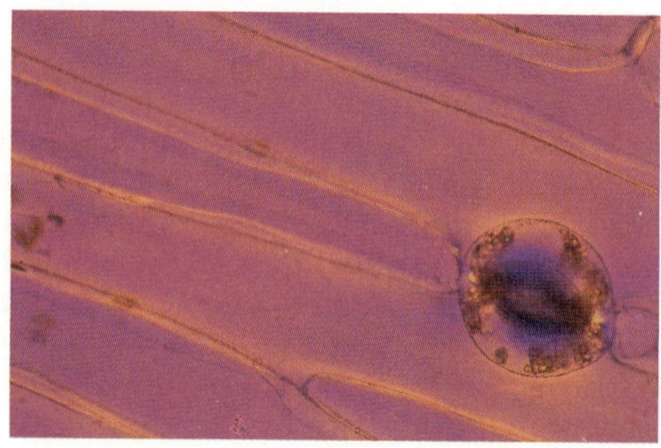

FIGURE 7 Image obtained with a polarizing microscope.

Dark-field

The basic design of the dark-field microscope is quite similar to that of the phase-contrast microscope. Again, the zero-order beam and the higher-order diffracted beams are separated, as in Fig. 4, but instead of introducing a phase difference between these beams, the zero-order beam is simply blocked (in Fig. 4, this could be done by filling the annular notch with black paint). The image is then formed with the higher-order beams only. Consequently, the background is dark (no higher-order beams arise from the background), and the image stands out sharply against this dark background (see Fig. 6).

This technique is especially useful for viewing very small, thin objects, such as the flagella of bacteria and the axons of nerve cells, which would be barely visible by other techniques.

Polarization

In the polarizing microscope, a polarizer (a sheet of Polaroid) is placed below the object, and an analyzer (a second sheet of Polaroid) is placed above the object. The preferential direction of the analyzer is arranged at 90^0 to the preferential direction of the polarizer. We know from Section 24.3 that such "crossed" polarizers by themselves will completely block an incident light wave — the first polarizer blocks light of one direction of polarization, and the second polarizer blocks the light of the perpendicular direction of polarization, leaving no light at all. However, when light passes through an object located between the two polarizers, its direction of polarization can be altered; many materials have preferential directions for polarized light, and they can slightly rotate the plane of polarization of the light that passes through them. If the object produces such a rotation, then the analyzer will not block the light completely, and the microscope will form an image of those parts of the object that have produced a rotation of the plane of polarization. The image shows up sharply against the dark background, where the light is blocked completely (see Fig. 7).

The polarization technique is best suited to specimens that contain crystals or structures with aligned chains of molecules, which produce strong polarization of the light. Hence this technique is used to view the shells of forams, radiolarians, and larval molluscs (crystals), and the details of muscle fibers and other fibers (aligned molecules).

Differential interference-contrast (DIC)

In the DIC microscope, each ray of light from the light source is split into two rays by a beam splitter. Both of these rays pass through the object, along adjacent paths (see Fig. 8). The rays then enter the microscope, where they are recombined by a beam combiner (similar to the beam splitter, but reversed). Before recombining the rays, the beam combiner introduces a phase difference of 180^0 between the rays. Hence, in the absence of the object, the two recombined rays would interfere destructively. But the object alters the phases of the rays that pass through it, and this alters the interference between the recombined rays, giving rise to some nonzero resultant ray which produces an image in the image plane. The image depends on the phase contrast between adjacent parts of the object, that is, it depends on the difference between the indices of refraction encountered by the split rays as they pass through adjacent parts of the object. The image is seen against a dark background (see Fig. 9).

Like the phase-contrast microscope, the interference-contrast microscope is often used to view living cells. It produces somewhat crisper images, with a three-dimensional appearance.

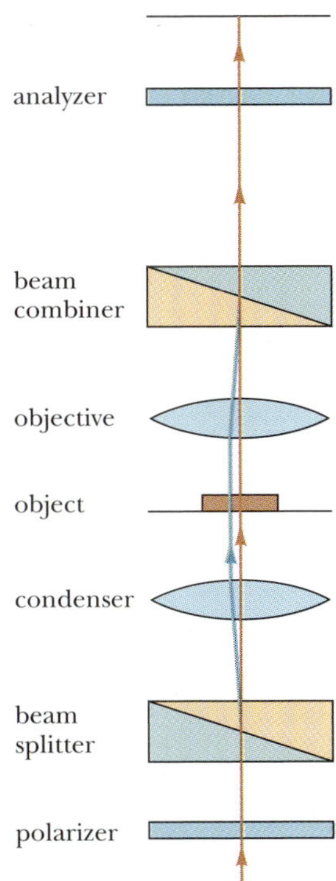

FIGURE 8 Schematic diagram of a DIC microscope. The ray of light from the source is split into two rays (colored red and blue in the diagram) by the beam splitter, consisting of a polarizer and a special prism. Both of these rays pass through the object along adjacent paths. They then enter the microscope, and are recombined by a beam combiner.

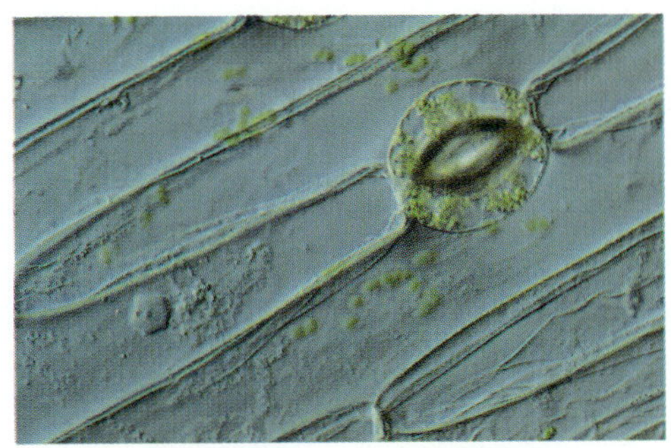

FIGURE 9 Image obtained with a DIC microscope.

Fluorescence

The fluorescence microscope exploits fluorescent dyes to stain the object. When illuminated with light of some wavelength, these dyes give off light of a different wavelength (generally, a longer wavelength). For example, Day-Glo orange paint contains such a fluorescent dye; when illuminated with sunlight, this dye gives off some extra orange light, which makes this paint look unusually bright. In the fluorescence microscope, the dyed object is illuminated with an intense beam of short-wavelength light so it fluoresces, giving off longer-wavelength light. The microscope is equipped with a filter that reflects the illuminating light (short wavelength), but allows the fluorescent light (long wavelength) to pass (see Fig. 10). The image shows up in the characteristic colors of the fluorescence against a dark background (see Fig. 11).

Fluorescent dyes are widely used as tracers to mark and track cells in studies of tissue development. Other fluorescent dyes are used to detect and track the nucleic acids RNA and DNA in the chro-

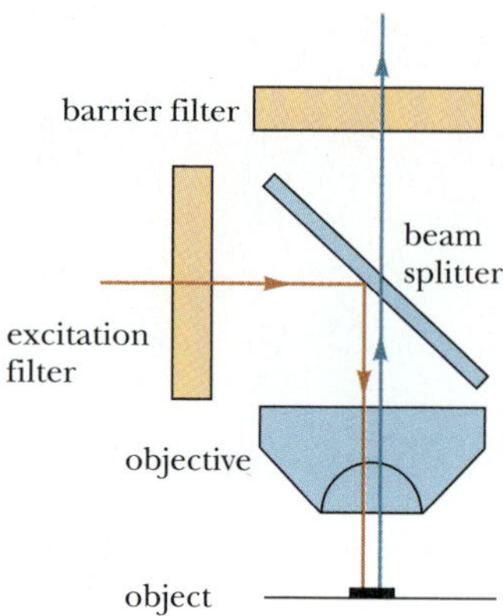

FIGURE 10 Schematic diagram of a fluorescence microscope.

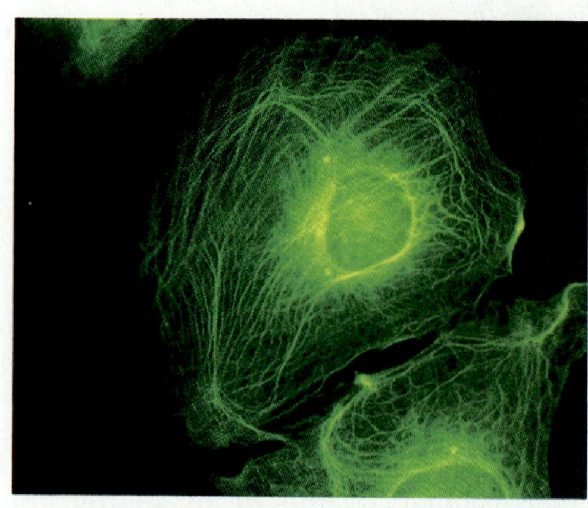

FIGURE 11 Image of human cheek cells obtained with a fluorescence microscope. The magnification is about 1200×.

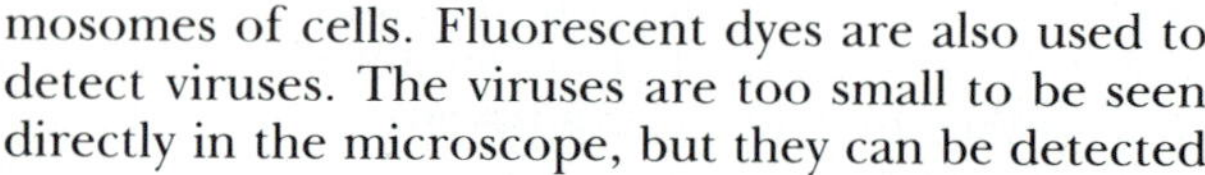

mosomes of cells. Fluorescent dyes are also used to detect viruses. The viruses are too small to be seen directly in the microscope, but they can be detected by tagging a virus-specific antibody with fluorescent dye; when the antibody seeks out and binds to the virus, it reveals the presence of the virus.

REFERENCES

Möllring, F. K., *Microscopy from the Very Beginning* (Zeiss, Oberkochen, 1981).

Spencer, M., *Fundamentals of Light Microscopy* (Cambridge University Press, Cambridge, 1982).

CHAPTER 27

The Stanford Linear Accelerator (SLAC) accelerates electrons to a final speed of 99.9999999948% of the speed of light.

THE THEORY OF SPECIAL RELATIVITY

As we saw in Chapter 3, Newton's laws of motion are equally valid in every inertial reference frame. Consequently, ***no mechanical experiment can detect any intrinsic difference between two inertial reference frames.*** This is the ***Newtonian principle of relativity.*** For a concrete illustration of this principle, consider the reference frame of a ship steaming away from the shore and consider the reference frame of the shore. Both of these reference frames are inertial, and the behavior of pucks used in a game of shuffleboard on the deck of the ship is not different from the behavior of similar pucks on shore.* Experiments with such pucks aboard the ship will

◀ **Newtonian principle of relativity**

* This assumes the ship moves steadily. If the ship pitches or rolls, it will cease to be an inertial reference frame, and the pucks will behave in a manner "inconsistent" with Newton's laws.

not reveal the uniform motion of the ship relative to the shore. To detect this motion, the crew of the ship must take sightings of points on the shore or use some other navigational technique that fixes the position and velocity of ship in relation to the shore. Hence, in regard to mechanical experiments, uniform translational motion of our inertial reference frame is always *relative* motion — it can be detected only as motion of our reference frame with respect to another reference frame. There is no such thing as *absolute* motion.

The question naturally arises whether the relativity of mechanical experiments also applies to electric, magnetic, optical, and other experiments. Do any of these experiments permit us to detect an absolute motion of our reference frame? Albert Einstein answered this question in the negative. He laid down a principle of relativity for *all* the laws of physics, according to which no experiment whatsoever can detect any intrinsic difference between two inertial reference frames. Before we deal with the details of Einstein's theory of relativity, we will briefly describe why nonmechanical experiments — and, in particular, experiments with light — might be expected to detect absolute motion which mechanical experiments cannot detect.

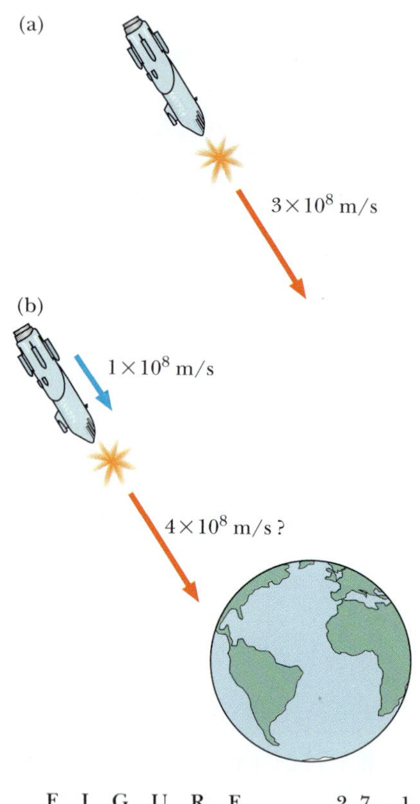

FIGURE 27.1
Addition of velocities according to the Galilean addition rule. (a) The light signal has a speed of 3.00×10^8 m/s in the reference frame of a spaceship speeding toward the Earth. (b) The light signal then has a higher speed in the reference frame of the Earth.

27.1 THE SPEED OF LIGHT; THE ETHER

Since the laws of mechanics are the same in all inertial reference frames, it might seem quite natural to assume that the laws of electricity, magnetism, and optics are also the same in all inertial reference frames. But this assumption immediately leads to a paradox concerning the speed of light. As we know from Chapter 24, light is an oscillating electric and magnetic disturbance propagating through space, and Maxwell's equations permit us to deduce that the speed of propagation of this disturbance must always be 3.00×10^8 m/s. The trouble with this deduction is that, according to the Galilean addition rule for velocities [Eq. (2.68)], the speed of light ought *not* to be the same in all reference frames. For instance, imagine that an alien spaceship approaching the Earth with a speed of, say, 1.00×10^8 m/s flashes a light signal toward the Earth; if this signal has a speed of 3.00×10^8 m/s in the reference frame of the spaceship, then the Galilean addition rule tells us that it ought to have a speed of 4.00×10^8 m/s in the reference frame of the Earth (see Fig. 27.1).

To resolve this paradox either we must give up the notion that the laws of electricity and magnetism (and the values of the speed of light) are the same in all inertial reference frames, or else we must give up the Galilean addition rule for velocities. Both alternatives are unpleasant: the former means that we must abandon all hope for a principle of relativity embracing electricity and magnetism, and the latter means that together with the Galilean addition rule we must abandon the transformation rule for position vectors measured in different reference frames [Eq. (2.65)] as well as the intuitively "obvious" notions of absolute time and length from which these rules are derived.

Since the failure of a relativity principle embracing electricity and magnetism seems to be the lesser of two evils, let us first explore this alternative. Let us assume that there exists a preferred inertial reference frame in which the laws of electricity and magnetism take their simplest form, that is, the form expressed in Maxwell's equations, Eqs. (24.4)–(24.7). In this reference frame, the speed of light has its standard value of $c = 3.00 \times 10^8$ m/s, whereas in any other reference frame it is larger or smaller according to the Galilean addition rule. The propagation of light is then analogous to the propagation of sound. There exists a preferred reference frame in which the equations for the propagation of sound waves in, say, air take their simplest form: the reference frame in which the air is at rest. In this reference frame, sound has its standard speed of 331 m/s. In any other reference frame, the equations for the propagation of sound waves are more complicated, but the speed of propagation can always be obtained directly from the Galilean addition rule. For instance, if a wind of 40 m/s (a hurricane) is blowing over the surface of the Earth, then sound waves have a speed of 331 m/s relative to the air, but their speed relative to the ground depends on direction — downwind the speed is 371 m/s, whereas upwind it is 291 m/s.

(a)

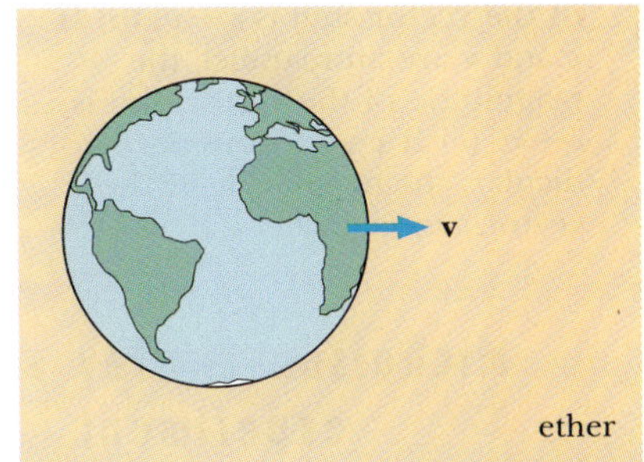

(b)

FIGURE 27.2

(a) The Earth moves relative to the ether. (b) In the reference of the Earth, the ether flows past the Earth, forming the ether wind.

This analogy between the propagation of light and of sound suggests that there exists some pervasive medium whose oscillations bring about the propagation of light, just like the oscillations of air bring about the propagation of sound. Presumably this ghostly medium fills all of space, even the interplanetary and interstellar space which is normally regarded as a vacuum. The physicists of the nineteenth century called this hypothetical medium the ***ether***, and they attempted to describe light waves as oscillations of the ether. The preferred reference frame in which light has its standard speed is then the reference frame in which the ether is at rest. The existence of such a preferred reference frame would imply that velocity is absolute — the ether frame would set an absolute standard of rest, and the velocity of any body could always be referred to this frame. For instance, instead of describing the velocity of the Earth relative to some other material body, such as the Sun, we could always describe its velocity relative to the ether.

◀ **Ether wind**

Presumably the Earth has some nonzero velocity relative to the ether. Even if the Earth were at rest in the ether at one instant, this condition could not last, since the Earth continually changes its motion as it orbits around the Sun. The motion of the ether past the Earth was called the ***ether wind*** by the nineteenth-century physicists (Fig. 27.2). If the Sun is at rest in the ether, then the ether wind would have a velocity opposite to that of the Earth around the Sun — about 30 km/s; if the Sun is in (steady) motion, then the ether wind would vary with the seasons — smaller than 30 km/s during one-half of the year and greater than 30 km/s during the other half.

Experimenters attempted to detect this ether wind by its effects on the propagation of light. A light wave in a laboratory on the Earth would have a greater speed when moving downwind and a smaller speed when moving upwind or across the wind. If the speed of the ether wind "blowing" through the laboratory is v, then the speed of light in this laboratory is $c + v$ for a light signal with downwind motion, $c - v$ for upwind motion,

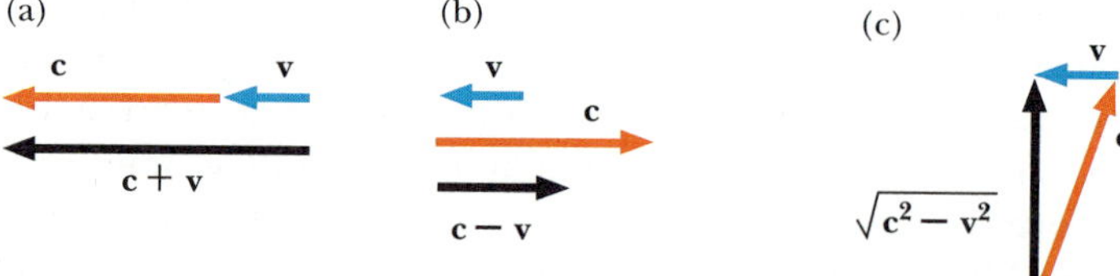

FIGURE 27.3
The velocity of light is **c** relative to the ether, and the velocity of the ether is **v** relative to the laboratory. The velocity of light relative to the laboratory is then the vector sum **c** + **v**. (a) If **c** and **v** are parallel, the magnitude of the vector sum is $c + v$. (b) If **c** and **v** are antiparallel, the magnitude of the vector sum is $c - v$. (c) If **c** and **v** are perpendicular, the magnitude of the vector sum is $\sqrt{c^2 - v^2}$.

and $\sqrt{c^2 - v^2}$ for motion perpendicular to the wind (see Fig. 27.3). With a value of 300,000 km/s for c and a value of approximately 30 km/s for v, the increase or decrease of the speed of light amounts to only about 1 part in 10,000 — a very sensitive apparatus is required for the detection of this small change.

Michelson-Morley experiment ▶

In a famous experiment first performed in 1881 and often repeated thereafter, A. A. Michelson and E. W. Morley attempted to detect small changes in the speed of light by means of an interferometer. As discussed in Section 26.2, they failed to detect any ether wind, even though the sensitivity of the original experiment of Michelson and Morley was such that an ether wind of 5 km/s would have produced a detectable effect. Since the expected wind is about 30 km/s, the experimental result contradicts the ether theory of the propagation of light. Later, more refined versions of the experiment established that if there were ether wind, its speed would certainly have to be less than 3 m/s. The experimental evidence therefore establishes conclusively that the motion of the Earth has no effect on the propagation of light. As the Earth moves around the Sun, its velocity changes continuously, and the Earth shifts from one inertial reference frame to another. But all these inertial reference frames appear to be completely equivalent in regard to the propagation of light. There is no preferred reference frame.

27.2 EINSTEIN'S PRINCIPLE OF RELATIVITY

Since both the laws of mechanics and the laws for the propagation of light fail to reveal any intrinsic distinction between different inertial reference frames, in 1905 Einstein took a bold step and proposed a general hypothesis concerning *all* the laws of physics. This hypothesis is the ***Principle of Relativity:***

Principle of Relativity ▶

All the laws of physics are the same in all inertial reference frames.

This deceptively simple principle forms the foundation of the theory of Special Relativity. Since Maxwell's equations and the laws for the propagation of light are included among the laws of physics, one immediate corollary of the Principle of Relativity is

Principle of Relativity for speed of light ▶

The speed of light (in vacuum) is the same in all inertial reference frames; it always has the value $c = 3.00 \times 10^8$ m/s.

As we pointed out in the preceding section, ***this invariance of the speed of light conflicts with the Galilean addition rule for velocities.*** We will therefore have to discard this rule, and we will also have to discard the transformation rule for position vectors or coordinates on which it is based (see Section 27.5).

The invariance of the speed of light also requires that we give up some of our intuitive, everyday notions of space and time. Obviously, the fact that a light signal always has a speed of 3.00×10^8 m/s, regardless of how hard we try to move toward it or away from it in a fast aircraft or spaceship, does violence to our intuition. This strange behavior of light is only possible because of a strange behavior of length and time in relativistic physics. As we will see later in this chapter, neither length nor time is absolute — they both depend on the reference frame in which they are measured and they suffer contraction or dilation when the reference frame changes.

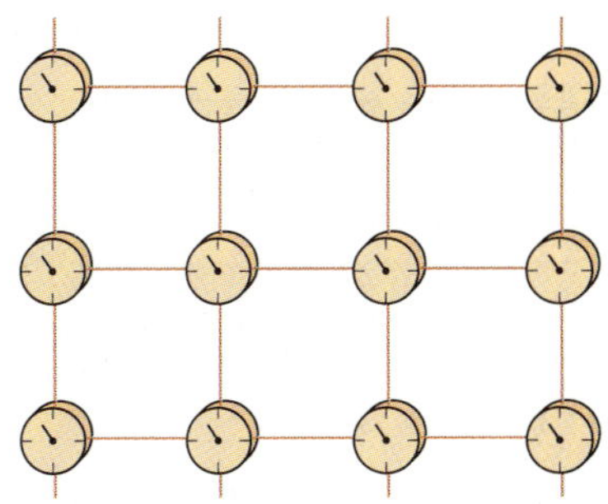

FIGURE 27.4

A reference frame consisting of a coordinate grid and synchronized clocks.

Before we can inquire into the consequences of the Principle of Relativity, we must carefully describe the construction of reference frames and the synchronization of clocks. ***A reference frame is a coordinate grid and a set of synchronized clocks*** (see Fig. 27.4) which can be used to determine the space and time coordinates of any ***event,*** that is, any point in space and time. One of the grid intersections is chosen as origin. The space coordinates of an event are then directly determined by the grid intersections nearest the event. The time coordinate of the event is the time registered by the clock at the event.

◀ **Reference frame**

Of course, the clocks of this reference frame must be synchronized with each other and with the master clock sitting at the origin of coordinates. Einstein proposed that this synchronization can be accomplished by sending out a flash of light from a point exactly midway between the clock at the origin and the other clock (see Fig. 27.5). The two clocks are synchronized if both show exactly the same time when the light from the midpoint reaches them. Note that this synchronization procedure hinges on the invariance of the speed of light. If the speed of light were not a universal constant, but were dependent on the reference frame and on the direction of propagation (say, faster toward the right in Fig. 27.5 and slower toward the left), then we could not achieve synchronization by the simple procedure with a flash of light from the midpoint.

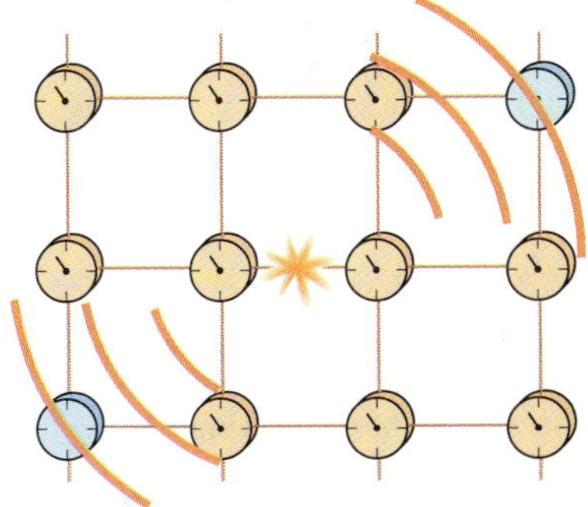

FIGURE 27.5

Synchronization procedure for a pair of clocks. A flash of light is sent from the midpoint to the clocks at the lower left corner and the upper right corner.

One immediate consequence of our synchronization procedure is that ***simultaneity is relative,*** that is, the simultaneity of two events depends on the reference frame. The following is a concrete example: Suppose that two bolts of lightning strike in Boston and in New York, respectively, at exactly 6:00 P.M. Eastern Standard Time. The emissions of brief flashes of light from the lightning bolts are then simultaneous in the reference frame of the Earth. However, in the reference frame of a fast spaceship passing by the Earth in the direction from Boston to New York, these two events are *not simultaneous* — as judged by the clocks on board the spaceship, the lightning in Boston occurs slightly later than the lightning in New York.

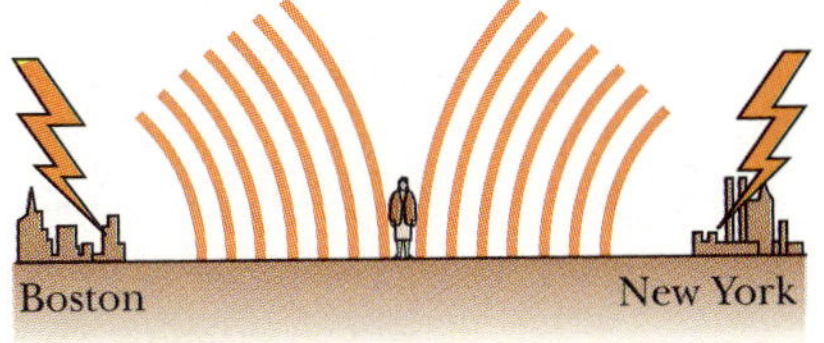

FIGURE 27.6

Observer on the ground at the midpoint between Boston and New York watches for the arrival of flashes of light.

To see how this difference between the two reference frames comes about, let us apply our procedure for testing simultaneity. In the reference frame of the Earth, an observer can test for simultaneity by placing herself exactly at the midpoint between Boston and New York (see Fig. 27.6); she will then receive flashes of light from the lightning in the two cities at the same instant. Thus, this observer will confirm that in the reference frame of the Earth, the lightning was simultaneous.

In the reference frame of the spaceship, an observer can likewise test for simultaneity by placing himself exactly at the midpoint between the

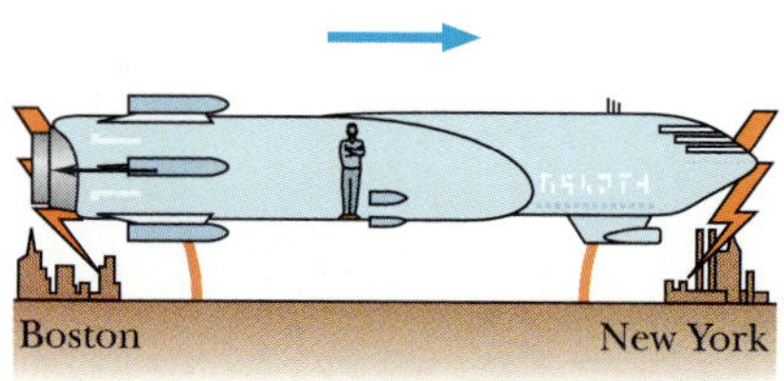

FIGURE 27.7
Observer in a fast spaceship at the midpoint between the front and rear ends of the spaceship, where lightning has perforated the skin.

places of origin of the two flashes of light. For him, the places of origin of the flashes of light are two grid points in his reference frame. To help us visualize the phenomenon as the observer in the spaceship perceives it, imagine that the front and the rear ends of the (large) spaceship are actually passing through New York and through Boston when the lightning strikes, and the lightning bolts penetrate through the ends of the ship, perforating its skin (see Fig. 27.7). The observer in the spaceship then regards the perforations as the places of origin of the flashes of light, and, to test for simultaneity, he places himself at the midpoint between these perforations and waits for the arrival of the light at his eyes. Will he receive the flashes of light from the front and the rear of the spaceship at the same instant?* The answer is obvious if we adopt the point of view of the observer on the ground: she sees the spaceship observer at the midpoint between Boston and New York when the lightning strikes, and she then sees him traveling toward New York, that is, she sees him traveling away from the flash of light trying to catch up with him from Boston and toward the flash coming to meet him from New York (see Fig. 27.7). Thus, from her point of view, it is quite clear that the flash of light from New York will encounter the spaceship observer before the flash of light from Boston can catch up. But if the spaceship observer receives the flash of light from New York before the flash from Boston, he will conclude the flashes were not emitted simultaneously — according to his reckoning, the lightning in Boston was late!

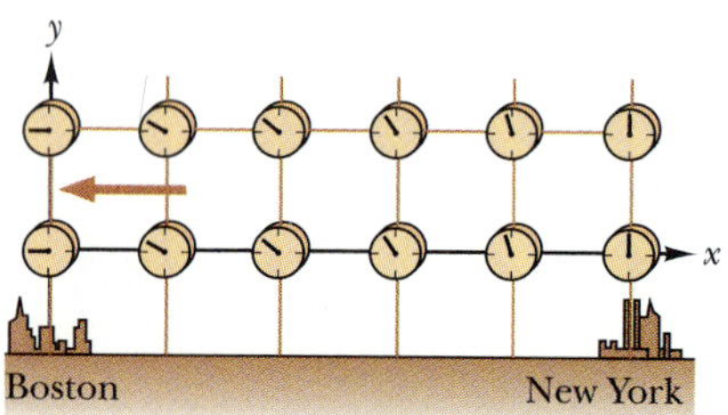

FIGURE 27.8
Clocks of the reference frame of the Earth as observed at one instant of spaceship time. Note that the clocks on the left part of the reference frame are late. (See Fig. 27.16 for a more accurate picture.)

Although this qualitative argument shows that simultaneity depends on the reference frame, it does not tell us by how much. A careful, quantitative calculation shows that for a spaceship traveling at, say, 90% of the speed of light, the lightning in Boston is late by about 0.001 s, as measured in the reference frame of the spaceship.

If simultaneity is relative, then the synchronization of clocks is also relative. In the reference frame of the Earth, clocks in Boston and New York are synchronized, that is, the hands of these clocks reach the 6:00 P.M. position simultaneously. But in the reference frame of the spaceship, the clock in Boston is judged to be late — in the same way as the lightning is late, the hands of the clock reach the 6:00 P.M. position later than in New York. Figure 27.8 shows the clocks belonging to the reference frame of the Earth as observed at one instant of time from the spaceship.

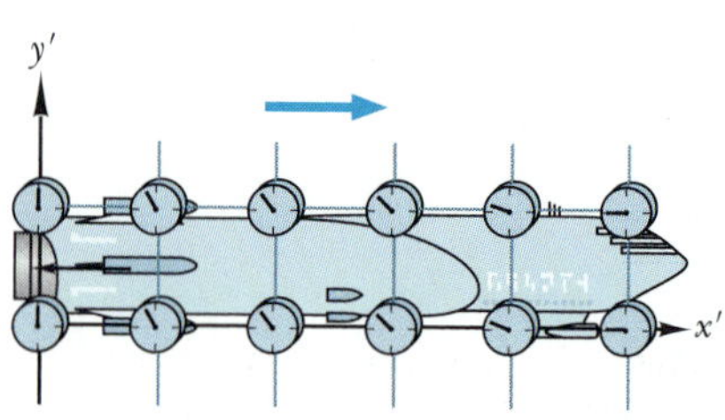

FIGURE 27.9
Clocks of the reference frame of the spaceship as observed at one instant of Earth time. Note that the clocks on the front part of the spaceship are late. (See Fig. 27.15 for a more accurate picture.)

The effect is symmetric. In the reference frame of the spaceship, all the clocks on board are synchronized. But, as observed from the reference frame of the Earth, the clocks on the front part of the spaceship are late. Figure 27.9 shows the clocks belonging to the reference frame of the spaceship as observed at one instant from the Earth. Note in Fig. 27.8 we are viewing the reference frame of the Earth moving past the spaceship, and in Fig. 27.9 we are viewing the reference frame of the spaceship moving past the Earth. In either case, the ***clocks on the leading edge of the reference frame are late.***

---

* Note that it makes no difference whether we consider the light as propagating outside or inside the (empty) spaceship. However, as a crutch to our imagination, we find it convenient to suppose that the relevant portions of the light waves in the reference frame of the spaceship actually travel inside the spaceship.

The relativity of synchronization is a direct consequence of the invariance of the speed of light, since our procedure for testing simultaneity depends crucially on the speed of light. The breakdown of absolute simultaneity implies that time is relative. There exists no absolute time coordinate; instead there exist only relative time coordinates associated with particular reference frames.

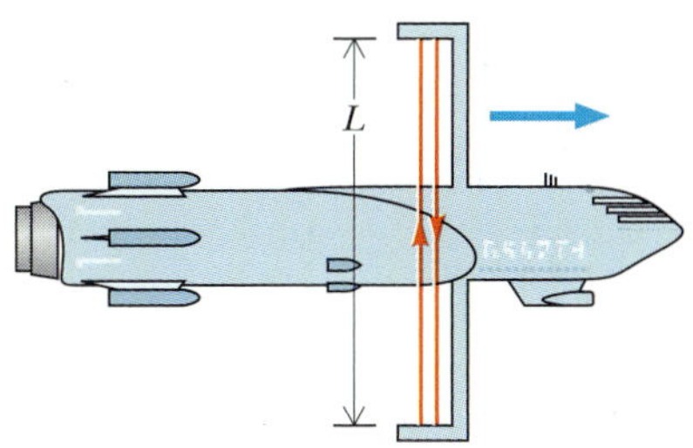

FIGURE 27.10
Spaceship with a "racetrack" for a light pulse.

27.3 TIME DILATION

The relativity of time shows up not only in the synchronization of clocks, but also in the rate of clocks. A clock on board the moving spaceship suffers ***time dilation*** relative to the clocks on Earth: ***the rate of the moving clock is slow compared with the rate of identically manufactured clocks at rest on Earth.*** To see how this comes about, imagine that the experimenters in the spaceship set up a racetrack of length L perpendicular to the direction of motion of the spaceship (see Fig. 27.10). If the experimenters use one of their clocks to measure the time of flight of a light signal that goes from one end of the track to the other and returns to its starting point, they will then find that the light signal takes a time of $\Delta t' = 2L/c$ to complete the roundtrip.

But experimenters on the Earth see that the light signal has concurrent vertical and horizontal motions (see Fig. 27.11). For the experimenters on Earth, the light signal travels a total distance *larger* than $2L$. Since the speed must still be the standard speed of light c, they will find that according to their clocks the light signal now takes a time Δt *longer* than $2L/c$ to complete the trip. Thus, a given time interval $\Delta t'$ registered by a clock on the spaceship is registered as a longer time interval Δt by the clocks on the Earth. This means that the clock on the spaceship runs slow when judged by the clocks on the Earth. Note that the experiment involves *one* clock on the spaceship (the clock at the starting point of the track), but several clocks on the Earth, because the light signal does not return to the point at which it started on Earth, and the observers on Earth will have to use one (stationary) clock at the starting point and another (stationary) clock at the end point to measure the time of flight.

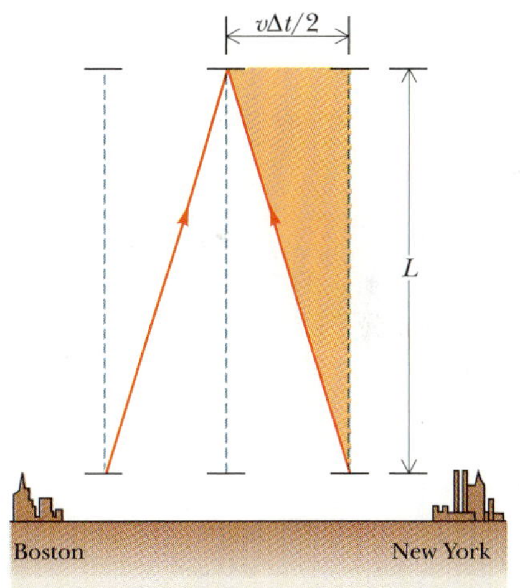

FIGURE 27.11
The trajectory of the light pulse as observed from the Earth.

For a quantitative evaluation of the time dilation, we note that in Fig. 27.11 the upward portion of the path of the light signal is the hypotenuse of a right triangle of sides L and $v\ \Delta t/2$, where v is the speed of the spaceship relative to the Earth. The total length of the path that the light signal has to cover in the reference frame of the Earth is therefore

$$2 \times \sqrt{L^2 + (v\ \Delta t/2)^2} \tag{1}$$

The time taken to cover this distance is

$$\Delta t = \frac{2 \times \sqrt{L^2 + (v\ \Delta t/2)^2}}{c} \tag{2}$$

If we square both sides of this equation, we obtain

$$(\Delta t)^2 = \frac{4L^2 + v^2(\Delta t)^2}{c^2} \tag{3}$$

which we can solve for $(\Delta t)^2$ and then for Δt:

$$(\Delta t)^2 = \frac{4L^2/c^2}{1 - v^2/c^2}$$

and

$$\Delta t = \frac{2L/c}{\sqrt{1 - v^2/c^2}} \tag{4}$$

Since in the reference frame of the spaceship $2L/c = \Delta t'$, this gives us

Time dilation ▶

$$\Delta t = \frac{\Delta t'}{\sqrt{1 - v^2/c^2}} \quad \text{for a clock at rest in spaceship} \tag{5}$$

This is the time-dilation formula. It shows that the time registered by the clocks on the Earth is longer than the time registered by the clock on the spaceship by a factor of $1/\sqrt{1 - v^2/c^2}$, that is, the clock on the spaceship runs slow when measured with the clocks on the Earth.

However, the time-dilation effect is symmetric: as measured by the clocks on the spaceship, a clock on the Earth runs slow by the same factor,

$$\Delta t' = \frac{\Delta t}{\sqrt{1 - v^2/c^2}} \quad \text{for a clock at rest on Earth} \tag{6}$$

The derivation of Eq. (6) can be based on an argument similar to that given above, with a racetrack for light at rest on the Earth. Incidentally: In these arguments we have implicitly assumed that the length of the racetrack is unaffected by the motion of the spaceship or the Earth, that is, we have assumed that the length is absolute. As we will see in the next section, this is true for lengths perpendicular to the direction of motion, although it is not true for lengths along the direction of motion.

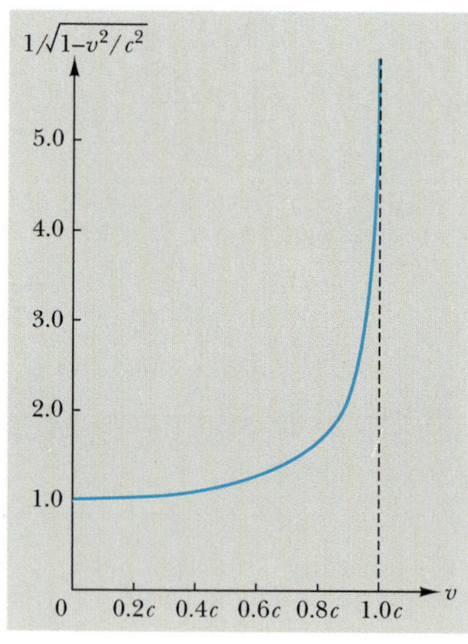

FIGURE 27.12
Time-dilation factor as a function of speed.

The slowing down of the rate of lapse of time applies to all physical processes — atomic, nuclear, biological, etc. At low speeds, the time-dilation effect is insignificant, but at speeds comparable to c, it becomes quite large. Figure 27.12 is a plot of the time-dilation factor $1/\sqrt{1 - v^2/c^2}$ for speeds in the range from 0 to c.

EXAMPLE 1. Very drastic time-dilation effects have been observed in the decay of short-lived elementary particles. For instance, a muon particle (see Chapter 32) usually decays in about 2.2×10^{-6} s; but if it is moving at high speed through the laboratory, then the internal processes that produce the decay will slow down and, as judged by the clocks in our laboratory, the muon lives a longer time. In accurate experiments performed at the Conseil Européen de Recherches Nucleaires (CERN) accelerator near Geneva, muons with a speed of 99.94% of the speed of light were found to have a lifetime 29 times as long as the lifetime of muons at rest. Is this dilation of the lifetime in agreement with Eq. (5)?

SOLUTION: We can regard the moun as a clock at rest in the reference frame of an (imaginary) spaceship with $v = 0.9994c$. This clock registers an elapsed time $\Delta t'$. According to Eq. (5) the time Δt registered by the clocks on the Earth is

$$\Delta t = \frac{\Delta t'}{\sqrt{1-(0.9994c)^2/c^2}} = \frac{\Delta t'}{\sqrt{1-(0.9994)^2}} = \frac{\Delta t'}{0.0346} = 28.9\ \Delta t'$$

This is in agreement with the experimental result.

Albert Einstein, *1879–1955, German (and Swiss, and American) theoretical physicist. Einstein was the most celebrated physicist of this century. He formulated the theory of Special Relativity in 1905 and the theory of General Relativity in 1916. Einstein also made incisive contributions to modern quantum theory, for which he received the Nobel Prize in 1921. Einstein spent the last years of his life in an unsuccessful quest for a unified theory of forces that was supposed to incorporate gravity and electricity in a single set of equations.*

At everyday speeds, the time-dilation effect is extremely small. For example, consider a clock aboard an airplane traveling at 300 m/s over the ground. The time-dilation factor is then

$$\frac{1}{\sqrt{1-v^2/c^2}} = \frac{1}{\sqrt{1-(300)^2/(3\times 10^8)^2}} \tag{7}$$

Evaluation of this gives 1.0000000000005, which means that a clock in the airplane will slow down by only 5 parts in 10^{13}! However, detection of such a small change is not beyond the reach of modern atomic clocks. In a recent experiment, scientists from the National Institute of Standards placed portable atomic clocks aboard a commercial airliner and kept them flying for several days, making a complete roundtrip around the world. Before and after the trip, the clocks were compared with an identical clock that was kept on the ground. The flying clocks were found to have lost time — in one instance, the total time lost because of the motion of the clock was about 10^{-7} s.

◂ Twin "paradox"

The time-dilation effect leads to the famous ***twin "paradox,"*** which we can state as follows: A pair of identical twins, Terra and Stella, celebrate their, say, twentieth birthday on Earth. Then Stella boards a spaceship that carries her at a speed of $v = 0.99c$ to Proxima Centauri, 4 light-years away; the spaceship immediately turns around and brings Stella back to Earth. According to the clocks on Earth, this trip takes about 4 years each way, so Terra's age will be 28 years when the twins meet again. But Stella has benefited from time dilation — relative to the reference frame of the Earth, the spaceship clocks run slow by a factor

$$\frac{1}{\sqrt{1-v^2/c^2}} = \frac{1}{\sqrt{1-(0.99)^2}} = \frac{1}{0.14} \tag{8}$$

(a)

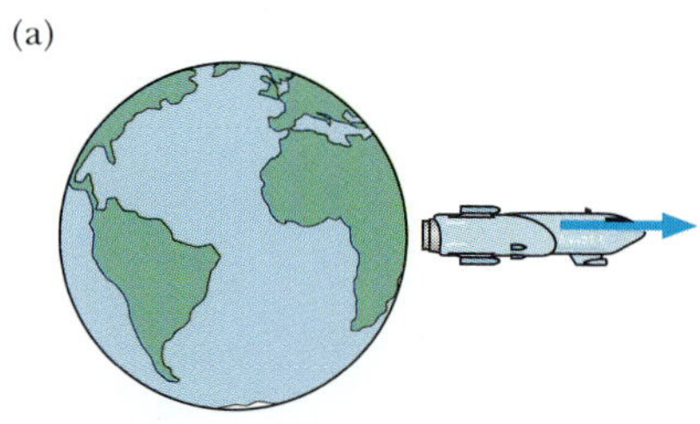
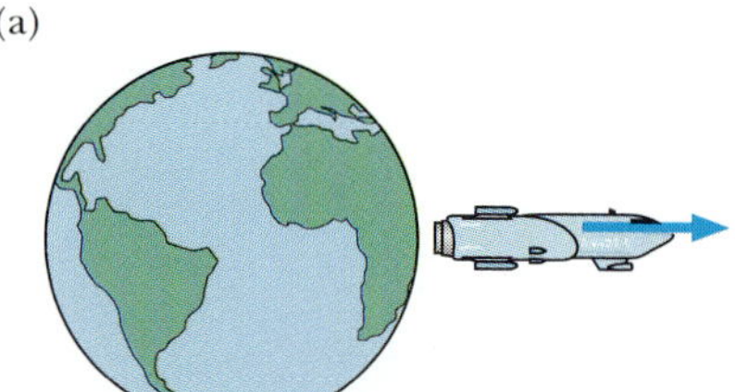

(b)

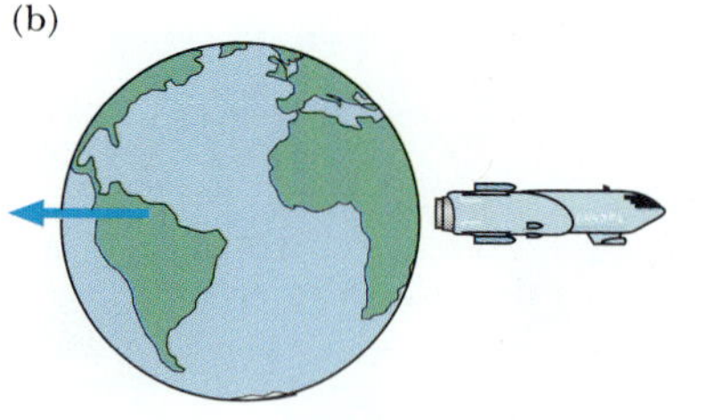

FIGURE 27.13

(a) In the reference frame of the Earth, the spaceship moves and its clocks experience time dilation. (b) In the reference frame of the spaceship, the Earth moves, and its clocks experience time dilation.

Hence 8 years of travel registered by the Earth clocks amount to only $8 \times 0.14 = 1.1$ years according to the spaceship clocks, and Stella's biological age on return will be only 21.1 years. Stella will be younger than Terra.

The paradox arises when we examine the elapsed times from the point of view of the reference frame of the spaceship. In this reference frame, the Earth is moving (see Fig. 27.13). Hence in this reference frame, the Earth clocks run slow — and Terra should be younger than Stella.

The resolution of this paradox hinges on the fact that our time-dilation formula is valid only if the time of a moving clock is measured from the point of view of an *inertial* reference frame. The reference frame of the Earth is (approximately) inertial, and therefore our calculation of the time dilation of the spaceship clocks is valid. But the reference frame of the spaceship is not inertial — the spaceship must decelerate when it reaches Proxima Centauri, stop, and then accelerate toward the Earth. Therefore, we cannot use the same time-dilation formula to find the time dilation of the Earth clocks from the point of view of the spaceship reference frame. The "paradox" results from the misuse of this formula.

A detailed analysis of the behavior of the Earth clocks from the point of view of the spaceship reference frame establishes that the Earth clocks run slow as long as the spaceship is moving with uniform velocity, but that the Earth clocks run *fast* when the spaceship is undergoing its acceleration at Proxima Centauri. The time that the Earth clocks gain during the accelerated portions of the trip more than compensates for the time they lose during the other portions of the trip. This confirms that Stella will be younger than Terra, even from the point of view of the spaceship reference frame.

27.4 Length Contraction

It turns out that because the synchronization of clocks is relative, length is also relative. A measuring rod, or any other body, on board the spaceship suffers ***length contraction*** along the direction of motion: ***the length of the moving measuring rod will be short when compared with the length of an identically manufactured measuring rod at rest on the Earth.*** The reason for this is that the length measurement of a moving body depends on simultaneity, and since simultaneity is relative, so is length.

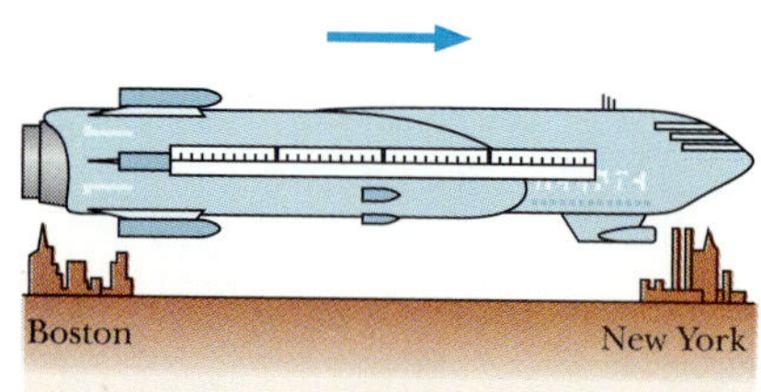

FIGURE 27.14

Spaceship with a measuring rod oriented along the direction of its motion.

Suppose that the spaceship, traveling from Boston toward New York, carries a measuring rod that has a length of, say, 300 km in the reference frame of the spaceship (see Fig. 27.14). To measure the length of this rod in the reference frame of the Earth, we station observers in the vicinity of New York and Boston with instructions to ascertain the positions of the front and the rear ends of the measuring rod at one instant of time, say, 6:00 P.M.

But when the observers on the Earth do this, the observers on the spaceship will claim that the position measurements were not done simultaneously, and that the observers in Boston measured the position of the rear end at a later time. In the extra time, the rear end moves an extra distance to the right, and hence the distance between the positions measured for the rear and the front ends will be reduced. From the point of view of the observers on the spaceship, it is therefore immediately obvious that the length measured by the observers on the Earth will be *short.* Figures 27.15 and 27.16 show the reference frame of the spaceship moving past the Earth and the reference frame of the Earth moving past the spaceship, respectively. In these figures the length contraction has been included (it was left out in Figs. 27.8 and 27.9). As illustrated in these figures, the length contraction effect is symmetric: a body at rest in the spaceship will suffer contraction when measured in the reference frame of the Earth, and a body at rest on the Earth will suffer contraction when measured in the reference frame of the spaceship.

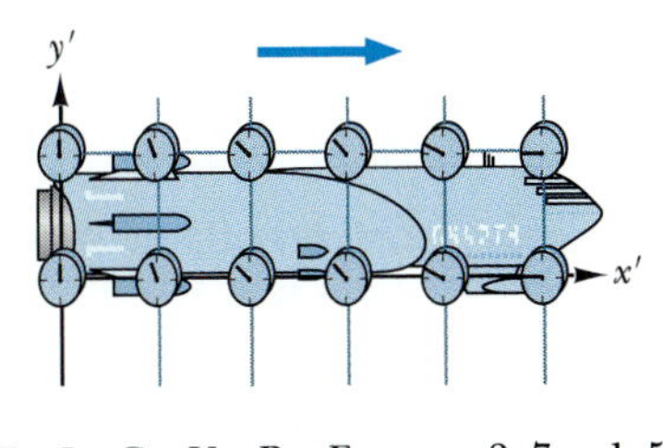

FIGURE 27.15 Reference frame of the spaceship as observed at one instant of Earth time (including length contraction).

We can obtain a formula for the length contraction by exploiting the formula for the time dilation. Consider a rod of length L at rest on the surface of the Earth. According to the Earth clocks, the nose of the spaceship takes a time $\Delta t = L/v$ to travel from one end of the rod to the other. Taking time dilation into account, we then conclude that the observers on the spaceship will judge that this takes a shorter time of only

$$\Delta t' = \sqrt{1 - v^2/c^2}\ \Delta t = \sqrt{1 - v^2/c^2}\ \frac{L}{v} \tag{9}$$

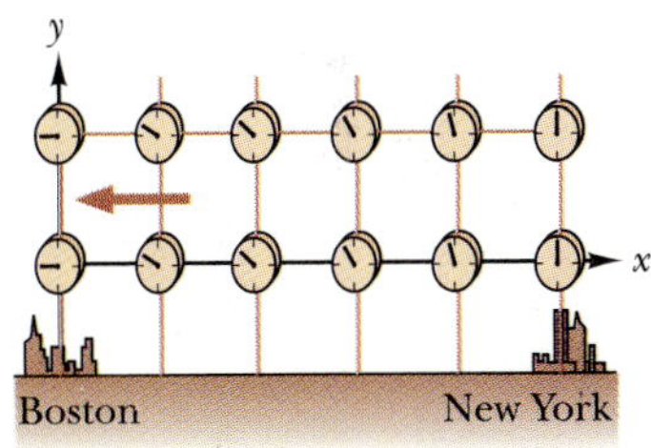

FIGURE 27.16 Reference frame of the Earth as observed at one instant of spaceship time (including length contraction).

However, the observers on the spaceship do not attribute this reduction of the travel time to a slowing of their clocks — in their own reference frame their clocks run at a normal rate. Instead, they attribute the reduction of travel time to a contraction of length of the rod. Measured in the reference frame of the spaceship, the rod on the Earth has some contracted length L', and it moves in the backward direction at speed v. Hence, the rod passes by the nose of the spaceship in a time L'/v. This time must agree with the time $\Delta t'$ calculated in Eq. (9); thus

$$\frac{L'}{v} = \sqrt{1 - v^2/c^2}\ \frac{L}{v} \tag{10}$$

or

$$L' = \sqrt{1 - v^2/c^2}\ L \qquad \text{for a rod at rest on Earth} \tag{11}$$

This is the formula for length contraction. According to this formula, the length of a body in motion relative to a reference frame is shortened by a factor of $\sqrt{1 - v^2/c^2}$. As already mentioned, this contraction effect is symmetric: if the rod is at rest in the spaceship and is measured in the reference frame of the Earth, the formula for the length contraction is

$$L = \sqrt{1 - v^2/c^2}\ L' \qquad \text{for a rod at rest in spaceship} \tag{12}$$

◀ **Length contraction**

The length contraction has not been tested directly by experiment. There is no practical method for a high-precision measurement of the

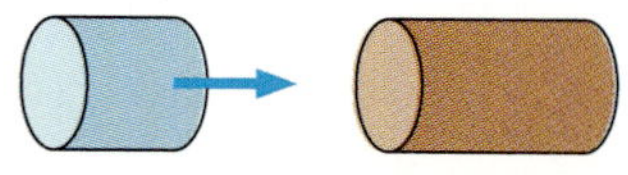

FIGURE 27.17
Two identical pieces of pipe. The piece on the right is at rest in the reference frame of the Earth; the piece on the left is at rest in the reference frame of the spaceship.

length of a fast-moving body. Our best bet might be high-speed photography, but this is nowhere near accurate enough, since the contraction is extremely small even at the highest speeds that we can impart to a macroscopic body. Note, however, that the experimental evidence for time dilation can be regarded as indirect evidence for length contraction, since, as we saw above, the former implies the latter.

The contraction effect applies only to lengths along the direction of motion of the body. ***Lengths perpendicular to the direction of motion are unaffected.*** The proof of this is by contradiction: Imagine that we have two identically manufactured pieces of pipe, one at rest on the Earth, one at rest on the spaceship (see Fig. 27.17). If the motion of the spaceship relative to Earth were to bring about a transverse contraction of the spaceship pipe, then by symmetry, the motion of the Earth relative to the spaceship would bring about a contraction of the Earth pipe. These contraction effects are contradictory, since in one case the spaceship pipe would fit inside the Earth pipe, and in the other case it would fit outside.

EXAMPLE 2. A proton is passing by the Earth at a speed of $0.50c$. In the rest frame of the proton, what is the length of the diameter of the Earth in a direction parallel to that of the motion of the proton? By how much is this shorter than the diameter in the rest frame of the Earth?

SOLUTION: The length contraction factor is

$$\sqrt{1 - v^2/c^2} = \sqrt{1 - (0.50c)^2/c^2} = \sqrt{1 - 0.25} = 0.87$$

The diameter of the Earth in its own rest frame is 1.3×10^7 m. Hence the length in the rest frame of the proton is

$$0.87 \times 1.3 \times 10^7 \text{ m} = 1.1 \times 10^7 \text{ m}$$

This is shorter than 1.3×10^7 m by 0.2×10^7 m.

COMMENTS AND SUGGESTIONS: The dimensions of the Earth perpendicular to the direction of motion do not contract. This implies that, in the reference frame of the proton, the Earth is not a sphere, but a somewhat flattened (pancaked) ellipsoid.

From the length contraction we can deduce the volume contraction. The volume of the Earth, which is calculated by taking a product of the dimension parallel to the motion and the two dimensions perpendicular to the motion, will be contracted by just one factor of $\sqrt{1 - v^2/c^2}$, that is, a factor of 0.87.

27.5* THE LORENTZ TRANSFORMATIONS AND THE COMBINATION OF VELOCITIES

Suppose that we measure the space and time coordinates of an event — such as the impact of lightning on a point on the ground — in the reference frame of the Earth and also in the reference frame of the spaceship. We will then obtain different values of these coordinates in the Earth and in the spaceship reference frames, but these different values of the coordinates are related by transformation formulas. In Einstein's physics, the transformation formulas for the coordinates are fairly complicated, because they are designed so as to keep the speed of light the same in all reference frames, and they incorporate the length contraction and the time dilation. Before dealing with these complicated formulas, let us examine the much simpler transformation formulas for coordinates in Newton's physics, where there is no length contraction and no time dilation.

Figure 27.18 shows the coordinate grids x–y and x'–y' of the first (Earth) and the second (spaceship) reference frames. The second reference frame is moving with velocity V_O along the x axis of the first reference frame. We assume that at time $t = 0$, the origins of the two coordinate grids coincide; at time t, the distance between the origins is then $V_O t$. The coordinates of the event are x, y in the first reference frame and x', y' in the second reference frame. By inspection of Fig. 27.18, we see that the distance x equals the sum of the distances x' and $V_O t$,

$$x = x' + V_O t \tag{13}$$

Hence

$$x' = x - V_O t \tag{14}$$

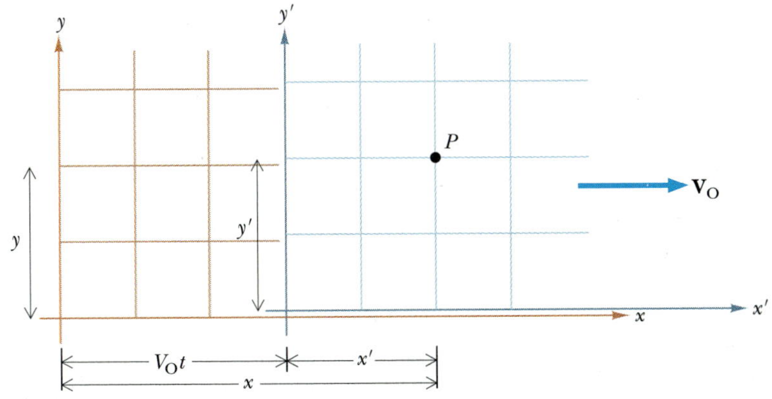

FIGURE 27.18
The coordinate grid x–y belonging to the reference frame of the Earth and the coordinate grid x'–y' belonging to the reference frame of the spaceship. The x'–y' grid moves with speed V_O along the x axis of the x–y grid.

* This section is optional.

Furthermore, the distance y equals the distance y',

$$y' = y \tag{15}$$

Equations (14) and (15) are the transformation equations for the x and y coordinates in Newton's physics. These two equations are merely the x and y components of the general vector equation $\mathbf{r}' = \mathbf{r} - \mathbf{V}_O t$ for the transformation of the position vector we found in Chapter 2 [see Eq. (2.65)]. We could have obtained our equations for the transformation of the x and y coordinates from the general vector equation; but it is just as easy to rederive these results by inspection of Fig. 27.18.

Note that although Fig. 27.18 makes the equations for the transformation of the x and y coordinates seem self-evident, these equations hinge on the absolute character of length in Newton's physics. Absolute length means that the observers in the two reference frames agree on the measurement of any length or any distance between two points. If the observers disagreed on the values of the distances x or x' — for example, if one observer claimed that the distance x was 3 m and the other observer claimed that this distance x was contracted to 2.5 m — then Eq. (13) would not be valid. The left side of Eq. (13) is a distance defined at one instant in the first reference frame, whereas the right side is a sum of a distance (x') defined at one instant in the second reference frame and a distance ($V_O t$) defined at one instant in the first reference frame, and such a sum makes no sense unless the observers agree on the values of these distances.

Furthermore, in Newton's physics time is absolute. This means that the times registered by the clocks in the two reference frames are always equal,

$$t' = t \tag{16}$$

Galilean transformation equations ▸

Taken together, Eqs. (14)–(16) are called the ***Galilean transformation equations;*** they relate the space and time coordinates in one reference frame to those in the other.

From these equations we can deduce the Galilean addition rule for the components of the velocity. For instance, if the x coordinate changes by Δx in a time Δt, then Eqs. (14) and (16) give us

$$\Delta x' = \Delta x - V_O \Delta t \tag{17}$$

$$\Delta t' = \Delta t \tag{18}$$

Dividing these two equations side by side, we obtain

$$\frac{\Delta x'}{\Delta t'} = \frac{\Delta x}{\Delta t} - V_O \tag{19}$$

Here, $\Delta x/\Delta t$ is the x velocity of the particle, light signal, or whatever, measured in the first reference frame, and $\Delta x'/\Delta t'$ is the x velocity measured in the second reference frame. Hence Eq. (19) says

Galilean addition rule for velocities ▸

$$v'_x = v_x - V_O \tag{20}$$

This, of course, is simply the Galilean addition rule for the x components of the velocities.

In Einstein's physics, the Galilean formulas for the transformation of the coordinates and for the addition of velocities must be replaced by more complicated formulas, designed in such a way as to keep the speed of light the same in all reference frames. The transformation equations that accomplish this trick are called the ***Lorentz transformations.*** If the new reference frame moves, again, with velocity V_O along the x axis of the first reference frame, and the origins coincide at time $t = 0$, the Lorentz transformations take the form

$$x' = \frac{x - V_O t}{\sqrt{1 - V_O^2/c^2}} \tag{21}$$

◀ **Lorentz transformation equations**

$$y' = y \tag{22}$$

$$t' = \frac{t - V_O x/c^2}{\sqrt{1 - V_O^2/c^2}} \tag{23}$$

These equations cannot be obtained by simple inspection of Fig. 27.18, because the distances displayed in this figure are not absolute in Einstein's physics, and they cannot be simply added as in Newton's physics. We will skip the derivation of the Lorentz transformation equations, but we can gain some understanding of the main features of these equations by comparing them with the Galilean transformation equations. Equation (21) differs from the Galilean equation only by the factor $1/\sqrt{1 - V_O^2/c^2}$; this factor represents the length contraction. Equation (22) is identical to the Galilean equation, because lengths transverse to the direction of motion remain unchanged. And Eq. (23) differs from the Galilean equation in two ways: it contains an extra factor $1/\sqrt{1 - V_O^2/c^2}$ representing the time dilation, and it contains an extra term $-V_O x/c^2$ representing the relativity of synchronization discussed in Section 27.2: from the point of view of the new reference frame, the clocks at the leading edge of the first reference frame are late — a clock at position x is late by an amount $V_O x/c^2$.

Hendrik Antoon Lorentz, *1853–1928, Dutch theoretical physicist. He investigated the relationship between electricity, magnetism, and mechanics. In order to explain the observed effect of magnetic fields on emitters of light, he postulated the existence of electric charges in the atom, for which he was awarded the Nobel Prize in 1902. He derived the Lorentz transformation equations, but he failed to recognize that these equations hinge on a new concept of space and time.*

Note that if the relative velocity between the two reference frames is small compared with the speed of light, then V_O/c in Eqs. (21) and (23) is small, and any term involving this quantity can be omitted in the equations. The Lorentz transformations then reduce to

$$x' \simeq x - V_O t \tag{24}$$

$$y' \simeq y \tag{25}$$

$$t' \simeq t \tag{26}$$

Thus, for low speeds, the Lorentz transformations reduce to the Galilean transformations.

But the crucial feature of the Lorentz transformations is that they leave the velocity of light unchanged. To verify this, we need to find the relativistic combination rule for velocity. If the x coordinate changes by Δx in a time Δt, then the Lorentz transformation equations tell us that

$$\Delta x' = \frac{\Delta x - V_O \Delta t}{\sqrt{1 - V_O^2/c^2}} \tag{27}$$

$$\Delta t' = \frac{\Delta t - V_O \Delta x/c^2}{\sqrt{1 - V_O^2/c^2}} \tag{28}$$

and dividing these two equations side by side, we find

$$\frac{\Delta x'}{\Delta t'} = \frac{\Delta x - V_O \Delta t}{\Delta t - V_O \Delta x/c^2} \tag{29}$$

On the right side we can divide both the numerator and the denominator by Δt, with the result

$$\frac{\Delta x'}{\Delta t'} = \frac{\Delta x/\Delta t - V_O}{1 - V_O(\Delta x/\Delta t)/c^2} \tag{30}$$

In this expression, $\Delta x/\Delta t$ is the x velocity of the particle or light signal measured in the first reference frame and $\Delta x'/\Delta t'$ is the x velocity measured in the second reference frame. Hence Eq. (30) may be written

Relativistic combination rule for velocities ▶

$$v'_x = \frac{v_x - V_O}{1 - v_x V_O/c^2} \tag{31}$$

This is the relativistic combination rule for the x components of the velocities (there are corresponding formulas for the combination of the other components of the velocities).

The relativistic combination rule for velocities is to be compared with the Galilean addition rule:

$$v'_x = v_x - V_O \tag{32}$$

It is the denominator in Eq. (31) that makes all the difference. For instance, suppose that v_x is the velocity of a light signal propagating along the x axis of the first reference frame. Then $v_x = c$, and Eq. (31) yields

$$v'_x = \frac{c - V_O}{1 - cV_O/c^2} = \frac{c(1 - V_O/c)}{1 - V_O/c} = c \tag{33}$$

Thus, as required by the Principle of Relativity, the velocity of the light signal in the second reference frame has exactly the same magnitude as in the first reference frame (see Fig. 27.19).

The relativistic combination rule for light velocities has been explicitly tested in an experiment at CERN involving a beam of very fast pions.

These particles decay spontaneously by a reaction that emits a flash of very intense, very energetic light (gamma rays). Hence, such a beam of pions can be regarded as a high-speed light source. In the experiment, the velocity of the pions relative to the laboratory was $V_O = 0.99975c$. The Galilean addition law for velocity would have predicted laboratory velocities of $1.99975c$ for light emitted in the forward direction and $0.00025c$ for light emitted in the backward direction. But the experiments confirmed the relativistic combination rule — the laboratory velocity of the light had the same magnitude in all directions.

(a)

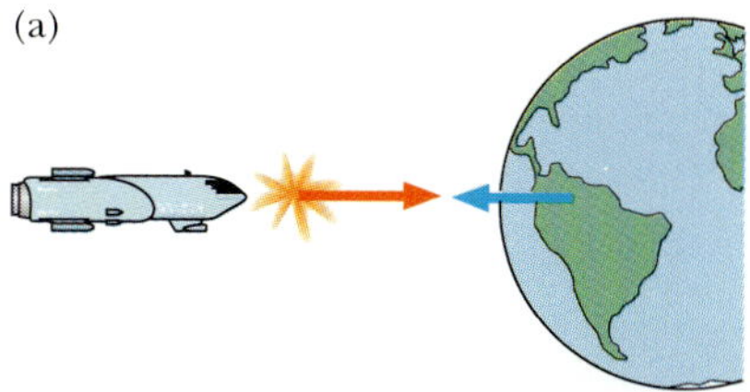

(b)

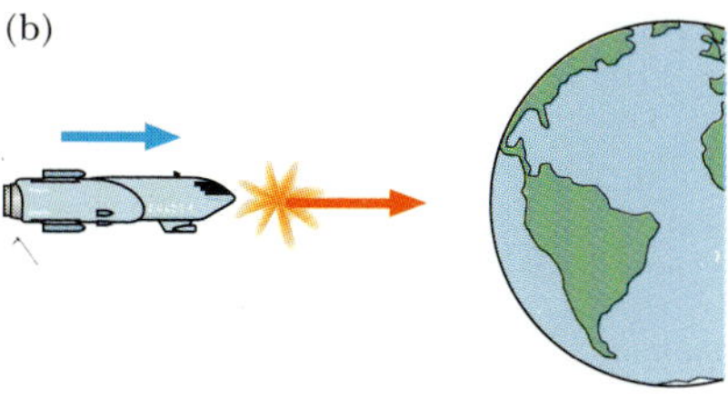

FIGURE 27.19

Addition of velocities according to the relativistic combination rule. (a) The light signal has a speed of 3.00×10^8 m/s in the reference frame of a spaceship speeding toward the Earth. (b) The light signal then also has this speed of 3.00×10^8 m/s in the reference frame of the Earth.

EXAMPLE 3. An alien spaceship approaching the Earth at a speed of $0.40c$ fires a rocket at the Earth. If the velocity of the rocket is $0.80c$ in the reference frame of the spaceship, what is its velocity in the reference frame of the Earth?

SOLUTION: For this calculation, it is convenient to regard the reference frame of the spaceship as the old reference frame, and the reference frame of the Earth as the new reference frame. The x axis is directed from the spaceship toward the Earth, and the velocity of the rocket in the reference frame of the spaceship is $v_x = 0.80c$. The velocity of the Earth relative to the spaceship is negative, $V_O = -0.40c$. Then Eq. (31) gives

$$v'_x = \frac{v_x - V_O}{1 - v_x V_O/c^2} = \frac{0.80c + 0.40c}{1 + (0.80c)(0.40c)/c^2}$$

$$= 0.91c$$

27.6 RELATIVISTIC MOMENTUM AND ENERGY

The drastic revision that the theory of Special Relativity imposes on Newton's concepts of space and time implies a corresponding revision of the concepts of momentum and energy. The formulas for momentum and energy and the equations expressing their conservation are intimately tied to the transformation equations of the space and time coordinates. To see that this is so, we briefly examine the Newtonian (nonrelativistic) case.

In Newton's physics, the momentum of a particle of mass m and velocity $\mathbf{v}$ is

$$\mathbf{p} = m\mathbf{v} \tag{34}$$

Since the Galilean transformation for velocity is

$$\mathbf{v}' = \mathbf{v} - \mathbf{V}_O \tag{35}$$

the transformation equation for the momentum is

$$\mathbf{p}' = m\mathbf{v}' = m\mathbf{v} - m\mathbf{V}_O = \mathbf{p} - m\mathbf{V}_O \tag{36}$$

From this we see that the momentum $\mathbf{p}'$ in the new reference frame differs from the momentum $\mathbf{p}$ in the old reference frame by only a constant quantity (a quantity independent of the velocity $\mathbf{v}$ of the particle). Hence, if the total momentum of a system of colliding particles is conserved in one reference frame, it will also be conserved in the other reference frame — and the law of conservation of momentum obeys the Principle of Relativity. This shows that the nonrelativistic formula for momentum and the nonrelativistic Galilean formula for the addition of velocities match in just the right way.

According to the relativistic physics of Einstein, we must replace the Galilean addition formula for velocities by the relativistic combination rule. If the law of conservation of momentum is to obey the Principle of Relativity, we must then design a new relativistic formula for momentum that matches the new relativistic combination rule for velocities. It turns out that the correct relativistic formula for momentum is

Relativistic ▶ momentum

$$\mathbf{p} = \frac{m\mathbf{v}}{\sqrt{1 - v^2/c^2}} \tag{37}$$

We will not give a proof of this formula.

If the speed of the particle is small compared with the speed of light, then $\sqrt{1 - v^2/c^2} \simeq 1$ and Eq. (37) becomes approximately

$$\mathbf{p} \simeq m\mathbf{v} \tag{38}$$

This shows that for low speeds, the relativistic and the Newtonian formulas for the momentum agree. We therefore see that although the Newtonian formula for the momentum is strictly false, we can regard it as a simple and useful approximation for low speeds; this approximation is quite adequate for the description of all the phenomena we encounter in everyday life and (almost) all the phenomena we encounter in the realm of engineering, such as the phenomena we dealt with in the earlier chapters of this book. But at high speeds, the formulas differ drastically. We must then abandon the Newtonian formula, and rely entirely on the relativistic formula. Note that the relativistic momentum becomes infinite as the speed of the particle approaches the speed of light. Figure 27.20 is a plot of the magnitude of the momentum vs. the speed.

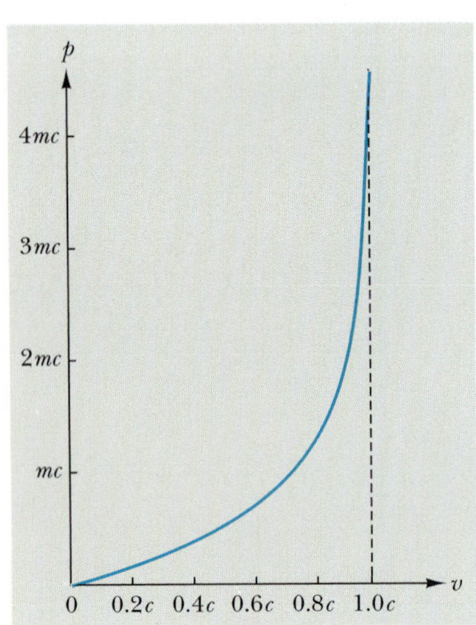

FIGURE 27.20
Momentum of a particle as a function of speed.

EXAMPLE 4. An electron in the beam of a TV tube has a speed of 1.0×10^8 m/s. What is the magnitude of the momentum of this electron?

SOLUTION: For this electron, $v/c = (1.0 \times 10^8 \text{ m/s})/(3.0 \times 10^8 \text{ m/s}) = 0.33$. According to Eq. (37), the magnitude of the momentum is then

$$p = \frac{mv}{\sqrt{1 - v^2/c^2}} = \frac{9.1 \times 10^{-31} \text{ kg} \times 1.0 \times 10^8 \text{ m/s}}{\sqrt{1 - (0.33)^2}}$$

$$= 9.7 \times 10^{-23} \text{ kg} \cdot \text{m/s}$$

COMMENTS AND SUGGESTIONS: Note that if we had calculated the momentum according to the nonrelativistic formula $p = mv$, we would have obtained 9.1×10^{-23} kg·m/s, and we would have been in error by about 6%.

We also need a new formula for kinetic energy. It turns out that this new relativistic formula for kinetic energy is

$$K = \frac{mc^2}{\sqrt{1 - v^2/c^2}} - mc^2 \tag{39}$$

◀ **Relativistic kinetic energy**

For low speeds, this relativistic formula for kinetic energy can be shown to agree approximately with the nonrelativistic formula $K = \frac{1}{2}mv^2$.

Figure 27.21 is a plot of the kinetic energy vs. the speed. The relativistic kinetic energy becomes infinite as the speed of the particle approaches the speed of light. This indicates that, for any particle (and for any body), ***the speed of light is unattainable, since it is impossible to supply a particle with an infinite amount of energy.***

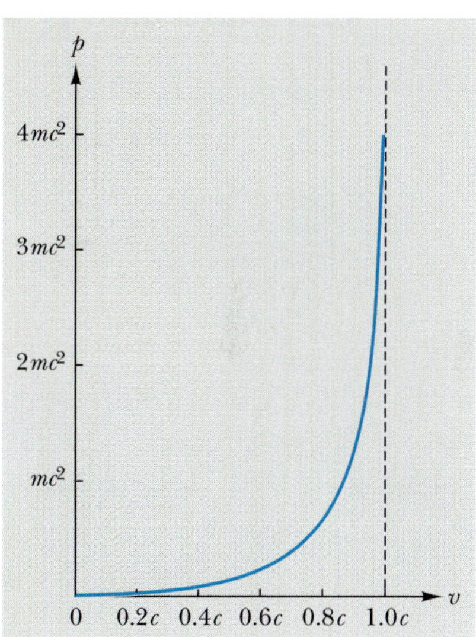

FIGURE 27.21

Kinetic energy of a particle as a function of speed.

EXAMPLE 5. The maximum speed that electrons achieve in the Stanford Linear Accelerator (SLAC) is $0.99999999967c$. What is the kinetic energy of an electron moving at this speed?

SOLUTION: The relativistic formula (39) contains a factor $\sqrt{1 - v^2/c^2}$. For v/c near 1, it is convenient to write

$$\sqrt{1 - v^2/c^2} = \sqrt{1 + v/c}\,\sqrt{1 - v/c} \simeq \sqrt{2}\,\sqrt{1 - v/c} \tag{40}$$

In our example, the value of $1 - v/c$ is $1 - 0.99999999967 = 3.3 \times 10^{-10}$. Hence

$$K = mc^2 \left(\frac{1}{\sqrt{1 - v^2/c^2}} - 1 \right)$$

$$= 9.1 \times 10^{-31} \text{ kg} \times (3.0 \times 10^8 \text{ m/s})^2 \times \left(\frac{1}{\sqrt{2}\,\sqrt{3.3 \times 10^{-10}}} - 1 \right)$$

$$= 3.2 \times 10^{-9} \text{ J}$$

Although the theory of Special Relativity requires a revision of the basic equations of mechanics, it does not require any revision of the basic equations of electricity and magnetism. Maxwell's equations are already relativistic, that is, they match the relativistic behavior of length and of time in just the right way. This concordance between Maxwell's equations and the requirements of relativity is no accident. When Einstein designed the theory of Special Relativity, he deliberately arranged it in such a way that Maxwell's equations would be preserved. In fact, Einstein's search for a theory of relativity was motivated by his faith in Maxwell's equations and his recognition that if Maxwell's equations were right then the Galilean coordinate transformations had to be wrong.

27.7 MASS AND ENERGY

One of the great discoveries that emerged from relativity is that energy can be transformed into mass, and mass can be transformed into energy. Thus, ***mass is a form of energy.*** The amount of energy contained in an amount m of mass at rest is given by Einstein's famous formula

Rest-mass energy ▶

$$E = mc^2 \tag{41}$$

This energy is called the ***rest-mass energy.**** The formula (41) can be derived from the theory of relativity, but, as with some other equations in this chapter, we will not give the derivation.

The most spectacular demonstration of Einstein's mass–energy formula is found in the annihilation of matter and antimatter (see Chapter 32). If a proton collides with an antiproton, or an electron with an antielectron, the two colliding particles react violently and they annihilate each other in an explosion that generates an intense flash of very energetic light. According to Eq. (41), the annihilation of just 1000 kg of matter and antimatter (500 kg of each) would release an amount of energy

$$E = mc^2 = 1000 \text{ kg} \times (3.00 \times 10^8 \text{ m/s})^2 = 9.0 \times 10^{19} \text{ J} \tag{42}$$

This is enough energy to satisfy the needs of the United States for a full year. Unfortunately, antimatter is not readily available in large amounts. On Earth, antiparticles can be obtained only from reactions induced by the impact of beams of high-energy particles on a target. These collisions occasionally result in the creation of a particle–antiparticle pair. Such pair creation is the reverse of pair annihilation. The creation process

* Throughout this section, *mass* means the mass that a body has when at rest or nearly at rest; to emphasize this, we use the term *rest mass.* The definition and the measurement of mass for a body in motion at high (relativistic) speeds are rather tricky, because Newton's equation $m\mathbf{a} = \mathbf{F}$ fails, and the direction of the acceleration is not necessarily the direction of the force. The only kind of mass that is unambiguously defined in Einstein's physics is the mass that the body has when at rest, and this is the only kind of mass we will consider.

transforms some of the kinetic energy of the collision into mass, and a subsequent annihilation merely gives back the original energy.

But the relationship between energy and mass in Eq. (41) also has another aspect. ***Energy has mass.*** Whenever the internal energy stored in a body is changed, its rest mass (and weight) is changed. The change in rest mass that accompanies a given change of energy is

$$\Delta m = \Delta E/c^2 \tag{43}$$

For instance, in the fission of uranium, the nuclear material loses energy, and correspondingly its mass (and weight) decreases. The complete fission of 1 kg of uranium releases an energy of 8.2×10^{13} J, and correspondingly the mass of the nuclear material decreases by $\Delta m = (8.2 \times 10^{13}\ \text{J})/c^2 = 0.0009$ kg, or about 0.1%.

The fact that energy has mass indicates that energy is a form of mass. Conversely, as we have seen above, mass is a form of energy. Hence mass and energy must be regarded as essentially the same thing. The laws of conservation of mass and conservation of energy are therefore not two independent laws — each implies the other. For example, consider the fission reaction of uranium inside the reactor vessel of a nuclear power plant (for details, see Chapter 31). The reaction conserves energy — it merely transforms nuclear energy into heat, light, and kinetic energy, but does not change the total amount of energy. The reaction also conserves mass — if the reactor vessel is hermetically sealed and thermally insulated from its environment, then the reaction does not change the mass of the contents of the vessel. However, if we open the vessel during or after the reaction, and let some of the heat and light escape, then the mass of the residues will not match the mass of the original amount of uranium. The mass of the residues will be about 0.1% smaller than the original mass of the uranium. This mass defect represents the mass carried away by the energy that escapes. Thus, the nuclear fission reaction merely transforms energy into new forms of energy and mass into new forms of mass. In this regard, a nuclear reaction is not fundamentally different from a chemical reaction. The mass of the residues in an exothermic chemical reaction is slightly less than the original mass. The heat released in such a chemical reaction carries away some mass, but, in contrast to a nuclear reaction, this amount of mass is so small as to be quite immeasurable.

The total energy of a free particle in motion is the sum of its rest-mass energy and its kinetic energy:

$$E = mc^2 + K = mc^2 + \frac{mc^2}{\sqrt{1 - v^2/c^2}} - mc^2 \tag{44}$$

This leads to a simple formula for the total relativistic energy of the particle:

$$E = \frac{mc^2}{\sqrt{1 - v^2/c^2}} \tag{45}$$

◂ **Relativistic total energy**

It is easy to verify (see Problem 50) that the relativistic energy can be expressed as follows in terms of the relativistic momentum:

$$E = \sqrt{c^2p^2 + m^2c^4} \tag{46}$$

For an ultrarelativistic particle, moving at a speed close to that of light, the first term (c^2p^2) within the square root is much larger than the second term (m^2c^4). Hence, for such a particle we can ignore the second term, and we then obtain the simple result

$$E \simeq \sqrt{c^2p^2}$$

or

$$E \simeq cp \tag{47}$$

Thus, the momentum and the energy of an ultrarelativistic particle are directly proportional.

EXAMPLE 6. Consider an electron of speed $0.99999999967c$, as in Example 5. What is the momentum of such an electron?

SOLUTION: Such an electron is ultrarelativistic. Its kinetic energy is much larger than its rest-mass energy, and the total energy is therefore approximately equal to the kinetic energy, which we have already calculated in Example 5:

$$E = mc^2 + K \simeq K = 3.2 \times 10^{-9}\ \text{J}$$

Hence Eq. (47) yields

$$P \simeq \frac{E}{c} = \frac{3.2 \times 10^{-9}\ \text{J}}{3.0 \times 10^8\ \text{m/s}} = 1.1 \times 10^{-18}\ \text{kg}\cdot\text{m/s}$$

SUMMARY

- **Principle of Relativity:** All the laws of physics are the same in all reference frames.
- **Time dilation:** $\Delta t = \dfrac{\Delta t'}{\sqrt{1 - v^2/c^2}}$ (clock at rest in spaceship)
- **Length contraction:** $L = \sqrt{1 - v^2/c^2}\, L'$ (rod at rest in spaceship)

- **Lorentz transformations:** $x' = \dfrac{x - V_O t}{\sqrt{1 - V_O^2/c^2}}$

$$y' = y$$

$$t' = \frac{t - V_O x/c^2}{\sqrt{1 - V_O^2/c^2}}$$

- **Combination of velocities:** $v'_x = \dfrac{v_x - V_O}{1 - v_x V_O/c^2}$

- **Relativistic momentum:** $\mathbf{p} = \dfrac{m\mathbf{v}}{\sqrt{1 - v^2/c^2}}$

- **Relativistic kinetic energy and total energy:**

$$K = \frac{mc^2}{\sqrt{1 - v^2/c^2}} - mc^2$$

$$E = \frac{mc^2}{\sqrt{1 - v^2/c^2}}$$

- **Rest-mass energy:** $E = mc^2$

QUESTIONS

1.–An astronaut is inside a closed space capsule coasting through interstellar space. Is there any way that the astronaut can measure the speed of the capsule without looking outside?

2.–Why did Michelson and Morley use *two* light beams traveling along perpendicular paths, rather than a single light beam traveling back and forth along one path, in their experiment?

3.–When Einstein was a boy he wondered about the following question: A runner holds a mirror at arm's length in front of his face. Can he see himself in the mirror if he runs at (almost) the speed of light? Answer this question both according to the ether theory and according to the theory of Special Relativity.

4.–Consider the piece of paper on which one page of this book is printed. Which of the following properties of the piece of paper are absolute, that is, which are independent of whether the paper is at rest or in motion relative to you? (a) The thickness of the paper, (b) the mass of the paper, (c) the volume of the paper, (d) the number of atoms in the paper, (e) the chemical composition of the paper, (f) the speed of light reflected by the paper, and (g) the color of the colored print on the paper.

5.–Two streetlamps, one in Boston and the other in New York City, are turned on at exactly 6:00 P.M. Eastern Standard Time. Find a reference frame in which the streetlamp in New York was turned on late.

6.–According to the theory of Special Relativity, the time order of events can be reversed under certain conditions. Does this mean that a sparrow might fall from the sky before it leaves the nest?

7.–Because of the rotational motion of the Earth about its axis, a point on the equator moves with a speed of 460 m/s relative to a point on the North Pole. Does this mean that a clock placed on the equator runs more slowly than a similar clock placed on the pole?

8.–According to Jacob Bronowski, author of *The Ascent of Man,* the explanation of time dilation is as follows: If you are moving away from a clock tower at a speed nearly equal to the speed of light, you keep pace with the light that the face of the clock sent out at, say, 11 o'clock. Hence if you look toward the clock tower, you always see its hands at the 11 o'clock position. Is this explanation correct? What is wrong with it?

9.–Suppose you wanted to travel into the future and see what the twenty-fifth century is like. In principle, how could you do this? Could you ever return to the twentieth century?

10.–According to the qualitative arguments of Section 27.3, a light signal traveling along a track placed perpendicular to the direction of motion of the spaceship (see Fig. 27.10) takes a longer time to complete a roundtrip when measured by the clocks on the Earth than when measured by the clocks on the spaceship. Would the same be true for a light signal traveling along a track placed parallel to the direction of motion? Explain qualitatively.

11.–A cannonball is perfectly round in its own reference frame. Describe the shape of this cannonball in a reference frame relative to which it has a speed of $0.95c$. Is the volume of the cannonball the same in both reference frames?

12.–Could we use the argument based on the two identical pieces of pipe (see Fig. 27.17) to prove that lengths *along* the direction of motion are not affected? Why not?

13.–A rod at rest on the ground makes an angle of $30°$ with the x axis in the reference frame of the Earth. Will the angle be larger or smaller in the reference frame of a spaceship moving along the x axis?

14.–In the charming tale "City Speed Limit" by George Gamow,* the protagonist, Mr. Tompkins, finds himself riding a bicycle in a city where the speed of light is very low, roughly 30 km/h. What weird effects must Mr. Tompkins have noticed under these circumstances?

15.–Suppose that a *very* fast runner holding a long horizontal pole runs through a barn open at both ends. The length of the pole (in its rest frame) is 6 m and the length of the barn (in *its* rest frame) is 5 m. In the reference frame of the barn, the pole will suffer length contraction and, at one instant of time, all of the pole will be inside the barn. However, in the reference frame of the runner, the barn will suffer length contraction and all of the pole will never be inside the barn at one instant of time. Is this a contradiction?

16.–Why can a spaceship not travel as fast as or faster than the speed of light?

17.–If the beam from a revolving searchlight is intercepted by a distant cloud, the bright spot will move across the surface of the cloud very quickly, with a speed that can easily exceed the speed of light. Does this conflict with our conclusion of Section 27.6, that nothing can move faster than the speed of light?

---

* George Gamow, *Mr. Tompkins in Wonderland.*

PROBLEMS

Section 27.3: Time Dilation

1.——If a moving clock is to have a time-dilation factor of 10, what must be its speed?

2.——Neutrons have an average lifetime of 15 minutes when at rest in the laboratory. What is the average lifetime of neutrons of a speed of 25% of the speed of light? 50%? 90%?

3.——Consider an unstable particle, such as a pion, which has a lifetime of only 2.6×10^{-8} s when at rest in the laboratory. What speed must you give such a particle to make its lifetime twice as long as when at rest in the laboratory?

4.——The speed of the Sun around the center of our galaxy is 200 km/s. Clocks in the Solar System will therefore run slow as compared with clocks at rest in the galaxy. By what factor are the Solar System clocks slow?

5.——A spaceship equipped with a chronometer is sent on a roundtrip to Proxima Centauri, 4 light-years away. The spaceship travels at $0.1c$, and returns immediately.

(a) According to clocks on the Earth, how long does this trip take?

(b) According to the chronometer on the spaceship, how long does this trip take?

6.*——Muons are unstable particles which — if at rest in a laboratory — decay after a time of only 2.2×10^{-6} s. Suppose that a muon is created in a collision between a cosmic ray and an oxygen nucleus at the top of the Earth's atmosphere, at an altitude of 20 km above sea level.

(a) If the muon has a downward speed of $v = 0.990c$ relative to the Earth, at what altitude will it decay? Ignore gravity in this calculation.

(b) Without time dilation, at what altitude would the muon have decayed?

7.**——To find the Doppler shift for a light source moving at a relativistic speed we must take into account the reduction of frequency produced by time dilation. By combining the familiar Doppler-shift formula for a moving source [Eq. (11.18)] with a time-dilation factor, show that the Doppler-shift formula for a receding light source becomes

$$f' = f\sqrt{\frac{1 - v/c}{1 + v/c}}$$

What is the Doppler-shift factor for $v/c = 0.5$, 0.6, 0.7, 0.8, and 0.9?

8.**——In a test of the relativistic time-dilation effect, physicists compared the rates of vibration of nuclei of iron moving at different speeds. One sample of iron nuclei was placed on the rim of a high-speed rotor; another sample of similar nuclei was placed at the center. The radius of the rotor was 10 cm and it rotated at 35,000 rev/min. Under these conditions, what was the speed of the rim of the rotor relative to the center? According to Eq. (5), what was the time-dilation factor of the sample at the rim compared with the sample at the center? (Hint: Use the approximation $1/\sqrt{1 - v^2/c^2} \simeq 1 + v^2/2c^2$ which can be shown to be valid if v/c is much smaller than 1.)

9.**——Suppose that a special breed of cat (*Felis einsteinii*) lives for exactly 7 years according to its own body clock. When such a cat is born, we put it aboard a spaceship and send it off at $v = 0.8c$ toward the star Alpha Centauri.

How far from the Earth (reckoned in the reference frame of the Earth) will the cat be when it dies? How long after the departure of the spaceship will a radio signal announcing the death of the cat reach us? The radio signal is sent out from the spaceship at the instant the cat dies.

10.** Because of the rotation of the Earth, a point on the equator has a speed of 460 m/s relative to a point at the North Pole. According to the time-dilation effect, by what factor do the rates of two clocks differ if one is located on the equator and the other at the North Pole? After 1 year has elapsed, by how many seconds will the clocks differ? Which clock will be ahead? (Hint: Use the approximation given in Problem 8.)

Section 27.4: Length Contraction

11.— A meter stick is moving by an observer at a speed of $0.5c$. What is the measured length of this meter stick in the reference frame of the observer?

12.— A cannonball flies through our laboratory at a speed of $0.3c$. Measurement of the transverse diameter of the cannonball gives a result of 0.20 m. What can you predict for the measurement of the length, or the longitudinal diameter, of the cannonball?

13.— According to the manufacturer's specifications, a spaceship has a length of 200 m. At what speed (relative to the Earth) will this spaceship have a length of 100 m in the reference frame of the Earth?

14.— What is the percent length contraction of an automobile traveling at 96 km/h? (Hint: Use the approximation given in Problem 8.)

15.* Suppose that a proton speeds by the Earth at $v = 0.8c$ along a line parallel to the axis of rotation of the Earth.

(a) In the reference frame of the proton, what is the polar diameter of the Earth? The equatorial diameter?

(b) In the reference frame of the proton, how long does the proton take to travel from the point of closest approach to the North Pole to the point of closest approach to the South Pole? In the reference frame of the Earth, how long does this take?

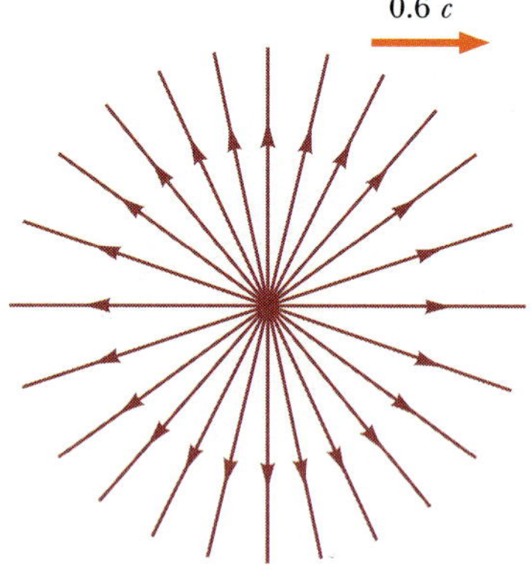

FIGURE 27.22

16.* Consider a cube measuring 1 m × 1 m × 1 m in its own rest frame. If this cube moves relative to the Earth at a speed of $0.6c$, what are its dimensions in the reference frame of the Earth? What are the areas of its faces? What is its volume? Assume that the cube moves in a direction perpendicular to one of its faces.

17.* Electric charge is uniformly distributed over a sphere; the charge density is 2.0×10^{-6} C/m^3. If this sphere is put in motion relative to the laboratory at a speed of $0.8c$, what will be the charge density? Keep in mind that the total amount of electric charge is unchanged by the motion of the sphere.

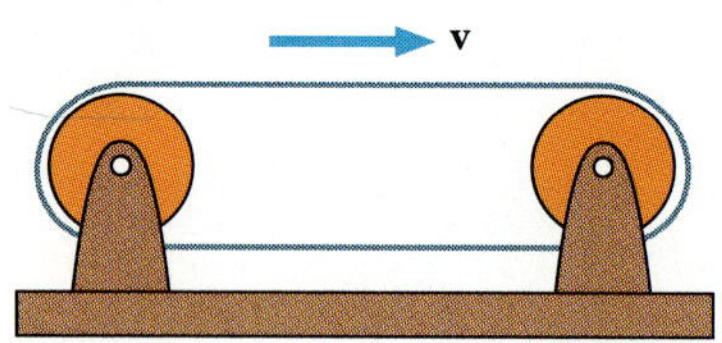

FIGURE 27.23
A drive belt and two flywheels.

18.* It can be shown that when a point charge moves at uniform velocity of relativistic magnitude, its pattern of electric field lines is contracted by the usual length contraction factor $\sqrt{1 - v^2/c^2}$ in the longitudinal direction and is unchanged in the transverse direction. Figure 27.22 shows the resulting pattern of field lines for a speed $v = 0.6c$. Draw a similar picture for a speed of $0.8c$.

19.** A flexible drive belt runs over two flywheels whose axles are mounted on a rigid base (see Fig. 27.23). In the reference frame of the base, the horizontal portions of the belt have a speed v and therefore are subject to a length contraction, which tightens the belt around the flywheels. However, in a reference frame moving to the right with the upper portion of the belt, the base is subject to length contraction, which ought to loosen the belt around the flywheels. Resolve this paradox by a qualitative argument. (Hint: Consider the lower portion of the belt as seen in the reference frame of the upper portion.)

20.** A spaceship has a length of 200 m in its own reference frame. It is traveling at $0.95c$ relative to the Earth. Suppose that the tail of the spaceship emits a flash of light.

(a) In the reference frame of the spaceship, how long does the light take to reach the nose?

(b) In the reference frame of the Earth, how long does this take? Calculate the time directly from the motions of the spaceship and the flash of light, and explain why you cannot obtain the answer by applying the time-dilation factor to the result of Part (a).

Section 27.5: The Lorentz Transformations and the Combination of Velocities

21. In the reference frame of the Earth, a firecracker is observed to explode at $x = 6.0 \times 10^8$ m, at $t = 4.0$ s. According to the Lorentz transformation equations, what are the x' and t' coordinates of this event as observed in the reference frame of a spaceship traveling in the x direction at a speed of $0.50c$? According to the Galilean transformation equations?

22. The captain of a spaceship traveling away from Earth in the x direction at $V_O = 0.80c$ observes that a nova explosion occurs at a point with spacetime coordinates $t' = -6.0 \times 10^8$ s, $x' = 1.9 \times 10^{17}$ m, $y' = 1.2 \times 10^{17}$ m as measured in the reference frame of the spaceship. He reports this event to the Earth via radio.

(a) What are the spacetime coordinates of the explosion in the reference frame of the Earth? Assume that the master clock of the spaceship coincides with the master clock of the Earth at the instant the spaceship passes by the Earth, and that the origin of the spaceship x', y' coordinates is at the midpoint of the spaceship.

(b) Will the Earth receive the report of the captain before or after astronomers on the Earth see the nova explosion in their telescopes? No calculation is required for this question.

23. Suppose that a spaceship is moving at a speed of $V_O = 0.2c$ relative to the Earth and a meteoroid is moving at a speed of $v_x = 0.1c$ relative to the Earth. What is the speed v'_x of the meteoroid relative to the spaceship according to Eq. (31)? What is the percent difference between this relativistic result and the Galilean result?

24. A spaceship is moving at a speed of $0.6c$ toward the Earth. A second spaceship, following the first one, is moving at a speed of $0.9c$. What is the speed of the second spaceship as observed in the reference frame of the first?

25. Find the inverse of Eq. (31), that is, express v_x in terms of v'_x.

26. A collision between two gamma rays creates an electron and an antielectron that travel away from the point of creation in opposite directions, each with a speed of $0.95c$ in the laboratory. What is the speed of the antielectron in the rest frame of the electron, and vice versa?

27.* A radioactive atom in a beam produced by an accelerator has a speed $0.80c$ relative to the laboratory. The atom decays and ejects an electron of speed $0.50c$ relative to itself. What is the speed of this electron relative to the laboratory, if ejected in the forward direction? If ejected in the backward direction?

28.* A spaceship traveling at $0.7c$ away from the Earth launches a projectile of muzzle speed $0.9c$ (relative to the spaceship). What is the speed of the projectile relative to the Earth if it is launched in the forward direction? In the backward direction?

29.** Show that the relativistic combination formula for the y component of the velocity is

$$v'_y = \frac{v_y\sqrt{1 - V_O^2/c^2}}{1 - v_x V_O/c^2}$$

30.** At $11^h0^m0^s$ A.M. a boiler explodes in the basement of the Museum of Modern Art in New York City. At $11^h0^m0.0003^s$ A.M. a similar boiler explodes in the basement of a soup factory in Camden, New Jersey, at a distance of 150 km from the first explosion. Show that in the reference frame of a spaceship moving at a speed greater than $V_O = 0.60c$ from New York toward Camden, the first explosion occurs *after* the second.

Section 27.6: Relativistic Momentum and Energy

31. Consider a particle of mass m moving at a speed of $0.1c$. What is its kinetic energy according to the relativistic formula? What is its kinetic energy according to the Newtonian formula? What is the percent deviation between these two results?

32. Suppose you want to give a rifle bullet of mass 0.01 kg a speed of 1% of the speed of light. What kinetic energy must you supply?

33. What is the kinetic energy of a spaceship of rest mass 50 metric tons moving at a speed of $0.5c$? How many metric tons of a matter–antimatter mixture would have to be consumed to make this much energy available?

34. A particle has a kinetic energy equal to its rest-mass energy. What is the speed of this particle?

35. The yearly energy expenditure of the United States is about 8×10^{19} J. Suppose that all of this energy could be converted into kinetic energy of an automobile of mass 1000 kg. What would be the speed of this automobile?

36. The speed of an electron in a hydrogen atom is 2.2×10^6 m/s. For this speed, does your electronic calculator show any difference between the kinetic energies calculated according to the relativistic formula and the Newtonian formula?

37. What is the speed of an electron if its kinetic energy is 1.6×10^{-13} J?

38. What is the momentum and what is the kinetic energy of an electron moving at a speed of one-half the speed of light?

39. Suppose that a spaceship traveling at $0.8c$ through our Solar System suffers a totally inelastic collision with a small meteoroid of mass 2.0 kg.
   (a) What is the kinetic energy of the meteoroid in the reference frame of the spaceship?
   (b) In the collision all of this kinetic energy suddenly becomes available for inelastic processes that damage the spaceship. The effect on the spaceship is similar to an explosion. How many tons of TNT will release the same explosive energy? One ton of TNT releases 4.2×10^9 J.

40.* At the Brookhaven AGS accelerator, protons of kinetic energy 5.3×10^{-9} J are made to collide with protons at rest.
   (a) What is the speed of one of these moving protons in the laboratory reference frame?
   (b) What is the magnitude of the momentum?

41.* At the SSC accelerator to be built in the United States, protons will be given kinetic energies of 3.2×10^{-6} J. What is the value of $c - v$ for such a proton, that is, by how many meters per second does the speed differ from the speed of light?†

42.*—The most energetic cosmic-ray particles have energies of about 50 J. Assume that such a cosmic ray consists of a proton. By how much does the speed of such a proton differ from the speed of light? Express your answer in meters per second.†

43.*—Consider the electrons of a speed $0.99999999967c$ produced by the Stanford Linear Accelerator. What is the magnitude of the momentum of such an electron?†

44.*—At the Fermilab accelerator, protons are given kinetic energies of 1.6×10^{-7} J. By how many meters per second does the speed of such a proton differ from the speed of light? What is the magnitude of the momentum of such a proton?†

45.*—Show that the velocity of a relativistic particle can be expressed as follows:

$$\mathbf{v} = \frac{c\mathbf{p}}{\sqrt{m^2c^2 + p^2}}$$

Section 27.7: Mass and Energy

46.—The atomic bomb dropped on Hiroshima had an explosive energy equivalent to that of 20,000 tons of TNT, or 8.4×10^{15} J. How many kilograms of rest mass must have been converted into energy in this explosion?

47.—How much energy will be released by the annihilation of one electron and one antielectron (both initially at rest)? Express your answer in electron-volts.

48.—Show that the momentum of a particle can be expressed in the concise form

$$\mathbf{p} = \frac{E\mathbf{v}}{c^2}$$

49.—The mass of the Sun is 2×10^{30} kg. The thermal energy in the Sun is about 2×10^{41} J. How much does the thermal energy contribute to the mass of the Sun? Express your answer in percent.

50.*—Show that the relativistic energy and the relativistic momentum are related by

$$E^2 = c^2p^2 + m^2c^4$$

51.*—Free neutrons decay spontaneously into a proton, an electron, and an antineutrino:

$$\mathrm{n} \rightarrow \mathrm{p} + \mathrm{e} + \bar{\nu}$$

The neutron has a rest mass of 1.6749×10^{-27} kg; the proton, 1.6726×10^{-27} kg; the electron, 9.11×10^{-31} kg; and the antineutrino, zero (or nearly zero). Assume that the neutron is at rest. What is the energy released in this decay?

52.**—A K^0 particle at rest decays spontaneously into a π^+ particle and a π^- particle. What will be the speed of each of the latter? The mass of the K^0 is 8.87×10^{-28} kg, and the masses of the π^+ and π^- particles are 2.49×10^{-28} kg each.

---

† Hint: Use the approximation given in Eq. (40).

CHAPTER 28

This thermogram reveals that warm human bodies emit thermal, infrared radiation. Note that the exposed faces radiate more than the clothed parts of the bodies, and that the hot cylinders of the engine of the motorcycle radiate the most.

QUANTA OF LIGHT

In Chapter 26 we examined the wave properties of light. We saw that light exhibits interference and diffraction, in agreement with Maxwell's theory, according to which light is a wave consisting of oscillating electric and magnetic fields with a continuous distribution of energy. In this chapter we will discover that light has particle properties. We will present experimental evidence that establishes that a light beam consists of a stream of discrete, particle-like energy packets. These energy packets are called ***quanta of light, or photons.***

Quanta of light, or photons ▶

The discovery of the quantization of the energy of light by Max Planck in 1900 initiated the modern era in physics. It was soon discovered that quantization of energy is a pervasive feature of the atomic and subatomic realm. As we will see in the next chapters, the energies of the atoms and the energies of the subatomic particles — electrons, protons, and neutrons — are quantized.

The fact that light has the dual attributes of a wave and of a particle indicates that neither the wave concept of classical physics nor the particle concept of classical physics gives an adequate description of light. We have to think of light as a wave–particle object, which we might call a ***wavicle.*** This object sometimes behaves pretty much like a classical wave, sometimes pretty much like a classical particle, and sometimes like a bit of both. Furthermore, electrons, protons, neutrons, and all the other known "particles" also exhibit such dual attributes of a wave and of a particle — we must regard all of them as wavicles.

28.1 BLACKBODY RADIATION

The first hint of a failure of classical physics emerged from the study of thermal radiation. When we heat a body to a high temperature, it gives off a glow; for instance, when we heat a bar of steel to 1200 or 1300 K, it glows in a deep red color (Fig. 28.1). This glow is thermal radiation ("radiant heat"). Thermal radiation is also emitted by bodies at lower temperatures, but the glow is then infrared, and not visible to the eye. ***The spectrum of thermal radiation is continuous*** — if we analyze the light emitted by a glowing body with a prism, we find that the energy is smoothly distributed over all wavelengths.

FIGURE 28.1

Bars of steel at a steel mill. The bar on the left, just out of the mill, is the hottest.

Classical physics cannot explain the details of the distribution of the energy over different wavelengths; in particular, it cannot explain the small amount of energy found at short wavelengths. For a quantitative description of the distribution of energy over different wavelengths, we plot the energy flux (or the power per unit area) radiated by the surface of the glowing body vs. the wavelength of the radiation. Such a plot may be regarded as giving the intensity distribution in the spectrum that an (ideal) prism produces when it is used to analyze the thermal radiation. Measurements of the thermal radiation emitted by glowing bodies show that ***the energy flux at very long and at very short wavelengths is quite small, and that the energy flux has a maximum at some intermediate wavelength.*** The location of this maximum depends on the temperature. For example, Fig. 28.2 gives plots of the distribution of the energy flux radiated by glowing

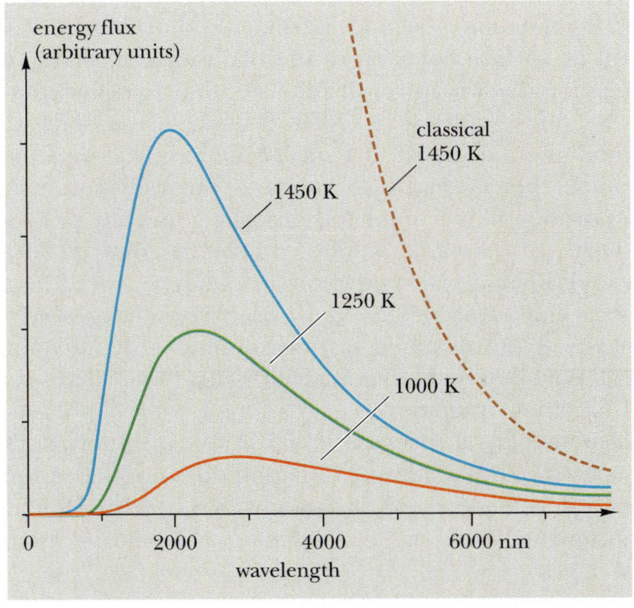

FIGURE 28.2

Distribution of the energy flux in the spectra of thermal radiation emitted by glowing bodies at temperatures of 1000 K, 1250 K, and 1450 K. The maxima (peaks) of these curves lie in the infrared region. The maximum is at 2900 nm for 1000 K, at 2300 nm for 1250 K, and at 2000 nm for 1450 K. The dashed curves gives the prediction of classical physics.

bodies at 1000 K, at 1250 K, and at 1450 K. By comparing these plots, we see that an increase of temperature produces more radiation at all wavelengths (the 1450-K curve is everywhere higher than the other curves); and we also see that an increase of temperature shifts the location of the maximum to shorter wavelengths (the peak of the 1450-K curve is located at a shorter wavelength than the peaks of the other curves).

The thermal radiation emerging from the surface of a glowing body is generated within the volume of the body by the random thermal motions of atoms and electrons. Before the radiation reaches the surface and escapes, it is absorbed and re-emitted many times and it attains thermal equilibrium with the atoms and electrons. This equilibration process shapes the continuous spectrum of the radiation, completely washing out all of the original spectral features that the radiation had when first emitted by the atoms in the body.

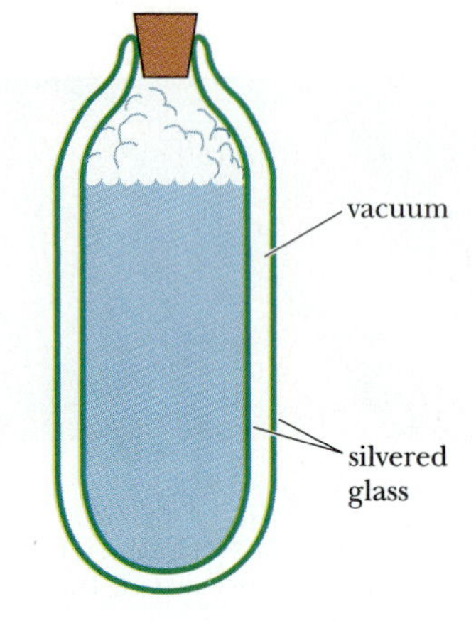

FIGURE 28.3
A thermos bottle.

The flux of thermal radiation emerging from the surface of a glowing body depends to some extent on the characteristics of the surface. The surface usually permits the escape of only a fraction of the flux reaching it from the inside of the body. Correspondingly, if the body is irradiated with an equal flux of thermal radiation from the outside, the surface permits the ingress of only an equal fraction of this flux, reflecting the rest. This equality of the emissive and absorptive characteristics of the surface can be deduced by an argument based on thermodynamics. Thus we are led to a general rule: ***A good absorber is a good emitter; and a poor absorber is a poor emitter.*** With this rule we can understand how the silvered, mirrorlike glass walls of thermos bottles or Dewars provide such excellent thermal insulation. These bottles are constructed with a double glass wall, and the space between these walls is evacuated (Fig. 28.3). Heat cannot flow across the evacuated space by conduction or convection; it can only flow by radiation. To inhibit radiation, the glass walls are silvered and thereby made into mirrorlike reflecting surfaces; these highly reflective surfaces are very poor absorbers and emitters of radiation. This keeps the heat transfer between the walls very small.

Blackbody ▶

A body with a perfectly absorbing (and emitting) surface is called a blackbody. When such a body is cold, it looks black because it does not reflect any of the illumination reaching it from the outside. But when a blackbody is hot, its surface emits more thermal radiation than any other hot body at the same temperature. In practice, the characteristics of an ideal blackbody are most easily achieved by a trick: take a body with a cavity, such as a hollow cube, and drill a small hole in one side of the cube (Fig. 28.4). The hole then acts as a blackbody — any radiation incident on the hole from outside will be completely absorbed, and any radiation incident from the inside of the cavity will be completely emitted. Because of this equivalence between a blackbody and a hole in a cavity, the terms *blackbody radiation* and *cavity radiation* are used interchangeably. The curves plotted in Fig. 28.2 are based on measurements of radiation emerging from a small hole in a body with a cavity; thus, these curves represent the spectra of blackbody radiation.

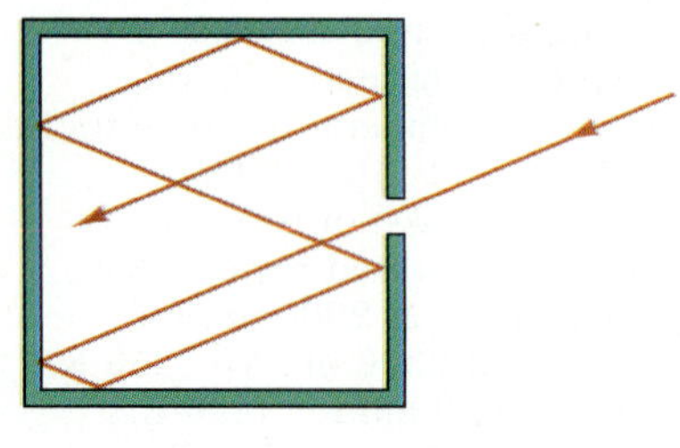

FIGURE 28.4
A cavity with a small hole. Any radiation entering the hole is trapped; it will suffer multiple reflections and it will ultimately be absorbed.

The blackbody plays a special role in the study of thermal radiation because the spectrum of its thermal radiation does not depend on the material of which it is made or on any other characteristics of the body. By another argument based on thermodynamics, it can be established that

the spectrum depends only on the temperature of the blackbody. The mathematical formula for the distribution of energy in this spectrum is therefore a universal law, and in the last years of the nineteenth century, physicists engaged in an intensive theoretical effort to find this universal law of blackbody radiation.

28.2 ENERGY QUANTA

One of the first attempts at a theoretical derivation of the distribution of energy in the spectrum of blackbody radiation was made by Lord Rayleigh. Since the energy flux of the radiation emerging from the hole in a cavity is directly proportional to the energy density of the radiation inside the cavity, Rayleigh decided to calculate the latter quantity. He began by noting that the radiation in a cavity is made up of a large number of standing waves; Fig. 28.5 shows some of these standing waves. Each of these standing waves can be regarded as a mode of vibration of the cavity. Rayleigh then appealed to the equipartition theorem, according to which, at thermal equilibrium, each mode of vibration has an average thermal energy of kT, where k is the Boltzmann constant (in Section 13.4 we stated a special case of the equipartition theorem for the free translational or rotational motion of a gas molecule). Thus, each of the standing waves of Fig. 28.5 ought to have an energy kT, and from this we can calculate the energy distribution in the spectrum of the radiation (the dashed curve in Fig. 28.2 shows the energy distribution obtained from this classical calculation). Although the classical calculation gave reasonable results at the long-wavelength end of the blackbody spectrum, it gave

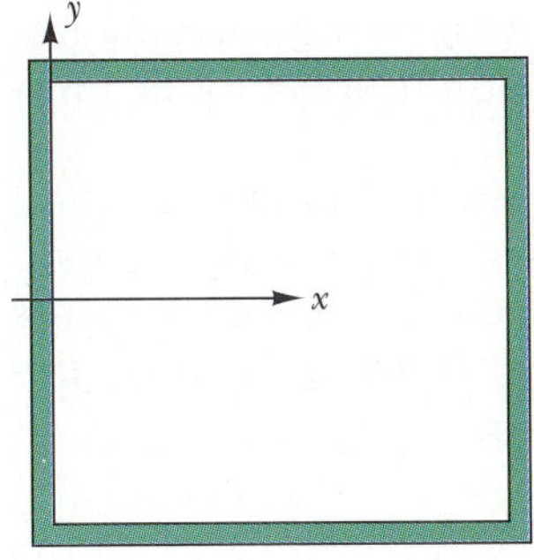

FIGURE 28.5

Some of the possible standing electromagnetic waves in a closed cavity. For the sake of simplicity, only waves with a horizontal direction of propagation are shown. The plots give the electric field as a function of x at one instant of time.

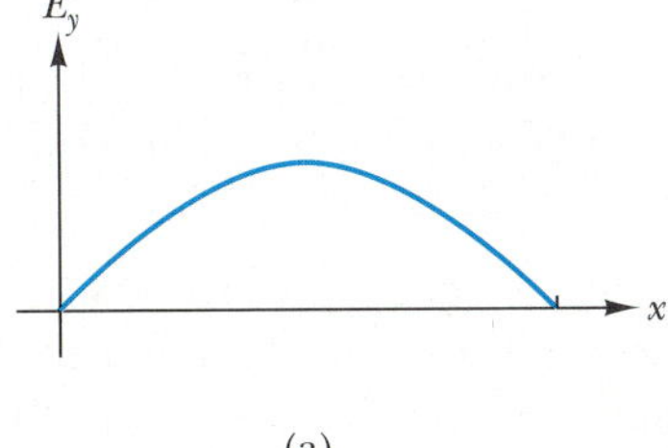

(a)

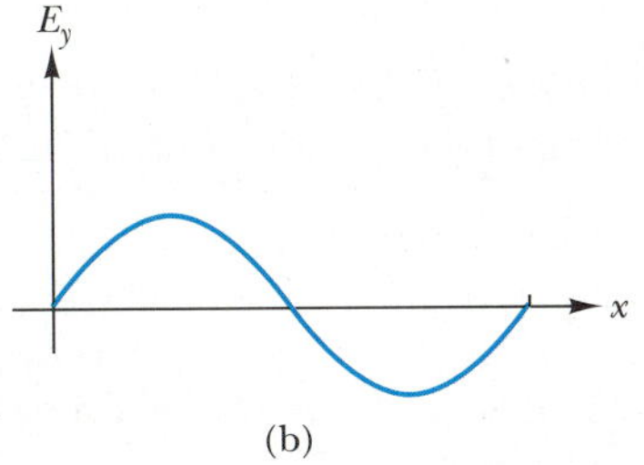

(b)

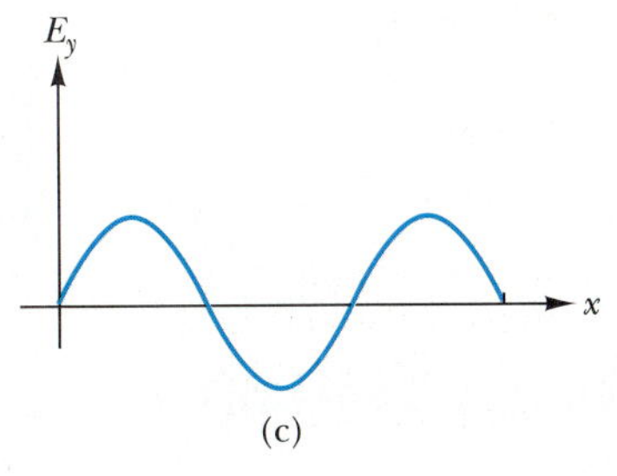

(c)

disastrous results at the short-wavelength end: the number of possible standing-wave modes of very short wavelength is infinitely large and, if each of these modes had an energy kT, the total energy in the cavity would be infinite! This disastrous failure of classical theory has been called the ***ultraviolet catastrophe*** (since short wavelengths belong to ultraviolet radiation).

Ultraviolet catastrophe ▶

The correct formula for the distribution of energy in the spectrum of blackbody radiation was finally obtained by Max Planck in 1900. By some inspired guesswork, based on thermodynamics, Planck found a mathematical formula that gave a precise fit to the experimental curves of blackbody radiation, such as those plotted in Fig. 28.2. He then searched for a theoretical explanation for his formula. The search led Planck to the discovery of the quantization of energy. This revolutionary discovery brought about the overthrow of classical Newtonian physics and the rise of quantum physics. For Planck, the postulate of the quantization of energy was "an act of desperation" which he committed because "a theoretical explanation had to be found at any cost, whatever the price."

Max Planck, *1858–1947, German physicist. Planck made significant contributions to thermodynamics before he became involved with the problem of blackbody radiation. He was a very conservative scientist, and he tried very hard to circumvent the quantization postulate, but without success. Planck received the Nobel Prize in 1918.*

Planck's derivation of the blackbody radiation formula involves some sophisticated statistical mechanics, and we cannot reproduce the details of his derivation here. We will merely give a sketchy outline of this derivation. Planck began by making a simple and convenient theoretical model of the walls of the cavity: he regarded the atoms in the walls as small harmonic oscillators, that is, small masses (with electric charges) attached to springs. Although this is a rather crude model of an atom, it was adequate for his purposes since, as described in the preceding section, the radiation in the cavity is known to be completely independent of the characteristics of the wall. The random thermal motions of the oscillators result in the emission of electromagnetic radiation. This radiation fills the cavity and acts back on the oscillators. When thermal equilibrium is attained, the average rate of emission of radiation energy by the oscillators matches the rate of absorption of radiation energy. Thus, the oscillators share their energy with the radiation in the cavity, and Planck was able to show that, under equilibrium conditions, the average radiation energy at some frequency f (or at a wavelength $\lambda = c/f$) is directly proportional to the average energy of an oscillator of frequency f.

These first steps of Planck's calculation involved nothing but classical mechanics. But in the next step of the calculation, Planck departed radically from classical physics. He proposed that the energy of the oscillators is quantized according to the following postulate: ***In an oscillator of frequency f, the only permitted values of the energy are***

$$E = 0,\ hf,\ 2hf,\ 3hf,\ \ldots \tag{1}$$

All other values of the energy are forbidden. The constant h in Eq. (1) is a new fundamental constant, called ***Planck's constant.*** The value of this constant is

Planck's constant ▶

$$h = 6.63 \times 10^{-34}\ \text{J} \cdot \text{s} \tag{2}$$

The energy hf is called an ***energy quantum.*** According to the quantization rule, the energy of an oscillator is always some multiple of the basic energy quantum hf,

◀ Energy quantum

$$E = nhf \qquad n = 0, 1, 2, 3, \ldots \tag{3}$$

◀ Energy quantization of oscillator

The integer n is called the ***quantum number*** of the oscillator.

◀ Quantum number

With this quantization condition, Planck calculated the average energy of the oscillators; and from that he derived his blackbody radiation formula. Although we cannot go into the details of this derivation, we can achieve a rough understanding of how Planck's postulate avoids the ultraviolet catastrophe. The thermal energy of the walls of the cavity is shared at random among all the oscillators in these walls. Some of these oscillators have high frequencies, some have low frequencies. For an oscillator of very high frequency, the energy quantum hf is very large. If this oscillator is initially quiescent ($n = 0$), it cannot begin to move unless it acquires one energy quantum; but since this energy quantum hf is very large, the random thermal disturbances will be insufficient to provide it — the oscillator will remain quiescent. Thus, the quantization of energy tends to inhibit the thermal excitation of the high-frequency oscillators. If the high-frequency oscillators remain quiescent, then they will not supply energy to the corresponding high-frequency standing waves in the cavity, and there will be no ultraviolet catastrophe.

Note that for an oscillator with a frequency of $f \simeq 10^{15}$ Hz, which is typical for atomic vibrations, the energy quantum is $hf = 6.6 \times 10^{-34}\ \text{J}\cdot\text{s} \times 10^{15}/\text{s} = 6.6 \times 10^{-19}$ J. Since this is a very small amount of energy, quantization does not make itself felt at a macroscopic level. But quantization plays a pervasive role at the atomic level.

Unfortunately, Planck could not offer any basic justification for this postulate of quantization of energy. His postulate gave him a blackbody radiation formula which was in complete agreement with the experimentally measured distribution of energy over wavelength (as displayed in Fig. 28.2), but his postulate brought him into conflict with classical physics. Superficially, the quantization of energy is analogous to the quantization of electric charge — we know from Chapter 16 that the electric charge of any particle or body is always some multiple of the fundamental charge e. However, whereas the quantization of charge is consistent with the laws of classical physics, the quantization of energy is inconsistent with these laws. The energy of any oscillator that obeys Newton's laws — such as a mass on a spring — can be changed by a small amount by pushing on the oscillator with a very weak force, and it should therefore be possible to change the energy of the oscillator continuously by any amount we please, and not just in discrete steps of one energy quantum. Thus, quantization of energy makes no sense in classical physics. Planck could justify his postulate only by its consequences; but, in theoretical physics, the end does not justify the means. A deeper explanation of the quantization of energy emerged only much later, with the development of quantum mechanics (see Chapter 29).

From Planck's formula for the distribution of energy in the spectrum of blackbody radiation it can be established that the energy flux has a maximum at a wavelength

$$\lambda_{\max} = \frac{1}{4.965}\frac{hc}{k} \times \frac{1}{T} \tag{4}$$

where c is, as always, the speed of light, k is Boltzmann's constant (see Section 13.1), h is Planck's constant, and T is the absolute temperature of the blackbody. Equation (4) is called ***Wien's Law.*** If we insert the numerical values of h, c, and k, Wien's Law takes the simple form

Wien's Law ▶

$$\lambda_{\max} = \frac{2.90 \times 10^{-3}\ \text{m} \cdot \text{K}}{T} \tag{5}$$

Wien's Law asserts that λ_{max} ***is inversely proportional to the temperature*** T. This means that an increase of temperature shifts the location of the maximum to shorter wavelengths, in agreement with the experimental results presented in Fig. 28.2. If the temperature is sufficiently high — 6000 K or so — the maximum of the spectrum lies in the visible region. For instance, the Sun, with a surface temperature of 5800 K, emits its largest flux of thermal radiation in the visible region.

EXAMPLE 1. According to Wien's Law, at what wavelength does the thermal radiation emitted by the Sun have its maximum? At what wavelength does the thermal radiation emitted by the tungsten filament in a light bulb have its maximum? Assume that the Sun and the tungsten filament are approximately blackbodies at temperatures of 5800 K and 3200 K, respectively.*

SOLUTION: With $T = 5800$ K, Wien's Law gives

$$\lambda_{\max} = \frac{2.90 \times 10^{-3}\ \text{m} \cdot \text{K}}{5800\ \text{K}} = 5.0 \times 10^{-7}\ \text{m}$$

and with $T = 3200$ K it gives

$$\lambda_{\max} = \frac{2.90 \times 10^{-3}\ \text{m} \cdot \text{K}}{3200\ \text{K}} = 9.1 \times 10^{-7}\ \text{m}$$

Thus, the maximum of the thermal radiation from the Sun lies in the visible region, but the maximum of the thermal radiation from the light bulb lies in the infrared.

* The blackbody approximation is fairly good for the Sun. But it is not good for an ordinary tungsten filament, because the tungsten surface is not a good absorber. The blackbody approximation becomes better if the filament is tightly coiled, like a solenoid, and if we examine the radiation in the interior of the coil (a cavity).

Furthermore, from Planck's formula we can establish that ***the combined energy flux for all wavelengths radiated from the surface of a blackbody is proportional to the fourth power of the temperature,***

$$S = \sigma T^4 \tag{6}$$

This is called the ***Stefan-Boltzmann Law.*** The constant of proportionality σ in this law can, again, be expressed in terms of h, c, and k. With the appropriate numerical values, the Stefan–Boltzmann Law then takes the form

$$S = \left(5.67 \times 10^{-8} \frac{\text{W}}{\text{m}^2 \cdot \text{K}^4}\right) \times T^4 \tag{7}$$

◀ **Stefan-Boltzmann Law**

Both the laws of Wien and of Stefan–Boltzmann had been discovered empirically many years before Planck supplied their theoretical foundation. Note that the energy flux S in Eq. (7) depends only on the temperature of the blackbody; it does not depend on the material the body is made of. Since S is the energy flux, or the power per unit area, we can obtain the net power radiated from the body by multiplying S by the surface area; thus, the net power depends on the size of the body.

EXAMPLE 2. On a clear night, the surface of the Earth loses heat by radiation. Suppose that the temperature of the ground is 10°C and that the ground radiates like a blackbody. What is the rate of loss of heat per square meter?

SOLUTION: The absolute temperature of the ground is 283 K. Hence, the Stefan–Boltzmann Law tells us that the radiated flux, or power per unit area, is

$$S = \sigma T^4 = 5.67 \times 10^{-8}\ \text{W}/(\text{m}^2 \cdot \text{K}^4) \times (283\ \text{K})^4 \tag{8}$$

$$= 364\ \text{W}/\text{m}^2$$

This amounts to 87 cal/m^2 per second.

COMMENTS AND SUGGESTIONS: This large radiative heat loss of the ground explains the sharp drop of temperature experienced during clear nights. The drop of temperature is much less severe if there is an overcast sky. The clouds then reflect most of the radiation back to the ground — they act like a blanket to keep the ground warm.

28.3 PHOTONS AND THE PHOTOELECTRIC EFFECT

In 1905, Einstein showed that Planck's formula could be understood much more simply in terms of a direct quantization of the energy of the radiation. Planck had postulated that the oscillators in the wall of the cavity have discrete quantized energies, but he had treated the electromagnetic radiation as a smooth, continuous distribution of energy, exactly as it is supposed to be according to classical electromagnetic theory.

In contrast, Einstein proposed that ***electromagnetic radiation consists of discrete particle-like packets of energy.*** He regarded a wave of some given frequency f as a stream of more or less localized energy packets, each with one quantum of energy hf. The wave then has an energy hf if it contains only one such quantum, $2hf$ if it contains two, and so on. ***The particle-like energy packets of magnitude*** hf ***are called photons.*** The thermal radiation in a cavity, with waves traveling randomly in all directions, can then be regarded as a gas of photons. Einstein applied statistical mechanics to calculate the energy spectrum of this gas, and he thereby obtained the energy spectrum of the blackbody radiation.

Photons ▶

The essential difference between Planck's and Einstein's views of the cavity radiation is that Planck quantized only the exchange of radiation with the walls of the cavity, whereas Einstein quantized the radiation itself. Thus, in Einstein's view, electromagnetic radiation is always quantized, regardless of where or how it is produced. Not only is the radiation quantized when it is produced by the oscillators in the walls of a cavity (as in the case of thermal radiation), but also when it is produced outside of a cavity, say, by the acceleration of electric charges on the antenna of a radio transmitter.

EXAMPLE 3. The energy flux of sunlight reaching the surface of the Earth is $1.0 \times 10^3\ \text{W/m}^2$. How many photons reach the surface of the Earth per square meter per second? For the purposes of this calculation assume that all the photons in sunlight have an average wavelength of 500 nm.

SOLUTION: The energy of a photon of wavelength 500 nm is

$$E = hf = h\frac{c}{\lambda} \tag{9}$$

$$= 6.63 \times 10^{-34}\ \text{J}\cdot\text{s} \times \frac{3.00 \times 10^8\ \text{m/s}}{5.0 \times 10^{-7}\ \text{m}} = 4.0 \times 10^{-19}\ \text{J}$$

The energy of sunlight incident per square meter per second is 1.0×10^3 J. To obtain the number of photons, we must divide this by the energy per photon, 4.0×10^{-19} J. This gives $1.0 \times 10^3\ \text{J}/4.0 \times 10^{-19}\ \text{J} = 2.5 \times 10^{21}$ photons per square meter per second.

COMMENTS AND SUGGESTIONS: Because the number of photons in sunlight and other common light sources is so large, at the macroscopic level we do not perceive the grainy character of the energy distribution in light.

◀ **Photoelectric effect**

With this concept of light as a stream of photons, Einstein was also able to offer an explanation of the ***photoelectric effect.*** In some early experiments on the production of radio waves by electric sparks, Hertz had noticed that light shining on an electrode tended to promote the formation of sparks. Subsequent careful experimental investigations demonstrated that the impact of light on an electrode can eject electrons, which trigger the sparks. The electrons emerge with a kinetic energy that increases directly with the frequency of the light.

Figure 28.6 is a schematic diagram of the apparatus used in the investigation of the photoelectric effect. Light from a lamp illuminates an electrode of metal *(C)* enclosed in an evacuated tube. Electrons ejected from this electrode travel to the collecting electrode *(A)*, and then flow around the external circuit. A galvanometer *(G)* detects this flow of electrons. The kinetic energy of the ejected photoelectrons can be determined by applying a potential difference between the emitting and the collecting electrodes by means of an adjustable source of emf *(V)*. With the polarity shown in the figure, the collector has a negative potential relative to the emitter, that is, the collector exerts a repulsive force on the photoelectrons. If the potential energy matches or exceeds the initial kinetic energy of the photoelectrons, then the flow of these electrons will stop. The corresponding potential is called the ***stopping potential;*** the measured value of this stopping potential gives us the kinetic energy of the electrons:

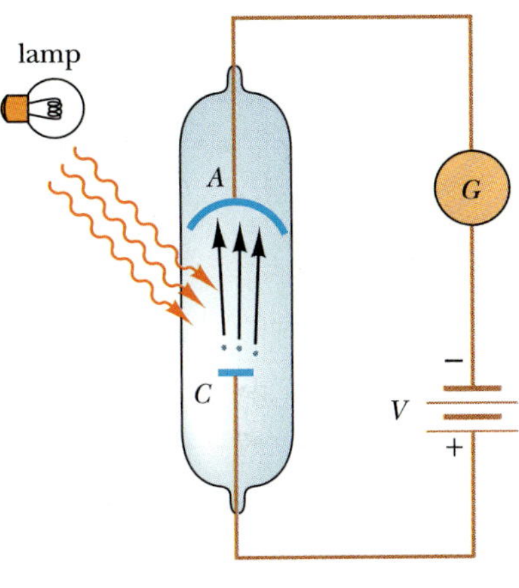

FIGURE 28.6

Schematic diagram of the apparatus for the investigation of the photoelectric effect. The light from the lamp ejects electrons from the electrode C (cathode), and they travel to the collecting electrode A (anode).

$$K = eV_{\text{stop}} \tag{10}$$

◀ **Stopping potential**

Experimentally, one finds that ***the kinetic energy determined in this way increases directly with the frequency of the incident light.*** For example, Fig. 28.7 is a plot of the kinetic energy vs. the frequency of the light for photoelectrons ejected from sodium. Note that, according to this plot, if the frequency is below 4.4×10^{14} Hz, then the light is incapable of ejecting electrons.

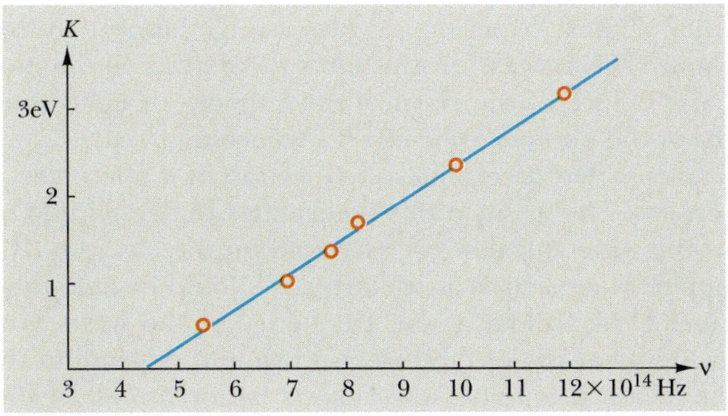

FIGURE 28.7

Kinetic energies (in electron-volts) of photoelectrons ejected from sodium by light of different frequencies.

Einstein's quantum theory of light accounts for these experimental observations as follows. The electrons in the illuminated electrode absorb photons from the light, one at a time. When an electron absorbs a photon, it acquires an energy hf. But before this electron can emerge from the electrode, it must overcome the restraining forces that bind it to the metal of the electrode. The energy required for this is called the ***work function*** of the metal, designated by ϕ. The energy retained by the electron is then $hf - \phi$, and this must be the kinetic energy of the emerging electron:

Work function ▶

$$K = hf - \phi \tag{11}$$

This is Einstein's photoelectric equation. It shows that the kinetic energy does indeed increase directly with the frequency of the absorbed photon, in agreement with the data of Fig. 28.7.

According to Eq. (11), ***a minimum frequency is required to achieve the ejection of an electron.*** This minimum frequency, called the ***threshold frequency,*** corresponds to the ejection of an electron of zero kinetic energy; such an electron is just barely ejected. The threshold frequency f_{thresh} is given by

$$0 = hf_{\text{thresh}} - \phi$$

or

$$f_{\text{thresh}} = \frac{\phi}{h} \tag{12}$$

When an electron absorbs a photon of this frequency, all of the energy of the photon is used to overcome the restraining forces that bind the electron to the metal, and nothing is left over for the ejected electron.

Einstein's photoelectric equation was verified in detail by a long series of meticulous experiments by Robert Millikan (the data in Fig. 28.7 are due to him). In order to obtain reliable results, Millikan found it necessary to take extreme precautions to avoid contamination of the surface of the photosensitive electrode. Since the surfaces of metals exposed to air quickly accumulate a layer of oxide, he developed a technique for shaving the surfaces of his metals in a vacuum, by means of a magnetically operated knife.

The results of these experiments gave strong support to the quantum theory of light. This success of Einstein's theory was all the more striking in view of the failure of the classical wave theory of light to account for the features of the photoelectric effect. According to the wave theory, the crucial parameter that determines the ejection of a photoelectron should be the intensity of light, since this determines the amplitude of the electric field in the wave and the force exerted on the electron. If an intense electromagnetic wave strikes an electron, it should be able to jolt it loose from the metal, regardless of the frequency of the wave. Furthermore, the kinetic energy of the ejected electron should depend on the intensity of the wave, since an intense wave exerts larger forces on the electron.

The observational evidence contradicts these predictions of the wave theory: ***A wave with a frequency below the threshold frequency never ejects an electron, regardless of its intensity. And, furthermore, the kinetic energy of the ejected electrons depends on the frequency*** [as specified by Eq. (11)], ***and not on the intensity.*** High-intensity light ejects more photoelectrons, but does not give the individual electrons more kinetic energy.

Today, the photoelectric effect finds many practical applications in sensitive electronic devices for the detection of light. For instance, in a photomultiplier tube, an incident photon ejects a photoelectron from an electrode at the faceplate of the tube. To convert this single electron into a measurable pulse of current, electric fields within the tube accelerate this single electron toward a second electrode (called a dynode; see Fig. 28.8) where its impact ejects several secondary electrons. These, in turn, are accelerated toward a third electrode where their impact ejects tertiary electrons, and so on. Thus, the single photoelectron from the first electrode generates an avalanche of electrons. In a high-gain photomultiplier tube, a pulse of 10^9 electrons emerges from the last electrode, delivering a measurable pulse of current to an external circuit. In this way, the photomultiplier tube can detect the arrival of individual photons. Some sensitive television cameras, such as the image orthicon, rely on the same multiplier principle to convert the arrival of a photon at a photosensitive faceplate into a measurable pulse of current. This permits these cameras to take pictures in faint light, where there are few photons.

Photomultiplier tubes.

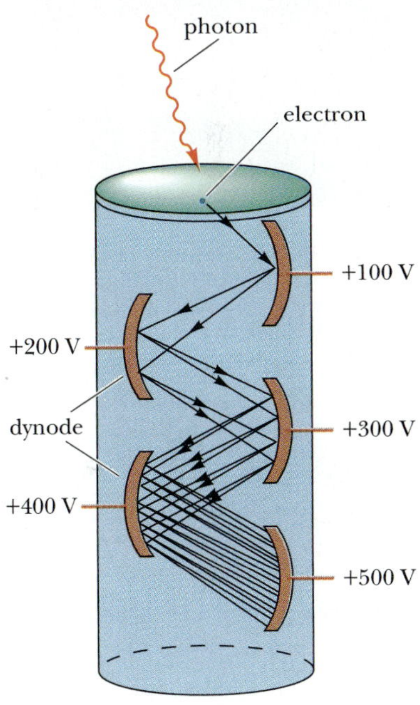

FIGURE 28.8

Schematic diagram of a photomultiplier tube. The secondary electrodes are called dynodes. For the purpose of this diagram, it has been assumed that each electron impact on a dynode releases two electrons. The arrows show an avalanche of electrons.

EXAMPLE 4. The work function for platinum is 9.9×10^{-19} J. What is the threshold frequency for the ejection of photoelectrons from platinum?

SOLUTION: If an electron absorbs a photon at the threshold frequency, it will just barely have enough energy to overcome the binding forces holding it in the metal, and it will emerge with zero kinetic energy. By Eq. (12) this corresponds to

$$f_{\text{thresh}} = \frac{\phi}{h} = \frac{9.9 \times 10^{-19}\ \text{J}}{6.63 \times 10^{-34}\ \text{J}\cdot\text{s}} = 1.5 \times 10^{15}\ \text{Hz} \tag{13}$$

28.4 The Compton Effect

Very clear experimental evidence for the particle-like behavior of photons was uncovered by Arthur Compton in 1922. Compton had been investigating the scattering of X rays by a target of graphite (pure carbon). According to Maxwell's theory, X rays are merely high-frequency light waves (see the next section). But according to quantum theory, they ought to contain photons; and since the frequency is much higher than that of ordinary light, the energy of such X-ray photons ought to be much larger than that of photons of ordinary light. When Compton bombarded the graphite with monochromatic X rays, he found that ***the scattered (deflected) X rays had a wavelength somewhat longer than that of the original X rays.*** Compton soon recognized that this effect could be understood in terms of collisions of X-ray photons with electrons, collisions in which the photons behave like particles.

Compton effect ▶

Arthur Holly Compton, *1892–1962, American experimental physicist. For his discovery of the Compton effect, he received the Nobel Prize in 1927.*

In this collision, the electron of a carbon atom can be regarded as free, because the force binding the electron to the atom is insignificant compared with the force exerted by the incident photon. When the photon bounces off the electron, the electron recoils and thereby picks up some of the photon's energy — the deflected photon is left with reduced energy. Since the energy of the photon is $E = hf = hc/\lambda$, a reduction of the energy of the photon implies an increase of wavelength. Qualitatively, we expect that those photons deflected through the largest angles should lose the most energy and therefore emerge with the longest wavelength. And this is just what Compton found in his experiments.

For a quantitative discussion of the photon–electron collision, we need an expression for the momentum of the photon. We can derive this expression from the relativistic formula for the momentum of a particle. Since the speed of the photon is the speed of light, the photon must be regarded as an ultrarelativistic particle. For such a particle, Eq. (27.47) tells us that

$$p = \frac{E}{c} \tag{14}$$

[Note that according to Eq. (27.45), the energy of a particle with a speed equal to the speed of light can be finite only if the mass m is zero; thus, photons must be regarded as particles of zero mass.] Since the energy of the photon is $E = hf = hc/\lambda$, the expression for the momentum can be written as

Momentum of photon ▶

$$p = \frac{h}{\lambda} \tag{15}$$

With the expression (15) for the momentum of the photon, Compton calculated the change of energy and the change of wavelength of a photon in an elastic collision with an electron, initially at rest. For a photon that emerges from the collision at an angle θ (the deflection angle; see Fig. 28.9), he calculated from the laws of conservation of energy and momentum that the change of wavelength is

Wavelength shift of photon ▶

$$\Delta\lambda = \frac{h}{m_e c}(1 - \cos\theta) \tag{16}$$

Thus, if the photon emerges from the collision at an angle of 90°, then $\cos\theta = \cos 90° = 0$, and the wavelength change is $\Delta\lambda = h/m_e c$. And if the photon emerges at 180°, in a direction opposite to that of its initial direction of motion, then $\cos\theta = \cos 180° = -1$, and $\Delta\lambda = 2h/m_e c$. Numerically, these wavelength changes amount to 0.243 nm and 0.485 nm, respectively. Such small wavelength shifts are difficult to detect unless the wavelength of the X rays is itself quite small, so $\Delta\lambda$ is an appreciable fraction of the wavelength.

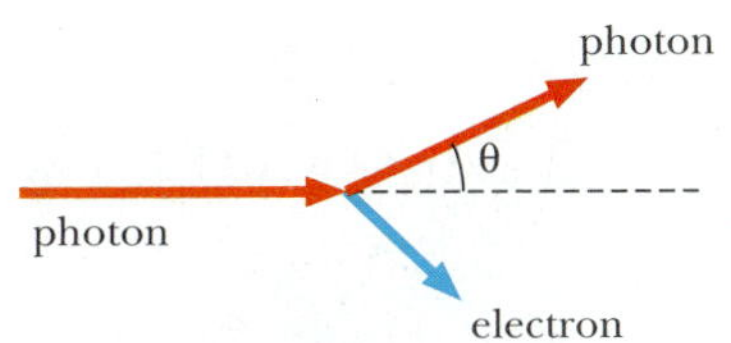

FIGURE 28.9

Directions of propagation, or directions of the momentum, of a photon before and after collision with an initially stationary electron. The deflection angle θ is the angle between the directions of propagation before and after the collision.

28.5 X RAYS

X rays were discovered in 1895 by Wilhelm Konrad Röntgen in experiments with beams of energetic electrons. Röntgen found that when such a beam of electrons moving in an evacuated glass tube struck the wall of the tube, some invisible, mysterious rays were emitted that caused a faint luminosity (fluorescence) on a nearby sheet of paper impregnated with chemicals. He also found that these rays could affect covered and wrapped photographic plates. These rays proved capable of penetrating through thick layers of opaque materials, and Röntgen immediately recognized the possible medical applications of these rays for making images of the tissues inside the human body, especially bones (see Fig. 28.10).

The distance that X rays can penetrate through a material depends on the density of the material. When X rays pass through atoms, they tend to be absorbed by the atomic electrons. Therefore ***materials of high density, such as lead, with a concomitant high density of electrons, strongly absorb and block X rays.*** In X-ray photographs of parts of the human body, bones throw sharp shadows because their density is higher than that of the surrounding tissues. Organs made of soft tissues of low density, such as the gastrointestinal tract, do not throw sharp shadows, and to enhance the contrast of the X-ray photograph it is advantageous to fill the organ with a barium solution, a high-density material which blocks the X rays (see Fig. 28.11).

Röntgen had named his rays ***X rays*** because their nature was unknown. Although he and his contemporaries suspected that they might be electromagnetic waves of extremely short wavelength, the conclusive experimental proof of this conjecture was not obtained until 1912, when Max von

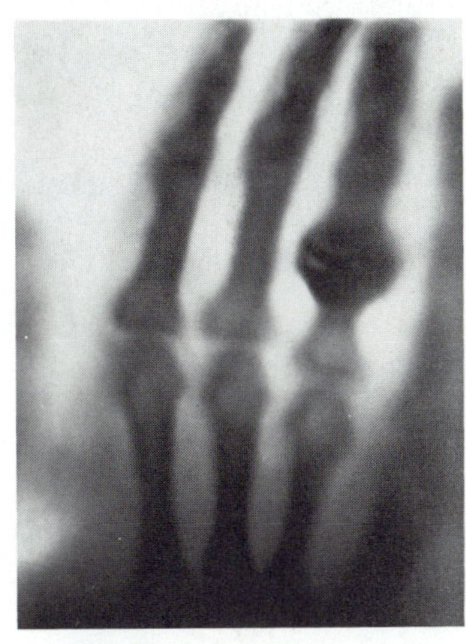

FIGURE 28.10

One of the first X-ray photographs prepared by Röntgen. It shows the hand of his wife; note the sharp image of the ring.

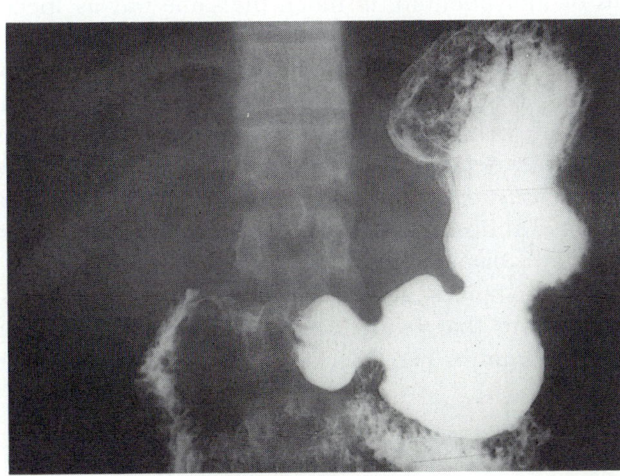

FIGURE 28.11

X-ray photograph of stomach with barium shadowing.

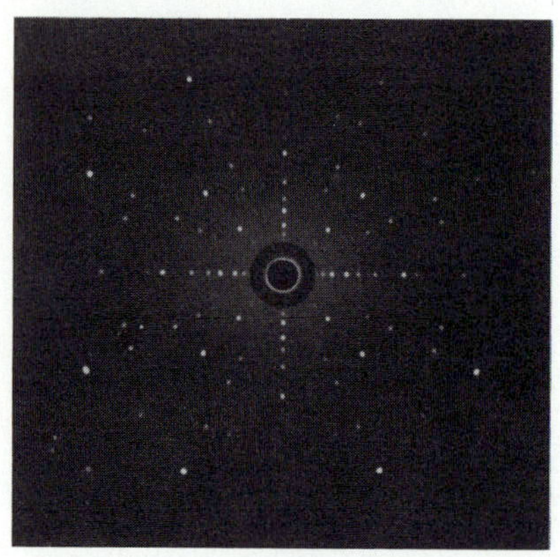

FIGURE 28.12
Interference pattern produced by X rays incident on a crystal.

Laue argued that if X rays are waves, they should display interference effects when passing through crystals. The distances between the rows of atoms in a crystal, such as rocksalt, are of the same order of magnitude as the wavelengths of X rays, and von Laue proposed that ***the crystal therefore can play the role of a "grating" for X rays,*** analogous to a multiple-slit grating used for interference experiments with light. Figure 28.12 shows a photograph of the interference pattern produced by X rays incident on a crystal. The wavelength of the X rays can be deduced from the angular positions of the interference maxima and the known size of the spacings in the crystal. The wavelengths of X rays are found to range from 0.001 nm to 10 nm.

Laue spots ▶

The bright spots in the photographs of X-ray interference patterns are called ***Laue spots.*** The beautiful symmetry of the pattern of Laue spots reflects the symmetry of the arrangement of the atoms in the crystal. In modern crystallography laboratories, the patterns of Laue spots produced by X rays incident on a crystal are often used to investigate the structure of the crystal and that of the molecules in it. For instance, such X-ray interference experiments played a crucial role in the determination of the structure of DNA.

Bremsstrahlung ▶

Since X rays are electromagnetic waves, their generation by the impact of energetic electrons on some sort of obstacle can be understood by the familiar mechanism of the emission of radiation by acceleration of electric charges. When the fast-moving electrons collide with the atoms in the obstacle, ***they suffer sudden decelerations and radiate intense electromagnetic waves of short wavelength.*** This kind of radiation is called ***Bremsstrahlung.**** Figure 28.13 shows an X-ray tube in which electrons emerging from a hot filament are accelerated through a potential difference of several kilovolts and then strike an obstacle, or target, made of a heavy metal, such as tungsten or molybdenum. The X rays produced in the violent decelerations of the electrons escape through the side of the tube.

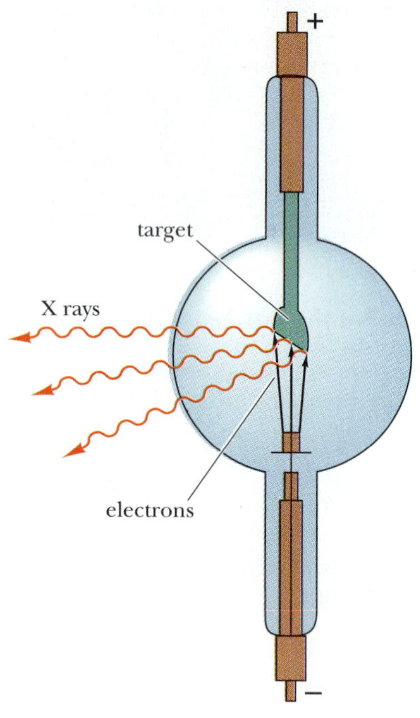

FIGURE 28.13
An X-ray tube.

Figure 28.14 is a plot of the distribution of energy of the X rays generated by electrons of 35 kilovolts striking a molybdenum target. Note that the X-ray energy is smoothly distributed over a wide range of wavelengths, but there also are two conspicuous spikes in the energy distribution. The smooth portion of the X-ray spectrum is due to Bremsstrahlung, whereas the discrete spikes are generated in the interior of the atoms of molybdenum, in much the same way as spectral lines of visible light are generated by the atoms. In the plot of the energy distribution, the shape and the position of the broad peak of the Bremsstrahlung portion of the spectrum depend on the energy of the incident electrons. But the positions of the spikes do not depend on the electron energy; instead they depend on the material of the target. The spikes are called the ***characteristic spectrum*** of the target material; each kind of target atom has its own distinctive characteristic spectrum, just as each kind of atom has its own distinctive spectrum of visible light.

Let us ignore the spikes, and concentrate on the smooth Bremsstrahlung spectrum. Note that below a certain wavelength — for instance, below about 0.036 nm in Fig. 28.14 — there is no Bremsstrahlung. ***The***

* German for *braking radiation.*

minimum wavelength emitted by the electrons is called the cutoff wavelength. Experimentally, the cutoff wavelength is found to be inversely proportional to the kinetic energy of the electrons. This cutoff makes no sense from the point of view of classical theory — if an electron suffers a sufficiently violent collision with an atom, it ought to be able to radiate waves of arbitrarily short wavelength (although the intensity of short-wavelength waves is expected to be low). But the cutoff is readily explained by the quantum theory of light, according to which the decelerating electron emits photons. Obviously, the maximum energy that a decelerating electron can give to a photon is *all* its energy. Thus, the maximum frequency f_{cutoff} of a photon is related to the kinetic energy of the incident electron by

◂ Cutoff wavelength

$$hf_{cutoff} = K \tag{17}$$

and

$$f_{cutoff} = \frac{K}{h}$$

But $\lambda = \frac{c}{f}$, and therefore

$$\lambda_{cutoff} = \frac{hc}{K} \tag{18}$$

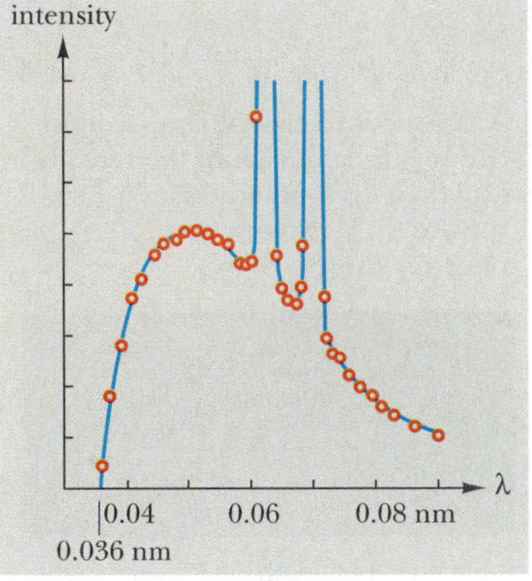

FIGURE 28.14
Distribution of energy in the spectrum of X rays produced by electrons of 35 kilovolts incident on a molybdenum target.

EXAMPLE 5. If an X-ray tube is operated at a potential of 35 kilovolts, the electron energy is 35 keV. What is the cutoff wavelength for the X rays emitted by electrons of this energy?

SOLUTION: With an electron energy $K = 35\text{ keV} = 5.6 \times 10^{-15}$ J, Eq. (18) gives

$$\lambda_{cutoff} = \frac{hc}{K} = \frac{6.63 \times 10^{-34}\text{ J}\cdot\text{s} \times 3.00 \times 10^{8}\text{ m/s}}{5.6 \times 10^{-15}\text{ J}}$$

$$= 3.6 \times 10^{-11}\text{ m} = 0.036\text{ nm}$$

This is an agreement with the cutoff wavelength indicated in Fig. 28.14.

28.6 Wave vs. Particle

From the photoelectric effect and the Compton effect we learned that photons have particle properties. On the other hand, we know that light

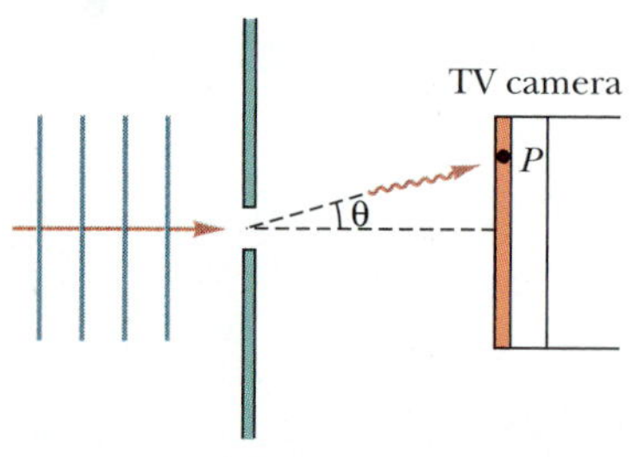

FIGURE 28.15

A diffraction experiment. A light beam is incident from the left on a plate with a narrow slit. A sensitive TV camera detects photons on the right.

(a)

(b)
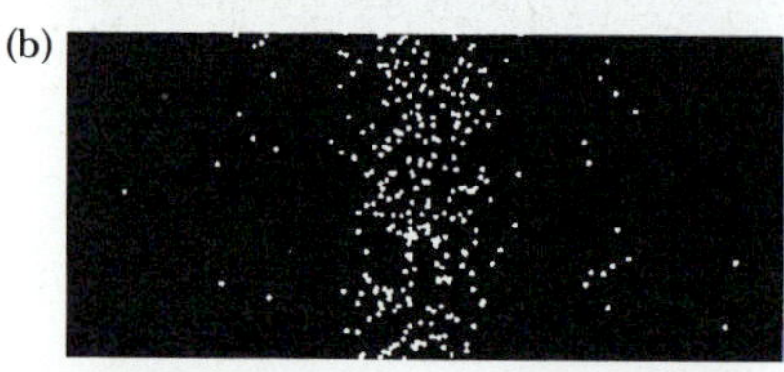

(c)

(d)

FIGURE 28.16

Patterns of impacts of photons on the faceplate in Fig. 28.15. (a) 30 photons; (b) 300 photons; (c) 3000 photons; (d) very large number of photons. The first three pictures are computer simulations of accumulated photon impacts; the last picture is a diffraction pattern obtained with a laser beam.

displays interference and diffraction phenomena, which prove that photons also have wave properties. Thus ***photons are neither classical particles nor classical waves.*** They are some new kind of object, unknown to classical physics, with a subtle combination of both wave and particle properties. Arthur Eddington has coined the name ***wavicle*** for this new kind of object.* It is difficult to achieve a clear understanding of the character of a wavicle because these objects are very remote from our everyday experience. We all have an intuitive grasp of the concept of a classical particle or of a classical wave from our experience with, say, billiard balls and water waves, but we have no such experience with wavicles.

We can gain some insight into the interplay between the particle behavior and the wave behavior of a photon by a new, detailed examination of a simple diffraction experiment. Figure 28.15 shows a light beam striking a plate with a narrow slit. In the usual diffraction experiments described in Chapter 26, we installed a screen or a photographic film at the far right, and with this we recorded the intensity of the light in the diffraction pattern. Instead, in our new arrangement we will install the faceplate of a very sensitive TV camera in place of the customary screen. With this, we can detect the individual photons of the light in the diffraction pattern.

If the incident light beam has a very low intensity, so there is only one photon passing through the slit at a time, we can watch the photons arriving one by one at the faceplate of our TV camera. Figure 28.16a shows a typical pattern of impacts of 30 photons. The pattern seems quite random. If the photons behaved like classical particles, they would travel along a straight line and they would reach only those points on the faceplate that are within the geometric image of the slit. The widely scattered impacts prove that the photons are certainly not traveling along such straight lines. Figure 28.16b shows the pattern of accumulated impacts for 300 photons; and Fig. 28.16c shows it for 3000 photons. In these figures, we can recognize a tendency of the photons to cluster in bandlike zones. These zones correspond to the maxima of the diffraction pattern predicted by the wave theory of light. Finally, Fig. 28.16d shows the pattern of accumulated impacts for a very large number of photons; this is simply the familiar intensity pattern of light diffracted by a slit (see also Fig. 26.25).

From this diffraction experiment we see that photons display both a wave aspect and a particle aspect. They behave like waves while passing through the slit; but they behave like particles when they strike the faceplate of the TV camera. Whether the wave aspect or the particle aspect predominates depends on the experimental equipment with which the photon is interacting. The slit brings out the wave aspect of the photon; the faceplate of the TV camera brings out the particle aspect.

The behavior of the photons is governed by a probabilistic law. The point of impact of an individual photon on the faceplate is unpredictable. Only the average distribution of impacts of a large number of photons is predictable: the distribution of photons matches the intensity distribution

* Other names that have been proposed are *quanticle* and *quon.* Most physicists prefer to avoid all such neologisms; if anything, they favor the descriptive phrase *quantum-mechanical particle* or *wave-mechanical particle.*

calculated from the wave theory of light. Thus, the probability that a photon arrives at a given point on the faceplate is proportional to the intensity of the wave at that point. Since the intensity of an electromagnetic wave is proportional to the square of its electric field, we can write the proportionality of probability and intensity as

$$[\text{probability for presence of a photon}] \propto E^2 \tag{19}$$

◀ **Probability interpretation of wave**

Here we have a connection between the wave and the particle aspects of the photon: ***the intensity of the photon wave at some point determines the probability that there is a photon particle at that point.*** This probability interpretation of the intensity of the wave was discovered by Max Born.

Max Born, *1882–1970, German, and later British, theoretical physicist. He was awarded the Nobel Prize somewhat tardily in 1954 for his discovery of the probabilistic interpretation of quantum waves in 1926.*

Our single-slit experiment can also teach us something about the limitations that quantum theory imposes on the ultimate precision of measurement of the position of a wavicle. Suppose we have a light wave consisting of one photon, and we want to measure the position of this photon. Of course, the position has x, y, and z components; we will concentrate on the y component, perpendicular to the direction of propagation. Figure 28.17 shows the wave propagating in the horizontal direction; the y direction is vertical. To determine this vertical position of the photon, we use a narrow slit placed in the path of the wave. If the photon succeeds in passing through this slit, then we will have achieved a determination of the vertical position to within an uncertainty

$$\Delta y = a \tag{20}$$

where a is the width of the slit. If the photon fails to pass through this slit, then our measurement is inconclusive and will have to be repeated.

By making the slit very narrow, we can make the uncertainty of our determination of the y coordinate very small. But this has a surprising consequence for the y component of the momentum of the photon: if we make the uncertainty in the y coordinate small, we will make the uncertainty in the y component of the momentum large. To see how this comes about, let us recall that according to our preceding discussion of the single-slit experiment, the photon suffers diffraction by the slit and emerges at some angle θ (Fig. 28.17). This angle θ is unpredictable; all we can say about the photon after it emerges from the slit is that it will be heading toward some point within the diffraction pattern. Thus, the direction of motion of the photon is uncertain. As a rough measure of the magnitude of this uncertainty in direction, we can take the angular width of the central diffraction maximum (most of the intensity of the photon wave is gathered within the region of this central maximum, and hence the photon is most likely to be found in this region). This estimate of the uncertainty of the angle gives us

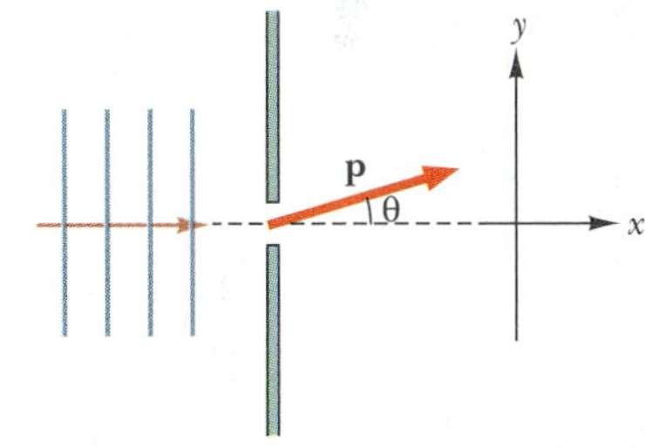

FIGURE 28.17

Photon passes through a narrow slit and emerges at an angle θ.

$$\Delta\theta \simeq \lambda/a \tag{21}$$

The y component of the momentum is $p_y = p \sin\theta$ (see Fig. 28.17); since we are concerned with a small angle, we can use the approximation

Werner Heisenberg, *1901–1976, German theoretical physicist. He was one of the founders of the new quantum mechanics, and received the Nobel Prize in 1932.*

$\sin\theta \simeq \theta$ and therefore $p_y \simeq p\theta$. The uncertainty in p_y is then

$$\Delta p_y \simeq p\Delta\theta \simeq p\lambda/a \tag{22}$$

But, according to Eq. (15), $p = h/\lambda$ and therefore

$$\Delta p_y \simeq h/a \tag{23}$$

Thus, if the slit is very narrow, the uncertainty in the y component of the momentum will be very large!

Comparing Eqs. (20) and (23), we find that the product of the uncertainty of position and the uncertainty of momentum is

$$\Delta y \Delta p_y \simeq h \tag{24}$$

This equation states that Δy ***and*** Δp_y ***cannot both be small; if one is small then the other must be large, so their product equals Planck's constant.***

Although we have obtained Eq. (24) by examining the special case of a position measurement by means of a slit, it turns out that this relation is actually of general validity for any kind of position measurement — the product of the uncertainty in the position and the uncertainty in the momentum is always roughly equal to or larger than Planck's constant. A more careful examination of a variety of arrangements for the simultaneous measurement of position and momentum shows that the product of the uncertainties rigorously obeys the inequality

Heisenberg's ▶ uncertainty relation

$$\Delta y \Delta p_y \geq \frac{h}{4\pi} \tag{25}$$

(The symbol $\geq$ means that the quantity on the left is bigger than or equal to the quantity on the right.) Equation (25) is one of ***Heisenberg's uncertainty relations.*** There are corresponding relations for the other components of position and momentum. The Heisenberg uncertainty relations tell us that ***there exist ultimate, insuperable limitations in the precision of our measurements.*** At the macroscopic level, the quantum uncertainties in our measurements can be neglected. But at the atomic level, these quantum uncertainties are often so large that it is completely meaningless to speak of the position or momentum of a wavicle.

EXAMPLE 6. Suppose we measure the vertical position of a photon by means of a narrow slit of width 10^{-5} m. With what vertical momentum uncertainty does the photon emerge from this slit?

SOLUTION: Equation (23) tells us that for such a position measurement

$$\Delta p_y \simeq \frac{h}{a} = \frac{6.6 \times 10^{-34}\,\text{J}\cdot\text{s}}{10^{-5}\,\text{m}} = 6.6 \times 10^{-29}\,\text{kg}\cdot\text{m/s}$$

SUMMARY

- **Energy quantization of oscillator:** $E = nhf$, $n = 0, 1, 2, \ldots$
- **Wien's Law:** $\lambda_{\max} = \dfrac{2.90 \times 10^{-3}\ \mathrm{m \cdot K}}{T}$
- **Stefan–Boltzmann Law:** $S = \left(5.67 \times 10^{-8} \dfrac{\mathrm{W}}{\mathrm{m^2 \cdot K^4}}\right) \times T^4$
- **Energy and momentum of a photon:** $E = hf$; $p = hf/c = h/\lambda$
- **Kinetic energy of photoelectron:** $K = hf - \phi$
- **Wavelength shift of photon (Compton effect):** $\Delta\lambda = \dfrac{h}{m_e c}(1 - \cos\theta)$
- **Probability interpretation of wave:**
  [probability for presence of photon] $\propto$ [intensity of wave]
- **Heisenberg's uncertainty relation:** $\Delta y \Delta p_y \geq h/4\pi$

QUESTIONS

1.–Is the light emitted by a neon tube thermal radiation? The light emitted by an ordinary incandescent light bulb?

2.–Does your body emit thermal radiation?

3.–The insulation used in the walls of homes consists of a thick blanket of fiberglass covered on one side by a thin aluminum foil. What is the purpose of these two layers?

4.–Black velvet looks much blacker than black paint. Why?

5.–For protection against the heat of sunlight, parts of the Lunar Lander (and some other spacecraft) were wrapped in shiny aluminum foil. Why is shiny foil useful for this purpose?

6.–If you look into a kiln containing pottery heated to a temperature equal to that of the walls of the kiln, you can scarcely see the pottery. Explain.

7.–Suppose that Planck's constant were much larger than it is, say, 10^{34} times larger. What strange behavior would you notice in a simple harmonic oscillator consisting of a mass hanging on a spring?

8.–Why do we not notice the discrete quanta of light when we look at a light bulb?

9.–Day-Glo paints achieve their exceptionally bright orange or red color by converting short-wavelength photons into long-wavelength (red) photons by inelastic interactions between the photons and the paint. Why can we not make such a paint in a blue or violet color?

10.–When light of a given wavelength ejects photoelectrons from the surface of a metal, why is it that not all of these photoelectrons emerge with the same kinetic energy?

11.–According to Eq. (16), a photon suffers a maximum change of wavelength in a collision with an electron if it emerges at an angle $\theta = 180°$, and a minimum change of wavelength (no change) if it emerges at an angle $\theta = 0°$. Is this reasonable?

12.–Can the Compton effect occur with visible light? Would it be observable?

13.–Give an example of an experiment in which photons behave like waves. Give an example of an experiment in which they behave like particles.

14.–If photons were classical particles, what pattern of impact points would we find in the diffraction experiment that led to the results described in Fig. 28.16a–d?

15.–In classical physics, the behavior of a system is deterministic, that is, from the initial positions of the particles in the system we can predict the positions at any later time. In quantum physics, according to Eq. (19), we can calculate only the *probabilities* that a photon or some other particle will be found at some given point at some time. Does this mean that the behavior of photons is unpredictable and undeterministic?

PROBLEMS

Section 28.1: Blackbody Radiation
Section 28.2: Energy Quanta

1.——Consider a seconds pendulum, that is, a pendulum that has a period of 2 seconds. What is the magnitude of one energy quantum for such a pendulum? Would you expect that quantum effects are noticeable in such a pendulum?

2.——In molecules, the atoms can vibrate about their equilibrium positions. For instance, in the H_2 molecule, the hydrogen atoms vibrate about their equilibrium positions with a frequency of 1.13×10^{14} Hz. What is the magnitude of the energy quantum for this oscillating system? Note that although there are two masses in this system, they must be regarded as a single oscillator because the vibrational motions of the two masses are always equal.

3.——In a solid, each atom is held in its position by elastic forces, which permit the atom to oscillate about its equilibrium point. The frequency of such oscillations of an aluminum atom in solid aluminum is 8×10^{11} Hz. The energy of this oscillatory motion is quantized. What is the magnitude of an energy quantum?

4.*——If we take Planck's model of the walls of a blackbody cavity seriously, we will have to assume that the oscillating masses are electrons. Imagine an electron oscillating with a frequency of 2×10^{15} Hz under the influence of a springlike force. What is the amplitude of oscillation of this electron if its energy of oscillation is one energy quantum? Two energy quanta?

5.*——An oscillator of frequency 2×10^{15} Hz consists of a mass of 9.1×10^{-31} kg attached to a spring. What is the amplitude of oscillation of this oscillator if its energy of oscillation is one energy quantum? Two energy quanta?

6.——Suppose that two stars have the same size, but the temperature of one is twice that of the other. By what factor will the thermal power radiated by the hotter star be larger than that radiated by the cooler star?

7.——The theoretical expression for the constant σ in Wien's Law is

$$\sigma = \frac{2\pi^5 k^4}{15h^3c^2}$$

Verify that the numerical value implied by this theoretical expression agrees with Eq. (7).

8.——The flux of thermal radiation from the star Procyon B is observed to have a maximum at a wavelength of 440 nm. Assuming the star radiates like a blackbody, what temperature can you deduce for this star?

9.——Melting iron has a temperature of 1808 K. At what wavelength does the iron radiate a maximum flux? Assume that the iron acts like a blackbody.

10.*—Interplanetary and interstellar space is filled with thermal radiation of a temperature of 2.7 K left over from the Big Bang.

(a) At what wavelength is the flux of this radiation maximum?

(b) What is the power incident on the surface of the Earth due to this radiation?

11.*—The tungsten filament of a light bulb is a wire of diameter 0.080 mm and length 5.0 cm. The filament is at a temperature of 3200 K. Calculate the power radiated by the filament. Assume the filament acts like a blackbody.

12.*—The temperature of the Sun is 5800 K and its radius is 6.96×10^8 m. If the Sun radiates like a blackbody, what can you predict for the thermal energy flux it emits? What can you predict for the energy flux of sunlight arriving at the Earth, at a distance of 1.5×10^{11} m? Compare with the measured value, 1.35×10^3 W/m^2.

13.*—The star Procyon B is at a distance of 11 light-years from Earth. The flux of its starlight reaching us is 1.7×10^{-12} W/m^2, and the surface temperature of the star is 6600 K. Calculate the size of the star.

14.**—At the Earth, the flux of sunlight per unit area facing the Sun is 1.35×10^3 W/m^2. The Earth absorbs heat from the sunlight and reradiates heat as thermal infrared radiation. For equilibrium, the power arriving from the Sun must equal the average power radiated by the surface of the Earth. This permits us to make a rough prediction for the average temperature of the Earth.

(a) Assume that the Earth absorbs all of the sunlight striking it. What is the power of the sunlight absorbed? (Hint: The relevant area is the cross-sectional area πR^2 of the Earth.)

(b) Assume that the surface of the Earth radiates like a blackbody. If the temperature of the surface is T, what is the expression for the power radiated? (Hint: The relevant area is the total surface area $4\pi R^2$.)

(c) By comparing (a) and (b), deduce the temperature of the Earth.

15.**—Deduce the surface temperature of Pluto by the method described in the preceding problem. Pluto is 39 times as far away from the Sun as the Earth, and hence the flux of sunlight at Pluto is $(39)^2$ times smaller than the flux at the Earth.

16.**—If you stand naked in a room, your skin and the walls of the room will exchange heat by radiation. Suppose the temperature of your skin is 33°C; the total area of your skin is 1.5 m^2. The temperature of the walls is 15°C. Assume your skin and the walls behave like blackbodies.

(a) What is the rate at which your skin radiates heat?

(b) What is the rate at which your skin absorbs heat? What is your net rate of loss of heat?

Section 28.3: Photons and the Photoelectric Effect

17.——Photons of green light have a wavelength of 550 nm. What is the energy and what is the momentum of one of these photons?

18.——You are lying on a beach, tanning in the sun. Roughly how many photons strike your skin in one hour? Assume that the energy flux of sunlight is as described in Example 3.

19.——For each of the following kinds of electromagnetic waves, find the energy of a photon: FM radio of wavelength 3 m, infrared light of 1×10^{-5} m, visible light of 5×10^{-7} m, ultraviolet light of 1×10^{-7} m, X rays of 1×10^{-10} m.

20.——A radio transmitter radiates 10 kW at a frequency of 8.0×10^{5} Hz. How many photons does the transmitter radiate per second?

21.——The energy flux in the starlight reaching us from the bright star Capella is 1.2×10^{-8} W/m^2. If you are looking at this star, how many photons per second enter your eyes? The diameter of your pupil is 0.70 cm. Assume that the average wavelength of the light is 500 nm.

22.——A laser emits a light beam with a power of 1 W. The wavelength of the light is 630 nm. How many photons per second does this laser emit?

23.——The energy density of starlight in intergalactic space is 10^{-15} J/m^3. What is the corresponding density of photons? Assume the average wavelength of the photons is 500 nm.

24.——An incandescent light bulb radiates 40 W of thermal radiation from a filament of temperature 3200 K. Estimate the number of photons radiated per second; assume that the photons have an average wavelength equal to the λ_{max} given by Wien's Law.

25.*——If you want to make a very faint light beam that delivers only 1 photon per square meter per second, what must be the amplitude of the electric field in this light beam? The wavelength of the light is 500 nm.

26.*——Show that if we express the energy of a photon in keV and the wavelength in nanometers, then

$$E = 1.24/\lambda$$

27.——According to Fig. 28.7, what is the work function of sodium? Express your answer in electron-volts.

28.——The work function of potassium is 2.26 eV. What is the threshold frequency for the photoelectric effect in potassium?

29.——The work functions of K, Cr, Zn, and W are 2.26, 4.37, 4.24, and 4.49 eV, respectively. Which of these metals will emit photoelectrons when illuminated with red light ($\lambda = 700$ nm)? Blue light ($\lambda = 400$ nm)? Ultraviolet light ($\lambda = 280$ nm)?

30.——The photons emitted by a sodium atom have a wavelength of 589 nm when at rest. If this atom is moving away from you at a speed of 2×10^{6} m/s, what is the energy that you measure for one of these photons?

31.*——By inspection of Fig. 28.7, find the slope of the line in eV/Hz. Convert these units into J · s, and verify that the slope is the same as Planck's constant.

32.*——The binding energy of an electron in a hydrogen atom is 13.6 eV. Suppose that a photon of wavelength 40 nm strikes the atom and gives up all of its energy to the electron. With what kinetic energy will the electron be ejected from the atom?

Section 28.4: The Compton Effect

33.——X rays emitted by molybdenum have a wavelength of 0.072 nm. What are the energy and the momentum of one of the photons in these X rays?

34.——In a collision with an initially stationary electron, a photon suffers a wavelength increase of 0.0022 nm. What must have been the deflection angle of this photon?

35.——For an experiment on the Compton effect, you want the X rays emerging at 90° from the incident direction to suffer an increase of wavelength by a factor of 2. What wavelength do you need for your incident X rays?

36.——A photon of wavelength 0.030 nm collides with a free electron at rest. Calculate the wavelength if the photon emerges from this collision with a deflection of 30°.

37.——X rays of wavelength 0.030 nm are incident on a graphite target. Calculate the wavelength of the X rays that emerge from this target with a deflection of 60°. Calculate the wavelength of the X rays that emerge from this target with a deflection of 120°.

38.*—What is the maximum energy that a free electron (initially stationary) can acquire in a collision with a photon of energy 4.0×10^3 eV?

39.*—In a collision with a free electron, a photon of energy 2.0×10^3 eV is deflected by 90°. What energy does the electron acquire in this collision?

40.*—A photon of energy 1.6×10^8 eV collides with a *proton* initially at rest. The photon is deflected by 45°. What is its new energy?

41.*—A photon of initial wavelength 0.040 nm suffers two successive collisions with two electrons. The deflection in the first collision is 90° and in the second collision it is 60°. What is the final wavelength of the photon?

Section 28.5: X Rays

42.——What is the energy of the photons in X rays of a wavelength of 0.050 nm?

43.——An X-ray tube is being operated with electrons of energy 25 keV. What is the cutoff wavelength of the emitted X rays?

44.——We want to use an X-ray tube to generate X rays of wavelength 1 nm. What is the minimum potential difference we must use to accelerate the electrons in the tube?

45.——The tube in a medical X-ray machine can be operated at potential differences in the range from 25 kV to 150 kV. What is the cutoff wavelength of the X rays emitted when the tube is operated at 25 kV? At 150 kV?

46.*—Figure 28.14 shows two of the discrete spectral lines of molybdenum. According to this figure, what are the wavelengths of these spectral lines? Would these spectral lines appear if instead of using electrons of 35 keV we were to use electrons of 19 keV in the X-ray tube?

Section 28.6: Wave vs. Particle

47.——A photon passes through a horizontal slit of width 5×10^{-6} m. What uncertainty in the vertical position will this photon have as it emerges from the slit? What uncertainty in the vertical momentum?

48.*—Consider a radio wave in the form of a pulse lasting 0.001 s. This pulse then has a length of $0.001\ \text{s} \times c = 3 \times 10^5$ m. Since an individual photon of this radio wave can be anywhere within this pulse, the uncertainty in the position of the photon is $\Delta x = 3 \times 10^5$ m along the direction of propagation.

(a) According to Heisenberg's relation, what is the corresponding uncertainty in the momentum of the photon?

(b) What is the uncertainty in the frequency of the photon?

PHYSICS IN ACTION

X-RAY IMAGES

X rays are exploited in medicine to make images of the internal organs of the human body. In an X-ray image, or ***radiograph,*** we see the shadows of the internal organs (Fig. 1). To make such an image, the X-ray tube is placed on one side of the body and a photographic plate or a fluorescent screen is placed on the opposite side. X rays propagate through the body along straight lines, without refraction. However, during their propagation through the tissues of the body, the X rays are gradually absorbed. Tissues that absorb strongly cast strong shadows; tissues that absorb weakly cast only weak shadows. Thus, the X-ray image is a map of the X-ray absorption of the tissues within the body.

The main absorption process for X rays arises from interactions between the X rays and the electrons in the atoms. Hence atoms with a large number of electrons are the strongest absorbers of X rays. Coincidentally, atoms with a large number of electrons are heavy atoms, and materials containing such atoms are dense. For instance, lead is an excellent absorber of X rays, and it is often used in protective shields to block X rays. The dense calcium in the bones of the human body also is a good absorber of X rays; but the less dense oxygen and hydrogen in the soft (watery) tissues are only weak absorbers of X rays. The difference between the X-ray absorption of bones and soft tissues is well illustrated in Fig 1, where the bones of the spine, lower ribs, and pelvis cast strong and sharp shadows; but the soft abdominal tissues cast only faint shadows, and the abdominal organs are barely visible. Note that the imaging of the soft tissues suffers not only from weak absorption, but also from a lack of contrast — the densities and absorption characteristics of all the kinds of soft tissues are quite similar, and therefore the boundaries of adjacent organs made of soft tissues tend to blur.

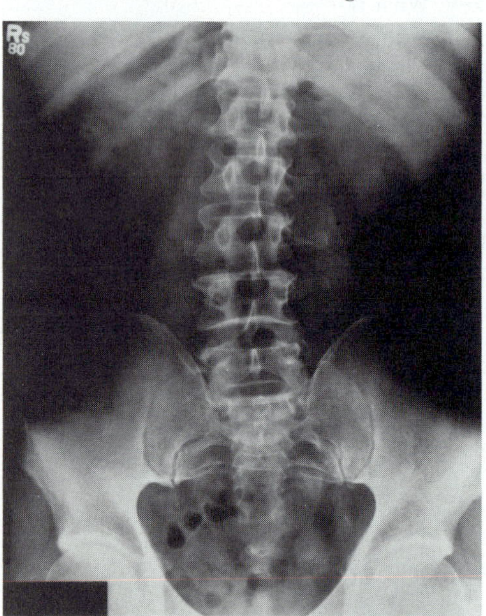

FIGURE 1 X-ray image of human abdomen. Like all X-ray images, this image is a negative — the bright areas indicate low exposure of the film (little blackening). The dense bones on the spine, lower ribs, and pelvis cast strong and sharp shadows, and they show up as bright areas in the X-ray image. Note that bubbles of gas in the intestine show up as dark areas (lower left quadrant), because they produce next to no shadowing (high exposure of the film, and heavy blackening).

Radiologists soon recognized that the visibility of soft tissues could be greatly enhanced by introducing ***contrast agents*** into the body, by ingestion or by injection. These contrast agents consist of chemical compounds that contain heavy atoms, such as barium or iodine, which are good absorbers of X rays. Barium is routinely used to enhance the contrast of X-ray images of the gastrointestinal tract. For example, Fig. 2 shows an X-ray image of the abdomen of a patient who has ingested a suspension of barium sulfate. The stomach and much of the small intestine, which were not visible in Fig. 1, are now strongly shadowed. Other contrast agents are used to enhance the imaging of the gallbladder (cholecystography), the bile ducts (cholangiography), the urinary tract (pyelography), and the blood vessels (angiography).

When using contrast agents, radiologists prefer to display the X-ray image on a fluorescent screen instead of a photographic plate. The impact of X rays on such a fluorescent screen produces pinpoints of visible light, which result in a glowing image, much like the image on a TV screen produced by the impact of electrons. With such a device, called a ***fluoroscope,*** the radiologist can view the X-ray image directly, and he can examine the motion of the contrast medium within the patient's body. In modern versions of the fluoroscope, a

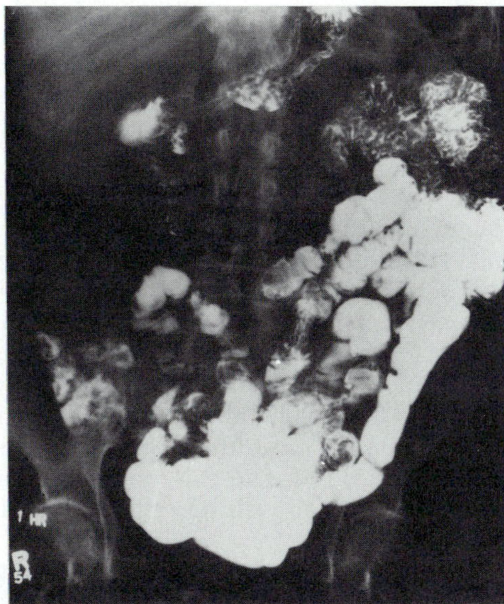

F I G U R E 2 X-ray image of the small intestine enhanced by barium. Some slight traces of barium remain in the stomach (specks in upper left quadrant), but most of the barium has passed into the small intestine, revealing the loops of the bowel.

video camera is attached to the fluorescent screen, and the radiologist can view the image on a video monitor, at a safe distance.

A further refinement of the contrast technique has been developed for the examination of the arteries of the brain or of other organs. In ***digital subtraction angiography (DSA),*** first an X-ray image of the organ is made without the contrast agent, and this image is converted to digital form and stored in the memory of a computer. Next an iodine contrast agent is injected into the arteries by means of a catheter, and a second image is made and converted to digital form. The computer then subtracts the first image from the second, and displays the result on a video monitor. In this subtraction, the portions of the two images that are identical cancel out, giving a uniform dark background; whereas the portions that were altered by the contrast agent stand out sharply. The subtraction eliminates all the irrelevant and murky clutter from the background and gives a sharp and clear picture of the arteries, revealing any obstructions or other abnormalities (see Fig. 3).

A severe disadvantage of conventional X-ray pictures, with or without contrast agents, is that such pictures are merely silhouettes — they merely present a collapsed, or projected, view of the body, as through the body had been flattened against the film or the fluorescent screen. For example, in the X-ray images of the abdomen (Figs. 1 and 2), organs lying at the front of the body are projected over those behind them, so their shadows overlap. This overlapping of shadows causes some difficulties for the physician attempting to interpret the image, and it is often necessary to make two or more images with different directions of projection, to get a better perception of the three-dimensional relationships among the organs within the body. The use of multiple X-ray projections is exploited to its fullest extent in a new technique called ***computed tomography (CT),*** which provides all of the missing three-dimensional information. Tomography gives us a view of a slice through the body (the word *tomography* comes from the Greek *tomos,* meaning slice). The CT scanner that constructs such a view uses a fan-shaped beam of X rays (see Fig. 4). After this X-ray beam passes through a narrow slice of the body, it reaches an array of detectors placed along an arc on the opposite side. The X-ray source and the detectors are rotated around the body, so as to scan the selected slice from many different directions, at intervals of one degree or less. By this procedure, a computer connected to the detectors

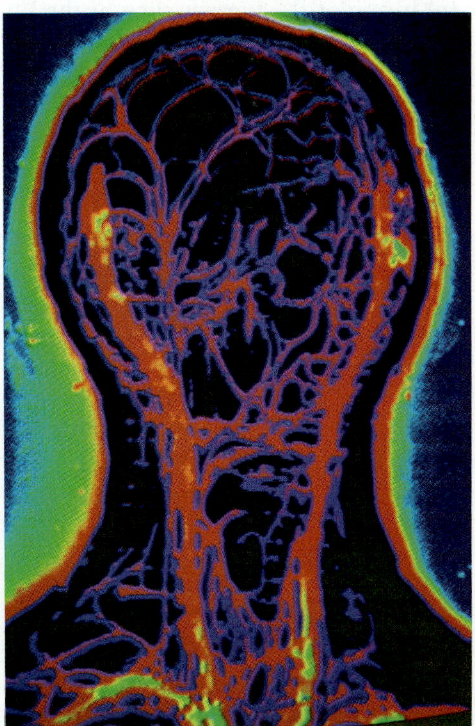

F I G U R E 3 Image of the arteries in the brain generated by DSA.

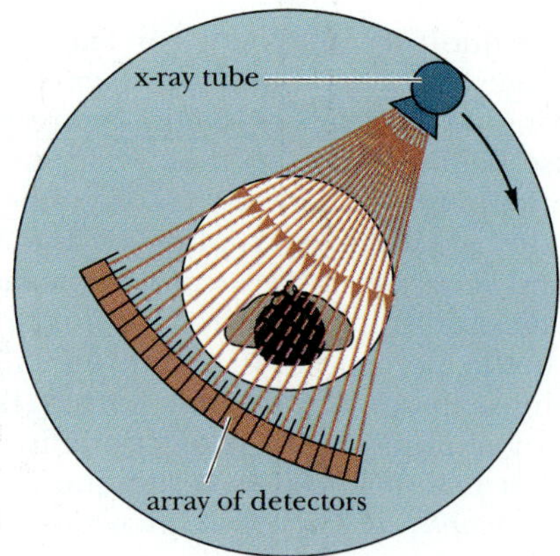

FIGURE 4 In this CT scanner, the X-ray source generates a fan-shaped beam of X rays which passes through one slice of the body and reaches an array of detectors placed along an arc on the opposite side. The source and the detectors rotate around the body, giving different views of the same slice from different angles.

acquires many views of the slice with many different directions of projection. Using a mathematical operation of "back projection," the computer then reverses the acquisition process and reconstructs the many closely spaced projections into a cross-sectional image of the narrow slice of the body.

Figure 5 is a tomographic image showing a slice through the skull, at eye level. This slice clearly displays the anatomical structures within the skull, without the troublesome overlap of shadows. By taking a sequence of CT images of adjacent slices though the body, we can construct a full three-dimensional view. Such three-dimensional CT images are especially useful in the examination of complicated fractures of the bones. For example, Fig. 6 shows a three-dimensional tomographic image of a broken vertebra.

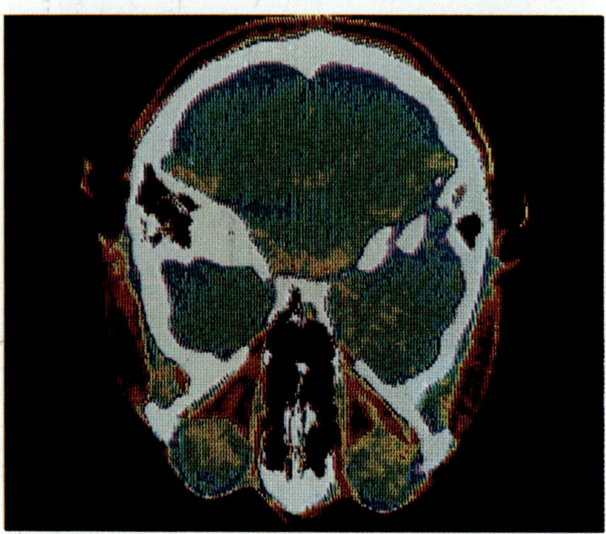

FIGURE 5 CT image of the skull.

The X rays used in all the above imaging techniques and also the gamma rays used in some other imaging techniques (see *Medical Imaging*, pp. 893A–893D) belong to a type of radiation called ***ionizing radiation.*** This kind of radiation also includes alpha particles, beta particles, and fast-moving neutrons (see Chapter 31). When any of these rays pass through some kind of material, their collisions with the atoms produce ions in the material. In living tissues, the ions produced in this way cause harm to the cells. The effect of ionizing radiation on the material depends on the amount of radiation energy absorbed by the material. The ***absorbed dose*** of ionizing radiation is measured in ***grays*** (Gy). The dose of radiation is 1 gray if the energy absorbed from the radiation is 1 joule per kilogram of tissue,

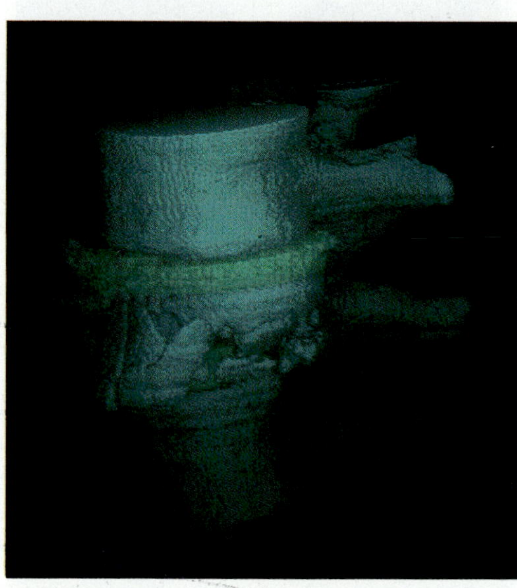

FIGURE 6 Three-dimensional CT image of a broken vertebra. This image was assembled out of several dozen CT images of adjacent slices through the vertebra.

$$1 \text{ gray} = 1 \text{ Gy} = 1 \text{ joule/kg} \qquad (1)$$

Some kinds of rays, such as alpha rays and neutrons, are more harmful to living tissues than others. To obtain a relative measure of the biological damage produced by different rays, the absorbed dose must be multiplied by a correction factor. The absorbed dose corrected in this way is called the ***equivalent absorbed dose.*** For X rays and gamma rays, the equivalent absorbed dose is the same as the absorbed dose (no correction factor); for alpha rays and neutrons, the equivalent absorbed dose is larger than the absorbed dose, in accord with the larger harm caused by these rays. To prevent confusion between these two kinds of doses, the unit for the equivalent dose has been given a new name; it is called a ***sievert*** (Sv). For the dose levels encountered in diagnostic procedures, smaller units called the milligray (mGy) and the millisievert (mSv) are often preferred.*

At very high doses, ionizing radiation can be fatal within a few days ("radiation sickness"). Moderately high doses cause delayed effects, such as cancers appearing years after the exposure and also genetic mutations in the offspring of the exposed individuals. The risk of radiation-induced cancers of all types is estimated at about 1 in 10,000 for an exposure of 10 mSv; thus, if each of 10,000 individuals is exposed to 10 mSv, there would be one additional case of cancer, in excess of cancers from other causes. For larger or smaller exposures than 10 mSv, the risk increases or decreases proportionately. The risk of genetic mutations is also estimated at about 1 in 10,000 for an exposure of 10 mSv; thus, if each of 10,000 individuals is exposed to 10 mSv, and each produces one child, there would be one additional case of genetic disorders. This contrasts with the natural rate of genetic disorders, estimated at about 10% of all live births.

* Two older and still widely used units are the rad, which equals 0.01 Gy, and the rem, which equals 0.01 Sv.

TABLE 1 RADIATION DOSES FROM DIAGNOSTIC X RAYS

Diagnostic procedure	Effective whole-body equivalent dose
CT scan of abdomen	15.0 mSv
CT scan of chest	7.1
CT scan of skull	2.2
Lumbar spine series (5 views)	0.90
Mammography (2 views)	0.75
Skull series (3 views)	0.05
Chest	0.02
Dental (whole mouth)	0.1–0.3

The radiation doses that a patient receives in some common diagnostic X-ray procedures are listed in Table 1. All the values are approximate; the dose that a patient receives in an X-ray procedure may vary by a factor of 2 or more, depending on machine settings, patient size, and other factors. Note that the values listed in the table are "effective whole-body doses." This is the dose of uniform whole-body exposure that produces the same risk as the actual localized exposure of the target organ and the surrounding organs (the actual localized exposure of the target organ is much higher than the value quoted in the table). It is of interest to compare the radiation doses from diagnostic procedures with the dose we receive from natural sources. We are all exposed to ionizing radiation contributed by cosmic rays incident from outer space and by radioactive elements in our environment. This "natural background radiation" contributes a dose of about 3 mSv per year for the average inhabitant of the United States.

REFERENCES

Hall, E. J., *Radiobiology for the Radiologist* (J. B. Lippincott, Philadelphia, 1988).

Laws, P. W., and The Public Citizen Health Research Group, *The X-Ray Information Book* (Farrar, Strauss, and Giroux, New York, 1984).

Putnam, C. E., and Ravin, C. E., eds., *Textbook of Diagnostic Imaging* (W. B. Saunders, Philadelphia, 1988).

Upton, A. C., "Health Effects of Low-Level Ionizing Radiation," *Physics Today*, August 1991.

Upton, A. C., "The Biological Effects of Low-Level Ionizing Radiation," *Scientific American*, February 1982.

Webb, S., ed., *The Physics of Medical Imaging* (Hilger, Bristol, 1988).

CHAPTER 29

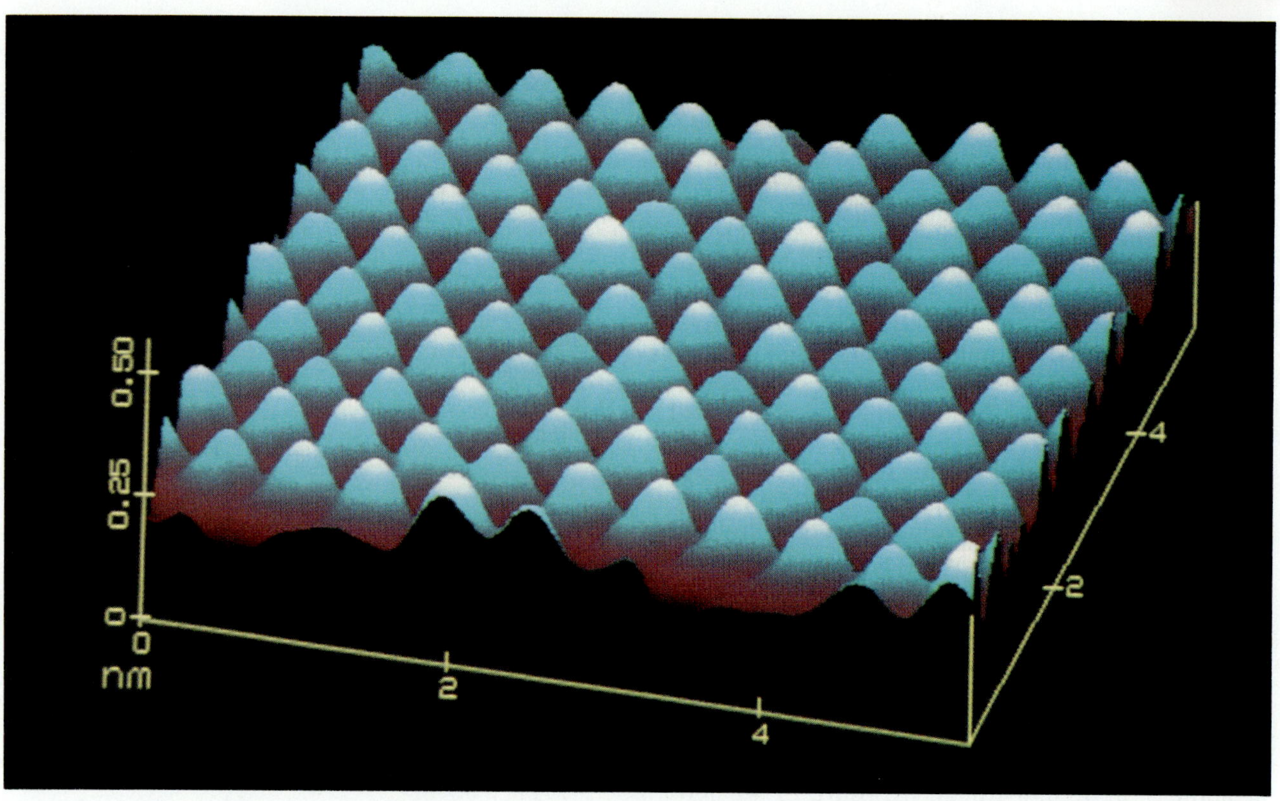

Atoms of sodium and chlorine on the surface of a crystal of ordinary table salt. This image was prepared with an Atomic Force Microscope (AFM). The scale is in nanometers.

SPECTRAL LINES AND BOHR'S THEORY

The photograph above gives convincing visual evidence that solids, liquids, and gases are made of atoms, small grains of matter with a diameter of about 10^{-10} m. This photograph was prepared with a powerful microscope of a special design. Unfortunately, not even this microscope is sufficiently powerful to reveal the inside of the atom. Hence, for the exploration of the internal structure of the atom, we still have to rely on the technique developed by Ernest Rutherford and his associates around 1910: bombard the atom with a beam of particles and use this beam as a probe to "feel" the interior of the atom.

By 1910, most physicists had come to believe that atoms are made of some combination of positive and negative electric charges, and that the attractions and repulsions between these electric charges are the basis for all the chemical and physical phenomena observed in solids, liquids, and gases. Since electrons were known to be present in all of these forms of matter, it seemed reasonable to suppose that each atom consists of a combination of electrons and positive charge. The vibrational motions of the electrons within such an atom would then result in the radiation of electromagnetic waves; this was supposed to account for the emission of light by the atom. However, both the arrangement of the electric charges within the atom and the mechanism that accounts for the characteristic colors of the emitted light remained mysteries until Rutherford's discovery of the nucleus and Niels Bohr's discovery of the quantization of atomic states. In this chapter we will look at these two momentous discoveries.

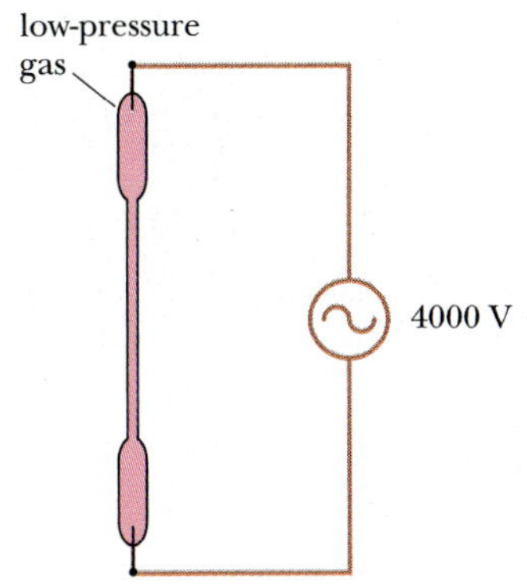

FIGURE 29.1

An electric-discharge tube. The tube contains gas at a very low pressure. When the terminals are connected to a high-voltage generator, an electric current flows through the gas and makes it glow.

The exploration of the internal structure of the atom led to the inescapable conclusion that ***in the atomic realm Newton's laws of motion are not valid.*** Electrons and other subatomic particles obey new equations of motion that are drastically different from the old equations of motion obeyed by planets, billiard balls, or shotgun pellets. ***The new theory of motion that rules the realm of the atom is called quantum mechanics.*** In contrast, the old theory of Newton is called ***classical mechanics.*** The discovery of an entirely new set of laws of motion was the greatest scientific revolution of this century.

◀ **Quantum mechanics vs. classical mechanics**

29.1 SPECTRAL LINES

The earliest attempts at a theory of atomic structure ended in failure — these early theories were not able to explain the characteristic colors of the light emitted by atoms. These colors show up very distinctly when a small sample of gas is made to emit light by the application of heat or of an electric current. For instance, if we put a few grains of ordinary salt into a flame, the sodium vapor released by the salt will glow with a characteristic yellow color. If we put neon gas into an evacuated glass tube and connect the ends of the tube to a high-voltage generator (Fig. 29.1), the gas will glow with the familiar orange-red color of neon signs (Fig. 29.2).

FIGURE 29.2

A neon sign.

The light emitted by an atom can be precisely analyzed with a prism (Fig. 29.3); this breaks the light up into its component colors. In the arrangement shown in Fig. 29.3, each discrete color generates a bright line. These are the ***spectral lines.*** Each kind of atom has its own discrete spectral lines. The color plate on page 834F shows the spectral lines of hydrogen, sodium, helium, and mercury; the numbers next to the spectral lines give the wavelengths in nanometers. Hydrogen has four spectral lines in the visible region (already mentioned in Chapter 25) and many ultraviolet and infrared lines not visible to the human eye. Obviously, the set of spectral lines, or ***spectrum,*** belonging to hydrogen is unmistakably different from those of sodium, helium, and mercury — the spectrum of an atom can serve as a fingerprint for its identification.

◀ **Spectrum**

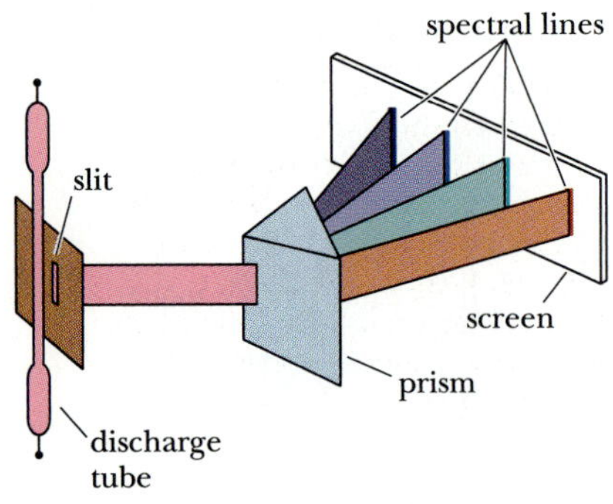

FIGURE 29.3

Analysis of light by means of a prism. In this arrangement, each separate color of the light emerging from the slit gives rise to a separate spectral line on the screen.

Spectroscopy provides us with a useful alternative to the traditional "wet" analysis familiar to students of chemistry. By heating a small sample of an unknown substance — preferably by means of an electric discharge — and inspecting the spectral lines, we can identify the kinds of atoms in the sample. In spectroscopy laboratories, scientists often perform the quantitative analysis of samples of atoms with absorption lines rather than emission lines. It so happens that an atom capable of emitting light of a given wavelength is also capable of absorbing light of that wavelength. When we illuminate a sample of atoms with white light (a mixture containing all colors or wavelengths), the atoms will absorb light of their characteristic wavelengths and, upon analyzing the remaining light with a prism, we find dark absorption lines in the continuous background generated by the white light. The last picture in the color plate shows such an absorption spectrum for sodium vapor. The dark lines of this absorption spectrum coincide with the bright lines in the emission spectrum (see the second picture in the color plate).

One advantage of spectroscopy over chemistry is that the analysis can be performed even on minuscule amounts of material. What is more, atoms can be identified at a distance. For example, we can identify the atoms on the surface of the Sun by careful analysis of the distribution of colors in sunlight — we do not need to pluck a sample of atoms from the Sun. The power of this technique is best illustrated by the story of the discovery of helium (the "Sun element"). In 1868, this gas was yet unknown to chemists when an astronomer discovered it on the Sun by means of its spectral lines; 30 years later chemists finally found traces of helium in minerals on the Earth. By spectroscopic techniques, astronomers can identify atoms in remote stars, clouds of interstellar gas, galaxies, and quasars. For example, Fig. 29.4 shows the spectrum of the star Caph in the constellation Cassiopeia; the spectral lines indicate the presence of hydrogen, calcium, iron, manganese, chromium, etc. Using spectroscopy, astronomers can perform a "chemical analysis" of the material on this star, even though it is 45 light-years away.

Note that only a small part of the light from a star is in the form of discrete spectral lines from individual atoms on the stellar surface. Most of the starlight is thermal radiation, a continuous mixture of all colors. This thermal radiation is produced in the stellar interior. As we know from the preceding chapter, this kind of light does not retain the fingerprint of the atoms that produced it. Light originating in the stellar interior cannot escape directly, but is first tossed back and forth (scattered) many times by the restless atoms of the hot stellar gas. The random motion of these atoms communicates random changes of wavelength to the light, and what finally emerges from the stellar interior is thermal radiation consisting of a continuous mixture of a wide range of wavelengths.

FIGURE 29.4

A portion of the spectrum of light from the star Caph (β Cassiopeiae), from 390 nm to 450 nm. The strong absorption lines on the left are due to ionized calcium. The other two strong lines (middle and right) are due to hydrogen.

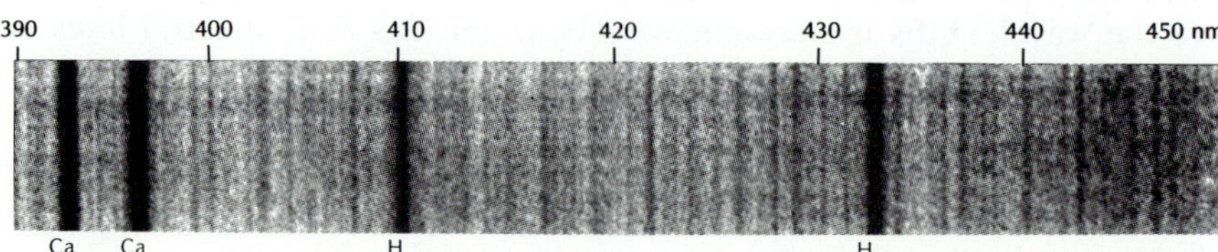

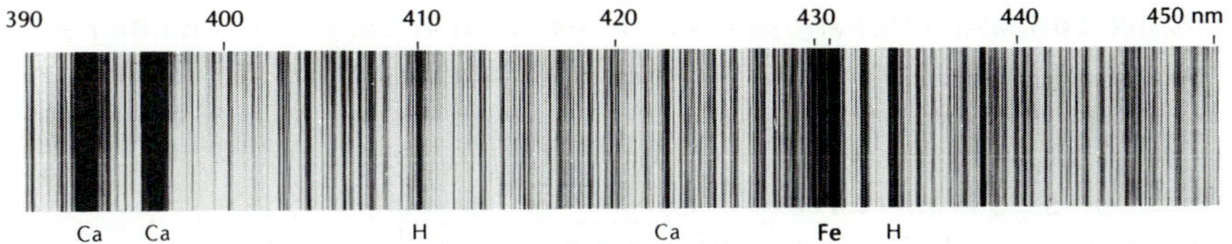

FIGURE 29.5

A portion of the spectrum of light from the Sun, from 390 nm to 450 nm. Besides the strong absorption lines of calcium and hydrogen, similar to those in Fig. 29.4, there are also strong lines of iron (close pair, center), and many other somewhat weaker lines.

Figure 29.5 shows a portion of the spectrum of the white light from our Sun. White light is a continuous and nearly uniform mixture of all colors. However, the high-resolution spectrum in Fig. 29.5 displays many dark lines in the spectrum of the Sun, caused by absorption in the gas surrounding the Sun. These are called the ***Fraunhofer lines.***

◀ Fraunhofer lines

29.2 THE BALMER SERIES

Careful examination of the spectral lines produced by an element reveals certain systematic regularities in the spacings of the lines. These regularities sometimes become especially striking if, instead of examining only the visible region of the spectrum, we examine both the visible region and the adjacent ultraviolet and infrared regions. For instance, Fig. 29.6 shows the spectrum of hydrogen in the visible and the near ultraviolet regions, as recorded on a photographic film. We notice immediately that the spacings of the lines and their intensities decrease systematically as we look at shorter and shorter wavelengths. The hydrogen lines in Fig. 29.6 are said to form a ***spectral series.*** In the spectra of other elements we find similar series; however, the spectra usually contain several overlapping series, and this makes it a bit harder to perceive the regularities in the spacings.

◀ Spectral series

The systematic pattern in the spacing of the spectral lines of hydrogen shown in Fig. 29.6 suggests that the wavelengths of these lines should be described by some simple mathematical formula. Table 29.1 lists the wavelengths of the first few of these spectral lines; there actually is an infinite number of spectral lines, the spacing between them becoming

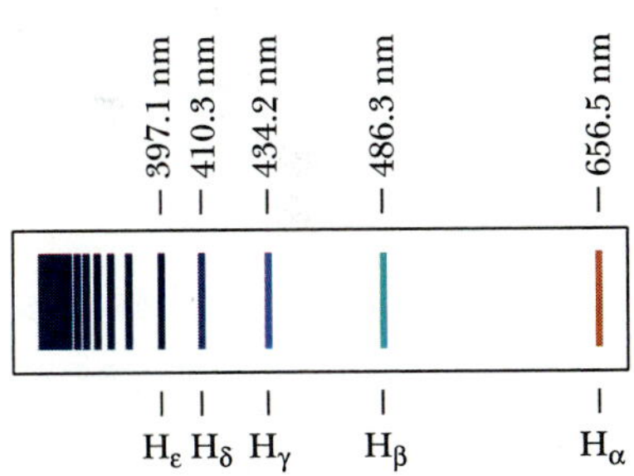

FIGURE 29.6

Spectrum of hydrogen in the visible and near ultraviolet regions. The spectral lines of this spectral series are consecutively labeled H_α, H_β, H_γ, and so on. The numbers give the wavelengths of the spectral lines in nanometers.

TABLE 29.1 THE BALMER SERIES IN THE HYDROGEN SPECTRUM

Wavelength λ[a]
656.47 nm
486.27
434.17
410.29
397.12
389.02
383.65
379.90
etc.

[a] Wavelengths are measured in vacuum.

smaller and smaller at shorter wavelengths. In 1855, Johann Balmer scrutinized the numbers in such a table and discovered that the wavelengths accurately fit the formula

$$\frac{1}{\lambda} = \frac{1}{91.176\ \text{nm}}\left(\frac{1}{4} - \frac{1}{n^2}\right) \tag{1}$$

Balmer series ▶

with $n = 3, 4, 5, \ldots$ This infinite series of spectral lines is called the ***Balmer series.*** Note that if n approaches infinity, the wavelength approaches the ultimate value $\lambda = 4 \times 91.176\ \text{nm} = 364.70\ \text{nm}$; this is called the ***series limit.***

Balmer's formula is usually written compactly as

$$\frac{1}{\lambda} = R\left(\frac{1}{2^2} - \frac{1}{n^2}\right) \tag{2}$$

where R is called the ***Rydberg constant,***

Rydberg constant ▶

$$R = \frac{1}{91.176\ \text{nm}} = 1.09678 \times 10^7\ \text{m}^{-1} \tag{3}$$

Balmer's formula was purely descriptive, or phenomenological; it did not explain the atomic mechanism responsible for the production of the spectral lines. Nevertheless, it proved very fruitful because it led to more general formulas describing other series of spectral lines. Balmer proposed that there might be other series in the hydrogen spectrum, with the 2^2 in the parentheses in Eq. (2) replaced by 1^2, or 3^2, or 4^2, etc. This yields the series of wavelengths

Lyman, Paschen, and Brackett series ▶

$$\frac{1}{\lambda} = R\left(\frac{1}{1^2} - \frac{1}{n^2}\right) \qquad n = 2, 3, 4, \ldots \tag{4}$$

$$\frac{1}{\lambda} = R\left(\frac{1}{3^2} - \frac{1}{n^2}\right) \qquad n = 4, 5, 6, \ldots \tag{5}$$

$$\frac{1}{\lambda} = R\left(\frac{1}{4^2} - \frac{1}{n^2}\right) \qquad n = 5, 6, 7, \ldots \tag{6}$$

These series of spectral lines were actually discovered many years after Balmer proposed them; they are called, respectively, the ***Lyman, Paschen, and Brackett series.*** The first of these series lies in the ultraviolet region of the spectrum; the other series lie in the infrared.

We can combine all the formulas for all the spectral series of hydrogen into a single general formula:

$$\frac{1}{\lambda} = R\left(\frac{1}{n_2^2} - \frac{1}{n_1^2}\right) \tag{7}$$

where n_1 and n_2 are positive integers, and n_1 is larger than n_2.

EXAMPLE 1. According to Eq. (7), what is the shortest wavelength that a hydrogen atom will emit or absorb?

SOLUTION: To find the shortest wavelength, we must choose n_1 and n_2 in Eq. (7) so as to obtain the largest possible value for the right side. This demands $n_1 = \infty$ and $n_2 = 1$, which gives

$$\frac{1}{\lambda} = R\left(\frac{1}{1} - \frac{1}{\infty}\right) = R \tag{8}$$

and

$$\lambda = 1/R = 91.176 \text{ nm}$$

Since $n_2 = 1$, this spectral line belongs to the Lyman series.

29.3 THE NUCLEAR ATOM

The regularity in the series of spectral lines of the atom must be due to an underlying regularity in the structure of the atom. We may think of an atom as analogous to a musical instrument, such as a flute. The atom can emit only a discrete set of spectral lines, just as the flute can emit only a discrete set of tones which make up a musical scale. The regularity in the spacing of tones in this musical scale is due to an underlying regularity in the structure of the flute — the tube of the instrument has regularly spaced blowholes that determine what kind of standing waves can build up within the tube and what kind of waves will be radiated.

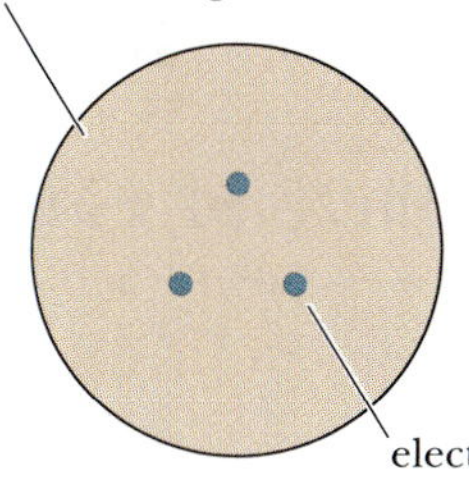

J. J. Thomson, the discoverer of the electron, made one of the first attempts at explaining the emission of light in terms of the structure of the atom. Having established that electrons are a ubiquitous component of matter, Thomson proposed the following picture: An atom consists of a number of electrons, say Z electrons, embedded in a cloud of positive charge. The cloud is heavy, carrying almost all of the mass of the atom. The positive charge in the cloud is $+Ze$, so it exactly neutralizes the negative charge $-Ze$ of the electrons. In an undisturbed atom, the electrons will sit at their equilibrium positions, where the attraction of the cloud on the electrons balances their mutual repulsion (Fig. 29.7). But if the electrons are disturbed by, say, a collision, they they will vibrate around their equilibrium positions, and this accelerated motion will cause the emission of electromagnetic radiation, that is, the emission of light. This model of the atom, called the ***"plum-pudding" model,*** does yield frequencies of vibration of the same order of magnitude as the frequency of light, but it does not yield the observed spectral series; for instance, on the basis of

FIGURE 29.7
The lithium atom according to the "plum-pudding" model. The three electrons sit at their equilibrium positions.

◀ **"Plum-pudding" model**

this model, hydrogen should have only one single spectral line, in the far ultraviolet. And in 1910, experiments by Ernest Rutherford and his collaborators established conclusively that most of the mass of the atom is not spread out over a cloud — instead, the mass is concentrated in a small kernel, or ***nucleus,*** at the center of the atom.

Nucleus ▶

Sir Ernest Rutherford, *1871–1937, British experimental physicist and director of the Cavendish Laboratory at Cambridge. Rutherford identified alpha and beta rays. He founded nuclear physics with his discoveries of the nucleus and of transmutation of elements by radioactive decay; he also produced the first artificial nuclear reaction. He was awarded the Nobel Prize in chemistry in 1908.*

Rutherford had been studying the emission of alpha particles from radioactive substances. These alpha particles carry a positive charge $2e$ and they have a mass of 6.64×10^{-27} kg, about four times the mass of a proton (alpha particles have the same structure as nuclei of helium atoms; see Section 31.3). Some radioactive substances, such as radioactive polonium and radioactive bismuth, spontaneously emit alpha particles with energies of several million electron-volts. These energetic alpha particles readily pass through thin foils of metal, thin sheets of glass, or other materials.

Rutherford was much impressed by the penetrating power of these alpha particles, and it occurred to him that ***a beam of these particles can serve as a probe to "feel" the interior of the atom.*** When a beam of alpha particles strikes a foil of metal, the alpha particles penetrate the atoms and are deflected by collisions with the subatomic structures; the magnitude of these deflections gives a clue about the subatomic structures. For example, if the interior of the atom had the "plum-pudding" structure proposed by J. J. Thomson, then the alpha particles would suffer only very small deflections, since neither the electrons, with their small masses, nor the diffuse cloud of positive charge would be able to disturb the motion of a massive and energetic alpha particle.

The crucial experiments were performed by H. Geiger and E. Marsden working under Rutherford's direction. They used thin foils of gold and of silver as targets and bombarded these with a beam of alpha particles from a radioactive source. After the alpha particles passed through the foil, they were detected on a zinc sulfide screen which registers the impact of each particle by a faint scintillation (Fig. 29.8). To Rutherford's amazement, some of the alpha particles were deflected by such a large angle that they came out backward. In Rutherford's own words: "It was quite the most incredible event that has ever happened to me in my life. It was almost as incredible as if you fired a 15-inch shell at a piece of tissue paper and it came back and hit you."

Rutherford's nuclear model ▶

Rutherford immediately recognized that the large deflection must be produced by a close encounter between the alpha particle and a very small but very massive kernel inside the atom. He therefore proposed the following picture: ***An atom consists of a small nucleus of charge $+Ze$ containing almost all of the mass of the atom; this nucleus is surrounded by a swarm of Z electrons, of charge $-Ze$.*** Thus, the atom is like a solar system — the nucleus plays the role of sun and the electrons play the role of planets.

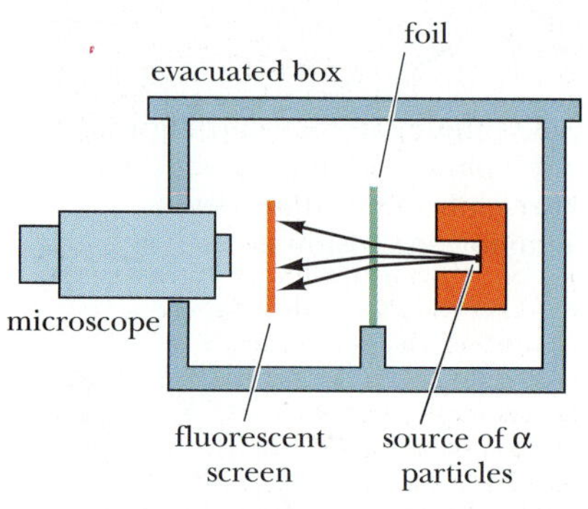

FIGURE 29.8
Rutherford's apparatus.

On the basis of this nuclear model of the atom, Rutherford calculated what fraction of the beam of alpha particles should be deflected through what angle. If an alpha particle passes close to the nucleus it will experience a large electric repulsion and it will be deflected by a large angle; if it passes far from the nucleus it will be deflected only by a small angle. Figure 29.9 shows the trajectories of several alpha particles approaching a nucleus; these trajectories are hyperbolas. The perpendicular distance between the nucleus and the original (undeflected) line of motion is called the ***impact parameter.*** In order to suffer a large deflection, the alpha particle must hit an atom with a very small impact parameter, 10^{-13} m or less; since the alpha particles in the beam strike the foil of metal at random, only very few of them will score such a close hit, and only very few will be deflected by a large angle.

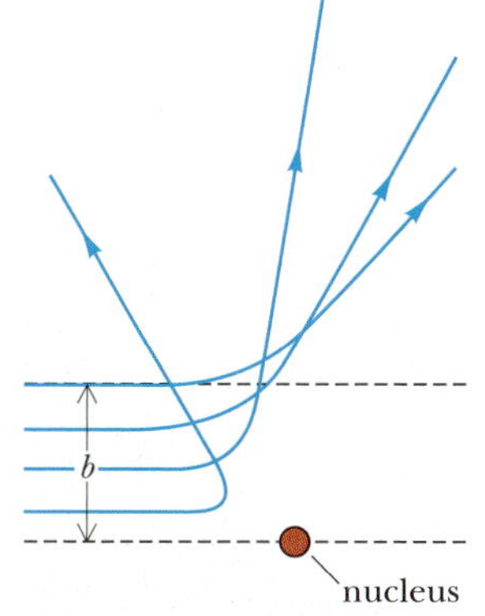

FIGURE 29.9
Four hyperbolic orbits of different impact parameters. The distance b is the impact parameter of one of these hyperbolic orbits.

29.4 Bohr's Theory

Rutherford's experiments did reveal the gross arrangement of the electrons in the atom, but not the details of their motion. Since the electrons make up the outer layers of an atom, their arrangement and motion should determine the chemical bonds of the atom and the emission of light, that is, ***their arrangement and motion should determine all the chemical and spectroscopic properties.*** But when physicists tried to calculate the electron motion according to the laws of classical mechanics and electromagnetism, they immediately ran into trouble.

To gain some insight into the source of this trouble, consider the case of the hydrogen atom. Suppose that the single electron of this atom is moving, according to the laws of classical mechanics, in a circular orbit of a radius of about 10^{-10} m around the nucleus. The electron would then have a centripetal acceleration which is very large, about 10^{23} m/s^2. Because of this acceleration, the electron would emit high-frequency electromagnetic radiation, that is, it would emit light. The energy carried away by the light must be supplied by the electron. Hence the emission process has the same effect on the electron as a friction force — it removes energy from the electron. This kind of "friction" would cause the electron to leave its circular orbit and gradually spiral in toward the nucleus, just like the residual atmospheric friction on an artificial satellite in a low-altitude orbit around the Earth causes it to spiral down toward the ground. A calculation using the laws of classical mechanics and electricity shows that the rate of emission of light by the orbiting electron in a hydrogen atom would be quite large. Correspondingly, the rate of energy loss of the electron would be large — the electron would spiral inward and collide with the nucleus within a time as short as 10^{-10} s!

Niels Bohr, *1885–1962, Danish theoretical physicist. He worked under J. J. Thomson and Rutherford in England and then became director of the Institute of Theoretical Physics in Copenhagen, for the foundation of which he was largely responsible. After formulating the quantum theory of the atom, he played a leading role in the further development of the new quantum mechanics. He received the Nobel Prize in 1922.*

Thus, our classical calculation leads us to the incongruous conclusion that hydrogen atoms, and other atoms, ought to be unstable — all the electrons ought to collapse into the nucleus almost instantaneously. Furthermore, the light that the electron emits during the spiraling motion ought to be a wave of continually increasing amplitude and increasing frequency (in musical terminology, crescendo and glissando); this is so because the closer the electron comes to the nucleus, the larger its acceleration and the higher its frequency of orbital motion. Hydrogen atoms do not behave as this classical calculation predicts. Hydrogen atoms are stable, and when they do emit light, they emit discrete frequencies (spectral lines) instead of a continuum of frequencies.

These irreconcilable disagreements between the observed properties of atoms and the calculated properties gave evidence of a serious breakdown of the classical mechanics of Newton and the classical theory of electromagnetism. Although these theories had proved very successful on the macroscopic scale, they were in need of some drastic modifications on the atomic scale.

In 1913, Niels Bohr took a bold step toward resolving these difficulties. He made the radical proposal that, ***at the atomic level, the laws of classical mechanics and of classical electromagnetism must be replaced or supplemented by other laws.*** Bohr expressed these new laws of atomic mechanics in the form of several postulates:

Bohr's postulates ▶

Stationary state ▶

Quantum jump ▶

1. ***The orbits and the energies of the electrons in an atom are quantized, that is, only certain discrete orbits and energies are permitted. When an electron is in one of the quantized orbits, it does not emit any electromagnetic radiation; thus, the electron is said to be in a stationary state. The electron can make a discontinuous transition, or quantum jump, from one stationary state to another. During such a transition from one stationary state to another stationary state of lower energy, the electron does emit radiation.***

2. ***The laws of classical mechanics apply to the orbital motion of the electron in a stationary state, but these laws do not apply during the transition from one state to another.***

3. ***When an electron makes a transition from one stationary state to another, the excess energy*** E ***is released as a single photon of frequency*** $f = E/h$.

4. ***The permitted orbits are characterized by quantized values of the orbital angular momentum*** L. ***This angular momentum is always an integer multiple of*** $h/2\pi$:

Quantization of angular momentum ▶

$$L = nh/2\pi \tag{9}$$

The number $n = 1, 2, 3, \ldots$ ***is called the angular-momentum quantum number.***

Let us now see how to calculate the stationary states and the spectrum of the hydrogen atom on the basis of these postulates. For the sake of simplicity, we will assume that the electron moves in a circular orbit around the proton, which remains at rest (Fig. 29.10). The centripetal acceleration of the electron is v^2/r. This centripetal acceleration is produced by the force of attraction between the electron and the proton, that is, the Coulomb force $e^2/4\pi\epsilon_0 r^2$. Thus, the equation of motion for the electron is

$$m_e \frac{v^2}{r} = \frac{1}{4\pi\epsilon_0} \frac{e^2}{r^2} \tag{10}$$

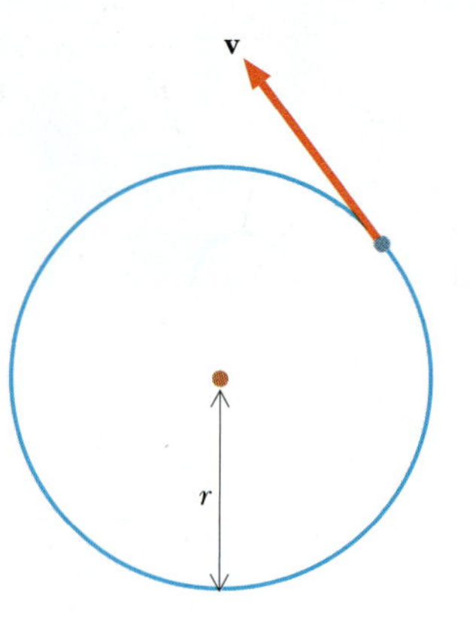

FIGURE 29.10
Electron in circular orbit around a proton.

The orbital angular momentum is $L = m_e vr$ [see Eq. (7.59)]. According to Bohr's postulate, this orbital angular momentum must be $h/2\pi$ multiplied by an integer,

$$m_e vr = nh/2\pi \tag{11}$$

It is convenient to rewrite this as

$$m_e v r = n\hbar \tag{12}$$

where $\hbar$ (pronounced "h bar") is Planck's constant divided by 2π,

$$\hbar = h/2\pi = 1.05 \times 10^{-34}\ \text{J}\cdot\text{s} \tag{13}$$

From Eq. (12), we obtain

$$v^2 = \frac{n^2\hbar^2}{m_e^2 r^2} \tag{14}$$

and, when we substitute this into Eq. (10), we obtain an equation for the radius of the orbit,

$$m_e \frac{n^2\hbar^2}{m_e^2 r^2}\frac{1}{r} = \frac{1}{4\pi\epsilon_0}\frac{e^2}{r^2} \tag{15}$$

Solving this for the radius, we find

$$r = \frac{4\pi\epsilon_0 n^2\hbar^2}{m_e e^2} \tag{16}$$

Thus, the radius of the smallest permitted orbit ($n = 1$) in the hydrogen atom is

$$a_0 = \frac{4\pi\epsilon_0\hbar^2}{m_e e^2} = 0.529 \times 10^{-10}\ \text{m} = 0.0529\ \text{nm} \tag{17}$$

◀ **Bohr radius**

This is called the ***Bohr radius.*** The radii of the other permitted orbits are multiples of the Bohr radius:

$$r = \frac{4\pi\epsilon_0 n^2\hbar^2}{m_e e^2} = n^2 a_0 \tag{18}$$

For $n = 2, 3, 4, \ldots$, this gives $r = 4a_0, 9a_0, 16a_0, \ldots$, respectively. Figure 29.11 shows some of the permitted circular orbits, drawn to scale.

The energy of the electron in one of these orbits is a sum of kinetic and potential energies. The kinetic energy is $\frac{1}{2}m_e v^2$, and the potential energy is the electrostatic potential energy $-(1/4\pi\epsilon_0)e^2/r$ [see Eq. (18.13)]. Hence the net energy is

$$E = \tfrac{1}{2}m_e v^2 - \frac{1}{4\pi\epsilon_0}\frac{e^2}{r} \tag{19}$$

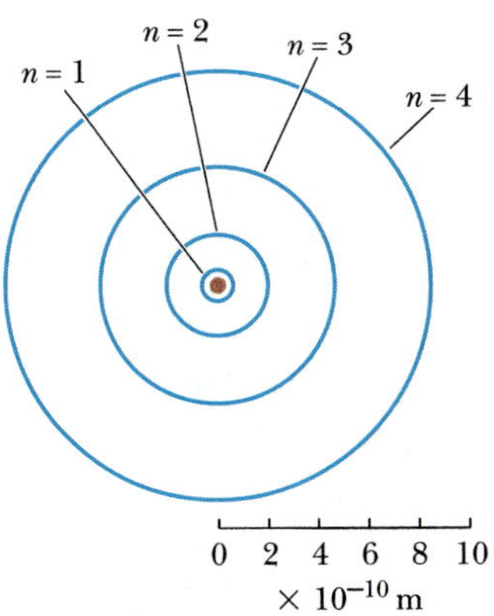

FIGURE 29.11

Possible Bohr orbits of an electron in the hydrogen atom.

Next, we substitute v^2 from Eq. (14) and then r from Eq. (16), and we obtain:

$$E = \tfrac{1}{2}m_e \frac{n^2\hbar^2}{m_e^2 r^2} - \frac{1}{4\pi\epsilon_0}\frac{e^2}{r} \tag{20}$$

$$= \tfrac{1}{2}m_e \frac{n^2\hbar^2}{m_e^2}\left(\frac{m_e e^2}{4\pi\epsilon_0 n^2\hbar^2}\right)^2 - \frac{e^2}{4\pi\epsilon_0}\left(\frac{m_e e^2}{4\pi\epsilon_0 n^2\hbar^2}\right)$$

$$= \frac{m_e e^4}{2(4\pi\epsilon_0)^2\hbar^2}\frac{1}{n^2} - \frac{m_e e^4}{(4\pi\epsilon_0)^2\hbar^2}\frac{1}{n^2} \tag{21}$$

or

$$E = -\frac{m_e e^4}{2(4\pi\epsilon_0)^2\hbar^2}\frac{1}{n^2} \tag{22}$$

We will label this energy with the subscript n,

$$E_n = -\frac{m_e e^4}{2(4\pi\epsilon_0)^2\hbar^2}\frac{1}{n^2} \tag{23}$$

According to this equation, the energies of all the stationary states are negative. The stationary state with the least energy, or the most negative energy, is the state with $n = 1$, for which:

$$E_1 = -\frac{m_e e^4}{2(4\pi\epsilon_0)^2\hbar^2} \tag{24}$$

$$= -\frac{9.11 \times 10^{-31}\ \text{kg} \times (1.60 \times 10^{-19}\ \text{C})^4}{2(4\pi \times 8.85 \times 10^{-12}\ \text{F/m})^2 \times (1.05 \times 10^{-34}\ \text{J}\cdot\text{s})^2}$$

$$= -2.18 \times 10^{-18}\ \text{J} = -13.6\ \text{eV}$$

And, by Eq. (23), the energies of the other stationary states are fractions of this energy:

Energy of stationary states of hydrogen ▶

$$E_n = -\frac{13.6\ \text{eV}}{n^2} \tag{25}$$

Thus, $E_1 = -13.6$ eV, $E_2 = -13.6/4$ eV, $E_3 = -13.6/9$ eV, and so on.

Energy-level diagram ▶

Figure 29.12 displays these quantized energies in an ***energy-level diagram.*** Each horizontal line represents one of the energies given by Eq. (25). According to Bohr's postulate, the electron emits a photon when it makes a quantum jump from one stationary state to a lower stationary state. Such quantum jumps have been indicated by arrows in Fig. 29.12. The stationary state of lowest energy is called the ***ground state;*** the next one is called the ***first excited state,*** and so on. Ordinarily, the electron of the hydrogen atom is in the ground state, that is, the circular orbit with radius a_0 and energy -13.6 eV. This is the configuration of least energy,

Ground state ▶

First excited state ▶

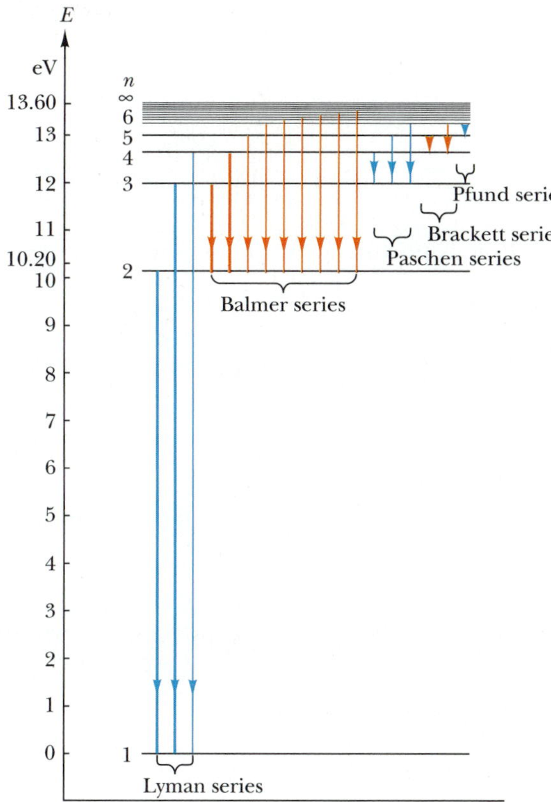

FIGURE 29.12

Energy-level diagram for the hydrogen atom. The red and blue arrows show the possible quantum jumps for the electron. Note that in this diagram the energies are given relative to the ground state, which is assigned an energy of zero.

and it is the configuration into which the atom tends to settle when it is left undisturbed. As long as the atom remains in the ground state, it does not emit light. To bring about the emission of light, we must first kick the electron into one of the excited states, that is, a circular orbit of larger radius and higher energy. We can do this by heating a sample of atoms or by passing an electric current through the sample. Collisions between the atoms will then disturb the electronic motions and occasionally kick an electron into a larger orbit. From there, the electron will spontaneously jump into a smaller orbit, giving off a quantum of light. Note that the quantum jumps indicated by colored arrows in Fig. 29.12 form several series: one series consists of all those jumps (indicated by blue arrows) that end in the ground state, another series consists of all those jumps (indicated by red arrows) that end in the first excited state, etc. These series of jumps give rise to the series of spectral lines: the Lyman series, the Balmer series, etc.

From our formula for the energies of the stationary states of the hydrogen atom we can calculate the frequency of the light emitted in a quantum jump from some initial state to a final state, as in the following example.

EXAMPLE 2. Calculate the frequency and the wavelength of the light emitted by the electron in a quantum jump from the second excited state ($n = 3$) to the first excited state ($n = 2$).

SOLUTION: The energy of the initial state is $E_3 = -13.6/9$ eV, and the energy of the final state is $E_2 = -13.6/4$ eV. Hence the electron releases an energy

$$E_3 - E_2 = -\frac{13.6}{9}\ \text{eV} - \left(-\frac{13.6}{4}\ \text{eV}\right) = 1.89\ \text{eV}$$

According to Bohr's postulate, this energy is radiated as a single photon of frequency

$$f = \frac{1.89\ \text{eV}}{h} = \frac{1.89\ \text{eV} \times 1.60 \times 10^{-19}\ \text{J/eV}}{6.63 \times 10^{-34}\ \text{J}\cdot\text{s}} = 4.56 \times 10^{14}\ \text{Hz}$$

and wavelength

$$\lambda = \frac{c}{f} = \frac{3.00 \times 10^{8}\ \text{m/s}}{4.56 \times 10^{14}\ \text{Hz}} = 6.58 \times 10^{-7}\ \text{m}$$

This wavelength agrees with the wavelength of the first spectral line of the Balmer series (see Table 29.1), except for a round-off error.

More generally, we can calculate the frequency of the light emitted in a quantum jump from some initial state of quantum number n_i to some final state of quantum number n_f. The initial energy of the electron is E_i and the final energy is E_f. Thus, the electron releases an energy $E_i - E_f$,

$$E_i - E_f = \frac{m_e e^4}{2(4\pi\epsilon_0)^2\hbar^2}\left(\frac{1}{n_f^2} - \frac{1}{n_i^2}\right) \tag{26}$$

This energy is radiated as a single photon of frequency

$$f = \frac{E_i - E_f}{h} \tag{27}$$

From the frequency of the light we can calculate the wavelength. Since the wavelength is inversely proportional to the frequency, $\lambda = c/f$, Eq. (27) implies the following expression for the wavelength of the emitted light:

$$\frac{1}{\lambda} = \frac{E_i - E_f}{ch} = \frac{E_i - E_f}{2\pi c\hbar} \tag{28}$$

or, with Eq. (26),

$$\frac{1}{\lambda} = \frac{m_e e^4}{4\pi(4\pi\epsilon_0)^2\hbar^3 c}\left(\frac{1}{n_f^2} - \frac{1}{n_i^2}\right) \tag{29}$$

Comparison of Eqs. (7) and (29) yields the following theoretical formula for the Rydberg constant:

$$R = \frac{m_e e^4}{4\pi(4\pi\epsilon_0)^2\hbar^3 c} \tag{30}$$

Upon insertion of the accurate values of the fundamental constants given in Appendix 7, we obtain

$$R = \frac{9.10953 \times 10^{-31}\ \text{kg} \times (1.602189 \times 10^{-19}\ \text{C})^4}{4\pi(4\pi \times 8.854178 \times 10^{-12}\ \text{F/m})^2 \times (1.054589 \times 10^{-34}\ \text{J}\cdot\text{s})^3 \times 2.997925 \times 10^8\ \text{m/s}}$$

$$= 1.09737 \times 10^7\ \text{m}^{-1} \tag{31}$$

This theoretical value of R agrees quite well with the experimental value quoted in Section 29.2.*

EXAMPLE 3. Suppose that the atoms in a sample of hydrogen gas are initially in the ground state. If we illuminate these atoms with light (from some kind of lamp), what frequencies will the atoms absorb?

SOLUTION: Absorption of light is the reverse of emission. When an electron in an atom absorbs a photon (supplied by the lamp), it jumps from the initial state to a state of higher energy. The energy of the photon must match the energy difference between the states. Thus, the frequencies of the photons that the electrons can absorb when jumping upward from the ground state are exactly those frequencies that they emit when jumping downward into the ground state, that is, the frequencies of the Lyman series.

With his theory Bohr attained the goal of explaining the regularities in the spectrum of hydrogen in terms of the regularities of the structure of the atom. By showing that this structure is based on a simple numerical sequence, he fulfilled the ancient dream of Pythagoras of a universe based on simple numerical ratios, a dream that arose from an analogy with musical instruments. Bohr's theory tells us how the atom plays its tune.

---

* The small disagreement between the theoretical value of R given in Eq. (31) and the experimental value given in Section 29.2 is due to the motion of the nucleus of the hydrogen atom, which we have neglected in our calculation. A careful calculation that takes into account the motions of electron and proton about their common center of mass eliminates the disagreement.

Louis Victor, prince de Broglie (de broy), *1892–1987, French theoretical physicist. He discovered the de Broglie wavelength by reasoning that if waves have particle properties, then maybe particles have wave properties. For his discovery of the wave properties of matter he was awarded the Nobel Prize in 1929, after the existence of these wave properties was confirmed experimentally.*

29.5 QUANTUM MECHANICS

Bohr's theory is a hybrid. It relies on some basic classical features (orbits) and grafts onto these some quantum features (quantum jumps, quanta of light). In the 1920s the cooperative efforts of several brilliant physicists — L. de Broglie, E. Schrödinger, W. Heisenberg, M. Born, P. Jordan, P. A. M. Dirac — established that the remaining classical features had to be eradicated from the theory of the atom. Bohr's semiclassical theory had to be replaced by a new quantum mechanics with an entirely different equation of motion.

The basis of the new quantum mechanics was laid by the discovery that electrons — as well as protons, neutrons, and all the other "particles" found in nature — have not only particle properties but also wave properties. When a beam of electrons is made to pass through an extremely narrow slit, the electrons exhibit diffraction. This means that electrons are neither classical particles nor classical waves. ***Electrons, just like photons, are a new kind of object with a subtle combination of particle and wave properties. Electrons are wavicles.*** As in the case of photons [see Eq. (28.15)], ***the wavelength associated with an electron or some other wavicle is inversely proportional to its momentum:***

$$\lambda = \frac{h}{p} \tag{32}$$

de Broglie ▶ wavelength

This wavelength was postulated by de Broglie, and it is called the ***de Broglie wavelength.*** This wavelength is quite small, even for electrons of the lowest energies attainable in experiments with beams of electrons.

EXAMPLE 4. What is the de Broglie wavelength of an electron of kinetic energy 1 eV, which is about the lowest energy that can be attained in experiments with beams of electrons?

SOLUTION: In joules, a kinetic energy of 1 eV is 1.6×10^{-19} J. This kinetic energy is related to the momentum as follows:

$$E = \frac{p^2}{2m_e}$$

Hence

$$p = \sqrt{2m_eE} = \sqrt{2 \times 9.1 \times 10^{-31}\ \text{kg} \times 1.6 \times 10^{-19}\ \text{J}}$$

$$= 5.4 \times 10^{-25}\ \text{kg} \cdot \text{m/s}$$

According to the de Broglie equation, the wavelength of the electron is then

$$\lambda = \frac{h}{p} = \frac{6.63 \times 10^{-34}\ \text{J}\cdot\text{s}}{5.4 \times 10^{-25}\ \text{kg}\cdot\text{m/s}}$$

$$= 1.2 \times 10^{-9}\ \text{m} = 1.2\ \text{nm}.$$

In the new quantum mechanics, or wave mechanics, the motion of an electron is described by a wave equation, the ***Schrödinger equation.*** This Schrödinger equation plays the same role for electrons as the Maxwell equations play for photons. The intensity of the electron wave at some point determines the probability that there is an electron particle at that point [compare Eq. (28.19)]. Furthermore, as a consequence of their wave properties, electrons obey the Heisenberg uncertainty relations for position and momentum [see Eq. (28.25)],

◀ Schrödinger equation

$$\Delta y \Delta p_y \geq \frac{\hbar}{2} \tag{33}$$

These quantum uncertainties are of crucial importance for the behavior of an electron inside an atom. For such an electron, ***the uncertainty in the position is very large*** — about as large as the size of the atom. This implies that the electron follows no definite orbit. It is therefore not surprising that the Bohr theory should have failed in all attempts at calculating the electron motions in the helium atom and in other atoms with several electrons; what is surprising is that this theory should have succeeded as well as it did in the case of the hydrogen atom.

EXAMPLE 5. Consider an electron in the ground state of hydrogen. Show that a well-defined orbit is inconsistent with the Heisenberg uncertainty relations.

SOLUTION: If the electron is to follow a well-defined orbit, the uncertainty in its momentum (in any direction) must be much smaller than the magnitude of the momentum. According to Eq. (14), the speed of the electron in the smallest circular orbit ($n = 1$) is $v = \hbar/m_e r$ and the magnitude of the momentum is $p = m_e v = \hbar/r$. For a well-defined orbit, the uncertainty of the momentum must be much smaller than this magnitude of the momentum,

$$\Delta p_y \text{ much smaller than } \hbar/r \tag{34}$$

Furthermore, we require that the uncertainty in the position must be much smaller than the size of the orbit,

$$\Delta y \text{ much smaller than } r \tag{35}$$

In view of the inequalities (34) and (35), the product $\Delta y \Delta p_y$ must be much smaller than $\hbar$,

$$\Delta y \Delta p_y \text{ much smaller than } \hbar \tag{36}$$

This is inconsistent with the Heisenberg uncertainty relation (33).

Erwin Schrödinger, *1887–1961, Austrian theoretical physicist. Another of the founders of the new quantum mechanics, he received the Nobel Prize in 1933.*

In wave mechanics, the quantization of the energy in the hydrogen atom and other atoms is an automatic consequence of the wave properties of the electron. The attractive electric force of the nucleus confines the electron wave to some region near the nucleus and causes the wave to reflect back and forth across the region, forming a standing wave. ***The different stationary states of the atom correspond to different standing-wave modes.*** As in the case of standing waves on a string, the standing electron waves in the atom have a discrete set of eigenfrequencies and, since $E = hf$, a discrete set of energies. The ground state of the atom is analogous to the fundamental mode of the string, the first excited state is analogous to the first overtone, and so on. However, whereas the determination of the eigenfrequencies of standing waves on a string is a quite trivial mathematical exercise, the determination of the eigenfrequencies of the electron waves in an atom is a formidable mathematical problem, which requires an investigation of the solutions of the Schrödinger wave equation.

Although we cannot deal here with the mathematical complexities of the Schrödinger equation, we can gain some insight into how electron waves determine the discrete energies in the hydrogen atom by means of the following simple calculation. Let us assume that the electron travels around the nucleus along an orbit of radius r, but instead of thinking of the electron as a particle, as in the Bohr theory, let us think of it as a wave. Figure 29.13 shows a "snapshot" of such an electron wave at one instant of time. If the wave is to have a well-defined amplitude at all points, it must repeat whenever we go once around the circumference — if it did not, then the wave amplitude would have two different values at a single point, which makes no sense. Hence, the wave is subject to the condition that ***some integer number of wavelengths must fit around the circumference***:

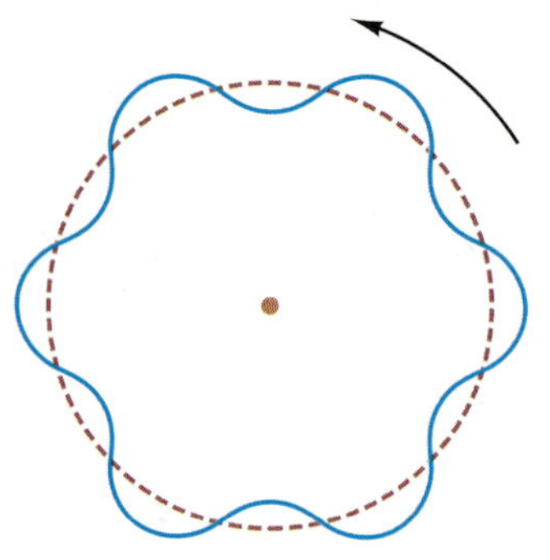

FIGURE 29.13

In this example, six wavelengths of the electron wave fit around the circumference of the orbit.

$$2\pi r = n\lambda \tag{37}$$

Although this equation looks like the condition for a standing wave on a string of length $2\pi r$, we are here dealing with a traveling wave, for which the entire wave pattern in Fig. 29.13 rotates rigidly around the center. With the de Broglie relation $\lambda = h/p$, Eq. (37) becomes

$$2\pi r = nh/p$$

or

$$rp = \frac{nh}{2\pi} \tag{38}$$

Since rp is the orbital angular momentum, this equation coincides with Bohr's quantization condition for the angular momentum. Thus, ***the wave picture of the electron implies the quantization of the angular momentum and, therefore, the quantization of the energy.*** But we must not take this calculation too seriously — its legitimacy is questionable, since it relies in part on the wave picture and in part on the particle picture. Furthermore, analysis of the Schrödinger equation shows that it is not enough to consider the behavior of the wave around the circumference; we must also consider the behavior of the wave outward along each radius. Thus, this simple calculation provides no more than a crude qualitative sketch of the role of the wave properties of electrons in the atom.

Within the atom, electrons always behave very much like waves. But outside the atom, they will sometimes behave pretty much like classical particles. Roughly, we can say that classical mechanics will be a good approximation whenever the quantum uncertainties are small compared with the relevant magnitudes of positions and momenta. For instance, for the electrons in the beam of a TV tube, the quantum uncertainty in the momentum is negligible compared with the magnitude of the momentum. Under these conditions, classical mechanics gives an adequate description of the motion of the electrons.

What we have said about the wave mechanics of electrons also applies to other "particles," or wavicles, found in nature — they all have wave properties and they all have quantum uncertainties in their position and momentum. Strictly speaking, even large macroscopic bodies have wave properties. For example, an automobile is a wavicle and it has some quantum uncertainty in its position. However, it turns out that the quantum uncertainties are very small whenever the mass of the body is large compared with atomic masses — the quantum uncertainty in the position of an automobile is typically no more than about 10^{-18} m, a number that can be ignored for all practical purposes. Hence, for automobiles and other macroscopic bodies, quantum effects are completely insignificant, and classical mechanics gives an excellent description of the motion of these bodies.

◀ **Electron microscopes (TEM, SEM, STM)**

The wave properties of electrons find practical application in the ***transmission electron microscope*** (TEM), which employs electron waves to form a highly magnified image of an object in the same way as an ordinary microscope employs light waves. The maximum magnification attainable by ordinary microscopes is limited by the wavelength of light. If we attempt to observe an object as small or smaller than the wavelength of light, the image becomes indistinct, because the light waves suffer severe diffraction when passing through and around such a small object, and the resulting diffraction fringes blur the image. The electron waves used in typical electron microscopes have wavelengths of 0.005 nm, which is about 10,000 times shorter than the wavelength of light; the electron waves are therefore much less susceptible to diffraction than is light.

The main "optical" elements in a transmission electron microscope are the same as in an ordinary microscope — an objective lens and an ocular, or projector, lens (see Fig. 29.14). The "lenses" are not made of glass, but of magnetic fields, carefully shaped so as to provide deflections similar to those experienced by light in a glass lens. The electron rays emerging from the projector lens are intercepted by a fluorescent screen, on which we can see and photograph the image.

Another kind of electron microscope is the ***scanning electron microscope*** (SEM). But the principle of operation of this microscope is entirely different, and bears no resemblance to the operation of ordinary microscopes. The scanning electron microscope relies on the particle properties of electrons, not on their wave properties. Instead of forming an image by means of electron waves, the scanning electron microscope uses a fine beam of electrons to bombard the object. The beam "scans" across the object line by line in a sweep pattern, like the sweep pattern of the electron beam in an ordinary TV tube, and a detector picks up the current of electrons that recoil from the object and of secondary electrons knocked out of the object by the incident primary electrons (Fig. 29.15). This detected current is amplified and fed into a video monitor, where it displays a picture. Scanning electron microscopes produce very crisp pictures with an exceptional depth of field and strong shadows that give a vivid three-dimensional impression, but they do not attain the extremely high magnifications of transmission electron microscopes.

(a)

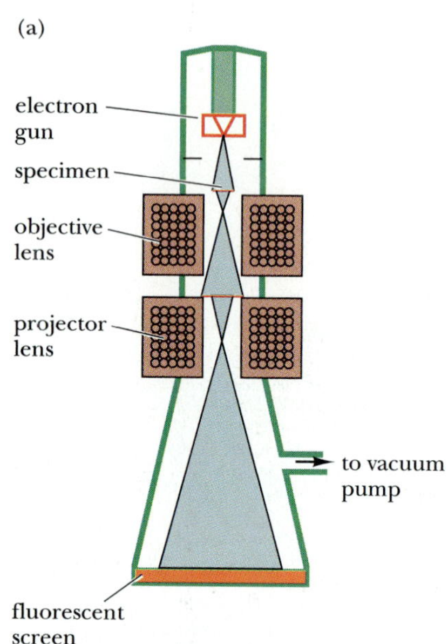

(b)

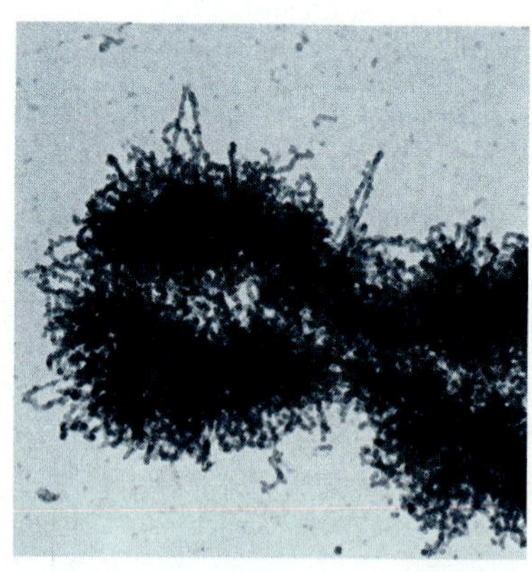

FIGURE 29.14

(a) Schematic diagram of a transmission electron microscope (TEM). (b) Picture of human chromosome prepared with TEM.

The newest and most powerful kind of electron microscope is the ***scanning tunneling microscope*** (STM). This microscope scans the surface of the object with a fine tungsten needle whose motion is precisely controlled by a delicate suspension system (see Fig. 29.16). A potential difference is applied between the needle and the surface of the object. The needle is not actually in contact with the surface; it merely sweeps over the surface line by line. The gap between the needle and the surface is effectively an insulator, which tends to block the motion of the electrons from the needle to the surface of the object. However, the wave properties of electrons permit them to spread for some distance into this gap, and leak across to the surface. This kind of leakage of an electron across a gap where its motion is actually forbidden is called ***tunneling.*** The probability for an electron to succeed in tunneling across the gap depends drastically on the size of the gap. Whenever the needle, during its sweep, comes near any peak in the surface — such as a bulging atom — the tunneling probability starts to increase drastically, and so does the current from the needle to the surface. This incipient increase of the current is detected by the electronic circuit connected to the needle, and an amplified feedback signal is sent to the suspension system to lift the needle, so as to keep it at constant height from the peak. Thus, the needle skims over the peaks and valleys of the surface, always maintaining a constant height above the "terrain." The amplified signal used to control the suspension is also sent to a computer, where it is processed and then fed into a video monitor which displays a relief picture of the surface. Modern scanning tunneling electron microscopes attain magnifications of up to 10^8, and they permit us to "see" individual atoms.

This technique for constructing a picture of the surface by sweeping a needle across it is also exploited in the ***atomic force microscope*** (AFM). But the needle of this kind of microscope is placed in direct contact with the surface — it is pressed against the surface with a force of, say, 10^{-8} N,

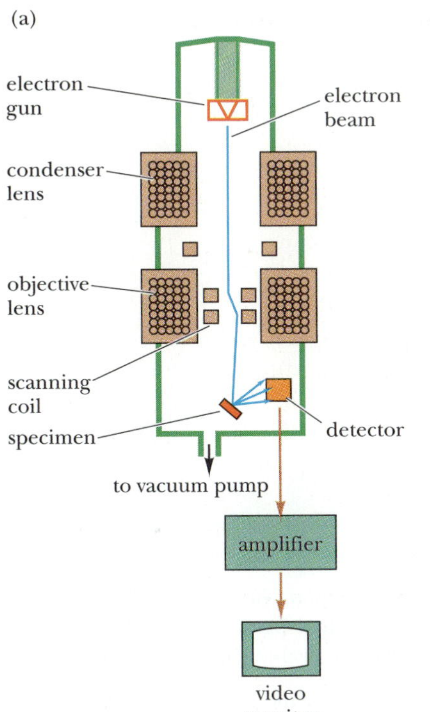

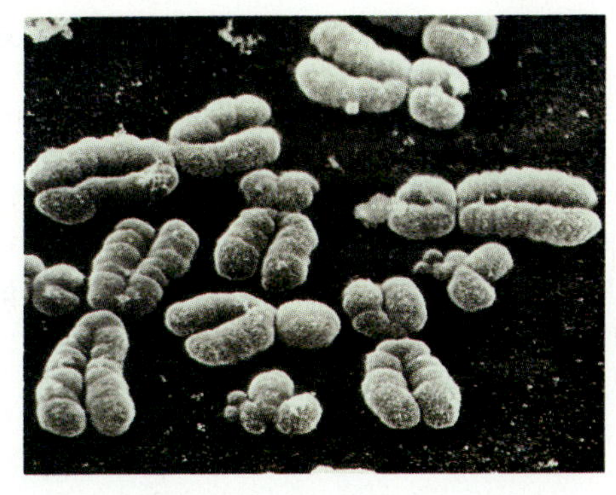

FIGURE 29.15

(a) Schematic diagram of a scanning electron microscope (SEM). (b) Picture of human chromosomes prepared with SEM. Figure 29.14b is a close-up view of one such chromosome.

and it is lifted or lowered by the suspension system so as to keep the force constant during the sweep. Thus, the needle actually explores the shape of the surface by directly feeling it, just as you might explore the shape of a surface by feeling it with the tip of your finger. Magnifications attained with atomic force microscopes are about the same as those attained with scanning tunneling microscopes. The picture displayed on the first page of this chapter was prepared with an atomic force microscope.

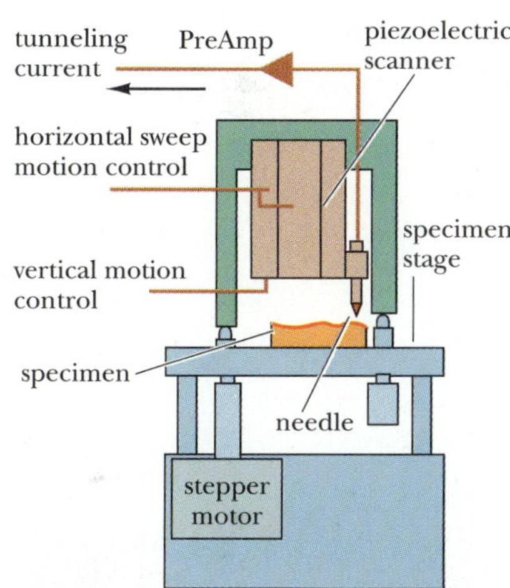

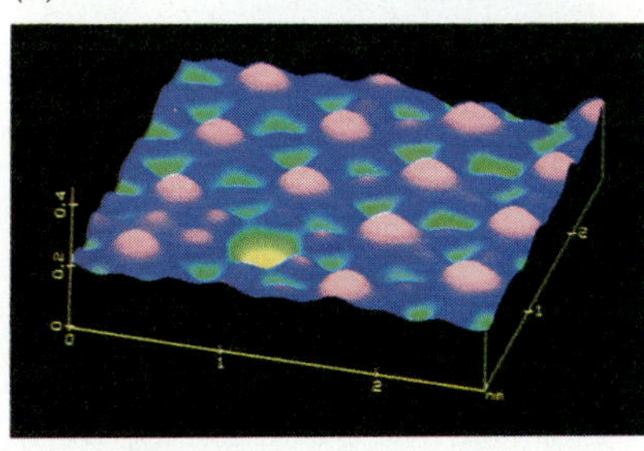

FIGURE 29.16

(a) Schematic diagram of a scanning tunneling electron microscope (STM). (b) Picture of iodine atoms prepared with STM.

SUMMARY

- **Spectral series of hydrogen:**

$$\frac{1}{\lambda} = R\left(\frac{1}{n_2^2} - \frac{1}{n_1^2}\right); \quad R = 1.09678 \times 10^7 \text{ m}^{-1}$$

- **Quantization of angular momentum:** $L = n\hbar$
- **Bohr radius:** $a_0 = \dfrac{4\pi\epsilon_0\hbar^2}{m_e e^2} = 0.0529$ nm
- **Energy of stationary states of hydrogen:**

$$E_n = -\frac{m_e e^4}{2(4\pi\epsilon_0)^2\hbar^2}\frac{1}{n^2} = -\frac{13.6\text{ eV}}{n^2}$$

▶ **Frequency and wavelength of photon emitted in transition:**

$$f = \frac{E_i - E_f}{h}$$

$$\frac{1}{\lambda} = \frac{f}{c} = \frac{E_i - E_f}{hc}$$

▶ **De Broglie wavelength of wavicle:** $\lambda = h/p$

QUESTIONS

1.–Do the spectral lines seen in a stellar spectrum (for example, Fig. 29.4) tell us anything about the chemical composition of the stellar *interior?*

2.–The spectrum of hydrogen shown in the color plate on page 834F displays all of the spectral lines simultaneously. Since a hydrogen atom emits only one spectral line at a time, how can all the lines be visible simultaneously?

3.–The target used in Rutherford's scattering experiment was a very thin foil of metal. What is the advantage of a thin foil over a thick foil in this experiment?

4.–How can Rutherford's experiment tell us something about the size of the nucleus?

5.–Why is Bohr's postulate of stationary states in direct contradiction with classical mechanics and electromagnetism?

6.–If an electron in a hydrogen atom makes a transition from some state to a lower state, does its kinetic energy increase or decrease? Its potential energy? Its orbital angular momentum?

7.–At low temperatures, the absorption spectrum of hydrogen displays only the spectral lines of the Lyman series. At higher temperatures, it also displays other series. Explain.

8.–The planets move around the Sun in circular orbits. Is their orbital angular momentum quantized?

9.–Would you expect that Bohr's theory of the hydrogen atom can be adapted to the singly ionized helium atom, that is, the helium atom with one missing electron? To what other ionized atoms can Bohr's theory be adapted?

10.–According to the ***Complementarity Principle,*** formulated by Bohr, a wavicle has both wave properties and particle properties, but these properties are never exhibited simultaneously: if the wavicle exhibits wave properties in an experiment, then it will not exhibit particle properties, and conversely. Give some examples of experiments in which wave or particle properties (but not both simultaneously) are exhibited.

11.–An electron and a proton have the same kinetic energy. Which has the longer de Broglie wavelength?

12.–Describe the interference pattern expected for an electron wave incident on a plate with two very narrow parallel slits separated by a small distance.

13.–Electron microscopes achieve high resolution because they use electron waves of very short wavelength, usually less than 0.01 nm. Why can we not build a microscope that uses *photons* of equally short wavelength?

PROBLEMS

Section 29.2: The Balmer Series

1.——Use Eq. (4) to calculate the wavelengths of the first four lines of the Lyman series.

2.——Show that the spectral lines of the Balmer series all have a higher frequency than the spectral lines of the Paschen series. Do the spectral lines of the Paschen series all have a higher frequency than those of the Brackett series?

3.*——When astronomers examine the light of a distant galaxy, they find that all the wavelengths of the spectral lines of the atoms are longer than those of the atoms here on Earth by a common multiplicative factor. This is the *red shift* of light; it is a Doppler shift caused by the motion of recession of the galaxy, away from the Earth. In the light of a galaxy beyond the constellation Virgo, astronomers find spectral lines of wavelengths 411.7 nm and 435.7 nm.

(a) Assume that these are two spectral lines of hydrogen, with the wavelengths multiplied by some factor. Identify these lines. What is the factor by which these wavelengths are longer than the normal wavelengths of the two spectral lines? What is the factor by which the frequencies are lower?

(b) If the speed of recession is low compared with the speed of light, the Doppler shift of light obeys a formula similar to that for the Doppler shift of sound [see Eq. (11.18)]. Calculate the speed of recession of the galaxy.

4.*——One of the spectral series of the lithium atom is the ***principal series,*** with the following wavelengths: 670.79 nm, 323.26 nm, 274.13 nm, 256.25 nm, 247.53 nm. Show that these wavelengths approximately fit the formula

$$\frac{1}{\lambda} = R\left[\frac{1}{(1+s)^2} - \frac{1}{(n+p)^2}\right]; \quad n = 2, 3, 4, \ldots$$

where $R = 1.09728 \times 10^7 \text{ m}^{-1}$ is the Rydberg constant for lithium, and s and p are constants, with $p = -0.040$ and $s = 0.59$.

Section 29.3: The Nuclear Atom

5.*——What is the distance of closest approach for a 5.5-MeV alpha particle in a head-on collision with a gold nucleus? With an aluminum nucleus?

6.*——The nucleus of platinum has a radius 6.96×10^{-15} m and an electric charge of $78e$. What must be the minimum energy of an alpha particle in a head-on collision if it is to just barely reach the nuclear surface? Assume the alpha particle is pointlike.

7.**——An alpha particle of energy 5.5 MeV is incident on a silver nucleus with an impact parameter 8.0×10^{-13} m. The distance of closest approach of the particle is 2.7×10^{-14} m. Find the speed of the point of closest approach. (Hint: The angular momentum is conserved.)

Section 29.4: Bohr's Theory

8.——What is the angular momentum of a vinyl record (a uniform disk) rotating at $33\frac{1}{3}$ revolutions per minute? The moment of inertia of the record is 1.3×10^{-2} kg · m^2. Express the answer as a multiple of $\hbar$.

9.——If you wanted to give an apple an angular momentum of $\hbar$, at what rate would you have to spin it about its axis? Treat the apple as a uniform sphere of mass 0.20 kg and radius 4.0 cm.

10.——What are the angular momentum, the kinetic energy, the potential energy, and the net energy of an electron in the smallest ($n = 1$) Bohr orbit in the hydrogen atom? Express your answers in SI units.

11.——What is the speed of an electron in the smallest ($n = 1$) Bohr orbit? Express your answer as a fraction of the speed of light.

12.——What is the centripetal acceleration of an electron in the smallest ($n = 1$) Bohr orbit in the hydrogen atom?

13.——What is the frequency of the orbital motion for an electron in the smallest ($n = 1$) Bohr orbit? In the next ($n = 2$) Bohr orbit? Do either of these frequencies coincide with the frequency of the light emitted during the transition $n = 2$ to $n = 1$?

14.——Hydrogen atoms in highly excited states with a quantum number as large as $n = 732$ have been detected in interstellar space by radio astronomers. What is the orbital radius of the electron in such an atom? What is the energy of the electron?

15.——Find the orbital radius, the speed, the angular momentum, and the centripetal acceleration for an electron in the $n = 2$ orbit of hydrogen.

16.——What is the ionization energy of hydrogen (that is, what energy must you supply to remove the electron from the atom when it is in the ground state)? Express the answer in electron-volts.

17.——If a hydrogen atom is initially in the ground state, what is the *longest* wavelength of light it will absorb?

18.——Calculate the frequency and the wavelength of the light emitted by the electron in a hydrogen atom in a quantum jump from the second excited state to the ground state. Identify the spectral line according to its spectral series.

19.——If a hydrogen atom is initially in the first excited state, what is the *longest* wavelength of light it will absorb? What is the *shortest* wavelength of light it will absorb?

20.——If you bombard hydrogen atoms in their ground state with a beam of particles, the collisions will (sometimes) kick atoms into one of their excited states. What must be the minimum kinetic energy of the bombarding particles if they are to achieve such an excitation?

21.*—A hydrogen atom emits a photon of wavelength 102.6 nm. From what stationary state to what lower stationary state did the electron jump?

22.*—The quantity $\hbar/(m_e c)$ is called the ***Compton wavelength.*** The quantity $e^2/(4\pi\epsilon_0 m_e c^2)$ is called the ***"classical electron radius."*** Show that the Bohr radius, the Compton wavelength, and the classical electron radius are in the ratio $1{:}\alpha{:}\alpha^2$, where $\alpha = e^2/(4\pi\epsilon_0\hbar c)$. The quantity α is called the ***fine-structure constant.*** What is the numerical value of this constant?

23.*—A hydrogen atom is initially in the ground state. In a collision with an argon atom, the electron of the hydrogen atom absorbs an energy of 15.0 eV. With what speed will the electron be ejected from the hydrogen atom?

24.*—Suppose that the electron in a hydrogen atom is initially in the second excited state ($n = 3$). What wavelength will the atom emit if the electron jumps directly to the group state? What two wavelengths will the atom emit if the electron jumps to the first excited state and then to the ground state?

25.*—An electron in a hydrogen atom is initially in the ground state. The electron absorbs a photon from an external light source and thereby makes a transition to the $n = 4$ state. What must have been the energy and the wavelength of the absorbed photon? If the electron now jumps spontaneously to the $n = 3$ state, what are the energy and the wavelength of the emitted photon?

26.*—The muon (or mu meson) is a particle somewhat similar to an electron; it has a charge $-e$ and a mass 206.8 times as large as the mass of the electron. When such a muon orbits around a proton, they form a ***muonic hydrogen atom,*** similar to an ordinary hydrogen atom, but with the muon playing the role of the electron. Calculate the Bohr radius of this muonic atom and calculate the energies of the stationary states. What is the energy of the photon emitted when the muon makes a transition from the $n = 2$ state to the $n = 1$ state?

27.*—Suppose that a sample of hydrogen atoms, initially in their ground states, is under bombardment by a beam of electrons of kinetic energy 12.2 eV. In an inelastic collision between a hydrogen atom and one of the incident electrons, the hydrogen atom will occasionally absorb all, or almost all, the kinetic energy of the electron and make a transition from the ground state to an excited state. If so, what excited state will the hydrogen atom attain? What are the possible spectral lines that the atom can emit subsequently?

28**—The singly ionized helium atom (usually designated HeII) has one electron in orbit around a nucleus of charge $2e$. Apply Bohr's theory to this atom and find the energies of the stationary states. What is the value of the ionization energy, that is, the energy that you must supply to remove the electron from the atom when it is in the ground state? Express the answer in electron-volts.

29.**—The doubly ionized lithium atom (usually designated LiIII) has one electron in orbit around a nucleus of charge $3e$. What is the radius of the smallest Bohr orbit in doubly ionized lithium? What is the energy of this orbit?

30.**—In principle, Bohr's theory also applies to the motion of the Earth around the Sun. The Earth plays the role of the electron, the Sun that of the nucleus, and the gravitational force that of the electric force.

(a) Find a formula analogous to Eq. (18) for the radii of the permitted circular orbits of the Earth around the Sun.

(b) The actual radius of the Earth's orbit is 1.50×10^{11} m. What value of the quantum number n does this correspond to?

(c) What is the radial distance between the Earth's actual orbit and the next larger orbit?

Section 29.5: Quantum Mechanics

31.—What must be the energy of an electron if its wavelength is to equal the wavelength of visible light, about 550 nm?

32.—Find the de Broglie wavelength for each of the following electrons with the specified kinetic energy: electron of 20 keV in a TV tube, conduction electron of 5.4 eV in a metal, orbiting electron of 13.6 eV in a hydrogen atom, orbiting electron of 91 keV in a lead atom.

33.—What is the de Broglie wavelength of an electron in the ground state of hydrogen? In the first excited state?

34.—What is the de Broglie wavelength of a tennis ball of mass 0.060 kg moving at a speed of 1.0 m/s?

35.—An electron microscope operates with electrons of energy 40 keV. What is the wavelength of such electrons? By what factor is this wavelength smaller than that of visible light?

36.——A photon and an electron each have an energy of 6.0×10^3 eV. What are their wavelengths?

37.——If the de Broglie wavelengths of two electrons differ by a factor of 2, by what factor must their energies differ?

38.——The "thermal" neutrons in a nuclear reactor typically have a kinetic energy of about 0.05 eV. What is the de Broglie wavelength of such a neutron?

39.*——Neutrinos of energy 10 MeV are emitted by the Sun. What is the de Broglie wavelength of such a neutrino? (Hint: Since the neutrino is a particle of zero mass, its momentum must be calculated according to the formula for the momentum of an ultrarelativistic particle.)

40.*——What is the de Broglie wavelength of an electron if its kinetic energy is equal to its rest-mass energy? (Hint: For a relativistic particle the momentum must be calculated according to the relativistic formula.)

41.*——What is the de Broglie wavelength of the Earth in its motion around the Sun?

42.*——What is the de Broglie wavelength of a N_2 molecule in air at room temperature (20°C)? Assume that the molecule is moving with the rms speed of molecules at this temperature.

43.*——Show that the de Broglie wavelength of an electron can be calculated from the formula

$$\lambda = 1.23/\sqrt{K}$$

where the wavelength λ is expressed in nm and the kinetic energy K is expressed in eV.

44.——Suppose that the velocity of an electron has been measured to within an uncertainty of ± 1 cm/s. What minimum uncertainty in the position of the electron does this imply?

45.——Interferometric methods permit us to measure the position of a macroscopic body to within $\pm 10^{-12}$ m. Suppose we perform a position measurement of such a precision on a body of mass 0.050 kg. What uncertainty in momentum is implied by the Heisenberg relation? What uncertainty in velocity?

46.——The nucleus of the aluminum atom has a diameter of 7.2×10^{-15} m. Consider one of the protons in this nucleus. The uncertainty in the position of this proton is necessarily less than 7.2×10^{-15} m. What is the minimum uncertainty in its momentum and velocity?

47.——If the position of a parked automobile of mass 2×10^3 kg is uncertain by $\pm 10^{-18}$ m, what is the corresponding uncertainty in its velocity?

PHYSICS IN ACTION

LASER LIGHT

In an ordinary light source, such as a light bulb, the atoms emit their light independently. The atom suffers concussions from thermal motions, and this excites an electron from a lower orbit to a higher orbit. At some later, unpredictable time, the electron spontaneously jumps back down, emitting a photon. Because the atoms emit at random, the light emerging from an ordinary light bulb is a confused combination of many different waves with no special phase relationships (see Fig. 1a). On the average, the total intensity of light is the sum of the intensities contributed by the individual atoms; therefore the total intensity is simply proportional to the number of atoms.

In a laser, the atoms emit their light in unison. The electrons in different atoms either jump down at the same time of else they jump with a time difference of one or several periods of oscillation of the light wave. Furthermore, the electrons emit their light waves in the same direction. The result is that the light emerging from a laser is a ***coherent*** combination of waves (see Fig. 1b). All the light waves from different atoms are in phase — the light wave contributed by each atom combines crest to crest with the light waves contributed by the other atoms. The total amplitude of the light wave is then proportional to the number of atoms, and the total intensity is proportional to the *square* of the number of atoms (remember that the amplitude of the light wave is its electric field, and the intensity is proportional to the square of the electric field). Since the number of atoms in even a fairly small light source is more than 10^{16}, the coherent emission from these atoms can be enormously stronger than the incoherent emission.

(a)

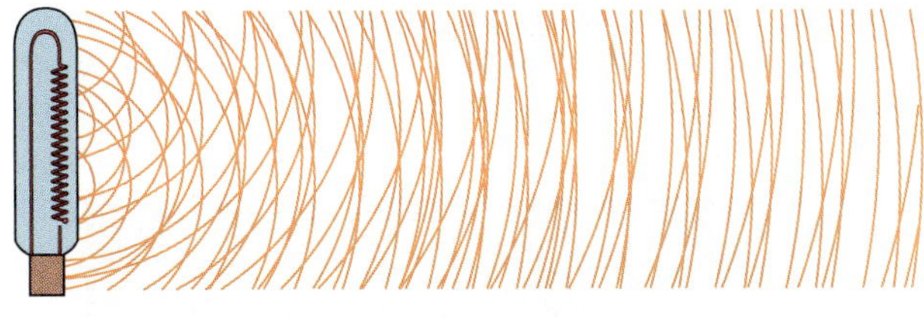

light bulb

(b)

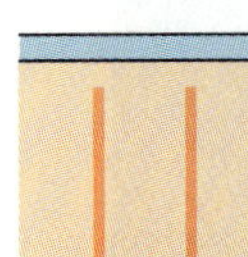

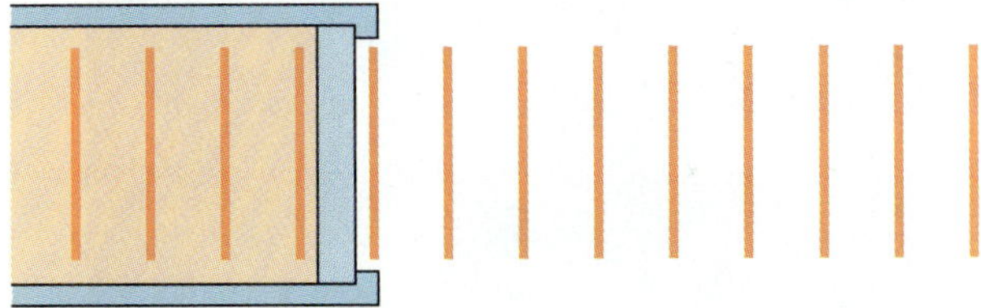

laser

FIGURE 1 (a) Light waves emitted by an ordinary light bulb. (b) Light waves emitted by a laser.

In the laser, the jumps of the electrons in different atoms are kept in step by a process of ***stimulated emission*** (the word *laser* is an acronym for *l*ight *a*mplification by *s*timulated *e*mission of *r*adiation). Imagine that an excited electron in one of the atoms in the laser jumps to its lower orbit, releasing some light. As this light wave passes by another atom with an excited electron, it jiggles that electron; this causes the electron to resonate and to jump down in unison with the light wave instead of waiting to jump down spontaneously. Thus, the passage of the light wave triggers, or stimulates, the emission of an additional, coherent light wave.

The key to the operation of a laser is the initial presence of a large number of atoms in the excited state. This is called a ***population inversion,*** because under normal conditions atoms tend to settle into an unexcited state. In the laser, there must be more atoms in the exited state than in the unexcited state — otherwise the light wave would lose energy by stimulated absorption rather than gain energy by stimulated emission. One method for achieving the required population inversion uses an intense flash of (ordinary) light to lift the electrons into excited states; this is called ***optical pumping.*** Unfortunately, the flash of light that pumps the electrons upward into an excited state can also pump them downward; hence direct pumping into an excited state does not produce a population inversion. This problem can be resolved by making the electrons jump upward by an indirect route: first a jump to some high excited state, and then a spontaneous jump to a somewhat lower excited state, where they remain for some time awaiting stimulated emission (see Fig. 2). The lasing action begins when a sufficiently large number of electrons have accumulated in these lower excited states.

The fundamental theoretical principles involved in the operation of the laser were first published by

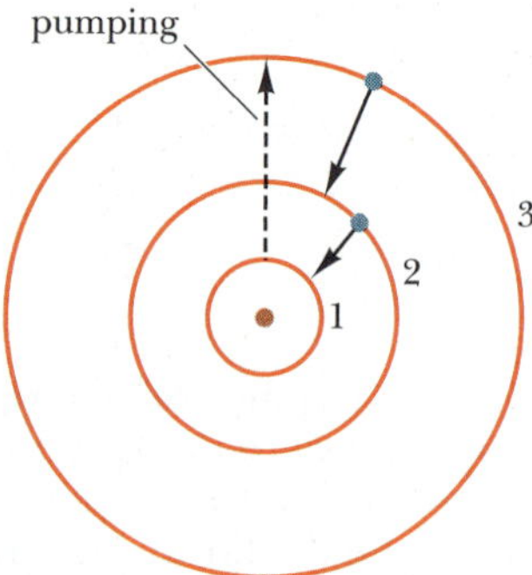

FIGURE 2. Upward and downward jumps of an electron in an atom subjected to optical pumping. A flash of light lifts the electron from its lowest orbit (1) to a high, excited orbit (3). From there the electron spontaneously jumps to a somewhat lower excited orbit (2). The lasing action occurs between this orbit and the lowest orbit.

C. H. Townes and A. L. Schawlow. The first laser designed according to these principles was a ***ruby laser*** built by T. H. Maiman (see Fig. 3). This kind of laser, still in common use today, generates light in a long cylindrical crystal of synthetic ruby. The crystal consists of corundum (an oxide of aluminum) with chromium impurities; the red color of ruby is due to the chromium impurities. In the ruby laser, only the chromium atoms lase. To pump these atoms into their excited states, the crystal is surrounded by a flash lamp (see Fig. 3); the pulse of high-intensity light from the flash lamp supplies the energy for the initial upward jumps of the electrons, and leads to a population inversion. The first electron that engages in the spontaneous emission of light will then trigger the stimulated emission of light by the other electrons.

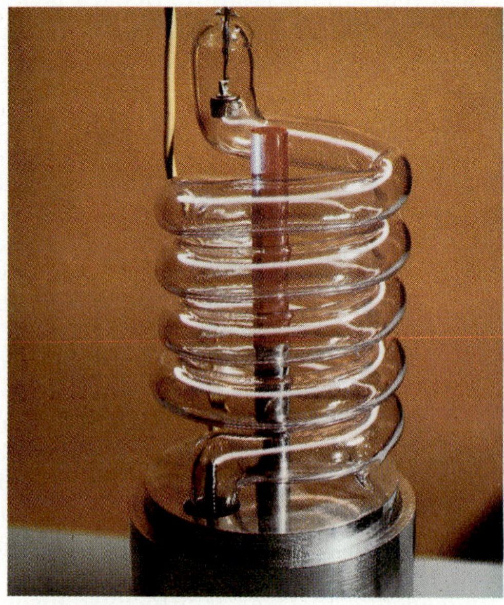

FIGURE 3 The ruby laser of Maiman.

In order to ensure that all, or almost all, of the excited atoms participate in stimulated emission, it is best to make the light wave traverse the ruby rod several times, so that any excited atoms not triggered on the first pass are triggered on a subsequent pass. For this purpose, the ends of the rod are polished flat and they are silvered, making them into mirrors. One end is only partially silvered, so that the light can ultimately escape. The back-and-forth reflections drastically enhance the lasing action for those light waves that are emitted exactly parallel to the axis of the rod and that have a wavelength that fits into the back-and-forth distance along the rod exactly an integral number of times. The chromium atoms in ruby will lase at one or another of several wavelengths in the red part of the spectrum, between 693 nm and 700 nm. The beam from a ruby laser emerges as a pulse lasting as long as excited atoms remain in the rod, about 10^{-3} s. When the excited atoms become exhausted, they must be pumped again, with another flash of light.

For many practical applications, it is desirable to build lasers with a continuous output of light. This requires that, on the average, the electrons in the atoms be pumped up into the excited state at the same rate as they jump down by stimulated emission. The most commonly employed laser with a continuous output of light is the ***helium–neon laser.*** It consists of a glass tube filled with a mixture of helium and neon at low pressure (see Fig. 4). The lasing action is due to the neon; the helium merely serves to pump the neon. When the electrodes at the ends of the tube are connected to a high-voltage supply, a current flows through the tube. This cur-

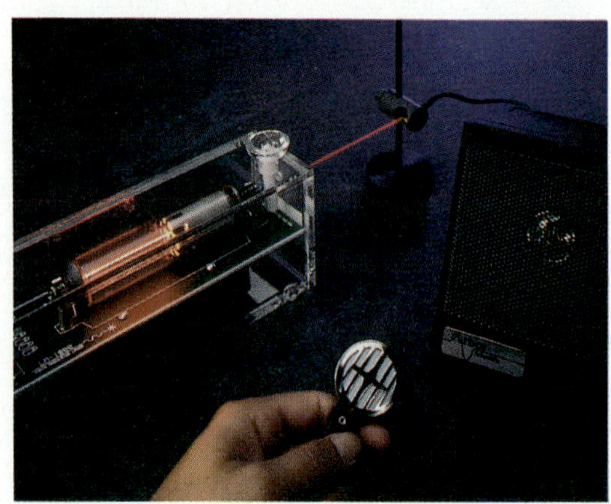

FIGURE 4 Helium–neon laser.

rent disturbs the helium atoms in the gas and excites these atoms. The excited helium atoms collide with the neon atoms; in these collisions, the neon atoms acquire energy from the helium atoms, and this pumps electrons into excited states, and produces a population inversion, suitable for stimulated emission. The neon atoms will lase at several wavelengths in the red and in the infrared parts of the spectrum.

Laser light is useful because of its directionality, high intensity, pure color, and coherence. The practical applications of laser light exploit one or several of these characteristics.

Because of its sharp directionality, the beam from a laser is a very convenient tool for laying out straight lines, and it is used for this purpose by surveyors and construction workers (see Fig. 5). For example, to dig a straight trench, we can aim a laser beam along the ground, and proceed to dig along this beam; we can check the depth of the trench by intercepting the laser beam with a vertical meter stick resting against the bottom of the trench. Laser beams can similarly be used in the construction of large aircraft and ships, to check the alignment of ribs and frames.

FIGURE 5 Laser being used for a survey of the Grand Canyon.

Laser pulses are used in rangefinders, especially for military purposes. The rangefinder consists of a laser and a light detector, both linked to a timing device. The laser sends a short pulse of light out to the target, and the target reflects part of this pulse back to the detector and the timing device. The travel time for this roundtrip indicates the distance.

In light-wave communication systems, light signals are piped through long, thin glass fibers. As we saw in Section 25.3, a light wave suffers total internal reflection within such an optical fiber, and it can be transmitted without loss over long distances. Telephone systems take advantage of this method of transmission by converting the sound of the human voice into a series of pulses of laser light. The lasers employed for this purpose are solid-state devices made of a sandwich of semiconductors, roughly the size of a grain of salt (see Fig. 6).

FIGURE 6 A solid-state laser, almost as small as the eye of the needle.

Many applications of lasers take advantage of the high concentration of energy in the laser light to melt, weld, cut, or vaporize materials. For industrial applications, a laser of a power of a few kilowatts makes an excellent welding torch, and such lasers are widely used in the manufacture of automobiles.

In medicine, lasers are used for surgery of the eye (see Fig. 7). A laser beam aimed through the (transparent) lens of the eye and focused on the ret-

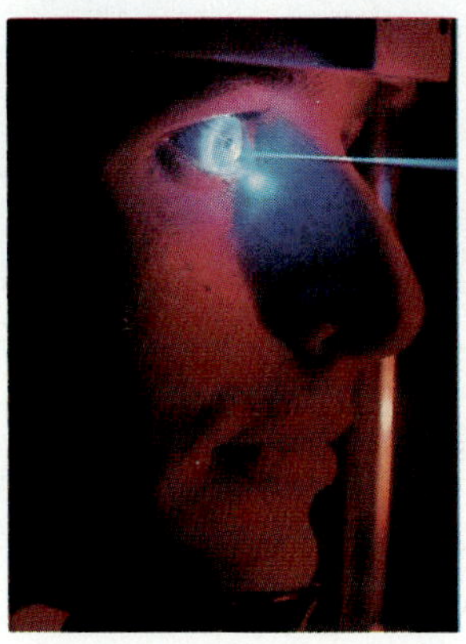

FIGURE 7 Eye surgery by laser.

(a)

(b)

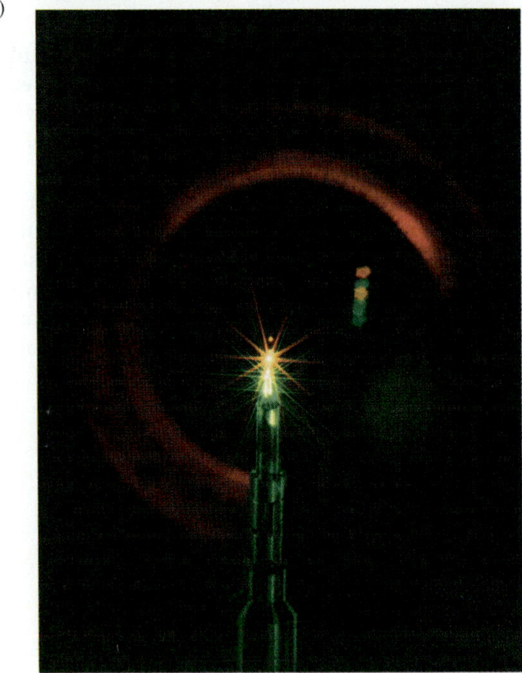

FIGURE 8 Target chamber of the NOVA laser at Lawrence Livermore Laboratory. The laser delivers a power of 2×10^{14} W in a pulse lasting 10^{-9} s. (b) Explosion of a heavy-hydrogen fuel pellet.

(a)

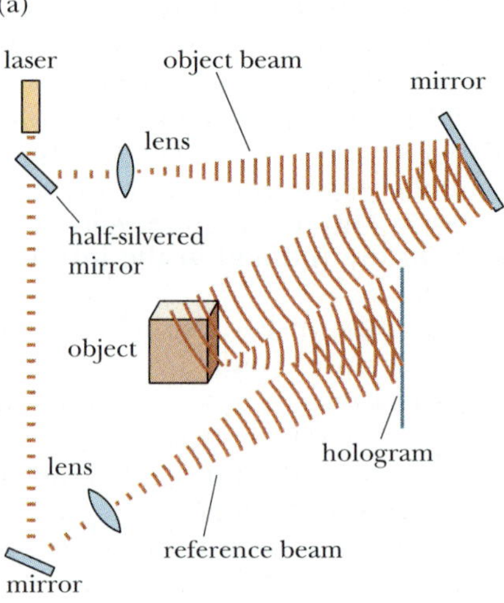

(b)

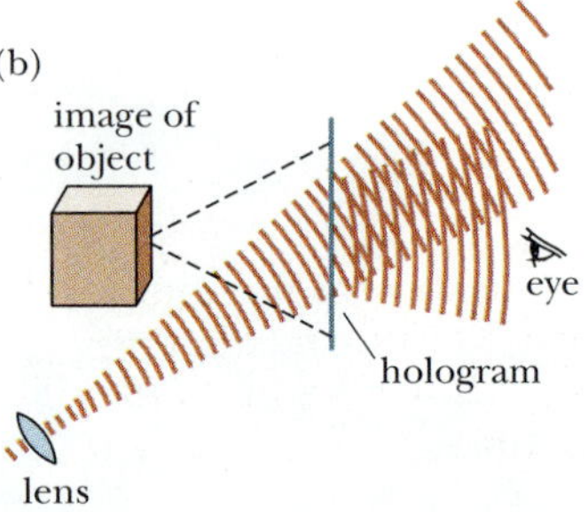

FIGURE 9 (a) To make a hologram, the beam from a laser is split into two beams by a half-silvered mirror. One of these beams is sent to the object, where it is reflected; the other beam is sent directly to the photographic plate. The two beams produce an interference pattern, which is recorded on the photographic plate. The two beams produce an interference pattern, which is recorded on the photographic plate. (b) To reconstruct the wave fronts of the light reflected by the object, the hologram is illuminated with a laser beam. The eye sees a three-dimensional replica of the object.

ina can "weld" a detached retina into its proper place. In other surgical operations, a laser beam serves as a scalpel; this is particularly suitable for operations on the blood-rich tissues, such as the liver, where the immediate cauterization of the tissues by the laser beam prevents excessive bleeding.

Extremely powerful lasers have been developed in an attempt to extract energy from thermonuclear reactions, similar to the reactions that occur at the center of the Sun. In these attempts, heavy hydrogen (deuterium and tritium) is heated to temperatures of about 10^8 K; the violent thermal collisions between the nuclei of heavy hydrogen then fuse them together into nuclei of helium (see Chapter 31 for details). The heavy hydrogen fuel is in the form of a small pellet which is placed in a combustion chamber where it is heated and compressed by an intense pulse of light from a large laser (see Fig. 8). The high temperatures and pressures ignite thermonuclear fusion in the pellet — the pellet acts as a miniature hydrogen bomb.

In scientific applications, lasers are widely used in diffraction and interference experiments. An interferometric application of laser light that has attracted much attention is ***holography,*** the making of three-dimensional pictures. An ordinary photograph merely records the intensity of the light reflected by an object on a photographic film; the image on the film contains no direct three-dimensional information. A hologram records the interference fringes produced by the interference of two laser beams, one of which arrives at the photographic film directly from the laser, and the other arrives after it is first reflected by the object (Fig. 9a). The interference fringes recorded on the photographic film contain a full description of the wave fronts reflected by the object, and therefore they contain information about the three-dimensional shape of the object. When such a hologram is later illuminated by a laser beam (Fig. 9b), it reconstructs the original wave fronts of the light reflected by the object, and therefore your eyes see the object just as though you were looking at the wave fronts reflected by the object itself — you see an exact three-dimensional replica of the object.

REFERENCES

Kock, W. E., *Lasers and Holography* (Doubleday, Garden City, 1969).

Schawlow, A. L., "Laser Light," *Scientific American,* September 1968.

Craxton, R. S., McCrory, R. L., and Soures, J. M., "Progress in Laser Fusion," *Scientific American,* August 1986.

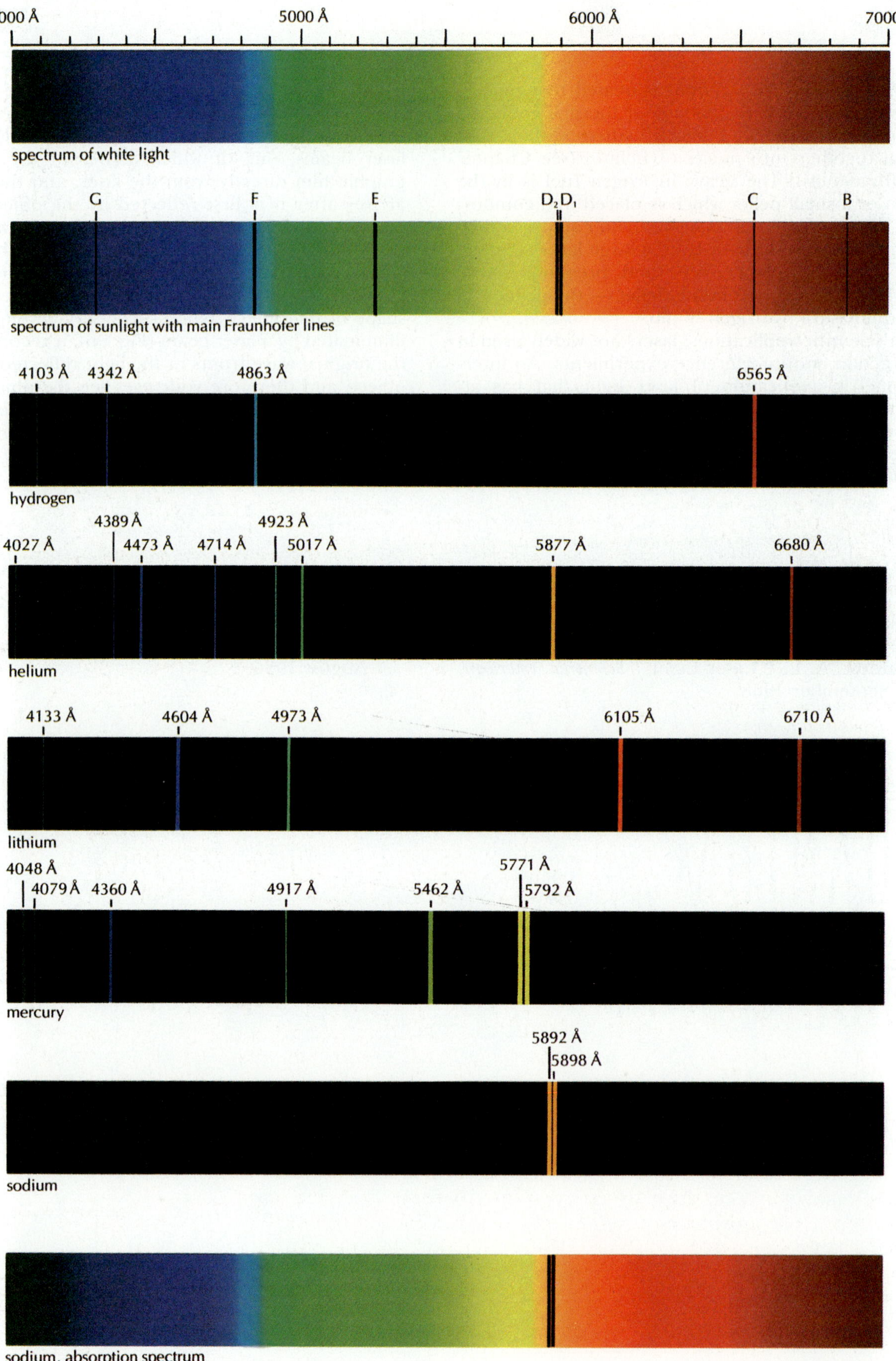

(All wavelengths are measured in vacuum.)

Courtesy Eastman Kodak Company

C H A P T E R 30

Spiral crystalline grains in a film of superconducting material. All the geometric patterns found in crystals hinge on the arrangement of the electrons of the atoms and molecules in the crystals.

QUANTUM STRUCTURE OF ATOMS, MOLECULES, AND SOLIDS

The physical and chemical properties of atoms and of molecules depend on the quantum behavior of their electrons. These electrons occupy most of the volume of the atom, and their arrangement in different orbits determines the size and the shape of the atom, the chemical bonds that the atom forms with other atoms, the energy required for ionization, the spectrum of light emitted and absorbed, and so on. Likewise, the physical properties of a solid — such as diamond, silicon, silver, copper — depend on the quantum behavior of the electrons in the solid. The geometry and the spacing of the crystal lattice of the solid, the electric and thermal conductivity, the magnetic properties, the mechanical properties all hinge on the arrangement of the electrons. For instance, we will see in Section 30.4 how the arrangement of the electrons determines the ability of the solid to conduct an electric current, and we will see how the differences among conductors, semiconductors, and insulators depend on what stationary states are available in the solid and which of these are occupied by electrons.

The arrangement of the electrons in the stationary states of an atom or a solid is subject to an important restriction: no more than two electrons can occupy the same orbital state. This is called the Exclusion Principle. In Section 30.1, we will become acquainted with the spin, or intrinsic angular momentum of the electron, and in Section 30.2, we will see that the Exclusion Principle is intimately linked to the spin. In later sections we will examine the implications of the Exclusion Principle for the arrangement of the electrons in atoms and in solids.

30.1 PRINCIPAL, ORBITAL, AND MAGNETIC QUANTUM NUMBERS; SPIN

In the preceding chapter, we saw that Bohr's theory characterizes the stationary states of the hydrogen atom and their energies by a single quantum number n. However, this simple version of Bohr's theory deals only with circular orbits. We know from the study of planetary orbits (Chapter 5) that the general orbit of a particle moving under the influence of an inverse-square force is an ellipse with one focus at the center of attraction. For an electron in such an elliptical orbit around a nucleus, the quantum number n, now called the ***principal quantum number,*** characterizes the overall size of the ellipse, that is, its major axis; this quantum number also characterizes the energy of the electron [the formula for the quantized energies of the elliptical orbits is the same as for the circular orbits, Eq. (29.25)]. But for the complete definition of the ellipse we need two extra quantum numbers l and m that characterize the elongation of the ellipse and the inclination of the ellipse, that is, its orientation in space. These two extra quantum numbers l and m are called, respectively, the ***orbital quantum number*** and the ***magnetic quantum number.***

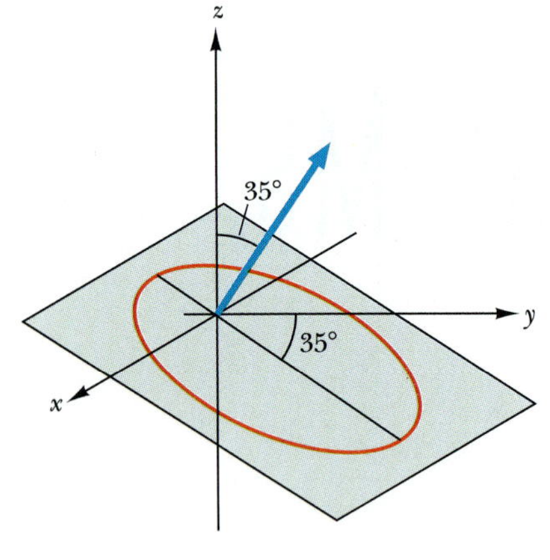

FIGURE 30.1
This elliptical orbit is inclined at an angle of 35° relative to the x–y plane. The angular-momentum vector **L** is perpendicular to the plane of the orbit; this vector is therefore inclined at an angle of 35° relative to the z axis. The z component of the angular-momentum vector is L_z.

As we will see below, the orbital quantum number l is related to the magnitude of the angular momentum. To understand how the magnetic quantum number m is related to the orientation of the orbit in space, we have to think of the angular momentum as a vector, of a direction perpendicular to the plane of the orbit (see Fig. 30.1). The orientation of the orbit in space can then be characterized by the direction of the angular-momentum vector; for instance, if the orbit is inclined at an angle of 35° relative to the x–y plane, as in Fig. 30.1, then the angular-momentum vector is inclined at an angle of 35° relative to the z axis. As we will see below, the magnetic quantum number m is related to the z component of the angular-momentum vector. A small inclination of the orbit implies a large value of this z component and a large value of the quantum number

m; a large inclination (near 90°) implies a small value of the z component and a small value of the quantum number m. (The quantum number m is called "magnetic" because it acquires a special significance when the atom is placed in a magnetic field; the energy of the orbit then depends not only on the size of the orbit and on the principal quantum number n, but also on the orientation of the orbit and, therefore, on the "magnetic" quantum number m.)

Although the two extra quantum numbers l and m were originally introduced on the basis of semiclassical considerations involving elliptical orbits, a later, more rigorous analysis based on wave mechanics and the Schrödinger equation confirmed that these quantum numbers are, indeed, required for the characterization of the stationary states. In wave mechanics ***a stationary state corresponds to a three-dimensional standing wave, and the quantum numbers*** n, l, ***and*** m ***characterize the "shape" of this standing wave.***

According to wave mechanics, the angular momentum is not directly specified by the principal quantum number n, but rather by the orbital quantum number l. From the study of the mathematical properties of the Schrödinger equation, it can be demonstrated that ***the magnitude*** L ***of the angular momentum obeys the quantization condition***

$$L = \sqrt{l(l+1)}\,\hbar \tag{1}$$

◂ **Quantization of magnitude of angular momentum**

The orbital quantum number l ***is restricted to integer values from*** 0 ***to*** $n-1$, that is,

$$l = 0, 1, 2, \ldots, n-1 \tag{2}$$

For example, if $n = 3$, then the possible values of the orbital quantum number are $l = 0$, $l = 1$, and $l = 2$; and the corresponding magnitudes of the angular momentum are $L = 0$, $L = \sqrt{2}\,\hbar$, and $L = \sqrt{6}\,\hbar$. Note that the smallest possible magnitude of the angular momentum is zero, in contrast to what is claimed by the Bohr theory, where the smallest magnitude of the angular momentum is $\hbar$.

If we think of the angular momentum as a vector, we must specify not only its magnitude, but also its direction. However, according to wave mechanics, the direction of the angular-momentum vector is determined only partially. The direction of the angular-momentum vector always has substantial quantum uncertainties, and, correspondingly, the x, y, and z components of this vector have substantial uncertainties. It can be demonstrated that at most *one* of the components is well determined — the other two components are completely uncertain. If we take the well-determined component to be the z component, then the quantization condition for this component is

◂ **Quantization of z component of angular momentum**

$$L_z = m\hbar \tag{3}$$

Here, ***the magnetic quantum number** m **is restricted to integer values from** $-l$ **to** $+l$*, that is

$$m = -l, -l + 1, \ldots, 0, \ldots, l - 1, l \tag{4}$$

Note that, for a given value of l, there are $2l + 1$ possible values of m. Each of these possible values of m corresponds to a possible direction of the angular-momentum vector relative to the z axis. For instance, if $l = 2$, then there are five possible values of m, namely, $m = -2$, $m = -1$, $m = 0$, $m = +1$, and $m = +2$; thus, there are five possible choices for the direction of the angular-momentum vector relative to the z axis. Figure 30.2 shows these possible directions.

From Eq. (3) we see that the z component of the angular momentum has a maximum possible value of $l\hbar$; this corresponds to the best attainable alignment of the angular momentum with the z axis. If we compare this maximum possible value of the z component with the magnitude of the angular momentum, we see that the former is always smaller than the latter, $l\hbar \leq \sqrt{l(l+1)}\,\hbar$. This means that the angular-momentum vector is never perfectly aligned with the z axis. The uppermost vector drawn in Fig. 30.2 indicates the best attainable alignment.

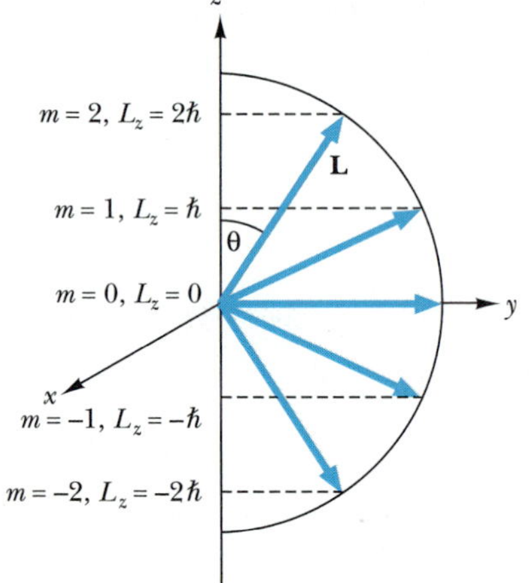

FIGURE 30.2

Possible direction of the angular-momentum vector for the case $l = 2$. In this case, there are five possible values of L_z, characterized by $m = -2, -1, 0, +1,$ and $+2$.

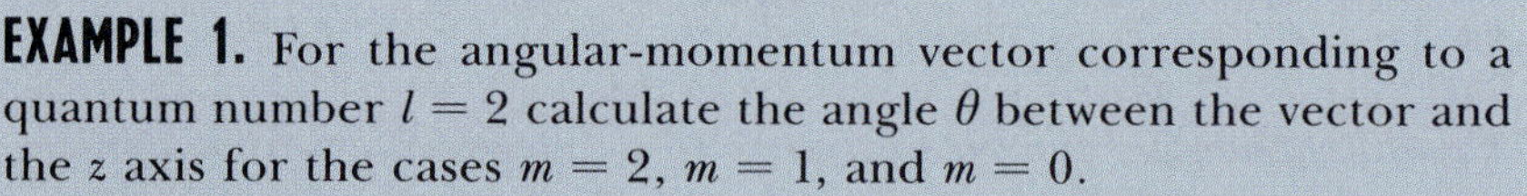

EXAMPLE 1. For the angular-momentum vector corresponding to a quantum number $l = 2$ calculate the angle θ between the vector and the z axis for the cases $m = 2$, $m = 1$, and $m = 0$.

SOLUTION: The directions of the angular-momentum vector for these cases are illustrated in Fig. 30.2. The magnitude of the angular-momentum vector is $L = \sqrt{l(l+1)}\hbar = \sqrt{2 \times 3}\hbar = \sqrt{6}\hbar$. In the case $m = 2$, the z component is $L_z = 2\hbar$, and the angle θ is given by

$$\cos\theta = \frac{L_z}{L} = \frac{2\hbar}{\sqrt{6}\hbar} = \frac{2}{\sqrt{6}} = 0.82$$

With our electronic calculator, we find that the angle with this cosine is 35°.

Likewise, in the case $m = 1$, the z component is $L_z = \hbar$, and

$$\cos\theta = \frac{\hbar}{\sqrt{6}\hbar} = 0.41$$

which gives us an angle of 66°.

Finally, in the case $m = 0$, the z component is $L_z = 0$, which gives us an angle of 90°.

Each possible set of values of n, l, and m corresponds to one kind of three-dimensional standing wave. The simplest of these waves is that for

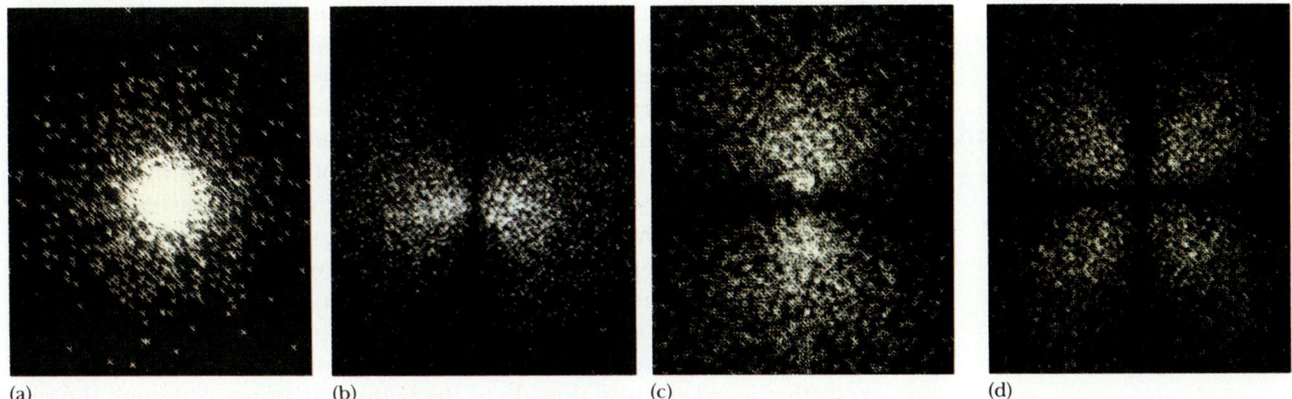

FIGURE 30.3 Intensities of possible standing waves in the hydrogen atom. The density of the dots is proportional to the intensity of the wave. (a) Ground state, $n = 1$, $l = 0$, $m = 0$. (b) First excited state, $n = 2$, $l = 1$, $m = 0$. (c) First excited state, $n = 2$, $l = 1$, $m = 1$. (d) Second excited state, $n = 3$, $l = 2$, $m = +1$ or -1. (Courtesy A. F. Burr and A. Fisher, New Mexico State University.)

the ground state, with $n = 1$, $l = 0$, and $m = 0$. Figure 30.3a shows a picture of the intensity of this wave for the hydrogen atom. In this picture, generated from the theoretical formulas of wave mechanics by a computer, the density of the dots indicates the intensity of the wave; thus, the wave is strongest near the center (near the nucleus of the atom), and gradually fades with increasing distance from the center. The intensity of the wave at any point is proportional to the probability for finding the electron at this point; thus, in the ground state of the hydrogen atom, the most probable position for the electron is right at the nucleus. Because of this probabilistic interpretation, the pictures of wave intensity shown in Fig. 30.3 are often called probability clouds.

Waves with other values of n, l, and m correspond to the excited states; all these waves are more complicated than the wave for the ground state, especially if n, l, and m are large. Figures 30.3b–d give some examples of waves corresponding to some excited states.

◀ **Spin quantum number**

In addition to these quantum numbers n, l, and m, one more quantum number is needed for the complete characterization of the stationary states of an electron in the hydrogen atom. ***This is the spin quantum number m_s that characterizes the spin, or intrinsic angular momentum, of the electron.*** The quantum number m_s was originally proposed by Wolfgang Pauli in an attempt to describe the "hyperfine" structure of the spectral lines: when the spectral lines of hydrogen and of other atoms are examined with a spectroscope of high resolving power, they are often found to consist of pairs, or doublets, of very closely spaced lines. This implies that one or both of the atomic energy levels involved in the transition must be a closely spaced pair of energy levels, and it implies that, besides n, l, and m, there must be another quantum number that distinguishes between the two energy levels in the pair. Pauli assigned the values $m_s = +\frac{1}{2}$ and $m_s = -\frac{1}{2}$ to the two energy levels in the pair, but offered no explanation of the physical significance of this new quantum number.

Shortly thereafter, S. Goudsmit and G. Uhlenbeck suggested that the electron has an intrinsic spin angular momentum, and that the two values of m_s correspond to different orientations of the axis of spin. They imagined the electron as a small ball of charge spinning about its axis, like the Earth spinning about its axis, with an angular momentum of magnitude $\sqrt{\frac{1}{2}(\frac{1}{2}+1)}\ \hbar$ and with an angular-momentum quantum number, or spin quantum number, of $\frac{1}{2}$. If we use the same rule for the possible directions of the spin as for the possible directions of the orbital angular momentum, we find that there are $2 \times \frac{1}{2} + 1 = 2$ possible directions for the spin. ***One of these directions is characterized by a magnetic quantum number*** $m_s = +\frac{1}{2}$, ***and the other by a magnetic quantum number*** $m_s = -\frac{1}{2}$; ***these two possible spin directions are called spin up and spin down, respectively.***

Spin up and spin down

Goudsmit and Uhlenbeck's simple picture of the spin as due to a rotation of the electron about its axis proved untenable. Instead, modern wave mechanics tells us that the spin is an angular momentum generated by a circulating energy flow in the electron wave. The orbital angular momentum is also generated by such a circulating energy flow in the electron wave, but the energy flow that generates the spin is distinguished in that it persists even in the electron wave of an isolated, free electron at rest (that is, an electron outside of the hydrogen atom). Although the simple picture of the electron as a rotating ball of charge is false, it can sometimes serve as a convenient crutch for our imagination. According to this simple picture, we expect that the electron has a magnetic moment, since a piece of charge rotating about an axis amounts to a current loop (see Fig. 30.4). The energy difference between the two energy levels that gives rise to doublets of spectral lines is, in fact, due to this magnetic moment, which interacts with the magnetic field generated by the motion of the nuclear charge seen in the reference frame of the electron. As already mentioned in Section 21.4, the magnetic moment of the electron also plays an important role in the behavior of ferromagnetic materials.

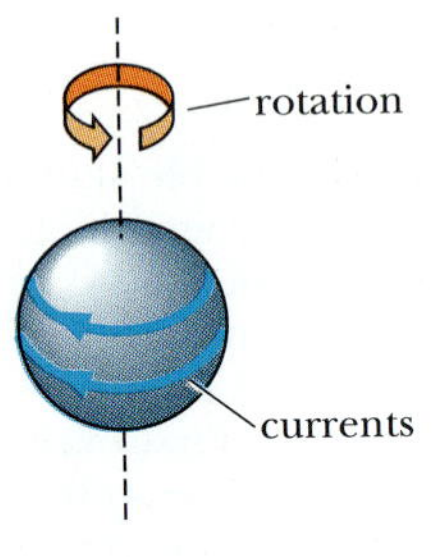

FIGURE 30.4
Simple picture of the electron as a rotating ball of charge. The motion of the negative charge in, say, counterclockwise circles is equivalent to a clockwise current of positive charge.

Many other elementary particles besides electrons have spin. For example, the proton and the neutron have spins of magnitude $\sqrt{\frac{1}{2}(\frac{1}{2}+1)}\hbar$ (spin quantum number $\frac{1}{2}$), the photon has spin of magnitude $\sqrt{1(1+1)}\hbar$ (spin quantum number 1), etc.

The quantum numbers n, l, m, and m_s provide a complete characterization of the stationary states of the hydrogen atom. The energies of the stationary state depend mainly on the principal quantum number n. According to Eq. (29.25), the energies are

$$E_n = -\frac{13.6 \text{ eV}}{n^2} \tag{5}$$

Thus, stationary states of the same value of n have (almost) the same energy, regardless of the values of the other quantum numbers. Classically, this means that the energy depends on the overall size of the elliptical orbit, but not on its elongation, or on its orientation in space. However, a more refined calculation of the energies of the stationary states of the hydrogen atom shows that the energies depend slightly on the orbital quantum number l and on the orientation of the spin relative to the plane of the orbit. This means that Eq. (5) for the energies of the stationary states of the hydrogen atom is not quite accurate. But the deviations from Eq. (5) are very small, and we can often ignore them.

Table 30.1 lists the permitted quantum numbers for the stationary states of the hydrogen atom. ***These quantum numbers can also be used to characterize the stationary states of atoms other than hydrogen,*** but the energies of other atoms are not given by the simple formula (1).

TABLE 30.1 QUANTUM NUMBERS OF ELECTRONIC STATES

Quantum Number	Symbol	Values
Principal	n	$1, 2, 3, \ldots$
Orbital	l	$0, 1, 2, \ldots n-1$
Magnetic	m	$-l, -l+1, -l+2, \ldots l-2, l, -1, l$
Spin	m_s	$-\frac{1}{2}, +\frac{1}{2}$

EXAMPLE 2. (a) What are the permitted quantum numbers for the case $n = 1$? (b) What are the possible quantum numbers for the case $n = 2$?

SOLUTION: (a) For $n = 1$, Table 30.1 tells us that the only permitted value of l is 0. Furthermore, if $l = 0$, then the only permitted value of m is also 0. The permitted values of m_s are $-\frac{1}{2}$ and $+\frac{1}{2}$, regardless of the values of the other quantum numbers. Hence, for $n = 1$, the list of quantum numbers is

$$n = 1;\ l = 0;\ m = 0;\ m_s = -\tfrac{1}{2} \text{ or } +\tfrac{1}{2}$$

(b) For $n = 2$, Table 30.1 tells us that the permitted values of l are 0 and 1. If $l = 0$, then the only permitted value of m is 0, and we find the same list as in Part (a):

$$n = 2;\ l = 0;\ m = 0;\ m_s = -\tfrac{1}{2} \text{ or } +\tfrac{1}{2}$$

However, if $l = 1$, then the permitted values of m are -1, 0, $+1$, and we find the following list:

$$n = 2;\ l = 1;\ m = -1, \text{ or } 0, \text{ or } +1;\ m_s = -\tfrac{1}{2} \text{ or } +\tfrac{1}{2}$$

Each alternative listed here represents a possible electronic state. For instance, $n = 2$, $l = 1$, $m = -1$, $m_s = +\frac{1}{2}$ represents one state; and $n = 2$, $l = 1$, $m = 0$, $m_s = +\frac{1}{2}$ represents another state.

30.2 THE EXCLUSION PRINCIPLE AND THE STRUCTURE OF ATOMS

The paramount question in atomic structure is the determination of the detailed arrangement of the electrons in their orbits around the nucleus. ***The electron arrangement, or configuration, determines all the physical and*** ◄ **Configuration**

TABLE 30.2 THE PERIODIC TABLE OF THE CHEMICAL ELEMENTS[a]

IA																	0
1 H 1.00794	IIA											IIIA	IVA	VA	VIA	VIIA	2 He 4.00260
3 Li 6.941	4 Be 9.01218											5 B 10.811	6 C 12.011	7 N 14.0067	8 O 15.9994	9 F 18.998403	10 Ne 20.1797
11 Na 22.98977	12 Mg 24.305	IIIB	IVB	VB	VIB	VIIB		VIII		IB	IIB	13 Al 26.98154	14 Si 28.0855	15 P 30.97376	16 S 32.066	17 Cl 35.4527	18 Ar 39.948
19 K 39.0983	20 Ca 40.078	21 Sc 44.9559	22 Ti 47.88	23 V 50.9415	24 Cr 51.996	25 Mn 54.9380	26 Fe 55.847	27 Co 58.9332	28 Ni 58.6934	29 Cu 63.546	30 Zn 65.39	31 Ga 69.723	32 Ge 72.61	33 As 74.9216	34 Se 78.96	35 Br 79.904	36 Kr 83.80
37 Rb 85.4678	38 Sr 87.62	39 Y 88.9059	40 Zr 91.224	41 Nb 92.9064	42 Mo 95.94	43 Tc (97.907)	44 Ru 101.07	45 Rh 102.9055	46 Pd 106.42	47 Ag 107.8682	48 Cd 112.411	49 In 114.82	50 Sn 118.71	51 Sb 121.757	52 Te 127.60	53 I 126.9045	54 Xe 131.29
55 Cs 132.9054	56 Ba 137.327	57–71 Rare Earths	72 Hf 178.49	73 Ta 180.9479	74 W 183.85	75 Re 186.207	76 Os 190.2	77 Ir 192.22	78 Pt 195.08	79 Au 196.9665	80 Hg 200.59	81 Tl 204.383	82 Pb 207.2	83 Bi 208.9804	84 Po (208.9804)	85 At (2109.9871)	86 Rn (222.0176)
87 Fr (223.020)	88 Ra (226.0254)	89-103 Acti- nides	104 Rf (261.11)	105 Ha (262.114)	106 (263.118)	107 (262.12)	108 (265)	109 (266)									

Rare Earths (Lanthanides)	57 La 138.9055	58 Ce 140.115	59 Pr 140.9077	60 Nd 144.24	61 Pm (144.913)	62 Sm 150.36	63 Eu 151.965	64 Gd 157.25	65 Tb 158.9253	66 Dy 162.50	67 Ho 164.9303	68 Er 167.26	69 Tm 168.93421	70 Yb 173.04	71 Lu 174.967
Actinides	89 Ac (227.0278)	90 Th 232.0381	91 Pa 231.0359	92 U 238.0289	93 Np (237.0482)	94 Pu (244.064)	95 Am (243.061)	96 Cm (247.070)	97 Bk (247.070)	98 Cf (251.080)	99 Es (252.083)	100 Fm (257.095)	101 Md (258.10)	102 No (259.101)	103 Lr (262.11)

[a] In each box, the upper number is the *atomic number*. The lower number is the *atomic mass*, that is, the mass (in grams) of one mole or, equivalently, the mass (in atomic mass units) of one atom. Numbers in parentheses denote the atomic masses of the most stable or best-known isotope of the element; all other numbers represent the average masses of a mixture of several isotopes as found in naturally occurring samples of the element.

chemical properties of the atom — if the electron configuration is known, all the properties of the atom can be deduced by theoretical considerations. For instance, the observed similarities of chemical properties among select groups of elements must be due to similarities in their electronic configurations. Chemists list similar elements in columns in the

Periodic Table ▸

Periodic Table of elements (see Table 30.2). Thus, the elements helium, neon, argon, krypton, etc., are listed in one column; these are the noble gases, which do not react chemically with anything. The elements fluorine, chlorine, bromine, iodine, etc., are listed in another column; these are the halogens, irritating, corrosive gases, all with quite noticeable and distinctive colors (pale yellow, greenish yellow, red, and blue-violet, respectively). And the elements lithium, sodium, potassium, rubidium, etc., are listed in the first column; these are the alkalis, silvery-white metals, which are extremely reactive. ***The pattern of the elements displayed in the Periodic Table can be explained by a study of the electron configurations.***

For the case of the hydrogen atom, the determination of the electron configuration is trivial: the single electron of this atom is in one or another of the stationary states characterized by the quantum numbers n, l,

m, and m_s. If the atom is in the ground state, the values of the quantum numbers of the electron configuration are $n = 1$, $l = 0$, $m = 0$, and $m_s = \pm\frac{1}{2}$; thus, everything is fixed, except the direction of the spin, which can be up or down.

But for atoms with several electrons, the determination of the electron configuration is not so trivial. It might be tempting to suppose that the ground state of the atom (the state of least energy) is attained by placing all the electrons in the lowest stationary state, with $n = 1$, $l = 0$, $m = 0$, $m_s = \pm\frac{1}{2}$, as for the hydrogen atom. But this would imply that all the atoms ought to have a spectral series similar to that of hydrogen, and it would also imply that atoms with a large number of electrons, or with a large atomic number Z, ought to be very small, since the Bohr radius for an atom of nuclear charge Ze is a_0/Z [if the nuclear charge is Ze instead of e, then in the denominator of the formula (29.17) for the Bohr radius of hydrogen, we must replace *one* of the factors of e by Ze]. These conclusions are in stark conflict with the observed properties of atoms: the spectra of most atoms are quite different, and the sizes of atoms of large Z — such as lead or bismuth — are considerably larger than hydrogen.

The rule that governs the configuration of the electrons in an atom is the ***Exclusion Principle,*** which was discovered by Wolfgang Pauli:

◀ **Exclusion Principle**

> ***Each stationary state of quantum numbers** n, l, m, m_s **can be occupied by no more than one electron.***

Since for each stationary orbital state of quantum numbers n, l, m, there are two possible spin states ($m_s = \pm\frac{1}{2}$), we can also rephrase the Exclusion Principle as follows: Each stationary orbital state of quantum numbers n, l, m can be occupied by no more than two electrons.

Pauli originally proposed this principle as an empirical rule, based on the observed features of atomic spectra. It was later established that the Exclusion Principle is intimately linked to the value of the spin of the electron; the Exclusion Principle can be shown to be a necessary consequence of the quantum theory of particles of half-integer spin. Thus, ***protons and neutrons also obey the Exclusion Principle,*** a fact of great importance for the configuration of these particles in the interior of the nucleus (see next chapter). ***In contrast, particles of integer spin, such as photons, do not obey the Exclusion Principle.*** There is no limit to the number of such particles that can be packed into a given stationary state, for instance, one of the standing-wave states in a cavity filled with blackbody radiation.

Wolfgang Pauli, *1900–1958, Austrian and later Swiss theoretical physicist. For his discovery of the Exclusion Principle, he was awarded the Nobel Prize in 1945. Pauli made an important contribution to the theory of beta decay by proposing that the emission of the beta particle is always accompanied by the emission of a neutrino (see Chapter 31).*

For our investigation of the electron configuration of atoms, we will find it convenient to start with a list of all the available states, in order of increasing energy (see Example 2). The states of lowest energy have $n = 1$; there are two such states:

States with $n = 1$ (K shell)

$n = 1$	$l = 0$	$m = 0$	$m_s = -\frac{1}{2}$
$n = 1$	$l = 0$	$m = 0$	$m_s = +\frac{1}{2}$

Next, consider $n = 2$; there are 8 available states:

States with $n = 2$ (L shell)

$n=2 \quad l=0 \quad m=0 \quad m_s=-\frac{1}{2}$
$n=2 \quad l=0 \quad m=0 \quad m_s=+\frac{1}{2}$
$n=2 \quad l=1 \quad m=-1 \quad m_s=-\frac{1}{2}$
$n=2 \quad l=1 \quad m=-1 \quad m_s=+\frac{1}{2}$
$n=2 \quad l=1 \quad m=0 \quad m_s=-\frac{1}{2}$
$n=2 \quad l=1 \quad m=0 \quad m_s=+\frac{1}{2}$
$n=2 \quad l=1 \quad m=+1 \quad m_s=-\frac{1}{2}$
$n=2 \quad l=1 \quad m=+1 \quad m_s=+\frac{1}{2}$

Shells (K, L, M, . . .) ▶

Likewise, for $n = 3$, there are 18 available states, and so on. ***The groups of states of a given value of n are called shells,*** and they are conventionally labeled with the letters K, L, M, etc. Thus, the two states with $n = 1$ form the K shell; the eight states with $n = 2$ form the L shell; the eighteen states with $n = 3$ form the M shell, etc.

According to the Exclusion Principle, ***each of the states listed above can accommodate one, and only one, electron.*** Thus, if an atom with Z electrons is in its ground state, the electrons will occupy the first Z of the states in the above list. We can therefore build up the configurations for all the atoms in the Periodic Table of elements by beginning with hydrogen and adding electrons one by one, sequentially filling the states in our list.

The second element in the Periodic Table is helium, which has two electrons. To obtain its electron configuration, we must add one electron to the hydrogen configuration; since the single electron of hydrogen occupies one of the states of the K shell, we can place the second electron in the other available state in the K shell. Helium therefore has a full K shell.

The third element is lithium, with three electrons. When we add one electron to the helium configuration, we must place this third electron in the L shell, in the state with quantum numbers $n = 2$, $l = 0$, $m = 0$, $m_s = -\frac{1}{2}$.

The next element is beryllium, with four electrons. Thus, we must add one more electron in the L shell, in the state with quantum numbers $n = 2$, $l = 0$, $m = 0$, $m_s = +\frac{1}{2}$.

We can continue in this way, filling up the states in our list one by one. With the tenth element, neon, we will have filled the L shell. And with the eleventh element, sodium, we must place one electron in the M shell, and so on.

This simple procedure for building up the electron configurations of the atoms provides us with an immediate explanation of the similarities of the elements in the columns of the Periodic Table. For instance, the similarity in the chemical behavior and the similarity in the spectra of helium and neon can be traced to a similarity of their electron configurations: both these atoms have full shells of electrons. Likewise, the chemical and spectroscopic similarities of hydrogen, lithium, and sodium can be traced to a similarity of their electron configurations: they all have a single electron outside of a full shell of electrons. This single, outer electron tends to come off the atom fairly easily, and in chemical reactions these atoms all tend to lose an electron. In contrast, flourine and chlorine are one electron short of a full shell, and in chemical reactions they tend to capture an electron to complete their shell.

Thus, the Exclusion Principle in conjunction with a simple counting procedure for the available stationary states is sufficient to explain the broad, qualitative features of the Periodic Table of elements. Detailed calculations, based on wave mechanics, provide quantitative theoretical results for ionization energies, spectral lines, atomic sizes, etc., in agreement with the observed atomic properties.

For atoms other than hydrogen, there is no simple formula for the energies of the stationary states. However, in an atom of fairly large atomic number, say $Z > 20$, the dominant force on the innermost electrons is the attractive Coulomb force exerted by the positive charge Ze of the nucleus, and the repulsive forces exerted by the other electrons can be neglected. Thus, ***these innermost electrons move on hydrogenlike orbits.*** The energy of an electron in such an orbit is given approximately by Eq. (29.23), with one modification: the product $e \times e$ of the electron and the proton charge must be replaced by the product $e \times Ze$ of the electron and the nuclear charge; hence e^4 in Eq. (29.23) must be replaced by e^4Z^2, which leads to the following approximate formula for the energy:

$$E_n = -\frac{m_e Z^2 e^4}{2(4\pi\epsilon_0)^2\hbar^2}\frac{1}{n^2} = -\frac{Z^2 \times 13.6\text{ eV}}{n^2} \tag{6}$$

From this formula, we can calculate the frequency and the wavelength of light emitted during a quantum jump from some initial state to a final state. However, such a quantum jump between the innermost orbits of the atom is not possible if the atom has its full complement of electrons — all the orbits are then occupied by electrons, and the Exclusion Principle forbids quantum jumps into an already occupied orbit. Thus, a jump is possible only if some external disturbance first removes one of the electrons from the atom, leaving a gap into which some other electron can jump. Such a process occurs when the target atoms in an X-ray tube are subjected to the impact of the electron beam. The target atoms are disturbed by this impact, and sometimes an electron in one of the innermost orbits is ejected, leaving a gap into which another electron can jump. The photon emitted in such a transition has a very short wavelength; it is an X ray. Thus, the quantum jumps of the innermost electrons of atoms give rise to the characteristic spectrum of X rays mentioned in Section 28.5.

EXAMPLE 3. Suppose that in an atom of molybdenum ($Z = 42$) in the target of an X-ray tube, one of the electrons in the $n = 1$ state is ejected during the impact of the electron beam on the target. Suppose that subsequently one of the other electrons in the atom jumps from the $n = 2$ state into this available empty $n = 1$ state. What are the energy and the wavelength of the photon emitted during this quantum jump?

SOLUTION: The initial energy of the electron is

$$E_2 = -\frac{Z^2 \times 13.6\text{ eV}}{4} = -\frac{42^2 \times 13.6\text{ eV}}{4} = -6.0 \times 10^3\text{ eV}$$

and the final energy is

$$E_1 = -\frac{Z^2 \times 13.6\text{ eV}}{1} = -42^2 \times 13.6\text{ eV} = -24.0 \times 10^3\text{ eV}$$

Hence the energy of the photon is

$$E_2 - E_1 = 18.0 \times 10^3\text{ eV} = 2.88 \times 10^{-15}\text{ J}$$

The frequency of the photon is

$$f = \frac{2.88 \times 10^{-15}\text{ J}}{h} = \frac{2.88 \times 10^{-15}\text{ J}}{6.63 \times 10^{-34}\text{ J}\cdot\text{s}} = 4.34 \times 10^{18}\text{ Hz}$$

and the wavelength is

$$\lambda = \frac{c}{f} = \frac{3.0 \times 10^8\text{ m/s}}{4.34 \times 10^{18}\text{ Hz}} = 6.9 \times 10^{-11}\text{ m} = 0.069\text{ nm}$$

30.3 ENERGY LEVELS IN MOLECULES

The chemical bonds that bind two or more atoms together in a molecule, such as O_2 or HCl, arise from a rearrangement of the outer electrons, or valence electrons, of the atoms. In some molecules (for example, O_2), the outer electrons are shared between the atoms, a sharing that produces an attractive force (covalent bond). In some other molecules (for example, HCl), one atom loses an electron to the other atom, and the atom with the missing electron is then electrically attracted by the atom with the extra electron (ionic bond).

The chemical bonds are elastic — they behave rather like springs tying the atoms together. The springs hold the atoms at an average equilibrium distance, but permit the atoms to oscillate back and forth about this average distance, with some kinetic and potential energy. This means that the energy of the molecule is the sum of the electronic energy of the atoms and the vibrational energy of the motion of the atoms in relation to each other. Let us focus on the vibrational energy of a molecule and examine its quantization.

The mass of the atom is concentrated in its center, in the nucleus, which is much smaller than the interatomic distances in a molecule. We can therefore schematically represent a molecule — for instance, a diatomic molecule — as a system of pointlike masses connected by a massless spring (see Fig. 30.5). The atoms oscillate in unison relative to the center of mass, which we can regard as fixed. Thus, the system is an oscillator, and the energy of this oscillator is subject to Planck's quantization condition, Eq. (28.3). If the frequency of oscillation is f, ***the energy of the vibrational motion is quantized according to***

◀ **Vibrational energies of molecule**

$$E = nhf \qquad n = 0, 1, 2, \ldots \tag{7}$$

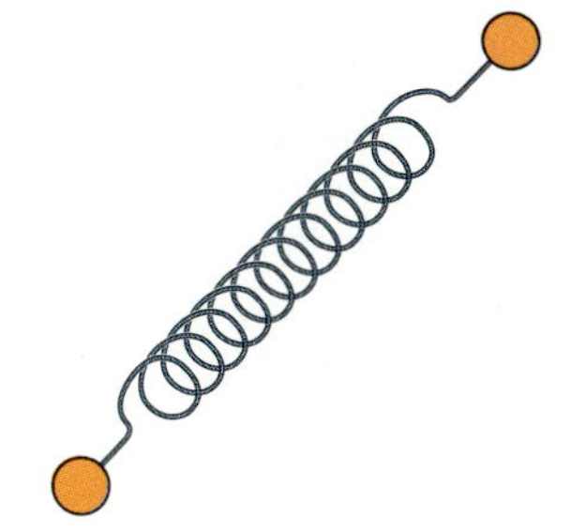

FIGURE 30.5

An oscillating diatomic molecule can be represented as two pointlike masses on the ends of a massless spring.

The corresponding energy-level diagram is shown in Fig. 30.6. The molecule will emit a photon if it makes a transition from an upper level to a lower level. The vibrational transitions in a molecule are restricted by a ***selection rule: the transition must proceed from one level to the next, that is, transitions spanning two or more levels in one jump are forbidden.*** This selection rule can be deduced from wave mechanics. The arrows in Fig. 30.6 indicate the permitted transitions. The frequencies of the radiation emitted during all these transitions are therefore the same,

$$f_{\text{rad}} = \frac{\Delta E}{h} = \frac{hf}{h} = f \tag{8}$$

Thus, the frequency of the radiation equals the frequency of vibration of the molecule. Typically, the frequencies of vibration of molecules are of the order 10^{13} Hz, and the wavelengths of the emitted radiation lie in the infrared.

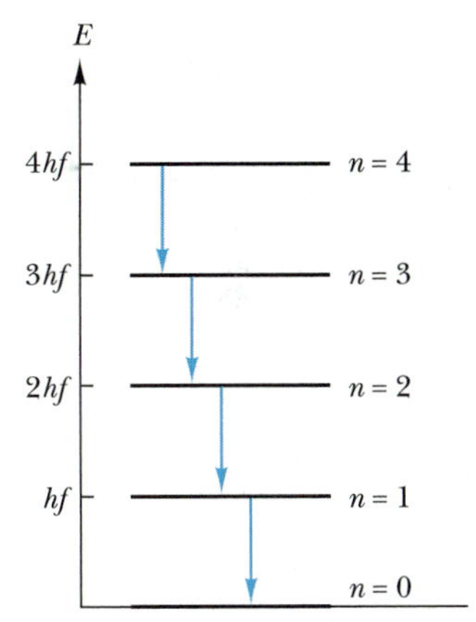

FIGURE 30.6

Energy-level diagram for the oscillating molecule. The arrows indicate the possible transitions.

Besides the vibrational motion, the molecule can also perform rotational motion. For the purposes of this rotational motion, we can regard the molecule as two pointlike masses linked by a massless rod, that is, a dumbbell (see Fig. 30.7). If the moment of inertia of the dumbbell about a perpendicular axis through the center of mass is I, then the kinetic energy of rotation is

$$E = \tfrac{1}{2} I\omega^2 \tag{9}$$

where ω is the angular frequency of the rotation. Let us express this in terms of the angular momentum. Since the angular momentum is $L = I\omega$, the angular frequency is $\omega = L/I$ and Eq. (9) becomes

$$E = \frac{L^2}{2I} \tag{10}$$

The angular momentum is quantized in the usual way, with $L = \sqrt{l(l+1)}\,\hbar$; this immediately implies that ***the energy of the rotational motion is quantized,***

$$E = \frac{l(l+1)\hbar^2}{2I} \qquad l = 0, 1, 2, \ldots \tag{11}$$

◀ **Rotational energies of molecule**

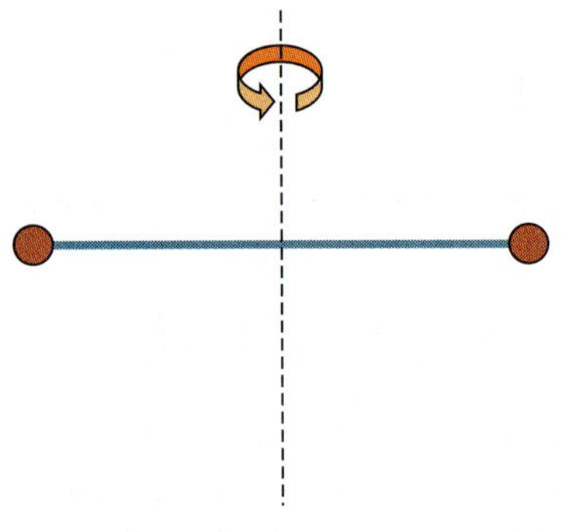

FIGURE 30.7
A rotating molecule can be regarded as two pointlike masses linked by a massless rigid rod.

Figure 30.8 displays the energy-level diagram for the rotational states of a molecule. The transitions are, again, subject to the selection rule that they must proceed from one level to the next. Such transitions are indicated by the arrows in Fig. 30.8.

EXAMPLE 4. The moment of inertia of the HCl molecule about its axis of rotation is $2.66 \times 10^{-47}\ \mathrm{kg \cdot m^2}$. What is the energy of the first excited rotational state of this molecule? The second excited rotational state?

SOLUTION: For the first excited state, $l = 1$ and Eq. (11) gives

$$E = \frac{2\hbar^2}{2I} = \frac{(1.05 \times 10^{-34}\ \mathrm{J \cdot s})^2}{2.66 \times 10^{-47}\ \mathrm{kg \cdot m^2}} = 4.14 \times 10^{-22}\ \mathrm{J}$$

For the second excited state, $l = 2$ and

$$E = \frac{2 \times 3 \times \hbar^2}{2I} = \frac{3 \times (1.05 \times 10^{-34}\ \mathrm{J \cdot s})^2}{2.66 \times 10^{-47}\ \mathrm{kg \cdot m^2}} = 1.24 \times 10^{-21}\ \mathrm{J}$$

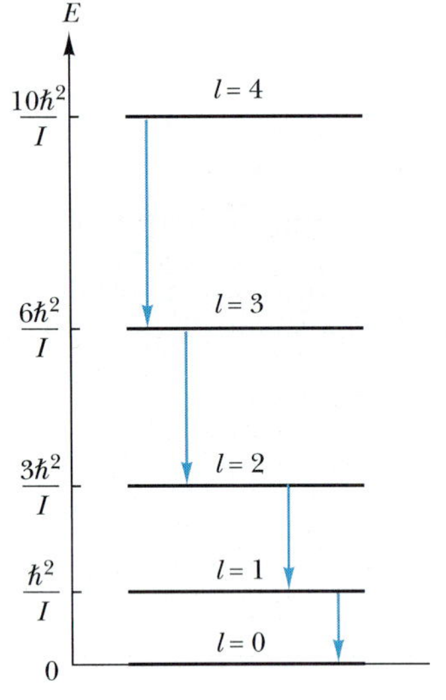

FIGURE 30.8
Energy-level diagram for the rotating molecule. The arrows indicate the possible transitions.

The rotational transitions in a molecule involve much smaller energies than the electronic transitions in an atom, such as the electronic transitions in a hydrogen atom. This means that the photons emitted in purely rotational molecular transitions are of rather low energy, and their wavelengths lie in the far infrared region of the spectrum. However, ***rotational molecular transitions are often observed in conjunction with a simultaneous electronic transition in one of the atoms of the molecule.*** This increases the energy of the transition, and reduces the wavelength of the radiation. As is obvious from Fig. 30.8, successive rotational transitions have slightly different energies and wavelengths; thus, they give rise to a group, or sequence, of adjacent spectral lines. This is called a ***band spectrum.*** Figure 30.9 is a photograph of several spectral bands in the spectrum of the NO molecule.

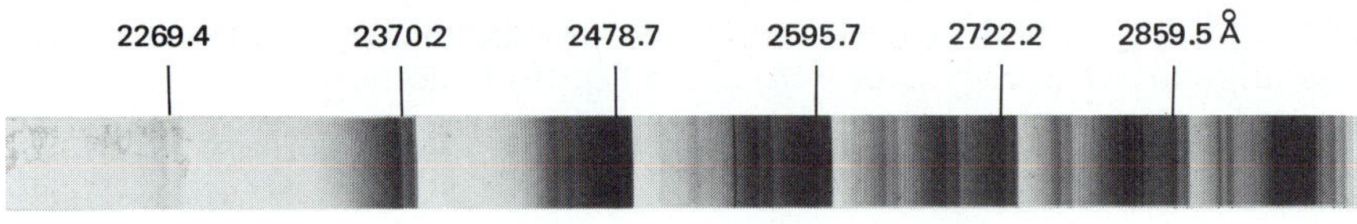

FIGURE 30.9
Bands of spectral lines emitted by the NO molecule.

30.4 Energy Bands in Solids

As discussed in Section 17.5, in a metal the outermost, or valence, electrons of the atoms are detached from their atoms, and they are free to wander all over the volume of the metal. However, whenever such a "free" electron passes by an atom, it experiences an attractive force. For an electron moving along a row of atoms in the crystal lattice of a metal the force will act repetitively, each time the electron passes by an atom (see Fig. 30.10). Under special conditions, such a repetitive action of a force can lead to a large cumulative effect. According to wave mechanics, we have to think of the electron as a wave, and an encounter with an atom scatters the wave: some fraction of the wave proceeds in its original direction of motion and some fraction is reflected. The repetitive scatterings at the atoms in the row will build up a large reflected wave by constructive interference if the scattering by one atom produces a reflected wave that is in phase with the reflected wave produced by the next atom. This will happen if the extra distance for a roundtrip from one atom to the next and back is equal to one de Broglie wavelength or a multiple of one wavelength. If the distance between the atoms is a, the distance for a roundtrip from one atom to the next is $2a$, and we can express the condition for constructive interference of all the reflected waves as

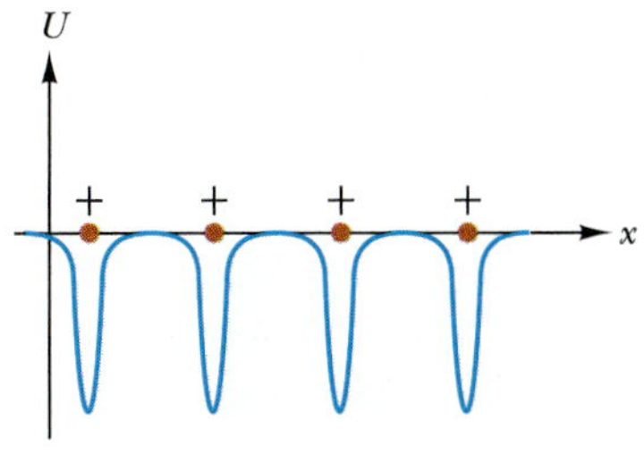

FIGURE 30.10

The plus signs mark the positions of the atoms along a row in a crystal. Whenever an electron comes near one of these atoms, it experiences an attractive force, which is here indicated by a dip in the potential.

$$2a = \lambda,\ 2\lambda,\ 3\lambda,\ \ldots \tag{12}$$

If this condition is satisfied, the reflected wave will gain more and more strength at each reflection at each atom, and finally match the strength of the incident wave. The result is a standing wave, which travels neither right nor left. This means that the electron cannot move through the lattice!

Since the de Broglie wavelength is related to the momentum of the electron by $\lambda = h/p$ [see Eq. (29.32)], we can express the condition for total reflection in terms of the momentum of the electron:

$$2a = \frac{h}{p},\ \frac{2h}{p},\ \frac{3h}{p},\ \ldots \tag{13}$$

or

$$p = \frac{h}{2a},\ \frac{2h}{2a},\ \frac{3h}{2a},\ \ldots \tag{14}$$

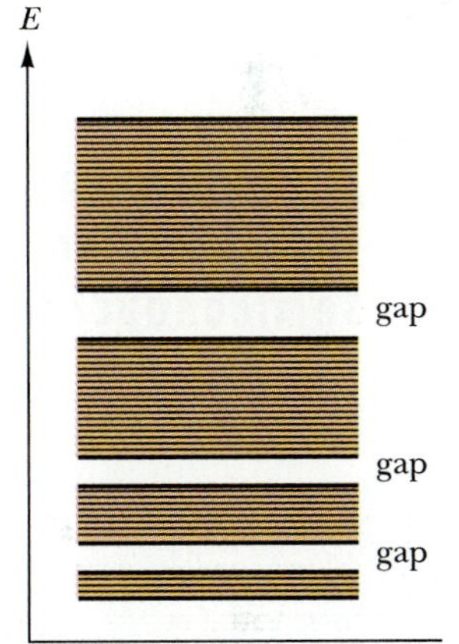

FIGURE 30.11

Energy-level diagram for an electron moving in a crystal. The permitted intervals of energies occur in bands, which are separated by forbidden intervals, or gaps.

These values of the momentum are forbidden values, that is, they are values at which the electron cannot move. Corresponding to the forbidden values of the momentum, there are forbidden values of the energy.

A more thorough analysis of the wave-mechanical motion of an electron through a crystal lattice reveals that the forbidden energies are actually forbidden energy gaps, that is, forbidden energy intervals. Figure 30.11 shows such forbidden energy gaps on an energy-level diagram. ***The permitted ranges of the energy, shown colored in the diagram, are called energy bands.*** The precise widths of the forbidden energy gaps and of the permitted energy bands depend on the details of the crystal lattice, but all crystals with "free" electrons have some kind of band pattern in their energy-level diagram.

◂ **Energy bands**

As in the case of the electron configuration of atoms, we can deduce the electron configuration of crystals by means of the Exclusion Principle.

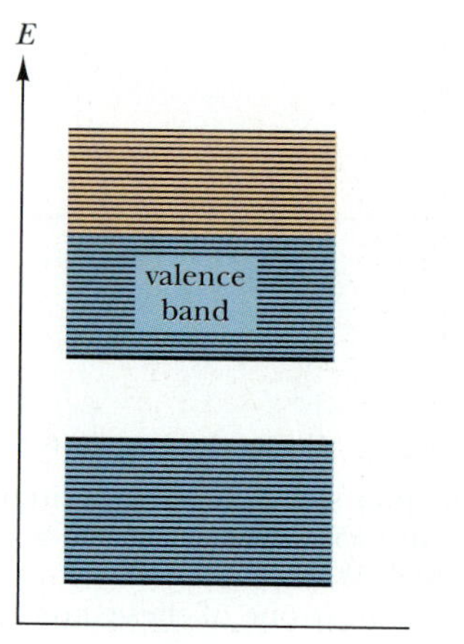

FIGURE 30.12
In a conductor, the upper energy band (valence band) is only partially filled with electrons. The portion of the band filled with electrons is shown in blue.

In a crystal in its ground state, the electrons settle in the available states of lowest energy. To discover the electron configuration, we proceed as before: we take all the "free" electrons and pack them, one by one, into the available energy bands. The lowest energy bands will then be completely filled, but the upper energy band will be either filled or partially filled, depending on the number of "free" electrons and the number of available states.

Conductor, Insulator, and Semiconductor

The differences among the electric properties of conductors, semiconductors, and insulators arise from the partial or complete filling of the upper energy band. ***In a conductor,*** such as copper or silver, ***the upper band is only partially filled with electrons*** (see Fig. 30.12). When the electrons in this partially filled band are subjected to an electric field, they absorb energy from the field and they make transitions to some of the slightly higher, empty states of the band. Thus, the electrons respond to the electric field, and they begin to carry an electric current.

In an insulator, such as diamond, ***the upper band is completely filled with electrons*** (see Fig. 30.13). When the electrons in this full band are subjected to an electric field, they cannot make transitions into other states in the band, because all of these states are already full, and the Exclusion Principle forbids transitions into already full states. This implies that the electrons in the full band cannot respond to the electric field; they therefore cannot accelerate in the direction of the electric force, and they cannot begin to carry a current. (The only way the electrons in the full band could respond to the electric field is by making transitions to the *next,* empty energy band; but this is too difficult, since it requires that an electron absorb a large amount of energy from the electric field.)

In a semiconductor, such as silicon or germanium, ***the upper band is completely filled with electrons,*** as in an insulator. ***However, the energy gap between this band and the next, empty band is much smaller than in an insulator*** (see Fig. 30.14). Typically, the width of the energy gap separating the full band from the next, empty band is of the order of only 1 eV, whereas in an insulator the gap is 6 eV or more. In a semiconductor at room temperature, the random thermal fluctuations of the energy will often permit an electron at the top of the full band to make a transition to the next, empty band. This means the electrons can respond to the electric field, and they can carry a current. The full band in an insulator or a semiconductor is called the ***valence band,*** and the empty band above it is called the ***conduction band*** (in a metal, the conduction band is the valence band, and this band is only partially filled).

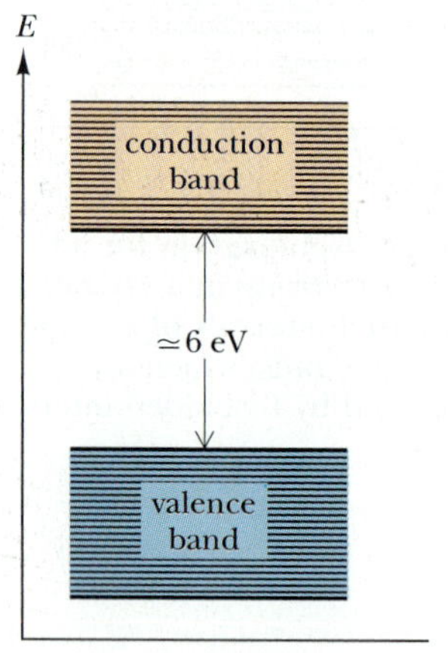

FIGURE 30.13
In an insulator, the upper energy band (valence band) is completely filled with electrons.

The values of the resistivities of semiconductors are between those of conductors and insulators. The resistivities of semiconductors vary over a wide range; the resistivities may be 10^4 to 10^{15} times as large as the resistivities of conductors. Semiconductors fall into two categories: n type and p type. ***In an n-type semiconductor, the carriers of current are free electrons that have reached the conduction band.*** Thus, the mechanism for conduction is the same as in a metallic conductor. However, the resistance of a semiconductor is higher than that of a metallic conductor, because the semiconductor has fewer free electrons in its conduction band than a metal in its partially filled upper band. Also, a (pure) semiconductor differs from a metal in that the resistivity *decreases* as the temperature increases. This curious behavior is due to an increase in the number of free electrons — as the temperature increases, more electrons are excited into the conduction band by random thermal fluctuations, and these extra free electrons more than compensate for the extra friction experienced by each at the higher temperature.

In a p-type semiconductor, the carriers of current are "holes" of positive charge. This type of semiconductor has a valence band that is almost, but not quite, filled with electrons. Thus, there are "holes" in the electron distribution, and if these holes move, they will transport charge. To see how such a transport of charge comes about, consider Fig. 30.15 showing an array of electrons and positive ions. In Fig. 30.15a, these electrons and ions form neutral atoms. Suppose that the right end of this array is connected to the positive pole of a battery (not shown) and the left end to the negative pole. If the battery pulls an electron out of the right end, it will leave the array with a hole, or missing electron, at the position of the last atom (Fig. 30.15b). The electrons will then play a game of musical chairs: the electron from the next-to-last atom will jump into this hole, leaving a hole at the position of the next-to-last atom (Fig. 30.15c); and then the electron from the next atom will jump, etc. The collective motion of the electrons from left to right can be conveniently described as the motion of a hole from right to left. The hole virtually carries positive charge from the right to the left. In essence, this is the mechanism for conduction in a p-type semiconductor. Instead of free electrons, this type of semiconductor has free holes. A flow of current is then a flow of holes, and the direction of the current is the same as the direction of motion of the holes.

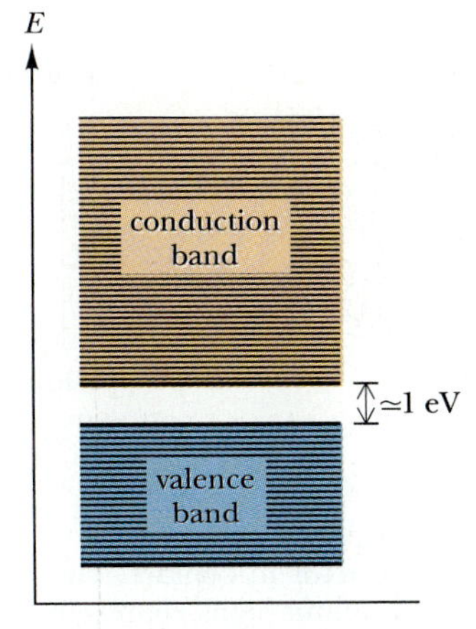

FIGURE 30.14

In a semiconductor, the upper energy band is also completely filled with electrons, but the gap between this band and the next is small.

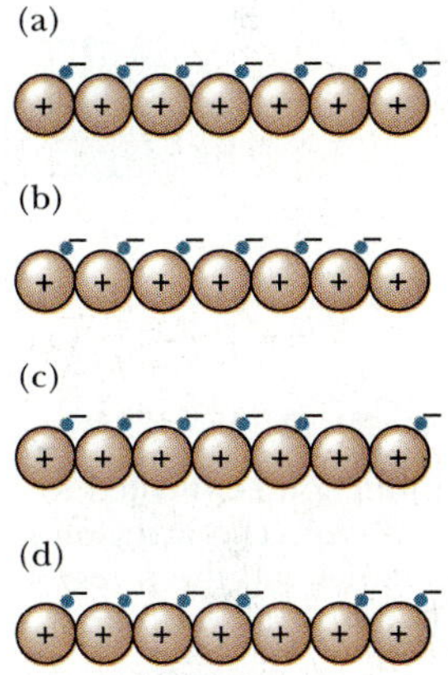

(a)

(b)

(c)

(d)

FIGURE 30.15

A row of positive ions (brown balls marked +) and electrons (blue dots marked −).

Semiconductors usually contain both free electrons and free holes. Whether a semiconductor is n type or p type depends on which kind of charge carrier dominates. The concentration of free electrons and of free holes is largely determined by the impurities that are present in the material. ***Donor impurities consist of atoms that release their valence electrons*** when placed in the semiconductor, and they thereby increase the number of free electrons. ***Acceptor impurities consist of atoms that trap electrons*** when placed in the semiconductor, and they thereby generate holes. Hence, a semiconductor with donor impurities will be n type and one with acceptor impurities will be p type. For instance, silicon with arsenic impurities is an n-type semiconductor, and silicon with boron impurities is a p-type semiconductor. Even though the added impurity atoms may only amount to a few parts per million, they completely change the conductivity because the semiconductor has so few current carriers to start with.

30.5* SEMICONDUCTOR DEVICES

The manipulation of the resistivity of semiconductor materials by intentional contamination with carefully selected impurities plays a crucial role

* This section is optional.

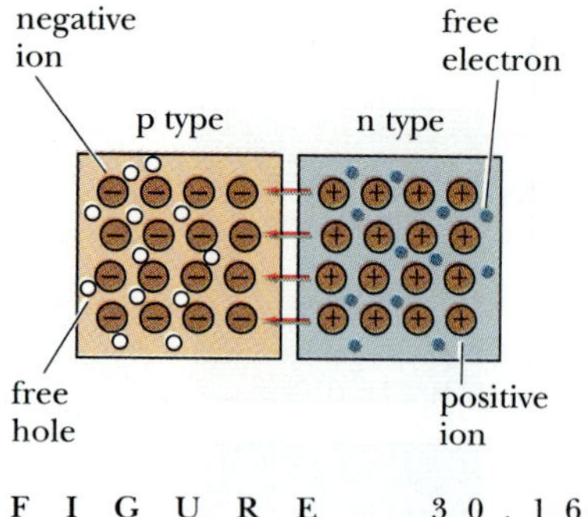

FIGURE 30.16

Pieces of p-type and n-type semiconductor in contact. The plus and minus signs represent the ions of the lattice. The black dots represent electrons and the white dots holes. The arrows indicate the electric field generated by the ions at the interface.

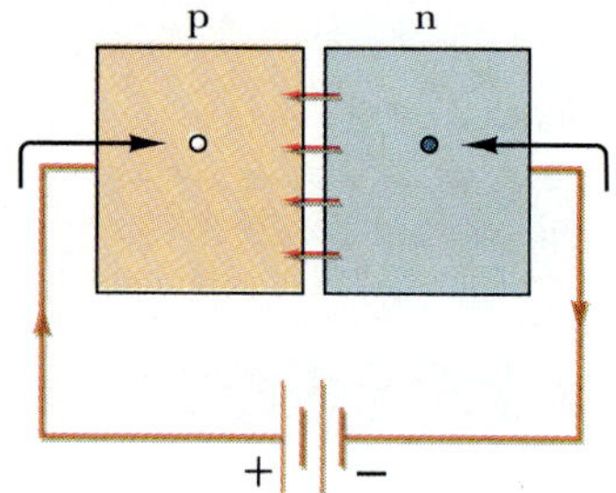

FIGURE 30.17

A p–n junction connected to a source of emf. The n region is at low potential and the p region at high potential (forward bias).

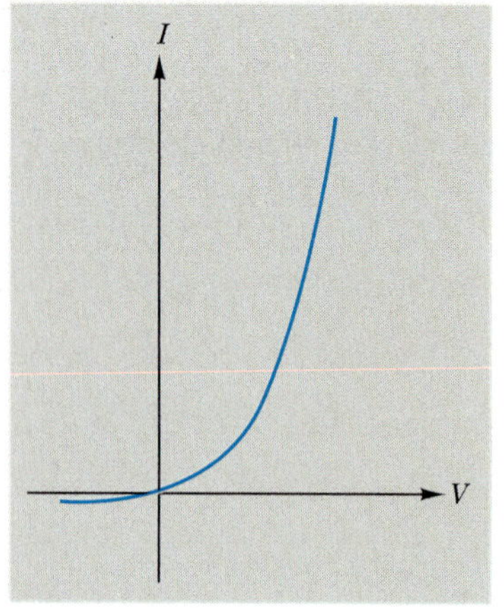

FIGURE 30.18

Plot of current vs. voltage for the p–n junction.

in the manufacture of semiconductor devices, such as diodes and transistors. It is a characteristic feature of semiconductor materials that the addition of impurities to the material has a drastic effect on the resistivity. For instance, the silicon used in electronic devices is usually contaminated, or "doped," with small amounts of arsenic or boron; the addition of just one part per million of arsenic will decrease the resistivity of silicon by a factor of more than 10^5. Pure semiconductor materials are hardly ever used in practical applications. In most cases, the presence of impurities is what gives the semiconductor materials their interesting and useful electric properties.

Rectifier (Diode)

A semiconductor rectifier consists of a piece of n-type and a piece of p-type semiconductor joined together. The n-type semiconductor has free electrons, and the p-type semiconductor has free holes; when they are joined, some of the free electrons will wander from the n region into the p region, and some of the holes will wander from the p region into the n region. Wherever the electrons and the holes meet, they annihilate each other — the electron falls into the hole and fills it, which means that both the electron and the hole disappear. The annihilation of some electrons and holes leaves residual positive and negative ions near the interface of the two regions, and the electric charges of these ions generate an electric field across the interface (see Fig. 30.16). This electric field opposes any further wandering of holes or electrons from one region into the other.

When such a p–n junction is connected to a battery or some other source of emf, it will permit the flow of current from the p region into the n region, but not in the opposite direction. Figure 30.17 shows the p–n junction connected to the source of emf so the p region is at high potential and the n region at low potential, a configuration called "forward bias." The source of emf pumps a steady flow of electrons into the n region and it removes electrons from the p region, which is equivalent to pumping holes into the p region. The electrons and the holes meet at the junction and they annihilate. This process can continue indefinitely, and therefore the source of emf can continue to pump current around the circuit indefinitely. Figure 30.18 is a plot of the current vs. the voltage applied to the p–n junction. The current increases steeply with the voltage, because the electric field associated with the applied voltage tends to cancel the internal electric field at the p–n junction, and this makes it easier for the electrons and holes to meet at the center. Evidently, the current is not simply proportional to the voltage, and the p–n junction does not obey Ohm's Law.

Now, consider what happens if the p–n junction is connected to the source of emf so the p region is at low potential and the n region is at high potential, as shown in Fig. 30.19. This configuration is called "reverse bias." The free electrons in the n region then flow away through the wire on the right, and the free holes in the p region flow away through the wire on the left. Consequently, each region is depleted of its charge carriers, and the flow of current stops almost immediately — the p–n junction blocks the current.

The p–n junction is called a rectifier because it can be used to convert an alternating current into a direct current. If the junction is connected to a source of alternating emf, ***it will pass current only during the "forward" part of the cycle.*** The alternating positive and negative emf then yields a periodic sequence of positive current pulses (see Fig. 30.20). Such solid-state rectifiers find many practical applications; for instance, they are used in the "alternators" that generate DC power in the electrical systems of automobiles.

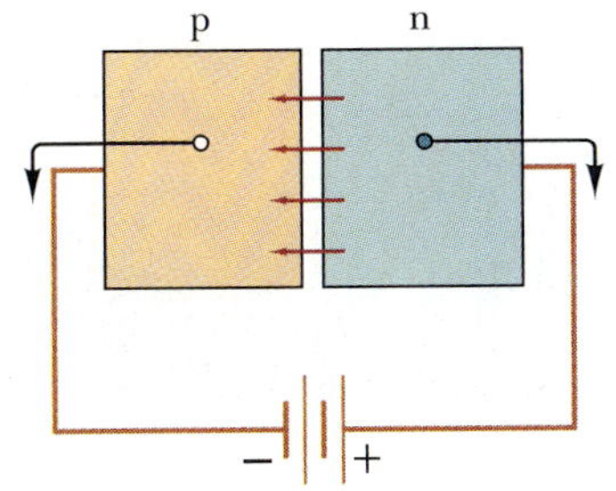

FIGURE 30.19

A p–n junction connected to a source of emf. The n region is at high potential and the p region at low potential (reverse bias).

Transistor (Triode)

A transistor consists of a thin piece of semiconductor of one type sandwiched between two pieces of semiconductor of the other type. Figure 30.21 illustrates an n–p–n junction transistor. The thin piece in the middle is called the base; and the pieces at the end are called the emitter and the collector, respectively. The transistor has three terminals which are connected to two sources of emf, V_B and V_C, so the emitter–base junction has a forward bias and the base–collector junction has a reverse bias. In this configuration, the emitter–base junction acts as diode with forward bias, and it permits the flow of electrons from the emitter into the base. However, the electrons that enter the p region fail to annihilate with holes, because the p region is quite thin and contains only a low density of holes, and the electrons pass through it before they have a chance to meet with a hole. The electrons wander to the base–collector junction, and the electric field across this junction (indicated by the arrows in Fig. 30.21) pulls the electrons into the collector. They leave the collector via the terminal connected to its end, and they continue around the external circuit, forming the external collector current I_C. Of the electrons that enter the base, a small fraction wander to the terminal connected to the base, and they leave via the wire connected there, forming the external base current I_B.

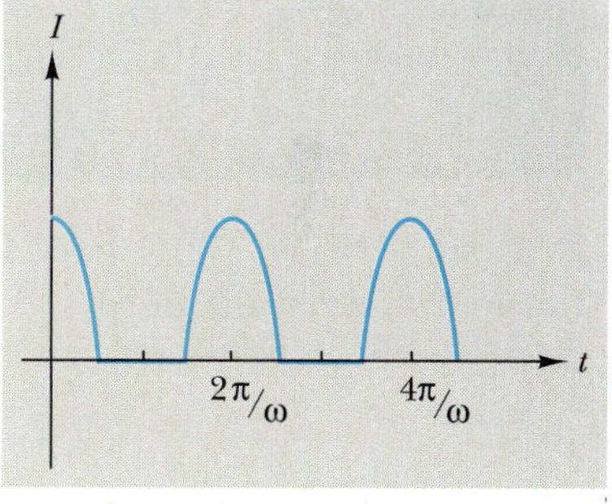

FIGURE 30.20

Current passed by the rectifier vs. time. The negative portions of the alternating current are blocked by the rectifier, and only the positive portions remain.

The use of the transistor as an amplifier for currents and voltages hinges on the relationship between the collector and the base currents: a small change in the base current I_B or the base voltage V_B leads to a quite large change in the collector current I_C. As in the case of the diode with forward bias, if we increase the base potential V_B, we cause a drastic increase in the flow of electrons entering the base from the emitter. But most of these electrons flow straight through to the collector, and only a small fraction flows to the terminal connected to the base. Thus, the result of an increase of base potential is a large increase of collector current, but only a small increase of base current. For a typical transistor, the ratio of the collector current increment and the base current increment is of the order of 100 or 200, and this ratio has a fixed value, over a wide range of currents. The ratio of these current increments is called the gain factor,

$$[\text{gain factor}] = \frac{\Delta I_C}{\Delta I_B} \tag{15}$$

This means that whenever we change the base current I_B by some amount

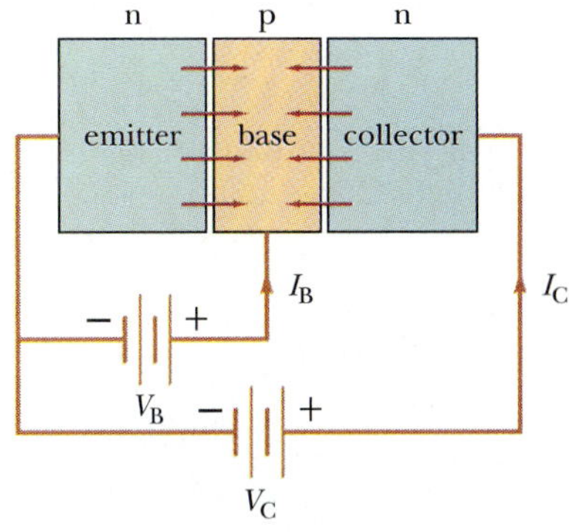

FIGURE 30.21

An n–p–n junction transistor. Two sources of emf V_B and V_C are connected to the base and the collector, respectively.

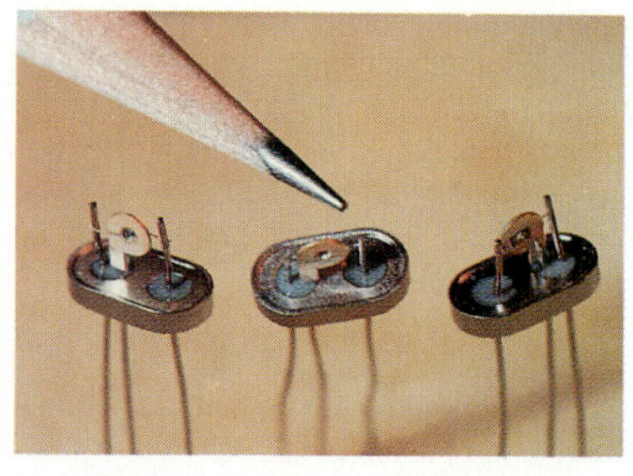

FIGURE 30.22
Transistors.

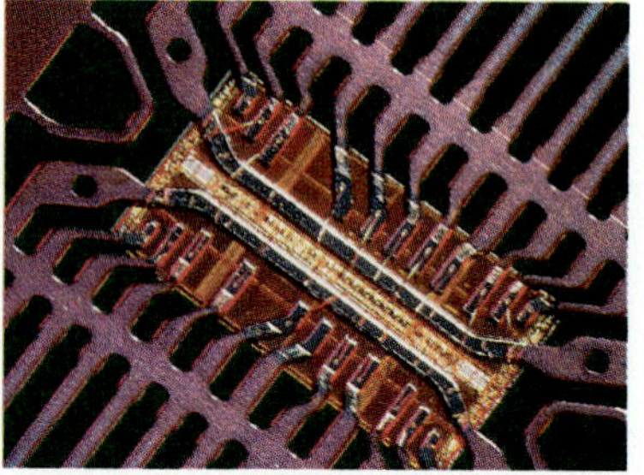

FIGURE 30.23
Integrated circuit with many miniature circuit elements.

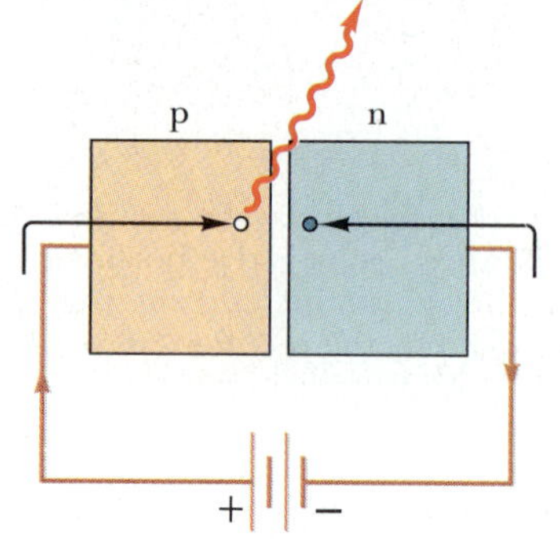

FIGURE 30.24
A p–n junction used as a light-emitting diode.

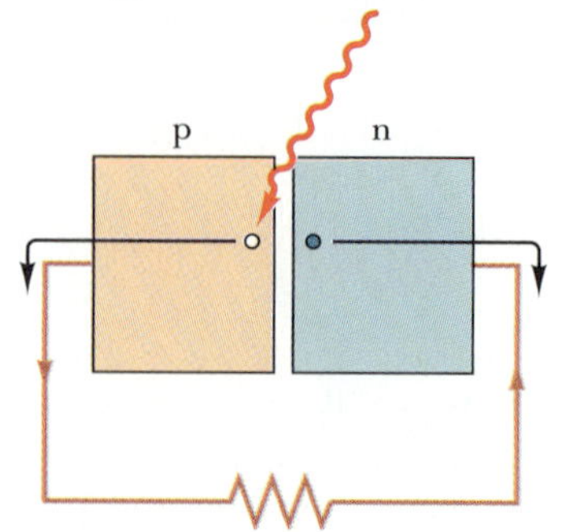

FIGURE 30.25
A p–n junction used as a solar cell.

(by adjusting the voltage V_B), we will change the collector current by an amount a hundred or so times larger. ***The transistor amplifies the current — a small input current at the base results in a much larger output current at the collector.*** For instance, in the "amplifier" in a radio receiver, the weak current picked up by the radio antenna is amplified by sending it into a transistor, as input current at the base. Further amplification can be achieved by connecting several transistors in tandem, so the output of each serves as input for the next.

Transistors are used in a wide variety of electronic circuits, to amplify and control currents. Figure 30.22 shows some ordinary transistors. In an integrated circuit, or chip, such as illustrated in Fig. 30.23, many transistors and other circuit elements of extremely small size are built up on a single, thin crystal of silicon. The small transistors are not manufactured by sticking together separate pieces of n- and p-type material, but by diffusing suitable concentrations of acceptor and of donor impurities into different layers of the silicon crystal.

Light-Emitting Diode (LED)

In principle, a light-emitting diode is simply a p–n junction operated with forward bias. At such a junction, electrons arriving from the n region meet holes arriving from the p region, and they annihilate, that is, the electrons jump into the holes. But this jump is a transition of the electron from a state of high energy in the conduction band to a state of lower energy in the valence band, a transition that releases energy. In gallium arsenide and some other semiconducting materials, the released energy takes the form of a photon of visible light (see Fig. 30.24). Thus, the p–n junction emits light when an electric current passes through it. Such light-emitting diodes have many practical applications in luminous displays in the dials of measuring instruments, watches, electronic calculators, clocks, automobile speedometers, and so on.

Solar Cell

A solar cell is simply a light-emitting diode operating in reverse. When sunlight is absorbed at the p–n junction, it excites an electron from the valence band to the conduction band, creating a free electron and a free hole (see Fig. 30.25). The electric field at the junction then pulls the electron toward the n region and the hole toward the p region. This means negative charge flows into the n region and from there into the external wire connected on the right, while positive charge flows into the p region and from there into the external wire connected on the left. Thus, sunlight striking the junction generates an electric current in the external circuit.

Solar cells are commonly manufactured out of p-type silicon and n-type silicon. The emf of such a silicon solar cell is only about 0.6 V, and the current it delivers is fairly small. For the solar cell shown in Fig. 30.26, the current delivered in full sunlight is about 0.1 A.

Solar cells are routinely used to generate electric power on communications satellites and other satellites in orbit around the Earth (see Fig.

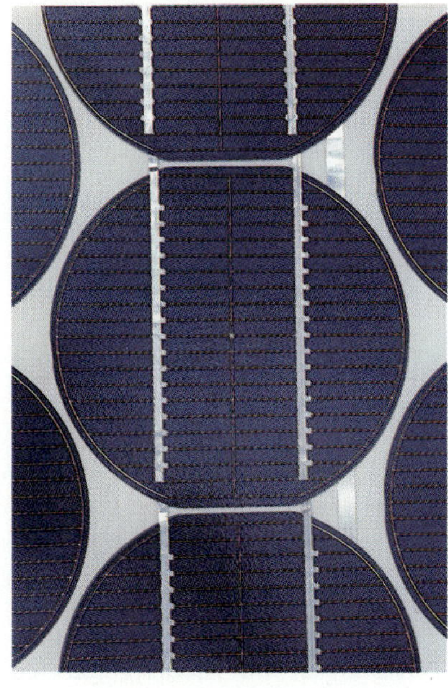

FIGURE 30.27
Solar panels on Skylab (above).

FIGURE 30.26
A solar cell (left).

FIGURE 30.28
The "Sunraycer," an experimental vehicle powered by several thousand solar cells covering its body. In full sunlight, these generate up to 1500 watts of electric power.

30.27). Solar cells have been used to generate power to drive experimental vehicles (see Fig. 30.28). Some attempts have also been made to use them to generate fairly large amounts of electric power for residential and industrial use. For instance, a solar power station at Carrisa Plains in California generates about 7000 kilowatts (see Fig. 30.29).

FIGURE 30.29
Solar power station at Carrisa Plains in California.

SUMMARY

- **Electron spin:** $\frac{1}{2}\hbar$
- **Quantum numbers of atomic states:** n, l, m, m_s (principal, orbital, magnetic, and spin quantum numbers)
- **Magnitude of angular momentum:** $L = \sqrt{l(l+1)}\,\hbar$
- **z component of angular momentum:** $L_z = m\hbar$
- **Exclusion Principle:** Each stationary state of quantum numbers n, l, m, and m_s can be occupied by no more than one electron.
- **Vibrational energies of molecule:** $E = nhf$
- **Rotational energies of molecule:** $E = \dfrac{l(l+1)\hbar^2}{2I}$
- **Conductor:** Partially filled conduction band
- **Insulator:** Empty conduction band, full valence band, large gap between bands
- **Semiconductor:** Empty conduction band, full valence band, small gap between bands

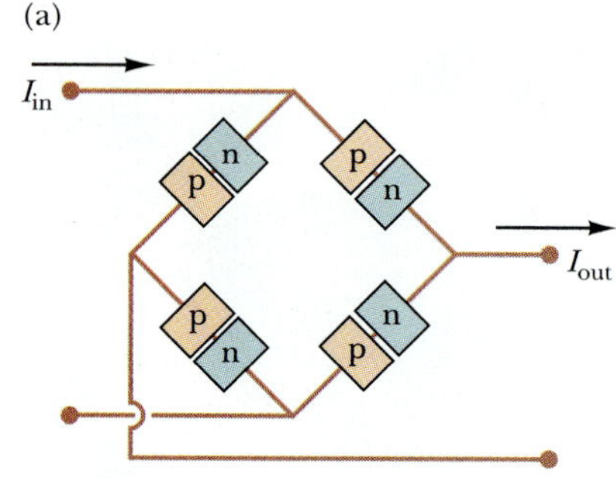

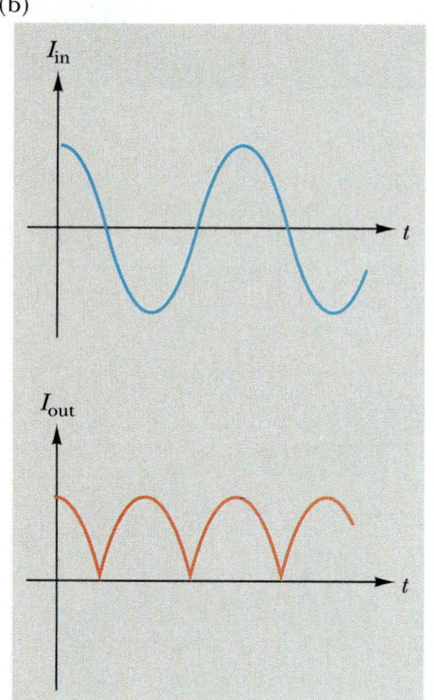

FIGURE 30.30

(a) Four diodes connected to form a full-wave rectifier. (b) The current going into the rectifier, and the current coming out.

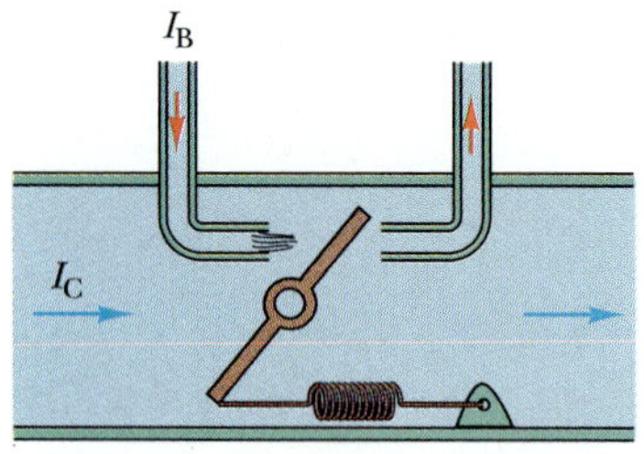

FIGURE 30.31

The butterfly valve in the large pipe is hinged at the center. The stream of water from the small pipe strikes the upper portion of the butterfly valve and pushes it open.

QUESTIONS

1.–According to classical mechanics, what kind of orbit would have zero angular momentum?

2.–If there were no Exclusion Principle, what would be the electron configuration of lithium?

3.–The bond in the NaCl molecule is similar to the bond in the KBr molecule. Explain this similarity.

4.–Rotational transitions in a molecule give a band spectrum, but vibrational transitions do not. Explain.

5.–How does n-type silicon differ from p-type?

6.–If you dope silicon with phosphorus impurities, will the silicon be n type or p type?

7.–What kind of valve in a hydraulic circuit is analogous to a diode rectifier? Draw a picture of a water pipe with such a valve.

8.–Figure 30.30 shows a full-wave rectifier consisting of four diodes connected together. This rectifier not only blocks the negative portion of an entering alternating current, but also reverses this portion, so the current is positive at all times. Describe the flow of current through the four diodes when the entering alternating current is positive and when it is negative.

9.–Do the current I_C and the voltage V_C in a transistor obey Ohm's Law?

10.–A spring-loaded butterfly valve in a large water pipe is controlled by a stream of water from a separate pipe (see Fig. 30.31). Is this a reasonable hydraulic analog of a transistor? How could you use such a device to amplify a water current?

PROBLEMS

Section 30.1: Principal, Orbital, and Magnetic Quantum Numbers; Spin

1.——According to wave mechanics, what are the possible values of the orbital quantum number l if the principal quantum number is $n = 1$? If $n = 2$? If $n = 3$?

2.——Suppose that a state in the hydrogen atom has orbital quantum number $l = 5$. What are the possible values of the magnetic quantum number m?

3.——Suppose that the magnitude of the orbital angular momentum vector is $\sqrt{20}\hbar$. What are the permitted values of the z component of the angular momentum?

4.——(a) How many of the stationary states (counting spin states) of the hydrogen atom have energy -13.6 eV? Energy -3.4 eV?

(b) How many of the stationary states (counting spin states) of the hydrogen atom have energy -3.4 eV and $l = 0$? Energy -3.4 eV and $l = 1$? Energy -3.4 eV and $l = 3$?

5.——Suppose that all you know about a state of the hydrogen atom is that the magnetic quantum number is $m = 3$. What conclusions can you draw about the value of the orbital quantum number? The principal quantum number? The energy of the state? (Hint: Is $l = 0$ compatible with the given value of m? Is $l = 1$? Is $l = 2$?)

6.——The circular orbits of orbital angular momentum $n\hbar$ in Bohr's theory roughly correspond to the wave-mechanical states of maximum orbital quantum number, that is, $l = n - 1$. If $n = 2$, compare the magnitude of the angular momentum given by Bohr's theory with the magnitude given by wave mechanics. Repeat for $n = 4$, $n = 10$, and $n = 500$.

7.——The ρ meson is a particle of spin quantum number 1. What is the magnitude of the spin of the ρ meson? What are the possible valus of m_s for this particle?

8.——In many atoms, the nucleus of the atom has spin. For example, the nucleus of one of the isotopes of magnesium (^{25}Mg) has spin, with a spin quantum number of 5/2. What is the magnitude of the spin of this nucleus? What are the possible values of m_s for this nucleus?

9.*——As stated in Section 30.1, the energies of the stationary states of the hydrogen atom depend slightly on the orbital angular-momentum quantum number l. An improved formula for the energy of the state of quantum numbers n and l is

$$E_{n,l} = -\frac{m_e e^4}{2(4\pi\epsilon_0)^2\hbar^2 n^2}\left[1 + \frac{e^4}{(4\pi\epsilon_0)^2\hbar^2 c^2 n}\left(\frac{1}{1 + \frac{1}{2} \pm \frac{1}{2}} - \frac{3}{4n}\right)\right]$$

where the term $\pm\frac{1}{2}$ corresponds to spin parallel and antiparallel, respectively, to the orbital angular momentum.

(a) For the case of the first excited state, $n = 2$, $l = 1$, and spin antiparallel to the orbital angular momentum, find the difference, in eV, between the energy calculated according to the Bohr theory and the energy calculated according to the improved formula.

(b) For $n = 2$, $l = 1$, find the difference, in eV, between the energies of the states of spin parallel and antiparallel to the orbital angular momentum calculated according to the improved formula. Which of these states has the lower energy?

10.*——Consider an angular-momentum vector with quantum number $l = 3$. What is the smallest possible angle that this angular momentum can make with the z axis?

11.*——The spin of the electron has a magnitude $\sqrt{\frac{1}{2}(\frac{1}{2}+1)}\hbar$ and a magnetic quantum number $m_s = +\frac{1}{2}$ or $m_s = -\frac{1}{2}$. For each of these two values of m_s, calculate the angle between the direction of the spin and the z axis.

12.*——Consider the possible directions of the orbital angular-momentum vector for an electron with orbital quantum number $l = 1$. What are the possible values of the magnetic quantum number m? For each value of m, calculate the angle between the angular-momentum vector and the z axis. Draw a diagram showing the possible orientations of the angular-momentum vector for all the different values of m.

13.*——Suppose that we regard the proton as a sphere of uniform density with a radius of 1×10^{-15} m rotating rigidly about its axis. According to classical mechanics, if the spin angular momentum of this sphere is to have a magnitude of $\sqrt{3/4}\hbar$, what must be the angular velocity of rotation? What must be the speed of a point on the equator?

14.*——For an electron in a state of gieveen orbital (l) and magnetic (m) quantum numbers the magnitude of the angular momentum and the z component of the angular momentum are well defined, but the x and y components are completely uncertain. Show that nevertheless the sum of the squares of the x and y components is well defined according to the formula

$$(L_x^2 + L_y^2) = l(l+1)\hbar^2 - m^2\hbar^2$$

With the additional assumption that, on the average, L_x^2 and L_y^2 are equal, show that the rms values of the x and y components of the angular momentum are

$$\sqrt{L_x^2} = \sqrt{L_y^2} = \sqrt{\frac{l(l+1) - m^2}{2}}\hbar$$

(This is similar to what happens in the case of the velocity components of a molecule of gas in a container. The, say, x component of the velocity is equally likely to be positive or negative, and therefore is completely unpredictable; nevertheless, the rms value $\sqrt{v_x^2}$ is well defined.)

Section 30.2: The Exclusion Principle and the Structure of Atoms

15.——How many possible electron states are there in the M shell ($n = 3$) of a hydrogen atom? Make a list of these states, like the list on page 844.

16.——List the quantum numbers of all the electrons of a boron atom in its ground state.

17.——List the quantum numbers of all the electrons of a carbon atom in its ground state.

18.——List the quantum numbers of all the electrons of a magnesium atom in its ground state.

19.——List the quantum numbers of all the electrons of a Na^+ ion in its ground state.

20.——What are the quantum numbers n and l for the outermost electron of the Li atom? The Na atom? The K atom? In what way are these quantum numbers of these different atoms similar?

21.——What are the quantum numbers n and l for the two outermost electrons of the Be atom? The Mg atom? The Ca atom? In what way are these quantum numbers of these different atoms similar?

22.——Suppose that the spin quantum number of the electron were $\frac{3}{2}$ instead of $\frac{1}{2}$. How many possible spin directions would the electron have in this case? What would be the possible permitted quantum numbers for the case $n = 1$ (K shell)? For the case $n = 2$ (L shell)? How many electrons could be placed in the K shell? The L shell? Compare the resulting Periodic Table of elements with the familiar Periodic Table in Section 30.2.

23.*—What energy is required to eject one of the electrons from the $n = 1$ orbit in molybdenum out of the atom? Express your answer in eV.

24.*—Calculate the energy and the wavelength of the X ray emitted during a quantum jump of an electron from the $n = 3$ to the $n = 1$ orbit in molybdenum.

25.*—An X-ray tube has a tungsten target ($Z = 74$). Calculate the energies and the wavelengths of the X rays emitted in the quantum jumps from the $n = 2$ to the $n = 1$ orbit and from the $n = 3$ to the $n = 1$ orbit.

26.*—For the molybdenum atom, the wavelength of the characteristic X ray emitted in the quantum jump from $n = 2$ to $n = 1$ is approximately 0.069 nm (see Example 3). For what atom is the wavelength of the corresponding characteristic X ray twice as large? Half as large?

27.*—A sample of an unknown element is being used as the target in an X-ray tube. It is found that the characteristic X-ray spectrum displays a strong spectral line at $\lambda = 0.0228$ nm. Assume that this spectral line results from the quantum jump of an electron from the $n = 2$ to the $n = 1$ orbit. Can you identify the unknown element?

Section 30.3: Energy Levels in Molecules

28.——The frequency of vibration of the H_2 molecule is 1.31×10^{14} Hz. What are the energies of the vibrational states? What is the frequency of the emitted radiation? The wavelength?

29.——The atoms of deuterium (D) and of hydrogen (H) have the same electron configuration (one electron), but deuterium is a heavier atom than hydrogen, because it has more mass in its nucleus. The mass of the deuterium atom is 2.014 u, whereas the mass of the hydrogen atom is 1.008 u. Given that the frequency of vibration of the H_2 molecule is 1.31×10^{14} Hz, deduce the frequency of vibration of the D_2 molecule.

30.——In the HF molecule, the chemical bonds holding the two atoms together behave like a massless spring of a spring constant $k = 9.7 \times 10^2$ N/m.

(a) Calculate the frequency of vibration of the molecule. For the sake of simplicity, assume that the fluorine atom remains at rest, and only the hydrogen atom moves.

(b) Calculate the energy of the first excited vibrational state of the molecule.

31.——Consider the HCl molecule described in Example 4. Suppose that this molecule is initially in the $l = 3$ rotational state. If this molecule sequentially makes purely rotational transitions to the $l = 2$ state, then to the $l = 1$ state, and finally to the $l = 0$ state, what are the energies and the wavelengths of the photons emitted in each of these transitions?

32.*——In the O_2 molecule, the distance between the two oxygen nuclei is 0.20 nm.

(a) What is the moment of inertia of the molecule for rotation about the perpendicular axis through the center of mass?

(b) What are the energies of the first, second, and third excited states? Express these energies in eV.

33.*——The distance between the K and the Br nuclei in the KBr molecule is 0.282 nm, and the center of mass is at a distance of 0.093 nm from the Br nucleus.

(a) What is the moment of inertia of a KBr molecule rotating about its center of mass?

(b) What are the energies of the first, second, and third excited states? Express these energies in eV.

34.*——For a rotating molecule, what are the ratios of the frequencies emitted in purely rotational transitions from the first excited state to the ground state, from the second excited state to the first, and from the third excited state to the second?

35.**——According to spectroscopic measurements, the energy difference between the first and the second excited rotational states of the N_2 molecule is 2.38×10^{-22} J. Deduce the moment of inertia of the molecule. Deduce the center-to-center distance between the N atoms.

Section 30.5: Semiconductor Devices

36.——Figure 30.21 shows a circuit diagram for an n–p–n transistor. Draw the analogous diagram for the p–n–p transistor, and explain how a small current I_B leads to a large current I_C.

37.——When a transistor is connected to a circuit such as shown in Fig. 30.32, it serves as an amplifier of voltage. The voltage gain factor is defined as the ratio of the output voltage (measured across the resistor R_C) to the input voltage V_B. Evaluate this ratio if $R_B = 3000\ \Omega$ and $R_C = 6000\ \Omega$. Assume that the current gain factor for this transistor is 100, and that the internal resistance of the emitter–base junction (a diode with forward bias) is negligible.

38.*——We want to connect two transistors in tandem, so that the net current amplification of the combination is the product of the individual current amplifications of the two transistors. Design a circuit that will accomplish this.

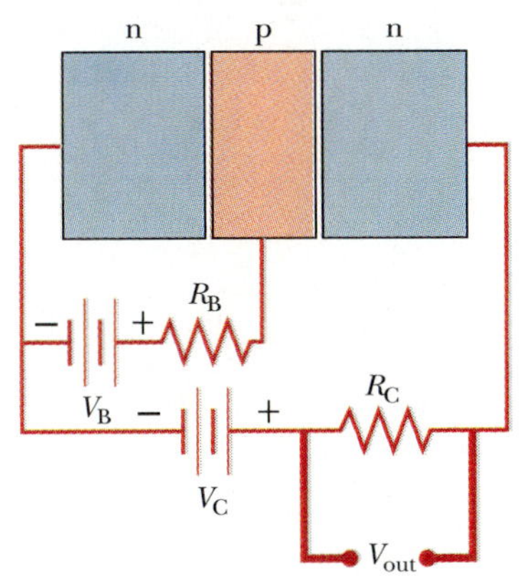

FIGURE 30.32

A transistor acting as a voltage amplifier. The source of emf V_B provides the input signal, and the voltage across the two free terminals constitutes the output signal.

39.*—A solar cell delivers 0.1 A at 0.6 V. How many such solar cells do you need, and how must you connect them, to obtain 2 A at 6 V for charging a battery?

40.*—A solar cell of area 5 cm^2 facing the Sun delivers 0.1 A at 0.6 V. What is the power delivered by this solar cell? Compare with the incident power of sunlight (1 kW/m^2) and deduce the effiency for the conversion of energy of light into electric energy.

Core of a small nuclear research reactor immersed in a pool of water. The water serves as catalyst, to promote the nuclear fission reactions, and also as shield to block nuclear radiations. The blue glow results indirectly when gamma rays emitted by the nuclear reactions strike the water.

NUCLEI

Rutherford's first experiments on the bombardment of atoms with a beam of alpha particles established that the nucleus of the atom is very small, but contains most of the mass of the atom. The nucleus is therefore very dense, and it must be made of massive particles packed very tightly together. In later experiments, Rutherford proceeded to explore the structure of the nucleus, again using a beam of alpha particles as a probe. He found that if the projectiles were energetic enough to penetrate the nucleus, they would often split it into two pieces, two smaller nuclei. The smallest such piece that could be split off was a proton, and Rutherford therefore conjectured that all nuclei contain protons. However, nuclei have more mass and less charge than expected if they contained nothing but protons — there must be some neutral particles in the nucleus or, al-

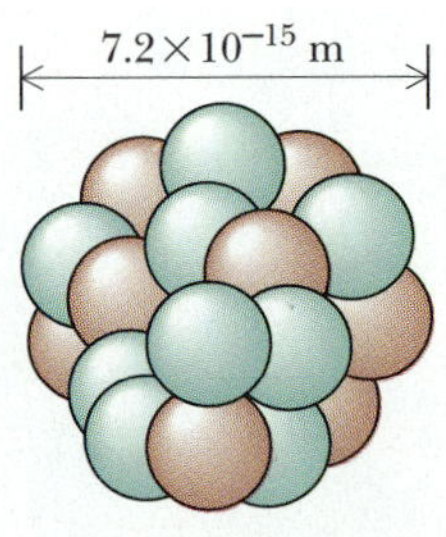

FIGURE 31.1
The nucleus of the aluminum atom, consisting of 13 protons (red) and 14 neutrons (green).

ternatively, some electrically neutral combination of particles of opposite charges. The mystery of the neutral constituent of the nucleus was not solved until 1932, when J. Chadwick discovered the neutron, a particle of about the same mass as the proton but of zero electric charge. This discovery led to the modern view of the nucleus as a tightly packed conglomerate of protons and neutrons (see Fig. 31.1).

Strong force ▶

Since the average distance between the protons in the nucleus is quite short, the repulsive electric force among the nuclear protons is very large. This force would burst the nucleus apart if there were not an extra, even larger, attractive force holding the protons and the neutrons together. This extra force is the ***nuclear force,*** or the ***"strong" force.*** Acting on two adjacent protons in a nucleus, this attractive force is about 100 times as large as the repulsive electric force. Thus, the strong force completely overwhelms the electric force. However, in heavy nuclei — such as uranium — with a large number of protons and a large total electric charge, the electric repulsion becomes important. The fission of uranium, as manifested in the explosion of a nuclear bomb, provides a spectacular demonstration of the electric force overpowering the strong force.

31.1 Isotopes

Nucleons ▶

Nuclei are made of protons and neutrons. Generically, these two kinds of constituents of the nucleus are called ***nucleons.*** Table 31.1 lists the main properties of protons and neutrons. The value of $\frac{1}{2}$ listed for the spin in this table is the spin quantum number; the magnitude of the spin is actually $\sqrt{\frac{1}{2}(\frac{1}{2}+1)}\,\hbar$. In tables of particles, physicists usually list the values of the spins of particles by giving the spin quantum number, rather than the magnitude of the spin.

It is instructive to compare these particles with the electron. Both the proton and the neutron have masses about 1800 times as large as that of the electron.* Their spin quantum numbers are $\frac{1}{2}$, the same as that of the electron. In contrast to the electron, which is a pointlike particle of no discernible size, both ***the proton and the neutron are small balls, of a radius of about 10^{-15} m.***

The number of protons in the nucleus of a (neutral) atom of a given element matches the number of its electrons, that is, it matches the atomic

TABLE 31.1 THE NUCLEONS

Nucleon	Mass	Spin	Radius
proton	$m_p = 1.00728$ u	$\frac{1}{2}$	1×10^{-15} m
neutron	$m_n = 1.00866$ u	$\frac{1}{2}$	1×10^{-15} m

* Expressed in atomic mass units, the electron mass is 5.49×10^{-4} u.

number of the element. For example, the carbon atom has six electrons and it has six protons in its nucleus.

All the atoms of a given chemical element, such as carbon, have exactly the same chemical properties, because they all have exactly the same number of electrons and the same electron configuration. However, the atoms of a chemical element can differ in mass, because their nuclei can have different numbers of neutrons. Thus, all carbon atoms have six protons in their nuclei, but some have six neutrons, some have seven, some have eight, and so on. ***Atoms with the same number of protons in their nuclei but different numbers of neutrons are called isotopes.*** Carbon has eleven known isotopes designated ^{8}C, ^{9}C, ^{10}C, ^{11}C, ^{12}C, ^{13}C, ^{14}C, ^{15}C, ^{16}C, ^{17}C, and ^{18}C (see Fig. 31.2). The superscript on the chemical symbol (for instance, the superscript "12" in "^{12}C") indicates the sum of the number of protons and the number of neutrons; this sum is called the ***mass number.*** If we designate the mass number by the symbol A, then

◀ **Isotopes**

◀ **Mass number, *A***

$$A = N + Z \qquad (1)$$

where N is the number of neutrons and Z is the number of protons, or the ***atomic number.*** Since the mass of each proton and each neutron is approximately one atomic mass unit (u), the mass number is approximately equal to the mass of the nucleus in atomic mass units.

◀ **Atomic number**

Natural samples of atoms of carbon or any other chemical element contain characteristic percentages of different isotopes. Natural carbon, as found in coal, is a mixture of 98.8% of the isotope ^{12}C and 1.11% of the isotope ^{13}C. Carbon dioxide, as found in air, contains not only the isotopes ^{12}C and ^{13}C, but also a very small amount (about $2.4 \times 10^{-10}\%$) of the isotope ^{14}C. The other isotopes of carbon do not occur naturally; they can only be produced artificially in transmutations of elements in a nuclear reactor or in an accelerator.

All chemical elements have several isotopes (see the excerpt from the chart of isotopes reproduced on the next page). Hydrogen has three isotopes (^{1}H, or ordinary hydrogen; ^{2}H, or deuterium; ^{3}H, or tritium). Helium has five isotopes, lithium has six, and so on. Some of these isotopes occur in nature, others can only be produced by artificial means.

Most of the isotopes listed in the chart of isotopes are unstable; they decay by a spontaneous nuclear reaction and transmute themselves into another element. The decay is accompanied by the emission of alpha rays, beta rays, or gamma rays. We will examine these decay processes in Section 31.3.

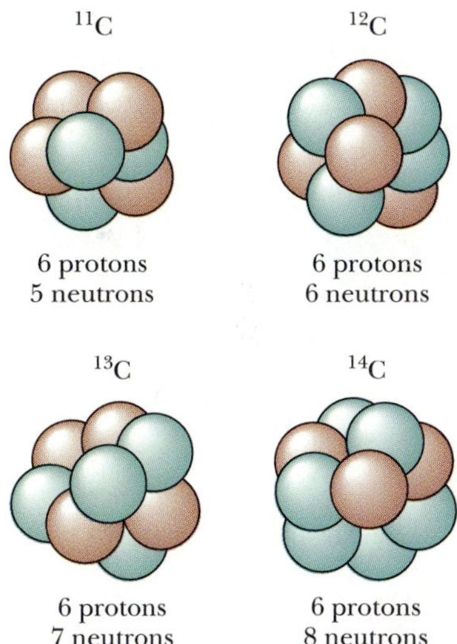

FIGURE 31.2
Some of the isotopes of carbon.

Experiments on the bombardment of nuclei with alpha particles and other particles indicate that the size of the nucleus is proportional to the cube root of the mass number. Specifically, these experiments indicate that the radius of a nucleus of mass number A is

$$R = (1.2 \times 10^{-15}\ \text{m}) \times A^{1/3} \qquad (2)$$

◀ **Radius of nucleus**

For example, the carbon-12 nucleus has a radius

$$R = 1.2 \times 10^{-15} \times (12)^{1/3}\ \text{m} = 2.7 \times 10^{-15}\ \text{m}$$

EXCERPT FROM THE CHART OF ISOTOPES[a]

Z \ N	Element	0	1	2	3	4	5	6	7	8	9	10	11	12	13	14	15	16	17
10	Ne 20.179								17 Ne 0.109 s β^+ 17.0177	18 Ne 1.67 s β^+	19 Ne 17.4 s β^+	20 Ne 90.5% 19.992439	21 Ne 0.27% 20.993847	22 Ne 9.22% 21.991384	23 Ne 37.6 s β^- γ	24 Ne 3.38 m β^- γ	25 Ne 0.61 s β^- γ	26 Ne 26.0005	27 Ne 27.0072
9	F 18.9984							15 F 15.0180	16 F $\sim 10^{-19}$ s p 16.011	17 F 66.0 s β^+ no γ	18 F 109.8 m β^+ EC	19 F 100% 18.998403	20 F 11.1 s β^- γ	21 F 4.36 s β^- γ	22 F 4.0 s β^- γ	23 F 2.2 s β^- γ	24 F 24.0093	25 F 25.0138	
8	O 15.9994						13 O 0.0089 s β^+	14 O 70.5 s β^+ γ	15 O 122 s β^+ no γ 15.003065	16 O 99.756% 15.994915	17 O 0.037% 16.99913	18 O 0.204% 17.999159	19 O 26.9 s β^- γ	20 O 13.6 s β^- γ	21 O 3.4 s β^- γ	22 O 23.0101	23 O 23.0193		
7	N 14.0067					11 N 11.0267	12 N 0.011 s β^+ γ	13 N 9.97 m β^+ no γ 13.005739	14 N 99.63% 14.003074	15 N 0.37% 15.000109	16 N 7.11 s β^- γ 16.006100	17 N 4.16 s β^- γ	18 N 0.63 s β^- γ	19 N 0.42 s 19.0176	20 N 20.0238	21 N 21.0289			
6	C 12.011				9 C 0.127 s β^+	10 C 19.4 s β^+ γ	11 C 20.4 m β^+ EC 11.011433	12 C 98.89% 12.00000	13 C 1.11% 13.003355	14 C 5730 y β^- no γ 14.003242	15 C 2.45 s β^- γ	16 C 0.74 s β^-	17 C 17.0226	18 C 18.0267	19 C 19.0370				
5	B 10.811				8 B 0.774 s β^+	9 B $\sim 8\times10^{-19}$ s p 9.01333	10 B 19.8% 10.012938	11 B 80.2% 11.009305	12 B 0.020 s β^- γ	13 B 0.017 s β^- γ	14 B 0.016 s β^- γ			17 B 17.0986					
4	Be 9.01218			6 Be $\gtrsim 3\times10^{-21}$ s p, α, Li^5 6.01973	7 Be 53.3 d EC γ 7.016930	8 Be $\sim 1\times10^{-16}$ s 2α 8.005305	9 Be 100% 9.012183	10 Be 1.6×10^6 y β^- no γ	11 Be 13.8 s β^- γ	12 Be 0.011 s β^-		14 Be 14.0440							
3	Li 6.941			5 Li $\sim 10^{-21}$ s p, α 5.0125	6 Li 7.5% 6.015123	7 Li 92.5% 7.016005	8 Li 0.85 s β^-	9 Li 0.17 s β^-		11 Li 0.009 s β^-									
2	He 4.00260		3 He 0.00013% 3.016029	4 He $\sim$100% 4.002603	5 He 2×10^{-21} s n, α 5.0122	6 He 0.802 s β^- no γ		8 He 0.122 s β^- γ											
1	H 1.0079	1 H 99.985% 1.007825	2 H 0.015% 2.014102	3 H 12.33 y β^- no γ 3.016049															
0			1 n 10.6m β^- 1.008665																

[a] The number Z, increasing vertically along the chart, is the number of protons in the isotope; it coincides with the atomic number. The number N, increasing horizontally, is the number of neutrons. In each box, the number directly below the symbol for the isotope gives the abundance in percent for naturally occurring isotopes, or else the half-life for unstable, artificially produced isotopes (the half-life is the time required for one-half of a sample of unstable isotope to decay). The Greek letters indicate the emissions that accompany the decay: β^- rays (electrons), β^+ rays (antielectrons), or γ rays. The bottom number gives the mass of the neutral atom (nucleus plus Z electrons) in atomic mass units.

whereas the uranium-238 nucleus has a radius

$$R = 1.2 \times 10^{-15} \times (238)^{1/3} \text{ m} = 7.4 \times 10^{-15} \text{ m}$$

We can gain some feeling for how small these radii are by comparing them with the radius of an atom, typically about 10^{-10} m. The comparison tells us that the size of the nucleus is less than 1/10,000 of the size of the atom.

The proportionality between R and $A^{1/3}$ implies that ***the number of nucleons per unit volume is the same for all nuclei.*** Since the volume of a sphere is $(4\pi/3)R^3$, the number of nucleons per unit volume is

$$\frac{A}{(4\pi/3)R^3} = \frac{A}{(4\pi/3)(1.2 \times 10^{-15} A^{1/3})^3 \text{ m}^3}$$

$$= 1.38 \times 10^{44} \text{ nucleons/m}^3 \qquad (3)$$

The mass of each nucleon is about 1.7×10^{-27} kg, and the mass density of the nuclear material is therefore

$$1.7 \times 10^{-27}\ \text{kg} \times \frac{1.38 \times 10^{44}}{\text{m}^3} = 2.3 \times 10^{17}\ \text{kg/m}^3 \qquad (4)$$

◀ **Density of nuclear material**

This means that one cubic centimeter, or 10^{-6} m^3, of nuclear material would have a mass of 230 million tons!

According to Eq. (3), the volume per nucleon is $1/(1.38 \times 10^{44}\ \text{m}^3)$. We can think of this volume as a cube enclosing the nucleon; the edge of the cube or, equivalently, the distance from one nucleon to its nearest neighbor, is therefore the cube root of the volume, $1/(1.38 \times 10^{44}\ \text{m}^3)^{1/3} \simeq 2 \times 10^{-15}$ m. By comparing this with the radius of a proton or neutron, about 1×10^{-15} m, we see that inside the nucleus the nucleons are so tightly packed together that they almost touch (see Figs. 31.1 and 31.2).

31.2 THE STRONG FORCE AND THE NUCLEAR BINDING ENERGY

Since the protons within a nucleus are at such short distances from one another, they exert very large repulsive electric forces on one another. Two neighboring protons, separated by a center-to-center distance of $\simeq 2 \times 10^{-15}$ m, experience an electric repulsive force of

$$F = \frac{1}{4\pi\epsilon_0}\frac{e^2}{r^2} = 8.99 \times 10^9\ \text{m/F} \times \frac{(1.6 \times 10^{-19}\ \text{C})^2}{(2 \times 10^{-15}\ \text{m})^2} \qquad (5)$$

$$= 58\ \text{N}$$

This is, roughly, four times the weight of this book; acting on a mass of only 10^{-27} kg, the magnitude of this force is colossal.

Obviously, some extra force must be present in the nucleus to prevent it from instantaneously bursting apart under the influence of the mutual electric repulsion of the protons. This extra force is the ***strong force,*** already mentioned in Section 3.5. ***This force acts equally between any two nucleons, regardless of whether they are protons or neutrons*** (the force is "charge independent").

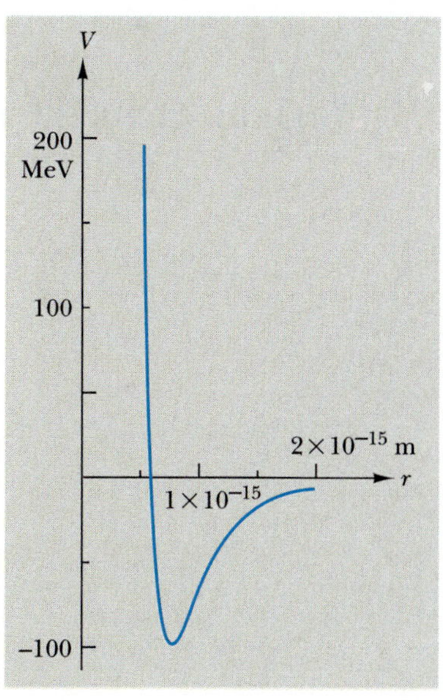

FIGURE 31.3

Potential energy for the strong force acting between two nucleons.

Figure 31.3 is a plot of the potential energy associated with the strong nucleon–nucleon force, calculated from experimental data on nuclear collisions. From Chapter 4, we know that an increasing potential energy means that the force does negative work when the nucleons separate; this corresponds to an attractive force. A decreasing potential energy means that the force does positive work when the nucleons separate; this corresponds to a repulsive force. We therefore see from the plot of the potential energy that the strong force is attractive over a range of internucleon distances from $\simeq 1 \times 10^{-15}$ m to $\simeq 2 \times 10^{-15}$ m. In this range of distances, the strong force is much larger than the electric force, as much as 100 times larger. The strong force is repulsive for internucleon distances less than $\simeq 1 \times 10^{-15}$ m; this means that the nucleons have a hard core

Range of strong ▸ force

that resists interpenetration. For distances larger than $\simeq 2 \times 10^{-15}$ m, the strong force decreases abruptly and finally vanishes. Thus, in contrast to the electric force, which fades only gradually and reaches out to large distances, ***the strong force cuts off sharply and has only a short range.*** In order to feel the strong force, the nucleons must be touching or almost touching, that is, the force acts only between nearest neighbors in the nucleus.

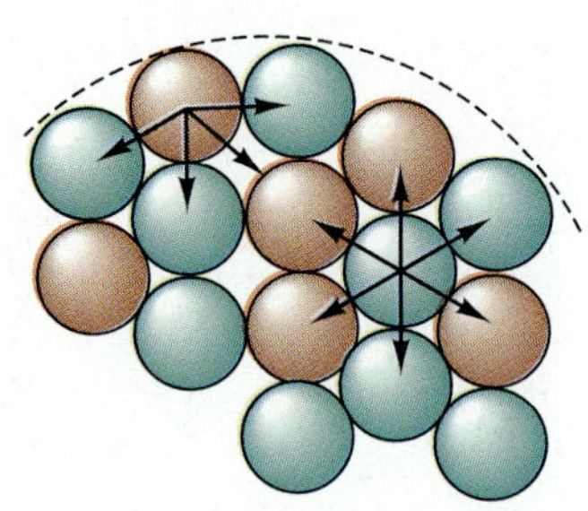

FIGURE 31.4
Forces on a nucleon at the nuclear surface, and forces on a nucleon in the nuclear interior.

In consequence of the short-range character of the strong force, a nucleon deep inside the nucleus does not experience any net force — the nucleon interacts only with its nearest neighbors, and since these pull it with equal force in almost all directions, the net force on the nucleon is zero or nearly zero (see Fig. 31.4). However, a nucleon at the nuclear surface has neighbors only on the side that is toward the interior, and hence these will exert a net force pulling the nucleon inward (see Fig. 31.4). Altogether this means that nucleons are more or less free to wander about the interior of the nucleus, but whenever they approach the nuclear surface, strong forces pull them back and prevent their escape.

This suggests that the nucleons in the nucleus behave somewhat like the molecules in a drop of water; such molecules are free to wander throughout the volume of the drop, but when they approach the water surface, intermolecular forces hold them back. This similarity between nuclei and drops of water rests on a similarity of the laws of force. The intermolecular force and potential have general features rather similar to those displayed in Fig. 31.3; the force is attractive over a short range and then becomes strongly repulsive when the molecules begin to interpenetrate. The hard repulsive core of the potential makes the water nearly incompressible, whereas the short-range attraction provides a cohesive force that prevents water droplets from falling apart. The balance of attraction and repulsion encourages water molecules to stay at a particular distance from one another, and this gives water a particular, uniform density.

Nucleus as liquid ▸ drop

Because of the similarities between a liquid and the nuclear material, ***the nucleus can be crudely regarded as a liquid drop of incompressible "nuclear fluid" of uniform density.*** The fluid is, of course, made of nucleons, but for some purposes we can ignore the individual nucleons, and we can calculate the properties of nuclei in terms of the gross properties of a liquid. For example, the spherical shape adopted by most nuclei can easily be understood as follows: Any nucleon located on the surface of a globule of nuclear fluid experiences an inward force pulling it back into the volume, and consequently the fluid tends to shrink its exposed surface to the smallest value compatible with its (fixed) volume. Since a sphere has the least surface area for a given volume, the globule of fluid will take the shape of a spherical droplet.

In a stable nucleus, the repulsive electric forces among the protons are held in check by the attractive strong forces. To achieve this balance of forces, the presence of neutrons is an advantage: a nucleus with more neutrons will have a larger size and, therefore, a larger average distance between pairs of protons — the neutrons in the nucleus dilute the repulsive effect of the electric force. Consequently, all stable nuclei, with the exception of hydrogen and one isotope of helium, contain at least as many neutrons as protons; heavy nuclei, such as uranium, contain substantially more neutrons than protons.

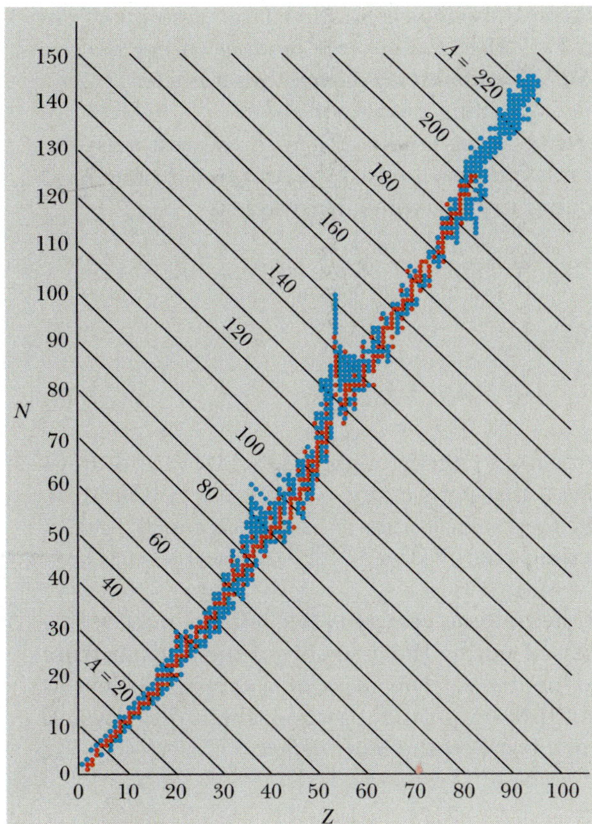

FIGURE 31.5

Number of protons (Z) vs. number of neutrons (N) for stable nuclei (blue dots) and unstable nuclei (red dots).

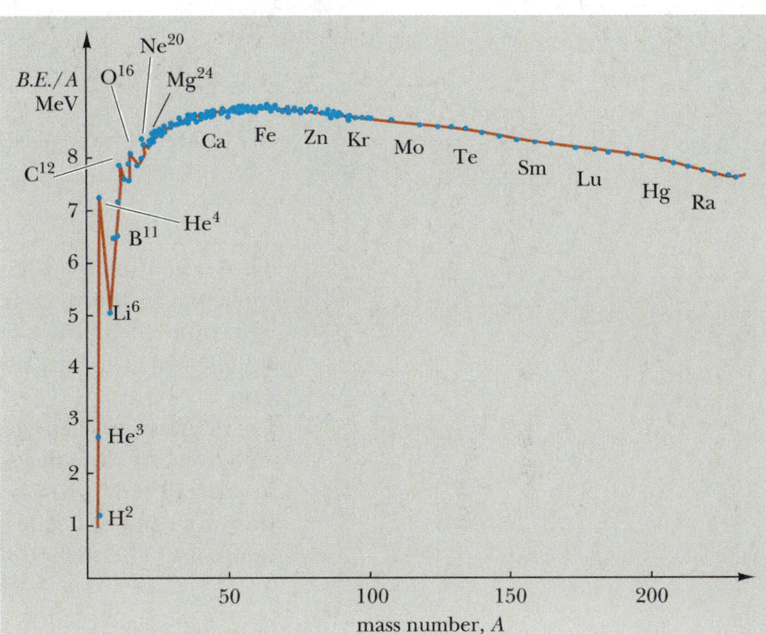

FIGURE 31.6

Average binding energy per nucleon vs. mass number.

Figure 31.5 is a plot of the number of neutrons vs. the number of protons (N vs. Z). On this plot, blue dots indicate the stable nuclei and red dots indicate unstable nuclei, that is, radioactive isotopes. Note that there is no stable nucleus beyond bismuth ($Z = 83$). However, several elements beyond bismuth have some isotopes with very long lifetimes; these are therefore almost stable and they occur naturally.

The energy stored in a nucleus is a sum of the potential energies contributed by the electric and the strong forces and the kinetic energies of the nucleons. The potential energy is negative, and its magnitude is larger than that of the kinetic energy. Thus, the "stored" energy is negative, and ***energy is released when the nucleus is assembled out of its constituent nucleons.*** Conversely, energy must be supplied to take the nucleus apart into its constituent nucleons. The negative of the "stored" energy is called the ***binding energy*** *(B.E.)*; this is the energy released during the assembly of the nucleus out of its constituent nucleons. Figure 31.6 is a plot of $B.E./A$, the binding energy divided by the number of nucleons, or the average binding energy per nucleon, vs. mass number. The curve plotted in Fig. 31.6 is called the ***curve of binding energy.*** The energy unit used in this plot is the MeV,

◂ **Binding energy** *(B.E.)*

◂ **Curve of binding energy**

$$1 \text{ MeV} = 10^6 \text{ eV} = 1.602 \times 10^{-13} \text{ J} \tag{6}$$

This unit is widely used in nuclear physics.

The binding energy of a typical nucleus is a rather large amount of energy. As may be seen from Fig. 31.6, ***the average binding energy per nucleon is in the vicinity of 8* MeV *for almost all nuclei;*** thus a nucleus with a mass number A typically has a binding energy of about $A \times 8$ MeV. To put this number in perspective, let us compare it with the rest-mass energy of the nucleons. Each nucleon has a mass of about one atomic mass unit. The energy corresponding to one atomic mass unit is

$$1 \text{ u} \times c^2 = 1.6606 \times 10^{-27} \text{ kg} \times (2.9979 \times 10^8 \text{ m/s})^2 = 1.492 \times 10^{-10} \text{ J}$$

or, in MeV units [see Eq. (6)],*

$$1 \text{ u} \times c^2 = 931.5 \text{ MeV} \tag{7}$$

Thus, each nucleon has a rest-mass energy of about 930 MeV, and the A nucleons in the nucleus have a rest-mass energy of about $A \times 930$ MeV. The ratio of binding energy to rest-mass energy is then about $8/930 \simeq 0.009$, which means the binding energy is nearly 1% of the rest-mass energy!

The mass associated with the binding energy is $B.E./c^2$; this mass is carried away by the energy released during the assembly of the nucleus from its constituent protons and neutrons. The mass of a typical nucleus is therefore about 1% less than the sum of the masses of these protons and neutrons. The mass difference is called the ***mass defect***,

Mass defect ▸

$$B.E./c^2 = [\text{mass defect}]$$
$$= [\text{mass of } N \text{ neutrons and } Z \text{ protons}] - [\text{mass of nucleus}] \tag{8}$$

Experimental values of the binding energy of a nucleus usually are obtained via the mass defect. A precise measurement of the mass of the nucleus is compared with the sums of the masses of the constituent protons and neutrons; the difference, or mass defect, then gives the binding energy according to Eq. (8). This is how the experimental points in the plot of the curve of binding energy were obtained.

EXAMPLE 1. What is the nuclear binding energy of the isotope ^{238}U? Express the energy in MeV. The mass of one atom of this isotope is 238.0508 u.

SOLUTION: Uranium-238 has 92 protons and 146 neutrons. According to Eq. (8),

$$B.E./c^2 = [92\ m_{\text{p}} + 146\ m_{\text{n}}] - [\text{mass of nucleus}] \tag{9}$$

* We have retained four significant figures in this calculation because we will be needing a precise value of the equivalence between energy and mass later on.

The mass of the uranium nucleus is the mass of the uranium atom minus the mass of the 92 electrons of this atom. Hence

$$B.E./c^2 = [92\ m_p + 146m_n] - ([\text{mass of U atom}] - 92m_e)$$

But 92 proton masses plus 92 electron masses equals 92 hydrogen masses, since a hydrogen atom consists of one proton and one electron. We therefore obtain the following convenient formula for the binding energy

$$B.E./c^2 = 92[\text{mass of H atom}] + 146m_n - [\text{mass of U atom}] \quad (10)$$

With [mass of H atom] = 1.007825 u and m_n = 1.00866 u, this yields

$$B.E./c^2 = 92 \times 1.007825\ \text{u} + 146 \times 1.00866\ \text{u} - 238.0508\ \text{u}$$

$$= 1.934\ \text{u}$$

The binding energy is therefore

$$B.E. = 1.934\ \text{u} \times c^2$$

Since u $\times\ c^2$ = 931.5 MeV [see Eq. (7)], the result is

$$B.E. = 1.934 \times 931.5\ \text{MeV} = 1802\ \text{MeV}$$

Besides its role as nuclear binding force, the strong force also plays a crucial role in nuclear reactions that occur when the nuclei in some target material are bombarded by a beam of protons, neutrons, or alpha particles. When such a projectile strikes the nuclear surface, the strong force pulls the projectile into the nucleus. The projectile then either remains bound within the nucleus, forming a heavier nucleus; or else the projectile disrupts the internal structure of the nucleus to such an extent that the nucleus ejects one or several fragments, that is, one or several protons, neutrons, alpha particles, or even other nuclei. Such nuclear reactions involve ***transmutation of elements: the original nucleus is transmuted into a new nucleus of different mass number and atomic number.***

◂ **Transmutation of elements**

The first such transmutation of elements was discovered in 1919 by Rutherford when he bombarded nitrogen with alpha particles. He found that this led to the transmutation of nitrogen into oxygen, according to the reaction

$$\alpha + {}^{14}\text{N} \rightarrow {}^{17}\text{O} + \text{p}$$

The proton is the nucleus of the hydrogen atom, ^{1}H; and, as we will see in Section 31.3, the alpha particle is actually the nucleus of a helium atom, ^{4}He. Hence the reaction can also be written

$$^4\text{He} + {}^{14}\text{N} \rightarrow {}^{17}\text{O} + {}^1\text{H} \quad (11)$$

FIGURE 31.7
The Cockcroft–Walton accelerator built in 1932.

In his early experiments, Rutherford used a naturally radioactive material as source of his beam of alpha particles. But in the 1930s, physicists began to build machines for the artificial acceleration of beams of charged particles. The first of these accelerators were electrostatic; they accelerated particles by means of strong static electric fields produced by a large amount of electric charge accumulated on a spherical capacitor (see Fig. 31.7). Many of the later accelerators use a combination of electric and magnetic fields. Thus, in the cyclotron (see Section 21.5), a uniform magnetic field holds protons in a circular orbit, while an electric field acts on them periodically, gradually increasing their energy step by step.

The investigation of nuclear reactions initiated by projectiles from accelerators led to the discovery of a multitude of new isotopes, most of them short-lived and highly radioactive. These investigations also led to a better understanding of the details of the strong force.

The energy of the projectile required to initiate a nuclear reaction or the energy released in a nuclear reaction can be calculated from the rest-mass energies of the isotopes that participate in the reaction, as illustrated by the following example.

EXAMPLE 2. Calculate the energy released in the reaction (11).

SOLUTION: The energy released in the reaction (11) is the difference between the total rest-mass energy before the reaction and the (smaller) total rest-mass energy after the reaction. The total rest-mass energy before the reaction is the rest-mass energy of the alpha particle (or ^{4}He nucleus) plus the rest-mass energy of the nitrogen nucleus, [mass of ^{4}He nucleus]c^2 + [mass of ^{14}N nucleus]c^2. The total rest-mass energy after the reaction is the sum of the rest-mass energies of the oxygen nucleus and the proton, [mass of ^{17}O nucleus]$c^2 + m_p c^2$. The energy released is therefore

$$Q = [\text{mass of } ^4\text{He nucleus}]c^2 + [\text{mass of } ^{14}\text{N nucleus}]c^2 - [\text{mass of } ^{17}\text{O nucleus}]c^2 - m_p c^2 \qquad (12)$$

Since the chart of isotopes lists the masses of the atoms, rather than the masses of nuclei, we want to express Eq. (12) in terms of atomic masses. For this purpose, we add the rest-mass energy of 18 electrons to the first two terms on the right side of Eq. (12), and we subtract the rest-mass energy of 18 electrons from the last two terms:

$$Q = [\text{mass of } ^4\text{He nucleus} + 4m_e]c^2 + [\text{mass of } ^{14}\text{N nucleus} + 14m_e]c^2 - [\text{mass of } ^{17}\text{O nucleus} + 17m_e]c^2 - [m_p + m_e]c^2$$

Each of the terms in brackets is now the mass of a complete atom:

$$Q = [\text{mass of } ^4\text{He atom}]c^2 + [\text{mass of } ^{14}\text{N atom}]c^2 - [\text{mass of } ^{17}\text{O atom}]c^2 - [\text{mass of } ^1\text{H atom}]c^2$$

Substituting the values of the masses listed on the chart of isotopes, we then find:

$$Q = 4.002603\ \text{u} \times c^2 + 14.003074\ \text{u} \times c^2 - 16.99913\ \text{u} \times c^2 - 1.007825\ \text{u} \times c^2$$

$$= 0.00128\ \text{u} \times c^2$$

Since $\text{u} \times c^2 = 931.5$ MeV, we can express our final result as

$$Q = 0.00128 \times 931.5\ \text{MeV} = 1.19\ \text{MeV}$$

COMMENTS AND SUGGESTIONS: Since the reaction releases energy, even alpha particles of very low (almost zero) kinetic energy can initiate this reaction. Most of the energy released in the reaction is carried away by the proton; the oxygen nucleus retains only a small fraction as recoil energy.

Note that if a reaction absorbs energy (which occurs if the total rest-mass energy after the reaction is larger than before), then the kinetic energy of the projectile must make up the deficit; thus, in such a reaction, the projectile needs a minimum energy to initiate the reaction.

31.3 RADIOACTIVITY

Radioactivity was discovered in 1896 by Henri Becquerel. Through an accident, he noticed that samples of uranium minerals emitted invisible rays which could penetrate through sheets of opaque materials and make an imprint on a photographic plate. Subsequent investigations established that uranium, and many other radioactive substances, emit ***three kinds of rays: alpha rays*** (α), ***beta rays*** (β), ***and gamma rays*** (γ). Of these, the alpha rays are the least penetrating; they can be stopped by a thick piece of paper. The beta rays are more penetrating; they can pass through a foil of lead or a plate of aluminum. The gamma rays are the most penetrating; they can pass through a thick wall of concrete. When these three kinds of rays are aimed into a magnetic field, the alpha and beta rays are deflected in opposite directions, whereas the gamma rays proceed without deflection (see Fig. 31.8). This simple experiment demonstrates that alpha rays are positively charged, beta rays negatively charged, and gamma rays are neutral.

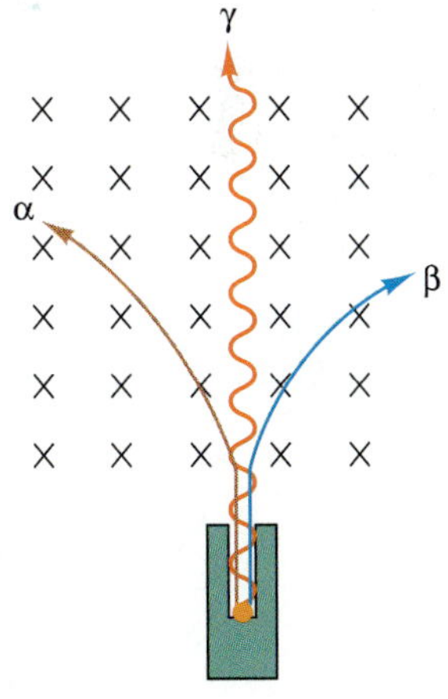

FIGURE 31.8

Alpha, beta, and gamma rays emitted by a radioactive source. When these rays enter a magnetic field (perpendicular to the plane of the page), the alpha and beta rays are deflected in opposite directions, and the gamma rays are not deflected.

By more detailed experiments, Becquerel identified the beta rays as high-speed electrons. Some years later, Rutherford investigated the nature of the alpha rays and demonstrated that they are identical to nuclei of helium. Thus, electrons and helium nuclei are, somehow, manufactured within the sample of uranium or other radioactive substance, and they are ejected at high speed. Gamma rays are very energetic photons emitted by the radioactive substance; the energy of these gamma-ray photons is typically a thousand times as large as that of X-ray photons.

The "manufacture" and ejection of these high-speed electrons and helium nuclei occur in the nuclei of the radioactive substance, by nuclear decay reactions. In the following, we will discuss the broad features of these nuclear reactions.

Alpha Decay

Alpha decay ▶

The alpha particle consists of two protons and two neutrons; it has the same structure as the helium nucleus, ^{4}He. When a radioactive nucleus ejects an alpha particle consisting of two protons and two neutrons, the radioactive nucleus does not create these protons and neutrons out of nothing — the nucleus merely takes two of its own protons and neutrons and spits them out. Since the nucleus loses two protons and two neutrons, both the atomic number and the mass number of the nucleus decrease. The atomic number decreases by two, and the mass number by four. Thus, the original isotope becomes a different isotope of a different chemical element. The following examples of alpha decays illustrate such transmutations of chemical elements:

$$^{238}\text{U} \rightarrow {}^{234}\text{Th} + \alpha \tag{13}$$

$$^{226}\text{Ra} \rightarrow {}^{222}\text{Rn} + \alpha \tag{14}$$

Parent and daughter ▶

In the first of these alpha-decay reactions, uranium is transmuted into thorium, and in the second, radium into radon. ***The original nucleus in a decay reaction is called the parent, and the resulting nucleus is called the daughter.***

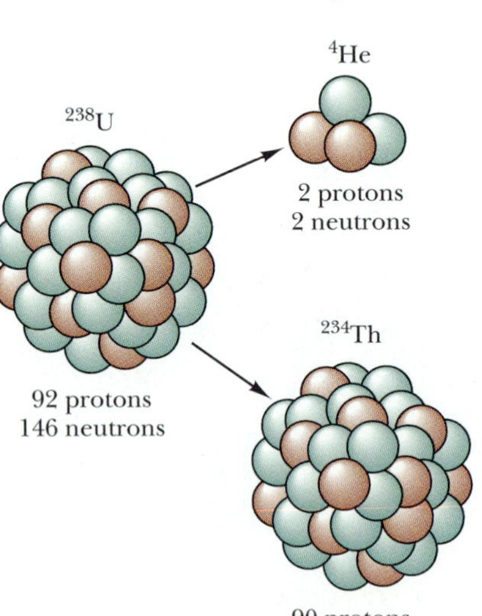

FIGURE 31.9

Fission of uranium into thorium and helium.

The alpha-decay reaction can be regarded as a fission, or splitting, of the nucleus into two smaller nuclei. The, say, uranium nucleus fissions into a thorium nucleus and a helium nucleus. The total number of protons and of neutrons is unchanged in this fission (see Fig. 31.9). The fission occurs spontaneously, because of an instability in the original nucleus. Large nuclei, such as uranium or radium, contain many protons which exert repulsive electric forces on each other. Although the repulsive electric force is balanced by the attractive strong force, the balance of these forces is rather precarious because the electric force easily reaches from one end of the nucleus to the other, whereas the strong force acts only between adjacent nucleons and therefore, in a large nucleus, cannot reach directly from one end to the other. Thus, any accidental, spontaneous elongation of the nucleus can shift the balance in favor of the electric force — the nucleus then elongates more and more, and ultimately bursts apart into two fragments. In true fission, the two fragments are of approximately equal size. In alpha decay, we are dealing with an extreme case of fission, with two fragments of very unequal size. The ejection of an alpha particle is strongly favored over ejection of, say, a hydrogen nucleus or a lithium nucleus, because the helium nucleus is an exceptionally tightly bound nucleus. In consequence of this large binding energy, the formation of an alpha particle, just before its ejection, makes more energy available for driving the fission reaction.

All the large, heavy nuclei beyond bismuth are unstable; they all are subject to alpha decay or other forms of spontaneous fission. Besides, many isotopes of somewhat smaller nuclei suffer from the same instability.

The kinetic energy of the alpha particles ejected in reactions such as (13) and (14) can be calculated from the masses of the participating nuclei and the mass of the alpha particle.

EXAMPLE 3. Calculate the energy released in the alpha decay of ^{238}U, and calculate the kinetic energy of the alpha particle ejected in this reaction. The atomic masses of the isotopes of uranium, thorium, and helium in this reaction are 238.0508 u, 234.0436 u, and 4.0026 u, respectively.

SOLUTION: The energy released in the reaction (13) is simply the difference between the total rest-mass energy before the reaction and the (smaller) total rest-mass energy after the reaction. The total rest-mass energy before the reaction is the rest-mass energy of the uranium nucleus, [mass of U nucleus]c^2. The total rest-mass energy after the reaction is the sum of the rest-mass energies of the thorium nucleus and the alpha particle, [mass of Th nucleus]c^2 + [mass of He nucleus]c^2. The energy released in the reaction is therefore

$$Q = [\text{mass of U nucleus}]c^2 - [\text{mass of Th nucleus}]c^2 - [\text{mass of He nucleus}]c^2 \qquad (15)$$

To express this in terms of the masses of the atoms rather than the masses of the nuclei, we add the rest-mass energy of 92 electrons to the first term on the right side of Eq. (15) and we subtract the same amount from the other two terms:

$$Q = [\text{mass of U nucleus} + 92m_e]c^2 - [\text{mass of Th nucleus} + 90m_e]c^2 - [\text{mass of He nucleus} + 2m_e]c^2 \qquad (16)$$

Now each of the terms in square brackets is the mass of a complete atom:

$$Q = [\text{mass of U atom}]c^2 - [\text{mass of Th atom}]c^2 - [\text{mass of He atom}]c^2 \qquad (17)$$

We know that these masses are 238.0508 u, 234.0436 u, and 4.0026 u, respectively; hence

$$Q = 238.0508\ \text{u} \times c^2 - 234.0436\ \text{u} \times c^2 - 4.0026\ \text{u} \times c^2$$

$$= 0.0046\ \text{u} \times c^2$$

or, with $u \times c^2 = 931.5$ MeV,

$$Q = 0.0046 \times 931.5 \text{ MeV} = 4.3 \text{ MeV}$$

Since the thorium nucleus is much heavier than the alpha particle, it suffers next to no recoil, and the alpha particle carries away almost all of the energy released in the reaction. Thus, the kinetic energy of the alpha particle will be 4.3 MeV.

Radioactive series ▸

In many cases of alpha decay, the daughter nucleus is also unstable, and decays some time after its formation, either by alpha decay or by beta decay. The daughter of the daughter then decays, and so on. ***The sequence of daughters descending from the original parent is called a radioactive series.*** The series ends when it reaches a stable isotope. For instance, the decay of uranium initiates a radioactive series, which ultimately ends with a stable isotope of lead.

Beta Decay

Beta decay ▸

The simplest beta-decay reaction is the decay of the neutron. ***The free neutron is unstable, and it decays into a proton, an electron, and an antineutrino:***

Decay of neutron ▸

$$n \rightarrow p + e^- + \bar{\nu} \tag{18}$$

In this equation, the symbol $\bar{\nu}$ (the Greek letter nu, with an added overbar) represents the antineutrino. Neutrinos and antineutrinos are particles of mass zero and spin quantum number $\frac{1}{2}$, which travel at the speed of light. Thus, they are somewhat similar to photons. However, in contrast to photons, which interact with electric charges, neutrinos and antineutrinos do not interact directly with electric charges — in fact, they hardly interact with anything at all, and they pass through the entire bulk of the Earth with little hindrance. As their names indicate, neutrinos and antineutrinos are ***antiparticles*** of each other; they can annihilate each other, producing a flash of light (gamma rays).

Antiparticles ▸

For a free neutron, the decay reaction (18), on the average, takes a time of 15 minutes.* For a neutron in a nucleus, the reaction may proceed at a faster rate or at a slower rate, depending on whether the nucleus promotes the reaction by supplying extra energy, or inhibits the reaction by withdrawing energy. When the reaction occurs in a nucleus, the net effect is the conversion of a neutron into a proton, which increases the atomic number of the nucleus by one, while leaving the mass number unchanged. Thus, the original isotope is transmuted into an isotope of the next chemical element. Two examples of beta decays are the following:

$$^{60}\text{Co} \rightarrow {}^{60}\text{Ni} + e^- + \bar{\nu} \tag{19}$$

* The average lifetime of free neutrons is 15.0 min, but the half-life (see next section) is 10.6 min.

$$^{14}C \rightarrow {}^{14}N + e^{-} + \bar{\nu} \quad (20)$$

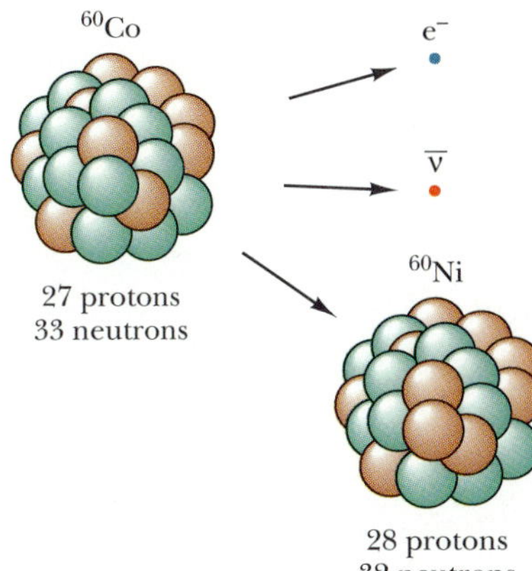

FIGURE 31.10
Beta decay of ^{60}Co.

In each of these reactions, the number of neutrons *(N)* decreases by one, and the number of protons *(Z)* increases by one (see Fig. 31.10).

The antineutrino emitted in these reactions is next to impossible to detect. Its existence, however, can be inferred from the conservation of energy. In the beta decay of, say, radioactive cobalt [Eq. (19)], sometimes the electron emerges with one energy, and sometimes with another. This is because the energy of the decay is shared between the electron and the neutrino, and sometimes one of these particles carries off most of the energy, sometimes the other. Hence the energy of the electron is unpredictable — it can be anything from zero up to a maximum amount that corresponds to the electron carrying away *all* the energy of the decay. In fact, it was on the basis of this variability of the energy of the ejected electron that Pauli first proposed the existence of the neutrino, since this was the only way to preserve the law of conservation of energy. When Pauli made this proposal in 1931, there was no direct evidence for the existence of an extra particle; neutrinos and antineutrinos were detected only much later, in experiments with nuclear reactors.

Besides the beta-decay reactions of the type (19) and (20), which involve the conversion of a neutron into a proton and the ejection of an electron and an antineutrino, there also are beta-decay reactions involving the conversion of a proton into a neutron and the ejection of an ***antielectron,*** or positron, and a neutrino. The antielectron has the same mass and spin as the electron, but the opposite electric charge. An example of a beta-decay reaction with ejection of an antielectron and a neutrino is

◀ **Antielectron (positron)**

$$^{22}Na \rightarrow {}^{22}Ne + e^{+} + \nu \quad (21)$$

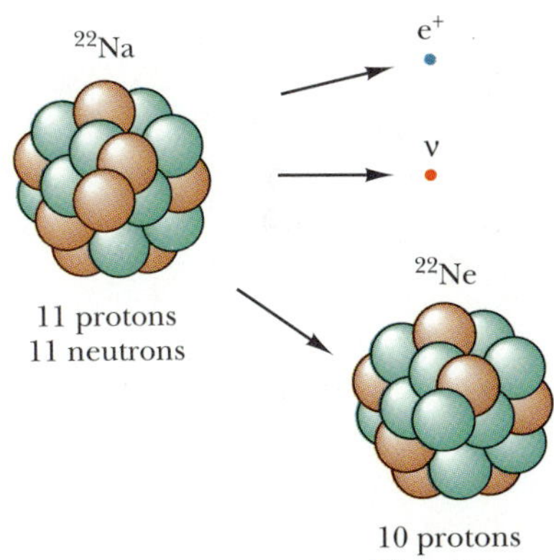

FIGURE 31.11
Beta decay of ^{22}Na, with ejection of an antielectron.

The antielectron ejected in this decay will sooner or later collide with one of the abundant atomic electrons in the surrounding environment and annihilate with it, producing two gamma rays. In the reaction (21), the number of protons *(Z)* decreases by one, and the number of neutrons *(N)* increases by one (see Fig. 31.11).

All the beta-decay reactions are instigated by a new kind of force, the ***"weak" interaction.*** This is one of the four fundamental forces found in matter (see Section 3.5). Whereas the electromagnetic and the strong forces play an important role in determining the structure of atoms and of nuclei, the weak force plays no role in this. Its only effect is to engender the beta-decay reactions (19)–(21) and other similar reactions.

EXAMPLE 4. What is the maximum kinetic energy of the beta rays emitted in the beta decay of ^{14}C? The atomic masses of ^{14}C and of ^{14}N are 14.00324 u and 14.00307 u, respectively.

SOLUTION: The energy released in the reaction (20) is, again, the difference between the total rest-mass energy before the reaction and the (smaller) total rest-mass energy after the reaction. The total rest-mass energy before the reaction is the rest-mass energy of the carbon nucleus, [mass of C nucleus]c^2. The total rest-mass energy after the reaction is the sum of the rest-mass energies of the nitrogen nucleus and the electron, [mass of N nucleus]$c^2 + m_e c^2$. The energy released in the reaction is therefore

$$Q = [\text{mass of C nucleus}]c^2 - [\text{mass of N nucleus}]c^2 - m_e c^2$$

If we add and subtract the rest-mass energy of 6 electrons on the right side of this equation, we obtain

$$\begin{aligned} Q &= [\text{mass of C nucleus} + 6m_e]c^2 \\ &\qquad - [\text{mass of N nucleus} + \text{mass of } 6m_e]c^2 - m_e c^2 \\ &= [\text{mass of C nucleus} + 6m_e]c^2 - [\text{mass of N nucleus} + 7m_e]c^2 \end{aligned}$$

Here the two terms in brackets are the masses of the complete atoms,

$$Q = [\text{mass of C atom}]c^2 - [\text{mass of N atom}]c^2$$

Substituting these masses and substituting $\text{u} \times c^2 = 931.5$ MeV, we find

$$\begin{aligned} Q &= 14.00324 \text{ u} \times c^2 - 14.00307 \text{ u} \times c^2 = 0.00017 \text{ u} \times c^2 \\ &= 0.00017 \times 931.5 \text{ MeV} = 0.16 \text{ MeV} \end{aligned}$$

The nitrogen nucleus, being much heavier than the electron, remains at rest, or nearly at rest. The decay energy is therefore shared between the electron and the antineutrino. If the antineutrino carries away none of the energy, then the electron will acquire all the energy. Thus, the maximum possible electron energy is 0.16 MeV.

Gamma emission ▶

Gamma Emission

Gamma rays are emitted by nuclei when nucleons make transitions from one stationary nuclear state to another. The emission of a gamma ray by a transition of a nucleon is similar to the emission of visible photons or of X rays by a transition of an atomic electron.

When a nucleus suffers alpha decay or beta decay, it is often left in an excited state, and it then eliminates the excitation energy in the form of a gamma ray. Thus, gamma emission is usually one step in a two-step process, with an alpha or beta decay preceding the gamma emission. For example, the beta decay of ^{60}Co leads to subsequent gamma emission according to the sequence of reactions

$$^{60}\text{Co} \rightarrow {}^{60}\text{Ni} + e^- + \bar{\nu}$$
$$\hookrightarrow {}^{60}\text{Ni} + \gamma \qquad (22)$$

The isotope ^{60}Co is commonly employed in high-intensity industrial irradiation cells, called cobalt cells. But, as shown by Eq. (22), the emitter of the gamma rays is actually nickel, not cobalt.

31.4 THE LAW OF RADIOACTIVE DECAY

In radioactive alpha or beta decay, the original isotope, or parent, becomes transmuted into another isotope, or daughter. If we have some given initial amount of parent material, say, 1 gram of radioactive strontium, some nuclei decay at one time, some at another, and the parent material disappears only gradually. In the case of radioactive strontium, one-half of the original amount disappears in 29 years, one-half of the remainder in the next 29 years, one-half of the new remainder in the next 29 years, and so on. Hence the amounts left at times 0, 29, 58, 87 years, etc., are 1, $\frac{1}{2}$, $\frac{1}{4}$, $\frac{1}{8}$ gram, etc. The time interval of 29 years is called the ***half-life,*** or $t_{1/2}$, of strontium.

◀ Half-life

Mathematically, we can represent the amount of strontium at different times by the formula

$$n = n_0\left(\frac{1}{2}\right)^{t/t_{1/2}} \qquad (23)$$

◀ Law of radioactive decay

where n is the number of strontium nuclei at time t, and n_0 is the number at the initial time. As expected, at $t = 0$, 29 years, 58 years, etc., the formula (23) yields

$$n = n_0\left(\frac{1}{2}\right)^0 = n_0 \times 1 = n_0 \quad \text{at } t = 0$$

$$n = n_0\left(\frac{1}{2}\right)^{29\text{ years}/29\text{ years}} = n_0\left(\frac{1}{2}\right)^1 = n_0 \times \frac{1}{2} \quad \text{at } t = 29 \text{ years}$$

$$n = n_0\left(\frac{1}{2}\right)^{58\text{ years}/29\text{ years}} = n_0\left(\frac{1}{2}\right)^2 = n_0 \times \frac{1}{4} \quad \text{at } t = 58 \text{ years}$$

and so on. However, the formula (23) is valid not only at these times, but also at intermediate times. For instance, at $t = 45$ years,

$$n = n_0\left(\frac{1}{2}\right)^{45\text{ years}/29\text{ years}} = n_0\left(\frac{1}{2}\right)^{1.55} = n_0 \times 0.34$$

Antoine Henry Becquerel, *1852–1908, French physicist. He was awarded the 1903 Nobel Prize for his discovery of radioactivity.*

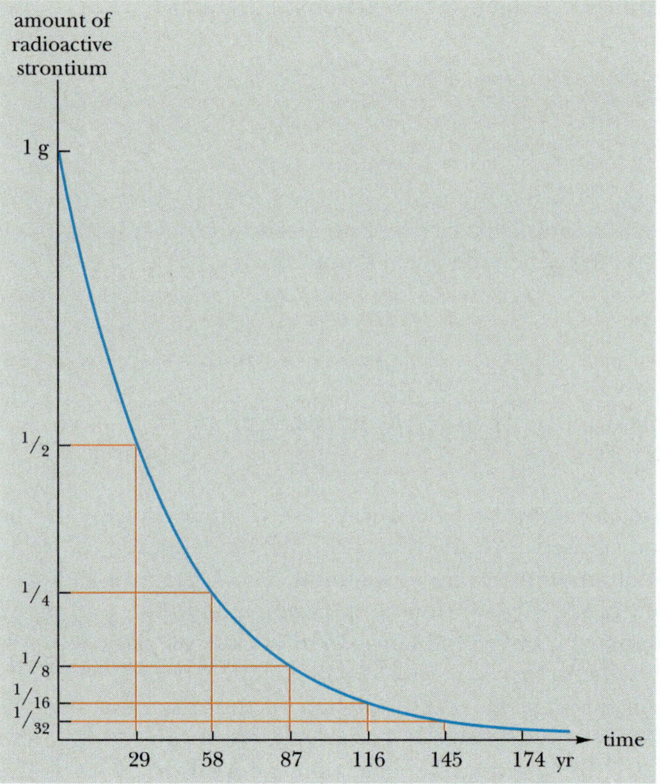

FIGURE 31.12

Amount of remaining strontium vs. time. In this plot, the amount of strontium is measured in grams. The initial amount is 1 gram, which corresponds to 6.7×10^{21} nuclei. After 29 years the remaining amount is $\frac{1}{2}$ gram, which corresponds to 3.3×10^{21} nuclei.

Figure 31.12 is a plot of the number of remaining strontium nuclei vs. time.

Equation (23) is called the ***law of radioactive decay.*** With appropriate values of the half-life, this law is, of course, applicable to the decay of any radioactive isotope. Lifetimes of different radioisotopes vary over a wide range. Some have half-lives of several billions of years, others have half-lives as short as a fraction of a second. Table 31.2 lists the half-lives of some radioisotopes. The chart of isotopes on page 864 lists some more half-lives.

TABLE 31.2 SOME RADIOISOTOPES

Radioisotope	Radioactivity	Half-life
^{14}C	β	5730 years
^{22}Na	β, γ	2.6 years
^{60}Co	β, γ	5.27 years
^{90}Sr	β, γ	28.8 years
^{131}I	β, γ	8.04 days
^{226}Ra	α, γ	1600 years
^{238}U	α, γ	4.5×10^9 years

Since each decay produces one alpha or beta ray, the rate of emission of rays by the parent material is equal to the ***decay rate,*** or the number of parent nuclei that decay per second. The instantaneous decay rate is given by the formula

$$\frac{\Delta n}{\Delta t} = 0.693 \frac{n}{t_{1/2}} \tag{24}$$

◀ **Decay rate**

Thus, ***the decay rate is directly proportional to the number of parent nuclei and inversely proportional to the half-life.*** Although the derivation of this formula for the instantaneous decay rate requires some advanced mathematics, we can obtain an approximate formula for the decay rate by simple mathematics. Let us look at the *average* rate of decay for a time interval of, say, one half-life. The average rate of change is simply the change in the number ($n_0/2$) divided by the time interval ($t_{1/2}$):

$$\frac{\overline{\Delta n}}{\Delta t} = \frac{n_0/2}{t_{1/2}} \tag{25}$$

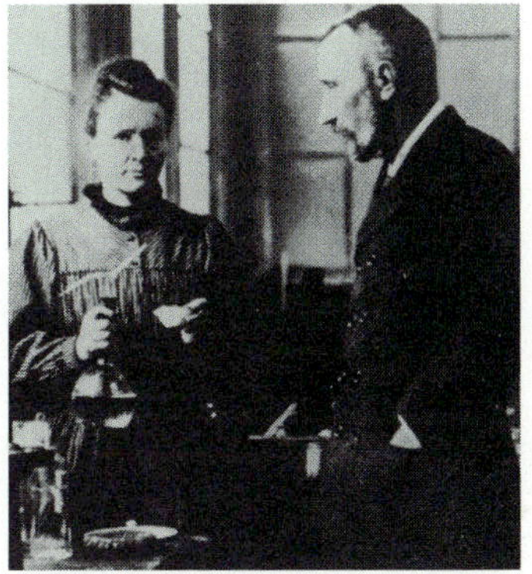

Marie Sklodowska Curie, *1867–1934, and* **Pierre Curie,** *1859–1906, French physicists and chemists. For their discovery of the radioactive elements radium and polonium, they shared the 1903 Nobel Prize with Becquerel. On the death of Pierre, Marie succeeded him in his professorship at the Sorbonne. She received a second Nobel Prize, in chemistry, for further work on radium. Her daughter, Irène Joliot-Curie, shared the 1935 Nobel Prize with Frédéric Joliot-Curie for their work on production of radioactive substances by bombardment with alpha particles.*

As is clear from Fig. 31.12, over a time interval of one half-life, the plot can be approximated quite well by a straight line, and the average number of nuclei is therefore simply the average of the initial number (n_0) and the final number ($n_0/2$):

$$\bar{n} \simeq \frac{n_0 + n_0/2}{2} = \frac{3}{4} n_0 \tag{26}$$

Hence, the average rate of change in Eq. (25) can be expressed in terms of the average number $\bar{n}$:

$$\begin{aligned} \frac{\overline{\Delta n}}{\Delta t} &\simeq \frac{4}{6} \frac{\bar{n}}{t_{1/2}} \\ &\simeq 0.666 \frac{\bar{n}}{t_{1/2}} \end{aligned} \tag{27}$$

This approximate equation for the average rate of change agrees quite closely with the exact equation (24) for the instantaneous rate of change, and provides some rough justification for it.

The radioactive decay rate of a sample, or the number of disintegrations per second, is usually called the ***activity*** of the sample. The SI unit for the radioactive decay rate is the ***becquerel*** (Bq),

$$1 \text{ becquerel} = 1 \text{ Bq} = 1 \text{ disintegration/s} \tag{28}$$

◀ **Becquerel (Bq)**

In practice, a larger unit is traditionally preferred; this larger unit is the ***curie*** (Ci),

$$1 \text{ curie} = 1 \text{ Ci} = 3.7 \times 10^{10} \text{ disintegrations/s} \tag{29}$$

◀ **Curie (Ci)**

EXAMPLE 5. What is the decay rate of 1 gram of radioactive strontium, ^{90}Sr?

SOLUTION: The atomic mass of this strontium isotope is 89.9 g; thus, 1 gram of strontium is 1/89.9 mole, or $1/89.9 \times 6.02 \times 10^{23}$ atoms $= 6.7 \times 10^{21}$ atoms. According to Eq. (24), the decay rate is

$$\frac{\Delta n}{\Delta t} = \frac{0.693}{29 \text{ years}} \times n = \frac{0.693}{29 \text{ years}} \times 6.7 \times 10^{21}$$

$$= \frac{0.693}{29 \times 3.16 \times 10^7 \text{ s}} \times 6.7 \times 10^{21}$$

$$= 5.1 \times 10^{12} \text{ /s} = 5.1 \times 10^{12} \text{ Bq}$$

Thus, 1 gram of strontium emits 5.1×10^{12} beta rays per second.

EXAMPLE 6. Atmospheric carbon dioxide contains the stable isotope ^{12}C and also the radioactive isotope ^{14}C, with a half-life of 5730 years (the latter isotope is produced by cosmic rays striking the upper atmosphere). While a plant lives, it absorbs carbon dioxide from the air and therefore maintains the same relative abundance of the isotopes ^{12}C and ^{14}C as in air. When the plant dies, it ceases to absorb carbon, and then the ^{14}C in its body decreases according to the law of radioactive decay. Thus, fresh wood or plant fibers have the same abundance of ^{14}C relative to ^{12}C as air, but ancient wood or plant fibers have less ^{14}C. Archeologists take advantage of this decrease of ^{14}C for the ***radioactive dating*** of materials found in ancient tombs. Suppose that a sample of cotton cloth from the wrappings of a mummy found in 1990 in an ancient Egyptian tomb has exactly half the abundance of ^{14}C relative to ^{12}C as air (the abundance of ^{14}C can be determined by measuring the beta activity of the sample). What age can you deduce for the sample?

Radioactive dating ▸

SOLUTION: If exactly one-half of the "normal" amount of ^{14}C has decayed, the age of the sample must be exactly one half-life, that is, 5730 years. This means the sample must be from the year 3740 B.C.

31.5 FISSION

In heavy, large nuclei — such as uranium — the balance between the attractive strong force and the repulsive electric force is quite precarious. Although the strong force between two adjacent nucleons easily exceeds the electric force, the strong force has only a short range, and it cannot bridge the distance from one end of a large nucleus to the other. Hence the interaction of the distant parts of a large nucleus is dominated by the electric force. In such a nucleus, the electric force becomes dominant if the nucleus suffers some extra elongation, either spontaneous or caused by an external disturbance, such as the impact of a neutron on the nucleus. The repulsive electric force then overpowers the attractive strong force, and the nucleus splits into two or more fragments. Such a splitting of a nucleus is called ***fission.*** Figure 31.13 illustrates the fission of a nucleus after it suffers a disturbance and elongation when hit by a neutron.

◀ **Fission**

Fission was first discovered by O. Hahn and F. Strassmann, who bombarded a sample of uranium with a beam of neutrons and found that the uranium fissions into barium and krypton, according to the reaction

$$\text{neutron} + {}^{238}\text{U} \rightarrow {}^{145}\text{Ba} + {}^{94}\text{Kr} \tag{30}$$

More generally, it was found that uranium can fission in several different ways, with the formation of a variety of fission fragments of about one-half of the mass of the original uranium nucleus. The fission fragments are usually radioactive, and they often emit one or several neutrons. Thus, the net reaction is actually

$$\text{neutron} + {}^{238}\text{U} \rightarrow \text{fission fragments} + 2 \text{ or } 3 \text{ neutrons} \tag{31}$$

While the fission fragments fly apart, the repulsive electric force does positive work on the fragments, giving them a large kinetic energy. This kinetic energy represents the energy released in the reaction. We can calculate the amount of energy released in the reaction in the usual way, by comparing the mass of the uranium nucleus with the (smaller) sum of the masses of the fission fragments. But even without such a calculation we can see by inspection of the curve of binding energy that ***fission of a large nucleus leads to a release of energy.*** Figure 31.6 tells us that a large nucleus has a somewhat lower amount of binding energy per nucleon than a medium-sized nucleus (the curve has a maximum at $A = 56$, corresponding to ^{56}Fe). Hence the fission of a large nucleus into two medium-sized nuclei results in an increase of the net binding energy, that is, it results in a release of energy.

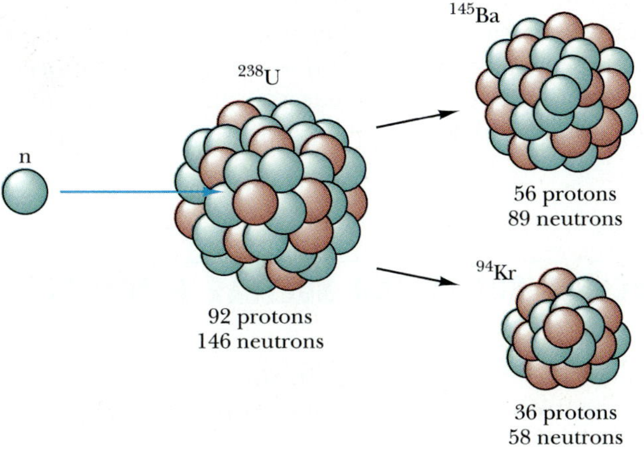

FIGURE 31.13
Fission of uranium triggered by the impact of a neutron.

On the average, the energy released in the fission of a uranium nucleus is about 200 MeV. We can gain a better appreciation of the magnitude of this energy by looking at the following numbers. The total energy released by the complete fission of 1 kg of uranium (a lump slightly larger than a golf ball) is 200 MeV times the number of atoms in 1 kg of uranium. Since 1 mole of uranium is 238 g, 1 kg of uranium is 1000/238 = 4.2 moles, or $4.2 \times 6.02 \times 10^{23}$ atoms = 2.5×10^{24} atoms, and the energy released by the fission of these atoms is

$$200\,\frac{\text{MeV}}{\text{atom}} \times 2.5 \times 10^{24}\ \text{atoms} = 5.1 \times 10^{26}\ \text{MeV}$$

$$= 8.1 \times 10^{13}\ \text{J} \tag{32}$$

This is equivalent to the energy released in the burning of 2 million liters of gasoline. It is also equivalent to the energy released in the explosion of about 20,000 tons of TNT.

To exploit the energy released by the fission of uranium we rely on the neutrons that emerge in the reaction (31). If one initial fission occurs in a lump of uranium, the 2 or 3 neutrons that emerge from this reaction can strike other uranium nuclei and trigger their fission, and the neutrons that emerge from these fissions can trigger further fissions, and so on. Thus, ***one initial neutron can initiate an avalanche of neutrons and an avalanche of fission reactions.*** Such an avalanche is called a ***chain reaction.*** If no neutrons, or very few neutrons, are lost from the chain reaction, the rate of fission and the rate of release of energy grow drastically with time. For instance, if on the average two of the neutrons released in each fission succeed in generating further fission reactions and further neutrons, then the numbers of fission reactions in successive steps of the chain will be 2, 4, 8, 16, ... If this multiplicative growth continues unchecked, the rate of release of energy will become explosive.

In the case of ^{238}U, conditions are not favorable for sustaining a chain reaction. Uranium-238 is a fairly stable nucleus, which does not readily fission when a neutron strikes it. Instead, the nucleus often merely absorbs the incident neutron, removing it from the chain reaction. This loss of neutrons blocks the chain reaction. However, naturally occurring uranium also contains a small amount (0.72%) of the isotope ^{235}U, which is unstable and much more susceptible to fission. This isotope is widely used as the nuclear fuel in practical applications of chain reactions in nuclear bombs and nuclear reactors. Another isotope used in such applications is ^{239}Pu, an isotope of plutonium. But this isotope does not occur naturally; it must be manufactured by nuclear transmutation reactions involving uranium, in a nuclear reactor.

Enrico Fermi, *1901–1954, Italian and later American physicist. He worked on experimental and theoretical investigations of beta decay, the artificial production of isotopes by neutron bombardment, for which he received the 1938 Nobel Prize, and the fission of uranium. Fermi was one of the leaders of the Manhattan Project, and he provided the first experimental demonstration of a chain reaction.*

31.6 Nuclear Bombs and Nuclear Reactors

In a given mass of ^{235}U or ^{239}Pu, neutrons produced by spontaneous fission or stray neutrons coming from elsewhere can initiate the first step in the chain reaction. Whether the reaction keeps going depends on how many neutrons are lost from the chain, by absorption without fission (as in ^{238}U) or by escape beyond the boundary of the mass. If the mass is large, few neutrons will reach its boundary before they are intercepted by a nucleus; thus, ***a large mass inhibits escape of the neutrons and favors the chain reaction.*** This mass is said to be ***critical*** if the number of neutrons lost from the chain reaction (by escape or by absorption) equals the number of neutrons released by the fissions. In this case the chain reaction merely proceeds at a constant rate — as in a nuclear reactor. The mass is said to be ***supercritical*** if the number of neutrons lost from the chain is smaller than the number of neutrons released in fission reactions. In this case the chain reaction proceeds at an ever-increasing, runaway rate leading to an explosion — as in a nuclear bomb. For pure ^{235}U arranged in a spherical configuration, the critical mass is about 50 kg.

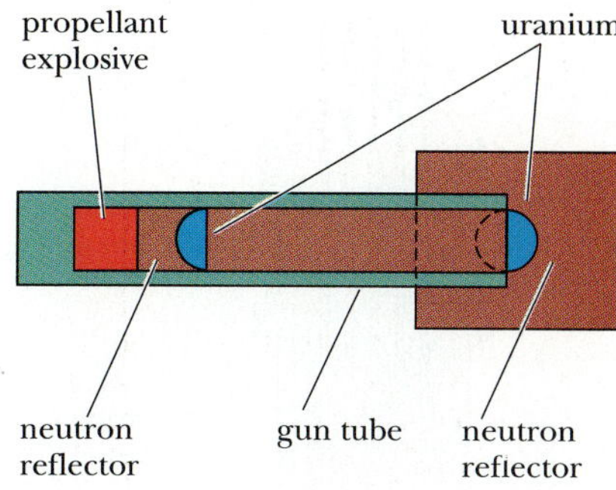

FIGURE 31.14
A fission bomb.

The simplest fission bomb, or ***A-bomb,*** consists of two pieces of ^{235}U such that separately their masses are less than the critical mass, but jointly their masses are more than the critical mass. To detonate such a bomb, the two pieces of uranium, initially at a safe distance from each other, must be suddenly brought close together. In the first such bomb (See Fig. 31.14), the device used for the assembly of the two pieces of uranium consisted of a gun which propelled one piece of uranium toward the other at high speed. More sophisticated fission bombs consist of a (barely) subcritical mass of ^{239}Pu; if this mass is suddenly compressed to higher than normal density, it will become supercritical. The sudden compression is achieved by the preliminary explosion of a chemical high explosive arranged in a layer around the mass of plutonium.

FIGURE 31.15
Explosion of a fission bomb.

◀ **A-bomb and H-bomb**

The energy released in the explosion of an A-bomb is typically equivalent to that released in the explosion of about 20 kilotons of TNT (see Fig. 31.15). Much higher explosive yields are achieved by ***H-bombs,*** in which an A-bomb is used to trigger fusion reactions similar to the fusion reactions that power the Sun (see Section 31.7). The energy released in the explosion of an H-bomb is typically one or several megatons. Such an explosion would level an entire city, with complete devastation and complete destruction of life by incineration and crushing out to a radius of 16 km from the center of the explosion.

◀ **Nuclear reactor**

For the peaceful exploitation of nuclear fission in a ***nuclear reactor,*** we must keep the chain reaction under control, so it releases energy at a steady rate. This means that the configuration of the uranium or other nuclear fuel must be critical rather than supercritical. The most common type of reactor operates with "enriched" uranium consisting of a few percent ^{235}U mixed with 90-odd percent of ^{238}U. Such a uranium mixture cannot by itself maintain a chain reaction — the ^{238}U soaks up too many of the neutrons. However, if the uranium is surrounded by a substance capable of slowing down the neutrons released in fissions, the chain reaction becomes viable. The substance that slows the neutrons is called the ***moderator.*** The role of the moderator in a fission reaction is analogous to that of a catalyst in a chemical reaction. The moderator enhances the chain reaction because slow neutrons are more efficient at producing fissions in ^{235}U than fast neutrons, and they are also less likely to be absorbed by ^{238}U.

◀ **Moderator**

◀ **Fuel rods**

Inside the reactor, the uranium is usually placed in long ***fuel rods,*** and

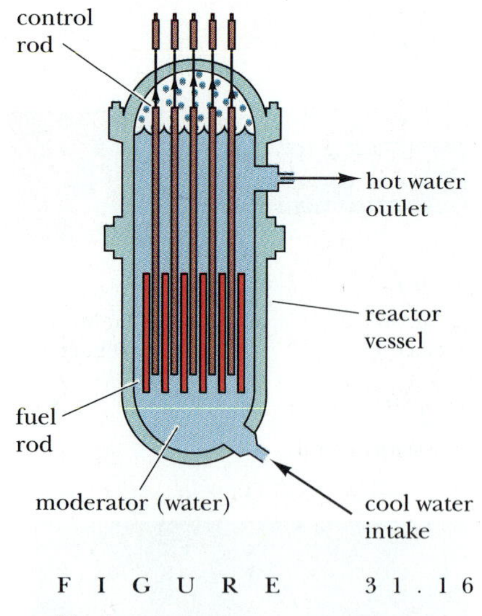

FIGURE 31.16
Nuclear reactor.

these are immersed in the bulk of moderator (see Fig. 31.16). Fast neutrons released by fissions travel from the fuel rods into the moderator; there the lose their kinetic energy by collisions with the moderator nuclei; and then they wander back into one or another of the fuel rods and trigger further fissions. The three best moderators are ordinary water (H_2O), heavy water (D_2O), and graphite (pure carbon).

The configuration of the reactor — the size, number, and location of the fuel rods and the shape of the moderator — must be designed so the reactor is nearly critical. Fine adjustments in the number of neutrons and the reaction rate are made by means of ***control rods*** of boron or cadmium. These substances greedily soak up neutrons, and by pushing the control rods in or pulling them out, the reaction rate can be decreased or increased.

The main application of reactors is for the generation of electric power. In the United States, most of the reactors used for this purpose have water-filled cores. The water acts simultaneously as moderator and as coolant. The water circulates through the core and removes the heat energy released by the fission reactions (see Fig. 13.17). The heat is transferred from the water to steam, which drives a steam turbine connected to an electric generator. Thus, the nuclear reactor plays the role of the furnace of a conventional steam engine — and uranium replaces coal or oil as fuel.

Nuclear power plants could meet all our energy requirements for the next several hundred years, or even the next several thousand years, if we exploit low-grade ores containing nuclear fuels (Fig. 31.18). Unfortunately, nuclear fission yields rather dirty energy — the fission reactions generate dangerous radioactive residues. Nuclear power plants must be carefully designed to hold these residues in confinement. The reactor core is enclosed in a massive reactor vessel, and as an extra precaution this vessel and the attached pumps and pipes are enclosed in a strong containment shell (see Fig. 31.19). The cumbersome safety features that are incorporated in the design of a nuclear power plant make the construction and the maintenance extremely expensive. Furthermore, when the load of fuel of the reactor has been spent, the residual radioactive

FIGURE 31.18
Nuclear power plant at Mt. Rainier, Oregon.

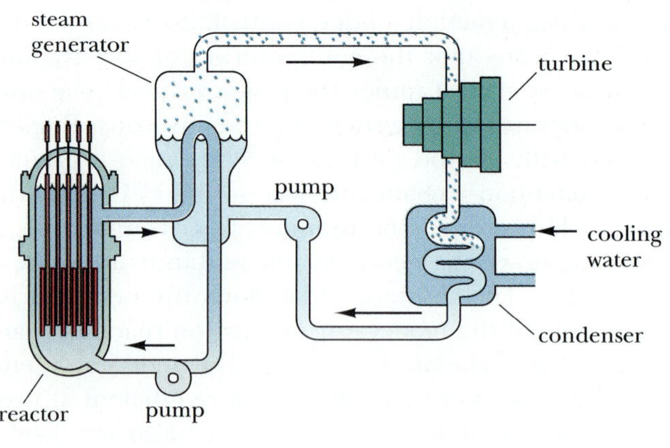

FIGURE 31.17
Schematic diagram of nuclear power plant.

wastes must be removed to a safe place to be held in storage for hundreds of years until their radioactivity has died away.

If some of the radioactive material contained in a nuclear reactor is released in the form of smoke or dust in an accidental explosion or fire, it can be carried away by winds, and it can then descend to the ground as lethal radioactive fallout. After the disastrous nuclear accident that happened at Chernobyl, near Kiev in the former U.S.S.R, in 1986, a plume of fallout spread north and west into Europe and produced hazardous contamination thousands of kilometers away. An area of 30-km radius around the reactor was so heavily contaminated it had to be permanently evacuated. The immediate cause of the Chernobyl disaster was operator error — through a series of almost incredible blunders, the operators lost control of the reactor, and its power level surged to 100 times normal values, which overheated the core, blew it apart, and started a fire in the graphite moderator.

FIGURE 31.19
The containment shell of a nuclear reactor.

The water-moderated reactors used in the United States are thought to be much safer than the Chernobyl reactor. But some experts question whether they are safe enough. Scenarios for conceivable reactor accidents have been analyzed in great detail. The worst that might happen is a loss-of-coolant accident, that is, the loss of the water that normally circulates through the core of the reactor, and which serves both as moderator and as coolant. Although such a loss of water would shut off the fission chain, the reactor contains a large amount of radioactive residues, and the heat from the radioactivity can by itself overheat the reactor vessel beyond safe limits. As a precaution, reactors are equipped with emergency cooling systems, and they are also encased in a containment shell.

Power reactors in the United States all use ^{235}U as fuel. Unfortunately, the supply of this nuclear fuel is rather limited, and it is expected to become exhausted early in the next century. There are, however, several other nuclear fuels in larger supply. One of these is ^{238}U. Although ^{238}U is incapable of supporting chain reactions, it can be converted into ^{239}Pu, which supports chain reactions. The manufacture of ^{239}Pu is an automatic side effect of the operation of nuclear reactors now in use; in all of these reactors, the fuel rods contain a mixture of ^{235}U and ^{238}U, and fission neutrons striking the ^{238}U gradually convert it into ^{239}Pu. A reactor fueled with ^{239}Pu not only makes good use of a material that would otherwise go to waste, but, if the reactor is surrounded by a blanket of ^{238}U, it can also manufacture extra ^{239}Pu. The number of neutrons released in the fission of ^{239}Pu is so large that in an efficiently designed reactor, neutrons can be diverted to the ^{238}U without hindering the chain reaction. Such a reactor can produce more ^{239}Pu (from ^{238}U) than it consumes (from its original supply). A reactor of this kind is called a ***breeder reactor.*** Because of worries over the safety of these reactors, they have not been adopted in the United States, but several are operating successfully in Europe.

◄ **Breeder reactor**

31.7 FUSION

Fusion is the merging of two small nuclei to form a larger nucleus. The fusion of two small nuclei, such as the nuclei of hydrogen, deuterium, or

◄ **Fusion**

tritium, releases energy. For instance, the heat of the Sun is generated by a fusion reaction called hydrogen burning, in which hydrogen nuclei fuse together to make helium nuclei. By inspection of the curve of binding energy, we can immediately recognize that the fusion of small nuclei leads to a release of energy. Figure 31.6 shows us that the smallest nuclei have exceptionally low binding energies per nucleon. Hence, the fusion of two such nuclei into a larger nucleus of higher binding energy results in a release of energy. Note that the process of fusion is the reverse of fission: ***small nuclei (such as hydrogen) release energy when they fuse; large nuclei (such as uranium) release energy when they split.***

The hydrogen burning in the Sun proceeds in three steps: first hydrogen nuclei fuse to make deuterium (^{2}H), then deuterium fuses with hydrogen to make tritium (^{3}He), and finally tritium fuses with tritium to make helium:

$$^1\text{H} + {}^1\text{H} \rightarrow {}^2\text{H} + e^+ + \nu \tag{33}$$

$$^1\text{H} + {}^2\text{H} \rightarrow {}^3\text{He} + \gamma \tag{34}$$

$$^3\text{He} + {}^3\text{He} \rightarrow {}^4\text{He} + {}^1\text{H} + {}^1\text{H} \tag{35}$$

Each of the first two reactions must occur twice for the last reaction to occur once. Each of these reactions releases energy; the net energy released when the first two reactions occur twice and the last occurs once (with the formation of one nucleus of ^{4}He) is 24.7 MeV. Four nuclei of ^{1}H are consumed in this process (6 are consumed when the first two reactions occur twice, but 2 are regenerated when the last reaction occurs); hence the amount of energy released per nucleon of "fuel" is 24.7 MeV per 4 nucleons, or 6.2 MeV per nucleon. This number is to be compared with the energy released in the fission of uranium, about 200 MeV per 235 nucleons of "fuel," or 0.85 MeV per nucleon. Thus, fusion of a given mass of ^{1}H in the Sun releases about 7 times as much energy as the fission of an equal mass of ^{235}U.

Thermonuclear reactions ▸

The fusion reactions (33)–(35) are called ***thermonuclear,*** because they can proceed only at extremely high temperatures and pressures, such as the temperature of about 15×10^6 K in the core of the Sun. The high temperature is needed to overcome the electric repulsion that the hydrogen nuclei experience whenever they come close together. At high temperatures, the hydrogen nuclei have high speeds, and their collisions are sufficiently violent to overcome the electric repulsion and bring the nuclei into the intimate contact required for fusion.

The thermonuclear reaction based on hydrogen fuel cannot be duplicated on Earth, because we cannot attain the pressure found in the core of the Sun. However, there are some fusion reactions based on deuterium and tritium that proceed at lower pressures, but require a temperature of about 10^8 K, even higher than in the center of the Sun. Such temperatures are attained in the explosions of H-bombs, where the preliminary explosion of a fission bomb heats and compresses a mixture of deuterium and tritium and thereby initiates fusion.

Experimental attempts at a peaceful exploitation of fusion power also rely on a deuterium–tritium fuel mixture, which is heated to the required temperature of about 10^8 K either by an intense electric current, or by a particle beam, or by a laser beam. At such a temperature, the deuterium and tritium will be in the form of a plasma, that is, a totally ionized gas, consisting of a mixture of independent nuclei and electrons. The plasma cannot be contained by the walls of a conventional reactor vessel, since any contact with the wall of the vessel would cool the plasma, and would also melt the wall. Instead, the plasma must be held suspended in the middle of the vessel, away from contact with the walls.

FIGURE 31.20
The Princeton Tokamak Test Reactor.

One scheme for controlled fusion, called ***magnetic confinement,*** attempts to suspend the plasma by means of magnetic fields. Figure 31.20 shows the Princeton Tokamak Test Reactor, in which the plasma is confined inside a large toroidal solenoid (a solenoid shaped like a doughnut). The plasma is heated by a combination of electric currents induced in the plasma and particle beams aimed into it.

◀ **Magnetic and Inertial confinement**

Another scheme for fusion attempts to extract energy by exploding small pellets of a deuterium–tritium mixture in a combustion chamber by hitting them with an intense laser beam; this scheme is called ***inertial confinement*** because the inertia of the pellet holds it together long enough for the reaction to occur. The beam heats the pellet so suddenly that it has no time to disperse before fusion begins. The pellet then explodes like a miniature hydrogen bomb. After the thermal energy has been extracted from the combustion chamber, the next pellet is placed in the chamber, and so on. Figure 31.21 shows the combustion chamber of the NOVA laser at Lawrence Livermore Laboratory. Although both of these schemes have been successful in initiating fusion, the amounts of energy released in fusion have remained far below the amount of energy that had to be fed into the reactor chambers to heat the plasma.

FIGURE 31.21
Combustion chamber of the NOVA laser.

Nuclear fusion is an attractive source of energy because it bypasses many of the safety problems associated with nuclear fission. It produces clean energy, without appreciable amounts of radioactive residues. Besides, we have available an enormous supply of deuterium. There is enough heavy water mixed with the ordinary water in the oceans of the Earth to satisfy our energy needs for millions of years.

SUMMARY

- **Radius of nucleus:** $R = (1.2 \times 10^{-15}\text{ m}) \times A^{1/3}$
- **Energy equivalent of atomic mass unit:** $1\text{ u} \times c^2 = 931.5\text{ MeV}$
- **Mass defect and binding energy:**
  [mass of N neutrons and Z protons] − [mass of nucleus] = $B.E./c^2$
- **Law of radioactive decay:** $n = n_0 \left(\frac{1}{2}\right)^{t/t_{1/2}}$
- **Decay rate:** $\frac{\Delta n}{\Delta t} = 0.693\,\frac{n}{t_{1/2}}$

- **becquerel:** 1 Bq = 1 disintegration/s
- **curie:** 1 Ci = 3.7×10^{10} disintegrations/s

QUESTIONS

1.–What is the average overall density of a typical atom, such as iron? Why is this much smaller than the density of a nucleus, as given by Eq. (4)?

2.–Naturally occurring magnesium has an atomic mass of 24.305. On the basis of this information, can you conclude that natural magnesium contains a mixture of several isotopes? Can you guess which isotopes?

3.–According to Fig. 31.6, which isotope has the largest binding energy?

4.–Why do alpha and beta emissions involve transmutation of elements, but gamma emission does not?

5.–If you irradiate a sample of material with alpha, beta, or gamma rays, is the sample likely to become radioactive? What if you irradiate it with neutrons?

6.–Radiation is sometimes used to sterilize food or surgical supplies. What are the advantages or disadvantages of this procedure over sterilization by heating?

7.–Can radioactive dating with ^{14}C be used to determine the age of a book printed during the last century? The age of deposits of mineral oil? The age of ancient Egyptian gold jewelry? Ancient Egyptian furniture?

8.–Tritium (^{3}H) is a radioactive isotope of hydrogen which occurs naturally in small concentrations in the ordinary water found in the environment. The half-life of this isotope is 12 years. Describe how you could take advantage of this isotope to determine the age of a bottle full of wine that your wine merchant claims is 25 years old.

9.–Alpha particles are more massive than beta particles, yet for equal kinetic energies, the alpha particles are stopped by a thinner layer of material than the beta particles. Can you explain why an alpha particle loses its energy more quickly? (Hint: Think of the electric forces that act on the particles in a collision with an atom.)

10.–All the best moderators consist of fairly lightweight nuclei. Why is a material with heavy nuclei, such as lead, not a good moderator?

11.–Why is it difficult to separate the isotopes ^{238}U and ^{235}U?

12.–Some artificial satellites carry small nuclear reactors as power supplies. What danger does this pose for people on Earth?

13.–Nuclear fission bombs produce a long-lived radioactive isotope of strontium (^{90}Sr), an element that is chemically similar to calcium. Explain why strontium poses a severe hazard to man and other vertebrates.

14.–Why is it more difficult to achieve controlled fusion in a reactor than controlled fission?

15.–A fusion reactor would produce a large amount of tritium (^{3}H), a radioactive isotope of hydrogen. If the tritium were accidentally released into the environment, it would be likely to contaminate the water supply. Explain.

PROBLEMS

Section 31.1: Isotopes

1.——How many protons and neutrons are there in the nucleus of the isotope ^{16}O? ^{56}Fe? ^{238}U?

2.——What isotope has 82 protons and 122 neutrons in its nucleus?

3.——What are the number of protons and the number of neutrons in each of the following isotopes: ^{24}Na, ^{27}Al, ^{52}Cr, ^{52}Mn, ^{63}Cu, ^{63}Zn, ^{124}Xe, ^{138}La?

4.——What isotope has 17 protons and 18 neutrons in its nucleus? What isotope has 18 protons and 17 neutrons? Such isotopes in which the numbers of protons and neutrons are exchanged are called ***mirror nuclei.*** Find another example of mirror nuclei in the chart of isotopes.

5.——Use the chart of isotopes on page 864 to make a list of all the isotopes of oxygen. What are the number of protons and the number of neutrons in the nucleus of each of these isotopes?

6.——By inspection of Fig. 31.5, determine the ratio of neutrons to protons in stable light nuclei ($Z \simeq 20$) and in stable heavy nuclei ($Z \simeq 80$).

7.——In any nuclear reaction, the nuclear electric charge must be conserved and the mass number must be conserved. Are the following reactions in accord with these conservation laws?

$$^{2}H + {}^{12}C \rightarrow {}^{4}He + {}^{9}B$$

$$^{4}He + {}^{10}B \rightarrow {}^{13}C + n$$

$$n + {}^{238}U \rightarrow {}^{121}Ag + {}^{118}Pd$$

8.——Naturally occurring boron is a mixture of 80.2% ^{11}B and 19.8% ^{10}B. From the masses of these isotopes (listed in the chart of isotopes) calculate the atomic mass of naturally occurring boron.

9.——According to Eq. (2), what is the nuclear radius of the smallest of the isotopes of carbon? The largest of these isotopes?

10.——The largest known nucleus is that of the isotope ^{266} of element 109, which has not yet been given a name. What is the radius of this nucleus?

11.*——Neutron stars are made (almost) entirely of neutrons, and they have approximately the same density as that of a nucleus. What is the radius of a neutron star of mass 0.5 times the mass of the Sun?

12.*——What fraction of the volume of your body is filled with nuclear material? The average density of your body is about 1000 kg/m^3.

13.*——A uranium nucleus of radius 7.4×10^{-15} m fissions into two equal spherical pieces. What is the radius of each piece?

14.**——Suppose you bombard a target of magnesium with alpha particles. If, in a head-on collision with a stationary magnesium nucleus, an alpha particle is to reach the nuclear surface just barely before being halted by the repulsive electric force, what must be the energy of the alpha particle? Express your answer in MeV.

15.**——Suppose that an alpha particle of energy 4.4 MeV collides head-on with a stationary gold nucleus. What is the distance of closest approach? Does the alpha particle make contact with the surface of the nucleus?

Section 31.2: The Strong Force and the Nuclear Binding Energy

16.——The binding energy of the electron in the ground state of the hydrogen atom is 13.6 eV. Calculate the corresponding mass defect of the hydrogen atom; express this in atomic mass units. (Since the mass defects associated with atomic binding energies are small, they are usually ignored in nuclear physics.)

17.——According to Fig. 31.6, what is the binding energy of ^{56}Fe? From this binding energy, calculate the mass of the isotope (in atomic mass units).

18.——What are the mass defect (in atomic mass units) and the nuclear binding energy (in MeV) of the isotope ^{4}He? The mass of ^{4}He is given in the chart of isotopes.

19.——What are the mass defect (in atomic mass units) and the nuclear binding energy (in MeV) of the isotope ^{12}C?

20.——Nuclei with 2, 8, 14, 20, 28, 50, 82, or 126 protons or neutrons are called ***magic nuclei*** because they have exceptionally large binding energies and they are exceptionally stable. Compare the binding energy of the magic nucleus ^{16}O with that of the nonmagic nucleus ^{16}F. The masses of these isotopes are given in the chart of isotopes.

21.*——How much energy (in MeV) is required to remove one neutron from the nucleus of the isotope ^{14}N? (Hint: Compare the binding energies of ^{14}N and ^{13}N; the masses of these isotopes are given in the chart of isotopes.)

22.——Verify that in the reaction (11), both sides have the same number of protons and neutrons.

23.*——When boron is bombarded with alpha particles, the following reaction is observed:

$$^{4}He + ^{10}B \rightarrow ^{13}C + ^{1}H$$

How much energy is released in this reaction?

24.*——Chadwick discovered the neutron when he bombarded boron with alpha particles, which resulted in the reaction

$$^{4}He + B \rightarrow ^{12}C + n$$

Calculate the energy released in this reaction.

25.*——Consider the following nuclear reaction:

$$^{2}H + ^{12}C \rightarrow ^{4}He + ^{10}B$$

What is the energy absorbed in this reaction? Can this reaction be initiated by bombarding carbon with a beam of protons of 1.2 MeV?

26.*——When ^{7}Li is bombarded with protons, the following reaction occurs:

$$^{1}H + ^{7}Li \rightarrow ^{7}Be + n$$

What is the minimum kinetic energy required for the proton to initiate this reaction? Neglect the recoil energy of the ^{7}Be nucleus.

27.*——Consider the reaction

$$^{1}H + ^{3}H \rightarrow ^{3}He + n$$

(a) What is the energy absorbed in this reaction?

(b) We can initiate this reaction either by bombarding a tritium target with protons, or by bombarding a hydrogen target with tritium ions. To initiate this reaction, the kinetic energy of the bombarding particle must be somewhat larger than the energy absorbed in the reaction, since the recoil of the reaction products retains some of the energy supplied by the bombarding particle. If tritium ions are used as bombarding particles, the required kinetic energy is larger than if protons are used as bombarding particles. Explain this difference.

28.**—The energy released in the reaction (11) is 1.19 MeV. Assume that the alpha particle has almost zero energy and zero momentum; the net momentum of the reaction products is then also zero. Calculate the separate energies of the oxygen nucleus and of the proton.

Section 31.3: Radioactivity

29.——What isotope is formed by the alpha decay of ^{210}Po? ^{239}Pu?

30.——What isotope is formed by the beta decay of ^{85}Kr? ^{63}Ni?

31.——What isotope is formed by the positron beta decay, or beta plus decay, of ^{22}Na? ^{64}Cu?

32.——The alpha decay of uranium [see Eq. (13)] is the first step in a radioactive series of decays. The next four steps are a beta decay, another beta decay, another alpha decay, and another alpha decay. What are the daughter nuclei produced in these four steps?

33.*—The alpha decay of ^{210}Po results in ^{206}Pb. Calculate the energy of the emitted alpha particle. The masses of these two isotopes are 209.9829 u and 205.9745 u, respectively.

34.*—What is the maximum kinetic energy of the beta ray emitted in the beta decay of a neutron, according to the reaction (18)? Ignore the recoil of the proton.

35.*—If neutrons had a somewhat smaller mass, then the (slow) electrons in atoms could combine with protons in the nucleus according to the reaction

$$e + p \rightarrow n + \nu$$

How much smaller would the mass of the neutron have to be to make this reaction viable? What consequences would this reaction have for the existence of atoms and the existence of life?

36.*—Find the maximum kinetic energy of the beta rays emitted in the decay

$$^{16}\mathrm{N} \rightarrow {}^{16}\mathrm{O} + e^- + \bar{\nu}$$

37.*—Nuclear physicists sometimes deduce the mass of an isotope from the energy of the beta rays emitted by the isotope. For instance, the maximum energy of the beta rays emitted in the decay of ^{27}Mg is 2.610 MeV. Given that the mass of ^{27}Al is 26.981541 u, what mass can you deduce for ^{27}Mg?

38.*—What isotope is formed in the beta decay of ^{14}C? Calculate the maximum energy of the beta rays emitted in this decay. The masses of the relevant isotopes are given in the chart of isotopes.

Section 31.4: The Law of Radioactive Decay

39.——According to Fig. 31.12, at what time will the remaining amount of radioactive strontium have fallen to 1/10 of the initial amount?

40.——According to the best available data, the half-life of ^{14}C is believed to be 5730 years. However, according to previous data, the half-life was thought to be 5570 years, and age determinations based on this value of the half-life were in error. What percentage error in an age determination does the error in the half-life introduce? For a sample from the year 3000 B.C., what is the error in years?

41.——Hydrogen bombs operate with tritium, a radioactive isotope of hydrogen with a half-life of 5.5 years. When a hydrogen bomb is held in storage, its tritium gradually decays, and must be replenished with fresh tritium every few years. According to a recent proposal, arms control could be achieved by halting the production of tritium. Every few years, the superpowers would then have to take some of their bombs out of service, extract the remaining tritium, and use it to replenish their other bombs. Suppose that the United States and the former U.S.S.R. each have 2000 hydrogen bombs now. Without fresh tritium, how many bombs will each superpower have in 11 years? In 22 years?

42.——The isotope ^{238}U found on the Earth was originally synthesized in nuclear reactions in supernovas that exploded in our Galaxy about 6.8×10^9 years ago and scattered debris through the Galaxy, some of which became trapped in the Earth during its formation, about 4.6×10^9 years ago. What fraction of the original amount of uranium synthesized in the Galaxy is still in existence? What fraction of the amount of uranium trapped in the Earth during its formation is still in existence? The half-life of ^{238}U is 4.5×10^9 years.

43.——The activity of an industrial ^{60}Co irradiation cell is 10^{16} Bq when the cell is new. What will be the activity after the cell has been in use for two years? The half-life of ^{60}Co is 5.3 years.

44.——What is the activity of 1 gram of pure radium, ^{226}Ra? The half-life of this isotope is listed in Table 31.2.

45.——What is the activity of 1 microgram of ^{22}Na? The half-life of this isotope is listed in Table 31.2.

46.*——The partial eradication of the thyroid in patients suffering from hyperthyroidism can be accomplished by injecting a compound containing radioactive iodine ^{131}I into the body; the iodine then concentrates in the thyroid and kill its cells. If the thyroid is to be subjected to an activity of 0.1 Ci, how many grams of ^{131}I should be injected? The half-life of ^{131}I is 8.04 days.

Section 31.5: Fission

Section 31.6: Nuclear Bombs and Nuclear Reactors

47.——The bomb exploded at Hiroshima had an explosive yield of about 20 kilotons of TNT, or 8.4×10^{13} J. How many kilograms of uranium actually underwent fission to release this amount of energy?

48.——A nuclear reactor in a power plant produces 1200 MW of heat. How many kilograms of ^{235}U does this reactor consume per year?

49.——Naturally occurring uranium contains 99.28% of the isotope ^{238}U and 0.72% of the isotope ^{235}U. How many kilograms of natural uranium must we process to extract the 50 kg of ^{235}U required for a bomb? Pretend that all of the ^{235}U can be extracted.

50.*——Consider the fission reaction

$$n + {}^{235}\text{U} \rightarrow {}^{144}\text{Ba} + {}^{92}\text{Kr}$$

Given that the masses of the isotopes ^{235}U, ^{144}Ba, and ^{92}Kr are 235.04394 u, 143.92285 u, and 91.92627 u, respectively, calculate the energy released in this reaction.

Section 31.7: Fusion

51.*——Consider the three fusion reactions (33)–(35) that occur in the core of the Sun. From the masses of the nuclei that participate in these reactions (see the chart of isotopes), calculate the energy released in each reaction. What is the net energy if the first two reactions occur twice, and the last reaction occurs once?

52.*——(a) The rate at which the Sun radiates heat and light is 3.9×10^9 W. If the energy for all of this radiation comes from the fusion reactions (33)–(35), what is the rate at which the Sun consumes hydrogen?

(b) The amount of hydrogen in the Sun is about 1.5×10^{30} kg. How long will this fuel last if the Sun goes on radiating at a steady rate?

53.* Whenever the first two of the fusion reactions (33)–(35) proceed twice and the last proceeds once, 24.7 MeV are released, and so are two neutrinos. Hence, the number of neutrinos released is 1 per 12.3 MeV. Assume that the fusion reactions (33)–(35) account for all of the 3.9×10^{9} W of heat and light radiated by the Sun. Calculate the rate at which the Sun releases neutrinos.

54.** A sequence of reactions that can be used in controlled fusion is the burning of deuterium "fuel," as follows:

$$^{2}\mathrm{H} + {}^{2}\mathrm{H} \rightarrow {}^{3}\mathrm{He} + \mathrm{n}$$

$$^{2}\mathrm{H} + {}^{2}\mathrm{H} \rightarrow {}^{3}\mathrm{H} + {}^{1}\mathrm{H}$$

$$^{2}\mathrm{H} + {}^{3}\mathrm{H} \rightarrow {}^{4}\mathrm{He} + \mathrm{n}$$

$$^{2}\mathrm{H} + {}^{3}\mathrm{He} \rightarrow {}^{4}\mathrm{He} + {}^{1}\mathrm{H}$$

The net result of this sequence of reactions is the transmutation of six nuclei of deuterium (^{2}H) into two nuclei of ^{4}He, two protons (^{1}H), and two neutrons. Calculate the energy released in each of these reactions. What is the net energy released per nucleon of fuel?

PHYSICS IN ACTION

MEDICAL IMAGING

We already have explored two of the imaging techniques used in medical diagnosis: ultrasound images (pp. 345A–345D) and X-ray images (pp. 809A–809D). Several other important imaging techniques exploit the properties of atomic nuclei to produce pictures of the tissues in the human body and also pictures of the metabolic activity in those tissues. These "nuclear" imaging techniques are ***radioisotope scans, single-photon emission tomography, positron-emission tomography,*** and ***magnetic-resonance imaging.*** Table 1 summarizes these imaging techniques.

TABLE 1 MEDICAL IMAGING TECHNIQUES

Imaging technique	What is detected
Sonography	Ultrasound waves reflected in body (echo)
Ordinary radiography	X rays transmitted through body
Computed tomography (CT or CAT)	X rays transmitted through body
Single-photon emission computed tomography (SPECT)	Gamma rays emitted by tracers
Positron emission tomography (PET)	Pairs of gamma rays emitted by tracers
Magnetic-resonance imaging (MRI)	Spin flip of hydrogen nuclei

Radioisotope imaging

In radioisotope imaging, some chemical compound that incorporates a radioactive "tracer" is injected into the body. Metabolic processes selectively accumulate the chemical compound in some organ, and the outline of the organ can then be detected by the radioactivity it emits. The tracers used in radioisotope scans are emitters of gamma rays, such as ^{123}I or ^{99}Tc. Like X rays, gamma rays have great penetrating power, and the rays emitted in an organ deep within the body can easily reach a detector placed outside the body — the detector can "see" the organ by the gamma-ray "light" it emits. The detector used to produce a picture of the intensity distribution of the gamma rays is called a ***gamma camera.*** However, this camera has little in common with an ordinary camera. The gamma rays cannot be focused by an ordinary lens, such as the lens of an ordinary camera — the gamma rays would penetrate through such a lens without refraction. Instead, the gamma camera uses a ***collimator,*** a lead plate with many fine parallel holes. Gamma rays that enter the collimator holes obliquely are blocked by the lead, while gamma rays aligned with the holes pass through; thus, each hole samples one spot of the organ (Fig. 1). A scintillator plate is placed directly behind the collimator, and the gamma rays that pass through the holes register as pinpoints of light, which form a picture of the organ. The pinpoints of light are detected by photomultiplier tubes, which convert the pinpoints of light into electric signals that are fed into a video monitor for display.

Different radioisotope tracers in conjunction with different chemical carriers are used to prepare pictures of different organs. For example, Fig. 2 shows a scan of a thyroid; this scan was made with an ^{131}I compound injected into the bloodstream.

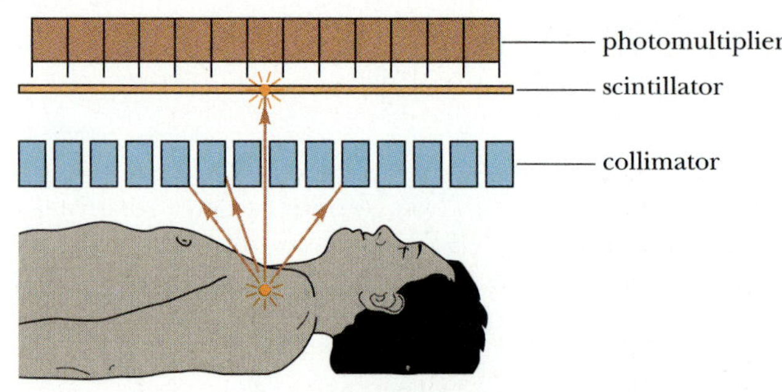

FIGURE 1 A gamma camera aimed at the neck. Only gamma rays aligned with the holes in the collimator can pass and reach the scintillator.

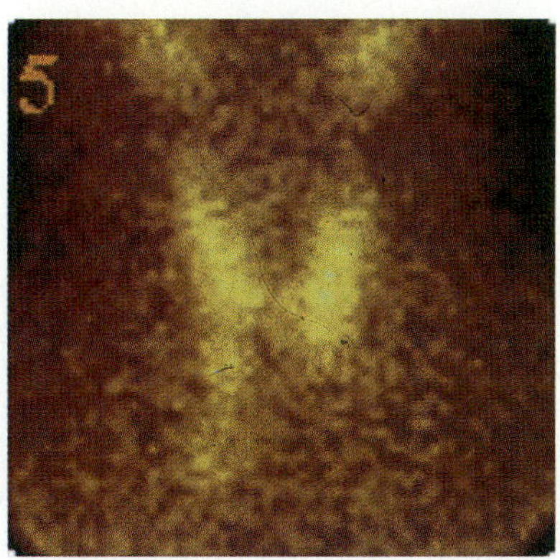

F I G U R E 2 Scan of a thyroid with an abnormally enlarged left lobe.

The thyroid metabolizes iodine, hence this element tends to concentrate there, and the thyroid then "glows" in its gamma-ray "light."

Single-photon emission computed tomography (SPECT)

In *X-ray Images* (pp. 809A–809D) we saw how computed tomography can generate a cross-sectional picture of the body from data obtained with X rays. A similar imaging technique can be applied to data obtained with gamma rays. As in tomography with X-rays, the detector of gamma rays is rotated around the body, and it records the intensity of gamma rays at different angles. However, in tomography with X-rays the source of the rays is external to the body, whereas in tomography with gamma rays the source is a radioisotope tracer, such as ^{99}Tc, in the interior of the body. This difference in the location of the source requires somewhat different computer processing.

Positron-emission tomography (PET)

This imaging technique is an elegant refinement of single-photon emission tomography. Ordinary gamma emitters, such as ^{99}Tc, emit only one gamma ray at a time. But positron emitters, such as ^{11}C, ^{13}N and ^{18}F, give rise to a pair of gamma rays in coincidence. When a positron emitter is embedded in tissue, the emitted positron, or antielectron, almost immediately collides with an electron in some nearby atom, and the consequent electron–antielectron annihilation produces two gamma rays in coincidence. Thus, positron-emission tomography is effectively "two-photon emission tomography."

The detectors for PET are arranged in a circle around the body (see Fig. 3). The detectors can identify the two gamma rays that make up a pair by their coincidence in time: the two gamma rays reach two opposite detectors at almost the same time. The big advantage of PET over SPECT is that no collimator is required to determine the direction of the gamma rays. The two gamma rays produced by one annihilation travel in almost exactly opposite directions, and therefore their directions of travel must lie along the straight line drawn from one detector to the other. Separate pairs of gamma rays yield separate straight lines, and the intersection of these lines reveals the location of the source. The absence of collimators means that the detectors can make use of a larger fraction of the total number of the emitted gamma rays, and that therefore PET can generate pictures of good quality with smaller amounts of radioactive tracer than SPECT. The disadvantage of PET is that the positron emitters ^{11}C, ^{13}N and ^{18}F have rather short half-lives (of the order of an hour), and they cannot be held in storage. The positron emitters must be produced in a cyclotron at or near the hospital, and they must be used immediately.

Figure 4 shows a picture of a brain scan prepared with PET. This picture displays activity in the brain under different conditions. This scan used a glucose-like compound tagged with ^{18}F. The active regions of the brain metabolize glucose and accumulate ^{18}F, and the scan reveals the distribution of ^{18}F.

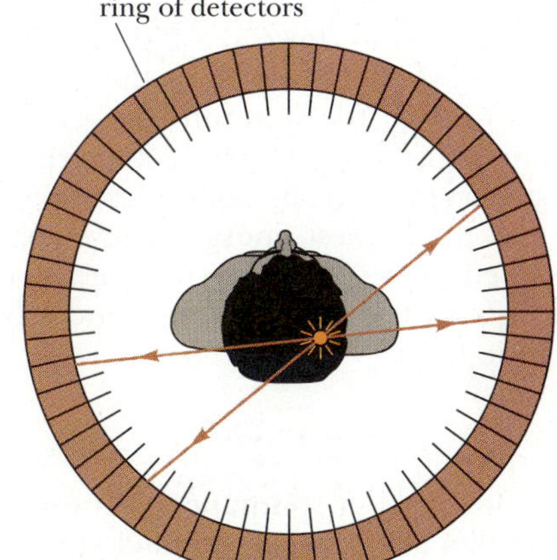

F I G U R E 3 Ring of detectors for PET. The directions of travel of the two gamma rays in a pair produced by electron–antielectron annihilation lie along a straight line from one detector to the other. The drawing shows the directions of travel of two separate pairs and the intersection.

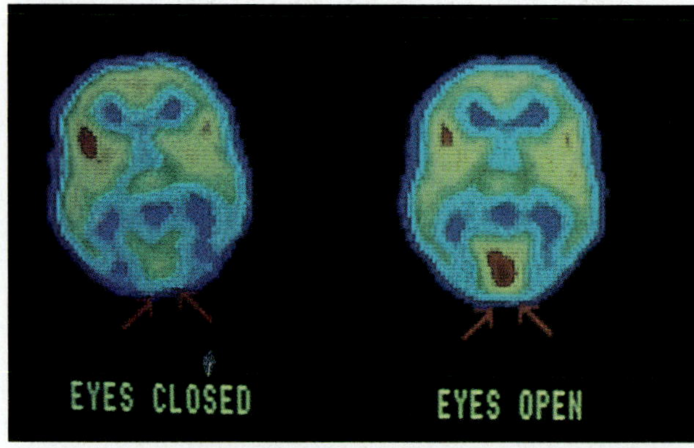

FIGURE 4 PET scan of the brain. Eyes of subject are closed on the left. Eyes of subject are open on the right. Note the differences in brain activity between the two.

Magnetic-resonance imaging (MRI)

Magnetic-resonance imaging exploits the spin and the magnetic moment of nuclei.* We know that electrons can be regarded as small balls of charge whose rotation gives them a magnetic moment. Likewise, nuclei can be regarded as small balls of charge, and their rotation also gives them a magnetic moment. When a nucleus with a spin and a magnetic moment is placed in a magnetic field, the orientation of the spin is quantized according to the usual quantization rules. For instance, a hydrogen nucleus, or proton, of spin $\frac{1}{2}$ has two possible orientations in the magnetic field: spin parallel to the magnetic field ("spin up") or spin antiparallel to the magnetic field ("spin down'see Fig. 5). The torque that the magnetic field exerts on the magnetic moment tends to twist the spin into the parallel orientation. Thus, this state of parallel orientation has the lowest energy, and the state of antiparallel orientation has the highest energy. The energy difference between these states is proportional to the magnetic moment μ of the proton and strength of the magnetic field,

$$\Delta E = 2\mu B \tag{1}$$

A proton can make a spontaneous transition from the antiparallel state to the parallel state, with the emission of a photon. This is called a ***spin flip.*** The frequency of the emitted photon is

$$f = \frac{\Delta E}{h} = \frac{2\mu B}{h} \tag{2}$$

For a magnetic field of 1 T, this frequency is 43 MHz, which is a radio frequency (RF). But a proton can also make a transition from the parallel state to the antiparallel state, provided that some electromagnetic disturbance "pumps" it at the natural frequency, as given by Eq. (2). Such a transition is a resonance phenomenon; if the frequency of the electromagnetic disturbance matches the natural frequency (2), the proton absorbs energy from the disturbance and flips from the parallel state to the antiparallel state. The electromagnetic disturbance used to pump the protons can be either a radio wave or simply an extra, oscillating RF magnetic field.

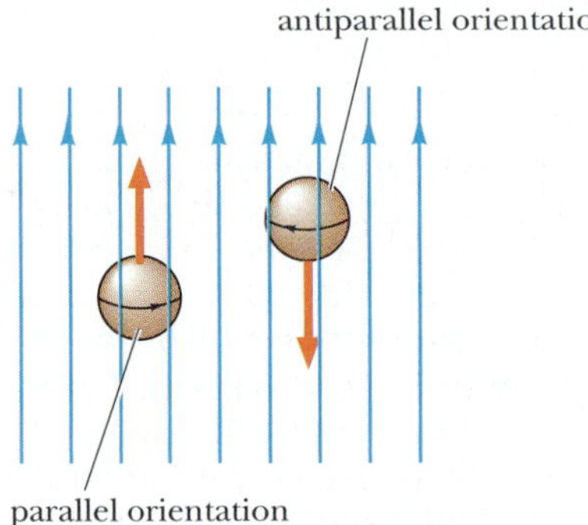

FIGURE 5 The two possible orientations of a proton in a magnetic field. The state of parallel orientation has a lower energy than the state of antiparallel orientation.

Magnetic-resonance imaging exploits this resonant spin flip of protons. The tissues of the human body contain an abundance of protons in the nuclei of hydrogen atoms. When the body is immersed in a strong magnetic field (see Fig. 6), a majority of these protons settle into the state that has spin parallel to the magnetic field. If now an extra RF magnetic field oscillating at the resonant frequency is briefly applied, the protons flip from the parallel state to the antiparallel state. After this resonant flip, they spontaneously flip back into their original state. These spontaneous spin flips can be detected by a search coil (a loop) placed near the body; when the spins flip, so do the magnetic fields associated with the magnetic moments of the protons, and this changes the magnetic flux in the search coil and thereby induces a detectable emf.

Such a detection of the spin flips of protons merely indicates the presence of hydrogen nuclei

* MRI was originally called nuclear magnetic resonance imaging, or NMR. But the salespeople of NMR equipment worried that the mention of anything "nuclear" might scare the public, and they decided to change the name to MRI.

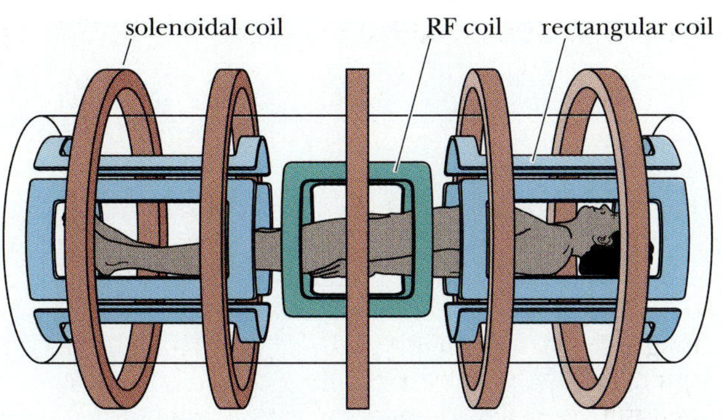

FIGURE 6 Human body inside a magnet used for MRI. The magnet is essentially a solenoid (red) which produces a more or less uniform magnetic field in the axial direction (head to toe). The rectangular coils (blue) permit alterations of this magnetic field, and the RF coil (green) generates a high-frequency oscillating magnetic field and also serves as search coil to detect the emf induced by the spontaneous spin flips.

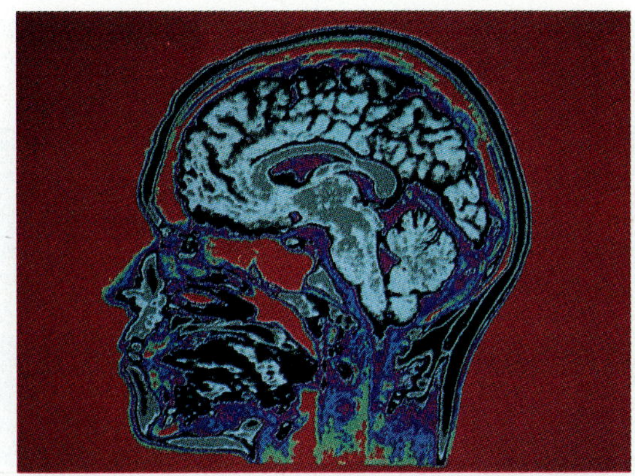

FIGURE 7 MRI scan of human head in longitudinal section.

within the body. To produce a picture of the distribution of hydrogen in the body, we need to determine the location of the hydrogen nuclei. This is done by varying the magnetic field B across the body. For instance, if the magnetic field B is stronger on the left side of the body than on the right side, then the resonant frequency is higher for the left side than for the right side, and if we "interrogate" the body with oscillating RF magnetic fields of several frequencies, we can determine the location of the hydrogen nuclei according to the frequency at which they respond. For a complete determination of the location of the hydrogen nuclei, we need to vary the magnetic field first in one direction and then in the two perpendicular directions, to obtain the x, y, z coordinates that describe the location.

A mere picture of the hydrogen distribution in the human body would not reveal a great deal of interesting detail, since hydrogen, in the form of water, is present in all the soft tissues of the body in about the same concentration. However, a refinement of the MRI technique permits us to discriminate among hydrogen atoms according to their chemical environment. After the hydrogen nuclei have been flipped into the antiparallel state by the briefly applied RF magnetic field, we leave them alone for a while, and we let them begin to flip back to the parallel state. The time that the spin of a nucleus takes to flip back (the rate of spontaneous transition) depends on disturbances that nearby atoms exert on the nucleus, that is, it depends on the chemical environment. After a while, we again apply the oscillating RF magnetic field. Then those nuclei that have flipped back will respond once more, but those that have not yet flipped back cannot respond. Hence, such repeated interrogations by the oscillating RF magnetic field reveal where nuclei have flipped back and where they have not, that is, they reveal details of the chemical environment.

Figure 7 is an MRI scan showing a longitudinal section through a human head. This image reveals many gross anatomical features, but its fine details also reveal chemical differences. The exact correlation between the fine details in such images and the chemical differences is not well understood, but even without a clear understanding of the details it is often easy to spot abnormalities by comparing the MRI image of a patient with the MRI image of a healthy person.

The magnetic fields required for MRI are very strong, of the order of 1 T. Such magnetic fields are produced by large, and expensive, magnets built with superconducting coils. An important advantage of MRI over radiography and over radioisotope scans is that the body is not subjected to hazardous X rays or hazardous radioactivity. The body is merely subjected to a strong magnetic field, which apparently has no harmful effects of any kind.

REFERENCES

Putnam, C. E., and Ravin, C. E., eds., *Textbook of Diagnostic Imaging* (W. B. Saunders, Philadelphia, 1988).

Webb, S., ed., *The Physics of Medical Imaging* (Hilger, Bristol, 1988).

C H A P T E R 32

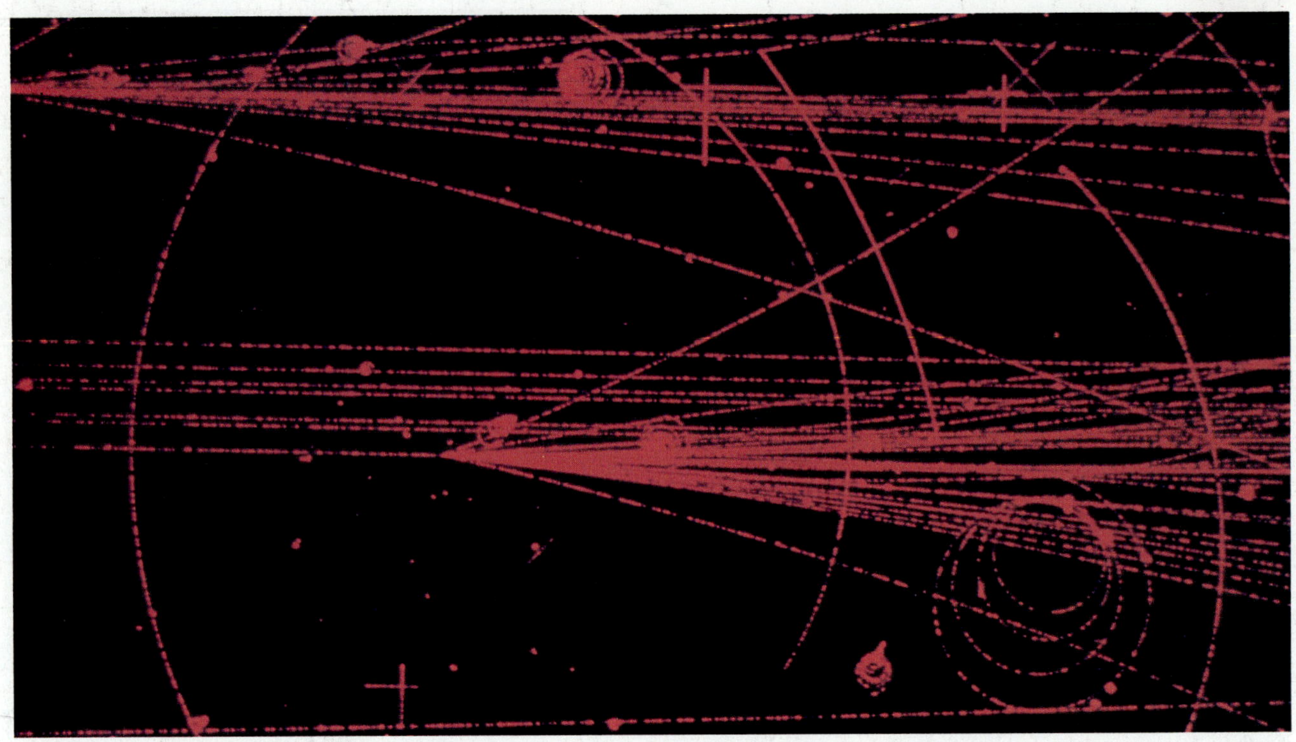

Trails of small bubbles reveal the passage of particles in a bubble chamber filled with liquid hydrogen. The parallel horizontal tracks show a beam of high-energy protons entering the chamber from the left. One of these protons collided with a proton in a hydrogen nucleus, and the collision produced a spray of several kinds of new particles.

ELEMENTARY PARTICLES

The search for the elementary, indivisible building blocks of matter is the most fundamental problem in physics. Early in this century physicists discovered that the atom is not an elementary, indivisible unit — each atom consists of electrons orbiting around a nucleus. And then they discovered that the nucleus is not an elementary, indivisible unit — each nucleus consists of protons and neutrons packed tightly together. In the 1930s, physicists began to build accelerating machines producing beams of energetic protons or electrons that could serve as projectiles; with these atom smashers physicists could split the nucleus. In the 1950s, they built much larger and more powerful accelerating machines; with these new machines they attempted to split the proton and the neutron. But the result of these attempts was chaos: when bombarded by very energetic projectiles, the proton and the neutron do not split into any simple subprotonic pieces. Instead, the violent collisions of protons and neutrons generate a multitude of new, exotic particles by the conversion of kinetic energy into mass. For want of a better name, these new particles were called "elementary particles." However, most of these elementary particles are more massive and more complicated than protons and neutrons — they are obviously not elementary, indivisible units. Only in recent years has some order emerged from this chaos. Physicists have found convincing circumstantial evidence that protons, neutrons, and other elementary particles are made of very small, compact subunits. The subprotonic building blocks are called ***quarks.***

◀ Quarks

32.1 THE TOOLS OF HIGH-ENERGY PHYSICS

Protons and neutrons are much "harder" than atoms or nuclei. A projectile of a bombarding energy of a few eV can shatter an atom, and a projectile of a bombarding energy of a few MeV can shatter a nucleus. But to make a dent in a proton, we need a bombarding energy of a few hundred or a thousand MeV. Elementary-particle physicists like to measure the energies of their projectiles in billion electron-volts (GeV) or in trillion electron-volts (TeV). Expressed in joules, these energy units are

$$1 \text{ GeV} = 10^9 \text{ eV} = 1.6 \times 10^{-10} \text{ J}$$

$$1 \text{ TeV} = 10^{12} \text{ eV} = 1.6 \times 10^{-7} \text{ J}$$

FIGURE 32.1
Panoramic view of Fermilab.

The acceleration of a projectile to such a high energy requires large, sophisticated accelerator machines. High energy — one or several GeV — characterizes the realm of particle physics. Collisions between particles at these high energies are capable of creating a large variety of new particles by the conversion of kinetic energy into mass.

◀ Accelerators

The two largest ***accelerators*** are the proton synchrotrons at the Fermi National Accelerator Laboratory (Fermilab, near Chicago) and at the Conseil Européen de Recherches Nucleaires (CERN, on the Swiss–French border near Geneva). Figure 32.1 shows an overall view of Fermilab; the accelerator is buried underground in a circular tunnel 6 km in circumference. The CERN accelerator is even larger, about 27 km in circumference. The Fermilab Tevatron accelerator produces a beam of protons with an energy of 1 TeV and a speed of 99.99995% of the speed of light. The protons travel in an evacuated circular beam pipe (Fig. 32.2). Large magnets placed along this pipe exert forces on the protons, preventing their escape — the protons move as if on a circular racetrack. At regular intervals the pipe is joined to cavities connected to high-voltage oscillators; in each of these cavities, oscillating electric fields impel the protons to higher energy. After several hundred thousand circuits around the racetrack, the protons reach their final energy of 1 TeV.

FIGURE 32.2
View of the underground tunnel housing the Tevatron accelerator at Fermilab. The long row of magnets encases the beam pipe.

Before the protons are allowed to enter the giant circular racetrack, they must pass through several smaller preliminary accelerators. At Fermilab there are three preliminary accelerators. The protons generated by a proton gun are first given an energy of about 1 MeV by an electrostatic generator; then their energy is raised to 200 MeV by a linear accelerator;

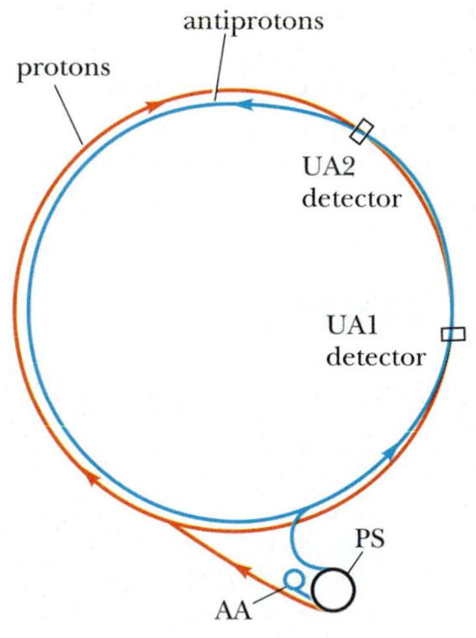

FIGURE 32.3

In the SPS accelerator at CERN, high-energy protons and antiprotons travel in opposite directions around a ring of circumference 7 km. The protons and antiprotons are produced by a medium-energy accelerator (PS, lower right). The antiprotons are temporarily stored in an accumulator ring (AA), while the protons are sent directly into the large ring. When enough antiprotons have accumulated, they are also sent into the large ring, and both kinds of particles are accelerated to their final, high energy. The orbits of the protons (red) and the antiprotons (blue) are slightly separated, but they intersect at two points, where the detectors UA1 and UA2 are located.

and then it is raised further to 8 GeV by a "small' circular accelerator. Only after the protons have passed through these preliminary stages do they enter the main ring.

An even larger accelerator, the Superconducting Super Collider (SSC), is being planned in the United States. This accelerator, to be built near Dallas in the 1990s, will have a circumference of 90 km, and it will accelerate protons and antiprotons to 20 TeV.

Once the particles have been given their maximum energy, they are guided out of the accelerator by steering magnets and made to crash against a target consisting of a block of metal or a tankful of liquid. Within the material of the target, the high-energy particles collide violently with the protons or neutrons of the nuclei. The reactions that take place in these collisions create new particles by conversion of energy into mass. Unfortunately, not all of the kinetic energy of the incident particles can participate in these reactions. As we saw in Section 6.6, the velocity of the center of mass remains constant during the collision, and therefore the particles must retain some kinetic energy. A relativistic calculation indicates that, in the collision of a high-energy proton and a stationary proton, the particles retain most of the kinetic energy, and only a small fraction becomes available for reactions. Such collisions are quite inefficient.

The efficiency improves drastically if two high-energy protons are made to collide head-on. The available energy is then the sum of the energies of the two protons. At CERN, a beam of protons and a beam of antiprotons of 450 GeV each are made to travel in opposite directions around the accelerator in slightly different orbits that cross at two places (Fig. 32.3). Where the two orbits cross, the two beams meet nearly head-on. Several other accelerator laboratories have recently installed similar storage rings with which they can bring about head-on collisions between particles. Head-on collisions will also be exploited by the Superconducting Super Collider.

When a beam of high-energy particles crashes into a stationary target or into an oncoming beam, the violent reactions create a wide variety of new particles. In order to observe the particles that emerge from the scene of these collisions, physicists use several kinds of particle detectors. Some of these — scintillation counters and Čerenkov counters — signal the passage of each electrically charged particle by giving off brief (and weak) flashes of light; these flashes of light are detected by sensitive photomultiplier tubes. Other detectors — bubble chambers and multiwire chambers — render the tracks of electrically charged particles visible, either on a photograph or on a computer-generated picture.

Bubble chamber ▶

A ***bubble chamber*** is a tank filled with a superheated liquid, usually liquid hydrogen, whose temperature is slightly above the boiling point. Such a liquid is unstable — it is about to start boiling but it will usually not start until some disturbance triggers the formation of the first few bubbles. A charged particle zipping through the chamber provides just the kind of disturbance the liquid is waiting for — a fine trail of small bubbles forms in the wake of the particle's passage. High-speed cameras can take a picture of these bubble tracks before they disperse and disappear in the turmoil of subsequent widespread bubbling and boiling of the liquid.

FIGURE 32.4

(a) The Big European Bubble Chamber (BEBC) at CERN, surrounded by the large magnet which almost completely hides it from view. (b) The BEBC before its installation in the magnet.

Figure 32.4 shows the BEBC bubble chamber at CERN and its ancillary equipment. The bubble chamber is surrounded by a large electromagnet, which aids in the identification of the particles passing through the bubble chamber. The magnetic field generated by this magnet pushes the particles into curved orbits as they pass through the chamber. The direction of the curvature depends on the sign of the electric charge, and the magnitude of the curvature depends on the momentum of the particles; thus, measurement of the curvature of a bubble track tells us the sign of the electric charge of the particle and the momentum of the particle.

The photograph in Fig. 32.5 was taken with a bubble chamber. It shows the track of a high-energy proton that entered the bubble chamber and collided with one of the protons at rest in the nucleus of one of the hydrogen atoms in the liquid filling the chamber. The collision destroyed one of the two protons, but produced seven positive pions, seven negative pions, a positive kaon, and a neutral lambda particle. The reaction can be summarized as follows:

$$\mathrm{p} + \mathrm{p} \rightarrow \mathrm{p} + 7\pi^+ + 7\pi^- + \mathrm{K}^+ + \Lambda \tag{1}$$

The Λ subsequently decayed into a proton and negative pion,

$$\Lambda \rightarrow \mathrm{p} + \pi^- \tag{2}$$

Note that the sum of the rest masses of the particles after the reaction (1) is much larger than the sum of the rest masses of two protons. The excess mass comes from the conversion of energy into mass — some of the kinetic energy of the incident proton has been converted into mass.

This conversion of kinetic energy into mass plays a crucial role in the discovery of new particles. Almost all the new particles discovered during the last 20 years are considerably heavier than protons and neutrons. Physicists need powerful accelerators to produce the large kinetic energies that must be supplied for the manufacture of these new heavy particles.

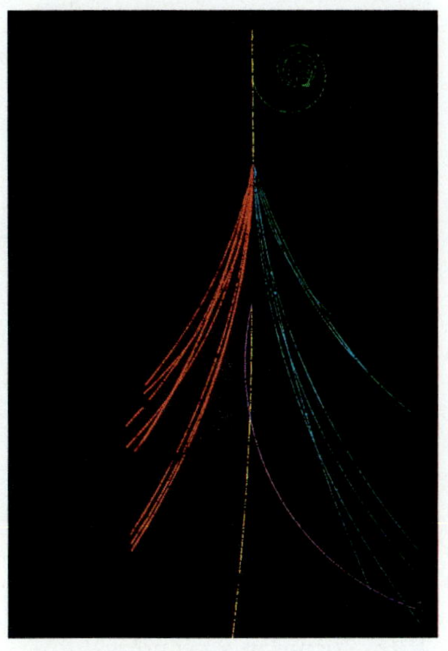

FIGURE 32.5

Tracks of particles in a bubble chamber (the tracks have been colored to distinguish different kinds of particles; in the original bubble-chamber photograph the tracks were all white). A proton (yellow track) enters the photograph from above and collides with one of the protons at rest in the nucleus of one of the hydrogen atoms in the liquid filling the chamber. The tracks of positively charged particles (red) and of negatively charged particles (blue) created in this collision can be seen emerging from the scene of the accident. The positive particles are one proton, one kaon, and seven positive pions; the negative particles are seven negative pions. Besides, a neutral lambda was created in the collision. Such a neutral particle leaves no track in a bubble chamber, but it reveals itself when it decays into a proton (yellow track, below) and a negative pion (purple track).

EXAMPLE 1. Elementary-particle physicists prefer to measure the masses of particles in MeV/c^2, using the conversion 1 u = 931.5 MeV/c^2. Thus a proton has a mass of 938 MeV/c^2, a pion has 140 MeV/c^2, a kaon has 494 MeV/c^2, and a lambda has 1115 MeV/c^2 (see Tables 32.2 and 32.3). Given these masses, calculate the minimum kinetic energy required for each proton if we want to initiate the reaction (1) in a head-on collision between two protons of equal kinetic energies.

SOLUTION: In a head-on collision between two protons of equal energies, all of the kinetic energy can be converted into rest-mass energy (this would *not* be true for the collision of a high-energy proton with a stationary proton, that is, it would not be true for the collision as photographed in Fig. 32.5). If the kinetic energy of each proton is K, the kinetic energy of both is $2K$, and this must equal the difference in the rest-mass energies before and after the reaction:

$$2K = m_{\rm p}c^2 + 14[\text{mass of pion}]c^2 + [\text{mass of kaon}]c^2 + [\text{mass of lambda}]c^2 - 2m_{\rm p}c^2$$

$$= 14 \times 140 \text{ MeV} + 494 \text{ MeV} + 1115 \text{ MeV} - 938 \text{ MeV}$$

$$= 2631 \text{ MeV} \tag{3}$$

Thus, the kinetic energy of each proton must be 2631/2 MeV, or 1316 MeV.

FIGURE 32.6
Wires strung in a cylindrical array for a multiwire chamber. The array is shown during construction, before it was placed inside the vacuum chamber.

Multiwire chamber ▶

Although bubble chambers yield the best pictures of particle tracks, they are very complex, very large, and very expensive machines, and they cannot take pictures fast enough to satisfy the needs of experimenters. In recent years, most bubble chambers have been replaced by ***multiwire chambers.*** These chambers are strung with thousands of fine wires, in an array that resembles a coordinate grid (Fig.32.6). The wires are connected to voltage supplies and to current sensors. When a charged particle passes through the gas in the chamber, it ionizes the gas along its track, and the electrons released in this ionization are attracted to the wires near the track, where they form miniature electric discharges, which are detected as current pulses by the sensors individually connected to the wires. These current pulses reveal which wires are near the track, and a computer can then immediately reconstruct the location of the track and draw a picture of the track. An important advantage of multiwire chambers over bubble chambers is that the raw data are in the form of current pulses, which can be immediately processed and stored in digitial form by a computer. In contrast, the raw data from a bubble chamber are in the form of photographs, and the tracks recorded on these photographs must be measured and converted into numbers before they can be processed and analyzed by a computer; this conversion of photographic data into digital data is very time-consuming.

The UA1 detector at CERN incorporates several multiwire chambers. Figure 32.7 shows a computer-generated picture of tracks of particles in the multiwire chambers of this detector.

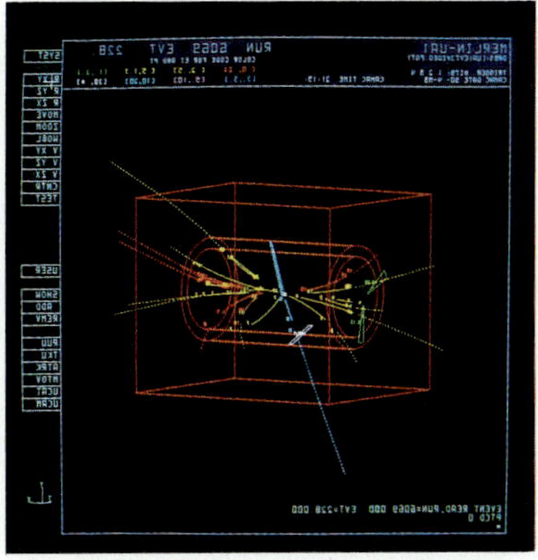

FIGURE 32.7
Computer-generated picture of particle tracks in the multiwire chamber of the UA1 detector at CERN. These are the tracks of particles produced in the collision of a proton and an antiproton that entered the chamber horizontally from opposite directions.

32.2 THE MULTITUDE OF PARTICLES

As physicists built accelerators of higher and higher energy, they discovered more and more new particles. By now, about 300 kinds of particles are known and a few extra particles are found every year. Whenever a new, more powerful accelerator makes available more energy for collisions, more particles and heavier particles are produced — there seems to be no end in sight.

The known particles fall into three groups: leptons, baryons, and mesons.

◄ Leptons

There are six different ***leptons*** (see Table 32.1). Among these, the electron (e) is the most familiar. The muon (μ) is very similar to an electron; it has the same electric charge as an electron but its mass is about 200 times as large. The tau (τ) is also similar to the electron, but its mass is even larger than that of the muon. The neutrinos (ν_e, ν_μ, and ν_τ) are particles of zero mass and zero electric charge.

In Table 32.1, the masses of the particles are expressed in units of MeV/c^2, and the electric charge and the spin are expressed in multiples of the proton charge e and in multiples of Planck's constant $\hbar$. The choice of MeV/c^2 for the unit of mass is very convenient in calculations involving the rest-mass energies. For instance, from the electron mass of 0.511 MeV/c^2 listed in the table, we immediately see that the electron rest-mass energy is 0.511 MeV.

Besides the leptons of Table 32.1, there are six antileptons: the antielectron, the antimuon, the antitau, and the three antineutrinos. These antiparticles have an opposite electric charge and some other opposite properties, but they have exactly the same mass and spin as the corresponding particles. (The notation for an antiparticle is a bar over the letter or, alternatively, a superscript indicating the electric charge; thus, the notation for the antielectron is $\bar{e}$ or e^+.)

◄ Baryons

The ***baryons*** are the most numerous group of particles. The most famil-

TABLE 32.1 THE LEPTONS

Particle	Symbol	Mass	Spin	Charge	Comments
electron	e	0.511 MeV/c^2	$\frac{1}{2}$	−1	constituent of atoms
muon	μ	105.6	$\frac{1}{2}$	−1	produced in decay of pion; abundant in cosmic rays
tau	τ	1784	$\frac{1}{2}$	−1	
electron neutrino	ν_e	0(?)	$\frac{1}{2}$	0	produced in beta decay; abundantly emitted by Sun
muon neutrino	ν_μ	0(?)	$\frac{1}{2}$	0	
tau neutrino	ν_τ	0(?)	$\frac{1}{2}$	0	

TABLE 32.2 SOME BARYONS

Particle	Symbol	Mass	Spin	Charge	Quark content	Comments
proton	p	938.3 MeV/c^2	$\frac{1}{2}$	+1	uud	constituent of nucleus
neutron	n	939.6	$\frac{1}{2}$	0	ddu	constituent of nucleus
lambda	Λ	1115	$\frac{1}{2}$	0	uds	has strangeness −1
sigma-plus	Σ^+	1189	$\frac{1}{2}$	+1	uus	has strangeness −1
sigma-minus	Σ^-	1197	$\frac{1}{2}$	−1	dds	has strangeness −1
sigma-zero	Σ^0	1192	$\frac{1}{2}$	0	uds	has strangeness −1
xi-minus	Ξ^-	1321	$\frac{1}{2}$	−1	dss	has strangeness −2
xi-zero	Ξ^0	1315	$\frac{1}{2}$	0	uss	has strangeness −2
omega-minus	Ω^-	1672	$\frac{3}{2}$	−1	sss	has strangeness −3
charmed lambda	Λ_c	2280	$\frac{1}{2}$	+1	udc	contains charmed quark

TABLE 32.3 SOME MESONS

Particle	Symbol	Mass	Spin	Charge	Quark content	Comments
pion-zero	π^0	135 MeV/c^2	0	0	uu and d$\bar{d}$	abundant in cosmic rays; carrier of nuclear force
pion-plus	π^+	140	0	+1	u$\bar{d}$	abundant in cosmic rays; carrier of nuclear force
pion-minus	π^-	140	0	−1	d$\bar{u}$	abundant in cosmic rays; carrier of nuclear force
kaon-zero	K^0	498	0	0	d$\bar{s}$	has strangeness +1
kaon-plus	K^+	494	0	+1	u$\bar{s}$	has strangeness +1
kaon-minus	K^-	494	0	−1	s$\bar{u}$	has strangeness −1
J/psi	J/ψ	3097	1	0	c$\bar{c}$	contains charmed quarks
D-zero	D^0	1865	0	0	c$\bar{u}$	contains charmed quark
D-plus	D^+	1869	0	+1	c$\bar{d}$	contains charmed quark
upsilon	Y	9460	1	0	b$\bar{b}$	contains bottom quarks

iar baryons are the proton and the neutron; these are the baryons of the least mass. Table 32.2 lists some of the other baryons. For every baryon in Table 32.2 there exists an antibaryon. As in the case of leptons, these antiparticles have an opposite charge but the same mass and spin as the corresponding particles.

Mesons ▶

Finally, the ***mesons*** are another numerous group of particles. Some of them are listed in Table 32.3. Note that all mesons have spin 0, or 1, or 2, etc., whereas all baryons have spin $\frac{1}{2}$, or $\frac{3}{2}$, or $\frac{5}{2}$, etc. This is the distinctive difference between mesons and baryons. For every meson there exists an antimeson; however, these antiparticles have already been included in Table 32.3. For example, the antiparticle to the π^+ is the π^- (and vice versa). The antiparticle to the π^0 is the π^0; this means that when two π^0 mesons meet, they can annihilate each other.

Most of the particles are unstable; they decay, spontaneously falling apart into several other particles. The only absolutely stable particles are the electron, the proton, and the neutrinos. However, in the terminology of elementary-particle physics, a particle is called "stable" if it lives long enough for physicists to do experiments on or with the particle. Thus, particles that live only 10^{-10} s or even 10^{-14} s are regarded as "stable," because such particles live long enough to be assembled in a beam and shot at a target. If this is stable, what is unstable? Most of the particles listed in Tables 32.2 and 32.3 have lifetimes of about 10^{-23} s. Such particles are regarded as unstable.

A particle that lives only 10^{-23} s is incapable of making a visible track in a bubble chamber. Thus, such an unstable particle cannot be detected directly, but its existence can be inferred from a careful study of the rates of reactions of stable particles engaged in collisions. The unstable particle participates in these reactions as an intermediate, ephemeral state, which causes a characteristic increase of the reaction rate whenever the energy of the stable particles coincides with the energy required for the production of the unstable particle. Because of their effects on reaction rates, the unstable particles are often called ***resonances.***

◂ Resonances

32.3 INTERACTIONS AND CONSERVATION LAWS

The reactions that occur among the particles in a high-energy collision are governed by the four fundamental forces: the "strong" force, the electromagnetic force, the "weak" force, and the gravitational force. However, at the microscopic level, particles are subject to quantum uncertainties in position and velocity, and the force acting on them is not well defined. Hence, physicists prefer to speak of four fundamental kinds of ***interactions,*** instead of forces. Mathematically, these interactions can be described by formulas that specify the amount of energy for each interaction — energies remain meaningful even at a microscopic level, and therefore energies are more relevant than forces. Nevertheless, for the purpose of the following qualitative discussion, let us continue to speak of forces, even though this intuitive concept is ambiguous and ought to be replaced by a more precise and sophisticated mathematical concept of interaction energies.

The ***strong force*** acts on baryons and mesons, but not on leptons. The particles that interact via the strong force are called ***hadrons;*** thus, baryons and mesons are hadrons, but leptons are not.

◂ Strong force

The ***electromagnetic force*** acts primarily on charged particles, but it also acts on neutral particles — such as the neutron — which contain an internal distribution of electric charge and are endowed with magnetic moments.

◂ Electromagnetic force

The ***weak force*** acts on leptons, baryons, and mesons. However, its effect on baryons and mesons is often hidden behind the much larger effects produced by the strong or electromagnetic forces. To see the purest

◂ Weak force

TABLE 32.4 THE FOUR FUNDAMENTAL FORCES

Force	Acts on	Relative strength	Range
Strong	Baryons and mesons (hadrons)	1	$\simeq 10^{-15}$ m
Electromagnetic	Particles with charge or magnetic moment	10^{-2}	Infinite
Weak	Leptons, baryons, and mesons	10^{-6}	$\simeq 10^{-18}$ m
Gravitational	All forms of matter	10^{-38}	Infinite

manifestation of the weak force, we have to examine reactions involving leptons. The weak force is deeply involved in many reactions that bring about the decay of unstable particles. For example, the weak force is responsible for the decay of the neutron.

Gravitational force ▶

The ***gravitational force*** is of no direct interest in particle physics. Although all particles and all forms of energy interact gravitationally, the gravitational effects produced by individual particles are too feeble to be of any significance at the energies available in laboratories on Earth.

Table 32.4 lists the strength of each of the fundamental forces and also the range, or the maximum distance over which this force can reach from one particle to another. In the table, the strengths of the forces are expressed relative to that of the strong force, to which a strength of 1 has been assigned arbitrarily.*

Baryon number ▶

All the forces, and all the reactions that they produce, obey the usual conservation laws for energy, momentum, angular momentum, and electric charge. The reactions also obey the conservation law for ***baryon number,*** which is a generalization of the conservation law for mass number familiar from nuclear physics. Each baryon has a baryon number of $+1$, each antibaryon -1, and all other particles have baryon number 0. The conservation law for baryon number then simply states that the net baryon number remains unchanged in any reaction. For example, consider the reaction (1). The baryon numbers for p, $\pi^{\pm}$, K^{+}, and Λ are 1, 0, 0, and 1, respectively; hence in the reaction (1), the net baryon number is $1 + 1$ before and $1 + 0 + 0 + 0 + 1$ after, that is, the net baryon number remains unchanged.

Besides these conservation laws for familiar quantities, experiments with high-energy particles have led to the discovery of new conservation laws involving several esoteric quantities, such as lepton number, isospin, strangeness, and parity.

Lepton number ▶

Lepton number is for leptons what the baryon number is for baryons. Each lepton has a lepton number $+1$, each antilepton -1, and all other particles have lepton number 0. The net lepton number remains unchanged in any reaction.

Isospin ▶

Isospin is somewhat more complicated because it is a quantity with several components; that is, isospin is a vector quantity. The conservation

* The strengths of the forces depend on the energies of the particles. The values in the table are appropriate for low energies.

law for isospin states that the net sum of the isospin vectors of all the particles involved in a reaction remains unchanged. Obviously, this is somewhat similar to the conservation law for the momentum vector or for the angular-momentum vector. However, the isospin is not a vector in our ordinary three-dimensional space but, rather, a vector in an abstract mathematical space that theoretical physicists invented as a computational tool for the description of the "strong" force.

◀ **Strangeness**

Strangeness is similar to the baryon and lepton numbers. Each hadron has a strangeness number: the proton has strangeness 0, the kaon-plus has $+1$, the lambda has -1, the pion has 0, etc. (values of the strangeness are given in the "Comments" column in Tables 32.2 and 32.3). The conservation law for strangeness states that the net strangeness number remains unchanged in any reaction. For example, in the reaction (1), the net strangeness is $0+0$ before and $0+0+0+1-1$ after; that is, there is no change.

◀ **Parity**

Parity characterizes the behavior of a quantum-mechanical wave under a reversal of the x, y, and z coordinates. Such a reversal is physically equivalent to forming a mirror image of the wave. It can be shown that the mirror image of the quantum-mechanical wave for a stationary state is either equal to the original wave (parity $+1$) or else equal to the negative of the original wave (parity -1). Conservation of parity means that the net parity (the product of all the individual parities) of all the particles participating in a reaction is unchanged.

The conservation laws for energy, momentum, angular momentum, electric charge, baryon number, and lepton number are absolute — no violation of any of them has ever been discovered. By contrast, ***the conservation laws for isospin, strangeness, and parity are approximate*** — they are valid for some reactions, but fail in some others. This, of course, raises the question of what possible meaning can be attached to a "law" that works sometimes and fails sometimes. The answer is that whether a conservation law is obeyed or not depends on the kind of interaction, or the kind of force, that drives the reaction. The reactions caused by the strong force obey all the conservation laws, but reactions caused by the electromagnetic or weak forces do not. It is usually easy to tell what force is involved in a reaction: reactions involving the strong force tend to be fast; reactions involving the other forces tend to be (relatively) slow. For example, reaction (1) is a fast reaction brought about by the strong force, whereas reaction (2) is a slow reaction brought about by the weak force.

Table 32.5 lists the conservation laws obeyed by the strong, electromagnetic, and weak forces.

TABLE 32.5 FORCES AND CONSERVED QUANTITIES

Force	Energy, momentum, and angular momentum	Charge	Baryon number	Lepton number	Strangeness	Parity	Isospin
Strong	✓	✓	✓	✓	✓	✓	✓
Electromagnetic	✓	✓	✓	✓	✓	✓	
Weak	✓	✓	✓	✓			

EXAMPLE 2. Do the following reactions conserve electric charge? Do they conserve baryon number? Do they conserve strangeness?

$$\Lambda \rightarrow n + \pi^0 \tag{4}$$

$$\Xi^- \rightarrow K^- + \pi^0 \tag{5}$$

SOLUTION: Both reactions conserve electric charge. Reaction (4) conserves baryon number, since the baryon number is 1 for the Λ and and $1+0$ for $n+\pi^0$. But Eq. (5) does not conserve baryon number, since the baryon number is 1 for Ξ^- and $0+0$ for $K^-+\pi^0$. This violation of an absolute conservation law implies that the reaction (5) is not a possible reaction.

Reaction (4) does not conserve strangeness, since the strangeness is -1 for the Λ and $0+0$ for $n+\pi^0$. Reaction (5) does not conserve strangeness, since the strangeness is -2 for the Ξ^- and $-1+0$ for $K^-+\pi^0$.

32.4 Fields and Quanta

According to classical theory, forces are mediated by fields. Distant particles do not act on one another directly; rather, each particle generates a field of force and this field acts on the other particles. As we saw in Section 17.1, the existence of fields is required by conservation of energy and momentum. Fields play the role of storehouses of energy and momentum; the energy and momentum stored in the fields balance any excess or deficit in the energy and momentum of the interacting particles engaged in (nonuniform) motion. The most spectacular example of the conversion of particle energy into field energy occurs in the annihilation of matter with antimatter: if an electron collides with an antielectron, the two particles annihilate one another, giving off a burst of very energetic light, or gamma rays. In this annihilation, the energy of the particles — including their rest-mass energy — is completely converted into field energy. The reverse reaction is also possible: if a gamma ray collides with a charged particle, it can create an electron–antielectron pair. In such a pair creation, the energy of the gamma ray is converted into the energy of the pair of particles (Fig. 32.8).

FIGURE 32.8
The two spiraling tracks in this bubble-chamber photograph were made by an electron (green) and an antielectron (red). These particles were created by a high-energy gamma ray in a collision with the electron of a hydrogen atom in the bubble chamber. The long, slightly curved down-ward track was made by the recoiling electron.

Each of the four fundamental forces is mediated by fields of its own. Hence there are gravitational fields, electromagnetic fields, strong fields, and weak fields. According to quantum theory, the energy stored in fields is not smoothly distributed; rather the energy is found in ***quanta,*** that is, small packets or lumps of energy.

In Chapter 28, we became acquainted with the quanta of the electromagnetic field; these quanta are the photons. Each of the other funda-

TABLE 32.6	FIELDS AND THEIR QUANTA	
Field	Quanta	Mass
Gravitational	Gravitons	0 MeV/c^2
Weak	W particles	80,000
	Z particles	91,000
Electromagnetic	Photons	0
Strong	Pions and some other mesons	140

mental fields also has quanta of its own. Table 32.6 lists the quanta of all the four fundamental fields. Like photons, the quanta of the gravitational field, or gravitons, are massless. In contrast, the quarta of the strong field, or pions, and the quanta of the weak field, or W and Z particles, are endowed with mass.

At the quantum level, we can picture the field of force generated by a particle as a swarm of quanta buzzing around the particle. For example, we can picture the electric field surrounding an electron, or any other charged particle, as a swarm of photons. The swarm is in a state of everlasting activity — the charged particle continually emits and reabsorbs the photons of the swarm. Emission is creation of a photon; absorption is annihilation of a photon. Hence we can say that the electric field arises from the continual interplay of three fundamental processes: creation, propagation, and annihilation of photons. The action of one charged particle on another involves a sequence of these three fundamental processes: a photon is emitted by one particle, propagates through the intervening distance, and is absorbed by the other particle. This exchange process can be represented graphically by a ***Feynman diagram*** (Fig. 32.9), invented by R. Feynman. The photon exchanged between the two electrons is called a ***virtual photon,*** because it lasts only a very short time and, being reabsorbed by an electron, is undetectable by any direct experiment. ***The steady attractive or repulsive force between two charged particles is generated by continual repetition of this photon exchange process.*** This is action-by-contact with a vengeance — at a fundamental level, all forces reduce to local acts of creation and destruction involving particles in direct contact.

In terms of a simple analogy, we can easily understand how the exchange of particles brings about forces. Imagine two boys tossing a ball back and forth between them (Fig. 32.10); it is intuitively obvious that this produces a net repulsive force between the boys due to the recoil they suffer when throwing or catching the ball. Our intuition suggests that no such exchange process can ever produce attraction. However, imagine

FIGURE 32.10

Two boys throw a ball back and forth.

Richard Phillips Feynman, *1918–1988, American physicist. His invention of the Feynman diagram revolutionized the computations of relativistic quantum processes. He shared the 1965 Nobel Prize with* **Julian Schwinger,** *1918–, American physicist, and* **Sin-Itiro Tomonaga,** *1906–1979, Japanese physicist, for work on the quantum theory of electrons.*

◀ **Feynman diagram**

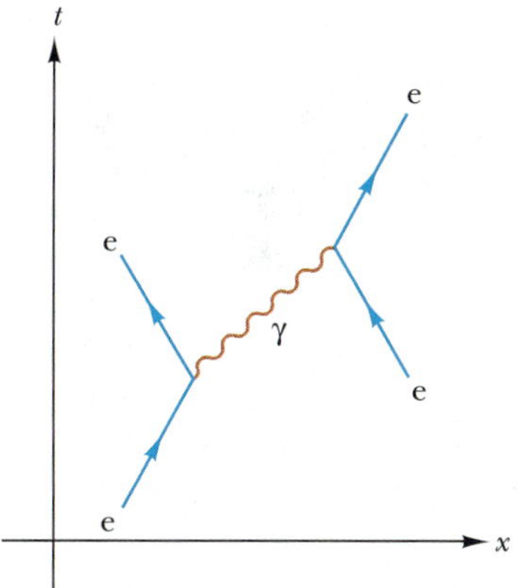

FIGURE 32.9

Feynman diagram representing the exchange of a virtual photon between two electrons. The solid blue lines indicate the motion of the two electrons, with the t axis plotted vertically and the x axis plotted horizontally. The wavy colored line indicates the motion of the photon. The electron on the left emits this photon and the electron on the right absorbs it.

FIGURE 32.11
Two boys throw a boomerang back and forth.

two Australian boys tossing a boomerang back and forth between them (Fig. 32.11); it is then obvious that this produces an attractive force between the boys. Whether a photon exchanged between two charges behaves like a ball or like a boomerang depends on the signs of the charges. Quantum calculations, which take into account the wave nature of all the particles involved, show that the net force is attractive for unlike charges and repulsive for like charges, as it should be.

The gravitational, weak, and strong forces are also generated by the exchange of virtual particles: gravitons, W and Z particles, and pions (and some other mesons). It is a general rule of quantum theory that the range of the force is inversely related to the mass of the particle that serves as the carrier of the force. Thus, the photons and the gravitons that are the carriers of the electromagnetic and the gravitational force have zero mass — the ranges of these forces are infinite. The pions that are the carriers of the strong force have a fairly large mass — the range of this force is short. The W and Z particles that are the carriers of the weak force have a very large mass — the range of this force is very short.

Unified field theory ▶

Although the four fundamental forces seem drastically different, theoretical physicists have sought to formulate a unified theory that treats several or all of these forces as aspects of a single, more fundamental force. Electromagnetism is the most familiar example of a ***unified field theory,*** that is, a theory that treats electric and magnetic forces as two aspects of a single, underlying force. To appreciate fully the unification of electricity and magnetism we would need to examine what relativity has to say about electric and magnetic fields; we could then see that electric and magnetic forces are merely two aspects of a single force called the electromagnetic force, just as space and time are two aspects of a single entity called spacetime. But even without adopting a relativistic point of view, we can see from Maxwell's equations that electricity and magnetism are intimately connected.

Electroweak force ▶

If we seek to unify the weak and electromagnetic forces, we must regard the carriers of these forces — the quanta whose exchange generates the forces — as closely related. This would seem to conflict with the large mass difference between these particles: the photon is massless, but the W and Z particles are the heaviest particles known. The unified theory of the ***electroweak force*** formulated by S. Weinberg, A. Salam, and S. Glashow attributes this mass difference to an imperfect symmetry (a "broken symmetry") between the photon and the W and Z particles. The theory asserts that perfect symmetry between these particles can be restored by giving them very high energies, in excess of 100 GeV; at such high energies, the photon and the W and Z particles should become essentially identical. Such high energies are difficult to achieve in our laboratories, but they were readily available during the very early stages of the Big Bang, when the Universe was younger than 10^{-10} s and had a temperature in excess of 10^{15} K. It is believed that at these early times, there was no difference between the photon and the W and Z particles, and there was no difference between the electromagnetic and weak forces.

The most impressive success of the unified theory was its prediction of the masses of the W and Z particles. The W and Z particles were detected in 1982 in experiments at the proton–antiproton collider at CERN. The experiments involved the observation of about a billion head-on collisions between protons and antiprotons of the same energy, 270 GeV. A few dozen W and Z particles were produced in these collisions (Fig. 32.12). The measured masses of the W and the Z particles are, respectively, 80 GeV/c^2 and 91 GeV/c^2. These measured values are within a fraction of a GeV/c^2 of the theoretically predicted values. This excellent agreement constitutes a brilliant confirmation of the unified theory of weak and electromagnetic interactions.

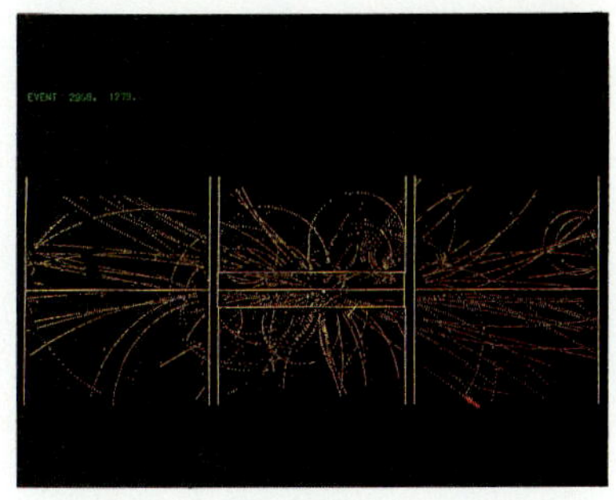

FIGURE 32.12

Track of particles produced in a very energetic head-on collision between a proton of 270 GeV that entered from the right and an antiproton of 270 GeV that entered from the left. In this collision a W^- particle was created. It immediately decayed into an electron and a neutrino. The track of the electron (red) emerges at the lower right. The track of the neutrino is not visible.

32.5 Quarks

Let us now return to our initial question. What are the ultimate, indivisible building blocks of matter? We know of about 300 particles. It is inconceivable that all these hundreds of particles are *elementary* particles. It is likely that most of them, or maybe all of them, are composite particles made of just a few truly elementary building blocks.

To discover these building blocks, physicists have tried to break protons into pieces by bombarding them with projectiles of very high energy. Unfortunately, if the energy of the projectile is large enough to make a dent in a proton, then it is also large enough to create new particles during the collision. This abundant creation of particles confuses the issue — we can never be quite sure which of the pieces that come flying out of the scene of the collision are newly created particles and which are fragments of the original proton. In fact, none of the pieces ever found in such collision experiments seems a likely candidate for an elementary building block. Typically, the particles that emerge from a collision between a high-energy projectile and a proton are pions, kaons, lambdas, deltas, and so on, all of which seem to be even less elementary than a proton.

Although brute-force collision experiments have failed to fragment protons into elementary building blocks, somewhat more subtle experiments have provided us with some evidence that distinct building blocks do indeed exist inside protons. At the Stanford Linear Accelerator (SLAC), very high-energy electrons were shot at protons; these electrons served as probes to "feel" the interior of the protons. The experiments showed that occasionally the bombarding electrons were deflected through large angles, bouncing off sharply from the interior of a proton. These deflections indicate the presence of some lumps or hard kernels in the interior of the proton, just as, in Rutherford's experiments, the large deflections of alpha particles by atoms indicated the presence of a hard kernel (nucleus) in the interior of the atom. Protons and all the other baryons and mesons seem to be composite bodies made of several distinct pieces. In contrast, electrons and the other leptons seem to be indivisible bodies with no internal structure. In recent experiments the electron was probed with beams of extremely energetic particles to within 10^{-18} m of its center. Even at these extremely short distances, no substructures of any kind were found. Thus, the electron seems to be a pointlike body with no size at all, a truly elementary particle.

Even before the experimental evidence for lumps inside protons became available, theoretical physicists had noticed that particles could be classified into groups or families of similar particles on the basis of their quantum numbers and their behavior in reactions. To explain these similarities, they had proposed theories in which all the known baryons and mesons are regarded as constructed out of a few fundamental building blocks. According to these theories, the similarity between particles in a given family reflects the similarity of their internal construction, just as the similarities between atoms in a group of the Periodic Table reflect the similarity of their internal construction.

The most successful of these theories is the quark model proposed by M. Gell-Mann and by G. Zweig. In this model, ***all particles are constructed of three kinds of fundamental building blocks called quarks.*** (Gell-Mann took the word *quark* from *Finnegans Wake,* a book by James Joyce.) The three quarks are usually labeled ***up, down,*** and ***strange,*** or simply u, d, and s. They all have spin $\frac{1}{2}$ and electric charges of $\frac{2}{3}$, $-\frac{1}{3}$, and $-\frac{1}{3}$, respectively (see Table 32.7). Of course, each quark — like any other particle — has an antiparticle, of opposite electric charge.

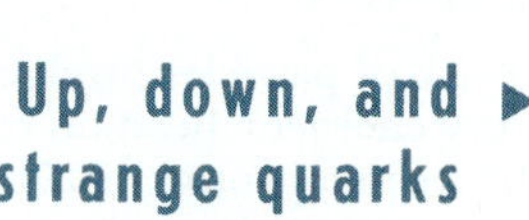

TABLE 32.7 THREE QUARKS

Quark	Mass[a]	Spin	Electric charge	Strangeness
u	5 MeV/c^2	$\frac{1}{2}$	$\frac{2}{3}$	0
d	10	$\frac{1}{2}$	$-\frac{1}{3}$	0
s	200	$\frac{1}{2}$	$-\frac{1}{3}$	+1

[a] Based on theoretical estimates.

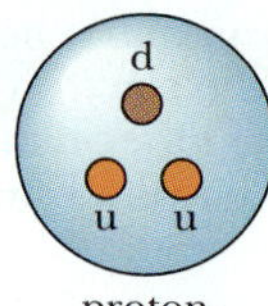

FIGURE 32.13 Structure of the proton: two u quarks and one d quark. The sizes of the quarks are not drawn to scale. Their sizes are probably much smaller than the size of the proton.

To make the ordinary particles out of the quarks, the latter must be glued together in diverse ways. For example, a proton is made of two u quarks and one d quark (Fig. 32.13). A neutron is made of two d quarks and one u quark (see Fig. 32.14). A positive pion is made of one u quark and one d antiquark (Fig. 32.15), and so on (Tables 32.2 and 32.3 list the quark composition of baryons and mesons). By gluing quarks together, we can build up all the known baryons and mesons and explain their quantum numbers and their similarities. Besides, we can predict some properties, such as magnetic moments and reaction rates, of the composite particles from the (assumed) properties of the quarks. The experimental confirmation of these predictions is strong circumstantial evidence for the quark model.

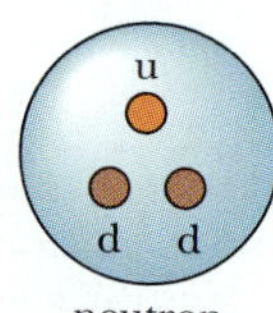

FIGURE 32.14 Structure of the neuton: two d quarks and one u quark.

There is only one snag: in spite of prodigious efforts, experimental searches for quarks have been unsuccessful. Physicists now suspect that quarks are permanently confined inside the ordinary particles so that there is no way to break a quark out of, say, a proton. Although the details of the confinement mechanism are still unclear, it seems that the quarks are held in place by an exceptionally strong force, which prevents their escape. This new force is called the ***"color" force.***

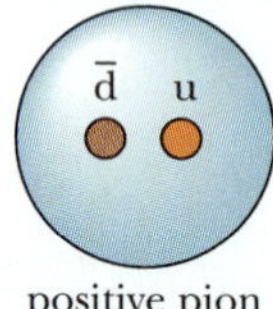

FIGURE 32.15 Structure of a positive pion (π^+): one u quark and one d antiquark.

The concept of "color" was first introduced to remedy an unacceptable violation of the Exclusion Principle. Quarks, like electrons and other particles of half-integer spin, ought to obey the Exclusion Principle. But investigations of the quantum states of quarks within protons, neutrons, sigmas, etc., disclosed that several apparently identical quarks were often found in the *same* quantum state. To avoid this apparent violation of the Exclusion Principle, physicists postulated that each of the quarks exists in three varieties, and that whenever two apparently identical quarks are found in the same quantum state, they actually are of different varieties. The varieties of quarks are characterized by a new property called ***color.*** Of course, this "color" has nothing to do with real color; it is merely a (somewhat unimaginative) name for a new property of matter. The different quark colors are ***red, green,*** and ***blue.*** Thus, there is a *red* u quark, a *green* u quark, and a *blue* u quark, and so on. The antiquarks have anticolors; the different antiquark colors are anti*red,* anti*green,* and anti*blue.*

◀ **"Colors" of quarks**

Color is a very subtle property of matter; it usually remains hidden inside the ordinary particles. ***All the normal particles are "colorless"*** — they consist of several quarks with an equal mixture of all three colors. For instance, one of the three quarks inside the proton is *red,* one is *green,* and one is *blue.* Nevertheless, color plays a crucial role in the theory of the forces that confine the quarks inside the ordinary particles.

The quarks are confined by extremely strong mutual attractive forces. These forces between quarks are called color forces because the source of these forces is color just as the source of electric forces is electric charge. Each of the three varieties of color (*red, green,* and *blue*) is analogous to a kind of electric charge. A body is color neutral, or "colorless," if it contains equal amounts of all three colors or if it contains equal amounts of color and anticolor, just as a body is electrically neutral if it contains equal amounts of positive and negative charge.

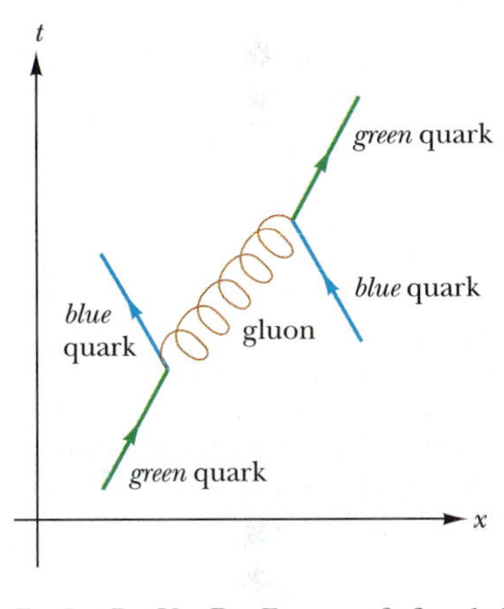

FIGURE 32.16

Exchange of a gluon between two quarks. Note that the quark on the left suffers a color change from *green* to *blue;* that on the right a color change from *blue* to *green.*

The color force is a fundamental force that should be included in our table of fundamental forces instead of the strong force (Table 32.4). The color force is closely related to the strong force — the latter is actually a special instance of the former. The relationship between the color force and the strong force is analogous to the relationship between the electric force and the intermolecular force. As we saw in Section 16.1, the force between two electrically neutral atoms or molecules is a residual electric force resulting from an imperfect cancellation among the attractions and repulsions of the charges in the two atoms or molecules. Likewise, the strong force between, say, two "colorless" protons is a residual color force resulting from an imperfect cancellation among the attractions and repulsions of the quarks in the two protons. Thus, the "strong" force between protons is no more than a pale reflection of the much stronger color forces acting within each proton.

At a fundamental level, the color force between quarks is due to an exchange of virtual particles between the quarks. The particle that acts as the carrier of the color force is the ***gluon.*** Figure 32.16 shows a Feynman diagram representing the exchange of a gluon between two quarks. Such an exchange of a gluon between two colored quarks is analogous to the exchange of a photon between two charged particles (see Fig. 32.9).

◀ **Gluon**

Another modification of the simple quark model with three quarks arose from the theory of the unification of the electromagnetic and weak forces. In order to make this theory fit some experimental data, physicists had to postulate the existence of a fourth quark, different from the u, d, and s quarks. This new hypothetical quark was labeled ***charmed.***

◀ **Charmed quark**

The hypothesis of the charmed quark soon received firm experimental support. In 1974, teams of experimenters at the Brookhaven accelerator and at the Stanford accelerator discovered the J/ψ meson and some other related mesons (see Table 32.3), all of which have exceptionally long lifetimes. It seems that these particles contain charmed quarks with an electric charge of $\frac{2}{3}$ and an (estimated) mass of 1500 MeV/c^2.

Unfortunately, the proliferation of quarks did not stop with four quarks. In 1977 a team of experimenters at Fermilab discovered the Υ mesons. These are by far the most massive mesons known (see Table 32.3). It seems that each of these contains a new and very massive quark.

Bottom and top quarks ▶

This fifth kind of quark has been labeled ***bottom.*** Theoretical considerations suggest that there should also exist a sixth quark, labeled ***top.*** But the existence of this sixth quark has not yet been confirmed experimentally.

Each of the six quarks (up, down, strange, charmed, top, bottom) comes in three varieties of color (*red, green, blue*); furthermore, for each quark there exists an antiquark, which comes in three varieties of "anticolor" (anti*red,* anti*blue,* and anti*green*). Altogether, this amounts to 36 quarks — and the proliferation of quarks will apparently not end with this. Some new theories postulate the existence of several dozen quarks (plus their color varieties plus their antiquarks). This continuing proliferation raises the question of whether matter really has such a large number of elementary building blocks. Pushing forward our search for the ultimate building blocks, we have uncovered layers of structures within layers of structure — nuclei within atoms, protons and neutrons within nuclei, and quarks within protons and neutrons. Is there another layer within quarks?

SUMMARY

▶ **Particles:**

- Leptons
- Baryons } hadrons
- Mesons } hadrons

▶ **Interactions:**

	Strengths:
Strong	1
Electromagnetic	10^{-2}
Weak	10^{-6}
Gravitational	10^{-38}

▶ **Conserved quantities:**

Absolute:	Approximate:
Energy	Isospin
Momentum	Strangeness
Angular momentum	Parity
Electric charge	
Baryon number	
Lepton number	

▶ **Quarks:** u, d, s, c, b, t

QUESTIONS

1.–Why are high-energy accelerators necessary for the production and discovery of new, massive particles?

2.–Physicists are planning to construct the SSC, a 20-TeV accelerator, which will cost some 4 billion dollars. Can such an expenditure be justified?

3.–Neutrons do not make tracks in a bubble chamber. Why not?

4.–Explain why some of the particle tracks in Fig. 32.5 curve upward and some downward.

5.–The names *baryon, meson,* and *lepton* come from the Greek *barys* (heavy), *mesos* (middle), and *leptos* (thin, slender). These names were originally intended to indicate the masses of the particles. According to the lists of particles and masses given in this chapter, is it true that the baryons have the largest masses and the leptons the smallest?

6.–How does the antiproton differ from the proton? The antineutron from the neutron?

7.–In principle, an antielectron and an antiproton can be put together to form an antihydrogen atom, and two of these atoms can form an antihydrogen molecule. Could we confine a sample of antihydrogen gas in an ordinary steel bottle? Can you think of any way of confining it?

8.–How would you refute the proposition that the Sun is made of antimatter?

9.–Why does a particle that lives only 10^{-23} s not make a track in a bubble chamber? (Hint: Suppose the particle moves at the maximum conceivable speed; how far will it travel in 10^{-23} s?)

10.–The strengths of the fundamental forces depend on the energies of the particles. In the case of the gravitational force, the strength increases with the energy. Why would you expect this to be true?

11.–The boomerang analogy described in Fig. 32.11 is defective in that the boomerang requires the presence of air. What would be the motion of a boomerang in vacuum?

12.–In Fig. 32.7, a large number of particles emerge in the longitudinal direction (toward the right and the left). Why is this expected, whereas the emergence of particles in the transverse direction (upward and downward) is surprising? (Hint: Consider a head-on collision between two aircraft; which way do you expect most fragments to spurt out?)

13.–According to some recent speculations, quarks are made of smaller constituents variously called prequarks, preons, or rishons (from the Hebrew word for first or primary). Can you think of a better name for the constituents of the quarks?

PROBLEMS

Section 32.1: The Tools of High-Energy Physics

——According to Eq. (21.27), the radius of the orbit of a charged particle of momentum p in a magnetic field is

$$r = \frac{p}{qB}$$

Expressed in this way, in terms of the momentum, this formula remains valid even if the particle is relativistic (although $p \neq mv$ for a relativistic particle).

(a) Show that for an ultrarelativistic particle the formula becomes

$$r = \frac{E}{qcB}$$

(b) At the CERN accelerator, protons of energy 450 GeV travel in a circular orbit of radius 1.1 km. Calculate the strength of the magnetic field required to achieve this orbital radius.

2.—The planned SSC accelerator will have a radius of 14 km and will produce protons of momentum 20 TeV/c, or 1.1×10^{-14} kg · m/s.

(a) What magnetic field is required to hold the protons in a circular orbit of this radius? (Hint: Use the formula given in Problem 1.)

(b) What is the period of the orbital motion? (Hint: The speed of the protons is nearly equal to the speed of light.)

3.—The relativistic formulas $\sqrt{2mc^2(2mc^2 + K)}$ gives the energy available for inelastic reactions when a particle of mass m and kinetic energy K is incident on a stationary particle of the same mass. Use this formula to calculate the available energy (in GeV) for an antiproton incident on a stationary proton for the following cases:

(a) The kinetic energy of the incident antiproton is 10 MeV.

(b) The kinetic energy of the incident antiproton is 1 TeV (as in the Tevatron).

(c) The kinetic energy of the incident antiproton is 20 TeV (as in the proposed SCC).

4.*—Show that for a particle of (total) energy E, the time-dilation factor can be expressed as E/mc^2. What is the time-dilation factor for a muon of energy 950 MeV? The lifetime of this muon is 2.2×10^{-6} s in its own rest frame. What is its lifetime in the laboratory reference frame?

5.*—Suppose we want to produce a Σ^+ particle by the head-on collision of two protons of equal energies according to the reaction

$$\mathrm{p} + \mathrm{p} \rightarrow \mathrm{p} + \mathrm{K}^0 + \Sigma^+$$

What is the minimum kinetic energy required for each proton to initiate this reaction? The masses of the particles are given in Tables 32.2 and 32.3.

6.*—In 1908, a meteoroid struck near Tunguska, in Siberia. The destructive effects of this impact have been estimated as equivalent to the explosion of 12.5 megatons of TNT, or 5.3×10^{16} J. One possible explanation for this explosion is that the meteoroid was made of antimatter, which annihilated with an equal amount of matter when it came in contact with the Earth's surface. If so, how much antimatter would account for the explosion?

7.*—The Ω^- particle decays in two alternative ways:

$$\Omega^- \rightarrow \Lambda + \mathrm{K}^-$$

and

$$\Omega^- \rightarrow \Xi^0 + \pi^-$$

Which of these reactions releases the most energy and gives the decay products the largest kinetic energy?

8.*—A Σ^- particle at rest decays into a pion and a neutron,

$$\Sigma^- \rightarrow \pi^- + n$$

What is the net kinetic energy of the decay products?

9.*—Consider the annihilation of an electron and an antielectron, resulting in two gamma rays,

$$e + \bar{e} \rightarrow \gamma + \gamma$$

Suppose that the electron and antielectron are initially at rest. What are the energies of the resulting gamma rays? What are their wavelengths?

10.*—The π^0 meson decays into two gamma rays,

$$\pi^0 \rightarrow \gamma + \gamma$$

If the pion is initially at rest, what are the energies of the two gamma rays? What are their wavelengths?

11.**—A K^0 particle at rest decays into two pions,

$$K^0 \rightarrow \pi^0 + \pi^0$$

What is the kinetic energy of each of these pions? What is the momentum of each? (Hint: Use the relativistic relation between energy and momentum.)

Section 32.2: The Multitude of Particles

12.—Which is the most massive particle listed in the tables of Section 32.2? Express the mass of this particle in atomic mass units and compare the mass with that of the helium atom and that of the lithium atom.

13.—Count the number of particles (including antiparticles) in the table of leptons.

Section 32.3: Interactions and Conservation Laws

14.—Verify that the reaction (2) conserves baryon number. Does the reaction conserve strangeness?

15.—Which of the following reactions are forbidden by an absolute conservation law?

$$\pi^+ + p \rightarrow \Lambda + K^0$$

$$K^- + p \rightarrow K^- + p + \pi^0$$

$$\pi^- + n \rightarrow \pi^- + \pi^0 + \pi^0$$

$$K^- + n \rightarrow \Sigma^- + \pi^0$$

$$e + \nu \rightarrow \pi^- + \pi^0$$

16.—Is strangeness conserved in the following reactions?

$$\pi^+ + \text{n} \rightarrow \text{K}^+ + \Lambda$$

$$\Lambda \rightarrow \text{p} + \pi^-$$

$$\text{K}^0 \rightarrow \pi^+ + \pi^-$$

17.—Show that the emission of a photon by a free electron is impossible because it conflicts with energy conservation. (Hint: Consider the emission process in the reference frame in which the electron is initially at rest.)

18.—Show that the annihilation of an electron and an antielectron into one single photon ($\text{e} + \bar{\text{e}} \rightarrow \gamma$) is impossible, because it conflicts with conservation of momentum. (Hint: Consider the reaction from the reference frame in which the two electrons have opposite velocities of equal magnitudes.)

19.—Consider the following reactions produced by a beam of K^- particles in a bubble chamber filled with liquid hydrogen:

$$\text{K}^- + \text{p} \rightarrow \Sigma^+ + \pi^-$$

$$\text{K}^- + \text{p} \rightarrow \Sigma^0 + \pi^0$$

$$\text{K}^- + \text{p} \rightarrow \Lambda + \pi^0$$

$$\text{K}^- + \text{p} \rightarrow \Lambda + \pi^+ + \pi^-$$

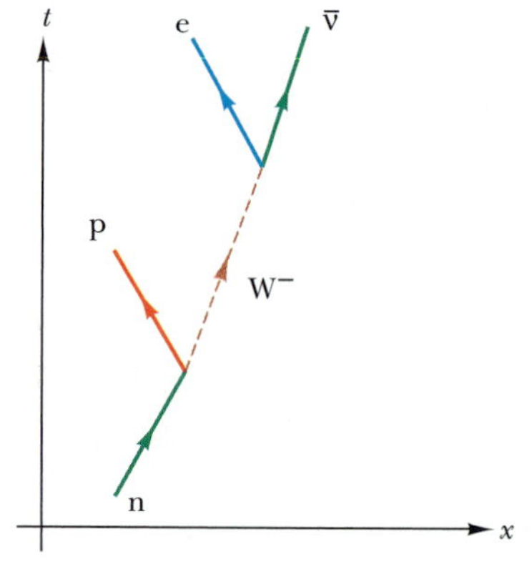

FIGURE 32.17
Feynman diagram for decay of a neutron.

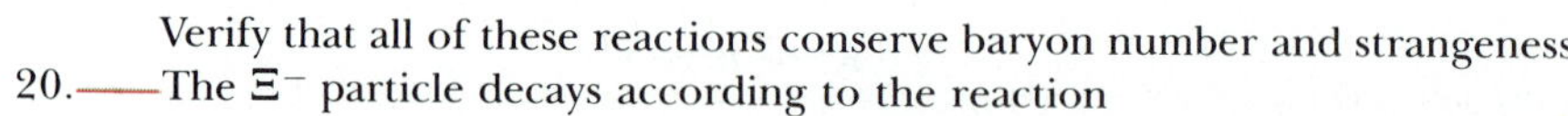

Verify that all of these reactions conserve baryon number and strangeness.

20.—The Ξ^- particle decays according to the reaction

$$\Xi^- \rightarrow \Lambda + \pi^-$$

Does this reaction conserve baryon number? Strangeness?

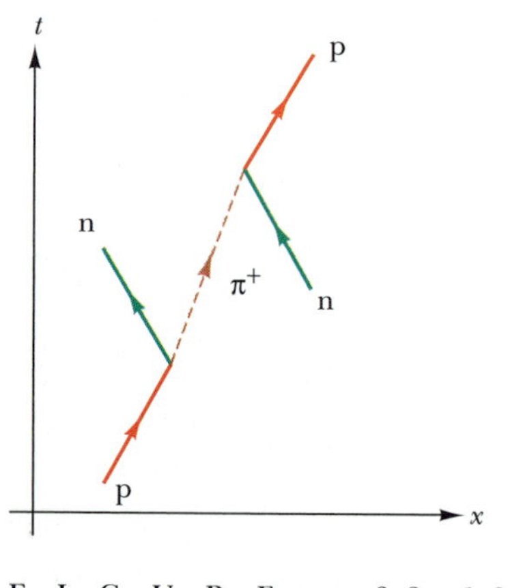

FIGURE 32.18
Feynman diagram for exchange of a π^+ between a proton and a neutron.

Section 32.4: Fields and Quanta

21.—The W particle can have either a positive charge (W^+) or a negative charge (W^-). Figure 32.17 shows the Feynman diagram for the decay of the neutron (n) via exchange of a W^-; the end products are a proton (p), an electron (e), and an antineutrino ($\bar{\nu}$). Can you guess the Feynman diagram for the decay of the antineutron?

22.—Figure 32.18 shows the Feynman diagram for the exchange of a π^+ between a proton and a neutron; note that the proton changes into a neutron, and vice versa, so electric charge is conserved at each vertex. Draw corresponding diagrams for the exchange of a π^- and of a π^0.

23.*—According to a speculative theory of the strong interactions, the proton should be unstable and decay with a lifetime of about 10^{33} years. Consider the protons in a mass of 10^6 kg of water. How many of these protons would decay in one year?

24.*—In the hot, early Universe, at a temperature of 10^{15} K, what was the average kinetic energy of the random thermal motion of gas particles? Compare this energy with the energy of about 100 GeV required to achieve symmetry between W particles and photons.

Section 32.5: Quarks

25.—Table 32.2 lists the quark composition of baryons. For all the Σ and the Ξ particles, verify that the listed quark composition gives the correct values of electric charge, baryon number, and strangeness.

26.——The antiproton ($\bar{p}$) is made of three quarks. What kinds of quarks are these?

27.——How many quarks are there in a hydrogen atom? In a water molecule? (The oxygen nucleus contains eight protons and eight neutrons.)

28.——How many quarks are created in the reaction (1)?

29.——A particle is made of one d quark and one u antiquark. What is the electric charge of this particle? What is this particle?

30.——According to a theoretical prediction based on the quark model, the masses of the nucleon (proton or neutron), the Ξ, the Λ, and the Σ should be related by

$$(m_N + m_\Xi) = \frac{1}{2}(3m_\Lambda + m_\Sigma)$$

Since this is intended as an approximate relation, the mass differences between the positive, negative, and neutral kinds of nucleon, Ξ, or Σ are to be neglected. Check this relation against the experimental values of the masses.

31.*——As described in Section 32.5, high-energy electrons produced by the Stanford Linear Accelerator were used to probe the internal structure of protons. To detect small lumps in the proton, the wavelength of the electrons must be smaller than the lumps. The electrons had an energy of 20 GeV (which is an ultrarelativistic energy). What is the de Broglie wavelength, $\lambda = h/p$, of these electrons?

APPENDICES

Appendix 1: Index to Tables

Appendix 2: Brief Review of Mathematics

A2.1 SYMBOLS

$a = b$ means a equals b
$a \neq b$ means a is not equal to b
$a > b$ means a is greater than b
$a < b$ means a is less than b
$a \geq b$ means a is not less than b
$a \leq b$ means a is not greater than b
$a \propto b$ means a is proportional to b
$a \simeq b$ means a is approximately equal to b
$a \gg b$ means a is much larger than b
$a \ll b$ means a is much less than b
$\pi = 3.14159\ldots$

A2.2 POWERS AND ROOTS

For any number a, the nth ***power*** of the number is the number multiplied into itself n times. This is written as a^n, and n is called the ***exponent.*** Thus,

$$a^1 = a, \quad a^2 = a \cdot a, \quad a^3 = a \cdot a \cdot a, \quad a^4 = a \cdot a \cdot a \cdot a, \text{ etc.}$$

For instance,

$$3^2 = 3 \times 3 = 9,\ 3^3 = 3 \times 3 \times 3 = 27,\ 3^4 = 3 \times 3 \times 3 \times 3 = 81, \text{ etc.}$$

A negative exponent indicates that the number is to be divided n times into 1; thus

$$a^{-1} = \frac{1}{a}, \quad a^{-2} = \frac{1}{a^2}, \quad a^{-3} = \frac{1}{a^3}, \text{ etc.}$$

A zero exponent yields 1, regardless of the value of a,

$$a^0 = 1$$

The rules for the combination of exponents in products, in ratios, and in powers of powers are

$$a^n \cdot a^m = a^{n+m}$$

$$\frac{a^n}{a^m} = a^{n-m}$$

$$(a^n)^m = a^{n \cdot m}$$

For instance, it is easy to verify that

$$3^2 \times 3^3 = 3^5$$

$$\frac{3^2}{3^3} = 3^{-1} = \frac{1}{3}$$

$$(3^2)^3 = 3^{2 \times 3} = 3^6$$

Note that for any two numbers a and b

$$(a \cdot b)^n = a^n \cdot b^n$$

For instance,

$$(2 \times 3)^3 = 2^3 \times 3^3$$

The nth ***root*** of a is a number such that its nth power equals a. The nth root is written $a^{1/n}$. The second root $a^{1/2}$ is usually called the square root, and designated by $\sqrt{a}$,

$$a^{1/2} = \sqrt{a}$$

As suggested by the notation $a^{1/n}$, roots are fractional powers, and they obey the usual rules for the combination of exponents:

$$(a^{1/n})^n = a^{n/n} = a$$

$$(a^{1/n})^m = a^{m/n}$$

A2.3 ARITHMETIC IN SCIENTIFIC NOTATION

The scientific notation for numbers (see page P-1 of Prelude) is quite handy for the multiplication and the division of very large or very small numbers, because we can deal with the decimal parts and the power-of-ten parts in the numbers separately. For example, to multiply 4×10^{10} by 5×10^{12}, we multiply 4 by 5 and 10^{10} by 10^{12}, as follows:

$$(4 \times 10^{10}) \times (5 \times 10^{12}) = (4 \times 5) \times (10^{10} \times 10^{12})$$

$$= 20 \times 10^{10+12} = 20 \times 10^{22} = 2 \times 10^{23}$$

To divide these numbers, we proceed likewise:

$$\frac{4 \times 10^{10}}{5 \times 10^{12}} = \frac{4}{5} \times \frac{10^{10}}{10^{12}} = 0.8 \times 10^{10-12} = 0.8 \times 10^{-2} = 8 \times 10^{-3}$$

When performing additions or subtractions of numbers in scientific notation, we must be careful to begin by expressing the numbers with the *same* power of ten. For example, the sum of 1.5×10^9 and 3×10^8 is

$$1.5 \times 10^9 + 3 \times 10^8 = 1.5 \times 10^9 + 0.3 \times 10^9 = 1.8 \times 10^9$$

A2.4 ALGEBRA

An equation is a mathematical statement that tells us that one quantity or a combination of quantities is equal to another quantity or combination. We often have to solve for one of the quantities in the equation in terms of the other quantities. For instance, we may have to solve the equation

$$x + a = b$$

for x in terms of a and b. Here a and b are numerical constants or mathematical expressions which are regarded as known, and x is regarded as unknown.

The rules of algebra instruct us how to manipulate equations and accomplish their solution. The three most important rules are:

1. Any equation remains valid if equal terms are added or subtracted from its left side and its right side.

This rule is useful for solving the equation $x + a = b$. We simply subtract a from both sides of this equation and find

$$x + a - a = b - a$$

that is

$$x = b - a$$

To see how this works in a concrete numerical example, consider the equation

$$x + 7 = 5$$

Subtracting 7 from both sides, we obtain

$$x = 5 - 7$$

or

$$x = -2$$

Note that given an equation of the form $x + a = b$, we may want to solve for a in terms of x and b, if x is already known from some other information but a is a mathematical quantity that is not yet known. If so, we must subtract x from both sides of the equation, and we obtain

$$a = b - x$$

Most equations in physics contain several mathematical quantities which sometimes play the role of known quantities, sometimes the role of unknown quantities, depending on circumstances. Correspondingly, we will sometimes want to solve the equation for one quantity (such as x), sometimes for another (such as a).

2. Any equation remains valid if the left and the right sides are multiplied by the same factor.

This rule is useful for solving

$$ax = b$$

We simply divide both sides by a, which yields

$$\frac{ax}{a} = \frac{b}{a}$$

or

$$x = \frac{b}{a}$$

Often it will be necessary to combine both of the above rules. For instance, to solve the equation

$$2x + 10 = 16$$

we begin by subtracting 10 from both sides, obtaining

$$2x = 16 - 10$$

or

$$2x = 6$$

And then we divide both sides by 2, with the result

$$x = \frac{6}{2}$$

or

$$x = 3$$

3. Any equation remains valid if both sides are raised to the same power.

This rule permits us to solve the equation

$$x^3 = b$$

Raising both sides to the power $\frac{1}{3}$, we find

$$(x^3)^{1/3} = b^{1/3}$$

or

$$x = b^{1/3}$$

As a final example, let us consider the equation

$$x = -\tfrac{1}{2}gt^2 + x_0$$

(as established in Chapter 1, this equation describes the vertical position x of a particle that starts at a height x_0 and falls for a time t; but the meaning of the equation need not concern us here). Suppose that we want to solve for t in terms of the other quantities in the equation. This will require the use of all our rules of algebra. First, subtract x from both sides, and then add $\frac{1}{2}gt^2$ to both sides. This leads to

$$0 = -\tfrac{1}{2}gt^2 + x_0 - x$$

and then to

$$\tfrac{1}{2}gt^2 = x_0 - x$$

Next, multiply both sides by 2 and divide both sides by g; this yields

$$t^2 = \frac{2}{g}(x_0 - x)$$

Finally, raise both sides to the power $\frac{1}{2}$, or, equivalently, extract the square root of both sides. This gives us the final result

$$t = \sqrt{\frac{2}{g}(x_0 - x)}$$

A2.5 EQUATIONS WITH TWO UNKNOWNS

If we seek to solve for two unknowns simultaneously, then we need two independent equations containing these two unknowns. The solution of such simultaneous equations can be carried out by the method of ***elimination:*** begin by using one equation to solve for the first unknown in terms of the second, then use this result to eliminate the first unknown from the other equation. An example will help to make this clear. Consider the following two simultaneous equations with two unknowns x and y:

$$4x + 2y = 8$$

$$2x - y = -2$$

To solve the first equation for x in terms of y, subtract $2y$ from both sides and then divide both sides by 4:

$$x = \frac{8 - 2y}{4}$$

Next, substitute this expression for x into the second equation:

$$2 \times \frac{8 - 2y}{4} - y = -2$$

To simplify this equation, multiply both sides by 4:

$$2 \times (8 - 2y) - 4y = -8$$

and combine the two terms containing y:

$$16 - 8y = -8$$

This is an ordinary equation for the single unknown y, and it can be solved by the methods we discussed in the preceding section, with the result

$$y = 3$$

It then follows from the above expression for x that

$$x = \frac{8 - 2y}{4} = \frac{8 - 2 \times 3}{4} = \frac{2}{4} = \frac{1}{2}$$

A2.6 LOGARITHMS

The logarithm of a (positive) number is the power to which 10 must be raised to obtain this number.* Thus, from $10 = 10^1$, and $100 = 10^2$, and $1000 = 10^3$, and $10{,}000 = 10^4$, we immediately deduce that

$$\log 10 = 1$$

$$\log 100 = 2$$

$$\log 1000 = 3$$

$$\log 10{,}000 = 4, \text{ etc.}$$

* Logarithms based on powers of ten are often called *common logarithms* to distinguish them from logarithms based on powers of other numbers, for instance, the so-called *natural logarithms* based on powers of 2.718. . . . But in this book we will use only logarithms based on powers of ten.

Likewise

$$\log 1 = 0$$

$$\log 0.1 = -1$$

$$\log 0.01 = -2$$

$$\log 0.001 = -3, \text{ etc.}$$

Obviously, the logarithm of a number between 1 and 10 is somewhere between 0 and 1, but to find the logarithm of such a number, we need the help of a table of logarithms or the help of a computer program (many electronic calculators have built-in computer programs that yield the value of the logarithm at the touch of a button).

The following list gives the logarithms of numbers from 1 to 10:

$$\begin{aligned}
\log 1 &= 0.000\\
\log 2 &= 0.301\\
\log 3 &= 0.477\\
\log 4 &= 0.602\\
\log 5 &= 0.699\\
\log 6 &= 0.778\\
\log 7 &= 0.845\\
\log 8 &= 0.903\\
\log 9 &= 0.954\\
\log 10 &= 1.000
\end{aligned}$$

The logarithm of the product of two numbers is the sum of the individual logarithms, and the logarithm of the ratio of two numbers is the difference of the individual logarithms. This rule makes it easy to find the logarithm of a number expressed in scientific notation. For example, the logarithm of 2×10^6 is

$$\log (2 \times 10^6) = \log 2 + \log 10^6 = 0.301 + 6 = 6.301$$

Note that the logarithm of any (positive) number smaller than 1 is negative. For example,

$$\log (5 \times 10^{-3}) = \log 5 + \log 10^{-3} = 0.699 - 3 = -2.301$$

Appendix 3: Perimeters, Areas, and Volumes

[perimeter of a circle of radius r] $= 2\pi r$
[area of a circle of radius r] $= \pi r^2$
[area of a triangle of base b, altitude h] $= hb/2$
[surface of a sphere of radius r] $= 4\pi r^2$
[volume of a sphere of radius r] $= 4\pi r^3/3$
[curved surface of a cylinder of radius r, height h] $= 2\pi rh$
[volume of a cylinder of radius r, height h] $= \pi r^2 h$

Appendix 4: Brief Review of Trigonometry

A4.1 Angles

The angle between two intersecting straight lines is defined as the fraction of a complete circle included between these lines (Fig. A4.1). To express the angle in ***degrees,*** we assign an angular magnitude of 360° to the complete circle; any arbitrary angle is then an appropriate fraction of 360°. To express the angle in ***radians,*** we assign an angular magnitude of 2π radian to the complete circle; any arbitrary angle is then an appropriate fraction of 2π. For example, the angle shown in Fig. A4.1 is $\frac{1}{12}$ of a complete circle, that is, 30°, or $\pi/6$ radian. In view of the definition of angle, the length of arc included between the two intersecting straight lines is proportional to the angle θ between these lines; if the angle is expressed in radians, then the constant of proportionality is simply the radius:

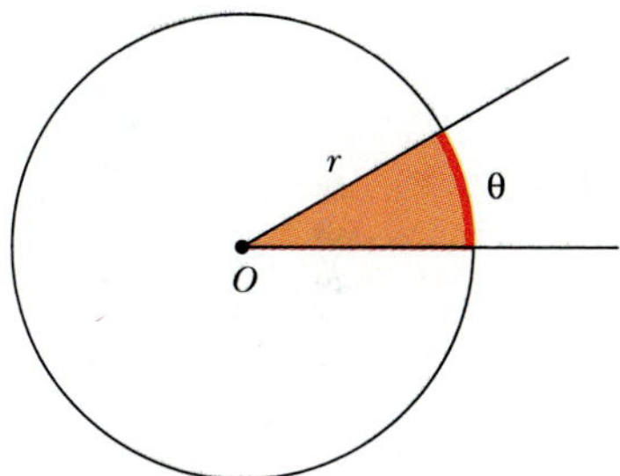

Fig. A4.1

$$s = r\theta \tag{1}$$

Since 2π radian $= 360°$, it follows that

$$1 \text{ radian} = \frac{360°}{2\pi} = \frac{360°}{2 \times 3.14159} = 57.2958° \tag{2}$$

Each degree is divided into 60 minutes of arc (arcminutes), and each of these into 60 seconds of arc (arcseconds). In degrees, minutes of arc, and seconds of arc, the radian is

$$1 \text{ radian} = 57° \ 17' \ 44.8'' \tag{3}$$

A4.2 THE TRIGONOMETRIC FUNCTIONS

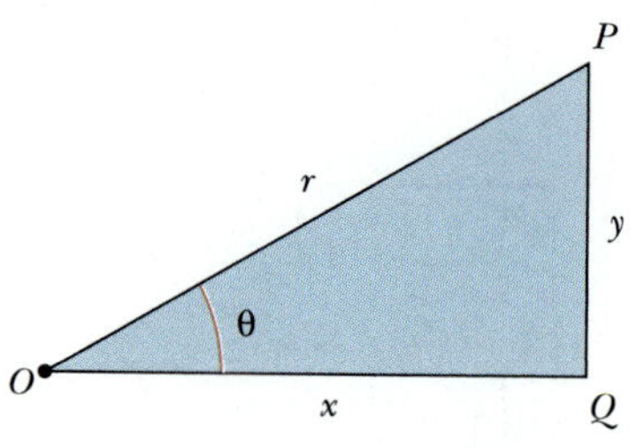

Fig. A4.2

The trigonometric functions of an angle are defined as ratios of the lengths of the sides of a right triangle erected on this angle. Figure A4.2 shows an acute angle θ and a right triangle, one of whose angles coincides with θ. The adjacent side OQ has a length x, the opposite side QP a length y, and the hypotenuse OP a length r. The ***sine, cosine, tangent, cotangent, secant,*** and ***cosecant*** of the angle θ are then defined as follows:

$$\sin\theta = y/r \tag{4}$$

$$\cos\theta = x/r \tag{5}$$

$$\tan\theta = y/x \tag{6}$$

$$\cot\theta = x/y \tag{7}$$

$$\sec\theta = r/x \tag{8}$$

$$\csc\theta = r/y \tag{9}$$

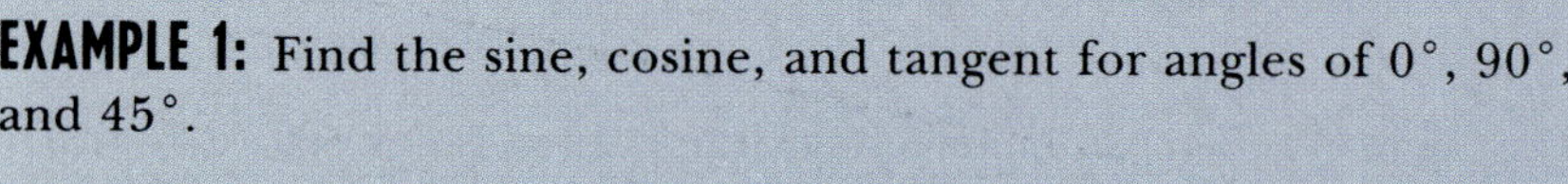

EXAMPLE 1: Find the sine, cosine, and tangent for angles of 0°, 90°, and 45°.

SOLUTION: For an angle of 0°, the opposite side is zero ($y = 0$), and the adjacent side coincides with the hypotenuse ($x = r$). Hence

$$\sin 0° = 0 \quad \cos 0° = 1 \quad \tan 0° = 0 \tag{10}$$

For an angle of 90°, the adjacent side is zero ($x = 0$), and the opposite side coincides with the hypotenuse ($y = r$). Hence

$$\sin 90° = 1 \quad \cos 90° = 0 \quad \tan 90° = \infty \tag{11}$$

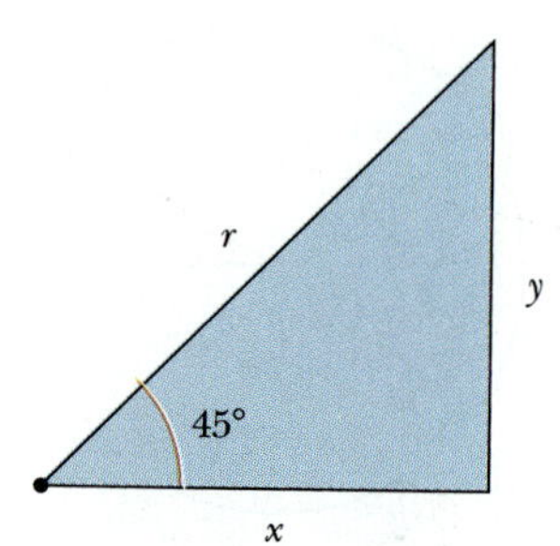

Fig. A4.3

Finally, for an angle of 45° (Fig. A4.3), the adjacent and the opposite sides have the same length ($x = y$) and the hypotenuse has a length of $\sqrt{2}$ times the length of either side ($r = \sqrt{2}x = \sqrt{2}y$). Hence

$$\sin 45° = \frac{1}{\sqrt{2}} \quad \cos 45° = \frac{1}{\sqrt{2}} \quad \tan 45° = 1 \tag{12}$$

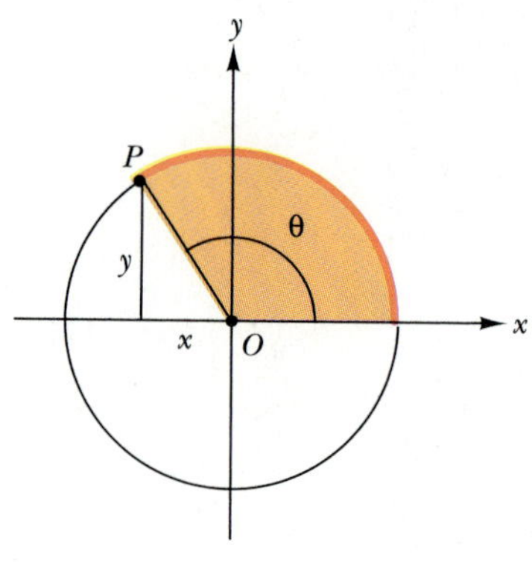

Fig. A4.4

The definitions (4)–(9) are also valid for angles greater than 90°, such as the angle shown in Fig. A4.4. In the general case, the quantities x and y must be interpreted as the rectangular coordinates of the point P. For any angle larger than 90°, one or both of the coordinates x and y are negative. Hence some of the trigonometric functions will also be negative. For instance,

$$\sin 135^\circ = \frac{1}{\sqrt{2}} \quad \cos 135^\circ = -\frac{1}{\sqrt{2}} \quad \tan 135^\circ = -1 \tag{13}$$

Figure A4.5 shows plots of the sine, cosine, and tangent vs. θ.

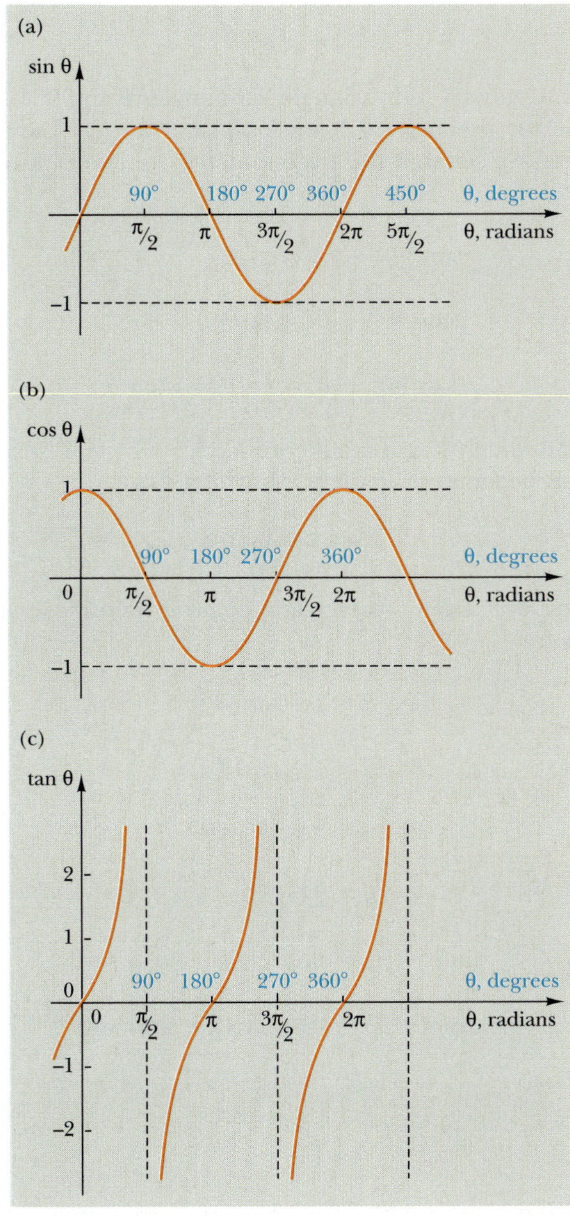

Fig. A4.5

A4.3 TRIGONOMETRIC IDENTITIES

From the definitions (4)–(9) we immediately find the following identities:

$$\tan\theta = \sin\theta/\cos\theta \tag{14}$$

$$\cot\theta = 1/\tan\theta \tag{15}$$

$$\sec\theta = 1/\cos\theta \tag{16}$$

$$\csc\theta = 1/\sin\theta \tag{17}$$

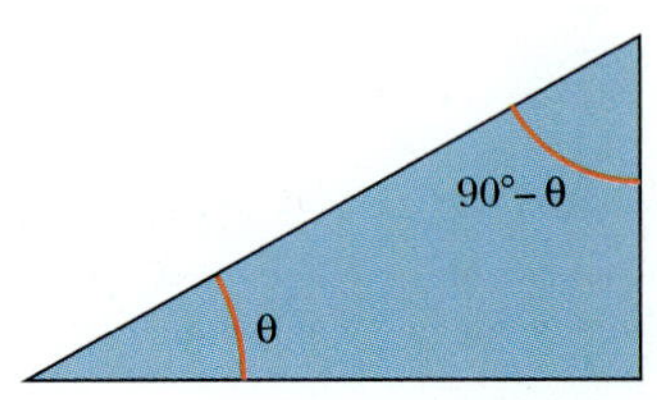

Fig. A4.6

Figure A4.6 shows a right triangle with angles θ and $90^\circ - \theta$. Since the adjacent side for the angle θ is the opposite side for the angle $90^\circ - \theta$ and vice versa, we see that the trigonometric functions also obey the following identities:

$$\sin(90^\circ - \theta) = \cos\theta \tag{18}$$

$$\cos(90^\circ - \theta) = \sin\theta \tag{19}$$

$$\tan(90^\circ - \theta) = \cot\theta = 1/\tan\theta \tag{20}$$

According to the Pythagorean theorem, $x^2 + y^2 = r^2$. With $x = r\cos\theta$ and $y = r\sin\theta$ this becomes $r^2\cos^2\theta + r^2\sin^2\theta = r^2$, or

$$\cos^2\theta + \sin^2\theta = 1 \tag{21}$$

The following are a few other trigonometric identities, which we state without proof:

$$\sec^2\theta = 1 + \tan^2\theta$$

$$\csc^2\theta = 1 + \cot^2\theta$$

$$\sin 2\theta = 2\sin\theta\cos\theta$$

$$\cos 2\theta = 2\cos^2\theta - 1$$

$$\sin(\alpha + \beta) = \sin\alpha\cos\beta + \cos\alpha\sin\beta$$

$$\cos(\alpha + \beta) = \cos\alpha\cos\beta - \sin\alpha\sin\beta$$

Appendix 5: The International System of Units (SI)

A5.1 BASE UNITS

The SI system of units is the modern version of the metric system. The SI system recognizes seven fundamental, or base, units for length, mass, time, electric current, thermodynamic temperature, amount of substance, and luminous intensity. The following definitions of the base units were adopted by the Conférence Générale des Poids et Mesures in the years indicated:

Meter (m) "The metre is the length of the path travelled by light in vacuum during a time interval of 1/299 792 458 of a second." (Adopted in 1983.)

Kilogram (kg) "The kilogram is . . . the mass of the international prototype of the kilogram." (Adopted in 1889 and in 1901.)

Second (s) "The second is the duration of 9 192 631 770 periods of the radiation corresponding to the transition between the two hyperfine levels of the ground state of the cesium-133 atom." (Adopted in 1967.)

Ampere (A) "The ampere is that constant current which, if maintained in two straight parallel conductors of infinite length, of negligible circular cross section, and placed one meter apart in vacuum, would produce between these conductors a force equal to 2×10^{-7} newton per meter of length." (Adopted in 1948.)

Kelvin (K) "The kelvin . . . is the fraction 1/273.16 of the thermodynamic temperature of the triple point of water." (Adopted in 1967.)

Mole "The mole is the amount of substance of a system which contains as many elementary entities as there are atoms in 0.012 kilogram of carbon-12." (Adopted in 1967.)

Candela (cd) "The candela is the luminous intensity, in a given direction, of a source that emits monochromatic radiation of frequency 540×10^{12} Hz and that has a radiant intensity in that direction of $\frac{1}{683}$ watt per steradian." (Adopted in 1979.)

Besides these seven base units, the SI system also recognizes two supplementary units of angle and solid angle:

Radian (rad) "The radian is the plane angle between two radii of a circle which cut off on the circumference an arc equal in length to the radius."

Steradian (sr) "The steradian is the solid angle which, having its vertex in the center of a sphere, cuts off an area equal to that of a [flat] square with sides of length equal to the radius of the sphere."

A5.2 DERIVED UNITS

The derived units are formed out of products and ratios of the base units. Table A5.1 lists those derived units that have been glorified with special names. (Other derived units are listed in the tables of conversion factors in Appendix A.6.)

TABLE A5.1 NAMES OF DERIVED UNITS

Quantity	Derived Unit	Name	Symbol
frequency	1/s	hertz	Hz
force	$kg \cdot m/s^2$	newton	N
pressure	N/m^2	pascal	Pa
energy	$N \cdot m$	joule	J
power	J/s	watt	W
electric charge	$A \cdot s$	coulomb	C
electric potential	J/C	volt	V
electric capacitance	C/V	farad	F
electric resistance	V/A	ohm	Ω
conductance	A/V	siemens	S
magnetic flux	$V \cdot s$	weber	Wb
magnetic field	$V \cdot s/m^2$	tesla	T
inductance	$V \cdot s/A$	henry	H
temperature	K	degree Celsius	°C
luminous flux	$cd \cdot sr$	lumen	lm
illuminance	$cd \cdot sr/m^2$	lux	lx
radioactivity	1/s	becquerel	Bq
absorbed dose	J/kg	gray	Gy
dose equivalent	J/kg	sievert	Sv

A5.3 PREFIXES

Multiples and submultiples of SI units are indicated by prefixes, such as the familiar *kilo, centi,* and *milli* used in *kilometer, centimeter,* and *millimeter,* etc. Table A5.2 lists all the accepted prefixes. Some enjoy more popularity than others; it is best to avoid the use of uncommon prefixes, such as *exa, peta,* and *atto,* since hardly anybody will recognize those.

TABLE A5.2 PREFIXES FOR UNITS

Multiplication Factor	Prefix	Symbol
10^{18}	exa	E
10^{15}	peta	P
10^{12}	tera	T
10^{9}	giga	G
10^{6}	mega	M
10^{3}	kilo	k
10^{2}	hecto	h
10	deka	da
10^{-1}	deci	d
10^{-2}	centi	c
10^{-3}	milli	m
10^{-6}	micro	μ
10^{-9}	nano	n
10^{-12}	pico	p
10^{-15}	femto	f
10^{-18}	atto	a

Appendix 6: Conversion Factors

The units for each quantity are listed alphabetically, except that the SI unit is always listed first. The numbers are based on "American National Standard; Metric Practice" published by the Institute of Electrical and Electronics Engineers, 1982.

Angle

1 **radian** = $57.30° = 3.438 \times 10^3\,' = \frac{1}{2\pi}$ rev = $2.063 \times 10^5\,''$
1 **degree** (°) = 1.745×10^{-2} **radian** = $60' = 3600'' = \frac{1}{360}$ rev
1 **minute of arc** (′) = 2.909×10^{-4} **radian** = $\frac{1}{60}° = 4.630 \times 10^{-5}$ rev = $60''$
1 **revolution** (rev) = 2π **radian** = $360° = 2.160 \times 10^4\,' = 1.296 \times 10^6\,''$
1 **second of arc** (″) = 4.848×10^{-6} **radian** = $\frac{1}{3600}° = \frac{1}{60}' = 7.716 \times 10^{-7}$ rev

Length

1 **meter** (m) = 1×10^{10} Å = 6.685×10^{-12} AU = 100 cm = 1×10^{15} fm = 3.281 ft = 39.37 in. = 1×10^{-3} km = 1.057×10^{-16} light-year = 1×10^6 μm = 5.400×10^{-4} nmi = 6.214×10^{-4} mi = 3.241×10^{-17} pc = 1.094 yd
1 **angstrom** (Å) = 1×10^{-10} m = 1×10^{-8} cm = 1×10^{-5} fm = 3.281×10^{-10} ft = 1×10^{-4} μm
1 **astronomical unit** (AU) = 1.496×10^{11} m = 1.496×10^{13} cm = 1.496×10^8 km = 1.581×10^{-5} light-year = 4.848×10^{-6} pc
1 **centimeter** (cm) = 0.01 m = 1×10^8 Å = 1×10^{13} fm = 3.281×10^{-2} ft = 0.3937 in. = 1×10^{-5} km = 1.057×10^{-18} light-year = 1×10^4 μm
1 **fermi** (fm) = 1×10^{-15} m = 1×10^{-13} cm = 1×10^5 Å
1 **foot** (ft) = 0.3048 m = 30.48 cm = 12 in. = 3.048×10^5 μm = 1.894×10^{-4} mi = $\frac{1}{3}$ yd
1 **inch** (in.) = 2.540×10^{-2} m = 2.54 cm = $\frac{1}{12}$ ft = 2.54×10^4 μm = $\frac{1}{36}$ yd
1 **kilometer** (km) = 1×10^3 m = 1×10^5 cm = 3.281×10^3 ft = 0.5400 nmi = 0.6214 mi = 1.094×10^3 yd
1 **light-year** = 9.461×10^{15} m = 6.324×10^4 AU = 9.461×10^{17} cm = 9.461×10^{12} km = 5.879×10^{12} mi = 0.3066 pc
1 **micron,** or **micrometer** (μm) = 1×10^{-6} m = 1×10^4 Å = 1×10^{-4} cm = 3.281×10^{-6} ft = 3.937×10^{-5} in.
1 **nautical mile** (nmi) = 1.852×10^3 m = 1.852×10^5 cm = 6.076×10^3 ft = 1.852 km = 1.151 mi
1 **statute mile** (mi) = 1.609×10^3 m = 1.609×10^5 cm = 5280 ft = 1.609 km = 0.8690 nmi = 1760 yd
1 **parsec** (pc) = 3.086×10^{16} m = 2.063×10^5 AU = 3.086×10^{18} cm = 3.086×10^{13} km = 3.262 light-years
1 **yard** (yd) = 0.9144 m = 91.44 cm = 3 ft = 36 in. = $\frac{1}{1760}$ mi

Time

1 second (s) = 1.157×10^{-5} day = $\frac{1}{3600}$ h = $\frac{1}{60}$ min = 1.161×10^{-5} sidereal day = 3.169×10^{-8} yr
1 day = 8.640×10^{4} s = 24 h = 1440 min = 1.003 sidereal days = 2.738×10^{-3} yr
1 hour (h) = 3600 s = $\frac{1}{24}$ day = 60 min = 1.141×10^{-4} yr
1 minute (min) = 60 s = 6.944×10^{-4} day = $\frac{1}{60}$ h = 1.901×10^{-6} yr
1 sidereal day = 8.616×10^{4} s = 0.9973 day = 23.93 h = 1.436×10^{3} min = 2.730×10^{-3} yr
1 year (yr) = 3.156×10^{7} s = 365.24 days = 8.766×10^{3} h = 5.259×10^{5} min = 366.24 sidereal days

Mass

1 kilogram (kg) = 6.024×10^{26} u = 5000 carats = 1.543×10^{4} grains = 1000 g = 1×10^{-3} t = 35.27 oz. = 2.205 lb = 1.102×10^{-3} short ton = 6.852×10^{-2} slug
1 atomic mass unit (u) = 1.6605×10^{-27} kg = 1.6605×10^{-24} g
1 carat = 2×10^{-4} kg = 0.2 g = 7.055×10^{-3} oz. = 4.409×10^{-4} lb
1 grain = 6.480×10^{-5} kg = 6.480×10^{-2} g = 2.286×10^{-3} oz. = $\frac{1}{7000}$ lb
1 gram (g) = 1×10^{-3} kg = 6.024×10^{23} u = 5 carats = 15.43 grains = 1×10^{-6} t = 3.527×10^{-2} oz. = 2.205×10^{-3} lb = 1.102×10^{-6} short ton = 6.852×10^{-5} slug
1 metric ton, or tonne (t) = 1×10^{3} kg = 1×10^{6} g = 2.205×10^{3} lb = 1.102 short ton = 68.52 slugs
1 ounce (oz.) = 2.835×10^{-2} kg = 141.7 carats = 437.5 grains = 28.35 g = $\frac{1}{16}$ lb
1 pound (lb)* = 0.4536 kg = 453.6 g = 4.536×10^{-4} t = 16 oz. = $\frac{1}{2000}$ short ton = 3.108×10^{-2} slug
1 short ton = 907.2 kg = 9.07×10^{5} g = 0.9072 t = 2000 lb
1 slug = 14.59 kg = 1.459×10^{4} g = 32.17 lb

Area

1 square meter (m^2) = 1×10^{4} cm^2 = 10.76 ft^2 = 1.550×10^{3} $in.^2$ = 1×10^{-6} km^2 = 3.861×10^{-7} mi^2 = 1.196 yd^2
1 barn = 1×10^{-28} m^2 = 1×10^{-24} cm^2
1 square centimeter (cm^2) = 1×10^{-4} m^2 = 1.076×10^{-3} ft^2 = 0.1550 $in.^2$ = 1×10^{-10} km^2 = 3.861×10^{-11} mi^2
1 square foot (ft^2) = 9.290×10^{-2} m^2 = 929.0 cm^2 = 144 $in.^2$ = 3.587×10^{-8} mi^2 = $\frac{1}{9}$ yd^2
1 square inch ($in.^2$) = 6.452×10^{-4} m^2 = 6.452 cm^2 = $\frac{1}{144}$ ft^2
1 square kilometer (km^2) = 1×10^{6} m^2 = 1×10^{10} cm^2 = 1.076×10^{7} ft^2 = 0.3861 mi^2
1 square statute mile (mi^2) = 2.590×10^{6} m^2 = 2.590×10^{10} cm^2 = 2.788×10^{7} ft^2 = 2.590 km^2
1 square yard (yd^2) = 0.8361 m^2 = 8.361×10^{3} cm^2 = 9 ft^2 = 1296 $in.^2$

Volume

1 cubic meter (m^3) = 1×10^6 cm^3 = 35.31 ft^3 = 264.2 gal. = 6.102×10^4 $in.^3$ = 1×10^3 liters = 1.308 yd^3
1 cubic centimeter (cm^3) = 1×10^{-6} m^3 = 3.531×10^{-5} ft^3 = 2.642×10^{-4} gal. = 6.102×10^{-2} $in.^3$ = 1×10^{-3} liter
1 cubic foot (ft^3) = 2.832×10^{-2} m^3 = 2.832×10^4 cm^3 = 7.481 gal. = 1728 $in.^3$ = 28.32 liters = $\frac{1}{27}$ yd^3
1 gallon (gal.)† = 3.785×10^{-3} m^3 = 0.1337 ft^3
1 cubic inch ($in.^3$) = 1.639×10^{-5} m^3 = 16.39 cm^3 = 5.787×10^{-4} ft^3
1 liter (l) = 1×10^{-3} m^3 = 1000 cm^3 = 3.531×10^{-2} ft^3
1 cubic yard (yd^3) = 0.7646 m^3 = 7.646×10^5 cm^3 = 27 ft^3 = 202.0 gal.

Density

1 kilogram per cubic meter (kg/m^3) = 1×10^{-3} g/cm^3 = 6.243×10^{-2} lb/ft^3 = 8.345×10^{-3} lb/gal. = 3.613×10^{-5} $lb/in.^3$ = 8.428×10^{-4} short ton/yd^3 = 1.940×10^{-3} $slug/ft^3$
1 gram per cubic centimeter (g/cm^3) = 1×10^3 kg/m^3 = 62.43 lb/ft^3 = 8.345 lb/gal. = 3.613×10^{-2} $lb/in.^3$ = 0.8428 short ton/yd^3 = 1.940 $slug/ft^3$
1 lb per cubic foot (lb/ft^3) = 16.02 kg/m^3 = 1.602×10^{-2} g/cm^3 = 0.1337 lb/gal. = 1.350×10^{-2} short ton/yd^3 = 3.108×10^{-2} $slug/ft^3$
1 pound per gallon (1 lb/gal.) = 119.8 kg/m^3 = 7.481 lb/ft^3 = 0.2325 $slug/ft^3$
1 short ton per cubic yard (short ton/yd^3) = 1.187×10^3 kg/m^3 = 74.07 lb/ft^3
1 slug per cubic foot ($slug/ft^3$) = 515.4 kg/m^3 = 0.5154 g/cm^3 = 32.17 lb/ft^3 = 4.301 lb/gal.

Speed

1 meter per second (m/s) = 100 cm/s = 3.281 ft/s = 3.600 km/h = 1.944 knot = 2.237 mi/h
1 centimeter per second (cm/s) = 0.01 m/s = 3.281×10^{-2} ft/s = 3.600×10^{-2} km/h = 1.944×10^{-2} knot = 2.237×10^{-2} mi/h
1 foot per second (ft/s) = 0.3048 m/s = 30.48 cm/s = 1.097 km/h = 0.5925 knot = 0.6818 mi/h
1 kilometer per hour (km/h) = 0.2778 m/s = 27.78 cm/s = 0.9113 ft/s = 0.5400 knot = 0.6214 mi/h
1 knot, or **nautical mile per hour** = 0.5144 m/s = 51.44 cm/s = 1.688 ft/s = 1.852 km/h = 1.151 mi/h
1 mile per hour (mi/h) = 0.4470 m/s = 44.70 cm/s = 1.467 ft/s = 1.609 km/h = 0.8690 knot

---

* This is the "avoirdupois" pound. The "troy" or "apothecary" pound is 0.3732 kg, or 0.8229 lb avoirdupois.

† This is the U.S. gallon; the U.K. and the Canadian gallon are 4.546×10^{-3} m^3, or 1.201 U.S. gallons.

Acceleration

1 meter per second squared (m/s^2) = 100 cm/s^2 = 3.281 ft/s^2 = 0.1020 gee

1 centimeter per second squared, or **Gal** (cm/s^2) = 0.01 m/s^2 = 3.281×10^{-2} ft/s^2 = 1.020×10^{-3} gee

1 foot per second squared (ft/s^2) = 0.3048 m/s^2 = 30.48 cm/s^2 = 3.108×10^{-2} gee

1 gee = 9.807 m/s^2 = 980.7 cm/s^2 = 32.17 ft/s^2

Force

1 newton (N) = 1×10^5 dynes = 0.1020 kp = 0.2248 lbf = 1.124×10^{-4} short ton-force

1 dyne = 1×10^{-5} N = 1.020×10^{-6} kp = 2.248×10^{-6} lbf = 1.124×10^{-9} short ton-force

1 kilopond, or **kilogram force** (kp) = 9.807 N = 9.807×10^5 dynes = 2.205 lbf = 1.102×10^{-3} short ton-force

1 pound-force (lbf) = 4.448 N = 4.448×10^5 dynes = 0.4536 kp = $\frac{1}{2000}$ short ton-force

1 short ton-force = 8.896×10^3 N = 8.896×10^8 dynes = 907.2 kp = 2000 lbf

Energy

1 joule (J) = 9.478×10^{-4} Btu = 0.2388 cal = 1×10^7 ergs = 6.242×10^{18} eV = 0.7376 ft · lbf = 2.778×10^{-7} kW · h

1 British thermal unit (Btu)* = 1.055×10^3 J = 252.0 cal = 1.055×10^{10} ergs = 778.2 ft · lbf = 2.931×10^{-4} kW · h

1 calorie (cal)† = 4.187 J = 3.968×10^{-3} Btu = 4.187×10^7 ergs = 3.088 ft · lbf = 1×10^{-3} kcal = 1.163×10^{-6} kW · h

1 erg = 1×10^{-7} J = 9.478×10^{-7} Btu = 2.388×10^{-8} cal = 6.242×10^{11} eV = 7.376×10^{-8} ft · lbf = 2.778×10^{-14} kW · h

1 electron-volt (eV) = 1.602×10^{-19} J = 1.602×10^{-12} erg = 1.182×10^{-19} ft · lbf

1 foot-pound-force (ft · lbf) = 1.356 J = 1.285×10^{-3} Btu = 0.3239 cal = 1.356×10^7 ergs = 8.464×10^{18} eV = 3.766×10^{-7} kW · h

1 kilocalorie (kcal), or **large calorie** (Cal) = 4.187×10^3 J = 1×10^3 cal

1 kilowatt-hour (kW · h) = 3.600×10^6 J = 3412 Btu = 8.598×10^5 cal = 3.6×10^{13} ergs = 2.655×10^6 ft · lbf

---

* This is the "International Table" Btu; there are several other Btus.

† This is the "International Table" calorie, which equals exactly 4.1868 J. There are several other calories; for instance, the thermochemical calorie, which equals 4.184 J.

‡ There are several other horsepowers; for instance, the metric horsepower, which equals 735.5 J.

Power

1 **watt** (W) = 3.412 Btu/h = 0.2388 cal/s = 1×10^7 ergs/s = 0.7376 ft · lbf/s = 1.341×10^{-3} hp

1 **British thermal unit per hour** (Btu/h) = 0.2931 W = 7.000×10^{-2} cal/s = 0.2162 ft · lbf/s = 3.930×10^{-4} hp

1 **calorie per second** (cal/s) = 4.187 W = 14.29 Btu/h = 4.187×10^7 erg/s = 3.088 ft · lbf/s = 5.615×10^{-3} hp

1 **erg per second** (erg/s) = 1×10^{-7} W = 2.388×10^{-8} cal/s = 7.376×10^{-8} ft · lbf/s = 1.341×10^{-10} hp

1 **foot-pound per second** (ft · lbf/s) = 1.356 W = 0.3238 cal/s = 4.626 Btu/h = 1.356×10^7 ergs/s = 1.818×10^{-3} hp

1 **horsepower** (hp)‡ = 745.7 W = 2.544×10^3 Btu/h = 178.1 cal/s = 550 ft · lbf/s

1 **kilowatt** (kW) = 1×10^3 W = 3.412×10^3 Btu/h = 238.8 cal/s = 737.6 ft · lbf/s = 1.341 hp

Pressure

1 **newton per square meter** (N/m^2), or **pascal** (Pa) = 9.869×10^{-6} atm = 1×10^{-5} bar = 7.501×10^{-4} cm-Hg = 10 dynes/cm^2 = 2.953×10^{-4} in.-Hg = 2.089×10^{-2} lbf/ft^2 = 1.450×10^{-4} lbf/$in.^2$ = 7.501×10^{-3} torr

1 **atmosphere** (atm) = 1.013×10^5 N/m^2 = 76.00 cm-Hg = 1.013×10^6 dynes/cm^2 = 29.92 in.-Hg = 2.116×10^3 lbf/ft^2 = 14.70 lbf/$in.^2$

1 **bar** = 1×10^5 N/m^2 = 0.9869 atm = 75.01 cm-Hg

1 **centimeter of mercury** (cm-Hg) = 1.333×10^3 N/m^2 = 1.316×10^{-2} atm = 1.333×10^{-2} bar = 1.333×10^4 dynes/cm^2 = 0.3937 in.-Hg = 27.85 lbf/ft^2 = 0.1934 lbf/$in.^2$ = 10 torr

1 **dyne per square centimeter** (dyne/cm^2) = 0.1 N/m^2 = 9.869×10^{-7} atm = 7.501×10^{-5} cm-Hg = 2.089×10^{-3} lbf/ft^2 = 1.450×10^{-5} lbf/$in.^2$

1 **inch of mercury** (in.-Hg) = 3.386×10^3 N/m^2 = 3.342×10^{-2} atm = 2.540 cm-Hg = 0.4912 lbf/$in.^2$

1 **kilopond per square centimeter** (kp/cm^2) = 9.807×10^4 N/m^2 = 0.9678 atm = 9.807×10^5 dynes/cm^2 = 14.22 lbf/$in.^2$

1 **pound per square inch** (lbf/$in.^2$, or psi) = 6.895×10^3 N/m^2 = 6.805×10^{-2} atm = 6.895×10^4 dynes/cm^2 = 2.036 in.-Hg = 7.031×10^{-2} kp/cm^2

1 **torr,** or **millimeter of mercury** (mm-Hg) = 1.333×10^2 N/m^2 = 0.1 cm-Hg

Electric Charge*

1 **coulomb** (C) ⇔ 2.998×10^9 statcoulombs, or esu of charge ⇔ 0.1 abcoulomb, or emu of charge

* The dimensions of the electric quantities in SI units, electrostatic units (esu), and electromagnetic units (emu) are different; hence the relationships among these units are correspondences (⇔) rather than equalities (=).

Electric Current

1 ampere (A) ⇔ 2.998×10^{9} statamperes, or esu of current ⇔ 0.1 abampere, or emu of current

Electric Potential

1 volt (V) ⇔ 3.336×10^{-3} statvolt, or esu of potential ⇔ 1×10^{8} abvolts, or emu of potential

Electric Field

1 volt per meter (V/m) ⇔ 3.336×10^{-5} statvolt/cm ⇔ 1×10^{6} abvolts/cm

Magnetic Field

1 tesla (T), or **weber per square meter** (Wb/m^2) ⇔ 1×10^{4} gauss

Electric Resistance

1 ohm (Ω) ⇔ 1.113×10^{-12} statohm, or esu of resistance ⇔ 1×10^{9} abohms, or emu of resistance

Electric Resistivity

1 ohm-meter ($\Omega \cdot$ m) ⇔ 1.113×10^{-10} statohm-cm ⇔ 1×10^{11} abohm-cm

Capacitance

1 farad (F) ⇔ 8.988×10^{11} statfarads, or esu of capacitance ⇔ 1×10^{-9} abfarad, or emu of capacitance

Inductance

1 henry (H) ⇔ 1.113×10^{-12} stathenry, or esu of inductance ⇔ 1×10^{9} abhenrys, or emu of inductance

Appendix 7: Best Values of Fundamental Constants

The values in the following table were taken from the report of the CODATA Task Group on fundamental constants by E. R. Cohen and B. N. Taylor, *The 1986 Adjustment of the Fundamental Physical Constants,* CODATA Bulletin No. 63, November 1986. The digits in parentheses are the one-standard deviation uncertainty in the last digits of the given value.

Quantity	Symbol	Value	Units	Relative Uncertainty (parts per million)
UNIVERSAL CONSTANTS				
speed of light in vacuum	c	299792458	ms^{-1}	(exact)
permeability of vacuum	μ_o	$4\pi \times 10^{-7}$	NA^{-2}	
		$=12.566370614\ldots$	$10^{-7}\ NA^{-2}$	(exact)
permittivity of vacuum	ϵ_o	$1/\mu_o c^2$		
		$=8.854187817\ldots$	$10^{-12}\ Fm^{-1}$	(exact)
Newtonian constant of gravitation	G	6.67259(85)	$10^{-11}\ m^3\ kg^{-1}s^{-2}$	128
Planck constant	h	6.6260755(40)	$10^{-34}\ Js$	0.60
in electron-volts		4.1356692(12)	$10^{-15}\ eVs$	0.30
	$\hbar = h/2\pi$	1.05457266(63)	$10^{-34}\ Js$	0.60
in electron-volts		6.5821220(20)	$10^{-16}\ eVs$	0.30
ELECTROMAGNETIC CONSTANTS				
elementary charge	e	1.60217733(49)	$10^{-19}\ C$	0.30
magnetic flux quantum, $h/2e$	Φ_0	2.06783461(61)	$10^{-15}\ Wb$	0.30
Josephson frequency–voltage ratio	$2e/h$	4.8359767(14)	$10^{14}\ HzV^{-1}$	0.30
quantized Hall conductance	e^2/h	3.87404614(17)	$10^{-5}\ \Omega^{-1}$	0.045
Bohr magneton, $e\hbar/2m_e$	μ_B	9.2740154(31)	$10^{-24}\ JT^{-1}$	0.34
in electron-volts		5.78838263(52)	$10^{-5}\ eVT^{-1}$	0.089
nuclear magneton, $e\hbar/2m_p$	μ_N	5.0507866(17)	$10^{-27}\ JT^{-1}$	0.34
in electron-volts		3.15245166(28)	$10^{-8}\ eVT^{-1}$	0.089
ATOMIC CONSTANTS				
fine-structure constant, $e^2/4\pi\epsilon_o\hbar c$	α	7.29735308(33)	10^{-3}	0.045
inverse fine-structure constant	α^{-1}	137.0359895(61)		0.045
Rydberg constant, $\frac{1}{2}m_e c\alpha^2/h$	R_∞	10973731.534(13)	m^{-1}	0.0012
Bohr radius, $4\pi\epsilon_o\hbar^2/m_e e^2$	a_o	0.529177249(24)	$10^{-10}\ m$	0.045
quantum of circulation	$h/2m_e$	3.63694807(33)	$10^{-4}\ m^2s^{-1}$	0.089
	h/m_e	7.27389614(65)	$10^{-4}\ m^2s^{-1}$	0.089
Electron				
electron mass	m_e	9.1093897(54)	$10^{-31}\ kg$	0.59
		5.48579903(13)	$10^{-4}\ u$	0.023
$m_e c^2$ in electron-volts		0.51099906(15)	MeV	0.30
electron–proton mass ratio	m_e/m_p	5.44617013(11)	10^{-4}	0.020
electron specific charge	$-e/m_e$	−1.75881962(53)	$10^{11}\ Ckg^{-1}$	0.30
Compton wavelength, $h/m_e c$	λ_C	2.42631058(22)	$10^{-12}\ m$	0.089
$\lambda_C/2\pi = \alpha a_0$	$\bar{\lambda}_C$	3.86159323(35)	$10^{-13}\ m$	0.089
classical electron radius, $e^2/4\pi\epsilon_o m_e c^2$	r_e	2.81794092(38)	$10^{-15}\ m$	0.13
Thomson cross section, $(8\pi/3)r_e^2$	σ_e	0.66524616(18)	$10^{-28}\ m^2$	0.27
electron magnetic moment	μ_e	928.47701(31)	$10^{-26}\ JT^{-1}$	0.34
in Bohr magnetons	μ_e/μ_B	1.001159652193(10)		1×10^{-5}
in nuclear magnetons	μ_e/μ_N	1838.282000(37)		0.020
electron magnetic moment anomaly, $\mu_e/\mu_B - 1$	a_e	1.159652193(10)	10^{-3}	0.0086
electron g-factor, $2(1 + a_e)$	g_e	2.002319304386(20)		1×10^{-5}

Quantity	Symbol	Value	Units	Relative Uncertainty (parts per million)
Muon				
muon mass	m_μ	1.8835327(11)	10^{-28} kg	0.61
		0.113428913(17)	u	0.15
$m_\mu c^2$ in electron-volts		105.658389(34)	MeV	0.32
muon–electron mass ratio	m_μ/m_e	206.768262(30)		0.15
muon magnetic moment	μ_μ	4.4904514(15)	10^{-26} JT^{-1}	0.33
in Bohr magnetons	μ_μ/μ_B	4.84197097(71)	10^{-3}	0.15
muon magnetic moment anomaly				
$[\mu_\mu/(e\hbar/2m_\mu)] - 1$	a_μ	1.1659230(84)	10^{-3}	7.2
muon g-factor, $2(1+a_\mu)$	g_μ	2.002331846(17)		0.0084
Proton				
proton mass	m_p	1.6726231(10)	10^{-27} kg	0.59
		1.007276470(12)	u	0.012
$m_p c^2$ in electron-volts		938.27231(28)	MeV	0.30
proton–electron mass ratio	m_p/m_e	1836.152701(37)		0.020
proton specific charge	e/m_p	9.5788309(29)	10^7 Ckg^{-1}	0.30
proton Compton wavelength, $h/m_p c$	$\lambda_{C,p}$	1.32141002(12)	10^{-15} m	0.089
proton magnetic moment	μ_p	1.41060761(47)	10^{-26} JT^{-1}	0.34
in Bohr magnetons	μ_p/μ_B	1.521032202(15)	10^{-3}	0.010
in nuclear magnetons	μ_p/μ_N	2.792847386(63)		0.023
Neutron				
neutron mass	m_n	1.6749286(10)	10^{-27} kg	0.59
		1.008664904(14)	u	0.014
$m_n c^2$ in electron-volts		939.56563(28)	MeV	0.30
neutron–electron mass ratio	m_n/m_e	1838.683662(40)		0.022
neutron–proton mass ratio	m_n/m_p	1.001378404(9)		0.009
neutron Compton wavelength, $h/m_n c$	$\lambda_{C,n}$	1.31959110(12)	10^{-15} m	0.089
neutron magnetic moment	μ_n	0.96623707(40)	10^{-26} JT^{-1}	0.41
in Bohr magnetons	μ_n/μ_B	1.04187563(25)	10^{-3}	0.24
in nuclear magnetons	μ_n/μ_N	1.91304275(45)		0.24
Deuteron				
deuteron mass	m_d	3.3435860(20)	10^{-27} kg	0.59
		2.013553214(24)	u	0.012
$m_d c^2$ in electron-volts		1875.61339(57)	MeV	0.30
deuteron–electron mass ratio	m_d/m_e	3670.483014(75)		0.020
deuteron–proton mass ratio	m_d/m_p	1.999007496(6)		0.003
deuteron magnetic moment	μ_d	0.43307375(15)	10^{-26} JT^{-1}	0.34
in Bohr magnetons	μ_d/μ_B	0.4669754479(91)	10^{-3}	0.019
in nuclear magnetons	μ_d/μ_N	0.857438230(24)		0.028
PHYSICO-CHEMICAL CONSTANTS				
Avogadro constant	N_A	6.0221367(36)	10^{23} mol^{-1}	0.59
atomic mass constant, $m_u = \frac{1}{12}m(^{12}C)$	m_u	1.6605402(10)	10^{-27} kg	0.59
$m_u c^2$ in electron-volts		931.49432(28)	MeV	0.30
Faraday constant	F	96485.309(29)	C mol^{-1}	0.30
molar gas constant	R	8.314510(70)	J $mol^{-1}K^{-1}$	8.4
Boltzmann constant, R/N_A	k	1.380658(12)	10^{-23} JK^{-1}	8.5
in electron-volts		8.617385(73)	10^{-3} eVK^{-1}	8.4
molar volume (ideal gas), RT/p				
$T = 273.15$ K, $p = 101{,}325$ Pa	V_m	22.41410(19)	liter/mol	8.4
Loschmidt constant, N_A/V_m	n_0	2.686763(23)	10^{25} m^{-3}	8.5
Stefan–Boltzmann constant,				
$(\pi^2/60)k^4/\hbar^3c^2$	σ	5.67051(19)	10^{-8} $Wm^{-2}K^{-4}$	34
Wien displacement law constant	b	2.897756(24)	10^{-3} mK	8.4

Appendix 8: The Chemical Elements and the Periodic Table

TABLE A8.1 THE PERIODIC TABLE OF THE CHEMICAL ELEMENTS[a]

IA																	0
1 H 1.00794	IIA											IIIA	IVA	VA	VIA	VIIA	2 He 4.00260
3 Li 6.941	4 Be 9.01218											5 B 10.811	6 C 12.011	7 N 14.0067	8 O 15.9994	9 F 18.998403	10 Ne 20.1797
11 Na 22.98977	12 Mg 24.305	IIIB	IVB	VB	VIB	VIIB		VIII		IB	IIB	13 Al 26.98154	14 Si 28.0855	15 P 30.97376	16 S 32.066	17 Cl 35.4527	18 Ar 39.948
19 K 39.0983	20 Ca 40.078	21 Sc 44.9559	22 Ti 47.88	23 V 50.9415	24 Cr 51.996	25 Mn 54.9380	26 Fe 55.847	27 Co 58.9332	28 Ni 58.6934	29 Cu 63.546	30 Zn 65.39	31 Ga 69.723	32 Ge 72.61	33 As 74.9216	34 Se 78.96	35 Br 79.904	36 Kr 83.80
37 Rb 85.4678	38 Sr 87.62	39 Y 88.9059	40 Zr 91.224	41 Nb 92.9064	42 Mo 95.94	43 Tc (97.907)	44 Ru 101.07	45 Rh 102.9055	46 Pd 106.42	47 Ag 107.8682	48 Cd 112.411	49 In 114.82	50 Sn 118.71	51 Sb 121.757	52 Te 127.60	53 I 126.9045	54 Xe 131.29
55 Cs 132.9054	56 Ba 137.327	57–71 Rare Earths	72 Hf 178.49	73 Ta 180.9479	74 W 183.85	75 Re 186.207	76 Os 190.2	77 Ir 192.22	78 Pt 195.08	79 Au 196.9665	80 Hg 200.59	81 Tl 204.383	82 Pb 207.2	83 Bi 208.9804	84 Po (208.9804)	85 At (209.9871)	86 Rn (222.0176)
87 Fr (223.020)	88 Ra (226.0254)	89-103 Acti- nides	104 Rf (261.11)	105 Ha (262.114)	106 (263.118)	107 (262.12)	108 (265)	109 (266)									

Rare Earths (Lanthanides)	57 La 138.9055	58 Ce 140.15	59 Pr 140.9077	60 Nd 144.24	61 Pm (144.913)	62 Sm 150.36	63 Eu 151.965	64 Gd 157.25	65 Tb 158.9253	66 Dy 162.50	67 Ho 164.9303	68 Er 167.26	69 Tm 168.93421	70 Yb 173.04	71 Lu 174.967
Actinides	89 Ac (227.0278)	90 Th 232.0381	91 Pa 231.0359	92 U 238.0289	93 Np (237.0482)	94 Pu (244.064)	95 Am (243.061)	96 Cm (247.070)	97 Bk (247.070)	98 Cf (251.080)	99 Es (252.083)	100 Fm (257.095)	101 Md (258.10)	102 No (259.101)	103 Lr (262.11)

[a] In each box, the upper number is the ***atomic number.*** The lower number is the ***atomic mass,*** that is, the mass (in grams) of one mole or, alternatively, the mass (in atomic mass units) of one atom. Numbers in parentheses denote the atomic masses of the most stable or best-known isotope of the element; all other numbers represent the average masses of a mixture of several isotopes as found in naturally occurring samples of the element.

TABLE A8.2 THE CHEMICAL ELEMENTS

Element	Chemical Symbol	Atomic Number	Atomic Mass[a]
Hydrogen	H	1	1.00794 u
Helium	He	2	4.00260
Lithium	Li	3	6.941
Beryllium	Be	4	9.01218
Boron	B	5	10.81
Carbon	C	6	12.011
Nitrogen	N	7	14.0067
Oxygen	O	8	15.9994
Fluorine	F	9	18.998403
Neon	Ne	10	20.179
Sodium	Na	11	22.98977
Magnesium	Mg	12	24.305
Aluminum	Al	13	26.98154
Silicon	Si	14	28.0855
Phosphorus	P	15	30.97376
Sulfur	S	16	32.06
Chlorine	Cl	17	35.453
Argon	Ar	18	39.948
Potassium	K	19	39.0983
Calcium	Ca	20	40.08
Scandium	Sc	21	44.9559
Titanium	Ti	22	47.88
Vanadium	V	23	50.9415
Chromium	Cr	24	51.996
Manganese	Mn	25	54.9380
Iron	Fe	26	55.847
Cobalt	Co	27	58.9332
Nickel	Ni	28	58.69
Copper	Cu	29	63.546
Zinc	Zn	30	65.38
Gallium	Ga	31	69.72
Germanium	Ge	32	72.59
Arsenic	As	33	74.9216
Selenium	Se	34	78.96
Bromine	Br	35	79.904
Krypton	Kr	36	83.80
Rubidium	Rb	37	85.4678
Strontium	Sr	38	87.62
Yttrium	Y	39	88.9059
Zirconium	Zr	40	91.22
Niobium	Nb	41	92.9064
Molybdenum	Mo	42	95.94
Technetium	Tc	43	(98)
Ruthenium	Ru	44	101.07
Rhodium	Rh	45	102.9055
Palladium	Pd	46	106.42
Silver	Ag	47	107.8682
Cadmium	Cd	48	112.41
Indium	In	49	114.82
Tin	Sn	50	118.69
Antimony	Sb	51	121.75
Tellurium	Te	52	127.60
Iodine	I	53	126.9045

(cont.)

[a] Numbers in parentheses denote the atomic masses of the most stable or best-known isotope of the element; all other numbers represent the average masses of a mixture of several isotopes as found in naturally occurring samples of the element.

TABLE A8.2 THE CHEMICAL ELEMENTS *(cont.)*

Element	Chemical Symbol	Atomic Number	Atomic Mass
Xenon	Xe	54	131.29
Cesium	Cs	55	132.9054
Barium	Ba	56	137.33
Lanthanum	La	57	138.9055
Cerium	Ce	58	140.12
Praseodymium	Pr	59	140.9077
Neodymium	Nd	60	144.24
Promethium	Pm	61	(145)
Samarium	Sm	62	150.36
Europium	Eu	63	151.96
Gadolinium	Gd	64	157.25
Terbium	Tb	65	158.9254
Dysprosium	Dy	66	162.50
Holmium	Ho	67	164.9304
Erbium	Er	68	167.26
Thulium	Tm	69	168.9342
Ytterbium	Yb	70	173.04
Lutetium	Lu	71	174.967
Hafnium	Hf	72	178.49
Tantalum	Ta	73	180.9479
Tungsten	W	74	183.85
Rhenium	Re	75	186.207
Osmium	Os	76	190.2
Iridium	Ir	77	192.22
Platinum	Pt	78	195.08
Gold	Au	79	196.9665
Mercury	Hg	80	200.59
Thallium	Tl	81	204.383
Lead	Pb	82	207.2
Bismuth	Bi	83	208.9804
Polonium	Po	84	(209)
Astatine	At	85	(210)
Radon	Rn	86	(222)
Francium	Fr	87	(223)
Radium	Ra	88	226.0254
Actinium	Ac	89	227.0278
Thorium	Th	90	232.0381
Protactinium	Pa	91	231.0359
Uranium	U	92	238.0289
Neptunium	Np	93	237.0482
Plutonium	Pu	94	(244)
Americium	Am	95	(243)
Curium	Cm	96	(247)
Berkelium	Bk	97	(247)
Californium	Cf	98	(251)
Einsteinium	Es	99	(252)
Fermium	Fm	100	(257)
Mendelevium	Md	101	(258)
Nobelium	No	102	(259)
Lawrencium	Lr	103	(260)
Rutherfordium	Rf	104	(261)
Hahnium	Ha	105	(260)
?		106	(263)
?		107	(262)
?		108	(265)
?		109	(266)

Appendix 9: Answers to Even Numbered Problems

Chapter 1

2. 8.3×10^{-3} cm
4. Virus = 8.3×10^{-7} in, Atom = 4×10^{-9} in, Fe Nucleus = 3×10^{-13} in, Proton = 8×10^{-14} in
6. 2.72 m
8. Distance from pole to equator = 1.00×10^{7} m
   Straight line distance, using the Pythagorean theorem = 9.0×10^{6} m
10. 6.9×10^{8} m
12. 8211 days
    (This excludes leap years and assumes average 30.5 day month)
    7.1×10^{8} s
14. 11.574 days
16. 3.7×10^{7} beats/year
18. (a) 4 significant figures 7.631×10^{3} sec
    (b) 2 significant figures 7.6×10^{3} sec
    (c) 1 significant figure 8×10^{3} sec
20. 3.532×10^{-3} ft³/m³
22. 195.7 m²
24. 60 s
26. 16.9 m/s
28. 23 mi/h
30. 4.604 km/hr
32. 5.867 h, 150.9 h
34. 1.9×10^{10} yr
36. (a) 17.5 km; (b) 42.5 km; (c) 60 km/h
38. 8 m/s, 3 s
40. 0.77 s, 13 m/s
42. (a) 12.5 s, 350 m; (b) 80 m
44. Avg. speed = 6.4 m/s, Avg. velocity = 0 m/s
46. 3.0×10^{3} m/s²
48. 12 m/s²
50. 29(km/h)/s
52. 2.4 m/s²
54. 270 km/h !! (170 mph)
56. 116.3 ft/s²
58. $t_{1/2} = 6.36 \times 10^{7}$ s
    Because the **magnitude** of the acceleration is the same for both parts of the trip, the time for the second half is identical to that of the first half. Thus, the total time for the trip, T is 1.3×10^{8} s $\approx$ 4.0 yr
    v = 6.2×10^{8} m/s (This exceeds the speed of light!!)
60. 15 m/s
62. (a) 65.6 m
    (b)

Speed, v_0		x_1	$\Delta x = v_0^2/2a$	$x = x_1 + \Delta x$
km/h	m/s	(m)	(m)	(m)
15	4.17	3.13	1.11	4
30	8.33	6.26	4.45	11
45	12.5	9.38	10.01	19
60	16.7	12.52	17.88	30
75	20.83	15.62	27.81	43

64. 44 m
66. 66 m
68. 13 m/s
70. 849 m/s
72. 34 m, 25l
74. (1 s) $\frac{g}{2}$, (2 s) $3\frac{g}{2}$, (3 s) $5\frac{g}{2}$
    $$\Delta S_{n,n-1} = (2n-1)\frac{g}{2} = (1,3,5,7\ldots)\frac{g}{2}$$
76. (a) 43.4 m/s
    (b) −260 m/s² (The minus sign denotes deceleration.)
78. 8.1 m, 11 m

Chapter 2

2. 45°
4. 6.5 m/s²
6. (a) 0.55 s; (b) 33 m
8. 0.95 m
10. 37.6 m/s (135 km/h)
12. (a) 1.72 s; (b) 14.6 m; (c) 58.1 m
14. 77.7 m/s, 13.3 m
16. 121.2 m/s, 375 m, 17.5 s
18. (a) 250 km; (b) 500 km
    (c) No. At such speeds air friction would be considerable. This is why the rocks do not rise to 250 km or cannot be thrown to distances approaching 500 km.
20. 85 m/s, 12 s
22. (a) 71 m/s; (b) 248 m; (c) 249 m
24. 43.1°, 9.8 m/s
26. 30° North of East
28. $|s|$ = 433 km, direction is due South
30. 11.2 km, 27.7° S of E
32. Analytical solution gives:
    A+B+C = 14.52 cm, θ = 7.9°
    A+B−C = 6.70 cm, θ = 17.4°
34. The straight-line distance is 12,750 km
    Distance around the equator is 19,810 km
36. A_x = 4.3 m, A_y = 2.5 m
38. A_x = 4.9 units, A_y = 6.3 units
40. 1.7
42. Magnitude = 5.8 units, $\theta_x = -31°$, $\theta_y = -121°$
44. (b) (2, 5) cm
46. (0.894, −0.447)
48. (a) 7.0 km, 5° E of N (see Problem 29, Chapter 2)
    (b) 5.6 km/h, 5° E of N; (c) 8.24 km/h
50. 0.52 m/s
52. 3.94×10^{6} m/s², 4.0×10^{5} gee
54. 8.99×10^{13} m/s² = 9.16×10^{12} gee
56. 5.9×10^{-3} m/s²
58. 0.0395, 0.0113, 5.95×10^{-3}
60. (a) +8.33 m/s; (b) 10.8 s

62. 13.0 m/s, 40°
64. 27 m/s, 68°
66. 12 m/s, 2.4° upward (i.e., mostly horizontal)
68. 20.3 km/h, 14.2°, from the W of North
70. (a) 8° W of N; (b) 213 km/h

Chapter 3

2. Total Mass $m_T = 1.99 \times 10^{30}$ kg
   Percentage in Planets = 0.133%
   Percentage in the Sun = 99.867%
4. 1.674×10^{-27} kg, 5.974×10^{26}
6. 3.8 m/s², 6.2×10^3 N
8. 6.6×10^3 N, which is 12 times the weight.
10. 1.2×10^4 N
12. 1.46×10^6 N
14. 9.6×10^3 N, 1.2×10^4 N, 1.5×10^3 N, 51°
16. 4.7×10^{20} N, 25°
18. 540 N(1, 0), 540 N(1, 0) + 720(sin 20°, 0)
20. By Newton's Third Law, the force on the small box is also 36 N.
22. 1.2×10^4 N(0.23, 0.64), 70°
24. 150 N, pointing away from the bank.
26. 2.4×10^2 N
28. (a) Force on diver = 736 N
    By the third law, the force on the Earth is 736 N
    (b) Acceleration of diver = 735 N/mass of diver
    Acceleration of the Earth = 735 N/mass of Earth
    $\approx 1.3 \times 10^{-22}$ m/s²
30. (a) 2.4×10^7 N; (b) 3.67 m/s²; (c) 12 m/s²
32. (b) $\dfrac{mg}{\sqrt{122}}$
34. (b) 602 N, 422 N; (c) 5.62 m/s²
36. 64 m, 5.1 s
38. 5717
40. 0.83
42. 180 m
44. 243 km/h
46. 0.17 m/s²
48. (b) $N = 1.7 \times 10^2$ N, $F = 1.0 \times 10^2$ N
    (c) $F_{max} = 1.5 \times 10^2$ N
    (d) $\mathbf{F}_{net} = 0$, as it must; $\mathbf{a} = 0$
50. 5.55 m/s², 6.5 s
52. 27°
54. 22.5 N, 11.2 N
56. Since k is not constant, the spring does not obey Hooke's law.
58. 2.0×10^{20} N
60. At $v = 22$ m/s the coins will start to slide (the frictional force will be inadequate to keep the coins accelerating towards the center with the car).
62. 22 m/s = 80 km/h
64. $v^2/2\mu_s g$ for stop without turning
    If the driver tries to turn, the force will be directed toward the center of the circle of turning, and the centripetal acceleration, as above will be $\mu_s g$. The radius of the circle is given by $v^2 r = a = \mu_s g$ or $r = v^2/\mu_s g$ which is also the minimum distance needed to clear the wall. Since $r = 2x_{min}$ *the best strategy is to brake.*

Chapter 4

2. 4928.6 J
4. 252 J
6. $W_H^{total} = W_H^{up} + W_H^{down} = mgh - mgh = 0$
   Must expend at least mgh to get the mass up to h. Must expend more energy on the way down (because she cannot recover the potential energy back into muscles on the way down).
8. (a) 1.36×10^4 N, 8.3×10^3 N; (b) 3.98×10^3 J
   (c) 1.81×10^4 J, 2.21×10^4 J
10. 2.1×10^3 J
12. 9.7×10^6 J, Total work = 3.2×10^7 J
14. 0.9 J, 2.5 J
16. 2.2×10^{-18} J
18.

Serious Accident Probability		
speed (km/h)		
95	110	125
$3\%\left(\frac{95}{80}\right)^2 = 4.2\%$	$3\%\left(\frac{110}{80}\right)^2 = 5.7\%$	$3\%\left(\frac{125}{80}\right)^2 = 7.3\%$

20. 1.27×10^7 J, 2.8 kg(6 lb − mass)
22. 196 m/s
24. (a) 6.8×10^4 J; (b) 3.4×10^4 J
26. 4.2×10^4 J, 3.5×10^7 J
28. (a) 0.74 J; (b) 0.49 J; (c) 0.25 J
30. 34.5 m/s = 124.2 km/h
32. 29.7 m/s
34. (a) 25.8 m/s; (b) 10.3 m/s
36. $\sqrt{2gh_0}$, $\sqrt{2g(h_0 - h_2)}$
38. (a) 2.5×10^4 N; (b) 1.2×10^7 J; (c) 15.3 m/s
40. 217 J
42. (a) 47 J; (b) 0.89 m
44. 0.26 m
46. Fusion of Uranium = 2.0×10^8 eV
    Annihilation of electron-positron = 1.0×10^6 eV
    Ionization energy hydrogen = 14 eV
48. 1.7×10^9 gal/day
50. 2.0×10^{-3} hp
52. 5.4×10^7 J
54. Rate of removal of momentum is 500 kg m/s²
    $\Delta E/\Delta t = 9000$ J/s
56. 560 slaves, 70 slaves
58. 5.6×10^{-3} m/s, 9.0×10^2 s
60. 3.24×10^{-2} m/s
62. 1.2×10^{-4} W
64. (a) 7.2×10^6 N; (b) 0.12 m/s
66. 140000 J, 6.5×10^4 W, 88 hp
68. 39.7 hp
70. 8.93 m/s
72. 1 kW
74. (a) 3.2×10^4 J/s; (b) 780 J/s; (c) 3.1×10^4 J/s
76. 8.9 m/s

Chapter 5

2. 4.7×10^{-35} N
4. 7.3×10^2 N, 5.5×10^2 N (at 1000 km)

6. 2.8×10^{-9} N
8. 3.41×10^{8} m
10. 0.15 m/s^2, 1.5% of W
12. 21 km/s, 42 km/s
14. 440 km/s, 5.6 hours
16. 2.8×10^{4} km/h
18. 2×10^{8} yr, 3.06×10^{3} m/s
20. 3.07×10^{3} m/s
22. 6.04×10^{24} kg
24. 1.90×10^{27} kg
26. 2.66×10^{7} m Only once/day.
28. In this time the earth would have spun around (360°/day × 0.0627 day) = 22.6°. Hence, the satellite is at the same latitude but 22.6° W, i.e., over ≈ Lincoln, Nebraska.
30. 9.08×10^{2} m/s
32. 1.034 or 3.4% greater
34. 2.04×10^{12} m, 3.2×10^{12} m
36. 5790 s = 96.5 min
38. $\frac{3}{2} \times 1.50 \times 10^{11}$ m
40. 1.41 years, 0.71 year to Mars
42. 7.8×10^{3} m/s, -1.4×10^{11} J
44. (a) 1.1×10^{4} m/s; (b) 1.2×10^{11} J (29 tons TNT)
    (c) 1.2×10^{5} m/s
46. The orbit is elliptical.
48. (a) 6500 s = 108 min
    (b) The orbit is closed (elliptical).
    (c) The orbit is not closed.
50. (a) 4.2×10^{4} m/s
    (b) Head-on collision: 7.2×10^{4} m/s
    Overtaking collision: 1.2×10^{4} m/s
52. 2.37×10^{3} m/s
54. (a) -3.3×10^{9} J; (b) 3.3×10^{9} J

Chapter 6

2. 1.8×10^{29} kg m/s (Earth)
   4.3×10^{7} kg m/s (jet airliner)
   3.8×10^{4} kg m/s (automobile)
4. $p_x = 1.65 \times 10^{-45}$ kg m/s, $p_y = 0.77 \times 10^{-45}$ kg m/s
6. -2.3×10^{5} J
8. −9.3 km/h, 34.3 km/h, 15.7 km/h
10. 150 bullets
12. −1.28 m/s
14. 2.7×10^{-9} N
16. 4.43×10^{3} N
18. 5×10^{-5} N
20. 13 m
22. 0, 0.43
24. $L/4$, $L/4$
26. 1.28 nm (from oxygen atom, between it and the nitrogen atom)
28. 4.1×10^{7} J
30. m_1 and m_2 collide at x_{cm}
32. 1080 kg
34. 1.05 m
36. 6.9×10^{6} m/s
38. 2.3×10^{4} N
40. 14.2 g (can't hold)
42. 3150 N
44. (a) 23 m/s; (b) 1.8×10^{3} m/s^2; (c) 9.4×10^{4} N
46. 39 m/s
48. (a) −70 m/s, 0; (b) +49 m/s, −21 m/s
    (c) +79 m/s, +9 m/s
50. 36 km/h
52. 29 J, which is enough to break the block.
54. 14 m/s
56. (a) 9.85 m/s; (b) 4.8×10^{5} J, 9.4×10^{4} J
58. +68 m/s (?) Inconsistent with statement of the problem. Tale was a TALL tale.
60. (a) 18.3 m/s; (b) 30 m/s
62. (a) It will rise to make an angle of θ with the vertical.
    (b) $l(1 - \cos\theta)/4$

Chapter 7

2. (a) 53 cm/s, 18 cm/s; (b) 1.8×10^{2} cm/s^2, 61 cm/s^2
4. ($2\pi \times 50$) rad/s, 2.2×10^{2} m/s
6. 3.5×10^{-4} m/s
8. 81.5 rad/s, 13 rev/s
10. (160, 0) km/h, (0, 0) km/h, (80, −80) km/h
    (Magnitude 113 km/h at 45° horizontal.)
12. −0.039/s^2
14. 12.3 rad/s^2
16. 3×10^{-4} kg m^2
18. 6.05×10^{-11} m. Thus, distance between atoms = $2R = 1.21 \times 10^{-10}$ m
20. (a) 225 kg m^2; (b) 4.4×10^{3} J
22. 2.2×10^{6} J
24. 4.9×10^{-5} kg · m^2
26. *With arms out:* 1.568 kg m^2, *With arms in:* 1.34 kg m^2
28. 3.59×10^{5} J, 3.73%
30. At middle: 80 Nm, At edge: 160 Nm
    Push at the edge of door to create the largest torque.
32. 1000 N
34. 3.1×10^{3} Nm
36. 3.1×10^{3} J
38. Power at maximum torque: 9.78×10^{4} W = 131 hp
    Torque at maximum power: 176 Nm
40. (a) $\omega_1 \frac{R_1}{R_2}$
    (b) On wheel 1: $(T - T')R_1$, On wheel 2: $(T - T')R_2$
    (c) $P_2 = (T - T')R_1\omega_1 = P_1$
42. 15 rad/s
44. 1.75×10^{-4} rad/s^2, 1.26×10^{4} N
46. 2.73×10^{4} Nm
48. 0.29 kg m^2/s
50. 110 J · s
52. 1019 J, 97.3 kg m^2/s^2
54. 1.6 kg m^2/s
56.

Radius, R(m)	velocity, v(m/s)	L = mrv(kg m^2/s)
0.529×10^{-10}	2.18×10^{6}	1.05×10^{-35}
2.12×10^{-10}	1.09×10^{6}	2.10×10^{-35}
4.76×10^{-10}	7.27×10^{5}	3.46×10^{-35}

58. (a) 2.4×10^{19} kg m^2/s; (b) 3.0×10^{-19} rad/s
60. −7.2°
    Thus, the rowboat turns 7.2° in a direction opposite to the direction of the turn of the woman.

Chapter 8

2. 3.6×10^2 N
4. 1.5 N
6. 780 N, 940 N
8. (a) 2.5×10^2 N; (b) 2.5×10^2 N at any angle
   (c) 2.1×10^2 N; 1.2×10^2 N; 2.1×10^1 N
10. $\pi r^2 l \rho g$
12. 30.7 ft
14. 49 N, 85 N
16. 6.88 m
18. $\theta_F = 10.1°$, $\theta_B = 4.45°$
20. $T = 14{,}210$ N, $F_H = 12{,}305$ N, $F_V = 25{,}235$ N
    Compressional force = 2.81×10^4 N
22. 2560 N (front wheel), 3350 N (back wheel)
24. 8.1×10^3 N, 2.8×10^4 N
26. The maximum protrusion = 1.04 L
    For an infinite number of books, the limiting protrusion is $\frac{L}{2}\left(1 + \frac{1}{2} + \frac{1}{3} + \frac{1}{4} + \frac{1}{5} + \ldots\right)$
28. 1.58×10^3 N, 1.33×10^3 N
30. 400 N
32. Mechanical Advantage is four. The two 200 N forces on the handles become two 800 N forces on the jaws resulting in a 1600 N compressive force.
34. 1200 N
36. 1000N, 2
38. 8.9
40. 2.26 m
42. 4×10^{-6} m
44. 0.52 mm
46. 1.8 cm
48. 5.9%
50. 0.25 m

Chapter 9

2. 4.0 cm, 10 rev/sec, 0.1 s, 20π rad/sec
4. 1.5 cm, 4000/60 Hz
6. (a) $t = 2$s equilibrium, $t = 4$s turning point
   (b) $t = 0$ or $t = 4$s equilibrium, $t = 2$s turning point
   (c) $\sin(\theta + \pi/2)$
8. 26 m/s, 1.7×10^4 m/s², 2.0×10^4 N
10. 1 m
12. (a) 3.0 m, 0.32 rev/s; (b) 12 m/s², 6 m/s²
14. 0.39 s
16. 81.12 kg
18. 1.9 kg/s²
20. 1.2/s
22. 9.26×10^{13} Hz
24. (a) 0.48/s, 54 N/m, 0.6 m/s; (b) $2\pi\omega'$, 1.0 m/s
26. $0.27 \cos 6t$
28. 3.6 J
30. 9.0 J
32. Amplitude: larger by factor of 2
    Frequency: No change
    Energy : larger by factor of 4
    Maximum speed: larger by factor of 2
    Maximum acceleration: larger by factor of 2
34. 8.75×10^{-19} J
    No, spring would have to compress to zero length.
36. 9.52 s
38. 15 m
40. Paris: $l = 0.9939$ m
    Buenos Aires: $l = 0.9926$ m
    Washington, D.C.: $l = 0.9930$ m
42. 35 s
44. 1.4 mm
46. 1.8×10^{-2} m/s²
48. (a) 0.32 m/s; (b) 3.9 N
50. f
52. (a) 50%; (b) 25%; (c) 8.4

Chapter 10

2. 4.3×10^{14} Hz, 7.5×10^{14} Hz
4. (a) 10.8 h; (b) 2.5/h
6. 7.8 m/s
8. 5.5×10 m, 4.2×10^7 m
10. (a) 4.4 m/s; (b) $v = 210$ m/s
12. (a) 0.03 m; (b) 5.2 m
    (c) Wavecrests: 0, 5.2 m, 10.5 m
    Wavetroughs: 2.6 m, 7.9 m, 13.1 m
14. 1.8×10^5 m
16. 7.7 m/s, 39 m
18. 0.010 m, 3.2 Hz
20. (a) 6.3 m/s; (b) 1 m
22. 64 m/s Crew will feel a bump.
24. 180 m/s
26. 28.9 m/s, 0.69 s
28. 110 m/s, 0.22 s
30. (a) 90°; (b) 0.042 m, 0.20 m
32. (a) up and down at a 45° slant; $2\sqrt{2}$ cm
    (b) up and down at a 45° slant; $2\sqrt{2}$ cm
    (c) circular motion radius $2\sqrt{2}$ cm
34. 31 Hz
36. Tension needs to be increased (or decreased) 2.8%, but we don't know which (increase or decrease) is required.
38. 392 Hz, 588 Hz, 784 Hz, 980 Hz
40. 1.58 Hz, 3.16 Hz
42. 71 N
44. 170 m/s, 28/s
46. 3.7 m/s, 6.8×10^3 m/s²
48. $f \alpha \mu^{-1/2}$
    Ratios of frequencies are 196: 294: 440: 659
    Ratios of densities are 1.00: 0.444: 0.198: 0.088
50. $(0.75 + 1.5\,n)$ m

Chapter 11

2. 55.0 Hz, 4187.2 Hz
4. 10 cm, 3300 Hz
6. (a) 0.652 m
   From the lowest to the highest frequencies, they are (i) 1.69 m, (ii) 1.13 m, (iii) 0.752 m, (iv) 0.502 m
   (b) First harmonic = 0.326 m
   $f = 2$ (fundamental frequencies) = 392 Hz, 588 Hz, 880 Hz, 1318 Hz
   From the lowest to the highest frequency they are (i) 0.844 m, (ii) 0.563 m, (iii) 0.376 m, (iv) 0.25 m

(c) 196 Hz = G, 294 Hz = D, 440 Hz = A
659 Hz = (329.5 × 2) = E, which are just the *octaves* of the four standard violin open string-notes

8. 3,0
10. 0.937×10^{-7} W/m²
12. 123 dB
14. 3.1 m
16. 130 times, 21 dB
18. 9.1 *s*; Use signals transmitted by light.
20. 0.41 m Low Frequencies
0.094 m High Frequencies
22. 250 m
24. $3f$
26. (a) 1.2×10^{-4} s; (b) The camera will focus on the glass.
28. 5 m/s, 3.2×10^8 m/s², a is 3.2×10^7 times larger than g
30. $n\frac{v}{2l}$; $n = 1, 2, 3$
32. $n(v/2l)$
34. 600 m/s
36. (a) 0.63 m
(b) 3.5 cm
3.3 cm
3.2 cm
3.0 cm
2.8 cm
2.7 cm
2.5 cm
2.4 cm
2.2 cm
2.1 cm
2.0 cm
1.9 cm
38. 150 m
40. 620 Hz
42. 21.5 m/s, 0.215 Hz
44. 594 Hz, 595 Hz
46. f'' = frequency heard by receiver $= f'(1 - v_t/v)$
$$= f\frac{(1 + v_t/v)}{(1 - v_t/v)} = f$$
If the position of the source/receiver is reversed, the minus signs become plus signs, and $f'' = f$.
48. 480 Hz
50. 660 m/s
52. (a) 33.5°; (b) 30.4 s

Chapter 12

2. 1.39 m/s, 22.3 m/s, 2.8 m/s, 25.1 m/s, 4.2 m/s, 23.9 m/s
4. 260 N
6. 1370 lb
Because the layer of air under the paper provides the same force upwards, "cancelling" this force.
8. 133 cm²
10. 3.6×10^4 N
12. 3.6×10^4 N
14. 52×10^5 N/m²
16. 130 m
18. 2×10^3 N
No, with 400 lbs of force on the chest the diver will not be able to breathe.
20. (a) 360 N; (b) 330 N
22. 0.86 m
24. 5 mm
26. 10.3 m
28. (a) 4.7×10^7 m³; (b) 4.3×10^{10} kg
30. (a) 0.1 m; (b) 0.05 m
32. The barrel floats.
34. 30.6 m
36. 3×10^5 kg
38. 0.066 kg
40. 28 N/m²
42. 1.4 l/s
44. (a) 15.3 m/s; (b) 1.05×10^5 N/m³
46. 1.8×10^5 N
48. 0.86 m/s, 61 l/s
50. (a) $\sqrt{2g(h_2 - h_1)}$; (b) $p_{atm} - \rho g h_2$; (c) $p_{atm}/\rho g$
52. 3.2 N/m²
54. 2.5 m
56. $0.84\, r$
58. $\frac{8\eta}{\pi\rho g r^4}\,\frac{\Delta V}{\Delta t}$

Chapter 13

2. 58.0°C = 331.2°K, −88.3°C = 184.9°K
4. 0.85 atm
6. 1.3 cm
8. 1.8×10^4 molecules per cm³
10. 1.05
12. 1.03×10^{23}
14. 3.34×10^{25} H_2O molecules, 3.34×10^{25} O atoms
6.68×10^{25} H_2 atoms (2/molecule)
16. −60.4°C, −13°C, −13.8°C, −91.1°C
18. 3.6×10^6 N/m²
20. 96.3 g
22. 0.008%
24. 95%
26. (a) 6.6% mass of air escapes.
(b) Net pressure on walls and windows is 0.07 atm = 7070 N/m². Force exerted on window is then 7070 N ≈ 1600 lb. *The window probably cannot stand the force.* It is equivalent to the weight of about ten 150-lb persons.
28. 6.4 kg
30. 21% O_2, 79% N_2
32. 3150 kg, 2.9×10^3 m³
34. (a) Water rises 1.2 m; (b) 2.5×10^5 N/m², 2.5 kg
36. 1.1×10^4 m/s, 6.1×10^5 m/s
38. (a) $\sqrt{2}$; (b) no change; (c) $\sqrt{2}$; (d) 1/2
40. 5.65×10^{-21} J
42. 1.822×10^5 m/s, 1.214×10^5 m/s, 8.87×10^{-16} J, 5.91×10^{-16} J
44. 1840 m/s
46. 8.9×10^{24}
48. 1.9×10^5 J, 0.6, 0.4
50. 284 K

Chapter 14

2. 97 kW
4. 25°C
6. 97 W
8. 4.3 cal/s
10. 0.024 l/s
12. 14.5%
14. (a) 9.6×10^5 W; (b) 0.019°C/s; (c) 6.9°C
16. 0.21 m
18. 136°C
20. 0.4 m, 0.06 m, 57 m^2
22. 2.7×10^{-2} J, 2.7×10^4 J, 10^6
24. 4.9×10^{-4} m, 17 N
26. (a) 710 kg/m^3
    (b) 43.3¢/kg, 44.4¢/kg. Buy gas on a cold day.
28. 1.2 cm
30. 20.5 cal/s = 86 W, 0.11, 0.89
32. 100.32°C
34. 0.028 m
36. 0.65 kg
38. (a) 1.0×10^{11} kcal; (b) 5
40. 22°
42. 41.2 kcal, 498 kcal/kg, 502 kcal/kg
44. 50 m/s
46. ≈ 50 kcal, 119 kcal
48. He = 750, A: = 75, CO = 180, N_2 = 180, CO_2 = 150, O_2 = 160, CH_4 = 406
    Highest value of specific heat per kilogram: Helium
    Lowest value: Argon
50. (a) 417 cal; (b) 298 cal; (c) 29 l, 0.77 atm; (d) 500 J
52. 4.94 cal/K mole, 6.93 cal/K mole
54. 14.6 kcal/hr

Chapter 15

2. 150 J
4. 0, 1.96 J, 1.96 J, 0.06°C
6. 9 J, 3.33×10^5 J
8. (a) $\Delta E = 150$ J, $Q = 150$ J, $W = 0$
   (b) $\Delta E = 150$ J, $Q = 300$ J, $W = 150$ J
10. −470 J (− sign means work done *on* gas)
12. (a) 4.9×10^{-3} m^3; (b) 90 J, 220 J, 310 J
14. 581 kcal/kg
16. 1.3×10^7 J, 0.60
18. 45%
20. 2.8 hp, 22%
22. 0.19 kg/s
24. 20%
26. 1.9×10^5 J, 1.4×10^5 J
28. 135 W, 85 W, 9×10^{-6} kg/s
30. (a) 0.61; (b) 2.5×10^3 J, 1.0×10^3 J
32. 0.88
34. 0.366 J
36. 2.1×10^2 J
38. (a) 0.067; (b) 1.39×10^7 W; (c) 180 kg/s
40. (a) 46.5 W; (b) 20.4 times larger
42. 7.1 kW
44. (8.2 kcal/K)/h
46. 1.8×10^3 J/Ks
48. 41 J/Ks
50. 1.56 kcal/K
52. 0.011 kcal/K, 0.096 kcal/K
54. 3.2 cal/°K per second, or 13.4 J/°K per second
56. 3×10^{17} J/K

Chapter 16

2. 9.109386×10^{-31} kg
4. 6.25×10^{12}
6. 8.4×10^{22} electrons
8. 8.1×10^{-10} N
10. 2.05×10^{-8} N, 0.91×10^{-8} N, 0.51×10^{-8} N
12. (a) 1.86×10^{-40} N, (b) 2.3×10^{-4} N, 1.24×10^{36}
    (c) 1.1×10^6 m
14. 2.89×10^{-9} N
16. 17 N, 2.6×10^{27} m/s^2
18. 8.125×10^{18} electron/s
20. 4.5×10^{-2} N, 4.9×10^{28} m/s^2
22. 4.2×10^{-3} N
    $F_{1x} = F \sin\theta = 2.3 \times 10^{-3}$; $F_{2x} = -F_{1x} = -2.3 \times 10^{-3}$
    $F_{1y} = F \cos\theta = 3.5 \times 10^{-3}$; $F_{2y} = -F_{1y} = -3.5 \times 10^{-3}$
24. N = 1.15×10^{-7} N, $F_{2x} = -F_{1x} = -5.1 \times 10^{-8}$ N
    $F_{2y} = -F_{1y} = -1.0 \times 10^{-7}$ N
26. 8.2×10^{-8} N, 2.2×10^6 m/s, 1.5×10^{-16} s
28. (a) 9.2×10^{-12} N; (b) 1.6×10^3 m/s
30. 3.57×10^{32} electrons on each
32. 8.7×10^{-4} C
34. $F_x = F_{1x} + F_{2x} = 2.0 \times 10^{-7}$ N
    $F_y = F_{1y} + F_{2y} = 1.7 \times 10^{-7}$ N
36. $\left[\left(1 + \frac{\sqrt{2}}{4}\right)F, \left(1 + \frac{\sqrt{2}}{4}\right)F\right]$
38. $(-3.1 \times 10^{-15}, 6.9 \times 10^{-16})$
40. 4 electrons
42. 5.6×10^{21} electrons

Chapter 17

2. -5.7×10^4 N/C
   $\mathbf{E} = (-5.7 \times 10^4 \text{ N/C}, 0)$
   Negative sign means **E** points in direction opposite to **a**.
4. 6.2 m/s^2 (i.e., going *up*.)
6. 1.3×10^{-13} N, 1.4×10^{17} m/s^2
8. 9000 N/C
10. 6.2×10^{-7} m
12. 5.1×10^{11} N/C
14. 2.61×10^{11} N/C
16. $(5.1 \times 10^{12}$ N/C, 0)
18. $\mathbf{E}_1 + \mathbf{E}_3 + \mathbf{E}_2 =$
    $$\frac{2Q}{4\pi\varepsilon_0(d^2 + y^2)}\left[0, \frac{-y}{(d^2 + y^2)^{1/2}}\right] + \frac{2Q}{4\pi\varepsilon_0 y^2}[0, 1]$$
20. 27.7 V/m (−0.83, 0.55)
22. 1.07×10^4 N/C, 337°
24. 5.1×10^2 N/C
26. 3.15×10^4 N/C, 189.8°
28. (a) $\mathbf{E} = 2.3 \times 10^6$ N/C, downward
    (b) Inside a metallic plate the electric field will be zero. Above and below the plate the electric field will be unchanged. The charge density on the upper sur-

face will be $-\sigma$, and the charge density on the lower surface will be σ.

30. 1.3×10^7 N/C, making angle 22.5° with each sheet
32. 9.4×10^{13} electrons
34. 8.3×10^{-9} C
36. 7.2 N/C ($r = 0.5$ m), 3.6 N/C ($r = 1.0$ m) 2.4 N/C ($r = 1.5$ m)
38. 3.3×10^{-9} C/m
40. $E = \dfrac{\lambda}{2\pi\varepsilon_0 r}$ (for $r > r_{\text{pipe}}$)
42. (a) $Q = \lambda h$; (b) Q/ε_0

    (c) $A = 2\pi rh \quad \rho_E = \dfrac{Q}{\varepsilon_0 A} = \dfrac{Q}{2\pi r\varepsilon_0 h} = \dfrac{\lambda}{2\pi r\varepsilon_0}$

    (d) $E = \rho_E = \dfrac{\lambda}{2\pi r\varepsilon_0}$
44. $\dfrac{\lambda}{2\pi\varepsilon_0 r}$, 1.1×10^7 N/C
46. $\dfrac{q}{2\varepsilon_0}$ (half of the flux goes upward!)
48. (a) $\dfrac{q}{8\varepsilon_0}$; (b) $\dfrac{3q}{8\varepsilon_0}$
50. 2.7×10^{-3} Vm, 2.3×10^{-3} Vm, 1.3×10^{-3} Vm
52.

d (m)	$2/(d^2 + 4)^{3/2}$	$\phi(\times 10^{-2}$ Vm)
0	.2500	1.35
0.5	.2283	1.23
1	.1789	0.97
2	.0884	0.48
3	.0427	0.23
4	.0223	0.12
5	.0128	0.07
6	.0079	0.04

54. $F = mg \propto \dfrac{mM}{A} \propto \dfrac{mM}{4\pi r^2}$ (in analog with $F = qE$)

    Thus the force is proportional to the product of the masses and inversely proportional to the square of the distance between them. (Universal Gravitation Law of Newton)
56. Pick a Gaussian Surface which is cube of side L. By Gauss' Law

    $EA = \Phi = \dfrac{Q}{\varepsilon_0}$; where $4\pi r^2$ = surface area (in on one side, out of the other)

    $= L^2(E) + L^2(-E)$

    $= 0$

    Therefore, $Q = 0$
58. $1.1 \times 10^3 \left(\dfrac{\text{Nm}^2}{\text{C}}\right)$
60. 2×10^5 Cm
62. (a) 2.1×10^{-29} Cm

    (b) The electron is not centered on the Cl nucleus but is on average shifted toward the H^+ (proton).
64. (a) 2×10^{-17} Cm; (b) 1.2×10^{-11} Nm

Chapter 18

2. -10 eV
4. 4500 V
6. 6.9×10^6 m/s
8. 0 (true for all points P).

   This is not true if both charges are positive.
10. (a) 2.1×10^6 m/s; (b) $v = 1.45 \times 10^6$ m/s
12. 2.1×10^{-14} m
14. (a) 5.5×10^{-12} J; (b) 4.0×10^7 m/s
16. $\sqrt{\dfrac{qQ}{3\pi m\varepsilon_0 d}}$
18. (a) 2.43×10^5 V; (b) 4.5×10^{18} V/m (c) 4.5×10^{18} V/m
20. -8.1×10^{-18} J
22. 5.8×10^6 eV
24. $\left[\dfrac{q^2}{8\pi\varepsilon_0 mL}\left(4 + \sqrt{2}\right)\right]^{1/2}$
26. 1.7 J
28. 5.1×10^{31} J/m³
30.

E (J)	$u = \frac{1}{2}\varepsilon_0 E^2$ (J/m³)
2×10^{21}	1.8×10^{31}
10^{14}	4×10^{16}
6×10^{11}	1.6×10^{12}
5×10^6	110

32. (a) 2.0×10^{-11} F (20 pF); (b) 4.0×10^{-6} C
34. (a) 1.6×10^{-10} F; (b) 380 V; (c) 7.5×10^4 V/m (d) 0.025 J/m³; (e) 1.1×10^{-5} J
36. 0.2 C, 2000 J
38. 9.9×10^4 V/m
40. 3.1×10^{-9} F
42. 15.5 μF, 1.5 μF
44. 8.9×10^{-4} J
46. $\dfrac{C}{2}$ for each pair in series
48. $Q_3 = Q = 8.9 \times 10^{-4}$ C, $Q_2 = 6.7 \times 10^{-4}$ C

    $Q_1 = 2.2 \times 10^{-4}$ C, $PE_1 = 1.2 \times 10^{-2}$ J

    $PE_2 = 3.7 \times 10^{-2}$ J, $PE_3 = 5.0 \times 10^{-2}$ J
50. In free space: $4\pi\varepsilon_0 R$

    In gas with dielectric constant $\kappa = 4\pi\kappa\varepsilon_0 R$
52. $V_2 = 48.05$ V, $\Delta V = 0.05$ V
54. 1.1×10^5 m³
56. $\dfrac{\text{Energy}}{\text{Volume}} = \dfrac{\frac{1}{2}Q\Delta V}{\text{Volume}}$

    $= \dfrac{\frac{1}{2}(\kappa\varepsilon_0 AE)(Ed)}{\text{Volume}} = \dfrac{\frac{1}{2}\kappa\varepsilon_0 E^2(Ad)}{\text{Volume}} = \dfrac{1}{2}\kappa\varepsilon_0 E^2$
58. (a) 1.6×10^5 N/C, 1.1×10^5 N/C (b) 3.0×10^5 N/C
60. $\dfrac{\varepsilon_0 A}{2d}(\kappa_1 + \kappa_2)$

Chapter 19

2. 2300 V/m
4. 1.4×10^5 C, 0.9×10^{24} electrons
6. 1.0×10^{-9} F
8. 3 Ω, 2.5 A

10. 400 A
12. $I' = 4I$
14. $R = \dfrac{\rho L}{m/\rho_m L} = \dfrac{\rho\rho_m L^2}{m} \Rightarrow$ inversely proportional to m at fixed L
16. 0.069 V/m
18. 0.32, 0.41, 0.51
20. 7.9 Ω
22. 2.24 cm
24. 0.88 Ω, 13.7 A
26. 0.018 V/m
28. 33°C
30. Minimum acceptable ≈ 14 gauge (extrapolated)
32. 8.0×10^{-5} A
34. Volume of aluminum = 0.071 m^3
Mass of aluminum = 191 kg
Resistance of aluminum wire = 4.0×10^{-3} Ω
Area of copper wire of same resistance = 4.3×10^{-4} m^2
Volume of copper wire = 0.043 m^3
Mass of copper = 383 kg
36. 5.4 A; $I_1 = 2.4$ A; $I_2 = 1.7$ A; $I_3 = 1.3$ A
38. 60 R
40. I_1 (iron) = 2.2 A, I_2 (brass) = 3.8 A
42. Fraction in iron = 0.999982; in water = 1.8×10^{-5}
44. 8.1×10^{-3} Ω, 4.9 V
46. 4.0 A, 2.4 A, 1.5 A, Total current = 7.9 A
48. 10.9 A, 7.1 A
50. $\dfrac{R_2R_3}{R_2 + R_3} = 3.43$ Ω; (a) 1.85 Ω; (b) 6.5 A
(c) 12.0 V; $I_1 = 3.0$ A; $I_2 = 2.0$ A; $I_3 = 1.5$ A

Chapter 20

2. 6.9×10^6 J
4. (a) Flashlight = 6500 J, automobile battery = 2.4×10^6 J
(b) Flashlight = 1.36×10^8 J/m^3
Automobile battery = 2.05×10^8 J/m^3
(c) Flashlight battery = 7.4×10^4 J/kg
Automobile battery = 10^5 J/kg
6. 20 cells
8. When R_2 is decreased, the voltage across R_1 is unchanged so its current is *unchanged.*
The voltage across $R_2 + R_3$ is also unchanged, so if R_2 decreases the current in R_2 and R_3 will *increase.*
10. (a) 53.3 Ω = R_{eq}
(b) R_1, R_2 = 0.23 A, R_3 = 0.08 A, R_4 = 0.15 A
12. 0.11 A
14. Current through R_1 = 1 A (down)
Through $R_2 = \dfrac{1}{4}$ A (up), Power = 3 W
16. 5/2 A, 11/4 A
18. $V_1 - (I_1' + I_2')R_1 - I_1'R_2 = 0$
$V_1 - (I_1' + I_2')R_1 - I_2'R_3 - V_2 = 0$
$(I_1 - I_2)R_2 - I_2'R_3 - V_2 = 0$ which is Eq. 20.13
20. 58¢
22. 10.4 A, 11.1 Ω
24. 13 A, 8.8 Ω
26. 10% increase in voltage.
28. 100 s
30. 6.2×10^{12} protons per s, 700 W
32. P'(alum.) = 0.6 P(copper)
34. (a) 47 A; (b) 71 km
36. 2¢
38. 0.12
40. (a) 2.0 W/m; (b) 0.08 V/m
42. (a) 3025 W for parallel arrangement
(b) 672 W for series connection
44. 19 l/min
46. R (for V = 110 V) = 12.1 Ω
R (for V = 220 V) = 48.4 Ω
Throwing the switches S_1 and S_2 (which can be ganged into a single sliding switch) will produce a 110 V operation. Sliding to the right will give a 220 V operation.
48. 433 Ω
50. 1.49936 V
52. 350 μs
54. 400×10^{-6} s
56. 10.7×10^{-3} s
58. (a) At $t = 0$, $Q = 0$ C, At $t = 0.002$ s, $Q = 76$ μC
(b) 120 μC; (c) 60×10^{-3} A

Chapter 21

2. 4.8×10^{-17} N, 2.8×10^{10} m/s^2
4. 2.9×10^{-15} N
6. 0.38 T
8. (a) 1.4×10^{-5} T; (b) 4.3 times
10. 4.2 T, DOWN
12. 9.3×10^{-7} m/s
14. 1.7×10^{-12} T, North
16. 1×10^{-17} N, $\theta_F = 16°$ above the y-axis, North and Upward.
18. (a) $\lambda v'$
(b) $B = \dfrac{\mu_0}{2\pi}\dfrac{I}{r} = \dfrac{\mu_0\lambda v'}{2\pi r}; \quad \dfrac{\lambda}{r} = 2\pi\varepsilon_0 E$
$= \dfrac{\mu_0 v'}{2\pi}\left(\dfrac{\lambda}{r}\right) = \dfrac{\mu_0 v'}{2\pi}(2\pi\varepsilon_0 E)$
$B = \mu_0\varepsilon_0 v'E$
20. (a) 6.4×10^{-4} T; (b) 4.3×10^{-5} T
22. $\dfrac{\mu_0}{2\pi}\dfrac{I}{d}\dfrac{3}{2}\sqrt{2}$
24. 6.4×10^{-16} N, 7.0×10^{14} m/s^2
26. $-\left(\dfrac{\mu_0}{2\pi}\right)\left(\dfrac{2Id}{d^2 + R^2}\right)$
B_{max}, the max. B is at $d = \pm R$ $\left[B = \left(\dfrac{\mu_0}{2\pi}\right)\left(\dfrac{I}{R}\right)\right]$
28. 8.4×10^{-5} T
30. 1.5×10^{13} m/s^2
32. 30 μA, 2.4×10^{-10} T
34. $0.34\ \mu_0 I/R$, $-x$
36. 26.5 A
38. (a) The poles attract. (b) These poles repel.
42. $\mu_0 I\left(\dfrac{-\sin\phi}{2\pi r}, \dfrac{\cos\phi}{2\pi r}, n\right)$
for $n >> \dfrac{1}{2\pi r}$ (i.e., at large r) the field is primarily in the z-direction with small components in the x and y directions which change orientation as one moves around the

wire. The magnetic field lines are not straight lines; they spiral slowly around the wire.

for $n << \frac{1}{2\pi r}$ (i.e., a very small r close to the wire) the magnetic field lines spiral rapidly around the wire. The pitch angle of the spiral is

$$\tan\alpha = \frac{B_z}{\sqrt{B_x^2 + B_y^2}} = \frac{n}{1/2\pi r} = 2\pi n r$$

44. 1.0×10^{-2} T, 1.6×10^5 Hz
46. 1.1×10^{-17} kg m/s
48. 3.43×10^{-17} kg m
50. 3.3 T
52. 3.84×10^{-19} kg m/s
54. 0.39 A
56. $\frac{4mv}{qd}$
58. (a) $F = 0$; (b) 0.18 N towards the North by the RHR.
    (c) 0.18 N towards the East by the RHR.
    (d) 0.16 N towards the East by the RHR.
60. 6.7×10^{-5} N
62. 0.12 N
64. 7.2×10^{-4} N; towards the long wire.
66. 8.0×10^{-27} Nm
68. 3.4×10^{-5} Nm

Chapter 22

2. 0.75 V
4. 2.2×10^{-3} V
6. 2.6×10^{-3} V
8. $\frac{\omega}{2} R^2 B$
10. $\frac{B_0 L^2 x_0}{mR} \frac{\Delta B}{\Delta t}$
12. 5.5×10^{-5} Tm^2
14. 4.8 V
16. -0.75 T/s
18. 6.7 V
20. -9.7×10^{-3} V
22. 300 A
24. $\frac{-\mu_0 La}{2\pi r} \frac{\Delta I}{\Delta t}$
26. $\frac{\Delta BAN}{R}$
28. (a) 90 Tm^2/s; (b) 90 V
    (c) At $t = 0$ flux in (2) is increasing to the *right*; induced emf will oppose this increase; therefore, direction of induced current (opposite to direction of current in coil 1).
30. 200 $\mu_0 nA$. (Shape of coil doesn't matter)
32. $\mu_0 \pi r_1^2 n_1 n_2$
34. (a) 6.3×10^{-3} H; (b) -1.9 V
36. 102 V
38. 1.2×10^{-3} J
40. 2.1×10^2 J
42. 4×10^{21}, 4.0×10^{11}, 3.6×10^8, 1.6×10^6
44. 3.4×10^7 J
46. (a) 0.024 T; (b) 220 J/m^3, 0.062 J
48. (a) 0 A; (b) 0.06 A; (c) 3 A/s
50. 0.6 H
52. (a) 0.75 A/s, 1.5 A/s; (b) $I_1 = I_2 = 0.25$ A

Chapter 23

2. 1.63 A, 1/4 cycle later = 0, 1/2 cycle later = -1.63 A
   3/4 cycle later = 0
4. (a) 10.4 A; (b) 14.75 A; (c) 11 Ω
6. 400 W
8. 3.2×10^{-2} A, 6.3×10^{-2} A
10. 2×10^6 Ω, 1.5×10^6 Ω
12. (a) 208 Ω; (b) 9.6×10^{-4} A
    (c) -8.3×10^{-4} A, -5.9×10^{-4} A
14. 0.1 μF
16. (a) 1.3×10^{-7} F; (b) 1.4×10^{-2} A
    (c) 1.6×10^{-6} C; (d) 9.4×10^{-6} J
18. 0.1 A
20. 1.5 kΩ, 4.5 kΩ
22. 5.6×10^4 Ω, 2.5×10^{-1} Ω, $X_C > X_L > R$
24. 9×10^{-1} A, 0, -9×10^{-1} A, 0
26. 5.0×10^{-3} A
28. (a) 10.8.6 V, 0; (b) 0 J, 26 J; (c) 1.95×10^4 W, $P = 0$
30. 3×10^3 Hz
32. 1.6×10^{-6} F, 6.6×10^{-2} H
34. $\frac{1}{\sqrt{LC}}$
36. 380 Hz
38. 670 Hz
40. (a) Frequency is unchanged $\omega = \frac{1}{\sqrt{2C}}$
42. 8 V
44. 2.8×10^4 V
46. 0.03 A
48. 0.16 A
50. 1.96×10^5 A, 4.35×10^5 A

Chapter 24

2. 380 A, 5.1×10^{-4} T
4. (a) 2.0×10^{-3} A; (b) 1 second
6. $\kappa\varepsilon_0 \frac{\Delta\Phi}{\Delta t}$, $\mu_0 I + \kappa\mu_0\varepsilon_0 \frac{\Delta\Phi}{\Delta t}$, 2.3×10^{11} Vm/s
12. 3.3×10^{-9} s, 1.5×10^8 Hz No.
14. 67 ps
16. 50×10^{-8} T, **B** points North, $C\frac{m}{s}$ T, $\frac{Vs}{m^2}$
18. **B** points North, 2.0×10^{-9} T
20. 39.2°
22. 0.25
24. 1.85 m
26. 120, 60, 30, 20, 15
28. 1.43 GHz
30. 2.7×10^{-2} m
32. 2.8×10^{-8} J/m^3
34. 1.6×10^{-14} W/m^2
36. 1.7×10^{-9} T, Vertically downwards, NORTH
    6.6×10^{-4} W/m^2

38. 2.8×10^{28} W
40. 4.0×10^{26} W
42. 2.1×10^{-13} W/m^2, 6.6×10^{-5} W
44. (a) 40 W/cm^2; (b) Yes!
46. (a) 4.7×10^6 V/m, 1.6×10^{-2} T; (b) 24 cal/sec
48. 6.7×10^{10} m
50. (a) 1.8×10^{-17} J/m^3, 1.8×10^{-17} J/m^3
    (b) 4×10^{-3} V/m, 0
    (c) 7.1×10^{-17} J/m^3, 0
    This is *not* a single electromagnetic wave!

Chapter 25

2. 85°
6. 0.21 m^2
8. 28.9°
10. 50°
12. 39°0′50″ or 39.0139°
14. 1.345
16. 56°
18. θ'_r does not exist.
20. $x \approx 2d\alpha/\theta$, $L + (2d/n)$
22. 48°
24. $\theta \geq 38.7°$
26. 41.1°, 61.0°
28. 5.7 m
32. −60 cm (behind mirror)
34. 200 cm
36. −14.5 m (behind the mirror)
38. 1.7 m, 0.24 m
40. $S > R/2$, $S < R/2$
42. Image is 1.96 cm behind doorknob. 0.13
44. Image forms 52.5 cm behind convex mirror.
46. (a) 25.9 cm; (b) 73.2 cm (image of candle 73 cm on opposite side of lens)
48. Incident parallel rays will diverge exciting the lens.
50. S′ = 75 cm, inverted, real, 1.5 times the size of the object.
52. Image is at 18.75 cm from the lens, virtual, erect, and smaller than the object.
54. Slide must be put 13.0 cm from lens.
56. 10 cm
58. (a) $S' = -60$ cm
    Virtual image formed by convex lens is 60 cm to *left* of lens.
    (b) $S' = -21.0$ cm
    Image at 21.0 cm to left of concave lens.
60. 48.7 cm
62. 0.022 second
64. 10.0
66. 1/64
68. 1344
70. Since $\theta \cong h/S$, then magnification $= 1 + \dfrac{25 \text{ cm}}{f}$

Chapter 26

2. $2d = \frac{1}{2}\lambda, \frac{3}{2}\lambda, \frac{5}{2}\lambda, \ldots$
4. $D' = 1.23\,D$ 23% increase in diameter
6. (a) 100 nm
   (b) 270 nm, 130 nm, 90 nm and less, all out of range of visible spectrum.
8. T is max. when $\lambda' = 2d$ ⎫ same formula for both, as
   R is min. when $\lambda' = 2d$ ⎭ expected
10. $2d/\cos\theta = 0, \lambda, 2\lambda, \ldots$
12. 7.4×10^{-3} m
14. *Average* value 5.5°, 20.5°, 35.5°, 51.5°
    Minima at 6.1°, 18.6°, 32.2°, 48.2°
16. 453 nm
18. 3.31 kHz, 6.62 kHz, ...
20. 22°, 61°, −7°, −39°
22. (a) $\sqrt{d^2+4h^2} - d = n\,\lambda$; (b) 97 MHz
24. 1.05×10^{-3}, 2.6×10^{-3} m
26. 3.4×10^{-4} m
28. $\theta_{water} < \theta_{air}$ and pattern compresses.
30. n = 1, $\theta \approx 25°$; $n = 2$, $\theta \approx 57°$
32. 0.82 rad
34. (a) 0.042 m; (b) 0.021 m
36. (a) The fringe pattern will not be resolved.
    (b) The fringe pattern can be resolved.
    (c) The diffraction pattern is so smeared as to be not observable.
38. 94°
40. 3.4×10^{-3} rad
42. 1.34×10^{-5} rad, 2.4×10^{-4} cm
44. (a) 3.1×10^{-7} rad; (b) 2.2 m
46. (a) 9.6×10^{-5} rad
    (b) 9.7×10^{-5} rad to 1.45×10^{-4} rad
    Somewhat improve resolution.
48. (a) 2.24×10^{-6} rad; (b) 0.34 m; (c) 5.4×10^{-6} m
50. (a) 1.95 km; (b) 850 W/m^2, 5.0×10^{-4} W/m^2
52. 1.08×10^{-5} rad, 6.2 cm

Chapter 27

2. 15.49 min, 17.32 min, 34.41 min
4. $1 + 2.5 \times 10^{-7}$
6. (a) 15.8 km; (b) 19.3 km
8. 370 m/s, 1.0000000000007
10. 3.7×10^{-5} s The clock at North Pole (stationary in internal frame) will be ahead.
12. 0.19 m
14. 4.0×10^{-15} (4.0×10^{-13}%)
16. 0.8 m; 1 m^2, (forward & backward);
    0.8 m^2; (front, back, top, bottom), 0.8 m^3
20. (a) 6.67×10^{-7} s
    (b) $\Delta t' = 2.14 \times 10^{-6}$ s, $t' = 2.13 \times 10^{-6}$ s
22. (a) 7.8×10^{16} m, -1.6×10^8 s
    (b) At the same time.
24. 0.65 c
26. 0.999 c
28. 0.98 c, −0.54 c
30. If $v > 0.6c$, the Camden explosion will occur first.
32. 45×10^9 J
34. 0.87 c
36. 2.66×10^{-5}
38. 1.58×10^{-22} kg − m/s, 9.46×10^{-14} J
40. (a) 0.999620 c; (b) 1.8×10^{-17} kg m/s
42. 1.4×10^{-15} m/s

44. 130 m/s, 5.3×10^{-16} kg · m/s
46. 9×10^{-2} kg
48. $\mathbf{p} = \gamma m\mathbf{v}$; $\gamma = \frac{E}{mc^2}$; Therefore, $\mathbf{p} = \frac{Em\mathbf{v}}{mc^2} = \frac{E\mathbf{v}}{c^2}$
52. 0.83 c

Chapter 28

2. 7.50×10^{-20} J
4. 0.136 nm, 0.192 nm
6. 16
8. 6.6×10^3 K
10. (a) 1.0×10^{-3} m; (b) 2.0×10^9 W
12. 1.4×10^3 W/m²
    Agrees with measured value of 1.38×10^3 W/m²
14. (a) 1.7×10^{17} W; (b) 3.3×10^2 W/m²
    (c) 3° C
16. (a) 746 W
    (b) 585 W Therefore, you lose heat at 161 W
18. 9×10^{24}
20. 1.9×10^{31} photons/s
22. 3×10^{19} photons/s
24. 2.2×10^{-19} J, 1.8×10^{20} photon/s
26. 1.24 keV nm/λ
28. 5.45×10^{14}/s
30. 593 nm
32. 17.5 eV
34. 85°
36. 0.0303 nm
38. Electron gains 62.5 eV
40. 1.52×10^8 eV
42. 25 keV
44. 1.25 kV
46. $\lambda_1 = 0.061$ nm; $\lambda_2 = 0.070$ nm
    Only λ_2 can be produced.
48. (a) 1.7×10^{-29} kg m/s; (b) 77 Hz

Chapter 29

2. Therefore, *not all* spectral lines of Paschen are higher than Brackett.
4. In close agreement with observed values.
6. 3.22×10^7 eV
8. $4.3 \times 10^{32}\,\hbar$
10. 2.20×10^{-18} J, -4.35×10^{-18} J, -2.15×10^{-18} J
12. 9.1×10^{22} m/s²
14. 2.8×10^{-5} m, 2.5×10^{-5} eV
16. 13.6 eV
18. 2.925×10^{15} Hz, 102.6 nm
20. 10.2 eV
22. $1 : \alpha : \alpha^2$ QED; $\alpha = \frac{1}{137}$
24 102.6 nm, 656.5 nm
26. 2.56×10^{-13} m, -2812 eV/n², 2109 eV
28. -54.4 eV
30 (a) $n^2 \frac{\hbar^2}{Gm^2M}$; (b) 2.6×10^{74}; (c) 1.2×10^{-63} m
32. 8.7×10^{-3} nm, 0.53 nm, 0.33 nm, 4.1×10^{-3} nm
34. 1.1×10^{-32} m
36. 2.07×10^{10} m, 1.59×10^{11} m
38. 1.28×10^{-10} m
40. 1.4×10^{-12} m
42. 0.039 nm
44. ± 0.07 m
46. 9.2×10^{-20} kg m/s, 5.5×10^7 m/s

Chapter 30

2. $m = 0, \pm 1, \pm 2, \pm 3, \pm 4, \pm 5$
4. (a) 2 states, 8 states
   (b) 2 states, 6 states, 0 states, 8 states in total
6. $2\hbar$, $1.4\hbar$; $4\hbar$, $3.5\hbar$; $10\hbar$, $9.5\hbar$; $500\hbar$, $499.5\hbar$
   $\Delta L = L_B - L_Q \approx 0.5\,\hbar$
8. $5/2\,\hbar$, $\pm 1/2$, $\pm 3/2$, $\pm 5/2$
10. 30°
12. $\sqrt{2}\,\hbar$
14. $L_x^2 + L_y^2 = L^2 - L_z^2 = l(l+1)\hbar^2 - m^2\hbar^2$
    $\sqrt{\frac{[l(l+1) - m^2]}{2}}\,\hbar$
16. 2 electrons $n = 1;\ l = 0;\ m = 0;\ m_s = \pm 1/2;$
    3 electrons $n = 2;\ l = 0;\ m = 0;\ m_s = \pm 1/2;$
    $n = 2;\ l = 1;\ m = 0;\ m_s = -1/2$
18. 2 electrons $n = 1;\ l = 0;\ m = 0;\ m_s = \pm 1/2;$
    2 electrons $n = 2;\ l = 0;\ m = 0;\ m_s = \pm 1/2;$
    6 electrons $n = 2;\ l = 1;\ m = 0, \pm 1;\ m_s = \pm 1/2;$
    2 electrons $n = 3;\ l = 0;\ m = 0;\ m_s = \pm 1/2$
20. Li, $n = 2;\ l = 0$
    Na, $n = 3;\ l = 0$
    K, $n = 4;\ l = 0$
    Single electron outside a closed shell.
22. $S = 3/2$ $S_z = \pm 3/2, \pm 1/2$ (4 states)
    (K shell)
    $n = 1;\ l = 0,\ m = 0,\ m_s = \pm 1/2, \pm 3/2$ (4 states)
    (L shell)
    $n = 2;\ l = 0,\ m = 0,\ m_s = \pm 1/2, \pm 3/2$ (4 states) } 16 states
    $l = 1,\ m = 0, \pm 1,\ m_s = \pm 1/2, \pm 3/2$ (12 states) }
24. 21 keV
26. 30 which is Zinc (Zn), 59 which is Praseodymium (Pr)
28. $n(8.69 \times 10^{-20}$ J); $n = 0, 1, 2\ldots$, $n(1.31 \times 10^{14}$ Hz)
    $\frac{2.29 \times 10^{-6}\ \text{m}}{n}$
30. (a) 7.6×10^{14} Hz, (b) 5.0×10^{-19} J
32. (a) 5.3×10^{-46} kg m² (Js²)
    (b) 1.3×10^{-4} eV; 3.9×10^{-4} eV; 7.8×10^{-4} eV
34. $\frac{\hbar^2}{I}$; $2\frac{\hbar^2}{I}$; $3\frac{\hbar^2}{I}$
36. A small change in the base voltage V_B near the value at which the Emitter-Base junction will start to conduct will cause a large current to flow in the Collector circuit.
40. 12% efficiency

Chapter 31

2. ^{204}Pb
4. ^{35}Cl, ^{35}Ar
6. $N/Z \approx 1.00$, $N/Z \approx 1.50$
8. 10.812024
10. 7.7×10^{-15} m
12. 4.3×10^{-15}

14. 7.5 MeV
16. 1.46×10^{-8} amu
18. 3.03×10^{-2} amu, 28.3 MeV
20. 16.13108, 0.0169 amu, 15.7 MeV
22. (2p, 2n) + (7p, 7n) $\rightarrow$ (8p, 9n) + (1p, 0n)
Both sides have 9p and 9n!
24. 3.0 MeV
26. 1.6 MeV minimum energy of proton neglecting the recoil energies.
28. 1.12 MeV, 0.07 MeV
30. ^{85}Rb, ^{63}Cu
32. ^{234}Pa, ^{234}U, ^{230}T, ^{226}Ra
34. 0.77 MeV
36. 10.4 MeV
38. 156 keV
40. 2.9%, 140 years
42. 0.351, 0.492
44. 3.65×10^{10} Bq
46. 8.2×10^{-7} g
48. 468 kg
50. 190 MeV
52. (a) 4.28×10^{-12} J/fusion cycle
(b) 4×10^{18} s $\approx 1.3 \times 10^{9}$ y
54. 3.27 MeV, 4.03 MeV, 17.50 MeV, 18.35 MeV
3.60 MeV/nucleon

Chapter 32

2. (a) 4.76 T; (b) 2.9×10^{-4} s
4. $\dfrac{E}{mc^2}$
6. 0.3 kg
8. 113 MeV
10. 1.8×10^{-14} m
12. 2.45 amu
14. Baryon # conserved Strangeness not conserved
16. Strangeness conserved Strangeness **not** conserved
Strangeness **not** conserved
18. Reaction cannot proceed.
20. Baryon number *is* conserved
Strangeness *is NOT* conserved
22. The exchange of a π^- exchanges the charge between the two nucleons. The exchange of a π^0 does not alter the identity of the scattering nucleons, protons, neutrons or even one of each.
24. 130 GeV
26. (uud)
28. 15 quarks and 15 antiquarks are created in the collision.
30. 2260 $\approx$ 2270 within 10 MeV or 0.5%

PHOTOGRAPHIC CREDITS

Biographical Portraits **Chapter 1: p.25** Galileo Galilei: Trinity Library, Cambridge **Chapter 3: p.79** Sir Isaac Newton: Royal Society **p.103** Leonardo da Vinci: Art Resource **Chapter 4: p.125** James Prescott Joule: Royal Society **p.136** Christiaan Huygens: Granger Collection **p.138** Joseph Louis, Comte Lagrange: Granger Collection **p.142** Hermann Von Helmholtz: American Institute of Physics **p.144** James Watt: National Portrait Gallery, London **Chapter 5: p.165** Henry Cavendish: Granger Collection **p.165** Nicholas Copernicus: Bettman Archives **p.168** Johannes Kepler: Granger Collection **Chapter 11: p.333** Christian Doppler: American Institute of Physics **p.335** Ernst Mach: American Institute of Physics **Chapter 12: p.354** Blaise Pascal: Art Resource **p.361** Daniel Bernoulli: Granger Collection **Chapter 13: p.382** William Thomson, Lord Kelvin: Hunterian Art Gallery, University of Glasgow **p.385** Robert Boyle: National Portrait Gallery, London **p.385** Ludwig Boltzman: American Institute of Physics **Chapter 14: p.403** Benjamin Thomson, Count Rumford: National Portrait Gallery, London **Chapter 15: p.434** Sadi Carnot: American Institute of Physics **p.438** Rudolph Clausius: Bettman Archives **p.442** Walther Hermann Nernst: Bettman Archives **Chapter 16: p.453** Benjamin Franklin: Bettman Archives **p.457** Charles Augustin de Coulomb: American Institute of Physics **Chapter 17: p.487** Karl Friedrich Gauss: American Institute of Physics **Chapter 18: p.504** Alessandro, Conte Volta: American Institute of Physics **Chapter 19: p.539** Georg Simon Ohm: Granger Collection **Chapter 20: p.561** Gustav Robert Kirchhoff: Granger Collection **Chapter 21: p.585** Hans Christian Oersted: Granger Collection **p.589** Nikola Tesla: American Institute of Physics **p.596** André Marie Ampére: Granger Collection **Chapter 22: p.620** Michael Faraday: National Portrait Gallery, London **p.631** Joseph Henry: By Henry Ulke; National Portrait Gallery, Smithsonian Institution/Art Resource, New York **Chapter 24: p.669** James Clerk Maxwell: American Institute of Physics **p.672** Heinriich Rudolf Hertz: American Institute of Physics **Chapter 25: p.699** Christiaan Huygens: Granger Collection **Chapter 26: p.734** Albert Abraham Michelson: University of Chicago **p.736** Thomas Young: Royal Society **Chapter 26: p.743** Augustin Fresnel: Granger **p.746** John William Strutt, 3rd Baron Rayleigh: Royal Society **Chapter 27: p.765** Albert Einstein: American Institute of Physics **p.771** Hendrik Antoon Lorentz: American Institute of Physics **Chapter 28: p.790** Max Planck: American Institute of Physics **p.798** Arthur Holly Compton: American Institute of Physics **p.803** Max Born: American Institute of Physics **p.804** Werner Heisenberg: American Institute of Physics **Chapter 29: p.816** Sir Ernest Rutherford: Cavendish Laboratory, Cambridge **p.817** Niels Bohr: American Institute of Physics **p.824** Louis Victor, Prince de Broglie: American Institute of Physics **p.826** Erwin Schrodinger: American Institute of Physics **Chapter 30: p.843** Wolfgang Pauli: American Institute of Physics **Chapter 31: p.877** Antoine Henri Becquerel: American Institute of Physics **p.879** Pierre & Marie Curie: American Institute of Physics **p.883** Enrico Fermi: American Institute of Physics **Chapter 32: p.905** Richard Philips Feynman: American Institute of Physics

Table of Contents **ix (center)** Runt/Schoenberger/Grant Heilman Photography **x (left)** NASA **(bottom)** ©Michael Tamborrino/Stock Market **xi (right)** NASA **(bottom)** Sandia National Laboratories **xii (left)** Jay W. Bruff **(bottom)** ©Brian Bielmann/ProFiles West **xiii** NASA **xiv (left)** ©Allsport USA/1989 **(bottom)** NCAR **xv** USCEA **xvi (top)** Siemens Solar Industries **xvi (bottom)** IBM Research **(top)** Tom Pantages **(bottom)** ©Matt Bradley/Tom Stack and Associates **xviii** NASA **xix** IBM Research **xx** ©Lawrence Berkeley Laboratory/Science Photo Library/ Photo Researchers

Prelude **P.2 (top)** Jay W. Bruff **(bottom)** Jay W. Bruff **P.3 (center)** Jay W. Bruff **(bottom)** James L. Mairs **P.4 (top)** General Electric Space Systems **(center)** NASA **(bottom)** NASA **P.5 (top)** NASA **(center)** NASA **(bottom)** NASA **P.9 (bottom)** ©Anglo-Australian Telescope Board **P.10 (top)** ©Anglo-Australian Telescope Board **(center)** Hale Observatories **(bottom)** Hale Observatories **P.11 (center)** J.R. Kuhn, Dept of Physics, Princeton Univ. **P.12 (center)** Jay W. Bruff **(bottom)** Courtesy of Carl Zeiss, Inc., Thornwood, NY **P.13 (top)** Dr. Ronald Radius The Eye Institute, Milwaukee, WI **(center)** Tissue and Organs: A Test Atlas of Scanning Electron Microscopy by Richard G. Kessel & Randy H. Kardon, W.H. Freeman & Co., San Francisco ©1979 **(bottom)** Tissue and Organs: A Test Atlas of Scanning Electron Microscopy by Richard G. Kessel & Randy H. Kardon, W.H. Freeman & Co., San Francisco ©1979 **P.14 (top)** Digital Instruments, Inc. **(center)** Lawrence Livermore Laboratory **(bottom)** Lawrence Berkeley Laboratory **P.15 (top)** Digital Instruments

Chapter 1 **CO** ©Tim Hoske 1990/Profiles West **1.4** NIS & T **1.5** H. Layer/NIS & T **1.6** NIS&T **Table 1.8a** ©Harold Edgerton 1992 Trust/ Palm Press **Table 1.8b** ©David Weintraub/Photo Researchers **Table 1.8c** Alaska Division of Tourism **1.9** FPG International **1-25** Lester Lefkowitz **1.28** ©Ellis Herwig/Stock Boston **Q-6** National Maritime Museum Greenwich **P.33** Wide World **P.46** ©Harold Edgerton 1992 Trust/Palm Press **1.30** NASA

Chapter 2 **CO** ©Robert Holland/The Image Bank **Ex.1** ©Harold Edgarton 1992 Trust/Palm Press **2.6a** ©Richard Megna 1990/ Fundamental Photos **2.9** ©Smithsonian Institution **2.10** ©Harold Edgarton 1992 Trust/Palm Press **2.11** ©Stephen Marks/Stockphotos Inc. **2.33** NASA **Q.20** ©E. Bordis/Leo de Wys **P.19** H. Armstrong Roberts **P.54** Fermilab

Physics In Action 1: Long Jumps. **1.1** ©Michel Hans Vandystadt 1992/Allsport USA **1.2**UPI/Bettmann Newsphotos **1.3** ©Gerard Vandystadt/Allsport USA **1.4** ©G. Vandystadt/Allsport USA

Chapter 3 **CO** ©Wayne Eastep 1991/Stock Market **3.3** NIS & T **3.4** Road and Track **3.5** ©Gianalberto Cigolini/Image Bank **3.6** Edmund Scientific Co. **3.7** NASA **S3.5a** ©Harold Edgerton Trust/Palm Press **b** ©Harold Edgerton Trust/Palm Press **c** Lawrence Livermore Laboratory **d** Brookhaven National Laboratory **3.21** ©Russell Cheyne 1988/Allsport USA **3.22** NASA **3.23** NASA **3.47** Beckman Instruments, Inc. **P.1** McDonnell Douglas **3.57** NIS & T

Chapter 4 **CO** ©Scott Markewitz 1990/FPG International **4.15** ©John Swart 1992/Allsport USA **Ex.6** ©Ian Tomlinson/Allsport USA **Table 4.3a** NASA **Table 4.3b** Delta Air Lines, Atlanta, GA **Table 4.3c** National Historical Photographic Agency **4.27** National Archives **P.31** ©Harold Edgerton Trust 1992/Palm Press **P.64** *Friction: An Introduction to Tribology,* by Bowden and Tabor, Bantam, Doubleday, Dell Publishing Group

Physics In Action 2: Muscular Contraction. **2.2** Courtesy of H.E. Huxley **2.4** Electrograph courtesy of G.F. Bahr, Beckenried 6375, Switzerland **2.6** ©Allsport USA/Seoul Olympics, 1988

Chapter 5 **CO** NASA **Ex 5** NASA **Table 5.1** NASA **5.12** Sovfoto/ Eastfoto **Ex 7** NASA **5.23** NASA **P.19** NASA **P.21** NASA **P.23** ©Anglo-Australian Telescope Board 1985 **P.26** **NASA**

Chapter 6 **CO** C. Powell, P. Fowler & D. Perkins/Science Photo Library/Photo Researchers **6.6** ©Richard Megna 1990/Fundamental Photos **6.11** Steven E. Sutton 1992/Duomo **6.14a** NASA **6.15** Mercedes-Benz of North America **6.20** ©Harold E. Edgerton/Palm Press **6.23** ©John Elk III/Stock Boston **6.26** ©Diane Schiumo 1991/Fundamental Photos **6.34** D.J. Roddy/U.S.G.S. Photo Library **P.59** ©Mike Powell/Allsport USA

Physics In Action 3: Automobile Collisions. **3.3** Insurance Institute for Highway Safety **3.4** Mercedes Benz of North America

Chapter 7 **CO** ©Andre Gallant/Image Bank **7.1** ©Harold Edgerton Trust 1992/Palm Press **7.2a)** ©Hank de Lespinasse 1988/Image Bank **b)** Jay W. Bruff **c)** Courtesy of Yamaha Electronics Corp. **Ex 3** Jay W. Bruff **7.21** ©Thomas Zimmerman 1991/FPG International **7.24** ©Herb Snitzer all rights reserved/Stock Boston **P.3** ©Guy Marche 1985/FPG International **P.22** Tom Pantages **P.44** Chicago Historical Society

Chapter 8 **CO** Jay W. Bruff **S8.1** Hans C. Ohanian **8.5** ©Gary Newkirk/Allsport USA **Ex. 7** Tom Pantages

Physics In Action 4: Animal Mechanics. **4.6** Aquino Productions

Chapter 9 **CO** Smithsonian Institution **9.7** NASA **9.9** Adapted from Energy, A Sequel to IPS by Uri Haber-Schaim, Prentice Hall, Englewood Cliffs, NJ, 1983 **9.15** Lester Lefkowitz **9.19** Smithsonian Institution **9.20** National Maritime Museum **9.22** Bibliotheque Nationale, Paris, France **9.23** Courtesy of the Time Museum, Rockford, IL **9.27** ©Ellis Herwig/Stock Boston **9.28** NASA

Chapter 10 **CO** Thomas Kitchin/Tom Stack & Associates **10.1** ©Richard Megna 1985/Fundamental Photographs **10.15** ©Richard Megna 1991/Fundamental Photographs **10.16** Wide World **10.17** Tom Pantages **10.18** National Automobile Museum **10.19** Jay W. Bruff **10.20** ©Daemmrich/Stock Boston

Chapter 11 **CO** Naval Research Laboratory **11.1** Winston E. Kock/Plenum Publishing Co. **11.4a–c** William Joyce **11.8** ©Howard Sochurek 1991/Stock Market **11.9** Courtesy of C.F. Quate and L. Lam, Hansen Laboratories, Stanford Univ. **11.14** ©Gregg Mancuso/Stock Boston **11.23** ©The Harold E. Edgerton 1992 Trust/Palm Press, Inc. **11.24** ©1991 British Airways Plc **11.25** Courtesy of John S. Shelton **11.26a,b** PSSC Physics, 2nd edition, 1965; D.C. Heath & Company and Educational Development Center, Inc., Newton, MA **11.27a,b** PSSC Physics, 2nd edition, 1965; D.C. Heath & Company and Educational Development Center, Inc., Newton, MA **11.28** B & W Loudspeakers of America, Buffalo, NY **11.29** ©Greig Cranna/Stock Boston **11.30** ©Moatil Gontier/The Image Works **11.31** Hummingbird **11.32** Tom Stack/Tom Stack & Associates **11.33** NCAR **11.34** Polaroid

Physics In Action 5: Ultrasound Images. **5.2** ©Will & Deni McIntyre/Photo Researchers **5.3** Ochsner Clinic, New Orleans **5.4** ©Howard Sochurek, 1985/The Stock Shop (Medichrome) **5.6** Hewlett-Packard **5.7** ©Current Medical Literature/Stephen Walton and Graham Leech

Chapter 12 **CO** ©Rob Catanach/Woods Hole Oceanographic Institution **12.7** A.D. Moore, Univ. of Michigan, from *Introduction to Electric Fields* by W.E. Rogers, McGraw-Hill Book Co., New York 12-8 D.C. Hazen and R.F. Lehnert, Subsonic Aerodynamics Laboratory, Princeton 12-9 D.C. Hazen and R.F. Lehnert, Subsonic Aerodynamics Laboratory, Princeton 12-10 D.C. Hazen and R.F. Lehnert, Subsonic Aerodynamics Laboratory, Princeton **12.10** ©Diane Schiumo/Fundamental Photographs **12.12** Courtesy of MCO Properties, Inc. **12.15** Naval Photographic Center **12.25** The Balloon Works **12.29** ©Greg Wenger/The Travel Image **12.42** H. Armstrong Roberts **12.43** National Archives **12.44** NASA **12.49** ©Robert Azzi 1983/Woodfin Camp and Associates

Chapter 13 **CO** NASA **Tbl 13.1a)** Princeton Plasma Physics Laboratory **b)** AISI; **c)** Runk/Schoenberger from Grant Heilman **d)** ©Dan McCoy/Rainbow **13.7** Weksler Instrument Corp. Freeport, NY **13.8** Tom Pantages **13.9** NIS & T **13.10** Tom Pantages **13.15** National Optical Astronomy Observatories **13.18** Physical Science Laboratory, National Scientific Balloon Facility

Chapter 14 **CO** ©Donald Dietz/Stock Boston **14.8** Jay W. Bruff **14.9** Wide World **S14.3a** Tom Pantages **S14.3b** ©David Hiser/Image Bank **S14.3c** Tom Pantages **14.16** ©Robert Azzi/Woodfin Camp **14.17** VANSCAM™ Thermogram by Daedalus Enterprises Courtesy of National Geographic Magazine **14.18** Tom Pantages **14.22** SLAC **14.23** Sears Roebuck, Inc.

Chapter 15 **CO** ©John Elk III/Stock Boston **15.5a** Tom Pantages 15.5b **15.7** JoAnn C. Schambier **15.13** Tom Pantages **15.15** Tom Pantages **15.19** ©1961 M.C. Escher/Cordon Art, Baarn, Holland **15.20** Scientific American **15.23** ©Gill C. Kenny 1981/Image Bank **15.25** ©New York State Department of Economic Development

Chapter 16 **CO** E.R. Degginger/H. Armstrong Roberts **16.8** NCAR **16.9** ©Electric Power Research Institute **16.10** Lightning Master **Q.10** ©Fundamental Photographs 1968

Chapter 17 **CO** ©Hank Morgan/Rainbow **17.29** Robert Matthews, Princeton University **17.36** Robert Matthews, Princeton University **17.37** Robert Matthews, Princeton University **P.10** Phil Jude/Science Photo Library/Photo Reseachers

Chapter 18 **CO** Lawrence Livermore Laboratory **Tbl 18-1a)** Runk/Schoenberger/Grant Heilman; **b)** New York Power Authority **c)** Tom Pantages; **d)** ©John Keating/Photo Researchers **18.30** Jay W. Bruff/Massachusetts Institute of Technology **Q.1** Jay W. Bruff **P.11** SLAC **P.27** Paul Brierley **P.32** 1988 Will/Deni McIntyre/Photo Researchers

Chapter 19 **CO** ©Dan McCoy/Rainbow **Table 19.1a)** NCAR **b)** Ewing Galloway **c)** ©Ronald Johnson/Image Bank/Stock Photos **d)** Delco Remy Division **e)** Canon, Inc **19.2** O. Jefimenko *American Journal of Physics* **30,** 19, 1962 **19.3** O. Jefimenko *American Journal of Physics* **30,** 19, 1962 **19.7** US National Bureau of Standards **P.18** Jay W. Bruff

Chapter 20 **CO** David Taylor/Science Photo Library/Photo Researchers **S20.2** Jay W. Bruff **20.7** NASA **S20.6** Tom Pantages **Q.11** Hans C. Ohanian **P.34** GM

Physics in Action 8: Thunderstorms and Lightning. **8.1** NCAR

Chapter 21 **CO** IBM **21.4** Tom Pantages **Tbl 21.1a)** Brookhaven National Laboratory, Upton, NY **b)** NASA **c)** ©David Hamilton/The Image Bank **d)** NOAO **21.14** ©Richard Megna 1990/Fundamental Photographs **21.16** ©Richard Megna 1990/Fundamental Photographs **21.22** ©Richard Megna 1990/Fundamental Photographs **21.27** ©Richard Megna 1990/Fundamental Photographs **21.30** ©Richard Megna 1986/Fundamental Photographs **21.31a** ©Lou Jones/The Image Bank **21.32** Zenith Electronics Corporation **21.35** Professor Lehman **21.63** Lawrence Berkeley Laboratory/Science Photo Library/Photo Researchers

Chapter 22 **CO** ©Chuck O'Rear/Westlight **P.5** ©Dave B. Fleetham/Tom Stack and Associates **22.30** IGC

Chapter 23 **CO** ©Alain Choisnet/Image Bank **23.18** Paul Brierly **P.1** Hans C. Ohanian **P.45** Hans C. Ohanian

Chapter 24 **CO** GM Hughes Electronic **24.14** ©Diane Schiumo 1992/Fundamental Photos **S24.4** ©Bill Gallery 1989/Stock Boston **24.17** ©Stacy Pick 1986/Stock Boston **24.23** Department of Defense Still Media Records Center **P.21** Hans C. Ohanian **P.26** ©Chuck O'Rear/Westlight **P.35** ©Dan Bundik/Woodfin **P.40** Courtesy of Carl Zeiss, Inc., Thornwood, NY

Chapter 25 **CO** Jan McCoy, University of Arizona **25.1** ©Marc Romanelli 1985/The Image Bank **25.2** ©Larry Keenan/The Image Bank **25.6** ©Richard Megna 1987/Fundamental Photographs **25.18** ©Richard Megna 1990/Fundamental Photographs **25.21b** Courtesy of AT&T Archives **25.22** ©Charles Lightdale/Photo Researchers **S25.3** ©David Parker/Science Photo Library/Photo Researchers **S25.6** ©Will/Demi McIntyre/Photo Researchers **25.57** ©Don Wong/Photo Researchers **25.60** Cal Tech Institute **Q.10** ©Paul Silverman 1990/Fundamental Photographs **25.62a** ©Scott Blackman/Tom Stack & Associates **25.75** Hughes Danbury Optics **P.31** JoAnn C. Schambier **P.60** Minolta **P.69** MMT Observatory

Chapter 26 **CO** NRAO/AUI **26.3** ©Eric Simmons/Stock Boston **26.5** Bausch & Lomb **26.8** Bureau of International des Poids et Mesures, Sevres, France **26.14** Educational Development Center, Newton, MA **26.15a** C.C. Jones, Union College **26.17** C.C. Jones, Union College **26.19** NRAO/AUI **26.20** Educational Development Center, Newton, MA **26.21** ©Ken Kay/Fundamental Photographs **26.25** Courtesy of C.C. Jones, Union College **26.26** Courtesy of C.C. Jones, Union College **26.27** Courtesy of C.C. Jones, Union College 26–28 STSI **26.29** The Arecibo Observatory is part of the National Astronomy and Ionosphere Center which is operated by Cornell University under a cooperative agreement with the National Science Foundation **26.30b** Bausch & Lomb **26.32** Sears, Zemansky & Young, *University Physics,* ©1982, Addison-Wesley, Reading, Mass. Reprinted with permission. Page 789, Fig. 41-18. **26.33** CNRI/Science Photo Library/Photo Researchers **26.38** Educational Development Center, Newton, Mass.

Physics in Action 9: Microscopy. **9.2** ©M.I. Walker/Photo Researchers **9.5** ©M.I. Walker/Photo Researchers **9.6** ©M.I. Walker/Photo Researchers **9.7** ©M.I. Walker/Photo Researchers **9.9** ©M.I. Walker/Photo Researchers **9.11** Courtesy of Carl Zeiss, Inc., Thornwood, NY

Chapter 27 **CO** SLAC

Chapter 28 **CO** VANSCAN™ Thermogram by Daedalus Enterprises, Inc. Courtesy of National Geographic Magazine **28.1** AISI **S28-3** Burle Photomultipliers **28.10** Deutsches Museum, München **28.11** ©Henry Przekop 1987/The Stock Shop (Medichrome) **28.12** Courtesy of General Electric **28.16a–c** E.R. Huggins **d)** C.C. Jones, Union College

Physics in Action 10: X-Ray Images. **10.1** Thomas Beck, Johns Hopkins Hospital **10.2** Thomas Beck, Johns Hopkins Hospital **10.3** ©Howard Sochurek 1981/The Stock Shop (Medichrome) **10.5** ©Science Photo Library/Science Source/Photo Researchers **10.6** ©Howard Sochurek 1986/The Stock Shop (Medichrome)

Chapter 29 **CO** Digital Instruments, Inc. **29.2** Unique Boutique **29.5** Hale Observatories **29.14** H.E. Huxley **29.15** ©Biophoto Associates/Science Source/Photo Researchers **29.16** Digital Instruments, Inc.

Physics In Action 11: Laser Light **11.3** GM Hughes Electronics **11.4** Clear Modulated Laser is distributed by Arbor Scientific, Ann Arbor, Michigan **11.5** ©Charles O'Rear/Westlight **11.6** Courtesy of AT&T Archives **11.7** ©Melchior Digiacomo/Image Bank **11.8** Lawrence Livermore Laboratory **p.834F** Eastman Kodak

Chapter 30 **CO** Digital Instruments, Inc. **30.3** A.F. Burr & R. Fisher, Physics Department, New Mexico State University **30.9** Hans C. Ohanian **30.22** IBM **30.23** IBM **30.26** ©John Cancalosi 1989/Stock Boston **30.27** NASA **30.28** GM Electric Vehicles **30.29** PE&G Photography

Chapter 31 **CO** Sandia National Laboratories **31.7** Cavendish Laboratory, University of Cambridge **31.15** Lawrence Livermore Laboratory **31.18** ©Earl Roberge/Photo Researchers **31.19** USCEA **31.20** Princeton Tokamak Test Reactor **31.21** Lawrence Livermore Laboratories

Physics in Action 12: Medical Imaging. **12.2** Siemens Medical Systems **12.4** ©John Mazziotta/Photo Researchers **12.7** ©Howard Sochurek/The Stock Shop (Medichrome)

Chapter 32 CO ©Dan McCoy/Rainbow **32.1** Fermilab **32.2** Fermilab **32.4** CERN Photo **32.5** Brookhaven National Laboratory/Science Photo Library/Photo Researchers **32.6** SLAC **32.7** ©David Parker/Science Photo Library/Photo Researchers **32.8** Lawrence Berkeley Laboratory/Science Photo Library/Photo Researchers **32.12** CERN Photo

INDEX

SPECIAL ▶ $23.95 ◀ OFFER

FOR WINDOWS . . .

The Norton Student Version of Pro Solv® by Problem Solving Concepts is an exceptional study tool that will help you develop problem-solving skills and strategies—just the sort of help that improves exam performance. The program contains the following special features:

- Expert tutorial help with 120 problems from *Principles of Physics.*
- Design and explore an almost limitless variety of problems covering basic physical situations; expert help available with the "computer solve" option.
- Simulations help you visualize the problems you are working on and cover the principles underlying basic concepts.
- Graphing and analysis of key equations and entered data.

System requirements: A 286 computer or better with at least 2 Mb of RAM and a VGA card; Microsoft Windows 3.0 or later

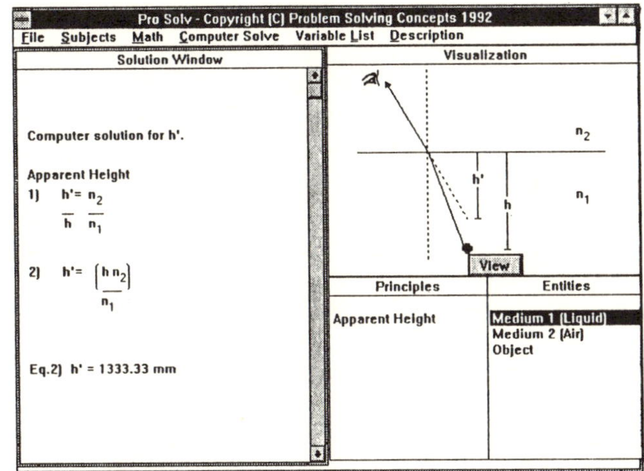

Waves

Waves

A **wave** is a transfer of energy. It travels with velocity **v**.
The **amplitude** is the maximum displacement of the medium.
The **frequency (f)** is the number of oscillations per unit time.
The **wavelength (λ)** is the distance between two highs or lows.

30 amplitude

$\lambda = (v)/(f) = 45$

4 frequency

FOR THE MACINTOSH . . .

Sensei Physics® by Broderbund Software is an innovative "personal tutor" that covers the basic material in your introductory course. This useful review contains the following special features:

- Over 300 problems for review and practice.
- Visualization of basic concepts through numerous interactively-designed animations.
- Hundreds of notecards covering basic physical principles.

System requirements: two 400K or one 800K disk drive.

YES, PLEASE SEND ME:

Choose one or both:
☐ The Norton Student Version of Pro Solv® for Windows [0-393-96580-5]–$23.95
☐ Sensei Physics® for the Macintosh [0-393-96581-3]–$23.95
(residents of CA, NY, PA, IL must add their local sales tax)

Phone orders to Customer Service at (800) 233-4830

☐ My check is enclosed or

Please charge to my: ☐ VISA ☐ MASTERCARD ☐ AMERICAN EXPRESS

Acct. # ____________ Exp. Date ________ Telephone ____________

Name ____________ Signature ____________

Ship to:
Address ____________
City ____________
State/Prov. ________ Zip/PC ________

Billing Address (if different):
Address ____________
City ____________
State/Prov. ________ Zip/PC ________

NO POSTAGE NECESSARY IF MAILED IN THE UNITED STATES

BUSINESS REPLY MAIL

FIRST CLASS PERMIT NO. 4008 NEW YORK, N.Y.

POSTAGE WILL BE PAID BY ADDRESSEE

W. W. NORTON & COMPANY, INC.
500 FIFTH AVENUE
NEW YORK, NEW YORK 10109-0145
ATTN: BEV BENZELESKI

▲ TAPE ▲

▲ TAPE ▲

▲ TAPE ▲

Fundamental Constants

Speed of light	$c = 3.00 \times 10^{8}$ m/s
Planck's constant	$h = 6.63 \times 10^{-34}$ J · s
	$\hbar = h/2\pi = 1.05 \times 10^{-34}$ J · s
Gravitational constant	$G = 6.67 \times 10^{-11}$ N · m²/kg²
Permeability constant	$\mu_0 = 1.26 \times 10^{-6}$ H/m
Permittivity constant	$\varepsilon_0 = 8.85 \times 10^{-12}$ F/m
Boltzmann constant	$k = 1.38 \times 10^{-23}$ J/K
Electron charge	$-e = -1.60 \times 10^{-19}$ C
Electron mass	$m_e = 9.11 \times 10^{-31}$ kg
Proton mass	$m_p = 1.673 \times 10^{-27}$ kg
Neutron mass	$m_n = 1.675 \times 10^{-27}$ kg
Atomic mass unit	$v = 1.66 \times 10^{-27}$ kg
Rydberg constant	$R_H = 1.10 \times 10^{7}$/m
Bohr radius	$4\pi\varepsilon_0\hbar^2/m_e e^2 = 5.29 \times 10^{-11}$ m

Miscellaneous Physical Constants

Acceleration of gravity	1 $g = 9.81$ m/s² $= 32.2$ ft/s²
Atomic mass unit	1 u $= 1.66 \times 10^{-27}$ kg
Avogadro's number	$N_A = 6.02 \times 10^{23}$/mole
Universal gas constant	$R = 8.31$ J/K
Molecular mass of air	28.98 g/mole
Density of dry air	1.29 kg/m³ (0°C, 1 atm)
Atmospheric pressure	1 atm $= 1.01 \times 10^{5}$ N/m²
Speed of sound in air	331 m/s (0°C, 1 atm)
Density of water	1000 kg/m³
Heat of vaporization of water	539 kcal/kg
Heat of fusion of ice	79.7 kcal/kg
Mechanical equivalent of heat	1 cal = 4.19 J
Index of refraction of water	1.33

The Planets

Planet	*Mean distance from Sun*	*Period of revolution*	*Mass*	*Equatorial radius*	*Surface gravity*	*Period of rotation*
Mercury	57.9×10^{6} km	0.241 year	3.30×10^{23} kg	2,439 km	0.38 g	58.6 days
Venus	108	0.615	4.87×10^{24}	6,052	0.91	243
Earth	150	1.00	5.98×10^{24}	6,378	1.00	0.997
Mars	228	1.88	6.42×10^{23}	3,397	0.38	1.026
Jupiter	778	11.9	1.90×10^{27}	71,398	2.53	0.41
Saturn	1,430	29.5	5.67×10^{26}	60,000	1.07	0.43
Uranus	2,870	84.0	8.70×10^{25}	25,400	0.92	0.65
Neptune	4,500	165	1.03×10^{26}	24,300	1.19	0.77
Pluto	5,890	248	1.5×10^{22}	1,500	0.045	6.39